MOLECULAR BIOLOGY

Organism	Gene name	Example	Mutant allele	Example	Protein name	Example
Bacteria	Three lowercase letters, followed by upper case letter, all italicized	*recA*	Same as gene name, followed by allele number (can have non-integer allele designations such as 'am' or 'ts' for amber- and temperature-sensitive mutants, respectively)	*recA11*	Same as gene name except first letter is upper case and gene name is not italicized	RecA
Saccharomyces cerevisiae	Letters (all uppercase if dominant, all lowercase if recessive) followed by an Arabic number, all italicized	*URA3*	Same as gene name followed by a hyphen and an Arabic number (can have additional information about how mutant was generated)	*ura3-52*	Uppercase first letter, followed by lowercase letters and number, not italicized	Ura3
Schizosaccharomyces pombe	Three lowercase letters followed by a number and superscript +, all italicized	$cdc2^{+}$	Same as gene name, followed by allele number (but no superscript +)	*cdc2-5*	Same as gene name except first letter is uppercase and gene name is not italicized and there is no superscript +	Cdc2
Caenorhabditis elegans	Three to four lowercase letters, followed by a hyphen and a number, all italicized	*dpy-5*	Same as gene name, followed by an allele name (one or two letters followed by a number) in parentheses	*dpy-5(e61)*	Same as gene name except all uppercase letters and gene name is not italicized	DPY-5
Drosophila melanogaster	Can be any word lowercase italicized (most genes also have a shorter unique symbol)	*dacapo* (*dap*)	Same as gene name followed by a superscript number(s) or letter(s) (for dominant mutants, the gene name is followed by a superscript D)	$dacapo^{4}$, $dacapo^{D}$	Same as gene name except first letter is uppercase and gene name is not italicized	Dacapo
Mus musculus	Usually three to five letters and Arabic numbers (maximum ten characters) begin with an uppercase letter (not a number), followed by lowercase letters and numbers, all italicized	*Grid2*	Same as the gene with the original mutant symbol added as a superscript to the gene symbol	$Grid2^{ho}$	Same as gene name except all uppercase letters and gene name is not italicized	GRID2
Homo sapiens	Maximum six characters: all uppercase letters or by a combination of uppercase letters and Arabic numbers, all italicized	*ATM*	Sequence variants are described by the specific sequence change in the DNA with sequence change, insertion, and deletions having specific nomenclature	c.1636C4G (p.Leu546-Val) (this example corresponds to a C to G change at position 1636 of the *ATM* coding sequence	Same as gene name except not italicized	ATM

Nomenclature table. Note that the names of some genes and proteins that have become accepted in the literature, such as the human Rb and p53 proteins, do not follow the conventions listed in this table.

Molecular Biology

Principles of Genome Function

SECOND EDITION

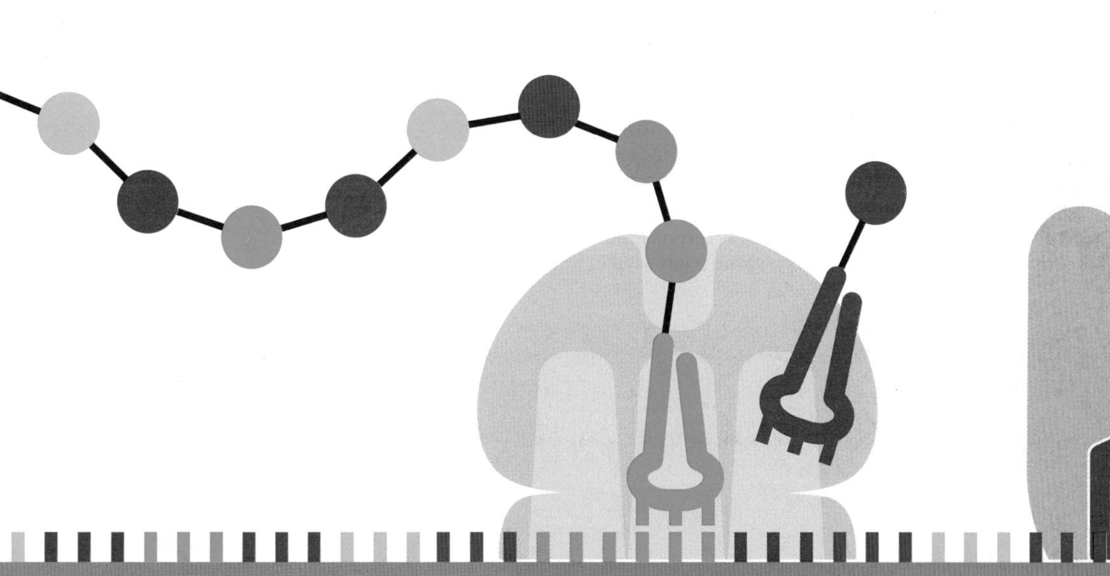

Nancy L Craig • Orna Cohen-Fix • Rachel Green
Carol Greider • Gisela Storz • Cynthia Wolberger

With end of chapter questions by Deborah Zies and Claire Burns

OXFORD
UNIVERSITY PRESS

OXFORD
UNIVERSITY PRESS

Great Clarendon Street, Oxford, OX2 6DP,
United Kingdom

Oxford University Press is a department of the University of Oxford.
It furthers the University's objective of excellence in research, scholarship,
and education by publishing worldwide. Oxford is a registered trade mark of
Oxford University Press in the UK and in certain other countries

First edition published 2010

Impression: 2

Published in the United States of America by Oxford University Press
198 Madison Avenue, New York, NY 10016, United States of America

British Library Cataloguing in Publication Data
Data available

Library of Congress Control Number: 2014934426

ISBN 978-0-19-870597-0
ISBN 978-0-19-965857-2 (pbk)
ISBN 978-0-19-871995-3 (instructors manual)

Printed in Great Britain by
Bell & Bain Ltd., Glasgow

To our teachers

ABOUT THE AUTHORS OF MOLECULAR BIOLOGY

Nancy L Craig received an A.B. in Biology and Chemistry from Bryn Mawr College in 1973 and a Ph.D. in Biochemistry in 1980 at Cornell University in Ithaca, New York, where she worked on DNA repair with Jeff Roberts. She then worked on phage lambda recombination as a postdoctoral fellow with Howard Nash at the National Institutes of Health. She joined the faculty of Department of Microbiology and Immunology at the University of California, San Francisco in 1984 and began her work on transposable elements. She joined the Department of Molecular Biology and Genetics at the Johns Hopkins University School of Medicine in 1991, where she is currently a Professor and a Howard Hughes Medical Institute Investigator, as well as the recipient of the Johns Hopkins University Alumni Association Excellence in Teaching Award. Nancy Craig is a Fellow of the American Academy of Microbiology, the American Academy of Arts and Sciences and the American Association for the Advancement of Science, and was elected to the National Academy of Sciences.

Orna Cohen-Fix received a B.A. from the Tel Aviv University in 1987 and a Ph.D. in Biochemistry with Zvi Livneh at the Weizmann Institute of Science in 1994. She did a post-doctoral fellowship with Doug Koshland at the Carnegie Institution of Washington in Baltimore, studying the regulation of chromosome segregation. In 1998, she moved to the National Institute of Diabetes and Digestive and Kidney Diseases in Bethesda, where she is now a Senior Investigator. Her research focuses on cell cycle regulation and nuclear architecture, using budding yeast and *C. elegans* as model organisms. She is also an Adjunct Professor at Johns Hopkins University and the Co-Director of the NIH/Johns Hopkins University Graduate Partnership Program. She is a recipient of a Presidential Early Career Award for Scientists and Engineers, and an Association of Women in Science Mentoring Award for her work on promoting the retention of women in science.

Rachel Green received a B.S. in chemistry from the University of Michigan in 1986 and a Ph.D. in Biological Chemistry from Harvard University in 1992, where she worked with Jack Szostak studying catalytic RNA. She then did postdoctoral work in the laboratory of Harry Noller at the University of California, Santa Cruz, studying the role played by the ribosomal RNAs in the function of the ribosome. She is currently a Professor in the Department of Molecular Biology and Genetics at The Johns Hopkins University School of Medicine, an Investigator of the Howard Hughes Medical Institute, and a member of the National Academy of Sciences. Her work continues to focus on the mechanism and regulation of translation in bacteria and eukaryotes. She is the recipient of a Johns Hopkins University School of Medicine Graduate Teaching Award.

Carol Greider received a B.A. from the University of California at Santa Barbara in 1983. In 1987, she received her Ph.D. from the University of California at Berkeley, where she and her advisor, Elizabeth Blackburn, discovered telomerase, the enzyme that maintains telomere length. In 1988, she went to Cold Spring Harbor Laboratory as an independent Fellow and remained as a Staff Scientist until 1997, when she moved to The Johns Hopkins University School of Medicine. She is currently a Professor and Director of the Department of Molecular Biology and Genetics, and her work focuses on telomerase and the role of telomeres in cell senescence, age-related disease, and cancer. She is a member of the National Academy of Sciences and is the recipient of numerous awards, including the Gairdner Foundation International Award, the Louisa Gross Horwitz Prize, and the Lasker Award for Basic Medical Research. In 2009, she was awarded the Nobel Prize in Physiology or Medicine together with Elizabeth Blackburn and Jack Szostak for the discovery of telomerase.

Gisela Storz graduated from the University of Colorado at Boulder in 1984 with a B.A. in Biochemistry and received a Ph.D. in Biochemistry in 1988 from the University of California at Berkeley, where she worked for Bruce Ames. After postdoctoral fellowships with Sankar Adhya at the National Cancer Institute and Fred Ausubel at Harvard Medical School, she moved to the National Institute of Child Health and Human Development in Bethesda, where she is now a Senior Investigator. Her research is focused on understanding gene regulation in response to environmental stress as well as elucidating the functions of small regulatory RNAs and very small proteins. She is a Fellow of the American Academy of Microbiology, American Academy of Arts and Sciences, and National Academy of Sciences, and received the American Society for Microbiology Eli Lilly Award.

Cynthia Wolberger received her A.B. in Physics from Cornell University in 1979 and a Ph.D. in Biophysics from Harvard University in 1987, where she worked with Stephen Harrison and Mark Ptashne on the structure of a phage repressor bound to DNA. She did postdoctoral work on eukaryotic protein–DNA complexes in the laboratory of Robert Stroud and the University of California, San Francisco and then in the laboratory of Carl Pabo at The Johns Hopkins University School of Medicine. She joined the faculty of the Department of Biophysics and Biophysical Chemistry in 1991, where she is now a Professor. Her research focuses on the structural and biochemical mechanisms underlying transcriptional regulation and ubiquitin-mediated signaling. She is a Fellow of the American Association for the Advancement of Science and a recipient of the Dorothy Crowfoot Hodgkin Award of the Protein Society.

Molecular Biologists of Fells Point, Baltimore: (L–R) Rachel Green, Gisela (Gigi) Storz, Orna Cohen-Fix, Nancy Craig, Cynthia Wolberger and Carol Greider. The photo-digital illustration was created by Robert McClintock, a Fells Point artist.

PREFACE

A contemporary perspective on molecular biology

Molecular Biology: Principles of Genome Function offers a distinctive approach to the teaching of molecular biology. It is an approach that reflects the challenge of teaching a subject that is in many ways unrecognizable from the molecular biology of the twentieth century – a discipline in which our understanding has advanced immeasurably, but about which many intriguing questions remain to be answered. Among the students being taught today are the molecular biologists of tomorrow; these individuals will be in a position to ask fascinating questions about fields whose complexity and sophistication become more apparent with each year that passes.

We have written the book with several guiding themes in mind, all of which focus on providing a faithful depiction of the key themes and challenges that surround molecular biology in the twenty-first century, and on communicating this reality to students in a way that will engage and motivate, rather than overwhelm and intimidate.

A focus on the underlying principles

Arguably one of the biggest challenges facing instructors and students of molecular biology today is the vast amount of information encapsulated by the field. It is impossible for an instructor to convey every last detail (and equally impossible for students to absorb everything that there is to know). Indeed, we believe that it is not necessary to delve into every fine detail in order to understand the main concepts. Therefore, our approach focuses on communicating the *principles* of the subject.

We believe it is better for students to truly understand the foundational principles rather than simply learn a series of facts. To this end, we do not try to be exhaustive in our coverage. In the digital age in which we live, it is easier than ever before for students to gather a vast amount of information on a particular topic of interest. This information is of little value, however, if the student lacks a conceptual framework within which to make sense of all the information to which they are exposed.

By focusing on key principles, we seek to equip students with a conceptual framework, which we believe will be invaluable to them during their later careers.

An emphasis on commonalities

Until relatively recently, much more was known about the molecular components and processes of bacterial systems than of their archaeal and eukaryotic counterparts. In recent years, however, our understanding of archaeal and eukaryotic systems has increased enormously. With this increased understanding has come the realization that bacterial, archaeal, and eukaryotic systems exhibit many commonalities – commonalities that point to the common ancestry of the three kingdoms of life.

Throughout this book, therefore, our emphasis is on the *common features* of bacterial, archaeal, and eukaryotic systems. Differences do exist, of course – an inevitable outcome of evolutionary processes generating biological diversity. However, we have strived where possible to present a single view of key topics based on conserved processes and components. We have then discussed key differences between bacterial processes and their archaeal and eukaryotic counterparts where they exist, and where they have helped to further our understanding.

We recognize that some may feel that the processes occurring in bacteria, and in eukaryotes and archaea, are best taught separately. However, our focus on principles – and on constructing an overarching conceptual framework – leads us strongly to believe that an emphasis on commonalities is a valuable educational approach.

Integration of key themes and concepts

One of the most startling realizations of recent years has been the widespread importance of certain molecular phenomena, such as chromatin modification, and regulatory RNAs, which have impacts on genome function in ways far more diverse than had previously been recognized. Rather than examining each of these phenomena in isolation, our approach reflects their diverse impacts by presenting them in the various contexts in which they function. We believe this overall approach reflects the reality of molecular biology, and helps students to appreciate molecular biology as a unified discipline, with many components and phenomena acting in concert, rather than as a series of isolated topics.

A demonstration of how we know what we know

At heart, molecular biology is an experimental science. Our understanding of the field is increased through the accumulation of experimental evidence, which leads to the gradual emergence of key ideas and paradigms. Therefore, a central element to the understanding of molecular biology is an appreciation of the approaches taken to yield the information from which concepts and principles are deduced.

However, as instructors, we face a potential conflict: a mass of experimental evidence can often be overwhelming for students, and can make it more challenging for them to grasp the central ideas and paradigms that the experimental evidence has allowed us to elucidate. On the other hand, ignoring the experimental evidence deprives students from fully understanding the fundamental aspects of molecular biology (and, indeed, of science in general). In response to this seeming conflict, our approach has been for the main body of the text to focus on the communication of key concepts, free from the layer of complexity that experimental evidence might introduce.

We have then complemented our coverage of key concepts in the main body of the text with separate 'Experimental approach' panels, which branch off from the text in a clearly signposted way. These panels describe pieces of research that have been undertaken and which have been particularly valuable in elucidating difference aspects of molecular biology.

Importantly, experimental research represents an ongoing journey of discovery, where the experimental approaches adopted develop as much as our understanding of the field. Uniquely, therefore, the experimental approach panels present, wherever possible, two approaches – one 'classic' and one 'contemporary'. Although all approaches have revealed valuable insights, regardless of whether they could be considered classic or contemporary, we believe that coupling the approaches in this way has additional educational value in terms of showing how both experimentation and the knowledge gained from such experimentation can evolve with time.

In addition to the experimental approach panels, further support for encouraging students to engage with experimental evidence is provided by an online Journal

Club, as described more fully in the description of the Online Resource Center, which follows.

The methods used in molecular biology

Many of the experimental approach panels (and the research work featured in the Journal Club papers) draw on particular laboratory techniques that are used in different contexts throughout molecular biology research. The final chapter of this book, 'Tools and techniques in molecular biology', provides an overview of the basic techniques that are exploited during the course of much experimental work in molecular biology. Rather than describing general methods in detail within the experimental approach panels, we have directed the reader to appropriate coverage in Chapter 19, where they can learn more about the methodological tools that are at a molecular biologist's disposal, how these tools work, and what they can tell us.

New for this edition

The preparation of this second edition has given us a welcome opportunity to refine a number of aspects of the text, particularly in the light of valuable feedback from those who have taught from the first edition. Beyond updating all topics in line with current research findings, the most notable changes are:

A new chapter on regulatory RNAs. We now include a broad survey of an area of molecular biology whose centrality to fully understanding genome function has come to the fore in recent years.

Individual chapters on the regulation of transcription and translation. Whereas the previous edition devoted one chapter each to transcription and translation, each of these topics is now covered in two chapters: one devoted to the core processes and another exploring how those processes are regulated.

New questions at the end of each chapter. Each chapter now includes a range of end-of-chapter questions, provided by Deborah Zies, University of Mary Washington, and Claire Burns, Washington and Jefferson College.

Additional Experimental approach panels. We have prepared a number of new Experimental approach panels, bringing the total throughout the book to 44. The format of these panels has also been refined, with new sub-headings to guide the reader, and links to explanations of relevant techniques in Chapter 19, 'Tools and techniques in molecular biology'.

Video animations. We have worked with the winner of OUP's Student Animation Award, Connor Hendrich of Colorado State University, to develop sixteen animations that illustrate some of the central processes in molecular biology, including those that are particularly challenging to visualize.

NLC
OCF
RG
CWG
GS
CW
Baltimore, Maryland, July 2013

LEARNING FROM THIS BOOK

Molecular Biology: Principles of Genome Function features a number of learning features to help students get the most out of their study of the subject.

The Experimental approach panels

As noted previously, molecular biology is an experimental science. To help you gain an understanding of how some of the key molecular processes and components described in this book were characterized, without overburdening the main text with a lot of experimental detail, many chapters feature 'Experimental approach' panels. These panels describe pieces of research that have been particularly valuable in elucidating different aspects of molecular biology. 'Related techniques' at the end of each panel direct you to relevant sections in Chapter 19, 'Tools and techniques in molecular biology', where you can learn more about the specific techniques mentioned in the panels.

6.2 EXPERIMENTAL APP

Discovery of the origin reco

Biochemistry and footprinting lead to the discovery of the eukaryotic replication initiation protein

While DNA sequences that function as origins of replication

End of chapter questions

A set of questions is presented at the end of each chapter, grouped by section to help instructors to assign questions according to specific topics taught. Included in these questions are *Challenge questions*, which are designed to stimulate students' thinking, and often encourage the use of data analysis skills. Some questions also relate to Experimental approach panels, to encourage students to engage more closely with the research explored in those panels.

8.1 OVERVIEW OF TRANSCRIPTION

1. Not all RNAs produced in the cell encode proteins. Explain.
2. Which of the following statements regarding RNA is NOT true?
   a. RNA contains the nitrogen base uracil.
   b. RNA is predominantly double stranded.
   c. RNA contains the sugar ribose.

Further reading

Each chapter ends with a list of further reading materials, typically review articles that we feel would make a good next step in exploring the topics covered in the book in more detail. Each further reading list is divided into chapter sections to help you pinpoint articles that are relevant to the particular topic you are interested in.

8.1 OVERVIEW OF TRANSCRIPTION

Cheung AC, Cramer P. A movie of RNA polymerase II transcription. *Cell*, 2012;**149**:1431–1437.

Lee TI, Young RA. Transcription of eukaryotic protein-coding genes. *Annual Review of Biochemistry*, 2000;**34**:77–137.

Orphanides G, Reinberg D. A unified theory of gene expression. *Cell*,

Glossary

Molecular biology, like many scientific disciplines, has its own particular vocabulary, and descriptions of molecular processes and procedures feature terms that may at first glance be unfamiliar. We have compiled an extensive glossary of all of the key terms featured in the book, which will be of value as you master the language of the subject.

GLOSSARY

(p)ppGpp ('magic spot'): a pentaphosphate guanosine anal an important role in signaling the stringent response systems.

–10: promoter sequence located ten nucleotides to the 5 of transcription and comprising part of the sequence dire

Cross-references

As we note previously, molecular biology comprises a range of interconnected topics, not a series of discrete, isolated ones. To help you make the connections between the topics presented in the book, and see how these topics come together to give a rounded picture of molecular biology, each chapter features numerous cross-references to other chapters in the book.

The use of nuclease digestion to probe DNA structure and interactions with proteins is covered in Section 19.12.

telomeres, the special DNA sequences at the ends of chromosomes. In addition, those sites in chromosomes that contain highly repetitive DNA sequences are also packaged into heterochromatin. We will learn much more about the nature of centromeres and telomeres later in this chapter.

The differences between heterochromatin and euchromatin are often detected experimentally by the **nuclease sensitivity** of the region. Heterochromatin is more resistant to digestion by DNase I than euchromatin. The relative resistance of heterochromatic DNA to digestion is often interpreted as indicating that heterochromatin is more condensed than euchromatin, although it may equally reflect differences in the ability of nucleases to interact with DNA bound by specific heterochromatin proteins.

The structural state of chromatin affects all processes directed by DNA

While most studies of the functional significance of chromatin structure have focused on gene transcription, the way in which DNA is packaged into chroma-

Scan here to watch a video animation explaining more about transcription initiation, elongation and termination, or find it via the Online Resource Center at www.oxfordtextbooks.co.uk/orc/craig2e/.

Links to video animations

The functioning of the genome involves a number of intricate processes operating in three dimensions. To help you visualize some of the key processes, a series of video animations are available for you to watch on YouTube. When you see a QR code image in the text, simply scan it with your phone or tablet camera to be taken directly to the animation that relates to the topic you are reading about.

PDB codes

Many of the molecular structures that appear throughout this book have been generated from data deposited in the Protein Data Bank (PDB). Each entry in the PDB is assigned a unique code; this code can be used to retrieve the data related to the entry in question, which often includes crystallographic data, and onscreen renderings of molecular structures in three dimensions. The PDB codes relating to many of the molecular structures in the book are given in the relevant figure captions. Visit the PDB website (http://www.rcsb.org/pdb/home/home.do) and enter the PDB codes related to molecules of interest to retrieve data related to those molecules.

Online Resource Center

Molecular Biology: Principles of Genome Function does not end with this printed book. Instead, additional resources for both instructors and their students are available in the book's Online Resource Center.

Go to http://www.oxfordtextbooks.co.uk/orc/craig2e/

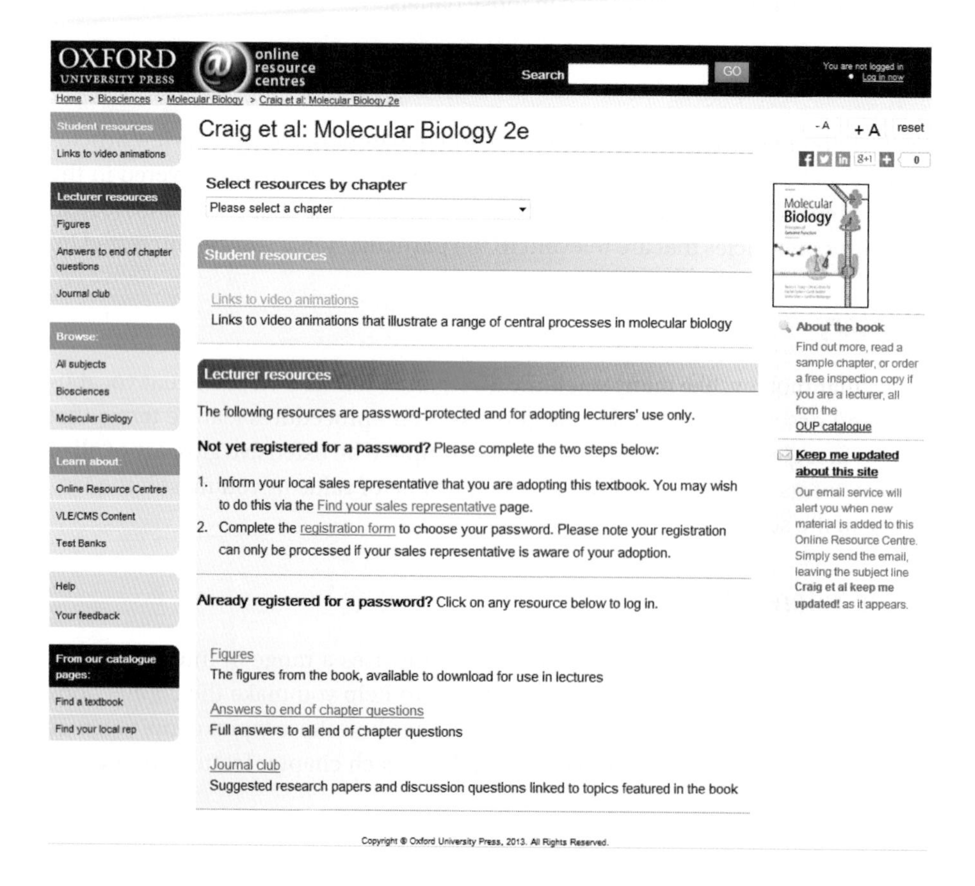

For instructors

Electronic artwork

Figures from the book are available to download, for use in lectures.

Answers to end of chapter questions

Full answers to all end of chapter questions are provided to help with the grading of your students.

Journal Club

Most chapters in the book are accompanied by an online Journal Club, which features suggested research papers and discussion questions linked to topics featured in the chapters. Understanding the details presented in primary literature articles can often be challenging; the purpose of the Journal Club is to guide students through some selected papers in a structured way, to build their confidence in reading and critically evaluating the work of others.

For students

Links to video animations

As an alternative to using the QR code images featured in the book, simply follow the links to view a series of video animations on YouTube, produced specifically to accompany the book.

ACKNOWLEDGMENTS

Many people made this textbook possible thanks to their advice and support.

We were fortunate to have worked under the guidance of two gifted individuals. The project was begun with Miranda Robertson at New Science Press, who convinced us that we could write a new textbook and helped show us the way. We benefited from her tremendous vision, advice, and insistence on clarity, as well as from her many visits to work with us in Baltimore and Washington. It was Jonathan Crowe at Oxford University Press who ushered us across the finish line by providing superb editorial advice, while teaching us how to work ever more effectively. He made outstanding contributions to the writing and organization of this book, and we are grateful for his efforts in helping us bring this project to completion as well as seeing us through the production of the second edition.

Our work was supported by many others, particularly Matthew McClements, who is responsible for the beautiful illustrations. At Oxford University Press, Joanne Hardern guided the book through the production process, while Elizabeth Farrell did an excellent job of copy editing. Our administrative assistants, particularly Patti Kodeck, helped to organize our meetings and carve out time for us to work on the book.

Our colleagues at Johns Hopkins and at the National Institutes of Health were generous, answering innumerable questions and providing comments on the text and figures. We are also grateful to the scientists outside our respective institutions for responding to our many emails and phone calls asking for information and clarifications. We owe special thanks to our students, postdoctoral fellows and laboratory staff, who gladly provided us with input while being patient when the writing took us away from the laboratory.

We extend our most profound thanks to our families for their support and encouragement throughout this project, which took far more time and work than any of us had imagined. We are grateful to our children, Rachel and Joshua Adams, Charles and Gwendolyn Comfort, Eric, Toby and Noel Cormack, Tal and Jonathan Fix, and Ella, Toby, and Felix Wu for their forbearance. We especially thank our spouses, Jeffrey Adams, Brendan Cormack, Alan Fix, Helen Lee McComas, and Carl Wu, for cheering us on while picking up the slack at home.

The authors and publisher extend their sincere thanks to the following individuals whose constructive comments either on the first edition or on draft chapters of the second have enhanced the book immeasurably.

David Angelini, American University
Daron Barnard, Worcester State University
Laura Baugh, University of Dallas
Prakash Bhuta, Eastern Washington University
Sue Biggins, Fred Hutchinson Cancer Research Center
Michael Botchan, University of California, Berkeley
Claire Burns, Washington and Jefferson College
Michael Carty, National University of Ireland, Galway
Karl Chai, University of Central Florida
Yijing Chen, Kent State University
Christopher Davis, Bethune-Cookman University
Brad Ericson, University of Nebraska Kearney
Dawn Franks, Loyola University of Chicago
Julia Frugoli, Clemson University
Susan Gottesman, Helix Systems, NIH
Erin Goley, Johns Hopkins University
Nigel Grindley, Yale University
Wolf-Dietrich Heyer, University of California, Davis
Babara Hoopes, Colgate University
James Hopper, The Ohio State University
Michael Ibba, The Ohio State University
Jim Kadonaga, University of California, San Diego
Craig Kaplan, Texas A&M University
Brett Keiper, East Carolina University
Thomas Kelly, Memorial Sloan Kettering Cancer Center
Nemat Keyhani, University of Florida
Nelson Lau, Brandeis University
Anthony Leung, Johns Hopkins University
Michael Lieber, University of Southern California
Boriana Marintcheva, Bridgewater State University
Manya Mascareno, State University of New York, College at Old Westbury
E. Stuart Maxwell, North Carolina State University
Mitch McVey, Tufts University
Paul Megee, University of Colorado Denver
Richard Myers, University of Miami
Jacqueline Nairn, University of Stirling

Luiza Nogaj, Mount St. Mary's College
Craig Peebles, University of Pittsburgh
Michael Persans, University of Texas Pan American
Jennifer Powell, Gettysburg College
Margaret Richey, Centre College
Daniel Schu, Helix Systems, NIH
Michael Seidman, National Institutes of Health
Mark Solomon, Yale University
Forrest Spencer, Johns Hopkins University
Dong Wang, University of California San Diego
Lauren Waters, University of Wisconsin Oshkosh
David Weinberg, University of California San Francisco
Sarah Wheelan, Johns Hopkins University
Daniel Williams, Gonzaga University
Lee Zou, Harvard University

We also continue to recognize those individuals whose feedback helped to shape the first edition, and whose influence continues to pervade the second:

Angelika Amon, MIT
John Atkins, The University of Utah
Brenda Bass, The University of Utah
Joel Belasco, New York University School of Medicine
Stephen D. Bell, Massachusetts Institute of Technology
Andrew Bell, University of Illinois
James Berger, University of California, Berkeley
Doug Black, University of California, Los Angeles
Michael Botchan, University of California, Berkeley
Chris Burge, Massachusetts Institute of Technology
Sean Burgess, University of California
Douglas Burks, Wilmington College
Mike Carey, University of California, Los Angeles
Mike Chandler, Laboratoire de Microbiologie et Génétique Moléculaire, CNRS Toulouse
Cris Cheney, Pomona College
Mitchell Chernin, Bucknell University
Christopher Cole, University of Minnesota
Joan Conaway, Stowers Institute for Medical Research
Victor Corces, Emory University
Jeffrey Corden, Johns Hopkins University School of Medicine
Don Court, Center for Cancer Research
Patrick Cramer, Gene Center Munich
James Dahlberg, University of Wisconsin-Madison
Seth Darst, The Rockefeller University
Stephanie Dellis, College of Charleston
Tom Dever, National Institutes of Health
Shery Dolhopf, Alverno College
Robert Drewell, Harvey Mudd College
Richard Ebright, Rutgers, The State University of New Jersey
Tom Eickbush, University of Rochester
Gary Felsenfeld, National Institutes of Health
Paul Foglesong, University of the Incarnate Word
Chris Francklyn, The University of Vermont
John Geiser, Western Michigan University
Raymond Gesteland, The University of Utah
Grace Gill, Tufts University
Rick Gourse, University of Wisconsin-Madison
Paula Grabowski, University of Pittsburgh
Carol Gross, University of California, San Francisco
Jim Haber, Brandeis University
Charlotte Hammond, Quinnipiac University
Phil Hanawalt, Stanford University
Frank Healy, Trinity University
Robert Heath, Kent State University
Tamara Hendrickson, Wayne State University
Alan Hinnebusch, National Institutes of Health
Ann Hochschild, Harvard Medical School
Paul Huber, University of Notre Dame
Dottie Hutter-Lobo, Monmouth University
Richard Jackson, University of Leeds
Steve Jackson, Gurdon Institute, University of Cambridge
Rudolph Jaenisch, Massachusetts Institute of Technology
Lisa Johansen, University of Colorado, Denver
Roland Kanaar, Erasmus Medical Center
Gary Karpen, University of California, Berkeley
Mikhail Kashlev, Center for Cancer Research
Paul Kaufman, University of Massachusetts Medical School
Tom Kelly, Memorial Sloan-Kettering Institute
Bob Kingston, Harvard University
Christopher Korey, College of Charleston
Roger Kornberg, Stanford University
Doug Koshland, University of California, Berkeley
Justin Kumar, Indiana University
John Kuriyan, University of California, Berkeley
Josef Kurtz, Emmanuel College
Michael Lichten, Center for Cancer Research
Lasse Lindahl, University of Maryland

Gary Linquester, Rhodes College
John Lis, Cornell University
Bob Lloyd, University of Nottingham
John Lorsch, Johns Hopkins University School of Medicine
Sue Lovett, Brandeis University
Karolin Luger, Colorado State University
Peter B. Moore, Yale University
Margaret Olney, Saint Martin's University
Jennifer Osterhage, Hanover College
Anthony Ouellette, Jacksonville University
William J Patrie, Shippensburg University
Tanya Paull, University of Texas at Austin
Stephen Pelsue, University of Southern Maine
Eric Phizicky, IBM Life Sciences Discovery Center, University of Rochester
Marie Pizzorno-Simpson, Bucknell University
Anna Marie Pyle, Yale University
Venki Ramakrishnan, MRC Laboratory of Molecular Biology
John Rebers, Northern Michigan University
Timothy J. Richmond, Swiss Federal Institute of Technology Zurich
Don Rio, University of California, Berkeley
Jeff Roberts, Cornell University
Peter Sarnow, Stanford University
Paul Schimmel, The Scripps Research Institute
Rey Sia, The College at Brockport, SUNY
Erik Sontheimer, Northwestern University
Forrest Spencer, Johns Hopkins University School of Medicine
Stephen Spiro, University of Texas at Dallas
Kevin Sullivan, National University of Ireland, Galway
Ben Szaro, State University of New York, Albany
Song Tan, Pennsylvania State University
David Tollervey, University of Edinburgh
Deborly Wade, Central Baptist College
Gabriel Waksman, University College London
Tracy Ware, Salem State College
Jon Warner, Albert Einstein College of Medicine, Yeshiva University
Alan Weiner, Washington University
Stephen West, London Research Institute
Jamie Williamson, The Scripps Research Institute
John Wilson, Michigan State University
Fred Winston, Harvard Medical School
Kate Leslie Wright, Rochester Institute of Technology
Stephen Wright, Carson-Newman College
Ramakrishna Wusirika, Michigan Tech University
Philip Zamore, University of Massachusetts Medical School
Ana Zimmerman, College of Charleston

CONTENTS IN OVERVIEW

CONTENTS IN FULL

ABBREVIATIONS

4E-BP	eIF4E-binding protein
53BP1	p53 binding protein 1
A	adenine
A	adenosine
A	aminoacyl
Å	angstrom
AAA+	ATPases associated with a variety of cellular activities
AAV	adeno-associated virus
aCGH	array comparative genomic hybridization
acetyl-CoA	acetyl coenzyme A
ADAR	adenosine deaminase that acts on RNA
ADP	adenosine diphosphate
AFM	atomic force microscopy
AIDS	acquired immunodeficiency syndrome
ALT	alternative lengthening of telomeres
AMP	adenosine monophosphate
APA	alternative polyadenylation site
APC	anaphase-promoting complex
ARE	AU-rich element
$Asn\text{-}tRNA^{Asn}$	asparagine-bound tRNA species
ATM	ataxia telangiectasia mutated
ATP	adenosine triphosphate
ATR	ATM related
ATRIP	ATR-interacting protein
attB	attachment site bacteria
attL	attachment site left
attP	attachment site phage
attR	attachment site right
BAC	bacterial artificial chromosome
bHLH	basic region-helix-loop-helix
BIR	break-induced replication
BLAST	Basic Local Alignment Sequence Tool
bp	base pairs
BRCA1	breast cancer type 1 susceptibility protein
BRCA2	breast cancer type 2 susceptibility protein
BRE	TFIIB recognition element
BSA	bovine serum albumin
bZIP	basic region-leucine zipper
C	cytidine
C	cytosine
C	carbon
Ca	calcium
Caf1	chromatin assembly factor
CAK	Cdk-activating kinase
cAMP	cyclic adenosine monophosphate
CAP	catabolite activator protein

CAR	cohesin-associated region
CBP	CREB-binding protein
CDE	centromere DNA element
Cdk	cyclin-dependent kinase
CENP-A	centromere protein-A
CFP	cyan fluorescent protein
ChIP	chromatin immunoprecipitation
CKI	cyclin kinase inhibitor
CLIP	cross-linking immunoprecipitation
CLIPS	chaperone-linked protein synthesis
CMCT	1-cyclohexyl-(2-morpholinoethylcarbodiimide metho-p-toluene sulfonate
CNV	copy number variant
CoA	Coenzyme A
CPE	cytoplasmic polyadenylation element
CRP	cAMP receptor protein
CsCl	cesium chloride
CSSR	conservative site-specific recombination
CTD	C-terminal domain
CTP	cytidine triphosphate
D	diversity segment
D-loop	displacement loop
Da	dalton
DAPI	4′-6′-diamidino-2-phenylindole
DARS	DnaA reactivating sequences
datA	DnaA titrator
ddNTP	dideoxynucleotide
DIC	differential interference contrast
DNA	deoxyribonucleic acid
DNA-PK	DNA-protein kinase
dNTP	deoxynucleotide triphosphate
DPE	downstream promoter element
dsRBD	double-stranded RNA-binding domain
DTT	dithiothreitol
E	exit
ECFP	enhanced cyan fluorescence protein
ECS	editing site complementary sequence
EFG	elongation factor
EGF	epidermal growth factor
eIF2	eukaryotic initiator
EJC	exon junction complex
EMS	ethyl methanesulfonate
EMSA	electrophoretic mobility shift assay
eRNA	enhancer RNA
ENU	ethylnitrosourea
ERAD	endoplasmic reticulum-associated degradation
ES	embryonic stem (cells)
ESE	exonic splicing enhancer
ESI	electrospray ionization
EYFP	enhanced yellow fluorescent protein
FACT	facilitates chromatin transcription

FAD	flavin adenine dinucleotide
FIG	field inversion gel
FIS	factor for inversion stimulation
FISH	fluorescence *in situ* hybridization
FLIP	fluorescence loss in photobleaching
FMN	flavin mononucleotide
FRAP	fluorescence recovery after photobleaching
FRET	fluorescence resonance energy transfer
G	guanine
G	guanosine
G0	G zero
G1	gap phase 1
G2	gap phase 2
GAP	GTPase-activating protein
GC	gas chromatography
GDP	guanosine diphosphate
GEF	guanine-nucleotide exchange factor
GFP	green fluorescent protein
GGR	global genomic repair
Gln-tRNA^{Gln}	glutamine-bound tRNA species
GlyRS	glycyl-tRNA synthetase
GPI	glycosylphosphatidylinositol
GST	glutathione-*S*-transferase
GTP	guanosine triphosphate
GWAS	genome-wide association studies
H	hydrogen
HAT	histone acetyltransferase
hcRNA	heterochromatic RNA
HCV	Hepatitis C virus
HDAC	histone deacetylase
hGH	human growth hormone
HGPS	Hutchinson–Gilford progeria syndrome
Hh-C	C-terminal domain of the Hedgehog protein precursor
Hh-N$_p$	N-terminal domain of the Hedgehog protein precursor
HisRS	histidyl-tRNA synthetase
HIV	human immunodeficiency virus
HLA	human leukocyte antigen
HMT	histone methyltransferase
HPLC	high-pressure or high-performance liquid chromatography
HSE	heat-shock element
HSF	heat-shock factor
IEF	isoelectric focusing
IF	initiation factor
IGF	insulin-like growth factor
IHF	integration host factor
IleRS	isoleucyl-tRNA synthetase
INR	initiator element
Int	integrase
iPS	induced pluripotent stem
IRE	iron response element
IRES	internal ribosome entry site

IRIF	ionizing-radiation-induced foci
IRP	iron regulatory protein
IS	insertion sequence
ISE	intronic splicing enhancer
ITAF	IRES-transacting factors
ITC	isothermal titration calorimetry
J	joining segment
JAK	Janus kinase
JmjC	Jumonji domain-containing
K48-linked	linked to lysine 48
K63-linked	linked to lysine 63
K_a	equilibrium association constant
K_d	equilibrium dissociation constant
K_{eq}	equilibrium constant
kb	kilobase pair (1000 base pairs)
kDa	kilodalton
KH	Hn RNP K homology
LC	liquid chromatography
LDL	low-density lipoprotein
Lk	linking number
lincRNA	long intergenic noncoding RNA
LINE	long interspersed element
LNA	locked nucleic acid
LOD	log of the odds
LTR	long terminal repeat
LUCA	last universal common ancestor
M	mitotic
MALDI	matrix-assisted laser desorption/ionization,
MASV	mobile element associated structural variant
Mbp	million base pairs
miRNA	microRNA
MRN	MRE 11-RAD50-NBS1
mRNA	messenger RNA
MRX	Mre11-Rad50-Xrs2
MS/MS	tandem mass spectrometry
MuLV	murine leukemia virus
myristoyl-CoA	myristoyl coenzyme A
N	nitrogen
NAC	nascent chain-associated complex
NaCl	sodium chloride
NAD	nicotinamide adenine dinucleotide
NAD^+	nicotinamide adenosine dinucleotide (oxidized form)
NADH	nicotinamide adenosine dinucleotide (reduced form)
NADP	nicotinamide adenosine dinucleotide phosphate
$NADP^+$	nicotinamide adenosine dinucleotide phosphate (oxidized form)
NADPH	nicotinamide adenosine dinucleotide phosphate (reduced form)
NCBI	National Center for Biotechnology Information
NER	nucleotide excision repair
NGD	no-go decay

NextGen	Next Generation
NF-κB	nuclear factor-κB
NHEJ	non-homologous end joining
NLS	nuclear localization signal
NMD	nonsense mediated decay
NMIA	*N*-methylisotoic anhydride
NMR	nuclear magnetic resonance
NO	nitric oxide
NOR	nucleolar organizer region
NPC	nuclear pore complex
NSD	non-stop decay
NTD	N-terminal domain
NTF2	nuclear transport factor 2
NTP	nucleoside triphosphate
NURF	nucleosome-remodeling factor
O	oxygen
***O*-GlcNAcase**	O-acetylglucosaminidase
OGT	*O*-GlcNAc transferase
OH	hydroxyl group
OMIM	Online Mendelian Inheritance in Man
ORC	origin recognition complex
ORF	open reading frame
P	peptidyl
PABP	poly(A)-binding protein
PADI4	peptidyl arginine deaminase
PAGE	polyacrylamide gel electrophoresis
PA-GFP	photoactivatable-GFP
PALB2	partner and localizer of BRCA2
PALM	photoactivated localization microscopy
PAR-CLIP	photoactivatable-ribonucleoside-enhanced cross-linking and immunoprecipitation
PARG	poly- adenosine diphosphate ribose glycohydrolase
PARP	poly-adenosine diphosphate ribose polymerase
PAZ	PIWI/Argonaute/Zwille
PCAF	CBP-associated factor
PCNA	proliferative cell nuclear antigen
PCR	polymerase chain reaction
PDB	Protein Data Bank
PDI	protein disulfide isomerase
PEP	phosphoenolpyruvate
PIKK	phosphoinositide-3-kinase–related protein kinase
piRNAs	Piwi-interacting RNAs
PITC	phenylisothiocyanate
PNPase	polynucleotide phosphorylase
pol I	DNA polymerase I
pol III	DNA polymerase III
pol α	DNA polymerase α
pol δ	DNA polymerase δ
pol ε	DNA polymerase ε
poly(A)	polyadenosine
POMC	proopiomelanocortin

PPB	preprophase band
PRC2	Polycomb repressive complex 2
pre-RNA	precursor RNA
preRC	prereplication complex
pri-miRNA	miRNA primary transcript
PSF	polyadenylation specificity factor
PTB	polypyrimidine tract binding protein
PTBP1	polypyrimidine tract-binding protein 1
PTBP2	polypyrimidine tract-binding protein 2
PTC	premature termination codon
PTGS	post-transcriptional gene silencing
qRT-PCR	quantitative reverse transcription-polymerase chain reaction
RAC	ribosome-associated complex
RACE	rapid amplification of complementary DNA ends
rasiRNA	repeat associated small interfering RNAs
RBD	RNA-binding domain
RdRp	RNA-dependent RNA polymerase
RF	release factor
RFC	replication factor C
RIDA	regulatory inactivation of DnaA
RIP	ribosome-inactivating protein
RISC	RNA-induced silencing complex
RITS	RNA-induced initiation of transcriptional silencing
RNA	ribonucleic acid
RNAi	RNA interference
RNP	ribonucleoprotein
RPA	replication protein A
RRM	RNA recognition motif
rRNA	ribosomal RNA
RSS	recombination signal sequence
rut	rho utilization
S	Svedberg
SAM	S-adenosylmethionine
SDS	sodium dodecyl sulfate
SDSA	synthesis-dependent strand annealing pathway
SDS-PAGE	sodium dodecyl sulfate-polyacrylamide gel electrophoresis
SECIS	selenocysteine insertion sequence
SEM	scanning electron microscope
SepRS	specialized class II synthetase
shRNA	small hairpin RNA
SILAC	stable isotope labeling with amino acids in culture
SINE	short interspersed element
SIR	silencing information regulator
siRNA	small inhibitory RNA
SKY	spectral karyotyping
SL RNA	spliced leader RNA
SMC	structural maintenance of chromatin
snoRNA	small nucleolar RNA
SNP	single nucleotide polymorphism
snRNA	small nuclear RNA
snRNP	small nuclear ribonucleoprotein

SPR	surface plasmon resonance
SRA	steroid receptor RNA activator
SRF	serum response factor
SRP	signal recognition particle
sRNA	small RNA
SSB protein	single-stranded binding protein
SSLP	simple sequence length polymorphism
STAT	signal transducers and activators of transcription
STED	stimulated emission depletion
STS	sequence tagged sites
T	thymidine
T	thymine
TAF	TBP-associated factor
TAL	transcription activator-like
TALE	transcription activator-like effector
TALEN	transcription activator-like effector nuclease
TBP	TATA binding protein
TCR	transcription-coupled repair
TEM	transmission electron microscopy
ter	terminus region
TERT	telomerase reverse transcriptase
TIR	Terminal Inverted Repeat
TLC	thin-layer chromatography
TLS	translesion synthesis
TPRT	target-primed reverse transcription
TR	terminal inverted repeat
tRNA	transfer RNA
Tw	twist (the number of turns in the DNA helix in a given fragment of DNA)
U	uridine
UAS_G	Upstream Activating Sequence for Gal
UBL	ubiquitin-like protein
UCE	upstream control element
UDP	uridine diphosphate
UEV	ubiquitin E2 variant
uORF	upstream open reading frame
UPR	unfolded protein response
UTP	uridine triphosphate
UTR	untranslated region
UV	ultraviolet
V	variable segment
VSR	viral suppressor of RNA silencing
Wr	Writhe
YAC	yeast artificial chromosome
YFP	yellow fluorescent protein
γ-TuRC	γ-tubulin ring complex

Genomes and the flow of biological information

INTRODUCTION

The biological world is a maze of interconnections on a multitude of scales: atoms join to form molecules – molecules cluster to form cells – cells interact to form tissues – and tissues aggregate to form an organism. The interplay continues beyond the organism, as organisms form populations, populations inhabit ecosystems, and ecosystems unite to form the world around us.

Through all the layers of connectivity runs a common thread: the communication and onward passage of information – from cell to cell, from organism to organism and, ultimately, from generation to generation. This information is stored, at the most fundamental level, as our **genome** in each living cell in our body. But how does this information – a static repository of data – come alive to direct the processes that constitute life?

The answer lies in the concerted action of molecular components, which cooperate in a series of ingenious processes to bring the genome to life. These components and processes lie at the heart of molecular biology, and are the key players in the chapters that follow.

Before embarking on our journey to discover these components and processes, however, we begin by considering some of the fundamental biological themes on which the study of molecular biology relies. You may be familiar with some of these themes, and the concepts and ideas introduced, from earlier studies. If so, look on this chapter as a refresher before later chapters lead you into more unfamiliar – but ultimately fascinating – topics.

1.1 THE ROOTS OF BIOLOGY

The diversity of life is unified by some common themes

Life on Earth is remarkably diverse, ranging from the relatively simple unicellular organisms, such as bacteria, to more complex multicellular organisms, such as plants and humans. Despite their diversity at the macroscopic level, where distinctive features of organisms are readily apparent with the naked eye (as illustrated by Figure 1.1), the core molecular features of all organisms are remarkably similar. Indeed, we can think of life on Earth as being unified by a number of key themes.

- First, a living organism must be distinct in some way from its environment, that is, it must be defined by a physical barrier, which serves to separate 'organism' from 'environment'. This separation allows the internal environment of an organism to be carefully regulated, and ultimately distinguishes 'self' from 'non-self'.

- Second, a living organism must be able to store its genome in a stable way, and also have a way of using this information to determine its characteristic features – its structural composition, how it functions, and so on.
- Third, a living organism must be able to reliably replicate and pass information from one generation to the next. **Replication** lies not only at the heart of the propagation of a **species** from generation to generation, but also at the heart of the growth and repair of a multicellular organism within a single generation. Without replication, not only would an organism be unable to reproduce to yield offspring, but a multicellular organism could not develop from an embryonic state to a fully formed individual. The processes of replication and transmission must remain faithful to the entity being replicated, just as a recipe – itself an information store – ensures that a given dish can be prepared in a consistent way time after time.
- Finally, living organisms require a source of energy from their surroundings to grow and reproduce; this energy is used to drive the biological processes that keep an organism alive. There are substantial differences in how energy is harvested by various organisms; for example, plants get their energy from the sun whereas humans get energy from food. Nevertheless, the core molecular mechanisms used by cells in all organisms to grow and propagate are remarkably similar (that is, they are evolutionarily conserved). For example, the same metabolic cycles are used in all organisms to transform sugars into the energy currency of the cell, adenosine triphosphate (ATP).

Figure 1.1 The life forms on Earth are diverse and interconnected. (a) A monarch butterfly on a milkweed plant, (b) a California Alligator Lizard carrying multiple ticks, (c) a clownfish surrounded by the tentacles of a sea anemone, (d) an oxpecker bird sitting on an African buffalo, and (e) a fungus on a tree.

In modern-day organisms, the physical barrier that separates individuals from their surroundings (and so defines 'self') is a lipid-based membrane, while a molecular species known as **nucleic acid** stores the biological information, and allows for ready replication of this information (thus conferring 'self' identity). Figure 1.2 depicts how these two basic molecular features – the lipid membrane surrounding the nucleic acid – come together to form the cell, which is the building block of organismal life. Some organisms are composed of a single cell (unicellular) whereas others are formed of many cells (multicellular). However, the cell is not the only fundamental building block of life; common building blocks exist at the molecular level too.

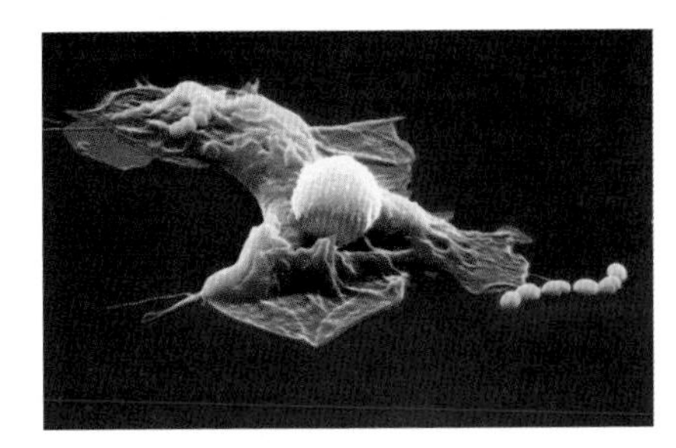

Figure 1.2 All cells share the same basic components. False-colored scanning electron micrograph showing chains of *Streptococcus* bacteria (yellow) and two human cells: a macrophage (gray) and a lymphocyte (pink). Both of the human cells are part of the immune response, and act together to eliminate the bacteria. Despite their different sizes, all three types of cell share the basic components of a cell: a membrane that isolates each individual cell; a genome that is the 'blueprint' of each cell; a means to divide and reproduce; and a means by which to utilize energy.

Electron micrograph courtesy of James A Sullivan.

Living organisms are constructed from common molecular building blocks

It is thought that the entire diversity of modern-day life derives from a common original life form often referred to as the last universal common ancestor (LUCA) or **progenote**, whose molecular components have been conserved from generation to generation, and from species to species, as different organisms have evolved. In support of this, when we compare modern organisms, we find that the core building blocks of all organisms are the same.

Figure 1.3 depicts the four basic classes of molecules from which living organisms are constructed. These are:

- nucleic acids
- proteins
- lipids
- carbohydrates.

These molecules reflect a further unifying theme – of large biological molecules being constructed from smaller, repeated subunits.

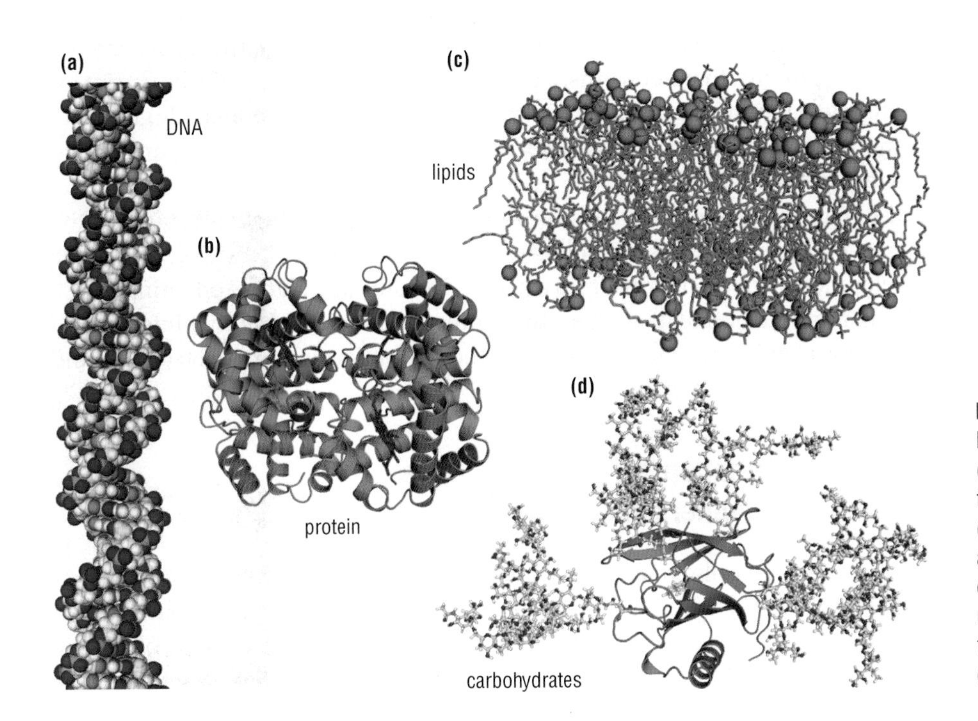

Figure 1.3 All cells are composed of four basic classes of molecules. (a) Nucleic acids (DNA and RNA) store information and serve as the interpreters of these instructions for life. (b) Proteins carry out most of the structural and functional tasks. (Protein Data Bank (PDB) code 1A00). (c) 'Water-fearing' lipids form the membranes that comprise the physical barrier to the outside of the cell. (d) Carbohydrates provide energy and often adorn extracellular proteins.

Nucleic acids typically have a simple, linear structure composed of monomer building blocks (the nucleotides); they contain the instructions for life, and thus for the propagation of life. The information stored in these molecules directs the synthesis of proteins (a 'coding' role), and, in so doing, specifies the structure and function of the cell, as noted later. In addition, nucleic acids play essential non-coding roles within the cell as functional components of molecular machines and by regulating how genomic information is used.

Proteins are also linear polymers, composed of distinct building blocks known as **amino acids**, and typically fold into complex three-dimensional structures. Proteins are often referred to as the workhorses of the cell, carrying out most of the critical structural and functional tasks.

Lipids are 'water-fearing' molecules that self-assemble to form the membranes that present a physical barrier to the exterior milieu, thus defining self. Lipids are also constructed from repeated subunits – multiple fatty acid molecules join together to form the lipid.

Carbohydrates are 'water-loving' molecules that play a wide range of roles in the cell, from providing energy to increasing the solubility of otherwise insoluble proteins. Carbohydrates continue to reflect the theme of large molecules being composed of smaller subunits: they are polymeric molecules constructed from simple sugar monomers.

➔ We learn more about the structure and chemical nature of these four classes of biological molecules in Chapter 2.

From these four components are built the macromolecular machines that are responsible for all cellular transactions – from copying the biological information stored in nucleic acids, to producing functional proteins, to harvesting energy from the environment.

The flow of biological information supports the maintenance of life

The title of this chapter refers to the flow of biological information – a phenomenon that is made possible by the components and processes that comprise 'molecular biology.' As we have mentioned earlier, storage and passage of information is one of the unifying characteristics of living organisms. But why is the flow of biological information necessary? Ultimately, a flow of information is needed to ensure that the offspring of an organism remains faithful to its parents in terms of the characteristics it exhibits. In other words, a parent must somehow act as the blueprint for its offspring. Without such a flow of information, a cat could not reliably give birth to another cat, and a human could not reliably give birth to another organism that was recognizably human.

The maintenance of life relies on the process of reproduction – both in the context of the growth of a specific tissue through rounds of cell division, and in the development of new offspring from a parent. In both cases, for the offspring to be a faithful reproduction of its parents, there must be a means of storing information and using it, and a mechanism through which this information can be copied and inherited by the next generation.

Heritable information is stored in a simple polymer of monomer building blocks

Information in all cells is stored in a repeating polymer structure called deoxyribonucleic acid (DNA), which is easily copied and transmitted to subsequent generations through processes we will explore later in the book. DNA is a type of nucleic acid, and is composed of four building blocks known as **nucleotides**: guanosine

(G), adenosine (A), thymidine (T), and cytidine (C). These nucleotides are strung together to form an extended linear polymer. The resulting strings of letters specify our genetic identity: they encode all functional molecules that must be synthesized for the cell - and ultimately the organism - to survive and propagate.

A particularly important chemical feature of these four distinct nucleotides is that the faces of A and T fit together, as do those of G and C. We say their shapes are complementary. When two different nucleic acids have complementary sequences - that is, where A on one strand is matched by T on another, and where C on one strand is matched by G on the other - the close association of these pairs of nucleotides allows the two sequences to form extended double-stranded structures with a characteristic double-helical conformation, as illustrated in Figure 1.4. This double-helical structure is the basic form in which the genetic information is stored.

Each DNA strand in a double helix has directionality: one end of the strand is denoted 3′ and the other 5′. When two complementary strands come together they do so in an antiparallel fashion - that is, the 3′ end of one strand aligns with the 5′ end of the other, and vice versa. (In essence, the two strands lie 'head to toe'.)

Duplication of the information stored in DNA is facilitated by the correspondence in complementary shapes - a complementary, fully informational strand can be made by separating the two strands and using the sequence of nucleotides in each of the two original strands to direct the synthesis of two new strands, as depicted in Figure 1.5. Such strand replication follows the same rules of complementarity described earlier: an A in one of the original strands directs the addition of a T in a new strand (and vice versa), and a C in the original strand directs the addition of a G in the new strand (and vice versa). It is important to note the word 'complementary': when a nucleic acid strand is replicated, the strand produced as a result is not identical to the original strand, but is complementary to it.

DNA is a very stable polymer, making it the nearly ubiquitous material for the storage of information in modern biology. However, the chemically similar but less stable nucleic acid, ribonucleic acid (RNA), is occasionally used for information storage. (This does not mean that RNA is a mere bit-player on the biological stage: throughout this book we will see that RNA has a myriad of functional roles in the workings of an organism.)

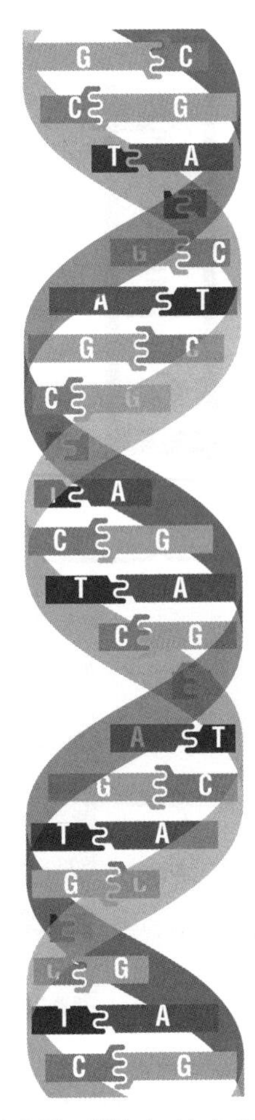

Figure 1.4 The DNA double helix is a repeating polymer that stores information in cells. DNA is composed of the nucleotide building blocks guanosine (G), adenosine (A), thymidine (T), and cytidine (C). The distinct feature that G pairs with C and A pairs with T allows a strand of DNA to be copied.

1.2 THE GENOME: A WORKING BLUEPRINT FOR LIFE

The sequence of nucleotides in DNA that comprises the genetic makeup of an organism is called the **genome**. This repository stores all information needed to specify cellular function and can be considered a blueprint for life. In the past 15 years, researchers across the world have determined the complete genome sequence of numerous organisms, from thousands of bacterial species to multiple individual humans. These approaches have generated great excitement in all areas of biology and medicine as analysis of these sequences has greatly increased our understanding of cellular functions and evolution.

Transmission and maintenance of the genome is essential for life

As we noted previously, reproduction by all organisms depends on their ability to make a copy of their blueprints for life to hand down to their progeny. At the cellular level, the copying of the genetic information (replication), the separation (or '**segregation**') of the copies, and transfer into the cells of the next generation is required even if the recipient is to remain part of the same organism - for example, a new skin cell, or muscle cell. We refer to this copying and segregation process as the replication and transmission of the genome. In molecular terms, this means that the nucleic acid, typically DNA, must be copied in its entirety, without many mistakes being made, in order to maintain the integrity of the genome - that is, the accuracy of the information contained within it. The process of segregation also must happen correctly, so that each daughter cell receives the complete genome.

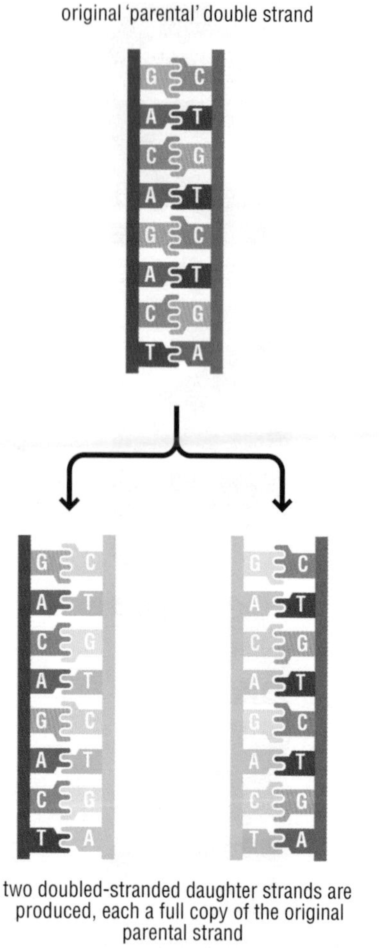

The molecular processes involved in transmission and maintenance of the genome are highly conserved from bacteria to humans. DNA is first copied or replicated by core machinery that separates the complementary strands of the double helix and simultaneously makes copies of each individual strand. The molecular machine, or enzyme, responsible for synthesizing the new DNA copies is known as **DNA polymerase**, and as we shall see, is very highly conserved across different organisms. Following this DNA replication process, the duplicated copies of the DNA are brought together to ensure that when the actual physical division of the cell occurs, the genome will be accurately partitioned (or segregated) between the two progeny cells; this process of segregation is illustrated in Figure 1.6. Throughout each of these processes, there are elaborate mechanisms in place to ensure that replication and segregation are extremely accurate.

original 'parental' double strand

two doubled-stranded daughter strands are produced, each a full copy of the original parental strand

Figure 1.5 The process of replication is common to all organisms. The complementary nature of the two DNA strands allows the DNA to be copied.

The genome is composed of genes and intergenic regions

The genome - the total DNA content of the cell - contains the instructions for cellular life. Within the genome, much of the critical information is found in discrete regions referred to as **genes**. A gene is typically defined as a region of DNA that controls a discrete, hereditary characteristic, and as such usually specifies the production of a functional product (a protein or RNA molecule). Different types of sequence are typically found within a given gene: some of the sequence directly encodes the product of the gene, whereas other sequences within a gene are important for specifying how and when to make the product. These regions are often referred to as coding and regulatory regions, respectively. Genes can vary in length from fewer than 100 nucleotides for certain functional RNAs, for example, to 2.4 million nucleotides for the human dystrophin gene (which is defective in muscular dystrophies). Typically, the DNA sequences between the genes are defined as intergenic regions.

It is worth noting that we give 'typical' definitions of a gene and intergenic region in the previous paragraph. In reality, giving a clear-cut definition for a gene - something that has never been as straightforward as it might appear - is becoming increasingly difficult. Why is this? As more genomes are sequenced and studied in detail, we are finding that far more of the genome appears to be expressed (that is, actively 'used' in some way) than previously anticipated. So, we may need to broaden our view of what a 'gene' really is to reflect the ever-expanding proportion of the genome that is active - beyond those areas that simply code for a single functional product.

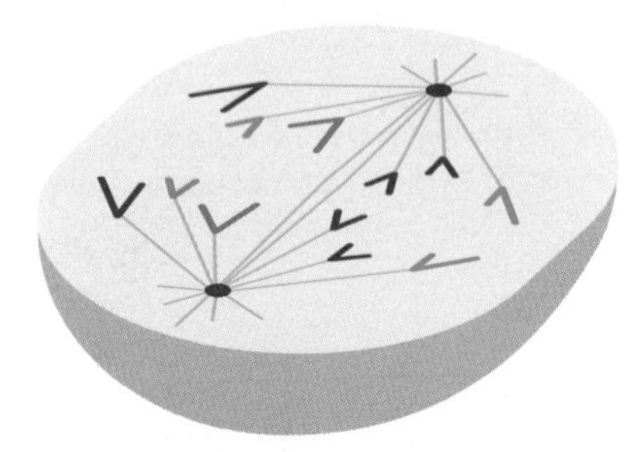

Figure 1.6 Replicated chromosomes are partitioned into two progeny cells by a process called segregation. To ensure accurate partitioning, the replicated chromosomes (having the same shape and color in this figure) are first brought together before they are pulled away into the progeny cells.

Going further, recent innovations are allowing for breathtaking amounts of sequence information to be rapidly generated and this new information coupled with functional analysis is changing our view of biology. We are now asking questions such as 'How discrete an element is a gene?' and 'Are all RNA transcripts worthy of being called a gene?' It is an exciting time to be a molecular biologist and this book sets the stage for understanding new discoveries and how they impact our understanding of evolution and biology as a whole.

➲ We learn about replication in more detail in Chapter 6 and about chromosomal segregation in Chapter 7.

The genome is physically organized into chromosomes

➲ Chromosomal features are the focus of Chapter 4.

The large amount of DNA that constitutes a genome must be packaged in an organized way that facilitates the various key events in the life of a cell, including the replication of the DNA and the expression of the genes. The individual genes are arranged in a linear array, either densely or less densely spaced with intervening intergenic DNA. All of this genomic DNA is packaged into one or more functional units called **chromosomes**. Some organisms have multiple, independent linear chromosomes, while others have a single, circular chromosome.

In eukaryotic chromosomes, the DNA is wrapped tightly around packing proteins (referred to as histones) to assume a highly ordered three-dimensional structure, as depicted in Figure 1.7. This highly compacted form of DNA is optimal for storage as it is very space efficient. However, highly compacted DNA does not facilitate information retrieval. Instead, specialized mechanisms drive the opening of this tightly wound package to allow access to the DNA strands by the various molecular components needed to mediate gene expression.

In addition to chromosomes, some cells have **plasmids**, which are extra-chromosomal pieces of DNA that can also carry important genetic information.

Given that the genome contains all of the genes of an organism, how then are the genes assembled? Are the genes placed in a particular order or in a particular location on the chromosome? Are they densely or sparsely distributed? What can we learn about an organism by examining overall DNA composition and structure? As we shall see, the overall composition of genomes varies widely across biology. For example, in small genomes, such as that of the bacterium *Escherichia coli*, which has 4.6 million base pairs (Mbp) of DNA, the genes are rather close together, sometimes even overlapping one another to minimize overall genome size. By contrast, the human genome comprises 3200 Mbp of DNA and is composed predominantly (98%) of intergenic regions that do not appear to contain 'typical' gene units. The composition of these intergenic regions is not uniform nor is the distribution of genes throughout the genome. As more genomes have been sequenced, and more studies of **gene expression** are performed, it has become very clear that we still have much to learn about the impact of the overall genome structure on genome function.

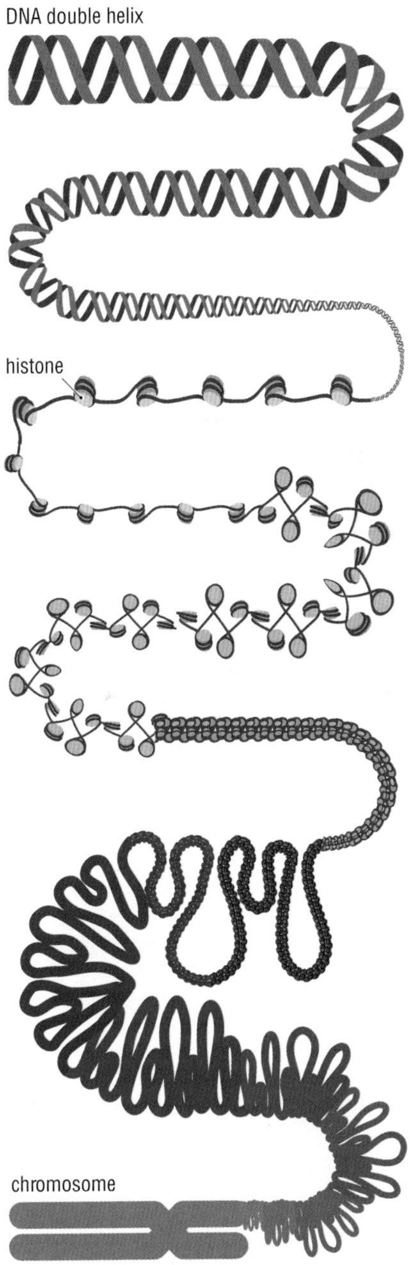

DNA double helix

histone

chromosome

Figure 1.7 DNA is packaged into chromosomes. The double-stranded DNA strands are wound around packing proteins (histones in eukaryotes) and then further wound to assume a highly ordered and compact three-dimensional structure.

Gene number and arrangement vary greatly among organisms

The number of 'typical' genes in an organism ranges widely – from 500 to 800 in some bacteria to a grossly estimated 30 000 in humans. Although single-celled bacteria generally have fewer genes than multicellular organisms, as shown in Figure 1.8, there is no simple correlation between genome size, gene number, and organism complexity. As an example, although zebrafish and lungfish appear to be very similar at the organismal level, their genome sizes are substantially different: the zebrafish has a genome of 2000 Mbp while that of the lungfish is 65 times larger with 130 000 Mbp. Furthermore, the nematode *Caenorhabditis elegans* and the fruit fly *Drosophila melanogaster* are certainly less complicated than humans, but possess only slightly fewer genes (around 20 000 in *C. elegans*, and around 15 000 in *D. melanogaster*). What are some potential explanations for this conundrum?

Approximate genome size and gene number for representative organisms

element	gene number	genome size
Mycoplasma genitalium	~480 genes	580 000 bp
Escherichia coli	4600 genes	4 600 000 bp
Saccharomyces cerevisiae	~ 6000 genes	12 100 000 bp
Schizosaccharomyces pombe	~ 5000 genes	14 000 000 bp
worm (*Caenorhabditis elegans*)	~23 000 genes	98 000 000 bp
fruit fly (*Drosophila melanogaster*)	~27 000 genes	130 000 000 bp
duckweed (*Arabidopsis thaliana*)	~29 000 genes	157 000 000 bp
zebrafish (*Danio rerio*)	~13 000 genes	2 000 000 000 bp
human (*Homo sapiens*)	~32 000 genes	3 200 000 000 bp
marbled lungfish (*Protopterus aethiopicus*)		130 000 000 000 bp
amoeba (*Amoeba dubia*)		670 000 000 000 bp

Figure 1.8 Approximate gene number and genome size for representative organisms.

First, it is clear that the number of different products (RNAs or proteins) that can be generated from a given gene is large. So, even though the genome of a given organism may feature a seemingly modest number of genes, the same genome may, in fact, encode a huge range of different functional products – more, perhaps, than a different organism that possesses a larger number of genes.

Further, it is clear that gene density varies widely from organism to organism, such that some genomes feature relatively more intergenic regions than others. But why is this important? Intergenic regions were previously thought to be little more than 'junk', acting solely as 'spacers' between genes. However, we now think that, far from being junk, many intergenic regions play critical roles in regulating gene expression. For example, we shall learn in Chapter 17 that much of the DNA of certain organisms is composed of **transposable elements** that can (or once could) move from place to place within a genome. While this DNA was for some time thought to be functionally unimportant, new information on its critical function in evolution and in gene expression is emerging. Further, despite earlier expectations that a majority of the human genome is functionally silent, recent studies indicate that a significant fraction of the human genome is actively being transcribed.

Therefore, it is becoming clear that the value of the genome, in terms of governing the processes that underpin biological complexity, comes from more than just the genes from which a genome is composed. Instead, it comes from a combination of gene number, the production of multiple gene products from the information carried in a single gene, and the regulatory activity of the genome's intergenic regions.

Molecular conservation is highlighted by DNA sequencing

We noted at the start of this chapter how the diversity of life is unified by some common themes, with commonalities extending from the molecular building blocks used to construct living organisms to the biological processes through which life is maintained. Direct proof of the remarkable extent of conservation in all cells comes from analysis of the genome of an organism.

With the advance of whole-genome sequencing has come new understanding of the extent to which commonalities exist between different organisms. The examination and comparison of the whole-genome sequences from different organisms has revealed marked similarities, which point, at least in part, to commonalities in the components and often the processes that operate at the molecular level. (This argument applies the logic that, if two genes have very similar sequences, then the gene products may also have a similar function.)

In this book we will focus on one particular area of cellular metabolism – on understanding key molecular mechanisms critical to the storage and copying of the genome and the expression of this information in its functional form.

Throughout the book, we will emphasize the commonalities of these mechanisms in all organisms, but also highlight instructive differences. The impact of whole-genome sequencing on our understanding of molecular evolution and function cannot be overstated, and promises to continue to reveal fundamental surprises about both biological commonality and diversity on Earth.

1.3 BRINGING GENES TO LIFE: GENE EXPRESSION

We have summarized how genetic information is captured by the sequence of nucleotides in DNA. But how is this information interpreted by the organism in which it is stored? How is it converted from a static sequence of nucleotides into dynamic biological activity, a process termed gene expression? We now begin our exploration of molecular biology to discover the molecular machinery and processes that bring the genome to life.

Genes encoded by DNA must be 'expressed'

While the blueprint of the cell is encoded in the DNA, the implementation of the blueprint is accomplished through the 'expression' of the genes into their functional forms. The gene products consist of two types of macromolecule, RNA and protein. These two molecules are the principal structural and functional components of the cell, and are responsible for reactions that harvest energy, synthesize cellular components, and facilitate the import and export of materials. In short, these molecules implement cellular function.

In order for the individual genes to be expressed, the packets of information need to be transformed into the specified products. It is this transformation that lies at the heart of this book, and is a process we will gradually unravel in the chapters that follow. Genomes are the blueprints for life, but it is through molecular biology – the array of molecular components and processes – that this blueprint is used to direct how life at the level of a given organism proceeds.

The first stage in the expression of all genes is the synthesis of a copy of the DNA region of interest in the form of RNA through a process called **transcription**. Once the RNA copy of the gene has been synthesized, it is either 'translated' into the corresponding protein sequence or used directly, such that the RNA molecule produced by transcription has its own specific biological function, as illustrated in Figure 1.9. Let us now consider each of these processes (transcription and **translation**) in a little more detail.

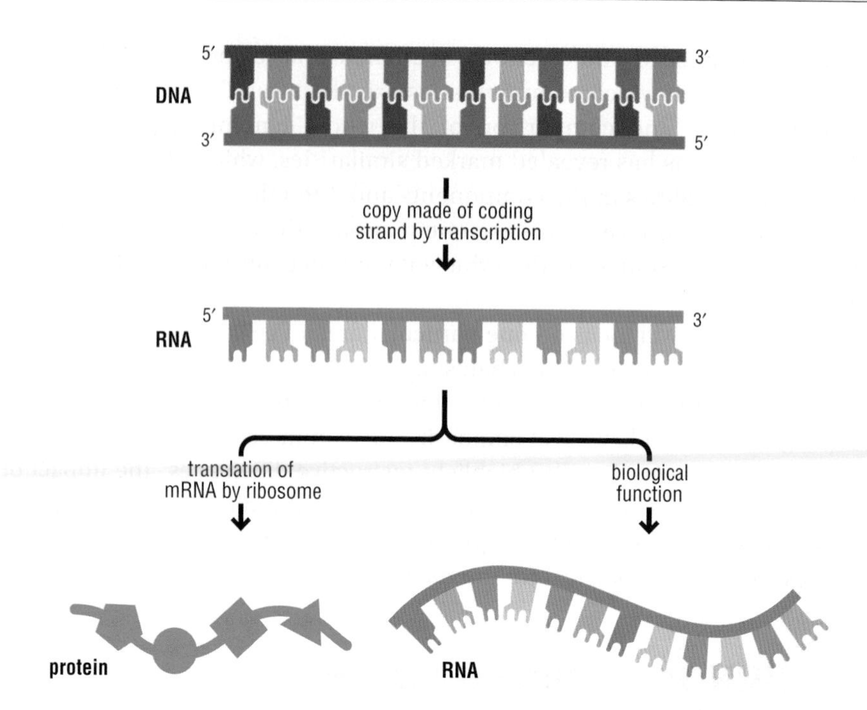

Figure 1.9 **Two functions of RNA.** RNA copies of DNA can serve as messenger RNAs (mRNAs) to direct the synthesis of a protein (left), or they can have intrinsic functions (right).

Transcription is the first step in gene expression for the production of both non-coding RNAs and proteins

The molecular details of the transcription process are described in greater detail in Chapter 8.

The process of copying the DNA sequence of interest into the corresponding RNA is referred to as transcription. The molecular machine that is responsible for carrying out transcription is called **RNA polymerase**, and is closely related to the DNA polymerase that is responsible for replicating the genome. The start point and end point of transcription along a particular DNA template strand are specified by sequences within the DNA template that are interpreted by RNA polymerase as a signal to start or stop.

The RNA polymerase catalyzes the synthesis of an RNA strand by joining together nucleotide monomers (single nucleotides) in a sequence dictated by the DNA template that the RNA polymerase is 'reading'. The process of joining together nucleotides to form a chain-like polymer is an example of a polymerization reaction. How are nucleotides added in the correct sequence to the growing RNA strand by RNA polymerase? The physical basis for the polymerization reaction is the complementarity in shape between a nucleotide on the template DNA and a nucleotide that becomes part of the growing RNA chain, as depicted in Figure 1.10. This complementarity mirrors that seen for the process of DNA replication, with one important difference: as in DNA replication, a C on the DNA strand directs the addition of a G on the growing RNA strand and vice versa; a T on the DNA strand directs the addition of an A on the growing RNA strand, but an A on the DNA strand directs the addition of a U (uridine, a nucleotide unique to RNA) on the growing RNA strand.

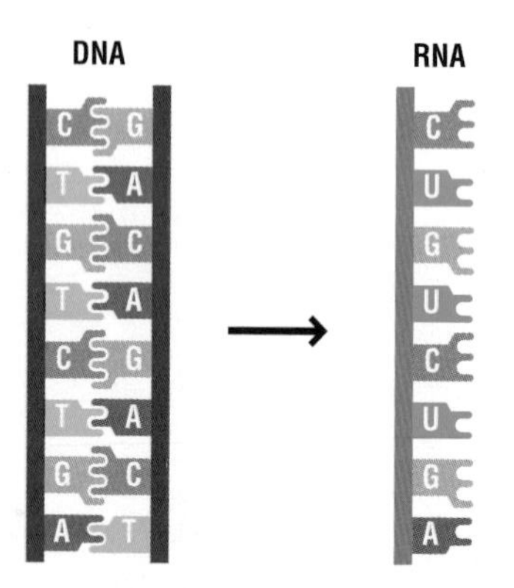

Figure 1.10 **The process of transcription is common to all organisms.** The complementary nature of the two DNA strands allows the DNA to be copied into a complementary strand of RNA. When DNA is copied into RNA, a uridine (U) (rather than a T) is incorporated opposite A.

The initial product of a transcription event is referred to as the primary transcript. For some genes, the final functional gene product is an RNA species derived from the primary RNA transcript; such RNA species often possess regulatory functions. For other genes, the RNA product is an intermediate that will be used to direct the synthesis of a protein product (called translation, as discussed in the following subsection). These two classes are commonly referred to as non-coding and coding

RNAs, respectively; the key distinction is that non-coding RNAs are not translated to give a protein product, whereas coding RNAs are.

This versatility of RNA is thought to have been central to the origin of life, and is still key to the functioning of modern-day cells. These ideas will be discussed more thoroughly in Chapters 2, 10, 11 and 13. For both non-coding and coding RNAs, the primary transcript RNA is typically subjected to a series of events that trim away and modify the RNA in a refinement process broadly referred to as RNA processing, to yield the final RNA molecule.

➔ The post-transcriptional RNA processing events are described in greater detail in Chapter 10.

The RNA sequence is next 'translated' to make functional proteins

Many genes encode proteins, which are composed of linear strings of 20 distinct building blocks, the amino acids. Proteins are encoded by the sequence of nucleotides comprising the RNA transcript template. Proteins built from this diverse set of building blocks (relative to the four quite similar nucleotide building blocks of DNA and RNA) can form a wide range of structures, as illustrated in Figure 1.11, that in turn perform quite diverse roles in the cell.

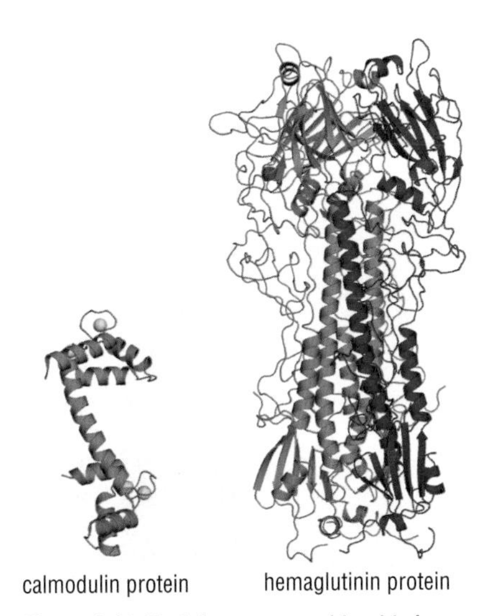

Figure 1.11 Proteins vary considerably in shape, size and function. Proteins can range in size from only a few amino acids to thousands of amino acids and have roles in every cellular process, ranging from roles as signaling molecules to receptors and structural scaffolds in the cell. In the examples shown here, the 148 amino acid calmodulin protein (PDB 1UP5) binds calcium (indicated by yellow balls), and the complex formed by the three proteins comprising the hemagglutinin protein (PDB 3HMG) mediates binding of influenza virus to the mammalian host cell and ultimately membrane fusion and entry.

The nucleic acid code for protein synthesis is composed of the four different nucleotides, which are assembled in long linear arrays that are 'read' three nucleotides at a time to specify which of the 20 different amino acids should be incorporated at a given point in the protein (see Figure 1.12), a process termed translation. As we note above, proteins are not directly synthesized from the DNA copy of the gene, but are instead synthesized from the transcribed RNA copy of the gene, which acts as an intermediate in the overall synthesis pathway. Such RNAs that function as coding molecules to direct protein synthesis are called messenger RNAs (mRNAs). Distinctive sequences in the mRNA specify the start and stop sites for protein synthesis, while the successive triplet nucleotide sequences in between encode the linear sequence of amino acids that comprise the final protein product.

The term 'translation' makes sense when one understands that this is the process by which the sequence of nucleotides within a nucleic acid is 'translated' into the sequence of amino acids in a protein. The code that correlates triplet nucleotides with the encoded amino acid is referred to as the genetic code, and is presented in Figure 1.13. The nucleotide triplets are called codons. We see, for example, that the codon UUU is translated as phenylalanine whereas AUG is translated as methionine. All codons dictate a particular outcome, either the incorporation of an amino acid or the halting of translation (a 'stop' site), and so we say that the code is non-ambiguous.

Translation is carried out by an elaborate macromolecular machine known as the **ribosome**, which comprises both RNA and protein molecules. At the core of the process of translation is a functional RNA called transfer RNA (**tRNA**), which acts as a physical link between the coding information (in the form of a triplet nucleotide sequence that reads the codons) and the appropriate amino acid. Figure 1.12 shows how the tRNA functions as the actual decoder that makes translation possible; in essence, a population of tRNAs sequentially 'reads' the codons in a processive march along the mRNA template, recruiting amino acids to the growing protein chain in the order specified by the mRNA template. These incoming amino acids are then chemically linked together as the march proceeds. The ribosome is the structural scaffold that facilitates this molecular march.

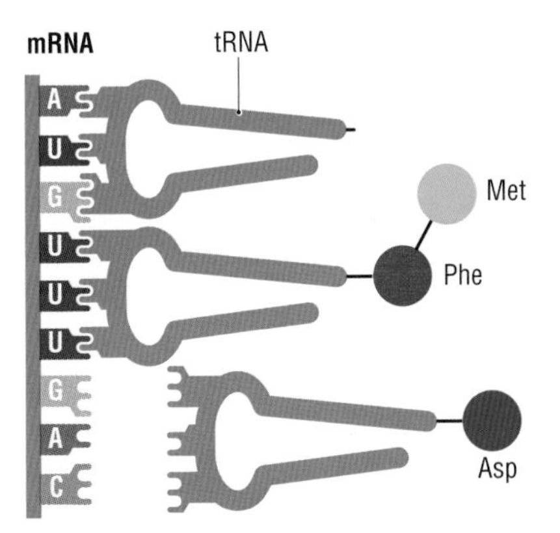

Figure 1.12 The process of translation is common to all organisms. The nucleic acids in RNA are read three at a time to specify which amino acid should be incorporated.

Most proteins that emerge from the ribosome are not fully functional. Different portions of the protein product may be removed (cleaved or spliced out) and additional moieties may be added. These processes are collectively referred to as post-translational processing.

Ist position (5′ end)	2nd position U	C	A	G	3rd position (3′ end)
U	Phe	Ser	Tyr	Cys	U
	Phe	Ser	Tyr	Cys	C
	Leu	Ser	stop	stop	A
	Leu	Ser	stop	Trp	G
C	Leu	Pro	His	Arg	U
	Leu	Pro	His	Arg	C
	Leu	Pro	Gln	Arg	A
	Leu	Pro	Gln	Arg	G
A	Ile	Thr	Asn	Ser	U
	Ile	Thr	Asn	Ser	C
	Ile	Thr	Lys	Arg	A
	Met	Thr	Lys	Arg	G
G	Val	Ala	Asp	Gly	U
	Val	Ala	Asp	Gly	C
	Val	Ala	Glu	Gly	A
	Val	Ala	Glu	Gly	G

Amino acids	Abbreviations		Codons
Alanine	Ala	A	GCA GCC GCG GCU
Cysteine	Cys	C	UGC UGU
Aspartic acid	Asp	D	GAC GAU
Glutamic acid	Glu	E	GAA GAG
Phenylalanine	Phe	F	UUC UUU
Glycine	Gly	G	GGA GGC GGG GGU
Histidine	His	H	CAC CAU
Isoleucine	Ile	I	AUA AUC AUU
Lysine	Lys	K	AAA AAG
Leucine	Leu	L	UUA UUG CUA CUC
Methionine	Met	M	AUG
Asparagine	Asn	N	AAC AAU
Proline	Pro	P	CCA CCC CCG CCU
Glutamine	Gln	Q	CAA CAG
Arginine	Arg	R	AGA AGG CGA CGC
Serine	Ser	S	AGC AGU UCA UCC
Threonine	Thr	T	ACA ACC ACG ACU
Valine	Val	V	GUA GUC GUG GUU
Tryptophan	Trp	W	UGG
Tyrosine	Tyr	Y	UAC UAU

Figure 1.13 The genetic code.

The functional repertoire of the cell is much more complex than the genome

➔ The features of the genetic code and process of translation are explored in more detail in Chapter 11 and post-translational processing is described in Chapter 14.

A direct consequence of the fact that RNA and protein products can be modified following their initial synthesis is that the complete set of functional RNAs and proteins is far more extensive than the set that is simply encoded by the genome. While the human genome encodes an estimated number of 30 000 genes, for example, the number of different human gene products is estimated in the range of one million and it is likely that this is an underestimate. Moreover, we are beginning to understand that certain classes of molecule have escaped notice for a variety of reasons, including low expression levels and sizes that are difficult to identify. However, recent technical advances are leading to more systematic approaches to gene identification. Thus, there has been a recent explosion in the discovery of several different classes of small, functional RNAs. These provide a compelling example of how the overall gene count is likely to remain in flux for the foreseeable future. These issues will be discussed in more detail in Chapter 18. It is the exquisite control of the production of an increasingly large number of cellular components that allows for the incredible complexity that we see in the living world.

1.4 REGULATING GENE EXPRESSION

Every step in the overall process of gene expression - be it transcription, RNA processing, translation, or protein modification - represents an opportunity for regulation, as depicted in Figure 1.14. This allows each gene to be expressed when and where the product is specifically needed by the organism, a process broadly referred to as **gene regulation.** As we shall see throughout this book, organisms commit tremendous cellular resources to gene regulation.

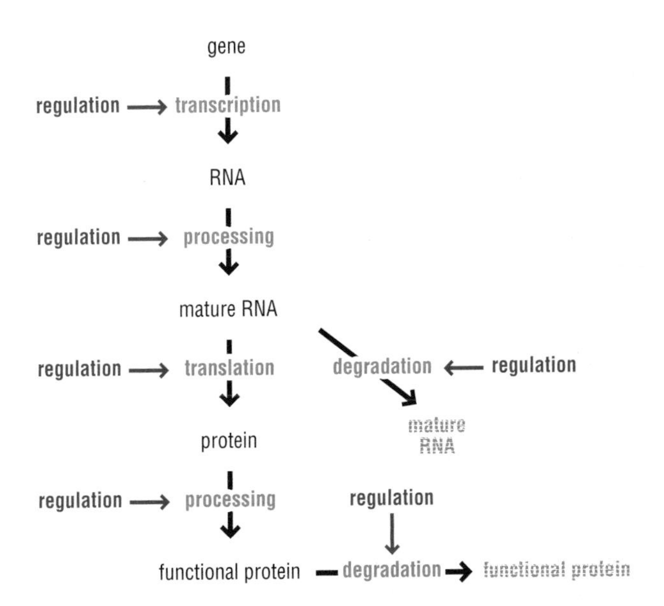

Figure 1.14 Expression of an RNA or a protein can be regulated at every level of synthesis, modification, or degradation.

Both the level and the activity of a gene product can be regulated

Gene expression can be controlled by simply regulating the level of the gene product. The level can be controlled by either promoting or preventing the *synthesis* of the RNA product at the level of transcription or by controlling the translation of a protein. The level of gene expression can also be modulated by either promoting or preventing the *degradation* of the RNA transcript or the protein. Thus a protein that is needed at only a particular stage in the development of an organism may only be synthesized just before it is needed and may then be rapidly degraded at a later stage in development. Such regulation can be illustrated by the levels of the Hy5 protein in *Arabidopsis*, illustrated in Figure 1.15. Under conditions of low light, the Hy5 protein is degraded and cannot regulate genes that limit the length of the stem.

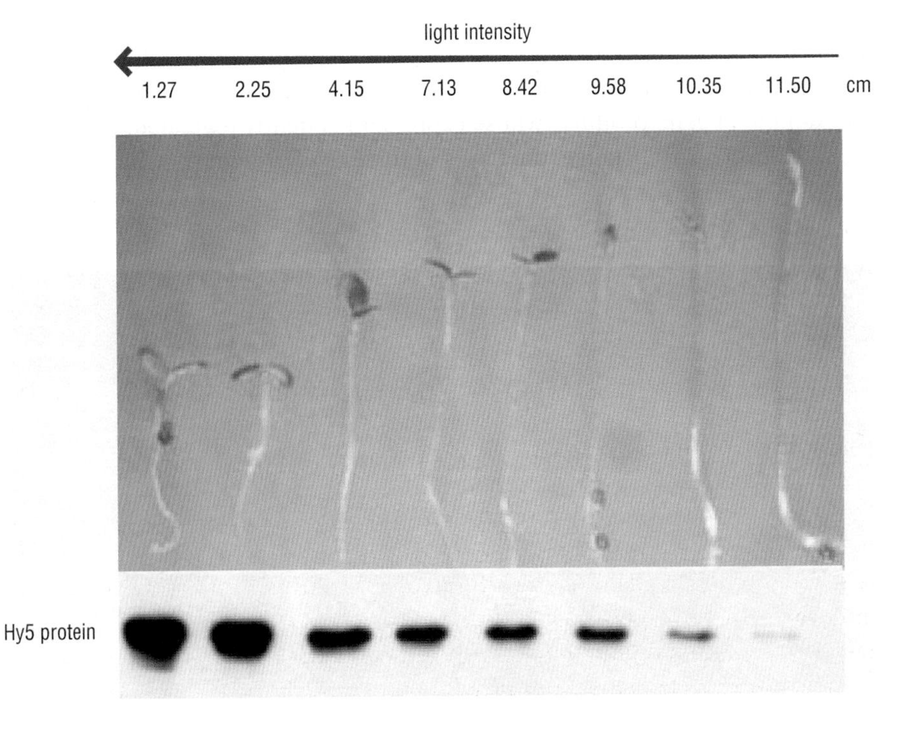

Figure 1.15 Targeted destabilization of HY5 during light-regulated development of *Arabidopsis*.

From Osterlund MT, Hardtke CS, Wei N, Deng XW, *Nature*, 2000; **405**: 462–466. Fig 5a.

Under conditions of high light, Hy5 protein levels are elevated, ultimately resulting in shorter seedlings.

The many modifications to which both RNA transcripts and protein products are subjected provide additional opportunities for regulating gene expression. For example, some RNAs and proteins are only active after a specific portion of the transcribed or translated product has been removed: the hormone insulin is initially synthesized as an inactive precursor protein and is only activated by the removal of 35 amino acid residues.

Gene expression is regulated in both time and space

Differences in cellular behavior or identity can be brought about by differences in when or at what level a given gene product is expressed. For example, different tissues are defined by the different cells from which they are formed, and different cells are defined by the particular elements of the genome - the particular subset of genes - they express. Therefore, a liver cell will express a different set of genes than a nerve cell; these differences in gene expression mean that the two cells are biologically distinct.

A view of the impact of spatial regulation of single genes on the correct development of an organism is provided by the segmentation of the *Drosophila* embryo during early development. Correct segmentation of the organism depends on the localized expression of several genes including *even-skipped*. Figure 1.16 shows how the expression of *even-skipped* is localized in discrete bands of cells along the length of the embryo, laying down the pattern for the segmentation of the embryo later in development. When this localized expression is disrupted, the proper patterning of the embryo fails to occur.

The expression of key gene products is regulated at multiple levels in response to multiple signals

The expression of many gene products, particularly gene products that are central to the functioning of cells, is regulated at multiple levels. Proteins that are frequently mutated in cancer cells, or represent the 'tipping point' in the cellular decision to initiate development or to initiate a dramatic change in cell metabolism, are an example of gene products whose expression is tightly regulated, often at all possible levels. For instance, the levels of the *E. coli* protein σ^S, which controls

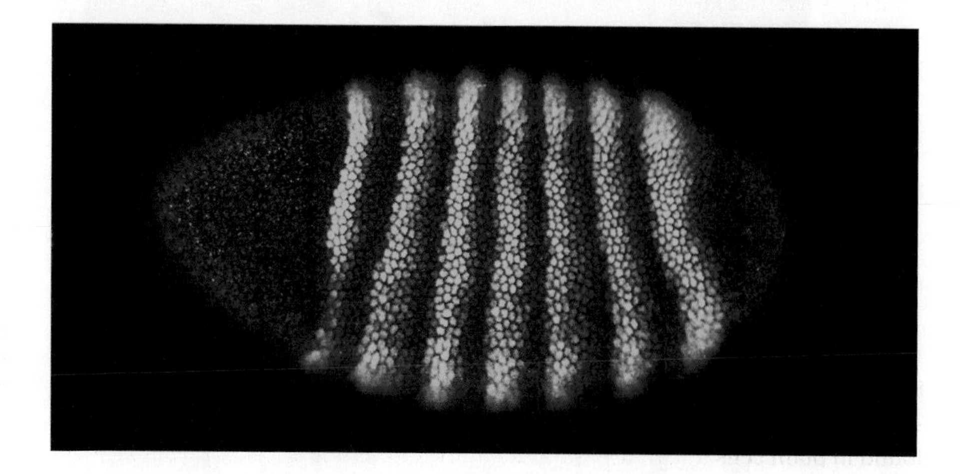

Figure 1.16 Expression of *even-skipped* gene in a *Drosophila* embryo. The *even-skipped* gene is expressed in discrete bands of cells along the length of the embryo at an early stage of *Drosophila* development and helps to lay down the pattern for the segmentation of the fly.

Image © Sabrina Desbordes.

the expression of genes important to surviving stress, are finely tuned to the environmental stresses encountered by the bacterium. Thus transcription of the gene encoding σ^S is regulated in response to available carbon sources; translation of the mRNA is regulated in response to temperature, levels of oxygen and changes to the cell membrane; and the stability of the protein is controlled by the availability of phosphate and magnesium. All of these environmental conditions - and others - impact expression of this gene product.

1.5 CELLULAR INFRASTRUCTURE AND GENE EXPRESSION

The cell is far from being a container in which molecular components tumble around freely. Instead, it is compartmentalized into distinct structures, each with its own function, which act in concert to mediate the processes essential to life. In this section we learn in broad strokes how the process of gene expression - from retrieval of information from the genome to its expression as gene products - is accommodated within the cell, and explore the cellular compartments in which these processes take place.

An internal membrane-bound compartment in the cell contains the genome in eukaryotes

At the beginning of this chapter, we discussed how the cellular membrane is key to life because it defines self and non-self. The membrane encapsulates the genetic information critical for self-propagation together with the functioning molecules of the cell. In addition to the exterior membrane shared by all organisms, more complex organisms have additional membrane-bound compartments in their interior. Many unicellular species, including members of the bacterial and the archaeal kingdoms, do not have internal membranes within their cells; because of this distinction they are referred to as **prokaryotes** (meaning 'before the kernel' or pre-nuclear). It should be noted, however, that these cells are not completely without compartmentalization. For example, the chromosomal DNA is sequestered in a region called the **nucleoid,** as shown in Figure 1.17.

By contrast, higher-complexity organisms (including those that are unicellular and multicellular) contain an internal membrane-bound structure called the **nucleus** where the genetic information is contained. The name eukaryote literally translates as 'well kernel,' and defines a kingdom of organisms having a nuclear compartment that contains their genetic information. The contents of the cell lying outside the nucleus are collectively referred to as the **cytoplasm**.

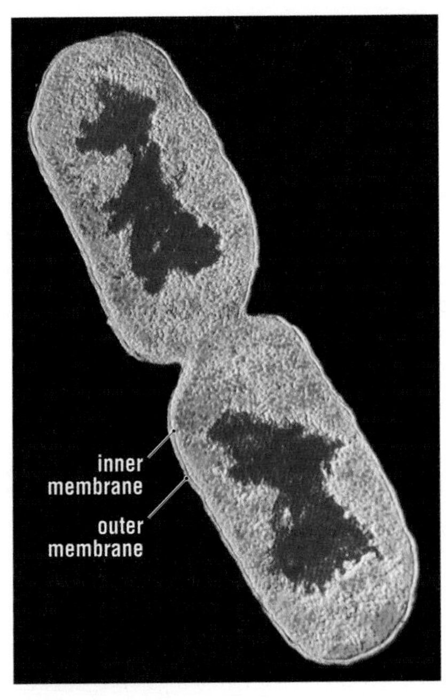

Figure 1.17 Electron micrograph of an *E. coli* cell dividing into two daughter cells by binary fission. The nucleoid containing the DNA appears in red.

Eukaryotic cells feature multiple membrane-bound compartments with specialized functions

While the nucleus is the defining compartment within eukaryotic cells, other membrane-bound compartments play equally important roles in cellular life. These compartments are generally referred to as organelles; the key organelles in eukaryotic cells are illustrated in Figure 1.18. The **mitochondria** are known as the powerhouses of the cell and contain the macromolecular machinery responsible for deriving chemical energy from food precursors. **Chloroplasts** are organelles found in plant cells that house the machinery for harvesting energy from sunlight

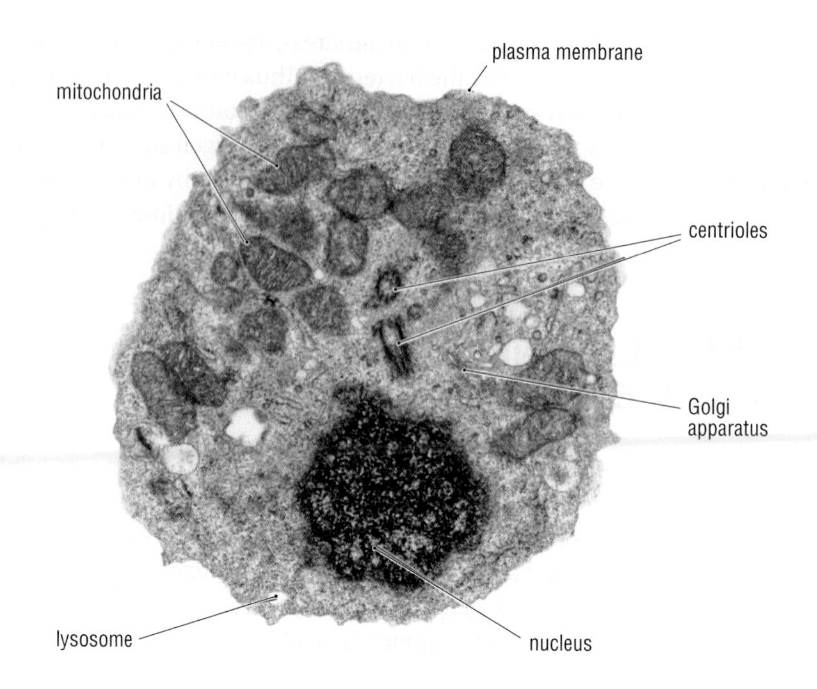

Figure 1.18 Electron micrograph of a human lymphocyte cell. This is the same type of cell that is shown in pink in Figure 1.2. The nucleus, Golgi apparatus, centrioles, mitochondria, lysosomes, and plasma membrane are labeled.
From Dr. Gopal Murti/Visuals Unlimited.

for the production of sugars within the cell. Both mitochondria and chloroplasts contain DNA unique to these organelles that encodes certain components critical to their function. This feature led scientists to hypothesize that these organelles originated as **endosymbionts**, organisms living within another cell that eventually became wholly dependent on their host for existence.

Other well-characterized membrane-bound compartments in the eukaryotic cell include the **endoplasmic reticulum** and the **Golgi apparatus**, which are essential for the production of proteins that are inserted into the membrane or secreted from the cell. Certain other organelles, which will not be discussed further in this book, are those devoted to the metabolic and catabolic needs of the cell. These include the lysosome and the peroxisome.

In addition to the membrane-bound compartments that we describe here, the eukaryotic cell also depends on other subcellular localization mechanisms both in the nuclear and cytoplasmic regions of the cell. These regions, which generally were first visualized by microscopy, include structures denoted P-bodies, stress granules, and neuronal particles in the cytoplasm, as well as the nucleolus and Cajal bodies in the nucleus.

What advantages does compartmentalization confer upon a cell? Why does it help biological processes occur in the way that they do?

Compartmentalization is important for facilitating chemical reactions

If we look at biological diversity, we see that overall cell size increases in the eukaryotic lineage when compared with simpler bacterial cells. All cellular reactions depend on interactions between molecules; by localizing various components together, their effective cellular concentrations are increased and so reactions can proceed more efficiently. These principles will be described in more detail in Chapter 2.

In addition to the chemical benefits of increased concentrations of cellular components, cells also benefit from the level of organization that accompanies compartmentalization. Just as we organize our own lives and make them more efficient by compartmentalizing our possessions – we store food items together in the kitchen ready for the preparation of meals, we gather books, class notes, and laptop on a desk ready for study – by co-localizing related processes with shared substrates, intermediates or macromolecular components, cellular efficiency is increased. Cell biological approaches provide striking images of detailed subcellular localization patterns for various cellular components that highlight the potential of such compartmentalization, as illustrated in Figure 1.19.

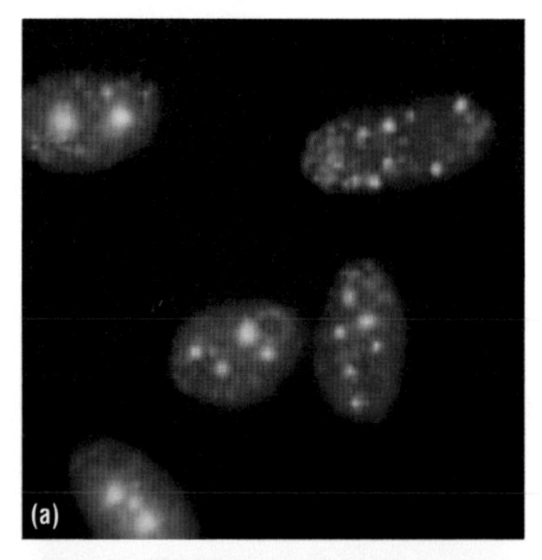

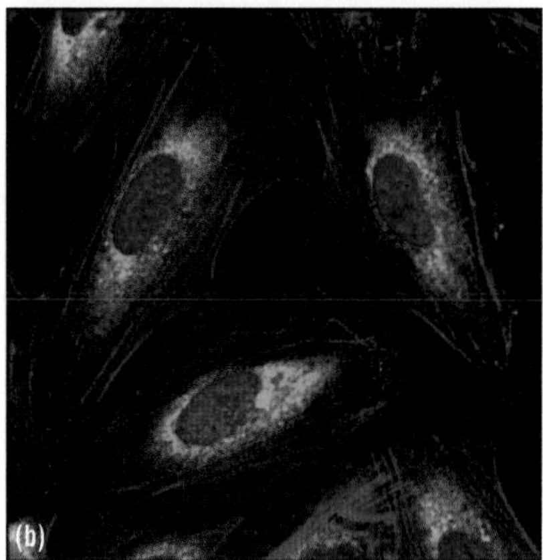

Figure 1.19 Subcellular localization allows for more efficient reactions and regulation. (a) An RNA polymerase (light blue) is localized to specific regions in the nucleus (stained dark blue with dye that binds to DNA) of a eukaryotic cell. (b) By contrast, ribosomes (a ribosomal protein is stained green) are localized to the cytoplasm surrounding the nucleus (blue). These differences in localization separate transcription and translation, providing an opportunity for regulation.

Compartmentalization is crucial for regulating gene expression in eukaryotic cells

The natural consequence of the fact that the DNA in eukaryotic cells is housed in a membrane-bound compartment, the nucleus, is that a large amount of regulatory control of gene expression has developed, which is based on this physical separation from the rest of the cell. Transcription of the DNA (to make RNA) takes place in the nucleus where the genome is housed, whereas translation of the RNA (to make proteins) takes place in the cytoplasm. This difference in location offers an immediate opportunity for regulation: just as you can regulate movement from one room to another in a house by locking (or unlocking) the door between them, so the physical barrier presented by the nuclear membrane provides an opportunity to regulate passage from the nucleus to the cytoplasm (and vice versa), such that passage can only occur if certain conditions are met.

For example, we know that in eukaryotic cells, mRNAs that are produced in the nucleus are significantly modified following transcription (RNA processing), and that this modification occurs before the transcript is exported from the nucleus to the cytoplasm, where it is translated. In fact, transcripts cannot be exported from the nucleus unless they have been fully processed. This feature thus provides the cell with an opportunity for quality control – the blocking of movement of the transcript from the nucleus to the cytoplasm across the nuclear membrane – which prevents inadequately processed mRNAs from encountering the translational machinery. In bacteria, where there is no nucleus to sequester the genetic information, transcription and translation are spatially coupled, leading to distinct forms of regulation in these organisms.

An additional complexity faced by gene expression is the production of proteins destined for different locations – for example, for the cytoplasm, for insertion into the cellular membranes, or beyond the cell itself. How can proteins be targeted to their correct locations?

As we will discover, proteins that are destined for the membrane or for export from the cell have physical properties that are incompatible with passage through the membrane without specific assistance. In bacteria, where there are no internal organelles, proteins destined for secretion are directly targeted to the extracellular membrane as they are translated. Eukaryotic cells appear to have elaborated on this plan, but instead of directly targeting proteins to the exterior of the cell, translation occurs on the membrane of an extensive organelle system known as the endoplasmic reticulum. This organelle is really a network of tubules where membrane and secreted proteins are translated, folded, modified and then sent on their way.

The final stop in the maturation of many proteins from genes is the Golgi apparatus, an organelle that modifies the nearly completed protein with sugars before packaging the protein for final export. Proteins are sent to the Golgi apparatus from the endoplasmic reticulum, and are sent from the Golgi to the cell membrane for export.

We describe the processing of RNAs and proteins in Chapters 10 and 14, respectively.

While this book will not focus on the cell biological aspects of gene expression, it is important to remember the critical role played by cellular structure in all biological processes.

1.6 EXPRESSION OF THE GENOME

In the preceding sections, we explored the expression of individual genes. However, genes do not operate in isolation. Just as individual humans in a community interact, and influence the behavior of others, so genes operate in networks, where the activity of one gene can quite often influence the activity of another. In this section we look beyond the expression of individual genes to ask how whole communities of genes – that is, genomes – are expressed.

Output of the genome is dictated by the interplay of genes

The expression of the information in a genome determines the physical properties of an organism. The visual features and properties of an organism are referred to as the **phenotype**, while the collective DNA sequence that determines the phenotype is referred to as the **genotype** of the organism. The output of the genome, the phenotype, is dictated by which genes are expressed, at what times, and at what levels. Each organism is thus the product of a combinatorial process dictated by a great many genes working together.

An additional complexity is that organisms can carry one or more copies of the individual genes, as depicted in Figure 1.20. For example, yeast can survive indefinitely as a **haploid** organism, carrying a single copy of its genome. By contrast, more complex multicellular organisms typically have two copies of their genome and are referred to as **diploid**. Some organisms have more than two versions of each of their genes and are said to have higher ploidy – for example, the frog species *Xenopus laevis*, commonly used for biochemical experiments, is **tetraploid** – it has four copies of each gene.

The importance of ploidy is that, in a haploid organism, a single version of a gene is expressed whereas in a diploid organism there is typically a mixed population of gene products derived from the expression of two different copies of the gene. Why can this be important? If one copy of a gene is defective, such that a functional gene product is not produced when the gene is expressed (or is produced in reduced amounts), a diploid organism has the advantage of a second copy of the gene to compensate for the defective one. By contrast, a haploid organism does not have the benefit of a 'back-up' gene, putting it at a greater risk of experiencing phenotypic defects if its only copy of a given gene is mutated.

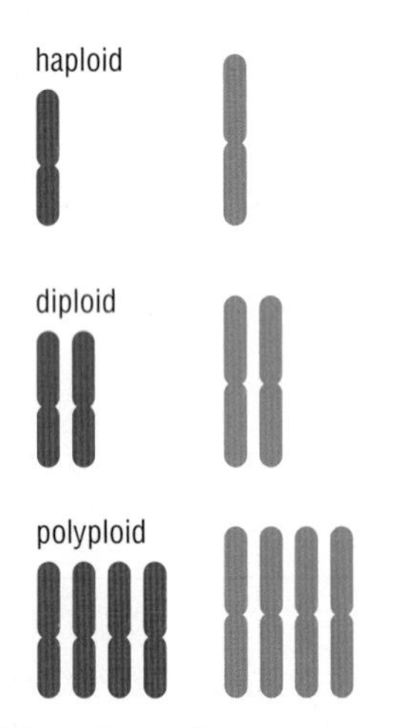

Figure 1.20 Cells can have one or more copies of their chromosomes. Haploid cells only contain a single copy of the two distinct red and blue chromosomes, and diploid cells have two copies. Some cells are polyploid and have multiple copies.

Genetic analysis has been essential in deciphering the natural function of genes

We see above how different genes are expressed to produce different RNA or protein products, and that it is the concerted action of all the genes in a genome that directs the development and function of an organism. But how do we know what individual genes actually do? How can we say that gene *x* drives process *y*?

Much of our understanding of biology comes from what are referred to as genetic approaches, which probe the natural function of genes by comparing so called **wild-type** (normal) properties, the typical or most common form, with non-wild-type (mutant) ones. In classical genetic approaches, scientists observed organisms under various environmental conditions following the imposition of DNA damage, looking for visible phenotypic changes. That is, they looked to see the consequence of different genes being rendered inactive or altered. When such mutant (variant) organisms were found, the specific changes in the genome that specified the difference between mutant and wild-type were pursued in order to provide information on potential gene function. So, for example, if a wild-type fruit fly had red eyes, but a defect in gene *a* produced a mutant with white eyes, one could deduce that the function of gene *a* was to determine eye color in some way. Examples of fruit flies carrying wild-type and mutant genes for eye color are shown in Figure 1.21.

Figure 1.21 ***Drosophila*** **mutant with a defective pigmentation gene.** Wild-type fruit flies have red eyes (left), while mutant flies carrying a defective *white* gene have white eyes (right).

Dr. Jeremy Burgess/Science Photo Library.

This general approach is typically referred to as '**forward genetics**'. For example, Figure 1.22 shows a single mutant *Arabidopsis thaliana* plant seedling surrounded by a group of wild-type plants. In the example shown, the phenotype of the mutant plant is an elongated stem while the genotype turns out to be a **mutation** in a gene encoding a blue light photoreceptor. A specific change in the gene for the blue photoreceptor is manifested as this obvious change in the growth properties of the plant. We will sometimes refer to these two different versions of the gene as the wild-type and mutant **alleles** of the gene. An allele is thus a version of a gene.

With a more complete understanding of genomes and their complete composition, and with sophisticated tools of gene manipulation, it has become possible to specifically disrupt a gene of interest in some organisms and to study the phenotypic consequences. This powerful approach is typically referred to as **reverse genetics** and has been used with wide success to decipher gene function.

Genotypic changes have phenotypic consequences

In forward genetic studies, the phenotype of the mutant organism is typically tracked by the scientist with the ultimate goal of identifying the specific changes in the DNA (mutation) that resulted in the phenotype in question. This type of analysis in principle establishes causality in that a certain change in the DNA results in a change in the physical properties being followed.

Mutations that cause differences from the wild-type organism are typically categorized as recessive or dominant. These terms take on meaning when one remembers that organisms do not always have just a single copy of a given gene. If, in a diploid, the phenotype of the mutant gene product is masked by the presence of the wild-type version, then the mutation is said to be recessive. A common example of a **recessive mutation** is a change in an enzyme that reduces its activity, but which is compensated for by the wild-type version that provides sufficient activity. By contrast, if a mutant gene product has obvious phenotypic consequences when the wild-type product is present, then the mutation is said to be dominant. A common example of a **dominant mutation** might be a change in a structural protein that disrupts higher-order protein structure whether or not the wild-type protein is present. Figure 1.23 shows schematically how recessive and dominant mutations impact mouse coat color.

Figure 1.22 ***Arabidopsis*** **mutant unable to sense blue light.** The stem (hypocotyl) extension of wild-type *Arabidopsis* seedlings is inhibited by blue light. The growth of a mutant carrying a defective blue light receptor *cry* gene is not inhibited by blue light so it stands out among the wild-type seedlings.

What is the physical nature of the mutations that can cause such phenotypic changes? Figure 1.24 shows how the mutations can be the result of very small changes in the gene (i.e. the alteration of a nucleotide), the rearrangement of large regions of the chromosome, the insertion of new segments of DNA or the **deletion** or excision of small or large sections of the genome. These changes can result in direct changes to the gene product encoded by a given gene, or in disruptions

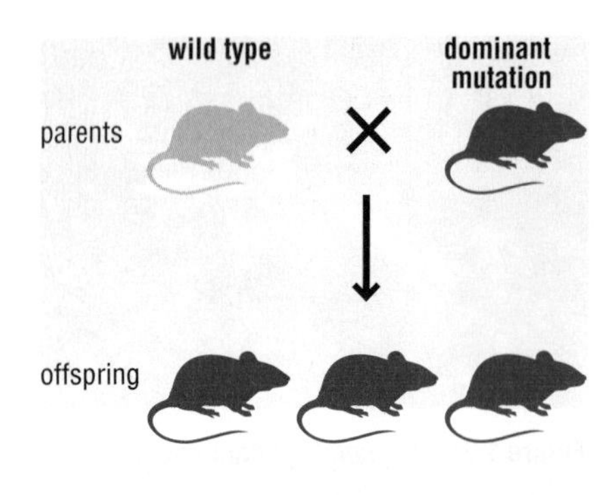

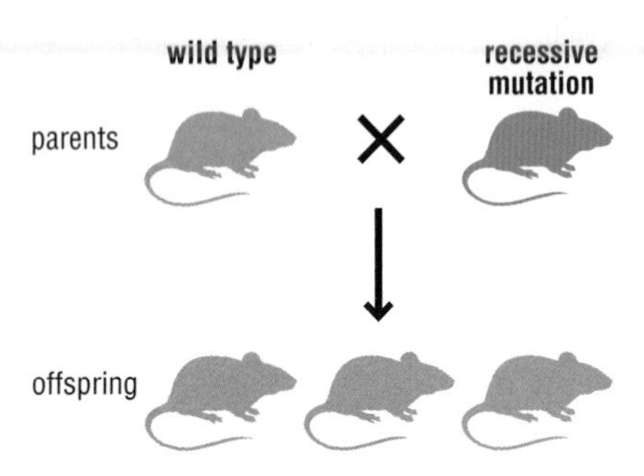

Figure 1.23 Effects of dominant and recessive coat color mutations in mouse. In the cross between one parent (gray) contributing the wild-type gene and another parent (pink) contributing a gene with a dominant mutation, all offspring will have the phenotype of the parent carrying the dominant mutation. In the cross between one parent (gray) contributing the wild-type gene and another parent (blue) contributing a gene with a recessive mutation, all offspring will have the phenotype of the parent carrying the wild-type gene.

wild type

point mutation

rearrangement

deletion

insertion

Figure 1.24 Types of mutation. The sequence of DNA can be mutated by a single nucleotide change (point mutation, pink), shuffling of different regions of DNA here indicated by the two shades of gray (rearrangement), excision of DNA (deletion) or insertion of DNA (blue).

of overall gene expression (where the actual identity of the gene product is not changed).

Mutations that eliminate the function of a gene are termed **null** or **loss of function mutations.** These types of mutation typically are associated either with a change in the product that completely eliminates function or a large disruption in the DNA sequence that eliminates gene expression.

Single nucleotide changes in the DNA sequence can directly result in changes in the identity of the encoded gene product (either RNA or protein) as illustrated in Figure 1.25. In protein coding genes a change of one nucleotide can be sufficient to change a codon from one which encodes a particular amino acid to one encoding a different one. Such changes in amino acid identity are referred to as **missense mutations.** If the amino acid substitution results in gene function being completely lost, the missense mutation can be thought of as a null mutation. By contrast, other substitutions may result in the production of partially functional gene products. Figure 1.25 also shows how, in some cases, there can be a change in the DNA sequence that introduces a premature stop signal such that only truncated species of the protein product are made. Such changes are typically referred to as **nonsense mutations.** Sometimes there are changes in the DNA sequence, such as a change in a codon encoding a particular amino acid to another encoding the same amino acid, which do not result in obvious phenotypic changes. Such changes are typically referred to as **silent mutations**.

Heritable phenotypic changes can arise from changes in chromosome packaging without changes in DNA sequence

In Section 1.2, we discussed how genomic DNA is packaged into a highly compacted structure. We also noted the existence of specialized mechanisms that change the opening of the compacted DNA. These modifications in chromosome packaging, which often are not associated with changes in the underlying DNA sequence, can be transmitted from one generation to another. Sometimes these modifications are associated with a phenotypic change - for example, the expression of a gene may be turned off (or 'silenced') by an increase in the compaction of the DNA, leading to a change in an organism without a change in DNA sequence. These heritable changes in chromosome packaging are called **epigenetic changes.** As we will learn in Chapter 18, great strides have recently been made in understanding the mechanisms of epigenetic change and how they influence phenotype.

The origin of disease can often be traced to mutations in the organism

In multicellular organisms, there are two general types of cell:

- **somatic cells** - these constitute the building blocks of the body of an organism and simply transmit their genetic information to daughter cells in the same organism
- **germline cells** - these are involved in the production of the progeny and transmit the genetic information to the next generation of the organism.

Consequently, mutations found in the germline cells typically affect the progeny of the organism whereas those found in somatic cells affect the organism itself.

One common consequence of gene mutation is the onset of disease, whereby normal biological function is disrupted in some way. Some diseases are the consequence of variation in a single gene and are said to be **monogenic**. A well-known example of a monogenic disease is phenylketonuria, in which a mutation in a gene encoding a specific liver enzyme can lead to severe mental retardation. In other cases, disease states are thought to result from variations in multiple genes and are said to be **polygenic** (though the term 'multifactorial' is also commonly used). Some of the most prevalent diseases, such as heart disease, Alzheimer's, arthritis, and diabetes are thought to be polygenic.

The presence of a specific mutation does not always lead to a disease state, however. For example, mutations in genes called *BRCA1* and *BRCA2* are associated with breast cancer, but not all individuals carrying these mutations develop cancer during their lifetime. The **penetrance** of a given mutation is defined as the percentage of individuals carrying a particular mutant genotype that exhibit the mutant phenotype. The risk of developing a mutant phenotype when carrying a mutant genotype is determined by both the genetic makeup of an organism and environmental influences, such as diet and exposure to different environmental conditions.

The relative extent to which the genotype (nature) and the environment (nurture) influence the phenotype associated with a specific mutation has long been debated and undoubtedly will continue to be a subject of controversy. In the example of the *BRCA1* and *BRCA2* mutations, other genetic factors clearly influence whether individuals are more or less susceptible to developing cancer, but environmental factors such as diet and smoking can also affect the penetrance of these mutations. As we shall in see Chapter 18, whole-genome sequencing is providing critical information for understanding the genetic basis of complex diseases such as cancer.

wild type	**ATG**	**AAT**	**ATT**	**CGA**	**GAT**
	Met	Asn	Ile	Arg	Asp
missense	**ATG**	**AAT**	**ACT**	**CGA**	**GAT**
	Met	Asn	Thr	Arg	Asp
nonsense	**ATG**	**AAT**	**ATT**	**TGA**	**GAT**
	Met	Asn	Ile	stop	Asp
silent	**ATG**	**AAT**	**ATC**	**CGA**	**GAT**
	Met	Asn	Ile	Arg	Asp

Figure 1.25 Multiple types of point mutation. Point mutations can lead to the incorporation of the wrong amino acid (missense), premature termination (nonsense), or not affect the sequence (silent) of the translated protein.

1.7 EVOLUTION OF THE GENOME AND THE TREE OF LIFE

The genome of a particular species is not a static, unchanging entity. If it were, there would not be the remarkable variety of life in the world around us. Instead, genomes are dynamic: they undergo gradual, incremental change over time, which results in progressive change in the organism whose development and function they are controlling. The progressive change in an organism's phenotype over many generations, underpinned by a gradual change in that organism's genotype, is described by the process of evolution.

The evolution and diversification of life is gradual

The argument in favor of a common ancestor for all living things is based most simply on the fundamental similarities in the molecular features of all organisms – for example, the usage of nucleic acids to store genetic information and the genetic code that specifies the amino acid sequence of proteins. It is highly unlikely that similar mechanisms would have evolved independently in many different organisms. Instead, it is far more likely that these common mechanisms were passed down from our ancestors into each new species, defined as a group of organisms capable of interbreeding and producing fertile offspring, as they evolved.

If common features have been conserved between species, how did the great diversity of life on Earth derive from a common ancestor? If we see such similarities between species, how could the differences that delineate different species arise at all? Evolution is the underlying principle of modern biology that explains this conundrum. It is the process by which organisms change over successive generations through a process of random change or mutation in the genome, coupled with natural selection. Let us consider a hypothetical example to illustrate this process.

Imagine a population that shares a common genome: all members of the population have the same genotype (genomic composition) and, hence, the same phenotype. This genome can only be passed on to future generations if an organism survives long enough to reproduce. In this case, however, all of this population will be equally healthy (because they have the same genome) and so all should be equally likely to reproduce, and have their genome transmitted to the next generation.

Now, imagine that one or two of the population develop a mutation in their germline cells, as depicted in Figure 1.26. This mutation would not affect them directly because it is in their germline and not their somatic cell line. However, the mutation *will* be expressed when the genome is transmitted to the next generation.

Let us assume the mutation confers upon the individual carrying it a particular resistance to disease. The next generation will therefore contain a minority of individuals who show unusually high disease resistance. If the population as a whole is then exposed to a disease, a form of natural selection, it is those individuals who carry the mutation conferring disease resistance who are most likely to survive - and are therefore most likely to reproduce. Consequently, there is an increased chance of this mutated gene being passed on to the *next* generation.

Over time, environmental influences, such as disease, will continue to 'select for' the mutant individuals and, over time, an ever-larger proportion of the population will contain the mutant gene, until it becomes the norm. This is the process of natural selection in action - and is illustrated in Figure 1.26 by the emergence of a new (pink) strain of fish. As a consequence of natural selection, as generations come and go, we see our original species evolving into one that is more disease-resistant than it was before. Notice how evolution is not a rapid process - it is a change in genotype which, if beneficial to the population, becomes increasingly established, driving a concomitant shift in the observed phenotype for the population.

So we see how mutations that result in increased success for the organism will be propagated or selected; such selection eventually leads to the emergence of a reproductively isolated group of organisms or species.

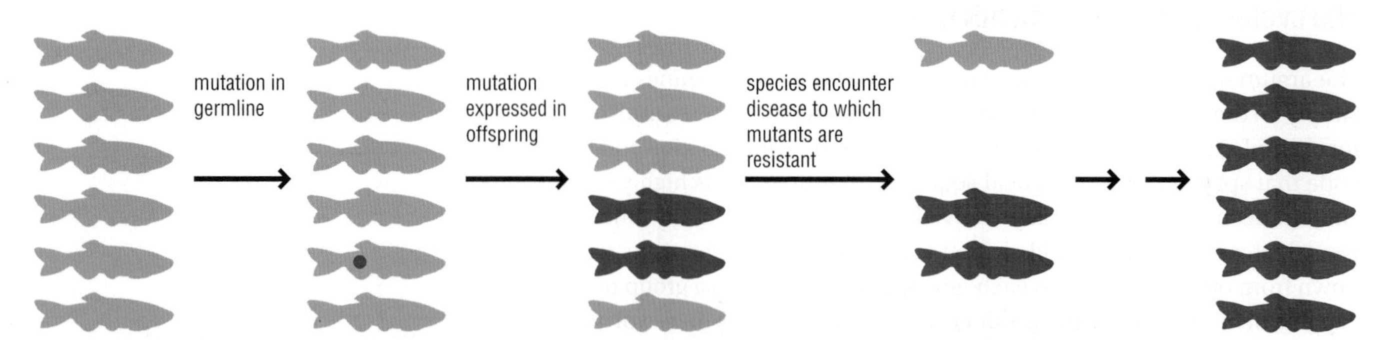

Figure 1.26 Ever-larger populations of individuals carrying specific DNA changes due to selection. Initially all the fish have the same genotype. Then a random mutation arises in the germline cells of one of the fish. By chance, the offspring (pink) of this fish have increased resistance to a particular disease. Since the offspring are more likely to survive, they are also more likely to reproduce, leading to an increase in the number of fish carrying the DNA change.

All organisms can be grouped into three domains of life based on sequence comparisons

We have learned that the collection of genes in an organism, and their expression, establishes the phenotype of an organism, in concert with environmental factors. By comparing the sequences of specific genes or whole genomes from different organisms, we can learn something about the likelihood that given gene products have similar functions, as well as how closely these organisms are related in evolutionary terms.

The sequences for individual genes can be very similar in certain regions while varying more extensively in others as shown in an alignment of the sequences of the same region of the ribosomal RNA genes from many different organisms depicted in Figure 1.27. The extent of relatedness can be quantified and these results then represented in a branched model called a **phylogenetic tree**. An example of a phylogenetic tree based on the sequence comparison of ribosomal RNA genes is shown in Figure 1.28. In these models, the length of the line reflects the relatedness of the gene at one end of the line to that at the other end, and thus by extension, the relatedness of these organisms. Therefore, an organism that is separated from its most recent ancestor by a short line is more closely related to that ancestor than a second organism that is separated from the same ancestor by a longer line.

Trees constructed from comparisons of different genes can differ, and so phylogenetic trees are typically based on comparisons of multiple genes. The precision of such classification approaches far exceeds more traditional methods of classification that depended on visually apparent physical features such as whether or not cells have nuclei.

Sequence comparisons based on many genes indicate that there are three distinct groups, or branches, of the tree into which all of organismal life can be split. These broad groups are referred to as the **three domains of life** and include the bacteria, archaea, and eukaryotes.

- The bacterial domain encompasses the largest number of organisms and continues to grow as more bacteria are identified by the sampling of an increasing number of environments. These organisms are single celled and have minimal subcellular compartmentalization.
- Archaeal organisms share features with organisms in the bacterial domain in that they are unicellular and do not have a nuclear structure. However, core archaeal proteins typically show more sequence similarity to the corresponding proteins found in eukaryotic organisms. Thus, early classification identified these organisms as bacterial, based on their physical features, but sequence comparisons allowed for the identification of this separate domain of life in 1977.

T. acidophilum	GGCCUUUAGUACGAGAGGAACAAGGG
E. coli	UGCUCCUAGUACGAGAGGACCGGAGU
B. subtilis	UGUCCUUAGUACGAGAGGACCGGGAU
E. histolytica	ACAACUCAGUACGAGAGGAACCGUUG
S. cerevisiae	UGAACUUAGUACGAGAGGAACAGUUC
C. elegans	CCUGCUUAGUACGAGAGGAACAGCGG
D. melanogaster	CCUGCGUAGUACGAGAGGAACCGCAG
R. norvegicus	GGUGCUCAGUACGAGAGGAACCGCAC
O. sativa	UCAACCUAGUACGAGAGGAACCGUUG
A. thaliana	UCAACCUAGUACGAGAGGAACCGUUG
H. sapiens	CCUGCUCAGUACGAGAGGAACCGCAG

Figure 1.27 Similarity between sequences from distantly related organisms. The alignment of sequences corresponding to ribosomal RNAs from bacteria to humans shows regions of similarity (pink) and variation (black).

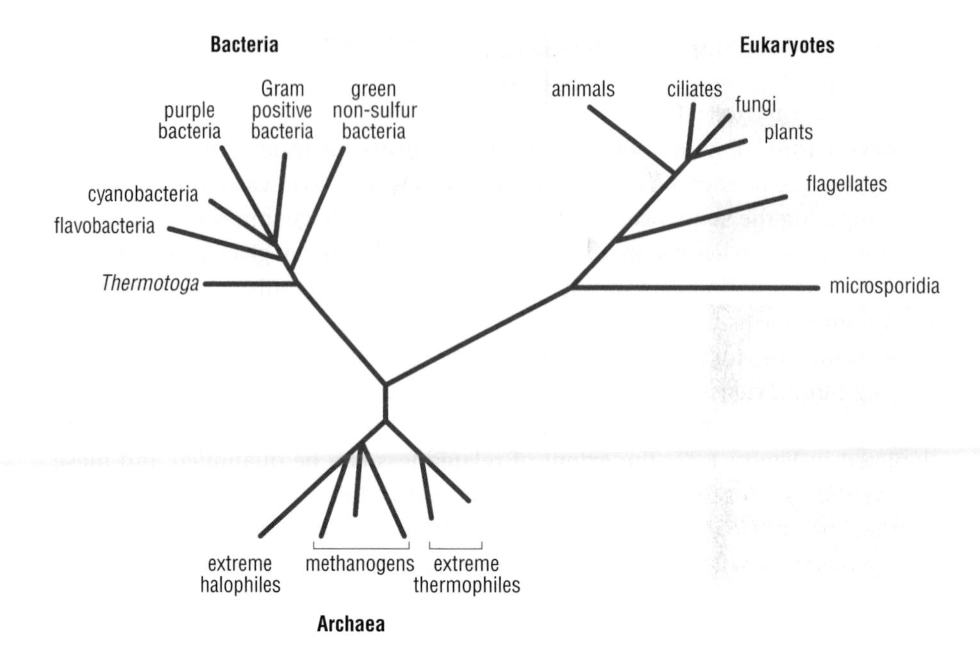

Figure 1.28 Relatedness of organisms depicted by a phylogenetic tree. The tree shows the relatedness of the sequences encoding ribosomal RNAs from bacteria, archaea, and eukaryotes. The length of each of the lines corresponds to the relatedness of the genes. From Woese, *Microbiological Reviews*, 1987; **51**: 221–271.

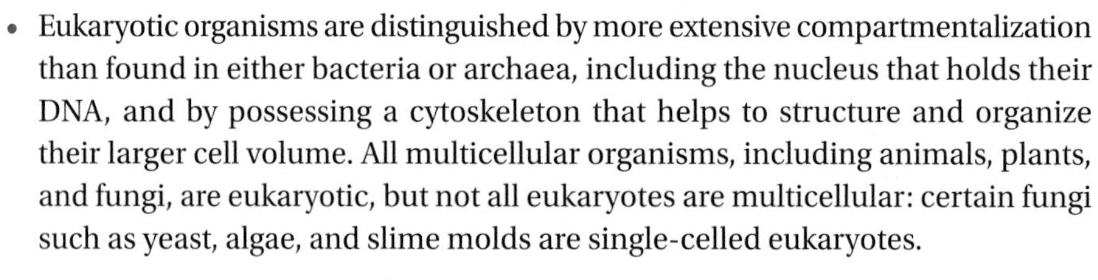

- Eukaryotic organisms are distinguished by more extensive compartmentalization than found in either bacteria or archaea, including the nucleus that holds their DNA, and by possessing a cytoskeleton that helps to structure and organize their larger cell volume. All multicellular organisms, including animals, plants, and fungi, are eukaryotic, but not all eukaryotes are multicellular: certain fungi such as yeast, algae, and slime molds are single-celled eukaryotes.

While both bacteria and archaea are prokaryotic, in the sense that they have no nucleus, the differences between them (as noted above) are the reason why we avoid using the term 'prokaryotic' throughout the rest of this book. It can be misleading to talk in general terms about prokaryotes when we see notable differences between bacteria and archaea.

The tree of life is an organizing principle for biological studies and we will see next how biologists have chosen a range of organisms covering some of this diversity to focus on explorations of molecular function.

Studies of model organisms have been key to understanding many biological processes

The study of organisms that have features of agricultural or medical importance, yet are easy to manipulate under laboratory conditions, has been critical to the understanding of the many processes in biology. These organisms, often referred to as model organisms, are typically directly evaluated and manipulated, subjected to mutational studies, or are used to prepare samples for biochemical analysis.

Two commonly studied bacteria, *E. coli* and *Bacillus subtilis*, belong to two distinct groups of the bacterial domain referred to as Gram-negative and Gram-positive, respectively, based on their differential staining with a particular violet dye (Gram stain). Two commonly studied fungi belonging to the eukaryotic domain are *Saccharomyces cerevisiae* and *Schizosaccharomyces pombe*, which are shown in Figure 1.29. While both are fungi, these two organisms exhibit significant differences in their biology – including differences in something as essential as their mode of cell division, where budding is seen in *S. cerevisiae* and fission in

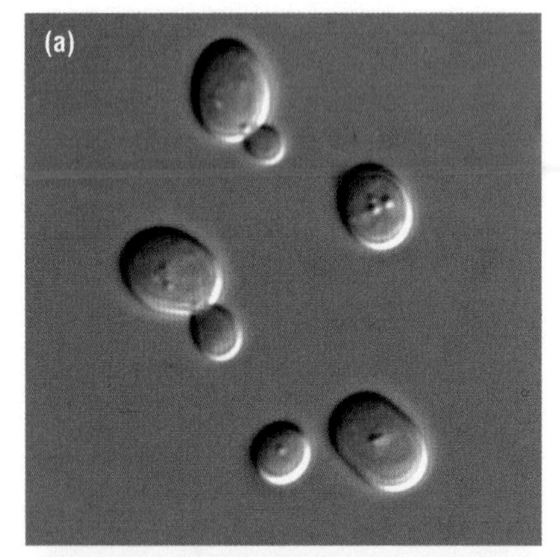

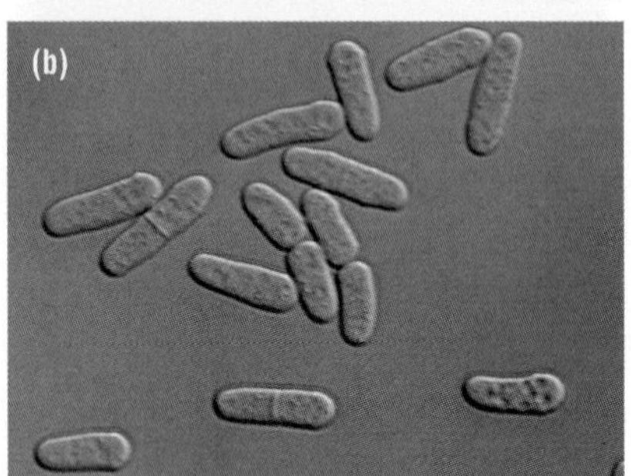

Figure 1.29 Commonly studied fungi *S. cerevisiae* and *S. pombe*. As seen in the micrographs, *S. cerevisiae* is oval shaped and divides by budding (a), whereas *S. pombe* is rod shaped and divides by fission (b).

Part (b) reproduced with permission of Kathleen Gould.

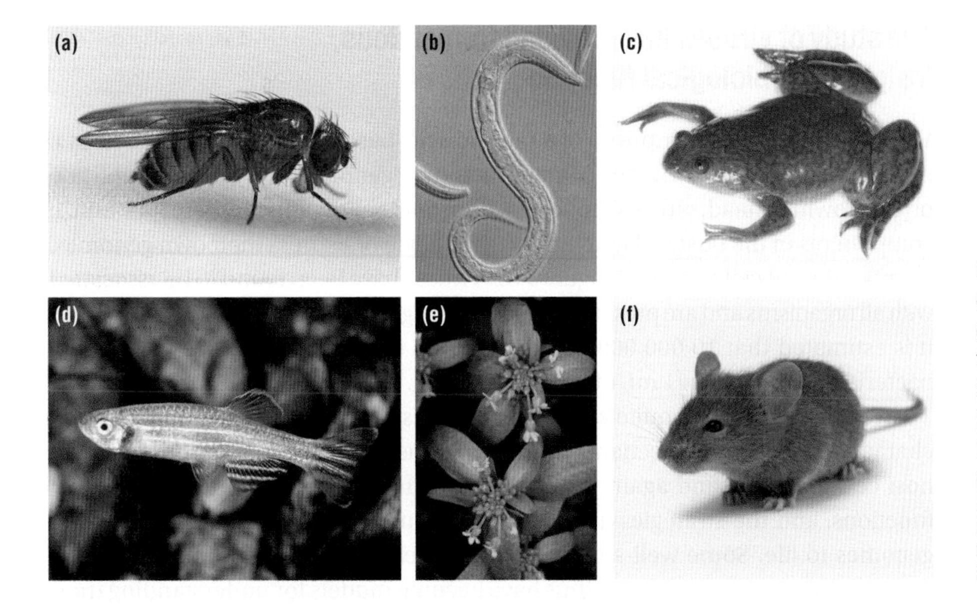

Figure 1.30 Commonly studied multicellular eukaryotes. Model multicellular eukaryotes commonly studied in biology include (a) the fruit fly *D. melanogaster;* (b) the worm *C. elegans*; (c) the frog *X. laevis;* (d) the zebrafish *D. rerio*; (e) the plant *A. thaliana*, (f) and the mouse *M. musculus*

(Image credits (a) Hermann Eisenbeiss/Science Photo Library; (b) Sinclair Stammers/Science Photo Library; (c) Zigmund Leszcynski/Photolibrary Group.com; (d) Oxford Scientific (OSF) Photolibrary Group; (e) Dr. Jeremy Burgess/Science Photo Library; (f) Oxford Scientific (OSF)/Photolibrary Group.)

S. pombe. Several aspects of these genomes suggest that *S. pombe* is more closely related to higher eukaryotes such as humans.

A number of multicellular eukaryotes that are commonly studied are shown in Figure 1.30; these include the fruit fly *D. melanogaster*, the worm *C. elegans*, the frog *X. laevis*, the zebrafish *Danio rerio* and the plant *A. thaliana*. The predominant mammalian model organism is the mouse *Mus musculus*. But what gives an organism its 'model' status? Most of the model organisms were chosen for practical reasons: they have advantages in terms of their biological composition, development, or behavior that make them particularly amenable to study. For example, many of the organisms have fast generation times: the average generation time of a fruit fly is two weeks, while an *Arabidopsis* plant is able to produce seed a month after germination. All of these model organisms are also relatively small and can thus be cultivated under typical laboratory conditions.

In addition, each of these organisms has different strengths and weaknesses regarding the types of analysis that can be readily performed. For example, *X. laevis* is a key organism for studying development in a biochemically tractable system since the developing oocyte is of large enough size to permit direct manipulation. However, genetic approaches are not very well developed in this system. By contrast, *D. melanogaster*, *C. elegans*, and *S. cerevisiae* are probably best known for their potential as genetically tractable organisms where genetic screens are readily performed.

It bears remembering that model organisms are simply that – they allow us to build models (or informed estimates) of how biological systems operate but do not allow us to formulate exhaustive descriptions of exactly what is going on in a single investigation. It is only by successive rounds of investigation, sometimes in different model organisms, each perhaps only revealing a small new insight, that our understanding is incrementally increased. As these insights build up, so we refine our understanding, and build gradually more detailed, and more refined, best guesses of the reality of life. Equally importantly, model organisms do not always provide a definitive indication of biological processes throughout the kingdoms of life; it may not be possible to extrapolate an observation made in one organism to a seemingly similar organism.

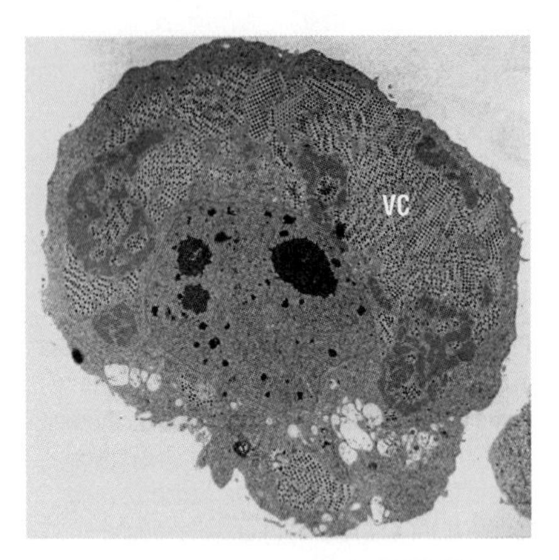

Figure 1.31 Electron micrograph of a virus-infected cell. The picture illustrates the relative size of the crystalline-like virus particles (VC) relative to the eukaryotic cell.

The study of viruses has provided tremendous insights into biological function

Viruses are particles composed of a nucleic acid-based genome surrounded by a protein coat. Although viruses contain their own genomes, they cannot replicate on their own. Instead, viruses replicate within another organism where they utilize components of the host cell machinery to replicate or express their own genomes. A virus-infected cell is show in Figure 1.31. Viruses have been found to be associated with all organisms and are present in an astounding variety and number. For example, it is estimated that 10 000 000 **bacteriophages** (the term for viruses that infect bacteria) are present in 1 mL of coastal seawater.

Given that viruses use quite varied mechanisms to usurp the host cell machinery, characterization of the mechanisms used by viruses to infect and replicate in their host cells has time and again provided fundamental insights into basic cellular functions, and the strategies and mechanisms used by organisms to bring their genomes to life. Some well-studied viruses, such as the bacteriophage Lambda and the Simian virus 40 (SV40), thus have become models for understanding transcription and replication in their own right.

The molecular components of different organisms are described using different nomenclature rules

As a consequence of the large body of work carried out with the model organisms mentioned earlier, many of the examples cited in the subsequent chapters will be from studies involving them. In citing such examples we need to be aware that, despite the common molecular components and processes that operate in different organisms, the ways in which these components are described varies. In particular, the conventions used for naming genes, proteins, and mutant alleles in different organisms - the nomenclature used - vary from organism to organism. For example, in some organisms, a wild-type gene name is written in uppercase letters, whereas the mutant gene name is written in lowercase letters. Sometimes, the gene name is in italics, and the name of the protein resulting from the expression of that gene is written in roman (non-italic) type. Throughout this book, we observe the most widely accepted nomenclature rules for each organism, and these are summarized in the table on page ii at the front of the book.

Also, we need to be aware that homologous genes and proteins in different organisms - that is, genes and proteins with conserved sequences (and, hence, conserved functions) - may have different names. For example, the eukaryotic protein Cdk1 (for cyclin-dependent kinase) that controls cell division is also called Cdc2 (for cell division control protein). These names came about historically as new genes discovered in different organisms were later found to be the same. It is difficult to go back and change the name of a gene that has been frequently used, although sometimes the community of scientists will get together and unify the nomenclature for specific genes. Such name differences can make it more challenging to see patterns and similarities, but they simply reflect the rich history and diversity of research that underpins our current knowledge of the field.

We have now completed our brief tour of biological systems, and the journey from genetic information to the living organism. We have seen how our genome directs our development and function, and how changes in our genome propagate changes in our physical character as evolutionary forces are brought to bear. We have also had our first glimpse of the molecular mechanisms that retrieve the information from our genome, and, by transforming information into activity, bring the genome to life.

In the next two chapters we move from the level of the cell to the level of atoms and molecules, to consider the chemical infrastructure of life, before beginning our more detailed exploration of molecular biology, and the principles of genome function.

THE UNIFYING FEATURES OF LIFE

- The diversity of life is unified by some common themes: a living organism must be distinct from its environment, have a way of storing biological information, be able to replicate and transmit the information to subsequent generations, and have a source of energy from the environment to grow and reproduce.
- The molecular building blocks of all modern organisms, nucleic acids, proteins, lipids, and carbohydrates, are conserved and are constructed from small, repeating subunits.

THE BLUEPRINT OF LIFE: THE GENOME

- Biological information is typically stored in the nucleic acid DNA referred to as the genome.
- A gene is typically defined as a region of DNA that controls a discrete hereditary characteristic. The number and arrangement of genes within a genome varies from species to species; gene number is a poor indicator of biological complexity.

GENE EXPRESSION

- The first stage of gene expression is transcription, in which an RNA copy of the gene is synthesized. The product of transcription may be a functional RNA, which does not encode a protein (a non-coding RNA), or a messenger RNA (mRNA), which does encode a protein (a coding RNA).
- mRNAs are translated to produce a protein product in a process mediated by the ribosome. During translation, the nucleotide sequence of the mRNA is deciphered in three-base triplets called codons; each codon represents a particular amino acid (or tells the ribosome to stop translation).
- A single gene may encode more than one RNA or protein product such that the estimated 30 000 human genes are estimated to encode more than one million gene products.
- The regulation of gene expression in both time and space is extensive and occurs at every level of expression.

THE CELLULAR BASIS OF GENE EXPRESSION

- Eukaryotic cells feature membrane-bound compartments called organelles, such as the nucleus which contains the genome. Bacteria and archaea lack organelles, though the genome is sequestered in a region of the cell called the nucleoid.
- Subcellular compartmentalization provides opportunities to increase the efficiency of cellular events and to regulate biological processes.

EXPRESSION OF WHOLE GENOMES

- The sequence of an organism's genome is its genotype. The overall expression of an organism's genome determines its phenotype (its physical features and properties).
- Null or loss of function mutations eliminate the function of a gene; missense mutations change the identity of a single amino acid within a protein product; and nonsense mutations result in the premature termination of translation to give a truncated protein product.
- Given that modifications in chromosome packaging can be inherited and can lead to phenotypic changes, these modifications, called epigenetic changes, may lead to a change in an organism without a change in DNA sequence.

EVOLUTION OF THE GENOME

- All modern-day life forms evolved from a single common ancestor.
- Random mutations that confer advantageous traits upon an organism will be selected for – that is, there will be an increased likelihood of such mutations being transmitted from generation to generation. This is the basis of the evolution of organisms through the process of natural selection.
- At the highest level, sequence similarities and differences allow us to divide all of organismal life into three domains – bacteria, archaea, and eukaryotes.

STUDYING GENOME FUNCTION

- The study of model organisms provides insights into molecular components and processes, helping us to understand how genomes function.
- Viruses exploit the host cellular machinery to ensure their own propagation. Consequently, the study of viruses has helped to illuminate how the host machinery operates.

1.1 THE ROOTS OF BIOLOGY

1. For each of the terms listed, fill in the table with the appropriate information.

Term	Definition	Importance to life
Cell membrane		
Nucleic acid		
Replication		
Energy		
Polymer		

2. What evidence exists to suggest that all life was derived from a last universal common ancestor (LUCA)?

3. Describe the meaning of each of the following terms with regard to the structure of DNA.

a. Polymer
b. Complementary base pairing
c. Directionality
d. Genome
e. Gene
f. Chromosome

1.2 THE GENOME: A WORKING BLUEPRINT FOR LIFE

1. What is meant by the term 'molecular conservation' and why is it important?

2. What relationship exists between genome size, gene number, and organism complexity? Give an example to support your answer.

Challenge Question

3. Explain the phrase 'biological complexity comes from more than just the genes of an organism'.

1.3 BRINGING GENES TO LIFE: GENE EXPRESSION

1. Which steps are necessary for the information contained in DNA to be used to direct the life of an individual organism?

2. Fill in the following table to highlight the similarities and differences in DNA replication versus transcription.

Feature	Replication	Transcription
Polymerase required		
Mechanism for adding the correct base		
Base complementary to C		
Base complementary to A		
Portion of genome involved		

3. Explain the difference between coding and non-coding RNA.

4. What is the significance of the table depicted in Figure 1.13?

5. It has been estimated that the human genome consists of approximately 30 000 genes. Describe some of the ways in which it is possible for the complete set of functional RNAs and proteins to be over one million.

1.4 REGULATING GENE EXPRESSION

1. What is meant by the phrase 'gene expression is regulated in time and space'?

Challenge Questions

2. Histone proteins are expressed in cells in the specific pattern that is depicted in this figure. The pattern is generated through multiple levels of gene regulation.

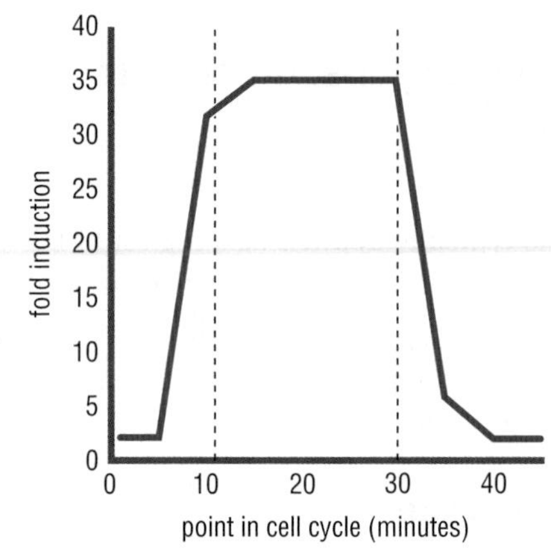

a. Describe the pattern shown in the graph and explain why it makes sense with regard to the timing of histone production.
b. Propose a basic regulatory mechanism by which the cell can control the level of histone protein in a way that produces the pattern shown in the graph.

3. How might the regulation of gene expression in time and space be important in the process of implanting an embryo during the female reproductive cycle?

1.5 CELLULAR INFRASTRUCTURE AND GENE EXPRESSION

1. Compare and contrast the nucleoid and the nucleus.

2. In addition to the nucleus, many cells contain a wide variety of other compartments. Describe some of the advantages of compartmentalization.

3. Compare the process by which proteins in bacteria and eukaryotes are made and secreted from cells.

Challenge Questions

4. In a leaf cell, both the production of sugars (photosynthesis) and the breakdown of sugars (cellular respiration) occur. Explain how the cell is capable of managing these two opposing biological processes.

5. In bacteria, transcription and translation occur in the same cellular compartment while in eukaryotes, they are separated. Give one advantage for each condition.

1.6 EXPRESSION OF THE GENOME

1. What is the relationship between ploidy and the genotype and phenotype of an organism?

2. Compare and contrast forward and reverse genetics.

3. Define the following terms with regard to their impact on human disease.

a. Somatic cell mutation
b. Germline mutation
c. Monogenic
d. Polygenic
e. Penetrance

4. Which of the following terms refers to a heritable change that alters the expression of a gene without altering the DNA sequence of a gene?

a. Compartmentalization
b. Epigenetics
c. Forward genetics
d. Recessive mutation
e. Somatic mutation

Challenge Questions

5. Figure 1.22 shows an *Arabidopsis* mutant that grows tall due to its inability to respond to inhibition by blue light.

a. What is the difference between a somatic cell mutation and a germline mutation?
b. How would you determine if the *Arabidopsis* mutation is somatic or germline?
c. What is the difference between a dominant and a recessive mutation?
d. How would you know if the *Arabidopsis* mutation is dominant or recessive?

6. If a normal zebrafish with a striped pattern on its body (wild-type) is crossed with a zebrafish that has a spotted pattern (leopard), all of the offspring are wild-type, but the second generation features a small number of leopard spotted offspring. Additionally, a comparison of the sequence of the gene responsible for stripes revealed that the leopard mutant had a premature stop codon in the amino acid sequence. Which of the following terms can be used to categorize and describe this mutation? Explain why you chose each answer.

a. Dominant
b. Recessive
c. Insertion
d. Deletion
e. Null
f. Missense
g. Nonsense
h. Silent
i. Somatic cell
j. Germline cell
k. Monogenic
l. Polygenic

7. Based on the following wild-type sequence, indicate if each of the mutations should be classified as either an insertion, deletion, missense, nonsense, or silent.

Wild-type:	AUG AUA CUA GAA AAC UGA	MILEN
a.	AUG AUA UUA GAA AAC UGA	*MILEN*
b.	AUG AUA CUA UAA AAC UGA	*MIL*
c.	AUG AUA CUA GAA AAC UAA	*MILEN*
d.	AUG AUC CUA AAC UGA	*MILN*
e.	AUG AUA CUA GAA AAA UGA	*MILEK*

8. For each disease state below, give a brief description of the disease and indicate which terms from Question 1.6.3 would apply.

a. Duchenne muscular dystrophy
b. Smith Magenis syndrome
c. Metabolic syndrome
d. Melanoma

1.7 EVOLUTION OF THE GENOME AND THE TREE OF LIFE

1. What is the relationship between DNA mutation, natural selection, and evolution?

2. What is a phylogenetic tree and how was the tree in Figure 1.28 generated?

3. Bacteria and Archaebacteria have been separated into two different domains of life.

a. Why were they originally put together?
b. What recent evidence separated them into two domains?

4. What is a genetic model organism? Give an advantage and a disadvantage to studies in model organisms.

Challenge Question

5. In a hypothetical forest, beetles of one species had an exoskeleton color that ranged from white to black. The color was dependent on the amount of black pigment produced by one gene. The amount ranged from none (white) to 100% (black). All of the beetles were able to find camouflage in the forest and were equally protected from bird predators. One day, a fire burned through the forest and turned the environment black.

a. Explain the mechanisms by which the variety of colored exoskeletons may have been produced and maintained in the population.
b. What natural selection is likely to occur due to the fire?
c. What is the likely outcome of natural selection in the next generation of beetles?
d. What would have happened in this scenario if there had been no variation in beetle exoskeleton color?
e. What is the relationship between this scenario and the development of the diversity of life?

Biological molecules

INTRODUCTION

Despite the extraordinary diversity of Earth's creatures, almost all processes necessary for life depend on a similar set of molecules. The transmission of genetic information, the manufacture of molecules that carry out all the many different tasks of a cell, the synthesis and breakdown of key nutrients, and the very structure of the cell itself depend on a variety of molecules that are uniquely suited to their tasks.

There are four major classes of biological molecules that play essential roles in all organisms:

- nucleotides
- amino acids
- carbohydrates
- lipids.

All of these molecules contain carbon atoms and are therefore called **organic** compounds. As shown in Figure 2.1, each of them can be found in cells both as individual small molecules or covalently linked to form larger molecules known as **polymers** or **macromolecules**. **Nucleic acids** are polymers of nucleotides that are responsible for carrying genetic information. Proteins, on the other hand, are polymers of **amino acids** that function as workhorses, carrying out most of the chemical reactions in the cell or giving cells their structure and shape. **Fatty acids** are key components of **lipids**, the molecules that form membranes that encapsulate cells, separating the cell contents from the environment and also helping to organize the cell's interior into specialized compartments. Polymers of sugars, called **carbohydrates**, serve diverse functions such as storing energy obtained from food, modulating the function of both proteins and lipids, and providing mechanical strength to cells. All of these molecules sometimes carry out other functions in addition to their principal roles.

In this chapter, we will review the basic properties of atoms and learn how they are joined together to form molecules. We will then see how the unique properties of the four major classes of biological molecules enable them to carry out distinct functions in the cell. Throughout, we will see how the behavior of all biological molecules is dominated by the properties of water (Figure 2.2). Cells consist of 70%–80% water by mass, which influences every aspect of a cell's biology. Water has also been a driving force throughout the evolution of life: life originated in the primordial seas, and the chemistry of modern-day organisms reflects their origins in that aqueous environment.

We begin our exploration of biological macromolecules by considering the individual components – the atoms – from which they are formed.

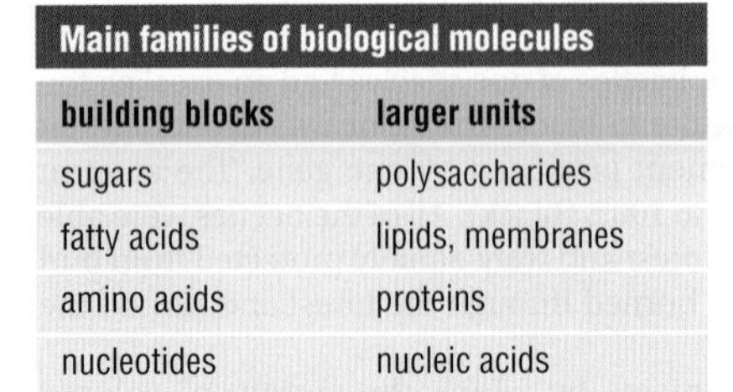

Main families of biological molecules

building blocks	larger units
sugars	polysaccharides
fatty acids	lipids, membranes
amino acids	proteins
nucleotides	nucleic acids

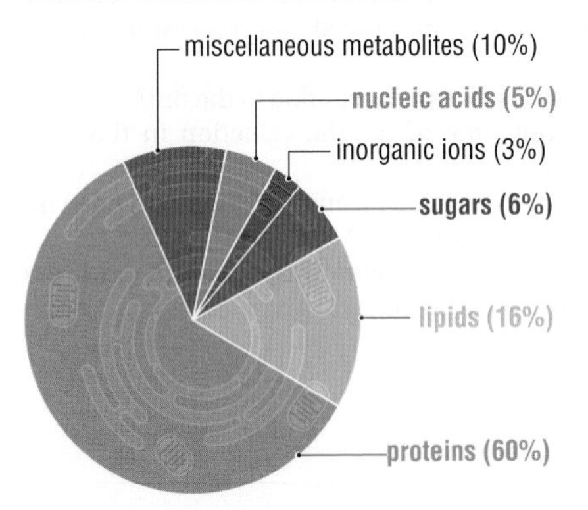

Figure 2.1 The four major classes of biological molecules. The table (top) lists the four major types of biological molecules, along with their building blocks. The pie chart (bottom) shows the relative percentage of the cell mass (excluding water) contributed by each class of molecules.

2.1 ATOMS, MOLECULES, AND CHEMICAL BONDS

The structure of atoms dictates their chemical reactivity

All matter in the universe is composed of a limited set of elements that obey the laws of physics and chemistry. The smallest unit of an element that retains its defining chemical properties is known as the atom. Atoms are, in turn, composed of subatomic particles: positively charged protons and electrically neutral neutrons, which are clustered in a central nucleus and surrounded by negatively charged electrons. The resultant atom is electrically neutral, with the positive protons and negative electrons balancing one another precisely. Different types of atoms (also known as elements) are distinguished by the number of protons they contain; this is known as the atomic number. For example, hydrogen (H), the lightest element, has a single proton and thus an atomic number of 1, whereas calcium (Ca) has 20 protons and thus an atomic number of 20. The atomic mass of an atom, given in units called daltons (Da), is equal to the number of protons plus neutrons in an atom. Carbon, for example, typically has six protons and six neutrons and therefore has an atomic mass of 12 Da. There are altogether 92 naturally occurring elements, of which just a handful predominate in living things, as listed in Figure 2.3. Four elements alone - carbon (C), oxygen (O), hydrogen (H), and nitrogen (N) - account for over 96% of the mass of an organism.

The arrangement of electrons in each atom determines its interactions with other atoms. Electrons are found in discrete states of varying distance from the nucleus, known as shells. The innermost shell contains up to two electrons, the second and third shells can accommodate eight electrons each, and the fourth shell contains up to 18 electrons. Figure 2.4 depicts the electron shells of hydrogen, carbon, nitrogen, and oxygen. An atom is most stable and chemically unreactive when its outermost electron shell is completely filled, that is, when it contains the maximum number of electrons for that shell.

Figure 2.2 Water influences the structure of cells and biological molecules. Jackson Lake, Grand Teton National Park, Wyoming.
Photo courtesy of CW.

Common elements in biology

element	symbol	atomic number	atomic mass (Daltons)	proportion of total body mass (%)
hydrogen	H	1	1	10
carbon	C	6	12	23
nitrogen	N	7	14	2.6
oxygen	O	8	16	61
sodium	Na	11	23	0.14
magnesium	Mg	12	24	0.027
phosphorus	P	15	31	1.1
sulfur	S	16	32	0.2
chlorine	Cl	17	35	0.12
potassium	K	19	39	0.2
calcium	Ca	20	40	1.4

Figure 2.3 Common elements in biology. The table lists the most common elements in biology, along with the symbol, atomic number (the number of protons), atomic mass of the most common isotope, and the proportion of the total body mass made up by each element.

Covalent bonds form when electrons are shared between atoms

An atom whose outer shell is not filled is chemically reactive, meaning that it can participate in chemical reactions with other atoms to form molecules. Why does having an unfilled outer shell make an atom chemically reactive? The answer lies in the way that an atom with an unfilled outer shell can share electrons with another atom. By sharing electrons, the atom can achieve a full outer shell, and becomes more stable as a result. This situation is depicted in Figure 2.5 for hydrogen and oxygen, which can combine to form a water molecule. Hydrogen has just one electron in its outer shell instead of the allowed two, while oxygen has six electrons in its outer shell instead of the allowed eight. When two hydrogen atoms share their electrons with an oxygen atom, each atom now has a complete outer shell and is therefore more stable.

Sharing of electrons between atoms is the basis of **covalent bonds**, whereby atoms are joined to one another to form **molecules**. The sharing of one pair of electrons gives rise to a **single bond** between two atoms (Figure 2.5a). Sharing of two pairs of electrons constitutes a **double bond** (Figure 2.5b). Covalent bonds are typically very stable; it takes a substantial amount of energy to break them. The precise amount of energy needed to break a covalent bond depends on the types of atoms involved and the number of bonds they form with one another, with double bonds generally being stronger than single bonds. The strength of covalent bonding makes biological molecules stable under the physiological conditions prevailing in cells.

A more complex bonding pattern can occur when an atom is joined to its neighbors by both single and double bonds. An important example in biology is the

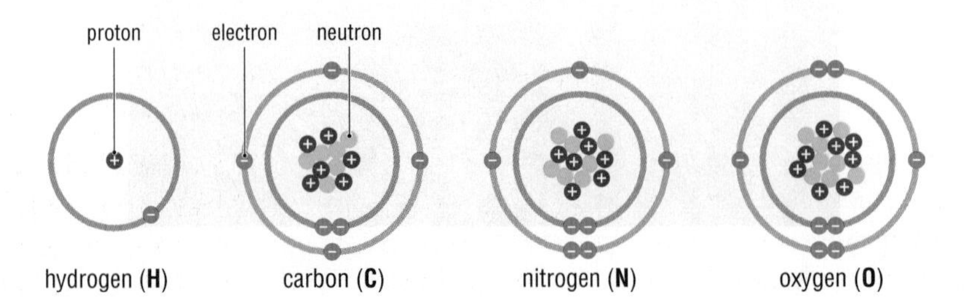

Figure 2.4 Atomic structure of hydrogen, carbon, nitrogen, and oxygen. Atoms are composed of a central nucleus of protons (red) and neutrons (green) surrounded by shells of electrons (blue) in which the first shell contains up to two electrons and the second up to eight electrons. Hydrogen has only one electron in its outer shell; carbon has four electrons; nitrogen has five electrons; and oxygen has six electrons instead of the allowed eight.

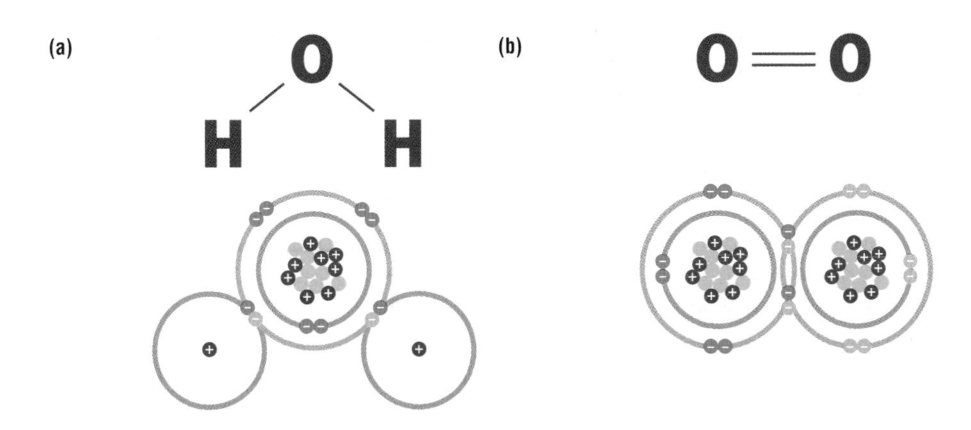

Figure 2.5 Covalent bonds. (a) A water molecule forms when each of two hydrogen atoms shares an electron with an oxygen atom. Each hydrogen atom forms a single covalent bond with the oxygen atom. (b) Oxygen is commonly found as the molecule O_2, in which two oxygen atoms share electrons. Each shared electron pair represents a single covalent bond. The two oxygen atoms therefore form a double covalent bond with one another.

peptide bond that, as we shall see later, joins amino acids together in a protein. Although we typically draw a double bond between the carbon and oxygen and a single bond between the carbon and nitrogen atom, there is actually a sharing of electrons between the three atoms that gives both the C–O and C–N bonds partial double bond character, as indicated in Figure 2.6. We shall see below how partial bonds constrain the shapes that a molecule containing them can adopt.

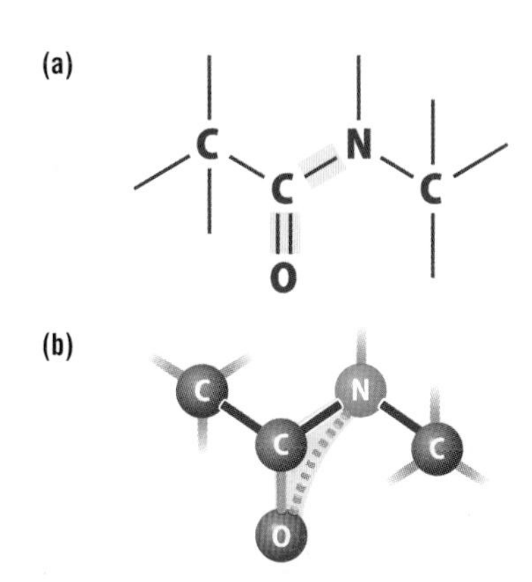

Figure 2.6 Partial double bonds. (a) Carbon atoms flanked by oxygen and nitrogen atoms are usually written as having one double bond and one single bond. However, there is actually a sharing of electrons as shown in (b), such that both bonds have partial double bond character.

Shape and flexibility of a molecule are determined by its electrons

The distribution of electrons among its component atoms determines the overall shape of a molecule. When an atom participates in a covalent bond with more than one other atom, the covalent bonds will be arrayed in a configuration that is characteristic of the particular atom, as illustrated in Figure 2.7. Carbon atoms, which can share all four of the electrons in their outer shell, make bonds that radiate outward and are equally spaced from one another, as if pointing towards the corners of a tetrahedron. The two bonds that can be formed with an oxygen atom form a V shape, whereas the unshared electron pairs are located on the opposing side of the oxygen atom.

The number of bonds that are formed between two atoms places an additional constraint on molecular geometry. Although atoms can rotate freely about a single bond, double bonds have properties that create a barrier to rotation. Even partial double bond character prevents rotation significantly, which means that electron sharing of the type shown in Figure 2.6 will also constrain the **conformation** of molecules that contain them. Together, bond order and the spatial arrangement of covalent bonds about each atom are important factors in determining the shapes of biological molecules. For example, the peptide bond, which joins adjacent **amino acid residues** in a **polypeptide** chain (Section 2.9), has partial double bond character (see Figure 2.6b). As a consequence, free rotation about two of the covalent bonds that comprise the peptide bond 'unit' is not possible, making the peptide bond itself planar, and placing restrictions on the three-dimensional conformations that a polypeptide can adopt.

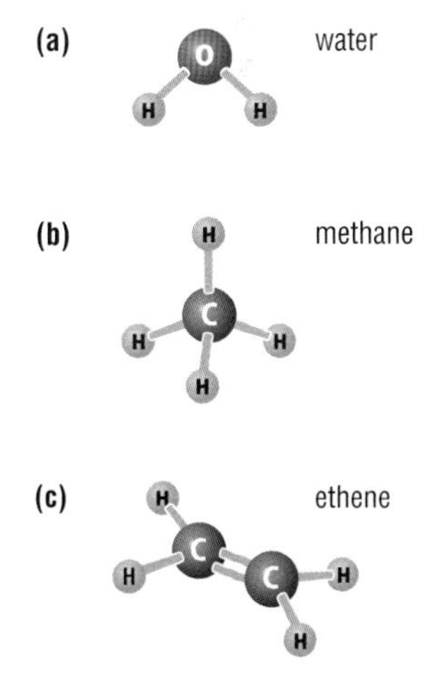

Figure 2.7 Geometry of bonds in oxygen and carbon atoms. (a) The bonds in an oxygen atom are positioned so that two single bonds have a V configuration, as shown for bonds with hydrogen in a water molecule. (b) Carbon can form four single bonds related by tetrahedral geometry. (c) Where two atoms participate in a double bond, other atoms bound to the same carbon atom are forced to lie in the same plane because the double bond prevents rotation around it.

The effective size of an atom is determined by its electrons

Atoms that are not joined to one another by covalent bonds behave in many ways like hard spheres with a defined radius. Just as two billiard balls cannot interpenetrate one another, non-bonded atoms can only come within a certain distance of one another before they begin to clash. The radius of the sphere is a characteristic

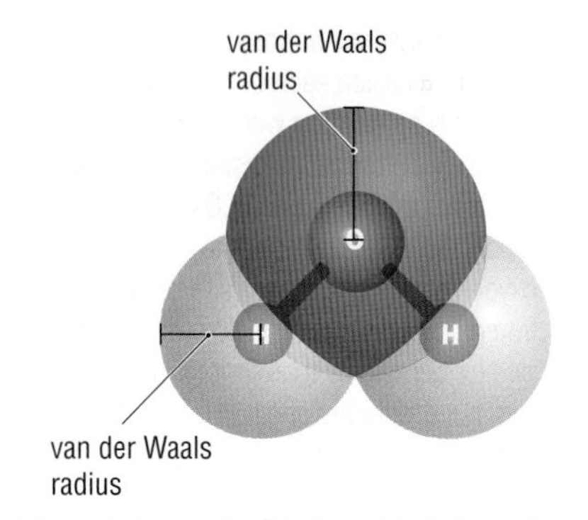

Figure 2.8 van der Waals radii of atoms in a water molecule.

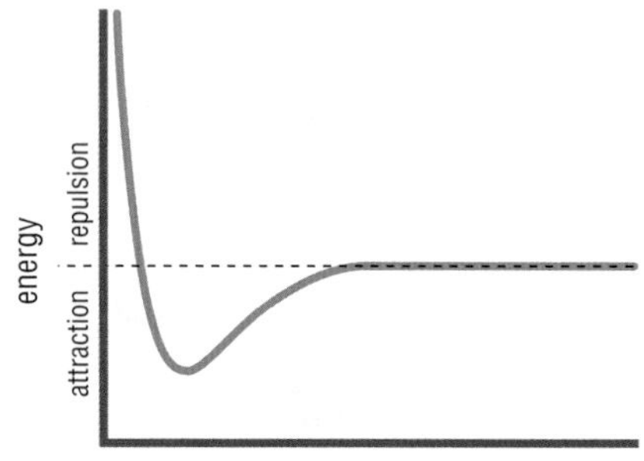

Figure 2.9 Interaction energy between two non-bonded atoms. The plot shows the interaction energy between two atoms as a function of the distance between them. The minimum in the plot occurs at a distance equal to the sum of the van der Waals radii, at which two atoms form the most favorable van der Waals interaction. At shorter distances, the atoms start to clash, and the energy increases greatly.

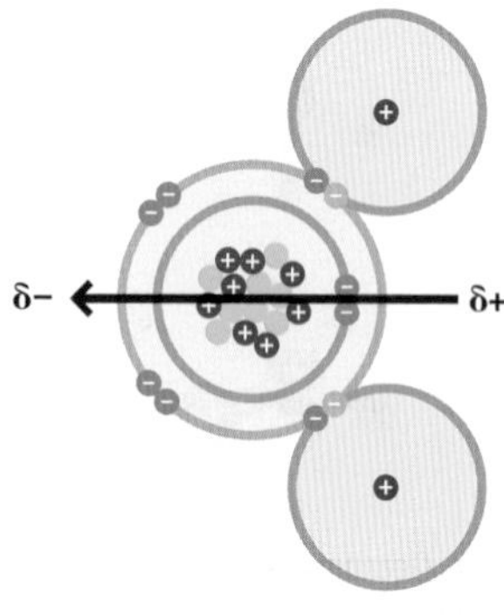

Figure 2.10 Unequal sharing of electrons confers molecular polarity. The oxygen atom (blue) in a water molecule takes a greater share of the shared electrons than each hydrogen atom (red). As a result, the oxygen atom has a partial negative charge (indicated as δ−) and each hydrogen atom has a partial positive charge (indicated as δ+).

of each type of atom, as depicted in Figure 2.8, and is called the **van der Waals radius.** When atoms come closer than the sum of their van der Waals radii, the interaction energy becomes highly unfavorable and they repel one another. When atoms are very far apart – that is, at distances equal to many times their van der Waals radii – they do not interact at all and the interaction energy is therefore nearly zero. However, atoms that are separated by just around the sum of their van der Waals radii weakly attract one another. Biologists call these attractions **van der Waals interactions**; they are also known as London dispersion forces. We will learn about the physical basis for this attractive force, as well as the role it plays in interactions between biological macromolecules, in the next section.

The way in which both attractive and repulsive forces vary with the distance between two atoms is shown in Figure 2.9. Look at this figure, and note how two atoms repulse each other when the nuclei are very close to one another, but then attract each other as the nuclei move slightly further apart. As the distance between the nuclei increases further, however, this attraction decreases until there is no net attraction or repulsion between the two nuclei.

Electrons are not always shared equally by atoms joined by covalent bonds

Electrons are often more closely associated with one of the two atoms in a covalent bond than the other. As a result, the atom with the greater share of the electrons has a partial negative charge and is therefore **electronegative** (denoted δ−), while the atom with the lesser share has a partial positive charge and is therefore **electropositive** (denoted δ+). This separation of charge is known as a **dipole**, and is represented in Figure 2.10; such covalent bonds are said to be **polar.** Molecules or chemical groups containing bonds in which electrons are shared more equally are **non-polar.**

Whether a chemical group is polar or non-polar has important consequences for its behavior in the aqueous environment of the cell, which consists predominantly of water molecules that are themselves polar. Polar chemical groups interact most favorably with polar molecules such as water, while non-polar groups prefer to interact with one another rather than with the polar solvent. Many molecules have both polar and non-polar regions. For example, the lipid molecules that form the double-layered cell membrane have a polar head and non-polar tail. It is this marked difference between the two regions of the lipid that drives the way multiple lipid molecules interact to form the cell membrane structure – with polar heads associating to form the outer faces of the cell membrane, and the non-polar tails associating to form the membrane interior. We will revisit lipids and membrane formation later in the chapter, when we see how the tendency of non-polar groups to associate with one another drives much of the behavior of biological macromolecules.

Atoms can become fully charged

As noted above, unequal sharing of electrons in polar bonds can give rise to a partial electrostatic charge on one or more atoms. However, because of their energetic preference to have a completely filled outer shell, atoms sometimes gain or lose electrons entirely and therefore become fully electrostatically charged. Atoms or chemical groups that become charged in this way are said to be **ionized** and are known as **ions.** Elements whose outer shells are just one electron short of being

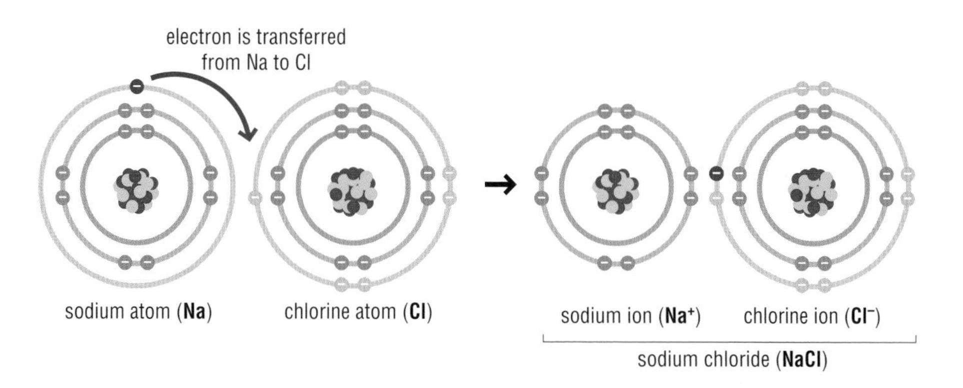

Figure 2.11 Ionic interaction between a sodium and a chlorine atom. The sodium atom has only one electron in its outer shell; the chlorine atom has seven. Loss of an outer shell electron from the sodium atom and gain of an electron by the chlorine atom leaves both atoms with filled outer shells. The result is a full positive charge on the sodium atom, which is thus a cation, and a full negative charge on the chlorine atom, which is thus an anion. The oppositely charged ions are strongly attracted to one another and form a salt.

filled can acquire an electron and thereby become negatively charged. Conversely, atoms that have just one or two outer shell electrons can lose these electrons and become positively charged. Atoms with opposite charges interact favorably with one another to form salts. A familiar example is the interaction between the positively charged sodium ions and negatively charged chloride ions in table salt as shown in Figure 2.11. This attraction between fully charged atoms, known as an ionic interaction, can also occur between chemical groups that bear full positive or negative charges. Full charges on chemical groups generally occur when a positively charged hydrogen ion associates with or dissociates from the group, as we shall see later in this chapter. Most biological molecules contain charged groups, and ionic interactions are therefore common in biology.

2.2 LIFE IN AQUEOUS SOLUTION

The chemistry of life - the biochemical reactions that keep us alive - occurs not in a test tube in the laboratory, but within the cells of our body. Our cells contain the cytoplasm, the water-based (aqueous) environment in which the biochemical reactions of life occur. For such a simple molecule, comprising just one oxygen atom and two hydrogen atoms, water has a number of important properties that make it particularly suited to being the medium of life. These include the ability to form hydrogen bonds (see Section 2.3) and the ability to become ionized. Without these properties, water would be unable to support life in the way that it does; it would be a poor solvent for many biological molecules, and would be a poor medium for biochemical reactions.

Water can become ionized

Any aqueous solution contains some water molecules that have dissociated into positively and negatively charged ions. As shown in Figure 2.12, a water molecule can lose a proton, denoted H^+, from one of its hydrogen atoms, leaving a negatively charged hydroxide ion, OH^-. The proton associates most favorably with the partial negative charge of the oxygen atom of another water molecule, forming the hydronium ion, H_3O^+. In a solution of pure water, only a very small percentage of water molecules will dissociate in this way; the concentration of OH^- and H_3O^+ ions is just 10^{-7}M. Importantly, the concentration of OH^- and H_3O^+ ions in a solution are intrinsically linked: the product of the concentration of these two ions, $[OH^-] \times [H_3O^+]$, is always equal to 10^{-14}. This relationship always holds true, even when the actual concentrations of OH^- and H_3O^+ ions change, as we will see next.

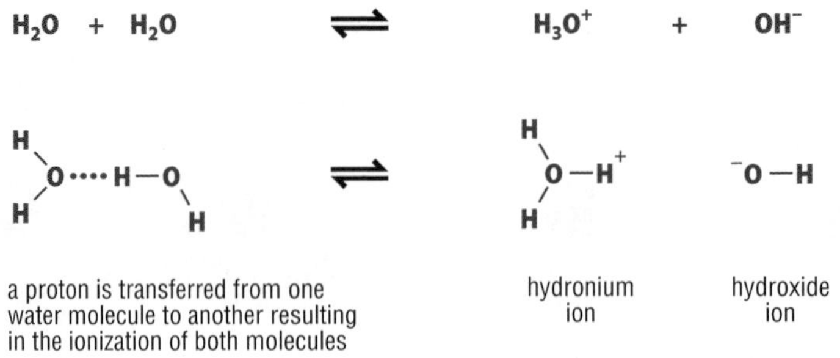

$$H_2O + H_2O \rightleftharpoons H_3O^+ + OH^-$$

a proton is transferred from one water molecule to another resulting in the ionization of both molecules

hydronium ion

hydroxide ion

Figure 2.12 The ionization of water. Water molecules can become ionized by gaining or losing a proton. Gain of a proton gives rise to the positively charged hydronium ion; loss of a proton gives rise to the negatively charged hydroxide ion.

pH denotes the concentration of hydrogen ions in solution

Hydrogen ions drive many reactions in biology, so their concentration is of great interest and importance. The concentration of hydrogen ions can change due to the ionization of molecules present in a given aqueous solution. A variety of compounds can become ionized in aqueous solution to release either OH^- or H^+ ions. The H^+ ions (protons) associate with water molecules to form H_3O^+ ions, whereas the OH^- ions remain 'free' in solution.

When there is an excess of either OH^- or H_3O^+ ions, there will be a decrease in the concentration of the other such that the product of the concentration of OH^- and H_3O^+ ions is preserved, as per the relationship $[OH^-][H_3O^+] = 10^{-14}$ noted above. Molecules that are able to release H^+ when they dissolve in a solution are called **acids**, while molecules that are able to absorb (or 'accept') H^+ to reduce the concentration of H^+ in a solution are called **bases.** Although virtually all free H^+ ions associate with water to form H_3O^+, it is a convention in biology to refer to the excess H^+ as if they were free in solution.

In water, the H^+ concentration can vary greatly. This range in H^+ concentration is conveniently expressed in terms of **pH**, which corresponds to the negative log of the H^+ concentration; this log scale is depicted in Figure 2.13. For example, if the concentration of H^+ in a given solution is 10^{-5} moles/L, the solution has a pH of 5 ($-\log 10^{-5} = 5$). In pure water, the H^+ concentration is equal to 10^{-7} moles/L. Pure water thus has a pH of 7, which we call 'neutral.' Solutions with a pH less than 7 are considered acidic. Conversely, solutions with a pH greater than 7 are called basic or alkaline. In the coming sections, we will see how the pH of a solution affects the ionization state of a variety of biological molecules.

The effect of pH on the properties of a biological molecule is important when we realize that the pH of the aqueous environments to which molecules are exposed in different parts of the human body can vary considerably. For example, the pH of the blood is 7.4 (almost neutral), whereas that of the gastric juices in the stomach is 1.6, which is very acidic. These variations in pH create challenges for pharmaceutical companies in the design of therapeutic drugs: a drug may have certain physical and chemical properties at one pH, but different properties at

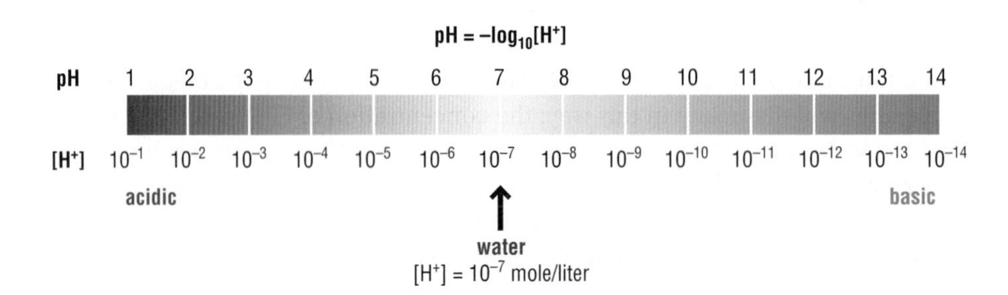

Figure 2.13 The pH scale. The concentration of H^+ in a solution is expressed as the $-\log_{10}$ of the H^+ concentration.

another pH. Drugs that are meant to be taken orally must be able to pass through the very acidic environment of the stomach without being destroyed, and therefore must be stable even at very low pH. A drug that is formulated for injection into the blood stream, however, simply needs to function at neutral pH.

2.3 NON-COVALENT INTERACTIONS

In Section 2.1, we explored the ways in which atoms come together to form molecules by forming covalent bonds, which hold atoms together as molecules. However, a number of weaker, non-covalent interactions are essential to the way different parts of a single molecule associate to give a molecule its defined three-dimensional shape, and to the way different molecules come together to form larger complexes. In this section, we explore non-covalent interactions, and discover their vital role in biological systems.

Non-covalent interactions allow dynamic interactions between molecules

The binding interactions that occur between biological molecules involve energies that are far lower than those of the covalent bonds, and can therefore be more readily disrupted and re-formed as they mediate dynamic processes in the cell. These weaker interactions, known as **non-covalent interactions**, underlie processes that require rapid or transient interactions, such as proteins binding to DNA to regulate gene expression, muscle proteins associating and then disengaging as the muscle fibers contract, and signaling proteins binding at cell surfaces. Non-covalent interactions also hold together more stable assemblages such as the molecular components of a ribosome or the two strands of a DNA molecule. These complexes have a lifetime of days or even years, but their components must be able to move apart and assemble to allow a protein chain to be synthesized, or the genetic code to be read.

All non-covalent interactions are fundamentally electrostatic in nature, which means that they depend on the attractive forces between opposite charges. There are three principal types of non-covalent interactions as shown in Figure 2.14:

- salt bridges, which are ionic interactions
- hydrogen bonds, which are interactions between polar atoms bearing partial charges
- van der Waals interactions, which arise from attractive forces between atoms of chemical groups.

Non-covalent interactions

type of interaction		distance between interacting atoms	energy of interaction (kJ/mole)
salt bridge	$—CH_2—COO^-$ ⋯ $H_3\overset{+}{N}—(CH_2)_4—$	0.27 nm	12.5–17
hydrogen bond	donor N—H •••••• O=C acceptor	0.27 nm	2–6
van der Waals	$—CH_3$ ⋯ $H_3C—$	0.38 nm	3–4

Figure 2.14 Non-covalent interactions. Three types of non-covalent interactions are shown. The distance over which each interaction operates is measured between the centers of the atoms involved.

As we will see later, biological molecules bind to one another through a combination of these interactions.

A salt bridge is an ionic interaction between charged chemical groups

We saw in the previous section that chemical groups within a molecule can become ionized, and that groups with opposite charges can interact with one another, as in ionic compounds such as salt. We therefore call an ionic interaction between chemical groups a **salt bridge** (see Figure 2.14). Since the electrostatic field of a fully charged atom extends at equal strength in all directions, the energy of a salt bridge does not depend on the precise orientation of the two oppositely charged groups and is only a function of the distance between them. This attractive energy between opposite charges falls off with increasing distance, *r*, between the two charges and is proportional to $1/r$. (By contrast, the force operating between the two charges is proportional to $1/r^2$.)

Hydrogen bonding between two dipoles is a common non-covalent interaction

Electrostatic interactions can also occur between atoms bearing partial charges, such as those that participate in polar covalent bonds. A **hydrogen bond** arises from electrostatic interactions between the positive end of one dipole and the negative end of another (see Figure 2.14). The positive end is located on a hydrogen atom that is joined by a polar bond to an electronegative atom, and the negative end is located on a different electronegative atom, usually oxygen or nitrogen.

A simple example of hydrogen bonding occurs between water molecules, with the electropositive hydrogen of one water molecule forming a hydrogen bond with the electronegative oxygen in another - a situation depicted in Figure 2.15. We refer to the hydrogen atom as the **hydrogen bond donor** and the electronegative partner - in this case the oxygen atom - as the **hydrogen bond acceptor**. Because hydrogen bonds depend on dipoles, whose electric field varies with angle about the dipole, the energy of a hydrogen bond is greatest when the three atoms involved lie in a straight line, as shown in Figure 2.15.

Hydrogen bonds can impose a high degree of specificity on the interactions between two binding partners, since the interaction energy is most favorable only if the hydrogen bond donors and acceptors on each macromolecule pair up at the binding interface so that the hydrogen bonds that form have the appropriate geometry and distance from one another. As is the case for all types of interactions, there is also an optimal distance at which the energy of hydrogen bonding is greatest.

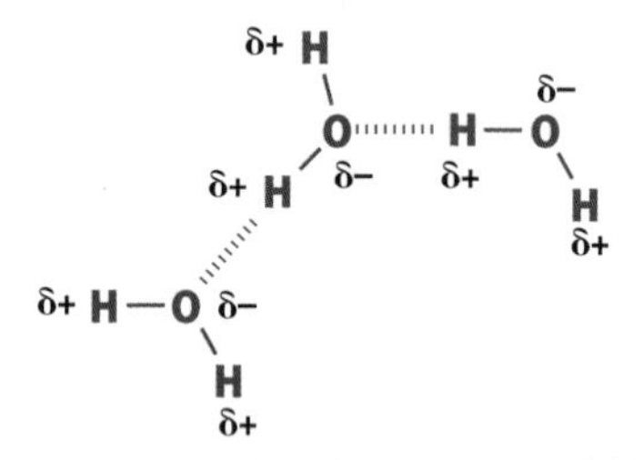

Figure 2.15 Hydrogen bonds between water molecules. Hydrogen bonds between water molecules with a partial negative charge on the oxygen atoms and a partial positive charge on the hydrogen atoms.

van der Waals interactions can occur between any two atoms

Any two atoms may attract one another, whether or not they bear a full or partial charge. We saw in the previous section that atoms separated by somewhat more than the sum of their van der Waals radii weakly attract one another. This attraction is the **van der Waals interaction**. The van der Waals interaction arises when the close approach of two atoms causes each atom to induce transient dipoles in the other atom. These dipoles occur because the electron distribution around each atom varies from moment to moment. The distribution is not equal, but varies in an uneven, asymmetrical way. Figure 2.16 shows how the uneven electron distribution produces a dipole, which in turn can induce a dipole of opposite polarity in a neighboring atom. The resulting favorable energy is due to attractive electrostatic forces between the opposing dipoles.

van der Waals interactions can occur between any two atoms, and are particularly important in interactions between non-polar atoms or chemical groups that cannot participate in other types of non-covalent interaction.

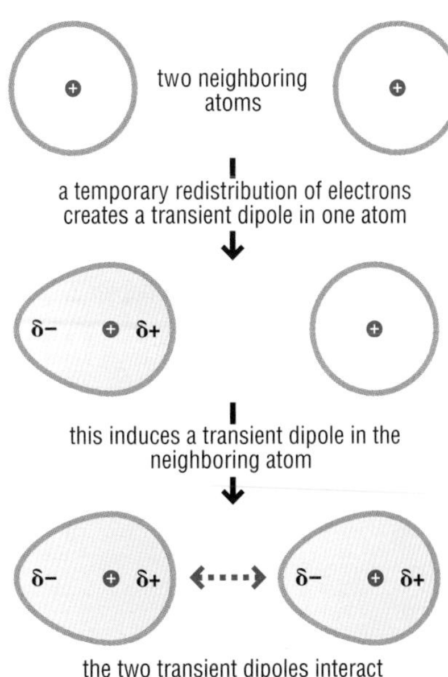
two neighboring atoms
a temporary redistribution of electrons creates a transient dipole in one atom
δ− δ+
this induces a transient dipole in the neighboring atom
δ− δ+
δ− δ+
the two transient dipoles interact favorably

Figure 2.16 van der Waals interactions. Non-polar atoms typically have a uniform distribution of electrons about the nucleus, and hence no dipole moment. From time to time, there is a spontaneous redistribution of electrons about the nucleus that gives rise to a partial negative charge to one side and a partial positive charge at the opposite pole. If another atom is nearby, the transient dipole in the first atom can induce a transient dipole of opposite polarity in the second, such that the two atoms form favorable electrostatic interactions.

The aqueous environment of the cell affects the energy of non-covalent interactions

The behavior of all biological molecules is dominated by the aqueous environment of the cell. The interactions we have described occur in the presence of many water molecules, as well as different types of ions within the cell. This affects the energy of the interactions, and also influences which types of interaction are most favorable. Salt bridges, for example, seldom have the large favorable energy one would expect from the attractive force between full opposite charges because the surrounding water molecules and ions also interact with charged groups, thereby reducing the attractive energy of the salt bridge interaction.

The aqueous environment in the cell influences not only the strength of interactions between molecules, but the types of interactions that occur. Polar molecules and highly charged molecules dissolve readily in water because of the many hydrogen bonding interactions they can make with water. By contrast, highly non-polar molecules do not make energetically favorable interactions with water and prefer, instead, to associate with one another. We say that these non-polar molecules and chemical groups are **hydrophobic**, or 'water-fearing,' because they separate from water. For example, oils are hydrophobic compounds that separate from water: when oil and water are mixed, the oil molecules associate with one another to form a discrete layer that floats on the surface of the water.

Polar and charged molecules that make favorable interactions with water are termed **hydrophilic**, or water-loving. The distinct behavior of hydrophilic and hydrophobic chemical groups in water dictate the three-dimensional structure of proteins and nucleic acids, the formation of cell membranes, and the way in which biological molecules interact with one another.

Hydrophobic interactions drive the folding of biological macromolecules

The terms 'hydrophilic' and 'hydrophobic' can also be applied to whole molecules. Instead, different portions of a single molecule that may contain chemical groups that have either hydrophilic or hydrophobic character. For example, one region of a polypeptide chain may be hydrophobic in character, whereas another region may be more hydrophilic.

In an aqueous environment, the hydrophobic portions of molecules preferentially cluster together, driven by the preference of hydrophobic groups to associate with one another. This 'clustering' usually results in the folding of hydrophobic regions towards the center of a macromolecule, such that these regions are shielded from their aqueous surroundings. By contrast, hydrophilic regions of a molecule can interact favorably with an aqueous environment, and will preferentially lie on the surface of a molecule, where they are exposed to the surrounding aqueous medium.

This partitioning – with hydrophobic regions of a molecule shielded from an aqueous environment, and hydrophilic regions directly exposed to it – is an important driving force behind the way that macromolecules such as nucleic acids and proteins **fold** into their characteristic three-dimensional shapes. For example, the characteristic three-dimensional structure of DNA – the double helix – has the hydrophobic bases located at the center of the molecule, and the hydrophilic sugar–phosphate backbone exposed more directly to the aqueous surroundings.

➔ We learn more about the thermodynamic driving forces behind hydrophobic interactions in Chapter 3.

Hydrophobic interactions are often grouped under the general umbrella of 'non-covalent interactions', although they refer more to the partitioning of hydrophobic groups in an aqueous environment.

Non-covalent interactions stabilize the binding of macromolecules at chemically complementary interfaces

Biological processes all depend on specific interactions between molecules in the cellular milieu. Enzymes must bind their substrates preferentially relative to other chemically similar species in the cell, while molecular machines, such as the ribosome, comprise a large number of molecules that interact in a precise way. The fundamental principles underlying all of these interactions are the same. Whether or not molecules bind to one another, and for how long, depends on the strength of the interaction and on the concentration of each component.

The strength of the interaction between macromolecules is determined by the shape and chemical nature of the contacting surfaces. Figure 2.17 shows how favorable chemical interactions arise when hydrophobic groups on one surface interact with hydrophobic groups on the other, when there is pairing of hydrogen bond donors and acceptors, or when positively and negatively charged side chains interact. If the two surfaces have complementary shapes, with the bumps on one surface fitting into depressions on the other, the two surfaces can fit very closely together, maximizing the number of favorable non-covalent interactions across the interface. In general, the strength of the interaction between two macromolecules increases with increasing size of the interaction interface. Preferential binding between certain molecules relative to others is dictated by relative binding strength, a concept generally referred to as **specificity**.

Specific biological interactions may be weak or strong, with their specificity determined by how well competing species bind relative to the interaction of interest. It is not the strength of a specific interaction, but rather the comparative strength of binding to the correct partner versus the incorrect partner that governs specificity. The greater the difference in binding strength, the more specific the interaction. This difference underpins the way that different proteins that bind DNA and regulate gene expression show specificity for particular DNA sequences, helping to ensure that the genes are not inappropriately activated by an array of different proteins. For example, if a protein shows specificity for a particular DNA sequence, it is more likely to bind to that sequence than to any other, because it binds to the 'preferred' DNA sequence particularly tightly.

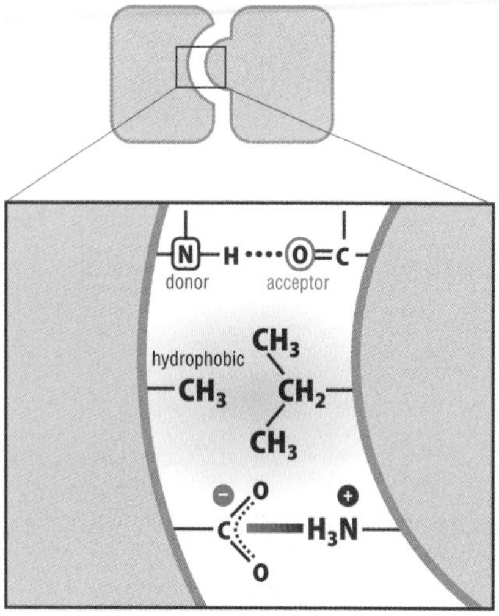

Figure 2.17 A chemically complementary interface. Two molecules bind to one another with surfaces that are complementary in shape. The interacting surfaces must also be chemically complementary: hydrogen donors (red) pair with acceptors (blue), hydrophobic patches (green) contact one another, and ionic interactions form between positively and negatively charged side chains.

2.4 NUCLEOTIDES AND NUCLEIC ACIDS

The physical laws we have described in the previous sections dictate how atoms are joined to make up the remarkable molecules in living things, as well as dictating the molecules' behavior. In the remainder of this chapter, we will learn about the major types of biological molecules and the properties that allow them to carry out their functions in the cell. We begin by considering the nucleic acids.

Nucleic acids play a fundamental role in the processes of life

Nucleic acids are responsible for storing and transmitting genetic information and therefore have a central role in biology. Cells contain two types of nucleic acid, deoxyribonucleic acid (DNA) and ribonucleic acid (RNA). The function of DNA is to store genetic information in all cells and in some viruses. RNA can also store genetic information and forms the chromosomes of other viruses, but has many other functions including decoding the genetic information in all forms of life and catalyzing the key chemical reactions during protein synthesis.

DNA and RNA are polymers that are synthesized from building blocks known as **nucleotides**. Other types of nucleotides also participate in a wide range of biochemical processes that include working together with proteins to catalyze chemical reactions and transfer chemical energy, most notably through the activity of adenosine triphosphate (ATP). Nucleotides are also commonly used as carriers of chemical groups that can be transferred to other molecules, to which they are covalently joined. We will see throughout this book several examples of the central role played by nucleotides in the many biochemical processes in cells.

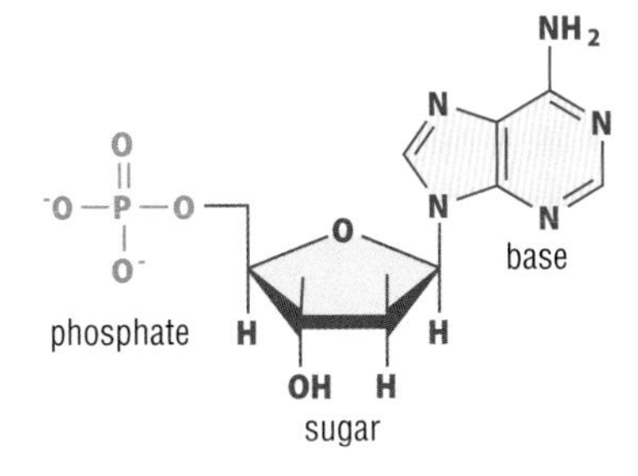

Figure 2.18 A nucleotide. Shown here is deoxyadenosine 5′-monophosphate (dAMP), which contains a base (adenine, pink), a sugar (deoxyribose, black), and a phosphate group (gold).

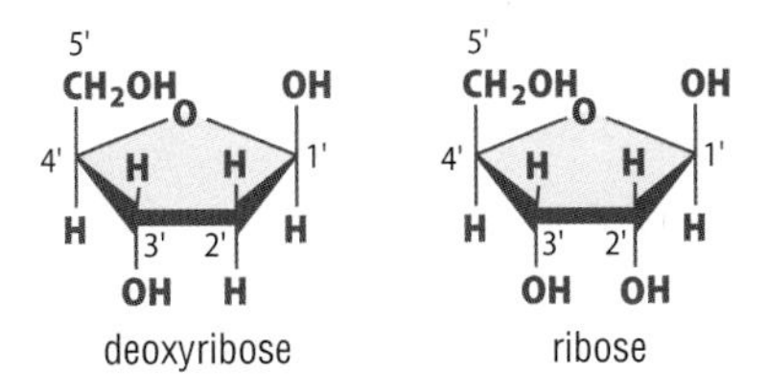

Figure 2.19 The sugars deoxyribose and ribose. According to convention, the five carbon atoms on the sugars are indicated by primed numbers. Deoxyribose differs from ribose only by the absence of an oxygen atom at the 2′ carbon (red).

Cells contain nucleotides of two general types

Nucleotides are small molecules composed of an aromatic group known as a **base** (also called a **nucleobase**), a sugar, and one or more phosphate groups; this structure is shown in Figure 2.18. RNA and DNA differ in the sugar their nucleotides contain. RNA contains the sugar **ribose**, while DNA contains the sugar **deoxyribose** as shown in Figure 2.19. DNA features two types of bases: the **purine** bases, adenine (A) and guanine (G), and the **pyrimidine** bases, cytosine (C) and thymine (T). The structure of these bases is illustrated in Figure 2.20. RNA also contains adenine, guanine, and cytosine, but with uracil (U) in place of thymine. The bases are planar, conjugated rings that are typically uncharged under physiological conditions, but can gain or lose a proton at near physiological pH. The **pK_a** values for each base, which are listed in Figure 2.20, correspond to the pH at which this transition generally occurs: a given atom is protonated at pH values below the pK_a and unprotonated at pH values above the pK_a. This property of bases is particularly important for the ability of RNA molecules to participate in chemical reactions, as we will see later in this book.

A base is joined to a sugar by a glycosidic bond, which is a covalent bond between the C1′ atom of the sugar and the N9 atom of the purine or the N1 atom of the pyrimidine (the numbering convention is shown in Figures 2.19 and 2.20). A base and sugar covalently linked in this way is called a **nucleoside**. Nucleosides are named for the base and are called adenosine (A), guanosine (G), cytidine (C),

(a) purines: adenine (**A**), guanine (**G**); pyrimidines: cytosine (**C**), thymine (**T**), uracil (**U**)

(b) pK_a valuesfor ionizable atoms in bases

base	atom	pK_a
adenine	N1	3.88
cytosine	N3	4.56
guanine	N7	3.60
guanine	N1	10.0
thymine	N3	10.5
uracil	N3	10.1

Figure 2.20 Purine and pyrimidine bases. (a) Adenine and guanine are purine bases, whereas thymine, cytosine and uracil are pyrimidine bases. The atoms in the rings are numbered according to the conventions for purine and pyrimidine rings. (b) The table shows the pK_a values of ionizable atoms in the bases.

thymidine (T), and uridine (U), and have a prefix indicating the linked sugar from which they are formed. Deoxyadenosine, for example, is composed of adenine covalently joined to deoxyribose. As it contains the deoxyribose sugar, deoxy-adenosine is found in DNA, and not in RNA. By contrast, adenosine contains the ribose sugar and is therefore found only in RNA.

A nucleotide is formed when one or more phosphate groups are attached by phosphate ester linkages to either the 3′ or 5′ hydroxyl groups of the nucleoside sugar as illustrated in Figure 2.18, giving rise to **ribonucleotides** and **deoxy-ribonucleotides**. The term 'phosphate ester' describes the chemical nature of the covalent bond that joins the phosphate group to the nucleoside. The phosphate groups are acidic (hence 'nucleic acid'), which means that they are negatively charged at physiological pH.

DNA and RNA are polymers of nucleotides

DNA is a nucleic acid polymer that is synthesized from deoxyribonucleotide building blocks. The nucleotides in DNA are joined by phosphodiester linkages between the 3′ hydroxyl of one nucleotide and the 5′ oxygen of the next, as shown in Figure 2.21. Each nucleotide is added with a 3′ to 5′ linkage in the same way, so that a DNA strand has defined directionality, with a 5′ phosphate group exposed at one end and a 3′ OH at the other end. The deoxyribose sugars and phosphodiester linkages comprise a monotonously repeating **sugar–phosphate backbone**, whereas the base attached to each sugar can vary from one nucleotide to the next along the DNA strand. By convention, the sequence of bases along a DNA strand is written starting at the 5′ end and proceeding to the 3′ end. It is this sequence of bases that we talk of when considering the 'sequence' of a strand of DNA.

A single strand of RNA is similar in most ways to DNA. The key difference is that RNA nucleotides contain ribose in place of deoxyribose (see Figure 2.19) in addition to containing the base uracil in place of thymine. We shall see later that

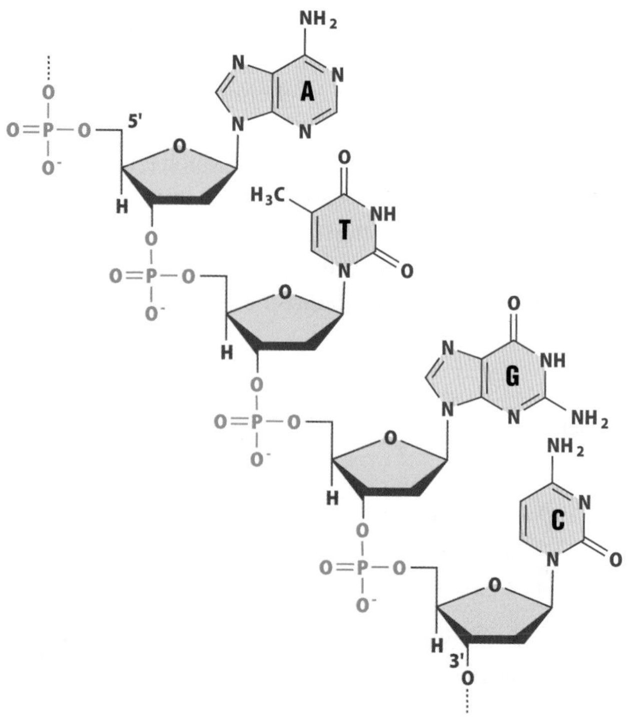

Figure 2.21 A single strand of DNA. Nucleotides are connected by phosphodiester linkages, which join the 3′ hydroxyl of one nucleotide to the 5′ oxygen of the next. The sequence of bases (red) is read from the 5′ end of the DNA. In this case, the sequence is ATGC.

the additional 2′ oxygen in the RNA sugar, ribose, makes RNA more chemically reactive than DNA, as well as enabling RNA to adopt a wider variety of three-dimensional structures. This seemingly small chemical difference is thereby responsible for the greater functional versatility of RNA.

Bases can exist as tautomeric isomers

In Figures 2.18 to 2.21 we have shown the arrangement of atoms in the bases in their most commonly found forms. However, these bases are in equilibrium with another form in which a proton has migrated to an alternative position; these forms of the base are known as **tautomeric isomers** or **tautomers** and are illustrated in Figure 2.22. In a typical population of nucleotides or nucleic acids, less than 0.01% of the bases are found in the rare tautomeric form. While seemingly uncommon, this still means that large genomes may contain a significant number of rare tautomers. If one considers the entire human genome of 3 billion base pairs (bp), about 100 000 bases may adopt the rare tautomeric form at any given moment. As we shall see in Chapter 15, this has important consequences for the fidelity of DNA replication during cell division, and provides one mechanism for generating genetic variation.

Nucleotide monomers have diverse important functions in cells

In addition to functioning as the building blocks for DNA and RNA, a large variety of nucleotides play essential roles in the cell. One of these is to store energy derived from the breakdown of food in high-energy covalent bonds within molecules such

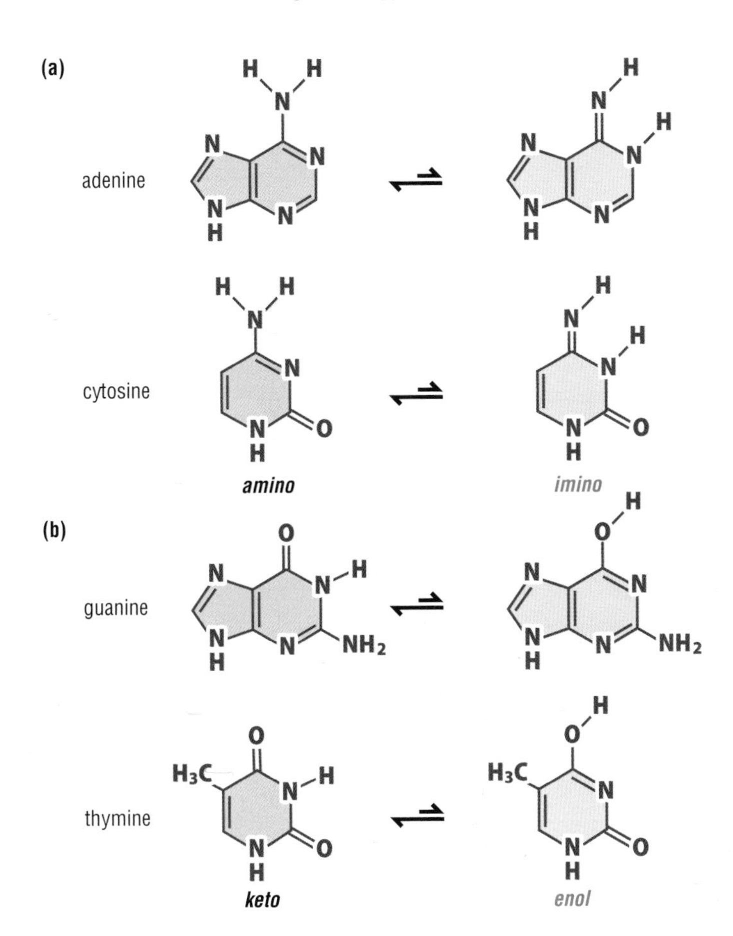

Figure 2.22 Base tautomers. Bases can adopt different arrangements of atoms called tautomers. (a) Adenine and cytosine normally adopt the common amino form (gray), but also occasionally adopt the imino form (blue). (b) Guanine and thymine are normally found in the keto form (gray), but can tautomerize to form much rarer enol forms (blue).

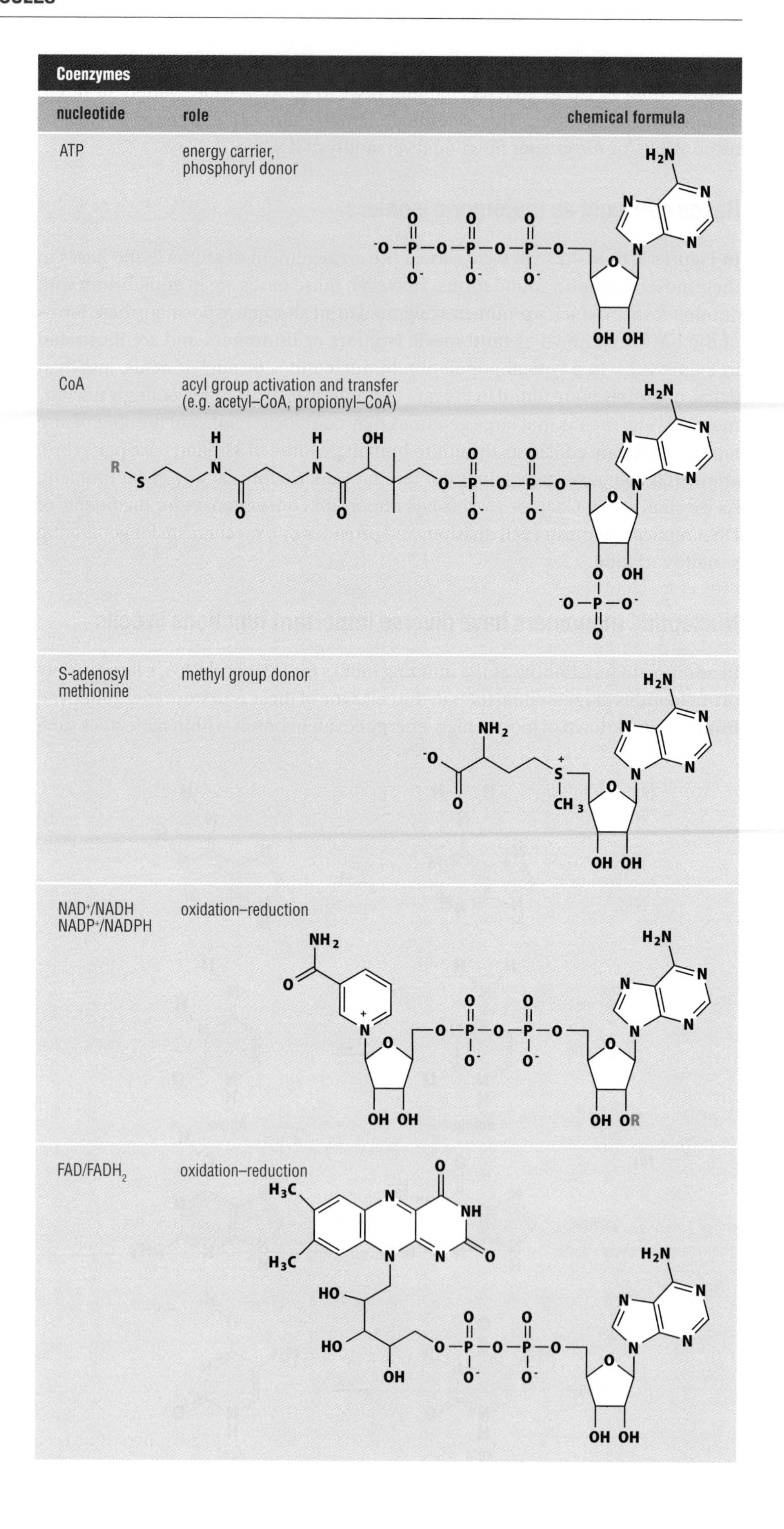

Coenzymes

nucleotide	role	chemical formula
ATP	energy carrier, phosphoryl donor	
CoA	acyl group activation and transfer (e.g. acetyl–CoA, propionyl–CoA)	
S-adenosyl methionine	methyl group donor	
NAD^+/NADH $NADP^+$/NADPH	oxidation–reduction	
FAD/$FADH_2$	oxidation–reduction	

Figure 2.23 Table of common nucleotides.

as ATP. These bonds can then be broken to release energy that can be used to drive chemical reactions and mechanical processes, such as the motion of flagella that allows bacteria to 'swim'. We will learn more in the next chapter about how various biological processes can be powered by the input of energy.

Nucleotides also carry chemical groups that can be transferred to other molecules, a process that occurs in a wide variety of biosynthetic reactions as well as in the regulation of biological molecules. Figure 2.23 shows some examples of single nucleotides and the essential roles they play in processes that we shall encounter throughout this book.

2.5 THE STRUCTURE OF DNA

Single strands of DNA associate non-covalently to form a double-helical structure

DNA is unique among biological macromolecules in the simple and direct way in which its structure is related to its biological function. In the cell, DNA molecules are double stranded. The bases on opposing strands form hydrogen bonds that cause them to pair with one another in a precise correspondence that is dictated by their geometry: cytosine on one strand makes three hydrogen bonds with the **complementary** surface of guanine on the other, whereas thymine on one strand makes two hydrogen bonds with the complementary surface of adenine on the other, as shown in Figure 2.24. We refer to these as **Watson–Crick base pairs** after the two scientists who deduced the structure of DNA (see Experimental approach 2.1). Since the sequence of bases in one strand dictates the sequence of bases in the other, we say that the two strands are complementary to one another. We shall see in Chapter 6 that this complementarity allows genetic information to be copied because each DNA strand is a **template** for synthesizing a copy of the complementary strand. We shall also see in Chapter 8 that the same type of base-pairing interactions are used to allow synthesis of a complementary strand of RNA – the messenger RNA molecule – as the first step in decoding the genetic information.

Two complementary DNA strands pair readily with one another when the strands are oriented in opposite directions, so that the 5′ end of one strand pairs with the 3′ end of the other. The two strands of DNA are therefore said to be **antiparallel.** The most energetically favorable conformation for double-stranded DNA is one in which the two strands wind around one another, forming a **double helix**. Such a double helix is illustrated in Figure 2.25. The helix shown is right-handed, meaning that the DNA strands spiral counterclockwise as you look down them.

The three-dimensional structure of double-stranded DNA can be explained by the physical properties of the nucleotides that make up the DNA strands. The hydrophilic sugar–phosphate backbone of each DNA strand lies on the outside of the helix where it can interact with water molecules, while the largely hydrophobic base pairs lie on the inside. The base pairs form a spiral stack at the interior of the double helix, with neighboring base pairs along the stack making favorable van der Waals interactions with one another (see Figure 2.25a). At the same time, the hydrophobic faces of the bases are sequestered from the surrounding aqueous environment. This arrangement, called **base stacking**, is energetically highly favorable and is a major factor in stabilizing the structure of double-stranded DNA and thus contributing to the durability required for the robust storage of genetic

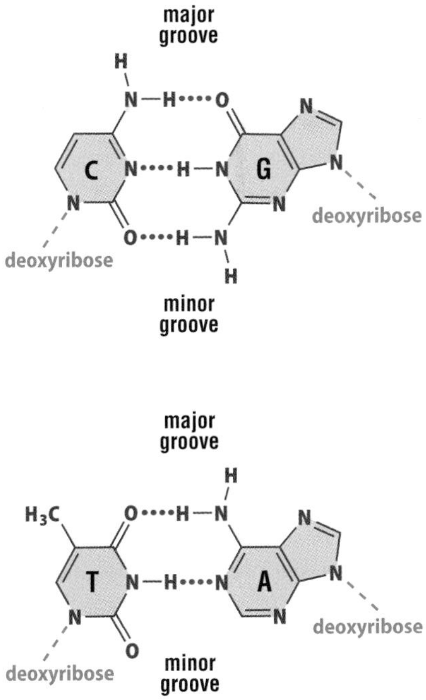

Figure 2.24 Watson–Crick base pairs. Adenine pairs with thymine while cytosine pairs with guanine. These pairings are due to the formation of hydrogen bonds (red dashed lines) between the bases. The deoxyribose sugars to which the bases are attached are indicated in blue.

2.1 EXPERIMENTAL APPROACH
The discovery of the DNA double helix

Chemical clues for DNA structure

The most renowned paper in molecular biology is undoubtedly the 1953 publication by James Watson and Francis Crick of their proposed structure of B-DNA. In just a single page, the authors presented the structure of the DNA double helix and, in doing so, began a new era in biology. This was because the discovery of base-pairing and complementarity between the two DNA strands immediately suggested how genetic information could be copied and transmitted.

How did Watson and Crick arrive at the structure of B-DNA? In 1953, methods for determining the crystal structures of macromolecules had not yet been devised, nor were there the means for isolating or synthesizing a uniformly pure sample of DNA that would have been suitable for single crystal studies. Instead, Watson and Crick relied on a combination of basic chemical information about nucleic acids, model building, some findings about DNA composition in different organisms, and x-ray diffraction data on DNA fibers produced by Rosalind Franklin.

By 1953, the chemical composition of single-stranded DNA was known, and it had been proposed that the bases were most likely to be oriented perpendicular to the ribose sugar. There had been some question about which tautomer the bases adopted, but a chemist named Jerry Donohue informed Watson and Crick that the bases were in the enol form, which proved to be key in determining how base pairs formed. Edwin Chargaff had recently found that adenine and thymine were always found in a 1:1 ratio, as were guanine and cytosine. All of these provided clues to the underlying structure of DNA.

X-ray diffraction from B-DNA fibers provides key information

The key information, however, came from x-ray diffraction studies. Long DNA fibers could be isolated from cells that had been disrupted. These fibers, which consisted of many DNA strands, could be lifted out of a solution and oriented in an x-ray beam. The x-rays scattered from the fiber were then recorded on film, as depicted in Figure 1; this scattering of x-rays is referred to as diffraction. The pattern of streaks and smears that results from x-rays striking the film is produced by x-ray waves that bounce off the DNA (or any other molecule) and combine with one another as they scatter in different directions. The particular pattern of constructive and destructive interference that is observed on the x-ray film depends upon the structure of the DNA fiber. Franklin

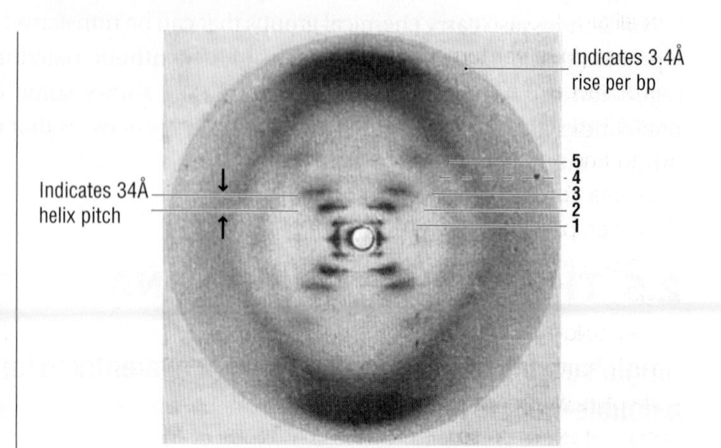

Figure 1 X-ray diffraction from B-DNA fibers. The appearance of the pattern is characteristic of diffraction from a helix, while the location of the smears and spots give some of the helix dimensions. Note that layer line 4 is missing, indicating that the two strands of the double helix are not evenly spaced along the helix axis, but instead are arranged to produce a larger (the major) and smaller (the minor) groove.

Reproduced from Franklin & Gosling, 1953. Molecular Configuration in Sodium Thymonucleate. *Nature* **171**, 740–741.

had discovered that the details of the diffraction pattern changed depending on how hydrated the DNA samples were. The pattern obtained at low humidity was dubbed the 'A' pattern, while that obtained at high humidity was dubbed the 'B' pattern. This is the origin of the names for the A-DNA and B-DNA helix types that produced these patterns.

The unusual pattern of scattered x-rays seen in Figure 1, with the characteristic cross pattern and the presence of individual streaks that fall on regularly spaced lines (known as layer lines), are characteristic of a helical molecule. Importantly, the details of the pattern, including the spacing between the layer lines and the location of the large smear along the vertical line, provided information on the dimensions of the helix and the spacing between each successive repeat. Diffraction from DNA fibers could even reveal that the DNA consisted of two intertwining helices, and that the strands were antiparallel to one another. These and other features could be determined from x-ray fiber diffraction patterns by taking advantage of the mathematical description of diffraction from helices.

Rosalind Franklin's famed diffraction image of B-DNA provided the high-quality data that was needed to deduce the parameters of the helix. From this photograph, it was possible to deduce that the sugar-phosphate backbone lay on the outside and was fairly extended. The location of the large smears along the vertical axis of the photograph meant that the vertical distance between the repeat unit of the helix was 3.4 Å, while the spacing between the layer lines and the number of layer lines observed (not all are readily apparent in Figure 1) meant that the helix had ten base pairs per turn and that the helix repeated every 34 Å. The absence of a layer line (note missing layer line 4 in Figure 1) meant that the two DNA strands were not evenly spaced relative to one another. It was also possible to determine that the helix was about 20 Å in diameter based on other features of the diffraction pattern.

Watson and Crick's leap of insight was to figure out how the base pairs fit together at the core of the helix. For this, they used model building and chemical intuition, along with knowledge of Chargaff's rule about base ratios, to arrive at the Watson–Crick base pairs we now know. The model was immediately appealing because it explained so much about the function of DNA, in addition to revealing its intrinsic beauty.

A crystal structure of B-DNA provides the final proof

Definitive proof of the proposed model for B-DNA did not arrive for another 27 years, when the first crystal structure of a B-DNA fragment was determined. In contrast with the model built by Watson and Crick, the crystal structure yielded an experimentally determined atomic model of DNA, based on the x-ray data alone without any assumptions other than basic information about covalent bonds and atomic radii. Dick Dickerson and his colleagues determined the atomic resolution structure of a 12 base pair oligonucleotide, dubbed the 'Dickerson dodecamer', in 1980; this structure is shown in Figure 2. They took advantage of a new method for synthesizing DNA in solution that made it possible to purify large quantities of DNA molecules of a defined length and sequence, which could then be crystallized. By 1980, there were also well-established methods for determining crystal structures of macromolecules – something that had not yet been possible back in 1953. (See section 19.14 for a description of x-ray crystal structure determination.)

The B-DNA crystal structure confirmed all of the essential features of the Watson and Crick model, with some interesting deviations. The DNA base pairs were not perfectly planar, but, rather, contained various degrees of propeller twist. In addition, the crystal structure had slightly more than ten base pairs per turn. Another key feature was that the helix axis was not perfectly straight, but rather was bent by about 19°. The authors duly noted that this degree of bending, when extended over longer DNA fragments, would be sufficient to allow DNA to wrap around a histone core to form a nucleosome, the fundamental unit of eukaryotic chromatin. Thus began a new era in the understanding of DNA structure and how it relates to its function.

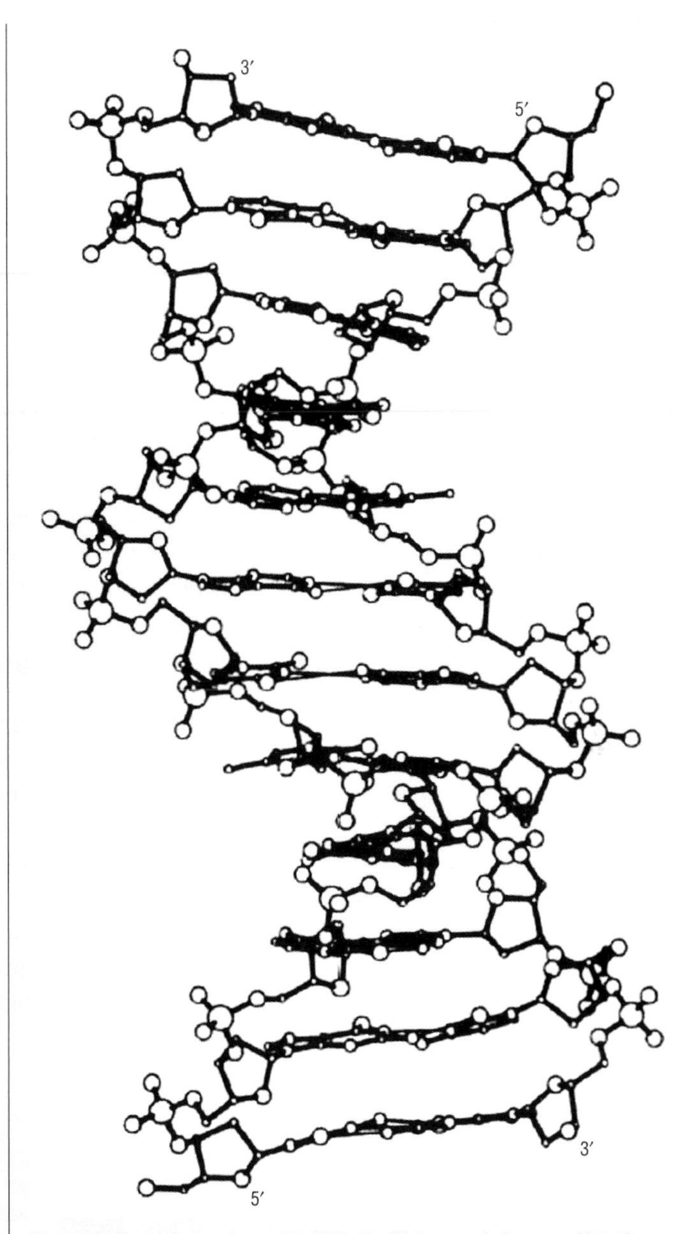

Figure 2 Crystal structure of B-DNA: the Dickerson dodecamer. Note the bend in the DNA axis.

Reproduced from Wing *et al.*, 1980. Crystal structure analysis of a complete turn of B-DNA. *Nature* 1980; **287**:755–758.

Find out more

Franklin RE, Gosling RG. Molecular configuration in sodium thymonucleate. *Nature*, 1953;**171**:740–741.

Watson JD, Crick FH. Molecular structure of nucleic acids; a structure for deoxyribose nucleic acid. *Nature*, 1953;**171**:737–738.

Wing R, Drew H, Takano T, *et al.* Crystal structure analysis of a complete turn of B-DNA. *Nature*, 1980;**287**:755–758.

Related techniques

DNA x-ray diffraction; Section 19.14

X-ray crystallography; Section 19.14

information. Since both A–T and G–C base pairs have similar widths, the helix has a fairly uniform diameter of 20 Å along its length.

DNA in the cell is predominantly B-form

The double helix can adopt several forms, and the differences between the forms have important functional consequences. Double-stranded DNA in cells generally adopts a conformation called **B-DNA**, whose helical structure repeats itself approximately every 10.5 bp as one looks along the double helix (see Figure 2.25a). Each successive base pair is separated by 3.4 Å. The two intertwined DNA strands produce a double helix with two grooves of unequal size: the wide **major groove** (~13 Å) and the narrower, shallower **minor groove** (~9 Å width); these grooves are highlighted in Figure 2.25b. The edges of the base pairs are exposed in these major and minor grooves. We see throughout this book how the structural features of the DNA helix, and in particular the atoms of the bases that are exposed in the grooves, govern the way in which proteins can interact with DNA in the processes of DNA replication, transcription, recombination, and repair.

DNA can adopt other conformations

Although the B-DNA helix is the dominant form in the cell, DNA can also adopt other helical conformations. One of these is the **A-DNA** helix shown in Figure 2.26, which has 11 bp per turn, with the base pairs tilted relative to the DNA axis as well as being displaced farther from the DNA axis than they are in B-DNA. As a result, the A-DNA helix has a narrower, deeper major groove and a shallower, wider minor groove than B-DNA. Although A-DNA differs in several details from B-DNA, they are both right-handed helices.

The A-helix conformation can be induced by some proteins that form a complex with DNA. It is therefore thought that some segments of chromosomal DNA

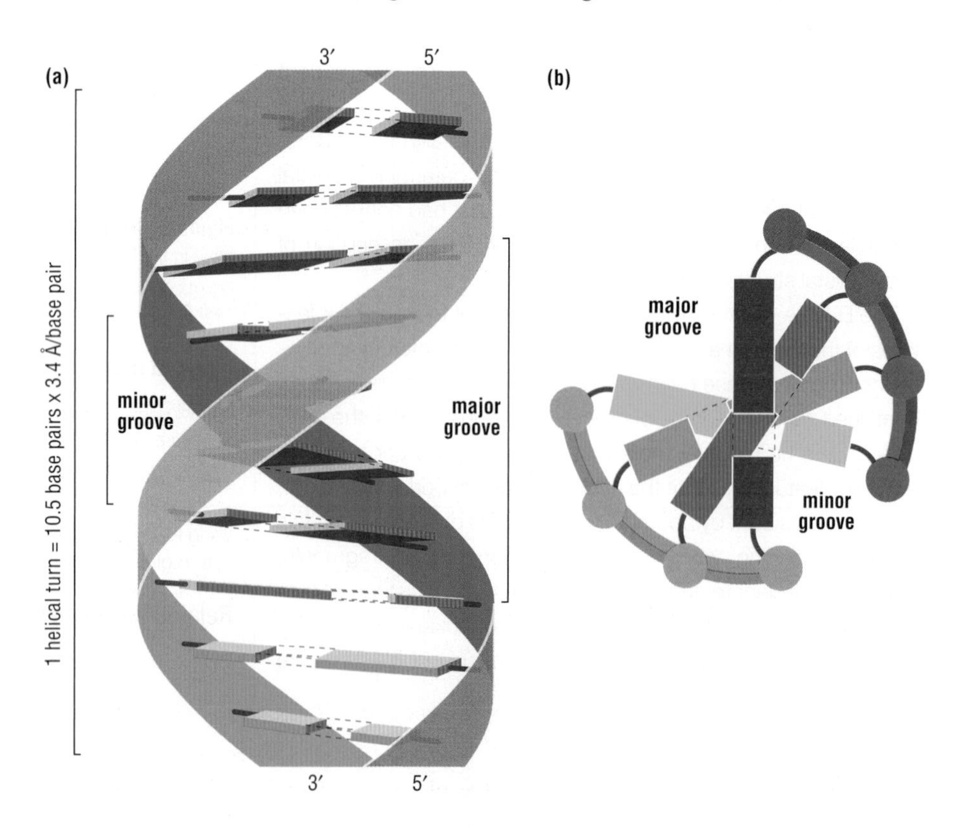

Figure 2.25 Double-helical DNA. (a) A side view of the helix showing that the base pairs (pink) are stacked perpendicular to the axis of the helix. The two DNA strands are antiparallel and wind around one another to form a helix with two different grooves of unequal width: the wider, major groove, and the narrower, minor groove. The helix winds around a full 360° turn about every 10.5 bp, with the base pairs about 3.4 Å apart; thus each turn of the helix is approximately 36 Å long. (b) A view looking down the B-DNA helix, showing how the wide major groove and the narrower, shallower minor groove are formed by the asymmetric attachment of the bases to the sugar-phosphate backbone (gray).

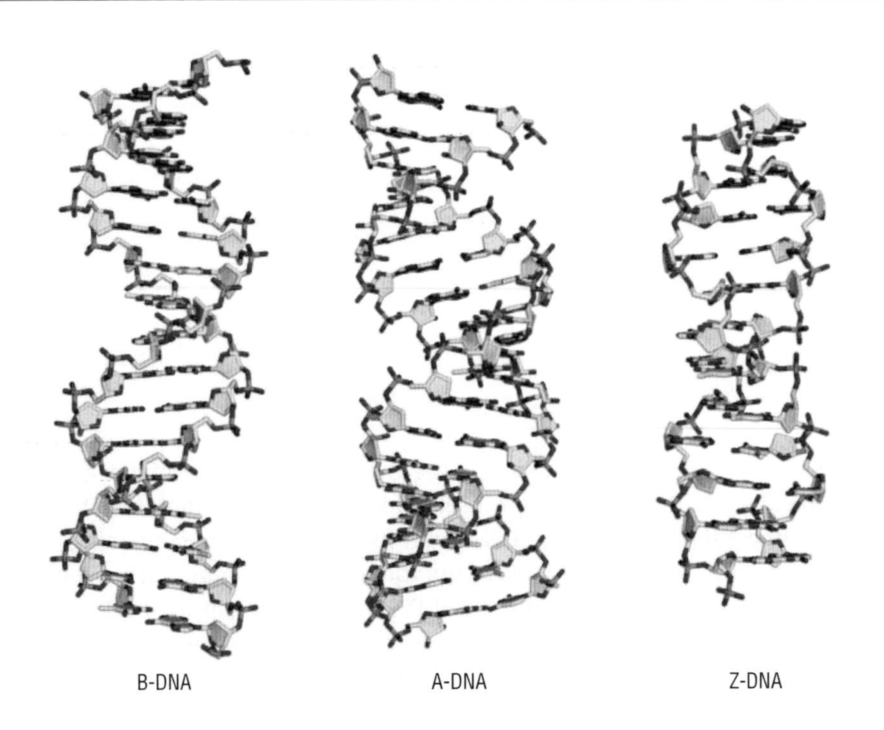

Figure 2.26 A-, B- and Z-DNA. A comparison of the three principal DNA helix types: B-DNA; A-DNA; and Z-DNA. In contrast with A- and B-DNA, which adopt right-handed helices, the Z-DNA helix spirals in the opposite direction and is therefore left-handed.

that are bound to these proteins may adopt a conformation closer to the A-DNA helix. In contrast to its minor prevalence in DNA, we see in Section 2.8 how the A-type helix is particularly important in RNA structure, since this is the only type of double helix that double-stranded RNA can form.

The particular sequence of bases along a DNA strand can influence the structure of the double helix in a number of ways. Although we often draw the DNA helix as following a straight path, DNA can bend somewhat, and certain base sequences are more bendable than others. In general, regions of the DNA double helix containing successive A–T followed by T–A base pairs tend to bend more readily than helices with G–C steps.

An unusual type of DNA helix that spirals in the opposite direction, and is therefore left-handed, can form under special conditions in DNA strands with alternating G and C bases. This helix, also shown in Figure 2.26, is known as **Z-DNA**. Formation of Z-DNA is favored by a chemical modification of cytosine, methylation (which we shall learn more about later) and high salt concentrations. It is also promoted by torsional stress on the DNA. While the ability of DNA to adopt the Z-type helix ***in vitro*** is well established, there is some debate about the role it might play in the cell.

DNA can contain topological variations known as supercoiling

The small amount of flexibility between DNA base pairs allows large fragments of DNA, containing thousands of base pairs, to appear like flexible ropes in electron micrographs. If the DNA is a closed circle, as is the case for many bacterial and viral chromosomes, it typically looks highly intertwined and tangled when it is isolated from a cell. This is because the DNA is **supercoiled**, which places the molecule under tension and causes it to twist on itself, as illustrated in Figure 2.27. In contrast, a **relaxed DNA** circle is open and not under tension.

To understand how supercoiled and relaxed circular DNA differ, imagine doing the following, as depicted in Figure 2.28: take a covalently closed, circular DNA molecule, cleave both strands in one place, and then rotate one DNA end around

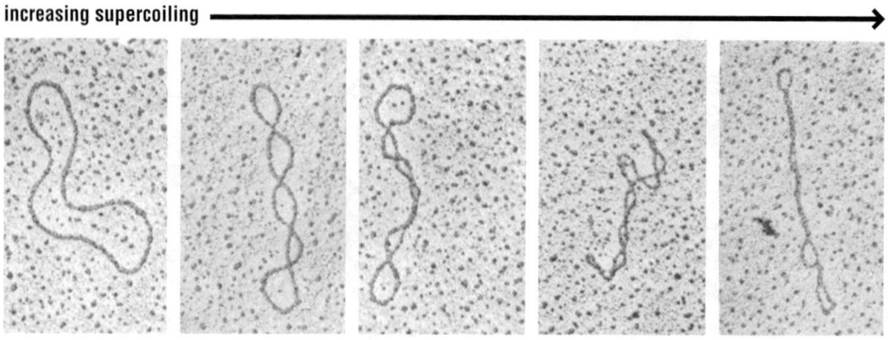

increasing supercoiling

Figure 2.27 Supercoiled and relaxed DNA. These electron micrographs show a closed, circular bacterial plasmid as the degree of supercoiling increases.

From Kornberg and Baker, *DNA Replication,* 2nd ed. New York: WH Freeman.

several times before covalently closing the circle again. The resulting DNA circle is supercoiled and will wind about itself. This is because rotating the DNA end changes the helical twist of the DNA from its preferred state containing about 10.5 bp per turn to one with a fewer number of base pairs per turn. When the DNA is ligated, the untwisted portion of DNA will tend to spring back to adopt its favored state, with 10.5 bp per turn. This, however, causes the plasmid to wrap around itself as shown in Figure 2.28. The closed DNA circle is now supercoiled.

There are two 'directions' of supercoiling - positive and negative. An initial unwinding of the DNA - which is in the clockwise direction and is equivalent to separating the DNA strands - leads to accumulation of **negative supercoils** as shown in Figure 2.28, whereas an initial twisting of the DNA in the counterclockwise direction leads to formation of **positive supercoils**.

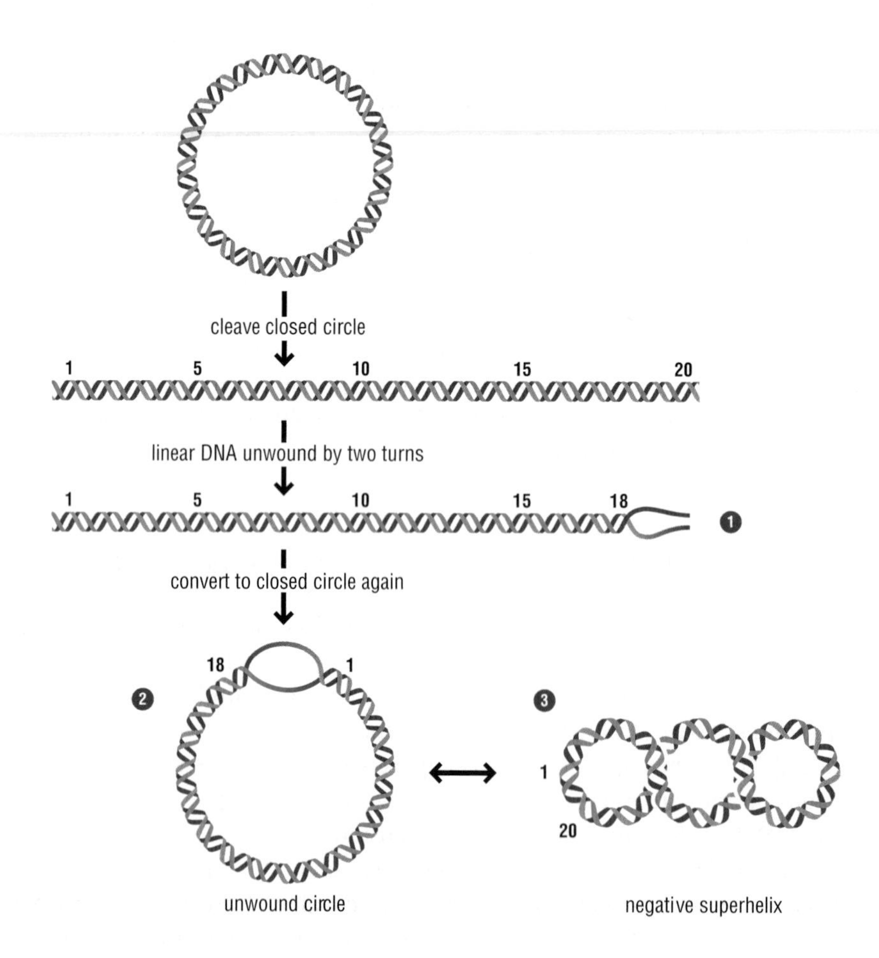

Figure 2.28 Introducing supercoils into DNA. Supercoiling can be introduced into DNA fragments whose ends are not free to rotate. In this example, the DNA in the linear fragment is unwound to remove two complete turns (1), then ligated to form a closed circle (2). The two unwound regions can twist up again to form a right-handed double helix, but this will cause two negative supercoils to form in the DNA (3).

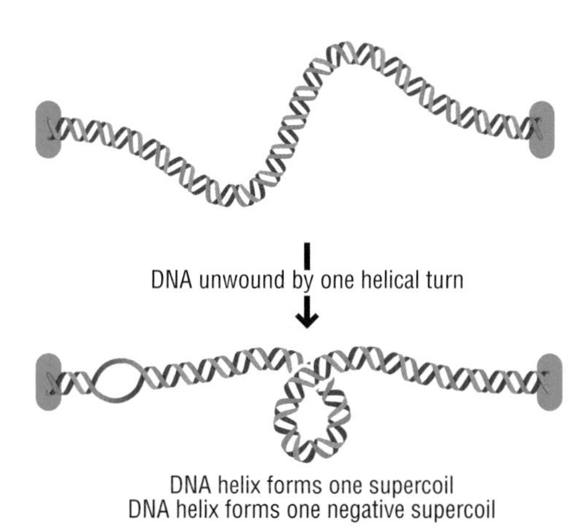

Figure 2.29 Supercoils can form in constrained linear DNA. A change in DNA twist can cause supercoils to form in any DNA fragment that is constrained at both ends.

Supercoiling can also occur in linear fragments of DNA in which the ends are immobilized and therefore not free to rotate and release superhelical tension, as shown in Figure 2.29. Thus, linear chromosomes or segments in the middle of a chromosome whose boundaries are constrained can also contain supercoils. Supercoiled DNA is under torsional strain (that is, there is tension in the molecule). If a nick or break is introduced into a supercoiled plasmid or chromosome, one end will swivel around the other to relax the supercoiling, and hence alleviate the strain on the DNA molecule.

DNA supercoiling is related to the linking number and twist

The relation between supercoiling and the winding or unwinding of the DNA helix can be described quantitatively. To understand this description, we must first define three quantities: linking number, twist, and writhe (which is another term for supercoiling).

Linking number (Lk) describes the number of times one strand of duplex DNA wraps around the other. Figure 2.30 shows how, for each turn of the DNA double helix (roughly every 10.5 bp of B-DNA), one single strand wraps around the other once. A closed, circular DNA plasmid of 1050 bp that is relaxed therefore has a linking number of 100. An important feature of the linking number is that it cannot change in closed circular DNA - or in a DNA fragment that is constrained at its two ends - without cleaving one or both of the DNA single strands. We say that the linking number is a topological invariant of closed circular or constrained DNA, as it is unaffected by changes in the overall shape of the DNA.

The twist is a property of the double-helical structure of DNA. **Twist (Tw)** represents the number of turns in the DNA helix in a given fragment of DNA; a relaxed plasmid or a straight fragment of DNA that lies in a plane has a twist of +1 for each turn of the double helix. (The positive sign is for a right-handed helix, whereas a negative sign would pertain for a left-handed helix.) For example, a 105 base pair fragment of relaxed B-DNA has a twist of 10 (= 105 bp ÷ 10.5 bp/turn), while a 1050 base pair fragment of relaxed B-DNA has a twist of 100.

Writhe (Wr) describes the supercoiling of closed circular DNA (Figure 2.28) or of DNA constrained at two ends (Figure 2.29). A relaxed plasmid has a writhe of zero; positively supercoiled DNA has a value of writhe greater than zero while negatively supercoiled DNA has a value of writhe that is less than zero. Figure 2.31 shows the different forms adopted by positively and negatively supercoiled DNA

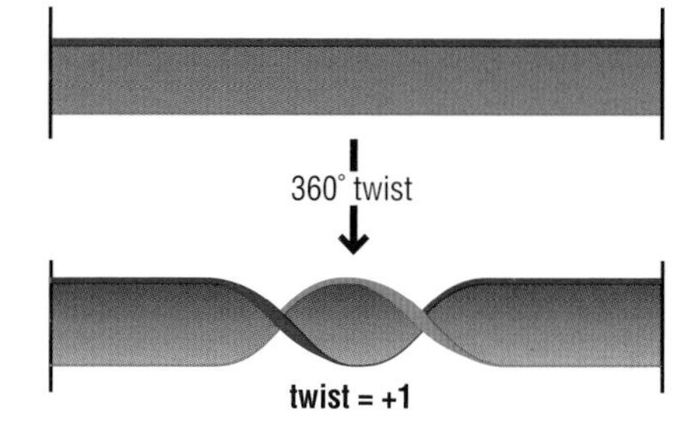

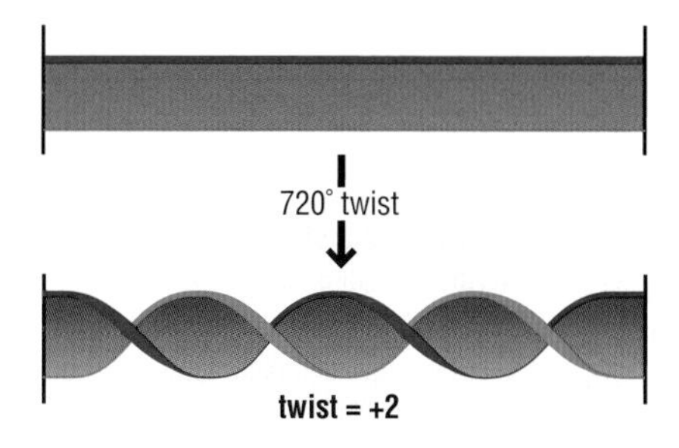

Figure 2.30 DNA twist. Two parallel DNA strands have a twist of 0. Each 360° rotation of the helix introduces a twist of +1. Relaxed DNA that is not supercoiled has a twist of +1 for every repeat of the DNA helix.

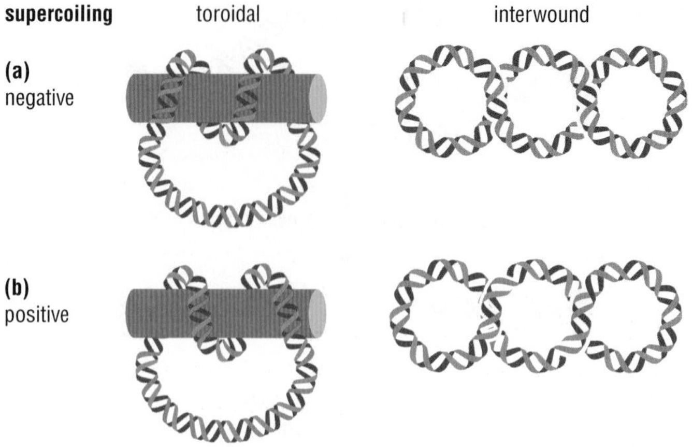

supercoiling toroidal interwound

(a) negative

(b) positive

Figure 2.31 The toroidal and interwound forms of supercoiled DNA. Supercoiled DNA can adopt either of two principal forms: toroidal or interwound. (a) Negatively supercoiled DNA can adopt the toroidal form, in which the DNA spirals in a left-handed manner, or an interwound form. (b) In the toroidal form of positively supercoiled DNA, the DNA spirals in a right-handed manner. Note how the interwound forms of positively and negatively supercoiled DNA are twisted in the opposite manner.

➔ We see how the toroidal form of negatively supercoiled DNA is similar to the way in which chromosomal DNA winds around histone proteins, which is how DNA is packaged in the nucleus, in Chapter 4.

circles. Note in the figure how supercoiled DNA can adopt either the toroidal form or the interwound form.

Having defined linking number, twist, and writhe, we will now see how changing the twist of a plasmid affects supercoiling. Although the linking number is constant for a closed circular or constrained DNA fragment, twist and writhe are free to vary according to the relation Lk = Tw + Wr. This means that any change in twist must be balanced by an opposite change in the value of writhe (supercoiling).

In the example depicted in Figure 2.28 and the one shown in Figure 2.32, the DNA has been unwound by two turns (a decrease in twist and linking number of 2) and the writhe is kept equal to zero before ligating the DNA circle. However, it is not favorable for the DNA to have a region in which two turns have been unwound, and this segment will tend to anneal and twist again to adopt a standard double helix.

If the twist goes up by +2, the writhe must change by -2; in other words, the DNA now contains two negative supercoils. As long as the plasmid is not nicked, such that the linking number cannot change, these two forms - one with no supercoils but a local region of unwinding (Tw = 18, Wr = 0) and the other with two negative supercoils (where the twist increases by 2, and the writhe decreases by 2, leading to Tw = 20, Wr = -2) - can interchange with one another.

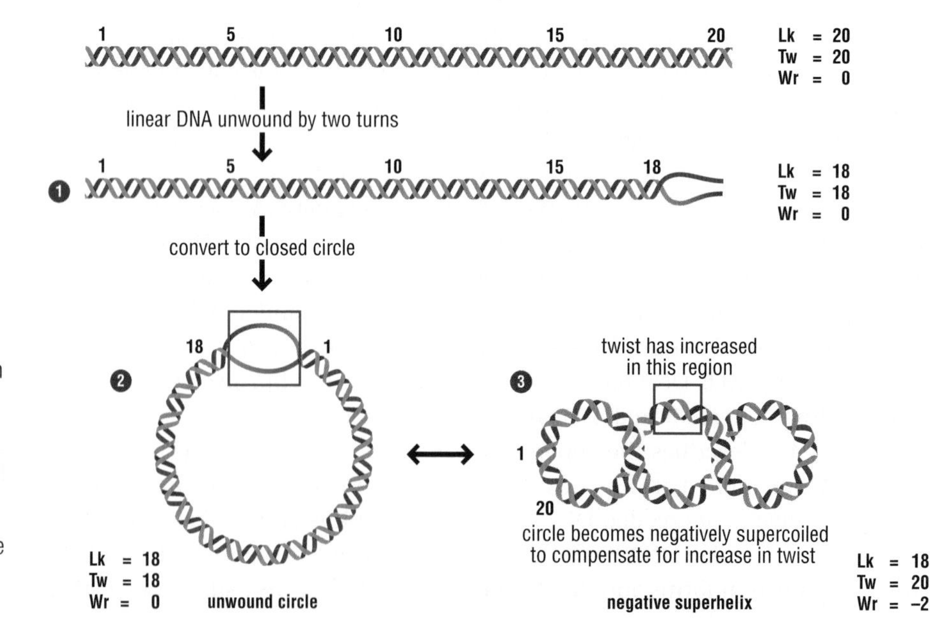

Figure 2.32 The effect of supercoiling on twist and writhe. Building on the example in Figure 2.28, DNA in the linear fragment is unwound to remove two complete turns (1), then ligated to form a closed circle (2) with a linking number (Lk) of 18 and a twist (Tw) of 18. If the two unwound regions twist up again (3) to form a right-handed double helix, we see a change in twist of +2. This causes two negative supercoils to form in the DNA, leading to a change in writhe of –2. Note that the linking number remains the same in both cases.

Negative supercoiling favors local unwinding of DNA

The balance between DNA twisting and supercoiling has important biological consequences. As shown in Figure 2.32, negative supercoils in DNA can be removed by unwinding one or more turns of DNA, which is equivalent to separating the strands. In this way, the unfavorable energy of strand separation is partially compensated for by the favorable energy due to relief of negative superhelical tension. This explains why negatively supercoiled DNA templates are needed for many biological processes that require separation of the DNA strands.

There are cellular enzymes called **topoisomerases** that introduce or remove supercoils from DNA in the cell. Topoisomerases change the supercoiling by introducing a transient break in either one or both DNA strands in the double-stranded DNA, depending on the type of enzyme. Since increasing the degree of supercoiling is energetically unfavorable, enzymes that introduce additional supercoils into DNA - usually negative - must use the energy provided by ATP hydrolysis to drive this process. Some topoisomerases can also separate chromosomes that have become tangled and play an important role during DNA replication.

➔ We look at topoisomerases in more detail in Chapter 6.

2.6 CHEMICAL PROPERTIES OF RNA

RNA is chemically distinct from the related nucleic acid, DNA

RNA is a nucleic acid that in many ways is similar to DNA, but with key chemical differences that give it far greater functional diversity. Like DNA, an RNA molecule is a polymer of nucleotides that are joined by phosphodiester linkages between the 3′ end of one nucleotide and the 5′ end of the next, as shown in Figure 2.33. As explained in Section 2.4, the principal chemical differences are that RNA nucleotides contain ribose in place of deoxyribose and uracil in place of thymine

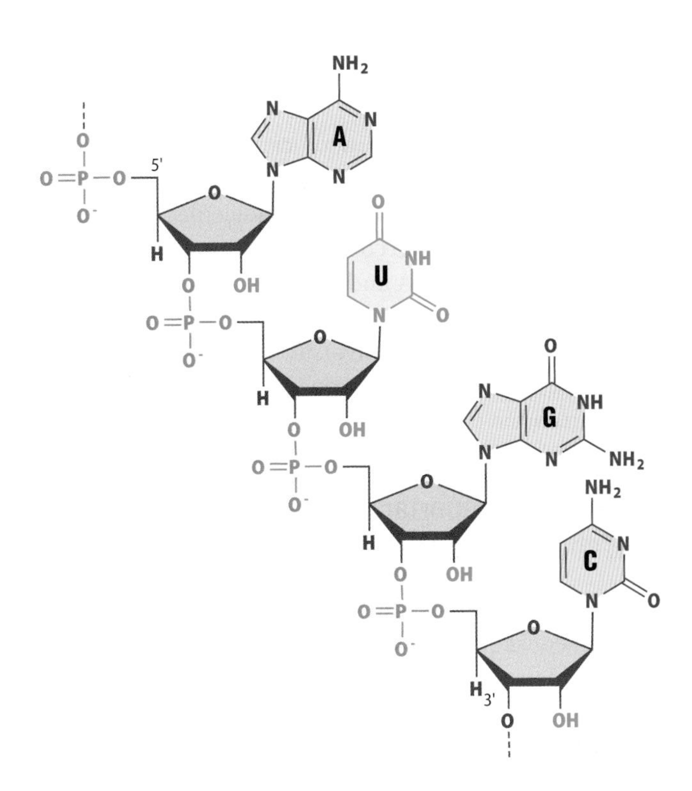

Figure 2.33 A single strand of RNA. RNA nucleotides are joined by single phosphodiester linkages, just as in DNA. The differences between RNA and DNA lie in the 2′ OH of ribose, rather than the 2′ H of deoxyribose found in DNA, and in the substitution of U in place of T. (See also Figures 2.19 and 2.20, which highlight these differences.)

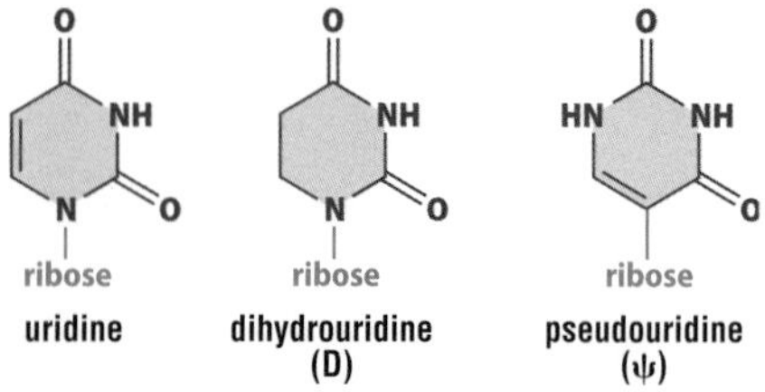

O NH N O ribose uridine
O NH N O ribose dihydrouridine (D)
O HN NH O ribose pseudouridine (ψ)

Figure 2.34 Examples of modified RNA bases. Left, the standard uridine nucleotide base; middle, dihydrouridine, found in the D-loop of tRNA; right, pseudouridine, found in abundance in eukaryotic rRNAs.

(see Figures 2.19 and 2.20). These differences do not affect the fundamental property of base-pairing; uracil can form base pairs with adenine just as thymine does. Furthermore, it is the ability of RNA to form base pairs with DNA, as well as with RNA, that allows cells to synthesize RNA copies of genes from a DNA or RNA template. However, the presence of the 2′ OH on the ribose and the distinct chemical structure of uracil, as compared with thymine, allows RNA to adopt diverse structures that are not readily formed by DNA.

RNA is often chemically modified after it is synthesized

Many RNA molecules require chemical modification of some of their nucleotides in order to become fully functional. These modifications, which alter the chemical structure of the bases, are introduced after the RNA is synthesized. These modifications are typically referred to as **post-transcriptional** modifications. Examples of modified uridine nucleotides are shown in Figure 2.34. Nearly 100 modified nucleosides have been identified to date, and there may well be additional modifications that have not yet been discovered. These modifications are generally not found in messenger RNA (mRNA, the RNA that acts as an intermediate in the flow of genetic information from DNA to proteins), but rather in RNA molecules that carry out other functions in the cell. A notable exception to this is the deamination of adenosine to produce inosine, which we shall learn about in Section 10.8. In contrast with DNA methylation, which plays a role in regulating whether a gene is transcribed or not and is reversible (Section 2.4), chemical modifications of RNA are largely irreversible and do not play a regulatory role.

The best-studied examples of modified nucleotides in RNA are found in the protein synthesis machinery. For example, the transfer RNA (tRNA) adaptor molecule must undergo numerous chemical modifications in order to be able to carry out its function. Around 10% of the roughly 75 nucleotides in tRNA are chemically modified after the RNA is transcribed. The ribosome, which is the large complex of protein and RNA molecules that synthesizes proteins, similarly contains many modified nucleotides in the ribosomal RNA (rRNA). In human rRNAs, approximately 200 nucleotides are modified after transcription, which represents about 3% of the 6500 nucleotides in rRNA.

➔ We explore post-transcriptional modifications in Chapter 10.

The position and nature of nucleotide modifications in a particular RNA molecule are often conserved among different species. This conservation suggests that these modifications play important roles in the molecules in which they are found (otherwise they would not have been conserved!). The chemical modifications expand the chemical diversity of RNA and extend its functional repertoire beyond that permitted by the four standard RNA nucleotides, A, U, C, and G. As we will see later, proteins have greater chemical diversity than RNA even without chemical modifications, since they are polymers constructed from 20 chemically distinct building blocks - the amino acids.

The 2′ OH of ribose has important structural and functional consequences

The presence of a 2′ OH group in the RNA ribose, as compared with deoxyribose in DNA, has important consequences for RNA structure and function. The chemical stability of RNA polymers is much lower than that of DNA because the oxygen in the 2′-OH of ribose facilitates a reaction that can cleave phosphodiester bonds. The chemical features of this simple reaction will be described in more detail in

Section 3.4, when we discuss catalysis. The instability of single strands of RNA, as compared with the far greater stability of DNA, is very likely the reason that DNA was selected during evolution for the long-term storage of genetic information. The less stable RNA molecule has been retained as the intermediate messenger, since mRNA is typically needed for only a limited period of time before it degrades chemically or is actively broken down by enzymes in the cell.

The additional hydroxyl group in the ribose also has important consequences for the three-dimensional structure of RNA molecules. Like DNA, RNA molecules can form double helices with base-pairing interactions between guanine and cytosine, and between adenine and uracil. However, the presence of the 2′ OH on the ribose causes double-stranded RNA to exclusively adopt the A-type helix (one of the helix types that DNA can adopt, as we saw earlier) for two reasons:

- First, the 2′ OH of ribose would clash with the sugar–phosphate backbone as depicted in Figure 2.35 if RNA were to form a B-type helix.
- Second, Figure 2.36 illustrates how the additional hydroxyl group on the 2′ position of ribose causes the sugar ring to buckle in a distinct fashion known as C3′ endo, which favors formation of an A-type helix, instead of the C2′ endo conformation typically found in B-type helices. These distinct conformations of the five-membered ribose ring are referred to as the sugar pucker.

Perhaps the most profound consequence of the 2′ OH of RNA derives from its ability to form hydrogen bonds. While DNA can also form hydrogen bonds, as we saw earlier, the 2′ OH of RNA confers on RNA the ability to form hydrogen bonds

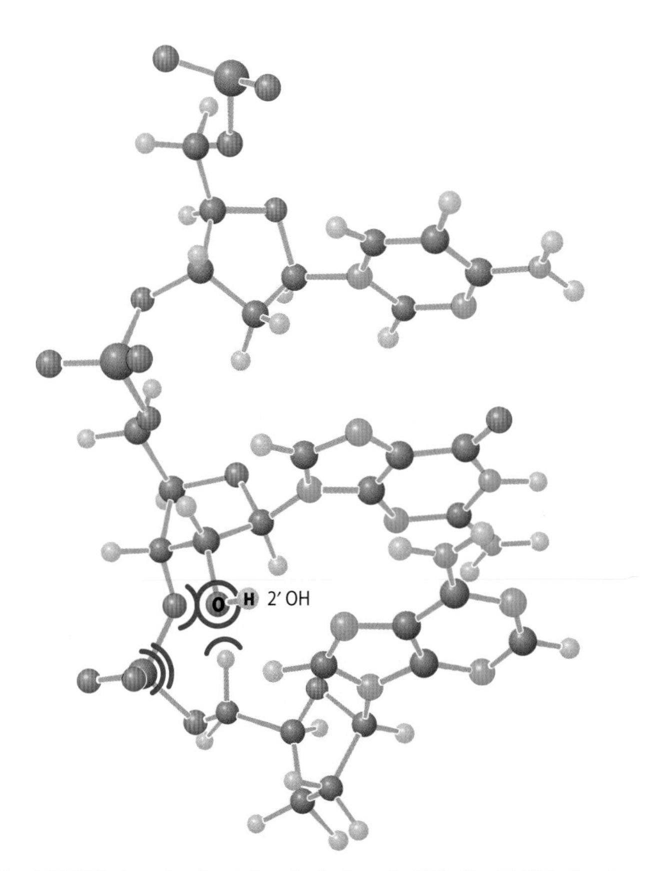

Figure 2.35 Role of 2′ OH in favoring the A-type helix form in RNA. The 2′ OH in the ribose would clash with other backbone atoms if RNA were to adopt the B-type helix.
From Berg *et al.*, *Biochemistry*, 5th ed. New York: WH Freeman.

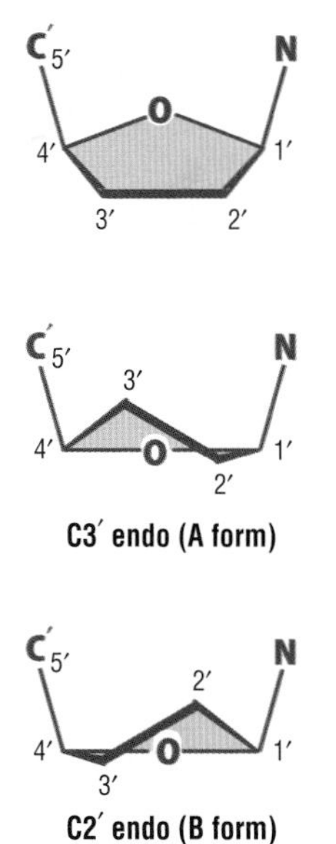

Figure 2.36 Sugar pucker in RNA and DNA. The sugar in RNA and DNA is non-planar and can adopt a variety of conformations referred to as sugar pucker. The top structure is energetically unfavorable and is shown here to highlight the chemical structure of the ribose sugar. The C3′ endo conformation is adopted in the A-type helix, whereas the C2′ endo prevails in the B-type helix.

more extensively than DNA. This added feature of RNA expands the potential for forming multiple interactions between nucleotides. Such hydrogen bonding by the 2′ OH promotes formation of complex structures that are unique to RNA molecules, as we discover in next section.

➔ We learn more about the role of regulatory RNAs in Chapter 13.

2.7 RNA FOLDING AND STRUCTURE

RNA can fold to form compact structures

Although DNA in cells is found overwhelmingly in a helical, double-stranded form, a far greater variety of RNA structures is found in biology. Many of these arise from single-stranded RNA molecules that fold to form compact three-dimensional structures. These RNA molecules do not typically code for proteins, and thus are termed 'non-coding'. Instead, they carry out structural, catalytic, and regulatory roles in many cellular processes including protein synthesis. The precise conformation, or **fold**, of an RNA molecule is determined by its nucleotide base sequence. Together, the fold and nucleotide composition determine the functional properties of the RNA. An ever-increasing number of genes in the human genome are known to be transcribed into non-coding RNA molecules, indicating the widespread involvement of these RNA molecules in many cellular processes.

RNA molecules can be described in terms of three levels of structural organization. The primary structure of RNA refers simply to the sequence of bases along an RNA strand, reading from the 5′ to 3′ direction. Portions of the RNA molecule can form short double-helical stretches, and we refer to these regions as RNA secondary structure, to distinguish them from the primary structure of nucleotides in the RNA strand. The single- and double-stranded portions of the RNA molecule come together to form a precise arrangement, which is its tertiary structure. We will expand upon each of these concepts in the sections that follow.

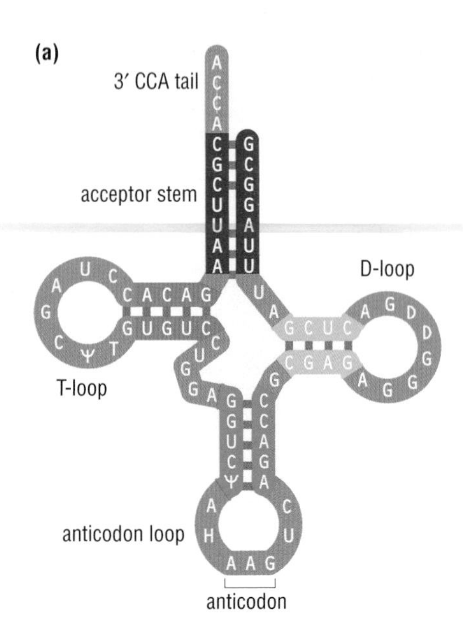

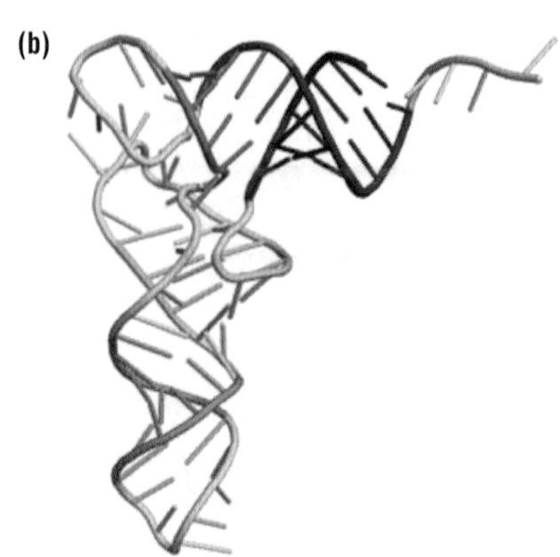

Figure 2.37 Double-helical regions of RNA. Single-stranded RNA molecules fold back on themselves to form short regions of double-stranded helices. (a) This RNA, called $tRNA^{Phe}$, has four helical regions, three of which are hairpin structures containing a double-helical region (blue, green, and orange) and a loop. The helix shown in red brings together the start and the end of the RNA strand to form a long-range pairing interaction. Several post-transcriptionally modified nucleotides are indicated: ψ is pseudouridine, D is dihydrouridine, T is ribothymidine, and H is a highly modified adenosine found proximal to the anticodon. (b) In the three-dimensional structure of this RNA, coaxial stacking of the helical segments to form the overall L-shape can be easily seen.

Short RNA helices are the building blocks of larger RNA structures

The fundamental structural unit of folded RNA molecules is the short, double-stranded helix, generally no longer than 6–8 bp in length, as shown in Figure 2.37. These double-stranded regions of RNA are A-type helices (see Figure 2.26) that form when two antiparallel nucleotide sequences form Watson–Crick base pairs with one another. If these complementary sequences are relatively close together in the linear or primary RNA sequence, the RNA forms what is called a **hairpin** structure (see Figure 2.37). The double-helical portion of the hairpin is called the stem and the unpaired bases are called the loop.

Complementary sequences that are widely separated from one another in the primary structure can also pair to form double-helical stretches, but they lack the hairpin. The collection of these base-paired elements (or helices) within a single RNA molecule is depicted in Figure 2.37a. Both types of helix are seen in this diagram of the secondary structure of a tRNA molecule – the D helix (in green) is formed from neighboring elements in the primary structure whereas the acceptor helix (in red) is formed from more distant elements.

Unlike DNA helices, RNA helices regularly contain base pairs that do not result from typical Watson–Crick interactions. Termed **non-canonical base**

pairs, these base pairs often feature bases that have been chemically modified in some way. A common modification is the addition of a methyl group in a process called methylation. Non-canonical base pairs are accommodated within double-helical regions, although they can cause distortions in the helix, as shown in Figure 2.38. These distortions are important for function. For example, there is a pairing interaction between a methylated G and an A in certain tRNAs that is critical for the formation of a functionally important kink in the structure.

One of the more commonly observed non-canonical base pairs is formed between G and U nucleotides as illustrated in Figure 2.39. This base pair is stabilized by two hydrogen bonds, just like a Watson–Crick A-U base pair, and is referred to as a 'wobble' base pair. We will discuss wobble interactions again in Section 11.2 when we learn about tRNA decoding interactions on the ribosome during protein synthesis.

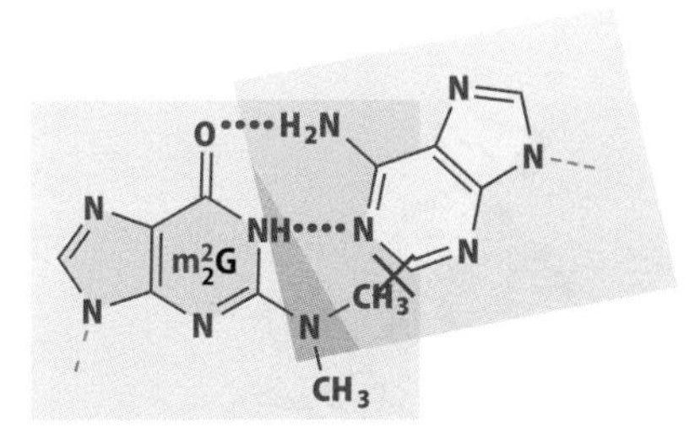

Figure 2.38 A base pair containing a methylated base. This base pair between a di-methylated guanine and adenine is non-planar because a methyl group in guanine clashes with the adenine ring ('X' marks the clash site).

Short helices assemble to form the final folded RNA structure

The full tertiary structure of RNA is formed when double-helical regions interact with one another, as well as with unpaired stretches of the RNA. The strong tendency of base pairs to stack upon one another gives rise to **coaxial stacking**, in which the helical segments stack with their central axes roughly lined up, as shown in Figure 2.37b. RNA structures typically contain several coaxially stacked helices. As seen in Figure 2.37b, which shows a tRNA molecule, each branch of the L-shaped molecule is composed of a distinct coaxial stack.

A second general feature of RNA structure is typified by interactions between nucleotides in unpaired and base-paired regions of a single RNA molecule. Hydrogen bonding interactions involving the 2′ OH of ribose or the exposed chemical groups on the bases are particularly common. An example is the **base triple** interaction, illustrated in Figure 2.40a, in which three bases interact with one another: two by Watson-Crick hydrogen bonds and the third, unpaired base by hydrogen bonding with the exposed edge of the base pair.

Another common example is the **A-minor motif**, where adenosine from one part of the molecule inserts into the minor groove of a double-helical region found elsewhere in the molecule. The unpaired adenosine hydrogen bonds with one or two 2′ OH groups flanking this groove as well as with the bases themselves (Figure 2.40b). These are just two examples of the many types of intramolecular interactions that stabilize a folded RNA structure.

A potential barrier to the folding of RNA into a compact three-dimensional shape is the high density of negative charge associated with the phosphodiester

Figure 2.39 Wobble base pair. The wobble base pair U–G, is shown next to the canonical Watson–Crick base pair, C–G.

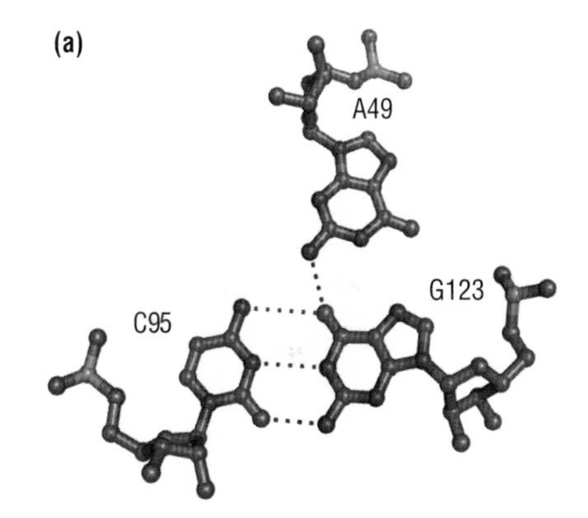

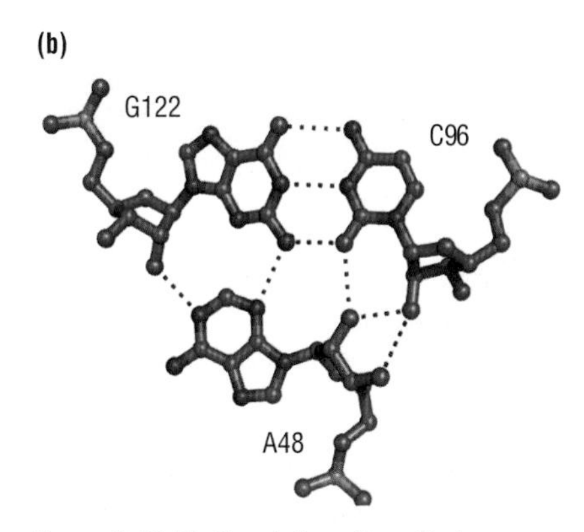

Figure 2.40 Tertiary interactions that stabilize RNA structure. (a) CGG base triple in which an isolated G residue interacts with a Watson–Crick G–C base pair in the major groove. (b) A-minor motif, in which an adenosine nucleoside forms hydrogen bonds with the ribose and bases in the minor groove of a helix.

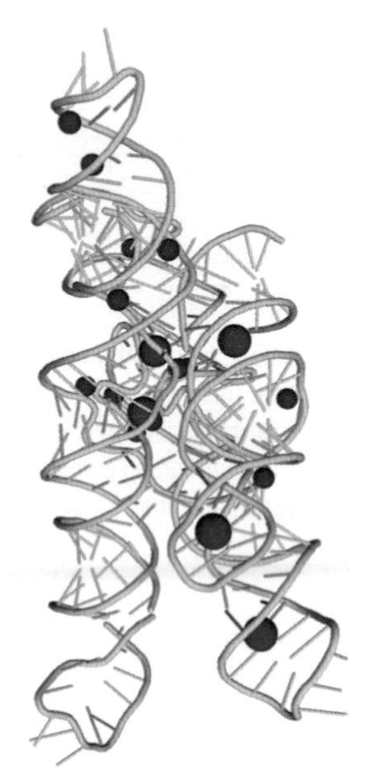

Figure 2.41 **Example of an RNA structure.** The tertiary structure of the group I intron provides an example of multiple sets of coaxially stacked helices and buried magnesium ions (red) that stabilize the overall fold of the molecule by minimizing charge repulsion of the phosphodiester backbone.

backbone, which would give rise to electrostatic repulsion when different segments of the backbone come together. RNA molecules overcome this problem by binding a large number of cations, most often magnesium. These positively charged ions balance the negative charges of the oxygen atoms in the sugar–phosphate backbone, allowing the backbones of different regions of the RNA to pack against one another as they are buried in the interior of the molecule. Figure 2.41 shows an example of a folded RNA structure with bound magnesium ions in the interior of the molecule that stabilize structure highlighted in red.

RNA secondary structures can be identified by phylogenetic analysis

The helical or base-paired regions of an RNA that assemble to form a complex tertiary structure can frequently be identified by searching in the RNA sequence for potential base-pairing interactions. This can be done with a computer program that searches for complementary sequences throughout the RNA molecule. Since this analysis can only identify regions with the potential to form helices, it will invariably pinpoint some short regions of complementarity that occur purely by chance, but that do not actually form helices in the RNA structure.

An approach that helps to identify which of the potential helix-forming regions indeed form helices in the folded RNA structure is known as **phylogenetic analysis**. This approach takes advantage of the fact that functionally equivalent RNA molecules are expected to have very similar secondary and tertiary structures, even if the actual nucleotide sequence is not conserved. Related RNA sequences (often from different species) can therefore be aligned and then analyzed for potential double-helical regions, as depicted in Figure 2.42. For a region of the RNA structure that is indeed double helical, the base-pairing interactions will be conserved even though the actual nucleotide sequence may differ among species. This is because there is selective pressure in the course of evolution to preserve the overall structure of the RNA, rather than the base sequence. Differences that conserve interactions are known as **covariations** because sequence changes at interacting bases are coupled in a way that preserves base-pairing.

	(1 ↓ at position 8; 2 ↓ at position 15)
A. laidlawii	CAACGCCCUUUUAAGGCGUGG
A. lipoferum	CAUCUGACUUUUAAUCAGAGG
B. burgdorferi	CAGCGCCCUUUUAAGGCGUUU
B. subtilis	CAUCUGACUUUUAAUCAGAGG
E. coli	CAGUUGACUUUUAAUCAAUUG
H. pylori	CAAUUCCCUUUUAAGGAAUGG
M. genitalium	CAUUUGACUUUUAAUCAAAGG
M. Pg50	CAACUGGCUUUUAACCAGUGG
M. pneumoniae	CAUUUGACUUUUAAUCAAAGG
S. aureus	CAUCUGACUUUUAAUCAGAGG
T. pallidum	CAACGCCCUUUUAAGGCGUGG

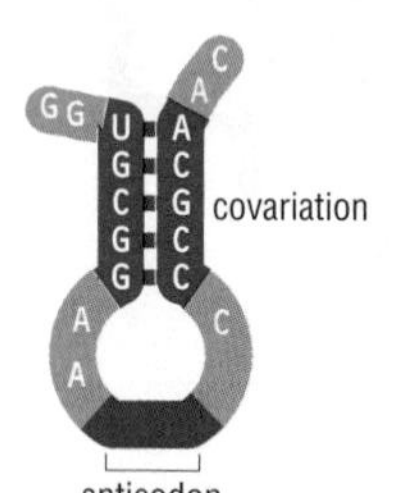

Figure 2.42 **Phylogenetic analysis in RNA structure.** Aligned tRNA[Lys] sequences are used to identify covarying regions (pink) that are predicted to form paired regions in the final folded RNA structure. For example, the identity of the base is not conserved at positions 1 or 2, but the base-pairing potential between these positions is.

2.8 THE RNA WORLD AND ITS ROLE IN THE EVOLUTION OF MODERN-DAY LIFE

RNA biology provides insights into the origins of life

Scientists have long been fascinated by the question of how life evolved on Earth. At some point in the distant past, there must have been a transition from a mixture of reactive molecules in the primordial seas – the early environment in which life evolved – to the formation of life. While defining life is a notoriously difficult task, the most fundamental properties of living organisms are that they are able to reproduce themselves and that they evolve as a result of chemical variations that arise as they reproduce. Although we cannot know what the first life forms were like, we can speculate about them based on our knowledge of chemistry and the various forms of life on Earth today.

Any autonomously replicating system that continually grows and divides requires a supply of small molecules and energy. Early life forms must have been chemically far simpler than modern organisms and are presumed to have depended upon a more limited set of biological molecules. As living things evolved to become more complex, they made use of additional classes of molecules and incorporated the necessary chemical reactions into their biological pathways. This led to the evolution of the common ancestor of all modern life (known as LUCA, the last universal common ancestor), which likely depended on RNA, DNA, proteins, carbohydrates and lipids to fulfill the requirements for life. Heritable information in this ancestor was stored as sequences of base pairs in a nucleic acid polymer, while other biochemical functions were, to a large extent, carried out by proteins. There are, as far as we know, no modern-day examples of organisms that resemble an earlier form of life from which all forms of life subsequently evolved, but it is interesting to speculate about the molecules these earlier life forms could have used. In particular, we would like to know how life evolved from simple chemical processes in the primordial world.

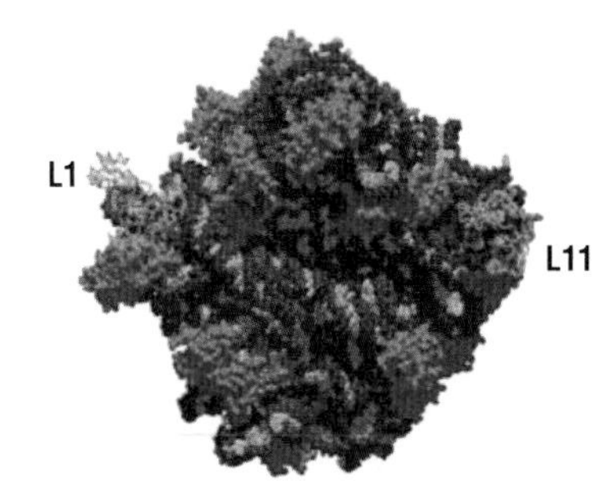

Figure 2.43 Remnant of the RNA world. The ribosome contains both RNA and proteins, with the RNA component carrying out most key functions during the synthesis of proteins.

There is modern-day evidence for the earlier existence of an RNA world

There is a great deal of circumstantial evidence that nucleic acids or similar molecules were the dominant biological macromolecules in the earliest forms of life, and were responsible for carrying out most biological functions before a sophisticated protein synthesis apparatus evolved. Nucleic acids are replicable by nature (since the base-pairing interactions allow one polynucleotide strand to serve as a template for another), while proteins are not. In addition to being able to store genetic information, RNA molecules in modern organisms can also catalyze chemical reactions, as well as adopt complex three-dimensional structures that carry out a variety of other biological functions. It is therefore possible to imagine an earlier RNA world in which RNA acted both as an information carrier and as the primary functional entity in cells, performing all the tasks needed by a primitive life form.

Some of the most compelling arguments for a putative RNA world derive from the diverse roles that RNA plays in modern-day cells. RNA plays a central role in processes including RNA processing, protein synthesis, and the regulation of gene expression. In these different roles, RNA can be a repository of information, provide the mechanism to decode the stored information, and sometimes be the catalyst to drive chemical reactions. Protein synthesis is the quintessential example of all these facets of RNA: the functional core of the ribosome, depicted in Figure 2.43, is composed predominantly of RNA; the molecule encoding a protein is contained in mRNA; and molecules that decode the information in the RNA message and orient amino acids for polypeptide synthesis are provided by RNA. When considering the evolution of protein synthesis, it is difficult to understand how a protein synthesis machine could be made of protein before such machinery existed - the classic chicken or egg first problem. But it is not difficult to imagine a protein synthesis machine composed entirely of RNA that emerged from a world in which RNA was the dominant macromolecule.

Another persuasive argument for life's origins in an RNA world lies in the composition of the many enzyme cofactors used by modern-day organisms. As we will learn in the next chapter, many protein enzymes that catalyze chemical reactions depend on a small molecule that is referred to as a coenzyme. Many of these coenzymes are ribonucleotides that are built from a core structure of the nucleoside, adenosine. For example, the coenzymes involved in oxidation and reduction reactions, including nicotinamide adenine dinucleotide (NAD) (Figure 2.44), nicotinamide adenine dinucleotide phosphate (NADP), flavin adenine dinucleotide (FAD), and flavin mononucleotide (FMN), contain a core adenosine. The adenosine serves as a carrier of the chemically active group but is not, itself, directly involved in the facilitated reactions. Similarly, adenosine-based nucleotides such as coenzyme A and S-adenosylmethionine (Figure 2.44) are the carriers

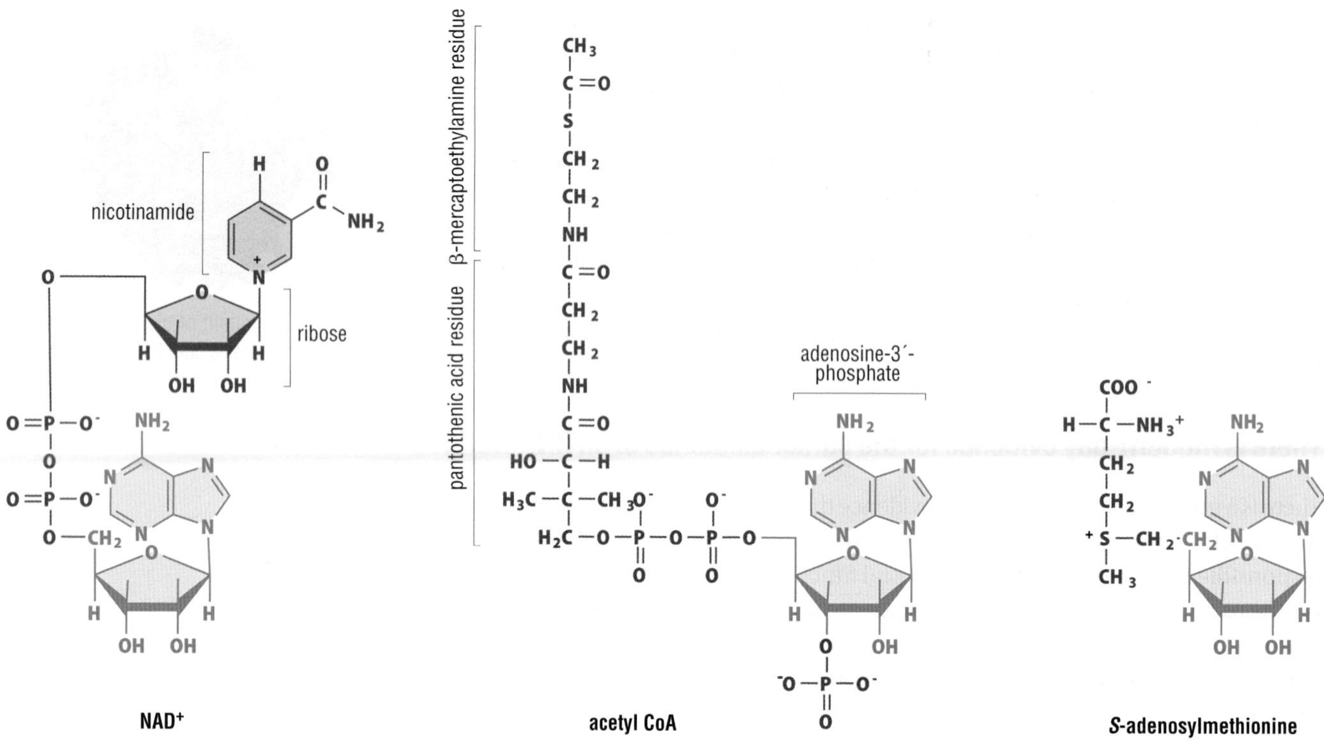

Figure 2.44 Remnant of the RNA world. Three examples of ribonucleotide cofactors that are all built from adenosine (blue): NAD, acetyl-coenzyme A (acetyl-CoA), and *S*-adenosylmethionine. The chemical groups that can be transferred to other proteins are shown in pink.

for different chemical moieties that are transferred to proteins or nucleic acids in a variety of cellular reactions.

Because each cofactor is used by many different enzymes to carry out similar functions, it has been suggested that the dependence of all current organisms upon these nucleotide cofactors became fixed at an early stage in evolution. To change a cofactor would compromise the function of all the enzymes using that cofactor, and the organism would struggle to survive such a deficit. Thus, the biological roles of nucleotide cofactors were locked into place during a time when RNA dominated the biochemical scene and cofactors were built from the molecular building blocks that were available.

Other arguments for the existence of an RNA world have rested on the ability of both natural RNA molecules and those derived from *in vitro* experiments to fold into complex three-dimensional structures that catalyze chemical reactions. Some researchers argue that the RNA molecules called riboswitches, which control gene expression throughout the bacterial world, might also be remnants of an RNA world. While these hypotheses are subject to considerable debate, they have helped guide our thinking about the types of molecules that may have been found in an early RNA world but whose functions have since been largely taken over by the more chemically diverse proteins.

Prebiotic chemistry and the feasibility of an RNA world

While the molecular evidence makes a strong case for the predominant role of nucleic acids early in the evolution of life, scientists have long struggled to understand how such an RNA world could have emerged from the primordial soup (a term that refers to the early environment in which life evolved). Early experiments to simulate the primordial soup, using spark chambers and presumed primitive earth components such as water, methane and ammonia, readily generated reasonably diverse mixtures of amino acids. However, identifying conditions that lead to the robust synthesis of nucleic acids has been much more difficult.

First, there are several chemically distinct components that are needed: the nucleotide bases, the sugar moieties, and the phosphate backbone. Although adenine is synthesized efficiently from mixtures of hydrogen cyanide and ammonia, the other bases (G, C, and U) are much less readily synthesized. Prebiotic synthesis of the sugar ring, ribose, presents another significant chemical challenge, as does formation of the glycosidic bond between the bases and the sugars. Despite these and other feasibility issues surrounding the synthesis of nucleic acid-like material on a primitive Earth, the concept of an RNA world is securely founded on the historical record provided by modern biology and it is here that we look to understand our origins.

This glimpse of the place of nucleic acids on the primitive Earth brings us to the end of our initial exploration of the nucleic acids. What we have seen here is just the foundation for what is to come in later chapters, where we investigate further the pivotal role that nucleic acids play in the biological processes that underpin life. Nucleic acids are only part of the story of genome function, however. Their position center-stage must be shared with another vitally important group of biological macromolecules, the proteins, which we explore in the next three sections.

2.9 FUNDAMENTALS OF PROTEIN STRUCTURE

Proteins are the most versatile macromolecules in biology

Almost every aspect of biology relies on proteins. Proteins can catalyze chemical reactions and form structural components of a cell, they can create channels in a membrane to allow molecules to pass in and out of cells, or they can recognize a vast array of infectious microorganisms. Proteins also form molecular machines that convert energy from one form into another, for example allowing muscles to contract and plants to harness the energy of sunlight. Proteins can carry out such an astonishing variety of functions because they are synthesized from a chemically diverse set of building blocks. This chemical diversity makes it possible for different proteins to adopt highly specialized structures that have been optimized through evolution to perform their particular tasks. We shall see here and in the following sections how the chemical makeup and structure of protein molecules has allowed them to evolve to be the dominant macromolecule in the cell.

Amino acids are the building blocks of proteins

Proteins are polymers composed of **amino acids** that are joined by covalent bonds. This is analogous to the way that nucleic acids are polymers of nucleotides, which are also joined by covalent bonds. Each free amino acid has an amino group, a carboxylate group, and a side chain, attached to a central carbon atom (denoted Cα); this generalized structure is shown in Figure 2.45. There are 20 different types of side chain, each with distinct chemical properties; their structures are shown in Figure 2.46. There are side chains that are largely hydrophobic in character, and those that contain polar chemical groups. A few side chains contain both

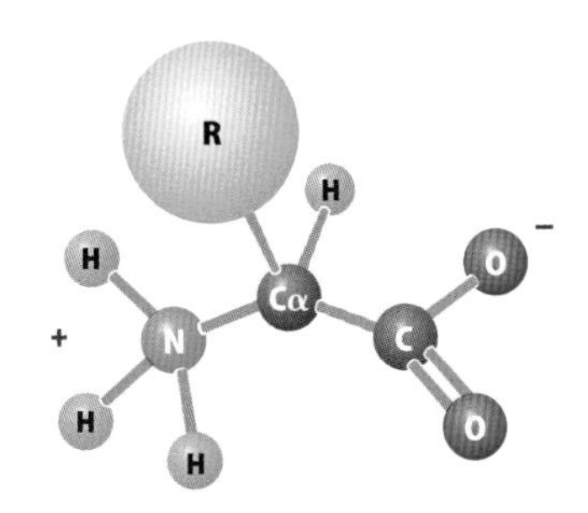

Figure 2.45 An amino acid. Chemical structure of an amino acid showing a central carbon atom (Cα) to which is attached an amino group (–NH_3^+), a carboxylate group (–COO^-), a hydrogen atom, and a side chain (which is by convention shown as R).

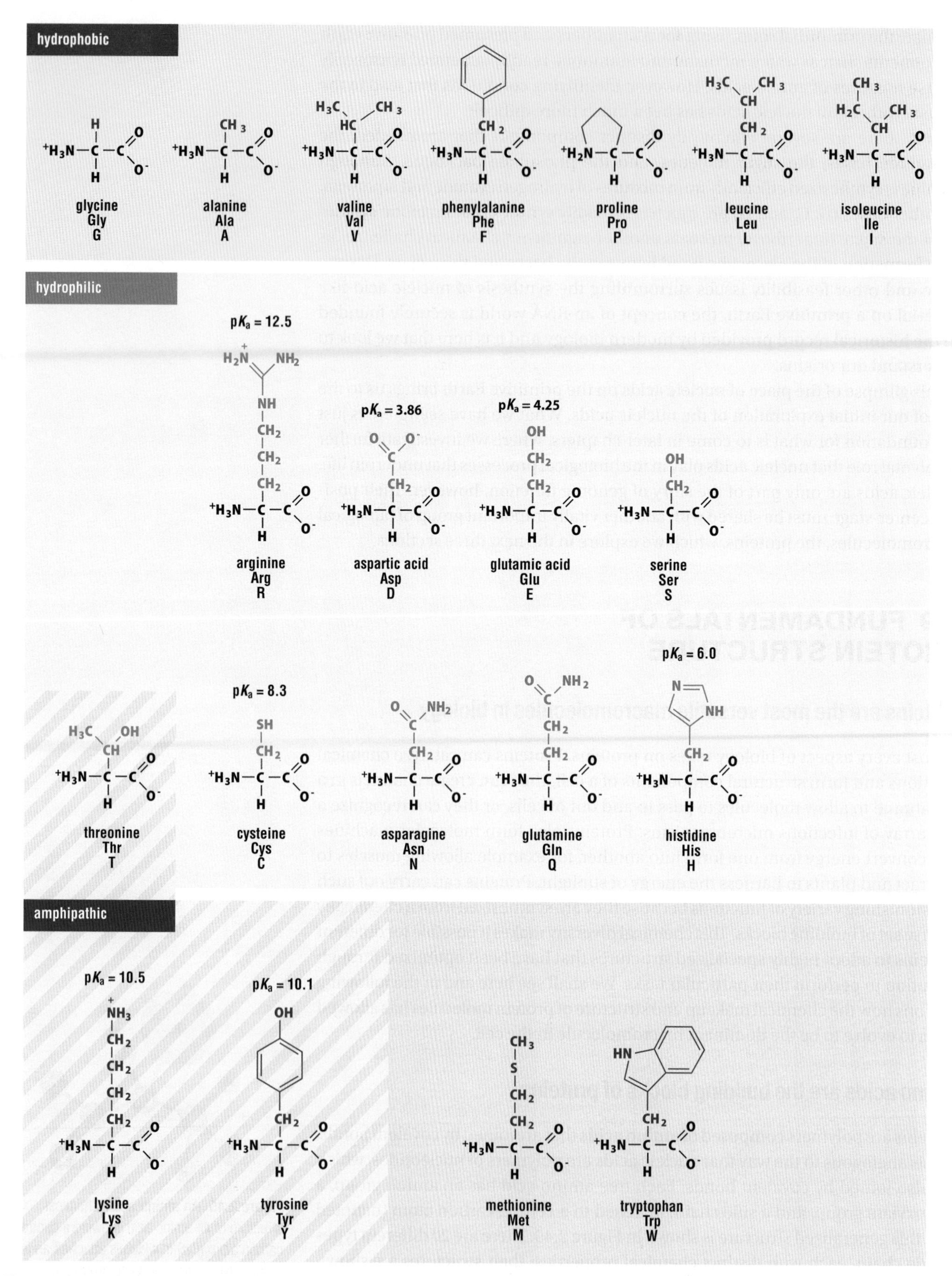

Figure 2.46 The 20 amino acids. Shown for each amino acid is the chemical structure, the full common name, and the three- and one-letter abbreviated name. The pK_a values are shown for ionizable side chains. Here we group the amino acids according to their hydrophobic (red), hydrophilic (cyan), or amphipathic (dark blue) character, but there are other ways to classify them.

hydrophobic and hydrophilic groups, and are as a result termed **amphipathic** side chains. Some side chains have a combination of properties; for example, lysine has a positively charged amino group at the end of a long hydrophobic side chain.

As we saw earlier for nucleic acid bases, Figure 2.47 illustrates how some amino acid side chains can gain or lose a proton and therefore carry an electrical charge at neutral pH. The basic amino acids, lysine and arginine, bear a full positive charge at neutral pH, while aspartate and glutamate bear a full negative charge. The typical pH at which the transition between the charged and uncharged species occurs is the pK_a.

The unique chemical properties of each side chain govern the interactions of side chains with one another, with other biological molecules, and with the aqueous environment of the cell. **Sulfhydryl** (-SH) groups on cysteine side chains can react with one another to form a covalent -S-S- linkage called a disulfide bond, which is an important bond for stabilizing protein structures. Importantly, however, disulfide bonds are only found in extracellular proteins (those that are secreted from the cell) because the interior of the cell is a reducing environment - an environment in which the -S-S- linkage is unable to form.

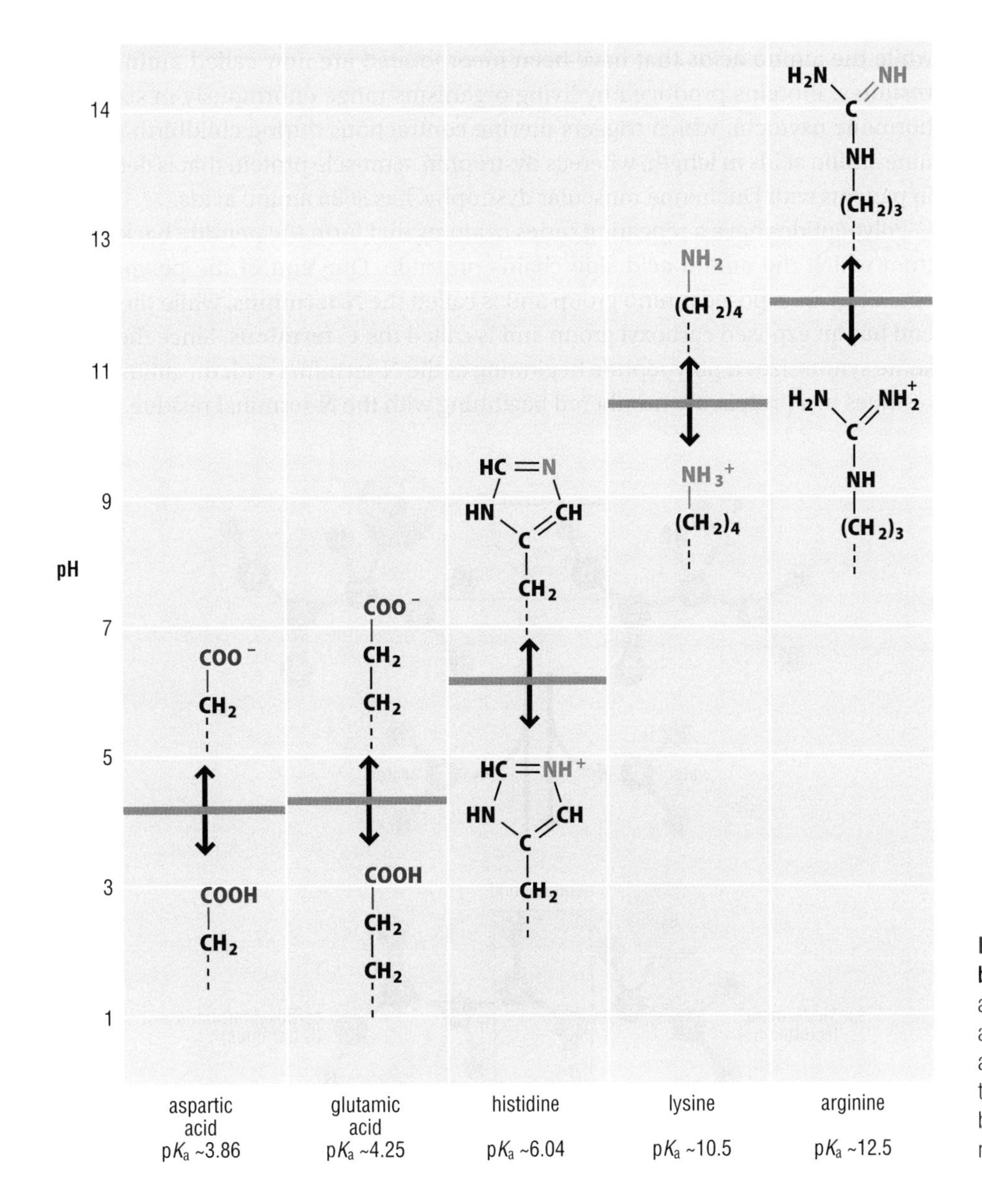

Figure 2.47 Amino acid side chains can become charged. The charge of any amino acid depends upon its pK_a (horizontal gray line) and the pH of the solution. The common charged amino acid side chains are shown, along with their chemical structures at pH values above and below the pK_a. The ionizable group is indicated in red (acidic) or blue (basic).

The chemical properties of side chains also allow them to play key roles in enzymatic reactions. The side chain of histidine has a pK_a around 6.5, which means that the transition between the protonated and unprotonated form of histidine occurs near the neutral pH of the cell. This property enables histidine to either donate *or* accept a proton during chemical reactions.

With the exception of glycine, the amino acids exist as two distinct stereoisomers that differ in the arrangement of atoms about Cα: the L- and D-amino acids. Like all stereoisomers, the L- and D-amino acids contain identical types of atoms and chemical bonds but are mirror images of one another, and are therefore not superimposable in space. Only L-amino acids are used by the main protein-synthesis apparatus of the cell, the ribosome.

Amino acids are joined by peptide bonds to form polypeptides

A protein is composed of amino acids that are joined by characteristic covalent bonds, as depicted in Figure 2.48. These linkages, called **peptide bonds**, form through a reaction between the carboxylate group of one amino acid and the amino group of the next, generating one water molecule for each peptide bond that is formed. A polymer formed from the joining of amino acids is called a **polypeptide**, while the amino acids that have been incorporated are now called **amino acid residues**. Proteins produced by living organisms range enormously in size; the hormone oxytocin, which triggers uterine contractions during childbirth, is just nine amino acids in length, whereas dystrophin, a muscle protein that is defective in patients with Duchenne muscular dystrophy, has 3685 amino acids.

Polypeptides have a repeating series of atoms that form the **peptide backbone**, from which the amino acid side chains protrude. One end of the polypeptide chain has an exposed amino group and is called the **N-terminus**, while the other end has an exposed carboxyl group and is called the **C-terminus**. Since the ribosome synthesizes a polypeptide beginning at the N-terminal end, the amino acid residues in a protein are numbered beginning with the N-terminal residue.

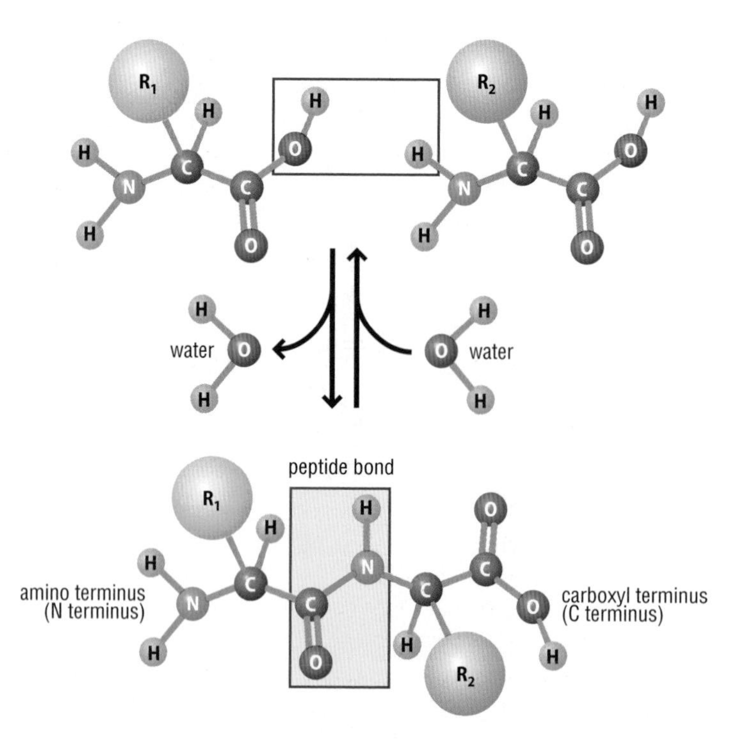

Figure 2.48 The peptide bond. Formation of a peptide bond between two amino acids, with the accompanying release of a water molecule. The addition of a water molecule is required to reverse the reaction and break the peptide bond.

Properties of the peptide backbone govern polypeptide conformation

The peptide bond has properties that limit the conformations that polypeptide chains can adopt. The atoms that make up the peptide bond all lie in a plane due to a phenomenon known as **resonance**, in which electrons are shared among several atoms of the peptide bond unit. This sharing is what we call delocalization. The delocalization of electrons in the peptide bond prevents free rotation of one of the bonds that form the overall peptide bond 'unit', locking the atoms into a planar (flat) conformation, as shown in Figure 2.49. Consequently, each amino acid unit in the polypeptide backbone has only two covalent bonds about which free rotation can occur: the N-Cα bond, which is defined by the rotation angle, ϕ and the Cα-C bond, which is defined by rotation angle ψ.

There are many combinations of ϕ and ψ angles that would give rise to collisions between atoms in the polypeptide chain or the side chains. These combinations of ϕ and ψ angles therefore do not occur. A **Ramachandran plot** provides a convenient graphical depiction of the allowable combinations of ϕ and ψ angles. An example of a Ramachandran plot is shown in Figure 2.50. While the planarity of the peptide bond restricts the possible conformations of the polypeptide backbone, it still allows proteins to adopt an enormous number of conformations and contributes to the stability of protein structures, as we shall see.

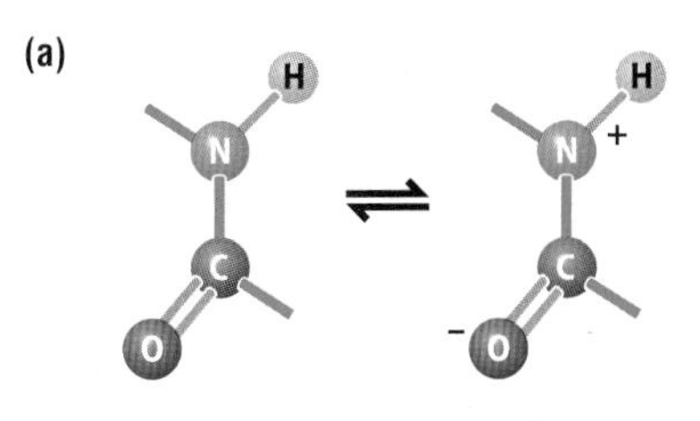

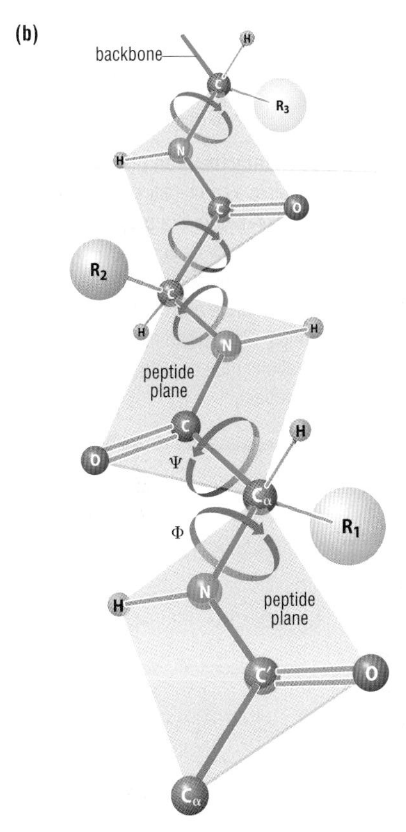

Figure 2.49 Bond rotations in polypeptide backbones. (a) The peptide bond has a partial double bond character due to the sharing of electrons between C–O and C–N atoms. As a result, the O–C–N–H atoms all lie in a plane. (b) A three amino acid polypeptide, showing the ψ and φ bonds that allow the amino acid residues to rotate about each other.

2.10 PROTEIN FOLDING

Proteins fold to adopt defined three-dimensional shapes

Each protein adopts a characteristic shape when the polypeptide folds upon itself to yield a much more compact three-dimensional arrangement of atoms. This process of collapse, referred to as **protein folding**, is driven by non-covalent interactions between atoms in the polypeptide chain. The resulting three-dimensional arrangement of the atoms is called the protein's structure, or **conformation**, and is determined by the sequence of amino acids along the polypeptide chain and by the chemical and physical properties of the polypeptide backbone.

There are four levels of organization that describe a protein's conformation; these are illustrated in Figure 2.51. The sequence of amino acids along a polypeptide chain is a protein's **primary structure**. Short regions of the polypeptide chain can form regular, repeating regions of structure stabilized by hydrogen bonds called **secondary structure**. When a protein folds, these regular elements come together, along with the other residues in the polypeptide chain, to form a defined shape called the **tertiary structure** of a protein. Many proteins are actually complexes formed by the association of several folded polypeptides, which together constitute the **quaternary structure** of a protein.

The most common forms of secondary structure are alpha helices and beta sheets

Polypeptide chains can form local regions of uniform structure known as **secondary structure**, which are characterized by repeating patterns of ϕ and ψ angles and of hydrogen bonds between neighboring atoms. There are two common forms of secondary structure, the **alpha helix** and the **beta sheet**. In an **alpha helix**, the polypeptide backbone curves in a right-handed helical path, allowing hydrogen

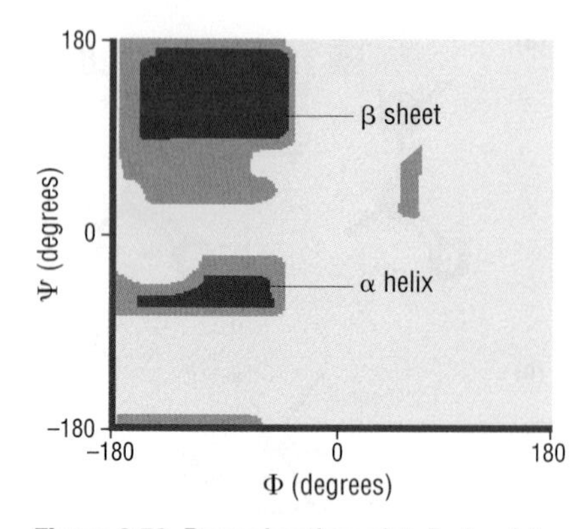

Figure 2.50 Ramachandran plot. Each point on the plot depicts a combination of φ and ψ angles about a single Cα in a polypeptide chain. The black regions show φ, ψ combinations that do not produce clashes between atoms, while the combinations in the gray regions produce minimal clashes. Glycine can adopt many more φ, ψ combinations than depicted here because it lacks a side chain.

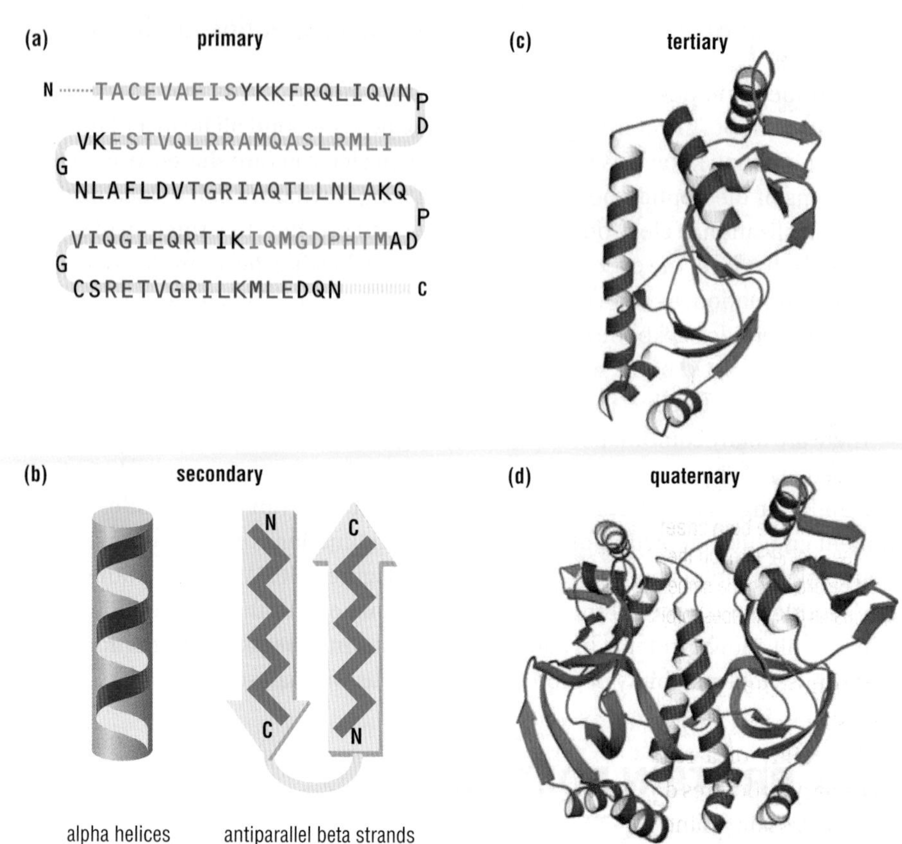

Figure 2.51 Levels of protein structure. The four levels of protein structure are illustrated for the catabolite activator protein (CAP), which regulates transcription in *E. coli*. (a) The amino acid sequence of a protein is its primary structure. (b) Short regions of regular, repeating structure in the protein are called secondary structure elements. The most common are alpha helices and beta sheets. (c) The three-dimensional structure of a polypeptide is its tertiary structure. (d) The arrangement of two or more polypeptide chains in a protein is its quaternary structure.

bonds to form between the carbonyl oxygen (C=O) of one residue and the amide nitrogen (N–H) four residues further along the polypeptide chain, as shown in Figure 2.52. The resulting structure is cylindrical, with the polypeptide backbone forming a well-packed interior and the side chains protruding on all sides.

Many alpha helices found in proteins have hydrophobic side chains exposed on one face and polar side chains exposed on the opposite face and are hence known as **amphipathic alpha helices.** Helices with this character frequently occur on the surfaces of proteins, where their polar faces are in contact with water while their hydrophobic faces interact with the tightly packed protein core, where water is excluded. The extensive hydrogen bonding in alpha helices also allows the polar atoms of the peptide bond to be buried in the hydrophobic core of a folded protein.

The **beta sheet** is formed when two or more segments of the polypeptide backbone, called **beta strands**, hydrogen bond to one another through their backbone carbonyl and amide groups, as illustrated in Figure 2.53. These aligned strands form a sheet-like structure that is slightly twisted. In contrast with the alpha helix, beta sheets can form between two or more non-contiguous segments of a polypeptide. The strands of the beta sheet can be oriented in the same direction (N- to C-terminus) and form a **parallel beta sheet**, or they can be oriented in opposite directions, forming an **antiparallel beta sheet** (see Figure 2.53). Mixed sheets, containing both parallel and antiparallel pairings, also occur. In all cases, the side

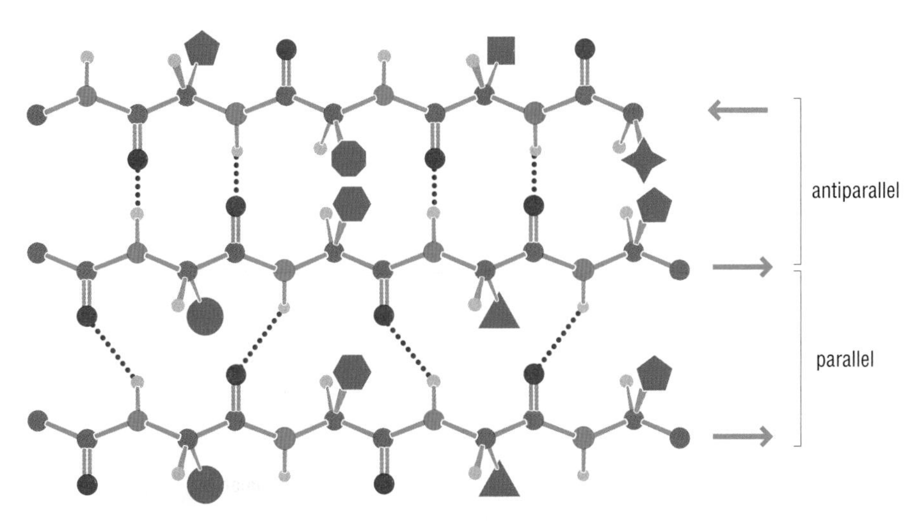

Figure 2.53 The beta sheet. In a beta sheet, two or more extended polypeptide chains hydrogen bond with one another through their backbone carbonyl and amide groups. Two strands of a beta sheet can be either parallel (with the same N- to C-terminus orientation) or antiparallel (with opposite orientation). The side chains (blue shapes) project above and below the beta sheet.

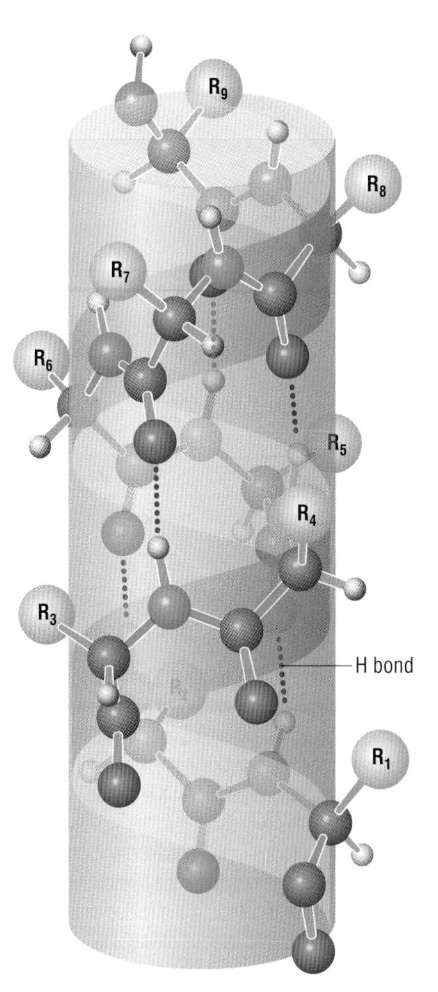

Figure 2.52 The alpha helix. The alpha helix is a coil in which the carbonyl oxygen of one residue hydrogen bonds with the amide hydrogen four residues further along the chain. Each residue in an alpha helix has a w angle of about –57° and a c angle of about –47°. The helix spirals in a right-handed direction, like a right-handed cork screw.

chains protrude above and below the beta sheet. Like alpha helices, amphipathic beta sheets are also found at protein surfaces.

The ability of a region of polypeptide chain to adopt one of the distinctive secondary structures described above depends on the primary structure of the polypeptide: certain amino acid residues favor the adoption of a given secondary structure more than others. For example, leucine, methionine, glutamine, and glutamic acid are often found in helices, while valine, isoleucine, and phenylalanine are more often found in beta sheets. By contrast, glycine and proline are often found in beta turns (which link helices or sheets that are adjacent in a polypeptide chain). Glycine shows exceptional conformational flexibility due its lack of a bulky side chain, so it is particularly well suited to forming sharp bends in the polypeptide chain.

Secondary structural elements condense to form the folded protein

The tertiary structure of a protein forms when the alpha helices and beta sheets, which are connected by less regular regions of polypeptide chain, associate in a precise manner to form a defined three-dimensional shape. The polar, aqueous environment of the cell drives protein folding by encouraging hydrophobic groups in the polypeptide chain to interact with one another inside the protein, while favoring the exposure of polar groups on the protein's surface, a behavior depicted in Figure 2.54. The final, folded protein has a core filled with predominantly non-polar side chains that form favorable van der Waals interactions with one another.

The formation of secondary structural elements helps to bury polar backbone groups in the protein's interior. Recalling that the secondary structure is stabilized by hydrogen bonding of the peptide backbone, it is energetically more favorable to bury hydrogen-bonded polar atoms in the hydrophobic interior of a protein than to bury unpaired hydrogen bond donor or acceptor atoms (that is, hydrogen bond donors or acceptors that are not participating in a hydrogen bond). The atoms in the interior of a protein fit together in a way that leaves very few empty spaces or gaps, which would be energetically unfavorable. The surface of a protein is primarily populated with charged and polar side chains that are exposed to the cell's aqueous environment, along with loops of polypeptide chain that connect the secondary structural elements.

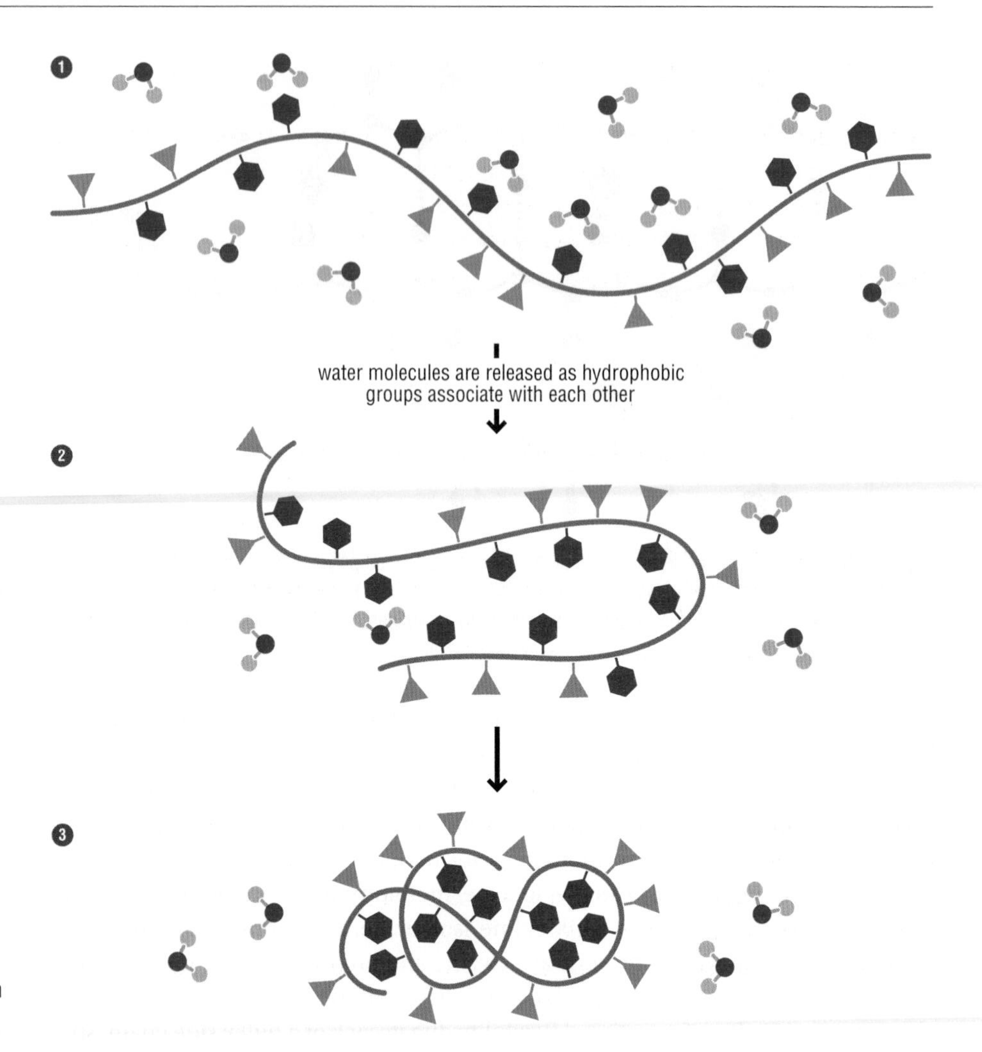
1
water molecules are released as hydrophobic groups associate with each other
2
3

Figure 2.54 Protein folding. The folding of a polypeptide chain is driven by the association of hydrophobic side chains (pink) with one another in the interior of the protein, with polar side chains (blue) exposed on the surface. Water molecules are ordered in the vicinity of hydrophobic groups in the unfolded polypeptide (step 1), but are released as the hydrophobic residues cluster together in the folded protein (steps 2 and 3). This release of water is energetically favorable and helps to drive protein folding.

The folded form of a polypeptide chain has more favorable energy in an aqueous environment than the unfolded chain. When hydrophobic side chains are exposed to water, the nearby polar water molecules tend to orient themselves in a way that cancels out one another's electric dipoles. By contrast, when the protein folds, the hydrophobic side chains associate with one another, freeing the water molecules to adopt a larger variety of orientations in their polar surroundings.

➔ We learn more about entropy in Chapter 3.

The number of conformations or positions a molecule can adopt is related to a thermodynamic quantity known as entropy – a measure of the disorder within a given system. An increase in entropy is energetically favorable. The increase in disorder of water molecules released by protein folding therefore makes the folding process energetically favorable. Even though the polypeptide chain itself has a decrease in disorder (and therefore lower entropy) as it changes from a flexible, freely moving polymer to one with a defined shape, the net change in entropy for the entire system – polypeptide and surrounding water – is positive. We shall see later in this chapter how the formation of cell membranes is driven by a similar process involving the partitioning of hydrophobic chemical groups from the polar environment of the cell.

Amino acid sequence determines protein structure

As we noted previously, the folded structure of a protein is determined by its amino acid sequence (that is, its primary structure), and most proteins adopt their particular three-dimensional structures spontaneously in an aqueous environment. There are many proteins that can be unfolded by a change in solution conditions – either by treatment with chemicals that destabilize the protein or by heating to high temperatures – and then spontaneously refold when the solution is returned to the original conditions. Some polypeptides require the assistance of proteins called chaperones to fold in the cell. The chaperones ensure that the folding process proceeds smoothly but do not determine the final structure. That said, there are some proteins that cannot refold properly; we say that they have been denatured. The white of an egg, which contains the protein albumin, is a classic example: when the egg white is heated, the albumin denatures, at which point the egg white essentially changes irreversibly from liquid to solid.

Since the process of protein folding is so complex and involves an intricate network of non-covalent interactions, we still cannot reliably predict the tertiary structure of a protein solely from its primary amino acid sequence. However, it is possible to infer whether two proteins are likely to adopt a similar structure by comparing their amino acid sequences: two polypeptides that have 50% of their amino acids in common generally have nearly identical structures. Even when two polypeptides have just 25% of their amino acids in common – also referred to as 25% sequence identity – their structures are generally fairly similar. Figure 2.55 shows an example of two proteins that regulate gene expression and are 27% identical in sequence, yet adopt virtually the same structure and carry out similar functions in the cell. Comparisons of amino acid sequence are commonly used to identify whether two proteins are predicted to have the same structure.

While it is not yet possible to predict tertiary structure on the basis of primary structure with any degree of accuracy, it is possible to predict secondary structural elements with some confidence. Indeed, we can now predict the occurrence of secondary structure on the basis of primary structure with an accuracy in excess of 75%.

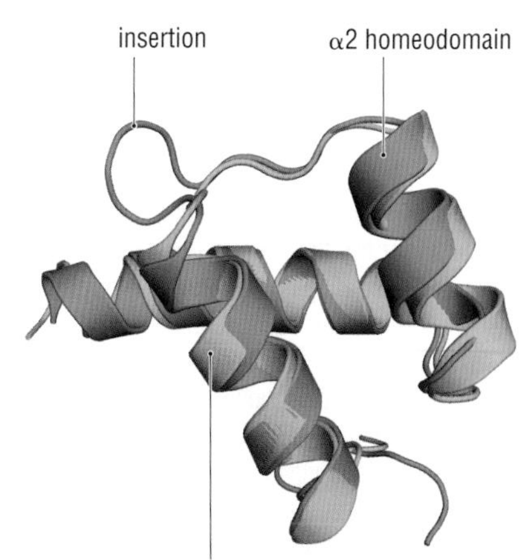

Figure 2.55 Sequence similarity and protein folding. The Matα2 (blue; Protein Data Bank (PDB) code 1APL) and engrailed (green; PDB code 1HDD) homeodomain proteins are only 27% identical in amino acid sequence, but adopt a nearly identical protein fold. Note the three additional amino acids inserted in MATα2 as compared with engrailed.

Amino acid sequence can reveal clues about protein function

Each protein can carry out its particular function in a cell because its structure and the chemical nature of the exposed amino acids are tailored for a particular task. We saw in the example above that amino acid sequence can diverge quite a bit while at the same time conserving overall structure and, generally, function. Since the amino acid sequence of a protein determines its structure and function, it is sometimes possible to predict the function of a newly identified gene based on the amino acid sequence of the protein it encodes. The sequence of this protein can be compared with the sequences of proteins whose functions are known, as represented schematically in Figure 2.56; if any proteins have at least 25% of their amino acids in common, they likely adopt the same fold. Often, although not always, this also means that the proteins perform a similar function.

Further clues to function may come from determining whether certain key amino acids – perhaps those required for catalysis, or binding to another protein – are also conserved in the newly identified protein. Ultimately, the protein's function must be verified with biochemical or genetic experiments. Sequence comparisons must be interpreted with caution, as there are many examples in which a particular protein fold has been adapted to perform a new function. Nevertheless, the ability to search databases of amino acid sequences is a powerful and efficient approach to obtaining important clues about a protein's function.

➔ We learn about searching databases in Section 19.15.

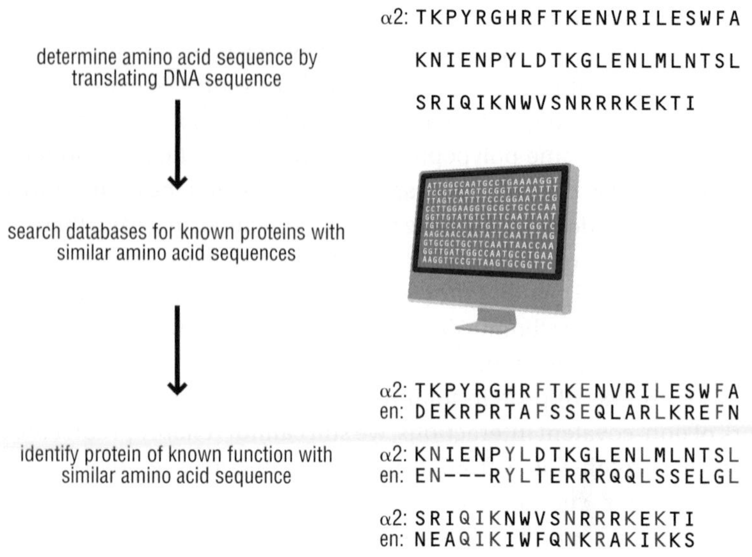

determine amino acid sequence by translating DNA sequence

α2: TKPYRGHRFTKENVRILESWFA
KNIENPYLDTKGLENLMLNTSL
SRIQIKNWVSNRRRKEKTI

↓

search databases for known proteins with similar amino acid sequences

↓

identify protein of known function with similar amino acid sequence

α2: TKPYRGHRFTKENVRILESWFA
en: DEKRPRTAFSSEQLARLKREFN

α2: KNIENPYLDTKGLENLMLNTSL
en: EN---RYLTERRRQQLSSELGL

α2: SRIQIKNWVSNRRRKEKTI
en: NEAQIKIWFQNKRAKIKKS

Figure 2.56 Comparing protein sequences can help identify the function of a protein. Flow chart illustrating how the function of an unknown protein can be identified by comparing its amino acid sequence to other proteins of known function. The sequence of a portion of the MATα2 protein (α2) is compared to a database of proteins whose functions are known, using a computer to align the sequence of α2 with each protein and obtain the best match. This identifies the protein engrailed (en), which is identical to α2 at 27% of its amino acids, and adopts a virtually identical structure, as shown in Fig. 2.55.

2.11 PROTEIN FOLDS

Protein folds have been conserved through evolution

Proteins can adopt an astonishing variety of tertiary structures. However, some structures look very much like one another in their overall shape, secondary structure content, and the way in which secondary structural elements pack together and are connected by loops. We refer to the arrangement of secondary structural elements that characterizes a particular protein as a **protein fold**. Proteins of identical structure - ones that have an identical arrangement of atoms in space - also have the same fold. However, two proteins that do not appear at first glance to have identical structures may, in fact, have identical folds if they contain essentially the same secondary structural elements that are connected to one another in the same way. As seen in the example in Figure 2.57, two proteins with the same fold can sometimes overlay rather poorly in space because of differences in the precise length of secondary structural elements and the way in which they pack against one another.

The great variety of protein folds is evident in the examples of common protein folds depicted in Figure 2.58. Some proteins, such as the enzyme triose phosphate isomerase (Figure 2.58b), are **globular**, meaning that they are roughly spherical and have a relatively low ratio of surface area to volume. Others are highly elongated, such as the protein, GCN4 (Figure 2.58d), which dimerizes through the association of two alpha helices that form a structure known as a **coiled coil**. The two helices gently coil around one another and interact through a set of hydrophobic side chains at the interface.

Figure 2.57 Two proteins with the same fold yet significant structural differences. Although these two small protein neurotoxins, NTII (blue; PDB code 1NOR) and Fasciculin I (red; PDB code 1FAS), have the same fold, there are many structural differences between them, especially in the length of the beta strands and in the conformation of the loops.

Folds also vary greatly in size and secondary structure composition; the homeodomain, which binds DNA, is small and contains only alpha helices (Figure 2.58a), while the much larger immunoglobulins are composed almost entirely of beta sheet (Figure 2.58c). We shall see throughout this book how the structural features of proteins are tailored to carry out specific functions.

While there is tremendous diversity in the types of protein folds that are known, it is also clear that the number of protein folds found in nature is limited. Most arbitrary amino acid sequences would fail to fold into a stable structure. The number of potential different sequences for a protein 100 amino acids long is 20^{100} or

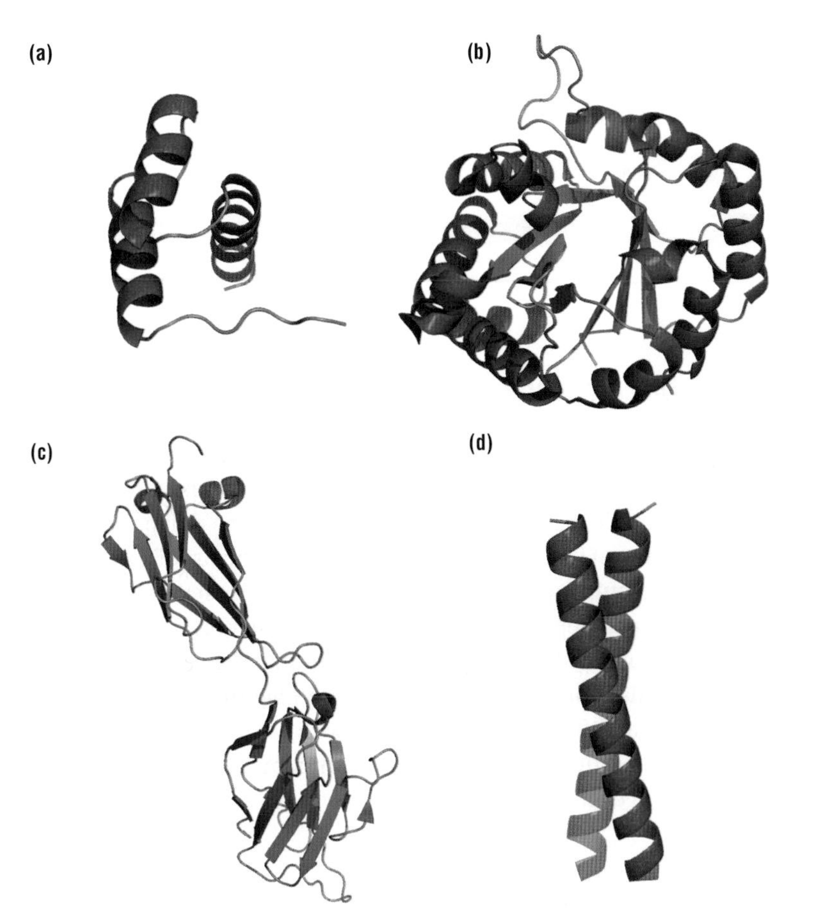

Figure 2.58 Examples of protein folds. A few examples of protein folds are (a) the homeodomain, which is entirely alpha helical, (b) triose phosphate isomerase, which consists of a TIM barrel, a mixed alpha helix, beta sheet structure, (c) immunoglobulin, which is almost entirely beta sheet, and (d) GCN4, which consists of a pair of long alpha helices that dimerize by forming a coiled coil.

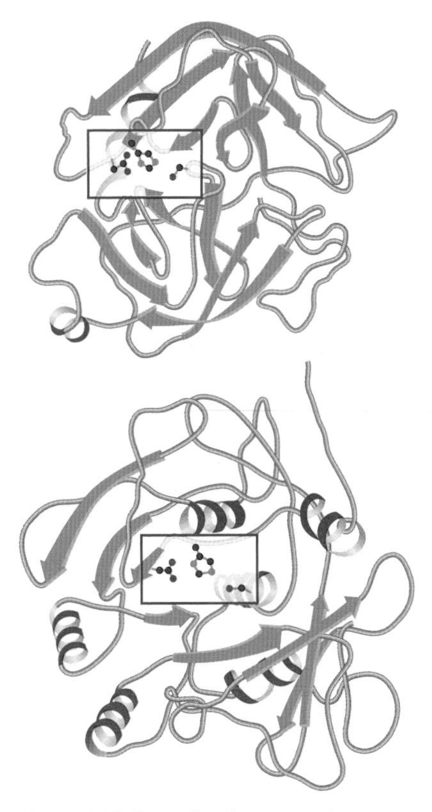

Figure 2.59 Example of convergent evolution. The enzymes chymotrypsin (top) and subtilisin (bottom) do not have the same fold, yet they have the same set of three amino acids known as the catalytic triad, which catalyzes peptide bond cleavage.

1.3×10^{130}! Yet of the >96 000 protein structures that were determined as of 2013, only a thousand or so folds have been observed. A variety of amino acid sequences can be accommodated by the same – or nearly the same – fold. Changes in the amino acid sequence (an amino acid substitution or other mutation) that do not alter the overall fold or properties of a protein may be tolerated – that is, they allow the protein to function as it did prior to the substitution. By contrast, those substitutions (or other mutations) that are more deleterious are not tolerated, and the protein no longer functions as it should. However, as it collects further mutations, or even insertions or deletions in the polypeptide chain, the protein may evolve a new, useful function.

Variants arise from a single ancestral protein through a process of **divergent evolution**, in which proteins with new characteristics, and eventually separate functions, evolve. Proteins whose primary structures may appear to be unrelated, but which have similar folds, are generally considered to have undergone divergent evolution. Occasionally, nature seems to have solved the same problem twice, producing two proteins that carry out a similar function but appear to have evolved independently. This less common process is referred to as **convergent evolution**. An example can be found in the enzymes chymotrypsin and subtilisin, which catalyze peptide bond cleavage using the same triad of amino acids but are otherwise dissimilar in structure, as shown in Figure 2.59. It is the continued selective pressure to carry out particular tasks that drives the evolution of proteins, and causes certain efficient solutions to be used again and again in biology.

➔ We learn about sequence variation and deleterious mutations in Chapter 18.

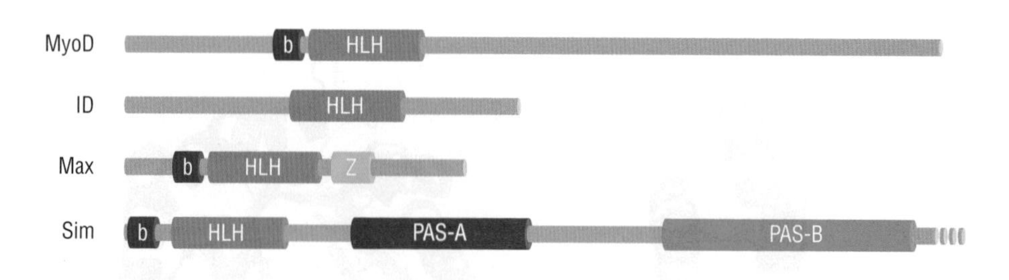

Figure 2.60 The modular nature of HLH proteins. The diagram shows how different proteins that bind DNA and regulate transcription can be built up from discrete domains. All of the proteins shown here contain a helix-loop-helix DNA-binding domain, denoted HLH. These proteins contain one or more other conserved domains: b, basic region, Z, leucine zipper, PAS-A domain, PAS-B domain.

Proteins can be composed of multiple domains

Many protein polypeptide chains contain more than one structurally distinct region. Each distinctive region is called a **domain**, which is a compact region of protein structure usually made up of a contiguous segment of the polypeptide chain that is capable of folding on its own. While some proteins consist of a single structural domain, most are built up in a modular fashion from several domains fused together in a single polypeptide chain. Each domain generally has a distinct fold and carries out a particular task. For example, proteins that regulate RNA transcription often have a modular composition, as depicted in Figure 2.60. Such proteins are composed of several different domains, each with a characteristic fold and a distinct function. The particular combination of functional domains gives rise to the unique character of each protein.

2.12 PROTEIN–DNA INTERACTIONS

Many of the processes covered throughout this book rely upon specialized classes of proteins that are able to bind to a particular DNA sequence, selecting that site from among the thousands or even millions of potential sites in the genome. There are particular features of the DNA double helix that are exploited by proteins that bind to specific DNA sequences via the non-covalent interactions that we learned about earlier in this chapter. Having now covered the principles of both protein and nucleic acid structure, we turn here to an overview of how proteins are able to recognize specific DNA sequences.

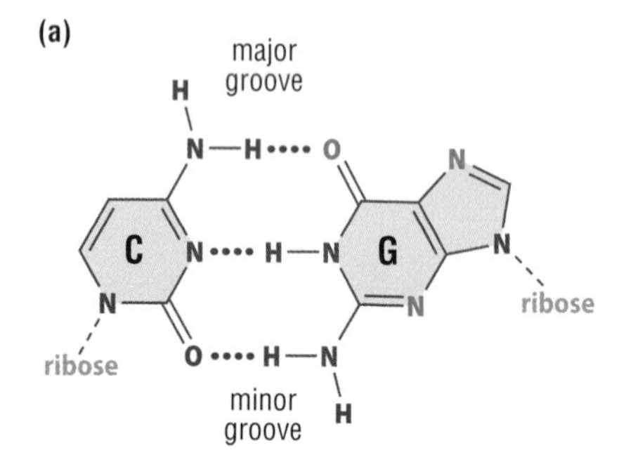

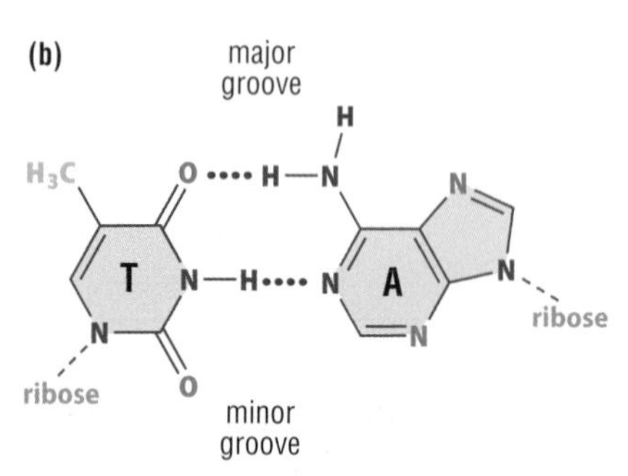

Figure 2.61 The chemical groups on the base pairs that are exposed in the major and minor grooves. The exposed hydrogen bond donors on Watson–Crick base pairs are in red and the acceptors are in blue. The hydrophobic methyl group of thymine is in green.

Each DNA sequence has a distinct chemical signature that can be recognized

As we learned earlier in this chapter, the DNA double helix is fairly uniform in structure, with a negatively charged sugar-phosphate backbone facing outward and the edges of the base pairs exposed to solvent in the major and minor grooves. Figure 2.61 illustrates how each base pair has a characteristic set of chemical groups that are thus available for interactions: hydrogen bond donors and acceptors and, in the case of thymine, a methyl group. A given DNA sequence is therefore characterized by a distinct array of hydrogen bond donors and acceptors, as well as by methyl groups that are exposed in the grooves, as depicted in Figure 2.62.

The pattern of functional groups exposed in the major groove is unique to each DNA sequence, whereas there is less variability in the chemical surface exposed in the minor groove. This limited variability in the minor groove occurs because the pattern of hydrogen bond donors and acceptors is the same for A–T and T–A base pairs, and for G–C and C–G base pairs, as seen in Figure 2.62. These therefore cannot be distinguished from one another through hydrogen bond interactions in the minor groove. Not surprisingly, most DNA-binding proteins that recognize a particular sequence of base pairs do so primarily through major groove contacts.

A protein recognizes a specific base pair sequence by forming a chemically complementary set of non-covalent interactions with the DNA. The surface of the protein that recognizes the DNA sequence contains a set of side chain and main chain atoms that can form favorable interactions with the particular array of DNA functional groups that are exposed. Thus, a hydrogen bond donor on the protein is paired with a hydrogen bond acceptor in the DNA, while a hydrophobic side chain might be in van der Waals contact with the methyl group of thymine. Figure 2.63 shows examples of side chain interactions with DNA bases. As in other examples of macromolecular interactions, there are often water molecules at the protein–DNA interface that mediate interactions with both the protein and the DNA via hydrogen bonds.

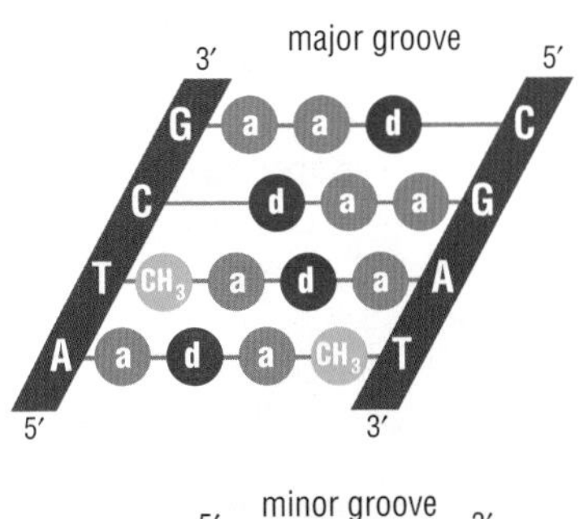

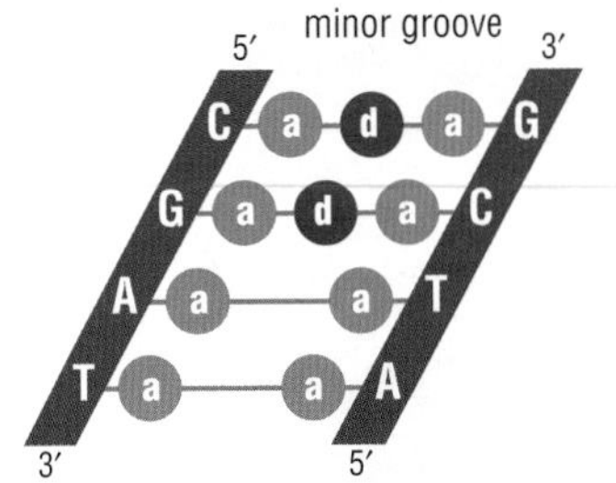

Figure 2.62 Pattern of chemical groups exposed in the major and minor groove. This schematic diagram shows the pattern of hydrogen bond donors and acceptors that are exposed in the major and minor grooves of DNA for the sequence, 5′–CGAT–3′. Note that each base pair has a unique chemical signature in the major groove, but not in the minor groove.

Shape and charge complementarity are important in protein–DNA interactions

A protein must form a sufficient number of non-covalent interactions with the DNA to ensure binding to the correct DNA sequence. (What constitutes a 'sufficient number' depends on many different factors, however, and will vary with both protein, DNA sequence, and the particular function of the proteins.) If the shape of the protein is complementary to that of the DNA, there is an increased chance that side chains on the surface of the protein will approach functional groups on the DNA closely enough to form favorable non-covalent interactions. Many proteins that contact bases in the major groove of the DNA therefore do so with an alpha helix, whose shape and dimensions allow it to fit in the major groove of B-DNA (see Figure 2.63). A two-stranded beta sheet can also fit readily into the major groove of B-DNA.

While the minor groove of B-DNA is normally too narrow to accommodate either a helix or a beta sheet, distortions can be induced by protein binding that open up the groove. Although there is an energetic penalty in distorting the DNA, it can be repaid by sufficient favorable interactions with the protein.

Any protein that interacts with DNA - whether or not it binds to a specific sequence - must contain chemical features that favor binding to the highly negatively charged sugar-phosphate backbone of DNA. For this reason, protein domains that interact with DNA generally contain many lysine and arginine side chains, which are positively charged and therefore interact favorably with the negatively charged phosphate groups, as depicted in Figure 2.64. Hydrogen bond donors such as the hydroxyl groups on serines or tyrosines, which bear a partial positive charge, can also form favorable electrostatic interactions with phosphate groups.

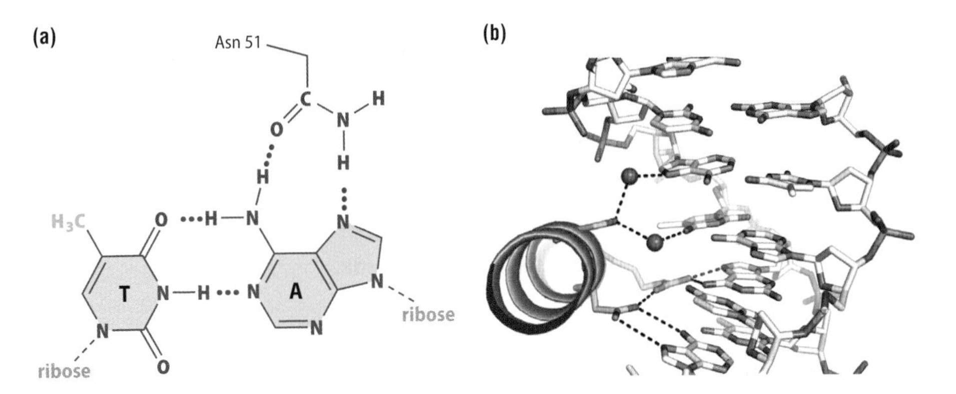

Figure 2.63 Examples of contacts between side chains and bases. (a) An asparagine side chain can form two hydrogen bonds with an A. (b) An alpha helix in the major groove with side chains contacting bases. Some side chains (N51 and R54) form direct hydrogen bonds with the base pairs, while others (S50) form water-mediated hydrogen bonds. PDB code 1YRN.

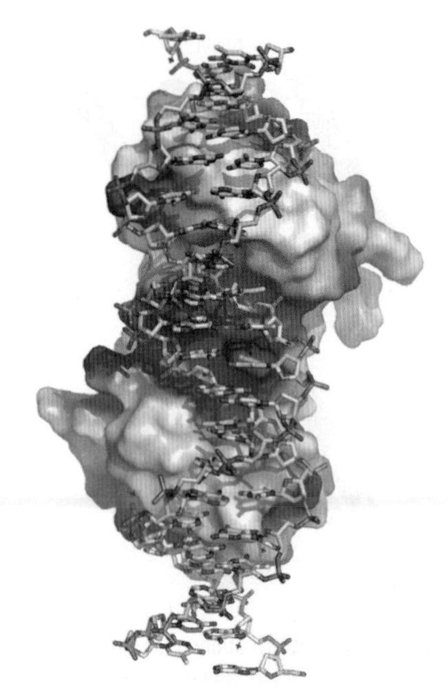

Figure 2.64 Electronegative DNA and electropositive protein surface. Surface rendering of DNA and protein is colored according to electrostatic potential. Electropositive surfaces are colored blue, electronegative surfaces are red and neutral light gray. The region of the protein surface that contacts the DNA is electropositive and therefore interacts favorably with the negatively charged DNA backbone.

During this section we have focused on the interactions that can exist between proteins and DNA. However, DNA isn't the only nucleic acid for which interactions with proteins are important. We see in Section 11.13 how proteins also interact extensively with RNAs via specific RNA-binding domains, and see examples of this interaction in practice when we consider the activity of small RNAs in Section 16.5.

Having now seen how proteins can recognize specific DNA sequences – a phenomenon that underpins numerous processes that we shall cover throughout this book – we finish this chapter by considering two other groups of biological macromolecules: carbohydrates and lipids.

2.13 SUGARS AND CARBOHYDRATES

Carbohydrates play a variety of roles in the cell

Carbohydrates comprise a diverse class of molecules that are involved in a wide range of biological processes. We have already encountered a simple type of carbohydrate: the ribose sugar that forms part of the backbone of RNA. Carbohydrates are also a major source of energy for all organisms, and are the form in which many cells story energy for future use. Polymers known as complex carbohydrates can form rigid structures that provide mechanical support, such as the cell walls of bacteria and plants or the rigid exoskeleton of insects. We shall see here how the chemical diversity of carbohydrates enables them to play myriad roles in biology.

Simple sugars can form linear or ring-like structures

Like the other major biological macromolecules, proteins and nucleic acids, **carbohydrates** are polymers, in this case built up from fundamental building blocks called **monosaccharides**. A monosaccharide is a simple organic molecule that generally has the chemical formula $(CH_2O)_n$, where n is greater than 3, and which cannot be hydrolyzed to form simpler saccharides. The nomenclature for naming monosaccharides indicates the number of carbon atoms in each molecule. For example, a six-carbon monosaccharide has the chemical formula $C_6H_{12}O_6$ and is therefore referred to as a **hexose**. A five-carbon monosaccharide has the formula $C_5H_{10}O_5$ and is called a **pentose**. We refer to these monosaccharides as sugars, but we also use the term sugar to refer to simple carbohydrates containing up to three monosaccharide molecules linked together.

Figure 2.65 shows some hexoses and pentoses that differ only in the location of their carbonyl (C=O) groups. Because the monosaccharides contain at least one

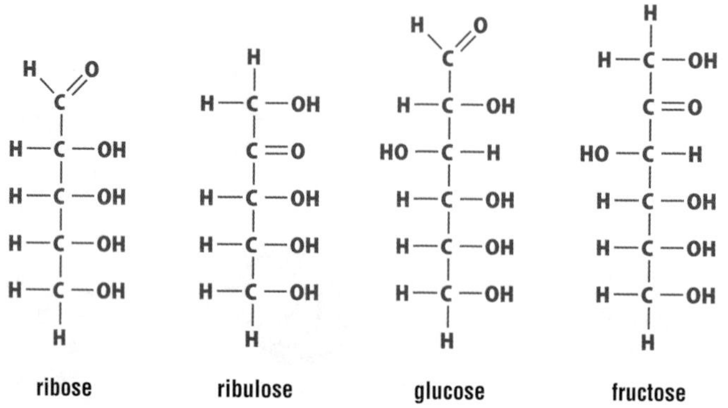

ribose ribulose glucose fructose

Figure 2.65 Simple linear sugars. Two five-carbon pentose sugars, ribose and ribulose, are shown, along with two six-carbon hexose sugars, glucose and fructose.

asymmetrical carbon, they can exist in two non-interconvertible forms, called isomers (D or L). The D- or L- designation is defined by the arrangement of atoms around the chiral carbon farthest from the aldehyde or ketone group as shown in Figure 2.66. Most naturally occurring pentoses and hexoses are found as the D form.

A structural feature of certain sugars is their ability to form covalently closed, ring-like structures. This occurs when a hydroxyl group reacts with a ketone or aldehyde carbonyl group, yielding a ring-like isomer. For pentose and hexose sugars dissolved in water, the cyclized form is more prevalent than the linear sugar. Examples of the linear and cyclic forms of a sugar are shown in Figure 2.67.

A hexose sugar can cyclize in one of two ways. A **pyranose** ring is formed when the C-1 aldehyde of a sugar such as glucose is attacked by the C-5 hydroxyl, forming a six-member ring (see Figure 2.67). A five-member **furanose** ring results when a C-2 keto group reacts with the C-5 hydroxyl. These relatively stable five- and six-member rings predominate in biology. Furanose and pyranose rings can form two different stereoisomers, called **anomers**, which differ in the orientation of the hydroxyl group at the C-1 carbon (illustrated in Figure 2.67). Ribose, when an integral part of nucleotides, is a furanose ring in the β anomeric form. RNA contains β-D-ribose, while DNA contains β-D-deoxyribose.

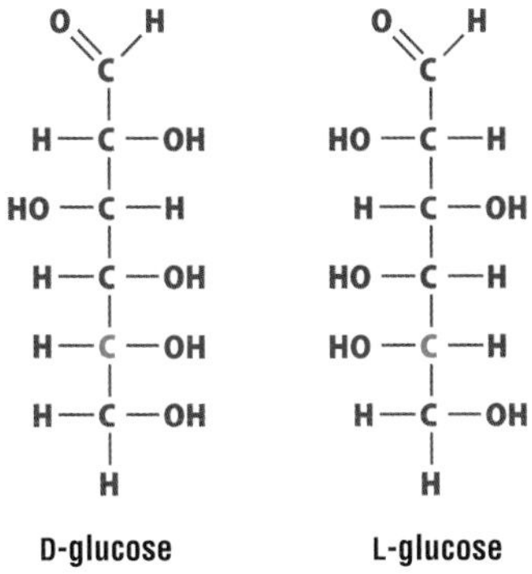

D-glucose L-glucose

Figure 2.66 Monosaccharide isomers. Monosaccharides can exist as two non-interconvertible isomers, denoted D and L. This designation is determined by the arrangement of atoms around the carbon furthest from the aldehyde or ketone group. D- and L-glucose are shown here, with the designations determined by the arrangement about the carbon atom highlighted in blue.

Monosaccharides can be linked together to form complex carbohydrates

Unlike proteins and nucleic acids, which are linear polymers, carbohydrates can form branched polymers. The building blocks of carbohydrates are cyclic monosaccharides, which can be covalently linked to form larger and more complex polymers. The hydroxyl of one monosaccharide can be joined to another in a neighboring molecule by a condensation reaction that releases a water molecule. The resulting **disaccharide** comprises two sugars joined by a linkage termed a **glycosidic bond**, as shown in Figure 2.68a. Disaccharides, in turn, can be linked with one another to form larger carbohydrates termed **oligosaccharides**. The largest carbohydrates, which consist of tens or hundreds of monosaccharides, are called **polysaccharides**.

Since sugars have many hydroxyl groups and are present in different anomeric forms, there are many different ways in which one sugar can be covalently joined with another. For example, there are 11 chemically distinct ways to join two D-glucose units. In addition, a single monosaccharide can be linked to two others. This allows the assembly of branched polysaccharides such as glycogen (represented schematically in Figure 2.68b), which is the form in which glucose is stored by mammalian cells until it is needed as a source of energy.

Because of the many possible ways in which sugars can be linked to form oligo- and polysaccharides, carbohydrates form a much more diverse family of polymers than proteins and nucleic acids, which have only one type of linkage joining one building block with the next. Not surprisingly, this also means that a much larger set of enzymes is needed to assemble or digest different carbohydrates, since a distinct enzyme is needed to catalyze formation or breakage of each type of linkage.

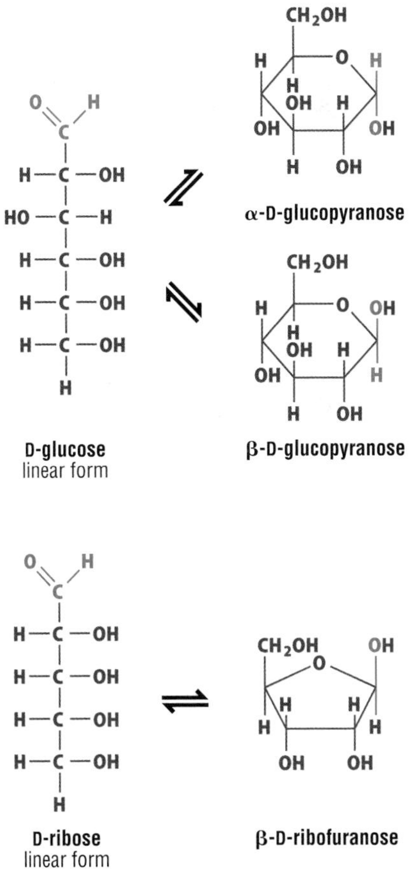

D-glucose linear form; α-D-glucopyranose; β-D-glucopyranose

D-ribose linear form; β-D-ribofuranose

Figure 2.67 Cyclized sugars. Glucose (top) can cyclize in one of two ways, producing ring structures that differ only in the orientation of the hydroxyl group attached to the carbon-1 atom. In the linear sugar, the aldehyde group is shown in blue and the hydroxyl is shown in red; the corresponding atoms in the cyclized sugars are colored accordingly. Ribose (bottom) can also cyclize in two ways, but only the b-form is shown.

Carbohydrates play structural roles in all organisms

Some polysaccharides form polymers that have great mechanical strength. Plant cell walls contain cellulose, a linear polysaccharide composed of glucose subunits that gives plants their rigid structure. The unbranched glucose polymers that comprise cellulose form long, linear chains that line up and hydrogen bond to

(a)

α-glucose + β-fructose → sucrose + H_2O

(b)

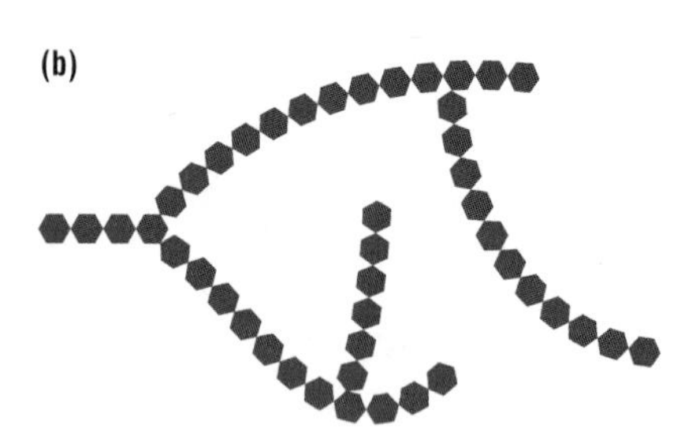

Figure 2.68 Polysaccharides. (a) Glucose and fructose can combine to form the disaccharide sucrose, with the concomitant release of water (red). (b) Glycogen is a complex, highly branched polysaccharide of linked glucose sugars (gray hexagons).

one another, thereby strengthening the plant cell wall. Bacteria have cell walls composed of **peptidoglycans**, which are linear polysaccharides that are joined to one another by peptide cross-links. The first specific antibiotic ever discovered, penicillin, acts by disrupting synthesis of the peptidoglycan bacterial cell wall. The hard outer shell of crustaceans and insects is composed of chitin, a linear polysaccharide. All multicellular organisms have carbohydrate polymers in the spaces between cells, which is called the extracellular matrix. The presence of structural carbohydrates in all organisms suggests that they appeared early in evolution.

Carbohydrates can be attached to proteins and lipids

Carbohydrates can be attached to other macromolecules, thereby altering the properties and behavior of the macromolecules in essential ways. Carbohydrates can be covalently attached to proteins to form **glycoproteins**, while lipids with covalently attached carbohydrates are called **glycolipids**. We shall learn more about covalent attachment of carbohydrates to proteins and lipids later in this chapter.

Animal cell surfaces are richly decorated with carbohydrates, due to the attachment of carbohydrates to cell surface proteins. The attachment of carbohydrates, termed **glycosylation**, stabilizes the structures of cell surface proteins and plays an important role in the interactions of cells with each other and with the free proteins surrounding cells. For example, the A, B, O and Rh blood types are distinguished by different patterns of protein glycosylation on the surface of red blood cells. Carbohydrates linked to proteins and lipids on the cell surface can be recognized by specialized proteins called **lectins**, which bind to particular carbohydrates and mediate interactions between cells. Carbohydrates on the cell surface can also be exploited by some pathogens to gain entry into the cell. Influenza virus binds to sialic acid, a glycosylation product found on certain cell surface proteins. *Vibrio cholerae,* which causes cholera, only invades epithelial cells with a particular type of glycolipid, G(M1) ganglioside, which contains a branched pentasaccharide linked to fatty acids. These infectious agents have clearly evolved strategies for infection that exploit the presence of carbohydrates on the surface of the cells they invade.

2.14 LIPIDS

Lipids are hydrophobic molecules

Lipids comprise a diverse class of biological molecules that are all strongly hydrophobic in character. The basic constituents of cell membranes, steroid compounds such as cholesterol, and the fats and oils that are used by cells to store energy are all lipids. Lipid molecules can be attached to proteins in order to tether the proteins to membranes, and lipids can be used to transmit signals within cells. We shall learn here about the basic properties of lipid molecules that are important for their biological functions.

Fatty acids are a key component in most types of lipid

The fundamental building block of most lipids in biology is the **fatty acid**. A fatty acid is a hydrocarbon chain with a carboxylic acid group at one end; several fatty acids are illustrated in Figure 2.69. The hydrocarbon chain is hydrophobic, whereas

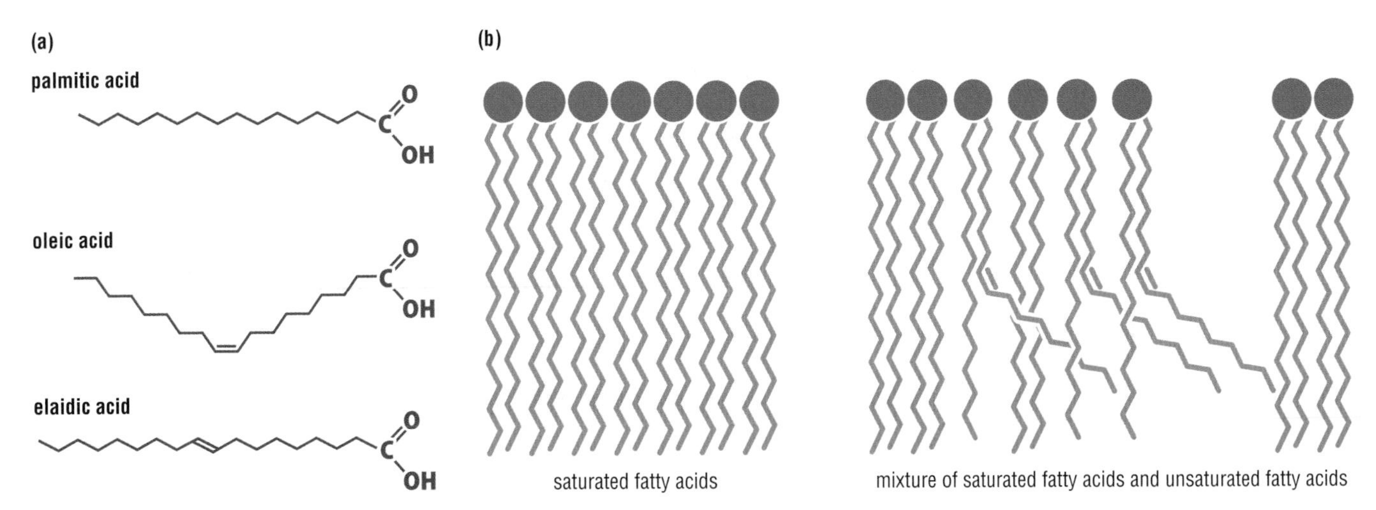

Figure 2.69 Saturated and unsaturated fatty acids. (a) Palmitic acid is a saturated fatty acid, meaning that there are single bonds connecting all the carbon atoms in the hydrocarbon chain. Oleic acid is an unsaturated fatty acid with a single *cis* double bond, which introduces a kink in the hydrocarbon chain. Elaidic acid is similar to oleic acid but has a *trans* double bond in place of the *cis* double bond. (b) *Trans* fatty acids can pack closely together and form extensive van der Waals interactions, while mixtures of *cis* and *trans* fatty acids cannot pack as tightly and hence form a less stable – and more fluid – array.

the negatively charged carboxylic acid is hydrophilic, which means fatty acids are amphipathic. A fatty acid can be **saturated**, meaning that all the carbon atoms are joined by single bonds, or they can be **unsaturated**, meaning that there are one or more double bonds joining the atoms in the hydrocarbon chain.

In biological systems, unsaturated bonds are generally found in the *cis* configuration, which introduces a kink in the hydrocarbon chain, rather than the *trans* configuration, which preserves the straight conformation found in saturated chains. For example, Figure 2.69a shows the structure of the **saturated fatty acid**, palmitic acid (which lacks double bonds in the hydrocarbon chain), and two unsaturated fatty acids, oleic acid and elaidic acid. Notice how the double bond in oleic acid adopts a *cis* conformation, while the double bond in elaidic acid adopts a *trans* conformation. By contrast, the bonds in palmitic acid are all *trans*, which is characteristic of saturated fatty acids. Fatty acids vary in length and in the number and position of double bonds, all factors leading to differences in their chemical properties.

One important consequence of saturation is its effect on how fatty acids associate with one another. In an aqueous environment, the hydrophobic hydrocarbon tails of fatty acids tend to pack against one another rather than with the surrounding water molecules. Saturated hydrocarbon tails are highly flexible molecules that can form extensive interactions with one another. Unsaturated tails, on the other hand, contain kinks caused by their double bonds and so do not generally pack together as tightly. (Look again at Figure 2.69a. Note how the hydrocarbon chain of oleic acid, which adopts a conformation, is kinked, whereas the hydrocarbon chain of elaidic acid, whose double bond is in the *trans* conformation, is straight.) As a result, the loosely packed unsaturated fatty acids are more fluid at room temperature than the tightly packed saturated fatty acids (Figure 2.69b). This is why butter, which is high in saturated fatty acids, is solid at room temperature, while vegetable oil, which is high in unsaturated fatty acids, remains liquid.

Plant oils can be hydrogenated *in vitro*, which reduces the number of double bonds per fatty acid chain and therefore makes them less fluid at room temperature. This process is used in the production of margarine. These chemically hydrogenated oils have been found to be harmful to human health, because the

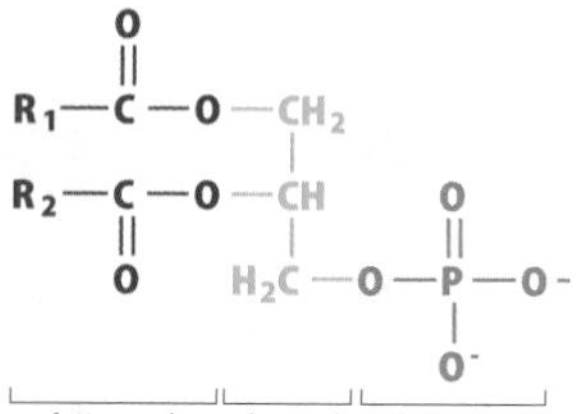

fatty acyl chains glycerol phosphate

Figure 2.70 Structure of a phospholipid. Two fatty acids (R_1 and R_2) are attached to a polar head group (blue) containing glycerol, a phosphate group, and an alcohol (X).

hydrogenation process yields fatty acids containing *trans* double bonds, which are known as *trans* fats. These *trans* fats do not resemble any of the naturally synthesized fats, and lead to the buildup of arterial plaques for reasons that are as yet poorly understood.

Fatty acid chain length also influences fluidity. Short-chain fatty acids exhibit a higher degree of fluidity than those with longer hydrocarbon chains because their intermolecular packing is less compact. Fatty acid lengths in biological systems can be up to 24 carbons in length, with those 16 and 18 carbons in length being the most common.

Many lipids are composed of fatty acids attached to a head group

Many lipid molecules consist of fatty acids attached to a polar moiety called a **head group**, as illustrated in Figure 2.70. The amphipathic nature of lipids, with their mixed hydrophobic and hydrophilic components, is a defining feature of lipids that allows them to form biological membranes, as we shall see below.

There are several different types of lipids common in biology, each distinguished by the type of head group to which the fatty acids are attached. The most common class of lipid found in cell membranes is the **phospholipid**. The head group of the most common phospholipids comprises glycerol, a phosphate group, and an alcohol (see Figure 2.70). Two fatty acids are covalently linked directly to the glycerol, forming the hydrophobic portion of the lipid molecule, while the head group is the hydrophilic part. There are many different types of glycerol-derived phospholipids, each distinguished by the type of alcohol attached to the phosphate group, and by the length and degree of saturation of the fatty acids. Glycolipids are another class of lipid molecules found in membranes, and are distinguished by the sugar moieties attached to their head groups.

Cholesterol, an abundant constituent of animal cell membranes, is a lipid whose chemical structure differs significantly from the lipids described above. It contains several linked hydrocarbon rings, a short hydrophobic tail, and a single polar hydroxyl group, as shown in Figure 2.71. Therefore, cholesterol lacks the more clearly defined head group of other lipids. Cholesterol is the metabolic precursor of a broader class of molecules called steroid hormones such as estrogen and progesterone. Its amphipathic character due to the polar OH group, combined with its relatively rigid fused ring system, results in important consequences for the fluidity of cell membranes, as we shall see in the next section.

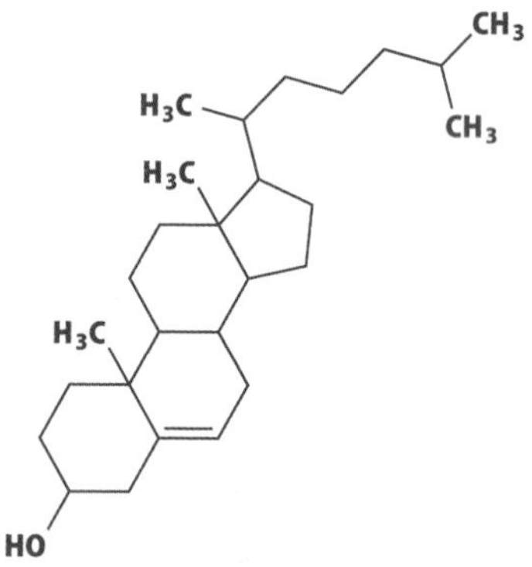

Figure 2.71 Cholesterol. This molecule contains the four-ring structure that is typical of steroids, as well as a hydrophobic tail (top right) and a hydrophilic hydroxyl group (bottom left).

Membranes contain a bilayer of oriented phospholipids

The unique properties of phospholipids enable them to form a protective sheath around cells called a **plasma membrane**, which allows the cell to maintain an intracellular environment that is different from its exterior. The amphipathic nature of phospholipids favors their arrangement in a **lipid bilayer**, with the polar head groups in contact with the surrounding aqueous medium and the hydrophobic fatty acid tails in the interior where they are shielded from the water. This arrangement is illustrated in Figure 2.72. van der Waals interactions between the fatty acid tails of the lipids help to stabilize the bilayer.

Lipid bilayers can be thought of as two-dimensional fluids because the lipids, along with other molecules in the membrane, can diffuse quite rapidly within each plane of the bilayer. The precise lipid composition of a membrane governs how readily this diffusion occurs. Fatty acid chain length and saturation influence

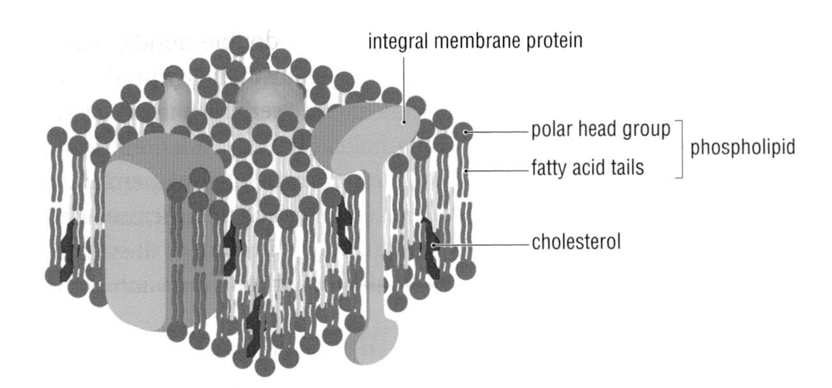

Figure 2.72 A cell membrane. The oriented phospholipids form a bilayer 5–10 nm thick, with the fatty acid tails pointing inward and the polar head groups facing the aqueous environment. Cholesterol molecules (red) can insert between the fatty acid tails. Proteins, shown in blue, can lie within the membrane.

membrane fluidity in the same way that they affect whether fats are solid or liquid at room temperature, as we learned earlier. Membrane fluidity can also be affected by the presence of cholesterol, which is found in most animal cell membranes. Membranes with significant amounts of cholesterol inserted into the bilayer have reduced fluidity, but also are less likely to become solid at very low temperatures. The ability to fine-tune membrane fluidity is important in regulating the diffusion of the proteins embedded within the membrane, as we will see next.

Proteins within the membrane allow communication between the cytoplasm and the cell's exterior

Cells must have the ability to take in a variety of molecules from the surrounding environment, as well as have a means to communicate with or respond to other cells or organisms. Lipid bilayers, however, form a barrier that is relatively impermeable to most molecules. Cell membranes therefore contain embedded proteins that allow the transfer of molecules and information across the lipid bilayer. The typical plasma membrane contains about 50% protein by mass, but protein content varies with membrane type. It ranges from a low of 18% in the membranes of nerve cells to 75% in the organelle membranes of mitochondria.

Figure 2.73 illustrates how membrane proteins vary a great deal in structure and in the degree to which they are embedded within the lipid bilayer. **Integral**

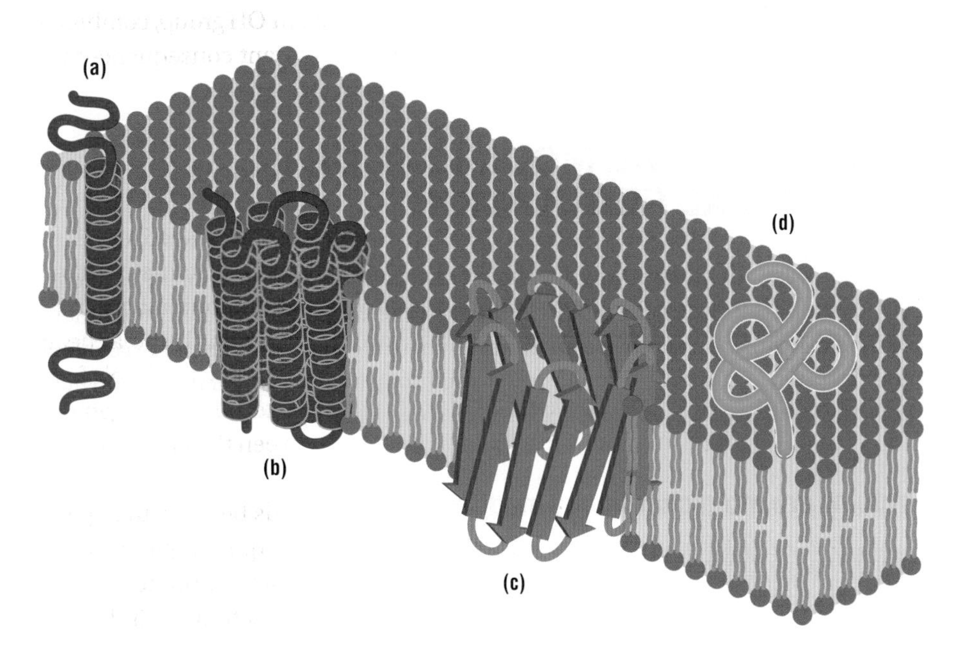

Figure 2.73 Membrane proteins. Integral membrane proteins can span membranes with one or more alpha helices as shown in (a) and (b), or with a beta sheet arrangement (c). Peripheral membrane proteins can be anchored to membranes by the covalent attachment of a fatty acid (d).

membrane proteins are embedded largely within the bilayer, whereas some proteins are anchored to the membrane by inserting just a small portion of the protein into the bilayer. Many membrane proteins have domains on the cell's interior and exterior joined by one or more alpha helices called **transmembrane helices** that span the membrane (Figure 2.73b). There are also integral membrane proteins that span the membrane and are composed primarily of structures rich in alpha helices (Figure 2.73b) or beta sheets (Figure 2.73c). Some of these proteins contain a central pore that allows ions or small molecules to pass across the membrane.

In all cases, the portion of the protein embedded in the membrane must be rich in hydrophobic side chains that interact favorably with the hydrophobic fatty acid chains within the lipid bilayer, with few, if any, polar or charged side chains. This is because of the unfavorable energy that results from burying a polar group within the non-polar lipid bilayer, where there is no water or other polar molecules with which it may form favorable hydrogen bonding interactions. For this reason, the peptide backbone that is buried within lipid bilayers is generally found in alpha-helical or beta-sheet form, because in these secondary structures the polar backbone atoms are hydrogen bonded with each other. **Peripheral membrane proteins** do not contain transmembrane regions, and can bind reversibly to the cell membrane.

Covalent attachment of lipids can target proteins to the membrane

Proteins that lack exposed hydrophobic regions can be targeted to membranes by becoming covalently modified with a lipid or fatty acid (see Figure 2.73d). Some peripheral membrane proteins, for example, have a covalently attached fatty acid that inserts into the lipid bilayer and anchors the protein to the membrane. The particular type of lipid that is attached can determine the intracellular membrane, and hence the organelle, with which the protein associates. Many different lipids, ranging from simple fatty acids such as palmitic acid (see Figure 2.69) to complex modifications containing carbohydrate and multiple fatty acids, can be attached to proteins. Cholesterol, in addition to being a membrane constituent that regulates the fluidity of a bilayer, can also be covalently attached to proteins and thereby anchor them in membranes. We will learn more about lipid modifications and their biological roles in Chapter 14.

2.15 CHEMICAL MODIFICATION IN BIOLOGICAL REGULATION

We see throughout biology how the chemical properties and function of a biological molecule are intrinsically linked. In the example above, the covalent attachment of a lipid can target a protein to a membrane. There are many other types of covalent modifications of proteins that are often used to activate or regulate a biological molecule's activity.

Proteins are often chemically modified after translation

After being synthesized, a protein may undergo chemical modifications of its amino acid side chains or of the polypeptide backbone itself in order to be able to carry

Post-translational modifications of proteins

modification	description	common function of modification	icon
lipid modification	attachment of lipid to protein (examples include palmitylation and myristoylation)	localization to the membrane	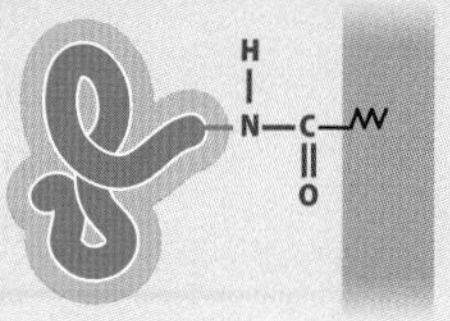
glycosylation	attachment of carbohydrates to side chains; typically Ser, Thr or Asn	protection from degradation, recognition by other proteins	
phosphorylation	attachment of phosphate group to Ser, Thr, His, Tyr, His, Asp	regulation of interactions or activity	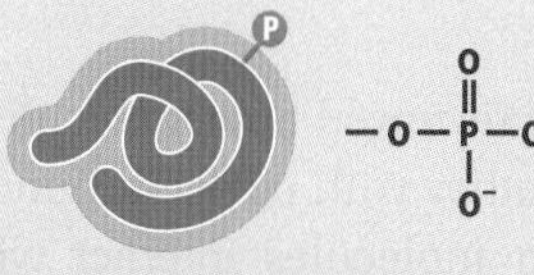
acetylation	attachment of an acetyl group to Lys side chains	regulation of interactions	
methylation	attachment of one or more methyl groups to Lys or Arg side chains	regulation of interactions	
ubiquitination	attachment of small protein, ubiquitin, to lysine side chains (one or more ubiquitins may be attached)	regulation of interactions, targeting of protein for degradation	 ubiquitin

Figure 2.74 Reversible modifications that alter the activity of proteins. At the right is a depiction of the protein (blue) along with the icon that represents the modification. Chemical structures are given for phosphate, acetyl, and methyl groups.

out its biological function. A range of such modifications is shown in Figure 2.74. In some cases, the chemical modification represents the final processing step that produces the fully functional proteins. Other modifications are transient and reversible, and are used to regulate an aspect of the protein's function. We refer to these chemical alterations as **post-translational modifications**, since they occur after the ribosome has translated the information in the messenger RNA.

➔ We learn more about post-translational modifications in Chapter 14.

Addition and subsequent removal of chemical groups is a major mechanism for regulating the activity of a protein. The covalent attachment of inorganic phosphate to specific side chains in a process called **phosphorylation** is a particularly common example of this. **Acetylation** and **methylation**, whereby acetyl and methyl groups are attached to amino acid side chains, also often occur in protein regulation. All of these chemical modifications exert their effect in one of two general ways: they can cause the protein to change conformation, perhaps activating or inhibiting the protein, or the chemically modified side chains can bind to other proteins. These events can then trigger changes in any of a wide variety of cellular processes. We shall see examples of such reversible chemical modifications throughout this book.

DNA methylation controls gene expression

Chemical modification of DNA also occurs and plays a variety of regulatory roles. The most common type of modification to DNA, and the only one we will encounter in this book, is methylation. In bacteria, both cytosine and adenine bases can be modified, whereas only cytosine methylation occurs in mammals and in plants. Bacteria use DNA methylation to distinguish the parental strand from the newly synthesized strand after DNA replication. DNA methylation is also used by bacteria to distinguish its own genome from that of invading viruses. In mammals and plants, cytosine methylation is used to turn off gene expression. We will learn about DNA methylation in detail in Chapter 4.

SUMMARY

In this chapter, we reviewed how molecules are built up by linking atoms together with covalent bonds, and explored the way in which molecules interact with one another non-covalently in the aqueous environment of the cell. We learned about the four classes of biological macromolecules – nucleic acids (RNA and DNA), proteins, carbohydrates, and lipids – and the way in which they are assembled from a set of chemical building blocks. Each of these macromolecules has a characteristic set of properties that makes it uniquely suited to its biological role. Many biological molecules can be covalently modified in ways that alter their chemical properties and allow their function to be regulated.

ATOMS AND MOLECULES

- Atoms are joined by covalent bonds to form molecules through the sharing of unpaired electrons.
- Electrons are not shared equally in some covalent bonds, with the result that some atoms have a partial positive or negative charge.
- Molecules with an overall unequal charge distribution are polar, whereas molecules with an overall even distribution of charge are non-polar.
- Water molecules are polar, a property that plays an important role in governing the behavior of molecules in aqueous solution.
- Polar chemical groups and molecules form favorable interactions with polar water molecules; we therefore call them hydrophilic.
- Non-polar molecules and chemical groups prefer to associate with one another, rather than with water; we therefore call them hydrophobic.
- Ions are atoms that become fully charged through the gain or loss of electrons.
- Water becomes ionized to form H^+ and OH^- ions; pH is a measure of the concentration of H^+ ions and is given by $pH = -\log_{10}[H^+]$.
- Molecules interact in the aqueous environment of the cell through three types of non-covalent interactions: salt bridges, hydrogen bonding, and van der Waals interactions.
  - Salt bridges are electrostatic interactions between chemical groups bearing opposite charges.
  - A hydrogen bond is an electrostatic interaction between a hydrogen atom with a partial positive charge (the hydrogen bond donor) and an atom with a partial negative charge (the hydrogen bond acceptor).
  - van der Waals interactions are attractive forces that arise between two atoms due to formation of transient dipoles that attract one another.

BIOLOGICAL MACROMOLECULES

Nucleic acids

- The family of molecules known as nucleic acids includes DNA, RNA, and a variety of nucleotide monomers.
- A nucleotide is composed of a base, a sugar, and one or more phosphate groups.
- In DNA, the sugar is deoxyribose, and the bases are adenine, guanine, cytosine, or thymine.
- A single strand of DNA is a polymer of covalently linked deoxyribonucleotides that are joined by phosphodiester linkages between the 3′ OH of one ribose and the 5′ OH of the next ribose.
- Two DNA strands associate with one another to form a double helix, with one strand spiraling around the other. The dominant type of DNA helix in cells is known as the B-helix, or B-DNA.
- The two strands of the double helix are antiparallel, with the sugar-phosphate backbone on the outside and the bases on the interior.
- The bases on opposite strands of the double helix hydrogen bond with one another to form Watson–Crick base pairs, with adenine pairing with thymine through

two hydrogen bonds and guanine pairing with cytosine through three hydrogen bonds.

- Long double-stranded DNA molecules can be under tension because of topological variations known as supercoiling. Negative supercoiling favors local unwinding of the DNA helix and separation of the two single strands.
- In RNA nucleotides, the sugar is ribose and the base is adenine, guanine, cytosine or uracil. Like DNA, RNA nucleotides are joined to one another by phosphodiester linkages.
- RNA nucleotides are subject to a wide variety of chemical modifications.
- The 2′ OH on the ribose confers on RNA additional structural and chemical diversity not found in DNA.
- Double-stranded RNA forms the A-type double helix.
- Some RNA molecules form complex structures containing short stretches of double-stranded regions.

Proteins

- Proteins are polymers of amino acids (also called polypeptides) that are joined to one another by peptide bonds.
- There are 20 amino acids, each consisting of an amino group, a carboxyl group, and a chemically distinct side chain.
- A polypeptide chain condenses in water to form a defined structure, with hydrophobic (non-polar) side chains buried on the inside and hydrophilic (polar) side chains on the outside.
- There are four levels of organization that describe protein structure:
  - primary structure - sequence of amino acids
  - secondary structure - formation of alpha helix and beta sheet, which contain characteristic regular, repeating patterns of hydrogen bonds between residues
  - tertiary structure - overall three-dimensional structure of a single polypeptide
  - quaternary structure - the arrangement of two or more polypeptides that form a complex with one another.
- The amino acid sequence dictates protein structure, which means that proteins that are similar in amino acid sequence adopt similar structures.

PROTEIN–DNA INTERACTIONS

- Functional groups on the edges of the bases are exposed in the major and minor grooves and can be contacted by DNA-binding proteins.
- The particular array of chemical groups exposed in the major groove is unique for all four base pairs (A·T, T·A, G·C, C·G). In the minor groove, A·T/T·A base pairs can be distinguished from G·C/ C·G.
- Proteins bind to specific DNA sequences by interacting with the array of chemical groups, unique to a particular base sequence, with a complementary set of hydrogen bonds and van der Waals interactions

CARBOHYDRATES AND LIPIDS

- Carbohydrates are polymers of sugar molecules known as monosaccharides.
- A monosaccharide has the general chemical formula $(CH_2O)_n$, where n >3. The five- and six-carbon sugars are the most common in biology.
- Monosaccharides are joined to one another in a condensation reaction that releases a water molecule, yielding disaccharides, oligosaccharides and polysaccharides, depending on the number of monosaccharide units involved.
- Monosaccharides can be linked to one another in a large variety of ways to form branched, as well as linear, polymers.
- Lipids are hydrophobic molecules, often composed of a non-polar fatty acid and a polar head group. This mixed hydrophobic and hydrophilic nature is referred to as amphipathic.
- Fatty acids are long hydrocarbon chains that are hydrophobic.
- The most common type of polar head group consists of glycerol covalently linked to a phosphate group and an alcohol. Lipids containing this type of head group are termed phospholipids.
- In water, phospholipids associate with one another to form a lipid bilayer, with the polar head groups facing the water and the hydrophobic lipids buried in the interior.

COVALENT MODIFICATION OF BIOLOGICAL MOLECULES

- All types of biological molecules can be modified by the covalent attachment of molecules.
- Carbohydrates can be covalently attached to protein, forming a glycoprotein, or to lipid, yielding a glycolipid.
- Lipids and fatty acids can be covalently attached to proteins.
- Methylation of DNA bases regulates a variety of processes.
- Proteins are subject to a wide variety of chemical modifications that include methylation, acetylation, and phosphorylation.

2.1 ATOMS, MOLECULES, AND CHEMICAL BONDS

1. van der Waals interactions are electrostatic attractions between atoms, even those that are not charged. Explain how this occurs.

2. Which four elements make up the vast majority of the mass of living organisms?

3. The following graph represents the repulsive and attractive forces on atoms as a function of distance. In which distance range – A, B or C – would you expect van der Waals forces to be most effective?

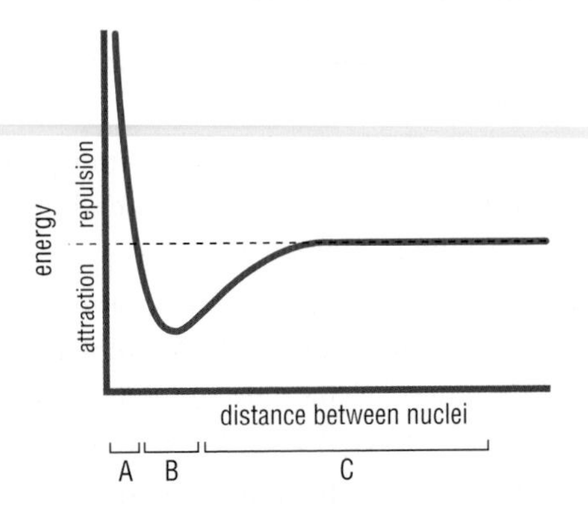

4. Explain why an oxygen atom makes two covalent bonds with other atoms but a hydrogen atom only makes a single covalent bond.

2.2 LIFE IN AQUEOUS SOLUTION

1. A solution with $[H^+] = 10^{-12}$ M is:
- a. Acidic
- b. Basic
- c. Neutral

2.3 NON-COVALENT INTERACTIONS

1. What is the pH of a 1 mM solution of KOH?

2. Why is salt (sodium chloride) soluble in water but not in oil?

2.4 NUCLEOTIDES AND NUCLEIC ACIDS

1. Two experimental tubes contain an equal number of nucleic acid molecules with the sequence 5′-AGCAGG-3′. The first tube contains single-stranded DNA and the second tube contains RNA. Which tube contains the greater mass? Explain your answer.

2. Which of the following correctly states the difference between RNA and DNA?
- a. RNA contains uracil instead of thymine.
- b. Ribose forms a phosphodiester bond in RNA through its 3′ carbon, deoxyribose in DNA through its 2′ carbon.
- c. Ribose has three phosphate groups, deoxyribose has none.
- d. Ribose cannot form phosphodiester bonds.

3. What is the difference between a nucleoside and a nucleotide?

2.5 THE STRUCTURE OF DNA

1. Which conformation does double-stranded DNA generally take in cells?

2. What type of interaction occurs between stacked bases in the DNA double helix and helps to stabilize the helix?
- a. Disulfide bridges
- b. Hydrogen bonds
- c. Ionic attraction
- d. van de Waals forces

3. Discuss how base-pairing occurs in a double-stranded DNA molecule.

4. Define linking number, writhe, and twist, and describe how they relate to one another without using an equation.

Challenge questions

5. The molecule shown in this figure represents relaxed, B-form DNA.

1 5 10 15 20

If you tightened this DNA molecule by three turns (winding the right-hand end in the counterclockwise direction as you look down the helix), then ligated the ends of the DNA and allowed the molecule to supercoil, what would the Lk, Tw, and Wr numbers be?

6. What aspect of nucleotide structure leads to the formation of major and minor grooves in B-form dsDNA?

2.6 CHEMICAL PROPERTIES OF RNA

1. RNA and DNA are chemically similar, yet RNA is much less stable than DNA. Explain why.

2.7 RNA FOLDING AND STRUCTURE

1. What are Watson–Crick and non-Watson–Crick interactions? Where do these types of interactions tend to occur?

2. RNA folds sometimes include associated metal ions. Explain why.

Challenge questions

3. The central dogma of biology (DNA is transcribed into RNA which is translated into protein) is not entirely accurate. Explain why, and give an example that illustrates this.

4. What is covariation, and how can it be used to help predict structurally important parts of RNA molecules? Use an example in your answer.

5. Discuss the rationales for and against the hypothesis that RNA was a precursor of other biological molecules during the development of life.

2.9 FUNDAMENTALS OF PROTEIN STRUCTURE

1. What type of interaction is a peptide bond?
- a. Covalent
- b. van der Waals
- c. Hydrogen bond
- d. Ionic bond
- e. Hydrophobic effect
- f. Peptide bond

2. What is the only type of covalent bond that stabilizes tertiary protein structure?
- a. A glycosidic bond
- b. A hydrogen bond
- c. A disulfide bond
- d. A phosphodiester bond
- e. A peptide bond
- f. A van de Waals interaction

3. Explain why atomic rotation around different types of covalent bonds is important for protein structure.

Challenge questions

4. Write the overall charge state (positive, negative, or neutral) of each amino acid residue in the following short peptide at pH 7.6. Do not forget to consider the termini of the peptide.

arginine-aspartate-proline-valine-lysine

5. You have a solution of a protein, but you don't know what the pH of the solution is. You are told that the protein contains glutamic acid and that half the glutamate side chains are charged and half are uncharged.

a. What is the approximate pH of the solution? Explain your answer.
b. You are told the protein in (a) contains only glutamic acid and aspartic acid. You run the protein on a native protein gel (one containing only polyacrylamide, with no sodium dodecyl sulfate (SDS) or other additives). The pH of the gel is 7. You load the protein on to the gel halfway between the positive and negative electrodes. Which way will the protein move? Explain your answer.

6. Many biological molecules have D and L isomers. Explain what this means and which molecules occur in biological systems. In addition, name a biological molecule that does not have these isomers and explain why not.

2.10 PROTEIN FOLDING

1. Name and briefly define the four levels of protein structure, and name the kinds of bonds that are important for holding that level of structure together.

Challenge questions

2. The Second Law of Thermodynamics states that for a spontaneous reaction to occur, the total entropy (disorder) of a system must increase. However, many proteins adopt an ordered fold in aqueous solution, in apparent conflict with this law. Explain this.

3. If given a primary amino acid sequence of a protein, how would you predict the secondary and tertiary structures that the mature protein would adopt?

2.11 PROTEIN FOLDS

Challenge question

1. Your lab partner discovers a two-subunit protein with one subunit containing a transmembrane protein and the other containing a pheromone receptor. A week later, you discover a single-subunit protein containing two domains: a pheromone receptor domain and transmembrane domain. These domains are identical to those present in your partner's protein, and she says that your protein is the same as hers.

a. Explain why you think that they cannot be the same proteins, being sure to note the differences and similarities between them and to define the terms 'subunit' and 'domain'.
b. Although you know that your protein and your friend's protein have different subunits, when you perform SDS-PAGE with the two proteins, both samples generate a single protein band. What was omitted from the polyacrylamide gel, and what does this tell you about the two-subunit protein and its likely cellular location? You may wish to refer to the description of SDS-PAGE in Chapter 19.

2.13 SUGARS AND CARBOHYDRATES

1. Carbohydrates frequently have more complex structures than nucleic acids and peptides. Explain why.

2.14 LIPIDS

1. Where on the surface of this transmembrane protein would isoleucine be more likely to occur? Explain your answer.

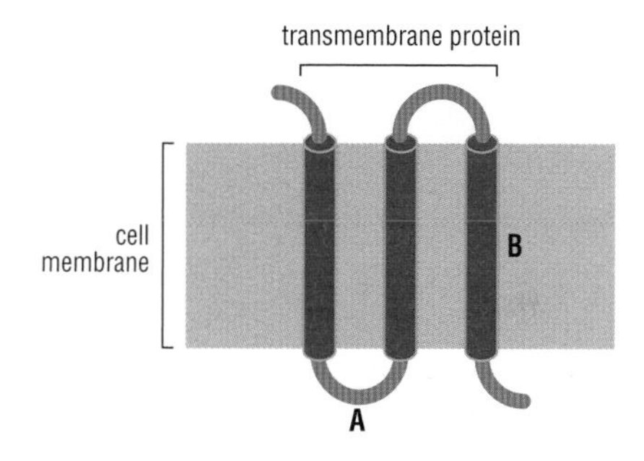

2. Explain how phospholipid fatty acid composition influences cell membrane fluidity.

3. Phospholipids are amphipathic. Explain why this is important for biological systems.

4. Given that cell membranes have a hydrophobic core, how do hydrophilic molecules pass through the membrane when needed?

FURTHER READING

2.1 ATOMS, MOLECULES, AND CHEMICAL BONDS

Pauling LC. *The Nature of the Chemical Bond and the Structure of Molecules and Crystals: An Introduction to Modern Structural Chemistry*, 3rd ed. Ithaca, NY: Cornell University Press, 1960.

2.8 THE RNA WORLD AND ITS ROLE IN THE EVOLUTION OF MODERN-DAY LIFE

Doudna JA, Cech TR. The chemical repertoire of natural ribozymes. *Nature*, 2002;**418**:222–228.

Robertson MP, Joyce GF. The origins of the RNA world. *Cold Spring Harbor Perspectives in Biology*, 2012;**4**:a003608.

White HB. Coenzymes as fossils of an earlier metabolic state. *Journal of Molecular Evolution* 1976;**7**:101–104.

2.9 FUNDAMENTALS OF PROTEIN STRUCTURE

Petsko GA, Ringe D. *Protein Structure and Function*. London: New Science Press, 2003.

3 The chemical basis of life

INTRODUCTION

The very definition of life requires that the molecules within a living organism be in constant flux. Different molecules are being continuously synthesized, broken down, chemically rearranged, and modified. One macromolecule binds to another to mediate a process and then becomes free again, to be available for yet another interaction in the cell. To understand the different processes that are required to maintain and propagate life, we must review the thermodynamic laws that govern how likely it is that a set of molecules will form a complex with one another, and how changing conditions in the cell affect that likelihood.

Similarly, it is important to know how likely it is that certain biochemical reactions will occur, and how those reactions are harnessed by the cell to carry out a variety of tasks. By studying the thermodynamics and chemistry underlying all of biology, we can understand the inner workings of the cell and how organisms respond to a changing environment. In this chapter, we will learn about the laws of thermodynamics and chemistry that underlie biological processes. We will also see how the cell is able to harness the energy released from one chemical reaction to power a second reaction that would otherwise not occur, much in the same way that a car can be propelled up a hill using the energy released by burning gasoline.

Most of the important chemical reactions in biology do not proceed rapidly enough to be useful to an organism. When, for example, a cell needs a particular protein, it cannot afford to wait for the free amino acids to react with one another and form a protein polymer. Even under very favorable conditions, in which all the free amino acids were available at high concentrations, it would take more than ten years for a 100 amino acid protein to form spontaneously. Cells therefore need molecules that can accelerate reactions so that they occur on a timescale that is useful to the organism. We shall learn in this chapter about some of the ways in which **enzymes** catalyze chemical reactions, and how enzyme activity is regulated in the cell.

3.1 THERMODYNAMIC RULES IN BIOLOGICAL SYSTEMS

Living things obey the laws of thermodynamics

All cellular functions depend on the synthesis and breakdown of macromolecules and on their intimate interactions. In the previous chapter, we learned how the remarkable physical and chemical properties of biological molecules govern their interactions with one another and allow them to adopt particular shapes. An understanding of these principles is not sufficient to predict macromolecular behavior, however. For example, proteins and DNA do not assemble spontaneously if their component building blocks are simply added to water; nor can the assembly of specific non-covalent complexes such as ribosomes or muscle fibers be guaranteed simply by mixing their components in an aqueous environment. Instead, to gain a full understanding, we need to consider the biochemical reactions that underpin macromolecular behavior by asking three key questions. First, is a reaction energetically favorable – that is, will it happen at all? Second, how far will the reaction proceed? And, third, how fast will the reaction happen?

Understanding how cellular macromolecules behave in the ordered way that they do, and what the answers to these questions are, requires an understanding of energetics as defined by the basic laws of thermodynamics. These laws provide us with a small number of rules that apply to any molecular process.

The energy associated with a system is indicated by its enthalpy

All biological systems rely on the flow of energy to drive forward the biochemical processes on which they rely: to digest and metabolize food; to power the contraction of muscles; to maintain the body at a constant, stable temperature. However, organisms cannot create energy. Instead, energy is merely converted from one form into another. This concept – the so-called conservation of energy – is known as the **first law of thermodynamics.** Energy can take many different forms: for example, kinetic energy (the energy associated with motion), chemical energy (the energy stored in chemical bonds), and thermal energy.

How much energy does an atom, a molecule, or an organism have? The measure of the internal energy that a given entity possesses is given by its **enthalpy** (represented by the symbol H). If an entity is of high energy, it has a large enthalpy; if it is of low energy, it has a small enthalpy. However, the enthalpy value in isolation is not all that useful in helping us to determine whether a biological process can happen or not. Instead, we are more interested in the *change* in enthalpy – that is, the difference between the enthalpy value at the start and end of a process. We represent the change in enthalpy by writing ΔH (where the symbol Δ (Greek uppercase delta) means 'change in').

If the final state of a process has a higher enthalpy than the starting state, the change in enthalpy is positive, as illustrated in Figure 3.1. Remembering that enthalpy is a measure of energy, this means that the end-state of the process has more energy than the starting state, so energy must be transferred *to* the system during the course of the process to make up this difference. We say that a process that exhibits a positive enthalpy change (and needs to have energy supplied to it to make it happen) is **endothermic**. By contrast, if the end-state of the process has a lower enthalpy than the starting state, the change in enthalpy is negative. Consequently, the end-point of the process has less energy than the starting state, so energy is

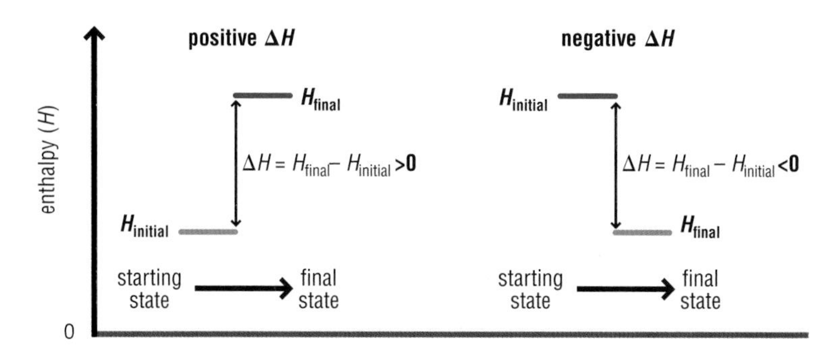

Figure 3.1 Enthalpy changes. If the final state of a process has a higher enthalpy than the starting state, ΔH (the change in enthalpy) is positive. If the enthalpy of the final state is lower than the enthalpy of the initial state, ΔH is negative.

released from the system during the course of the process. A process with a negative enthalpy change – that releases energy to the surroundings – is **exothermic**.

So why is this significant? Biological systems comprise many interlinked reactions and processes. The enthalpy change for some of these processes is positive; for others it is negative. For example, plants carry out the process of photosynthesis, in which they synthesize the sugar, glucose, from carbon dioxide and water according to the equation:

$$6CO_2 + 6H_2O \rightarrow C_6H_{12}O_6 + 6O_2.$$

This reaction, which also releases oxygen, has an enthalpy change of +2870 kJ/mol (685 kcal/mol) and is therefore endothermic, meaning that it requires an input of energy. That energy comes from sunlight. By contrast, the breakdown of glucose in the reverse reaction

$$C_6H_{12}O_6 + 6O_2 \rightarrow 6CO_2 + 6H_2O$$

has an enthalpy change of –2870 kJ/mol (–685 kcal/mol). This negative number tells us that the breakdown of glucose releases energy; it is highly exothermic.

Entropy gives the measure of disorder in a system

Not all of the energy associated with a change of enthalpy during the course of a reaction or process is free to be channeled into doing something useful (what we describe as being free to do 'work'). Why is this? To find the answer, we have to introduce the concept of entropy.

As well as being described in terms of its enthalpy, the amount of energy something has can also be described in terms of its **entropy**, which is represented by the symbol *S*. Entropy is something of an abstract concept, but essentially gives a measure of disorder: something with high entropy is more disordered than something with lower entropy, as depicted in Figure 3.2a. For example, let us consider the population of water molecules shown in Figure 3.2b. If these water molecules have very little energy, they will tend to adopt a very organized, low-energy

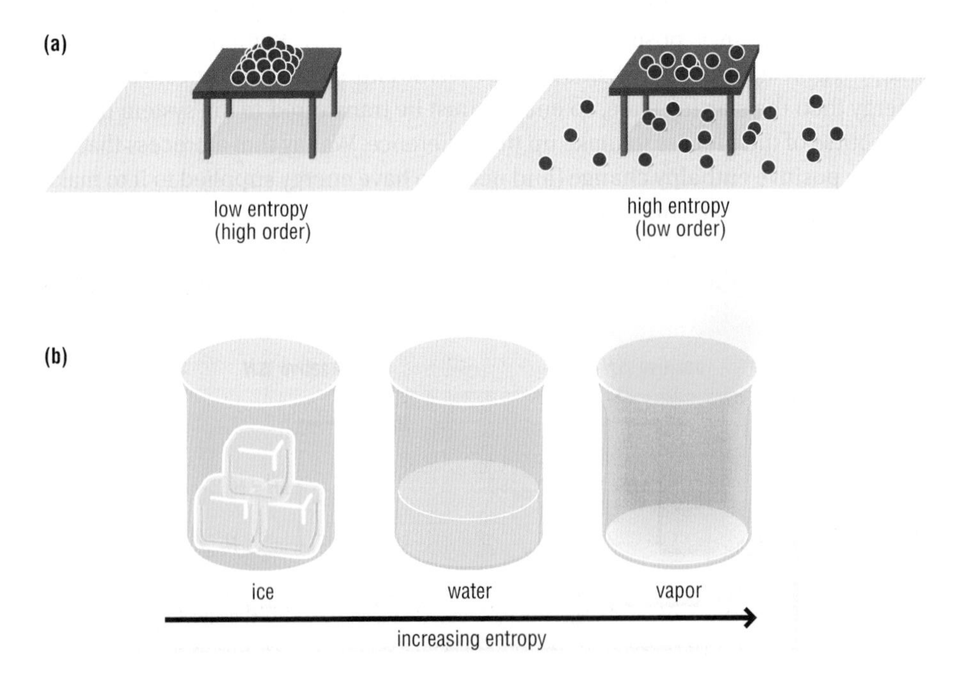

Figure 3.2 Entropy. (a) Entropy, denoted by the quantity *S*, gives a measure of order. The table on the left has low entropy because the balls are stacked neatly in a pile. On the right, the stack has fallen and the balls lie in random positions. This configuration for the balls is of higher entropy than the one on the left. It would take the input of energy to stack the balls up again in a neat pile, whereas a slight jostling of the table will cause them to disperse in a random way. (b) Water is in a low entropy state when it is in the solid state as ice, which contains water molecules aligned and immobilized within a crystal lattice. Liquid water has higher entropy than ice, as the water molecules are free to tumble and diffuse around the beaker. Water vapor has even higher entropy than liquid water, as the water molecules in vapor are free to leave the beaker and diffuse around the room.

structure – namely, ice. Ice has low entropy. If we transfer energy to the molecules in the form of heat, however, they will start to become more disorganized: the ordered array of molecules in ice will fall apart, and the molecules will become liquid water, which is of higher entropy than ice. If we add more energy still, the molecules will become even more disorganized, drifting apart from each other to form water vapor, which is of yet higher entropy. The link between disorder and entropy holds true even when we consider biological molecules: a folded protein, which represents a highly organized structure, has lower entropy than an unfolded protein, which is disorganized.

Just as we see a change in enthalpy during the course of a process (written as ΔH), so we also see a change in entropy, written as ΔS. A positive ΔS means that the final state has less order than the initial state, whereas a negative ΔS means that the final state has greater order. Therefore, the freezing of water – from liquid to solid – corresponds to a decrease in entropy, so the entropy change, ΔS, is negative. By contrast, the vaporization of water – from liquid to gas – corresponds to an increase in entropy, so the entropy change is positive.

For a process to happen spontaneously, the overall entropy change must be positive

Why is entropy important? The answer lies in the **second law of thermodynamics** which states, in its simplest form, that disorder in the universe can only increase. Put another way, for a process to happen spontaneously – without outside influence – the change in entropy must be positive.

How, then, do biological systems continually build ordered structures – macromolecules, cells, complex tissues – out of simple building blocks without defying the second law of thermodynamics? How does the folding of proteins occur if this process – from a disorganized to an organized polypeptide – represents a *decrease* in entropy? This apparent contradiction is resolved if we consider the cell in the context of the universe as a whole (that is, the cell and all of its surroundings). As we shall see, the processes and reactions that generate increased order in biology, such as the synthesis of macromolecules or the assembly of cell membranes, are coupled to other reactions (just as in our discussion of enthalpy above). Together, the net effect is to *increase* the disorder in the surroundings and, hence, the universe.

In the case of protein folding, the decrease in entropy associated with the organized folding of a protein is offset by the *increase* in entropy stemming from an increase in the disorder of water molecules surrounding the protein. When the protein is unfolded, water molecules surround the molecule in a relatively organized way. Upon folding, however, this organized interaction is disrupted, leading to an increase in the entropy of the water molecules.

Let us now come back to our earlier comment that not all of the energy associated with a change in enthalpy is free to do work. The reason for this is that some of the energy is used to drive an increase in the entropy of the system – that is, it is used to satisfy the second law of thermodynamics. The proportion of energy that is used to drive an increase is entropy is therefore not available to do work. The amount of energy that is actually free to do something useful is called the **Gibbs free energy**, *G*.

The free energy measures how favorable a process is

The difference in the Gibbs free energy of the initial state and the final state of a reaction or process, represented by ΔG, governs how readily two molecules will bind

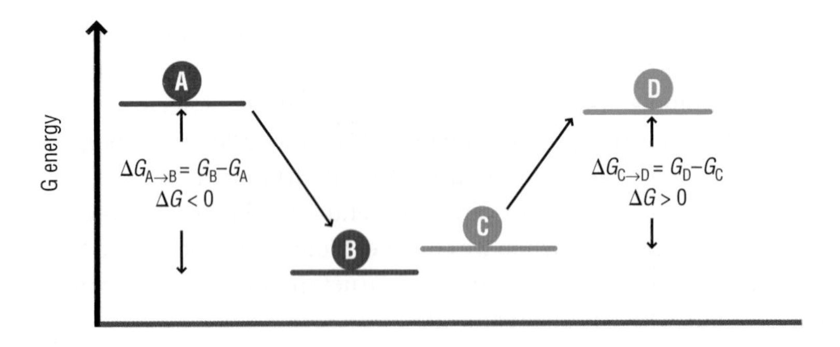

Figure 3.3 Free energy changes. A reaction converting molecule A into B is favorable because the free energy of the final state is lower than the free energy of the initial state. The free energy change, given by $\Delta G = G_B - G_A$, is therefore negative. A reaction such as that converting molecule C to D is unfavorable because the free energy change is positive ($G_D - G_C$). The situation is analogous to rolling a ball up a hill, which requires energy (C→D), while a ball will roll freely down a hill (A→B) and decrease its energy.

to one another or whether a chemical reaction will occur spontaneously. Figure 3.3 shows that, for a process to happen spontaneously, the change in free energy, ΔG, must be negative, such that the final state has a lower free energy than the initial state. Such a process gives out energy rather than requiring energy to be supplied. By contrast, a process with a positive ΔG requires energy and is not spontaneous. These processes can be compared to that of a ball rolling freely down a hill or requiring an input of energy for rolling up the hill. When an ice cube is placed in a glass of warm water, it will rapidly melt because the free energy change associated with the ice melting is negative. The synthesis of sugars by plants has a positive ΔG and therefore requires an input of energy, which is supplied by sunlight.

The free energy is related to both the enthalpy and entropy by the equation:

$$\Delta G = \Delta H - T\Delta S$$

where T is the temperature at which the reaction occurs, expressed in degrees Kelvin. This equation tells us that negative changes in enthalpy, when heat is released, or positive changes in entropy, when there is an increase in disorder, can both contribute to the value of ΔG being negative and, hence, to a reaction being spontaneous.

An important consequence of the composite nature of the free energy change is that a process that involves an unfavorable change in either enthalpy or entropy can still occur spontaneously as long as the equation relating the two terms yields a negative overall free energy change. A simple example is the spontaneous dissolving of table salt, sodium chloride (NaCl), in water. The breaking of the ionic lattice holding the Na^+ and Cl^- ions together in the salt crystals requires energy, so this process draws heat from the solution. Consequently, the enthalpy change is positive, and hence unfavorable. However, there is a large favorable change in entropy that arises from freeing into solution the many ions that had been part of an organized crystal lattice, offsetting the amount of energy needed to break apart the lattice; the change in free energy for this process is negative.

Generally speaking, then, where biological systems continually require unfavorable reactions for the synthesis of macromolecules and the regulation of their behavior, the price of these unfavorable changes is paid for by favorable changes in a second reaction that thereby drives the first. In so doing, we see the second law of thermodynamics in action, with the *overall* entropy - the level of disorder in the system - increasing.

Energy is needed to drive unfavorable reactions

Many reactions that are necessary for life are energetically unfavorable. For example, a DNA strand will not form spontaneously from free nucleoside monophosphates because the free energy of forming phosphodiester linkages is +25 kJ/mol

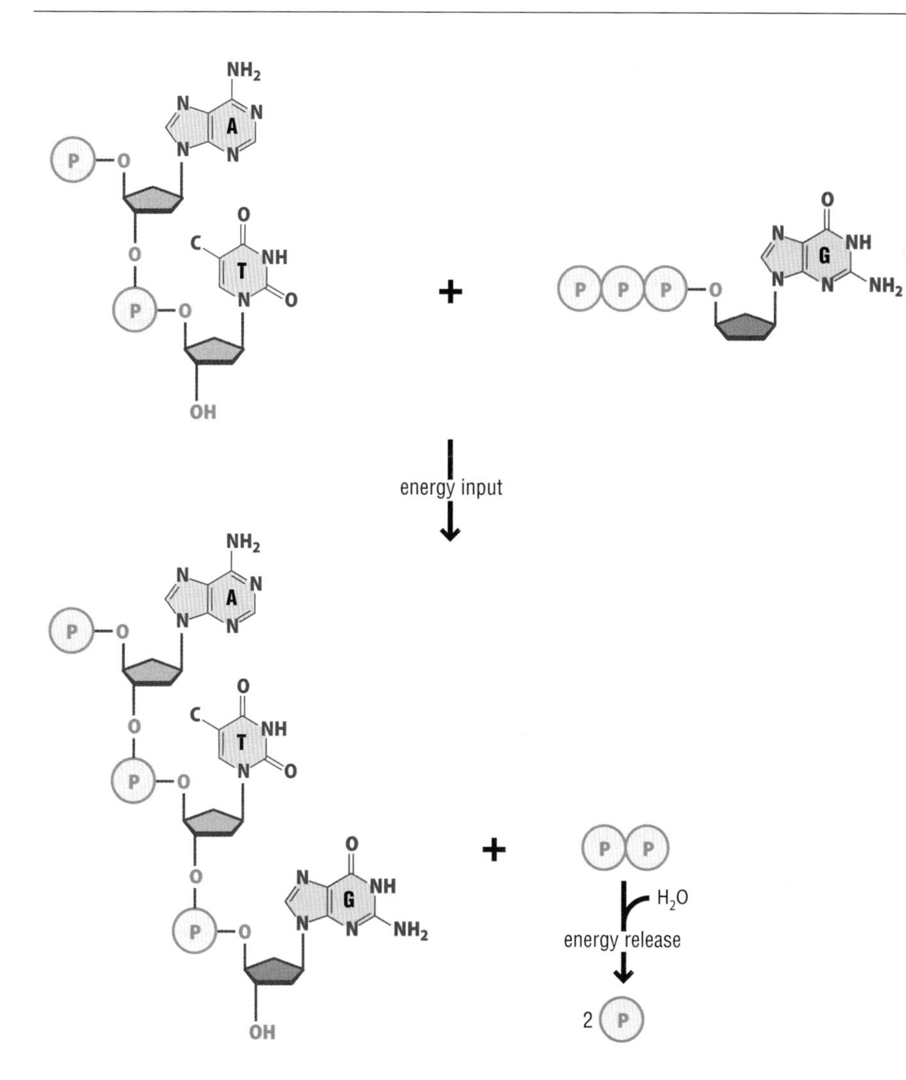

Figure 3.4 Coupling of favorable and unfavorable reactions in DNA synthesis. Synthesis of DNA from nucleoside monophosphates is unfavorable. NTPs as the precursors in the reaction allow the coupling of NTP hydrolysis to the unfavorable condensation reaction, thereby making the net reaction of DNA synthesis favourable

(+6.0 kcal/mol). This unfavorable process can, however, be driven by coupling it with an energetically favorable process. If the sum of the free energy of the two coupled reactions is negative, the overall reaction can proceed.

The coupling of an unfavorable chemical reaction with a highly favorable one is illustrated by the way in which cells synthesize DNA. Instead of sequentially adding nucleoside monophosphates in a simple condensation reaction, cells use nucleoside triphosphates (NTPs) as precursors in the reaction. These NTPs are referred to as high-energy molecules because hydrolysis of the bonds that join the phosphate groups releases a large amount of energy. As illustrated in Figure 3.4, NTPs are used to synthesize DNA in a reaction that releases the last two phosphate groups, producing **pyrophosphate** (PPi), and adds a nucleoside monophosphate to the growing DNA chain. This reaction is still somewhat unfavorable, with a ΔG of +2 kJ/mol (+0.5 kcal/mol), but notably much more favorable than the simple condensation of monophosphates.

Subsequent hydrolysis of pyrophosphate to organic monophosphate by pyrophosphatase, with a ΔG of –31 kJ/mol (–7.3 kcal/mol), is coupled to the bond forming reaction. Thus the total free energy change of the coupled reactions is –29 kJ/mol (–6.8 kcal/mol), which is quite favorable. There are many reactions in biology in which the energy released by cleaving a high-energy molecule is used to drive an otherwise unfavorable reaction.

The free energy of a reaction is influenced by the concentration of reactants

The free energy change of a reaction depends on both the inherent properties of the participating components and on their concentrations in solution at a given point in time. The concentration of individual proteins, nucleic acids, and various small molecules can vary greatly among different types of cells and under different conditions. This means that the particular conditions in the cell or in the test tube will determine whether a particular reaction will proceed spontaneously, and ultimately what proportion of the molecules will undergo the reaction.

To determine how molecules will react with one another under a set of conditions that may be of interest to us, we start by considering the energy change for the reaction in which the reactants and products are in their **standard states**. Compounds in solution are in their standard state when they are present at a concentration of 1M. Standard states can be defined at any temperature but, if it is not specified, we usually assume the temperature to be 298 K (25 °C).

Measurements made under standard conditions can act as reference values, which we can use to characterize a system when it exists under other, non-standard conditions. The **standard free energy change**, denoted by ΔG^o, describes the inherent tendency of a specific reaction to proceed when the reactants and products are in their standard states. This quantity is the result of the particular chemical and physical properties of the reacting molecules and of the products. (The superscript, °, tells us that a thermodynamic quantity – such as ΔG – has been measured when the reactants and products are in their standard states.)

In general, however, we are interested in determining the free energy change when the concentration of the reactants differs from the standard state. To illustrate how concentration affects the overall free energy change, we will start by considering a very simple reaction in which X is converted into Y – for example, a *cis* proline amino acid converting to a *trans* proline in a polypeptide chain. This reaction can be written in the following way:

$$X \rightleftharpoons Y.$$

The arrows are shown pointing in both directions to denote that this is a reversible reaction. So how can we determine ΔG for this reaction under non-standard conditions? The free energy change for the formation of complex Y can be calculated from the equation:

$$\Delta G = \Delta G^o + RT \ln[Y]/[X]$$

where R is the gas constant (8.31 J K^{-1} mol^{-1} or 1.987×10^{-3} kcal $mol^{-1}deg^{-1}$), T is the temperature expressed in degrees Kelvin, ln is the natural logarithm, and the square brackets denote concentration in units of mol/L. The first term, ΔG^o, provides information about the specific reaction being considered when the reactants and products are in their standard states; we can often look up these values from reference tables rather than having to calculate them ourselves. By contrast, the value of the second term, $RT\ln[Y]/[X]$, is determined by the relative concentrations of reactant (X) and product (Y) in the reaction system that we are interested in.

An important consequence of this equation is that the reaction becomes more spontaneous as the relative concentration of X increases, as this causes the $\ln[Y]/[X]$ term to become increasingly negative. This means that increasing the amount of reactants in solution initially can drive the reaction forward towards conversion to Y. If, on the other hand, there is a great excess of the product, Y, the term ln[Y]/[X] will be positive. If the net value of $RT\ln[Y]/[X]$ is large enough such that adding it to ΔG^o gives a net positive value, the reaction will be unfavorable and will not proceed as written. Instead, the reverse reaction will occur, with Y converting to X.

This example illustrates an important point: the initial concentration of both products and reactants determines the direction of the reaction. A sufficiently large excess of reactants will drive the reaction forward, while a large enough excess of product will drive the reaction in the reverse reaction. This is true for all types of reactions - chemical reactions and binding reactions. This fundamental point is central to understanding how biological interactions and reactions are driven throughout the cell.

The ratio of products and reactants at equilibrium is determined by the standard free energy

We have seen how the standard free energy change, ΔG^o, and the initial concentration of products and reactants determines whether or not a reaction is initially favorable, and in which direction the reaction will proceed. But what happens after that initial time, as reactants are converted into products (or vice versa) and the balance of reactants and products changes? When hemoglobin in red blood cells carries oxygen to tissues, how much oxygen will it deposit there? For any chemical reaction, how much product will be produced under a particular set of conditions?

To see what happens as a reaction proceeds, we will again consider the very simple reaction in which X is converted into Y. This is a reversible reaction comprising two separate reactions happening in parallel: a forward reaction, $X \rightarrow Y$, and a reverse reaction, $Y \rightarrow X$.

Figure 3.5 illustrates what happens as this reaction proceeds. Let us imagine that we start out with a solution containing only X molecules and a favorable (i.e. negative) ΔG^o for the conversion of X to Y. As the reaction proceeds, reactants are consumed and the concentration of product, Y, increases. This change causes the [Y]/[X] term in the equation to increase, which will make the overall free energy less negative, as given by the equation $\Delta G = \Delta G^o + RT\ln[Y]/[X]$. In other words, the forward reaction will become less favorable than under the initial conditions as more and more reactant is converted into product.

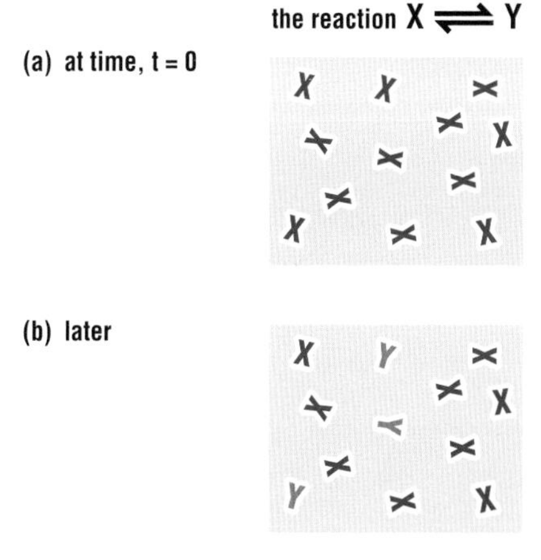

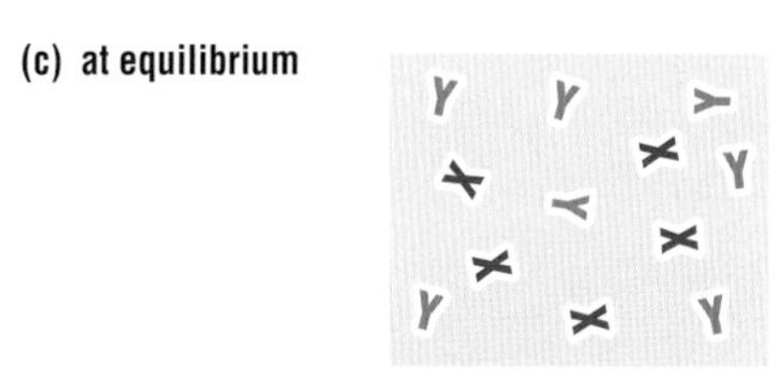

Figure 3.5 The conversion of X to Y. The reaction in which X is converted to Y is a reversible reaction. (a) Initial conditions: all X, no Y. (b) As the reaction proceeds, X is converted into Y, and Y begins to accumulate. Some Y is also converted back into X, but the net rate of Y formation exceeds the rate of the reverse reaction. (c) At equilibrium, the relative amount of X and Y remains constant, although some X continues to be converted to Y while an equal amount of Y is converted back to X.

As the product, Y, accumulates, the rate of the reverse reaction - namely the conversion of Y back to X - will increase. As long as the net free energy is negative, this reverse reaction will occur relatively less often than the forward reaction, and Y will continue to accumulate overall.

As the reaction proceeds still further, the ratio of product to reactant (Y to X) will reach a point at which ΔG reaches 0. At this point, there is no net change in the system: the rate of formation of product from reactants exactly balances the rate of conversion of product into reactants. We say that the reaction has reached **equilibrium**. The reaction has proceeded as far as is favorable, and the relative amounts of reactant and product will change no further.

At equilibrium:

$$\Delta G = 0 = \Delta G^o + RT\ln([Y]/[X])$$

$$\text{or} \qquad \Delta G^o = -RT\ln([Y]/[X]).$$

We can see from this expression that, under equilibrium conditions, the standard free energy change, ΔG^o, can tell us the ratio of product to reactant. In the coming sections, we will see in greater detail how this relationship can be used to tell us how much product and reactant will be present in a chemical reaction, and what proportion of binding partners in a solution will form a complex.

3.2 BINDING EQUILIBRIA AND KINETICS

Virtually all of the biological processes covered in this book depend on the binding of two or more molecules to one another. A protein that regulates whether or not a gene is expressed must bind to DNA. The ribosome, which synthesizes proteins, must bind to messenger RNA and transfer RNA charged with amino acids before it can form a polypeptide chain. We can use the thermodynamic principles we have discussed thus far to understand binding reactions: how likely it is that a binding reaction will occur, and how much complex will form under a given set of conditions. Since conditions in living things frequently change, it is also important to understand how quickly complexes form and come apart and how long it takes for a binding reaction to reach equilibrium. As we will see, the proportion of molecules that form a complex at equilibrium is directly related to the rates governing binding and dissociation: the binding kinetics.

The ratio of bound and free molecules at equilibrium is determined by the standard free energy

The general equation describing the free energy of any reaction can be adapted to describe a binding interaction as follows: Consider the binding reaction of two molecules, A and B, to form a complex denoted AB. This reaction can be written in the following way:

$$\mathrm{A+B \rightleftharpoons AB.}$$

The arrow is pointing in both directions to denote that this is a reversible reaction: just as A associates with B to form a complex, the AB complex can dissociate to release the individual A and B components.

As we saw above, the free energy change for the formation of complex AB is given by the equation:

$$\Delta G = \Delta G^{\mathrm{o}} + \mathrm{R}T\ln([\mathrm{AB}]/[\mathrm{A}][\mathrm{B}]).$$

In this case, the ΔG^{o} term provides information about the change in energy when A and B form a complex, relative to their being free in solution. If the two molecules form energetically favorable interactions with one another such that the formation of the complex is favored, the value of ΔG^{o} will be negative. At equilibrium (when $\Delta G = 0$):

$$\Delta G^{\mathrm{o}} = -\mathrm{R}T\ln([\mathrm{AB}]/[\mathrm{A}][\mathrm{B}]).$$

The ratio of complex AB to reactants at equilibrium is determined by the standard free energy change, ΔG^{o}. If ΔG^{o} is negative, telling us that complex formation is energetically favorable, the numerator in the above equation (the concentration of AB at equilibrium), must be larger than the denominator (the concentration of free A molecules multiplied by the concentration of free B molecules that are left in solution at equilibrium). In other words, the complex predominates at equilibrium, and binding is favored.

Reaction rates are influenced by the concentrations of the reactants

Although thermodynamic quantities tell us whether a reaction is spontaneous and how much product will be formed, these constants do not reveal how quickly – or how slowly – the process occurs. For example, once the hormone, estrogen, enters a cell, how long will it take for certain genes to be activated in response? To answer this question, we must know the rate at which the underlying processes occur. The description of chemical reaction rates is known as kinetics.

Let us begin by considering how long it will take for the binding reaction that we have been considering to reach equilibrium. For a simple binding reaction during which molecules A and B associate to form the AB complex, Figure 3.6 shows how the rate of binding is given by the **rate constant**, k_{on}, multiplied by the concentration of the binding partners: k_{on}[A][B]. (Notice how the rate constant carries the subscript 'on'; this indicates that the process represents a binding reaction, or the association of two (or more) species.)

The rate constant for this particular reaction has units of $M^{-1}s^{-1}$, and is termed a bimolecular (second-order) rate constant because it describes the rate at which two molecules bind to one another. For a rate constant of 10^7 $M^{-1}s^{-1}$ and initial reactant concentrations of 10^{-3} M for both A and B, the initial rate at which A and B form a complex will be 10 M s^{-1}.

While the rate at which a complex forms can vary quite a bit, the k_{on} cannot be greater than the diffusion limit, which is the maximum rate at which two molecules can move through an aqueous environment and eventually collide with one another. For a typical interaction between proteins, the diffusion limit is on the order of 10^8 $M^{-1}s^{-1}$. When k_{on} approaches this value, essentially all encounters of the two species result in productive (that is, successful) binding. However, most biological binding reactions exhibit k_{on} values that are considerably lower than this upper limit. The diffusion limit affects all types of chemical reactions and also places an upper limit on the rate of enzyme-catalyzed reactions, as we see below.

So far, we have only considered the association of two species – that is, their binding. Of equal importance in biology is the reverse process, that of dissociation. For example, the process by which oxygen is delivered to our tissues depends on both binding and dissociation: oxygen must participate in a binding reaction with hemoglobin in the lungs, but must then undergo dissociation once hemoglobin has reached the target tissue, such that the oxygen can be taken up by that tissue.

The rate at which the AB complex dissociates is given by a rate constant, k_{off}, multiplied by the concentration of the complex [AB]: k_{off}[AB] (see Figure 3.6). Since the rate describes the dissociation of a single species, the AB complex, it is a unimolecular (first-order) rate constant and has units of per second (s^{-1}). This value tells us how likely it is that the AB complex will dissociate in a given time frame, and thus permits us to calculate the average lifetime of a biological complex.

The kinetics of a binding process can have important biological consequences. Consider an example of a drug that must bind to a particular protein in the cell in order to be effective. If the k_{on} is too slow, the drug may be excreted from the body before it engages its target. By contrast, a too-slow k_{off} may mean that it may take a very long time for the drug to be cleared from the body to the extent that it might accumulate in potentially toxic concentrations.

The time it takes for half of all complexes in a given population of identical complexes to come apart is called the half-life, $t_{1/2}$, as illustrated in Figure 3.7. For simple binding equilibria, the half-life of a complex is related to the rate of dissociation of the complex by the expression $t_{1/2} = 0.693/k_{off}$. So, for example, for a

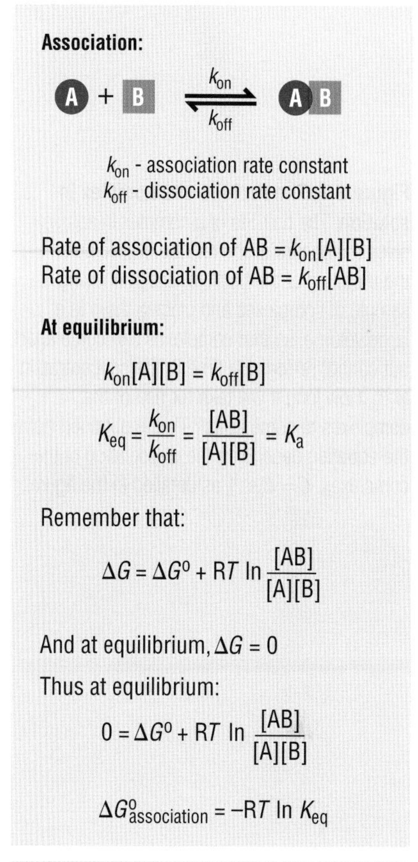

Association:
A + B ⇌ AB
k_{on}
k_{off}
k_{on} - association rate constant
k_{off} - dissociation rate constant
Rate of association of AB = k_{on}[A][B]
Rate of dissociation of AB = k_{off}[AB]
At equilibrium:
$k_{on}[A][B] = k_{off}[B]$
$K_{eq} = \frac{k_{on}}{k_{off}} = \frac{[AB]}{[A][B]} = K_a$
Remember that:
$\Delta G = \Delta G^o + RT \ln \frac{[AB]}{[A][B]}$
And at equilibrium, $\Delta G = 0$
Thus at equilibrium:
$0 = \Delta G^o + RT \ln \frac{[AB]}{[A][B]}$
$\Delta G^o_{association} = -RT \ln K_{eq}$

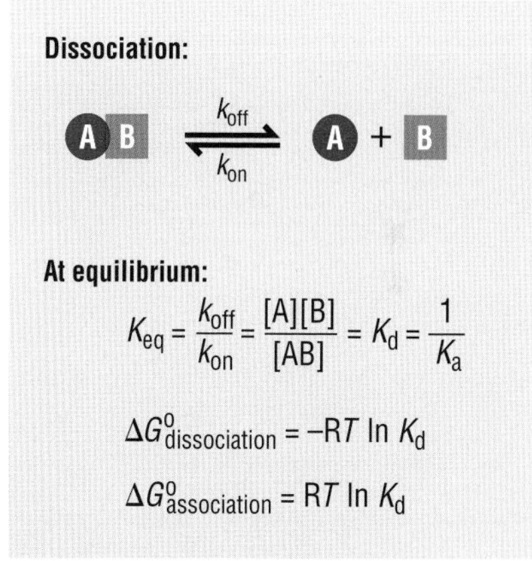

Figure 3.6 Kinetics of binding and equilibrium constants. Formation of a complex between molecules A and B is governed by the rate constants, k_{on} and k_{off}, which in turn determine the equilibrium binding constant, K_{eq}. Note that the rate constants are denoted by a lower case k. k_{on} is a second-order rate constant with units of M^{-1} s^{-1} while k_{off} is a first-order rate constant with units of per second. Equilibrium binding constants are indicated by an upper case K and, in this example, K_a has units of M^{-1} whereas K_d has units of M.

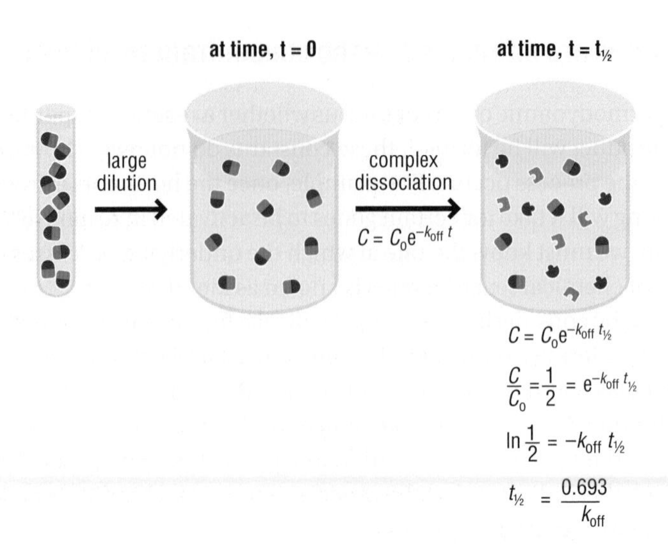

Figure 3.7 The half-life of a complex in solution. The half-life of a complex in solution gives a measure of the off rate, k_{off}. To see the meaning of the half-life, imagine taking a sample of complexes and placing them in a large volume, so that complexes can come apart but cannot re-form. At time 0, the concentration is C_0. How long it will take for half of the complexes to come apart? This is obtained from the equation describing the dissociation of the complexes, $C = C_0e^{-kt}$, as detailed in the figure.

complex with a dissociation rate of 0.1/s, half of all the complexes in the population will come apart in 0.693/0.1 s, or 6.93 seconds.

Unlike the rate of complex formation, the rate at which a complex dissociates is not limited by the speed at which molecules diffuse through liquid, since a single complex can come apart at any time without needing to interact with any of the other molecules in solution. There is therefore no inherent limit on k_{off}.

The equilibrium constant is a ratio of the forward and reverse reaction rates

At equilibrium, the rate at which the complex, AB, dissociates must equal the rate at which A and B associate. In terms of the rates we defined above, it is the point at which

$$k_{on}[A][B] = k_{off}[AB].$$

Rearranging, we obtain

$$k_{on} / k_{off} = [AB]/[A][B].$$

This ratio of forward and reverse rate constants is defined as the **equilibrium constant (K_{eq})**:

$$K_{eq} = k_{on} / k_{off} = [AB]/[A][B].$$

Since the particular reaction we are describing is the association of two molecules with one another to form a complex, we also call this the **equilibrium association constant (K_a)**. (Note that we use the upper case K to denote equilibrium constants, and lower case k to denote rate constants.)

The equilibrium constant gives a measure of the proportion of the molecules in solution that are bound in a complex at equilibrium. It is not the absolute value of either rate constant that matters; it is the ratio of the on and off rates that determines K_{eq}. The magnitude of the forward rate, k_{on}, will simply determine how long it takes to reach equilibrium once the individual components are mixed in solution, while the k_{off} will determine how long it takes the complex to fall apart (see Figure 3.6).

The equilibrium constant, K_{eq}, relates to a general reversible reaction and is always the ratio of the forward and reverse rate constants. When considering the specific case of a binding reaction such as $A + B \rightleftharpoons AB$, however, we saw that the equilibrium constant for the forward reaction is known as K_a, the equilibrium association constant. Since the equilibrium association constant is equal to [AB]/[A][B], it has units of $M/(M \times M)$, which simplify to M^{-1}. A large value for K_a tells us that the products (in this case, the complex AB) predominate over reactants at equilibrium, whereas a small value for K_a tells us that reactants predominate over products. We can therefore say that two species exhibiting a large value for K_a show a greater degree of binding (they bind more readily) than two species exhibiting a smaller K_a.

By contrast, the equilibrium constant for the reverse reaction, namely the dissociation of complex AB:

$$AB \rightleftharpoons A + B$$

is known as the **equilibrium dissociation constant (K_d)**, and is simply the inverse of the association constant, K_a:

$$K_d = k_{off}/k_{on} = [A][B]/[AB] = 1/K_a.$$

The K_d, which is expressed in units of concentration (M), gives us a measure of the proportion of molecules bound up in a complex at equilibrium. A large value for K_d indicates that the products (in this case the single, dissociated species A and B) predominate at equilibrium, whereas a smaller value suggests that reactants (the complex) predominate, as depicted in Figure 3.8. The K_d is used nearly universally in biology to describe binding reactions.

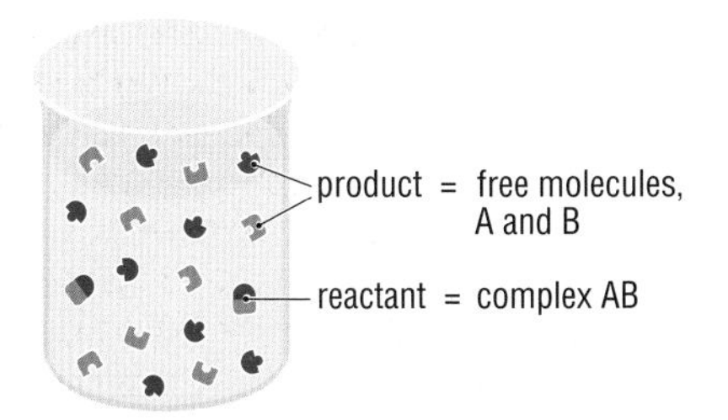

products dominate at equilibrium
$[AB] < [A][B]$
K_d is large

reactants dominate at equilibrium
$[AB] > [A][B]$
K_d is small

Figure 3.8 Relation of K_d value to the equilibrium concentrations of complex and free molecules. A large value for K_d indicates that the products (in this case the single, dissociated species A and B) predominate at equilibrium, whereas a smaller value suggests that reactants (the complex) predominate.

The equilibrium constant is related to the free energy of the interaction

The equilibrium constant has a thermodynamic meaning because the relative amounts of free components and of the complex present at equilibrium are directly related to differences in the free energy, ΔG^o, of the free and complexed components. Combining the expressions shown above for the free energy change at equilibrium and the equilibrium constant for the binding reaction $A + B \rightleftharpoons AB$ gives the equation:

$$\Delta G^o = -RT \ln([AB]/[A][B]) = -RT \ln K_{eq}.$$

This important equation gives the relationship between the equilibrium constant and the free energy of a reaction. In general, if the equilibrium constant is greater than 1, ΔG^o is negative, and complex formation is favored. By contrast, if the equilibrium constant is less than 1, ΔG^o is positive, and most of the A and B molecules will not be part of a complex at equilibrium. Notice how this mirrors our discussion of the free energy of binding reactions.

When considering a given reaction in biology we are often particularly interested to know how favorable it is - that is, what is the ΔG^o - and how far the reaction proceeds - that is, what the ratio of products to reactants is at equilibrium. For a binding reaction, the ratio of products to reactants at equilibrium is given by K_d, which is the equilibrium constant for the dissociation reaction. Combining the information above - that $\Delta G^o = -RT \ln K_{eq}$ and $K_a = 1/K_d$ - we see for the binding reaction $A + B \rightleftharpoons AB$ that

$$\Delta G^o = -RT \ln K_a = -RT \ln(1/K_d) = RT \ln K_d.$$

Figure 3.9 lists some values for equilibrium dissociation constants along with the corresponding free energy changes for the association reaction. As can

Dissociation Constants (K_d) and Corresponding Free Energy of Association

$K_d = \left(\frac{[A][B]}{[AB]}\right)$	free energy (kJ/mole)	free energy (kcal/mole)
10^3 M	17.1	4.09
1 M	0	0
10^{-3} M (mM)	–17.1	–4.09
10^{-6} M (μM)	–34.2	–8.17
10^{-9} M (nM)	–51.4	–12.3
10^{-12} M (pM)	–68.5	–16.4
10^{-15} M (fM)	–85.6	–20.5

Figure 3.9 Table of equilibrium dissociation constants and free energy change of association. The table lists free energy changes of binding and their corresponding equilibrium dissociation constants, as defined by the equation $\Delta G_{association} = RT \ln K_d$.

(a)

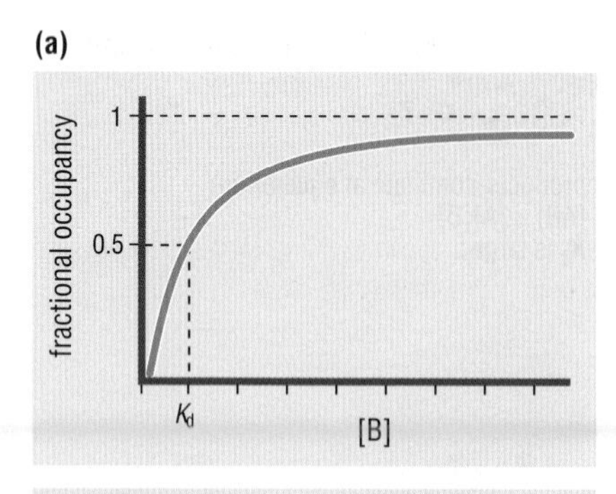
1
0.5
fractional occupancy
K_d
[B]

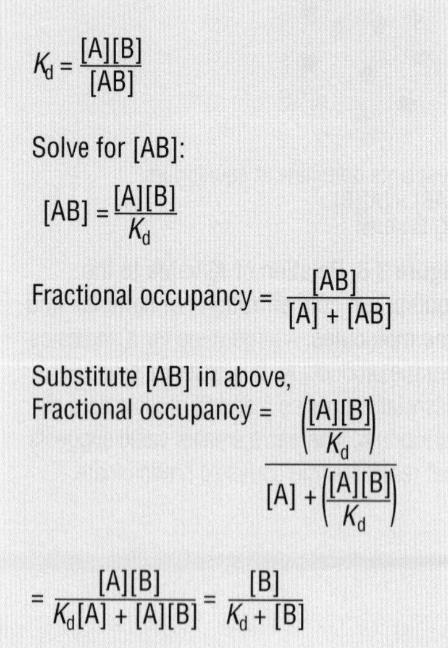
$K_d = \frac{[A][B]}{[AB]}$
Solve for [AB]:
$[AB] = \frac{[A][B]}{K_d}$
Fractional occupancy = $\frac{[AB]}{[A] + [AB]}$
Substitute [AB] in above,
Fractional occupancy = $\frac{\left(\frac{[A][B]}{K_d}\right)}{[A] + \left(\frac{[A][B]}{K_d}\right)}$
$= \frac{[A][B]}{K_d[A] + [A][B]} = \frac{[B]}{K_d + [B]}$

(b)

concentration of B as a function of K_d (M)	fractional occupancy of A
0.001 K_d	0.001
0.01 K_d	0.01
0.1 K_d	0.091
K_d	0.5
10 K_d	0.91
100 K_d	0.99

Figure 3.10 Fractional binding as a function of concentration. (a) This graph shows the fractional occupancy of A as a function of [B] for the simple binding reaction $A + B \rightleftharpoons AB$ under conditions where B is in excess. The concentration of B where 50% of A is bound (fractional occupancy of A = 0.5) corresponds to the dissociation constant (K_d) for the binding reaction. (b) The table shows specific examples of the fraction of A bound to B as [B] is increased relative to K_d.

be seen from the values, a negative ΔG^o corresponds to a K_d that is less than 1, which means that the concentration of the complex at equilibrium exceeds that of the uncomplexed molecules. A biological interaction of reasonable affinity, characterized by a K_d of 10^{-9} M, has a corresponding ΔG^o of roughly –51.5 kJ/mol (–12.3 kcal/mol). Keep in mind that we are comparing the energy of the *association* reaction with the equilibrium constant for the *dissociation* reaction, K_d.

A very important implication of equations relating free energy and the equilibrium constant is that seemingly small differences in binding energy, which contribute to the overall free energy, will result in large differences in K_d. For example, the free energy value of a single hydrogen bond is in the order of 4.2–13 kJ/mol (~1–3 kcal/mol), which means that adding or subtracting a hydrogen bond from a binding interface between two proteins has the potential to change the K_d by several orders of magnitude! This fundamental relationship between binding energy and the dissociation constant is the basis for understanding biological interactions in the cell, and how seemingly similar molecular species can have quite distinct binding partners and biological consequences.

3.3 BINDING PROCESSES IN BIOLOGY

Until now, we have been discussing the simple example of binding in which just two molecules bind to one another, such as the binding of a substrate to an enzyme. However, much of biology involves many more molecules coming together in a network of interactions. In this section, we explore some of the characteristic features of the binding interactions that are so important in biological systems, and the way they are underpinned by the thermodynamic concepts we have considered so far.

A binding curve shows how complex formation increases with concentration

One of the most central questions about the interaction between any two biological molecules is the proportion that will form a complex at a given concentration of each molecule, A and B. The answer will depend on the concentration of A and B, as well as on the equilibrium dissociation constant, K_d.

The importance of using the K_d to describe a binding reaction is that, under certain conditions, the K_d tells us the concentration at which half of a set of molecules are bound up in a complex while the other half are free in solution. For the example we have been considering, $AB \rightleftharpoons A + B$, we will consider conditions where there is much more B than A in solution (or vice versa). By plotting the amount of complex, AB, formed as a function of the total amount of B added to a reaction system, we obtain a hyperbolic curve that is known as a **binding curve**. The binding curve provides a simple depiction of how much complex is formed at different ligand concentrations. The equilibrium dissociation constant, K_d, corresponds to the concentration of molecule B at which half of the A molecules are bound in a complex and half are free. Therefore, the value of K_d can be readily determined from such a hyperbolic plot, as illustrated in Figure 3.10a.

The reason that K_d gives the concentration of B at which half-maximal binding occurs can be explained as follows: if the concentration of B is very much greater than A, then when some B is 'used up' to form a complex with A, the impact on the concentration of B is negligible. Under such conditions at equilibrium, we can

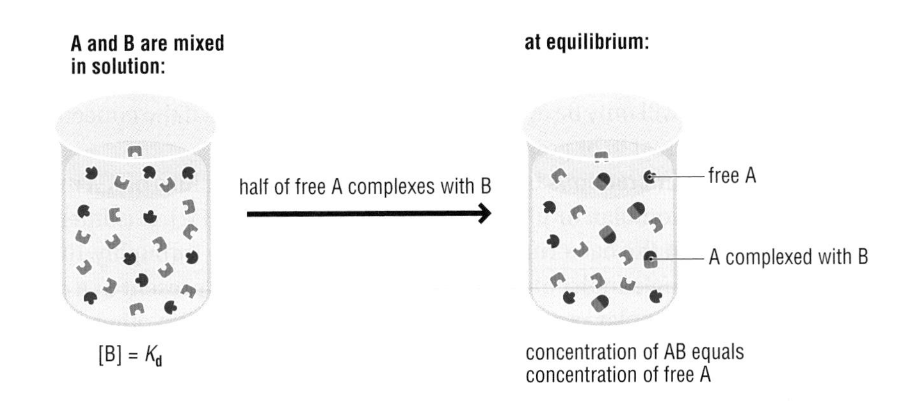
A and B are mixed in solution:
at equilibrium:
half of free A complexes with B
free A
A complexed with B
[B] = K_d
concentration of AB equals concentration of free A

Figure 3.11 K_d and the concentration of complex at equilibrium. Molecules A and B mixed in solution (left) will start to bind to one another. At equilibrium (right), the K_d gives the concentration of B (denoted as [B]) at which half of its binding partners, A, will form a complex with B, with the other half free in solution.

write that [A] = [AB]. We learned earlier that K_d = [A][B]/[AB]. So, when [A] = [AB], [A]/[AB] = 1 and, therefore, K_d = [B]. In other words, when the concentration of B is equal to the K_d, there is as much of A bound to B as there is free A in solution. This situation is depicted in Figure 3.11.

Taking this one step further, by comparing the concentration of B to the value of K_d for the complex being examined, we can estimate the number of molecules of A and B that are likely to be bound to one another at a given concentration of B, as depicted in Figure 3.12. If [B] in a reaction mixture is much lower than K_d then we infer that only a small proportion of A will have formed a complex with it. (If [B] is less than the K_d then less than half of the available A will have formed a complex.) By contrast, if [B] is much higher than the K_d, then we infer that a large proportion of A will have formed a complex with B.

Comparing the concentration of interacting species to the K_d can help us predict the proportion of the species that are likely to form a complex at a given concentration. The values listed in Figure 3.10b show that when the concentration of B is much higher than the K_d, almost all of A is complexed with B, while at concentrations of B much lower than the K_d, almost all of A is free (uncomplexed). It is very important to note that it does not matter how *many* molecules of A there

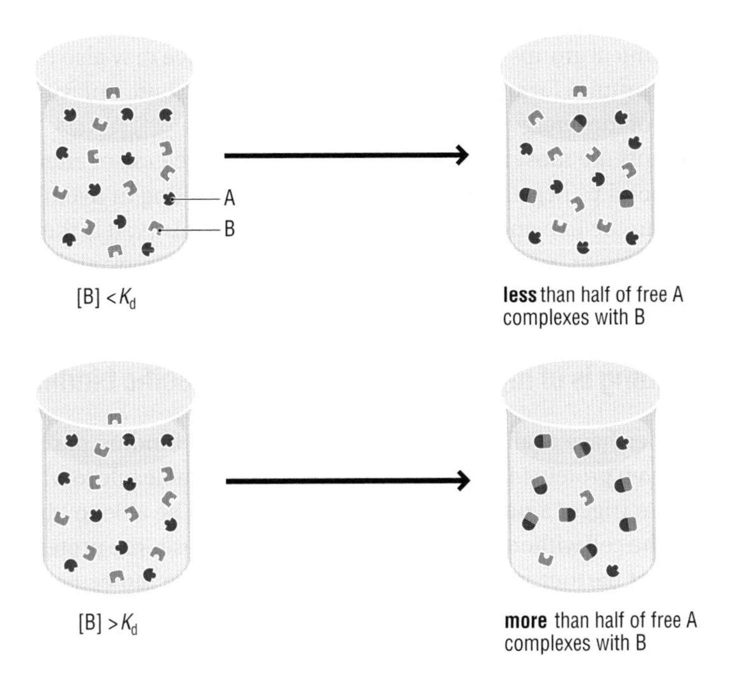
A
B
[B] < K_d
less than half of free A complexes with B
[B] > K_d
more than half of free A complexes with B

Figure 3.12 The ratio of [B] to K_d and the proportion of complex formed at equilibrium. We perform the same experiment as in Figure 3.11, but with different starting concentrations of B. When [B] is less than K_d, less than half of the free A molecules will form a complex with B. When [B] is greater than K_d, more than half of free A molecules will form a complex with B.

are as compared to B; it is the *concentration* of A and B and the K_d for the binding interaction that matter. Even when there are a million molecules of B for every molecule of A, there will only be appreciable complex formation if the concentration of B is roughly equal to or greater than the K_d.

Typical biological interactions have dissociation constants (values of K_d) in the range of 10^{-6} M (micromolar, or μM) to 10^{-9} M (nanomolar, or nM). It is interesting to consider that in the bacterium, *E. coli*, whose cell volume is roughly 10^{-15} L (a femtoliter, fL), a single molecule expressed in the cell will be present at a concentration in the nanomolar range. By contrast, one molecule within other, larger cell types would be present at a much lower concentration. It stands to reason that most cellular components will be present at higher concentrations than this when they are synthesized in an *E. coli* cell (otherwise there would be none present). In general, the K_d corresponding to a particular biological interaction – as dictated by the strength of that interaction – is close to the cellular concentration at which the molecules must bind in the cell to achieve a particular biological outcome.

The chemical details of the bound and free molecules govern the energy of interaction

The net binding energy for two interacting partners represents the energy difference of the entire system before and after the molecules form a complex. The total energy of complex formation will be the net change in enthalpy and entropy summed over all the different interactions, according to the equation $\Delta G = \Delta H - T\Delta S$. When two molecules bind to one another, interactions between the surrounding water and the molecules are broken and replaced with hydrogen bonds, salt bridges, and van der Waals interactions between the two molecules at the binding interface. The water molecules are thus released when the two interacting surfaces come together, contributing positively to the entropy of the system.

Sometimes, water molecules are trapped at an intermolecular interface, where they participate in networks of hydrogen bonding interactions, as shown in Figure 3.13. Immobilized water molecules have less entropy than free water molecules, and thus contribute unfavorably to the entropy term of the energy equation. Chemical groups located at the binding interface may also become less mobile upon complex formation, which would also contribute unfavorably to the entropic term of the energy equation.

Binding events are most typically associated with the release of energy as heat, which correspond to negative changes in enthalpy, resulting in overall favorable contributions to the free energy. As in all aspects of chemistry, it is the *net* energy change in the system, summed over all the changes in entropy and enthalpy in going from the free state to the bound, that determines the overall energy of binding.

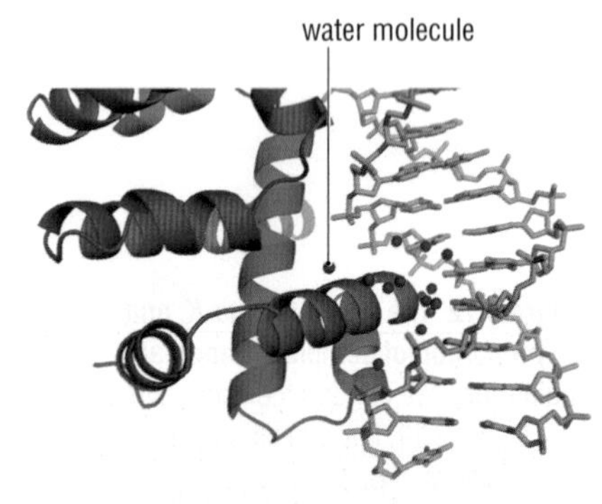

Figure 3.13 Water molecules immobilized at a binding interface. Solvent molecules can form bridges between two molecules, as in this example of Trp repressor bound to DNA (Protein Data Bank (PDB) code 1TRO).

Specific binding is of higher energy than non-specific binding

Macromolecules have available to them many potential binding partners in the cell. While a given protein may have one particular macromolecule to which it must bind in order to trigger a particular response, it may also be able to bind to other molecules in the cell without yielding a productive response. The protein is able to select its correct partner from among the other competing molecules because it binds more tightly to its correct partner than to any other molecule in the cell (see Experimental approach 3.1). Binding to the correct partner is often termed **specific binding**, whereas lower-energy interactions with other molecules are termed

3.1 EXPERIMENTAL APPROACH

Insights from kinetics into how DNA-binding proteins find their target sites

The kinetics of a given binding process can give important insights into how the two binding partners locate each other in the cell. A fundamental problem for all proteins that interact with specific DNA sequences is how they locate these relatively small sites (6–30 base pairs) within the vast excess of non-specific genomic DNA. In a classic study by Riggs, Bourgeois, and Cohn of the kinetics of *Escherichia coli* lac repressor binding to its operator sites, the authors found that binding to specific DNA sites occurred very rapidly. Indeed, the observed rate constant for the association of lac repressor with the lac operator on a 50 kb piece of DNA was 7×10^9 M^{-1} s^{-1}, a value considerably higher than the generally accepted figure for the diffusion limit for protein–ligand association reactions, 10^8 M^{-1} s^{-1}. This observation led to the view that lac repressor did not locate its binding sites by simply diffusing randomly through solution before binding to its correct DNA site, but instead bound first to a random site in the chromosome and then slid along the DNA until it reached its correct binding site, as depicted in Figure 1. This type of binding mechanism was termed facilitated diffusion or one-dimensional diffusion, because the protein was confined to sliding in one dimension along a DNA strand.

Experiments on a number of different proteins appeared to support the view that proteins located their DNA sites by one-dimensional diffusion. For example, it was found that the restriction enzyme, EcoRI, could locate its target sequence more rapidly when the site was located within a long DNA fragment as compared with a short one (Jack *et al.*, 1982). In the experiment depicted in Figure 2, the authors compared binding of EcoRI to a 4361 bp fragment and a 34 bp fragment. The enzyme was mixed with both fragments at time 0 and complex formation was monitored. While the protein initially bound overwhelmingly to the long DNA fragment, binding to the two fragments was equalized as the complexes were allowed to equilibrate. This suggested that the equilibrium dissociation constant (K_d) was the same for the two DNA fragments, but that the on rate (and hence also the off rate) was more rapid for the longer fragment.

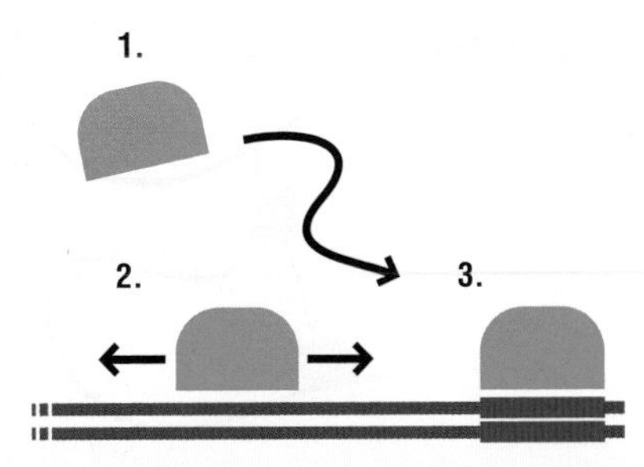

Figure 1 Model for one-dimensional diffusion along DNA. The protein (1) diffuses through solution and binds to a non-specific site (2). It then randomly diffuses back and forth along the DNA until it locates its specific binding site (3).

Using modern approaches that permit one to manipulate individual DNA molecules, evidence for additional contributions to DNA binding has accumulated. According to this updated view, a protein rapidly 'searches' large amounts of non-specific DNA using a combination of one-dimensional diffusion and three-dimensional 'hops' of the protein between DNA segments, as shown in Figure 3a . van den Broek and colleagues (2008) came to this conclusion by directly measuring the association rate for EcoRV endonuclease binding to DNA, but comparing the rates for the same DNA fragment as a function of how compact it is.

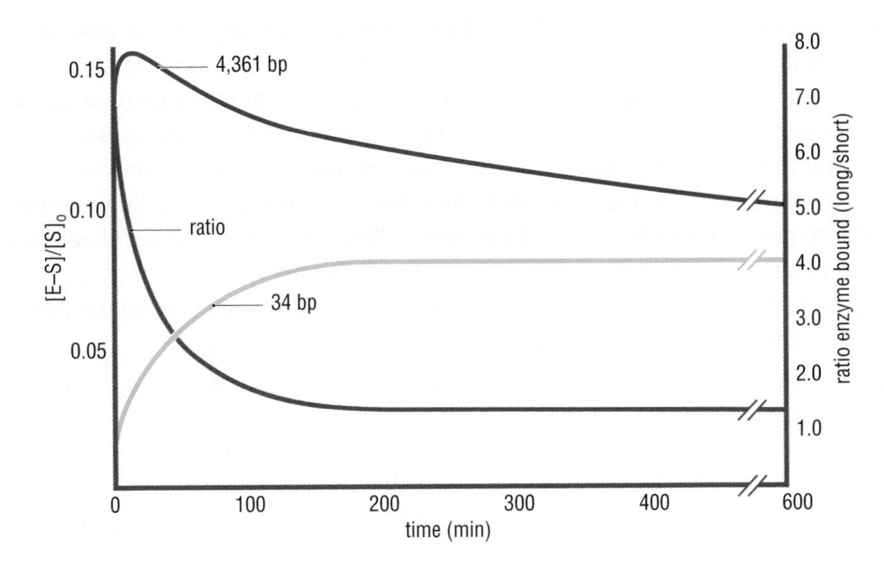

Figure 2 Binding of EcoRI to long and short DNA fragments. At time 0, EcoRI was mixed with two DNA fragments, one long and one short, and the binding to each fragment was then monitored as a function of time. While the protein bound very rapidly to the long DNA fragment at first, with time the binding to the two fragments was nearly equal.

Reproduced from Jack *et al.*, *Proceedings of the National Academy of Sciences of the U S A*, 1982:**79**:4010–4014.

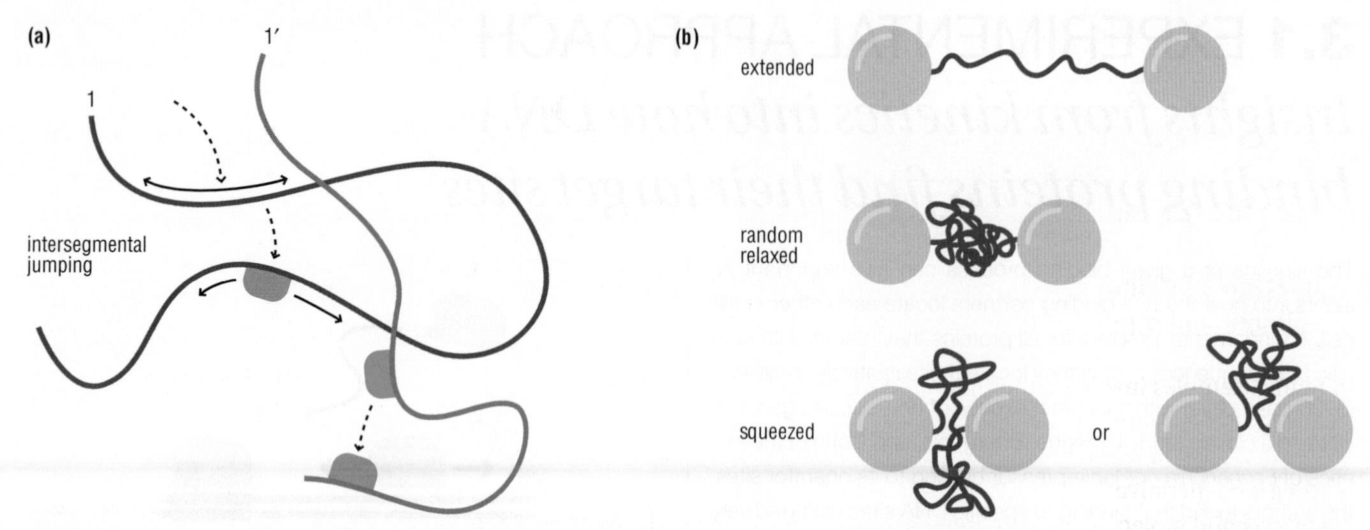

Figure 3 DNA binding by diffusion and intersegmental hopping. (a) A protein binds to DNA and can then either diffuse along for short distances or hop from one segment to another. The latter process is shown to be facilitated by DNA compaction in a single molecule analysis. (b) Illustration of the experiment, in which each end of a long DNA fragment is attached to a bead that can be manipulated to condense or extend the DNA. Protein association is facilitated when DNA is condensed (random relaxed), but not too compressed (squeezed), which makes hopping from strand to strand more likely.

Reproduced from van den Broek *et al.*, *Proceedings of the National Academy of Sciences of the USA*, 2008:**105**:15738–15742.

DNA compaction was varied by attaching a bead to each end of the DNA fragment and manipulating the distance between them using a device known as optical tweezers (see Figure 3b). This approach is generally referred to as 'single molecule' analysis since it is based on direct observation of individual molecules through a microscope. By studying the binding process as the distance between the two beads was varied, the authors found that the rates of association (the on rates) were somewhat higher when the degree of DNA compaction was higher. In thinking about their results, the researchers argued that in the crowded intracellular environment where DNA density is extremely high, the contributions of such hopping to the search algorithm of a protein could be quite significant for biochemical function.

Find out more

Jack WE, Terry BJ, Modrich P. Involvement of outside DNA sequences in the major kinetic path by which EcoRI endonuclease locates and leaves its recognition sequence. *Proceedings of the National Academy of Sciences of the U S A*, 1982;**79**: 4010–4014.

Riggs AD, Bourgeois S, Cohn M. The lac repressor-operator interaction. 3. Kinetic studies. *Journal of Molecular Biology*, 1970;**53**: 401–417.

van den Broek B, Lomholt MA, Kalisch SMJ, Metzler R, Wuite GJL. How DNA coiling enhances target localization by proteins. *Proceedings of the National Academy of Sciences of the U S A*, 2008;**105**:15738–15742.

Related techniques

Protein-DNA binding; Section 19.12

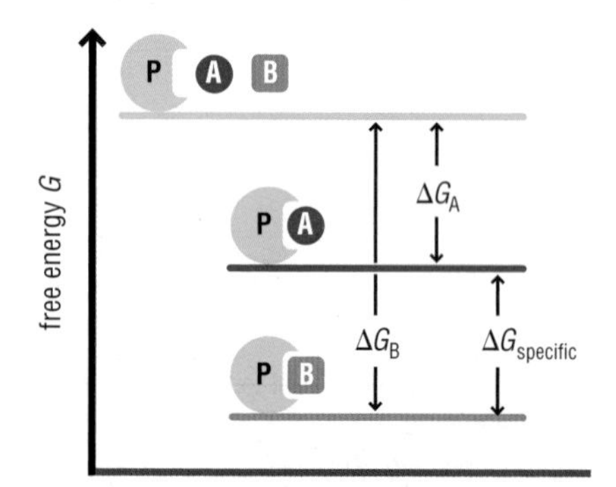

Figure 3.14 Binding specificity. Protein P binds preferentially to protein B, with which it forms a more favorable complex than with protein A. The net free energy change upon binding is greater (i.e., more negative, and therefore more favorable) when P binds to B as compared with binding to A.

non-specific binding. The specificity of a given interaction depends on how likely it is that the specific binding partner is selected from among competing ligands.

The specificity of a given binding reaction can be determined by comparing the thermodynamic constants of binding to the correct partner with the constants for binding to other, competing molecules. The free energy of association with the correct partner will be more negative than that associated with binding to other, random molecules in the cell, as illustrated in Figure 3.14. This will be reflected in the equilibrium dissociation constants, as the K_d for the specific interaction will be lower (reflecting tighter binding) than the K_d for the non-specific interaction.

The ratio of these two constants,

$$K_d^{specific}/K_d^{non\text{-}specific},$$

gives a measure of the specificity of a given interaction. Specificity can vary widely: a ratio of 10 or less typifies an interaction with limited specificity, that is, the two molecules only show limited preference for one another. By contrast, ratios of 1000 or greater mean that the binding interaction is highly specific. The bacterial lac repressor protein, for example, binds to DNA of arbitrary sequence with a K_d of around 10^{-7} M, while it binds to its specific DNA site with a K_d of around 10^{-11} M. This 10 000-fold preference for the correct site over random sites ensures that the lac repressor will regulate the correct set of genes in the bacterial genome. The specificity of different interactions has been fine-tuned in the course of evolution to ensure that they can take place in the crowded cell milieu.

Cooperative binding makes an important contribution to the sensitivity and regulation of biological systems

In many examples involving multiple binding species, the free energy change as each successive subunit is added to the larger complex can depend on the molecules that have previously been incorporated into the complex. We say that the binding is cooperative if the binding of one molecule affects the likelihood of binding of a second molecule. If the binding of one molecule makes the binding of the next molecule more favorable, it is described as **positive cooperativity. Cooperative binding** can also be negative, whereby the binding of one molecule makes the binding of the next less favorable. However, the examples we will encounter in this book will be almost exclusively those of positive cooperativity. Cooperative binding can be exploited by the cell to fine-tune responses to small changes in the concentration of individual binding partners.

A simple example of cooperativity - and one that we shall encounter later in this book - occurs when two proteins bind to two adjacent sites on a double-helical strand of DNA, a situation depicted in Figure 3.15. If there are no interactions between the two proteins when they are each bound to the neighboring DNA sites, the binding affinity for one site is unaffected by the binding of another protein to an adjacent site. However, if proteins bound to neighboring sites can form favorable interactions with one another, the binding of protein to the first DNA site can make the binding of a second protein to the neighboring site more favorable. This additional energy makes the free energy of binding to the second site more negative; in other words, more favorable.

The effect of cooperative binding, and its importance in fine-tuning regulation, can be readily seen in the changes that it produces in a **binding curve**. Instead of the hyperbolic curve that we described above for a simple binding interaction, cooperative binding produces an S-shaped, **sigmoidal curve**, in which complex formation at first rises slowly with increasing protein concentration and then rises steeply over a narrow concentration range; such a sigmoidal curve is shown in Figure 3.15b. The sharper increase occurs because, as one DNA site is filled, the effective affinity for the next site increases; that site will become occupied at a lower protein concentration than in the absence of cooperative interactions. Cooperativity has the effect of narrowing the concentration range over which the binding partners make the transition from being largely free to being mostly bound up in a complex.

Cooperative interactions are critical to many biological switches because they enable the cell to respond to smaller changes in the concentration of a particular ligand. If the cellular concentration of a macromolecule or its ligand is near the midpoint of the sigmoidal response curve, where the binding reaction is half-saturated, small fluctuations in concentration can result in large changes in the extent of its binding. This poises a system to rapidly switch processes on and off when ligand concentrations fluctuate in small but biologically significant ways.

A variety of mechanisms can give rise to cooperative binding besides the example described. In the example of cooperative binding of protein to DNA,

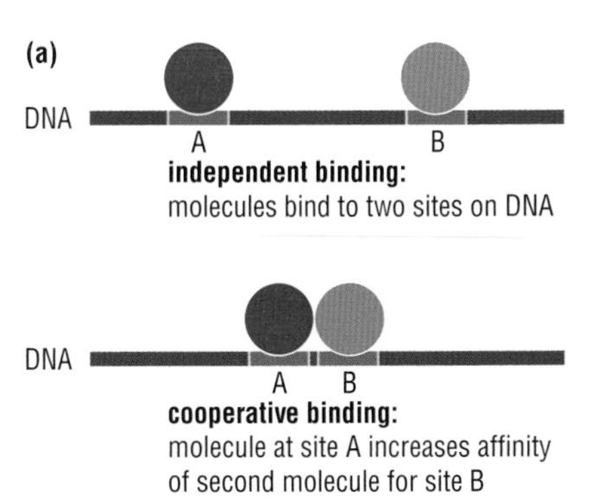

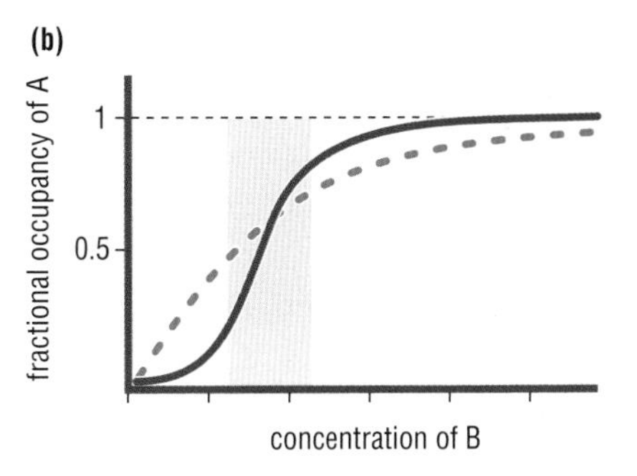

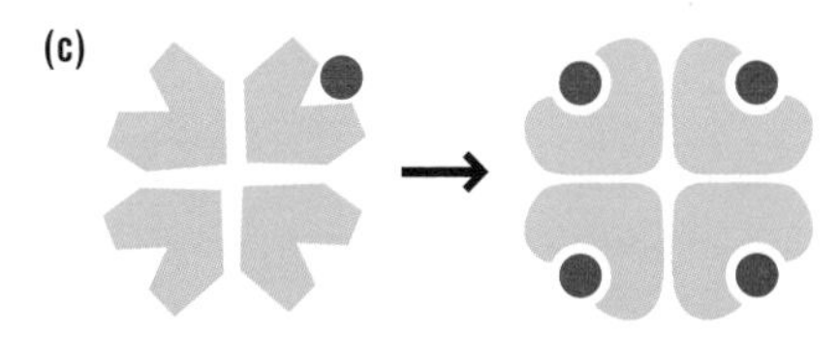

Figure 3.15 Effect of positive cooperativity on binding curves. (a) Above, two proteins are shown binding independently to two binding sites on DNA; below, two proteins bind cooperatively to adjacent sites, with the binding of one protein to site A increasing the affinity of the second protein for site B. (b) The solid line shows the sigmoidal binding curve characteristic of cooperativity. Note that within a critical range of concentration (shaded area), relatively small differences in concentration of B have large effects on binding. Simple binding with no cooperativity gives the hyperbolic curve shown as a blue dotted line. (c) Binding of one molecule of ligand B (red circle) to one subunit of protein A (green) causes a conformational change that both increases the affinity of the bound subunit for the ligand, and induces the same change in the other subunits, increasing their affinity for ligand B.

favorable interactions between protein bound to adjacent sites on the DNA are sufficient to make the binding cooperative. In another type of mechanism, the binding of a ligand can induce a conformational change in protein structure, which can then alter the affinity for binding the next ligand. The conformational change can be within a single polypeptide, or in the quaternary arrangements of subunits in a complex. The binding of oxygen to the hemoglobin tetramer in blood cells is a classic example of cooperativity: once a single oxygen molecule binds to one protein subunit, the affinity of oxygen for the remaining three subunits is significantly increased. Protein folding is also cooperative: once the first turn of an alpha helix forms, the remainder folds much more quickly to form the final helical form.

3.4 ENZYME CATALYSIS

An activation energy must be overcome for a reaction to proceed

We note above that the free energy change associated with a reaction determines whether a reaction happens spontaneously or not. However, the spontaneity (or otherwise) of a reaction tells us nothing about how fast a reaction will occur. A reaction may exhibit a favorable (i.e. negative) change in Gibbs free energy, but may still proceed very slowly indeed. For example, the chemical breakdown of a polypeptide chain into individual amino acids through hydrolysis is spontaneous, but happens slowly over months to years. What makes such reactions happen so slowly?

The answer is that every reaction has associated with it an **activation energy** – an energy threshold that must be overcome along the reaction path from reactants to products, as depicted in Figure 3.16. If the activation energy for a reaction is very high at room temperature, as denoted in Figure 3.16, only very few reactant molecules possess the energy required to overcome the activation energy barrier. Consequently, the reaction happens very slowly.

Many important biological reactions have activation energies that are so high that the rates at which they would proceed under 'normal' conditions would be so slow as to prohibit life: few molecules would possess energies higher than the activation energy, and only these few molecules could undergo chemical change. One way to overcome this problem would be to heat the reactant molecules so that more of them possessed energies greater than the activation energy barrier. However, biological systems cannot be exposed to extremes of heat – not least because many essential molecules would become denatured and cease to function.

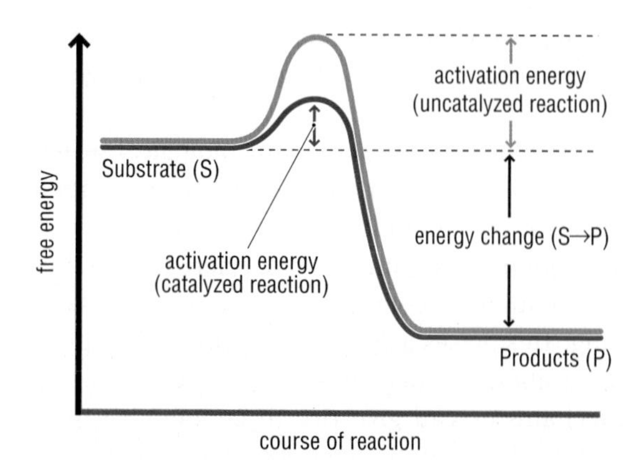

Figure 3.16 Activation energy. For each chemical reaction, an activation energy barrier must be overcome in the process of converting substrates into products. Even if the net energy change of the reaction is very favorable, few molecules at room temperature may have the energy to overcome the activation energy (blue). An enzyme can increase the rate of product formation by lowering the activation energy for the reaction (red), so a greater proportion of molecules will have the energy needed to overcome the barrier. Note that enzyme catalysis does *not* alter the energy difference between substrates and products.

So, how does nature circumvent this problem? The answer lies in the activity of a vital class of biological molecules, the enzymes.

Enzymes accelerate chemical reactions

Cells must continually carry out a remarkable assortment of chemical reactions, which must proceed at a rate that is fast enough to be useful to the cell. All living things therefore depend on molecules called **enzymes** that accelerate the rate of chemical reactions, a process known as **catalysis**. Enzymes catalyze chemical reactions by lowering the activation energy barrier (see Figure 3.16). Most enzymes are proteins, although some are composed of RNA and are therefore called **ribozymes**. Enzymes vary in the degree to which they accelerate reaction rates. Some accelerate a chemical reaction by just several orders of magnitude, while some enhance reaction rates by a remarkable factor of 10^{17}.

There are several features that are common to all enzymes. An enzyme binds to one or more molecules known as **substrates**, promotes a chemical reaction in the substrate molecules, and then releases the products of the reaction. The enzyme is not permanently modified in the process, so it is then able to bind another substrate molecule and perform the reaction again. Enzymes generally catalyze just one type of chemical reaction and are usually highly specific for the types of substrates on which they act. For this reason, cells produce many different types of enzymes, each uniquely designed for a particular chemical task.

A chemical reaction with a negative free energy change should occur spontaneously, given enough time. However, in the course of the chemical reaction, the substrates must undergo transient changes in stereochemistry, charge, or covalent structure before completing the necessary rearrangements to produce the final products. These changes are energetically unfavorable, which means that the first steps in the chemical reaction generally involve a positive free energy change. These are then followed by chemical steps with a negative free energy change of even greater magnitude, resulting in a chemical reaction whose net free energy change, when summed over all the steps, is negative, as indicated in Figure 3.16.

A consequence of this energetic pathway is that there is a free energy barrier between the initial and final states of the chemical reaction. The **transition state**, which is the highest point in free energy on the reaction pathway from substrate to product, marks the top of the free energy barrier. This species exists only transiently (for perhaps as little as 10^{-15} seconds) and is generally different in stereochemistry and charge configuration from either the substrate or the product. It is the energy difference between the ground state and the transition state that constitutes the activation energy.

Without the aid of an enzyme, few substrate molecules at any one time typically have sufficient energy to reach the transition state; as a result, the uncatalyzed reaction proceeds slowly. Enzymes achieve their catalytic effects by lowering the activation energy barrier using several different strategies that we will discuss below.

The enzyme active site contains residues that catalyze the chemical reaction

Each enzyme has an **active site**, which is a pocket or region in the enzyme containing precisely positioned chemical groups that catalyze the chemical reaction, as illustrated in Figure 3.17. In addition, there may be other molecules bound in the active site that either assist or actively participate in the chemical reaction. These include metal ions and water, as well as larger molecules such as the nucleotide cofactors we learn about later in this chapter. The great majority of enzymes are proteins, and the key chemical groups in their active sites typically are amino acid side chains, although the peptide backbone itself can occasionally participate. By contrast, the active sites of ribozymes are composed of chemical groups found on the nucleotide bases or on the sugar–phosphate backbone.

The chemical nature and shape of the active sites are uniquely tailored to the reaction that the enzyme catalyzes. Some enzymes also have an additional pocket that binds to more remote regions of the substrate molecule to aid in specific substrate recognition. Evolutionarily divergent enzymes that catalyze the same reaction may vary a great deal in sequence and structure, but are usually very similar in the active site.

Enzymes lower the activation energy in a variety of ways

Enzymes use a number of different mechanisms to lower the activation energy barrier of a reaction. Many enzymes lower the energy barrier by specifically stabilizing the transition state, through favorable free energy of binding between the enzyme and the transition state, as depicted in Figure 3.18. If the enzyme binds more tightly to the transition state than to the substrate or the product, the barrier to the reaction will be lowered, making the reaction more likely to proceed. Some enzymes are thought to catalyze a reaction by simply bringing two reactive substrates together, thereby pre-organizing them for catalysis. This is often referred to as proximity catalysis and is perhaps the simplest way to stabilize the transition state of the reaction.

Another way of reducing the size of the energy barrier is to destabilize the original substrate, raising its free energy as compared with the molecule free in solution. This is called ground state destabilization. A substrate bound to the enzyme in a less stable, higher-energy conformation starts out farther up the free energy hill towards the transition state, and the reaction will therefore proceed at a higher rate.

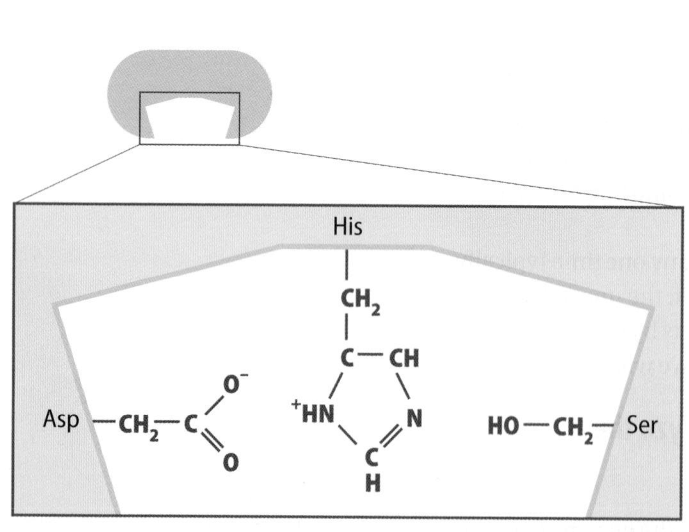

Figure 3.17 Enzyme active site. The chemical reaction catalyzed by an enzyme takes place in the active site, which contains side chains that bind the substrate molecules, catalyze the chemical reaction, and promote the subsequent release of reaction products. The three side chains shown here are found in the active sites of many proteases that cleave polypeptide chains.

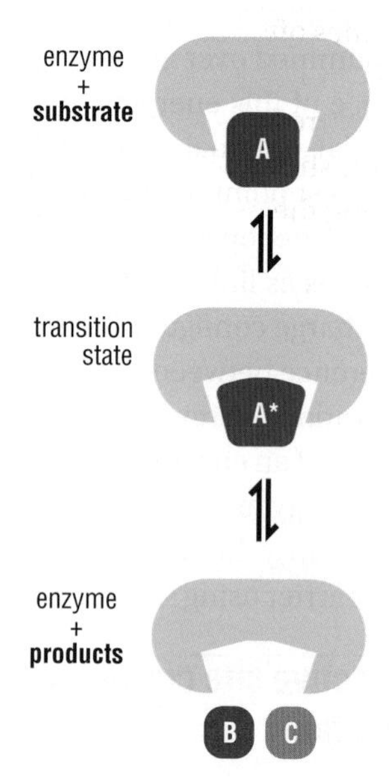

Figure 3.18 Transition state stabilization. Enzymes generally bind the transition state (A*) more tightly than either the substrate (A) or products (B and C). If an enzyme bound more tightly to the substrate, the reaction would be inhibited. If the enzyme bound more tightly to the products, the products would not be rapidly released from the enzyme following catalysis.

Finally, an enzyme can lower the barrier to a chemical reaction by causing it to proceed through a different path composed of a series of steps, each with a smaller energy barrier than that of the uncatalyzed reaction. We refer to each of the chemical species that is transiently formed during such a reaction as a **reaction intermediate**. Enzymes commonly employ more than one strategy to catalyze a reaction.

A few catalytic strategies are common among biological enzymes

Having now seen the general methods through which enzymes catalyze reactions, we need to consider another question: what is happening in chemical terms - at the atomic level - when a reaction is catalyzed? Although the human genome encodes many thousands of enzymes, there are a few conserved reaction mechanisms that enzymes adopt to lower the activation barrier for reactions in order to promote catalysis.

The most common type of reaction that we will observe throughout this book is a nucleophilic displacement reaction, in which an atom that is rich in electrons, called a **nucleophile**, donates electrons to an atom that is deficient in electrons, known as an **electrophile**. Another term used for this process is nucleophilic attack. As with any movement of electrons within and between molecules, this reaction - and the transfer of electrons it represents - leads to a change in chemical bonding, as depicted in Figure 3.19. This nucleophilic displacement reaction lies at the heart of DNA, RNA and protein synthesis as well as most reactions involved in nucleic acid and protein transformations.

In each of these polymerization reactions - the joining of subunits to form a larger molecule - the nucleophile is activated to initiate a nucleophilic attack through removal of a proton by a **general base** on the enzyme (see Figure 3.19). A molecule that is able to donate (or lose) a proton is called an acid, while a molecule that is able to accept (or gain) a proton is called a base. Acids and bases must act in tandem: for an acid to donate a proton, it must have a base to donate the proton to. Enzymes often facilitate such proton movements by providing moieties that act as acids and bases during the reaction cycle. This type of participation by the enzyme is referred to as general acid-base catalysis.

In the reaction shown in Figure 3.19, a base removes a proton from the 3′ OH of ribose, allowing the oxygen to carry out a nucleophilic attack on the innermost

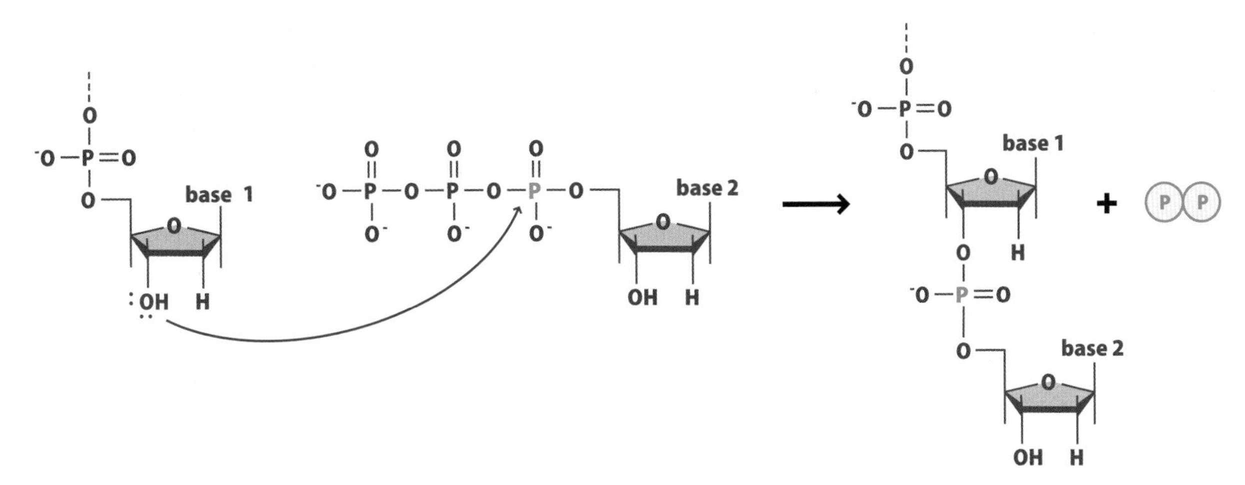

Figure 3.19 Nucleophilic attack in a DNA polymerization reaction. In the synthesis of a DNA strand, the 3′ OH at the end of the growing strand carries out a nucleophilic attack on the innermost phosphate of the incoming nucleoside triphosphate. A general base withdraws a proton from the 3′ OH, making the oxygen a better nucleophile.

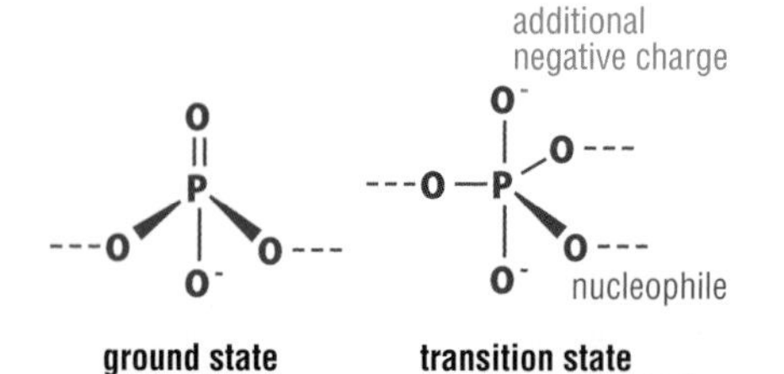

Figure 3.20 Pentacovalent phosphate intermediate. During the course of the polymerization reaction catalyzed by DNA and RNA polymerase, a transition state intermediate is formed where the phosphate atom is coordinated by five ligands. This is the pentacovalent intermediate.

phosphate of the incoming deoxynucleotide triphosphate. The side chain of histidine is particularly useful in mediating acid–base catalysis; since its pK_a is near 7, it can act as both an acid and a base at neutral pH to both accept and donate protons during the course of an acid–base reaction.

Another common feature of these reactions is that the transition state structure contains an additional bond. For example, during the course of DNA and RNA synthesis a pentacovalent phosphate intermediate is formed (rather than the ground state tetrahedral structure); both the tetrahedral ground state and transition state structures are shown in Figure 3.20. This transition state structure forms favorable interactions with the DNA and RNA polymerase enzymes that catalyze the polymerization reaction. Similarly, protein synthesis involves formation of a tetravalent carbon intermediate (rather than the ground state trivalent structure). These transition state structures have charge configurations and covalent bonding arrangements that are quite distinct from those of the ground state. The exquisite specificity of enzymes permits their recognition of these differences in structure to stabilize specific transition states and thus promote catalysis.

Active sites position chemical moieties to promote catalysis

The cleavage of RNA by the enzyme, RNAse A, provides a well-understood example of nucleophilic catalysis that also highlights some of the ways in which enzymes catalyze chemical reactions. The RNAse A reaction proceeds by acid–base catalysis, which means that the enzyme must provide chemical moieties that function as acids and/or bases to facilitate proton movement. As with all other enzymes, RNAse A contains features that stabilize the transition state structure.

The reaction, which is diagrammed in Figure 3.21, begins with the nucleophilic attack on the phosphate backbone of the RNA strand by the 2′ OH of ribose. Since this chemical group, on its own, is not sufficiently nucleophilic, a proton must first be removed in order for it to function as a nucleophile, thereby enabling catalysis to occur. RNAse A facilitates this deprotonation with a histidine side chain, which removes a proton from the 2′ OH and thereby activates the oxygen atom (top panel). This histidine is therefore acting as a general base. The nucleophilic attack results in the formation of a negatively charged pentacovalent phosphate intermediate in the transition state. Positively charged lysine side chains in the enzyme active site surround this group, stabilizing the transition state through favorable ionic interactions between the negatively charged phosphate and the positive charges on the lysines. At the same stage (top panel), another histidine side chain donates a proton to the other part of the cleaved strand, which is the 5′ oxygen of the leaving group. The enzyme remains bound to the remaining strand that contains the cyclic phosphate intermediate.

In the second step of the reaction, the cyclic intermediate is hydrolyzed to yield the final product containing a 3′ phosphate group. The two active site histidine side chains, one that gained a proton and the other that lost a proton, now reverse their roles. A water molecule is activated when one histidine removes a proton, and then attacks the cyclic intermediate. At the same time, the second histidine donates a proton, leading to formation of the final product (middle panel). The use of histidine in this enzyme as both a general acid and a general base highlights its special utility in enzymatic reactions.

Transesterification reactions involve the breakage and re-forming of a phosphodiester linkage

There are many reactions in which the phosphodiester linkage is broken and then re-formed with a different nucleic acid fragment in a concerted reaction that maintains the same sort of chemical linkage. This type of reaction is called a **transesterification** reaction, since an ester linkage is broken and then re-formed with a different fragment. An important feature of this type of reaction is that there is no energetic difference between the substrate and the product. Figure 3.22 outlines the key steps of a transesterification reaction that is central to the RNA splicing that occurs when RNA is cleaved to remove an intron. During this reaction, the 3′ end of one exon (labeled exon 1) is joined to the 5′ end of another (labeled exon 2).

Many enzymes require an additional molecule to catalyze a chemical reaction

Not every biological reaction can be carried out efficiently using the somewhat limited chemical properties of amino acids. Some enzymes require an additional molecule that participates in the reaction, called a **cofactor** (or **coenzyme**). An enzyme cofactor may be covalently attached to the enzyme, but is more generally simply tightly bound to the enzyme near the active site. Cofactors typically act as donors or acceptors of chemical groups, ranging from electrons to slightly larger groups such as acetyl groups. Once a cofactor loses or gains atoms during the course of the reaction, it must be regenerated before it can participate again in subsequent rounds of catalysis.

Figure 3.23 lists some of the common cofactors and their chemical function in reactions. All organisms use a universal set of cofactors, indicating that their use developed very early in evolution. Most of these cofactors are derived from ribonucleotides, probably because the chemistry of early life forms was dominated by RNA.

Several cofactors are essential for enzyme-catalyzed **oxidation** and **reduction** reactions. Reduction is the process of gaining electrons, while the loss of electrons is referred to as oxidation. Oxidation and reduction must always occur together, since whenever electrons are lost by one molecule, another must receive them. Reactions in which one species is reduced and another oxidized are therefore called oxidation–reduction, **redox reactions** or electron transfer reactions. This type of reaction, which is one of the most common in biology, typically depends upon high-energy cofactors such as nicotinamide adenine dinucleotide (NAD) and nicotinamide adenine dinucleotide phosphate (NADP). NAD is found in the cell in its oxidized form, **NAD$^+$**, and in a higher-energy reduced form, **NADH.** NAD$^+$ is converted to NADH by accepting two electrons and a proton, H$^+$, which is equivalent to accepting a hydride ion, H$^-$. NADH can then participate in reactions as a donor of H$^-$, as illustrated in Figure 3.24.

NADP$^+$ and **NADPH**, the oxidized and reduced forms of NADP, respectively, play similar roles, but the additional phosphate group generally prevents NADP$^+$

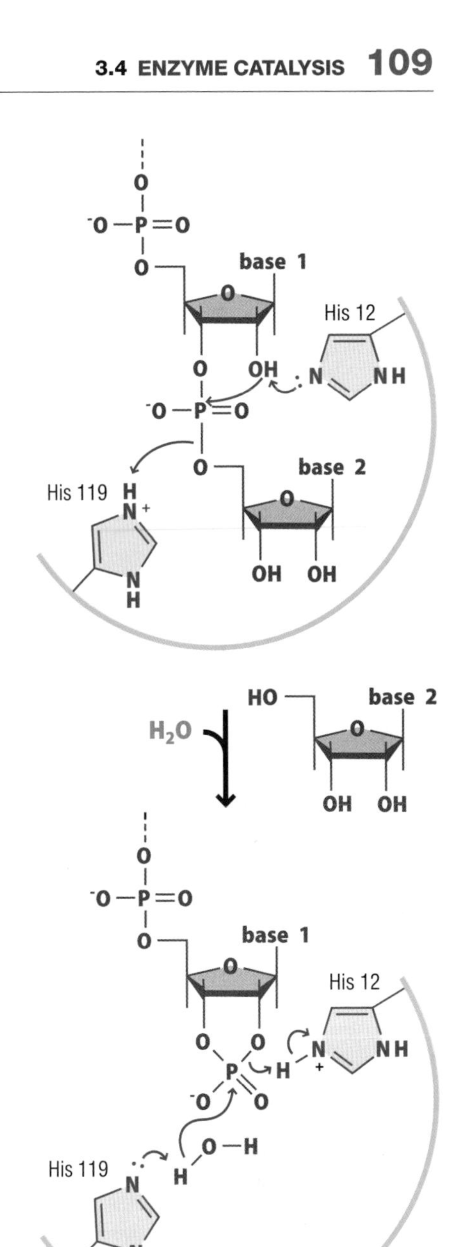

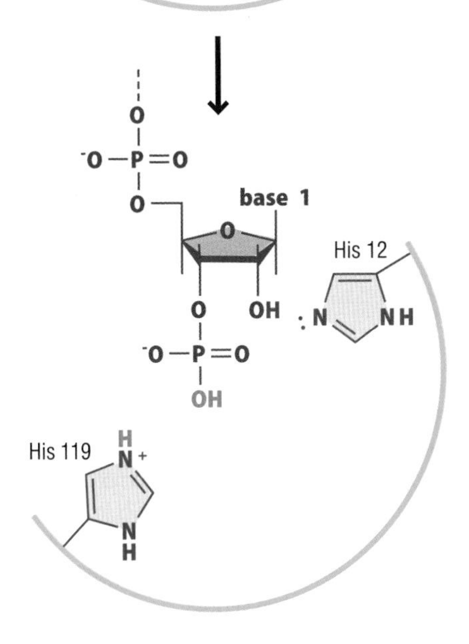

Figure 3.21 Mechanism of RNA cleavage by RNAse. His 12 acts as a general base to remove a proton from the 2′ OH to promote a nucleophilic attack by the oxygen on the adjacent phosphate group. His 119, acting as a general acid, donates a proton to the 5′ oxygen of the leaving group, leading to strand cleavage and formation of a 2′, 3′ cyclic intermediate. In the second step, a water molecule attacks the cyclic intermediate with His 12 and His 119 reversing their roles, with His 119 acting as a general base and His 12 as a general acid.

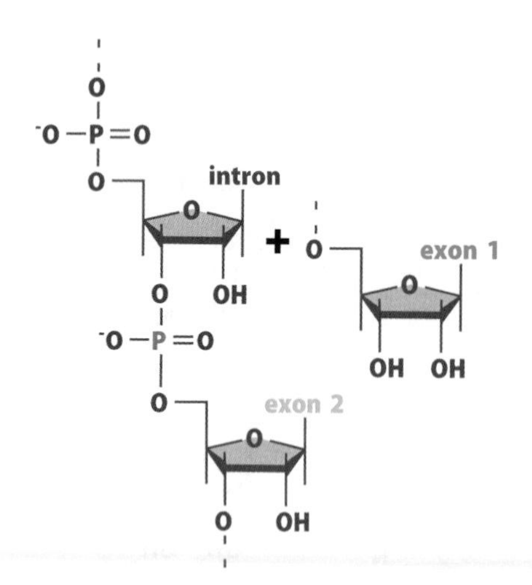

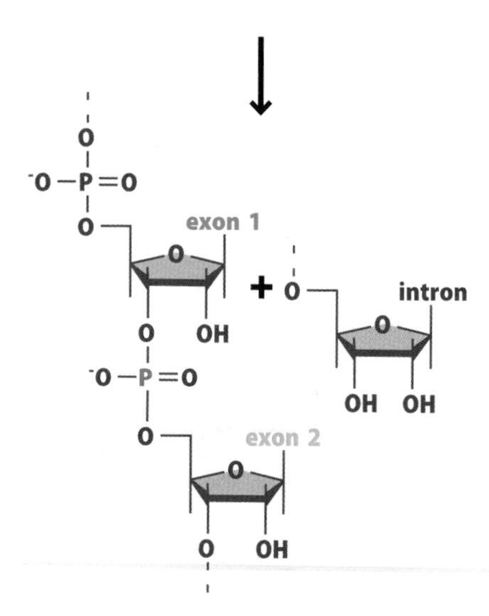

Figure 3.22 Transesterification. The schematic diagram shown here illustrates the final steps in exon ligation, in which one RNA fragment is cleaved and its 3′ end joined to the 5′ end of a new fragment.

Common Coenzymes

coenzyme	role	example
NAD^+/NADH	oxidation-reduction	dehydrogenases (NAD^+)
$NADP^+$/NADPH	oxidation-reduction	reductases (NADPH)
FAD/$FADH_2$	oxidation-reduction	dehydrogenases (FAD)
CoA	acyl-group activation and transfer	histone acetyltransferases
Biotin	CO_2 transfer	carboxylases

Figure 3.23 Common coenzymes.

and NADPH from binding to enzymes designed to function with NAD^+ and NADH. This provides the cell with two different pools of electron transfer molecules, and each can be maintained at different levels and can function in independent pathways. In our cells, these two pools of coenzymes, NAD^+/NADH and $NADP^+$/NADPH, are generally used for catabolic and anabolic reactions, respectively. Cells must maintain adequate and appropriately balanced pools of these and related cofactors in order to drive the many redox reactions needed for life.

Some enzymes require metal ions for catalysis

Many enzymes require a bound metal ion to catalyze chemical reactions. These metal ions are typically tightly bound to the enzyme and may even be covalently attached, such as the iron-carrying heme cofactor. Such enzymes are referred to as **metalloenzymes**. The most common metal ions used by enzymes are iron, copper, zinc and magnesium. Not coincidentally, these metals are among the most abundant in the Earth's crust. Like the coenzymes described above, metal ions can participate in electron transfer reactions through changes in the oxidation state of the metal. In addition, metal ions can assist reactions by providing positive charge density that stabilizes negative charge and, in doing so, can orient substrates and promote catalysis. Finally, metal ions can directly participate in catalysis by lowering the pK_a of bound water, thus providing a ready source of nucleophilic OH^- ions. We will see many examples throughout this book of both protein and RNA enzymes (ribozymes) that use metal ion cofactors in catalysis.

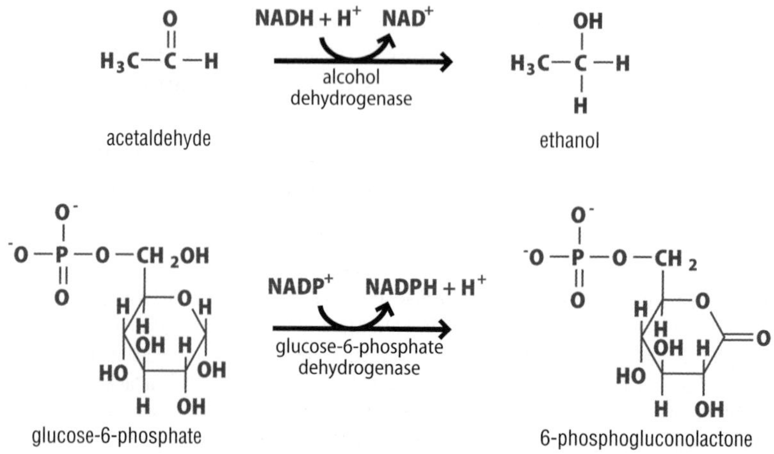

acetaldehyde — $NADH + H^+$ → NAD^+ (alcohol dehydrogenase) → ethanol

glucose-6-phosphate — $NADP^+$ → $NADPH + H^+$ (glucose-6-phosphate dehydrogenase) → 6-phosphogluconolactone

Figure 3.24 Oxidation–reduction of NAD and NADP. Acetaldehyde is reduced to ethanol by alcohol dehydrogenase with the concomitant oxidation of the reduced form of NADH to NAD^+. Conversely, glucose-6-phosphate is oxidized by glucose-6-phosphate dehydrogenase with the concomitant reduction of $NADP^+$.

Other enzymes bind metal ions more loosely; these metal ions are usually cations of the alkali or alkaline earth metals like sodium, potassium, magnesium or calcium. In these cases, the metal ions more typically play a structural rather than a catalytic role – they stabilize charges to promote proper polypeptide folding and, hence, proper function. Therefore, the corresponding proteins are not referred to as metalloenzymes.

3.5 ENZYME KINETICS

We see above how enzymes increase the rate at which a process occurs, most particularly by lowering the activation energy barrier. Understanding the rate at which a given process will happen is often of huge importance: if we administer a drug to a patient we need to know how long the drug will remain in the body. Will it remain long enough for it to elicit a therapeutic effect, or will it be metabolized too quickly? In this section, we learn more about how we can describe the rates of reactions as we explore the field of enzyme kinetics.

Enzyme activity can be described by Michaelis–Menten kinetics

The properties of enzymes are well suited to the cellular conditions in which they operate. Enzymes bind to their substrates and catalyze their reactions at rates that allow for the continued life of the cell. **Enzyme kinetics** is the study of how quickly enzymes carry out reactions. Understanding the kinetics of enzymatic reactions can shed a great deal of light on the molecular details of the catalytic mechanism and, potentially, on the conditions in the cell under which the enzyme has evolved to function.

The kinetic behavior of many enzymes can be described according to the Michaelis-Menten model, which can be used to derive useful parameters that describe the activity of an enzyme. Many enzymes carry out a relatively straightforward reaction that begins when the enzyme binds to a single substrate, S, forming an enzyme–substrate complex, ES. This first step is analogous to the simple binding reactions that we saw earlier in this chapter, and is therefore associated with an association (k_{on}) and dissociation (k_{off}) rate constant, as shown in Figure 3.25. Once the enzyme–substrate complex is formed, the enzyme catalyzes a reaction that converts the substrate into product. The rate constant describing this reaction is denoted k_{cat} and is also referred to as the turnover number of the enzyme. The overall rate at which substrate is converted into product is called the reaction velocity and is expressed in units of mol s^{-1}.

According to this model for enzyme behavior, when substrate concentration is sufficiently high that all enzyme is found in the ES form, then the reaction is rate limited by k_{cat}, the rate at which the enzyme-substrate complex is converted into enzyme and product, and is insensitive to higher substrate concentrations. We say that the enzyme is saturated.

Enzymes whose activity can be described in terms of these simple assumptions are generally characterized by the **Michaelis–Menten equation** (see Figure 3.25). The Michaelis–Menten equation can be used to describe how the rate of the reaction changes as the concentration of substrate is varied. When studying enzymatic reactions, it is most useful to look at the initial velocity at which the enzyme catalyzes a reaction, before a significant amount of product has accumulated. This is because the rates of some enzymatic reactions are altered – usually decreased – by

Many enzymatic reactions proceed according to the following simple scheme:

$$E + S \underset{k_{off}}{\overset{k_{on}}{\rightleftharpoons}} ES \xrightarrow{k_{cat}} E + P$$

The initial reaction rate, v_0, can be described by the Michaelis-Menten equation:

$$v_0 = V_{max} [S] / (K_M + [S])$$

where the maximum rate at which the reaction proceeds is given by:

$$V_{max} = k_{cat} [ES]$$

We define the constant K_M as:

$$K_M = \frac{k_{off} + k_{cat}}{k_{on}}$$

Figure 3.25 Michaelis–Menten kinetics. Enzymatic reactions that follow Michaelis–Menten kinetics are characterized by this simple equation, in which E stands for enzyme, S for substrate, ES the enzyme–substrate complex, and P the product. The Michaelis–Menten equation assumes that there is no conversion of product into substrate, which is true if initial reaction rates are measured before the product has accumulated to any significant degree.

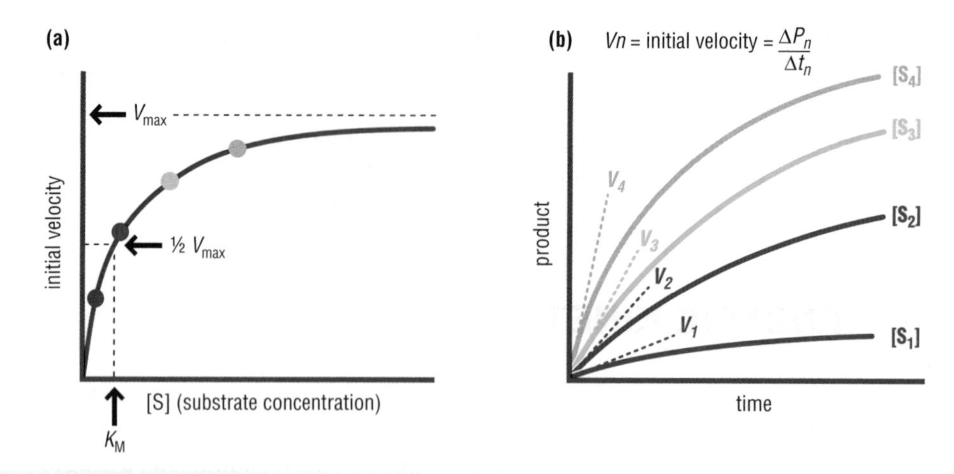

Figure 3.26 Initial velocity of a Michaelis–Menten reaction. (a) The plot shows how substrate concentration affects the initial velocity of a reaction that follows Michaelis–Menten kinetics. At low substrate concentrations, the velocity increases proportionally with substrate concentration. At higher substrate concentrations, the reaction velocity approaches a maximum velocity, V_{max}. The K_M is the substrate concentration at which the velocity is half that of V_{max}; it is a characteristic feature of each enzyme and it does not vary with the enzyme concentration. (b) The individual points in a Michaelis-Menten plot are derived from a set of curves as depicted here. Each line shows product formation as a function of time at a particular substrate concentration. The initial velocity is the slope of the line in the early part of the reaction. The colored data points in panel (a) are derived from the correspondingly colored curves in (b).

the presence of large amounts of product, so examining the initial velocity provides a measure of the intrinsic catalytic power of the enzyme.

In order to evaluate a specific reaction in terms of the Michaelis–Menten equation, a series of reactions are monitored over time with a range of different substrate concentrations. For each of these reactions, the amount of product generated at each time point is plotted and the initial velocity is determined at each concentration, as shown in Figure 3.26b. When only the early time points of the reaction are included in the analysis, before very much substrate is consumed, the assumption that the substrate concentration remains roughly the same over the period of measurement for each data point is valid. Since very little product will have been generated in the initial stages, we can neglect the rate of the back reaction, namely the conversion of product back into substrate.

The hyperbolic plot shown in Figure 3.26a shows how the initial reaction velocity increases as substrate concentration is increased according to the Michaelis–Menten equation. The rate at which the product is formed is equal to k_{cat} multiplied by the concentration of enzyme–substrate complex. If the substrate concentration is sufficiently high that all the enzyme has substrate bound to it, then the reaction rate is as high as it can be. This maximal velocity is called V_{max}; it is determined by k_{cat} multiplied by the total enzyme concentration.

The concentration of substrate at which the initial velocity is half that of V_{max} is called the Michaelis constant, K, and is defined as $(k_{off} + k_{cat})/k_{on}$. For enzymes where the k_{cat} is slow relative to k_{off}, then the K_M provides a fairly good measure of the equilibrium dissociation constant, $K_d = k_{off}/k_{on}$, for substrate binding to the enzyme.

More generally, the K_M provides a measure of how the reaction rate of an enzyme varies with substrate concentration. A low K_M means that the enzyme reaches its maximal reaction velocity at low substrate concentrations, whereas an enzyme with a higher K_M requires higher concentrations of substrate to approach its maximal reaction velocity. The K_M of a particular enzyme is generally close in value to the physiological concentration of its substrate. This ensures that the enzyme is catalyzing the reaction at a robust rate, while ensuring that a change in the concentration of substrate will result in a significant change in reaction velocity. When the substrate concentration is substantially above the K_M, the reaction velocity changes little in response to modest variations in substrate concentration (see Figure 3.26).

Another parameter that derives from the Michaelis–Menten model for enzyme behavior is k_{cat}/K_M, which is often referred to as the specificity factor. This ratio is related to how often the encounter of an enzyme and substrate results in a productive reaction (to generate product, P) and is therefore used to describe an enzyme's efficiency with a particular substrate.

Enzymes can be inhibited by small molecules

An inhibitor is a molecule that binds to an enzyme and interferes with the enzymatic reaction. Inhibitors can be small molecules or large proteins, and they can be found naturally in the cell where they regulate the activity of an enzyme or can be supplied exogenously. Many drugs are enzyme inhibitors. Drugs used to treat human immunodeficiency virus (HIV) infection, such as azathioprine (brand name Retrovir) and indinavir (brand name Crixivan), inhibit the enzymatic activity of the HIV reverse transcriptase and protease enzymes, respectively. They are both examples of reversible inhibitors, which can bind to the enzyme and then dissociate. Other less commonly seen inhibitors covalently react with the enzyme, causing permanent inactivation, and are commonly referred to as suicide inhibitors.

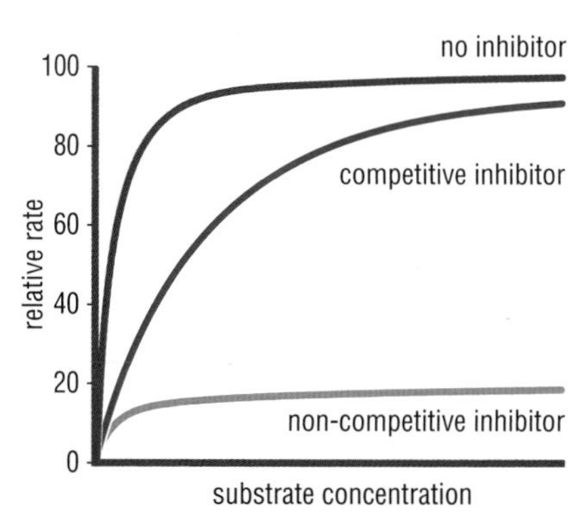

Figure 3.27 Enzyme inhibition. A competitive inhibitor (pink line) competes with the substrate for binding to the enzyme, increasing the apparent K_M. A non-competitive inhibitor (blue line) binds to the enzyme–substrate complex and prevents catalysis from occurring at its maximal rate, lowering the apparent V_{max}.

There are several ways in which a reversible inhibitor can interfere with an enzymatic reaction. A **competitive inhibitor** binds to an enzyme in a way that directly competes with substrate binding; it often closely resembles the substrate and binds within the substrate-binding site, but it does not engage in the chemical reaction catalyzed by the enzyme. The effects of a competitive inhibitor, shown in Figure 3.27, can be overcome by simply adding more substrate. A competitive inhibitor therefore does not affect the V_{max} of an enzyme. Sometimes the products of an enzymatic reaction can be competitive inhibitors, since they also bind to the substrate-binding site. This form of competitive inhibition is known as product inhibition, and can occur if the accumulating product reaches a concentration sufficient to favor significant binding to the enzyme.

A **non-competitive inhibitor** reduces the rate of an enzymatic reaction without interfering with the ability of the substrate to bind to the enzyme. These inhibitors typically bind in a location that is removed from the substrate-binding site. An example is the class of anti-HIV drugs known as non-nucleoside reverse transcriptase inhibitors, which bind to the HIV reverse transcriptase enzyme and inhibit synthesis of viral RNA without interfering with substrate binding. Non-competitive inhibitors lower the V_{max} of an enzyme, as depicted in Figure 3.27, and their effects cannot be overcome by increasing the concentration of substrate.

While there are other forms of enzyme inhibition, competitive and non-competitive inhibition are the two principal types that will be encountered in this book.

SUMMARY

The behavior of molecules in living cells is governed by the laws of chemistry and physics. The thermodynamic and kinetic principles underlying chemical reactions in the cell can be used to describe biological processes and predict how cells will respond to different conditions. We learned in this chapter how to apply these principles to understand and predict binding interactions and enzyme catalysis.

THERMODYNAMIC PRINCIPLES

- The enthalpy, *H*, is a measure of the internal energy of a system.
- The entropy, *S*, is a measure of disorder in the system.
- The energy that is available to do useful work is called the Gibbs free energy, *G*.

- Every process in biology – the binding of two or more molecules, a biochemical reaction – has associated with it a *change* in the Gibbs free energy, ΔG.
- The free energy change is related to enthalpy and entropy by the relation, $\Delta G = \Delta H - T\Delta S$.
- A process must have a negative ΔG in order to be spontaneous.
- An unfavorable reaction (with positive ΔG) can be made favorable by coupling it to a favorable reaction such that the overall free energy change of the coupled reactions is negative.
- Every process has associated with it a standard free energy change, ΔG^o, which is the free energy change occurring when the reactants and products are in their standard states.
- The overall free energy change for the reaction $X \rightleftharpoons Y$ is influenced by the concentrations of reactant and product and is given by:

$$\Delta G = \Delta G^o + RT\ln[Y]/[X], \text{ where } R = 8.31\ \text{J mol}^{-1}\text{K}^{-1}$$
$$\text{or } 1.987 \times 10^{-3}\ \text{kcal mol}^{-1}\text{deg}^{-1} \text{ and } T \text{ is the temperature in degrees Kelvin.}$$

- At equilibrium, the rate of the forward reaction equals the rate of the reverse reaction and the overall free energy change is equal to zero. At this point,

$$\Delta G^o = -RT\ln[Y]/[X].$$

BINDING INTERACTIONS

- For the binding reaction, $A + B \rightleftharpoons AB$, the ratio of complex to free molecules at equilibrium is given by:

$$\Delta G^o = -RT\ln[AB]/[A][B].$$

- The rate of complex formation is given by $k_{on}[A][B]$, where k_{on} is a rate constant.
- The rate of complex dissociation is given by $k_{off}[AB]$, where k_{off} is a rate constant.
- The time it takes for half of all complexes in a given population of identical complexes to come apart is called the half-life:

$$t_{1/2} = 0.693/k_{off}.$$

- The equilibrium constant for the binding reaction, $A + B \rightleftharpoons AB$, is the ratio of k_{on} to k_{off}, and is also related to the ratio of complex to free molecules as follows:

$$K_{eq} = k_{on}/k_{off} = [AB]/[A][B].$$

- This equilibrium constant for this reaction is also written K_a because is describes the association reaction (binding of A and B).
- The free energy change and the equilibrium constant for any reaction are related by the equation

$$\Delta G^o = -RT\ln K_{eq}.$$

- In biology we often refer to the equilibrium constant describing the dissociation reaction, $AB \rightleftharpoons A + B$. This equilibrium dissociation constant relates to the reverse of the association reaction, and is given by:

$$K_d = k_{off}/k_{on} = [A][B]/[AB] = 1/K_a.$$

- The equilibrium dissociation constant, K_d (with units of M), gives the concentration at which half of the molecules are bound in a complex while the other half are free.
- The binding curve depicts how much complex is formed at different ligand concentrations.
- Cooperative interactions can occur when two or more ligands bind to the same molecule, and the binding of one molecule makes the binding of the next molecule more (or less) favorable. The resulting binding curve is S-shaped, or sigmoidal.

ENZYME CATALYSIS

- The activation energy is an energy threshold that must be overcome before a reaction can proceed.
- Enzymes accelerate the rate of a chemical reaction without affecting the free energy change between reactants and products.
- Enzymes catalyze reactions by lowering the activation energy barrier.
- Enzymes use a variety of strategies to lower the activation energy barrier, including binding preferentially to the transition state of the chemical reaction and binding simultaneously to two substrates, orienting them favorably for reaction.
- Binding of substrate occurs at the active site, a pocket or region of the enzyme containing precisely positioned chemical groups that bring about catalysis by participating in the chemical reaction.
- Many enzymatic reactions obey Michaelis–Menten kinetics, in which the initial rate of product formation is given by:

$$v_o = V_{max}[S]/(K_M + [S]).$$

- The maximum reaction velocity is given by $V_{max} = k_{cat}[E]$, where k_{cat} is the rate at which the enzyme–substrate complex is converted into product and [E] is the enzyme concentration.
- Inhibitors are molecules that bind to enzymes and interfere with the chemical reactions they catalyze. Competitive inhibitors increase the apparent K_M of an enzyme, while non-competitive inhibitors reduce the V_{max}.

3.1 THERMODYNAMIC RULES IN BIOLOGICAL SYSTEMS

1. Where is the potential chemical energy of a molecule stored?

2. What is the standard state of a chemical in solution, and for a gas?

3. You dissolve a salt in water. How would you determine if the reaction is exothermic or endothermic?

3.2 BINDING EQUILIBRIA AND KINETICS

1. What is the equilibrium constant?

2. What limits the maximum rate at which two molecules in solution can bind? Does this also limit k_{off}?

3. At a concentration of ligand that is 10-fold higher than the K_d value, what fraction of the target molecule has bound ligand?

Challenge questions

4. A concentrated solution of the molecular complex AB is rapidly diluted to cause the complex to dissociate.
   a. What does the half-life of the reaction mean?
   b. If the half-life of the reaction is 5 s, calculate the k_{off} of the reaction.
   c. Calculate the time it takes for the concentration of AB to reduce to ¼ of its original concentration

5. In water, hydrofluoric acid (HF) dissociates to form H^+ and F^- ions. This reaction has a K_d value of 6.3×10^{-4} M. Calculate the Gibbs free energy change of the reaction under standard conditions.

6. An amount of AB is dissolved in 1 liter of water. AB has a dissociation constant (K_d) of 4 M. The reaction is left to reach equilibrium, at which point the concentration of A is 2 M.
   a. What is the concentration of AB at equilibrium?
   b. How much AB was dissolved?
   c. If an additional 2 M of B is added to the solution what will happen to the concentration of AB?

3.3 BINDING PROCESSES IN BIOLOGY

1. The binding of two molecules is typically associated with:
   a. An increase in energy and a lowering of enthalpy
   b. A release of heat energy and an increase in enthalpy
   c. A release of heat energy and a decrease in enthalpy
   d. An increase in energy and an increase in enthalpy

2. The K_d of the protein complex AB is 10^{-7} M, and the K_d for the protein complex AC is 10^{-3} M. What is the binding specificity of protein A for protein B relative to protein C?

3. What is negative cooperative binding?

4. How does positive cooperativity alter the shape of a binding curve?

Challenge question

5. Draw the binding curve for the reaction AB $\leftrightarrow$ A + B. Identify the value of K_d on the graph.

3.4 ENZYME CATALYSIS

1. Even though some reactions have a favorable change in Gibbs free energy they may still proceed slowly. Why is this?

2. Which of these is not a property of enzymes?
   a. Enzymes increase the rate of reactions by reducing their activation energy.
   b. Enzymes are very specific to a certain type of reaction.
   c. Enzymes are permanently modified during the reactions they catalyze.
   d. Enzymes can bind one or more substrate molecules.

3. What is a nucleophile, and how does it function in a chemical reaction? How does an enzyme typically initiate this type of reaction? Give an example of this type of reaction.

4. RNAse A catalyzes the cleavage of RNA molecules by breaking the phosphate backbone, but how does the histidine side chain in the active site function in the earliest stage of the reaction?
   a. It acts as a nucleophile.
   b. It acts as an electrophile.
   c. It acts as a general acid.
   d. It acts as a general base.

5. Transesterification results in a product that has
   a. A higher-energy state than the starting substrates.
   b. A lower-energy state than the starting substrate.
   c. The same energy state as the starting substrate.

6. Many proteins contain metal ions. What roles do they perform?

7. Which of the following statements is true?
   a. Enzymes bind the substrate more tightly than the transition state.
   b. Enzymes bind the transition state more tightly than the products but less tightly than the substrate.
   c. Enzymes bind the transition state more tightly than the substrate but less tightly than the products.
   d. Enzymes bind the transition state most tightly.

Challenge questions

8. A reaction occurring under standard conditions has a change in enthalpy of -100 kJ mol^{-1} and a change in entropy of -200 J mol^{-1} K^{-1}. Calculate the Gibbs free energy. Will the reaction occur spontaneously?

9. Enzymes can lower the activation energy of a reaction in a number of ways. Describe the different mechanisms.

3.5 ENZYME KINETICS

1. What three rate constants are used to describe simple enzyme kinetics, and to what do they refer?

2. Under what conditions is the rate of an enzymatic reaction limited by k_{cat}?

3. Enzyme A has a K_M value of 10 nM and Enzyme B has a K_m value of 1000 nM for the same substrate. Which enzyme functions most rapidly (assuming their k_{cat}s to be equivalent) at low substrate concentration?

4. How does the K_M of an enzyme relate to the rate of the reaction?

5. Reaction rates can be estimated by measuring the change in absorptivity of a substrate or reaction product over time, as described in Chapter 19. NAD+ is a cofactor in many enzymatic reactions, in which is it reduced to NADH. NADH absorbs substantially at 340 nm, whereas NAD+ absorption at this wavelength is minimal. Thus, the change in absorption at 340 nm is often used to measure enzymatic reactions for which NAD+ is a cofactor and in which NADH is produced.
   a. You measure a reaction over time and obtain the following data.

time (seconds)	absorbance at 340 nm
0	0.15
5	0.35
10	0.49
15	0.65
20	0.87
25	1.00
30	1.23

Plot a graph, determine the equation of the line, and use this to estimate the reaction rate in terms of the change in absorbance per second.

b. As described in Chapter 19, Beer's Law can be used to determine the concentration of the measured product based on the absorbance and the known molar absorptivity of the substance being measured, using the equation $A = \varepsilon cl$, where A = absorbance, ε = molar absorptivity (M/cm), c = concentration (M), and l = path length (cm). The ε of NADH at 340 nm is 6220 M/cm.

What is the reaction rate from part (a) expressed in NADH M/s produced? Assume a path length of 1 cm.

c. Reaction rate is influenced by initial substrate concentration. The substrate concentration you used for the measurements in part (a) was 17 μM. You also then perform the experiment using substrate concentrations of 10 μM and 24 μM and obtain the following line equations from plotting time(s) against absorbance. Calculate the reaction rates in NADH M/s for these substrate concentrations.

10 μM	$y = 0.0216x + 0.1493$
24 μM	$y = 0.0481x + 0.1521$

6. Draw a graph showing the influence of substrate concentration on enzyme reaction rate for a reaction mixture containing no inhibitor, a competitive inhibitor, and a non-competitive inhibitor.

EXPERIMENTAL APPROACH 3.1 – INSIGHTS FROM KINETICS INTO HOW DNA BINDING PROTEINS FIND THEIR TARGET SITES

Challenge question

1. Proteins that bind DNA often have a higher association rate constant than would be expected for randomly diffusing molecules. How do they achieve this higher rate?

GENERAL

Berg JM, Tymoczko JL, Stryer L. *Biochemistry*, 7th ed. New York, NY: WH Freeman, 2012.

Voet DJ, Voet JG. *Biochemistry*, 4th ed. Hoboken, NJ: Wiley-Blackwell, 2011.

Chromosome structure and function

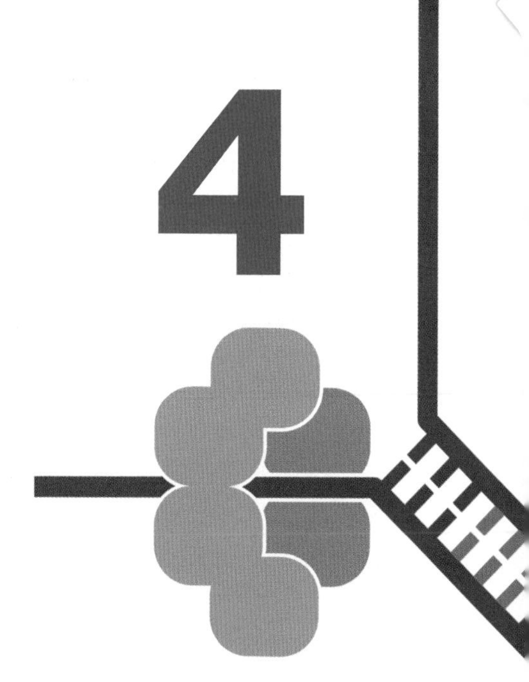

INTRODUCTION

Chromosomes are the vehicles for the transmission of genes

The genetic information for each organism comprises a set of instructions that enables them to live, grow, reproduce, and pass the information along to their progeny. These instructions are in the form of genes, which are organized into very long DNA molecules. Each individual DNA molecule is packaged into what we call a **chromosome**, a compacted form of DNA that acts as the vehicle that ensures stable maintenance of the genetic information and facilitates its transmission to the next generation. Chromosomal DNA is associated with many proteins that help to package the DNA inside the cell; the proteins also help the stored genetic information to be read and used to synthesize the vast array of molecules that are necessary for life.

The central subject of this book is how genetic information is read, transmitted and preserved. In this chapter, we describe how the vehicle for that information – the chromosome – is organized, and give an overview of the changes that take place in chromosomes as cells grow, divide, and respond to changing conditions. We will learn about special features of the DNA and proteins in chromosomes that make it possible for an organism to make faithful duplicates of each chromosome as cells grow and divide. Since key elements of the chromosome are important when cells divide and DNA is replicated, we will briefly review the basic stages of the **cell cycle**, which are the series of events that lead to the replication and segregation of chromosomal DNA during cell division. We will learn about the packaging of eukaryotic DNA into a protein–DNA complex known as **chromatin**, and how modifications of these proteins and of the DNA molecule can give rise to changes in chromatin structure. Then, we will learn about special sequence elements within the DNA that are required for key processes in the cell cycle, and for preserving the integrity of chromosomes. Finally, we will see how chromosomes are organized within the eukaryotic nucleus, and how and why chromosomes change their organization during the cell cycle.

Variations in chromosome structure and chromatin modifications impact all processes that involve chromosomal DNA, including DNA replication, transcription, and DNA repair – all topics that we describe in later chapters. Therefore, while this chapter provides a general introduction to chromosome structure and function, it is not the complete story of how chromosomes influence the molecular processes that determine genome function. Instead, we will gradually build this complete picture during later chapters, when we revisit many of the themes and topics of this chapter in more detail.

➔ Topics introduced in this chapter feature particularly strongly in Chapters 5, 6, 7, and 8.

4.1 ORGANIZATION OF CHROMOSOMES

Chromosomes contain genetic information in double-stranded DNA molecules

A cell's genome – the total DNA content of a cell – is divided among one or more chromosomes that organize, store, and transmit information for producing protein and RNA molecules. Each chromosome contains one double-stranded DNA molecule, along with proteins that help to condense and organize the DNA within the cell. The DNA molecule can be linear or circular: all eukaryotes have linear chromosomes, whereas most bacteria and archaea have circular chromosomes.

A single chromosome contains many genes embedded within a single DNA molecule. Each gene consists of a linear string of DNA bases that contain the information for producing the RNA and protein molecules needed by the cell. Interspersed between the genes are regions of DNA known as **intergenic DNA** that do not, to our knowledge, contain genes. Although once considered by some scientists to be 'junk' DNA with no function, ongoing studies have revealed many regulatory functions for these stretches of DNA that lack apparent protein-coding sequences. We will return to this intriguing issue later in this book.

There is great variation in the relative amount of gene versus intergenic DNA in chromosomes from different organisms. The genes of less complex, single-celled organisms, such as bacteria or budding yeast, are located fairly close to one another in the chromosome, with little intergenic DNA. By contrast, more complex organisms have a lower density of genes within their chromosomes and more intergenic DNA. Human chromosome 18, for example, contains one gene for each million base pairs, whereas a budding yeast chromosome contains 500 genes per million base pairs. Even within species, the gene density on different chromosomes is not the same: thus, human chromosome 1, which is the largest chromosome, has twice the gene density of human chromosome 18.

Each species has a characteristic set of chromosomes

Every organism maintains a fixed number of chromosomes in each cell. The number of chromosomes per cell, along with the size of each chromosome, is a defining characteristic of each species. Most bacteria and archaea have a single chromosome in each cell, while eukaryotic cells have multiple chromosomes within a membrane-bound compartment known as the nucleus.

The number of chromosomes per cell varies greatly among different species. Human cells contain 46 chromosomes in the nucleus, half of which are inherited from the mother and half from the father. The fission yeast, *Schizosaccharomyces pombe*, has six chromosomes, while the kingfisher, a bird, has 132. Archaea usually have a single, circular chromosome, as do most bacteria. There are, however, a few examples of bacteria with linear chromosomes or more than one circular chromosome.

Another defining characteristic of each type of cell is the number of sets of identical chromosomes it contains, a characteristic described by the term **ploidy**. Human cells, like those of most eukaryotes, are **diploid**, which means that they contain two copies of each chromosome. (In the case of humans, there are two copies each of chromosomes 1 to 22, and either two X chromosomes, or one X and one Y.) The egg and sperm cells that are involved in sexual reproduction are **haploid**, meaning that they contain a single set of chromosomes. When an egg is fertilized by a sperm, the two sets of genetic material come together in one nucleus to produce a diploid, fertilized egg cell that goes on to develop into a diploid individual. The copy of each chromosome inherited from the mother is **homologous** to the copy inherited from the father, meaning that it contains the same genes and gene order, but may contain different versions of each gene.

Exceptions to this simple alternating pattern between haploid and diploid do occur. **Polyploid** cells, which contain more than two sets of chromosomes, are found in some organisms during certain stages of their lifecycles. **Polyploidy** is particularly common among cultivated plants such as wheat, coffee, and oats, and also occurs in wild maize. However, it is not confined to plants, and is also observed in a small number of specific human cell types. Two examples are hepatocytes, which are cells in the liver that produce proteins, lipids, and bile salts, and bone marrow megakaryocytes, which produce blood platelets.

Other classes of DNA can be found within cells

There is additional DNA found within cells besides the chromosomal DNA contained in the nucleus. Mitochondria, the membrane-enclosed organelles in eukaryotic cells that are the primary source of cellular adenosine triphosphate (ATP), contain a single chromosome that is usually circular. The presence of distinct mitochondrial DNA is one of the reasons that mitochondria are thought to have arisen from independently living bacteria that were ingested by another single-celled organism, within which the bacteria developed a symbiotic relationship. In this scenario, the single-celled organism and ingested bacteria would have had separate, unique genomes, a difference still reflected by mitochondria and the cells they occupy to this day. Chloroplasts, the plant organelles where photosynthesis occurs, also have their own circular chromosome and are thought to have originated from blue-green algae, incorporated into another cell by a similar but independent event during evolution.

In addition to their main genomic DNA, bacterial cells can contain double-stranded, circular DNA molecules known as **plasmids**, which contain a small number of genes. Plasmids are much smaller than bacterial chromosomes and typically encode a few proteins that confer on the bacterium a trait that provides a selective advantage, such as drug resistance. Since plasmids can be readily introduced into some types of bacteria, they have been enormously useful tools in the laboratory for genetic engineering.

➔ We explore how plasmids are used to introduce novel or altered genes into bacteria in Section 19.1.

Although this chapter focuses on the chromosomes in cells, it is useful to contrast cellular genomes with the genetic information found in viruses. Viruses are infectious agents rather than free-living organisms, and cannot propagate on their own. Their genetic information is stored, replicated, and delivered to progeny in ways that sometimes differs from the mechanisms used by cells. Viruses carry genetic information in the form of either DNA or RNA, and their chromosomes may be either single- or double-stranded. Like cellular chromosomes, viral DNA can be either circular or linear. Viruses propagate by taking over a host cell and using its machinery to produce many new viral genomes and viral proteins, which assemble into active viral particles that can be dispersed. Studies of viruses and the way they harness the host cell machinery have provided key insights, not just into disease, but also into the basic workings of the cell.

4.2 THE CELL CYCLE AND CHROMOSOME DYNAMICS

Chromosomes are duplicated and segregated to daughter cells during the cell cycle

The faithful transmission of genetic information from one generation to the next depends on a cell's ability to make copies of each chromosome and then distribute the complete set of chromosomes to the two daughter cells. The incorrect segregation of chromosomes - for example, where one daughter cell receives an extra chromosome, and the other cell receives none - may have a serious, negative impact on the viability of either or both daughter cells. Chromosomes therefore have many features that ensure the proper copying and distribution of chromosomes during the many rounds of cell division that occur during the lifetime of a single individual. To see how these features function and when they are needed, we will first review the key events that occur during the process of cell division.

Before a cell can divide, it must replicate its DNA and segregate the duplicated chromosomes equally between what will become the two daughter cells. The highly coordinated sequence of events that occurs during this division process is called the cell-division cycle, or **cell cycle**. We focus here on somatic cells, which are the cells that form the body of an organism; these are distinct from germline cells that take part in sexual reproduction, such as those cells that produce the haploid sperm and eggs, and which exhibit a modified form of chromosome duplication and segregation compared to somatic cells. The cell cycle and its regulation will be discussed in detail in Chapter 5 and chromosome segregation will be discussed in Chapter 7. Here we outline the stages of the cell cycle and describe the changes in the chromosomes - the **chromosome cycle** - that occur at each stage.

The eukaryotic cell cycle is divided into four distinct phases

The cell-division cycle of a somatic cell has four distinct phases that happen in sequence: **G1**, **S**, **G2**, and **M**. These four phases are illustrated in Figure 4.1. In the first phase, **G1** (for gap phase 1), cell growth occurs until cells attain a minimum size that is required to progress to the next phase. During this next phase, **S phase** (for DNA synthesis), the DNA is copied, thereby duplicating all of the chromosomes. The two identical copies of each chromosome, called **sister chromatids**, remain physically associated with one another. S phase is followed by **G2** (for gap phase 2), during which the cell prepares for **mitosis**.

In the final cell-cycle stage the **mitotic (M) phase**, sister chromatids are separated and a complete set of chromosomes is delivered to each pole of the cell. This process is known as chromosome segregation.

G1, S, and G2 are often collectively referred to as **interphase**, since cells in these stages are not easily distinguished in the microscope. By contrast, cells undergoing mitosis can be readily distinguished due to the changes in chromosome and nuclear structures that are clearly evident in the microscope. Mitosis is followed by **cytokinesis**, which completes the division of the mother cell into two daughter cells, each containing the same number of chromosomes.

During mitosis, the cell must separate all sister chromatid pairs to form two identical chromosome sets for distribution to daughter cells. This process occurs in the series of steps illustrated in Figure 4.1. Here we will give a brief overview of the steps.

➔ The steps in mitosis are described in greater detail in Chapters 5 and 7.

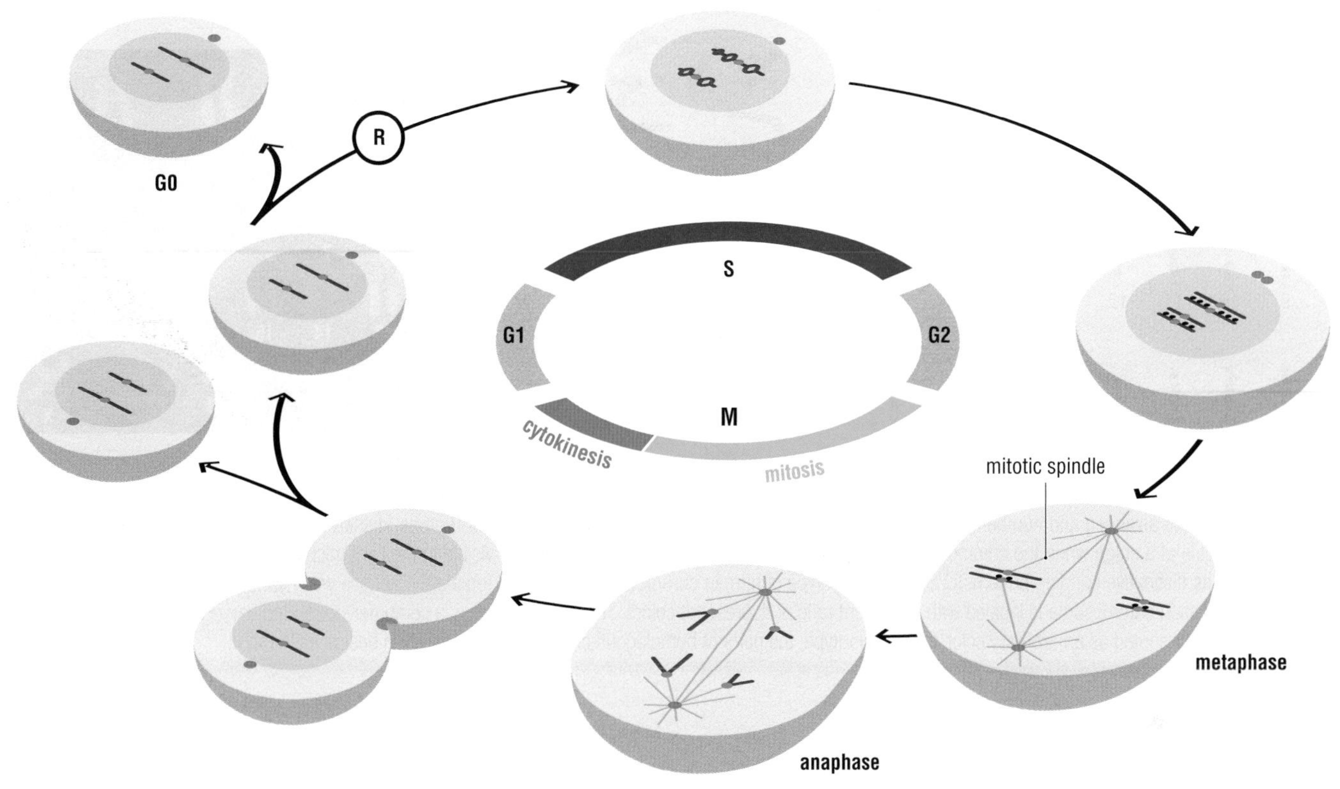

Figure 4.1 The chromosome cycle describes the behavior of the chromsomes during the cell cycle. The cell cycle is divided into four phases: G1 for gap phase 1, S for DNA synthesis, G2 for gap phase 2, and M for mitosis. During S phase, the chromosomes (red lines) are duplicated, yielding two sister chromatids per chromosome. At the end of G2, the chromosomes condense to prepare for mitosis. During mitosis, the chromosomes are highly condensed to facilitate their distribution to daughter cells. Microtubules (green lines) emanate from the centrosomes (blue circles), bind to the centromere (orange circles) found on each chromosomes, and the sister chromatids are pulled to opposite poles. The cells then divide, producing two daughter cells, each with an identical set of chromosomes. Chromosomes condense during M phase and later decondense during G1 to allow proteins access to the DNA for transcription and repair.

Before M phase, sister chromatids are held together by protein complexes that bind along the length of the chromosome. As the cell enters M phase, the nuclear membrane is disassembled, and sister chromatids condense into compact structures that remain bound together. Human chromosomes become condensed by a factor of 10, which allows them to segregate without becoming entangled. The **centromeres** are specialized regions that bind microtubules which emanate from centrosomes. The centrosome is a structure that defines each pole of the mitotic spindle. The spindle organizes the sister chromatid pairs at the center of the cell; this portion of M phase is known as metaphase. The ties that bind the sister chromatids together are then removed, and each sister chromatid moves to its pole powered by the spindle. This movement ensures that each pole receives one full set of chromosomes. The sets of chromosomes at each pole become encapsulated by a nuclear membrane, and the cell divides to yield two daughter cells.

Chromosomes have a distinctive appearance at mitosis

Condensed mitotic chromosomes have been studied under the microscope for many years. The most familiar picture of human chromosomes comes from a **karyotype**, a display of the chromosome set of an individual, lined up from the largest to smallest, like a collection of butterflies; an example of a karyotype is

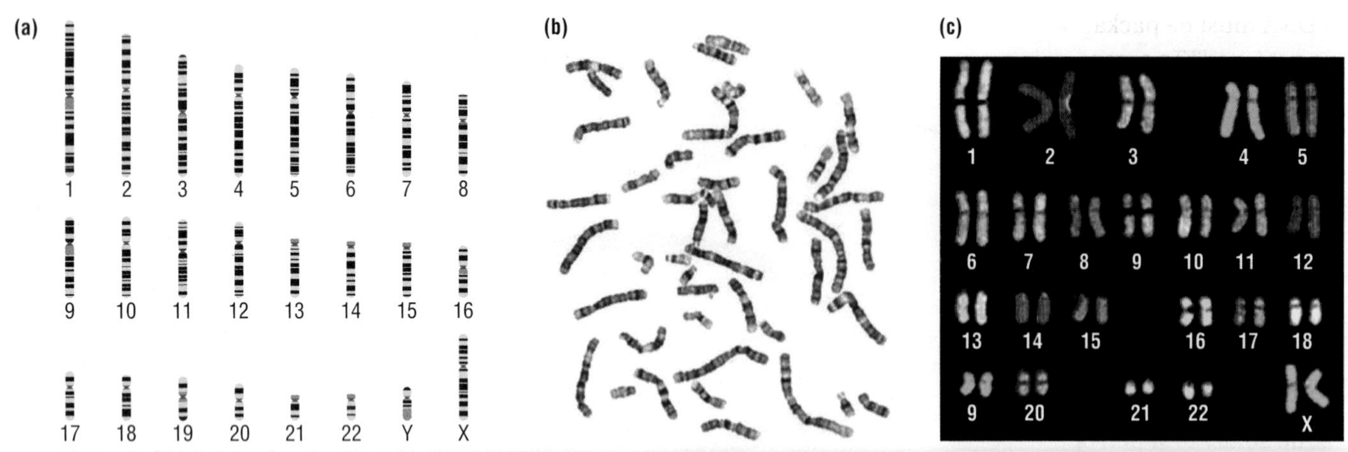

Figure 4.2 Karyotype of human chromosomes. (a) Diagram of a human male karyotype. The banding pattern arises when the highly condensed chromosomes are stained with Giemsa dye. Each chromosome has been duplicated and is still attached at the centromere. As dictated by karyotype convention, the chromosomes are arranged and numbered from largest to smallest. (b) Photomicrograph of a spread of Giemsa-stained chromosomes from a diploid cell from a human male at a stage when the chromosomes have become sufficiently condensed to become visible in the light microscope. Each chromosome can be identified by its distinctive banding pattern. (c) Spectral karyotype of human chromosomes. A normal female metaphase was hybridized with specific DNA probes such that each chromosome is labeled with a different ratio of fluorescent colors. The image shown is a computer-generated karyotype in which each chromosome is represented as an arbitrary color. In this karyotype, the pairs of homologous chromosomes are shown beside each other.

Courtesy: National Human Genome Research Institute.

shown in Figure 4.2a. The chromosomes typically seen in a karyotype are taken from microscope images of metaphase chromosomes, when the sister chromatids are maximally condensed but have not yet separated (see Figure 4.2b). We will learn in the next section how the binding of special proteins helps to package and condense DNA. When these proteins are removed from mitotic chromosomes, the DNA appears as large loops that are apparently anchored near the midline of the chromosome as illustrated in Figure 4.3. Organisms with smaller genomes do not have such a high level of condensation. *Saccharomyces cerevisiae*, for example, performs only a twofold compaction of the DNA, and its metaphase chromosomes cannot be seen under the microscope.

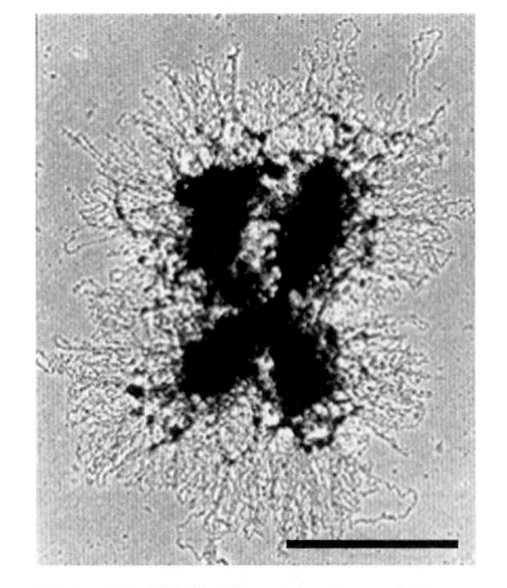

Figure 4.3 Mitotic chromosomes contain large looped domains of DNA. A mitotic chromosome has been treated to release proteins. Large loops of DNA released from the condensed chromosome can be seen as a halo extending outward.

4.3 PACKAGING CHROMOSOMAL DNA

We noted at the start of this chapter that a chromosome is composed of DNA packaged with specific proteins, which help to condense the DNA into a relatively tiny volume of space. This packaging also helps to protect the genome from mutation and rearrangements. The DNA–protein complex that comprises chromosomes is called chromatin. In this section, we explore how chromatin is formed, and learn about the distinct three-dimensional structures it adopts in order to organize the cell's genome within the nucleus and regulate the access of other proteins to the DNA.

DNA condensation occurs in multiple steps

The length of DNA packed into a cell nucleus is astounding. In a single human cell, over 2 m of DNA must fit inside a nucleus that is 10^{-5} m (10 μm) in diameter. This is like taking 30 miles of string, cutting it into 46 pieces, and then packing the pieces neatly into your backpack. Bacteria have a similar problem: 1500 μm of DNA must be packaged into a cell that is only 2 μm long. How is this packaging achieved? To organize DNA into such a small space, cells use multiple levels of packing.

The first level of DNA packaging in all organisms is the binding of small, basic (positively charged) proteins directly to the DNA along its length. The many positively charged side chains on these proteins help to counterbalance the negatively charged DNA. In bacteria, the binding of small basic proteins bends the DNA as depicted in Figure 4.4, which helps pack it into a more compact structure. The compacted bacterial chromosome is called the **nucleoid**. The DNA within this DNA–protein complex becomes supercoiled as a result of its packaging; that is, it is wrapped around itself like a twisted rubber band, and thus is much more compact than non-supercoiled DNA.

➔ We describe supercoiling in more detail in Section 2.5.

The ability to supercoil DNA is essential for bacterial viability, and supercoiled DNA plays a part in many chromosomal functions in eukaryotes as well. Many of the enzymes that act on DNA, such as recombinases, only act on supercoiled DNA, as we shall see in later chapters.

Figure 4.4 The bacterial DNA-binding protein IHF bends DNA. The mainly alpha-helical IHF protein binds to DNA as a dimer. The two monomers are shown as red and blue ribbons, respectively. IHF binding bends the DNA, which facilitates its packaging and helps to recruit other proteins and enzymes that act on the bent DNA. Protein Data Bank (PDB) code 1IHF.

Eukaryotic DNA and histones are packaged into nucleosomes

In eukaryotes, the basic DNA-binding proteins that package DNA into chromatin are called **histones**. There are four core **histone proteins**, H2A, H2B, H3, and H4, which evolved early in the history of eukaryotes. Because core histones are essential and are required to interact with many chromosomal proteins, they represent some of the most highly conserved proteins in eukaryotes. All four of the core histones have a common protein fold known as the histone fold, and are rich in lysine and arginine residues. The side chains of the amino acids lysine and arginine carry positive charges, which help to counterbalance the many negative charges on phosphate groups in the DNA backbone. This electrostatic interaction - between positive and negative charges - helps to stabilize the overall histone–DNA assembly.

DNA wraps around a complex of histone proteins to form a **nucleosome**, which is the fundamental unit of eukaryotic chromatin. The structure and composition of the nucleosome is illustrated in Figure 4.5. At the center of the nucleosome is a disk-shaped protein complex, the **histone octamer**, which contains two copies of each of the four core histone proteins. Two monomers each of histones H3 and H4 come together to form a tetramer, while histones H2A and H2B form heterodimers. The full histone octamer is assembled on the DNA in an active process that is facilitated by a number of different proteins.

The H3·H4 tetramer is assembled on the DNA followed by the association of two H2A·H2B heterodimers to give the complete histone octamer. Then, DNA is wrapped around the histone octamer to form a nucleosome (shown in Figure 4.5). About 146 base pairs of DNA wrap around the histone octamer; this amounts to somewhat less than two complete windings of the DNA around the histone

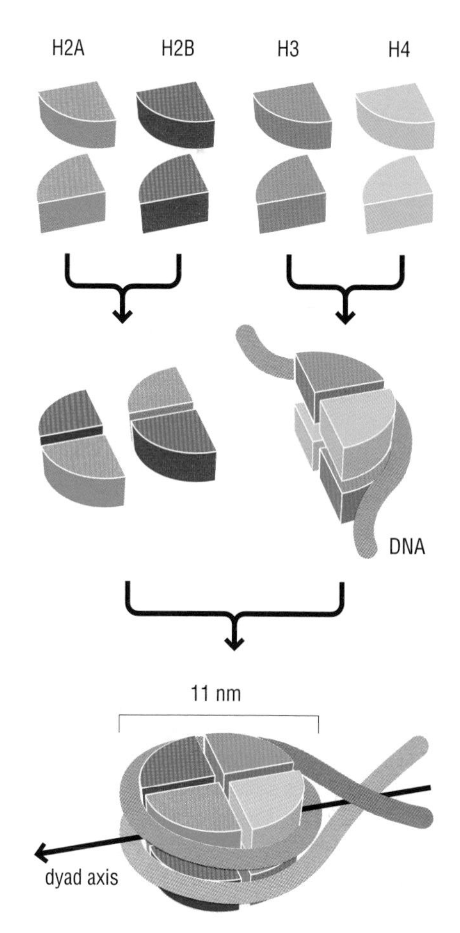

Figure 4.5 Histones assemble on DNA to form nucleosomes. Assembly of the histone octamer. Histones H3 and H4 assemble to form a heterotetramer containing two copies of H3 and two copies of H4, while histones H2A and H2B form heterodimers. The H3•H4 heterotetramer binds to DNA first (possibly assembling on the DNA from two H3•H4 dimers), followed by two H2A•H2B heterodimers, forming a nucleosome. DNA is wrapped around the histone octamer in a left-handed manner to form a nucleosome.

The role of negative supercoiling in favoring strand separation is discussed in Section 2.5.

octamer (1.76 turns, to be precise). This number has been deduced from two different kinds of experiments: the three-dimensional atomic structure of the nucleosome core particle (shown in Figure 4.6), and from nuclease digestion experiments, which are described in Experimental approach 4.1. The DNA winds around the histone octamer in a left-handed manner as depicted in Figures 4.5 and 4.6. The direction of this winding is important: when a histone octamer is stripped away from the DNA, it leaves behind negatively supercoiled DNA. But why is this important? Negative supercoiling makes it easier for double-stranded DNA to be drawn out into separate strands (a requirement for DNA replication and transcription). By contrast, positive supercoiling makes it *more* difficult to separate the two strands.

Each core histone protein has an N-terminal 'tail' that extends away from the nucleosome core between the coils of DNA (see Figure 4.6). These tails are stretches of up to 25 amino acids that do not adopt a defined structure. The histone tails interact with adjacent nucleosomes to condense the chromatin structure further, although their precise role in chromatin compaction remains an area of active study. In addition, as we will see later in the chapter, the histone tails are subject to covalent modifications that alter chromatin structure and function.

Although the four core histones – H2A, H2B, H3, and H4 – are the most common histones in nucleosomes, there are other histone proteins in the cell that can be incorporated into a histone octamer instead. These **histone variants**, illustrated in Figure 4.7, are highly similar to core histones in amino acid sequence, but are encoded by distinct genes. Histone variant proteins are found in place of the common core histones at different locations in chromatin, where they play special functions. For example, CENP-A (centromere protein-A) is a variant of histone H3 that is incorporated into nucleosomes at centromeres and is essential for their formation and function. H2A.Z, a variant of H2A, is more commonly found upstream of genes and has been implicated in transcriptional regulation. Another H2A variant, H2A.X, is phosphorylated at sites of DNA double-strand breaks and is thought to recruit DNA repair machinery. We will learn later in this book about the role that these and other histone variants play.

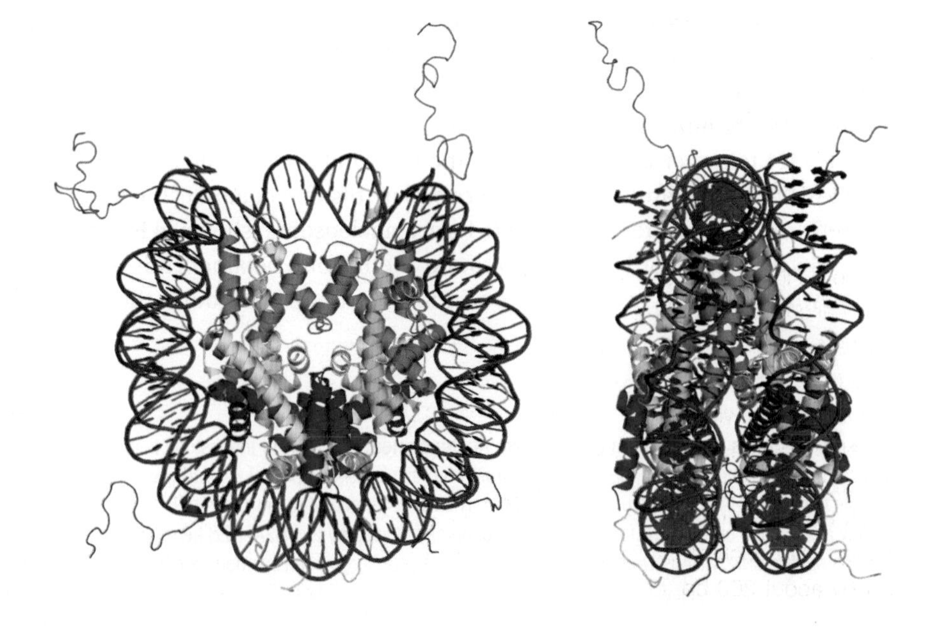

Figure 4.6 Crystal structure of the nucleosome core. The histone proteins are shown in colors corresponding to those in Figure 4.5. The DNA wrapped around the histone octamer is shown in gray. The left view is of the top of the disk-like octamer; the right view is from the side. The N terminal tails of the histones project outward from the nucleosome core (see text). PDB code 1KX5.

4.1 EXPERIMENTAL APPROACH
Nuclease probes of chromatin organization

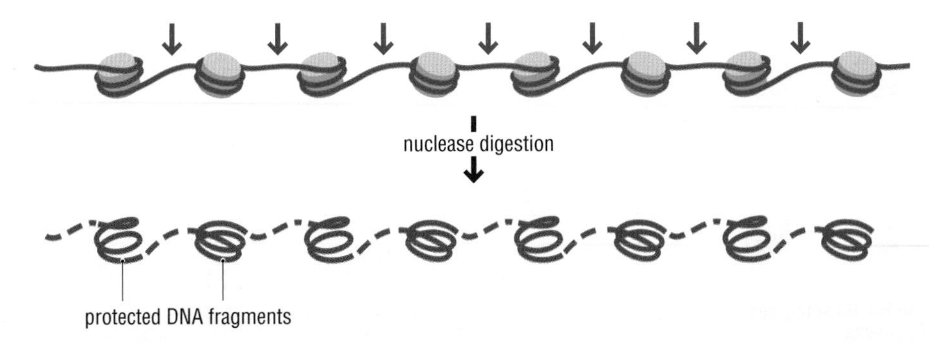

Figure 1 Protection of nucleosomal DNA from micrococcal nuclease digestion. The DNA that is not in contact with protein is digested into nucleotides, while the nucleosomal DNA is protected from digestion.

Nuclease digestion to examine chromatin structure

In the early 1970s, it was known that chromatin consisted of DNA packaged in some way with histone proteins, but the nature of the packaging was not at all clear. In 1974, Roger Kornberg proposed that chromatin consisted of a regular 'repeating unit of two each of the four main types of histone and about 200 base pairs of DNA'. Empirical evidence for this model of the histone octamer and its role in packaging DNA was reported in the same year by Markus Noll, in the form of nuclease digestion experiments.

The Noll group used nucleases to probe the way eukaryotic DNA is packaged into chromatin. Micrococcal nuclease cleaves accessible double-stranded DNA, and thus can differentiate DNA wrapped around a histone octamer from 'naked' unpackaged DNA. After DNA cleavage the protected regions are ultimately visualized by stripping the histone proteins away from the DNA and analyzing the resulting DNA fragments by gel electrophoresis (see cartoon in Figure 1).

The nucleosome protects about 200 base pairs of DNA from digestion

We see in Figure 2 the banding pattern of DNA that results from a nuclease treatment of rat chromosomes. The pattern of DNA fragments is consistent with the model of a regular repeating unit as proposed by Kornberg. Chromatin that was incubated for a short time with micrococcal nuclease yielded a variety of discrete fragments of DNA, corresponding to the protection by one, two, three, or more nucleosomes. The smallest DNA fragments were about 170 bp and 205 bp in length (two discrete bands), while each of the longer fragments was larger than the next by about 200 bp, consistent with the proposed model. It turns out that the 205 bp fragment is the first one typically generated by the nuclease, when it nicks a site more or less halfway between two sequential octamers. Further digestion by the nuclease nibbles closer to the octamer, yielding the smaller 170 bp fragment.

Histone precipitation allows genome-wide mapping of nucleosome positions

Nuclease digestion of DNA can now be used in conjunction with modern DNA sequencing methods to determine the location of individual nucleosomes on each chromosome. This was done by Mavrich and colleagues in their 2008 study of nucleosome positions in the yeast genome. In this approach, yeast chromosomes were first treated with a chemical cross-linking agent that causes covalent bonds to form between the histone proteins

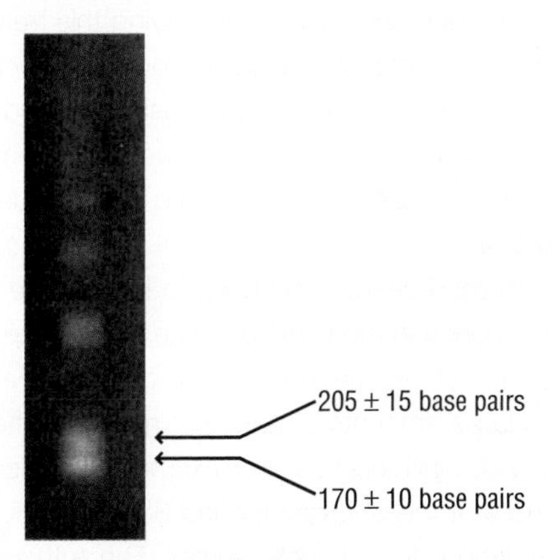

Figure 2 Nuclease digestion pattern of chromatin. The electrophoretic pattern shows the fragments of DNA that were produced from limited digestion of rat liver chromosomes with micrococcal nuclease.
Reproduced from Noll. Subunit structure of chromatin. *Nature*, 1974;**251**:249–251.

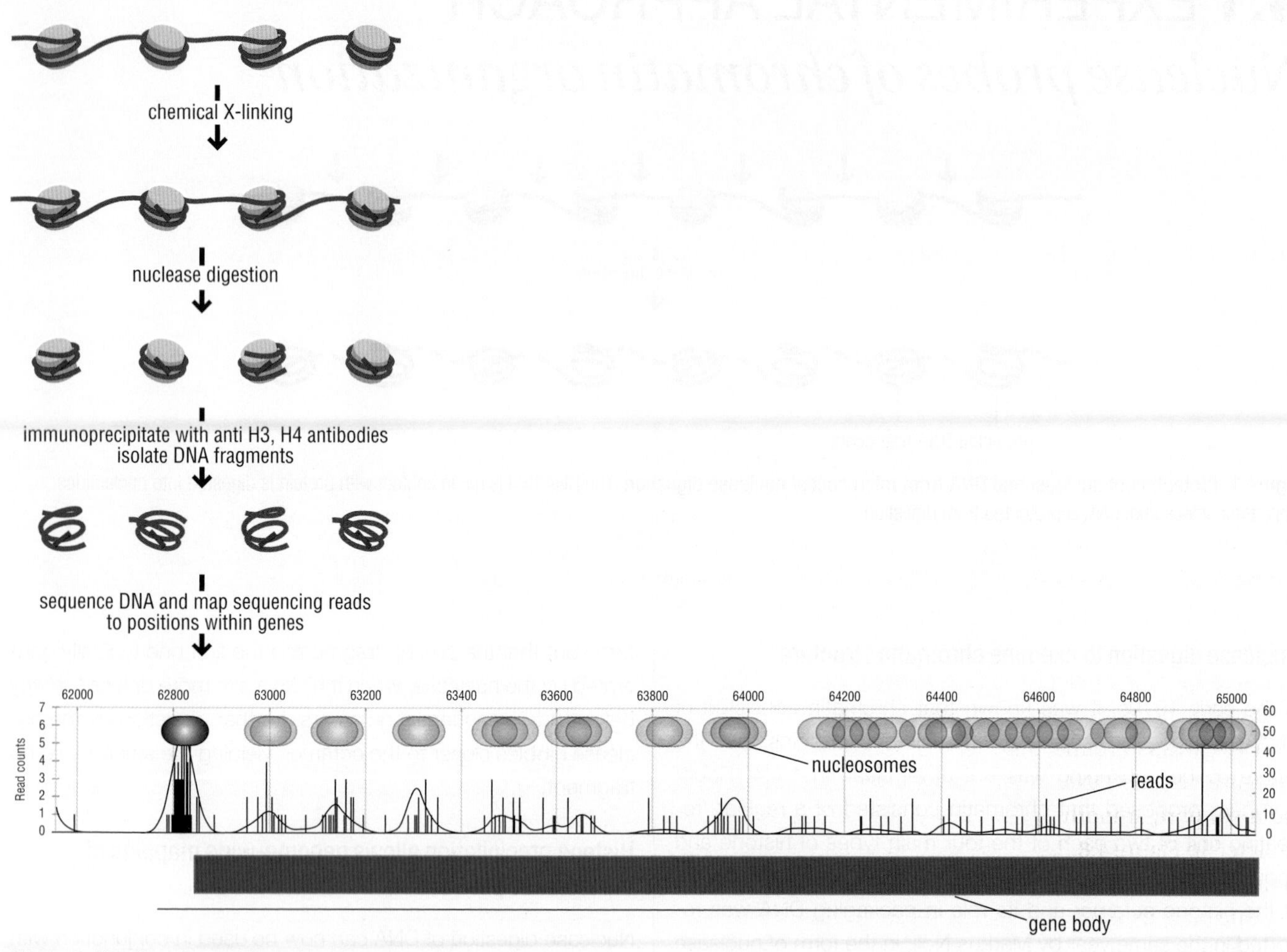

Figure 3 Mapping the positions of nucleosomes. The diagram shows the procedure used to identify the positions of nucleosomes in the yeast genome. Once the DNA wrapped around each nucleosome has been sequenced, a computer search of the yeast genome is done to map the sequence to its correct position and the location relative to specific genes can then be analyzed.

and the DNA in each nucleosome. Following this treatment, the chromosomes were digested with micrococcal nuclease, and the nucleosome particles were purified using antibodies against histones H3 and H4. The chemical links were then removed, and the different DNA fragments in the mixture were sequenced (outlined in Figure 3).

A computer-aided search was used to identify the location of each DNA sequence in the yeast genome, thereby giving the location of each nucleosome in the yeast genome. From such a study comes large amounts of data that must be sifted through to identify trends and insights. For example, Mavrich and colleagues found that the 5' promoter and 3' ends of *S. cerevisiae* genes were relatively free of nucleosomes. The authors proposed that such nucleosome-free regions help proteins involved in transcription initiation and termination to gain access to their target locations on the DNA. This study thus provided immediate insights into genome function and gene expression.

Find out more

Kornberg RD. Chromatin structure: a repeating unit of histones and DNA. *Science*, 1974;**184**:868–871.

Mavrich TN, Ioshikhes IP, Venters BJ, *et al.* A barrier nucleosome model for statistical positioning of nucleosomes throughout the yeast genome. *Genome Research*, 2008;**18**:1073-1083.

Noll M. Subunit structure of chromatin. *Nature*, 1974;**251**:249–251.

Related techniques

Nuclease digestion and DNA gel electrophoresis; Section 19.7

Nucleosome positioning via sequencing of nuclease protected fragments; Section 19.12

Variant histones		
histone class	variant histones	proposed function
H2A	H2A.X H2A.Z MacroH2A1 MacroH2A2 H2A-Bbd	DNA damage signaling transcriptional activation binds inactive X chromosome binds inactive X chromosome
H2B	H2B.1	
H3	CENP-A H3FA3 H3.3B H3FT	defines centromere chromatin
H4		

Figure 4.7 Variant histones and their proposed functions, where known.

Nucleosomes form preferentially at certain DNA sequences

Since all eukaryotic DNA is packaged by histones, the way in which histone octamers bind to DNA is relatively insensitive to the DNA sequence. However, nucleosomes do form preferentially along certain types of DNA sequences. When yeast chromosomes are examined, for example, there are certain regions that contain nucleosomes at particular DNA sequences, whereas nucleosome positions in other regions appear to be more variable.

How do the preferred locations arise? The answer lies in the way in which the DNA wraps around the histone octamer to form a nucleosome. As we see in Figure 4.6, the DNA that is wrapped around the core histone octamer is, for the most part, smoothly bent, but contains sharper bends at some locations as depicted in Figure 4.8. The DNA sequence at the sharp bends generally consists of pyrimidine–purine base steps (the sequence thymine/adenine (TA) or cytosine/guanine (CG) as viewed along one strand), which are more highly bendable than other base steps. The minor groove narrowing that must occur in order for the DNA to wrap smoothly around the nucleosome core is energetically more favored in AT-rich sequences. A favored nucleosome binding regions sequence would contain AT-rich sequences alternating with stiffer GC-rich sequences spaced roughly half a helical turn apart from one another. In this way, the base sequence helps to determine the relative location of the nucleosome.

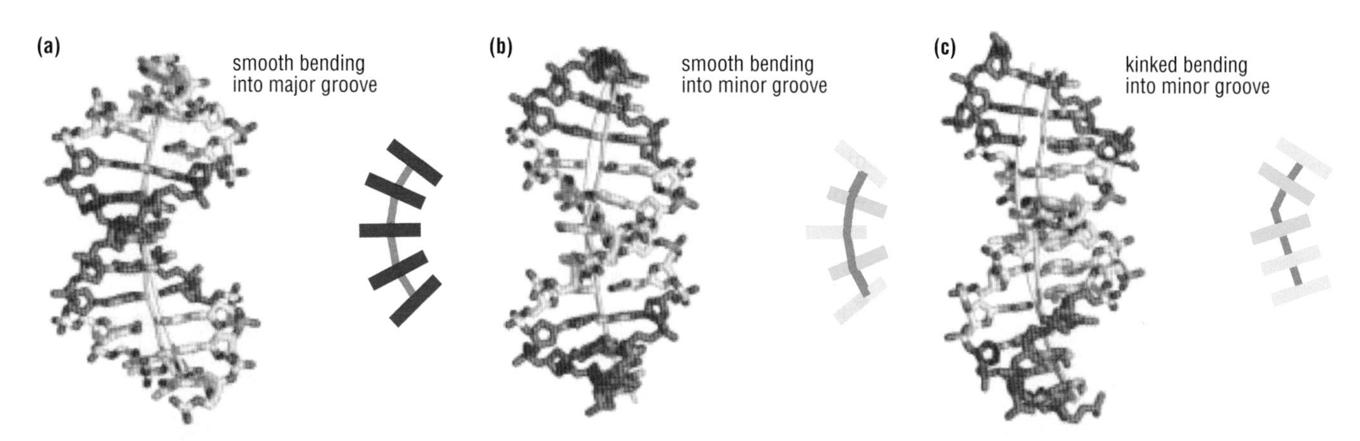

Figure 4.8 Different types of DNA bends allow the DNA to wrap around the histone octamer. The DNA in the nucleosome is smoothly bent in some regions (a and b) and more sharply bent at some points (c). The cartoon to the right of each image depicts the bending of the helix. The gray line represents the path of the center of the DNA helix, and the red and yellow bars represent how the bases fit on the bent DNA. In the smooth bend in panel (b), some of the base pairs are shifted from their more typical positions relative to the helix axis. PDB code 1KX5.

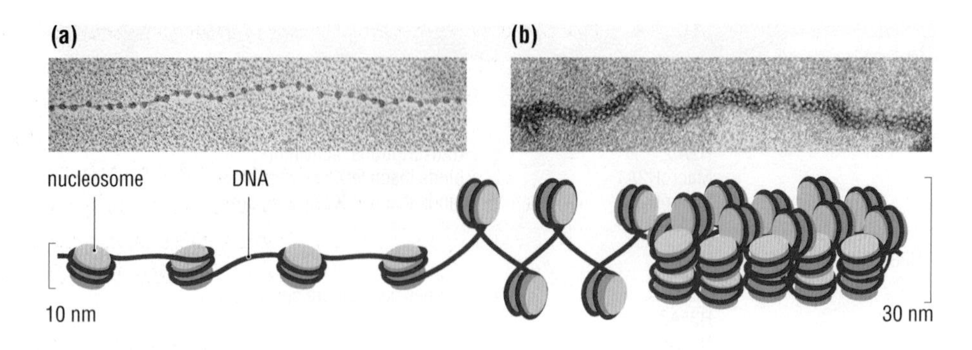

Figure 4.9 Nucleosomes become packaged into higher-order structures in eukaryotic chromosomes. (a) Electron micrograph of the 10 nm fiber of chromatin (top) and schematic (bottom), showing the 'beads on a string' organization of nucleosomes along DNA. (b) Electron micrograph (top) and schematic (bottom) of the 30 nm fiber that is formed by further packing. The less compact zigzag model is shown on the left and a more condensed form on the right. Condensation occurs through interactions between nucleosomes that are not yet fully understood. H1 and other proteins are present in the 30 nm fiber but are not shown here for simplicity.

Chromatin is organized into 30 nm filaments and higher-order structures

The packaging of chromatin occurs at several successive levels, working up from the small to large scale, as illustrated in Figure 4.9. These levels are visible in electron micrographs as the 10 nm fiber, the 30 nm fiber, and looped structures. Let us now consider each of these forms in turn.

The first stage of packaging of eukaryotic DNA into chromatin is the 10 nm fiber. This level of packaging arises from the way that histone octamers associate with double-stranded DNA to form nucleosomes, which are separated from one another by stretches of DNA to form a fiber that appears in electron micrographs like 'beads on a string' (Figure 4.9a). In this state, which occurs in the test tube at low ionic strength, the individual nucleosomes (the 'beads') are clearly visible and are separated by linker DNA (the 'string').

The 10 nm fiber can be further folded up into a more condensed filament 30 nm in diameter (Figure 4.9b) with the help of a protein, histone H1. In the 30 nm fiber, the nucleosome core particles form a regular, alternating arrangement that brings the nucleosomes in contact with one another. Our current understanding of how chromatin is organized into the 30 nm fiber is shown in Figure 4.9. Histone H1 binds to the linker DNA that connects successive nucleosomes, facilitating the compaction of the chromatin as shown in Figure 4.10. Histone H1 is therefore known as a linker histone.

The tails of the core histone proteins that protrude from each nucleosome are also important in formation of the 30 nm fiber, although the precise way in which they stabilize the fiber remains an open question. One of the salient questions that have occupied researchers has been that of how the linker DNA connects one nucleosome to the next, and how chromatin with linkers of varying length is organized into a 30 nm fiber. As new data emerge from a variety of physical measurements, our understanding of the structure of the 30 nm fiber will need to be modified accordingly.

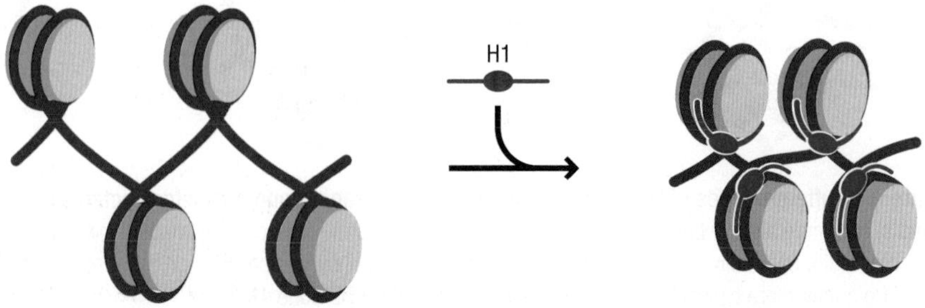

Figure 4.10 The binding of linker histone H1 to the nucleosome. Histone H1 (shown in red) binds to the linker DNA in the nucleosome, facilitating formation of the 30 nm fiber (shown in Figure 4.9b).

In the cell, the 30 nm fiber is further organized to become part of the compact forms of chromosomes that are seen in electron micrographs (see Figure 4.3). We know little about the precise nature of how chromatin is further compacted to form these structures. However, it is clear that large loops of chromatin, which has been organized into 30 nm fibers, are anchored to a central scaffold whose precise composition is still not completely understood. Nevertheless, many studies have clarified the role of the histone proteins both in organizing the genetic material and in allowing access to the DNA during transcription, DNA replication, and other processes, as we shall see throughout this book.

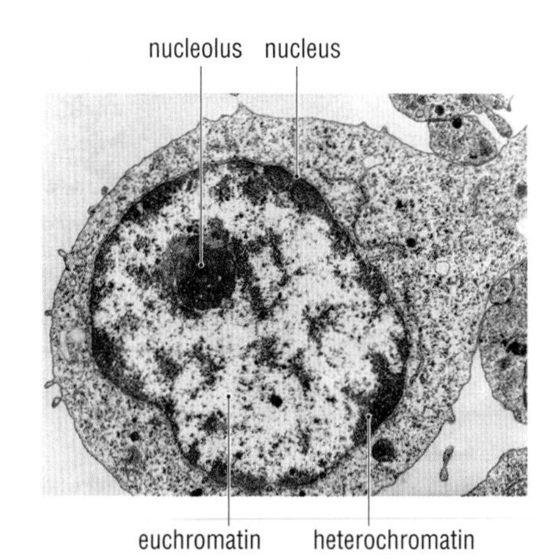

Figure 4.11 Electron micrograph of a cell showing the heterochromatin localized to the nuclear periphery. The large dark-rimmed circle in the middle of the cell is the nucleus. The white regions within the nucleus represent euchromatin and the dark-staining regions heterochromatin. The heterochromatin is concentrated at the periphery of the nucleus. Many nuclei also contain spherical structures called nucleoli (the nucleus shown has one visible nucleolus). The nucleolus will be discussed further towards the end of this chapter.

4.4 VARIATION IN CHROMATIN STRUCTURE

The chromosomes of a cell undergo a variety of changes in the course of the cell cycle, with marked changes taking place in the packaging and organization of DNA. During interphase in particular, when chromosomes are relatively decondensed and various genes may be actively transcribed, there is great variation in DNA compaction in different regions of the chromosome. We begin by describing some of these variations in chromosome structure, after which we will explore some of the basic mechanisms underlying the variations in chromatin that regulate access of proteins to the DNA at all stages of the cell cycle.

Interphase chromatin is not uniformly packaged

During interphase, chromosomes are in an extended, relatively decondensed form and genes are actively transcribed. However, within the relatively decondensed chromosomes, we find significant variations in structure. Our earliest understanding of these differences came from studies of whole nuclei, such as the electron micrograph shown in Figure 4.11. In these images, some of the chromatin stained lightly with the dye used to visualize DNA while other regions stained more darkly. The light staining regions were interpreted as being relatively decondensed regions, and were named **euchromatin**, and the strongly staining regions appeared to be more compacted regions and were named **heterochromatin**. The term chromatin, which has as its root the Greek word for color, chroma, derives from these early observations of how dyes bound to different regions of DNA.

Interphase chromosomes contain regions of both heterochromatin and euchromatin. Genes that are actively transcribed typically lie in regions of euchromatin. Euchromatin is relatively decondensed, and it is this open structure that presumably allows ready access to DNA by the enzymes and proteins that must bind to it for transcription, repair and recombination to occur. By contrast, heterochromatin appears to be more condensed and is typically concentrated near the periphery of the nucleus (see Figure 4.11). While it was originally thought that genes in heterochromatin were not transcribed, we now know that there is indeed some transcription in these regions. At the same time, there is well-established evidence that the **translocation** of certain genes to the vicinity of heterochromatin actively prevents transcription, as well as other processes, such as DNA repair and recombination.

➲ DNA repair is covered in Chapter 15 and DNA recombination in Chapter 16.

Beyond the general occurrence of heterochromatin described above, certain specific regions of chromosomes also form heterochromatin in a characteristic way. These regions include the DNA at and near the centromeres, which is the segment of the chromosome to which the spindle attaches for segregation, and at

The use of nuclease digestion to probe DNA structure and interactions with proteins is covered in Section 19.12.

telomeres, the special DNA sequences at the ends of chromosomes. In addition, those sites in chromosomes that contain highly repetitive DNA sequences are also packaged into heterochromatin. We will learn much more about the nature of centromeres and telomeres later in this chapter.

The differences between heterochromatin and euchromatin are often detected experimentally by the **nuclease sensitivity** of the region. Heterochromatin is more resistant to digestion by DNase I than euchromatin. The relative resistance of heterochromatic DNA to digestion is often interpreted as indicating that heterochromatin is more condensed than euchromatin, although it may equally reflect differences in the ability of nucleases to interact with DNA bound by specific heterochromatin proteins.

The structural state of chromatin affects all processes directed by DNA

While most studies of the functional significance of chromatin structure have focused on gene transcription, the way in which DNA is packaged into chromatin affects all processes that involve interactions with chromosomes. The role of chromatin structure in the regulation of transcription will be discussed in detail in Chapter 8. Here we briefly describe some examples of how chromatin structure influences transcription, replication, recombination, and chromosome transmission.

As described above, telomeres are often packaged into heterochromatin. If a gene that is active in its normal location in euchromatin - let us say *URA3* from yeast - is artificially placed adjacent to a telomere, that gene becomes silenced. The chromatin structure in the telomeric heterochromatin does not allow the recruitment of transcription factors that is necessary for gene expression.

We describe replication origins in detail in Section 6.7.

In a similar manner, the effects of chromatin structure on DNA replication can be demonstrated by experiments in which an origin of replication normally found in euchromatin is placed into heterochromatin. An origin of replication is a location in the chromosome at which DNA replication begins.

Replication origins located at different positions along a chromosome initiate replication at different times during S phase. Chromosome rearrangements that place an early-replicating origin into heterochromatin result in late replication of that origin. The functional importance of regulating replication timing is not entirely clear; however, observations on the budding yeast *S. cerevisiae*, suggest that late-replicating origins may allow cell division to be delayed if DNA is damaged or other steps in DNA replication are incomplete.

Differences in chromatin structure can also control which parts of the chromosomes are able to undergo recombination, a process by which a strand of DNA is broken and then joined to a different DNA molecule. In yeast, the tandem array of genes encoding ribosomal RNA (rDNA genes) is packaged into heterochromatin and normally has a reduced recombination rate, which helps prevent the loss and rearrangement of genes that can occur between highly repetitive sequences. When heterochromatin is not properly maintained, however, there is increased mitotic recombination in the rDNA gene cluster. The packaging of repetitive DNA into heterochromatin therefore helps stabilize the genome.

We learn in detail about recombination in Section 16.8.

Chromatin structure also plays a role in chromosome segregation and cell division. In many species, disruption of normal heterochromatin at centromeric DNA causes major defects in chromosome segregation at mitosis, and chromosome pairing at meiosis. Thus in the regulation of chromatin structure is critical for centromere function and chromosome stability.

4.5 COVALENT MODIFICATIONS OF HISTONES

Chromatin is far from being a permanently static entity, in which sections of chromosomal DNA are irreversibly 'frozen' into euchromatin or heterochromatin. Neither, as we noted earlier, is the biological role of chromatin purely structural - merely a convenient means of storing DNA in the cell. Instead, there are dynamic changes in chromatin structure that are mediated by various modifications to the chromatin. These modifications - of both the physical location of the nucleosomes, within a chromosome and the biochemical composition of the histone proteins that comprise the nucleosomes - are central to regulation.

Modification of histones affects chromatin structure and function

The histone proteins that package DNA into chromatin are subject to a variety of post-translational modifications. These modifications, which add chemical groups or small proteins to histone side chains, serve as important signals in regulating virtually all processes that involve interactions with chromatin. Acetylation, methylation, phosphorylation, and **ubiquitination**, are covalent modifications that are commonly found in histones.

➔ We discuss the post-translational modification of proteins in detail in Chapter 14. A brief overview can be found in Section 2.15.

Post-translational modifications occur both in the histone tails, as shown in Figure 4.12, and in the globular core of the histone proteins, shown in Figure 4.13. Once modified, these sites are thought to recruit specific proteins to the chromatin, which then affect chromatin structure and function. The covalent modifications also constitute chemical changes that can alter chromatin structure directly; acetylation, for example, removes the positive charge from lysine side chains and thereby alters interactions with the negatively charged DNA. While **histone modifications** are relatively minor in terms of the changes they introduce at the chemical level, the biological impact of such changes can be very dramatic. For example, the addition of just one simple methyl group to an arginine residue in a

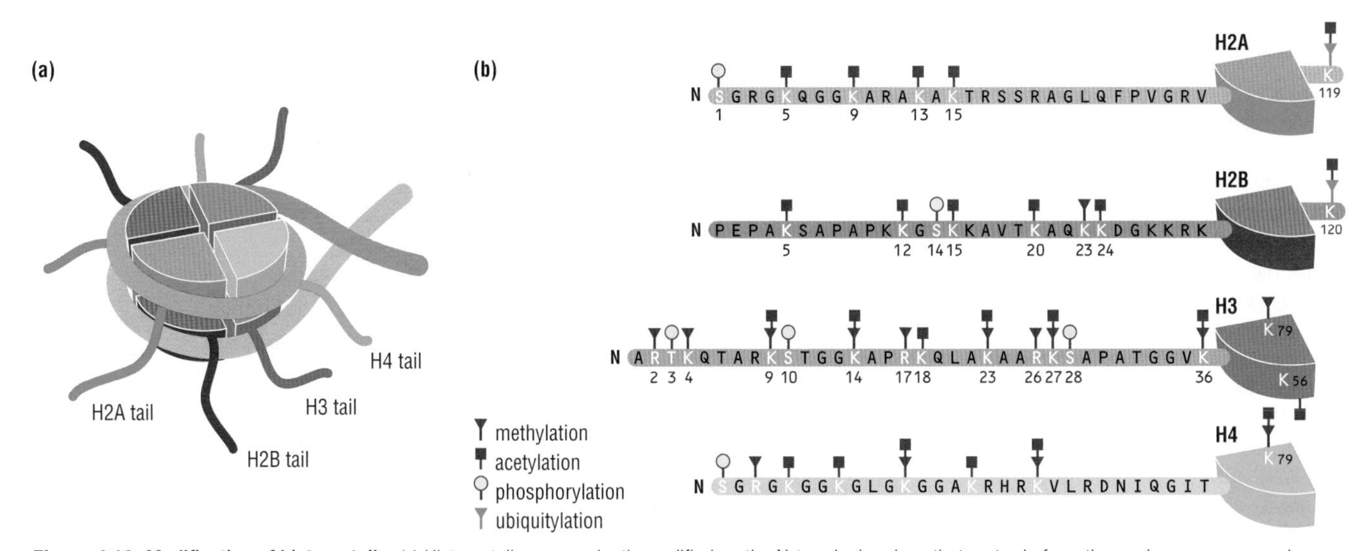

Figure 4.12 Modification of histone tails. (a) Histone tails are covalently modified on the N-terminal regions that protrude from the nucleosome core region. (b) The tails are methylated, acetylated, phosphorylated, or ubiquitinated at specific amino acids. The combination of specific modifications on a given histone tail can influence the function or accessibility of the associated region of DNA. This is often referred to as the histone code.

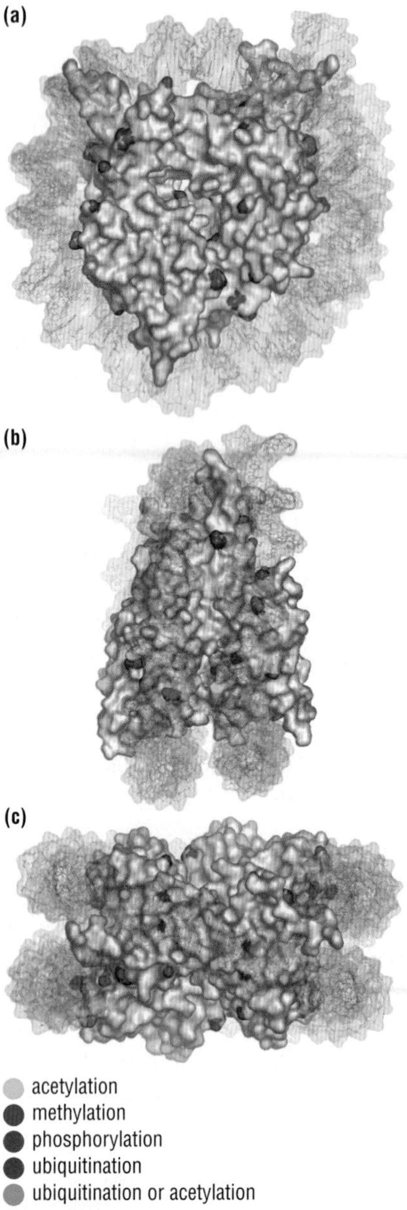

(a)

(b)

(c)

acetylation
methylation
phosphorylation
ubiquitination
ubiquitination or acetylation

Figure 4.13 Residues in the core of the histone octamer are also modified. Residues that lie on the surface of the globular core of the histone octamer are also subject to post-translational modifications. The DNA wrapped around the histone octamer is shown in blue. The sites of acetylation, methylation, phosphorylation, ubiquitination, and either ubiquitination or acetylation are shown as colored residues. (a) A view of the nucleosome down the DNA double-helix axis; (b) As in panel a, but rotated 90° around the 'North-South' axis (c) As in panel a, but rotated 90° around the 'East-West' axis.

From Cosgrove, S, Boeke, Jef D, and Wolberger, C. Regulated nucleosome mobility and the histone code. *Nature Structural and Molecular Biology*, 2004;**11**: 1307–1043.

histone can have a major effect on the fate of the DNA bound to the nucleosome with that modification. Thus, histone modifications are an important means of modulating any process that involves accessing the DNA.

An important feature of histone modifications is that they are reversible. There are specialized enzymes that add particular chemical modifications to histone side chains, and other types of enzymes that remove the modification. Figure 4.14 shows some examples of different histone modifications and the enzymes that either add or remove them. The dynamic addition and removal of particular modifications - the biological equivalent of an on/off switch - is used by all eukaryotic cells to regulate a wide array of processes.

Histone acetylation is associated with active chromatin

Virtually all regions of eukaryotic chromosomes contain histones with acetylated lysine residues. However, the relative levels of acetylation vary substantially: actively transcribed regions in euchromatin contain highly acetylated histones, whereas heterochromatin, which contains few transcribed genes, has very low levels of acetylation. The primary targets of acetylation are histones H3 and H4, which contain numerous lysines that can become acetylated (see Figure 4.12). Enzymes called **histone acetyltransferases (HATs)** covalently attach acetyl groups to lysine side chains, while **histone deacetylases (HDACs)** remove the acetyl groups (see Figure 4.14). The isolation of the first histone acetyltransferase is described in Experimental approach 4.2.

Histone acetylation generates sites for the binding of proteins containing domains known as **bromodomains**. These domains bind specifically to particular acetyl-lysines, as illustrated in Figure 4.15 - for example, acetylated lysine 14 on histone H3. The bromodomain protein may then recruit enzymes that act on DNA. The effect of acetylation may also derive from the resulting change in electrostatic charge, since unmodified lysines are positively charged, while acetyl-lysine is electrostatically neutral.

Different Classes of Histone-Modification Enzymes

enzyme family	species	complex
histone acetyltransferase (HAT) complexes		
Gcn5 family	*S. cerevisiae*	SAGA, ADA
TAF250	human	TFIID
PCAF	human	PCAF
histone deacetylase (HDAC) complexes		
HDAC1, HDAC2 family	*S. cerevisiae*	Sin3
	human	Sin3. NURD
HDAC5	human	N-CoR/SMRT
Sir2	*S. cerevisiae*	Sir
histone methyltransferase (HMT) complexes		
PRMT family	*S. cerevisiae*	RMT1/HMT1
	human	PRMT1
Set domain family	human	Suv39h1
Jumonji family	*S. cerevisiae*	JHDM1
LSD1 family	human	LSD1

Figure 4.14 Examples of different classes of histone-modifying enzymes in yeast and humans.

4.2 EXPERIMENTAL APPROACH

Discovery of a histone acetyltransferase

Histone acetylatransferase identified from *Tetrahymena*

Although it was known that histones were acetylated and that histone acetylation was associated with transcriptionally active chromatin, the enzyme responsible for this post-translational modification was not known. David Allis and colleagues were able to crack this problem by developing an assay in which acetyltransferase activity could be detected in proteins that were separated on a gel impregnated with histones.

This special gel was prepared by adding histone proteins to the polyacrylamide-SDS mixture before the gel polymerized. Protein samples could then be electrophoresed on the gel and separated by molecular weight, as on a standard protein gel. The gel was then treated to remove the SDS, allowing the proteins to adopt their native fold, and the gel was then soaked in radiolabeled acetyl-CoA (the acetyl group donor). Active histone acetyltransferases in the gel could then catalyze transfer of the labeled acetyl group onto histone proteins. Since the histone proteins were everywhere in the gel, while the acetyltransferase proteins were only located in particular bands on the gel, an autoradiograph of the gel would show which protein band on the gel contained histone acetyltransferase activity. This approach allowed the researchers to directly analyze still-heterogeneous column fractions and identify which protein in that particular mixture was the active protein.

Taking advantage of the fact that the macronuclei of *Tetrahymena* are an enriched source of the enzyme, Brownell and Allis fractionated extracts using various protein purification techniques and used the histone-impregnated gel to monitor the activity in each fraction. In this way, the researchers observed a 55 kDa protein (which they called p55) that had histone acetyltransferase activity, as illustrated in Figure 1. Brownell *et al.* then identified the protein by eluting it from the gel and determining the amino acid sequence of several peptide fragments of the purified protein. The partial amino acid sequence made it possible to clone the entire cDNA encoding the protein using PCR methods (see Section 19.3). The resulting DNA sequence of the cDNA matched a known gene called GCN5 in yeast that was required for full transcriptional activity of some genes. This match unambiguously linked histone acetylation and transcriptional activation. (Today, the partial amino acid sequence would lead to direct identification of the gene (as described in Section 19.8) without the intervening step of cDNA cloning, because the complete genome sequence of *Tetrahymena* is available.)

Antibodies against modified histones allow genome-wide analysis of chromatin modifications

Since the first histone acetyltransferase was identified, numerous enzymes that acetylate and methylate histones at specific residues have been isolated. Since the identity of the modified residue and the location of the modified histone within the chromosome has important functional significance, techniques have been developed to find where in the genome particular modified histone proteins are located. These methods rely upon antibodies that recognize very specific modifications, such as

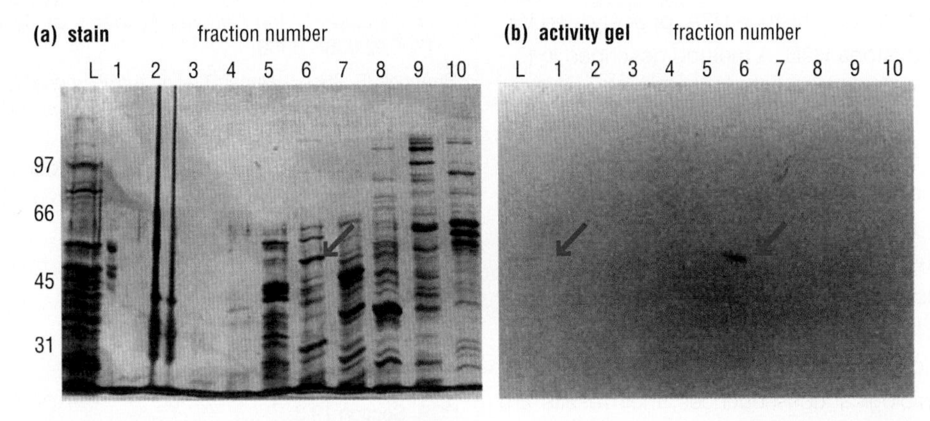

Figure 1 Activity gel showing identification of histone acetyltransferase. (a) The gel is stained with Coomassie dye and shows the proteins in the partially purified fractions. (b) Autoradiograph of the same gel as in (a) after the proteins were renatured and the gel was incubated with radiolabeled acetyl-coenzyme A (acetyl-CoA), followed by washing with unlabeled acetyl-CoA to remove the unincorporated radiolabel. The bands indicated a single 55 kDa protein that acetylated histones within the gel matrix. Lane 'L' contains all the proteins in the extract (namely, before fractionation). The arrows point to the protein band that exhibits acetyltransferase activity (in L and fraction 6).

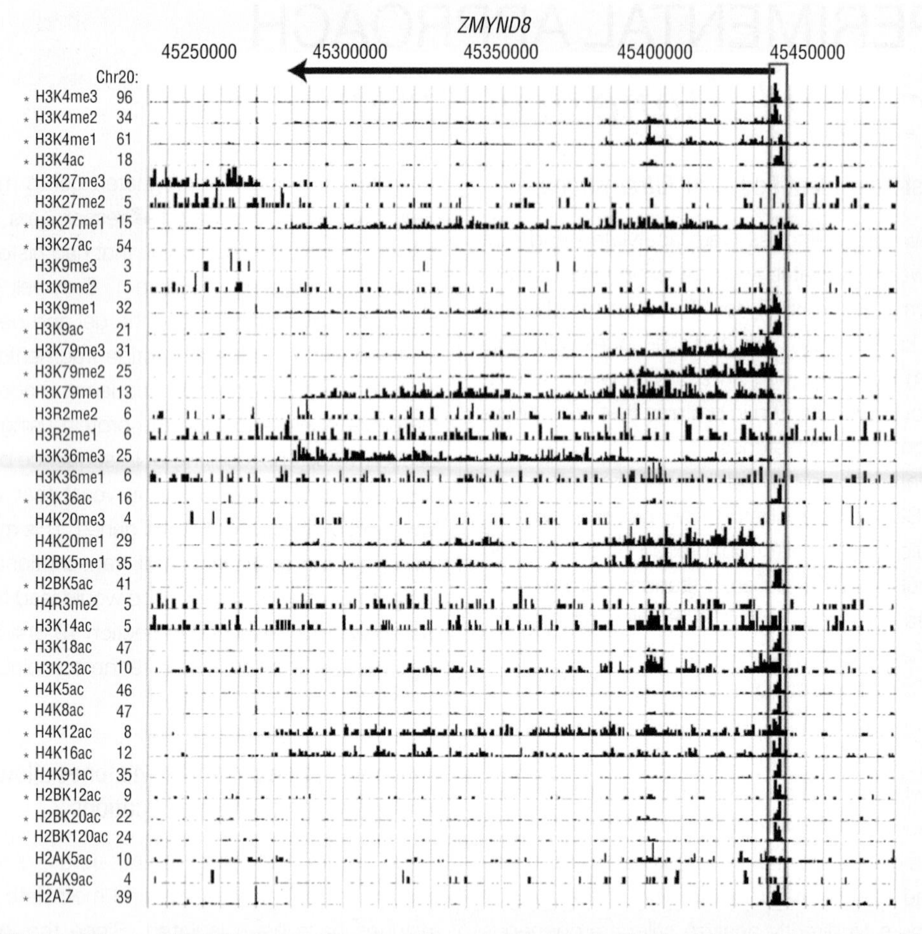

Figure 2 Mapping of different histone modifications across a region of human chromosome 20. Each line shows the location of the indicated modified histone. The column marked by red lines indicates the transcription start site of a gene residing in this region. Modifications that were determined experimentally to be statistically significant are labeled with red asterisks on the left.

Reproduced from Wang *et al.* Combinatorial patterns of histone acetylations and methylations in the human genome. *Nature Genetics*, 2008; **40**:897–903.

acetylated lysine 5 (K5) or K9 in histone H2A, or acetylated K5, K12, K20, or K120 in histone H2B. A method described in Experimental approach 4.1 can then be used to identify where in the genome histones with particular modifications are located: the histones can be cross-linked to the DNA, after which the DNA sheared, and the histones of interest immunoprecipitated with the appropriate antibody (which binds specifically to the modification of interest). Using this approach, Wang and colleagues mapped histone acetylation and methylation sites across large regions of the human genome, yielding detailed information that provides insights into the roles of histone modification in dictating biological consequences; their results are shown in Figure 2

Find out more

Brownell JE, Allis CD. An activity gel assay detects a single, catalytically active histone acetyltransferase subunit in *Tetrahymena* macronuclei. *Proceedings of the National Academy of Sciences of the U S A*, 1995;**92**:6364-6368.

Brownell JE, Zhou J, Ranalli T, *et al. Tetrahymena* histone acetyltransferase A: a homolog to yeast Gcn5p linking histone acetylation to gene activation. *Cell*, 1996;**84**:843-851.

Wang Z, Zang C, Rosenfeld JA, *et al.* Combinatorial patterns of histone acetylations and methylations in the human genome. *Nature Genetics*, 2008;**40**:897-903.

Related techniques

Protein gel electrophoresis; Section 19.7

Protein purification techniques; Section 19.7

PCR; Section 19.3

Tetrahymena (as a model organism); Section 19.1

Protein mass spectrometry; Section 19.8

Chromatin IP; Section 19.12

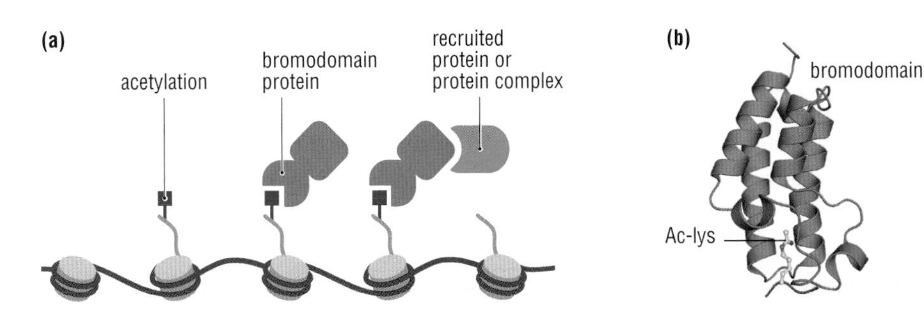

Figure 4.15 Histone tail acetylation allows recruitment of proteins that act on chromatin. (a) Acetylation of histones in chromatin recruits bromodomain-containing proteins that can, in turn, specifically recruit other proteins that may, for example, activate nearby genes. (b) Structure of the bromodomain from GCN5 bound to acetyl-lysine in a histone H4 peptide. PDB code 1E6I.

Histone methylation can be associated with either active or inactive chromatin

Histones H3 and H4 can be methylated at both lysine and arginine residues (see Figure 4.12). Up to three methyl groups can be covalently attached to a single lysine, while up to two can be attached to an arginine side chain. The different degrees of methylation - namely mono-, di-, or trimethyl lysine, and mono- or dimethyl arginine - can have different consequences by recruiting different proteins that recognize the different methylated side chains. Histones are methylated by enzymes known as **histone methyltransferases (HMTs)** and are removed by **histone demethylases**. Some examples of these enzymes are listed in Figure 4.14.

Unlike acetylation, which is uniformly associated with active transcription, methylation is associated with either activation *or* repression of transcription. The precise effect of histone methylation depends upon which residue is methylated. For example, methylation of lysine 9 in the tail of histone H3 is associated with transcriptionally silent chromatin - including heterochromatin - while methylation of H3 lysine 4 is associated with active chromatin. It is, therefore, the recognition of specific modifications on histones by other proteins, not the chemical change on the histone itself, that determines the functional consequence of the methylation.

Like acetyl-lysine, methylated lysine and arginine side chains are recognized by proteins that bind specifically to methylated residues. **Chromodomains**, for example, bind to specific methylated lysines as depicted in Figure 4.16. Chromodomains are typically found in proteins that mediate transcriptional **silencing**; an example is the protein HP1, which helps initiate heterochromatin formation in mammalian cells.

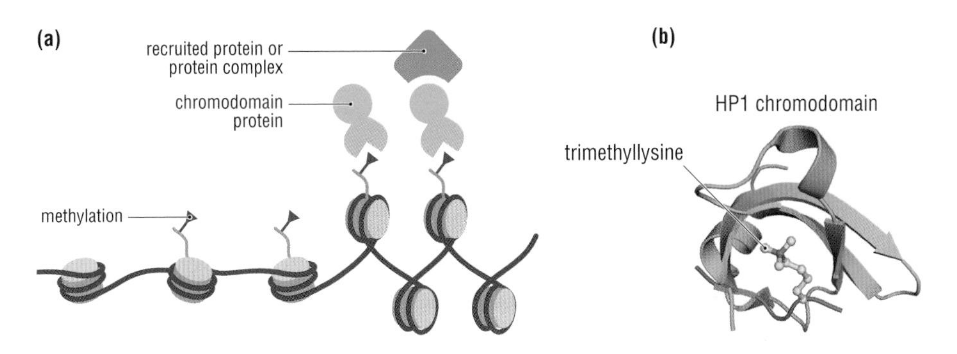

Figure 4.16 Histone tail methylation recruits proteins that initiate heterochromatin formation. (a) Methylation of lysine 9 on H3 provides a site for recruitment of proteins containing a chromodomain (green). An example of a chromodomain protein is the protein HP1, which helps initiate heterochromatin formation. Some chromodomain proteins interact with other proteins (orange) and recruit them to sites of H3 Lys 9 methylation. (b) The structure of the HP1 chromodomain bound to trimethyl-lysine is shown. (PDB code 1KNE).

Histone phosphorylation and ubiquitination serve as additional regulatory signals

Histone tails can be phosphorylated at either serine or threonine residues, as depicted in Figure 4.12. The phosphate groups are added by kinases and removed by phosphatases. Histone phosphorylation plays a variety of roles: phosphorylation of histone H3 at serine 10 alone allows the transcription of genes required for cell growth. In contrast, chromosome condensation at mitosis is tightly correlated with the phosphorylation of histone H3 at both serines 10 and 28. Phosphorylated residues can be recognized by specialized protein domains, although the precise mechanism by which phosphorylation of histone exerts an effect remains to be elucidated.

The covalent attachment of ubiquitin to lysine side chains is the largest histone modification, since ubiquitin is a 76-amino acid protein, a notable contrast to the handful of atoms added by each of the previously mentioned histone modifications. Ubiquitin is attached to histones in a series of enzymatic steps, and can be removed by a deubiquitinating enzyme (DUBs). Histones can be similarly modified by the covalent attachment of Sumo (small ubiquitin-like modifier), which is similar in structure to ubiquitin. Ubiquitination and sumoylation of histones play roles in regulating different steps of transcription as well as DNA repair.

Different patterns of histone modifications have distinct consequences

Post-translational modification of one histone protein can trigger a cascade of events that leads to further chromatin modifications, either in another histone protein or in the DNA. For example, phosphorylation of serine 10 promotes the acetylation of H3 lysine 14, which, as we saw above, is associated with active chromatin. Figure 4.17 shows how the acetylation of H3 lysine 14 then inhibits methylation of lysine 9, which, as we saw earlier, is a docking signal for proteins associated with heterochromatin formation.

The concerted interaction of histone modifications, together with the large number of possible combinations of modifications, has led to the suggestion that there is a histone code, in which unique combinations of modifications specify particular functional states of chromatin. According to this model, the particular pattern of histone modifications found in the vicinity of a gene influences whether or not that gene is transcribed. The precise nature of functional information defined by the proposed histone code is an area of active investigation.

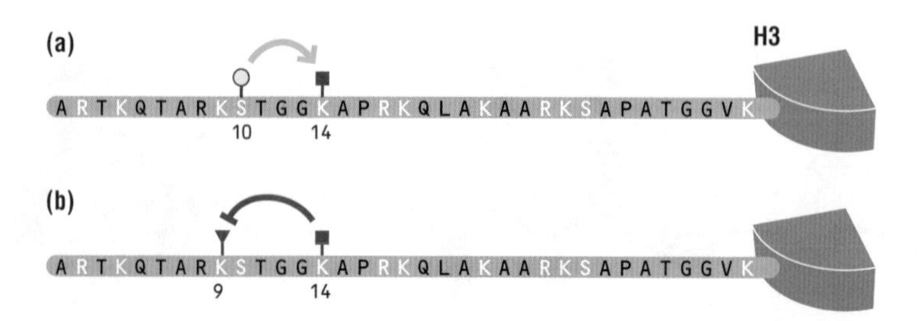

Figure 4.17 Modification of one site on histone influences the modification of a second site. (a) The phosphorylation of serine 10 on H3 promotes the acetylation of lysine 14. (b) The acetylation of H3 lysine 14 inhibits the methylation of lysine 9.

4.6 NUCLEOSOME-REMODELING COMPLEXES

ATP-dependent nucleosome remodeling alters the location of nucleosomes on DNA

While the packaging of eukaryotic DNA into chromatin helps to organize the genetic material in the nucleus, it presents a potential barrier to different proteins and enzymes that must access the DNA. For example, the wrapping of DNA around a histone core can block binding of the RNA polymerase enzyme, thereby repressing transcription. To overcome this barrier, the location of nucleosomes along the DNA can be altered in a regulated manner so that transcription can proceed, as illustrated in Figure 4.18. This alteration is mediated by **ATP-dependent nucleosome-remodeling complexes**, which are multisubunit enzyme complexes that use the energy provided by ATP hydrolysis to change the positions of nucleosomes on DNA. Nucleosome-remodeling enzymes can either slide the histone octamer along the DNA or remove the histone octamer entirely and transfer it from one segment of the DNA molecule to another. In addition, nucleosome- remodeling complexes can increase accessibility of DNA by introducing loops into the portion of DNA that is wrapped around a single histone core, thus increasing the accessibility of the DNA.

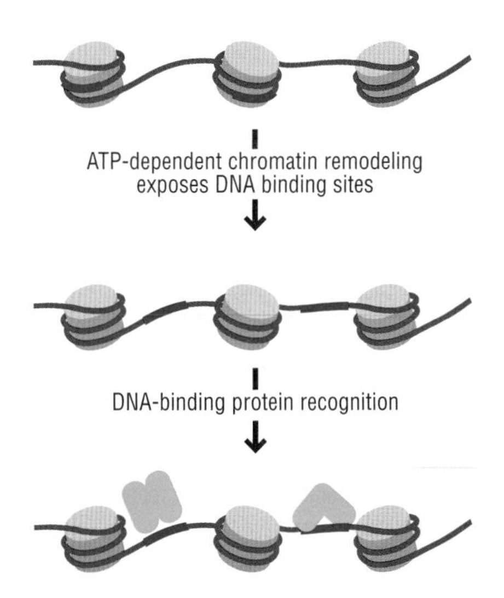

Figure 4.18 ATP-dependent chromatin remodeling allows access to specific sites on DNA. The binding sites for specific proteins in DNA (pink) may be inaccessible to DNA binding proteins owing to association of the region with the nucleosome. ATP-dependent chromatin remodeling results in the movement of the nucleosome along the DNA, exposing the binding site. Specific DNA-binding proteins (green) can now bind to their sites.

Figure 4.19 shows how there are several classes of nucleosome-remodeling complexes, which are distinguished by their subunit composition as well as by their effect on chromatin. The SWI/SNF complex, for example, disrupts nucleosome positioning, while the ACF complex (the ISWI family) helps to position nucleosomes during chromatin assembly. Despite their diversity, all nucleosome-remodeling enzymes have in common a conserved ATPase domain that hydrolyzes ATP. Some complexes also contain subunits that bind to chemically modified residues in histone proteins. The SWI/SNF complex, for example, contains 11 subunits, including an ATPase domain that catalyzes DNA-dependent ATP hydrolysis and a bromodomain that binds to acetylated lysine residues in histones or in other proteins associated with the DNA.

One of the first chromatin-remodeling complexes to be studied, SWI/SNF, was identified in genetic screens for yeast defective in mating-type switching (SWI) and sucrose fermentation (SNF, sucrose non-fermenter). The mutations found to be responsible for the observed defects, termed *swi/snf*, could be suppressed by mutations in histone genes, suggesting that the wild-type SWI/SNF proteins might mediate function by altering chromatin structure.

A number of different models have been proposed to explain how ATP-dependent nucleosome-remodeling complexes alter the positions of nucleosomes on DNA; these models are depicted in Figure 4.20. One proposal is that the enzymes propagate a local distortion in the DNA around the octamer, which

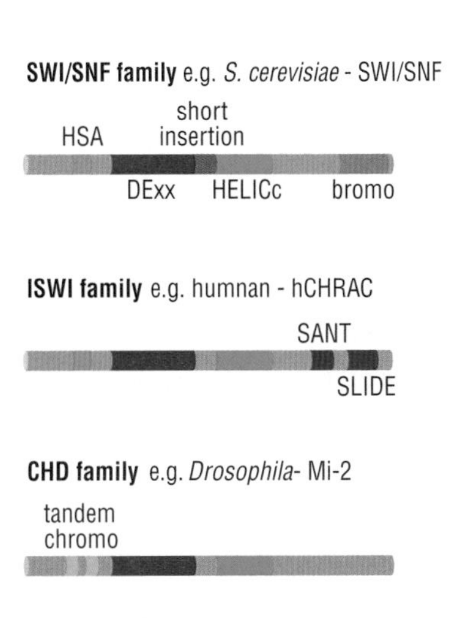

Figure 4.19 The four families of ATP-dependent nucleosome-remodeling complexes. All nucleosome remodeling complexes contain a subunit with a ATPase domain that is split into two parts called DExx and HELICc. These two domains can be separated by either a short or long insertion. Different families are distinguished from one another by the other domains contained within the ATPase subunit. The domains are called HSA (helicase-SANT), bromodomain (binds acetyl-lysine), chromodomain (binds methyl-lysine), and SANT-SLIDE (binds to histone tails).

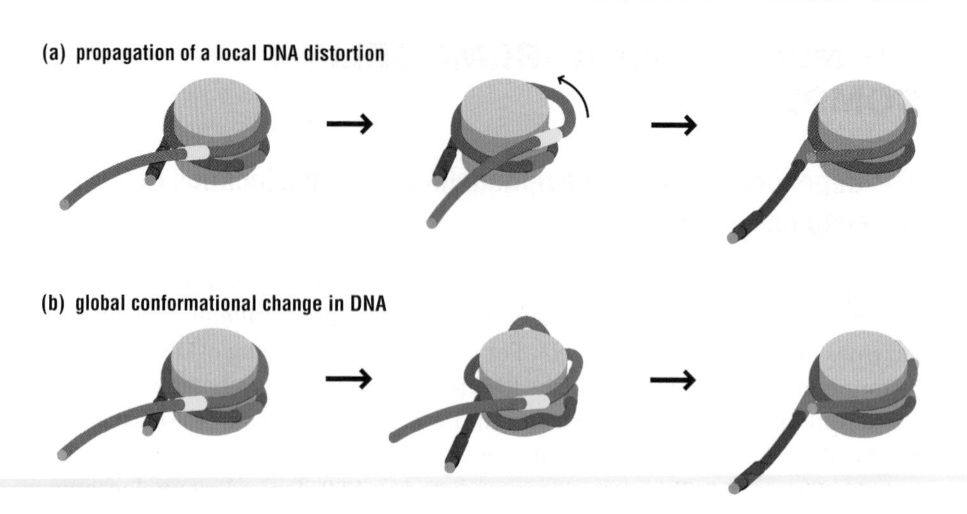

Figure 4.20 Different ways in which nucleosomes might be repositioned along the DNA by the actions of ATP-dependent nucleosome-remodeling enzymes. A number of models have been proposed for how nucleosome remodeling enzymes reposition nucleosomes: (a) propagation of a local DNA distortion around a histone octamer can lead to 'sliding' of the octamer along the DNA; and (b) a global conformational change in the DNA around the histone octamer could trigger repositioning of the nucleosome on the DNA.

gives rise to a nucleosome in a new position on the DNA strand (Figure 4.20a). It is also possible that a global distortion in the DNA, perhaps one that puts the DNA under torsional stress, could trigger repositioning of the nucleosome on the DNA (Figure 4.20b). The precise mechanism by which nucleosome-remodeling complexes alter nucleosome position is an active area of inquiry.

Nucleosome-remodeling complexes can be targeted to specific sites by interactions with other proteins

Since nucleosome remodeling is typically required in precise locations on the DNA, the cell must have a mechanism for recruiting nucleosome-remodeling complexes to specific sites on the DNA where they are needed. The most common way of recruiting these complexes is through association with proteins that bind to specific sequences in the DNA, as shown in Figure 4.21a. A wide variety of eukaryotic transcriptional regulators can recruit nucleosome-remodeling complexes through direct, non-covalent binding between proteins. Post-translational modifications in histone proteins can also recruit nucleosome-remodeling enzymes (see Figure 4.21b). As mentioned earlier, some nucleosome-remodeling complexes contain specialized domains that recognize specific histone modifications. In this way, chemical modification of histones can trigger the binding of nucleosome remodeling complexes, which in turn alter transcription or other processes involving access to the DNA.

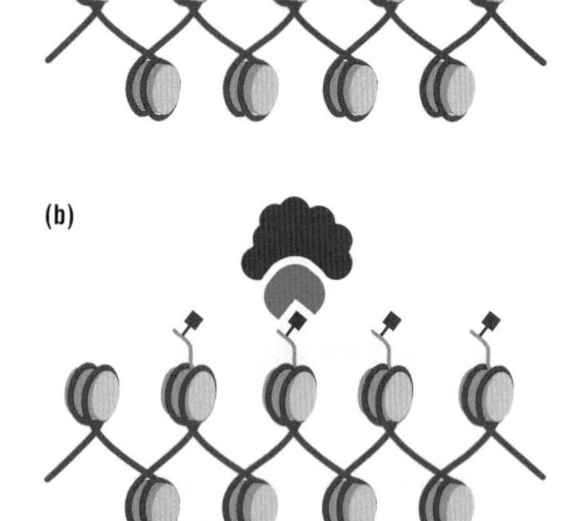

Figure 4.21 Recruitment of nucleosome-remodeling complexes to chromatin. (a) Proteins (orange) bound directly to a particular DNA sequence can specifically recruit nucleosome-remodeling enzymes (dark blue). (b) Acetylated lysine residues (pink squares) in histone proteins may be specifically recognized by a bromodomain (blue circle) that is part of a nucleosome-remodeling complex.

Adapted from Becker, P., and Horz, W. ATP-dependent nucleosome remodelling. *Annual Review of Biochemistry*, 2002; **71**: 247–273.

4.7 DNA METHYLATION

Methyltransferases add methyl groups to cytosine in DNA

Not all chromatin modifications involve modification of proteins; chemical modification can also occur on the DNA itself. In both bacteria and eukaryotes, methyl groups can be added to cytosine residues to generate 5-methyl cytosine, whose structure is shown in Figure 4.22. This modification, known as **DNA methylation**, is carried out by enzymes known as **DNA methyltransferases**(or **DNA methylases**). The basic mechanism of methylation is similar in bacteria and eukaryotes; however, the biological function of DNA methylation differs in each system, as we shall see.

DNA methylation carries some risk to the cell, however, of leading to permanent alterations to the DNA. Methylated cytosines are chemically less stable than unmethylated cytosines, and undergo rare deamination. This deamination of methyl cytosine generates a thymine in the DNA (Figure 4.22c), and thus a methylated CpG dinucleotide becomes TpG. This residue change can lead to a permanent change in the DNA base sequence, as sometimes the single nucleotide mismatch is not recognized by the DNA repair machinery prior to replication. In contrast, the less frequent deamination of unmethylated cytosine generates a uracil residue. This is efficiently recognized as damage to the DNA and is correctly repaired to restore the original thymine moiety.

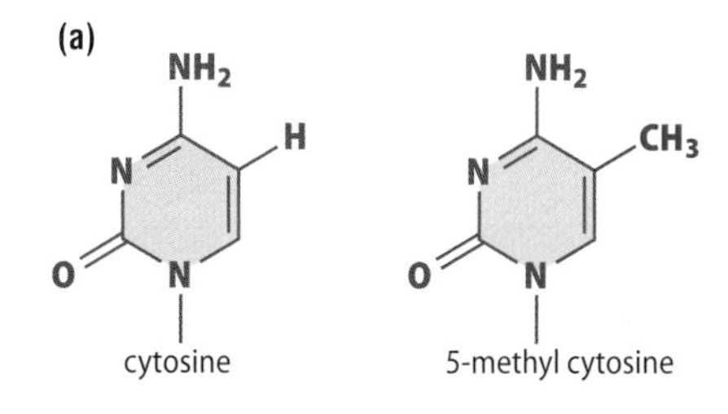

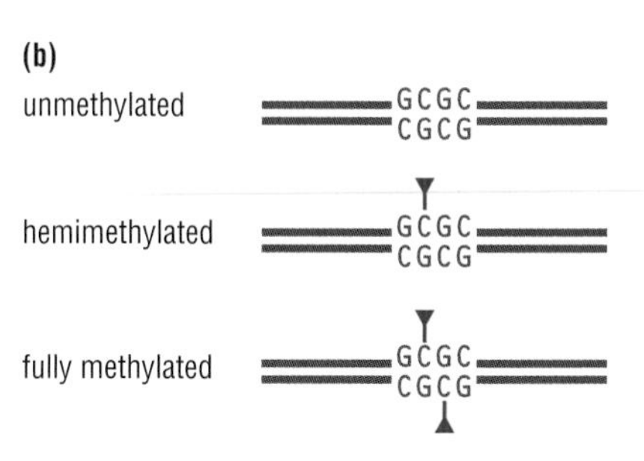

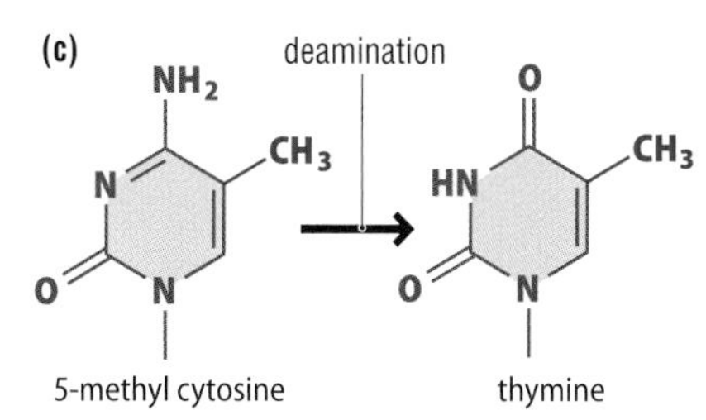

Figure 4.22 DNA methylation. (a) The chemical structure of unmethylated and methylated cytosine. (b) DNA is preferentially methylated at cytosine in CG pairs. When only one of the two strands are methylated, it is termed hemi-methylated DNA. (c) Deamination of methyl cytosine gives thymine, leading to a mutation of C to T.

➔ We describe how DNA lesions are repaired in Chapters 8 and 16.

Base flipping is used to access bases to be methylated

The stable structure of DNA makes enzyme-driven base modification reactions, such as cytosine methylation, difficult. The bases are tightly packed inside the double helix, which is stabilized by base-pairing and base-stacking (see Figure 2.25). DNA methyltransferases therefore use an elegant mechanism called base flipping to gain access to cytosine bases, as was first discovered for the bacterial methyltransferase, HhaI. During base flipping, the cytosine base to be modified is flipped completely out of the DNA helix and an amino acid side chain from the enzyme is inserted in its place to stack temporarily with the adjacent nucleotides as illustrated in Figure 4.23. In the meantime, the flipped-out cytosine is modified by the methyltransferase before being reinserted into the double helix.

Having now seen how this somewhat modest change in the chemical structure of cytosine is brought about, we now need to ask: what is the biological impact of DNA methylation?

Bacteria use methylation to distinguish newly synthesized DNA and to mark and protect their own DNA

In *Escherichia coli*, adenine methylation can be used to distinguish the newly replicated DNA strand from the old strand in repair processes. While both strands of DNA are methylated in the growing cell, immediately after DNA replication, only

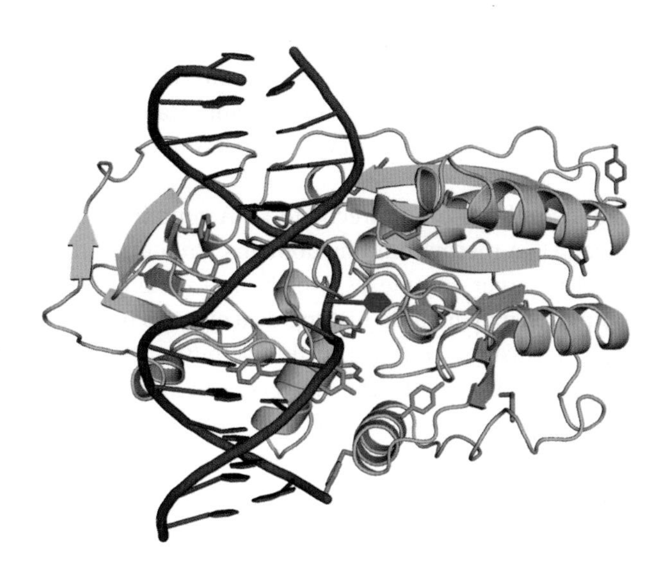

Figure 4.23 Methylation of DNA requires that the base to be modified is flipped out from the double helix. The structure shown is of HhaI methyltransferase (green) bound to DNA (gray). The single cytosine residue to be modified (red) is flipped entirely out of the DNA helix and placed in the methyltransferase active site. This rather drastic distortion of the DNA is the mechanism used by most DNA methyltransferases studied. PDB code 3MHT.

the parental strand remains methylated. We say that such a half-methylated site on a chromosome is **hemi-methylated**. The methylation status of the two strands is 'read' by DNA replication and repair enzymes . Repair proteins search for nearby methylated adenines to identify the parental strand so that they can replace the mismatched, incorrect base in the newly synthesized strand using the correctly identified parental strand as the template.

DNA methylation is also used by bacteria to distinguish their genomic DNA from invading bacteriophage DNA. Bacteria have specific endonucleases called restriction enzymes that cut DNA at sites containing particular recognition sequences; these enzymes are used to destroy foreign DNA that gets into the cell. To avoid destroying their own DNA, bacteria methylate cytosine or adenine residues in the recognition sites in their own genomes, which prevents cleavage by their own restriction enzymes. For example, in some bacteria, the cytosines in all HhaI recognition sites with the sequence GCGC are methylated in the host so that the endogenous HhaI endonuclease will not cleave the DNA at those sites. Foreign DNA from an invading virus will not be methylated, however, and thus will be cleaved at every GCGC site and consequently destroyed.

DNA methylation is associated with gene silencing in plants and mammals

In many eukaryotes, particularly plants and mammals, DNA methylation is used to silence transcription of genes. In other words, the presence of methylated cytosines along a DNA strand is interpreted by the cell to signal that the DNA that is modified in this way should not be transcribed. Genes that are tagged in this way usually become switched off.

Cytosine methylation in eukaryotes occurs most commonly at the sequences, **CpG** and **CpXpG**. Up to 60% of CpG base steps (that is, adjacent C and G nucleotides joined by a phosphodiester bond) in the human genome are methylated in somatic cells. An important feature of this form of transcription repression is that the methylated - and hence silenced - state can be passed on to daughter cells, thereby maintaining the silenced state over many generations of cell division. That is, when the DNA is replicated, it is not only the DNA sequence that is preserved, but also the patterning of methylation along that DNA strand. The pattern can be conserved through successive cell generations by fully methylating the hemi-methylated sites created during replication. This type of gene regulation is referred to as **epigenetic silencing**, which means that the information inherited by daughter cells is not due to changes in the DNA nucleotide sequence. (The term 'epigenetic' refers to any type of heritable state or change that is not encoded in the DNA base sequence, but which affects the expression state of that DNA.)

Certain protein complexes bind directly to DNA containing methylated cytosine via specialized DNA methyl-binding domains. Proteins with these domains are often subunits of larger complexes, which are thus directed to methylated DNA. These complexes may include histone deacetylases or methyltransferases that then modify the histone proteins in the vicinity of the methylated DNA, as depicted in Figures 4.24a and b. The converse can also occur: there are proteins that bind to modified histones, which in turn recruit DNA methyltransferases that methylate DNA in the vicinity of the modified histones (Figure 4.24c). Nucleosome-remodeling complexes can also be recruited to DNA through specific interactions with methylated DNA (Figure 4.24d). Since some eukaryotes,

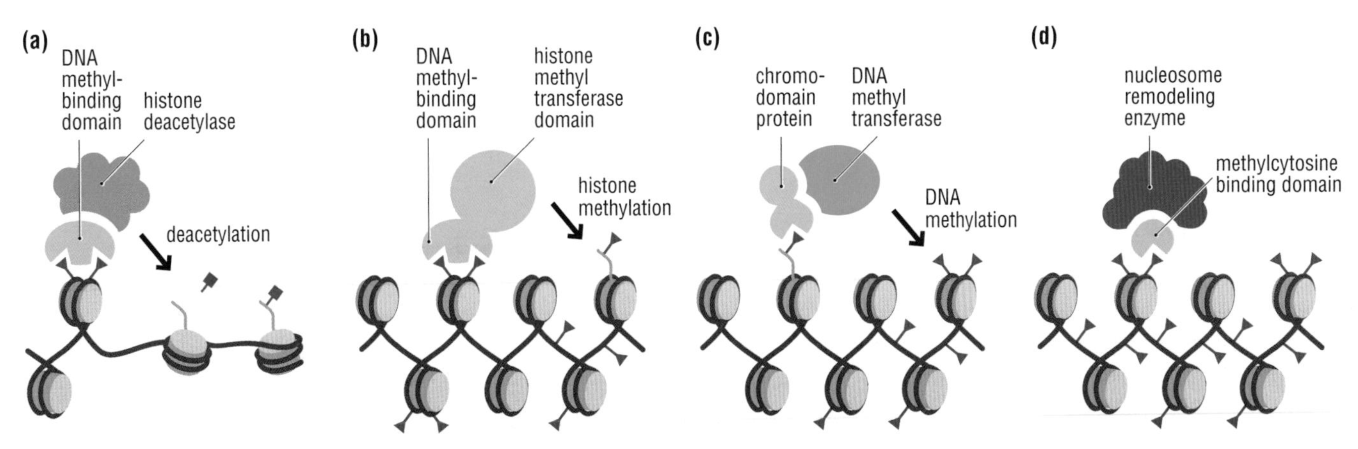

Figure 4.24 DNA methylation can recruit proteins involved in histone modification and vice versa. Methyl groups (blue triangles) can be added to both DNA and histone tails. (a) Methyl groups on DNA recruit methyl-binding proteins (green), which in turn can recruit histone deacetylase (orange) which modifies chromatin by removing acetyl groups (pink squares) from histone tails and thus initiates silencing. (b) Methyl-binding proteins that have histone methyltransferase activity (green) can bind to sites of DNA methylation and methylate histones (blue triangle), which can be a signal for chromatin silencing depending on the site methylated. (c) Methylated histone tails can bind chromodomain-containing proteins, which in turn can recruit DNA methyltransferases (aqua) that methylate adjacent DNA. (d) Nucleosome remodeling complexes (dark blue) can bind specifically to regions of methylated DNA, using subunits that specifically recognize methylated cytosine.

such as yeast and fruit flies, silence specific genes using histone modification but lack DNA methylation, it is thought that DNA methylation is a special mechanism present in mammals and some other species to lock in a silent chromatin state.

DNA methylation is involved in X-chromosome inactivation and genomic imprinting in mammals

X inactivation is one of the best examples of the interrelation between DNA methylation, formation of heterochromatin and the silencing of chromosome functions. In female mammals, one of the two X chromosomes in each cell is permanently inactivated early in development so that dosage of the gene products encoded on the X is the same as in males, who have only one X chromosome. What do we mean by the 'dosage of gene products'? We can imagine a female carrying two copies of a given gene - one on each X chromosome. By contrast, a male would carry just one copy of the same gene, because they only have one X chromosome. If both X chromosomes in a female were transcriptionally active, the female would express double the amount of gene product relative to the male, because the female has two copies of the gene being transcribed, rather than just the one. X inactivation helps to correct the imbalance by switching off one of the female's X chromosomes such that both sexes have the same number of X chromosomes in an active state. The inactive X bears all the hallmarks of mammalian heterochromatin: the DNA is heavily methylated, the histones are modified as in heterochromatin, and most regions are completely inactive.

DNA methylation is also linked to a form of gene inactivation called **imprinting**, which is the silencing of one copy of a gene on either the maternal or the paternal chromosome, while the homologous genes on the other chromosome are expressed. Imprinting, which occurs in mammals, is thought to control the activity of critical paternal and maternal genes that regulate growth of the embryo. Figure 4.25 shows how, for some genes, the copy inherited from the mother is silenced, while that inherited from the father is active. For others, the copy from the father is silenced. The inactive copy of the gene is usually highly methylated.

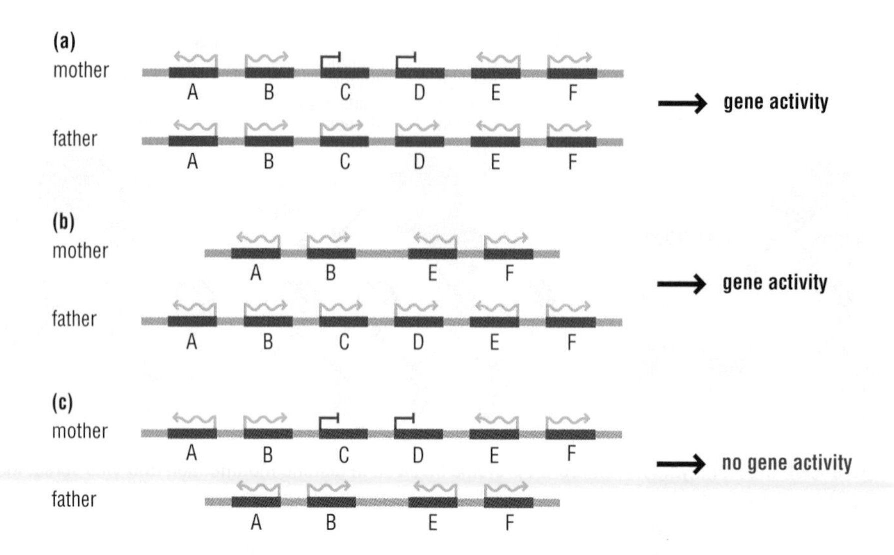

Figure 4.25 The activity of some genes is affected by genomic imprinting. (a) For loci where imprinting occurs, the copy of the gene that was inherited from the mother may be packaged differently and silenced, while the copy inherited from the father is active (green arrows indicate 'active' or expressed, while red bars indicate 'inactive' or silenced). (b) If a deletion removes a region containing genes C and D from the mother, there is still a functional gene product from the father. (c) If a deletion occurs in the copy from the father, however, no functional gene product for genes C and D will be made. Although this example shows the maternal allele being silenced, which copy is imprinted depends on the particular locus.

In normal individuals, expression of only one copy of these genes is needed. However, if the active gene is deleted, there is no expression of the gene at all, which usually leads to developmental abnormalities or death of the embryo. An example of an imprinted gene is *IGF2* (*insulin-like growth factor 2*), which encodes a peptide hormone that plays a role during embryonic development. Normally, only the paternal copy of the gene is expressed because the maternal copy is silenced. In some types of tumor cells, the imprinting of the maternal gene is lost, leading to aberrant expression of both copies of the gene. Imprinted genes are often found in clusters, suggesting that extensive regions of chromatin may be imprinted coordinately.

4.8 THE SEPARATION OF CHROMATIN DOMAINS BY BOUNDARY ELEMENTS

Special DNA sites establish the transition between euchromatin and heterochromatin

The different chemical modifications we have learned about in the preceding sections help to differentiate actively transcribed chromatin – the euchromatin – from transcriptionally silent heterochromatin. Cells appear to have specific boundaries between euchromatin and heterochromatin, implying that there is a mechanism for maintaining the separation between these two distinct forms of chromatin and preventing one from encroaching on the other. These sites on the chromosome are known as **insulators** or **boundary elements**.

Evidence for specific boundaries between euchromatin and heterochromatin came first from classical experiments on **position-effect variegation** in fruit flies. Wild-type fruit flies have red eyes, which comes from the expression of a gene called *white*$^+$. (Mutant flies that lack an intact copy of this gene, or which cannot produce a functional protein product, lack the red pigment and therefore have white eyes.) However, an unusual pattern was sometimes observed in which the compound eye of the fruit fly had a mottled appearance, with some facets of the eye colored red while the rest were white. These differences are illustrated in Figure 4.26a.

How could the expression of the *white*$^+$ gene differ from facet to facet in the eye of a single fly? It turns out that there is a rearrangement in the chromosome of

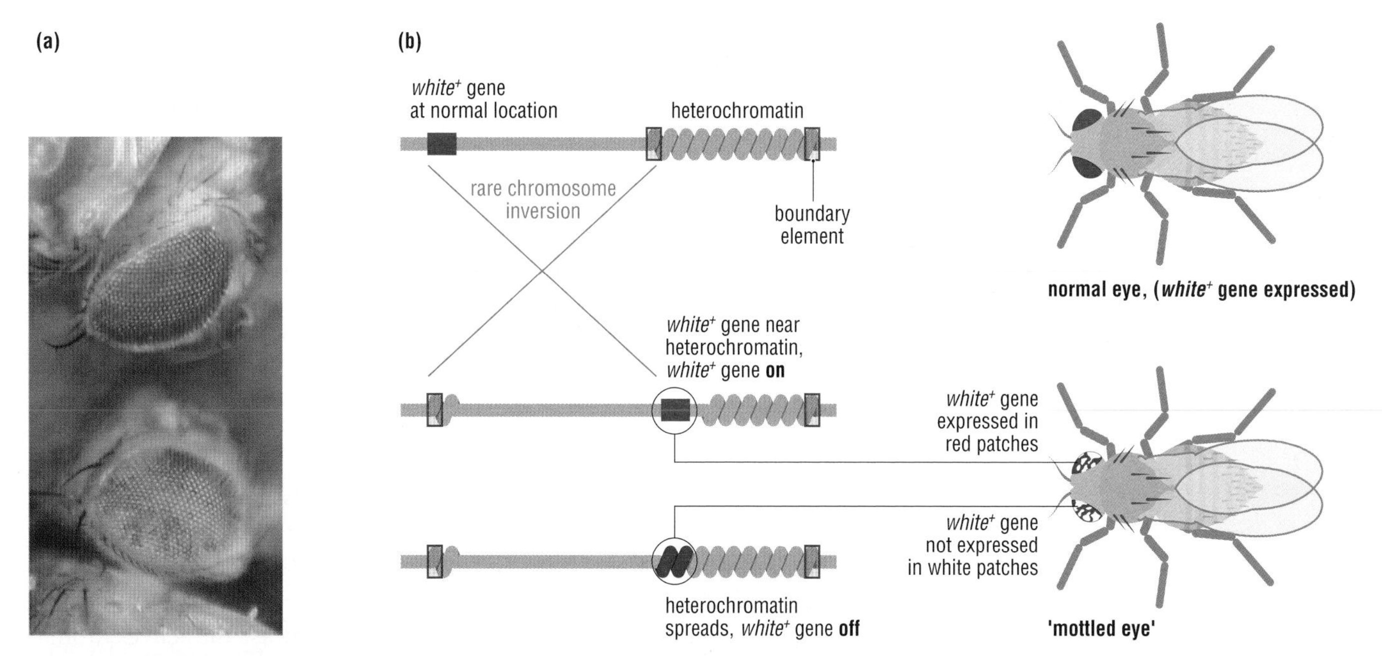

Figure 4.26 Position-effect variegation in *Drosophila*. (a) The normal *Drosophila* eye (top) and a mottled eye due to position-effect variegation (bottom). (b) Wild-type fruit flies that express the *white*+ gene have red eyes. The *white*+ gene is normally located in euchromatin. There is a nearby region of heterochromatin that is flanked by boundary elements (open boxes). In some mutants, a chromosome inversion occurs that places the *white*+ gene adjacent to the heterochromatin and removes the boundary element between them. In some eye cells, the *white*+ gene continues to be expressed and the cells are red, but in other cells the heterochromatin encroaches on the *white* locus due to loss of the boundary element. In these cells, *white*+ is silenced, giving rise to patches of white cells in the eye.

Figure 4.26a reproduced from Lippman and Martienssen. The role of RNA interference in heterochromatic silencing. *Nature*, 2004; **431**: 364–370.

the mutant fly that moves the white gene from its normal location in euchromatin to a location immediately adjacent to a heterochromatic region, as shown in Figure 4.26b. This rearrangement, known as a chromosomal inversion, also likely removes a barrier element that normally separates the euchromatic and heterochromatic regions, thus allowing the heterochromatin to spread into sequences that are usually euchromatic.

The spreading of the heterochromatin silences genes that are normally expressed – in this case, causes silencing of the *white*+ gene in some cells in the compound eye. The degree of heterochromatic spreading varies from cell to cell, such that some cells in which spreading is minimal will continue to express the *white*+ gene and will be red, while others will not (and will be white). Further, because the new heterochromatin states are heritable, the descendants of a cell still expressing the *white*+ gene will give rise to a small patch of red cells (see Figure 4.26b). By contrast, clones of cells in which the heterochromatin has spread into the *white*+ gene will not express the gene and will be white. The result is red and white mottled, or variegated, eyes.

This phenomenon is called position-effect variegation because it depends on the position of the gene on the chromosome. The inherited alteration in gene expression is an example of an epigenetic effect, since, unlike a mutation, the DNA sequence is no different in the two cell types although the expression state of the gene is.

Boundary elements can affect chromatin modification

Boundary elements that apparently stop the spreading of heterochromatin along the chromosome have been identified in a variety of different settings. A direct

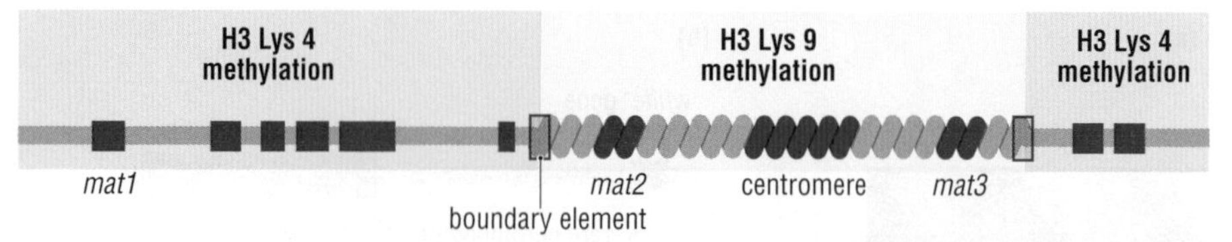

Figure 4.27 Boundary elements mark transitions in different states of histone modification. The *S. pombe* heterochromatic centromere region (dark blue) on chromosome 3 is bounded by two boundary elements. Outside this region, the expressed mating-type locus (*mat1*; in red) and several other expressed genes (red) are located in euchromatin, as indicated by the methylation of lysine 4 on histone H3. Between the boundary elements, shown as open boxes, the DNA is packaged into heterochromatin. The *mat2* and *mat3* loci located in this region are silenced and histone H3 is modified by methylation at lysine 9.

connection between the spreading of heterochromatin and post-translational modification of histone proteins has been found in the fission yeast, *S. pombe*. In this organism, two boundary elements border a region of silent heterochromatin around the centromere. The chromatin between the two boundary elements contains a high level of histone H3 methylated at lysine 9 (see Section 4.4), indicating that this region is silenced, as shown in Figure 4.27. In contrast, the active chromatin to the left and right of the silent region has a high level of lysine 4-methylated histone H3. Deletion of the left boundary element results in the leftward spreading of H3 lysine 9 methylation and chromatin silencing, but does not affect the chromatin to the right of the other boundary. Similarly, deletion of the right boundary element results in spreading of the heterochromatin and H3 lysine 9 methylation to the right. Knowing that boundary elements mediate these effects, the obvious next question is: how?

Boundary elements act by binding specific proteins

There are a number of models that can explain heterochromatin spreading and how the propagation of heterochromatin along the chromosome can be delimited by a boundary element. The best-understood boundary elements are specific DNA sequences that bind those proteins that regulate the spreading of histone-modification activity. Figure 4.28a shows how heterochromatin can spread by the progressive modification of one nucleosome after the other until a boundary element is reached. Heterochromatin spreading can be initiated by deacetylation of histone H3, followed by methylation of H3 at lysine 9 and subsequent binding of a chromodomain-containing chromatin-silencing protein, such as Swi6 in yeast or HP1 in animals, and/or recruitment of other silencing factors. Boundary elements act as physical barriers to this progressive histone modification, inhibiting modification beyond the boundary element itself.

Another way in which boundary elements could function is to separate large defined looped domains of chromatin, as depicted in Figure 4.28b. By anchoring the domains at the base of the loops, they could delimit distinct, separate regions of the chromosome that may differ in chromatin structure and, possibly, in the relative amount of transcription that occurs in that region. There is evidence that, in some cases, proteins that specifically bind to boundary elements anchor that region of the chromosome to the nuclear periphery, which is followed by silencing of the genes in that region. Heterochromatin is often found at the nuclear periphery, which suggests that the relocation of a chromosome segment to the nuclear periphery triggers heterochromatin formation (see Figure 4.28c and Figure 4.11).

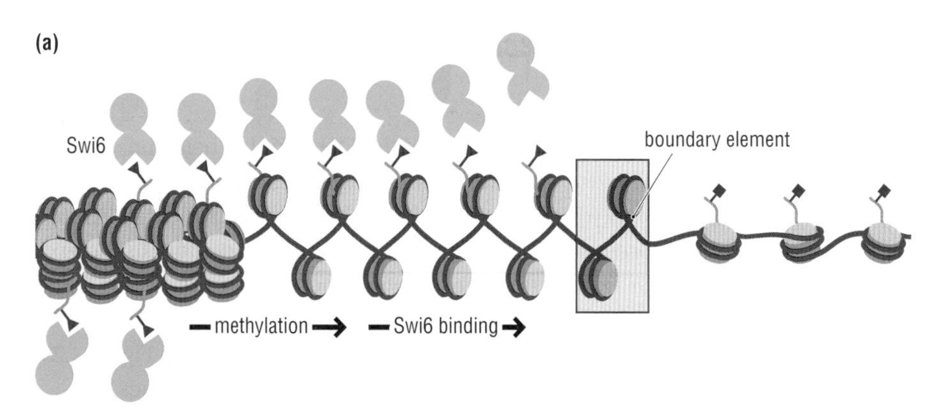

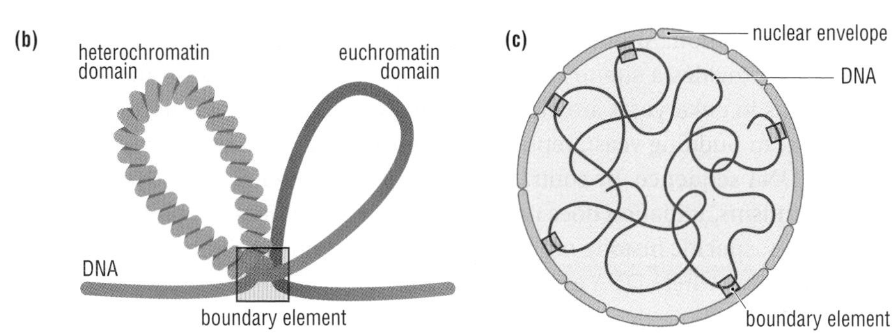

Figure 4.28 Possible ways in which boundary elements might function. (a) Some boundary elements are sites that stop the propagation of progressive histone modification. Histones in euchromatin can become methylated (dark blue triangles), which then allows binding of heterochromatin-inducing proteins such as Swi6 (green balls). Binding of Swi6 facilitates the binding of more Swi6, and thus a region of DNA can become progressively incorporated into heterochromatin. At the boundary element, this propagation stops. The chromatin on the other side can be active euchromatin, with acetylated histones (pink squares). (b) Boundary elements might be sites that organize chromatin into large looped domains that differ in structure. All of the chromatin in one loop will be heterochromatin, and in a separate loop it will all be euchromatin. (c) Another suggestion is that boundary elements are sites at which DNA is attached to the nuclear envelope. This is supported by the fact that heterochromatin is often found around the periphery of the nucleus.

4.9 ELEMENTS REQUIRED FOR CHROMOSOME FUNCTION

In earlier sections, we saw how chromosomes comprise two key components - DNA and histone proteins. These two components assemble to form the nucleosome, the fundamental 'unit' of chromosome packing. However, there are other structural elements of a chromosome that are central to its correct behavior. In the next sections, we explore these elements - the origin of replication, the centromere, and the telomere.

All chromosomes contain specialized regions that direct essential chromosomal functions

Chromosomes contain special features that allow them to be duplicated during DNA replication, and to ensure that the resulting copies of the chromosomal DNA are segregated into the two daughter cells at cell division.

Replication in both bacteria and eukaryotes initiates at the **origins of replication** (*ori*) as illustrated in Figure 4.29a. In bacteria, sequences near the origin of replication also govern the distribution of the duplicated chromosomes to the two daughter cells, while a specific **terminus region** (*ter*) specifies termination of replication. In eukaryotes, the **centromere** directs chromosome segregation. In addition, specialized regions called **telomeres** at the ends of eukaryotic chromosomes stabilize the ends and assure the maintenance of the most terminal portions of the DNA (see Figure 4.29b). We will briefly introduce replication origins and terminus regions here, and will discuss telomeres and centromeres in detail in the next two sections.

➔ Replication origins and the mechanism and regulation of DNA replication are described in more detail in Chapter 6.

DNA replication initiates at specific chromosomal sites termed origins

For DNA replication to begin, the two strands of the DNA double helix must be unwound and separated to allow the replication machinery access to the DNA. Origins of replication (*ori*) are the sites at which this unwinding occurs. Once initiated, DNA replication is bidirectional, proceeding away from the origin in both directions. In bacteria, DNA replication initiates at a single site. In eukaryotes, whose chromosomes are linear and typically much larger than bacterial chromosomes, DNA replication initiates at multiple sites along the chromosome; these differences are illustrated in Figure 4.29. Origins of replication are typically found in regions that do not contain genes.

➔ The reason the very end of the chromosome cannot be copied is discussed in Section 6.11.

In *E. coli* and other bacteria, a distinctive DNA sequence specifies the single origin of replication. Although the exact sequence differs in different bacteria, in all cases it provides a site to which specialized initiator proteins bind. Replication origins in eukaryotes are often, but not always, defined by a particular DNA sequence. In budding yeast, replication origins are defined by a relatively short, specific DNA sequence. By contrast, in *Drosophila*, humans, and other multicellular organisms, initiation does not always occur at a unique origin sequence. In these cases, specific histone modifications, rather than a unique DNA sequence, may determine where DNA replication begins.

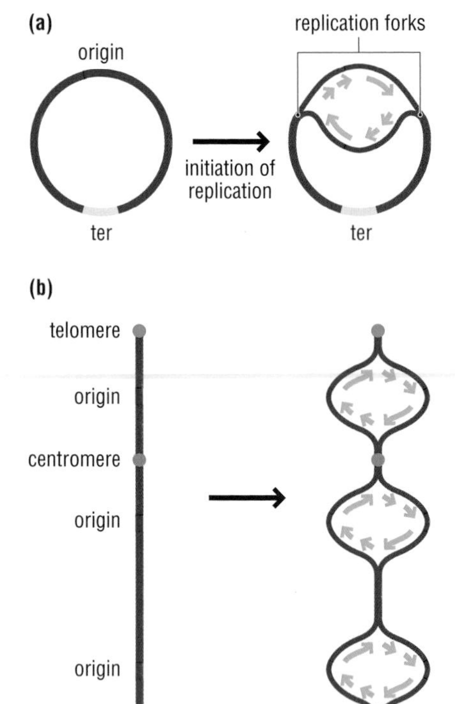

Figure 4.29 Components of chromosomes in bacteria and eukaryotes. (a) A circular bacterial chromosome has a single origin (red), which initiates replication that proceeds in both directions around the chromosome and terminates at the *ter* region (yellow). (b) The multiple origins on a linear eukaryotic chromosome initiate bidirectional replication independently at many sites along the chromosome. Chromosomes also contain a centrosome (in orange) needed for chromosome segregation, and telomeres at chromosomes end (turquoise circles) that protect the ends from attrition.

DNA replication of linear chromosomes poses a special problem

In linear chromosomes, bidirectional replication from internal origins allows copying of all of the sequences in the internal part of the chromosome (Figure 4.29b). However, when the replication fork gets to the very end of the DNA molecule, there is a small region on one strand of DNA that cannot be copied. (This end-replication problem is discussed in detail in Section 6.11.)

Thus, when there is no mechanism to compensate for this, the chromosome gets shorter every time the DNA is replicated. Telomeres provide a mechanism of overcoming this end-replication problem, and also protect the chromosomes from degradation and from fusing end-to-end with other chromosomes, as we discover in Section 4.11.

4.10 THE CENTROMERE

The centromere is a region of DNA necessary for chromosome segregation

Centromeres are components of all eukaryotic chromosomes that are necessary for correct chromosome segregation at cell division. In most species, each chromosome has a single centromere, which is packaged into heterochromatin and whose structure is stably inherited from one cell generation to the next. The centromeric DNA binds specific proteins that form a structure called the **kinetochore**. In mitotic chromosomes, in which duplication of DNA strands has occurred but sister chromatids have yet to separate, the kinetochore appears as the site where paired sister chromatids are constricted (Figure 4.30).

During mitosis, kinetochores of sister chromosomes attach to microtubules emanating from opposite spindle poles. This attachment allows the spindle to pull sister chromatids apart and distribute them to the two daughter cells, as shown in Figure 4.31. The segregation of chromosomes through the attachment of centromeres to a mitotic spindle is an essential conserved function in all eukaryotes. Yet, remarkably, centromeres of different species have distinct DNA sequences. Instead, it is the protein components of the centromeric chromatin and kinetochores that are conserved. Here we will discuss properties of centromeres.

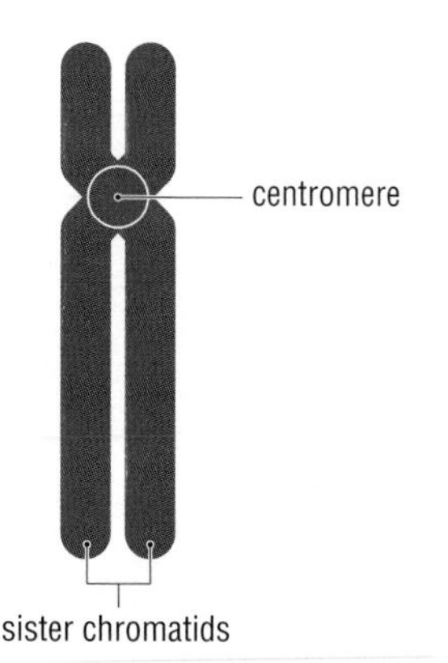

Figure 4.30 Duplicated chromosomes consist of two sister chromatids still joined together. Diagram of two sister chromatids at metaphase after duplication but before the identical duplicated copies have separated.

➔ We discuss kinetochore structure and function in Section 7.4.

➔ The identification of individual chromosomes through the method of fluorescence *in situ* hybridization (FISH) is covered in Section 19.9.

Centromere sequence and size vary widely between species

Figure 4.32 shows how centromeres in different organisms vary not only in sequence but also in size. *S. cerevisiae* and some other yeasts have very small, defined DNA sequences at the centromere; these are termed **point centromeres** because of their small size and simple organization. The way in which yeast centromeres were first isolated is described in Experimental approach 4.3. The *S. cerevisiae* centromeric DNA is 125 bp long and is divided into three distinct elements: centromere DNA elements (CDE) I, II, and III. These elements are sufficient to provide centromere function when incorporated into another DNA molecule. CDE I and III each consist of a unique conserved sequence. Mutations in these elements disrupt centromere function. CDE II consists of 80–90 bp of AT-rich DNA but, unlike CDE I and III, the exact sequence is not important to its function. In other yeast species with point centromeres, the DNA sequence differs from that in *S. cerevisiae*.

Regional centromeres, in contrast, consist of hundreds of kilobases (kb) of repetitive DNA sequence. The fission yeast *S. pombe*, the fruit fly *Drosophila*, and mammals all have regional centromeres. Each of the three distinct chromosomes of *S. pombe* has a slightly different centromere DNA sequence, ranging in size from 40 kb to 100 kb. Each centromere has a large inverted repeat sequence, which itself consists of smaller repetitive sequences, surrounding a unique central core sequence of 5–6 kb (see Figure 4.32). The core sequence differs on each chromosome, while the repetitive sequence is similar. The core alone is not sufficient to establish centromere function.

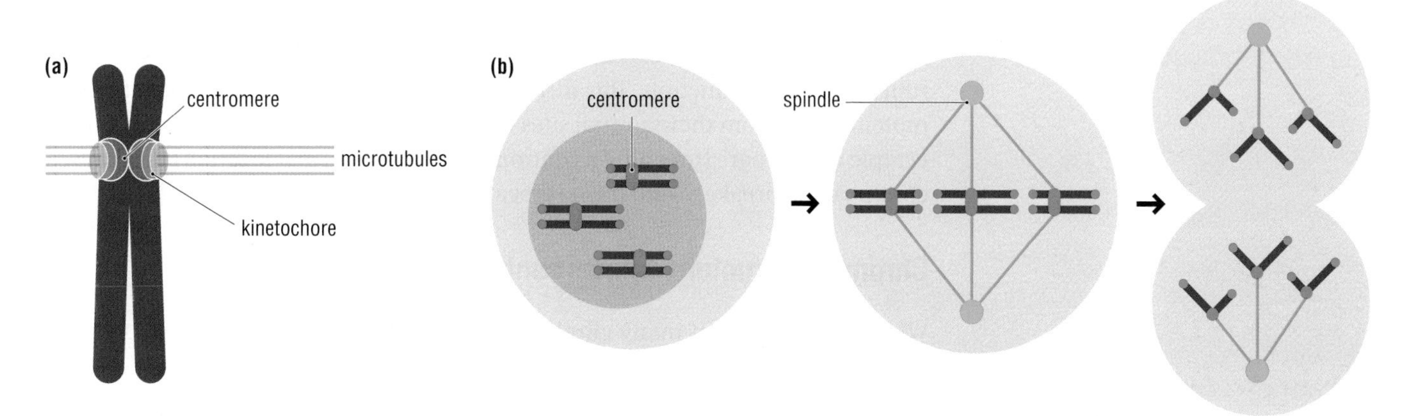

Figure 4.31 Chromosome segregation in eukaryotes. (a) The kinetochore mediates the attachment of spindle fibers to the centromere. (b) At mitosis, the centromere of each duplicated chromosome becomes attached to the mitotic spindle (here represented in highly schematic form in green), and the two sister chromatids are pulled to opposite poles of the cell. Telomeres are shown as turquoise circles.

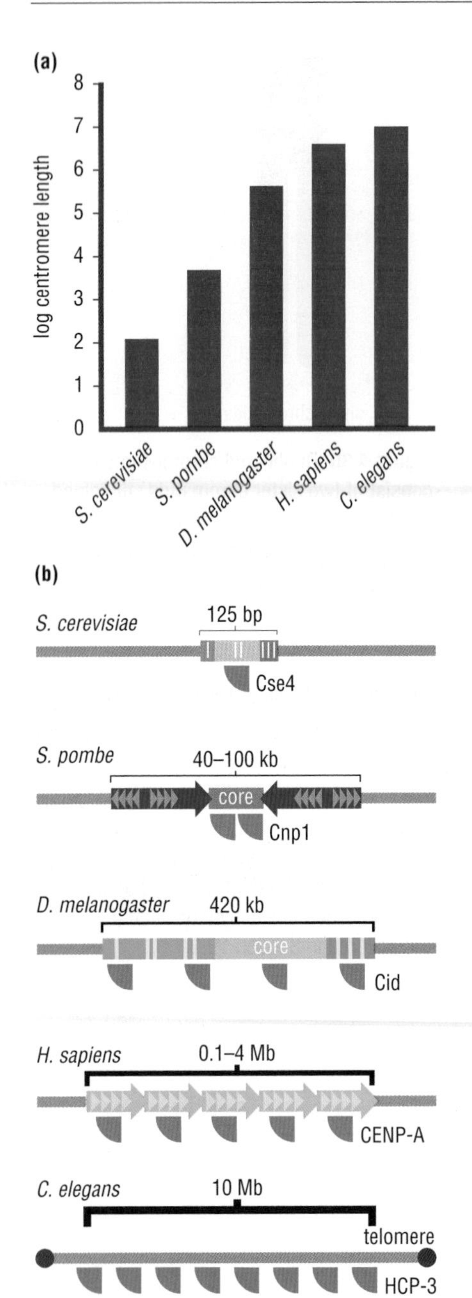

Figure 4.32 Centromeres differ in size and DNA sequence in different species. *S. cerevisiae* centromeres are about 125 bp long and consist of three DNA elements: CDE I, II, and III (dark/light blue). *S. pombe* centromeres are 40–100 kb in size and are made of many repeated sequence elements (small purple arrows) that make up larger repeats (large dark blue arrows). Between the two large inverted repeats is a core region of unique sequence, which differs between chromosomes. *Drosophila* centromeres are around 420 kb in size and are made of repetitive sequence; they include many transposable elements (yellow lines). Human centromeres are 100 kb to 4 Mbp in size and consist of tandem arrays of alpha-satellite repeats (small light green arrows) grouped into higher-order repeats (large green arrows). *C. elegans* chromosomes are holocentric: a centromere can be established at any point along the chromosome. In all organisms, a homolog of the mammalian histone H3 variant CENP-A (blue quarter-circles) binds centromeric DNA.

Drosophila centromeres are about 400 kb in size and also consist mostly of repetitive DNA. Human and mouse centromeric DNA consists of about a megabase (1 million bases; 1Mbp) of repetitive sequence termed **alpha (α)-satellite** repeats. It is unrelated to the sequence in *S. pombe* or *Drosophila*, and consists of tandem repeats of 171 bp, which are organized into an array of higher-order repeats as depicted in Figure 4.32. Although the sequences of alpha-satellite repeats are all very similar, each chromosome contains its own specific variant and, indeed, can be identified by it.

Some organisms, including the nematode (worm) *Caenorhabditis elegans*, do not have defined centromeres (and so lack defined centromeric DNA sequences); their chromosomes are said to be **holocentric**. Rather than attaching at a unique site, multiple spindle microtubules attach at numerous sites along the length of the chromosome.

Centromeres bind a histone H3 variant called CENP-A

Despite dramatic sequence differences, centromeres of all organisms have a defined region in which nucleosomes contain a histone H3 variant, known as CENP-A (or CENH3, in place of histone H3). CENP-A shows specificity for centromeric regions, and particularly the AT-rich domains. The region of the centromere demarked by the presence of CENP-A is the domain onto which kinetochore proteins assemble. The presence of CENP-A is crucial for centromere function: cells that lack CENP-A fail to recruit most known kinetochore components to the kinetochore. We do not yet know how this histone variant gets incorporated only at centromeric sites, but it is clear that the levels of CENP-A are carefully regulated, as abnormally high CENP-A protein levels can lead to binding at non-centromeric sites and disrupt kinetochore function. This disruption may result from drawing kinetochore proteins away from their normal sites of assembly, or by directing assembly of an ectopic centromere (meaning a centromere in an abnormal location) that causes chromosome breakage and mis-segregation during mitosis.

Chromatin structure defines centromere function

While the function of many chromosomal elements depends on their DNA sequence, surprisingly it is chromatin structure, not DNA sequence, that defines a functional centromere. This was initially suggested by two observations. First, in human cells, not all large tracts of alpha-satellite sequence form a functional centromere. Human chromosomes that are rearranged to generate two centromeres, and chromosomes into which an extra-large tract of alpha-satellite DNA

4.3 EXPERIMENTAL APPROACH

The isolation of a yeast centromere

Development of an assay for centromere function

Although centromeres had been seen in cytological studies over a hundred years ago and had long been studied using genetic approaches, their molecular structure was first uncovered by Clarke and Carbon in 1980. These scientists used a combination of genetics and molecular biology to isolate the first centromere sequence from the yeast *S. cerevisiae*. The key to the success of this experiment was their development of an assay for centromere function, illustrated in Figure 1. Their assay relied on a type of plasmid known as an ARS plasmid (for autonomously replicating sequence) that has a replication origin. When ARS plasmids are introduced into yeast, they can replicate, but do not segregate when the cell divides. As a consequence, the plasmids accumulate in the mother cell, and the daughter cells do not receive the plasmid at cell division. Clarke and Carbon reasoned that if a DNA element that conferred centromere function was added to the plasmid, it would segregate at mitosis so that all of the cell progeny would have the plasmid. This assay turned out to work tremendously well.

Genetic mapping and chromosome walking allowed isolation of the first centromere sequence

To start the hunt for centromere DNA, Clarke and Carbon focused on a region of chromosome III where the approximate location of the centromere was known from genetic studies. They had in their collection DNA clones for two genes that flanked the centromere: LEU2 mapped to the left and CDC10 mapped to the right of the centromere on the genetic map, as illustrated in Figure 2. These experiments were done long before the genome sequence was known, so they had to find a way to isolate DNA fragments between these two known markers. They employed a technique called chromosome walking, in which successive overlapping clones are isolated that span the region of the DNA between the two known markers. Once they had a set of clones that they suspected might contain a centromere, they tested each clone in their plasmid segregation assay (see Figure 1). Indeed, three independent clones conferred mitotic segregation on the usually unstable ARS plasmids, these clones were then further characterized.

To identify the sequence responsible for 'centromere' function, they tested smaller and smaller pieces of the initial DNA candidates that they had cloned, finally narrowing in on a 1.6 kb sequence that was sufficient to confer mitotic stability. When this region was deleted from the plasmid, the plasmid no longer segregated to both mother and daughter cells. Moreover, in addition to conferring plasmid segregation at mitosis, this piece of DNA also conferred proper segregation at meiosis. We see how the careful development of a sensitive assay, coupled with the relatively straightforward procedure of cloning, paid off to yield results of longstanding significance.

Find out more

Clarke L, Carbon J. Isolation of a yeast centromere and construction of functional small circular chromosomes. *Nature*, 1980; **287**: 504–509.

Related techniques

S. cerevisiae (as a model organism); Section 19.1

Plasmid libraries; Section 19.4

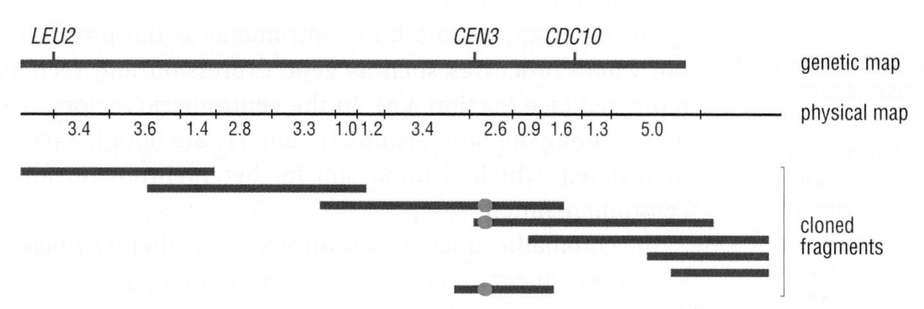

Figure 1 Genetic and physical map of the centromere region of chromosome 3. The top line represents the genetic map near CEN3. The line below represents the physical map of the region. The lines below the physical map show the DNA fragments that were assayed to test for centromere activity. The orange circle represents the region carrying functional centromere sequences.

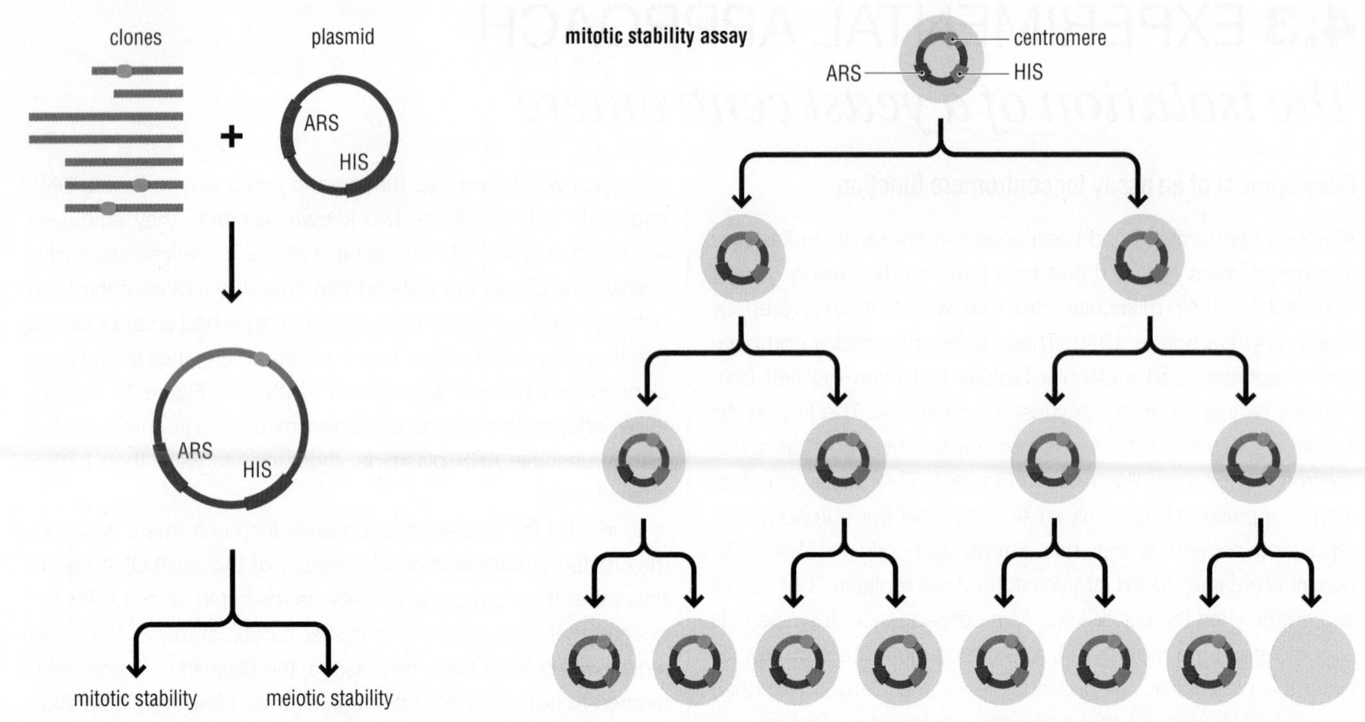

Figure 2 Assay for identification of a functional centromeric DNA. Fragments of yeast DNA were cloned into a plasmid containing an ARS (red) and a selectable marker (HIS (blue), which allows certain yeast cells to grow on growth media lacking histidine). Each clone was tested for its ability to segregate in a mitotic stability assay. For those clones that have a centromere (orange), the plasmid segregates at mitosis and all of the progeny contain the plasmid. Occasionally, the plasmid fails to replicate or segregate, resulting in cells lacking the plasmid (right-most cell in the bottom row). In contrast, if the plasmid lacked a centromere sequence the plasmid would accumulate in the mother cells and most cells would not have the HIS genetic marker (Not shown in figure)

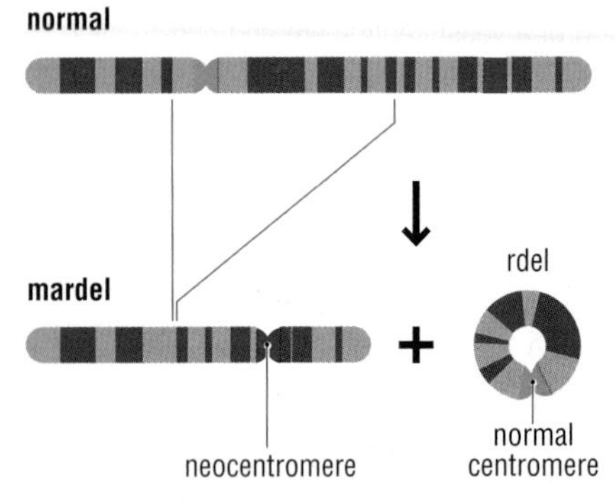

Figure 4.33 Neocentromeres can form away from the usual centromere region. A normal human chromosome 10 is shown at the top with the centromere marked in orange. A deletion removes the centromere, and the region marked in pink becomes a neocentromere; the new chromosome is known as mardel (marker deletion). The region of DNA containing the centromere that was deleted can circularize and form a separate chromosome known as rdel (ring deletion). Neocentromeres appear to be generated very rarely. The formation of neocentromeres is considered an epigenetic phenomenon, as the function of the DNA is altered in a heritable manner without any alteration in its DNA sequence.

is introduced, would be expected to have two centromeres, but in fact often are found to contain just one active centromere. The second potential centromere is inactivated, and essential centromere proteins are found to no longer be bound there. This **centromere inactivation** prevents the chromosome breakage that would occur if two centromeres on the same chromosome were pulled to opposite poles at cell division.

The second observation that chromatin structure defines centromeres was the discovery of new functional centromeres, or **neocentromeres**, that can form on a DNA fragment that has no other active centromere, as depicted in Figure 4.33. The formation of neocentromeres can occur in a chromosome that has lost its centromere through a deletion of that segment of DNA. Neocentromeres can form at sites that do not contain alpha-satellite DNA, but can nevertheless assemble a functional kinetochore.

An important hallmark of centromeres is the presence of heterochromatin, where processes such as gene expression and recombination are often repressed (see Section 4.4). In the centromeric regions that flank the CENP-A-containing domain, histone H3 and H4 are both hypoacetylated and hypermethylated, which is important for heterochromatin formation and proper centromere function.

The chromatin structure of centromeres is inherited epigenetically, that is, the sites at which centromeres are formed on newly replicated DNA are not solely determined by the DNA sequence of the parent chromosome. Instead, the new centromeres form at sites determined by the location of centrosomes on chromosomes in the parent cell, as illustrated in Figure 4.34.

How then are centromeres established at a *new* site? It has been proposed that repetitive, AT-rich DNA, such as alpha-satellite DNA, has a high probability of binding CENP-A to form a centromere-specific chromatin structure. Thus, alpha-satellite sequences found at all centromeres are most likely to form centromeres. At a low probability, however, CENP-A can bind to a DNA sequence other than alpha-satellite and establish a centromere, thus forming a neocentromere. This newly established functional centromere is self-propagating, irrespective of its DNA sequence, which is why neocentromeres are stably inherited at each cell division. Self-propagation of centromere structure is achieved by epigenetic mechanisms for maintaining heterochromatin, such as histone modification and the recruitment of specific proteins described in Section 4.4. We discuss this inheritance of chromatin structure in more detail in Section 6.12.

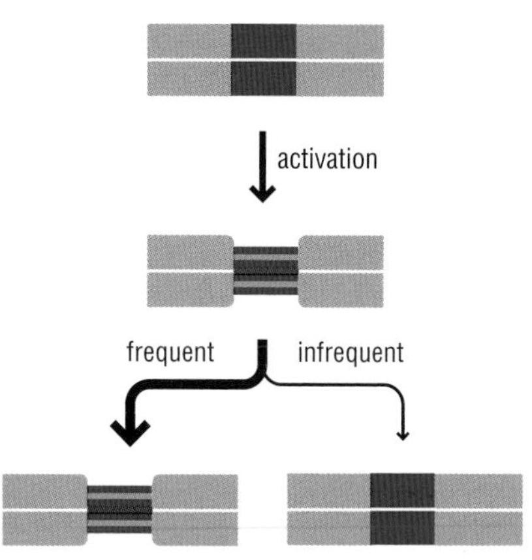

Figure 4.34 Centromeres are formed and lost by epigenetic mechanisms. A region of DNA, usually comprising alpha-satellite sequence, can become activated to form a centromere (striped blue and gray), probably by binding CENP-A. With each cell division there is a high probability that this centromere will self-propagate and a new centromere will form at that location on the daughter chromosomes. Very infrequently, however, the centromere will be inactivated (solid blue box), self-propagation will not occur, and centromere function will be lost.

4.11 THE TELOMERE

Having now considered centromeres, the specialized structures that appear within the chromosome, let us move on to consider a specialized element that appears at the end of the chromosome - the telomere.

Telomeres protect the ends of linear chromosomes

Telomeres are specialized structures that define the ends of eukaryotic chromosomes and are essential for their stability. Telomeric DNA consists of simple tandem repeated sequences that have one strand that is G-rich (containing many G residues) and the complementary strand that is C-rich. Representative repeat sequences from different organisms are listed in Figure 4.35; human telomeres, for example, consist of many repeats of the sequence TTAGGG. The **G-rich strand** in all eukaryotes is the DNA strand that extends 5' to 3' towards the chromosome end and terminates in a small single-stranded region, as depicted in Figure 4.36a. Thus, while most of the telomere repeats are double-stranded, there is a small G-rich overhang that is important for telomere function. The total length of this repetitive DNA region varies from about 50 bp to 30 000 bp, depending on the species.

In some organisms with long telomeres, the G-rich overhang may be processed into a t-loop (Figure 4.36b). This loop is formed by base-pairing of the single-stranded G overhang with the double-stranded region of the telomere DNA. These t-loop structures are thought to protect the very end of the chromosome, though their precise role is not yet understood.

The telomere repeats are binding sites for specific proteins that mark the chromosome ends as natural ends and thereby distinguish them from DNA breaks. This prevents the inappropriate activation of the cell's DNA-damage response, which often leads to cell death. If either the telomeric DNA sequences or the proteins that bind them are defective, the cell will either die or undergo end-to end chromosomal fusion. Telomeres are also the means of overcoming the end-replication problem, which, as we have already mentioned (Section 4.9), arises from the inability of DNA polymerase to copy the ends of a linear DNA molecule completely. While the bulk of the double-stranded telomere sequence is replicated by conventional DNA replication machinery, the extreme ends are maintained by a specialized enzyme called **telomerase**, which adds new telomere sequences to the ends of chromosomes.

Telomere sequence repeats

group	organism	repeat sequence
ciliates	*Tetrahymena*	TTGGGG
flagellates	*Trypanosoma*	TTAGGG
fungi	*S. cerevisiae*	TG_{1-3}
	S. pombe	$TTAC(A)G_{2-5}$
nematodes	*C. elegans*	TTAGGC
plants	*Arabidopsis*	TTTAGGG
vertebrates	Human	TTAGGG

Figure 4.35 Telomere repeat sequences in various organisms. The telomeric DNA in most eukaryotes consists of tandem arrays of a simple repeat. In some organisms the repeat is irregular such as *S. cerevisiae* and *S. pombe*. The telomeres are made of double stranded DNA, but only the G-rich strand of the DNA sequence is shown for simplicity. The G-rich strand always runs 5' to 3' toward the end of the chromosome. At the very end of the telomere is a 30–100 bp overhang on the G-rich strand.

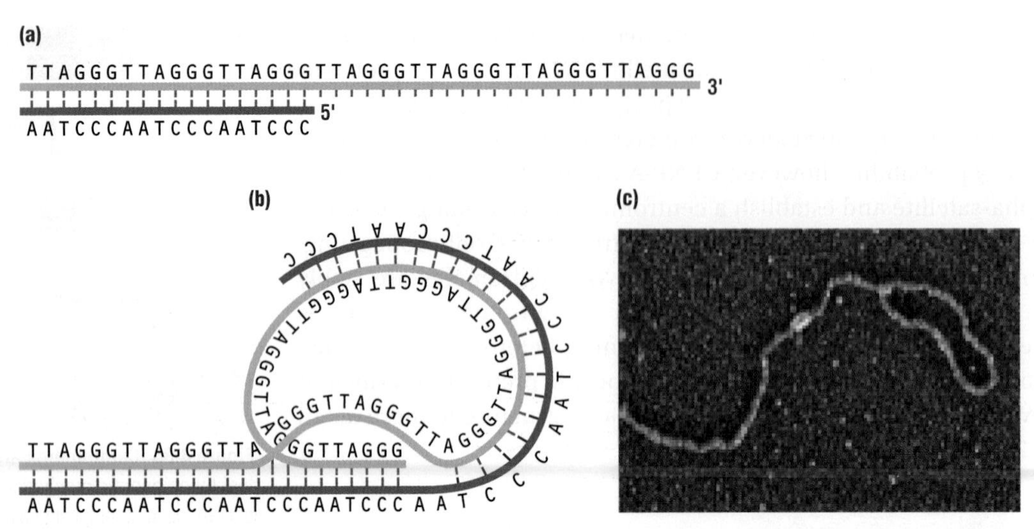

Figure 4.36 Linear and t-loop structure of human telomeres. Telomere DNA may be linear or may form in some organisms a structure known as a t-loop. (a) In the linear state there is a G-rich overhang that is bound and protected by end-binding proteins (see text). (b) The t-loop forms when the single-stranded G-rich tail is base-paired with the adjacent double-stranded DNA. (c) Electron micrograph showing t-loop.

Figure 4.36c reproduced with permission from Jack Griffith.

Telomeres are maintained by the enzyme telomerase

Telomerase is an unusual DNA polymerase that has both a protein component and an essential RNA component. The RNA contains a short template region that specifies the sequence of the telomere repeats that are added onto the chromosome ends. The protein component synthesizes the telomeric DNA, using the RNA component as a template.

In humans, telomerase is active - and telomere length is maintained - in stem cells that divide frequently to maintain tissues in the body and in germline cells that produce the sperm and eggs for the propagation of the species (Figure 4.37a). Similarly, telomerase is active in unicellular eukaryotes such as yeast, which divide indefinitely.

In many human tissues, however, there is not enough telomerase to maintain telomere length and progressive telomere shortening therefore occurs. The importance of telomere maintenance is highlighted by the fact that mutations in telomerase cause inherited diseases in humans such as bone marrow failure, because short telomeres limit the number of divisions cells can undergo. Inappropriate activation of telomerase, on the other hand, can allow cells to continue to grow when they should not, as occurs in cancer. Indeed, telomerase is active in more than 90% of human tumors, and is often required to enable tumor cells to divide indefinitely.

Although telomerase activity is the principal mechanism for telomere maintenance, some species use alternative mechanisms. *Drosophila*, for example, has two transposable elements that compensate for sequence loss by repeatedly transposing new elements onto chromosome ends. Some other insects maintain their chromosome ends through a special form of DNA recombination, which can copy or transfer sequences between two chromosomes.

DNA recombination is also used as a backup mechanism in yeast and humans when telomerase is absent. For example, when telomerase is artificially inactivated in yeast, the telomeres become shorter at each cell division until the cells

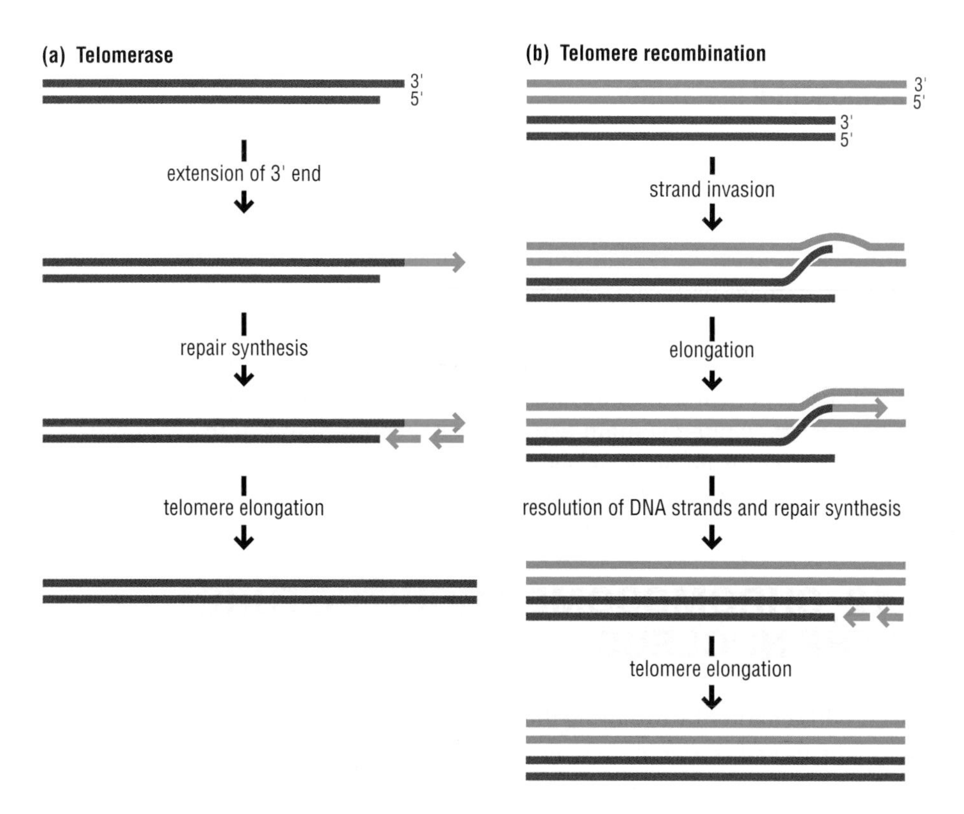

Figure 4.37 Telomerase elongation of telomeres is most frequent, but recombination can elongate telomeres as well. (a) Telomeres contain a single-stranded overhang region on the G-rich 3' strand. The 3' end can be elongated by telomerase. The C-rich strand can then be filled in by repair DNA polymerases. (b) Recombination can lead to the net elongation of telomeres. The sequence at the ends of chromosomes is highly repetitive and is identical on all telomeres in a given species. The 3' end of one chromosome can invade a second chromosome and use the sequence of the second chromosome as a template for telomere elongation by DNA polymerase. The lagging strand can then be filled in by conventional replication mechanisms as detailed in chapter 6.

stop dividing. A few cells, however, continue to grow by using a specialized form of recombination to elongate the telomeres, as illustrated in Figure 4.37b. These cells can therefore grow indefinitely despite the absence of telomerase. This pathway also allows some tumor cells to grow in the absence of telomerase.

Telomere-binding proteins are essential for telomere function

Telomeres contain a variety of proteins that not only physically protect the ends of chromosomes, but also help maintain the telomere at a particular length. There are three classes of telomere-binding proteins, which are illustrated in Figure 4.38. End-binding proteins bind to and protect the G-rich overhang at the extreme terminus; double-stranded DNA-binding proteins bind along the length of the telomere sequence; and there are also proteins that associate with the telomeric DNA-binding proteins.

Telomere-binding proteins regulate the length of telomeres; overexpression of the double-stranded DNA-binding telomere proteins results in shorter telomeres than normal, whereas decreasing their binding results in telomere lengthening. These telomere-binding proteins are thought to regulate the access of telomerase to the telomere, although we do not yet fully understand how this occurs.

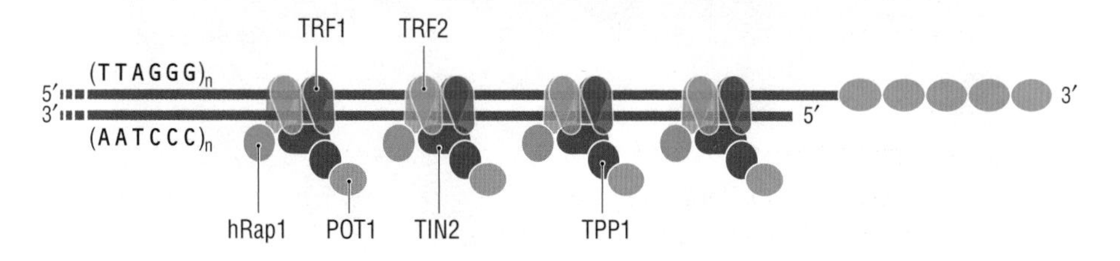

Figure 4.38 Human telomere-binding proteins. Most organisms have three classes of telomere-associated proteins that carry out telomere function: end-binding proteins, double-stranded DNA-binding proteins, and telomere-associated proteins that do not bind the DNA directly but are associated with other bound proteins. The exact proteins involved vary from organism to organism: the human telomere proteins that comprise the shelterin complex are shown here. Pot1 (turquoise ovals) is bound to the single-stranded G-rich tail DNA and is also associated with TPP1. TRF1 (dark blue elongated ovals) and TRF2 (orange elongated ovals) are bound along the length of the double-stranded telomeric DNA. Rap1 (orange circle), Tin2 (red ovals) bind indirectly through their association with TRF1 and TRF2. The double-stranded DNA is also organized into nucleosomes, not shown here for simplicity.

Loss of telomere function leads to a DNA-damage response and chromosomal instability

Telomere function can be disrupted in a variety of ways. Shortening the telomere, for example, leads to the loss of binding sites for proteins. The direct inhibition of protein binding will also leave the chromosome ends unprotected. In all cases, the unprotected end can be recognized by the cell as a DNA break, which activates cellular DNA-damage pathways, causing arrest of the cell cycle or cell death. On rare occasions, the exposed end may recombine with DNA on another chromosome, leading to translocation of genetic material and genetic abnormalities; or it may fuse with another chromosome end. Chromosomes fused end-to-end are unstable because they often contain two functional centromeres. If the centromeres are pulled to opposite poles at cell division, the chromosome will break and the daughter cells will inherit broken chromosomes.

➔ We discuss cellular DNA-damage pathways in more detail in Chapter 15.

4.12 CHROMOSOME ARCHITECTURE IN THE NUCLEUS

We end this chapter by exploring the location of eukaryotic chromosomes in the context of the cell as a whole. It is tempting to view the nucleus as a vessel in which chromosomes move around freely to occupy random locations. However, the reality is rather different.

Chromosomes occupy specific territories in the eukaryotic nucleus

During interphase, genes are being actively transcribed in the nucleus to provide the cell with the macromolecules it needs to carry out a myriad of functions. Within the interphase nucleus, each chromosome occupies a characteristic, discrete space termed a **chromosome territory** as depicted in Figure 4.39. This organization is remarkable because each interphase chromosome, which is in a relatively extended state, is hundreds of times longer than the diameter of the nucleus. One might imagine that in packing this much DNA into a nucleus, the chromosomes would be jumbled together like worms in a bag, instead of the organized distribution that is actually found.

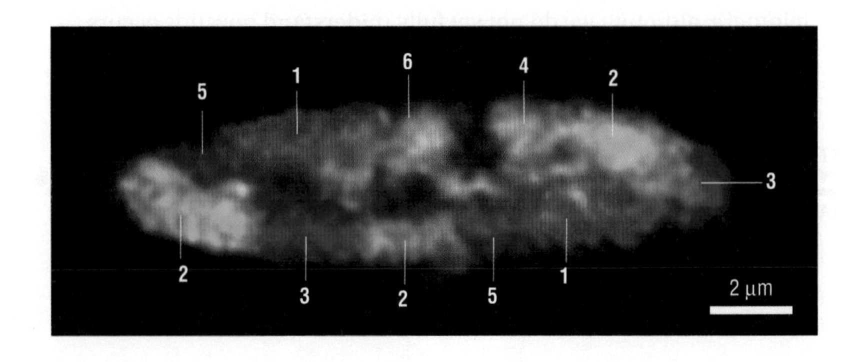

Figure 4.39 Chromosomes are localized to defined territories in the interphase nucleus. The chromosomes in a nucleus from a cultured chicken cell are stained with chromosome-specific fluorescently labeled probes, as in the karyotype shown in Figure 4.2c. An optical section through the nucleus shows each chromosome stained with a different color. Each copy of the five chromosomes is localized to a distinct domain within the nucleus (as the cell is diploid, there are two copies of each chromosome).

From Cremer and Cremer. Chromosome territories, nuclear architecture and gene regulation in mammalian cells. *Nature Reviews Genetics*, 2001; **2**:292–301.

The mechanism by which chromosome territories are established and maintained is not yet clear, although their locations may represent where each chromosome came to rest after mitosis and decondensed, with little further movement. Indeed, in some tissues the three-dimensional arrangement of chromosomes during mitosis is quite clearly maintained in the resulting nuclei, with all the centromeres located at one end and the telomeres at the other. This reflects the arrangement the chromosomes had during segregation as they were pulled to a pole of the cell, with the centromeres in front and the chromosome ends trailing behind.

Even though the actual location of a chromosome in the nucleus during interphase appears not to change markedly, the size of the territory can vary. When transcription of a gene on a given chromosome is activated, for example, the volume of the chromosome's territory increases. This probably reflects further unpacking of the chromatin to allow the transcription machinery to access the DNA.

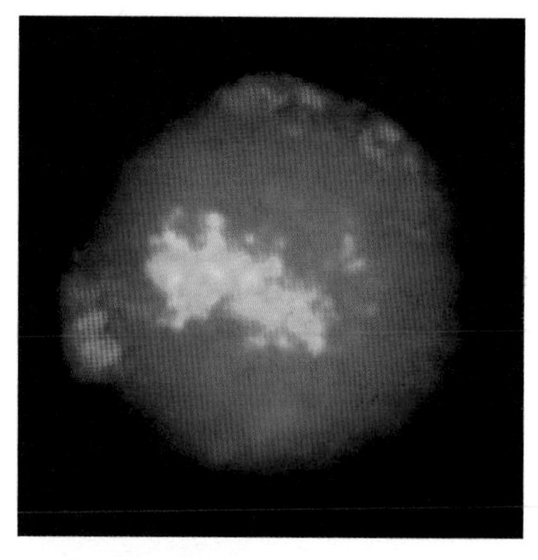

Figure 4.40 The location of human chromosomes 18 and 19 within the nucleus. Human chromosomes 18 and 19 are of a similar size but chromosome 18 has many fewer genes. Chromosome 18, shown colored in red, is reproducibly located at the periphery of the nucleus, while chromosome 19, shown in green, is located more centrally.

From Chubb JR and Bickmore WA. Considering nuclear compartmentalization in the light of nuclear dynamics. *Cell*, 2003; **112**, 4, p. 403–406.

The location of a chromosomal region is correlated with transcriptional activity

It is not just the location of a chromosome, as a complete entity, within the nucleus that follows a characteristic pattern. In addition, particular regions *within a single chromosome* can be found in characteristic locations. Specifically, the nuclear location of a chromosomal region is correlated with its transcriptional activity. Heterochromatic regions, in which genes are silenced, are often associated with the nuclear envelope (see Section 4.7). Electron micrographs first revealed the dense-staining heterochromatin around the periphery of the nucleus (see Figure 4.11), and many subsequent studies have reinforced this observation. In female mammals, for example, the heterochromatic inactive X chromosome is always located at the nuclear periphery. In yeast, the telomeres, which are heterochromatic, cluster together in distinct foci at the nuclear periphery. In human cells, late-replicating DNA, another hallmark of heterochromatin, is also located at the nuclear periphery.

Certain intrinsic chromosomal properties are also correlated with their location. In human cells, for example, the density of genes on a chromosome appears to correlate with its nuclear location. Human chromosomes 18 and 19 are of similar size, but differ greatly in their number of genes. Figure 4.40 shows how the gene-rich chromosome 19 is reproducibly located in the interior of the nucleus, while the gene-poor chromosome 18 is located near the nuclear periphery. The significance of the central location of gene-rich chromosomes is not yet clear.

Chromosomal regions can be silenced if placed adjacent to an area of the nucleus occupied by centromeric heterochromatin

A gene does not need to be physically integrated into heterochromatin for it to be silenced. Simply bringing a gene into the proximity of a region of centromeric heterochromatin, that is, *in trans*, can have the same effect. This has been demonstrated in *Drosophila*, in which homologous chromosomes remain paired with one another throughout the mitotic cell cycle, unlike in mammals.

The *brown* locus in *Drosophila* is at the opposite end of the chromosome from the centromeric heterochromatin. An inactive mutant form of this gene, known as *brown*^{D}, arises from an inappropriate insertion of a block of heterochromatin in the vicinity of the *brown* locus. In cells that contain one wild-type *brown* gene

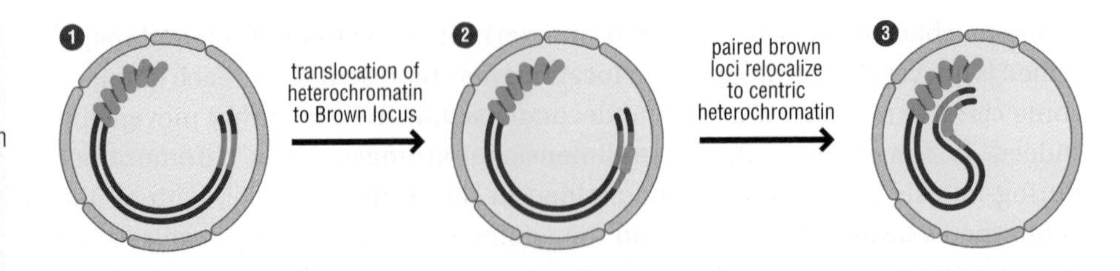

Figure 4.41 Proximity to centromeric heterochromatin in the nucleus can silence genes in *Drosophila*. (1) The paired *brown* gene loci (blue) are usually located far away from the nuclear region occupied by the centromere heterochromatin on the same chromosome (orange). (2) Insertion of a small region of heterochromatin (orange) into one of the two *brown* genes inactivates this allele and also causes movement of the chromatin containing the paired loci to a new nuclear region adjacent to the centromere heterochromatin (3). This relocation results in the silencing of the active copy of the *brown* gene.

on one homolog and *brown*^{D} on the other, both are inactive. This inactivation of the wild-type copy seems to be due to the relocation of the chromosomal regions containing wild-type and mutant *brown* loci to a site in the nucleus adjacent to the centromeric heterochromatin, as shown in Figure 4.41. The heterochromatic region inserted near the mutant gene directs this relocation in some way. Maternally and paternally inherited homologs are generally thought to control gene expression independently of one another in mitotic cells. Thus, *trans* silencing and concerted relocation of both loci reveals an interaction between homologs that is poorly understood.

Human cells show a similar phenomenon. When human B lymphocytes switch from a resting state to active cell division, certain genes move from their typical nuclear locations in resting cells to a location adjacent to centromeric heterochromatin, and are thereby silenced. It is not yet clear whether silencing causes the relocation or the relocation causes the silencing. Conversely, activation of the beta-globin locus in developing human red blood cells is associated with its relocation away from centromeric heterochromatin. We are some way from developing a clear, unambiguous picture of the mechanism behind this silencing.

The nucleolus is a nuclear substructure that forms around the ribosomal RNA genes

Electron micrographs show a prominent feature inside the nucleus called the **nucleolus** (see Figure 4.11), which is where actively transcribed rRNA genes are found. The ribosome is the large nucleoprotein complex that synthesizes proteins and contains several large RNA subunits termed rRNA, which are encoded by rDNA genes. In addition to containing the rDNA, the nucleolus is rich in other macromolecules required for assembly of ribosomes and other RNA containing complexes. These include enzymes that process the RNA molecules after they are transcribed, as well as ribosomal proteins. Although the nucleolus appears in electron micrographs to have distinct boundaries (see Figure 4.11), it lacks a membrane. Instead, it is the way the proteins and RNA that are specific to the nucleolus are stained in electron microscopy that gives the nucleolus its distinctive appearance.

➔ We discuss the structure of the ribosome in Chapter 11.

In order to meet the cell's need for new ribosome synthesis, many organisms have multiple copies of each gene encoding the ribosomal RNA subunit. The total number of repeats – termed rDNA repeats – varies among organisms and ranges from several hundred to 10 000. The rDNA repeats are arranged in clusters in the eukaryotic chromosome; in human chromosomes, there are ten sets of ribosomal gene repeats (five homologous pairs). At the end of M phase, transcription of the rDNA commences and the nucleolus forms as a visible, distinct structure around these highly transcribed sets of genes. All ten sets of genes coalesce in one or more nucleoli, where the rRNA subunits are transcribed and then assembled into ribosomes.

Figure 4.42 shows how the nucleolus is sometimes seen physically associated with the region of the DNA that contains the rRNA genes. This region of the chromosome where the nucleolus is associated is termed the nucleolar organizer region (NOR). The nucleolus usually disappears in early M phase and re-forms once the cell has divided, the chromosomes have decondensed, and transcription of various genes commences. In some cell types such as amphibian oocytes, the nucleolus remains visible during the early stages of chromosome condensation.

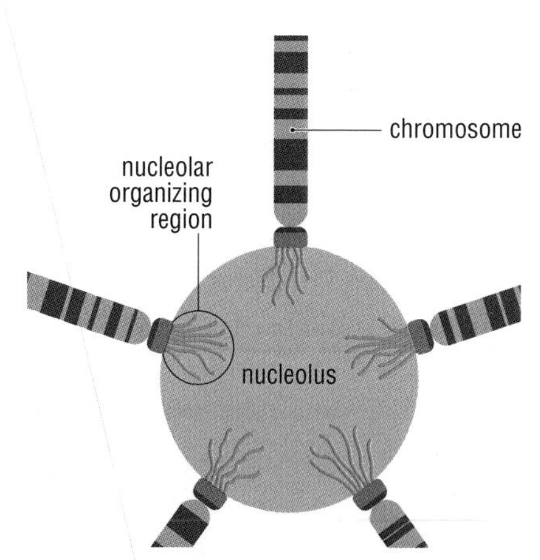

Figure 4.42 rDNA repeats and the nucleolus. In humans, there are rDNA repeats near the ends of five different chromosomes. These regions (called nucleolar organizing regions (NOR)) localize to the nucleolus, where the rDNA genes are actively transcribed. The figure shows five chromosomes: in a diploid cell there are two copies of each, giving a total of ten.

Reproduced from Chubb, JR and Bickmore, WA. Considering nuclear compartmentalization in the light of nuclear dynamics. *Cell*, 2003; **112**: 403–406.

Despite the high level of transcription in the nucleolus, a subset of the rDNA repeats are transcriptionally silent and have the features of heterochromatin. These heterochromatic regions are found at the periphery of the nucleolus, while the actively transcribed rDNA are found within the nucleolus. In addition to silencing transcription, heterochromatin formation also prevents fragmentation of the rDNA repeats, whose many repeated DNA sequences make them prone to recombination and chromosome instability. Indeed, mutations in proteins that maintain rDNA silencing give rise to instability of the rDNA.

In later chapters, we will see how the activity of a gene can be controlled by sophisticated mechanisms that modulate transcription and translation. However, we saw throughout this chapter how there are numerous ways of controlling this activity at a more fundamental level - the physical structure of the chromosome in which the gene is located. This control is mediated using some elegantly simple cues, from the addition or removal of methyl groups and other chemical entities, to the physical location of the chromosome within the nucleus. These mechanisms show the ingenious ways in which biological systems have evolved to carefully regulate their activity, seemingly with no opportunity for control overlooked.

In this chapter, we learned about the basic features of chromosomes, including how genomic DNA is packaged by proteins into chromatin and how special regions of the chromosomes play a role in ensuring the transmission of genetic information to daughter cells. In subsequent chapters, we will explore in detail how these features impact all processes that involve the chromosomes.

SUMMARY

CHROMOSOME BASICS

- An organism's genome is divided into one or more chromosomes.
- Each chromosome contains many genes embedded within a single DNA molecule.
- Between the genes lie stretches of intergenic DNA. There is great variation in the relative amount of intergenic DNA found in a given chromosome.
- A chromosome can be linear or circular.
- Each organism contains a characteristic number of chromosomes in each cell.
- Ploidy describes the number of identical sets of chromosomes in a cell. Diploid cells contain two sets of chromosomes in each cell, while haploid cells contain just a single set of chromosomes.
- Chromosomes are duplicated and distributed to daughter cells in the course of the cell cycle, which is divided into four phases: G1, S, and G2 (known collectively as interphase), and M (the mitotic phase).
- Mitotic chromosomes are highly condensed and have a distinctive appearance, whereas chromosomes have a more diffuse appearance during interphase.

CHROMATIN

- Chromosomes in all organisms are associated with proteins that help to condense and organize the DNA molecules inside the cell.
- The compacted bacterial chromosome is known as the nucleoid.

- Eukaryotic DNA is packaged by histone proteins into chromatin.
- The basic building block of chromatin is the nucleosome, which consists of around 146 bp of DNA wrapped twice around the histone octamer in a left-handed manner.
- The histone octamer consists of two copies each of the histone proteins, H2A, H2B, H3, and H4. The histone proteins have flexible N-terminal extensions, known as 'tails,' that extend between the turns of DNA.
- DNA packaged into nucleosomes adopts a 'beads on a string' appearance known as the 10 nm fiber. This is further compacted into a 30 nm fiber with the assistance of the linker histone, H1.
- Histone proteins are subject to numerous post- translational modifications, including acetylation, methylation, phosphorylation, ubiquitination, and sumoylation, which play key roles in regulating numerous processes that involve interactions with chromatin.
- Nucleosome remodeling complexes can alter the positions of nucleosomes using the energy from ATP hydrolysis.
- DNA can be methylated at cytosine bases and in bacteria also at adenine bases.
- Interphase chromosomes contain regions of heterochromatin, which appear compact and stain vividly in the microscope and euchromatin, which appears more diffuse in the microscope and is more easily digested by nucleases.
- While transcribed genes lie primarily in euchromatic regions, it is now clear that there is some transcription in heterochromatic regions, as well.
- Boundary elements separate heterochromatic and euchromatic regions and can limit the spreading of heterochromatin along the chromosome.

ELEMENTS REQUIRED FOR CHROMOSOME FUNCTION

- DNA replication begins at the origin of replication. Bacterial chromosomes have a single origin of replication, while eukaryotic chromosomes can have multiple origins.
- The centromere is a region in the eukaryotic chromosome that is involved in chromosome segregation.
- The centromere is heterochromatic and contains a specific variant histone, CENP-A in place of histone H3.
- Telomeres are special structures at the ends of eukaryotic chromosomes that protect the chromosomes from degradation or end-to-end fusion, and prevent the shortening of chromosomes after many rounds of DNA replication.
- Telomeres consist of many repeats of a short DNA sequence (TTAGGG in human chromosomes). Each telomere is mostly double-stranded but also contains a single-stranded overhanging extension.
- The telomeric repeats are added to the ends of chromosomes by the enzyme telomerase, which has an RNA subunit containing the template for synthesizing the telomeric repeats.

CHROMOSOME ARCHITECTURE IN THE NUCLEUS

- During interphase, each chromosome occupies a distinct chromosome territory in the nucleus.
- Actively transcribed regions of the chromosome tend to be near the center of the nucleus, whereas heterochromatin is often associated with the nuclear envelope.
- Transcription of the rRNA subunits (encoded in the rDNA) takes place in the nucleolus, a distinct substructure in the nucleus. The nucleolus forms around the actively transcribed rDNA repeats.
- The repeated rDNA genes are found in heterochromatin, which prevents recombination among the repeats.

INTRODUCTION

1. What is the function of an organism's chromosomes?

2. Which of the following definitions best describes the term 'chromatin'?

a. The basic repeating unit of DNA packaging in eukaryotes
b. The circular DNA of bacteria
c. The highly condensed DNA structure formed during cell division
d. The complex of DNA and proteins that forms the eukaryotic chromosome
e. The two copies of DNA that are held together at the centromere

4.1 ORGANIZATION OF CHROMOSOMES

1. Define each of the following terms and put them in order from least inclusive to most inclusive: chromosome, DNA, gene, genome.

2. In the domestic dog, *Canis familiaris,* there are 39 different chromosome types.

a. How many chromosomes are in a diploid cell of this organism?
b. How many chromosomes are in a haploid cell of this organism?
c. How many chromosomes are in a single gamete of this organism?
d. How many chromosomes are in a skin cell of this organism?

3. Fill in the table below with information regarding typical bacterial and eukaryotic cells.

Feature	Bacteria	Eukaryote
Number of chromosomes		
Shape of chromosomes		
Type of extrachromosomal DNA		
Main life stage is haploid or diploid?		

4.2 THE CELL CYCLE AND CHROMOSOME DYNAMICS

1. What relationship exists between the cell cycle and the chromosome cycle?

2. Fill in the table below with information regarding the cell cycle.

Cell cycle stage	Main events	Sister chromatids present?	Sister chromatids attached? (put NA for stages that do not have sister chromatids)
G1			
S			
G2			
M			

Challenge question

3. Propose a reason for the maintaining attachment of sister chromatids in the transmission of genes to the next generation.

4.3 PACKAGING CHROMOSOMAL DNA

1. What is the difference in structure and function of the histone core versus the histone tail?

2. What two factors facilitate the formation of the 30 nm fiber?

3. What is the difference between heterochromatin and euchromatin in terms of:

a. Gene expression
b. Dye staining
c. Nuclease sensitivity
d. Level of acetylation
e. Chromosome territory location

Challenge question

4. Propose a theory that explains why the histone protein sequences are among the most highly conserved sequences across species.

4.5 COVALENT MODIFICATION OF HISTONES

1. What is the significance of the fact that histone modifications are reversible?

2. Which of the following histone modifications is associated with the condensation of chromatin at mitosis?

a. Acetylation of lysine 14 on histone H3
b. Methylation of lysine 4 on histone H3
c. Phosphorylation of serines 10 and 28 on Histone H3
d. Sumoylation of lysine 4 on histone H3
e. Ubiquitination of lysine 14 on histone H3

Challenge question

3. The Flowering Locus C (*FLC*) gene in *Arabidopsis* is a regulator of plant flowering. When *FLC* is expressed, it produces a protein that reduces the expression of genes that promote flowering. Therefore, when *FLC* is active, the plant is in the vegetative growth phase and when *FLC* is silenced, the plant flowers. The expression of *FLC* is regulated in part by chromatin structure. Give three examples of mechanisms by which the plant might modify nucleosomes to promote the expression of *FLC* and explain how each of these modifications can be altered, or new modifications made to silence the expression of *FLC*.

4.6 NUCLEOSOME-REMODELING COMPLEXES

1. What relationship exists between nucleosome remodeling and histone modifications?

Challenge question

2 Angiotensin converting enzyme (ACE1) is a key regulatory enzyme in the pathway that regulates blood pressure. ACE1 activation leads to the constriction of blood vessels and the retention of salt and water. In comparing normal to hypertensive rats, it was determined that there was a difference in the histone modifications H3Ac, H3K4, and H3K9 in aorta, heart, kidney and liver.

a. What is hypertension more commonly known as?
b. Would you expect ACE1 to be more expressed in the normal or the hypertensive rats? Why?
c. For each modification, is it associated with increased or decreased expression? Which rat type would you expect to be enriched for the modification?

Modification	Rat with enrichment for this modification (normal or hypertensive)	ACE1 expression in rat with modification (higher or lower)
H3Ac		
H3K4(me3)		
H3K9(me2)		

4.7 DNA METHYLATION

1. What main physical barrier must be overcome for DNA to be successfully methylated?

2. Compare and contrast the use of DNA methylation in bacteria versus eukaryotes.

Challenge question

3. Describe one similarity and one difference between epigenetic silencing and DNA mutation.

4.8 THE SEPARATION OF CHROMATIN DOMAINS BY BOUNDARY ELEMENTS

1. What is the function of a boundary element? Describe the fission yeast experiment that suggested this function.

2. Which of the following statements best describes the basic mechanism by which boundary elements carry out their function?

a. The DNA that makes up a boundary element is always actively transcribed thereby preventing the formation of heterochromatin.
b. Nucleosomes are positioned at boundary elements and prevent the formation of heterochromatin.
c. Specific proteins bind the boundary element sequence and physically block the formation of heterochromatin.
d. Variant histones are incorporated into the nucleosomes at boundary elements and prevent the formation of heterochromatin.

Challenge questions

3. Why was the mottled appearance of the fruit fly eye termed 'position-effect variegation'?

4. Use your own words to describe the difference between Figures 4.28a and 4.28b.

4.9 ELEMENTS REQUIRED FOR CHROMOSOME FUNCTION

1. Briefly describe the main differences between the elements required to initiate DNA replication in bacteria and eukaryotic organisms.

4.10 THE CENTROMERE

1. Briefly describe the evidence that suggests chromatin structure rather than DNA sequence is most important in determining centromere function.

4.11 THE TELOMERE

1. What is the end-replication problem?

2. How do telomere-binding proteins determine the length of the telomere?

Challenge question

3. The catalytic subunit of human telomerase is transcribed from the *hTERT* gene. It has been shown that DNA methylation plays two opposing roles in regulating the transcription of this gene. At the core promoter, hypomethylation is necessary for transcription while the binding site for the CTCF repressor must be hypermethylated for transcription to occur.
   a. Based on your understanding of this chapter, how does hypomethylation of the core promoter lead to transcription?
   b. Propose a mechanism by which hypermethylation of the regulatory sequence might also lead to transcription.
   c. In many cancer cell types, telomerase is highly transcribed and active. What is the likely methylation status of the core promoter and the regulatory sequence in this cell type?
   d. It is now known that in addition to its role in regulating individual gene transcription, CTCF also acts as an insulator that affects the regulation of entire domains of gene transcription. Propose a mechanism by which DNA methylation of the insulator sequence may play a role in activating and repressing expression of gene regions.

4.12 CHROMOSOME ARCHITECTURE IN THE NUCLEUS

1. What is a chromosome territory? What is one hypothesis for the mechanism by which chromosome territories are established?

EXPERIMENTAL APPROACH 4.1 – NUCLEASE PROBES OF CHROMATIN ORGANIZATION

1. Why can micrococcal nuclease distinguish between unpackaged DNA and DNA that is wrapped around a histone?

2. In the experiment by Kornberg, how were the repeating units of DNA that have an approximately 200 bp difference in size (200, 400, 600, etc.) formed?

3. Once the position of each nucleosome was determined and analyzed by computer, what insight was learned about the positioning of yeast nucleosomes?

Challenge questions

4. In the experiment by Kornberg, how were the 170 bp fragment and the 205 bp fragment of DNA formed?

5. In the methods of the two experiments (Figures (1) and (3)) how did their opening stages differ? What was the purpose of the difference in the experiment by Mavrich?

EXPERIMENTAL APPROACH 4.2 – DISCOVERY OF A HISTONE ACETYLTRANSFERASE

1. In the experiment by Allis, what special composition of gel was developed for the assay?

2. How was this special gel used to identify a histone acetyltransferase?

Challenge question

3. In Figure 1, (a) and (b) represent the same gel viewed in two different ways. Why are there many bands in (a) and only one band in (b)?

FURTHER READING

4.1 ORGANIZATION OF CHROMOSOMES

Casjens S. The diverse and dynamic structure of bacterial genomes. *Annual Review of Genetics*, 1998;**32**: 339–377.

Duret L, Bucher P. Searching for regulatory elements in human noncoding sequences. *Current Opinion in Structural Biology*, 1997;**7**:399–406.

4.2 THE CELL CYCLE AND CHROMOSOME DYNAMICS

Blow JJ, Tanaka TU. The chromosome cycle: coordinating replication and segregation. Second in the cycles review series. *EMBO Reports*, 2005;**6**:1028–1034.

Diffley JF, Labib K. The chromosome replication cycle. *Journal of Cell Science*, 2002;**115**: 869–872.

Morgan DO. *The Cell Cycle*. London: New Science Press, 2006.

4.3 PACKAGING CHROMOSOMAL DNA

Thadani R, Uhlmann F, Heeger, S. Condensin, chromatin crossbarring and chromosome condensation. *Current Biology*, 2012;**22**:R1012–1021.

Wu C, Bassett A, Travers A. A variable topology for the 30-nm chromatin fibre. *EMBO Reports*, 2007;**8**:1129–1134.

Wyman C, Kanaar R. Chromosome organization: reaching out to embrace new models. *Current Biology*, 2002;**12**:R446–R448.

4.4 VARIATION IN CHROMATIN STRUCTURE

Bailis JM, Forsburg SL. It's all in the timing: linking S phase to chromatin structure and chromosome dynamics. *Cell Cycle*, 2003;**2**:303–306.

Rusche LN, Kirchmaier AL, Rine J. The establishment, inheritance, and function of silenced chromatin in *Saccharomyces cerevisiae*. *Annual Review of Biochemistry*, 2003;**72**:481–516.

4.5 COVALENT MODIFICATIONS OF HISTONES

Beisel C, Paro R. Silencing chromatin: comparing modes and mechanisms. *Nature Reviews Genetics*, 2011;**12**:123–135.

Berger SL. The complex language of chromatin regulation during transcription. *Nature*, 2007;**447**:407–412.

Cedar H, Bergman Y. Linking DNA methylation and histone modification: patterns and paradigms. *Nature Reviews Genetics*, 2009; **10**, 295–304.

Ruthenburg AJ, Li H, Patel DJ, Allis CD. Multivalent engagement of chromatin modifications by linked binding modules. *Nature Reviews*, 2007;**8**:983–994.

Suganuma T, and Workman JL. Signals and combinatorial functions of histone modifications. *Annual Review of Biochemistry*, 2011; **80**, 473–499.

Taverna SD, Li H, Ruthenburg AJ, Allis CD, Patel DJ. How chromatin-binding modules interpret histone modifications: lessons from professional pocket pickers. *Nature Structural and Molecular Biology*, 2007;**14**:1025–1040.

4.6 NUCLEOSOME-REMODELING COMPLEXES

Becker PB, Hörz W. ATP-dependent nucleosome remodeling. *Annual Review of Biochemistry*, 2002;**71**:247–273.

Clapier CR, Cairns BR. The biology of chromatin remodeling complexes. *Annual Review of Biochemistry*, 2009;**78**:273–304.

Rando OJ, Winston F. Chromatin and transcription in yeast. *Genetics*, 2012;**190**:351–387.

4.7 DNA METHYLATION

Cedar H, Bergman Y. Programming of DNA methylation patterns. *Annual Review of Biochemistry*, 2012;**81**;97–117.

Ideraabdullah FY, Vigneau S, Bartolomei MS. Genomic imprinting mechanisms in mammals. *Mutation Research*, 2008;**647**:77–85.

Law JA, Jacobsen SE. Establishing, maintaining and modifying DNA methylation patterns in plants and animals. *Nature Reviews Genetics*, 2010;**11**:204–220.

Smith ZD, Meissner A. DNA methylation: roles in mammalian development. *Nature Reviews Genetics*, 2013; **14**;204–220.

Suzuki MM, Bird A. DNA methylation landscapes: provocative insights from epigenomics. *Nature Reviews Genetics*, 2008;**9**:465–476.

4.8 BOUNDARY ELEMENTS SEPARATE CHROMATIN DOMAINS

Barkess G, West, A.G. (). Chromatin insulator elements: establishing barriers to set heterochromatin boundaries. *Epigenomics*, 2012;**4**:67–80.

Bushey AM, Dorman ER, Corces VG. Chromatin insulators: regulatory mechanisms and epigenetic inheritance. *Molecular Cell*, 2008;**32**:1–9.

Wallace JA, Felsenfeld G. We gather together: insulators and genome organization. *Current Opinion in Genetics and Development*, 2007;**17**:400–407.

Wei GH, Liu DP, Liang CC. Chromatin domain boundaries: insulators and beyond. *Cell Research*, 2005;**15**:292–300.

4.9 ELEMENTS REQUIRED FOR CHROMOSOME FUNCTION

Bergmann JH, Martins NM, Larionov V, Masumoto H, Earnshaw WC. HACking the centromere chromatin code: insights from human artificial chromosomes. *Chromosome Research*, 2012;**20**:505–519.

Mechali M. DNA replication origins: from sequence specificity to epigenetics. *Nature Reviews Genetics*, 2001;**2**:640–645.

Murray AW, Szostak JW. Artificial chromosomes. *Scientific American*, 1987;**257**:62–68.

4.10 THE CENTROMERE

Amor DJ, Choo KH. Neocentromeres: role in human disease, evolution, and centromere study. *American Journal of Human Genetics*, 2002;**71**:695–714.

Bloom K. Centromere dynamics. *Current Opinion in Genetics and Development*, 2007;**17**:151–156.

Burrack LS, Berman, J. Flexibility of centromere and kinetochore structures. *Trends in Genetics*, 2012;**28**;204–212.

Kalitsis P, Choo, KH. The evolutionary life cycle of the resilient centromere. *Chromosoma*, 2012;**121**:327–340.

Maddox PS, Corbett KD, Desai A. Structure, assembly and reading of centromeric chromatin. *Current Opinion in Genetics and Development*, 2012;**22**:139–147.

Nechemia-Arbely Y, Fachinetti D, Cleveland DW. Replicating centromeric chromatin: spatial and temporal control of CENP-A assembly. *Experimental Cell Research*, 2012;**318**:1353–1360.

Stimpson KM, Matheny JE, Sullivan BA. Dicentric chromosomes: unique models to study centromere function and inactivation. *Chromosome Research*, 2012;**20**:595–605.

4.11 THE TELOMERE

Baumann P, Price C. Pot1 and telomere maintenance. *FEBS Letters*, 2010;**584**:3779–3784.

Lewis KA, Wuttke DS. Telomerase and telomere-associated proteins: structural insights into mechanism and evolution. *Structure*, 2012;**20**:28–39.

Palm W, de Lange T. How shelterin protects mammalian telomeres. *Annual Review of Genetics*, 2008;**42**:301–334.

4.12 CHROMOSOME ARCHITECTURE IN THE NUCLEUS

Cremer T, Cremer M. Chromosome territories. *Cold Spring Harbor Perspectives in Biology*, 2010;2:a003889.

Geyer PK, Vitalini, MW, Wallrath, LL. Nuclear organization: taking a position on gene expression. *Current Opinion in Cell Biology*, 2011;**23**:354–359.

Gilbert N, Gilchrist S, Bickmore WA. Chromatin organization in the mammalian nucleus. *International Review of Cytology*, 2005;**242**:283–336.

McKeown PC, Shaw PJ. Chromatin: linking structure and function in the nucleolus. *Chromosoma*, 2009;**118**:11–23.

Nemeth A, Langst G. Genome organization in and around the nucleolus. *Trends in Genetics*, 2011;**27**:149–156.

The cell cycle

INTRODUCTION

Cell division is the process by which one cell divides to produce two daughter cells. Single-celled organisms divide in order to reproduce themselves, while cells of multicellular organisms divide in order to increase in mass, to generate distinct tissues, to form gametes, and to allow tissue renewal. To undergo cell division, cells must first increase in mass and duplicate their chromosomes and other cellular components. They then must properly distribute these components to the two daughter cells. This process is summarized in Figure 5.1. The events that lead to the formation of two daughter cells from one parental cell are part of the **cell cycle**, which encompasses a series of highly coordinated processes that ensure cell duplication in a precise and timely manner.

The overall principles of cell division apply to all kingdoms of life. However, as we will see in this chapter, the mechanisms of cell cycle regulation differ between bacteria and eukaryotes. Much of our understanding of the cell cycle is based on experiments that were initially done in model organisms, such as yeast, flies, and mice, because they are less complex than human cells and are more amenable to experimental manipulations. Nonetheless, since the eukaryotic cell cycle is evolutionarily conserved, what we learn from model organisms is often applicable to humans.

In this chapter we introduce the events that occur as cells progress through the cell cycle, and we will learn about the regulatory mechanisms, triggered by internal cues or by signals from outside the cell, that affect cell cycle progression. First, we will describe the cell cycle in eukaryotes; we will then discuss a few different types of cell cycles utilized by bacteria. We will learn more about specific events and processes that make up the cell cycle, such as DNA replication and chromosome segregation, in Chapters 6 and 7.

5.1 STEPS IN THE EUKARYOTIC CELL CYCLE

Our exploration of the molecular processes and components that underpin the cell cycle first focuses on how the process takes place in eukaryotes. We begin with a general overview of the cell cycle as a whole, before considering the underlying mechanisms and components in more detail.

The eukaryotic cell cycle is divided into four distinct phases

As described in Chapter 4, the eukaryotic cell cycle can be divided into four stages, or cell cycle phases, according to the major event that takes place during that time. These four major cell cycle phases are G1, S, G2, and M, as depicted in Figure 5.2:

- **G1 phase** is a period of growth before chromosome duplication.
- **S phase** is when DNA synthesis occurs, duplicating each chromosome to form two identical sister chromatids.

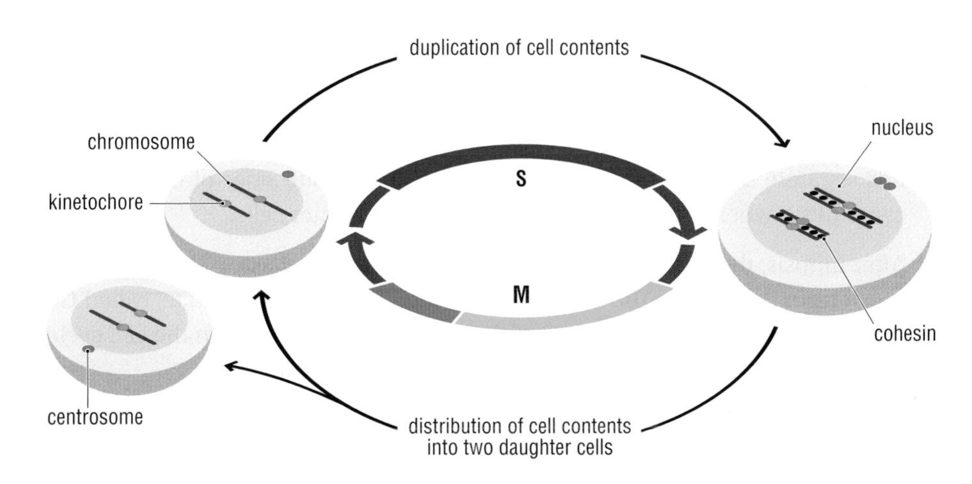

Figure 5.1 The cell cycle. The process of cell division generates two cells from one parental cell (shown with two chromosomes, in red). An essential part of this process is the duplication of certain cellular components, including the exact duplication of each chromosome during a specific part of the cell cycle called S phase. Chromosomes are divided equally between two daughter cells in M phase in a process that involves chromosome segregation and cytokinesis (M-phase sections marked in green and blue, respectively). Other cellular components that are important for this process are the centrosomes (shown in blue), centromeres and their associated kinetochores (shown in orange), and cohesin (shown in black). These will be discussed later in this chapter and in the following chapters.

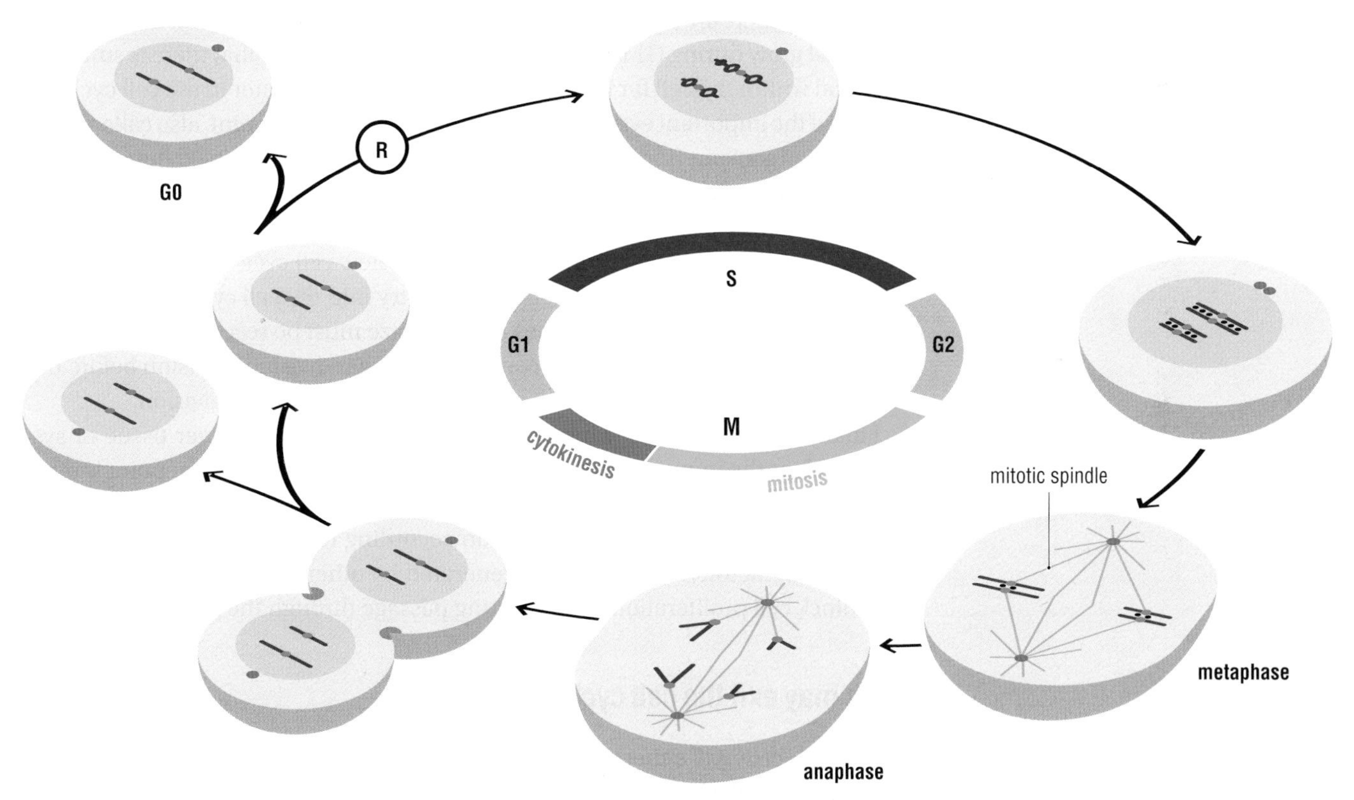

Figure 5.2 The events of the eukaryotic cell cycle. A newly born cell begins its life cycle in G1 phase. The cell may exit mitosis and become quiescent (in a G0 phase) or it may increase in size and pass through the restriction point (designated as R in the figure) and commit to a new round of cell division. Once the cell reaches the appropriate size it enters S phase, where DAN replication takes place (designated by red chromosomes with replication bubbles). At the end of S phase, each chromosome is made of two sister chromatids that are held together by cohesin (black dots). Another gap phase, G2, follows after S phase. During this time the centrosome (blue dot), which resides in the cytoplasm, duplicates. The two centrosomes then begin to form a spindle (not shown). At the end of G2, as cells enter mitosis, the double membrane that surrounds the nucleus (called nuclear envelope) breaks down (note the disappearance of the nucleus, shown in darker gray). This removes the barrier that existed between the spindle and the chromosomes, allowing the spindle (in green) to contact chromosomes at a site called the centromere (in orange). The mitotic spindle moves the chromosomes such that they are aligned in the spindle midzone, at metaphase. At this point, cohesin, which linked the two sister chromatids, is removed and the sister chromatids are pulled away from each other by the spindle in anaphase. Once anaphase is complete the cell divides in a process called cytokinasis, giving rise to two daughter cells, each with a full complement of chromosomes and a single centrosome.

- **G2 phase** is a gap phase that allows both growth and preparation for the separation of the sister chromatids. The mitotic spindle, the apparatus that mediates chromosome segregation, begins to form during this phase.
- **M phase** includes two major events: mitosis, during which the sister chromatids separate and are moved by the spindle to opposite sides of the cell, and cytokinesis, when the cell divides to create two daughter cells.

After mitosis, the two daughter cells then return to G1 phase. Cells may pause and exit from the cycle in the G1 phase and enter a non-dividing state, called **G0** (G zero). The timing and precise execution of each of these steps in the cell cycle is necessary for the generation of two fully functional daughter cells.

Cell growth occurs in the G1 phase

The G1 phase is the stage during which cell growth occurs. The full name of this phase is Gap phase 1 – so-called because there is no major observable change in the shape or size of the chromosomes during this stage. This is in contrast to S phase and M phase where noticeable changes in the structure and appearance of the chromosomes do occur. While the term 'gap' may seem to imply that nothing is happening, there are in fact essential processes occurring throughout this stage: the cell must grow during G1 to increase its cellular contents so that there is sufficient material within the cell for it to divide into two daughters later in the cell cycle.

One of the important events in G1 phase is the **restriction point**, also called 'Start' in certain organisms (indicated in Figure 5.2): once cells have passed the restriction point they are committed to go through the entire cell cycle, although they may be paused along the way if certain types of cellular defects arise. Because of this irreversible transition, the decision to embark on a new cell cycle depends, to a large extent, on nutrient availability. In addition, entry into the cell cycle is often tightly linked to cell size: in many cell types a critical size must be reached before the cell can pass the restriction point and commit to another division. Division before this critical size is reached would result in ever-smaller cells at each division.

Finally, in animal cells, where dividing cells are part of a larger tissue or system (such as the immune system), the transition past the restriction point is often regulated by external signals. These include **growth factors**, which are generated by other cells and stimulate cell division according to the organism's needs, and anti-growth factors, which are also generated by other cells and whose function is to block cell proliferation by preventing passage through the restriction point.

Cells may exit the cell cycle and enter G0

During G1, cells will either commit to a new round of division or will enter a resting stage termed G0, also known as **quiescence**, prior to passing the restriction point. For example, unicellular organisms will exit the cell cycle under conditions of poor nutrient availability, while cells in multicellular organisms become quiescent in response either to negative growth signals from neighboring cells or to lack of positive growth signals. The exit from the cell cycle also occurs during development: initially, during embryogenesis, there is rapid cell division to form a large number of cells. Later, cells become specialized, or differentiated, to perform a defined task as part of a specific tissue, and they stop dividing.

Some cells in our body, such as adult stem cells, keep dividing to replenish a particular tissue throughout our lifetime; for example, adult skin stem cells are constantly dividing to provide new skin cells. However, most human cells are naturally arrested in G0. Some can be stimulated to re-enter the cell cycle; others, however, are permanently arrested in G0. For example, human nerve cells carry out essential functions and yet they cannot resume cell division.

It is important to note that quiescent cells are not just in a very long G1 phase; rather, cells in G0 have certain markers that distinguish them from G1 cells. For example, the levels of certain cell cycle inhibitors are higher in quiescent cells than in G1, and certain DNA replication proteins that are present in G1 cells in preparation for DNA replication are absent in quiescent cells. Some types of quiescent cells also have lower metabolic rates and reduced nucleotide synthesis than their actively dividing counterparts. It is important to remember, however, that while quiescent cells are not actively dividing, they can still grow in size. For example, certain motor neurons stop dividing during early development, but their axons may continue to elongate as the organism increases in size.

DNA is replicated during S phase

Once a cell passes the restriction point, it enters S phase, when DNA replication takes place. The purpose of DNA replication is to produce an accurate copy of all the chromosomes, thereby creating two complete copies of the genome. Each chromosome is duplicated to form two identical sister chromatids, which remain associated with each other until mitosis. The sister chromatids are held together by a protein complex called cohesin.

➲ The mechanisms of DNA replication and its regulation are discussed in more detail in Chapter 6. The action of cohesin is described in more detail in Section 7.2.

To ensure accurate duplication, there are regulatory mechanisms that stop cell cycle progression until the DNA is fully replicated, and other mechanisms that block replication in a region of the chromosome that has already been duplicated. Once the DNA is completely replicated, the cell can move to the next cell cycle stage, termed the G2 phase.

The G2 phase allows cells to prepare for chromosome segregation

After the completion of DNA synthesis, cells enter the G2 phase, during which they continue to grow and prepare for mitosis. During G2 there is a further increase in cellular content in order to generate the cellular components that will be distributed between the two daughter cells. Other events that occur in G2 include the formation of the **mitotic spindle**, the apparatus that drives chromosome segregation.

➲ The structure and function of the mitotic spindle is described in detail in Section 7.3.

In addition, G2 phase is a point at which cell cycle progression can be stopped if chromosomes experience DNA damage. For example, if a chromosome has a double-stranded DNA break, there is a regulatory mechanism that will prevent cells from continuing to mitosis until the chromosome is repaired. Progression through mitosis with a broken chromosome could result in the loss of genetic material.

Regulatory mechanisms, often called **checkpoint** pathways, monitor the integrity of different cellular structures. If a problem is detected, these checkpoint pathways prevent cell cycle progression, thereby allowing the cell to fix the problem before it results in permanent damage that may be passed on to the daughter cells. (We discuss checkpoint pathways further in Section 5.7.) Thus, although passage through the restriction point commits a cell to complete the cell cycle, a cell may pause its progression at different cell cycle stages to repair certain types of damage.

Chromosome segregation occurs during M phase

Mitosis (M phase) is the stage at which chromosomes are segregated into two new daughter cells. The progression through M phase is characterized by specific changes in chromosome organization.

➲ The process of chromosome segregation and cytokinesis are discussed in greater detail in Sections 7.4 through 7.6.

A key event in mitosis is the attachment of the chromosomes to the mitotic spindle, which then pulls the sister chromatids apart to create two identical clusters of chromosomes at opposite poles of the cell, each containing the same chromosome composition as that of the parent cell. Errors that occur during this stage can lead to chromosome mis-segregation, and consequently give rise to cells with too few or too many chromosomes. Either of these situations can be deleterious to the cell or the organism.

Just like in S phase and in G2 phase, there is also a checkpoint pathway in mitosis that stops cell cycle progression if, for example, the chromosomes are not attached properly to the spindle. After chromosome segregation is completed, the cell divides between the two separated sets of chromosomes in a process called cytokinesis to give rise to two daughter cells, which are in G1 phase.

The length of the cell cycle varies between different cell types

The general description of the cell cycle given above is accurate for many single-celled organisms and human somatic cells. However, it is important to remember that the total length of the cell cycle, and the relative lengths of the different phases, can vary greatly between cell types, as illustrated in Figure 5.3a. For example, the entire cell cycle in budding yeast takes about 90 minutes, while in many human cells it takes 18 hours or more. Even different yeasts exhibit different types of cell cycle: budding yeast have a long G1 and an almost non-existent G2, while fission yeast has a short G1 and a long G2.

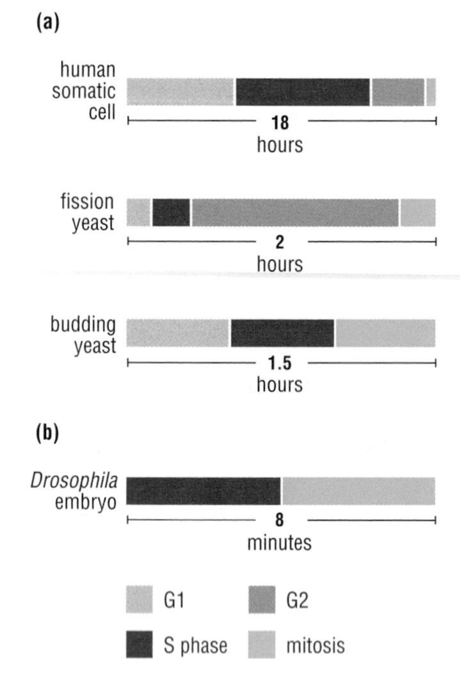

Figure 5.3 Variations can occur in the cell cycle of different cell types. (a) The cell cycle of a human somatic cell follows the outline described in Figure 5.1, and often lasts many hours. In other cell types, the overall length of the cell cycle and the relative length of each phase can vary. Shown are the cell cycles of fission and budding yeasts, which are much shorter than the typical human cell cycle, and differ in the relative length of G2. (b) The first 13 *Drosophila* embryonic divisions are much more rapid than those of human or yeast cells, and they lack gap phases.

Some cell types have a specialized cell cycle organization

Not only can the relative lengths of the cell cycle phases differ between cell types, but in certain cell types some cell cycle phases may be absent altogether. For example, in certain organisms, early embryonic development requires rapid cell divisions. In these early embryos there is no new protein synthesis and the embryo relies instead on proteins that were deposited in the oocyte by the mother. In the cell cycle of these embryonic cells there are no gap phases (G1 and G2): the cells shuttle between S phase and M phase instead (Figure 5.3b).

Nonetheless, in all dividing cells, there is a common engine that drives the cell cycle forward regardless of the length of the various phases or even the types of phases that make up their unique cell cycle. This engine is based on the activity of a family of kinases (enzymes that catalyze the transfer of a phosphate group from a donor, such as adenosine triphosphate (ATP), to a target protein) that are regulated by proteins called **cyclins**. In the next section we see how these **cyclin-dependent kinases (Cdks)** move the cell cycle forward, from one cell cycle phase to the next.

5.2 CYCLINS AND Cdks

Cdks are key regulators of cell cycle progression

The central components of the machinery that drives and regulates the cell cycle are enzymes called Cdks. Cdks belong to a family of serine/threonine protein kinases that phosphorylate specific serine and/or threonine residues on their protein substrates. Some organisms, such as mammals, have multiple Cdks, while other organisms, such as yeast, have a single Cdk. Read Experimental approach 5.1 to

5.1 EXPERIMENTAL APPROACH

Cyclin-dependent kinases: of yeast and men

Identification of the human cyclin-dependent kinase using yeast

The first Cdks that were identified were those of the budding yeast *Saccharomyces cerevisiae* (a protein called Cdc28) and the fission yeast *Schizosaccharomyces pombe* (called Cdc2). The genes coding for these proteins were identified in genetic screens for mutants causing cell division cycle defects (hence the name cdc; see Experimental approach 5.2). At the time, it was not known that the activity of these kinases was dependent on cyclins, and hence the term Cdk had not yet been coined. It was known, however, that these proteins play key roles in cell cycle progression, and thus it was of interest to identify the human homolog.

To accomplish this, Paul Nurse and colleagues exploited the likely evolutionary conservation between Cdk genes in different species. Although *S. cerevisiae* and *S. pombe* are both yeasts, they are only distantly related from an evolutionary standpoint. Nurse and colleagues recognized a remarkable 60% amino acid identity between the *S. cerevisiae* and *S. pombe* proteins and, based on this, reasoned that the human Cdk is likely to be similar to the yeast Cdks. They further reasoned that, this being the case, they could identify the human Cdk homolog by its ability to complement the growth defect of an *S. pombe cdc2* mutant strain.

The *cdc2* mutant strain used by the Nurse group carried a temperature-sensitive allele of *cdc2* (called *cdc2-33*) that only allowed cell division at a permissive temperature. (For a description of how *cdc* mutants were isolated, see Experimental approach 5.2.) At the non-permissive temperature, the mutant cells could not divide and became elongated. The investigators then introduced a human cDNA library (a large collection of plasmids each carrying a different human cDNA, described in Chapter 19) into the *cdc2-33* mutant strain, as illustrated in Figure 1a. As they had predicted, the investigators found a subset of cells that now could divide at the non-permissive temperature. Several examples are shown in Figure 1b where cells carrying the 'suppressing' plasmid appear normal (pink arrow), while cells that have a different plasmid cannot divide and appear elongated (blue arrow). The investigators isolated the

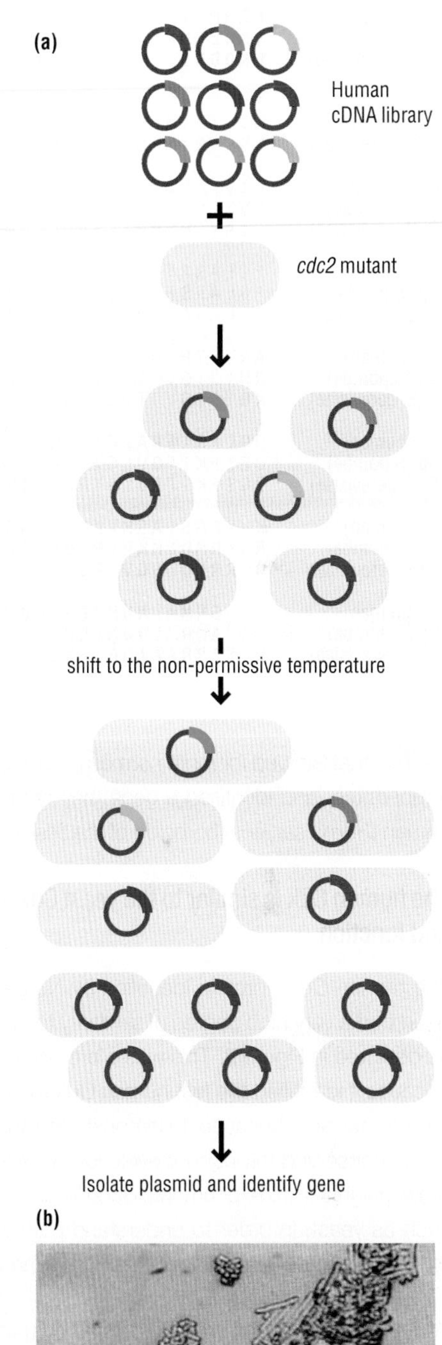

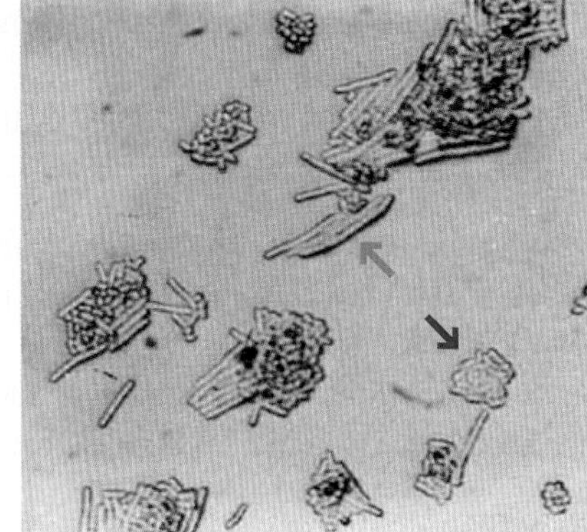

Figure 1 Cloning of human Cdk. (a) The strategy used by Lee and Nurse to clone human CDC2 (from the top): A library of human cDNA clones, containing many plasmids with different cDNA segments, and indicated by the different colors, was transformed into a fission yeast *cdc2* mutant strain to yield a collection of yeast cells, each transformed with a different cDNA clone. These cells were then shifted to the non-permissive temperature. Cells that contained a plasmid that complemented the *cdc2* defect (the plasmid with a pink insert) could divide and form colonies at the non-permissive temperature. All other cells contained non-complementing plasmids and thus could not divide, and instead formed elongated cells. (b) An image of the *cdc2* strain transformed with the human *CDC2* clone. Cells that contained the plasmid divided and had a normal shape (pink arrow). Cells that do not contain the Human Cdc2 clone cannot divide and instead have an elongated shape (blue arrow).

```
2Hs (human)               MEDYTKIEKIGEGTYGVVYKGRHK  TT
2Sp (S.pombe)             MENYQKVEKIGEGTYGVVYKARHK  LS
28 (S. cerevisiae)   MSGELANYKKLEKVGEGTYGVVYKALDLRPGQ

2Hs (human)          GQ VVAMKKIRLESEEEGVPSTAIREISLLKE
2Sp (S.pombe)        G RIVAMKKIRLEDESEGVPSTAIREISLLKE
28 (S. cerevisiae)   GQRVVALKKIRLESEDEGVPSTAIREISLLKE

2Hs (human)          LRHPNI    VSLQDVLMQD SRLYLIFEFLS
2Sp (S.pombe)        VNDENNRSNCVRLLDILHAE SKLYLVFEFLD
28 (S. cerevisiae)   LKDDNI    VRLYDIVHSDAHKLYLVFEFLD

2Hs (human)          MDLKEYLDSIPPGQYM  DSSLVKSYLYQILQ
2Sp (S.pombe)        MDLKKYMDRISETGATSLDPRLVQKFTYQLVN
28 (S. cerevisiae)   LDLKRYMEGIPKDQPL  GADIVKKFMMQLCK

2Hs (human)          GIVFCHSRRVLHRDLKPQNLLIDDKGTIKLAD
2Sp (S.pombe)        GVNFCHSRRIIHRDLKPQNLLIDKEGNLKLAD
28 (S. cerevisiae)   GIAYCHSHRILHRDLKPQNLLINKDGNLKLGD

2Hs (human)          FGLARAFGIPIRVYTHEVVTLWYRSPEVLLGS
2Sp (S.pombe)        FGLARSFGVPLRNYTHEIVTLWYRAPEVLLGS
28 (S. cerevisiae)   FGLARAFGVPLRAYTHEIVTLWTRAPEVLLGG

2Hs (human)          ARYSTPVDIWSIGTIFAELATKKPLFHGDSEI
2Sp (S.pombe)        RHYSTGVDIWSVGCIFAEMIRRSPLFPGDSEI
28 (S. cerevisiae)   KQYSTGVDTVSIGCIFAEMCNRAPIFSGDSEI

2Hs (human)          DQLFRIFRALGTPNNEVWPEVESLQDYKNTFP
2Sp (S.pombe)        DEIFKIFQVLGTPNEEVWPGVTLLQDYKSTFP
28 (S. cerevisiae)   DQIFKIFRVLGTPNEAIWPDIVYLPDFKPSFP

2Hs (human)          KWKPGSLASHVKNLDENGLDLLSKMLIYDPAK
2Sp (S.pombe)        RWKRMDLHKVVPNGEEDAIELLSAMLVYDPAH
28 (S. cerevisiae)   QWRRKDLSQVVPSLDPRGIDLLDKLLAYDPIN

2Hs (human)          RISGKMALNHPYFNDLDNQIKKM
2Sp (S.pombe)        RISAKRALQQNYLRDFH
28 (S. cerevisiae)   RISARRAAIHPYFQES
```

Figure 2 Comparison of the amino acid sequences of human and yeast Cdc2. 2Hs is human *CDC2*, 2Sp is *S. pombe cdc2*, and 28 is the *S. cerevisiae CDC28*. Identities between the human and yeast proteins are shown in pink.

With permission from Lee and Nurse. Complementation used to clone a human homologue of the fission yeast cell cycle control gene cdc2. *Nature*, 1987;**327**:31–35.

plasmid that allowed for more normal growth at the non-permissive temperature, and identified a gene they named *CDC2*Hs, for the human (*Homo sapiens*) homolog of the fission yeast *cdc2*.

The human Cdk is similar to the yeast Cdk in both sequence and function

The human Cdc2 protein identified by Nurse and colleagues indeed had a very high degree of similarity to its yeast counterparts, as depicted in Figure 2. This was a breakthrough finding not only because it identified the first human Cdk (other Cdks would soon follow), but also because it demonstrated that cell cycle regulation is conserved throughout evolution. This allowed scientists to draw insights from experiments done in simple model organisms, such as yeast, in order to understand pathways of regulation in more complicated organisms, such as humans.

Yeast Cdk substrates were identified by using a modified Cdk and an ATP analog

Following the discovery of Cdks and the realization that these kinases regulate cell cycle progression by phosphorylating other proteins, investigators wanted to identify Cdk substrates. This was not a simple task: one could not simply identify all the phosphorylated proteins in the cell because cells have many different kinases, and only a fraction of the phosphorylated proteins are Cdk substrates. Instead, there was a need to develop a method to distinguish Cdk substrates from other phosphorylated proteins. This problem was addressed in a clever way by David Morgan, Kevan Shokat, and colleagues who devised a mutant form of the *S. cerevisiae* kinase Cdc28 (also called Cdk1) that had an altered ATP binding pocket (see Figure 3). The mutant Cdc28, called Cdc28-as1 (for analog sensitive), could bind an ATP analog that was too bulky to fit into the ATP binding pockets of other kinases, thus allowing its activity to be specifically followed in a sea of other kinases. By incubating this mutant kinase (Cdc28-as1) and an ATP analog radioactively labeled at the gamma phosphate in a cellular extract, the scientists could immediately observe that many proteins are targeted by Cdc28-as1 in a phosphorylation reaction.

Specific candidates were next identified using a yeast collection in which each individual open reading frame (ORF) was fused to an affinity tag (glutathione S-transferase, or GST) that allowed the corresponding protein to be purified from the extract (Figure 3c, lower panel). The 'semi-purified' proteins were then subjected to phosphorylation by Cdc28-as1 again in the presence of the radioactive ATP analog (Figure 3c); of the 700 proteins tested, 360 were found to be Cdc28 substrates, many of them novel targets. Follow up experiments subsequently confirmed many of these proteins as valid *in vivo* Cdk substrates. In Experimental approach 15.2, we will see another approach used to identify the substrates of a protein kinase.

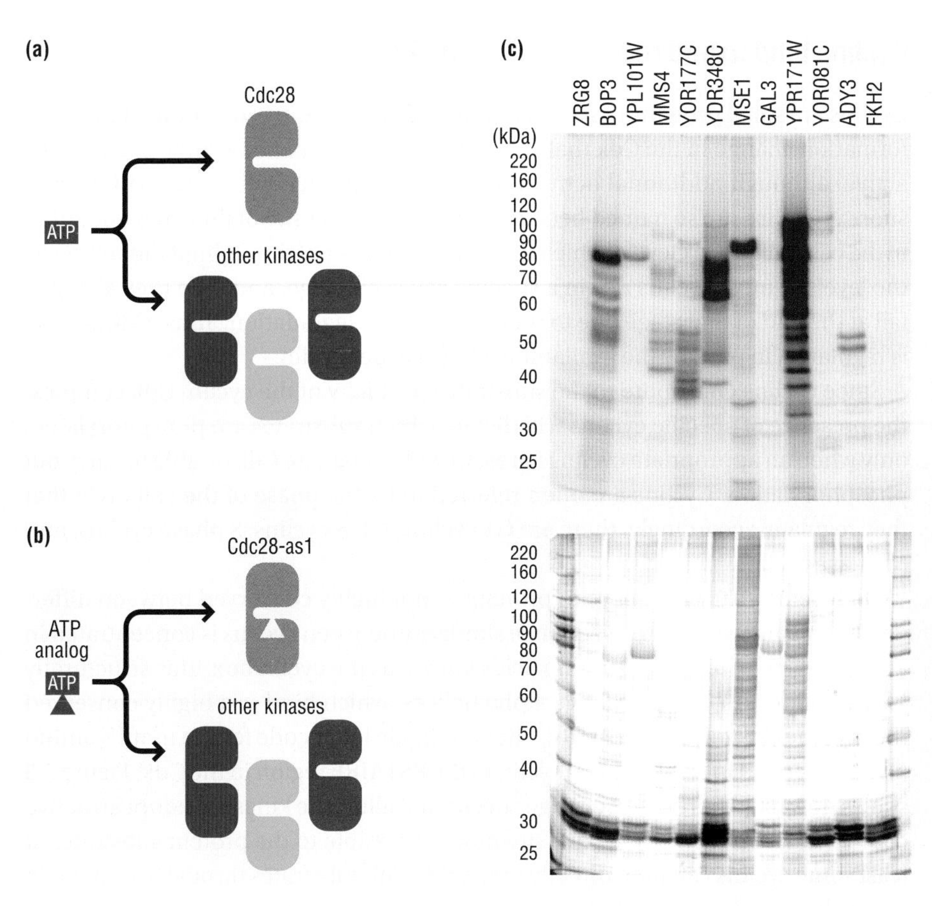

Figure 3 An analog specific kinase. (a) Various kinases, including the *S. cerevisiae* Cdc28 (sometimes also called Cdk1), share a similar binding pocket for ATP (shown in pink, binding pocket is shown in white). (b) Kevan Shokat and colleagues devised a Cdc28 mutant protein (Cdc28-as1) with an altered ATP binding site, which can accommodate both ATP and an ATP analog. The ATP analog does not fit into the ATP-binding pocket of other kinases. (c) A representative sample of yeast proteins that were tested for phosphorylation by Cdc28-as1. Strains carrying the indicated genes fused to a tag were grown individually and subjected to affinity purification to enrich for the tagged protein. The extracts were then incubated with purified Cdc28-as1 bound to the Clb2 cyclin and the radioactive ATP analog. At the end of the reaction the extract was run on a gel. The top image is an autoradiograph showing the various radioactive proteins on the gel. Note that some proteins, such as those encoded by *ZRG8*, *GAL3*, and *FKH2* were not phosphorylated, while others, such as those encoded by the ORFs *YOR177C* or *TPR171W*, were heavily phosphorylated. The bottom image is the silver-stained gel from which the autoradiograph was obtained. Note that most lanes contain many proteins, which is typical of this type of affinity purification.

Find out more

Lee MG, Nurse P. Complementation used to clone a human homologue of the fission yeast cell cycle control gene cdc2. *Nature*, 1987;**327**:31–35.

Ubersax JA, Woodbury EL, Quang PN, *et al*. Targets of the cyclin-dependent kinase Cdk1. *Nature*, 2003;**425**:859–864.

Related techniques

S. pombe (as a model organism); Section 19.1

cDNA libraries; Section 19.4

in vitro phosphorylation (radioactive); Section 19.6

Affinity tagging; Section 19.5

find out about the discovery of the first human Cdk and the identification of Cdk substrates.

The activity and specificity of Cdks is determined by regulatory proteins called cyclins. Different cyclins are present during different cell cycle phases, as shown in Figure 5.4. This leads to the activation of distinct cyclin–Cdk complexes at different cell cycle stages. As the cell progresses through the cycle, specific cyclin–Cdks become active and phosphorylate specific substrates that carry out particular cell cycle events. This phosphorylation may change the substrate's activity, its stability, or its interaction with other proteins.

Not only do cyclin–Cdks activate the events in their cell cycle phase, they also lead to the activation of the cyclin–Cdk of the next cell cycle phase (for example, by promoting the expression of the next cyclin), thereby pushing the cell cycle forward. Interestingly, despite the presence of multiple Cdks that are active at distinct cell cycle phases, studies in mice and other organisms have revealed that there is a significant functional overlap between the Cdks, with one Cdk being able to compensate for another that has been deleted. The exception is Cdk1, which is essential for viability.

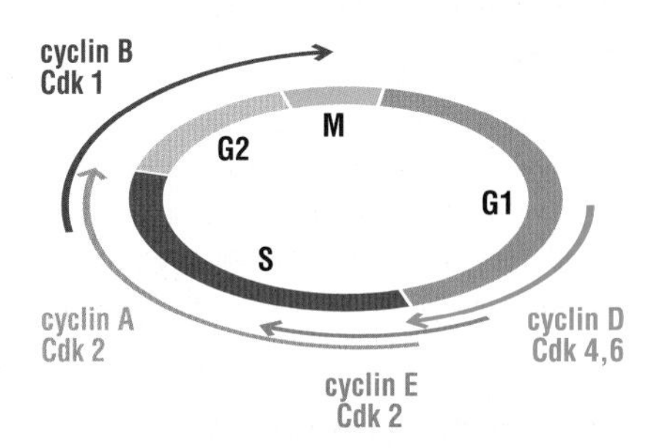

Figure 5.4 Different cyclins and Cdks function in different cell cycle stages. This diagram illustrates the different cyclins and Cdks that act during the cell cycle of a mammalian somatic cell. Different cyclins are present during different phases of the cell cycle and pair with different Cdks. Cyclin D (a G1 cyclin) can activate either Cdk4 or Cdk6, cyclin E (a G1/S cyclin) can activate Cdk2, cyclin A (an S-phase cyclin) can also activate Cdk2, and cyclin B (an M-phase cyclin) can activate Cdk1.

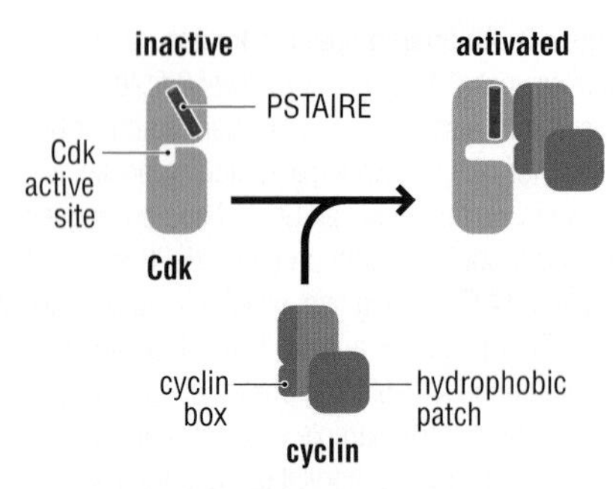

Figure 5.5 Cdk activation. The activation of Cdks at specific cell cycle stages requires binding to a specific cyclin. Cyclins contain a region called the cyclin box and a region called a hydrophobic patch (both shown in dark blue), which are important for cyclin function (see text for further detail). A Cdk can be represented as a bi-lobed protein with its active site in a cleft between the two lobes. The binding of cyclin to Cdk results in a conformational change in the Cdk, including movement of the PSTAIRE region (shown in red), opening the active site to potential substrates. For simplicity, this diagram does not include changes in Cdk phosphorylation, which also contribute to Cdk activation (see Figures 5.7 and 5.8).

Cyclins bind to and regulate Cdk activity

Cyclins are small proteins that bind tightly to Cdks. In doing so, they lead to a conformation change in the Cdk that is necessary for the Cdk to become catalytically active, although additional activation steps are still needed, as we will discuss shortly. Cyclins are so named because of the cyclical nature of their accumulation in the cell: unlike Cdk levels, which are relatively constant throughout the cell cycle, the levels of most cyclins rise in a specific cell cycle stage, and then they abruptly fall at the end of that stage due to their proteolytic degradation. Thus, Cdk activity is regulated, in part, by the availability of its partner cyclin.

Since cyclins contribute to the substrate specificity of the cyclin–Cdk complex, the presence of specific cyclins will dictate which substrates are phosphorylated: only when an appropriate cyclin is present will its cognate Cdk be able to carry out phosphorylation. Cyclins are often referred to by the phase of the cell cycle that they regulate: accordingly, there are G1 cyclins, G1/S cyclins, S phase cyclins, and M phase cyclins.

The overall structure of cyclin proteins is not highly conserved between different organisms. Instead, the sequence similarity between cyclins is concentrated in a small region of about 100 amino acids known as the **cyclin box**. This structurally conserved region consists of five alpha helices, which bind to a highly conserved motif in the Cdk termed **PSTAIRE**, after the single letter code for the motif's amino acid sequence. Upon binding of cyclin to the PSTAIRE region in the Cdk, Figure 5.5 shows how major structural changes occur that allow the kinase to adopt an active conformation, making the active site more accessible to the protein substrate. At least some cyclins can bind directly to specific Cdk substrates through a domain in the cyclin called a **hydrophobic** patch. Thus, the binding of cyclin to the Cdk not only activates the Cdk but also promotes recruitment of Cdk substrates.

Cyclin–Cdks move the cell cycle forward

The function of different cyclin–Cdk complexes is to promote cell cycle progression. This means that cyclin–Cdks not only promote stage specific processes, but they also promote the transition to the next cell cycle stage. The actual network of interactions that drive cell cycle progression is very complex, involving dozens of proteins. A simplified example taken from the cell cycle regulatory mechanism in budding yeast is shown in Figure 5.6. Note that there are both positive interactions, such as those that promote protein activity or gene expression, and negative interactions, which inhibit activity or lead to protein degradation.

When cells are in S phase, S-phase cyclin–Cdk1 activates DNA replication, as depicted in Figure 5.6. In order to move on to mitosis, however, M-phase cyclin–Cdk1

Figure 5.6 A regulatory cell cycle circuit acting at the beginning of M phase. Entry into M phase is dependent on the activity of M-phase cyclin–Cdk. However, prior to M phase, M-phase cyclins are unstable due to proteolysis (step 3 in the regulatory circuit). To allow entry into M phase, S-phase cyclin–Cdk, which is needed to promote S-phase processes (step 1) inhibits M-cyclin proteolysis (step 2). This allows M-phase cyclins to accumulate and activate its Cdk (step 4), thereby promoting entry into M phase (step 5). In addition, M-phase cyclin–Cdk also inhibits the expression of S-phase cyclins (step 6), thereby extinguishing processes that occur earlier in the cell cycle. Keep in mind that the actual regulatory circuit is much more complex. Another example of a cell cycle regulatory circuit is shown in Figure 5.10. The relationships between the various players in this scheme are denoted by green arrows, which represent activation processes, or by red bars, which represent inhibitory processes.

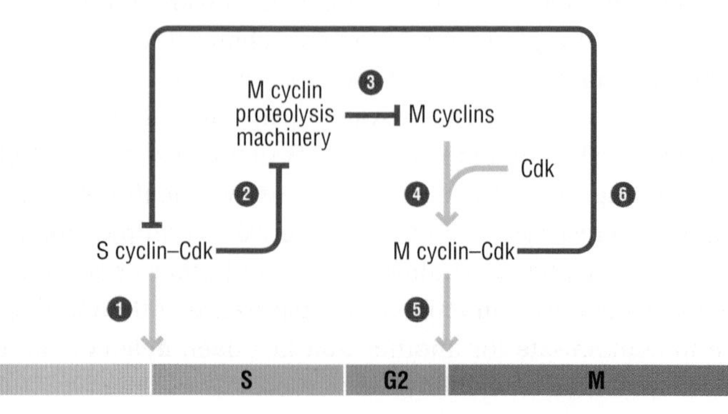

has to be active. Prior to mitosis – in G1 and early S phase – an M-cyclin degradation process causes M cyclins to be unstable. Consequently, the M cyclin–Cdk complex cannot be formed. Part of S-phase cyclin–Cdk1 activity is to counter this instability and promote entry into mitosis. It does so by inactivating the M-phase cyclin proteolysis machinery, hence preventing the M cyclins from being degraded. This type of regulation – where two negative reactions result in a positive outcome – is very common in cell cycle regulation. (In this instance, the two 'negative' reactions are the proteolysis of M cyclin and the inhibition of this proteolysis, while the positive outcome is the stabilization of M cyclin.)

The cell cycle circuitry is obviously more complex, because not only does the cell cycle machinery have to activate the next cell cycle step, it also has to turn off the previous one. In the example shown in Figure 5.6, the association of M-phase cyclins with the Cdk not only drives progression through mitosis but also inhibits the transcription factors responsible for S-phase cyclin expression; this activity turns off processes that preceded M phase and force the cell cycle forward. Such processes, and many others like them, make up the cell cycle engine that drives cell cycle progression.

5.3 REGULATION OF Cdk ACTIVITY

Cdk activity is regulated by multiple mechanisms

The fundamental importance of Cdks is underscored by the regulation of their activity at multiple levels. We have seen in Section 5.2 that Cdk activity is regulated by the presence or absence of a particular cyclin. However, this is not the only way by which Cdk activity is regulated. Cdks are also regulated, both positively and negatively, by phosphorylation. There are specific amino acid residues in the Cdk that must be phosphorylated and different residues that must be dephosphorylated for the enzyme to be active. Thus, the careful control of both the kinases that add these phosphate groups, and the phosphatases that remove them, allows the cell to finely tune the activity of the Cdks. In addition, there are small proteins termed cyclin kinase inhibitors that bind to and inhibit Cdk activity. In this section we consider these processes in more detail as we discuss some of the mechanisms that regulate Cdk activity.

Cdks can be activated by phosphorylation and dephosphorylation

The binding of cyclin to a Cdk changes the conformation of the Cdk such that its active site is now more accessible to its substrates. However, even with the cyclin bound, there is still a structural hindrance that prevents cyclin–Cdk from phosphorylating its substrates. For a Cdk to be active, a specific threonine residue adjacent to the kinase active site must be phosphorylated. Phosphorylation at this site is catalyzed by enzymes called **Cdk-activating kinases (CAKs),** as illustrated in Figure 5.7. CAK activity is usually high throughout the cell cycle. Therefore, although CAK-dependent phosphorylation is necessary for Cdk activation, it can be viewed as a post-translational modification rather than as a cell cycle-regulated mode of activation. Thus, both cyclin binding and phosphorylation by CAK are needed to fully activate the cyclin–Cdk complex.

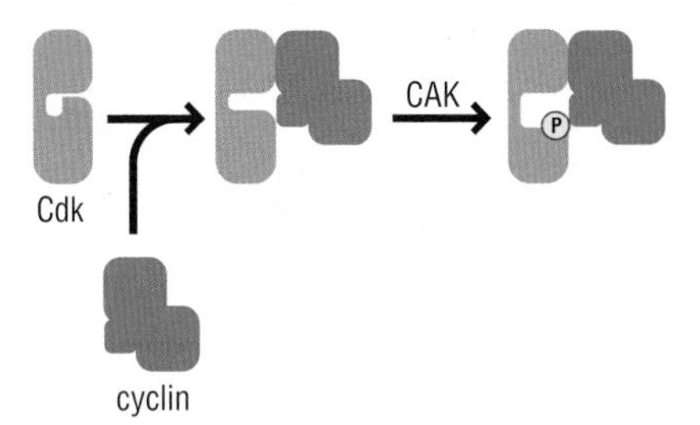

Figure 5.7 Two steps in Cdk activation. Cyclin binding alone causes partial activation of Cdks, but complete activation also requires phosphorylation by CAK. In animal cells, CAK phosphorylates the Cdk subunit only after cyclin binding, and so the two steps in Cdk activation are usually ordered as shown here, with cyclin binding occurring first. CAK tends to be in constant excess in the cell, so that cyclin binding is the rate-limiting step in Cdk activation. Cdk is shown in gray and cyclin is shown in blue. Phosphorylation is in yellow.

In the case of the M-phase cyclin–Cdk complex, not only does CAK have to activate the Cdk, two specific phosphate groups on Cdk must also be *removed* for full Cdk activity. One site of inhibitory phosphorylation is a tyrosine residue (Tyr 15 in

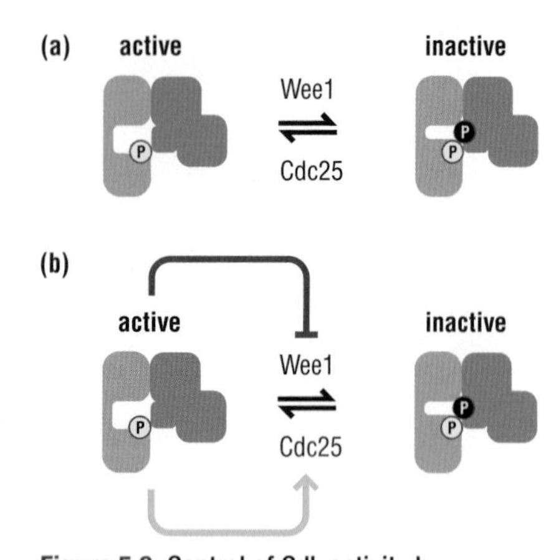

Figure 5.8 Control of Cdk activity by inhibitory phosphorylation. (a) The fully active cyclin–Cdk complex (phosphorylated by CAK, as indicated by the yellow P) can be inhibited by Wee1 phosphorylation at the active site of the Cdk (inhibitory phosphorylation is indicated by the black P). Dephosphorylation of the inhibitory sites by the phosphatase Cdc25 leads to reactivation of the cyclin–Cdk complex.
(b) Wee1 and Cdc25 are regulated by M-phase cyclin–Cdk: phosphorylation of Wee1 by M-phase cyclin–Cdk inhibits the activity of Wee1 (shown as a red bar) while phosphorylation of Cdc25 by M-phase cyclin–Cdk stimulates the activity of Cdc25 (shown as a green arrow). This leads to a positive feedback loop, where as more active M-phase cyclin–Cdk is formed, Cdc25 becomes more active while Wee1 is further inhibited.

➔ We discuss the response to DNA damage in section 5.6 and throughout Chapter 15.

human Cdks) that is found in all major Cdks. The second site of inhibitory phosphorylation is found only in animal cell Cdks, and is at the adjacent threonine residue (Thr 14). These two residues, Thr 14 and Tyr 15, are located in the ATP-binding site of Cdk; their phosphorylation interferes with the orientation of ATP in the binding pocket. Therefore, as long as Thr 14 and Tyr 15 are phosphorylated, M-phase cyclin–Cdk cannot phosphorylate its substrates.

The phosphorylation state of Thr 14 and Tyr15 is determined by the opposing activities of a kinase and a phosphatase acting at these sites: Figure 5.8a shows how Thr 14 and Tyr15 are phosphorylated by the **Wee1** kinase, and dephosphorylated by the **Cdc25** phosphatase. Wee1 was identified as a mutation that caused cells to be smaller (from the Scottish term 'wee' meaning small) because they initiated mitosis prematurely, at a smaller than normal cell size. Wee1 is active during G1, S, and G2 phases, while Cdc25 becomes active in G2. Moreover, the activities of both Wee1 and Cdc25 are regulated by their substrate, the M-phase cyclin–Cdk complex: Wee1 is inhibited and Cdc25 is activated by M-phase cyclin–Cdk phosphorylation (see Figure 5.8b). Thus, M-phase cyclin–Cdk inhibits its inhibitor and activates its activator. This type of regulation is called a **positive feedback loop**; in this instance it leads to a sudden increase in Cdk activity.

How does this sudden increase in Cdk activity come about? Imagine that a small fraction of M-phase cyclin–Cdk becomes active. This active cyclin–Cdk inhibits Wee1, making it easier for Cdc25 to activate more cyclin–Cdk. Cdc25 itself becomes more active through phosphorylation by Cdk, further increasing the amount of active cyclin–Cdk. This type of regulatory circuit leads to an abrupt rise in Cdk activity. In a system where there is no positive feedback, such that Cdk activity would solely depend on cyclin binding, the increase in cyclin–Cdk activity would be gradual and directly proportional to the levels of cyclin in the cell, as illustrated in Figure 5.9a. However, the presence of a positive feedback loop creates an ON/OFF switch, as depicted in Figure 5.9b: the levels of cyclin–Cdk complex increase in proportion to the level of cyclin, but the complex remains inactive. The activity of Cdc25, however, removes a phosphate and, in so doing, leads to an abrupt increase in cyclin–Cdk activity. This positive feedback loop allows a sharp transition from G2 into mitosis.

Cdc25 is also involved in the regulation of Cdk activity in the presence of DNA damage. When DNA damage is detected, a regulatory pathway called the DNA damage checkpoint pathway is activated and leads to an inhibitory phosphorylation of Cdc25, which inactivates Cdc25. As a result, the inhibitory phosphorylations at Thr14 and Tyr15 cannot be removed and the cell cycle progression is blocked until the damage is repaired.

Cdk inhibitors negatively regulate cyclin–Cdk activity

The binding of specific inhibitory proteins called **cyclin kinase inhibitors (CKIs)** provides yet another level of regulation of cyclin–Cdk complexes. These small

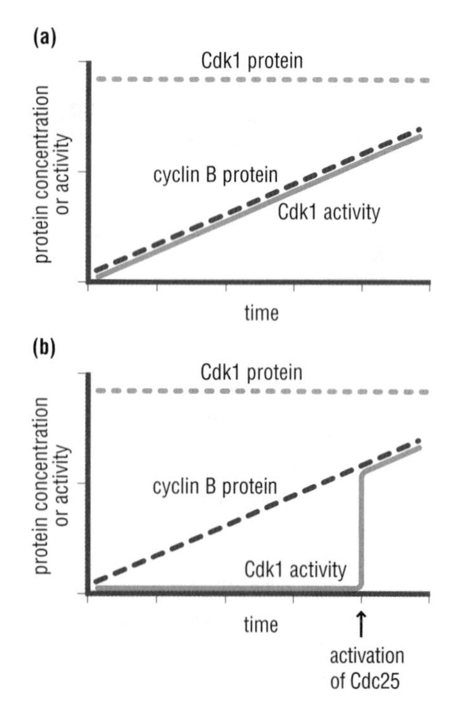

Figure 5.9 The contribution of Wee1/Cdc25 to the switch-like behavior of M-phase cyclin-Cdk. (a) The levels of Cdk in the cell (dashed gray line) are relatively constant and are in excess compared to the levels of M-phase cyclin (in this example cyclin B, shown as a dashed dark blue line). Furthermore, the activity of CAK is also constant throughout the cell cycle (not shown) and is not rate limiting. Thus, if the only requirement for Cdk activation was the binding of M-phase cyclin, the activity of M-phase cyclin–Cdk (orange line) would be directly proportional to the levels of the cyclin, and it would increase gradually as the cyclin is being synthesized. (b) However, cyclin–Cdk is also regulated by the Wee1/Cdc25 positive feedback loop (see Figure 5.8). As a result, the activity of cyclin–Cdk is repressed until Cdc25 is activated, when cyclin-Cdk is further stimulated by the positive feedback loop, leading to a burst of cyclin–Cdk activity.

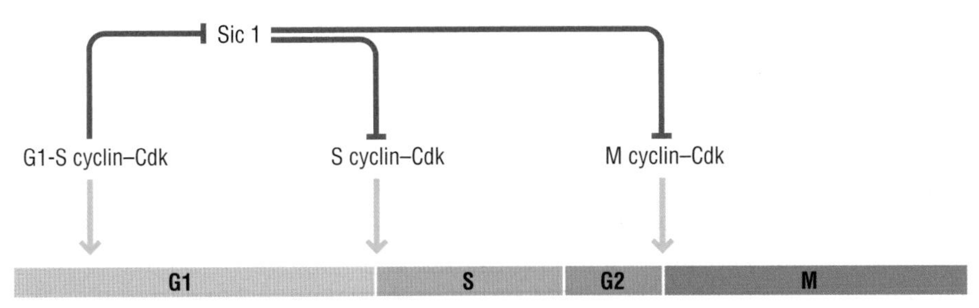

Figure 5.10 A regulatory cell cycle circuit acting at the beginning of S phase. In budding yeast, progression beyond G1 is regulated by the Cdk1 inhibitor Sic1 (budding yeast have only a single Cdk). Sic1 is an inhibitor of S-phase cyclin–Cdk1 and of M-phase cyclin–Cdk1, but not of G1-S cyclin–Cdk1. As G1-S cyclins accumulate, they phosphorylate Sic1 and target it for degradation. This alleviates the inhibition on S-phase cyclin–Cdk1 and allows entry into S phase. Note that at this point M-phase cyclin–Cdk cannot be activated due to the instability of M-phase cyclins (Figure 5.6). Thus, Sic1 and protein degradation are two mechanisms that down-regulate the activity of M-phase cyclin–Cdk during early stages of the cell cycle. Activating processes are shown as green arrows while inhibitory processes are shown as red bars.

inhibitory proteins fall into two main classes that differ at the amino acid sequence level, but both of which bind and inactivate cyclin–Cdk complexes. Mutations in CKIs are of clinical importance because they may lead to uncontrolled cellular proliferation and in doing so are sometimes associated with cancer.

One class of CKIs inhibitors binds to both the cyclin and the Cdk of G1-S or S-phase cyclin–Cdk complexes. CKIs of this class bind the hydrophobic patch of the cyclin (see Figure 5.5) and to a large domain on the Cdk, thereby distorting Cdk's structure and directly inhibiting ATP binding. In contrast, inhibitors of the second class bind to G1 Cdk monomers on the face of the protein opposite the active site. This binding induces a global change in the Cdk structure, which reduces cyclin binding and alters the Cdk active site.

An example of a CKI is the budding yeast Sic1 protein. Although Sic1 inhibits cyclin–Cdks, it is not similar in structure to any of the animal CKIs. Sic1 ensures that cells remain in G1 by inhibiting the major S and M-phase cyclin–Cdk1 complexes. Sic1, however, does not inhibit G1 and G1-S cyclin–Cdk1 complexes, so G1 processes and passage through start (the yeast restriction point equivalent) are not inhibited. At the beginning of S phase, Sic1 is phosphorylated by G1-S cyclins and targeted for destruction, thereby allowing S- and M-phase cyclin–Cdks to become active, as depicted in Figure 5.10. Thus, CKIs and other regulatory processes play an important role in ensuring the timely execution of the various cell cycle processes.

5.4 CELL CYCLE REGULATION BY Cdks

In the previous section, we discussed how the activity of Cdks is regulated. But what is the outcome of Cdk activity? What effects do Cdks actually have in the cell once active? We consider the answers to these questions in this section.

Cdks have many cellular targets

Cyclin–Cdks regulate the cell cycle through the phosphorylation of specific substrates. Once phosphorylated, these substrates have an altered activity or altered stability. This, in turn, affects the processes in which these proteins participate. But what are the processes that respond to Cdk-dependent phosphorylation? We know of over 400 proteins in yeast that are specific targets of the cyclin–Cdk kinase, and recent studies in animal cells have begun to systematically uncover an even greater

➔ We discuss a way in which Cdk substrates in yeast have been identified in Experimental approach 5.1.

number of cyclin-Cdk substrates. The identified substrates play roles in many cellular processes. Some proteins have roles in processes that are directly related to cell cycle progression, such as DNA replication, chromosome condensation, mitosis, and cell division. Other protein substrates, such as those involved in protein synthesis, may contribute to cell cycle regulation in a less direct way.

Many Cdk substrates have multiple phosphorylation sites. The method for elucidating how Cdk phosphorylation contributes to protein function often involves the identification of phosphorylated serine or threonine residues, mutating the corresponding DNA sequence so that these sites can no longer be phosphorylated, and then determining the consequences caused by the presence of an 'unphosphorylatable' protein. For some substrates, the role of Cdk phosphorylation has been elucidated, but for many others the role of phosphorylation is not yet known. In the sections that follow we will discuss several examples, out of many, of how Cdk phosphorylation affects cell cycle progression.

Cyclin–Cdks regulate the transcription of specific genes

One mechanism by which cyclin–Cdks regulate cell cycle events is through the regulation of transcription. Indeed, many of the identified cyclin–Cdk targets are transcription factors. In budding yeast, over 1200 genes - nearly 20% of the genome - are transcribed in a cell cycle-regulated manner, and are responsible for carrying out the various cell cycle processes. This cell cycle-regulated pattern of expression is controlled by cyclin–Cdk activity.

For example, in mammalian cells, entry into G1 is under transcriptional regulation: prior to the restriction point, the transcription factor **E2F** is inhibited through its association with one of the **Rb** family of proteins, as shown in Figure 5.11. Phosphorylation of Rb by the G1 cyclin–Cdks, cyclin D/Cdk4 or cyclin D/Cdk6, releases E2F from Rb, allowing E2F to activate transcription of a specific set of genes required for the G1/S transition. Mutations in the Rb gene that abolish its ability to bind E2F lead to inappropriate gene activation, which drives cell division and gives rise to an increased risk of cancer. In fact, the Rb gene gets its name from retinoblastoma, a type of cancer that occurs in the retina of the eye and is often associated with a mutation in the Rb gene. The regulatory pathway controlling Rb function is one of the most frequently altered pathways in many different kinds of cancer in addition to the rare retinoblastoma, as we will discuss towards the end of this chapter.

Cyclin–Cdks regulate many distinct cellular processes

Many cellular organelles and structures change during cell cycle progression. For example, the nuclear envelope disassembles during mitosis. Many of the details of this process are still to be elucidated, but we know that various proteins associated with the nuclear envelope, including components of the nuclear pore complex (the multiprotein structures that allow selective passage of proteins and other molecules between the cytoplasm and the nucleoplasm) and the nuclear lamina (a structural

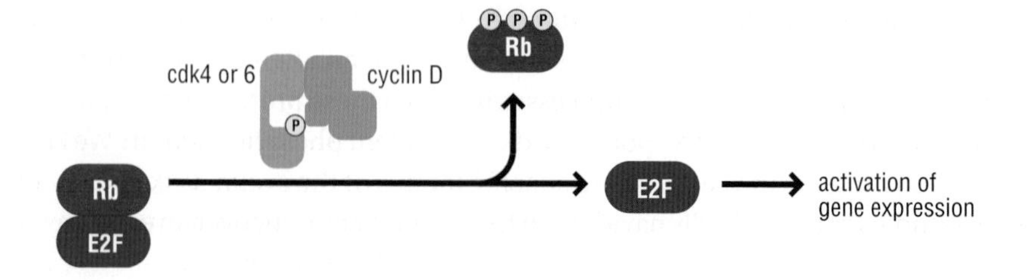

Figure 5.11 Regulation of gene expression by cyclin–Cdk phosphorylation. In G0 and early G1 of metazoan cells, the transcription activator E2F is bound to Rb. This association prevents E2F from activating transcription. In order to enter the cell cycle, cyclin D associates with either Cdk4 or Cdk6, and the resulting cyclin–Cdk phosphorylates Rb. This causes Rb to dissociate from E2F, freeing E2F to activate the expression of genes necessary for entering the cell cycle. Note that if Rb is inactivated, due to a mutation, for example, E2F may promote unscheduled cell cycle entry, leading to uncontrolled cell division, such as in cancer cells.

component of the nuclear envelope made of intermediate filaments that lies beneath the inner nuclear membrane) are phosphorylated by M-phase cyclin–Cdks; this phosphorylation contributes to the disassembly of the nuclear envelope during mitosis.

As cells divide, each daughter cell needs to receive a complete set of chromosomes; they also need to inherit other cellular structures and organelles. We now know that the inheritance of at least some of these structures is dependent on Cdk activity. For example, in yeast, the movement of a cellular structure called the vacuole during cell division is controlled by Cdk phosphorylation. Likewise, the reorganization of the Golgi apparatus, an organelle involved in protein sorting, that allows its segregation to the daughter cells is also dependent on Cdk phosphorylation. Thus, by controlling multiple cellular processes, Cdk phosphorylation ensures the timely execution of events that must be tightly coordinated.

5.5 REGULATION OF PROTEOLYSIS BY Cdks

Cyclin–Cdks regulate proteolysis of specific substrates

A way to ensure that the cell cycle moves in a forward direction is to eliminate proteins needed for the previous cell cycle stage by degrading them. Indeed, cyclin–Cdks drive the cell cycle forward by promoting the proteolytic destruction of key regulatory proteins. The main mechanism for cell cycle-regulated protein degradation is by the **ubiquitin**-mediated proteolysis machinery. Ubiquitin is a small protein that is highly conserved in evolution. Proteins can be degraded by being modified by poly-ubiquitin chains, which target them for degradation by the **proteasome**. The process of ubiquitin attachment is called ubiquitination or ubiquitylation, and involves a cascade of three enzymes, between which the ubiquitin is passed until it is finally attached to the protein that is targeted for degradation. The last step in this process, namely the attachment of a ubiquitin molecule to the substrate, is catalyzed by the enzyme **ubiquitin ligase**.

➔ The process of ubiquitination and protein degradation by the proteasome are discussed in greater detail in Chapter 14.

There are two major ways in which cyclin–Cdks regulate protein degradation. First, some cell cycle regulators have to be phosphorylated in order to be recognized by certain ubiquitin ligases. Thus, cyclin–Cdk targets these substrates for degradation by phosphorylating them. Second, cyclin–Cdk directly phosphorylates a number of ubiquitin ligases, leading to their activation. In this case, cyclin–Cdk targets proteins for degradation by activating their ubiquitination machinery. To illustrate this, let us discuss two ubiquitin ligases that are central to cell cycle progression.

The SCF ubiquitin ligase complex catalyzes ubiquitination of phosphorylated substrates

The **SCF** complex is a ubiquitin ligase that regulates the transition from G1 into S phase by degrading specific proteins that are phosphorylated by G1-S cyclin–Cdks. SCF is named for three of its central components – **Skp1**, **cullin**, and the **F-box protein**; its structure is illustrated in Figure 5.12. The complex also contains a RING domain protein, often found in ubiquitin ligases. The F-box protein is the subunit that binds the protein substrate that will be ubiquitinated. In the cell there are many F-box proteins, each with specificity for a distinct set of substrates.

The activity of SCF is constant throughout the cell cycle. However, the substrate must be phosphorylated in order to bind the F-box protein. It is this step that is regulated by cyclin–Cdk phosphorylation. An example of a protein that is targeted for degradation by SCF-dependent ubiquitination is the yeast Cdk inhibitor Sic1,

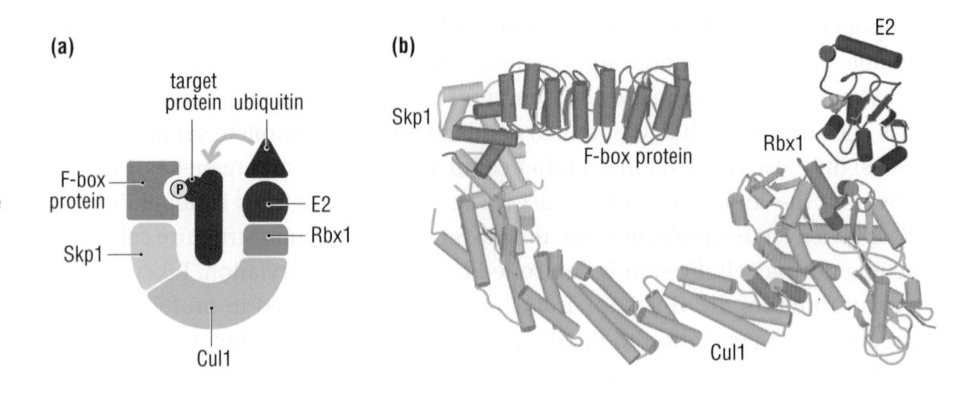

Figure 5.12 SCF is a multisubunit enzyme. (a) SCF contains several subunits: the subunits that it is named after, which are Skp1, Cul1, and an F-box protein, and a fourth subunit, Rbx1, which contains a RING domain characteristic of ubiquitin ligases and binds another component of the ubiquitination process, an E2 enzyme. The phosphorylated target protein binds to the F-box subunit, after which ubiquitin is transferred from the E2 to a lysine side chain on the target protein. (b) A composite model structure for human SCF derived from multiple crystal structures. Protein Data Bank (PDB) codes 1LDK, 1FBV, 1FQV.

Adapted from Zheng, N, Schulman, BA, Song, L *et al.* Structure of the Cul1-Rbx1-Skp1-FboxSkp2 SCF ubiquitin ligase complex. *Nature*, 2002;**416**: 703–709.

which we encountered in Section 5.3. Recall that Sic1 is an S-phase cyclin–Cdk inhibitor (CKI) (see Figure 5.10). Sic1 phosphorylation by G1-S cyclin–Cdks leads to its ubiquitination by the SCF and its subsequent degradation by the proteasome. As a result of the decline in Sic1 levels, S-phase cyclin–Cdks become active.

Other proteins that are degraded by the SCF are the G1-S cyclins themselves; by contrast, S-phase cyclins are not SCF targets. Thus, through Cdk activity, cyclins that are needed for early cell cycle events are degraded, allowing the cell to move forward in the cell cycle.

The anaphase-promoting complex ubiquitin ligase, which drives anaphase and mitotic exit, is activated by Cdk phosphorylation

A second mechanism by which Cdk activity regulates proteolysis is by directly activating a ubiquitin ligase known as the **anaphase-promoting complex (APC)** early in mitosis. The APC, as its name implies, is essential for promoting the anaphase stage of mitosis, during which sister chromatids separate. The importance of the APC was shown genetically in yeast, where the inactivation of the APC blocks the cell's ability to progress through mitosis, as illustrated in Figure 5.13. Although the APC catalyzes the ubiquitination of many proteins, it has three major substrates whose destruction is particularly critical for the precise execution of mitosis: securin and shugoshin, whose destruction is necessary for the removal of cohesin from sister chromatids at the onset of anaphase, and cyclin B, whose destruction is necessary for timely exit from mitosis.

➔ The mitotic substrates of the APC are discussed in greater detail in Section 7.5.

In most organisms, the APC is a complex of 11–13 subunits, including a cullin subunit and a RING domain subunit, much like SCF. Look at Experimental approach 5.2 to see how some of the genes coding for APC subunits were identified. Unlike SCF, however, the APC does not have an F-box protein; its substrate specificity is instead

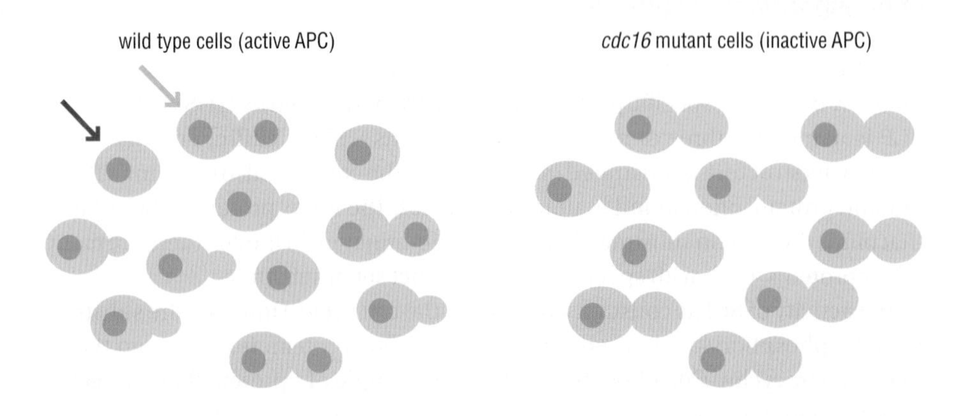

Figure 5.13 The APC is needed for progression to anaphase. The different cell cycle stages of budding yeast can be assessed by the size of the bud (from no bud at all in G1 to a large bud in mitosis) and by the state of the DNA (one nucleus before anaphase; two nuclei – one in the mother and one in the bud – after anaphase). A culture of wild-type cells (left panel) contains a mixture of cells in various cell cycle stages (see, for example, an unbudded cell indicated by the red arrow, and a cell in anaphase (large bud, two DNA masses), indicated by the green arrow). In contrast, in a culture of cells where one of the APC subunits was inactivated (for example, cells carrying a temperature-sensitive allele of the APC subunit Cdc16 that are grown at the non-permissive temperature), cells cannot enter anaphase, and they accumulate in mid-mitosis as large budded cells with a single nucleus.

5.2 EXPERIMENTAL APPROACH

The cell division cycle: from genes in yeast to function in animal cells

Inactivation of a cell cycle protein leads to a uniform population of cells

A major breakthrough in studies of the cell cycle came in 1970 with the isolation of *S. cerevisiae* mutants that were defective in cell cycle progression – these were referred to as cell division cycle (*cdc*) mutants. Lee Hartwell and colleagues based their genetic screen to find cell cycle mutants on three considerations. First, the group assumed that the cell cycle of budding yeast could be followed by cellular and nuclear morphology. Second, they thought that genes that contribute to cell cycle progression would likely be essential. Thus they would need to isolate temperature-sensitive (*ts*) mutants that are viable at the permissive temperature (23 °C) but are non-viable at some other non-permissive temperature (36 °C in this case). Third, they assumed that cell cycle mutants would arrest with a uniform cell cycle morphology.

With these criteria in mind, the Hartwell group screened a yeast *ts* mutant collection, looking for those strains that failed to progress beyond a particular cell cycle stage at the non-permissive temperature. In real terms, this meant that they looked in the microscope at many different yeast cell cultures and identified those where all the cells in the population had a similar morphology (i.e., all might be 'unbudded' or 'large budded' (see Figure 5.13)).

Of the 100 initial temperature-sensitive strains that were screened, 15 exhibited a uniform cell cycle arrest. These were called *cdc* genes and were assigned a number according to their order of isolation (*cdc1*, *cdc2*, etc). One such example is the budding yeast cdc2 mutant, which, when grown at the non-permissive temperature, causes cells to arrest as large budded cells as depicted in Figure 1.

The terminal phenotype of a mutant may not reflect the point in the cell cycle where the underlying defect has occurred.

We now know that the budding yeast Cdc2 protein is involved in DNA replication, and that the phenotype observed by Harwell and colleagues was probably due to a cell cycle arrest caused by DNA replication checkpoint activation. This raises an important point: inactivation of certain proteins (for example, the anaphase promoting complex) will lead to a cell cycle arrest at the point where this activity is needed, namely mitosis. In this case, it is said that the execution point, which is the stage in a process where the activity is needed, is the same as the terminal phenotype, which is where the cells arrest in the absence of the activity. However, this is not always the case. For example, if the assembly of the spindle is blocked in late S phase/early G2, cells will progress through the cell cycle and will stop in mitosis due to spindle checkpoint activation. In this case, the execution point (S/G2) is different from the terminal phenotype (mitosis).

Combination of genetics and biochemistry led to the identification of the anaphase promoting complex

Genetic studies such as those carried out by the Hartwell group could generally pinpoint the cell cycle process that was disrupted by a mutation, but often could not identify the specific molecular function of the proteins encoded by the genes. Twenty years after

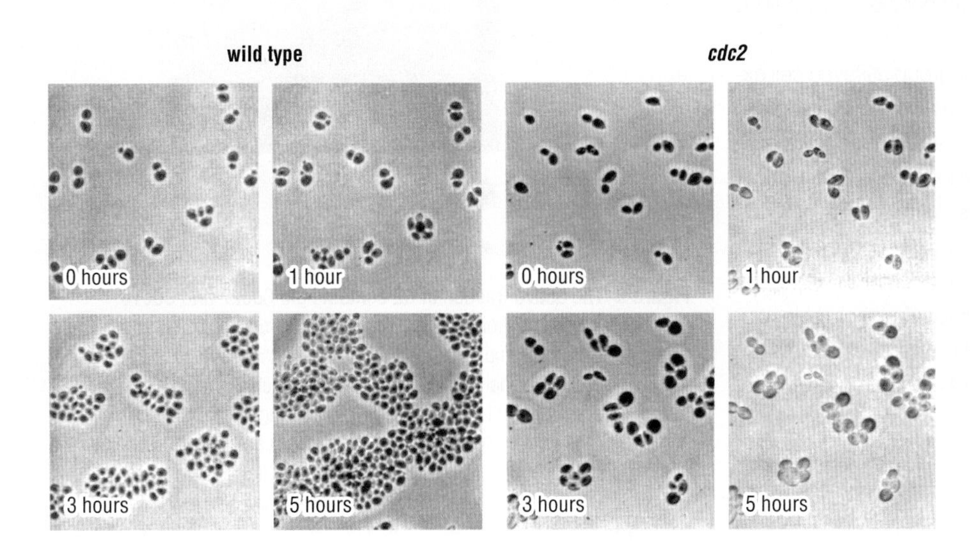

Figure 1 The isolation of *cdc* mutants. Wild-type control cells (left panels) and mutant cells (right panels) were plated on agar plates containing yeast media, and incubated at 36 °C, which is the non-permissive temperature for the mutant strain. Cell growth was then monitored at 1, 3, and 5 hours after the temperature shift. The wild-type cells grew continuously, while the mutant cells arrested as cells with large buds.

Figure 1 is adapted from Hartwell *et al.* Genetic control of the cell division cycle in yeast, 1. Detection of mutants. *Proceedings of the National Academy of Sciences of the U S A*, 1970;**66**:352–359.

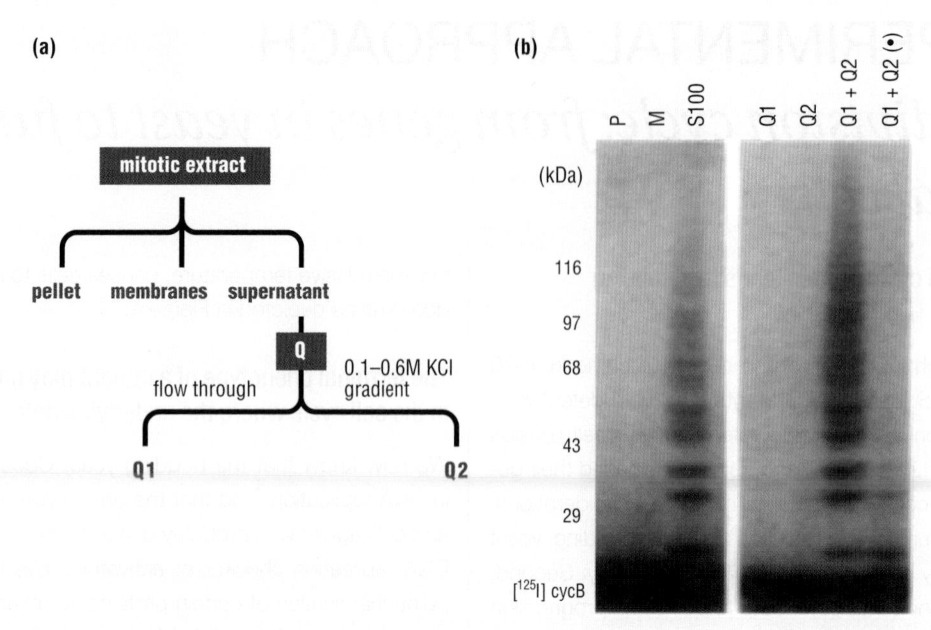

Figure 2 Identification of a cell cycle-regulated activity for cyclin B ubiquitination. (a) Fractionation scheme of *Xenopus* egg extract. The investigators used either mitotic egg extract (meaning that the extract was derived from oocytes that were in mitosis), an interphase egg extract (not shown). After subjecting the extract to centrifugation, which separated the insoluble material (pellet, P), membrane (M) fraction and supernatant (S100 in panel (b)), the supernatant was subjected to further fractionation on an anion exchange column (Q), resulting in two fractions, Q1 (the flow through, namely the material that did not bind the column) and Q2, which contained material eluted from the column with a gradient of 0.1–0.6 M KCl, and then combined. (b) Ubiquitination activity was followed by the appearance of a ladder on SDS-polyacrylamide gel electrophoresis (SDS-PAGE) of radiolabeled cyclin B ((125I)-cyclin B). Only the supernatant fraction could promote ubiquitination. Neither fraction Q1 alone, nor Q2 alone could promote cyclin B ubiquitination, but the mixture of both was active. The investigators showed in this study that Q2 was active only when prepared from mitotic extracts, but not from interphase extracts (denoted (•) in the figure). The mitotic activity of Q2 was important because it suggested that the cell cycle-regulated activity of oocyte extract was present in this fraction.

From King *et al.* A 20S complex containing CDC27 and CDC16 catalyzes the mitosis-specific conjugation of ubiquitin to cyclin B. *Cell*, 1995;**81**:279–288.

the initial discovery of the *cdc* genes, Marc Kirschner and colleagues used a biochemical approach to examine the mechanism of cyclin B degradation during mitotic progression. At the time, it was known that cyclin B destruction occurred in a cell cycle dependent manner and that it involved ubiquitination of cyclin B. The role of ubiquitination, the attachment of the small ubiquitin protein, in protein degradation is discussed in Section 14.9. It was also understood that cyclin B destruction was necessary for exit from mitosis, though the machinery responsible for cyclin B ubiquitination was not yet known.

To approach this question, the Kirschner group assayed cell extracts from *Xenopus laevis* (frog) oocytes, which readily progress through nearly every cell cycle process in a test tube, including DNA replication, spindle formation, chromosome condensation, and nuclear envelope breakdown and assembly. To identify the machinery involved in cyclin B destruction, Kirschner and colleagues labeled cyclin B with radioactive iodine and searched for an enzymatic activity that resulted in a ladder of higher molecular weight cyclin B, which is typical of ubiquitination; their results are shown in Figure 2. Upon fractionating the *Xenopus* egg extracts, they discovered that at least two of the required components were present in fractions Q1 and Q2. Size fractionation of Q2 revealed that the activity of interest eluted as a very large 20S complex that contained an E3 ubiquitin ligase. Previous genetic and biochemical studies in budding yeast suggested that a complex of several proteins, including Cdc16, Cdc23 and Cdc27, is needed for mitotic progression. The Kirschner group next showed that these very same proteins were indeed present in the active complex that they had identified; see Figure 3. Based on

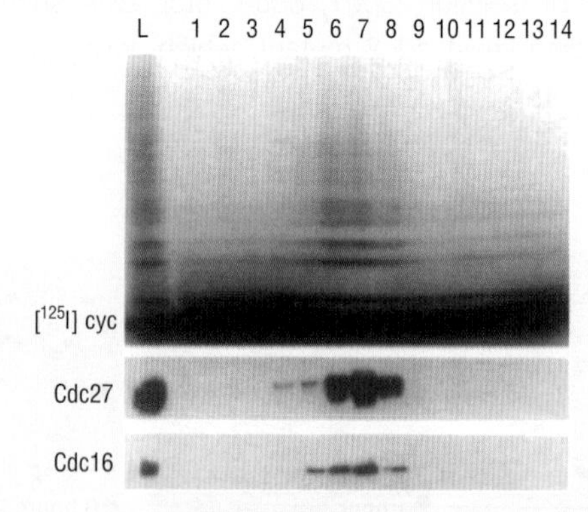

Figure 3 The active fraction in Q2 contains the Cdc27 and Cdc16 proteins. Fraction Q2 from Figure 2 was subjected to a sucrose gradient, where proteins and complexes sediment according to their density. A sample from the starting material was loaded in the lane labeled L and the sucrose fractions were loaded in lanes numbered 1–14. Using antibodies against Cdc27 and Cdc16, the investigators found that the fractions that were active in cyclin B ubiquitination (upper panels) contained Cdc27 and Cdc16 (lower panels).

From King *et al.* A 20S complex containing CDC27 and CDC16 catalyzes the mitosis-specific conjugation of ubiquitin to cyclin B. *Cell*, 1995;**81**:279–288.

its activity in cell cycle progression, they called the complex the anaphase-promoting complex (or APC).

Both yeast genetics and the fractionation of frog oocytes were critical to advancing our understanding of the cell cycle. These genetic screens and biochemical purifications illustrate how different experimental approaches can complement each other to provide a more complete understanding of the biological process at hand.

Find out more

King RW, Peters JM, Tugendreich S, *et al.* A 20S complex containing CDC27 and CDC16 catalyzes the mitosis-specific conjugation of ubiquitin to cyclin B. *Cell* 1995;**81**:279–288.

Hartwell LH, Culotti J, Reid B. Genetic control of the cell division cycle in yeast, I. Detection of mutants. *Proceedings of the National Academy of Sciences of the U S A* 1970;**66**:352–359.

Related techniques

S. cerevisiae (as a model organism); section 19.1

Using temperature sensitive mutants; section 19.5

Microscopy (phase); section 19.13

Protein fractionation; section 19.7

Using *Xenopus* egg extracts; section 19.1

determined by activator subunits. These activator subunits function in much the same way as F-box proteins in that they serve as adaptors that link the substrate with the catalytic site of the ubiquitin ligase. Two activator proteins, **Cdc20** and **Cdh1**, are particularly important. Cdc20 targets proteins for ubiquitination by the APC early in mitosis, while Cdh1 targets proteins for ubiquitination by the APC in late mitosis and G1. Figure 5.14a depicts how multiple APC subunits are phosphorylated early in mitosis by mitotic cyclin–Cdk, a step that is necessary for APC activation. This phosphorylation promotes Cdc20 binding to the APC, which, in turn, promotes the ubiquitination, and subsequent degradation, of securin and shugoshin - proteins that inhibit sister chromatid separation. Cdc20 also recruits mitotic cyclins for ubiquitination, thereby initiating the process of mitotic cyclin–Cdk inactivation. Thus, mitotic cyclin–Cdks promote their own demise by activating the APC.

In yeast, Cdh1 is also phosphorylated by M-phase cyclin–Cdk, but unlike the APC itself, this phosphorylation is inhibitory. Cdh1 remains phosphorylated until after anaphase, when a particular phosphatase becomes active (see Figure 5.14b). Thus, by phosphorylating Cdh1, the cell cycle machinery ensures that Cdh1 will become active only after anaphase. Once the phosphatase removes the inhibitory phosphate from Cdh1, Cdh1 becomes active and, in association with the APC, targets M-phase cyclins for degradation. This further reduces the activity of M-phase cyclin–Cdk1, thereby making the removal of Cdh1 phosphate more efficient, and enabling APC–Cdh1 to target more M-phase cyclins for degradation. Ultimately, this facilitates mitotic exit.

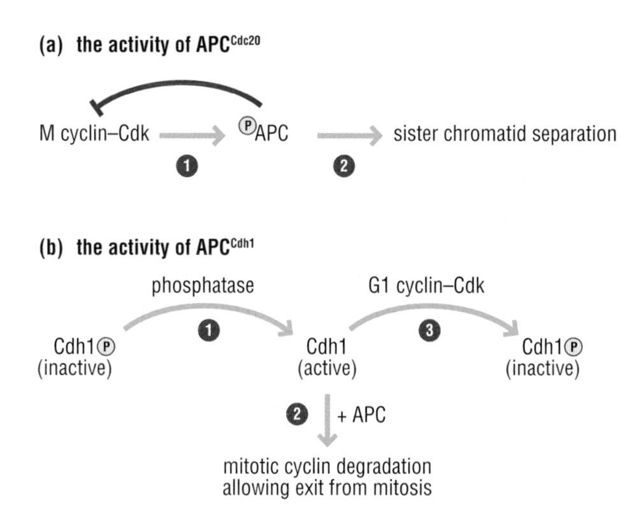

Figure 5.14 The APC and its activator subunits, Cdc20 and Cdh1, regulate progression through, and exit from, mitosis. (a) The APC is activated through its phosphorylation by M-phase cyclin–Cdk (step 1 in the regulatory circuit). Once activated, the APC promotes progression through mitosis by targeting anaphase inhibitors for degradation, which promotes sister chromatid separation (step 2). The APC also targets M-phase cyclins for degradation (step 3), a process that is necessary for mitotic exit. (b) In the beginning of mitosis, Cdh1 is still phosphorylated by M-phase cyclin–Cdk, which prevents its association with the APC. The APC^{Cdc20}-dependent initiation of anaphase activates a phosphatase that dephosphorylates Cdh1 (step 1). This allows Cdh1 to bind to the APC and target M-phase cyclins for degradation (step 2). M-phase cyclin degradation, in turn, promotes the transition from M to G1. Once cells are in G1, G1 cyclins begin to accumulate, because they are not APC substrates. This results in the activation of the G1 cyclin–Cdk that phosphorylates Cdh1 (step 3), thereby inactivating the APC^{Cdh1} and allowing S-phase cyclins to accumulate (not shown). Activating processes are shown as green arrows while inhibitory processes are shown as red bars.

APC–Cdh1 remains active in G1, thereby preventing the accumulation of M-phase cyclin. This ensures that the cell cycle will not go back into mitosis. However, there must come a point at which the cell cycle has to move past G1 – but if Cdh1 remains active, S-phase cyclins will not accumulate. G1 cyclins are not targeted for degradation by APC–Cdh1, and when their level of activity rises, they phosphorylate and inactivate Cdh1. This allows the S-phase cyclins to accumulate and promote the transition from G1 to S. Therefore, by regulating protein degradation, the Cdk machinery can move the cell cycle forward.

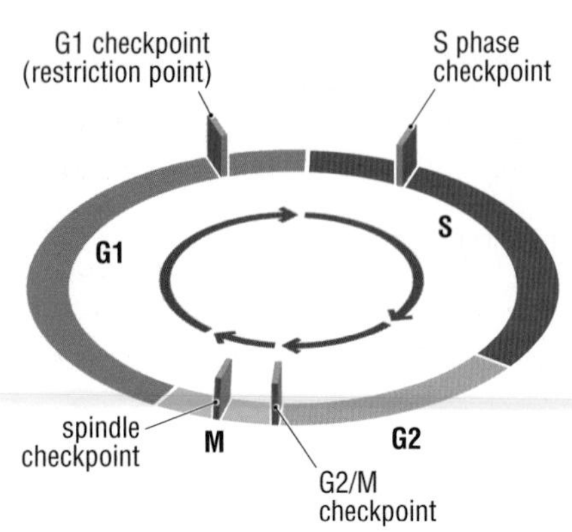

Figure 5.15 Key cell cycle transitions can be blocked by checkpoint pathways. Checkpoint mechanisms prevent cell cycle progression in the presence of different kinds of insults (e.g., DNA damage, spindle defects) until the damaged structure is repaired. In eukaryotes, different checkpoint pathways can halt, or delay, cell cycle progression at various key transitions, as described later in this section. These delay points are shown in the figure as gates that can be traversed only if there is no inhibitory signal from the respective checkpoint pathway. (Note that the S phase checkpoint pathway can lead to a cell cycle block anywhere within S phase; only one S phase checkpoint 'gate' is shown for simplicity.) The DNA damage checkpoint pathway can arrest cell cycle progression at the restriction point (the G1–S boundary) or in G2/M. The DNA replication checkpoint pathway can inhibit further DNA replication during S phase and can delay entry into mitosis. The spindle assembly checkpoint pathway blocks the initiation of anaphase and the exit from mitosis.

5.6 CHECKPOINTS: INTRINSIC PATHWAYS THAT CAN HALT THE CELL CYCLE

The cell monitors certain cell cycle processes and can arrest at specific transition points

In addition to the mechanisms described earlier in this chapter that drive the cycle forward, the cell also has intrinsic surveillance mechanisms, often called checkpoint pathways, that can stop the cell cycle when certain things go wrong. Thus, checkpoint pathways can be viewed as quality control mechanisms that contribute to the fidelity of the cell division cycle. We discuss the discovery of the first checkpoint gene, and the rationale behind the checkpoint concept, in Experimental approach 5.3.

Different checkpoint pathways monitor various essential cellular processes, and activate specific responses when errors are detected. These responses include the activation of repair or correction pathways, and the induction of a cell cycle delay, thereby providing the cell with time to correct the errors. In eukaryotes, cell cycle progression can be stopped at different points in the cell cycle – the G1-S transition, S phase, the G2–M transition and mitosis – depending on the checkpoint pathway that is activated. The different points in the cell cycle at which checkpoint pathways act are illustrated in Figure 5.15.

The checkpoint mechanism is not unique to eukaryotes: bacteria have checkpoint mechanisms as well. Both bacteria and eukaryotes have checkpoint pathways that monitor DNA integrity, and eukaryotes also have a checkpoint pathway that monitors the chromosome segregation process. In this section we will briefly touch upon the main properties of different checkpoint pathways in eukaryotes. The bacterial checkpoint, called the SOS response, will be described when we discuss the bacterial cell cycle in Section 5.9. We also discuss the eukaryotic checkpoints in greater detail in later chapters in the context of the processes that they monitor.

➔ The checkpoint that is activated in response to DNA damage is discussed extensively in Chapter 15, and the checkpoint that is activated due to spindle malfunction is described in Section 7.5.

Checkpoint pathways are composed of sensors, transducers, and effectors

The activation of a checkpoint pathway in response to the presence of a cellular defect requires proteins, often referred to as **sensors**, which sense the cellular defect and activate the checkpoint response. For example, Figure 5.16 illustrates how DNA damage is often associated with the formation of long stretches of single-stranded DNA. In this case, the activation of the checkpoint pathway is dependent on the binding of sensor proteins to the single-stranded DNA. Different types of cellular damage recruit distinct sensor proteins, and this recruitment is necessary for the appropriate cellular response. Specifically, the involvement of damage-type specific sensors ensures that the correct checkpoint pathway is triggered, and that other checkpoint pathways remain dormant.

5.3 EXPERIMENTAL APPROACH

Checkpoint pathways: delay and conquer

Radiation sensitive mutant uncover genes involved in the DNA damage checkpoint pathway

The presence of DNA damage poses a serious threat to cells because it can affect chromosome integrity. Scientists had noted that cells exposed to DNA damaging agents exhibited delayed cell cycle progression in G2, but until 1988 the mechanism responsible for this delay was not known. To understand the nature of this cell cycle delay, Ted Weinert and Lee Hartwell decided to use budding yeast *S. cerevisiae* to identify genes involved in regulating cell cycle progression in response to DNA damage. They reasoned that mutants in this process would exhibit a greater sensitivity to DNA damaging agents because they would not arrest but instead continue through the cell cycle even in the presence of DNA damage.

At the time, there already existed a large collection of yeast mutants that were hypersensitive to radiation and thus termed *rad* mutants. Weinert and Hartwell decided to look through this collection to identify mutants in which cell cycle progression failed to arrest after irradiation. The assay they used was simple: cells from asynchronous populations of wild-type or mutant strains were placed on solid growth media, the cells were then exposed to x-rays, and cell cycle progression was followed in the microscope (by examining how many times each cell divided). And indeed, a mutation in one of the genes, *RAD9*, seemed to meet their criteria for a checkpoint gene.

Mutants defective in checkpoint activation fail to arrest cell cycle progression in the presence of DNA damage

Since budding yeast cells divide by budding, Weinert and Hartwell scored the appearance of new buds to mark cells that were actively dividing. Haploid cells were used for this assay because, if damaged

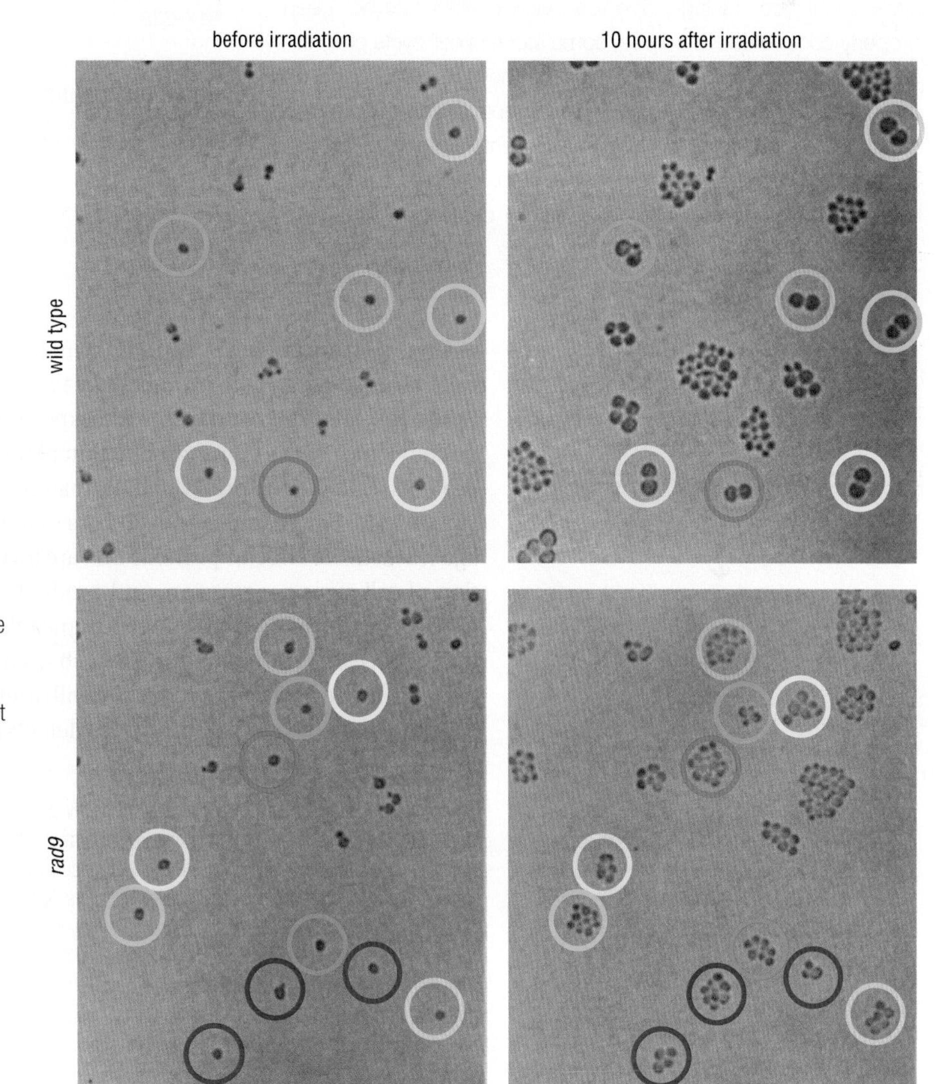

Figure 1 No DNA damage-induced arrest in *rad9* mutant cells. Wild-type and *rad9* cells were plated, x-ray irradiated at a dose of 2 kilorads, and then allowed to grow for 10 hours at 23 °C. At this temperature, the cell division cycle is about three hours long. Note that although some of the wild type cells grew in the irradiated sample, most of the cells that were unbudded at the time of irradiation arrested as large-budded cells (see color matched cells before and after irradiation). In contrast, most of the unbudded *rad9* cells continued to divide despite being irradiated, indicating that they are missing the regulatory pathway that would normally arrest cell cycle progression in the presence of DNA damage.

From Weinert and Harwell. The RAD9 gene controls the cell cycle response to DNA damage in *Saccharomyces cerevisiae*. *Science*, 1988;**241**:317–322.

in G1, they should exhibit the greatest cell cycle delay since they have neither a homologous chromosome nor a sister chromatid that can serve as a template for repair. Indeed, most unbudded cells on the plates exposed to radiation formed a bud (indicative of progression to G2), but then remained arrested as large budded cells for at least 10 hours (see Figure 1, top panels).

In contrast, the investigators found one mutant strain, *rad9*, in which unbudded cells failed to arrest after irradiation and divided multiple times within the 10-hour time span (Figure 1, bottom panels). When the results shown in these images were quantified, the investigators found that 80% of unbudded wild-type cells arrested as large budded cells after irradiation, whereas less than 8% of unbudded *rad9* mutant cells arrested as large budded cells under the same conditions.

Checkpoint pathways are regulatory mechanisms that prevent the execution of later cell cycle stages until earlier ones have been completed successfully

These findings led Weinert and Hartwell to propose the idea of cell cycle checkpoints, regulatory mechanisms that prevent a cell cycle event from taking place if an earlier event has not been properly completed. Of course, some steps in cell cycle progression cannot take place simply because they directly depend on the product of an earlier event. For example, inactivation of the anaphase promoting complex (APC) prevents anaphase initiation because APC activity is critical for this step. In other cases, however, a regulatory mechanism prevents a cell cycle event from taking place before an earlier event has been completed. In the case of DNA damage, the DNA damage checkpoint pathway prevents progression through mitosis when there is, for example, a double-stranded DNA break. This break, by itself, does not prevent the chromosomes from binding to spindle microtubules, nor is the break an impediment for chromosome segregation. Rather, the presence of the DNA break activates a regulatory mechanism, the DNA damage checkpoint pathway, which prevents cells from activating the processes that are necessary for mitosis.

Checkpoint-based control of the cell cycle, first elucidated in *S. cerevisiae*, is critical in all eukaryotes, and components of these checkpoints are conserved throughout evolution. It is interesting to note that mutations that affect the p53 protein, a major regulator of checkpoint function in human cells, are found in over 50% of all human cancers. Why is this? Mutations in p53 result in a failure of cells to arrest in response to DNA damage, and the continued growth of these cells leads to tumor formation.

Find out more

Weinert TA, Hartwell LH. The *RAD9* gene controls the cell cycle response to DNA damage in *Saccharomyces cerevisiae*. *Science*, 1988;**241**: 317–322.

Related techniques

S. cerevisiae (as a model organism); Section 19.1

Microscopy (phase); Section 19.13

Irradiation; Section 19.5

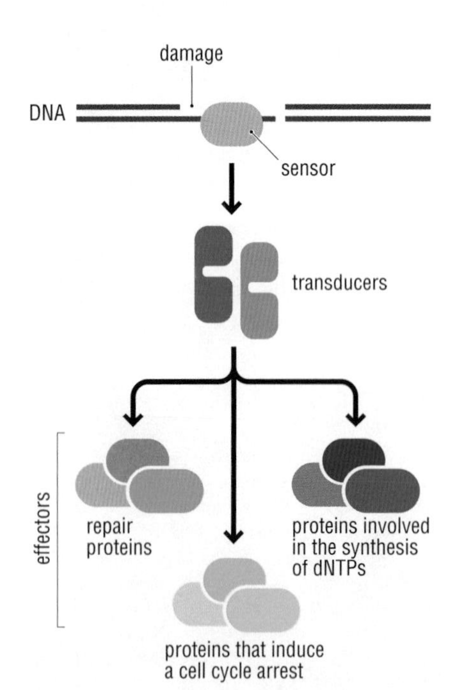

Once sensors associate with the damage, they activate a second set of proteins, called **transducers**, whose function is to launch the checkpoint pathway response. In the case of the DNA damage checkpoint pathway, some of the transducers are protein kinases. The common function of transducers is to activate a third group of proteins, called **effectors**, which ultimately carry out the checkpoint response (see Figure 5.16). For example, transducers that are activated in response to DNA damage will activate effector proteins that are involved in repair of DNA damage, inhibition of cell cycle progression, and production of nucleotides, the building blocks for DNA synthesis. By activating multiple effectors, transducer proteins can initiate many processes simultaneously, which then act in parallel to effectively repair the damage. In the next two sections, we will briefly discuss some of the main features of checkpoint pathways that respond to defects in DNA integrity and spindle function.

Figure 5.16 The structure of a checkpoint pathway. The example shown is of a checkpoint pathway induced by the presence of DNA damage in the form of a double-stranded DNA break. The sensor protein associated with the damaged site activates a transducer protein. The transducers, which in this case are protein kinases, activate multiple substrates, called effectors. In the case of DNA damage, some effectors are involved in DNA damage repair (orange spheres), others may induce a cell cycle arrest (green spheres), while yet others contribute to the synthesis of deoxyribonucleotide triphosphates (dNTPs) that are required for DNA synthesis after the damage is repaired (blue spheres). (dNTP is a general term referring to dATP, dCTP, dGTP, and dTTP.) Note that if the DNA damage persists in animal cells, the checkpoint pathway may activate a cell death process (called apoptosis) that will eliminate cells that may carry potentially detrimental DNA lesions (not shown).

Insults to DNA integrity can activate checkpoint pathways that signal repair and stop cell cycle progression

Both bacteria and eukaryotes have mechanisms to respond to different types of DNA insults, although the sensors, transducers, and effectors involved in these processes are quite distinct. There are two main forms of DNA insults that are monitored by the cell: stalled DNA replication forks and DNA double strand breaks.

During DNA replication, regions of the double-stranded DNA are separated into single strands, with each strand serving as the template for the synthesis of a new daughter strand. During this process, the DNA replication machinery usually moves along the DNA without perturbation, unwinding the DNA into single strands as it goes. Occasionally, however, replication is impeded - for example, due to a physical block on the DNA strand that is being copied, or due to insufficient levels of nucleotides from which to synthesize the DNA. When this happens, DNA replication stops and single-stranded DNA accumulates. This single-stranded DNA activates the S-phase checkpoint.

➔ The process of DNA replication is discussed in Chapter 6.

A different kind of DNA damage can occur when the cell is not replicating its DNA. For example, a DNA double strand break can occur due to exposure to chemicals or radiation. This damage is also detected by specific sensor proteins. The ends of the broken DNA are sometimes processed by exonucleases in the cell, resulting in the formation of single-stranded DNA. DNA breaks can be repaired by various mechanisms, but if these breaks linger through mitosis they can result in permanent damage to the cell due to loss of genetic material.

➔ The process of DNA repair is discussed in Chapters 15 and 16.

With both stalled replication forks and DNA breaks, one of the triggers that activate the checkpoint pathway is the presence of the single-stranded DNA that is recognized by certain sensor proteins. In the case of DNA double strand breaks, at least one sensor protein can also be activated by a non-resected DNA double-stranded end. Note, however, that different sensor proteins get recruited to single-stranded DNA in the case of a replication block versus DNA damage such as a double-stranded DNA break. The association of sensor proteins with these damaged DNA sites leads to their activation, which initiates the checkpoint response. The sensor proteins in turn activate transducers, in the form of protein kinases, which then can activate a variety of effectors that lead to a range of cellular responses, including DNA repair and cell cycle arrest.

The stage at which cell cycle arrest occurs depends on when the damage is detected. For partially replicated DNA, cell cycle progression will stop in S phase, and the checkpoint will inhibit mitotic cyclin–Cdks to ensure that mitosis does not take place with partially replicated DNA. When the DNA damage is detected in G1, the checkpoint pathway will block passage through the restriction point by activating a Cdk inhibitor, thereby inhibiting G1 cyclin–Cdk activity. For DNA damage that is detected in late S phase or in G2, the checkpoint pathway will block entry into mitosis by inhibiting the activation of mitotic cyclin–Cdk and arresting at G2-M transition (for example, by preventing Cdc25 activation; see Figure 5.8).

In mammalian cells, if the perturbation to DNA integrity persists for a long time, checkpoint pathways will induce programmed cell death (apoptosis) to eliminate cells whose DNA cannot be properly repaired. Thus, in response to defects in the DNA, checkpoint pathways activate repair mechanisms, block cell cycle progression, and may trigger cell death.

The importance of checkpoint pathways that respond to DNA damage or defects in DNA replication is underscored by the fact that mutations in some checkpoint genes are associated with disease. For example, mutations in the gene encoding for the DNA damage checkpoint protein **ATM** are the underlying cause of the disorder ataxia telangiectasia. Ataxia telangiectasia has multiple manifestations, one of which is an increased risk of cancer. ATM is a kinase that is recruited by DNA damage sensor proteins to the site of DNA damage, where ATM becomes active. Some of ATM's targets are well-known cancer-associated genes, such as p53 and BRCA1. Thus, failure to properly respond to perturbations in DNA integrity can be deleterious to the cell and organism.

For more on ATM, see Section 15.9.

The spindle assembly checkpoint pathway inhibits progression through mitosis

In addition to mechanisms that monitor DNA integrity, eukaryotic cells also monitor the processes required to correctly segregate chromosomes in mitosis. In Chapter 4 we discussed the mitotic spindle that attaches to chromosomes and moves them to the daughter cells. We will discuss the segregation process, as well as the structure and function of the spindle, in more detail in Chapter 7. A crucial step in the chromosome segregation process is the attachment of the chromosomes to the spindle: lack of attachment or an incorrect attachment can result in the segregation of both copies of a chromosome to one daughter cell and no copies of that chromosome to the other daughter cell.

To ensure proper chromosome segregation, the spindle assembly checkpoint pathway monitors the attachments between chromosomes and the spindle. When a chromosome is not attached to the spindle, the spindle checkpoint pathway is activated, although the exact nature of this pathway's sensor is not yet known. Numerous proteins, some of which are protein kinases, then participate in the transduction of the signal to the main effector of this pathway, the APC. As mentioned earlier, APC activity is required for progression through mitosis. The spindle assembly checkpoint pathway blocks the activity of the APC, thereby preventing both anaphase initiation and mitotic exit. Thus, as long as there are chromosomes that are not properly attached to the spindle, the spindle assembly checkpoint pathway will inhibit progression through mitosis, thereby providing more time for the chromosome to correctly attach to the spindle. Once proper spindle attachments are formed, the checkpoint signal is extinguished and progression though mitosis can resume.

Thus, checkpoint pathways ensure the integrity of the cell cycle process, and failure in checkpoint pathway activity can have dire consequences.

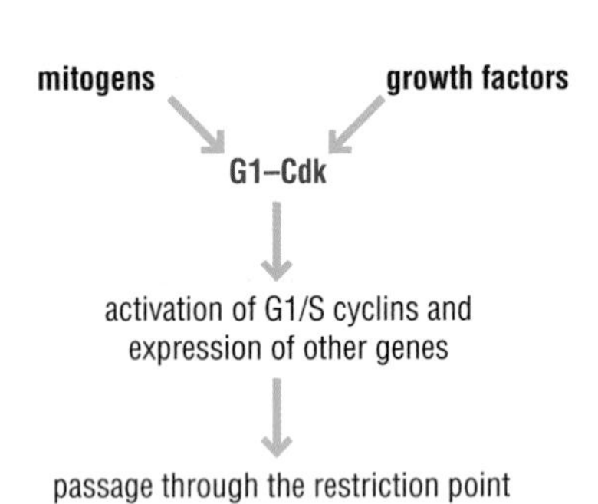

Figure 5.17 External regulation of cell proliferation. External regulation of entry into G1 can be triggered by growth factors and mitogens. Activation by extracellular growth factor or mitogens at the cell surface leads to a signaling cascade (not shown) that stimulates G1 cyclin–Cdk activity. G1 cyclin–Cdk complexes then promote the activation of gene regulatory proteins that stimulate the expression of numerous G1/S genes, including genes encoding G1/S cyclins, S cyclins, and the proteins that carry out the events of the early cell cycle. The resulting G1/S– and S–Cdk activities leads to passage through the restriction point and irreversible commitment to cell division.

5.7 EXTRINSIC REGULATORS OF CELL CYCLE PROGRESSION

In the previous sections, we discussed how the monitoring of events within the cell leads to the regulation of cell cycle progression. However, it is not just the internal environment of the cell that influences this progression: factors in the environment surrounding the cell are also important regulators. We consider some of the key environmental factors - so-called extrinsic regulators - in this section.

Cell growth and division are regulated by extracellular mitogens and growth factors

In addition to internal signals, cells also respond to signals from their environment. These signals include **mitogens** and growth factors, both of which promote the passage through the restriction point, as depicted in Figure 5.17. Mitogens stimulate cells to divide by activating the cell cycle machinery that drives cells past the restriction point. Growth factors, as their name implies, are molecules that stimulate cells to increase their mass, for example by stimulating protein synthesis. In some cell types this increase in mass is necessary in order to pass through the restriction point (in which case the growth factor is said to have mitogenic activity); in others, however, growth factors stimulate cell growth without cell division. In addition, cells also respond to anti-proliferative signals (also known as anti-growth factors) that block the activation of the cell cycle machinery so that cells cannot go through the restriction point.

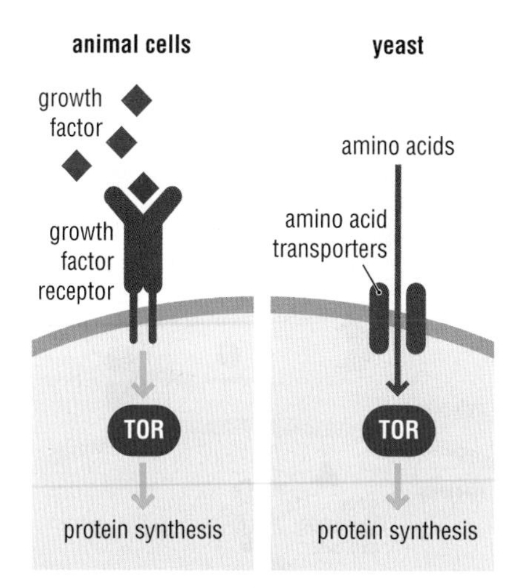

Figure 5.18 Growth factors and nutrients regulate cell division though increasing growth rate. In both unicellular eukaryotes and metazoans, cell growth is sensitive to stimuli from outside the cell. In organisms such as yeast (right panel), one of the environmental cues that is translated into growth signals is nutrient availability, such as the abundance of amino acids. These amino acids are transported into the cell by specific amino acid transporters. In animal cells (left panel), cell growth is often dependent on growth factors, which bind growth factor receptors that have both an extracellular and an intracellular component. Upon growth factor binding to the extracellular part of the receptor, a signaling cascade is triggered by the intracellular part of the receptor. In both animal cells and yeast, signals from nutrients and growth factors converge on the TOR kinase, which controls cell growth in part by regulating protein synthesis.

Growth factors increase growth rate and stimulate division indirectly

In certain cells, where progression through the restriction point requires that cells reach a minimum size threshold before division can occur, growth factors can stimulate cell division, and thus increase proliferation, by promoting an increase in cell mass. Many single-celled eukaryotes monitor the levels of various nutrients in the environment and adjust their rates of growth and metabolism accordingly. If the concentration of important nutrients declines, cells respond by decreasing their growth rate.

The dependence of cell division on cell growth results from links between the cell's metabolic machinery and the cell cycle control system. In budding yeast, for example, changes in the overall rate of protein synthesis affect the activities of both the G1 cyclin–Cdk1 and G1–S cyclin–Cdk1. There is also a link between growth rate and division in mammalian cells. However, the precise mechanism that links growth and G1 entry is not yet clear.

Growth rate is regulated by a protein kinase called **TOR**, which is central to the response to nutrients and growth factors in all eukaryotes. In animal cells, growth factors activate a receptor by binding to it at the cell surface, and this leads to the initiation of a signaling cascade. In this cascade, one regulator stimulates the activity of the next in a series of steps that can be modulated at many levels. A simplified version of the cascade is shown in Figure 5.18, with the TOR kinase being activated at the end of the cascade. Activation of TOR leads to the subsequent activation of many processes, chief among them being an increase in the rate of protein synthesis, which drives an increase in cell mass.

Mitogens control the division of mammalian cells

Cells of mature multicellular organisms usually divide only when the maintenance or repair of tissues is required. Many mammalian cells regulate this kind of cell proliferation by responding to external mitogenic signals. Mitogens are generally soluble small proteins secreted by neighboring cells. These proteins directly signal cell division by activating cyclin–Cdks. Some are highly specific regulators of division in just one cell type, whereas others have much broader action on many cell types.

An example of a mitogen is **epidermal growth factor (EGF)**, a soluble polypeptide that controls the initiation of division in many different cell types. Though the name that was originally given to this polypeptide was 'growth factor', we now know that it acts as a mitogen that directly stimulates entry to the cell cycle at G1. An example of how a mitogen activates cell proliferation is shown in Figure 5.19.

Mitogens bind to a receptor that initiates a signal transduction cascade involving multiple protein kinases. The presence of multiple steps gives the cell numerous opportunities to regulate this process. Through this cascade, the signal is transduced

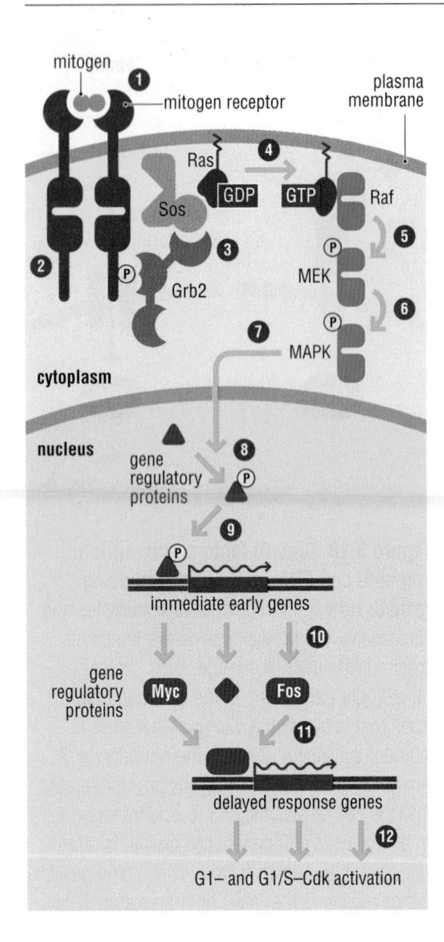

Figure 5.19 Mitogens stimulate cell division through a signaling cascade. The mitogen receptor has an extracellular component, which can bind a mitogen, and an intracellular component, which, upon mitogen binding, can activate a signal transduction cascade. Mitogen binding to its receptor (step 1 in this signaling cascade) results in receptor dimerization (step 2). This triggers the recruitment of several proteins (in the example shown: Sos and Grb2) to the plasma membrane (step 3) that activate the membrane-associated Ras protein (by exchanging Ras-bound guanosine diphosphate (GDP) with guanosine triphosphate (GTP)) (step 4). Activated Ras, in turn, activates the first of a series of cytoplasmic kinases, each of which phosphorylates the next kinase in the cascade (Raf phosphorylates mitogen-activated or extracellular signal-regulated protein kinase (MEK, step (5)), which phosphorylates mitogen-activated protein kinase (MAPK, step (6)). The last kinase in this cascade, MAPK, is transported into the nucleus (step 7), where it activates, through phosphorylating a transcription factor (step 8), the expression of a series of genes (step 9), including the genes coding for Myc and Fos (step 10). Myc, Fos, and other proteins then activate the expression of another set of genes (step 11), including those encoding G1 and G1–S cyclins (step 12). Thus, through this cascade, the mitogen signal is transmitted from outside the cell, through the plasma membrane, to the cytoplasm and into the nucleus, where it induces the expression of genes required for cell proliferation.

from outside the cell all the way into the nucleus. This cascade results in the activation of various G1- and S-phase genes, including cyclins. Thus, this cascade links cell surface signaling mediated by the mitogen to entry into the cell cycle. For example, following mitogen stimulation, the G1 cyclin, cyclin D, binds and activates Cdk4, which phosphorylates Rb. As we saw in Section 5.5, Rb phosphorylation frees E2F to activate G1 gene expression and this, in turn, promotes cell cycle progression.

In some cases, a single extracellular protein acts both as a growth factor and as a mitogen, resulting in a coordinated increase in the rates of both growth and division. As we will see in the next section, dysregulation of various steps in this pathway can lead to uncontrolled cell division and cancer.

Extracellular anti-proliferative factors can inhibit cell division

In addition to growth factors and mitogens, cells also respond to soluble factors that inhibit cell division. One such factor is TGFβ, which is secreted by a variety of cell types. TGFβ binds to the TGFβ receptor on its target cell, triggering a signal transduction cascade that, among other things, blocks cell proliferation in at least two ways: First, activation of the TGFβ inhibits the phosphorylation of Rb thereby keeping Rb bound to E2F and preventing E2F from activating transcription that is necessary for cell cycle progression (see Section 5.4). Second, TGFβ activates G1 and G1/S Cdk inhibitors through its signal transduction cascade (see Section 5.3), thereby blocking the transition through the restriction point. Cancer cells often lose the ability to respond to TGFβ – for example, by acquiring a mutation that renders the TGFβ receptor or components of the signal transduction pathway non-functional; this allows these cancer cells to evade TGFβ anti-proliferative signals and to divide in an uncontrolled fashion.

5.8 THE CELL CYCLE AND CANCER

Tumor cells lack growth controls and accumulate mutations

Cancer is a disease that derives primarily from uncontrolled cell division. This inappropriate accumulation of cells can result from either inappropriate cell growth or from insufficient cell death. It is of little surprise, therefore, that the mechanisms that regulate cell cycle progression are frequently altered in cancer. There are three primary changes occurring in cells that fuel cancer growth: dominant activation of

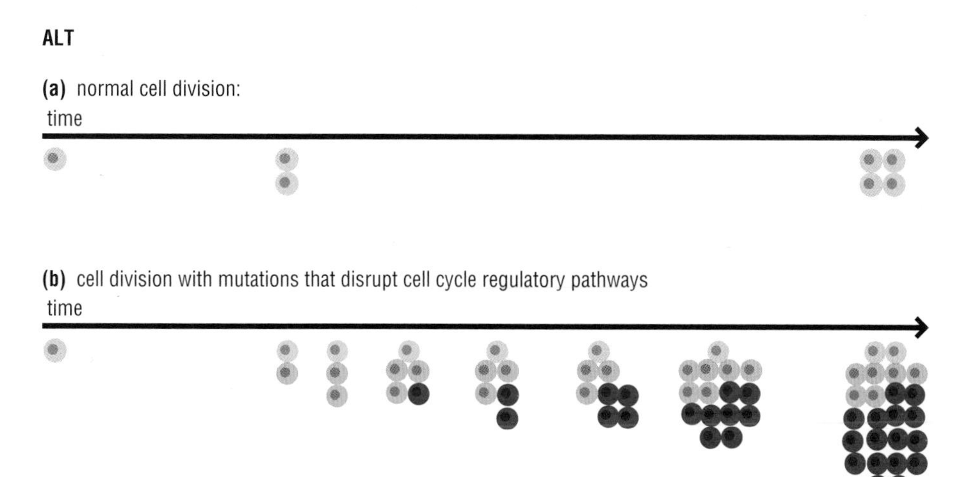

Figure 5.20 Evolution of a tumor. (a) Under normal growth conditions, a cell (shown in gray) may divide infrequently within a given period of time. (b) During the first division, one of the daughter cells (in light pink) acquired a mutation that causes it to divide more rapidly than the parent, blue cell. A second mutation (dark pink cell) causes this cell and its descendants to overcome additional cell cycle inhibitory signals and divide even faster. As a result, in the same time where a normal cell will divide twice to give rise to four cells (panel a), the acquired mutations cause hyper-proliferation, resulting in a much larger mass of cells.

extrinsic signaling pathways, loss of regulatory mechanisms (such as checkpoint pathways) that normally limit cell growth, and genetic alterations that allow the accumulation of multiple mutations.

As we have seen throughout this chapter, there are multiple overlapping controls to the cell cycle. In order for a cell to become a tumor cell and grow uncontrolled, often more than one event that alters cell growth has to take place. The multiple mutations required for a cell to become cancerous do not occur all at once but rather accumulate over time. For example, Figure 5.20 illustrates how one mutation may happen in one cell that allows it to divide more frequently and outgrow the cells that surround it. These early cancer cells now represent a significantly larger fraction of the cells and, if additional growth control pathways are mutated, these cells will have a continued advantage. In this way, the cancer cells accumulate mutations and rapidly outgrow the other cell and thus take over the population.

The role of accumulated mutations in the evolution of cancers is highlighted by the increased incidence of cancer as people get older. The accumulation of multiple genetic changes that promote cancerous growth can take years. When cancer appears in relatively young individuals, it is suspected that these individuals have a genetic predisposition for cancer.

An example of a gene that leads to a genetic predisposition for cancer when mutated is the *BRCA1* gene. The BRCA1 protein has several functions, one of which is to repair DNA damage; women with mutations in the *BRCA1* gene develop breast cancer at a younger age and at a much high frequency than the general population.

While there are many factors that are needed for cancer cells to become a tumor, including changes in the surrounding tissue, here we will focus on the changes in the cancer cell itself that alter cell cycle regulation and lead to increased growth. The genes that are altered in cancer are often classified into two general groups: **oncogenes**, which are genes that when mutated or overexpressed promote abnormal growth, and **tumor suppressors**, which are genes that normally halt the cell cycle but fail to do so when mutated.

Oncogenes are often aberrant versions of positive regulators of the cell cycle

Cells of multicellular organisms normally proliferate in response to growth factors and mitogens, which are essential extrinsic regulators of cell cycle progression,

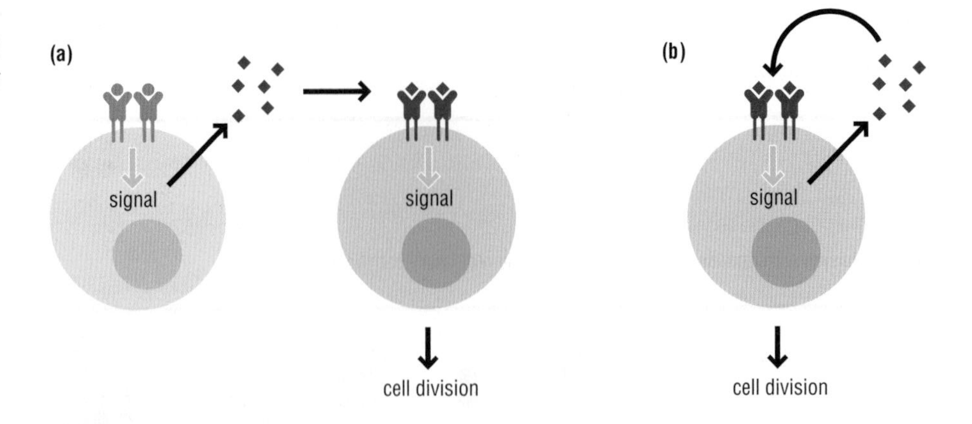

Figure 5.21 Cell signaling. (a) Under normal conditions, cell proliferation would be stimulated only when appropriate signals are generated, for example, by neighboring cells. In the example shown, the gray cell will only divide when it is exposed to a secreted stimulating signal (pink diamonds) from the blue cell. (b) Under certain conditions, cancer cells generate their own inducing signals, generating an autocrine loop. These signals occur irrespective of the neighboring cells and, thus, regardless of whether the tissue requires additional cell divisions. In this case, the cell divides uncontrollably, stimulated by its own signals.

and usually stimulate entry into the cell cycle at G1. The uncontrolled growth of cancer cells is frequently due to defects in the mitogenic signaling pathway (see Figure 5.19). Oncogenic proteins may function at any level in this pathway. In some cases, cells aberrantly make and secrete a stimulatory protein that signals its own cell division. This **autocrine loop** stimulates a cell to promote its own division as well as that of its surrounding cells, as depicted in Figure 5.21.

An oncogenic protein may also stimulate cell division by constitutively activating one of the effectors in a mitogenic pathway. A gene that normally would be activated by a signal from the outside may acquire a mutation that makes it active in the absence of that signal. An example of such an oncogene is *ras*, which codes for a component of the pathway that induces cell cycle entry following mitogenic stimulation (see Figure 5.19). In the normal setting, the activation of Ras requires the association of a mitogen with its receptor, followed by the recruitment of proteins that are necessary for Ras activation to the plasma membrane. However, the mutant version of the *ras* gene is constitutively active - that is, it can induce the protein kinase cascade in the absence of mitogen. Consequently, the cell responds as if there is mitogen present, even though there is none, and activates the cell division pathway.

Similarly, the machinery of the cell cycle itself may be altered such that the cell enters the cell cycle at G1 constitutively without waiting for the appropriate signal. For example, in some cancers, mutations in Cdk4 render it insensitive to the cyclin-dependent inhibitors described in Section 5.3, which regulate the activity of cyclin–Cdk kinases.

Another kind of oncogene is one that codes for a protein that normally protects cells against cell death. As we discuss in detail in Chapter 15, cells often undergo apoptosis, or cell death, when inappropriate or conflicting signals are present. However, overexpression of some genes will block the cell death signals and allow continued growth of cells that would otherwise die. If this occurs in cells that are proliferating, cancer can result.

Tumor suppressor genes are often defective negative regulators of the cell cycle

The uncontrolled growth of tumor cells is often due to their failure to respond to normal signals that inhibit the cell cycle. In such cases, the defect is in proteins that would normally stop growth. These tumor suppressor mutations can be in checkpoint genes that normally signal for repair, and which also arrest cell division when things go amiss in the cell cycle. For example, **p53** is regulated by checkpoint kinases that are activated in response to DNA damage. In the presence of DNA damage, p53 activates the expression of genes that are required for cell cycle arrest or cell death. In the absence of p53, cells continue to proliferate despite the presence of DNA damage, greatly increasing the chance of accumulating deleterious mutations. For many cancers, loss of p53 function occurs during the multistep mutational process that results in tumor growth. In fact, greater than 50% of human cancers have mutations in p53.

➔ We see further in Chapter 15 that p53 plays an essential role in the response to DNA damage by triggering either cell cycle arrest or apoptosis.

In addition to mutations in checkpoint genes, tumor suppressors can also be normal negative regulators of cell cycle progression. One example is the Rb transcriptional regulator that we described in Section 5.4. A mutation in Rb that blocks its ability to inhibit the E2F transcription factor (see Figure 5.11) can promote inappropriate entry into the cell cycle in G1. The pathway that regulates G1 entry is so central to cell cycle progression that the vast majority of cancers have mutations in one of the components of this regulatory cascade.

The link between cell cycle regulation and cancer underscores the importance of understanding the cell cycle machinery. In the next two chapters, will we explore various aspects of cell cycle regulation in the context of specific processes: DNA replication (Chapter 6) and chromosome segregation (Chapter 7). As you read through these chapters notice the intricate relationship between the cyclin-Cdk machinery and the proteins that it regulates. It is then easy to understand how mis-regulation can lead to dire consequences for the cell.

5.9 THE BACTERIAL CELL CYCLE

Like single-celled eukaryotes, the rate of bacterial growth is determined by environmental conditions. For example, bacteria will accumulate faster in chicken soup, which is nutrient rich, than in tap water, which is nutrient poor. Much like eukaryotes, where passage through the restriction point occurs only when conditions are right, bacteria also have pathways that translate internal and environmental cues, such as nutrient availability, into signals that regulate the cell cycle. Moreover, certain bacteria inhibit cell cycle progression in the presence of DNA damage, again in a similar manner to eukaryotes. Most importantly, however, both eukaryotes and bacteria have cell cycle machineries that are made up of signaling pathways that drive cell cycle progression.

Unlike eukaryotes, however, in which the cell cycle machinery is conserved across species, different bacteria utilize distinct mechanisms to regulate the cell cycle. This is likely the result of evolutionary diversity caused by the broad range of ecological niches that bacteria inhabit, which may have posed distinct selective pressures that require different cell cycle paradigms.

As researchers began to study cell cycle regulation in different bacteria, they uncovered a menagerie of intricate pathways and processes, many of which had no parallels in other bacteria. In the study of the eukaryotic cell cycle, researchers often rely on discoveries made in one organism to decipher similar processes in another. However, this is not always possible when studying the bacterial cell cycle. As a result, our knowledge of cell cycle regulation in bacteria is less complete than that of eukaryotes, and there is still a wealth of information relating to pathways, proteins and processes remaining to be discovered. In this section we will discuss general features of the bacterial cell cycle and examine specific examples from several bacteria species.

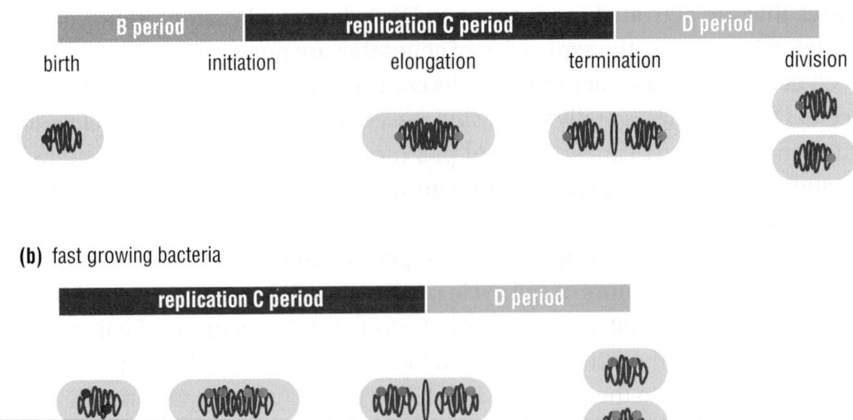

Figure 5.22 The periods of the bacterial cell cycle. (a) Under slow growth conditions, the bacterial cell cycle can be divided into three periods: the 'B period', which is from when the cell is born until DNA replication begins; the 'C period', which encompasses DNA replication and chromosome segregation; and the 'D period', which is the interval from the end of the C period until cell division takes place. Under these conditions, the initiation of DNA replication from the single replication origin (red dot) duplicates the origin region (forming the blue and orange dots) and eventually the entire chromosome. This is followed by cell division, which gives rise to two cells, each with a single origin and a completely replicated chromosome. (b) Under rapid growth conditions, some types of bacteria can divide while replication is still ongoing. This is possible because the cell is born with DNA that has already begun to duplicate in the parent cell, as seen by the presence of separated origins of replication (in red). In this cell, the existing replication forks will complete the duplication of the chromosome, and in the meantime a new round of replication is initiated during the C period: the red origin on the left is duplicated to become the two blue origins, while the red origin on the right is duplicated to become the two orange origins. This gives rise to two cells with origins of replications that have already been duplicated. Under these growth conditions, DNA replication is ongoing as the cell is born, and there is no B period.

The main landmarks of most bacterial cell cycles are DNA replication and cell division

Under slow growth conditions, such as in poor nutrient conditions, the bacterial cell cycle can be divided into three main periods as illustrated in Figure 5.22a:

- 'B period' – during which the bacterium increases in mass and size
- 'C period' – during which DNA replication and chromosome segregation take place
- 'D period' – the time between the end of DNA replication and the subsequent cell division, during which chromosome segregation and division are completed.

Under conditions that support rapid growth (Figure 5.22b), such as nutrient-rich growth conditions, certain types of bacteria grow and divide so fast that the length of the cell cycle is shorter than the time it takes to complete DNA replication. For example, under rich nutrient conditions, *E. coli* can divide every 20 minutes, but it takes over 30 minutes to complete the replication of the entire *E. coli* chromosome. This rapid division is possible because, in these cells, a new round of replication begins before the previous one has been completed – a phenomenon known as **multifork replication**. Under these conditions, the 'B period' is eliminated and the newborn cells, which inherit partially duplicated chromosomes, continue with DNA replication as soon as they are formed.

This situation is different from eukaryotic cells, where the presence of partially replicated DNA will stop cell cycle progression. Moreover, not all bacteria undergo multifork replication; in some bacteria, cell division will take place only after a single round of DNA replication has been completed.

➔ The regulation of DNA replication is described in Chapter 6.

Bacterial cell growth and division are sensitive to nutritional cues

The mechanistic details of bacterial DNA replication, chromosome segregation, and cell division will be discussed in the coming chapters. Here we will focus on the regulation of cell cycle events. For example, what dictates when a bacterial cell divides or when it initiates DNA replication?

Bacteria growing in nutrient-rich growth media tend to be larger than those grown in nutrient-poor media, suggesting that the cell division machinery is sensitive to nutritional cues. For example, in *Bacillus subtilis*, cell division is inhibited

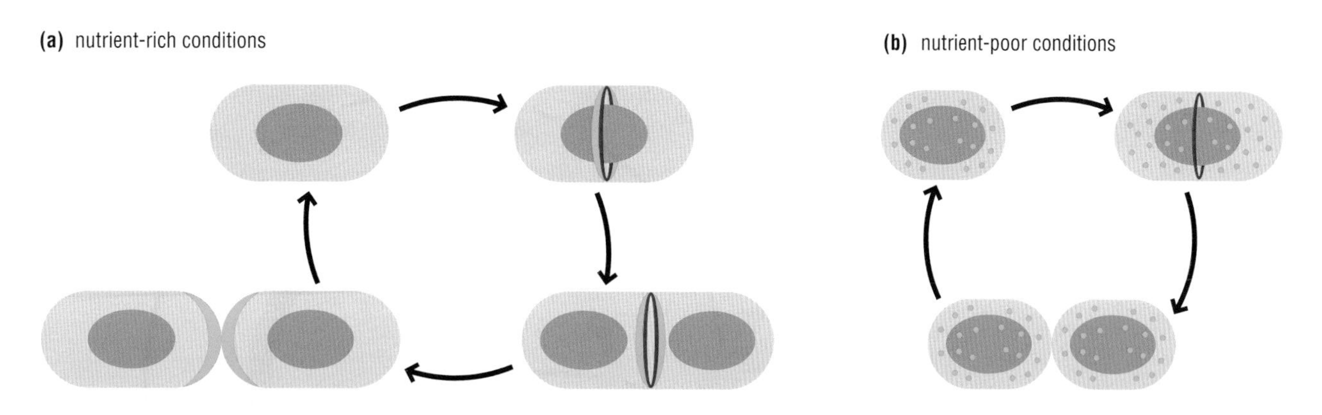

Figure 5.23 The regulation of cell division by the *Bacillus subtilis* UgtP. Cell division is mediated by the FtsZ ring, shown in red (the nucleoid, containing the bacterial chromosome, is shown in blue). Under nutrient-rich conditions (a), cell division is delayed to allow the cell to increase in size. This delay in cell division is mediated by the protein UgtP (in green), which localizes to the cell midzone (the future site of cell division) and binds to FtsZ to inhibit its activity. When the cell reaches the appropriate size, inhibition by UgtP is alleviated, and cell division takes place. Under nutrient-poor conditions (b), UgtP is sequestered in foci away from the midcell where it cannot inhibit cell division. Hence, under nutrient-poor conditions, division occurs at a substantially smaller cell size than during growth in a nutrient-rich medium.

by a metabolic pathway that senses the amount of available glucose: under rich growth conditions, where glucose is abundant, an enzyme called UgtP inhibits cell division, thereby forcing the cell to reach a larger size before the cell divides, as depicted in Figure 5.23.

The UgtP protein inhibits cell division through its interaction with another protein, FtsZ, which forms a ring-like structure and is needed for cell division in bacteria. UgtP can associate with itself or with FtsZ, though the binding to FtsZ increases in presence of uridine dephosphate (UDP)-glucose. The binding of UgtP to FtsZ also prevents the formation of a functional FtsZ ring, thereby inhibiting cell division. Under poor growth conditions, both UDP-glucose and UgtP levels are reduced, and the remaining UgtP self-assembles in foci away from the FtsZ ring. As a result, cell division takes place at a smaller cell size. Thus, UgtP and UDP-glucose link cell division to nutrient availability.

➔ FtsZ is discussed in detail in Section 7.8.

Caulobacter crescentus has defined cell cycle stages

The study of cell cycle regulation in any organism or cell type is often facilitated by the existence of visible landmarks, such as the size of the bud in dividing budding yeast, or the organization of chromosomes in animal cells. These landmarks change as the cell progresses through the cell cycle, allowing researchers to easily determine specific cell cycle stages. Determining the cell cycle stage of bacteria is often much harder because many lack easily detectable features. For example, the only morphological cell cycle change in the rod-shaped *E. coli* is the length of the organism. Perhaps for this reason, many advances in our knowledge of cell cycle regulation in bacteria have come from *Caulobacter crescentus*, a crescent-shaped bacterium, which has several clearly visible structures, including a stalk, a flagellum and several pili. The key morphological changes exhibited by *C. crescentus* as the cell cycle proceeds are illustrated in Figure 5.24. Importantly, the appearance and disappearance of these structures is tightly linked to cell cycle events such as chromosome replication and segregation.

Another reason *C. crescentus* emerged as a powerful model organism for cell cycle studies in bacteria is that it initiates one round of DNA replication in each cell cycle. Many types of bacteria, including *E. coli*, can restart DNA replication before the previous replication cycle has been completed when growth conditions

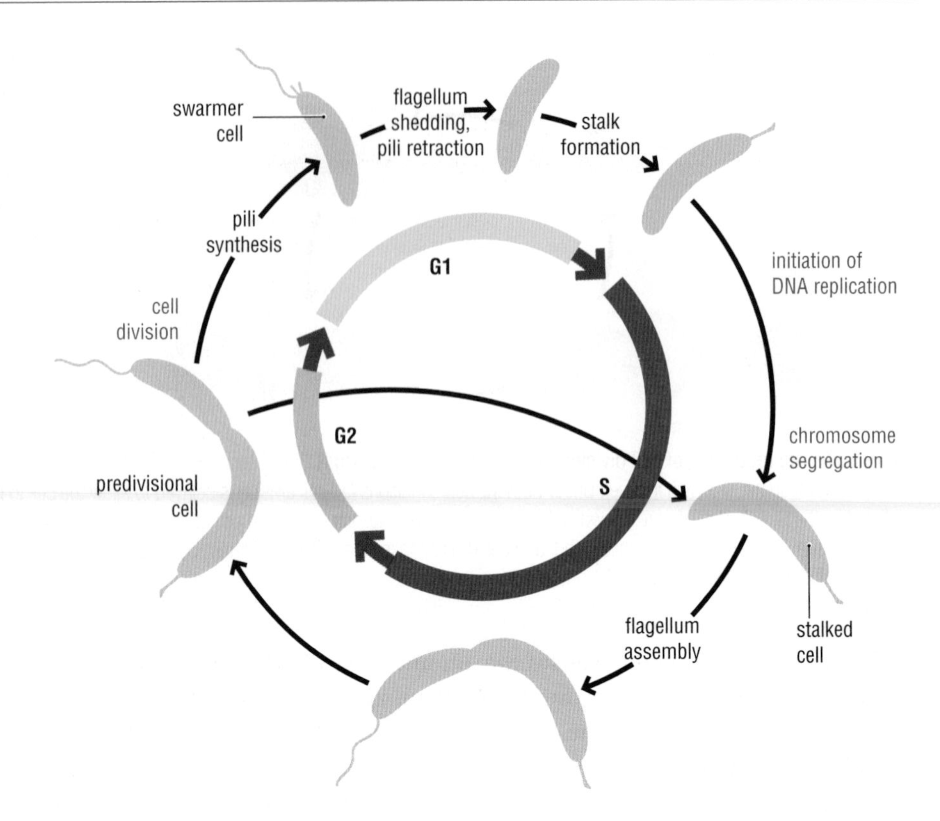

Figure 5.24 The cell cycle phases of *C. crescentus*. The cell cycle phases (G1, S, G2), key morphological changes, and cell cycle events (pink text). See text for further details.

are favorable (the multifork replication we mentioned earlier). This multifork replication makes it difficult to study the relationship between the initiation of DNA replication and other cell cycle events, and limits the experiments to those done in poor growth conditions. In contrast, *C. crescentus* initiates and completes one cycle of DNA replication in the same cell cycle, allowing the demarcation of an S phase, much like in eukaryotes.

During the *C. crescentus* S phase, chromosomes are replicated and segregated concomitantly. This S phase is preceded by a gap phase (called G1), during which the flagellum and pili are lost and the stalk begins to form, and it is followed by a second gap phase (called G2), during which cell division takes place (see Figure 5.24).

➲ The mechanism of chromosome segregation in *C. crescentus* and other bacteria is discussed in Section 7.8.

The *C. crescentus* cell cycle stages are characterized by the expression of distinct sets of genes. Analysis of gene expression has revealed that over 550 *C. crescentus* genes (out of roughly 3800) are expressed in a cell cycle-regulated fashion. Three key regulators involved in the cell cycle are CtrA, GcrA, and DnaA, as shown in Figure 5.25. Phosphorylated CtrA is found in swarmer cells (which are in G1) and predivisional cells (which are cells before cell division), where it inhibits the initiation of DNA replication by binding to the origin of replication. The binding of phosphorylated CtrA to the origin prevents the binding of DnaA to the same location – an interaction that is necessary for the initiation of DNA replication.

CtrA also inhibits cell division, and other morphological changes that should occur later in the cell cycle, by inhibiting the expression of the *gcrA* gene. GcrA is needed for the expression of genes involved in late stages of the cell cycle. When the cell enters S phase, corresponding with the transition from a swarmer to a stalked cell, CtrA is degraded and DnaA begins to accumulate, facilitating the initiation of DNA replication.

DnaA is not only needed for DNA replication, but also controls the transcription of a variety of genes in *C. crescentus* (such as *gcrA* and other genes) needed

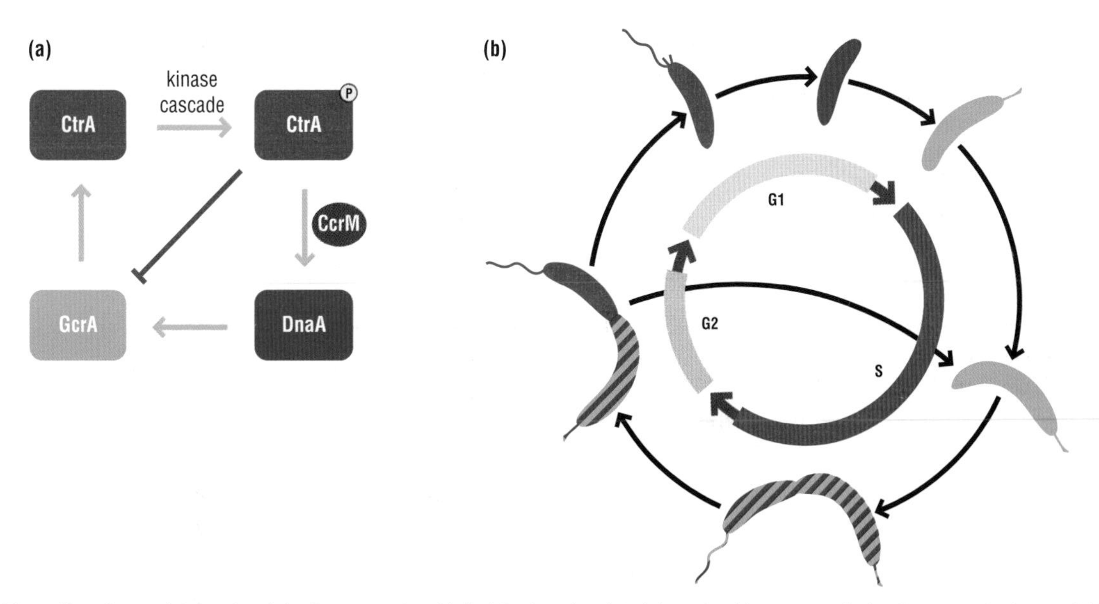

Figure 5.25 The cell cycle regulatory circuit in *C. crescentus*. (a) CtrA is phosphorylated through a kinase cascade that is present in the predivisional cell and swarmer cell. Phosphorylated CtrA inhibits initiation of DNA replication and the expression of genes needed for late stages in the cell cycle, including *gcrA* (illustrated by red bar). CtrA induces the expression of DnaA indirectly by inducing the expression of an enzyme, CcrM, that methylates the *dnaA* gene, enabling its expression. Once the cell develops a stalk, CtrA is degraded (not shown), and DnaA induces both DNA replication and the expression of numerous genes including *gcrA*. The GcrA protein, in turn, activates the expression of genes needed for S phase and G2. It also induces the expression of *ctrA*, thus completing the cycle. (b) The presence of CtrA (in red), DnaA (in dark blue), and GcrA (in blue) throughout the cell cycle. The hatched marks indicate the presence of two proteins.

for growth, DNA replication, and cell division. GcrA activates additional genes required for S phase and G2, and it also activates the expression of the gene coding for CtrA, which negatively regulates DNA replication, as described above. Late in the cell cycle, CtrA is phosphorylated, allowing it to bind the origin of replication and to prevent further cycles of DNA replication until the next cell cycle. This regulatory circuit, and many other additional processes that impinge on this circuit, drive cell cycle progression forward in a manner similar to eukaryotic regulatory loops.

➔ The role of DnaA in the initiation of replication is discussed in Section 6.7.

C. crescentus is not only a good model for bacterial cell cycle studies, but also provides insights into processes associated with asymmetric cell division. As we mentioned earlier, following cell division, two different cell types are formed: a swarmer cell, which can move thanks to its flagella, and a stalked cell, which is immobile (see Figure 5.24). The degradation of CtrA occurs only in cells that have a stalk due to the accumulation of the CtrA degradation machinery near the stalked pole of the cell (see Figure 5.25b). Thus, immediately after cell division, the stalked cell contains no CtrA and can initiate DNA replication immediately. By contrast, the swarmer cell has stable and active CtrA, and the cell must first differentiate into a stalked cell before CtrA is degraded and the cell can progress into S phase.

The bacterial SOS response induces transcription of DNA repair genes and inhibits cell division in response to DNA damage

We have now seen how DNA replication and cell division are controlled during normal cell growth. In some cases, however, cell cycle progression needs to be halted to permit repair of cellular damage, such as DNA damage, so that the transmission of damaged genetic material to daughter cells is prevented. In eukaryotes, this type of regulation is mediated by checkpoint pathways, such as the DNA damage and DNA replication checkpoint pathways (see Section 5.7). In many, but not all, bacteria, the presence of DNA damage induces a cellular response called the SOS response.

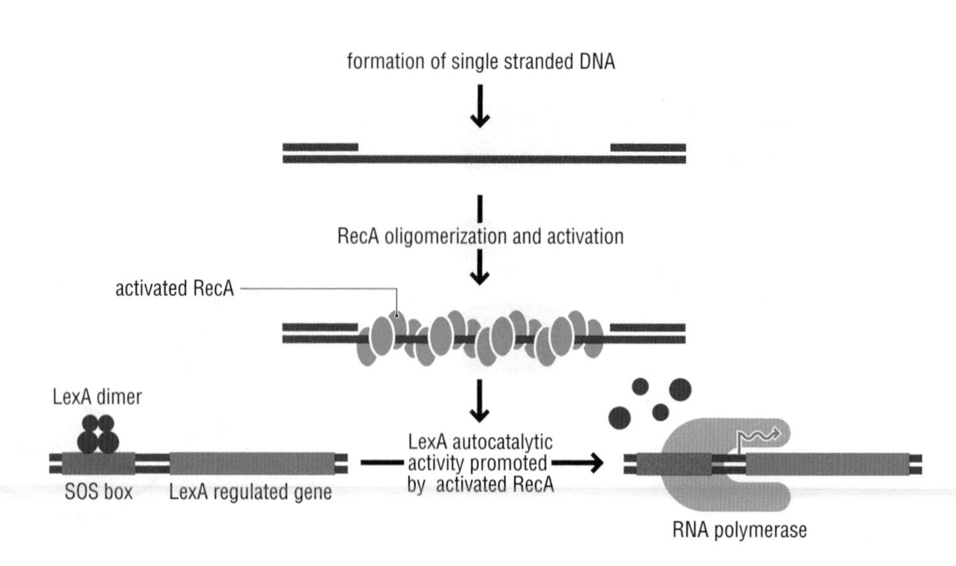

Figure 5.26 The SOS response. In the presence of single-stranded DNA (which can form as a result of a block to DNA replication or resection of one of the strands from a double-stranded DNA break), the RecA protein (in orange) oligomerizes on the single-stranded DNA. This process activates RecA. Activated RecA then promotes the autocatalytic activity of LexA, which leads to LexA auto-cleavage and consequently inactivation. This allows RNA polymerase to transcribe LexA-controlled genes, some of which participate in DNA repair and inhibition of cell cycle progression.

The molecular aspects of the SOS response will be described in more detail in Section 15.8; here we will discuss how the SOS response affects the cell cycle.

The SOS response was the first checkpoint pathway to be discovered, although its discovery preceded the coining of the term 'checkpoint'. In the SOS response, the presence of DNA damage leads to the formation of single-stranded DNA. The binding of a protein called RecA to this single-stranded DNA promotes the autocatalytic inactivation of a transcriptional repressor called LexA, as illustrated in Figure 5.26. This inactivation leads to the transcription of genes involved in repair and cell cycle arrest. If we compare the SOS response to eukaryotic checkpoint pathways (Section 5.7), RecA serves as both sensor and transducer in the SOS response, while LexA is the effector protein.

The mechanism of the SOS response is best characterized in *E. coli*. LexA inhibits the expression of roughly 40 genes, including the *sulA* gene, which regulates cell division. When induced, the SulA protein binds to the FtsZ protein, which (as we note earlier) is part of a contractile ring necessary for the formation of the septum that separates the two daughter cells. The binding of SulA to FtsZ inhibits the formation of the FtsZ ring, thereby blocking cell division. Taken together, the presence of DNA damage leads to the activation of the SOS response, which, in turn, inhibits cell division, thereby preventing the segregation of damaged DNA molecules to the daughter cells.

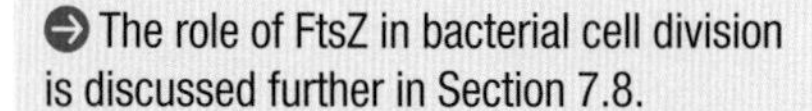
➔ The role of FtsZ in bacterial cell division is discussed further in Section 7.8.

C. crescentus also has an SOS response that controls over 30 genes, but *C. crescentus* does not contain a homolog of the *E. coli sulA* gene. Instead, inactivation of *C. crescentus* LexA allows the expression of a small membrane protein, sidA, which inserts next to the FtsZ ring and prevents its constriction. Thus, while the cellular consequences of SOS response induction are conserved among bacteria, the proteins that mediate the effect are often not.

SUMMARY

In this chapter we learned about the machinery that drives and regulates cell cycle progression in both eukaryotes and bacteria. We saw how in eukaryotes cyclin–Cdks promote not only the processes that take place during a particular phase, but how cyclin–Cdks activate the next phase and extinguish the previous one. We also learned that regulatory pathways called checkpoints are superimposed on the cyclin–Cdk machinery, and inhibit cell cycle progression in the presence of damage or certain types of errors in the cell. These checkpoint pathways ultimately stop cell cycle progression by affecting cyclin–Cdks. The multiple levels of regulation illustrate the importance in carrying out cell cycle processes with the utmost accuracy. Indeed, we saw that failure to do so can result in cancer. This theme will be further elaborated in the coming chapters.

We also learned that bacteria are more diverse in terms of their cell cycle schemes: some bacteria can carry out multifork replication, whereas others do not. The major regulatory steps in many bacteria are the initiation of DNA replication, which is often coupled to chromosome segregation, and cell division. Although the bacterial cell cycle differs substantially from the eukaryotic cell cycle, they both have regulatory mechanisms that inhibit progression when errors or certain types of damage are detected. In the case of bacteria, the SOS response inhibits cell division in the presence of DNA damage.

THE EUKARYOTIC CELL CYCLE

- The cell cycle is divided into four phases: G1, S, G2, and M. Cells can exit the cell cycle to a G0 phase.
- Cell cycle progression is driven by cyclin–Cdks. Cdks are protein kinases that are active only when bound to cyclins.
- Different cyclins associate with different Cdks at various cell cycle stages. The specificity of the cyclin–Cdk complex is determined by the cyclin.
- Cyclin–Cdks are regulated at multiple levels, including the expression and degradation of cyclins, activating and inhibitory phosphorylation of the Cdk, and association with Cdk inhibitors.
- Cyclin–Cdks have many targets, and can promote cell cycle progression by activating other kinases, promoting gene expression, and promoting protein degradation.

CHECKPOINT PATHWAYS

- Checkpoint pathways are processes that become activated in response to certain cellular insults. Different checkpoint pathways respond to a different type of cellular damage or defect.
- In general, checkpoint pathways feature sensors, transducers, and effectors.
- Checkpoint pathways can lead to the induction of repair processes, cell cycle arrest and, in multicellular organisms, apoptosis.

EXTRINSIC REGULATORS

- The cell cycle is regulated in response to extrinsic signals such as growth factors and mitogens.
- Growth factors stimulate the cell to increase its mass.
- Mitogens directly stimulate cells to divide and proliferate without stimulating growth.

CANCER

- The three primary changes that fuel cancer growth are: dominant activation of extrinsic signaling pathways, loss of regulatory mechanisms (including checkpoint pathways that normally limit cell growth), and genetic instability.
- Oncogenic proteins are often positive regulators that are being activated in an aberrant way.
- Tumor suppressors are negative regulators of cell cycle progression and are often inactive in cancer cells.

THE BACTERIAL CELL CYCLE

- Among bacteria, there are many different cell cycle paradigms, often involving non-conserved proteins.
- Key regulatory points in many bacterial cell cycles are the initiation of DNA replication and cell division.

5.1 STEPS IN THE EUKARYOTIC CELL CYCLE

1. Match each of the following statements to the appropriate cell cycle stage; G0, G1, S, G2, or M.

a. Human nerve cells are in this phase.
b. Cell cycle progression can be stopped in this phase if DNA damage is detected.
c. Chromosomes are separated into daughter cells in this phase.
d. Cells committed to cell division pass the restriction point in this phase.
e. The cytoplasm is divided in this phase.
f. Sister chromatids are formed in this phase.
g. Mitotic spindle formation begins in this phase.
h. Cells in this phase have increased expression of cell cycle inhibitors.

2. Some cell types have a modified cell cycle in which they alternate between S phase and M phase without ever entering G1 or G2.

a. What is the main result of this modification?
b. What types of cells would benefit from this modification?

Challenge question

3. Increased cellular ploidy is widespread during developmental processes of multicellular organisms, especially in plants. Elevated ploidy levels are typically achieved by two mechanisms: failure to transition from S phase to M phase and failure to exit M phase. The plant genetic model organism *Arabidopsis thaliana* is diploid and has ten chromosomes in its somatic cells (two sets of 5 chromosomes). However, certain cells in this plant have a greater ploidy.

a. What would be the main difference between the chromosomes of cells that failed to transition from S phase to M phase and those that failed to exit M phase?

b. UV Insensitive 4 (UVI4) is a newly identified protein in *Arabidopsis*. Mutations in the gene block mitosis and promote endoreplication (that is, another round of DNA replication without mitosis). Based on your knowledge of the cell cycle, propose a possible mechanism by which UVI4 would normally allow cells to progress through to M phase and explain why mutations would lead to a failure to endoreplication.

c. GIG1 is a newly identified protein in *Arabidopsis*. Mutations in this gene lead to cells that fail to exit M phase. Based on your knowledge of the cell cycle, propose a possible mechanism by which GIG1 would normally allow cells to exit M phase and explain why mutations would lead to a failure to exit M phase.

5.2 CYCLINS AND CDKS

1. Describe the roles of cyclins and Cdks in the control of the cell cycle.

2. With regard to cell cycle progression, what does it mean to say that 'two negatives make a positive'?

5.3 REGULATION OF CDK ACTIVITY

1. Changes in the activity of a variety of Cdks are essential for accurate progression through the cell cycle and yet the levels of Cdk expression are fairly constant during the cell cycle. Briefly describe three mechanisms by which the activity of Cdks is regulated.

Challenge questions

2. What would be the effect on the cell cycle if a cell acquired a mutation that rendered each of the following domains inactive? Explain your reasoning.
   a. The PSTAIR region of an M-phase Cdk
   b. The Wee1 kinase
   c. The Sic1 cyclin-Cdk inhibitor
   d. The kinase that normally phosphorylates Sic1

3. Pancreatic β cells are the principal source of the hormone insulin, which is required for the regulation of blood glucose and general metabolic homeostasis. It was once thought that these cells were incapable of proliferation, but it is now known that they can respond to a variety of extrinsic and intrinsic factors. INK4 is an example of a protein that has been shown to regulate the cell cycle in β cells. INK4 is highly expressed in adult β cells. Mice that are deleted (knocked out) for INK4 expression have an increase in proliferation of β cells. Propose a mechanism by which INK4 may regulate β cell proliferation.

5.4 CELL CYCLE REGULATION BY CDKS

1. In yeast alone, over 400 proteins that are phosphorylated and therefore likely regulated by the action of cyclin-Cdks have been identified. Briefly describe one example from each of the following categories of cyclin-Cdk targets.
   a. Transcription factors
   b. Changes in cellular structures
   c. Proteolysis (Section 5.5)

5.5 REGULATION OF PROTEOLYSIS BY CDKS

1. APC is an ubiquitin ligase that is activated by M-cyclin Cdk.
   a. What is the function of APC when bound to Cdc20?
   b. What is the function of APC when it is bound to Cdh1?
   c. How is the binding of APC^{Cdh1} to its substrates regulated?

2. Which of the following statements best describes the mechanism by which the ubiquitination of target proteins by the SCF complex is regulated?
   a. SCF complex phosphorylation by specific cyclin-Cdks determines when target proteins are ubiquinated.
   b. SCF complex is bound by cyclins and then becomes active.
   c. SCF complex is degraded when it is not needed and stabilized when it is needed.
   d. Target protein phosphorylation by cyclin-Cdks allows it to be recognized by the SCF complex.
   e. When target proteins reach specific cellular concentrations, they are recognized by the SCF complex.

3. Which of the following statements best describes the mechanism by which the APC complex ubiquitination of target proteins is regulated?
   a. The association of the APC with Cdc20 or Cdh1 determines when target proteins are ubiquinated.
   b. APC complex is bound by specificity factors that recognize specific APC substrates and facilitate their ubiquitination.
   c. APC complex is degraded when it is not needed and stabilized when it is needed.
   d. Target protein phosphorylation by cyclin-Cdks allows it to be recognized by the APC complex.
   e. When target proteins reach specific cellular concentrations, they are recognized by the APC complex.

Challenge question

3. The function of the APC complex:
   a. In Figure 5.13, the APC complex was rendered inactive, and yeast cell were allowed to grow. They all got stuck at the point at which they had large buds and one nucleus. Explain this result.
   b. In terms of the cell cycle, describe the most direct consequence of a non-function mutation in the Cdc20 protein.
   c. Yeast cells that carry a mutation that inactivates Cdh1 can still exit mitosis. Explain how that might happen.

5.6 CHECKPOINTS: INTRINSIC PATHWAYS THAT CAN HALT THE CELL CYCLE

1. The two major intrinsic pathways that halt the cell cycle act in response to DNA damage and spindle assembly during chromosome separation. Describe the basic components of checkpoint pathways.

5.7 EXTRINSIC REGULATORS OF CELL CYCLE PROGRESSION

1. What is the basic difference between a growth factor and a mitogen?

Challenge question

2. Estrogens are a group of naturally occurring steroid hormones that are best known for their actions in the development of the human female secondary sex characteristics. One additional action of estrogens bound to the estrogen receptor is to activate the transcription of cyclin D.
   a. With regard to the cell cycle, what effect would estrogen have on cells that contain the estrogen receptor? Explain your answer.
   b. About 80% of breast cancers are estrogen sensitive. What effect would estrogen have on these cancer cells?
   c. Mushrooms have been shown to contain anti-aromatase compounds, and certain studies suggest that females that eat mushrooms have a reduced risk of developing breast cancer. What is the action of aromatase and how might anti-aromatase compounds reduce the risk of breast cancer?

5.8 THE CELL CYCLE AND CANCER

1. Most cancers appear as individuals grow older. Why is that the case, and what is the likely reason for cancer to develop in younger individuals?

2. With regard to the regulation of the cell cycle, explain why the retinoblastoma protein is considered a tumor suppressor and *ras* is considered an oncogene.

5.9 THE BACTERIAL CELL CYCLE

1. Match each of the following statements to the appropriate cell type.
   A Bacteria
   B Eukaryotes
   C Both

a. Highly conserved cell cycle machinery
b. Cell cycle is stimulated or inhibited by environmental cues
c. Cell cycle can increase in speed by allowing a new round of replication to begin before the original is completed
d. Cell cycle can increase in speed by eliminating growth phase
e. Can inhibit the cell cycle in response to DNA damage

2. *C. crescentus* has emerged as a model organism for the study of the cell cycle in bacteria. Describe two of the unique characteristic of this organism that make it suitable for cell cycle study.

Challenge question

3. Based on your knowledge of the cell cycle in *C crescentus*, fill in the table below regarding the results of the mutations given. Assume that the cells were growing normally and that the mutation was introduced as the cell started a new G1 phase.

Mutation	Effect on the cell cycle	Phenotype of the cell
CtrA that cannot be phosphorylated		
Non-functioning Ccrm		
Non-functioning GcrA		
Non-functioning DnaA		

EXPERIMENTAL APPROACH 5.1 – CYCLIN-DEPENDENT KINASES: OF YEAST AND MEN

1. What was the purpose of the experiments carried out by Nurse and colleagues?

2. How did Nurse and colleagues use the idea of evolutionary conservation in their experiments?

3. What was the basis for the assay that was used by Nurse and colleagues?

4. Explain the significance of Figure 2.

5. Why was it difficult to determine targets of the Cdks?

6. How was this difficulty overcome?

EXPERIMENTAL APPROACH 5.2 – THE CELL DIVISION CYCLE: FROM GENES IN YEAST TO FUNCTION IN ANIMAL CELLS

1. Describe the basic approach used by Hartwell and colleagues to identify the original cell division cycle mutants (cdc).

2. Figure 1a depicts examples of wild type (normal) and cdc mutant yeast as observed by Hartwell and colleagues. Explain how the mutants are different from the wild-type cells over the time in which they were observed.

3. What was the purpose of the biochemical assay performed by Kirschner and colleagues?

4. What do the large number of high molecular weight bands seen in the S100 lane and the Q1 + Q2 lane of Figure 2b represent, and what can be concluded from this result?

Challenge question

5. Figure 3 represents a combination of results from genetic screens, like those in Figure 1, and biochemical assays, like those in Figure 2. Explain why this result is a combination, and what conclusion could be made from the result.

EXPERIMENTAL APPROACH 5.3 – CHECKPOINT PATHWAYS: DELAY AND CONQUER

1. What observation was at the heart of the experiments designed by Weinert and Hartwell?

2. What hypothesis did Weinert and Hartwell make based upon the observation?

3. What was the general approach to test the hypothesis?

Challenge question

4. What result was obtained, and what conclusion was drawn based upon the result?

5.1 STEPS IN THE EUKARYOTIC CELL CYCLE

Blomena VA, Boonstraa J. Cell fate determination during G1 phase progression. *Cellular and Molecular Life Sciences*, 2007;**64**:3084–3104.

Morgan DO. *The Cell Cycle Principles of Control*. London: New Science Press, 2007.

5.2 CYCLINS AND CDKS

Bloom J, Cross FR. Multiple levels of cyclin specificity in cell-cycle control. *Nature Reviews Molecular Cell Biology*, 2007;**8**:149–160.

Hochegger H, Takeda S, Hunt T. Cyclin-dependent kinases and cell-cycle transitions: does one fit all? *Nature Reviews Molecular Cell Biology*, 2008;**9**:910–916.

Kaldis P, Aleem E.Cell cycle sibling rivalry: Cdc2 vs. Cdk2. *Cell Cycle*, 2005;**4**:1491–1494.

5.3 REGULATION OF CDK ACTIVITY

Besson A, Dowdy SF, Roberts JM. CDK inhibitors: cell cycle regulators and beyond. *Developmental Cell*, 2008;**14**:159–169.

Kaldis P. The cdk-activating kinase (CAK): from yeast to mammals. *Cellular and Molecular Life Sciences*, 1999;**55**:284–296.

5.4 CELL CYCLE REGULATION BY CDKS

Manning AL, Dyson NJ. pRB, a tumor suppressor with a stabilizing presence. *Trends in Cell Biology*, 2011;433–441.

van den Heuvel S, Dyson NJ. Conserved functions of the pRB and E2F families. *Nature Reviews Molecular Cell Biology*, 2008;**9**:713–724.

5.5 PROTEOLYTIC REGULATION BY CDKS

Harper JW, Burton JL, Solomon MJ. The anaphase-promoting complex: it's not just for mitosis any more. *Genes and Development*, 2002;**16**:2179–2206.

Peters JM. The anaphase-promoting complex: proteolysis in mitosis and beyond. *Molecular Cell*, 2002;**9**:931–943.

5.6 CHECKPOINTS: INTRINSIC PATHWAYS THAT CAN HALT THE CELL CYCLE

Harper JW, Elledge SJ. The DNA damage response: ten years after. *Molecular Cell*, 2007;**28**:739–745.

Lew DJ, Burke DJ. The spindle assembly and spindle position checkpoints. *Annual Review of Genetics*, 2003;**37**:251–282.

5.7 EXTRINSIC REGULATORS OF CELL CYCLE PROGRESSION

Hallstrom TC, Nevins JR. Balancing the decision of cell proliferation and cell fate. *Cell Cycle*, 2009;**8**:4, 532–535.

5.8 THE CELL CYCLE AND CANCER

Jones RG, Thompson CB. Tumor suppressors and cell metabolism: a recipe for cancer growth. *Genes and Development*, 2009;**23**: 537–548.

Luo J, Solimini NL, Elledge SJ. Principles of cancer therapy: oncogene and non-oncogene addiction. *Cell*, 2009;**136**: 823–837.

Malumbres M, Barbacid M. Cell cycle, CDKs and cancer: a changing paradigm. *Nature Reviews Cancer*, 2009;**9**:153–166.

5.9 THE BACTERIAL CELL CYCLE

Collier J. Regulation of chromosomal replication in *Caulobacter crescentus. Plasmid*. 2012;76–87.

Erill I, Campoy S, Barbe J. Aeons of distress: an evolutionary perspective on the bacterial SOS response. *FEMS Microbiology Reviews*, 2007;**31**: 637–656.

Leonard AC, Grimwade JE. Regulation of DnaA assembly and activity: taking directions from the genome. *Annual Reviews in Microbiology*, 2011;**65**:19–35.

Lutkenhaus J. The ParA/MinD family puts things in their place. *Trends in Microbiology*, 2012;**20**:411–418.

Mott ML, Berger JM. DNA replication initiation: mechanisms and regulation in bacteria. *Nature Reviews Microbiology*, 2007;**5**: 343–354.

DNA replication

6

INTRODUCTION

Before cells divide, they must make a complete and faithful copy of the DNA in their chromosomes. This process of copying is called DNA replication. During DNA replication, each strand of the DNA double helix is copied to make a new strand, thus producing two new daughter DNA double helices. Replication is a carefully orchestrated process, requiring the activity of many separate enzymes all working together in a coordinated fashion. It is also a process for which accuracy is paramount: the viability of successive cell generations is dependent on faithful DNA replication so that the daughter cells each receive an exact copy of the parent cell's DNA. Failure to correctly copy the DNA sequence from generation to generation can have serious consequences both for the cell itself and for the organism of which it is a part.

In this chapter we explore DNA replication by first considering the various stages that make up the overall process of replication, building a broad picture of how two double-stranded helices are generated from one initial double-stranded helix. We then consider the key components of the molecular machinery that makes replication happen. Finally, we revisit each stage of replication in more detail to discover how the different molecular components come together to ensure that DNA replication happens as it should, and how regulatory mechanisms ensure that DNA is replicated only once in each cell-division cycle.

6.1 OVERVIEW OF DNA REPLICATION

We begin the chapter by looking at DNA replication as an overall process, identifying its key features and key components.

Each strand of the DNA double helix serves as a template for the synthesis of a complementary new strand

DNA replication is a fundamental process carried out by all living organisms. It is therefore not surprising that the basic mechanism of replication had an early evolutionary origin and is the same in bacteria, archaea, and eukaryotes. According to this basic mechanism of replication, both strands of the parental DNA molecule are copied during replication, each strand serving as a template for the synthesis of a new daughter strand, as illustrated in Figure 6.1. We call this process **semi-conservative**: when replication is complete, two daughter helices have been produced, each containing one old strand (which has been conserved from the parent) and one new strand. The experiments that first showed this to be the case are described in Experimental approach 6.1.

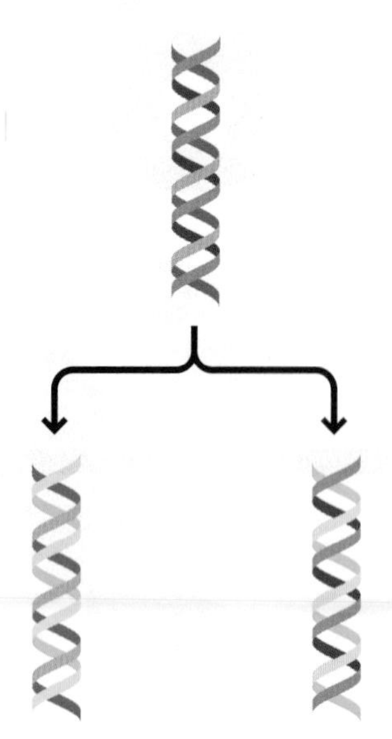

Figure 6.1 Semi-conservative replication. The parental double helix is shown in light and dark gray. After replication, each daughter has one of the two old parental strands (gray) and one newly synthesized strand (green).

Scan here to watch a video animation explaining more about the replication fork and how it moves, or find it via the **Online Resource Center** at www.oxfordtextbooks.co.uk/orc/craig2e/.

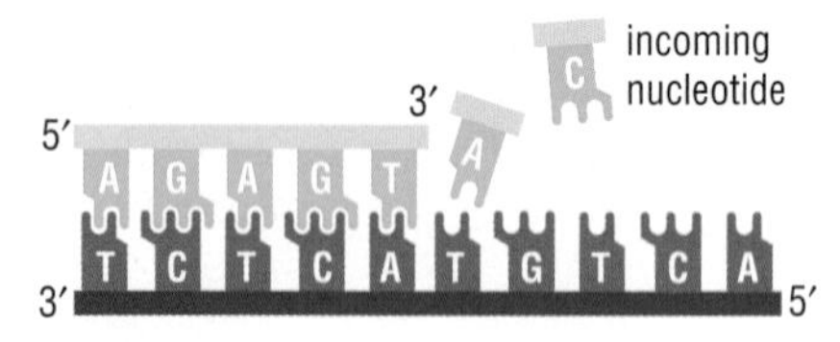

Figure 6.2 Template-directed DNA synthesis. A parental DNA strand (dark gray) acts as a template for the synthesis of an exactly complementary new strand (green) by the successive base-pairing of incoming nucleotides to the template. The base adenosine (A) pairs with thymine (T), and cytosine (C) pairs with guanine (G). Correctly paired nucleotides are joined to the new strand one at a time by the enzyme DNA polymerase. The new strand is always synthesized in the 5′ to 3′ direction.

Figure 6.2 shows how the process of copying the template strand involves the recognition of a base in the template strand and the addition to the new daughter strand of the complementary base. This method of replication ensures not only that the daughter molecules are identical to the parent DNA, but that they are also identical to each other. Consequently, when the cell divides and one copy is passed to each daughter cell, both cells receive exactly the same genetic material.

DNA synthesis occurs at replication forks that move outwards from origins of replication

In chromosomes, DNA replication starts at specialized sites called origins of replication (see Section 4.9) and moves away from the origin in both directions, creating a structure known as a **replication bubble**, as depicted in Figure 6.3. The DNA double helix is opened at the origins of replication and then unwound on both sides of the origin to form structures called **replication forks**. The replication forks are the sites at which single-stranded DNA is exposed and at which DNA synthesis occurs. As replication continues, the two replication forks that have originated from a single origin of replication move away from each other, proceeding along the DNA template strand in opposite directions.

DNA replication comprises three distinct phases

The process of DNA replication can be divided into three consecutive phases: initiation, elongation, and termination. The overall process of DNA replication is summarized in Figure 6.4. We will first develop an overview of the process of DNA replication, identifying the key components in the molecular machinery that we will examine in more detail later. Towards the end of the chapter, we will further discuss each step in the process.

Initiation of replication occurs at origins

During initiation, the first phase of DNA replication, the helix of the double-stranded parent DNA is opened up to give the replication enzymes and other proteins access to the single strands that form the templates for the daughter strands that will be synthesized (Figure 6.4, step 1). We see in Section 4.9 that initiation occurs at specific origins of replication termed *ori* (see Figure 4.29). Bacterial chromosomes typically have a single origin while eukaryotic chromosomes have multiple origins.

The origin is recognized by a specific **initiator protein** that facilitates opening of the double-stranded DNA and recruits a class of enzymes called helicases (Figure 6.4, step 1). **DNA helicases** act to unwind the double-stranded DNA to form the single-stranded templates required for DNA replication (see Section 6.5). As soon as single-stranded DNA is formed, it is coated with single-stranded binding proteins, which prevent the single-stranded DNA from forming secondary structures or re-pairing with the other strand, and protect the single strand from endonucleases. These single-stranded DNA-binding proteins are then removed as DNA synthesis proceeds. As we will explore later in the chapter, initiation is under strict control to ensure that DNA replication is initiated only once per cell division.

6.1 EXPERIMENTAL APPROACH

Discovery of semi-conservative DNA replication

Development of density gradient centrifugation allowed visualization of newly synthesized DNA.

When James Watson and Francis Crick proposed the double-helical structure of DNA (see Experimental approach 2.1), they realized that the structure suggested a potential way to make exact duplicates of the double-stranded DNA during cell division. They wrote in a now famous understatement at the end of their paper 'it has not escaped our attention that the specific pairing we have postulated immediately suggests a possible copying mechanism for the genetic material'. From casual examination, it would seem that simply unzipping the helix and copying each separate strand would be an obvious solution. Others, however, pointed out that the helical nature of DNA itself posed a problem to simply opening and copying the helix. This led to much debate about whether DNA replication was indeed carried out by such a strand separation and copying mechanism.

Matthew Meselson and Franklin Stahl resolved this theoretical argument by designing an elegant set of experiments showing that the DNA is indeed copied in a semi-conservative manner; that is, one parental strand remains intact while a copy of that complete strand is generated by the synthesis of new DNA. In their experiments, the researchers labeled DNA from *Escherichia coli* by growing the cells in a medium enriched in the heavy isotope of nitrogen, N^{15}, which imparts a different density to DNA than the natural isotope, N^{14}. To examine the distribution of heavy and light DNA in their samples, an experimental approach called density gradient centrifugation was developed. As they discovered, when a cesium chloride (CsCl) solution is spun at high speeds in an ultracentrifuge, a density gradient is established in the salt solution. DNA added to this tube will migrate to and form a band at a position in the gradient that is equivalent to its own buoyant density.

Using this method, Meselson and Stahl first showed that DNA isolated from bacteria grown in N^{15}-containing media formed a band at a different position in the gradient than DNA from bacteria grown in the standard 'light' nitrogen source, N^{14}. Having established the method, the researchers then grew a culture of bacteria in N^{15} and shifted it to N^{14}-containing media; they then followed the density of DNA in the cultures closely through four cell divisions. At several time points (reported here as 'generations'), they removed a sample of the culture, isolated DNA from the sample, and centrifuged it in a CsCl density gradient; their results are shown in Figure 1. At the earliest time-point (0 generations, before the addition of the N^{14} media), the entire population of DNA ran as 'heavy' band N^{15} DNA, while after one cell division, all of the DNA was located at an intermediate density between heavy and light. After two full cell divisions, there was a band of all light DNA and a band of intermediate density.

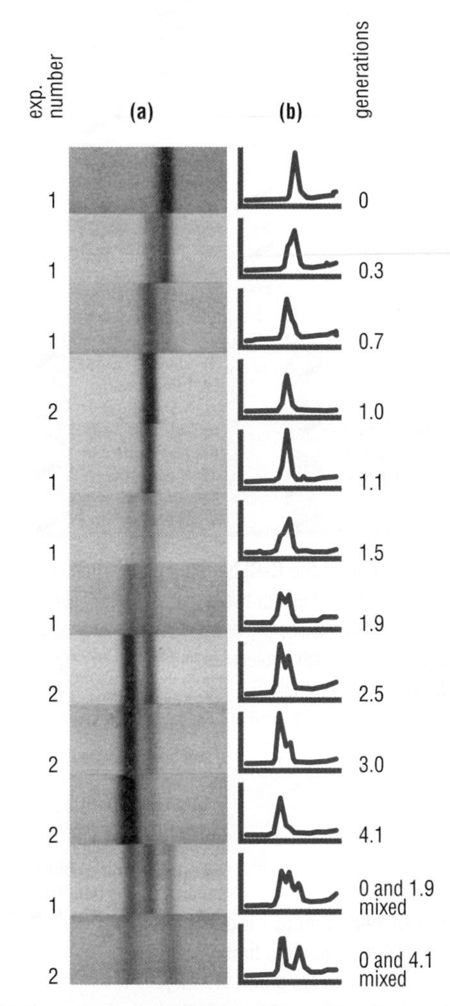

Figure 1 CsCl gradient centrifugation differentiates heavy and light DNA peaks. (a) The image of the DNA bands taken from the ultracentrifuge camera is shown for 12 different time points. (b) The tracing of the bands shown in (a) and the estimated number of generations is shown. The generations were calculated based on the growth rate and doubling time determined for the cells in this experiment. The DNA moved from heavy:heavy to heavy:light to light:light peaks as the cells grow in N^{14} light media. The last two panels show a mixing of 0 and 1.9 generations and 0 and 4.1 generations to show the overlap in the bands.

Reproduced from Meselson and Stahl. The replication of DNA in *Escherichia coli*. *Proceedings of the National Academy of Sciences of the U S A*, 1958; **44**: 671–682.

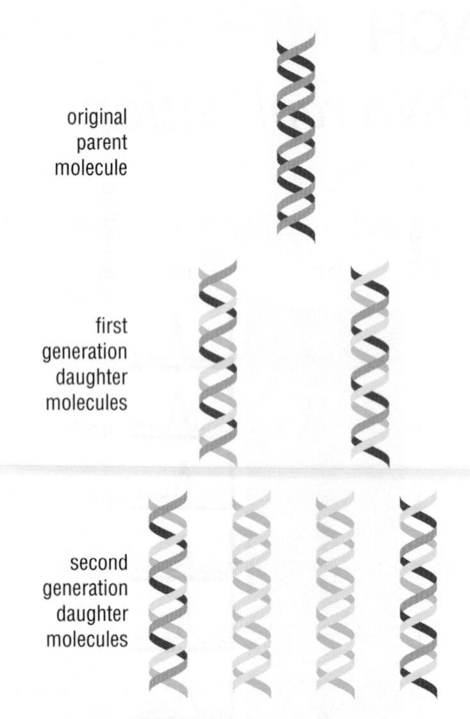

Figure 2 Interpretation of the CsCl banding experiment shown in Figure 1. The initial helix is made of two strands of heavy DNA. After one cell division, both daughter molecules are heavy:light, but after two rounds of cell division, there are two light:light molecules and two heavy:light molecules; no heavy:heavy molecules will remain.

Reproduced from Meselson and Stahl. The replication of DNA in *Escherichia coli*. *Proceedings of the National Academy of Sciences of the U S A*, 1958; **44**: 671-682

Interpretation of bands at different densities revealed replication is semi conservative

The authors interpreted the results shown in Figure 1 with the diagram shown in Figure 2. After the first round of DNA synthesis, all of the DNA was of intermediate density because the old strand was 'heavy' (containing N^{15}) and the new strand was 'light' (containing N^{14}). During the second round of DNA synthesis, the parental light strand was replicated with light DNA giving rise to light:light DNA (which exhibited a light density after centrifugation); and the parent heavy strand was also replicated with light DNA, giving rise to heavy:light DNA, which exhibited the intermediate density. After two more rounds of synthesis, most of the DNA was light:light, with a small amount at the intermediate (light:heavy) density. As a marker of the density for each type, they mixed the 0 time point and the 2 generation time point and saw bands of all three densities.

Density gradient centrifugation applied genome-wide shows timing of origin firing is not uniform.

This elegant set of simple experiments not only clearly showed the semi-conservative nature of DNA replication but also provided a powerful tool for analysis (CsCl density gradient centrifugation) that became a mainstay for separation and analysis of DNA in molecular biology. Indeed, CsCl-based density gradient centrifugation is still widely used for studying replication. With this method, and knowing

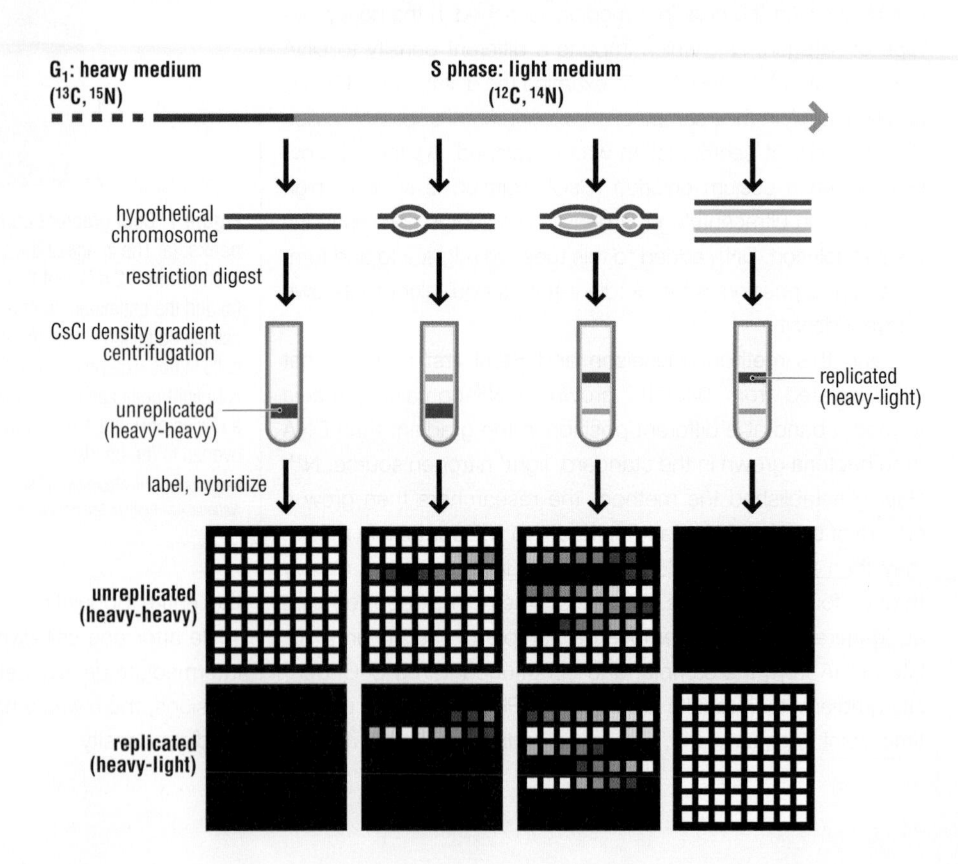

Figure 3 Experimental strategy for genome-wide analysis of DNA replication timing. Top: timeline of experiment: cells are grown in heavy medium for several generations and a sample of DNA is taken at time 0. Samples are then taken at several intervals over the course of one cell cycle. Middle: the replication of a hypothetical region of the chromosome is depicted and the expected banding location in CsCl gradients is shown. Bottom: the heavy:heavy DNA is hybridized to a microarray representing the entire yeast genome, and in a separate experiment, DNA from the heavy:light region is hybridized to a duplicate microarray. At the beginning of the experiment all of the regions of the genome are represented in the heavy:heavy fraction. At the end of the experiment, all of the DNA is in the heavy:light fraction. At the intermediate time point, the origin regions, the earliest regions to replicate, show up on the array probed with the heavy:light fraction and at the same time disappear from the heavy:heavy fraction.

Adapted from Raghuraman *et al. Science*, 2001;**294**: 115–121.

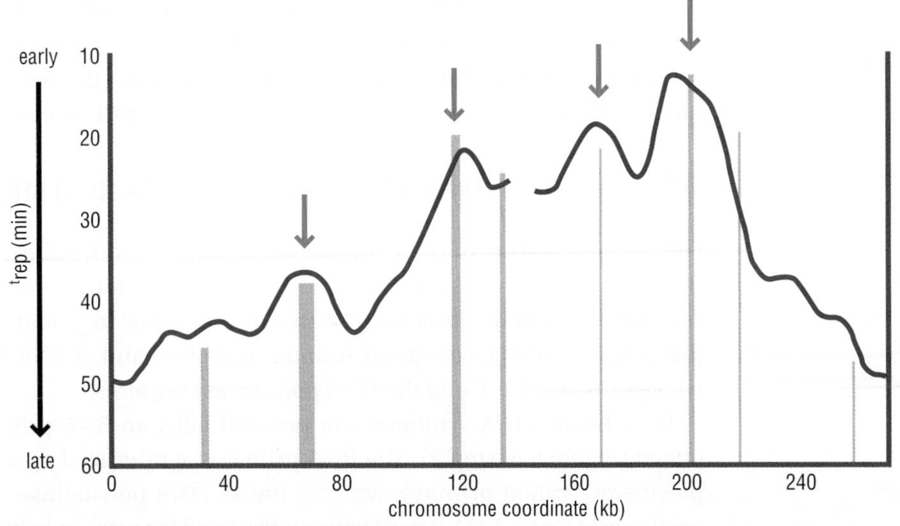

Figure 4 Replication timing across all of yeast chromosome VI. The data from the experiment outlined in Figure 3 is plotted for the length of one chromosome, yeast Chromosome VI. The Y-axis shows the timing of replication for a given chromosomal region. Dark gray bars correspond to DNA regions known to contain replication origins. Blue arrows indicate the four origins used in more than 50% of the cell cycles.

Adapted from Raghuramen *et al.* Replication dynamics of the yeast genome. *Science*, 2001;**294**:115–122

the sequence of the entire yeast genome, M. K. Raghuraman *et al.* simultaneously mapped the timing of replication of all replication origins in yeast. As we discuss in this chapter, some origins fire early in the replication cycle while others fire later. Raghuraman *et al.* again used density labeling, growing yeast first in a 'heavy' media (containing both N^{15} and carbon C^{14}) and then shifting the cells to a 'light' media (with N^{14} and C^{12}). Replication was then followed in a manner similar to the Meselson and Stahl experiment; but in this case, the authors added an isolation step wherein they directly purified DNA from the heavy:heavy and heavy:light bands at different time points. This purified DNA was then sheared and hybridized to a microarray chip with the whole genome arrayed (microarrays are discussed in Section 16.9) (Figure 3).

The time at which a sequence disappears from the heavy:heavy fraction and appears in the heavy:light fraction represents the time at which that sequence starts to replicate. Thus, early firing origins, and the sequences near them, are the first to move to the heavy:light fraction, while late firing origins, and the sequences near them, move much later. Using this approach, the timing of replication along each individual chromosome was thus determined (Figure 4).

Find out more

Meselson M, Stahl FW. The replication of DNA in *Escherichia coli*. *Proceedings of the National Academy of Sciences of the U S A*, 1958;**44**:671–682.

Raghuraman MK, Winzeler EA, Collingwood D, *et al.* Replication dynamics of the yeast genome. *Science*, 2001;**294**:115–121.

Related techniques

CsCl gradient centrifugation; Section 19.7

Density labeling; Section 19.6

Microarray chips; Sections 19.9, 19.10

During elongation each base in the parent DNA strand is read by DNA polymerase to direct synthesis of a daughter strand in a 5′ to 3′ direction

After the double-stranded DNA has been opened up during initiation, the enzymes that are needed to copy the DNA assemble and we enter the elongation phase of DNA replication. During this phase, the replication enzymes move along the parent DNA template strands, and the daughter strands are gradually elongated.

The synthesis of a new DNA strand is catalyzed by an enzyme called **DNA polymerase**. This enzyme moves along the template strand, reading the nucleotides one at a time and adding the complementary nucleotide to the end of the new

We discuss the polarity of the DNA strand in Section 3.4.

daughter strand. The DNA polymer has a distinct polarity: the chemistry of the nucleotide-addition reaction only allows DNA polymerase to add new nucleotides to the 3′ end of the growing DNA strand. This has the very important consequence that DNA synthesis can only proceed in a 5′ to 3′ direction.

DNA replication starts with the synthesis of a short stretch of RNA

One unusual feature of DNA polymerase is that it cannot build a new DNA strand from the very start of its parent strand; it can only add nucleotides to the 3′ end of a nucleotide fragment that already exists. The nucleotide fragment on which the DNA polymerase builds its daughter strand is a short strand of RNA termed a **primer**, which provides the 3′ end the DNA polymerase requires.

Thus, before DNA synthesis can proceed fully, an RNA primer must be generated (Figure 6.4, step 2). The RNA primer is synthesized by a specialized RNA polymerase called **primase**, which - unlike DNA polymerase - can initiate the synthesis of a new RNA strand without the need for a pre-existing 3′ end.

Later in the process of replication, the RNA primer is degraded and replaced with DNA, as we will discuss in detail in Section 6.8.

DNA polymerase is recruited to the DNA at primer–template junctions

After the primer is synthesized, the stage is now set for the recruitment of DNA polymerase. This recruitment - and the stable association of DNA polymerase with the template strand - is mediated by two specific protein complexes, the **clamp loader**, and the **sliding clamp**, which we learn about further in Section 6.6.

As the name suggests, the clamp loader acts to load the sliding clamp on to the DNA template. The sliding clamp then recruits DNA polymerase to the template, ready for DNA synthesis to commence. Again, as indicated by its name, the sliding clamp remains clamped to the DNA template, sliding along it as DNA synthesis occurs, and keeping the DNA polymerase tethered to the template.

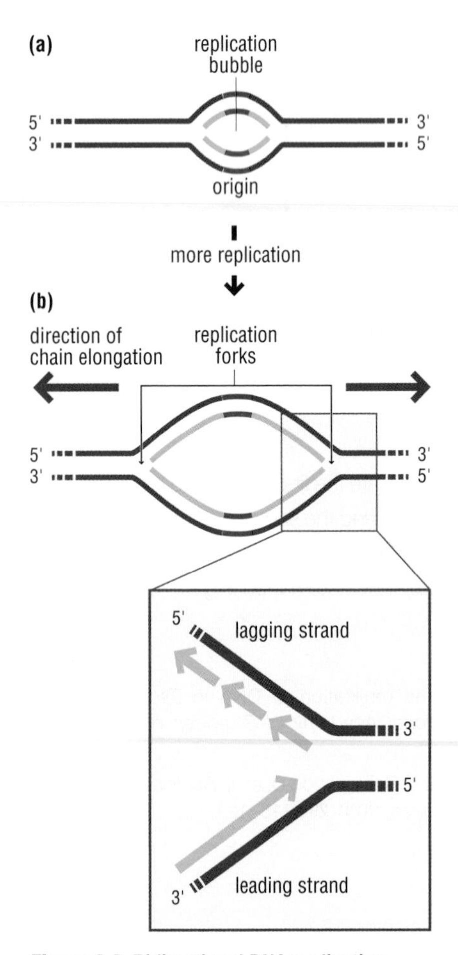

Figure 6.3 Bidirectional DNA replication. (a) Replication initiates at an origin (red) and proceeds by copying both strands of DNA in both directions away from the origin. This creates what is known as a replication bubble. (b) At either end of the bubble are moving replication forks at which the DNA helix is being continuously unwound and DNA is being synthesized. Newly synthesized DNA is shown in green. The yellow box shows a close-up of one replication fork. Note that both strands are being synthesized in the 5′ to 3′ direction. However, one strand, called the leading strand, is being synthesized continuously (solid green arrow on the bottom strand) while the other strand, called the lagging strand, is being synthesized discontinuously (short green arrows on the top strand of the fork). These fragments of the lagging strand will be eventually joined to form a continuous DNA strand.

Figure 6.4 Steps in DNA replication. Shown are the steps as they occur in the bacteria *Escherichia coli*. DNA replication can be described as a series of discrete steps. Initiation: (1) Helicase loading: helicases (light orange rings) are recruited to the origin (red) by initiator proteins (not shown); the DNA double helix is opened to allow both strands to be copied. The *E. coli* helicase binds to single-stranded DNA and moves in a 5′ to 3′ direction on the template of the lagging strand (see below). In contrast, the eukaryotic replicative helicase binds to the template of the leading strand and moves in a 3′ to 5′ direction (see Figure 6.11). The small replication bubble, formed by the actions of the bound helicases, is now ready for DNA replication to start. (2) Priming: primases (light orange) are recruited by the helicases and initiate the synthesis of RNA primers (turquoise) on both strands. The primers allow DNA polymerases to synthesize the DNA strands. Bacterial primase is shown here for simplicity. In eukaryotes, pol α-primase synthesizes an RNA primer followed by a short stretch of DNA. Elongation: (3) Loading of the sliding clamp: the clamp loader (not shown) binds to the 3′ primer–template junction and places the sliding clamp (light orange) around the DNA. (4) Recruitment of the replicative polymerase (purple): interaction with the sliding clamp places the polymerase in the correct position to elongate the 3′ end of the primer. (5) Bidirectional fork movement and strand elongation: the two forks move away from the origin in opposite directions and new DNA synthesis occurs on both strands. DNA synthesis on the lagging strand is discontinuous, producing short segments of DNA. Successive segments are each started at the replication fork and are elongated in the 5′ to 3′ direction. (6) Termination: once replication is complete, DNA ligase (not shown) seals any remaining gaps in the DNA backbone between DNA segments, resulting in two complete daughter molecules.

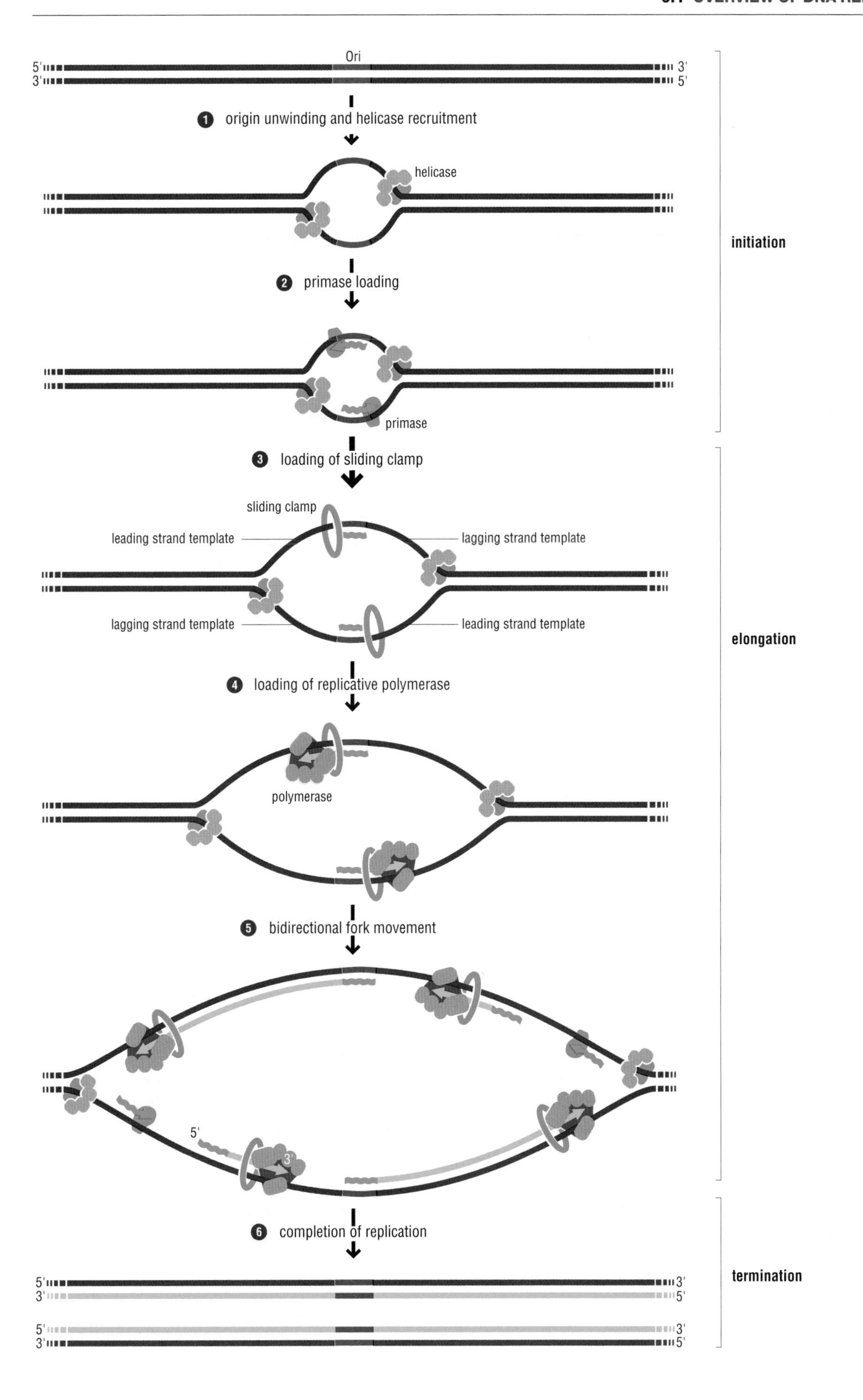
Ori
5'
3'
3'
5'
1 origin unwinding and helicase recruitment
helicase
2 primase loading
primase
3 loading of sliding clamp
sliding clamp
leading strand template
lagging strand template
lagging strand template
leading strand template
4 loading of replicative polymerase
polymerase
5 bidirectional fork movement
5'
3'
6 completion of replication
5'
3'
3'
5'
5'
3'
3'
5'
initiation
elongation
termination

The clamp loader and sliding clamp play an important role in recruiting the DNA polymerase to the appropriate location on the DNA template: they localize specifically at the 3′ primer–template junction, the region where DNA synthesis needs to commence (Figure 6.4, step 3). In eukaryotes, the first few nucleotides of the daughter strand are synthesized by DNA polymerase α (pol α), which acts in tight association with primase (forming a complex called pol α-primase). The replicative polymerase is then recruited to begin processive elongation (Figure 6.4, step 4). In eukaryotes, this involves a switch from pol α-primase to the replicative polymerases δ (pol δ) or ε (pol ε). We learn more about the different types of polymerases in Section 6.4.

DNA synthesis is continuous on one strand and discontinuous on the other strand

The fact that DNA can only be synthesized in the 5′ to 3′ direction poses a problem at the replication fork: both strands of the DNA molecule must be copied but yet the two template strands have opposite orientations - one strand runs in the 5′ to 3′ direction, while its complementary strand runs in the 3′ to 5′ direction. To solve this problem, DNA synthesis on the two strands is different, as illustrated in Figure 6.3. The synthesis of one strand is continuous: the daughter DNA can be synthesized as a single, uninterrupted strand from 5′ to 3′. This strand is known as the **leading strand**. By contrast, the other strand, known as the **lagging strand**, is synthesized discontinuously: a series of short DNA fragments are synthesized from 5′ to 3′, which are then subsequently joined into a continuous strand. Primer synthesis and polymerase loading occur once on the leading strand, but must occur repeatedly on the lagging strand.

Polymerases, helicases, and primases travel with the replication fork

The helicases that unwind the DNA at the origin travel ahead of the polymerases to unwind the DNA at the replication fork. Directly behind the helicase, the leading strand polymerase synthesizes DNA in the 5′ to 3′ direction, while a substantial region of single-stranded DNA is generated on the lagging strand and is bound by single-stranded binding proteins, which bind and protect the single-stranded DNA. Primase at the fork binds this single-stranded region of DNA and synthesizes a new primer. Primer synthesis enables the initiation of further DNA synthesis on the lagging strand to yield a relatively short segment of new DNA (Figure 6.4, step 5).

DNA synthesis on the leading and lagging strands is carried out by a separate DNA polymerase on each strand; the two polymerases move together in a coordinated fashion as the replication fork progresses along the DNA. This coordinated movement entails the looping of the lagging strand to allow simultaneous 5′ to 3′ synthesis on both strands. So it is important to note that Figure 6.4 is somewhat artificial as it does not show the coupling of the two polymerases. However, it illustrates in a simplified manner the principles and the order of the steps that occur.

Termination of replication occurs either when two forks meet or when the ends of linear chromosomes are reached

Termination of DNA replication occurs when the two forks moving in opposite directions meet and the **replication complexes** are disassembled. In bacteria this occurs at a specific site, *ter* (see Section 4.9); by contrast, no specific termination sequence is required in eukaryotes; the replication of a particular DNA segment is terminated when adjacent replication forks meet. A regulated mechanism presumably exists to disassemble the two large fork complexes that meet between adjacent origins. However, little is known about the details of this final step.

In eukaryotes, when synthesis of a large genome segment is complete, a fork coming from one origin will meet the previously initiated strand of the neighbor origin. The RNA primer is then removed and replaced with DNA, and the two adjacent newly synthesized DNA strands are connected by DNA ligase (Figure 6.4, step 6). Finally, at the very end of a linear chromosome, termination occurs when the fork reaches the end and, presumably, falls off.

6.2 DNA POLYMERASES: STRUCTURE AND FUNCTION

Having reviewed the key aspects of DNA replication, let us now examine the proteins that are key players in the replication machinery. After we have learned about the proteins themselves, we will bring both process and machinery together to see in more detail how the cell achieves the remarkable feat of replicating DNA with as few errors as possible.

DNA polymerases make a complementary copy of one DNA strand

The central enzymes in DNA replication are DNA polymerases, which move along the template strand of DNA, reading the bases and adding the correct complementary nucleotide to the new strand. At each position, DNA polymerase selects the correct nucleotide to be added, catalyzes its linkage to the 3′ OH end of the chain, and then moves to the next template residue, maintaining contact with the template during the process. The accuracy or **fidelity** of DNA replication is essential for correct transmission of the genetic material from one generation to the next. If errors creep in, mutations occur that could affect a specific gene product or regulatory region, neither of which is desirable.

Much of the accuracy of replication is due to the activity of DNA polymerases: in addition to selecting the correct nucleotide for addition, many polymerases also proofread for errors. If the wrong nucleotide is added, it is recognized and hydrolyzed by an **exonuclease** activity in the polymerase, and the correct nucleotide is then inserted. We will discuss the mechanisms that ensure the accuracy of replication in Section 6.3.

DNA polymerases carry out a number of different functions in the cell

All cells have a number of different DNA polymerases that perform distinct functions. Some polymerases are responsible for replicating the genome, some for the repair of DNA, and some for continuing replication across damaged DNA. Others have even more specialized functions, such as the replication of telomeres.

➔ We discuss telomeres in more detail in Sections 4.11 and 6.11.

Some polymerases comprise just a single protein, which exhibits a range of functions, including nucleotide addition, proofreading and substrate specificity. By contrast, other polymerases are made of multiple subunits, with each subunit having a distinct function, such that full polymerase activity is only achieved when all subunits come together to form a complex. In this section, we will discuss the common features of polymerase structure and function, focusing on the major

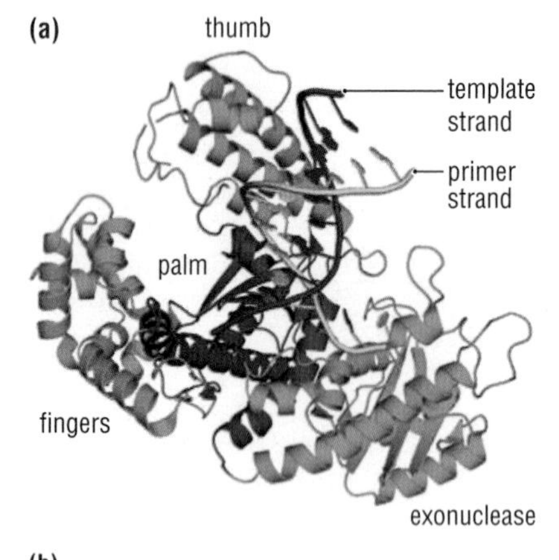

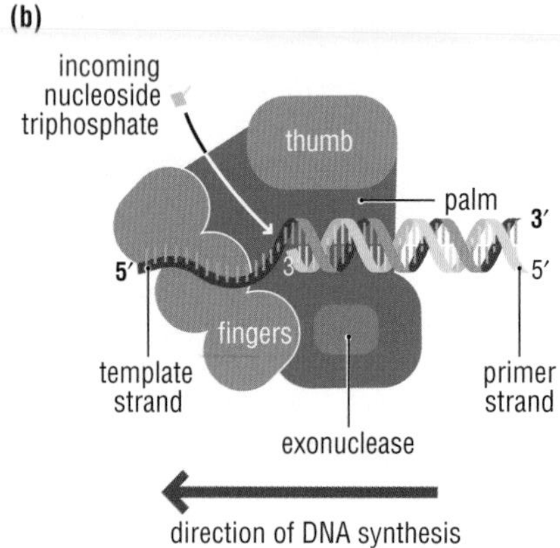

Figure 6.5 DNA polymerase resembles a right hand. (a) Crystal structure of the Klenow fragment of *E. coli* pol I, which comprises the polymerase domains and the 3′ to 5′ exonuclease activity but lacks the 5′ to 3′ exonuclease domain of pol I. The polymerase domain resembles a right hand with the fingers on the side, the palm domain forming a cleft where the DNA will lie and the thumb on the top. The exonuclease domain for this class of polymerases is to one side of the three core polymerase domains (Protein Data Bank (PDB) code 1KLN). (b) Schematic representation of the polymerase structure in (a). The palm domain is in the interior of the structure. The single-stranded template DNA is fed past the fingers and into the active site in the palm where nucleotide addition is catalyzed. The incoming nucleotide is joined to the free 3′ OH in the polymerase active site in the palm. The newly synthesized double helix exits from the back of the polymerase domains.

replicative polymerases; some of the more specialized polymerases are described in Section 6.4 and their role in DNA repair is discussed in Chapter 16.

In bacteria, one major replicative polymerase, termed DNA polymerase III (pol III), carries out genome duplication. In eukaryotes, there are two replicative polymerases at the replication fork: pol δ synthesizes the lagging strand, while pol ε synthesizes the leading strand. These enzymes synthesize DNA rapidly and with high fidelity. The high degree of sequence conservation of the replicative polymerases in bacteria, archaea, and eukaryotes reflects the early evolutionary origin of their specialized function.

While we note above that some polymerases comprise just a single protein, the major replicative polymerases are composed of multiple protein subunits that help them achieve speed while also enabling them to maintain the fidelity of replication. Bacterial DNA replication proceeds at a rate of around 1000 nucleotides per second, while eukaryotic chromosomes are replicated more slowly, at the rate of around 50 nucleotides per second. To achieve this speed, polymerases have specialized subunits that carry out specific functions. Some polymerase subunits are involved in proofreading - monitoring for errors in the way one would do when proofreading text - while others help attach the polymerase to the DNA template so that it can replicate long stretches of DNA before it dissociates. This ability to remain associated with the template is referred to as the **processivity** of a polymerase.

The structure of a DNA polymerase resembles a right hand

The most highly conserved component of the polymerase complex is the subunit that contains the catalytic site for nucleotide addition. The structure of this core polymerase subunit is exemplified by that of DNA polymerase I (pol I) from *E. coli*. Pol I is a specialized polymerase that is essential for finishing DNA replication and removing the RNA primers that initiated replication. It is composed of a single protein chain that folds into three structural domains as illustrated in Figure 6.5; taken together, these domains look something like your right hand. Because of this resemblance, the three domains are called the palm, the fingers, and the thumb.

The spatial relationship of the three domains to each other and the structure of the finger and thumb regions vary among different polymerases, but the overall structure and catalytic mechanism is conserved in polymerases from all organisms. The most highly conserved domain is the palm, which forms a cleft into which the growing double-stranded DNA fits. Meanwhile, the single-stranded template strand wraps through the fingers.

During the elongation phase of DNA replication, nucleoside triphosphates (NTPs) are added to the growing daughter strand in a reaction catalyzed by the polymerase. In most polymerases, the finger domain positions the incoming NTP in relation to the template DNA. The thumb domain helps to hold the elongating duplex DNA and also maintains the contact with the template that permits processive synthesis. In pol I, the 3′ to 5′ exonuclease function that removes incorrect bases is present as an additional domain of the same protein (see Figure 6.5).

The active site of the enzyme, where the reaction catalyzed, lies deep in the cleft of the beta sheet comprising the palm domain. At the active site, carboxylate groups of two aspartate residues coordinate two magnesium ions that participate in catalysis, as we'll now discuss in more detail.

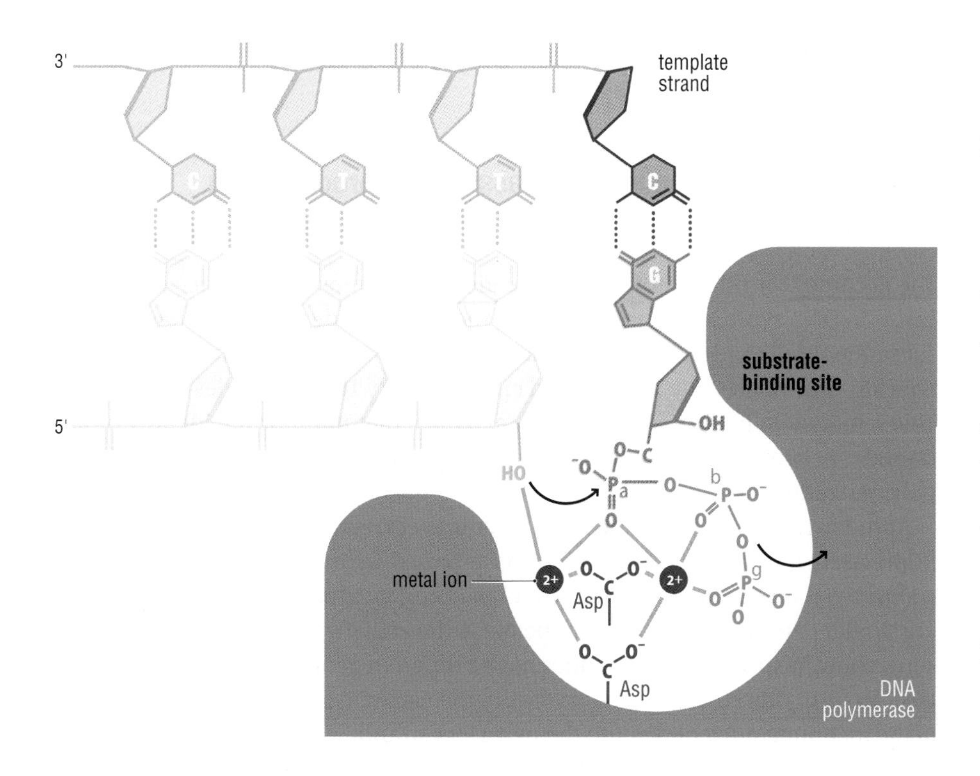

Figure 6.6 Two-metal-ion mechanism of catalysis. The carboxylate groups of two conserved aspartate residues in the polymerase active site coordinate two metal ions (typically magnesium ions [Mg^{2+}] in DNA polymerases) and hold them in the correct orientation to participate in catalysis. One Mg^{2+} interacts with the 3′ OH group of the growing strand (light green), and the other interacts with the incoming nucleoside triphosphate (dark green and orange). The Mg^{2+} bound to the growing strand facilitates the attack of the 3′ OH on the α -phosphoryl group of the incoming nucleoside triphosphate, and a new O–P bond is formed. The two phosphate groups highlighted in orange are released as pyrophosphate and are subsequently hydrolyzed. Both metal ions stabilize the structure of the transition state of the reaction. This two-metal-ion catalyzed phosphoryl transfer is similar to that used in all polymerases, including RNA polymerases.

The polymerase active site catalyzes the addition of nucleotides onto the growing DNA chain

The active site of the polymerase catalyzes a phosphoryl transfer reaction that adds a nucleotide to the nascent DNA strand by linking the 5′ phosphoryl group on the nucleotide to the 3′ hydroxyl on the end of the chain to form a phosphodiester bond (see Section 2.6). The continuous series of phosphate bonds linking adjacent nucleotides in the DNA molecule comprises the so-called DNA backbone, which lies on the outside of the double helix. Let us now explore the mechanism of this reaction in a little more detail.

Incoming nucleotides are in the form of NTPs. Their addition to a growing strand proceeds through nucleophilic attack by the nascent chain 3′ OH on the α-phosphate of the incoming NTP, releasing pyrophosphate (PPi), as illustrated in Figure 6.6. Critical to this reaction are two Mg^{2+} ions present in the active site that are positioned by carboxylate moieties on two conserved aspartate residues. One of these Mg^{2+} ions functions primarily to activate the 3′ OH of the terminal nucleotide, whereas the second Mg^{2+} interacts with the incoming NTP and stabilizes a developing negative charge on the leaving oxygen during the reaction. The incoming triphosphate is also stabilized by interactions with side chains on an alpha helix in the finger domain. Subsequent hydrolysis of the released pyrophosphate provides the free energy that drives the reaction forward.

The general chemical features of this polymerization reaction - including the nucleophilic attack, the use of two metal ions in the active site to promote the reaction, and the use of pyrophosphate hydrolysis to drive the reaction forward - are similar to RNA polymerase and other biological reactions, which we encounter a number of times in this book.

purine–pyrimidine mispairs

G–T
G–C

purine–purine mispairs

G–A
T–A

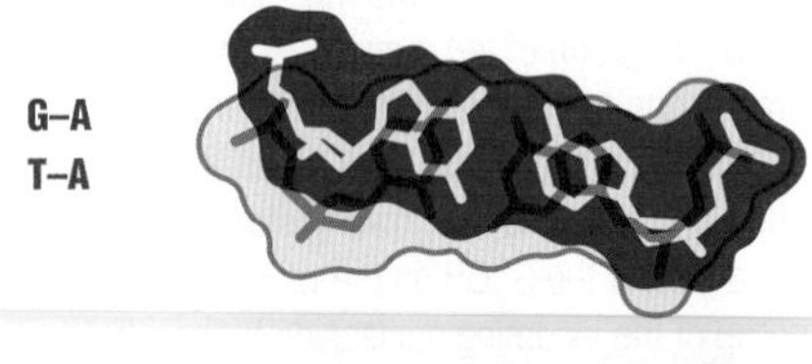

G–G
G–C

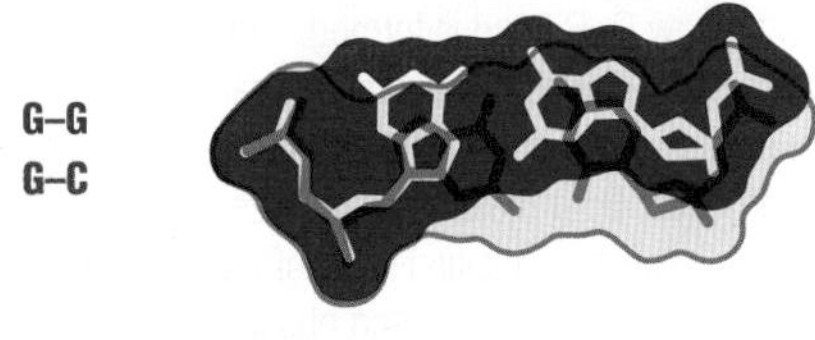

incorrect
correct

Figure 6.7 Structure of mismatched base pairs. Incorrectly base-paired nucleotides have a distinctly different molecular shape from that of a correct base pair. The top row shows the shape of a correct G–C base pair outlined in gray. Superimposed in pink is the shape of an incorrect G–T pair. The difference in the two shapes is obvious. The two lower rows illustrate the same point for G–A and G–G mispairings. The correct base pair is outlined in gray in each case; note how similar the shapes of G–C and T–A base pairs are to each other.

Johnson SJ, Beese LS. Structures of mismatch replication errors observed in a DNA polymerase. *Cell*, 2004;**116(6)**: 803-16.

6.3 DNA POLYMERASES: FIDELITY AND PROCESSIVITY

The fidelity of DNA replication is assured by proofreading mechanisms

The accuracy, or **fidelity**, of DNA replication is of utmost importance to the cell. Errors in DNA synthesis cause changes in the genetic code that will be passed to all subsequent cells and can alter essential cellular functions. A replicative polymerase typically only makes one uncorrected error for every 10^7 nucleotides synthesized. This remarkable fidelity is achieved by activities that monitor the identity of nucleotides at two distinct stages: during the polymerization reaction itself, and after polymerization has occurred.

At the first level, the polymerase recognizes correct nucleotides because they fit precisely into the active site when base-paired with the template strand. The geometry of an adenine–thymine (A–T) pair and a cytosine–guanine (C–G) pair are similar to each other and will fit neatly in the active site. By contrast, Figure 6.7 shows how mismatched nucleotides have a different geometry that does not fit as well. In turn, this correct geometry favors the catalytic reaction and the addition of the incoming nucleotide onto the new DNA strand.

Despite these spatial constraints, an incorrect nucleotide is added to the new DNA strand about once in every 10^5 nucleotides. This error can be corrected by **proofreading**, the second level at which fidelity is monitored. Most polymerases have an exonuclease activity that hydrolyzes the final nucleotide from a DNA strand, if it is incorrect, before the next one is added. This activity is referred to as 3′ to 5′ **exonuclease** activity.

The polymerase active site and the exonuclease active site are spatially separated on the polymerase such that, when a correct base is added, the polymerase simply moves on to the next template position and proceeds with polymerization. However, when a mispaired nucleotide is incorporated, the rate of incorporation of the next nucleotide by the polymerase is substantially slowed, thus giving the exonuclease time to function. For the excision reaction associated with proofreading, the 3′ end of the newly generated DNA is moved from the polymerase active site into the exonuclease active site, where the terminal nucleotide is removed, as depicted in Figure 6.8. The 3′ end is then repositioned in the polymerase active site and elongation can proceed. These events are all accomplished while the polymerase enzyme remains associated with the template DNA.

The recognition of correct nucleotides based on their geometry derives from the inherent energetic differences between correct and incorrect nucleotides. As such, proofreading at this level does not need an input of energy for recognition to occur because the enzyme does not actually get as far as adding the wrong nucleotide to the growing chain. By contrast, if an incorrect nucleotide *is* incorporated, the cell does need to expend energy - in this case, to excise the incorrect nucleotide. Such processes that allow an enzyme to achieve greater discrimination in a second step than permitted by the inherent energetic differences in the interactions of the correct and incorrect nucleotide are referred to as kinetic proofreading, and typically are associated with energetic expenditure (the removal of incorrect nucleotide(s) in this case). As we shall see when we discuss the translation of RNA into protein (Chapter 10), two-step mechanisms for ensuring accuracy are a common feature of fundamental molecular processes.

The overall accuracy of DNA transmission from one cell to the next is assured by three steps - two that involve DNA polymerase as described here, and one that involves post-replication DNA repair. The selection of the correct base by the DNA polymerase accounts for an error rate of only of one wrong base out of 10^5 bases synthesized. However, the addition of the editing or kinetic proofreading step brings an additional 100-fold accuracy, lowering the total error rate to only one mistake in every 10^7 bases synthesized. As we will see in Chapter 15, a final step that helps assure fidelity of the transmission of the DNA sequence between generations is DNA repair, which occurs after replication. This process lowers the error rate by an additional 10^3 nucleotides, making the overall error rate that is propagated during each cell division to one base in about 10^{10} total base pairs (bp).

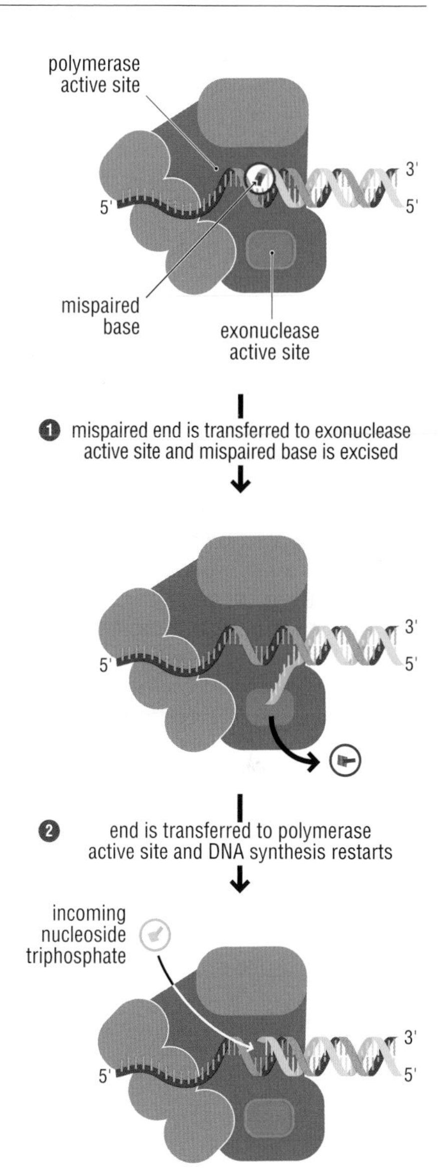

Figure 6.8 Steps in proofreading. When an incorrect base pair is incorporated despite the discrimination step described in Figure 6.7, it can be removed from the end by editing. The mispaired base at the 3′ end of the growing strand (shown in pink) is detected, and the 3′ end is moved to the 3′ to 5′ exonuclease site on the polymerase (step 1). This editing mode involves breaking several base pairs between the nascent strand and the template. In the editing site, the 3′-most nucleotide is removed from the DNA chain, and the new 3′ end is then allowed to return to the polymerase active site in the palm (step 2). Now a new round of nucleotide addition can occur in the polymerase active site.

Processive and distributive polymerases have specific roles in DNA replication

After each addition of a nucleotide, there is a chance that the polymerase will dissociate from the template. If this were to occur too often, replication would be slow and inefficient. As noted in Section 6.2, the polymerases engaged in general replication have high processivity: that is, they remain associated with the template through many rounds of nucleotide addition. A processive polymerase may copy many thousands of nucleotides before dissociating, thus allowing long regions of a chromosome to be copied efficiently. The main replicative polymerases are intrinsically processive, but also have several accessory proteins that tether them to the DNA or to other proteins at the fork, as we shall see later in the chapter.

Not all polymerases are inherently processive, however. Some dissociate frequently from the template, polymerizing only tens or hundreds of nucleotides at a time, depending on the particular polymerase. After a polymerase dissociates, a new polymerase molecule then binds the 3′ end of the growing strand and continues elongation. In these cases, repeated rounds of polymerase dissociation and reassociation are required for the template strand to be fully copied, as depicted in Figure 6.9. This is known as a **distributive** mechanism of elongation.

Examples of distributive polymerases include the polymerases involved in DNA repair. These polymerases only need to copy short stretches of DNA at locations where single-stranded gaps are left by the repair process. Another example is pol α, which in eukaryotes synthesizes a short stretch of DNA onto the RNA primer made by the primase. Pol α is not very processive, adding only 30–100 nucleotides before dissociating from the template.

6.4 SPECIALIZED POLYMERASES

Distinct DNA polymerase families have specialized roles in DNA replication

The replicative DNA polymerases - pol III in bacteria and pol δ and pol ε in eukaryotes - are responsible for copying the bulk of the DNA in order to duplicate the genome. In addition to these replicative DNA polymerases, however, a number of other important DNA polymerase enzymes are found in both bacterial and eukaryotic cells. The overall structure and mechanism of these different DNA polymerases is remarkably conserved, but they differ from each other in subtle ways that allow them to carry out distinct functions in the cell.

➔ We learn more about DNA repair in Chapter 16.

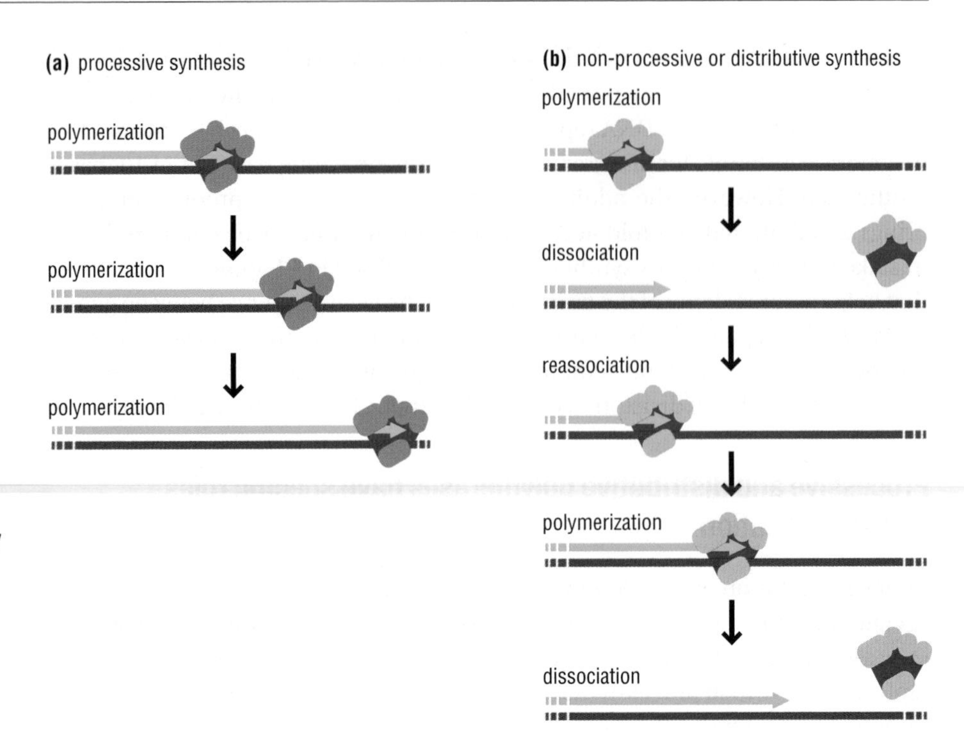

Figure 6.9 Processive and non-processive polymerases. (a) In processive DNA synthesis, the polymerase remains associated with the template DNA strand for many thousands of nucleotides. Some DNA polymerases require additional factors, such as a sliding clamp, in order to be processes. (b) Polymerases that synthesize only a short region of DNA are usually non-processive or distributive. The polymerase is bound to the template and synthesizes DNA; however, with a certain probability, it will dissociate from the DNA. A new polymerase molecule can then bind to the primer template DNA, and further polymerization will occur.

DNA polymerases can be grouped into functional families according to their evolutionary relatedness

The active site and the catalytic mechanism for nucleotide polymerization are conserved among polymerases. Beyond the active site, however, other regions of the polymerase can vary considerably. Figure 6.10 shows how DNA polymerases are grouped into six different families – A, B, C, D, X, and Y – on the basis of structural similarity and conserved sequence domains. Strikingly, in any given organism the similarities in both structure and function within a family are greater than the similarities between the different polymerases, indicating that these polymerase families diverged early in evolution.

The A family of polymerases includes the pol I from *E. coli*, which is involved in finishing DNA replication and removing the RNA **primers** (see Section 6.1). Pol I contains a 5′ to 3′ exonuclease domain that allows it to remove DNA or RNA ahead of it; it also features the 3′ to 5′ proofreading exonuclease activity already noted. The B, C, and D families contain the replicative polymerases of eukaryotes, bacteria, and archaea. These polymerases have high fidelity and possess a 3′ to 5′ proofreading exonuclease. The X family of polymerases is specialized for DNA repair, filling in the gaps generated in DNA during repair. **Y-family polymerases** are specialized to replicate past bulky adducts in damaged DNA; other polymerases will stop polymerization when they encounter such lesions.

➔ We learn more about DNA repair and the polymerases involved in Chapter 16.

Reverse transcriptases are DNA polymerases that use RNA templates to make DNA

In addition to the DNA polymerases that are involved in DNA replication and repair, eukaryotic cells and some viruses have DNA polymerases that copy RNA into

DNA polymerase families			
family	polymerase	source	function
A	pol 1 Taq T7	*E. coli* *Thermus aquaticus* phage T7	gap repair replication replication
B	pol α pol δ pol ϵ pol ζ (Rev3) pol II	eukaryotes eukaryotes eukaryotes eukaryotes *E. coli*	primase and repair replication replication translesion synthesis
	T4 RPB69 pol B	phage phage archaea	 replication
C	pol III	*E. coli*	replication
D	pol D	archaea	replication
X	pol β pol λ pol μ pol σ	eukaryotes eukaryotes eukaryotes eukaryotes	gap repair gap repair gap repair gap repair
Y	pol η (Rad 30) pol ι pol κ REV 1	eukaryotes eukaryotes eukaryotes eukaryotes	translesion synthesis translesion synthesis translesion synthesis translesion synthesis
	Din B (Pol IV) UmuCD (Pol V)	*E. coli* *E. coli*	translesion synthesis translesion synthesis
	Dbh Dpo4	archaea archaea	translesion synthesis translesion synthesis
RT	reverse transcriptase	retrovirus eukaryotes	copy genome copy retrotransposons
	telomerase	eukaryotes	elongate telomeres

Figure 6.10 DNA polymerase families.

DNA. These are called **reverse transcriptases.** Unlike all the other DNA polymerases described so far (which use single-stranded DNA as their template), reverse transcriptases use single-stranded RNA as a template to synthesize a complementary DNA strand. The structure of reverse transcriptase is very much like that of other polymerases: it has thumb and finger domains and a palm domain. The active site of reverse transcriptase is in the palm domain and it uses the same two metal ion catalytic mechanism for adding nucleotides as described above for DNA polymerase.

Reverse transcriptases are typically encoded by viruses as well as DNA elements called retrotransposons, which are found in eukaryotic genomes. These are a class of the transposable genetic elements, which we will learn more about in Chapter 17. We will later discuss telomerase, a specialized form of a reverse transcriptase that is needed to synthesize chromosome ends, in Section 6.11.

6.5 DNA HELICASES: UNWINDING OF THE DOUBLE HELIX

DNA helicase unwinds double-stranded DNA for copying

Having now considered the key enzymes involved in the actual synthesis of DNA during replication, let us now turn to an enzyme that has an important function before synthesis can actually occur – the unwinding of the double-stranded helix of the parent DNA.

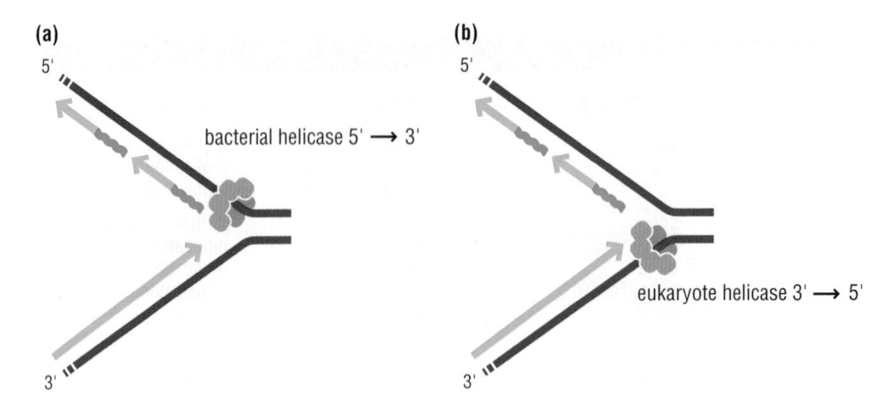

Figure 6.11 Bacterial vs eukaryotic replicative helicases. The replicative helicases are six-membered rings that encircle a single strand of DNA as they move with the replication fork. The helicases in bacteria and eukaryotes, however, encircle opposite strands at the replication fork. The direction of movement of a helicase is described by the polarity of the DNA that it is moving along. a) The *E. coli* helicase binds to single-stranded DNA and moves in a 5′ to 3′ direction on the template strand; we can therefore say that it is traveling on the template of the lagging strand. b) The eukaryotic replicative helicase binds to the template of the leading strand and moves in a 3′ to 5′ direction.

When base-paired in the double-stranded helical DNA molecule, nucleotides are inaccessible to polymerases, making copying impossible: both primase and polymerases need access to single-stranded DNA. To achieve this, the internal base-pairing must be broken and the helix unwound.

As we discuss in Section 6.1, the first step in this unwinding process is the initial opening of the helix, a step performed by the initiator protein at the origin of replication. Once opened, the unwinding of the double helix to expose single-stranded DNA for copying can begin. This unwinding process is catalyzed by an enzyme called **DNA helicase**. As we discuss in Chapter 16, the cell also needs to open the helix for DNA repair and recombination, and there are unique DNA helicases for these repair processes.

Helicases open the double-stranded DNA and then travel with the replication fork, continuously unwinding the DNA to provide the single-stranded template for the polymerase to copy. The helicases involved in replication in both bacteria and eukaryotes are composed of six subunits that form a ring structure, which surrounds one strand of the DNA. Curiously, while the structure is conserved from bacteria to eukaryotes, the helicases translocate on different strands of the replication fork. In bacteria it moves along the leading strand in the 5′–3′ direction (relative to the strand that it is bound to). By contrast, the eukaryotic helicase MCM travels on the lagging strand, moving in the 3′–5′ direction, as depicted in Figure 6.11.

The structure of the replicative DNA helicase from papillomavirus provides insights into the mechanism of helicase action. This viral helicase is a hexamer of one protein, E1. Crystallographic studies have revealed that a single strand of DNA fits into the center of the channel formed by the hexamer ring, as depicted in Figure 6.12. The other strand of the DNA is displaced by the helicase and is bound by single-stranded binding proteins.

In *E. coli*, the helicase associated with DNA replication is known as **DnaB** and is also a hexamer of six identical subunits. In eukaryotes and archaea, the replicative helicase is called **MCM** and is a complex of six different proteins, Mcm2–7, which assemble to form a ring. The eukaryotic replicative helicase is conserved throughout evolution and requires accessory factors for helicase activity.

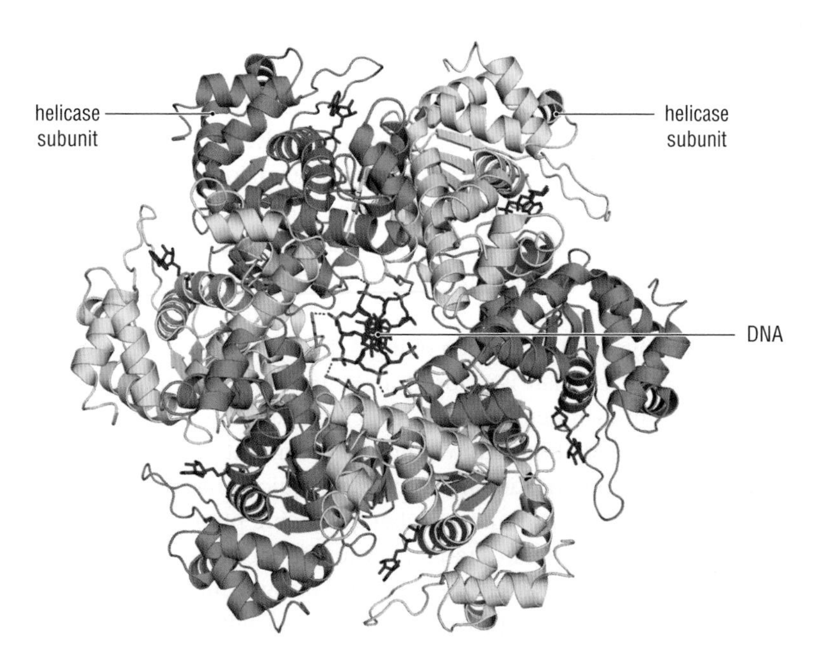

Figure 6.12 Structure of the papillomavirus helicase E1 protein and DNA. The helicase protein is a ring composed of six identical subunits (PDB code 2GXA). The structure shows that single-stranded DNA fits through the center on the helical structure. There are five contact points between the phosphates on the single-stranded DNA and specific helices within the helicase that allow the movement of the helicase along DNA.

Single-stranded DNA-binding proteins keep DNA strands separated

The helicase exposes a region of single-stranded DNA that must be kept open for copying to proceed. This is achieved by coating the strand with single-stranded binding proteins as depicted in Figure 6.13. In bacteria, a monomeric protein called **single-stranded binding protein (SSB protein)** associates to form tetramers around which the DNA is wrapped in a manner that significantly compacts the single-stranded. In eukaryotes, the single-stranded binding protein is a complex of three different subunits called **replication protein A (RPA).**

The SSB and RPA proteins both stabilize the single-stranded DNA and interact specifically with other proteins needed for replication. Coating of single strands is particularly important on the lagging strand, because long stretches of single-stranded DNA are generated as a result of the discontinuous nature of replication on this strand. (Remember that replication of the lagging strand requires the synthesis of a series of short fragments of DNA, which are later joined to form a continuous strand. The regions of single-stranded DNA are protected by a coat of SSB proteins before they are copied.)

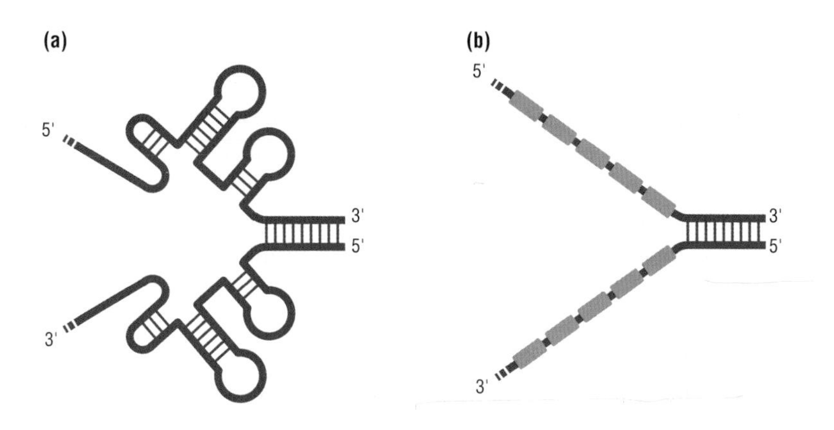

Figure 6.13 (a) Single-stranded DNA can adopt many secondary structures that would make copying by polymerase difficult. (b) To protect single-stranded DNA and present it to the polymerase for copying (and to other proteins such as repair proteins), the single-stranded binding proteins coat single-stranded DNA and melt any secondary structures. In bacteria, the single-stranded binding protein is a single polypeptide called SSB. In eukaryotes, it is a three-subunit protein termed RPA.

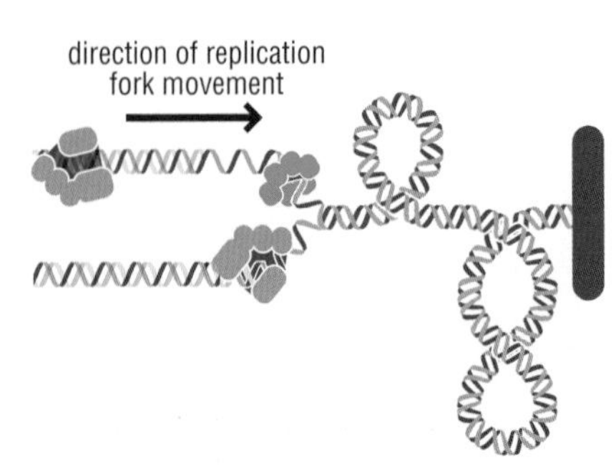

Figure 6.14 Supercoiling of DNA ahead of the replication fork. As the helicase at the fork unwinds the DNA, the helical twist from the DNA is translated into supercoils ahead of the fork as it moves forward. The circular DNA of a bacterial chromosome or eukaryotic chromosomal DNA packaged into chromatin is not free to swivel, and thus behaves as if it were anchored (bar at right). The supercoils must be removed by either topoisomerase I or II for the fork to continue its movement.

➜ The relationship between twist, supercoiling, and linking number is discussed in Chapter 2.

Topoisomerases assist helicases by removing supercoils from DNA

One full turn of the double helix is unwound for every 10 bp of DNA that the replication machinery travels along. As helicases and polymerases move along DNA, unwinding and copying it as they go, they generate torsional stress in the double-stranded helical DNA in front of them. Since the ends of the chromosome are not free to rotate, strand separation during DNA replication is accompanied by an overwinding of the DNA ahead of the replication fork, as illustrated in Figure 6.14. As we learned in Chapter 2, changes in the winding of the two DNA strands - defined as twist - lead to changes in DNA supercoiling if the ends of the DNA are prevented from rotating or if the template is a closed circle. The supercoils that build up ahead of the replication fork generate tension, which makes strand separation increasingly more difficult. Indeed, with unwinding happening at such a pace, the tension generated ahead of the replication machinery can quickly become considerable.

To envision how this tension accumulates, imagine phone charger cords that are twisted around each other in a tangle. If you find the ends of the cords and just pull in an attempt to unwind them, the cords will become increasingly twisted and tangled the more you pull. The process of unwinding will be met with increasing resistance, making it more difficult.

To deal with the DNA-unwinding problem, the cell has enzymes called **topoisomerases**, which can cut either single or double-stranded DNA. Topoisomerases remove supercoils ahead of the replication fork by transiently breaking the DNA backbone, allowing the strands to 'swivel' past each other, and thereby allowing the supercoils to relax. They then reconnect the DNA once tension has been relieved. In principle, you could use a similar solution to deal with entwined phone charger cords but it would not be practical: you would need to splice the two ends of the cord back together with each cut you make.

The name of this family of enzymes denotes the fact that the substrate of a topoisomerase and the product of the reaction are chemically identical, yet differ in topology.

There are several different ways that topoisomerases couple strand cleavage with a change in supercoiling; Figure 6.15 shows how these enzymes are grouped according to the particular mechanism they use. Type IA and IB topoisomerases break one of the two DNA strands and do not require adenosine triphosphate (ATP); in contrast, type II topoisomerases cleave both strands in a reaction that is powered by ATP hydrolysis. Both types of topoisomerases are present at replication forks in both bacteria and eukaryotes. Because they break the DNA backbone, topoisomerases can also untangle DNA that is inappropriately linked or knotted.

Type I topoisomerases change supercoiling by introducing transient breaks in one strand of DNA

How can an enzyme change the topology of DNA? While each class of topoisomerase carries out a distinct set of steps to achieve this change, all topoisomerase reactions share some common features. A central event is an attack on the phosphodiester backbone of DNA by a tyrosine hydroxyl on the topoisomerase enzyme, leading to formation of a phosphoester linkage between the tyrosine and either the 5′ or 3′ end of the cleaved strand. This bond holds one end of a DNA strand in place while the DNA molecule as a whole undergoes a change in topology. The cleavage reaction is then reversed, and the two ends of the cleaved strand re-ligated. Since the substrate and product in the reaction are chemically identical, no net energy input is needed.

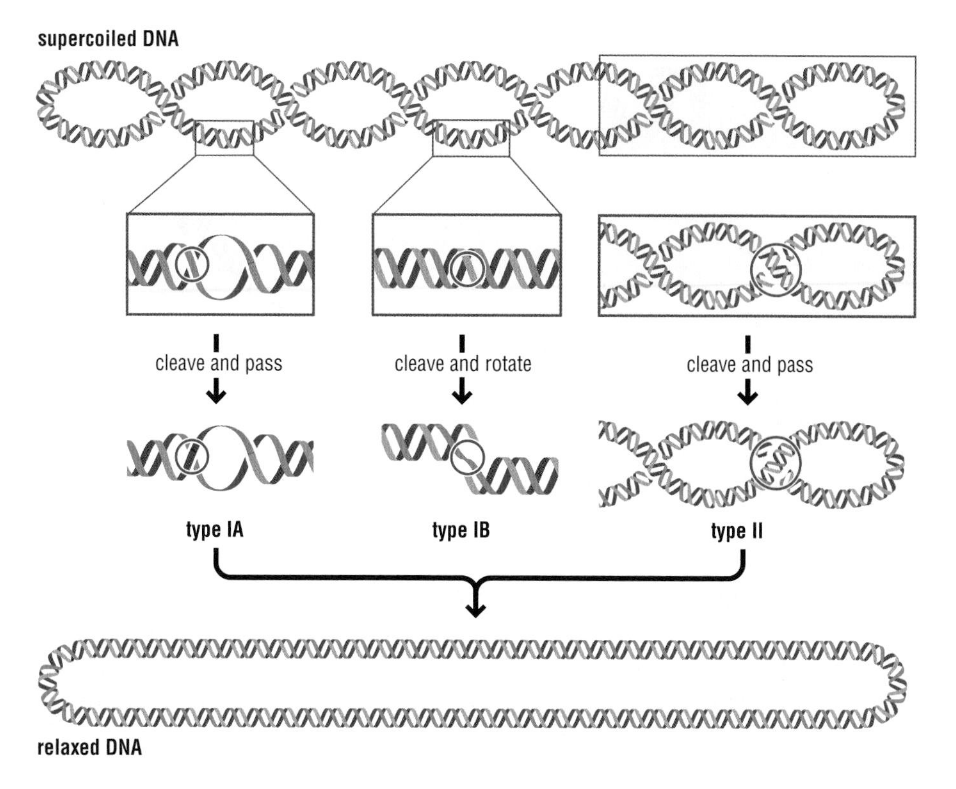

Figure 6.15 Three different mechanisms by which topoisomerases relax supercoils in DNA. (a) Type IA topoisomerases cleave single-stranded DNA and facilitate strand passage of one single strand past the other. (b) Type IB topoisomerases also cleave single-stranded DNA and allow the free end to swivel to release supercoils before religation. (c) Type II topoisomerases cleave the double strand of DNA and pass one strand behind the other. The mechanism by which type II topoisomerases relax supercoils is shown in more detail in Figure 6.16.

From Figure 1 of Corbett KD, Berger JM. Structure, molecular mechanisms, and evolutionary relationships in DNA topoisomerases. *Annual Review of Biophysics and Biomolecular Structures.* 2004;**33**:95–118.

Type IB enzymes rely upon the superhelical tension in the DNA to drive the relaxing of supercoils ahead of the replication fork (see Fig. 6.14). Specifically, these type IB enzymes cleave one DNA strand and form a transient covalent bond between an active site tyrosine residue in the enzyme and the 3′ end of the cleaved DNA strand. The superhelical tension in the DNA causes the free 5′ end to swivel about the uncleaved strand, with each complete rotation changing the linking number of the DNA in steps of one. After one or more rotations (up to five rotations have been observed), the free 5′ end attacks the bond between the enzyme, and the 3′ end and the DNA strand is re-ligated, generating double-stranded DNA with fewer supercoils.

Type IA and type II topoisomerases change supercoiling via strand passage

In contrast to the type IB enzyme described above, the type II and type IA enzymes change the supercoiling of DNA in a different way, by a process known as strand passage. Figure 6.16 illustrates how type II topoisomerases use strand passage to change the linking of double-stranded DNA. The dimeric enzyme cleaves both strands of the DNA duplex, forming a covalent bond between a tyrosine and the 5′ end of each cleaved strand. Conformational changes in the enzyme separate the two ends of the cleaved strand while allowing another double-stranded segment to pass through the gap in the DNA. The cleavage reaction is then reversed and the DNA is re-ligated. This reaction requires ATP hydrolysis, which presumably helps to power the conformational change in the enzyme. The resulting DNA has fewer supercoils and its linking number is changed by two for each cycle.

While most type II enzymes relieve supercoiling, an important exception is bacterial DNA gyrase, which introduces negative supercoils into DNA. DNA gyrase actively maintains the circular bacterial chromosome and plasmids in a negatively supercoiled state. This negative supercoiling has the opposite effect of the

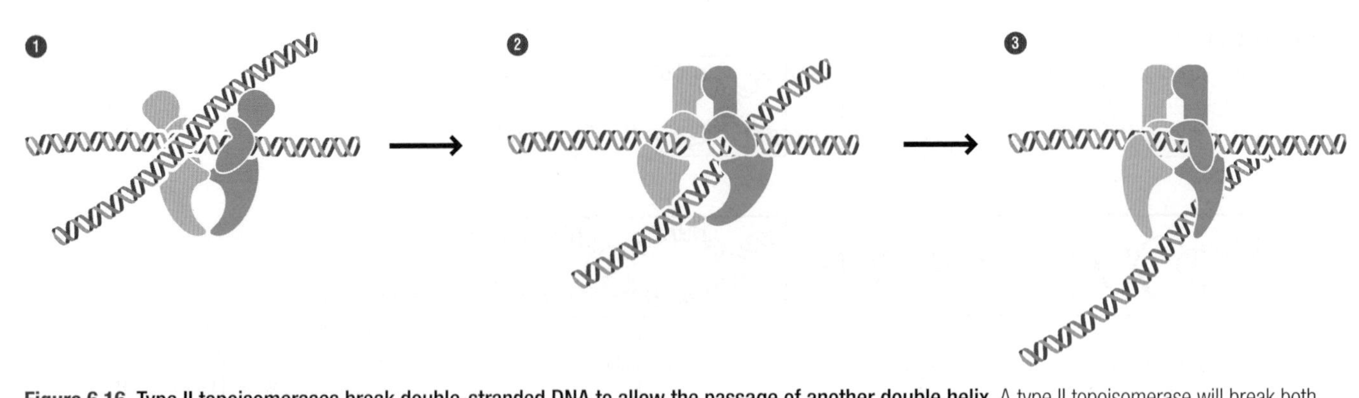

Figure 6.16 Type II topoisomerases break double-stranded DNA to allow the passage of another double helix. A type II topoisomerase will break both strands of one DNA molecule and allow one double helix to pass through another. This allows two intertwined helices to be resolved and ultimately separated. (1) The topoisomerase binds both double-stranded helices; at this point the blue helix is above the pink one. (2) The double-stranded DNA of the pink helix is broken, and the blue helix passed though the broken ends. (3) The pink helix is religated, and the blue helix is now below the pink one. Strand passage in this manner allows supercoils to be released.

positive supercoiling generated by movement of the replication fork: while the positive supercoils hinder strand opening, negative supercoils allows for much easier strand opening even compared with a relaxed DNA (which lacks supercoiling). This energetically favorable state is important in opening the DNA for the initiation of both DNA replication and transcription.

The final class of topoisomerases comprises the type IA enzymes. Like the type II topoisomerases, the type IA enzymes use a strand passage mechanism to remove supercoils from DNA. However, an important difference is that type IA enzymes introduce a transient break in just *one* of the DNA strands, and linking number changes in steps of one. So, instead of passing a double-stranded segment of DNA through a transient break in the DNA duplex, type IA enzymes pass a single strand of DNA through a transient break in another single strand. Like the other topoisomerases, a transient break is introduced into the DNA and the 5′ end of the single DNA strand forms a covalent bond with a tyrosine. This DNA is then re-ligated after the strand passage.

➔ Read about the relationship between supercoiling (writhe), twist, and linking number in Section 2.5.

The activities of type IA and type II topoisomerases in allowing the passage of DNA strands past one another facilitates the untangling of daughter chromosomes that are intertwined after replication. This untangling allows the replicated chromosomes to segregate at cell division, the mechanism of which will be discussed in more detail in Section 6.10 when we discuss termination of replication. In addition, both type I and type II topoisomerases are important in removing supercoils that accumulate during replication. For this reason, a variety of drugs have been identified that target toposiomerase, because blocking these enzymes causes cell death. Cipro is an antibiotic that specifically targets bacterial topoisomerases, while eukaryotic topoisomerase inhibitors are used as chemotherapeutic agents to block the growth of rapidly dividing cells in a tumor.

6.6 THE SLIDING CLAMP AND CLAMP LOADER

A sliding clamp and associated clamp loader increase the processivity of DNA polymerase

Let us now move on to consider some other proteins that play important roles during DNA replication.

We noted earlier that the replicative polymerases are highly processive - that is, they remain associated with the DNA template strand for a considerable length of time before dissociating. The high processivity of the replicative polymerases is largely due to an important multisubunit protein called the **sliding clamp** that keeps the polymerase tethered to the DNA. This clamp is assembled by another multisubunit protein known as the **clamp loader**. The structures of these proteins are highly conserved from bacteria to archaea and eukaryotes, indicating that they arose early in evolution and are of fundamental importance to efficient replication.

The clamp loader, as well as other proteins associated with replication that we shall learn about later in the chapter, are members of the **AAA+ family** of ATPases. (AAA+ stands for **ATPases associated with a variety of cellular activities.**) These proteins all share a conserved region of 220 amino acids that includes an ATPase domain. Many of the actions of AAA+ ATPases associated with replication undergo conformational changes; these changes, which are required for their functions, are driven by ATP hydrolysis.

The ring-shaped sliding clamp keeps DNA polymerase attached to the DNA

The sliding clamp is a donut-shaped protein complex that completely encircles the double-stranded DNA; its structure is illustrated in Figure 6.17. The clamp is a very stable structure and, once clamped onto the DNA, will remain associated with it (and will ensure that the DNA polymerase is tethered to the DNA template) while thousands of nucleotides are copied. Sliding clamp proteins are present in bacteria, archaea and eukaryotes. In bacteria they are known as the **β protein** and in eukaryotes as **PCNA**. The name PCNA, which stands for proliferating cell nuclear antigen, reflects the history of discovery of this protein, which was first identified by specific antibodies that recognized a protein present in the nuclei of dividing cells but not resting cells. Because antibody recognition of PCNA changed in the cell cycle, it was thought that the protein levels changed during the cell cycle. However, this protein is actually present at a constant level; it is only antibody accessibility that varies throughout the cell cycle. Thus the protein itself does not undergo the periodic destruction we saw for cell cycle proteins in Chapter 5.

Overall, the sliding clamp has a very similar structure in all organisms, reflecting conservation of the mechanism of DNA replication throughout evolution. In *E. coli* the ring is a homodimer of β proteins, as illustrated in Figure 6.17(a). Each β protein has three similar domains, so dimerization generates a structure with six symmetrically arranged domains. In eukaryotes, an analogous structure is made by the interaction of three copies of the PCNA protein, each of which is composed of two domains with structural similarity to those of bacterial β protein (Figure 6.17(b)). Thus, the *E. coli* and eukaryotic sliding clamps have a very similar ring structure consisting of six domains surrounding a central cavity of 35 Å that can accommodate duplex DNA.

During replication, the clamp binds DNA polymerase and slides along the DNA with the polymerase, thus keeping the polymerase tethered to the DNA. PCNA interacts with pol δ through a motif of eight amino acids present in each PCNA subunit, a motif that is conserved in eukaryotic and archaeal sliding clamps. The β protein has a similar short peptide motif that interacts with pol III, further indicating the conservation of the β protein and PCNA. We will see in Chapter 16 that a similar ring-shaped structure is formed by a complex of proteins that are involved in detecting sites of DNA damage, and is similarly thought to track along the DNA.

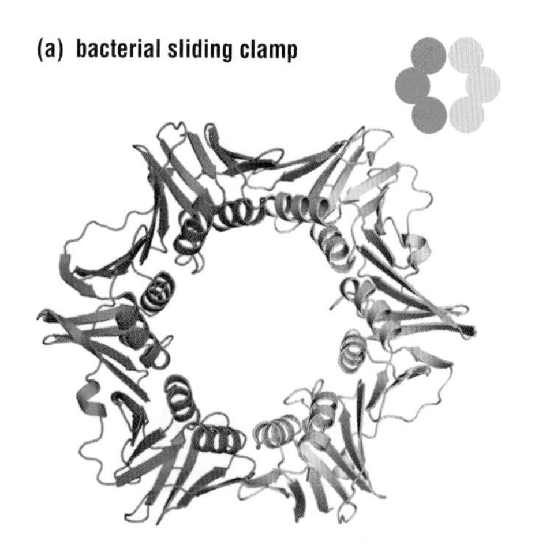

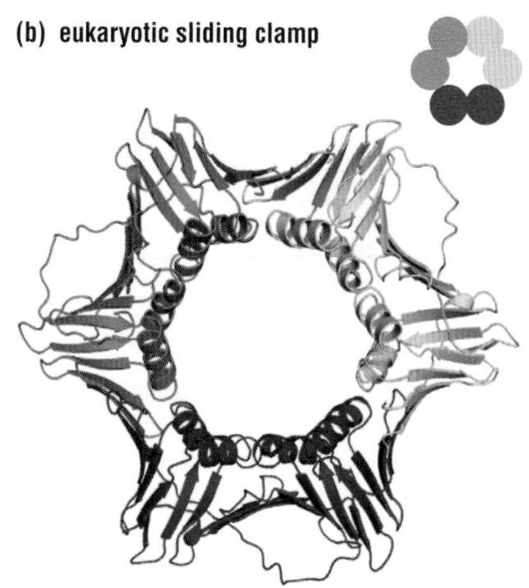

Figure 6.17 Structures of the bacterial and eukaryotic sliding clamps. (a) The β protein sliding clamp from *E. coli* is composed of two identical β proteins, each of which has three similar domains. Dimerization forms a ring that has six similar domains (PDB code 2POL). (b) The eukaryotic sliding clamp is made of three molecules of the PCNA protein. Each PCNA molecule has two similar domains, and the trimer of PCNAs forms a ring with six domains (PDB code 1AXC). Note that the structures of the two sliding clamps are remarkably similar despite differences in protein sequence and the number of protein subunits. The DNA is threaded through the hole in the center of the ring.
From Bruck I, O'Donnell M. The ring-type polymerase sliding clamp family. *Genome Biology*, 2001;**2(1)**:reviews3001.1–3001.

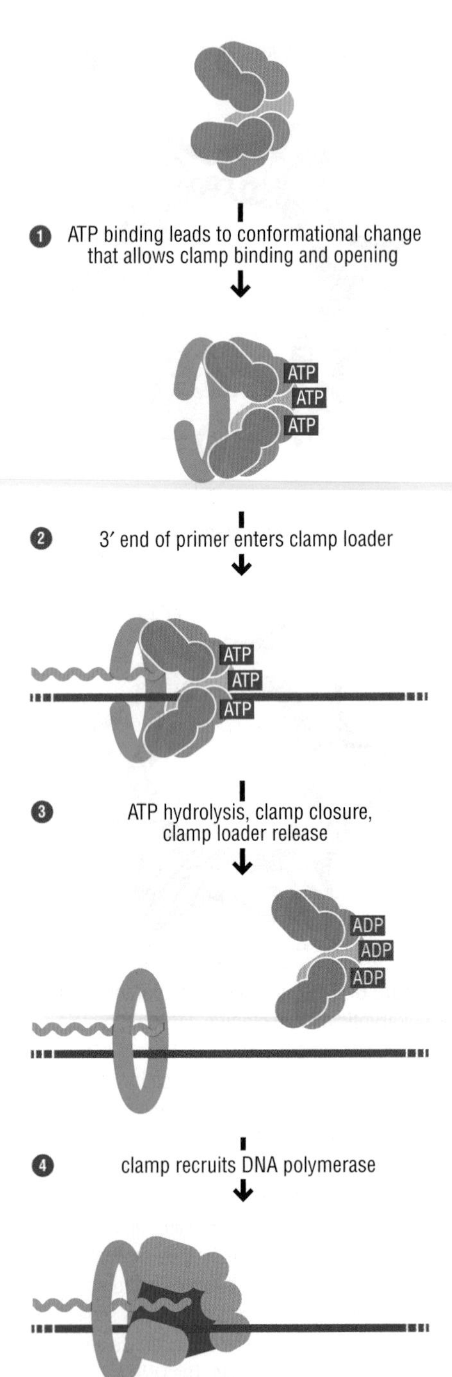

Figure 6.19 **Assembly of the sliding clamp by the clamp loader onto primer–template DNA.** (1) The binding of ATP to the clamp loader (blue) leads to its association with the sliding clamp (orange) and opening of the sliding clamp. (2) The opened clamp loader–sliding clamp has high affinity for the 3′ end of a primer–template junction in the DNA. (3) DNA binding triggers ATP hydrolysis, leading to clamp closure and the release and recycling of the clamp loader. (4) The sliding clamp is now loaded on the primer–template junction and can recruit DNA polymerase (purple) to the site to begin elongation.

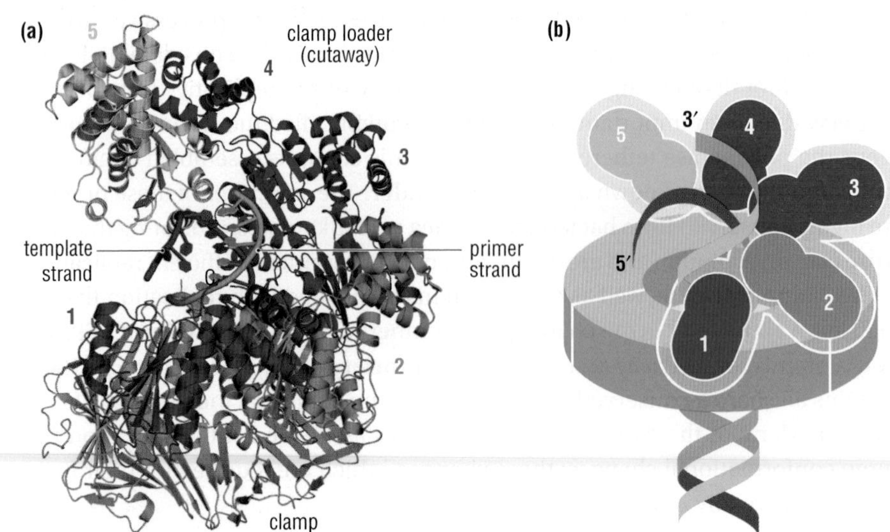

Figure 6.18 **Structure of the eukaryotic clamp loader.** (a) Structure of a co-crystal of the RFC clamp loader and the PCNA sliding clamp, with DNA modeled into the structure (PDB codes 1SXJ and 3GLF). PCNA is shown in orange, and each subunit of RFC is shown in a different color. The RFC clamp forms a helix that matches the twist of the DNA. (b) The RFC/PCNA/DNA complex, showing each of the five RFC subunits and the manner in which the DNA is threaded through the complex.

From Bowman GD, O'Donnell M, Kuriyan J. Structural analysis of a eukaryotic sliding DNA clamp–clamp loader complex. *Nature*, 2004;**429**(6993):724–30.

The clamp loader assembles the sliding clamp onto DNA

The sliding clamp, which exists as a ring, must somehow be loaded onto the DNA when needed for replication. This requires that the sliding clamp be opened, DNA be inserted into the center, and the clamp be allowed to close again. This is accomplished by the clamp loader, which also has a multiprotein ring structure. The core of the clamp loader in both bacteria and eukaryotes consists of five protein subunits as illustrated in Figure 6.18. In *E. coli* it is composed of three copies of a protein subunit called γ, one δ subunit, and one δ' subunit. In eukaryotes, the clamp loader is known as **replication factor C (RFC)** and consists of five different protein subunits, called Rfc1–5.

Some subunits of both the bacterial and eukaryotic clamp loader are AAA+ ATPases, and the cycle of sliding clamp loading to DNA is driven by conformational changes brought about by the binding of ATP and its subsequent hydrolysis. In the absence of ATP, the clamp loader has low affinity for the sliding clamp. Upon binding ATP, however, the clamp loader undergoes a conformational change, which allows it to bind and force open the sliding clamp, as depicted in Figure 6.19. In bacteria, the δ subunit interacts with and opens the sliding clamp, while, in eukaryotes, the Rfc1 subunit is involved in ring opening.

The clamp loader-sliding clamp complex, which assumes a spiral shape rather than a closed circle, has high affinity for the primer template junction - the region encompassing the 3′ end of the newly synthesized primer as it associates with the template DNA. In eukaryotes, the complex engages with 3′ end of the short sequence of DNA made by the pol α-primase complex. Binding of the clamp loader-sliding clamp complex to this DNA 3′ end stimulates ATPase activity, which leads to another conformational change that closes the sliding clamp and releases the clamp loader from the DNA. The sliding clamp is thus ready to bind DNA polymerase at exactly the right position for the polymerase to begin elongation (see Figure 6.19).

On dissociation of the adenosine diphosphate (ADP; the product of ATP hydrolysis), the clamp loader can be recharged with ATP and the cycle repeated. We shall see later that although the clamp loader is released from the sliding clamp and the DNA, it remains associated with the polymerase.

6.7 ORIGINS AND INITIATION OF DNA REPLICATION

The proteins required for DNA replication must be assembled into a replication complex and be loaded onto the DNA in a highly coordinated fashion. Each step in DNA replication must occur at the proper time and only when the previous steps are completed. This is accomplished by the successive recruitment of proteins to DNA and close regulation of their activity, which also helps guarantee the accuracy of replication. In this and following sections, we discuss in more detail the molecular mechanisms adopted to make DNA replication happen, before moving on to consider how regulation is achieved.

Replication begins at discrete sites on the chromosome

The initiation of DNA replication requires the binding of initiator proteins to replication origins – the sites at which the DNA helix is initially opened (see Section 4.8). Although origins of replication in some organisms contain specific DNA sequence elements, in other organisms it is the binding of initiator proteins to chromatin that marks a site as a DNA replication origin.

The initiator proteins in both bacteria and eukaryotes are members of the AAA+ family of ATP-binding proteins (see Section 6.6). In *E. coli*, the initiator protein is a single-subunit protein called **DnaA**, which self-associates into a helical multisubunit complex when bound to ATP. In eukaryotes, the initiator is a protein complex called the **origin recognition complex (ORC)**. As described in Experimental approach 6.2, this complex was first identified in the budding yeast *Saccharomyces cerevisiae*, where it has six essential subunits, Orc1–6. Five of these (Orc1–5) are members of the AAA+ family. There are homologs of ORC proteins in all eukaryotes; despite the differences in origin structure, it appears that all eukaryotes require the ORC for origin activation.

As we will describe later, the binding and hydrolysis of ATP is used to regulate the initiation process in both bacteria and eukaryotes. ATP binding allows DnaA to oligomerize and bind to the bacterial origin. Although the mechanism for ORC is different, its activity is also dependent upon ATP binding.

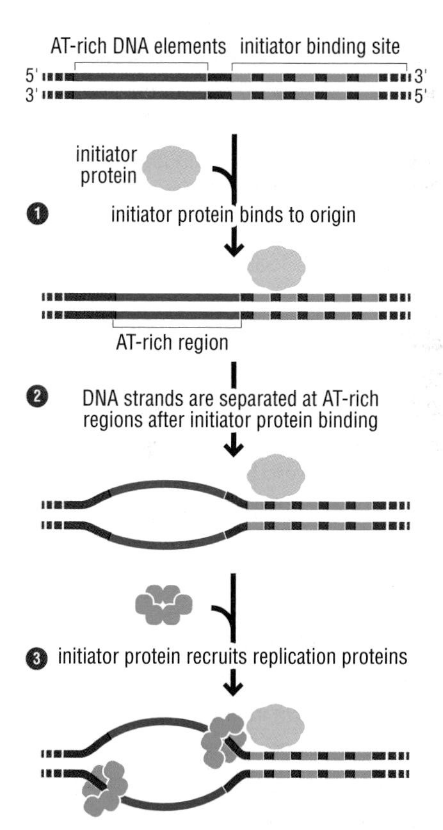

Figure 6.20 Initiation of replication occurs at specific sites in DNA. Origins of replication typically have a specific DNA sequence (orange dashes) that binds the initiator protein (green) with high affinity, adjacent to an AT-rich DNA sequence (pink). Binding of the initiator protein to its high-affinity binding site facilitates unwinding of the AT-rich region. Replication proteins, such as DNA helicase (orange), can then bind to the unwound DNA and begin to recruit other proteins for DNA replication.

The initiation of replication requires DNA unwinding and recruitment of helicases

The binding of initiator proteins to DNA specifies where replication begins. In many cases these initiator proteins bind to a well-defined DNA sequence, as depicted in Figure 6.20. Typically, an origin of replication has recognizable binding sites for the initiator proteins and an adjacent AT-rich region that facilitates unwinding. Unwinding is facilitated because A–T base-paired DNA requires less energy to separate than C–G base pairs due to there being fewer hydrogen bonds between the bases, as discussed in Chapter 1. These AT-rich regions are sometimes referred to as DNA-unwinding elements. Organisms with well-defined, specific origin

6.2 EXPERIMENTAL APPROACH
Discovery of the origin recognition complex

Biochemistry and footprinting lead to the discovery of the eukaryotic replication initiation protein

While DNA sequences that function as origins of replication had been identified in a number of systems using a variety of approaches, it was not clear what factors were localized to these regions to coordinate the activities of replication initiation. The ORC was discovered by Stephen Bell and Bruce Stillman when they set out to look for proteins that bind specifically to a DNA fragment containing a replication origin from yeast. Bell and Stillman used a DNA footprinting assay (described in Section 19.12) to identify cellular components that bound to a well-characterized yeast origin, ARS1.

Briefly, nuclear protein extract from yeast cells was incubated with a ^{32}P-labeled DNA fragment containing ARS1, and a DNase1 protection assay was used to see whether any proteins in the extract bound to the DNA. As initial footprinting results were promising, the researchers began to fractionate their nuclear extract on different column matrices to identify a fraction enriched in the binding activity of interest. Following each fractionation step, the various column fractions were evaluated for ARS1 binding activity using the footprinting assay. After four column fractionation steps, the active fraction was further subdivided by ultracentrifugation through a glycerol gradient; this approach distinguishes various proteins and protein complexes by size and shape. The proteins residing in the final fractions from the glycerol gradient were resolved on polyacrylamide gels and visualized (Figure 1a). Substantial ARS1 binding activity can be seen in fractions 18-20 (Figure 1b), correlating with the visible proteins present in those two fractions.

It turns out that the proteins found in these fractions represent the six subunits of ORC with sizes 120 kD, 72 kD, 62 kD, 56 kD, 53 kD, and 50 kD. In subsequent publications, Bell and Stillman identified these proteins by cutting them out of a protein gel like the one shown in Figure 1a, and subjecting them to peptide sequencing (described in Section 19.8). The gene encoding each of the subunits was ultimately identified, and cloned, by inferring the codon sequence from the sequence of amino acids in the protein.

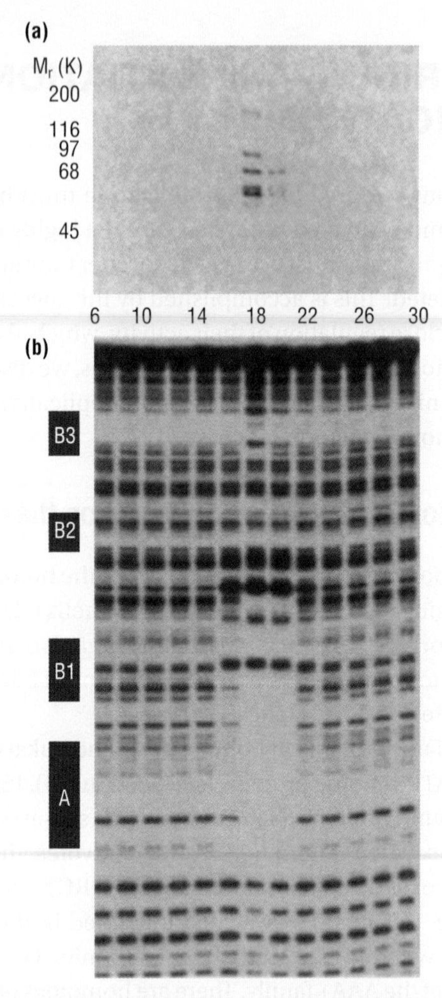

Figure 1 Footprinting approach used to identify ORC binding activity. (a) Protein gel showing the protein complex that was purified after four chromatography steps and glycerol gradient centrifugation. The even fractions from the gradient numbered 6–30 are shown; fraction 18 contains the highest level of a set of proteins that corresponds to origin recognition complex (ORC). (b) Nuclease protection (footprinting) assay showing that fractions 16 through 20 contain a DNA binding activity that protects the ARS1 sequence in a DNA fragment. The proteins are not visible in fraction 16 in the gel above because there is not enough protein present to detect by staining in the protein gel, even though there is enough to detect in the sensitive footprinting assay.
Reproduced from Bell and Stillman, *Nature*, 1992;**357**:128–134.

Recombinant protein expression allowed functional analysis of ORC complex

The final proof that this complex of six proteins is the ORC that footprints across the ARS1 sequence came from reconstitution experiments. Each of the six cloned yeast proteins was simultaneously expressed in insect cells using a recombinant protein expression system. In this way the entire ORC was able to assemble inside the cells before purification and subsequent activity assays. The researchers found that ARS1 footprint could be reconstituted when all six proteins were simultaneously expressed. If just one component was omitted, however, no footprint was seen. This reconstitution verifies that the six ORC proteins together make up the active ORC.

Subsequent experiments from Bell and Stillman and a number of other groups showed that the ORC protein complex that was identified in yeast is conserved in many other organisms. The human ORC has been reconstituted and has many of the same properties of the ORC isolated from yeast. Indeed, related ORCs function in the initiation of DNA replication in all eukaryotic species examined.

Find out more

Bell SP, Stillman B. ATP-dependent recognition of eukaryotic origins of DNA replication by a multiprotein complex. *Nature*, 1992;**357**:128-134.

Bell SP, Mitchell J, Leber J, Kobayashi R, Stillman B. The multidomain structure of Orc1p reveals similarity to regulators of DNA replication and transcriptional silencing. *Cell*, 1995;**83**:563-568.

Related techniques

DNA footprinting; Section 19.12

Protein purification; Section 19.7

SDS-PAGE; Section 19.7

Peptide sequencing; Section 19.8

sequences include *E. coli*, bacteriophage lambda, *S. cerevisiae*, and mammalian viruses.

The study of *E. coli* has revealed much about the steps by which initiation occurs. The *E. coli* origin, termed *oriC* for 'origin of chromosomal replication', consists of a 245-base sequence. This sequence contains seven 9 bp elements called DnaA boxes, which bind seven or more molecules of the initiator protein DnaA with high affinity. When fully loaded with ATP, the DnaA protein AAA+ domains multimerize into a spiral, or filament, as illustrated in Figure 6.21. This filament interacts with the DnaA boxes and wraps around the DNA strand. This interaction distorts the DNA, causing a bend in the helix with an angle of up to 40°. The DNA is then further bent by the binding of additional bacterial histone-like proteins. This binding of DnaA, and perhaps also the bending of the DNA, facilitates the local unwinding of DNA in the AT-rich region to yield single-stranded DNA. The DnaA protein may then interact directly with the single-stranded DNA to facilitate strand opening.

As well as generating single-stranded DNA at the origin, DnaA also recruits the bacterial DNA helicase, DnaB, to the origin. Since the DNA helicase forms a ring structure around a single strand of DNA, DnaB needs to be loaded onto the DNA in a manner similar to the loading of the sliding clamp discussed above. The loader for DnaB is called the DnaC protein complex. To accomplish loading, a complex of two hexameric DnaB molecules together with the helicase loader

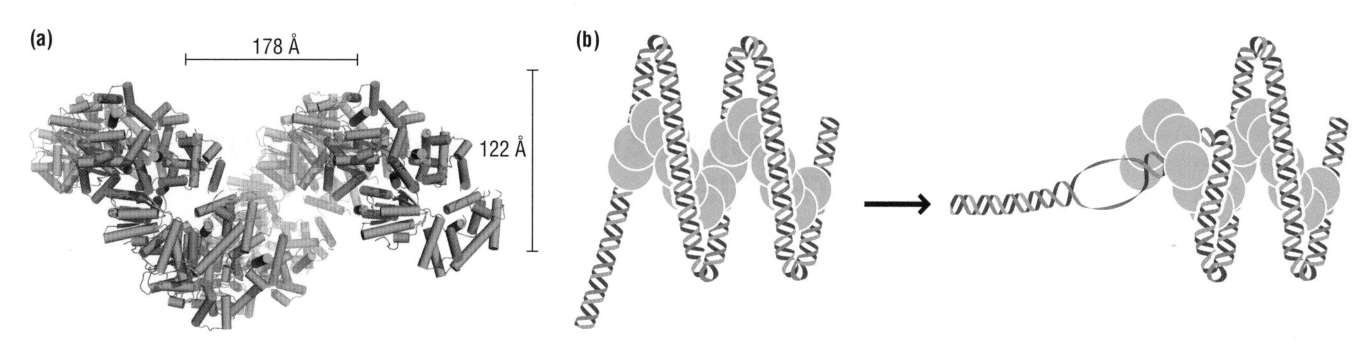

Figure 6.21 DNA wrapped around initiator protein DnaA. (a) Structure of DnaA oligomers forming a spiral around which the DNA will wrap (PDB code 2HCB). (b) Cartoon of the wrapping of the DNA around the ATP-bound DnaA and the subsequent unwinding of the origin sequence.
From Erzberger JP 2006.

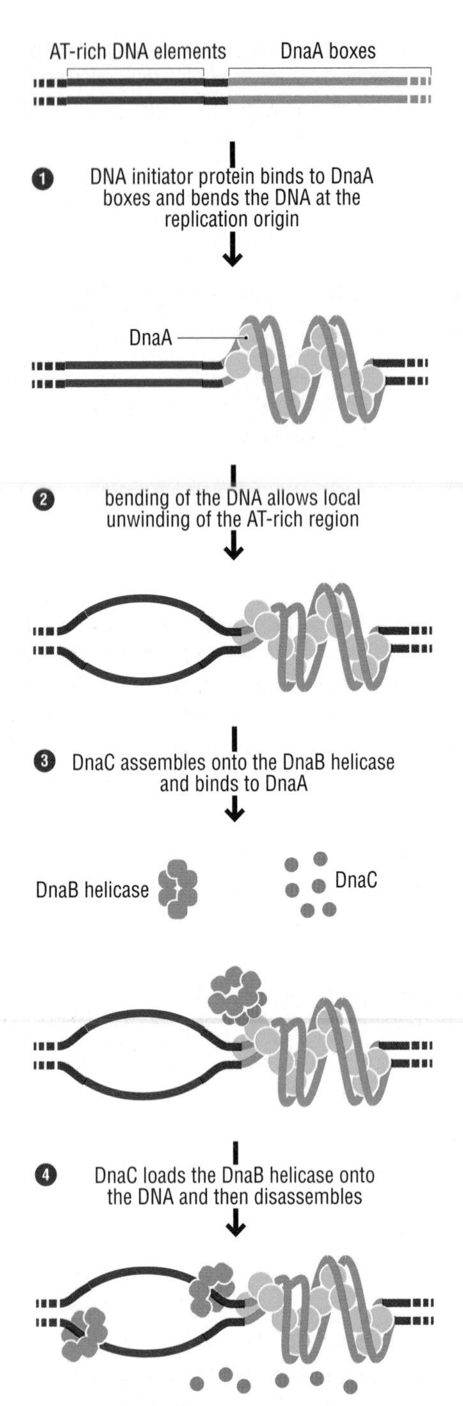
AT-rich DNA elements
DnaA boxes
1 DNA initiator protein binds to DnaA boxes and bends the DNA at the replication origin
DnaA
2 bending of the DNA allows local unwinding of the AT-rich region
3 DnaC assembles onto the DnaB helicase and binds to DnaA
DnaB helicase
DnaC
4 DnaC loads the DnaB helicase onto the DNA and then disassembles

Figure 6.22 Initiation of DNA replication at *E. coli oriC*. (1) The initiator protein DnaA binds to the high-affinity DNA elements called DnaA boxes at the *E. coli* origin. The binding of the first DnaA proteins allows the binding of additional DnaA proteins. When complexed with ATP, these proteins oligomerize into a right-handed helical oligomer. (2) The wrapping of the DNA around the protein complex and the DnaA itself facilitate local unwinding of the adjacent AT-rich region. (3) The helicase loader DnaC binds the helicase DnaB and loads it onto the single-stranded DNA. (4) DnaC then dissociates, and the origin is ready for recruitment of primase and other replication proteins.

protein DnaC interacts with DnaA bound at the origin (see Figure 6.22, steps 3 and 4). These interactions with DnaA and DnaC are essential for loading the DnaB helicase onto the single-stranded DNA at the origin. Once the helicase is loaded, primase and polymerase can be recruited.

DNA replication origins in the yeast *S. cerevisiae* have a similar general structure to that of *oriC*. A typical origin in this yeast is about 100 bp long and contains two sites, A1 and B1, at which the ORC binds; the structure of such an origin is illustrated in Figure 6.23. Adjacent to these sites are AT-rich regions B2 and B3, which also interact with the ORC and may facilitate unwinding.

As in *E. coli*, the origin DNA is wrapped around the ORC complex. ORC recruits two helicase-loading proteins Cdc6 and Cdt1, which in turn recruit the helicase Mcm2–7 (see Section 6.5). The complex of all these proteins is called the **prereplication complex** or **preRC**. As we will discuss in more detail below, assembly of this complex is not sufficient to initiate replication (unlike assembly of the initiation complex at the bacterial origin). Instead, it is loaded onto DNA well before DNA replication is actually due to begin, and the complex must be specifically activated to initiate replication. This regulation ensures that replication occurs at the correct time in the cell cycle, as we learned in Section 5.1.

Chromatin structure determines replication origins in some organisms

Whereas origins are determined by unique DNA sequences in many unicellular organisms and viruses, chromatin structure appears to play an important role in multicellular organisms. In humans and other animals, replication is reproducibly initiated in specific chromosomal regions, and yet these regions do not contain discrete, highly conserved DNA sequences as found in origins in yeast and bacteria, and mammalian viruses. Despite the lack of a specific origin sequence, however, the ORC is still required for the initiation of replication.

How does chromatin structure have its effect? In *Drosophila*, the ORC is recruited to origins where histone tails are hyperacetylated. We know that chromatin structure can determine origin activity because an artificial increase in the overall level of histone acetylation leads to an increase in origin firing (the initiation of replication from an origin). The regulation of transcription initiation also involves acetylation of histone tails. Thus, proteins such as histone acetyltransferases, which are known to regulate transcription through their effect on chromatin structure, may also play a role in regulating the usage of replication origins.

Other organisms that do not have specific, high-affinity binding sites for ORC may also make use of aspects of chromatin structure, such as histone modification, to specify origin location and usage.

6.8 LEADING AND LAGGING STRAND SYNTHESIS

DNA primase synthesizes the RNA primers that start all new DNA strands

After an origin of replication has been opened and helicases have been recruited, DNA replication can begin. As discussed in Section 6.1, the first step in DNA replication is the synthesis of primers, which are required on both the leading and lagging strands. Because of the discontinuous nature of DNA synthesis, however, priming occurs much more frequently on the lagging strand (see Figure 6.3).

A primer is a short stretch of RNA or RNA followed by a short stretch of DNA, which is synthesized by a specialized polymerase known as **primase**, as illustrated in Figure 6.24(a). Primase binds single-stranded DNA that is coated with SSB proteins. Unlike DNA polymerases, which require a 3′ OH group onto which the DNA polymerase can add nucleotides, RNA polymerases such as primase can start RNA synthesis *de novo*. On the leading strand, priming occurs only at the origin. By contrast, during discontinuous DNA replication on the lagging strand, primase must synthesize a primer for each new DNA fragment.

In bacteria, a short RNA primer of around 10–30 bases is synthesized by a primase that then hands off replication directly to pol III, the bacterial replicative polymerase. In eukaryotes, however, the primase complex contains four subunits: two subunits that function as a primase, bound in a complex with the pol α catalytic subunit and an accessory subunit 6.23(b). This pol α-primase complex first synthesizes a stretch of 10–30 nucleotides of RNA, which the poly α subunit of the complex elongates with a short stretch of DNA. After this primer synthesis, the replicative polymerases pol δ or pol ε take over and carry out the rest of the elongation phase of DNA replication. This handover from pol α-primase to pol δ or pol ε is known as **polymerase switching** and is illustrated in Figure 6.25. A polymerase switch occurs each time a new DNA strand is started on both the leading and the lagging strands.

The polymerase switch ensures that the replicative DNA polymerase is loaded at the right place. As described above, the replicative polymerase is recruited by the sliding clamp, which can only be loaded onto DNA at the primer-template junction by the clamp loader (see Section 6.6). Only after clamp-loader release can synthesis occur, because pol δ or pol ε binds to the same face of the clamp where the loader had bound. This sequence of events ensures that the DNA polymerase is loaded onto the DNA only after a primer has been made and is also in the right place to begin DNA elongation.

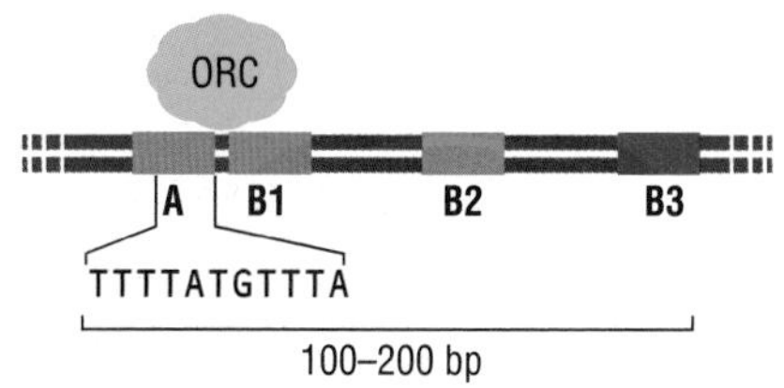

Figure 6.23 Budding yeast origin of replication. Origins of replication in *S. cerevisiae* contain two core sequences (A and B1) that together bind the ORC. The A element includes an 11 bp region called the ARS consensus, whose sequence is highly conserved among all budding yeast origins. The replication origin also contains a less conserved region (B2) that serves as the major site of DNA unwinding.

➲ We learn more about histone acetylation in Section 4.5 and more about the regulation of transcription in Chapter 9.

The lagging strand is synthesized in short stretches of DNA called Okazaki fragments

As the replication fork progresses along the chromosome, DNA is synthesized on both template strands, a process that requires two polymerases. As the two strands have opposite chemical polarities, this creates a problem: for both strands to be synthesized simultaneously, the polymerases would apparently have to synthesize one strand in the 3′ to 5′ direction and the other in the 5′ to 3′ direction. However, as described in Section 6.1, DNA polymerases cannot synthesize DNA in the 3′ to 5′ direction. Instead, the lagging strand is synthesized in the 5′ to 3′ direction, but in short discontinuous segments. This discontinuous synthesis is coupled to the movement of the replication fork, as we discuss later.

The initiation of new DNA synthesis on the lagging strand is carried out in an identical manner to that described above for the leading strand, but occurs much more frequently. The short stretches of DNA synthesized on the lagging strand are called **Okazaki fragments** (named after their discoverer); once each fragment is completed, it is joined to the previous fragment in a process called **Okazaki fragment maturation** to make one continuous DNA strand.

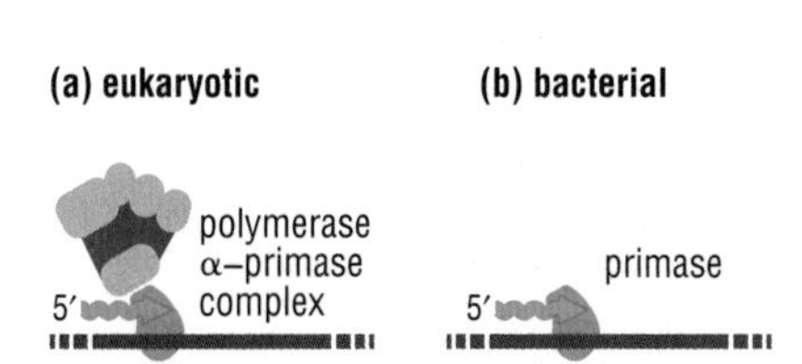

Figure 6.24 Primases catalyze the synthesis of a short RNA primer for DNA replication. (a) The eukaryotic primase is part of a pol α–primase complex. This complex is shown here in simplified form: the native eukaryotic primase (orange) is a two-subunit enzyme, as is pol α (pink). (b) Bacterial primases (orange) are single-subunit RNA polymerases. The primase from *E. coli* is also known as DnaG.

Okazaki fragment maturation generates a continuous DNA strand

Okazaki fragment maturation takes place in several steps, as illustrated in Figures 6.26 and 6.27. First, the RNA primer is removed and replaced by DNA, and then the two adjacent DNA fragments are ligated - that is, joined together (Figure 6.27).

The mechanism of Okazaki fragment maturation differs between bacteria and eukaryotes. In bacteria, pol III synthesizes DNA up to the beginning of the next RNA primer, then stops and dissociates from the sliding clamp and the DNA. The sliding clamp remains on the DNA and is involved in recruiting the proteins for Okazaki fragment maturation. Primer removal is the task of another polymerase, pol I, which possesses a 5′ to 3′ exonuclease activity that removes the RNA primer of the previous Okazaki fragment. As it is removing the RNA in front of it, pol I synthesizes

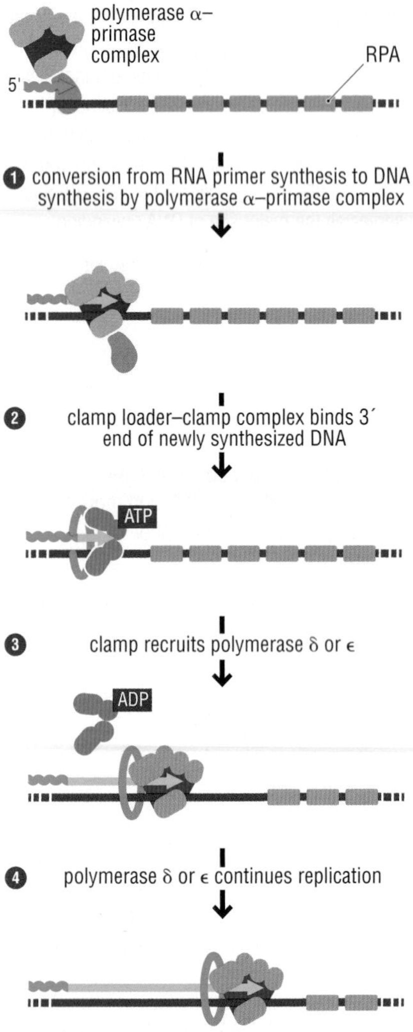

polymerase α–primase complex

RPA

5′

1 conversion from RNA primer synthesis to DNA synthesis by polymerase α–primase complex

2 clamp loader–clamp complex binds 3′ end of newly synthesized DNA

ATP

3 clamp recruits polymerase δ or ϵ

ADP

4 polymerase δ or ϵ continues replication

Figure 6.25 The polymerase switch. In eukaryotes, the RNA primer is synthesized by the pol α–primase complex (pink), which binds to single-stranded DNA coated with the single-stranded binding protein RPA (light purple). It synthesizes an initial RNA primer and then changes to DNA synthesis (step 1). Pol α is soon replaced by one of the more processive polymerases, pol δ or pol ε. The polymerase switch occurs as a result of the binding of the clamp loader (blue) onto the junction between the 3′ end of the primer and the single-stranded DNA template (step 2). The clamp loader loads the sliding clamp onto DNA (orange), which in turn recruits pol δ or pol ε (purple) (step 3), the DNA polymerases that continue DNA synthesis (step 4).

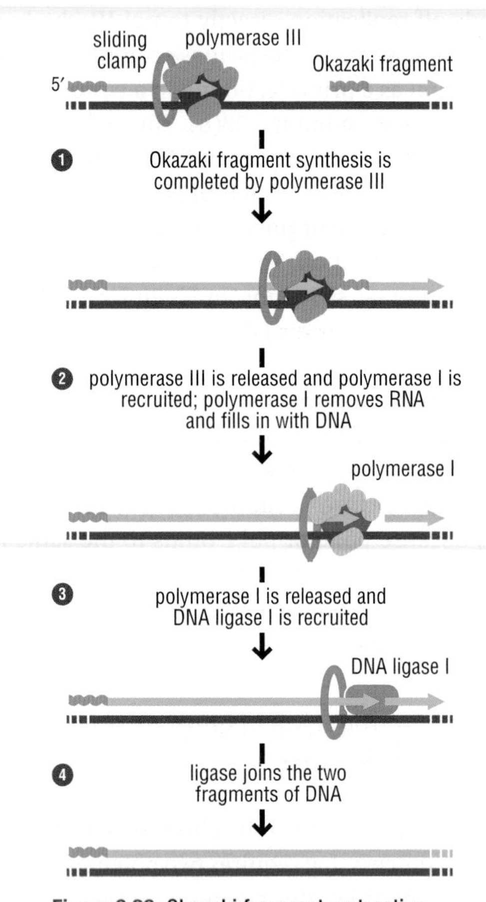
sliding clamp
polymerase III
Okazaki fragment
5′
1 Okazaki fragment synthesis is completed by polymerase III
2 polymerase III is released and polymerase I is recruited; polymerase I removes RNA and fills in with DNA
polymerase I
3 polymerase I is released and DNA ligase I is recruited
DNA ligase I
4 ligase joins the two fragments of DNA

Figure 6.26 Okazaki fragment maturation in *E. coli*. DNA polymerase III synthesizes new DNA from the 3′ end of a primer until it reaches the beginning of the primer of the previous Okazaki fragment (step 1). When polymerase III encounters the primer, it leaves the DNA and polymerase I is recruited (step 2). The 5′ to 3′ exonuclease activity of DNA polymerase I removes the RNA primer, and its nucleotide-polymerizing activity fills in the gap with DNA (step 3). The gap in the DNA backbone between the end of this fragment and the beginning of the next is joined by DNA ligase, which is initially recruited to the site by binding to the sliding clamp (step 4).

DNA to fill the gap the RNA has left. When it reaches the beginning of the next DNA segment fragment, the DNA polymerase cannot join the two adjacent DNA strands; this ligation is done by an enzyme called **DNA ligase**, which catalyzes the formation of a covalent bond between adjacent 5′ and 3′ DNA ends. This ligation reaction links the two fragments together into one continuous DNA strand (see Figure 6.26).

Okazaki fragment maturation in eukaryotes is similar in principle, but removal of the RNA primer and the filling-in with new DNA are carried out in a slightly different way. When it reaches the RNA of the adjacent fragment, pol δ or pol ε continues synthesizing DNA rather than stopping. This results in the end of the strand that was ahead of the polymerase being lifted like a flap and displaced as the polymerase runs into it. This strand displacement occurs with the assistance of several polymerase accessory proteins. A protein called Fen1 then uses its endonuclease activity to cleave the partly displaced single-stranded DNA, or flap, as depicted in Figure 6.27. DNA ligase is then recruited to join the two strands.

Under some circumstances, Fen1 cleavage does not occur right away, and a long, displaced single strand is created. In this situation, Dna2, which has both helicase and endonuclease activity, first cleaves the long flap, which is followed by further processing by Fen1 and finally ligation.

DNA elongation and Okazaki fragment maturation all take place in the context of the replication complex at the replication fork, which is moving along the DNA and synthesizing both the leading strand and lagging strand simultaneously. In the next several sections, we will examine how this coordinated synthesis is achieved.

➔ We learn more about pol I in Section 6.2.

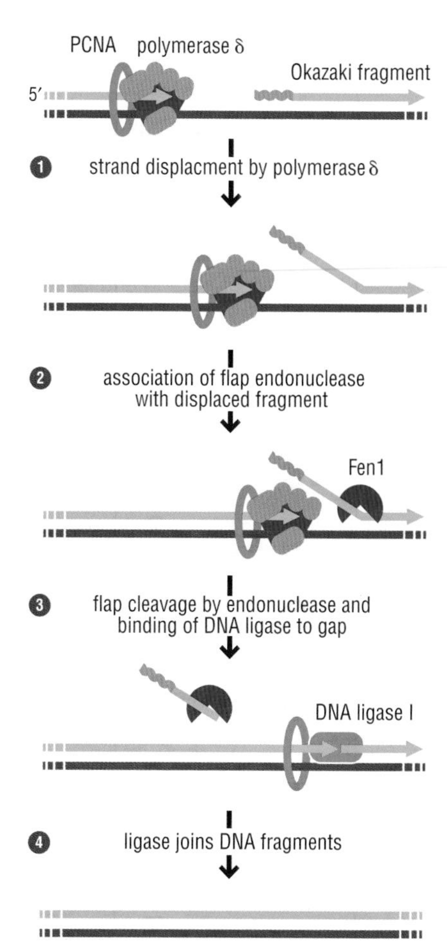

Figure 6.27 Okazaki fragment maturation in eukaryotes. Pol δ synthesizes DNA up to and past the 5′ end of the adjacent Okazaki fragment, displacing a single-stranded 5′ tail that contains the RNA primer (step 1). The Fen1 endonuclease then cleaves the flap (steps 2 and 3), leaving a simple nick in the DNA backbone that is sealed by DNA ligase (step 4). PCNA is the eukaryotic sliding clamp.

6.9 THE REPLICATION FORK

Leading and lagging strand synthesis is coupled at the replication fork

We see in Section 6.1 how DNA synthesis occurs at replication forks that move outwards from origins of replication. The general structure of a replication fork is depicted in Figure 6.28. While leading strand synthesis proceeds continuously from 5′ to 3′, with the strand elongation in the same direction as the moving fork, lagging strand synthesis is discontinuous, with strand elongation in the direction opposite to that of the fork (Figure 6.28). Yet, strikingly, the movement of the leading and lagging strand polymerases are coordinated at the fork. To achieve this coordination, the lagging strand must loop around to allow both polymerases to travel together in the direction of fork movement (Figure 6.29). The discovery of this looping is described in Experimental approach 6.3.

The coordination of leading and lagging strand polymerases allows the replication of both strands to be regulated together. If leading strand replication stalls, lagging strand replication will also be halted. For example, if damaged DNA is encountered on one strand, the entire fork complex will halt and wait for repair.

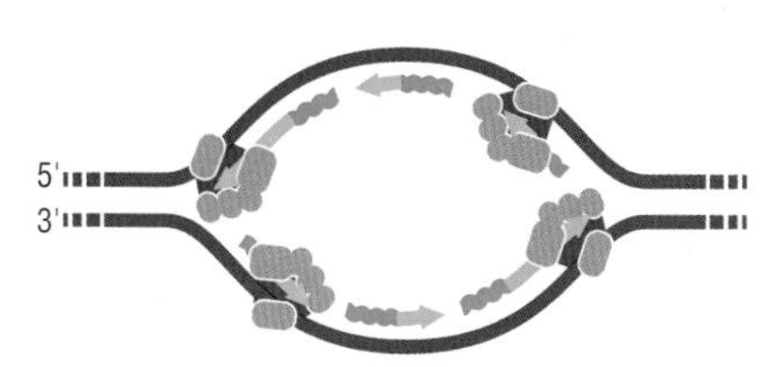

Figure 6.28 The replication fork. A replication bubble showing the two polymerases on the leading and lagging strands. In fact the polymerases on the leading and lagging strands travel together in the same directions as shown in Figure 6.29.

Leading and lagging strand synthesis in *E. coli* is coordinated by the replisome

The coupling of leading and lagging strand polymerases is best understood for bacterial replication complexes, as depicted in Figure 6.29. In *E. coli,* the multisubunit assembly of DNA polymerases, sliding clamps, and clamp loader is called the **polymerase III holoenzyme** or the **replisome**, which moves as a whole as the fork

6.3 EXPERIMENTAL APPROACH

Coupling of polymerases at the replication fork: the trombone mode

Functional studies reveal coupling of two polymerases at the replication fork

The architecture of the DNA replication fork is both elegant and complex. Early biochemical studies by Bruce Alberts and colleagues showed that leading and lagging strand polymerase activities are coupled at the fork, raising a topological conundrum: how can polymerization go in two directions at the fork? On the basis of the biochemical data, a model was proposed whereby the template DNA of the lagging strand is looped out, allowing the two polymerases to effectively proceed together in the same physical direction, with the lagging strand polymerase continually reinitiating the extension reaction on an available template. This model was referred to as the trombone model because of the similarity between the looped-out DNA as imagined by this model and the slider on a trombone; the model is illustrated in Figure 1.

Electron microscopy allows visualization of the proposed trombone model

Electron microscopy and a biochemically reconstituted system eventually allowed for direct visualization of this proposed structure. To set up such an experiment, all of the proteins known to be required for DNA synthesis and fork movement were purified and incubated *in vitro* with DNA template in conditions where active replication would occur. The preparation was then fixed and concentrated, and spread on a special carbon grid on which the DNA and protein complexes could lie flat. The grid was then coated with fine tungsten powder, which allowed the contrast in different surfaces to be visualized.

In the micrograph in Figure 2a, a circle is being replicated to generate a long double-stranded copy (so-called rolling circle replication). The double helix of DNA of the template circle can be seen as a wavy garden hose winding around on the grid. The polymerase complex, or replisome, sits at the junction between this helix and the circle; there is a looped region that represents the trombone structure adjacent to the circle. This loop is generated by the coupled polymerase as depicted in Figure 2b. Subsequent microscopy work in which specific components were individually labeled and imaged demonstrated unambiguously that each of the expected fork proteins are found in the large visible protein complex.

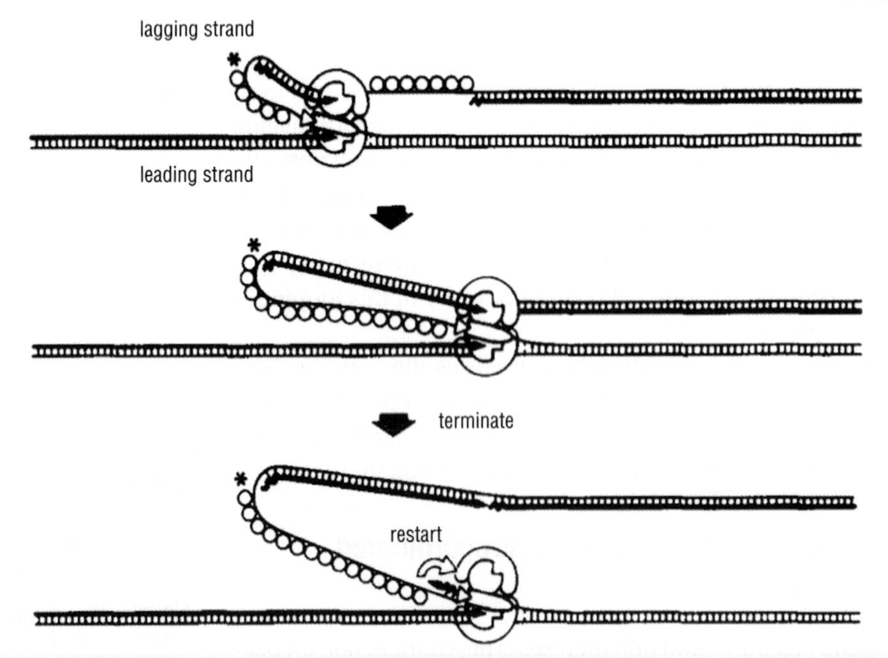

Figure 1 Trombone model for replication forks. Biochemical experiments indicated that both the leading and lagging strand polymerases are coupled together during replication. Such coupling of the leading and lagging strands was predicted in this model to cause looping of the lagging strand DNA to allow both polymerases to travel together.

Reproduced from Alberts, *Trends in Genetics*, 1985;**1**:26–30.

(a)

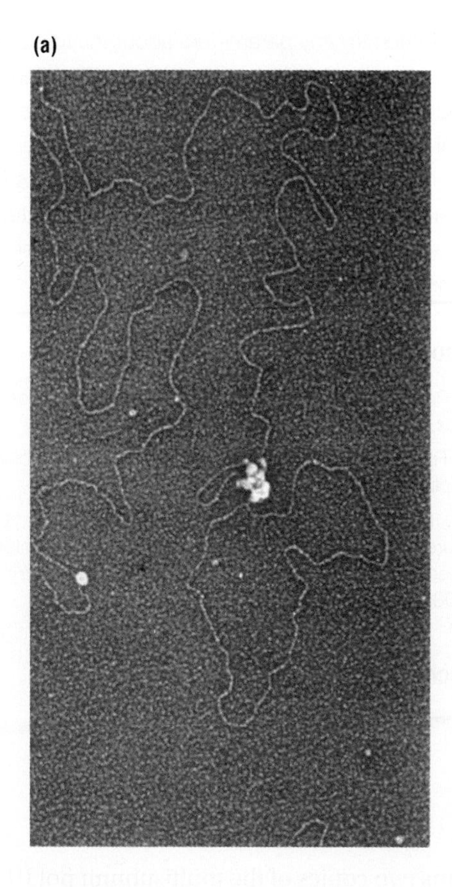

(b)

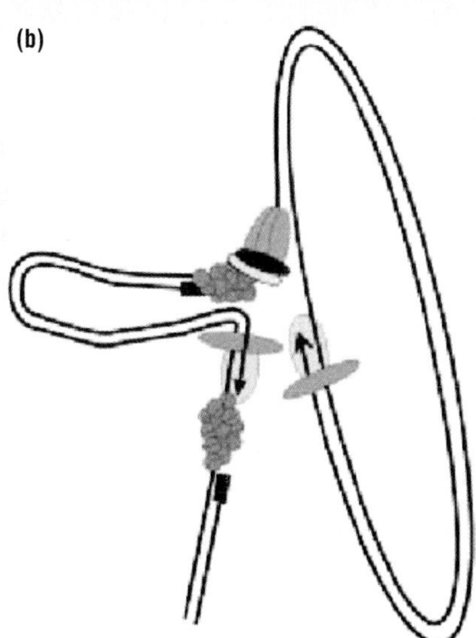

Figure 2 Visualization of trombone looping. (a) Electron micrograph of rolling circle replication showing the replisome and the looped out lagging strand of DNA. (b) Interpretation of the electron micrograph.

Reproduced from Chastain *et al.* Architecture of the Replication Complex and DNA Loops at the Fork Generated by the Bacteriophage T4 Proteins. *Journal of Biological Chemistry*, 2003;**278**:21276–21285

(a)

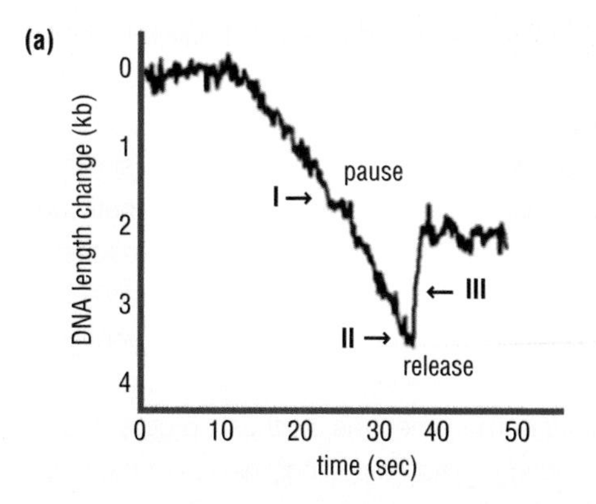

(b)

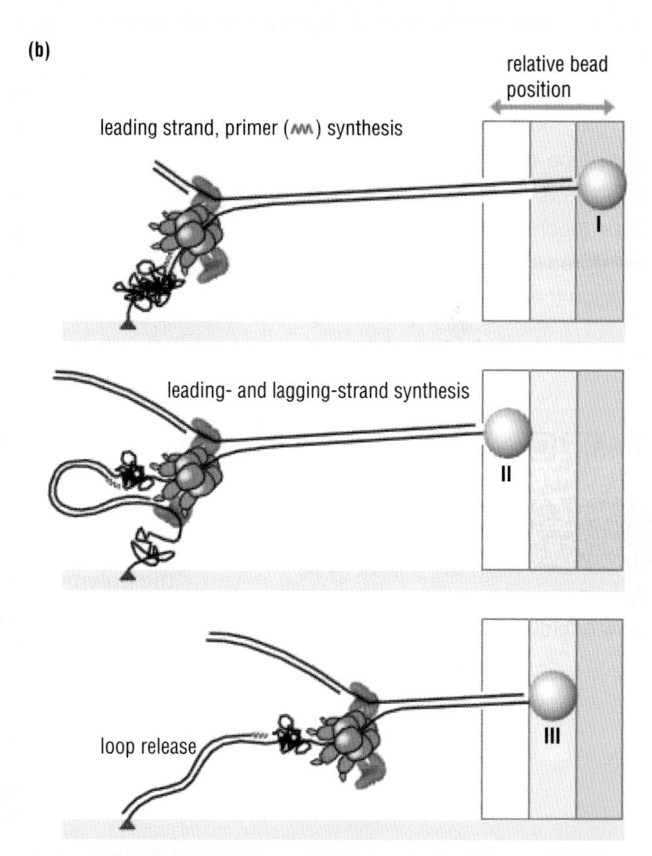

Figure 3 Single molecule experiments show dynamics of replication fork progression. (a) A tracing of the position of a single molecule of DNA with an assembled replication fork. The DNA is tethered to a slide at one end and the position of the other end is monitored over time. As the polymerization reaction proceeds initially there is no change in the length of the DNA. As a loop grows, however, the length of the DNA shortens as some of the DNA is incorporated into the loop. However upon release of the looped out strand the DNA abruptly gets longer. (b) Diagram of the experimental set up used and interpretation of the data shown in (a).

Reproduced from Lee *et al. Nature*, 2006;**439**:621–624.

Single molecule experiments can measure the forces at replication fork and further verify the polymerase coupling

The power of such techniques that visualize individual molecules has recently been adapted to answer detailed functional questions about replication fork activity. Using powerful optical microscopes it is possible to visualize single DNA molecules in an ongoing biochemical reaction. In one study, a DNA molecule was chemically tethered at one end to a glass while the other end was attached to a magnetic bead. Replication fork components were next assembled on the DNA (by adding the required purified proteins), and buffer was flowed across the DNA—protein solution, effectively stretching the DNA and pushing the bead end away.

Up to 50 individual DNA molecules could be followed simultaneously by precisely measuring the position of each magnetic bead over time. The replication reaction was initiated simultaneously on all molecules at a particular time (by the addition of dNTPs, for example) and the position of the bead was followed over time as each fork synthesized DNA (see Figure 3b). The change in bead position represents fork movement as synthesis progresses; the *abrupt* changes in bead position represent release of the looped lagging strand DNA. By following the movement of individual molecules, these investigators obtained a representative picture of how the process is occurring at the molecular level.

The stochastic nature of specific steps such as loop release can be measured as differences in the timing of the release step. One example of such a single molecule tracing is shown in Figure 3. If all molecules were averaged, as they are in typical 'bulk' biochemical experiments, instead of single molecules being monitored it would not be possible to see the specific steps in the reaction that may vary.

With this system, many parameters about the role of each protein component of the fork can be assayed by altering the reaction conditions. For example, omitting the helicase blocks the extension of the fork, indicating that unwinding is essential for fork movement. In the experiment shown here, the authors found that lagging strand release from the replisome is a regulated process and that the loop length is regulated by the primase-mediated initiation of lagging strand synthesis.

Find out more

Alberts BM. Protein machines mediate the basic genetic processes. *Trends in Genetics*, 1985;**1**:26–30.

Lee JB, Hite RK, Hamdan SM, *et al*. DNA primase acts as a molecular brake in DNA replication. *Nature*, 2006;**439**:621–624.

Nossal NG, Makhov AM, Chastain PD 2nd, Jones CE, Griffith JD. Architecture of the bacteriophage T4 replication complex revealed with nanoscale biopointers. *Journal of Biological Chemistry*, 2007;**282**: 1098–1108.

Related techniques

Electron microscopy; Section 19.13

Scan here to watch a video animation explaining more about how the leading and lagging strands are coordinated during replication, or find it via the Online Resource Center at **www.oxfordtextbooks.co.uk/orc/craig2e/**

elongates. The replisome contains two copies of the multisubunit pol III together with sliding clamps, and a clamp loader that continually reloads sliding clamps on the lagging strand. One polymerase replicates the leading strand, and one replicates the lagging strand as the fork progresses along the double helix.

The polymerases are linked together via two subunits of a protein called tau (τ), which is associated with the clamp loader, and which also links the complex to the helicase. One of the functions of this large replication complex is to keep a lagging

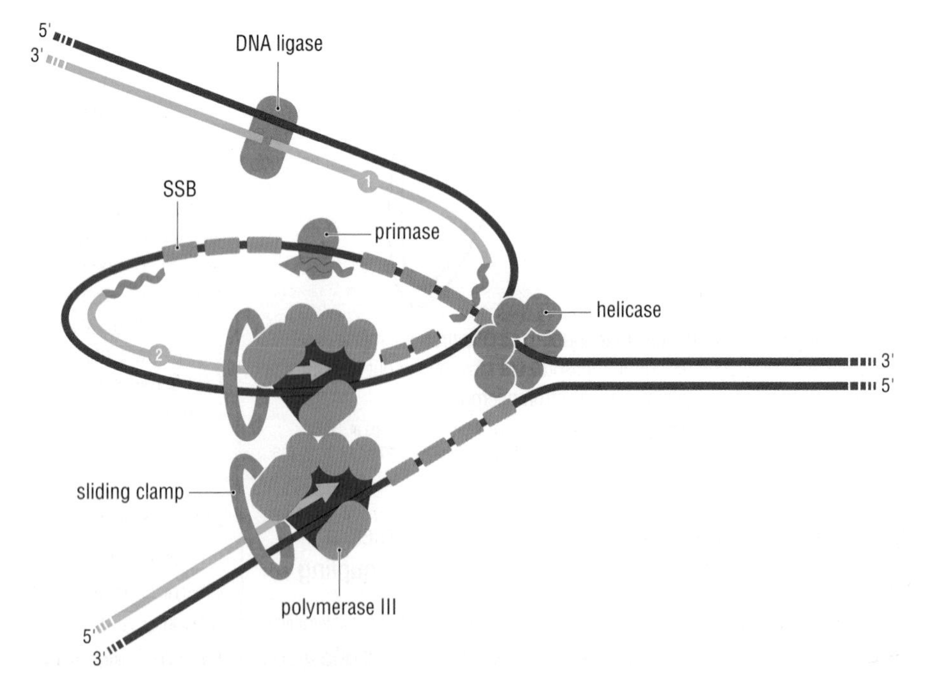

Figure 6.29 Replication fork coupling in *E. coli*. The leading and lagging strands are coupled by looping. The lagging strand is shown looped with a completed Okazaki fragment (1) and one pol III molecule almost completing the synthesis of the next Okazaki fragment (2). A new primer is being synthesized by primase on the unwound lagging strand. The leading strand is being synthesized continuously, and the whole fork is moving to the right. The replisome that moves with the fork contains two associated pol III molecules bound to the sliding clamp and linked by the tau protein (not shown).

strand polymerase associated with the fork even though it is being released from the DNA at the end of each Okazaki fragment. This association of polymerase, clamp loader, and sliding clamp means that the loading of a new clamp and a polymerase onto DNA at the beginning of each Okazaki fragment can be efficiently coordinated with the progress of the polymerase on the leading strand.

A bacterial replication fork moves at a rate of around 750–1000 bp per second, with a new Okazaki fragment being produced every 1–2 kb; so every second or two, the clamp loader is putting a new clamp onto a newly primed 3′ end and a polymerase is re-engaging with a clamp and DNA.

The eukaryotic replication complex shares many features of the *E. coli* complex

The basic mechanism of replication fork progression in eukaryotes is very similar to that described above for *E. coli*. Both of the replicative polymerases – pol δ and pol ε – are present at the replication fork, with pol ε synthesizing the leading strand and pol δ synthesizing the lagging strand. It is thought that leading and lagging strand synthesis are coupled in eukaryotes, as they are in *E. coli*. However, unlike the replisome of *E. coli*, the eukaryotic replication proteins do not seem to exist in a single, large complex: biochemical experiments show that the components do not co-purify as the *E. coli* proteins do. In addition, no protein that links the two polymerases, equivalent to bacterial tau, has been identified. Therefore, when we refer to the eukaryotic replication complex, we are really considering a collection of proteins working in concert, but not necessarily as a tightly bound complex.

6.10 TERMINATION OF DNA REPLICATION

We have explored the process by which DNA replication is initiated, and how two daughter strands are synthesized using the original double-stranded parent as a template. We now consider what must happen to complete replication – the process of termination.

Termination of replication occurs at a specific site in bacterial chromosomes

For the circular chromosomes of bacteria, termination of DNA replication occurs when the two replication forks moving away from the single origin meet on the opposite side of the circle. As the forks approach each other, the replication complex must disassemble, and the two growing strands must be joined.

In *E. coli* this termination occurs at a DNA region called *ter* (see Section 4.9) on the opposite side of the circular chromosome from the replication origin, *oriC*. As the two replication forks that initiate at *oriC* progress at the same pace around the chromosome, they will meet near *ter*, which contains specific protein binding sites. The *ter* element is bound by a replication terminator protein known as Tus in *E. coli*. When the forks approach each other, one fork arrests when it is stopped by the bound Tus protein. When the other fork arrives it too arrests. At this point, disassembly of the forks presumably occurs, leaving a short length of single-stranded DNA uncopied. This gap is assumed to be filled in by a specialized polymerase, pol I, which is also used in DNA repair, as discussed in Chapter 16. The molecular details of how the fork is disassembled to allow repair synthesis are not yet clear.

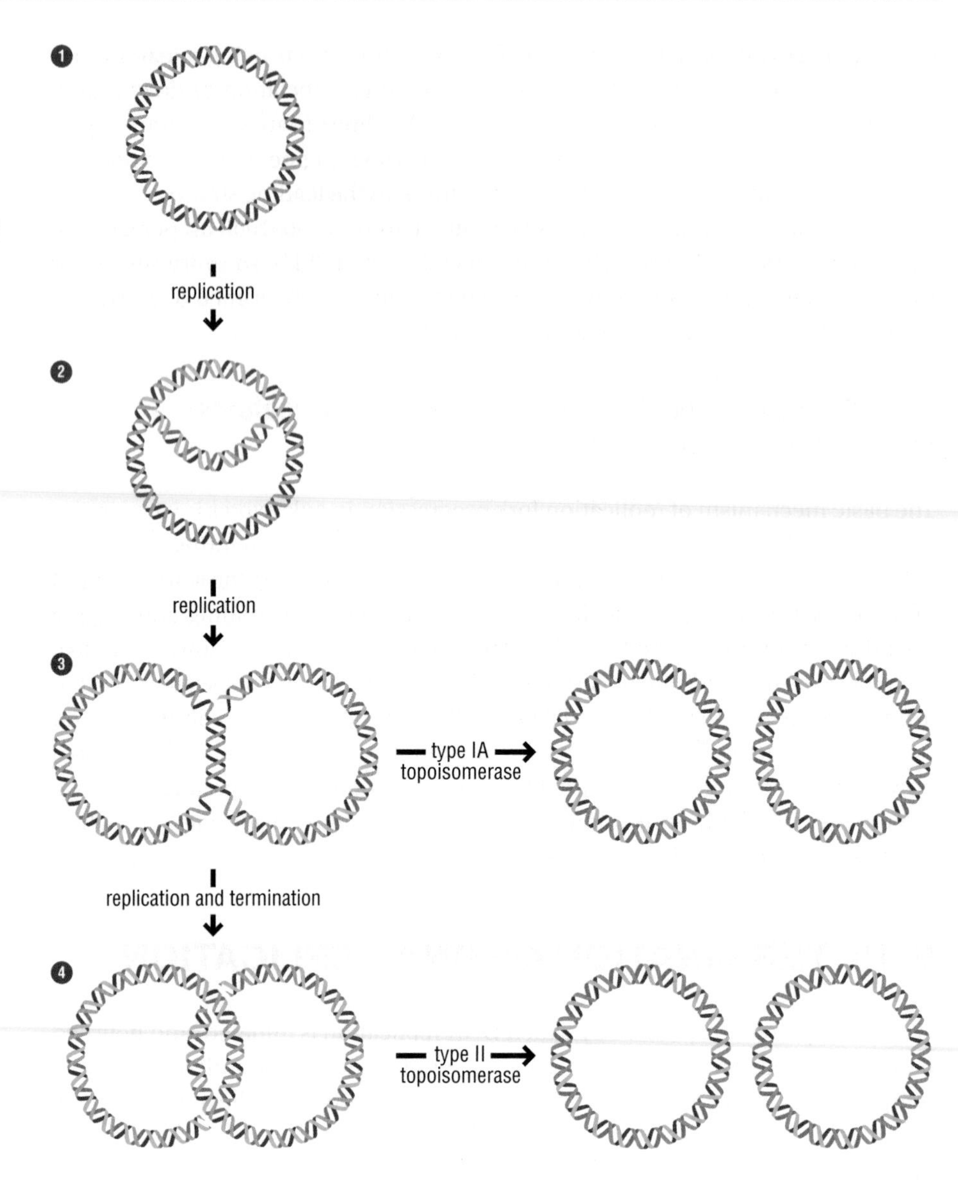

Figure 6.30 Replication of circular chromosomes results in daughter molecules that are still linked to each other. Replication of a circular DNA double helix results in two interlinked or catenated double-stranded daughter molecules. The circular chromosomes must be unlinked or decatenated by type 1A topoisomerases at step 3 or type II topoisomerases at step 4 to allow separation.

In circular bacterial chromosomes or circular viral genomes, the intertwining of the double helix results in the interlinking of the two new daughter DNA molecules, as illustrated in Figure 6.30. The two circular DNA molecules must then be resolved to allow the duplicated chromosomes to separate. These daughters are separated in a process called **decatenation**, which involves strand breakage and passage by either type IA or type II topoisomerases (see Section 6.5). Breakage and strand passage of the single-stranded DNA by type IA topoisomerase before replication is complete will unlink the almost-complete molecules shown in Figure 6.30, step 3. By contrast, double strand DNA breakage and passage by type II topoisomerases can unlink the fully replicated daughter molecules shown in Figure 6.30, step 4.

Replication does not usually terminate at specific sites in eukaryotic chromosomes

The existence of multiple replication origins in eukaryotes means that there is no one specific site where termination occurs, in contrast to the situation in bacteria. Instead, termination can occur anywhere on the chromosome. We do not yet know how the two converging replication forks interact to allow their disassembly and assure the completion of DNA synthesis between them. There are, however, some regions in eukaryotic chromosomes where replication is blocked in one direction by a specialized structure. In budding yeast, for example, there is a specific termination site termed the **replication fork barrier** in the cluster of repeated genes encoding ribosomal DNA (rDNA). A protein called FOB1 binds to a specific DNA sequence at the fork barrier and blocks progression of replication forks coming from one direction. This ensures that the repeated rRNA genes are always replicated in the same direction; if a fork approaches from the other direction, it is stopped. This mechanism may be specific to sites of repeated genes and may help to prevent errors in copying the repeated DNA sequences. Alternatively it may limit polymerase to replication in one direction to prevent collisions between DNA polymerase and RNA polymerase on the heavily transcribed rRNA genes.

Due to the double-helical nature of DNA, replication of linear chromosomes is also translated into helical intertwining of the two finished daughter molecules as we described earlier for *E. coli.* Although eukaryotic chromsomes are linear, the long bulky nature of the chromosome does not allow for passive untangling from the ends; so, topoisomerases are also required for the daughter chromosomes to be separated, as depicted in Figure 6.31.

In contrast to internal origins, replication termination at the *end* of a linear chromosome can occur by the polymerase simply copying to the end of the molecule. When this happens, however, it poses a problem: how can the polymerase replicate all of the nucleotides at the very end of the DNA sequence? We explore the answer in the next section.

Figure 6.31 Converging forks create interlinked DNA on linear chromosomes. As two forks converge and replicate the region of DNA between them, the double-helical nature of DNA is translated into helical intertwining of the two finished daughter molecules. This intertwining is then resolved by topoisomerase before the daughters are separated.

6.11 THE END-REPLICATION PROBLEM AND TELOMERASE

The end of a linear DNA molecule cannot be completely replicated by the replication complex

The mechanism of DNA replication described so far in this chapter allows faithful copying of the vast majority of chromosomal DNA. For organisms with linear chromosomes, however, it results in the non-replication and eventual loss of some sequence from the ends of the DNA molecules. This loss occurs because lagging strand synthesis cannot complete the copying of the very end of a linear DNA molecule. This is known as the **end-replication problem** and, after repeated rounds of replication, can lead to loss of sequence from chromosome ends.

Incomplete replication is often described as arising from RNA primer removal. However, in principle, it can occur in several ways. The removal of the last Okazaki fragment synthesized before the replication complex runs off the end of a DNA template strand will leave a single-stranded region of DNA, as depicted in Figure 6.32. Alternatively, the dissolution of the replication fork when it reaches the

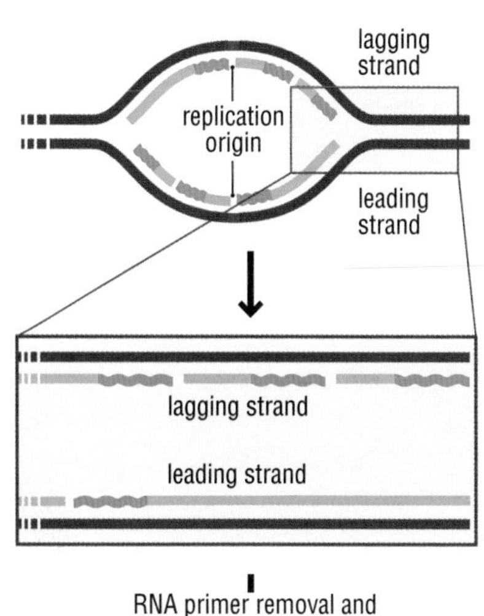

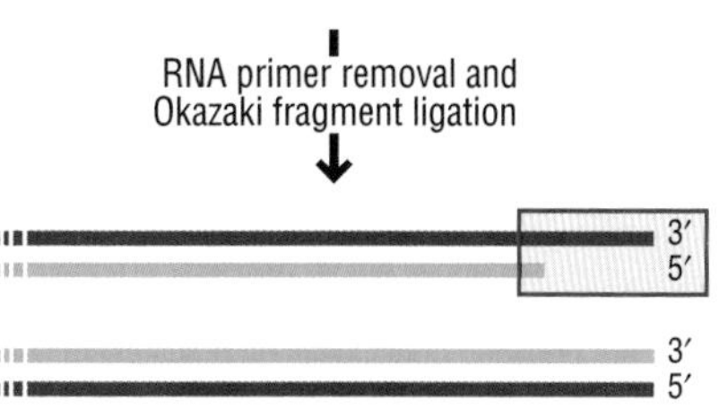

Figure 6.32 Inability to fill the gap left after the removal of the last RNA primer is one cause of the end-replication problem. After initiation of replication, leading and lagging strand synthesis proceeds to the end of the chromosomal DNA. During replication, the RNA primers between adjacent Okazaki fragments are removed and replaced with DNA, and the fragments are ligated to eventually produce one long continuous strand. The terminal RNA primer on the lagging strand has, however, no adjacent fragment to ligate to, and so once it is removed, the lagging strand is shorter. When this shortened strand is replicated at the next round of DNA synthesis, the sequence at the very end of the chromosome will be permanently lost.

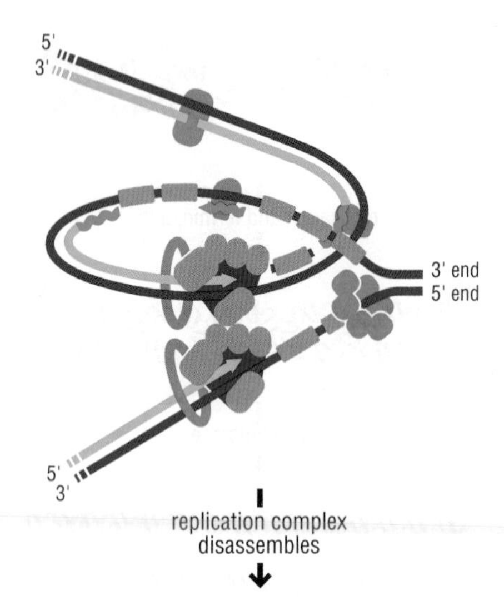

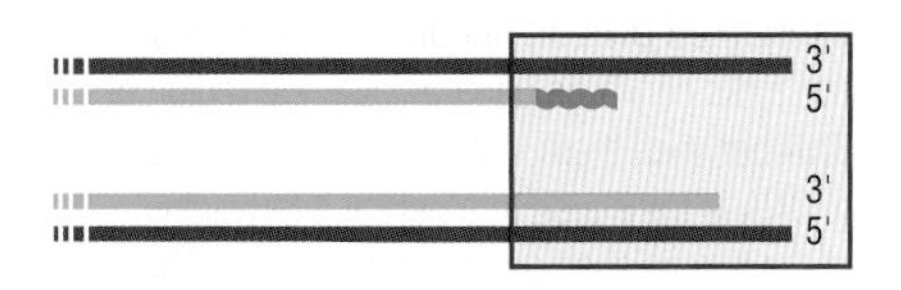

Figure 6.33 Dissolution of the replication complex may lead to the loss of the terminal Okazaki fragment. A eukaryotic replication fork is shown with the single-stranded DNA bound by SSB. The replication fork machinery disassembles when it reaches the end of a DNA molecule. Because the lagging strand is looped out to allow coordinated leading and lagging strand synthesis, the last Okazaki fragment begun may not be completed before the fork disassembles. This leads to loss of sequence on the lagging strand. A small region on the leading strand may also not be copied if the fork collapses before it reaches the end of the chromosome. Note that because this is a eukaryotic replication fork, the helicase is on the leading strand.

end of the leading strand may result in premature termination of DNA synthesis on the looped-out lagging strand and may also lead to incomplete copying of the leading strand, as shown in Figure 6.33.

The end-replication problem can be solved by several different mechanisms

Cells and viruses have several different mechanisms for overcoming the end-replication problem, as summarized in Figure 6.34. In circular chromosomes, such as those of most bacteria, the problem is avoided altogether. Viruses with linear genomes have evolved several different mechanisms to replicate the ends of their DNA. For example, bacteriophage T4 strings several copies of its genome together before replication to reduce the number of ends. Another mechanism is found in the vaccinia virus. Its double-stranded DNA genome has the two strands covalently linked to each other at each end of the DNA, creating hairpins that are replicated; and then the daughter chromosomes are processed to generate two complete genomes. In bacteriophage ϕ29 and mammalian adenovirus, replication is primed at the very end of the molecule by a terminal protein rather than an RNA primer. By initiating from a protein, there is no terminal RNA primer to remove, so the full DNA sequence can be maintained.

Most eukaryotes, in contrast, overcome the end-replication problem in an entirely different way. As described in Section 4.11, eukaryotic chromosomes have telomeres at their ends, which consist of simple sequence repeats (see Figure 4.35). The length of the whole tract of telomere repeats varies from less than 100 bp to more than 20 000 bp, depending on the species. The bulk of the telomere sequence is replicated by the replicative polymerases as they move along the double-stranded telomere DNA, and, for the reasons described above, the ends of the telomeres grow shorter at each round of replication.

Telomeres can, however, be elongated by the enzyme telomerase independently of the standard DNA replication machinery. This enzyme adds specific telomere DNA sequences *de novo* onto the chromosome ends: when the telomeres shorten, telomerase is recruited to add telomere repeats. This results in net telomere elongation and allows telomeres to be maintained around an equilibrium length, as described in Section 4.11 (see Figure 4.37). In *Drosophila*, telomere length is maintained by yet another mechanism:

Mechanisms of overcoming the end-replication problem	
mechanism	**example**
circular chromosome	bacterial chromosome, plasmids
end recombination	mitochondrial genomes, bacteria
transposition to telomeres	*Drosophila* telomeres
genome concatamerization	T4 phage
hairpin ends	vaccinia virus
terminal protein primer	adenovirus, phage ϕ29
telomere repeat addition	eukaryotic chromosomes

Figure 6.34 Mechanisms for overcoming the end-replication problem. The table lists several different mechanisms that are used by a variety of organisms to avoid loss of sequence at the ends of a DNA molecule when it is replicated.

Transposable elements called Het and TART transpose onto the ends of the telomeres to maintain their length. Transposable elements are described further in Chapter 17.

Telomerase contains a specialized reverse transcriptase and an essential RNA component

Telomerase is a unique DNA polymerase that is made up of a catalytic protein and an integral RNA molecule. The RNA contributes to enzyme function and also provides the template for the synthesis of telomere repeats. In addition to these two universal conserved subunits, there are also a variety of species specific accessory proteins in different organisms that are part of telomerase.

The catalytic protein of telomerase is called telomerase reverse transcriptase (TERT) and is conserved between distantly related eukaryotes. Telomerase has similar catalytic domains to those in other polymerases, with conserved motifs that make up the palm, fingers and thumb domains. Mutation of the conserved residues that coordinate the two magnesium molecules in the palm domain inactivates the enzyme, indicating that it uses a similar catalytic mechanism to that of other polymerases.

➔ We learn more about other specialized polymerases in Section 6.4.

The essential RNA component of telomerase varies in size from 150 nucleotides in *Tetrahymena*, to 450 in humans, to over 1300 nucleotides in yeast. Though the sequence between these different RNAs is not conserved, its overall secondary structure is. In addition to providing a small stretch of template RNA, specific structures within the telomerase RNA are required for telomerase activity. Mutations in the RNA in sites distant from the template region render the enzyme inactive, indicating that these residues play an as yet undefined role in catalysis.

Telomerase adds telomeric repeats processively

Telomerase binds to single-stranded telomeric DNA at the very end of the chromosome in a manner that allows base-pairing of the DNA 3′ end with the template sequence of the telomerase RNA, as shown in Figure 6.35. The enzyme then uses the RNA as a template to add telomeric sequences to the 3′ end of the telomeric DNA. Once the end of the template is reached, the newly synthesizes DNA 3′ end is realigned with the beginning of the RNA template while the enzyme remains associated with the telomere DNA in a translocation step. The telomerase then makes another copy of the template, but this time adding to the 3′ end of the new DNA. As such, the translocation event, and the repeated synthesis from a single template, allows telomerase to processively add hundreds of nucleotides onto a telomere before it dissociates.

6.12 CHROMATIN REPLICATION

Chromosome replication requires duplication of both the DNA and the protein components of chromatin

In all organisms, the chromosomes contain not only a DNA molecule but also specific proteins bound to the DNA. In eukaryotes, DNA is packaged into nucleosomes

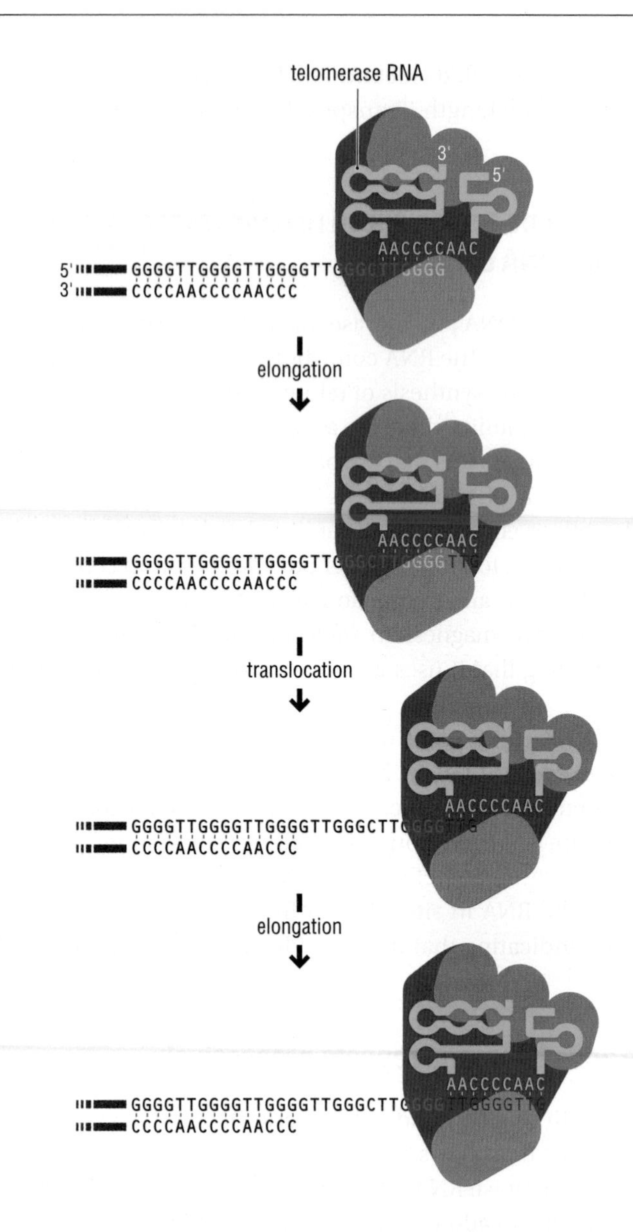

Figure 6.35 Elongation of telomeres by telomerase. Telomerase binds to the single-stranded GT-rich DNA at the end of a telomere. The 3′ OH end of the DNA is base-paired with the RNA template region (the sequence in green) in telomerase. In the elongation step, telomere repeats are copied using the template as a guide and are added to the DNA (shown as red bases). Once the end of the template is reached, the telomerase translocates forward to reposition the new DNA 3′ end with the RNA template for a second round of elongation. Telomerase does not dissociate from the DNA during translocation and can processively add hundreds of base pairs of new telomere sequence.

as described in Section 4.3. Therefore when we consider replication of the DNA we need to consider the context of how that DNA is packaged.

To maintain chromosome function, not only must the DNA be copied, but also specific proteins must be recruited to both daughter molecules after the replication fork passes, to generate faithful copies of both the DNA and the protein components of the parental chromosome. In addition, modifications to the DNA such as methylation must be re-established on both daughter molecules after replication. The re-establishment of the chromatin structure allows the epigenetic inheritance of heterochromatic and euchromatic regions in the chromosomes, as introduced in Chapter 4 and described further in Chapter 8.

➔ We discuss the packaging of DNA into nucleosomes in Section 4.3 and DNA methylation in Section 4.7.

Old histones present ahead of the replication fork are recycled onto newly replicated DNA

To replicate chromosomal DNA in eukaryotes, the replication fork must disrupt the nucleosomes on the parental DNA, separate the DNA strands and then reassemble

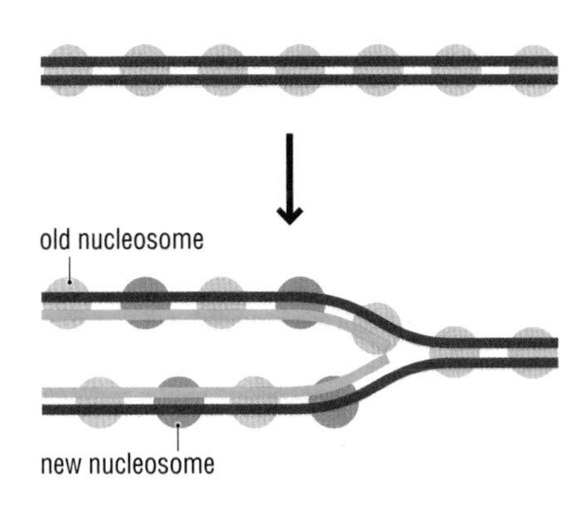

Figure 6.36 Histone segregation. Chromosome replication in eukaryotes proceeds through the duplication of chromatin as well as DNA. As the replication fork passes though chromatin, half of the old H3 and H4 core histones (light balls) go to one daughter DNA helix and half to the other; the full nucleosome is then reconstituted behind the fork. Newly synthesized H3 and H4 core histones (dark balls) are then deposited on both daughter molecules to maintain a ratio of one nucleosome approximately every 200 bp. The inheritance of old histones H3 and H4 by each DNA daughter molecule allows for the modification state of the histones in any given region of DNA to be inherited by both daughter chromosomes. The new nucleosomes are presumed to be modified according to the modification state of their neighbors.

nucleosomes on both daughter DNA molecules behind the fork. Remarkably, the histones that were bound in front of the replication fork rebind behind it, maintaining the modification state of the chromosomal region. As the fork proceeds, the modification state of the parental chromatin is thus passed on by a redistribution of the old nucleosomes between the two daughter DNAs in a process called **parental histone segregation**, which is depicted in Figure 6.36.

As we saw in Section 4.5, histones are subject to a number of modifications including methylation, acetylation and phosphorylation, and modifications differ between heterochromatin and euchromatin. During chromatin replication, the old H3 and H4 core histones are segregated to each daughter, and a reconstituted nucleosome is assembled with H2A and H2B. This segregation of the old, modified histones to both daughter strands is thought to provide a guide to the reconstruction of the parental chromatin state in each daughter chromosome.

The distribution of old histones to both new DNA molecules is carried out by the active transfer of H3 and H4 by a protein complex that contains the histone chaperone Asf1. Asf1 binds a half-nucleosome containing a tetramer of two H3 and two H4 histones and guides its addition onto DNA. This segregation of old nucleosomes means that, immediately behind the fork, each of the two daughter DNA strands has only half the number of nucleosomes required to package the DNA. New nucleosomes must be assembled, and these are added directly behind the fork by the **chromatin assembly factor (Caf1)** together with Asf1. The newly synthesized and deposited histones are surrounded by old, modified histones, which are thought to recruit histone-modification enzymes that rapidly modify the new histones according to the state of the old histone neighbors to which they are adjacent.

➔ We discuss histone-modification enzymes in Section 4.5.

Replication-dependent histones are synthesized and modified during replication just before their assembly onto DNA

To generate the large number of histone proteins needed during replication to make new nucleosomes, histone synthesis is tightly linked to DNA synthesis. The cell has a number of different histone variants (see Section 4.3), but only the major histones - H2A, H2B, H3, and H4 - are synthesized during replication to provide the building blocks for new nucleosome assembly. They are therefore known as the **replication-dependent histones** or **S-phase histones**.

Histone assembly into nucleosomes occurs stepwise by the formation of subcomplexes, which then assemble into the full nucleosome on the DNA. First, histones H2A and H2B form dimers, and independently two subunits each of histones H3 and H4 assemble into a stable tetrameric complex. The H3–H4 tetramer is first deposited on the DNA, after which two H2A–H2B dimers are added to complete the histone octamer, as described in Section 4.3.

➔ We learn more about PCNA in Section 6.6.

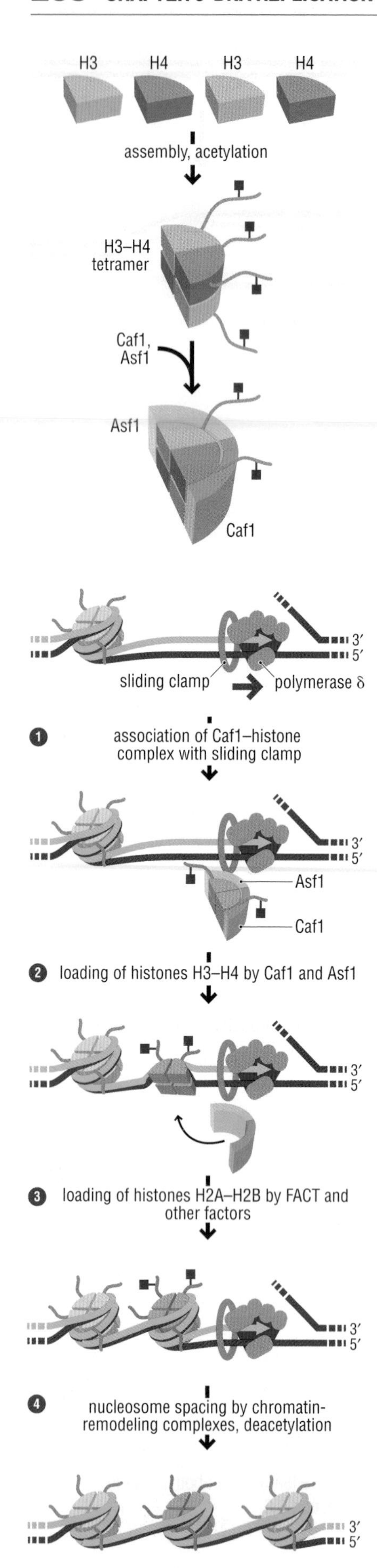

Figure 6.37 Caf1 assembly of nucleosomes. The newly synthesized and acetylated histone H3 and H4 proteins, in association with Asf1, assemble into a half-nucleosome. This acetylated half-nucleosome binds to the Caf1 protein. Caf1 together with Asf1 then facilitates histone placement on DNA.

Histones are deposited behind the replication fork by the chromatin assembly factor

Newly synthesized S-phase histones H3 and H4 are marked by the acetylation of lysine residues in their N-terminal tails. The specific lysine that is acetylated differs from those acetylated in areas of active chromatin: in humans, *Drosophila* and yeast, lysine 5 and lysine 12 in newly synthesized histone H4 are acetylated.

These acetylated histones are bound by the chaperone proteins Asf1 and Caf1 as described above; Figure 6.37 depicts how the latter then deposits them on DNA. Caf1 is a complex of three protein subunits that binds tightly to newly synthesized, acetylated histone H3–H4 tetramers and interacts with the sliding clamp protein PCNA to target the H3–H4 tetramer to the DNA behind the replication fork, and, together with Asf1, facilitate assembly of the H3–H4 tetramer into a nucleosome behind the replication fork (Figure 6.38). After loading of H3–H4 tetramers, the addition of newly synthesized histone H2A–H2B dimers into the nucleosome is assisted by additional histone chaperone proteins. Once the histones are deposited on DNA, they are deacetylated, ready to be modified by enzymes thought to be recruited by the adjacent old modified histones.

6.13 REGULATION OF INITIATION OF REPLICATION IN *E. COLI*

The concluding sections of this chapter brings together all that we have seen in the chapter so far to examine DNA replication as a carefully coordinated, tightly regulated sequence of events.

The initiation of replication is tightly regulated

DNA replication must be coordinated with cell growth and cell division to ensure accurate cell duplication. The regulation of DNA replication is controlled at initiation in both bacteria and eukaryotes. Eukaryotes contain multiple replication origins along the chromosome, whereas bacteria contain a single origin. In addition, the processes controlling cell division differ in the two types of organisms. So we can see how the regulation of DNA replication in these two groups differs significantly. In this section, we discuss the mechanisms that regulate the firing of the

Figure 6.38 Caf1-mediated incorporation of H3–H4 tetramers into replicating DNA. To incorporate nucleosomes with newly synthesized H3 and H4 histones (1), Caf1 and Asf1 bound to an acetylated H3–H4 half-nucleosome interact with the sliding clamp protein, PCNA (2), and deposits the H3–H4 tetramer on the DNA. Other chromatin-assembly factors including FACT (for facilitates chromatin transcription; not shown) then assemble the H2A–H2B tetramer together with the H3–H4 tetramer to make a full nucleosome (3). Nucleosome spacing is adjusted by chromatin-remodeling complexes, and the histones are subject to other modifications such as deacetylation (4). Caf1 together with Asf1 only loads H3–H4 tetramers onto the newly synthesized DNA behind the replication fork. Only one strand of the fork is shown for simplicity.

E. coli origin, *oriC*, as an example of regulation in bacteria; in the following Section 6.14 we will discuss the regulation of eukaryotic origins.

The initiation of replication in *E. coli* is regulated in several ways that not only activate the origin and cause it to fire, but also block it from firing again until sufficient time has passed for cell division to occur. However, in *E. coli* these latter mechanisms can be bypassed under certain circumstances to allow very rapid reinitiation and faster cell division.

There are several independent mechanisms that serve to control the initiation of DNA replication in bacteria: regulatory inactivation of DnaA (RIDA); binding of SeqA to *oriC* to block both DNA methylation and DnaA binding; the existence of the *datA* DNA site that titrates out initiator protein; and a DARS DNA sequence that mediates recycling of DnaA-ADP to DnaA-ATP. We consider each of these in turn.

ATP binding and hydrolysis regulates the binding of initiator proteins to the origin of replication

The RIDA mechanism is central to the blocking of replication because it involves the hydrolysis of DnaA-bound ATP to DnaA-ADP and the consequent inactivation of the DnaA initiator activity (see Section 6.7). So why is the hydrolysis of ATP important? As discussed in Section 6.6, the *E. coli* initiator protein DnaA is a member of the AAA+ ATPase family and will only multimerize to initiate replication at *oriC* when it is bound with ATP, as illustrated in Figure 6.39. The binding and hydrolysis of ATP helps to establish the irreversibility of the initiation process. After binding to *oriC*, the ATP-bound DnaA unwinds the DNA for replication initiation. After DNA unwinding, the β sliding clamp recruits another AAA+ protein, called Hda, which stimulates the ATPase activity of DnaA, causing the ATP bound by DnaA to be hydrolyzed to ADP. This ADP-bound form of the molecule cannot multimerize and thus cannot initiate origin firing, effectively blocking reinitiation.

SeqA binding to hemi-methylated DNA helps to block the reinitiation of replication

While the RIDA mechanism described above is considered to be the major negative regulatory step that prevents initiation, there are also other mechanisms that prevent reinitiation: one mechanism involves the methylation of the DNA sequences at *oriC*. Bacterial chromosomal DNA is methylated on the A residue at **GATC sites** by the enzyme DNA adenine methylase (Dam methylase). As described in Section 4.7, both strands are usually methylated. Immediately after replication, however, only one strand, the parental strand, will be methylated until Dam methylase can modify the new strand (see Figure 4.23). Hence, the DNA is said to be **hemi-methylated**.

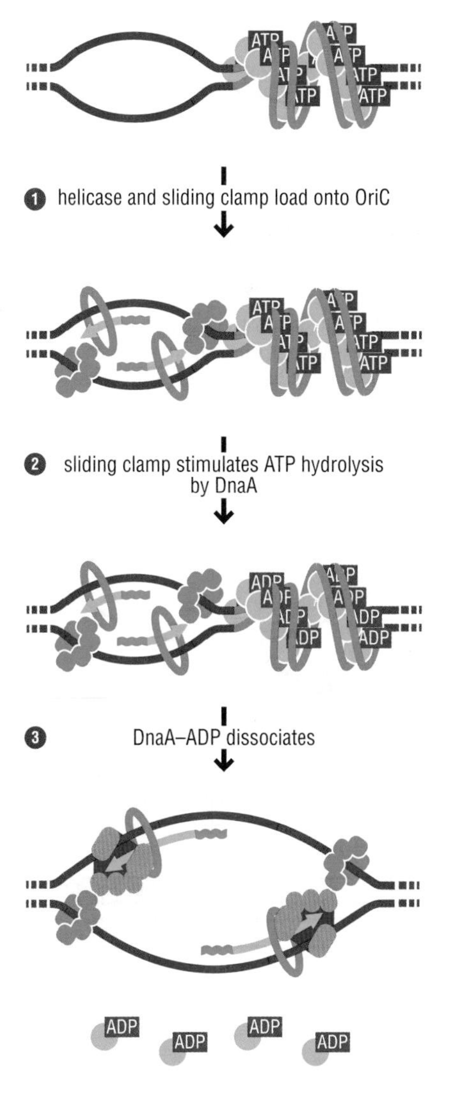

1 helicase and sliding clamp load onto OriC
2 sliding clamp stimulates ATP hydrolysis by DnaA
3 DnaA–ADP dissociates
ADP
ADP
ADP
ADP

Figure 6.39 ATP binding of DnaA regulates *oriC* firing. DnaA protein bound to ATP oligomerizes and binds to DnaA sites (orange) in *oriC*. The DNA wraps around the oligomer to initiate unwinding as described in Figure 6.21. The DnaB helicase is loaded by the DnaC protein (not shown) onto the unwound DNA. Primer synthesis occurs, and the sliding clamp and pol III are recruited. The interaction of the sliding clamp, pol III, and the Hda protein (not shown) with ATP-bound DnaA stimulates hydrolysis of ATP. The ADP-bound DnaA no longer oligomerizes nor binds DNA and thus dissociates from the DNA. This is one of several levels of regulation that prevent premature refiring of *oriC*.

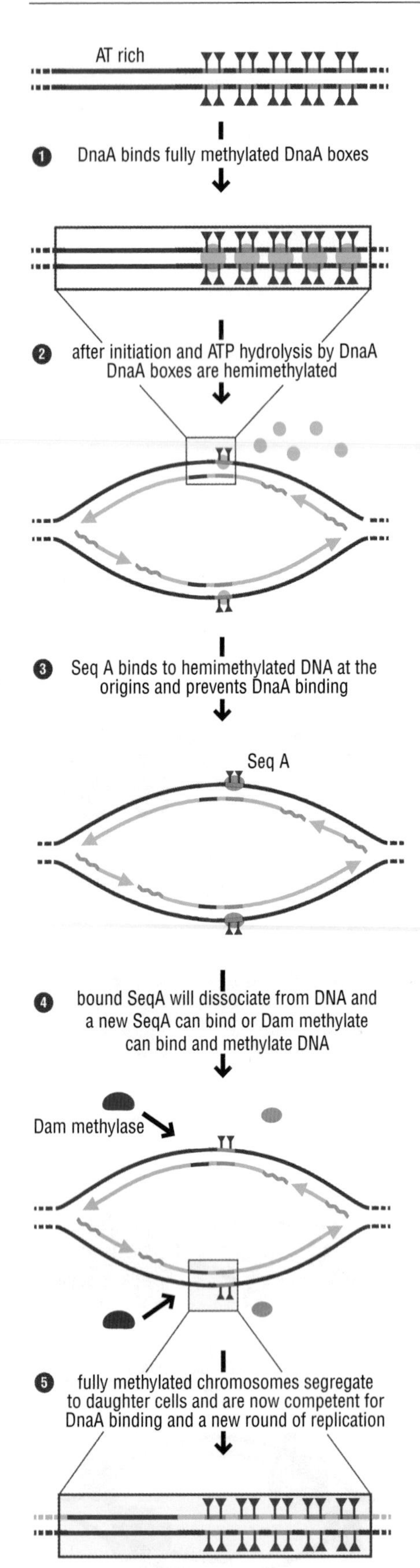

AT rich
1 DnaA binds fully methylated DnaA boxes
2 after initiation and ATP hydrolysis by DnaA DnaA boxes are hemimethylated
3 Seq A binds to hemimethylated DNA at the origins and prevents DnaA binding
Seq A
4 bound SeqA will dissociate from DNA and a new SeqA can bind or Dam methylate can bind and methylate DNA
Dam methylase
5 fully methylated chromosomes segregate to daughter cells and are now competent for DnaA binding and a new round of replication

Figure 6.40 *oriC* contains GATC sites for methylation by Dam methylase. Before replication, the GATC sites overlapping DnaA boxes within the *oriC* DNA are fully methylated (blue triangles). ATP-bound DnaA (green balls) binds to these sites (orange) and allows initiation, as described in Figure 6.39. For a short time after initiation, the GATC sites in the newly replicated DNA are in the hemi-methylated state, as the newly synthesized DNA strand has not yet been methylated. The SeqA protein binds to these hemi-methylated sites and prevents DnaA rebinding. After some time, the SeqA will dissociate and allow full remethylation of the GATC sites by Dam methylase so that replication initiation can occur again in the new daughter cell. SeqA also binds the hemi-methylated promoter of the *dnaA* gene and prevents synthesis of new DnaA (which is rapidly converted to DnaA-ATP) until the *dnaA* promoter is fully remethylated.

There are 11 GATC sites in the *oriC* DNA sequence that overlap the DnaA binding sites. Figure 6.40 shows how the DnaA protein will bind to the fully methylated binding sites. By contrast, a protein known as SeqA specifically binds only to the hemi-methylated, but not fully methylated, GATC sites. Immediately after replication begins, SeqA can bind to the hemi-methylated DNA in the origin and blocks the binding of DnaA. It also blocks the access of methylase to these sites and thus inhibits the methylation of the new strand. This block only lasts a short time, however, as SeqA dissociates from the hemi-methylated DNA, allowing Dam methylase to regain access and fully methylate the new strand; full methylation then prevents rebinding of SeqA. The replicated origin is usually fully methylated about ten minutes after initiation, so SeqA binding can only temporarily block DnaA binding and origin re-firing.

DnaA binding to *datA* DNA sequences sequesters the initiator protein after initiation

A third mechanism that prevents reinitiation is the binding of the initiator protein DnaA to a 1 kb region of DNA called *datA* (for DnaA titrator) to the right of *oriC*. This region has sites that bind DnaA with very high affinity, and can sequester about 25% of the free DnaA molecules in the cell. After replication - but before cell division - there will be two copies of this region in the cell, which together sequester excess DnaA, making the protein unavailable for rebinding to *oriC*. When segregation of chromosomes occurs, there is just one copy of *datA* per cell again, and new DnaA is synthesized, promoting initiation once more.

The DARS DNA sequence mediates recycling of DnaA-ADP to DnaA-ATP

Although the three processes just mentioned prevent replication and promote the formation of DnaA-ADP, DnaA-ATP must again become available for the next round of initiation. This is achieved by the new synthesis of DnaA, which is rapidly converted to DnaA-ATP. In addition, the accumulated DnaA-ADP can be recycled to DnaA-ATP. Two DnaA binding sites outside of *oriC*, called DARS (DnaA reactivating sequences), serve as cofactors to stimulate the exchange of ADP bound to DnaA to ATP.

Initiation can take place before cell division is complete when *E. coli* cells are growing rapidly

In *E. coli*, the timing of initiation is tightly coordinated with growth rate. Under different nutrient conditions, cell division may occur as rapidly as every 20 minutes or as slowly as every 150 minutes. The elongation rate for copying DNA is constant, and it takes about 50 minutes to replicate the bacterial chromosome, whatever the growth conditions. To enable a more rapid rate of cell division in good growth conditions, *E. coli* initiates a second round of chromosome replication before the first round is complete. Therefore, when cells are growing slowly, there is one copy of *oriC* per cell before initiation and two copies after initiation. In contrast, when cells are dividing rapidly, there can be four to eight copies of *oriC* in a single cell.

Initiation occurs on every origin in the cell synchronously, and so the number of origins is always a power of two. This synchronous initiation is due to the four levels of regulation of initiation described above. Perturbation of any one of these mechanisms results in reinitiation, leading to an odd number of origins per cell. The timing of initiation is proportional to cell size, and it is thought that the ratio of DnaA molecules bound to ATP versus those bound to ADP is the primary regulator of reinitiating in fast-growing cultures. Thus, with adequate nutrients there will be a higher level of ATP-bound DnaA, which promotes initiation. Although this origin reinitiation in *E. coli* is well regulated, it is in contrast to the situation in eukaryotes where the firing of origins more than once per cell division is tightly blocked, as we will see in Section 6.14, below.

➔ We discuss the regulation of bacterial origin firing further in Chapter 7.

6.14 REGULATION OF REPLICATION INITIATION IN EUKARYOTES

Eukaryotic chromosomes are duplicated only once per cell cycle

Eukaryotes have multiple origins of replication along the chromosomes and the regulation of initiation is critically important. Each origin must fire once - and only once - in each cell cycle. If origin re-firing occurred at multiple origins along the chromosome, the colliding forks might collapse, causing damaged and unreplicated DNA. The mechanism that assures that each origin fires only once is tightly linked to the eukaryotic cell cycle that was described in Chapter 5.

The firing of each origin once per cell division is achieved, in part, by dividing the initiation of DNA replication into tightly regulated steps that occur at different times in the cell cycle. First, in the G1 phase, replication origins are selected and prepared for activation. Second, in the S phase of the cell cycle, the preselected origins are activated. Finally, once an origin has fired in S phase, it cannot be re-used until it is prepared for activation in the daughter cell in the next G1 phase. Thus, each origin is activated only once in each cell cycle.

The origin recognition complex is the central component of the prereplication complex

As described in Section 6.13, the initiation of replication at an origin depends on the binding of specific initiator proteins to origin DNA. The eukaryotic initiator protein, called the origin recognition complex (ORC), consists of six tightly associated subunits (Orc1–6). In *S. cerevisiae*, the ORC binds to a clearly defined origin DNA sequence; in organisms in which origins are not so well-defined by sequence, however, the ORC recognizes elements of chromatin structure that specify replication

➔ The cell cycle is described in Chapter 5.

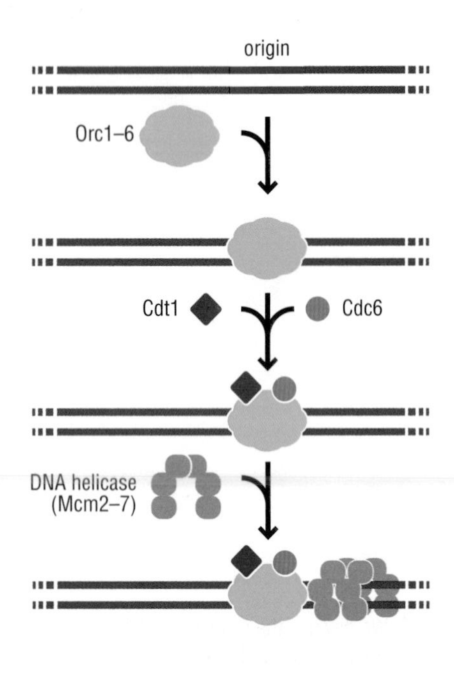

Figure 6.41 **Assembly of the preRC.** When bound to an origin, the ORC recruits two additional proteins, Cdc6 and Cdt1, to form a larger complex that is then responsible for loading the six-subunit Mcm2–7 complex onto the DNA. Two MCM complexes are loaded onto the DNA as a double hexamer, in a head-to-head configuration. This, presumably, precedes the separation of the two MCM complexes so that they can promote bidirectional DNA replication. When they are loaded, the MCM helicases are inactive, and additional steps are required to initiate replication (see Figure 6.42).

origins. The ORC is the central component of a large assembly of proteins - called the **prereplication complex** or **preRC** - that forms at the replication origin in G1, as illustrated in Figure 6.41

After the ORC has bound the origin, it recruits a number of other factors that make up the preRC. Two proteins, Cdc6 and Cdt1, facilitate loading of the MCM helicase onto the origin. As described in Section 6.5, MCM is the six-membered (Mcm2-7) DNA helicase that unwinds the DNA at the origin. In the preRC, however, the MCM helicase is not active; only after origin activation in S phase does MCM become activated to unwind the DNA.

➲ We discuss the ORC in more detail in Section 6.7.

Once the MCM helicase is loaded onto the DNA in the preRC, it recruits Cdc45, Sld3/7, Sld2, and the GINS complex. GINS, whose acronym Go-Ichi-Ni-San, stands for five-one-two-three in Japanese, is a complex of four proteins needed for replication fork progression.

Activation of the preRC is regulated by cell cycle kinases

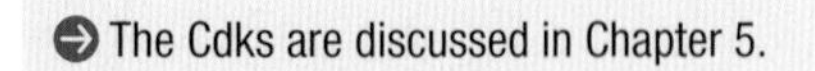
➲ The Cdks are discussed in Chapter 5.

After the preRC has been established in G1 phase of the cell cycle, the transition to the S phase causes activation of the preRC by two specific kinases, as shown in Figure 6.42. First, the S-phase specific cyclin-dependent kinase (Cdk) phosphorylates Sld3; the Cdc7-Dbf4 kinase (also known as Dbf4 dependent kinase or DDK) then phosphorylates the MCM helicase. Cdk also phosphorylates Ctd1 and Cdc6, which leads to their proteolytic destruction (via ubiquitin mediated proteolysis, as discussed in Chapter 5). This destruction ensures that Ctd1and Cdc6 are not available in the S phase, G2 phase, or M phase to build new preRCs. Upon origin activation by Cdk and Cdc7 kinase activity, the replicative polymerases are recruited, and Cdc45, the Mcm proteins, and GINS form a larger complex called CMG. It is this CMG complex that is the activity helicase that travels with the fork, allowing single-stranded template DNA to be copied by the polymerase.

Formation of the preRC occurs only in G1

One important mechanism that limits origin firing in eukaryotes to once per cell cycle is the fact that a preRC can only form in G1. As described above, when Cdk levels are high - in S phase, G2 phase, and M phase - Cdc6 and Ctd1 are phosphorylated and proteolytically destroyed. However, when the cell returns

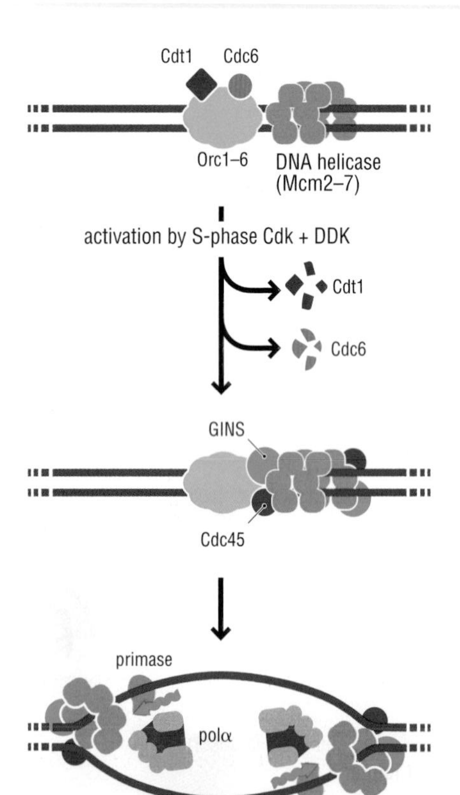

Figure 6.42 **The preRC is activated to initiate replication by the activity of Cdk and DDK.** The S-phase Cdks and the Dbf4-dependent kinase Cdc7, also known as DDK, trigger origin activation by phosphorylating several proteins, including Cdc6, components of ORC, and Mcm subunits. Several proteins, including GINS and Cdc45 (and additional proteins not shown) get recruited to the origin, and both Cdc6 and Cdt1 are removed from the origin and then destroyed. Origin activation results in activation of the Mcm2–7 helicase: the two MCM complexes, along with GINS and Cdc45 (known as the CMG complex), separate and begin to unwind the DNA helix. RPA proteins (not shown) bind the single-stranded DNA to prevent it from reannealing, and pol α-primase then initiates primer synthesis and DNA replication.

to G1 and Cdk activity is low, Cdc6 and Ctd1 accumulate and can now establish new preRCs.

A second mechanism that restricts origin firing in metazoans is the activity of Geminin protein. Geminin binds to Ctd1, preventing it from loading MCM to the origin to generate the preRC. Geminin is only present in the cell - and hence only blocks Ctd1 function - in S phase, G2 phase, and M phase. At the metaphase to anaphase transition, Geminin is phosphorylated by the anaphase-promoting complex (APC), which was described in Section 5.5. The phosphorylation by APC triggers the proteolytic destruction of Geminin, allowing Cdt1 to establish the preRC in the G1 phase of the cell cycle.

Replication must be completed before chromosome segregation occurs

It is important that chromosome replication is finished before the cell proceeds to separate the daughter chromosomes. If the cell attempts to segregate partially duplicated chromosomes it could result in broken chromosomes in the daughter cells. In most eukaryotic cells, this order of events is ensured by the DNA damage response checkpoint system, which monitors progression through S phase. If replication fails, then replication forks will stall. Stalled forks trigger the DNA damage response, and a signal is sent to the cell cycle control system to block progression into mitosis, as described in Section 5.6. Thus, the cell is prevented from attempting to segregate chromosomes that are not fully duplicated.

➔ The DNA damage response is discussed further in Chapter 15.

SUMMARY

PRINCIPLES

- DNA is duplicated via semi-conservative replication, whereby each original strand serves as a template for the synthesis of a new complementary daughter strand.
- DNA replication occurs in three distinct phases: initiation, elongation, and termination.

INITIATION OF REPLICATION

- Initiation of replication usually occurs at a unique chromosomal location called an origin, which is recognized by specific proteins.
- DNA strands must be separated by a DNA helicase to allow the replication machinery to gain access to the two parental strands.
- New strands are initiated by RNA or combined RNA and DNA primers, which are synthesized by primase in bacteria and the pol α-primase complex in eukaryotes. After their initial synthesis, the primers are then elongated by the replicative DNA polymerase.

ELONGATION DURING REPLICATION

- One half of a replication bubble where parental DNA strands are separated and new DNA is synthesized on both strands is called the replication fork.
- Since DNA polymerase can only add nucleotides to the elongating chain in a 5′ to 3′ direction, the two strands are replicated differently.
- The leading strand is synthesized as one long stretch of DNA while the lagging strand is made as short pieces called Okazaki fragments, which need to be stitched together by DNA ligase.
- DNA polymerases replicating the leading and lagging strand are coupled so that they move together at the replication fork.

PROCESSIVITY AND FIDELITY OF DNA REPLICATION

- Sliding clamps facilitate the processivity of the replicative DNA polymerase.
- The assembly of the sliding clamp onto the DNA strand is mediated by the clamp loader.

- Accurate DNA replication is critical and is ensured through two distinct mechanisms mediated by DNA polymerase: the insertion of the nucleotide with the correct geometry and proofreading, and the removal of incorrectly incorporated nucleotides.

TERMINATION OF REPLICATION

- The termination of replication occurs when two replication forks traveling in opposite directions meet. In bacteria this happens at a specific chromosomal position called ter.
- Replicated chromosomes are linked together and must be unlinked or decatenated by topoisomerases.
- DNA polymerases are unable to completely replicate the ends of linear DNA. This is called the end-replication problem.
- For many eukaryotic chromosomes the end-replication problem is solved by telomerase, which adds telomere sequences onto the ends of chromosomes.

SPECIALIZED POLYMERASES

- Specialized DNA polymerases carry out a variety of functions including DNA repair and the bypass of damaged DNA. One specialized polymerase, reverse transcriptase, copies RNA into DNA.

REPLICATION OF CHROMATIN

- During the replication of eukaryotic chromosomes, the parental nucleosomes are segregated such that half go to each daughter strand, and new nucleosomes are added to each strand by chromatin assembly factors. In this way, histone modifications are equally segregated to each daughter strand and can direct modification of the new nucleosomes.

REGULATION OF DNA REPLICATION

- The initiation of replication is tightly regulated by multiple mechanisms.
- In bacteria, where there is a single origin, replication initiation is regulated by the availability of the initiation factor DnaA in its ATP-bound state, and the availability of the origin sequence.
- In eukaryotes, where there are multiple origins, replication initiation is tightly linked to the cell cycle.
- In G1 phase, the preRC is formed at the origins, but is inactive. After initiation of S phase, the preRC is activated, allowing initiation.
- PreRCs cannot reform until the next G1 due to the inactivation of the components by phosphorylation and degradation in all other phases.

QUESTIONS

6.1 OVERVIEW OF DNA REPLICATION

1. In the context of the process of replication, which of the following enzymes unwinds DNA at the replication fork?
  a. Primase
  b. Polymerase
  c. Helicase
  d. Topoisomerase

2. Which of the following statements correctly completes this sentence?

DNA replication is
  a. continuous on both the leading and lagging strands.
  b. discontinuous on both the leading and lagging strands.
  c. continuous on the leading strand and discontinuous on the lagging strand.
  d. discontinuous on the leading strand and continuous on the lagging strand.

3. Briefly describe the role of each of the following components during DNA replication.
  a. DNA helicase
  b. DNA polymerase
  c. Sliding clamp
  d. Primase
  e. Topoisomerase

6.2 DNA POLYMERASES: STRUCTURE AND FUNCTION

1. Briefly explain the function of the following conserved domains in DNA polymerase.
  a. Palm
  b. Thumb
  c. Fingers

Challenge question

2. Replicative DNA polymerases have a much higher rate of fidelity than RNA polymerases. Explain why this is necessary.

6.3 DNA POLYMERASES: FIDELITY AND PROCESSIVITY

1. Some DNA polymerases have proofreading activity in which erroneously added nucleotides are removed. What type of activity is represented by this proofreading?
  a. 5′ to 3′ endonuclease
  b. 5′ to 3′ exonuclease
  c. 3′ to 5′ endonuclease
  d. 3′ to 5′ exonuclease

Challenge question

2. Explain why it might be beneficial for some polymerases (like pol III) to be highly processive but not for others (like pol I).

6.4 SPECIALIZED POLYMERASES

1. Which of the following is true?
  a. Pol I uses 5′ to 3′ exonuclease activity to remove RNA primers, but pol ε and pol δ do not.
  b. Pol I has 3′ to 5′ exonuclease activity, but pol III does not.
  c. Flap endonuclease and pol I use 5′ to 3′ exonuclease activity to remove RNA primers.
  d. Both eukaryotes and bacteria need a non-processive polymerase in order to remove RNA primers.

2. Which of these polymerases can replace pol α during replication?
   a. Pol I
   b. Pol III
   c. Pol β
   d. Pol ε

Challenge questions

3. Y-family DNA polymerases have a lower fidelity than replicative polymerases; explain what Y-family polymerases do, and why this necessitates a lower fidelity rate (you may need to refer to Chapter 15).

6.5 DNA HELICASES: UNWINDING OF THE DOUBLE HELIX

1. Topoisomerases remove supercoils to relieve torsional stress. In this context, what is unusual about DNA gyrase?
   a. It does not require ATP.
   b. It is a type III topoisomerase.
   c. It is found only in eukaryotes.
   d. It induces negative supercoiling.

2. Supercoiling is described by the equation $Lk = TW + Wr$ (see Chapter 2). If topoisomerase were not active during replication, which of the following statements would be true?
   a. DNA ahead of and behind the polymerase has $Wr = 0$.
   b. DNA ahead of and behind the polymerase has $Wr < 0$.
   c. DNA ahead of and behind the polymerase has $Wr > 0$.
   d. DNA behind the polymerase has $Wr < 0$, and DNA ahead of it has $Wr > 0$.
   e. DNA behind the polymerase has $Wr > 0$, and DNA ahead of it has $Wr < 0$.

6.6 THE SLIDING CLAMP AND CLAMP LOADER

1. Explain the purpose of sliding clamps and how they are loaded onto DNA.

6.7 ORIGINS AND INITIATION OF DNA REPLICATION

1. Explain how ATP and its hydrolysis are important in the initiation of DNA replication and the regulation of that initiation in *E. coli*.

2. Thinking in terms of the function of the proteins listed below, which is found at the replisome throughout replication?
   a. DnaA
   b. DnaB
   c. DnaC
   d. DnaD

3. What is the major mechanism preventing early re-firing of replication origins in *E. coli?*
   a. Cyclin-dependent kinase regulation
   b. Hydrolysis of ATP to ADP on DnaA
   c. DnaA sequestration close to the origin
   d. Delay in methylation at *oriC*

Challenge question

4. Design an experiment to discover where the origin of replication is located on the chromosome of a newly discovered bacterium. It has already been established that *E. coli* replication proteins are effective at stimulating replication in this particular bacterium.

6.8 LEADING AND LAGGING STRAND SYNTHESIS

1. Which of the following statements about Okazaki fragment maturation is true?
   a. A flap of RNA and DNA is displaced during maturation in both eukaryotes and bacteria.
   b. Both eukaryotes and bacteria use a non-processive DNA polymerase during maturation.
   c. Both eukaryotes and bacteria use DNA ligase I to seal nicks during maturation.
   d. Okazaki fragments only occur in eukaryotes, not bacteria.

2. DNA replication is not fully completed at the end of which strand(s) in chromosomes?
   a. Leading strand in circular chromosomes
   b. Leading strand in linear chromosomes
   c. Lagging strand in circular chromosomes
   d. Lagging strand in linear chromosomes
   e. Neither strand

Challenge question

3. What is polymerase switching, and how is this different in eukaryotic DNA synthesis as opposed to bacterial DNA synthesis?

6.9 THE REPLICATION FORK

1. What is a replication bubble, and how does it relate to replication forks?

2. Explain why it is necessary that DNA is replicated continuously on the leading strand but discontinuously on the lagging strand.

Challenge question

3. You wish to carry out an experiment on a mutated bacterial pol III. The mutation is in a region of the protein that you suspect weakens the binding of the polymerase to the replisome. Devise a hypothesis for how this mutation might affect replication, and explain your reasoning.

6.10 TERMINATION OF DNA REPLICATION

1. In the context of the process of replication, which of the following enzymes separates the strands of the double helix ahead of the replication fork?
   a. Polymerase
   b. Helicase
   c. Ligase
   d. Topoisomerase

2. *E. coli* has a single origin of replication on its chromosome, whereas eukaryotes have many. Explain why.

6.11 THE END-REPLICATION PROBLEM AND TELOMERASE

1. What is the end-replication problem? What mechanisms, other than the use of telomerase, have evolved to address this problem?

2. Explain the purpose and function of telomerase.

Challenge question

3. Telomerase has functional and structural similarities to several other enzymes. Choose one of these (other than DNA polymerase or RNA polymerase) and explain why you think they are similar.

6.12 CHROMATIN REPLICATION

1. Synthesis of which histone protein is not replication dependent?
   a. CenH3 (H3 variant)
   b. H2A
   c. H2B
   d. H3
   e. H4

2. Explain the importance of histones H3 and H4 undergoing parental histone segregation (as opposed to simply adding completely new histones to newly synthesized DNA).

6.13 REGULATION OF INITIATION OF REPLICATION IN *E. COLI*

1. Replication in *E. coli* is initiated at *oriC*. List the mechanisms that regulate replication initiation so that the origin does not fire too frequently. Choose one of these mechanisms and describe how it functions.

6.14 REGULATION OF REPLICATION INITIATION IN EUKARYOTES

1. During which phase of the eukaryotic cell cycle are replication complexes assembled at origins?
   a. M
   b. G1
   c. S
   d. G2

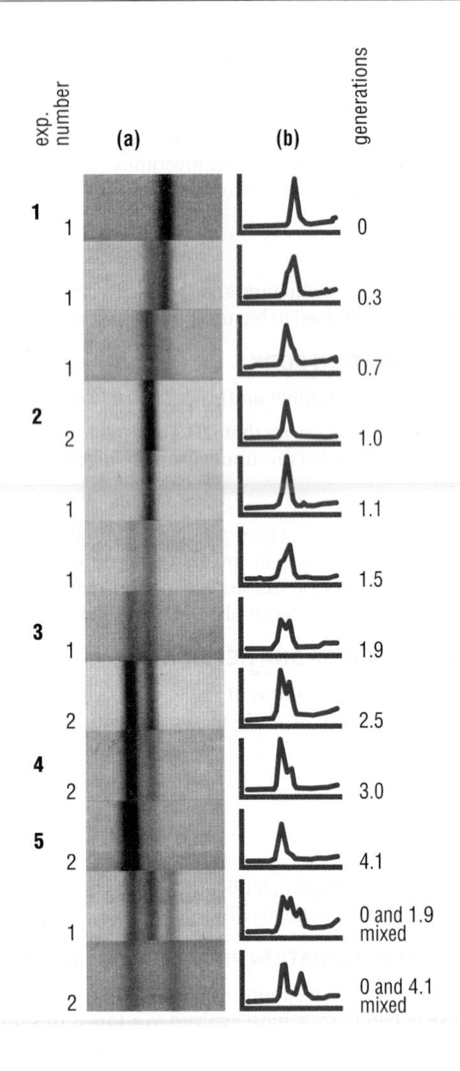

Figure A

Challenge question

2. Explain how the cell cycle control system restricts origin firing to S phase, and the role of Cdks in this process.

EXPERIMENTAL APPROACH 6.1 – DISCOVERY OF SEMI-CONSERVATIVE DNA REPLICATION

Challenge questions

1. Figure A, modified from Experimental approach 6.1, shows data from the Meselson and Stahl experiment on DNA replication. Which numbered part of the figure (numbers on left in bold) represents the situation in which half of the total DNA contains N^{14} and half contains N^{15}? Explain your answer.

2. In a classic experiment to figure out whether DNA replication in a replication bubble is unidirectional or bidirectional, researchers labeled replicating DNA with radioactive nucleotides (Gyurasits, Wake. *J Mol Biol*, 1973;**73**:55–63). The DNA was labeled with a low level of radioactivity followed with a high level of radioactivity (after a specific period of time that allowed for some replication to have occurred). Draw what you would expect the autoradiograph to look like for a) unidirectional replication and b) bidirectional replication. Explain how this experiment proves bidirectional replication.

EXPERIMENTAL APPROACH 6.3 – COUPLING OF POLYMERASES AT THE REPLICATION FORK: THE TROMBONE MODEL

1. Figure B depicts the trombone model of DNA replication. Use this to answer the following questions.

a. Label the replication components in the diagram, assuming this is bacterial replication.
b. With reference to the stages shown, describe how DNA is synthesized on the lagging strand according to the trombone model. Note the relative "trombone loop" sizes at the different stages.
c. Some important proteins that are involved in DNA replication are NOT shown on this diagram. List as many as you can think of, then choose one and briefly describe its role in replication.
d. Compare and contrast Okazaki fragment maturation in eukaryotes with that in bacteria.

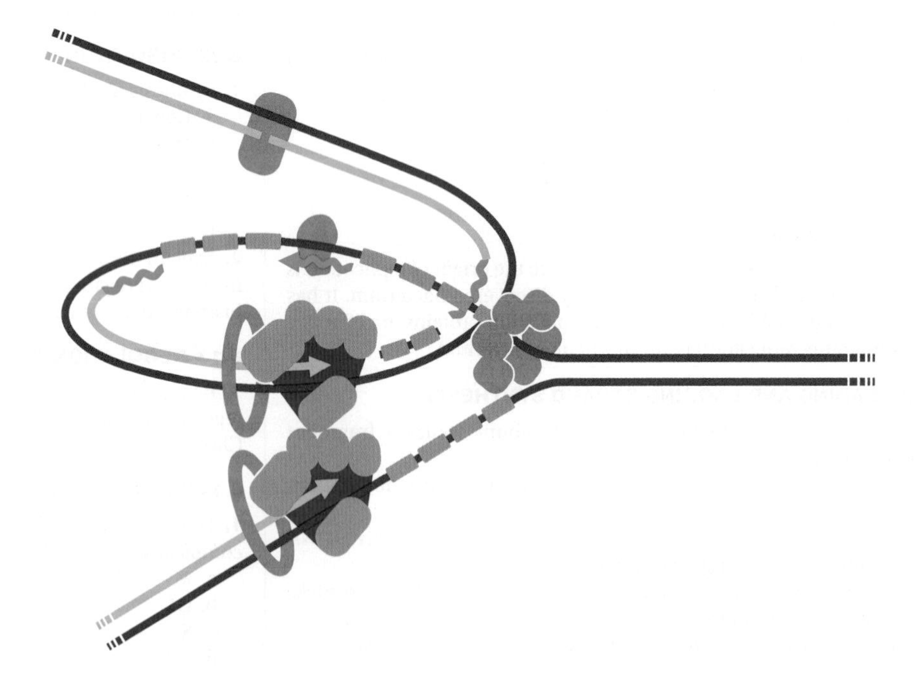

Figure B

6.1 OVERVIEW OF DNA REPLICATION

Davey MJ, Jeruzalmi D, Kuriyan J, O'Donnell M. Motors and switches: AAA+ machines within the replisome. *Nature Reviews Molecular Cell Biology*, 2002;**3**:826–835.

Kunkel TA, Burgers PM. Dividing the workload at a eukaryotic replication fork. *Trends in Cell Biology*, 2008;**18**:521–527.

Waga S, Stillman B. The DNA replication fork in eukaryotic cells. *Annual Review of Biochemistry*, 1998;**67**:721–751.

6.2 DNA POLYMERASES: STRUCTURE AND FUNCTION; AND 6.3 DNA POLYMERASES: FIDELITY AND PROCESSIVITY

Francklyn CS. DNA polymerases and aminoacyl-tRNA synthetases: shared mechanisms for ensuring the fidelity of gene expression. *Biochemistry*, 2008;**47**:11695–11703.

Joyce CM, Benkovic SJ. DNA polymerase fidelity: kinetics, structure, and checkpoints. *Biochemistry*, 2004:**43;**14317–14324.

Joyce CM, Steitz TA. Function and structure relationships in DNA polymerases. *Annual Review of Biochemistry*, 1994;**63**:777–822.

Kunkel TA, Bebenek K. DNA replication fidelity. *Annual Review of Biochemistry*, 2000;**69**:497–529.

6.4 SPECIALIZED POLYMERASES

Friedberg EC, Wagner R, Radman M. Specialized DNA polymerases, cellular survival, and the genesis of mutations. *Science*, 2002;**296**:1627–1630.

Prakash S, Johnson RE, Prakash L. Eukaryotic translesion synthesis DNA polymerases: specificity of structure and function. *Annual Review of Biochemistry*, 2005;**74**:317–353.

Sharma S, Helchowski CM, Canman CE. The roles of DNA polymerase ζ and the Y family DNA polymerases in promoting or preventing genome instability. *Mutation Research*, 2012;**743–744**:97–110.

6.5 DNA HELICASES: UNWINDING OF THE DOUBLE HELIX

Enemark EJ, Joshua-Tor L. On helicases and other motor proteins. *Current Opinion in Structural Biology*, 2008;**18**:243–257.

Forsburg SL. The MCM helicase: linking checkpoints to the replication fork. *Biochemical Society Transactions*, 2008;**36**:114–119.

Gai D, Chang YP, Chen XS. Origin DNA melting and unwinding in DNA replication. *Current Opinion in Structural Biology*, 2010;**20**:756–762.

Masai H, You Z, Arai K. Control of DNA replication: regulation and activation of eukaryotic replicative helicase, MCM. *IUBMB Life*, 2005;**57**:323–335.

Soultanas, P. Loading mechanisms of ring helicases at replication origins. *Molecular Microbiology*, 2012;**84**:6–16.

Wang JC. Cellular roles of DNA topoisomerases: a molecular perspective. *Nature Reviews Molecular Cell Biology*, 2002;**3**:430–440.

6.6 THE SLIDING CLAMP AND CLAMP LOADER

Bloom LB. Dynamics of loading the *Escherichia coli* DNA polymerase processivity clamp. *Critical Reviews in Biochemistry and Molecular Biology*, 2006;**41**:179–208.

Ellison V, Stillman B. Opening of the clamp: an intimate view of an ATP-driven biological machine. *Cell*, 2001;**106**:655–660.

Indiani C, O'Donnell M. The replication clamp-loading machine at work in the three domains of life. *Nature Reviews Molecular Cell Biology*, 2006;**7**:751–761.

Johnson A, O'Donnell M. Cellular DNA replicases: components and dynamics at the replication fork. *Annual Review of Biochemistry*, 2005;**74**:283–315.

Prives C, Gottifredi V. The p21 and PCNA partnership: a new twist for an old plot. *Cell Cycle*, 2008;**7**:3840–3846.

6.7 ORIGINS OF REPLICATION AND THE INITIATION OF DNA REPLICATION

Bell SP, Dutta A. DNA replication in eukaryotic cells. *Annual Review of Biochemistry*, 2002;**71**:333–374.

Boos, D, Frigola, J, and Diffley, JF. Activation of the replicative DNA helicase: breaking up is hard to do. *Current Opinion in Cell Biology*, 2012;**24**, 423–430.

DePamphilis ML. Cell cycle dependent regulation of the origin recognition complex. *Cell Cycle*, 2005;**4**:70–79.

Diffley JF. Regulation of early events in chromosome replication. *Current Biology*, 2004;**14**:R778–786.

Méchali M. DNA replication origins: from sequence specificity to epigenetics. *Nature Reviews Genetics*, 2001:**2**:640–645.

Prasanth SG, Mendez J, Prasanth KV, Stillman B. Dynamics of pre-replication complex proteins during the cell division cycle. *Philosophical Transactions of the Royal Society of London Series B, Biological Sciences*, 2004;**359**:7–16.

6.8 LEADING AND LAGGING STRAND SYNTHESIS

Balakrishnan L, Bambara RA. Eukaryotic lagging strand DNA replication employs a multi-pathway mechanism that protects genome integrity. *Journal of Biological Chemistry*, 2011;**286**, 6865–6870.

Corn JE, Berger JM. Regulation of bacterial priming and daughter strand synthesis through helicase-primase interactions. *Nucleic Acids Research*, 2006;**34**:4082–4088.

Kunkel TA. Balancing eukaryotic replication asymmetry with replication fidelity. *Current Opinion in Chemical Biology*, 2011;**15**:620–626.

6.9 THE REPLICATION FORK

Burgers PM. Polymerase dynamics at the eukaryotic DNA replication fork. *Journal of Biological Chemistry*, 2009;**284**:4041–4045.

Kunkel TA, Burgers PM. Dividing the workload at a eukaryotic replication fork. *Trends in Cell Biology*, 2008;**18**:521–527.

Kurth, I., and O'Donnell, M. New insights into replisome fluidity during chromosome replication. *Trends in Biochemical Sciences*, 2012;**38**:195–203.

O'Donnell M. Replisome architectureand dynamics in *E. coli*. *Journal of Biological Chemistry* 2006:**281**;10653–10656.

Waga S, Stillman B. The DNA replication fork in eukaryotic cells. *Annual Review of Biochemistry* 1998:**67**:721–751.

6.10 TERMINATION OF DNA REPLICATION

Dalgaard JZ, Eydmann T, Koulintchenko M, Sayrac S, Vengrova S, Yamada-Inagawa T. Random and site-specific replication termination. *Methods in Molecular Biology*, 2009;**521**:35–53.

MacNeill SA. Structure and function of the GINS complex, a key component of the eukaryotic replisome. *The Biochemical Journal*, 2010;**425**:489–500.

Mohanty BK, Bastia D. Binding of the replication terminator protein Fob1p to the Ter sites of yeast causes polar fork arrest. *Journal of Biological Chemistry*, 2004;**279**:1932–1941.

Neylon C, Kralicek AV, Hill TM, Dixon NE. Replication termination in *Escherichia coli*: structure and antihelicase activity of the Tus-Ter complex. *Microbiology and Molecular Biology Reviews*, 2005;**69**:501–526.

Wang JC. Cellular roles of DNA topoisomerases: a molecular perspective. *Nature Reviews Molecular Cell Biology*, 2002;**3**:430–440.

6.11 THE END-REPLICATION PROBLEM AND TELOMERASE

Collins, K. Single-stranded DNA repeat synthesis by telomerase. *Current Opinion in Chemical Biology*, 2011;**15**:643–648.

Lingner J, Cooper JP, Cech TR.Telomerase and DNA end replication: no longer a lagging strand problem? *Science*, 1995;**269**: 1533–1534.

Podlevsky JD, Chen JJ. It all comes together at the ends: telomerase structure, function, and biogenesis. *Mutation Research*, 2012;**730**:3–11.

6.12 CHROMATIN REPLICATION

Altaf M, Auger A, Covic M, Cote J. Connection between histone H2A variants and chromatin remodeling complexes. *Biochemistry and Cell Biology* 2009;**87**:35–50.

Burgess RJ, Zhang Z. Histone chaperones in nucleosome assembly and human disease. *Nature Structural and Molecular Biology*, 2013;**20**:14–22.

Corpet A, Almouzni G. Making copies of chromatin: the challenge of nucleosomal organization and epigenetic information. *Trends in Cell Biology*, 2009;**19**:29–41.

Hondele M, Ladurner AG. The chaperone-histone partnership: for the greater good of histone traffic and chromatin plasticity. *Current Opinion in Structural Biology*, 2011;**21**:698–708.

Li Q, Burgess R, Zhang Z. All roads lead to chromatin: Multiple pathways for histone deposition. *Biochimica et Biophysica Acta*, 2012;**1819**:238–246.

Martin C, Zhang Y. Mechanisms of epigenetic inheritance. *Current Opinion in Cell Biology* 2007;**19**:266–272.

Mermoud JE, Rowbotham SP, Varga-Weisz PD. Keeping chromatin quiet: how nucleosome remodeling restores heterochromatin after replication. *Cell Cycle*, 2011;**10**:4017–4025.

6.13 REGULATION OF INITIATION OF REPLICATION IN *E. COLI*

Leonard AC, Grimwade JE. Regulation of DnaA assembly and activity: taking directions from the genome. *Annual Review of Microbiology*, 2011;**65**:19–35.

Mott ML, Berger JM. DNA replication initiation: mechanisms and regulation in bacteria. *Nature Reviews Microbiology*, 2007;**5**:343–354.

Waldminghaus T., Skarstad K. The *Escherichia coli* SeqA protein. *Plasmid*, 2009;**61**:141–150.

6.14 REGULATION OF REPLICATION INITIATION IN EUKARYOTES

Aparicio OM. Location, location, location: it's all in the timing for replication origins. *Genes and Development*, 2013;**27**:117–128.

Bell SP. The origin recognition complex: from simple origins to complex functions. *Genes and Development* 2002;**16**:659–672.

Caillat C, Perrakis A. Cdt1 and geminin in DNA replication initiation. *Sub-cellular Biochemistry*, 2012;**62**:71–87.

Labib, K. How do Cdc7 and cyclin-dependent kinases trigger the initiation of chromosome replication in eukaryotic cells? *Genes and Development*, 2010;**24**:1208–1219.

Sclafani RA, Holzen TM. Cell cycle regulation of DNA replication. *Annual Review of Genetics*, 2007;**41**:237–280.

Chromosome segregation

7

INTRODUCTION

Cell proliferation is the process by which one cell gives rise to two daughter cells. This process is critical for the survival of single-celled organisms such as bacteria and fungi. It is also essential for the development of multicellular organisms, during the replenishment of renewable tissue such as blood and skin, and for wound healing. The key challenge in this process is to ensure that each daughter cell receives one, and *only* one, copy of each chromosome. Indeed, failure to meet this goal can have dire consequences. For example, Down syndrome, in which chromosome 21 appears in three copies instead of just two, is due to a **chromosome segregation** error that took place during the formation of an egg or sperm. Cancer cells also tend to exhibit an abnormal number of chromosomes.

Successful cell proliferation depends on a flawless transmission of chromosomes from one generation to the next. This process, known as chromosome segregation, involves many structural and regulatory components that will be discussed in this chapter.

Most of this chapter will be dedicated to **mitosis**, the cell-division program of somatic cells (all the eukaryotic cells other than those involved in sexual reproduction). We will also discuss chromosome segregation in **meiosis**, the specialized cell-division program required to generate the **gametes** that are involved in sexual reproduction. Finally, we will discuss the process of chromosome segregation in bacteria, whose principles are similar to those that govern mitosis, but with different structural and regulatory machineries.

Before considering the specific cellular and molecular processes that underpin mitosis and meiosis, we begin with a broad overview of the stages of mitosis, to give us a general framework to which we can add more detail later.

7.1 THE STAGES OF MITOSIS

Processes that take place early in the cell cycle set the stage for mitosis

To understand the mechanism and regulation of chromosome segregation in eukaryotes, we must first become familiar with cellular events that lead up to and take place during mitosis. These events are part of the eukaryotic cell cycle, which we discussed in Chapter 5. Dividing cells spend most of their lifecycle in **interphase**, which encompasses the G1, S, and G2 phases of the cell cycle, as illustrated in Figure 7.1, step 1. Two events critical for chromosome segregation take place during interphase – DNA replication and the initial stages of spindle formation (see Section 5.1). As we learned in Chapter 6, DNA replication leads to the duplication of the cell's chromosomal DNA, resulting in the formation of

Scan here to watch a video animation explaining more about chromosome segregation, or find it via the **Online Resource Center** at **www.oxfordtextbooks.co.uk/orc/craig2e/**.

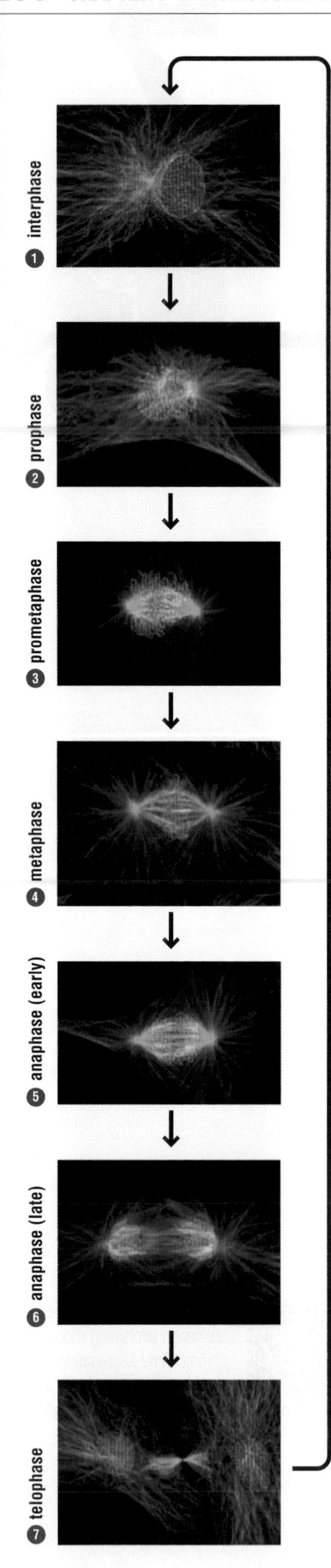

two identical chromosomes, or **sister chromatids**. After DNA replication, sister chromatids remain associated with each other by the presence of a protein complex called **cohesin** (introduced in Section 4.2), which links the two sister chromatids to each other. This association, or '**cohesion**', is important for directing the segregation of sister chromatids to different daughter cells during mitosis (see Section 7.2).

During mitosis of animal cells, the movement of chromosomes to the daughter cells occurs via the attachment of the chromosomes to a structure called the **mitotic spindle** (or 'spindle', for short), which is made of filaments called **microtubules** (see Section 7.3). The first steps in spindle formation take place during interphase in the cytoplasm. Because during interphase the nucleus is surrounded by a membranous structure called the **nuclear envelope**, the chromosomes in the nucleus cannot bind to the spindle, which is in the cytoplasm. To allow the chromosomes to attach to the spindle microtubules, the nuclear envelope disassembles early in mitosis, and re-forms only after chromosome segregation is completed, at the end of mitosis.

Early mitosis is marked by changes in chromosome structure and nuclear integrity

Mitosis has been divided into distinct steps, based on cytological observations. The first visible signs of mitosis occur in **prophase** (Figure 7.1, step 2), when the duplicated chromosomes begin to condense and individual pairs of sister chromatids become apparent. Prophase is also the time when the spindle grows in length, ultimately forming a structure made of parallel microtubules, which tapers at both ends (hence the name 'spindle'). Immediately after prophase, in a stage called **prometaphase** (Figure 7.1, step 3), the nuclear envelope breaks down and the spindle microtubules can now make contact with the chromosomes.

The attachment of chromosomes to microtubules leads to chromosome alignment at the spindle mid-zone

The spindle microtubules contribute to chromosome segregation by binding to pairs of sister chromatids and pulling them apart, such that each sister chromatid moves to a different end – or 'pole' – of the spindle. This means that each sister chromatid in a given pair has to bind spindle microtubules that are connected to a different spindle pole. Following nuclear envelope breakdown, the attachments between the chromosomes and spindle microtubules are initially random, meaning that both sister chromatids may bind microtubules that are connected to the same spindle pole. Eventually, however, regulatory proteins help each sister chromatid to bind stably to microtubules that are associated with opposite spindle poles, a state called **bi-orientation**. At this point, the spindle microtubules exert pulling forces on the sister chromatids, but the chromatids do not separate because

Figure 7.1 The stages of mitosis. The cell cycle can be divided into interphase (1), which spans the phases G1, S, and G2 (see Section 5.1), and mitosis (2-7), which is followed by cytokinesis (not shown). The stages of mitosis are (2) prophase, (3) prometaphase, (4) metaphase, (5, 6) anaphase, and (7) telophase. The chromosomes are shown in pink and microtubules are shown in green. Note the changes in microtubule organization and chromosome condensation between interphase and mitosis. The nuclear envelope cannot be seen in these images, but it is present during interphase, prophase, and telophase. Images courtesy of Tamara A. Potapova and Gary J. Gorbsky, Oklahoma Medical Research Foundation.

they are still connected to each other by cohesin complexes. However, because the paired sister chromatids are subjected to equivalent forces from microtubules attached to two *different* spindle poles, the chromatids move to the spindle mid-zone equidistant between the two poles, much like in a tug-of-war game between equally strong teams. The stage at which all chromosomes are bi-oriented and aligned on the spindle mid-zone is called **metaphase** (Figure 7.1, step 4) - a stage that we consider in more detail in Section 7.4.

Separation of sister chromatids at anaphase is a point of no return

For chromosome segregation to take place, sister chromatid cohesion has to be dissolved, which occurs only when *all* chromosomes have bi-oriented. Cohesion dissolution is accomplished by the cleavage of a subunit of the cohesin complex, thereby inactivating the complex. This occurs during the **anaphase** stage of mitosis and allows the sister chromatids to be pulled to opposite spindle poles by forces acting through the spindle microtubules, as illustrated in Figure 7.1, steps 5 and 6. Anaphase is a critical point in mitosis because it is irreversible: once sister chromatids separate, it is no longer possible to correct erroneous chromosome–spindle attachments that could lead to inaccurate chromosome segregation. Therefore, multiple regulatory processes control the onset of anaphase, and in particular the timing at which sister chromatid cohesion is dissolved (see Section 7.5).

Events in late mitosis set the stage for the return to interphase

Once anaphase is completed, cells begin the journey back to interphase. During **telophase**, which is the last stage of mitosis (Figure 7.1, step 7), the spindle disassembles, the chromosomes decondense and the nuclear envelope reassembles to create two daughter nuclei. All that is left to complete cell division is the physical separation of the two daughter cells, in a process called **cytokinesis**. The plane of cell division is determined in anaphase, and it corresponds to the location of the spindle mid-zone. This localization ensures that the plane of cell division is situated between the two daughter nuclei, and that the division event itself will result in two cells, each with a full complement of chromosomes. Upon completion of cytokinesis, the cells are back in interphase.

7.2 CHROMOSOME CONDENSATION AND COHESION

Having discussed mitosis in general terms, and having identified the key stages that make up the process as a whole, we can now move on to examine in more detail the molecular and cellular events that characterize each stage. We begin by considering cohesion and **condensation** of chromosomes.

Mitosis is associated with changes in chromosome structure

Chromosome segregation is accompanied by major changes in chromosome organization. During interphase, the chromosomes are in an extended, intertwined state that is unlikely to yield orderly chromosome movement. The compaction that occurs as cells enter mitosis protects the chromosomes from getting their trailing arms caught in the plane of cell division during cytokinesis, and is necessary for the untangling of the two sister chromatids. This process of compaction is called **chromosome condensation**, and it requires a protein complex called **condensin**. We learn more about this condensation process later in this section.

Another structural feature of chromosomes in dividing cells is **sister chromatid cohesion.** As mentioned above, cohesion is mediated by cohesin complexes, which link the two sister chromatids to each other prior to segregation and ensure that they bi-orient on the mitotic spindle. The cohesin complex is related in structure and protein composition to condensin.

Condensin and cohesin form ring-like structures

Condensin and cohesin belong to the **structural maintenance of chromatin (SMC)**protein complex family. In eukaryotes, SMC protein complexes, such as condensin and cohesin, are composed of a heterodimer of SMC proteins, each containing an ATPase domain, a long coiled-coil arm, and several associated subunits, as illustrated in Figure 7.2. The SMC complex is not unique to eukaryotes: bacteria also contain SMC complexes, but those identified so far contain a homodimer of SMC proteins rather than a heterodimer. The binding of adenosine triphosphate (ATP) and its subsequent hydrolysis, together with the actions of the associated subunits, is believed to result in conformational changes in the SMC heterodimer that lead to changes in chromosome structure. For example, ATP hydrolysis may open and close the SMC coiled-coil arms, allowing the complex to associate with or dissociate from chromosomes. We do not yet know how these complexes contribute to chromosome cohesion and condensation, but it is thought that the SMC arms are long enough to encircle one, and possibly two, chromosomes.

Of particular note is the kleisin subunit (from the Greek word for 'closure'), which bridges the ATPase domains of the SMC subunits (see Figure 7.2). The kleisin subunit of cohesin is cleaved at the onset of anaphase (as discussed further in Section 7.5). However, we don't yet know whether the kleisin subunit of condensin regulates condensin function.

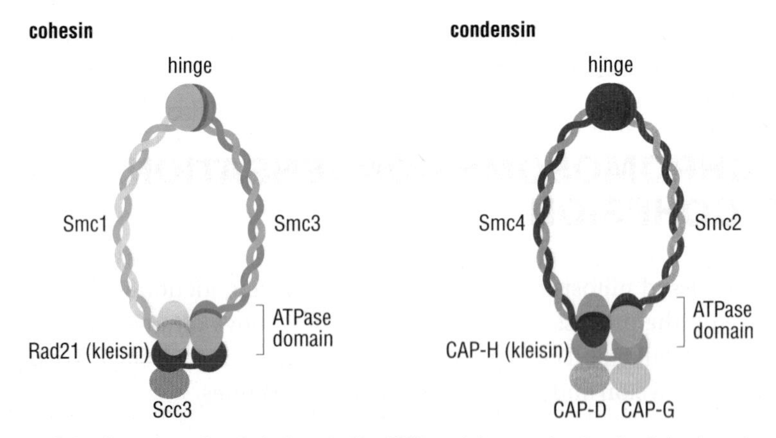

Figure 7.2 Cohesin and condensin belong to the SMC protein complex family. Cohesin and condensin are composed of two SMC proteins: cohesin is made of Smc1 and Smc3 while condensin is made of Smc2 and Smc4. Both cohesin and condensin contain additional subunits, including a subunit that is referred to as kleisin, which is thought to bind near the ATPase domain of the SMC subunits. In many organisms, the kleisin subunits of cohesin and condensin are called Rad21 and CAP-H, respectively. In both complexes, the kleisin subunit bridges between the ATPase domains of the two Smc molecules. Shown are the predicted structures of individual cohesin and condensin complexes; the structure of these complexes when associated with chromosomes is not known.

Chromosome condensation requires the condensin complex

Despite the striking visible changes that accompany chromosome condensation, which are illustrated in Figure 7.3, we do not know much about the molecular mechanisms that drive chromosome condensation. The condensin complex is needed for stable chromosome condensation. However, additional proteins such as topoisomerases are likely also involved in mitotic chromosome condensation; we know this because depletion of condensin results in condensation being only partially defective. It is thought that condensin can compact chromosomes only when transcription is shut off during mitosis. Chromosome condensation occurs because condensins associate with widely separated chromosomal regions and bring them closer together to compact the chromosome - a process that requires functional ATPase domains in the SMC proteins. However, we don't yet know how ATP binding or hydrolysis contribute to condensation, or whether condensin complexes are involved in additional processes outside mitosis.

We discuss topoisomerases further in Section 6.5.

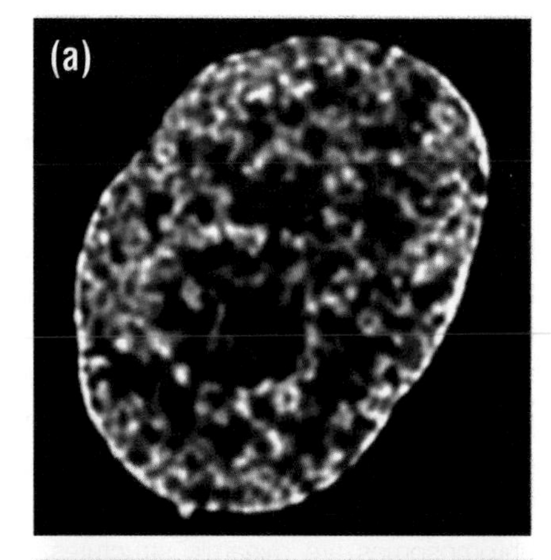

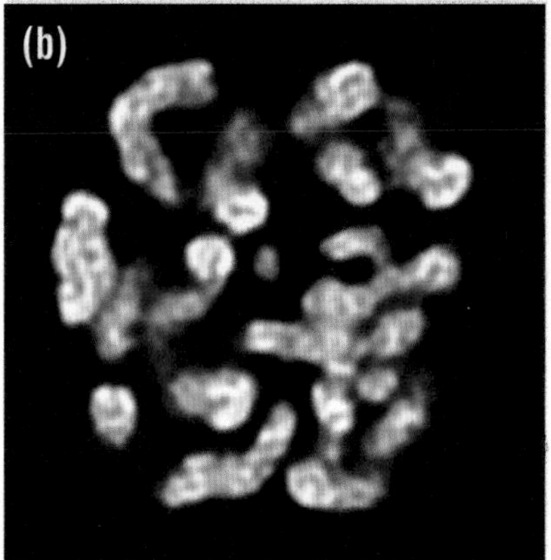

Figure 7.3 Chromosome condensation. During interphase, chromosomes are decondensed, and it is difficult to distinguish between individual chromosomes. In prophase, chromosomes are condensed: they become thicker and appear as discrete units. Shown are light microscopy images of chromosomes from a Chinese hamster ovary cell in (a) interphase and (b) late prophase/early prometaphase.

Adapted from Kireeva N, Lakonishok M, Kireev I, Hirano T, Belmont AS. Visualization of early chromosome condensation: a hierarchical folding, axial glue model of chromosome structure. *Journal of Cell Biology*, 2004;**166**:775–85.

Cohesin is essential for accurate chromosome segregation

By contrast with what little we know about the mechanism of condensin, we have a better understanding of the cohesion complex. Sister chromatid cohesion is necessary for accurate chromosome segregation during mitosis; in the absence of sister chromatid cohesion, there would be no way of coordinating the binding of the two sister chromatids to different spindle poles. The importance of cohesion in the segregation process is illustrated by the observation that, if cohesin is inactivated, neither of the daughter cells receives a full complement of chromosomes. As a result, chromosomes grossly mis-segregate and the cell dies. We learn how sister chromatid separation has been studied experimentally in Experimental approach 7.1.

Cohesin complexes assemble on chromosomes in interphase (from telophase to the G1/S boundary, depending on the organism), well before mitosis. Some cohesin complexes will ultimately be involved in cohesion, while others will contribute to the regulation of gene expression by a yet unknown mechanism. Cohesion is established in coordination with DNA replication, so that the sister chromatids are held together from the time they are formed in S phase. Linking the establishment of cohesion to DNA replication circumvents the difficulty that would arise if sister chromatids were allowed to separate after replication and then had to be identified out of a mixture containing all the other chromosomes.

Cohesins cannot associate with chromosomes on their own, but are loaded onto chromosomes by a specific loading complex. Although cohesins show a preference for localizing to AT-rich regions on DNA, known as cohesin-associated regions (CARs), they do not bind to a specific DNA sequence. It is possible that cohesins (or their loading factors) favor particular types of chromatin modification that are abundant at CARs. Alternatively, CARs often coincide with intergenic regions, and it is therefore possible that cohesins are excluded from active sites of transcription.

The binding of cohesin to chromosomes is not sufficient to generate cohesion. Once cohesins are loaded onto the chromatids, the action of yet another protein, called Eco1, is required for the establishment of cohesion, namely the physical linking of the two sister chromatids. In other words, in the absence of Eco1, chromosomes are decorated with cohesin complexes but the two sister chromatids

7.1 EXPERIMENTAL APPROACH

The advantage of being able to see: following chromosome movement in single-celled organisms

Fluorescence in situ hybridization (FISH) provides a snapshot of chromosome positioning

Often, significant leaps in our understanding of a cellular process have come from a new development in microscopy, which gives us a newfound ability to see structures or detect specific proteins, DNA or RNA sequences in the cell in a way that was not previously possible. Such leaps forward have been instrumental for our understanding of chromosome movement. It was known from the observation of large animal cells by light microscopy that chromosomes are dynamic structures, changing conformation and position throughout the cell cycle. The monitoring of individual chromosomes by light microscopy even allowed the detection of some chromosomal abnormalities. However, small changes in chromosome structure cannot be detected by light microscopy, and it is not possible to follow individual chromosomes in eukaryotic microorganisms such as yeast.

A powerful development in both research and diagnostics was fluorescence *in situ* hybridization, or FISH (see Section 19.9), wherein DNA is visualized through hybridization to a fluorescently labeled DNA probe. In this method, cells are immobilized on a slide, and the DNA is denatured. The fluorescent DNA probe is then added to the treated cells, where it anneals only to the complementary sequence located within the chromosome. With this technique, one can monitor any locus in the cell, in both mitotic and interphase cells, and one can detect even small deletions and chromosomal rearrangements if the probe is designed carefully.

Doug Koshland and colleagues modified the FISH method to directly observe yeast chromosomes in strains that were suspected to have defects in chromosome segregation. Unlike animal cells, individual yeast chromosomes cannot be distinguished by light microscopy. The investigators knew from genetic analysis that when *pds1-1* mutant cells were stalled in mitosis with the use of drugs, they lost viability when subsequently allowed to continue through the cell cycle. (By contrast, under the same conditions, wild-type cells remained viable.)

To determine the cause of this phenotype, they used FISH to examine sister chromatid cohesion in wild-type cells and in *pds1-1* mutant cells that were held in prometaphase. They found that in haploid wild-type cells in prometaphase, a FISH probe to a particular region on one of the chromosome arms gave only a single fluorescent signal because the sister chromatids were held very close together (see Figure 1a (top panel)). However, in *pds1-1* mutant strains, two fluorescent signals were observed, indicating that the sisters drifted apart (Figure 1a (bottom panel)). To substantiate their findings, the researchers looked at many wild-type and *pds1-1* mutant cells, and quantified the number of signal per nucleus (Figure 1b). Based on this and other results, they concluded

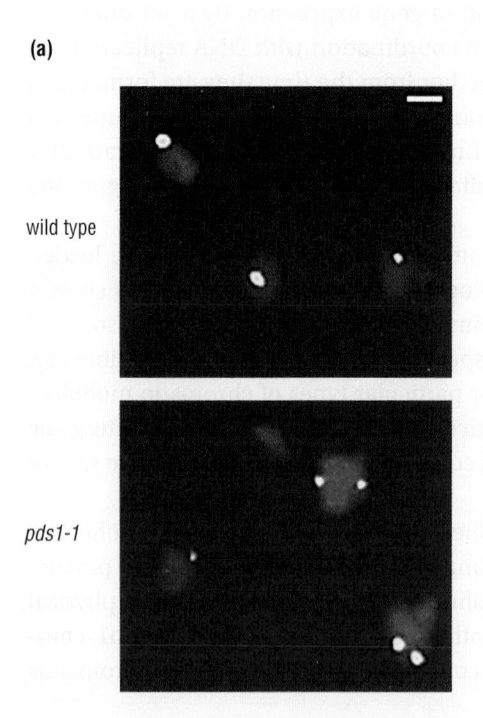

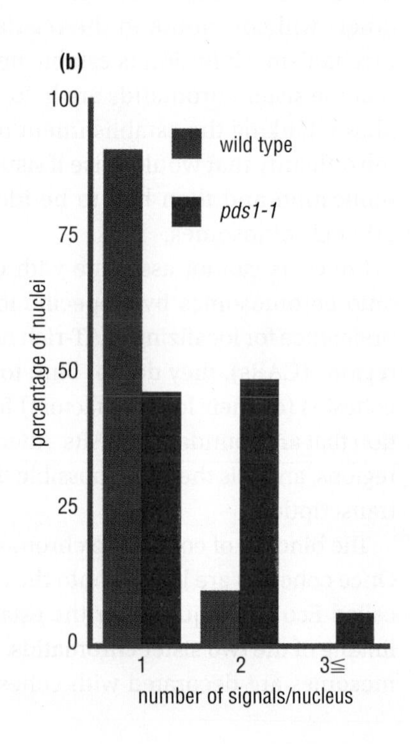

Figure 1 Fluorescence *in situ* hybridization (FISH). (a) Wild-type and *pds1-1* mutant cells were treated with the drug nocodazole, which depolymerizes microtubules. This results in the activation of the spindle checkpoint pathway and leads to a cell cycle arrest in prometaphase. The cells were then exposed to a FISH probe against a region of chromosome 16. In wild-type cells (top panel), sister chromatids maintain their cohesion, and consequently these cells mostly have one FISH signal. *pds1-1* mutant cells, on the other hand, fail to maintain cohesion, and a large fraction of cells exhibit two FISH signals (bottom panel). (b) A quantification of 200 nuclei such as those shown in (a).

From Yamomoto *et al.* Pds1p, an inhibitor of anaphase in budding yeast, plays a critical role in the APC and checkpoint pathway(s). *Journal of Cell Biology*, 1996; **133**:99–110.

that Pds1 (which stands for precocious dissociation of sisters) is required to maintain sister chromatid cohesion in metaphase.

Chromosome movement in live cells can be followed using a DNA binding protein fused to a fluorescent protein

Although the FISH method can be used to follow almost any chromosomal locus in any cell type, it cannot be used to follow chromosome movement in live cells. Live cell imaging is of particular importance when studying the dynamics of chromosome segregation. To follow a particular chromosomal region in live cells, Andrew Murray, Andy Belmont, and colleagues established a method where *in vivo* expressed fluorescent fusion proteins are targeted to specific loci in the genome through a DNA binding motif that recognizes an array of binding sites in the chromosome. This group took advantage of the tight binding between the *Escherichia coli* LacI transcriptional repressor and its DNA binding site, *lacO*. As shown in Figure 2a, an array containing many tandem *lacO* binding sites is integrated at a particular chromosomal locus such that when a LacI-GFP (green fluorescent protein) fusion protein is expressed in these cells, multiple LacI-GFP proteins bind to the *lacO* array. We note that in such an experiment, the signal from the many bound LacI-GFP proteins coalesces to a single fluorescent spot; a single LacI-GFP generates a signal that is too weak to detect by conventional fluorescence microscopy. This approach provides a powerful tool that enables researchers to follow chromosome movement in live cells.

In one nice example, Kerry Bloom and colleagues used this approach to follow a haploid budding yeast cell undergoing mitosis, as shown in Figure 2b. The yeast in this case were engineered to have a *lacO* array near the centromere of chromosome 3. As we see in Figure 2b, the yeast have undergone DNA replication, but because the sister chromatids remain next to each other until anaphase, only a single GFP spot is observed (left-most image). When anaphase begins, the sister chromatids separate, resulting in the appearance of two GFP spots, each corresponding to the centromeric region of the duplicated chromosome 3. As anaphase continues, the spindle (which cannot be seen) elongates, and the centromeres of the sister chromatids move away from one other, such that each cell (the mother and the bud) receives one copy of chromosome 3. This approach is useful for many applications, but it requires the insertion of the *lacO* array into the desired place in the genome.

Find out more

Straight AF, Belmont AS, Robinett CC, Murray AW. GFP tagging of budding yeast chromosomes reveals that protein-protein interactions can mediate sister chromatid cohesion. *Current Biology*, 1996;**6**:1599–1608.

Yamamoto A, Guacci V, Koshland D. Pds1p, an inhibitor of anaphase in budding yeast, plays a critical role in the APC and checkpoint pathway(s). *Journal of Cell Biology*, 1996;**133**:99–110.

Related techniques

FISH; Section 19.9

Light microscopy; Section 19.13

Use of GFP; Sections 19.10, 19.11

S. cerevisiae (as a model organism); Section 19.1

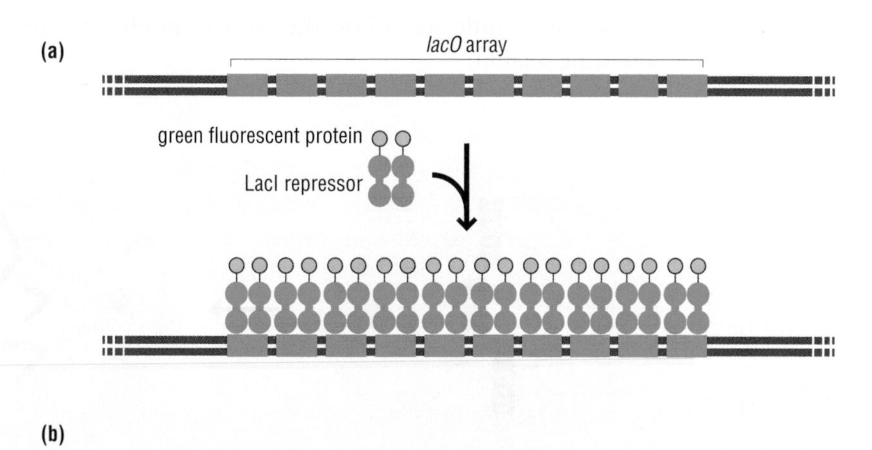

(b)

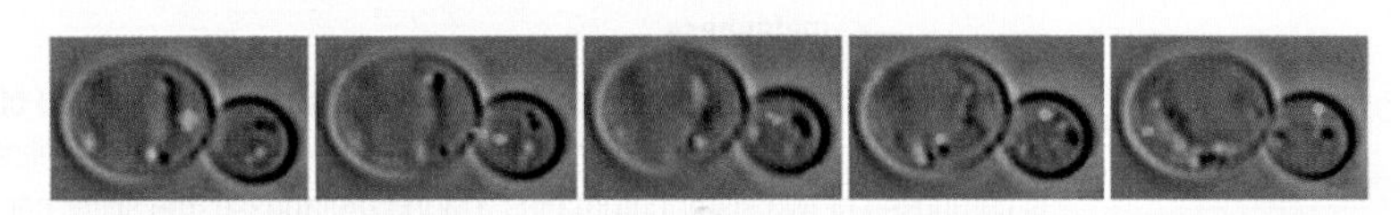

Figure 2 Following a specific chromosomal locus in live cells. (a) A schematic drawing of the *lacO* array/LacI-GFP system. (b) A haploid budding yeast cell expressing LacI-GFP (in green) and carrying a *lacO* array (256 repeats) near the centromere of chromosome 3. The first image on the left shows a live cell prior to sister chromatid separation. The following images (from left to right) show sister chromatid segregation during anaphase. Note that each cell received one copy of chromosome 3.

Figure 2b is from Thrower *et al*. Dicentric chromosome stretching during anaphase reveals roles of Sir2/Ku in chromatin compaction in budding yeast. *Molecular Biology of the Cell*, 2001;**12**:2800–2812.

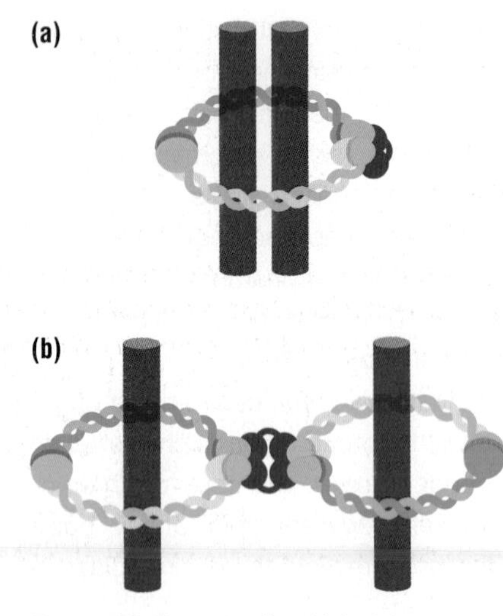

Figure 7.4 Two ways in which sister chromatids may be held together by cohesin. (a) Cohesin complexes may mediate sister chromatid cohesion by encircling two sister chromatids simultaneously (chromatids are represented by the gray rods). (b) An alternative proposal is that cohesin encircles a single chromatid and that chromatids are held together by interactions between cohesin complexes on different sister chromatids. In this representation, we show the two cohesin complexes interacting through their kleisin subunits, but other types of cohesin–cohesin interactions are possible.

are not associated with each other. Eco1 has acetyltransferase activity (that is, it transfers an acetyl group from one compound to another) and it acetylates one of cohesin's SMC subunits. This activity is essential for cohesion, but the exact mechanism is unknown.

While we don't yet know the mechanism by which cohesin complexes hold the sister chromatids together, the size of the SMC protein arms suggests the possibility that a single cohesin complex traps both sister chromatids (Figure 7.4a). Another possibility is that each cohesin complex encircles a single sister chromatid, and cohesion is mediated by the interactions between cohesin complexes on different sister chromatids (Figure 7.4b). In either case, it is thought that ATP binding and hydrolysis by the ATPase domain of the SMC heterodimer is responsible for conformational changes in the cohesin complex that allow it to open, encircle one or both chromatids, and then close again. Because cohesion is established as chromosomes duplicate, the physical proximity of the newly formed sister chromatids dictates that cohesin will link sister chromatids rather than unrelated chromosomes.

The exception to the coupling between cohesion establishment and DNA replication is when DNA breakage occurs after DNA replication. In this case, cohesin is loaded during G2/early mitosis in the vicinity of the break, as well as throughout the genome, and Eco1 then establishes cohesion at these sites. This extra cohesion is thought to facilitate the repair of the break by tethering the broken sister chromatid to its intact twin, which can serve as a template for repair. Indeed, a role for cohesin in repair of DNA damage is evident from the name of one of cohesin's subunits, Rad21, which stems from its initial identification as a gene that can lead to radiation sensitivity when mutated.

We will learn later in this chapter that the separation of sister chromatids during anaphase is induced by the proteolytic cleavage of the kleisin subunit of cohesin (see Section 7.4). This event inactivates the cohesin complex and destroys cohesion, allowing the sister chromatids to be pulled apart by the spindle microtubule, as shown in Figure 7.5. Thus, the next step in understanding chromosome segregation is to understand how the mitotic spindle functions, as described in the following section.

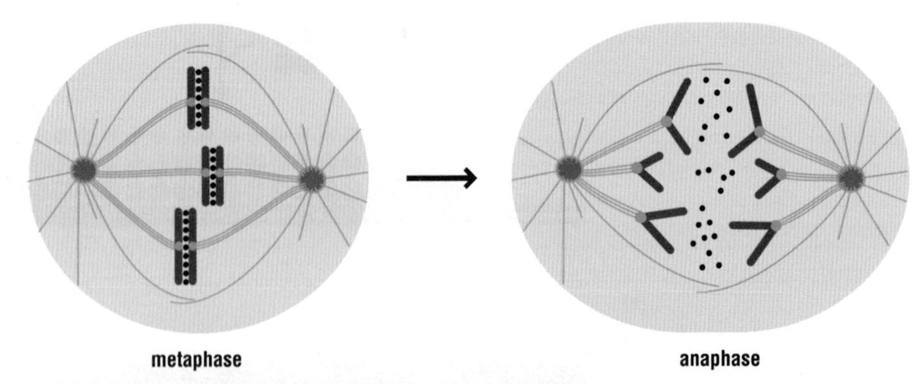

Figure 7.5 Sister chromatid cohesion is necessary for accurate chromosome segregation. Shown is a hypothetical cell with three different chromosomes after DNA replication, so that each chromosome is composed of two sister chromatids. During metaphase, the sister chromatids are held together by the cohesin complex (black dots). The spindle microtubules, shown in green, are pulling on the chromosomes but cohesion prevents the sister chromatids from separating. At anaphase, the cohesin complex dissociates from the chromosomes, allowing the sister chromatids to separate and be pulled away from each other by microtubules of the mitotic spindle. Had cohesin been absent, the sister chromatids would have separated immediately following DNA replication and would have attached at random to the spindle microtubules.

7.3 THE MITOTIC SPINDLE

Chromosome segregation is mediated by the mitotic spindle

A key feature of chromosome segregation is chromosome movement: first as the paired sister chromatids align during prometaphase and metaphase, and then as sister chromatids separate in anaphase (see Figure 7.1). Chromosome movement occurs through the attachment of chromosomes to the mitotic spindle. Therefore, to understand the process of chromosome movement we need to understand how chromosomes attach to spindle microtubules and how microtubules move chromosomes to the right place at the right time. In this section we will learn about the components of the spindle. In the coming sections we will then learn how chromosomes attach to the spindle and how the movement of chromosomes is regulated.

The spindle of most animal cells has two main structural components: microtubules, which are protein filaments, and two large structures called **centrosomes**. Figure 7.6 shows how one centrosome is located at each pole of the spindle. Spindle microtubules are anchored at one of their ends to a centrosome. At their other end, different types of microtubules can interact with different targets: a chromosome, a microtubule emanating from the opposite spindle pole, or the plasma membrane at the cell periphery. All three types of microtubules are important for proper chromosome segregation.

Microtubules that interact with chromosomes and facilitate their movement are called **kinetochore microtubules**, named after the chromosome-associated structure with which these microtubules interact (see Section 7.4). Microtubules that interact with other microtubules emanating from the opposite spindle pole are called **polar microtubules**, and they serve to elongate the spindle during anaphase (see Section 7.5). Finally, microtubules that extend toward the cell periphery are called **astral microtubules**, and they play a role in positioning the spindle within the cell throughout mitosis.

Microtubules are dynamic structures, undergoing periods of shortening and lengthening throughout their lifetime. Chromosome movement, spindle elongation, and spindle positioning are made possible by this dynamic behavior of microtubules and by the presence of specialized proteins, called **motor proteins**, which associate with microtubules. Motor proteins, as the name implies, have the ability to move proteins and other structures along microtubules, which serve as racetracks. Motor proteins utilize energy, in the form of ATP hydrolysis, to mediate this movement. Different motor proteins have different functions: some motor proteins

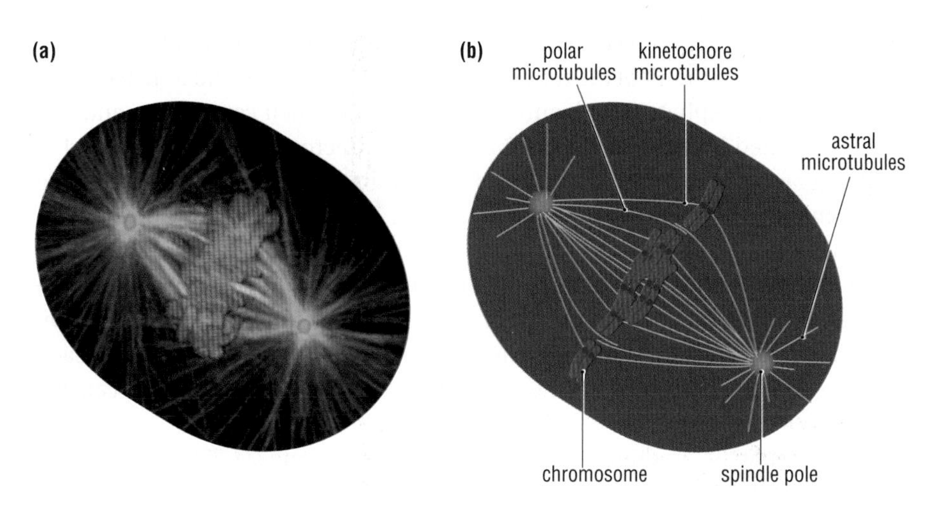

Figure 7.6 The mitotic spindle. (a) An image of a spindle from a cell in metaphase. The DNA is shown in pink, microtubules in green and γ-tubulin, a component of a structure that is located at the spindle pole, called the centrosome, is shown in blue. (b) A schematic diagram of a mitotic spindle. Microtubules (in green) are anchored at one of the spindle poles, to the centrosome (in blue). Microtubules that bind to chromosomes (in pink) are called kinetochore microtubules (see Section 7.5). Microtubules interacting with other microtubules emanating from the opposite spindle pole are polar microtubules. Microtubules emanating from the spindle poles toward the cell cortex are called astral microtubules.

Figure 7.6a adapted from Wong C, Stearns T. Mammalian cells lack checkpoints for tetraploidy, aberrant centrosome number, and cytokinesis failure. *BMC Cell Biology*, 2005;**6**:6.

facilitate the movement of chromosomes along kinetochore microtubules, some help slide polar microtubules relative to each other, some affect microtubule dynamics (namely lengthening and shortening), and some move astral microtubules along the cell cortex. We next describe the special properties of microtubules that promote their dynamic nature, and we will see how microtubules, centrosomes, and motor proteins create a functional spindle that can promote chromosome segregation.

Figure 7.7 Microtubules are hollow. A three-dimensional reconstruction of a section of a microtubule based on data from cryoelectron microscopy, showing the 13 side-by-side protofilaments that form the hollow microtubule.
Adapted from Li H, DeRosier DJ, Nicholson WV, Nogales E, Downing KH. Microtubule structure at 8Å resolution. Structure, 2002;**10**:1317–28.

Microtubules are made of tubulin dimers

Microtubules are hollow protein filaments, as shown in Figure 7.7. The building blocks of microtubules are **tubulin heterodimers** made of one **β-tubulin** and one **α-tubulin** subunit. Figure 7.8 shows how tubulin heterodimers join head to tail to form linear **protofilaments**; 13 protofilaments are aligned side by side to form a microtubule. Because the tubulin heterodimer is asymmetrical, microtubules have two distinct ends: the face of one end (called the **microtubule plus end**) is made of β-tubulin subunits whereas the face at the other end (called the **microtubule minus end**) is made of α-tubulin subunits. In the spindle, the minus ends of microtubules are usually anchored at the centrosome whereas the plus ends extend toward chromosomes, the cell periphery, or microtubules from the opposite spindle pole.

In addition to their role in the spindle, microtubules are involved in other various cellular processes, such as intracellular transport, cell motility, and cell shape. In animal cells during interphase (Figure 7.1, part 1), microtubules form an extensive network that serves as a highway system along which motor proteins move macromolecular assemblies and vesicles within the cell. The interphase microtubule network disassembles before mitosis, and the tubulin subunits become dedicated to spindle assembly.

Microtubules are dynamic structures

Microtubules grow and shrink rapidly in the early stages of mitosis. This property is referred to as **dynamic instability**. Since microtubules find their targets by chance encounters, the dynamic nature of microtubules allows them to probe a large volume of the cell, increasing the likelihood that they will encounter their target. The changes in microtubule length are due to the addition or removal of tubulin heterodimers at microtubule ends, with tubulin exchanging faster in the plus end than the minus end. It is important to note that tubulin heterodimers can only be added or removed from microtubule ends. The process of rapid microtubule shortening is called **catastrophe**, and the reverse process, during which microtubules grow again, is called **rescue**. These two processes are depicted in Figure 7.9.

The β-tubulin subunit of the tubulin heterodimer is able to bind and hydrolyze guanosine triphosphate (GTP), and the transition between catastrophe and rescue is governed by whether the β subunits at the microtubule end are bound to GTP or guanosine diphosphate (GDP). Free tubulin heterodimers that are added to the microtubule end contain β-tubulin in a GTP-bound state. Once bound, this GTP is gradually hydrolyzed to GDP. GTP-bound tubulin heterodimers have a high affinity toward each other; at the microtubule end, they tend to associate and recruit additional GTP-bound tubulin, causing the microtubule to grow. On the other hand, GDP-bound tubulin heterodimers have a low affinity toward each other; and, when present at the microtubule end, they tend to dissociate, causing the microtubule to shorten.

When tubulin heterodimers are added faster than the rate of GTP hydrolysis, the microtubule will have a cap of GTP β-tubulin, which protects the microtubule from

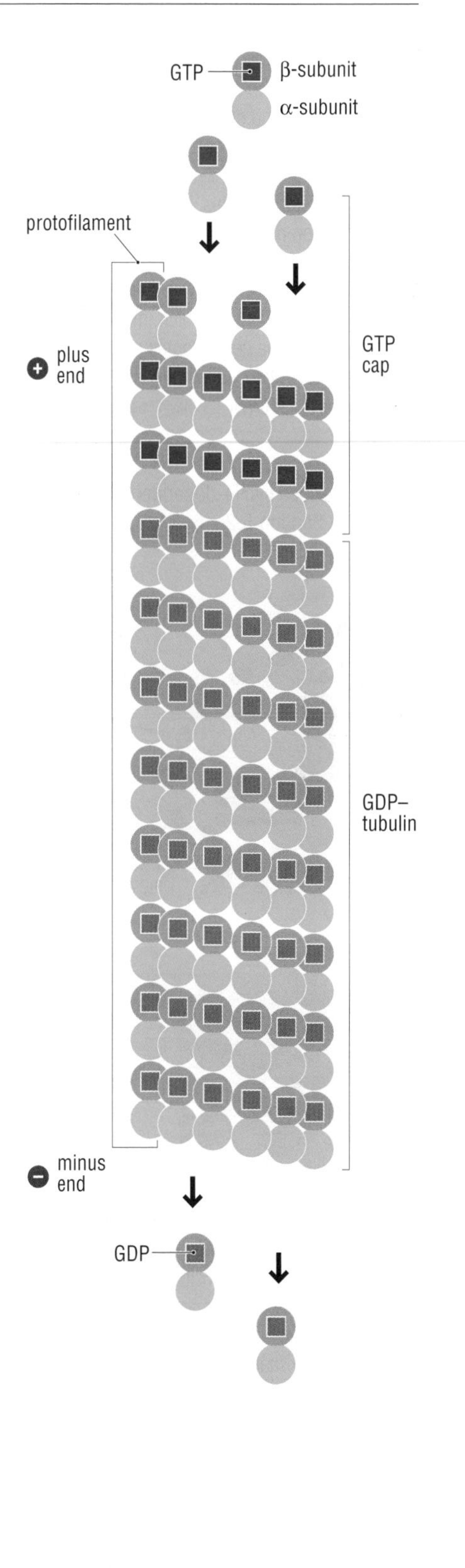

Figure 7.8 Microtubules are made of tubulin dimers. Tubulin heterodimers are made of α-tubulin (light-green spheres) and β-tubulin (dark-green spheres). The β-tubulin subunit can bind guanosine triphosphate (GTP; pink square) and hydrolyze it to guanosine diphosphate (GDP; gray square). Tubulin heterodimers bind head to tail (namely the α-tubulin subunit of one dimer binds to the β-tubulin subunit of another) to form long protofilaments. The protofilaments line up next to each other to form a cylindrical, hollow microtubule. The protofilaments are slightly off-register, creating a helical pitch. The end of the microtubule that has the α-tubulin subunits exposed is designated the minus end, and the end of the microtubule that has the β-tubulin subunits exposed is designated the plus end. Tubulin heterodimers can only be added or removed from the ends of the microtubule, with the minus end being more stable (namely tubulin heterodimers are added or removed more slowly) than the plus end.

disassembly and allows further heterodimers to be added (see Figure 7.9). However, if the rate of tubulin heterodimer addition is slow, such that GTP is hydrolyzed to GDP faster than new GTP–tubulin heterodimers are added, then heterodimers at the microtubule end will be GDP bound, causing these heterodimers to dissociate from the microtubule end so that the microtubule shortens. (Remember that tubulin heterodimers can only be added or removed from the microtubule end.) When the rate of GTP-tubulin addition increases again, the process of microtubule shortening will be rescued, and the microtubule will once again grow.

These cycles of catastrophe and rescue are responsible for the dynamic nature of microtubules and play an important role in chromosome movement, as we will see in the next section.

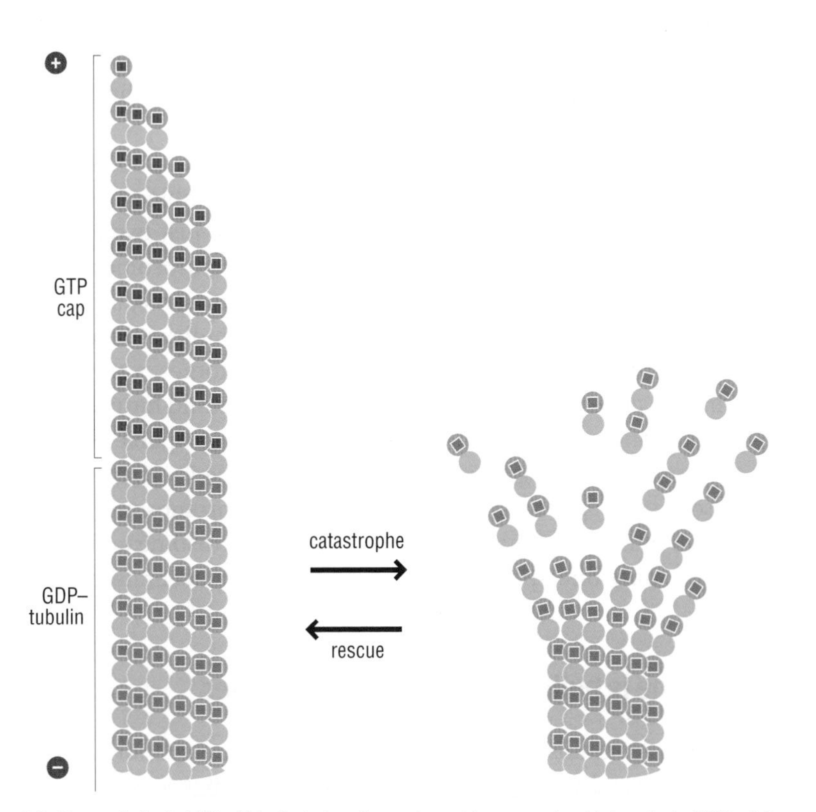

Figure 7.9 Dynamic instability. Tubulin heterodimers bound to guanosine triphosphate (GTP; pink squares) form stable interactions within the microtubule, whereas tubulin heterodimers bound to guanosine diphosphate (GDP; gray squares) form weak interactions, causing the microtubule to depolymerize. Microtubule depolymerization involves the dissociation of tubulin heterodimers as well as the splaying apart of the protofilaments. When the rate of GTP-bound tubulin addition is slower than that of GTP hydrolysis, the GTP cap is lost and the microtubule begins to depolymerize (catastrophe). When the GTP cap re-forms, growth can resume (rescue). Repeated episodes of catastrophe and rescue contribute to the dynamic instability of microtubules.

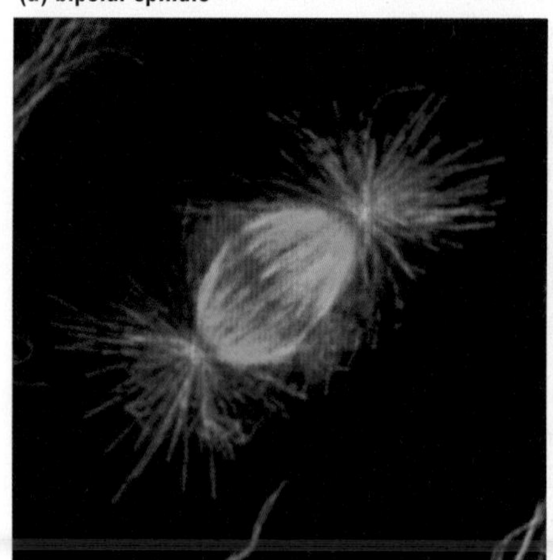

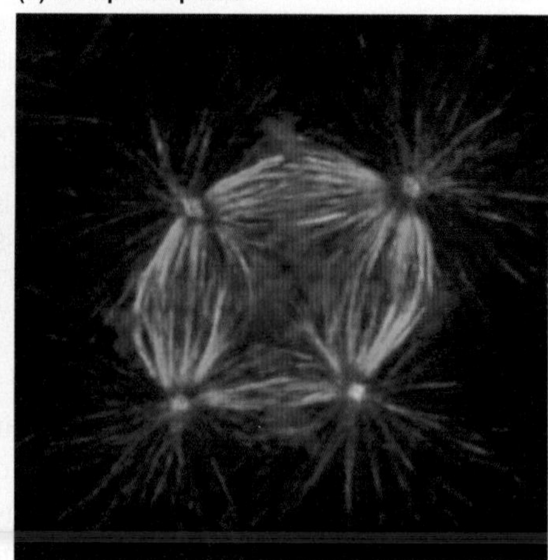

Figure 7.10 Centrosomes dictate the structure of the spindle. Errors in cell division, drug treatments, and defects in centrosome duplication can result in cells with an abnormal number of centrosomes. Since centrosomes nucleate microtubules, this, in turn, can lead to the formation of an abnormal spindle. (a) A normal bipolar spindle with two centrosomes and (b) a multipolar spindle with four centrosomes. The DNA is in blue, microtubules are in green and the centrosomes are marked by a centrosomal protein, called pericentrin, in red.

Images courtesy of Neil Ganem and David Pellman.

We discuss cyclin-dependent kinases (Cdks) in Section 5.2.

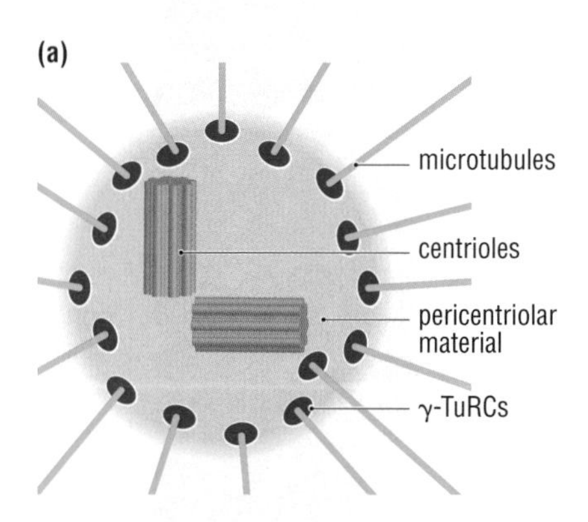

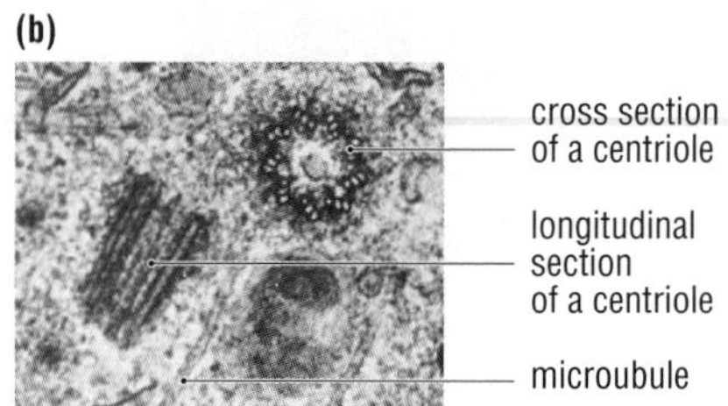

Figure 7.11 The structure of the centrosome. (a) Each centrosome contains a pair of centrioles (shown in green) that are positioned at right angles to each other. The centrioles are surrounded by pericentriolar material (shown in gray). Within the pericentriolar material are multiple γ-tubulin ring complexes (γ-TuRCs) (shown in pink), which nucleate microtubules (green). γ-tubulin serves as a template for microtubule formation and caps the microtubule at the minus end. (b) Electron micrograph of centrioles within a centrosome. Because of the orientation of the centrioles relative to each other, one centriole is seen in cross-section whereas the other is seen in a longitudinal section. Each centriole is composed of a cylindrical array of nine groups of three microtubules, as can be seen in the cross-section.

Micrograph in Figure 7.11b from McGill M, Highfield DP, Monahan TM, Brinkley BR. Effects of nucleic acid specific dyes on centrioles of mammalian cells. *Journal of Ultrastructure Research*, 1976;**1**:43–53.

The centrosome is a microtubule-nucleating structure that duplicates once per cell cycle

The centrosome is a cytoplasmic structure in animal cells that provides a platform on which microtubules form, or 'nucleate'. Cells normally inherit a single centrosome from their parent cell. During interphase, this centrosome serves as a cytoplasmic microtubule-organizing center for the cell-wide microtubule network (Figure 7.1, part 1). Toward the end of G1, the action of G1 and S phase cyclin-dependent kinases triggers centrosome duplication, marking the beginning of spindle formation.

Upon entry into prophase, the interphase network of microtubules disassembles, and the two centrosomes separate and begin to assemble a mitotic spindle (see Figure 7.1). Normally, at this point, a cell will have just two centrosomes. However, if a cell ends up with an aberrant number of centrosomes, the spindle could adopt an abnormal shape. If there is a single centrosome, a situation that can arise from a failure in centrosome duplication, a spindle will form with only one pole. Conversely, if there are more than two centrosomes, a situation that is often seen in cancer cells, a spindle with multiple poles could form, as illustrated in Figure 7.10. Monopolar and multipolar spindles are incapable of supporting accurate chromosome segregation; only a bipolar spindle is capable of supporting chromosome segregation such that two identical chromosome clusters will form at the end of mitosis. Thus, inheriting a single centrosome and duplicating the centrosome only once per cell cycle are crucial for accurate chromosome segregation.

The centrosome is built around a pair of centrioles

Each centrosome contains a pair of **centrioles**, which are cylindrical structures made of short microtubules that are positioned at right angles to each other; this structure is depicted in Figure 7.11. Centrioles perform two functions. First, they are required for centrosome duplication. Centrosome duplication begins with the separation of the centrioles, followed by centriole duplication. The mechanisms that promote centriole duplication and restrict centrosome duplication to once per cell cycle are not yet understood, although they clearly are a crucial step in the formation of a proper spindle.

Centrioles are also required to recruit proteins that form the pericentriolar matrix, a protein-dense region that surrounds the centrioles. One of the components of the pericentriolar matrix is a type of tubulin called γ-tubulin, which, together with several other proteins, makes up γ-tubulin ring complexes (γ-TuRCs). These complexes provide templates onto which the α/β tubulin heterodimers assemble to form the cylindrical microtubule - in other words, γ-TuRCs act as the foundations on which the microtubules are constructed.

It is important to note that spindles can also form independently of centrosomes. For example, in female meiosis (the chromosome segregation process that generates oocytes), spindles are formed entirely without centrosomes. In this case, spindle self-assembly is attributed to the ability of chromosome-associated proteins to nucleate microtubules and to the activity motor proteins that cross-link microtubules to form a stable bipolar array of microtubules. Centrosome-independent spindle assembly is subjected to cell cycle regulatory processes, such that spindle self-assembly does not happen before nuclear envelope breakdown. It is thought that even in cells that do have centrosomes, both centrosome-dependent and centrosome-independent microtubule nucleation events contribute to spindle assembly.

The spindle is organized by the action of motor proteins

In addition to centrosome duplication, the formation of a bipolar spindle also requires the activity of microtubule-associated proteins, commonly referred to as MAPs, which include motor proteins. The motor proteins that are associated with the spindle have two functional domains: a motor domain that promotes the movement of the motor along the microtubule, and a non-motor domain, called a cargo-binding domain, which binds to particular cellular structures depending on the cellular context in which they are operating. These two domains are depicted in Figure 7.12a.

During interphase, the cargo domain of certain motor proteins may bind proteins that are to be transported along microtubules. By contrast, in mitosis, motor proteins contribute to chromosome segregation in a variety of ways: the cargo domains of some motor proteins bind to chromosome-associated proteins and facilitate chromosome movement along microtubules. The cargo-binding domain of other motor proteins can bind proteins associated with the plasma membrane, while their motor domain binds a microtubule. In this configuration, motor proteins can reel in microtubules toward the cell periphery, helping to position the spindle within the cell. In some cases, motor proteins do not have a cargo-binding domain but rather have motor domains at both ends (see Figure 7.12b). Such motor proteins can bind microtubules at both ends and they can help slide microtubules relative to each other.

The motor domains of particular motor proteins can move in one direction only - some motor proteins move toward the microtubule plus end whereas others move toward the minus end. One way to determine the direction of movement of a particular motor protein is to mark it with a fluorescent probe and to follow its movement on purified microtubules under the microscope. A combination of different motor proteins, some that move to the plus end and others that move to the minus end, is needed for the variety of microtubule movements to take place. The two main types of motor protein involved in spindle assembly and function are the kinesin-like motor proteins and dynein.

Kinesin-like motor proteins are typically plus-end-directed motors, while dynein is a minus-end-directed motor. Kinesin motor proteins that have motor

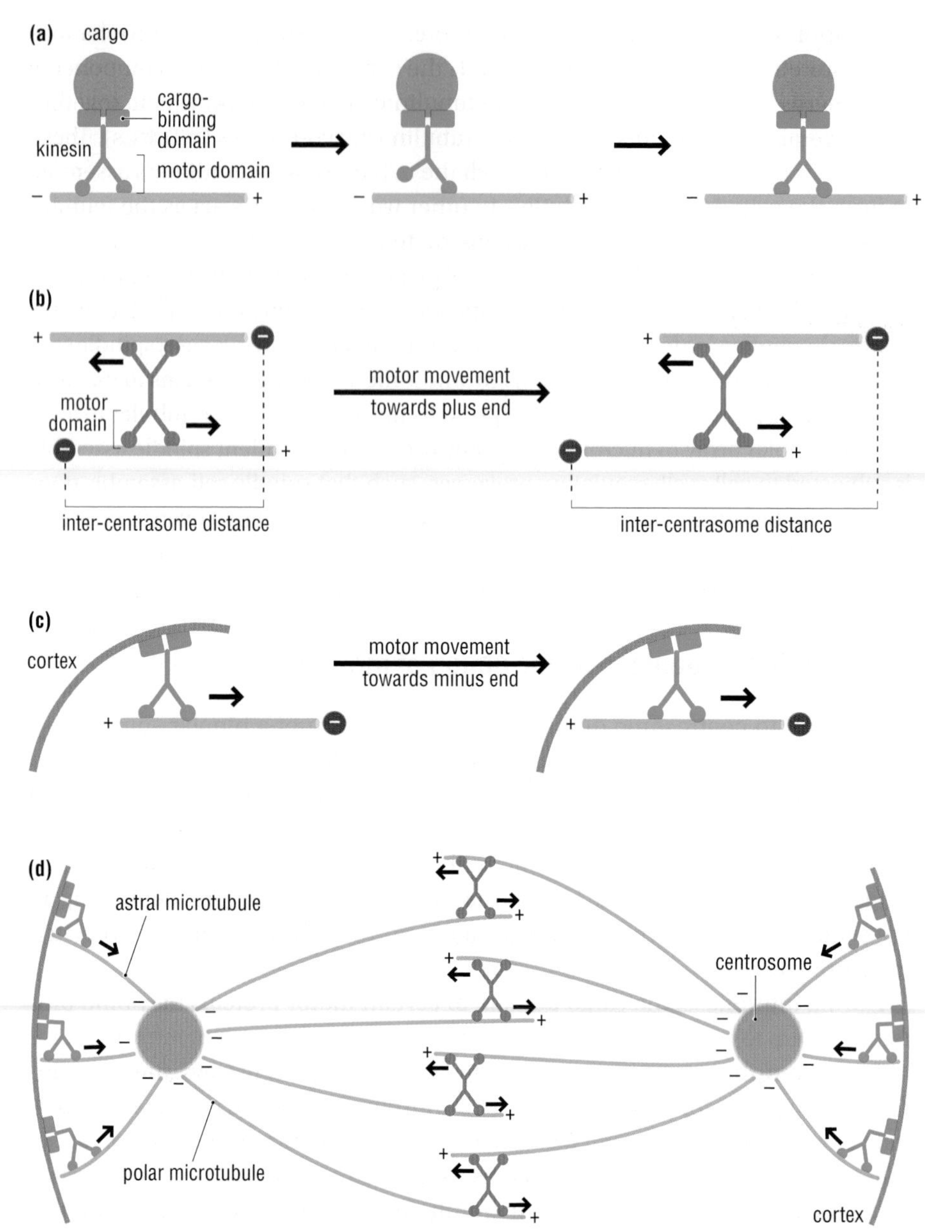

Figure 7.12 Motor proteins. (a) The motor protein shown (turquoise) is a homodimeric plus-end-directed motor containing two motor domains that are associated with a microtubule (green), and two cargo-binding domains associated with a cargo (orange). Movement of the motor protein is mediated by the ATPase activity of its motor domains, which leads a conformation change in the motor domain, causing the motor to 'walk' along the microtubule toward the plus end. As the motor moves along the microtubule, so does its cargo. (b) The motor protein shown (in blue) has plus-end-directed motor domains at both its ends. Therefore, it can cross-link polar microtubules. These microtubules are attached to a centrosome at their minus end (in gray). When the motor walks toward the microtubule plus end, its activity moves the microtubules relative to each other, thereby increasing the distance between the two centrosomes. (c) The cargo-binding domain of the motor protein shown (in turquoise) is bound to the cell cortex. The motor domain has minus-end-directed motor activity. The motor protein is anchored to the cell cortex and is therefore stationary. As this motor protein moves toward the microtubules minus end, it reels in the microtubule, along with its associated centrosome. As a result the centrosome gets closer to the cortex. Motors that have this kind of activity include the dynein family of motor proteins. (d) The actions of different motor proteins in the spindle (the chromosomes are not shown). A plus-end-directed motor protein with motor domains at both ends cross-links the polar microtubules and pushes the spindle poles apart. Cytoplasmic dynein anchors the plus ends of microtubules at the cell cortex and reels in the centrosome toward the cortex. Not shown are motor proteins associated with chromosomes that help move chromosomes along microtubules, as well as motor proteins that affect microtubule depolymerization.

domains on both ends can bind simultaneously to two microtubules from opposite spindle poles. Because these microtubules are in opposite orientations, the movement of each motor domain toward the plus ends pushes the spindle poles apart. This action is important both in separating the centrosomes after their duplication and in elongating the spindle during anaphase. When dynein is localized to the cell cortex, its minus end motor activity reels in the microtubule toward the cell cortex, pulling the whole spindle with it. This action helps position the spindle in the cell, as shown in Figure 7.12c.

The actions of motor proteins in shaping and positioning the spindle are summarized in Figure 7.12d. Many other non-motor proteins participate in spindle assembly, including some that provide structural support by bundling microtubules, and some that act in a regulatory capacity to affect microtubule dynamics.

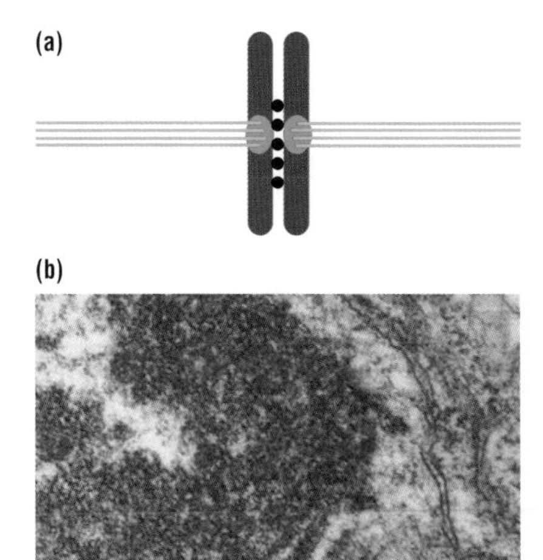

Figure 7.13 Structure of the kinetochore. (a) Microtubules (green) attach to chromosomes (gray rods) through a multisubunit protein structure called the kinetochore (orange). Kinetochores of animal cells can bind more than 20 microtubules. In metaphase, the pulling forces exerted by the microtubules are opposed by the presence of cohesin complexes (in black) that link the two sister chromatids. (b) The inner and outer layers of an animal cell kinetochore can be distinguished in the electron micrograph, which shows a kinetochore bound to microtubules.

Figure 7.13 b courtesy of Jeremy Pickett-Heaps.

7.4 PROMETAPHASE AND METAPHASE

The initial steps of spindle assembly occur in the cytoplasm, before nuclear envelope breakdown (see Section 7.1). However, the completion of spindle assembly, which happens when the spindle microtubules engage with the chromosomes once the nuclear envelope disassembles, occurs in prometaphase, as we describe next.

Chromosomes attach to the spindle in prometaphase

At the end of prophase, the nuclear envelope breaks down, allowing spindle microtubules to gain access to the chromosomes. This marks the beginning of prometaphase, during which chromosomes become associated with microtubules of the mitotic spindle. Prometaphase ends when all the chromosomes are properly attached to spindle microtubules and the chromosomes have moved to the spindle mid-zone. At this point, the cell is said to be in metaphase.

Microtubules interact with chromosomes through a complex called the kinetochore, illustrated in Figure 7.13, which assembles at a specified chromosomal locus called the centromere. In most cell types each chromosome has one centromere. However, in some cell types, such as somatic cells of the nematode *Caenorhabditis elegans*, chromosomes have multiple centromeres distributed along the entire length of the chromosomes (See Section 4.10). We will first discuss the structure and function of centromeres before considering the properties of kinetochores.

Centromeres specify the site at which kinetochores assemble

Centromeres are the chromosomal loci at which kinetochores, the complex that links chromosomes to the mitotic spindle, assemble. Centromeres can be classified into two types based in part on their size and complexity: point centromeres and regional centromeres. (Look back at Figure 4.32 to see the distinction.) Point centromeres are typical of certain fungi, such as the budding yeast *S. cerevisiae*; they are small in size (the budding yeast centromere is 125 base pairs (bp)), have a defined sequence, and lead to the association of a single microtubule with the chromosome. In contrast, regional centrosomes, which are characteristic of most other eukaryotes including fission yeast, flies, and mammals, range in size from tens to hundreds of kb. They are not defined by a specific sequence but rather contain repetitive DNA sequences, and they promote the assembly of more complex kinetochores that lead to the association of multiple microtubules with the chromosome.

➔ We previously learned about centromeres in Chapter 4 when we discussed the various structural elements of chromosomes.

Centromeres have a distinct chromatin structure

The centromere is defined by the presence of CENP-A, a histone H3 variant that replaces histone H3 in centromeric nucleosomes. (CENP-A is also called CenH3 in more recent publications.) Point centromeres have a single nucleosome containing CENP-A, while in regional centromeres nucleosomes with CENP-A are interspersed between nucleosomes containing the canonical histone H3. CENP-A is critical for centromere function: cells depleted of CENP-A fail to assemble functional kinetochores. The structure of the CENP-A containing nucleosome is still a subject of much debate, and thus is it not clear how CENP-A, but not the canonical histone H3, promotes kinetochore assembly.

In regional centromeres, the region containing CENP-A is defined as the centromere's core. The chromatin structure of this region is unique and is unlike euchromatin or heterochromatin: the histone H3 in this region is mostly devoid of both acetylation and methylation on lysine 9 (which are markers of euchromatin and heterochromatin, respectively). This unique pattern may help distinguish this region from the rest of the chromosome, dictating the assembly of the kinetochore at only one specific locus. The centromere's core is surrounded by pericentromeric heterochromatin with unique modifications to histone H3 and H4. The histone modifications are typical of silenced chromatin, and indeed genes placed at heterochromatic regions tend to be silenced.

➔ We discuss histone modifications in Section 4.5.

Since CENP-A association with chromatin can confer centromere function onto that region, it is important to restrict CENP-A deposition to only one specific region of the chromosome. In budding yeast, this is accomplished by limiting the amount of CENP-A through protein degradation. Moreover, CENP-A deposition requires to the activity of a specific histone chaperone, called Scm3 in yeast and HJURP in mammals. However, it is not entirely clear how these CENP-A chaperones are targeted specifically to sites designated for centromere function.

➔ We learn more about the RNAi machinery in Chapter 13.

The RNAi machinery is required for pericentric heterochromatin

In addition to the presence of CENP-A at the centromere core, the presence of heterochromatin in the pericentric region is critical for centromere function. How does this heterochromatin state persist? Recent studies in fission yeast have revealed that the RNAi machinery plays a major role in this process. Briefly, although the heterochromatin state in the pericentric regions can silence genes, the DNA repeats are transcribed during S phase, resulting in the formation of double-stranded RNA. These RNAs are processed by the RNAi machinery to produce short double-stranded RNAs; these dsRNAs interact with additional proteins to direct heterochromatin formation, as shown in Figure 7.14. Thus, the transient transcription of DNA sequences at pericentric regions is responsible for directing the heterochromatic state at these sites. A similar mechanism is used at the fission yeast mating type locus whereby RNAi can induce heterochromatin formation (Section 13.6). Thus, the combination of CENP-A at the centromere's core, along with the pericentric heterochromatin, provide the appropriate chromatin structure for kinetochore assembly.

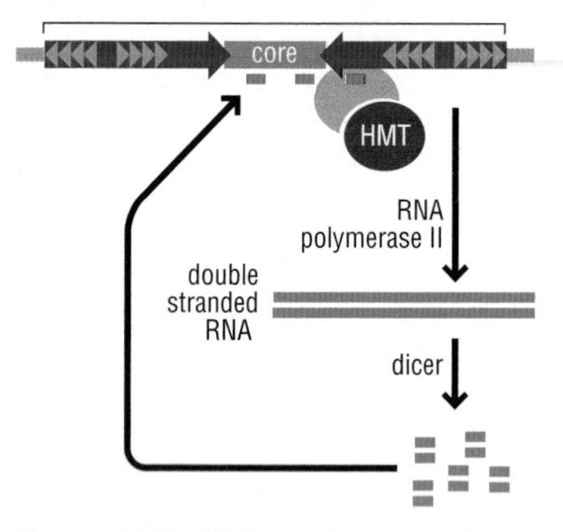

Figure 7.14 The RNAi machinery is required for pericentric heterochromatin. The repetitive sequences at the pericentric region of the *S. pombe* centromere are transcribed by RNA polymerase II. This results in the formation of double-stranded RNA, which is processed by an enzyme called Dicer to yield short double-stranded RNA segments, about 20 base pairs (bp) in length. One strand of these RNAs associated with a protein complex (called RITS, in gray) and targets it back to the pericentric region, where the protein complex recruits a histone methyltransferase (HMT, in red). The methyltransferase then methylates the appropriate histone residues, reestablishing a heterochromatin state.

The kinetochore links chromosomes to spindle microtubules

The kinetochore, the large protein structure that functions at the centromere, has multiple functions. As well as interacting with microtubules, it associates with motor proteins that help move chromosomes along microtubules. It also ensures that chromosomes are oriented properly with respect to the spindle. Sister chromatids need to move away from each other during chromosome segregation, so the binding of microtubules to chromosomes must be coordinated such that sister chromatids bind microtubules that originate from *opposite* spindle poles. Proteins associated with the kinetochore have the ability to distinguish between correct and incorrect kinetochore–microtubule attachments and detach improperly attached kinetochores from microtubules. In addition, certain kinetochore proteins have the ability to induce a cell cycle arrest if kinetochores are not attached to microtubules, as we discuss further in Section 7.5. Next we will discuss the properties of the kinetochore and how correct kinetochore–microtubule attachments are formed.

Kinetochores vary in size but have conserved proteins and functions

Like centromeres, which vary in size from around 100 bp in budding yeast to hundreds of kb in animal cells, kinetochores also vary greatly in size. The budding yeast kinetochore binds a single microtubule, whereas animal cell kinetochores are much larger and can bind more than 20 microtubules. Nonetheless, kinetochore proteins are highly conserved in species from yeast to humans, suggesting that the basic mechanism for microtubule binding by kinetochores is evolutionarily conserved.

The identification of the exact protein composition of a kinetochore has been very challenging because even the simplest known kinetochore, that of budding yeast, is over 5 megadaltons (MDa) in size, which is twice as large as a ribosome. Despite these challenges, many kinetochore proteins have been identified in simple model organisms through genetic screens for mutations that disrupt chromosome segregation. With the advent of protein identification techniques such as mass spectrometry, some kinetochore proteins were identified biochemically, for example, by purifying known kinetochore proteins and identifying other proteins with which they associate. In all cases, the designation of a protein as a kinetochore protein is done only after confirming that it associates with the centromeric region of the chromosome.

To date, more than 80 proteins that participate in building a kinetochore of animal cells have been identified; some have been shown to have structural, motor, or regulatory functions, but the function of most is unknown. We explore how kinetochore proteins have been identified in more detail in Experimental approach 7.2.

Many kinetochore-associated proteins are called CENP-X, where CENP stands for centromere protein, and X represents the order in which the proteins were discovered: the list of CENP proteins currently runs from A through U. (We encountered CENP-A when discussing centromere structure.) Kinetochore proteins can be roughly assigned to two groups based on their physical proximity to the centromere: inner kinetochore proteins, which are physically associated with the DNA or located very close to it, and outer-kinetochore proteins, which are further away from the centromere. Several of the proteins defined as inner kinetochore proteins are constitutively associated with the centromere, including the histone H3 variant CENP-A. Others, including all of the outer-kinetochore proteins, only join the kinetochore during mitosis.

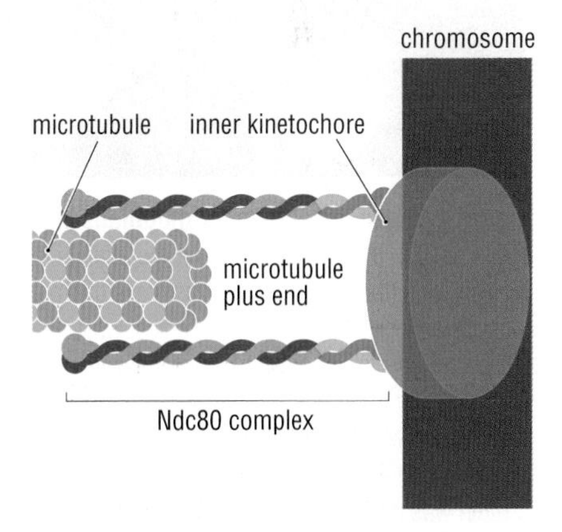

Figure 7.15 Proposed organization of the protein complexes that attaches kinetochores to microtubules. The Ndc80 complex (in blue and red), which is part of the outer kinetochore, binds directly to microtubules (in green). Note that the end of the microtubule is available for the addition of tubulin. When microtubules depolymerize, kinetochore proteins slide further toward the minus end, thus maintaining an association between the kinetochore and the microtubule.

Microtubules are captured by proteins of the outer kinetochore

Kinetochores interact with microtubule plus ends. The association between the kinetochore and microtubules is mediated by the conserved rod-shaped outer-kinetochore subcomplex, called the Ndc80 complex, as illustrated in Figure 7.15. However, microtubules can grow and shrink at the plus end. So how can kinetochores maintain an attachment to dynamically unstable microtubules?

7.2 EXPERIMENTAL APPROACH

The hunt for kinetochore proteins

A genetic screen in yeast identifies genes needed for faithful chromosome segregation

Kinetochore structures in metazoans have long been observed by microscopy, but the molecular composition of these structures was not known. To identify kinetochore components, Phil Hieter and colleagues conducted a genetic screen in budding yeast. As we saw in the screen for cell cycle mutants in Experimental approach 5.1, key in the design of any genetic screen are (1) the predicted phenotype(s) that might arise from a mutation in a genes involved in the process of interest, and (2) the means to detect the predicted phenotype. When screening for mutations in genes involved in kinetochore function, the Hieter group predicted that such mutations would lead to increased rates of chromosome loss. As a strategy for detecting this predicted loss, the investigators followed the presence (or loss) of a non-essential truncated short chromosome (called a chromosome fragment) containing a marker gene that affects the color of the colony. When the chromosome fragment, and with it the marker gene, is lost, red sectors appear in otherwise white yeast colonies, as depicted in Figure 1.

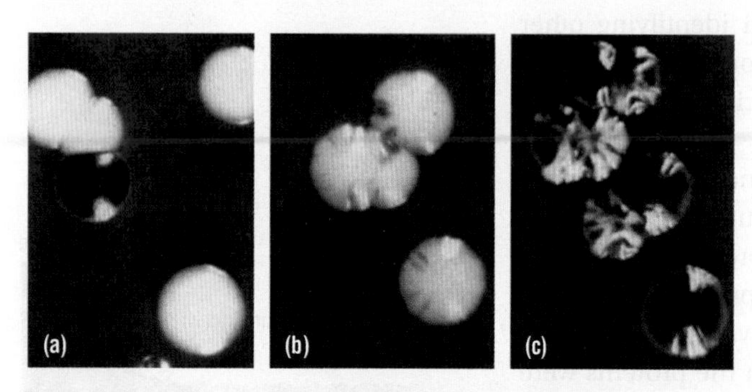

Figure 1 A genetic screen for mutations in kinetochore genes. The yeast strain used in this screen carried a mutation in a gene involved in adenine biosynthesis (the *ADE2* gene). Yeast cells that are proficient in adenine biosynthesis give rise to white colonies, while yeast carrying the *ade2-101* allele give rise to red colonies. In this assay, cells also carried a non-essential chromosome fragment that had a centromere and a gene for a tRNA suppressor, which allowed the synthesis of the complete Ade2 protein from the *ade2-101* allele. Therefore, when the tRNA suppressor was present, the Ade2 protein was made and the colonies were white. When the suppressor was missing, for example due to the loss of the chromosome fragment, the Ade2 protein could not be made, and the colonies were red. Loss of chromosome fragment during the growth of the colony resulted in the appearance of a red sector (this appears as a black sector in the black and white images). Wild-type cells (panel a) give rise to mostly white colonies, unless the chromosome fragment was lost before the cells were plated. Mutations that affected kinetochore function (b – a mild defect; c – a severe defect) increased the loss rate of the chromosome fragment, resulting in the appearance of red sectors. Adapted from Spencer *et al.* Mitotic chromosome transmission fidelity mutants in *Saccharomyces cerevisiae. Genetics,* 1990;**124**:237–249.

The use of the chromosome fragment that contained a functional centromere was a critical aspect of this screen. Since kinetochores are essential for chromosome segregation, mutations that severely diminish kinetochore function would likely result in massive chromosome loss, and thus in cell death. However, as a chromosome fragment is less stable than a normal chromosome, the group was able to isolate phenotypically mild mutations that had a more pronounced effect on the chromosome fragment compared to an endogenous chromosome, thus having a minimal effect on cell viability. We might refer to this strain carrying the chromosome fragment marker as a 'sensitized' background for conducting the screen.

To conduct the screen, Hieter and colleagues treated yeast cells with a chemical that induced point mutations, and then isolated cells that showed a high rate of chromosome fragment loss, as evidenced by the appearance of red sectors within the white colonies. Through this screen, the researchers isolated over 130 independent mutants that corresponded to mutations in 11 different genes. After their initial identification, evidence that these proteins were *bona fide* kinetochore proteins was still needed. Subsequent studies have shown that while some of the genes identified in this screen code for kinetochore proteins that are conserved through the vertebrate lineage, others are involved in processes such as sister chromatid cohesion. This result was not surprising given that defects in sister chromatid cohesion would also be expected to increase chromosome loss rate.

Biochemical isolation of novel kinetochore proteins that physically interact with a previously identified centromeric protein, CENP-A

Kinetochore proteins were also identified through biochemical approaches. Investigators suspected the kinetochore to be a large complex, but the identity of only a few kinetochore proteins was known. Knowing that the kinetochore assembles at the centromere, Don Cleveland and colleagues turned to biochemistry in order to isolate proteins that physically interacted with the centromeric protein CENP-A. The researchers designed a CENP-A gene fused to a protein tag and expressed it in mammalian cells, as depicted in Figure 2. Immunoprecipitation was then used to purify the tagged CENP-A and its associated proteins from a protein extract prepared from these cells. The purified proteins were separated on a gel and identified by mass spectrometry.

In this analysis, the Cleveland group found a number of proteins that were associated with CENP-A. The group then pursued two approaches to determine if these proteins were indeed kinetochore proteins. First, the researchers generated gene fusions

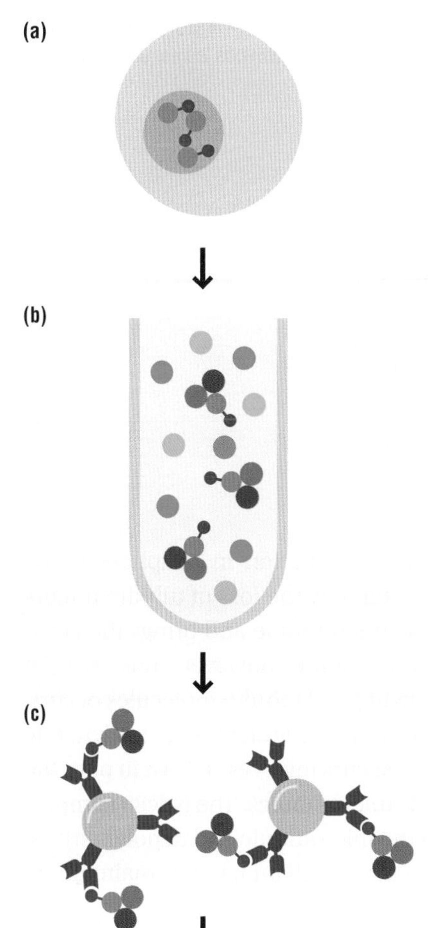

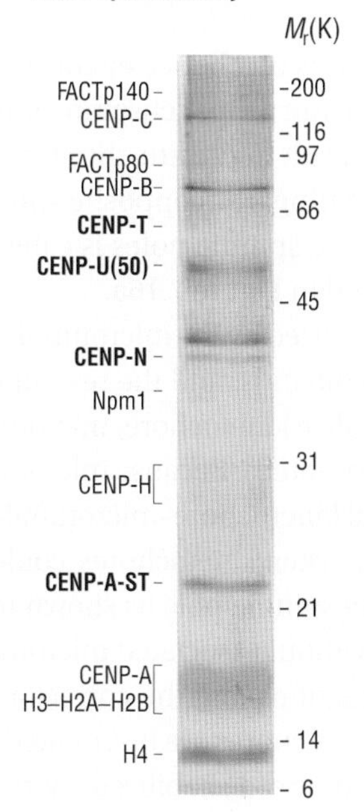

Figure 2 Identifying kinetochore proteins through a biochemical approach. To purify kinetochore components, researchers engineered cells to synthesize CENP-A (shown in blue) that was fused to a protein tag (in red) for which antibodies are available (a). Under the right conditions, the only protein that these antibodies will recognize is the tagged CENP-A. To purify tagged CENP-A, a protein extract is prepared (b) and mixed with beads that are bound to the antibodies (c). When the beads are removed from the protein mixture they will bring with them the tagged CENP-A, and any proteins to which CENP-A is bound (in gray), leaving behind the rest of the proteins (in green and orange). To identify the proteins to which CENP-A is bound, the beads are stripped of the proteins, which are then separated according to their size on a polyacrylamide gel (d). The identity of each protein band, as indicated on the left side of the stained gel, is determined by mass spectrometry. The protein names shown in bold are those that were not previously known to be associated with CENP-A. CENP-A-ST is the tagged version of CENP-A.

Adapted from Foltz *et al.* The human CENP-A centromeric nucleosome associated complex. *Nature Cell Biology*, 2008;**8**:458–469.

of the newly identified proteins (designated CENP-K through T) with GFP and showed that these proteins localized to the same place as previously identified kinetochore proteins (an example for CENP-N is shown in Figure 3a). Second, the group hypothesized if these proteins were involved in kinetochore function, inactivation of these genes would lead to chromosome segregation defects. To examine this, the researchers down-regulated the levels of each of the identified proteins using small inhibitory RNAs (siRNA, see Sections 13.3 and 19.5), and confirmed the participation of a number of the identified proteins in kinetochore

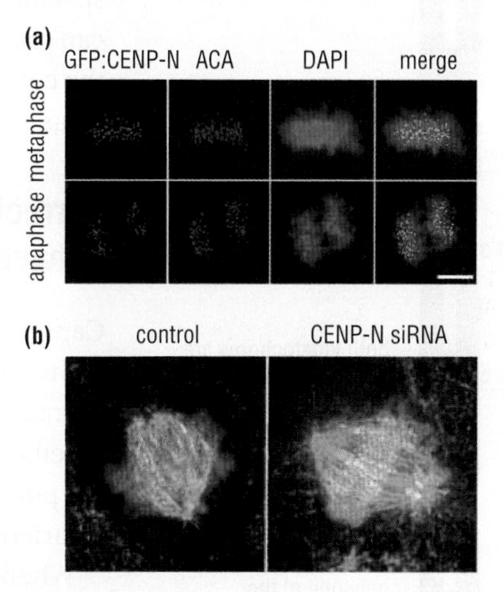

Figure 3 CENP-N is a kinetochore protein. (a) To determine if the proteins that were identified by mass spectrometry were indeed kinetochore proteins, Foltz *et al.* made protein fusions with the green fluorescent protein (GFP), expressed them in cells, and determined where they localize (panel a). Shown is an example of the co-localization (that is, localization in the same place) of GFP:CENP-N and the target of the anti-centromere antibodies (ACA), which was previously shown to identify kinetochore proteins. The merged images on the far right show the co-localization of the two proteins. (b) To confirm that the proteins identified by mass spectrometry have a role in kinetochore function, Foltz *et al.* examined what happens when the levels of these proteins is reduced by the use of siRNA (see Section 19.5). For example, whereas in the control cell the chromosomes are all aligned on the metaphase plate (left image), down-regulation of CENP-N leads to the appearance of lagging chromosomes that have not properly attached to the spindle and are not on the metaphase plate (right image).

Adapted from Foltz *et al.* *Nature Cell Biology*, 2008;**8**:458–469.

function. For example, in the case of CENP-N (Figure 3b), down-regulation resulted in defects in chromosome movement towards the metaphase plate, as would be expected from a defect in kinetochore–microtubule binding.

These examples nicely demonstrate how genetic and biochemical approaches often yield complementary information. Neither approach typically yields a complete picture, but through a combination of approaches, scientists can get increasingly detailed pictures of biological processes.

Find out more

Foltz DR, Jansen LE, Black BE, *et al*. The human CENP-Acentromeric nucleosome associated complex. *Nature Cell Biology*, 2006;**8**:458–469.

Spencer F, Gerring SL, Connelly C, Hieter P. Mitotic chromosome transmission fidelity mutants in *Saccharomyces cerevisiae*. *Genetics*, 1990;**124**:237–249.

Related techniques

S. cerevisiae (as a model organism); Section 19.1
Chemical mutagenesis; Section 19.4
Tissue culture; Section 19.2
Epitope tagging; Sections 19.4, 19.7, 19.11
Co-immunoprecipitation; Section 19.12
Mass spectrometry; Section 19.8
Use of siRNAs; Section 19.4
Fluorescence microscopy; Section 19.13

One of the driving forces of sister chromatid separation in anaphase is the depolymerization of the microtubule plus end (namely the loss of tubulin dimers from the microtubule end), which shortens the microtubule and brings the chromosome closer to the spindle pole. The chromosome, however, must remain attached to the microtubule even when this shedding of tubulin molecules occurs. Therefore the kinetochore must somehow accommodate attachment to a dynamic microtubule end. The mechanism by which this occurs involves, at least in part, the specific way in which the Ndc80 complex binds microtubules. The Ndc80 complex can track along microtubules, such that when the microtubule end depolymerizes, the complex will move to a microtubule region that is still intact, thus maintaining an association between the kinetochore and the microtubule.

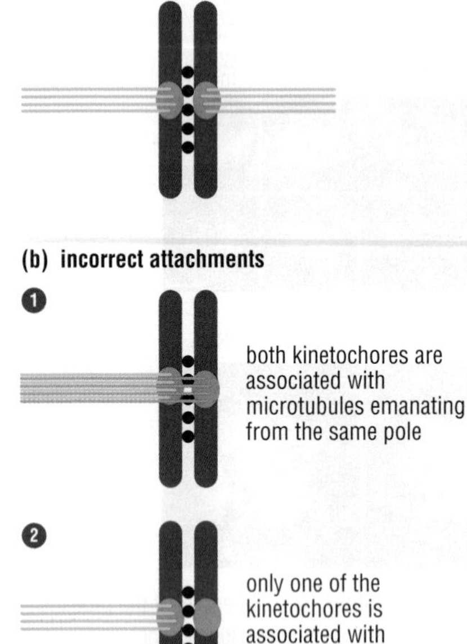
(a) amphitelic attachment
(b) incorrect attachments
1
both kinetochores are associated with microtubules emanating from the same pole
2
only one of the kinetochores is associated with microtubules

Figure 7.16 Correct and incorrect kinetochore–microtubule attachments. (a) In the correct attachment of microtubules (green) to kinetochores (orange), sister chromatids (gray rods) are bound to microtubules emanating from opposite spindle poles. This attachment is called an amphitelic attachment, and the sister chromatids are bi-oriented. (b) Examples of incorrect attachments include the association of kinetochores with microtubules emanating from the same spindle pole (1), or when one of the kinetochores is unattached (2).

Correct kinetochore–microtubule attachment creates tension between sister kinetochores

Capture of chromosomes by spindle microtubules depends on chance encounter of the plus end of a microtubule with a kinetochore. To ensure proper chromosome segregation, these chance encounters must ultimately result in the attachment of kinetochores of sister chromatids to microtubules emanating from opposite spindle poles. The attachment of sister chromatids to opposite spindle poles is called bi-orientation or **amphitelic attachment**, and is depicted in Figure 7.16a.

When kinetochores are bi-oriented, the pulling forces exerted by the microtubules are opposed by the cohesion between the two sister chromatids, and the resulting tension between the sister kinetochores is thought to stabilize kinetochore–microtubule attachments and promote the binding of more microtubules. Because microtubules encounter their target by chance, many of the initial kinetochore–microtubule attachments are incorrect or incomplete. For example, sister kinetochores could both attach to microtubules that emanate from the same spindle pole as shown in Figure 7.16b, or one of the sister kinetochores could be without associated microtubules (Figure 7.16c). The difference between these interactions and the correct attachment is that tension is generated only when sister kinetochores are bi-oriented.

Incorrect attachments, where no tension is generated, are destabilized by the activity of a protein kinase called aurora protein kinase, which detects the lack of tension by a yet unknown mechanism. In the absence of tension, aurora protein kinase phosphorylates several kinetochore proteins, including the region of the Ndc80 complex that interacts with microtubules. Aurora phosphorylation reduces

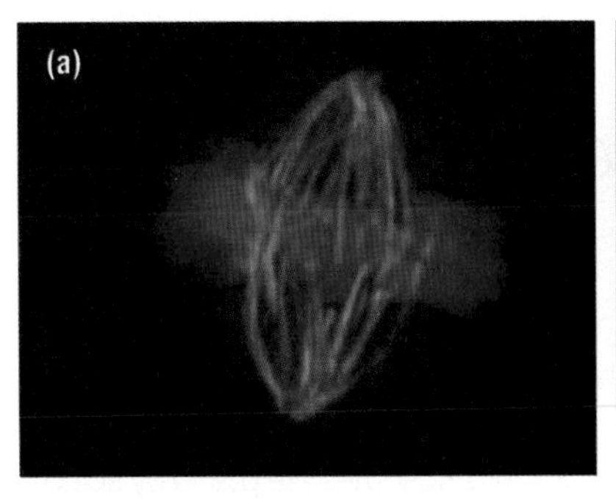

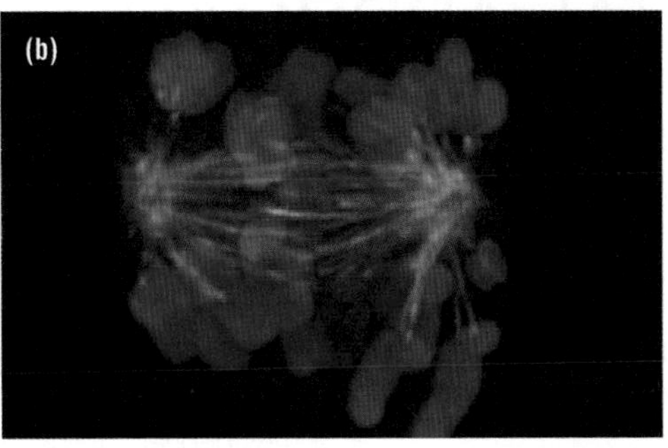

Figure 7.17 Inactivation of aurora protein kinase leads to the persistence of incorrect kinetochore–microtubule attachments. (a) When aurora protein kinase is active, incorrect kinetochore–microtubule attachments do not persist, and all chromosomes eventually bi-orient. This type of binding leads to a balance of forces acting on the chromosomes (blue), such that all the chromosomes align at the spindle mid-zone. (b) When aurora protein kinase is inactivated by drug treatment, the microtubules are still arranged in a bipolar spindle, but incorrect microtubule–kinetochore attachments cannot be corrected. As a result, sister chromatids may form attachments to microtubules from only one spindle pole. Consequently, pulling forces on the paired sister chromosomatids are greater toward that spindle pole, resulting in a failure to align at the spindle mid-zone.

Adapted from Lampson MA, Renduchitala K, Khodjakov A, Kapoor TM. Correcting improper chromosome-spindle attachments during cell division. *Nature Cell Biology*, 2004;**3**:232–7.

the binding affinity between the kinetochore and microtubules, causing the kinetochore to release the associated microtubules and creating another opportunity for microtubule binding. It is possible that in the presence of tension, the aurora kinase substrates are pulled away from the kinase, and their phosphorylation is thus prevented, thereby preserving kinetochore-microtubule attachments. We learn more about the investigation of kinetochore tension in Experimental approach 7.3.

This process of attachment and release mediated by aurora protein kinase is repeated until the sister chromatids bi-orient and tension is produced. Since the persistence of incorrect kinetochore-microtubule interactions leads to chromosome mis-segregation, scientists have targeted this process for drug therapy to disrupt cell division in cancerous cells. Indeed, inhibition of aurora protein kinase by chemical inhibitors leads to massive chromosome mis-segregation, which is accompanied by cell death. Such a situation is depicted in Figure 7.17.

Incorrect kinetochore-microtubule attachments lead to the activation of a regulatory pathway, called the **spindle checkpoint pathway** (also known as the spindle assembly checkpoint). This pathway inhibits further progression through mitosis as long as incorrect attachments exist. Indeed, it will be activated by even a single unattached kinetochore. The mitotic delay induced by this checkpoint pathway will be discussed in the next section.

➔ We introduced the spindle checkpoint pathway in Chapter 5.

At the end of prometaphase, the bi-oriented chromosomes are held in alignment under tension along the spindle mid-zone. When this occurs, the cell is in metaphase, and the chromosomes are poised for segregation, as we will discuss next.

7.5 ANAPHASE: AN IRREVERSIBLE STEP IN CHROMOSOME SEGREGATION

At anaphase sister chromatids separate and move to opposite spindle poles

Anaphase is the stage in mitosis when sister chromatids separate and are pulled to opposite poles of the spindle by kinetochore-associated microtubules. As we learned in Section 7.2, sister chromatids are associated with each other from the time of DNA replication through metaphase by the presence of cohesin complexes. At the onset of anaphase, however, these cohesin complexes must be removed to allow the spindle microtubules to pull the sister chromatids apart, as illustrated in Figure 7.18. In the first part of this section we will discuss how cohesin is removed.

The metaphase-to-anaphase transition marks a point of no return in mitosis: mistakes in chromosome-microtubule attachments must be corrected before this stage, or the daughter cells will inherit an abnormal number of chromosomes.

7.3 EXPERIMENTAL APPROACH

Chromosome bi-orientation: how to generate and detect tension

Chromosomes of praying mantis spermatocytes reveal the requirement for tension during chromosome segregation

One of the most important steps in the cell division cycle is the segregation of one copy of each sister chromatid to each of the daughter cells. In preparing for chromosome segregation, the two sister chromatids initially bind microtubules at random, but eventually they bi-orient. The spindle checkpoint pathway prevents progression to anaphase until *all* paired sister chromatids have bi-oriented. A similar situation occurs in meiosis, where homologous chromosomes (rather than sister chromatids) also need to bi-orient (see Section 7.7).

What is the mechanism that signals that bi-orientation has been accomplished? Based on his observations of chromosome segregation using microscopy, Richard McIntosh was the first to propose that bi-orientation is distinguished from any other kind of chromosome-microtubule attachment by the presence of tension between the two sister chromatids (or homologous chromosome in meiosis) due to pulling forces, in opposite directions, exerted by the microtubules.

To test this hypothesis, Xiatong Li and Bruce Nicklas took advantage of the ability to directly micro-manipulate chromosomes in praying mantis spermatocytes. These spermatocytes are unusual in that they normally contain three sex chromosomes: one Y and two X chromosomes, designated X_1 and X_2. The X and Y chromosomes can pair thanks to a short region of homology. In order to bi-orient in meiosis, these sex chromosomes have to form a trivalent complex, with the Y chromosome facing one pole, and the X_1 and X_2 facing the other. In other words, when these chromosomes bi-orient, the X_1 and X_2 chromosomes are pulled by spindle microtubules in one direction, while the Y chromosome is pulled in the other direction, as shown in Figure 1a. Sometimes one of the X chromosomes fails to associate with the X–Y pair, resulting in a spindle checkpoint-dependent cell cycle arrest (Figure 1b).

Li and Nicklas argued that if the delay was due to a lack of tension, because this lone X chromosome was not paired with any other chromosome and therefore could not bi-orient, artificially creating tension by pulling on this lone X chromosome with a micro-manipulator should alleviate the checkpoint-induced delay. Indeed, when they pulled on the lone X chromosome in the appropriate direction, the cell cycle delay was considerably shortened (Figure 1c). This was an elegant demonstration that tension plays a direct role in signaling bi-orientation.

Intra-kinetochore stretching leads to tension that signals chromosome bi-orientation

Once the link between tension and bi-orientation was established, researchers began to search for the actual source of the tension and a mechanism by which that tension regulates anaphase. One possibility was that tension in mitosis was created between sister chromatids when their kinetochores were pulled apart (referred to as inter-kinetochore tension, Figure 2b). Alternatively, the source of tension could be within the kinetochore itself (intra-kinetochore

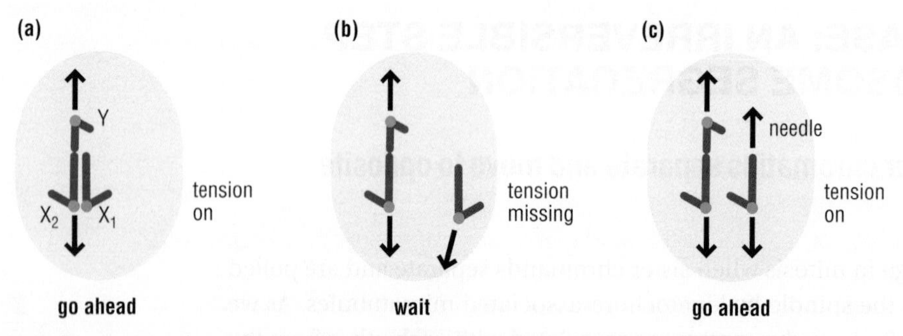

Figure 1 Tension between sex chromosomes in praying mantis spermatocytes. Praying mantis spermatocytes contain three sex chromosomes: two X chromosomes, designated X_1 and X_2, and one Y chromosome. During the first meiotic division, the three chromosomes need to form a trivalent species, with the X chromosomes being pulled by microtubules from one spindle pole and the Y chromosome being pulled by microtubules from the opposite pole (Figure 1a, microtubules are not shown, the arrows point to the direction at which the chromosomes are pulled. Orange circles are kinetochores). This configuration generates tension and the cells can proceed through meiosis ('go ahead' situation). When one of the X chromosomes is not attached to the other X and Y chromosomes, the lone chromosome is pulled by microtubules in only one direction and therefore does not experience tension, resulting in spindle checkpoint activation and inhibition of cell cycle progression (Figure 1b, 'wait' situation). If, however, the lone X chromosome is pulled by a micro-needle in the opposite direction to that of the microtubules, this creates tension within the chromosome and its kinetochore, thereby satisfying the tension requirement and allowing the cell cycle to proceed.

Adapted from Li and Nicklas, Mitotic forces control a cell cycle checkpoint. *Nature*, 1995;**373**:630–632.

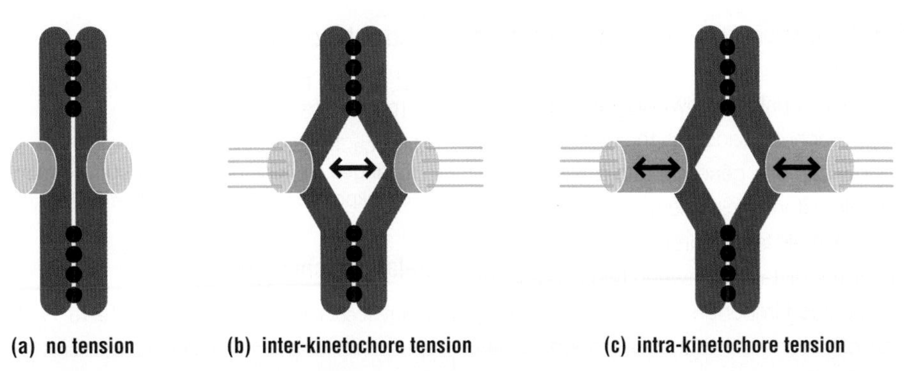

Figure 2 Inter- and intra-kinetochore tension. In mitosis, chromosomes bi-orient when sister kinetochores (in orange) bind microtubules (in green) from opposite poles. Cohesion between the sister chromatids (illustrated as black dots) opposes the pulling forces of the microtubules and can create two types of tension (black arrows): between sister kinetochores (inter-kinetochore) (b) or within each kinetochore (intra-kientochore) (c). Note that both types of tension can be present concomitantly (such as in (c)). There is no tension when microtubules are not attached (a).

tension; see Figure 2c). To distinguish between these models, Thomas Maresca and Ted Salmon constructed a *Drosophila melanogaster* (fruit fly) cell line in which they could follow both inter- and intra-kinetochore tension. In this cell line, the centromere closest to the inner part of the kinetochore was labeled with a fly homolog of the CENP-A protein, CID, fused to a red fluorescent protein (mCherry), whereas the outer kinetochore was labeled with an outer kinetochore protein, Ndc80, fused to GFP (Figure 3a). The use of these color markers enabled the investigators to directly measure the separation between the two proteins by microscopy.

When there was no tension, such as in prophase, the signals from the inner and outer kinetochore proteins nearly overlapped, as depicted in Figure 3b, top panels. However, during metaphase when the tension is known to be high, the two Ndc80-GFP signals were clearly separated from the two CID-mCherry signals (Figure 3b, bottom panels). These data argued that the kinetochore itself was being stretched, and thus that intra-kinetochore tension could be critical for dictating the checkpoint control. Indeed, when the researchers examined the structure of the kinetochore immediately before sister chromatid separation, they found that the signal for tension correlated with intra-kinetochore,

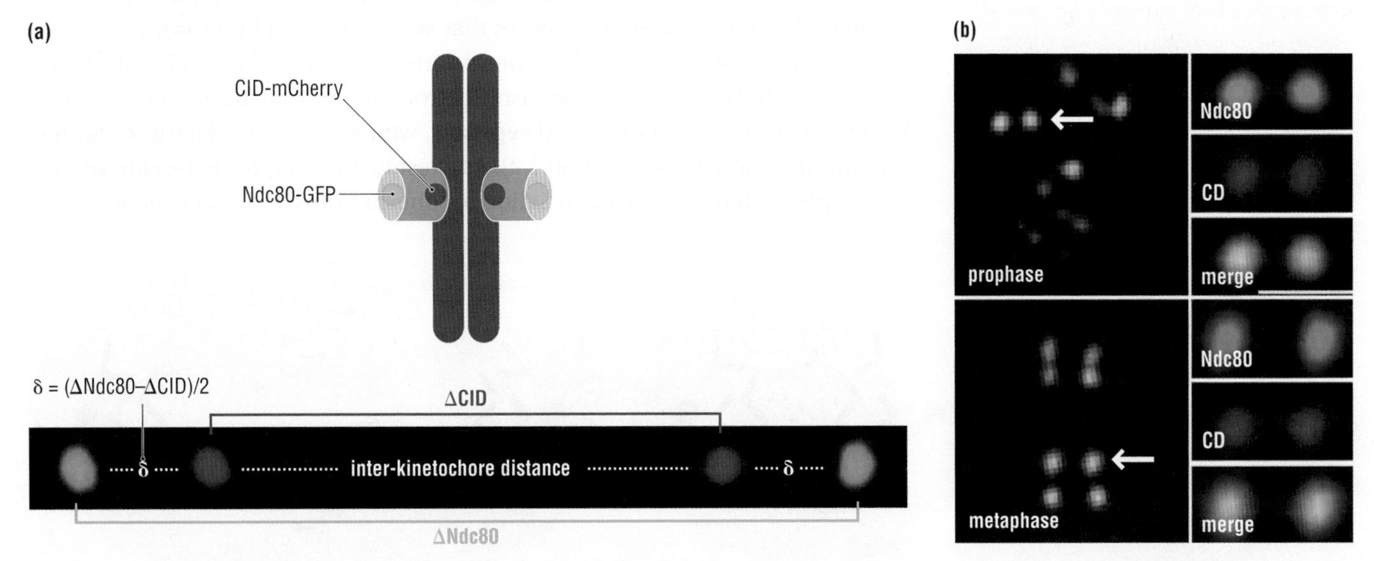

Figure 3 Creating a tension monitor. (a) To measure both inter- and intra-kinetochore tension in live cells, Maresca and Salmon constructed a fly cell line where a centromere-associated protein (CID) and outer kinetochore protein (Ndc80) were labeled with mCherry and GFP, respectively (illustrated in red and green in panel a). The inter-kinetochore distance was measured between the two mCherry signals (ΔCID). The intra-kinetochore distance, designated as ∂, was determined by measuring the distance between the two Ndc80-GFP signals (ΔNdc80), subtracting the ΔCID value, and dividing the result by two, as there are two kinetochores between the two Ndc80-GFP signals. The measurements were done from the center of each fluorescent signal. (b) Top panel: In prophase the kinetochores have not yet bound microtubules, and therefore there is no tension. Consequently, the signals from CID-mCherry and Ndc80-GFP overlap almost entirely. In metaphase (bottom panel), in contrast, the pulling forces of the microtubules cause the kinetochores to extend, resulting in a partial displacement of the Ndc80-GFP signal from the CID-mCherry signal.

rather than inter-kinetochore, tension. Thus, the tension sensor is likely to be the kinetochore itself.

The studies described here illustrate how investigators often need to adopt creative approaches to address the questions they are seeking to answer. How can one generate and measure tension in a single cell and follow it in a microscope? Here, praying mantis spermatocytes, with their large chromosomes, and the ability to mark and monitor the distance between two fluorescent protein spots in fly cells provided important insights. It is striking how often in biology an organism with unusual properties has allowed the field to propel itself forward.

Find out more

Li X, Nicklas RB. Mitotic forces control a cell cycle checkpoint. *Nature*, 1995;**373**:630–632.

Maresca TJ, Salmon ED. Intrakinetochore stretch is associated with changes in kinetochore phosphorylation and spindle assembly checkpoint activity. *Journal of Cell Biology*, 2009;**184**:383–381.

Related techniques

Fluorescence microscopy; Section 19.13

Drosophila (as a model organism); Section 19.1

Use of GFP; Sections 19.7, 19.11

Thus, the dissolution of sister chromatid cohesion must be delayed until proper kinetochore-microtubule attachments are formed. This is the function of the spindle checkpoint pathway that will be discussed further in the second part of this section.

Once cohesin is removed, chromosome movement occurs in two steps as shown in Figure 7.18: during **anaphase A**, sister chromatids move toward their respective spindle poles, and during **anaphase B**, the spindle poles move away from each other, further increasing the distance between the sister chromatids. This microtubule-dependent movement will be discussed in the last part of this section.

Dissolution of cohesin at the beginning of anaphase depends on separase and is regulated by the anaphase-promoting complex

The separation of sister chromatids at the onset of anaphase depends on the complete removal of the cohesin complexes that hold sister chromatids together. In animal cells, most cohesin complexes that associate with chromosome arms are removed during prophase, while cohesin complexes located around centromeres persist until the beginning of anaphase. Centromere-associated cohesin is removed by the activity of a protease called **separase**, which cleaves the kleisin subunit of the cohesin complex (see Section 7.2). As a result, the integrity of the cohesin ring is disrupted, leading to the dissociation of cohesin from the chromosomes.

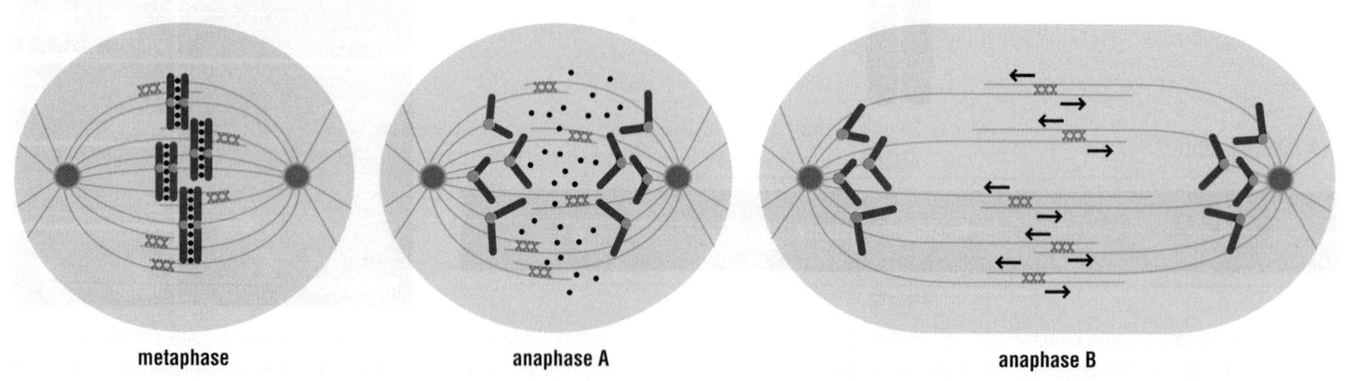

Figure 7.18 Anaphase. At metaphase, sister chromatids have bi-oriented. The cohesin complexes (black dots) oppose the pulling forces of the spindle microtubules (green) on kinetochores (orange). At this point the microtubule pulling forces acting on each of the bi-oriented sister kinetochores are roughly the same, causing the chromosomes to align on the spindle mid-zone. At the metaphase-to-anaphase transition, cohesin is removed from chromosomes, allowing the spindle microtubules to pull the sister chromatids apart. At first, the chromosomes move toward the poles (in gray), in what is known as anaphase A. Subsequently, motor proteins associated with polar microtubules (shown as blue crosses) cause the spindle to elongate, concomitantly with cell elongation. This, along with motor proteins associated with the cortex, which pull on astral microtubules (not shown), cause the spindle poles to move apart, in what is known as anaphase B.

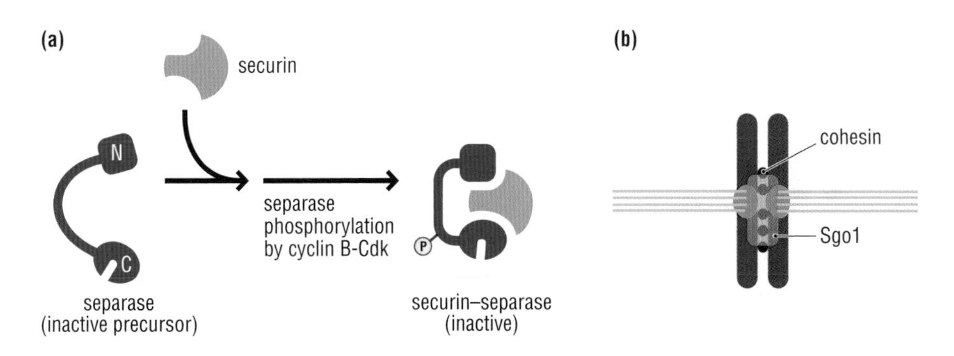

Figure 7.19 Sister chromatid separation is regulated by multiple proteins. (a) Separase (shown in dark blue) is inhibited by securin (in light blue), which binds to the N and C termini of separase and inhibits the protease activity. Separase is also inhibited by cyclin–Cdk phosphorylation (yellow P in black circle). (b) Cohesin (black circles) at centromeres (marked in orange) is protected from the cleavage by the presence of shugoshin (Sgo1; light blue), which blocks the access of separase to centromere-bound cohesin.

The proper timing of kleisin cleavage by separase is crucial: if sister chromatid cohesion is dissolved before chromosomes bi-orient, there will be no mechanism to ensure the proper attachment of sister chromatids to microtubules from the opposite spindle pole. Instead, chromosomes will attach randomly to the spindle, leading to aberrant chromosome segregation. Thus, separase activity must be under tight regulation. Indeed, until anaphase, separase is kept inactive by multiple regulatory mechanisms, thereby ensuring that cohesin does not dissociate prematurely. Three proteins participate in regulating separase activity: cyclin B, **securin**, and **shugoshin** (a Japanese word that means guardian spirit; also called Sgo1).

Prior to anaphase, cyclin B–Cdk phosphorylates separase, thereby inhibiting separase's catalytic activity, as illustrated in Figure 7.19a. In addition, separase binds to a protein called securin, which prevents separase's catalytic site from gaining access to cohesin's kleisin subunit. Finally, cohesin complexes located at centromeres are protected from separase's proteolytic action by the kinetochore-associated protein shugoshin (Figure 7.19b). These three proteins, cyclin B, securin, and shugoshin, are degraded at the onset of anaphase, thereby lifting the inhibition of separase activity and allowing it to promote sister chromatid separation.

➔ We discuss ubiquitination further in Chapter 14.

The degradation of cyclin B, securin, and shugoshin occurs through a ubiquitin-mediated protein degradation process in which the proteins are targeted for degradation by the addition of polyubiquitin chains to lysine side chains on the proteins in question. Ubiquitination of these proteins is carried out by the anaphase-promoting complex (APC), a ubiquitin ligase that was described in Chapter 5. At the metaphase-to-anaphase transition, the APC acts in conjunction with the substrate-recognition subunit **Cdc20**, which recognizes numerous cell cycle proteins, including cyclin B, securin, and shugoshin.

The chain of events that results in anaphase initiation is summarized in Figure 7.20. It begins with the APC^{Cdc20}-dependent ubiquitination and subsequent degradation of cyclin B, securin, and shugoshin. The active separase then cleaves cohesin's kleisin and leads to the dissolution of sister chromatid cohesion. Cdc20 is therefore a key player in promoting anaphase and is one of the main targets of the spindle checkpoint pathway.

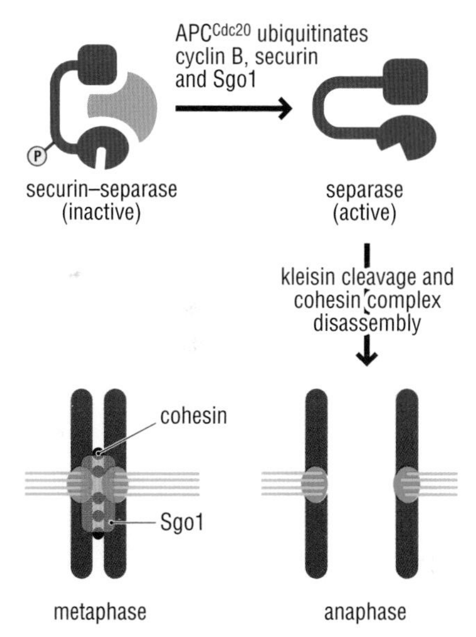

Figure 7.20 The APC promotes the degradation of proteins that inhibit separase. At the onset of anaphase, the APC, in conjunction with Cdc20, promotes the ubiquitination and subsequent degradation of shugoshin (Sgo1), cyclin B, and securin (light blue). Cyclin B degradation facilitates the removal of the inhibitory phosphorylation from separase (dark blue) by phosphatases (not shown). Securin degradation leads to further activation of separase, which can then cleave the kleisin subunit of cohesin, which is no longer blocked by shugoshin. This, in turn, leads to the dissolution of cohesion, which allows the sister chromatids to separate at anaphase.

The spindle checkpoint inhibits anaphase initiation by inactivating APC^{Cdc20}

The spindle checkpoint is activated by kinetochores that are not associated with microtubules. Unattached kinetochores are present in early mitosis, before nuclear envelope breakdown, when microtubules have yet to gain access to chromosomes. Unattached kinetochores also result from the activity of aurora protein kinase, which disassembles incorrect kinetochore–microtubule associations, as discussed in Section 7.4. The presence of an unattached kinetochore causes several spindle

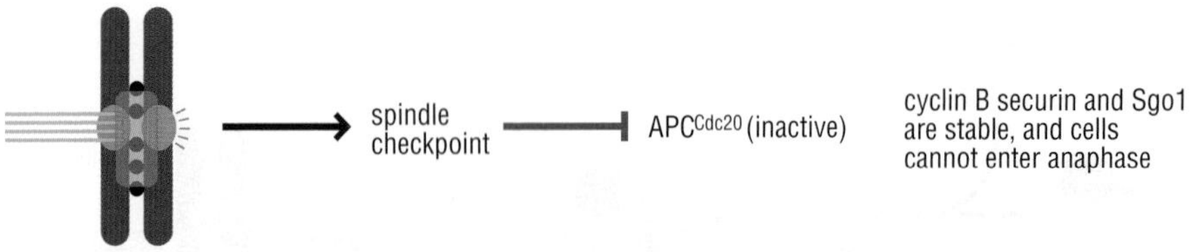

Figure 7.21 Unattached kinetochores inhibit anaphase by activating the spindle checkpoint pathway. An unattached kinetochore generates a signal that activates the spindle checkpoint pathway. Proteins of this pathway bind and inactivate Cdc20. Since Cdc20 is necessary for APC-mediated ubiquitination of shugoshin (Sgo1), cyclin B, and securin, the presence of an unattached kinetochore effectively inhibits the initiation of anaphase.

checkpoint proteins to bind and inhibit Cdc20, thereby inactivating the APC^{Cdc20} complex, as depicted in Figure 7.21. Therefore, as long as there are unattached kinetochores in the cell, the spindle checkpoint inhibits the APC^{Cdc20}, and anaphase is delayed. Once all the chromosomes successfully bi-orient, the signal for checkpoint activation is extinguished, the APC^{Cdc20} becomes active, and anaphase can begin. Thus, the spindle checkpoint pathway links the state of kinetochore attachment to cell cycle progression, and ensures that anaphase initiation does not occur prematurely.

The spindle checkpoint pathway is crucial for ensuring proper chromosome segregation. Indeed, components of this pathway are essential for viability; mice in which spindle checkpoint genes are deleted die *in utero*. The importance of the spindle checkpoint pathway in the mouse model system is further underscored by the association between mutations in spindle checkpoint genes and the predisposition to cancer. One of the hallmarks of cancer cells is an abnormal number of chromosomes, referred to as **aneuploidy**. It is thought that when the spindle checkpoint pathway is inactive, the cell cycle proceeds aberrantly despite the presence of unattached kinetochores, and this leads to chromosome mis-segregation and aneuploidy. Changes in chromosome copy number cause many cells to die, but in a small fraction of cells the combination of chromosomes will be such that it could allow selection for cancer cells that have lost tumor suppressor genes or gained oncogenes, as described in Chapter 5.

Chromosome segregation in anaphase is driven by microtubule flux, microtubule depolymerization, and spindle-associated motor proteins

Once separase is active and cohesion has been dissolved, spindle microtubules pull the sister chromatids apart. As we mentioned in the beginning of this section, sister chromatids first move toward their respective spindle poles (known as anaphase A), and then the spindle poles begin to move away from each other, further increasing the distance between the sister chromatids (known as anaphase B). The movement of the chromosomes to the spindle poles at anaphase A is driven by a combination of **microtubule flux** and depolymerization of the kinetochore microtubules at their plus ends.

Microtubule flux is the gradual movement of tubulin heterodimers from the plus end of a microtubule, where tubulin is being added, toward the minus end at the spindle pole, where tubulin is being removed, as illustrated in Figure 7.22. Microtubule flux exerts a pulling force on sister chromatids toward the poles to which they are connected. Imagine that the kinetochore is associated with a particular site on a microtubule: as tubulin subunits are removed from the minus end, this site gets closer to the spindle pole, and so does the chromosome.

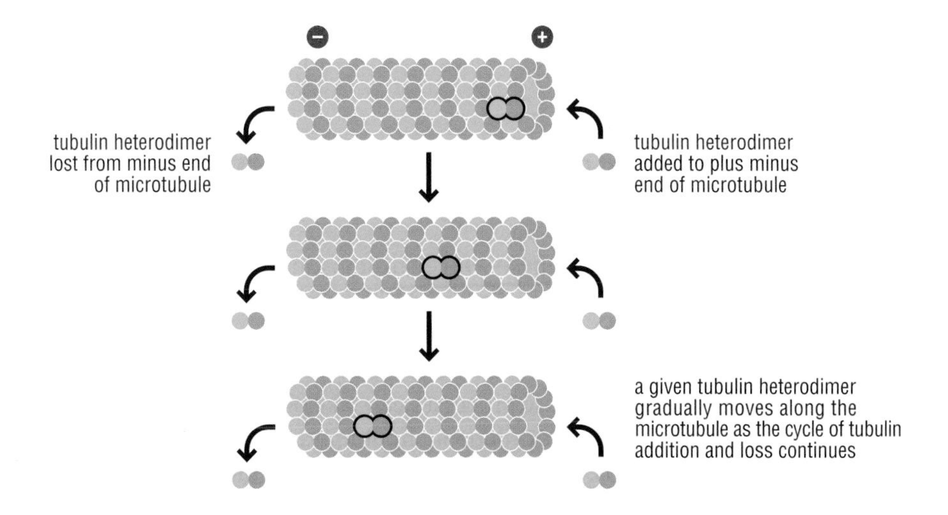
tubulin heterodimer
lost from minus end
of microtubule
tubulin heterodimer
added to plus minus
end of microtubule
a given tubulin heterodimer
gradually moves along the
microtubule as the cycle of tubulin
addition and loss continues

Figure 7.22 Microtubule flux. Under conditions where tubulin heterodimers are added at the microtubule plus end and removed from the microtubule minus end, the incorporated tubulin heterodimers gradually move closer to the microtubule's minus end. If a protein were to be tethered to a newly incorporated tubulin heterodimer, it would gradually more toward the minus end (imagine a person standing on an escalator step at one end of the escalator and being carried with it as the step moves toward the escalator's other end).

Kinetochore-associated proteins, such as motor proteins and the Ndc80 complex, also contribute to the movement of chromosomes toward the microtubule minus end. Minus-end-directed motor proteins actively move their chromosome cargo toward the minus end of the microtubule. Other motor proteins induce microtubule depolymerization at the microtubule plus end; because kinetochore proteins, such as the Ndc80 complex, ensure that the kinetochore remains bound to the part of the microtubule that has yet to disassemble, the combined action of motor activity and microtubule depolymerization moves the sister chromatids away from each other and closer to the spindle poles, as illustrated in Figure 7.23.

After anaphase A, during which chromosomes start moving toward the poles, anaphase B begins, during which the poles move away from each other. There are two driving forces for anaphase B: the first is the activity of plus-end-directed motors that can bind simultaneously to two polar microtubules from opposite poles of the spindle, and which push the poles apart (as we saw in Figure 7.12b). The second force is a pulling force exerted by minus-end-directed motor proteins that bind to the cell cortex and pull on the astral microtubules (Figure 7.12c). By the end of anaphase B, the two chromosome clusters are at the poles of the elongated spindle. If there were no errors in chromosome attachments to the spindle, these two clusters will have an identical chromosome composition. The next step will be to physically separate these DNA masses by cell division, creating two cells that are genetically identical to their parent.

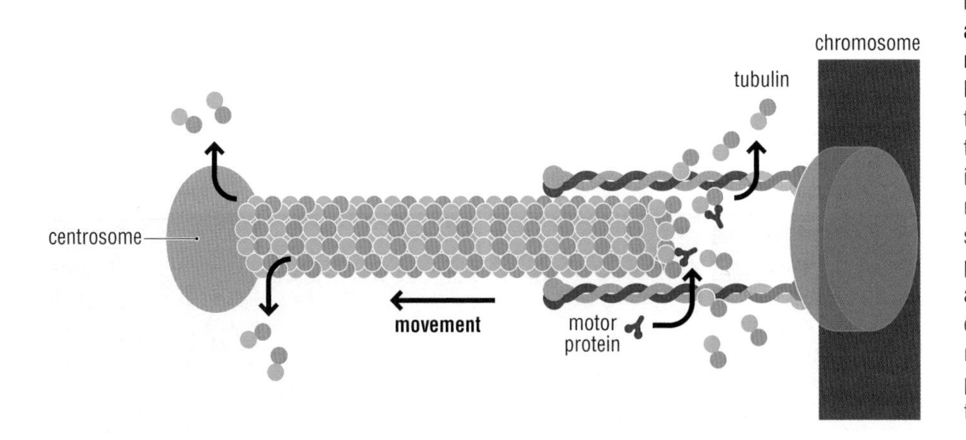
chromosome
tubulin
centrosome
movement
motor
protein

Figure 7.23 Anaphase A is driven by a combination of microtubule flux and microtubule plus end depolymerization. Microtubule flux results in the movement of tubulin within the microtubule from the plus end toward the minus end. During anaphase, there is a loss of tubulin dimers from the microtubule minus end, leading to microtubule flux. At the same time, depolymerization of the microtubule plus end results in splaying of the protofilaments and microtubule shortening. The Ndc80 complex (red-blue) maintains contact with the microtubule as the microtubule shortens from its plus end, causing the chromosome to be pulled toward the spindle pole.

7.6 THE COMPLETION OF MITOSIS AND CYTOKINESIS

Mitosis ends with spindle disassembly and nuclear envelope reassembly

The final phase of mitosis is telophase. Once chromosome segregation is completed, the spindle disassembles (see Figure 7.1). The chromosomes, which have been condensed throughout mitosis, decondense, and the nuclear envelope reforms around the two chromosome clusters; the re-formation of the nuclear envelope is shown in Figure 7.24. These processes can take place only when mitotic Cdk has been inactivated by the APC-dependent ubiquitination and subsequent degradation of mitotic cyclins. Mitotic cyclin degradation begins at the onset of anaphase and is completed by telophase. The inactivation of Cdk allows for the dephosphorylation of proteins that were phosphorylated earlier in the cell cycle by Cdk; this, in turn, triggers mitotic exit and cytokinesis.

Cytokinesis in animal cells, but not in plant cells, is driven by the contraction of an actomyosin ring

Cytokinesis, which happens after telophase, is the process by which a parent cell divides to give rise to two daughter cells. Cell division requires the placement of a membrane between the two newly formed nuclei. Animal cells and plant cells do this in different ways. Figure 7.25 shows how animal cells pull the plasma membrane inward, toward the cell center, whereas plant cells form a membrane at the cell center and expand it toward the cell cortex.

In animal cells, cytokinesis is mediated by a **contractile ring**, called the **actomyosin ring**, which is made of actin filaments bound to motor proteins called myosin. This ring binds to the plasma membrane at a site called the **cleavage furrow**, the site where cell division will take place. (The question of how the site of the cleavage furrow is determined is an interesting one, which we will discuss later in this section.) Ring contraction is facilitated by the motor activity of myosin, which pulls the actin filaments toward each other, thereby pulling the plasma membrane inward (Figure 7.25a). Many proteins besides actin and myosin are involved in this process, and some act by controlling the motor activity of myosin and the polymerization or depolymerization of the actin filaments. The contraction of the actomyosin ring, and the invagination of the plasma membrane with it, proceeds until a small hole is left. Cytokinesis is completed when this hole is filled. This filling occurs by the fusion of membrane vesicles that are targeted to the site of cytokinesis, forming an intact membrane that separates the cytoplasms of the two daughter cells.

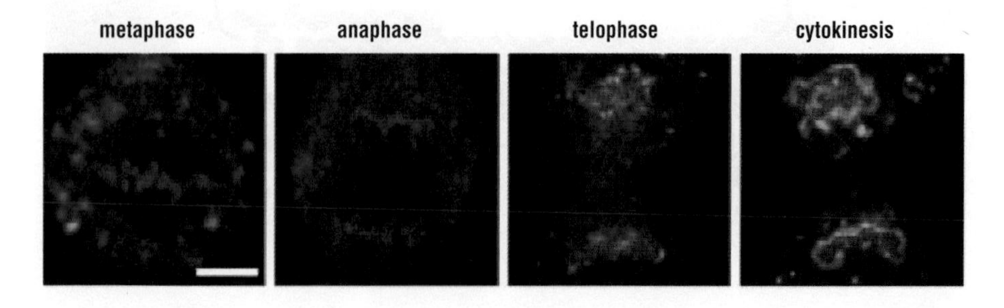

Figure 7.24 Nuclear envelope formation at the end of mitosis. The images show a cell from a human cell line synthesizing histone H2B fused to a red fluorescent protein (shown in red), which marks the chromosomes, and a nuclear envelope protein (Sun1) fused to a green fluorescent protein (in green). During metaphase, nuclear envelope proteins localize to the cytoplasm or the endoplasmic reticulum. These proteins are recruited to the chromosomes as cells exit mitosis (telophase and cytokinesis). The bar depicts 10 μM in length.

Figure 7.24 adapted from Anderson, DJ and Hetzer, MW. Reshaping of the endoplasmic reticulum limits the rate for nuclear envelope formation. *Journal of Cellular Biology*, 2008;**182**:911–924.

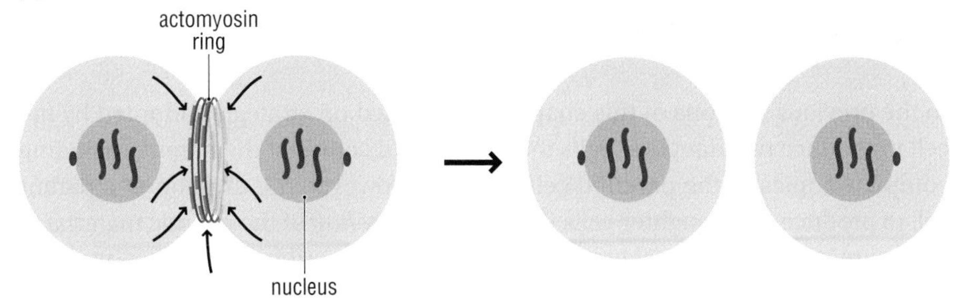

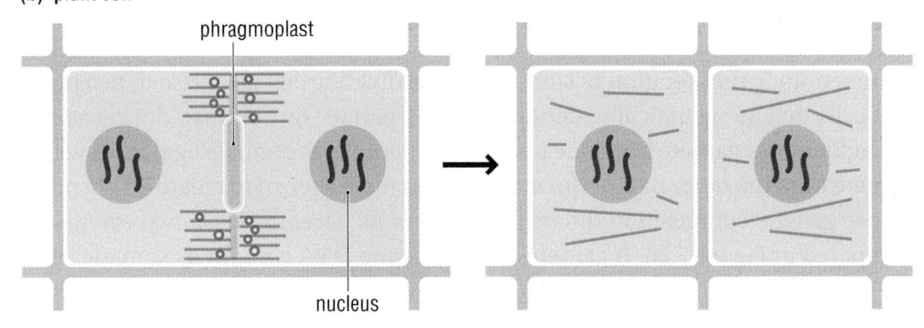

Figure 7.25 Cytokinesis in animal cells occurs from the cell surface inward whereas cytokinesis in plant cells occurs from the cell center outward. (a) During cytokinesis in animal cells, the contraction of an actomyosin ring associated with the cell membrane leads to the invagination of the membrane. At the end of this process the cytoplasm of the two daughter cells is separated by a membrane. (b) During cytokinesis in plant cells, a phragmoplast forms in the middle of the cell and recruits, via microtubules (green), membrane vesicles (blue) that form a membranous structure extending toward the cell cortex. Once the membrane is fully extended, the cytoplasm of the parent cell has been divided into two daughter cells.

In plants, the membrane that separates the two daughter cells is formed independently of the plasma membrane. In this case, a cytoplasmic structure called the **phragmoplast** is responsible for delivering membrane vesicles to the plane of cell division by targeted secretion. Successive rounds of vesicle fusion generate a membrane that extends from the cell center toward the cell cortex to separate the daughter cells (Figure 7.25b). In plants, cytokinesis does not involve a contractile ring.

Cytokinesis is regulated both spatially and temporally

For the successful completion of chromosome segregation, the timing of cytokinesis must be regulated so that the cell divides only when the two sets of chromosomes are fully separated. In addition, the position of the cleavage furrow, or phragmoplast, must be such that it is situated between the two daughter nuclei. In animal cells, the position of the cleavage furrow corresponds to the spindle midzone, and is thought to be determined by a signal from the spindle, although the nature of the signal is not known. In plant cells, the place of cell division is determined in G2 by the preprophase band (PPB), a cortical array of cytoskeletal filament (including microtubules and actin). Unlike animal cells, the position of the PPB is determined not by the spindle but by the nucleus, through an unknown mechanism. As we will see later in this chapter, bacteria have yet another process that defines where cells divide. Thus, different species have evolved different ways of placing the plane of cell division, but all have some relationship to the position of the nucleus or the chromosomes.

The temporal regulation of cytokinesis is also poorly understood. In animal cells we know that cytokinesis depends on proteins associated with the mid-zone of the elongated anaphase spindle, which comprises bundled polar microtubules. This structure forms only *after* chromosome segregation is completed. Linking cytokinesis to the spindle mid-zone therefore ensures that cell division does not take place before the end of anaphase. This combination of spatial and temporal cues ensures that cell division results in the formation of two daughter cells, each containing a full set of chromosomes.

7.7 MEIOSIS: GENERATING HAPLOID GAMETES FROM DIPLOID CELLS

In the previous sections of this chapter, we focused on strategies adopted by the cell to produce two daughter cells that are faithful copies of the parent, including complete copies of the parental cell genome. However, the division of a parent cell to produce two daughter cells that contain only *half* of the genetic material of their parent is also a central process in biology. We describe this process, meiosis, in this section.

Meiosis generates cells that are involved in sexual reproduction

During mitotic proliferation of somatic cells, a dividing cell gives rise to two progeny cells that are genetically identical to their parent. By contrast, during sexual reproduction, a gamete from one parent (for example, the female egg) fuses with a gamete from the other parent (for example, the male sperm) to create an offspring whose genetic composition differs from that of its parents. These two events are compared in Figure 7.26. If gametes had the same DNA content as somatic cells, the total DNA content, or ploidy, of the progeny would double with each successive generation. To circumvent this problem, the ploidy of gametes is half that of somatic cells and, therefore, the fusion of two gametes results in a cell with the same number of chromosomes as the cells of the parent organism. In most animals, gametes are

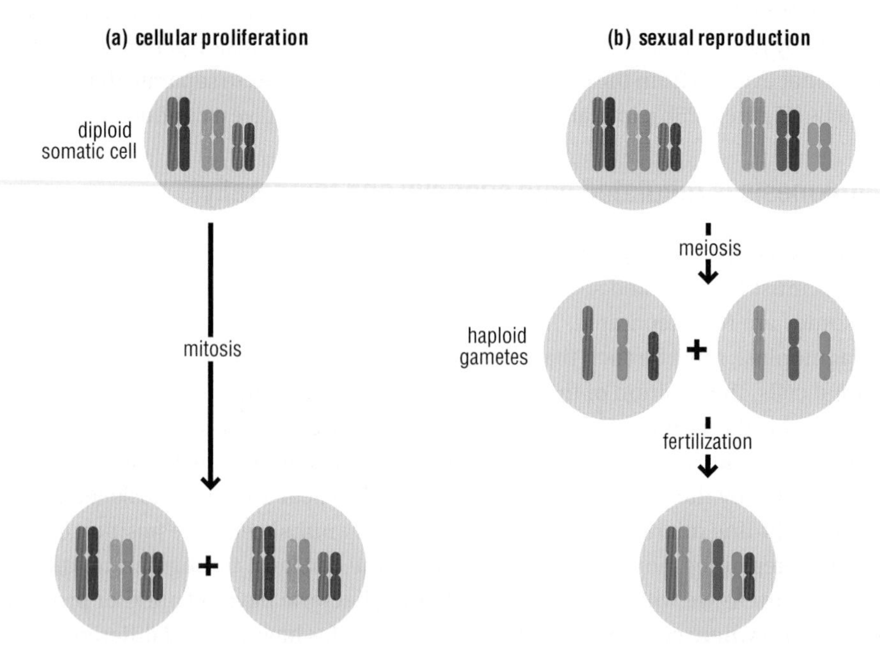

Figure 7.26 Somatic cell proliferation versus sexual proliferation. (a) Somatic cells proliferate by undergoing mitosis, during which cell division results in two daughter cells that are identical to their parent. (b) During sexual reproduction, two parents (whose genotypes are represented as the two diploid cells on the top), form haploid gametes by a meiotic division, which in humans would be the egg and sperm. Gametes are fused to give rise to a progeny cell that differs in chromosome composition from both its parents. The haploid gametes shown for each parent are just one of many possible combinations. Not shown are the exchanges between homologous chromosomes that occur during meiosis. Those will be explained later in this section. Note that some organisms have a higher ploidy (for example, tetraploid organisms have four homologs of each chromosome). Nonetheless, while in mitosis the number of chromosomes is the same in the parent and daughter cells, during sexual reproduction, the number of chromosomes in the gametes is always half that of the parent.

haploid (having one copy of each chromosome), and the fusion of male and female gametes gives rise to a diploid offspring (Figure 7.26b). The specialized cell-division process that creates gametes is called meiosis. It is also referred to as a reductional division because it leads to a reduction in ploidy from parent cell to gamete.

Meiosis involves two successive rounds of chromosome segregation

As in mitosis, meiosis is preceded by a single round of DNA replication, so that each chromosome is made of two identical sister chromatids that are associated with each other by cohesin. Unlike mitosis, however, Figure 7.27 depicts how DNA replication in meiosis is followed by *two* successive rounds of chromosome segregation: in the first round, called meiosis I, **homologous chromosomes** first pair, exchange pieces of DNA, and then segregate away from each other. This is unlike mitosis, where the homologous chromosomes do not interact. In the second round of chromosome segregation, called meiosis II, sister chromatids segregate away from each other, much like in mitosis.

The meiotic segregation process has several requirements: homologous chromosomes need to associate with each other before chromosome segregation begins, sister chromatids have to remain attached during the first meiotic division

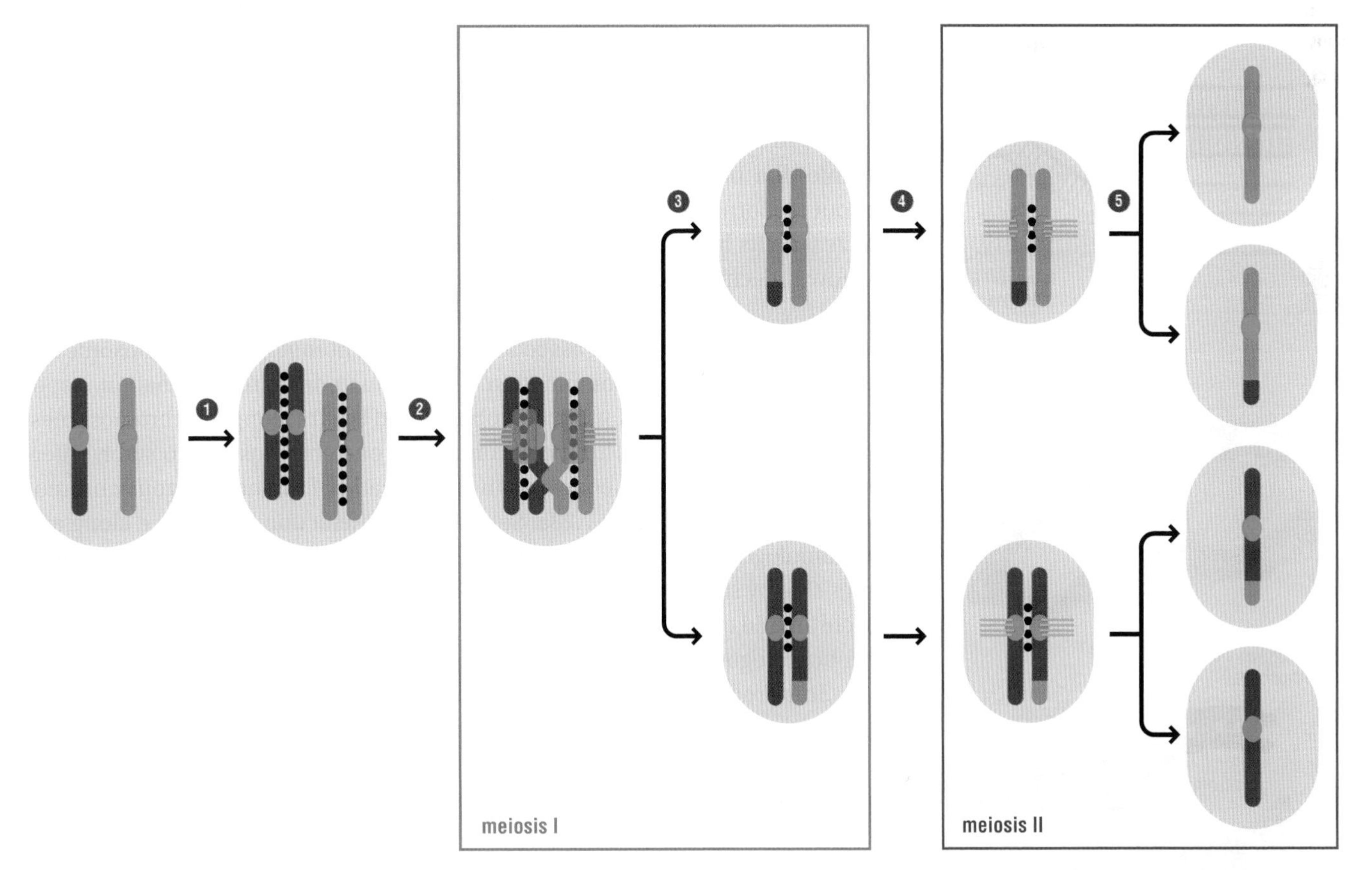

Figure 7.27 Chromosome segregation during meiosis. Animal cells that undergo meiosis are typically diploid, containing one set of chromosomes from the mother and one from the father (for simplicity, only one homologous pair is shown). During DNA replication (1) sister chromatids become attached to each other by cohesin complexes (shown as black circles). Prior to the first meiotic division, homologous chromosomes align by the action of homologous recombination, creating recombination crossovers that result in exchange of DNA between homologous sister chromatids (2). The homologous chromosomes, which are attached to each other through the crossover, bi-orient on the meiosis I spindle (green). Note that at this stage, kinetochores (in orange) of sister chromatids co-orient (namely, attached to microtubules from the same spindle pole). During meiosis I, cohesin is removed from chromosome arms but not around centromeres due to the presence of shugoshin (light blue). This allows homologous chromosomes to separate while sister chromatids remain associated (3). During meiosis II, kinetochores of sister chromatids bi-orient on the meiosis II spindle (4). This is followed by cohesion dissolution and chromosome segregation, resulting in four gametes, each containing a single copy of each chromosome (5).

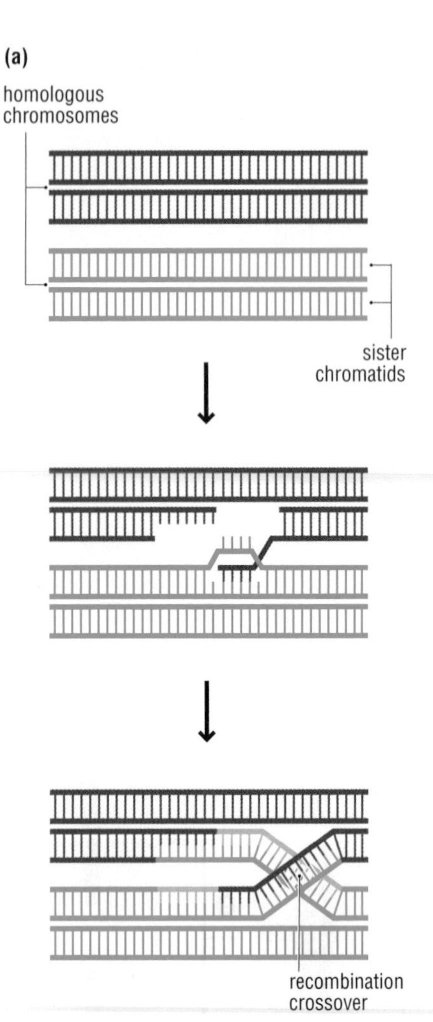

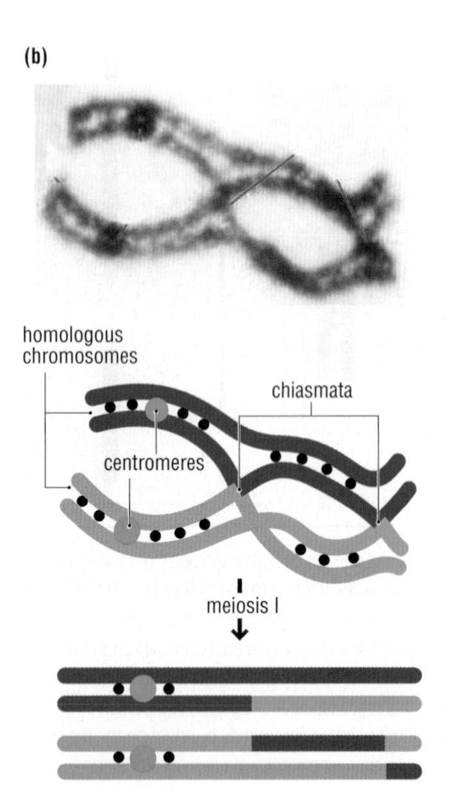

but separate during the second meiotic division, and the kinetochores of sister chromatids have to co-orient during meiosis I but bi-orient during meiosis II. The details of these processes are described in the following text.

Homologous chromosomes become physically attached before the first meiotic division through homologous recombination

A prerequisite for chromosome segregation in both meiosis and mitosis is the physical association between the chromosomes that need to be separated from each other. (Imagine sorting a basket full of unpaired socks of different colors into two identical piles by first matching each sock with its partner.) Earlier in this chapter we learned that, in mitosis, sister chromatids associate with each through cohesin complexes before being separated by spindle microtubules. In meiosis I, homologous chromosomes need to associate with each other before segregation. However, unlike sister chromatids, which are bound to each other by cohesin complexes as they are formed, homologous chromosomes must find one another in a nucleus full of other, non-homologous chromosomes.

The feature that distinguishes two homologous chromosomes from all other chromosomes is their high degree of sequence similarity, and this is the basis for their selective pairing. This sequence similarity allows the formation of complementary base-pairing between DNA strands of homologous chromosomes. It is not entirely clear how DNA strands from homologous chromosomes initially interact: one idea is that DNA breaks are necessary for strands from one chromosome to probe other chromosomes for complementary sequences. It is also possible that the initial search for complementary sequences may take place without the formation of DNA breaks, and that an intact chromosomal region interacts with its homologous sequence on the homologous chromosome by a yet unknown mechanism. Either way, homologous chromosomes rely on the presence of highly similar (though not identical) sequences in order to pair.

The basis for the physical association between homologous chromosomes in meiosis is homologous recombination. In order to undergo recombination, a specialized nuclease creates multiple double-strand breaks along the chromosomes, allowing a strand of DNA from a chromatid of one chromosome to stably interact with complementary sequences in a chromatid of the homologous chromosome, as depicted in Figure 7.28a. Several such interactions take place along any two homologous chromosomes, thereby aligning the two homologous chromosomes side by side. These breaks are then repaired. In the process, a strand from one chromatid can become permanently joined to a DNA strand from a chromatid of its homologous chromosome, creating a continuous **recombinant DNA** molecule. This process, which is known as **homologous recombination**, also plays an important role in the accurate repair of broken chromosomes in somatic cells.

Figure 7.28 Homologous chromosomes interact through homologous recombination. (a) The diagram shows two homologous chromosomes (gray and turquoise) each composed of two sister chromatids. A DNA double-strand break in one of the chromatids of the gray chromosome allows a DNA segment near the break to anneal to its homologous sequence on a chromatid of the homologous (turquoise) chromosome. As cells progress through meiosis, some of these recombination intermediates are converted to covalent attachments between chromatids of homologous chromosomes, such that a turquoise chromatid is attached to the end of the gray chromatid and vice versa. This type of exchange is called a recombination crossover. (b) A recombination crossover as visualized by light microscopy. Shown is a homologous pair of grasshopper chromosomes in prophase of meiosis I. The X-shaped structure that is the crossover is called a chiasmata. Below is a schematic representation of the chromosomes and their resolution after meiosis I.
Figure 7.28b courtesy of James Kezer, University of Oregon.

These physical connections between chromatids of homologous chromosomes are called **recombination crossovers**, and their cytological manifestations, as viewed by microscopy, are called **chiasmata**; a microscopy image showing chiasmata is shown in Figure 7.28b. Since the sister chromatids are themselves held together by cohesin, recombination crossovers create a stable connection between the two homologous chromosomes. Note that, despite the proximity and sequence identity between sister chromatids, meiotic homologous recombination takes place preferentially between complementary sequences of homologous chromosomes, not between sister chromatids. The mechanism that dictates this preference is not fully understood.

➔ We discuss the process of homologous recombination further in Chapter 16.

Recombination crossovers in meiosis occur at least once per homologous chromosome pair, ensuring that all chromosomes are stably associated with their homolog. Keep these recombination crossovers in mind: they are also responsible for genetic variation, which we will discuss at the end of this section.

The synaptonemal complex provides an additional link between homologous chromosomes

Homologous recombination is essential for the initial interaction between homologous chromosomes. However, the creation of stable interactions also requires a protein structure called the **synaptonemal complex**, which forms along the entire length of the paired homologous chromosomes. The structure of the synaptonemal complex is illustrated in Figure 7.29. A subset of the proteins that make up the synaptonemal complex form two structures, called **lateral elements**, along each of the paired sister chromatids, which organize the sister chromatids within each homolog into discrete loops. The lateral elements are connected by other proteins of the synaptonemal complex that form two rows of a coiled-coil protein known as the **transverse filament**.

There is a tight link between synaptonemal complex formation and the recombination process: mutants that fail to undergo recombination are also defective in synaptonemal complex formation, and the synaptonemal complex helps to stabilize recombination intermediates. This massive structure is dissolved prior

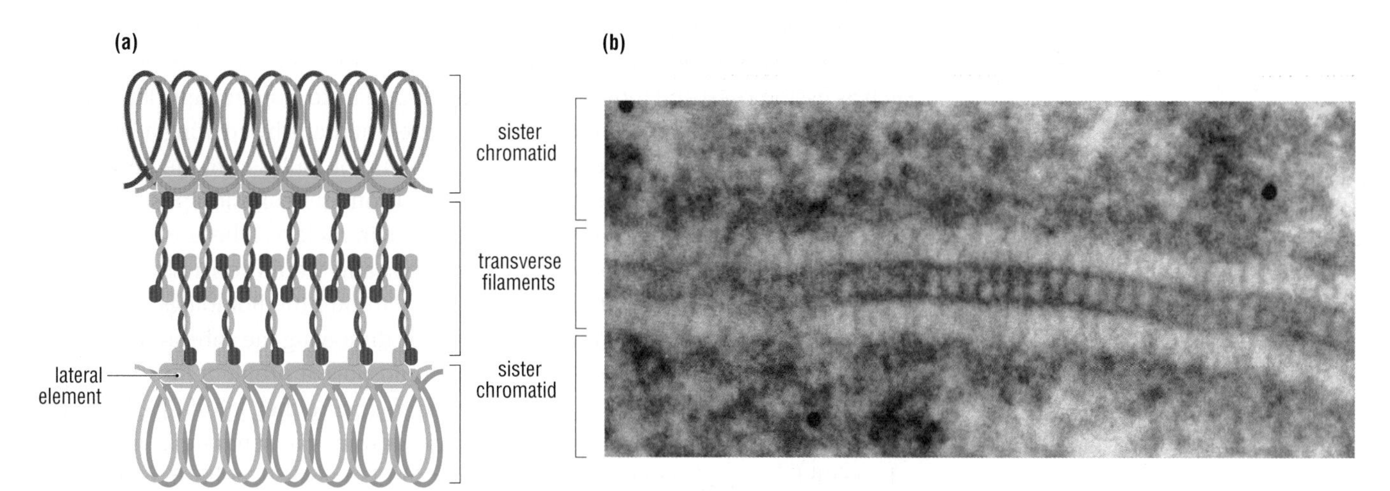

Figure 7.29 Components of the synaptonemal complex. (a) During early meiosis, the paired sister chromatids (one shown in gray and the other in turquoise) are compacted to create a series of loops held together at their base by a protein structure called the lateral element (green). Lateral elements of homologous chromosomes are linked via the transverse filament. Within the transverse element the proteins (in pink) form homodimers that interact with each other via their N-terminal domain and with the lateral element via the C-terminal domain. (b) An electron micrograph showing the synaptonemal complex. Figure 7.28b from Schmekel K, Daneholt B. Evidence for close contact between recombination nodules and the central element of the synaptonemal complex. *Chromosome Research*, 1998;**6**:155–9.

to chromosome segregation in meiosis I, leaving behind the DNA crossovers and sister chromatid cohesion to maintain the physical attachment between homologous chromosomes. It is possible that another role for the synaptonemal complex is to direct recombination events between homologous chromosomes rather than between sister chromatids.

The completion of meiosis I can take many years

After the synaptonemal complex disassembles, the first of the two chromosome segregation events (namely meiosis I) is ready to begin. During sperm formation, chromosome segregation follows immediately after synaptonemal complex disassembly. Mammalian oocytes, on the other hand, remain arrested prior to the completion of meiosis I, awaiting the appropriate hormonal trigger. For women, this arrest can last over 40 years. Loss of cohesion over time may lead to subsequent mis-segregation, for example, because the two homologous chromosomes separate prematurely and consequently segregate to the same spindle pole during meiosis I. In fact, this age-dependent deterioration of cohesion is thought to be one of the contributing factors to the increase in the incidence of trisomy, such as Down syndrome, in offspring of older women. In cells of individuals with Down syndrome, there are three copies of chromosome 21. In this case, one of the gametes of the parents, most likely the egg, had two copies of chromosome 21, rather than just one.

In meiosis I, kinetochores of sister chromatids co-orient and cohesion around centromeres remains intact

We just learned that in meiosis I, sister chromatids do not separate. This means that kinetochores of sister chromatids must face the same spindle pole, or co-orient, unlike in mitosis, where kinetochores of sister chromatids bi-orient. The mechanism that allows co-orientation of sister kinetochores during meiosis I has been deciphered in budding yeast, where it was found that co-orientation is achieved by the binding of a protein complex, called monopolin, to sister kinetochores in early meiosis I. How monopolin affects co-orientation is not known. However, it is assumed that either it inactivates one of the sister kinetochores, or it fuses them so that they effectively work as a single kinetochore. Further, monopolin is not active during meiosis II when sister kinetochores need to bi-orient. In other organisms, such as fission yeast and mouse, cohesin itself appears to play an important role in mono-orientation of homologous chromosomes, perhaps by creating a structure that favors the binding of sister kinetochores to microtubules from the same pole.

The existence of both crossovers and sister chromatid cohesion poses a problem when separating homologous chromosomes in meiosis I, because these two links effectively lock the two homologs together and would resist chromosome segregation if spindle microtubules were to try to pull the homologs apart (see Figure 7.27). In theory, this could be resolved by dissolving the cohesion between all sister chromatids in meiosis I, but that would pose a problem for meiosis II, in which sister chromatid cohesion is needed to ensure they faithfully segregate from one another. To allow homologous chromosomes to separate in meiosis I, while still holding sister chromatids together until meiosis II, cohesin complexes are removed in two steps. First, during meiosis I, cohesin complexes on chromosome arms are removed while cohesin complexes near centromeres remain in place. Then, during meiosis II, all the remaining cohesin is removed.

In both meiosis I and meiosis II, cohesin removal is mediated by separase, the protease that is also responsible for cohesin removal in mitosis as described earlier

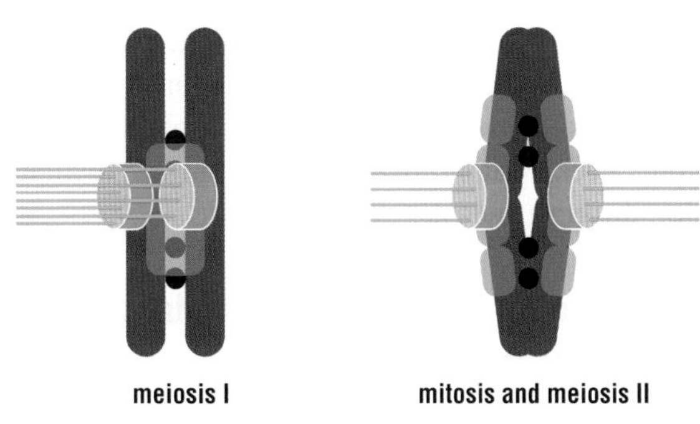

Figure 7.30 Shugoshin protects centromeric cohesion during meiosis I. During meiosis I, when there is no tension between the co-oriented kinetochores (shown in orange) of sister chromatids, shugoshin (shown in light blue) is present near cohesin (black circles), and this presumably prevents separase from accessing cohesin. However, in mitosis and meiosis II, when sister kinetochores bi-orient, pulling by spindle microtubules (in green) in opposite directions displaces shugoshin from cohesin, thereby exposing cohesin to separase's action.

in the chapter. How is it, then, that cohesin close to the centromere is spared from the action of separase during meiosis I? At least two proteins participate in this process: the shugoshin protein and a meiosis-specific kleisin subunit of cohesin called Rec8.

We learned earlier that in mitosis, shugoshin prevents separase from cleaving centromeric cohesin prior to anaphase (see Figure 7.26, steps 2 and 3). Shugoshin has the same role in meiosis I. In order to inactivate the cohesin complex, separase has to cleave Rec8, but it can only do so when Rec8 is phosphorylated. Shugoshin forms a complex with a phosphatase that keeps Rec8 in a dephosphorylated state, thus protecting it against cleavage by separase. In meiosis II, as in mitosis, shugoshin is degraded prior to sister chromatid separation, allowing sister chromatids to separate.

The kinetochore also plays a role in determining whether cohesin will be removed or spared from degradation as illustrated in Figure 7.30: in mitosis and meiosis II, the tension between kinetochores of sister chromatids could displace shugoshin from cohesin, allowing separase to infiltrate and cleave kleisin. On the other hand, during meiosis I, where there is no tension between the kinetochores of sister chromatids, shugoshin remains in close proximity to cohesin and protects kleisin from cleavage.

Recombination crossovers and sister chromatid cohesion are needed for homologous chromosome bi-orientation

In previous sections, we saw that bi-orientation of sister kinetochores leads to tension during mitotic chromosome segregation, and that the absence of tension is indicative of incomplete or incorrect kinetochore–microtubule attachments (see Sections 7.4 and 7.5). An analogous situation exists in meiosis I, where tension is created as kinetochores of homologous chromosomes bi-orient. In this case, it is the recombination crossovers that counteract the pulling forces of the spindle microtubules (see Figure 7.27, meiosis I). This reveals another role of homologous recombination: not only is it involved in identifying homologous chromosomes and holding them together, but it is also necessary for the tension which signals that the homologous chromosomes have bi-oriented.

Note that crossovers alone are not sufficient to counteract the pulling forces of the spindle; sister chromatid cohesion between the crossover and the chromosome ends must be in place to prevent the exchanged chromosomes from being pulled apart. As in mitosis, if kinetochores of homologous chromosomes attach to microtubules from the same spindle pole so that no tension is created, regulatory proteins reverse these attachments. This generates unattached kinetochores that activate the spindle checkpoint pathway (see Section 7.5), which in turn inhibits further progression through meiosis.

Chromosome segregation in meiosis II is similar to chromosome segregation in mitosis

At the end of meiosis I, the homologous chromosomes, each still comprising a pair of joined sister chromatids, have separated from each other. The cell now moves on to meiosis II, the ultimate goal of which is to separate the sister chromatids to produce haploid gametes.

Once chromosome segregation in meiosis I is complete, the meiosis I spindle disassembles and meiosis II can begin. Meiosis II is very much like mitosis, except that meiosis II is not preceded by DNA replication. Initiation of DNA replication can occur only after Cdk activity has declined at the end of mitosis to levels low enough to allow the components at the replication origin to assemble, as we saw in Section 6.7. During meiosis I, Cdk activity declines to a level low enough to allow completion of meiosis I and dissolution of the meiosis I spindle, but it is maintained at a level too high for re-initiation of DNA replication. In this way, replication between meiosis I and meiosis II is prevented.

At the onset of meiosis II, kinetochores of sister chromatids bi-orient on the meiosis II spindle (Figure 7.27, step 4). At anaphase of meiosis II, the remaining cohesin is removed by separase, and sister chromatids segregate away from each other (Figure 7.27, step 5). Thus, following two rounds of segregation without an intervening replication step, each gamete has half the number of chromosomes compared with somatic cells.

Meiosis generates diversity

Mitotic divisions generate progeny that are identical to their predecessors. In meiosis, however, the resulting gametes are all different. Two aspects of meiosis contribute to the distinct genetic makeup of gametes. First, consider that, prior to meiosis, the diploid parental cell has two set of chromosomes, each originally contributed by one of the organism's two parents. (For the purpose of this explanation we will designate them maternal and paternal sets.) During meiosis I, when homologous chromosomes separate, some maternally contributed chromosomes will go to one pole while the rest will go to the other. The same is true for the paternally contributed chromosomes. Therefore, at the end of meiosis I, not only does each spindle pole contain a distinct combination of maternal and paternal chromosomes, but two identical cells undergoing meiosis are unlikely to give rise to genetically identical gametes. The way in which meiosis I generates diversity is illustrated in Figure 7.31. Look at this figure and notice how meiosis can generate genetically diverse gametes despite starting with the same chromosomes initially.

Second, homologous recombination creates unique chromosome combinations because of the exchange between maternal and paternal chromosomes during homologous recombination. Different cells have recombination crossovers at different chromosomal loci so that the chromosomes resulting from meiosis are likely to possess allele combinations that are different from the 'parent' chromosomes. This explains why gametes of an individual are likely to have a different

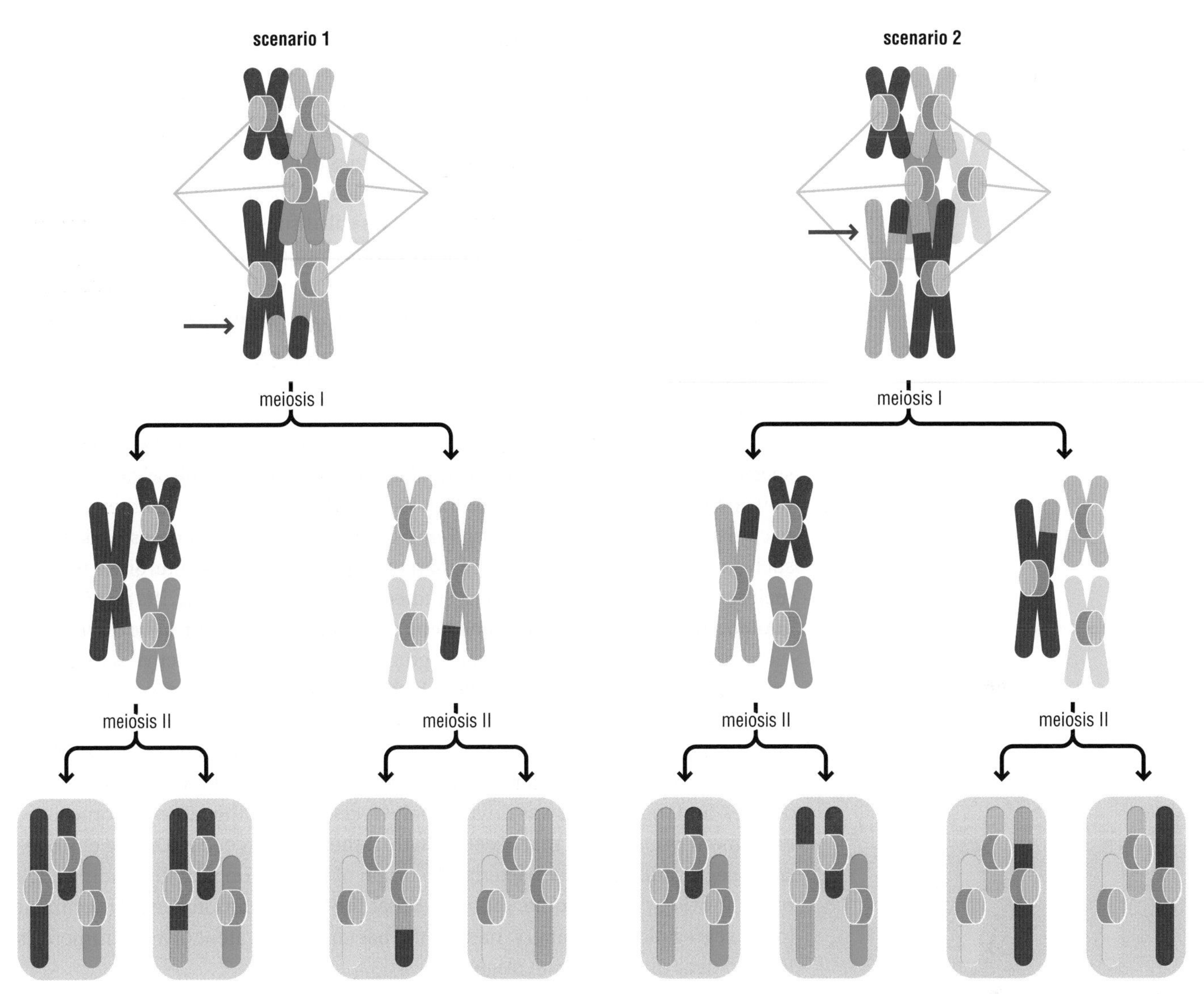

Figure 7.31 Meiosis as a way of generating diversity. The hypothetical diploid organism shown has three chromosomes (pink, turquoise, and blue), each represented by two homologous chromosomes (dark and light). As the cells of this organism undergo meiosis, different gametes are generated, depending on the orientation of the homologous chromosomes on the meiosis I spindle. In scenario I, all the dark chromosomes are facing one pole while the light chromosomes face the other. In scenario II, the dark pink, dark turquoise, and light blue chromosomes face one pole, whereas the light pink, dark turquoise, and dark blue chromosomes face the other pole. Note that a recombination event happened between the dark blue and light blue chromosomes, but in different places in each of the two scenarios. As we follow these chromosomes through meiosis, we can see that the gametes generated in scenario I are different from the gametes generated in scenario II, due to both chromosome orientation in meiosis I and recombination crossovers.

genetic makeup to one another. Thus, the combination of a random assortment of maternal and paternal chromosomes, combined with homologous recombination, is the genetic reason for why siblings look alike but are not identical.

➔ We discuss homologous recombination in more detail in Section 16.8.

7.8 CHROMOSOME SEGREGATION IN BACTERIA

The need for cellular machinery that drives chromosome segregation is not unique to eukaryotes: organisms from all kingdoms rely on regulated chromosome segregation to ensure the proliferation of the species. While the nature of the segregation process differs between bacteria and eukaryotes, all cells must deliver one copy of each chromosome to the future daughter cells. So far we have discussed the nature of the eukaryotic segregation machinery. Less is known about the process of chromosome segregation in bacteria. Only recently have scientists begun to understand the bacterial chromosome segregation process at a molecular level, thanks in part to technical advances in cell biology methodologies that allow observations in cells as small as bacteria. We now know that multiple distinct chromosome segregation processes have evolved among bacteria. Here we will discuss the general principles of bacterial chromosome segregation and provide examples from specific organisms.

The bacterial chromosome is segregated via an active mechanism that differs from the eukaryotic chromosome segregation machinery

We should begin our consideration of bacterial chromosome segregation by stressing that our current knowledge is derived from only a small number of bacterial species. Most of these have a single circular chromosome with one origin of replication and a defined replication terminus (see Sections 6.7 and 6.10), although some bacteria, such as *Borrelia* and *Streptomyces*, have linear chromosomes, and others, such as *Vibrio cholerae*, have more than one circular chromosome. Since the vast majority of bacteria have never been studied in a laboratory setting, it is possible that additional mechanisms for chromosome segregation are waiting to be discovered.

In certain bacteria, chromosome segregation is driven by an active segregation machinery, as in eukaryotes, although the mechanisms themselves differ. This active segregation is evident from the rate of chromosome movement following chromosome duplication, which exceeds the rate of bacterial cell growth: if chromosomes were simply tethered to a site in the cells - the cell membrane, for example - and segregation was driven solely by cell growth, the rate of chromosome movement would equal that of cell growth, rather than exceed it. During chromosome segregation of most bacteria, the origin of replication is at the leading edge of the moving chromosome, indicating that the site of chromosome attachment to the segregation machinery, namely the bacterial equivalent of a centromere, is located near the replication origin.

There are several features that distinguish bacterial chromosome segregation from eukaryotic segregation. In eukaryotes, DNA replication and chromosome segregation occur at distinct cell cycle phases, but in bacteria the two processes are coupled; chromosome segregation initiates well before the completion of DNA replication, and each chromosomal locus is segregated sequentially soon after it is replicated.

Also, although most bacteria, like all eukaryotes, have mechanisms that regulate the initiation of DNA replication so that it occurs only once per cell cycle, some bacteria can complete a round of cell division faster than the time required to replicate the chromosome under favorable growth conditions. This is possible because these bacteria undergo **multifork replication** under conditions of rapid growth, whereby the single origin of replication fires before the previous round of replication is completed (see Section 5.9). Thus, in rapidly dividing bacteria, the daughter cell is 'born' with a chromosome that is already partially replicated. This is in contrast to eukaryotic DNA replication, where each chromosome has multiple origins of replication, and each origin is strictly regulated to fire only once per cell division.

Finally, while chromosome segregation in eukaryotes relies on motor proteins associated with a microtubule-based spindle, no such motor proteins have been implicated in chromosome segregation in bacteria, and the structures that promote bacterial chromosome movement are not made of tubulin-like molecules.

➔ We learn more about chromosome duplication in Section 6.12.

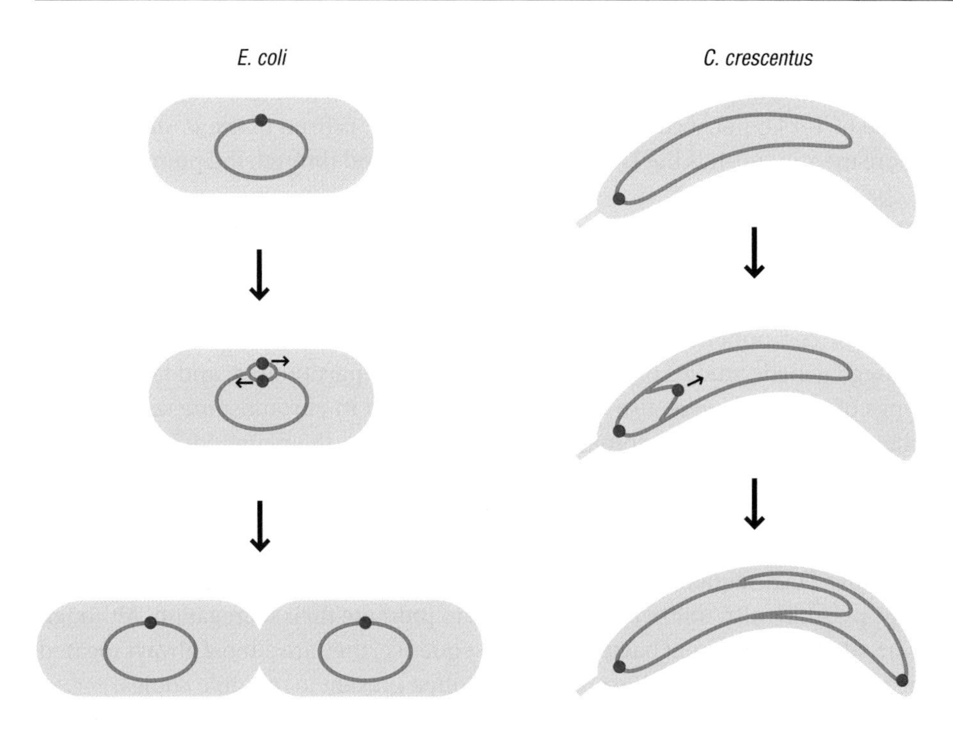

Figure 7.32 Chromosome segregation in *E. coli* and *C. crescentus*. The *E. coli* origin of replication (red) is situated near the cell center. As the chromosome duplicates, the origins move apart and end up near the centers of the two new daughter cells. In contrast, the *C. crescentus* origin of replication is located near one of the poles (the stalked pole). When DNA replication occurs, one origin stays at that pole (the old pole), while the other moves to the opposite pole (the new pole).

Different bacteria adopt different chromosome segregation strategies

The geometry of the segregation process differs among bacteria. For example, in *Escherichia coli,* the replication machinery is positioned near the cell mid-zone, and the duplicated chromosomes move in a symmetrical fashion away from the mid-zone, toward the cell poles (Figure 7.32, left). In contrast, in *Caulobacter crescentus,* DNA replication occurs at one end of the cell, and one of the replicated chromosomes migrates to the opposite cell end (Figure 7.32, right). In either case, the origin of replication is at the leading edge of the moving chromosome(s). In the case of *C. crescentus,* this is because the *parS* site, the bacterium's centromere equivalent, is positioned near the origin of replication. As we will see below, *parS* associates with the ParB protein that is needed for attachment to the segregation machinery, itself composed of ParA. (Chromosome segregation in bacteria is often referred to as chromosome partitioning; thus many of the genes involved are termed Par to designate their role in this process.)

The FtsK DNA transporter is needed for complete segregation of the chromosome into the *Bacillus. subtilis* forespore

In Chapter 5 we discussed the asymmetrical cell-division process in *C. crescentus.* The *Bacillus subtilis* system provides another example of asymmetrical cell division that poses an interesting problem during chromosome segregation. During sporulation in *B. subtilis,* cells divide asymmetrically, generating a large mother cell and a smaller forespore, as illustrated in Figure 7.33. However, as the replicated chromosomes move away from each other, they adopt an elongated shape, with the origins of replication at the two poles and the replication termini near the cell center. Therefore, at this point, most of the DNA destined to be in the forespore still resides within the mother cell.

This DNA is actively pumped into the forespore by a DNA transporter called **FtsK** (or SpoIIIE in *B. subtilis*). This is the same transporter that moves the DNA

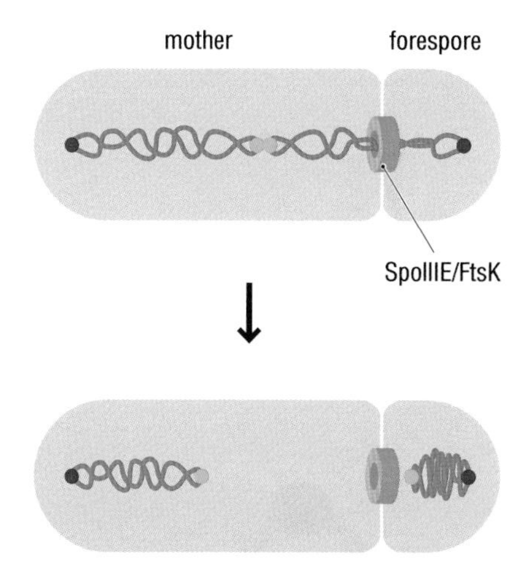

Figure 7.33 Chromosome segregation during *B. subtilis* sporulation. During sporulation, *B. subtilis* cells divide asymmetrically, forming a large mother cell (on the left) and a small forespore (on the right). The DNA segregates such that the origins (red) are each near a cell pole, and the rest of the chromosome stretches between the origins, with the chromosome termini (in green) situated near the cell's center. As a result, most of the DNA that should be in the forespore is in the mother. FtsK (orange) is a DNA translocase that pumps the chromosome into the forespore.

away from the plane of cell division during a regular cell cycle in *B. subtilis* and other bacteria (see following text). FtsK-type transporters can move along DNA at a rate of 4–7 kb per second, but since FtsK/SpoIIIE is fixed in the *B. subtilis* forespore membrane, it is the DNA that is actively moved through the pump.

Chromosome segregation in many bacteria involves the binding of the ParB protein to the *parS* site

The first components of the bacterial segregation machinery were identified through genetic studies on plasmid segregation or partitioning, and later similar types of elements were also shown to be involved in chromosome segregation. These genes and chromosomal loci were often given the name *parX*, where X is a different letter for each gene/protein (such as **ParA**, **ParB**, etc.). (Homologous proteins in different bacteria may have different names. To simplify the discussion, here we will use the ParX nomenclature.) Many, but not all, bacterial chromosomes contain a DNA sequence called **parS**, which functions as a centromere. Some low copy plasmids also utilize *parS* sequences to promote their segregation. Although *parS* sites from different bacteria vary in sequence, they are almost always located near the origin of replication and they are often present in multiple copies.

In several bacteria, including *C. crescentus* and *B. subtilis*, *parS* recruits a protein called **ParB**, which assembles into a nucleoprotein filament along the DNA flanking *parS*, as shown in Figure 7.34. ParB, in turn, recruits a second protein, ParA, to the *parS* site. ParA proteins have the ability to bind and hydrolyze ATP. In *C. crescentus*, the *parS*, ParA, and ParB equivalents are all essential for viability, and their inactivation results in chromosome segregation defects. In *B. subtilis*, on the other hand, inactivation of the ParA homolog has few consequences, but inactivation of the ParB homolog results in chromosome disorganization. Thus, *B. subtilis* is likely to have a chromosome segregation machinery that acts in parallel to ParA.

How do ParA and ParB promote chromosome segregation? Several hypotheses have been proposed by comparing them to functionally similar proteins found in plasmid partitioning systems. The ParA-like protein involved in the R1 plasmid partitioning mechanism, called **ParM**, can form a long polymer that associates with two replicated plasmids at its ends, as shown in Figure 7.35. As the ParM polymer grows, it pushes the plasmids away from each other toward the cell poles, thereby increasing the likelihood that the two plasmids will end up in two different cells when the bacterium divides.

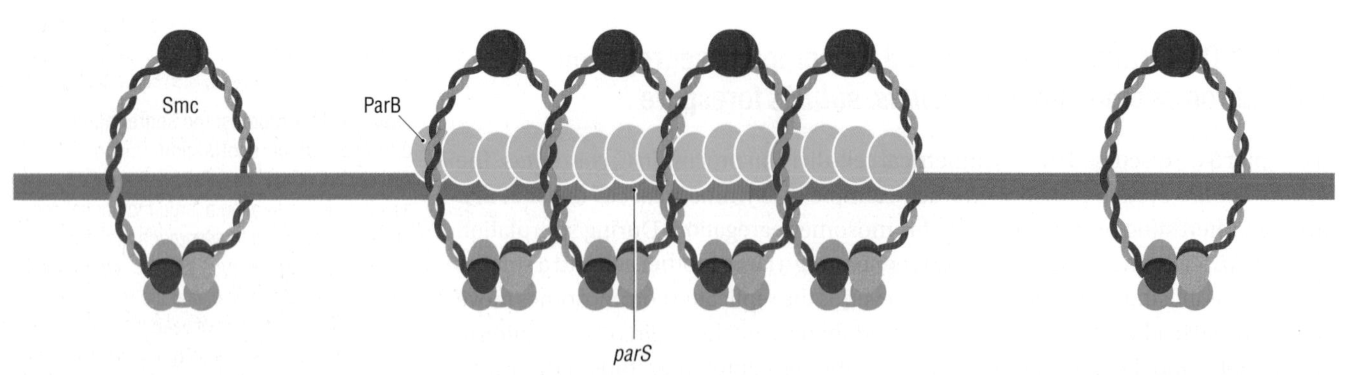

Figure 7.34 The bacterial centromere. The *parS* site (blue), present in many plasmids and bacterial chromosomes, is situated close to the origin of replication (not shown). *parS* recruits ParB (green), which assembles into a nucleoprotein filament. ParB, in turn, recruits ParA (not shown) and the SMC complex (red/pink). Although purified SMC complex forms ring structures, the structure of the SMC complex when it is bound to DNA is not known, nor is it known how the binding of the SMC complex affects DNA structure. Although the SMC complex concentrates around the *parS*/ParB site, it can also be found throughout the bacterial chromosomes.

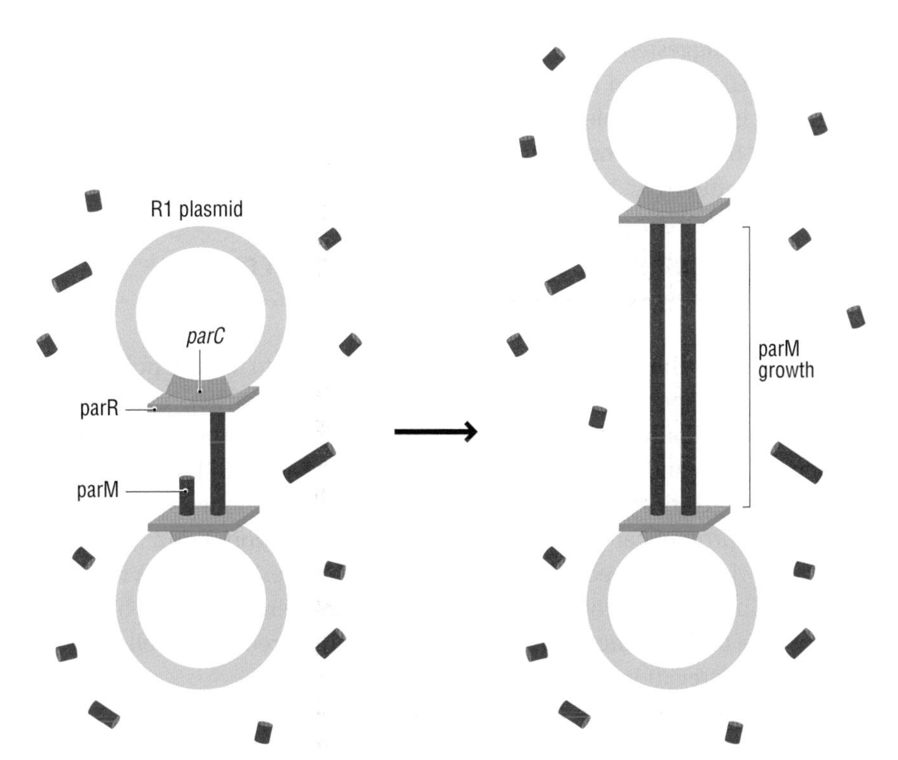

Figure 7.35 The segregation machinery of the R1 plasmid. The R1 plasmid (green) requires two proteins for its segregation: ParR (orange), which binds to the centromere site, called *parC*, on the plasmid (blue), and ParM (red), which can form filaments and associates with the R1 plasmid through its binding to ParR. The ParM filaments grow by incorporating free ParM near the attachment site to ParR. The elongation of the ParM filament pushes the R1 plasmids away from each other.

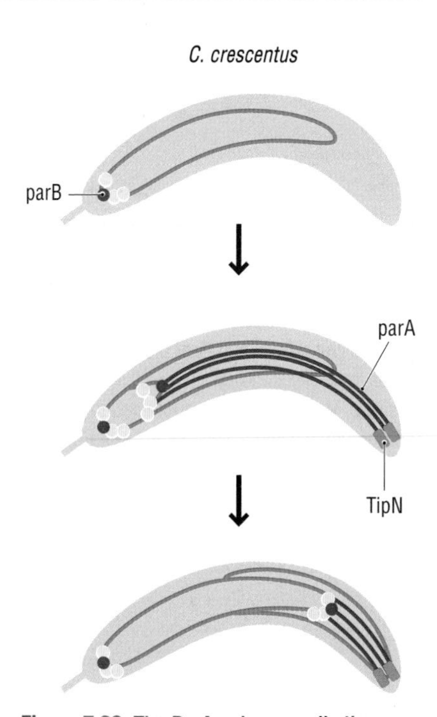

Figure 7.36 The ParA polymer pulls the *C. crescentus* chromosome. After DNA replication begins, ParB (in yellow) assembles on the *parS* regions that are around the origin of replication (in red). One of the origins remains near the cell pole, while the other moves to the opposite pole. This movement is due to the binding of ParA (in purple, anchored at the opposite pole through TipN, in orange) to ParB, followed by ParA polymer shrinkage, which pulls the chromosome toward the opposite pole.

The ParA protein involved in the segregation of C. *crescentus* chromosome also forms filaments that associate with the chromosome through ParB bound to the *parS* site. But unlike the pushing mechanism proposed for plasmid R1, in *C. crescentus,* ParA appears to pull the chromosome to the cell pole, as shown in Figure 7.36. It is thought that after binding to ParB at the parS site, the ParA polymer depolymerizes, pulling the chromosome with it. This is reminiscent of microtubule instability in animal cells (see Section 7.3), but the mechanism promoting ParA polymerization and depolymerization is currently not known.

The SMC complex assists in chromosome segregation by contributing to chromosome compaction

In addition to recruiting ParA to the *parS* region, ParB is also involved in recruiting a complex that belongs to the **SMC** family of proteins (see Figure 7.34). Bacteria have only one type of SMC complex, which is similar in structure and function to the eukaryotic condensin (see Section 7.2). As in eukaryotes, loss of bacterial SMC function leads to chromosome decondensation and defects in chromosome organization and segregation. It is interesting that although the eukaryotic and bacterial segregation machineries are quite distinct, they both involve proteins of the SMC family.

The SMC complex is bound throughout the bacterial chromosome but localizes at high concentrations around *parS*. Even bacteria that lack *parS* sites, such as *E. coli,* still have SMC-like complexes that bind near the origin of replication. We don't currently know how the SMC complex promotes chromosome segregation,

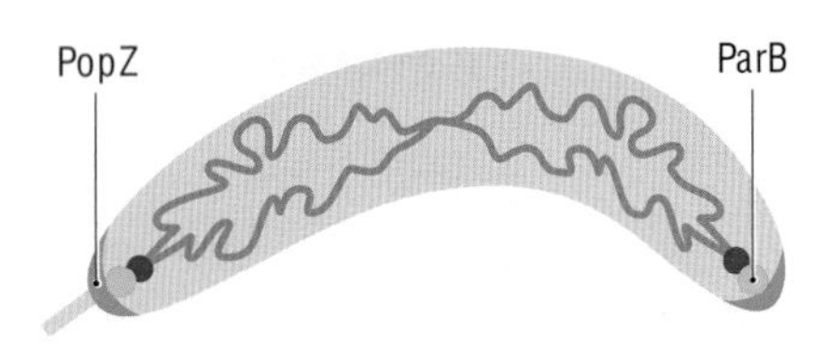

Figure 7.37 PopZ anchors the origin-proximal centromere to the poles in *C. crescentus*. In *C. crescentus*, as the DNA is being replicated, one origin of replication (red), with its associated *parS* site (not shown) and ParB protein (green), moves from the old pole (with the stalk) to the new pole. The origin becomes tethered to the new pole through the interaction of ParB with PopZ (orange). PopZ is also responsible for tethering of the origin to the old pole.

but it has been suggested that SMC complexes at the two newly replicated origins compact the DNA and, in so doing, move the two sister chromatids away from each other. It is also possible that, by compacting the two new copies of the chromosome, the SMC proteins contribute to the partitioning of the chromosomes away from one another.

During chromosome segregation, centromeres of certain bacteria become tethered to the cell's pole

In certain bacteria, the origin of replication that is moved to the cell pole becomes anchored there by specialized machinery. As we discussed earlier, in *C. crescentus*, one copy of the chromosome moves from one cell pole to the other, led by *parS* and its associated ParA and ParB proteins (see Figures 7.34 and 7.37). When it reaches the new pole, ParB becomes associated with a polymeric network composed of a proline-rich protein called PopZ. Prior to DNA replication, PopZ is present at the old pole, where it is needed to anchor the centromere. When DNA replication begins, PopZ also appears at the new pole, such that there is PopZ at both poles. In the absence of PopZ, the origin regions are not anchored to the cell poles, and chromosome segregation is severely disrupted.

A similar type of anchoring mechanism also exists in *B. subtilis* during sporulation (albeit with other types of proteins), and so it is probably a common feature of bacteria that segregate their *parS* sites to the cell poles. Anchoring of the chromosome to the poles may be important for setting up the proper cell-division site, as we will discuss next.

Bacterial cells divide by the contraction of an FtsZ ring

In both bacteria and eukaryotes (but not plants), cell division is mediated by a contractile ring. However, the nature of this ring differs greatly between the three kingdoms. In eukaryotes the ring is made of actin and myosin, whereas in bacteria and some archaea, it is made of **FtsZ** (a homolog of tubulin) and numerous associated proteins. The bacterial contractile ring, called the **Z ring**, is tethered to the inner face of the membrane at the cell mid-zone. As the ring contracts, it pulls the membrane with it, forming a septum that separates the two daughter cells.

A potential problem during cell division is that the septum, which is formed as the Z ring contracts, could bisect lagging chromosomes. To avoid this problem, the DNA in some bacteria is condensed into a compact structure by specialized proteins, including the conserved complex containing SMC proteins discussed above. Eukaryotes have a further regulatory mechanism that is linked to mitotic cyclin–CDK levels and which inhibits cell division until telophase, when the two DNA masses are well apart. Bacterial chromosomes, on the other hand, often remain catenated (that is, interlinked) after DNA replication (see Section 6.10). As a result, linked chromosomes could get stuck in the cell's mid-zone.

To avoid these linked chromosomes getting bisected by the FtsZ contractile ring, bacteria actively decatenate and remove lagging DNA from the plane of cell division by an ATP-dependent DNA translocase, called FtsK, that associates with the FtsZ ring. We encountered FtsK (or SpoEIII in *B. subtilis*) earlier in this section when we discussed the segregation of the chromosome into the *B. subtilis* forespore. As in the forespore case, the tethering of FtsK to the FtsZ ring means that ATP hydrolysis does not move the protein along the DNA but rather moves the DNA through the protein, thereby pumping the ends of the chromosomes away from the contractile ring.

It is important that the FtsK translocates the DNA at the terminus of the chromosome toward the pole and not the DNA at the origin end back toward the midline, because reeling in the origin will result in one daughter cell with two chromosomes and one daughter cell with none. The directionality of FtsK action is determined by DNA sequences called the FtsK-orienting polar sequences. These sequences are dispersed throughout the chromosome and serve as DNA arrows that point from the origin of replication toward the terminus. FtsK recognizes these sequences and reels in the DNA in the appropriate direction. Thus, the action of FtsK serves as a mechanism that couples cell division to the completion of chromosome segregation.

The placement of the cell-division site depends on positional information from both the chromosome and the cell poles

Once the sister chromosomes have segregated, the bacterial cell is ready to divide. A universal challenge faced by all dividing cells is how to position the division site correctly, so as to bisect the cell with the segregated chromosomes on either side. We saw in Section 7.6 that, in eukaryotes, the spindle mid-zone guides the placement of the cleavage furrow. Two of the best-understood mechanisms that cooperate to specify the cell-division site in bacteria are called **nucleoid occlusion** and the **MinCDE system**, which operate in *E. coli* and other bacteria.

The bacterial chromosome is organized into a protein-containing structure called a nucleoid, which was introduced in Section 4.1. The nucleoid occlusion mechanism involves DNA binding proteins, such as Noc in *B. subtilis* and SlmA in *E. coli*, that bind DNA and inhibit FtsZ polymerization by an unknown mechanism. Nonetheless, because of this inhibitory activity, the FtsZ ring can only assemble, in principle, either between the two nucleoids or between one nucleoid and the poles of the bacterium, as illustrated in Figure 7.38a.

The Min system also inhibits the assembly of the cell-division apparatus. *In E. coli*, the MinC and MinD proteins form a gradient of inhibitory signal that is highest at the cell poles and lowest at the cell mid-zone. MinC directly antagonizes FtsZ ring stability, and this allows the FtsZ ring to assemble in a wide region around the cell's center (Figure 7.38b). By combining the Min system with the nucleoid occlusion process, the FtsZ ring can only assemble in a narrow region in the center of the cell, between the segregated chromosomes (Figure 7.38c).

In *C. crescentus*, which lacks the Min system and nucleoid occlusion proteins, the assembly of the FtsZ ring is inhibited by a protein called MipZ. MipZ interacts with ParB, and therefore the accumulation of MipZ at the cell poles and away from the cell mid-zone is dependent on PopZ, the protein that forms a filamentous network at the cell pole. Prior to chromosome segregation, PopZ is absent from the new pole so MipZ can inhibit FtsZ throughout the cell. However, once chromosome segregation begins and PopZ accumulates at the new pole, MipZ concentrates at the two cell poles as well, clearing the mid-zone for FtsZ assembly. Thus, the association of MipZ and ParB with PopZ couples the assembly of FtsZ to the chromosome partitioning processes, thereby ensuring that cell division will take place only when chromosome segregation is completed.

(a) nucleoid occlusion

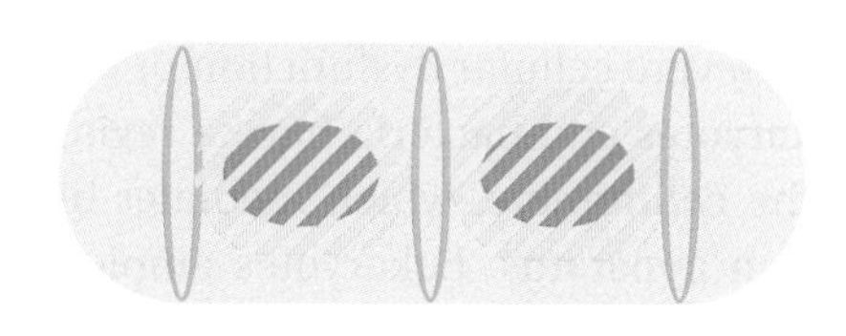

(b) Min system

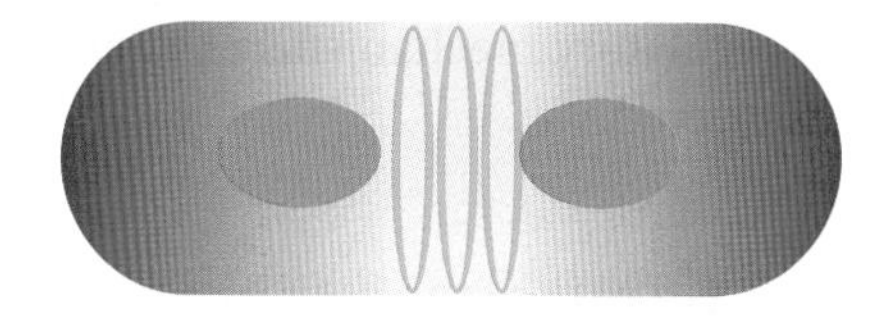

(c) both mechanisms

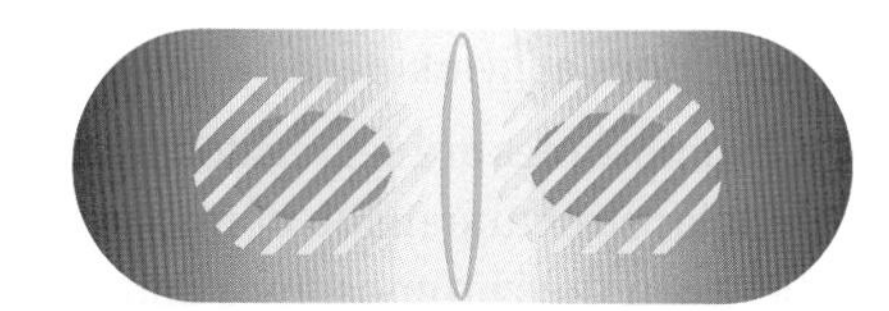

Figure 7.38 Determining the site of cell division. Two mechanisms act in concert to define the site at which cell division will take place: (a) the nucleoid occlusion pathway and (b) the MinCDE system. In the nucleoid occlusion pathway, the nucleoid (blue oval shapes) emits a negative signal (yellow) that blocks the assembly of the FtsZ ring (green) in the vicinity of the nucleoid. When only this pathway is working, the FtsZ ring can assemble either between the nucleoids or near the cell poles. The MinCDE system generates a gradient of negative signal (shown in red), that is highest at the cell poles and lowest at the cell mid-zone, allowing FtsZ ring assembly in a wide area between the two nucleoids. (c) The combination of the two signals directs the assembly of the FtsZ ring to a narrow region between the two nucleoids.

In this chapter we learned about the mechanism and regulation of chromosome segregation in mitosis and meiosis of eukaryotes, as well as in bacteria. In all cases chromosomes attach to a segregation apparatus, which separates the chromosomes to form two identical copies of the genome. The one overarching principle of chromosome segregation is that it must be accurate: cells must receive one, and only one, copy of each chromosome. (Remember that in meiosis, each gamete must receive one, and only one, copy of either of the two homologous chromosomes.) Deviation from this rule will have dire consequences: an abnormal number of chromosomes can lead to severe cellular dysfunction or even cell death. Both eukaryotes and bacteria have a region of the chromosome, the centromere, which mediates binding to the segregation apparatus. Eukaryotes employ a microtubule-based spindle, associated with motor proteins, to move chromosomes, while some bacteria use ParA polymers, but no motor proteins. Eukaryotes insure accurate chromosome segregation through checkpoint pathways. Whether similar regulatory mechanisms exist in bacteria remains to be discovered.

MITOTIC CHROMOSOME SEGREGATION IN EUKARYOTES

- Mitosis includes five main steps: prophase, prometaphase, metaphase, anaphase, and telophase. Mitosis concludes at cytokinesis, when the parent cell divides.
- Chromosomes are segregated by the spindle apparatus, a bipolar structure made of microtubules. The attachment between the chromosomes and the microtubules is mediated by the kinetochore, which assembles at the chromosome's centromere.
- Sister chromatid cohesion plays an essential role in chromosome segregation: it allows the cell to recognize the two identical chromosomes as sister chromatids and it enables the generation of tension when proper microtubule attachment (bi-orientation) is achieved.
- Sister chromatid cohesion is dissolved at the onset of anaphase by the protease separase. Separase is negatively regulated in multiple ways, including by the spindle checkpoint pathway, to ensure that sister chromatids do not separate precociously.
- The spindle checkpoint pathway inhibits progression into anaphase until all chromosomes bi-orient on the mitotic spindle.

MEIOTIC CHROMOSOME SEGREGATION

- Meiosis is the chromosome segregation process that generates gametes. The products of meiosis have half the ploidy of the parent cell.
- Meiosis I involves the separation of homologous chromosomes while meiosis II involves the separation of sister chromatids.
- Recombination crossovers, one of the products of homologous recombination, are needed in order to align homologous chromosomes and facilitate the generation of tension when homologous chromosomes bi-orient during meiosis I.
- In meiosis, cohesin is removed in two steps: cohesin complexes associated with the chromosome arms are removed during meiosis I while cohesin complexes associated with centromeres are removed during meiosis II.
- Kinetochores of sister chromatids co-orient during meiosis I and bi-orient during meiosis II.
- Homologous recombination and random assortment of maternal and paternal chromosomes during meiosis I create genetic variation.

BACTERIAL CHROMOSOME SEGREGATION

- Different bacteria utilize different segregation machineries.
- Some bacterial chromosomes are segregated by a machinery that involves a sequence that serves as a centromere (parS) and centromere binding proteins called ParA and ParB. ParB recruits the SMC protein complex that promotes chromosome compaction.
- Certain plasmids (e.g., R1) are segregated by ParM filaments, which bind plasmids at the filament ends and push the plasmids apart as the filament elongates. The ParA polymer may move chromosomes by a pulling mechanism that involved binding to ParB followed by polymer depolymerization.
- In most bacteria studied to date, the origins of replication are at the leading edge of the segregating chromosomes. In some bacteria, the parS locus, which is near the origin, becomes tethered to the cell pole. In *C. crescentus*, this tethering is mediated by the PopZ protein.
- The FtsK protein (SpoEIII inB. subtilis) acts as a DNA pump that moves the DNA away from the plane of cell division. The directionality of this movement is determined by FtsK-orienting polar sequences.

- Bacterial cell division is mediated by a contractile ring made of the FtsZ protein, which is similar to the eukaryotic tubulin.
- In certain bacteria, the site of cell division is determined by two mechanisms: nucleoid occlusion, whereby proteins associated with the nucleoid inhibit the assembly of the FtsZ ring, and the MinCDE system, which forms a FtsZ inhibitory gradient that is highest at the cell poles and lowest at the cell mid-zone. The combination of these two mechanisms confines the assembly of the FtsZ ring to a narrow region at the cell's mid-zone.

QUESTIONS

7.1 THE STAGES OF MITOSIS

1. During which cell cycle stage does chromosome segregation occur?
   a. G1
   b. G2
   c. S
   d. M
2. Which is the first stage in mitosis after which faulty attachments in chromosome-microtubule attachment can no longer be corrected?
   a. Prophase
   b. Metaphase
   c. Anaphase
   d. Telophase
3. Briefly describe what happens during each of the following cell cycle stages.
   a. Prophase
   b. Prometaphase
   c. Metaphase
   d. Anaphase
   e. Telophase
   f. Cytokinesis
4. Briefly define/describe the following:
   a. Mitotic spindle
   b. Kinetochore
   c. Centromere
   d. Sister chromatid

7.2 CHROMOSOME CONDENSATION AND COHESION

1. What are the functions of cohesin and condensin?
2. Briefly explain why cohesion is crucial for proper chromosome segregation.

Challenge question

3. Cohesin:
   a. Propose a hypothesis for the outcome of mitosis if cohesin could not assemble properly. Explain your reasoning.
   b. You have a temperature sensitive haploid strain of *Saccharomyces cerevisiae* that has a defect in cohesin assembly. 'Temperature sensitive' means that the yeast will grow normally at low temperatures, but is defective when grown at higher temperatures. Utilizing this strain, design an experiment to test your hypothesis from part (a).

7.3 THE MITOTIC SPINDLE

1. What are the three different types of microtubule in a cell, and where do they attach?
2. Describe the structure of a microtubule.
3. Microtubules constantly polymerize and depolymerize. How does this occur?

Challenge question

4. Explain why it is important that spindles are bipolar rather than monopolar or multipolar.

7.4 PROMETAPHASE AND METAPHASE

1. Kinetochore vs centromere:
   a. What is the difference between the kinetochore and the centromere?
   b. If a kinetochore protein is associated with the centromere throughout the cell cycle, what type of kinetochore protein is this and why?
2. Kinetochores attach to microtubules at random. Explain the main ways that microtubules can attach, and the one that is necessary for correct chromosomes segregation.

Challenge question

3. Explain why random attachment of kinetochores does not usually lead to chromosome mis-segregation.

7.5 ANAPHASE: AN IRREVERSIBLE STEP IN CHROMOSOME SEGREGATION

1. Anaphase cannot proceed until the spindle checkpoint is satisfied. Explain the purpose of this checkpoint and describe how it functions.
2. Microtubules remain attached to chromosomes even though polymerization and depolymerization occurs. Explain how this attachment is maintained.
3. What are the roles of the following in regulating cohesin cleavage?
   a. Separase
   b. Securin
   c. Shugoshin
   d. Spindle checkpoint proteins

7.6 THE COMPLETION OF MITOSIS AND CYTOKINESIS

1. Briefly describe how cytokinesis occurs:
   a. In plant cells
   b. In animal cells
2. How is cytokinesis regulated in animal cells such that daughter cells with uneven sizes or uneven numbers of chromosomes are not generated?

7.7 MEIOSIS: GENERATING HAPLOID GAMETES FROM DIPLOID CELLS

1. Briefly define/describe the following:
   a. Homologous chromosome
   b. Recombination crossover
   c. Chiasmata
   d. Synaptonemal complex

2. Draw a diagram of the synaptonemal complex in association with DNA.

Challenge questions

3. Explain why this diagram depicting crossovers in meiosis is incorrect and draw a more accurate diagram.

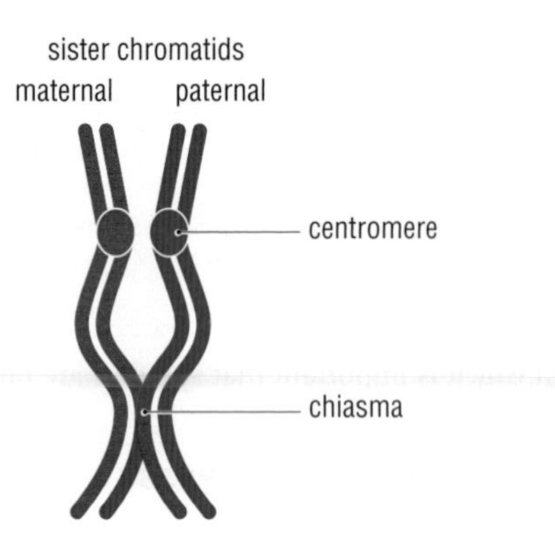

4. Compare kinetochore–microtubule attachment in mitosis with that in meiosis. Explain why any differences are important for proper segregation.

5. Compare cohesin cleavage in mitosis with that in meiosis and explain why any differences are important for proper segregation.

6. Meiosis involves one round of replication followed by two rounds of cell division. Which meiotic division most resembles mitosis? Explain your answer.

7.8 CHROMOSOME SEGREGATION IN BACTERIA

1. Briefly describe the hypothesized roles of the following during *C. crescentus* chromosome segregation.
  a. *parS*
  b. ParA
  c. ParB

2. How does bacterial cytokinesis resemble animal cell cytokinesis? How does it differ?

3. Briefly explain how the following processes regulate the site of bacterial cell division.
  a. Nucleoid occlusion
  b. MinCDE

EXPERIMENTAL APPROACH 7.1 – THE ADVANTAGE OF BEING ABLE TO SEE: FOLLOWING CHROMOSOME MOVEMENT IN SINGLE-CELLED ORGANISMS

Challenge questions

1. Experimental approach 7.1 describes the development and use of the FISH technique to observe chromosome behavior. Using FISH to follow a single chromosomal locus, how many fluorescent spots would you expect to see in a single wild-type cell in the following conditions? Explain your answers.
  a. Just prior to prophase in a diploid cell
  b. At metaphase just prior to the first meiotic division
  c. After the second meiotic division but before cytokinesis

2. Experimental approach 7.1 describes use of GFP to follow chromosome movements in live haploid yeast cells.
  a. In Figure 2b, what is indicated by the presence of the two green dots in the far-right panel?
  b. If the far-right panel in Figure 2b showed two green dots, but they were both in the mother cell, what might this suggest biologically?

EXPERIMENTAL APPROACH 7.2 – THE HUNT FOR KINETOCHORE PROTEINS

Challenge question

1. Experimental approach 7.2 describes the identification of several kinetochore proteins. Why was it necessary to use a small, non-essential chromosome fragment for this experiment?

EXPERIMENTAL APPROACH 7.3 – CHROMOSOME BI-ORIENTATION: HOW TO GENERATE AND DETECT TENSION.

Challenge questions

1. Figure 1 of Experimental approach 7.3 indicates how the importance of tension in mitosis was discovered.
  a. Explain why this experiment would have been difficult to perform using human spermatocyte cells rather than praying mantis spermatocyte cells.
  b. Wild-type cells arrest, as shown in Figure 1b, occurs when the second X chromosome is not paired. Would arrest be expected if the experiment were repeated with a mutant containing a non-functional spindle checkpoint pathway? Explain your answer.

FURTHER READING

7.2 CHROMOSOME CONDENSATION AND COHESION

Hirano T. Condensins: universal organizers of chromosomes with diverse functions. *Genes and Development*, 2012;**26**:1659–1678.

Onn I, Heidinger-Pauli JM, Guacci V, Unal E, Koshland DE. Sister chromatid cohesion: a simple complex with a complex reality. *Annual Review of Cell and Developmental Biology*, 2008; **24**:105–129.

7.3 THE MITOTIC SPINDLE

Walczak CE, Heald R. Mechanisms of mitotic spindle assembly and function. *International Review of Cytology*, 2008;**265**:111–158.

7.4 PROMETAPHASE AND METAPHASE

Cheeseman IM, Desai A. Molecular architecture of the kinetochore-microtubule interface. *Nature Reviews Molecular Cell Biology*, 2008; **9**:33–46.

O'Connell CB, Khodjakov A, McEwen BF. Kinetochore flexibility: creating a dynamic chromosome-spindle interface. *Current Opinion in Cell Biology*, 2012;**24**:40–47.

7.5 ANAPHASE: AN IRREVERSIBLE STEP IN CHROMOSOME SEGREGATION

Holland AJ, Cleveland DW. Losing balance: the origin and impact of aneuploidy in cancer. *EMBO Reports*, 2012;**13**:501–514.

Musacchio A. Spindle assembly checkpoint: the third decade. *Philosophical Transactions of the Royal Society*, 2011;**366**:3595–3604.

Pines J. Mitosis, a matter of getting rid of the right protein at the right time. *Trends in Cell Biology*, 2006;**16**:55–63.

Sullivan M, Morgan DO. Finishing mitosis, one step at a time. *Nature Reviews in Molecular and Cell Biology*, 2007;**8**:894–903.

7.6 THE COMPLETION OF MITOSIS AND CYTOKINESIS

Fededa JP, Gerlich DW. Molecular control of animal cell cytokinesis. *Nature Cell Biology*, 2012;**14**:440–447.

Kutay U, Hetzer MW. Reorganization of the nuclear envelope during open mitosis. *Current Opinion in Cell Biology*, 2008;**20**:669–677.

Müller S, Wright AJ, Smith LG. Division plane control in plants: new players in the band. *Trends in Cell Biology*, 2009;**19**:180–188.

7.7 MEIOSIS: GENERATING HAPLOID GAMETES FROM DIPLOID CELLS

Jones KT. Meiosis in oocytes: predisposition to aneuploidy and its increased incidence with age. *Human Reproduction Update*, 2008; **14**:143–158.

Marston AL, Amon A. Meiosis: cell cycle controls shuffle and deal. *Nature Reviews Molecular Cell Biology*, 2004;**5**:983–997.

Watanabe Y. Geometry and force behind kinetochore orientation: lessons from meiosis. *Nature Reviews in Molecular and Cell Biology*, 2012;**16**:370–382.

7.8 CHROMOSOME SEGREGATION IN BACTERIA

Erickson HP, Anderson DE, Osawa M. FtsZ in bacterial cytokinesis: cytoskeleton and force generator all in one. *Microbiology and Molecular Biology Reviews* 2010;**74**:504–528.

Ramamurthi KS, Losick R. Grasping at origins. *Cell*, 2008;**134**:916–918.

Schumacher MA. Bacterial plasmid partition machinery: a minimalist approach to survival. *Current Opinion in Structural Biology*, 2012. **22**:72–79.

Thanbichler M. Synchronization of Chromosome Dynamics and Cell Division in Bacteria. *Cold Spring Harbor Perspectives in Biology*, 2010;**2**:a000331.

Toro E, Shapiro L. Bacterial chromosome organization and segregation. *Cold Spring Harbor Perspectives in Biology*, 2010;**2**:a000349.

8 Transcription

INTRODUCTION

The information stored in chromosomes contains instructions for synthesizing all of the macromolecules needed by an organism to live and reproduce. This information is encoded in the sequence of bases in the chromosomal DNA. We covered in the previous chapters how genetic information is copied and distributed to newly divided cells. In this chapter, we consider how the information in a linear sequence of adenine (A), thymine (T), guanine (G), and cytosine (C) bases in DNA is used to produce a functional protein or RNA molecule in the cell.

The flow of genetic information from DNA to RNA to protein (Figure 8.1) is a universal feature of all cells. Although important differences and adaptations have evolved among different organisms, we shall see that the basic mechanisms underlying gene expression are remarkably conserved from bacteria to humans, suggesting that these essential processes evolved well before the divergence of the three kingdoms of life.

In this chapter, we will describe the individual steps that must occur to produce an RNA copy of a gene. We shall see that, despite the many differences between bacteria, archaea, and eukaryotes, there are common principles underlying the mechanism by which RNA is synthesized. We begin by identifying the key features of transcription in general terms, and describe the important molecular components that are involved in this process.

8.1 OVERVIEW OF TRANSCRIPTION

Transcription produces a copy of the coding strand of DNA using ribonucleotides

To produce a functional RNA molecule, the cell must make an RNA copy of a DNA sequence. This process is known as **transcription**, because a copy of one strand of the DNA known as the **coding strand** is transcribed into a single strand of ribonucleotides. Figure 8.1 shows how, to produce a protein, the sequence of ribonucleotides in the RNA copy of a gene must then be read by the ribosome, which translates the information encoded by the bases into a polypeptide chain. The RNA is therefore named **messenger RNA (mRNA)** for its role in carrying a copy of the genetic information to the ribosome.

➔ The process of translation is described in detail in Chapter 11.

Not all RNA transcripts encode proteins. In some cases, the RNA molecules play a structural role or form a part of an enzyme complex. The ribosome itself, for example, is composed partly of RNA. There are also RNA transcripts that play regulatory roles, which we shall learn about in Chapter 13. The overall process by which these different classes of RNA are transcribed is largely the same.

DNA is transcribed into RNA by an **RNA polymerase** enzyme, which makes a copy of the coding strand by using the complementary strand of DNA, known as the **template strand** (or **non-coding strand**). RNA polymerase must first separate the two strands of DNA to expose the template DNA strand. This makes it possible for the precursors of RNA - the **ribonucleoside** triphosphates - to form Watson-Crick base pairs with the template strand, a critical step in ensuring the faithful transfer of information from DNA to RNA.

The bases in the ribonucleoside triphosphates are the same as those in DNA, with one exception: RNA contains uracil (U) in place of thymine (T). RNA also has the sugar, ribose, whereas DNA has the sugar, deoxyribose. The RNA polymerase enzyme catalyzes formation of phosphodiester linkages as successive ribonucleotides are added to the 3′ end of the growing RNA chain. The chemical steps required to synthesize an RNA molecule are thus quite similar to those in DNA replication. The final product of transcription, however, is a single-stranded RNA molecule called the **transcript**, whose base sequence is a copy of that of the coding strand.

➔ We introduce the building blocks of RNA in Chapter 2.

Transcription proceeds through three key stages

Scan here to watch a video animation explaining more about transcription initiation, elongation, and termination, or find it via the Online Resource Center at www.oxfordtextbooks.co.uk/orc/craig2e/.

The process of transcription, outlined in Figure 8.2, can be divided into three main stages: **initiation**, **elongation**, and **termination**. Transcription begins when RNA polymerase locates and binds to a DNA sequence immediately preceding the gene. This sequence is called the **promoter** and signals where transcription should start. The transcription start site is the first base of the coding strand that is transcribed and is denoted +1. Since transcription proceeds in the 5′ to 3′ direction, sites located 3′ to this or any given site are said to be located **downstream**, whereas those located to the 5′ side are said to be **upstream**.

In the **initiation** phase, the RNA polymerase enzyme separates the two strands of DNA, creating a short unpaired region known as a **transcription bubble**. This separation makes the template strand available for base-pairing with incoming ribonucleoside triphosphates, so that RNA polymerase can now catalyze formation of the RNA transcript. RNA polymerase begins by synthesizing the first few ribonucleotides of the RNA transcript while remaining at the promoter. When RNA polymerase succeeds in synthesizing an RNA product of sufficient size, the enzyme moves past the promoter and changes conformation, becoming more stably associated with the DNA.

After this transition, the polymerase enzyme continues to move along the template strand and elongate the RNA transcript. In this **elongation** phase, the enzyme unwinds the DNA ahead of it to expose the single-stranded template DNA. Behind the enzyme, the template strand pairs again with the coding strand to re-form the DNA duplex. In this way, RNA polymerase maintains the transcription bubble as it proceeds along the chromosome. Within this open region, a short stretch of the DNA template strand is base-paired with the most recently synthesized RNA, while the rest of the growing RNA strand is extruded through RNA polymerase.

Elongation of the RNA transcript continues until RNA polymerase encounters a DNA sequence known as a **terminator** that signals where RNA synthesis should stop (the **termination** phase). The polymerase enzyme can then dissociate from the DNA template and is free to initiate a fresh round of transcription.

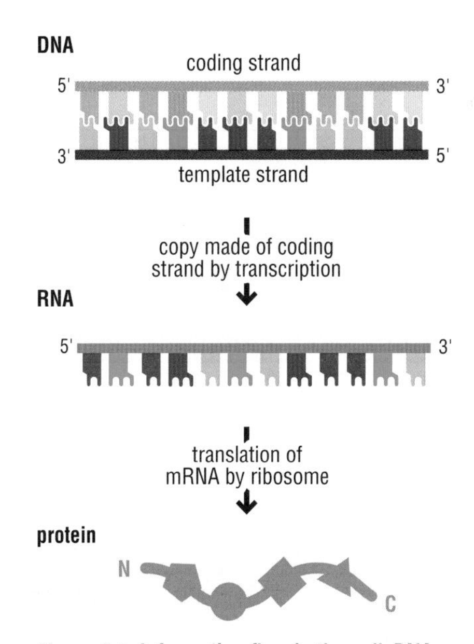

Figure 8.1 Information flow in the cell: DNA → RNA → protein. Genetic information in all cellular organisms is stored in the base sequence of double-stranded DNA. The cell harnesses this information by making an RNA copy of the coding strand of the DNA, using the complementary non-coding strand as a template. For protein-coding genes, this RNA is called messenger RNA (mRNA). In the process of translation, the ribosome synthesizes a protein whose amino acid sequence is specified by the sequence of bases in the mRNA, with each three consecutive bases specifying a single amino acid.

The packaging of eukaryotic DNA into chromatin poses a barrier to transcription

The packaging of eukaryotic DNA into chromatin affects all aspects of transcription from initiation through elongation. We saw in Chapter 4 that DNA in

➔ We learn more about histones and histone proteins in Section 4.3.

eukaryotic cells is packaged by histones and other proteins, with the nucleosome as the fundamental building block of chromatin. Recall that each nucleosome consists of 147 base pairs (bp) of DNA wrapped around a histone octamer (two each of histone H2A, H2B, H3, and H4). This wrapping can present a barrier to transcription because it can prevent RNA polymerase and other proteins from binding to DNA.

Eukaryotic cells overcome the chromatin barrier by using three general classes of enzymes that must also be recruited to the transcribed gene in addition to RNA polymerase. Nucleosome-remodeling enzymes, which reposition histones and histone octamers from one location on the DNA to another, help remove the barrier to transcription imposed by the structure of chromatin and allow binding of the necessary proteins to the promoter. (These enzymes can also be marshaled to block transcription by relocating nucleosomes to positions that interfere with transcription.) Histone chaperones, which disassemble and reassemble the histone octamer, also help to remove the chromatin barrier from promoters and from the path of the elongating RNA polymerase enzyme. Histone chaperones are also needed to insert variant histones at particular locations in a gene. Finally, there are enzymes that catalyze reversible chemical modification of histone proteins. These modifications play a central role in modulating chromatin structure and in governing the recruitment of proteins that bind to particular histone modifications.

Acetylation, methylation, ubiquitination, sumoylation, phosphorylation, and adenosine diphosphate (ADP) ribosylation are all post-translational modifications found in histone proteins. These modifications are important in all processes that involve DNA and chromatin, including replication and DNA repair, as well as transcription. The structure of the nucleosome, ATP-dependent nucleosome remodeling, and post-translational modification of histones are discussed in Chapter 4. We will see throughout this chapter how enzymes that target chromatin play a central role in facilitating transcription in eukaryotes.

Transcription is regulated in all cells

Although the genome encodes every protein and RNA molecule needed by a cell, not all of these molecules are needed at any given moment. Instead, the needs of a cell vary significantly with time, developmental stage, and changes in the environment. The amount of each protein and RNA molecule in the cell can be adjusted by controlling the number of RNA transcripts made of each gene. Transcription is therefore a highly regulated process. We explore the regulation of transcription in Chapter 9.

Before examining each stage of transcription in more detail, we need to consider the key molecular component, RNA polymerase, without which transcription could not occur.

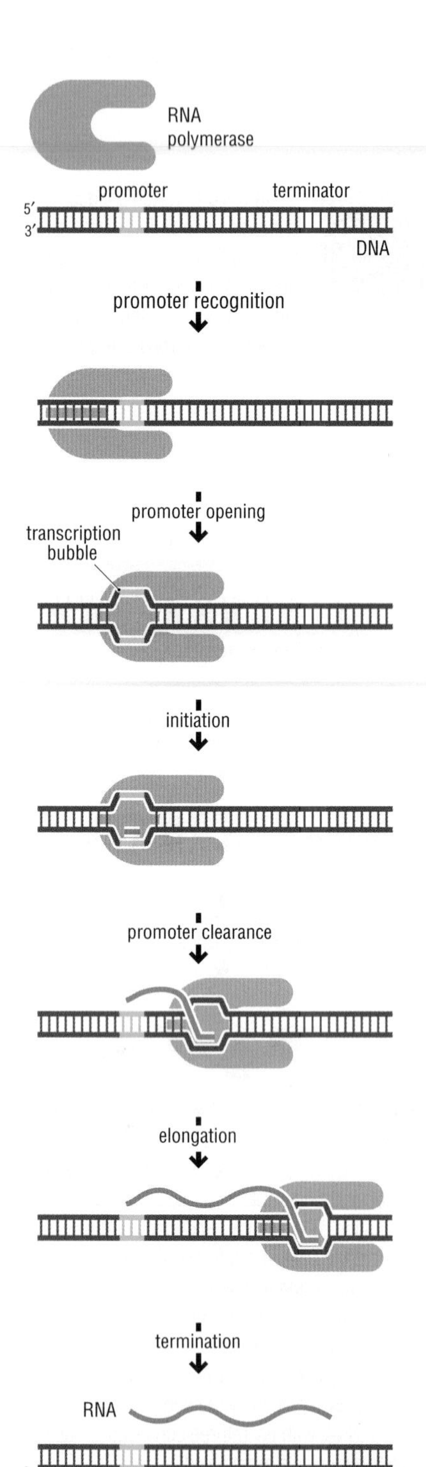

Figure 8.2 Transcription consists of a number of distinct stages. RNA polymerase recognizes and binds to the promoter (green) The enzyme separates the two DNA strands (promoter opening) and initiates polymerization of the first ribonucleotides of the RNA transcript. Once it has moved past the promoter (promoter clearance), the conformation of RNA polymerase changes. The enzyme then elongates the transcript until it encounters a terminator sequence (red) where it is released from the DNA template along with the RNA transcript.

Functions of the eukaryotic RNA polymerases

RNA polymerase	Genes transcribed
RNA polymerase I	Ribosomal RNA (rRNA)
RNA polymerase II	mRNA, small regulatory RNAs
RNA polymerase III	tRNA, 5S RNA, sn RNA
RNA polymerases IV, V (plants only)	si RNA

Figure 8.3 Functions of the eukaryotic RNA polymerases.

8.2 RNA POLYMERASE CORE ENZYME

RNA polymerase is a multisubunit enzyme with a conserved core

The RNA polymerases that transcribe cellular DNA are large, multisubunit enzymes. Bacteria and archaea each possess a single type of RNA polymerase, while eukaryotes typically have three RNA polymerase enzymes that are each responsible for transcribing a different class of genes (Figure 8.3). RNA polymerase I directs transcription of large ribosomal RNA (rRNA) genes, RNA polymerase II directs transcription of all genes encoding mRNAs, and RNA polymerase III directs transcription of a variety of RNA molecules including transfer RNA (tRNA) and 5S rRNA. Plants are distinct in having two additional RNA polymerases (IV and V) that transcribe regulatory RNA molecules.

RNA polymerase II has been studied most extensively because of its role in transcribing protein-coding genes and regulatory RNAs, and will therefore be covered in greatest detail in this chapter. However, it is important to note that RNA polymerase I and III together are responsible for more than 80% of the total RNA synthesized in growing eukaryotic cells.

➔ We learn about the function of regulatory RNAs in Chapter 13.

At the heart of all RNA polymerase enzymes is a set of subunits called the **core enzyme**, which catalyzes the synthesis of RNA using a DNA template. The core enzyme cannot operate in isolation, however; it depends on additional proteins that target it to promoter sequences. The actual number of protein subunits in the core enzyme and the sequences of the individual proteins vary among the three domains of life: bacteria have the smallest enzymes, whereas eukaryotes and archaea have larger, more complex core enzymes. Despite these variations, Figure 8.4 shows how the RNA polymerase core enzymes from bacteria, archaea, and eukaryotes share the same basic architecture, reflecting a common mechanism for synthesizing RNA using a DNA template. This conservation of the enzyme core indicates that all cellular RNA polymerases evolved from a common ancestor.

The bacterial core enzyme illustrates key features of RNA polymerases

Bacterial RNA polymerase is the smallest and simplest of the cellular RNA polymerases. The bacterial core enzyme is a complex of roughly 400 kD, comprising five polypeptides: two copies of the α subunit and one each of β, β', and ω (Figure 8.4a). In some bacteria, the large β and β' subunits are produced as a single polypeptide. The α subunit is composed of an N-terminal domain (known as the **α NTD**) and a C-terminal domain (known as the **α CTD**) joined by a flexible linker. The five core subunits assemble to form a large complex with two jaw-like lobes separated by a central cleft, as illustrated in Figure 8.4b.

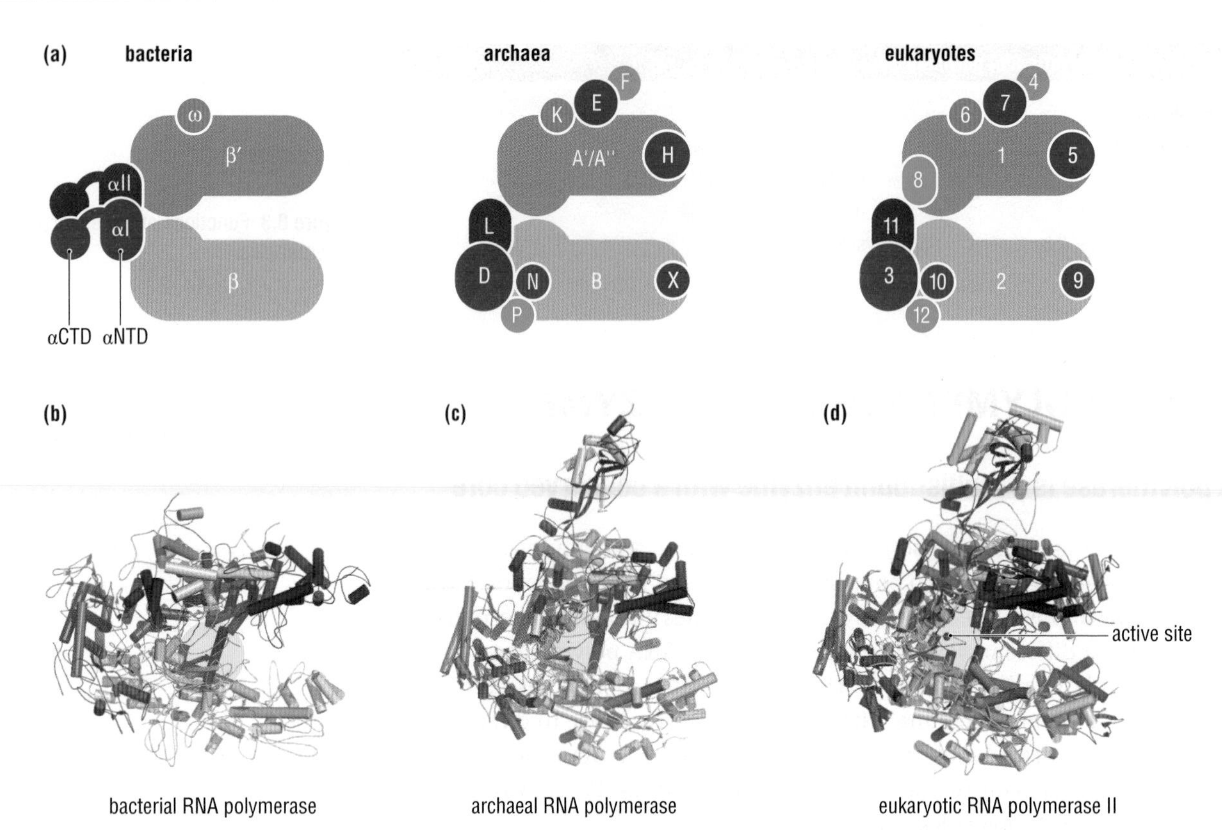

Figure 8.4 Common features of bacterial, archaeal, and eukaryotic RNA polymerases. (a) The enzymes from all three domains of life contain a conserved core corresponding to the five subunits of bacterial RNA polymerase, colored in purple (β′), light purple (β), red (α), yellow (second α subunit), and blue (ω). The corresponding core subunits in archaeal and eukaryotic RNA polymerases are shown in the same colors. The numbers on the eukaryotic RNA polymerase subunits denote Rpb1, Rpb2, etc. Archaeal and eukaryotic polymerases contain subunits in addition to the conserved core. Structures of (b) *Thermus aquaticus* (bacterial, Protein Data Bank (PDB) code 1HQM), (c) *Sulfolobus solfataricus* (archaeal, PDB code 2PMZ) and (d) yeast (eukaryotic, PDB code 1Y1W) RNA polymerase with subunits color-coded as in (a). The Mg^{2+} ion at the active site is shown as a pink sphere.

The upper and lower jaw (as viewed in Figure 8.4b) are formed by the β and β′ subunits. At the base, where the β and β′ subunits meet, are two α subunit N-terminal domains. The ω subunit completes the core polymerase structure. The C-terminal domain of the α subunit does not form part of the globular core enzyme, but rather extends outward, where it is available for interactions with the DNA template and with transcriptional regulators.

The active site of the RNA polymerase enzyme lies at the base of the long cleft separating the two jaws. The active site is also accessible to the outside of the jaw through an opening in the β′ subunit. These two points of access to the active site are key structural elements that position the components of the chemical reaction – the ribonucleotides that are joined together during the synthesis of an RNA molecule – in an optimum way for the reaction to occur.

Eukaryotic and archaeal RNA polymerases share a common central architecture with the bacterial enzyme

All three eukaryotic RNA polymerases and the archaeal RNA polymerases contain a central core that is remarkably similar to that of bacterial RNA polymerase (Figure 8.4), even though the eukaryotic and archaeal enzymes are larger and contain more subunits, as shown by the data in Figure 8.5. RNA polymerase enzymes from all three domains of life contain five central subunits that are similar in amino acid sequence and

RNA polymerase subunits					
RNA polymerase	**Pol I**	**Pol II**	**Pol III**	**bacterial**	**archaeal**
ten-subunit core					
	A190	Rpb1	C160	β'	A'/A''
	A135	Rpb2	C128	β	B
	AC40	Rpb3	AC40	α	D
	AC19	Rpb11	AC19	α	L
	AC12.2	Rpb9	C11		X
	Rpb5 (ABC27)	Rpb5	Rpb5		H
	Rpb6 (ABC23)	Rpb6	Rpb6	ω	K
	Rpb8 (ABC14.5)	Rpb8	Rpb8		
	Rpb10 (ABC10α)	Rpb10	Rpb10		N
	Rpb12 (ABC10β)	Rpb12	Rpb12		P
Rpb4/7 subcomplex					
	A14	Rpb4	C17		F
	A43	Rpb7	C25		E
Additional subunits					
	A49		C37		
	A34.5		C53		
			C82, C34, C31		

Figure 8.5 RNA polymerase subunits. The table shows the components of eukaryotic RNA polymerase II (Pol II) that can collectively transcribe DNA, and their similarities to the eukaryotic polymerase I and III enzymes as well as the archaeal and bacterial RNA polymerases.

in three-dimensional structure. The similarities are highest in the region containing the active site, where the amino acid sequence is nearly identical in bacterial, eukaryotic, and archaeal enzymes. This invariance of key catalytic residues means that all cellular RNA polymerases use an identical mechanism for synthesizing RNA.

The correspondence between the core subunits of the bacterial enzyme and their counterparts in the eukaryotic and archaeal enzymes can be seen in Figures 8.4 and 8.5. The eukaryotic RNA polymerase II subunits, Rpb1 and Rpb2, correspond to the large bacterial β′ and β subunits, Rpb6 corresponds to ω, and Rpb3 and Rpb11 correspond to the two α subunits.

Looking again at Figure 8.4a, we see how the additional subunits found in eukaryotic and archaeal RNA polymerases are arranged around the five central subunits. Eukaryotic RNA polymerase II, for example, contains seven additional subunits, giving a total of 12 subunits for the enzyme. Most of the additional subunits are conserved in the three eukaryotic polymerases and, with the exception of Rpb8, in archaea.

The three types of eukaryotic RNA polymerase share a structurally conserved core of ten subunits (see Figure 8.5). There are additional subunits that are particular to RNA polymerase I, II, or III. RNA polymerase I has a total of 14 subunits, and RNA polymerase III has 17 subunits, in comparison with the 12 subunits of RNA polymerase II. The subunits that are unique to a particular eukaryotic polymerase are presumably responsible for functions specific to each type of polymerase.

The RNA polymerase II CTD couples transcription with pre-mRNA processing

The RNA polymerase II enzyme, alone of the three eukaryotic polymerases, carries out an additional function: it couples transcription to processing of the RNA transcript to produce the mature mRNA. The mRNA synthesized by RNA polymerase II is rarely in its mature, fully functional form. Instead, it is termed a pre-mRNA, and must be subjected to various processing events that yield the final mRNA. These chemical modifications include the addition of a guanosine cap to the 5′ end of the nascent transcript as well as the attachment of a poly-adenosine (polyA) tail to the 3′ end.

The coupling of transcription to RNA processing depends on an important feature unique to RNA polymerase II called the C-terminal domain (**CTD**). This additional domain comprises the C-terminus of the Rpb1 subunit, which forms part of the conserved enzymatic core. The CTD consists of many repeats of seven amino acids (a heptad) with the consensus sequence Tyr-Ser-Pro-Thr-Ser-Pro-Ser; this heptad is repeated 26 or 27 times in the yeast enzyme and 52 times in the human enzyme. Selected residues in the CTD are phosphorylated at different stages of the transcription cycle. The phosphorylated CTD then recruits RNA processing enzymes, which modify the RNA while it is being transcribed. The CTD has a central role in the transition from initiation to elongation and in processing of the RNA transcript, as we shall see in later sections.

➔ We explore mRNA processing in depth in Chapter 10.

Having examined transcription in general terms and learned the basic features of the enzyme that is central to this process, we can now begin to explore each of the stages of transcription in more detail. We start with promoter recognition.

8.3 PROMOTER RECOGNITION IN BACTERIA AND EUKARYOTES

The RNA polymerase core enzyme can synthesize an RNA copy of any DNA template, but is unable to locate the transcription start site on its own. Additional proteins are needed to guide the core enzyme complex to the promoter, which is the DNA sequence adjacent to the transcription start site where RNA polymerase first binds. Although the RNA polymerase core enzyme that we learned about in Section 8.2 has been conserved through evolution across all three domains of life, bacteria and eukaryotes have evolved distinct solutions to the problem of locating the promoter sequence. We begin by describing how the bacterial enzyme is brought to the transcription start site.

Sigma factors direct bacterial RNA polymerase to promoters

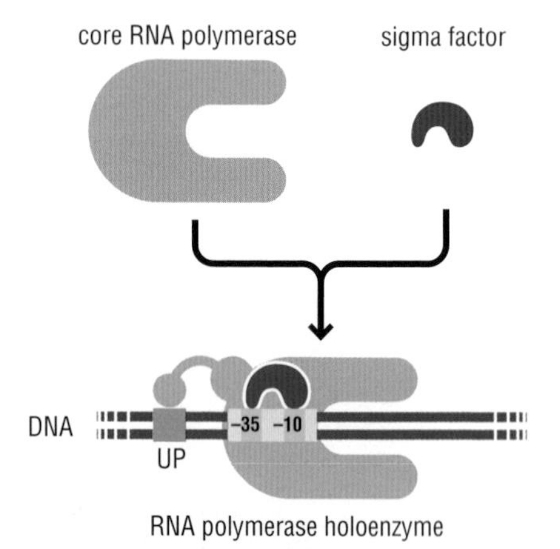

Figure 8.6 Promoter recognition by the bacterial RNA polymerase holoenzyme. The core RNA polymerase associates with a sigma factor that binds to a specific promoter sequence (green) in the DNA, contacting sequences centered at −10 and −35 region. The αC-terminal domains contact the UP element (orange), which is located upstream of the −35 region of some promoters.

Bacterial RNA polymerase core enzymes can transcribe RNA on their own, but they do not recognize promoter sequences. Instead, the bacterial core enzyme requires a specialized subunit called a **sigma factor** (σ), which directly contacts the promoter sequence, as illustrated in Figure 8.6. This contact thereby ensures that transcription begins at the proper position in the DNA. The complex formed by the sigma subunit and the polymerase core enzyme constitutes the bacterial **holoenzyme**. Bacteria contain a variety of sigma factors that specifically recognize different promoter sequences. It is therefore the particular sigma factor utilized by a given RNA polymerase that determines which genes will be transcribed by that polymerase holoenzyme.

All cells have a **primary sigma factor**, which directs transcription from the promoters of genes encoding essential proteins that are needed by growing cells. These genes are therefore often referred to as housekeeping genes. In addition, bacteria have a variety of **alternative sigma factors** whose levels or activities are increased in response to specific signals or stress conditions. The alternative sigma factors direct transcription of genes that are only required under certain conditions, or whose expression must be increased or decreased in response to changing conditions. For example, *Escherichia coli* responds to heat shock (a sudden increase in temperature) by synthesizing an alternative sigma factor (called

σ^H or σ^{32}) that it needs to transcribe genes encoding proteins that help the cells respond to heat shock.

Bacterial species differ significantly in the number of sigma factors they encode, reflecting the different environmental conditions to which they must respond. *E. coli*, a bacterium that leads a relatively sheltered life in the gut of other organisms (including humans), has only seven sigma factors. In contrast, the genome of the soil bacterium, *Bacillus subtilis*, encodes 17 sigma factors, while *Streptomyces coelicolor*, which is also free living in the soil, has 65. The large number of sigma factors enables these organisms to initiate transcription from different promoters at different times, depending on the specific sigma factors produced in the cell at a given moment. This allows the cell to synthesize additional proteins encoded by the newly transcribed genes, thus enabling these microorganisms to respond to changing conditions in their environment and undergo major morphological changes when threatened with starvation.

Sigma factors bind to specific sequences that define bacterial promoters

The sigma subunit guides the bacterial RNA polymerase holoenzyme to the promoter by binding to specific sequences upstream of the start site of transcription, whose position along the DNA sequence is denoted as +1. Bacterial promoters generally contain two key elements that are recognized by a sigma factor: a conserved sequence most commonly centered at 35 nucleotides upstream of the start of transcription and therefore termed the **−35** (minus 35) element, and a second conserved sequence centered at 10 nucleotides upstream of the start site known as the **−10** (minus 10) element. These elements are illustrated in Figure 8.7a. Each of these promoter elements is contacted by the sigma factor protein, as we shall see shortly.

A given sigma factor will favor binding to promoters with particular DNA base sequences in the −35 and −10 regions. In addition, there is a preferred spacing between the −35 and −10 promoter elements that favors avid binding. Some examples of different sigma factors and the optimal promoter elements that they recognize are shown in Figure 8.7b. Variations in both promoter element spacing and sequence allow for differential regulation of genes, since these DNA sequence variations will have distinct effects on the relative binding of different classes of sigma factors. The more closely the promoter sequence matches the optimal sequence for a given sigma factor, the more tightly sigma will bind. Tighter binding to a particular promoter will lead to higher relative levels of transcription, which

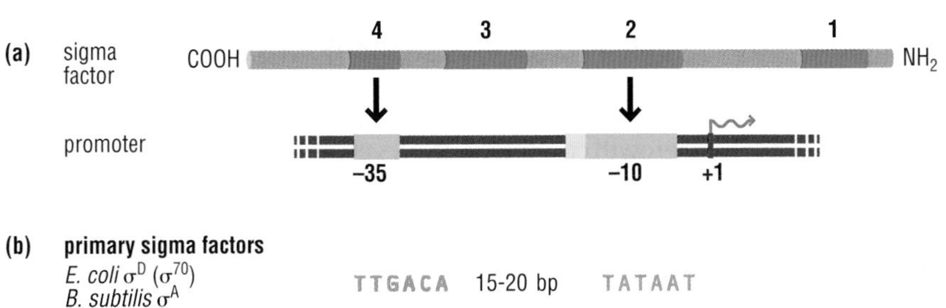

Figure 8.7 Structure of bacterial consensus promoters and regions of sigma subunits that contact DNA. (a) The four major regions of conservation are indicated on the diagram of a sigma factor. Regions of contact between the sigma factor and the promoter are indicated by arrows. (b) The consensus sequences for promoters recognized by the major *E. coli* and *B. subtilis* sigma factors and the *E. coli* σ^H, σ^N, and σ^F alternative sigma factors. N denotes, any nucleotide; W denotes, A or T. Sigma factors are referred to either by a letter designation or by a number indicating the molecular weight in kilodaltons.

in turn will lead to higher levels of protein expressed, whereas weak binding to another promoter will result in lower rates of transcript production. Promoter sequence variations also influence the rates of promoter opening and transcription initiation, as we explore further below.

Some bacterial promoters contain sequences in addition to the –35 and –10 regions that can further modulate RNA polymerase binding and initiation of transcription. At some promoters that lack an optimal –35 sequence, transcription initiation depends on sigma binding to an **extended –10 region** consisting of an additional two bases located on the upstream side of the –10 region, as illustrated in Figure 8.7a. These bases provide additional sites for favorable contacts with the sigma subunit, partly making up for the absence of an optimal –35 region.

Some bacterial genes that are transcribed at a high rate contain an additional AT-rich sequence called the **UP element** located upstream side of the –35 region (see Figure 8.6). The UP element serves as a binding site for the C-terminal domain of the α subunit of RNA polymerase. These additional contacts with the UP element strengthen the binding of RNA polymerase to the DNA, thus helping to promote high rates of transcription of housekeeping genes such as those encoding rRNA.

The different promoter elements are recognized by distinct domains of the sigma subunit. The sigma subunit contains a poorly structured region at the N-terminus, followed by three structural domains (domains 2–4) (see Figure 8.7a). When sigma binds to the RNA polymerase core enzyme to form the RNA polymerase holoenzyme, the three structural domains of sigma are positioned to recognize specific elements of the promoter sequence, as illustrated in Figure 8.8. Domain 2 (σ_2) binds to the –10 region and promotes promoter melting, in which the two strands of the DNA duplex separate. This promoter melting is favored by domain 2 of sigma binding to a single unpaired DNA strand of the opened duplex. The immediately adjacent sigma domain 3 (σ_3) recognizes the two bases comprising the extended –10 region. Domain 4 (σ_4) recognizes the –35 region and is bound to a flexible flap of the polymerase β subunit that may allow movement of domain 4 to accommodate variably spaced –35 and –10 elements.

An interesting feature that can be found between domains 3 and 4 is a loop called region 3.2, which extends into the active site. Residues in this loop help to position incoming nucleotides in the active site.

An N-terminal region of sigma called region 1.1, which is found in a subset of sigma factors (for example, σ^{70}), prevents DNA binding by a free sigma subunit (that is, sigma that is not part of the holoenzyme); it also accelerates the rate of DNA strand separation at certain promoters. Region 1.1 is unstructured when it is part of the holoenzyme and thus is not depicted in Figure 8.8.

Sigma factor activity can be regulated in many ways

While the amounts of the core subunits of RNA polymerase remain relatively constant under different growth conditions, the levels and activities of many sigma factors can vary significantly in response to environmental or developmental signals. This enables the cell to alter its pattern of gene expression to meet changing needs. The relative amount of sigma factor in the cell can be regulated by modulating the transcription or translation of the sigma factor, or by modulating the stability of the sigma factor protein or of its mRNA transcript.

The incorporation of a sigma factor into a holoenzyme can also be regulated. Some sigma factors are synthesized as pro-sigma factors, meaning that they

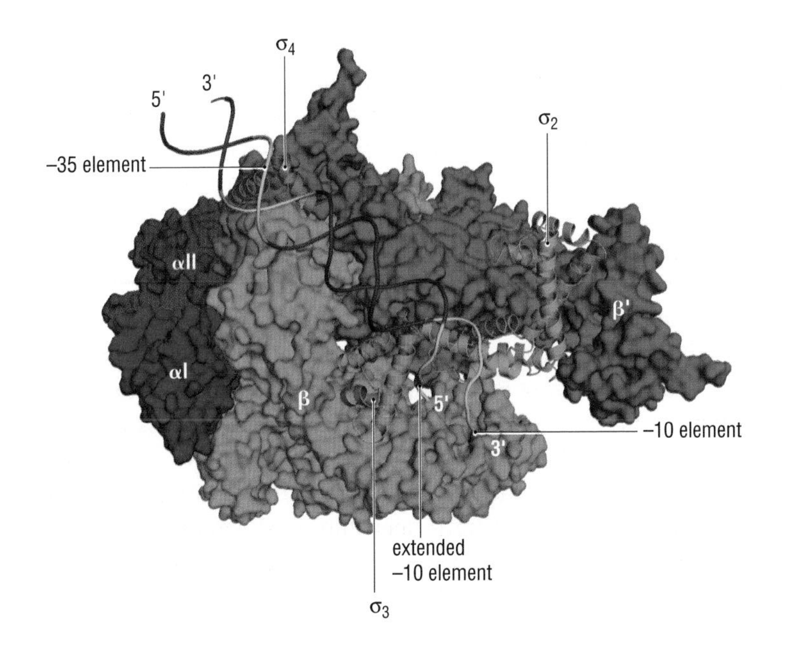

Figure 8.8 Structure of a bacterial RNA polymerase holoenzyme bound to a promoter. The structure of the *T. aquaticus* holoenzyme bound to promoter DNA shows how the three structural domains of the sigma subunit bind to the core enzyme in a position to recognize the promoter elements. The core component of the holoenzyme is shown as a molecular surface, with the αI, αII, and β′ subunits labeled accordingly. The sigma factor is shown as a Cα cartoon depiction. The DNA is numbered relative to the transcription start site at +1. The σ2 domain (orange) recognizes the −10 region of the promoter (green), while the σ3 domain (blue) binds to the flanking base pairs of the extended −10 region. The σ4 domain (orange) binds to the −35 element (green) (hybrid of PDB codes 1L9Z and 1IW7).

contain inhibitory domains that must be cleaved off before the sigma factor can associate with the core RNA polymerase. There are also regulatory proteins called **anti-sigma factors** that bind to certain sigma factors and inhibit their function, for example by preventing sigma factor from interacting with the RNA polymerase core enzyme. The activities of anti-sigma factors, in turn, can be regulated by controlling their expression in the cell or by sequestering them through interactions with other cellular proteins (called anti-anti-sigma factors).

In the example of sigma factor regulation illustrated in Figure 8.9, the anti-sigma factor FlgM forms a complex with the sigma factor, σ^F (also called σ^{28}), which directs the transcription of genes required to complete the assembly of flagella in the bacterium *Salmonella typhimurium*. (Flagella are tail-like structures that allow the cell to propel itself.) While the proteins that form the base of the flagella are being synthesized, FlgM is bound to σ^F, preventing incorporation of the sigma subunit into the holoenzyme and thereby preventing expression of the genes whose transcription depends upon σ^F. During the late stages of flagellar synthesis, however, FlgM is exported from the cell through the incomplete flagellar apparatus, releasing σ^F to **activate** transcription of the genes necessary for final assembly of the flagella.

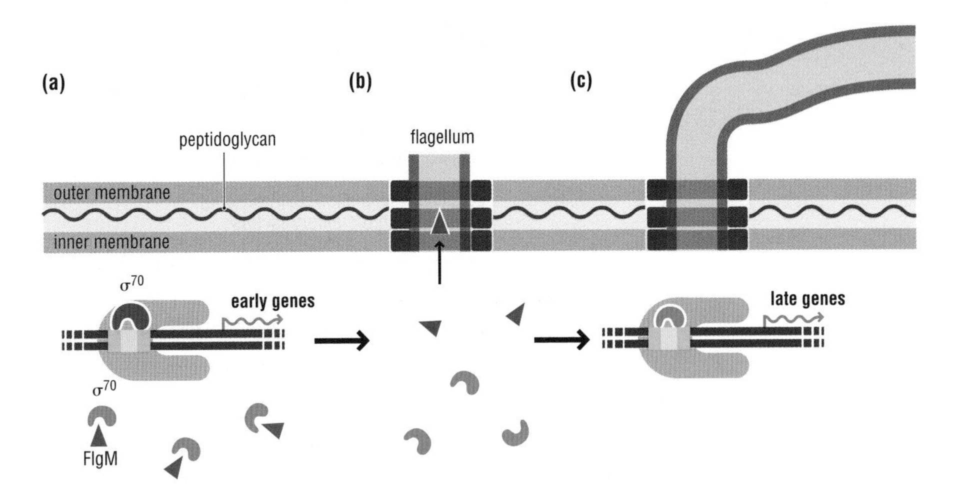

Figure 8.9 Regulation of *S. typhimurium* sigma factor activity during flagellum biosynthesis. σ^F is needed for the expression of genes late in the assembly of the flagellar motility motor. The genes required for the synthesis and assembly of the initial hook–basal body complex of the flagellum (gray), as well as the genes for σ^F and the anti-sigma factor, FlgM (pink triangles), that inhibits σ^F, are transcribed by an RNA polymerase containing the housekeeping sigma factor σ^{70} (blue). FlgM binds σ^F and prevents it from interacting with the RNA polymerase core enzyme. Once the intermediate hook and basal body structure of the flagellum has been completed, however, FlgM, which carries signals that cause its export at this stage, releases σ^F and is secreted from the cell. The released σ^F then can interact with the RNA polymerase core enzyme and directs transcription of genes required for the completion of the flagellum.

Eukaryotic and archaeal RNA polymerases require general transcription factors for promoter recognition

Having explored the initiation of transcription in bacteria, let us now consider how the process occurs in eukaryotes and archaea. Like the bacterial polymerase described in Section 8.2, the RNA polymerase core enzymes in eukaryotes and archaea also depend on additional proteins to target them to specific sites in DNA where transcription is initiated. These proteins are called **general transcription factors** and come together to form a large complex that binds RNA polymerase and helps it initiate transcription at specific promoters. But while the bacterial sigma factors target the polymerase to the promoter and mediate promoter melting, the eukaryotic and archaeal general transcription factors also catalyze additional steps in the initiation of transcription, as we discuss further below.

The general transcription factors that are required to initiate transcription by eukaryotic RNA polymerase II have been the most extensively studied. These factors, which are either single proteins or protein complexes, are named for the distinct, biochemically purified species that were required to reconstitute promoter-specific transcription initiation *in vitro*. Denoted TFIIA (Transcription Factor for RNA polymerase II, species A), TFIIB, TFIID, TFIIE, TFIIF, and TFIIH, these proteins (or protein complexes) assemble at promoters together with the 12-subunit polymerase core enzyme to form the **pre-initiation complex**. Figure 8.10 presents an overview of their functions.

The eukaryotic RNA polymerase I and III enzymes require distinct sets of proteins to form a transcription initiation complex. However, just one protein is required for transcription initiation by *all* eukaryotic polymerases; this protein is called **TATA binding protein (TBP)**, which is also a subunit of the TFIID complex mentioned above. TBP is also conserved in archaea, where it is also required for transcription initiation by the single archaeal RNA polymerase.

Eukaryotic and archaeal promoters are often defined by a TATA box and other core promoter elements

The promoters of eukaryotic genes contain a variety of sequences called **core promoter elements** that direct the binding of the transcription initiation complex,

General transcription factors associated with RNA polymerase II

factor	number of subunits in human	key subunits	functions
TFIIA	3		stabilizes binding of TBP and TFIIB
TFIIB	1		promoter recognition stabilize early transcribing complex
TFIID	15	TBP, TAFs	promoter recognition, DNA bending interacts with regulatory factors
TFIIE	2		recruits TFIIH
TFIIF	2		suppresses non-specific DNA binding captures non-template strand upon melting
TFIIH	9	p62, p52 XPD, XPB MAT1, Cdk7 Cyclin H p34, p44	unwinds promoter DNA, phosphorylates CTD

Figure 8.10 Basal transcription factors required for transcription initiation by eukaryotic RNA polymerase II.

thereby determining the start site and direction of transcription. Promoters for each of the three eukaryotic RNA polymerases typically contain a characteristic set of promoter elements, although these vary significantly in sequence and conservation among individual promoters.

The typical structure of the promoters of genes transcribed by RNA polymerase II is illustrated in Figure 8.11. These promoters commonly contain a **TATA box**, named for its TATA(T/A)A(T/A) consensus sequence, to which the TBP protein binds. The TATA box is centered approximately 25–30 bp to the upstream side of the transcription start. A promoter element called BRE (for TFIIB recognition element) is the binding site for the transcription factor, TFIIB. Additional elements that may be present are the initiator element (INR) and the downstream promoter element (DPE), the latter of which is located downstream of the start site of transcription. RNA polymerase II promoters can contain any combination of these elements, which are identified based on the nucleotide sequence of the promoter. The promoters of actively transcribed genes are generally free of nucleosomes, leaving the DNA available for binding to the transcription initiation complex.

In some cases, the sequence of the promoter element does not closely match the consensus sequence, yet still serves the same function; this is the case for so-called 'TATA-less promoters' in yeast, which still contain a binding site for TBP. Experimental approach box 8.1 describes how the binding sites for TBP were located in yeast promoters lacking a strongly conserved TATA box sequence. Understanding how transcription is initiated from promoters that lack well-conserved promoter elements is an ongoing area of research.

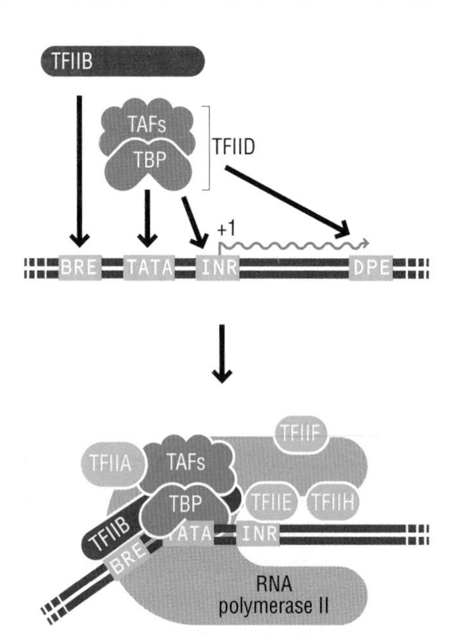

Figure 8.11 Structure of RNA polymerase II promoters and the regions contacted by TFIIB and TBP. Conserved DNA sequences called core promoter elements (green) are found in many eukaryotic promoter regions. The TATA box is the binding site for TBP (blue) which promotes binding of TFIIB (red) to the adjacent BRE element. These help nucleate assembly of the rest of the transcription initiation complex.

TFIID nucleates assembly of the RNA polymerase II pre-initiation complex

The first step in the assembly of the RNA polymerase II transcription initiation complex occurs when TFIID binds to the TATA box. TFIID plays the same role as the bacterial sigma factor in locating the promoter. TFIID was first identified as a purified biochemical fraction that was needed for transcription *in vitro*, but turned out to consist of a complex of proteins. The subunit of the TFIID complex that binds to DNA and recognizes the TATA box sequence is **TBP (TATA binding protein)**. The structure of TBP is depicted in Figure 8.12. It is a saddle-shaped protein that binds to the minor groove of the DNA, with its concave face inducing profound bending and partial unwinding of the DNA double helix. The remaining components of TFIID, called **TBP-associated factors (TAFs)**, include at least a dozen proteins with various functions. Some of the TAFs mediate recognition of other core promoter elements, including the INR and DPE sequences (see Figure 8.11). Other TAFs

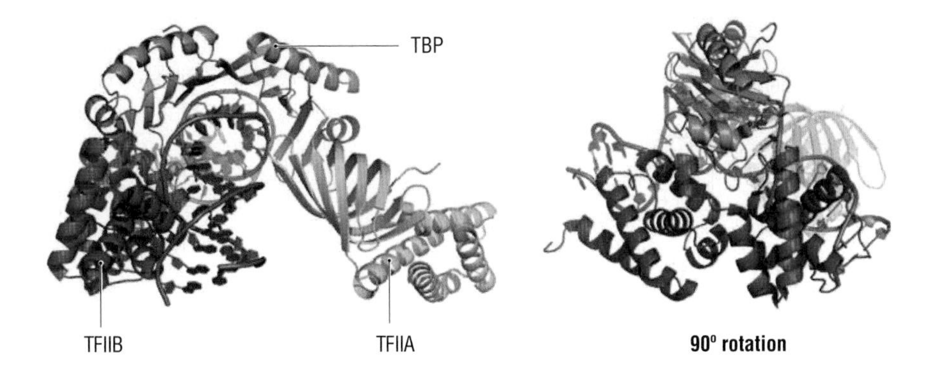

Figure 8.12 Structure of TBP bound to the minor groove of a TATA promoter element. TBP (blue) binds the TATA box and induces a profound distortion in the DNA, while the TFIIB (red) and TFIIA (green) bind to either side. Hybrid of PDB codes 1NVP and 1D3U.

8.1 EXPERIMENTAL APPROACH

Mapping the locations of the transcription pre-initiation complex at promoters with and without TATA boxes

How does TFIID recognize TATA-less promoters?

The proteins required for transcription initiation in eukaryotes have been known for well over a decade through a combination of studies carried out *in vitro* and *in vivo*. One of the remaining puzzles centered on how transcription was initiated from promoters lacking an apparent TATA box, whose consensus sequence is TATAWAWR (where W denotes A or T, and R denotes A or G). This sequence is recognized by the TATA-binding protein (TBP) subunit of the general transcription factor, TFIID. It was known that TFIID was required for transcription from TATA-less promoters. How then could TBP help to direct TFIID to the promoter if there was no TATA box in the promoter?

Determining the precise location of transcription pre-initiation complexes throughout the yeast genome

Frank Pugh and Ho Sung Rhee utilized a new method they had developed for precisely mapping the locations throughout the genome where a particular protein is bound to DNA (Rhee and Pugh, 2011). This method, illustrated in Figure 1, is called ChIP-exo and is similar in some respects to a method known as ChIP-seq, in which proteins are immunoprecipitated along with the DNA to which they are bound, after which the DNA is sequenced. The authors' innovation was to treat the cross-linked protein–DNA complexes with 5' exonuclease that digests the DNA until it encounters the first protein–DNA crosslink, which prevents further digestion. By reversing the cross-links to remove the protein and then sequencing the DNA, the authors can precisely map the location of the complex on the DNA to within a few base pairs. The sequences of the different binding sites can then be compared using sophisticated computer algorithms, thus making it possible to arrive at a consensus binding sequence.

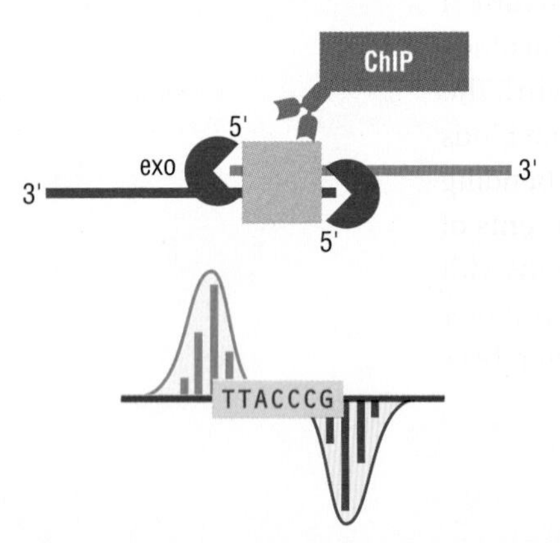

Figure 1 Illustration of the ChIP-exo method for mapping the location of complexes bound to DNA. A protein complex (grey square) is cross-linked to DNA and the protein–DNA complex is immunoprecipitated with antibodies (ChIP stands for chromatin immunoprecipitation). A 5′ exonuclease then digests the DNA, stopping when the first cross-link is encountered. The cross-links are reversed, and DNA sequencing is used to identify the 5′ end on each strand, thus defining where the protein was bound to near-base pair resolution. The histograms depict the distribution of 5′ ends that results from analyzing multiple isolates of the same gene.

Reproduced from Rhee and Pugh, *Cell*, 2011;**147**:1408–1419.

Using the ChIP-exo method, Rhee and Pugh mapped the precise locations of all of the general transcription factors and of RNA polymerase II in yeast. Figure 2 shows the average occupancy for these proteins at genes whose promoters contain TATA boxes. Note the high occupancy near the TATA box and the rapid falloff near the transcription start site, indicating that the general transcription factors remain at the promoter during the transcription cycle.

'TATA-less' promoters actually have a poorly conserved TATA box

Analyzing the data for promoters lacking a TATA box was more difficult. Attempts to identify a match in each promoter to the TATAWAWR consensus sequence failed. They therefore decided to relax the stringency of their computer search, instead looking for sequences in the promoters that deviated from the TATA consensus at one or two positions. This approach succeeded: now the authors could identify a 'TATA-like' sequence in each gene that was an imperfect match, deviating at one or two positions from the TATAWAWR sequence (see Figure 3). The distribution of general transcription factors such as TFIIB around the TATA-like elements also matched that observed for TATA-containing promoters. The finding that all yeast promoters contain a sequence that at least partly matches the TATA consensus sequence is consistent with the universal requirement that TATA-binding protein (TBP) bind to the promoter.

Insights into pre-initiation complex binding throughout the yeast genome

In addition to information on TATA-less promoters, this study yielded a wealth of information on the location of pre-initiation

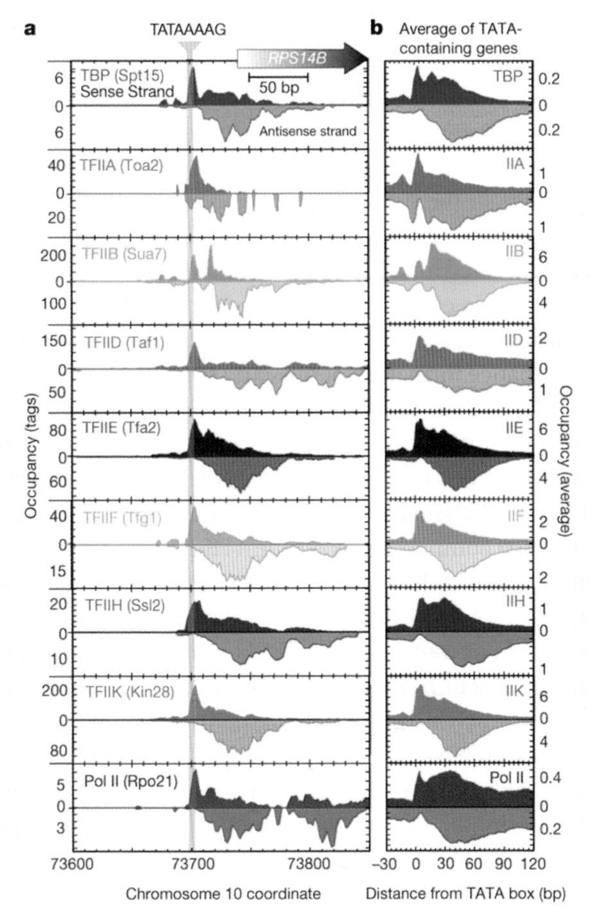

Figure 2 Average occupancy of the general transcription factors and of RNA polymerase II in genes with TATA boxes. (a) Data for a single gene, *RPS14B*. (b) Average occupancy for 676 genes that contain TATA boxes. Note that the kinase subunit of TFIIH is denoted TFIIK.

Reprinted from Rhee and Pugh, *Nature*, 2012;**483**:295–301.

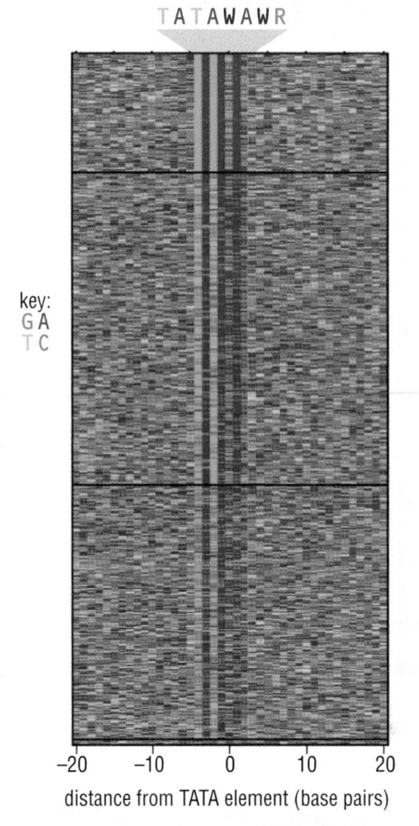

Figure 3 Alignment of TATA-like elements at TATA-less genes. Each line represents a single promoter sequence, with bases A, T, C, and G colored as shown. The sequences are aligned by their TATA-like sequences.

Reprinted from Rhee and Pugh, *Nature*, 2012;**483**:295–301.

complex assembly relative to nucleosomes flanking the promoter. One of the more intriguing observations was that a large number of preinitiation complexes are responsible for transcribing non-coding RNAs from divergent promoters. The interested student is encouraged to read about this and other findings in the Rhee and Pugh 2012 paper. We will learn in Chapter 13 about non-coding RNA and its role in regulating gene expression.

Find out more

Rhee HS, Pugh BF. Comprehensive genome-wide protein-DNA interactions detected at single-nucleotide resolution. *Cell*, 2011;**147**: 1408–1419.

Rhee HS, Pugh BF. Genome-wide structure and organization of eukaryotic pre-initiation complexes. *Nature*, 2012;**483**:295–301.

Related techniques

ChIP-exo (similar to ChIP-seq); Section 19.12

Genome-wide sequence alignment; Section 19.15

have enzymatic activities and are required for activation of transcription by gene-specific transcriptional regulators.

Once TFIID is bound to the promoter region, assembly of the pre-initiation complex proceeds with the recruitment of TFIIB, a single polypeptide that recognizes the adjacent BRE promoter element (see Figure 8.11). The asymmetrical binding of TFIIB helps to establish the correct direction of transcription, since TFIIB forms direct contacts with the core enzyme. When TFIIB binds to the RNA polymerase core enzyme, an extended linker called the B reader that connects the N-terminal and C-terminal domains of TFIIB extends deep into the active site of the enzyme. This loop, which is reminiscent of a portion of bacterial sigma domain 3 (Figure 8.8), is thought to enhance formation of the early transcribing complex. The additional binding of TFIIA further stabilizes the TBP–DNA interaction.

Assembly of the pre-initiation complex proceeds with the recruitment of the core RNA polymerase II enzyme, which is in a complex with TFIIF. This is then followed by the binding of TFIIE and TFIIH, thus yielding the complete transcription pre-initiation complex. The TFIIH complex catalyzes the ATP-powered unwinding of DNA to expose the template strand. This is aided by a subunit of TFIIH called XPB, an enzyme that promotes DNA unwinding (also known as a helicase). In addition to facilitating the initiation of transcription, TFIIH also plays a role in the repair of damaged DNA; mutations in XPB, as well as in another TFIIH subunit, XPD, give rise to human diseases such as xeroderma pigmentosum and Cockayne syndrome, which are characterized by defects in DNA repair. TFIIH also contains enzymatic subunits that phosphorylate the CTD of the polymerase subunit, Rpb1, which occurs later in the transcription cycle during the transition to the elongating complex.

Additional protein complexes are required for transcription by RNA polymerase II

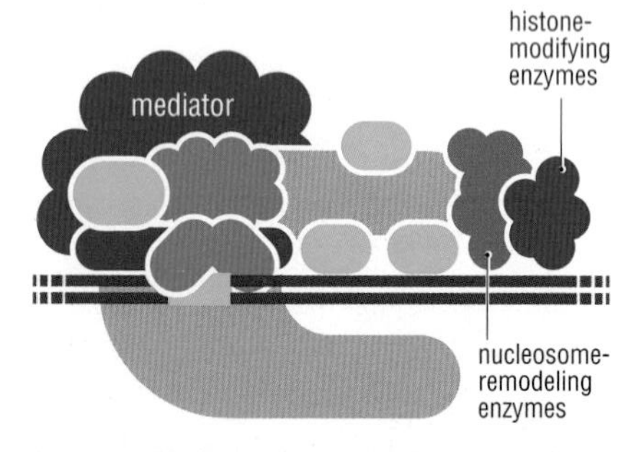

Figure 8.13 Proteins required for regulated RNA polymerase II transcription. The RNA polymerase II holoenzyme consists of three sets of proteins: the multisubunit core polymerase enzyme (light purple), which catalyzes synthesis of the RNA transcript, the general transcription factors (blue, green and red), which help RNA polymerase recognize specific promoters, and the Mediator complex (purple), which is required for regulated transcription at most promoters. This large complex of proteins assembles at the promoter region of a gene before transcription is initiated. Histone modifying enzymes and nucleosome remodeling enzymes that are recruited to particular promoter regions further stimulate transcription initiation and elongation.

The pre-initiation complex, consisting of the RNA polymerase II core enzyme and the general transcription factors, is sufficient to initiate transcription at specific promoters *in vitro*. Additional protein complexes (depicted in Figure 8.13), however, are required for transcription of most genes *in vivo*. These include a variety of enzymes that act on chromatin, as well as a complex known as **Mediator**.

Mediator is a large protein complex that is required to activate many genes transcribed by RNA polymerase II. As we will learn in Chapter 9, genes transcribed by RNA polymerase II are regulated by proteins that bind upstream of the promoter region. Mediator interacts with transcriptional activator proteins that are bound to upstream regulatory regions, promoting assembly of the transcription initiation complex at the promoter. The Mediator complex contains more than 20 individual subunits that are largely conserved between yeast and humans; it was isolated after the general transcription factors had been identified, when it was discovered that there appeared to be a missing component that was needed for transcriptional regulation. The experiments that led to the discovery of Mediator are described in Experimental approach 8.2.

The packaging of the template DNA into chromatin presents a potential obstacle to RNA polymerases in general. Changes in chromatin structure therefore are needed for transcription initiation, thus clearing the way for the assembly of the pre-initiation complex and allowing RNA polymerase to transcribe through chromatin. Histone-modification enzymes, which covalently modify histones, and nucleosome-remodeling enzymes, which reposition histones and histone

8.2 EXPERIMENTAL APPROACH
Identification of Mediator

'Squelching' of gene activation by Gal4 suggests existence of a missing factor

The existence of Mediator was first proposed on the basis of indirect evidence suggesting that there was an unidentified component in the yeast nucleus that was needed for transcription activation. Experiments had shown that when excess amounts of the activator protein, Gal4-VP16 (a genetically engineered hybrid protein containing the Gal4 DNA-binding domain fused to the VP16 activation domain), were added to yeast nuclear extracts, transcription decreased from fully activated levels. This phenomenon was dubbed 'squelching'. The hypothesis was that the decrease in transcription was the result of the binding of Gal4-VP16 to factors in the extract that were needed for full activation of transcription. The addition of excess RNA polymerase II or general transcription factors (such as TFIID, TFIIA, etc.) did not overcome squelching, suggesting that those proteins were not the ones being sequestered by the excess Gal4-VP16.

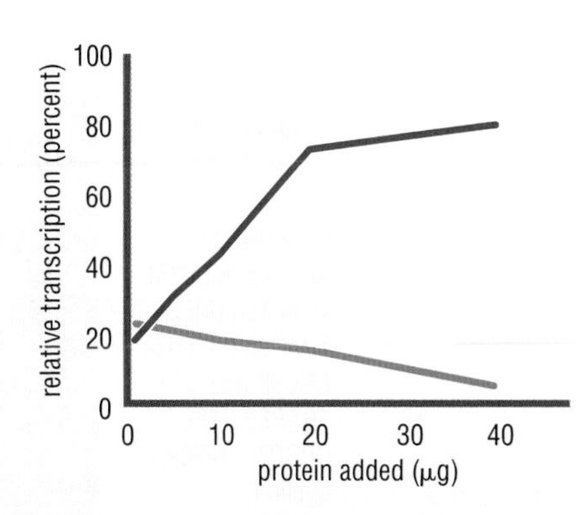

Figure 1 Reversal of activator squelching by yeast fraction isolated from an ion exchange column. The fraction containing the putative Mediator (pink) stimulates transcription by overcoming the squelching effect, while a control fraction (blue) fails to do so.

Adapted from Kelleher *et al.* A novel Mediator between activator proteins and the RNA polymerase II transcription apparatus. *Cell*, 1990:**61**:1209–1215.

Fractionating yeast extracts to identify the missing factor, Mediator

If RNA polymerase and the general transcription factors were not the target of squelching, could there be other undiscovered proteins required for transcriptional activation? To address this question, Roger Kornberg and colleagues fractionated yeast nuclear extracts by ion exchange chromatography, and added those fractions (eluted from the column at different salt concentrations, see Section 19.7) to an *in vitro* transcription assay. They found that addition of one of the fractions stimulated transcription and overcame the squelching effect in a dose-dependent manner (see Figure 1). Without knowing what was in the fraction, they dubbed the unknown factor 'Mediator'. Subsequent studies have shown Mediator to be a complex containing around 20 proteins with a mass of about 1 MDa. The individual components of Mediator eventually were identified through a combination of biochemistry and genetics.

Using affinity tags and mass spectrometry to identify the human Mediator complex

Mediator complexes from other organisms can now be analyzed rapidly using modern mass spectrometry methods that make it possible to identify the individual proteins found in a single enriched sample (see Section 19.8). An example can be found in a report by Joan and Ron Conaway and their co-workers, who identified the subunits in human Mediator complexes isolated from HeLa cells. Several groups had independently purified Mediator from mammalian cells, but each reported a different subset of proteins in their complexes. This suggested either that not all Mediator proteins had been identified, or that there were several different Mediator complexes. The Conaways took advantage of a highly sensitive mass spectrometry method known as MudPIT (Multidimensional Protein Identification Technology) to resolve this confusion.

To attack the problem, the researchers transfected HeLa cells with a DNA vector expressing one of the putative Mediator subunits fused to an N-terminal FLAG epitope (a type of tag used for affinity purification; see Section 19.7). Antibodies that recognized the FLAG epitope were used to immunoprecipitate the Mediator complex, and the individual subunits were then directly identified by mass spectrometry. To ensure that the search was not biased by the particular subunit that was tagged, or by the position of the tag, the experiment was repeated with several different putative Mediator subunits. Figure 2 shows the polypeptides that had been reported previously (Figure 2a: red boxes) along with the subunits identified in the Conaway study (Figure 2b: red boxes).

By using this very sensitive detection method, a consistent set of human Mediator subunits could be identified. Importantly, all the subunits that had previously been found in one or another experiment were present in the complexes isolated in this study. Differences between the sets of proteins identified in this and previous studies may reflect differences in the sensitivity of the approach, or in the purity or homogeneity of the particular preparation characterized.

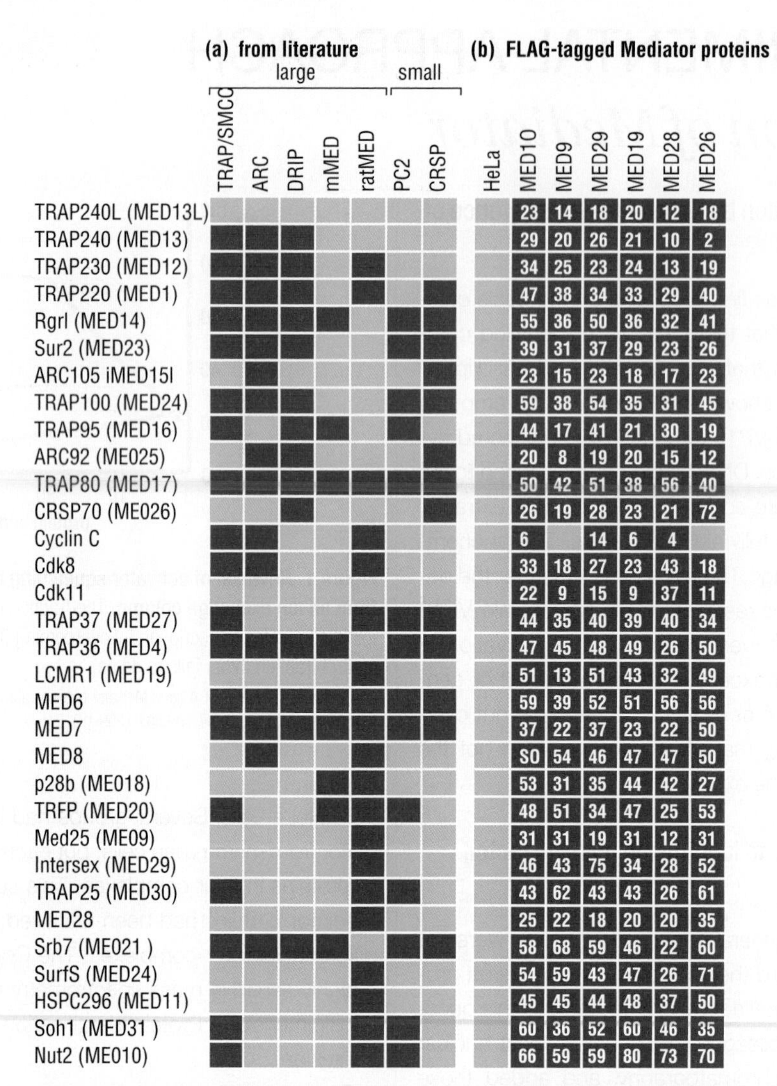

(a) from literature
large
small
(b) FLAG-tagged Mediator proteins
TRAP/SMCC
ARC
DRIP
mMED
ratMED
PC2
CRSP
HeLa
MED10
MED9
MED29
MED19
MED28
MED26
TRAP240L (MED13L) 23 14 18 20 12 18
TRAP240 (MED13) 29 20 26 21 10 2
TRAP230 (MED12) 34 25 23 24 13 19
TRAP220 (MED1) 47 38 34 33 29 40
Rgrl (MED14) 55 36 50 36 32 41
Sur2 (MED23) 39 31 37 29 23 26
ARC105 iMED15I 23 15 23 18 17 23
TRAP100 (MED24) 59 38 54 35 31 45
TRAP95 (MED16) 44 17 41 21 30 19
ARC92 (ME025) 20 8 19 20 15 12
TRAP80 (MED17) 50 42 51 38 56 40
CRSP70 (ME026) 26 19 28 23 21 72
Cyclin C 6 14 6 4
Cdk8 33 18 27 23 43 18
Cdk11 22 9 15 9 37 11
TRAP37 (MED27) 44 35 40 39 40 34
TRAP36 (MED4) 47 45 48 49 26 50
LCMR1 (MED19) 51 13 51 43 32 49
MED6 59 39 52 51 56 56
MED7 37 22 37 23 22 50
MED8 SO 54 46 47 47 50
p28b (ME018) 53 31 35 44 42 27
TRFP (MED20) 48 51 34 47 25 53
Med25 (ME09) 31 31 19 31 12 31
Intersex (MED29) 46 43 75 34 28 52
TRAP25 (MED30) 43 62 43 43 26 61
MED28 25 22 18 20 20 35
Srb7 (ME021) 58 68 59 46 22 60
SurfS (MED24) 54 59 43 47 26 71
HSPC296 (MED11) 45 45 44 48 37 50
Soh1 (MED31) 60 36 52 60 46 35
Nut2 (ME010) 66 59 59 80 73 70

Figure 2 Identification of human Mediator subunits by MudPIT. On the left is a list of Mediator subunits found in one or more independent isolates. (a) The names of complexes purified by various groups are listed across the top; a red square indicates that the subunit is present. (b) Mediator complexes isolated by immunoprecipitation from the indicated FLAG-tagged Mediator protein were characterized by MudPIT. Protein subunits that were identified are indicated with a red square (with the number corresponding to the percent amino acid sequence coverage by the MudPIT approach). The column labeled 'HeLa' corresponds to control cells that did not express a FLAG-tagged protein.

Reproduced from Sato *et al.* A set of consensus mammalian Mediator subunits identified by multidimensional protein identification technology. *Molecular Cell*, 2004;**14**:685–691.

Find out more

Kelleher RJ 3rd, Flanagan PM, Kornberg RD. A novel Mediator between activator proteins and the RNA polymerase II transcription apparatus. *Cell*, 1990;**61**:1209–1215.

Sato S, Tomomori-Sato C, Parmely TJ, *et al.* A set of consensus mammalian Mediator subunits identified by multidimensional protein identification technology. *Molecular Cell*, 2004;**14**:685–691.

Related techniques

Use of GAL4 hybrid system; Section 19.12

Protein extract fractionation; Section 19.7

Ion exchange chromatography; Section 19.7

S. cerevisiae (as a model organism); Section 19.1

Mass spectrometry; Section 19.8

Use of HeLa cells; Section 19.2

Epitope tagging (FLAG); Section 19.4

Co-immunoprecipitation; Section 19.12

octamers, help remove the barrier to transcription imposed by the structure of chromatin. Many of these enzymes are recruited to promoter regions through the action of proteins that regulate transcription, as we shall see in the next chapter.

➲ We learn about histone-modification enzymes and nucleosome remodeling enzymes in Chapter 4.

➲ We learn about tRNAs and the ribosome in Chapter 11.

RNA polymerases I and III use TBP to recognize their characteristic promoters

Eukaryotic RNA polymerases I and III are very similar to RNA polymerase II, with several of the core enzyme subunits identical in all three eukaryotic RNA polymerases (see Figure 8.5). RNA polymerase I and III, however, recognize quite distinct promoter sequences and have unique general transcription factors, as depicted in Figure 8.14. However, one feature in common among all three RNA polymerase enzymes is their dependence upon TBP to initiate transcription. In each case, TBP is a subunit of a larger complex: SL1, TFIID, and TFIIIB, complexes that are required for transcription by RNA polymerases I, II, and III, respectively.

RNA polymerase I, which transcribes rRNA genes, binds to a promoter containing a core promoter element and an upstream control element (UCE). TBP, which is part of a complex called SL1, helps RNA polymerase I to recognize the core promoter, as shown in Figure 8.14a.

Genes transcribed by RNA polymerase III, which include a variety of RNA genes such as tRNA and the 5S RNA subunit of the ribosome, have an unusual promoter structure: some of the key promoter elements are located *downstream* of the transcription start site. Figures 8.14b and c show the different pre-initiation complexes formed by RNA polymerase III upstream of tRNA genes and the 5S RNA gene. Promoter recognition by RNA polymerase III is mediated by TBP - which in this case is a subunit of TFIIIB - mirroring promoter recognition by polymerases I and II.

(a) rRNA gene

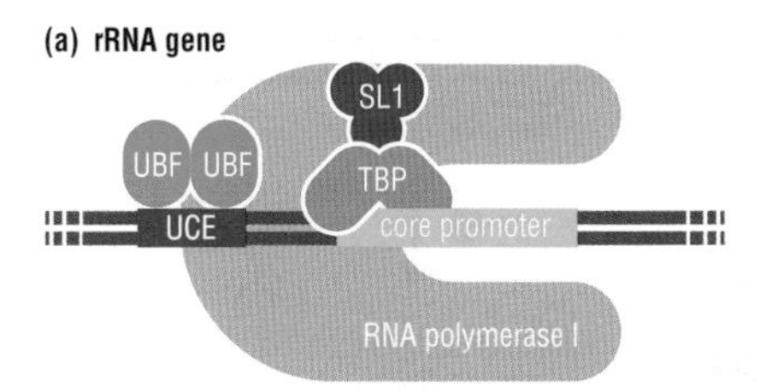

(b) tRNA gene

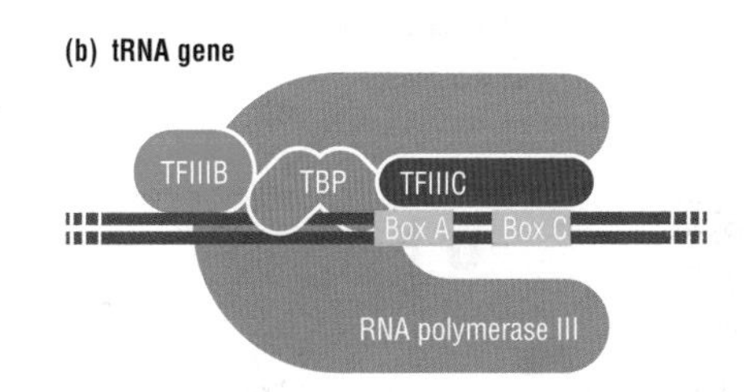

(c) 5S RNA gene

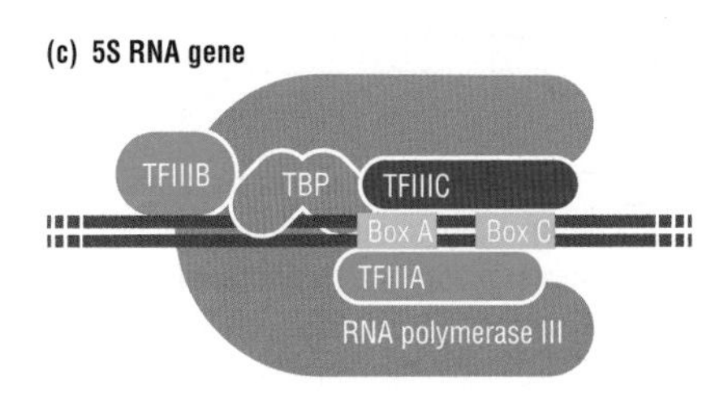

Figure 8.14 Pre-initiation complexes formed by RNA polymerase I and III. (a) RNA polymerase I promoters contain an upstream control element (UCE) and a core region, which are contacted by the general transcription factors, UBF and SL1, in human cells. The latter contains a TBP subunit. (b) RNA polymerase III pre-initiation complex bound to the promoter of a yeast tRNA gene. (c) The promoter region of the 5S RNA gene transcribed by RNA polymerase III has a different sequence from tRNA genes and also requires an additional protein, TFIIIA. Note that, in both (b) and (c), some of the general transcription factors bind downstream of the transcription start site.

8.4 INITIATION OF TRANSCRIPTION AND TRANSITION TO AN ELONGATING COMPLEX

In the preceding section, we discussed how bacterial, archaeal, and eukaryotic RNA polymerases bind to the promoter of a gene in the correct position and orientation to start transcription. At this stage, the complex of RNA polymerase and the promoter is called a **closed complex.** The enzyme is now ready to begin the cascade of events required to initiate transcription.

RNA polymerase must separate the two strands of DNA

The first step in transcribing a gene is to separate the two strands of DNA in the region of the transcription start site to expose the template strand (see Figure 8.15). RNA polymerase opens up a short region of around 14 bases known as the **transcription bubble.** The unpaired template strand binds to the polymerase near the active site and is kept apart from the non-template strand by regions of the RNA polymerase called the 'rudder', 'lid', and 'zipper', as illustrated in Figure 8.16a. The complex of RNA polymerase that is bound to an opened region of DNA and poised to initiate transcription is called the **open complex.**

The transition to the open complex can either occur spontaneously or may require the input of energy, depending upon the class of polymerase enzyme. Eukaryotic RNA polymerase II depends on the helicase subunits of the general

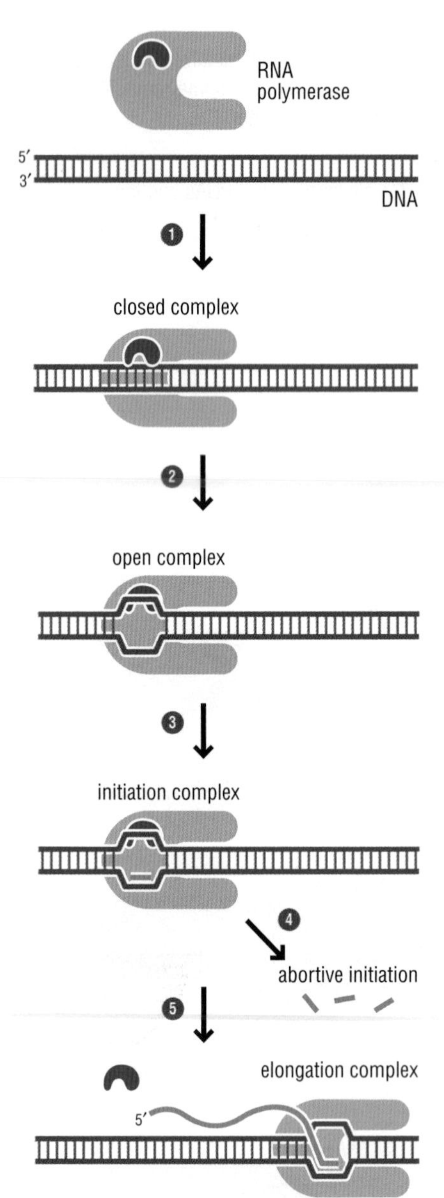

Figure 8.15 Steps in promoter recognition, promoter opening, transcription initiation, and promoter clearance by bacterial RNA polymerase. (1) The RNA polymerase holoenzyme binds to the promoter region, forming the closed complex. (2) The enzyme then separates the template and non-template strand, forming the open complex. (3) Synthesis of an RNA strand is initiated, using base-pairing interactions with the template strands. (4) RNA polymerase frequently synthesizes only a few base pairs at first and then releases the product, in a process known as abortive initiation. (5) When an RNA product of around 9–11 nucleotides is successfully synthesized, RNA polymerase undergoes a conformational change and loosens its grip on the sigma subunit to become an elongating complex that continues to synthesize RNA as it translocates along the DNA template.

transcription factor, TFIIH, to unwind and separate the two DNA strands in a process that is fueled by hydrolysis of ATP. The requirement of ATP-derived energy to separate the DNA strands is unique to RNA polymerase II; it is not required by RNA polymerases I or III, or by the archaeal enzyme, for whom the change from closed to open complex occurs spontaneously. For bacterial RNA polymerase, the change from the closed to the open complex is usually also spontaneous and does not require additional energy. A special exception to this is the transcription of genes that depend on σ^{54}, which requires energy from ATP hydrolysis.

RNA polymerase initiates synthesis of the nascent RNA chain

Once the template strand is exposed and positioned near the active site of the RNA polymerase, the enzyme can begin synthesis of RNA. Figure 8.16b shows how free ribonucleoside triphosphates enter through the funnel region of RNA polymerase and form complementary base pairs with bases on the exposed template strand. The first base in mRNA is typically a purine (A or G). Once base-paired with the template strand, successive nucleotides are added to the 3′ end of the growing RNA chain, as outlined in Figure 8.17 . The 3′ OH of the last base in the growing RNA chain initiates a nucleophilic attack on the innermost phosphate of the incoming ribonucleoside triphosphate. This is an energetically favorable reaction that leads to formation of a phosphodiester bond with the 5′ carbon of the newly added nucleotide and release of pyrophosphate. The 5′ end of the growing RNA chain exits through a special exit channel while a loop within the lid helps to separate the RNA from the DNA template (see Figure 8.16b).The catalytic mechanism for ribonucleotide addition is remarkably similar to that in DNA replication.

See Section 6.2 (in particular, Figure 6.6) to review the mechanism of DNA synthesis.

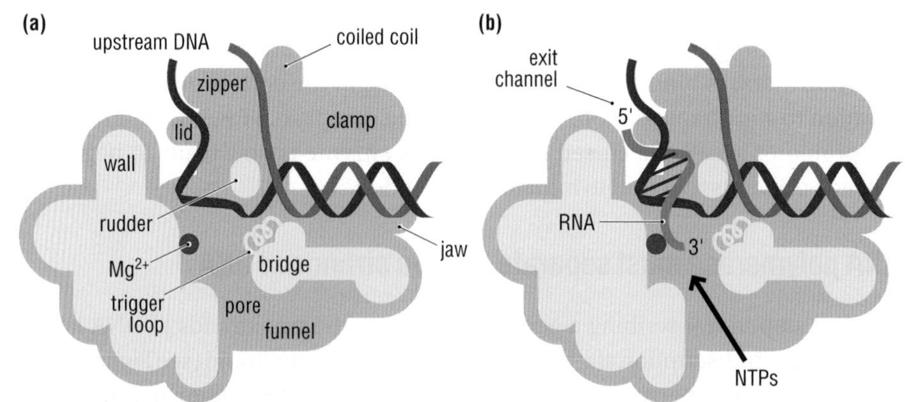

Figure 8.16 The path of the DNA and RNA in RNA polymerase. This is a cutaway view of the interior of the RNA polymerase enzyme. (a) The open complex consists of the RNA polymerase bound to the promoter DNA (gray), with the transcription bubble opened and the template strand positioned near the Mg^{2+} ion (pink) in the enzyme's active site. The rudder, lid, and zipper help keep the two strands of DNA apart. The downstream DNA is held in the cleft between the two large subunits. (b) Transcription is initiated, and incoming ribonucleoside triphosphates (NTPs) enter the active site and are added to the growing RNA chain (blue). A conformational change that closes the clamp head around the upstream DNA helps prevent dissociation of the enzyme–DNA–RNA complex, thereby helping to ensure processivity.
Adapted from Carmer, P. *Current Opinion in Structural Biology*, 2002;**12**:89–97.

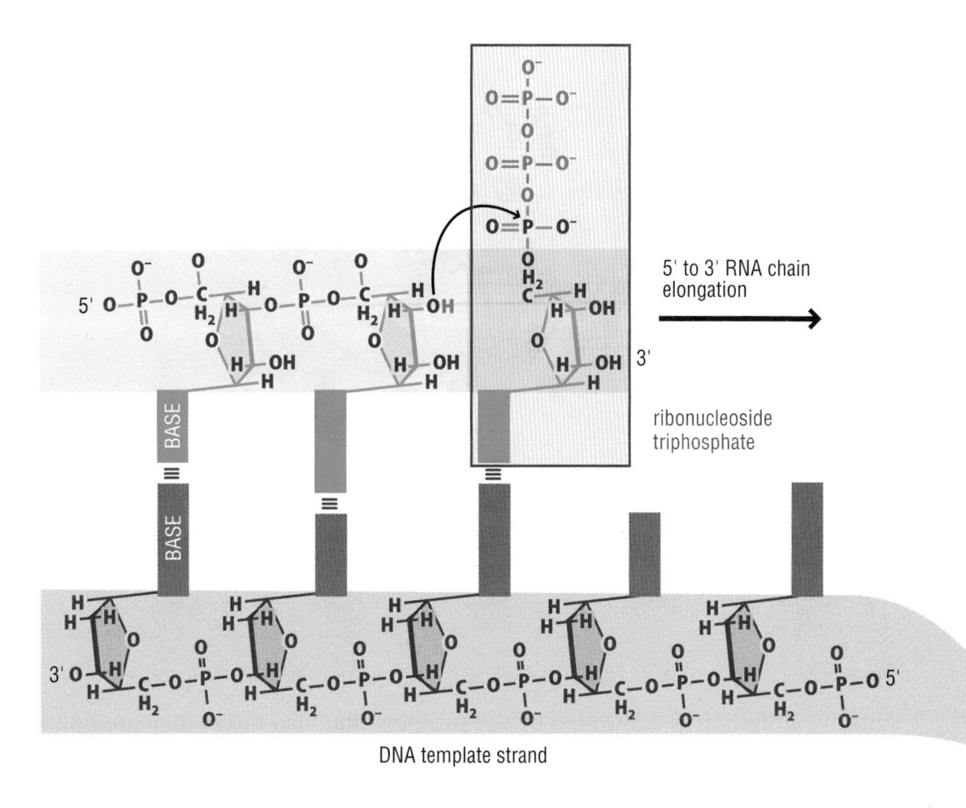

Figure 8.17 The biochemical reaction catalyzed by RNA polymerase. An incoming ribonucleoside triphosphate is selected for its ability to base pair with the exposed DNA strand. The innermost phosphate of this ribonucleotide (the α phosphate) is then subjected to nucleophilic attack by the 3′ OH group of the last nucleotide of the growing RNA chain. The release and further cleavage of the resulting pyrophosphate generates energy for the reaction.

Just like DNA polymerase, RNA polymerase has two Mg^{2+} ions in the active site that play key roles in activating the 3′-OH for nucleophilic attack and stabilizing a developing negative charge on the leaving oxygen; the locations of these ions are depicted in Figure 8.18a. One Mg^{2+} ion is stably bound in the RNA polymerase active site (see Figure 8.18), whereas the second is brought in by the incoming ribonucleoside triphosphate. A feature in the active site known as the trigger loop helps to position the incoming ribonucleoside triphosphate and ensure that it is a correct match to the template strand.

Once the new phosphodiester bond is formed, pyrophosphate is released and then further broken down by hydrolysis, releasing additional energy and thereby favoring the polymerization reaction. The trigger loop and bridge helix help to translocate the RNA out of the active site and make room for the next incoming dNTP (see Figure 8.16).

➲ We learn more about the nature of nucleophilic attack as a reaction mechanism in Chapter 2.

Surprisingly often, RNA polymerase fails to synthesize the full-length RNA on the first attempt and instead enters into a series of abortive cycles of synthesis (see step 4 in Figure 8.15). These cycles, referred to as **abortive initiation**, result in the production and release of short RNA products (2–9 nucleotides in length), which are subsequently degraded. The amount of abortive initiation varies from one promoter to another and can be regulated by sequence-specific DNA-binding proteins or by the availability of particular nucleotides. If insufficient cellular components are available to support transcription, the process will be aborted. Read Experimental approach 8.3 to learn about how the products of abortive initiation were first detected *in vitro*, and how they have more recently been visualized *in vivo*.

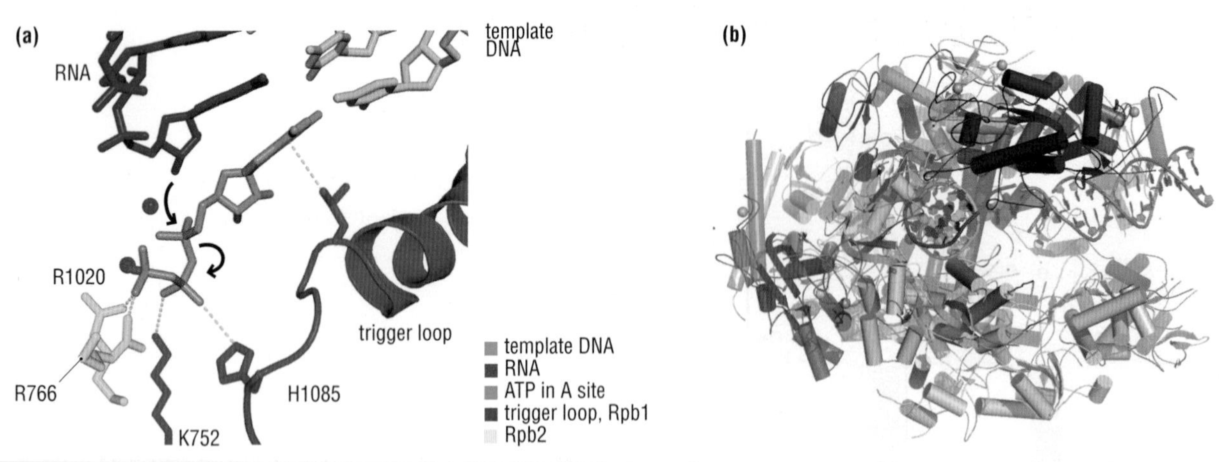

Figure 8.18 Two metal ions and the trigger loop in the RNA polymerase II active site. (a) The active site contains two bound Mg^{2+} ions (pink spheres): one that is always bound to the enzyme, and a second that is brought in by the incoming ribonucleoside triphosphate. The trigger loop (purple) positions the ribonucleoside triphosphate and helps to discriminate whether the correct base has been inserted. (b) RNA polymerase II elongation complex, with the template DNA (gray) and RNA (blue). PDB code 2E2H.

Abortive initiation is controlled by the proteins that also direct the core polymerase to the promoter. Both the bacterial sigma protein and the eukaryotic TFIIB basal transcription factor play similar roles in abortive initiation. But how do they exert their effect? We learned in Section 8.3 that loop 3.2 of sigma and the B finger loop of TFIIB extend into the active site of the core enzyme (Figure 8.8). Each of these loops is in a position to block the elongating transcript after a short initial transcript has been synthesized. In order for transcription to continue unimpeded, the loop must somehow move out of the way so that the transcript can continue to grow. If the loop does not move, nucleotide addition ceases and the short transcript is released (as shown in Figure 8.19, step 1). However, if the loop is displaced, nucleotide addition can continue (Figure 8.19, step 2). Displacement of the loop (or B finger) is thought to help the polymerase enzyme break away from the promoter by disrupting contacts with proteins that helped to recruit the core polymerase, setting the stage for the next step in the transcription cycle: **promoter clearance**.

RNA polymerase undergoes a conformational change as it leaves the promoter

When RNA polymerase succeeds in synthesizing an RNA about 9–11 nucleotides long, the enzyme breaks free from the promoter and becomes a productive 'elongating complex'. During this process, termed **promoter clearance** (or **promoter escape**), the polymerase enzyme undergoes a conformational change that closes the clamp region around the DNA and helps make the association of the enzyme with the DNA very stable (Figure 8.19, step 3). At the same time, RNA polymerase loosens its grip on initiation factors – the bacterial sigma subunit or the eukaryotic general transcription factors – leading to their eventual release. The enzyme is now free to translocate along the DNA, allowing the growing RNA chain to exit through a special channel (see Figure 8.19, step 4).

Eukaryotic RNA polymerase II becomes phosphorylated as it converts from the initiating complex to the elongating complex. Specifically, the general transcription factor, TFIIH, phosphorylates the CTD of the large subunit, Rpb1

8.3 EXPERIMENTAL APPROACH

Abortive initiation by E. coli RNA polymerase

In vitro transcription produces a mix of very short and very long transcripts

The phenomenon of abortive initiation was discovered by examining the transcripts produced by *E. coli* RNA polymerase at the *lac UV5* promoter. Earlier experiments had hinted that the progression from transcription initiation to elongation was not necessarily smooth. Some of those experiments, however, were conducted under conditions in which one or more nucleotide precursors were limiting. In addition, the detection methods that had previously been used – such as paper chromatography – could not resolve many different short nucleotide products. Agamemnon Carpousis and Jay Gralla (1980) decided to use the relatively new method of high-resolution gel electrophoresis to see what nucleotide products were produced *in vitro* when all four ribonucleoside triphosphates were present.

In the experiment shown in Figure 1, RNA polymerase holoenzyme was incubated with DNA containing the *lac UV5* promoter, along with radiolabeled nucleotide precursors, ATP, GTP, CTP, and UTP. The products were then separated by gel electrophoresis and visualized by autoradiography on film. The authors found that many short RNA transcripts, in the range of two to six nucleotides in length, were produced at each concentration of nucleotide tested (see Figure 1). Near the top of the gel are the much longer transcripts that resulted from productive elongation. The relative absence of transcripts of an intermediate length supported the idea that cycles of abortive initiation were followed by highly processive transcription once RNA polymerase entered into the elongation phase.

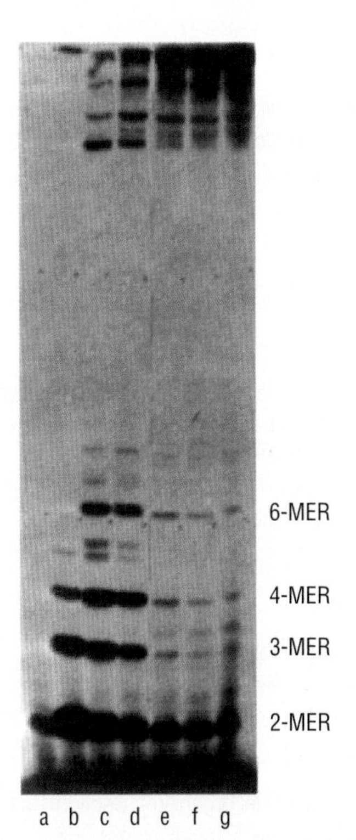

Figure 1 Separation of abortive initiation products by gel electrophoresis. Lane a shows a control in which DNA was omitted from the reaction. The remaining lanes show the products of abortive initiation in the presence of different nucleotide concentrations.

Reproduced from Carpousis and Gralla, Cycling of ribonucleic acid polymerase to produce oligonucleotides during initiation in vitro at the lac UV5 promotor. *Biochemistry*, 1980;**19**:3245–3253.

Could abortive initiation be an *in vitro* artifact?

Despite longstanding observations of abortive initiation by both bacterial and eukaryotic RNA polymerase enzymes, the phenomenon had until recently only been observed in a test tube. While it was speculated that abortive initiation was a universal feature of transcription initiation that also took place *in vivo*, there had been no direct evidence that abortive transcripts were generated *in vivo*, much less that these transcripts accumulated in the cell. Was abortive initiation an *in vitro* artifact? Could there be factors in cells that suppressed abortive initiation? Even if abortive initiation occurred in cells, would the products not be rapidly degraded?

Confirming that abortive initiation occurs in live bacterial cells

To answer these questions, Seth Goldman, Richard Ebright, and Bryce Nickels (2009) searched for abortive initiation products in *E. coli* using methods that had been recently devised to detect small RNA molecules in the cell. Instead of studying the *lac UV5* promoter, they chose a bacteriophage promoter (the phage T5 N25anti- promoter) that was known to exhibit particularly high rates of abortive initiation *in vitro*. Since the cell could express abortive initiation products from many different promoters, the researchers used oligonucleotide hybridization (described in Section 19.10) to detect the specific products of this promoter. The probe used was an oligonucleotide complementary to the first 15 nucleotides of the putative abortive transcripts, but made with special nucleotides (to yield 'locked nucleic acid' or 'LNA') that hybridize with exceptionally high affinity and specificity.

Experiments were first performed *in vitro* to compare detection by oligonucleotide hybridization with the radiolabeling method used by Carpousis and Gralla (Figure 2a, right and left, respectively). The hybridization method with the LNA probe successfully identified abortive transcripts of 11–15 bp in length. The same

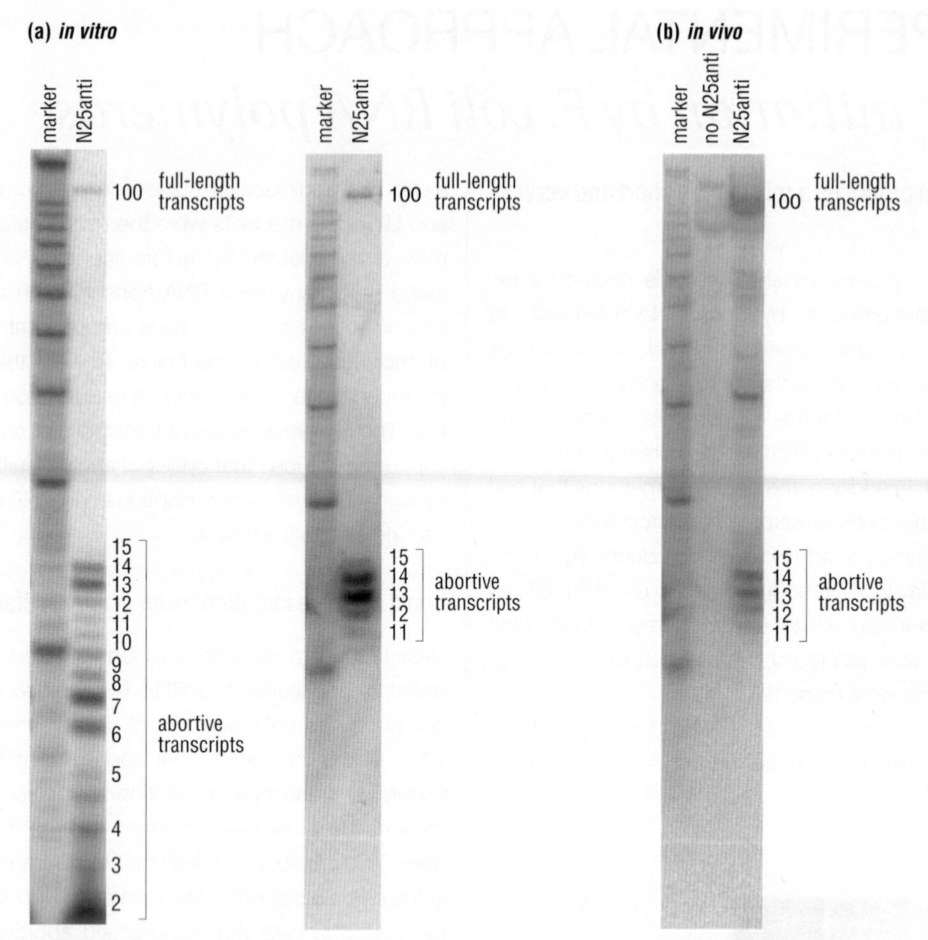

Figure 2 Abortive initiation occurs *in vivo.* (a) Abortive initiation *in vitro*. Left panel: Gel electrophoresis followed by autoradiography shows transcripts generated in 'hot' (radioactive) reactions. Right panel: Detection of abortive initiation products by hybridization of locked nucleic acids (LNA; non-radioactive) reactions. (b) Abortive initiation *in vivo*. Detection was performed by hybridization of LNAs to transcripts generated *in vivo*.

Goldman *et al.* Direct detection of abortive RNA transcripts in vivo. *Science*, 2009; **324**:927–928.

experiments were then carried out on lysates prepared from cells containing a plasmid with the T5 N25anti-promoter and the gene complementary to the LNA probe; Figure 2b shows that the same set of short transcripts is detected *in vivo*.

These experiments confirmed that abortive initiation takes place *in vivo*, and that its products accumulate – at least in *E. coli*. An important question that remains is whether the abortive products play regulatory roles *in vivo*. Among the many possibilities is that these small nucleic acid products could serve as sequence-specific primers for RNA synthesis or lagging-strand DNA synthesis, or perhaps could stabilize other RNA molecules through base-pairing interactions.

Find out more

Carpousis AJ, Gralla JD. Cycling of ribonucleic acid polymerase to produce oligonucleotides during initiation in vitro at the lac UV5 promoter. *Biochemistry*, 1980;**19**:3245–3253.

Goldman SR, Ebright RH, Nickels BE. Direct detection of abortive RNA transcripts in vivo. *Science*, 2009;**324**:927–928.

Related techniques

E. coli (as a model organism); Section 19.1

Gel electrophoresis; Section 19.7

Radiolabeling; Section 19.6

Autoradiography; Section 19.5

Bacteriophage T5 (as a model organism); Section 19.1

Oligonucleotide hybridization; Section 19.9

1 abortive initiation

sigma factor

RNA

bridge

NTPs

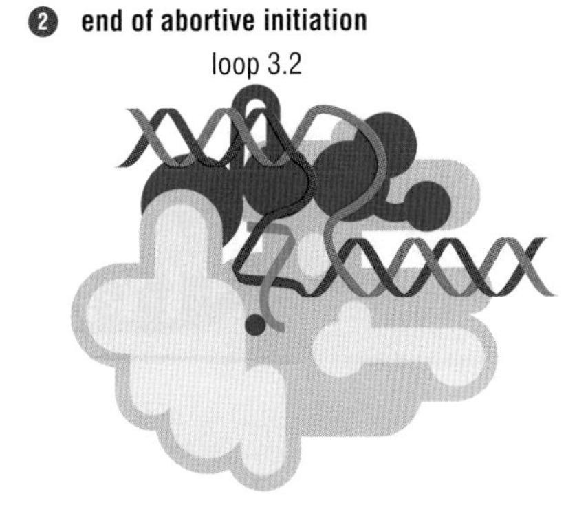

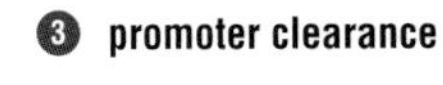

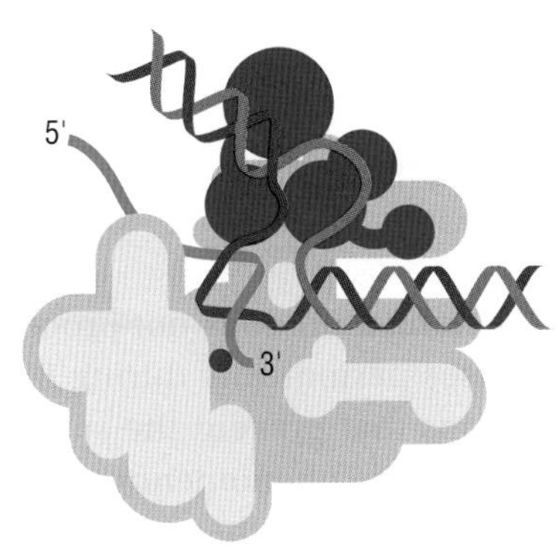

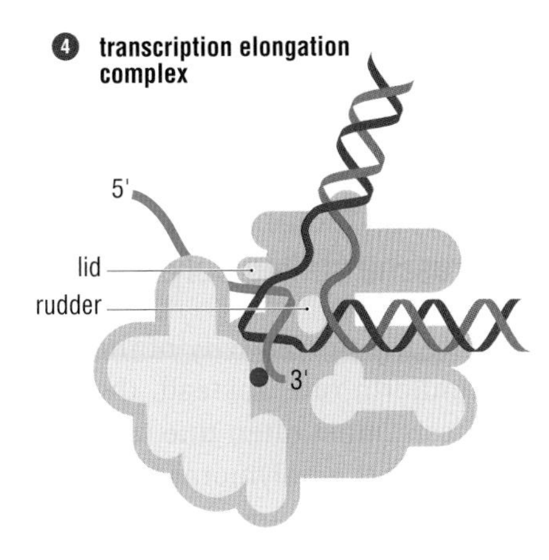

Figure 8.19 Structural transitions from abortive initiation through elongation for bacterial RNA polymerase. Shown are cutaway views of the RNA polymerase holoenzyme (with σ, blue; rest of RNA polymerase gray; catalytic Mg^{2+}, pink sphere), promoter DNA (template strand, dark gray; non-template strand, light gray) and the RNA transcript (blue). (1) Abortive initiation, with sigma factor loop 3.2 blocking path of RNA. (2) End of abortive initiation, with loop 3.2 swung out of the way. (3) Promoter clearance. (4) Transcription elongation complex. Adapted from Murakami KS, Darst SA. *Current Opinion in Structural Biology*, 2003;**13**:31-9.

(see Section 8.1), at the serine that is the fifth residue of the heptad repeat. This is the first in a series of CTD phosphorylation events that are needed to recruit proteins required for processing the mRNA transcript, as we shall learn next.

8.5 TRANSCRIPTION ELONGATION

Transcription proceeds in a highly processive manner

Once RNA polymerase has cleared the promoter region and made the transition to an elongating complex, transcription proceeds in a processive manner, such that one nucleotide after another is added in quick succession by a single RNA polymerase enzyme for hundreds or even thousands of bases. This processivity indicates that the elongating polymerase enzyme is very stably bound to the DNA. The stability results from the way that the two large subunits of RNA polymerase close around the DNA like a set of jaws, preventing the elongation complex from falling off the DNA. The energy released by cleavage of the incoming nucleoside triphosphates and subsequent pyrophosphate hydrolysis also favors the polymerization process, such that the polymerase enzyme can synthesize the growing transcript at a rate of about 20–50 nucleotides per second.

A transcription bubble containing around 10–14 unpaired bases is maintained by the enzyme as it moves along the DNA, as illustrated in Figure 8.20. RNA polymerase separates the downstream DNA into unpaired strands, while base pairs re-form behind the translocating enzyme. Within the transcription bubble are the nine most recently added ribonucleotides of the RNA transcript, which remain

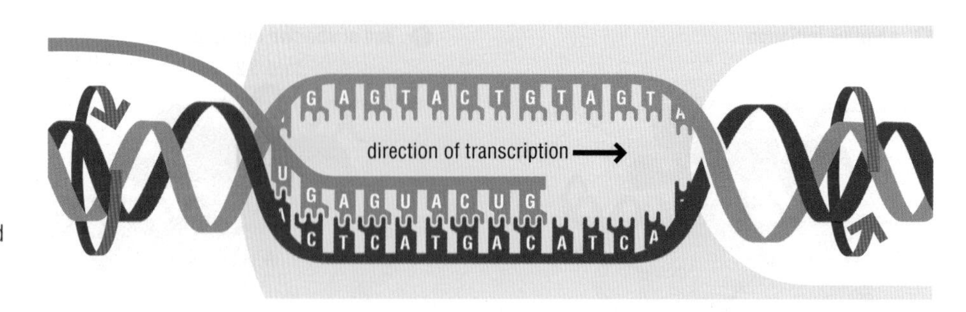

Figure 8.20 Transcription bubble. The transcription bubble consists of 10–14 bp of unpaired DNA, with the nine most recently added RNA nucleotides base-paired with the template strand.

base-paired to the template strand. These characteristics of the transcription bubble persist as successive ribonucleotides are added to the 3′ end of the RNA transcript (see Figure 8.16). The association of the polymerase with the RNA–DNA hybrid within the transcription bubble contributes to the stability of the elongation complex.

Elongation factors suppress the pausing of transcription and rescue stalled polymerase enzymes

Despite the processivity of the reaction, chain elongation does not proceed at a uniform rate. Instead, the polymerase pauses periodically during RNA synthesis, most often due to physical obstructions that prevent the smooth passage of the polymerase along the DNA strand. This phenomenon, called **transcriptional pausing**, can be caused by physical features that arise due to the way in which the template DNA sequence and complementary RNA transcript interact.

One type of sequence-dependent pausing occurs when the RNA transcript contains short complementary stretches that form a transient hairpin. This short stretch of RNA containing base-paired nucleotides interacts with the polymerase and temporarily halts RNA synthesis; the precise mechanism for this pausing remains to be determined. Transcriptional pausing can also be brought about by the presence of a weak DNA–RNA hybrid in the transcription bubble, the result of an AU-rich sequence or a misincorporated base.

The polymerase can eventually resume RNA synthesis on its own, although pausing can be either relieved or enhanced by proteins called **elongation factors. Transcriptional arrest** occurs when the enzyme is unable to resume RNA synthesis without the assistance of elongation factors. As we shall see in the next chapter, transcriptional pausing and arrest can be regulated in bacteria and in eukaryotes, providing another point at which cells can modulate gene expression.

The critical importance of elongation factors is clear from the many human diseases that arise from defects in one or more of these proteins, as summarized in Figure 8.21. For example, the ELL protein is one of a class of elongation factors that suppress pausing of RNA polymerase and thereby stimulate the overall rate of elongation. ELL is part of a complex called the Super Elongation Complex (SEC) that also includes the RNA polymerase II CTD kinase, pTEFb, along with additional subunits (AFF4, ENL, AF9). Chromosomal translocations that result in the expression of a fusion protein containing portions of the ELL or pTEFb proteins can give rise to acute myeloid leukemia.

In many genes transcribed by RNA polymerase II, the enzyme stalls after transcribing around 35–50 base pairs and only resumes efficient transcription

Functions of eukaryotic elongation factors and their roles in human disease		
factor	function	disease/note
TFIIS	transcript cleavage, suppresses pausing	
CSB	suppresses pausing	Cockayne syndrome (associated with a defect in nucleotide excision repair)
elongins ABC	suppress pausing, associate with VHL protein	von Hippel-Lindau disease (familial cancer predisposition syndrome)
ELL	disrupts interaction with TBP and TFIIB	acute myeloid leukemia
p-TEFb	phosphorylates CTD of Rpb1 subunit of RNA PolII (functions in concert with HIV Tat)	acute myleloid leukemia HIV infection
NELF, DSIF	negative elongation factor promotes pausing	implicated in Wolf-Hirschhorn syndrome
elongator complex	aids in elongation (Elp1-6) through chromatin; contains a histone acetyltransferase subunit	neurodegenerative disease
FACT complex (Spt16, Pob3)	H2A/H2B chaperone; nucleosome disassembly/reassembly	
Asf1	H3/H4 chaperone; nucleosome disassembly/reassembly	

Figure 8.21 Examples of eukaryotic elongation factors, their functions and their roles in human disease.

with the assistance of other proteins. Pausing is promoted by the negative elongation factors, NELF and DSIF, which are thought to stabilize the paused polymerase on the DNA template. Termed **promoter proximal pausing**, this phenomenon was originally thought to be restricted to a handful of genes. More recent findings, however, indicate that many eukaryotic genes contain a paused RNA polymerase enzyme just a few dozen base pairs from the start site of transcription, awaiting the action of proteins that can restart transcription. These observations suggest that regulation of promoter proximal pausing may be more important to eukaryotic gene regulation than had been previously thought.

➔ We learn about promoter proximal pausing in the regulation of heat shock genes in Chapter 9.

Phosphorylation of the RNA polymerase II CTD couples elongation with pre-mRNA processing

We have already seen that phosphorylation at the fifth serine in the CTD heptad repeats takes place when RNA polymerase II clears the promoter region. As outlined in Figure 8.22, phosphorylation of serine 5 of the CTD repeats triggers recruitment of RNA-processing enzymes that add the 5′ guanosine cap to the nascent transcript. The temporary pause triggered by negative elongation factors that we learned about above is thought to allow sufficient time to recruit the RNA processing enzymes and add the guanosine cap. Capping of the mRNA triggers recruitment of a kinase, P-TEFb, which phosphorylates the CTD at the second serine in each heptad repeat and causes RNA polymerase to resume elongation. Phosphorylation of serine 2 also enables the CTD tail to recruit additional RNA-processing enzymes that are required for later processing of the 3′ end of the complete transcript.

➔ We explore RNA processing in depth in Chapter 9.

Pausing and arrest of transcription can be alleviated by cleavage of the RNA strand

When there is a pause in RNA synthesis, Figure 8.23 shows how the elongation complex may **backtrack**, reversing direction on the DNA. This causes

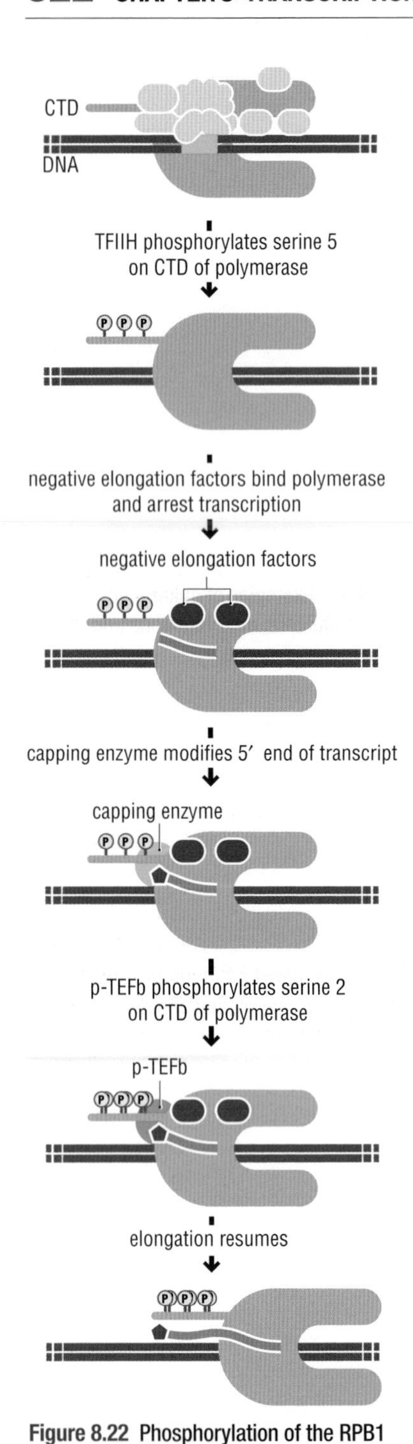

Figure 8.22 Phosphorylation of the RPB1 CTD couples pre-mRNA processing with transcription elongation by RNA polymerase II. Phosphorylation of the fifth serine residue in the heptad repeats in the CTD of the RNA polymerase II large subunit leads to binding of several proteins (pink) that cause transcriptional arrest. This then allows recruitment of enzymes (green) that catalyze formation of a 5′ cap on the nascent RNA. The capping of the mRNA is followed by further phosphorylation of the CTD at serine 2 residues, which leads to resumption of transcription elongation.

the most recently synthesized portion of the RNA to separate from the template strand and protrude through the funnel region of RNA polymerase as the transcription bubble moves backward. Transcription can be restarted with the assistance of **transcript cleavage factors**, which promote cleavage of the protruding 3′ end of the RNA transcript and allow transcription to resume. Transcript cleavage has been postulated to have a role in error correction: if mismatched bases are incorporated into the growing mRNA, they may destabilize the RNA–DNA hybrid and produce distortions in the duplex region, which causes the RNA polymerase to stall. By cleaving this mismatched region, the RNA polymerase can then correct this error by resynthesizing the portion that has been cleaved.

Transcript cleavage factors promote the hydrolysis of the phosphodiester bond, thus releasing the unpaired RNA nucleotides at the 3′ end of the transcript. By contrast, hydrolysis proceeds at a negligible rate in the absence of transcript cleavage factors. GreA and GreB are the transcript cleavage factors in *E. coli*, while TFIIS is the transcript cleavage factor for the eukaryotic RNA polymerase II. These proteins bind in the funnel region of RNA polymerase, where the 3′ end of the mRNA protrudes. Figure 8.24 shows how GreB and TFIIS bind to RNA polymerase and extend into the funnel, where they position a magnesium ion in the active site. The metal activates a water molecule for hydrolysis of the phosphodiester bond, thereby promoting cleavage of the last two to ten nucleotides of the transcript. This exposes a new 3′ OH at the end of the RNA transcript that is correctly positioned in the active site for a fresh round of nucleotide addition. This reaction is reminiscent of proofreading by DNA polymerases, which hydrolyze the phosphodiester bond to remove misincorporated bases.

Nucleosomes must be removed from the path of the elongating polymerase for transcription to proceed

Just as nucleosomes can interfere with the assembly of the transcription preinitiation complex on DNA, they also pose a barrier to the elongating RNA polymerase enzyme. The DNA must be unwrapped from the histone core in order to open a transcription bubble, yet this has an energetic cost that could cause the elongating polymerase to stop transcribing. Eukaryotic cells therefore have proteins that actively remove nucleosomes from the path of the transcribing polymerase and reassemble them in its wake, as shown in Figure 8.25. This is accomplished with the help of histone chaperones, which aid in the temporary disassembly and reassembly of histone octamers. Histone chaperones such as FACT (Facilitates Chromatin Transcription), Asf1, and Spt6 each bind to a subset of the histones belonging to a partially disassembled histone octamer; for example, FACT binds to H2A–H2B dimers while Asf1 binds to H3–H4 tetramers.

There is some evidence that the eviction of histones is either preceded by or coupled to a wave of histone modifications, such as acetylation and H2B ubiquitination. Precisely how all these events are coupled during the process of transcription remains an exciting area of research.

The need to dismantle and then reassemble nucleosomes during transcription has been established primarily for RNA polymerase II. However, there is evidence that similar changes in chromatin are also required for transcription elongation by the other eukaryotic RNA polymerases, I and III.

Transcription elongation produces supercoiling that must be relieved by topoisomerases

As the transcription bubble moves through a double-stranded DNA template, Figure 8.26 shows how RNA polymerase is continually unwinding the DNA template at the downstream end of the enzyme. At the same time, the unpaired strands at the upstream end must twist around one another to re-form the DNA double helix. As we learned in Chapter 2, changes in the winding of the two DNA strands - defined as twist - lead to changes in DNA supercoiling if the ends of the DNA are prevented from rotating or if the template is a covalently closed circle. As transcription proceeds, the DNA flanking the transcribed region is not free to rotate even in linear chromosomes due to the packaging of the DNA into chromatin. Unwinding the downstream DNA therefore reduces the twist of the DNA, which is compensated for by an increase in positive supercoiling ahead of the transcribing polymerase. Similarly, the reannealing DNA behind the polymerase enzyme increases in twist as the double helix re-forms, leading to an accumulation of negative supercoils in the upstream region.

The build-up of superhelical tension could cause the elongating complex to stall and therefore must be relieved. This is achieved through the action of topoisomerase enzymes. In *E. coli*, DNA gyrase removes the positive supercoils ahead of the transcribing polymerase, and DNA topoisomerase I removes the negative supercoils behind the enzyme.

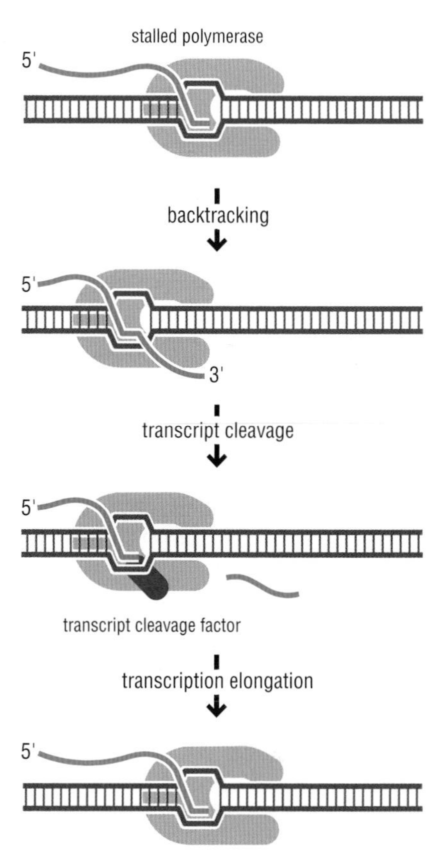

Figure 8.23 Transcript cleavage factors rescue a backtracked polymerase. A stalled RNA polymerase can backtrack, meaning that the transcription bubble moves backwards and the 3′ end of the RNA protrudes through the enzyme. By promoting cleavage of the exposed 3′ end of the RNA, transcript cleavage factors can allow RNA synthesis to resume.

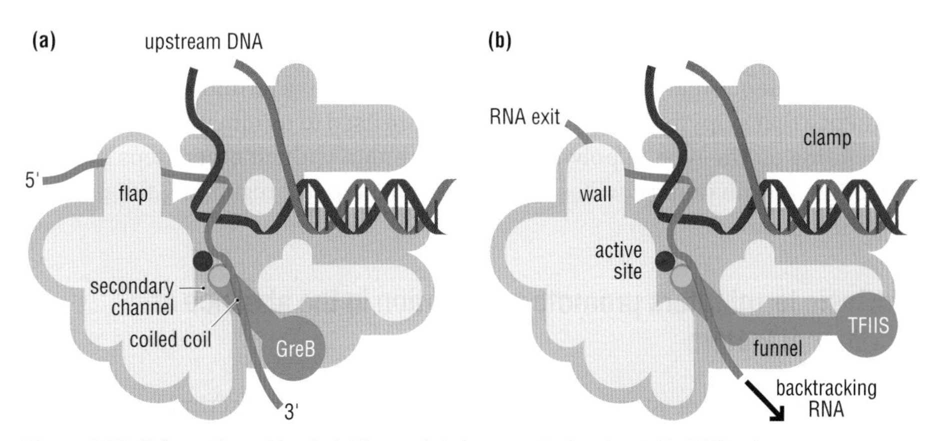

Figure 8.24 Eukaryotic and bacterial transcript cleavage factors bound to RNA polymerase. (a) GreB, a bacterial transcription cleavage factor, extends into the active site of bacterial RNA polymerase and positions a second metal ion (green). (b) TFIIS, the eukaryotic transcript cleavage factor, similarly extends into the active site of RNA polymerase II, where it positions a second metal ion (green).
Adapted from Cramer, *Current Opinion in Genetics and Development*, 2004; **14**: 218–226.

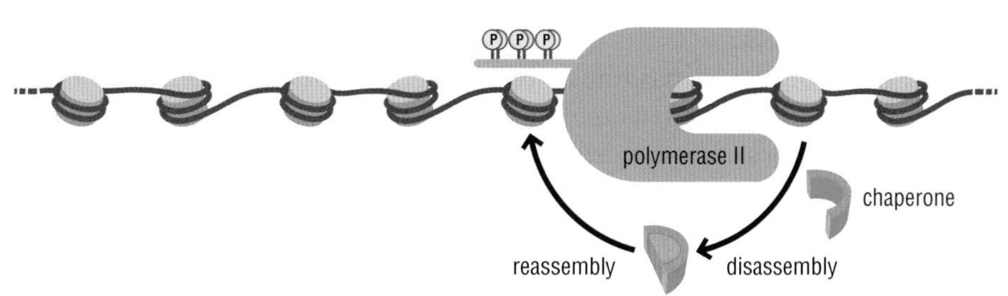

Figure 8.25 Nucleosome disassembly and reassembly during elongation. Nucleosomes are disassembled ahead of the elongating polymerase and reassembled in its wake with the help of histone chaperones.

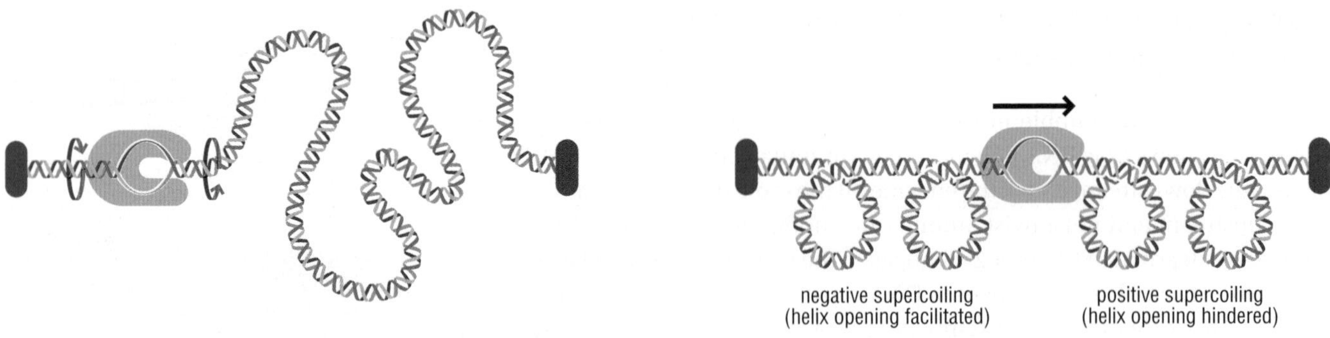

Figure 8.26 DNA supercoiling results from transcription elongation. RNA polymerase moving along the template. DNA gives rise to the accumulation of negative supercoiling behind the enzyme and positive supercoiling ahead of it.

➔ Histone chaperones also play a role in DNA replication, as we saw in Chapter 6.

➔ We learn about histone chaperones and topoisomerases in Chapter 6.

8.6 TRANSCRIPTION TERMINATION

Transcription is terminated by a variety of mechanisms

Once RNA polymerase has succeeded in transcribing a complete gene or set of genes, the enzyme must stop transcribing at the proper position and release both the completed transcript and the DNA. This prevents inappropriate read-through transcription of genes further downstream and frees the polymerase to initiate transcription again. As described earlier, RNA polymerase stops transcription at defined DNA sequences called terminators. In essence, the effect of terminators on transcription is opposite to that of promoters: a promoter is a defined sequence that promotes transcription, whereas a terminator is a defined sequence that stops it. Various mechanisms are used to terminate transcription, depending on the type of polymerase. In common with all other steps in transcription, termination provides another opportunity for regulating the process of gene expression.

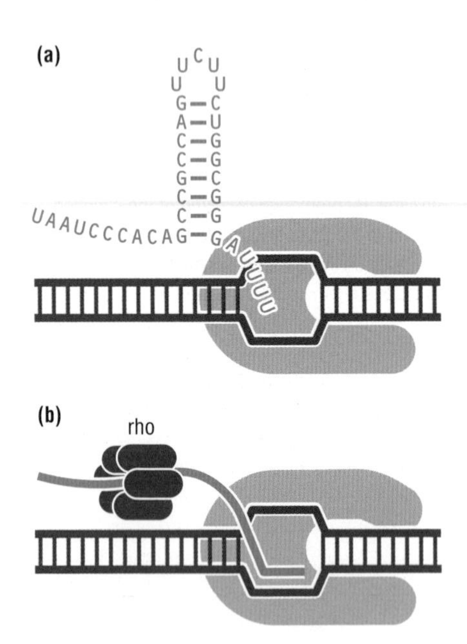

Figure 8.27 Terminators in bacteria. (a) Intrinsic, Rho-independent terminators encode an RNA stem–loop hairpin and a stretch of eight to ten U residues. The poly(U)–poly(A) RNA–DNA hybrid that is generated on transcribing this region has an unusually weak base-paired structure, and the RNA hairpin helps to pull the RNA out of the active site. (b) At enzymatic, Rho-dependent terminators, the Rho hexamer (pink) uses the energy from adenosine triphosphate (ATP) hydrolysis to pull the transcript away from RNA polymerase.

Intrinsic terminators promote termination in the absence of additional factors

At certain terminator sites, RNA polymerase terminates transcription in the absence of any other factors. These DNA sites are called **intrinsic** or **simple terminators.** In bacteria, intrinsic terminators have two characteristic features. One is an inverted repeat sequence that specifies a secondary structural element in the RNA in which the RNA strand folds back on itself and base-pairs to form a hairpin-like structure called a **stem-loop**, as illustrated in Figure 8.27a. The second is a series of around eight to ten adenosine residues (A) (**polyadenosine** or poly(A)) immediately downstream that encodes a stretch of uridine residues (U) (poly(U)) in the RNA.

Once RNA polymerase has transcribed the terminator sequence, the complementary halves of the inverted repeat in the RNA transcript base-pair with one another and form the stem-loop structure shown in Figure 8.27a. At the same time, the poly(U)-poly(A) RNA-DNA hybrid that results from transcribing the poly(A) region has an unusually weak base-paired structure and is hence unstable. The opening up of this region within the transcription bubble is thought to arrest transcription while the RNA hairpin part of the terminator sequence interacts with the polymerase and somehow helps to pull the RNA out of the active site, causing the elongation complex to dissociate.

In eukaryotes, specific terminator sequences are found in genes transcribed by RNA polymerase III. These terminators are defined by a stretch of U residues in the RNA transcript, which again may destabilize the RNA–DNA hybrid.

Termination of some transcripts requires additional protein factors

In many cases, the DNA sequence alone is not sufficient to terminate transcription. At these sites, denoted **enzymatic terminators**, additional protein factors are required for RNA polymerase to terminate transcription and be released from the DNA template.

Certain *E. coli* genes depend on a protein called Rho to terminate their transcription. Terminators that require Rho for transcription termination are called **Rho-dependent termination** sites, while those that do not are called **Rho-independent terminators**. The terminators of Rho-dependent genes do not contain the characteristic hairpin and run of U residues that is found in intrinsic terminators.

Rho is a hexameric ATPase that binds to C-rich sites on the nascent RNA chain termed *rut* sites (for *r*ho *ut*ilization). The Rho hexamer, whose structure is depicted in Figure 8.28, can exist as an open ring, which facilitates its loading onto the RNA. Once RNA is bound inside the Rho hexamer, it is thought that the ring closes, and concerted cycles of ATP hydrolysis by Rho drive structural changes that pull the RNA through the center of the ring and away from the RNA polymerase (see Figures 8.27b and 8.28b). This activity of Rho also helps to displace the RNA from the DNA and prevent formation of RNA–DNA hybrids known as R-loops, in which several base pairs of the RNA hybridize with the template strand, thus displacing the complementary strand of DNA.

Transcription termination by eukaryotic RNA polymerase I also requires additional proteins. The terminators recognized by RNA polymerase I contain U-rich sequences just as intrinsic terminators do, but termination requires the action of a DNA-binding protein (Reb1p in yeast and TTF1 in mouse) that binds to sequences downstream of the **transcription unit**. While the instability of the RNA–DNA hybrid is sufficient to lead to transcript release and termination in yeast once the DNA-binding protein has triggered termination, an additional release factor is required in animals.

(a)

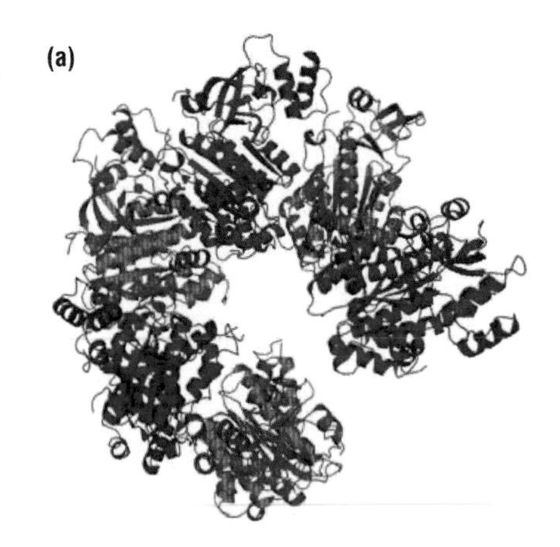

(b)

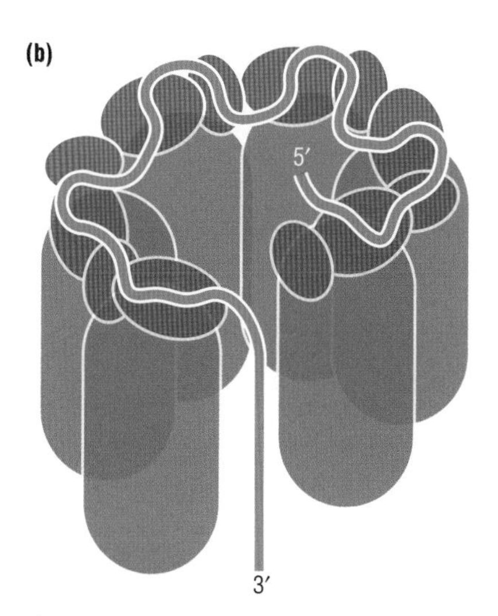

Figure 8.28 Rho termination factor. (a) The six subunits of the hexameric Rho helicase form a ring that is open at one end. PDB code 1PV4. (b) Proposed binding of RNA to Rho. The red ovals are the RNA-binding domains that are not present in the crystal structure shown in (a). Upon binding of the mRNA, Rho forms a closed ring, and sequential rounds of ATP hydrolysis by Rho drive structural changes that pull the RNA along the channel of the ring and away from RNA polymerase.

Figure 8.28b is adapted from Skordalakes E, and Berger JM. Structure of the Rho Transcription Terminator: Mechanism of mRNA Recognition and Helicase Loading, *Cell*, 2003;**114**:135-146.

Termination by RNA polymerase II is coupled to RNA processing

Transcription termination by RNA polymerase II is coupled to RNA processing. Just as the early stages of transcription by this enzyme are coupled to processing the 5′ end of the nascent transcript, termination of transcription is coupled to processing of the 3′ end. Most mRNAs transcribed by RNA polymerase II have a string of adenosines called a poly(A) tail that is added to the 3′ end of the transcript without the use of a DNA template. RNAs that are not polyadenylated (for example, small non-coding RNAs or the mRNA encoding histones) are processed at their 3′ ends nonetheless. The information that signals what type of processing should occur is found in the template DNA, which is in turn transcribed into RNA.

Polyadenylation of mRNA is linked to transcription termination

RNA polymerase II continues to transcribe beyond the nucleotide sequence that signals polyadenylation. As the nascent transcript exits the polymerase enzyme, the mRNA is cleaved at the poly(A) signal and the 3′ end of the mRNA is processed, yielding a polyadenylated transcript. At the same time, however, RNA polymerase II continues to transcribe. What then signals polymerase to stop and dissociate from the template?

Two models for how transcription is terminated have been proposed, and are illustrated in Figure 8.29. Recall that the RNA-processing proteins associate with both the processing 'signals' in the RNA and with the phosphorylated CTD of the RNA polymerase II large subunit. According to the 'allosteric' model shown in Figure 8.29a, recognition of the poly(A) or 3′-end processing signals, along with RNA cleavage at the processing sites, could lead to a conformational change in the polymerase enzyme that causes it to dissociate from the DNA template and release the remaining 3′ RNA fragment. ('Allostery' is a general term used to describe the way that the activity of a protein may be modulated through a change in its conformation, whereby the conformational change is stimulated by some kind of 'signal' – often the binding of a signaling molecule to the protein in question.)

A second model for how transcription is terminated, however, has emerged from the finding that termination is linked to degradation of the second RNA fragment by a 5′ to 3′ ribonuclease. This is known as the 'torpedo' model and

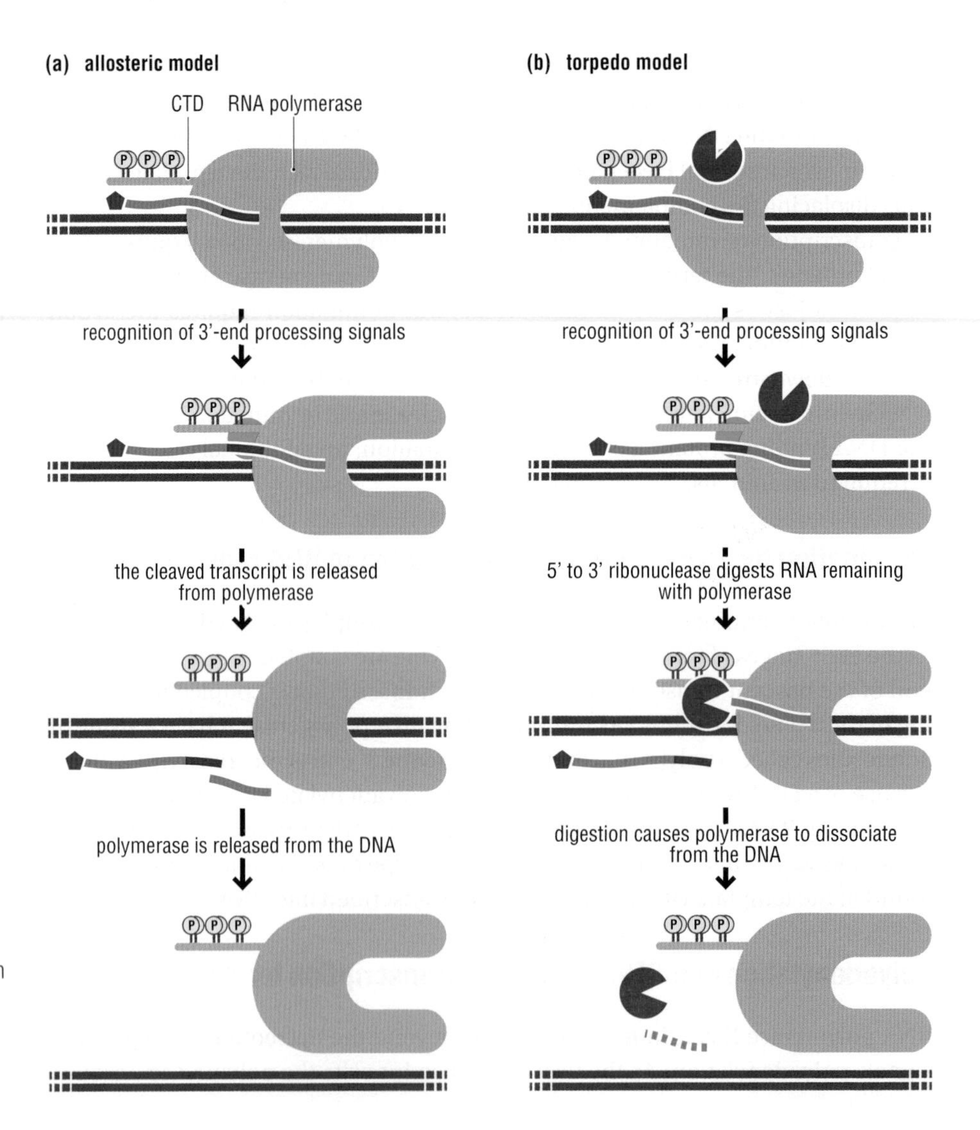
(a) allosteric model
CTD
RNA polymerase
recognition of 3'-end processing signals
the cleaved transcript is released from polymerase
polymerase is released from the DNA
(b) torpedo model
recognition of 3'-end processing signals
5' to 3' ribonuclease digests RNA remaining with polymerase
digestion causes polymerase to dissociate from the DNA

Figure 8.29 Transcription termination by eukaryotic RNA polymerase II is coupled to RNA processing. Two proposed models for transcription termination are illustrated.(a) Allosteric model: RNA polymerase II transcribes through the polyadenylation and 3′-end processing signals (pink), and RNA-processing complexes (turquoise) associate with both the processing signals and the phosphorylated CTD of the large subunit. Recognition of the processing signals and/or cleavage at the processing sites causes termination. The cleaved RNA is released and RNA polymerase dissociates from the DNA template. (b) Torpedo model: RNA downstream of the poly(A) cleavage site (blue line) is digested by a 5′ to 3′ ribonuclease (the 'Torpedo', shown in red), which remains with RNA polymerase II throughout the length of the gene. After poly(A) site cleavage, the exonuclease torpedo causes polymerase to dissociate from the DNA template.

is shown in Figure 8.29b. After the mRNA is cleaved at the poly(A) signal, a ribonuclease called Rat1 begins to degrade the nascent transcript emerging from the still-transcribing RNA polymerase II. The ribonuclease tracks along the RNA, degrading it in the 5′ to 3′ direction until it runs into the polymerase like a torpedo, disrupting polymerization and causing the enzyme to dissociate from the DNA template. It is quite possible that a combination of the torpedo and allosteric models may be responsible for transcription termination.

SUMMARY

In order for the information in the genome to be expressed, the cell must synthesize an RNA copy of the DNA - a process called transcription. Since this is the key step in gene expression, transcription is highly regulated in all cells. The RNA polymerase enzyme catalyzes the synthesis of the RNA from ribonucleotide triphosphate precursors, using the DNA template strand to assemble a copy of the coding strand. RNA is synthesized from the 5′ to the 3′ end, just as in DNA replication, in a defined series of events known as the transcription cycle: initiation, elongation, and termination. Cells can regulate the transcription of individual genes by targeting any of the steps in the transcription cycle, as we discover in the next chapter.

RNA POLYMERASES

- RNA polymerase enzymes in all three domains of life share a structurally similar core and synthesize RNA using a common mechanism.
- Bacteria and archaea have a single RNA polymerase enzyme, while eukaryotes have three:
  - RNA polymerase I - transcribes genes encoding rRNAs.
  - RNA polymerase II - transcribes all protein-coding genes as well as some non-coding regulatory RNA molecules.
  - RNA polymerase III - transcribes tRNA and other RNA molecules needed for RNA splicing.
- Plants are an exception among eukaryotes in possessing two additional RNA polymerases (IV and V) that transcribe regulatory RNA molecules.
- Transcription by RNA polymerase II is closely coupled to RNA processing. RNA polymerase II has a C-terminal domain that recruits proteins required for processing the mRNA. This recruitment is triggered by changes in the phosphorylation of residues in the CTD.
- Bacterial RNA polymerase is targeted to its promoter by the sigma subunit, which recognizes the −35 and −10 regions of the promoter.
- Eukaryotic and archaeal RNA polymerases require a set of proteins called general transcription factors that recognize a set of promoter elements, and promote some of the later events in the transcription cycle. All of these polymerases depend on a subunit called TBP for promoter recognition.
- The complex of RNA polymerase with the double-stranded DNA is called the closed complex.

STEPS IN TRANSCRIPTION

- Transcription begins with isomerization of RNA polymerase to form the open complex, in which the DNA strands are separated and synthesis of the RNA begins, using the template strand to position the incoming ribonucleotide triphosphates.
- RNA polymerase may first go through several cycles of abortive initiation, producing very short transcripts, before breaking free of the promoter and entering into the elongation phase.
- The elongating polymerase maintains an open transcription bubble consisting of a melted region of the DNA (around 10–14 base pairs) and the 8–9 most recently added nucleotides of the RNA transcript base-paired with the template DNA.
- During transcription elongation, RNA polymerase occasionally stalls to correct an error due to a misincorporated base, or because of features of the DNA or RNA that cause the polymerization to pause.
- If the stalled RNA polymerase backtracks, transcript cleavage factors promote cleavage of the most recently added ribonucleotides and thereby help to restart transcription.
- Transcription termination in bacteria can occur by two different mechanisms. In rho-independent termination, the

DNA contains an intrinsic terminator sequence that gives rise to a stem–loop in the transcribed RNA that causes transcription to terminate. Rho-dependent termination requires the action of the ATP-dependent enzyme, Rho.

- Termination of transcription by eukaryotic RNA polymerase II is coupled to the final steps in mRNA processing.
- Since eukaryotic DNA is packaged into chromatin, chromatin-modifying and nucleosome-remodeling enzymes are required for efficient transcription initiation and elongation, as well as regulation of transcription in eukaryotes.

8.1 OVERVIEW OF TRANSCRIPTION

1. Not all RNAs produced in the cell encode proteins. Explain.

2. Which of the following statements regarding RNA is NOT true?
a. RNA contains the nitrogen base uracil.
b. RNA is predominantly double stranded.
c. RNA contains the sugar ribose.
d. RNA nucleotides are connected by phosphodiester bonds.

3. Match each of the following stages of transcription with the appropriate statements.

A: Initiation; B: Elongation; C: Termination

a. RNA polymerase unwinds the DNA ahead of the transcription bubble.
b. The transcription bubble is formed during this stage.
c. RNA polymerase encounters a terminator DNA sequence.
d. RNA polymerase dissociates from the DNA template.
e. RNA polymerase synthesizes a few ribonucleotides of the RNA transcripts while remaining at the promoter.

4. What feature of eukaryotic chromosomes presents a barrier to transcription? In general, how do eukaryotic cells overcome this barrier?

Challenge question

5. Draw a picture of a DNA molecule featuring a transcription bubble that is approximately midway through transcribing a gene.
a. Add labels to your diagram for the RNA polymerase and the RNA transcript being produced.
b. Label the following positions on the DNA: +1, downstream DNA, the non-coding (template) strand of DNA, and upstream DNA.

8.2 RNA POLYMERASE CORE ENZYME

1. The subunits that make up the core RNA polymerases are capable of synthesizing RNA molecules using a DNA template. Why do they need accessory proteins to function properly?

2. Why is it significant that bacteria, archaea, and eukaryotic RNA polymerases have the highest similarity in the amino acid sequence that contains the active site?

Challenge question

3. Propose an explanation for the fact the RNA polymerase II has a C-terminal domain but RNA polymerases I and III do not.

8.3 PROMOTER RECOGNITION IN BACTERIA AND EUKARYOTES

1. In bacteria, how does RNA polymerase get oriented on a gene for transcription?

2. In bacteria, what is the relationship between sigma factors and gene regulation?

3. How do differences among promoter sequences play a role in gene regulation?

4. In eukaryotes, how does RNA polymerase get oriented on a gene for transcription?

5. Once the pre-initiation complex is formed on DNA, initiation will proceed *in vitro* but not *in vivo*. Why is this the case, and what must be added to overcome the *in vivo* problem?

6. What are the functions of the three eukaryotic RNA polymerases?

7. In addition to RNA polymerase, what protein is required for transcription initiation by all eukaryotic polymerases?

8. Match each of the following general transcription factors with their functions.

A: TFIIA; B: TFIIB; C: TFIID; D: TFIIH

a. Assist in establishing the correct direction of transcription
b. Recognition of the TATA box sequence
c. Recruitment of RNA polymerase
d. Unwinding of DNA to expose the template strand

Challenge questions

9. A hypothetical *E. coli* σ^{70} promoter has the following sequence:

TTGACA – 15 bases – TATAAT

a. What change in the level of transcription would you predict if the sequence were mutated to:

TTCCCA – 15 bases – TATAAT?

b. What additional mutations might you make to return transcription to its original level?

10. In eukaryotes, how might the level of transcription change if the Mediator were deleted?

8.4 INITIATION OF TRANSCRIPTION AND TRANSITION TO AN ELONGATING COMPLEX

1. Compare and contrast the transition from a closed to an open complex for eukaryotic RNA polymerases I, II, and III.

2. Describe the molecular events that must occur to enable RNA polymerase to transition from abortive initiation to promoter clearance.

8.5 TRANSCRIPTION ELONGATION

1. Transcriptional pausing:
a. What is transcriptional pausing?
b. Why does it happen?
c. How is it overcome?

2. What relationship exists between the phosphorylation of the CTD tail of RNA polymerase and transcriptional pausing?

Challenge questions

3. The protein eleven-nineteen lysine rich leukemia (ELL) is a known elongation factor that facilitates transcription elongation and reduced transcriptional pausing. Misregulation of ELL is also known to be associated with several forms of cancer. Propose a mechanism by which this association might occur.

4. In bacteria grown in normal conditions, the coupling of transcription and translation, along with the action of topoisomerase A (encoded by the *TopA* gene), prevents excessive supercoiling of DNA and transcriptional pausing. *TopA* mutants can grow relatively normally in optimal growth conditions, but can be easily killed by stresses such as heat shock and exposure to chemicals. Additionally, the *TopA* gene is upregulated by the increased expression of an alternative sigma factor. Propose a mechanism by which *TopA* may protect cells against stress and suggest why *TopA* mutants cannot tolerate the same stress.

8.6 TRANSCRIPTION TERMINATION

1. What is an intrinsic terminator? Briefly describe the mechanism by which intrinsic terminators end transcription.

2. Compare and contrast the allosteric and torpedo models for eukaryotic transcription termination.

EXPERIMENTAL APPROACH 8.1 – IDENTIFICATION OF MEDIATOR

1. What observation led to the hypothesis of a Mediator?

2. Briefly describe the experiment that was performed to prove the existence of Mediator.

3. What is MudPIT and how was MudPIT used to resolve inconsistency in the literature regarding the human Mediator?

EXPERIMENTAL APPROACH 8.2 – ABORTIVE INITIATION BY *E. COLI* RNA POLYMERASE

1. What was the limitation of the original experiments that showed abortive initiation?

2. What is a locked nucleic acid, and how did it help to resolve the problem above?

Challenge question

3. Explain how the three gels shown in Figure 2 differ from each other and what conclusion can be drawn based upon the results.

FURTHER READING

8.1 OVERVIEW OF TRANSCRIPTION

Cheung AC, Cramer P. A movie of RNA polymerase II transcription. *Cell*, 2012;**149**:1431–1437.

Lee TI, Young RA. Transcription of eukaryotic protein-coding genes. *Annual Review of Biochemistry*, 2000;**34**:77–137.

Orphanides G, Reinberg D. A unified theory of gene expression. *Cell*, 2002;**108**:439–451.

8.2 RNA POLYMERASE CORE ENZYME

Cramer P, Armache KJ, Baumli S, *et al.* Structure of eukaryotic RNA polymerases. *Annual Review of Biophysics*, 2008;**37**:337–352.

Murakami KS, Darst SA. Bacterial RNA polymerases: the whole story. *Current Opinion in Structural Biology*, 2003;**13**:31–39.

Paule MR, White RJ. Survey and summary: transcription by RNA polymerases I and III. *Nucleic Acids Research*, 2000;**28**: 1283–1298.

8.3 PROMOTER RECOGNITION IN BACTERIA AND EUKARYOTES

Gruber TM, Gross CA. Multiple sigma subunits and the partitioning of bacterial transcription space. *Annual Review of Microbiology*, 2003; **57**:441–466.

Hughes KT, Mathee K. The anti-sigma factors. *Annual Review of Microbiology*, 1998;**52**:231–286.

Juven-Gershon T, Hsu JY, Theisen JW, Kadonaga JT. The RNA polymerase II core promoter — the gateway to transcription. *Current Opinion in Cell Biology*, 2008;**20**:253–259.

Kornberg RD. Mediator and the mechanism of transcriptional activation. *Trends in Biochemical Sciences*, 2005;**30**:235–239.

Österberg S, del Peso-Santos T, Shingler V. Regulation of alternative sigma factor use. *Annual Reviews Microbiology*, 2011;**65**:37–55.

8.4 INITIATION OF TRANSCRIPTION AND TRANSITION TO AN ELONGATING COMPLEX

Nechaev S, Adelman K. Pol II waiting in the starting gates: Regulating the transition from transcription initiation into productive elongation. *Biochim Biophys Acta*, 2011;**1809**:34–45.

Saunders A, Core LJ, Lis JT. Breaking barriers to transcription elongation. *Nature Reviews Molecular Cell Biology*, 2006;**7**:557–567.

8.5 TRANSCRIPTION ELONGATION

Borukhov S, Lee J, Laptenko O. Bacterial transcription elongation factors: new insights into molecular mechanism of action. *Molecular Microbiology*, 2005;**55**:1315–1324.

Gilmour DS. Promoter proximal pausing on genes in metazoans. *Chromosoma*, 2009;**118**:1–10.

Selth LA, Li T, Sigurdsson S, Svejstrup JQ. Transcript Elongation by RNA Polymerase II. *Annual Review of Biochemistry*, 2010;**79**:271–293

Workman JL. Nucleosome displacement in transcription. *Genes and Development*, 2006;**20**:2009–2017.

8.6 TRANSCRIPTION TERMINATION

Kuehner JN, Pearson EL, Moore C. Unravelling the means to an end: RNA polymerase II transcription termination. *Nature Reviews. Molecular Cell Biology*, 2011;**12**:283–294.

Santangelo TJ, Artsimovitch I. Termination and antitermination: RNA polymerase runs a stop sign. *Nature Reviews. Microbiology*, 2011;**9**: 319–329.

9 Regulation of transcription

INTRODUCTION

In the previous chapter, we explored the mechanisms by which genes are transcribed into RNA: messenger RNA (mRNA), which contains the information that is translated into a polypeptide chain by the ribosome, or other RNA molecules that serve a variety of functions in the cell. Yet, while every gene in a genome plays a role in the life cycle of the organism, the need for a particular gene product – whether it is a protein or an RNA molecule – can vary greatly with time, stage of development, environment, and cell type. The reason a skin cell is different from a neuron is that each cell type expresses a different subset of genes, giving each tissue type its unique characteristics. Regulation of transcription, and thus of gene expression, therefore lies at the heart of cell differentiation and development.

The importance of transcriptional regulation in determining cell identity is highlighted by the remarkable finding that just four proteins that regulate transcription are needed to 'reprogram' B lymphocytes (which play a role in the immune response) into pluripotent stem cells, which are capable of differentiating into any tissue type. The importance of transcription regulation also is highlighted by the dramatic phenotypic consequences of mutating a single gene encoding the transcription factor agamous, which dramatically alters the appearance of the flowering plant, *Arabidopsis thaliana* (Figure 9.1). In this chapter, we will learn about a remarkable array of mechanisms for regulating transcription of individual genes, thereby ensuring that appropriate levels of protein and RNA molecules are available in the cell.

9.1 PRINCIPLES OF TRANSCRIPTION REGULATION

In this chapter, we begin by exploring the principal ways in which transcription is regulated in bacteria, eukaryotes, and archaea, identifying both the differences between the three branches of life, and the common themes and strategies that unite them. We cover the key principles that underpin the process of transcriptional regulation, and the key players that participate in this process. We go on to explore specific mechanisms by which transcription is regulated, in both bacteria and eukaryotes, later in the chapter.

Any step in the transcription cycle can be regulated

The RNA polymerase holoenzyme can transcribe any gene that has an intact promoter. Yet at any given time, the relative rate at which a complete transcript is generated varies tremendously among different genes. While some of these variations are due to intrinsic differences in promoter strength – and hence how avidly

Figure 9.1 A mutation in a transcription factor alters a flower's appearance. The shape and arrangement of flower petals in the plant, *Arabidopsis thaliana*, change dramatically when a transcription regulator called agamous is mutated or deleted.

From Weigel J. Flower development: repressing reproduction. *Current Biology*, 1997;**7**: r373–r375.

RNA polymerase binds to DNA and changes conformation to give the elongating complex – most of the differences are due to targeted gene regulation.

Any step in the transcription cycle, from binding of the RNA polymerase enzyme to DNA to release of the complete transcript, can be regulated. Cells of all types contain a remarkable array of proteins that control the relative amount of transcription from each promoter, increasing or decreasing the rate at which RNA transcripts of selected genes are produced. Regulatory proteins that decrease the rate of transcription initiation are termed **repressors**, while proteins that increase transcription are denoted **activators.**

Most gene regulation occurs at the level of transcription initiation. However, we will also see examples in which later steps in the transcription cycle, such as elongation and termination, are regulated to either increase or decrease synthesis of a complete transcript.

The transcribed RNA itself can also play a role in regulating transcription. Some RNA transcripts fold to adopt structured regions that can inhibit or promote transcription. The structures adopted can be influenced by the binding of proteins or small metabolites that exert control over transcription through their effect on RNA structure. In addition, there are RNA transcripts whose sole function is to regulate gene expression rather than to encode a functional protein or RNA molecule. We will see several examples of RNA-mediated regulation later in this chapter.

Proteins that bind to DNA target specific genes for regulation

Most genes are regulated by proteins that bind to the chromosome in the vicinity of the transcribed gene. Specific regions in the chromosome called **regulatory sequences** are the sites in the DNA to which regulatory proteins bind. Regulatory sequences are most commonly located upstream of the promoter, and can be immediately adjacent to the promoter (or even overlapping) or up to many kilobases away. A gene might be regulated by a single protein, or by several different proteins that bind to distinct sites within the regulatory sequence.

In bacteria, the DNA sequences recognized by transcriptional regulators are termed **operator** sites. This term was originally used exclusively for repressors but has come to refer to DNA sites to which any transcriptional regulator binds. Bacterial operators are located upstream of the transcription start site, typically close to or overlapping the promoter, whereas sites to which activators bind lie just upstream; these differences are illustrated in Figure 9.2a and b. This arrangement allows the regulatory

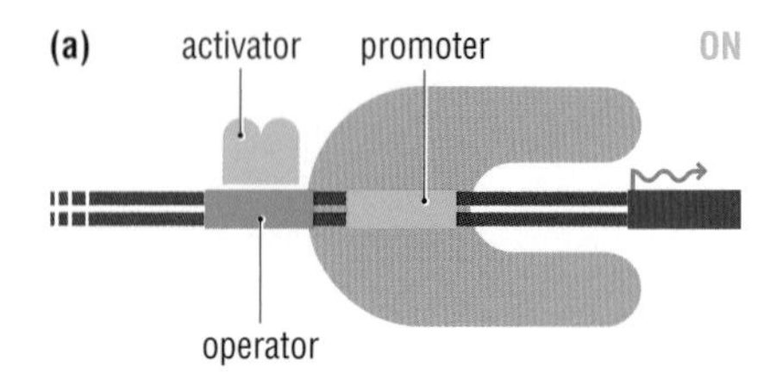

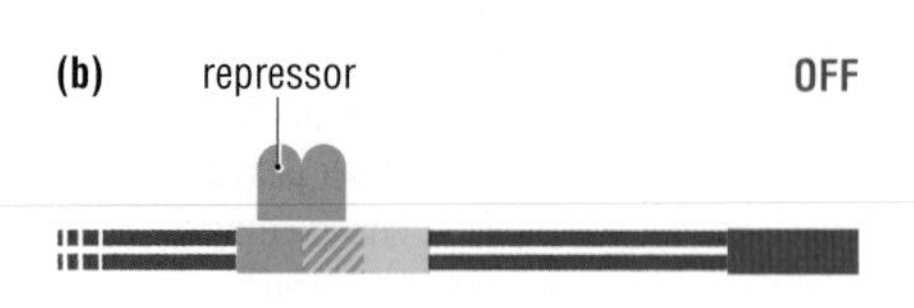

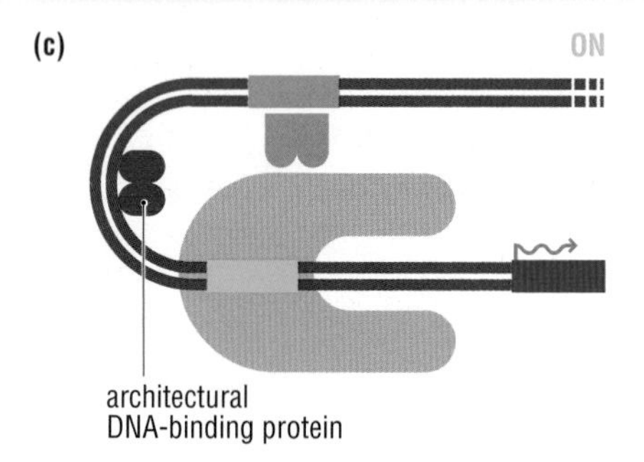

Figure 9.2 The position of different regulatory sequences relative to bacterial promoters. (a) A regulatory sequence located just upstream (5′) of the promoter region is recognized by a DNA-binding protein that activates transcription from the promoter by helping RNA polymerase to bind. (b) The regulatory sequence overlaps the promoter. In this case, binding of the regulatory protein to its DNA site can block binding of RNA polymerase to the promoter, thereby repressing transcription. (c) Regulatory proteins that bind several hundred or more base pairs upstream influence transcription of the downstream promoter as the result of DNA looping, which brings the regulatory proteins in contact with the downstream polymerase enzyme. An architectural DNA-binding protein can promote looping by bending the DNA.

proteins to interact with the RNA polymerase enzyme at the promoter site. However, some bacterial regulatory sequences are located several hundred or more base pairs away from the promoter. In this case, the intervening DNA must loop out to allow the regulatory protein to interact with the polymerase enzyme, as illustrated in Figure 9.2c. Proteins called **architectural DNA-binding proteins** sometimes assist in the process of looping out DNA by promoting the bending of the intervening sequence.

Complex regulatory sequences located far from the gene are common in higher eukaryotes

Regulatory sequences that control eukaryotic genes are typically located farther from the promoter than in bacteria. It is also far more common for a eukaryotic gene to be regulated by multiple proteins. In multicellular organisms, regulatory sequences can encompass binding sites for six or more individual proteins and may be located at great distances from the promoter region – sometimes several thousand base pairs or more away, as illustrated in Figure 9.3. These sequences, called **enhancers**, bind proteins that activate transcription. An enhancer can be located either upstream or downstream of the target gene. The region of the chromosome over which an enhancer acts can be delimited by an insulator sequence, which we learned about in Chapter 4.

Enhancers are able to regulate transcription at distant promoters because the DNA forms a loop that allows proteins bound to the enhancer region to contact proteins bound to the promoter region (see Figure 9.3c). In higher eukaryotes, there are even more complex regulatory regions known as **locus control regions**, which contain a combination of enhancer and insulator elements. For example, transcription of the globin genes, which encode different subunits of hemoglobin during the course of development, is under control of a locus control region, as illustrated in Figure 9.4.

Eukaryotic DNA-binding proteins recruit protein complexes that regulate transcription

In eukaryotic cells, DNA-binding proteins typically do not control transcription directly, but rather exert their function by recruiting additional protein complexes that either activate or repress transcription. These complexes, denoted **co-activators** or **co-repressors**, are not able to bind DNA on their own and must therefore be brought to the regulatory region by the DNA-binding proteins, as denoted in Figure 9.5. Most DNA-binding proteins selectively recruit particular co-activators or co-repressors. These intermediary proteins or protein complexes can interact directly with the transcription machinery or harbor enzymatic activities that modify chromatin.

Proteins that regulate transcription in eukaryotes are typically modular in nature, with a DNA-binding domain that recognizes specific DNA sequences and a separate activation or repression domain that recruits a co-activator or co-repressor, respectively (see Figure 9.5). We know the roles of the two domains are

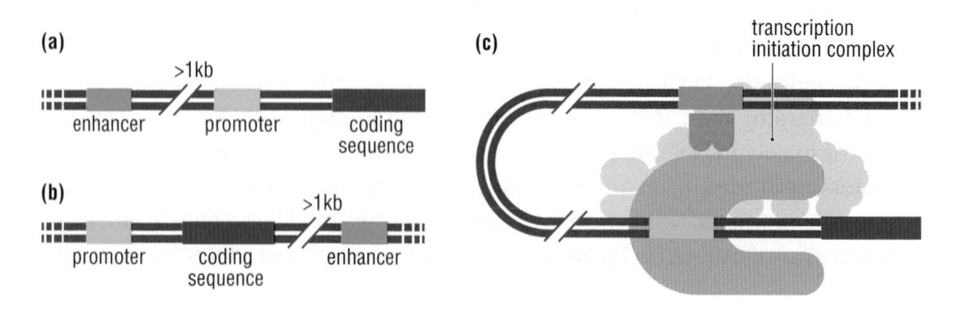

Figure 9.3 Eukaryotic enhancers can be located far from the promoter. Enhancers can be located several thousand base pairs (a) upstream or (b) downstream of the promoter and coding regions. (c) Looping of the DNA that separates the enhancer region from the promoter enables the enhancer-bound regulatory proteins to contact the transcription initiation complex.

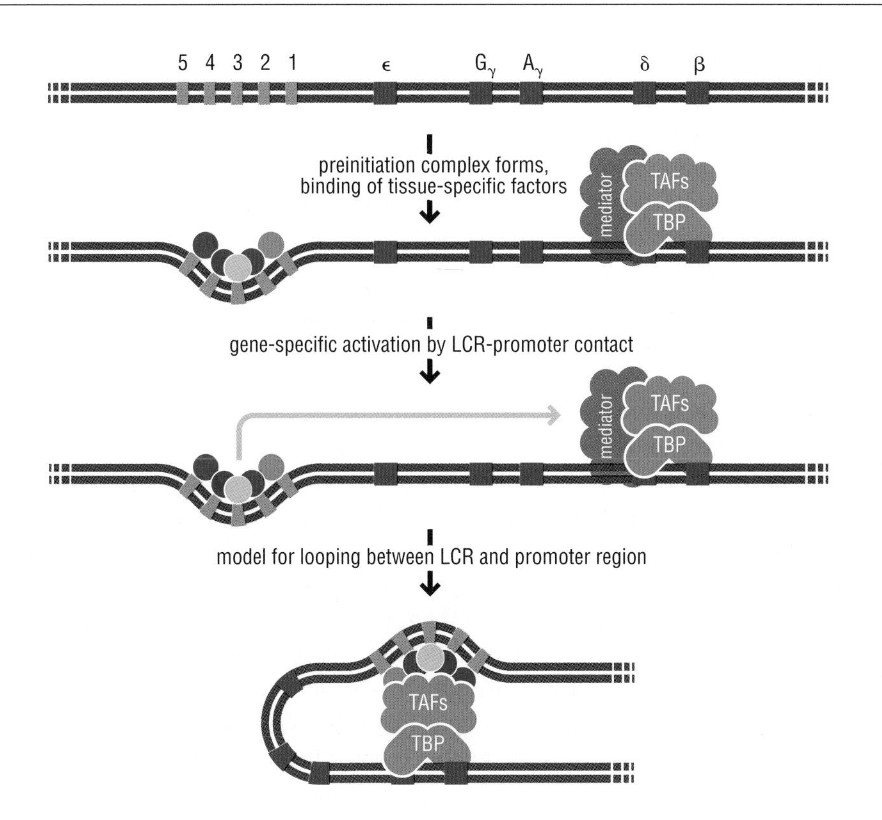

Figure 9.4 Locus control region (LCR) of the beta-globin genes. Model for activation of beta-globin genes, ε, Gγ, Aγ, δ and β, by proteins bound to the locus control region. Several different proteins bind to the locus control region, leading to recruitment of additional proteins and activating transcription of the downstream gene. One model is that DNA looping brings the LCR near the promoter region, where the bound proteins can promote assembly of a transcription initiation complex.

Adapted from Spector, D.L., and Gasser, S.M. A molecular dissection of nuclear function conference on the dynamic nucleus: questions and implications. *EMBO Reports*, 2003;**4**:18–23.

separable because it is possible to generate functional chimeric proteins with the DNA-binding domain from one protein and the activation domain from another. Additional distinct domains that serve additional functions, such as dimerization or binding to additional regulatory proteins, can also be found in some proteins that regulate eukaryotic transcription.

Chromatin modifications play a central role in the regulation of eukaryotic transcription

All aspects of transcription in eukaryotes are affected by the fact that the DNA is packaged into chromatin. We learned in Section 8.1 that the packaging of eukaryotic DNA into chromatin affects transcription initiation and elongation, and that chromatin structure must therefore be altered through the combined action of adenosine triphosphate (ATP)-dependent nucleosome remodelers, enzymes that mediate post-translational modification of histones, and histone chaperones. The activities of all of these enzymes must be coordinated in order to transcribe a gene. An important mechanism by which genes are activated or repressed in eukaryotes is the recruitment of enzymes that modify chromatin.

We learned in Section 4.4 that regions of chromatin that are **hyperacetylated** (that is, are heavily acetylated) are sites of active transcription, whereas low rates of transcription are found in regions of **hypoacetylated** chromatin (those sites that are acetylated at low levels). The activation of many eukaryotic genes is linked to recruitment of histone acetyltransferase enzymes, also known as HATs, which acetylate lysine residues within specific histone proteins. Histone acetylation, in turn can help to recruit proteins that contain bromodomains, which bind specifically to acetyl-lysine. Thus, for example, the SWI/SNF nucleosome-remodeling enzyme, which contains a bromodomain, is recruited to acetylated regions of chromatin. Similarly, genes can be repressed by recruitment of histone deacetylase enzymes, known as

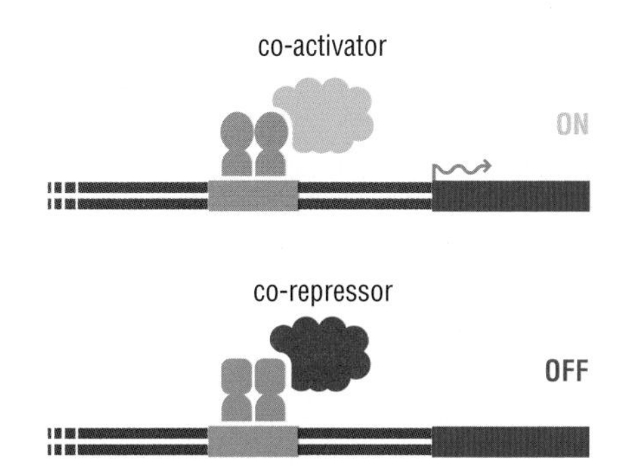

Figure 9.5 Recruitment of co-activator and co-repressor proteins to the DNA. Some sequence-specific DNA-binding proteins can recruit co-activator proteins to the DNA, which activate transcription of downstream genes. The proteins that recruit co-activators typically have a separate DNA-binding domain and an activation domain that recruits the co-activator. Other DNA-binding proteins recruit co-repressors, which repress transcription of downstream genes.

➲ We give a comprehensive discussion of chromatin structure, histone modifications, and nucleosome remodeling in Chapter 4.

HDACs. Since HATs and HDACs also modify lysines in other proteins, these enzymes are also referred to as KATs and KDACs, where K is the one-letter symbol for lysine.

Post-translational modifications other than acetylation also play important roles in transcription regulation. Such modifications include methylation, phosphorylation, ubiquitination, and sumoylation. Like acetylation, these histone modifications both alter the chemical properties of side chains and create binding sites for specialized proteins that bind specifically to the modified side chains. However, there is no simple correlation between any of these modifications and transcriptional activity. We learned in Section 4.5 that methylation of histone tails, for example, is correlated with both repressed *and* activated transcription depending on the lysine residue that is modified.

Although there is much yet to be discovered about the precise set of chromatin modifications needed to transcribe any given gene, a picture has been emerging thanks to modern methods for screening whole genomes for the presence of particular histone modifications, variant histones, and DNA methylation. The observed connection between particular histone modifications and subsequent outcomes has led to the proposal that there is a 'histone code' consisting of particular combinations of covalent modifications that specify precise outcomes. Indeed, characteristic patterns of histone modifications have been observed in actively transcribed genes. As an example, Figure 9.6 shows the histone modifications found across a typical yeast gene that is actively transcribed.

The activity of regulatory proteins can be modified

The transcription of many genes is altered in response to a variety of changing conditions. One way in which gene expression can be altered in response to an external signal is through a direct change in the activity of the DNA-binding protein. The transcription of some genes can be altered by small molecules that bind directly to regulatory proteins and act as **allosteric effectors**. Allosteric effectors trigger a conformational change in the proteins to which they bind, altering the protein's DNA-binding properties or its interactions with other proteins. The nucleotide cyclic adenosine monophosphate (cAMP) and the hormone estrogen are two examples of allosteric effectors that bind to transcription regulators and alter the way in which they modulate transcription.

Covalent modification of proteins that regulate transcription, such as phosphorylation or acetylation, can also be used to alter the DNA-binding properties of a protein or its effect on transcription. These modifications may be the end result of a signaling cascade, a series of biochemical events triggered by an extracellular signal that we describe in more detail in Section 9.8. Whether or not a transcriptional regulator is localized in the nucleus and is thus available to bind DNA and regulate

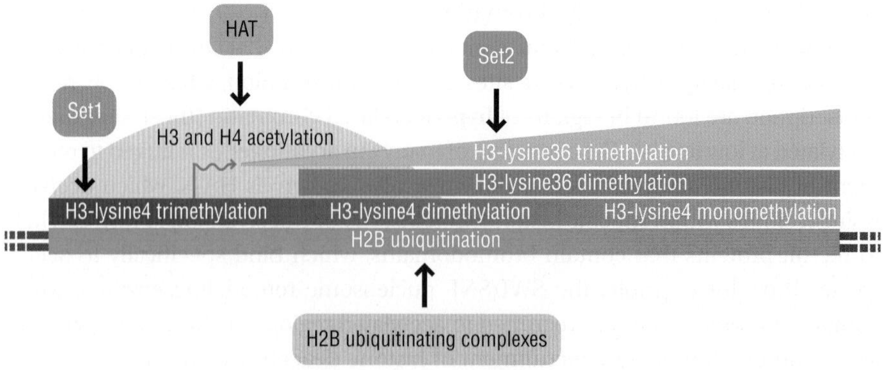

Figure 9.6 The pattern of histone modifications across a eukaryotic gene. The pattern of histone modifications found across the promoter and transcribed region of a yeast gene is shown, along with the enzymes responsible for the modifications.

Reproduced from Saunders, A. Core, LJ, and Lis, JT. Breaking barriers to transcription elongation. *Nature Reviews Molecular Cell Biology*, 2006;**7**: 557–567.

transcription can be controlled by regulating nuclear import or export. As we will see, multiple strategies are often used to fine-tune regulation of a single gene.

9.2 DNA-BINDING DOMAINS IN PROTEINS THAT REGULATE TRANSCRIPTION

Proteins that regulate transcription contain DNA-binding domains that recognize specific DNA sequences, thus targeting them to particular regulatory sites. The specificity of this interaction – between DNA-binding protein and target DNA sequence – is central to the appropriate regulation of gene expression. It is therefore worth pausing to consider the range of DNA-binding domains that are found in transcriptional regulators, and the commonalities and differences that exist between them.

Sequence-specific DNA-binding proteins can be classified broadly by different DNA-binding motifs

A remarkable variety of domains that recognize specific DNA sequences are found among transcription regulators. Called **DNA-binding motifs**, these protein folds are tailored to accommodate the structural features characteristic of B-DNA. Some of these motifs are found universally in all organisms, while others appeared later in evolution and may be found just in eukaryotes, or only in metazoans (multicellular organisms). Here we touch upon some of the most common DNA-binding motifs and how they interact with DNA.

➔ We learn more about the principles governing the way in which proteins bind selectively to particular DNA sequences in Section 2.12.

The helix-turn-helix is a common DNA-binding motif found in all organisms

The **helix-turn-helix** was the first DNA-binding motif to be identified. It consists of two alpha helices related by a fixed angle and connected by a tight bend (the 'turn'), as illustrated in Figure 9.7. The second, the C-terminal helix, is the **recognition helix** that fits in the major groove of the DNA and forms contacts with the base pairs that

➔ We learned about how proteins recognize specific DNA sequences in Section 2.12.

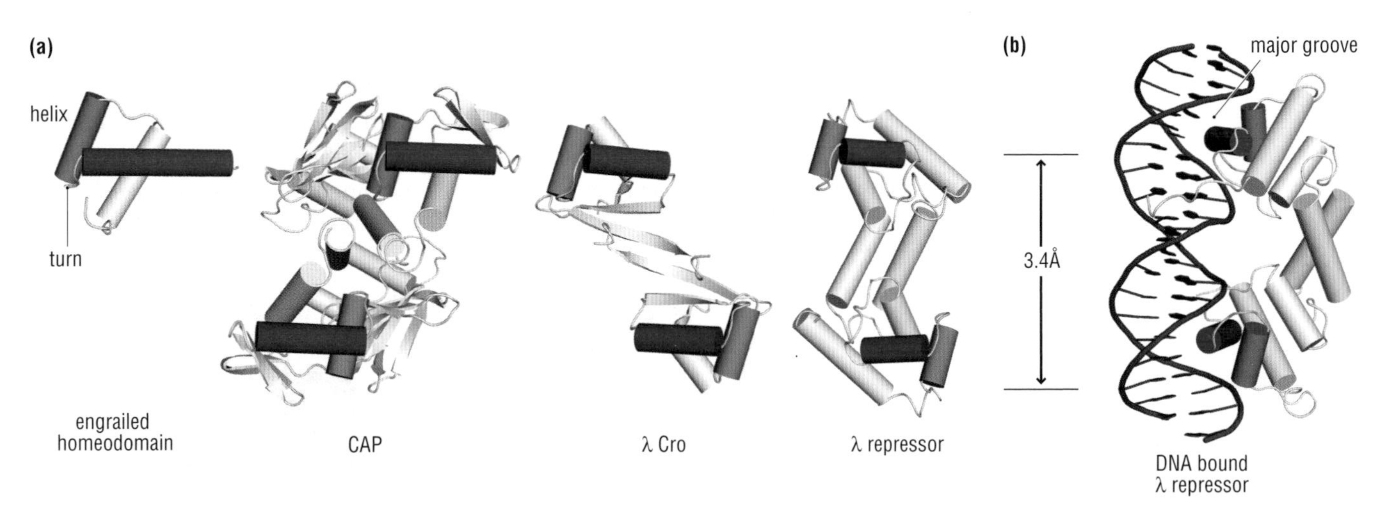

Figure 9.7 Examples of helix-turn-helix proteins. (a) The helix-turn-helix in each protein is shown in blue (first helix) and red (second, recognition helix). The *E. coli* CAP protein (also called CRP) (Protein Data Bank (PDB) code 1CGP), phage λ Cro (PDB code 1CRO) and phage λ repressor (or cI protein; PDB code 1LMB) are dimers in which the respective recognition helices are separated by 34 Å (3.4 nm). The engrailed homeodomain protein from *Drosophila* (PDB code 1HDD), which also contains a helix-turn-helix, can bind DNA as a monomer. (b) The structure of λ repressor bound to DNA. In addition to the helix-turn-helix, each monomer has an N-terminal 'arm' that binds in the major groove on the opposite face of the DNA.

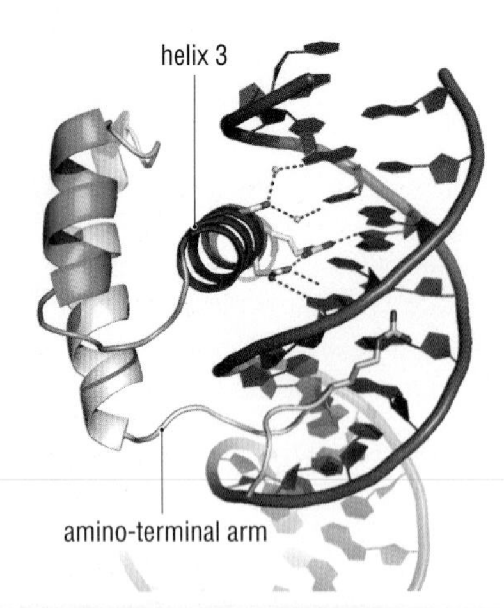

Figure 9.8 The MATα2 homeodomain bound to DNA. Helix 3 (red) of the MATα2 homeodomain lies in the major groove of the DNA, where side chains form contacts with individual base pairs. An N-terminal arm (green) forms additional contacts in the minor groove of the DNA (PDB code 1APL).

'read' the DNA sequence. As can be seen in Figure 9.7, the helix-turn-helix is found in a variety of protein folds that otherwise have little in common. Many helix-turn-helix proteins are dimers that position the recognition helices of the two monomers 34 Å apart, which matches the spacing of successive major grooves in B-DNA. The DNA sequence that is recognized has a symmetry that roughly matches that of the dimer. In this way, each monomer forms very similar side chain contacts with each half of the binding site in DNA.

The **homeodomain**, one of the most common types of DNA-binding domains in eukaryotes, is a monomeric helix-turn-helix protein whose recognition helix is somewhat longer than its bacterial counterparts; its structure is shown in Figure 9.7a. An N-terminal arm that inserts into the adjacent minor groove allows homeodomain proteins to form additional contacts with DNA bases (Figure 9.8). We will see later that some homeodomain proteins actually bind DNA cooperatively with other DNA-binding proteins, thereby forming a complex that binds DNA very tightly and with high specificity for particular DNA sequences.

The zinc finger is the most common DNA-binding domain in the human genome

A number of different DNA-binding folds are organized around one or more zinc ions whose role is to stabilize the overall structure of the protein. Most prominent of these is the **zinc finger**, a small domain of about 30 amino acids whose single alpha helix and two antiparallel beta strands are held together by a central zinc ion, as depicted in Figure 9.9a. The zinc is coordinated by two cysteine and two histidine residues that are universally conserved in this family of proteins; these domains are therefore known as Cys_2His_2 zinc fingers.

Figure 9.9b illustrates how a single protein will typically contain several zinc fingers arranged in tandem, with each domain recognizing two to three base pairs. Each zinc finger inserts its alpha helix end-on into the major groove, where side chains form base contacts. Variation in the number of zinc fingers per protein, as well as the amino acid sequence of each finger, allows different proteins to recognize DNA sequences of differing length and sequence. It is perhaps for this reason that the zinc finger is the most common DNA-binding domain encoded in the human genome. The modular nature of these proteins has been exploited to engineer 'designer' zinc finger proteins with different DNA sequence specificities.

A very different class of DNA-binding protein that contains structural repeats, the TALE (transcription activator-like effector) proteins found in certain plant pathogens, has proven even more amenable to being re-engineered to recognize

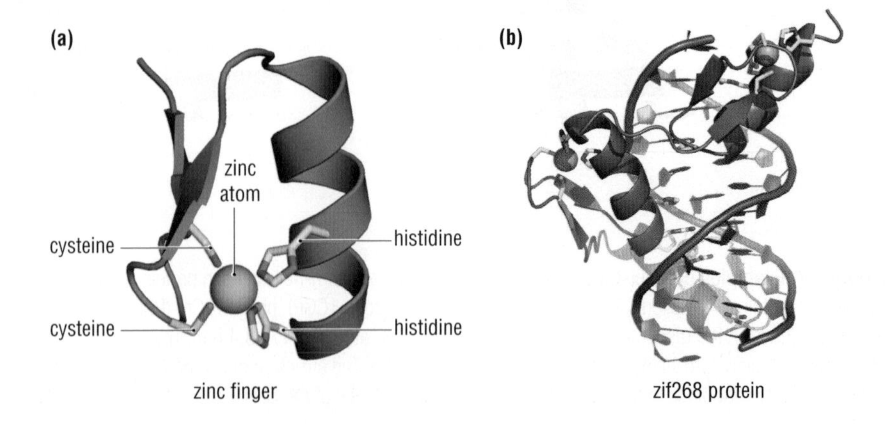

Figure 9.9 The zinc finger domain. (a) A zinc finger contains a central zinc atom, coordinated by two histidine and two cysteine side chains, which stabilizes the entire structure. (b) Zinc finger domains typically occur in multiple copies, with each domain recognizing a short DNA sequence. In zif268, it is the end of each alpha helix that inserts into the major groove, with each zinc finger contacting two to three base pairs (PDB code 1ZAA).

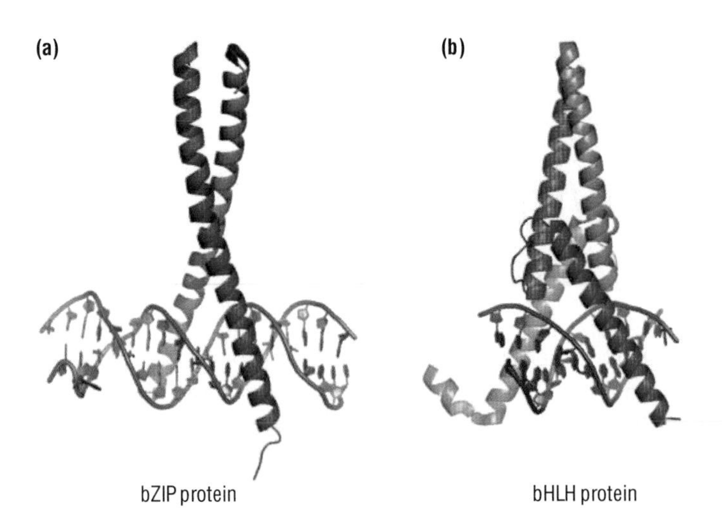

Figure 9.10 Coiled coils in leucine zipper and helix-loop-helix proteins. Both these classes of DNA-binding domains have in common a coiled coil, which consists of two alpha helices that gently coil around one another and interact through a series of hydrophobic residues. (a) In bZIP proteins, each monomer is a single helix that extends into the major groove of the DNA. (b) In bHLH proteins, each monomer consists of two helices joined by a loop. One portion of the dimerization interface consists of a four-helix bundle, in addition to the coiled coil interface. (PDB codes 1YSA, 1HLO).

particular sequences. TALE proteins contain many repeats of a small structural element, each of which recognizes a single base pair. These proteins are described in Experimental approach 9.1. The one-to-one correspondence between structural repeats and contacted base pairs makes it possible to engineer TALE proteins to recognize a particular DNA sequence.

Basic region-leucine zipper and helix-loop-helix motifs mediate DNA binding and dimerization of eukaryotic regulators

Coiled coils, which are pairs of parallel alpha helices that gently wind around each other, form an integral part of some DNA-binding domains. For instance, basic region-leucine zipper (bZIP) proteins consist of long, uninterrupted alpha helices of about 60 residues. These proteins associate via their C-terminal halves, forming a parallel coiled coil with leucine residues at the hydrophobic dimer interface, as illustrated in Figure 9.10. (The leucine residues lie along one face of each alpha helix and 'zip' together when the helices associate, stabilizing the coiled coil structure.) The N-terminal portions of the dimerized helices splay out and insert into the major groove on either side of the DNA.

A striking feature of these proteins is that the helical structure of the DNA-binding region is coupled to DNA binding: these residues are unstructured in the absence of DNA and only form the organized DNA-binding motif in the presence of DNA.

Basic region-helix-loop-helix (bHLH) proteins also contain a coiled coil that helps to mediate dimerization, and an alpha helix that lies in the major groove of DNA, as shown in Figure 9.10b. Each bHLH monomer consists of two helices joined by a loop. Portions of the four helices – two from each monomer – meet to form what is known as a four-helix bundle, which forms part of the dimer interface. The coiled coil that extends past the four-helix bundle mediates additional dimer contacts. The basic region-leucine zipper and helix-loop-helix protein families have many members that form both homodimers and heterodimers.

DNA recognition can also be mediated by beta sheets or loops

All the structures described so far recognize DNA through alpha-helical elements. As mentioned above, the major groove of B-DNA can also accommodate a pair of beta strands. The *Escherichia coli* MetJ repressor protein binds DNA using two antiparallel beta strands, which fit into the major groove of B-DNA, as

9.1 EXPERIMENTAL APPROACH

DNA sequence recognition by TAL effector proteins is mediated by tandem structural repeats

Plant pathogens use TAL effectors to regulate host genes

A variety of pathogens can infect plants and alter host cell gene expression, activating or repressing transcription of genes that favor growth and propagation of the pathogen. One such pathogen is the bacterium, *Xanthomonas campestris* pv. *vesicatoria* (*Xcv*), which infects pepper and tomato plants. *Xanthomonas* expresses a distinctive type of protein called a transcription activator-like (TAL) effector, which is injected into plant cells, where it activates genes that promote colonization and spread of the bacteria. These proteins turn out to have a structurally distinctive manner of recognizing particular DNA sequences that has made TAL effector proteins attractive targets for protein engineering aimed at designing proteins that recognize a particular DNA sequence.

A one-to-one correspondence between TAL repeats and DNA base pair recognition

TAL effector proteins consist of multiple tandem repeats of a ~34 amino acid sequence, with most containing roughly 1–2 dozen of these repeats. Bonas and colleagues studied the DNA-binding properties of these proteins and noted that the amino acid sequence of each repeat is highly conserved, with the notable exception of two amino acids. Using a bioinformatics approach, they were able to align sequences of TAL effector proteins as well as the DNA sequences of known *in vivo* binding sites to arrive at a likely correspondence between different amino acid pairs and particular DNA bases. A graphical depiction of the correspondence is shown in Figure 1. The investigators were able to verify their observations by examining the effects of mutations

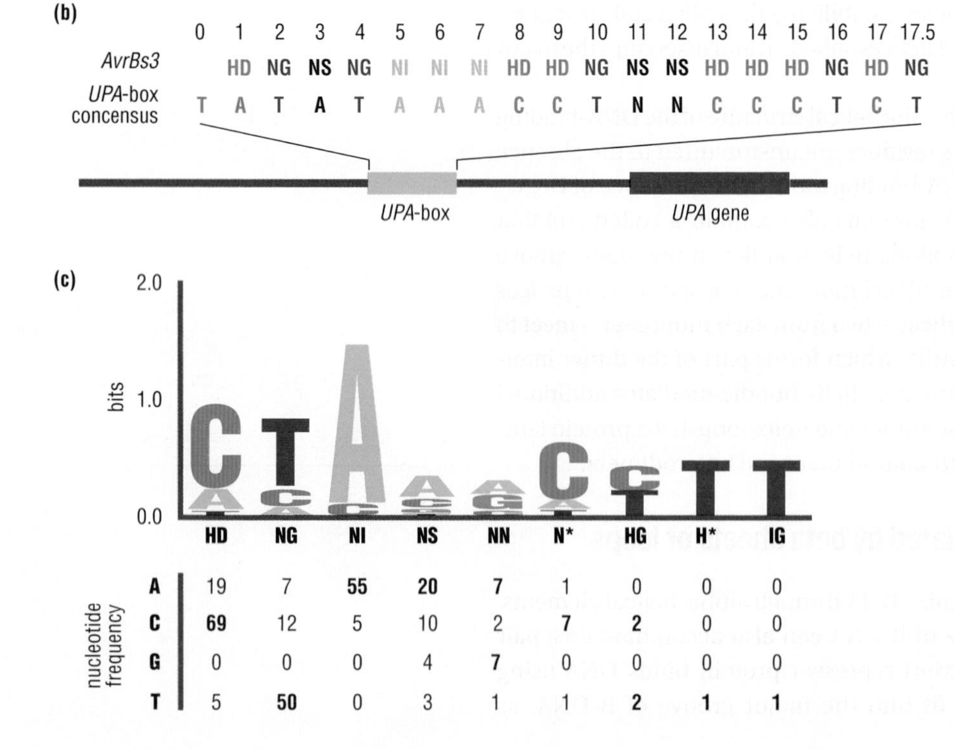

Figure 1 Model for DNA-target specificity of TAL effectors. (a) TAL effectors contain central tandem repeats, nuclear localization signals (NLSs), and an activation domain (AD). Shown is the amino acid sequence of the first repeat of AvrBs3. Hypervariable amino acids 12 and 13 are shaded in yellow. (b) Hypervariable amino acids at position 12 and 13 are aligned to a consensus DNA sequence. (c) Repeats of TAL effectors and predicted target sequences in promoters of induced genes were aligned manually. The size of the colored letter (A, T, G, and C) indicates how likely it is that the given base occurs at that position in the TAL effector DNA-binding sites. Below each position is the identity of the hypervariable amino acids predicted to contact each base pair. Asterisks indicate a TAL repeat missing amino acid 13 of the hypervariable repeat. Highest nucleotide frequencies are in bold.

Reproduced Boch *et al.* Breaking the code of DNA binding specificity of TAL-type III effectors. *Science* 2009; **326**:1509–1512.

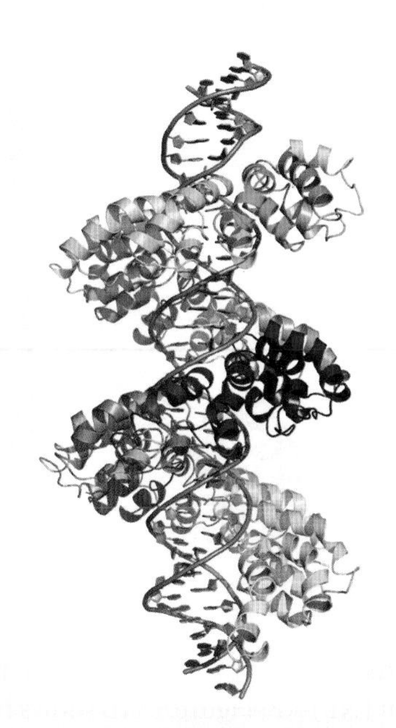

Figure 2 Structure of a TAL effector protein bound to DNA. (a) Side view. (b) Top view.

From Mak *et al.* The crystal structure of TAL effector PthXo1 bound to its DNA target. *Science*, 2012;**335**:716–719.

on DNA binding, thus arriving at a TAL effector DNA recognition code: HD = C; NG = T; NI = A; NS = A, C, G, or T; NN = A or G; and IG = T.

Structures show that TAL repeats spiral around the DNA double helix

Crystal structures of TAL effector proteins bound to DNA revealed the unique way in which TAL effector proteins recognize DNA base pairs. Barry Stoddard and colleagues determined a crystal structure showing the remarkable and beautiful spiral arrangement of the TAL repeats (Figure 2) and how they mediate interactions with DNA. Each structural repeat forms a pair of alpha helices shown in Figure 3a. These repeats are arranged in a right-handed helical arrangement whose dimensions track the major groove of B-DNA (Figure 2a). Within each repeat, the variable residues responsible for DNA recognition are located in the bend separating the two helices. These residues are inserted in the major groove of the B-DNA helix, with each TAL repeat contact a single base pair (Figure 3b).

The modular nature of TAL effector proteins naturally lends itself to protein engineering aimed at designing proteins that will recognize specific DNA sequences. Given a particular DNA sequence, genetic engineering can be used to generate a protein with a given number of TAL repeats corresponding to the length of the site to be recognized. By introducing the appropriate side chains in the DNA-contacting bend of each repeat, a 'designer' TAL effector protein can be engineered. These sequence-specific DNA-binding domains can then be fused to another domain, for example an endonuclease, that can be targeted to specific DNA sequences in a cell. We learn in Section 19.4 how designer TAL proteins are used to target endonucleases to specific sites in the genome.

Find out more

Boch J, Scholze H, Schornack S, *et al.* Breaking the code of DNA binding specificity of TAL-type, III, effectors. *Science*, 2009;**326**:1509–1512.

Mak AN, Bradley P, Cernadas RA, *et al.* The crystal structure of TAL effector PthXo1 bound to its DNA target. *Science*, 2012;**335**:716–719.

Deng D, Yan C, Pan X, *et al.* Structural basis for sequence-specific recognition of DNA by TAL effectors. *Science*, 2012;**335**:720–723.

Related techniques

Genome editing; Section 19.5

Sequence alignment; Section 19.15

Crystal structure determination; Section 19.14

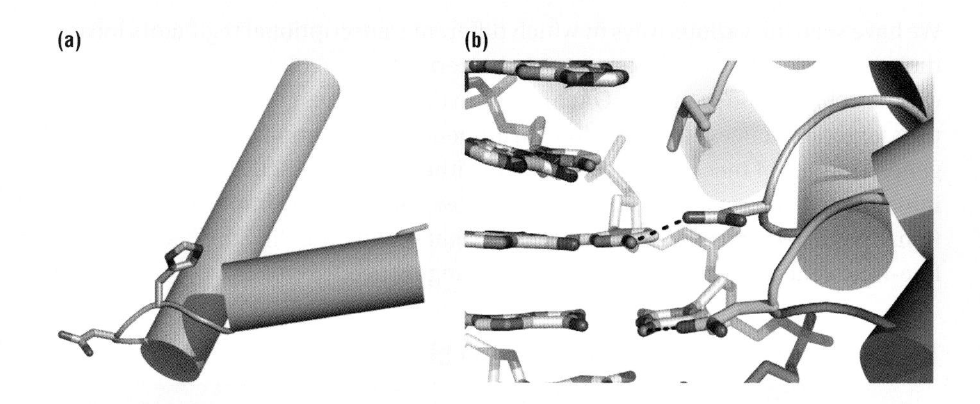

Figure 3 Topology and contacts between TAL effector repeats and DNA bases. (a) A single TAL repeat. (b) Three successive TAL repeats contact three bases in the major groove. Each repeat is colored differently. (PDB code 3UGM).

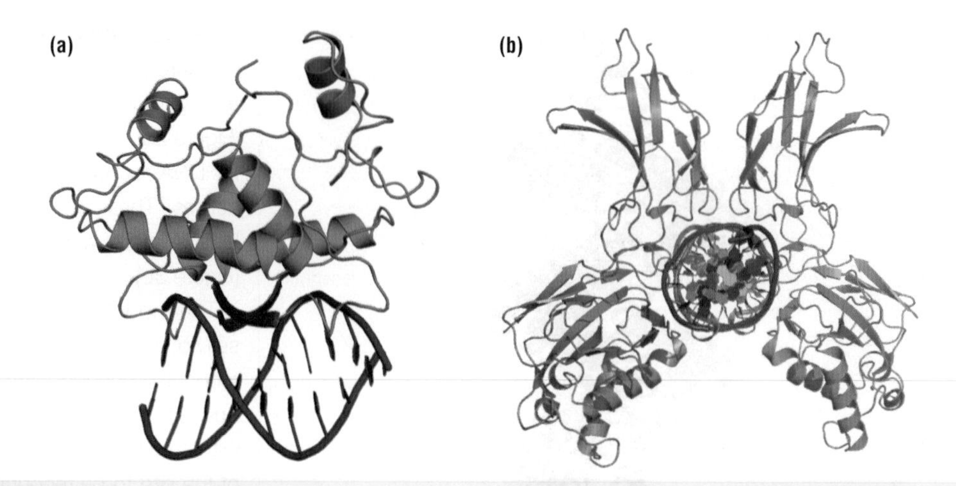

Figure 9.11 Beta sheets in DNA recognition. (a) The MetJ repressor binds DNA by inserting two antiparallel β strands in the major groove. (PDB code 1CMA) (b) NF-κB belongs to the Rel homology domain class of proteins, which contact DNA via a set of loops. (PDB code 1SVC)

shown in Figure 9.11a. The two proteins, p50 and p65, comprise the mammalian transcriptional regulator, nuclear factor (NF)-κB; Figure 9.11b shows how they are both composed mostly of beta sheet. The two proteins belong to what is denoted the Rel homology domain family of DNA-binding proteins, which all resemble an immunoglobulin protein. Immunoglobulin proteins are composed entirely of beta sheet and connecting loops, and play a variety of roles in the immune response. Rel homology domains use the loops connecting successive beta strands to contact the DNA.

We have now reviewed the key structural features that mediate protein–DNA interactions, which underpin the regulation of transcription. But how do regulatory proteins effect regulation in practice? What mechanisms are used by different organisms to ensure that transcriptional regulation occurs as it should? We discover the answers to these questions in the next sections of this chapter.

9.3 MECHANISMS FOR REGULATING TRANSCRIPTION INITIATION IN BACTERIA

We have seen the various ways in which different transcriptional regulators locate their target sequences in the genome. In bacteria, these regulatory sequences generally lie close to the promoter. Once a protein that regulates gene expression binds to its target sequence, it is in a position to influence transcription of the downstream gene. A variety of mechanisms are used to either repress or activate transcription initiation in bacteria. The examples that follow illustrate how transcription initiation is regulated in response to a variety of signals, thus enabling the bacterium to fine-tune gene expression in response to changing needs.

Trp repressor blocks RNA polymerase binding

One of the simplest ways in which a protein can repress transcription is by preventing RNA polymerase from binding to the promoter. This type of gene regulation is illustrated by the *E. coli* Trp repressor, a helix-turn-helix protein whose DNA-binding activity is controlled by the level of tryptophan in the cell. The genes encoding the enzymes needed to synthesize the amino acid tryptophan are clustered together in the *E. coli* chromosome and are under the control of a single promoter and operator, as illustrated in Figure 9.12a. These genes are transcribed as a single mRNA transcript called a **polycistronic** mRNA. Such an arrangement of genes, whereby they are regulated and transcribed as a unit, is known as an **operon**.

When cellular levels of the amino acid tryptophan are low, genes that encode enzymes required for the synthesis of tryptophan are expressed, as shown in Figure 9.12b. When levels of tryptophan are sufficiently high, transcription from the *trp* operon is repressed. This is accomplished by the Trp repressor, which binds to a site directly overlapping the promoter and thereby blocks access by RNA polymerase.

The Trp repressor is able to respond directly to the concentration of tryptophan in the cell because tryptophan is an allosteric effector of Trp repressor's DNA-binding activity. When sufficient amounts of tryptophan are present, one molecule of tryptophan binds to each monomer in the Trp repressor dimer and orients the helix-turn-helix motif of the repressor so that it is presented properly to the DNA major groove. The Trp repressor then binds DNA and prevents RNA polymerase from binding to the promoter.

In the absence of tryptophan binding, however, which occurs when intracellular concentrations of tryptophan are low, the helix-turn-helix unit collapses inward. With the recognition helices no longer properly aligned to fit in successive major grooves of the DNA, the protein cannot bind to the operator or repress transcription and the genes in the *trp* operon are transcribed (see Figure 9.12b).

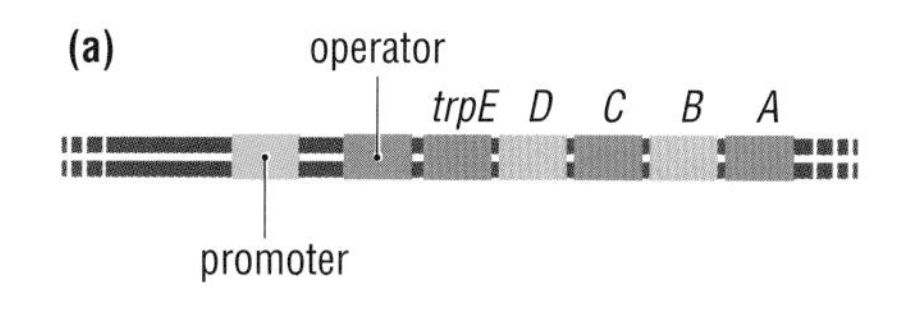

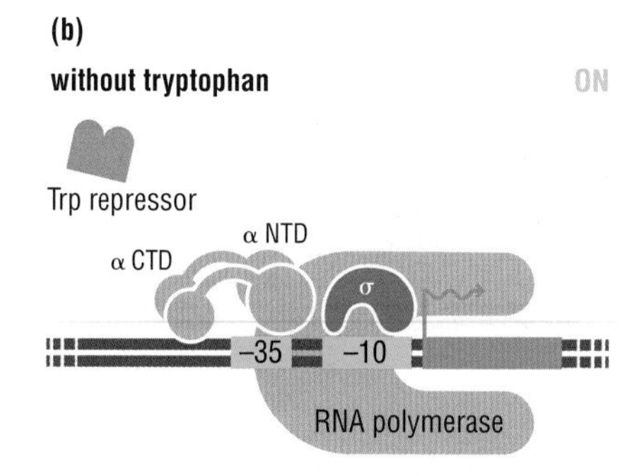

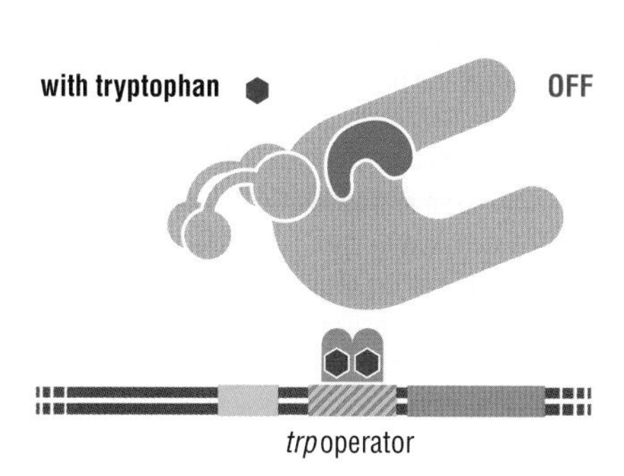

Figure 9.12 Regulation of transcription by the Trp repressor. (a) The *trp, trpE, D, C, B, A* operon. The genes *trpE, D, C, B,* and *A* are transcribed as a single polycistronic mRNA. (b) In the absence of tryptophan, Trp repressor does not bind DNA, and the genes of the *trp* operon are transcribed. When tryptophan is present, it binds to the Trp repressor and renders the protein capable of binding DNA. The Trp repressor binds to regulatory sites that overlap the promoter region, thereby preventing RNA polymerase binding and repressing transcription of the downstream genes.

The catabolite activator protein recruits RNA polymerase by contacting the C-terminal domain of the RNA polymerase α subunit

Just as blocking the binding of RNA polymerase to the promoter represses transcription, enhancing the binding of polymerase to the promoter can activate transcription. Most bacterial regulators activate transcription by helping to recruit RNA polymerase to a promoter. One example is the catabolite activator protein (CAP; also known as the cAMP receptor protein or CRP). CAP activates transcription of more than 100 promoters in *E. coli* under conditions in which the levels of carbon sources such as glucose are too low to allow rapid metabolism.

Glucose depletion in bacteria leads to increased synthesis of cAMP, which binds to the dimeric CAP and increases its affinity for DNA. The CAP dimer binds to specific DNA sites overlapping or near target promoters and enhances the ability of the RNA polymerase holoenzyme to bind to its promoter and initiate transcription, as shown in Figure 9.13.

CAP can activate transcription in several ways, and promoters can be grouped according to the mechanism by which CAP stimulates transcription at that particular promoter. At class I promoters, CAP binds DNA upstream of the promoter and recruits RNA polymerase by binding directly to the C-terminal domain (CTD) of the RNA polymerase α subunit (see Figure 9.13). By helping the RNA polymerase holoenzyme to bind DNA, CAP promotes formation of the closed complex. At class II promoters, however, the CAP-binding site overlaps the binding site for RNA polymerase. At these promoters, interactions between CAP and the α subunit CTD also facilitate binding of RNA polymerase to promoter DNA. In addition, interactions between CAP and the α subunit N-terminal domain promote isomerization of the closed complex to the open complex.

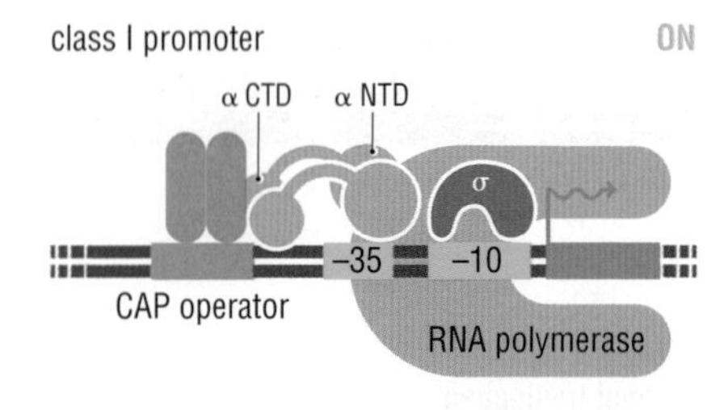

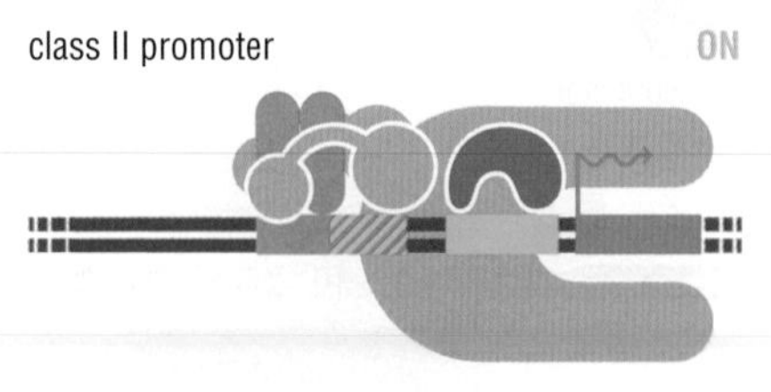

Figure 9.13 Mechanism of CAP activation of class I and class II promoters. Class I promoters (top) have a CAP-binding site just upstream of the promoter, allowing CAP to contact the CTD of the α subunit. At class II promoters (bottom), the CAP operator overlaps the –35 region and allows CAP to interact simultaneously with the CTD and N-terminal domain (NTD) of the α subunit of RNA polymerase. The positioning of the –35 region and the CAP-binding site allow RNA polymerase and CAP to bind simultaneously to the DNA.

CAP and LacI repressor regulate the *lac* operon

Many genes are regulated in response to multiple signals. The regulation of the *lac* operon is a classic example of how transcription of a set of genes can be modulated in response to two distinct signals: the relative availability of glucose and lactose. The simple sugar, glucose, is a preferred carbon source for *E. coli*, but the bacterium can also metabolize lactose, a disaccharide. Lactose utilization is a more complex process and therefore costs the cell more energy. Consequently, when both sugars are available, the bacterium will preferentially metabolize glucose over lactose, so the genes required for lactose utilization are only expressed at a low level. Only when glucose levels are low and lactose levels are high will the genes required for lactose utilization be switched on.

Figure 9.14 shows how the *lac* operon encodes the three genes required for lactose uptake and utilization. The key to regulating the *lac* operon in response to two separate signals is the use of two different regulatory proteins: CAP, which indirectly responds to glucose levels, and the *lac* repressor (or LacI), which senses lactose levels. These two proteins, which are both members of the helix-turn-helix family, have opposite effects on transcription and also respond differently to the binding of an allosteric regulator. As we learned above, CAP binds DNA and *activates* transcription only when it is bound to cAMP, which is found in abundance when glucose levels are low. The *lac* repressor, by contrast, binds DNA and *represses* transcription only in the *absence* of its allosteric regulator, allolactose. The binding of allolactose, a metabolite of lactose that is only present when lactose is being actively metabolized by the bacterium, inactivates the *lac* repressor.

We can now see how the *lac* repressor and CAP regulate transcription in response to the availability of glucose and lactose; the process is illustrated in

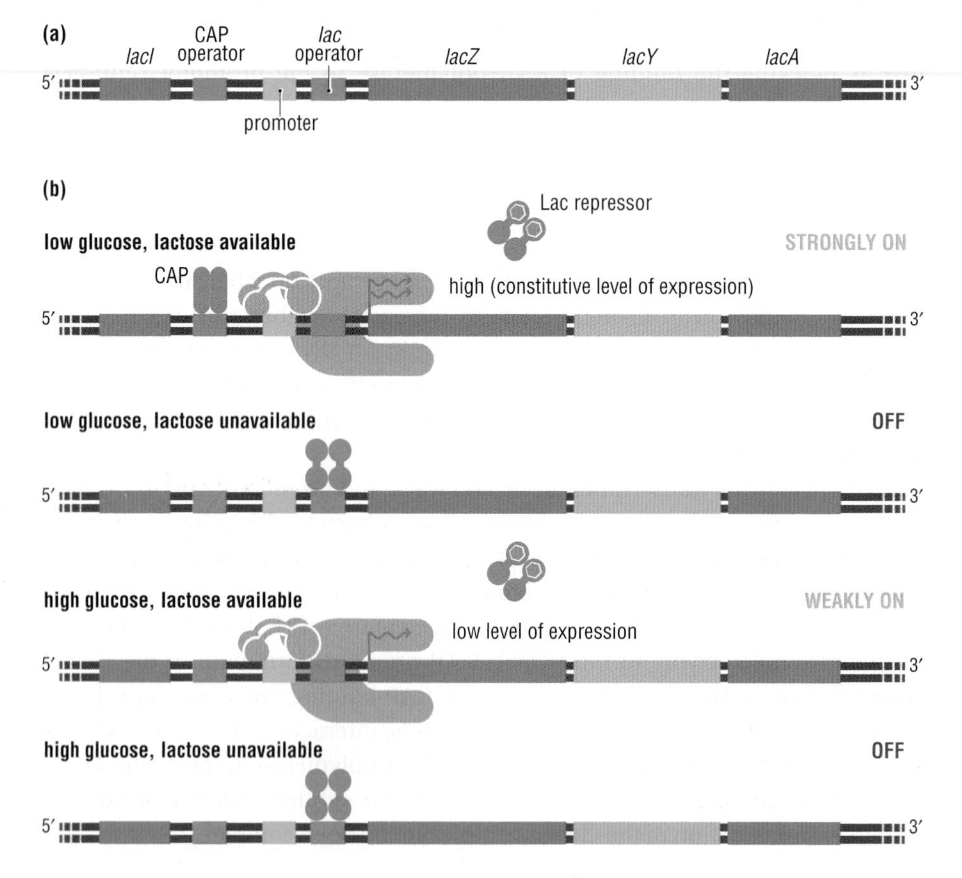

Figure 9.14 Regulation of transcription by the Lac repressor and CAP activator. (a) The *lac* operon. The *lacZ, lacY,* and *lacA* genes, which encode proteins needed for lactose utilization, are transcribed from a single promoter as a polycistronic mRNA transcript. The *lacI* gene encodes the lac repressor. (b) Diagram of how transcription of the *lac* operon is regulated by CAP and the Lac repressor (LacI) in response to different levels of glucose and lactose. CAP binding activates transcription of the *lac* genes, while LacI binding blocks polymerase binding and represses transcription. The maximum level of expression occurs when LacI does not bind DNA (high lactose concentration) and glucose levels are low (CAP binds DNA).

Figure 9.14. The *lacZ*, *lacY*, and *lacA* genes encode proteins required for lactose metabolism and uptake, and are transcribed as a single transcript (just as we saw in the case of the *trp* operon). The *lacI* gene, which encodes the *lac* repressor, is constitutively expressed (meaning that it is always transcribed), so the *lac* repressor is always present in the cell. In the absence of lactose, the *lac* repressor binds to a DNA site (*lacO*, the *lac* operator) that overlaps the promoter and represses transcription. What happens when lactose is present? In this case, the *lac* repressor is inactivated by the binding of allolactose, which means it no longer represses transcription, causing it to dissociate from DNA.

Scan here to watch a video animation explaining more about the regulation of transcription, or find it via the **Online Resource Center** at **www.oxfordtextbooks.co.uk/orc/craig2e/**.

The relative *level* of transcription, however, will depend on whether glucose is also present. If glucose levels are low, CAP will bind upstream of the operon and activate transcription of the *lac* operon, since cAMP binding enables CAP to bind DNA. The proteins encoded by the *lac* operon will therefore be synthesized at high levels and the bacterium will be able to make efficient use of lactose. If glucose levels are high, however, CAP does not bind DNA (because cAMP levels are low) and the *lac* genes are transcribed at a much lower level. This unstimulated level of transcription is referred to as **basal transcription.** The bacterium will make less efficient use of lactose since there is adequate glucose to serve the cell's metabolic needs.

MerR-family activators alter promoter structure

The variable spacing between −10 and −35 sequences provides another opportunity for regulating transcription regulation in bacteria. If the spacing between these two sites is much larger than is optimal, RNA polymerase binding to the promoter is significantly weakened. The MerR family, named for a protein that activates expression of genes required for mercury detoxification, forms a unique group of transcriptional regulators that can change the way the promoter elements are organized in space relative to each other. The −35 and −10 sequences of promoters regulated by MerR are spaced an unusually long 19 bp apart, as compared with the ideal spacing of 17 bp, as illustrated in Figure 9.15. The two additional base pairs not only place the −35 and −10 promoter elements farther apart, but also affect the relative orientation of the two promoter elements about the DNA axis. This is

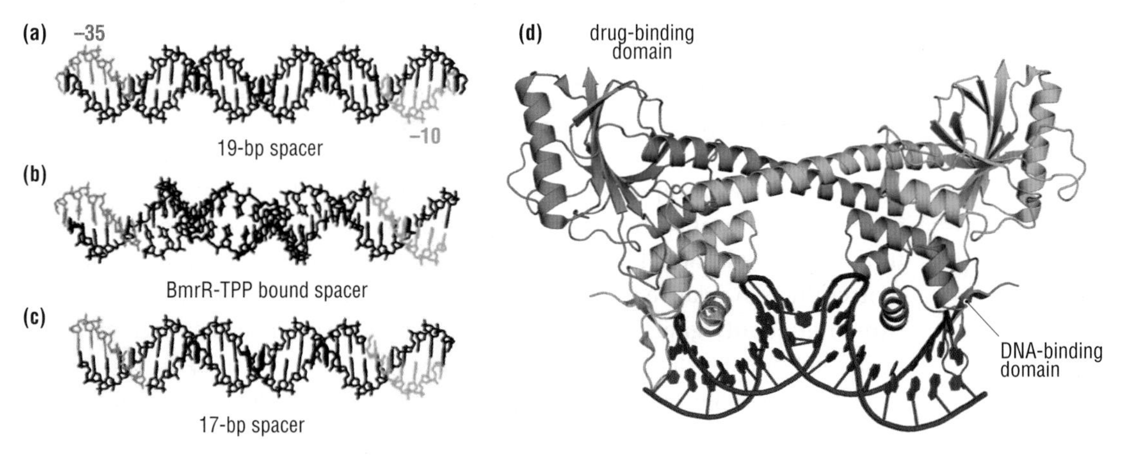

Figure 9.15 Mechanism of promoter opening by MerR-family members. The structure of BmrR explains how MerR-family members facilitate the opening of promoters in which the −10 (green) and −35 (pink) sequences are spaced 19 base pairs (bp) apart. (a) Promoter with a 19 bp spacer. (b) BmrR binding leads to a distortion of the DNA such that the −10 and −35 elements now have the same spacing and orientation as the promoter element with −35 and −10 spaces, 17 bp apart, shown in (c). (d) Structure of BmrR bound to DNA (PDB code 1EXJ).

From Heldwein, EE, and Brennan, RG. Crystal structure of the transcription activator BmrR bound to DNA and a drug. *Nature*, 2001;**409**:328.

because each additional base-pair step is rotated roughly 34° about the helix axis and translated a distance of about 0.34nm along the helix.

MerR activates transcription by altering the relation between the –10 and –35 regions to more closely resemble an ideal promoter (Figure 9.15b). The binding site for the MerR protein overlaps the promoter region. As seen in the structure of a related protein, BmrR, shown in Figure 9.15d, proteins in this family are able to change the distance and angular spacing between the –10 and –35 regions by distorting the DNA structure between the two promoter elements. With the alignment of the two promoter regions now closer to that found in optimally spaced promoters, RNA polymerase binds more readily. Notice how BmrR binds on the opposite face of the DNA and therefore does not interfere with RNA polymerase binding.

➔ The initial characterization of NtrC is described in Experimental approach 11.2.

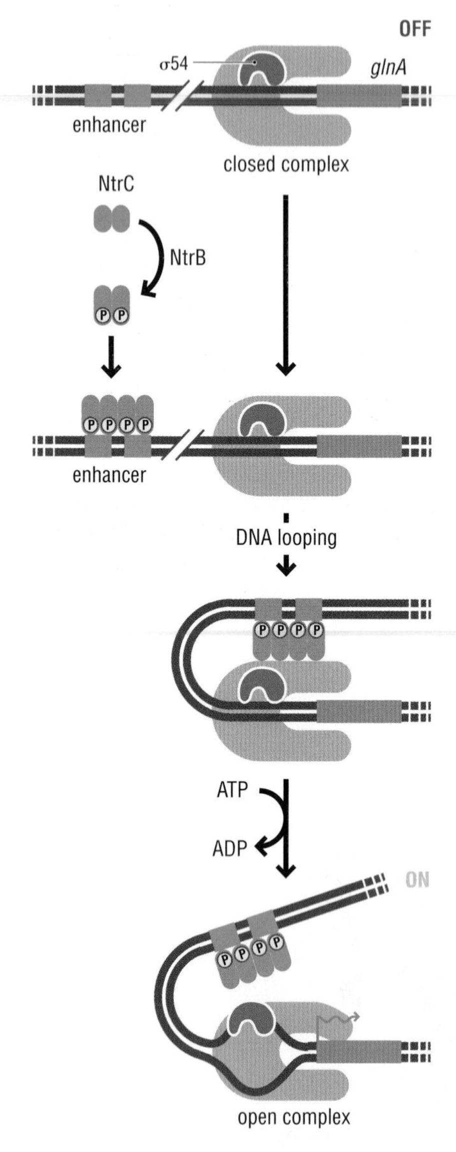

Figure 9.16 Mechanism of action of NtrC. The NtrB kinase phosphorylates NtrC, causing NtrC to oligomerize and bind to its regulatory region in the DNA. The intervening DNA loops out, allowing NtrC to contact σ^{54} RNA polymerase. Oligomerization also triggers the hydrolysis of ATP to adenosine diphosphate (ADP) by NtrC, which helps promote open-complex formation by the σ^{54} RNA polymerase.

Activation by NtrC involves DNA looping and ATP hydrolysis

Promoters that are transcribed by RNA polymerase holoenzyme containing the σ^{54} (also known as σ^{N}) subunit must be activated by protein complexes that use ATP hydrolysis to drive promoter opening. A well-characterized example of this class of activators is the *E. coli* NtrC protein, which activates genes that are important for nitrogen metabolism. The mechanism of action of NtrC is illustrated in Figure 9.16. Genes regulated by NtrC generally contain two or more binding sites for NtrC well upstream of the promoter. NtrC can bind to these sites but it does not activate transcription until a regulatory domain within the protein is phosphorylated. This phosphorylation event triggers oligomerization of NtrC, which has two effects on the protein's activity. First, oligomerization of NtrC promotes interaction of the NtrC protein with the RNA polymerase enzyme that is bound to the promoter, causing the intervening DNA to loop out. At some promoters, the architectural DNA-binding protein integration host factor (IHF) helps promote the looping by binding to the intervening sequences and bending the DNA.

Oligomerization of NtrC also stimulates its rate of ATP hydrolysis, which in turn helps to promote open-complex formation at σ^{54}-dependent promoters. Somehow, the σ^{54} holoenzyme bound to a promoter is unable to isomerize to the open complex in the absence of ATP hydrolysis. NtrC uses the energy released by ATP hydrolysis to help the σ^{54} holoenzyme overcome this block.

Two-component signal transduction pathways are composed of histidine kinases and response regulators

There are many examples in which the first response to an extracellular signal is the phosphorylation of a cytoplasmic protein. The phosphorylated protein, in turn, triggers additional biochemical events that eventually lead to a change in transcription. A relatively simple example of this type of signaling cascade is the **two-component signal transduction pathways** found in bacteria, fungi and plants. Two-component systems have a **sensor kinase**, which detects the relative amounts of specific molecules in the cell's environment, and a **response regulator**, which is activated when it is phosphorylated by the sensor protein; this composition is illustrated in Figure 9.17.

The sensor protein is a histidine kinase, typically membrane-bound, that becomes autophosphorylated on a histidine residue when it detects a certain signal. This phosphate is then transferred to an aspartate residue on the response regulator, which often contains a DNA-binding domain. Phosphorylation of the response regulator changes its DNA-binding properties, thus leading to a downstream effect on gene regulation.

There are many variations of this type of signaling. There are sensor kinases that are cytoplasmic, response regulators that have functions other than transcriptional regulation, and more complex networks that involve more than two proteins. However, the common feature of these signaling systems is the transfer of a phosphate from the histidine kinase sensor to a downstream effector protein.

The response of *E. coli* to changing levels of inorganic phosphate is a simple example of two-component signal transduction. As shown in Figure 9.17, the PhoR sensor kinase is a transmembrane protein that binds a free phosphate in the periplasmic space of the bacterial cell. At low levels of phosphate, the phosphate dissociates from the periplasmic sensor domain, triggering a change in the cytoplasmic domain. This change activates the sensor kinase, causing it to catalyze transfer of the γ-phosphate of ATP to a histidine in the sensor kinase domain. The PhoR sensor kinase then transfers the phosphate from its histidine residue to an aspartate residue in PhoB, the response regulator. Phosphorylated PhoB then activates transcription of a number of genes required by the cell when phosphate concentrations are low.

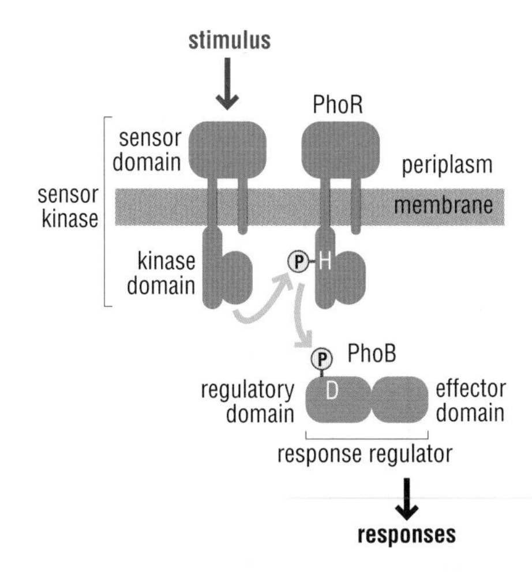

Figure 9.17 Two-component signaling mechanisms in response to extracellular stimuli. The basic two-component phosphotransfer system, found in bacteria, consists of a dimeric transmembrane receptor histidine kinase and a cytoplasmic response regulator. Information flows between the two proteins in the form of a phosphoryl group that is transferred from the histidine kinase to the response regulator. Histidine kinases catalyze ATP-dependent autophosphorylation of a specific conserved His residue (H). The phosphoryl group (P) is then transferred to a specific aspartic acid residue (D) located within the conserved regulatory domain of a response regulator. Phosphorylation of the regulator typically activates an associated (or downstream) effector domain, which ultimately elicits a specific cellular response.

9.4 COMPETITION BETWEEN cI AND Cro AND CONTROL OF THE FATE OF BACTERIOPHAGE LAMBDA

It is frequently necessary for an organism to respond to a complex set of signals by adjusting transcription of multiple genes. This type of regulation is often the result of the concerted action of a network of proteins, which may exert antagonistic effects on one another. The outcome – whether transcription happens or not – then depends on the balance of power between regulators that have opposing effects on transcription. In this section we consider a prime example of how transcription can be regulated by a network of proteins: the regulation of bacteriophage lambda genes.

Bacteriophage lambda can undergo lytic growth or lysogeny

Bacteriophage lambda presents a well-studied example of how the competition between several different sequence-specific DNA-binding proteins can lead to very different physiological outcomes. This bacterial virus infects *E. coli* by injecting its short chromosome into the bacterium's cytoplasm, where the **phage** can have one of two fates, as depicted in Figure 9.18. Phage lambda can undergo **lytic growth**, reproducing many virus particles within the cell, leading to its eventual lysis (bursting of the host cell). Alternatively, the phage can undergo **lysogeny**, whereby the phage DNA integrates into the bacterial chromosome and becomes dormant.

The integrated phage is denoted a **prophage** and a bacterial cell whose chromosome contains an integrated copy of the bacteriophage chromosome is called a **lysogen**. In the lysogen, the viral DNA is replicated along with the bacterial chromosome. Conditions that threaten the viability of the bacterial host cell, such as DNA damage, can trigger the prophage to excise itself from the chromosome, undergo lytic growth, and produce viral progeny that burst out of the damaged host cell.

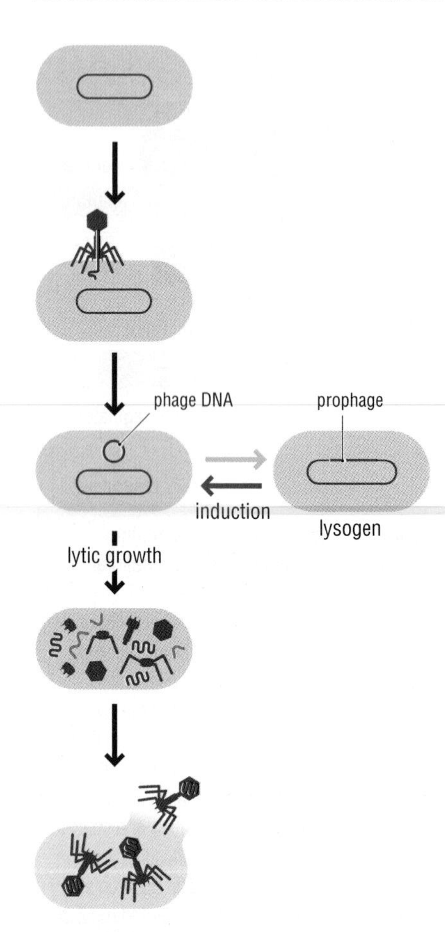

Figure 9.18 The lifecycle of bacteriophage lambda. Bacteriophage lambda infects a cell by injecting its linear chromosome (pink) into a host *E. coli* cell. The lambda chromosome circularizes and has one of two fates: it can direct lytic growth of the phage, directing synthesis of new phage particles that are released by bursting the host cell, or it can incorporate itself into the host cell's chromosome and become a prophage, remaining in this quiescent state indefinitely. Damage to the host cell's chromosome can induce the integrated prophage to excise itself from the host cell's chromosome, circularize, and commence lytic growth.

The choice between lysis and lysogeny is controlled by the levels of the cI, cII, and Cro proteins

The choice between lysis and lysogeny is determined by the relative levels of three DNA-binding proteins encoded by the phage: cI (also called the lambda repressor), cII, and Cro. These three regulators, which are the first proteins to be synthesized after phage lambda infects an *E. coli* cell, bind to specific sites in the phage chromosome and control transcription from the four promoters illustrated in Figure 9.19: P_R (rightward promoter), P_L (leftward promoter), P_{RE} (promoter for repressor establishment), and P_{RM} (promoter for repressor maintenance). As we shall see next, whether the phage grows lytically or establishes a lysogen depends upon the rate at which these proteins build up in the infected cell and occupy their DNA-binding sites.

After a lambda phage injects its genetic material into a bacterial cell, the linear phage chromosome circularizes, and the host cell RNA polymerase initiates basal transcription from P_L and P_R. The first genes in the P_R transcript encode the Cro and cII regulators, while the first gene in the P_L transcript is the N protein (see Figure 9.19). As we will learn in Section 9.5, N prevents premature termination of the P_{RM} transcript and thereby allows transcription of both Cro and cII. If sufficient cII protein accumulates, it binds DNA and stimulates transcription from the P_{RE} promoter, producing an mRNA transcript encoding the cI regulator. The newly synthesized cI protein, in turn, stimulates transcription from the weak P_{RM} promoter, yielding another transcript encoding cI. As cI builds up in the cell, the protein binds to sites in the P_L and P_R promoter regions and represses transcription of the lytic genes encoded by the P_L and P_R transcripts. cI also activates transcription of the gene that encodes integrase, which mediates lambda DNA integration into the bacterial chromosome (see Experimental approach 9.2). Thus, lysogeny is established and maintained by high levels of cI.

By contrast, if cII does not accumulate in sufficient amounts and too little cI protein is synthesized, the Cro protein will bind the P_{RM} promoter and repress transcription of the cI gene. By preventing cI synthesis, Cro ensures expression of the lytic genes from P_L and P_R.

The choice between lysis and lysogeny depends upon whether or not conditions in the host and its environment favor propagation of the phage. The lysogenic state is favored in a nutrient poor environment, when the host cell is growing slowly; by contrast, lysis is favored when the host is growing rapidly in a nutrient-rich environment. If conditions are favorable to the host, it is likely that there are many other hosts around to infect, so lytic growth would generate more phage, which could in turn infect many more hosts. If conditions are not favorable to the host, it is less likely that there will be other potential hosts in the vicinity, so lysogeny makes more sense for the survival of the phage.

The mechanism by which the choice between lysis or lysogeny is made after the bacteriophage infects a host cell hinges on the activity of cII. cII is sensitive to the host cell proteases, making it unstable in the cell. If the environmental factors are such that cII accumulates and directs expression of enough cI, a lysogenic state is established. If cII is degraded, cI is not synthesized and all other bacteriophage genes are expressed, allowing autonomous viral replication, packaging, and, ultimately cell lysis.

The DNA-binding affinities of cI and Cro are a key component in the choice between lysis and lysogeny

A closer look at the P_L and P_R promoters reveals that a key component of the molecular switch controlled by cI and Cro is the relative affinities of these proteins

9.2 EXPERIMENTAL APPROACH

Cooperative binding and DNA looping in gene regulation by the lambda cI repressor

The cI repressor binds to the O_R1, O_R2, and O_R3 sites with different affinities

One of the key components involved in regulating the switch between lysis and lysogeny in bacteriophage lambda depends on cooperative binding of the cI protein to DNA. This property of cI was discovered in 1979 by Alexander Johnson, Barbara Meyer, and Mark Ptashne, who measured the binding affinity of cI to the three sites within O_R: O_R1, O_R2, and O_R3. It had already been established that cI dimers bound to these three sites, as could an N-terminal fragment containing just the DNA-binding domain. Without the C-terminal domain, the DNA-binding domains could not dimerize efficiently and therefore bound DNA more weakly. Ptashne and colleagues uncovered a second property of the C-terminal domain: the ability to mediate cooperative binding between cI dimers.

The way in which the bacteriophage lambda cI repressor fills successive binding sites in O_R was determined from nuclease protection experiments (see Section 19.10). As seen in Figure 1, these experiments show vividly where in the operator DNA cI binds. Lane 1 shows the pattern of DNA fragments produced when a radiolabeled DNA fragment containing O_R is subjected to limited digestions by DNase I and the products are then separated by gel electrophoresis. As cI repressor is added in successively higher concentrations (lanes 2–8), the protein dimers bind to each of the individual DNA sites – O_R1, O_R2, and O_R3 – protecting those base pairs from digestion. This causes the bands in that region to fade and eventually disappear, as more and more DNA fragments are protected from DNase I digestion by the bound protein.

By noting how the sites on the DNA filled as a function of cI concentration, the equilibrium dissociation constants (K_d) could be determined for the three sites (O_R1, O_R2, and O_R3) within the rightward operator. Recall that the K_d is the protein concentration at which half-maximal binding occurs. (Turn to Section 3.2 to review the principles of equilibrium binding.) As shown in Figures 1 and 2a, cI fills O_R1 and O_R2 at roughly the same concentration in the titration; this results from the fact that cI dimers bind cooperatively to these two sites (see Section 3.3 for a discussion of cooperativity). By contrast, the O_R3 site was only filled when the protein concentration was tenfold higher (a tenfold higher K_d). However, if a mutation was introduced into O_R1 that abrogated DNA binding (O_R1^-), cI bound to O_R2 and O_R3 with the *same* affinity (Figure 2a). In addition, the affinity of cI for O_R3 became higher (~ fivefold) such that it matched the affinity for O_R2. This observation suggested that, in the absence of cI binding to O_R1, there were now cooperative interactions between dimers bound to O_R2 and O_R3.

cI dimers bind cooperatively to pairs of neighboring sites

When the same experiments were performed with the N-terminal domain of cI (missing the C-terminal domain), cooperative binding was no longer observed; indeed, this fragment bound more tightly to O_R1 than to O_R2 (Figure 2b). In addition, mutations in O_R1 had no effect on binding of the N-terminal fragment to the other two sites (Figure 2b).

The results of these experiments could be explained by the model shown in Figure 3a, dubbed 'alternate pairwise cooperativity.' The C-terminal domain of cI was proposed to mediate cooperative interactions between adjacently bound dimers. The nature of the interaction was such that one dimer could interact with

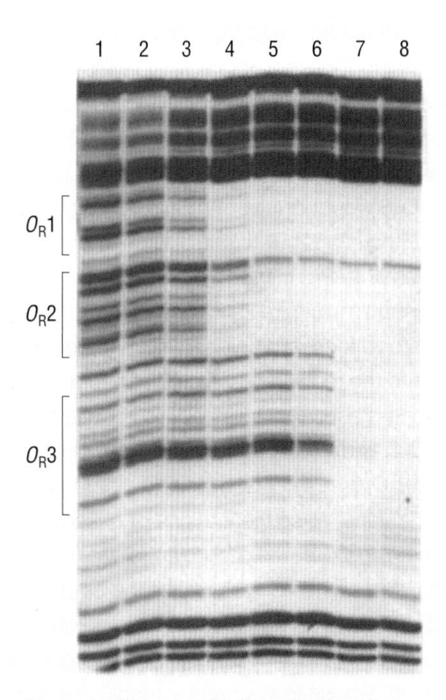

Figure 1 DNase I protection experiment showing the binding of the bacteriophage lambda cI repressor to O_R1, O_R2, and O_R3. The concentration of cI increases in each successive lane.

Reproduced from Johnson *et al.* Interactions between DNA-bound repressors govern regulation by the lambda phage repressor. *Proceedings of the National Academy of Sciences of the U S A*, 1979;**76**:5061–5065.

Figure 2 Relative concentration of cI repressor required for half-maximal binding to wild-type and mutant O_R. The concentration of lambda repressor and the N-terminal domain is expressed as a multiple of the concentration of each protein needed to fill O_R3. (a) Intact cI. (b) N-terminal domain of cI.

Reproduced from Johnson *et al.* Interactions between DNA-bound repressors govern regulation by the lambda phage repressor. *Proceedings of the National Academy of Sciences of the U S A*, 1979;**76**:5061–5065.

	relative repressor concentration					
	intact repressor			N-terminal domain		
DNA	O_R3	O_R2	O_R1	O_L3	O_L2	O_L1
O_R^+	25	2	1	25	25	1
O_R1^- (*vs326*)	5	5	-	-	25	1
O_R2^- (*virC23*)	25	-	2	25	-	1
O_R1^- O_R2^- (*Δ265 virC23*)	25	-	-			
O_R1^- O_R3^- (*Δvc1 vc3*)	-	25	-	-	25	-

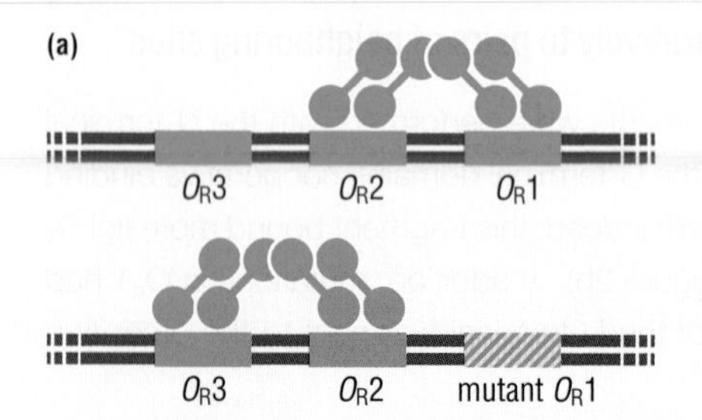

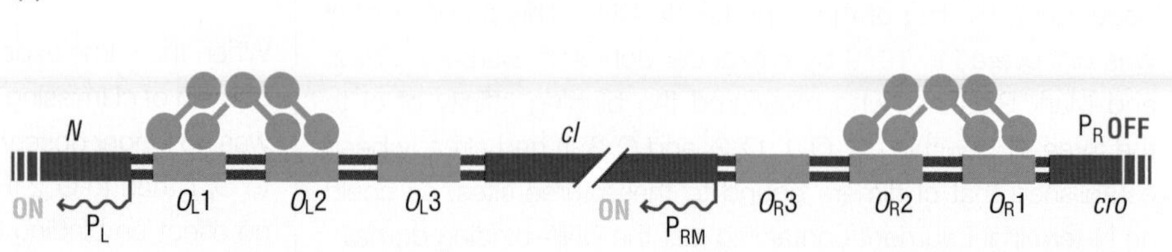

Figure 3 Alternate pairwise cooperativity model. (a) Figure illustrating that just two dimers at a time could interact with one another. (b) Model for how cI bound cooperatively to the O_R and O_L operators. O_R3 was predicted to fill at much higher cI concentrations, due to the absence of cooperativity at that site in the wild-type operator. The DNA looping shown in Figure 8.43 was discovered over 20 years later.

Adapted from Hochschild, The lambda switch: cI closes the gap in autoregulation. *Current Biology*, 2002;**12**:R87–R89.

only one other dimer on the DNA, giving the pairwise interactions shown in Figure 3a. This meant that, in the wild-type operator, a cI dimer bound to O_R3 would not bind DNA cooperatively because the dimer bound to the adjacent O_R2 site was already interacting with a dimer at O_R1 and was therefore not available for additional interactions. O_R3 would only fill at much higher concentrations of cI (Figure 2a), thereby regulating its own expression by shutting off transcription from P_{RM} (Figure 3b).

The model shown in Figure 2b for how cI regulated transcription in a lysogen stood for over 20 years until another study revealed a new aspect to the regulatory switch that had gone undetected. One of the long-standing puzzles about lambda had to do with the roles of O_R3 and O_L3 in autoregulation of cI expression. The affinity of cI for O_R3 seemed too weak for cI to shut off its own transcription, at least at concentrations of cI known to exist in a lysogen; moreover, there was no known role for O_L3 in a lysogen. In 2001, Ian Dodd and colleagues made the surprising discovery that binding of cI to O_R3 was aided by cooperative interactions with O_L3. They replaced the *cI* gene with the gene expressing β-galactosidase, whose levels can be readily measured in solution, while the cI protein itself was encoded on a separate plasmid whose synthesis could be controlled independently. This system made it possible to measure transcription from P_{RM} as a function of increasing cI concentration. As shown in Figure 4, repression of P_{RM} by cI requires an intact O_L operator (compare the level of transcription in the absence and presence of O_L as cI concentration increases).

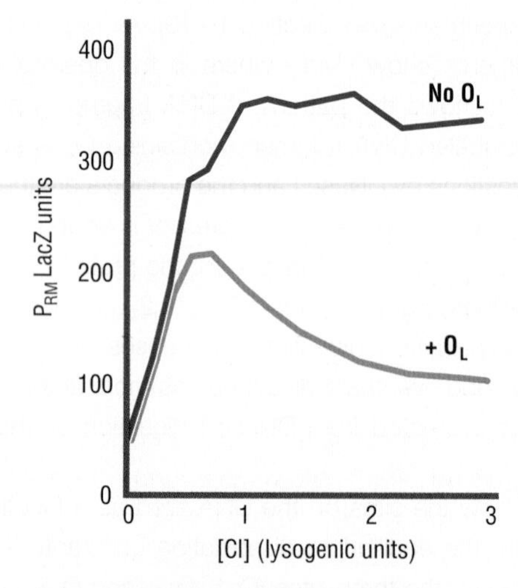

Figure 4 Transcription from P_{RM} in the presence and absence of O_L. In the absence of O_L, as the concentration of cI increases, there is little repression of transcription from P_{RM} with a wild-type O_R3. In the presence of O_L, high concentrations of cI repress transcription from P_{RM}.

Figure kindly supplied by Ian Dodd.

cI dimers binds cooperatively to O_R3 and O_L3 and loop out the intervening DNA

How could binding to O_L affect binding to a site in O_R several thousand base pairs away? The model that explains these observations is that the intervening DNA loops out as depicted in Figure 9.20c, allowing cI dimers bound to O_L to directly contact dimers bound to O_R. This idea did not come out of thin air – it had been recently discovered that cI bound to distant sites (>3.5 kb) could induce a loop in the DNA, although it was not clear what the physiological significance of the looping could be. Independently, biochemical experiments had shown that cI could form octamers at high concentrations. The model in Figure 9.20c brings together all of these observations: a pair of cI dimers bound to O_R1 and O_R2 interacts with a pair of cI dimers bound to O_L1 and O_L2, forming an octamer, which is mediated by the C-terminal domains and looping out of the intervening DNA. The looping then facilitates cooperative binding between dimer bound to O_R3 and O_L3, as shown in Figure 9.20. Cooperativity between these two distant sites ensures that P_{RM} is actually repressed 50% of the time in a lysogen, resulting in a cI concentration that perfectly poises the switch for induction.

Find out more

Dodd IB, Perkins AJ, Tsemitsidis D, Egan JB. Octamerization of lambda CI repressor is needed for effective repression of PRM and efficient switching from lysogeny. *Genes and Development*, 2001;**15**:3013–3022.

Johnson AD, Meyer BJ, Ptashne M. Interactions between DNA-bound repressors govern regulation by the lambda phage repressor. *Proceedings of the National Academy of Sciences of the U S A*, 1979;**76**: 5061–5065.

Related techniques

Bacteriophage lambda (as a model organism); Section 19.1

Nuclease protection studies; Section 19.12

Gel electrophoresis; Section 19.7

Radiolabelling of DNA; Section 19.6

LacZ assay; Section 19.6

for six DNA-binding sites. The dimeric cI protein is composed of two discrete domains, as illustrated in Figure 9.20a: an N-terminal DNA-binding domain with a helix-turn-helix motif and a C-terminal dimerization domain. cI dimers further oligomerize to form tetramers, which in turn can associate to form an octamer. These higher-order interactions are required for cooperative cI binding to DNA. In contrast, the Cro protein consists of a single, small domain (also with a helix-turn-helix) and binds DNA as a dimer.

The operators, O_L and O_R, each contain three sites to which either cI or Cro can bind, as depicted in Figure 9.20b. The three sites in O_L – O_L1, O_L2, and O_L3 – overlap the P_L promoter, while O_R1, O_R2, and O_R3 overlap the divergent P_R and P_{RM} promoters. At both O_L and O_R, cI binds first to its highest-affinity sites, O_L1 and O_R1, and recruits a second cI dimer to the weaker O_L2 and O_R2 binding sites through cooperative interactions. The pairs of cI dimers bound to O_R and O_L interact with one another to form an octamer as shown in Figure 9.20c, looping out the intervening DNA. This arrangement of cI proteins on the DNA prevents RNA polymerase from binding to P_R and P_L and stimulates transcription from P_{RM} as a result of interactions between RNA polymerase and the cI dimer bound to O_R2. The increased transcription from P_{RM} leads to increased cI levels.

As cI builds up, the protein can bind cooperatively to the weakest binding sites, O_R3 and O_L3, which now are close in space due to the DNA looping (see Figure 9.20c). Occupancy of O_R3 occludes P_{RM} and suppresses synthesis of cI. This **autoregulation**, whereby cI regulates its own synthesis, ensures there is sufficient cI in the cell to maintain the lysogenic state, but not so much that prophage induction is prevented.

A lysogen can persist for many generations, with the phage stably incorporated in the bacterial genome. With each cell division, the daughter cells receive a copy of the circular chromosome containing the integrated prophage, along with enough cI protein to stimulate fresh cI synthesis, maintaining the lysogenic state. This inheritance of the lysogenic state is a form of epigenetic inheritance, since the information that the prophage remain in the 'off' state is transmitted by the cI protein and not by differences in the DNA sequence.

What if, early in the stage of phage infection, there is insufficient cI protein to establish lysogeny and ensure sufficient transcription of the cI gene? In this case, the

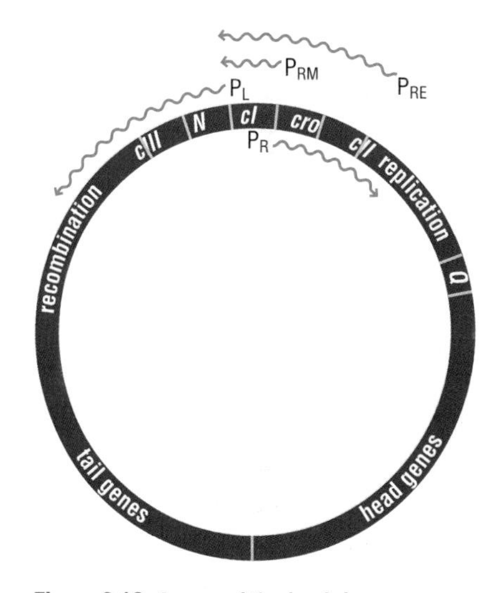

Figure 9.19 A map of the lambda chromosome. Map of the 48.5 kilobase (kb) bacteriophage lambda chromosome showing some of the genes and the promoters that are important in the early stages of the phage lifecycle. The P_L and P_R promoters direct transcription of genes required for lytic growth, whereas the P_{RM} transcript encodes cI, which is required for establishment and maintenance of the lysogenic state. Following infection of a host cell, transcription from P_R leads to synthesis of the Cro and cII proteins. The cII protein stimulates transcription from P_{RE}, which leads to synthesis of cI.

➔ We introduce the concept of epigenetic inheritance in Section 4.7.

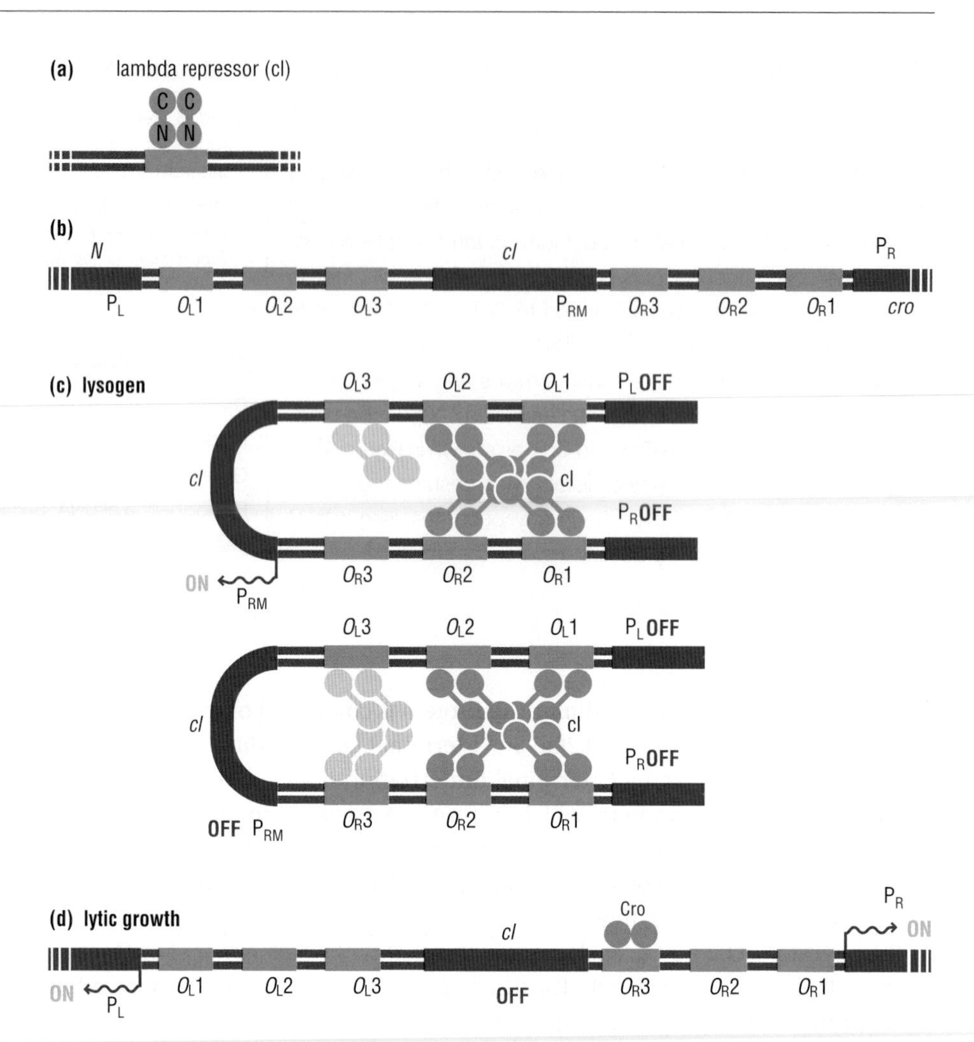

Figure 9.20 Binding of cI and Cro to the rightward and leftward operators. (a) Each monomer of cI contains an N-terminal DNA-binding domain and a C-terminal dimerization domain. (b) The arrangement of DNA-binding sites and promoters in the leftward (O_L) and rightward (O_R) operators. Each operator region contains three DNA sequences that can be recognized by both cI and Cro. (c) In a lysogen, cI dimers bind cooperatively to O_L1 and O_L2 in the leftward operator and to O_R1 and O_R2 in the rightward operator. The four cI dimers bound to these sites interact via their C-terminal domains, looping out the intervening DNA. This arrangement of cI proteins on the DNA represses transcription from the P_R and P_L promoters, while stimulating transcription from P_{RM}. Binding of cI to O_R3 (light orange), which occurs at higher concentrations of cI, turns off transcription of the *cI* gene. The DNA loop promotes cooperative interactions between cI dimers bound to O_R3 and O_L3. (d) The Cro protein binds to the same DNA sites as cI, but with different relative preferences. At the start of lytic growth, the Cro protein binds first to O_R3 and represses transcription from P_{RM}, turning off synthesis of cI. As lytic growth proceeds and concentrations of Cro build up in the cell, Cro binds to the remaining sites in O_L and O_R, turning down expression of the early viral genes.

Cro protein will build up and bind to the same operator sites as cI, as illustrated in Figure 9.20d. A key difference is that, in contrast to cI, Cro binds with greatest affinity to O_R3, where it blocks RNA polymerase binding to P_{RM} and represses cI synthesis. As the concentration of Cro builds up in the cell later during lytic growth, it binds to the remaining operator sites in O_L and O_R (see Figure 9.20d). This leads to the autorepression of *cro* as well as the repression of all other early lytic genes, whose expression is not needed late in the lytic cycle when the phage is expressing proteins that will package the phage DNA into viral particles that are released when the cell lyses.

The prophage is induced on cI cleavage

The prophage is stably integrated into the *E. coli* chromosome and can be maintained through many rounds of DNA replication and cell division. However, under conditions that threaten survival of the host cell and hence of the integrated phage DNA, the prophage can excise itself from the host chromosome and undergo lytic growth, thereby producing new phage and escaping the compromised host. This switch from lysogeny to lytic growth can be brought about by a response to DNA damage that disables the cI repressor.

➔ We learn about the response to DNA damage in Chapter 15.

Damage of the host cell DNA – for example by chemical agents or by ultraviolet light – triggers the bacterial **SOS response**, which leads to activation of the RecA protein. RecA promotes autocatalytic cleavage of the CTD of cI, separating it from the N-terminal DNA-binding domain, as depicted in Figure 9.21. Without the CTD, cI can no longer dimerize efficiently or form highly cooperative interactions between adjacent cI dimers. On its own, the N-terminal domain has relatively low affinity for the operator sites and therefore dissociates from the DNA. Now transcription of the early lytic genes from P_R and P_L can occur; there is little further synthesis of cI because P_{RM} is a relatively weaker promoter than P_L.

One of the first lytic genes to be transcribed from the P_L promoter is *cro* (see Figure 9.19). The cro protein binds to OR3 and represses transcription of the *cI* gene from P_{RM}, preventing re-establishment of the lysogenic state and allowing lytic growth to occur. The resulting phage particles lyse the damaged host cell and are free to infect fresh cells.

The types of transcription feedback loops we see in the bacteriophage lambda underlie complicated environmental sensing and developmental decisions from systems as 'simple' as a bacteriophage to systems as 'complicated' as humans.

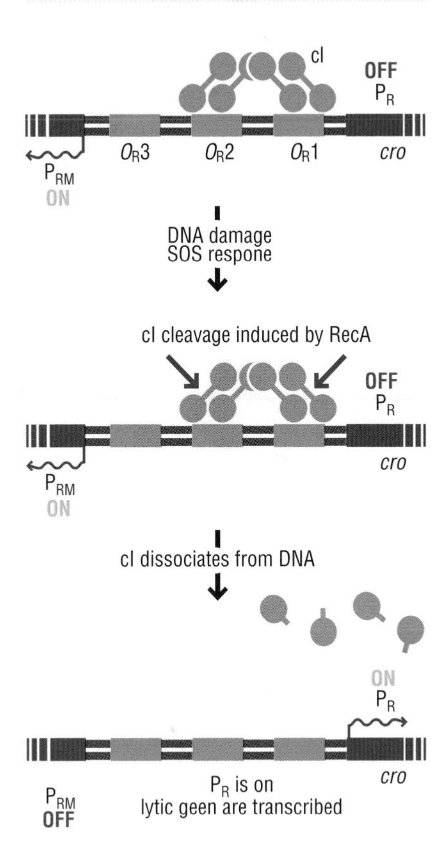

Figure 9.21 Induction of the lambda prophage. The presence of single-stranded DNA, a consequence of DNA damage, induces the cellular SOS response, which includes activation of the RecA protein. RecA induces autocatalytic cleavage in cI that separates the DNA-binding domain from the domain that mediates dimerization and cooperativity. cI dissociates from the DNA, allowing transcription from PR and thus initiating the lytic program. The first gene transcribed is *Cro* (see Figure 9.19), whose product binds to O_R3 and prevents re-establishment of the lysogenic state.

9.5 REGULATION OF TRANSCRIPTION TERMINATION IN BACTERIA

Most of the mechanisms for regulating transcription that we have discussed up to this point have the effect of controlling whether or not transcription of a given gene is initiated. Yet, even after RNA polymerase has initiated transcription and broken free of the promoter, the subsequent steps in elongation and termination provide additional opportunities for the cell to regulate whether the full-length transcript is made. Here we discuss further mechanisms for regulating whether a full-length transcript is synthesized by the RNA polymerase. We will see that, in some cases, the transcribed RNA folds into a defined structure that plays a central role in the regulation of elongation and termination: that is, the RNA performs a function, rather than merely serving as a template for protein synthesis.

The bacteriophage lambda N and Q proteins prevent stalling and termination by altering the properties of bacterial RNA polymerase

In Section 9.4, we learned how the expression of some bacteriophage lambda genes is tightly controlled at the level of transcription initiation. Other bacteriophage genes are transcribed only when premature termination is prevented. This type of regulation is therefore denoted **anti-termination**. As we saw in Section 9.4, the transcription of the phage lambda lytic genes begins at the divergent P_L and P_R promoters. Initially, the transcripts produced are short because RNA polymerase

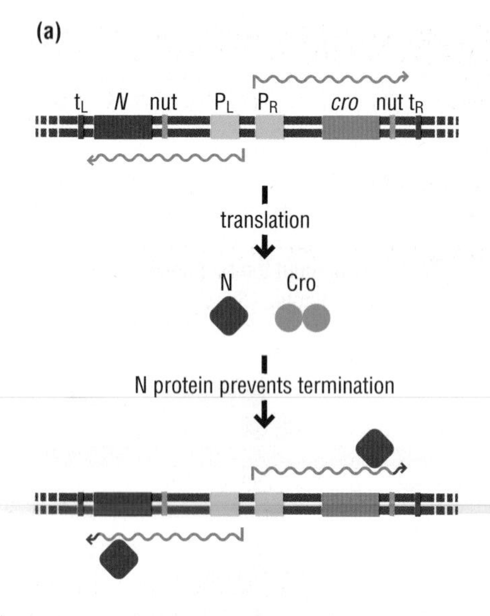

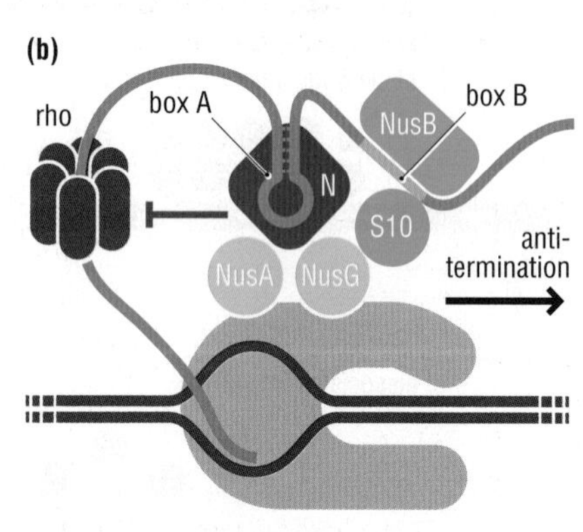

Figure 9.22 Anti-termination by the phage lambda N protein. (a) In the lytic phase, transcription initiated at both the P_R and P_L promoters initially terminates at the t_R and t_L terminators, respectively. This leads to synthesis of the Cro and N proteins encoded by these two transcripts. The N protein prevents termination at the two terminator sites, allowing production of longer transcripts. (b) A number of proteins are required along with N to relieve termination. The N protein binds to a stem–loop structure formed by the *nut* site (shown in (a)). This RNA–protein complex interacts with several other cellular proteins, including NusA, NusG, NusB, and S10, as well as RNA polymerase. In some way that is not yet understood, this relieves termination by the cellular Rho protein and allows polymerase to transcribe through this region.

encounters sequences in the RNA, t_L and t_R, that lead to termination of transcription, as depicted in Figure 9.22a. Thus, only the Cro and N proteins are synthesized at the early stages of lytic growth. We previously described the role of Cro in binding DNA and preventing re-establishment of the lysogenic state (Section 9.4); the other important lytic gene product is the N protein, which also begins to build up in the cell. At sufficient concentrations, the N protein alleviates termination at the t_L and t_R sites (see Figure 9.22a). This allows RNA polymerase to transcribe genes needed early in the lytic cycle, including cII.

The N protein causes anti-termination by interacting with special features in the transcribed RNA termed *nut* sites (for *N utilization*), which are depicted in Figure 9.22a. Within each *nut* site are two RNA sequences called Box A and Box B. Figure 9.22b shows how Box B forms an RNA stem–loop to which N binds, while Box A is a binding site for the *E. coli* protein NusB. The binding of N protein, together with NusB and other cellular proteins shown in Figure 9.22b (NusA, NusE, and NusG), alters the properties of RNA polymerase such that it gains the ability to transcribe efficiently through the transcription termination site. The precise mechanism by which these proteins act in concert to relieve termination is not yet understood.

With termination suppressed, RNA polymerase can now continue to transcribe the downstream genes, which encode several proteins including those needed for phage chromosome replication. Another of the downstream genes encodes the lambda Q protein, which overrides transcription termination at the $P_{R'}$ promoter and allows transcription of the late lytic genes, as shown in Figure 9.23.

In contrast to the N protein, the Q protein prevents termination by binding directly to DNA. The binding site for the Q protein, called QBE, is located between the –35 and –10 regions of the $P_{R'}$ promoter, which is transcribed by RNA polymerase holoenzyme containing the σ^{70} subunit. In the absence of Q, transcription terminates just downstream of the –10 region. When sufficient Q protein is present, it binds to the QBE site and directly contacts the σ^{70} subunit, causing elongation to resume. The Q protein remains associated with the elongation complex as the downstream sequences are transcribed.

Attenuation is a mechanism for controlling the effect of terminator sequences

One form of regulation in bacteria takes advantage of the ability of certain mRNA sequences to form two alternative RNA secondary structures – one that leads to transcription termination and another to productive gene expression. Called **attenuation**, this mechanism is used to control transcription of genes encoding enzymes required for amino acid biosynthesis. Consequently, under certain conditions, expression of the full-length transcript is promoted, whereas, under other conditions, termination of transcription is favored.

An example of attenuation is found in the regulation of the *E. coli trp* operon, as illustrated in Figure 9.24. Transcription of the *trp* operon is governed by the levels of tryptophan present in the cellular environment, as noted in Section 9.3. Different levels of tryptophan also cause the transcript to adopt different secondary structures that directly influence how transcription proceeds. So how does this regulation occur in practice?

We learned in Section 9.3 that the genes of the *trp* operon are transcribed as a single **polycistronic message**. This mRNA contains a leader sequence near its 5′ end that contains an intrinsic terminator (also called an **attenuator**) along with two different additional RNA sequences capable of forming stem–loop structures.

The leader sequence, which encodes a small open reading frame that is rich in codons for the amino acid, tryptophan, contains four key blocks of ribonucleotides – regions 1, 2, 3, and 4 – which can form alternative pairing arrangements. Regions 1 and 2 are complementary and can form a stem–loop structure, while regions 3 and 4 can also base-pair and form a Rho-independent terminator. Regions 2 and 3 can form an alternative stem–loop structure, which then leaves regions 1 and 4 unpaired.

Figure 9.24 shows how these different possible pairings are exploited to regulate transcription. Once RNA polymerase transcribes the beginning of the leader sequence, the ribosome binds to the message and begins to translate the leader peptide while RNA polymerase continues to transcribe the downstream sequences. When levels of tryptophan are low, however, the ribosome stalls within the short open reading frame while waiting for a tRNA to be charged with tryptophan. This occludes region 1 in the leader sequence of the RNA, which then cannot base-pair with region 2. If region 4 has not yet been transcribed, regions 2 and 3 form an alternative stem–loop known as an anti-terminator. This prevents the terminator (the 3–4 pair) from forming and thereby ensures that transcription elongation will continue. At high levels of tryptophan, however, there is enough charged tRNA to allow the ribosome to move rapidly to the end of the small open reading frame. This allows the terminator stem–loop to form, leading to transcription termination.

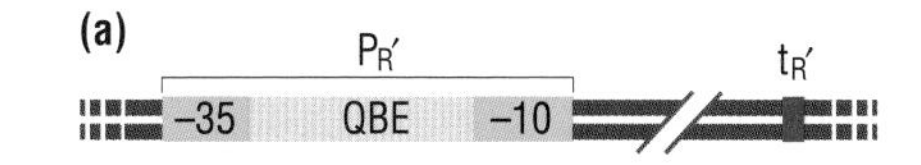

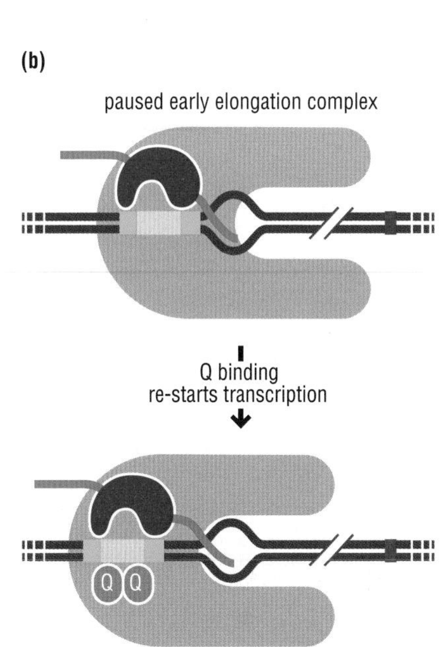

Figure 9.23 Anti-termination by the phage lambda Q protein. (a) The late lytic genes are under control of the P_R promoter, which is transcribed by the RNA polymerase holoenzyme containing σ^{70}. In the absence of Q, transcription pauses upstream of the lytic genes. (b) Q protein relieves termination by binding directly to a DNA sequence located between the –10 and –35 sequences in the promoter. Q interactions with σ^{70} relieve the pause, and the holoenzyme complex with Q resumes elongation.

➡ The process of tRNA charging is described in Chapter 11.

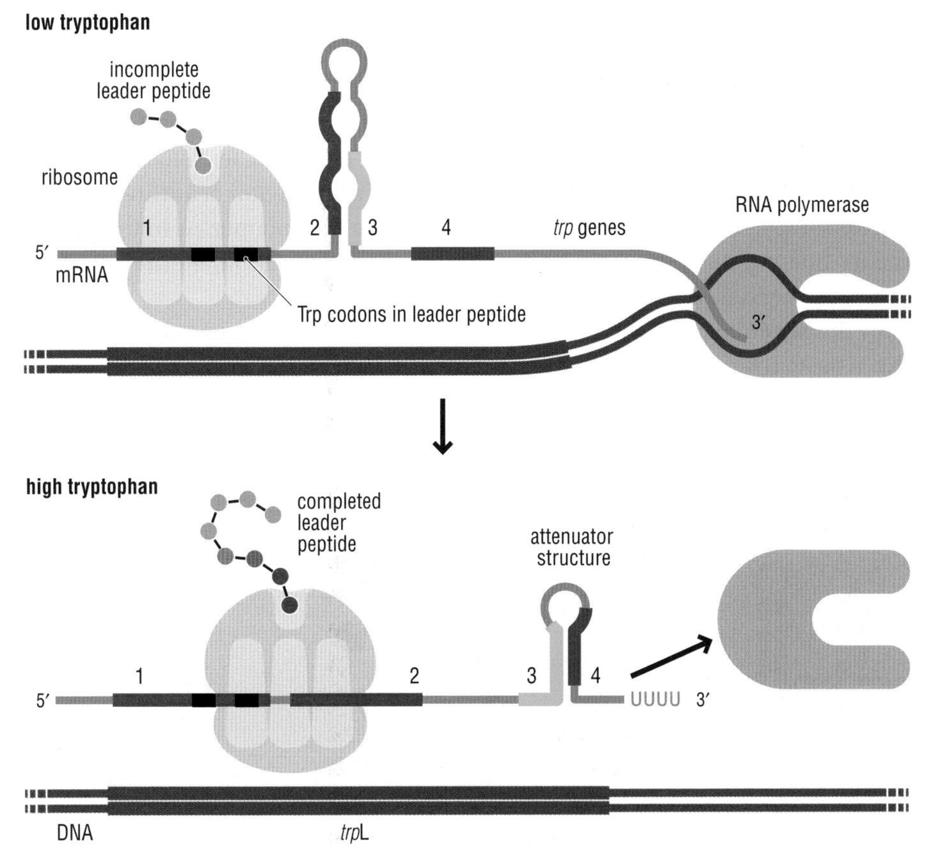

Figure 9.24 Transcription attenuation regulates amino acid biosynthesis in bacteria. The *trp* operon leader transcript can fold into two alternative RNA secondary structures. At low levels of tryptophan (top), the structure with the attenuator stem–loop 2:3 is formed, and transcription proceeds through the entire operon. At high levels of tryptophan (bottom), the structure with a stem–loop formed between RNA sequences 1:2 and a terminator stem–loop between sequences 3:4 is formed, leading to termination.

Direct binding of metabolites to RNA riboswitches can control transcription termination

The formation of alternative RNA structures can be regulated even more directly, for example, by the binding of small metabolites to specific regions of RNA transcripts denoted riboswitches. A **riboswitch** binds directly to a small molecule that controls the secondary structure that the RNA transcript adopts, which in turn determines whether a gene is expressed. Riboswitches can regulate a variety of processes; we focus here on their role in regulating transcription. Riboswitches are found primarily in bacteria – indeed, 2% of the genes in *Bacillus subtilis* are regulated by riboswitches – but have also been identified in plants and fungi and are thought to exist in archaea.

Riboswitches regulate transcription by controlling whether or not a terminator or anti-terminator stem–loop forms in the mRNA, which determines whether transcription is completed appropriately. All riboswitches (which we will learn more about in Chapter 13) contain two key regions, as illustrated in Figure 9.25a: the aptamer, a sometimes complex RNA structure that binds to the metabolite being sensed, and an expression platform, which consists of segments of the RNA that control the output, such as transcription. The binding of metabolite to the aptamer alters the structure of the riboswitch and thereby regulates whether or not the downstream gene is expressed. Examples of riboswitches include those regulated by purine nucleotides, thiamine pyrophosphate, *S*-adenosyl methionine, glycine, and lysine.

Purine riboswitches provide a simple illustration of how metabolite binding can regulate transcription. The adenine riboswitch of *B. subtilis* regulates the transcription of genes required for adenine synthesis and transport. Expression of these genes depends upon whether a terminator or an anti-terminator forms in the 5′ end of the mRNA transcript; the formation of a terminator or anti-terminator is determined, in turn, by the binding of adenine to the aptamer. As in the case of the *trp* operon, the RNA contains four regions that can pair in alternative ways; these four regions are shown in Figure 9.25b.

In the absence of adenine, the downstream genes are transcribed because the riboswitch forms the RNA secondary structure shown in Figure 9.25b,

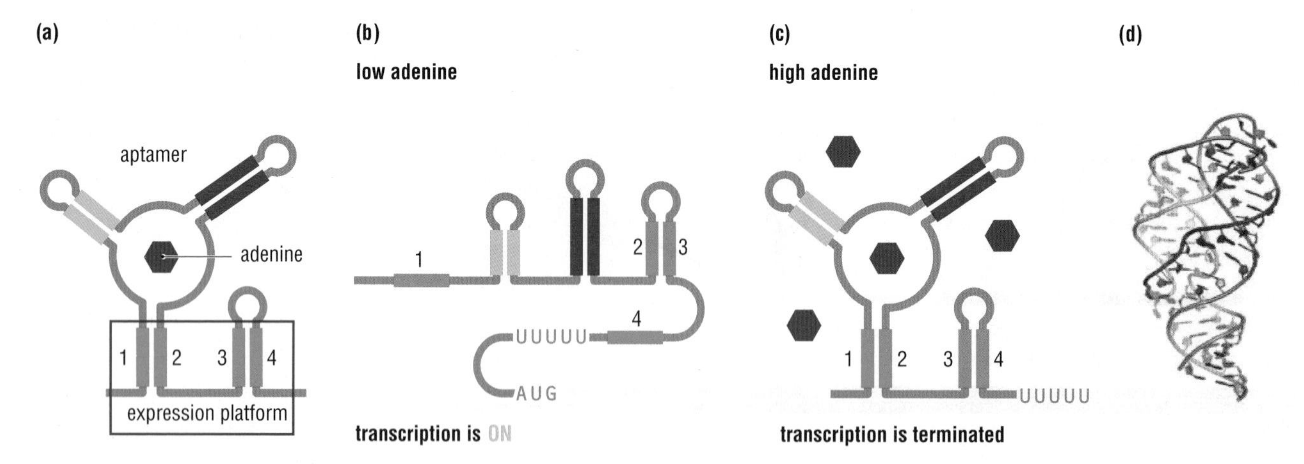

Figure 9.25 Riboswitch regulation of transcription termination. (a) All riboswitches contain an aptamer, which binds to the metabolite, and an expression platform. In the example shown, regions 3 and 4 of the RNA form a terminator. (b) The adenine riboswitch. When concentrations of adenine are low, the RNA forms the secondary structure shown, and the downstream gene is transcribed. The pairing of regions 2 and 3 form an anti-terminator, which prevents regions 3 and 4 from forming a terminator. (c) When the concentration of adenine is high, adenine binds directly to the aptamer and triggers formation of an alternative RNA structure. Regions 3 and 4 pair to form a terminator. (d) The structure of the adenine riboswitch (PDB code Y26).

with regions 2 and 3 forming an anti-terminator. By contrast, when the adenine concentration is high, adenine binds directly to the aptamer and triggers formation of an alternative structure shown in Figure 9.25c, in which regions 1 and 2 pair. This allows regions 3 and 4 to form a terminator, thereby suppressing transcription of the downstream genes. Figure 9.25d shows how the aptamer folds around the bound adenine nucleotide to form a globular three-dimensional structure.

➔ We give a comprehensive discussion of chromatin structure, histone modifications, and nucleosome remodeling in Chapter 4.

9.6 REGULATION OF TRANSCRIPTION INITIATION AND ELONGATION IN EUKARYOTES

Having covered key examples of gene regulation in bacteria, let us now consider several examples that illustrate some of the diverse ways in which transcription is regulated in eukaryotes. As in bacteria, transcription of most genes is regulated at the point of initiation. However, we will also see an example in which elongation is the step that is targeted to regulate synthesis of the complete mRNA.

We begin by exploring how transcription initiation by RNA polymerase II is regulated in eukaryotes.

ELK1 activates transcription by recruiting the mediator complex

We learned in Section 8.3 that the multisubunit Mediator complex is generally required to activate transcription by RNA polymerase II. Mediator is thought to stimulate transcription either by promoting assembly of the transcription pre-initiation complex, or by stimulating an increase in the rate of transcription initiation.

One way in which transcriptional activators can function is by recruiting the Mediator complex directly. The mammalian ELK1 protein activates transcription in response to extracellular signals known as mitogens, which are molecules that cause a cell to enter mitosis. The process by which Mediator is recruited by ELK1 is illustrated in Figure 9.26. In the absence of a mitogen, ELK1 binds DNA together with another protein (called serum response factor (SRF)), but does not activate transcription. However, mitogen binding to a cell surface receptor activates protein kinases in the cytoplasm, which subsequently phosphorylate the activation domain of ELK1. The phosphorylated activation domain recruits the Mediator complex by binding to a specific subunit in the complex (called Med23). A histone acetyltransferase complex is also recruited. This increases the rate of transcription initiation, thereby stimulating an increase in the overall level of transcription. Note that the DNA binding by these activators is not the critical aspect of their function that is regulated, nor does the binding of small molecules play a role in the same way as some of the bacterial regulators we discussed earlier. Instead, the response to an extracellular signal is communicated to the transcriptional regulator via an enzyme that phosphorylates ELK1, thereby altering its ability to recruit co-activator complexes.

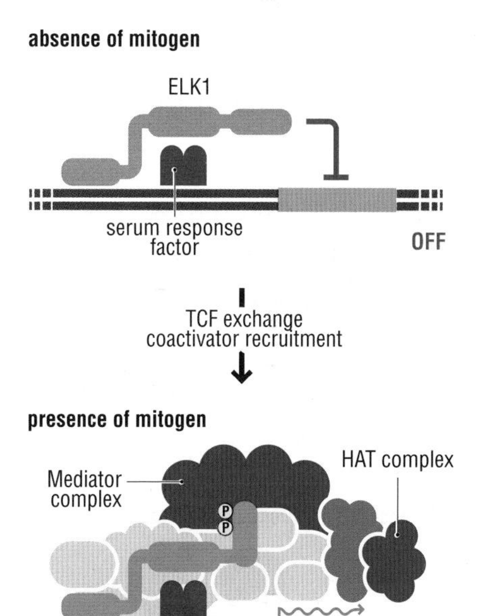

Figure 9.26 Recruitment of Mediator by the activation domain of ELK1. The ELK1 protein binds to DNA along with a second protein called serum response factor (SRF; magenta). When mitogens bind to receptors on the cell surface, this leads to activation of mitogen-activated protein (MAP) kinase, which phosphorylates the activation domain of ELK1. The phosphorylated ELK1 recruits the Mediator complex. A HAT complex is also recruited. As a result, the rate of transcription initiation is increased.

Gal4 recruits RNA polymerase through a co-activator complex

The regulation of transcription by the yeast Gal4 protein provides an example of a more complex interplay between several proteins that regulate transcription in response to the presence of a particular signal. Gal4 regulates genes that encode

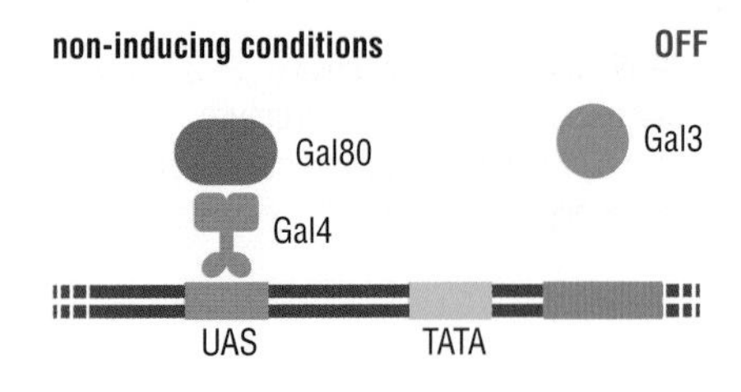

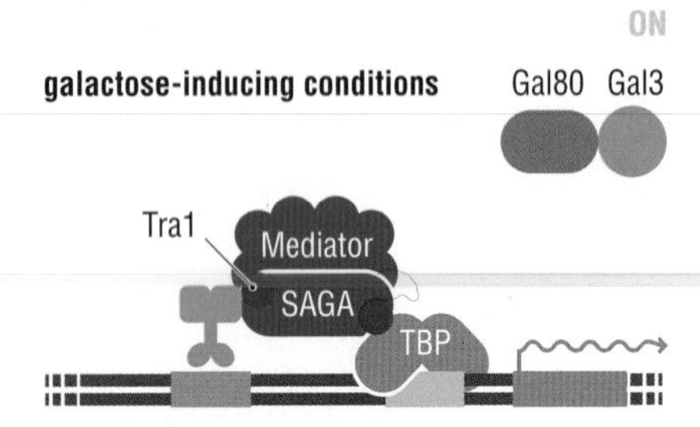

Figure 9.27 Mechanism of Gal4 recruitment of RNA polymerase II. In the absence of galactose (top), Gal4 is bound to a UAS, but Gal80 binds to Gal4 and blocks its ability to activate transcription. In the presence of galactose (bottom), Gal3 binds to Gal80 and transports it to the cytoplasm, relieving the inhibitory effect on Gal4. Gal4 is now able to recruit the SAGA co-activator complex, which in turn promotes assembly of the transcription initiation complex.

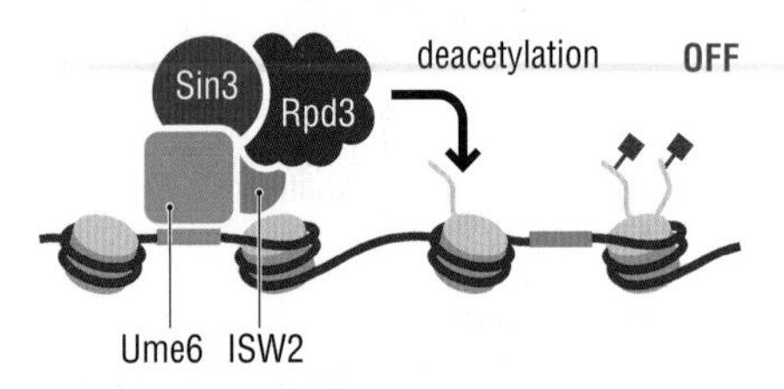

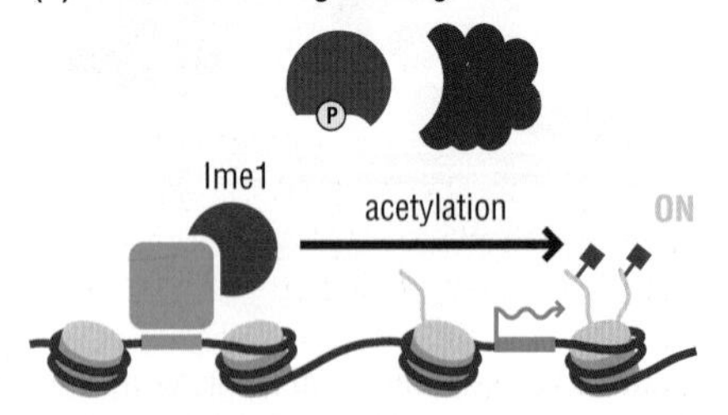

Figure 9.28 Transcriptional repression through recruitment of a histone deacetylase. The Ume6 protein binds DNA and recruits the Sin3/Rpd3 co-repressor complex, which deacetylates histone tails and helps cause transcriptional repression (top). Starvation triggers phosphorylation of Ume6, which causes Sin3/Rpd3 to dissociate and be replaced by a co-activator, Ime1 (bottom).

proteins required for galactose metabolism. When galactose levels are elevated, Gal4 activates transcription of genes encoding enzymes that convert galactose into glucose-1-phosphate (Figure 9.27). As is typical for most eukaryotic activators, Gal4 has separate DNA-binding and activation domains. The activation domain is rich in acidic amino acids, a property common to many, though not all, activation domains. The separation of the DNA-binding and activation functions of Gal4 into discrete domains has been exploited to produce chimeric proteins that are useful tools to molecular biologists, as will be discussed in Section 19.12.

Gal4 forms a dimer that binds to a target sequence known as the UAS_G (upstream activating sequence for Gal) found in the promoter regions of genes whose transcription is induced in the presence of galactose. Unlike the bacterial Trp repressor and CAP activator, Gal4 does not bind directly to the allosteric effector, galactose. Instead, the ability of Gal4 to activate transcription is controlled by two proteins, Gal80 and Gal3, as depicted in Figure 9.27.

In the absence of galactose, Gal80 binds to the activation domain of Gal4 and prevents it from activating transcription. Under these conditions, Gal80 can be found both in the nucleus and in the cytoplasm, whereas Gal3 is found only in the cytoplasm. When cellular galactose levels are elevated, galactose binds to Gal3, along with ATP. Binding of these ligands to Gal3 somehow alters its properties, causing Gal3 to bind to Gal80. Gal3 thus sequesters Gal80 in the cytoplasm, freeing the Gal4 activating region to recruit a multiprotein co-activator complex called SAGA.

The SAGA complex, which is conserved across all eukaryotes, has multiple functions in activating transcription. One of these functions it to recruit Mediator, which, together with SAGA, promotes assembly of the transcription initiation complex by contacting the basal transcription factors, which in turn help to recruit RNA polymerase II. SAGA also acetylates histones and removes monoubiquitin from histone H2B, both of which are thought to help make the chromatin template more accessible to RNA polymerase. SAGA and Mediator also recruit the SWI/SNF nucleosome-remodeling enzyme, which helps to clear nucleosomes from the promoter region and facilitates transcription through the chromatin template.

Ume6 represses transcription by recruiting a co-repressor complex with histone deacetylase activity

The majority of transcriptional regulators in eukaryotes activate transcription, but some proteins actively *repress* transcription of their target promoters. One protein that can both activate and repress transcription is the yeast Ume6 protein. Ume6 ensures that cells respond appropriately to their nutritional environment, binding upstream of genes whose expression is modulated in response to changes in the levels of metabolites such as glucose, nitrogen, and inositol.

When there is adequate nitrogen and a source of carbon, Ume6 represses transcription by binding DNA and recruiting co-repressor complexes, as illustrated in Figure 9.28a. One of these complexes contains a protein called Sin3 and a histone deacetylase called Rpd3. In addition, Ume6 recruits a nucleosome-remodeling enzyme, Isw2. The deacetylase targets the N-terminal tails of adjacently bound histones, removing acetyl groups from the lysine residues. Deacetylation of residues of the tail of histone H4, in particular, allows more compact chromatin states to form, and consequently represses transcription. At the same time, the nucleosome-remodeling enzyme helps to establish an altered chromatin structure that further contributes to repression of transcription. We see from this example that nucleosome-remodeling enzymes can either repress transcription or, as in the case of Gal4 above, activate transcription, depending upon how it repositions nucleosomes.

In the absence of nitrogen and glucose, however, Ume6 activates transcription, as shown in Figure 9.28b. The lack of these specific nutrients triggers phosphorylation of Ume6, which causes Sin3 and Rpd3 to dissociate. Ume6 instead recruits a transcriptional co-activator, Ime1, which now activates transcription.

➲ We learn about molecular chaperones in Chapter 14.

Heat-shock factor regulates transcription elongation by relieving promoter proximal pausing

We have now considered several examples in which mRNA levels are regulated by controlling whether or not transcription of a given gene is initiated. Yet, even after RNA polymerase has initiated transcription and broken free of the promoter, there are additional points at which the cell regulates whether the full-length transcript is made.

We learned in Section 8.5 that RNA polymerase sometimes stalls after transcribing a few dozen nucleotides, a phenomenon known as promoter proximal pausing. Restarting a paused polymerase is thus another way in which transcription can be activated. An example can be found in the transcription of the *Drosophila hsp70* gene. The Hsp70 protein is a molecular chaperone that plays a variety of protective roles when cells are subject to abnormally high temperatures, a condition known as heat shock. The *hsp70* gene is one of several genes needed under these conditions, which are collectively known as heat-shock genes.

One of the functions of Hsp70 is to prevent aggregation of partially synthesized or unfolded proteins, which can accumulate rapidly under conditions of heat shock. The cell thus needs to turn on Hsp70 synthesis rapidly to avoid the toxic effects of protein aggregation. To enable rapid activation, *hsp70* and other heat-shock genes have a paused RNA polymerase that is activated upon heat shock, thus leading to a rapid transcriptional response. The discovery of the transcriptional response to heat shock in *Drosophila* is described in Experimental approach 9.3.

The regulation of the *hsp70* gene under normal and heat-shock conditions is illustrated in Figure 9.29. Upstream of the *hsp70* TATA elements, there are multiple binding sites for two sequence-specific DNA-binding proteins, GAGA factor and heat-shock factor (Hsf). Under normal conditions, GAGA factor binds to the *hsp70* promoter and recruits the ATP-dependent nucleosome-remodeling factor (NURF). NURF helps to maintain the promoter in a nucleosome-free open configuration that allows binding of the RNA polymerase II basal transcription machinery. The polymerase enzyme can initiate transcription and synthesizes transcripts of 21–35 nucleotides in length. However, at normal temperatures of around 25°C, the polymerase enzyme pauses due to the action of negative elongation factors, NELF and DSIF, because it fails to undergo sufficient phosphorylation to be converted into a productive elongating complex (see Section 8.5).

The stalled polymerase can be restarted by a sudden rise in temperature, leading to expression of the heat-shock gene and synthesis of Hsp70. When subjected to heat shock, in which the temperature is shifted to 37°C, the previously monomeric Hsf protein forms trimers that are able to bind to DNA sequences called heat-shock elements (HSEs) in the *hsp70* promoter, as illustrated in Figure 9.29b. The conversion of Hsf from the monomeric to trimeric form is thought to involve a conformational change, likely caused by heat-induced redirection of chaperone

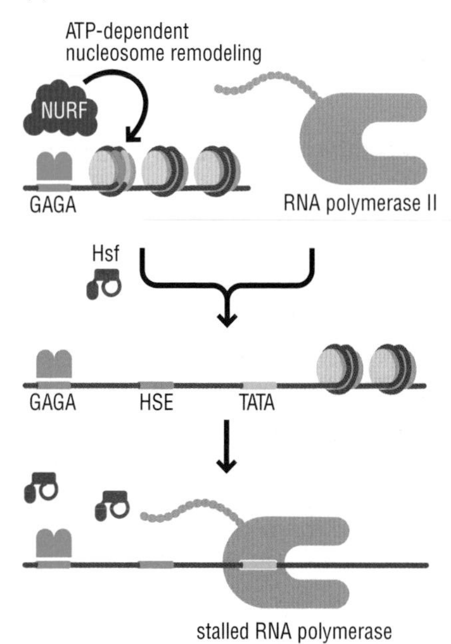

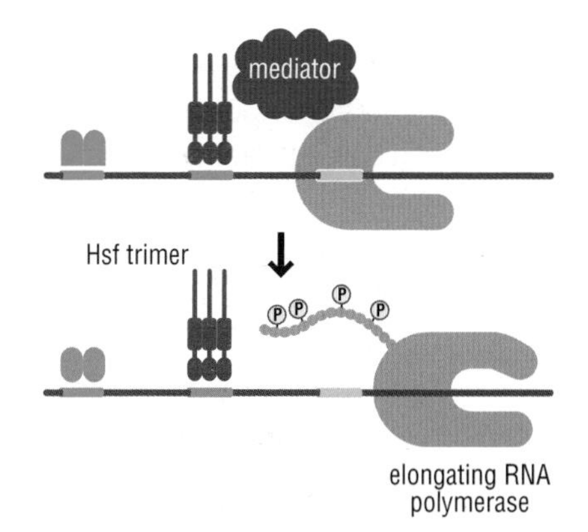

Figure 9.29 Mechanism of heat-shock factor (Hsf) conversion of RNA polymerase II into a productive elongating complex. The *Drosophila hsp70* gene contains upstream binding sites for the GAGA and Hsf regulatory proteins. (a) Under normal conditions, Hsf monomers are present in the cell but cannot bind DNA. GAGA binds DNA and recruits the nucleosome-remodeling factor (NURF). RNA polymerase II synthesizes a short stretch of RNA and then stalls. (b) Under conditions of heat shock, the Hsf monomers trimerize to form an active DNA-binding complex. The DNA-bound Hsf trimer contacts the Mediator complex, helping to rescue the stalled polymerase and convert it into a productive elongating complex.

9.3 EXPERIMENTAL APPROACH

Serendipitous discovery of the heat-shock response opened the door for numerous studies of transcriptional regulation

A laboratory mishap leads to discovery of the Drosophila heat-shock response

Sometimes new directions in research are brought about by serendipity. This is the case for studies of the heat-shock response, which have yielded many important insights into transcriptional regulation in both bacteria and eukaryotes because the response is rapid and easy to induce. That an increase in temperature led to a rapid change in transcription was discovered by Ferruccio Ritossa in 1962 when someone in the laboratory changed the temperature of the incubators in which he was maintaining his stocks of *Drosophila*. At the time, Ritossa was examining the polytene chromosomes from the salivary glands of *Drosophila* larvae (which are easy to see under the microscope because they are polyploid). He and others were finding that different regions of the chromosomes 'puffed up' during different developmental stages and that these 'puffs' corresponded to regions where radioactive uridine was incorporated, indicating that RNA was being made. As shown in Figure 1, Ritossa reported that if the larvae were shifted from 25°C to 30°C for 30 minutes, the puffing pattern changed markedly at a defined number of chromosomal positions.

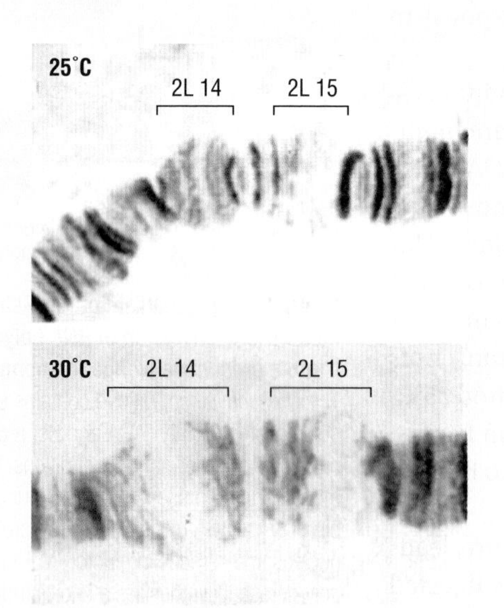

Figure 1 Increased transcription of heat shock puffs. The 2L 14 and 15 regions of the salivary gland chromosome of *Drosophila buschii* larvae reared at 25 C (a) and 30°C (b).

Reproduced with permission from Figures 1 and 2 in Ritossa, A new puffing pattern induced by temperature shock and DNP in Drosophila. *Experimentia*, 1962;**18**:517–573 with kind permission from Springer Science & Business Media B.V.

This exciting discovery, brought about because of a change in incubator temperature, spawned many important research directions that led to the discovery of protein chaperones (discussed in Chapter 14), and produced many insights into transcriptional regulation. A key question that arose immediately was how the transcription of heat-shock genes is regulated, aspects of which are still under investigation today.

Heat shock induces a strong transcriptional response

Hugh Pelham made contributions to answering this question by defining the promoter regions of the *Drosophila hsp70* promoter required for the heat-shock induction. He cloned portions of the *hsp70* gene into an SV40 vector (which contains the replication origin from the SV40 virus) and transfected this vector into COS cells (monkey cells) (see Section 19.4 for a discussion of cloning technology). By following the end of the *hsp70* transcript using 5′ S1 mapping (described in Section 19.10), he showed that transcription of the *hsp70* promoter could be induced in a heterologous system; his results are shown in Figure 2. By deleting different regions of the *hsp70* promoter, he further showed that induction was lost when sequences downstream of –66 were deleted. These experiments indicated that sequences between –10 and –66 were sufficient for the heat-shock induction.

The very strong and clear transcriptional induction of the specific genes upon heat shock, as is shown in Figure 2, has been exploited in a wide range of experiments, including the development of microarrays to examine the transcript levels of thousands of genes in parallel. When Patrick Brown, Ronald Davis, and colleagues first set out to develop DNA arrays by attaching (printing) 1046 human cDNAs to glass slides, heat shock was one of the first conditions they tested. Even though only a limited number (by today's standards) of cDNAs were printed on the slides, they discovered new genes whose expression was induced when human T (Jurkat) cells were subjected to heat shock, as depicted in Figure 3. In the experiment shown, the cDNAs prepared from samples grown under two conditions (37°C and 43°C) were labeled through the incorporation of two different fluorescent nucleotides, fluorescein-dCTP or Cy5-dCTP. When the resulting two cDNA populations were hybridized to the array, differences in the ratio of the two dyes indicated whether a given gene was induced or repressed upon the heat shock treatment.

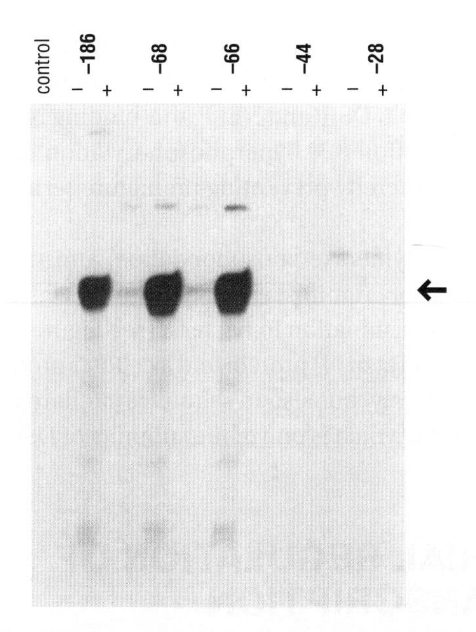

Figure 2 Promoter regions required for heat shock induction. S1-mapping of *hsp70* transcripts (indicated by the arrow) synthesized in COS cells for different deletion mutants of the *hsp70* promoter without (–) and with heat (+).

Reproduced with permission from Pellham, A regulatory upstram promoter element in the *Drosophila hsp 70* heat-shock gene. *Cell*, 1982;**30**:517–528.

Find out more

Pelham HR. A regulatory upstream promoter element in the *Drosophila hsp 70* heat-shock gene. *Cell*, 1982;**30**:517–528.

Ritossa F. A new puffing pattern induced by temperature shock and DNP in *Drosophila*. *Experientia*, 1962;**18**:517–573.

Schena M, Shalon D, Heller R, *et al*. Parallel human genome analysis: microarray-based expression monitoring of 1000 genes. *Proceedings of the National Academy of Sciences of the U S A*, 1996;**93**: 10614–10619.

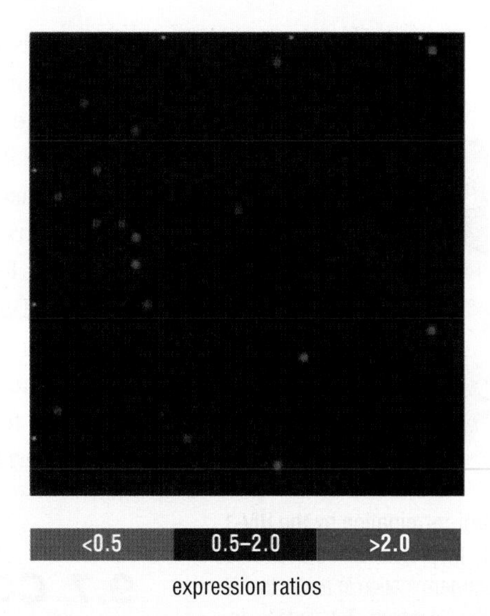

Figure 3 Identification of new heat shock genes by microarray analysis. Fluorescein-labeled probes from Jurkat cells subjected to heat shock were compared with Cy5-labeled probes from untreated cells. Visible spots represent genes where differences in expression are readily detected.

Reproduced with permission from Schena *et al.*, Parallel human genome analysis: microarray-based expression monitoring of 1000 genes. *Proceedings of the National Academy of Sciences of the U S A*, 1996;**93**:10614–10619 with kind permission from Springer Science & Business Media B.V.

Related techniques

Drosophila (as a model organism); Section 19.1

5′ S1 mapping; Section 19.10

DNA microarrays; Section 19.10

Use of cDNA; Section 19.3

proteins, that leads to the formation of the active trimer. Once bound to DNA, Hsf contacts the Mediator complex, recruiting it to the promoter. This complex in turn interacts with other factors, including a kinase (p-TEFb) that phosphorylates the CTD of the Rpb1 subunit of RNA polymerase II and the pausing complexes of NELF and DSIF, thereby converting the paused enzyme into a productive elongating complex. The *hsp70* gene is transcribed and the encoded protein synthesized, allowing the cell to mount a protective response.

The human immunodeficiency virus-1 anti-termination protein Tat functions similarly to lambda N

Transcription of the human immunodeficiency virus (HIV)-1 genome depends upon a viral anti-termination factor whose function is similar to that of the phage lambda N protein. The HIV-1 genome contains a sequence called *tar* that, much like the phage lambda *nut* site, causes formation of an RNA stem–loop in the transcribed RNA called the TAR element. Formation of the stem–loop leads to premature termination of transcription without the action of Tat, a small viral protein

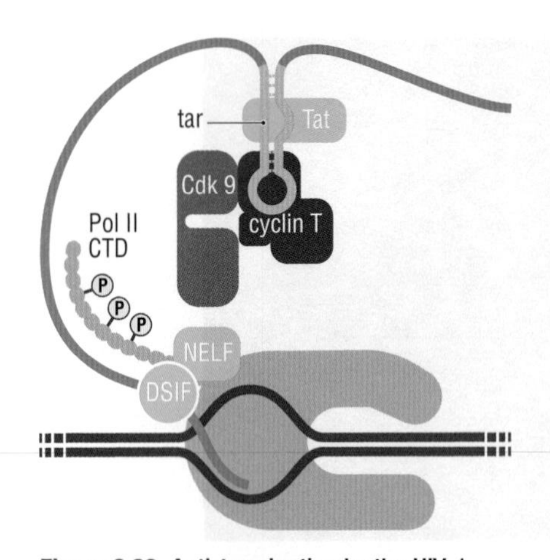

Figure 9.30 Anti-termination by the HIV-1 Tat protein. The Tat protein from the retrovirus HIV-1 causes anti-termination in a way that is reminiscent of the l N protein. Tat binds to an RNA sequence in the HIV-1 genome called *tar*, causing premature termination. The association of the Tat–*tar* complex with several cellular proteins allows RNA polymerase to overcome the block and continue transcribing. CTD, C-terminal domain.

that binds to the *tar* sequence. Tat binds cooperatively to *tar* together with the cellular kinase P-TEFb, which phosphorylates the CTD of RNA polymerase II as well as the two regulators of elongation, DSIF and NELF; this binding is illustrated in Figure 9.30. As we learned in Section 8.5, hyperphosphorylation of the RNA polymerase CTD relieves pausing, thereby preventing premature termination of the viral transcript.

Except for the involvement of polymerase phosphorylation, anti-termination by the HIV-1 Tat protein is remarkably reminiscent of the way in which lambda N protein overcomes transcription termination in bacteria (see Figure 9.22). This extends not only to the use of a viral (Tat or N) protein and a cellular protein (Cdk9 or NusB) to bind the RNA cooperatively, but also to sequence similarity between the bacterial NusG protein and Spt5, which is one of the subunits of DSIF.

9.7 COMBINATORIAL REGULATION OF EUKARYOTIC TRANSCRIPTION

We saw in Section 9.4 how the bacteriophage lambda uses a combination of regulatory proteins to modulate transcription. In this section we see how a similar network of regulators is often employed in eukaryotic systems.

The complexity of higher organisms depends on elaborate regulatory networks that can respond differently to a variety of signals. A response to multiple signals can be accomplished by regulating a single gene with several different proteins, each of which is controlled by a different parameter. Whether or not the target gene is transcribed depends upon the particular combination of proteins that binds to the regulatory sequences. This type of regulation, termed **combinatorial control**, is a hallmark of eukaryotic gene regulation. Regulating a gene with multiple proteins also makes it possible to activate (or repress) a gene only when a certain set of conditions are met. Regulating transcription with more than one protein also provides a simple way to regulate genes that specify cell type.

Yeast cell type is determined by different combinations of four transcriptional regulators

A remarkably elegant combinatorial circuit controls the genes that determine cell type in the budding yeast, *Saccharomyces cerevisiae*. These single-celled eukaryotes can exist as one of three cell types: **a**, α, or **a**/α cells, as depicted in Figure 9.31. The **a** and α cell types are haploid, meaning that they each have just a single complement of chromosomes. The **a**/α cell type is diploid, and forms when an **a** cell fuses with an α cell in a process that is known as **mating**. Each haploid cell type produces a distinct type of signaling molecule known as a pheromone that is secreted into the surrounding medium. These **a** cells and α cells also express pheromone receptors on their surfaces that detect the presence of an opposite cell type. Only haploid cells of opposite **mating type** can mate with one another – that is, **a** and α cells can mate, but not two **a** cells, or two α cells. The diploid **a**/α cell that results from mating of an **a** and an α cell is incapable of mating but, under conditions of starvation, can undergo meiosis to form haploid spores.

➔ We learn about meiosis in Section 7.7.

The different phenotypes of **a**, α, and **a**/α cells result from distinct patterns of gene expression in each cell type; these expression patterns are summarized in Figure 9.32. The expression patterns in each cell type are specified by four

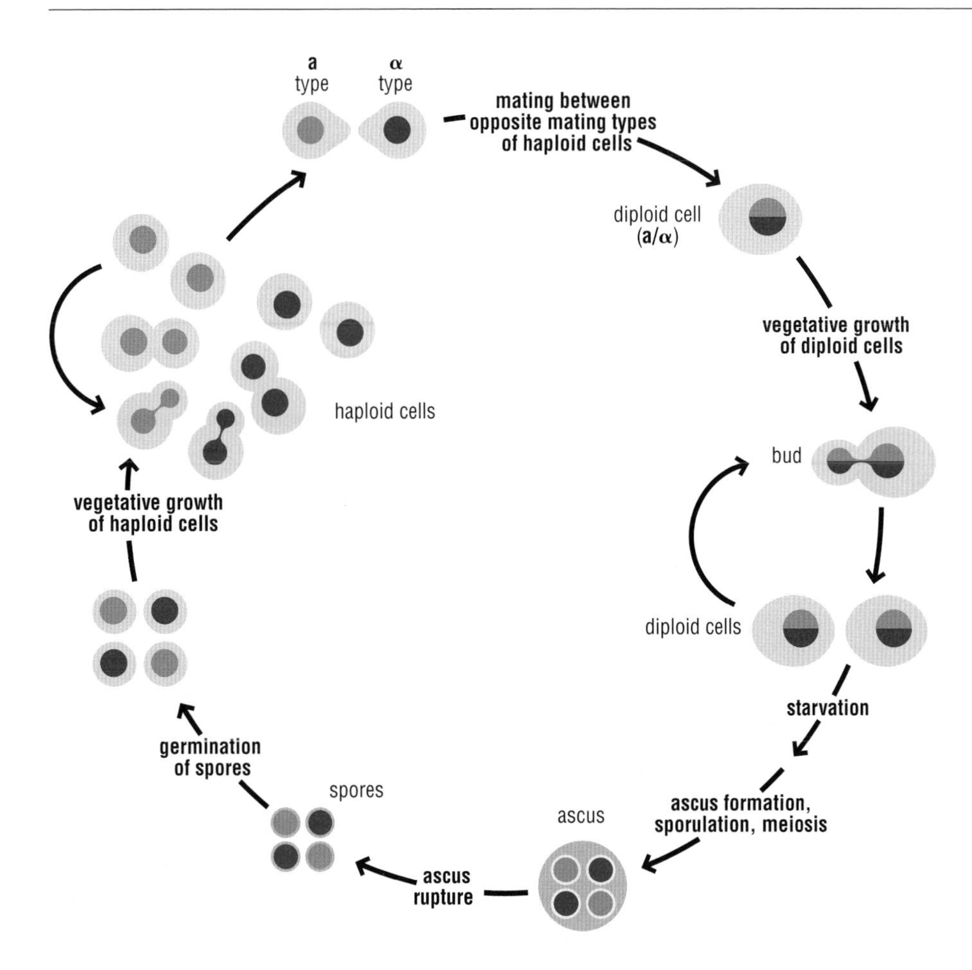

Figure 9.31 The three yeast cell types. Budding yeast can exist as one of three cell types: the a or α haploid cell types, or the a/α diploid. a and α cells mate by fusing with one another, forming diploid a/α cells, which are incapable of mating but can continue to bud and yield mother and daughter cells. When deprived of adequate nutrients, a diploid cell will undergo meiosis to produce four haploid cells (spores). The spores can commence vegetative growth under favorable conditions, thus resuming the cycle.

proteins, a1, α1, α2, and MCM1, that bind DNA in different combinations. The a1, α1, and α2 regulatory proteins are encoded at the ***MAT* locus**, which is located on yeast chromosome III. The DNA sequence of this segment of chromosome III differs with cell type: **a** cells encode the a1 protein at the *MAT* locus, while α cells encode the α1 and α2 proteins. The MCM1 protein is expressed in all cell types.

In **a** cells, the genes whose transcription confers the **a** phenotype – the **a**-specific genes, which encode proteins that confer the **a** phenotype, are activated by a cellular protein called Mcm1, as depicted in Figure 9.33a. The α-specific genes, which require an activator to be expressed, are not transcribed in **a** cells. The a1 protein, although present, does not bind DNA in **a** cells but plays an important role in diploid cells, as explained below.

Figure 9.33b shows how the pattern of activation and repression is reversed in α cells: the α1 protein activates transcription of α-specific genes while α2 represses transcription of **a**-specific genes by recruiting a co-repressor complex called Tup1/Ssn6. Interestingly, both α1 and α2 exert their effect in concert with the cellular Mcm1 protein, which activates transcription when recruited by α1 but helps to repress transcription when in complex with α2.

The fusion of an **a** and an α cell during mating gives rise to a new regulatory scenario in the resulting **a**/α diploid cell, as shown in Figure 9.33c. The a1 and α2 proteins combine to form a heterodimer that binds to a different set of DNA sites than either α2 or a1 alone. The a1/α2 heterodimer represses transcription of a variety of genes that are needed only in haploid cells. One of the genes repressed by a1/α2 encodes α1; without this protein, the α-specific genes are not activated. As in α cells, the α2 protein also represses transcription of **a**-specific genes. With no expression of either **a** or α genes, the diploid cells are incapable of mating.

	a-specific genes	α-specific genes	haploid-specific genes
a-cell (haploid)	ON	OFF	ON
α-cell (haploid)	OFF	ON	ON
a-cell (diploid)	OFF	OFF	OFF

Figure 9.32 Expression patterns in the three yeast cell types. Three classes of genes are important in specifying yeast cell type.

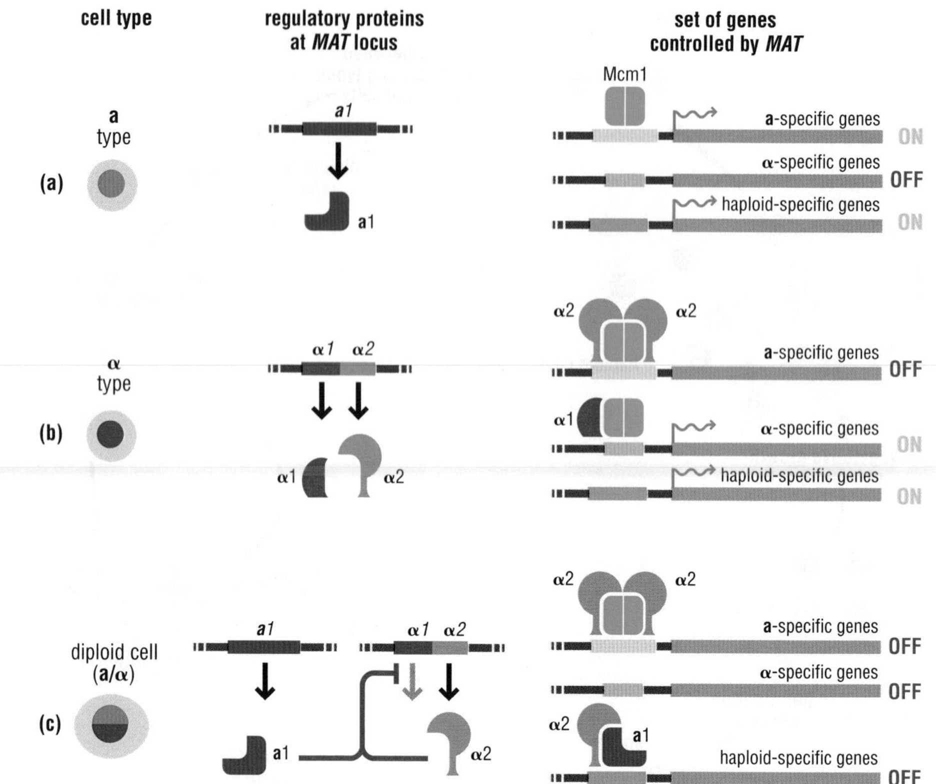

Figure 9.33 Control of cell type in budding yeast. The budding yeast *S. cerevisiae* can exist as one of three cell types: the a or α haploid cell types, or the a/α diploid. Each cell type is established by regulating three groups of genes: the a-specific genes, the α-specific genes, and the haploid-specific genes. These genes are regulated by proteins expressed by the MAT locus.

One property that distinguishes haploid from diploid cells is that only diploid cells can undergo meiosis. The genes required for meiosis are therefore not transcribed in haploid cells. One of these haploid specific genes is *RME1*, which encodes a DNA-binding protein that represses genes required for meiosis. When a1/α2 repress transcription of *RME1*, the Rme1 repressor protein is no longer synthesized, and the genes required for meiosis can be expressed. Diploid cells are thus able to undergo meiosis to produce haploid spores when the yeast cells are starved.

Regulation of the human interferon-α enhancer integrates signals from several pathways

The regulation of the human interferon-β gene provides an example of how synergy between several regulatory proteins makes it possible to regulate transcription of a single gene in response to multiple signals. The interferon protein encoded by the interferon-β gene helps to defend cells from viral infection by inhibiting synthesis of the viral genome and stimulating the host immune response. When viruses initially infect human cells, genes that encode interferon proteins are induced, allowing the cells to mount an antiviral response before the immune system has had time to respond.

Induction of the interferon-β gene depends on the binding of a number of different proteins to the interferon-β enhancer. As shown in Figure 9.34, these proteins bind to precisely positioned sites within the enhancer to form a complex of DNA-bound transcriptional regulators called the **enhanceosome**. Cooperative interactions between neighboring proteins are required for stable complex assembly. These interactions are facilitated by binding of an architectural DNA-binding protein called HMG-I(Y) (see Figure 9.34a), which helps to nucleate assembly of the complex of DNA-bound transcriptional regulators, although it is not present

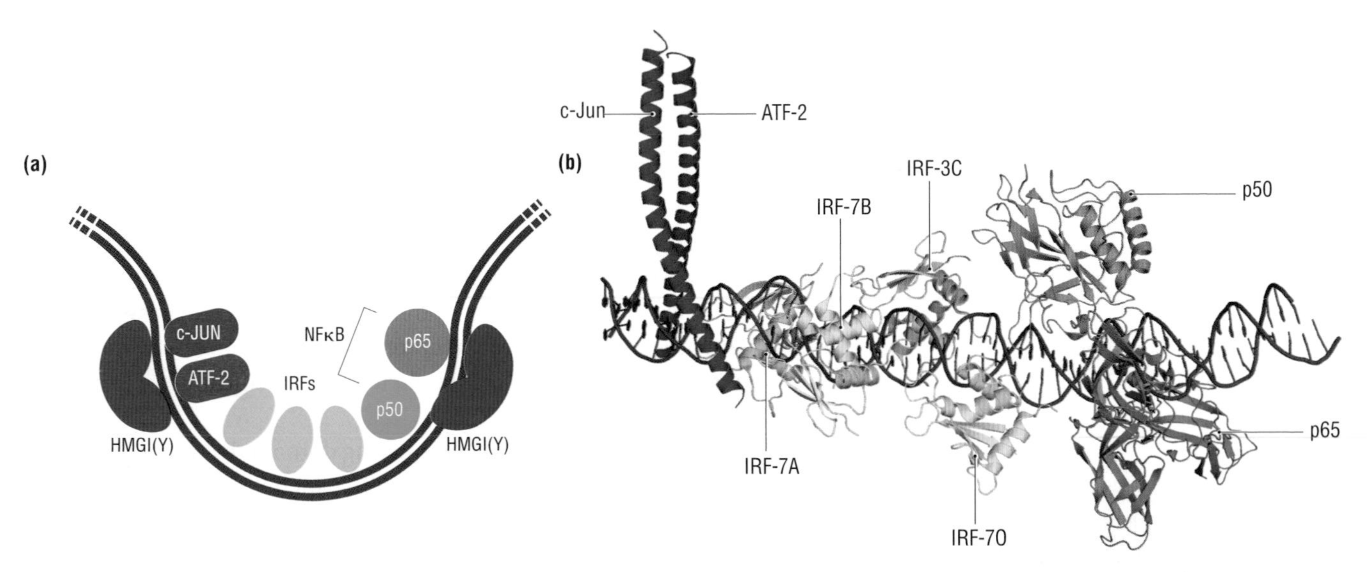

Figure 9.34 Proteins bound to the interferon-β enhanceosome. (a) Multiple proteins are required to bind cooperatively to the interferon-β enhancer to trigger activation of the interferon-β gene. The architectural DNA-binding protein HMG-I(Y) (dark blue) introduces pronounced bends into the DNA, which facilitates the binding of the other transcriptional regulators. (b) A model of the regulatory proteins bound to the interferon-β enhanceosome based on crystal structures. C-Jun and ATF2 are bZIP DNA-binding domains, p65 and p50 belong to the Rel homology domain family, and IRFeA, 3C, 7B, and 7D all belong to the IRF family. Model a hybrid of PDB codes 2O6G, 2O61, 1T2K.

from Panne, D, Maniatis, T, and Harrison, SC. An Atomic Model of the Interferon-β Enhanceosome. *Cell*, 2007;**129**:1111–1123.

in the final assembled complex, whose structure is shown in Figure 9.34b. When HMG-I(Y) binds DNA in the first step in enhanceosome assembly, it introduces bending of the DNA, which facilitates binding of the other transcriptional regulators. In the final complex, however, the DNA is not bent.

All of the proteins bind to the DNA in a highly cooperative manner, with the binding of certain proteins early in complex formation facilitating the binding of further components of the enhanceosome. Once the enhanceosome is complete, the interferon-β gene is activated. The interferon-β gene is not transcribed in cells that synthesize only a subset of the activator proteins. In general, the requirement for using a large number of proteins to activate a single gene ensures that the gene is activated only under a very precise set of conditions.

9.8 THE ROLE OF SIGNALING CASCADES IN THE REGULATION OF TRANSCRIPTION

Cells are continually altering their patterns of gene expression in order to respond to changes in their surroundings. Shifting hormonal levels within an organism, alterations in the environment, and responses to infections are all examples of changes that trigger alterations in gene expression. We have already seen several examples in which changes in the level of sugars or amino acids within the cellular environment can lead to increased or decreased transcription of particular genes. However, extracellular signals can also trigger a variety of biochemical events whose end result is a change in gene expression. These signaling cascades can range from the simple to the elaborate and can change the activities of gene-specific regulators through a variety of mechanisms. In this section, we describe several examples that illustrate how cells modulate transcription patterns in response to signals from the environment.

Nuclear receptor proteins respond to effector molecules that diffuse through the membrane

One of the simplest ways in which an extracellular signal can be transmitted to the nucleus is through steroid hormones, which are allosteric effectors of regulatory proteins. These signaling molecules include the sex hormones, estradiol (commonly known as estrogen), and testosterone, as well as thyroid hormone, which controls metabolism. These hormones are lipid soluble and can therefore diffuse through membranes into the cell, where they bind directly to **nuclear receptor proteins**. At least 50 members of the nuclear receptor superfamily have been identified in the human genome, each of which is thought to bind specifically to a particular signaling molecule.

Scan here to watch a video animation explaining more about the regulation of transcription, or find it via the **Online Resource Center** at www.oxfordtextbooks.co.uk/orc/craig2e/.

Nuclear receptor proteins contain a DNA-binding domain that recognizes specific DNA sequences and a ligand-binding domain, which binds the hormone or other small ligand. The binding of ligand induces a conformational change in the ligand-binding domain of the receptor. Rather than affecting the binding of the receptor to DNA, as in the case of the bacterial Trp repressor, the ligand-induced conformational change instead affects the ability of the nuclear hormone receptor to recruit different co-repressor and co-activator complexes to the DNA, as shown in Figure 9.35a.

While most nuclear receptor proteins bind DNA as homodimers, some, such as the thyroid hormone receptor–retinoid X receptor, can bind DNA as heterodimers. The ability to heterodimerize makes more complex regulatory networks possible, since two proteins that are capable of forming either homodimers or heterodimers can give rise to three different types of dimers (two different homodimers and one heterodimer). Moreover, if the level or activity of each protein is regulated in response to different signals, the heterodimer will of necessity by regulated by both sets of signals. In the absence of thyroid hormone, for example, the thyroid hormone receptor–retinoid X receptor heterodimer recruits two complexes, N-CoR and SMRT, which are co-repressors that contain histone deacetylase subunits. In contrast, the thyroid-hormone–bound heterodimer recruits two other proteins, CREB-binding protein (CBP) and P300/CBP-associated factor (PCAF), which are activators with histone acetyltransferase activity.

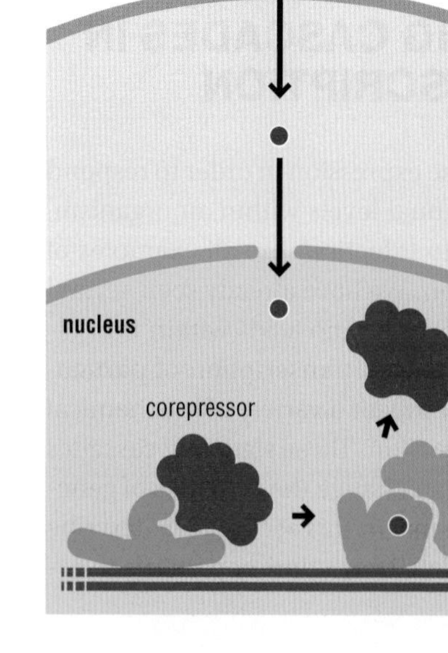

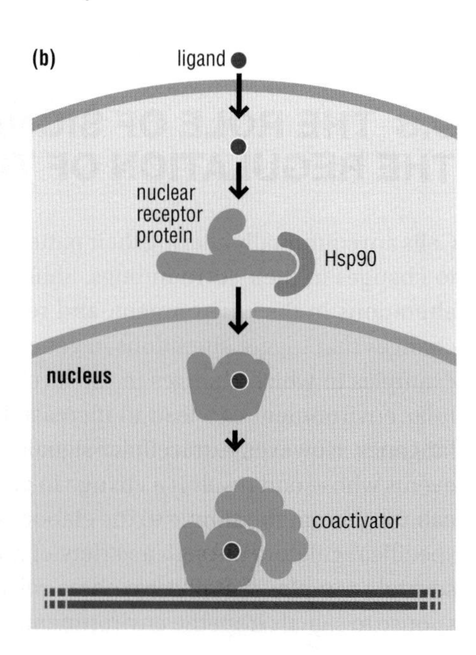

Figure 9.35 Co-activator recruitment and nuclear translocation of steroid hormone receptors upon hormone binding. (a) Some nuclear receptor proteins bind to DNA in the absence of ligand. These receptors recruit co-repressor complexes in the absence of ligand, and co-activator complexes in the presence of ligand. (b) Some nuclear receptor proteins are found in the cytoplasm, sequestered by the heat-shock protein HSP90. Ligand binding leads to dissociation of HSP90, allowing the receptor to enter the nucleus, bind to DNA, and recruit a co-activator complex.

In the case of other nuclear receptor proteins such as the glucocorticoid receptor, ligand binding also triggers translocation of the protein to the nucleus. In the absence of hormone, glucocorticoid receptor is in a complex with an inhibitory protein, HSP90, which causes the glucocorticoid receptor to remain in the cytoplasm. The binding of hormone causes HSP90 to dissociate and allows the receptor to enter the nucleus, as depicted in Figure 9.35b.

The very direct effect of ligand binding on transcription makes it possible to design drugs that have potent effects on nuclear receptor signaling pathways. Tamoxifen, for example, is highly successful in treating some types of breast cancer in which the tumor cells require estrogen to proliferate. Tamoxifen competes with estradiol for binding to the estrogen receptor protein but its different molecular structure prevents the ligand-binding domain from adopting its normal estradiol-bound conformation. The altered protein conformation in the presence of tamoxifen prevents the estrogen receptor from recruiting the co-activator complex, thereby blocking activation of estrogen-responsive genes and preventing cell proliferation.

➔ Ubiquitin and the proteasome are described in detail in Chapter 14.

A cascade of events involving protein phosphorylation, proteolysis, and nuclear translocation regulates NF-κB activity

Many signaling cascades involve multiple steps and may take advantage of a number of different mechanisms for transmitting a signal from one molecule to the next. This is particularly common in metazoans, which require more complex regulatory networks than unicellular organisms. NF-κB, a DNA-binding dimer that plays a key role in mammalian inflammatory and immune responses, provides a well-characterized example of a multistep cascade of events that leads to transcriptional activation. Key events in this cascade are summarized in Figure 9.36.

NF-κB is a heterodimer of the p50 and p65 proteins. As mentioned earlier, these DNA-binding domains belong to the Rel homology domain family. In unstimulated cells, the NF-κB heterodimers are held in the cytoplasm by an inhibitory protein called I-κB, which binds to a specialized region of NF-κB (the nuclear localization signal) that is responsible for targeting NF-κB to the nucleus. However, extracellular signals that are induced by infection trigger a cascade of events that lead to the activation of the I-κB kinase (IKK). Activated IKK phosphorylates two serine residues in the regulatory domain of I-κB. The phosphorylated I-κB protein is now recognized by an enzyme called an E3 ubiquitin ligase, which attaches a chain of ubiquitin proteins to I-κB. This ubiquitin chain targets IκB for degradation by a large enzymatic complex called the **proteasome**. Thus the ubiquitin-tagged I-κB is degraded, exposing the nuclear localization signal of NF-κB and freeing the protein to translocate to the nucleus, where it binds DNA and activates gene expression.

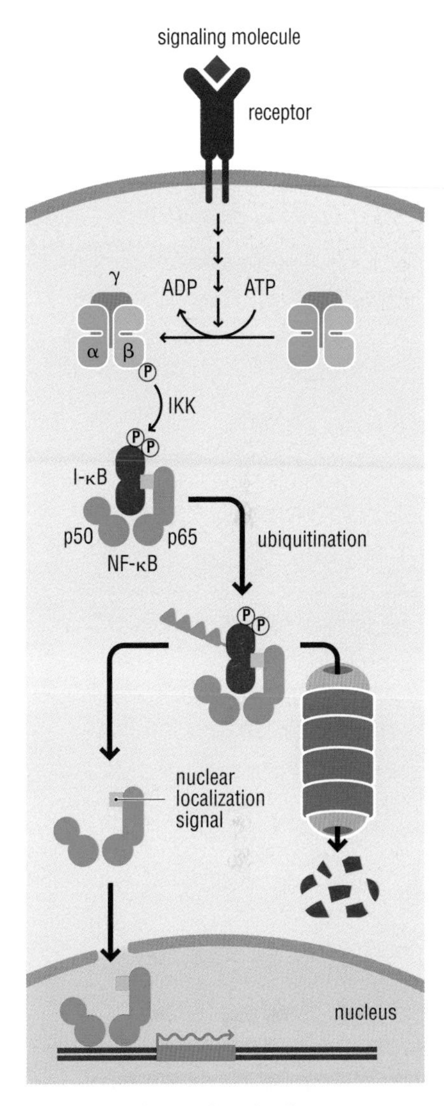

Figure 9.36 Change in subcellular localization of nuclear factor (NF)- κB in response to a diverse set of extracellular signals. In response to extracellular signals, a cascade of events (arrows) leads to the phosphorylation and thus the activation of the inhibitory (I)- κB kinase (IKK) (blue). Activated IKK then phosphorylates the regulatory domain of I-κB (red). Phosphorylated I-κB is modified with a K48-linked polyubiquitin chain (orange triangles), targeting I-κB for degradation by the proteasome. This releases NF-κB dimers (orange) and exposes the nuclear localization signal (green), causing NF-κB to translocate to the nucleus and activate gene expression.

9.9 GENE SILENCING

So far, the regulatory mechanisms we have discussed are those controlling synthesis of an individual mRNA transcript. We now turn to a different type of regulatory mechanism found in eukaryotes, whereby transcription in large regions of the chromosome containing multiple genes may be suppressed for long periods, often over the course of many cell divisions. This mechanism is referred to as silencing, to distinguish it from other modes of repression that are transient and can generally be quickly reversed. The silenced state, by contrast, is stable and can be inherited through many cell generations. This mode of inheritance is known as epigenetic inheritance.

➔ We discuss epigenetic inheritance further in Section 4.7.

Gene silencing is brought about by changes in chromatin structure

Gene silencing is brought about by an alteration in chromatin structure that can often be detected by assays that measure differences in nuclease accessibility and staining. In Chapter 4, we saw that expressed genes usually lie in euchromatic regions of the chromosome, while heterochromatic regions are condensed in a way that generally leads to gene silencing. Another feature that distinguishes silencing from simple repression is that silenced regions can spread along a chromosome. Spreading can be blocked by boundary elements, as we saw for example in Section 4.8.

In this section, we describe several mechanisms by which silencing is established. We shall see that each of these examples involves chemical modifications of either histone proteins or DNA, but that these modifications are established and maintained in rather different ways. Notably, the final example we will consider shows a heterochromatic state that can be established by transcription within the heterochromatic region itself, through a mechanism known as RNA interference, or RNAi. We will cover RNAi in greater detail in Chapter 13.

For our first example we turn again to the budding yeast mating-type loci. This organism is one of the most amenable experimental systems for exploring transcriptional regulatory mechanisms, yet the same basic mechanisms also appear to function in animal cells. In all cases, the silent state must first be established at a specific chromosomal location from which the silent state then often spreads. Subsequently, the silent state must be maintained over an extended period of time.

The budding yeast mating-type loci and telomeric regions are silenced by establishing a heterochromatic state

We saw in Section 9.7 how cell type in budding yeast is determined by the DNA-binding proteins that are expressed in a region of chromosome III called the *MAT* locus. Yeast **a** and α cells differ only in the genes encoded by the *MAT* locus: **a** cells contain the a1 regulatory gene at the *MAT* locus, and α cells have in its place the α regulatory genes α1 and α2, as illustrated in Figure 9.37. In addition to the actively transcribed sequence at the *MAT* locus, chromosome III contains an additional copy of the **a** and of the α gene sequences that are, however, *not* expressed. As shown in Figure 9.37, to the right of the *MAT* locus is a silenced copy of the **a** sequence called *HMR*a and to the left is a silenced copy of the α sequences called *HML*α. These silent copies are present in all yeast cells, irrespective of which genes are expressed at the *MAT* locus.

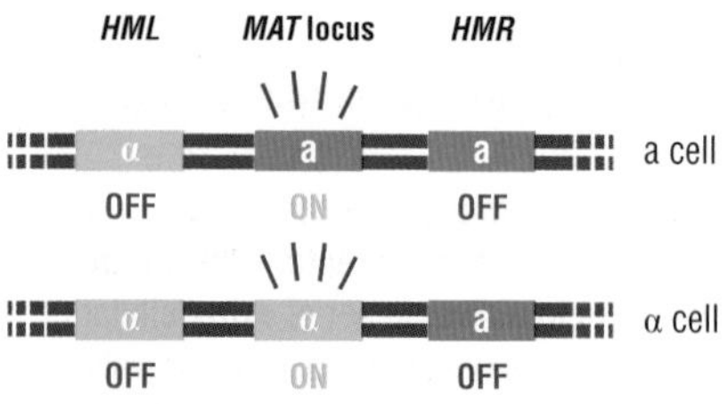

HML *MAT* locus *HMR*

α a a — a cell

OFF ON OFF

α α a — α cell

OFF ON OFF

Figure 9.37 Silencing of the mating-type cassettes in budding yeast. Yeast chromosome III contains a silent copy of the α cassette, which contains genes encoding α1 and α2, at the HML locus and a silent copy of the a cassette, which contains the gene encoding a1, at the HMR locus. The *MAT* locus contains an actively transcribed copy of either the a or the α cassette.

Silencing of the *HMR*a and *HML*α **mating-type cassettes** (also called the *HM* loci) depends upon the silencing information regulator (SIR) proteins, which establish and maintain a region of heterochromatin at *HML* and *HMR*. The SIR proteins are recruited by several proteins including Abf1 which, along with two other proteins called Rap1 and Orc1, binds to regions within *HMR* and *HML* (Figure 9.38). These proteins help establish silencing by recruiting the other three SIR proteins, Sir2, Sir3, and Sir4, which spread along the chromatin fiber through a network of interactions between one another and with the N-terminal tails of histones H3 and H4. The silencing activity depends upon the enzymatic activity of Sir2, a nicotinamide adenine dinucleotide (NAD^+)-dependent deacetylase. Sir2 maintains the chromatin in a hypoacetylated state, deacetylating Lys16 in histone

H4. Sir2 belongs to a special class of lysine deacetylases distinguished by their dependence on NAD^+, which makes silencing in yeast sensitive to intracellular levels of NAD^+.

Genes that are near the yeast telomeres are also silenced, in a mechanism that is very similar to the silencing of the HM loci. The Rap1 protein, which helps to nucleate silencing at the HM loci, also binds to the telomeric repeat sequences. Rap1 recruits the Sir2/Sir3/Sir4 complex, which spreads along the chromatin and establishes a heterochromatic state that silences gene expression. When localized to the genes near the telomere, the silencing is referred to as telomeric silencing.

Sir2 is also involved in a third form of silencing, in which it represses RNA polymerase II transcripts in the intergenic regions between rDNA genes. This helps to establish a heterochromatic state that prevents recombination between the multiple copies of the rDNA genes. This form of silencing relies on a different set of partner proteins that act along with Sir2 to inhibit transcription.

➲ We learn about the rDNA repeats and the role of heterochromatin in preventing recombination among them in Sections 4.4 and 4.12.

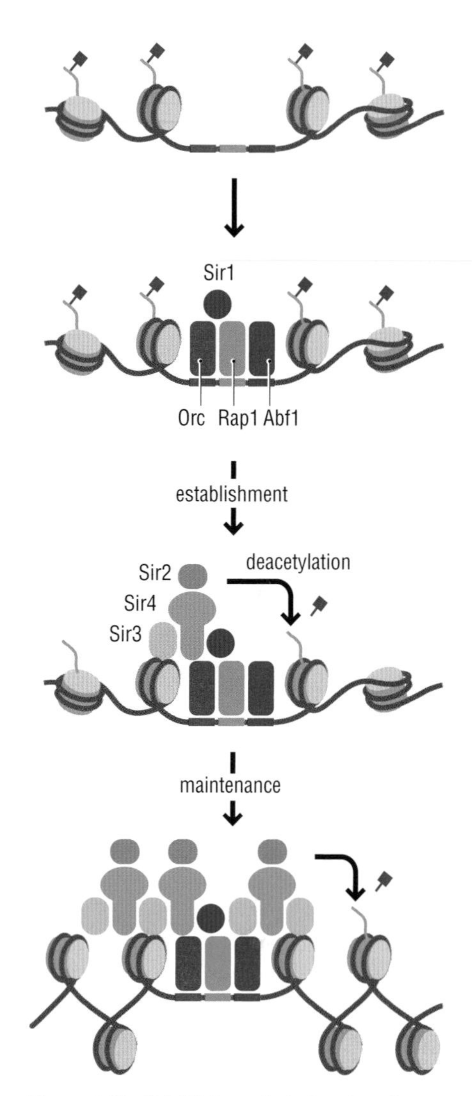

Figure 9.38 Establishment of silencing of the *HML* and *HMR* loci in budding yeast. Silencing in *S. cerevisiae* is initiated by the recruitment of SIR proteins to specific regions of the chromosome by the DNA-binding proteins Rap1, Orc1, and Abf1. The Sir2, Sir3, and Sir4 proteins then spread along the chromatin, and the Sir2 histone deacetylase maintains the chromatin in a hypoacetylated state. Sir1 plays a role in establishing silencing, although its precise function is not well understood.

Telomeric silencing in a pathogenic yeast is linked to niacin levels in the urinary tract

Telomeric silencing is exploited by *Candida glabrata*, a pathogenic yeast that causes urinary tract infections, to respond to the host environment. An important step in infection occurs when this organism adheres to the host urinary tract, through the action of adhesion proteins. Expression of one of the genes encoding an adhesion protein is controlled by telomeric silencing and is therefore switched off when there is adequate NAD^+ for Sir2 deacetylase activity. Importantly, while other yeast species can synthesize NAD^+ from a variety of precursor molecules, *C. glabrata* requires niacin (nicotinic acid) as a precursor, which it obtains from its surroundings. In the urinary tract, however, there is typically inadequate niacin. *C. glabrata* therefore cannot make enough NAD^+, which reduces Sir2 deacetylase activity and relieves silencing of the adhesion gene. The pathogenic yeast can now adhere to the urinary tract tissues and infect the host.

Note how the level of NAD^+ indirectly controls adhesion of the yeast by determining whether the Sir2 deacetylase is active and, hence, whether the required adhesion genes are expressed. More generally, the dependence of the entire Sir2 family of enzymes on NAD^+ allows the activity of these enzymes to be coupled to cellular metabolism, in which NAD^+ plays a critical role.

DNA methylation of an enhancer-blocking insulator regulates silencing of an imprinted gene

Transcriptional silencing can be mediated by modifications to DNA, as well as to histone proteins. As we learned in Chapter 4, mammalian chromosomes can be methylated at cytosine residues, and DNA methylation can mediate transcriptional silencing by blocking the binding of proteins that bind preferentially to unmethylated DNA. This mechanism underlies the regulation of the imprinted genes, *IGF2* and *H19*. As we learned in Chapter 4, gene imprinting is a mechanism by which only one of the parental copies of a gene is expressed while the other is silenced.

➲ We introduce DNA methylation, imprinting, and insulator elements in Chapter 4.

The *IGF2* gene encodes an insulin-like growth factor hormone that is required by the developing embryo, while *H19* encodes a long RNA transcript that is thought to play a role in cancer. These genes are located near one another on human chromosome 11 and can be activated by the same enhancer element. However, only one of these genes is expressed in each chromosome: the *IGF2* gene is expressed in the paternal chromosome, while the *H19* gene is expressed in the maternal chromosome. What gives rise to this pattern of expression?

The difference between maternal and paternal gene expression arises because of different patterns of DNA methylation in the insulator control region (ICR), which lies between the *IGF2* and *H19* genes. The ICR is the binding site for a protein called CCCTC-binding factor (CTCF), which contacts the ICR with a set of zinc finger domains. CTCF can only bind to this DNA sequence when the cytosine bases are *not* methylated. Figure 9.39a illustrates that, in the paternal chromosome, cytosine bases in the ICR are methylated and CTCF cannot bind. By contrast, this DNA sequence is not methylated in the maternal chromosome, and CTCF is able to bind.

It is the binding of CTCF to the ICR that helps to control which gene the enhancer acts upon. So how does this happen? In the maternal chromosome, proteins bound to the enhancer activate transcription of the *H19* gene. However, CTCF bound to the ICR blocks the ability of the enhancer to activate the *IGF2* gene, which remains inactive.

In paternal chromosomes, however, both the ICR and the promoter of the adjacent *H19* gene are methylated. Methylation of the ICR means that CTCF can no longer bind to it. Consequently, the effect of the enhancer on the *IGF2* gene is not blocked, and transcription of *IGF2* is activated. Methylation of the promoter region of the *H19* gene, however, compromises the binding of the transcriptional machinery, and the *H19* gene remains inactive. As shown in Figure 9.39b, CTCF is thought to control formation of separate chromosome loops, which in turn influence whether transcription of one or the other gene is activated by proteins bound to the enhancer.

The importance of imprinting at this locus is seen in the developmental disorder called Beckwith–Wiedemann syndrome. Children with this syndrome are much larger than normal at birth and are prone to a variety of cancers in childhood. The syndrome arises when the *IGF2* gene is expressed in both copies of chromosome 11,

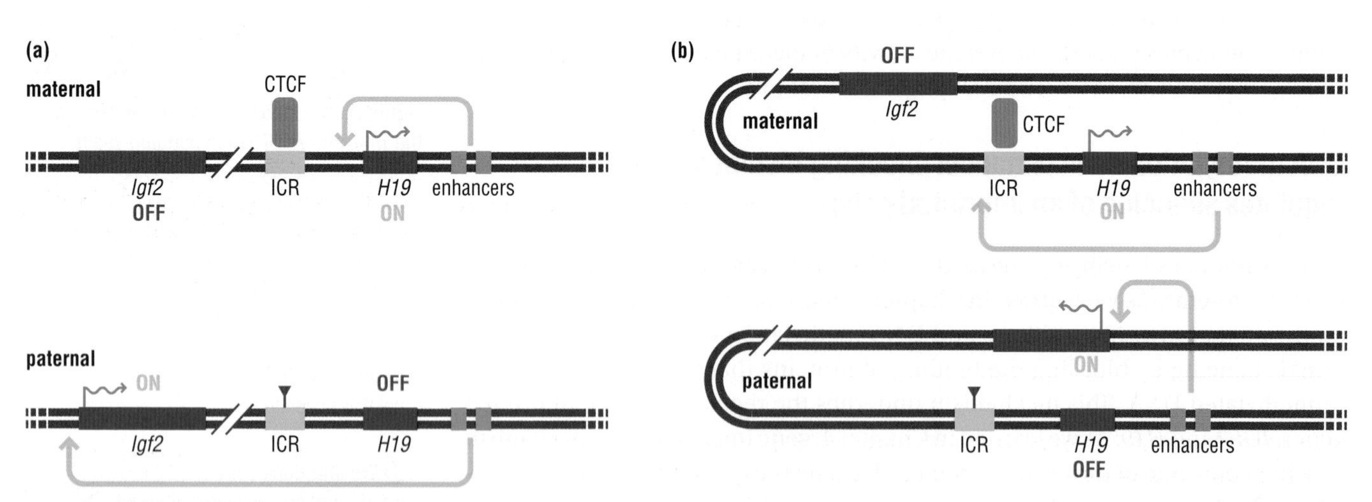

Figure 9.39 Imprinting and silencing. (a) The maternal chromosome (top) is not methylated, so CTCF can bind DNA. The enhancer activates transcription of the *H19* gene but CTCF blocks activation of *IGF2*. In the paternal chromosome (bottom), the CTCF binding site and *H19* are methylated. CTCF does not bind to DNA and the enhancer activates transcription of *IGF2*. (b) Model of proposed looping that accounts for the patterns of expression in part (a). It has been proposed that CTCF mediates formation of a loop that represses transcription of *IGF2* and allows transcription of *H19*.

while the *H19* gene is inactive – a consequence of defects in methylation of the paternal chromosomes, or chromosomal rearrangements that disrupt the ICR.

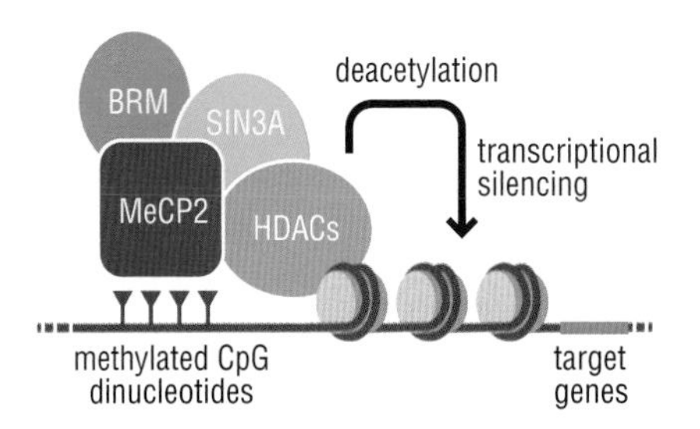

Figure 9.40 MeCP2 binds to methylated DNA and recruits a co-repressor complex. MeCP2 binds specifically to methylated CpG and recruits a complex containing a histone deacetylase (HDAC) that represses transcription of the target genes.

Methylated DNA can recruit enzymes that modify chromatin and repress transcription

DNA methylation also can have an opposite effect on regulatory protein binding. Instead of preventing proteins from binding to DNA, as is the case for CTCF, methylation can in some cases promote binding of proteins. Methyl-binding domains are protein domains that bind specifically to methyl cytosine. Some proteins that contain a methyl-binding domain recruit nucleosome-remodeling or histone-modifying enzymes, which repress transcription. The DNA methylation is the stable epigenetic signal that is maintained through cell division, while the methyl-binding domain helps to 'interpret' the signal to repress transcription.

The protein MeCP2 is a methyl-binding domain that binds to methylated DNA and recruits the Sin3A transcriptional co-repressor, which contains a histone deacetylase; the structure of this co-repressor complex is depicted in Figure 9.40. MeCP2 represses a variety of genes in humans, including genes in neuronal cells. Mutations in the gene encoding MeCP2 can lead to a devastating disease known as Rett syndrome, which affects 1 in 10 000 girls. Girls with this disorder appear normal at birth but soon develop autism-like behaviors, repetitive hand movements, and mental retardation, along with physical symptoms such as scoliosis and seizures. The syndrome is found primarily in girls because the gene for MeCP2 is located in the X chromosome. As in the case of other X-linked traits, this means that affected girls will have one mutant and one wild-type copy of MeCP2, whereas an affected boy has only a single mutant copy. Without a normal copy of the gene, most male fetuses do not survive to term, and the few that are born die within the first two years of life.

The synthesis of small RNA molecules triggers silencing of the fission yeast mating-type loci and other loci with repetitive sequences

Like the budding yeast, *S. cerevisiae*, the fission yeast, *Schizosaccharomyces pombe,* also silences its mating-type cassettes by establishing a heterochromatic state. However, the mechanism of silencing used in *S. pombe* is quite different from that of its evolutionarily distant cousin, the budding yeast. Establishment of the silenced chromatin in fission yeast exploits a more recently discovered mechanism that depends upon small regulatory RNA molecules; this mechanism is known as **RNA interference**.

As we saw in Section 4.8, the silent mating-type cassettes in *S. pombe* lie in a region of heterochromatin that flanks the centromere. The establishment of this heterochromatic region is outlined in Figure 9.41; it begins with transcription of

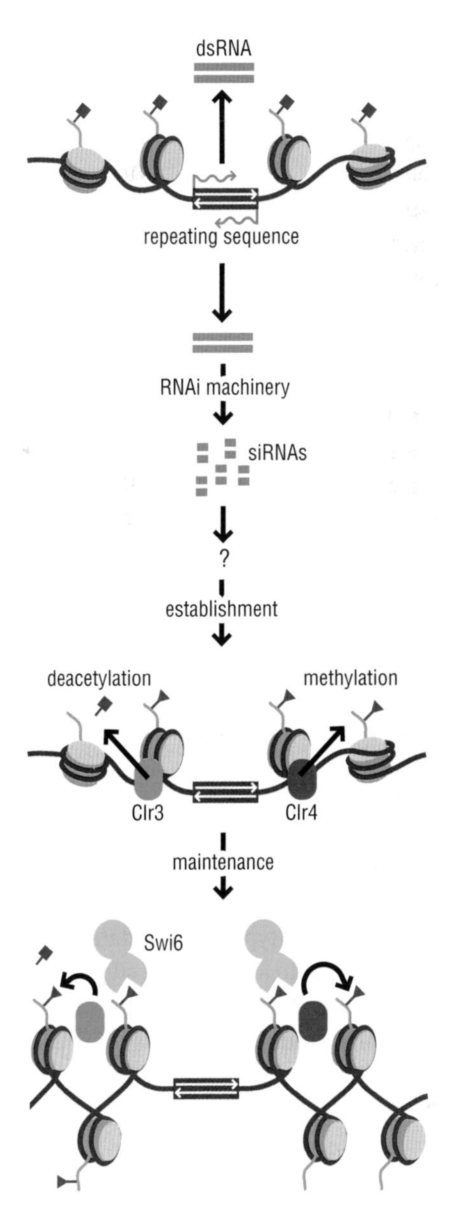

Figure 9.41 Establishment of heterochromatic silencing in the fission yeast. Silencing of the mating-type loci in *S. pombe* requires transcription of the coding and non-coding strands of a set of repeated sequences. These two transcripts form double-stranded RNA (dsRNA, shown in blue) that is processed into short fragments by the RNAi machinery. The resulting short siRNAs lead, through unknown mechanisms, to formation of a repressive chromatin structure that involves the Clr3 histone deacetylase, Clr4 histone H3 lysine 9 methyltransferase and Swi6, which binds to H3 tails methylated at lysine 9.

the centromere, which consists of a series of repeated nucleotide sequences. Although transcription of silenced regions seems counterintuitive, *some* transcription is required at many loci where synthesis of small RNAs leads to silencing of the locus. Both strands of the centromeric DNA are transcribed, which gives rise to complementary RNA strands that can base-pair with one another. These double-stranded RNA molecules are then cleaved into 21 bp fragments that are incorporated into a protein complex (called RITS, which stands for RNA-induced initiation of transcriptional silencing), which then associates with the actively transcribed centromeres. The RITS complex, in turn, recruits a histone deacetylase and a histone methyltransferase that specifically target lysine 9 of histone H3. The histone-modifying enzymes ensure that the histones in this region have the chemical hallmarks of heterochromatin in *S. pombe*, namely, hypoacetylated histones and methylated lysine 9 of histone H3.

The methylated lysine 9 in the N-terminal tail of histone H3 recruits additional histone-modifying enzymes to the chromatin template and facilitates the spreading of heterochromatin. The protein, Swi6, binds to the methylated histone tail via its chromodomain, a specialized protein domain that binds to methylated lysine (which we learned about in Section 4.8). Swi6, in turn, recruits additional histone methyltransferase and deacetylase enzymes, which target the neighboring histones. As additional histones are modified, additional Swi6 proteins bind and the cycle continues, spreading the heterochromatin further down the chromatin fiber. As we learned in Section 4.8, the spreading of heterochromatin continues through the mating-type genes, thereby silencing them. The heterochromatin spreads until a boundary element is reached.

RNA interference has been found to similarly silence the transcription of many other eukaryotic genes, particularly repeated genes, although small regulatory RNA molecules also target steps subsequent to transcription. The important role played by RNA in regulating gene expression was not appreciated until recently, when several observations pointed to a role for small non-coding RNA molecules in regulating protein levels. We learn more about this exciting aspect of biology in later chapters.

➔ We discuss the discovery of RNA interference and the mechanisms by which the small regulatory RNAs are generated and how they function in a variety of contexts in Chapter 13.

SUMMARY

- Any step of the transcription cycle can be regulated to control the rate at which RNA is transcribed.
- Proteins that decrease the level of transcription are called repressors, whereas those that increase the level of transcription are called activators.
- Genes are typically regulated by one or more proteins that bind to a regulatory site in the DNA. A regulatory site may lie near the promoter (as is typical in bacteria) or many thousands of base pairs upstream or downstream of the gene (as is more common in eukaryotes).
- Many genes in eukaryotes are regulated by multiple proteins that bind to one or more regulatory regions.
- The binding of an allosteric effector molecule can alter the ability of a protein to bind DNA or regulate transcription. In this way, proteins such as the Trp repressor in bacteria and the glucocorticoid receptor protein in humans can regulate transcription in direct response to the level of a ligand present in the cell.
- A variety of mechanisms are used to repress or activate transcription initiation in bacteria. Transcription can be repressed by blocking the binding of RNA polymerase (Trp repressor, cI). Transcription can be activated by recruiting the polymerase holoenzyme (CAP), stimulating open-complex formation (CAP, NtrC), or by altering the structure of promoter DNA (MerR).

- Eukaryotic genes are regulated by co-activator and co-repressor complexes, which are recruited to the DNA by sequence-specific DNA-binding proteins. These complexes typically contain enzymes that reposition nucleosomes or that add or remove post-translational modifications from histones and other transcription factors.
- A complex called Mediator is required for appropriate gene regulation by RNA polymerase II.
- Histone acetyltransferases are used to activate transcription in eukaryotes, while histone deacetylases repress transcription. Other enzymes, such as histone methyltransferases and ATP-dependent nucleosome-remodeling complexes, can be used either to activate or repress transcription.
- The expression of some bacterial and eukaryotic genes is regulated by controlling transcription elongation or termination. The bacterial N and Q proteins, and the HIV Tat protein, regulate transcription in this way.
- Small molecule metabolites can regulate elongation and termination by binding to a region of the mRNA called a riboswitch, which adopts different secondary structural elements depending upon whether or not the metabolite is bound. Many bacterial genes and some eukaryotic and archaeal genes are regulated in this way.
- Transcription in large regions of eukaryotic chromosomes can be repressed or silenced by a variety of mechanisms that alter chromatin structure and sometimes depends on small regulatory RNAs (RNA interference).

QUESTIONS

9.1 PRINCIPLES OF TRANSCRIPTION REGULATION

1. If every cell in an individual human contains the same DNA, how is it that cells of different tissues have unique characteristics?

2. Why regulate protein production at every step in the process rather than just at the level of transcription?

3. What is the connection between the histone code and gene regulation?

4. Why do most DNA-binding domains interact with the major groove as opposed to the minor groove of DNA?

5. Explain the terms enhancer, co-activator, and co-repressor.

6. Eukaryotic DNA is associated with histone proteins and packaged as chromatin. Explain why this is important for transcriptional regulation.

7. Which of the following statements most accurately describes chromatin state and transcription?

a. Open chromatin is associated with hypoacetylation and has high levels of transcription.
b. Histone deacetylases promote closed chromatin and result in repressed transcription.
c. Hyperacetylation is associated with closed chromatin and low levels of transcription.
d. Histone acetyltransferases result in hypoacetylation and high levels of transcription.

Challenge question

8. What is the meaning of the phrase 'eukaryotic regulatory proteins are typically modular in nature'?

9.2 DNA-BINDING DOMAINS IN PROTEINS THAT REGULATE TRANSCRIPTION

1. What is the purpose of DNA-binding domains, and which DNA conformation do DNA-binding domains usually interact with?

2. Briefly describe the features of the following DNA-binding domains and how they interact with DNA.

a. Helix-turn-helix
b. Homeodomain
c. Zinc finger
d. Basic region-leucine zipper
e. Basic region-helix-loop-helix

9.3 MECHANISMS FOR REGULATING TRANSCRIPTION INITIATION IN BACTERIA

1. The *E. coli lac* operon is regulated both negatively and positively by the binding of the *lac* repressor (LacI) and the CAP respectively. Propose a hypothesis to explain why both negative and positive regulation have evolved for this operon.

Challenge questions

2. Consider *E. coli* cells with each of the following mutations. Complete the table to indicate the extent of transcription in the presence and absence of lactose (no glucose, CAP is present and active).

Mutation	Transcription in the presence of lactose	Transcription in the absence of lactose
No mutation		
Mutant lac operator (cannot bind repressor)		
Mutant lac repressor (cannot bind the operator)		
Mutant lac repressor (cannot bind lactose)		
Mutant lac promoter (cannot bind CAP)		

3. In *E. coli*, the galactose operon is responsible for transcribing the enzymes necessary to convert galactose into glucose so that it can be used in cellular respiration. The operon is regulated by both a repressor (galR) and the CAP. Additionally, it has been shown that the repression by galR requires HU, an architectural histone-like binding protein.

a. Make a diagram of the galactose operon that includes the likely positions of the binding sites for RNA polymerase, galR, HU, and CAP.
b. Make two additional diagrams that show the activation of the gal operon in response to low glucose and high galactose and

the repression of the gal operon in response to low galactose. Be sure to show the appropriate proteins bound and any structural changes to the region of DNA.

c. The galactose operon has an extended –10 region that is well conserved and a –35 sequence that weakly matches the consensus sequence. What is the significance of these sequences (what is the consequence of matching versus divergence from the consensus sequence)?

d. It was hypothesized that the weakly conserved –35 region was necessary to block galR/HU repression and allow transcription to take place. Propose a mechanism by which this sequence could be responsible for blocking repression.

9.4 COMPETITION BETWEEN cI AND CRO AND CONTROL OF THE FATE OF BACTERIOPHAGE LAMBDA

1. Which of the following situations leads to lysogeny during phage lambda infection of *E. coli*?

a. Digestion of cII by host protease
b. cII binding at P_L and P_R
c. cI binding at P_{RM}
d. Cro binding at O_R3

Challenge question

2. A cell infected with lambda bacteriophage can follow one of two pathways: the lytic or lysogenic pathway.

a. Describe the similarities and differences between the structure of cI and Cro, paying particular attention to the features that allow them to carry out their different functions.

b. It was hypothesized that the cooperative binding and subsequent folding was not necessary for repression and that strong binding of the lambda repressor, cI, to its binding site alone could lead to repression. Therefore, a mutant phage was constructed in which the lambda repressor could still bind DNA but could not form tetramers and octamers. These mutants, however, could not establish lysogenic colonies. Explain the mechanism for *establishing* the lysogenic pathway and give a possible explanation for why this is an expected result of the mutation.

c. These phage were allowed to grow in the lytic cycle for several generations and new mutants were isolated that contained a second mutation that allowed these phage to establish lysogeny. The secondary mutation was found in the P_{RM} promoter. Based on your knowledge of the mechanism for establishing lysogeny, give a possible explanation for why this type of mutation would help overcome the lambda repressor mutation.

d. The phage with two mutations could establish lysogeny, but at a lower level than the wild type; it also still had defects in switching from lysogeny to the lytic stage. So, another set of mutants were recovered that now contained three mutations. These mutants had a lambda repressor that could not cooperatively bind DNA, an altered P_{RM} promoter, and a mutation in the cI binding site, O_R2. The mutant phage could establish lysogeny at wild-type levels and they could switch to the lytic cycle, but at a lower level than the wild type. Based on your knowledge of the mechanism for establishing lysogeny, give a possible explanation for why this set of mutations would help overcome the problem of switching to the lytic stage.

e. Based on the fact that these triple mutants were able to establish lysogeny and were able to switch from the lysogenic to the lytic stage, what conclusion would you draw regarding the need for cooperativity in the life cycle of the lambda phage?

9.5 REGULATION OF TRANSCRIPTION TERMINATION IN BACTERIA

1. Several regulatory mechanisms control gene expression during phage lambda infection of *E. coli*. Explain how the N and Q proteins affect transcription.

2. Explain how riboswitches function. Use an example to further illustrate your explanation.

Challenge question

3. What would happen to the transcription of the bacterial *trp* operon if each of the four RNA nucleotide blocks were deleted individually?

9.6 REGULATION OF TRANSCRIPTION INITIATION AND ELONGATION IN EUKARYOTES

1. Transcription of heat-shock protein genes can be regulated by promoter proximal pausing.

a. What is promoter proximal pausing?
b. Explain why regulation by promoter proximal pausing of heat-shock genes is beneficial to the cell.
c. Give a brief overview of how promoter proximal pausing occurs and is regulated at the *Drosophila hsp70* gene.

Challenge question

2. Gal4 is involved in the regulation of galactose metabolism.

a. Which common molecular biology technique uses Gal4, and what features of the Gal4 protein make this possible?
b. Describe how transcription would be affected in the presence of the following mutations.
   i. A mutation that resulted in an inability of Gal80 to enter the nucleus.
   ii. A mutation that resulted in a lack of ability of Gal3 to bind galactose.
   iii. A mutation that resulted in Gal4 being unable to bind the UAS_G.

9.7 COMBINATORIAL REGULATION OF EUKARYOTIC TRANSCRIPTION

1. Briefly explain combinatorial control and give an example of its occurrence.

2. Which transcriptional activator turns on genes in haploid **a** cells in *S. cerevisiae?*

a. Mcm1
b. a1
c. α1
d. α2

Challenge question

3. Consider yeast cell with the following inactivating mutations. Complete the following table to show the effect of these mutations on the α, **a**, and haploid specific genes in a diploid yeast cell?

Mutation	α-specific genes	a-specific genes	Haploid specific genes
No mutation			
Mutation in a1			
Mutation in α1			
Mutation that allowed α1 to be expressed			
Mutation in α2			
Mutation in Mcm1			

9.8 THE ROLE OF SIGNALING CASCADES IN REGULATION OF TRANSCRIPTION

1. Compare the mechanisms by which the cell can respond to an extracellular signal that can diffuse through the cellular membrane versus a signal that cannot.

9.9 GENE SILENCING

1. In what way is transcriptional silencing distinguished from transcriptional repression?

2. Briefly describe the two main mechanisms by which genes can become silenced.

3. *S. cerevisiae* can switch mating types. Which of the following statements correctly summarizes where the information comes from for it to do this?
   a. A simple one -nucleotide mutation is all that is needed for the switch, so no extra information is needed.
   b. *S. cerevisiae* can conjugate with other yeast cells to acquire the new mating type.
   c. If *S. cerevisiae* is in low N and C conditions, the transcribed RNA loops and forms an activator to coordinate expression of the other mating type.
   d. The mating locus is flanked by inactive copies of the two mating-type loci that can be used as information to replace the active copy at the mating locus.

Challenge question

4. You have a mutant mouse strain in which cytosines at the ICR have all been changed to thymines. Devise a hypothesis for how the expression of the IGF2 and H19 genes might be affected. Explain your answer.

EXPERIMENTAL APPROACH 9.2 – COOPERATIVE BINDING AND DNA LOOPING IN GENE REGULATION BY THE LAMBDA REPRESSOR

1. What is the basic principle of a nuclease protection assay?

2. What role was later discovered for O_R3? Briefly describe how this hypothesized role for O_R3 was tested.

Challenge question

3. What conclusions could be drawn from the nuclease protection assay shown in Figure 1? Use the results shown to explain your answer.

EXPERIMENTAL APPROACH 9.3 – SERENDIPITOUS DISCOVERY OF THE HEAT-SHOCK RESPONSE OPENED THE DOOR FOR NUMEROUS STUDIES OF TRANSCRIPTIONAL REGULATION

1. What evidence was presented to suggest that chromosome puffs correspond to regions of transcription?

2. What is the relationship between chromosome puffs and heat shock?

3. As part of the experimental design for Figure 2, the *hsp70* promoter was cloned into a vector. Why?

4. Explain the experimental design for the result that is shown in Figure 3.

Challenge questions

5. In Figure 2:
   a. What do the numbers at the top of the figure represent?
   b. What do the symbols '+' and '–' at the top of the figure represent?
   c. What conclusion can be drawn from this figure? Be sure to explain how the figure is used to draw the conclusion.

6. Explain the results in terms of how red, green, and black spots were found on the microarray.

FURTHER READING

9.1 PRINCIPLES OF TRANSCRIPTION REGULATION

Li B, Carey M, Workman JL. The role of chromatin during transcription. *Cell*, 2007;**128**:707–719.

9.2 DNA-BINDING DOMAINS IN PROTEINS THAT REGULATE TRANSCRIPTION

Garvie C, Wolberger C. Recognition of specific DNA sequences. *Molecular Cell*, 2001;**8**:937–946.

9.3 MECHANISMS FOR REGULATING TRANSCRIPTION INITIATION IN BACTERIA

Lawson CL, Swigon D, Murakami KS, *et al*. Catabolite activator protein: DNA binding and transcription activation. *Current Opinion in Structural Biology*, 2004;**14**:10–20.

Schumacher MA, Brennan RG. Structural mechanisms of multidrug recognition and regulation by bacterial multidrug transcription factors. *Molecular Microbiology*, 2002;**45**:885–893.

Xu H, Hoover TR. Transcriptional regulation at a distance in bacteria. *Current Opinion in Microbiology*, 2001;**4**:138–144.

9.4 COMPETITION BETWEEN cI AND CRO AND CONTROL OF THE FATE OF BACTERIOPHAGE LAMBDA

Hochschild A, Lewis M. The bacteriophage lambda CI protein finds an asymmetric solution. *Current Opinion in Structural Biology*, 2009; **19**:79–86.

Ptashne M. *A Genetic Switch*, 3rd ed. Cold Spring Harbor, New York: Cold Spring Harbor Laboratory Press, 2004.

9.5 REGULATION OF TRANSCRIPTIONAL TERMINATION IN BACTERIA

Henkin TM, Yanofsky C. Regulation by transcription attenuation in bacteria: how RNA provides instructions for transcription termination/antitermination decisions. *BioEssays*, 2002;**24**:700–707.

Roberts JW, Shankar S, Filter JJ. RNA polymerase elongation factors. *Annual Review of Microbiology*, 2008;**62**:211–233.

Roth A, Breaker RR. The structural and functional diversity of metabolite-binding riboswitches. *Annual Review of Biochemistry*, 2009;**78**:305–334.

9.6 REGULATION OF TRANSCRIPTON INITIATION AND ELONGATION IN EUKARYOTES

Adelman K, Lis JT. Promoter-proximal pausing of RNA polymerase II: emerging roles in metazoans. *Nature Reviews Genetics*, 2012;**13**: 720–731.

Buchwalter G, Gross C, Wasylyk B. Ets ternary complex transcription factors. *Gene*, 2004;**324**:1–14.

Traven A, Jelicic B, Sopta M. Yeast Gal4: a transcriptional paradigm revisited. *EMBO Reports*, 2006;**7**:496–499.

Weake VL, Workman, JL. Inducible gene expression: diverse regulatory mechanisms. *Nature Reviews Genetics*, 2010;**11**:426–437.

9.7 COMBINATORIAL REGULATION OF EUKARYOTIC TRANSCRIPTION

Herskowitz I. A regulatory hierarchy for cell specialization in yeast. *Nature*, 1989;**42**:749–757.

Kassir Y, Adir N, Boger-Nadjar E, *et al.* Transcriptional regulation of meiosis in budding yeast. *International Review of Cytology*, 2003; **224**:111–117.

Panne D. The enhanceosome. *Current Opinion in Structural Biology*, 2008;**18**:236–242.

9.8 THE ROLE OF SIGNALING CASCADES IN THE REGULATION OF TRANSCRIPTION

Bain DL, Heneghan AF, Connaghan-Jones KD, Miura MT. Nuclear receptor structure: implications for function. *Annual Review of Physiology*, 2007;**69**:201–220.

Stock AM, Robinson VL, Goudreau PN. Two-component signal transduction. *Annual Review of Biochemistry*, 2000;**69**:183–215.

Vallabhapurapu S, Karin M. Regulation and function of NF-kappaB transcription factors in the immune system. *Annual Review of Immunology*, 2009;**27**:693–733.

Zhang J, Lazar MA. The mechanism of action of thyroid hormones. *Annual Review of Physiology*, 2000;**62**:439–466.

9.9 GENE SILENCING

Cam HP, Chen ES, Grewal SIS,. Transcriptional scaffolds for heterochromatin assembly. *Cell*, 2009;**136**:610–614.

Moazed D. Small RNAs in transcriptional gene silencing and genome defense. *Nature*, 2009;**457**:413–420.

Rusche LN, Kirchmaier AL, Rine J. The establishment, inheritance, and function of silenced chromatin in Saccharomyces cerevisiae. *Annual Review of Biochemistry*, 2003;**72**:481–516.

RNA processing

10

INTRODUCTION

RNAs have diverse roles in the flow of information from DNA to protein. Messenger RNAs (mRNAs) are informational molecules, which specify the amino acid sequence of proteins: they carry the information encoded by genes and direct ribosomes to synthesize proteins from this information. Other RNAs participate directly as players in different cellular functions – as parts of the machinery for translating the mRNA into protein (which we discuss in Chapters 11 and 12) or as regulators of gene expression (as we will see in Chapter 13). For example, ribosomal RNAs (rRNAs) provide structural and catalytic components of the protein synthesis machinery, and transfer RNAs (tRNAs) function as adaptors between the nucleotide sequence of an mRNA and the amino acid sequence of the encoded protein.

We saw in Chapter 8 how RNAs are synthesized from their DNA template by RNA polymerases. However, many RNA molecules that emerge from RNA polymerase are not fully functional. Instead, these **precursor RNAs (pre-RNAs)** first need to be modified in one or more ways to generate the functional or **mature RNA**. In this chapter, we consider the great variety of different alterations, collectively known as **RNA processing**, to which an RNA molecule may be subjected, and the machines that execute them. We also discover how these modifications play an important part in generating biological diversity, enabling single RNA molecules to encode multiple end products.

We begin this chapter by briefly reviewing the key forms of modification that occur. We will then revisit each of these approaches in more detail throughout the rest of the chapter.

10.1 OVERVIEW OF RNA PROCESSING

RNA is processed in many different ways

Figure 10.1 summarizes the many ways in which RNAs can be processed. In the simplest case, the ends of the transcript are removed from the larger precursor RNA by RNA-cleaving enzymes called **ribonucleases.**

The coding sequence of many genes, particularly eukaryotic mRNAs, is not continuous but is interrupted by non-coding sequences known as **introns**. Despite their seemingly disruptive nature, introns are included in the precursor RNA produced by RNA polymerase along with the coding segments, or **exons**. However, introns must be removed to generate a functional species so that, for example, an mRNA will be translated into a functional protein. The removal of the introns and rejoining of the exons that remain is known as **RNA splicing**.

Eukaryotic pre-mRNAs also undergo a modification known as **5′ capping** whereby the 5′ end becomes modified with a 7-methyl guanosine; they also undergo **polyadenylation** in which a long stretch of adenosine nucleotides is added to the 3′ end as part of the transcription termination process.

➔ We describe the termination of transcription in Section 8.6.

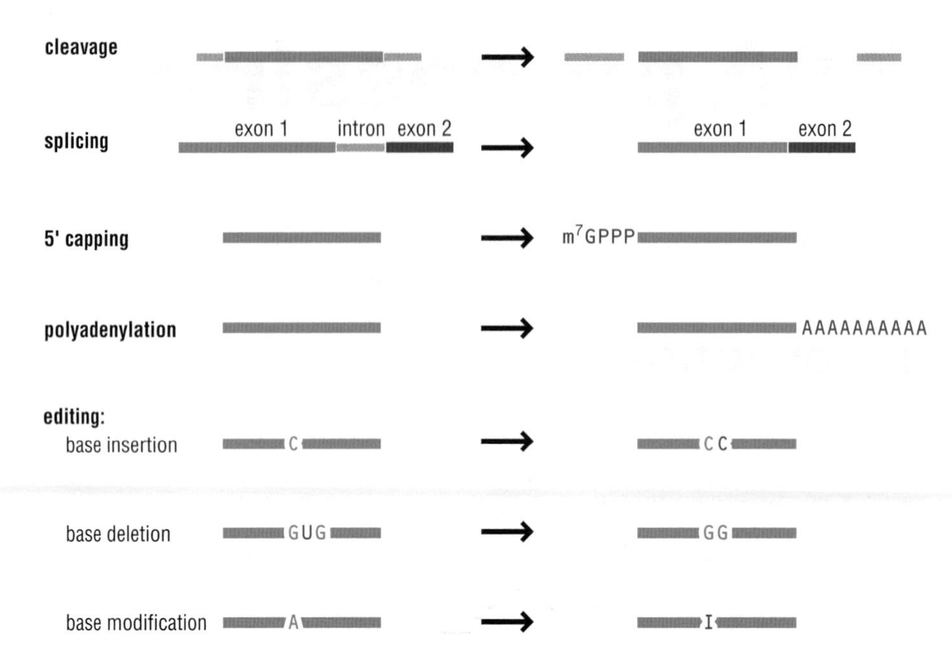

Figure 10.1 Types of RNA processing. The regions that make up the mature RNA are indicated by blue lines. Some RNAs are cleaved from longer precursors; for others, introns are removed and exons are spliced together. The 3′ ends of RNAs can also be modified by polyadenylation, and eukaryotic mRNAs undergo 5′-end capping. Finally, some transcripts are edited by the insertion, deletion, or modification of a base. m^7G, 7-methyl guanosine.

RNA processing can also involve the insertion or deletion of bases or the chemical modification of bases: these changes are collectively known as **RNA editing**. Finally, during the last stage of their cellular life, RNA molecules are degraded by nucleases into individual nucleotides that can be recycled by the cell.

RNA processing events are points for regulation and quality control, and are sources of diversity

Fundamentally, RNA processing is required to produce functional RNAs, but processing also brings with it at least three associated benefits. First, processing events contribute to the regulation of gene expression: the efficiency with which a particular RNA is spliced, capped, polyadenylated, and edited governs how quickly the RNA can function or be translated. In addition, the concentration of any RNA in a cell is regulated by the balance between the transcription of the RNA (the rate at which a given RNA is synthesized) and the degradation of the RNA by nucleases. For example, if the transcription of a gene is induced by an environmental signal such as a steroid hormone, the synthesis of the protein can be switched off by the degradation of the RNA when the hormone signal is no longer present.

Second, as well as providing a means for regulating gene expression levels, RNA processing can generate many different RNAs from just one gene. Many genes contain multiple introns and exons. As illustrated in Figure 10.2, exons can be spliced together in a variety of combinations to produce different proteins. An extreme example of the multiplicity of products that can be generated from a single gene is provided by the *Drosophila melanogaster* gene *dscam*. Proteins produced from this gene help to guide nerve cells in the central nervous system. The *dscam* gene has 115 exons; of these, 95 exons are in four clusters that are alternatively spliced. Thus there is the potential to produce 38 016 different proteins. This diversity contributes to the exquisite specificity observed for neuronal connections.

Alternative splicing determines whether a given sequence is present or not in the synthesized product: each exon encodes an amino acid sequence that

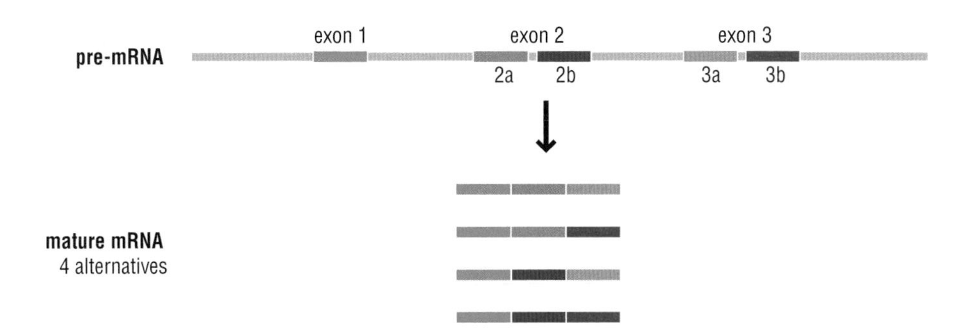

Figure 10.2 Protein diversity brought about by alternative splicing. The putative pre-mRNA shown has one exon 1, two possible exon 2s (2a and 2b) and two possible exon 3s (3a and 3b). Alternative splicing of this example could lead to the production of four different RNA products (exons 1+2a+3a, 1+2a+3b, 1+2b+3a, and 1+2b+3b).

may or may not be included in a protein synthesized from the precursor RNA once splicing has occurred. RNA editing, on the other hand, involves removing, inserting, or chemically modifying individual nucleotides, all events that can change the amino acid sequence of the encoded protein. Indeed, we will see in Section 10.7 an example where up to half the nucleotides in a fully processed RNA are the result of editing.

Third in our list of associated benefits, RNA processing also serves as a quality control step. mRNAs that are defective in some way (for example, as a result of errors during transcription or splicing) are detected and degraded by cellular machinery. As we will explore in more detail in Chapter 13, in addition to these systems that monitor endogenous RNA quality, eukaryotic cells possess a mechanism referred to as **RNA interference (RNAi)** to detect and degrade foreign RNAs produced by parasites such as viruses. The RNAi machinery shares enzymes with the small RNA-based systems involved in transcriptional silencing, which we discussed in Chapter 9. All of the quality control steps prevent the accumulation of RNAs (or eventually proteins) that could be detrimental to the cell.

Many RNA processing machines have RNA components

Most RNA processing steps are performed by multicomponent macromolecular machines. In many cases these machines interact with and influence one another. In addition, many complexes are localized to specific compartments in eukaryotic cells: some RNA processing complexes are found in the nucleus, either distributed throughout or localized to specific regions such as the nucleolus, whereas others are found in the cytoplasm. Specific interactions and localization allow the processes to occur in the appropriate sequential manner and provide checks to ensure the proper processing has occurred. Thus, for example, mRNAs are not exported from the nucleus to the cytoplasm until they are appropriately modified.

Many RNA processing machines contain both protein and RNA components and are therefore referred to as **ribonucleoproteins (RNPs)**. While a subset of the RNAs function only as structural components of the RNP, many RNAs play active parts in the functions of an RNP. For example, as shown schematically in Figure 10.3, some RNAs have catalytic activity and are termed ribozymes. Others are called **guide RNAs** because they base-pair with the precursor RNAs and guide the catalytic components of an RNP to where the processing events should take place. In the course of this chapter, and in the following chapters on protein synthesis and regulatory RNAs, we shall see many examples of structural, catalytic, or guide functions of RNAs in diverse RNP complexes.

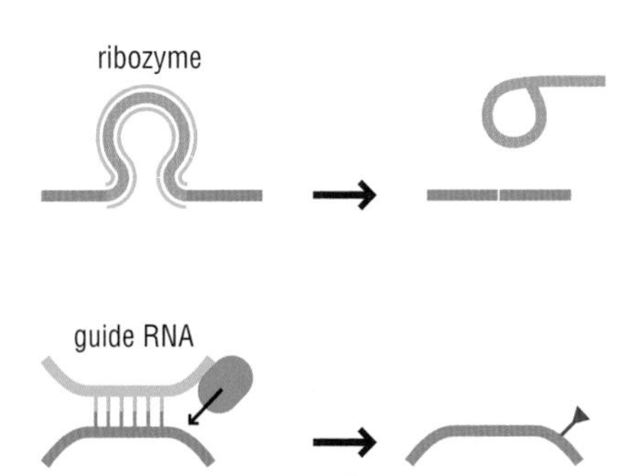

Figure 10.3 Functions of RNAs in RNA processing machines. Some RNAs, called ribozymes, have intrinsic enzymatic activity. Although these RNAs can often catalyze reactions in the absence of proteins, proteins generally increase the activity of ribozymes. Other RNAs, referred to as guide RNAs (green), direct proteins (orange) to carry out reactions at specific sequences by base-pairing with the target mRNAs.

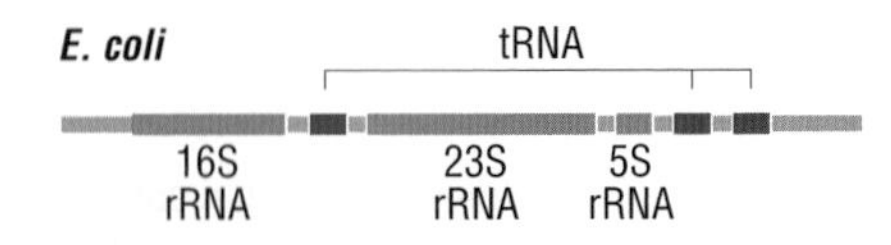

Figure 10.4 *E. coli* and *S. cerevisiae* precursors encoding tRNAs and rRNAs. The mature, functional tRNAs and rRNAs in bacteria and eukaryotes are usually cleaved from longer precursor RNAs.

10.2 tRNA AND rRNA PROCESSING

Having reviewed the main modifications that occur to RNA molecules, and some of the key components involved in the modification processes, we will now consider a range of specific processing events in more detail. We begin with the processing associated with the synthesis of mature tRNAs and rRNAs.

tRNAs and rRNAs are synthesized as longer precursors

The transcripts for most tRNAs and rRNAs in all organisms are synthesized as long precursor molecules that must be processed to produce functionally active RNAs. The precursor might simply need one or both ends trimmed; alternatively, more extensive processing may be required. Often, a precursor RNA includes two or more functional RNAs that are cut, or excised, from the precursor by enzymatic cleavage. Two examples of this are shown in Figure 10.4. An *Escherichia coli* rRNA precursor encodes three rRNAs as well as several tRNAs, whereas the *Saccharomyces cerevisiae* precursor encodes three rRNAs. Synthesis of a single long precursor ensures that similar amounts of the functional RNAs, such as all three rRNAs, are produced, simply because the synthesis of one is intrinsically linked to another as they come from the same precursor molecule.

Ribonucleases, the enzymes that cleave the RNAs into smaller parts, can operate in one of two ways, as illustrated in Figure 10.5. Some ribonucleases, denoted exonucleases, digest the RNA strand by successively removing nucleotides from the end. Most exonucleases recognize the 3′ end and digest in the 3′ to 5′ direction, although some bacterial and eukaryotic exonucleases begin digestion at the 5′ end and proceed in the 3′ direction. Generally these enzymes act on single-stranded ends and do not exhibit sequence specificity. Other ribonucleases, termed endonucleases, cleave the RNA internally (that is, within a strand, and not at an exposed end). Endonucleases differ in their specificities for RNA targets; some endonucleases cleave double-stranded RNA while others cleave single-stranded RNA. Features of two representative and well-characterized endonucleases, RNase III and RNase P, are discussed below.

The processing of tRNAs and rRNAs can involve many steps and can involve more than one pathway, as illustrated in Figure 10.6 for the processing of the rRNA precursor from *S. cerevisiae*. The reason why multiple processing pathways have evolved is not known. In general, the various processing steps do not occur in isolation, but instead are coordinated with each other and with other cellular processes. For example, in eukaryotes, rRNA transcription and early processing steps both occur in the nucleolus and are tightly coupled. Further rRNA processing is tied to the early steps of ribosomal protein assembly on the rRNA as well as, in some cases, export to the cytoplasm.

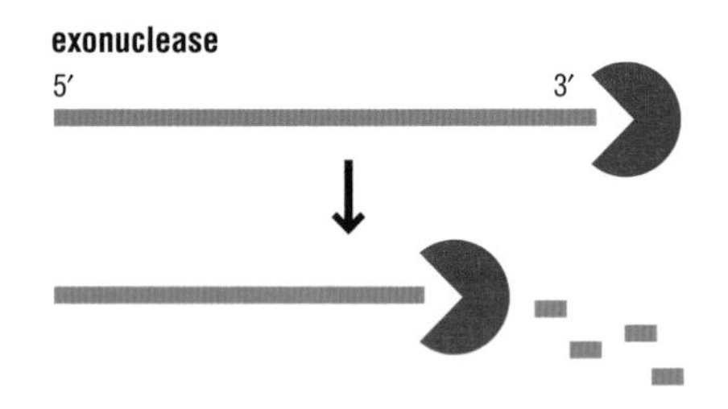

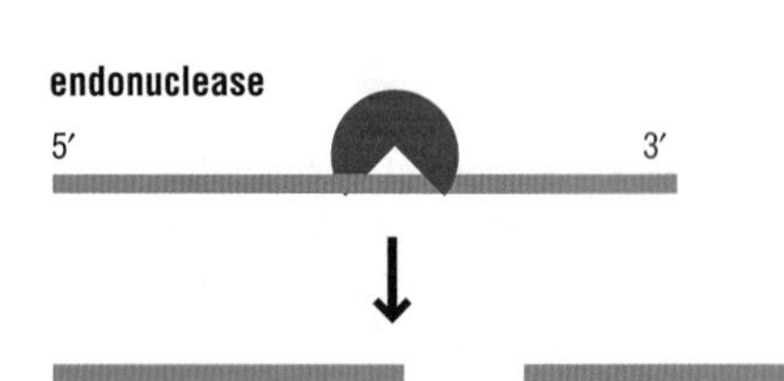

Figure 10.5 The action of exonucleases and endonucleases. Exonucleases sequentially digest RNA from the 3′ end in a 5′ direction or from the 5′ end in a 3′ direction (not shown), while endonucleases cleave at internal sites.

The endonuclease RNase III cleaves and trims rRNA precursors

Excision of the bacterial rRNAs from longer precursors and the trimming of some rRNAs and tRNAs is carried out by the endonuclease RNase III, which recognizes double-stranded RNA. This class of ribonuclease binds stem structures in precursor RNAs and then cleaves the double-stranded RNA on one or both sides of the stem.

Endonucleases similar to RNase III are involved in many processes in the cell. In Chapter 13, we will see how these enzymes in eukaryotes are important in

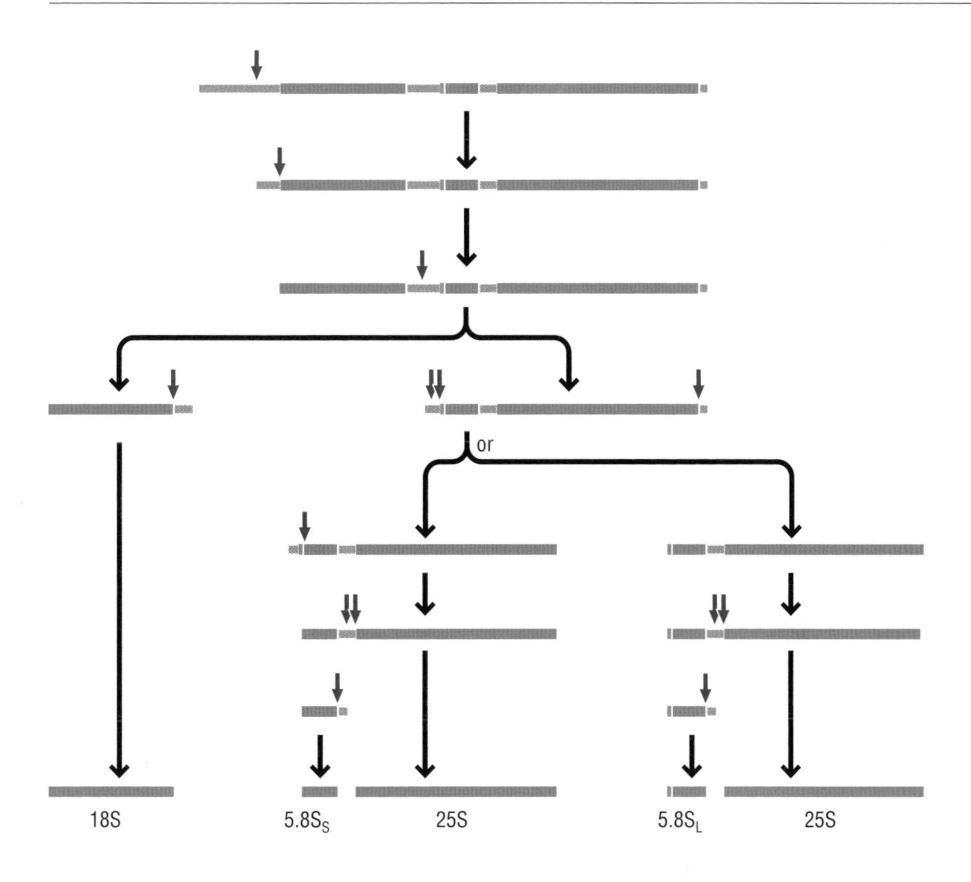

Figure 10.6 Cleavage steps during the processing of mature *S. cerevisiae* rRNAs from the precursor RNA. Multiple cleavage steps (indicated with pink arrows) are required for the processing of mature rRNAs from the precursor RNA. The products of this processing – two large rRNAs (labeled 18S and 25S) and the short rRNA (5.8S) – are discussed in Chapter 11. At various stages during the processing of the precursor, one of two different steps is carried out, for example, leading to the production of either the short (5.8SS) or long form (5.8SL) of the 5.8S rRNA. This diagram does not show the additional processing events in which nucleotide bases are chemically modified.

generating both microRNA (**miRNA**) regulators and the small interfering RNAs (**siRNAs**) that inhibit the expression of detrimental genes.

The RNA-containing endonuclease RNase P cleaves the 5′ ends of tRNAs

The precise trimming of the 5′ end of tRNAs is accomplished by the endonuclease ribonuclease P (RNase P), which cleaves single-stranded RNA. Unlike RNase III, the bacterial, archaeal, eukaryotic, and mitochondrial RNase P enzymes all contain an RNA component in addition to one or more protein components. The RNA component of bacterial RNase P alone can act as an enzyme to cut the RNA, and was one of the first RNAs shown to act as a ribozyme (as we discuss in Experimental approach 10.2). The single protein subunit of the bacterial enzyme enhances performance relative to that achieved by the RNA component alone, and may broaden the range of substrates the enzyme can cleave. By contrast, the isolated RNA component of archaeal, eukaryotic, and mitochondrial RNase P does not act as a ribozyme, although the RNA is essential for catalysis, and is probably the functional core of these enzymes.

Several tRNAs and some rRNAs have introns that require splicing

The precursor RNAs for several tRNAs and some rRNAs are interrupted by introns, which must be removed to produce the functional RNA; this removal provides an additional step for regulation and quality control. The tRNA precursors undergo splicing that is catalyzed by protein factors. In contrast, some introns found in rRNA

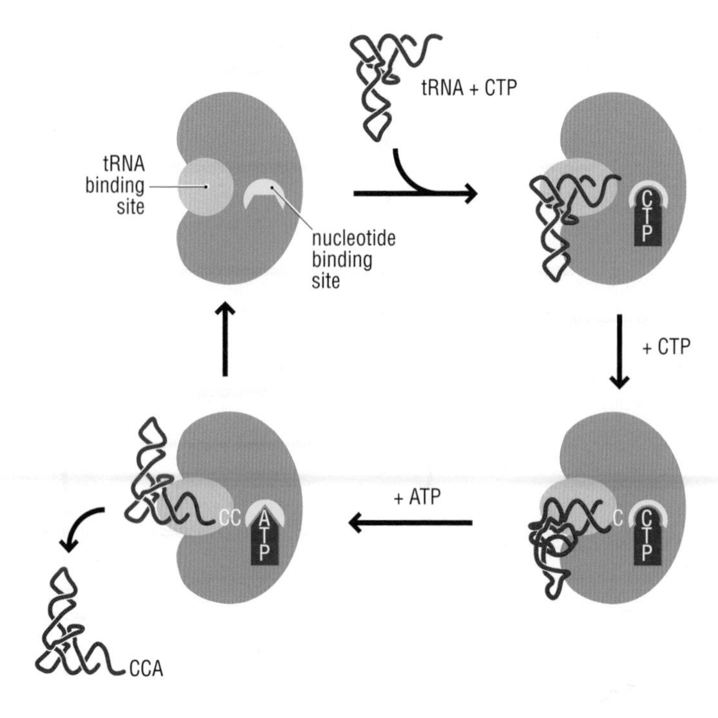

Figure 10.7 Specificity of nucleotide addition by CCA adding enzymes is determined by the conformation of the nucleotide binding pocket (red). This binding pocket can have three different conformations depending on which form of the tRNA is bound. These conformations determine whether a C or an A residue is added.

Augustin MA, Betat H, Huber R, Mörl M, Steegborn C. Crystal structure of the human CCA-adding enzyme: insights into template-independent polymerization. *Journal of Molecular Biology*, 2003;**328**:985–994.

precursors catalyze their own removal and are thus **self-splicing**. In these cases, the precursor rRNAs themselves act as ribozymes that catalyze the removal of introns and the joining of the exons that remain (referred to as intron excision and exon ligation, respectively). We consider the details of the different splicing reactions in subsequent sections.

CCA is added to tRNAs in the absence of a nucleic acid template

The 3′ termini of tRNAs contain a universally conserved CCA sequence, which is the site of attachment for the amino acid that the tRNA carries to the ribosome during protein synthesis.

➔ We describe the attachment of amino acids to tRNAs in more detail in Section 11.3.

For some bacterial tRNAs, this CCA sequence is encoded by the tRNA gene, but for most tRNAs this sequence is added post-transcriptionally by a CCA-adding enzyme. Indeed, even in organisms in which the CCA sequence is encoded by the tRNA gene, the CCA-adding enzyme is required for the repair and maintenance of the CCA sequence because the single-stranded 3′ termini of these tRNAs are particularly susceptible to digestion by exonucleases.

Although the CCA-adding enzyme shares sequence similarity with a DNA repair polymerase, the enzyme does not use a nucleic acid template to direct the addition of two C residues followed by a single A residue to the tRNA. Instead, the enzyme contains a binding pocket that can accommodate either a cytidine triphosphate (CTP) or an adenosine triphosphate (ATP). As illustrated in Figure 10.7, both the size and shape of this binding pocket change depending on the form of tRNA bound – the tRNA without the CCA tail, or the tRNA with a single C added, and so on. These changes to the binding pocket determine which nucleotide is added to the 3′ end of the tRNA at each step, and hence drive the synthesis of the correct CCA tail. As we discuss in Experimental approach 10.1, the CCA-adding enzyme has an additional role wherein it targets unstable tRNAs for degradation through the addition of even longer CCACCA or CCACC tails.

10.1 EXPERIMENTAL APPROACH

CCA-adding enzyme plays a specialized role in targeting unstable tRNAs for rapid degradation

Classical genetic analysis revealed genes that work together in a common pathway to remove undermodified tRNAs

Genetic approaches can lead to the identification of important, unanticipated biological connections. Eric Phizicky and colleagues used such an approach to evaluate potential interactions between yeast genes encoding tRNA modification enzymes, which target various positions in the tRNA, including, for example, the acceptor stem. (The different parts of a tRNA are shown in Figure 2.36. We will learn more about the roles of the different regions in Chapter 11.) In the process, they uncovered a novel form of tRNA quality control referred to as rapid tRNA degradation or RTD. In this case, while none of these tRNA modification genes was essential for viability, when the m^7G_{46} methyl transferase *TRM8* or *TRM82* gene deletions were combined with a variety of other tRNA modification gene deletions (including *TRM4*, *PUS7* and *DUS1-3*), cell viability was severely compromised, as seen in Figure 1. Such an approach is commonly known as a synthetic lethal screen and can identify genes that act together in a pathway.

In the same study, the growth defects in the double deletion strains were correlated with decreases in the abundance of several specific tRNA species in yeast: $tRNAVal^{(AAC)G73}$, $tRNAVal^{(AAC)A73}$, and $tRNA^{Cys(GCA)}$, as shown in Figure 2. Each of these tRNA species happens to be modified by several of the key enzymes identified in the genetic screen. By contrast, the levels of other tRNAs not modified by this set of modification enzymes (i.e., $tRNA^{Arg(ACG)}$) were relatively high and unaffected in strains lacking these modification enzymes. The loss of specific tRNA species explains the observed loss of cell viability, since overexpression of these specific tRNAs restores viability to the cells. These genetic studies point to a crucial role for tRNA modifications in tRNA stability and cell survival.

Studies on the stability of non-coding RNAs in mammalian cells reveal insights into a tRNA quality control pathway

Clues to how hypomodified tRNAs are targeted for decay came from the study of two tRNA-like non-coding RNAs in mammalian cells (mascRNA and MEN β tRNA-like small RNA). In this study, Eric Phizicky, Phil Sharp, and colleagues found that the CCA-adding enzyme, which is normally responsible for the post-transcriptional addition of CCA to tRNAs, plays a key role in tagging certain RNA species (including one of the two non-coding RNAs). This discovery came from the sequencing of these different non-coding RNAs, one of which was stable and the other not; the

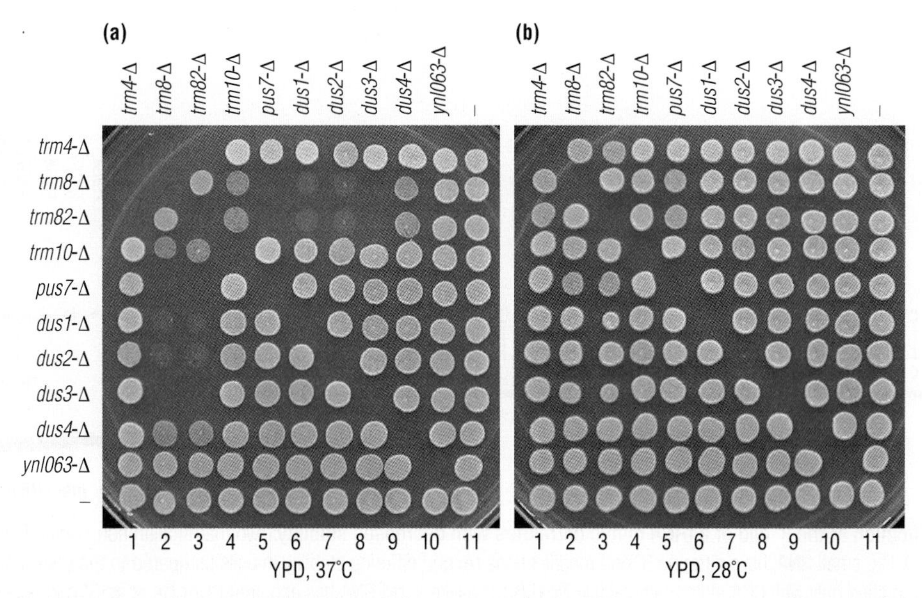

Figure 1 Synthetic growth defects are evident at high temperature in yeast carrying multiple temperature sensitive mutations in tRNA modifying enzymes. Cultures of strains carrying listed pairs of double gene deletions were spotted on plates and grown at (A) 37°C and (B) 28°C. The lack of growth for strains carrying **trm8 or trm82** mutations in combination with the *trm4*, *pus7*, *dus1*, *dus2*, and *dus3* mutations at 37°C indicate these combinations are synthetic lethal and are thereby proposed to function in the same biological pathway.

From Alexandrov *et al.* Rapid tRNA decay can result from lack of nonessential modifications. *Molecular Cell*, 2006;**21**:87–96.

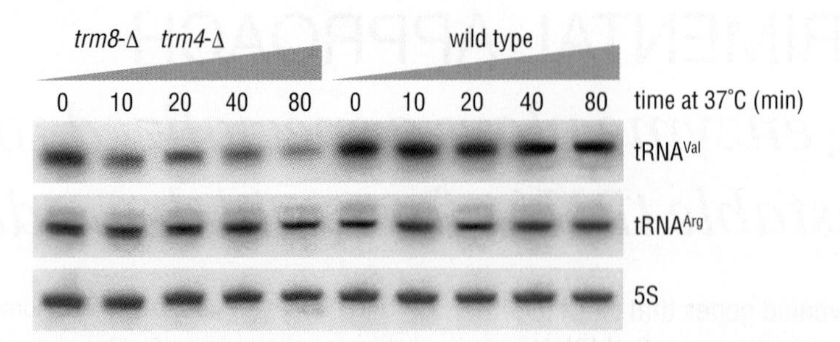

Figure 2 Northern analysis shows loss in stability of certain tRNA species in yeast strains lacking multiple modification enzymes. Here, $tRNA^{Val}$ is destabilized in the *trm8-Δ trm4-Δ* background while $tRNA^{Arg}$ is not.

From Alexandrov *et al.* Rapid tRNA decay can result from lack of nonessential modifications. *Molecular Cell,* 2006;**21**:87–96.

unstable RNA appeared to carry additional uncoded CCACCA or CCACC sequences at the 3′end while the more stable RNA had only a CCA addition (like a canonical tRNA). These additional extensions at the 3′ tail of the tRNA-like species (and tRNAs themselves) correlated with lowered stability (see Figure 3).

The specific features that target a tRNA for tagging by additional CCACCA or CCACC sequences are found in its acceptor stem. The key feature that is recognized is overall decreased base pairing of the acceptor stem (or its mimic), which is influenced by nucleotide modifications, correlating with the earlier observations from the genetic studies in yeast. Taken together, these data yield a mechanism for tRNA quality control that is similar from yeast to mammals: destabilized tRNAs are recognized as such by the CCA-adding enzyme, and are thereby targeted for rapid degradation by 3′ to 5′ exonucleases in the cell (including Xrn1 and Rrp44 in mammals). CCA-adding enzyme thus plays a dual role, first in generating and maintaining the integrity of the 3′ end of the tRNA, and second, in targeting troubled tRNAs for degradation. Conservation of this quality control system from yeast to man emphasizes the critical role played by post-transcriptional modifications across all kingdoms of life.

Find out more

Alexandrov A, Chernyakov I, Gu W, Hiley SL, Hughes TR, Grayhack EJ, Phizicky EM. Rapid tRNA decay can result from lack of nonessential modifications. *Molecular Cell,* 2006;**21**:87–96.

Wilusz JE, Whipple JM, Phizicky EM, Sharp PA. tRNAs marked with CCACCA are targeted for degradation. *Science*, 2011;**334**:817–821.

Related techniques

Gene mutation; Section 19.5

S. cerevisiae (as a model organism); Section 19.1

Northern blots; Section 19.10

RNA sequencing; Section 19.10

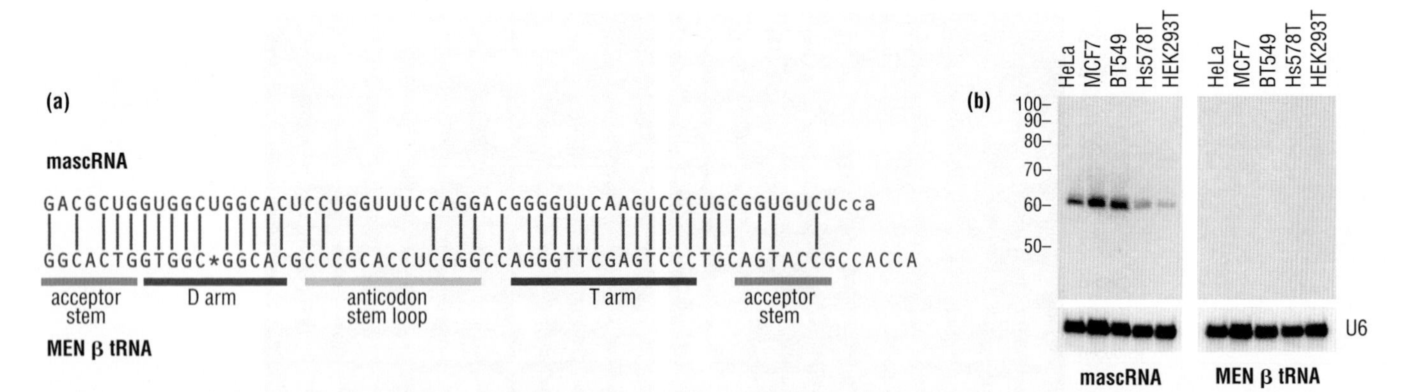

Figure 3 CCACCA modification at the 3′ end of a tRNA mimic correlates with decreased stability. Two mammalian non-coding RNAs that mimic tRNAs, mascRNA and MEN β tRNA-like small RNA, have differing 3′ end modifications (A) and differing stability in cells compared to the control U6 RNA (B). The more stable RNA, mascRNA, is modified only with CCA at its 3′ end, while MEN β tRNA-like small RNA has acquired CCACCA at its 3′ end (see red sequences at end).

From Wilusz *et al.* tRNAs marked with CCACCA are targeted for degradation. *Science*, 2011;**334**:817–821.

10.3 tRNA AND rRNA NUCLEOTIDE MODIFICATIONS

Nucleotides in tRNAs and rRNAs are modified in numerous ways

Many of the nucleotides in tRNAs and rRNAs are chemically modified after transcription. The modifications can be found on the nucleotide base or the ribose sugar ring, and can be relatively small: they may comprise the addition of hydrogen atoms, the methylation of nitrogen or oxygen, or the addition of selenium as seen in Figure 10.8a. The modifications also can be quite large, such as the incorporation of the amino acid threonine. Some nucleotides even carry multiple independent modifications. Typically, particular positions in different tRNA and rRNA species are specifically modified and, for those modifications that have been studied in detail, they often appear to be required for the optimal survival and growth of organisms. Some modifications do not have any obvious functional role, yet are well conserved during evolution, suggesting that they do nevertheless confer a selective advantage to the organism.

More than 80 different modifications have been seen in tRNAs. Some of these modifications seem to contribute to the overall structural stability of tRNAs and to their ability to interact with other molecules. It has been argued that the chemical diversity of the modifications adds to the limited repertoire of the four standard ribonucleotides, thus allowing for higher order function. For example, we will see in Chapter 11 that some modifications allow tRNA to interact properly with mRNA during protein synthesis to ensure that this process occurs with high accuracy.

The most common rRNA modifications are shown in Figure 10.8b; these include **ribose 2′-O-methylation** (to produce methylribose) and **pseudouridylation** (the conversion of uridine to pseudouridine). Many of the rRNA modifications are found in regions of the ribosome that are known to be important for ribosome function. It is thought that the extra hydrogen-bonding capacity of pseudouridine relative to uridine could allow this modified nucleotide to help stabilize the structure of the ribosome.

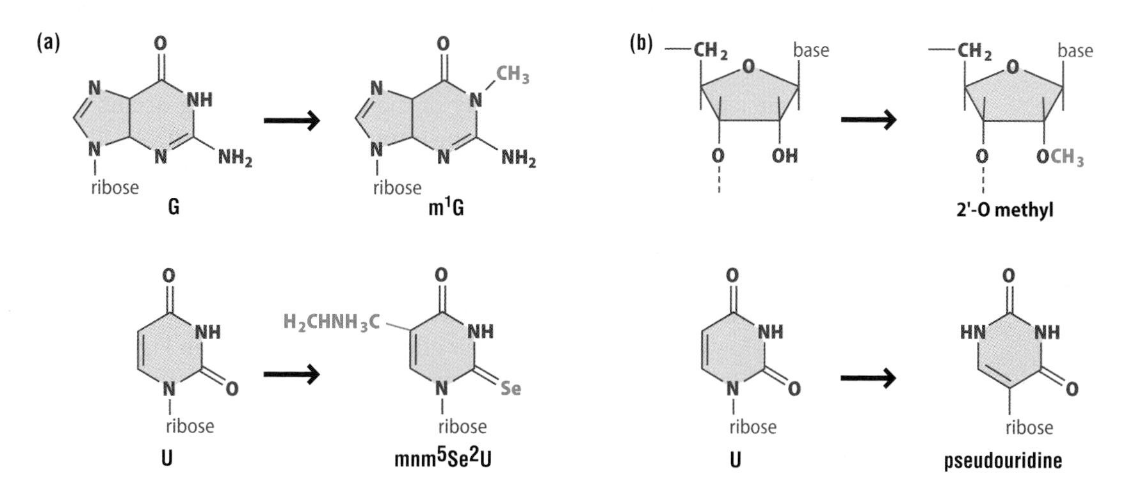

Figure 10.8 Examples of nucleotide modifications found in tRNA and rRNA. Modifications can occur to the ribose sugar or the nucleotide base, and they can consist of the addition of a single methyl group or be more extensive. Shown here are (a) two tRNA modifications, the methylation of guanosine, and the addition of a methyl amino methyl group and selenium to uridine, as well as (b) the two most common rRNA modifications, the methylation of ribose to produce 2′-O methylribose and the conversion of uridine to pseudouridine.

Mutations in the protein dyskerin, the enzymatic component of a molecular complex involved in pseudouridylation, lead to a human disease of skin and bone marrow failure known as dyskeratosis congenita, underlining the importance of the modified nucleotide pseudouridine in biological function.

Specific enzymes catalyze individual tRNA and some rRNA modifications

In bacteria, the modifications found on tRNAs and rRNAs are added by site-specific enzymes that catalyze a given modification at a specific site in the RNA. Some of the more complex modifications require the actions of several enzymes to synthesize the final modified nucleotide. In a few cases, such as that of the pseudouridine synthases (which catalyze the conversion of uridine to pseudouridine) and several of the methylases (which add methyl groups to a target molecule), we have a detailed understanding of the catalytic mechanisms of the modifying enzymes. For most of the modifications, however, we know little about the enzymes involved and how specific sequences are recognized. Despite this, the large number and diversity of modifications and the large number of proteins that are committed to their synthesis indicate that these modifications must play a biologically significant role.

snoRNAs guide methylation and pseudouridylation of eukaryotic and archaeal rRNAs and some tRNAs

We mentioned earlier that two particularly common modifications to rRNA are ribose methylation and the conversion of uridine to pseudouridine. Although the enzymes that carry out these modifications in all organisms are similar, the mechanism by which most sites are recognized in archaea and eukaryotes is very different from the mechanism found in bacteria. In particular, the archaeal and eukaryotic machineries make use of short RNA molecules called **small nucleolar RNAs (snoRNAs)** to guide the enzymes to the site in the RNA that is to be modified. The snoRNAs associate with specific sets of proteins, which include the enzyme that carries out the modification reaction, to form an **RNP** known as a **snoRNP**.

SnoRNAs range in length between 60 and 300 nucleotides and are found in the nucleolus, where the majority of the rRNA processing occurs. Most vertebrate snoRNAs are processed from the introns of precursor mRNAs, illustrating the fact that introns can themselves encode functional RNAs. The snoRNAs fall into two major classes that possess distinct conserved sequence elements and direct different nucleotide modifications: the RNAs that direct ribose methylation are referred to as the box C/D snoRNAs, and the RNAs that direct pseudouridylation are referred to as the box H/ACA snoRNAs.

How do snoRNAs exert their effects? As illustrated in Figure 10.9, the snoRNA within the ribonucleoprotein complex base-pairs with the target RNA and thus directs the modification enzymes to particular positions. While the function of most of the snoRNAs is to guide modifications of the eukaryotic rRNAs and tRNAs, a subset have been found to direct modifications of specific mRNAs. Similar RNA-guided reactions will be seen again in Section 10.6 where we describe how RNPs containing guide RNAs direct the precise removal of introns from the mRNA.

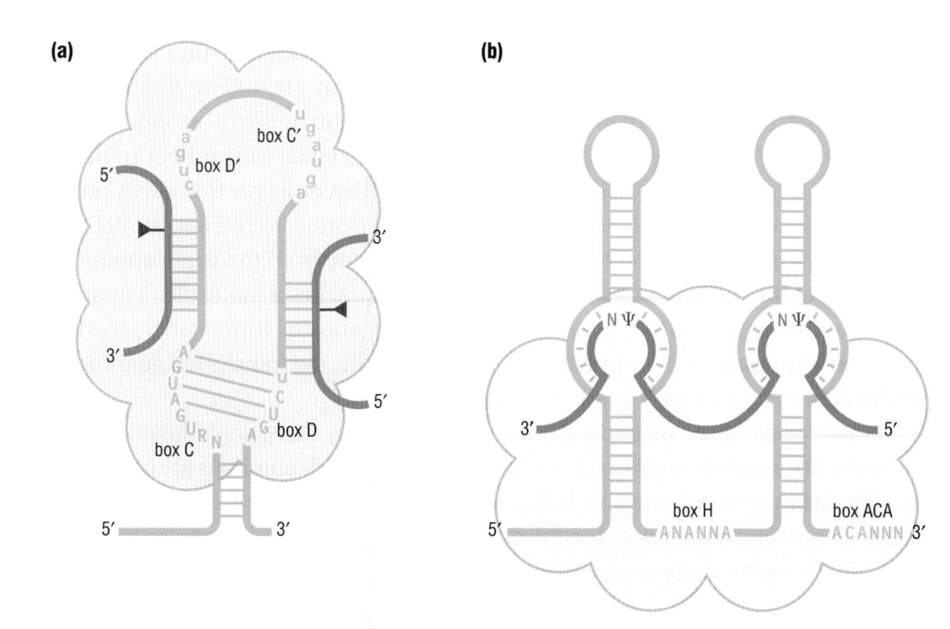

Figure 10.9 Ribose methylation and pseudouridylation guided by box C/D and box H/ACA snoRNAs, respectively. (a) snoRNAs containing conserved C/D box motifs (green) base-pair with a 10–21-nucleotide region of a target RNA (blue) and direct the methylation (blue triangle) of specific bases by a fibrillarin-containing protein complex (pale orange). (b) snoRNAs containing conserved H/ACA box motifs (green) base-pair with two 3–10-nucleotide regions of a target RNA (blue) and direct the conversion of uridines to pseudouridines (Ψ, red) by a dyskerin-containing protein complex (pale yellow).

10.4 mRNA CAPPING AND POLYADENYLATION

We now move on from tRNAs and rRNAs to consider the modification of mRNAs. Although mRNAs are typically not functional molecules in their own right, they play a vital part in the way biological information is transferred from genome to protein, and their processing both during and after transcription is key to their correct activity.

Eukaryotic mRNAs are capped at the 5′ end in a multistep process

Both the 5′ and 3′ ends of eukaryotic mRNAs are modified during transcription by RNA polymerase II. The 5′ ends of the mRNAs are 'capped' with a special 7-methylguanine nucleotide that is attached to the end of the message via a unique 5′-5′ triphosphate linkage, as shown in Figure 10.10. This structure, called the **5′ cap**, is essential for the efficient elongation and termination of transcription as well as the subsequent processing of the mRNA. The cap structure also functions as

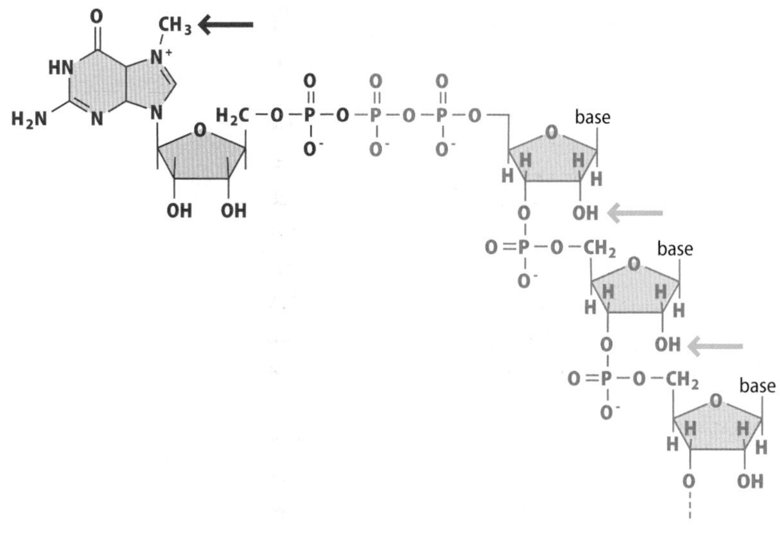

Figure 10.10 Structure of the 5′cap of eukaryotic mRNAs. A guanine nucleotide (pink) is added to the 5′ end of an mRNA (blue) via a 5′–5′ linkage. The guanine is subsequently methylated at the N7 position (pink arrow). In higher eukaryotes, the 2′-O position of the second base and sometimes the third base are also methylated (green arrows).

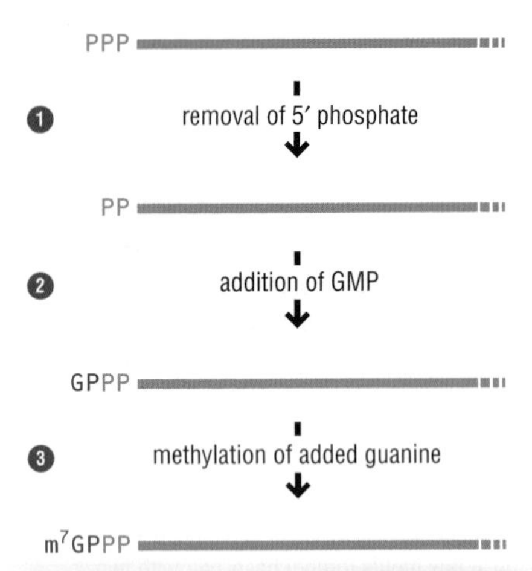

Figure 10.11 The 5′ cap is added in three successive steps. The 5′ capping process begins as soon as 20–30 nucleotides have been synthesized by RNA polymerase II. First, one phosphate is removed from the triphosphate at the 5′ end of the mRNA. A molecule of GMP is then attached to the end by a guanyl transferase. Finally, the transferred guanine is methylated by a guanine-7-methyltransferase.

➔ We learn more about the initiation of protein synthesis in Section 11.7.

a binding site for proteins that export the mRNA from the nucleus to the cytoplasm, and for directing the initiation of protein synthesis from the mRNA. In addition, the cap protects the 5′ end of the mRNA from degradation by 5′ to 3′ exonucleases.

As illustrated in Figure 10.11, the 5′ cap is added to eukaryotic mRNA in three successive steps that occur soon after the growing RNA emerges from RNA polymerase II, when only 20–30 nucleotides have been synthesized. First, an RNA 5′ triphosphatase catalyzes the removal of one phosphate from the triphosphate at the 5′ end of the mRNA. A guanyl transferase then attaches a molecule of guanosine monophosphate (GMP; in a reaction utilizing guanosine triphosphate (GTP)) to this end through a 5′–5′ triphosphate linkage. Finally, the transferred guanine is methylated by a guanine-7-methyltransferase.

In yeast, these three reaction steps are performed by three separate enzymes, while the first two reactions in *Caenorhabditis elegans* and mammals are catalyzed by a single bifunctional enzyme. In higher eukaryotes, the 2′ O positions of ribose in the second and sometimes the third nucleotides of the mRNA are also methylated, which seems to allow more efficient translation of these mRNAs.

Eukaryotic mRNAs are polyadenylated at the 3′ end

The 3′ ends of all eukaryotic mRNAs, except those encoding metazoan histones, are modified by the addition of a tail of approximately 200 adenosine residues, called a **polyadenosine**, or **poly(A) tail**. Like the 5′ cap, the 3′ poly(A) tail protects the mRNA from degradation by exonucleases and plays a role in the initiation of protein synthesis.

Some mRNAs have more than one site, called the **polyadenylation site**, at which the precursor mRNA is cleaved and the poly(A) tail added. These alternative polyadenylation sites (APAs) contribute to the regulation of protein synthesis and expand the range of protein products made from a single mRNA species. For example, as we will discuss in more detail in later sections and subsequent chapters, mRNA stability and translation in eukaryotes are often regulated by factors, both proteins and regulatory RNAs, which bind to the 3′ **untranslated regions (UTRs)** of a given mRNA. Thus, the variable 3′ UTR lengths that can be specified by different polyadenylation site selection can determine what regulatory sequences will be included. Many mRNAs produced in human proliferating cells are present in multiple lengths, while typically shorter versions of the mRNAs are preferentially expressed in transformed cancer cell lines. In Figure 10.12, we see an example of such an mRNA, cyclin D1, where the use of a proximal alternative polyadenylation site leads to a shorter mRNA that lacks binding sites for several different miRNAs (miR-15, miR-16, and let-7) that normally modulate expression of this gene.

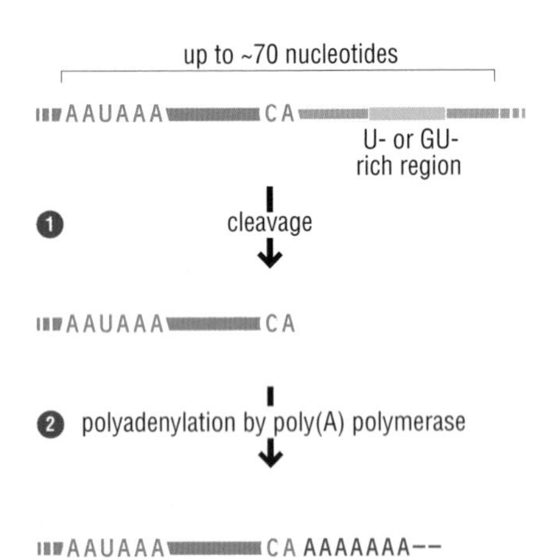

Figure 10.13 Polyadenylation occurs between a conserved AAUAAA sequence and a downstream U- or GU-rich region. The precursor mRNA is cleaved after a CA dinucleotide, which is found somewhere between the AAUAAA hexanucleotide sequence and a downstream U- or GU-rich region. Following cleavage, up to 200 adenine nucleotides are added to the new 3′ end of the mRNA by poly(A) polymerase.

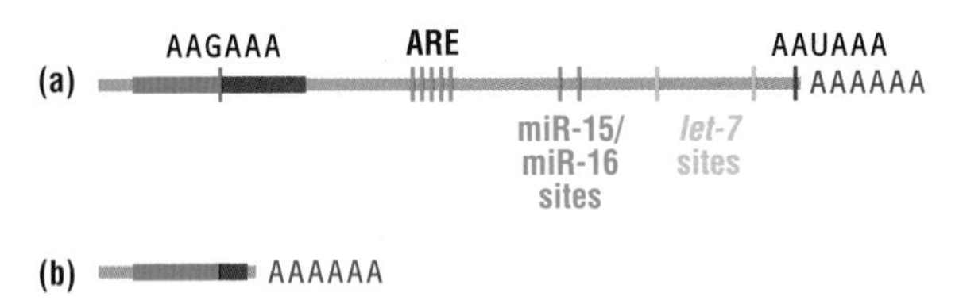

Figure 10.12 Alternative polyadenylation removes regulatory sequences. (a) Polyadenylation at the distal polyadenylation site (AAUAAA, pink) in the cyclin D1 mRNA retains AU-rich regulatory sequences (ARE, yellow) and sequences targeted by small regulatory RNAs (two shades of green). (b) Polyadenylation at the proximal alternative polyadenylation site (AAGAAA, pink) eliminates the ARE and small RNA target sequences.

Adapted from Figure 1B of Mayr C, Bartel DP. Widespread shortening of 3′UTRs by alternative cleavage and polyadenylation activates oncogenes in cancer cells. *Cell*, 2009; **138**:673–684.

The steps in the polyadenylation of the 3′ end of eukaryotic mRNAs are presented in Figure 10.13. The process begins with an initial cleavage of the mRNA. For most transcripts, the cleavage occurs after a CA nucleotide pair that lies somewhere between a conserved AAUAAA hexamer sequence and a U- or GU-rich region further downstream. Following this cleavage, a tail of approximately 200 adenosines is added by poly(A) polymerase. In general, a larger complex of proteins is required for this sequence of reactions than is required for the 5′ capping reactions: it is speculated that the larger complex is required because the polyadenylation enzymes must be able to perform the difficult job of recognizing and processing the different polyadenylation sites in different mRNAs.

The 3′ ends of mRNAs encoding metazoan histones are not polyadenylated, but instead carry highly conserved stem–loop structures that engage specialized RNA-binding proteins with functions similar to the proteins that bind the poly(A) tails. Bacteria also contain poly(A) polymerases, and bacterial RNAs have been shown to be polyadenylated. However, the fraction of bacterial RNA that is polyadenylated is lower than in eukaryotic cells, and polyadenylation in bacteria is generally associated with mRNA destabilization.

Capping and polyadenylation are intimately coupled to each other, and to transcription, by RNA polymerase II

There is an intimate association between 5′ capping and 3′ polyadenylation, and also with other processes that occur during transcription via RNA polymerase II. As we saw in Chapter 8, capping is required during transcription to allow RNA polymerase II to continue elongation of the mRNA (see Figure 8.24), and polyadenylation is required for the efficient termination of transcription (see Figure 8.29). In between capping and polyadenylation, RNA polymerase II recruits proteins that remove introns from the mRNA, as we will see in the next few sections on RNA splicing.

The region of RNA polymerase II that is responsible for orchestrating all these mRNA processing events is the C-terminal domain (CTD) of RPB1, the largest subunit of the polymerase. Figure 10.14 shows how there is a sequential interaction between the CTD and the different processing complexes. RNA polymerases I and III do not have a domain equivalent to the RPB1 CTD (of RNA polymerase II), and RNAs that are synthesized by these polymerases are not capped and polyadenylated.

Transcription and polyadenylation are also coupled to appropriate mRNA localization

Although the transcription and processing of eukaryotic mRNAs occurs in the nucleus, the translation of mRNAs occurs in the cytoplasm. Thus, mRNAs must be

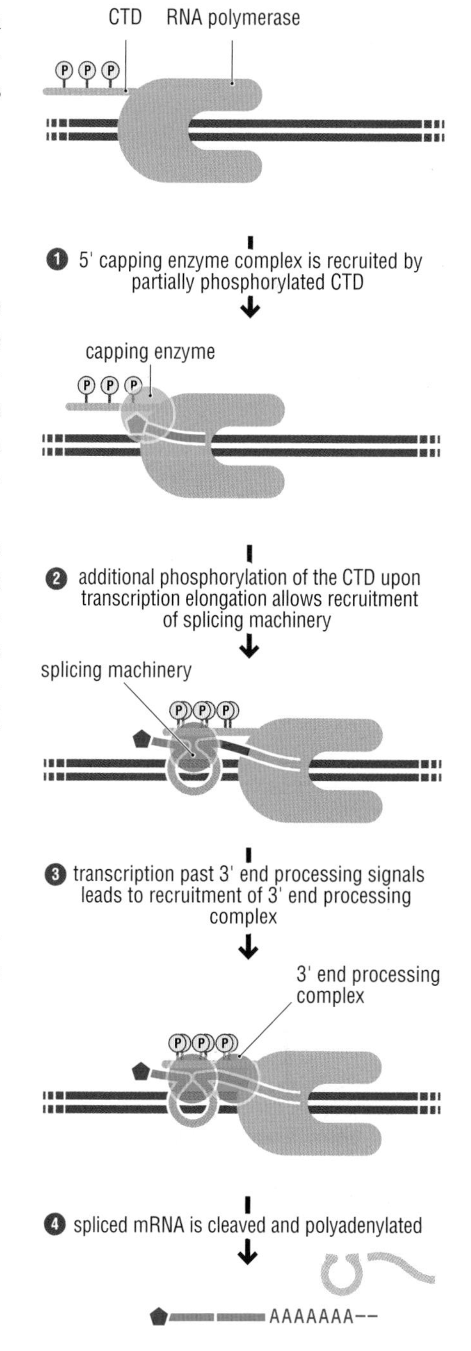

Figure 10.14 The C-terminal domain of RNA polymerase II coordinates many steps in the processing of mRNAs. The C-terminal domain of RNA polymerase II (CTD) plays a critical role in assembling and coordinating the activities of mRNA processing complexes. On the initiation of transcription, repeated amino acid sequences in the RNA polymerase CTD become partially phosphorylated (indicated by yellow balls). This form of the CTD recruits the enzymes (green) that add the 5′-cap (pink pentagon) to the growing RNA. After the RNA is capped, RNA polymerase can continue elongation, which is associated with further phosphorylation of the CTD. This additional phosphorylation in turn allows the recruitment of RNA splicing machinery (orange), which removes introns (gray lines) from the RNA. The additional phosphorylation also leads to the recruitment of the complex (blue) that cleaves and polyadenylates the 3′ end of the mRNA.

exported out of the nucleus before translation can occur. The downstream steps of mRNA export and translation are tightly linked to both transcription and processing of the mRNA; the protein factors required for nuclear export are loaded onto the precursor mRNA during transcription, but the export-competent mRNA-protein complexes are not released from the transcription complex until they receive signals from polyadenylation enzymes.

Beyond merely being transported out of the nucleus, some mRNAs are localized to very specific regions of the cytoplasm, allowing synthesis of the encoded proteins in defined regions of the cell. For example, a number of neuronal mRNAs are only translated in the distant synapses of the neuron. This specific localization requires complete processing of the mRNA as well as the presence of 'localization elements', generally found at the 3′ end of the transcript. These localization sequences are bound by specific RNA-binding proteins that often have dual functions – affecting both localization and regulating translation. For example, in the developing *Drosophila* oocyte, *bicoid* mRNA is localized to the anterior pole through binding to Staufen, an RNA-binding protein that interacts with sequence elements in the 3′ UTR of the *bicoid* mRNA as well as with components of the cell cytoskeleton and the translation complex.

The biochemical modifications made to mRNA molecules, which we have explored throughout this section, cause somewhat subtle changes to the primary structure of the molecules involved, changing only the ends of the molecules. However, we now go on to explore how many mRNAs are also subjected to a much more invasive form of processing, called splicing, in which entire sections of primary structure are simply removed from the transcribed molecule.

10.5 RNA SPLICING

Introns must be removed from precursor RNAs

As we note in Section 10.1, the mature transcript for many genes is encoded in a discontinuous manner as a series of discrete exons, which are separated from each other along the DNA strand by non-coding introns. mRNAs, rRNAs, and tRNAs can all contain these encoded introns, which must be removed from precursor RNAs to produce functional molecules. There are a number of different classes of intron, whose relative distribution and abundance depends on the kingdom in which they are found. Some are excised by proteins, some by RNPs, and some excise themselves; yet all use some variation of a series of transesterification reactions that we describe below.

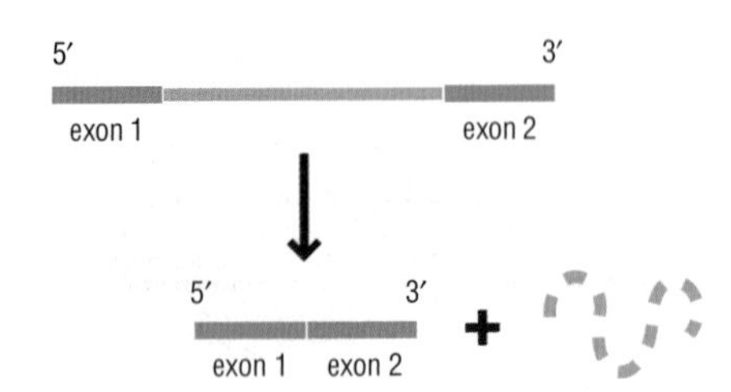

Figure 10.15 RNA splicing. RNA splicing involves cutting an intron (gray) from RNA and joining together (splicing) the two neighboring exons (blue). The spliced exons form the functional RNA and the intron is usually degraded.

Most introns do not themselves contain genes and are simply removed from the precursor RNAs and degraded, as depicted in Figure 10.15. Exceptions to this rule are the snoRNAs (see Section 10.3) and certain miRNAs in invertebrates and mammals (which will be discussed in Chapter 13), both of which are commonly found in the introns of mRNAs.

While introns are widespread, they are far more common in eukaryotes than bacteria. Indeed, the existence of introns has allowed for a process referred to as **exon shuffling**, wherein the DNA encoding exons can be exchanged and reordered through genetic recombination between intronic DNA. This process allows for the emergence of novel genes and is thought to have played a major part in the evolution of eukaryotic genomes. As we will see in Section 10.7, the differential removal of introns from an RNA transcript can result in the production of different spliced RNAs from a single gene, thereby increasing the number of gene products encoded in an organism's genome.

➔ We discuss recombination between DNA sequences in more detail in Chapters 16 and 18.

RNA splicing occurs by two transesterification reactions

There are several different pathways by which introns are removed from genes, but most of them involve the same underlying chemical reactions. The two exons that surround an intron are not directly joined to one another in a single step. Instead, a two-step reaction occurs in which the intron is first detached from exon 1, freeing this exon to react with exon 2. Both of these reaction steps are **transesterifications** in which a single phosphodiester bond between two nucleotides is broken and replaced with an energetically equivalent phosphodiester bond as illustrated in Figure 10.16.

The replacement of one chemical bond with a similar chemical bond has two implications: first, the reaction can occur without the net input of energy from an energy source such as ATP; and, second, the reaction can be readily reversible. In fact, reverse splicing reactions associated with certain classes of intron, which allow them to insert into duplex DNA, are thought to have played a significant role in dispersing introns and thus in shaping the evolution of the genome.

Most introns in eukaryotes are removed by a complex ribonucleoprotein machine called the spliceosome through a mechanism that we will discuss in Section 10.6. However, a few introns in eukaryotes, bacteria, and organelles are able to catalyze their own removal from RNA and are therefore called **self-splicing introns**. The discovery of self-splicing introns is described in Experimental approach 10.2. There are two groups of self-splicing introns, which illustrate some general features of RNA splicing; we consider these self-splicing introns first, before discussing the more common but complex case of splicing by spliceosomes.

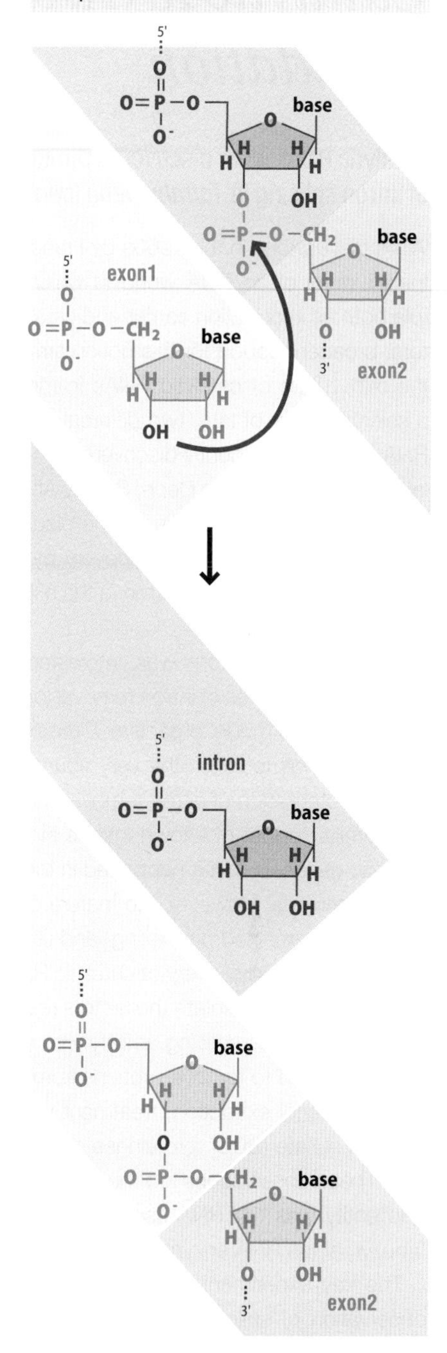

Figure 10.16 The transesterification reaction of RNA splicing. The 3′ OH of an incoming nucleotide (or RNA) (pink) attacks and breaks the 3′–5′ phosphodiester bond between two nucleotides in the target RNA (orange). As a result, a new 3′–5′ phosphodiester bond is formed between the incoming nucleotide (or RNA) and the target RNA. Alternatively, the 2′ OH of the incoming RNA can attack the target RNA, in which case the incoming and target RNA become joined though a 2′–5′ phosphodiester bond (not shown).

Group I introns can be self-splicing

Group I introns are a class of introns found in a number of genes in bacteria, viruses, lower eukaryotes and plants. Many of these introns are autocatalytic RNAs, or ribozymes, which can catalyze their own excision from a primary transcript in a splicing reaction involving two sequential transesterifications; this splicing reaction is depicted in Figure 10.17a. The first transesterification reaction has the unusual feature that a free guanosine nucleotide (rather than a specific nucleotide within the RNA molecule) is the attacking species that detaches the 5′ end of the intron from its neighboring exon 1. In the second transesterification reaction, the released terminus of exon 1 attacks at the intron–exon 2 junction, thus joining the two exons together and releasing the free intron.

The splice sites are precisely defined by the three-dimensional structure of the intron and by the recognition of a conserved G-U wobble pairing interaction found within a paired helical region.

The activity of other group I introns is dependent on the assistance of cellular and intron-encoded proteins, and so these are not referred to as self-splicing.

10.2 EXPERIMENTAL APPROACH

Catalytic RNA: discovery and tertiary structure elucidation

Catalytic RNAs were discovered through the characterization of intron splicing in *Tetrahymena thermophila*

While discussions in the 1960s by Francis Crick and Leslie Orgel fueled ideas of an RNA world in which RNA played a primary role both as information carrier and catalyst in our earliest ancestors, broader support for this notion arrived with the discovery in the mid-1980s of catalytic RNAs in modern-day organisms. In a scientific twist of fate, two different examples of such catalytic RNA were independently discovered around the same time in the laboratories of Thomas Cech, Sidney Altman, and Norman Pace, forever changing our views on the plausibility of an RNA world. Here we will focus on the discovery by Thomas Cech and colleagues of a self-splicing intron in the rRNA gene of *Tetrahymena thermophila*.

The Cech laboratory was interested in studying the basic mechanistic features of intron removal (or splicing); the group was working on the model organism *T. thermophila,* where the intron of interest was found in the very abundant rRNA transcript. Initial characterization of intron removal suggested that the process was similar to that of known introns in the mRNAs of other eukaryotic genes, in that it happened in the nucleus by a cleavage-ligation process. However, the 'nature of the activity responsible for splicing remained perplexing' and ultimately led the scientists to the surprising discovery of catalytic RNAs.

In their efforts to isolate the factors responsible for splicing, the authors found the splicing activity was able to withstand procedures intended to remove proteins from the splicing reaction, including phenol extraction, treatment with the detergent sodium dodecyl sulfate (SDS), proteinase digestion, and heat denaturation. These observations led the group to propose that the catalytic entity was the RNA itself, for which they coined the term 'ribozyme' (an RNA enzyme).

The key experiment that supported this hypothesis was the observation of splicing activity with a transcript of the *T. thermophila* rRNA gene precursor synthesized in a test tube by a heterologous RNA polymerase enzyme from *Escherichia coli* in the presence of radioactive GTP. The original figure from the paper is shown in Figure 1, where we focus on specific lanes. Lanes 1 and 3 were loaded with the transcription reactions for two different intron-containing constructs. The two bands that migrate at the sizes of the linear form of the excised intron (L IVS) and the circular form of the excised intron (C IVS) provide evidence that a splicing reaction has taken place. Lanes 4 and 6 were loaded with similar transcription reactions that were further incubated with free GTP, Mg^{+2}, and salt – all components found to stimulate the splicing reaction, as seen here by the diminished amount of full length transcript.

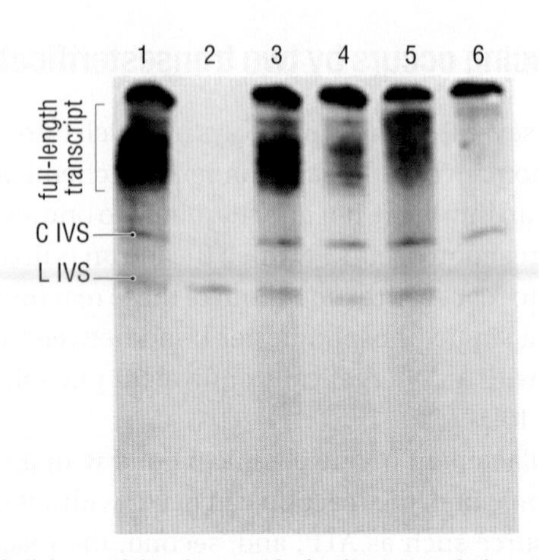

Figure 1 Splicing activity observed in *in vitro* transcription reactions with purified *E. coli* RNA polymerase. Various plasmid DNA templates were transcribed in reactions containing a-[^{32}P]-GTP, and the products loaded directly on a 4% polyacrylamide, 8 M urea gel. C IVS is the circular form of the excised intervening sequence RNA. L IVS is the linear form of the excised intervening sequence RNA.

Reproduced with permission from Kruger *et al.*, *Cell*, 1982;**31**:147–157.

As there was no possibility that *T. thermophila* proteins were present in any of these reactions, it became clear that the catalytic activity must derive from the RNA itself. While many subsequent experiments more elegantly defined how GTP stimulates the splicing reaction, what the reaction products are, and how they are made, this experiment was simple and the results sufficiently clear to conclude that RNAs can be catalytic.

A year later, Sidney Altman, Norman Pace, and colleagues reported that the RNA component of the ribonuclease P (RNase P) enzyme responsible for processing the 5′ end of the precursor tRNA transcript is also a ribozyme. Since that time, a number of other natural ribozymes have been identified, and there is now considerable support for the idea that key RNPs in modern-day biology (in particular the spliceosome and the ribosome) are at their functional core driven by RNA. Together, these discoveries changed the way biologists view evolution and added to our understanding of the ever-expanding roles of RNA throughout biology as we explore in this book.

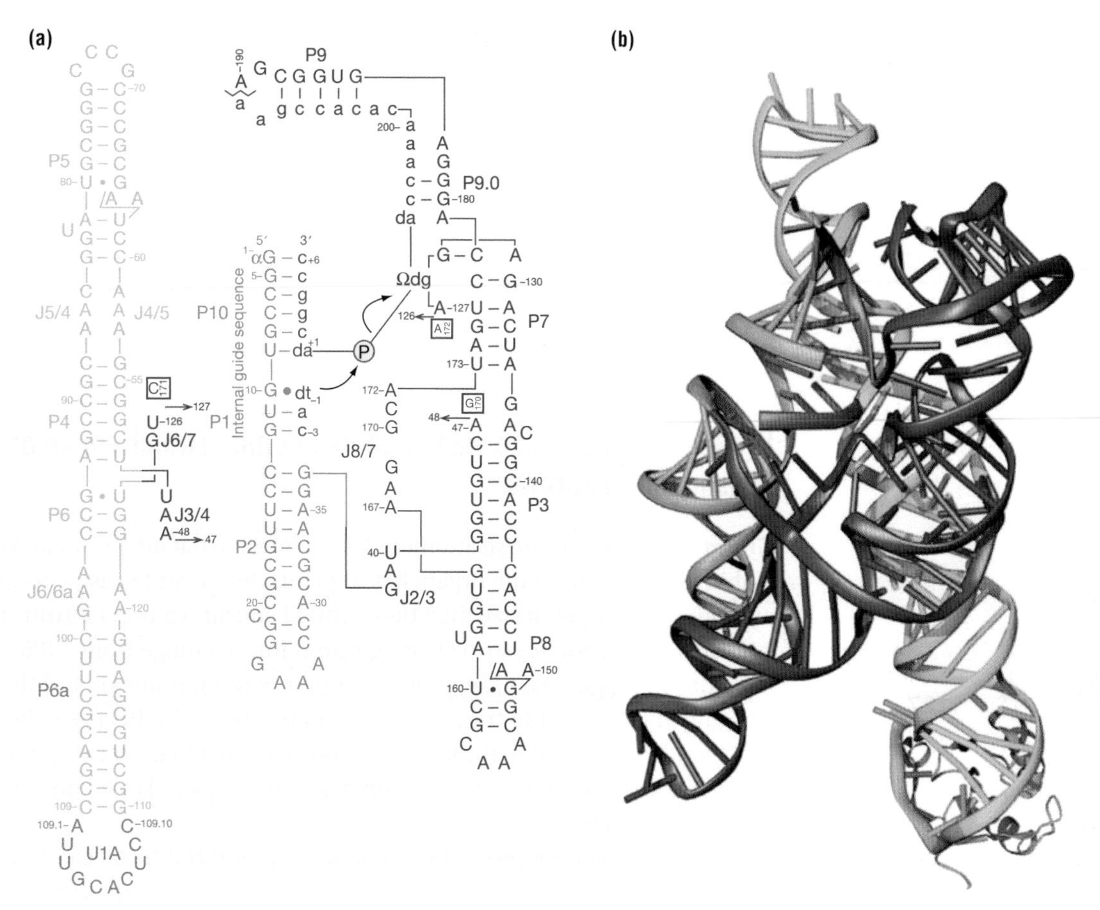

Figure 2 Overall secondary and tertiary structure of the group I intron. (a) Secondary structure of the group I intron from the pre-tRNA^{Ile} from the purple bacterium *Azoarcus sp. BH72*. The principal features of secondary structure, and many tertiary interactions, were determined through phylogenetic analysis. Paired regions are sequentially numbered (P1 through P10) as are joining segments (e.g., J4/5 connects P4 and P5). (b) Tertiary structure of the same intron with coloring corresponding to the secondary structure in (a). The gray U1A feature represents an RNA feature added to allow binding of a protein used to aid crystallization.

Adapted from Figure 1 of Adams *et al.* Crystal structure of a self-splicing group I intron with both exons. *Nature*, 2004;430:45–50.

Phylogenetic and structural studies determined the secondary and tertiary structural features of the group I intron

As biochemical studies established the catalytic prowess of the rRNA intron in *T. thermophila*, phylogenetic analyses by Francois Michel and Bernard Dujon identified a class of introns in mitochondrial, chloroplast, and nuclear-encoded genes in various fungi that appeared to share characteristic sequence features with the *Tetrahymena* intron. These introns were referred to as 'group I' and could easily be distinguished from the more standard eukaryotic pre-mRNA introns as they lacked the canonical GU…AG splice site recognition elements. Comparative sequence analysis of these introns eventually led to the secondary structural model predictions for the group I class on which the field relied for many years to guide biochemical studies (see Figure 2a).

To determine the tertiary structure, Scott Strobel and colleagues obtained crystals of the group I intron, with both exons attached, which diffracted to a resolution of 3.1 Å. The structure solved for these crystals, shown in Figure 2b, provides a view of the global three-dimensional structure of this catalytic RNA, with several co-axially stacked helices (seen in green and blue) converging to provide a binding site for the substrate helix (orange) involved in the splicing reaction. Even in the absence of a hydrophobic core, the RNA is able to form a secure binding site where specific catalysis can be promoted.

The structure also revealed key features of the intron that allow for efficient catalysis – the 3′ oxygen of the 5′ exon is beautifully poised for in-line nucleophilic attack on a conformationally constrained phosphate at the intron–3′ exon junction. In this structure, metal ions were also visible in the active site. (In fact, there were two of them.) This finding corroborated much biochemistry over the years arguing for the existence of several Mg^{+2} ions important for catalysis. This structure of an intact group I intron precursor RNA provided both an unprecedented view of a catalytic RNA, and a wealth of information on the structural motifs that underpin the complex three-dimensional structures of RNA throughout biology.

Find out more

Adams PL, Stahley MR, Kosek AB, Wang J, Strobel SA. Crystal structure of a self-splicing group I intron with both exons. *Nature*, 2004;**430**:45–50.

Guerrier-Takada C, Gardiner K, Marsh T, Pace N, Altman S. The RNA moiety of ribonuclease P is the catalytic subunit of the enzyme. *Cell*, 1983;**35**:849–857.

Kruger K, Grabowski PJ, Zaug AJ, *et al*. Self-splicing RNA: Autoexcision and autocyclization of the ribosomal RNA intervening sequence of *Tetrahymena*. *Cell*, 1982;**31**:147–157.

Related techniques

Tetrahymena (as a model organism); Section 19.1

In vitro transcription studies; Section 19.6

RNA gel electrophoresis; Section 19.7

Crystal structure determination; Section 19.14

Group II introns splice via the same chemical mechanism as spliceosomal introns

Group II introns comprise another class of introns that are found in bacteria and within genes found in the organelles of plants, fungi, and yeast. Group II introns are somewhat larger in size than their group I cousins (group I introns range from ~120 to 450 nucleotides in size and group II introns range from ~400 to 1000 nucleotides in length, excluding internal open reading frames), and they perform a slightly different set of transesterification reactions. In this case, the 2′ OH of a specific adenosine within the intron functions as the initial attacking nucleophile in a two-step series of transesterification reactions, depicted in Figure 10.17b, which link exon 1 with exon 2.

➔ We discuss base-pairing and tertiary interactions in RNA in Section 2.7.

As we have seen for group I introns, some group II introns can function without the assistance of any cellular proteins, whereas others do depend on protein cofactors. As was true for the group I intron, the splice sites of the group II intron are defined by the three-dimensional structure of the intron itself and by specific

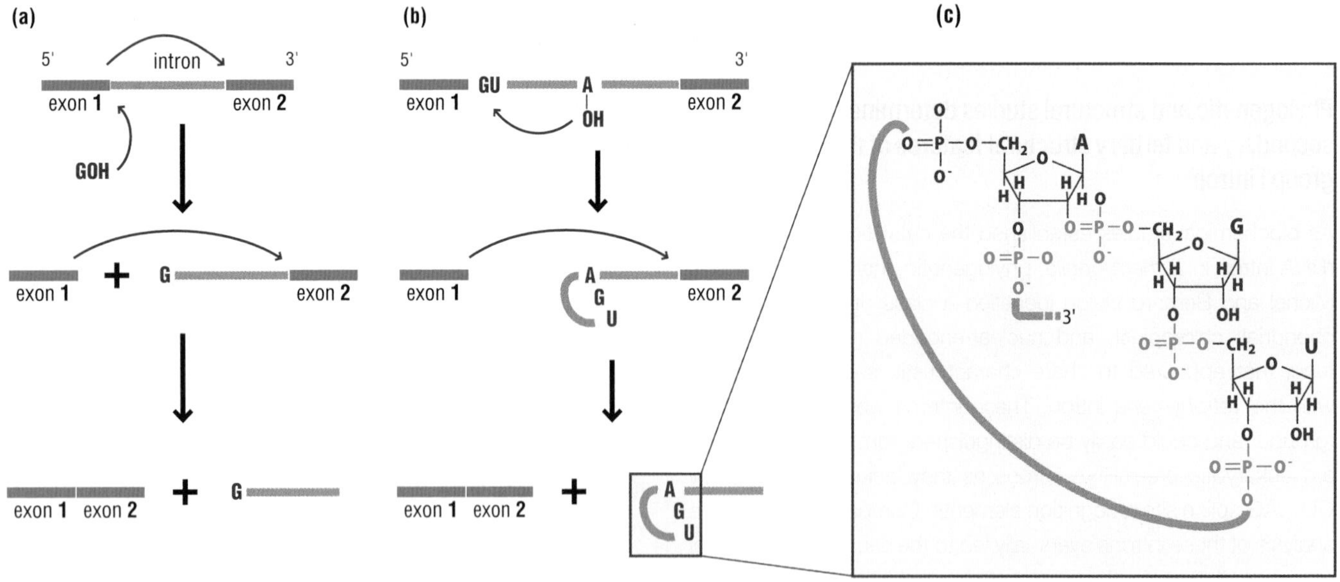

(a) 5′ intron 3′ exon 1 exon 2 GOH exon 1 + G exon 2 exon 1 exon 2 + G

(b) 5′ 3′ exon 1 GU A OH exon 2 exon 1 A G U exon 2 exon 1 exon 2 + A G U

(c) O=P–O–CH$_2$ A G U OH 3′

Figure 10.17 Splicing schemes for group I and group II introns. (a) The group I intron first uses the 3′ OH of an exogenous guanosine nucleotide (pink) to attack the boundary between exon 1 and the intron, releasing the intron from the exon. Next, the 3′ end of exon 1 attacks the boundary between the intron and exon 2, detaching the intron and joining the two exons together. (b) The group II splicing reaction uses an internal adenosine 2′ OH (pink) to attack the exon 1–intron boundary. The second splicing reaction occurs as it does in group I introns, with exon 1 attacking the intron–exon 2 boundary to join the two exons and release the intron. The resulting branched structure of the intron (pink box) is called a lariat and is shown in more chemical detail in (c).

base-pairings and tertiary interactions between intron and exon segments that position the critical sites within the catalytic center of the ribozyme.

The group II introns are of particular interest in furthering our understanding of the evolution of splicing: the utilization of an intron-internal adenosine (the branch-point adenosine) in the chemistry of the reaction results in the formation of a branched **lariat** intermediate (see Figures 10.17b, c), as is observed for the more common eukaryotic spliceosomal mechanism (see Section 10.6). These parallels speak to likely common evolutionary origins and inform our mechanistic understanding of both processes.

➲ We discuss transposition by group II introns further in Chapter 17.

Group II introns are also interesting in that they can act as mobile genetic elements that attack and insert themselves into duplex DNA by a reversal of the splicing mechanism. Also of interest is the fact that group II introns often contain large open reading frames, which encode 'maturase' proteins that assist structurally in splicing and directly in the catalysis of intron mobility reactions.

10.6 EUKARYOTIC mRNA SPLICING BY THE SPLICEOSOME

Splicing of eukaryotic mRNAs is carried out by a large RNP machine called the spliceosome

Genes in higher eukaryotes usually contain several introns. These introns are often of considerable length, sometimes extending to many thousands of bases. As such, they can account for greater than 90% of the length of a typical precursor mRNA. For example, exons correspond to only 14 kb of the 2400 kb human dystrophin gene shown in Figure 10.18. By comparison, lower eukaryotes such as yeast have fewer introns, and they are generally short (<300 nucleotides). The vast majority of eukaryotic introns are not self-splicing, and the formidable task of identifying and splicing together exons among all the intronic RNA is performed by a large ribonucleoprotein machine, the spliceosome. The spliceosome is composed of several individual **small nuclear ribonucleoproteins (snRNPs)**, pronounced 'snurps', which comprise both RNA and protein components) and many more additional proteins that come and go during the splicing reaction.

The splicing reaction catalyzed by the spliceosome is similar to that of the group II introns we discussed in the preceding section (see Figure 10.17b). In both cases, the 2′ OH of an adenosine residue located within the intron itself attacks the exon 1-intron boundary, detaching the intron from the exon and producing a branched intron structure (the lariat). Next, the terminal 3′ OH of the newly released exon attacks the intron-exon 2 junction, splicing together the two exons and releasing the lariat intron.

The high similarity in mechanism between the transesterification reactions catalyzed by the group II intron and the spliceosome has led to the idea that the simpler self-splicing intron is the evolutionary predecessor of the more complicated multicomponent eukaryotic machinery.

Figure 10.18 Exon–intron structure of the human dystrophin gene. The dystrophin gene contains 78 introns (gray), which comprise most of the length of the pre-mRNA. As a result, 79 exons (blue) can be many thousands of bases apart.

From Roberts, RG. Dystrophin, its gene, and the dystrophinopathies. *Advanced Genetics*, 1995;**33**:177–231.

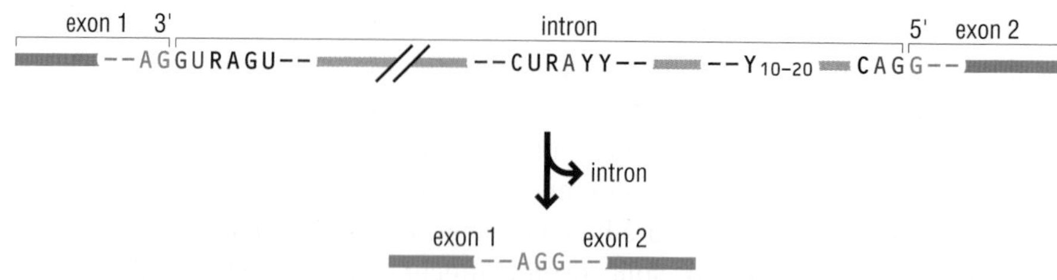

Scan here to watch a video animation explaining more about RNA splicing, or find it via the Online Resource Center at www.oxfordtextbooks.co.uk/orc/craig2e/.

Figure 10.19 Sequences in precursor mRNAs recognized by spliceosomes. Spliceosomes recognize a 5′ splice site motif (at the exon 1–intron border), a 3′splice site motif (at the intron–exon 2 border), the branch-point sequence (with the branch-point adenosine in pink) and a polypyrimidine (Y) tract upstream of the 3′ exon. This diagram shows the consensus sequences of mammals; other eukaryotes have their own functionally equivalent sequences.

Splice site recognition and spliceosome assembly are dominated by RNA–RNA interactions

Before splicing can occur, the spliceosome must identify the splice sites - the sites at which exons are separated from their neighboring introns, and at which two exons are subsequently attached. In contrast to the group I and II introns, where the splice sites are defined by the three-dimensional structures of the introns themselves, the spliceosome identifies splice sites by recognizing short sequence motifs found in each pre-mRNA. Key sequences indicated in Figure 10.19 are the **5′** and **3′ splice sites**, a **branch-point nucleotide** within the intron (equivalent to the branch-point adenosine we saw in the preceding section), and a **polypyrimidine tract** before the 3′ splice site.

In addition to these essential sequence elements, there are other sequences within the exons and in nearby intronic regions that help to define the exon–intron junctions. These additional sequence determinants allow cells to produce different mature RNAs from a precursor RNA depending on which introns and exons are utilized by spliceosomes, a process we explore further in Section 10.7.

The spliceosome is composed of five snRNPs (U1, U2, U4, U5, and U6) each containing an RNA molecule called a **small nuclear RNA (snRNA)** that is usually 100–300 nucleotides long and several protein factors. The snRNAs perform many of the spliceosome's mRNA recognition events by forming Watson–Crick base pairs with the precursor mRNA and one another. Sm proteins, which form ring-like structures, are common core constituents of the snRNPs. Other protein factors recognize specific sequences in the mRNA or promote the conformational rearrangements in the spliceosome that are required for the splicing reaction to progress.

Figure 10.20 shows a basic outline of the steps that are thought to occur during splicing. First, the 5′ splice site is recognized by the U1 snRNP (Figure 10.20, step 1). Other splice site consensus sequences are recognized by non-snRNP factors; the branch-point sequence is recognized by the branch-point-binding protein (BBP), and the polypyrimidine tract and 3′ splice site are bound by two specific protein components of a splicing complex referred to as U2AF (U2 auxiliary factor), U2AF65 and U2AF35, respectively.

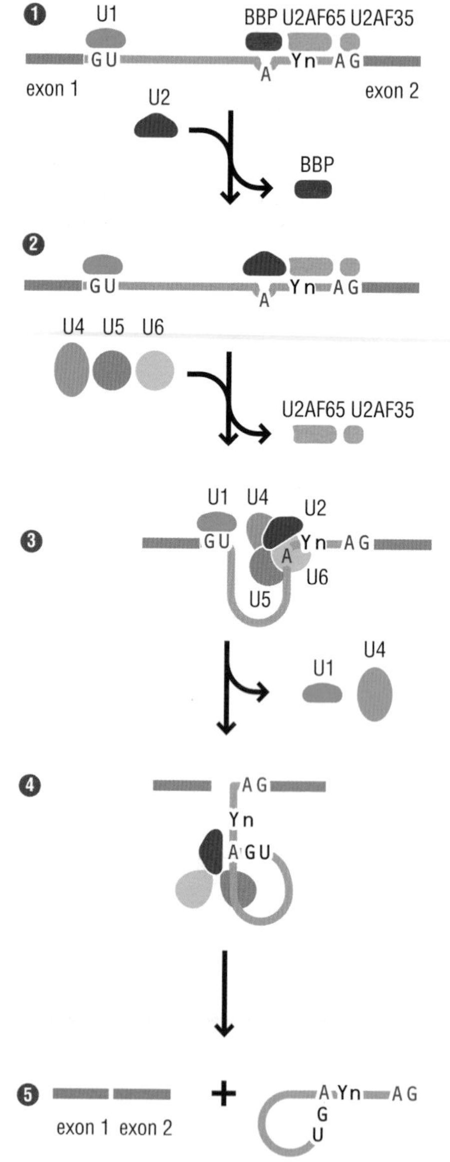

Figure 10.20 Spliceosome assembly and the splicing reaction. This schematic diagram shows the assembly of the spliceosome and the subsequent splicing reactions on an mRNA transcript. (1) The 5′ splice site (at the exon 1–intron border) is bound by the U1 snRNP, the branch-point sequence is bound by the branch-point-binding protein (BBP), and the polypyrimidine tract (Y(n)) and the 3′ splice site (at the intron–exon 2 border) are bound by a complex of the U2AF proteins, U2AF65, and U2AF35, respectively. (2) The U2 snRNP displaces BBP from the branch point, and a complex of the U4–U5–U6 snRNPs displaces the U2AF complex. (3) A rearrangement of the snRNPs releases the U1 and U4 snRNPs and (4) the initial transesterification reaction occurs to form the lariat intermediate. (5) Further rearrangements bring the two splice sites together, and the two exons are spliced together and the lariat intron is released.

Following the binding of these initial components, the remainder of the splicing apparatus assembles around them, in some cases displacing some of the previously bound components (Figure 10.20, steps 2 and 3). The precursor mRNA then undergoes a structural rearrangement, the first transesterification step takes place, and the lariat forms (Figure 10.20, step 4). Other rearrangements in the spliceosome next bring together the newly freed exon 1 and the intron–exon 2 junction – the two RNA pieces that will be joined to create the final mRNA product – and the second transesterification reaction completes the splicing (Figure 10.20, step 5).

While the molecular details of the individual steps are not fully understood, it is clear that the spliceosome is a dynamic ribonucleoprotein machine, and that conformational changes in the RNA components themselves are central to the spliceosome's function. Indeed, the known ATP-dependence of the splicing reactions is thought in part to reflect the requirements of the many helicases that promote and regulate these RNA-based conformational rearrangements. Based on similarities in mechanism to the group II intron, it is also generally believed that the snRNAs themselves (the U2 and U6 snRNAs being the primary candidates) play a central functional role in catalysis.

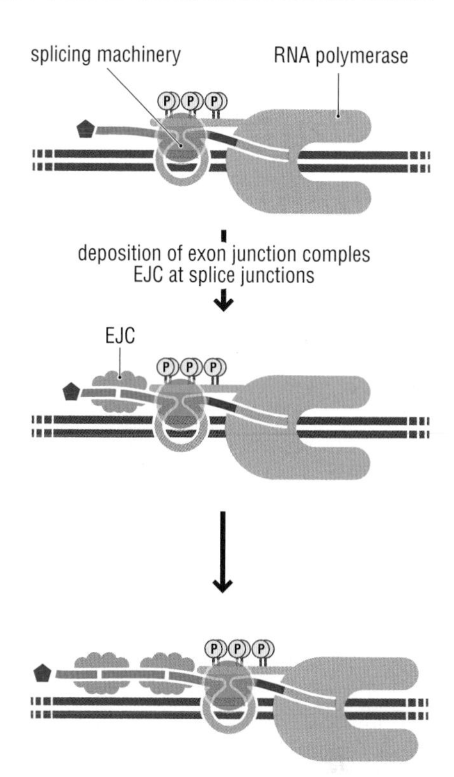

Figure 10.21 **Coordination of mRNA splicing with transcription.** This expansion of the steps that occur after step 2 in Figure 10.14 shows the deposition of exon junction complexes (EJCs) at splice junctions after splicing is completed. The EJCs mark the transcript as processed and interact with the export and translation machineries.

Protein factors contribute in multiple ways to pre-mRNA splicing

In addition to the snRNPs and proteins we have already mentioned, there are hundreds of other proteins that are associated with the splicing machinery and play important roles in the process. For example, PRP8, an essential, large protein found near the active site of the assembled spliceosome, has long been thought to be well positioned to play a critical role in catalysis. A number of ATPases probably facilitate structural rearrangements of the snRNAs during the sequential splicing steps as well as promoting mRNA and intron lariat release after the reaction is complete. It also has been proposed that these same ATPases are important for 'proofreading' mechanisms that promote fidelity in splice site selection. Mis-splicing in the cell can have dire consequences as the desired product is not produced, and often the wrong products can be toxic for the cell. Such an expenditure of energy for the purpose of promoting fidelity is a common feature of biological reactions.

Other proteins appear to regulate the splicing reactions. For example, the SR proteins (named after the serine(S)/arginine(R) dipeptide repeats that they possess) bind to the pre-mRNA where they are thought to help recruit other components of the spliceosome to the 5′ and 3′ splice sites. Some SR proteins play an important constitutive role in splicing reactions, whereas others may be produced in a cell-type specific manner where they are involved in controlling alternative exon splicing reactions (as we will discuss in Section 10.7). SR proteins can be heavily modified by phosphorylation on their abundant serine residues, and the extent of this phosphorylation may be altered throughout the splicing cycle to modulate splicing events.

Extensive study of the human spliceosome has shown that some of its integral components are also involved in transcription, polyadenylation, nuclear mRNA export, and translation, broadly highlighting the intimate coordination of biochemical events in mRNA biogenesis (see Figure 10.14). In one well-understood example of such coordination shown in Figure 10.21, the process of splicing leaves behind a set of proteins, the **exon junction complex (EJC)**, at splice junctions, thus 'marking' the transcript as processed and ready for export and translation. These EJCs are eventually recognized by the translational machinery in a process that evaluates whether the mRNA represents a full-length product (a form of quality control). The identification and characterization of the EJC is described in Experimental approach 10.3.

10.3 EXPERIMENTAL APPROACH

Definition of an EJC and a higher order mRNP structure

Protein components (denoted Exon Junction Complex or EJC) that specifically associate with spliced mRNAs provide a 'memory' of splicing

The life cycle of an mRNA in the cell is in large part specified by its complement of RNA binding proteins that, together with the mRNA, form the messenger RNA protein particle (mRNP). The mRNP proteins have important functions in determining the subcellular localization, translational efficiency, and the half-life of the mRNA. In eukaryotic cells, the 5′ cap structure of a pre-mRNA is bound in the nucleus by cap binding proteins, CBP20/80, which associate with the transcript during transcription and which are ultimately replaced by the translation initiation factor eIF4E in the cytoplasm. Similarly, the 3′ poly(A) tail of an mRNA transcript is bound initially by a nuclear version of poly(A) binding protein, PABP2, which ultimately is replaced by PABP1 in the cytoplasm.

Other mRNP components, such as hnRNP and SR proteins, recognize specific sequences in the mRNA and 'landmark' features, such as where pre-mRNA processing events have occurred. For example, introns are typically bound by the hnRNP proteins (hnRNP A, hnRNP B, and hnRNP C) while exons are bound by SR proteins. During splicing, a set of proteins referred to as the exon junction complex (EJC) is specifically deposited upstream of exon-exon junctions, marking the site where an intron was previously located. In higher eukaryotes, EJCs are important both for increasing the translation efficiency of newly made mRNAs in the cytoplasm and for recognizing premature termination codons (PTCs), a feature that targets the transcript for degradation by nonsense mediated decay (NMD).

The identification of the components of the EJC and its binding site in the mRNA, and the gradual elucidation of the biological role of the EJC have been exciting stories to follow. Early studies identified a number of protein components differentially associated with spliced and unspliced mRNA transcripts. Other studies argued that termination codons were recognized as premature when they were found in exons that preceded the terminal exon, placing them upstream of an exon-exon junction on the mRNA.

These ideas came together beautifully in a biochemical study by Melissa Moore, Lynne Maquat, and colleagues (2000), where the authors used an RNase H footprinting approach to identify regions of the RNA that were protected following a splicing reaction. In the study, the RNA was incubated with a series of oligonucleotides. Where these oligonucleotides base-pair with the mRNA, the duplex becomes a substrate for RNase H; where these oligonucleotides are not able to base-pair (i.e., because of an obstruction), the mRNA is protected. We see in Figure 1 a small protected region of mRNA just upstream of the exon-exon junction specifically on transcripts that have undergone splicing. The site of protection was found to be sequence independent, and the footprint was identical in spliced transcripts generated from either HeLa cell or *Xenopus* splicing extracts, establishing the generality of the observation.

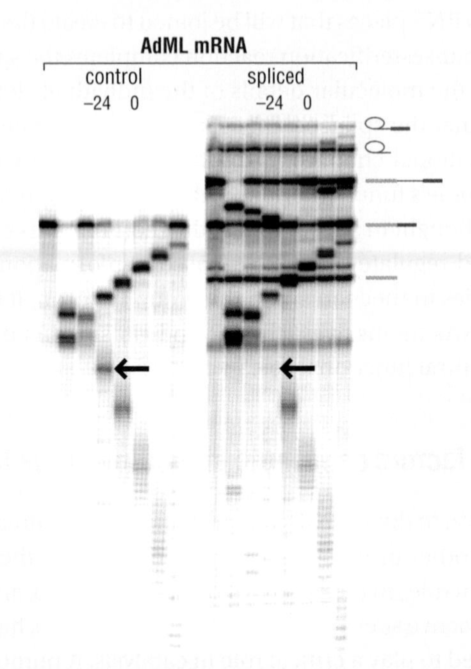

Figure 1 Splicing-dependent protection of localized region of mRNA from targeted RNase H cleavage. On left, the control mRNA (that looks like the product of splicing) is exposed to RNase H targeted cleavage using a series of DNA oligonucleotides that anneal along the length of the transcript; in this case, each oligo triggers cleavage by RNase H as indicated. On right, a pre-mRNA version of the same transcript (unspliced) is incubated in a splicing reaction, and then subjected to the same RNase H-targeted cleavage. Here we see that a single site (indicated by the arrow) along the length of the transcript is protected from RNase H-targeted cleavage. This site is protected by a bound EJC.

Reproduced with permission from Le Hir *et al.* The spliceosome deposits multiple proteins 20-24 nucleotides upstream of mRNA exon-exon junctions. *EMBO Journal*, 2000;**19**;6860–6869.

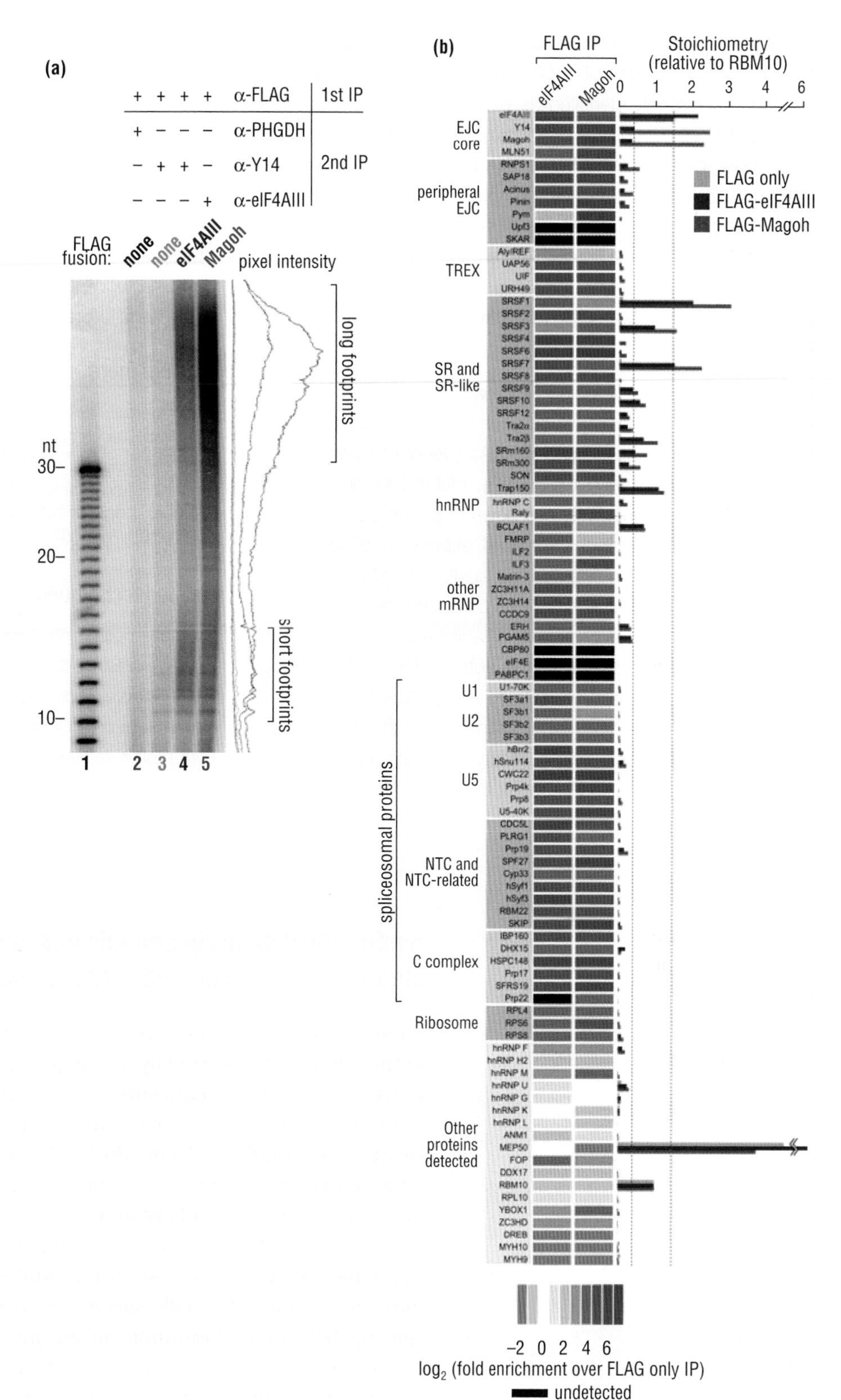

Figure 2 Biochemical analysis of the mRNP reveals a large protein complex. (a) A dual affinity pull down of components of the EJC shows large RNA fragments associated with the EJC complex. The mRNA complexes were purified using FLAG antibodies targeting a tagged version of either eIF4AIII or Magoh followed by an antibody directed against Y14, eIF4AIII, or the control protein PHGDH. RNA associated with these pull downs was labeled with radioactivity and analyzed on a denaturing gel. Inspection of this gel reveals two broad footprints (darker bands) for the lanes with the FLAG-tagged eIF4AIII or Magoh proteins, one short and one long, that map to the original EJC footprint site determined by RNase H mapping (above) and to the bulk of the exon, respectively. (b) Mass spectrophotometric analysis of the protein components of the FLAG-tagged eIF4AIII or Magoh protein complexes reveals an extensive set of proteins previously implicated in RNA biogenesis and function, including in particular a large variety of SR proteins.

From Singh *et al.* The cellular EJC interactome reveals higher-order mRNP structure and an EJC-SR protein nexus. *Cell*, 2012;**151**(4):750–764.

The authors further established that the RNP complex bound to that site had a mass of ~350 kD and contained proteins that had been previously identified as being differentially associated with spliced transcripts. This study gave physical meaning to a 'memory' of splicing for the mRNA transcript: the memory of splicing was a residual protein complex deposited at the exon-exon junction, the EJC. Subsequent studies established eIF4AIII, Y14, and Magoh as the key EJC core components.

High throughput sequencing approaches expanded our definition of an EJC

High throughput sequencing approaches have recently extended our understanding of the composition and structure of RNPs associated with spliced mRNAs (denoted mRNPs). In a seemingly straightforward experiment, Melissa Moore and colleagues performed an affinity pull-down of the core EJC proteins (eIF4AIII, Y14, and Magoh) from cell extracts and sequenced the associated RNA fragments. This analysis confirmed that the major 'canonical' EJC site *in vivo* lies 24 nucleotides upstream of exon-exon junctions, as previously established *in vitro*.

More interestingly, however, the authors found that endogenous EJCs do not exist as isolated complexes but rather are predominantly found in megadalton-sized complexes composed of multiple EJCs and a variety of additional RNA binding proteins. As shown in Figure 2a, this large complex protects extended segments of RNA (of the size of the average exon) from extensive treatment with nucleases. Mass spectrometric analysis revealed a large numbers of proteins associated with the EJC core, as enumerated in Figure 2b. Together, these findings suggest that another function of the EJC is to help package spliced mRNAs into highly condensed mRNPs suitable for transport out of the nucleus and to their ultimate destinations in the cytoplasm. These intriguing ideas about higher order mRNP particles will undoubtedly be subjected to experimental testing in the near future.

Find out more

Le Hir H, Izaurralde E, Maquat LE, Moore MJ. The spliceosome deposits multiple proteins 20-24 nucleotides upstream of mRNA exon-exon junctions. *EMBO Journal*, 2000;**19**:6860–6869.

Singh G, Kucukural A, Cenik C, *et al*. The cellular EJC interactome reveals higher-order mRNA structure and an EJC-SR protein nexus. *Cell*, 2012;**151**:750–764.

Related techniques

RNAse H footprinting; Section 19.12

Protein extraction; Section 19.7

Xenopus (as a model organism); Section 19.1

Use of HeLa cells; Section 19.2

RNA-protein immunoprecipitation + sequencing; Section 19.12

Mass spectrometry; Section 19.8

A minority of genes in animals and plants are processed by an alternative set of splicing snRNPs (the AT–AC spliceosome)

While the majority of pre-mRNA introns have GU and AG dinucleotides at their termini and are recognized by the 'major' spliceosome described above, some introns in multicellular organisms are processed by a spliceosome that includes the same U5 component as above, but four different snRNPs (U11, U12, U4atac, and U6atac) functionally replacing the U1, U2, U4, and U6 snRNPs. This machinery is referred to as the minor or U12-dependent spliceosome and is far less abundant in cells than the major spliceosome.

Certain introns spliced by the U12-dependent spliceosome have unusual splice site sequences with AU and AC dinucleotides at the intron termini. Consequently, the spliceosome was initially referred to as the AT–AC spliceosome. In general, the splicing pathway for this minor spliceosome appears to be similar to that followed by the major spliceosome. It is not clear why this secondary splicing pathway was retained during the evolution of metazoans.

Trans splicing joins exons from two separate RNA transcripts

The different splicing processes described above are all examples of ***cis* splicing** – that is, the splicing events occur between parts of the *same* molecule. Another unusual form of splicing is ***trans* splicing**, which results in exons from two separate RNA molecules becoming joined together. *Trans* splicing has mostly been explored in nematode worms and trypanosomes (the single-celled animals that cause African sleeping sickness), where these events are reasonably common (trypanosomes splice all of their mRNAs in this way). Similar *trans* splicing has also been observed in *Drosophila*, in which it occurs in a few genes.

In *trans* splicing, illustrated in Figure 10.22, the 5′ exon is a short RNA called the spliced leader RNA (**SL RNA**) that is generally 20–50 nucleotides long. During the *trans* splicing process, this SL RNA is joined to a variety of different mRNA transcripts. Following the *trans* splicing reaction, the SL RNA represents the first exon in the mature mRNA and is typically located just upstream of the AUG initiation site.

During the *trans* splicing reaction, the SL RNA is bound to proteins to form the SL snRNP, effectively replacing the U1 snRNP bound to the 5′ splice site during the more standard splicing reactions. This complex then interacts with the introns and exons within the mRNA and the U2, U4, U5, and U6 snRNPs to carry out a splicing reaction essentially analogous to the ***cis* splicing** reactions described above. The sequence and cap structure of the SL RNA are thought to allow the highly efficient recognition of these *trans*-spliced mRNAs by the ribosome and critical translation factors, thus resulting in overall robust translation.

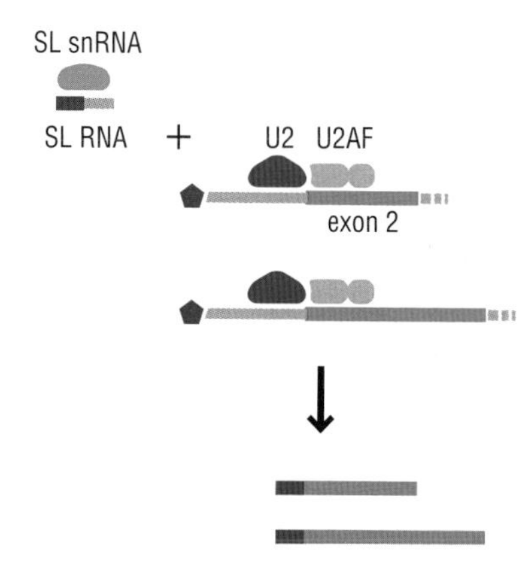

Figure 10.22 Mechanism of *trans*-splicing. In contrast to *cis*-splicing where splicing events take place between parts of the same RNA, in *trans*-splicing, a short RNA called the spliced leader RNA (SL RNA) is joined to a variety of different mRNA transcripts. In the case of *trans*-splicing, a SL snRNP, which replaces the U1 snRNP bound to the 5′ splice site in *cis*- splicing reactions, interacts with the U2, U4, U5, and U6 snRNPs at the 3′ splice site to carry out a reaction essentially analogous to ***cis*-splicing** reactions.

10.7 EXON DEFINITION AND ALTERNATIVE SPLICING

Diversity in gene expression in eukaryotes derives from alternative splicing

In a gene comprising multiple introns and exons, the process of splicing may yield more than just a single mature mRNA in which the exons are joined in only one way. Instead, at least 75% of genes in higher eukaryotes undergo alternative splicing, in which different combinations of exons in a gene are spliced together to produce different mature mRNAs. There are many ways in which the splicing pattern of a multi-exon mRNA can be altered. Although most exons are constitutive, meaning that they are always included in the final mRNA, regulated exons that may or may not be included are also common.

Some of the possible outcomes of alternative splicing are shown schematically in Figure 10.23. In addition to the potential exclusion or inclusion of a particular exon, the splicing pattern of an mRNA can be altered by the utilization of alternative 5′ or 3′ splice sites. Alternative splicing combined with the utilization of alternative transcriptional start or polyadenylation sites can lead to even further diversity. What alternative splicing occurs depends on a variety of factors such as the cell type, the developmental stage of the cell or organism, or cues from the cellular environment, and are typically mediated through interactions with a large group of auxiliary proteins involved in the splicing reactions that we will describe below.

Alternative exons represent an immense source of potential genetic diversity. As mentioned in Section 10.1, an extreme example is the *Drosophila dscam* pre-mRNA that can in principle be processed to generate more than 38 000 different mature transcripts. Indeed, some of the most extensive regulation through alternative splicing has been observed in the mammalian nervous system, which is not surprising given the complexity of neuronal interactions.

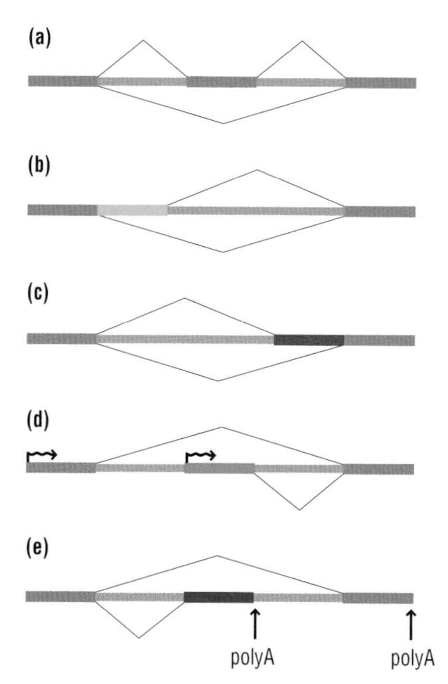

Figure 10.23 Possible outcomes of alternative splicing. A particular exon may be included or excluded (a) or alternative 5′ (b) or 3′ (c) splice sites may be used. Additional variation can come from the utilization of alternative transcriptional start (d) or termination (polyadenylation) (e) sites.

In Section 10.6 we learned about the sequence motifs - the 5′ and 3′ splice sites, the branch-point consensus and the polypyrimidine tract - which are the primary determinants used by the spliceosome to identify splice sites in precursor RNAs. We now look in more detail at how spliceosomes distinguish and discriminate between potential splice sites in an mRNA, and how cells can regulate alternative splicing.

Exon and intron definition models help to explain how the ordered recognition of splice sites can occur

The fundamental task of the spliceosome is to recognize splice junctions and accurately join together the appropriate exons. The sequences that define the exon–intron junctions are, as we have seen, relatively simple. Consequently, similar sequences commonly occur in introns purely by chance: such accidental splicing signals are known as **cryptic splice sites**. The spliceosome must therefore not only reliably detect true splice sites, but must also reliably ignore cryptic splice sites.

There are two models for how the spliceosome correctly identifies splice sites. These models are called **exon definition** and **intron definition**. Experiments that have explored the alternative models are discussed in Experimental approach 10.4.

The relative contributions of each model to splice site identification are thought to be somewhat species-dependent. In mammals, the likely general explanation for how the spliceosome correctly identifies splice sites is through **exon definition**, in which the exon is 'defined' for the spliceosome to recognize (Figure 10.24a). According to this model, the 5′ and 3′ ends of an exon are brought together by interactions between the U1 snRNP and the U2 complex (composed of the U2 and U2AF snRNP). Figure 10.25 shows how the exon is typically marked by SR proteins that are needed to 'define' the exon and assemble the activated splicing complex, while the introns are typically bound by a heterogeneous group of proteins (referred to as hnRNPs), which can help in masking the numerous potential

(a) exon definition - predicted consequence is exon exclusion

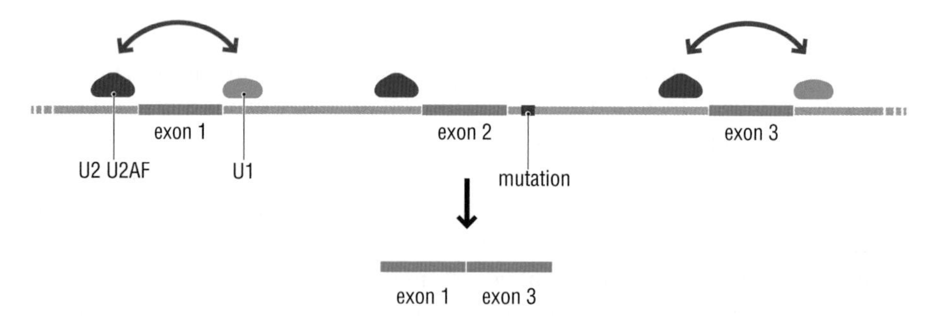

(b) intron definition - predicted consequence is intron inclusion

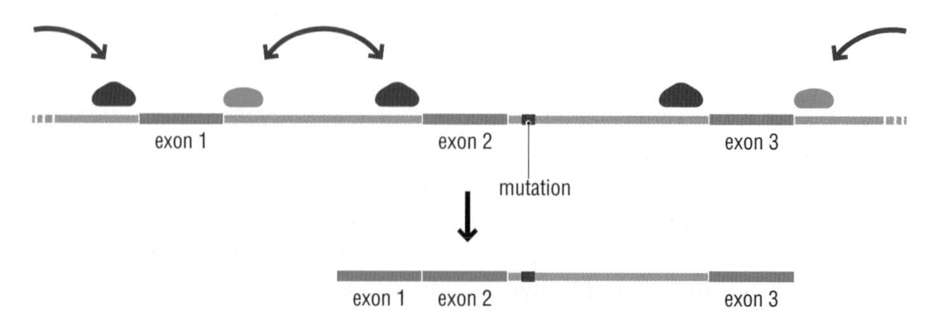

Figure 10.24 Two possibilities for how correct splice sites are identified. (a) In the exon definition hypothesis, splice sites are brought together across the short exon sequences by direct interactions between the U2–U2AF and the U1 snRNP complexes (or between these components and the cap binding complex at the 5′ end of the mRNA and the poly(A)-binding proteins at the 3′ end of the mRNA). This idea predicts exon exclusion when mutation of the 5′ splice site occurs. (b) Alternatively, interactions between the U2–U2AF and the U1 snRNP complexes might occur across the relatively longer intron sequences. This idea predicts intron inclusion when mutation of the 5′ splice site occurs.

cryptic splice sites. This hypothesis nicely explains several general observations from mammalian splicing systems: that exons tend to be short relative to introns and thus simpler to define; and that mutation of 5′ splice sites tends to result in exon exclusion rather than intron inclusion.

The alternative to the exon definition model is that introns are defined through interactions between sequential 5′ and 3′ splice site bound factors, denoted **intron definition** (see Figure 10.24b). The prediction of this model is that intron inclusion results from the mutation of a 5′ splice site, as has been observed in certain systems.

While both models help us to understand how sequential splice sites are identified to promote ordered and accurate splicing, it is clear that the final active splicing complex must be one in which 5′ and 3′ splice site protein complexes interact with one another across the intron (to facilitate its removal). In the cases where exons are 'defined,' this requires reorganization of the splicing intermediates, whereas in the cases where introns are 'defined,' no such reorganization is required. The question of how such reorganization might take place in the case of exon definition remains open.

An important feature of both models (exonic vs. intronic) for splice site definition is that the 5′ and 3′ splice sites are marked co-transcriptionally (that is, while transcription is still in progress). These events are facilitated by specific interactions between the unusual C-terminal tail of RNA polymerase (see Section 10.4) and splicing components including the U1 snRNP. Co-transcriptional splicing allows sequential splice sites to 'see' one another as they emerge from the polymerase, thus minimizing the chance that distant exons are mistakenly joined together. Moreover, after the splicing reactions are complete, the EJC (the residual protein 'mark' we mention earlier) is left behind, effectively telling the translational machinery that the splicing machinery has completed its task.

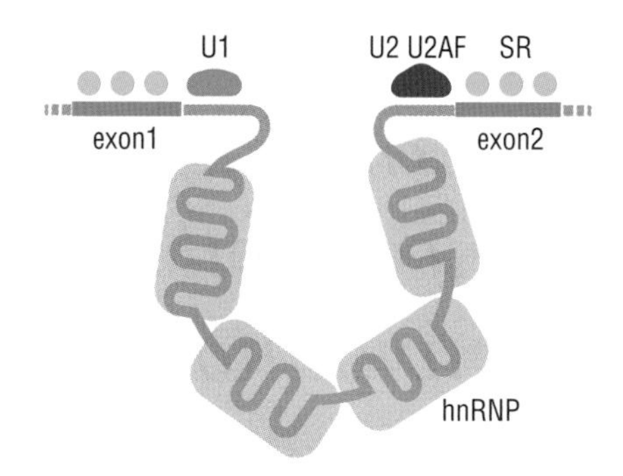

Figure 10.25 Definition of exon and intron sequences. Binding of the SR proteins to exons and hnRNP complexes to introns helps the spliceosome to differentiate between exon and intron sequences.

Alternative splicing is regulated by a complex set of pre-mRNA sequences in exons and introns that are recognized by diverse cellular factors

Exon or intron definition and the co-transcriptional assembly of the splicing machinery explain in general terms how exons can be identified by the spliceosome and joined to the next exon in the pre-mRNA transcript. However, the splice site consensus sequences do not provide sufficient information to determine whether a site will be identified as functional by the spliceosome. Instead, other information and interactions are required to stimulate their use. Moreover, the selection of particular splice sites is often tightly regulated in order to include or omit specific exons, or portions of them, from the mature mRNA (as described in Figure 10.23).

In addition to the exon or intron 'defining' interactions described above, there are many non-splice site regulatory sequences that strongly affect spliceosomal function. RNA elements that positively affect splicing are referred to as enhancers. Such **intronic** and **exonic** splicing **enhancer sequences** (ISE and ESE, respectively) typically bind proteins that increase the ability of spliceosomes to identify nearby splice sites in a pre-mRNA. The SR family of proteins appears to be broadly involved in binding ESE sequences (and recruiting other components) and can be thought of as important determinants of exon definition.

Intronic and **exonic** splicing **silencer sequences** (ISS and ESS, respectively) have the opposite effect, often masking splice sites or blocking spliceosome assembly, thus causing exons to be omitted from the mature mRNA. A well-characterized

10.4 EXPERIMENTAL APPROACH

Determining the roles of exon and intron definition in eukaryotic splicing though mutational and computational analyses

Early evidence for exon definition came from splicing reactions reconstituted *in vitro*

Early studies of eukaryotic pre-mRNA splicing identified short consensus splice site recognition sequences that appeared to be rather ubiquitously used to define complex splicing patterns. These rules were relatively simple: GU at the 5′ splice site and AG at the 3′ splice site, with several adjacent positions that are biased toward particular bases and determine the intrinsic strength of the site. However, the low information content of these sites immediately suggested that there must be many other determinants that are involved in defining the appropriate and sequential splice sites with high precision; these additional determinants have come to be known as enhancers and silencers and are found in both exons and introns (see Section 10.7 for further discussion of these elements).

As discussed in Section 10.7, there have been two dominant models for how splice site recognition occurs, through exon definition (wherein interactions between the 5′ and 3′ splice sites span an exon) or through intron definition (wherein interactions between the 5′ and 3′ splice sites span an intron). The rules appear to be different in different organisms, and these differences appear to correlate with critical features in the pre-mRNAs themselves. While insights into these models and where they apply have come from numerous approaches, certainly some of the early biochemical studies by Susan Berget and colleagues were instrumental in guiding our thinking about this problem. Berget realized that exon and intron definition models could be distinguished by following the efficiency of spliceosome assembly and actual splicing in *in vitro* reconstituted mammalian splicing extracts with a carefully chosen set of pre-mRNA reporters carrying different combinations of complete and partial exons and introns.

If we first look at Figure 1, we see the authors compare the formation of a spliceosomal assembly intermediate (α) on a native gel using two sets of pre-mRNA constructs (radioactively labeled). The constructs have either a complete exon (with both 5′ and 3′ splice sites) or a partial exon (with only the 3′splice site present) (a vs b) followed by variable length 'intronic' sequences. We see in Figure 1a (the E2 series) that, when a complete exon sequence is

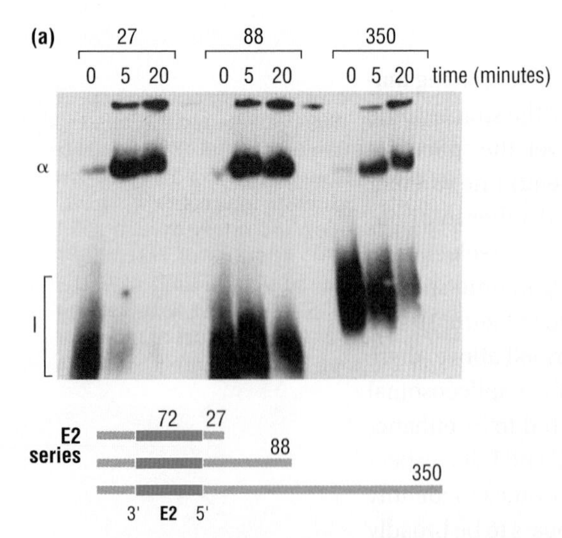

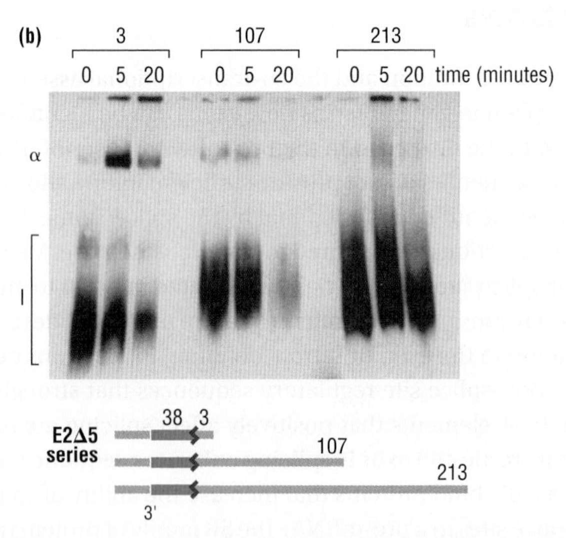

Figure 1 Enhanced complex assembly for mRNA substrates containing a full exon. Complex assembly in mammalian splicing extract was followed on 3.5% native acrylamide-agarose gels following incubation with [^{32}P]-radiolabeled mRNA substrates. Cartoons of pre-mRNA constructs are shown below the gel with complete (a) or incomplete (b) blue exons and gray introns (of variable length). Equal amounts of substrate were incubated in the extracts for the indicated times before gel electrophoresis. I represents the uncomplexed mRNA, whereas α represents a known splicing intermediate. 5′ and 3′ refer to the splice sites (not directionality of the mRNA).

present, an intermediate (α) forms in each case, though with diminishing efficiency as the intronic sequence length is increased. By contrast, we see in Figure 1b (the E2Δ5 series) that formation of the assembly intermediate with an incomplete exon (indicated by squiggly gray line) is substantially less efficient, and the stability of the complex over time is compromised.

These data immediately suggest that the presence of both consensus splice sites and the exon between them are important determinants of spliceosome assembly. However, these experiments do not directly compare the efficiency of spliceosomal assembly on a complete exon with that on a complete intron (since there is no complete intron in these pre-mRNAs).

In the next experiment, the researchers again compared spliceosomal assembly efficiency on a set of pre-mRNAs containing a full exon (E1) and intron, as well as portions of another exon (E2). In Figure 2a (the E1–E2 series), complex assembly was quite efficient (now forming three distinct complexes, A, A′, and B) on all constructs carrying two complete exons surrounding an intron. Splicing is also efficient with each of these constructs (not shown). In Figure 2b (the E1-E2Δ5 series), complex assembly was severely compromised when the second exon was incomplete (missing the 5′ splice site recognition sequence).

Since there is a complete intron present in these constructs in both parts of Figure 2, and yet the efficiency of assembly is very different, the immediate conclusion was that recognition of splice sites across the exon (but not the intron) is critical to the assembly process. While the experiments represented in Figures 1 and 2 followed the assembly of the splicing machinery, the results correlate very well with overall splicing activity observed in the same *in vitro* extracts. From these experiments, the exon definition model emerged as a robust framework for thinking about splicing in mammals.

Bioinformatic analysis of covariation between splice sites supports exon definition model in mammals (and an intron definition model in other organisms)

The exon definition model (and its role in mammals) has been supported by bioinformatic analyses by Christopher Burge and colleagues using the vast wealth of information available in the annotated genomes. These studies modeled the coevolution of splicing elements in orthologous exon pairs in several different eukaryotic taxa using network formalisms, called Principal Variation-based Subset Analysis (PVSA), more commonly applied to the analysis of protein interaction networks.

The basic idea behind the analysis was to use comparative genomics to ask whether covariations (or signs of co-dependence) were apparent between the 5′ and 3′ splice sites located across exons from one another or across introns. In essence, the method asked whether large decreases in the strength of the 5′ splice sites of exons in one organism relative to another were associated with compensating increases in the strength of the upstream (cross-intron) or downstream (cross-exon) 3′ splice site. This analysis revealed major trends in splicing element coevolution that are different between mammals and other eukaryotic taxa.

In terms of the exon versus intron definition models, these computational studies observed clear cross-exon (but not cross-intron) compensatory interactions between the 5′ and 3′ splice sites in human and mouse genes, whereas cross-intron compensatory interactions were more commonly observed in plants, fungi, and invertebrates (indicated with green arrows in Figure 3a). These observations are strikingly consistent with the length variability of introns and exons across these diverse taxa (depicted in Figure 3b) and even with the relative extents of exon exclusion versus intron retention observed experimentally in these systems

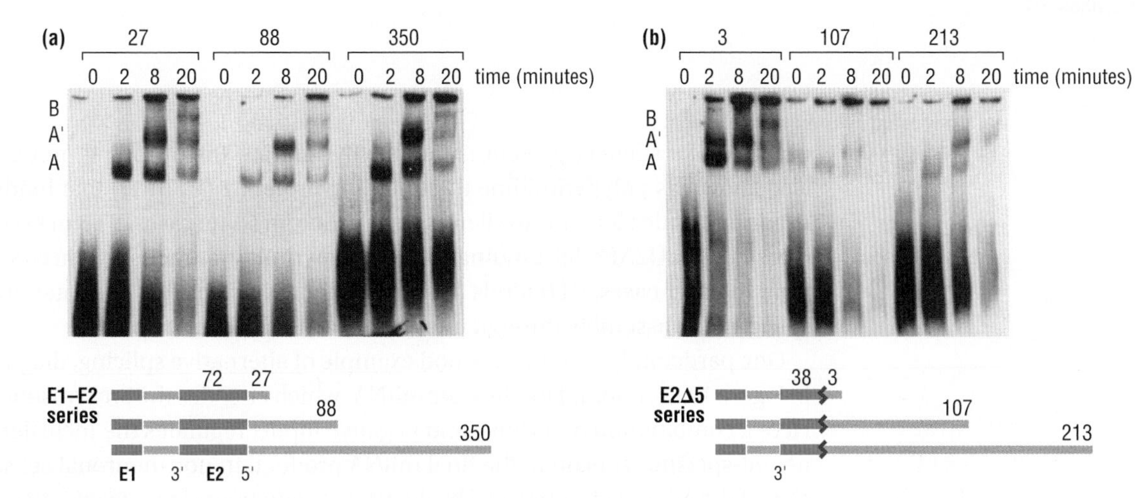

Figure 2 Enhanced complex assembly on two-exon precursor mRNAs with a 5′ splice site following exon 2. Splicing assembly reactions were followed as in Figure 1. Cartoons of pre-mRNA constructs are shown below the gel with complete (a) or incomplete (b) blue exons and gray introns (of variable length) A, A′, and B represent productive splicing complexes for two-exon containing pre-mRNAs. Again, 5′ and 3′ refer to the splice sites (not directionality of the mRNA).

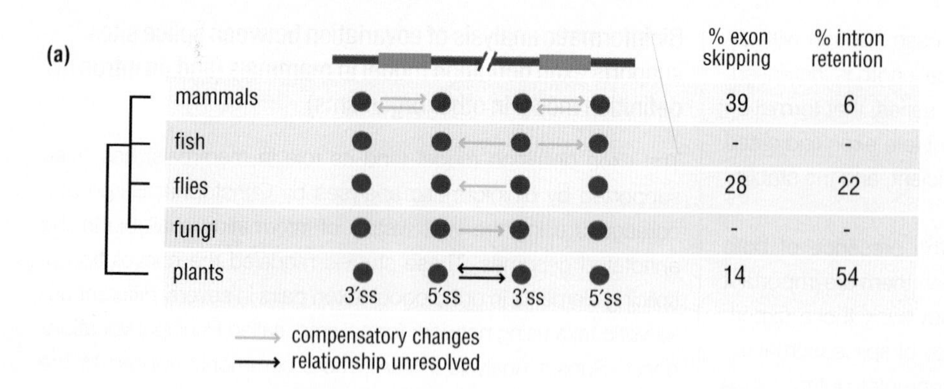

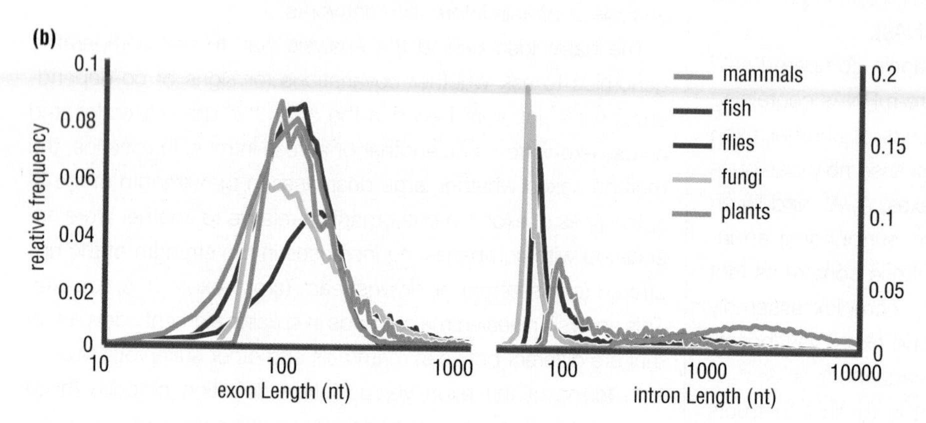

Figure 3 Coevolution of splice sites in diverse eukaryotic taxa. (a) Using PVSA, the coevolution of four sequence elements within the pre-mRNA were modeled (the 3′ and 5′ splice sites of two consecutive exons and intron length). An arrow from one node to another (the ovals) indicates significant covariation seen in the analysis, with compensatory relationships shown in green, and relationships with unresolved directionality in black. The percentages of exon skipping and intron retention events among observed alternative splicing events are also listed. (b) Semilog plot of exon and intron length distributions in five representative genome classes.

Reproduced from Xiao *et al.* Coevolutionary networks of splicing cis-regulatory elements. *Proceedings of the National Academy of Sciences of the U S A*, 2007;**104**:18583–18588.

(see columns in Figure 3a). These data provide experimental support (in the terms of evolutionary history) for distinct modes of splice site definition across the eukaryotic lineage, and highlight the power of phylogenetic analysis for identifying functionally relevant biological interactions.

Find out more

Robberson BL, Cote GJ, Berget SM. Exon definition may facilitate splice site selection in RNAs with multiple exons. *Molecular and Cellular Biology*, 1990;**10**:84–94.

Xiao X, Wang Z, Jang M, Burge CB. Coevolutionary networks of splicing cis-regulatory elements. *Proceedings of the National Academy of Sciences of the U S A*, 2007;**104**:18583–18588.

Related techniques

Native polyacrylamide gels; Section 19.7

Comparative genomics; Section 18.2

example of a silencer protein is the polypyrimidine tract binding protein (PTB; also known as polypyrimidine tract binding protein 1 (PTBP1)) that binds to intronic or exonic elements to silence weak exons. In some cases, PTB may compete directly with U2AF^{65} for binding to the polypyrimidine tract sequence of the intron. In other cases, PTB binds to other pyrimidine-rich sites and interferes with spliceosome assembly through interactions with the U1 snRNP.

One particularly well-understood example of alternative splicing, diagrammed in Figure 10.26, is found for the c-*src* mRNA, which encodes the Src tyrosine kinase. Here a combination of positive and negative inputs regulates the inclusion of the neural-specific N1 exon in the final mRNA product. In non-neuronal cells, inclusion of the N1 exon is repressed by the PTB protein bound to intronic silencer elements on both sides of the exon. One or more of the RNA-binding domains of this PTB protein interact with the U1 snRNA and prevent its assembly with the downstream U2 complex to form a pre-spliceosome. In neuronal cells, PTB is replaced

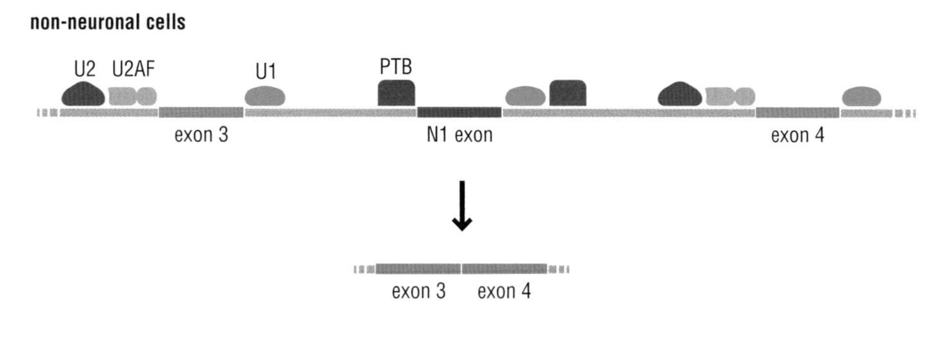

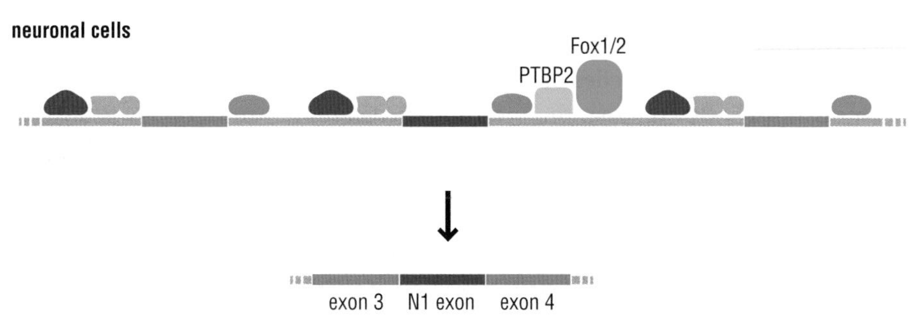

Figure 10.26 Alternative splicing of the c-*src* mRNA in neuronal cells. In non-neuronal cells, inclusion of the N1 exon is repressed by the PTB protein bound to intronic silencer sequences. In neuronal cells, the PTB protein is replaced by a related PTBP2 protein such that nearby sequences can function as intronic enhancer sequences.

by the related polypyrimidine tract binding protein 2 (PTBP2) protein that binds to the same elements as PTB, but does not form a repressive complex. Instead, a nearby enhancer sequence is activated that stimulates N1 splicing through binding to regulatory proteins of the Rbfox family. There are additional regulatory proteins that also bind to the region of the N1 exon to affect its splicing. This combination of positive and negative control affecting the same exon is seen in most examples of tissue-specific splicing.

We see from this example how the overall regulation of alternative splicing involves a complex interplay of pre-mRNA sequences with a variety of protein factors that may be expressed at different levels in different cell types. Further, mutations that emerge in the genes themselves can also affect the recognition of splice sites, leading to the activation of cryptic splice sites and the creation of new exons, generating even more genetic diversity.

10.8 RNA EDITING

The splicing pathways explored in the previous sections provide an important means of generating diversity from a specific pool of genes. In this section, we see how the pool of gene products yielded from a single gene can be broadened still further through a series of additional modifications, which are collectively termed RNA editing.

mRNAs can be edited by the modification, insertion, and deletion of specific nucleotides

In addition to splicing, capping, and polyadenylation, some mRNAs also undergo RNA editing, whereby specific nucleotides are modified, inserted, or deleted. This editing can affect only one or two nucleotides (akin to the correction of a few typographical errors) or it can be surprisingly extensive (akin to the complete

```
UGAUACAAAAAAACAUGACUACAUGAUAAGUA
UCAUUUUAUGUUAUUUUUGGUAGUUUUUUUAC
AUUUGUAUCGUUUUACAUUUG*GUCCACAGCA
UCCCG***CAGCACAUG**GUGUUUUAUGUUG
UUUAUUGUAUUUUUGUGGUGA*AUUUAUUGUU
UA**UAUUGAUUGUAUUAUA***GGUUAUUUG
CAUCGUGGUACAGAAAAGUUAUGUGAAUAUAA
AAGUGUAGAACAAUGUCUUCCGUAUUUCGACA
GGUUAGAUUAUGUUA*GUGUUUGUUGUAAUGA
GCAUUUGUUGUCUUUA***UGUUUUGAGUAUA
UGUUGCGAUGUUGUUUGUCGUUACGUUGUGCA
UUUAUUAAUUG****GAAUUUAC***CCGUAG
UUUUAAUGGUUUGUUGUGUAUAUCAUGUAUGG
UUUUGG*AUUUAGGUUGUUUGUCUCCGUUG*U
UAUGAUCAUUUGAGGAA***CG*UGACAAAUU
GAUGACAUUUUUUGAUUUAUG**UUGUGGUUG
UCGUAUGCAUUUGGCUUUCAUGGUUUUAUUA*
GGUAUUCUUGAUGAUUUUGUUUUUGGUUUUGU
UGAUUUUUUGUUGUUGUUGA***UAAUAUCAU
GUUUGUUUGUUAUGGAUUGUUAUGAUUUGUUA
UUUGUGGGUAAUCGUUUAUUUUAUUUUAUUUG
CGUUUGC***GUGGUUUGUCAUUUUUUGAUUU
AUAUGAUUUA**GUUUUUA**A**UAGUUUAA
GUGGUGUUUUGUCUCGUUCGUUAGGUAUGGUG
UGAGAUUGUCGUUUAUUUAGUUGUUA****UG
A****GUUGUAUUUUAUGUUUUGUUAUGAUUA
UUGUUUUUGUUUUAUAGGUGAUGCAUUUGA*U
CGUUUAUUUUUACGUUUGUUUGAUAUUUG***
**GUUUUGUUUGUUUGUUUG**AUUAUUUAUA
UUGUGAUAUUACCAUUG****AGACCAUUAUU
AUGUUAUUUUAUAGUUUGUGGUGUUGUUGUUU
GCCGGGUAUA*UCAUUUGC*UUGUGUUGAACA
CCCCAAAGGUGA***GUAUUGUUUGUUAUUA*
***UGUUUUUGUGUUGGUUUAUGUUCUCGUUU
GCGUUGUGCGGAUUUUUUGCA*UAUUUGUUUA
UUGGAUGUUUGUUUGCGUGGUUUUUUAUUGCA
UGAUUUAGUUGC***C*GUUUUAGGUAAUAUU
GAUGUUGUUUUUGGAUCCGUAGAUCGUUA*GU
UUUAUAUGUG**A******GGUUAUUGUAGGA
UUGUUUAAAAUUGAAUAAAAA-poly(A)
```

Figure 10.27 An example of extensive RNA editing. The sequence of the NADH dehydrogenase 7 gene of *Trypanosoma brucei* after RNA editing. Black nucleotides are encoded by mitochondrial DNA, blue asterisks indicate uridines that are deleted and red nucleotides indicate uridines that are inserted.

rewriting of a text): it can involve as many as half of the nucleotides in the mRNA, as the example in Figure 10.27 shows. Editing is another way for a cell to vary the amino acid sequence of a protein that is encoded by a given mRNA, and makes it extremely difficult for us to predict the protein sequence encoded by a gene whose precursor RNA is processed in this way. In some cases, RNA editing provides cells with a mechanism to produce more than one form of a protein from a single gene. In this way, RNA editing can function like the alternative splicing mechanism that we discussed in Section 10.7 in that it increases the coding capacity of the genome.

Two widespread forms of editing shown in Figure 10.28 are the deamination of adenosine to form inosine and the deamination of cytidine to form uridine, the effects of which we discuss below. These nucleotide conversions have been seen in RNA viruses, the organelles of a variety of eukaryotes, and the nuclei of metazoan organisms. By contrast, the insertion or deletion of nucleotides is far less widespread. Uridine insertion or deletion has been found almost exclusively in the mitochondrial RNAs of a few single-celled eukaryotes such as trypanosomes (the parasites that cause Chagas disease and African sleeping sickness). Cytidine insertion has thus far only been seen in the mitochondrial RNAs of slime molds, such as *Physarum polycephalum*. The origins of these unusual mechanisms for gene control are not well understood.

Adenosine deaminases that act on RNA convert adenosine to inosine

The deamination of adenosine to inosine is the most widespread form of mRNA editing in higher eukaryotes and can occur in coding sequences, introns, and the 5′ and 3′ UTR of transcripts. The replacement of adenosine with inosine has the same effect on coding as a change to guanosine - that is, inosine is interpreted as guanosine by ribosomes. Consequently, changes of this type within coding regions generally result in changes in the sequence of the protein translated from the mRNA. Indeed, there are several well-studied examples in the nervous system in which adenosine deamination allows more than one form of a protein to be expressed from the same gene.

We discuss RNA-binding motifs in Section 10.11.

The enzymes that catalyze adenosine to inosine editing are referred to as **ADARs** (adenosine deaminases that act on RNA). These enzymes contain one or more double-stranded **RNA-binding motifs** and are known to target double-stranded regions of RNA. Insights into how such double-stranded sites in the mRNA are targeted for editing can be seen in Experimental approach 10.5. Interestingly, ADARs deaminate as many as half of the adenosines in some RNAs while in others they target a specific adenosine among hundreds. These differences in specificity are not well understood, but undoubtedly derive from differences in the structures of the RNA substrates.

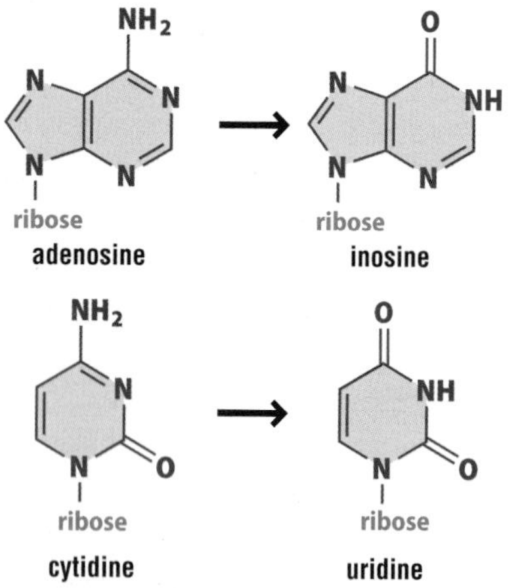

NH_2 N N N N ribose adenosine → O N NH N N ribose inosine

NH_2 N N O ribose cytidine → O NH N O ribose uridine

Figure 10.28 Deamination of adenosine and cytidine. The deamination of adenosine and cytidine produces inosine and uridine, respectively.

Sequence elements in mRNA direct cytidine to uridine deamination

Numerous plant mitochondrial and chloroplast mRNAs, as well as the mRNA encoding mammalian apolipoprotein, are edited by the deamination of cytidine to uridine. Whether this type of editing is more widespread in mammalian systems is a subject of current study. Apolipoprotein B is involved in the binding and transport of lipids throughout the human body. Figure 10.29 illustrates how the deamination of a single cytidine in the apolipoprotein B mRNA introduces a stop codon into the transcript, causing premature termination of translation;

10.5 EXPERIMENTAL APPROACH
Identification of RNA editing sites through serendipity and comparative genomics

Discrepancies between genomic and cDNA sequences led to the discovery of site-specific A to I editing.

The discovery of RNA editing emerged initially from routine comparison of the sequences of cDNAs (DNA copies of mRNAs) with their genomic equivalent. Typically, cDNA samples are prepared from poly(A) selected mRNAs (mRNAs that are fully processed and are found in the cytoplasm) that have been reverse transcribed and prepared for sequence analysis. In the case of certain ionotropic glutamate channel genes in mammalian cells, these approaches identified single nucleotide discrepancies in the sequences of the cDNA and genomic DNA copies of the gene (an apparent A to G change) that would result in the exchange of a glutamine (Q) residue for an arginine (R) in the protein.

Other studies had shown that these amino acid changes are found in a transmembrane segment of the protein and profoundly alter the properties of ion flow in membranes in which these proteins are found. These sequence changes are thus important for protein function. Of course, to be certain that such changes in the mRNA sequence result from editing, the potential existence of multiple copies of the sequence within the genome needed to be eliminated as a possibility through other forms of analysis.

In a study by Peter Seeburg and colleagues, the authors demonstrated that, for several glutamate channels (GluR-B, GluR-5, and GluR-6) where an arginine codon had been observed in at least some fraction of the cDNA clones, the genomic copy clearly specifies glutamine (Q), and there is only one copy of the exon that encodes this transmembrane segment. Further analysis of a number of cDNA clones of these three genes (*GluR-B*, *GluR-5*, and *GluR-6*), prepared using the polymerase chain reaction (PCR), provided clear evidence for the specificity of the editing phenomenon as compared with close, apparently unedited, homologs (*GluR-C* and *GluR-D*). This comparison is shown in Figure 1. For example, for the *GluR-B* gene, all 61 sequenced clones carry the edited version (R) of the gene; for *GluR-C* and *GluR-D*, all sequenced clones carry the non-edited version (Q); and *GluR-5* and *GluR-6* appear to be partially edited.

These observations provided compelling support for the idea of RNA editing and motivated experiments that ultimately identified the enzymes responsible for what turned out to be A to I editing activity (the adenosine deaminases acting on RNA or ADARs) and the particular sequence features that these enzymes depend on for the selection of editing sites.

Compilation of glutamine and arginine codons in transmembrane II of glutamate receptor subunit cDNAs

	GluR-B	GluR-C	GluR-D	GluR-5	GluR-6
Q	0	150	426	122	89
R	61	0	0	78	260

Figure 1 Analysis of independent cDNA clones of the indicated glutamate receptor genes provide evidence for highly specific RNA editing. The cDNA constructed from rat brain RNA was used as a template for PCR to amplify the coding region for the relevant transmembrane segment in these proteins. Sequence identity of the various clones was determined through 'plaque' hybridization (each phage plaque on a plate derives from a single cDNA sequence) to Q- and R-specific radioactive probes.

The targeted A to I editing sites were identified by comparative genomics

Biochemical studies conducted on the ADAR enzymes during subsequent years found that the pre-mRNA editing substrate is typically an imperfect RNA duplex. This duplex is formed between the exon containing the adenosine that is to be modified and an intronic non-coding element called the editing site complementary sequence (ECS). Because the ECS can be rather distantly located relative to the edited site (from hundreds to thousands of nucleotides away), and because the pairing interactions are not precise, it is notoriously difficult to identify RNA editing sites by simple inspection of sequences.

In an effort to further identify editing sites, Robert Reenan and colleagues took a phylogenetic approach and analyzed the characteristics of the editing site of the *paralytic* gene (a Na^+ channel) in *Drosophila*. The researchers hypothesized that if RNA editing of this particular site were conserved between species, then conservation might be seen in the intronic ECS sequences that are critical for editing. Intronic sequences typically are not conserved, but conservation of nucleotides would be predicted if the sequences have functional significance. To the researchers' surprise, however, the most dramatic conservation observed for an alignment of the *paralytic* gene sequences from 18 distinct *Drosophila* species was in the exonic sequence surrounding the edited site, as shown in Figure 2. In the alignment shown, we see that third position degeneracy, typically found even within highly conserved protein coding sequences, is completely absent. In Figure 3, we see how exonic conservation emerges from such analysis; in the case of the *DSC1* gene, exon 15 is seen to be more highly conserved than any others, a clue

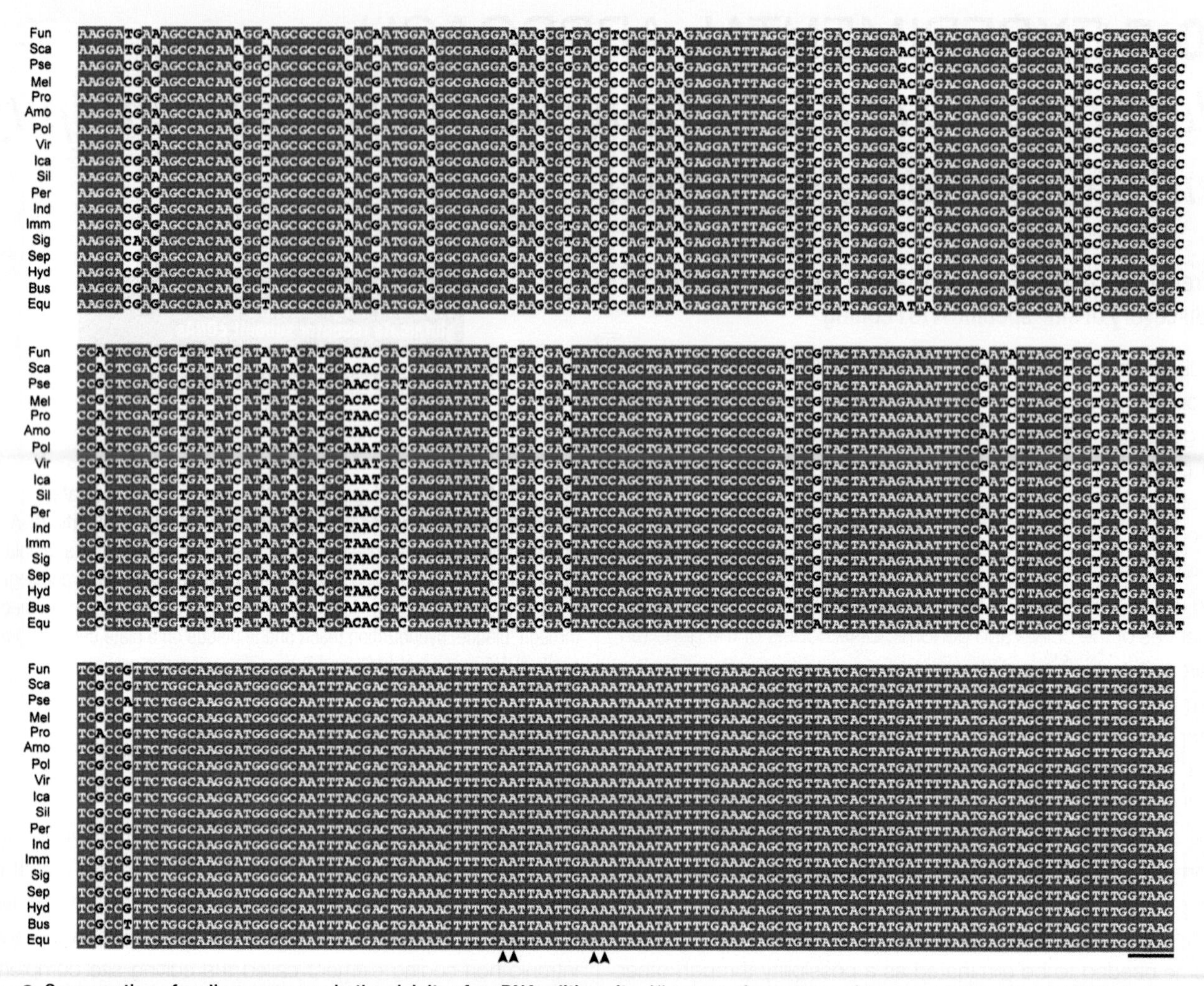

Figure 2 Conservation of coding sequence in the vicinity of an RNA editing site. Alignment of sequences of the *paralytic* Na+ channel T/M site from 18 *Drosophila* species spanning 60–80 million years divergence time. Sequence of the entire exon containing the editing site is shown, with the 5′ splice site underlined. Nucleotides shaded in red are conserved among all 18 species. Arrowheads indicate the position of the adenosine residues that are modified by dADAR (d for *Drosophila*). Reproduced with permission from Hoopengardner *et al.*, *Science*, 2003;**301**:832–836.

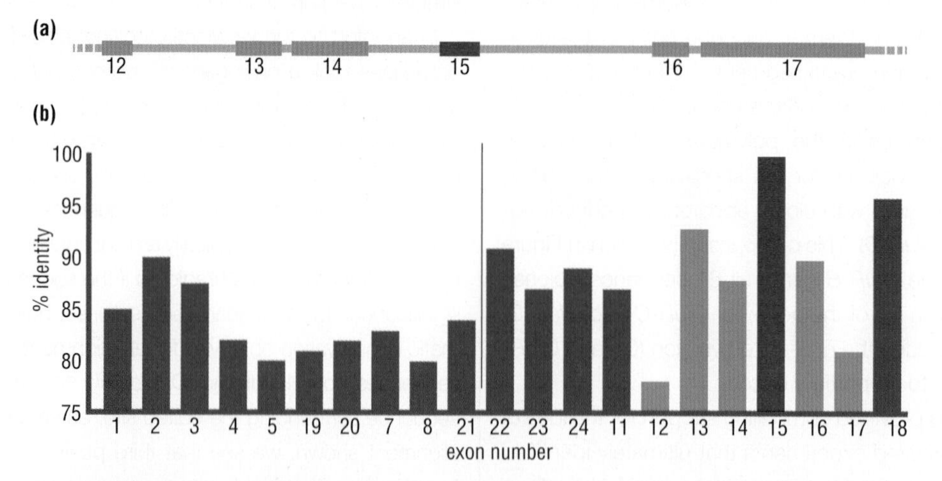

Figure 3 Experimental identification of novel edited site through analysis of exonic conservation. Bar graph showing percent sequence identity for the exons of *DSC1* transcript in *Drosophila*. Red bar indicates the edited exon, blue bars correspond to unedited neighboring exons shown in (a), and black bars represent the unedited, more distant exons.

Reproduced with permission from Hoopengardner *et al.*, *Science*, 2003;**301**:832–836.

that ultimately led to its identification as an authentic editing site. Indeed, similar comparative analysis of exonic conservation in 914 genes (annotated as potentially relevant for neurological behavior) identified 16 previously unknown ADAR target genes that could be experimentally confirmed.

The success of this particular study highlights the power of comparative genomics to identify biologically essential events that might escape identification through other more targeted, organism-specific approaches.

Find out more

Hoopengardner B, Bhalla T, Staber C, Reenan R. Nervous system targets of RNA editing identified by comparative genomics. *Science*, 2003;**301**:832–836.

Sommer G, Köhler M, Sprengel R, Seeburg PH. RNA editing in brain controls a determinant of ion flow in glutamate-gated channels. *Cell*, 1991;**67**:11–110.

Related techniques

Comparative genomics; Section 18.2

cDNA synthesis; Section 19.3

PCR; Section 19.3

Drosophila (as a model organism); Section 19.1

this results in the formation of a short form of apolipoprotein B called APOB48. In humans, the editing occurs in the small intestine but not in the liver. Thus, APOB48 is produced in the small intestine, where it is required for the absorption of lipids from food, while the long form of the protein (APOB100) is synthesized in the liver, where it acts to transport cholesterol from the liver to the rest of the body.

RNA editing allows these two specialized forms of the protein to be produced from a single gene. The site of cytidine deamination in the apolipoprotein B mRNA is determined by specific 'mooring' sequences within the mRNA: several sequence elements surrounding the cytidine base-pair with one another and are thought to form a double-stranded element which positions the cytidine such that it is recognized and modified by a cytidine deaminase enzyme.

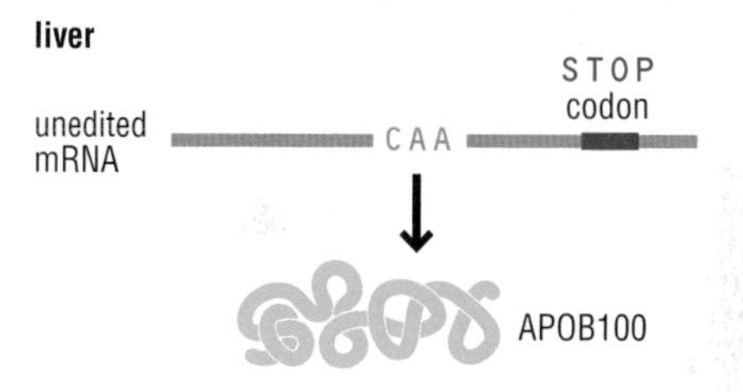

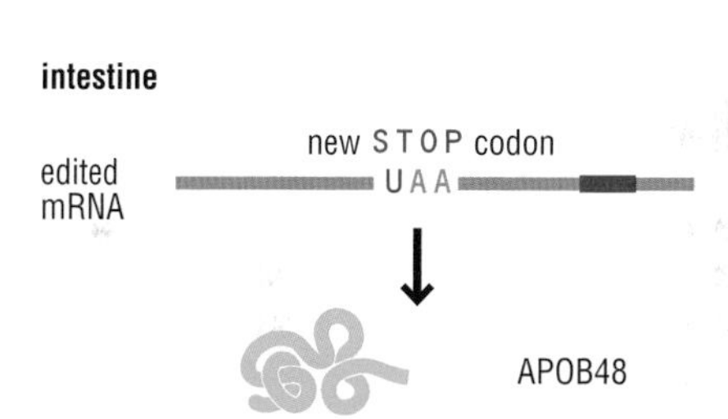

Figure 10.29 Cytidine deamination changes the protein expressed by the Apolipoprotein B mRNA. In the liver, the APOB mRNA is not edited and produces a long form of the protein. In the small intestine, the deamination of a single cytidine to uridine introduces a translation stop codon resulting in the synthesis of a shorter protein.

Uridine addition and deletion is directed by guide RNAs

One of the most unusual forms of editing is that of uridine addition and deletion in trypanosomes. Figure 10.27 shows an example of the numerous uridine nucleotides that are inserted into or deleted from a single mRNA encoded by trypanosome mitochondria. Why trypanosomes edit their mRNAs so extensively is a mystery, but the editing mechanism itself has been well characterized. The editing depicted in Figure 10.30 involves guide RNAs approximately 20–50 nucleotides long that bind to the mRNA and define where the insertions or deletions of uridines occur. The guide RNAs are partially complementary to the unedited RNA but are more extensively complementary to the fully edited mRNA.

The process of editing begins when the guide RNAs first base-pair with part of the mRNA. The editing is then guided by the base-pairing interaction. During each editing step, an endonuclease cuts the mRNA at a mismatch between the guide RNA and mRNA, and the guide RNA then serves as a template to direct the addition or deletion of uridines. The addition of uridines, which base-pair with adenosines or guanosines in the guide RNA, is catalyzed by a 3′ terminal uridylyl transferase, as illustrated in Figure 10.30a. Alternatively, a 3′ to 5′ exonuclease *removes* uridines that are unable to base-pair with the guide RNA (Figure 10.30b). The final step in both the addition and deletion steps of uridine editing is the ligation of the resulting edited fragments of the mRNA.

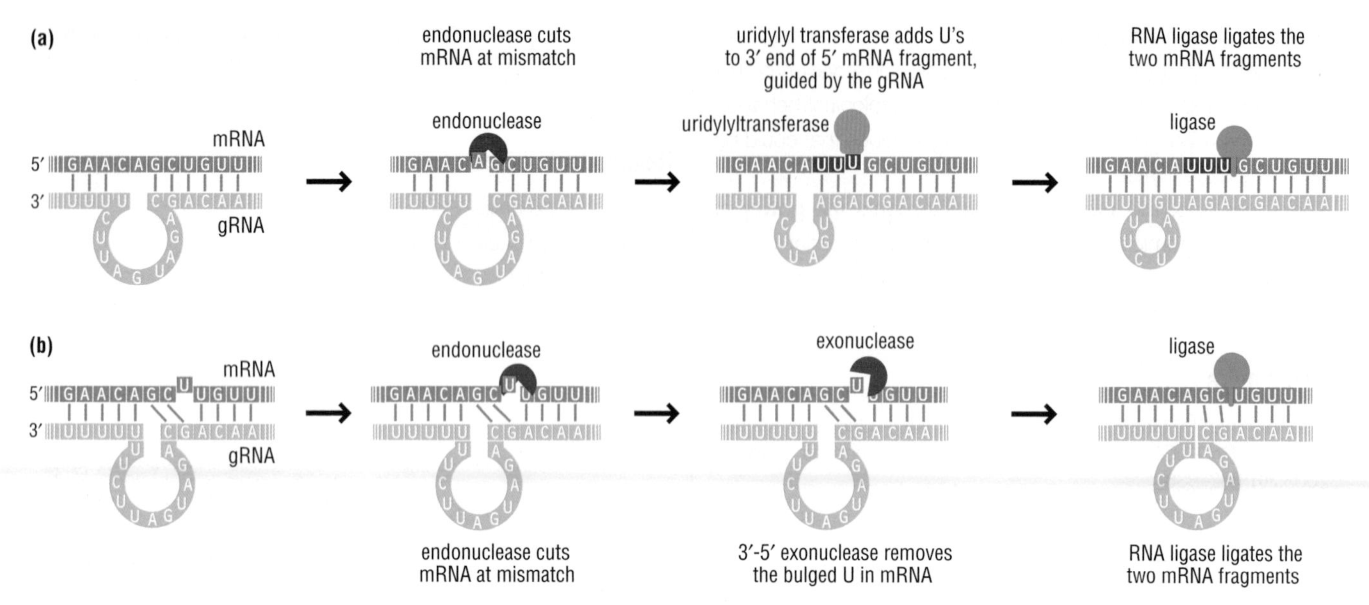

Figure 10.30 The insertion and deletion of uridine nucleotides is directed by guide RNAs. During both the insertion (a) and deletion (b) of uridine nucleotides, guide RNA molecules (gRNA, green) base-pair with the target mRNA. RNA editing occurs where there are mismatches between the guide RNA and the mRNA. The guide RNA acts as a template for the insertion or deletion of uridines, as directed by the base-pairing.

10.9 DEGRADATION OF NORMAL RNAs

In the preceding sections, we have considered ways in which the properties of an RNA molecule can be modified through chemical modifications to the molecule. However, a much starker way of modulating the activity of an RNA is simply to remove it from a biological system completely. In this section, we explore such controlled removal of RNA – the process of degradation.

RNA degradation can influence gene expression

All cells have mechanisms to degrade RNA, digesting the RNA back to individual nucleotides. This degradation allows cells to remove RNAs that are no longer required or which might be present at too high a level for the current needs of the cell. The precious nucleotides in the RNA can then be recycled in new RNA transcripts. Cells have methods to degrade 'normal' RNAs – those RNAs that perform useful functions to cells – and different methods to efficiently destroy foreign or defective RNAs, which we will discuss in the next section.

Some RNAs, such as rRNAs, are needed for long periods of time and are comparatively stable. In contrast, other RNAs, such as some mRNAs, are only needed briefly and are rapidly degraded. The stability of a given RNA species in the cell is described by the **RNA half-life**, the time in which the amount of the RNA is reduced by one half. RNA half-life varies widely: *E. coli* transcripts have half-lives of less than one minute to one hour, and the half-lives of vertebrate mRNAs range from 20 minutes to more than 24 hours.

A variety of determinants impact RNA stability

Various factors can affect the stability of an RNA, including the sequences and structures at both the 5′ and 3′ ends of RNAs. As we mentioned above, the 5′ cap structure present on eukaryotic mRNAs protects the 5′ end from degradation by 5′ to 3′ exonucleases. Similarly, bacterial RNAs with a 5′ triphosphate are more stable than mRNAs with a 5′ monophosphate. Further protection is provided by stem–loop structures present at the very 5′ end of some bacterial RNAs.

Stem–loop structures found at the 3′ end of bacterial RNAs, including the stem-loops corresponding to the Rho-independent terminators described in Section 8.6, also generally stabilize the transcript by protecting it from 3′ to 5′ exonucleases. On the other hand, the presence of a poly(A) tail in bacteria is associated with *decreased* stability. By contrast, in eukaryotes, the poly(A) tail generally lends *increased* stability to RNA transcripts whereas other elements, including the AU-rich elements (AREs) found in the 3′ UTRs of some mRNAs, cause instability by directing poly(A) tail removal.

A dramatic example of the importance of AREs is illustrated in Figure 10.31 by c-*fos*, a short-lived mRNA with an ARE, which encodes a transcriptional regulator involved in promoting the growth of vertebrate cells. Some tumor-causing viruses contain a mutated form of c-*fos*, called v-*fos*, which lacks an ARE. When these viruses infect a host cell, the absence of an ARE prevents the v-*fos* mRNA from being rapidly degraded, leading to high levels of the v-*fos* protein, which promote the uncontrolled growth of infected cells.

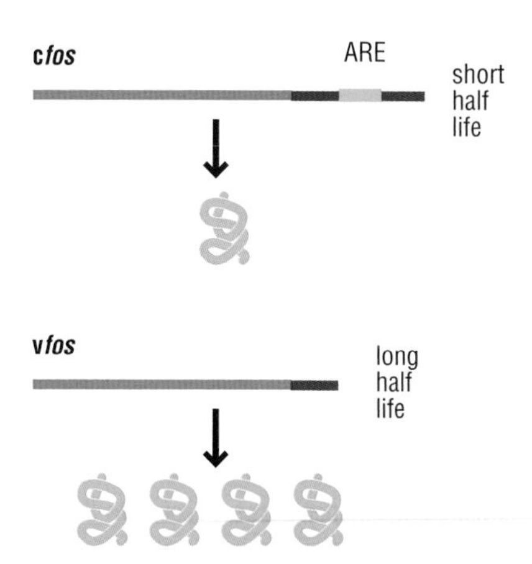

Figure 10.31 Consequences of a lack of an mRNA stability element for the *fos* mRNA. Vertebrate cells express the c-*fos* mRNA, which carries an ARE that leads to rapid turnover of the mRNA. Some tumor-causing viruses express v-*fos*, a mutated form of the mRNA lacking the ARE. The elevated levels of the v-*fos* mRNA lead to abnormally high levels of the Fos transcription regulator and uncontrolled growth of the virus infected cells.

In all organisms, processes that act on RNAs - such as RNA splicing, transport of the RNA through cellular compartments, or the translation of an mRNA into protein - also impact RNA half-life. In addition, specific proteins and non-coding RNAs can bind to RNA transcripts to regulate their stabilities, as we shall see in examples that follow here and in Chapter 13. In all of these cases, a site for digestion by the degrading enzymes can either be blocked or become more accessible.

RNA degradation in eukaryotes and bacteria involves multiple steps

Although the degradation processes for bacterial and eukaryotic RNAs were originally thought to be distinct, the discovery of an increasing number of common degradation machinery components points to some similarity between the two pathways. Figure 10.32 illustrates how, for many *E. coli* RNAs, degradation is initiated by an endonuclease. The products of the initial endonuclease digestion are then degraded by 3′ to 5′ exonucleases. Most often, the initiating endonuclease is RNase E, which is part of a protein complex called the **degradosome**. The degradosome contains a number of proteins likely to facilitate degradation, including an RNA helicase that uncoils double-stranded regions of the RNA and a 3′ to 5′ exonuclease denoted polynucleotide phosphorylase (PNPase). Other bacterial species that lack RNase E instead use an enzyme (RNase J) with both endonuclease and 5′ to 3′ exonuclease activity to degrade RNA.

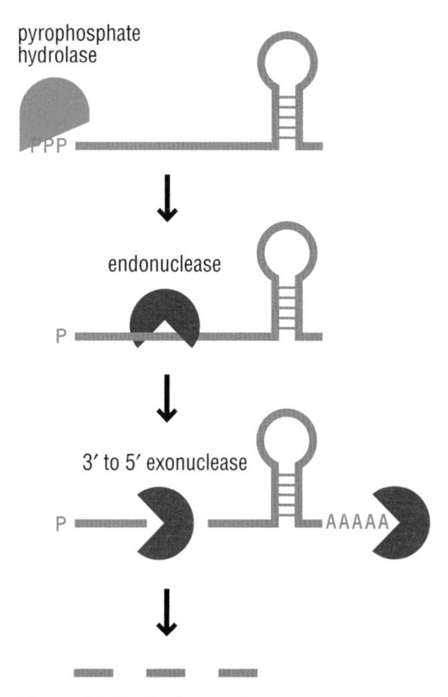

Figure 10.32 Pathways for degradation of bacterial mRNAs. The degradation of most bacterial mRNAs is initiated by an endonucleolytic cleavage. The RNA fragments are then degraded to individual nucleotides by 3′ to 5′ exonucleases. The removal of 5′ triphosphates stimulates the endonucleolytic cleavage, and the polyadenylation of the 3′ ends facilitates 3′ to 5′ exonuclease degradation, particularly of those RNAs with 3′ structures (such as stem–loops) that are resistant to exonuclease cleavage.

The presence of a triphosphate at the 5′ ends of an RNA inhibits RNase E activity and protects against 5′ to 3′ exonuclease degradation. Thus, the conversion of a 5′ triphosphate to a 5′ monophosphate by the RppH pyrophosphate hydrolase enzyme can stimulate degradation. All 3′ ends can eventually be digested by 3′ to 5′ exonucleases. However, RNA stem–loop structures can block the accessibility of the 3′ end, making RNAs resistant to exonuclease action. The digestion of such RNAs is facilitated by polyadenylation of the 3′ ends, which provides an accessible, unstructured end for the action of exonucleases.

In eukaryotes, RNA degradation generally is not initiated by the action of endonucleases. Instead, mRNA decay involves a series of exonucleolytic digestions. Moreover, poly(A) tails usually prevent rather than facilitate degradation. Thus, as

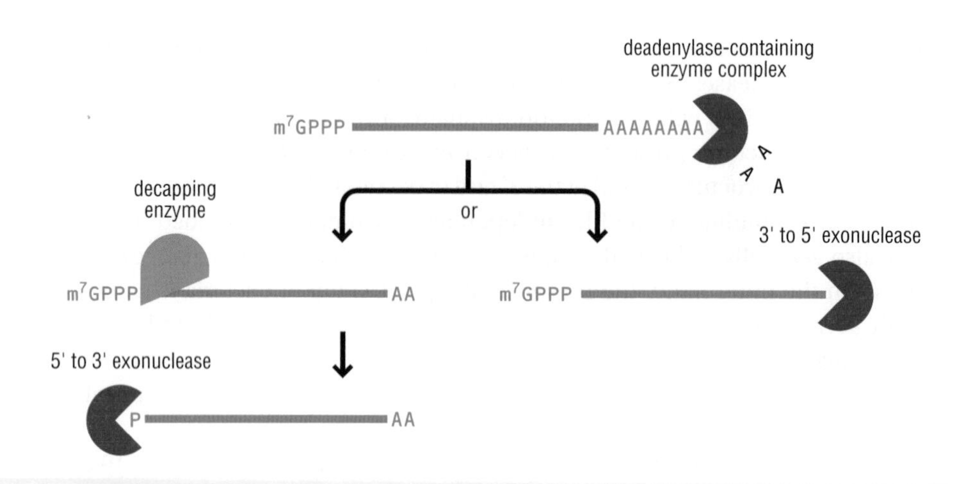

Figure 10.33 Pathways for the degradation of eukaryotic mRNAs. While the degradation of some eukaryotic mRNAs can be initiated by endonuclease cleavage, the first step in the degradation of most eukaryotic mRNAs is the shortening of the poly(A) tail, catalyzed by a deadenylase. Subsequently, degradation can proceed by further digestion by 3′ to 5′ exonucleases, or by the removal of the 5′ cap structure, which then allows digestion by 5′ to 3′ exonucleases. m^7G, 7-methyl guanosine.

shown in Figure 10.33, the first step in the degradation of many normal eukaryotic mRNAs is the shortening of poly(A) tails in a process referred to as **deadenylation**. Deadenylation is catalyzed by a multiprotein complex in which the key factor is the 3′ to 5′ exonuclease CCR4. While deadenylation is a necessary first step in eukaryotic mRNA decay, it does not necessarily commit an mRNA for destruction. In certain biological situations, deadenylation is reversible, and post-transcriptional readenylation is possible.

Committing an mRNA to decay after deadenylation involves one of two distinct mechanisms: the mRNA can be destroyed from either the 5′ or the 3′ end. Destruction of an mRNA from the 5′ end involves the assembly of a complex of proteins that removes the 5′ cap structure of the mRNA. The catalytic subunit of this complex, DCP2, has pyrophosphatase activity and is distantly related to the bacterial RppH enzyme. The action of the decapping enzyme liberates 7-methyl-GDP and leaves a 5′ monophosphate on the mRNA body. Decapping is a highly regulated step in mRNA decay and commits the message for destruction. The 5′ monophosphate is recognized by a highly processive 5′ to 3′ exonuclease called XRN1, which rapidly destroys the mRNA. An mRNA also can be degraded from the 3′ end following deadenylation. In these cases, the mRNA is digested by a complex of proteins called the exosome, a 3′ to 5′ exonuclease related to bacterial PNPase.

Of the two mRNA decay pathways described here, the one that predominates is determined by the situation. In organisms such as yeast, decapping appears to be the major decay pathway. In humans, both decapping and exosome-mediated degradation have been observed on specific transcripts. Thus, mRNA decay is organism, tissue, cell type, and mRNA specific.

Degradation of RNAs can be regulated in response to specific signals

The steps in the degradation of RNAs can all be promoted or blocked in response to specific environmental signals, at different points in the cell cycle, in different cell types, or in different stages of development.

In the mammalian example shown in Figure 10.34, the intracellular concentration of iron is regulated in part by the transferrin receptor, a protein that imports transferrin-bound iron into the cell. (Transferrin itself is a glycoprotein that binds iron very tightly.) The mRNA encoding the transferrin receptor is rapidly degraded by nucleases when iron levels in a cell are high and the cell has no need to import more iron. However, when iron levels are low, the transferrin receptor mRNA is protected from degradation. This protection is provided by two iron

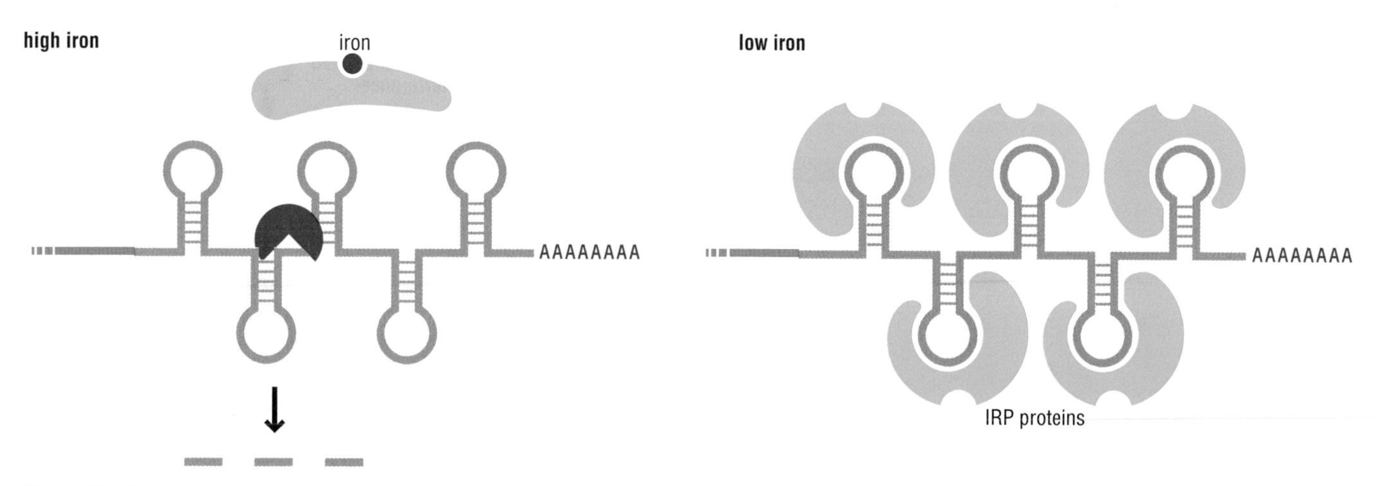

Figure 10.34 Degradation of mammalian transferrin receptor mRNA is inhibited by low cellular iron levels. When intracellular iron levels are high, the 3′ end of the transferrin receptor mRNA is susceptible to cleavage by endonucleases (pink). When intracellular iron levels are low, the mRNA is protected by iron regulatory proteins (green, representing either IRP1 or IRP2), which block nuclease access.

regulatory proteins, IRP1 and IRP2, that bind to stem–loop structures called iron response elements (IREs) found in the 3′ UTR of the transferrin receptor mRNA. When iron is plentiful in the cell, the IRPs are no longer able to bind RNA, thus allowing the transferrin receptor mRNA to be degraded. By regulating the stability of the mRNA in this way, mammalian cells are able to promote or prevent the synthesis of the transferrin receptor in response to cellular iron levels.

As we shall see in Chapter 12, IRPs also directly regulate the synthesis of proteins from other mRNAs by binding to IRE stem–loop structures located in 5′ UTRs. In addition, in Chapter 13, we discuss a plethora of small, regulatory RNAs found in almost all organisms, whose expression is induced in response to a wide range of environmental inputs or in specific cell types, and which base-pair with mRNAs to affect their stability.

Sometimes, under conditions of starvation or extreme stress, it might be advantageous to the cell to degrade a large proportion of RNA molecules in a last ditch effort to scavenge nucleotides and survive. This area of research is just beginning to receive more attention, but in bacteria it is becoming clear that many proteins initially identified as toxins have endoribonuclease activity. The synthesis and activity of these proteins, denoted 'interferases', are tightly regulated, but their activity is thought to allow a small subset of cells to persist under conditions of infection.

10.10 DEGRADATION OF FOREIGN AND DEFECTIVE RNAs

Foreign and defective RNAs can be harmful to cells

Several types of RNA molecules can have harmful effects on a cell. For example, 'foreign' RNAs, derived from foreign DNA such as viruses or transposons, can be detrimental to the cell if the RNAs are translated into proteins that inhibit critical cell functions or promote transposition. In addition, an RNA that is defective in some way, due to errors during transcription or RNA processing, has the potential to disrupt normal cellular processes by encoding a product with altered function.

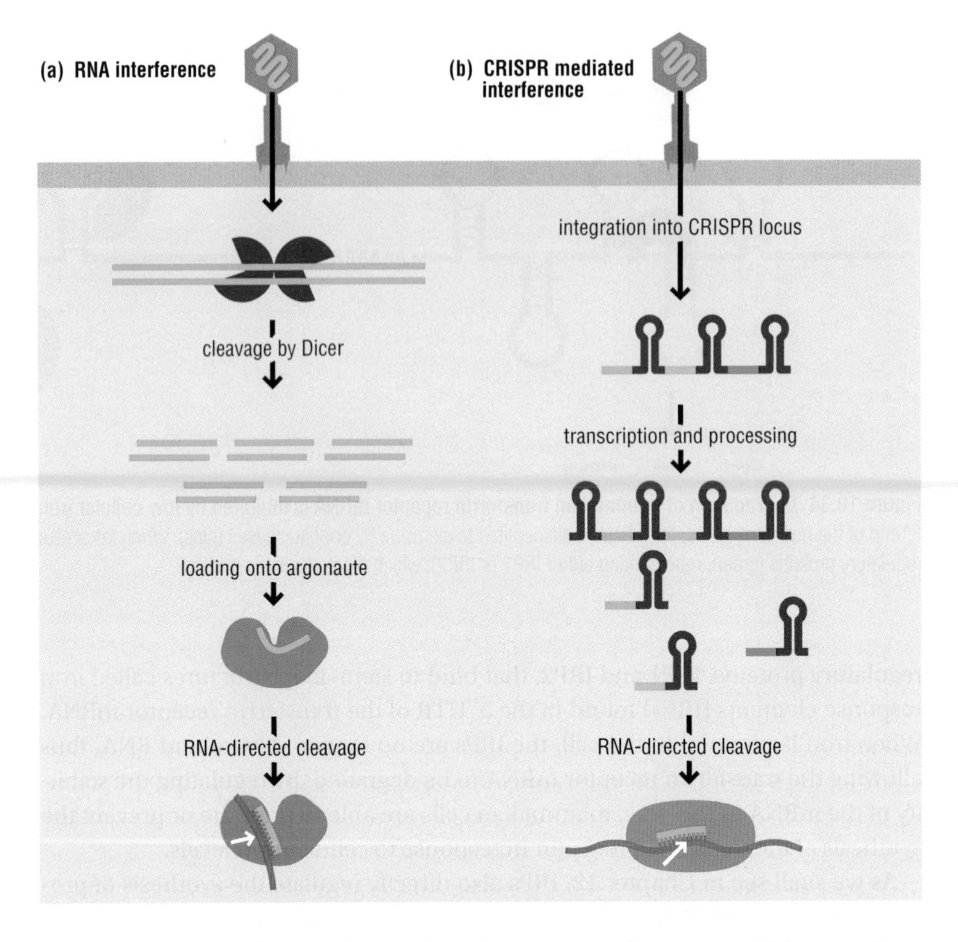

Figure 10.35 RNA-directed cleavage of nucleic acids protects cells against foreign RNA and DNA. (a) In eukaryotic RNA interference, long double-stranded RNA recognized as foreign is cleaved by RNase III family Dicer proteins into short interfering RNAs. These siRNAs are loaded onto Argonaute proteins, where they direct the cleavage of the foreign RNA. (b) In bacterial and archaeal CRISPR systems, foreign DNA is integrated into the CRISPR locus. Long transcripts from these loci are processed into shorter fragments, which are loaded onto specialized proteins to direct the cleavage of homologous foreign DNA or RNA.

Adapted from Fig 1 in Wiedenheft *et al.* RNA-guided genetic silencing systems in bacteria and archaea. *Nature*, 2012;**482**:7382.

Just as for the degradation of normal RNA transcripts, however, cells have various means to identify and destroy these potentially harmful RNAs.

Foreign RNAs are removed by an RNAi mechanism in eukaryotes and CRISPR systems in bacteria

The RNA molecules derived from the transcription of many viruses that enter eukaryotic cells or from the transcription of transposons or other repetitive sequences in the chromosome are typically double-stranded. As summarized in Figure 10.35a, eukaryotic cells possess defense mechanisms that detect and destroy these RNAs. This phenomenon has been observed in a number of different organisms; it was first called **RNA interference (RNAi)** in *C. elegans* and **post-transcriptional gene silencing (PTGS)** in plants.

We now have a detailed understanding of how the double-stranded RNAs are cleaved into 20–30 nucleotide fragments by the RNase III-like Dicer enzyme before being loaded onto specialized Argonaute proteins with nuclease domains. The short RNAs bound by the Argonaute complexes ultimately guide the cleavage of the deleterious RNAs.

Bacteria and archaea similarly possess an RNA-guided system to target and degrade foreign DNA and RNA, as depicted in Figure 10.35b. This defense system was denoted **CRISPR** based on the first evidence for this system: clustered regularly interspaced short palindromic repeats identified by genome sequencing. As for RNAi, the CRISPR system involves multiple nucleases, accessory protein factors, and RNA-guided cleavage. We explore the mechanisms of RNAi- and CRISPR-mediated RNA and DNA degradation in detail in Chapter 13.

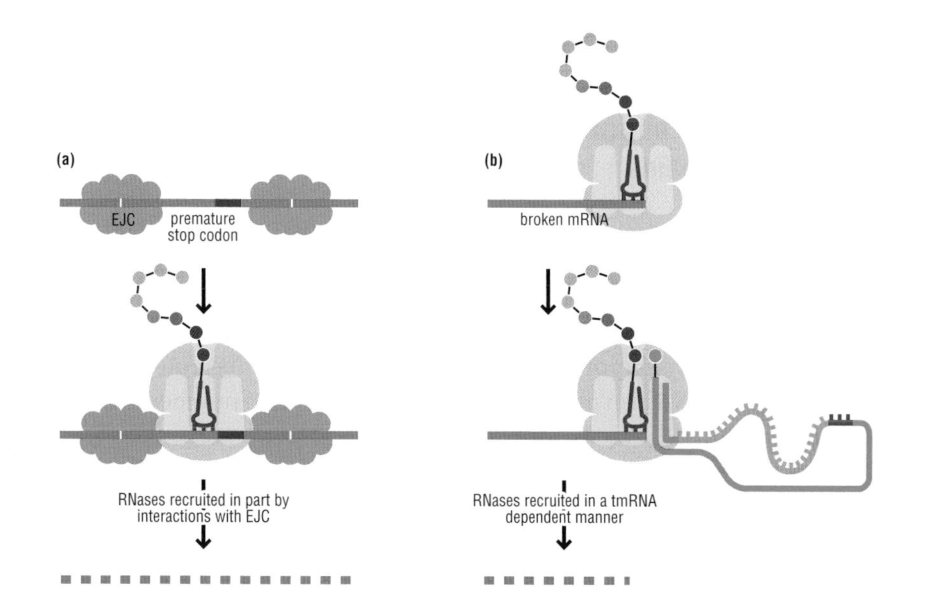

Figure 10.36 Complexes that recognize stalled ribosomes recruit RNases to degrade damaged mRNAs. (a) In eukaryotes, ribosomes transcribing defective mRNAs are marked by interactions with complexes such as the EJC, which recruit specific RNases to degrade the mRNAs. (b) In bacteria, stalled ribosomes are recognized by a specialized tmRNA-containing complex, which releases the ribosome and targets the mRNA for degradation.

Defective mRNAs are recognized and removed by specific decay mechanisms

In Chapter 11, we will discuss how a molecular machine called the ribosome synthesizes a protein molecule by reading the sequence of nucleotides in an mRNA. Normally, when the protein has been fully synthesized, the ribosome recognizes a particular sequence of three nucleotides in the mRNA called a *stop codon* (or *nonsense codon*), which causes the ribosome to terminate protein synthesis. However, some defective mRNAs contain a *premature* translation termination codon that causes the ribosome to stop synthesis before the natural protein is completed; others completely lack a termination codon that prevents the termination reaction from occurring; while still others have problematic coding sequences that can promote stalling during translation. At the very least, the production of a truncated or aberrant protein is a waste of a cell's energy and chemical resources; at worst, it results in the generation of a protein product that disrupts cellular functions. As illustrated in Figure 10.36, cells have evolved means to specifically degrade such defective RNAs.

In eukaryotic cells, processes termed **nonsense-mediated mRNA decay (NMD),** non-stop decay (NSD), and no-go decay (NGD) degrade the different classes of defective mRNA discussed above. As we will discuss in more detail in Chapter 11, each of these decay mechanisms depends on specific proteins and molecular signals. In the case of NMD, the signals are well understood, and include the EJC protein complex deposited at splice junctions during pre-mRNA splicing (see Section 10.6).

In bacterial cells, ribosomes stalled on the mRNAs - and particularly truncated mRNAs lacking stop codons - are recognized by a unique RNA that functions as both a tRNA and mRNA (and is thus called tmRNA). The tmRNA binding allows the stalled ribosome to finish the process of translation using the tmRNA coding sequence while simultaneously promoting the degradation of the defective mRNA as we will also discuss in Chapter 11.

Throughout this chapter, we have considered a range of processes, many of which depend on the interaction of an RNA molecule with one or more proteins. Often, these interactions are specific and are mediated by characteristic features of both RNA and protein, which facilitate specific interactions. We end this chapter by exploring structural features of some of the RNA-binding proteins essential to the processes described.

10.11 RNA-BINDING DOMAINS IN PROTEINS

RNA structure can be specifically recognized by a variety of RNA-binding domains

Most of the processes described in this chapter depend on precise interactions between the RNAs being processed and the various protein machines involved in the processing reactions. For these interactions to be specific, a protein must only recognize the RNA to be modified. Just as proteins that bind to specific DNA sequences possess a variety of different DNA-binding motifs, the proteins that specifically recognize RNA also possess one or more of a limited number of different **RNA-binding motifs**.

There are, however, fundamental differences in DNA and RNA structure that lead to differences in the way that proteins bind to these two types of nucleic acid. DNA is generally found in a regular B-form helical structure, and proteins most often interact with nucleotide bases in the wider major groove. By contrast, double-stranded RNA is generally found in an A-form helical structure in which the deep major groove is relatively less accessible compared with the shallower minor groove. Moreover, RNAs rarely assume an extended helical structure and instead are composed of a series of shorter interrupted helices that pack together to form more complicated three-dimensional structures (see Figure 2.41 and Experimental approach 10.2).

➔ We learn more about protein-DNA interactions in Section 2.12.

RNA bases in single-stranded regions of an RNA molecule can also interact via extensive stacking interactions with the aromatic side chains of proteins. While some RNA-binding motifs in proteins recognize single strands by their sequence (and the constellation of hydrogen-bonding interactions that form between them, for example), others recognize specific loops or bulges associated with a unique backbone configuration (a configuration that typically differs from standard A-form double-stranded helix). In general, RNA-binding motifs exhibit a greater structural variety than DNA-binding motifs, reflecting the greater structural versatility of RNA. Here we discuss the most prevalent of these motifs.

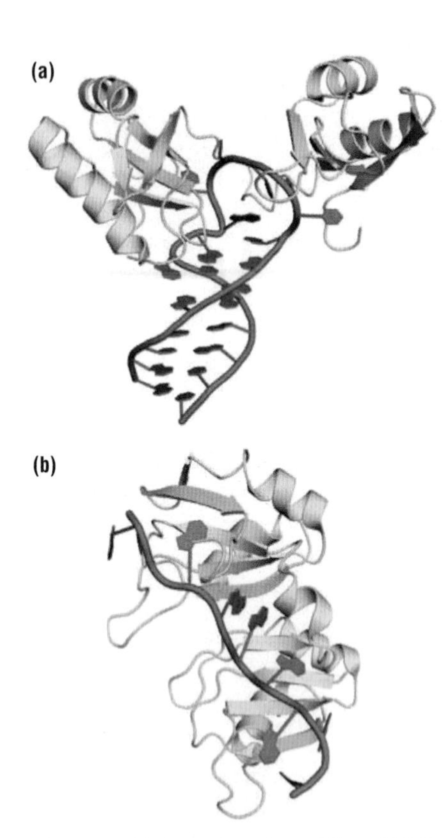

Figure 10.37 Examples of two different RRM platforms binding to single-stranded RNA. The pairs of RRM domains in hamster nucleolin (a) and human PABP protein (b) in complex with RNA illustrate the flexibility with which the RRM platform of four beta-strands supported by two alpha helices can interact with RNA targets. Protein Data Bank (PDB) codes 1FJE, 1CVJ.

Single-stranded RNA is bound by RNA-recognition motif domains in a variety of configurations

The most abundant **RNA-binding motif** is the **RNA-recognition motif (RRM)**, also called the **RNA-binding domain (RBD)** or **ribonucleoprotein (RNP) domain**. The RRM domain was first identified in the poly(A) binding protein (PABP) and in one of the hnRNP proteins (discussed in Section 10.6); it is now known to occur in many other RNA-binding proteins.

The motif provides a versatile platform for binding RNA that is used to direct RNA processing factors to many different RNAs in the cell. The structures of two RRM domains binding to RNA are shown in Figure 10.37. In these domains, alpha helices and beta strand elements come together in a sandwich structure such that a four-stranded antiparallel beta sheet is packed against two alpha helices, forming a platform that predominantly binds single-stranded RNA. A feature of this platform is the adaptability of its interactions with RNA, allowing it to bind to RNAs that can be very different in structure.

There are also two other relatively common single-stranded RNA-binding motifs: the KH (for hnRNP K homology) and the PAZ (for PIWI/Argonaute/Zwille) domains. In KH domains, such as those also found in hnRNP proteins (described in Section 10.6), an RNA-binding groove is formed by two alpha helices and one beta strand. The PAZ domain is found in a number of proteins involved in the RNAi processing pathway, which will be described in Chapter 13, including Dicer, the key enzyme in siRNA generation, and Argonaute, the key catalytic component in initiating RNAi-mediated RNA degradation. In these proteins, illustrated in Figure 10.38, a pocket formed by beta strands and an alpha helix of the PAZ domain specifically interacts with the single-stranded overhangs of siRNAs generated by Dicer cleavage.

Figure 10.38 The PAZ domain forms a pocket for single-stranded RNA ends. In the PAZ domain of human Argonaute 2, four beta strands and an alpha helix form a pocket for the single-stranded overhangs of siRNAs. PDB code 1SI3.

Double-stranded RNA-binding domains distinguish between double-stranded RNA, DNA, and RNA–DNA hybrids

The RNase III family enzymes that are central to both rRNA processing and RNAi have **double-stranded RNA-binding domains (dsRBDs)**. Examples of proteins carrying this RNA-binding motif also include the ADAR protein involved in adenosine deamination in eukaryotic cells (which we discussed in Section 10.8). These proteins have an alpha helix and two loop regions that contact 2′ OH and phosphate groups in three adjacent minor, major, and minor grooves of an RNA helix as shown in Figure 10.39. These interactions allow this class of proteins to recognize double-helical RNA and to discriminate against DNA or DNA–RNA hybrids.

Some RNAs are recognized by proteins with multiple, repeated motifs

Several RNA-binding proteins contain more than one RNA-binding motif in tandem. For example, some proteins contain two or more RRM domains that form more extensive contacts with their target RNAs and thus allow for increased binding affinity (or specificity). A subset of RNA-binding proteins even have multiple, repeated copies of an individual binding motif. These motifs can be present in a single protein or in oligomers, where each subunit contributes a binding motif. Striking examples are provided by members of the eukaryotic PUF family of proteins (named after the founding *Drosophila* Pumilio and *C. elegans* FBF proteins), of which one is shown in Figure 10.40.

These proteins contain eight imperfect repeats of 36 amino acids and bind to single-stranded sequences in the 3′ ends of mRNAs to repress translation. Three conserved amino acids in each of the repeats in the protein contact an individual RNA nucleotide base, with one residue in particular making extensive stacking interactions with the base. The eukaryotic Sm proteins that form the core of eukaryotic splicing complexes (and related proteins found in archaea and bacteria) present another example. These proteins form oligomeric rings comprised of six or seven protein subunits, where each subunit can provide stacking interactions with uridine residues found in a variety of single-stranded RNAs.

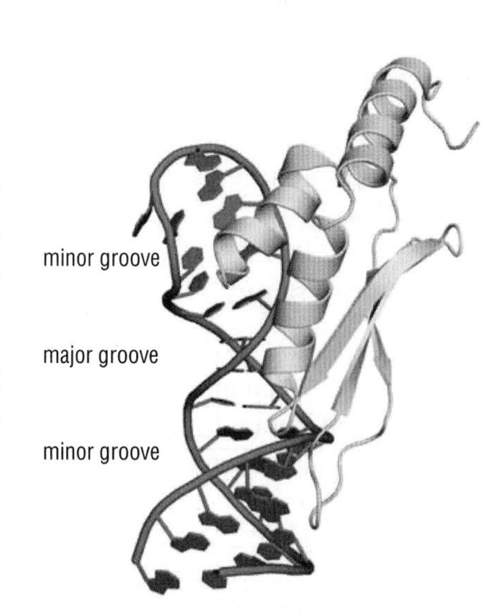

Figure 10.39 dsRBM domains make multiple contacts with adjacent grooves in double-stranded RNA. Helix 1, loop 4, and loop 2 of the dsRBM of *Saccharomyces cerevisiae* RNase III make contacts with sequential minor, major, and minor grooves, respectively. PDB code 1T4L.

RNA recognition can also be mediated by small basic motifs or zinc finger domains

For a subset of proteins, RNA binding requires the protein to possess small, basic stretches that are rich in arginine and lysine residues. While these regions, which can fold into beta hairpins or alpha helices, are required for RNA recognition, they do not bind RNA in isolation and often do not fold into secondary structures in the absence of RNA. Two proteins encoded by the human immunodeficiency

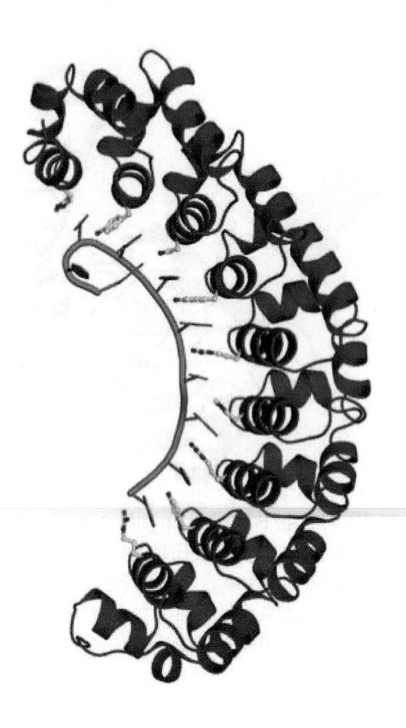

Figure 10.40 An example of RNA binding by a protein with a repeated domain. The human Pumilio 1 homology domain contains multiple repeats of a two-helix motif, each of which interacts with a single nucleotide in the RNA and contributes a histidine, tyrosine, or arginine residue that stacks with the nucleotide. The side chains for these stacking amino acids are shown in yellow. PDB code 1M8Y.

virus (HIV), Tat and Rev, which bind the TAR and RRE sequence elements of the HIV RNA, respectively, are well-studied examples of proteins with such small basic regions.

Other proteins containing the zinc finger DNA-binding domain described in Section 9.2 also bind to RNA. Unlike the binding of a zinc finger to DNA, where the finger inserts its alpha helix end-on into the major groove and side chains form base contacts, the interactions between zinc fingers and RNA are much more varied and include contacts with both single- and double-stranded regions of RNA.

In the next chapter, we will learn how RNA-binding proteins, together with the processed tRNAs and rRNAs, act to translate the processed mRNAs described in this chapter.

SUMMARY

OVERVIEW

- RNA processing events are points for regulation and quality control, and are sources of diversity.
- Many RNA processing reactions are directed by RNA components.

tRNA AND rRNA PROCESSING

- tRNAs and rRNAs are processed out of longer precursor transcripts, and the nucleotides are post-transcriptionally modified.

mRNA PROCESSING

- The maturation of eukaryotic mRNAs requires the addition of a 5′ cap and a poly(A) tail, processes that are closely tied to transcription, splicing, transport out of the nucleus, and ultimately to translation.

RNA SPLICING

- RNA splicing allows the generation of great diversity in RNA products and can be catalyzed by the RNA itself (self-splicing) or by a large protein and RNA-containing complex called the spliceosome.
- Splice site recognition depends on many protein factors that help in defining which exons are brought together. These choices can be regulated to generate alternatively spliced products.

RNA EDITING

- The sequence of mRNAs can be edited (nucleotides deleted, inserted, or changed) such that it might not be possible to predict the protein sequence from the DNA sequence of a gene.
- The sites of editing are specified by diverse mechanisms involving the direct recognition of RNA sequences or structures, or through interactions with guide RNAs.

RNA DEGRADATION

- The degradation of RNAs is another mechanism by which gene expression can be regulated and can be initiated by endonuclease cleavage or by removal of 5′ or 3′ end RNA features that lead to further exonucleolytic degradation.
- Cells have mechanisms to sense and destroy foreign and defective RNA.

RBDs

- The structures of RBDs, which include the RRM domain, the PAZ domain, and the dsRBD domain, exhibit more variety than DNA-binding domains.

QUESTIONS

10.1 OVERVIEW OF RNA PROCESSING

1. What benefits are associated with RNA processing?

10.2 tRNA AND rRNA PROCESSING

Challenge question

1. Compare the addition of RNA nucleotides to mRNA by RNA polymerase to the addition of the CCA nucleotides to tRNAs by CCA-adding enzyme.

10.3 tRNA AND rRNA NUCLEOTIDE MODIFICATIONS

1. What is the main purpose of tRNA and rRNA nucleotide modifications?

2. What is a snoRNP, and why is this type of complex necessary for many tRNA and rRNA modifications?

10.4 mRNA CAPPING AND POLYADENYLATION

1. What are the main functions of the 5′cap?

2. What are the main functions of polyadenylation?

3. Explain how 5′capping and polyadenylation are coordinated with transcription.

Challenge question

4. In *Arabidopsis*, the onset of flowering requires a functional FCA protein. Sequencing of the *FCA* gene revealed that the gene had two alternative polyadenylation sequences: one close to the promoter of the gene and one after the last exon. Propose a mechanism by which these alternative polyadenylation sequences may be used to regulate flowering.

10.5 RNA SPLICING

1. What is transesterification, and what transesterification reactions are needed to splice introns?

2. Compare group I and group II introns.

10.6 EUKARYOTIC mRNA SPLICING BY THE SPLICEOSOME

1. In addition to the proteins that make up the basic spliceosome, hundreds of other proteins are thought to be required for efficient and accurate splicing. Describe some of the proposed functions of these additional proteins.

10.7 EXON DEFINITION AND ALTERNATIVE SPLICING

Challenge questions

1. The *presenilin-2* pre-mRNA can be alternatively spliced in the presence of the protein HMGA1a. Under hypoxic conditions, HMGA1a is produced and leads to the skipping of exon 5 in the transcript. The variant form of presenilin-2 (PS2V) produced is associated with the development of Alzheimer's disease. It has been suggested that the development of an oligonucleotide matching the 3′ end of exon 5 that could be delivered to cells could be a treatment for Alzheimer's disease by preventing exon 5 skipping and the accumulation of PS2V. Propose a mechanism by which HMGA1a may lead to exon skipping that also explains why the oligonucleotide would prevent exon skipping.

2. Neurexin2 (NRXN2) is a protein expressed in the brain that has been implicated in synaptic plasticity and memory processing. Depolarization of rat brain neurons grown in culture induces exon 11 skipping and the production of an alternative *NRXN2* transcript. A bioinformatics search of the *NRXN2* gene found a potential enhancer protein binding site. Researchers hypothesized that in polarized cells (exon 11 included), the enhancer protein is bound and increases the use of the U1 splice site required for exon 11 inclusion.

In order to test the hypothesis, a minigene that contained only the exons and introns beginning at exon 10 and ending after exon 12 of the *NRXN2* gene was constructed. This was considered the wild-type (WT) minigene. Additionally, four mutant minigenes were created. In M1, the first U1 binding site was mutated; in M2, the enhancer protein binding site was mutated; in M3, the first U2 binding site was mutated; and in M4, the next U2 (required for exon 12 splicing) was mutated. They injected the minigenes into cells and performed real-time PCR when the cells were polarized. They determined the amount of RNA produced that contained exon 11 vs the amount produced in which exon 11 was skipped.

a. M1, M3, and M4 were controls for the experiment. Explain why they are controls and what result is expected for each.
b. M2 is the experimental mutant. Explain what result is expected if the enhancer hypothesis is true.
c. Provide an alternative hypothesis to explain exon 11 skipping in depolarized cells.

10.9 DEGRADATION OF NORMAL RNAs

1. Compare and contrast the normal degradation mRNAs in E. coli and eukaryotes? What aspects are different? What aspects are similar?

Challenge question

2. In bacteria, the addition of a poly(A) tail decreases RNA stability and makes it more susceptible to degradation while, in eukaryotes, the addition of the poly(A) tail increases RNA stability and protects the RNA from degradation. Explain this difference.

10.10 DEGRADATION OF FOREIGN AND DEFECTIVE RNAs

1. Describe what type of RNA is targeted by
a. RNA interference
b. CRISPR-mediated interference
c. Nonsense-mediated decay
d. tmRNA-mediated decay

EXPERIMENTAL APPROACH 10.2 – CATALYTIC RNA: DISCOVERY AND TERTIARY STRUCTURE

1. What were the properties of the intron cleavage activity that made it 'perplexing'?

2. Why was the gel shown in Figure 1 considered to be sufficient evidence that the introns were self-splicing?

3. How did bioinformatics add to the information regarding self-splicing introns?

EXPERIMENTAL APPROACH 10.4 – DETERMINING THE ROLES OF EXON AND INTRON DEFINITION IN EUKARYOTIC SPLICING THROUGH MUTATIONAL AND COMPUTATIONAL ANALYSES

1. What is covariance, and how did studies on the covariance of 5′ and 3′ splice sites add to our understanding of splicing?

Challenge questions

2. Look at Experimental approach 10.4 Figure 1.
a. What was the purpose of the experiment performed to produce the result shown in Figure 1?

b. What do the numbers 27, 88, and 350 at the top of the gel in (a) represent?c.What do the numbers 0, 5, and 20 at the top of the gel in (a) and (b) represent?
d. What does α represent?
e. Explain the main difference between the gel in (a) and the gel in (b). What conclusion could be drawn based on the result?

3. In what way is Figure 2 different from Figure 1, and what conclusion could be drawn based upon the results?

EXPERIMENTAL APPROACH 10.5 – IDENTIFICATION OF RNA EDITING SITES THROUGH SERENDIPITY AND COMPARATIVE GENOMICS

1. What observation serendipitously led to the hypothesis of an RNA editing mechanism?

2. What characteristic of RNA editing made it difficult to identify authentic editing sites, and how was this difficulty overcome?

10.1 OVERVIEW OF RNA PROCESSING

Hattori D, Millard SS, Wojtowicz WM, Zipursky SL. Dscam-mediated cell recognition regulates neural circuit formation. *Annual Review of Cell and Developmental Biology*, 2008;**24**:597–620.

10.2 tRNA AND rRNA PROCESSING

Phizicky EM, Hopper AK. tRNA biology charges to the front. *Genes & Development*, 2010;**24**:1932–1860.

Weiner AM. tRNA maturation: RNA polymerization without a nucleic acid template. *Current Biology*, 2004;**14**:R883–L R885.

10.3 tRNA AND rRNA NUCLEOTIDE MODIFICATIONS

Decatur WA, Fournier MJ. RNA-guided nucleotide modification of ribosomal and other RNAs. *Journal of Biological Chemistry*, 2003; **278**:695–698.

Kiss T. Small nucleolar RNAs: an abundant group of noncoding RNAs with diverse cellular functions. *Cell*, 2002;**109**:145–148.

Ofengand J. Ribosomal RNA pseudouridines and pseudouridine synthases. *FEBS Letters*, 2002;**514**:17–25.

10.4 mRNA CAPPING AND POLYADENYLATION

Jensen TH, Dower K, Libri D, Rosbash M. Early formation of mRNP: license for export or quality control? *Molecular Cell*, 2003;**11**:1129–1138.

Moore MJ, Proudfoot NJ. Pre-mRNA processing reaches back to transcription and ahead to translation. *Cell*, 2009;**369**:688–700.

Tian B, Manley JL. Alternative cleavage and polyadenylation: the long and short of it. *Trends in Biochemical Sciences*, 2013;**38**:312–320.

10.5 RNA SPLICING

Doudna JA, Lorsch JR. Ribozyme catalysis: not different, just worse. *Nature Structural and Molecular Biology*, 2005;**12**:395–402.

Toor N, Keating KS, Pyle AM. Structural insights into RNA splicing. *Current Opinions in Structural Biology*, 2009;**19**:260–266.

10.6 EUKARYOTIC mRNA SPLICING BY THE SPLICEOSOME

Hoskins AA, Moore MJ. The spliceosome: a flexible, reversible macromolecular machine. *Trends in Biochemical Sciences*, 2012;**37**:179–188.

Michaeli S. Trans-splicing in trypanosomes: machinery and its impact on the parasite transcriptome. *Future Microbiology*, 2011;**6**:459–474.

10.7 EXON DEFINITION AND ALTERNATIVE SPLICING

Kornblihtt AR, Schor IE, Alló M, Dujardin G, Petrillo E.Muñoz MJ. Alternative splicing: a pivotal step between eukaryotic transcription and translation. *Nature Reviews in Molecular Cell Biology*, 2013;**14**:153–165.

Li Q, Lee JA, Black BL. Neuronal regulation of alternative pre-mRNA splicing. *Nature Reviews in Neuroscience*, 2007;**8**:819–831.

Matlin AJ, Clark F, Smith CW. Understanding alternative splicing: towards a cellular code. *Nature Reviews Molecular Cell Biology*, 2005;**6**:386–398.

10.8 RNA EDITING

Horton TL, Landweber LF. Rewriting the information in DNA: RNA editing in kinetoplastids and myxomycetes. *Current Opinion in Microbiology*, 2002;**5**:620–626.

Hundley HA, Bass BL. ADAR editing in double-stranded UTRs and other noncoding RNA sequences. *Trends in Biochemical Sciences*, 2010;**35**:377–383.

Keegan LP, Gallo A, O'Connell MA. The many roles of an RNA editor. *Nature Reviews Genetics*, 2001;**2**:869–878.

Simpson L, Sbicego S, Aphasizhev R. Uridine insertion/deletion RNA editing in trypanosome mitochondria: a complex business. *RNA*, 2003;**9**:265–276.

10.9 DEGRADATION OF NORMAL RNAs

Belasco JG. All things must pass: contrasts and commonalities in eukaryotic and bacterial mRNA decay. *Nature Reviews in Molecular Cell Biology*, 2010;**11**:467–478.

Houseley J, Tollervey D. The many pathways of RNA degradation. *Cell*, 2009;**136**:763–776.

10.10 DEGRADATION OF FOREIGN AND DEFECTIVE RNAs

Isken O, Maquat LE. Quality control of eukaryotic mRNA: safeguarding cells from abnormal mRNA function. *Genes Development*, 2007;**21**:1833–1856.

Malone CD, Hannon GJ. Small RNAs as guardians of the genome. *Cell*, 2009;**136**:656–668.

Wiedenheft B, Sternberg SH, Doudna JA. RNA-guided genetic silencing systems in bacteria archaea. *Nature*, 2012;**482**:331–338.

10.11 RNA-BINDING DOMAINS IN PROTEINS

Chen Y, Varani G. Protein families and RNA recognition. *FEBS Journal*, 2005;**272**:2088–2097.

Maris C, Dominguez C, Allain FH. The RNA recognition motif, a plastic RNA-binding platform to regulate post-transcriptional gene expression. *FEBS Journal*, 2005;**272**:2118–2131.

Translation

INTRODUCTION

In the preceding three chapters, we have examined how the information stored in a specific region of a genome is extracted in the form of messenger RNA (mRNA), and how this mRNA is processed to reveal the information required to direct the synthesis of the protein products needed by the cell. How does the cell decode this information? How can a sequence of nucleotides be used to determine the composition of a biochemically distinct molecule, a polypeptide?

In this chapter, we conclude our journey from gene to protein by exploring the process of **translation**, the series of events through which the cell uses the information in an mRNA molecule to direct the synthesis of a protein. We explore the key molecular components that play vital roles in the process of translation; we describe what happens in each individual step of this overall process. In the following chapter, we then see how – as we have for each of the other processes we have encountered throughout this book – the process of translation is regulated, to ensure that it happens at the right time, in the right amount, with an end result that meets the required standards.

While we consider individual aspects of translation in some detail later in the chapter, we begin by painting an overall picture of the process, to give a framework on which the subsequent detail can be placed.

11.1 OVERVIEW OF TRANSLATION

Translation is the final step in decoding the genetic information

Translation is the process by which a polypeptide chain is synthesized according to the instructions in an mRNA template molecule; it is thus the final step in the flow of genetic information from DNA to protein. Translation transforms genotype into phenotype. The earlier steps of this information flow - DNA replication and transcription - are essentially two similar processes in which a new DNA or RNA molecule is synthesized by forming complementary nucleotide base-pairings with a pre-existing DNA template, and joining these nucleotide bases together to form a continuous complementary strand. The synthesis of protein from mRNA is an altogether more complex task: the language of proteins has an alphabet of 20 amino acids, as opposed to the four-nucleotide alphabet of mRNA, and there are no simple amino acid–nucleotide interactions that substitute for the direct Watson–Crick base-pairings between nucleotides.

Transfer RNA molecules provide the physical link that decodes the mRNA into the encoded polypeptide

How can a polypeptide chain, whose amino acid sequence is determined by the sequence of nucleotides in an mRNA molecule, be synthesized? The solution to the

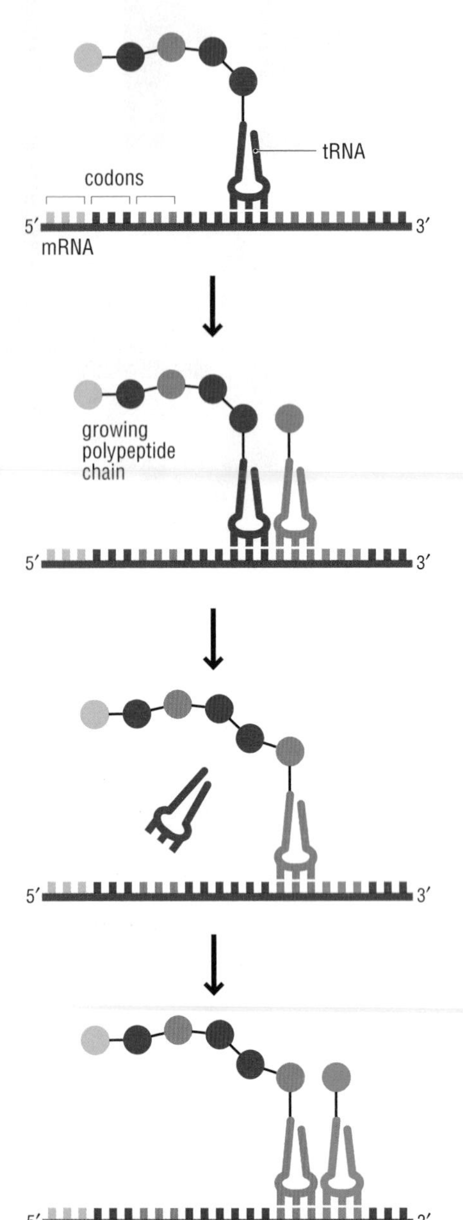

Figure 11.1 tRNA links the mRNA sequence with the amino acid sequence. Distinct tRNA molecules bind to different three-nucleotide long sequences (called codons) in mRNA. Each tRNA also carries a particular amino acid (balls) that is incorporated into a polypeptide chain. Thus, tRNAs are able to specify the sequence of amino acids that are 'read' from an mRNA sequence during protein synthesis. In the figure, several tRNAs are shown aligned with the mRNA to emphasize how tRNAs link nucleotide and amino acid sequences. Inside cells, the tRNAs only bind to mRNA found inside the ribosome, two or three at a time.

➲ We learn about the chemical structure of peptide bonds in Chapter 2.

problem of coupling nucleotide and amino acid sequence is provided by a class of RNA called **transfer RNA (tRNA)**. These tRNAs are bifunctional: they decode the mRNA through direct base-pairing interactions, and participate in the synthesis of the polypeptide as they carry a chemically reactive amino acid at their 3′ end. These amino acids are the building blocks for the polypeptide that is synthesized during translation. The enzymes responsible for attaching the appropriate amino acids to the various tRNAs are the **aminoacyl-tRNA synthetases**; this attachment event plays a key role in deciphering the **genetic code**.

Each tRNA molecule recognizes a sequence of three nucleotides (collectively called a **codon**) in the mRNA through direct pairing interactions with a region within the tRNA called the **anticodon**, and each tRNA molecule carries a particular amino acid that corresponds to the codon that it recognizes and to which it binds, as illustrated in Figure 11.1. By having these two functions - binding to mRNA and carrying an amino acid - tRNAs can decode the information stored in mRNA and ensure that the correct amino acid encoded by this information is inserted into the growing polypeptide. The fidelity of decoding depends therefore on the loading of tRNAs with the appropriate amino acids (a job performed by the aminoacyl-tRNA synthetases) and the specific interaction of the anticodon with the codon during translation.

Translation is carried out by a macromolecular RNA–protein complex called the ribosome and is facilitated by a number of protein factors

The complex task of protein synthesis is carried out by a large molecular machine called the **ribosome**. Ribosomes are composed of RNA and protein and, in all organisms, have two subunits, the 'small' and the 'large', which are responsible for the two principal events of translation. The small subunit is responsible for deciphering the mRNA, whereas the large subunit mediates the formation of peptide bonds between the amino acids; this overall process of synthesis is depicted in Figure 11.2. The mass of the large subunit is about twice that of the small subunit, and the overall mass of the ribosome is composed roughly of two-thirds RNA relative to one-third protein. The overall diameter of an intact bacterial ribosome with both large and small subunits together is about 220 Å, about twice as large as that of the *Escherichia coli* RNA polymerase holoenzyme.

The ribosome moves processively (5′ to 3′) along an mRNA molecule, forging peptide bonds between amino acids along the way to synthesize a polypeptide chain. In bacterial cells, a ribosome synthesizes protein at a rate of about 15 amino acids per second, and amino acids are incorporated with an error rate of around 10^{-3} to 10^{-4} per residue. To put this another way, roughly nine out of ten proteins that are 1000 amino acids long will be synthesized accurately (i.e., without any mistakes) if fidelity is 10^{-4} (0.9999^{1000}), whereas one in three will be accurately synthesized if fidelity is 10^{-3} (0.999^{1000}). While neither the rate nor the fidelity of protein synthesis come close to those reported for DNA replication (which proceeds at a rate of almost 1000 nucleotides per second with an astonishing accuracy of 10^{-10}), they are similar to those reported for transcription (which proceeds at a rate of 50–100 nucleotides per second and an accuracy of around 10^{-4}). As we have seen in other systems and will see again for translation, fidelity is energetically costly (and can slow down the overall process) and speed is often important for cellular robustness. Ribosomes have evolved to provide a suitable compromise for the cell between high speed and accuracy during protein synthesis.

Many of the proteins that contribute to translation are not permanent components of the ribosome. These proteins, referred to as **translation factors**, temporarily associate with the ribosome and leave when their task is complete. Protein factors help to coordinate the orderly progression of events during translation, ensuring that the right events happen at the right time. Many of these translation factors are GTPases (or G proteins) that couple the energy of guanosine triphosphate (GTP) hydrolysis to events on the ribosome (movement or dissociation, for example). Translation factors are also important for regulating protein synthesis: cells can modify the activity of certain factors to increase or decrease the amount of protein the cells synthesize to suit their needs.

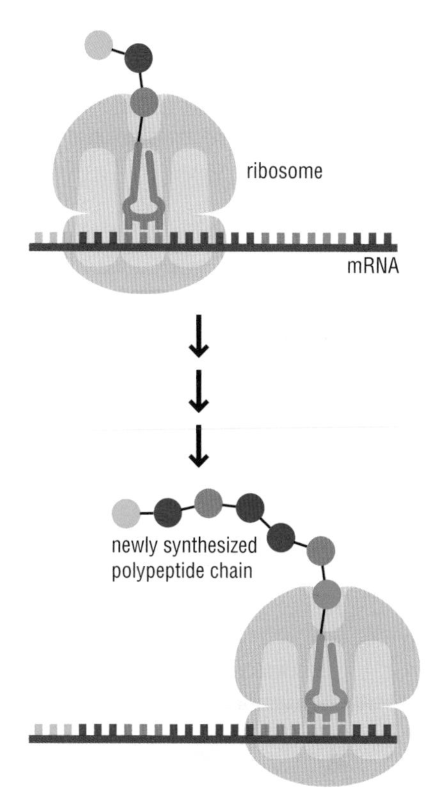

Figure 11.2 Ribosomes are protein synthesis machines. A ribosome moves along an mRNA molecule, synthesizing a protein as it goes. The sequence of amino acids in the polypeptide chain is determined by the sequence of nucleotides in the mRNA.

The translation cycle is composed of four basic stages

The translation cycle is composed of four basic stages: **initiation**, **elongation**, **termination**, and **ribosome recycling**, which we explore in detail in later sections. At each of these stages, RNA elements in the ribosome itself play critical functional roles in the process; in addition - and with the exception of peptide bond formation - all steps in protein synthesis depend on the actions of auxiliary protein factors.

Let us now consider in more detail the three core components that lie at the heart of translation: tRNA, aminoacyl-tRNA synthetases, and the ribosome. We begin with tRNA.

11.2 tRNA AND THE GENETIC CODE

With a broad picture of translation now in our minds, we need to consider a question that is central to what happens during translation: how is a sequence of nucleotides in a nucleic acid converted into a sequence of altogether different molecules, the amino acids that comprise a polypeptide? The answer lies in a remarkable molecule that straddles the worlds of both nucleic acids and proteins: tRNA.

tRNA is a bifunctional molecule that links gene sequence to protein sequence

Two processes are essential to translation: the deciphering of triplet **codons** in mRNA, and the incorporation of amino acids encoded by the triplets into a growing polypeptide chain. **tRNAs** link these two processes by virtue of their structure. We will first look at some general features of tRNA structure and then see how these features allow tRNAs to function during translation.

tRNA molecules are small RNAs ranging from 75 to 94 nucleotides in length. There are many different tRNAs in an organism, each specific to a particular amino acid; the specificity is determined by the enzymes that attach the amino acids to the tRNA. (This topic will be discussed in Section 11.3.) Many of the nucleotides in the tRNAs form base pairs that are best visualized by drawing a tRNA in a cloverleaf pattern in two dimensions, as depicted in Figure 11.3a. In this representation, tRNA can be seen to consist of four base-paired regions referred to as stems or arms that are interspersed with single-stranded regions called loops.

The 5′ and 3′ ends of the tRNA meet in a double-stranded region called the **acceptor stem**. The acceptor stem ends with a short, universally conserved sequence of bases called the **3′ CCA tail**, which is the attachment point for amino acids that will later be incorporated into the polypeptide chain. At the other end

➲ We learn about CCA tail addition and post-transcriptional modification in Section 10.2.

of the cloverleaf, in a loop called the **anticodon loop**, are three nucleotides collectively called the **anticodon** that base-pair in an antiparallel fashion with mRNA codons. (That is, the 5′ end of the mRNA codon base-pairs with the 3′ end of the anticodon, and vice versa.) In this way, the tRNA provides a physical link between the amino acid and the triplet codon that specifies it.

The nucleobases of the anticodon (typically positions 34–36) are typically stacked on one another in a structure that follows a well-defined turn in the RNA backbone known as a U-turn. This anticodon loop conformation positions the nucleotides of the anticodon to effectively base-pair with the mRNA. A hypermodified purine residue occurs just after the three nucleotides of the anticodon (typically position 37). For example, the wybutosine base (Y) found in certain eukaryotic tRNAs is shown in Figure 11.4. The site of modification on the Watson–Crick face of this base prevents it from base-pairing with the mRNA and thus helps in aligning the codon–anticodon base-pairing interactions. Modification at this site is known to be critical for ensuring the high fidelity of **decoding**.

Two other regions of tRNA, the **DHU (D) loop** and **TpsiC (T) loop**, are named after the modified bases they contain: dihydrouridine (D) in the D loop, and ribothymidine (T) and pseudouridine (ψ) in the T loop. Figure 11.4 shows the chemical structures of several post-transcriptionally modified bases that are found in the D-, T-, and anticodon loops of various tRNAs.

About 10% of the nucleotides in a given tRNA are post-transcriptionally modified. The group of such modified nucleotides is specific to a given tRNA species. The many different post-transcriptional modifications are proposed to promote tRNA function in a number of ways, which include the aiding of folding, the stabilization of overall structure, and the increase in the specificity of interactions with various components of the translation system (including the codons themselves).

The overall structure of a tRNA in three-dimensional space is an L, as illustrated in Figure 11.3b. The elbow region of this structure (the 'bend' in the L shape) is stabilized by a number of unusual base–base interactions between nucleotides in the D and T loops of the molecule. For example, there are several places where three nucleotides interact together (base triple interactions), and a conserved methylated G–A base pair (such as that shown in Figure 2.37) is critical for kinking the D-stem of the tRNA into a particular conformation. The L shape of the tRNA appropriately spaces the anticodon and acceptor stems with respect to one another to interact both with the aminoacyl-tRNA synthetases (that attach the amino acids to the tRNA) and with the two subunits of the ribosome where protein synthesis takes place.

(a)

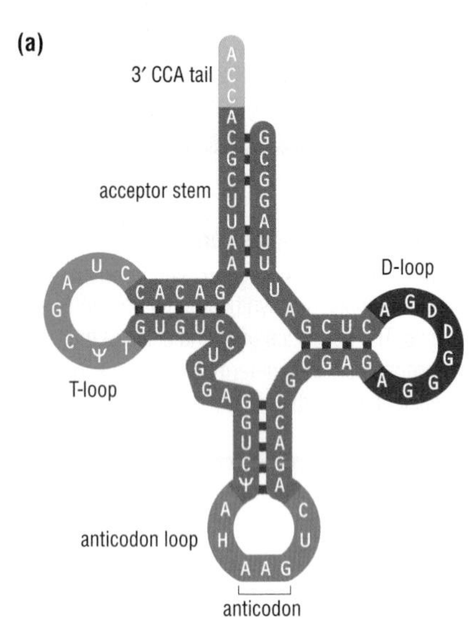

(b)

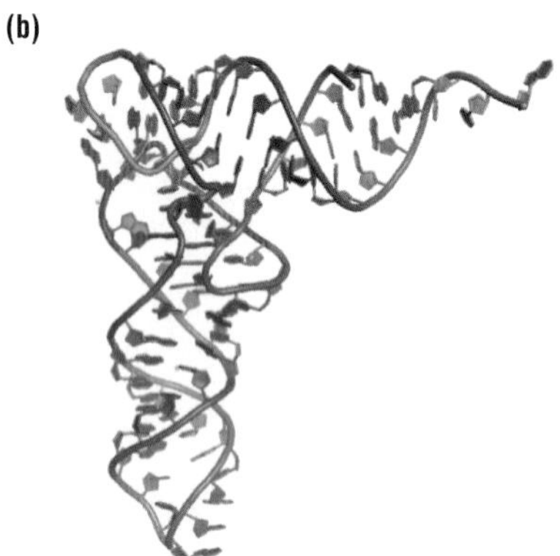

Figure 11.3 The structure of tRNA. (a) A cloverleaf depiction of the pattern of base-pairing in tRNA. The structures of some of the modified nucleotides indicated as D (dihydrouridine), H (hypermodified purine), and ψ (pseudouridine) are provided in Figure 11.4. (b) The L-shaped three-dimensional structure of a folded tRNA. PDB code 1TRA.

Triplet nucleotide codons specify the genetic code

We have discussed how two regions of tRNA – the 3′ CCA tail to which amino acids are attached and the anticodon that base-pairs with mRNA – enable tRNA to link the **genetic code** to protein sequences. The code that defines the relationship between mRNA sequences and protein sequences is explained in Section 1.3 and is summarized in Figure 11.5. Each triplet mRNA codon specifies a single amino acid (a **sense codon**) or no amino acid (a **nonsense codon**). A triplet code is the simplest code that can specify as many as 20 amino acids with only four nucleotides. (By contrast, a singlet code – whereby one base = one amino acid – could only specify four amino acids, 4^1; a doublet code 16, 4^2; and the triplet code 64, 4^3.) **Stop codons**, often referred to as nonsense codons, signal the end of the protein-coding region of an mRNA molecule.

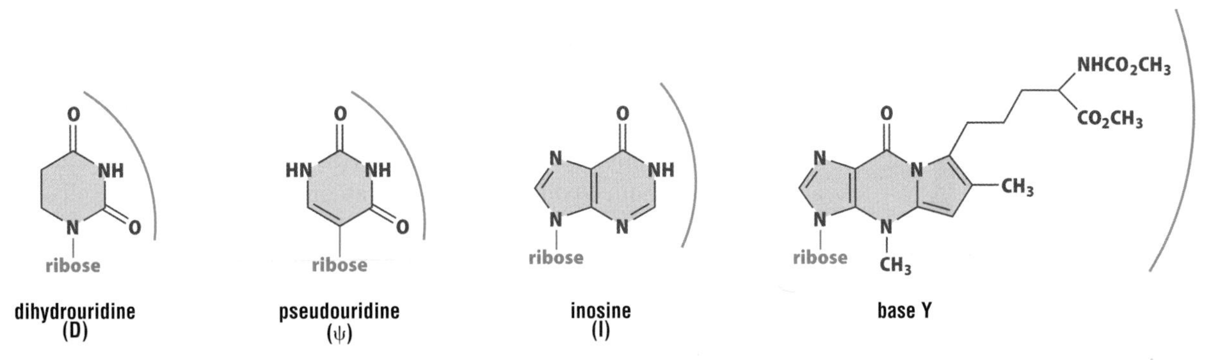

Figure 11.4 Examples of post-transcriptionally modified nucleotides found in tRNA. Dihydrouridine and pseudouridine are modifications of uridine and are found in conserved positions in the D and T loops of tRNA, respectively. Inosine and base Y (an example of a hypermodified purine) are modifications of purines (A or G) and are found in the anticodon loop, where they play critical roles in decoding mRNA. The Watson–Crick face of each nucleotide is indicated (orange arc).

	AGA									UUA					AGC					
	AGG									UUG					AGU					
GCA	CGA						GGA			CUA				CCA	UCA	ACA			GUA	
GCC	CGC						GGC		AUA	CUC				CCC	UCC	ACC			GUC	UAA
GCG	CGG	GAC	AAC	UGC	GAA	CAA	GGG	CAC	AUC	CUG	AAA		UUC	CCG	UCG	ACG		UAC	GUG	UAG
GCU	CGU	GAU	AAU	UGU	GAG	CAG	GGU	CAU	AUU	CUU	AAG	AUG	UUU	CCU	UCU	ACU	UGG	UAU	GUU	UGA
Ala	Arg	Asp	Asn	Cys	Glu	Gln	Gly	His	Ile	Leu	Lys	Met	Phe	Pro	Ser	Thr	Trp	Tyr	Val	stop
A	R	D	N	C	E	Q	G	H	I	L	K	M	F	P	S	T	W	Y	V	

Figure 11.5 Codon table. The amino acids that are decoded from each trinucleotide codon in the mRNA are indicated. At the bottom are listed their standard three-letter and single-letter abbreviations.

Some amino acids, for example tryptophan and methionine, are specified by a single codon, whereas others are specified by multiple codons. (Arginine, for example, typically is specified by six.) If we examine the codon table in Figure 11.5 we see that, in many cases, the first two nucleotides of a codon are sufficient to specify an amino acid while the third nucleotide is irrelevant. For example, the four codons GCA, GCC, GCG, and GCU (which we can write more simply as GCN, where N is any nucleotide) encode alanine, while the GGN family encodes glycine. These observations do not necessarily mean that a single tRNA can decode all four of the GCN alanine codons. Indeed, the multiple codons in such a four-member group are typically decoded by more than one species of tRNA; the set of tRNAs that are loaded with the same amino acid are referred to as **isoacceptors**.

How then does recognition of the codon take place? The first two positions of the codon (reading from the 5′ to 3′ direction in the mRNA) are read by strict Watson-Crick pairing interactions with the third and second positions of the anticodon (typically positions 36 and 35, respectively, numbering sequentially from the 5′ end of the tRNA); the mRNA and tRNA anticodon effectively form an antiparallel mini RNA helix. Complexities in deciphering the code are seen at the third position of the codon where Watson–Crick pairing, and certain deviations from Watson–Crick pairing, are permitted. This promiscuity in reading of the third codon position by the first position of the anticodon (the 5′ proximal nucleotide, position 34) is called **wobble pairing.** Wobble pairing allows certain non-Watson–Crick pairing interactions to be accepted by the translational machinery as shown in Figure 11.6a. A summary of allowed wobble pairings in bacteria is presented in Figure 11.6b.

Wobble pairing most often involves G–U base pairs, with the G found either in the codon or in the anticodon. More seldomly, the unusual post-transcriptionally modified purine nucleotide inosine is found in the anticodon (at position 34), where it effectively pairs with U, C, or A in the mRNA.

(a)

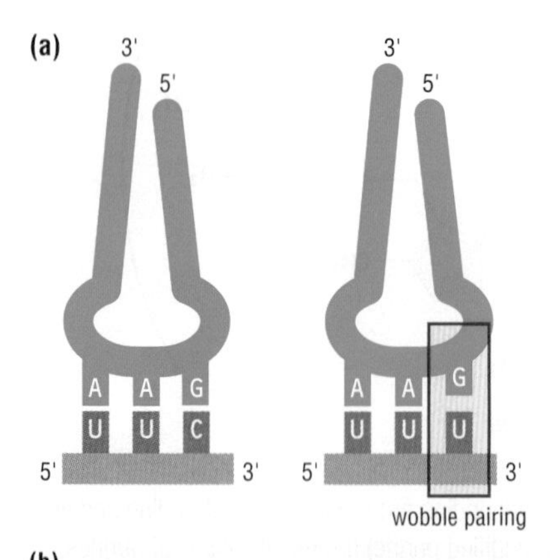

(b)

1st base of anticodon	3rd base of codon
C	G
A	U
U	A or G
G	U or C
I	U, C or A

Figure 11.6 Wobble pairing. (a) The first two nucleotides of the mRNA codon (gray) must interact in a strict Watson–Crick pairing interaction with the anticodon (blue), while the third position may interact through certain non-canonical pairing interactions. A specific example is shown indicating how tRNA^{Phe} can decode both UUC and UUU codons by taking advantage of 'wobble' pairing. (b) The rules for allowed wobble pairings seen in bacteria. For example, a tRNA with inosine in the wobble position can pair with three different mRNA codons that have U, C, or A at the third position.

As a result of this relaxation in the stringency of the pairing rules, tRNAs can sometimes decode several different codons by forming non-Watson–Crick base-pairing interactions with mRNA at the third position. Because of wobble pairing, the number of tRNAs needed to decode all mRNA codons is less than 61 (the number of sense codons), yet still more than 20 (the number of amino acids) - generally, there are about 40 different tRNAs in an organism that decode all 61 codons. It is interesting to note that not all tRNAs are equally abundant and that some codons are used less frequently than others. (We call these rare codons.) Not surprisingly, the rare codons tend to be decoded by the less abundant tRNA species.

The genetic code is nearly universal across all life forms. AUG, for example, codes for methionine in all organisms, whereas CCC codes for proline. In addition, there are typically three stop codons - UAA, UAG, and UGA. Generally speaking, the genetic code has evolved to minimize the deleterious effect of genetic mutations: single nucleotide changes tend to result in the insertion of chemically similar amino acids into a protein. For example, mutation of the first nucleotide of the phenylalanine codon (UUU) will result in the replacement of phenylalanine with other hydrophobic amino acids such as leucine (CUU), isoleucine (AUU), or valine (GUU).

Codon reassignment does happen occasionally

There are exceptions to the universality of the genetic code. Mitochondria, which have much of their own protein translation machinery, decode AUA as methionine rather than as isoleucine, thus providing two distinct codons for specifying methionine: AUG and AUA. In *Mycoplasma* species, the stop codon UAG is decoded as tryptophan, in addition to the standard tryptophan codon UGG. In some organisms, stop codons are programmed to incorporate non-standard amino acids, thus expanding the coding potential of the DNA. For example, we see in Section 11.13 that selenocysteine is incorporated in a number of systems (bacterial, archaeal, and eukaryotic) through recognition of a UGA codon found in a specialized mRNA context.

Although exceptions to the basic rules of the genetic code are of considerable biological interest, the universal features of the code indicate that a sophisticated protein translation apparatus (and the basic code) evolved before the divergence of the three kingdoms of life.

11.3 AMINOACYL-tRNA SYNTHETASES

Transfer RNAs are not the only vital components of the molecular machinery that make translation happen. Vital, too, are the enzymes that couple tRNAs with the appropriate amino acids for addition to the growing polypeptide chain. Only if the right amino acid is coupled to the right tRNA will the reading of a codon by the translational machinery result in the insertion of the correct amino acid. In this section, we discuss the enzymes that are responsible for this essential coupling step, the aminoacyl-tRNA synthetases.

Aminoacyl-tRNA synthetases are responsible for the first step in decoding

Attachment of the appropriate amino acids to the specified tRNAs is the first key step in the decoding of genetic information. This attachment, termed **aminoacylation**, is carried out by enzymes called **aminoacyl-tRNA synthetases**. (We often

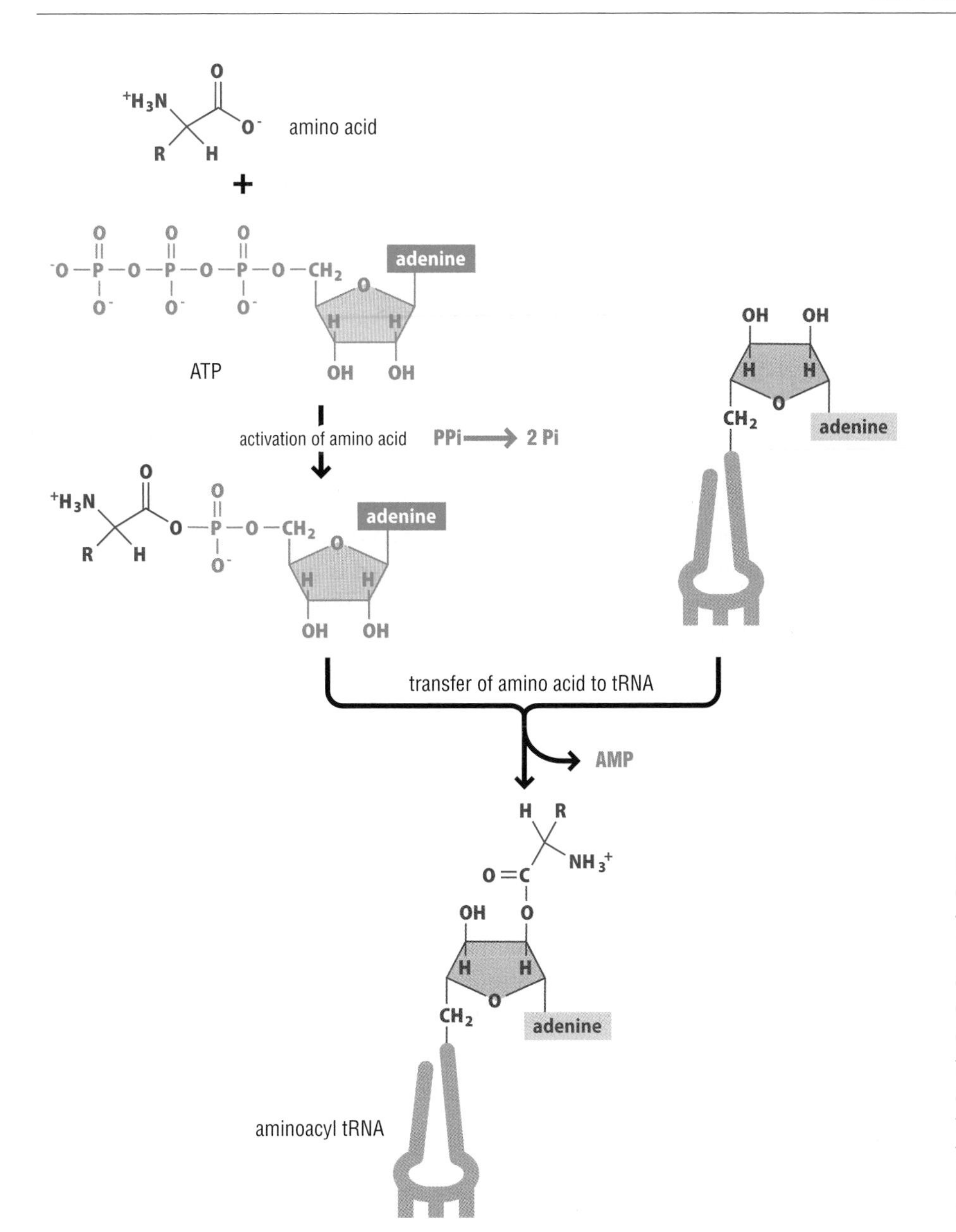

Figure 11.7 Aminoacyl-tRNA synthetases catalyze the aminoacylation of tRNA in a two-step process. Amino acids are activated by the attachment of AMP (yellow) to form an aminoacyl adenylate in which the carboxyl group of the amino acid is linked to the phosphoryl group of AMP. The high-energy bond between the amino acid and AMP can be easily broken, allowing transfer of the aminoacyl group to the 2′ or 3′ hydroxyl group of the ribose of the terminal adenosine of the tRNA (gray). The hydrolysis of pyrophosphate (PPi), producing two phosphates (2Pi), drives the reaction forward.

refer to them as aaRS. For example, GlyRS refers to the glycyl-tRNA-synthetase.) Aminoacyl-tRNA synthetases are specific both for the amino acid and the tRNA species to which the amino acid is attached. As such, this class of enzymes establishes the rules of the genetic code. How these enzymes recognize particular tRNAs and amino acids is central to translation of the nucleotide code into protein.

The direct attachment of an amino acid to a tRNA is not energetically favorable, so aminoacyl-tRNA synthetases use a two-step process that is driven by the hydrolysis of adenosine triphosphate (ATP), as illustrated in Figure 11.7. The amino acid is first activated by the attachment of adenosine monophosphate (AMP) with release of pyrophosphate; this chemical reaction produces an aminoacyl-adenylate that remains bound to the synthetase. As in many biological reactions, hydrolysis of the pyrophosphate moiety to free phosphate by pyrophosphatases in the cell provides critical energy that drives the overall process.

The activated amino acid is then directly transferred to the 2′ or 3′ OH of the ribose of the terminal adenosine of tRNA (the nucleophile in the reaction) with the release of AMP. The resulting **aminoacyl-tRNA** is composed of a chemically reactive amino acid attached to the 3′ CCA tail of the tRNA. The amino acid is chemically reactive because of the high energy bond that joins it to the tRNA. The chemical lability of the bond between the amino acid and the tRNA is a useful property, since the amino acid will next be transferred to a growing peptide chain. The resulting aminoacyl-tRNA is protected from spontaneous hydrolysis by immediately binding to an abundant cellular protein known as EFTu in bacteria (and eEF1A in eukaryotes and archaea), which plays a key role in shepherding aminoacyl-tRNAs through the translation process.

Aminoacyl-tRNA synthetases recognize specific tRNA species

Each tRNA species must be loaded with the amino acid that corresponds to the mRNA codon to which the tRNA binds – this amino acid is referred to as **cognate**. For example, the codon AUG codes for methionine, so the tRNA recognizing this codon must be loaded with methionine and not a different amino acid. The specific amino acid with which a given tRNA is loaded is indicated by a superscript three-letter shorthand for the amino acid – $tRNA^{Met}$, $tRNA^{Gly}$, and so on. This loading reaction is performed by the aminoacyl-tRNA synthetases with a remarkable accuracy of less than one error per 10^4 aminoacylation events. There is generally one aminoacyl-tRNA synthetase for each of the standard 20 amino acids in a cell.

How do the synthetases select their tRNAs and amino acids? Aminoacyl-tRNA synthetases recognize tRNAs on the basis of specific sequence and structural features (termed **identity elements**) found primarily in the anticodon loop and acceptor stems. A notable example of a well-defined identity element is the G3:U70 base pair in the acceptor stem of $tRNA^{Ala}$. The G3:U70 base pair serves as the principal determinant for recognition of alanine tRNAs in all kingdoms, and can confer alanine identity on most tRNAs when the element is simply transplanted to the acceptor stem. (That is, when a tRNA molecule is engineered to carry the G3:U70 base pair, it will become charged with alanine.) Thus, while the overall three-dimensional structure of all tRNAs is highly similar, there are sufficient differences among them to allow synthetases to specifically recognize their cognate tRNAs.

Aminoacyl-tRNA synthetases select the correct amino acid in a two-step process

Aminoacyl-tRNA synthetases use various chemical features, such as charge, hydrophobicity, size, and shape, to discriminate between the different amino acids. It is easy to understand how synthetases can discriminate against amino acids larger than their cognate amino acid. For example, the synthetase for glycine tRNA (GlyRS) would be unlikely to bind phenylalanine or other large amino acids in its active site simply because they are so much larger than glycine. Consequently, these substrates are readily excluded. However, it is more difficult for synthetases to discriminate against amino acids that are similar to or slightly smaller than the cognate amino acid, as these amino acids are often able to enter the active site. Indeed, certain **non-cognate** amino acids are activated by the synthetase and in some cases are even transferred to the tRNA.

For these errors, a quality control step plays an important role in increasing the fidelity of aminoacylation. Most synthetases have a secondary site (the **editing site**) that has evolved to specifically accommodate the amino acids that tend to be mis-activated in the aminoacylation site – that is, the non-cognate amino acids. The location of the editing site within the overall structure of one of the tRNA synthetases is shown in Figure 11.8a. This site, depending on the synthetase, can

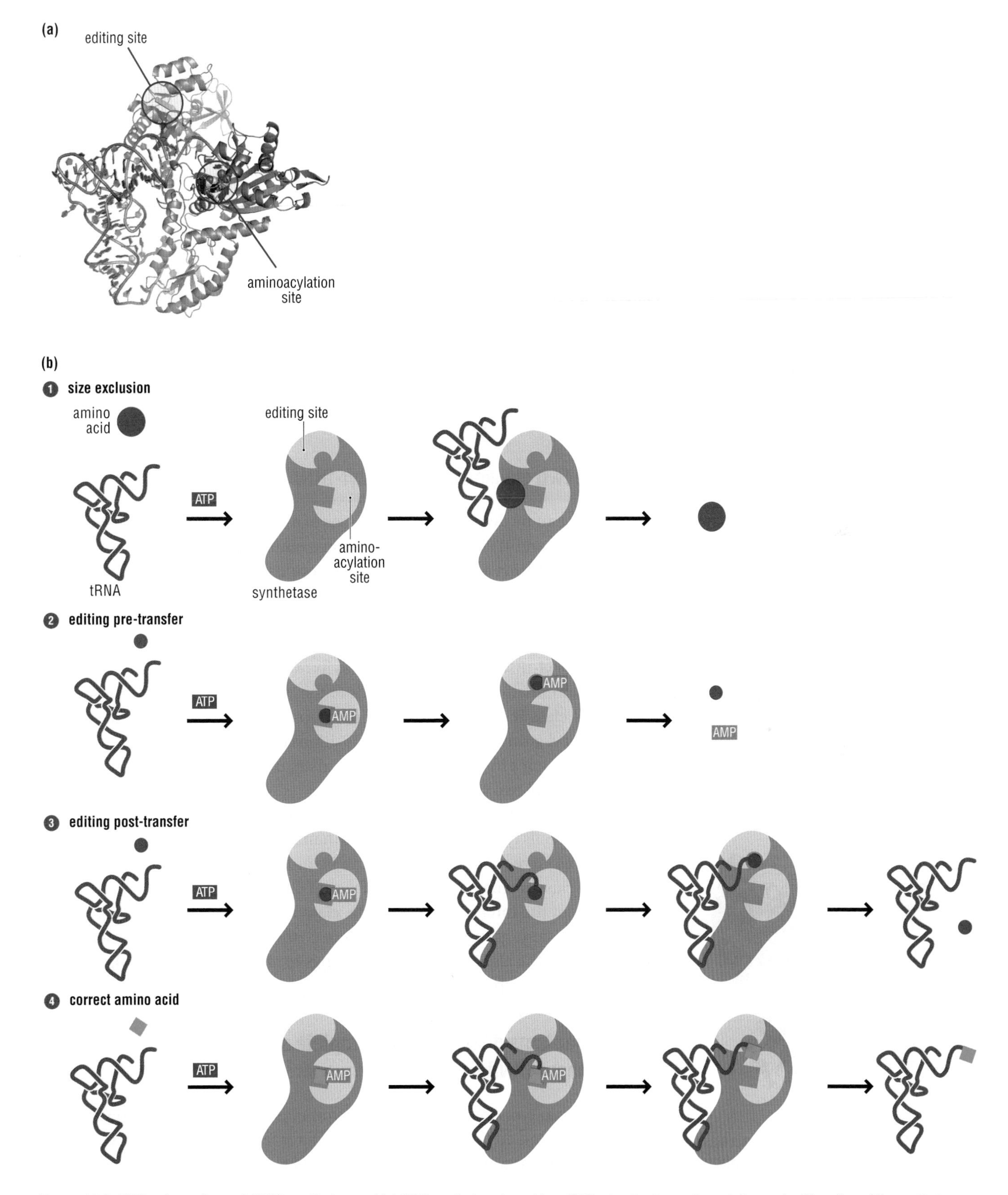

Figure 11.8 Editing by aminoacyl-tRNA synthetases. (a) A tRNA synthetase bound to a tRNA, showing the aminoacylation and editing sites of the synthetase. PDB code 1QF6. (b) The editing site increases the fidelity of the aminoacylation reaction: (1) Large non-cognate amino acids are excluded from the aminoacylation site (top); (2) smaller non-cognate amino acids are occasionally activated (to form the aminoacyl adenylate), but are recognized by the editing site and are hydrolyzed (pre-transfer); or (3) are transferred to the tRNA in the aminoacylation site, and then enter the editing site and are hydrolyzed from the tRNA (post-transfer); (4) the cognate amino acid for the tRNA cannot enter the editing site and therefore remains attached to the tRNA.

recognize either the aminoacyl-adenylate species (the activated amino acid) or the aminoacyl end of the tRNA, and promote a hydrolytic reaction. If the hydrolytic reaction takes place prior to transfer of the activated amino acid to the tRNA, the aminoacyl-adenylate itself is hydrolyzed (pre-transfer); if the hydrolytic reaction occurs after transfer of the amino acid to the tRNA, the aminoacyl linkage between the amino acid and the tRNA is hydrolyzed (post-transfer). These quality control steps are referred to as editing, and both pre- and post-transfer versions are outlined in Figure 11.8b (steps 2 and 3). Typically, the cognate aminoacyl-adenylates or aminoacyl-tRNAs cannot enter the editing site and so are spared from the editing activity (Figure 11.8b, step 4).

Let us consider one well-documented example: valine is smaller, but chemically similar to the isoleucine side chain, and as such is readily activated and transferred onto $tRNA^{Ile}$ by the isoleucyl-tRNA synthetase (IleRS). This non-cognate aminoacyl-tRNA species is recognized as mismatched and the aminoacyl linkage is rapidly hydrolyzed in the editing site of the synthetase (post-transfer, as in Figure 11.8b, step 3), thus increasing the overall fidelity of the aminoacylation process.

There are two classes of aminoacyl-tRNA synthetases

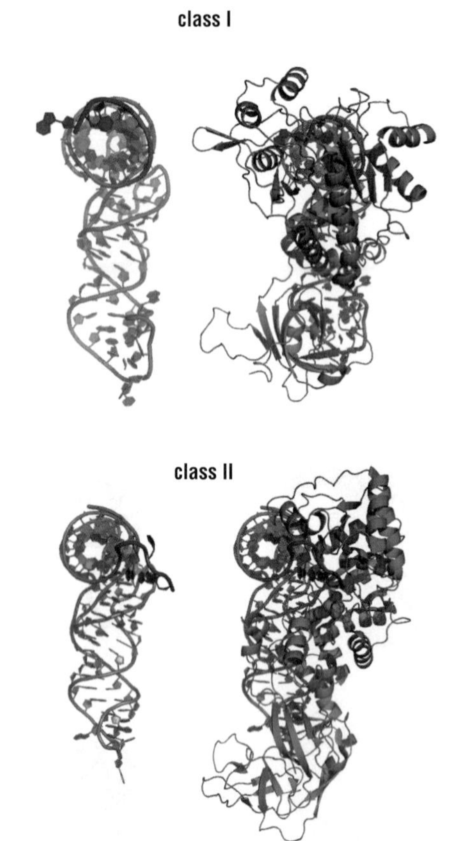

Figure 11.9 The evolutionarily distinct class I and class II synthetases recognize different faces of tRNA. The synthetases are shown in blue, the tRNA in gray, and the CCA end of the tRNA in red. The protein structures of these classes of synthetase are completely different, and recognize distinct faces of the tRNA. PDB code 1EXD, 1ASY.

Synthetases can be divided into two classes of approximately ten members each, termed class I and II. These two classes have entirely different structures and so evolved from distinct ancestors. Each binds opposite faces of tRNA as illustrated in Figure 11.9, allowing them to explore somewhat different molecular features of tRNA – class I synthetases typically recognize the minor groove of the acceptor stem, whereas the class II synthetases recognize the major groove. In addition, each aminoacyl-tRNA synthetase class has its own regio-specificity such that class I synthetases initially attach amino acids to the 2′ OH of the terminal ribose of the tRNA while class II synthetases attach them to the 3′ OH. However, since migration of the acyl group between the two positions of the ribose sugar is rapid, there is no real difference in the attachment chemistry of the aminoacyl-tRNA products released from the two synthetase classes.

It is not clear why two synthetase classes have evolved. Since we saw in Section 11.2 that cells typically produce more than 40 different tRNA species, it is possible that this large number of tRNAs can be more easily distinguished if the two synthetase classes recognize different faces of the tRNA.

Throughout the three kingdoms of life, class I and class II synthetases typically recognize the same groups of amino acids (e.g., the arginine synthetase belongs to class I whereas the alanine synthetase belongs to class II, in all species), although there are noted exceptions.

Some organisms do not have a complete set of 20 aminoacyl-tRNA synthetases

While most eukaryotic organisms have 20 synthetases, one for each standard amino acid, bacterial and archaeal organisms often have fewer than 20. Most typically in these cases, the asparaginyl and the glutaminyl synthetases are missing. A more complicated example can be found for *Methanococcus jannaschii*, where genome sequencing at first revealed only 16 of the 20 known aminoacyl-tRNA synthetase homologs. The lysyl synthetase, normally a class II enzyme, was ultimately identified as a class I synthetase (representing the first example of class switching among the synthetases) – truly missing were the asparaginyl, glutaminyl, and cysteinyl synthetases.

In a variety of organisms missing the genes encoding the asparagine and glutamine tRNA synthetases (AsnRS and GlnRS), the aspartate and glutamate tRNA synthetases (AspRS and GluRS) have dual specificity for tRNA^{Asp} and tRNA^{Asn}, or tRNA^{Glu} and tRNA^{Gln}, respectively. And, following misacylation of tRNA^{Asn} with aspartate or tRNA^{Gln} with glutamate, a **transamidase** enzyme chemically transforms the side chain of the bound amino acid from an acid to an amide, thus producing the desired asparagine- and glutamine-bound tRNA species (Asn-tRNA^{Asn} and Gln-tRNA^{Gln}), as illustrated in Figure 11.10.

What about the missing cysteine tRNA synthetase (CysRS) in *M. jannaschii*? It is now known that in *M. jannaschii*, a very unusual adaptation has occurred: a phosphorylated serine is first loaded onto the tRNA^{Cys} by a specialized class II synthetase (SepRS); and then a second enzyme, a **cysteine desulfurase**, directly converts the phosphorylated serine (on the tRNA) to a cysteine. This mode of formation of Cys-tRNA^{Cys} appears to be relatively common in archaea, as many are missing the CysRS. These exceptions to the rule of at least one synthetase per amino acid reveal interesting clues about the early evolution of tRNAs and the genetic code.

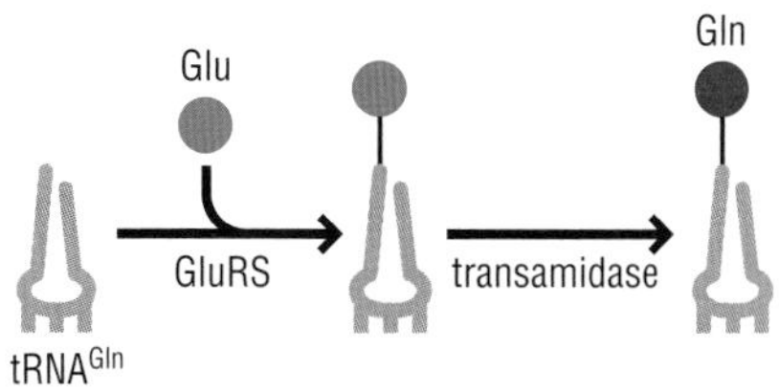

Figure 11.10 Mischarging of tRNA^{Gln} with glutamate is recovered by a transamidase enzyme. In certain organisms missing the gene encoding GlnRS, GluRS has dual specificity for both tRNA^{Glu} and tRNA^{Gln}. The product of the reaction with tRNA^{Gln}, Glu-tRNA^{Gln} is then transformed into Gln-tRNA^{Gln} through the actions of a transamidase enzyme.

11.4 STRUCTURE OF THE RIBOSOME

In the previous sections, we have seen how the link between nucleic acid and amino acid is forged by tRNA molecules, which act as mediators between the mRNA template and the growing polypeptide chain. While tRNAs are clearly the heart and soul of translation – in transducing the flow of biological information from a gene to its protein product – tRNA molecules depend for their correct function on a remarkable molecular assembly that orchestrates the overall process of translation. In the following sections, we explore this assembly: the ribosome. We begin by describing key molecular and structural features of the ribosome, before considering how it achieves the remarkable feat of coordinating the individual steps of translation.

Ribosomes are composed of two conserved subunits

The ribosome is the macromolecular machine that facilitates the interaction between the tRNAs and the mRNA and forms peptide bonds between amino acids carried by the tRNAs to yield the end product of translation: a polypeptide. The ribosome is thus responsible for protein synthesis. It is a large machine, ranging in mass from 2.5 MDa in bacteria to more than 4 MDa in eukaryotic cells, which reflects the complexity of the tasks it must carry out. In both eukaryotes and bacteria, approximately two-thirds of the ribosome's mass is composed of RNA, termed **ribosomal RNA (rRNA)**, and about one-third is composed of a collection of proteins referred to as **ribosomal proteins (r-proteins)**. We will see that the rRNAs are central to all of the activities of the ribosome. Given the ribosome's fundamental role in synthesizing protein, it is not surprising that ribosomes are highly conserved across all life forms both in structure and in function.

All ribosome particles are composed of a large subunit and a small subunit, each of which contains both rRNA and protein (detailed in Figure 11.11). The small subunit mediates the interactions between mRNA and tRNA, and the large subunit catalyzes peptide bond formation. These distinct events are integrated at the interface between the two subunits, where gross movements of the interface can shift the position of the tRNA–mRNA complex as amino acids are sequentially added to the growing polypeptide chain. These movements are coordinated by specific bridge elements between the two subunits, which are composed of RNA and proteins.

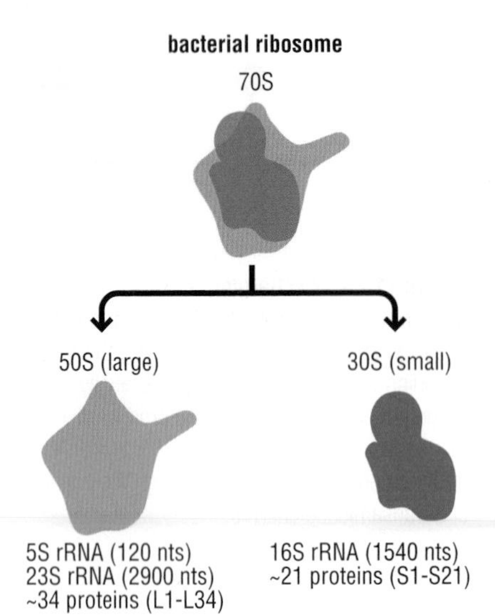

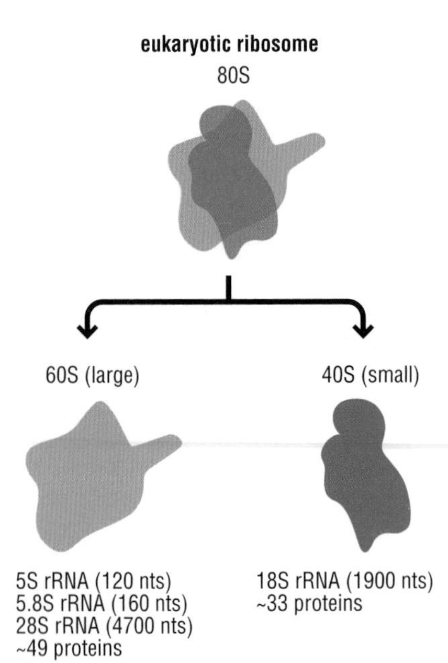

Figure 11.11 Composition of the ribosome. All ribosomes (bacterial, archaeal, and eukaryotic) are composed of one large and one small subunit. The RNA and approximate number of protein components are listed beneath each subunit for the bacteria and eukaryotes. The subunits and their components are named historically according to how fast they sediment during centrifugation: the large bacterial 50S subunit sediments faster than the small 30S subunit (where S is the Svedberg unit of sedimentation speed).

Another important structural feature of the ribosome is the exit tunnel in the large subunit through which the polypeptide is extruded as protein synthesis proceeds. We will see in Section 11.14 how many **antibiotics** function by blocking the exit tunnel and in Section 14.1 how external factors involved in protein secretion and folding assemble at the site on the exterior of the large subunit where the tunnel opens to the cytosol.

Protein and rRNA are differentially distributed on the exterior and interface surfaces of the ribosome

The interface between the small and large subunits of the ribosome, where the tRNA substrates bind and function, is rich in rRNA elements and relatively poor in proteins, while ribosomal proteins are more evenly distributed on the exterior. Figure 11.12 shows the distribution of protein and rRNA in the small and large subunits. A number of the ribosomal proteins have unusual structures, consisting of globular domains embedded in the exterior face of the subunit particle with long extended arms that snake their way into the core of the rRNA structure, as illustrated in Figure 11.13. These arms are generally enriched in basic amino acids and are thought to act as mortar in packing the negatively charged rRNA phosphate backbone into a compact tertiary structure.

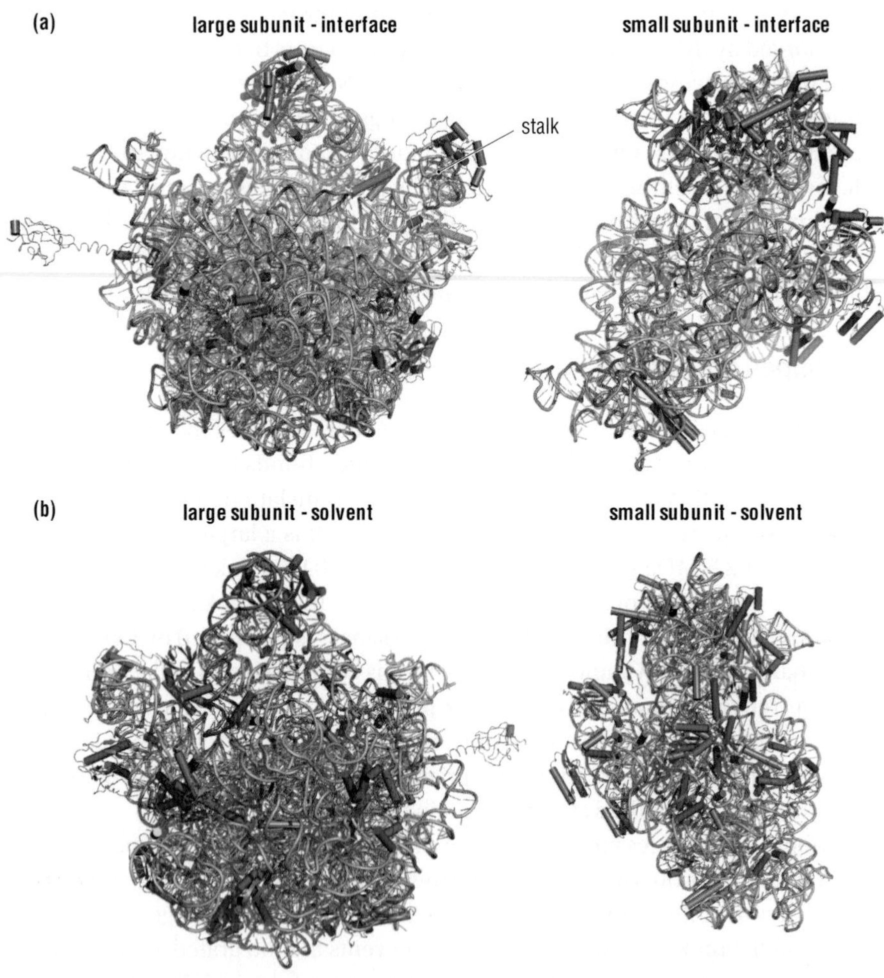

Figure 11.12 Distribution of rRNA and protein in the small and large ribosomal subunits. The interface (a) of the small and large subunits, which interact with one another and the tRNAs, is composed mainly of rRNA, while the solvent face (b) of both subunits has a greater concentration of protein. RNA is shown in gray and proteins in turquoise. PDB codes 1GIX, 1GIY.

What we have recently learned from X-ray and cryoelectron microscopy structures of diverse eukaryotic ribosomes is that additional layers of RNA and protein appear to increase the size of the ribosomes as the complexity of the organism's life style increases. We see the simplest core bacterial structure in Figure 11.14, and then increasingly adorned ribosomes as we move up the evolutionary ladder. In archaea, an additional protein layer is added, in yeast still more protein and RNA are added, and finally, in human ribosomes, so-called RNA expansion segments extend outward from the surface of the ribosome. While discrete functions for these additional layers of protein and RNA are not well defined, these additions likely add molecular handles that can facilitate increasingly complex translational regulation in these cells.

The large rRNAs from the two ribosomal subunits can be subdivided into distinct regions, or domains, based on their secondary structures. The 16S (or 18S in eukaryotes) rRNA is divided into three major and one minor domain whereas the 23S (or 28S) rRNA is divided into six different domains; these domains are illustrated in Figure 11.15a and b. The organization of the rRNA domains in the two ribosomal subunits is strikingly different. In the small subunit, each rRNA domain folds independently into a structure that meets in a central pseudoknot region, whereas the rRNA domains in the large subunit are intricately interwoven, as shown in Figure 11.15c. It

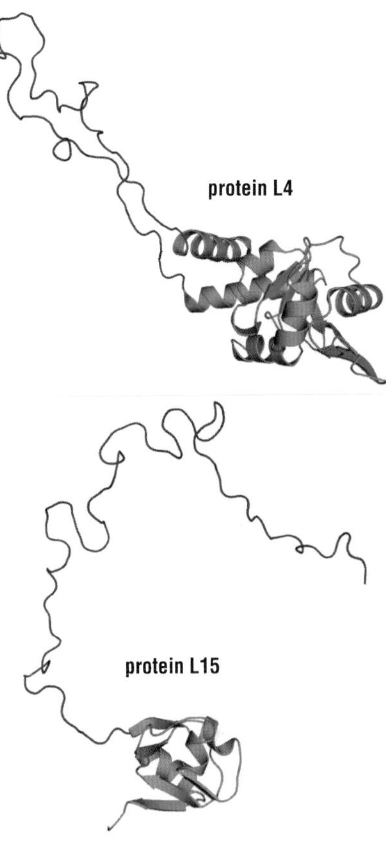

Figure 11.13 The unusual structure of ribosomal proteins. Several large subunit proteins show unusual structures with globular domains attached to extended positively charged arms. PDB code 1FIN.

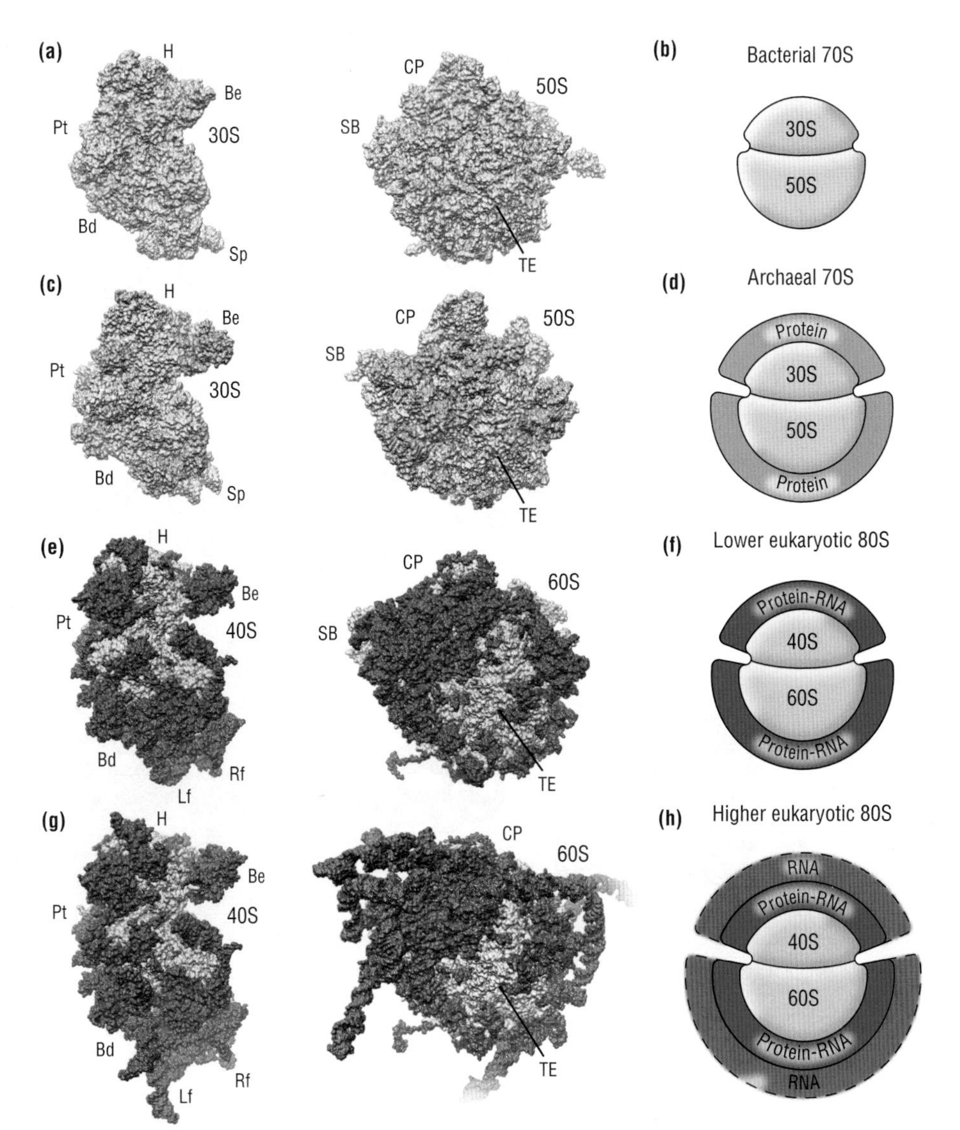

Figure 11.14 The changing composition of ribosomes throughout phylogeny. Bacterial ribosomes are in gray at top and represent the core of all ribosomes. Archaeal ribosomes have an additional layer of proteins, lower eukaryote (yeast) ribosomes have more protein and RNA in a shell around the subunit, and higher eukaryote (human) ribosomes have yet another layer of RNA that extends as 'tentacles' from the surface of both subunits.

has been suggested that this organization reflects functional differences between the subunits: interdomain flexibility is suited to large-scale movements performed by the small subunit, whereas stability provided by the interwoven RNA domains is suited to protection of the active site for protein synthesis by the large subunit.

In addition to the large rRNAs in both subunits (16S/18S and 23S/28S), the large subunit in bacteria and eukaryotes has a smaller RNA known as 5S RNA. In eukaryotes, there is a second small RNA, the 5.8S rRNA; this latter species really represents the 5′ end of the large subunit large RNA that has undergone additional processing in these cells (see Section 10.2).

The crucial roles played by rRNAs in the function of the ribosome are reflected in their remarkably conserved secondary and tertiary structure and even in the conservation of the primary structure. Certain nucleotide stretches in the rRNA are conserved across all species: for example 16 of 17 contiguous nucleotides in the large rRNA are more than 95% conserved across the three kingdoms of life. A great many ribosomal proteins also are highly conserved, and these probably are those most critical to ribosome function; a number of other ribosomal proteins are not present in all species and thus likely play less essential roles.

rRNAs are central to each of the ribosomal functions

At first glance, the observation that ribosomes are made of RNA and protein appears to raise an intriguing 'chicken-and-egg' problem: how could a machine that synthesizes

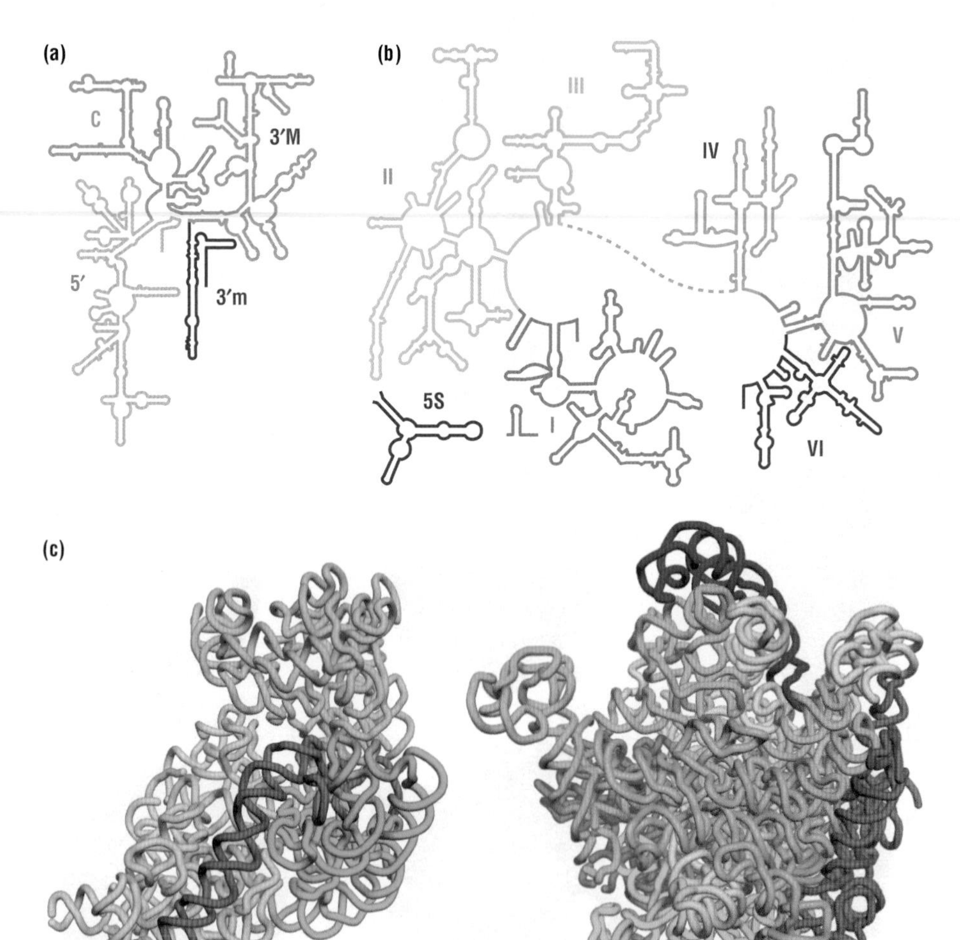

Figure 11.15 Organization of rRNA in the large and small subunits. Secondary structure of (a) the small bacterial subunit rRNA (16S) with domains labeled 5′ for 5′ domain, C for central, 3′M for 3′ major, and 3′m for 3′ minor; (b) the large bacterial subunit rRNA (23S and 5S), with the six domains labeled (5′ to 3′) I–VI; and (c) their three-dimensional organization in the two subunits, respectively. PDB codes 1GIX, 1GIY. The rRNA elements have corresponding colors in the secondary structure and three-dimensional representations. The domains of the 23S (and 5S) rRNA are more intricately interwoven than the domains of the 16S rRNA.

proteins be made of proteins before proteins actually existed? The answer to this apparent conundrum lies in the composition of the ribosome's functional center.

Importantly, the primary molecular component in the functional centers of the ribosome is RNA. In the small subunit, the so-called 'decoding' region of 16S rRNA directly mediates the interaction between tRNA and mRNA. In the large subunit, the 23S rRNA in the catalytic or **peptidyl transferase** center directly interacts with the aminoacyl end of the tRNAs. And, as there are no protein elements within 18 Å of the region where peptide bond formation occurs, it seems likely that rRNA, and not protein, catalyzes this reaction. In Chapter 10, we saw that catalytic RNAs are involved in the processing of introns in mRNA, providing clear evidence for the potential of RNA catalysis in an RNA world early in evolution. The RNA-driven ribosome arguably provides the strongest evidence for this RNA world, explaining how proteins could be synthesized before they had themselves evolved into the components of a protein synthesis machinery.

It is believed that the protein components of ribosomes were late additions to the pre-existing protein synthesis machinery, allowing protein synthesis to become increasingly efficient and accurate, and more finely regulated. Ribosomes certainly evolved early in the evolution of life because they are remarkably similar in each of the three kingdoms of life – bacteria, archaea, and eukaryotes.

11.5 THE TRANSLATION CYCLE: THE RIBOSOME IN ACTION

We have now encountered all of the key components of the cellular machinery that mediate translation, and have seen in general terms how this machinery – with the ribosome as its centerpiece – deciphers the information in an mRNA molecule to synthesize a polypeptide chain. In this and the sections that follow, we explore each step in the translation cycle at a higher resolution to reveal the molecular mechanisms that underpin the processes we have described in only quite general terms so far.

Ribosomes have three binding sites for tRNAs

We saw earlier in this chapter how the small ribosomal subunit is responsible for binding the mRNA and facilitating interactions between the mRNA codons and the anticodons of aminoacyl-tRNAs, and how the large subunit of the ribosome is responsible for catalyzing peptide bond formation between amino acids. A polypeptide chain is synthesized as the ribosome moves from codon to codon along an mRNA template. The ribosome binds an aminoacyl-tRNA at each step along the way and transfers its attached amino acid to the end of a growing polypeptide chain. The mRNA is threaded through the ribosome in the 5′ to 3′ direction, and the polypeptide chain is synthesized from the N-terminus to the C-terminus, leaving the ribosome through the exit channel or tunnel. The subunit interface region of each ribosomal subunit has three binding sites for tRNA as they proceed through the cycle, shown nicely through cryoelectron microscopy in Figure 11.16.

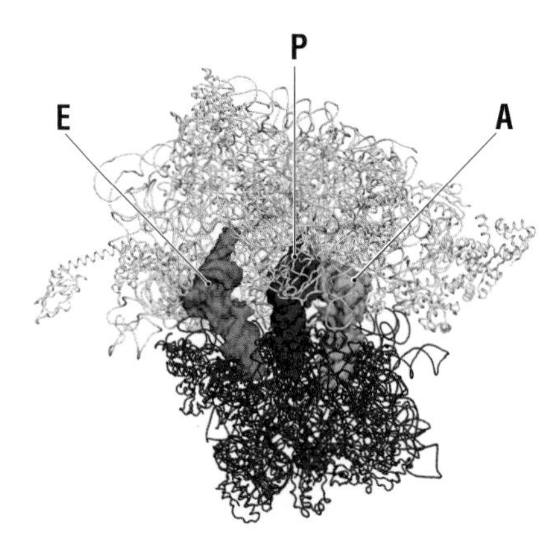

Figure 11.16 Ribosomes have three binding sites for tRNA. The tRNAs are bound to ribosomes in the RNA-rich interface between the large (light gray) and small (dark gray) subunits. This molecular model derived from cryoelectron microscopy experiments shows three tRNAs (orange, red and green) bound in the exit (E), peptidyl (P), and aminoacyl (A) sites of the ribosome. PDB codes 1GIX, 1GIY.

Once translation is underway, each aminoacyl-tRNA first binds in the **aminoacyl (A) site** before moving to the **peptidyl (P) site** after a peptide bond is formed between its amino acid and the growing polypeptide chain. A third site, the **exit (E) site**, is occupied by each tRNA (now without an amino acid) as it leaves the ribosome. In the A and P sites, but possibly not in the E site, there is a requirement for complementary base-pairing between the tRNA and mRNA for active and accurate ongoing translation.

With a picture of the structure of a ribosome – its two subunits and three tRNA binding sites – in mind, we can now look at the basic steps of translation from the beginning to the end of protein synthesis, a process called the translation cycle.

Translation of a message involves initiation, elongation, termination, and ribosome recycling

Scan here to watch a video animation explaining more about translation initiation, elongation, and termination, or find it via the **Online Resource Center** at www.oxfordtextbooks.co.uk/orc/craig2e/.

Figure 11.17 shows the four basic steps of translation: initiation, elongation, termination, and ribosome recycling. These steps are briefly outlined here and will be considered in more detail in Sections 11.7–11.11. Each step in translation is facilitated by protein factors that help to ensure high accuracy and efficiency – translation can proceed in the absence of these factors on certain simple messages but at greatly reduced speed and accuracy.

During the first step, **initiation**, the AUG codon that signals the start of an open reading frame (ORF; the protein-coding region of the mRNA) is identified by a process that involves the interaction of the ribosome, a specialized initiator methionine tRNA, and proteins called **initiation factors (IFs)**. Initiation effectively decodes the AUG start codon and generates a ribosome with a methionine-loaded $tRNA^{Met}$ bound in the P site; the ribosome is now poised to move along the ORF as shown in the first image of Figure 11.17. While the early steps of initiation involve only the small subunit of the ribosome, the final step in the process involves the joining of the large ribosomal subunit. At this point elongation proceeds, freeing up the AUG **initiation codon** for the next round of initiation. Indeed, initiation does typically occur again and again, thus piling up ribosomes on a single mRNA, which become distributed along the ORF. These large ribosome-loaded mRNA complexes are referred to as polysomes and are easily visualized by electron microscopy, as demonstrated in Figure 11.18. The details of initiation vary substantially between bacteria and eukaryotes. For example, only three initiation factors appear to be critical in bacteria while as many as 28 polypeptides are critical in a mammalian eukaryotic system.

Following initiation, the **elongation** cycle can begin wherein amino acids are sequentially added to a growing polypeptide chain as the ribosome moves directionally (5′ to 3′) along the mRNA. The elongation cycle consists of three basic steps. First, the codon of the mRNA located in the A site of the ribosome specifies which aminoacyl-tRNA species is to be loaded next into the ribosome (Figure 11.17, step 1). In Section 11.3, we mentioned that free aminoacyl-tRNAs in cells are always complexed with a specific protein, which protects the linkage between the amino acid and tRNA from hydrolysis. This **elongation factor**, called EFTu in bacteria and eEF1A in eukaryotes, plays an essential role in loading the aminoacyl-tRNAs into the ribosome as they use the energy of GTP hydrolysis to 'evaluate' the interaction between the codon and anticodon (to determine whether it is a match), and deposit the appropriate aminoacyl-tRNA in the A site.

The second step of elongation is the catalysis of peptide bond formation, which joins the amino acid brought into the A site by its cognate tRNA with the growing polypeptide chain. Before formation of the peptide bond, the polypeptide chain being synthesized is attached to the tRNA located in the P site; this polypeptide–tRNA species is called a peptidyl-tRNA. During peptide bond formation, this growing polypeptide chain is transferred to the amino acid attached to the aminoacyl-tRNA bound in the A site. This peptide bond forming reaction is catalyzed by the peptidyl transfer center (or active site) of the large ribosomal subunit (Figure 11.17, step 2). The energetic driving force for peptide bond formation comes from the ATP that was used in the activation of the amino acids by the aminoacyl-tRNA synthetases.

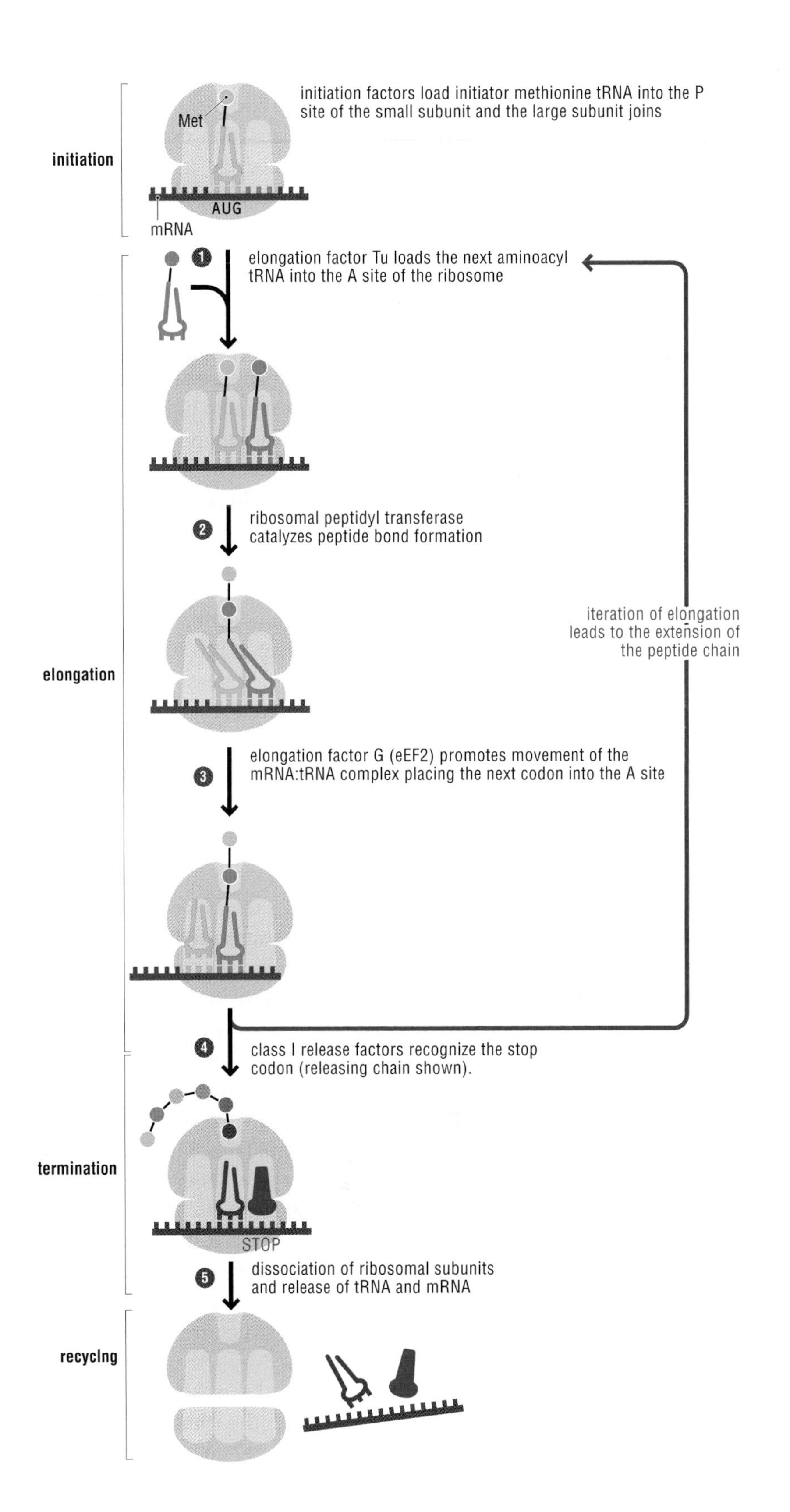

initiation
Met
AUG
mRNA
initiation factors load initiator methionine tRNA into the P site of the small subunit and the large subunit joins
elongation
1 elongation factor Tu loads the next aminoacyl tRNA into the A site of the ribosome
2 ribosomal peptidyl transferase catalyzes peptide bond formation
iteration of elongation leads to the extension of the peptide chain
3 elongation factor G (eEF2) promotes movement of the mRNA:tRNA complex placing the next codon into the A site
4 class I release factors recognize the stop codon (releasing chain shown).
termination
STOP
5 dissociation of ribosomal subunits and release of tRNA and mRNA
recycling

Figure 11.17 The translation cycle. The basic steps of translation are initiation, elongation (1–3), termination (4), and recycling (5), which are described in the main text. Briefly, the first amino acid of a polypeptide chain is generally methionine, and it is found attached to a tRNA that is loaded into the P site of the ribosome (initiation). This methionine is then attached to the subsequent amino acid at step 2, extending the polypeptide chain. The ribosome moves one codon further along the mRNA, and the elongation cycle then repeats (pink arrow). When a stop codon is reached in the mRNA, termination takes place (4) wherein the completed polypeptide chain is released from the ribosome. Translation ends when the ribosomal subunits are dissociated during ribosome recycling (5), and are thus ready to begin the synthesis of a new protein.

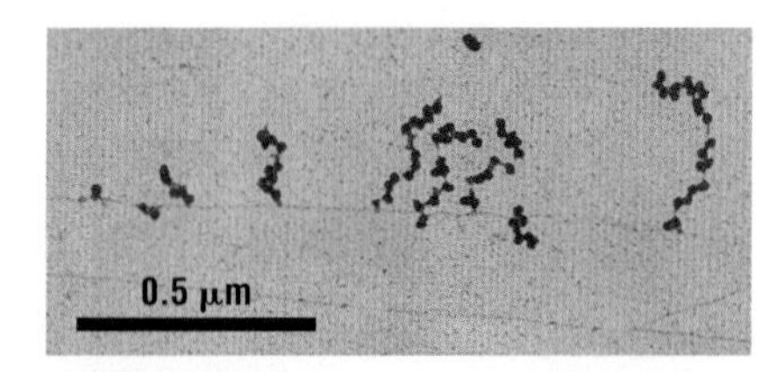

Figure 11.18 Polysomes form when multiple ribosomes initiate and elongate multiple mRNAs from the same gene. The polysomes here are visualized by electron microscopy: the thin strand is the DNA, and the blobs emerging from it are ribosomes bound to newly synthesized mRNA transcripts. Multiple ribosomes are found on each extending transcript.

From Miller, OL, Hamkalo, BA, and Thomas, CA. Visualization of bacterial genes in action. *Science*, 1970;**169**:392–395. Reprinted with permission from AAAS.

Finally, another GTP-dependent elongation factor (EFG in bacteria and eEF2 in eukaryotes) promotes the three-nucleotide movement of the mRNA–tRNA complex through the ribosome (Figure 11.17, step 3), bringing the newly minted peptidyl-tRNA fully into the P site and the next codon into the A site. This process, termed **translocation**, effectively opens the A site for the next incoming aminoacyl-tRNA. The elongation cycle then repeats.

At the end of the ORF, the ribosome reaches a stop codon (UAA, UAG, or UGA) in the mRNA. This signals the next step in the translation cycle, **termination**, when the fully synthesized polypeptide chain is released from the ribosome. Stop codons are not recognized by specific tRNAs but by protein factors known as **class 1 release factors** (Figure 11.17, step 4). In bacteria, there are two class 1 release factors; RF1 recognizes the codons UAA and UAG, and RF2 recognizes the codons UAA and UGA. In eukaryotes, there is a single class 1 release factor, eRF1, which recognizes all three stop codons. The interaction between the release factor and the ribosome promotes release of the completed peptide chain from the final peptidyl-tRNA, allowing the peptide chain to diffuse away from the ribosome. There is an additional GTPase involved in termination both in bacteria and eukaryotes, RF3 and eRF3, respectively. These GTPases are referred to as **class 2 release factors** though their mechanisms of action appear to be rather distinctive. (We explore these mechanisms of action in more detail in Section 11.10.)

In the final step of translation, **ribosome recycling**, the large and small ribosomal subunits dissociate from each other, releasing the remaining peptide-less tRNA and mRNA (Figure 11.17, step 5). In bacteria, a protein known as ribosome recycling factor (RRF) works together with EFG to promote ribosome recycling; in eukaryotes recycling is promoted by an ATPase known as Rli1 in yeast (or ABCE1 in humans) working together with the still bound class 1 release factor eRF1. After subunit dissociation, the liberated ribosomal subunits can then bind again to initiation factors in preparation for new rounds of protein synthesis.

The hybrid states model explains how tRNAs move through the ribosome

How do tRNAs move through the three tRNA binding sites on the ribosome during the translation cycle? One idea for this is the **hybrid states model**, in which tRNA molecules ratchet their way through the subunit interface region of the ribosome, always maintaining contact with one subunit while moving with respect to the other. In this model, the tRNAs move first with respect to the large subunit directly following peptide bond formation. EFG then promotes the movement of the anticodon ends of the tRNAs (still associated with the mRNA!) with respect to the small subunit of the ribosome (Figure 11.17, steps 2 and 3). We refer to a ribosome with tRNAs straddling different binding sites on the ribosomal subunits as a hybrid state (the panel between steps 2 and 3 in Figure 11.17); when tRNAs are bound in this state, the ribosome is also in a different state that we refer to as 'rotated'. After these two independent movements of the tRNAs have taken place, the A site is fully accessible for the next round of elongation to continue. As the tRNAs move, the ribosomal subunits themselves move with respect to one another, oscillating between a 'classic' and a 'ratched' state, using their two subunit structure to great advantage.

The ordered process of translation is maintained by the oscillating movements of the ribosomal subunits and the two ends of the tRNAs, during which the tRNAs are never fully released from contact with the ribosome. Initial support for the hybrid state of tRNA binding came from chemical modification experiments as outlined in Experimental approach 11.1. These results have been

11.1 EXPERIMENTAL APPROACH

Chemical modification analysis to study RNA structure and function without high-resolution structures

Chemical modification analysis can be broadly used to study the structure and function of RNA molecules

Chemical modification analysis has been extensively used to define molecular features of a large number of biologically interesting RNAs and ribonucleoproteins (RNPs). These approaches were especially useful for early analysis of the ribosome structure and function because more high-resolution approaches, such as x-ray crystallography, were not then technically feasible.

The basic idea behind chemical modification analysis is to bombard an RNA species of interest with chemical reagents that attack and chemically modify RNA elements such that the sites of modification can be easily identified. Accessible RNA sites will be modified, whereas those found within the core of the macromolecule, or those that are protected by various ligands, will not. The sites of modification are typically identified as pauses or stops on a sequencing gel that follows the progress of a reverse transcription reaction as it extends primers along the chosen (rRNA) template. The strength of a given band in the primer extension gel is dependent on whether that site has been modified, and how extensively. If a given RNA molecule escapes modification, reverse transcriptase should fully extend the primer, yielding a full-length complementary DNA strand. If the RNA is modified, reverse transcription will be halted at the modified site, yielding a visibly shorter band on the gel. This general approach is described more thoroughly in Section 19.8.

The hybrid state of tRNA binding on the ribosome is defined by chemical modification analysis

How do these methods allow the researcher to identify positions of functional importance? Well, when different ligands are bound to the ribosome, for example, specific regions are protected from chemical modification, thus identifying potential ligand interaction sites. In an informative study by Harry Noller and colleagues in 1989, structural probing was performed on a series of ribosomal complexes containing various tRNA substrates stalled at different points in the translational cycle. These experiments provided a structural view of tRNA movements within the ribosome, and, surprisingly, indicated that tRNAs can move independently with respect to the large and small subunits of the ribosome. The intermediate tRNA binding state (where tRNA has moved with respect to the large, but not the small, subunit) was referred to as the 'hybrid' state of binding (as discussed in Section 11.5).

Representative data from these experiments are shown in Figure 1. In the primer extension sequencing gels, we see that several different nucleotide modification patterns can be followed within lanes documenting the modification patterns of the 16S and 23S rRNAs found in the small and large ribosomal subunits, respectively. Lanes 1 through 5 contain a set of ribosome complexes with tRNAs bound during different stages in the elongation cycle, and we see that the patterns of accessibility change in response to distinctly bound ligands. When empty ribosomes – with nothing bound to them – are bombarded with chemical reagents (lane 1) we see a number of potential ligand interaction sites being modified. In lanes 2–5, ribosomes with tRNA ligands bound show different modification patterns – for example, nucleotide 926 (Figure 1b) is protected from modification when any tRNA substrate is bound to the ribosome (lanes 2–5), while the modification patterns for U2585 and A1492/93 (Figure 1a and c, respectively) show variability in modification that is dependent on whether peptide bond formation and/or translocation have occurred. These different modification patterns ultimately provided Noller and colleagues with experimental support for the hybrid states model of tRNA movement.

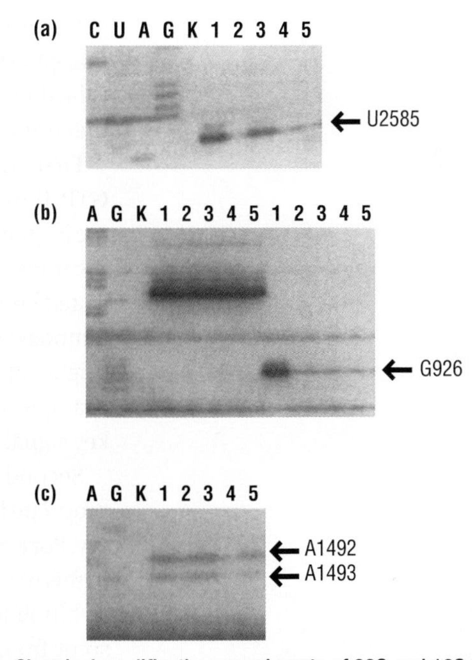

Figure 1 Chemical modification experiments of 23S and 16S rRNA with CMCT (U specific, panel (a)), kethoxal (G specific, panel (b)), and DMS (A and C specific, panel (c)). Panel (a) examines a region in 23S rRNA while (b) and (c) examine two regions in the 16S rRNA. Lane 1, 70S ribosomes and polyU; lane 2, 70S, polyU, N-Ac-Phe-tRNA^{Phe}; lane 3, 70S, polyU, deacylated tRNA^{Phe}; lane 4, 70S, polyU, deacylated tRNA^{Phe}, N-Ac-Phe-tRNA^{Phe}; lane 5, 70S, polyU, deacylated tRNA^{Phe}, N-Ac-Phe-tRNA^{Phe} plus EFG:GTP. C, U, A, and G refer to sequencing lanes, K to unmodified control.

Reproduced from Moazed and Noller, Intermediate states in the movement of transfer RNA in the ribosome. *Nature*, 1989;**342**:142–148.

This study provides an excellent example of the power of chemical modification analysis to assess properties of functional RNAs at the nucleotide level. For the ribosome, such approaches successfully identified most of the functionally critical rRNA nucleotides long before high-resolution structures clarified their three-dimensional location and function. Even now, when high-resolution structures of the ribosome provide increasingly detailed atomic insight into core molecular features, chemical modification approaches remain useful as relatively complicated biological questions can be addressed in the absence of a perfectly behaved homogeneous (and crystallize-able) sample. Recent studies have further expanded the application of these approaches by combining structural probing with kinetic analysis to watch in real time the assembly of large macromolecular complexes. Such studies are particularly useful still as they can address questions where high resolution structures remain elusive.

Find out more

Moazed D, Noller HF. Intermediate states in the movement of transfer RNA in the ribosome. *Nature*, 1989;**342**:142–148.

Related techniques

Chemical modification analysis; Section 19.8
Sequencing gels; Section 19.10
Primer extension; Section 19.10

further supported by a number of independent approaches, including cryoelectron microscopy to visualize the 'hybrid' state and detailed bulk and single-molecule based kinetic studies.

11.6 PROTEIN FACTORS CRITICAL TO THE TRANSLATION CYCLE

The translation cycle is facilitated by the actions of protein factors

In the brief overview of the translation cycle in the preceding section, we have seen that each step in the cycle, with the possible exception of peptide bond formation, is facilitated by the action of protein factors. Broadly speaking, there are two ways that translation factors act:

First, many factors are GTPases (or G proteins), which catalyze the hydrolysis of GTP. The conformational changes that these factors undergo during GTP hydrolysis are intimately linked with the orderly, stepwise progression of the ribosome through the translation cycle. EFTu, for example, depends on GTP hydrolysis (and the associated conformational rearrangements) in its role as guardian and deliverer of the aminoacyl-tRNAs during elongation. Such G proteins represent the evolutionary origins of many important players in cellular signal transduction and cancer. For example, the oncogene protein Ras, a well-characterized, small GTPase involved in key signal transduction events in eukaryotic cells, is homologous with EFTu.

Second, some translation factors simply bind to the ribosome and prevent inappropriate interactions with tRNAs or other components of the translational machinery. For example, in bacteria, the initiation factors IF1 and IF3 are believed to function in this manner by preventing binding of the **initiator tRNA** to the A site (thus favoring its binding to the P site) and by preventing the large and small subunits of the ribosome from associating before the initiator tRNA has been loaded into its binding site.

An elongation factor that we introduced in Section 11.5, EFG (or eEF2 in eukaryotes), triggers the movement of the mRNA–tRNA complex through the ribosome during the elongation phase of translation. EFG both mimics the structure of tRNA and is a GTPase. In later sections, we look in more detail at how translation factors act during the translation cycle; but first we will outline further the general principles by which GTPase activities and molecular mimicry are thought to facilitate different steps in the translation cycle.

The evolutionarily related GTPases interact with the protein-rich flexible stalk region of the ribosome

GTPases act at several steps during translation. During initiation, eukaryotic eIF2 is required for loading Met-tRNA$_i^{Met}$ into the AUG-coded P site of the ribosome while bacterial IF2 (or eIF5B in eukaryotes) is required for subunit joining. During elongation, EFTu (eEF1A) is critical for depositing the cognate aminoacyl-tRNAs into the A site, EFG (eEF2) promotes the directional movement of the mRNA–tRNA complex through the ribosome and is critical for ribosome recycling, while RF3 (eRF3) is involved in the termination of protein synthesis.

These GTPases are all related and contain a **P loop motif** (Gly-X-X-X-X-Gly-Lys) common to many nucleotide-binding proteins, including Ras. This motif specifically interacts with the phosphate groups of GTP and undergoes substantial conformational changes on GTP hydrolysis, as depicted in Figure 11.19.

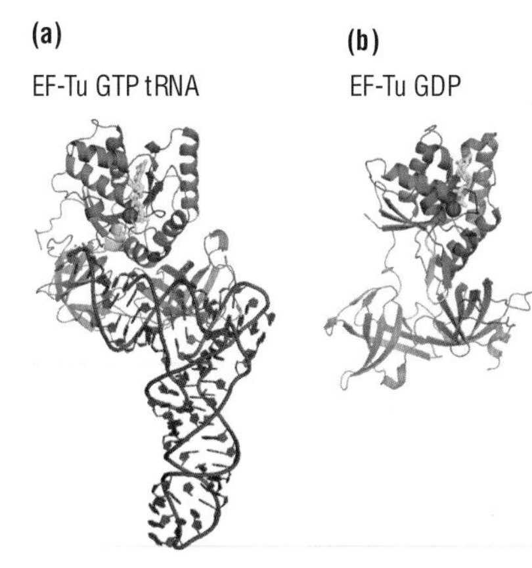

Figure 11.19 The P loop in GTPases undergoes a large structural rearrangement upon GTP hydrolysis. The GTPase in this example is EFTu (blue), shown with bound GTP (a) and GDP (b). An aminoacyl-tRNA (gray) that is bound to EFTu (in its GTP bound form) interacts with the P loop (green). The shift in position of the P loop following GTP hydrolysis causes EFTu to dissociate from the aminoacyl-tRNA. PDB codes 1D2E, 1TTT.

As with other G proteins, the identity of the bound nucleotide (guanosine triphosphate (GTP) or guanosine diphosphate (GDP)) dictates the conformation of the protein, and thus the ligands with which it interacts. In translation, the ligands that the GTPases recognize are the distinct states of the ribosome during the translation cycle. So, for example, EFG in its GTP-bound (active) state recognizes pre-translocation state ribosomes (that is, ribosomes ready to move to the next mRNA codon) whereas eIF2 in its GTP-bound state recognizes small ribosomal subunits ready to accept a P-site tRNA. The GDP forms of translation factors typically dissociate from the ribosome after they have performed their functions.

As with related GTPases, both the background rate of GTP hydrolysis and the rate of exchange of GDP for GTP can be slow. In situations where GTP hydrolysis or exchange is slow, the reactions are typically promoted by **GTPase-activating proteins (GAPs)** and **guanine-nucleotide exchange factors (GEFs)** respectively, as represented schematically in Figure 11.20. A number of additional translation factors play essential roles in translation by acting as GAPs or GEFs. For example, bacterial EFTs (related to eukaryotic eEF1Balpha) is the GEF for EFTu (or eEF1A) while eukaryotic eIF5 is the GAP for eIF2 (that loads tRNA into the P site during initiation in eukaryotes).

In some cases, the ribosome itself functions as a GAP (or potentially even as a GEF). The best-documented example of this is with EFTu: once a cognate aminoacyl-tRNA is base-paired with the codon in the ribosome's A site, portions of the ribosome promote the hydrolysis of GTP by EFTu, thus promoting full acceptance of the aminoacyl-tRNA into the A site and release of EFTu-GDP.

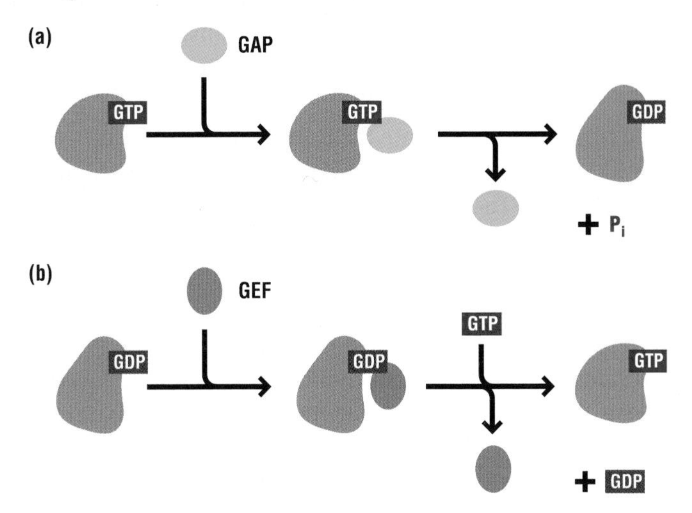

Figure 11.20 GAPs and GEFs. Various protein factors and the ribosome itself are thought to function as the (a) GAPs and (b) GEFs for various GTPases involved in translation.

The GTPases involved in translation appear to interact extensively with the protein-rich flexible stalk region of the large ribosomal subunit. The rRNA portion of the stalk region is seen as the prominent extension on the right in Figure 11.12; the full extent of the protein-rich component of the stalk is not seen since the protein components are not well resolved in the various atomic resolution structures determined to date. As a group, the GTPases probably communicate with the interior of the ribosome via extended protein domains, associated protein factors, or tRNA substrates, and ultimately bring about distinct functional consequences (e.g., translocation, tRNA loading, class 1 release factor dissociation).

Protein factors are not absolutely required for very minimal translation *in vitro*; such reactions can proceed slowly in their absence on certain unstructured mRNAs such as polyuridine (encoding poly-phenylalanine). As a whole, the factors are highly conserved and it is thought that they evolved to contribute speed, precision, and processivity to a crude but already functioning translation apparatus that initially depended only on the coordinated interactions of tRNAs, mRNAs, and the ribosome itself. Once a single GTPase became integrated into the translation cycle, related GTPases could be derived from this original template molecule allowing for the evolution of increasingly complex regulation of the process. We will see examples of such regulation later in this chapter.

Other nucleotide triphosphate hydrolyzing proteins promote different steps in the translational cycle

In addition to the extensive family of translational GTPases, a number of different nucleotide triphosphate hydrolyzing enzymes have also been implicated in other steps of the translation cycle. For example, eIF4A is a 'DEAD-box helicase' (named for conserved amino acids that are important for the nucleotide hydrolysis activity) that is important during eukaryotic translation initiation, likely unwinding inhibitory mRNA structures that impede the ribosome. The ABC-type ATPase Rli1 in yeast (or ABCE1 in humans) plays an essential role in eukaryotic ribosome recycling, using the energy of ATP hydrolysis to promote subunit dissociation.

Molecular mimicry of tRNA may be used by some translation factors to access the interior of the ribosome

One striking feature of several translation factors, including some GTPases, is their apparent mimicry of the shape and size of tRNA substrates of translation. Indeed, looking like a tRNA seems to be an efficient way for translation factors to access the functionally critical interface region of the ribosome. Perhaps the most compelling example of tRNA mimicry is seen with the elongation factor (EFG), which binds in the A site of the ribosome to promote the translocation of the mRNA–tRNA complex. Figure 11.21 shows how there is marked similarity in the shape of EFG and the shape of the tRNA–EFTu–GTP complex, both of which are known to function bound within the A site of the ribosome. In functional and structural terms, the class 1 release factors also mimic the A site tRNA in their role in termination; we will explore these similarities more thoroughly in Section 11.11.

Having seen how the translation cycle is facilitated by various protein factors, let us now consider the molecular mechanisms that are employed during each stage in this cycle. We start by considering the first stage, that of initiation.

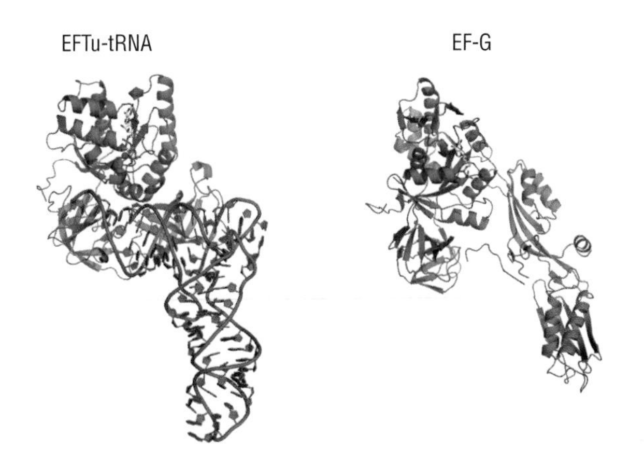

Figure 11.21 Molecular mimicry of tRNAs by translation factors. EFG shares overall structural features with the aminoacyl-tRNA–EFTu complex. In the view shown, domain IV of EFG is colored in gray to highlight its similarity to the tRNA bound to EFTu (also in gray). This structural similarity is consistent with the fact that both molecules bind in the A site of the ribosome. PDB codes 1TTT, 1DAR.

11.7 TRANSLATION INITIATION – SHARED FEATURES IN BACTERIA AND EUKARYOTES

During initiation, a methionyl-tRNA and the AUG start site of the mRNA engage one another in the P site of the small ribosomal subunit. The initiation step of translation is the process by which a ribosome identifies the translation start site in an mRNA and traps it in the P site with an initiating methionyl-tRNA in preparation for the beginning of the elongation cycle (see Figure 11.17). The consequence of these steps is that the first codon of the ORF is decoded as methionine and polypeptide synthesis can begin when the next aminoacyl-tRNA is loaded into the A site of the ribosome.

There are three main tasks that take place during translation initiation. First, the small ribosomal subunit must identify the first codon of an ORF, the **initiation codon**, within the mRNA. Second, a methionyl-tRNA must be loaded into the P site of the ribosome where it base-pairs with the initiation codon. Third, the large ribosomal subunit must join the small subunit that now has the methionyl-tRNA positioned in the P site. Both bacteria and eukaryotes must achieve these three tasks, but bacteria and eukaryotes have differently structured genes, and the details of how initiation occurs in these organisms differ accordingly.

Initiator tRNAs have distinctive features

In both eukaryotes and bacteria, the initiation codon is typically AUG (which specifies methionine) and is decoded by a specialized tRNA species called the **initiator tRNA**. These initiator molecules, $tRNA^{fMet}$ in bacteria and $tRNA_i^{Met}$ in eukaryotes, are distinct from other tRNAs in several ways that prevent them from associating with EFTu (or eukaryotic eEF1A), the protein which escorts aminoacyl-tRNAs to the A site during elongation. The lower case 'f' on the bacterial initiator tRNA refers to the 'formyl' moiety found on the amine function of the methionine of the aminoacylated tRNA species (formyl-Met-$tRNA^{fMet}$) as illustrated in Figure 11.22a. The formyl group is added to the methionine by a transformylase enzyme that recognizes Met-$tRNA^{fMet}$ following the aminoacylation reaction. This formyl group functions to stabilize the ester linkage between the amino acid and the tRNA body. In so doing, the stabilizing formyl group found in the bacterial system may help to compensate for protection usually afforded to aminoacylated tRNAs by EFTu/eEF1A (for elongator tRNAs) or by eIF2 (for eukaryotic initiator tRNA).

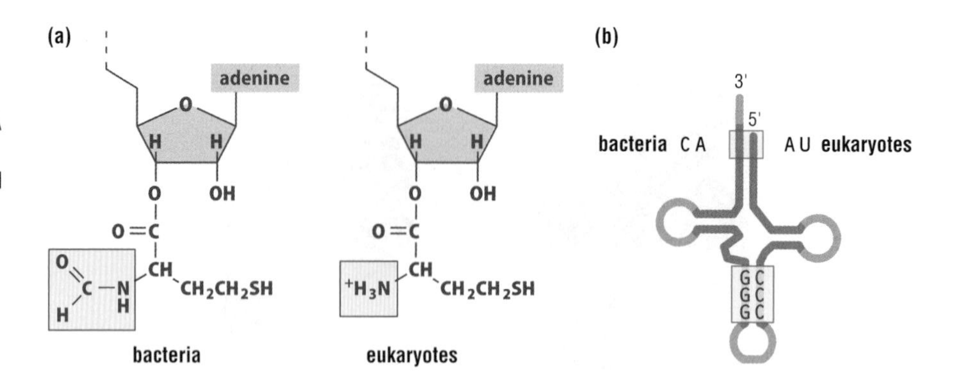

Figure 11.22 Features of the initiator tRNAs in bacteria and eukaryotes. (a) Bacterial and eukaryotic initiator tRNAs are different in their chemical composition, as the bacterial Met-tRNA species is formylated on the amine functionality while the eukaryotic Met-tRNA is not (highlighted boxes). (b) At the primary sequence level, both bacterial and eukaryotic initiator tRNAs have three consecutive GC base pairs in the anticodon stem, and unusual (for tRNAs) base pairs at the top of the acceptor stem (CA and AU, respectively) that prevent binding to EFTu (eEF1A).

In bacteria, the identity elements of the initiator tRNA^{fMet} include a C–A wobble mismatch at the start of the acceptor stem and a stretch of three consecutive G–C base pairs in the anticodon stem. Eukaryotic initiator tRNA identity elements include the same consecutive G–C base pairs in the anticodon stem, and an A–U base pair at the start of the acceptor stem, as depicted in Figure 11.22b. In each system, these identity elements are important anti-determinants for binding of the initiator tRNAs to EFTu (eEF1A), so these tRNAs are not loaded into the A site by the normal pathway for elongator tRNAs.

How then are initiator tRNAs loaded directly into the P site in these two systems? In eukaryotes, the Met-tRNA$_i^{Met}$ is bound by a GTPase, eIF2, that plays a direct role in start codon recognition and loading of the P site. In bacteria, there is no known homolog of this protein, though a different (i.e., non-homologous) GTPase, IF2, likely plays an important role in directing binding of f-Met-tRNA^{fMet} to the AUG-programmed P site.

In the subsequent section (Section 11.8), we shall describe what is known of the features of the initiation process in bacteria, and in Section 11.9, we shall see how in eukaryotes initiation proceeds in a similar yet distinct way. In the following chapter (Chapter 12), we will focus on the many layers of regulation that target translation initiation in bacteria and eukaryotes.

11.8 BACTERIAL TRANSLATION INITIATION

Locating the AUG start site in bacteria is relatively simple

The first step in translation initiation is to identify the AUG that signals the start of the intended ORF. Figure 11.23 shows the basic organization of a typical bacterial mRNA. Bacterial mRNA molecules are often **polycistronic**, meaning that they contain multiple ORFs, each encoding a different protein. Each ORF in the mRNA has an initiation and a termination codon; we will see in Section 11.11 how ribosomes are able to process each reading frame either independently, or, sometimes, in a coupled fashion. For now, we will simply consider the means by which bacterial ribosomes identify the initiation codons within these multicistronic mRNAs.

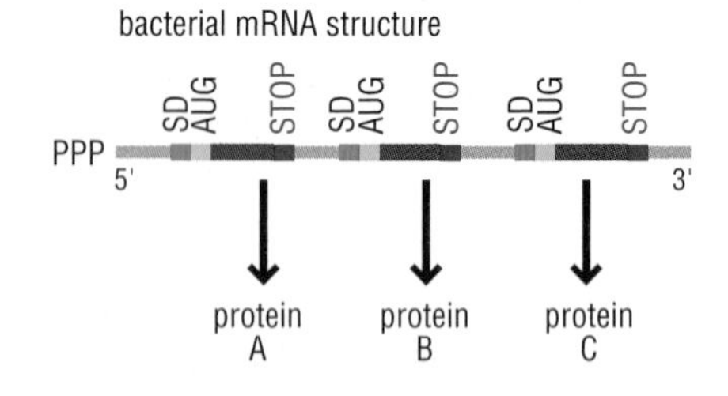

Figure 11.23 General features of bacterial mRNAs. Bacterial transcripts often encode more than one ORF (that is, they are polycistronic). Each ORF begins with an AUG initiation codon (green) and ends with a stop codon (red). The initiation codons are typically preceded by a Shine–Dalgarno sequence (blue) that is recognized by the 16S rRNA (anti-Shine–Dalgarno sequence) of the small ribosomal subunit during initiation.

Initiation codons (AUG, or sometimes GUG) in bacteria are most typically accompanied by a sequence element in the mRNA called a **Shine–Dalgarno sequence** (or **ribosome-binding site**), a polypurine tract located six to eight nucleotides upstream of the AUG initiation codon. The Shine–Dalgarno sequence has the consensus sequence AGGAGGU and base-pairs with a polypyrimidine tract in the 3′ end of the bacterial 16S rRNA (the **anti-Shine–Dalgarno sequence**) of the small ribosomal subunit, as shown in Figure 11.24. Deviations from the

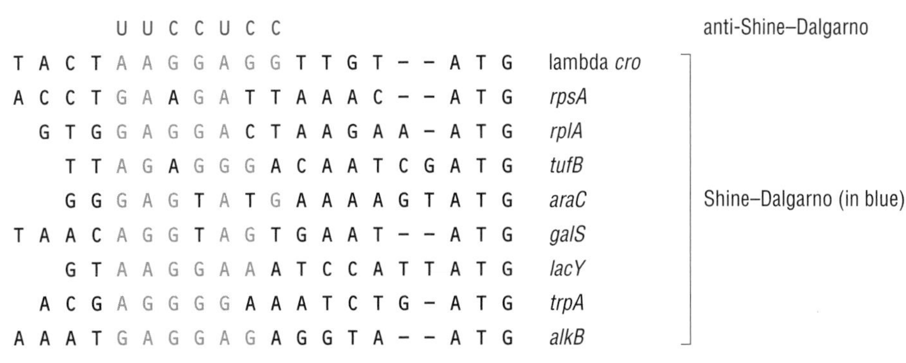

U U C C U C C
anti-Shine–Dalgarno
T A C T A A G G A G G T T G T – – A T G lambda *cro*
A C C T G A A G A T T A A A C – – A T G *rpsA*
G T G G A G G A C T A A G A A – A T G *rplA*
T T A G A G G G A C A A T C G A T G *tufB*
G G G A G T A T G A A A A G T A T G *araC*
T A A C A G G T A G T G A A T – – A T G *galS*
G T A A G G A A A T C C A T T A T G *lacY*
A C G A G G G G A A A T C T G – A T G *trpA*
A A A T G A G G A G A G G T A – – A T G *alkB*
Shine–Dalgarno (in blue)

Figure 11.24 The purine-rich Shine–Dalgarno sequence specifically interacts with a pyrimidine-rich region of the 16S rRNA. Shown here are a number of bacterial genes with varying levels of complementarity to the anti-Shine–Dalgarno region of the 16S rRNA.

consensus sequence result in weaker interactions with the ribosome and are important in dictating how efficiently a given mRNA is translated in the cell. Experimental approach 11.2 describes some of the early observations of Shine and Dalgarno and how they led to the identification of this crucial interaction involved in bacterial translation initiation.

Base-pairing between the Shine–Dalgarno sequence and the anti-Shine–Dalgarno sequence on the small ribosomal subunit occurs and is positioned approximately six to eight nucleotides upstream of the AUG initiation codon, and effectively guides this AUG into the region of the P site of the ribosome where it is poised to engage with the initiator f-Met-tRNA^{fMet}. There is also an additional substantial class of mRNAs in bacteria (and archaea) that has no leader sequence upstream of the AUG start site; these mRNAs appear to depend on somewhat different mechanisms to initiate translation (so-called 'leaderless initiation'), although the mechanism is not yet well understood.

Initiator tRNA binding is guided by a small set of protein factors in bacteria

The initiator f-Met-tRNA^{fMet} is guided to the P site of the ribosome with the help of three initiation factors, IF1, IF2, and IF3. In the absence of mRNA or f-Met-tRNA^{fMet}, IF1 and IF3 independently associate with the small ribosomal subunit in the regions where the A and E site tRNAs usually bind, leaving primarily the P site unfilled, as shown in Figure 11.25a. These factors also appear to block sites on the small subunit that are important in forming intersubunit interactions. As such, these two proteins effectively direct the initiator tRNA to the P site and prevent the premature association of the two ribosomal subunits. The third factor, the GTPase IF2, also seems to play an important role in this early stage of initiation, helping to direct f-Met-tRNA^{fMet} to the AUG in the P site.

The large ribosomal subunit joins the small subunit in the final step in bacterial translation initiation

Once the initiator tRNA has been deposited in the P site, the large subunit joins the small subunit so that elongation can begin. The steps in this assembly process are outlined in Figure 11.25b. The GTPase IF2 couples the energy of GTP hydrolysis to this subunit joining, ultimately resulting in the displacement of all three initiation factors (IF1, IF2, and IF3) and formation of the initiation complex. At this stage, the

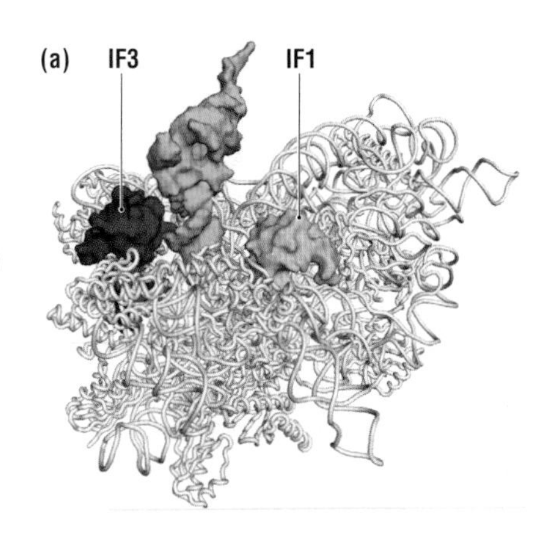

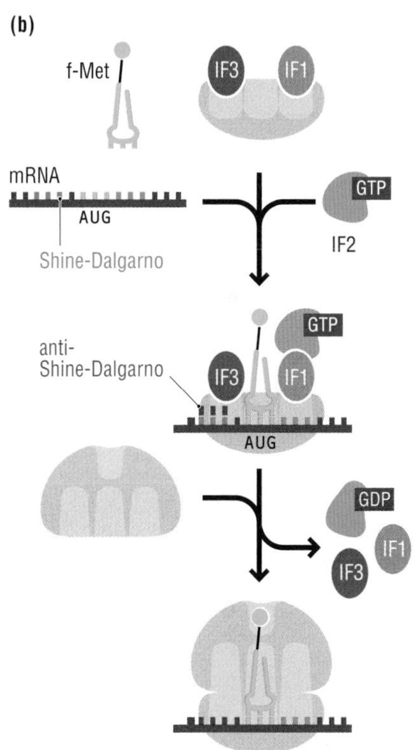

Figure 11.25 Initiation factors orchestrate bacterial initiation complex assembly. (a) IF1 and IF3 (gray and dark gray, respectively) block regions of the small ribosomal subunit where the A and E site tRNAs usually bind, in addition to blocking sites that are involved in intersubunit contacts in the 70S ribosome. These initiation factors thus help in guiding the initiator tRNA (green) into the P site. PDB codes 1GIX, 1HRO. (b) Diagram showing the ordered steps of assembly of bacterial initiation. IF1 and IF3 initially bind to the small subunit A and E sites, guiding binding of the initiator fMet-tRNA^{fMet} to the P site (with the help of IF2-GTP). The Shine–Dalgarno sequence of the mRNA directly interacts with the anti-Shine–Dalgarno sequence of the 16S rRNA to position the AUG in the P site. Once the initiator tRNA finds the AUG, IF2 facilitates subunit joining (using the energy of GTP hydrolysis), and the three IFs are released from the complex.

11.2 EXPERIMENTAL APPROACH
Phylogenetic covariation analysis

Covariation analysis is widely used to define the secondary and tertiary structural interactions in RNA species

Phylogenetic covariation analysis has been used extensively in the study of functional RNAs to identify potential interacting regions, allowing for the elucidation of secondary structures and even longer-range tertiary interactions. In general terms, this approach uses comparative sequence analysis – the analysis of related sequences, often from different organisms – to identify patterns of variation within the RNA that are non-random and instead reflect a co-dependence during evolution. For example, if two sequences interact closely through base-pairing, and one of the two sequences undergoes change, their interaction can only be sustained if the second sequence undergoes a compensatory change. If such concerted changes are observed, the two positions in the RNA are said to covary. The process of evolution will select for such concerted changes if the interaction is important for the function of the RNA.

Covariation analyses have been used with great success to determine the secondary structure of numerous RNA species beginning with the cloverleaf structure of tRNA in the 1960s (Figure 10.3), the 16S rRNA in the 1970s, and the growing body of non-coding RNAs. Such analyses even permit the identification of long-range interactions within these RNAs that stabilize their three-dimensional structures. For example, Michael Levitt saw covariations within the tRNA sequences that led him in the 1960s to propose the existence of the known base triple interactions that stabilize their elbow region. And, in perhaps the most stunning of such analyses, Francois Michel and Eric Westhof in 1990 used phylogenetic analysis to identify a number of long-range interactions in the group I intron. Their predictions were remarkably accurate; interactions that they identified in these studies represent some of the most common and important core structural motifs in large RNAs (for example, the A minor motif). Indeed, these motifs have since then been observed repeatedly in atomic resolution structures of the group I and II introns and the ribosome.

Discovery of the bacterial ribosome binding site through covariation analysis

In an example pertinent to this chapter on translation, two researchers, John Shine and Lynn Dalgarno, were exploring sequence variations in the pyrimidine-rich 3′ end of the 16S rRNA of the small subunit of the ribosome, whose structure is depicted in Figure 1, when they noticed that variation in this sequence (in ribosomes from different organisms) correlated with the level of expression of certain mRNA cistrons (or genes). Shine and Dalgarno further noted that there was a purine-rich region just upstream of the AUG initiation codon in these genes, and that there was sequence variation in this purine-rich region that correlated

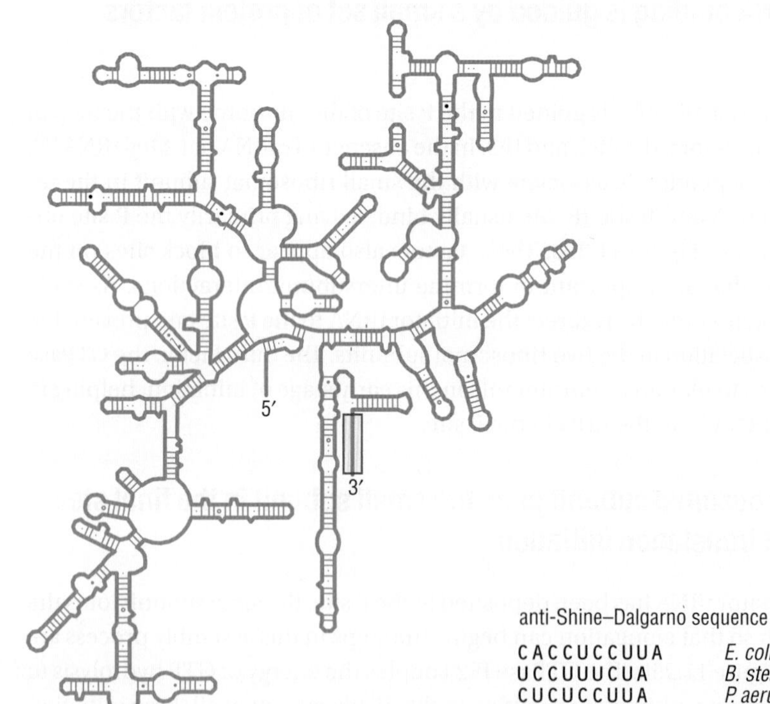

Figure 1 Secondary structure of 16S rRNA with the anti-Shine–Dalgarno region at the 3′ end highlighted in red. Listed on the right are several of the sequences from different bacteria determined by Shine and Dalgarno, exhibiting variability that they noted correlated with the varying levels of protein expression for certain phage genes.

From Shine and Dalgarno. Determinant of cistron specificity in bacterial ribosomes. *Nature* 1975;**254**:34–38.

with the variation in the pyrimidine-rich sequences in the 16S rRNAs. Based on these patterns, Shine and Dalgarno proposed that the pyrimidine-rich 3′ end sequence of the 16S rRNA paired directly with the purine-rich upstream leader sequence of mRNA cistrons, and that the more extensive the pairing (and, hence, the more direct complementarity between the two sequences), the more effectively the mRNA would be translated.

The role of these base-pairing interactions was tested directly when Herman de Boer and colleagues constructed a 'specialized' ribosome system. In these experiments, the human growth hormone (hGH) gene was cloned in a plasmid where the mRNA, carrying one of several highly mutated Shine–Dalgarno sequences, was constitutively expressed in a number of different *Escherichia coli* strains carrying either wild-type or mutant 16S rRNA genes (in addition to wild-type chromosomally expressed ribosomes). The basic idea behind these experiments was to switch the sequences in the Shine–Dalgarno and anti-Shine–Dalgarno motifs, and ask whether expression of the hGH protein depended on the presence of ribosomes with a complementary region in the 16S rRNA. The results were straightforward: the hGH protein was efficiently expressed only in cells where ribosomes with a complementary anti-SD sequence are found (see Figure 2). We broadly refer to this as a 'specificity swap' experiment – an approach that works especially well for RNA because of the simplicity of the Watson–Crick pairing interaction and the fact that all base pairs are more or less isosteric with one another.

Identification of a specific interaction leads to a general rule for bacterial translation initiation

We now know that the so-called Shine–Dalgarno sequence found upstream of the start site in a majority of bacterial genes plays a key role in the initiation of translation, directly guiding the ribosome to the appropriate AUG. Moreover, the Shine–Dalgarno sequence upstream of the AUG start site remains an important predictor for identifying start sites in gene finder programs in bacteria (and archaea), as depicted in Figure 3.

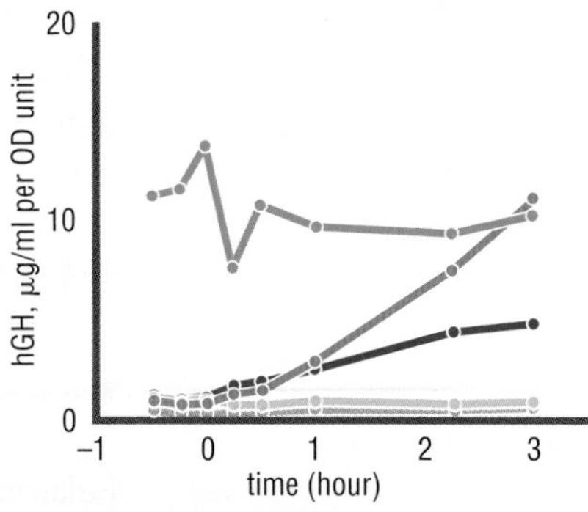

Figure 2 Specificity swap Shine–Dalgarno experiment. Human growth hormone (hGH) accumulation in various cells measured at various times. Orange line, wild-type ribosomes and an *hGH* mRNA with a standard *E. coli* upstream sequence; green and purple lines, mRNAs carrying altered upstream sequences but no specialized ribosomes; pink and blue lines, have mRNAs with altered upstream sequences and specialized ribosomes that match.

Reproduced from Hui and de Boer. Specialized ribosome system: Preferential translation of a single mRNA species by a subpopulation of mutated ribosomes in *Escherichia coli*. *Proceedings of the National Academy of Sciences of the U S A*, 1987;**84**:4762–4766.

Phylogenetic approaches continue to play a critical role in defining the secondary structure (and occasionally tertiary interactions) of RNA species of interest. New computational tools and computing power have increased the sophistication of phylogenetic analysis, allowing models to be evaluated in more systematic and robust ways. The relatively recent discovery of riboswitches has provided the field with a new set of challenges to solve using such approaches.

Find out more

Hui A, de Boer HA. Specialized ribosome system: Preferential translation of a single mRNA species by a subpopulation of mutated ribosomes in *Escherichia coli*. *Proceedings of the National Academy of Sciences of the U S A*, 1987;**84**:4762–4766.

Michel F, Westhof E. Modelling of the three-dimensional architecture of group I catalytic introns based on comparative sequence analysis. *Journal of Molecular Biology*, 1990;**216**:585–610.

Shine J, Dalgarno L. Determinant of cistron specificity in bacterial ribosomes. *Nature*, 1975;**254**:34–38.

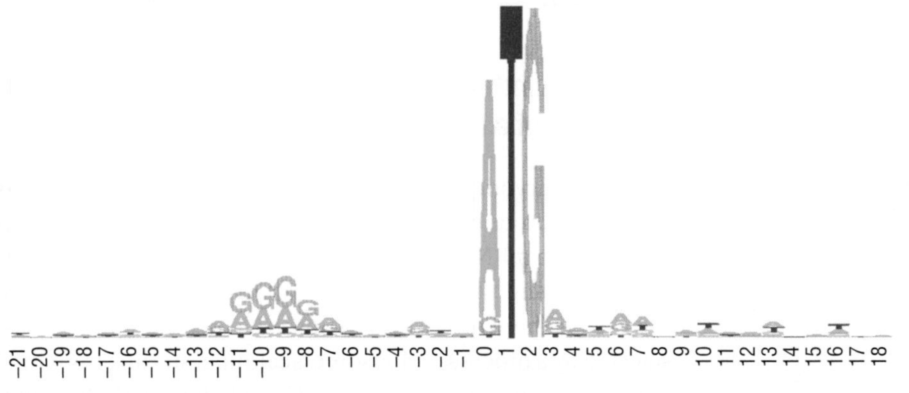

Figure 3 Nucleotide conservation plot for *E. coli* genes in the region just upstream of the AUG start site. The conserved purine-rich Shine–Dalgarno sequence located six to eight nucleotides upstream of the start site is readily apparent.

Adapted from Tom Schneider 'A Gallery of Sequence Logos'; http://schneider.ncifcrf.gov/sequencelogo.html.

complex is composed of both ribosomal subunits, a bound mRNA and an initiator f-Met-tRNA^{fMet} poised in the P site; initiation is complete and translation elongation can proceed.

11.9 EUKARYOTIC TRANSLATION INITIATION

Translation initiation in eukaryotes is more complex than in bacteria

Eukaryotic initiation of translation resembles bacterial initiation in that it occurs at an AUG codon that is decoded by a special initiator Met-tRNA$_i^{Met}$ (albeit distinct from the formyl-Met-tRNA^{fMet} of bacterial initiation; see Figure 11.22). However, there are radical differences between the two systems; in fact, no other aspect of translation is so extremely different. Broadly speaking, eukaryotic ribosomes identify the AUG start site using a scanning mechanism that begins at the 5′ end of the mRNA at the eukaryotic specific 'cap' structure and depends on distinct factors and functions.

Unlike the bacterial 30S small subunit, eukaryotic 40S small ribosomal subunits do not bind directly to the mRNA upstream of the AUG start site. Indeed, eukaryotic AUG initiation sites do not have an equivalent of the oligopurine Shine–Dalgarno motif that is the key to start site recognition in bacteria; consistent with this fact is the observation that, despite the highly homologous nature of the bacterial 16S and eukaryotic 18S rRNAs, the anti-Shine–Dalgarno polypyrimidine tract is wholly missing in the 18S rRNA. (Exceptions where 40S subunits bind directly to the mRNA near the AUG start site are known as internal ribosome entry sites or IRESs, and are discussed in Section 12.4)

Without direct base-pairing interactions to guide the ribosome to the appropriate AUG start site, eukaryotic ribosomes depend on a distinctive mechanism known as 'scanning' for identification of the AUG start site. The greater complexity of this initiation process is evident from the fact that eukaryotic mammalian initiation requires no fewer than 11 distinct factors: eIF1, 1A, 2, 2B, 3, 4A, 4B, 4E, 4G, 5, and 5B. Some of these factors are composed of several polypeptide chains: eIF2 has three polypeptide chains, eIF2B has five, and eIF3 no fewer than 12 (with a mass about half the size of the small ribosomal subunit). All told, the minimal number of polypeptides involved in the mammalian system is 28. Compare this to initiation in bacterial systems, which requires just three distinct initiation factor proteins (IF1, IF2, and IF3), each of which is a single polypeptide.

Eukaryotic mRNAs have unique features that lead to the formation of a closed loop initiation complex

Mature eukaryotic mRNAs are generally monocistronic (that is, they encode a single protein product), and initiation usually occurs at the first AUG triplet from the 5′ end of the mRNA. Occasionally, for some mRNAs, initiation at this first AUG is inefficient, and there is more initiation at the second (or even third) AUG from the 5′ end. These observations suggest a mechanism whereby the initiation site is selected by a process of ribosome 'scanning', starting from the 5′ end. The explanation for recognition of the first AUG not being 100% efficient is that recognition is sensitive to the sequence context of the AUG. The preferred sequence context in mammals is (A/G)XXAUGG and is commonly referred to as the **Kozak sequence**.

The extent to which the context of a given AUG differs from the consensus in part dictates how efficiently it is recognized by the scanning ribosome.

Other distinctive features of eukaryotic mRNA that help guide initiation are the 7-methyl guanosine cap (at the 5′ end) and the poly(A) tail (at the 3′ end); these features are depicted in Figure 11.26. As we saw in Chapter 10, both of these features are added to the mRNA cotranscriptionally. The 7-methyl guanosine cap and poly(A) tail allow for several preparatory steps, which depend on the eIF4 family of initiation factors, to prime the mRNA for the subsequent scanning stage. The 5′ cap structure is specifically bound by a protein factor eIF4E (known as the cap binding protein) and the 3′ poly(A) tail is bound by a protein called the poly(A) binding protein (PABP). These factors interact with eIF4A/4B (an ATPase complex that assists in the unwinding of mRNA structure) and eIF4G (thought to act as a scaffold for assembly of the initiation complex).

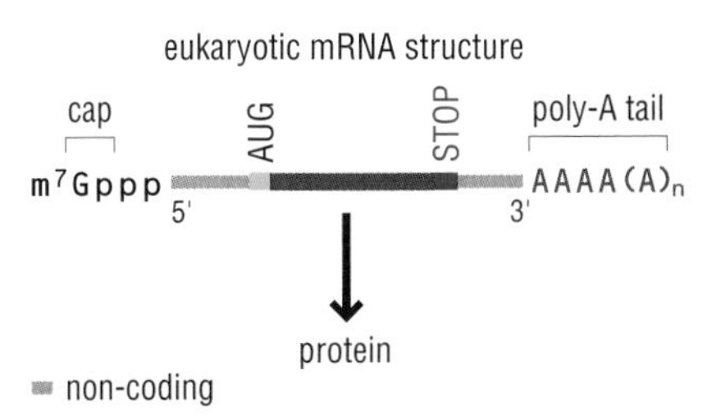

Figure 11.26 General features of eukaryotic mRNAs. Eukaryotic mRNAs are usually monocistronic, where the open reading frame begins with an AUG initiation codon (green) and ends with a stop codon (red). At the 5′ end is the 7-methyl guanosine cap (m7Gppp), and at the 3′ end is the poly(A) tail.

The consequence of the engagement of these multiple binding interactions is the formation of a **closed loop complex** on the mRNA that seems to be important for the most efficient translation. This circular form of the mRNA also appears to serve as a useful 'quality control' feature, allowing the system to immediately distinguish between competent and incompetent mRNA substrates (i.e., complexes lacking a cap or a poly(A) tail are not efficiently translated, and are deemed 'incompetent'). The structure of this closed loop complex is depicted in step 1 of Figure 11.27.

The initiation codon in a eukaryotic mRNA is identified by a ribosome scanning mechanism that begins at the 5′ cap structure

If formation of the closed loop complex is a first step required for the identification of the AUG start site, how then is the actual start site found? This process depends on a second set of preparatory steps during which most other initiation factors (including eIF1, eIF1A, eIF2, eIF3, and eIF5) prime the 40S ribosomal subunit for scanning. The eukaryotic homolog of IF1, eIF1A, similarly binds in the area of the A site of the 40S subunit and blocks premature subunit joining, while eIF1, clearly related to IF3, positions itself around the E site of the 40S subunit.

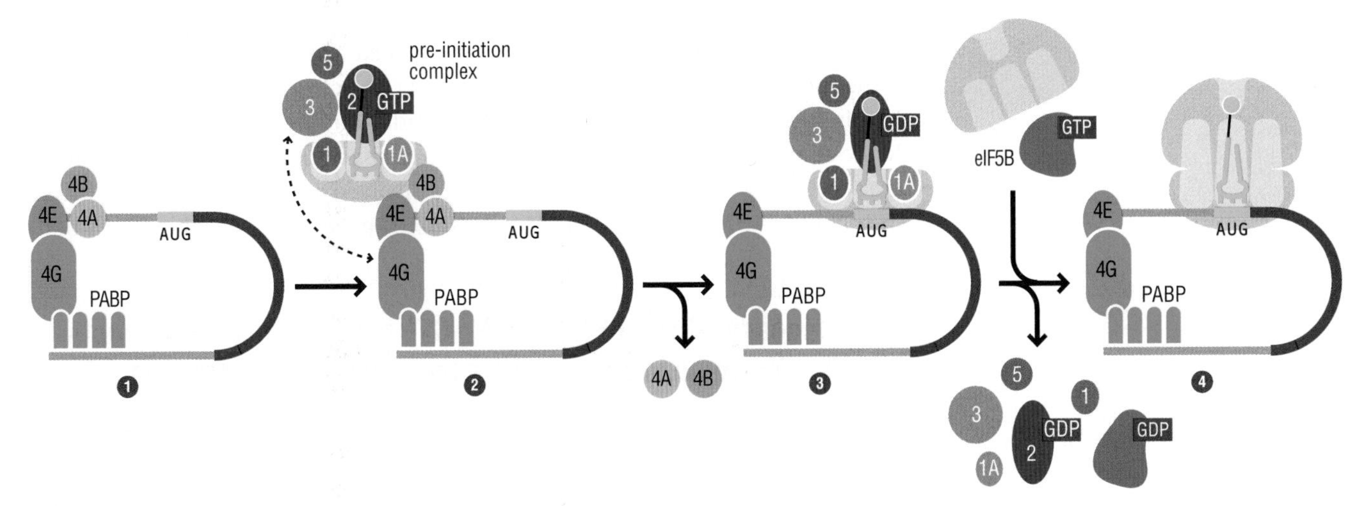

Figure 11.27 The scanning model for eukaryotic translation initiation. (1) The closed loop of mRNA results from the interactions of eIF4G with eIF4E (which binds the mRNA 5′ cap) and with PABP (which binds the mRNA poly(A) tail). eIF4A/4B are also present and are important for unwinding the structured mRNA. (2) After formation of the mRNA loop, the pre-initiation complex associates with the mRNA complex. (3) This initiation complex scans along the mRNA in search of the AUG initiation codon (green). Once the start codon is found, eIF5 acts as GAP to stimulate GTP hydrolysis on eIF2. (4) Finally, eIF5B-GTP promotes the final step in initiation, called subunit joining, and triggers the departure from the complex of many of the bound initiation factors.

These two proteins thus guide the initiator tRNA to the P site, as was the case for bacterial initiation (see Figure 11.25). An important distinction is that the initiator tRNA, Met-$tRNA_i^{Met}$, is bound to a protein, eIF2, in a ternary complex with GTP, and this protein is required for the loading of the P site. The primed complex is generally referred to as '48S' based on its slightly faster migration in a sucrose gradient (relative to a bare 40S subunit) – a more general term would be the 'pre-initiation complex'.

At this stage, the complex is ready for scanning and is brought to the eIF4 factor-bound mRNA through direct interactions between eIF3 and the 'scaffold protein' eIF4G (Figure 11.27, step 2). Scanning now proceeds, with eIF4A (the ATPase) and its associated factor (eIF4B) unwinding secondary structural blocks in the mRNA, and with the initiator tRNA evaluating triplet codons as they go by to determine whether they are complementary to its anticodon. Once the scanning ribosome complex identifies an appealing AUG codon (subject to the context features described above and with the specific help of eIF1 in evaluating the pairing interaction), eIF5 functions as a GAP and triggers GTP hydrolysis on eIF2. This GTP hydrolysis associated step serves as a fidelity checkpoint for translation, only rapidly firing on recognition by the complex of an authentic start site (Figure 11.27, step 3).

With start site recognition now complete, initiation factors must dissociate (though the precise order of these departures is unknown), and subunit joining must take place. In this final step, the eukaryotic homolog of IF2, the GTPase eIF5B, promotes large subunit (60S) joining in a GTP hydrolysis-dependent step (Figure 11.27, step 4). Once the large subunit has been recruited to join the smaller subunit, the completed initiation complex (now 80S) is prepared to begin translation elongation.

11.10 TRANSLATION ELONGATION: DECODING, PEPTIDE BOND FORMATION, AND TRANSLOCATION

The initiation step of translation, described in the previous two sections, essentially primes the ribosome for peptide synthesis: it locates the site on the mRNA from which the sequential decoding of codons will occur (in the 5′ to 3′ direction), and brings both ribosomal subunits together, ready for active synthesis to commence. In this section, we describe the core events underpinning peptide synthesis itself. In the elongation cycle, the parallels between the bacterial and eukaryotic systems are very strong – both in terms of the factors involved and their functions – and so the details we present here should be considered generally applicable across both systems.

The aminoacyl-tRNA–EFTu(eEF1A)–GTP ternary complex interacts with the ribosome to ensure high accuracy during translation

Following initiation, the ribosome sits poised on the mRNA with an initiator methionyl-tRNA in the P site and an empty A site ready to begin the elongation cycle. The elongation cycle comprises three basic steps in all organisms: decoding, peptide bond formation, and translocation.

During the decoding step, the ribosome selects an aminoacyl-tRNA with an anticodon that is complementary to the mRNA codon in the A site. This aminoacyl-tRNA is bound to elongation factor EFTu-GTP (or eEF1A-GTP in eukaryotes), which interacts with the ribosome to ensure high fidelity in the selection process. In its role as an escort, EFTu acts as another guardian of high accuracy in protein synthesis, the first guardian being the aminoacyl-tRNA synthetases that load tRNAs with only their cognate amino acids (see Section 11.3). The role of EFTu is thus to increase the accuracy of decoding beyond the levels brought about by the simple differences between the binding energy of **cognate**, **near-cognate** (single mismatches), and **non-cognate** (more than one mismatch) **interactions** between the codon and the anticodon. Remember that the rules of the genetic code demand Watson–Crick pairing interactions between the codon and anticodon at the first and second codon positions, but certain non-standard pairing interactions are permitted at the third codon position. The overall fidelity of this EFTu-facilitated selection process is thought to be of the order of 10^{-3}, meaning that an incorrect aminoacyl-tRNA is selected by the ribosome only once in a thousand times.

How are cognate tRNAs recognized and selectively incorporated by the ribosome and EFTu ternary complex? First, cognate aminoacyl-tRNAs associate more tightly with the ribosome complex than near-(or non-)cognate ones, favoring their ultimate acceptance into the ribosome. Moreover, these binding differences can be utilized twice by the ribosome, before and after GTP hydrolysis on EFTu. This provides the ribosome with two opportunities to discriminate between cognate and near-cognate tRNAs and, hence, increase the fidelity of selection. These two different phases of the selection process are referred to as '**initial selection**' and '**proofreading**'. We think of these two cognate-discriminated binding steps as thermodynamic contributors to fidelity (red arrows in Figure 11.28).

In addition to the thermodynamic contributions to fidelity, the geometry of the small helix of RNA formed by the codon and anticodon is carefully evaluated in its minor groove by specific ribosomal elements located in the 'decoding center', the region of the small subunit surrounding the codon–anticodon complex.

➔ We learn about the difference between the major and minor grooves of a double-stranded nucleic acid in Chapter 2.

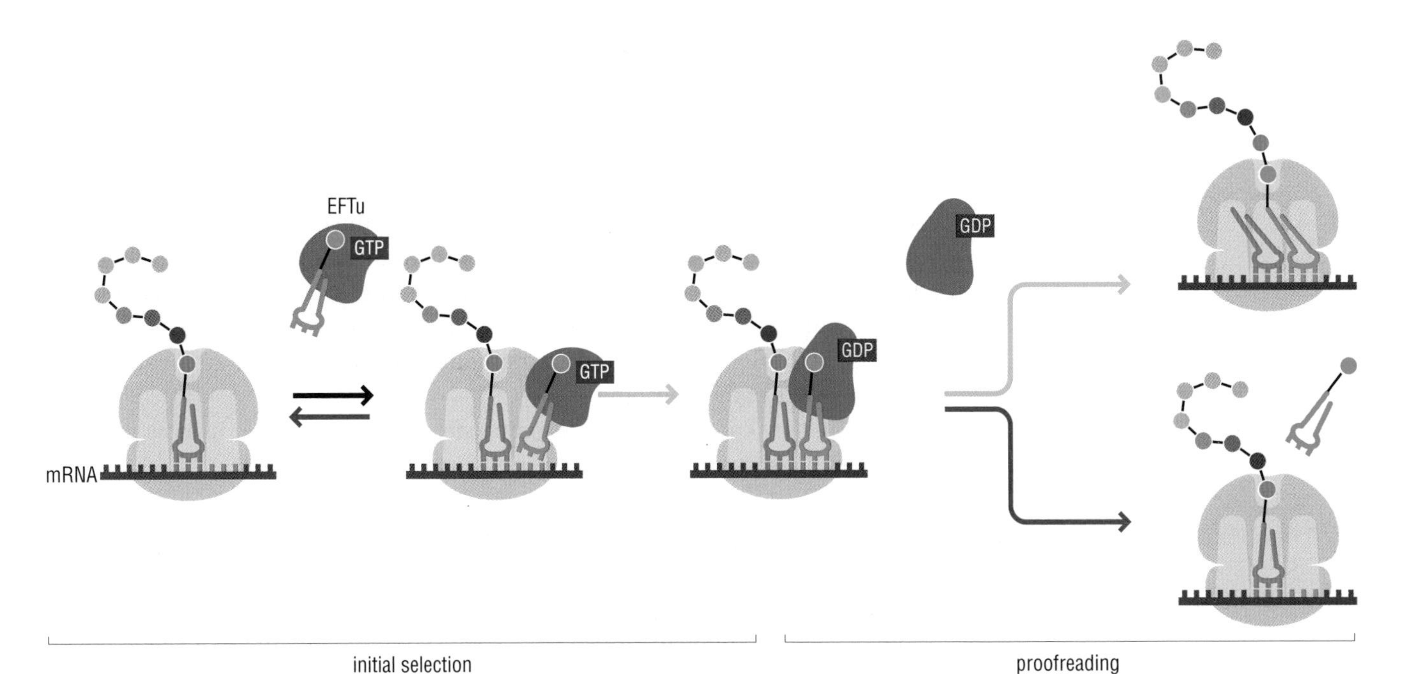

Figure 11.28 Decoding (tRNA selection) by the ribosome and EFTu. The selection of a cognate aminoacyl-tRNA in the A site of the ribosome involves two distinct steps separated by the hydrolysis of GTP by EFTu. In the 'initial selection' phase, an aminoacyl-tRNA complexed with EFTu can be rejected from the ribosome before EFTu hydrolyzes GTP. Following GTP hydrolysis and the departure of EFTu, ribosomes still can reject near-cognate tRNA in a 'proofreading' phase. Both thermodynamic rejection steps are shown with red arrows. Additionally, cognate tRNAs promote more rapid GTPase activation and accommodation (shown in green arrows), allowing for kinetic discrimination mechanisms to contribute to the overall fidelity of tRNA selection.

Recognition of a cognate helix promotes conformational changes in proximal regions of the subunit (see below), and these structural changes are in turn propagated to more distant regions of the ribosome. These transmitted signals ultimately tell the ribosome to act as a GAP for EFTu (eEF1A) to increase the rate of GTPase activation and to promote more rapid '**accommodation**' of the aminoacyl-tRNA fully into the A site (there is a rather large movement of the aminoacyl-tRNA following dissociation from EFTu). We think of these two cognate-stimulated forward steps as kinetic contributors to fidelity (green arrows in Figure 11.28). These ideas are discussed from an experimental perspective in Experimental approach 11.4.

As a result, cognate tRNAs move more rapidly than near-cognate tRNAs through the initial selection process - first because their ternary complexes are less likely to fall off the ribosome complex than near-cognate tRNAs and, second, because the GTPase activation step is more rapid. These cognate tRNAs are also more efficiently accepted by the ribosome during the second 'proofreading' phase - first, because they fall off less often than near-cognate tRNAs and, second, because the accommodation step is more rapid. So, while near-cognate tRNAs do bind to the ribosome, they are readily discriminated against by this two-step selection process.

As we discuss in Section 11.11, there are other mechanisms in place that monitor the fidelity of protein synthesis following the misincorporation of an incorrect amino acid into the polypeptide chain or following a frameshifting event. These mechanisms depend on premature termination as a mechanism to increase the overall fidelity of protein synthesis.

Conformational changes in the rRNA near the codon–anticodon pairing are critical for tRNA selection (decoding)

We now know in detail some of the conformational changes that take place in the ribosome when a cognate tRNA is recognized. In the regions surrounding the anticodon-codon complex, the decoding center, there are several universally conserved nucleotides in the16S rRNA that respond in a dramatic way to cognate tRNA recognition. The nucleotides A1492, A1493, and G530 undergo substantial conformational rearrangements on binding a cognate tRNA; in this rearranged (or induced) state they stably interact with the minor groove of the cognate codon-anticodon helix, as illustrated in Figure 11.29. These structural changes, which are not stabilized in the presence of near- and non-cognate tRNAs, are thought to be responsible for inducing more remote structural changes in the ribosome that stimulate GTPase activation on EFTu and accommodation of the cognate aminoacyl-tRNA into the A site of the ribosome.

Having now considered the decoding step of elongation, let us now go on to consider the second step - peptide bond formation.

Peptide bond formation is catalyzed in an active site containing only RNA

After a cognate aminoacyl-tRNA has been accommodated in the A site of the ribosome, the aminoacyl- and peptidyl-tRNAs are ready for the catalysis of peptide bond formation wherein the growing polypeptide chain is transferred to the aminoacyl-tRNA. The process of peptide bond formation is represented schematically in Figure 11.30a. As we discussed earlier, the peptidyl transferase active site is tightly packed with universally conserved rRNA elements that surround the two

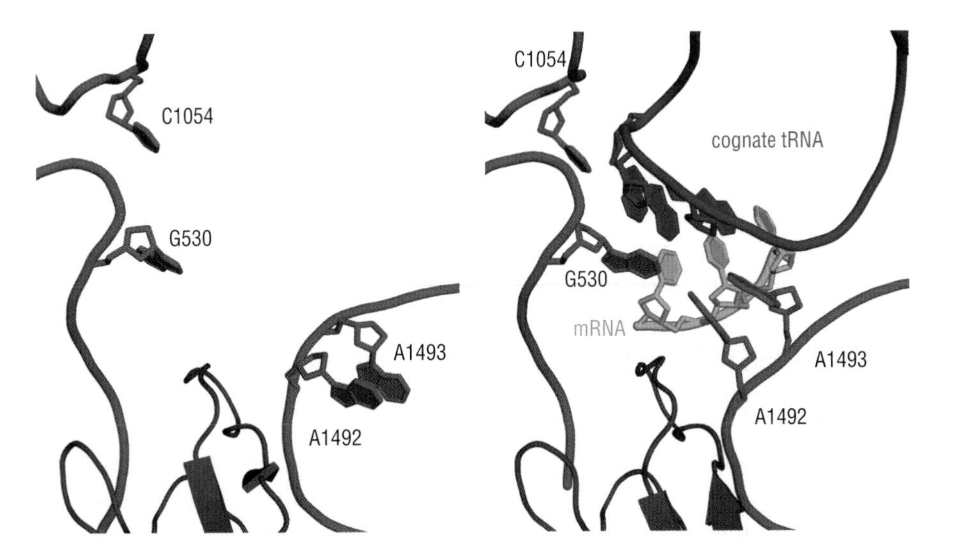

Figure 11.29 Structural changes in the ribosome induced by the binding of a cognate tRNA. Three universally conserved nucleotides in the 16S rRNA in the A site of the ribosome (G530, A1492, and A1493) change conformations on recognition of a cognate codon (blue):anticodon (red) helix. The structure of the nucleotides in the absence of the mRNA–tRNA interaction is shown on the left. In their rearranged state, these nucleotides directly interact with the minor groove of the decoding helix. These structural changes are thought to be critical for triggering GTPase activation by EFTu and for accommodation of cognate tRNA in the A site. PDB codes 2WDG, 1J5E.

tRNA substrates and position them for catalysis. (There are almost no nearby ribosomal proteins, with the exception of L27 in bacteria.)

How is catalysis promoted in this active site? The chemistry of the reaction is relatively simple and consists of a nucleophilic attack of the amino group of the aminoacyl-tRNA on the electron deficient carbonyl of the peptidyl-tRNA substrate, as depicted in Figure 11.30b.

Prior to accommodation of the aminoacyl-tRNA, the peptidyl-tRNA is bound to EFTu in an unreactive configuration for nucleophilic attack. However, when an aminoacyl-tRNA is accommodated in the A site through the tRNA selection process, conformational rearrangements occur in the large subunit catalytic center, which orient the peptidyl-tRNA substrate in a position appropriate for nucleophilic attack. At this stage, the two substrates are poised for catalysis with the conserved 3′ CCA tails forming specific Watson–Crick base pairs with universally conserved guanosine residues in characteristic loops of the 23S rRNA (or the 28S rRNA in eukaryotes) (the A and P loops in Figure 11.30b). A set of universally conserved nucleotides within the large subunit rRNA are positioned nearby, though a role for them in catalysis *per se* has not been identified.

➲ We learn more about the nature of nucleophilic attack in Chapter 2.

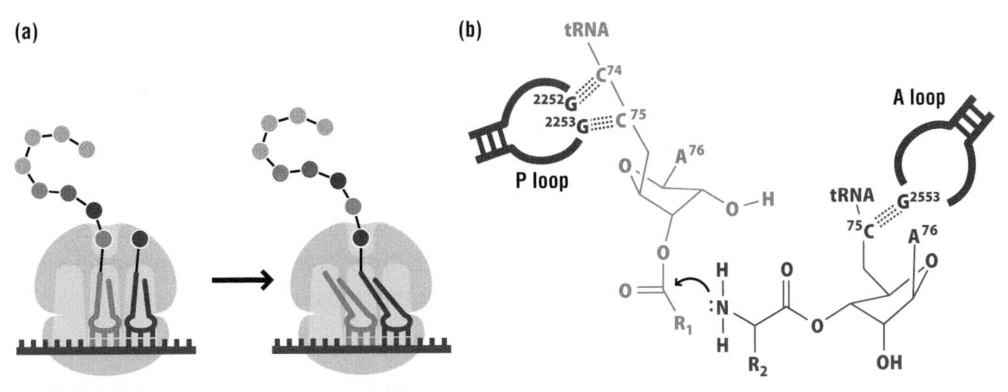

Figure 11.30 Peptide bond formation. (a) Schematic showing the tRNA substrates on the ribosome before and after peptidyl transfer. As a result of catalysis, the growing polypeptide chain is transferred from the P-site (peptidyl) to the A-site (aminoacyl) tRNA. At some point following peptidyl transfer, the tRNAs are thought to assume a hybrid state of tRNA binding in preparation for translocation. (b) A more detailed chemical view of the two substrates prior to peptidyl transfer. Watson–Crick pairing interactions between the ribosome (the P and A loops) are indicated. The 2′ OH of the peptidyl-tRNA substrate is nearby and facilitates proton transfer to promote the nucleophilic displacement reaction. R_1 refers to the growing polypeptide chain and R_2 to the side group of the amino acid on the A-site tRNA.

Adapted from Parnell and Strobel HIS & HERS, magic magnesium and the ballet of protein synthesis. *Current Opinion in Chemical Biology*, 2003;**7**:528–533.

(a)

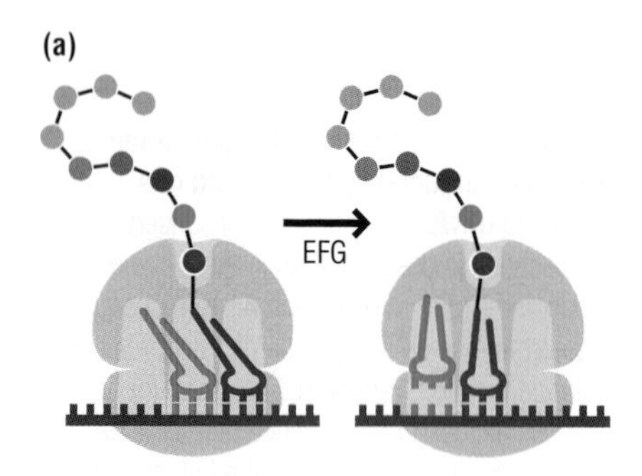

(b)

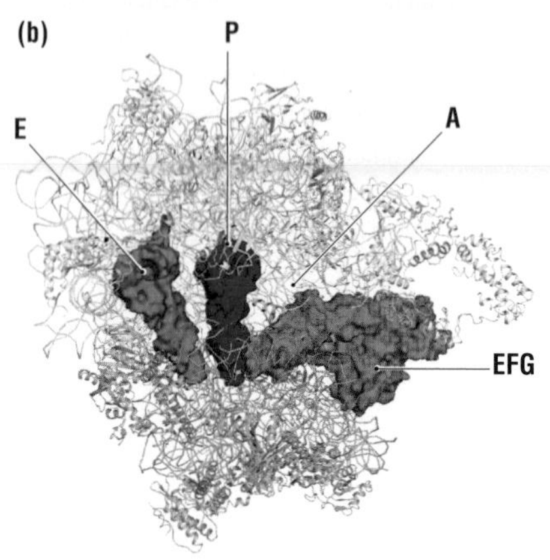

Figure 11.31 Translocation of the mRNA–tRNA complex. (a) Schematic showing the movements of the mRNA–tRNA complex with respect to the ribosome that occur during the process of translocation. EFG interacts with pre-translocation state ribosomes, with tRNAs bound in the 'hybrid' state, promoting the translocation of the mRNA:tRNA complex. A domain of EFG mimics the anticodon of a tRNA, and interacts with the small subunit decoding center, biasing forward movement. (b) Cryoelectron microscopy image of EFG stalled on the ribosome (orange). The image clearly shows how domain IV of EFG directly engages the normal A-site tRNA binding site on the small subunit. PDB codes 2WRK, 2WRL.

From Valle *et al.* Locking and Unlocking of Ribosomal Motions. *Cell*, 2003;**114**:123–124.

In addition to these required positioning elements of the ribosome (the A and P loops and highly conserved active site nucleotides), the 2′ OH of the peptidyl-tRNA substrate, which is critically positioned in the active site (see Figure 11.30b), appears to play an important role in catalysis. This chemical group has been proposed to be involved in facilitating the movement of protons during the reaction. Such a tRNA-centric mechanism is consistent with the idea that this chemical reaction emerged very early in the evolution of protein synthesis, before the arrival of a highly sophisticated ribosome.

Translocation of the mRNA–tRNA complex is promoted by the GTPase EFG

In the final step of elongation – translocation – the complex of mRNA bound to tRNAs in the A and P sites moves through the ribosome, opening up the A site to receive the next incoming aminoacyl-tRNA (Figure 11.31a). This translocation step is specifically promoted by the GTPase EFG (or eEF2 in eukaryotes). Translocation entails large-scale movements of the tRNAs (and associated mRNA), and conformational changes in the ribosome wherein numerous tertiary contacts must be disrupted and reformed. We have seen how each aminoacyl-tRNA first binds in the A site initially and then must move to the P site (after a peptide bond is formed), and we have learned that each tRNA occupies the exit (E) site as it leaves the ribosome (Section 11.5). We also discussed the hybrid states of tRNA binding as a model to explain how tRNA movement seems to occur in two motions with respect to the large and small ribosomal subunits. First, the acceptor end of the tRNA (carrying the amino acid) moves with respect to the large subunit as the peptide bond forms (see Figure 11.31a). Next, the anticodon end of the tRNA (still base-paired with the codon) moves with respect to the small subunit in a reaction catalyzed by EFG. In Section 11.6 we discussed how the GTPase EFG (eEF2 in eukaryotes) mimics the shape and structure of tRNA bound to EFTu and apparently uses this mimicry to promote structural rearrangements involved in the translocation process. Figure 11.31b gives evidence for this tRNA mimicry, showing a cryoelectron microscopy view of EFG bound to the A site of the ribosome

This view argues that EFG in part functions like a 'pawl' in a motor, using a domain that mimics tRNA to bind in the small subunit decoding center and promote forward movement of the mRNA:tRNA complex. The reaction is driven in the forward direction by the hydrolysis of GTP by EFG. At the completion of the translocation reaction, tRNAs are found in a post-translocation or 'classic' state. Structural studies indicate that the ribosome itself also assumes several different states that are critical to the translocation process, one state favoring 'hybrid' pre-translocation-bound tRNAs and another favoring non-hybrid ('classical') post-translocation-bound tRNAs.

Despite these many insights, we still have much to learn about how EFG directly couples the energy of GTP hydrolysis to discrete three-nucleotide movements of the mRNA–tRNA complex in the ribosome.

11.11 TRANSLATION TERMINATION, RECYCLING, AND REINITIATION

As the three-step cycle of elongation described in the previous section continues, the ribosome glides along its mRNA template, continually extending the polypeptide chain as it moves. Though it might seem plausible that elongation should continue until the ribosome runs out of mRNA to decipher, the biological solution is different. The end-point of the process of translation is signaled by three specific stop (or nonsense) codons that are recognized by specific protein factors. This process of termination, like other steps in translation, is subject to various levels of regulation and control, as we shall discover in Chapter 12.

Class 1 and class 2 release factors are critical to termination in bacterial and eukaryotic systems

In most organisms, the end of a protein-coding region of mRNA is signaled by one of three **stop codons**: UAA, UAG, and UGA. There are usually no tRNAs that recognize these stop codons (though certain exceptions to this rule will be described in Section 11.13); instead, they are recognized by proteins known as **class 1 release factors (RFs)**. In bacteria, there are two different class 1 release factors, RF1 and RF2, with overlapping specificity - both recognize UAA, while RF1 recognizes UAG and RF2 recognizes UGA. In eukaryotes, one class 1 release factor, eRF1, recognizes all three stop codons.

In addition to recognizing stop codons in the A site of the ribosome, release factors promote a hydrolytic reaction in the peptidyl transferase center that releases the completed peptide chain from the peptidyl-tRNA in the P site. Thus the class 1 release factors are bifunctional molecules, like tRNAs, linking an interpretation of the genetic code in the small subunit decoding center with a catalytic event in the large subunit catalytic center. In this case, however, it is a protein factor rather than an RNA that facilitates specific recognition of codons and stimulates the catalytic event that results in the release of the peptide chain.

Figure 11.32a shows how the structures of a tRNA and a class 1 release factor (from bacteria), both in their ribosome-bound state, superimpose quite nicely on one another. Notice how there are protein domains of the class 1 release factor that correspond to the functional ends of the tRNA, the anticodon loop and the acceptor CCA end. Moreover, Figure 11.32b shows how the release factor bound to the ribosome is very clearly positioned in the A site, spanning the decoding center in the small subunit and the catalytic center in the large subunit. It is interesting to note that the class 1 release factors in bacteria and eukaryotes are unrelated to one another, as illustrated in Figure 11.33. However, both have evolved to perform the same core functions, and even have the same universally conserved GGQ motif that is essential for catalysis.

In addition to the class 1 release factors, a set of GTPases (known as the **class 2 release factors**) is important for termination in both the bacterial and eukaryotic

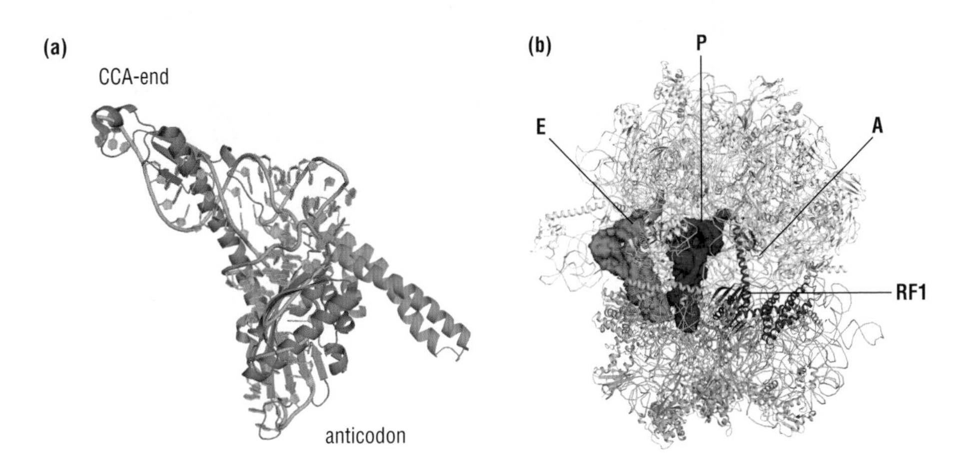

Figure 11.32 Class 1 release factors are protein mimics of functional tRNAs. (a) Bacterial RF1 structure (blue) superimposed on tRNA^{Phe} structure (gray), both in their ribosome-bound conformation. (b) Atomic resolution structure of RF1 bound to the ribosome, showing how the release factor occupies the A site of the small and large subunits as it deciphers the codon and facilitates peptide release. PDB codes 1PNS, 3D5C, 3D5D.
Figure 11.32a from Youngman, EM, McDonald, ME, and Green, R. Peptide release on the ribosome: mechanism and implications for translational control. *Annual Review of Microbiology*, 2008;**62**:353–373.

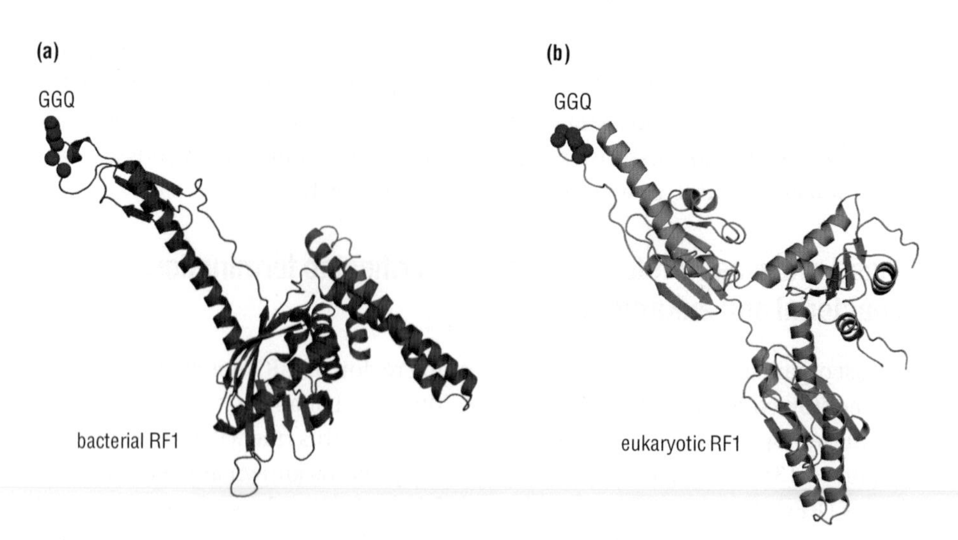

Figure 11.33 Bacterial and eukaryotic class 1 release factors are evolutionarily distinct. Atomic resolution structures of class 1 release factors from bacteria (a) and eukaryotes (b) indicate that the structures are distinct, though both rely on the conserved GGQ motif for efficient peptide release. PDB codes 3D5C, 1DT9.

systems. While both proteins are GTPases, RF3 derives from EFG while eRF3 derives from EFTu. And, indeed, these factors appear to play somewhat distinct roles in these two systems; these differences are depicted in Figure 11.34. In bacteria, the GTPase RF3 appears to function after peptide release to promote the dissociation of the class 1 release factor, coupling GTP hydrolysis to this event (Figure 11.33a). In eukaryotes, the GTPase eRF3 appears to behave as an escort

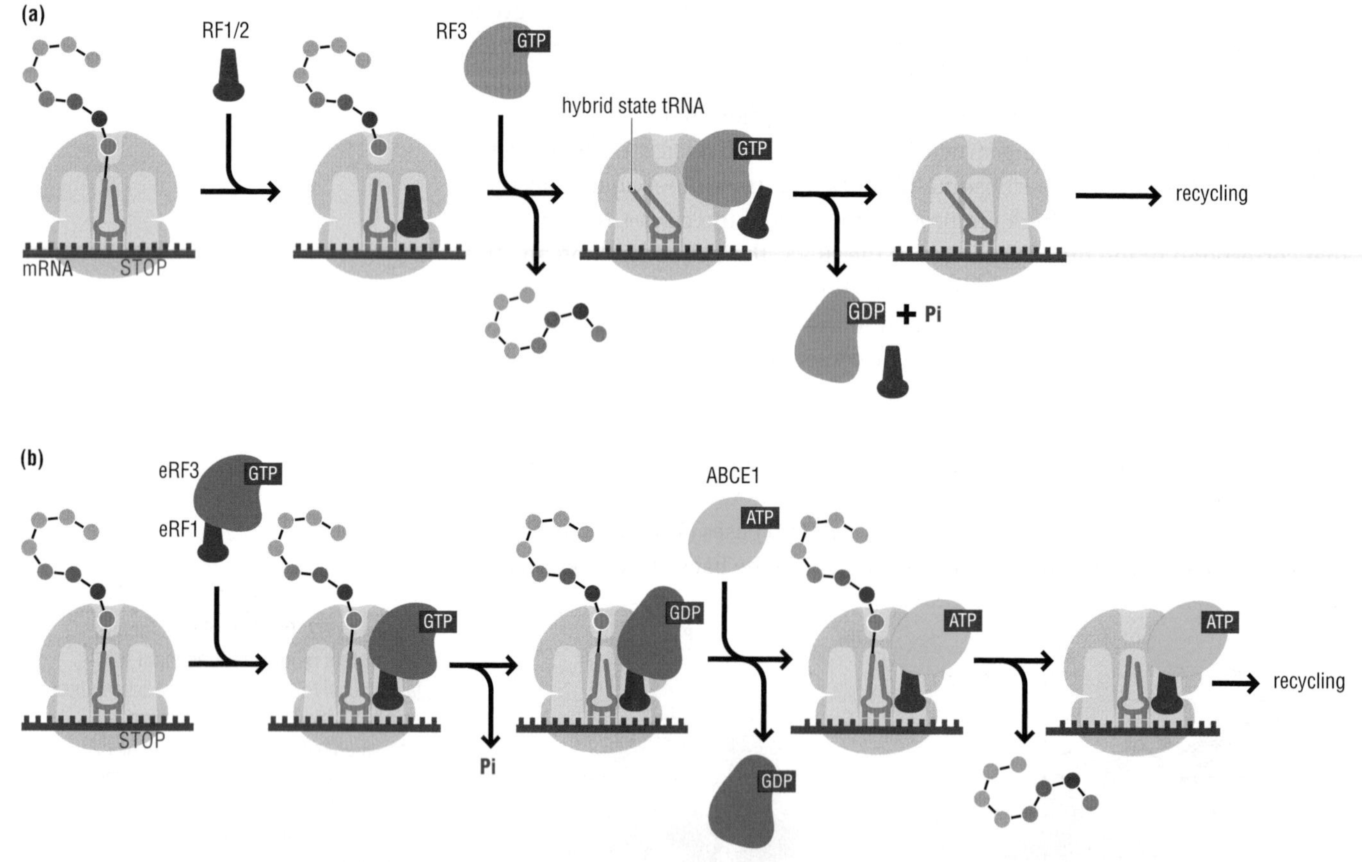

Figure 11.34 Roles of class 1 and 2 release factors in bacteria and eukaryotes. (a) In bacteria, a class I release factor (RF1 or RF2) recognizes the stop codon and triggers the release of the synthesized polypeptide chain from the protein. RF3-GTP then associates with the termination complex (or GTP is acquired on binding to the ribosome), GTP hydrolysis occurs and the class 1 release factor and RF3-GDP dissociate. The remaining ribosome complex contains an mRNA and a tRNA bound in a hybrid state. (b) In eukaryotes, eRF1 and eRF3 are associated with one another prior to binding to the ribosome, with eRF3 functioning similar to EFTu as it chaperones tRNAs to the A site. GTP hydrolysis then occurs, and eRF3 is released. ABCE1 (or Rli1 in yeast) then engages the ribosome in an ATP-bound state and promotes the peptide release reaction. For this step, ATP hydrolysis is not required.

for eRF1, akin to the role played by EFTu (eEF1A) for the aminoacyl-tRNAs, and ultimately using the energy of GTP hydrolysis to promote its own departure from the pre-termination complex (see Figure 11.34b). At this stage, another factor, an AAA+ ATPase ABCE1 (or Rli1 in yeast) enters the scene, binding near where the departed eRF3 was previously bound, and promotes the peptide release reaction in the absence of ATP hydrolysis. We also see that in eukaryotes, the class 1 release factor eRF1 remains bound to the ribosome following the peptide release reaction where, as we shall see, it will play an important role in the subsequent recycling reaction. These substantial differences are not surprising given that the class 1 release factors in these two kingdoms evolved independently, and thus their interaction with their class 2 GTPase must also have evolved separately.

Stop codon recognition by class 1 release factors is highly accurate

Release factor selection in the A site takes place with extremely high fidelity, mirroring tRNA selection. Studies in bacteria suggest that premature peptide release on non-stop (or sense) codons occurs only once in every 100 000 events. Indeed, class 1 release factors appear to have exquisite mechanisms in place to distinguish between stop and sense codons, depending again on strategies of induced conformational rearrangements in the small and large subunits to promote catalysis only when true stop codons are recognized.

As for tRNA selection, specific conserved nucleotides in the decoding center (including nucleotides G530 and A1492) undergo distinct conformational rearrangements on recognition of the stop codon by the class 1 release factor. And, as for peptide bond formation, the 2′ OH of the peptidyl-tRNA substrate appears to be a key component in the peptide-releasing hydrolytic reaction, working in concert with the conserved GGQ motif on the class 1 release factor. Such coupling of recognition of stop codons in the decoding center with conformational rearrangements in the remote catalytic center (~ 80 Å away in the large subunit) allows for an extremely low rate of premature termination during protein synthesis.

Abortive termination on the bacterial ribosome increases the overall fidelity of protein synthesis

Although the processes of tRNA and release factor selection have evolved to take place with very high accuracy, mistakes nevertheless do happen. Indeed, such decoding errors are increasingly likely during times of stress or starvation when, for example, a particular aminoacyl-tRNA might be depleted. Additionally, frameshifting events where the tRNAs become repositioned with respect to the mRNA are another class of potential 'miscoding' in the cell.

➔ We discuss frameshifting in more detail in Section 11.13.

How are such miscoding events detected and dealt with? Misincorporation events on the bacterial ribosome appear to be recognized because the codon:anticodon helix, positioned in the P site following translocation, carries a mismatch. When such a perturbation of the P-site decoding helix is detected by the bacterial ribosome, the fidelities of tRNA and release factor selection on the next codon in the A site are substantially decreased. This decrease in fidelity initially leads to an iteration of errors, which eventually leads to premature termination of the error-containing polypeptide through promiscuous recruitment of the class 1 release factor, as illustrated in Figure 11.35.

Interestingly, this specialized premature termination reaction depends on direct contributions of RF3 (the class 2 release factor) to the actual rate of peptide

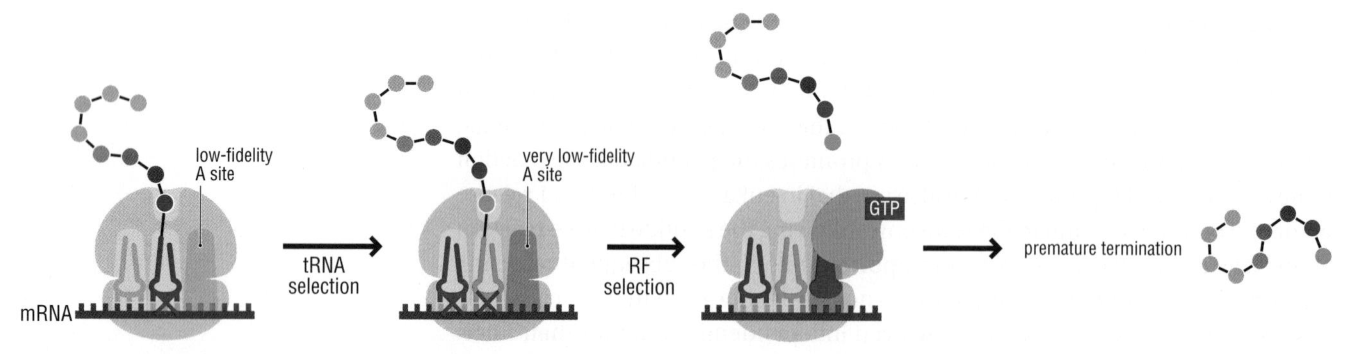

Figure 11.35 Quality control on the ribosome following a misincorporation event. When a misincorporation takes place due to an error during tRNA selection, a mismatched decoding helix is found in the P site following translocation (indicated by a red X; notice that the tRNA color does not match the codon). This mismatch triggers structural changes in the ribosome complex that substantially decrease the fidelity of tRNA and RF selection in the A site, leading most immediately to an iteration of errors, and ultimately to premature termination at sense codons.

release, a role for RF3 distinct from its usual relatively minor role in termination. The prematurely truncated peptide is likely degraded by cellular peptidases and proteases, while the ribosome and mRNA are then liberated for additional rounds of translation. This quality control system on the bacterial ribosome can be compared with the proofreading or editing functions of both DNA polymerases and aminoacyl-tRNA synthetases discussed earlier in the book. Whether a similar mechanism of post-peptidyl transfer quality control exists in eukaryotes is not yet established.

➔ We discuss the proofreading/editing functions of DNA polymerase and aminoacyl-tRNA synthetases in Sections 6.3 and 11.3, respectively.

Ribosome recycling is the final step in protein synthesis

Following the release of the completed polypeptide chain, the ribosome complex must be disassembled to allow the ribosomal subunits and remaining bound components to take part in another round of protein synthesis. Minimally, in both bacteria and eukaryotes, the substrate of recycling is a complex consisting of the two subunits of the ribosome, the mRNA, and a deacylated tRNA that once carried the growing peptide chain. There are substantial differences in recycling in these systems however that connect directly to the very substantial differences that we observed for termination.

The process of ribosome recycling in bacteria is summarized in Figure 11.36a. In bacteria, the class 1 release factor dissociates from the ribosome following peptide release, assisted by the action of RF3 (though this factor is not present in all bacteria). At this point, a protein factor known as ribosome recycling factor (RRF) wedges itself between the subunits, and in concert with EFG (and GTP hydrolysis) promotes complex disassembly. The initiation factor IF3 binds to the small subunit to stabilize the dissociated state. This event completes a full cycle of translation by a given ribosome in bacteria.

The process of recycling in eukaryotes is quite distinct from that seen in bacteria, as summarized in Figure 11.36b. First, eRF1 remains ribosome-bound after the peptide release reaction, in the A site, from where it helps to promote ribosomal subunit dissociation. This eRF1-promoted subunit dissociation reaction is further enhanced by the AAA+ ATPase ABCE1 in mammals (or Rli1 in yeast) through the utilization of the energy of ATP hydrolysis. Finally, as in the bacterial system, core initiation factors such as eIF1, 1A, and 3 appear to trap the dissociated subunits in preparation for the next round of initiation. These fundamental differences are consistent with the fact that there is no RRF homolog in eukaryotes

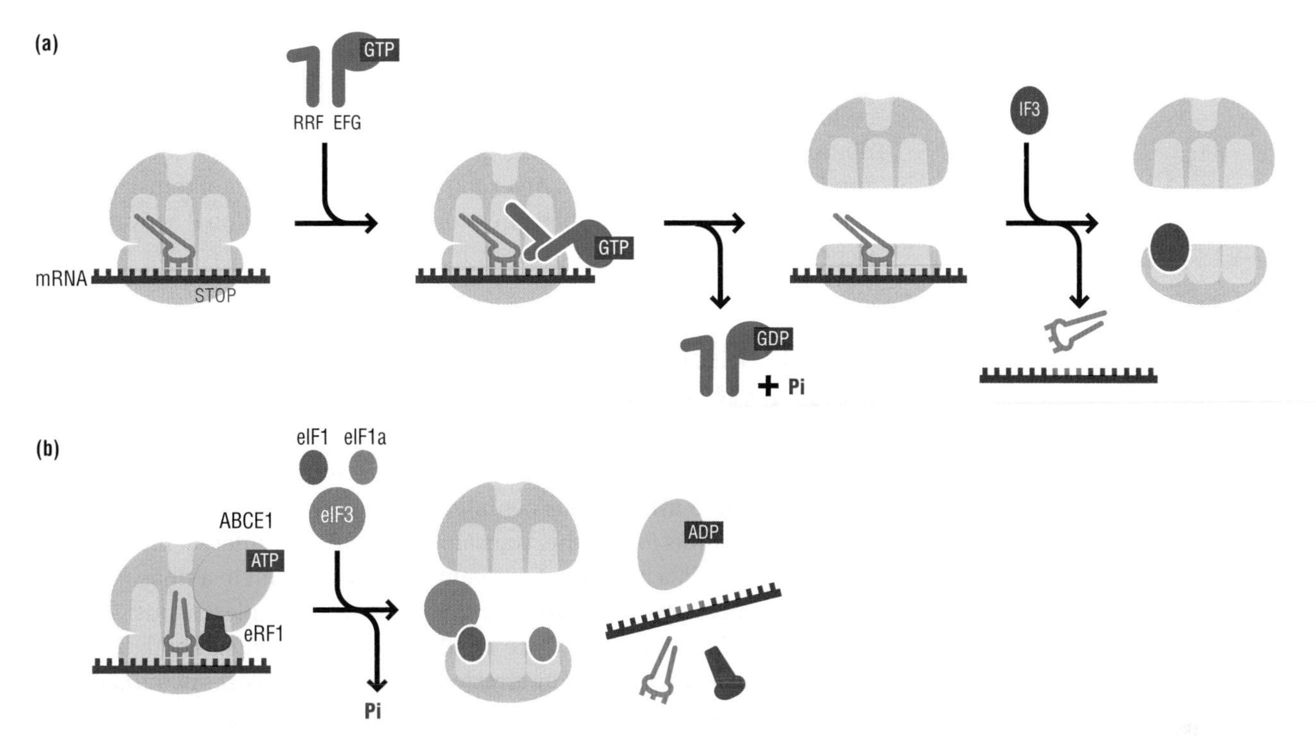

Figure 11.36 Ribosomal recycling in bacteria and eukaryotes. (a) In bacteria, post-termination ribosome complexes are found in a rotated state. The combined actions of RRF and EFG-GTP cause dissociation of the two subunits of the ribosome. The tRNA and mRNA eventually fall off of the small subunit, and IF3 stabilizes their departure. The small ribosome subunit–IF3 complex is then free to begin the initiation process with another mRNA. (b) In eukaryotes, post-termination ribosome complexes are dissociated by still bound eRF1 and ABCE1. This dissociation reaction depends on ATP hydrolysis by ABCE1. Finally, as in bacteria, the dissociated state of the ribosomal subunits is stabilized by bound initiation factors eIF1, eIF1A, and eIF3.

that might play a similar role in the recycling reaction. We will see in Section 12 how related mechanisms are utilized in eukaryotes to impose quality control on translation.

Reinitiation of translation on downstream ORFs in polycistronic mRNAs is routine in bacteria

In bacteria, genes are typically transcribed into mRNAs that carry more than one ORF; such transcripts are called polycistronic mRNAs. In these mRNAs, each ORF is typically preceded by a Shine–Dalgarno sequence that can independently guide the small subunit to the appropriate AUG for loading of the initiator f-Met-tRNA^{fMet} (as we described in Section 11.6). Following initiation, translation elongation, termination, and ribosome recycling can then take place such that ribosomal subunits, tRNA and mRNA are fully released from the actively translating complex. In some cases, however, there appears to be coupling between closely spaced ORFs, in a **polycistronic message** such that ribosome recycling is somewhat inefficient. In this situation, the small ribosomal subunit is thought to remain associated with the mRNA and scan in both the forward (3′) and backward (5′) direction in search of another AUG to initiate the next round of protein synthesis. The rate at which the small subunit dissociates from the mRNA is a crucial factor in determining how much 'scanning' can occur before the subunit completely departs from the mRNA.

While a vast majority of eukaryotic genes are monocistronic, a few are polycistronic in the sense that there are multiple sites within the transcript where translation initiation and termination take place, though such 'cistrons' are often extremely small (with just a few codons) and the product that they encode may not have any cellular function. In some of these cases, the **upstream ORFs (uORFs)** can be used to regulate gene expression. In brief, the efficiency of ribosomal recycling impacts the likelihood of reinitation occurring at downstream ORFs - a ribosome that is still partially engaged on an mRNA can subsequently search for another AUG for reinitiation. An example of how this form of translational regulation can be used to control expression of the transcriptional activator protein, Gcn4, will be examined more closely in Section 12.2.

11.12 RIBOSOME RESCUE IN BACTERIA AND EUKARYOTES

When all goes well, ribosomes initiate, elongate, terminate and are recycled from a given template mRNA, thus completing a round of translation, and releasing ribosomal subunits for another translational cycle on a different mRNA template. However, in a cell, a number of different events can result in a ribosome pausing, stalling, or most severely, arresting; these arrested complexes must be resolved by cells in order to survive. There are three general events in both bacteria and eukaryotes that are triggered by arrested ribosome complexes in the cell: first, the mRNA sequence on which the arrested ribosome sits is targeted for mRNA decay; second, the protein being synthesized, which is not full length, is targeted for proteolysis; and, finally, the ribosomes themselves are recycled, allowing the precious ribosomes to be used again in subsequent rounds of translation. These three events are summarized in Figure 11.37. The very distinct evolutionary solutions to these problems in bacteria and eukaryotes are outlined in this section.

Deleterious effects from truncated bacterial mRNAs are minimized by the tmRNA protein tagging system

Truncated mRNAs pose a significant threat to bacteria: when ribosomes encounter the physical end of an mRNA rather than terminating at a stop codon, the complex is effectively trapped, taking a valuable ribosome out of circulation, and at best generating a truncated polypeptide. Truncated mRNAs arise from incomplete transcription reactions and from mRNA decay. Bacteria cannot distinguish truncated from full-length mRNAs during translational initiation, and have evolved a system

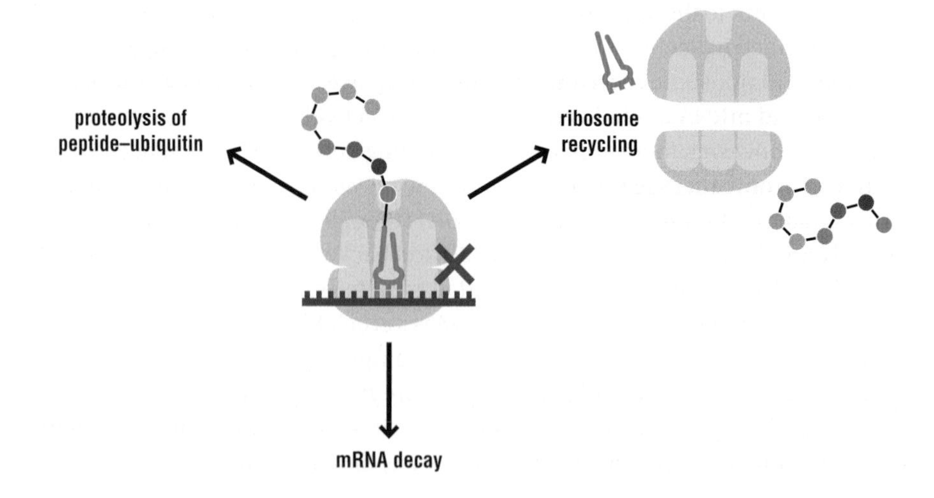

Figure 11.37 Three consequences of a stalled ribosome. When ribosomes arrest during elongation, for a variety of reasons (red X), the mRNA is degraded (mRNA surveillance), the incomplete polypeptide is targeted for decay, and the ribosome complex is rescued. These pathways are mechanistically distinct in bacteria and eukaryotes, but the broad outcomes outlined here are similar.

to rescue stalled ribosomes, as described below. In contrast, truncated mRNAs are less problematic in eukaryotes because translation is separated from transcription both spatially and temporally, allowing quality control of mRNA prior to translation. In particular, translational initiation in eukaryotes depends strongly on the presence of both the 5′ cap and the 3′ poly(A) tail structures.

Bacterial ribosomes stalled on truncated mRNAs are released by an unusual RNA species known as **tmRNA**. Its name reflects the fact that tmRNA has features that resemble both tRNA and mRNA, as depicted in Figure 11.38. A tmRNA acts first as a tRNA, adding an alanine residue to the growing peptide chain; it then acts as an mRNA, encoding an additional ten amino acids before terminating translation at a stop codon. These additional amino acids form a degradation 'tag' (as shown in Figure 11.38), which targets the truncated protein to cellular proteases for degradation. In this manner, tmRNA recovers stalled ribosomes and stimulates the degradation of potentially toxic polypeptides.

Given its ability to enter the ribosome and hijack translation, tmRNA could in principle abort translation reactions that are proceeding properly. However, this does not happen: tmRNA reacts selectively with ribosomes that are stalled on truncated mRNAs. This selectivity arises from way that tmRNA and its protein partner SmpB bind in the ribosomal A site. The tRNA-like domain of tmRNA and the SmpB protein form a structure that mimics the shape of a canonical tRNA, as illustrated in Figure 11.39. Notice how SmpB is positioned where the anticodon stem-loop of the tRNA would otherwise be. When the tmRNA-SmpB complex is delivered to the A site by EFTu, SmpB interacts with the decoding center on the 30S subunit. Before peptidyl transfer to tmRNA occurs, the ~30 residue C-terminal tail of SmpB inserts into the mRNA channel. While this occurs rapidly in ribosomes stalled on truncated mRNAs, where there is no mRNA in the A site or mRNA channel, peptidyl transfer to tmRNA is blocked with intact mRNAs. In effect, SmpB monitors the length of mRNA in the ribosome to discriminate stalled ribosomes from actively translating ones. Finally, once the degradation tag is added to the polypeptide, the ribosome then reaches a stop codon, and termination and recycling occur through standard mechanisms. In this way, the bacterial cell effectively eliminates aberrant protein products and recovers stalled ribosomes.

Ribosome rescue in eukaryotes shares mechanistic features with the termination and recycling steps of normal translation

In eukaryotes, only full-length mRNAs are effectively translated because formation of the initiation complex depends so strongly on the presence of both the 5′ cap and the 3′ poly(A) tail structures. Even so, there are thought to be multiple signals that trigger pausing by the ribosome in eukaryotes, and ribosome rescue pathways are activated if these pauses are sufficiently long.

Ribosomal arrest can be triggered by substantial mRNA structure, by specific peptide stalling sequences, or even by codons for which the corresponding aminoacyl-tRNA is in low abundance. Such stalling events on the ribosome are generally thought to trigger a rescue pathway referred to as no-go decay (NGD).

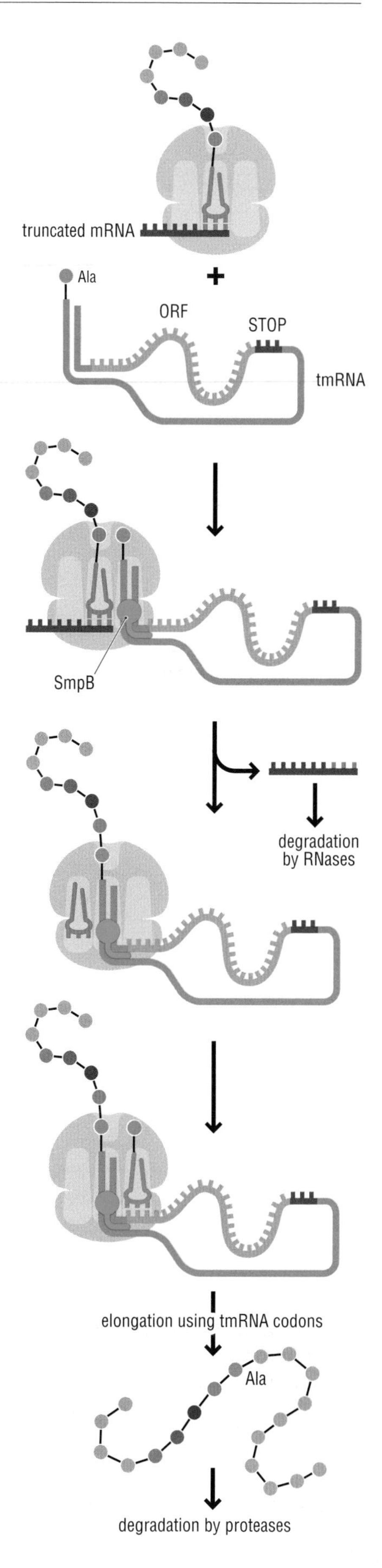

Figure 11.38 tmRNA:SmpB directs a quality control step on stalled ribosomes in bacteria. Bacterial ribosomes occasionally stall on truncated mRNAs that lack a stop codon. This stalled complex is recovered by tmRNA, which acts first as an aminoacyl-tRNA (with a $tRNA^{Ala}$-like moiety) and second as an mRNA encoding an 11-residue peptide terminated with a stop codon. Following the tmRNA-mediated events, free ribosomes are liberated for further protein synthesis while the truncated proteins carry an 11-residue peptide that targets them for degradation (to avoid potential dominant effects on the cell).

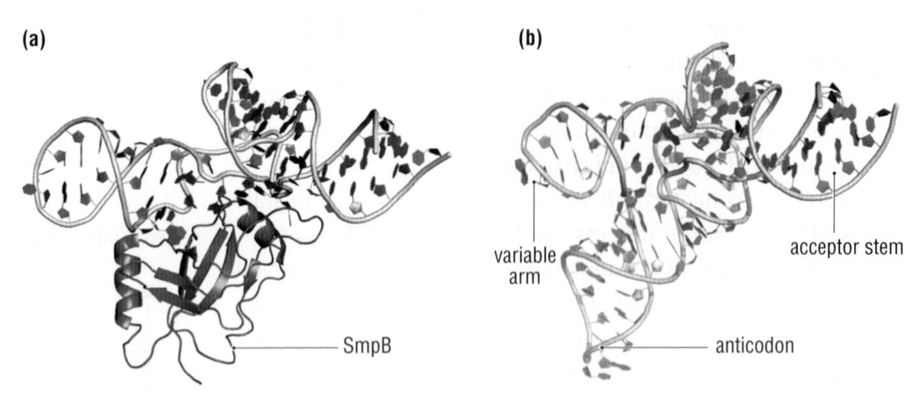

Figure 11.39 tmRNA:SmpB mimics an intact tRNA to engage the A site of the ribosome (a) tmRNA bound to SmpB and (b) tRNASer with its variable arm indicated. See how the SmpB domain (color) mimics the structure of the anticodon of the tRNA, thus allowing the tmRNA:SmpB complex to effectively bind the A site of empty ribosomes.

➔ We will explore the degradation of proteins in Chapter 14.

➔ We discuss the NGD pathway in more detail in Experimental approach 11.3.

While the details are not fully defined, what is known is that, under some stalling circumstances, the mRNA and incomplete protein product are degraded (the latter through a ubiquitin-mediated proteasomal process), and the ribosomes are rescued through the actions of termination factor homologs known as Pelota (or Dom34 in yeast) and Hbs1; these proteins are related to eRF1 and eRF3, respectively. These factors promote a recycling reaction on the ribosome that dissociates the small and large ribosomal subunits, releasing the mRNA and the peptidyl-tRNA for recovery by the cell for subsequent translation events. This recycling reaction is depicted in Figure 11.40.

Another related rescue pathway is referred to as non-stop decay (NSD), which occurs either when an mRNA is truncated prior to the stop codon (and the ribosome reads to the end of the message and stops) or when there is no stop codon at the end of the ORF and the ribosome instead reads into the polyA tail sequence. These stalls are also likely rescued through the actions of Pelota (Dom34) and Hbs1. Both NSD and NGD in yeast depend on the actions of another eRF3 homolog Ski7 that also interacts with the exosome complex. There are likely a number of other factors involved in detecting stalled ribosomes, as well as in targeting the mRNA and the truncated protein product for degradation. In this section, we have focused on the ribosome rescue events that are most relevant to this chapter on translation.

Nonsense-mediated decay targets mRNAs with premature termination codons for decay

Nonsense mediated decay (NMD) is an mRNA surveillance pathway that targets for degradation mRNAs that encode a premature termination codon (PTC) located

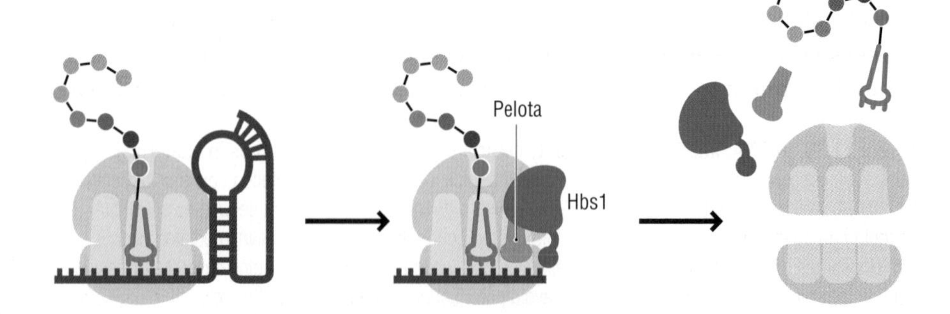

Figure 11.40 Pelota:Hbs1 complex rescues stalled ribosomes in eukaryotic cells. When ribosomes encounter a sufficient stall, such as a very stable mRNA structural element, Pelota and Hbs1 mimic eRF1 and eRF3 and promote a recycling reaction, with the help of ABCE1. The result of this rescue reaction is the dissociation of the ribosomal subunits and release of peptidyl-tRNA and mRNA.

11.3 EXPERIMENTAL APPROACH
mRNA surveillance factors for no-go decay reveal insights into mechanism of recycling on eukaryotic ribosomes

No-go decay is an mRNA surveillance pathway that reports on the translational capacity of an mRNA in the cell

Aberrant mRNA transcripts are recognized in the eukaryotic cell through a number of specialized RNA decay pathways that include nonsense-mediated decay (NMD), no-go decay (NGD), and non-stop decay (NSD). The mRNA surveillance by NGD and NSD appears to be triggered by translational stalls that are subsequently resolved by a set of specialized protein factors. Two such factors, Dom34 and Hbs1, were implicated in NGD in a study by Roy Parker's lab that used a series of **reporter constructs** that allowed them to follow the fate of the encoded mRNA species in response to various obstacles placed within the open reading frame (ORF).

Reporter constructs are powerful tools that allow for the analysis of a given phenomenon on a non-essential gene that is typically easily visualized (e.g., β-galactosidase, green fluorescent protein (GFP), or luciferase). Typically, many related variants can be directly compared to one another to decipher molecular features that are critical to the behavior being followed. In this study, the insertion of a 34 base pair (bp) stable stem–loop into the PGK1 gene led to endonucleolytic cleavage of the mRNA encoding the stem–loop in a fashion that directly depended on translation of the ORF. Other sequences proposed to cause stalling (including strings of rare codons or pseudoknots) also resulted in similar endonucleolytic cleavage events and eventual mRNA decay.

The intellectual leap made in this study was the observation that the mRNA fragments generated at a pausing sequence were only seen in the presence of the factors Dom34 and Hbs1 – conserved proteins that were known to resemble the canonical eukaryotic termination factors eRF1 and eRF3, respectively, at the sequence and structural levels; see Figure 1.

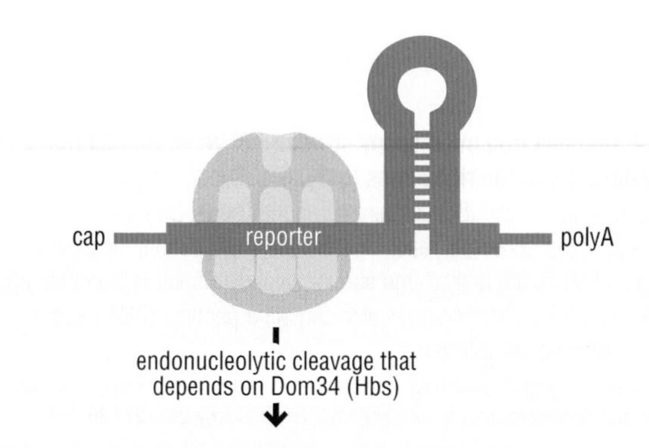

Figure 1 Cartoon model detailing reporter gene with inserted stem–loop mRNA structure that leads to Dom34:Hbs1-dependent endonucleolytic cleavage.

Structural studies provide immediate clues on the ribosome-targeted role of Dom34 and Hbs1

Further insights into how Dom34 and Hbs1 trigger NGD came from subsequent structural and biochemical studies. As already suspected from similarities between the structures of Dom34 and eRF1 and those between Hbs1 and eRF3, cryoelectron microscopy revealed that Dom34 binds to the A site of the eukaryotic ribosome such that one arm mimics the GGQ region of eRF1 and extends towards the peptidyl transferase center of the large subunit while another arm extends towards the decoding center of the small subunit, as depicted in Figure 2. Further, Hbs1 binds where other translational GTPases bind, interacting with the stalk region of the large subunit. Interestingly, Hbs1 has an N-terminal extension not found in eRF3 that is seen to localize near the mRNA entrance channel, quite a distance from the majority of the Hbs1 protein binding site (see Figure 2). It is speculated that this domain somehow monitors mRNA length to aid in the identification of optimal ribosomal complexes for targeting for NGD.

Biochemical studies provide insights into the function of Dom34 and Hbs1

But what do these factors actually do on binding to the ribosome? The answer to this question emerged from biochemical studies where native gel electrophoresis (and sucrose gradient analysis) was used to follow the fate of radioactively labeled ribosome complexes. In this experiment, the addition of Dom34 and Hbs1 to ribosome complexes resulted in the release of peptidyl-tRNA; it was ultimately determined that peptidyl-tRNA was released because the ribosomal subunits had been dissociated by these factors (see Figure 3). Through the use of a purified component system and some well-defined assays, the direct physical outcome of engagement of Dom34 and Hbs1 on the ribosome was clearly defined. These reactions are A-site codon independent but exhibit significant Hbs1-conferred sensitivity to the length of the mRNA species: ribosomes bound to shorter mRNA fragments are a better target for the dissociative activity of the Dom34/Hbs1 protein complex. The mRNA length sensitivity documented by the biochemistry connected beautifully to several other observations in the literature: (1) the observation that ribosome stalling typically

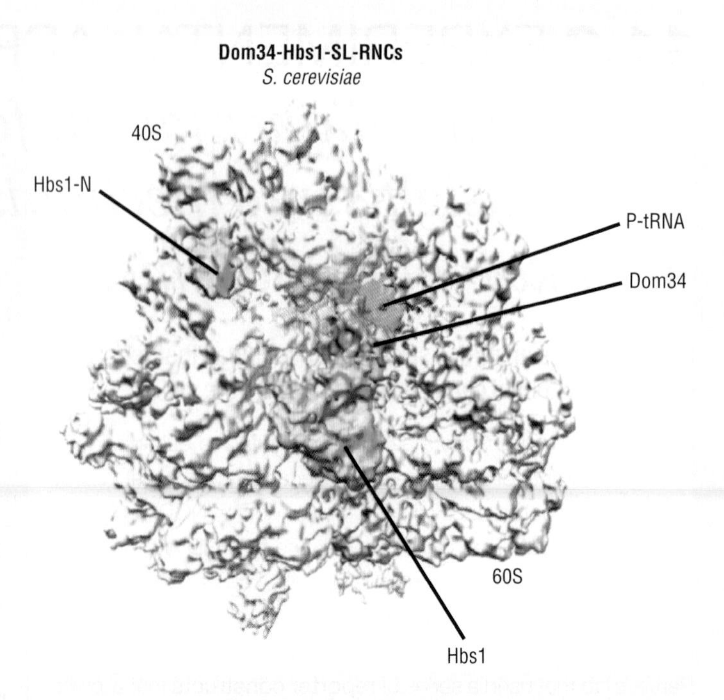

Figure 2 Cryoelectron microscopy structure of yeast Dom34 (blue) and Hbs1 (yellow) bound to ribosomes (gray). Ribosome complexes (RNCs) prepared on a stem-loop (SL) containing reporter gene. Dom34 is positioned in the A site of the ribosome, reaching from the small subunit decoding center (albeit not quite) to the large subunit peptidyl transferase center (again not quite). These stalled ribosomes also contain a peptidyl-tRNA (green) bound in the P site, as anticipated.

Frackenberg *et al.* Structural view on recycling of archaeal and eukaryotic ribosomes after canonical termination and ribosome rescue. *Current Opinion in Structural Biology*, 2012;**22**:786–796.

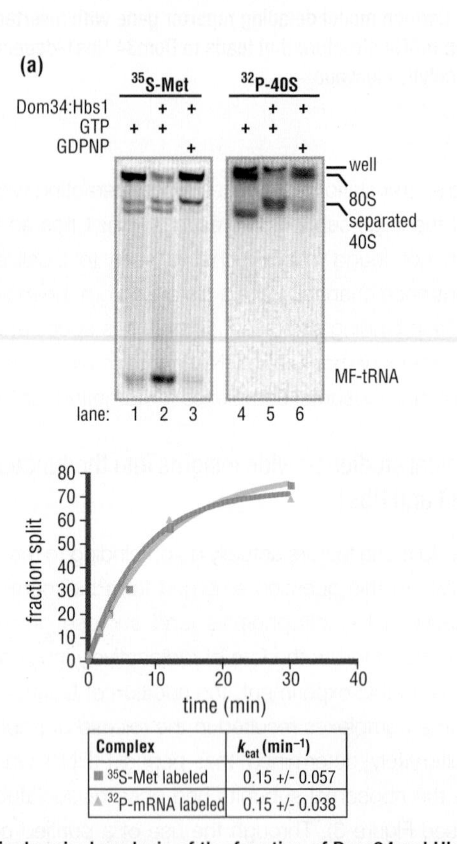

Complex	k_{cat} (min^{-1})
^{35}S-Met labeled	0.15 +/- 0.057
^{32}P-mRNA labeled	0.15 +/- 0.038

Figure 3 Biochemical analysis of the function of Dom34and Hbs1 on yeast ribosomes. (a) Native gel electrophoresis of radioactively labeled ribosome complexes shows that addition of Dom34:Hbs1:GTP leads to release of peptidyl-tRNA (left, when radioactive methionine is found on the peptide) and to separation of the ribosomal subunits (right, when the subunits themselves are radioactively labeled). (b) The rates of these two different reactions are indistinguishable from one another, consistent with a model arguing that subunit separation precedes peptidyl-tRNA release.

Shoemaker *et al.* Dom34:Hbs1 promotes subunit dissociation and peptidyl-tRNA drop-off to Initiate No-Go Decay. *Science*, 2010;**330**, 369.

leads to endonucleolytic cleavage of the mRNA in the RNase-rich cellular milieu; and (2) the observation of unanticipated electron density attributable to Hbs1 at the mRNA entrance channel of the ribosome (noted previously in 'Structural studies provide immediate clues on the ribosome-targeted role of Dom34 and Hbs1').

Taken together, the biochemical and structural studies suggest a model wherein Dom34 and Hbs1 scan the cytoplasm for ribosomes with a truncated mRNA species (and any codon in the A site!), perhaps in this way identifying unproductive ribosomes that need to be 'rescued'. Engagement of these ribosome complexes may lead to effective ribosome recycling, thus recovering the stalled and precious translational components. As we read in Section 11.11, the parent termination factor homologs (eRF1 and eRF3) also participate directly in ribosome 'recycling'.

Find out more

Becker T, Armache JP, Jarasch A, *et al.* Structure of the no-go mRNA decay complex Dom34-Hbs1 bound to a stalled 80S ribosome. *Nature Structural & Molecular Biology*, 2011;**18**:715–720.

Doma MK, Parker R. Endonucleolytic cleavage of eukaryotic mRNAs with stalls in translation elongation. *Nature*, 2006;**440**:561–564.

Pisareva VP, Skabkin MA, Hellen CU, Pestova TV, Pisarev AV. Dissociation by Pelota, Hbs1 and ABCE1 of mammalian vacant 80S ribosomes and stalled elongation complexes. *The EMBO Journal*, 2011;**30**:1804–1817.

Shoemaker CJ, Eyler DE, Green R. Dom34:Hbs1 promotes subunit dissociation and peptidyl-tRNA drop-off to initiate no-go decay. *Science*, 2010;**330**:369–372.

Related techniques

Reporter constructs; Section 19.10

Cryoelectron microscopy; Section 19.14

Native gel electrophoresis; Section19.7

Use of sucrose gradients; Section 19.7

upstream of the normal termination codon (i.e., within the normal ORF of the mRNA), as depicted in Figure 11.41. The PTC may have arisen through genomic mutations, a transcriptional error, a splicing error, or through some other post-transcriptional event. These errors are important for the cell to recognize since the protein produced from a PTC-containing message will be truncated and may have dominant effects within the cell.

➔ We learn more about the EJC and how its role in facilitating NMD was elucidated in Experimental approach 10.3.

➔ We learn more about the mechanism of splicing in Section 10.5

How does the cell distinguish between a natural stop codon (UAA, UGA, or UAG) and a similar sequence located at a different upstream position within the mRNA transcript? In higher eukaryotes, the authentic stop codon is almost always located within the terminal exon of a gene, and so stop codons found in earlier exons are defined as premature. As the splicing machinery leaves a protein mark at the exon-exon junction (known as an exon junction complex or EJC) during splicing, a ribosome that encounters a stop codon upstream of an EJC tags this transcript as faulty and triggers recruitment of specialized components that target the mRNA and protein for degradation (and almost certainly the ribosome for rescue!). In lower eukaryotes like *Saccharomyces cerevisiae*, NMD is known to occur, and yet few genes are spliced; other mechanisms in this organism must contribute to the identification by the ribosome of a PTC.

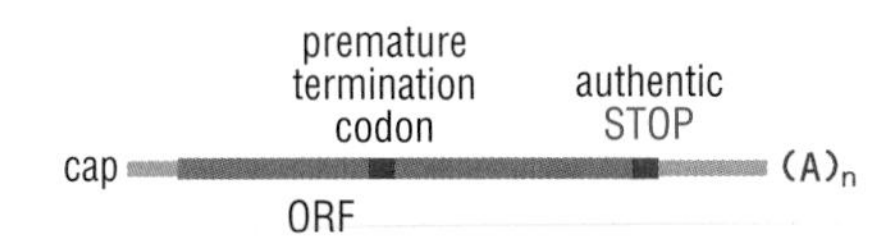

Figure 11.41 PTCs are stop codons found within the ORF. PTCs can arise by a variety of different mechanisms, including, for example, genomic mutation, transcription error, or splicing error.

Specialized cellular factors implicated in NMD include Upf1, Upf2, and Upf3. (The name UPF derives from the emergence of these genes in genetic screens for 'up' frameshifting.) Upf1 possesses a helicase domain and interacts directly with the termination factor eRF3, while Upf2 and Upf3 bridge the interaction between this ribosome complex and the EJC, as shown in Figure 11.42. Although we have information regarding these protein-protein interactions, the exact roles of these factors in promoting mRNA and proteolytic decay and ribosome rescue remain unknown.

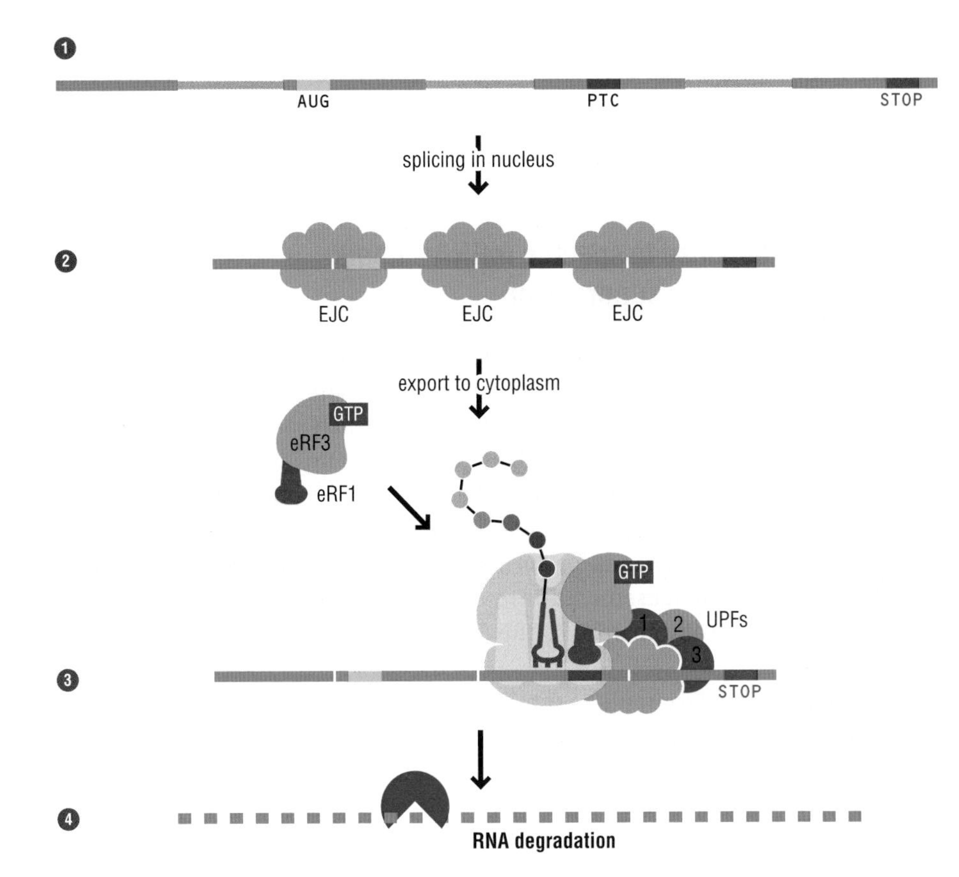

Figure 11.42 NMD is triggered by recognition of exon–exon junction complexes downstream of the PTC in higher eukaryotes. (1) Unspliced transcript in the nucleus showing normal start (green) and stop (red) codons as well as PTC (pink). (2) Spliced transcript carries EJCs at each exon–exon junction, marking the sites that have undergone splicing. (3) Ribosomes translating the transcript displace EJCs until a stop codon is encountered. Upon recognition, interactions between eRF1 and eRF3 of the termination complex, the UPFs and the EJC ultimately (4) recruit factors responsible for degradation of the faulty mRNA transcript in the cell. The molecular consequences of the UPFs interacting with the terminating ribosome complex are not known.

Recent studies reveal a broad and critical role for NMD in dictating gene expression throughout the eukaryotic kingdom. This system serves as a beautiful example of how events occurring in the nucleus and cytoplasm (splicing, and translation and mRNA degradation, respectively) are integrated to specify the 'lifetime' of an mRNA.

11.13 RECODING: PROGRAMMED STOP CODON READ-THROUGH AND FRAMESHIFTING

In the preceding sections, we discussed the molecular details of the translation cycle, from initiation to ribosome recycling, and saw the general principles governing this process, including the nature of the genetic code from which successive three-letter nucleotide sequences are transformed into the primary sequence of a protein. However, there are few hard and fast rules in biology; the genetic code is just one more example of a rule that fails to hold absolutely. In this section, we discover how biological systems can often flex the rules of the genetic code, introducing more diversity to the way the nucleic acid sequence of a messenger RNA is interpreted.

In contrast with codon reassignment described in Section 11.2, the rules of the genetic code are superseded in a different way in **recoding**, wherein readout is reprogrammed in an mRNA-specific fashion. In recoding, the meaning of a codon is not globally reassigned such that all translation events in the organism or organelle interpret the codon in this different way. Instead, the codon is reinterpreted in the context of a specific mRNA, *in competition with* a standard reading of the mRNA. Such recoding can lead to the production of multiple distinct proteins from a single gene and can be used for regulatory purposes. We will first look at several well-characterized examples of the recoding of stop codons (depicted conceptually in Figure 11.43a) and then at some more complex examples of recoding wherein the frame in which an mRNA is read is altered during the process of translation (as represented by Figure 11.43b).

(a) nonsense suppression

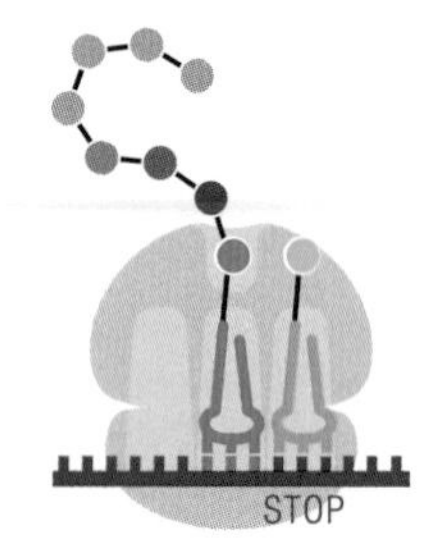

(b) frameshifting (–1)

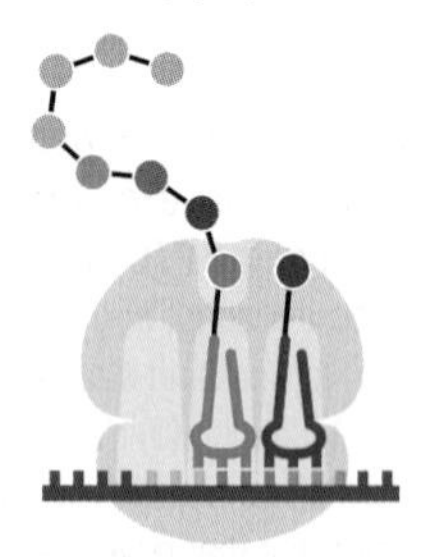

Figure 11.43 Nonsense suppression and frameshifting are relatively common recoding events. (a) Nonsense suppression happens when a nonsense codon is misread by a near-cognate tRNA species, thus reading past the stop codon to extend the polypeptide sequence. (b) Frameshifting happens when the mRNA slips in register, such that protein synthesis continues in a different reading frame.

Recoding in response to specific mRNA signals can lead to nonsense suppression

When any of the three normal stop codons are misread by aminoacyl-tRNA species such that the termination of translation fails to occur, we refer to the process as nonsense suppression (Figure 11.43a). Although such read-through normally happens with very low frequency, the likelihood of it occurring can be significantly affected by specific sequence elements in the mRNA. These phenomena have been studied most deliberately in viral mRNAs where a relatively simple hexanucleotide motif (CARYYA, where R stands for purines and Y for pyrimidines) found 3′ to the stop codon triggers increased read-through by near-cognate aminoacyl-tRNAs. UAG codons are typically misread by $tRNA^{Gln}$ while UGA codons are typically misread by $tRNA^{Trp}$.

In one well-studied example in the Murine Leukemia Virus (MuLV), the Gag-Pol precursor polypeptide required for the viral lifecycle is synthesized through a nonsense suppression mechanism. Some fraction of the time, the UAG codon of the *gag* ORF is read as 'stop', producing the *gag* gene product. However, some fraction of the time, the UAG codon is misread by Gln-$tRNA^{Gln}$ (a near-cognate

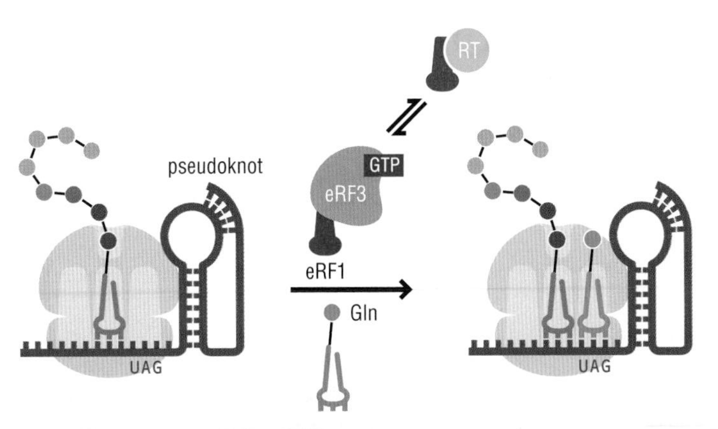

Figure 11.44 Programmed nonsense suppression. In MuLV A, a pseudoknot structure in the RNA downstream of the stop codon (UAG) triggers misreading by Gln-tRNA^{Gln} and thus the production of the Gag-Pol viral gene product. This miscoding must compete effectively with the normal termination factors (eRF1-eRF3), though eRF1 is apparently sequestered through interactions with another viral gene product, RT.

aminoacyl-tRNA for the UAG codon) in an event that is specifically promoted by a 3′ proximal structured **pseudoknot** element in the mRNA: nonsense suppression thus results in translation of the Gag-Pol fusion gene product.

An additional contributor to nonsense suppression in this case is the partial sequestration of eRF1, the natural reader of the UAG codon in the cell, by a virally encoded gene product, reverse transcriptase (RT). This mechanism is outlined in Figure 11.44. Because the nonsense suppression event must compete with normal stop codon recognition by release factors, nonsense suppression is not 100% efficient. The extent of misreading in this system plays a key role in regulating the levels of several critical viral gene products.

Of medical interest, many genetic diseases are caused by mutations within ORFs that insert a premature stop codon, leading to the premature termination of protein synthesis, an absence of full-length protein product, and NMD (see Section 11.12). In an interesting twist on nonsense suppression, researchers have discovered mechanisms for promoting the 'read-through' of such PTCs to sometimes ameliorate disease symptoms; we find out more about these discoveries in Experimental approach 11.4.

Stop codons can be recognized by systems that incorporate non-standard amino acids into proteins

Stop codons can also be recoded to allow a non-standard amino acid (not one of the 20 common amino acids) to be inserted into a protein, often at positions of functional importance for the activity of the protein. The best-characterized example of this mechanism is the insertion of selenocysteine, an amino acid similar to cysteine that carries a selenium atom instead of a sulfur atom in its side chain, as shown in Figure 11.45a. This so-called 'twenty-first amino acid' is incorporated into several different enzymes, where it appears to play a role as a powerful reducing agent in the active sites of these proteins.

In organisms ranging from *E. coli* to humans, there are structural elements in the mRNA that are critical for specifying the insertion of the selenocysteine residue. In *E. coli*, an RNA stem-loop known as a **selenocysteine insertion sequence (SECIS)** is found immediately 3′ to the UGA codon that will be recoded. In

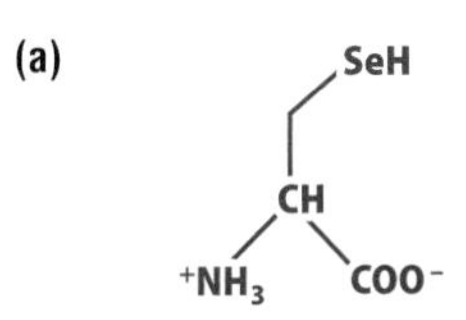

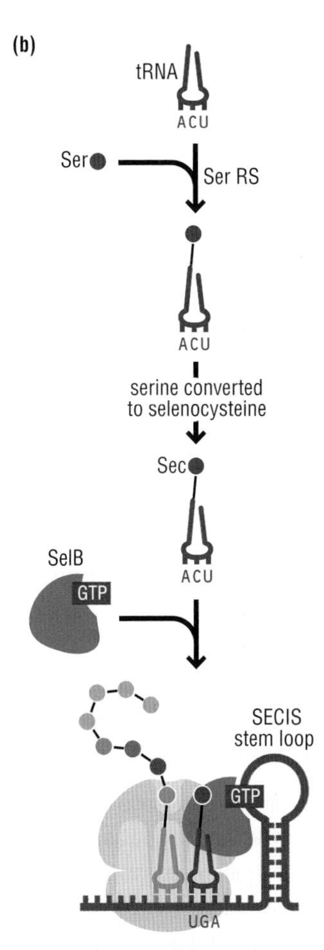

Figure 11.45 The recoding of UGA stop codons for selenocysteine insertion. (a) The chemical structure of the amino acid selenocysteine. (b) Schematic showing how a specialized tRNA in bacteria is first loaded with a serine that is then enzymatically converted to selenocysteine. The resulting Sec-tRNA^{Sec} is delivered to the ribosome by an EFTu homolog, SelB, that simultaneously recognizes the selenocysteine insertion sequence (SECIS) RNA element downstream of the UGA codon. This mechanism allows for the insertion of selenocysteine only at certain UGA codons in the cell.

11.4 EXPERIMENTAL APPROACH

Mechanism of action for aminoglycosides and treatment of PTC-triggered genetic diseases

It has long been appreciated that aminoglycoside antibiotics function through their interaction with the ribosome in bacterial targets to promote miscoding (and inhibition of bacterial growth). However, a mechanistic understanding of this phenomenon only came with high-resolution crystal structures of the decoding center of the ribosome with bound aminoglycosides, and detailed biochemical studies.

Initial insights into small subunit architecture came from chemical modification experiments

Chemical modification analysis of the ribosome indicated that there were a few key highly conserved nucleotides in the 16S rRNA (among them A1492, A1493, and G530) which underwent conformational rearrangements that were likely critical to function. These clues were rationalized with high-resolution x-ray crystal structures of the small subunit showing that when tRNA ligands bind in the aminoacyl (A) site, residues A1492 and A1493 are repositioned from a stacked conformation within helix 44 to a 'flipped out' conformation that allows them to directly engage the minor groove of the codon:anticodon mini-helix. Additionally, in the presence of aminoacyl-tRNA, G530 of the 16S rRNA undergoes a conformational rearrangement from a syn to an anti conformation of the purine base relative to the ribose sugar.

X-ray crystal structures of the small ribosomal subunit revealed the binding site of aminoglycoside antibiotics (and a likely mechanism of action)

The significance of these conformational rearrangements to aminoglycoside function was revealed when Venki Ramakrishnan and colleagues used crystallography to identify the binding site for several different aminoglycoside antibiotics. Paromomycin, a known miscoding agent, was seen to bind in helix 44 of the small subunit, in the position normally occupied by A1492 and A1493 (as depicted in Figure 1), thus effectively displacing A1492 and A1493 from their interhelical position. In this way, the aminoglycoside stabilizes an 'activated' state of the decoding center, making misincorporation increasingly likely.

Kinetic approaches provided new insights into aminoglycoside function

Marina Rodnina and colleagues used pre-steady kinetic approaches to further define how cognate and near-cognate tRNAs are distinguished on the ribosome during tRNA selection. Contrary to earlier models, these scientists found that a non-equilibrium process allowed the ribosome to actively promote the acceptance of cognate tRNA (and not near-cognate tRNA) – such a process is generally referred to as 'kinetically driven'. Interestingly, two distinct steps are specifically accelerated in the tRNA selection pathway – activation of the GTPase center in EFTu and movement of aminoacyl-tRNA into a fully accommodated state in the large subunit.

Nicely correlating with these observations, the aminoglycoside paromomycin accelerates these same two key steps, GTPase activation and accommodation, even when near-cognate tRNAs are bound, as shown in Figure 2. The establishment by the Rodnina group of a kinetic and thermodynamic framework for the process

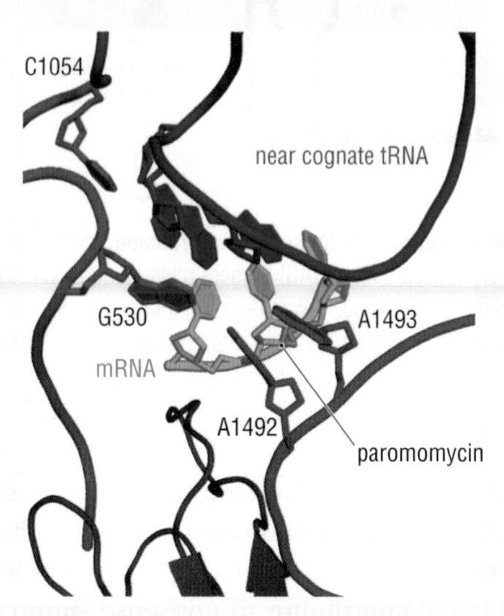

Figure 1 Atomic resolution structure of the decoding center of the small subunit of the ribosome bound by paromomycin reveals conformational rearrangement. Here we see nucleotides A1492 and A1493 in a 'flipped out' conformation while paromomycin is bound within helix 44. G530 is also found in an anti configuration in the presence of the antibiotic.

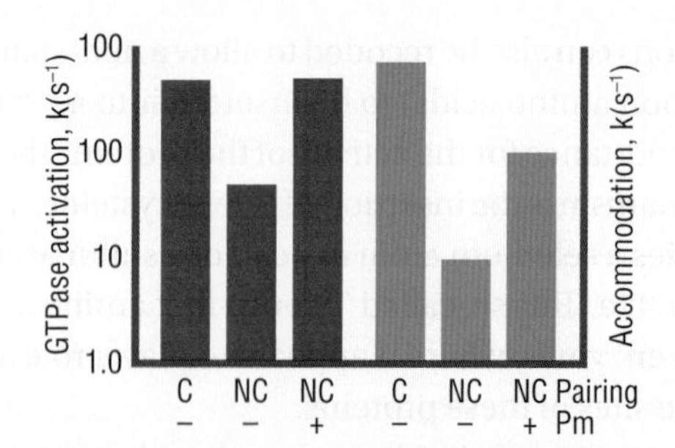

Figure 2 Pre-steady state kinetic studies indicate that rate constants for two forward steps, GTPase activation (red bars) and accommodation (blue bars), in the tRNA selection process are accelerated by paromomycin for the near-cognate tRNA. Pm for paromomycin, C for cognate, NC for near-cognate.

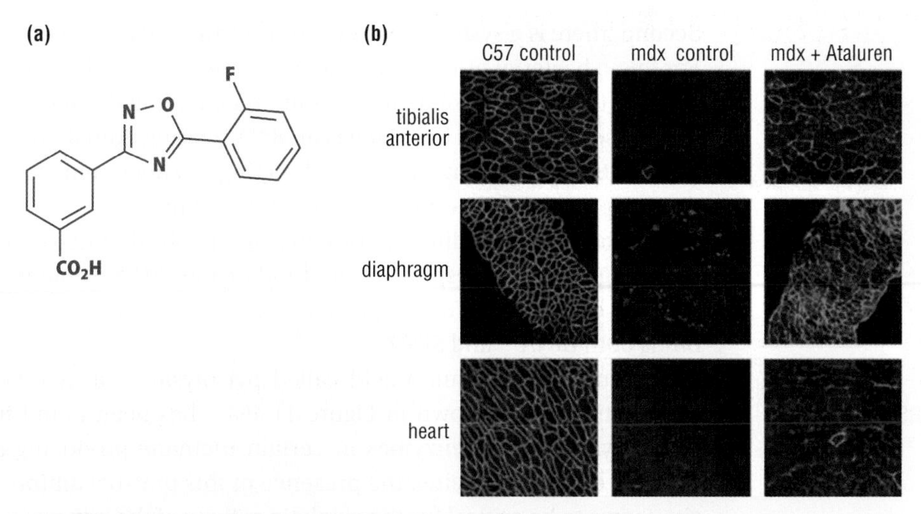

Figure 3 Ataluren as a therapeutic treatment for nonsense promoted diseases. (a) Chemical structure of PTC124 (Ataluren) that emerged from screen to identify compounds that increase misreading of stop codons by aminoacyl-tRNA. (b) Tissues from mdx mouse before and after treatment with Ataluren. Presence of green staining is indicative of expression of *mdx* gene in different tissues.

Peltz *et al.* Ataluren as an agent for therapeutic nonsense suppression. *Annual Review of Medicine*, 2013;**64**:407–425.

of tRNA selection has allowed for a much deeper mechanistic understanding of this process.

Therapeutic application of miscoding reagents for the treatment of genetic diseases

A practical application of these mechanistic insights has been the treatment of certain genetic diseases that result from mutations that place a premature stop codon within genes and thereby eliminate protein expression. Allan Jacobson and Stuart Peltz reasoned that drugs with properties comparable to aminoglycosides might promote the misreading of stop codons by near-cognate tRNA species and would thus allow for some expression of the mutated gene. Because aminoglycosides are broadly toxic to humans with extended use, drug screens were performed to identify compounds that might behave similarly. PTC124 or Ataluren (the trade name) is a small molecule that emerged from such a screen and very effectively promotes nonsense suppression in animal models for cystic fibrosis and muscular dystrophy; its structure is shown in Figure 3. Clinical trials on this compound in humans with either of these two diseases look quite promising.

Find out more

Carter AP, Clemons WM, Brodersen DE, *et al*. Functional insights from the structure of the 30S ribosomal subunit and its interactions with antibiotics. *Nature*, 2000;**407**:340–348.

Moazed D, Noller HF. Transfer RNA shields specific nucleotides in 16S ribosomal RNA from attack by chemical probes. *Cell*, 1986;**47**:985–994.

Pape T, Wintermeyer W, Rodnina MV. Conformational switch in the decoding region of 16S rRNA during aminoacyl-tRNA selection on the ribosome. *Nature Structural Biology* 2000;**7**:104–107.

Welch EM, Barton ER, Zhuo J, *et al*. PTC124 targets genetic disorders caused by nonsense mutations. *Nature*, 2007;**447**:87–91.

Related techniques

X-ray crystal structure determination; Section 19.14

Cell imaging; Section 19.13

Activity assays; Section 19.6

mammals, the regulatory RNA structure is more complex and seems to function when found anywhere within the 3′ untranslated region (UTR) of the mRNA (much as transcriptional enhancers can act by binding to DNA somewhat remote from the gene's promoter). In addition to these RNA elements, a number of other factors are required for selenocysteine incorporation, as illustrated in Figure 11.45b. First, there is a system for generating the selenocysteine-tRNA. This system includes a specialized tRNA (tRNA^{Sec}) that is charged with serine (by SerRS) and has an anticodon that matches the UGA stop codon; additionally, there are enzymes for converting the serine attached to this specialized tRNA to selenocysteine.

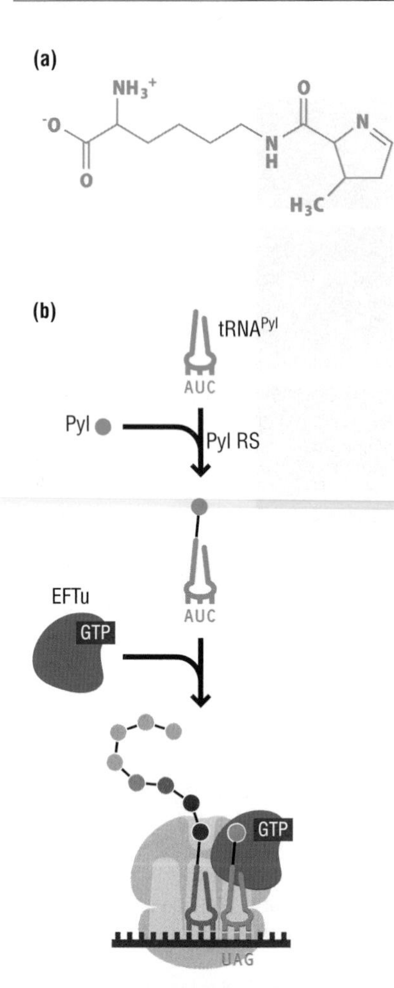

Figure 11.46 The recoding of UAG stop codons for pyrrolysine insertion. (a) The chemical structure of the non-standard amino acid pyrrolysine. (b) Schematic showing how a specialized tRNA is directly aminoacylated with pyrrolysine by a pyrrolysine specific aminoacyl-tRNA synthetase (PylRS). EFTu is thought to load this tRNA like a standard elongator tRNA, albeit in competition with release factor present in the cell that normally recognizes UAG.

Second, there is a system that determines which UGA codons will be recognized. For this, a bacterial protein, SelB, that is homologous to EFTu is key. SelB in *E. coli* appears to be bifunctional: in addition to delivering the selenocysteinyl-tRNA to the ribosome (as EFTu does) it has an RNA binding domain that allows it to interact with the SECIS stem-loop, thus directing selenocysteine to be incorporated in the polypeptide chain only at specific stop codons.

In eukaryotes, two different proteins are involved in the actual selenocysteine decoding event, one responsible for binding the SECIS element in the RNA (SBP2) and the other an EFTu-like protein with a C-terminal extension (eEFsec) that binds both $tRNA^{Sec}$ and SBP2.

A twenty-second amino acid called pyrrolysine – a modified lysine residue whose structure is shown in Figure 11.46a – has been found in the active sites of methyltransferase enzymes in certain methane-producing organisms. As in the case of selenocysteine, the presence of this unusual amino acid in the active site seems to be critical for the catalytic activity of the enzymes where it is found. Figure 11.46b shows how pyrrolysine is incorporated into these enzymes at UAG stop codons in somewhat the same way as selenocysteine, with both a specialized tRNA ($tRNA^{Pyl}$) and aminoacyl-tRNA synthetase (PylRS) being required; interestingly, in this case, tRNA loading is thought to be promoted by EFTu through the standard pathway. It is not yet known whether RNA elements that function like the SECIS element are also required to ensure that pyrrolysine is inserted into proteins only at certain UAG codons.

Programmed frameshifting is triggered by local mRNA elements

During translation, ribosomes move sequentially from one mRNA codon to the next by advancing three nucleotides along the mRNA. If the ribosome moves some different number of nucleotides (not comprising a multiple of 3), then this wholly changes the protein sequence decoded from the mRNA, and the reading frame of the mRNA is said to be shifted; several such shifts are depicted in Figure 11.47. By contrast, when the mRNA:tRNA complex moves in steps comprising integers of three (so steps of 6, 9, 12, etc.) the frame is maintained, though some number of amino acids encoded by the gene may be omitted (or duplicated).

The change of the reading frame, called **frameshifting**, has most commonly been seen to occur as a single nucleotide shift in the +1 or –1 direction. Such events can take place in very specific mRNA contexts (leading to the term 'programmed frameshift') and can control gene expression in bacterial, mammalian, and viral systems.

An illustrative example of frameshifting is in the expression of the bacterial translation termination factor, RF2, as shown in Figure 11.48. In 70% of bacterial species, the complete RF2 protein is encoded in two ORFs: a short one immediately followed by a longer one in the +1 frame. Almost without exception, the site of frameshifting within the ORF of the mRNA is demarked by the sequence CUU

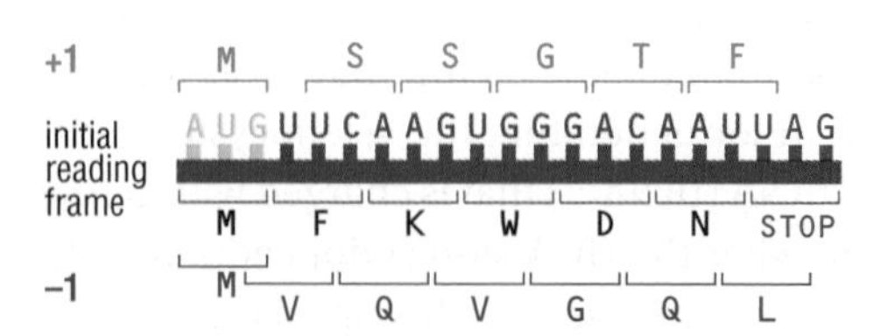

Figure 11.47 Frameshifting disrupts the reading frame of protein synthesis. The initial reading frame of mRNA can be shifted either in the forward or backward direction, here by +1 (blue) or –1 (pink) nucleotide, which changes the sequence of the protein (shown in single-letter code) synthesized from the mRNA.

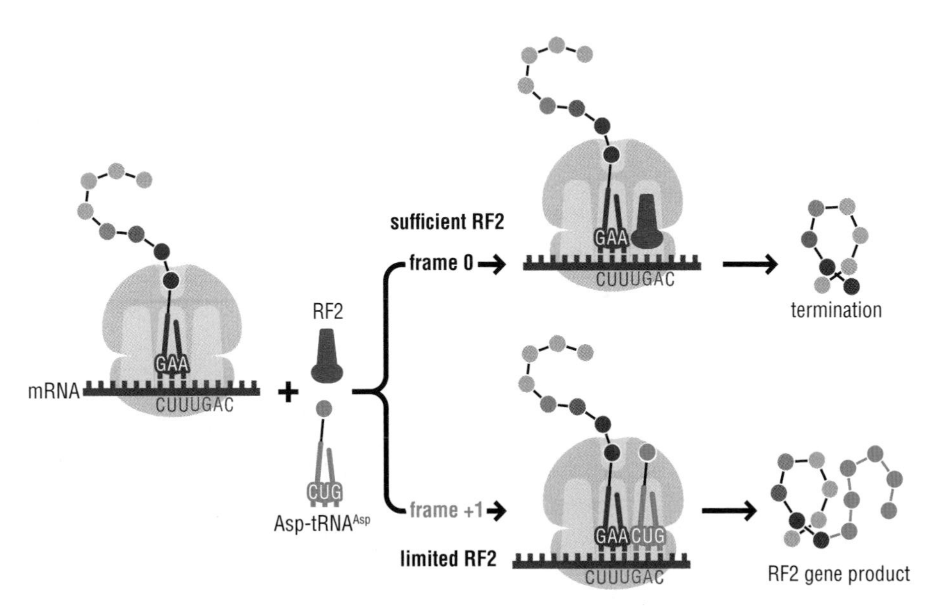

Figure 11.48 Programmed +1 frameshifting controls the amount of RF2 produced in bacteria. The initial reading frame (frame 0) of the mRNA encoding RF2 contains a stop codon (red) that prematurely terminates protein synthesis. When RF2 levels in the cell are low, termination is inefficient. Given sufficient time, the aminoacyl-tRNA in the P site of the ribosome can pair with the mRNA codon in the +1 reading frame, thus allowing sampling of the A site by a cognate Asp-tRNA^{Asp}, resetting the frame to generate full-length RF2 gene product.

UGA C, with the stop codon (shown in red in Figure 11.48) being recognized by RF2. When there is sufficient RF2 in the cell, the UGA codon is efficiently recognized by RF2 that is already present and translation is terminated. When the level of RF2 in the cell is low, this termination event is slow and +1 frameshifting occurs, leading to the production of the complete RF2 protein. A +1 frameshift is permitted because Leu-tRNA^{Leu} in the P site of the ribosome can pair with either of two different codons - most effectively with the CUU codon in the initial frame and less effectively with the near-cognate UUU codon in the second frame that is shifted in the +1 direction.

Following the frameshift, translation proceeds normally and RF2 is synthesized. Thus, the establishment of competition between somewhat slow termination (on UGA) and the sampling of a new frame (and a different tRNA selection event on GAC) results in the regulated expression of the RF2 gene. Other documented examples of +1 frameshifting are typically stimulated by kinetically sluggish decoding, for example, at codons decoded by tRNAs that are of low abundance in the cell (rare tRNAs).

Another well-known example of frameshifting is the -1 event that is necessary for the production of the polymerase gene product (Pol) in many **retroviruses** and retrotransposons. During expression of the viral genome, 90% of ribosomes terminate translation at the end of the *gag* gene to produce the main structural protein of the viral capsid. However, a small percentage of ribosomes undergo a -1 frameshift that results in the synthesis of the Gag protein fused to the *pol* gene product (Gag-Pol). As for the case of MuLV described above (which uses nonsense suppression to regulate read-through between the *gag* and *pol* genes), this -1 frameshifting provides a mechanism for the virus to modulate the relative amounts of the same two proteins, as depicted in Figure 11.49. In this case, the -1 frameshifting is thought to occur by slippage of the mRNA in the ribosome, facilitated by a downstream RNA pseudoknot element that seems to perturb interactions between the ribosome and the mRNA. The mRNA sequence where recoding occurs is said to be 'slippery' because the tRNA in the P site of the ribosome can bind reasonably well to both the initial and the -1 reading frames, thus allowing the reading frame to slip backward one nucleotide along the mRNA during the recoding event.

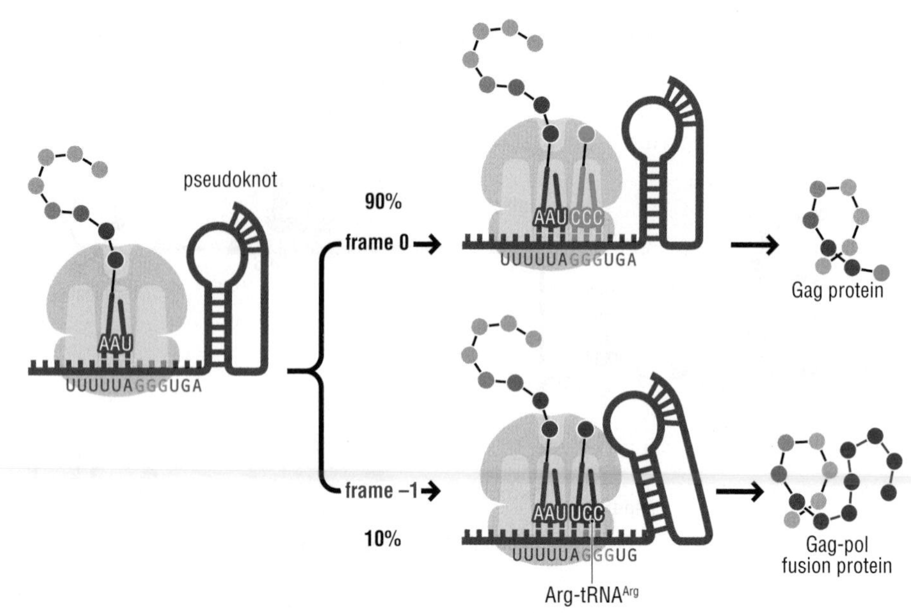

Figure 11.49 Programmed −1 frameshifting controls the relative amounts of Gag and Gag-Pol gene products for retroviruses. The viral Gag protein of the Rous sarcoma virus is produced from reading frame 0 of the mRNA. A pseudoknot element in the mRNA downstream of the programmed frameshift site stimulates ~10% of ribosomes to undergo a −1 frameshift, resetting the frame to generate sufficient full-length Gag-Pol fusion protein.

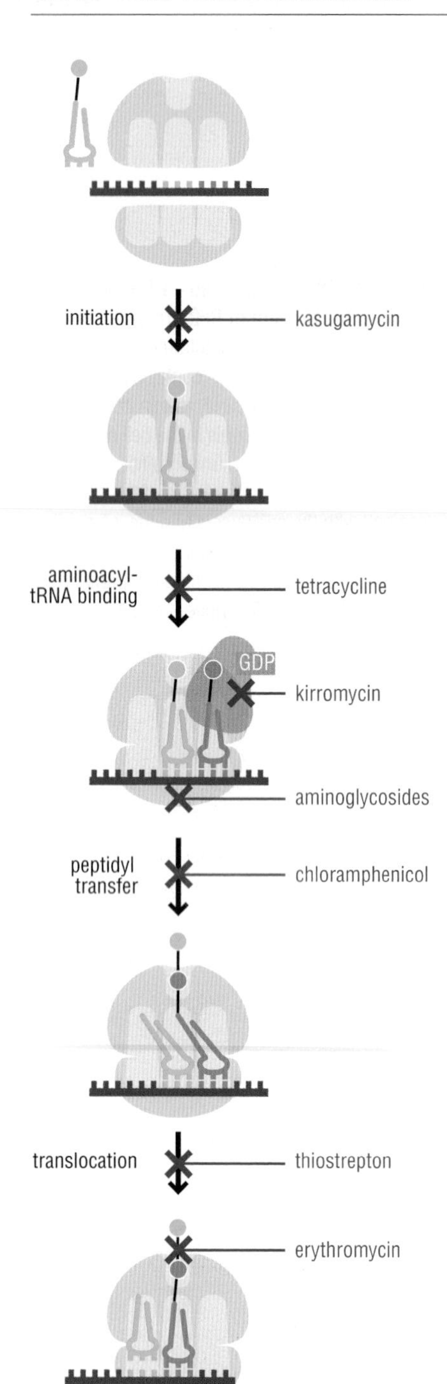

Figure 11.50 Distinct antibiotics inhibit the various steps in translation. Kasugamycin inhibits formation of the initiation complex. During elongation, tetracycline destabilizes binding of aminoacyl-tRNA to the A site, and kirromycin prevents the dissociation of EFTu from the ribosome. Aminoglycosides (such as streptomycin and kanamycin) are miscoding agents, which stimulate the ribosome to accept near-cognate tRNAs. Chloramphenicol inhibits the peptidyl transferase reaction and thiostrepton blocks the translocation reaction. Erythromycin blocks the exit channel through which synthesized peptide chains leave the ribosome.

11.14 ANTIBIOTICS THAT TARGET THE RIBOSOME

In the previous section, we saw how biological systems can flex the genetic code, and the way in which mRNAs can contribute to this flexing. However, other more invasive ways of interrupting the 'standard' process of translation exist. Primary among these is the use of antibiotics to disrupt translation - or even prevent it altogether - as we discover in this section.

Many antibiotics target the ribosome or translation factors

Antibiotics are small molecules that kill bacteria (bactericidal) or fungi (fungicidal) or stop these organisms from growing (bacteristatic or fungistatic). Antibiotics can be extremely effective drugs if they disrupt a critical cellular process in the bacteria or fungi without disrupting the equivalent process in mammals, if there is one. Mammalian cells, for example, do not have a cell wall like that of bacteria, and the antibiotic penicillin stops bacterial growth by blocking the synthesis of peptidoglycans required to build this cell wall. The antibiotic rifampicin selectively inhibits bacterial transcription by targeting a distinctive pocket on the bacterial RNA polymerase. Similarly, the seemingly subtle structural differences between the active sites of ribosomes from eukaryotes and bacteria can also be discriminated between by small molecules, and indeed antibiotics that disrupt many steps in protein synthesis have been identified, as denoted in Figure 11.50. Many of the antibiotics that we discuss throughout this book are produced in a variety of microbes, while others can be man-made; we discuss both classes in this section.

How do antibiotics target protein synthesis? Some antibiotics bind directly to the ribosome while others interact with translation factors to inhibit their function. Since antibiotics typically have a molecular weight of around 10^3 Da, 1000 times smaller than the size of a ribosome (2.5×10^6 Da) (see two examples in Figure 11.51), antibiotics typically must target functionally critical regions of the ribosome to be effective. Figure 11.52 shows how erythromycin (and others in this so-called macrolide class, such as telithromycin) sits neatly in the exit tunnel of the large subunit, blocking the growing peptide chain from entering the tunnel, and thus preventing translation from proceeding. Chloramphenicol binds more centrally in the active site of the large ribosomal subunit, blocking access to critical ribosomal nucleotide residues that are thought to be important for binding tRNAs and for the peptide bond forming and release reactions catalyzed by this active site. A number of antibiotics appear to inhibit more than one step in the translation cycle (e.g. the aminoglycoside paromomycin affects decoding, termination, and recycling), suggesting that functionally critical regions of the ribosome may be close together or that movement common to multiple steps might be targeted.

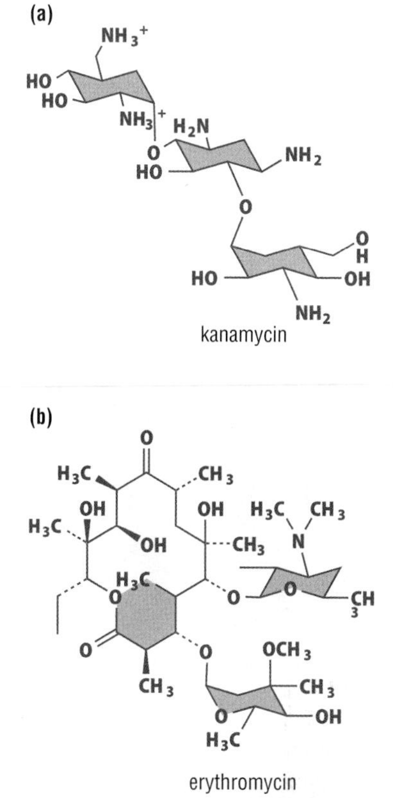

(a)

kanamycin

(b)

erythromycin

Figure 11.51 The chemical structures of antibiotics from two important classes. Two naturally occurring classes of antibiotics are the (a) aminoglycosides such as kanamycin and the (b) macrolides such as erythromycin.

Antibiotic resistance mutations can reveal how antibiotics and the ribosome function

Bacteria can become resistant to antibiotics; consistent with this, genetic mutations conferring resistance to antibiotics that disrupt translation are often found in components of the ribosome. By identifying these mutations, we can gain insights into how an antibiotic acts and also about how the ribosome functions more generally.

For example, streptomycin is an antibiotic that acts as a **miscoding agent** - it causes ribosomes to misread mRNA with the result that synthesized proteins have errors in their amino acid sequence. Streptomycin exerts its effect by binding to the small ribosomal subunit and, in the presence of near-cognate tRNAs, induces conformational changes that are normally seen only in the presence of cognate tRNA. These rearrangements trick the ribosome into incorporating an incorrect amino acid; more mechanistic insight into the function of such aminoglycoside antibiotics are found in Experimental approach 11.4. Streptomycin-resistant bacteria are routinely found to have mutations in the small subunit protein S12 that is located near the streptomycin binding pocket proximal to the decoding center on the interface of the subunit. In the absence of streptomycin, these resistant bacteria show **restrictive** protein synthesis, meaning they have unusually accurate and slow protein synthesis. Thus, the resistance mutations in the S12 protein appear to counteract the effects of streptomycin by increasing the inherent fidelity of the ribosome. In the absence of streptomycin, the resistant bacteria grow more slowly than their wild-type relatives.

Additional mutations can be isolated that increase the growth rates of these streptomycin-resistant bacteria (carrying S12 mutations). These additional mutations are typically found in two small subunit proteins, S4 and S5, which interact with one another on the solvent side of the small subunit. These mutations on their own normally cause bacteria to have highly error-prone translation (referred to as a ***ram*** phenotype, for **ribosomal ambiguity**). The analysis of this series of variant ribosomes has over many years yielded some of the most important insights into how the ribosome balances the demands of high speed and fidelity during protein synthesis.

A general observation that has emerged from the analysis of antibiotic resistance is that mutations tend to be found in the ribosomal proteins rather than the rRNA. Given that the ribosome is fundamentally an RNA machine, why might this

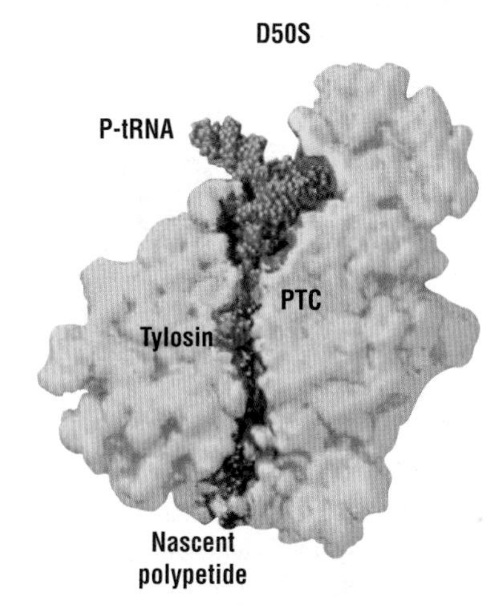

Figure 11.52 Many antibiotics function by sterically obstructing the peptide exit tunnel of the ribosome. This cross-section view of the ribosome (cut through the interior of the large subunit) shows how the exit tunnel is blocked by tylosin (an erythromycin derivative). PDB code 1K9M.

(a)

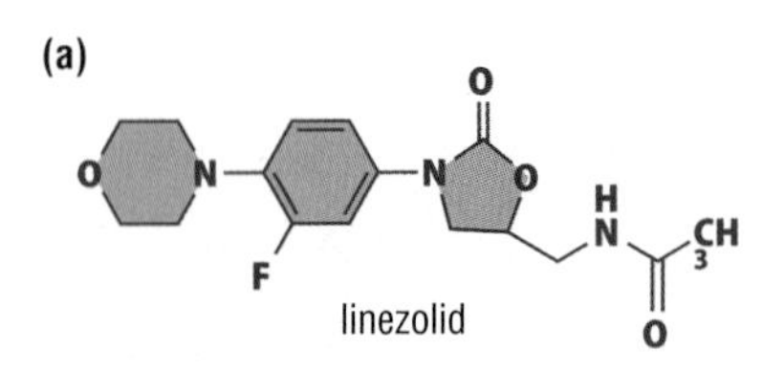

(b)

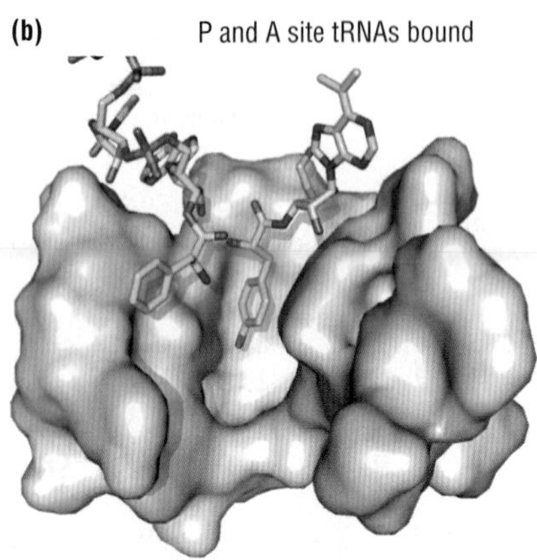

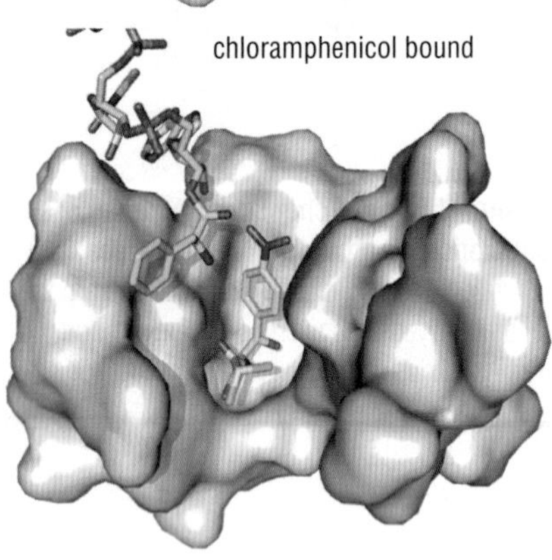

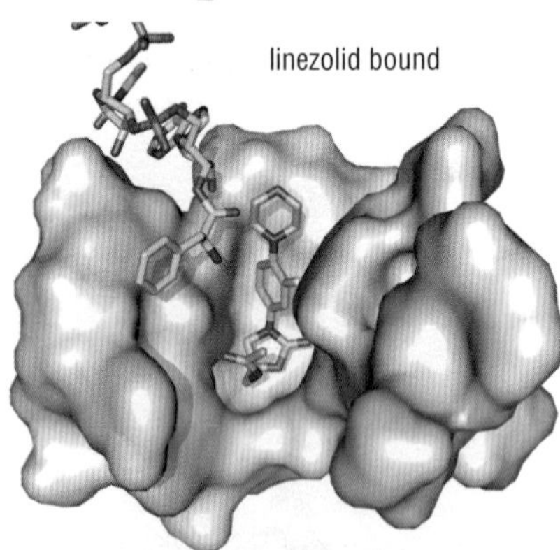

Figure 11.53 Structure and function of new class of synthetic antibiotics (the oxazolidinones). (a) The chemical structure of linezolid, the founding member of a new class of approved synthetic antibiotics that do not resemble any naturally occurring ones. (b) X-ray structure comparing the P (green) and A (pink) site tRNAs, chloramphenicol, and linezolid bound in similar positions in the peptidyl transferase center of the ribosome.

be the case? The answer seems to be that although bacteria have a single copy of the gene for each ribosomal protein, they have multiple copies of the rRNA genes. Therefore resistance is only observed when a reasonable fraction of the rRNA genes acquire the requisite mutation.

An interesting exception to this general rule is seen in bacterial resistance to the macrolide antibiotics such as erythromycin. In this case, the most common mechanism for resistance is the acquisition by the bacteria of a methylase gene that specifically modifies the binding site for this class of antibiotic. Methylation of a particular adenosine (A2058) in the active site of the large subunit prevents binding of macrolide antibiotics and thus confers resistance. While resistance to erythromycin can also be achieved by direct mutation of the same nucleotide (A2058) in multiple 23S rRNA genes, the acquisition of the methylase gene through horizontal transfer of DNA from other bacteria is apparently an easier means of acquiring resistance.

New classes of antibiotics continue to be discovered

The most prolific antibiotic producers are organisms from the genus *Streptomyces* (Gram-positive bacteria) that use these small molecules (often referred to as secondary metabolites) to challenge the microbial competition in their environment. Over time, the targeted bacteria have evolved a variety of mechanisms to combat the antibiotics, and these mutations determine the clinical efficacy of the antibiotic. Indeed, there is currently an acute need for the development of new classes of antibiotics for which resistance has not yet emerged. A recent success has been the identification of a new synthetic antibiotic class, the oxazolidinones, whose founding member, linezolid, was rather recently approved for use in humans for last line of defense antibacterial therapy. This antibiotic, whose structure is depicted in Figure 11.53a, is structurally unrelated to naturally occurring antibiotics and so resistance has been relatively slow to develop. Mechanistically, it appears to target peptide bond formation by binding directly to the large subunit catalytic center, much like chloramphenicol and others. This binding is depicted in Figure 11.53b.

Ribotoxins and ribosome-inactivating proteins target a highly conserved loop in eukaryotic 28S rRNA and inactivate interactions with the elongation factors

Ribosome-inactivating proteins (RIPs) and ribotoxins are cytotoxic enzymes that inhibit translation in eukaryotes by modifying a highly conserved and functionally important loop in the 28S rRNA. RIPs are *N*-glycosidases most often produced by plants; the best known is the highly poisonous ricin from *Ricinus communis*. Ribotoxins are ribonucleases secreted by certain fungi with the most noted, alpha-sarcin, produced by *Aspergillus giganteus*. Both cytotoxins target a surface feature of the large subunit rRNA aptly named the sarcin/ricin loop. Ribotoxins cleave a specific phosphodiester bond and RIPs depurinate a specific nucleotide in this region, inactivating interactions with the elongation factors eEF1A and eEF2 and leading to a complete shutdown of protein synthesis. Studies of these toxins and their interactions with the ribosome have led to important insights into the role of this highly conserved rRNA element in overall ribosome function.

SUMMARY

- Translation is the process by which the genetic information, stored as polymers of four nucleotides, gets transformed into proteins composed of amino acids.
- The genetic code specifies how each three-nucleotide codon corresponds to a specific amino acid, or stop signal, in the mRNA transcript.

MOLECULAR COMPONENTS

- tRNAs are bifunctional molecules that decipher the genetic information through Watson–Crick pairing interactions with the mRNA, bringing the activated amino acid along for peptide bond formation.
- Aminoacyl-tRNA synthetases are responsible for the first, and primary, event in decoding as they determine which of the 40 or so tRNAs in the cell get attached to which amino acids.
- The ribosome is the macromolecular ribonucleoprotein machine that coordinates the process of translation through events in two distinct subunits, one primarily responsible for decoding and the other for making peptide bonds.
- Many extra-ribosomal protein factors are GTPases that facilitate ribosome-based events through GTP hydrolysis-dependent conformational rearrangements. Other translation factors act by simply binding and occluding specific sites on the ribosome, sometimes by mimicking tRNAs.

THE MECHANISM OF TRANSLATION

- Translation can be divided into multiple stages: initiation, elongation, termination, and ribosome recycling.
- Each stage of protein synthesis depends on the action of the ribosome, tRNAs, mRNA, and a variety of extraribosomal protein factors.
- Bacteria identify the AUG start site through direct interactions between the small subunit rRNA and the Shine–Dalgarno region of the mRNA.
- Eukaryotes identify the AUG start site through a more complex 5′ to 3′ scanning process.
- Elongation is composed of three basic steps: tRNA selection (or decoding), peptide bond formation, and translocation of the mRNA–tRNA complex.
- Termination involves recognition of stop codons by specific protein factors that promote peptide release.
- Ribosome recycling involves dissociation of the mRNA and tRNAs, and dissociation of the two ribosomal subunits and is quite different in bacteria and eukaryotes.
- Ribosome rescue pathways help to recover arrested ribosome complexes in bacteria and eukaryotes.
- Recoding events allow for an alternative interpretation of the standard genetic code in an mRNA-specific 'programmed' fashion.
- Selenocysteine and pyrrolysine are two unusual amino acids that are specifically incorporated by recoding stop codons in a variety of organisms.
- Frameshifting is often used by viruses to regulate the expression levels of key viral proteins.
- Antibiotics target different key steps in protein synthesis.

QUESTIONS

11.1 OVERVIEW OF TRANSLATION

1. Define genotype and phenotype. Explain the statement 'translation transforms genotype into phenotype.'

2. Describe the role of tRNA in the flow of genetic information from genotype to phenotype.

3. Which of the following statements accurately reflects a property or function of the ribosome?

a. The large subunit deciphers the RNA code.
b. The ribosome does not require an input of energy to carry out protein synthesis.
c. RNA elements play a crucial role in ribosome function.
d. The small subunit mediates the formation of peptide bonds.
e. Translation factors are permanent components of the ribosome that coordinate the progression of translation events.

Challenge question

4. The error rate for DNA polymerase in DNA replication is one to two orders of magnitude lower than that of RNA polymerase in transcription and the ribosome in translation. Propose an explanation for this difference.

11.2 TRANSFER RNA AND THE GENETIC CODE

1. The genetic code consists of 64 possible codons, and yet there are typically only 40 different tRNAs. Explain how the cell deals with this discrepancy.

11.3 AMINOACYL-tRNA SYNTHETASES

1. Although tRNAs have very similar three-dimensional shapes, they are still recognized by specific aminoacyl-tRNA synthetases and become charged with the appropriate amino acid with high accuracy. What characteristics of the tRNAs and the aminoacyl-tRNA synthetases account for this apparent discrepancy?

2. Explain how some bacteria are able to survive without an aminoacyl-tRNA synthetase for asparagine and glutamine.

11.4 STRUCTURE OF THE RIBOSOME

1. The components of the ribosome can be classified as being RNA and protein, or as belonging to large and small subunits. For each classification, compare and contrast the function of the components.

2. What evidence suggests that the rRNAs play a crucial role in the function of the ribosome?

11.5 THE TRANSLATION CYCLE: THE RIBOSOME IN ACTION

1. Match each of the following events with the appropriate stage in the translation cycle:

A: Initiation; B: Elongation; C: Termination; D: Ribosome recycling

a. Peptide bond formation
b. Decoding the AUG codon
c. Translocation of the ribosome
d. Uncharged tRNA and the mRNA are released from the ribosome.
e. Most steps during this event involve only the small subunit of the ribosome.
f. Release factors recognize the stop codon.

11.6 PROTEIN FACTORS CRITICAL TO THE TRANSLATION CYCLE

1. In general, describe the function and the significance of each of the following factors:

a. GTPases
b. GAPs
c. GEFs

11.7 TRANSLATION INITIATION – SHARED FEATURES IN BACTERIA AND EUKARYOTES

1. What makes the bacterial and eukaryotic initiator tRNAs different from other tRNAs, and how do these differences help them to carry out initiation rather than elongation?

11.8 BACTERIAL TRANSLATION INITIATION

1. What is the function of the anti-Shine–Dalgarno sequence found in the bacterial small subunit rRNA?

2. Which of the following correctly describes the sequence of events in the initiation of translation in bacteria?

a. Displacement of initiation factors, IF1 and IF3 bind to small ribosomal subunit, initiator tRNA binds to P site, large subunit and small subunit join.
b. IF1 and IF3 bind to small ribosomal subunit, initiator tRNA binds to P site, large subunit and small subunit join, displacement of initiation factors.
c. Initiator tRNA binds to P site, IF1 and IF3 bind to small ribosomal subunit, large subunit and small subunit join, displacement of initiation factors.
d. IF1 and IF3 bind to small ribosomal subunit, initiator tRNA binds to P site, displacement of initiation factors, large subunit and small subunit join.

11.9 EUKARYOTIC TRANSLATION INITIATION

1. Which eukaryotic translation initiation factors are thought to promote circularization of the mRNA and why might this to be useful for the cell?

2. Compare the factors involved in deposition of the initiator tRNA in the P site in bacteria and eukaryotes.

Challenge question

3. Compare the mechanism by which bacteria and eukaryotes identify the first AUG for translation.

11.10 TRANSLATION ELONGATION: DECODING, PEPTIDE BOND FORMATION, AND TRANSLOCATION

1. Explain the role of accommodation is peptide bond formation within the ribosome.

2. Explain the role of molecular mimicry in the translation process.

3. What two classes of proteins act as guardians against errors in the amino acid sequence during translation?

Challenge question

4. Tetracycline is a broad spectrum antibiotic that blocks bacterial protein synthesis by blocking recognition of the cognate aminoacyl tRNA by the A site of the ribosome. Interestingly, bacterial cells that are sensitive to tetracycline are first suppressed from growth (not directly killed) but then die due to the inability to grow. This is significant because cells treated with tetracycline can be studied while in the suppressed state prior to cell death. In order test the hypothesis that tetracycline inhibits the proofreading stage rather than the initial selection stage, measurements of GTPase activity were taken in cells in the suppressed state.

a. Why was GTPase activity a good choice of activity to measure to test this hypothesis?
b. What result would you expect if the proofreading stage is the one that is inhibited?

11.11 TRANSLOCATION TERMINATION, RECYCLING, AND REINITIATION

1. Compare the mechanisms by which bacteria and eukaryotic cells minimize translation of incomplete or degraded RNAs.

Challenge question

2. The mitochondria of eukaryotic cells contain two forms of the EF G protein. EF G1mt acts an elongation factor as expected, but EF G2mt acts as a ribosome recycling factor. Individuals with certain single nucleotide polymorphisms (SNPs) in the EF G2mt gene are known to have an increased number of side effects associated with taking statin drugs that treat high cholesterol. The most common side effect is muscle weakness, but they can be as severe as muscle breakdown and liver failure. Propose a mechanism by which SNPs in the EF-G2mt gene may lead to these side effects.

11.12 RIBOSOME RESCUE IN BACTERIA AND EUKARYOTES

1. Draw a picture of a tmRNA used to rescue ribosomes stalled at the end of incomplete RNAs in bacteria. Label each important domain of the RNA and explain why it is important in the rescue process.

2. Explain the mechanism by which eukaryotic cells ensure that incomplete RNAs are never translated.

11.13 RECODING: PROGRAMMED STOP CODON READ-THROUGH AND FRAMESHIFTING

Challenge question

1. Compare and contrast recoding a stop codon with reassignment of a stop codon.

11.14 ANTIBIOTICS THAT TARGET THE RIBOSOME

1. Explain why the ribosome is a good target for antibiotic development.

2. Compare the general mechanisms that signal poor nutrient status and reduction of global translation in bacteria and eukaryotes.

Challenge question

3. Tetracycline was once among the most widely used antibiotics in the United States. It was effective, relatively cheap, taken orally, and had very few side effects. Today, however, tetracycline has only limited use in treatment of clinical infections because resistance has appeared in many groups of bacteria. One mechanism for tetracycline resistance involves the acquisition of a new gene (tetO) from environmental transfer. This gene encodes a known ribosomal protection protein (RPP). RPPs display sequence similarity to the elongation factors EFG and EFTu (EFs).
  a. Why would sequence similarity to these two elongation factors be a benefit in conferring tetracycline resistance?
  b. If the RPP just looked like an elongation factor then, upon exposure to tetracycline, it would be blocked from entering the ribosome just like the EFG and EFTu. Therefore, describe one additional feature (activity) that might you see in a RPP that is not a function of EFs.
  c The sequence similarity between RPPs and EFs suggests that the RPPs may be evolutionarily derived from the EFs. Keeping in mind that the functions of the elongation factors are critical for cell survival, explain one possible mechanism by which RPPs could have been derived from the EFs.

EXPERIMENTAL APPROACH 11.1 – CHEMICAL MODIFICATION ANALYSIS TO STUDY RNA STRUCTURE AND FUNCTION WITHOUT HIGH-RESOLUTION STRUCTURES

1. What was the purpose of the chemical modification analysis represented in Figure 1?

2. What does the change in the band intensity in lanes 2–5 represent in terms of the elongation process?

3. What conclusion could be drawn based on the type of results represented in Figure 1?

Challenge question

4. In lanes 1–5 on each gel, different substrates were incubated with the ribosome prior to the chemical modification. Based upon the substrates added, what stage in translation elongation do each of the lanes represent?

EXPERIMENTAL APPROACH 11.2 – PHYLOGENETIC COVARIATION ANALYSIS

1. With regard to rRNA structure, what is covariance?

2. In this article, what was the covariance used to locate?

3. Briefly describe what Shine and Dalgarno noticed and what they proposed.

4. What is a specificity swap experiment?

5. What do the colored lines in Figure 2 represent, and what conclusion could be made based on the figure?

FURTHER READING

11.1 OVERVIEW OF TRANSLATION

Ramakrishnan V. Ribosome structure and the mechanism of translation. *Cell*, 2002;**108**:557–572.

11.2 tRNA AND THE GENETIC CODE

Chapeville F, Lipmann F, Von Ehrenstein G, *et al.* On the role of soluble ribonucleic acid in coding for amino acids. *Proceedings of the National Academy of Sciences of the U S A*, 1962;**15**:1086–1092.

Nirenberg MW, Matthaei JH. The dependence of cell-free protein synthesis in *E. coli* upon naturally occurring or synthetic polyribonucleotides. *Proceedings of the National Academy of Sciences of the U S A*, 1961;**47**:1588–1602.

11.3 AMINOACYL-TRNA SYNTHETASES

Fersht AR. Sieves in sequence. *Science*, 1998;**280**:541.

Hou YM, Schimmel P. A simple structural feature is a major determinant of the identity of a transfer RNA. *Nature*, 1988;**333**:140–145.

Ling J, Reynolds N, Ibba M. Aminoacyl-tRNA synthesis and translational quality control. *Annual Review of Microbiology*, 2009;**63**:61–78.

11.4 STRUCTURE OF THE RIBOSOME

Ban N, Nissen P, Hansen J, *et al.* The complete atomic structure of the large ribosomal subunit at 2.4Å resolution. *Science*, 2000;**289**:905–920.

Wimberly BT, Brodersen DE, Clemons WM Jr, *et al.* Structure of the 30S ribosomal subunit. *Nature*, 2000;**407**:327–339.

Yusupov MM, Yusuova GZ, Baucom A, *et al.* Crystal structure of the ribosome at 5.5Å resolution. *Science*, 2001;**292**:883–896.

11.5 THE TRANSLATION CYCLE: THE RIBOSOME IN ACTION

Moazed D, Noller HF. Intermediate states in the movement of transfer RNA in the ribosome. *Nature*, 1989;**342**:142–148.

11.6 PROTEIN FACTORS CRITICAL TO THE TRANSLATION CYCLE

Brodersen DE, Ramakrishnan V. Shape can be seductive. *Nature Structural Biology*, 2003;**10**:78–80.

Rodnina MV, Stark H, Savelsbergh A, *et al.* GTPases mechanisms and functions of translation factors on the ribosome. *Biological Chemistry*, 2000;**381**:377–387.

Valasek LS. 'Ribozoomin'—translation initiation from the perspective of ribosome-bound initiation factors. *Current Protein & Peptide Science*, 2012;**13**:305–330.

11.7 TRANSLATION INITIATION – SHARED FEATURES IN BACTERIA AND EUKARYOTES

Mayer C, Stortchevoi A, Köhrer C, Varshney U, RajBhandary UL. Initiator tRNA and its role in initiation of protein synthesis. *Cold Spring Harbor Symposia on Quantitative Biology*, 2001;**66**:195–206.

Simonetti A, Marzi S, Jenner L, *et al.* A structural view of translation initiation in bacteria. *Cellular and Molecular Life Sciences*, 2009;**66**:423–436.

11.8 BACTERIAL TRANSLATION INITIATION

Shine J, Dalgarno L. Determinant of cistron specificity in bacterial ribosomes. *Nature*, 1975;**254**:34–38.

11.9 EUKARYOTIC TRANSLATION INITIATION

Kapp LD, Lorsch J. The molecular mechanics of eukaryotic translation. *Annual Review of Biochemistry*, 2004;**73**:657–704.

Sonenberg N, Hinnebusch AG. Regulation of translation initiation in eukaryotes: mechanisms and biological targets. *Cell*, 2009;**136**: 731–745.

Hinnebusch AG, Lorsch JR. The mechanism of eukaryotic translation initiation: new insights and challenges. *Cold Spring Harbor Perspectives in Biology*, 2012;**4**:a 011544.

11.10 TRANSLATION ELONGATION: DECODING, PEPTIDE BOND FORMATION, AND TRANSLOCATION

Beringer M, Rodnina MV. The ribosomal peptidyl transferase. *Molecular Cell*, 2007;**26**:311–321.

Daviter T, Gromadski KB, Rodnina MV. The ribosome's response to codon-anticodon mismatches. *Biochimie*, 2006;**88**:1001–1011.

Moore PB, Steitz TA. The structural basis of large subunit function. *Annual Review of Biochemistry*, 2003;**72**:813–850.

Shoji S, Walker SE, Fredrick K. Ribosomal translocation: One step closer to the molecular mechanism. *ACS Chemical Biology*, 2009;**4**:93–107.

11.11 TRANSLATION TERMINATION, RECYCLING, AND REINITIATION

Dever TE, Green R. The elongation, termination, and recycling phases of translation in eukaryotes. *Cold Spring Harbor Perspectives in Biology*, 2012;**4**:1–16.

Youngman E, McDonald ME, Green R. Peptide release on the ribosome: mechanisms and implications for translational control. *Annual Review of Microbiology*, 2008;**62**:353–373.

Zaher HS, Green R. Fidelity at the molecular level: Lessons from protein synthesis. *Cell*, 2009;**136**:746–762.

11.12 RIBOSOME RESCUE IN BACTERIA AND EUKARYOTES

Karzai AW, Roche ED, Sauer RT. The ssrA-smpB system for protein tagging, directed degradation and ribosome rescue. *Nature Structural Biology*, 2000;**7**:449–455.

Moore SD, Sauer RT. The tmRNA system for translational surveillance and ribosome rescue. *Annual Review of Biochemistry*, 2007;**76**: 101–124.

Shoemaker CJ, Green R. Translation drives mRNA quality control. *Nature Structural & Molecular Biology*, 2012;**19**:594–601.

Stalder L, Muhlemann O. The meaning of nonsense. *Trends in Cell Biology*, 2008;**18**:315–321.

11.13 RECODING: PROGRAMMED STOP CODON READ-THROUGH AND FRAMESHIFTING

Craigen WJ, Caskey CT. Expression of peptide chain release factor 2 requires high-efficiency frameshift. *Nature*, 1986;**322**:273–275.

Farabaugh PJ. Programmed translational frameshifting. *Annual Review of Genetics*, 1996;**30**:507–528.

Gesteland RF, Atkins JF. Recoding: dynamic reprogramming of translation. *Annual Review of Biochemistry*, 1996;**65**:741–768.

Sheppard K, Yuan J, Hohn MJ. From one amino acid to another: tRNA-dependent amino acid biosynthesis. *Nucleic Acids Research*, 2008;**36**:1813–1825.

11.14 ANTIBIOTICS THAT TARGET THE RIBOSOME

Carter AP, Clemons WM, Brodersen DE, *et al.* Functional insights from the structure of the 30S ribosomal subunit and its interactions with antibiotics. *Nature*, 2000;**407**:340–348.

Elseviers D, Gorini L. Direct selection of mutants restricting efficiency of suppression and misreading levels in *E. coli* B. *Molecular and General Genetics*, 1975;**137**:277–287.

Hansen JL, Ippolito JA, Ban N, *et al.* The structures of four macrolide antibiotics bound to the large ribosomal subunit. *Molecular Cell*, 2002;**10**:117–128.

Schlunzen F, Zarivach R, Harms J, *et al.* Structural basis for the interactions of antibiotics with the peptidyl transferase centre in eubacteria. *Nature*, 2001;**413**:814–821.

Vicens Q, Westhof E. RNA as a drug target: the case of aminoglycosides. *Chembiochem*, 2003;**4**:1018–1023.

Wilson DN. On the specificity of antibiotics targeting the large ribosomal subunit. *Annals of the New York Academy of Sciences*, 2011;**1241**:1–16.

Wilson DN, Schluenzen F, Harms JM, Starosta AL, Connell SR, Fucini P. The oxazolidinone antibiotics perturb the ribosomal peptidyl-transferase center and effect tRNA positioning. *Proceedings of the National Acadamy of Sciences of the U S A*, 2008;**105**:13339–13344.

Regulation of translation

INTRODUCTION

The importance of regulation in ensuring the correct and appropriate execution of the processes on which life depends is a recurring theme throughout this book. The process of translation is no exception. In this chapter, we explore the key strategies that are employed by both bacteria and eukaryotes to ensure that the translation machinery operates as it should, and that translation as a whole yields the correct end-products.

During the chapter, we will consider the ways in which the cell regulates translation both globally and at the level of specific messenger RNAs (mRNAs) to achieve protein synthesis at the right time, in the right place as demanded during an organism's development and ongoing existence. We will see how the diverse strategies employed to achieve regulation in both bacteria and eukaryotes have conceptual similarities – mirroring the way that the molecular machinery that conducts translation is conserved across all organisms. Let us begin by considering how protein synthesis in both bacteria and eukaryotes is regulated at a global level – at the level of an entire protein population.

12.1 GLOBAL REGULATION OF INITIATION IN BACTERIA AND EUKARYOTES

Translation is globally regulated in response to amino acid starvation

Protein synthesis in the cell needs to be globally regulated in response to external conditions and stimuli. Protein synthesis rates should be high when nutrients are plentiful, to allow maximal cell growth, and low in times of nutrient deprivation. Both bacteria and eukaryotes respond generally to the levels of precursors in the cell; for translation, the precursors are the amino acids. Under normal circumstances, transfer RNAs (tRNAs) are immediately aminoacylated once released from a round of elongation on the ribosome, and are bound in the cell by EFTu (or eEF1A) once they are loaded with the appropriate amino acid. As such, uncharged tRNA – a tRNA without an amino acid attached – is an indicator of low levels of amino acid and a trigger for global shutdown of protein synthesis in both bacterial and eukaryotic cells. However, the *mechanisms* of response are completely different in these two systems.

In bacteria, the uncharged tRNA induces a stress response when it begins to compete with decreasing concentrations of ternary complex (EFTu-guanosine triphosphate (GTP)-aa-tRNA) and binds directly to the ribosomal A site, as illustrated in Figure 12.1a. This event immediately blocks overall translation as the A site of the cellular population of ribosomes is now occupied with unproductive pseudo-substrate. More importantly, however, this binding event on the ribosome recruits a protein called RelA to the complex. Once recruited, RelA synthesizes

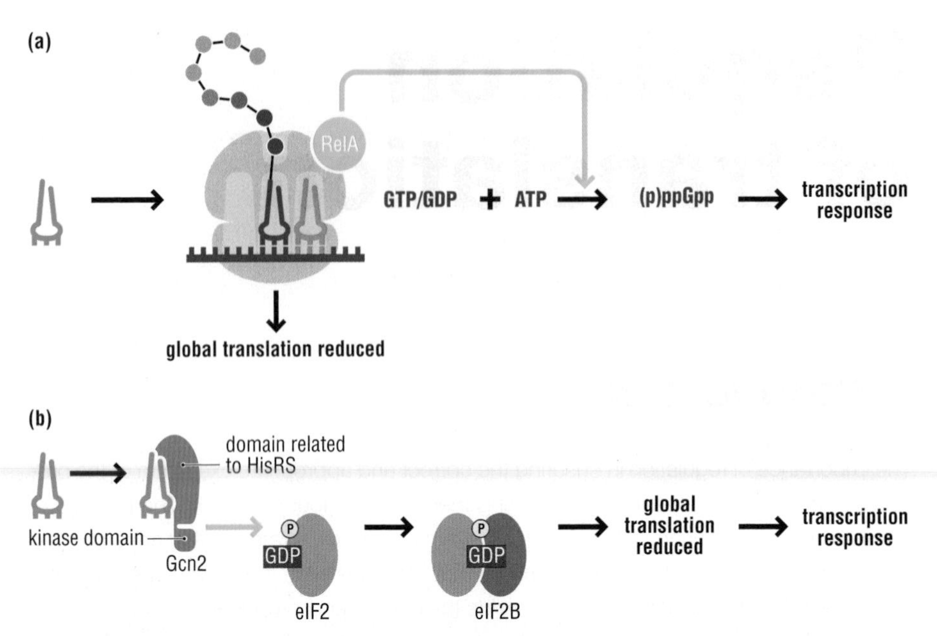

Figure 12.1 Amino acid starvation triggers distinct responses in bacteria and eukaryotes for global translational shutdown. (a) In bacteria, when amino acids are depleted, uncharged tRNAs become increasingly abundant and eventually compete for binding to the ribosomal A site. RelA protein recognizes this ribosomal complex, in turn stimulating the production of (p)ppGpp by RelA, and activation of the stringent response pathway. (b) In eukaryotes, amino acid depletion also leads to increased amounts of uncharged tRNA. In this case, these tRNAs are bound by a protein factor, Gcn2, thus activating a kinase domain on Gcn2 which targets eIF2. Once phosphorylated, eIF2 binds with high affinity to its GEF, eIF2B, thus effectively depleting cellular eIF2 and shutting down overall translation.

high (mM) concentrations of (p)ppGpp (a pentaphosphate guanine nucleotide known as 'magic spot') from GTP/GDP and ATP precursors. The nucleotide (p) ppGpp itself has profound effects on transcription, which include a decrease in the overall quantity of stable RNAs (such as ribosomal RNA (rRNA) and tRNA) and the induction of stress response factors that go about the task of replenishing the cell's declining resources (for example, inducing the expression of genes involved in amino acid biosynthesis and transport). This process is broadly referred to as the 'stringent response' and is the best understood example in bacteria of a global translational response.

The response in eukaryotes is quite distinct, as depicted in Figure 12.1b. There, the uncharged tRNA binds to a protein Gcn2 (in yeast), and more specifically to a domain in this protein related to the histidyl-tRNA synthetase (HisRS). Gcn2 appears to bind uncharged tRNAs with higher affinity than the corresponding charged tRNAs. Once Gcn2 binds to the uncharged tRNA, its kinase domain is activated and phosphorylates the translational initiation factor eIF2. We remember from Section 11.9 that eIF2 is a GTPase that is responsible for binding initiator $tRNA_i^{Met}$ and guiding it into the P site during the initiation process. Phosphorylation of eIF2 increases its affinity for its GEF, eIF2B, to the point where active eIF2 is effectively depleted from the cell.

The result of this phosphorylation event is thus an overall shutdown of protein synthesis. However, as we saw in bacteria, the cells also respond to this stress by inducing the expression of certain genes that are important for restoring cellular resources. While some of this induction takes place at the level of transcription, we will see in Section 12.2 how translation of the protein Gcn4 (a transcriptional activator of many amino acid biosynthetic genes) is up-regulated during this global translational shutdown.

Much global translational control is achieved through modulation of eIF2 and its GEF partner eIF2B

The example above shows how a general translation factor (eIF2) can be directly modified to globally affect the efficiency of cellular translation. This type of mechanism is broadly utilized in eukaryotic cells, with proteins involved in multiple different steps in the initiation process serving as targets of regulation.

Phosphorylation of eIF2, as illustrated in Figure 12.2, is a common means of affecting the equilibrium between free eIF2 and eIF2B:eIF2; the specific phosphorylation of serine 51 on the alpha subunit of eIF2 (eIF2alpha) results in the constitutive binding of eIF2–GDP to eIF2B. This binding interaction sequesters eIF2 in its GDP-bound state, thus preventing the protein from participating in initiation. In this role, phosphorylated eIF2 acts as a competitive inhibitor of the GEF function of eIF2B. The overall outcome of phosphorylation of eIF2 is the global down-regulation of translation.

At least three different kinases that phosphorylate serine 51 of eIF2alpha in response to stress have been identified in a variety of systems. PKR encodes a mammalian double-stranded RNA-activated kinase that is important in the cellular response to viral invasion; GCN2, as mentioned above, is a member of a kinase family that is responsive to amino acid and glucose starvation, and high salinity; and the PERK kinase family responds to unfolded proteins in the endoplasmic reticulum. Each of these kinases, activated by diverse stresses, phosphorylates serine 51 of eIF2alpha and affects global translation. GCN2 homologs have been identified in yeast, flies, worms, and mammals, suggesting that this protein may be the founding member of this family of related kinases.

The other partner in the functionally important equilibrium with eIF2 is the GEF protein eIF2B. This protein can also be phosphorylated under certain stress conditions, a modification that typically prevents it from interacting with eIF2 and thus from promoting GDP/GTP exchange. The effect of this modification is again global translational down-regulation.

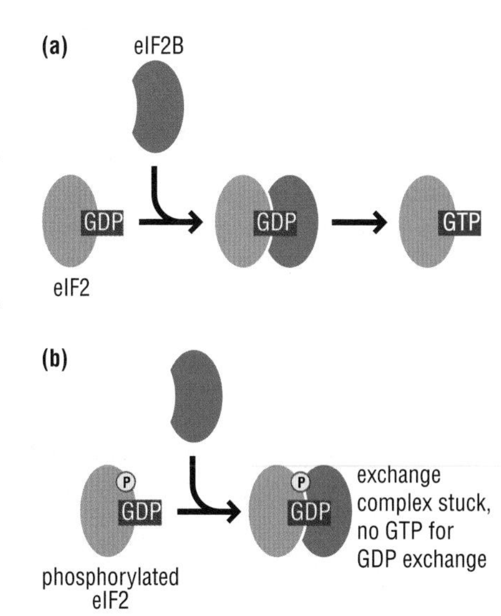

Figure 12.2 Phosphorylation status of eIF2 determines the amount that is free in solution. (a) Under normal circumstances, eIF2B acts as a GEF for eIF2, stimulating eIF2 to exchange GDP for GTP. (b) Phosphorylated eIF2 binds more tightly to eIF2B, thus preventing GTP/GDP exchange and the release of eIF2 from the complex.

The mRNA-bound cap complex initiation factors are also commonly targeted to modulate translation in eukaryotes

Generally, mRNAs are at a higher concentration than the components of the translational machinery and therefore mRNAs must compete for the available initiation factors if they are to be translated. We learned in Section 11.9 that formation of the initiation complex in eukaryotic cells begins with a closed loop mRNA structure where eIF4E is bound to the cap, and PABP to the poly(A) tail; these two proteins are physically linked through interactions with eIF4G (see Figure 11.27). Each of the components in this complex offers a potential target for the regulation of initiation.

How do cells typically regulate this group of factors? First, the transcription of eIF4E is regulated in response to external stimuli. Second, eIF4E itself is phosphorylated in response to external stimuli, which leads to an increase in its affinity for the 5′ cap of mRNA, leading to globally increased initiation. Finally, eIF4E can be sequestered by binding proteins known as **4E-BPs** (eIF4E-binding proteins), which effectively sequester eIF4E from the active population (decreasing overall translation) as shown in Figure 12.3. Not surprisingly, the binding of eIF4E to 4E-BPs can be further modulated by post-translational modifications: phosphorylation of the 4E-BPs can occur in response to different stimuli, decreasing affinity of eIF4E for 4E-BP and leading to a global increase in translational initiation.

➜ We learn more about post-translational modifications in Chapter 14.

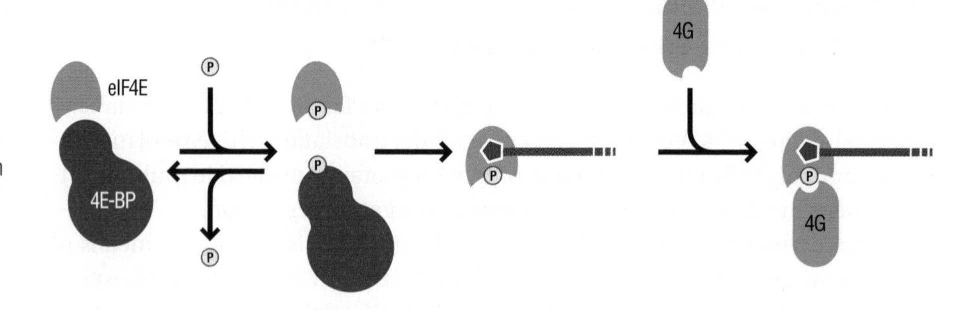

Figure 12.3 eIF4E availability is regulated by 4E-BPs. eIF4E can bind to both 4E-BPs (left) and eIF4G (right). The phosphorylation of either eIF4E or the 4E-BPs (center) prevents them from interacting with one another as effectively, thus increasing the overall effective concentration of eIF4E. In addition, phosphorylated eIF4E has higher affinity for the 5′ cap structure.

Multiple distinct 4E-BPs have been identified in mammalian cells, suggesting that this is a diverse family of proteins that enables cells to increase or decrease their protein synthesis rates in response to a number of different stimuli.

12.2 REGULATION OF INITIATION BY *CIS* ACTING SEQUENCES IN THE 5′ UNTRANSLATED REGION IN BACTERIA AND EUKARYOTES

In Section 12.1, we considered how organisms regulate translation at a general, global level. We now explore some examples of mRNA-specific regulation conferred by particular sequences in the mRNA.

Gene-specific translational regulation depends on sequence features in the mRNA

The rate at which any particular protein in the cell is synthesized is regulated according to the needs of a cell. The rate of synthesis of a protein depends on many factors – the rate of transcription of the gene, the stability of the mRNA, and the rate of its translation, to name the most obvious. Arguably, the most rapid changes in the rates of protein synthesis can be achieved by regulating the rate of translation of an mRNA population that is already present in the cell; in this case, there is no requirement to produce new mRNA to increase the downstream step of protein synthesis. Indeed, cells often respond rapidly to environmental stimuli by controlling the rates of translation initiation and, more occasionally, elongation.

As we see below, the alternative structural conformations that mRNAs can assume under differing conditions can be critical for the regulation of gene expression. In bacteria, the best-known examples of the regulation of initiation involve direct obstruction of the Shine–Dalgarno sequence in the 5′ untranslated region (UTR) of an mRNA, preventing formation of competent initiation complex. Translation initiation in eukaryotes involves both the 5′ and 3′ UTRs of the mRNA, thus allowing for more diverse mechanisms of control.

Shine–Dalgarno sequestration is a common mechanism for the regulation of translation in bacteria

We saw in Section 11.8 that a critical feature in the initiation of translation in bacteria is the presence and composition of the Shine–Dalgarno sequence upstream of the AUG start site in an mRNA. If the Shine–Dalgarno sequence is not accessible to

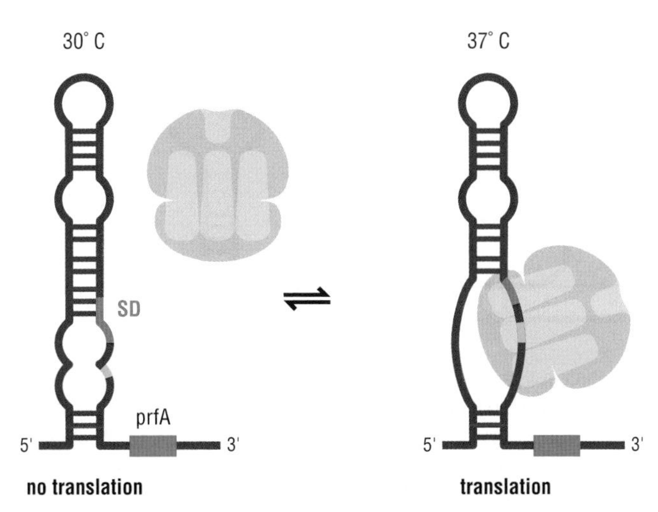

Figure 12.4 Translation of prfA is regulated by temperature dependent changes in RNA structure. At low temperature (30°C), the prfA 5′ UTR RNA structure is quite stable and the Shine-Dalgarno (SD) sequence is obstructed from the ribosome; at higher temperature (37°C), breathing of the RNA structure allows the ribosome to access this region critical for ribosome binding.

form interactions with the small subunit of the ribosome then initiation cannot take place. Indeed, bacteria commonly sequester the Shine–Dalgarno sequence using various strategies, including the binding of small regulatory RNAs, RNA binding proteins and small molecules. We explore each of these strategies below. These broad examples will also be discussed in Chapter 13, as part of a broader exploration of regulatory RNAs.

Likely the simplest mechanism for sequestering the Shine–Dalgarno sequence depends on the mRNA 5′ leader sequence itself and its three-dimensional structure. One such example is seen in *Listeria monocytogenes*, a bacterium that infects warm-blooded animals. The expression of PrfA, a key transcriptional activator of genes required for infection by this organism, is regulated by temperature-dependent changes of the structure of the 5′ UTR, as illustrated in Figure 12.4. At low temperatures outside host animals, the Shine–Dalgarno sequence of the *prfA* mRNA is locked in an RNA structure that renders it inaccessible to the ribosome, thus blocking translation. Once a bacterium enters a host, however, temperature-dependent structural reorganization of this region of the mRNA makes the Shine–Dalgarno sequence accessible to the ribosome, and translation of the *pfrA* mRNA can occur.

There are a growing number of examples where the expression of biosynthetic genes is subject to autoregulation - that is, where the end product of the pathway directly regulates its own synthesis. In bacteria, the small molecule product often binds to the mRNA sequence upstream of the open reading frame (ORF) and changes the structure of the mRNA. Such elements are generally referred to as **riboswitches**. As we discussed in Chapter 9, the binding of these small regulatory molecules to some transcripts can lead to transcription termination. For other mRNAs, the binding of the small molecule can directly block the initiation of translation.

Scan here to watch a video animation explaining more about the regulation of translation, or find it via the Online Resource Center at **www.oxfordtextbooks.co.uk/orc/craig2e/**.

For example, the conserved 5′ regions of the mRNAs of the *thiM* and *thiC* thiamine biosynthetic genes in *Escherichia coli* fold into a structure that binds thiamine pyrophosphate (TPP), an intermediate in the synthesis of thiamine. When TPP is bound to the mRNA, the Shine–Dalgarno region is sequestered in the tertiary structure of the mRNA, as shown in Figure 12.5a, and translation is inhibited. By contrast, when TPP levels are low, the free mRNA adopts a structure in which the Shine–Dalgarno sequence is no longer sequestered and expression of the biosynthetic genes is induced (Figure 12.5b). It is plausible (and even likely)

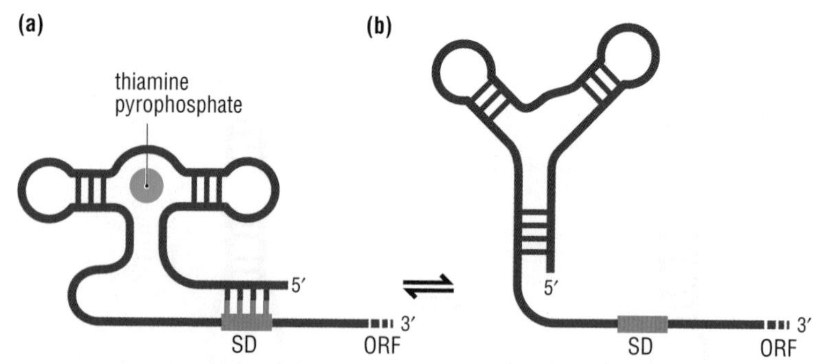

Figure 12.5 Translation of thiamine biosynthetic genes is regulated by the binding of thiamine pyrophosphate (TPP) to the 5′ UTR of the mRNA. (a) In the presence of TPP, the Shine–Dalgarno sequence (SD) in the mRNA of thiamine biosynthetic genes is sequestered through interactions with portions of the 5′ UTR, thus preventing ribosome binding and translation. (b) In the absence of TPP, the SD is revealed through conformational rearrangements in the 5′ UTR of the mRNA, thus increasing translation of the downstream gene.

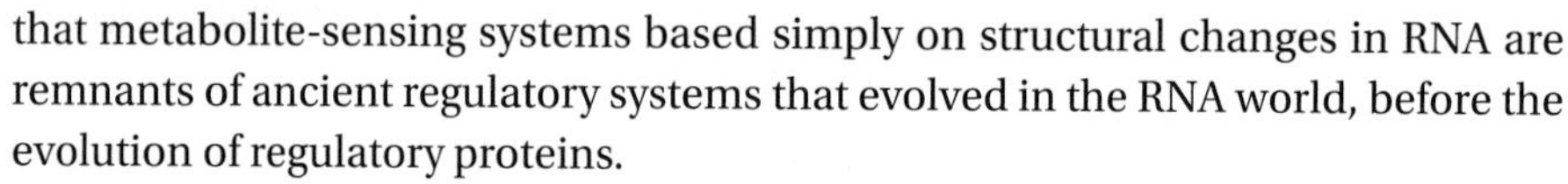

that metabolite-sensing systems based simply on structural changes in RNA are remnants of ancient regulatory systems that evolved in the RNA world, before the evolution of regulatory proteins.

A class of small, regulatory RNAs (generally 100–150 nucleotides in length) in bacteria has also been shown to either promote or limit translation by directly base-pairing with the 5′ UTR such that the Shine–Dalgarno sequence is either sequestered or revealed. The establishment of these base-pairing interactions is often dependent on the RNA binding protein, Hfq, which is synthesized in high levels in many of these organisms. It is also believed that base-pairing interactions between small RNAs and the initial codons of a gene (up to five codons) can act to inhibit translation initiation.

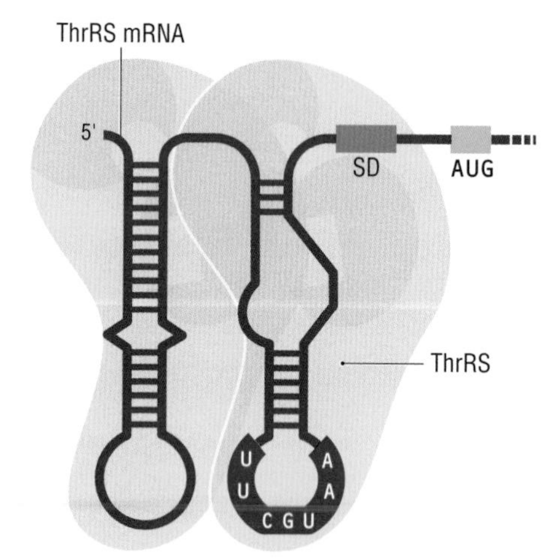

Another specific example of regulated Shine–Dalgarno sequestration is seen in the control of the expression of the gene for threonine-tRNA synthetase (*thrS*) in *E. coli*. In this case, sequestration is mediated by a protein: ThrRS binds to its own mRNA, sterically blocking the Shine-Dalgarno sequence and thus preventing the translation of more synthetase. Otherwise, ThrRS binds to unacylated $tRNA^{Thr}$ (when protein synthesis levels are high and unacylated $tRNA^{Thr}$ is being rapidly generated) and performs its usual task of aminoacylation. Thus *thrS* mRNA and unacylated $tRNA^{Thr}$ directly compete with one another for binding to threonine-tRNA synthetase, depending on the overall rates of protein synthesis (or really, the levels of unacylated $tRNA^{Thr}$).

The ability of ThrRS to bind both its own mRNA and unacylated $tRNA^{Thr}$ is a consequence of sequence similarities between the ThrRS mRNA and the unacylated $tRNA^{Thr}$. Figure 12.6 shows how the ThrRS mRNA and the unacylated $tRNA^{Thr}$ share a common sequence (identical mRNA, highlighted in pink).

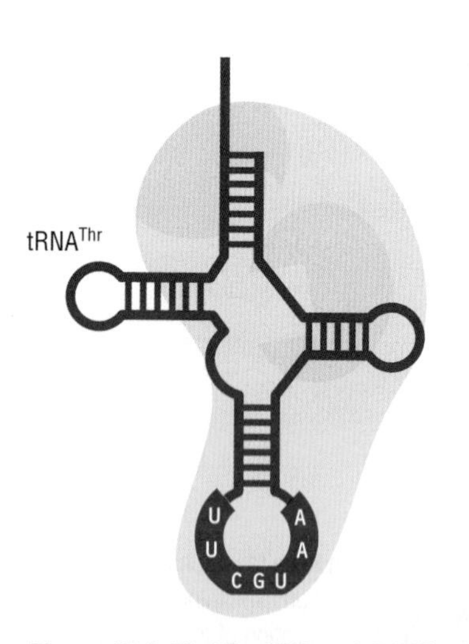

Figure 12.6 ThrRS mRNA and $tRNA^{Thr}$ share molecular features in their RNA structures. These similarities allow ThrRS to bind either species, depending on the amount of deacylated $tRNA^{Thr}$ available in the cell. Binding to the 5′ UTR by ThrRS blocks access to the Shine–Dalgarno sequence, and thus translation.

Initiation in eukaryotes can be regulated by blocking access to the ribosome-binding site

While the sequestration of 5′ UTRs seems to be the rule in bacterial translational control, this mode of regulation has been less commonly observed in eukaryotes. That said, there are examples in eukaryotes where protein binding in the 5′ UTR does regulate translation.

Iron is an essential nutrient whose levels in eukaryotic cells must be tightly controlled as high levels can be toxic. Inside cells, iron is bound to a storage protein

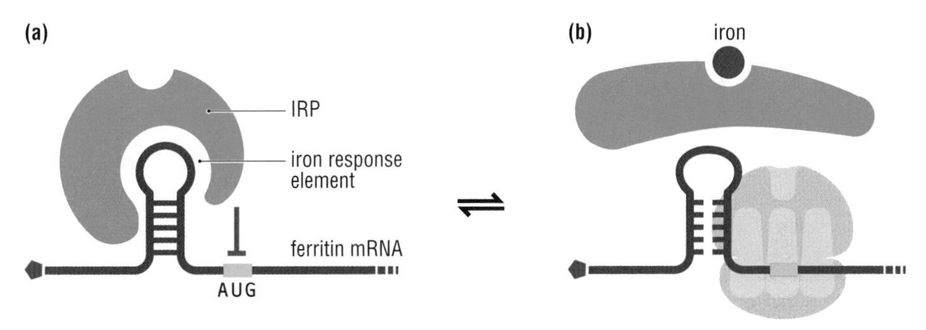

Figure 12.7 Initiation of translation of ferritin mRNA is regulated through the 5′ UTR. (a) When iron is scarce in the cell, ferritin translation initiation is prevented by the binding of IRP1/2 to iron-response elements (IREs) in the 5′ UTR of the mRNA. (b) When iron is plentiful, the IRPs (IRP1/2) no longer bind the IREs, and ferritin translation initiation proceeds.

known as ferritin; the higher the concentration of iron in a cell, the more ferritin the cell requires. In the 5′ UTR of ferritin mRNA, there are several stem-loop structures called iron response elements (IREs), which bind to proteins known as the iron-regulatory proteins (IRPs). When iron is scarce in the cell, the IRPs binds to the IRE as illustrated in Figure 12.7a. This binding prevents the initiation of translation by preventing ribosome access to the AUG start site, probably by blocking the scanning ribosome.

When iron is plentiful, however, the IRPs bind to iron as shown in Figure 12.7b, which prevents them from binding to the IREs. This, in turn, allows for efficient initiation of ferritin translation. We recall that the same IRPs also regulate the expression of a different gene, the transferrin receptor, modulating mRNA stability through IREs located in the 3′ UTR (see Section 10.9). This example illustrates how feedback control can be used to exquisitely tune the amount of a protein produced under a specific set of conditions.

Upstream ORFs regulate the synthesis of Gcn4 protein in yeast

One of the best-characterized examples of translational control in eukaryotes is the sophisticated regulated expression of the Gcn4 protein in yeast, a transcriptional activator of many amino acid biosynthetic genes; much has been learned about this mechanism of gene regulation from analysis in the yeast system. As we discussed in Section 12.1, when amino acid levels in eukaryotic cells are low, the eIF2 kinase Gcn2 is activated, eIF2alpha is phosphorylated, and overall protein synthesis in the cell is reduced. However, the rate of translation of the gene *GCN4* is specifically increased under these conditions.

This upregulation occurs by an unusual mechanism that depends on the presence of multiple upstream ORFs (uORFs; uORF1-4) found in the transcript upstream of the actual Gcn4 coding region. Under all conditions - whether the cell is starved for amino acids or not - uORF1 is effectively translated. What happens beyond uORF1, however, differs depending on the cellular levels of eIF2 available, as illustrated in Figure 12.8. Central to all possible outcomes is that the uORF1 stop codon (and the sequence context in which it is located) favors the resumption of scanning following termination, rather than full ribosome drop-off and recycling - that is, the 40S subunit remains associated with the mRNA and travels further along the mRNA to encounter downstream sequence elements; this process of reinitiation was briefly discussed in Section 11.11

Under conditions of amino acid abundance (non-starved), eIF2alpha is largely unphosphorylated, and the ternary complex (eIF2:Met-tRNA$^{:Met}$:GTP) is

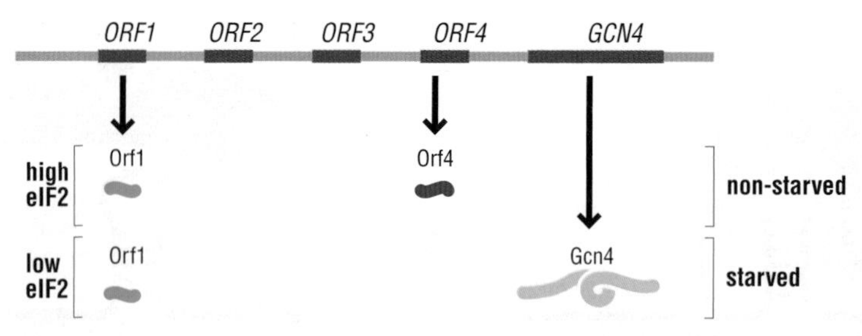

Figure 12.8 Model explaining how Gcn4 synthesis is regulated by uORFs and the availability of eIF2. Translation of the polycistronic GCN4 mRNA begins with the translation of uORF1 (left). When amino acids are abundant in the cell (top pathway), eIF2 is largely unphosphorylated, and as a result, eIF2 activity is high. Efficient reinitiation following the translation of uORF1 is thought to lead to the translation of uORF4 (and possibly also uORF2 and uORF3) (top pathway) and subsequent full recycling (ribosome drop-off). When amino acids are not abundant (bottom pathway), eIF2 is largely phosphorylated and is therefore less available to participate in certain initiation events. In this case, reinitiation at uORF4 is inefficient, and reinitiation at the *GCN4* open reading frame becomes more likely.

maximally available. Hence, virtually all 40S subunits scanning downstream from uORF1 are thought to rebind initiator tRNA before reaching uORF 2, 3, or 4, and these ribosome complexes reinitiate translation at one of these start codons. Unlike uORF1, however, the sequence context of the stop codons for these uORFs (2–4) does not favor the reinitiation of scanning, and complete ribosome recycling (i.e., including dissociation of the 40S subunit) happens more efficiently. As a result, the 40S subunit is not able to reinitiate translation again, and thus rarely makes it to the actual *GCN4* ORF. As a result, overall expression of the Gcn4 protein is low under these conditions.

By contrast, under conditions of amino acid depletion (starved), the concentration of the eIF2-GTP-Met-$tRNA_i^{Met}$ ternary complex is low. As a consequence, a reasonably large fraction of the 40S ribosomes scanning downstream from uORF1 fails to rebind the ternary complex until reaching the interval between uORF4 and the *GCN4* ORF. As a result, these ribosomes reinitiate translation after uORF4. Incidentally, the *GCN4* gene start site is thought to be a relatively robust one, thus ensuring that ribosomes don't easily scan past it as well. Consequently, Gcn4 protein is translated at relatively high levels. This model for *GCN4* regulation, shown in Figure 12.8, was developed largely on the basis of *in vivo*-based experiments using *GCN4* expression as a readout.

In Experimental approach 12.1, we see how ribosome profiling has revealed new insights into this complex form of translational control. While much of what was learned from more traditional yeast genetic and biochemical approaches holds true, we see how a new genomic approach for observing ribosomes in action in the cell is revealing new features of regulation that we have yet to fully understand. As anticipated, we see that globally decreased levels of protein synthesis can lead to increased expression of certain genes found in an appropriate mRNA context (*GCN4*). In this case, 'mRNA context' includes the series of uORFs in the 5′ UTR as well as differing stop codon sequence contexts that affect the balance between ribosome recycling (dissociation) and reinitiation. However, we fail to observe certain signatures that were predicted from earlier approaches, including for example increased ribosome occupancy of uORFs2–4 during non-starvation conditions. Moreover, we observe completely unanticipated ribosome occupancy 5′ of uORF1 whose origin and function remains poorly understood. Similar examples of translational control that depend on uORFs, although a relatively rare form of regulation, have been observed in mammals, maize, and certain fungi.

12.1 EXPERIMENTAL APPROACH
Studying translational control using ribosome profiling

Sucrose gradients can be used to explore ribosome density on the mRNA

The translational activity of ribosomes on a given mRNA has traditionally been followed using a technique referred to as polysome analysis. In a cell, an mRNA that is translationally active tends to be loaded up with ribosomes all along its length, with initiation occurring again (and again) once the previous ribosome has elongated beyond the initiating AUG codon (see Figure 11.18). Such 'ribosome-loaded' mRNAs migrate as large particles in a sucrose (or glycerol) gradient. The absorbance of the sample at around 260 nm is proportional to the amount of nucleic acid present. Therefore this measure directly reports on the predominant RNA in the cell, the rRNA found within the ribosomes. With this read-out, the various peaks in the sample can be easily identified – at the top of the gradient we find mRNPs, then the first real peak (starting from the top of the gradient) corresponds to small ribosomal subunits (30S or 40S), the second peak to large ribosomal subunits (50S or 60S), the next to full ribosomes (70S or 80S, or monosomes), then disomes (two ribosomes attached to a single mRNA), trisomes (three attached ribosomes) and so on; such a profile is shown in Figure 1. mRNAs are sometimes associated with isolated small subunits (presumably in a pre-initiation phase), but primarily are found associated with one or more intact ribosomes (monosomes and higher).

When the various fractions collected from the sucrose gradient are analyzed for their mRNA content (for example, by northern analysis), then clues can be obtained about the translational activity of an mRNA of interest in the cell, and even the type of regulation that might be occurring. Genes that are actively expressed tend to be found in the heavy polysomal fractions where mRNAs are loaded with multiple ribosomes, while genes that are translationally repressed at the level of initiation tend to be excluded from the polysomal fractions, migrating instead as smaller particles closer to the top of the sucrose gradient. Genes that are translationally repressed, but at a stage following initiation (for example, during elongation or termination) are predicted to be found within the polysomal fractions, but in a state that is stalled or where movment is slow. In such a situation, we might expect to observe even more ribosomes than normal loaded on a repressed mRNA. We see from this discussion that polysomal analysis can be used to explore how the translation of a given mRNA might be regulated.

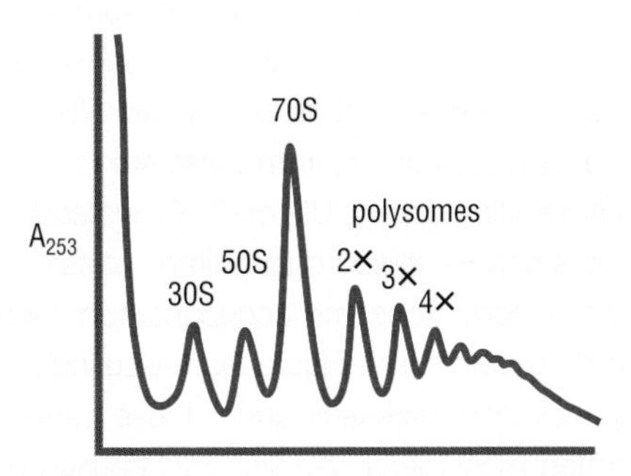

Figure 1 Polysome profile using A_{260} to follow the fractionation of ribosomes in a sucrose gradient following centrifugation. Ribosomal subunits (30S and 50S), 70S, disomes, trisomes and larger are clearly resolved. The UV-absorbing material at the top of the gradient is mRNP and other small RNAs.

Ribosomes protect a small fragment of mRNA that can be mapped

Joan Steitz first showed in 1969 that ribosomes protect discrete fragments of mRNA using the R17 bacteriophage and Sanger sequencing methods to identify the mRNA fragments. Some years later, Sandra Wolin and Peter Walter built on these results and developed a different experimental approach to map the specific sites of ribosome occupancy on an mRNA (found in polysomes). Like Steitz, Wolin and Walter understood that mRNA bound to a ribosome would be protected from exogenously added nuclease, and could be isolated and identified following such a treatment (effectively an RNase protection assay). The real challenge of the experiment was in developing a means to identify small amounts of a gene product of interest.

In the study by Wolin and Walter, the authors chose to hybridize the isolated ribosome protected fragments to a single-stranded template carrying the anti-sense strand of the gene of interest. The template complex was then analyzed by reverse transcription (see Section 19.10) using a downstream primer to yield a cDNA strand, and the products of the reaction were resolved on a sequencing gel. In this analysis, the authors identified a pause site within the bovine preprolactin mRNA where ribosome occupancy was increased in the presence of exogenously supplied signal recognition particle (SRP) (see Figure 2, band b). As we have come to understand, the ribosome occupancy site in this case corresponded to pausing by the ribosome as the signal peptide emerges from the exit channel of the ribosome (on the solvent side of the large subunit) and engages machinery involved

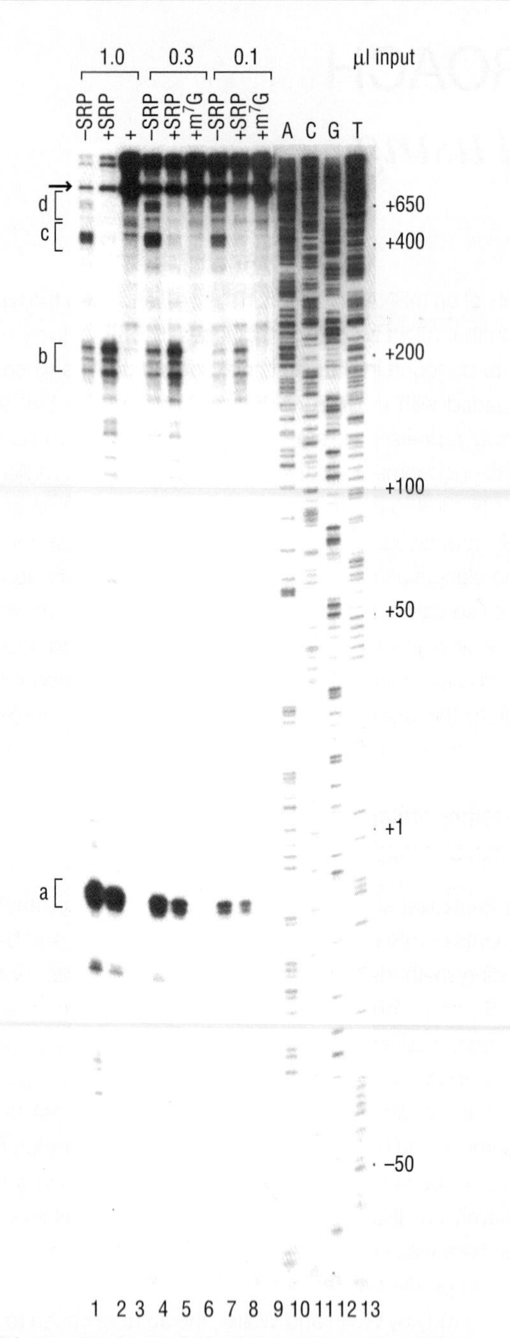

Figure 2 Primer extension analysis of ribosome pausing during translation of the preprolactin mRNA in a wheat-germ translation extract. Ribosome-protected fragments of mRNA were prepared from an *in vitro* translation reaction either in the presence or absence of signal recognition particle (SRP) or 7-methyl guanosine cap (m^7G) (to compete with cap dependent translation). Four specific stalls can be identified (a–d) that correspond to distinct sites in the mRNA. Band a represents the initiating population of ribosomes at the AUG start site while band d corresponds to the terminating population at the UAA stop codon. Bands b and c are both centered on rare GGC codons, but the intensity of band b is increased in the presence of SRP. The m^7G competes with translation of the mRNA, and so none of the pauses are seen when it is present. Sequencing lanes on the right are labeled A, C, G, T. Three different concentrations of ribosome-protected fragment were used in the primer extension reactions (1.0, 0.3, and 0.1 μl input, top of gel).

Reproduced from Wolin and Walter, Ribosome pausing and stacking during translation of a eukaryotic mRNA. *The EMBO Journal*, 1988;**7**:3559–3569.

in downstream steps in the protein translocation process (described in Section 14.2).

In addition to this primary pause site, the authors identified other ribosome protected fragments that appeared to be centered on rare glycine codons (GGC) in the mRNA template; the authors argued that these sites resulted from the ribosome spending a bit more time paused at these positions during the decoding step of elongation. Both examples show how detailed mapping of ribosome positioning on the mRNA template can provide information on translational events within the cell and their timing. Moreover, these results foreshadow more recent high-throughput approaches described briefly below.

Ribosome profiling allows for genome-wide analysis of ribosome occupancy

Two different high throughput approaches have since tackled this same general problem. In the first example, Daniel Herschlag, Patrick Brown, and colleagues coupled microarray approaches with polysomal profiling to globally assess the translational status of mRNAs in the cell. To do this, polysomal fractions were isolated, intact mRNA was extracted from them, and the distribution of each mRNA across the gradient assayed through hybridization to a microarray. The approach yielded a number of insights into broad features of translational activity in the cell. For example, most mRNAs were seen to be somewhat underpopulated with ribosomes, relative to their anticipated capacity based on length, suggesting that initiation might be the rate-limiting step during translation. Despite the broad success of this approach, its resolution was limited because whole mRNAs were followed rather than the individual 'footprints' of a ribosome on the mRNA.

More recently, ribosome mRNA occupancy has been more precisely assessed by Jonathan Weissman and colleagues, using high-throughput sequencing to take a comprehensive look at mRNA status in the cell. (They call the approach 'ribosome profiling'.) In these experiments, ribosome footprints ~28 nucleotides in length were isolated from yeast ribosomes, cloned, and sequenced, with tens of millions of individual sequence reads being obtained for a given set of conditions. New insights into translation are rapidly emerging from these studies. For example, non-canonical initiation sites (UUG or GUG, instead of AUG) appear to be more heavily utilized during times of cellular stress, as indicated by increased ribosome occupancy on such sites during amino acid starvation (see Figure 3); reassuringly, *GCN4* ribosome occupancy also increases under these same conditions, as demonstrated most clearly by Alan Hinnebusch during years of *in vivo*-based studies. We can already see that the level of resolution that comes from such a deep sequencing approach will yield an unprecedented view of translational complexity both at the level of the single gene, and of the genome.

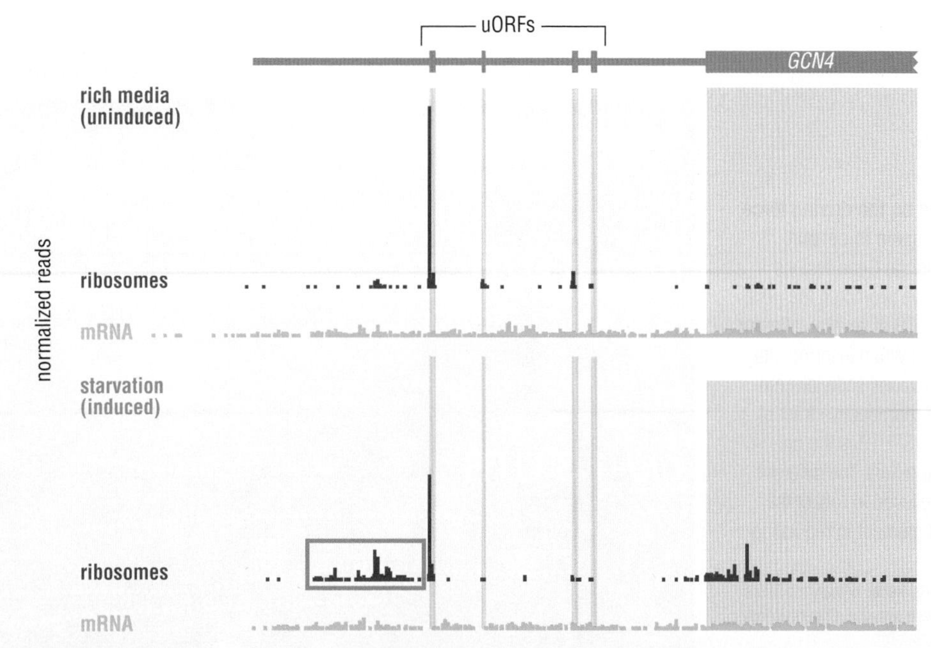

Figure 3 Ribosome profiling explores translation activity at the *GCN4* locus in yeast under normal (top, red) and starved (bottom, yellow) conditions. The number of reads of a particular region in the gene (the x-axis) are tabulated on the y-axis both for a total yeast fragmented mRNA population (green) and for ribosome-protected fragments (pink). As expected (see Section 12.2), *GCN4* expression is induced (more 'reads' within the *GCN4* open reading frame (ORF)) under conditions of amino acid starvation. Boxed in orange is another upstream location (containing no AUG codons) where there is clear increased ribosome occupancy during starvation; within this region there are several proposed non-canonical initiation sites.

Find out more

Arava Y, Wang Y, Storey JD, *et al*. Genome-wide analysis of mRNA translation profiles in *Saccharomyces cerevisiae*. *Proceedings of the National Academy of Sciences of the U S A*, 2003;**100**:3889–3894.

Hinnebusch AG. Evidence for translational regulation of the activator of general amino acid control in yeast. *Proceedings of the National Academy of Sciences of the U S A*, 1984;**81**:6442–6446.

Ingolia NT, Ghaemmaghami S, Newman JR, Weissman JS. Genome-wide analysis in vivo of translation with nucleotide resolution using ribosome profiling. *Science*, 2009;**324**:218–223.

Steitz JA. Polypeptide chain initiation: Nucleotide sequences of the three ribosomal binding sites in bacteriophage R17 RNA. *Nature*, 1969;**224**: 957–964.

Wolin SL, Walter P. Ribosome pausing and stacking during translation of a eukaryotic mRNA. *The EMBO Journal*, 1988;**7**:3559–3569.

Related techniques

Reverse transcription; Section 19.3
Sequencing gels; Section 19.8
Ribosome profiling; Section 19.12
Deep sequencing; Section 19.10

12.3 REGULATION OF TRANSLATION THROUGH *CIS* ACTING SEQUENCES IN THE 3′ UTR IN EUKARYOTES

In Section 12.2, we discussed ways in which translation of specific genes in bacteria and eukaryotes can be regulated by structural changes or by interactions with the 5′ UTR of an mRNA that disrupt translation initiation. While direct obstruction of the Shine–Dalgarno sequence is probably the most likely means of regulating translation in bacteria, the 5′ UTRs of eukaryotic genes are typically shorter than the 3′ UTR and are less commonly involved in such regulatory processes. Instead, the 3′ UTRs of eukaryotic genes can be very large and are the driving force for the regulation of a number of developmentally important genes. This concept is illustrated in

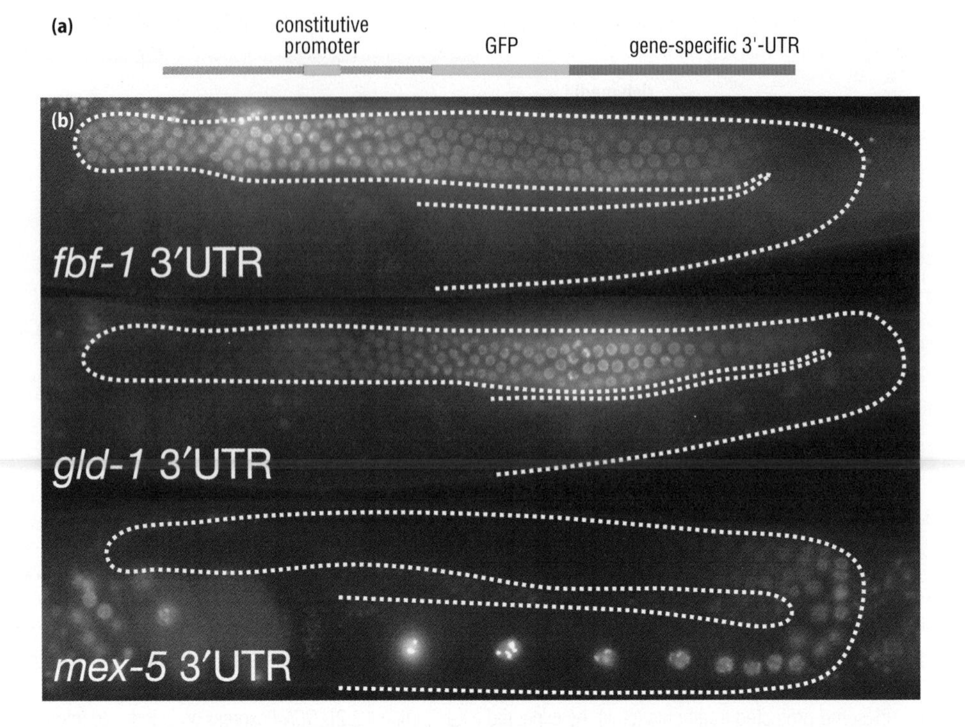

Figure 12.9 3′ UTRs can be the driving force in regulating gene expression in certain developmental programs. (a) Gene fusions carrying a strong constitutive promoter, the gene encoding GFP, and the 3′ UTR of several different *C. elegans* genes (color). (b) When inserted into the worm genome, these fusions show diverse expression patterns in the reproductive tract of the organism. Because the 3′ UTR is the only portion of the reporter that differs, the different expression patterns are believed to reflect the inherent properties of this element in regulating translation.

Taken from Merritt, C *et al.* 3′ UTRs are the primary regulators of gene expression in the *C. elegans* germline. *Current Biology*, 2008;**18**:1476–1482

Figure 12.9, which shows that the developmentally specific expression of reporter proteins in the *Caenorhabditis elegans* reproductive tract can be specified simply by the inclusion of the gene-specific 3′ UTR on the green fluorescent protein (GFP) reporter construct.

The general understanding of such regulation is that elements within the 3′ UTR recruit different cellular factors that impact downstream events, for example disrupting formation of the closed loop mRNA initiation complex and thus down-regulating translation initiation. Not surprisingly, these mechanisms target many of the same steps as we saw for global translational repression in Section 12.1; in these cases, however, the effects are gene-specific as they are mediated through factors specifically recruited to the 3′ UTR of the regulated mRNA.

Polyadenylation levels can be regulated by 3′ UTR binding proteins

One mechanistically well-understood example of 3′ UTR-mediated translational control is seen during development in the *Xenopus* oocyte. The oocyte contains maternally derived mRNAs that are not immediately translated into protein – these mRNAs are said to be dormant or translationally repressed. Dormant mRNAs have short poly(A) tails, which only during maturation are extended to allow the mRNAs to be translated into protein.

How does this repression and subsequent activation of translation occur? The 3′ UTRs of the dormant mRNAs have sequences referred to as cytoplasmic polyadenylation elements (CPEs), which play a key role in controlling their expression. Translation is repressed when the CPEs are bound by a protein known as CPEB. CPEB interacts with a protein called Maskin, which interacts in turn with eIF4E (Figure 12.10, step 1). The interaction of eIF4E with Maskin directly precludes the binding of eIF4G to eIF4E, thus 'masking' the expression of an mRNA by preventing formation of the closed circle initiation complex.

The region of Maskin that directly interacts with eIF4E contains a conserved sequence element YXXXXLphi (where Y corresponds to tyrosine, X to any amino

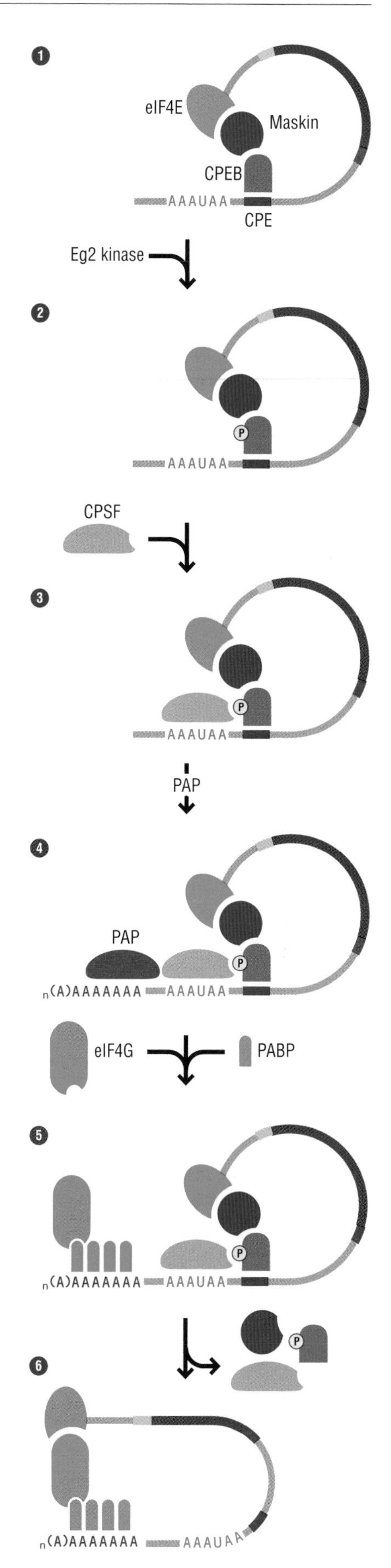

Figure 12.10 Translational regulation of *Xenopus* oocyte dormant mRNAs through RNA elements found in the 3′ UTR. (1) Dormant mRNAs contain short poly(A) tails and CPEs. Translation of these mRNAs does not proceed because eIF4E is blocked from binding to eIF4G by Maskin. (2) Following the phosphorylation of CPEB by the kinase Eg2, (3) CPEB recruits CPSF to the hexanucleotide sequence AAUAAA in the mRNA. (4) CPSF in turn recruits poly(A) polymerase (PAP) to the mRNA, and the poly(A) tail of the mRNA is extended. (5) When PABP (the poly(A) binding protein) binds to the newly extended poly(A) tail, eIF4G can bind and (6) is ultimately able to displace Maskin from eIF4E such that translation can proceed.

acid, L to leucine, and phi to a hydrophobic residue), which mimics a related element in eIF4G. Many cellular factors that regulate translation initiation use this same trick to compete with natural interactions in the cell between eIF4E and eIF4G; these proteins have been described in Section 12.1 and are broadly referred to as 4E-BPs. Maskin is a type of 4E-BP with additional specificity for the protein CPEB that allows for gene-specific regulation through 3′ UTR recruitment.

Activation of the dormant mRNAs requires Maskin to be displaced from eIF4E. The first step in this process is the phosphorylation of CPEB by a kinase, Eg2 (Figure 10.56, step 2), which increases the affinity of CPEB for a protein known as CPSF (cytoplasmic polyadenylation specificity factor) (Figure 12.10, step 3). Once recruited by CPEB, CPSF binds to the standard AAUAAA polyadenylation signal and recruits poly(A) polymerase to the mRNA. The poly(A) polymerase then extends the poly(A) tail of the mRNA. Once extended (Figure 12.10, step 4), PABP (the poly(A) binding protein) can bind to the 3′ tail of the mRNA (Figure 12.10, step 5). Having associated with the poly(A) tail, PABP recruits eIF4G, which then displaces Maskin from eIF4E. Once eIF4E and eIF4G interact, the initiation complex is formed and translation of the activated mRNAs can proceed (Figure 12.10, step 6).

This rather complicated sequence of events allows mRNAs that are held dormant by the CPE sequence element to be activated by a single event - the activation of the kinase Eg2. This method of translation control is probably more broadly utilized; related proteins appear to be important in specifying critical functions in developing neurons.

RNAs or proteins bound to the 3′ UTR can regulate translation at multiple steps following polyadenylation

We see above how the cytoplasmic polyadenylation element in the 3′ UTR of a dormant mRNA in *Xenopus* oocytes regulates translation of the mRNA before the mRNA is fully polyadenylated. The expression of many genes in eukaryotic systems can be controlled by other 3′ UTR regulatory sequences, and often the regulation is applied to different stages in the initiation (or even elongation) process. The basic theme is the same as for the CPEB story described above. Sequence elements in the 3′ UTR recruit a variety of different proteins that impinge on different steps in the translational cycle.

In one example, expression of the *lox* mRNA that encodes a lipoxygenase in erythroid precursor cells is controlled via specific sequences (the differentiation control element or DICE) in the 3′ UTR that bind the hnRNP K and E1/E2 proteins. These specific interactions in the 3′ UTR prevent the large ribosomal subunit from joining the small subunit during the initiation process, as illustrated in Figure 12.11a. How such interactions block this step, which is normally catalyzed by the GTPase eIF5B, is not well understood.

In another example, depicted in Figure 12.11b, the expression of a protein called Nanos (which is involved in the development of *Drosophila* embryos) is regulated through a 3′ UTR element known as a TCE. In the case of Nanos, the

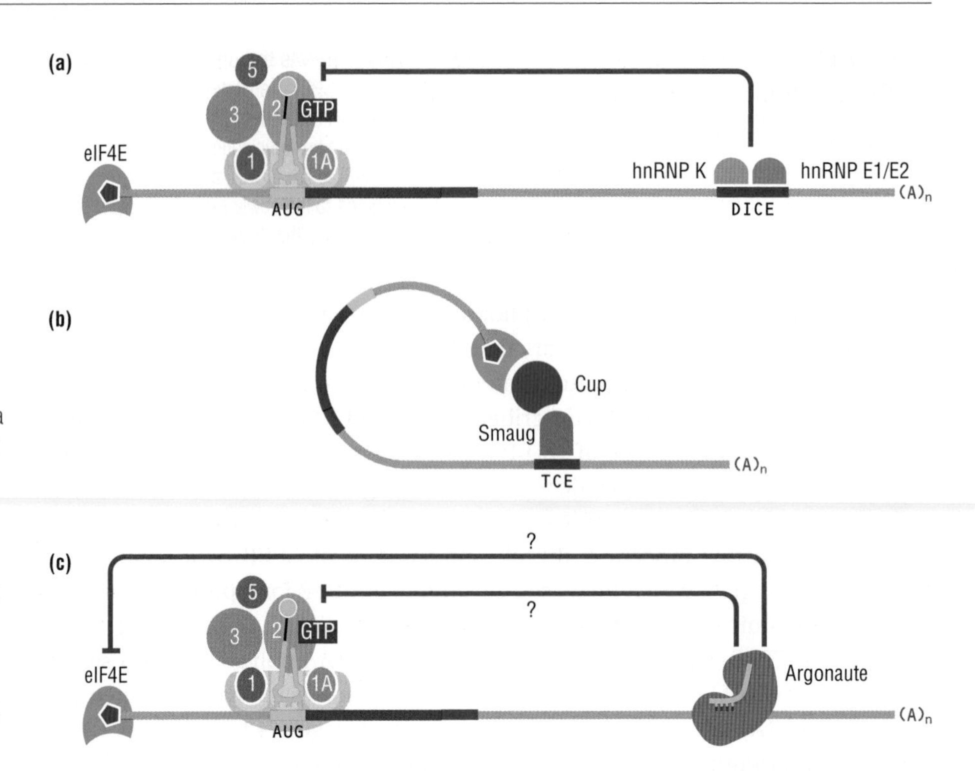

Figure 12.11 Translation can be regulated by the binding of proteins and small RNAs to the 3′ UTR of genes. (a) Expression of the *lox* mRNA in erythroid precursor cells is blocked through a binding element in the 3′ UTR (DICE) that recruits hnRNP K and E1/E2 to repress the subunit joining step of initiation. (b) Expression of the Nanos protein in *Drosophila* embryos is blocked by the binding of Smaug to a translational control element (TCE) in the 3′ UTR of the mRNA. TCE-bound Smaug then interacts with Cup, a 4E binding protein (4E-BP), which directly binds eIF4E and prevents productive interactions with eIF4G. (c) Expression of *lin14* and *lin28* mRNA during *C. elegans* development is blocked during certain larval stages through interactions with the miRNA lin-4. The binding of the lin-4-miRISC complex to the 3′ UTR of the gene somehow blocks translation, through a mechanism that has not yet been defined.

TCE is bound by a protein known as Smaug, which in turn interacts with Cup, an eIF4E-binding protein (4E-BP). This complex of proteins represses translation by competing with the normal eIF4G interaction. This example represents a gene-specific version of what we described in Section 12.1 as a mechanism for global translational shutdown, and a variant of the mechanism just described for the regulation of polyadenylation by the CPE element.

Finally, the microRNA (miRNA) class of small RNAs in eukaryotes is proposed to regulate the translation of specific mRNAs. These small RNAs are known to act through binding interactions with partially complementary sequences in the 3′ UTRs of genes. Through these binding interactions, miRNAs are thought to repress gene expression by interfering with translation initiation (as shown in Figure 12.11c), and this initial translational repression leads eventually to mRNA degradation through standard pathways (as discussed in 13.3). The molecular mechanism through which initiation is down-regulated remains unknown.

➔ We discuss the biogenesis of miRNAs and their lifecycle more thoroughly in Section 13.3.

While the mechanistic details of translational repression in some of these systems remain cloudy, these examples serve to illustrate how gene-specific translational regulation can be specified through sequence elements located within the 3′ UTRs of different genes.

12.4 VIRAL CORRUPTION OF THE TRANSLATIONAL MACHINERY

The previous sections have illustrated the often sophisticated mechanisms used by the cell to ensure that translation proceeds when it should and with appropriate accuracy, either through the implementation of general mechanisms that have a broad effect on translation, or through more specific mechanisms that target individual mRNAs. The machinery that is so central to the survival of a cell is not immune to exploitation from intruders, however. We end this chapter by describing how the translation machinery is hijacked quite spectacularly by viruses, which depend on the machinery for survival every bit as much as the host cell.

Viruses utilize the host cellular protein synthesis machinery in order to propagate

The goal of a virus is to enter a host cell and take advantage of the host's cellular machinery to promote its own replication and propagation. As viral genomes do not encode the basic components of the translation machinery, they rely on their host to provide these resources. As such, the viral mRNAs must directly compete with the host mRNAs for access to the translational machinery, and many of the battles between host and virus over protein synthesis machinery are fought in the initiation step of translation. Some viruses have evolved specific mechanisms for selectively turning off host protein synthesis while others have simply developed mechanisms for outcompeting the host mRNAs. Exploring the mechanisms by which viruses co-opt the host translation machinery has taught us much about how the normal process of translation can be controlled.

➔ We discuss the viral lifecycle briefly in Section 1.7.

Viruses can disrupt cap-dependent translation initiation

As described in Section 11.9, mRNAs that are efficiently translated in eukaryotes form a closed loop structure in which a 5′ cap structure and a 3′ poly(A) tail are brought together by a set of protein factors including eIF4E, eIF4G, and PABP. Translation of these mRNAs is said to be **cap-dependent** because formation of the closed loop initiation complex depends on recognition of the 5′ mRNA cap. A number of viruses disrupt cap-dependent initiation in order to redirect the host's translation machinery to their own mRNAs.

For example, picornaviruses (such as poliovirus) produce proteases that target eIF4G and PABP for cleavage, disrupting the formation of the closed loop initiation complex. Influenza viruses disrupt the formation of closed loop complexes by producing an endonuclease that cleaves the host mRNA 5′ caps (but not the viral 5′ caps). In this latter case, cap-dependent translation can still be utilized by the cleavage-resistant viral mRNAs. These different points of inhibition are outlined in Figure 12.12a. Yet another picornavirus, encephalomyocarditis (EMCV), produces viral proteins that dephosphorylate the 4E-BPs, which results in the sequestration of eIF4E (Figure 12.12b). Again, global cellular cap-dependent translation is inhibited.

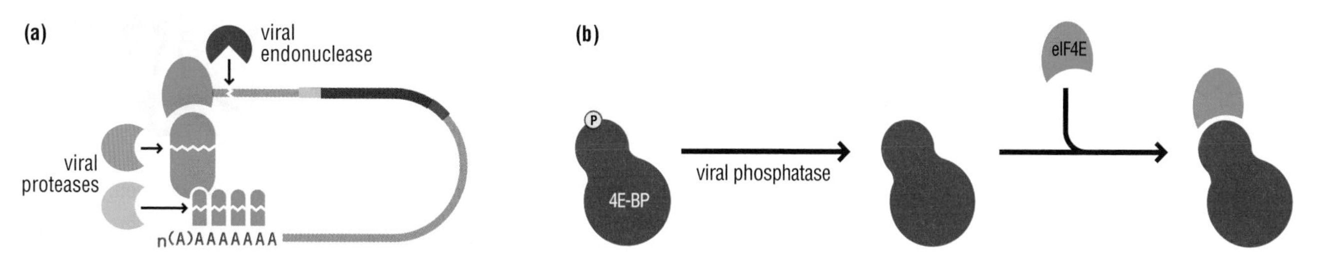

Figure 12.12 Schematics indicating various targets of translational control as executed by different viruses. (a) Closed circle initiation complex indicating several targets for viral proteases and RNases. (b) Viral phosphatases can dephosphorylate 4E-BPs (eIF4E-binding proteins), thus favoring sequestration of eIF4E by the 4E-BPs.

Viral mRNAs must be effectively translated despite disruptions to the host's translational machinery

Whatever mechanism a virus utilizes to disrupt translation of a host's own mRNAs, the virus must also have mechanisms in place that allow for its own mRNAs to be translated. For example, the picornaviruses, which inhibit cap-dependent translation by cleavage of eIF4G and PABP, have evolved a cap-independent method of initiating translation of their mRNAs; this method involves RNA elements known as **internal ribosome entry sites (IRES)** found in the 5′ UTR of their mRNAs. These sometimes large and complex IRES RNA structures (including the example shown in Figure 12.13a) allow the viral mRNAs to directly recruit small ribosomal subunits, typically using a reduced set of initiation factors.

Cryoelectron microscopic structures of IRES RNAs bound to the ribosome, such as that shown in Figure 12.13b, show that they bind just where the Shine–Dalgarno interaction would be located in bacterial ribosomes. Indeed, they function in much the same way, positioning the mRNA appropriately, without scanning, so that the appropriate AUG start site is positioned in the P site and ready for action. For example, picornavirus IRES-dependent initiation does not depend on the eIF4 factors that it has destroyed, but instead utilizes distinct factors referred to as **IRES-transacting factors (ITAFs)**.

It is interesting to note that the discovery of viral IRESs eventually led researchers to identify IRESs in eukaryotic genes that are thought, for example, to function during times of cellular stress. These molecular species have been less well characterized, but probably use strategies similar to those of the sophisticated viral IRESs.

In other examples, the viral RNA appears simply to compete for limiting translation factors. The rotaviruses, for example, express mRNAs that carry a 5′ cap structure but lack a 3′ poly(A) tail. Instead, the 3′ end of the viral RNA contains a conserved tetranucleotide motif that binds specifically to a virally encoded protein, NSP3, that itself binds to eIF4G with higher affinity than PABP. In this manner, the viral RNA effectively competes with cellular mRNAs for available eIF4G.

One viral RNA obviates the need for the cellular factors that promote closed loop complex formation by simply utilizing direct RNA–RNA interactions between the 5′ and 3′ UTRs, effectively forming a circularized RNA that promotes efficient translation.

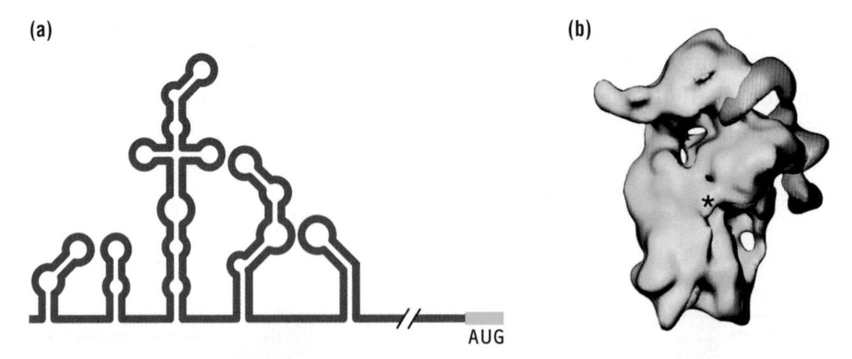

Figure 12.13 IRES elements in the 5′UTR allow for cap-independent translation initiation. (a) A typical complex picornavirus IRES with complex RNA structure. (b) Cryoelectron microscopic view of the hepatitis C virus (HCV) IRES (purple) directly interacting with the small ribosomal subunit in the region where mRNA is known to bind. Protein Data Bank (PDB) codes 2AGN, 2ZKQ.

From Spahn, MT, Kieft, JS, Grassucci, RA *et al.* Hepatitis C virus IRES RNA-induced changes in the conformation of the 40S ribosomal subunit. *Science*, 2001;**291**:1959–1962. Reprinted with permission from the American Association for the Advancement of Science (AAAS).

The hepatitis C virus (HCV) and the cricket paralysis virus (CrPV) have been of particular interest since they both depend on a minimal set of cellular initiation factors to initiate – HCV has dispensed with all but eIF3 and the eIF2 ternary complex, whereas CrPV has even figured out how to initiate translation independent of the eIF2 ternary complex, as we discuss below.

eIF2 dependent Met-tRNAi^{Met} loading is also a critical control point

As we discussed in Section 12.1, the availability of eIF2 for loading initiator met-tRNA in the ribosome is a critical control point for protein synthesis. Recall that various environmental stimuli can affect the phosphorylation state of the eIF2alpha subunit, and thus the availability of eIF2. Of particular interest in this respect is PKR, a cellular protein that phosphorylates eIF2alpha when activated by the presence of double-stranded viral RNA. This phosphorylation promotes an overly stable interaction between eIF2 and eIF2B, slowing down global protein synthesis.

As might be expected, viruses have evolved specific responses to this global attenuation of host cell protein synthesis by acquiring mechanisms for initiation that are independent of eIF2-mediated loading of Met-tRNA$_i^{Met}$ into the ribosome. In two remarkable examples, the viral RNAs themselves appear to have RNA elements either in the 5′ or 3′ UTR that mimic the function and/or shape of a tRNA, thus allowing initiation to take place at non-standard initiation codons (non-AUG).

In the turnip yellow mosaic virus (TYMV), an RNA element in the 3′ UTR of the viral transcript mimics a tRNA^{Val} and is directly aminoacylated by cellular ValRS. This aminoacylated RNA element appears to function to initiate translation on an AUG codon (albeit with a non-canonical first codon position pairing mismatch), as illustrated in Figure 12.14a. The subsequent alanine codon (GCC) is then normally decoded, and the elongation cycle continues.

In CrPV, the initiation of translation depends on base-pairing interactions between three nucleotides in the 5′ UTR of the viral transcript and three nucleotides directly upstream of an alanine codon, as shown in Figure 12.14b. These interactions appear to mimic a codon–anticodon interaction in the P site of the ribosome. The true initiation of protein synthesis (where the first amino acid is incorporated) then begins at the alanine codon (GCU) positioned in the A site of the ribosome.

Thus, for both viruses, translation initiation is cap- and eIF2-independent (no Met-tRNA$_i^{Met}$ is involved). As eIF2alpha is heavily phosphorylated during infection by these viruses, these mechanisms of initiation allow them to bypass the global cellular translational shutdown triggered by viral RNA in the cell.

Although there are numerous other examples of elegant mechanisms through which viruses take advantage of cellular machinery to propagate, the examples described here capture the key points.

Ultimately, translation in the cell is a competitive process: different mRNAs are in competition with each other to recruit the molecular components required for translation; the translation assemblies (the ribosome and associated protein factors) on different mRNAs must compete with each other for the building blocks of translation – the tRNAs and amino acids – if translation is to proceed. Beyond this, a cell must compete with external scavengers and viruses, to secure the resources it needs. Any changes in the relative balance of reagents (initiation factors, capped mRNAs, tRNAs etc.) will affect the outcome of these battles for cellular resources; translation, like all the other processes that mediate genome function, is a sensitive, finely balanced operation.

(a) Turnip yellow mosaic virus

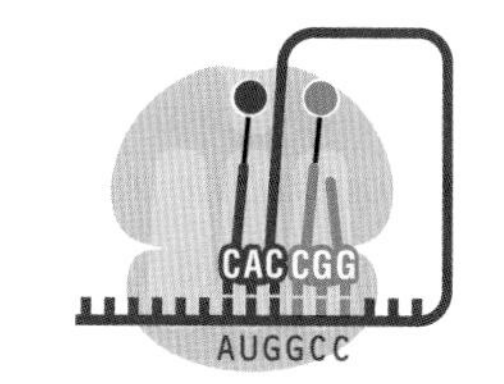

(b) Cricket paralysis virus initiation

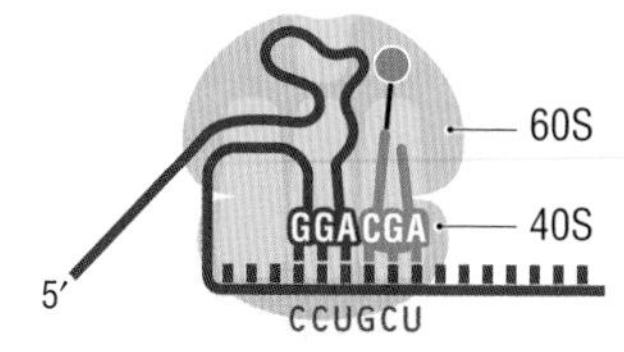

Figure 12.14 Unusual strategies used by viral RNA elements to promote eIF2-independent translation initiation. (a) TYMV initiates translation in an eIF2-independent manner by utilizing a tRNA^{Val} mimic found in its 3′ UTR to directly fill the P site. (b) CrPV initiates translation in an eIF2-independent manner by utilizing sequence in the 5′ UTR to create an artificial 'decoding helix' that somehow bypasses normal initiation requirements in the P site.

SUMMARY

- Translation initiation can be globally affected through the modification of core translational factors in eukaryotes.
- Both bacterial and eukaryotic organisms respond to amino acid limitation by shutting down overall protein synthesis, although the mechanisms differ.
- Initiation of specific bacterial mRNAs can be regulated through obstruction of the Shine–Dalgarno sequence by intrinsic RNA structure, metabolites, small RNAs, or proteins.
- Eukaryotic initiation is most often regulated through interactions of various factors (proteins and RNAs) with the 3′ UTR of the transcript.
- Viral systems have evolved ingenious mechanisms for slowing down host protein synthesis, so that their own proteins can be more efficiently synthesized.

QUESTIONS

12.1 GLOBAL REGULATION OF INITIATION IN BACTERIA AND EUKARYOTES

1. When the amino acid levels in eukaryotic cells are low, general protein synthesis is reduced. Gcn4 translation, however, is increased.
  a. Why?
  b. In general, what is the mechanism by which Gcn4 levels are increased?
  c. What would happen under high and low amino acid conditions if only one of the upstream ORFs were deleted from *GCN4*?
  d. What would happen under high and low amino acid conditions if the upstream ORFs were deleted from *GCN4*?

12.2 REGULATION OF INITIATION BY *CIS* ACTING SEQUENCES IN THE 5′ UNTRANSLATED REGION IN BACTERIA AND EUKARYOTES

1. Iron is an essential nutrient for eukaryotic cells, but high levels of free iron can be toxic. Mechanisms to regulate iron levels were discussed in both Chapter 10 (RNA processing) and Chapter 11 (translation). Compare and contrast these mechanisms.

12.3 REGULATION OF TRANSLATION THROUGH *CIS* ACTING SEQUENCES IN THE 3′ UTR IN EUKARYOTES

1. Discuss the similarities between the global regulation of translation described in Section 12.1 and the specific regulation described in Section 12.3.

Challenge question

2. Figure 12.9 shows the results of an experiment using the reproductive tract of *C. elegans* (outlined in white dots) to demonstrate the importance of the 3′ UTR in the regulation of translation.
  a. Figure (a) depicts the general reporter gene construct used for this experiment. What is GFP and why was it used as a reporter gene?
  b. How do the specific constructs that are used to produce each of the three results differ?
  c. Look up the function of each of the three genes in Wormbase (http://www.wormbase.org/#01-23-6) and explain the pattern of expression seen for each.
  d. What conclusion can be drawn from these results?

12.4 VIRAL CORRUPTION OF THE TRANSLATIONAL MACHINERY

1. Viruses that infect eukaryotic cells have developed mechanisms for reducing host cell translation, including the removal of the 5′ cap from host cell mRNA. The virus must then also have mechanisms to promote their own mRNA translation without a 5′ cap and under conditions in which the host cell is reducing overall translation. Describe two such mechanisms.

EXPERIMENTAL APPROACH 12.1 – STUDYING TRANSLATIONAL CONTROL USING RIBOSOME PROFILING

1. What is the purpose of translational profiling?

2. Compare and contrast the experimental designs for the results that are shown in Figure 2 and Figure 3.

Challenge question

3. What is the relationship between Figure 3 in this Experimental approach and Figure 12.8 in the text?

FURTHER READING

12.1 GLOBAL REGULATION OF INITIATION IN BACTERIA AND EUKARYOTES

Potrykus K. (p) ppGpp: still magical? *Annual Review of Microbiology*, 2008;**62**:35–51.

Sonenberg N, Hinnebusch AG. Regulation of translation initiation in eukaryotes: mechanisms and biological targets. *Cell*, 2009; **136**:731–745.

12.2 REGULATION OF INITIATION VIA *CIS* ACTING SEQUENCES IN THE 5′ UNTRANSLATED REGION IN BACTERIA AND EUKARYOTES

Hinnebusch AG. Translational regulation of yeast G CN4. A window on factors that control initiator tRNA binding to the ribosome. *Journal of Biological Chemistry*, 1997;**272**:21661–21664.

Johansson J, Cossart P. RNA-mediated control of virulence gene expression in bacterial pathogens. *Trends in Microbiology*, 2003;**11**: 280–285.

Pantopoulos K. Iron metabolism and the IRE/IRP regulatory system: an update. *Annals of the New York Academy of Sciences*, 2004;**1012**:1–13.

Storz G, Opdyke JA, Zhang A. Controlling mRNA stability and translation with small, noncoding RNAs. *Current Opinion in Microbiology*, 2004; 7:140–144.

Vitreschak AG, Rodionov DA, Mironov AA, Gelfand MS. Riboswitches: the oldest mechanism for the regulation of gene expression? *Trends in Genetics*, 2004;**20**: 44–50.

12.3 REGULATION OF TRANSLATION VIA *CIS* ACTING SEQUENCES IN THE 3′ UTR IN EUKARYOTES

Dever TE. Gene-specific regulation by general translation factors. *Cell*, 2004; **108**: 545–556.

Djuranovic S, Nahvi A, Green R. A parsimonious model for gene regulation by miRNAs. *Science*, 2011:**331**;550–553.

Lasko P, Cho P, Poulin F, Sonenberg N. Contrasting mechanisms of regulating translation of specific Drosophila germline mRNAs at the level of 5′-cap structure binding. *Biochemical Society Transactions*, 2005;**33**:1544–1546.

Richter JD. CPEB: a life in translation. *Trends in Biochemical Sciences*, 2007;**32**:279–285.

Szostak E, Gebauer F. Translational control by 3′-UTR-binding proteins. *Briefings in Functional Genomics*, 2013:**12**;58–65.

12.4 VIRAL CORRUPTION OF THE TRANSLATIONAL MACHINERY

Jackson RJ. Alternative mechanisms of initiating translation of mammalian mRNAs. *Biochemical Society Transactions*, 2005;**33**: 1231–1241.

Kieft JS. Viral IRES RNA structures and ribosome interactions. *Trends in Biochemical Sciences*, 2008;**33**:274–283.

Martinez-Salas E, Pacheco A, Serrano P, Fernandez N. New insights into internal ribosome entry site elements relevant for viral gene expression. *Journal of General Virology*, 2008;**89**:611–626.

Sarnow P, Cevallos RC, Jan E. Takeover of host ribosomes by divergent IRES elements. *Biochemical Society Transactions*, 2005:**33**;1479–1482.

13 Regulatory RNAs

INTRODUCTION

The regulation of gene expression is central to all aspects of biology and has been a focus throughout this book. Thus far, most of the regulators that we have encountered have been proteins, which have been extensively characterized since their discovery during the earliest days of molecular biology. However, the last decade has seen RNA molecules become recognized as ubiquitous regulators of gene expression from bacteria to eukaryotes, and considerable effort has been focused on understanding the underlying mechanisms by which these regulatory RNAs control gene expression.

The principal function of regulatory RNAs is to control the fate of other molecules. In this chapter, we will see examples of regulatory RNAs that directly modulate transcription, messenger RNA (mRNA) degradation, and translation in bacteria and in eukaryotes. In many cases, these RNAs identify their targets through simple complementary base-pairing interactions. In other cases, however, the RNA molecules fold into structures that are recognized by small molecules or proteins. We will describe in some detail the unifying principles of RNA-mediated regulation in bacteria and eukaryotes, as well as specific mechanisms and examples of these processes.

An important difference between the RNAs that we will describe in this chapter and those that we discussed previously is that their primary function is to regulate other processes. Regulatory RNAs thus perform an analogous function to those proteins whose sole function is to regulate processes such as transcription. If we consider the telomerase RNA involved in eukaryotic telomere maintenance (Chapter 6), the small nuclear RNAs (snRNAs) of eukaryotic splicing (Chapter 10), and the transfer RNAs (tRNAs) and ribosomal RNA (rRNAs) of the translational apparatus (Chapter 11), each of these RNAs is directly involved in bringing about a function, such as elongating a telomere, splicing an mRNA, or translating a protein, respectively. By contrast, the regulatory RNAs can be thought of as akin to the conductor of an orchestra: not responsible for producing the actual sound, but for guiding it into a coherent form.

Regulatory RNAs control biological processes across the kingdoms of life. For example, in bacteria, regulatory RNAs are critical in the organism's response to environmental stresses such low iron, low oxygen, or low glucose. In eukaryotes, regulatory RNAs specify embryonic development by determining which proteins are expressed at particular stages. In the nematode worm *Caenorhabditis elegans*, for instance, mutations in the small RNA (sRNA) lin-4 ('lin' for lineage) lead to striking delays in larval development (as described in Experimental approach 13.1). The sRNAs are now known to be critical for controlling gene expression throughout the life of an organism; for this reason, regulatory RNA defects are correlated with illnesses that range from heart disease to cancer.

As we saw in Chapter 1, increasingly sensitive techniques for detecting RNA have revealed that, in all organisms, far more of the genome is transcribed than was initially anticipated from the annotation of all protein-coding genes. Indeed, greater than 60% of the human genome is currently thought to be transcribed, while less than 2% of the human genome is composed of known open reading frames (ORFs), which code for proteins.

13.1 EXPERIMENTAL APPROACH

Analysis of the lin4–lin14 *genetic interaction and the development of miRNA target prediction approaches*

The first miRNA *lin-4* was discovered in genetic screens.

miRNAs are ubiquitously involved in the regulation of gene expression in plants and animals. The first miRNA was identified through genetic analyses in the experimentally tractable worm *C. elegans*. Genetic screens identified a number of worms with a class of mutation that resulted in a particular type of recognizable developmental defect; with these so-called lineage (lin) mutations, specific cell fates occur either earlier or later than they normally would. Several of the mutations that caused the developmental defect were mapped in the genome to two distinct genes, *lin–4* and *lin–14*.

Further analysis of the *lin–4* and *lin–14* genes led to the discovery of miRNAs and the first ideas about how they might work. Understanding the *lin–14* mutations proved to be relatively straightforward: the *lin–14* gene encodes a protein whose synthesis varies during development, with high levels seen in larval stage 1 (L1) and lower levels throughout subsequent stages (L2/L3). The mutations causing lineage defects yielded unregulated expression patterns, with inappropriately high levels of protein detected in the larvae at post-L1 stages. Gary Ruvkun's group found that most of the interesting mutations in *lin–14* were located in the 3′ UTR, immediately suggesting the presence of regulatory elements that are responsible for down-regulating *lin–14* expression under normal circumstances.

Interpretation of the *lin–4* mutations was less straightforward but more exciting. Like the *lin–14* variant worms, the *lin–4* variant worms exhibited a loss of regulation of LIN-14 protein levels, failing to down-regulate expression of LIN-14 in the post-L1 stages of development. The observed phenotype argued for the existence of a *lin–4* gene product, assumed to be a protein, that acts as a repressor of LIN-14 protein synthesis. However, when Victor Ambros' group identified the *lin–4* gene, they discovered that it did not encode a protein. Instead, they showed that the gene specifies the production of an RNA transcript that is ultimately processed into an sRNA (around 22 nucleotides in length) now classified as a miRNA; their results are shown in Figure 1.

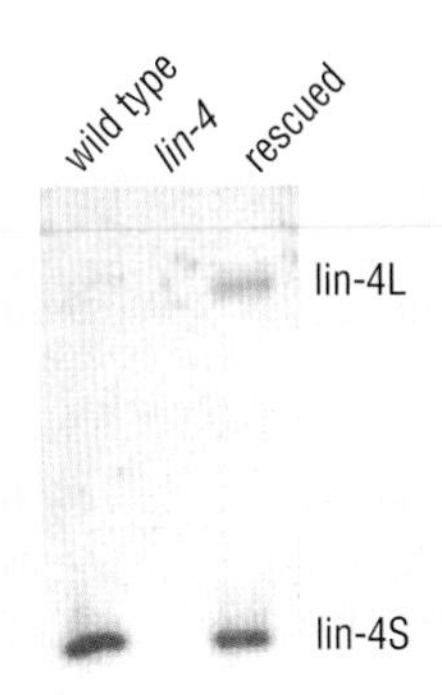

Figure 1 Identification of two small *lin-4* transcripts. Northern blot of total RNA from a *C. elegans* wild-type strain, a *lin-4* mutant strain, and from a strain with a rescuing plasmid carrying the *lin-4* wild-type gene. Two RNA products are seen, a less abundant longer product (*lin-4L*, 61 nucleotides in length) and a more abundant shorter product (*lin-4S*, ~22 nucleotides long); we now know these correspond to pre-miRNA and fully processed miRNA, respectively.

Reproduced from Lee *et al.* The *C. elegans* heterochronic gene encodes small RNAs with antisense complementarity to *lin-14*. *Cell*, 1993;**75**:843–854.

The *lin-4* RNA acts at a post-transcriptional level by base pairing with the *lin-14* mRNA

The levels of the *lin–14* mRNA were shown to be roughly equivalent in larval stages L1 and L2, whereas the amount of LIN-14 protein present during these same stages was very different (see results in Figure 2). Consistent with the genetics, an increase in

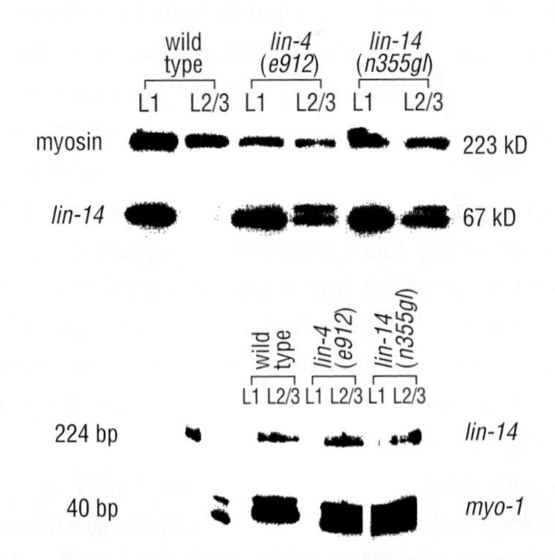

Figure 2 Analysis of mRNA and protein levels in different larval stages during *C. elegans* development. (a) Western analysis of Lin-14 levels (and control myosin) in synchronized preparations of wild-type worms. Lin-14 is down-regulated in wild-type embryos at late stages, but this regulation is lost in the mutant worms. (b) RNase protection analysis of *lin-14* mRNA levels (and control myosin). In this analysis, the *lin-14* mRNA levels are essentially the same in the early and late extracts. L1 refers to early stage larva and L2/L3 to late stage.

Wightman *et al.* Posttranscriptional regulation of the heterochrinic gene *lin-14* by lin-4 mediates temporal pattern formation in *C. elegans*. *Cell*, 1993;**75**:855–862.

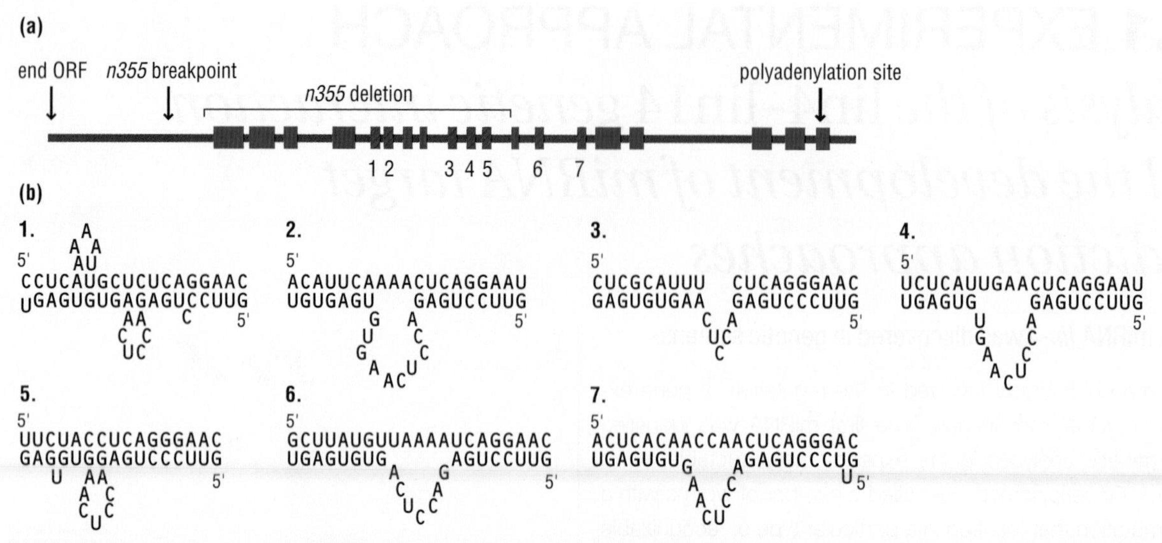

Figure 3 Conserved sequences in the *lin-14* 3′ UTR are complementary to the *lin-4* RNA. (a) Representation of regions in the *C. elegans* 3′ UTR that are conserved in between *C. elegans* and *C. briggsae* (blue). Shown in pink (overlaid on blue) are conserved regions that are complementary to the *lin-4* RNA. The mutant breakpoints for several lineage mutants and polyadenylation sites are indicated. (b) Predicted pairings between *lin-4* RNA and the proposed *lin-4* target sites in the *lin14* 3′ UTR.

Wightman *et al.* Posttranscriptional regulation of the heterochrinic gene *lin-14* by lin-4 mediates temporal pattern formation in *C. elegans*. *Cell*, 1993;**75**:855–862.

lin–4 RNA correlates with the down-regulation of LIN-14 protein levels in larval stage L2, suggesting that *lin-4* regulates (and, in fact, represses) LIN-14 synthesis. Together, these observations led to the proposal that the mechanism of regulation by miRNAs is post-transcriptional, potentially affecting mRNA stability, some step in translation, or protein stability.

Perhaps the most informative insight coming from the early analyses of these genes (and noted by both the Ambros and Ruvkun groups) was that the *lin-4* RNA product could form direct pairing interactions with as many as seven potential complementary regions within the 3′ UTR of the *lin-14* mRNA (see Figure 3). These targeted regions in the *lin–14* mRNA are evolutionarily conserved, and fall within the region that is affected in the *lin–14* mutants. Collectively, these observations suggested that the *lin–4* RNA exerts its effect by specifically binding to complementary regions of the *lin–14* 3′ UTR. These initial predictions about how targets are identified by the miRNAs turn out to be broadly accurate: miRNAs indeed recognize their mRNA targets through direct base-pairing interactions.

These early genetic discoveries opened a new and exciting area of biology that has been well served by the existence of fully sequenced genomes. When a second sRNA (*let–7*) with similar genetic and biochemical properties was discovered, and was seen to be conserved from *C. elegans* to humans, researchers began to wonder if these small regulatory RNAs were not more widespread throughout nature. The floodgates then opened as the sequencing of sRNAs isolated from worms, flies, and human cells revealed the existence of hundreds of miRNA genes. High-throughput sequencing approaches, many which took advantage of the specialized 5′ monophosphates and 3′ hydroxyl groups found on processed miRNAs, were used to explore the numbers and diversity of this class of biological macromolecule in multiple organisms – current numbers confidently identify more than 200 miRNA genes in *C. elegans*, and in *Drosophila melanogaster*, and more than 1000 in humans.

Approaches to identify miRNA targets on a genome wide basis are being developed

The next task has been to identify the biological targets of these many small RNAs. Several complementary approaches have been taken. First, bioinformatic approaches have been used to identify potential mRNA pairing partners for miRNAs, with particular emphasis on those potential targets that are conserved between organisms. Second, such data sets have been compared with the output from high-throughput assays – microarray, deep sequencing, and proteomic analyses – to evaluate the impact of miRNAs on RNA and protein levels in the cell. These approaches have exhibited a marked overlap in potential mRNA target sets – RNA and protein levels are both affected by a given miRNA, as depicted in Figure 4 – and together have been used to develop target prediction software that can identify many of the significant targets of a given miRNA.

As a third approach, denoted cross-linking immunoprecipitation (CLIP), groups stabilize the interactions between Argonaute proteins and associated miRNAs and target mRNAs by treating cells with ultraviolet (UV) light or other cross-linking agents to generate protein–RNA cross links. The Argonaute proteins are then purified. Upon partial digestion of the cross-linked RNA and, in some protocols reversal of the cross links, the RNA species that

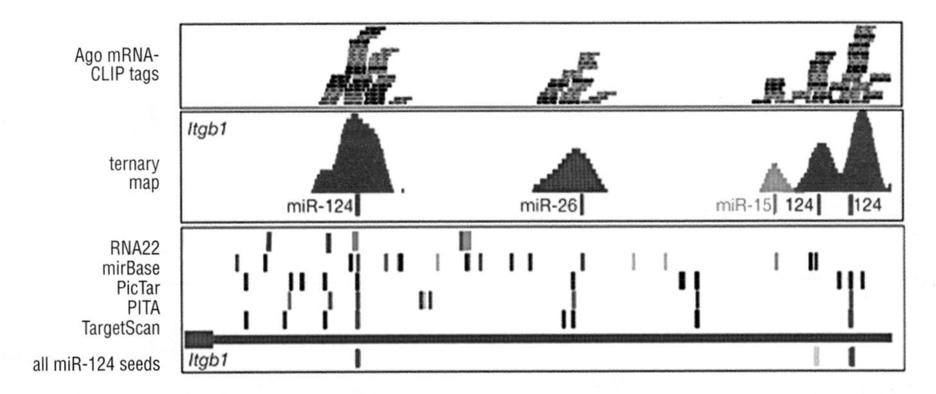

Figure 4 Sequences obtained in an Argonaute -mRNA CLIP experiment reveal target sites of miR-124 in the 3′ end of *Itgb1* mRNA. The top panel shows the positions of all raw sequence reads, which are represented in normalized clusters in the ternary map (middle panel). The bottom panel shows miRNA sites predicted by programs indicated on left with different colored bars corresponding to different miRNAs.

Chi *et al.* Argonaute HITS-CLIP decodes microRNA-mRNA interactions maps. *Nature*, 2009;**460**:479–486.

co-purify with the Argonaute protein can be subjected to high-throughput sequencing.

In one application of this approach, Robert Darnell and colleagues identified targets for miR-124 and 20 other miRNAs in mouse brain. As shown in Figure 4, the mRNA fragments associated with the Eif2c Argonaute protein for the 3′ UTR of the *Itgb1* mRNA map to five positions that show good overlap with target sites predicted for the miR-124, miR-26, and miR-15 miRNAs.

As each conserved miRNA has evolutionarily conserved interactions with hundreds of mRNAs in the cell, it is clear that miRNAs play an important and broad role in metazoan biology. While there remain gaps in our molecular understanding of how miRNAs regulate gene expression, target identification has placed the field in a position to pursue interesting biology.

Find out more

Baek D, Villén J, Shin C, *et al.* The impact of microRNAs on protein output. *Nature*, 2008;**455**:64–71.

Chi SW, Zang JB, Mele A, Darnell RB. Argonaute HITS-CLIP decodes microRNA-mRNA interaction maps. *Nature*, 2009;**460**:479–786.

Lee RC, Feinbaum RL, Ambros V. The *C.elegans* heterochronic gene encodes small RNAs with antisense complementarity to *lin-14*. *Cell*, 1993;**75**:843–854.

Lewis BP, Shih IH, Jones-Rhoades MW, Bartel DP, Burge CB. Prediction of mammalian microRNA targets. *Cell*, 2003;**115**:787–798.

Lim LP, Lau NC, Garrett-Engele P, *et al.* Microarray analysis shows that some microRNAs downregulate large numbers of target mRNAs. *Nature*, 2005;**433**:769–773.

Selbach M, Schwanhäusser B, Thierfelder N, *et al.* Widespread changes in protein synthesis induced by microRNAs. *Nature*, 2008;**455**:58–63.

Wightman B, Ha I, Ruvkun G. Posttranscriptional regulation of the heterochronic gene *lin-14* by *lin-4* mediates temporal pattern formation in C. *elegans*. *Cell*, 1993;**75**:855–862.

Related techniques

C. elegans (as a model organism); Section 19.1

Northern blots; Section 19.10

Western blots; Section 19.11

RNAse protection; Section 19.10

Mouse (as a model organism); Section 19.1

Microarrays; Section 19.10

Proteomic analysis; Section 19.8

These discrepancies – and a growing appreciation of the potential of RNA to play regulatory roles – have led scientists to believe that much of the genome that was thought to be functionally inert may, in fact, be vital for regulating expression of the genome from bacteria to man.

➔ We discuss the concept of genome annotation in Section 18.2.

13.1 OVERVIEW OF REGULATORY RNAs

Despite the variety of ways in which regulatory RNAs operate, there are some recurring principles. First, regulatory RNA molecules are often enzymatically processed from a longer primary transcript to yield the final functional RNA species. Second,

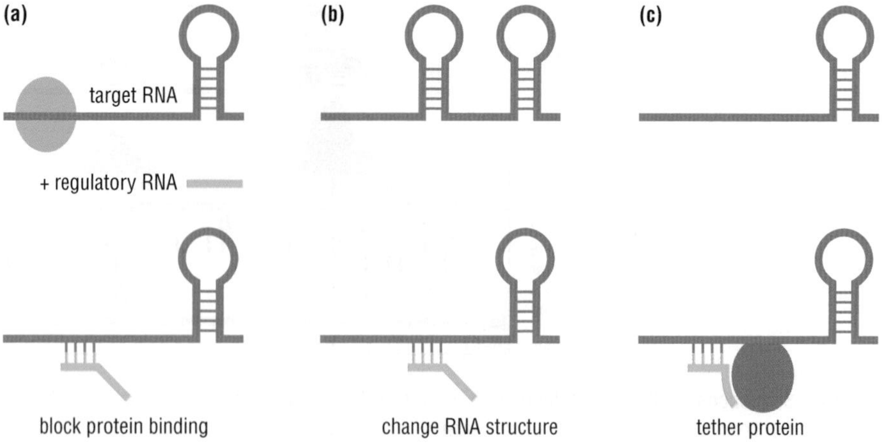

Figure 13.1 Possible ways RNAs can act as regulators by base-pairing. Base-pairing between the regulatory RNA (green) and the target RNA (blue) can (a) block a protein-binding site, (b) change the structure of the target, RNA or (c) bring a protein or proteins into the proximity of the target RNA.

regulatory RNAs typically use base-pairing interactions to bind to their DNA or RNA targets. Third, regulatory RNAs frequently interact with other components, typically proteins, to perform their regulatory function. We will explore each of these features in the rest of this chapter.

It is important to keep in mind that the study of regulatory RNAs is still an emerging area, about which we are continuing to learn more and more with every year that passes. Throughout this chapter, we discuss the general categories of regulatory RNAs that are best understood - but it is not always easy to put these RNAs into neatly defined categories, not least because a given RNA molecule has the potential to have multiple functions. For example, a given transcript may regulate gene expression through more than one mechanism, or may also encode a protein. Nonetheless, a number of general principles are emerging from studies of RNA regulators, as we discuss below.

RNA is ideally suited to regulate through base-pairing interactions

RNAs are particularly well suited to regulate other nucleic acid species through simple complementary base-pairing interactions. Base-pairing interactions between a regulatory RNA and the RNA molecule it is targeting typically either directly obstructs function (by blocking interactions with other partners), alters the target RNA structure (and thus potential interactions and functions), or recruits other factors that, in turn, mediate downstream events. These outcomes are illustrated in Figure 13.1. RNA regulators that work by base-pairing have been extensively characterized in both bacteria and eukaryotes. As depicted in Figure 13.2a, the base pairing RNAs can be encoded on the strand of DNA that is antisense to the one encoding their target RNA; these antisense RNAs typically have extensive complementarity with their target. Regulatory RNAs can also be encoded at a site by a region that is entirely separate (in *trans*) from the genes encoding the target RNA

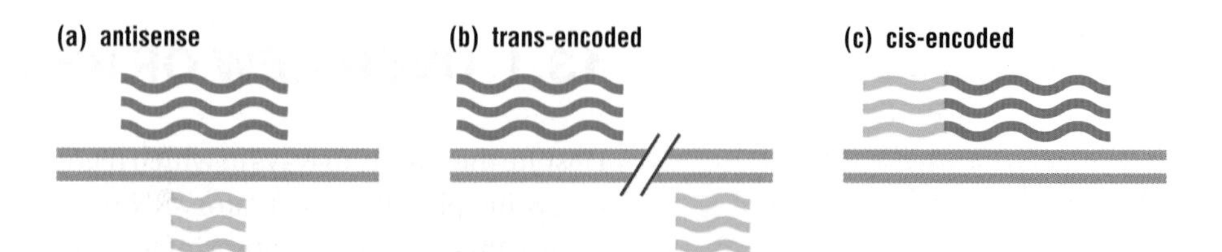

Figure 13.2 Regulatory RNAs can be encoded relative to their targets in a variety of configurations. (a) Base-pairing RNAs (green) can be encoded on the strand opposite or antisense to their target RNA (blue) or (b) at a distance from their target. Although the *trans*-encoded sRNA in (b) is encoded on the strand opposite the target it could just as well be encoded on the same strand. (c) Regulatory RNAs can also be encoded as part of (in *cis* to) their target.

or RNAs with which they base-pair, in which case they typically have more limited complementarity with their target or targets (Figure 13.2b).

Flexibility in RNA structure allows for regulation via metabolite or protein binding

Given the inherent flexibility of RNA molecules which enables them to adopt diverse structures in the cell, it is not surprising that RNAs also can act as regulators by providing binding sites for metabolites or proteins. (Recall the impressive three-dimensional structures that RNA can assume – for example, the tRNAs and rRNAs mediating translation that we encounter in Chapter 11). Figure 13.2c illustrates how highly structured RNA regulators often act in *cis*, meaning that they are directly connected to the transcript that they regulate. Structured RNAs can also act in *trans* and provide binding sites for proteins and influence their activities. In addition, RNA can act as a regulator by effectively serving as a scaffold to bring two or more proteins into proximity with one another or with DNA.

Regulatory RNAs in bacteria and eukaryotes have similar biological effects but achieve them in different ways

Regulatory RNAs in all organisms can impact the transcription, stability, or translation of the target RNA. However, the way in which the RNAs are processed and the mechanisms and machinery through which they manifest their effects are different in organisms from the different kingdoms of life. We will highlight both the similarities and differences throughout this chapter.

13.2 BACTERIAL BASE-PAIRING sRNAs

Let us begin exploring the principles that govern how base-pairing RNAs can act by considering their activity in bacteria. Regulatory RNAs in bacteria are generally synthesized as independent transcripts in the range of 100 to 300 nucleotides in length and thus are commonly designated as 'small' RNAs (sRNAs). A subset of these primary transcripts are processed to shorter fragments but the majority act as full-length RNAs, in contrast to what we will see for eukaryotic base-pairing RNAs. Most of the base-pairing sRNAs characterized to date in bacteria are encoded in *trans* and have only limited complementarity with their mRNA targets, but a limited number of antisense sRNAs encoded on the complementary strand relative to their target RNAs have also been studied.

Trans-encoded sRNAs can regulate ribosome binding or target mRNAs for degradation

Trans-encoded base-pairing sRNAs in bacteria are commonly expressed in response to specific environmental conditions such as limiting iron or a shift from aerobic to anaerobic conditions. As we will see in an example below, many of the sRNAs base-pair with sequences in the mRNA in the vicinity of the site to which ribosomes must bind in order to begin translation (the 'Shine–Dalgarno' motif discussed in Section 11.8). The base-paired RNA duplex prevents the ribosome from binding, resulting in decreased synthesis of the protein encoded by the mRNA. In selected circumstances, the sRNAs can have the opposite effect, actually facilitating

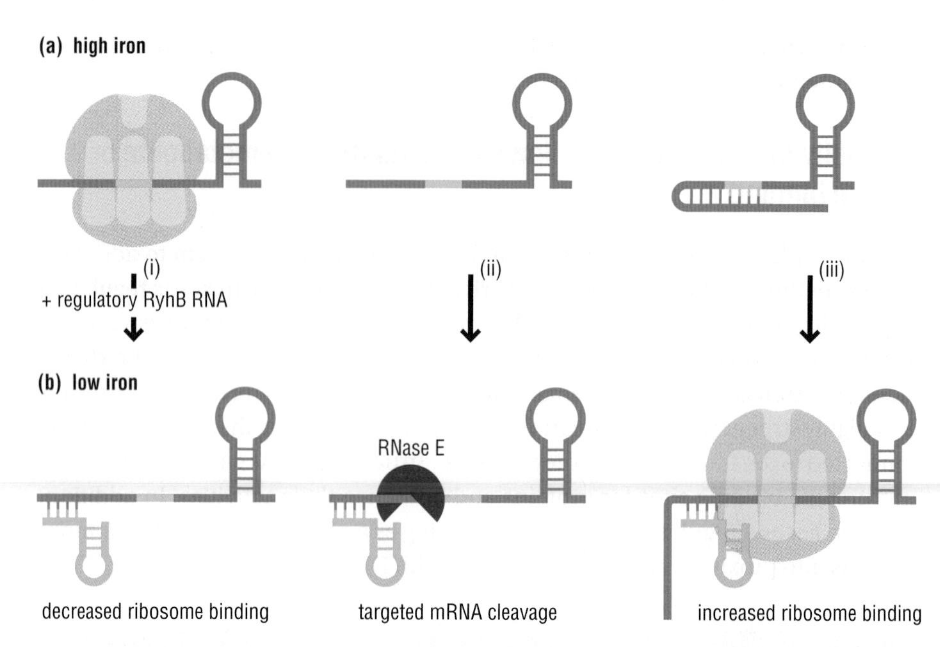

Figure 13.3 Low iron-induced RyhB RNA illustrates mechanisms by which bacterial sRNAs can act. (a) When intracellular iron levels are high, expression of the RyhB RNA (green) is repressed, and the mRNAs encoding negatively-regulated mRNA targets are translated (i) or are stable (ii), while the secondary structure of a positively regulated mRNA target prevents translation (iii). (b) When intracellular iron levels are low, expression of the RyhB RNA is induced, and RyhB RNA base-pairing with negatively-regulated mRNAs (i) blocks ribosome binding and/or (ii) targets the mRNAs for degradation or, in a case of positive regulation (iii), RyhB RNA base-pairing induces a change in the target mRNA secondary structure to facilitate ribosome binding.

translation of target mRNAs. In these cases, the sRNA may base-pair with the 5′ untranslated region (UTR) of an mRNA and thereby prevent formation of RNA secondary structure that would otherwise block ribosome binding and inhibit translation.

In combination with modulating translation, bacterial sRNAs can stimulate the degradation of the target mRNAs by recruiting ribonucleases. In some cases, the sRNA itself is degraded together with the mRNA. There also are examples where sRNAs base-pair with sequences within the coding region and stimulate RNase E-dependent cleavage of the target mRNA, independent of effects on translation.

A well-studied sRNA that illustrates the different mechanisms by which these regulators can modulate gene expression is the 90-nucleotide RyhB RNA of *Escherichia coli*, whose effects on gene expression are illustrated in Figure 13.3. Expression of RyhB is induced when cells encounter low iron levels. Under these conditions, RyhB represses the synthesis of non-essential iron-containing enzymes, allowing the limited iron to be more effectively utilized by essential iron-requiring enzymes. For many mRNA targets, RyhB base-pairs at or near the translational start codon of the mRNA, thereby blocking ribosome binding. For one target, the base pairing of RyhB near the ribosome-binding site blocks translation while also directing RNase E to cleave within the downstream coding sequence. On another target, RyhB base-pairs with part of the structured 5′ UTR of the mRNA to promote ribosome binding and increase synthesis of a gene product that is important for the production of siderophore molecules that facilitate iron uptake. Thus, a single sRNA acts in three different ways on three different targets. Recall our discussion in Chapter 10 and Chapter 12 of the eukaryotic IRPs, which also modulate mRNA stability and translation in response to cellular iron levels.

Trans-encoded sRNAs in many bacteria require the chaperone protein Hfq for base-pairing

The base complementarity between the *trans*-encoded sRNAs and their mRNA targets is usually limited to a length of 10–20 base pairs. While the overall base-pairing

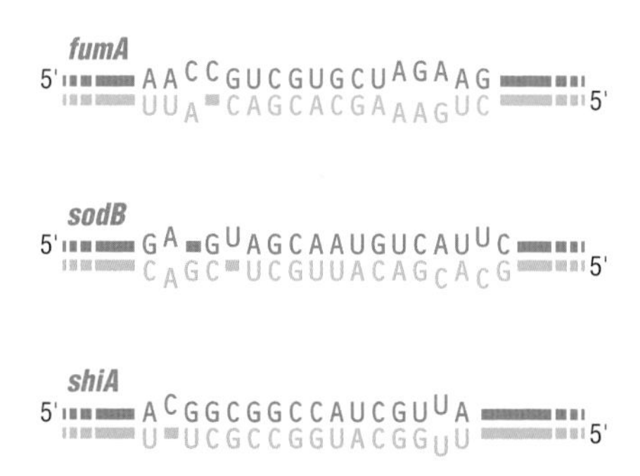

Figure 13.4 Examples of base-pairing that can take place between bacterial sRNAs and their targets. This figure shows regions of base-pairing between the RyhB RNA and three targets, with different regions of the sRNA involved in each case. The longest regions of contiguous base-pairing are likely to be the initial site of base-pairing (sometimes referred to as a 'seed' region but less defined than the 'seed' regions delineated for small eukaryotic RNAs).

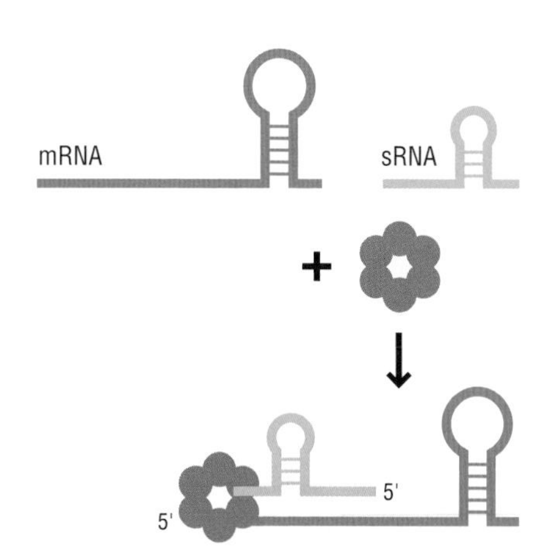

Figure 13.5 Model for Hfq action. The Hfq protein, which has been shown to have RNA binding on the proximal, rim, and distal faces of the hexameric ring, is thought to facilitate sRNA (green) and mRNA (blue) base-pairing by bringing the RNAs into proximity and affecting their structures.

is often somewhat discontinuous, Figure 13.4 shows how there is usually a region of continuous base-pairing that is most critical for the interaction. The region of the sRNA involved in base-pairing with the target RNA is generally single-stranded and more conserved than the rest of the molecule. This single-stranded region can be found at different positions within a particular sRNA - at the 5′ end, in an internal loop, or in single-stranded regions.

Given the limited base-pairing interactions and the need for the sRNA to locate its specific target mRNAs among thousands of other transcripts, many bacterial sRNAs require the assistance of the RNA chaperone protein, Hfq. Intriguingly, this hexameric, ring-shaped protein is the homolog of the Sm proteins that are found in the eukaryotic splicing complexes, which we discussed in Chapter 10. The sRNAs and target mRNAs appear to bind to different surfaces of the Hfq, and so it is thought that Hfq functions to bring two regions involved in base-pairing into proximity with one another, as depicted in Figure 13.5.

The action of *trans*-encoded sRNAs can be modulated in multiple ways

Sometimes, the levels of Hfq or the sRNAs in the cell can be limiting relative to the mRNA targets. In these situations, competition for Hfq binding can have a direct impact on regulation. For example, elevated levels of a particular sRNA that is induced upon oxidative stress can saturate the Hfq binding sites. Hfq is thus not available to bind to other sRNAs that also depend upon Hfq for their activity, thereby preventing these sRNAs from carrying out their functions. In another twist on this theme, an mRNA can act as a 'sponge' or 'decoy' to sequester a given sRNA, preventing this sRNA from acting on a different target. In the example shown in Figure 13.6, the base-pairing of an sRNA with an abundant mRNA also can lead to the degradation of the sRNA and thus its removal from the active pool of sRNAs. Similar nuances in regulation are observed for small eukaryotic base-pairing RNAs, which we will discuss next.

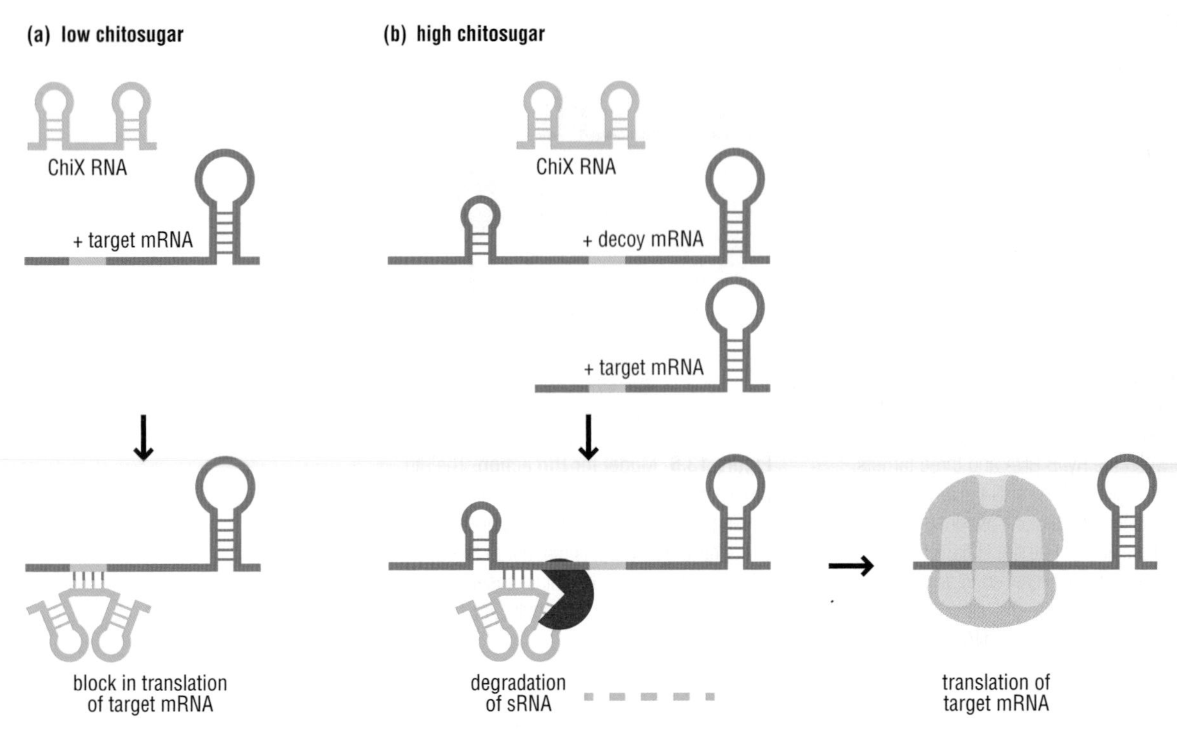

Figure 13.6 Consequence of competition between mRNA targets. (a) When cells are exposed to low levels of chitosugar (oligosaccharides derived from the ubiquitous chitin polymer), the ChiX sRNA (green) can base-pair with target mRNAs (dark blue) to repress translation. (b) When the levels of chitosugars increase, transcription of a decoy mRNA (light blue) is induced. Base-pairing between ChiX and the decoy leads to degradation of the sRNA. As a consequence, translation of the target mRNA is no longer blocked, and the corresponding porin, which allows the uptake of chitosugars, can be synthesized.

Antisense sRNAs repress the synthesis of potentially toxic proteins by modulating translation and mRNA stability

Fewer antisense sRNAs have been studied in bacteria, but one theme that has emerged is that these RNAs tightly regulate potentially detrimental proteins such as transposases and toxins. These antisense RNAs are continuously expressed, preventing synthesis of the gene product encoded on the sense RNA. Although sense and antisense RNAs by definition share extensive complementarity, base-pairing between the RNAs typically initiates through limited contact between two single-stranded regions in what has been termed a 'kissing' complex. The region of pairing can subsequently expand, ultimately blocking translation or serving as a substrate for ribonucleases such the double-strand specific endonuclease RNase III.

13.3 EUKARYOTIC sRNAs: miRNAs, siRNAs, AND rasiRNAs

Let us now focus on the sRNAs from eukaryotes that are broadly involved in gene silencing or 'RNA interference' (RNAi). The first evidence for this type of interference came from studies of unusual post-transcriptional silencing observed in plants and in *Neurospora crassa* (where it is known as 'quelling'). The characterization of the RNAi pathways has revolutionized our thinking about the roles of RNA in biology. We now know that sRNAs silence gene expression at the level of transcription, RNA stability, or translation by base-pairing with their targets in the cell. Unlike the bacterial sRNAs discussed in Section 13.2, small regulatory RNAs in eukaryotes are rather similar in length, ranging from 20 to 30 nucleotides. These RNAs are processed from longer primary transcripts that derive from multiple sources, and thus commonly have 5′phosphate groups and 3′ hydroxyl termini.

The eukaryotic sRNAs ubiquitously associate with a family of proteins known as the **Argonautes**, which facilitate pairing interactions between the bound regulatory RNA and cellular RNA targets. This interesting name derives from the squid-like appearance of leaves of *Arabidopsis thaliana* mutants with defects in this protein. The particular Argonaute protein with which the regulatory RNA interacts dictates, in part, the biological consequence. For example, some Argonaute proteins catalyze a cleavage reaction that leads to RNA degradation, some trigger chromatin modifications at specific genomic loci, while others repress translation and promote mRNA decay. Given these distinctions in mechanism, the 'sorting' of the various sRNAs in the cell onto the appropriate Argonaute protein can impact their biological function.

In the sections that follow, we will focus on the different steps in the life cycle of eukaryotic small regulatory RNAs: their origin, their processing, their loading into an Argonaute protein, and, finally, their functional output.

The sRNAs in eukaryotes can be grouped into broad classes

At least three broad classes of small regulatory RNAs can be categorized in eukaryotes based on differences in each step of the sRNA life cycle. We'll begin by considering the three classes in overview before moving on to discuss the stages of their life cycles in more detail.

MicroRNAs (**miRNAs**) are specific short RNA sequences which are derived from a primary transcript that is generated in the nucleus from a specific gene encoded in the genome; these RNAs generally down-regulate cytoplasmic mRNAs through translational repression and mRNA decay and are integral to many regulatory networks.

Small interfering RNAs (**siRNAs**) are short RNA sequences derived from longer duplex RNAs found in the cytoplasm, including viral RNAs or various endogenous duplex RNAs. Endogenous duplex RNA can arise in RNA transcripts that have fold-back structures, two RNA molecules that are transcribed convergently, or antisense RNAs synthesized by RNA-dependent RNA polymerases from coding RNA templates. The siRNAs are thought to target RNAs for cleavage in the cytoplasm and as such can serve as a cellular defense system.

Repeat-associated small interfering RNAs (**rasiRNAs**) are short sequences specifically derived from repetitive regions of the genome (including centromeres and transposon-rich regions); these RNAs down-regulate transcription from diverse repetitive regions of the genome. The rasiRNAs fall into several subclasses. Two well-studied subclasses are the heterochromatic RNAs (hcRNAs) of *Schizosaccharomyces pombe* and the mammalian Piwi-interacting RNAs (piRNAs) named for their preferred binding partners, the Piwi family of Argonaute proteins.

13.4 PROCESSING OF EUKARYOTIC sRNAs

The eukaryotic small regulatory RNAs are generally derived from primary transcripts that must be further processed by specialized machinery. However, each class of sRNA differs in terms of the source of RNA that provides the substrate for the RNA processing machinery – endogenous or exogenous, single-stranded or

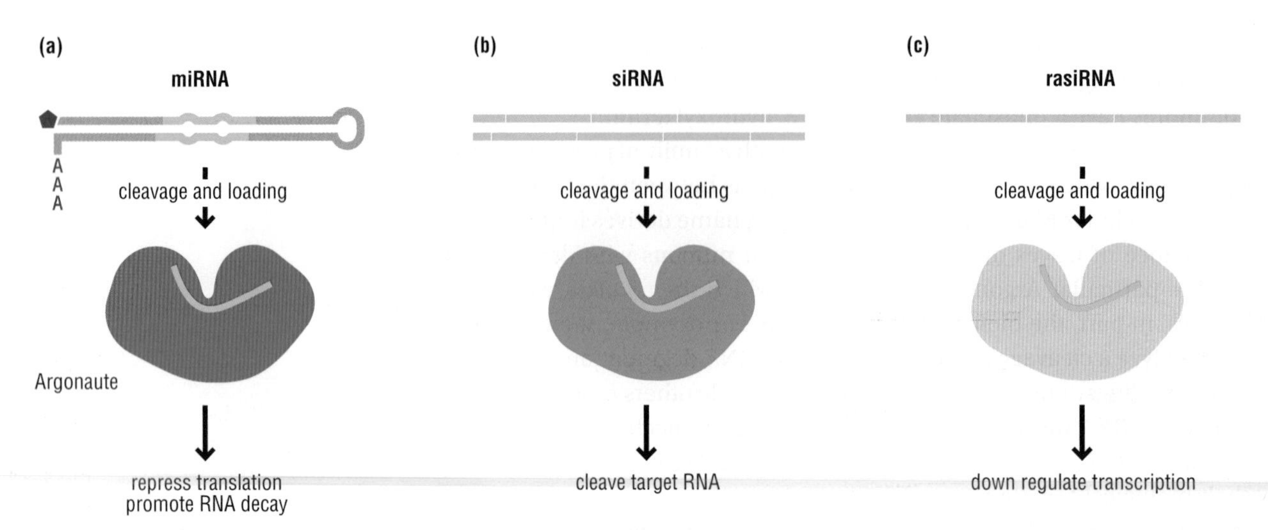

Figure 13.7 Three general classes of eukaryotic small RNAs are (a) miRNAs, (b) siRNAs, and (c) rasiRNAs. All are processed from longer precursor RNAs: miRNA from endogenous primary transcripts, siRNAs from endogenous or exogenous sources of double-stranded RNA, and rasiRNAs from single-stranded RNA transcribed from repeat sequences. They are all processed by different mechanisms and associate with different Argonaute proteins to carry out the indicated functions.

double-stranded, specific (discrete) or non-specific. These differences are summarized in Figure 13.7. The miRNAs, siRNAs, and hcRNAs in yeast share a number of features in their processing pathway. By contrast, the processing pathway for piRNAs is quite distinct. What unifies all of these small RNAs, however, is that for each class, a longer RNA precursor is processed into appropriately sized pieces of RNA, which are loaded into the Argonaute proteins to function.

RNase III family enzymes generate miRNAs, siRNAs, and some rasiRNAs of defined length by endonucleolytic cleavage

RNase III family enzymes are key in the processing of both miRNAs and siRNAs. The domain structures of the RNase III family are depicted in Figure 13.8. Broadly speaking, the RNase III family endonucleases use a dimeric catalytic domain to promote the cleavage of double-stranded RNA at two sites. Some RNase III enzymes only possess one catalytic domain and so must dimerize to catalyze cleavage (for example, enzymes involved in the processing of rRNA described in Chapter 10). In contrast, Drosha and Dicer, two enzymes involved in miRNA and siRNA processing, both possess two distinct RNase III domains and so function as monomers.

Another critical feature is the way in which these enzymes set the length for their RNA cleavage products. Each enzyme digests double-stranded RNA to produce short duplex fragments of a particular length. The mechanisms by which

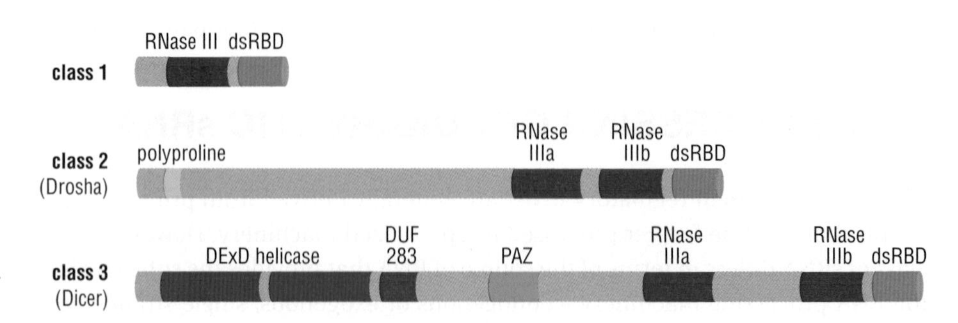

Figure 13.8 Domain structure of different classes of RNase III enzymes. Pink regions correspond to the ribonuclease domain of RNase III enzymes. Other functional domains include a double-stranded RNA binding domain (blue), a proline-rich region (brown), a DExD helicase domain (purple), a DUF283 domain (red), and a PAZ domain (yellow).

Based on Fig. 1 in MacRae & Doudna, Ribonuclease revisited: structural insights into ribonuclease III family enzymes. *Current Opinion in Structural Biology*, 2007;**17**:1–8.

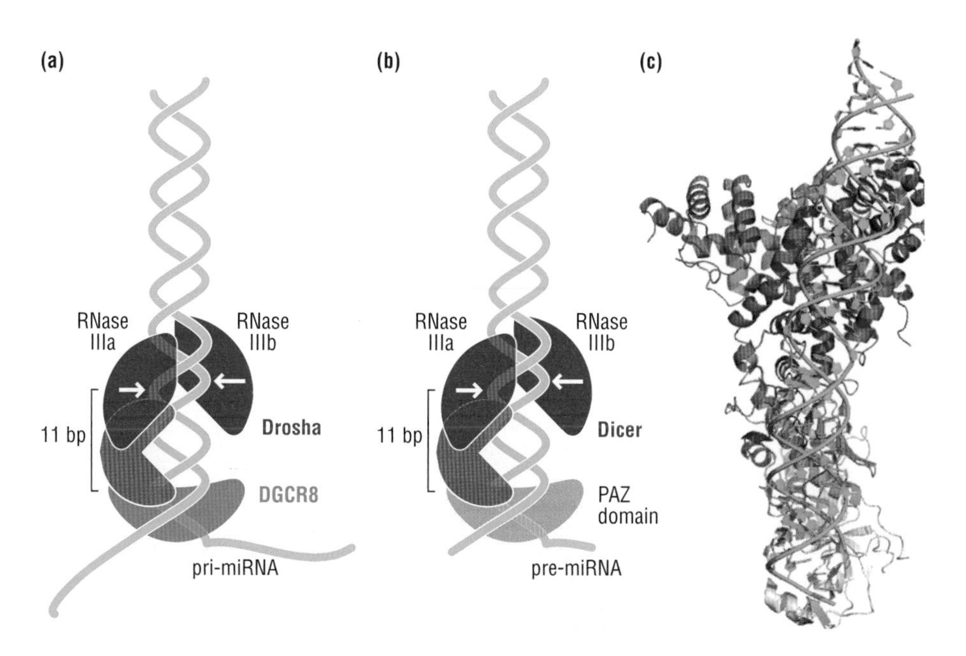

Figure 13.9 Different mechanisms by which Drosha and Dicer proteins determine position of cleavage. (a) The Drosha interaction with the DGCR8 proteins determines the position of cleavage while (b) the PAZ domain of the Dicer proteins determines the position of cleavage. (c) Structure of the Dicer protein.

Based on Fig. 4 in MacRae & Doudna, Ribonuclease revisited: structural insights into ribonuclease III family enzymes. *Current Opinion in Structural Biology*, 2007;**17**:1–8.

different RNase III enzymes determine the position of cleavage is outlined in Figure 13.9. Some RNase III enzymes (like Dicer) can 'measure' a fragment of the requisite length by the nature of their binding to RNA, typically securing the 3′ end of the RNA in a protein domain and having the catalytic domain located a defined distance away. Other RNase III enzymes (like Drosha) rely on accessory factors to define the distance between one end of the RNA and the site of cleavage.

miRNAs are encoded in the genome and are commonly processed sequentially by two distinct RNase III enzymes

The miRNAs derive from specific genes found in genomes across the lineages of multicellular animals and plants. As such, these RNAs begin as primary transcripts (pri-miRNAs), usually transcribed by RNA polymerase II. For example, the precursor for the miR-35 through miR-41 miRNAs in *Caenorhabditis elegans* is depicted in Figure 13.10. Like other RNA polymerase II products, the pri-miRNA transcripts are capped and polyadenylated and often contain introns, but do not necessarily encode a protein product.

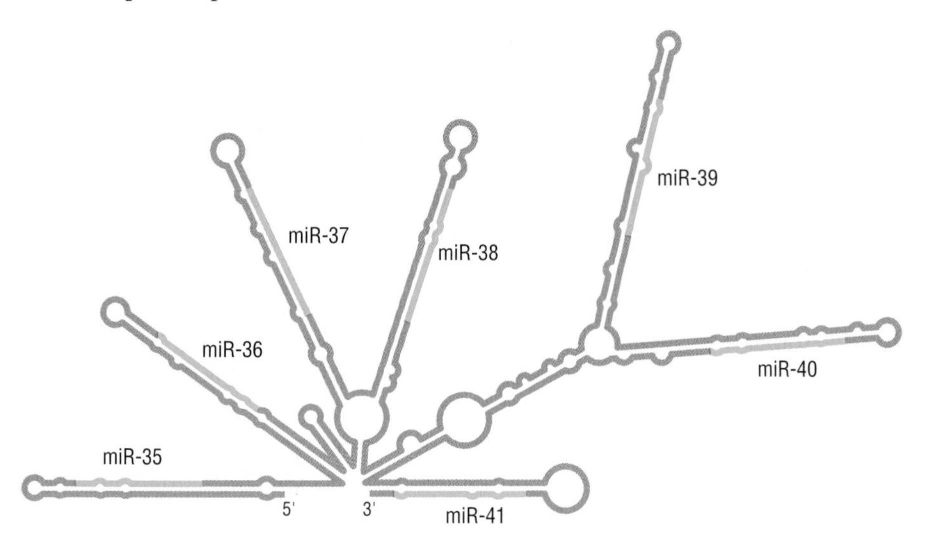

Figure 13.10 Precursor for the miR-35 through miR-41 miRNAs in *C. elegans*. The seven miRNAs (indicated in green) are cleaved out of the longer precursor RNA by the Drosha and Dicer enzymes. Note how all are cleaved from the right side of a characteristic stem–loop structure.

Adapted from Lau, NC, Lim, LP, Weinstein, EG, and Bartel, DP. An abundant class of tiny RNAs with probable regulatory roles in *Caenorhabditis elegans*. *Science*, 2001;**294**:858–862, Figure 1D.

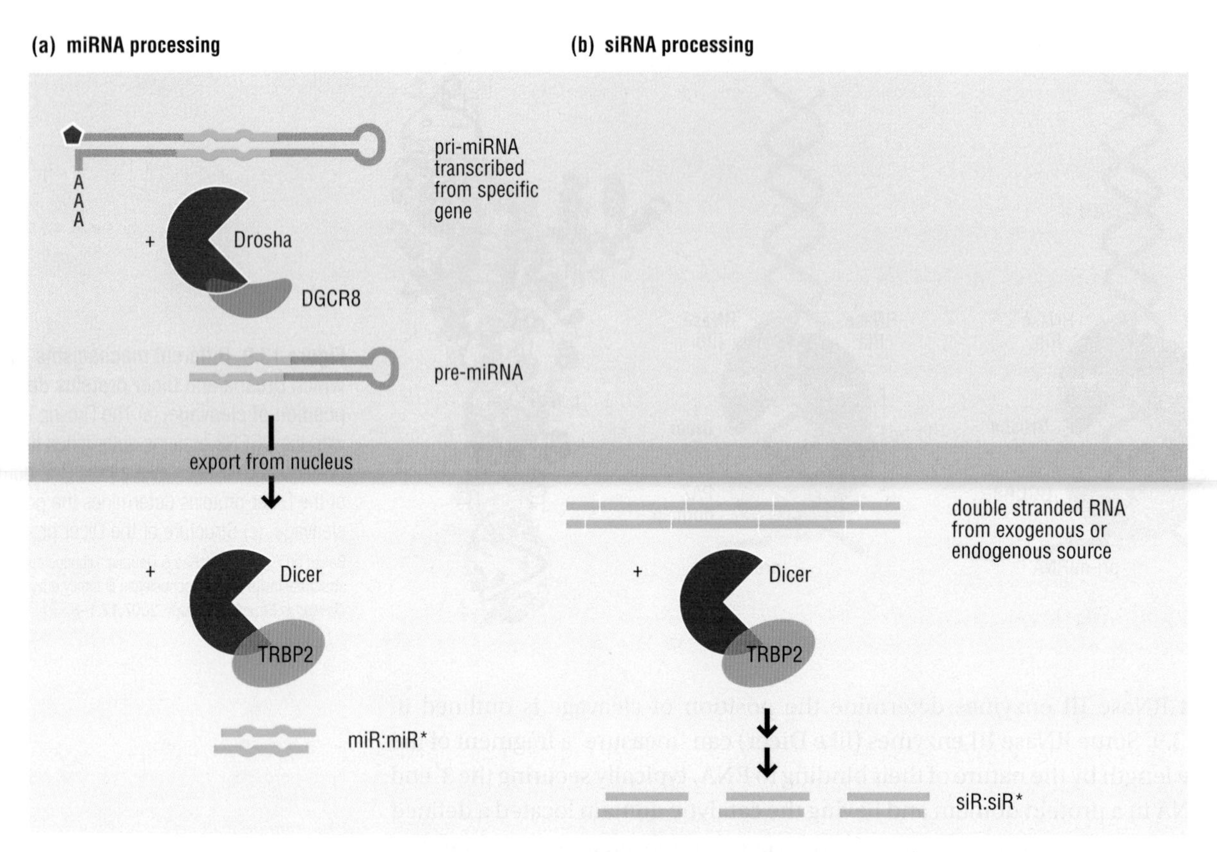

Figure 13.11 Mechanism of (a) miRNA and (b) siRNA biogenesis. For miRNAs, a primary pri-miRNA transcript with a 5′-end cap and 3′-end poly(A) tail, is cleaved by the Drosha ribonuclease in the nucleus to give the pre-miRNA, which is exported to the cytoplasm, where it is further processed by the Dicer ribonuclease. For siRNAs, double-stranded RNAs of either exogenous or endogenous origin are randomly cleaved by the Dicer ribonuclease in the cytoplasm. Proteins associated with the Drosha and Dicer enzymes, such as DGCR8 and TRBP2, provide specificity.

Scan here to watch a video animation explaining more about the generation and action of siRNAs and miRNAs, or find it via the **Online Resource Center** at **www.oxfordtextbooks.co.uk/orc/craig2e/**.

We will first describe the most common pathway for miRNA biogenesis in animals, illustrated in Figure 13.11a. The pri-miRNA transcript folds into a stem loop structure (with an internal region of unpaired nucleotides) surrounded by 5′ and 3′ single-stranded extensions; this structure serves as the initial substrate for processing by Drosha, a nuclear RNase III enzyme. Drosha interacts with a specialized RNA-binding protein, DGCR8 (in humans), to form what is termed the Microprocessor complex. This complex performs a cleavage reaction that removes the 5′ and 3′ extensions, liberating a 60–70 nucleotide transcript known as a pre-miRNA that adopts the form of a hairpin. Importantly, like other RNase III enzymes, the RNAs generated by this processing carry monophosphates at their 5′ termini and dinucleotide overhangs with hydroxyl groups at their 3′ termini.

The hairpin structure generated by the Microprocessor complex is then recognized by a nuclear export factor (Exportin 5) that delivers the pre-miRNA to the cytoplasm. There, a second endonucleolytic cleavage reaction, generally referred to as 'dicing', is catalyzed by Dicer, another RNase III enzyme, in complex with a double-stranded RNA-binding protein (TAR RNA-binding protein 2 or TRBP2). Dicer possesses a PAZ domain, which directly binds to the 3′ end of double-stranded RNAs, thus positioning one end of the duplex RNA. The double-stranded RNA extends along the remainder of the Dicer protein such that the dual cleavage sites of the RNase III domains are positioned at a distance appropriate to generate sRNAs of a precise length. The product of this cleavage reaction (miR:miR*)

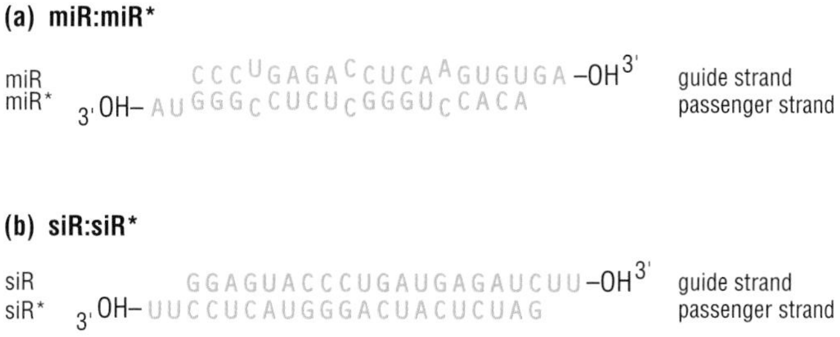

(a) miR:miR*

```
miR            CCCUGAGACCUCAAGUGUGA-OH3'   guide strand
miR*   3'OH-AUGGGCCUCUCGGGUCCACA           passenger strand
```

(b) siR:siR*

```
siR            GGAGUACCCUGAUGAGAUCUU-OH3'  guide strand
siR*   3'OH-UUCCUCAUGGGACUACUCUAG          passenger strand
```

Figure 13.12 Examples of (a) miR:miR* and (b) siR:siR* duplexes. As a result of the Dicer-dependent processing, both types of duplexes carry monophosphates at their 5′ termini and two-nucleotide overhangs with hydroxyl groups at their 3′ termini. miR:miR* duplexes are usually not fully complementary, in contrast to siR:siR* duplexes.

shown in Figure 13.12a again carries 5′monophosphates and 3′ overhangs of two nucleotides. It is these two products – the miR and miR* RNAs – that are ultimately loaded into the Argonaute protein. It is noteworthy that the duplex usually is not fully complementary, but instead has interruptions in the base-pairing between the miR and miR* RNAs.

The processing of some miRNAs involves variations such as methylation

While the pathway detailed in Figure 13.11a is thought to be the most common for miRNA processing in animals, there are less typical paths through which miRNA processing can occur, including pathways in which ribonucleases other than Drosha carry out the initial cleavages. (For example, some pri-miRNAs are processed into pre-miRNAs known as 'mirtrons' using the endogenous splicing machinery instead of Drosha.) In plants, both of the major processing steps are performed in the nucleus by a Dicer-type enzyme known as Dicer-like 1 (DCL1), and then the miR:miR* duplex is exported for loading directly into the Argonaute protein.

It is interesting to note that, in addition to the core cleavage steps described here, post-transcriptional modification of sRNAs can take place during the course of the RNA processing reaction. For example, a methyltransferase (HEN1) methylates the 3′ ends of miRNAs in plants. The specific function of such post-transcriptional modifications is largely unknown, although they do appear to protect the 3′ ends from trimming (by exonucleases) and tailing (by polymerases). These differences in processing of miRNAs in animals and plants likely reflect their independent evolution from extant siRNA and piRNA pathways.

siRNAs are processed from long double-stranded RNAs from a range of sources

In contrast to miRNAs, siRNAs are not generally encoded in the genome as specific genes (i.e., as RNA products of a specific sequence that are precisely excised from their primary transcript). Instead, siRNAs are derived from double-stranded RNA that comes from several different sources; these sources may be endogenous or exogenous. Endogenous duplex RNA can arise from the normal transcription of genomic loci that have extensive hairpin structures, or from the annealing of sense and antisense RNAs that have both been transcribed from a given locus. The increased ability to detect transcribed RNAs mentioned in the introduction to this chapter is making it clear that there is abundant transcription from both strands of the DNA throughout the genome; this transcription is the typical source of duplex RNA in mammals and in flies. In worms and plants, RNA-dependent RNA polymerases further contribute to the generation of sRNAs, typically by synthesizing complementary copies of single-stranded precursor RNAs (which are normally transcribed) and thus generating duplex RNA.

Exogenous sources of double-stranded RNA include viral RNAs and duplex structures that have been synthetically introduced into cells by scientists for experimental purposes. While siRNA-mediated RNA interference is very clearly a cellular process that has evolved to perform a variety of functions (including cellular defense against viral invasion), this endogenous mechanism has been widely exploited to inactivate genes in a variety of reverse and forward genetic experiments. Indeed, key insights into the role of duplex RNA in triggering RNA interference came from experimental manipulations of various organisms including fungi (*Neurospora crassa*), plants (the petunia), and worms (*C. elegans*). The discovery of RNA interference as a result of such experimental manipulation is described in Experimental approach 13.2.

➔ We discuss forward and reverse genetic approaches in Section 19.5.

Figure 13.11b depicts how duplex RNAs, irrespective of their origins, become substrates for an RNA processing reaction akin to the one described above for the miRNAs, but lacking a Drosha-dependent cleavage reaction. Instead, a Dicer enzyme 'dices' the long duplex structures, sequentially cleaving along the duplex approximately every 20–25 base pairs. The cutting frame is set by interactions between the double-stranded RNA and the PAZ region of the Dicer protein. As shown in Figure 13.12b, the products of this reaction are short duplex RNAs, similar to the miR:miR* duplexes, but fully base-paired along their length. Additionally, as for the miRNAs, post-transcriptional modification of the siRNAs occurs in some organisms.

The rasiRNAs are produced from longer precursor RNAs that are transcribed from repetitive regions of the genome

As mentioned above, the rasiRNAs fall into several classes. These sRNAs are typically somewhat longer than the miRNAs and siRNAs and share the property of being transcribed from repetitive sequences in genomes, which include retrotransposons, DNA transposons, and microsatellite DNA sequences. The most general view is that the rasiRNAs all interfere with gene expression from these repetitive sequences, in some cases by targeting nascent mRNA transcripts for degradation by the Argonautes and in other cases by directly modulating chromatin structure and transcription at these loci.

➔ We discuss repetitive sequences such as transposons in Chapter 17 and microsatellite DNA sequences in Chapter 18.

Some rasiRNAs are processed in a manner similar to the siRNAs

In fission yeast, the repeat-associated RNAs known as hcRNAs help to maintain the integrity of centromeres. In this system, bidirectional transcription of pericentromeric repeat elements yields double-stranded RNA that is cleaved by a Dicer endonuclease (SpDcr1) into hcRNAs, similar to siRNAs in Figure 13.11b. These hcRNAs are loaded into a fungal Argonaute protein (SpAgo1), forming the RNA-induced transcriptional silencing (RITS) complex. The RITS complex is then targeted to nascent pericentromeric transcripts through base-pairing interactions with the hcRNA. This interaction, in turn, recruits an RNA-dependent RNA polymerase (SpRdp1) to the nascent transcript, which synthesizes further antisense transcripts that are processed by SpDcr1 to produce more hcRNAs to be loaded onto SpAgo1.

We will see below that hcRNAs establish silent heterochromatin at the centromere. Paradoxically, the silencing requires low levels of transcription for continued production of hcRNAs. The transcription of the pericentromeric repeats

13.2 EXPERIMENTAL APPROACH

The discovery of RNA interference and its broad application to genetic analysis

Silencing of *C. elegans* genes is brought about by the injection of double-stranded RNA

While in the 1990s it was known that the introduction of RNA into cells could sometimes interfere with the function of endogenous genes, a mechanistic understanding of such interference was lacking. The consensus at the time was that such interfering RNAs (typically designed as antisense to the gene of interest) hybridize directly with their target to somehow block subsequent gene expression. However Andrew Fire, Craig Mello, and colleagues made the surprising observations that both sense and antisense RNA preparations could cause 'gene silencing' in the nematode *C. elegans*, and that the observed interference could persist into the next generation of organisms.

A major breakthrough in understanding the process of RNA interference (or RNAi) came with the insightful experiments of Fire and Mello in 1998, when they established that the most potent trigger of gene silencing was neither a sense nor an antisense RNA transcript, but instead a double-stranded RNA species. These scientists reasoned that such duplex RNA would be naturally generated by phage RNA polymerases during the transcription reactions used to prepare the samples, and might explain the surprising potency of the sense transcripts in their silencing experiments.

To address this hypothesis, the authors injected 'deliberately prepared' double-stranded RNA (as well as purified single-stranded RNA) targeting a non-essential gene, the myofilament protein encoding *unc-22*. Reductions in *unc-22* gene expression are easily measured in the progeny organism since loss of function *unc-22* worms exhibit a rather obvious 'twitching' phenotype. As shown in Figure 1, targeting of several different exons (covering exons 21–22 or 27) in *unc-22* with double-stranded RNA (sense + antisense, S + AS) resulted in dramatic loss of function phenotypes (100% strong twitchers), whereas injection of the single-stranded RNAs (either S or AS) did not.

Two other genes, *fem-1* and *hlh-1*, were also targeted by these approaches, as their elimination from *C. elegans* also conferred an easily measured, but distinct, phenotype. Again, the injection of double-stranded RNA targeting exons of the genes of interest resulted in the expected loss of function phenotypes, but the injection of the equivalent single-stranded RNAs did not. Importantly, in no case did non-specific RNAs produce the very particular phenotypes being assayed. Finally, in a visually appealing experiment, green fluorescent protein (GFP)-synthesizing worms were targeted with double-stranded RNAs to GFP, and the amount of GFP protein expressed in these organisms was dramatically reduced; these results are shown in Figure 2.

RNA interference in *C. elegans* targets RNA and can be amplified

The mechanism of action of the observed RNA interference was more fully explored when the authors compared the efficacy of

Effects of sense, antisense and mixed RNAs on progeny of infected animal

gene fragment	segment	size (kb)	injected RNA	F1 phenotype
unc22A	exon 21-22	742	S	wild type
			AS	wild type
			S + AS	strong twitchers (100%)
unc22B	exon 27	1033	S	wild type
			AS	wild type
			S + AS	strong twitchers (100%)
fem1A	exon 10	531	S	hermaphrodite (98%)
			AS	hermaphrodite (>98%)
			S + AS	female (72%)
fem1B	intron 8	556	S + AS	hermaphrodite (98%)
hlh1A	exons 1-6	1033	S	wild type (<2% lpy-dpy)
			AS	wild type (<2% lpy-dpy)
			S + AS	lpy-dpy larvae (>90%)
hllh1D	intron 1	697	S + AS	wild type (<2% lpy-dpy)

Figure 1 Effects of sense, antisense, and mixed RNAs on progeny of infected animals. The indicated RNA was injected into six to ten adult hermaphrodite worms, and the eggs were collected after an appropriate time. Progeny phenotypes were scored upon hatching and at appropriate intervals. Adapted from Fire A *et al*. Potent and specifi c genetic interference by double-stranded RNA in Caenorhabditis elegans . *Nature*, 1998;391:806–811.

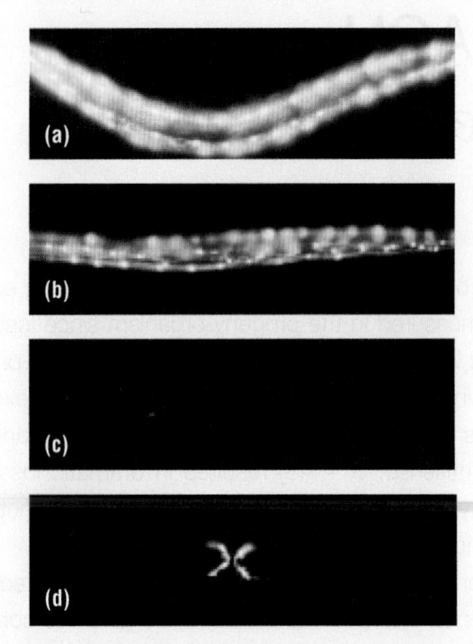

Figure 2 Direct visualization of RNA interference effects in the whole *C. elegans* animal. Fluorescence micrographs show progeny of injected animals from GFP-reporter strain PD4251. a, b Progeny of animals injected with a control RNA (double-stranded *unc22A* RNA (ds-*unc22A*)). c, d Progeny of animals injected with double-stranded GFP RNA (ds-*gfpG*). a, c: young larva; b, d: adult.

Reproduced from Fire *et al.* Potent and specific genetic interference by double-stranded RNA in *Caenorhabditis elegans*. *Nature*, 1998;**391**:806–811.

double-stranded RNAs targeting either the introns or the promoters of the genes, and observed no phenotypic changes as a consequence (see Figure 2). These observations suggested that the mechanism of action of gene silencing involved the targeting of an mRNA species that has already been transcribed and processed, and thus was most likely to take place in the cytoplasm following export of the mRNA from the nucleus. Moreover, the fact that the knock-down phenotypes could be discerned in progeny organisms, where the initial input RNA concentrations would be very dilute, indicated that the mechanism of action included features that allowed for an amplification of the input signal.

Many of the mechanistic features uncovered in this landmark publication have turned out to be broadly accurate in a wide range of organisms, and these same features have allowed for a revolution in genetic studies. By simply delivering double-stranded RNAs to a given organism, loss of function variants can be directly evaluated without the necessity of directly manipulating the genome through more complicated approaches, thus greatly expediting genetic analysis. In the fifteen years that have passed since these earliest observations, RNA interference has arguably become the single most commonly used tool for probing gene function in eukaryotic systems.

Even more importantly, these initial discoveries foreshadowed and dovetailed with the discovery of multiple classes of sRNAs (including miRNAs, piRNAs, and siRNAs) now known to play ubiquitous and central roles in modulating gene expression throughout the eukaryotic lineage. The implications of these earliest insightful experiments will be far-reaching for the foreseeable future.

Find out more

Fire A, Xu S, Montgomery MK, *et al.* Potent and specific genetic interference by double-stranded RNA in *Caenorhabditis elegans*. *Nature*, 1998; **391**:806–811.

Related techniques

RNAi; Section 19.5
C. elegans (as a model organism); Section 19.1
GFP protein tagging; Sections 19.10, 19.11
Fluorescence microscopy; Section 19.13

during a particular time in the cell cycle is thought to provide the hcRNAs that establish the centromeric heterochromatin. We see in the phylogenetic tree of Argonaute proteins below that SpAgo1 is closely related to the plant Ago proteins, which also likely bind RNAs from repetitive regions and are involved in heterochromatin formation.

Mammalian piRNAs are generated from single-stranded precursors by a 'ping-pong' mechanism

In animal cells, piRNAs are derived from long transcripts of specific genomic loci known as piRNA clusters. Many of these regions are enriched in fragmented transposon remnants and are located in the most transposon-rich areas of the genome. Processing of these long single-stranded RNAs is quite distinct from that of the miRNAs and siRNAs.

The biogenesis of piRNAs was first described in *Drosophila*, where primary piRNAs, which are antisense to transposon mRNA, derive from single-stranded

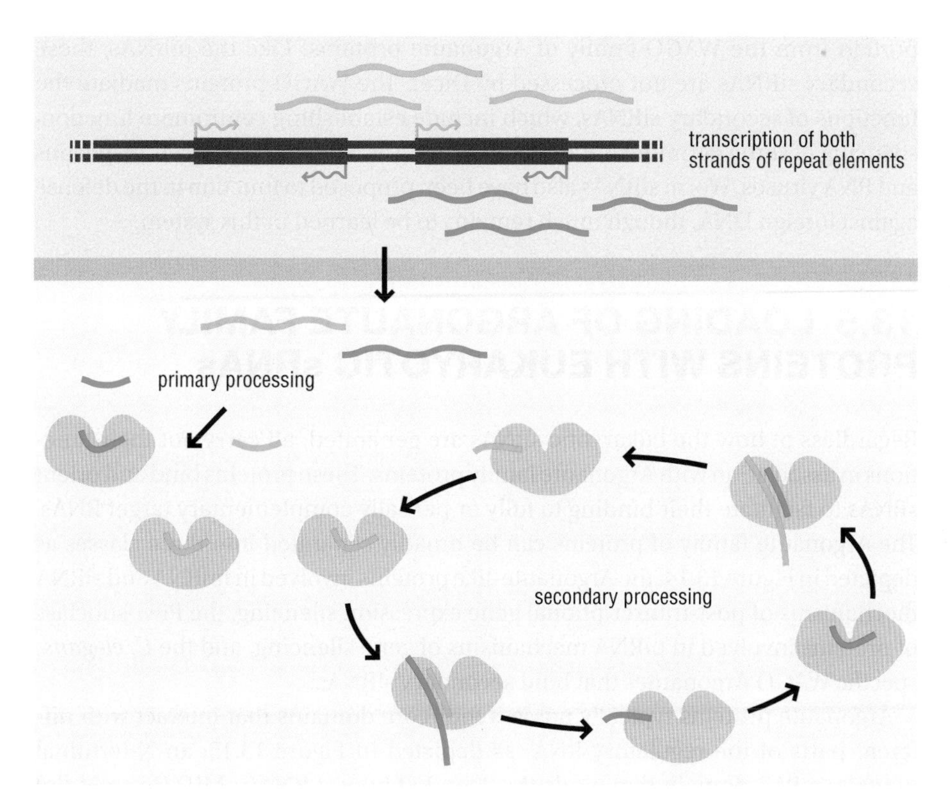

Figure 13.13 Ping-pong mechanism of piRNA biogenesis. piRNAs are generated from single-stranded RNAs transcribed from repetitive elements such as transposons in the genome. The piRNAs are not generally processed by Dicer but rather are cleaved by the PIWI domain of the Piwi protein. A transcript can be cleaved directly in a primary processing step that is not well understood or via secondary processing whereby a piRNA loaded onto a Piwi protein directs the cleavage of a complementary strand to generate a new piRNA. Given the back-and-forth nature, secondary processing has been said to occur via a ping-pong mechanism.

Based on Fig 3 of Kim *et al.* Biogenesis of small RNAs in animals. Nature Reviews *Molecular Cell Biology*, 2009;**10**:126–139.

precursor RNAs. These primary piRNAs direct the cleavage of transposon RNAs and thereby lead to the production of secondary piRNAs. Secondary piRNAs, in turn, direct production of further primary piRNAs by promoting cleavage of long transcripts antisense to transposon mRNAs. Both classes of piRNA (primary and secondary) are loaded into several Piwi proteins (a subclass of the Argonaute family) by a mechanism that is poorly understood. These piRNA-loaded Piwi proteins ultimately function as endonucleases for the processing reaction. Importantly, the Dicer-type proteins are not required for these reactions.

This iterative process of generating piRNAs, illustrated in Figure 13.13, is termed a 'ping-pong mechanism', where existing piRNAs bound to a Piwi protein define the cleavage position within a target RNA to form a new piRNA (which will subsequently bind to another Piwi protein, and so on). A similar ping-pong mechanism generates piRNAs in mammals, though some details differ. The piRNAs that are specifically produced via this ping-pong reaction typically have a uracil at the 5′ end (as do many miRNAs) which leads to an adenine bias at position 10 in the secondary piRNA. These features undoubtedly help direct these RNAs to bind the appropriate Argonaute proteins.

As we discuss shortly, the piRNAs are thought to be important in the germline and during early embryo development, where they silence expression from repetitive regions of the genome that might cause significant problems during the most sensitive stages of organismal life.

Secondary siRNAs in *C. elegans* are abundant but are particular in their properties relative to other sRNA classes

Worms express an abundant class of sRNAs known as secondary siRNA that is not found in other animals. RNA-dependent RNA polymerases use target mRNAs as templates to produce these siRNAs, which are directly loaded onto a worm-specific protein from the WAGO family of Argonaute proteins. Like the piRNAs, these secondary siRNAs are not processed by Dicer. The WAGO proteins mediate the functions of secondary siRNAs, which include establishing centromere function, silencing of some germline mRNAs, and serving as a defense against transposons and RNA viruses. Worm siRNAs also have been proposed to function in the defense against foreign DNA, though much remains to be learned in this system.

13.5 LOADING OF ARGONAUTE FAMILY PROTEINS WITH EUKARYOTIC sRNAs

Regardless of how the eukaryotic sRNAs are generated, all carry out their functions in association with Argonaute family proteins. These proteins bind and orient sRNAs to facilitate their binding to fully or partially complementary target RNAs. The Argonaute family of proteins can be broadly separated into three classes as depicted in Figure 13.14: the Argonaute-like proteins involved in miRNA and siRNA mechanisms of post-transcriptional gene expression silencing, the Piwi-subclass of proteins involved in piRNA mechanisms of gene silencing, and the *C. elegans*-specific WAGO Argonautes that bind secondary siRNAs.

Argonaute proteins typically possess four core domains that interact with different parts of the regulatory RNA, as depicted in Figure 13.15: an N-terminal domain; a PAZ domain that binds the 3′ end of bound RNA; a MID (for middle) domain that interacts with the 5′ end of bound RNA; and an RNase H-related

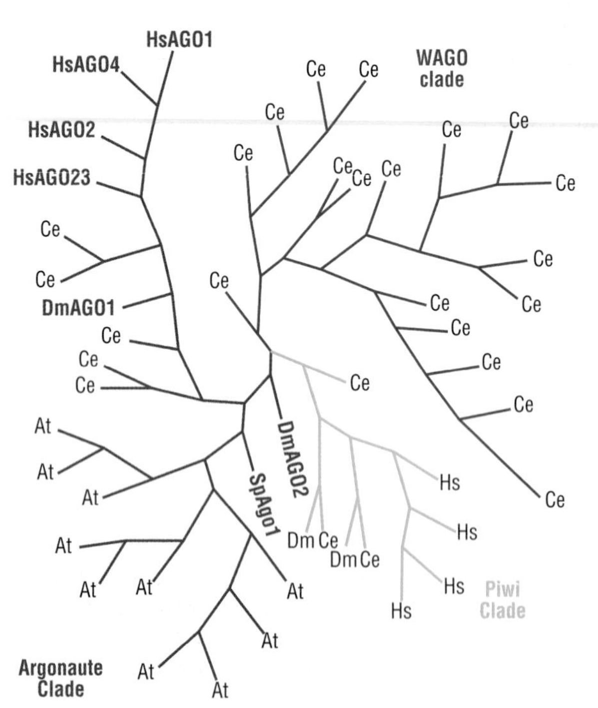

Figure 13.14 Tree of Argonaute proteins. The Argonaute group found in plants (*A. thaliana*), animals (*C. elegans*, *Drosophila melanogaster* and *Homo sapiens*), and fungi (*S. pombe*) is indicated in black. Piwi-family proteins are indicated in green. A separate clade of WAGO-family proteins indicated in red is only found in *C. elegans*.

Based on Box 1 of Hutvagner G, Simard MJ. Argonaute proteins: key players in RNA silencing. *Nature Reviews Molecular Cell Biology*, 2008;**9**:22–32.

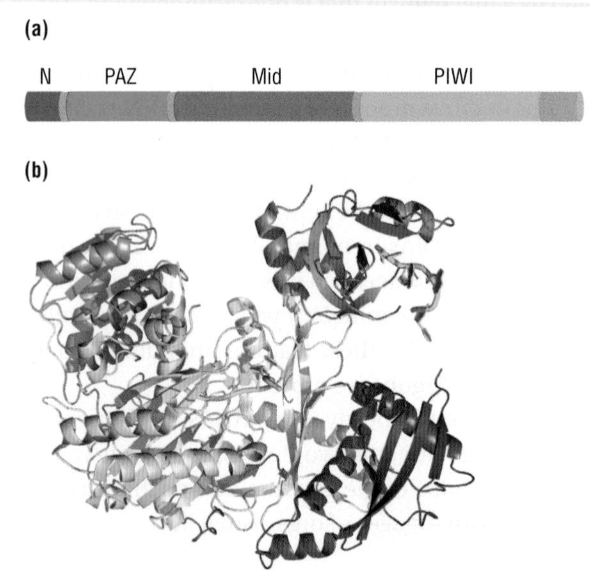

Figure 13.15 Domain structure of Argonaute proteins. (a) Human AGO2 protein has a PAZ domain (orange), a PIWI domain (yellow), and a MID domain (tan). (b) Structure of *Pyrococcus furiosus* Argonaute protein with RNA (green). The colors of the Argonaute protein correspond to the domains delineated in (a).

Based on Figure 1 of Hutvagner G, Simard MJ. Argonaute proteins: key players in RNA silencing. *Nature Reviews Molecular Cell Biology*, 2008;**9**:22–32.

PIWI domain that interacts along the length of the sRNA. The PIWI domain often possesses the residues DEDH, which are mostly acidic in nature and are essential for catalytic activity.

The PAZ, MID, and PIWI domains together make extensive sequence-independent contacts with the backbone of the 'seed' region, thus pre-orienting this portion of the guide RNA in a helical conformation that facilitates the scanning of cellular RNAs for complementary sequences. In addition, the MID domain of the Argonaute binds to (and thereby flips out) the first nucleotide of the guide RNA to coordinate its 5′ phosphate and to sometimes recognize this first nucleotide. These recognition features help explain the partial preference of piRNAs and some miRNAs for a uracil at the 5′ end, and explains why the first nucleotide of the guide does not participate in identifying cellular targets. We will first describe the process by which small RNAs of the various classes are loaded into the appropriate Argonaute protein and then discuss the actions of the loaded Argonaute proteins in the next section.

miRNA and siRNA sorting into the Argonaute proteins involves identification of the guide strand and elimination of the passenger strand

As described in the previous section, the outcome of the RNA processing pathway for miRNAs and siRNAs is a duplex RNA, which is either partially or fully self-complementary. The two strands of this duplex RNA are called the guide and passenger strands. Once the duplex species has been generated, it is loaded into the appropriate Argonaute protein. One of the RNA strands (the miR or siR guide strand) is retained, while other strand (the miR* or siR* passenger strand) is selectively removed (Figure 13.16). The guide strand will ultimately be involved in directing the silencing of the target RNA. This process of selecting one strand and removing the other is called 'sorting'.

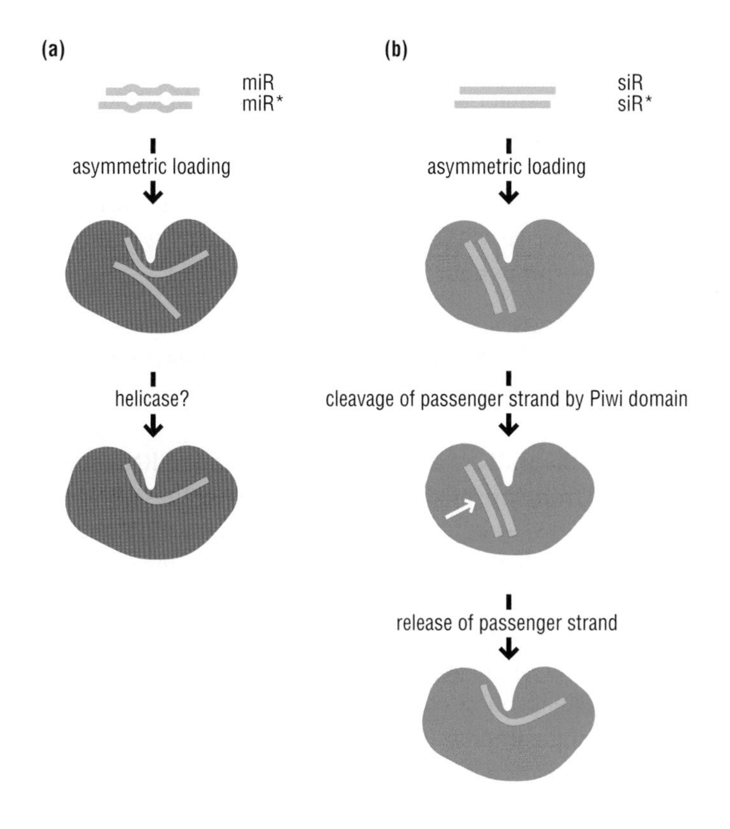

Figure 13.16 Loading of (a) miRNAs and (b) siRNAs into Argonaute complexes. The partially complementary miR:miR* and fully complementary siR:siR* duplexes are loaded onto the appropriate Argonaute protein after Dicer cleavage. For miR:miR* duplexes, it is not completely clear how the miR* passenger strand is selectively removed. For siR:siR* duplexes, the endonuclease activity of the PIWI domain of the Argonaute protein catalyzes cleavage of the siR* passenger strand, allowing its unwinding and release.

This 'sorting' problem is complex. It appears that a number of factors help to determine which of two strands is the guide strand and is thus loaded onto the Argonaute protein, and which is the passenger strand that is released and degraded. Sorting appears to be affected by the Dicer protein (and its interacting protein partners) involved in the final RNA cleavage step, by the thermodynamic properties of the duplex RNA (where the guide strand has the less stable 5′ end pairing), and by the destination Argonaute protein.

For example, in *Drosophila*, there are two Argonaute proteins, DmAgo1 and DmAgo2; the former principally binds miRNAs while the latter binds siRNAs. Sorting of the different sRNAs into these two proteins is likely affected by the preceding processing step, in which two distinct Dicer proteins, DmDcr1 and DmDcr2, separately process miRNAs and siRNAs, respectively, although the details are not fully understood. By contrast, there are four Argonaute proteins in humans; all are thought to be involved in miRNA-mediated silencing mechanisms under normal circumstances, and there appear to be few specialized sequence preferences for miRNA binding to specific Argonaute proteins. Thus, when duplex RNAs are provided exogenously to the cell, they are loaded into the same set of Argonaute proteins but instead act through an siRNA-mediated cleavage pathway. While siRNA-mediated cleavage of target RNAs is an extremely powerful experimental tool that is widely used in human systems, it is not clear whether the human Argonaute proteins are normally utilized by the cell for the prototypical RNAi defense mechanism. In this regard, it is noteworthy that humans have other, more sophisticated immune defense systems against foreign RNA including, for example, the interferon system, which responds to duplex RNAs in the cell that result from viral infection.

Once the duplex RNA is bound in an appropriate orientation on the Argonaute protein, the second (passenger) strand must be eliminated. For imperfect duplexes (miR:miR*, the duplex composed of the intended miRNA (miR) and its nearly complementary sequence (miR*)), it is not certain how the passenger strand is selectively removed, though the action of a helicase may well be helpful. For perfect duplexes (siR:siR*, the duplex of the intended guide siRNA (siR) and its complementary passenger sequence (siR*)), the endogenous endonuclease activity of the PIWI domain of the Argonaute protein catalyzes cleavage of the passenger strand. The cleaved strands are then unwound and released. This mechanism makes use of the RNase H fold of the PIWI domain and acidic residues that are poised for such catalysis; we will see that this nuclease activity of the PIWI domain is used multiple times throughout the life cycle of the sRNA.

Single-stranded piRNAs are loaded onto Piwi proteins as they are generated

As we noted earlier, the piRNAs are short single-stranded RNA species that are generally processed from longer single-stranded RNAs. These RNAs are thought to bind directly to the Piwi proteins after processing via the ping-pong mechanism (see Figure 13.13). As such, these RNAs do not depend on the strand dissociation and discard step during Argonaute loading.

Independent of the route taken, the output of the loading process for all classes of eukaryotic sRNAs is a single-stranded guide RNA bound to an Argonaute-type protein: this ribonucleoprotein complex is commonly referred to as **RISC**, for RNA-induced silencing complex. This complex is the central component in the gene silencing mechanisms we will discuss next.

13.6 GENE SILENCING BY SMALL EUKARYOTIC RNAs

Once Argonaute proteins are charged with their 'guide' RNA, they are ready to bind their target RNA and promote gene silencing. The multiple RNA-protein contacts along the length of the guide RNA poise it to identify and engage potential target RNAs. This task is a major challenge given the many RNAs in a cell that must be evaluated by these Argonaute-containing complexes. Some target RNAs (usually targets for the miRNAs) are incompletely complementary while other targets (usually those for the siRNAs and the piRNAs) are fully complementary to the small guide RNA. Argonaute proteins bound to partially or fully complementary duplexes assume different conformational states, which are likely important in determining the downstream function of the RISC complex. Target RNAs with only partial complementarity to the guide RNA are generally not cleaved by the Argonaute protein but instead are translationally silenced and subsequently destabilized by other enzymes. By contrast, fully complementary target RNAs are often cleaved by the PIWI domain of the Argonaute protein. This cleavage event is referred to as 'slicing'. In addition, the subcellular localization and expression pattern of the RISC complex further dictate target specificity and mode of action.

We will first discuss how the different conformations of the Argonaute protein define these distinct biochemical outcomes - whether or not cleavage occurs - and will then examine how these outcomes, in turn, affect cellular events.

Partially complementary and fully complementary pairing to target RNAs induce the Argonaute protein to adopt distinct conformations

Argonautes bound to miRNAs typically identify targets with imperfect complementarity, as depicted in Figure 13.17a. The most important pairing region for the miRNA is referred to as the 'seed' sequence, which generally encompasses nucleotide positions 2–8 of the miRNA species and defines its principal cellular targets. (As mentioned earlier, the 5′ nucleotide of the guide RNA is sequestered in the MID domain of the Argonaute protein and is thus not available to engage in base-pairing interactions with the target RNA.) In these cases, extensive base-pairing with downstream portions of the miRNA do not typically occur; the absence of extended duplex formation allows the 3′ end of the miRNA to remain securely bound to the PAZ domain.

By contrast, in the presence of fully complementary target RNA, the Argonaute:guide siRNA complex engages the target all along its length to form extended duplex nucleic acid, as shown in Figure 13.17b. When such extended duplex forms, the 3′ end of the guide RNA is released from the PAZ domain of the Argonaute protein, leading to conformational changes that are thought to activate the cleavage activity of the RNase H PIWI domain of the Argonaute protein.

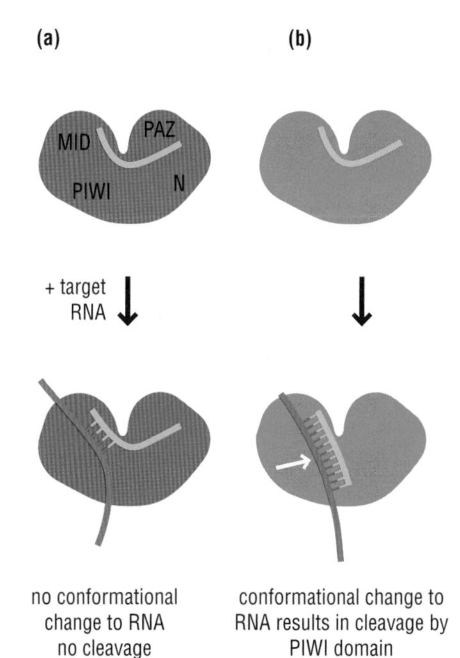

Figure 13.17 Conformational changes to Argonaute proteins brought about by target RNA binding to (a) miRISC and (b) siRISC complexes. The miRNA loaded in a RISC complex usually base-pairs through the 'seed' sequence corresponding to nucleotides 2–8. This configuration allows the 3′ end of the miRNA to tightly bind the PAZ domain, leaving the PIWI domain inactive. In contrast, extended base-pairing along the length of the siRNA releases the 3′ end of this RNA from the PAZ domain, inducing a conformational change that activates the catalytic activity of the RNase H PIWI domain.

An miRISC typically targets 3′ UTRs to promote translational repression and mRNA decay

An miRISC is the complex formed between an miRNA and its favored Argonaute pairing partner. These complexes are generally found in the cytoplasm of the cell, where mRNAs are targeted for silencing. Functional miRNA binding sites are predominantly found in the 3′ UTRs of the targeted mRNAs, as illustrated in Figure 13.18a.

As mentioned previously, pairing between the miRNA and the target RNA is typically imperfect, with mismatches minimally found at positions 10 and 11. As a

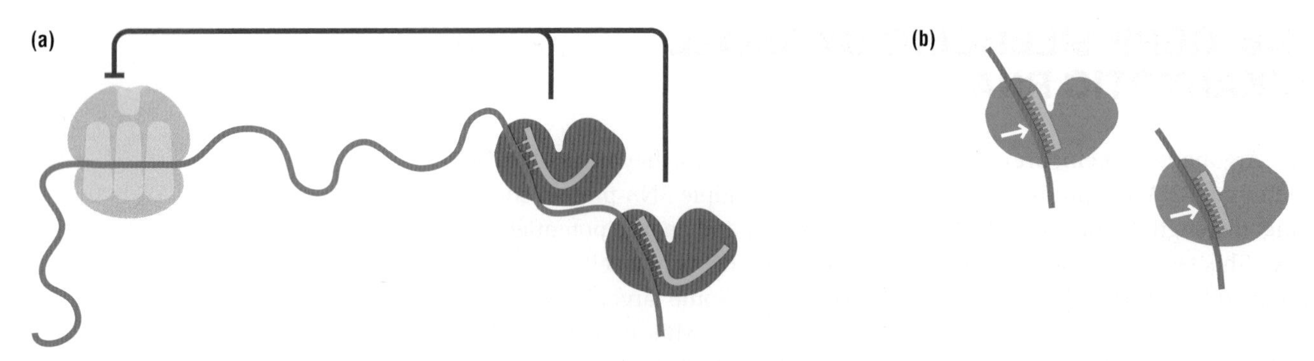

Figure 13.18 Functional consequences of (a) miRNA-loaded and (b) siRNA-loaded RISC complexes. miRISC often targets multiple sequences in 3′ UTRs and represses translation by a mechanism that is not yet fully understood but involves additional proteins. Frequently, as shown in Experimental approach 13.1, there are multiple target sequences for a particular miRNA or even multiple miRNAs. The activated RNase H activity in siRISC cleaves target RNA, leading to its degradation.

Scan here to watch a video animation explaining more about the generation and action of siRNAs and miRNAs, or find it via the **Online Resource Center** at www.oxfordtextbooks.co.uk/orc/craig2e/.

consequence of this particular mismatch, the PIWI domain is generally not able to cleave mRNAs targeted by miRNAs. The most critical pairing region for the miRNA is referred to as the 'seed' sequence, which generally encompasses nucleotide positions 2–8 of the miRNA species; binding to this short region is sufficient for repressing translation of the target. The extent of base-pairing between miRNA and target varies across kingdoms, with more extensive complementarity commonly seen in plants. As described in Experimental approach 13.1, bioinformatic approaches can be used to predict potential targets for a given miRNA by identifying regions complementary to seed sequences whose location is conserved among different organisms. Crosslinking between the Argonautes and bound miRNAs or their target RNAs also is increasingly being used as a strategy to identify cellular targets.

Once bound to targets, how do miRISCs silence gene expression? Argonaute proteins appear to recruit additional factors, including in particular a GW-rich protein (that is, a protein that contains multiple glycine-tryptophan repeats) known as TRNC6 in humans (or GW182 in *Drosophila*). This protein, and likely others, appear to be involved in repressing translation and destabilizing the mRNA through mechanisms that remain poorly understood. What is clear is that the repression mechanism does not depend on the 'slicer' activity of the Argonaute protein. Degradation of the target mRNA is thought to occur subsequent to translational repression via normal RNA decay pathways, which involve the deadenylation, decapping, and exonucleolytic degradation steps that we learned about in Section 10.9. This rough ordering of events is consistent with the idea that translation and mRNA decay are tightly coupled processes in all cells.

The miRNA-mediated gene regulation has only been documented in multicellular plants and animals and is notably absent in fungi; as such, this pathway is considered to be a more recent evolutionary adaptation of existing sRNA regulatory pathways involved in genome surveillance. Through their actions, miRNAs appear to play key roles in organismal development by, for example, preventing synthesis of inappropriate proteins during specific stages of development, as we see in Experimental approach 13.1.

An siRISC targets cytoplasmic RNAs for endonucleolytic cleavage and degradation

An siRISC, composed of an Argonaute protein and an siRNA, is often found in the cytoplasm, where it silences gene expression of complementary targets, although there is increasing evidence for nuclear roles for these Argonaute complexes. In contrast to miRISC, the RNAs targeted by siRISC tend to be fully complementary to the siRNA, which potently activates the complex to cleave ('slice') the target mRNA, as depicted in Figure 13.18b. The cleaved mRNA is thought to immediately become a target for the exosome and other standard RNA decay pathways in the cell.

What are the normal cellular targets for an siRISC? Double-stranded viral RNAs are processed to generate siRNAs, which subsequently bind to Argonaute and then eventually target their own RNAs for silencing. This antiviral response obviously protects the cell. Additionally, duplex RNAs generated naturally by transcription in the nucleus can lead to the generation of siRNAs, which in turn may target these transcripts for silencing. In this way, small amounts of transcript can lead to subsequent silencing of the locus through a post-transcriptional mechanism; such a pathway could, in principle, function in the cytoplasm or in the nucleus.

Of particular interest to scientists is the broad utility of the RNAi pathway to perform genetic experiments by knocking out the function of particular genes in an organism. In most cells, researchers are now able to introduce specific duplex RNA species that then target endogenous mRNAs for silencing. In this way, the roles of specific mRNAs (or genes) can be studied without more complicated manipulation of the eukaryotic genome. These techniques have had a tremendous impact on biological exploration during the past decade, as we explore further in Experimental approach 13.2.

An rasiRISC can target RNAs for degradation and affect chromatin structure by associating with genomic DNA

The piRNAs associated with Piwi-family Argonaute proteins are generally present in the nucleus of cells, where they silence the activity of repetitive regions in the genome. More specifically, piRNAs are predominantly found in the germline of their host organism, where they are presumed to down-regulate transposon activity during development. The piRNA-bound Piwi proteins appear to function in multiple ways. First, the piRNA:Piwi complexes are thought to recognize targets in a fashion similar to the canonical siRNA pathway, as illustrated in Figure 13.19a. In this model, piRNA:Piwi complex binds to the nascent mRNAs and represses

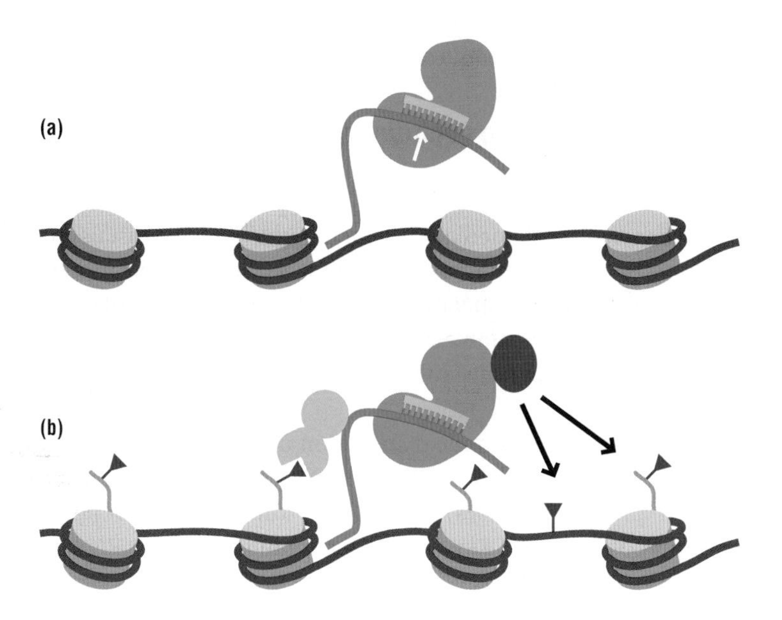

Figure 13.19 Multiple actions of rasi-loaded RISC complexes. (a) As for siRISC, rasiRISC complexes can cleave target RNA. (b) In addition, rasiRISC complexes can associate with chromatin and other proteins to bring about DNA or histone-tail methylation, resulting in a repressive chromatin structure.

accumulation of the transcript, which may or may not include cleavage of the transposon-derived mRNAs through a 'slicer' mechanism.

The piRNAs also appear to bring about epigenetic changes in the DNA and chromatin structure, as shown in Figure 13.19b. For example, piRNA:Piwi complexes facilitate the methylation of repetitive DNA in mammals. In *Drosophila*, certain Piwi-like Argonautes interact with HP1a, a non-histone chromosomal protein that plays a role in gene silencing at the chromosome. These latter roles for the Piwi proteins suggest that they help to modulate transcription directly at repetitive loci. The pathways all could be important in limiting the expression of potentially detrimental sequences in the genome, especially during the very sensitive early phase of organismal development.

➔ We discuss the concept of epigenetic changes in Section 4.7.

Similar to the piRNAs, sRNAs called hcRNAs in the fission yeast, *S. pombe*, appear to be involved in regulating heterochromatin structure in the centromeric repetitive regions of the genome. In this case, a set of proteins known as the RITS complex, containing the Argonaute protein (SpAgo1), a chromodomain protein (Chp1), and a GW-repeat protein (TAS3), triggers patterns of histone H3 lysine 9 methylation, which are characteristic of heterochromatin. The establishment of these marks allows for proper chromosome segregation and general maintenance of gene silencing at the centromere. As for certain piRNAs, these effects are likely manifested primarily at the level of transcription, preventing the transcriptional machinery from expressing RNA from these loci.

➔ We learn more about the structure of centromeres in Section 4.10

In other fungi and ciliates, small RNA systems also appear to be important in managing repeated elements in the genome through diverse, but probably related, mechanisms. For example, in *Tetrahymena thermophila*, the repetitive regions of the genome are eliminated from the somatic macronucleus during the sexual cycle through an RNA-based mechanism. This process also depends on a Piwi-family protein. The degree to which the underlying mechanisms resemble the better-studied piRNAs remains to be discovered.

13.7 VIRAL DEFENSE ROLE OF BACTERIAL, ARCHAEAL, AND EUKARYOTIC sRNAs

The function of base-pairing sRNAs is not restricted to regulating expression of the host genome. Indeed, defense against foreign DNA or RNA is a ubiquitous role for base-pairing sRNAs in all kingdoms of life.

Viral defense was likely the evolutionary origin of sRNAs in eukaryotes

The different classes of RNA-mediated silencing mechanisms discussed here are widely distributed throughout the eukaryotic lineage. However, the most conserved functions for this machinery appear to be related to the cleavage of cognate RNAs and the transcriptional repression of homologous DNA sequences, rather than the more nuanced regulation of translation and decay imposed by miRNA-mediated silencing mechanisms.

Careful analysis of the taxonomic distribution of key components of the RNA-mediated silencing pathways – including the RNase III endonuclease Dicer, the Argonaute/Piwi protein effectors of silencing, and the RNA-dependent RNA polymerases that often play a key role in generating the guide RNAs that drive silencing – has revealed a consistent view of how these pathways emerged. Each of these components is widely (albeit not ubiquitously) distributed throughout the eukaryotic kingdom, indicating that RNAi can be traced back to the common ancestor of eukaryotes that possessed, at a minimum, one each of an Argonaute-/Piwi-like protein, a Dicer, and an RNA-dependent RNA polymerase protein. The deduced early origins of these components suggest that RNA-mediated silencing played an early and important role in defending the genome against genomic parasites such as transposable elements and viruses.

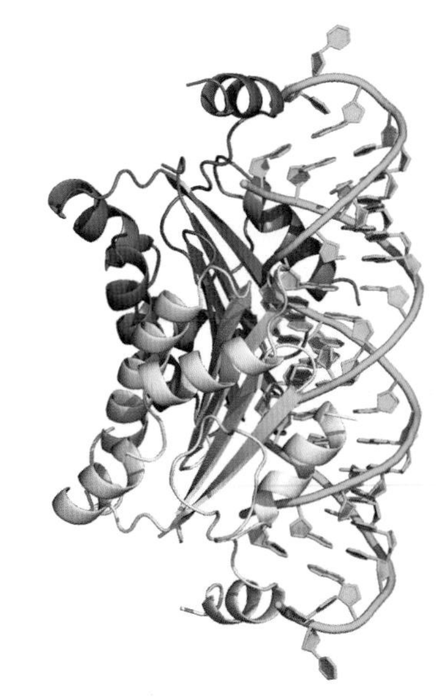

Figure 13.20 Structure of the carnation Italian ringspot virus p19 protein. The p19 protein interferes with the anti-viral silencing by binding to 21-nucleotide siRNA molecules (green) in a caliper-like fashion. Protein Data Bank (PDB) code 1RPU.

Viruses express a variety of anti-silencing defense mechanisms

Not surprisingly, viruses have evolved a wide variety of mechanisms to suppress RNAi surveillance (often denoted VSRs for viral suppressors of RNA silencing). For example, a plant virus called the carnation Italian ringspot virus produces a caliper-like protein that specifically binds double-stranded siRNAs and prevents their incorporation into the RISC complex; the structure of this protein is shown in Figure 13.20. As another example, a protein of the cucumber mosaic virus binds to the PAZ domain of the AGO1 Argonaute protein to inhibit the activity of the already assembled RISC complex in plants.

Viral defense is provided by CRISPR systems in bacteria and archaea

While the bacterial sRNAs that act by base-pairing (described in Section 13.2) do not appear to have a prominent role in defense against foreign DNA, an intriguing viral defense mechanism involving RNA has been recently uncovered in bacteria and archaea. (Bacterial viruses are often referred to as phage.) The sequences encoding this defense were identified as clustered regularly interspaced short palindromic repeat (CRISPR) sequences. The repeats are 20 to 50 nucleotide sequences that are frequently palindromes interspaced with unique spacer sequences of similarly short lengths, as illustrated in Figure 13.21a. The observation that many of the spacer sequences match sequences found in viral genomes led to the hypothesis that these sequences serve as a defense against foreign viral or plasmid DNA. This hypothesis has been borne out with direct tests as well as an increasingly detailed understanding of the mechanism of CRISPR-mediated immunity.

The CRISPR sequences are found adjacent to several conserved sets of CRISPR-associated (or *cas*) genes, which encode proteins containing domains likely to be useful in different nucleic acid based transactions (for example, endonuclease domains for cleaving and helicase domains for unwinding). These groups of genes fall into several broad classes, defined by the quite diverse sets of genes associated with the CRISPR systems in different bacterial and archaeal species.

Generally speaking, the CRISPR systems carry out two main processes: adaptation, in which foreign DNA is recognized and new spacers are integrated into the genome; and interference, in which the CRISPR locus is expressed, foreign nucleic acids are cleaved, and immunity is conferred. Given the different complements of associated proteins factors, it is not surprising that there are mechanistic distinctions in how these processes are carried out by the different classes of CRISPR systems. Experimental approach 13.3 describes some of the experiments that led to insights into how CRISPR loci interfere with the expression of foreign DNA through different mechanisms.

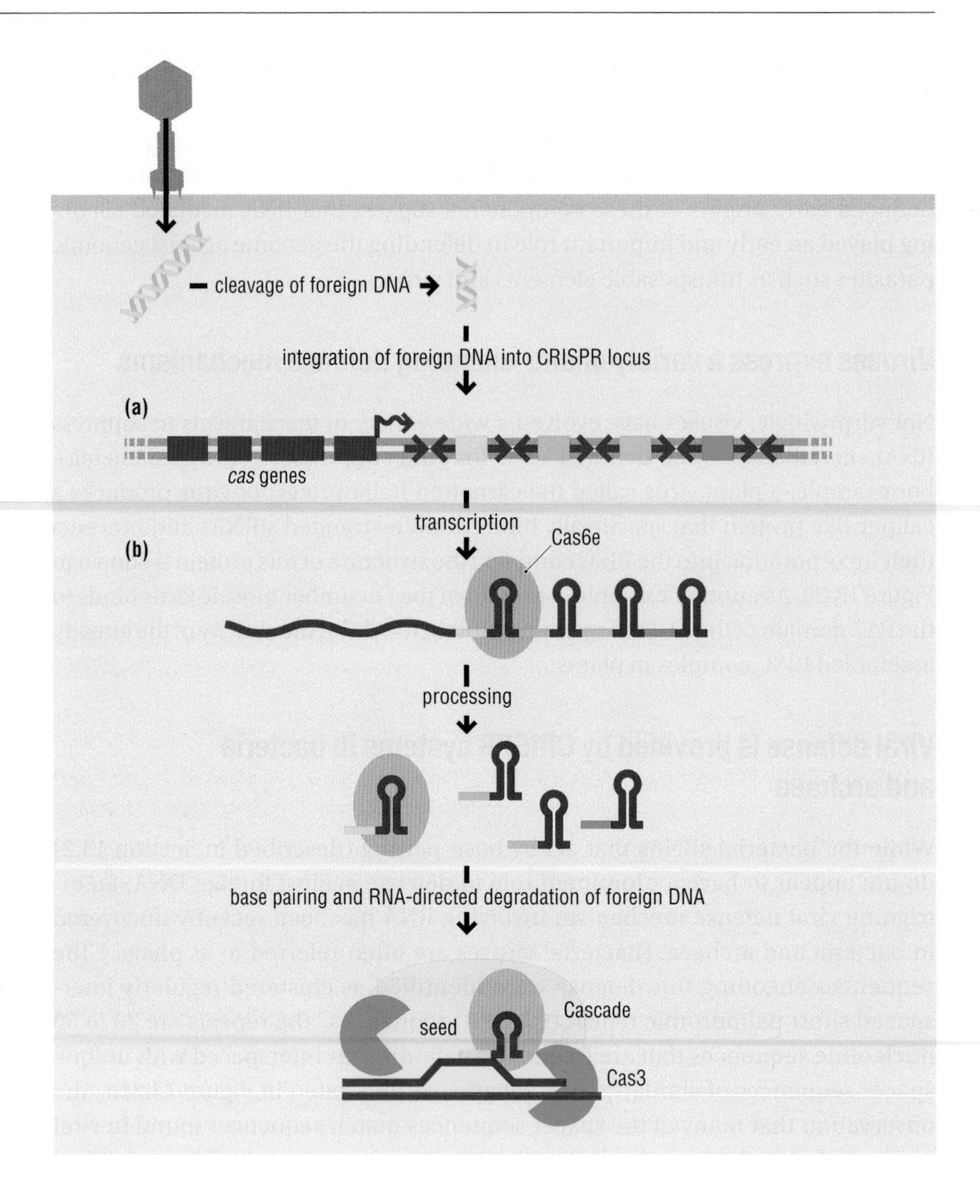

Figure 13.21 Viral defense provided by CRISPR systems. (a) Diagram of CRISPR locus with unique spacers of integrated fragments of foreign DNA (green) interspaced between short repeat sequences (arrows). A number of genes encoding Cas proteins with exonuclease and helicase domains are usually found adjacent to the repeats. (b) The region encompassing the repeat is transcribed as a long RNA from an upstream promoter. This long precursor is then processed by Cas proteins to give rise to short crRNAs that each contain one unique sequence. In one mechanism of CRISPR mediated silencing, the crRNAs direct cleavage of the complementary DNA.

Based on Fig 2 of Wiedenheft B, Sternberg SH, Doudna JA. RNA-guided genetic silencing systems in bacteria and archaea. *Nature*, 2012;**482**:331–338.

For many CRISPR systems, the first step in interference is the transcription of a long RNA that encompasses the CRISPR repeat sequences. This long transcript is then processed into shorter fragments termed crRNAs, each containing a spacer sequence, by one of several different mechanisms. In perhaps the most extensively studied pathway from *E. coli*, which is illustrated in Figure 13.21b, adjacent palindromic repeat sequences base-pair, thus providing a double-stranded substrate for the Cas6e endonuclease, which cleaves within each repeat sequence. In the other systems, different endonucleases are involved, including an endogenous bacterial RNase III enzyme in one system. There are also examples in which the repeats are transcribed as independent transcripts.

The resulting short spacer-containing RNA fragments next bind to a set of Cas proteins – we broadly refer to these complexes as the crRNA:Cas complexes. How do these complexes then function? In some CRISPR systems, the crRNA:Cas complex is thought to cleave viral or plasmid DNA sequences that are homologous to the spacer sequences; in other CRISPR systems, the crRNA-Cas complexes target homologous RNA transcripts generated from the foreign DNA sequences. In the model *E. coli* system, the crRNAs are bound by five different

13.3 EXPERIMENTAL APPROACH

CRISPR-mediated silencing is carried out by multiple mechanisms

Does CRISPR-mediated silencing target DNA or RNA?

During the sequencing of bacterial and archaeal genomes, investigators noted unusual sequences they denoted clustered regularly interspaced short palindromic repeats (CRISPR). The realization that many of the sequences interspaced between the palindromic repeats corresponded to sequences found in viruses and also some plasmids led to the hypothesis that CRISPR sequences represented a form of defense system against these external viral and plasmid threats. This was subsequently borne out in various direct tests. It was also shown that the CRISPR sequences were transcribed and cleaved into short RNAs. The conceptual similarities between RNA interference found in eukaryotes and the CRISPR systems in archaea and bacteria led to questions about whether the CRISPR RNAs target the foreign DNA directly or instead target the RNAs transcribed from the foreign DNA.

CRISPR interference in Staphylococci targets DNA

Luciano Marraffini and Erik Sontheimer addressed the target question in the bacterial pathogen *Staphylococcus epidermidis.* There they took advantage of a CRISPR system carrying a spacer sequence homologous to the nickase (*nes*) gene found on plasmids that enter staphylococcal species by conjugation (whereby DNA is transferred from a donor cell to a recipient cell by cell-to-cell contact). Marraffini and Sontheimer found that the *S. epidermidis* CRISPR system very effectively prevented the conjugative plasmid from being maintained in the recipient cell.

With this system in hand, the scientists asked whether the CRISPR system was targeting the plasmid DNA or a plasmid-derived transcript, using a clever experiment that depended on the incorporation of a self-splicing intron sequence into the target region of the *nes* gene, as illustrated in Figure 1a. This integrated sequence disrupts the CRISPR target sequence on the DNA but not the RNA, since excision of the self-splicing intron in

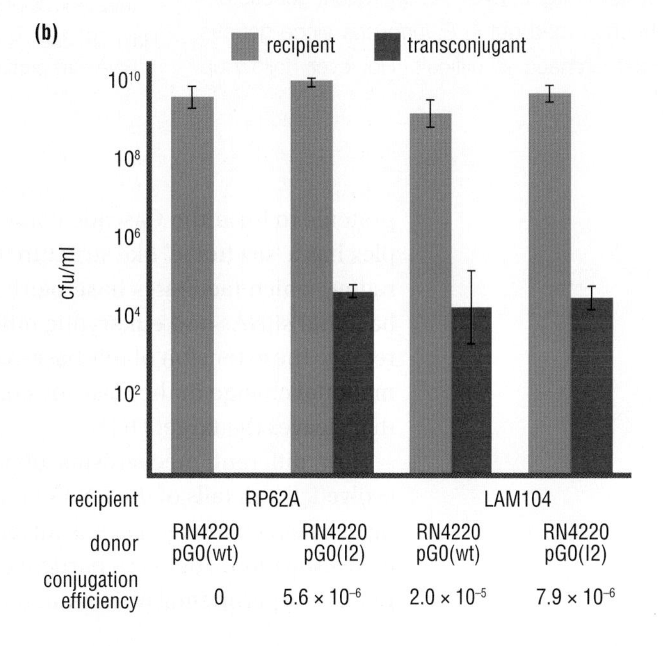

Figure 1 Effects of an interrupted target sequence in DNA but not RNA. (a) Position of the insertion of the sequence encoding a self-splicing intron in the *nes* gene of the conjugative plasmid pG0400. (b) Conjugation efficiency as assayed by the number of colony forming units observed after conjugation of the wild type-pG0(wt) or mutant pG0(I2) into the recipient RP62a strain carrying an intact CRISPR locus and the corresponding strain LAM104 lacking a CRISPR locus.

Fig 2 in Marraffini LA, Sontheimer EJ. CRISPR interference limits horizontal gene transfer in Staphylococci by targeting DNA. *Science*, 2008;**322**:1843–1845.

the transcript would regenerate the target sequence. As shown in Figure 1b, the insertion inhibited conjugation by six orders of magnitude in a manner dependent on an intact CRISPR system in the recipient cell. These data argue that DNA is the target of the *S. epidermidis* CRISPR system.

CRISPR interference in Pyrococcus targets RNA

At the same time, the group of Michael Terns used a biochemical approach to examine the target specificity of the CRISPR system in the archaeal species *Pyrococcus furiosus*. In this study, as shown in Figure 2, purified *P. furiousus* Cas proteins (the proteins encoded by genes associated with CRISPR loci) in complex with CRISPR RNA fragments were incubated with labeled RNA containing the target sequence (lanes marked 1), labeled RNA containing the reverse of the target sequence (lanes marked 2) or labeled DNA containing the target sequence (lanes marked 3). The investigators only detected cleavage products for the labeled RNA containing the target sequence. This experiment together with other controls, led Terns and colleagues to conclude that the *P. furiousus* CRISPR system targets RNA transcripts derived from the invading DNA species.

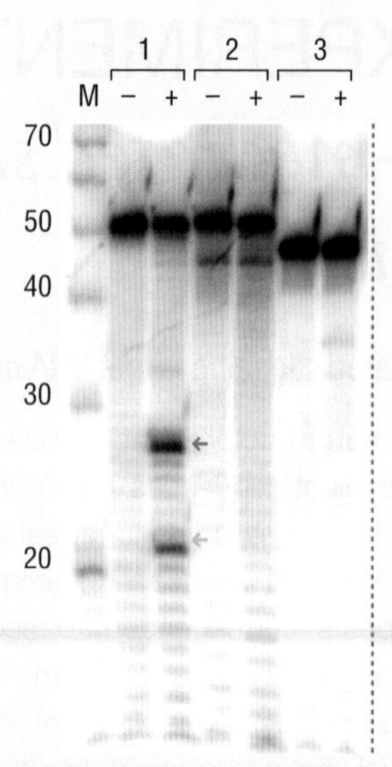

Figure 2 Cleavage of labeled target substrates by a purified CRISPR RNA containing ribonucleoprotein complex. 5′end-labeled single-stranded RNA with perfect complementary to the CRISPR RNA (1), RNA lacking complementary to the CRISPR RNA (2) or DNA with perfect complementary to the CRISPR RNA were incubated in the presence (+) or absence (–) of CRISPR RNA containing RNP complexes. Arrows indicate two primary cleavage products.

How can different conclusions be reconciled?

Such a situation wherein different groups reach different conclusions is not uncommon, but usually means that scientists do not have a complete understanding of a system. In the case of CRISPR targeting, the two groups used very different approaches (genetic versus biochemical) as well as very different organisms (bacteria versus archaea). Conceivably, one of the approaches has hidden flaws. Alternatively, both groups may be correct.

Since the first studies described here, purified bacterial Cas proteins in complex with CRISPR RNAs also have been shown to cleave DNA, lending credence to the finding that the *S. epidermidis* CRISPR system targets DNA. An equivalent genetic experiment has not been carried out in *P. furiousus* since genetic manipulation of most archaea is difficult. However, follow-up studies on a wide range of organisms are showing there are multiple distinct CRISPR systems that rely on different types of Cas proteins; it is conceivable that these systems act through distinct mechanisms. These complexities illustrate the importance of studying related systems in different organisms. What is found in one species may not always be true in another.

Find out more

Marraffini LA, Sontheimer EJ. CRISPR interference limits horizontal gene transfer in Staphylococci by targeting DNA. *Science*, 2008;**322**:1843–1845.

Hale CR, Zhao P, Olson S, *et al*. RNA-guided RNA cleavage by a CRISPR RNA-Cas protein complex. *Cell*, 2009;**139**:945–956.

proteins to form the Cascade complex. As illustrated in Figure 13.22, this complex has a 'sea horse' like structure that presents the crRNA in a helical configuration, which facilitates base-pairing with the target DNA. As we have seen for bacterial sRNAs and eukaryotic miRNAs, base-pairing initiates through a 'seed' region. The extension of this base-pairing interaction is associated with a conformational change in the Cascade complex and recruitment of the Cas3 nuclease that cleaves the target DNA.

How different mechanisms of action using the same CRISPR RNAs have evolved, the details of the RNA and DNA targeting, and, in particular, the mechanism by which foreign DNA is integrated into a CRISPR array, are all active areas of investigation. There is a particular focus on modifying these systems for use as tools for experimental manipulation of genomes, such as cleaving DNA at specific

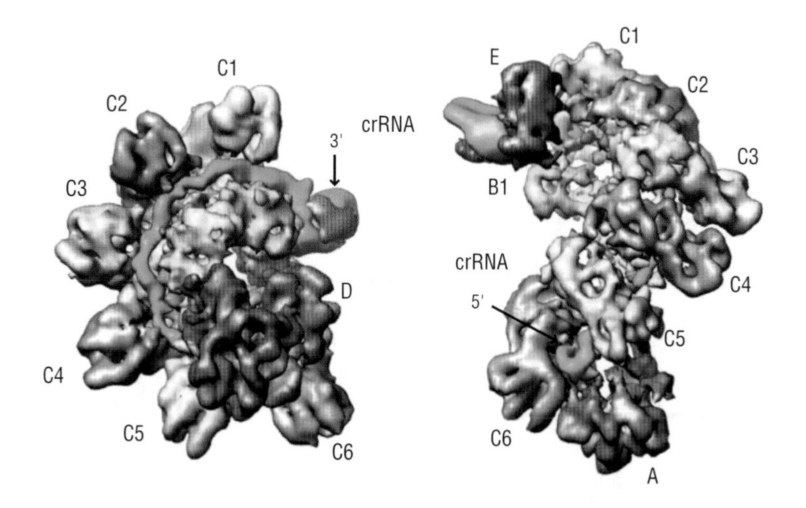

Figure 13.22 (a) Top-down and (b) side view of a single-particle cryoelectron microscopy structure of the *E. coli* crRNA-Cascade complex showing the crRNA displayed on the side of the structure in green.

Based on Fig 1 of Wiedenheft *et al.* (2011) Structures of the RNA-guided surveillance complex from a bacterial immune system. *Nature*. Sep 21;**477**(7365):486–489.

sites allowing for the insertion of DNA sequences of interest in eukaryotes, as well as for silencing specific genes in bacteria.

13.8 RNA-MEDIATED REGULATION IN *cis*

The sRNAs that we have discussed until now are necessarily somewhat limited in what they can do given their small size. Indeed, these 'small' RNAs are generally thought to do little more than identify targets through simple base-pairing interactions. There are, however, considerably larger, and thus structurally more complex, regulatory RNAs that have been identified in bacteria and eukaryotes - the list of which appears to be growing as rapidly as the list of small RNAs. While their large size does not mean that these RNAs do not act through simple base-pairing interactions, their function seems to be more complex, consistent with their greater information content and thus functional potential.

During this and the next few sections, we will focus on larger RNAs that appear principally to dictate whether certain regions of the genome are expressed, acting either at the level of transcription, or afterwards (post-transcriptionally). While we still have much to learn about their functions, there are certain features that appear to be shared and that provide some themes for discussion.

cis-encoded RNA elements regulate transcription, translation, or splicing of mRNAs

We discuss in Section 9.5 how sequences at the 5′ end of bacterial mRNAs can impact transcription elongation by a phenomenon known as attenuation, in which short self-complementary regions in the RNA base-pair with one another and thereby impact transcription. The first 5′ sequences found to modulate transcription contained small ORFs encoding proteins that are responsive to the levels of particular amino acids. In recent years it has become clear that the structure of *cis* regions in the RNA can be responsive to many different signals. For example, the RNA structure can be directly altered by temperature, or by the binding of proteins, tRNAs, or small molecules. These changes in structure can impact whether downstream transcription or translation can proceed.

In perhaps the simplest case, a segment of RNA functions like a 'thermometer', sensing temperature and reporting the results in an interpretable fashion. In this

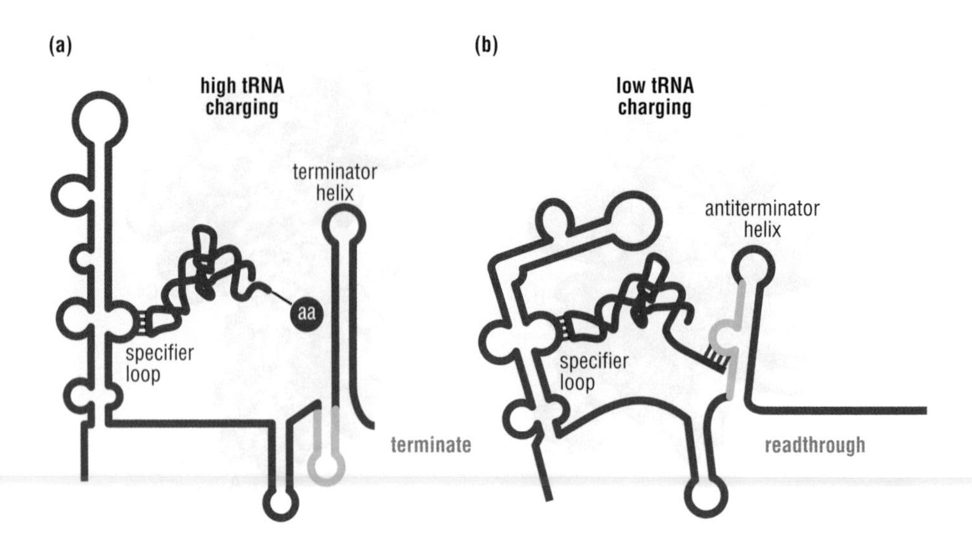

Figure 13.23 T-box mediated regulation of the transcription of a gene encoding aminoacyl tRNA synthetase. (a) At high levels of tRNA charging, the mRNA folds into a terminator structure that prevents transcription into the coding sequence. (b) At low levels of charging, the tRNA interacts with two regions of the mRNA, triggering a conformational change that allows read-through into the synthetase gene.

Based on Figure 2 of Green NJ, Grundy FJ, Henkin TM. The T box mechanism: tRNA as a regulatory molecule. *FEBS Letters*, 2010;**584**:318–324.

case, a secondary structure that inhibits ribosome binding (by sequestering the Shine–Dalgarno sequence) is folded at lower temperature, but is unfolded at an elevated temperature. In the pathogenic bacteria in which they are found, RNA thermometers make it possible to turn on synthesis of virulence regulators and chaperone proteins when the bacteria are exposed to elevated temperatures in the host during infection.

In another permutation, uncharged tRNA molecules bind specifically to the 5′ untranslated sequence of an mRNA sequence that encodes their cognate aminoacyl tRNA synthetase. When the amount of a given synthetase in the cell is limiting, deacylated tRNA is abundant and free in solution (since charged tRNAs are generally bound by EFTu). Binding of the tRNA in *cis* to the mRNA triggers a conformational change that promotes antitermination, and thus increases gene expression, as depicted in Figure 13.23.

In addition to these specific examples - in which RNA regulates gene expression on its own - there is an increasing appreciation for the wide range of regulatory proteins that bind to mRNAs, both upstream and downstream of the ORF, to modulate all steps in the lifetime of an mRNA (see Figures 10.34 and 12.11).

Riboswitch RNA elements respond to a wide range of small molecules

One particularly well characterized class of *cis*-encoded RNA regulatory elements folds into exquisite secondary structures that bind to small molecules and can regulate downstream events including transcription, translation, and even RNA splicing. These rather elegant *cis* motifs are broadly referred to as riboswitches. While riboswitches appear to be primarily found in bacteria, examples have also been identified in archaea and eukaryotes.

These RNA regulatory elements typically comprise two parts, as shown in Figure 13.24: an aptamer domain (from the Latin *aptus*, to fit) and an effector domain. In general, the metabolite binds to the aptamer domain and induces changes in the structure of the effector domain, which brings about changes in gene expression. The effector domain can encompass a transcription terminator sequence whose activity can either be enhanced or diminished when a ligand binds to the aptamer domain (Figure 13.25a). Alternatively, the effector domain can effectively sequester or release the ribosome-binding site, thus either blocking or facilitating

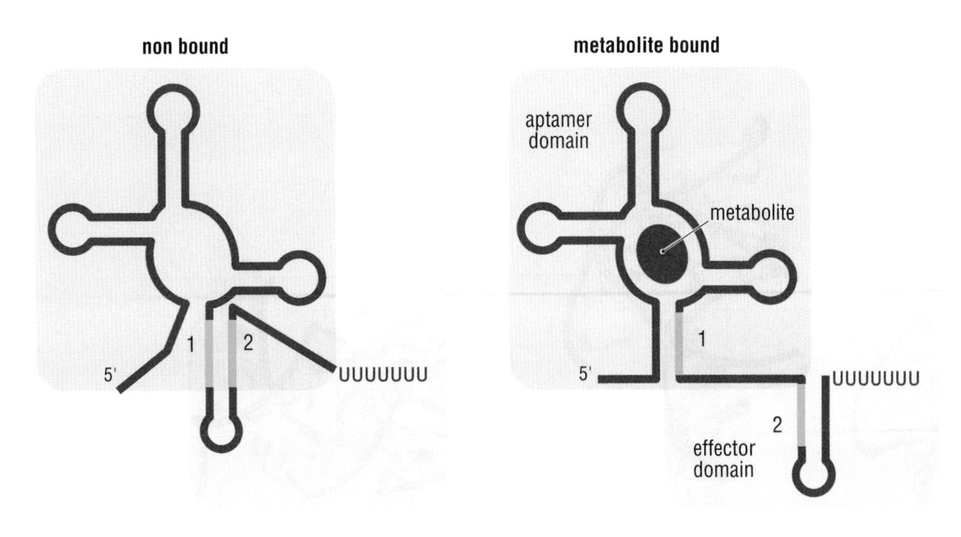

Figure 13.24 General structure of a riboswitch. Riboswitches typically comprise an 'aptamer' domain and an 'effector' domain. Upon binding a metabolite, the aptamer domain induces a change in the conformation of the effector domain to bring about a change in the transcription, translation, or splicing of the downstream RNA.

translation initiation (Figure 13.25b). Metabolite binding can also promote changes in RNA structure that, for example, enhance or prevent splicing. Certain riboswitches even occur in tandem such that their action is enhanced by cooperativity.

In a typical scenario, the association of a particular metabolite with a riboswitch regulates (either negatively or positively) the synthesis of a gene product encoded by the downstream ORF that is involved in metabolizing or transporting the same metabolite. For example, a riboswitch located upstream of a gene encoding a vitamin B_{12} transporter binds to B_{12} itself; when intracellular B_{12} levels are high, B_{12} binds to the riboswitch, turning off transcription of the downstream gene.

The range of molecules recognized by riboswitch elements and the specificity of binding are both impressive. Molecular species ranging from atoms like fluoride and magnesium ions to phosphorylated sugars, purines, vitamins, and amino acids are all recognized by riboswitches. More than three different classes of riboswitch recognize S-adenosyl methionine alone. Yet, riboswitch binding can be so specific that a guanine-binding riboswitch can easily discriminate against

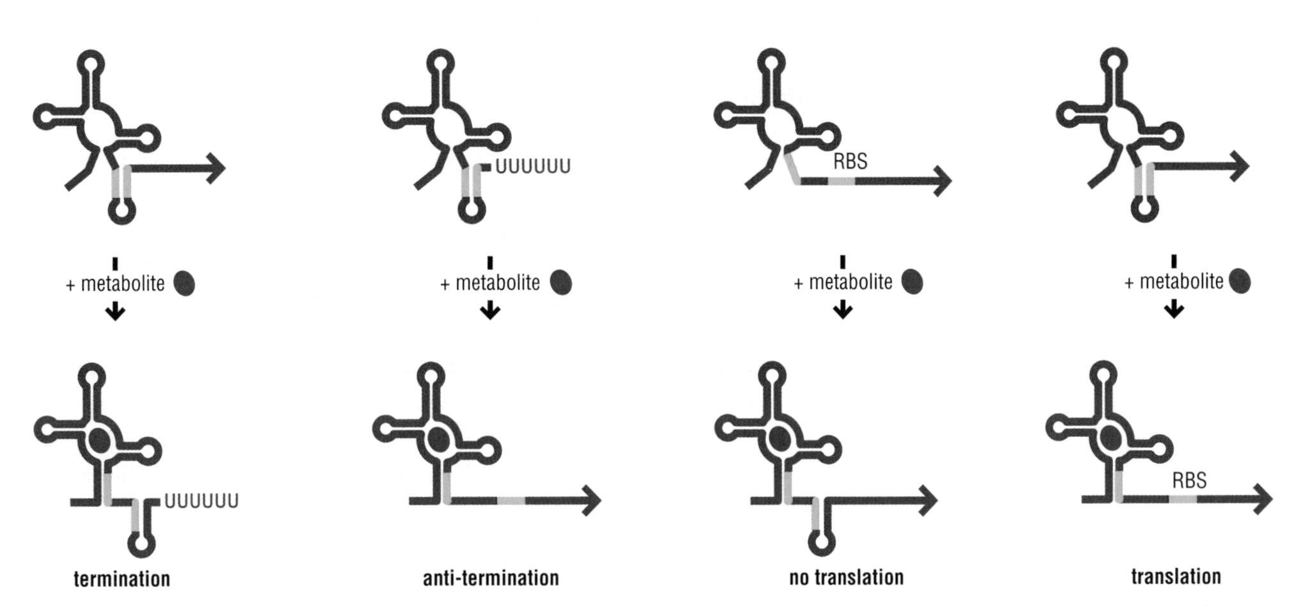

Figure 13.25 Outcomes of metabolite binding to riboswitches. Metabolite binding to the aptamer domain of a riboswitch can bring about a conformational change in the effector domain to lead to or prevent the formation of a transcription terminator, or block or increase access to a ribosome binding site.

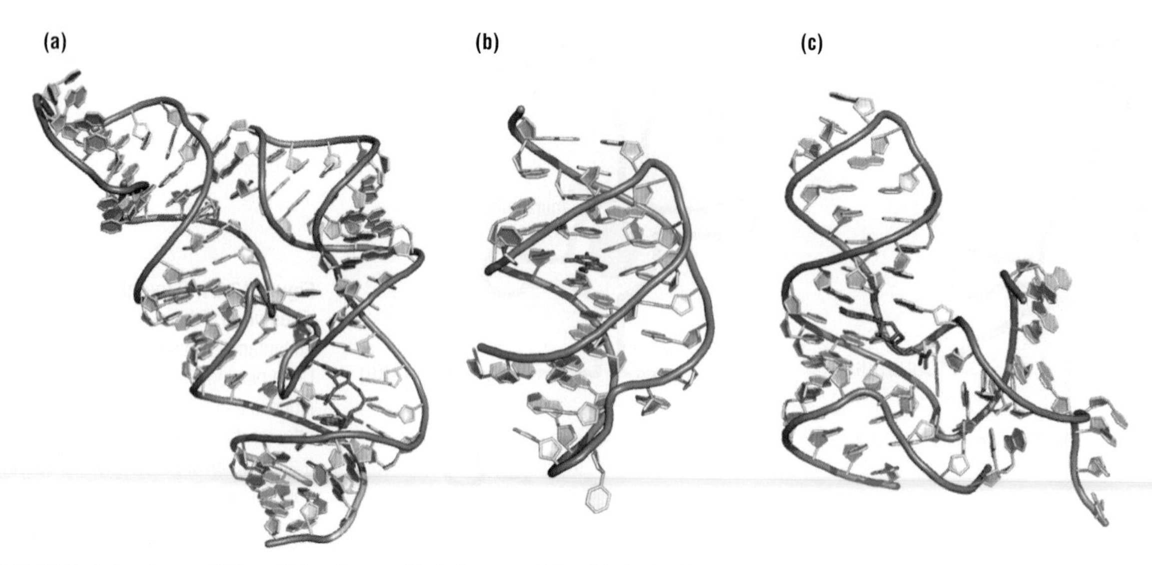

Figure 13.26 Varied structures of riboswitch aptamers that allow exquisite discrimination among molecules. This figure shows riboswitch aptamers binding the following metabolites (a) cyclic-di-GMP (c-di-GMP; PDB code 3IRW), (b) queuosine (preQ$_1$; PDB codes 2KFC and 3FU2), and (c) S-adenosylhomocysteine (SAH; PDB code 3NPQ).

all other nucleotide bases, even deoxyguanine. In general terms, as illustrated in Figure 13.26, riboswitch elements illustrate the amazing versatility of RNA to adopt complex, functionally diverse, three-dimensional structures that rival those formed by proteins.

13.9 PROTEIN-BINDING REGULATORY RNAs

The simple binding of proteins to a regulatory RNA also can broadly impact the biological outcome. For example, an RNA can mimic the structure of another nucleic acid and thus titrate a protein away from its normal target. Alternatively, the RNA can bring individual proteins or protein complexes together with each other or with genomic DNA. Let us now consider some examples in more detail.

Protein-binding regulatory RNAs can titrate a protein away from its target

Given the hundreds of RNA-binding proteins and the ability of RNA to fold into a wide range of structures, it is not surprising that some regulatory RNAs act by sequestering proteins away from their usual target sites. A straightforward example of such regulation is the CsrB family of bacterial RNAs (denoted RsmY in some organisms), as illustrated in Figure 13.27. These RNAs typically have multiple repeats of a sequence that is recognized by the RNA-binding protein, CsrA. When CsrB is not induced, CsrA binds to the 5′ untranslated regions of target mRNAs and modulates their stability or translation. However, when CsrB levels increase during specific stages of growth, the CsrA protein is titrated away from these target mRNAs by the many binding sites present in CsrB RNA. In this situation, CsrA-mediated regulation of target mRNAs is lost.

In another bacterial example, the ~180-nucleotide 6S RNA mimics the structure of DNA in an open transcription bubble, as shown in Figure 13.28. Not surprisingly, the 6S RNA effectively sequesters one form of RNA polymerase that is expressed during stationary phase growth, thus limiting overall transcription at this

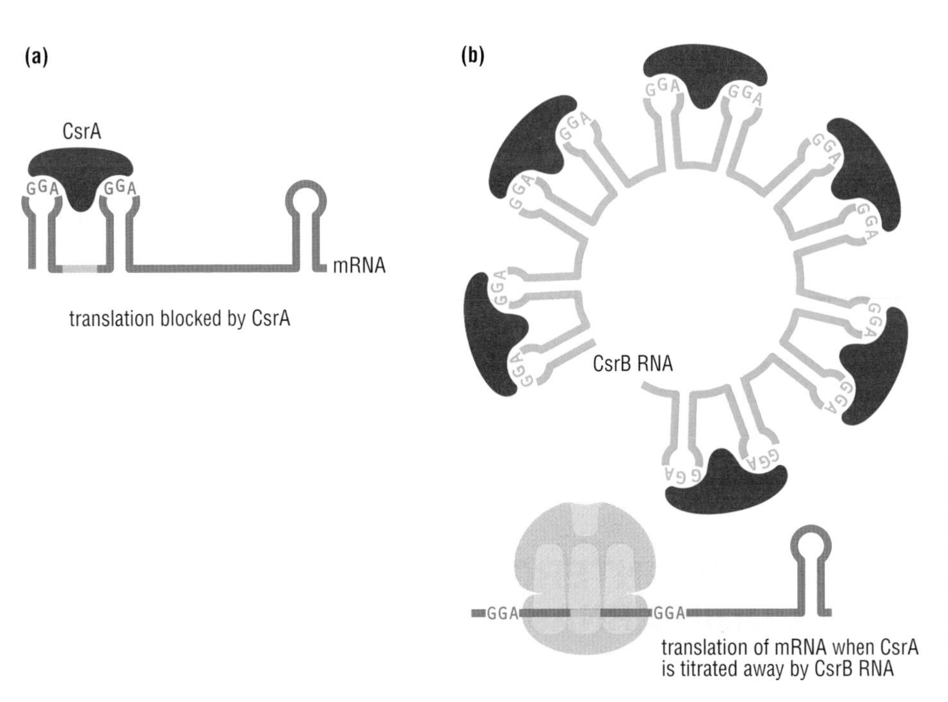

Figure 13.27 Mechanism of bacterial CsrB action. The CsrB RNA (green) has multiple repeats of the 'GGA' binding motif bound by the CsrA protein (red). (a) In the absence of CsrB, CsrA binds to mRNA targets (blue) to modulate translation and stability. (b) When CsrB is induced, it titrates CsrA away from the mRNA targets.

stage. Intriguingly, this mimic can even be subsequently used as a substrate for RNA polymerase: during exponential growth - when 6S-mediated inhibition is no longer advantageous - the regulatory RNA transcript is extended by several nucleotides by the bound RNA polymerase. This extension releases the polymerase from inhibition as the transcribed 6S RNA no longer binds effectively.

The ~330-nucleotide 7SK RNA provides an example of a eukaryotic RNA that sequesters a protein, also affecting transcription. In this case, the kinase activity of the RNA polymerase II transcription elongation factor, P-TEFb, is inhibited when it associates with the 7SK ribonucleoprotein (RNP) complex. In the presence of specific proteins such as the human immunodeficiency virus (HIV) activator protein, Tat, P-TEFb is released to promote transcription elongation.

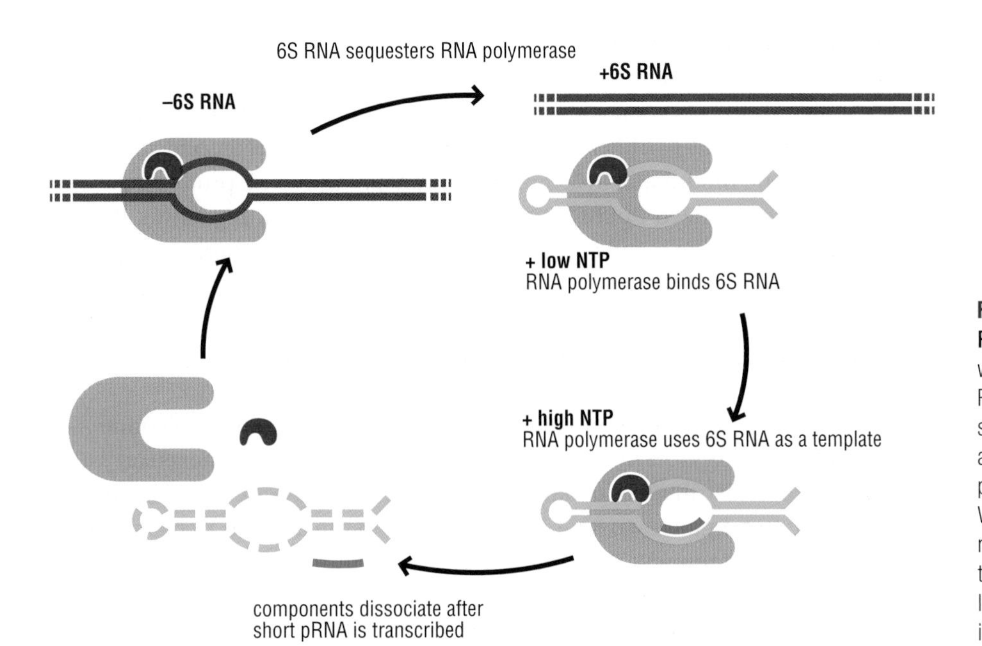

Figure 13.28 Mechanism of bacterial 6S RNA action in bacteria. The 6S RNA competes with DNA promoters (gray) for binding to the RNA polymerase holoenzyme (blue) during the stationary phase when 6S RNA (green) levels are high and nucleotide levels are low, thus preventing transcription of some promoters. When nucleotide levels become high as cells re-enter the exponential phase, the 6S RNA is transcribed. Synthesis of the short pRNA (aqua) leads to the release of the 6S RNA, lifting the inhibition of transcription.

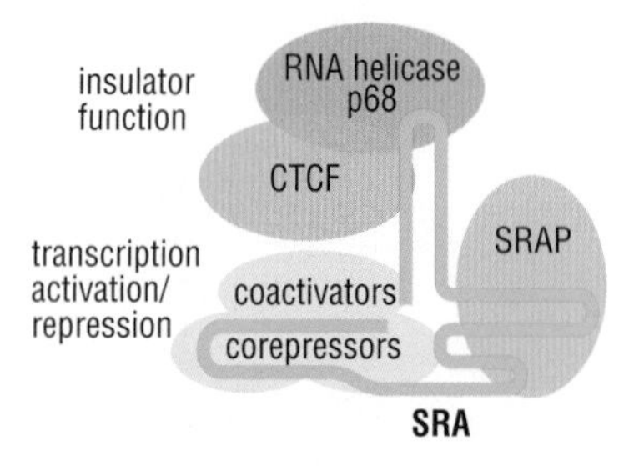

Figure 13.29 Mechanism of SRA RNA action. The SRA RNA (green) serves as a tether to bring multiple protein complexes into proximity at various promoters to modulate transcription.
Based on Fig 2e of Kugel JF, Goodrich JA. Non-coding RNAs: key regulators of mammalian transcription. *Trends in Biochemical Sciences*, 2012;**37**:144–151.

Protein-binding regulatory RNAs can act as tethers

Somewhat longer regulatory RNAs can also serve as scaffolds to bring proteins into close proximity. An example is the ~870-nucleotide mammalian steroid receptor RNA activator (SRA) RNA. As its name implies, this RNA was first serendipitously identified in a screen for protein co-activators of the steroid hormone receptor. While the precise mechanism of SRA RNA action is still under investigation, it is clear that the RNA binds multiple co-regulator proteins (each with RNA-binding domains), as illustrated in Figure 13.29. The precise configuration of these recruited SRA-binding proteins impacts the expression of hormone receptor-regulated genes. The importance of such protein-binding RNA regulators is underscored by the finding that the levels of SRA are altered in certain breast cancers.

13.10 LONG INTERGENIC NON-CODING RNAs

Throughout this chapter, we have focused primarily on the regulatory functions of sRNAs – those RNAs that comprise tens to a few hundreds of nucleotides. However, in recent searches to identify all RNAs in eukaryotes, researchers have found extensive evidence for transcription in non-protein-coding regions – and in some cases these transcripts are extremely long, ranging from thousands to hundreds of thousands of nucleotides in length. These RNAs, a few of which have been shown to have regulatory functions, are generically referred to as long intergenic non-coding RNAs (lincRNAs) or long non-coding RNAs (lncRNAs).

As in the case of the miRNAs, these putative lincRNAs typically possess the hallmarks of protein-coding RNA polymerase II transcripts: they are spliced, capped, and polyadenylated. Despite these features and their length, most lincRNA species have limited protein-coding potential. Short potential ORFs can be identified throughout the RNA sequence, but these relatively short ORFs are typically not conserved in related organisms, and there is little evidence that functional proteins are made from them.

The characterization of these longer regulatory RNAs is an exploding area of research. Given their length, it is likely that the RNAs will be able to function by using combinations of the mechanisms outlined in the preceding sections – for example, by both base-pairing interactions and through more complicated structures that may be modulated by ligands or may serve to recruit proteins. Several of the best-characterized lincRNAs to date are important in regulating the expression of the genome in eukaryotic cells, and so we discuss these examples in more detail.

X-inactivation is carried out by lincRNAs

The best known of the long regulatory RNAs is surely the 17 000 nucleotide *Xist* RNA, which plays a key role in inactivating the X chromosome in higher eukaryotes. Despite its broad distribution in mammals, *Xist* RNA shows poor overall sequence conservation. In humans, *Xist* RNA is expressed from the inactive X chromosome during female development. The primary transcript is processed like a normal RNA polymerase II transcript but is then retained in the nucleus, where it surrounds or 'coats' the X chromosome from which it has been transcribed; this coating is

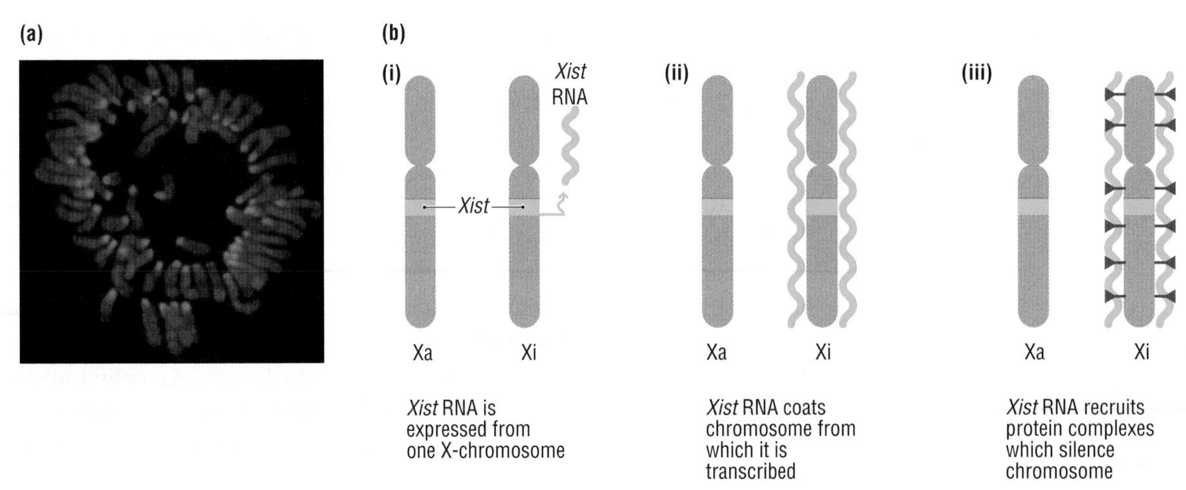
(a)
(b)
(i)
Xist RNA
Xist
Xa
Xi
Xist RNA is expressed from one X-chromosome
(ii)
Xa
Xi
Xist RNA coats chromosome from which it is transcribed
(iii)
Xa
Xi
Xist RNA recruits protein complexes which silence chromosome

Figure 13.30 X-inactivation by *Xist*. (a) RNA-fluorescence in situ hybridization (FISH) to detect the *Xist* RNA on condensed mouse chromosomes. (b) General mechanism for the inactivation of one X chromosome (Xi) involves (i) transcription of the *Xist* RNA from one of a pair of X chromosomes. This *Xist* RNA then (ii) coats the chromosome from which it is transcribed, and (iii) recruits the Polycomb repressive complex, which trimethylates lysine 27 of histone H3. Transcription of other genes on this chromosome is silenced as a result of this trimethylation.

(a) Fig 1 in Ng *et al.* Xist and the order of silencing. *EMBO Reports*, 2007 January;**8**(1):34–39.

illustrated in Figure 13.30a. Exactly how the *Xist* RNA is recruited there is not well understood.

Once localized, *Xist* RNA recruits a protein complex known as Polycomb repressive complex 2 (PRC2), as depicted in Figure 13.30b. This complex is known to trimethylate histone H3 at lysine 27, marking the associated chromatin region as transcriptionally silent. Consistent with this view, the coat of *Xist* RNA surrounding the chromosome creates an almost visually distinct nuclear compartment that is thought to exclude RNA polymerase II. This silent state of the X chromosome is stably inherited by progeny cells.

As if this story were not complex enough, the plot is further complicated by the role of another non-coding RNA, *Tsix*, which is an antisense RNA transcribed from the same region of the X chromosome. *Tsix* is generally thought to antagonize the role of *Xist* RNA, perhaps allowing for an additional layer of regulation. Another RNA, RepA, that is transcribed from a somewhat conserved portion of the *Xist* RNA gene also appears to contribute to X-chromosome inactivation in ways that are still poorly understood.

Long non-coding RNAs are abundant

Some of the lessons learned in studying *Xist* RNA appear to have broader applicability. For example, the *Air* RNA plays a critical role in imprinting at the *Igf2r* locus on the paternal chromosome; in this case, expression of *Air* RNA from the paternal chromosome leads to a repressed chromatin state, allowing the maternal copy of the locus to be the only one expressed. *HOTAIR* and *HOTTIP* RNAs contribute to imprinting at yet other loci. All of these RNAs, like *Xist*, appear to localize to certain domains of the chromosome, where they recruit repressive chromatin-modifying factors. While in some cases these lincRNAs directly impact the locus from which they are transcribed, in other cases, they appear to regulate distant loci to which they are somehow recruited. The molecular details that distinguish these behaviors are not yet well established.

➔ We discuss imprinting in more detail in Section 18.9.

Other lincRNAs have recently been proposed to function as 'enhancers' of transcription. These large RNAs are transcribed in the opposite direction from another RNA polymerase II transcribed gene, and are referred to as enhancer RNAs (eRNAs). If, as speculated, these RNAs turn out to be essential in regulating gene expression at these loci, their impact is paradoxically the opposite of the repressive lincRNAs such as *Xist*. Still other lincRNAs are cytoplasmically localized and thus likely function in a distinctive manner.

In broad terms, it is clear that far more of the genome is transcribed than ever before suspected. How many of these transcripts have function and how and where do they act? How does a cell distinguish between the transcript at a given locus that is destined for processing and export to the cytoplasm for translation by the ribosome, and the regulatory transcript that is destined for regulatory roles in the nucleus. We expect to learn much more about RNA regulators in the coming years.

The discovery of expanding roles for RNA in genome function has evolutionary implications

An expanding role for RNA in genome function may at first seem disconcerting and then reassuring. It has been argued that the ever-expanding complexities of eukaryotic evolution (i.e., the multicellularity, the specialization of cell types, and the increasing complexity of cognition) would have required a substantial increase in the number of regulatory features when compared with relatively simpler organisms. The human species is certainly more complex in most ways than the nematode, and yet the gene count in gross terms is roughly equivalent (~20 000). While these seeming discrepancies might be explained by the complexities of alternative splicing, alternative polyadenylation, and post-transcriptional and post-translational processing, it might also be that these discrepancies are partially reconciled by the arrival in the genome of increasingly complex control by RNA.

The reality is that the evolution of a new protein fold is a rare (and remarkable) event. This point is made by the observation that there is a relatively limited number of protein folds, with re-use of existing folds being a dominant mechanism for developing new functional roles. For example, consider that there are hundreds of proteins in the human genome that contain a Rossman fold, likely because this relatively simple fold can be repeatedly adapted to evolve proteins with new functions, not because it has arisen more than once during evolution. By contrast, 'functional' RNAs are more simply evolved *de novo*. For example, target recognition by miRNAs involves as few as six specific nucleotide residues (the 'seed' sequence). As such, the evolution of a miRNA to target a set of different transcripts in the cell may involve the substitution of just one nucleotide with another in an existing miRNA sequence, or the *de novo* insertion of a simple stem–loop miRNA precursor structure. The arrival of RNA-based regulation may have provided a competitive edge that could address the escalating costs of protein-based gene regulation, and allowed increasingly complex life forms to evolve.

OVERVIEW

- RNA molecules can act as regulators by base-pairing with target RNAs or even DNA, or by binding metabolites or proteins, to block or facilitate the binding of other RNAs or proteins and bring molecules into proximity.

BASE-PAIRING RNAS IN BACTERIA

- Most base-pairing RNAs in bacteria are expressed in response to specific environmental conditions, and many require the RNA chaperone Hfq for function.
- Many bacterial RNAs repress or induce translation or modulate target RNA stability via limited base-pairing.

miRNAs, siRNAs, AND rasiRNAs

- Eukaryotic sRNAs, which are involved in a range of regulatory processes, act in association with Argonaute-containing complexes. These sRNAs include:
  - miRNAs, which are encoded by distinct genes and modulate translation and mRNA destabilization
  - siRNAs, which are generated through excision along the length of a double-stranded RNA of either external origin (e.g., viruses) or endogenous origin (e.g., repeat sequences), and direct RNA interference and transcriptional silencing
  - rasiRNAs, which are derived from repetitive regions of the genome and repress the expression of diverse repeat sequences in the nucleus. Subsets of rasiRNAs include the piRNAs associated with Piwi-family Argonaute proteins and hcRNAs associated with heterochromatin.

PROCESSING OF miRNAs, siRNAs, AND rasiRNAs, AND LOADING ONTO ARGONAUTE PROTEINS

- miRNAs are processed from primary transcripts by the ribonuclease Drosha in the nucleus followed by the ribonuclease Dicer in the cytoplasm, resulting in the generation of an imperfectly paired double-stranded RNA, miR:miR*.
- After the double-stranded miR:miR* complex is loaded onto an Argonaute enzyme in the cytoplasm, one strand is chosen as the guide strand miR to target complementary RNAs, while the passenger miR* strand is removed, possibly by a helix activity.
- siRNAs are processed from long double-stranded RNAs by Dicer in the cytoplasm.
- After the perfectly complementary siR:siR* complex is loaded onto an Argonaute enzyme, the passenger strand is cleaved by the PIWI domain of the Argonaute protein.
- piRNAs are processed by a 'ping-pong' mechanism whereby an existing piRNA bound to a Piwi-family Argonaute protein defines the site of cleavage in a target RNA to form a new piRNA, which will bind another Piwi protein and repeat the cycle.

FUNCTIONS OF mi-, si-, AND rasiRISCs

- miRNAs generally base-pair imperfectly with sequences found in the 3′ UTRs of endogenous mRNA to repress translation and destabilize mRNAs.
- siRNAs are fully complementary with their target RNA sequences, a feature which activates the RNase H cleavage of the PIWI domain of the Argonaute protein leading to cleavage of the target RNA.
- rasiRNAs can be involved in cleaving target RNAs transcribed from repetitive regions of the genome or in bringing about changes in DNA or chromatin modifications by associating with chromosomes.

VIRAL DEFENSE ROLE OF sRNAs

- Sequence comparisons indicate RNA silencing by siRNAs evolved as a viral defense in eukaryotes.
- Viruses have evolved a wide range of mechanisms to suppress RNA silencing.
- In bacteria and archaea, viral defense is provided by CRISPR systems in which an RNA complementary to viral sequences is processed into short RNA fragments.

cis-ENCODED REGULATORY RNA ELEMENTS

- Transcription, splicing, or translation of an mRNA can be regulated by the intrinsic structure and the binding of proteins, tRNAs, and small molecules to *cis*-encoded RNA elements.
- The characterization of small molecule-binding elements called riboswitches has illustrated the exquisite secondary structures into which RNAs can fold.

PROTEIN-BINDING RNAs

- RNAs can act as regulators by mimicking another RNA or DNA structure to titrate a protein from other binding sites.
- RNAs can act as regulators by acting as tethers to bring proteins together.

lincRNAs

- lincRNAs, of which the *Xist* RNA required for X-chromosome inactivation is an example, modulate chromatin domains and regulate transcription.
- Increasing numbers of regulatory RNAs are being discovered in eukaryotes. The increasing complexity in regulation that can be facilitated by these RNAs might explain the evolution of increasingly complex life forms.

QUESTIONS

INTRODUCTION

1. In what way are the regulatory RNAs that are discussed in this chapter fundamentally different from the functional RNAs (such as tRNA and rRNA) discussed in previous chapters?

2. What is the significance of the finding that 60% of the human genome is transcribed and yet only 2% of the human genome is known to contain ORFs?

13.1 OVERVIEW OF REGULATORY RNAs

1. Which of the following statements does NOT describe a feature of regulatory RNAs?

a. They are processed after transcription.
b. They have different biological effects in bacteria and eukaryotes.
c. They identify targets using complementary base-pairing.
d. They interact with other cellular components to carry out their function.

2. In what way does the flexibility structure facilitate the function of regulatory RNAs?

13.2 BACTERIAL BASE-PAIRING sRNAs

1. Explain how it is possible that bacterial sRNAs can both inhibit and enhance translation of their target mRNA.

2. What is the normal function of the bacterial Hfq protein? How might Hfq participate in the modulation of sRNA activity?

Challenge question

3. Consider this figure in which gene A is transcribed from one DNA strand and a small antisense RNA is produced from the complementary strand.

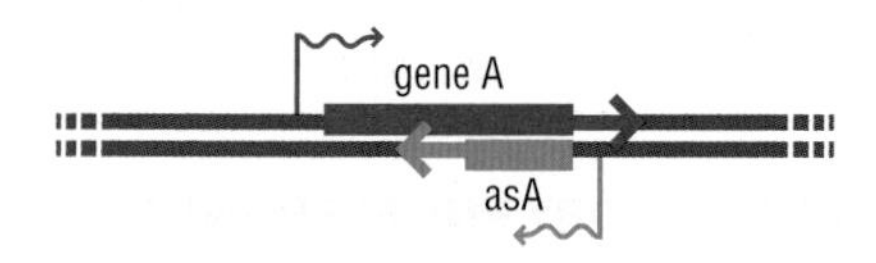

a. If gene A were responsible for enhancing a cellular response to low iron, propose a mechanism by which transcription of the antisense RNA might regulate the activity of gene A.
b. If gene A were responsible for inhibiting the low iron response under normal iron conditions, propose a mechanism by which transcription of the antisense RNA might regulate gene A's activity.

13.4 PROCESSING OF EUKARYOTIC sRNAs

1. Which of the following is not a source of siRNA?

a. Genes that have extensive hairpin structures
b. The product of specific genes
c. Transcription of antisense RNA
d. Viral RNA

2. Compare and contrast the processing of miRNAs and siRNAs.

3. What is the basis for classification as a rasiRNA?

4. What is unique about the class of rasiRNAs found in worms?

Challenge question

5. Compare and contrast the Microprocessor and Dicer complexes.

13.5 LOADING OF ARGONAUTE FAMILY PROTEINS WITH EUKARYOTIC sRNAs

1. Name and describe the function of the three core domains of the Argonaute family proteins.

2. What are the two strands of the duplex RNA from miRNA and siRNA called, and what are their fates during the loading into the Argonaute protein? How do the fates of miRNAs and siRNAs differ?

3. What is the end result of the loading process for all eukaryotic sRNAs?

13.6 GENE SILENCING BY SMALL EUKARYOTIC RNAs

1. Briefly describe the mechanism by which imperfect complementary base-pairing between the eukaryotic sRNA and its target can lead to a different outcome than a perfect complementary base-pairing.

2. What is the significance of the discovery of the RNAi pathway for forward genetic experiments?

Challenge question

3. Compare and contrast the function of piRNA:Piwi complexes with mi- or siRISCs.

13.7 VIRAL DEFENSE ROLE OF BACTERIAL, ARCHAEAL, AND EUKARYOTIC sRNAs

1. Give one example of a mechanism by which a virus developed an RNAi defense.

13.8 RNA-MEDIATED REGULATION IN *cis*

1. How is RNA-mediated regulation in *cis* different from the other types of regulation described above?

2. Describe the meaning of the term 'RNA thermometer'.

Challenge question

3. Consider two hypothetical genes, A and B, identified in bacteria. Each has a 5′ UTR region that can fold into two different complex secondary structures dependent upon the presence of a specific metabolite. These 5′ UTR regions were therefore hypothesized to act as riboswitches.

a. What downstream pathways are known to be affected by the use of riboswitches?
b. In one experiment, the 5′ UTR sequence of genes A and B were connected to the reporter gene luciferase. Each construct was assayed for production of luciferase in the presence and absence of their specific metabolites. The results of this experiment are shown in Figure A, top panel. For each gene, what was the effect of the metabolite?

c. In a second experiment, the same constructs were assayed in the presence and absence of their metabolite and the presence and absence of BCM, a known Rho Terminator inhibitor. The results of this experiment are shown in Figure A, bottom panel.
   i. What is the function of the Rho Terminator protein? (You may find it helpful to refer to Chapter 8.)
   ii. Based on the results in Figure A, bottom panel, what conclusions can be drawn regarding the function of the riboswitches located in the 5′UTRs for gene A and gene B?

Figure A

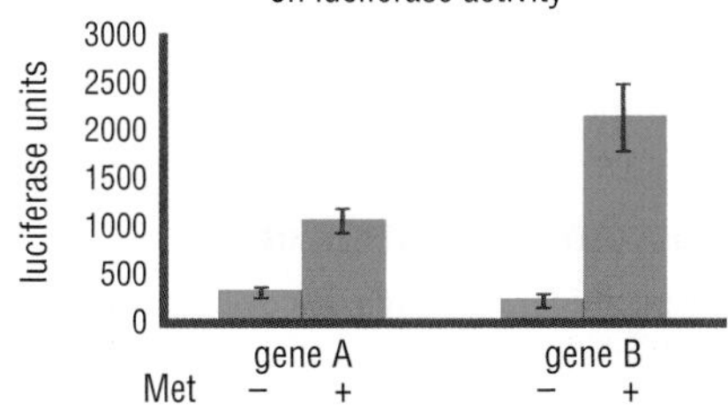

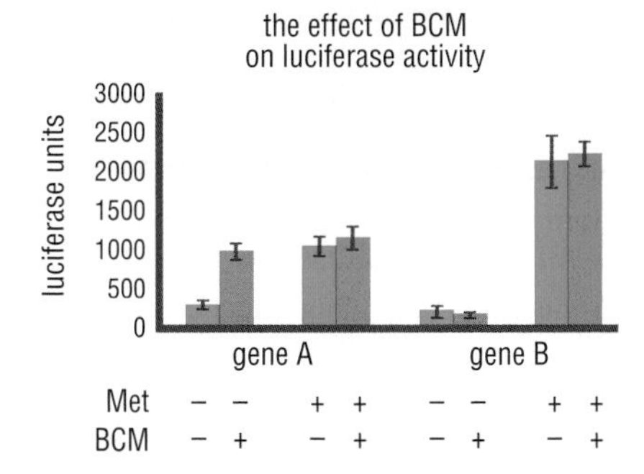

13.9 PROTEIN-BINDING REGULATORY RNAs

Challenge question

1. A newly identified protein in bacterial cells was shown to inhibit translation of its target genes by binding to the 5′ UTR of the mRNA and preventing ribosome binding. Propose a mechanism by which this type of inhibition may be relieved by an sRNA.

13.10 LONG INTERGENIC NON-CODING RNAs

1. In what way are lincRNAs fundamentally different from the other regulatory RNAs discussed in this chapter?

2. Which three lincRNAs are involved in X-inactivation, and what is known about their function?

3. Describe the significance of regulatory RNAs in creating the increased complexity of organisms.

EXPERIMENTAL APPROACH 13.1 – ANALYSIS OF THE *LIN4-LIN14* GENETIC INTERACTION AND THE DEVELOPMENT OF miRNA TARGET PREDICTION APPROACHES

1. In the *C. elegans* genetic screen described, *lin-14* and *lin-4* mutations resulted in similar lineage mutation phenotypes. Why was the understanding of the *lin-14* mutation considered straightforward but the *lin-4* mutations less so?

2. What hypothesis regarding the mechanism of action of the *lin-4* gene could be made after sequence analysis of both the *lin-14* and *lin-4* genes?

Challenge question

3. Look at Experimental approach 13.1 Figure 2.
   a. What is the main difference between the procedures used to create Figures 2a and 2b?
   b. What is the main difference between the results of Figures 2a and 2b?
   c. What conclusion can be drawn based upon the results of Figure 2?

4. How was the Argonaute protein used to help identify RNA targets of miRNAs?

EXPERIMENTAL APPROACH 13.2 – THE DISCOVERY OF RNA INTERFERENCE AND ITS BROAD APPLICATION TO FORWARD GENETICS

1. In 1998, Fire and Mello established that the most potent trigger of RNA interference was double-stranded RNA.
   a. What is the *unc22* gene?
   b. What phenotype appears if the *unc22* gene is not functioning?
   c. In Figure 1, why is there both *unc22A* and *unc22B*?
   d. What were the three types of RNA that were injected into the infected animal?
   e. How do the results of the experiment support the statement that double-stranded RNA is the most potent trigger of RNA interference?

2. Why is it significant that the RNAi knock-down phenotype is observable in the progeny (offspring) of the targeted organism?

3. Why is RNAi one of the most commonly used tools for probing gene function in eukaryotic systems? (That is, without RNAi, what would you have to do to study gene function?)

Challenge questions

4. Why is it significant that RNAi only works if you target exons and not if you target introns and promoters of genes?

5. Many cancers exhibit an overexpression of the *MYC* gene. This type of cancer had been considered to be untreatable using drugs because *MYC* serves an essential role in normal cells and inhibiting its function would have intolerable side effects. It has, however, been discovered that cells that overexpress *MYC* are highly dependent upon other genes and pathways for their survival, but some of these pathways may not be essential for normal cells. Propose an experiment in which you could use RNAi to identify genes in these pathways.

EXPERIMENTAL APPROACH 13.3 – CRISPR-MEDIATED SILENCING IS CARRIED OUT BY MULTIPLE MECHANISMS

1. Why were CRISPR sequences originally hypothesized to be part of a defense mechanism against viral infection?

Challenge questions

2. Look at Experimental approach 13.3, Figure 1.
   a. Describe the model system that Marraffini and Sontheimer developed to address the CRISPR target question.
   b. How did the investigators alter the model system described in (a) in order to determine if the CRISPR sequence targeted the nickase DNA or the nickase RNA?
   c. What would be the expected result if the CRISPR system targeted the DNA? The RNA? What was the result?

3. Look at Experimental approach 13.3, Figure 2.
   a. Describe the assay system used by Terns to address the same question in archaeal species.
   b. In Figure 2, what do the 1, 2, and 3 represent?
   c. In Figure 2, what do the symbols '-' and '+' represent?
   d. What conclusion could be drawn based on the results shown in this figure?

13.1 OVERVIEW OF REGULATORY RNAs

Atkins JF, Gesteland RF, Cech TR Eds. *RNA Worlds: From Life's Origins to Diversity in Gene Regulation*. Cold Spring Harbor, NY: Cold Spring Harbor Laboratory Press, 2011.

13.2 BACTERIAL BASE-PAIRING sRNAs

Storz G, Vogel J, Wassarman KM. Regulation by small RNAs in bacteria: expanding frontiers. *Molecular Cell*, 2011;**43**:880–891.

13.3 EUKARYOTIC sRNAs: miRNAs, siRNAs, AND rasiRNAs

Shirayama M, Seth M, Lee H-C, *et al.* piRNAs initiate an epigenetic memory of nonself RNA in the *C. elegans* germline. *Cell*, 2012;**150**:65–77.

13.4 PROCESSING OF EUKARYOTIC sRNAs

Kim VN, Han J, Siomi MC. Biogenesis of small RNAs in animals. *Nature Reviews*, 2009;**10**:126–139.

Hutvagner G, Simard MJ. Argonaute proteins: key players in RNA silencing. *Nature Reviews Molecular Cell Biology*, 2008;**9**:22–32.

13.5 LOADING OF ARGONAUTE AND PIWI PROTEINS WITH EUKARYOTIC sRNAs

Czech B, Hannon GJ. Small RNA sorting: matchmaking for Argonautes. *Nature Reviews Genetics*, 2011;**12**:19–31.

Schalch T, Job G, Shanker S, Partridge JF, Joshua-Tor L. The Chp1-Tas3 core is a multifunctional platform critical for gene silencing by RITS. *Nature Structural & Molecular Biology*, 2011;**18**:1351–1357.

13.6 GENE SILENCING BY SMALL EUKARYOTIC RNAs

Djuranovic S, Nahvi A, Green R. A parsimonious model for gene regulation by miRNAs. *Science*, 2011;**331**:550–553.

Ishizu H, Siomi H, Siomi MC. Biology of PIWI-interacting RNAs: new insights into biogenesis and function inside and outside of germlines. *Genes and Development*, 2012;**26**:2361–2373.

Joshua-Tor L, Hannon GJ. Ancestral roles of small RNAs: An Ago-centric perspective. *Cold Spring Harbor Perspectives in Biology*, 2011;**3**:1–11.

13.7 VIRAL DEFENSE ROLE OF BACTERIAL, ARCHAEAL, AND EUKARYOTIC sRNAs

Cerutti H, Casas-Mollano JA. On the origin and functions of RNA-mediated silencing: from protists to man. *Current Genetics*, 2006;**50**:81–99.

Ding SW, Voinnet O. Antiviral immunity directed by small RNAs. *Cell*, 2007;**130**:413–426.

Wiedenheft B, Sternberg SH, Doudna JA. RNA-guided genetic silencing systems in bacteria and archaea. *Nature*, 2012;**482**:331–338.

13.8 RNA-MEDIATED REGULATION IN *cis*

Breaker RR. Prospects for riboswitch discovery and analysis. *Molecular Cell*, 2011;**43**:867–879.

Serganov A, Nudler E. A decade of riboswitches. *Cell*, 2013;**152**:17–24.

13.9 PROTEIN-BINDING REGULATORY RNAs

Kugel JF, Goodrich JA. Non-coding RNAs: key regulators of mammalian transcription. Trends in Biochemical *Sciences*, 2012;**37**:144–151.

13.10 LONG INTERGENIC NON-CODING RNAs

Guttman M, Rinn JL. Modular regulatory principles of large non-coding RNAs. *Nature*, 2012;**482**:339–346.

Wutz A. RNA-mediated silencing mechanisms in mammalian cells. *Progress in Molecular Biology and Translational Science*, 2011;**101**:351–376.

Mattick JS. Deconstructing the dogma: A new view of the evolution and genetic programming of complex organisms. *Annals of the New York Academy of Science*, 2009;**1178**:29–46.

Protein modification and targeting

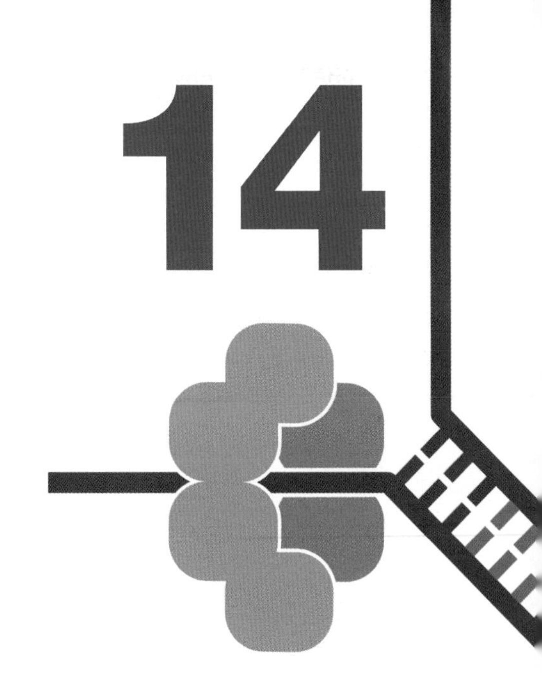

INTRODUCTION

In Chapters 11 and 12, we learned how proteins are synthesized, and how this process is regulated to determine when a given gene product is produced. However, many polypeptide chains are not fully functional upon their release from the ribosome. The proper three-dimensional folding of numerous polypeptides is only achieved with the help of accessory proteins. In addition, a protein may only be needed in a specific location in the cell and thus must be transported to this site after synthesis. The covalent addition of lipids may also be required for a protein to be associated with a membrane. Furthermore, many proteins need to be cleaved or modified by the addition of sugars to become active.

As we have already seen in a number of examples in this book, the activity of a protein can also be modulated by the addition and subsequent removal of one or more small molecule adducts such as phosphate. The covalent additions of lipids, sugars, and small molecule adducts are termed **post-translational modifications**. Because so many chemical changes can occur in a polypeptide chain after it has been synthesized, simply knowing the complete genome sequence of an organism, and hence the amino acid sequence of all translated proteins, is not enough to provide a complete picture of the true diversity of proteins in a cell.

Post-translational modifications do not always enhance a protein's activity. Proteins can undergo detrimental covalent modifications as well as misfolding, which in some cases can lead to disease. Degradation of proteins to yield their component amino acids is also utilized to maintain the proper concentration of a particular protein in the cell, a process that is generally regulated. In this chapter, we explore the range of fates for which different proteins may be destined, and the mechanisms by which they occur.

14.1 CHAPERONE-ASSISTED PROTEIN FOLDING

Chaperone proteins assist the proper folding of many proteins

A protein must adopt the correct three-dimensional structure in order to carry out its function. While the primary sequence of a protein contains all of the information required for it to adopt the correct three-dimensional structure, the sequence alone usually is not enough to direct proper folding in an appropriate time span within the context of a cell. Instead, protein folding in the cell often requires the assistance of a class of proteins known as **chaperones** which, like their adult Victorian counterparts, prevent illicit interactions. Chaperone proteins have multiple functions: they prevent a polypeptide from folding too early, aid the correct folding at the proper time, promote the correct folding of a misfolded polypeptide, and assist the unfolding of certain proteins.

Protein misfolding is associated with several neurodegenerative diseases

Several features of protein synthesis and the cellular environment make chaperone-assisted protein folding necessary. The final fold of some proteins contains interactions between amino acid residues that are widely separated in the polypeptide chain. However, the directional nature of protein synthesis - from the N- to the C-terminus - means that folding of the polypeptide chain may begin as the N-terminus of the protein is extruded from the ribosome, long before the protein has been completely synthesized. Thus, if improper associations occur between neighboring regions of the polypeptide chain before the protein is fully synthesized, the protein will be trapped in a misfolded state.

Another factor impacting protein folding is the remarkably crowded nature of the intracellular environment, with protein concentrations of approximately 300 mg/mL. These crowded conditions increase the likelihood of inappropriate interactions between partially folded polypeptides. For example, we learned in Section 2.3 that the driving force for protein folding is the sequestration of hydrophobic residues in the core of the folded protein, where these residues form favorable interactions with one another and are protected from water. However, such hydrophobic regions can readily interact with one another in inappropriate ways early in the folding process - that is, before they have folded to form the intended hydrophobic core - instead forming undesirable, insoluble protein aggregates.

Insoluble protein aggregates are observed in a number of neurodegenerative diseases, such as Alzheimer's disease and Parkinson's disease. Another example is the group of spongiform encephalopathies, such as 'mad cow' disease, which are caused by the prion protein. In these neurodegenerative diseases, a misfolded, abnormal form of the prion protein induces the conversion of normal forms of the protein into the abnormal state, again leading to aggregates.

Given the extremely detrimental consequences of protein aggregation, how is misfolding prevented?

Co-translational folding is facilitated by ribosome-associated chaperones

Nascent polypeptide chains begin to emerge from the exit tunnel of the ribosome after approximately 30 amino acids have been covalently joined to one another. The ribosomal exit tunnel is composed of both ribosomal RNA (rRNA) and ribosomal proteins. It is thought that this composition makes the tunnel chemically diverse, and thus inert enough to prevent most peptide sequences from associating with it too strongly. The tunnel has a width of about 15 Å and can potentially accommodate an alpha helix, but not more globular protein structures. So, while it is thought that some local structures, such as alpha helices, begin to assemble within the exit tunnel, proteins generally emerge from the ribosome in an extended conformation. As the peptides emerge, they immediately encounter a number

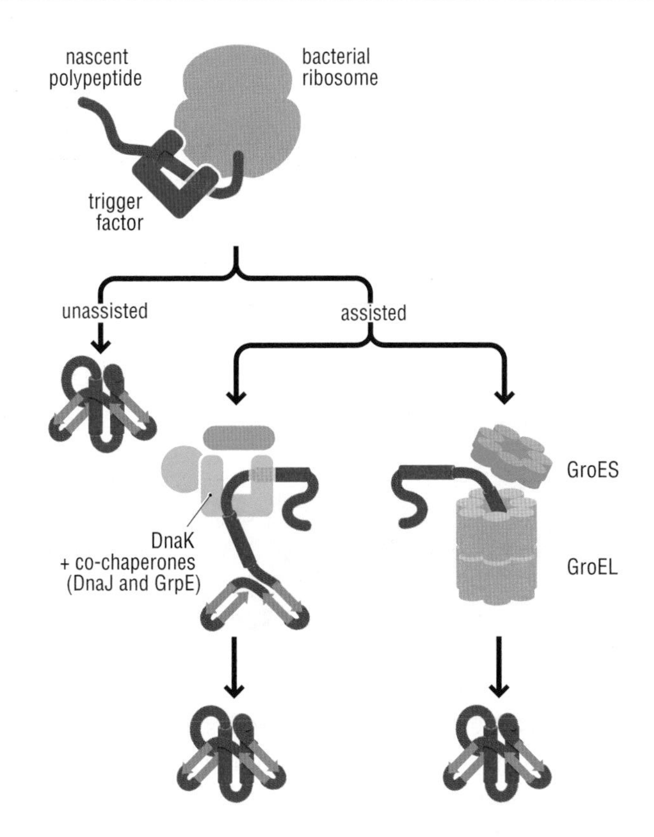

Figure 14.1 Protein chaperones in bacteria. The trigger factor chaperone interacts with and cradles the nascent polypeptide as it emerges from the ribosome. Although some proteins fold autonomously after release from the ribosome, most require the actions of the ATP-dependent Hsp70 or Hsp60 chaperones. The Hsp70 proteins (DnaK) act like a clamp to hold the hydrophobic part of the protein whereas the Hsp60 proteins (GroES + GroEL) form a barrel that provides a protected environment for folding. Examples are shown for bacteria, but homologs or analogs of all proteins are present in eukaryotic cells.

Adapted from Deuerling, E, and Bukau, B. Chaperone-assisted folding of newly synthesized proteins in the cytosol. *Critical Reviews in Biochemistry and Molecular Biology*, 2004; **39**:261–277.

of chaperone proteins, which in some cases are physically associated with ribosomes. Binding of the emerging nascent chain by these chaperones, as illustrated in Figure 14.1, prevents immediate (and potentially inappropriate) folding.

In bacteria, the nascent chain makes contact with the ribosome-associated chaperone known as trigger factor after about the first 100 residues of the polypeptide chain have been synthesized. Trigger factor protects the emerging polypeptide against misfolding and aggregation by using a hydrophobic region on its surface to interact with hydrophobic sequences in the nascent polypeptide, thus providing an environment that shields these regions from inappropriate interactions with other hydrophobic regions of the poly peptide or in other proteins. Eukaryotes have multiple chaperones linked to protein synthesis (in a process denoted as chaperone-linked protein synthesis (CLIPS)) including both adenosine triphosphate (ATP)-dependent and ATP-independent chaperones. One of these, the nascent chain-associated complex (NAC), is an ATP-independent chaperone that associates with most nascent chains and protects them against misfolding. Another ATP-dependent chaperone belonging to the Hsp70-class, which we will learn about next, associates with aggregation-prone, nascent chains in a manner that is regulated by the ribosome-associated complex (RAC), itself comprised of Hsp70- and Hsp40-class chaperones. The specificities and functions of these eukaryotic ribosome-associated chaperones are less well understood than for the bacterial trigger factor, but the additional complexity reflects the need to facilitate the folding of many more, sometimes extremely large, proteins.

Post-translational folding is assisted by Hsp70 and Hsp60 chaperones

Proteins usually have to pass through several distinct stages of folding to be able to adopt their final three-dimensional shape. Consequently, after the early

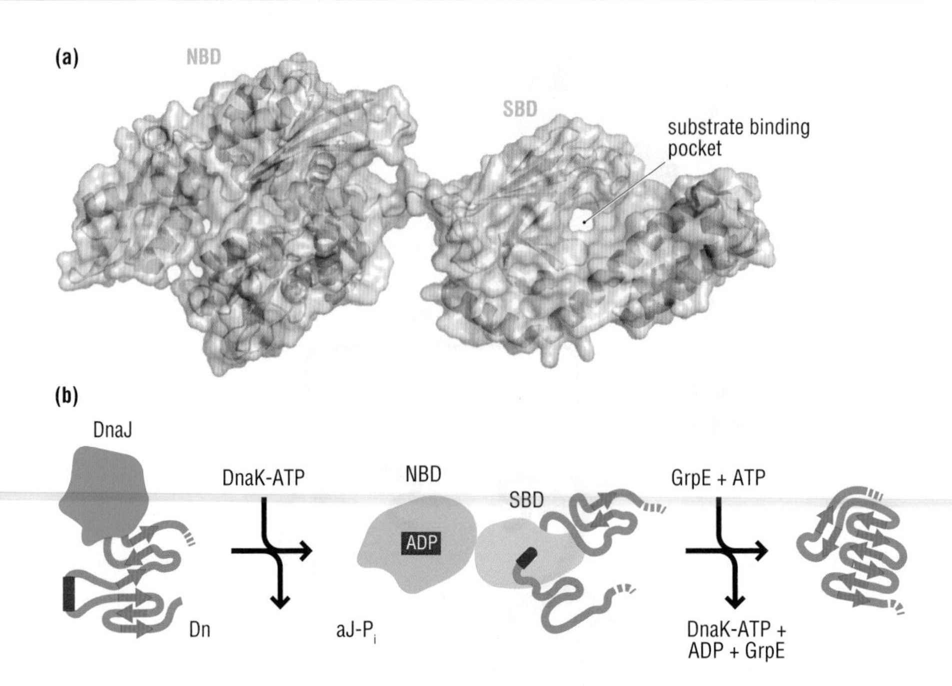

Figure 14.2 Unfoldase function of Hsp70-family chaperone DnaK. (a) Structure of DnaK showing the nucleotide-binding domain (NBD), which binds and hydrolyzes ATP, and substrate-binding domain (SBD), which clamps the substrate (Protein Data Bank (PDB) code 2KHO). (b) Mechanism of DnaK-mediated unfolding. DnaJ binds to the misfolded region (red) of a substrate protein (gray) and delivers the protein to ATP-loaded DnaK thereby stimulating ATP hydrolysis. As a result, the SBD tightly clamps the misfolded region of the structure eliminating secondary structure. Upon binding of GrpE, ADP release and rebinding of ATP, the substrate protein is ejected from DnaK and allowed to refold into its proper native conformation.

Adapted from Baneyx F and Nannenga BL. *Nature Chemical Biology*, 2010;**6**:880–1.

engagement with ribosome-associated chaperones, many proteins still require the assistance of other chaperones – which promote productive over unproductive folding steps – to reach their final folded state (see Figure 14.1). Several conserved classes of chaperone are found in all organisms. Most of these chaperones are known as heat-shock proteins because many of them were first identified as proteins that are strongly induced when cells are exposed to elevated temperature. This heat-shock induction suggests that chaperones are needed to prevent proteins from becoming unfolded or denatured by heat and to refold the proteins if they do become denatured. The best-known classes are Hsp70 and Hsp60, where the number corresponds loosely to the molecular weight of the heat-shock protein.

In general terms, the workings of the Hsp70 and Hsp60 classes of chaperone are quite similar: the chaperones recognize and bind transiently to hydrophobic regions on the non-native polypeptide, with ATP driving the conformational changes required for the chaperone to bind and then release the polypeptide. As illustrated in Figure 14.2, members of the Hsp70 class of proteins, which include the bacterial DnaK protein, are monomeric and act like a clamp that firmly holds an extended hydrophobic segment of polypeptide chain. Upon ATP binding, the chaperone releases the polypeptide chain, giving it a chance to fold properly. In bacteria, these steps are assisted by co-chaperone proteins such as DnaJ and GrpE.

The Hsp60 chaperones, called chaperonins, comprise a structurally distinct class of proteins that form oligomeric barrel-like structures with detachable 'lids' (depicted in Figure 14.1). The cavity of the barrel is lined with hydrophobic residues, which capture the non-native protein through the exposed hydrophobic surfaces. The binding of ATP and the 'lid' changes the character of the cavity wall from hydrophobic to hydrophilic. As a result, the protein captured in the closed cavity is released from the wall and can now fold as a monomeric protein, protected from aggregating with other cellular components. Upon ATP hydrolysis, the lid opens and the folded protein is released. Examples of Hsp60 chaperones include the bacterial GroEL-ES proteins and TRiC or CCT chaperones in eukaryotes.

Once a protein reaches its stable, folded state, the hydrophobic surfaces are largely buried and the protein is thus no longer recognized by the chaperones or by the degradation machinery of the cell. Deciphering the complex patterns that

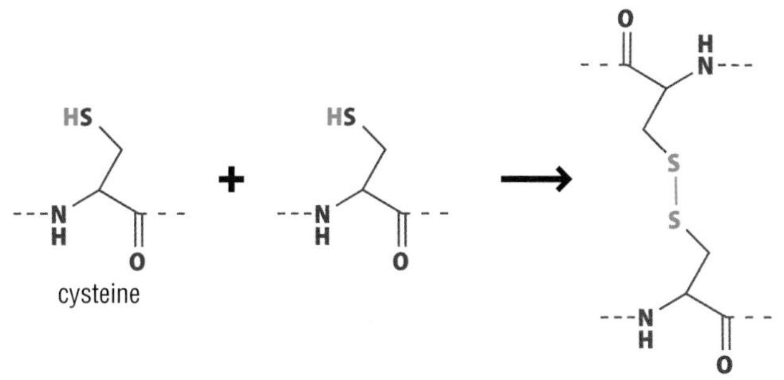

Figure 14.3 Disulfide bond formation. The thiol groups of two cysteine residues are oxidized to form a covalent bond between the two sulfur atoms.

specify the interactions between chaperones and nascent or unfolded peptides remains a significant challenge.

Disulfide bond formation depends on oxidative reactions

A critical step in the folding of some proteins is the formation of one or more **disulfide bonds** that covalently link pairs of sulfur-containing cysteine residues. Cysteine residues can exist stably either in a reduced form known as a thiol (-SH) or in an oxidized form in which one cysteine residue forms a covalent S-S bond, a disulfide bond, with another cysteine. The formation of a disulfide bond is shown in Figure 14.3. Since the cytosol of the cell tends to be a reducing environment, disulfide bonds are rarely found in cytosolic proteins. However, the bacterial periplasm and the eukaryotic endoplasmic reticulum are oxidizing compartments that favor the formation of disulfide bonds. Thus disulfide bonds are very common in secreted proteins, and the correct formation of these bonds is part of proper folding.

Not surprisingly, cells contain enzymes and organic molecules that catalyze the formation or rearrangement of disulfide linkages to favor the folded native state of the protein, as depicted in Figure 14.4. Functionally similar catalysts have been discovered in both bacteria and eukaryotes. Specifically, two different types of

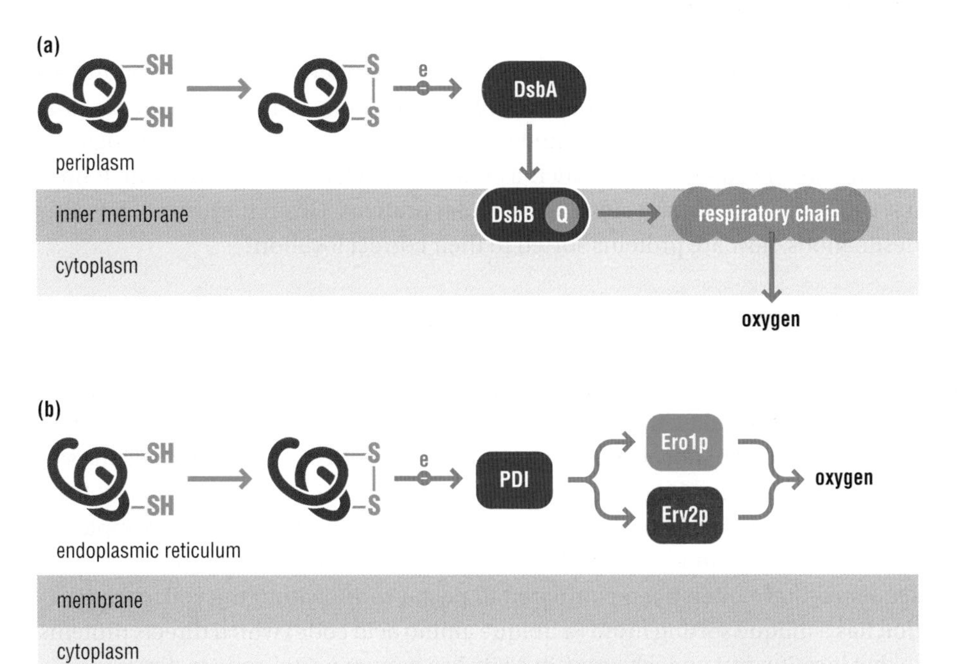

Figure 14.4 Proteins required for protein disulfide bond formation. (a) In bacteria, the oxidoreductase DsbA catalyzes disulfide bond formation in the periplasm. DsbA is reoxidized by DsbB, which transfers the reducing equivalents (electrons) to coenzyme Q. Coenzyme Q transfers the electrons to the respiratory chain and ultimately to oxygen during aerobic growth. (b) In eukaryotes, PDIs catalyze disulfide bond formation in the endoplasmic reticulum. In yeast, the PDI is reduced by FAD-dependent oxidases Ero1p and Erv2p. Again the reducing equivalents are ultimately transferred to oxygen.

➔ We learn more about oxidation and reduction in Section 3.4.

oxidative reaction are carried out: sulfhydryl oxidation results in the *de novo* formation of disulfide bonds, while the isomerization of disulfide bonds allows for the swapping of disulfide bonds until the correct folded state is achieved. In bacteria, the protein DsbA catalyzes sulfhydryl oxidation to facilitate *de novo* disulfide bond formation, while a second protein, DsbC, is responsible for the isomerization of disulfides. In eukaryotic cells, protein disulfide isomerases (PDIs) have been shown to carry out both reactions.

In order to carry out the cysteine oxidation reactions, the thiol-disulfide oxidoreductases DsbA and PDI must themselves be reoxidized after each reaction. This is achieved by the transfer of reducing equivalents (electrons) from the oxidoreductase to other oxidases such as DsbB in bacteria and Ero1p and Erv2p in yeast. The reducing equivalents are ultimately transferred from these proteins to the electron transport chain.

14.2 TARGETING OF PROTEINS THROUGHOUT THE CELL

In Chapter 1, we learned that the cell has many different compartments. In this section, we discuss how the subcellular localization of a protein can affect its function, and how proteins are targeted to the proper location.

Many proteins need to be targeted to specific regions of the cell

Subcellular targeting by the cell allows for proteins to be involved in distinct processes in different areas of the cell. Localization of proteins also increases their effective concentration by confining many copies of a single protein to a defined part of the cell, thus helping to drive the reactions and interactions that the protein is involved in by mass action. Distinct isoforms of a protein can be differentially localized, thus increasing the number of cellular functions for a gene, without the burden of evolving novel genes.

In bacterial cells, certain proteins must be targeted to the membrane of the cell or to the extracellular environment to perform their function. In the more complicated eukaryotic cells, proteins are targeted to a number of membrane-bound organelles, such as the nucleus, the endoplasmic reticulum, and the mitochondria. In addition to the membrane-bound structures, eukaryotes contain cytologically distinguishable substructures, such as nuclear speckles and cytoplasmic P bodies, which contain distinct subsets of cellular proteins. Given the array of possible destinations, how are proteins sorted to their correct location?

Sorting motifs direct proteins to the proper subcellular compartments

Cells contain a number of different sorting and targeting machineries that dispatch proteins to specific subcellular locations. **Sorting motifs**, also denoted sorting signals or, in some cases, signal sequences, are sequence identifiers within proteins that target the proteins to the appropriate localization machinery. Examples of sorting motifs found in bacteria and eukaryotes are given in Figure 14.5. These sequences have often been compared to postal codes within the cell: each location has a unique sorting motif (a unique amino acid code) which directs proteins to that location just as each street in a city has its own postal code to direct mail.

Examples of sorting motifs		
bacteria		
inner membrane	amino-terminus	positively charged/hydrophobic/polar
eukaryotes		
endoplasmic reticulum	amino-terminus	positively charged/hydrophobic/polar
nuclear import	amino-terminus	positively charged (PKKKRKV)
nuclear export	amino-terminus	leucine-rich (LQLPPLERLTL)
mitochondria	amino-terminus	positively charged amphiphilic α helixes
peroxisomes	carboxy-terminus	SKL
ER retention	carboxy-terminus	NPXY or YXXØL; [DE]XXXXL[LI] or DXXLL
lysosomes + endosomes	carboxy-terminus	KDEL

Figure 14.5 Examples of sorting motifs. Example sequences are given in parentheses. The motifs that direct proteins to the endoplasmic reticulum in eukaryotes and the inner membrane in bacteria are often referred to as signal sequences.

Sorting motifs may be present at either the N- or C-terminus of a protein (or both) and can comprise specific amino acids, or a patch of amino acids with specific features. For example, the related secretion machineries in bacteria and eukaryotes recognize a series of hydrophobic amino acids. Often the N-terminal signal sequences are cleaved from proteins during their translocation through the membrane, so that the sorting motif is not present in the mature protein. In addition, some proteins have multiple sorting motifs. Thus a protein may have both a sequence that targets the protein for insertion into the endoplasmic reticulum as well as a second sequence that signals retention in the endoplasmic reticulum upon its arrival there.

The importance of the cellular sorting process can be illustrated by looking at the consequences of mutations in the sorting motifs themselves. For example, in one form of familial hypercholesterolemia, an inherited condition in which an accumulation of abnormally high levels of low-density lipoprotein (LDL) in the bloodstream leads to premature atherosclerosis, a cysteine replaces tyrosine in the sorting motif of the LDL receptor. This single mutation impairs transport of the LDL receptor to its proper subcellular location. As a consequence, the cholesterol-containing lipoprotein particles cannot be cleared from the bloodstream. The resulting deposits that accumulate on artery walls can lead to premature heart attacks.

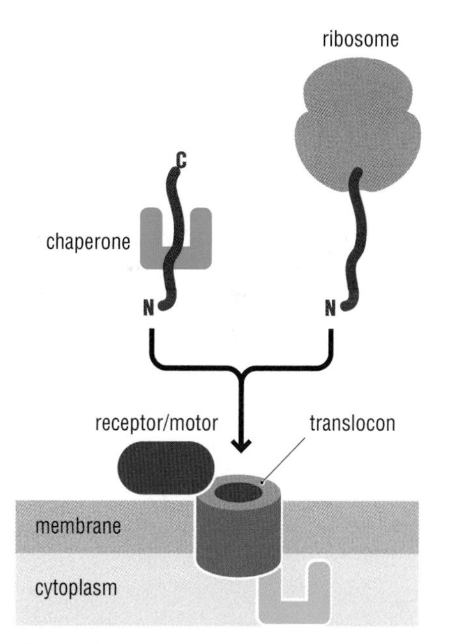

Figure 14.6 General mechanism of protein translocation across a membrane. The sorting motif (red) of the protein to be translocated can be recognized by the receptor as the protein is being translated or after synthesis is complete. If the protein is folded, it may need to be unfolded by a chaperone protein. After the signal sequence is recognized and the protein is brought to the membrane, it is threaded through the translocon. The protein can be pushed across the membrane as it is being extruded from the ribosome or it may be pulled across the membrane by the refolding promoted by chaperones on the other side of the membrane.

Protein targeting is determined by multiple distinct machineries in the cell

The trafficking of proteins into the correct subcellular compartment requires two processes to occur. First, the sorting motif must be recognized and the protein must be transported to the proper membrane. Then, the protein must be translocated - moved into or across - the membrane. The composition of the transport and translocation complexes varies from one compartment to another, but all carry out the same basic steps as illustrated in Figure 14.6.

Targeting can be initiated either as the protein is being synthesized (co-translationally) or following synthesis (post-translationally). After the protein with the sorting motif is recognized and transported to the proper membrane, it is transferred to the proteins carrying out the translocation step. In many cases, this transfer step is associated with a conformational change that opens the channel or pore through which the protein is threaded. Finally, the protein must be inserted into the membrane. This can be accomplished by 'pushing' the protein across the membrane, with the energy provided by the extrusion of the nascent polypeptide from the ribosome. Alternatively, the protein may be 'pulled' across the membrane, with the energy provided by the chaperone proteins that refold the protein as it is extruded on the other side of the membrane.

Protein targeting in bacteria is relatively simple and principally involves the secretion apparatus, which directs most membrane-localized and secreted proteins to their final destinations. It is estimated that up to 40% of proteins synthesized in the bacterial cell are targeted using this basic machinery. By contrast, eukaryotic protein targeting mechanisms are considerably more diverse, requiring trafficking of proteins from their site of synthesis in the cytoplasm to a much wider range of membrane-enclosed compartments. Let us now consider some examples of the strategies used to transport proteins into these different destinations.

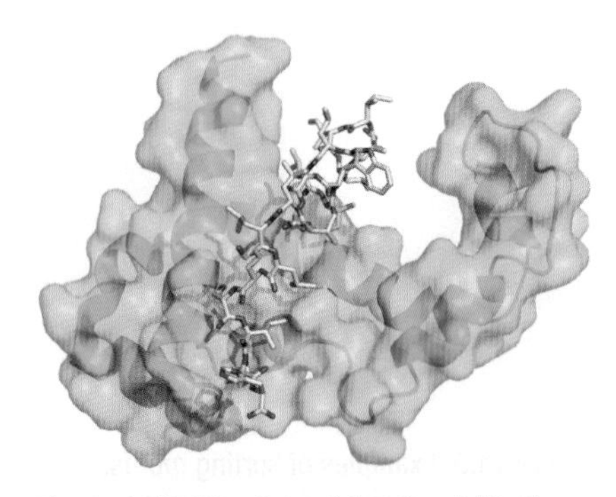

Figure 14.7 Structure of the signal-binding component of the SRP bound to the signal peptide. Surface depiction of *Thermus aquaticus* SRP54 protein shows groove where the signal recognition sequence (gray sticks) is bound. PDB code 3KL4.

Secreted proteins are recognized and targeted by related machineries in bacteria and eukaryotes

The machinery that targets many proteins to the endoplasmic reticulum and beyond in eukaryotes is evolutionarily related to the secretion machinery in bacteria. Both systems depend on a hydrophobic signal sequence that is recognized by signal recognition particles (SRPs) that greet the polypeptide chain as it emerges from the exit tunnel of the ribosome. As shown in Figure 14.7, these SRPs have a pocket lined with hydrophobic residues, thus making this binding pocket especially suited to broad recognition of hydrophobic amino acids. The SRP in both systems is composed of both RNA and protein and depends on guanosine triphosphate (GTP) hydrolysis for downstream steps in the translocation pathway.

Once it binds to the signal sequence, the SRP interacts with the SRP receptor located either on the inner membrane of bacteria or on the surface of the endoplasmic reticulum where the nascent polypeptide is handed over to the ATP-driven translocase - SecA in bacteria and Sec61 in eukaryotes. This translocase then guides the polypeptide through a channel into the periplasm of the bacteria or the lumen of the endoplasmic reticulum.

Other membrane-embedded translocases operate post-translationally to transport unfolded proteins

Unlike the secretion machinery described above, many cellular targeting machineries recognize and target proteins upon release from the ribosome. For example, the mitochondrial targeting machinery recognizes an amphipathic helix, with positively charged amino acids on one face and hydrophobic residues on the other, in a fully synthesized protein. The targeting machinery then engages the protein with the energy-dependent translocation machinery, transporting the proteins into the mitochondrion. A key feature of such membrane-embedded translocases is that they typically transport *unfolded* proteins. The chaperone proteins that we described in the previous section (the Hsp70 family) seem to play an important role both in unfolding the proteins before they are engaged by the translocation machinery and in refolding the proteins within the mitochondria.

Movement into and out of the nucleus depends on a gated transport process

Transport of proteins into and out of the nucleus is highly regulated. In the case of ribosome assembly, ribosomal proteins that are synthesized in the cytoplasm must be brought together with the rRNA, which is located in a nuclear sub-region known as the nucleolus. The assembled ribosome must then be exported to the cytoplasm, where translation takes place. It is worth pausing for a moment to reflect on the level of movement needed here: the protein-based ribosome components are synthesized in the cytoplasm, after which they must be transported to the nucleus to assemble with their RNA counterparts, before being transported *back* to the cytoplasm where the complete ribosome has its activity. Such shuttling back and forth between cellular compartments demonstrates just how extensive a process the trafficking of proteins (and other biological molecules) is.

Macromolecules are transported into and out of the nucleus through structures referred to as nuclear pore complexes (NPCs) that are more elaborate than the translocases discussed above. While the nuclear pore is relatively permeable to small molecules, it is quite selective as to which large molecules it lets through. Protein import depends strictly on the recognition of nuclear localization sequences, which are rich in positively charged amino acids. The nuclear localization sequence (NLS) is recognized by soluble nuclear import receptors that associate with the nucleoporin proteins that extend into the cytoplasm from the nuclear pore complex, as depicted in Figure 14.8. Once bound by the receptors, the imported proteins pass through a gated aqueous pore, without unfolding. Nuclear export appears to work in a similar fashion, although it depends on a different set of signals and receptors. The nuclear pore can be as large as 26 nm in

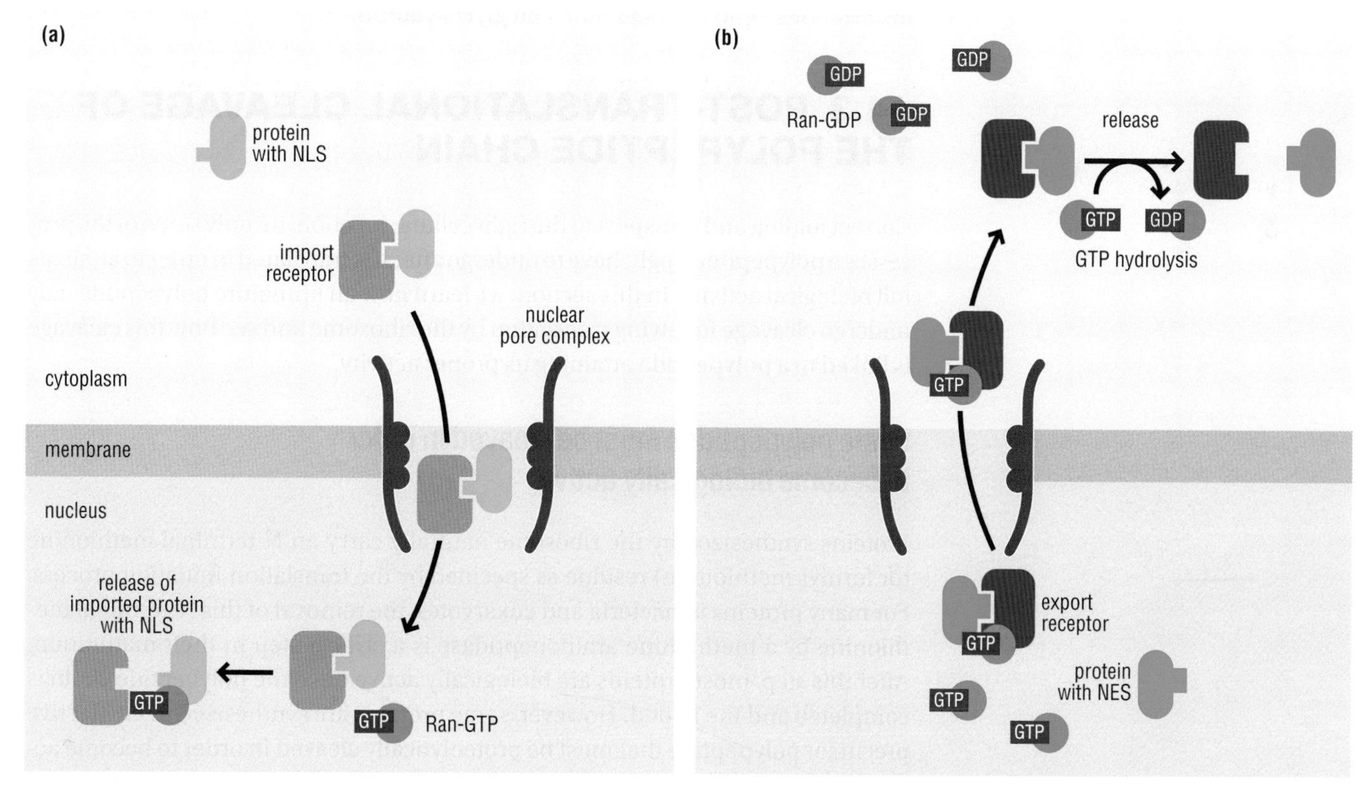

Figure 14.8 Transport in and out of the eukaryotic cell nucleus. (a) For proteins trafficking into the nucleus, the NLS is recognized by soluble nuclear import receptors (light blue) that also bind to components of the nuclear pore complex. After passing through the aqueous pore, the GTP-bound Ran protein stimulates release of the imported protein from the receptor. (b) For proteins trafficking out of the nucleus, binding of an export receptor (dark blue) to a protein with a nuclear export signal (NES) is promoted by Ran-GTP. This tri-partite complex passes through the pore, whereupon release of the exported protein from the receptor is stimulated by hydrolysis of the GTP bound to Ran. Ran-GDP then is transported back into the nucleus by the specialized transport factor NTF2 (not depicted).

diameter, a dimension that can readily accommodate the ribosomal particles that must pass through in their assembled state.

Since molecules are transported through the nuclear pore in both directions, how is directionality established? This is the function of the GTPase Ran, a molecular switch that is found predominately in the GTP-bound form in the nucleus and in the guanosine diphosphate (GDP)-bound form in the cytoplasm. In the nucleus, Ran-GTP promotes the release of imported proteins from their receptors and the binding of proteins to be exported to their receptors, as illustrated in Figure 14.8. That is, Ran-GTP facilitates the drop-off of imported proteins once they reach their intended destination, and the pick-up of proteins for export, preparing them to begin their journey out of the nucleus. In the cytoplasm, Ran hydrolyzes GTP to GDP, promoting the release of the exported protein from the export receptor. Another protein called nuclear transport factor 2 (NTF2) transports Ran-GDP to the nucleus, where the GDP on Ran is exchanged for GTP, thus allowing the cycle to repeat itself.

Covalent modifications can be used to target proteins to different locations

Though the molecular machineries involved in these different protein localization events are different, each system depends on intrinsic signals encoded within the primary sequence of the protein itself. These distinct signals allow the cellular machineries to accurately select and target the wide range of proteins that must reach specialized cellular locations. In Sections 14.4 and 14.5, we will learn about two other strategies utilized by the cell to localize proteins to different subcellular localizations, lipid modification, and **glycosylation**.

14.3 POST-TRANSLATIONAL CLEAVAGE OF THE POLYPEPTIDE CHAIN

Correct folding and transport to the right cellular location are only some of the processes a polypeptide might have to undergo after it is translated in order to attain its full biological activity. In this section, we learn how an immature polypeptide may undergo cleavage following translation by the ribosome and see how this cleavage is linked to a polypeptide attaining its proper activity.

Some polypeptides must be cleaved in order to become biologically active

Proteins synthesized by the ribosome naturally carry an N-terminal methionine (or formyl-methionine) residue as specified by the translation initiation process. For many proteins in bacteria and eukaryotes, the removal of this N-terminal methionine by a methionine aminopeptidase is a critical step in their maturation. After this step, most proteins are biologically active once the polypeptide chain is completed and the folded. However, some proteins are synthesized as an inactive precursor polypeptide that must be proteolytically cleaved in order to become active. The insertion of such a processing step into the lifetime of a protein provides another point at which the cell can regulate gene expression. The following examples illustrate why it is beneficial to produce certain proteins in an inactive state, and to specifically activate them only when they reach an appropriate location or when their activity is specifically required.

Proteolytic processing is widely used for the timely activation of polypeptides

There are many examples of enzymes that cleave polypeptide chains but must themselves first be proteolytically processed to become active. Chymotrypsin is a digestive enzyme in the gut, but must be prevented from digesting the pancreatic cells that produce it. To accomplish this, the enzyme is translated as an inactive precursor called chymotrypsinogen in pancreatic cells. The precursor lacks enzymatic activity because the uncleaved portions of the enzyme prevent the precursor from folding into a catalytically active configuration. Once released from pancreatic cells, chymotrypsinogen is transported to the gut where a protease cleaves the inactive precursor at specific amino acid positions to generate the active chymotrypsin enzyme. This cleavage process is illustrated schematically in Figure 14.9.

The hormone insulin provides another example of how proteolytic processing is required for activation. Insulin triggers glucose uptake from the blood into tissues such as liver or muscle. Figure 14.10 shows how an inactive preproinsulin precursor contains a signal sequence that directs its co-translational import into the endoplasmic reticulum, where the signal sequence is cleaved off and three disulfide bonds are formed. Subsequent cleavage by three additional proteases gives rise to active, secreted insulin. The many steps required to generate the hormone provide multiple opportunities for regulation. In patients with type 1 (juvenile) diabetes, insulin is no longer produced and processed because of immune-mediated destruction of insulin-secreting pancreatic beta cells.

A third example of proteolytic processing is found in the retrovirus family that includes the human immunodeficiency virus (HIV). In these viruses, a number of essential proteins are synthesized as a single precursor **polyprotein**. In Chapter 11, we learned how the translation of the polyprotein is regulated by a programmed frameshift. For this discussion, it is important to know that the precursor polypeptide contains domains corresponding to a number of distinct proteins, including a reverse transcriptase that uses viral RNA as a template to synthesize a DNA copy, and an integrase that integrates the viral genome into that of the host cell. After the polyprotein is synthesized, the retroviral-encoded protease cleaves the polypeptide chain to release the individual functional domains, as shown in Figure 14.11. This process is particularly relevant to molecular pharmacology, since a number of successful anti-retroviral therapies depend on drugs that inhibit the protease and thus prevent the release of the individual, active viral proteins.

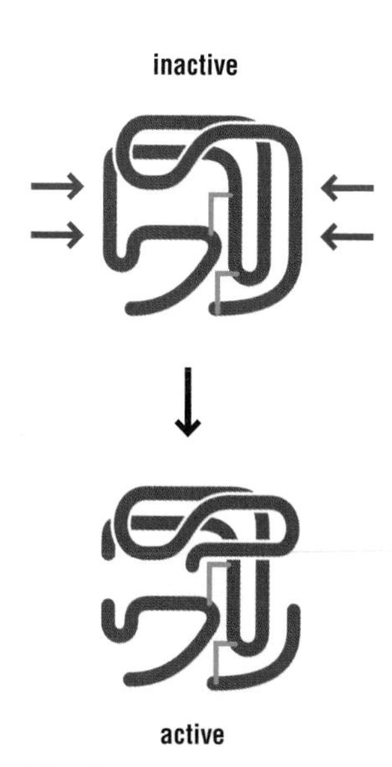

Figure 14.9 Chymotrypsin activation by specific cleavage. The chymotrypsin enzyme is translated as an inactive precursor called chymotrypsinogen, which is cleaved by a protease at specific amino acid positions to generate the active enzyme. The three fragments generated by the proteolysis remained connected by disulfide bonds.

Some proteins contain self-excising domains called inteins

Several proteins found in bacteria, archaea, and single-celled eukaryotes contain an internal domain referred to as an **intein**, which must be excised from the final expressed polypeptide to yield an active protein product. Like intron removal by self-splicing RNA, the intein catalyzes its own excision from the translated polypeptide and ligates the flanking polypeptides to form a modified protein product; this excision and ligation is illustrated in Figure 14.12a. The flanking protein sequences are referred

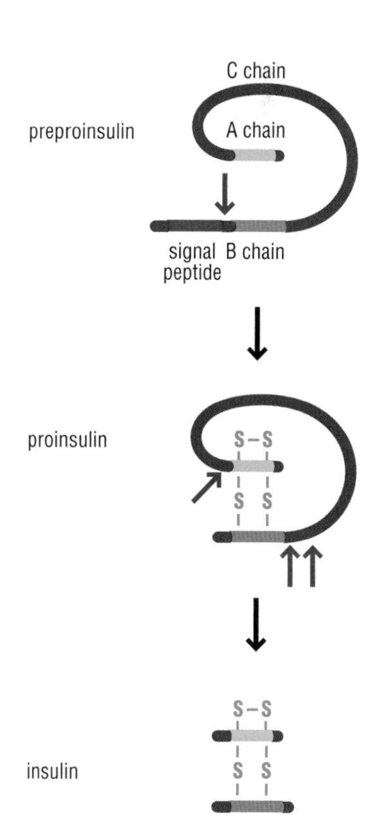

Figure 14.10 Insulin maturation by specific cleavage and disulfide bond formation. The inactive preproinsulin precursor contains a signal sequence (red), which directs the co-translational import of the protein into the endoplasmic reticulum. There the signal sequence is cleaved off, and three disulfide bonds are formed. Subsequent, cleavage by three additional proteases gives rise to active insulin.

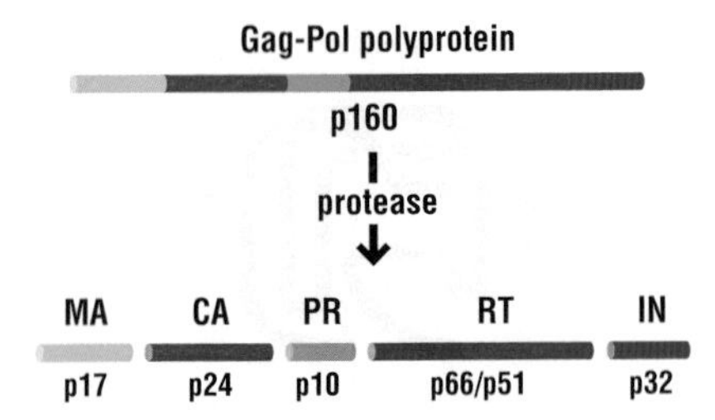

Figure 14.11 Proteolytic processing of the HIV polyprotein. Five viral proteins (MA, matrix; CA, capsid; PR, protease; RT, reverse transcriptase; and IN, integrase) synthesized as part of a single precursor polyprotein are released by specific proteolytic cleavage.

➔ We discuss the self-splicing of RNA in Section 10.5.

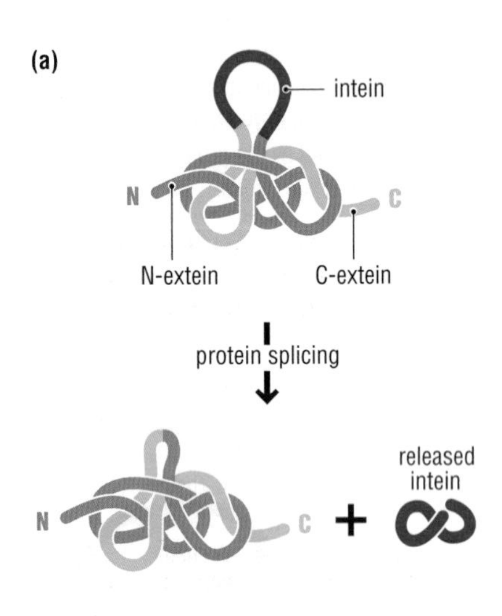

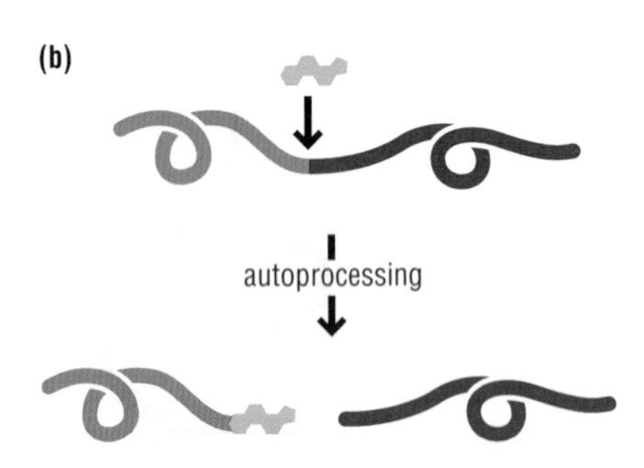

Figure 14.12 Autoprocessing reactions catalyzed by proteins. (a) Self-excision catalyzed by inteins. Key nucleophilic attacks are catalyzed by conserved serine, cysteine, or threonine residues. (b) Autoprocessing reaction catalyzed by the Hedgehog protein. In this case, the key nucleophilic attack is catalyzed by an exogenous cholesterol molecule.

to as **exteins** (with analogy to exons). In this way, two proteins, the ligated exteins and the excised intein polypeptide, can be derived from a single polypeptide chain.

The intein is not always junk that is discarded following excision. It often encodes an endonuclease domain that specifically excises the DNA sequence that encodes the endonuclease and inserts this DNA fragment into genes that lack the intein sequence. This feature thus allows the intein to propagate throughout genomes as a mobile genetic element. While the joined exteins can, in principle, be any protein, approximately 70% of inteins reside in proteins involved in DNA replication and repair. The reason for this is not clear.

As is the case for self-splicing introns, conserved residues located within the intein itself catalyze the proteolytic autoprocessing reactions through which the intein is excised. The processing reaction is initiated by the residue at the N-terminus of the *intein* (a serine, cysteine, or threonine), which attacks the carbonyl group of the preceding peptide bond as illustrated in Figure 14.13. The same carbonyl group is then subject to a nucleophilic attack by the first residue (again a serine, cysteine, or threonine) in the C-terminal extein. Finally, the last residue of the intein internally cyclizes.

This set of reactions yields a branched intermediate that leads to release of the intein and the formation of a genuine peptide bond between the N- and C-terminal extein moieties. Although the reactions involved in intein excision are chemically somewhat more complex than those involved in RNA splicing, the ultimate outcomes of the two processes are quite comparable.

Autoprocessing of the metazoan Hedgehog protein is chemically similar to intein splicing

The chemical reaction carried out by inteins is remarkably similar to that performed by the eukaryotic Hedgehog protein, which plays a key role in developmental patterning in vertebrates and invertebrates. Central to the mechanism of action of the Hh protein is its ability to catalyze a series of self-processing reactions. As with intein-containing proteins, Hh is expressed as a precursor protein that undergoes autocatalytic cleavage to yield two protein fragments, in this case, N-terminal and C-terminal domain products.

Both Hh autoprocessing and the intein reactions begin with chemical rearrangement of the peptide backbone. However, in the case of the Hh protein, this rearrangement is followed by a nucleophilic attack from an exogenous cholesterol molecule. As a result, cleavage of the peptide bond between the two domains of Hh occurs with concomitant covalent attachment of a cholesterol molecule to the last residue of the N-terminal domain, as depicted in Figure 14.12b. The modified N-terminal signaling domain is then freed to diffuse to adjacent cells where it can mediate its patterning effect. Although inteins are found only in single-celled organisms and Hh proteins are found only in metazoans, the similarities in the sequences of their self-splicing domains suggest a common evolutionary origin.

14.4 LIPID MODIFICATION OF PROTEINS

We have explored a number of ways in which the primary sequence of a protein may need to be altered before the protein can adopt its final, fully functional form. However, changes to many proteins do not stop once this final length is reached. A number of covalent additions or modifications occur to modulate localization and function. In this section, we will discuss lipid modifications and their properties.

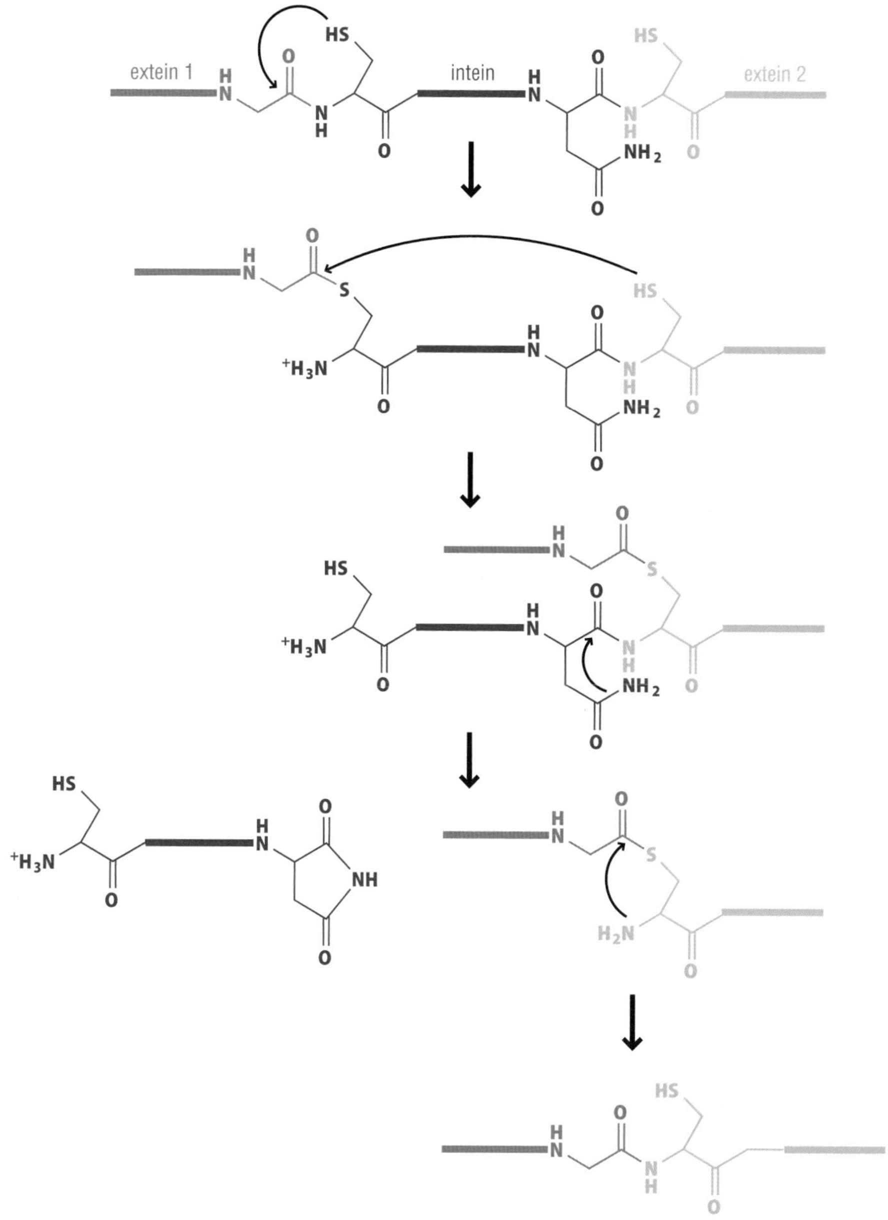

extein 1 intein extein 2

Figure 14.13 Intein cleavage reaction. The intein processing reaction is initiated by the residue (a serine, cysteine, or threonine) at the N-terminus of the intein (red) attacking the carbonyl group of the preceding peptide bond linking the N-terminal extein (blue) with the intein. The same carbonyl group is then subject to a nucleophilic attack by the first residue (again a serine, cysteine, or threonine) in the C-terminal extein (green). Finally, the last residue of the intein internally cyclizes, leading to the release of the intein and the formation of a genuine peptide bond between the N- and C-terminal extein moieties.

Covalent attachment of lipids targets proteins to membranes

Lipid modifications generally target proteins to the cell membrane, enabling the proteins to insert themselves and reside in the hydrophobic lipid bilayer. Different types of lipid modifications are thought to target eukaryotic proteins to distinct cellular membranes with different compositions, such as the plasma membrane, the Golgi, or lysosomes, as well as to particular regions within a given type of membrane. Lipid modifications are not specific to eukaryotic cells and are used for similar purposes in bacterial cells.

More than one type of lipid can be covalently attached to a protein, thereby allowing the cell to use a combination of lipids to fine-tune the localization of a protein in a membrane, just as a postal address features a combination of items – street name, city name, country name – to direct an item of mail to a very specific location. In some cases, modification with a lipid is a reversible process, thus enabling dynamic changes in the targeting of proteins to membranes. Lipid modification is therefore a strategy used by cells both to dispatch proteins to the appropriate intracellular membrane, and to regulate cellular processes that are mediated by these membrane-associated proteins.

Lipids are covalently attached to specific amino acid residues, typically at or near either the N-terminus or the C-terminus of the protein. In this section, we shall learn about three broad types of lipid modification that have somewhat distinct functional properties, and that are classified according to the type of attached lipid:

- **acylation** by fatty acyl groups such as myristoyl or palmitoyl
- **prenylation** by isoprenoid groups such as farnesyl or geranylgeranyl
- addition of glycoinositol phospholipids (referred to as **glycosylphosphatidylinositol (GPI) anchors**).

These three modifications are illustrated in Figure 14.14.

Acylation processes target the N- and C-termini of proteins

Two different fatty acid modifications, myristoyl and palmitoyl groups, typically have different attachment points. The relatively rare 14-carbon saturated fatty acid myristoyl is attached to proteins at the N-terminus via a stable amide linkage to an N-terminal glycine residue, yielding an *N*-myristoylated protein (*N* refers to the resulting linkage to nitrogen; see Figure 14.14a). The myristoyl moiety is donated by the carrier, myristoyl coenzyme A (myristoyl-CoA), and is attached to the protein co-translationally before folding of the polypeptide chain occurs. The 14-carbon myristoyl chain allows the modified protein to associate with cell membranes and

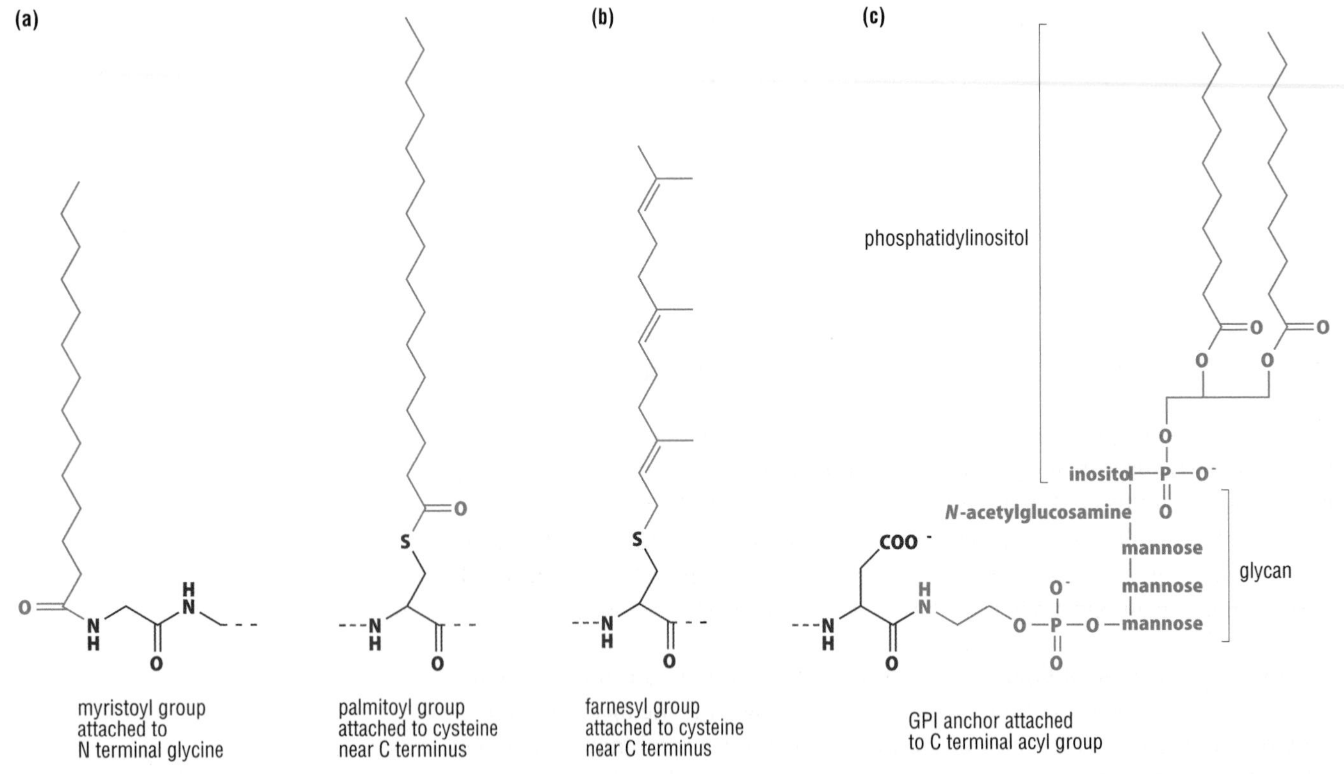

Figure 14.14 Major classes of lipid modifications. Proteins can be (a) acetylated by the addition of a fatty acid such as a myristoyl or a palmitoyl group or (b) prenylated by the addition of an isoprenoid group such as farnesyl. The third type of lipid modification (c) involves the addition of a GPI anchor, which is composed of both carbohydrates and lipids.

is generally a stable modification throughout the life of the protein. Proteins that are *N*-myristoylated include certain alpha subunits of heterotrimeric G proteins and the Src family of tyrosine kinases, both of which are involved in mediating the transmission of signals across cell membranes.

The more common 16-carbon saturated palmitoyl is generally attached post-translationally to C-terminal cysteines by a covalent bond with the sulfur side chain, as depicted in Figure 14.14a. The thioester linkage that attaches the palmitoyl moiety can be broken with relative ease (and can thus be considered to be chemically labile). This ease of modification appears to be used by the cell to reversibly modulate the localization of target proteins and, in some cases, to affect protein stability. For example, lymphoma proprotein convertase, an enzyme that activates proproteins in the secretory pathway, is found in both palmitoylated and unmodified forms. Interestingly, the palmitoylated form of the protein is degraded much faster than the unmodified one, allowing the lifetime of this enzyme to be regulated by lipidation.

Other fatty acid chains, such as the 18-carbon stearoyl, can sometimes substitute for the palmitoyl group, leading to a more general term for this modification known as ***S*-acylation** (*S* refers to the resulting linkage to sulfur).

Some proteins are both *N*-myristoylated and *S*-palmitoylated. This dual modification appears to have the important consequence of targeting the dually modified species to distinct membrane subdomains termed lipid rafts: the presence of both the myristoyl and palmitoyl groups targets the modified protein to a location distinct from that to which a protein carrying only a myristoyl or palmitoyl group alone might be directed.

Prenylation is a stable modification of C-terminal cysteine residues

Prenylation is the modification of a protein through the addition of a group belonging to the isoprenoid family – generally either a C_{15} farnesyl or C_{20} geranylgeranyl group – via a thioether linkage to the sulfur of a cysteine residue; this modification is shown in Figure 14.14b. The prenylation machinery recognizes several different cysteine-containing motifs at the C-terminus of the protein to be modified. One of these motifs is CaaX, w here 'a' stands for any aliphatic amino acid residue and where 'X' determines which prenyl group will be added. When X is glutamine, methionine, or serine, a farnesyl group is attached to the protein and when X is leucine, a geranylgeranyl group is attached.

As illustrated in Figure 14.15, following modification with the prenyl group, the C-terminus is proteolytically processed, removing the three terminal amino acids to yield a protein terminating in the lipidated cysteine residue. Finally, the newly exposed C-terminal carboxyl group is modified with a methyl group that neutralizes its negative charge. Other prenylation motifs include C-terminal CXC or CC amino acid residues, which can be modified with either one or two prenyl groups. The distinctions between these modifications may seem subtle, but probably have important functional consequences that have not yet been completely deciphered. In all examples of prenylation, the modification is stable and so is not believed to serve in a regulatory role.

Almost all prenylated proteins are small GTP-binding proteins. A number of *S*-prenylated proteins also appear to be subject to *S*-acylation at a nearby cysteine residue. This dual tagging has been observed to drive the variable targeting of proteins – for example, determining the subcellular localization of the signaling protein, Ras. Ras is a GTPase that, when mutated, gives rise to a variety of human cancers. All Ras proteins are farnesylated, and can in addition be acylated with

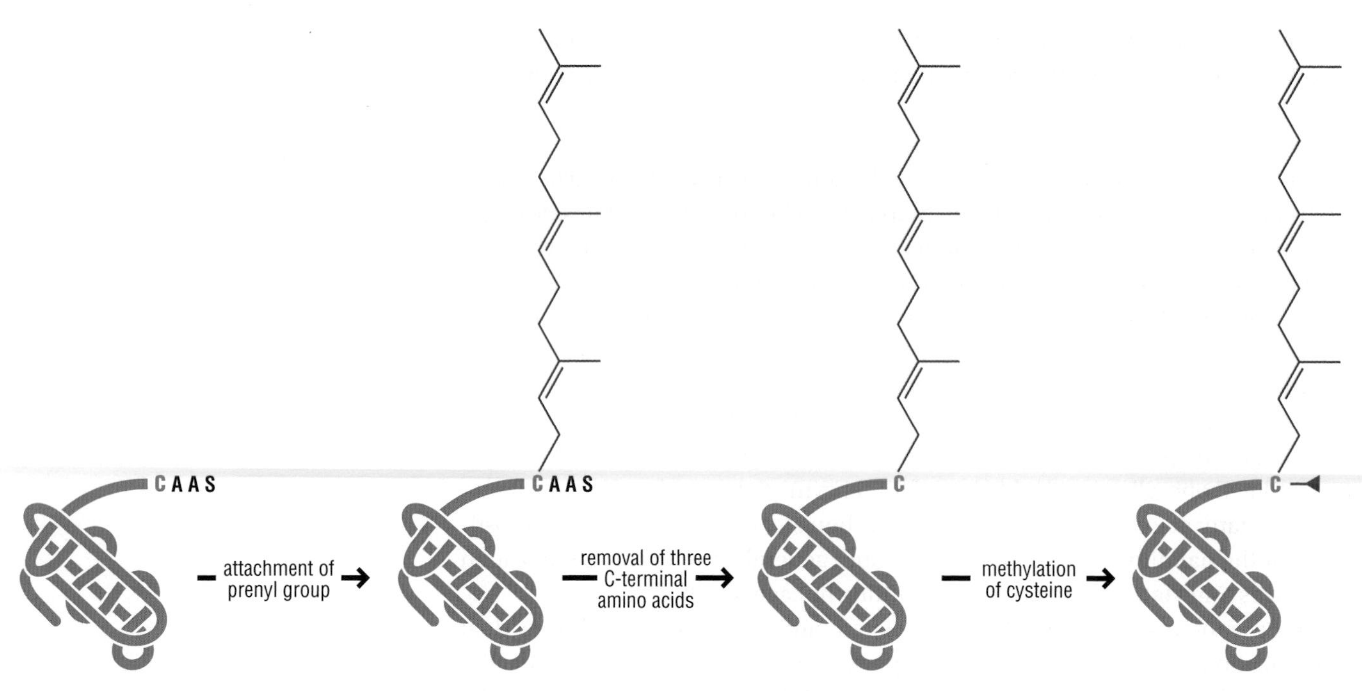

Figure 14.15 Protein prenylation. The prenyl group is first attached to a cysteine residue close to the C-terminus of the protein. All C-terminal amino acids following the cysteine are then removed by proteolytically processing, after which the newly exposed C-terminal carboxyl group of the lipidated cysteine is modified with a methyl group.

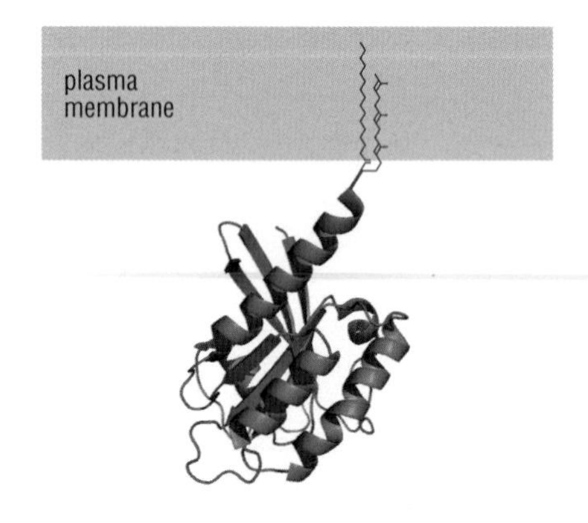

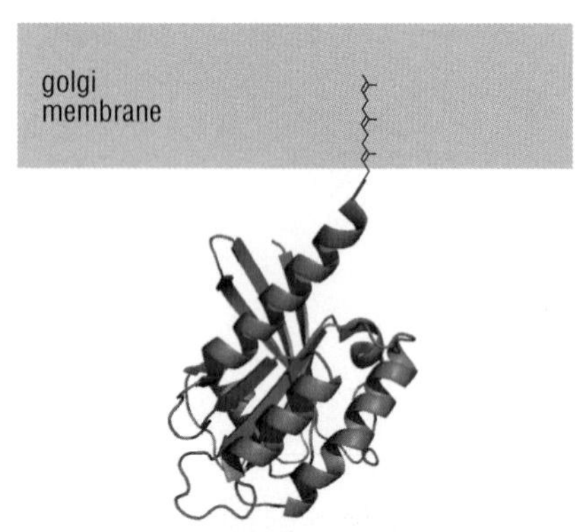

Figure 14.16 Subcellular localization of Ras is regulated by farnesylation and palmitoylation. Ras proteins bearing both farnesyl and palmitoyl groups are localized to the plasma membrane whereas Ras proteins with only a farnesyl group are localized to the Golgi membrane. PDB code 1Q21.

palmitoyl groups. Ras proteins bearing both lipid modifications associate with the plasma membrane, whereas removal of the palmitoyl group to leave just the farnesyl group causes Ras to be targeted to the Golgi membrane instead. By cycling between palmitoylated and depalmitoylated states, the cell can carry out Ras signaling in different cellular compartments, as depicted in Figure 14.16.

GPI anchors comprise both carbohydrates and lipids

The modifications we have just been considering involve the addition of lipid groups to a protein. The attachment of a C-terminal GPI anchor to a protein is distinct from the lipid modifications described above, both in chemical composition and in the manner of its attachment. The GPI anchor is composed of a carbohydrate moiety that is tethered to the membrane by two fatty acid chains, as shown in Figure 14.14c. The identity of the fatty acids and of the core tetrasaccharide can vary, lending diversity to this modification with likely functional consequences.

The GPI anchor is assembled in the endoplasmic reticulum through a complex series of enzymatic steps. The final step involves attachment of the GPI anchor to the C-terminus of the protein with simultaneous cleavage of the protein's transmembrane segment so that the protein is only tethered to the membrane via the GPI anchor. The presence of the carbohydrate moiety directs GPI-anchored proteins to the cell surface via transport through the secretory pathway.

Since the proteins are only anchored to the membrane via the GPI anchor, the protein is readily released from the membrane by the action of phospholipases, which cleave the lipid. GPI-anchored proteins participate in a wide variety of processes in the cell including nutrient uptake, cell adhesion, and membrane signaling. A number of human parasites have GPI-anchored enzymes on their cell surfaces that are inactive while tethered to the membrane but which become activated when the parasite encounters host phospholipases.

Multiple lipid modifications on the Hh protein are important for specifying its subcellular localization

There are still other lipid modifications that play critical roles in targeting proteins to membranes. The Hh protein, which plays a central role in regulating embryonic development, must be modified with cholesterol in order to carry out its biological function. This modification is shown in Figure 14.17. We learned earlier in this chapter that Hh undergoes an autocatalytic cleavage that separates the protein into two fragments, and the cleavage is accompanied by the attachment of a covalent cholesterol adduct to the newly exposed C-terminus of the N-terminal domain. The discovery of this modification is described in Experimental approach 14.1. Of particular interest here is that the N-terminus of the N-terminal domain is also modified with a palmitoyl residue, suggesting that multiple lipid modifications may help to specify the localization of Hh to particular membranes within the cell.

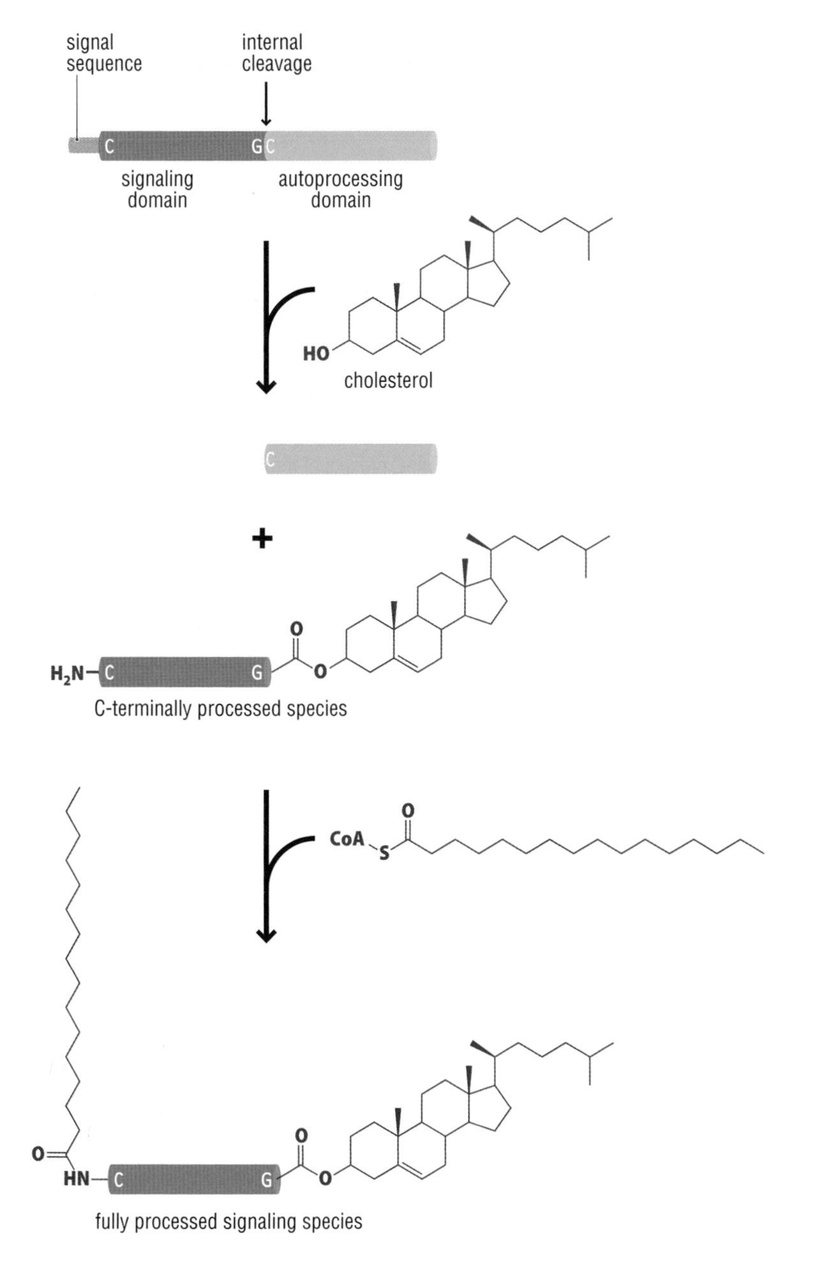

Figure 14.17 Lipid modifications of the Hh protein. The Hh signaling protein undergoes autocatalytic cleavage that is accompanied by the attachment of a cholesterol molecule to the exposed C-terminus of the N-terminal domain. A second lipid modification, the palmitoylation of the N-terminal cysteine residue, completes the processing of this signaling domain.

14.1 EXPERIMENTAL APPROACH

The discovery of cholesterol modification for membrane tethering during signaling

Autocatalytic cleavage is required for activation of the Hedgehog protein

The Hedgehog family of secreted proteins is required for developmental patterning of a wide variety of embryonic structures in diverse organisms. Work by Phil Beachy's group in the early 1990s established much about how this protein is produced and what its *in vivo* roles might be. The group found that the Hedgehog protein undergoes several proteolytic processing steps to yield the final active species. First, the signal sequence at the N-terminus of the protein is removed by standard processing. In addition, the 45 kilodalton (kDa) precursor undergoes an autocatalytic cleavage event to yield two products – a 20 kDa N-terminal domain (Hh-N_p) and a 25 kDa C-terminal domain (Hh-C). The required molecular components for this autocleavage event are principally contained within Hh-C (a particular histidine found there is critical), while Hh-N_p constitutes the extracellular signal responsible for dictating essential tissue-patterning effects during development. The mechanisms for Hh and intein cleavage ultimately turned out to be related, consistent with sequence similarities between these proteins, in particular the conservation of an essential cysteine at the cleavage site.

Cholesterol is covalently attached to the C-terminus of Hh-N_p

The Beachy group noticed that the internal thioester intermediate of the reaction (akin to that of intein cleavage) yielded a covalent modification of Hh-N_p that increased its 'hydrophobic character' and then sought a molecular explanation. It turns out that the hydrophobic modification that is a consequence of autoprocessing is at the heart of Hedgehog function, specifying where the signaling molecule Hh-N_p goes and thus where it exerts its effect.

The study referred to here focused on identifying the hydrophobic covalent modification and its chemical linkage, and thus how the movement of this protein through tissues is regulated. Since previous studies had determined that the intermediate of the autoprocessing reaction was a thioester carbonyl, it was already clear that high concentrations of thiol reagents (or strong nucleophiles) could attack the intermediate, resulting in the release of Hh-C and the formation of an adduct between the nucleophile and Hh-N_p. Given the hydrophobic nature of Hh-N_p, it was proposed that a lipid might function as the nucleophile *in vivo*, thus producing the more hydrophobic adduct. Beachy and colleagues took this idea as a starting point in their quest to identify this adduct. Bulk

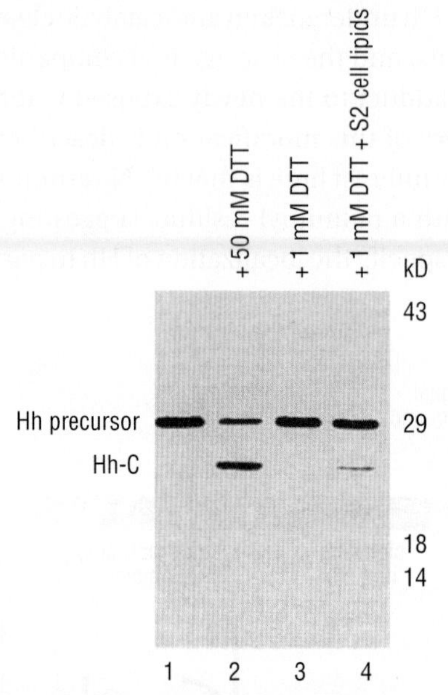

Figure 1 Lipid stimulation of Hedgehog autoprocessing *in vitro.* Coomassie blue-stained sodium dodecyl sulfate (SDS)-polyacrylamide gel showing *in vitro* autocleavage reactions of the Hedgehog (Hh) precursor protein (~29 kDa) incubated for three hours at 30°C with no additions (lane 1), 50 mM DTT (lane 2), 1 mM DTT (lane 3), or 1 mM DTT plus bulk S2 cell lipid extract (lane 4). The Hh-C product of the reaction migrates as an ~25 kDa species (lanes 2 and 4), and the ~5 kDa Hh-N_p product is not seen in this gel.

Reproduced from Porter *et al.* Cholesterol modification of hedgehog signaling proteins in animal development. *Science*, 1996;**274**:255–259.

lipids isolated from *Drosophila* S2 cultured cells clearly stimulated the processing reaction (much like the strong reducing agent dithiothreitol (DTT)), yielding the visible 25 kDa Hh-C fragment (Figure 1; the construct used in these studies had a truncated N-terminus yielding only a 5 kDa Hh-N_p product not visible in this gel system).

The authors next fractionated the lipid extract by thin-layer chromatography (TLC), as shown in Figure 2a, and found that one specific excised band from the TLC (spot B) stimulated the *in vitro* processing reaction (Figure 2b). Spot B was ultimately identified as cholesterol, first through comparisons to lipid standards, and ultimately through *in vivo* labeling experiments. The chemical linkage to the cholesterol was also determined. Given the potential intein type reaction, it was proposed that the reactive nucleophile on cholesterol would be the 3′ hydroxyl group; the base-labile nature of the predicted ester adduct of this reaction was readily confirmed.

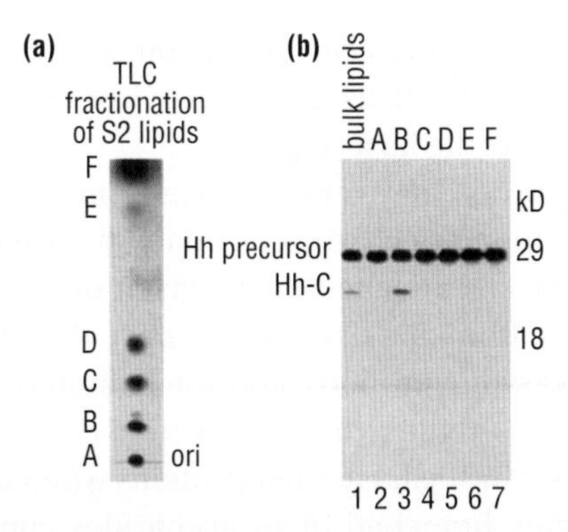

Figure 2 Identification of cholesterol as the stimulatory lipid in the Hedgehog autoprocessing reaction. (a) TLC plate coated with silica gel G (Merck) showing the fractionation of bulk S2 lipids with a heptane:ether:formic acid solvent (80:20:2). Six major spots are visualized by acid charring and are labeled A through F. (b) Coomassie blue-stained SDS-polyacrylamide gel showing *in vitro* autocleavage reactions of the Hedgehog (Hh) precursor protein incubated with 1 mM DTT plus either unfractionated S2 cell lipids (lane 1) or lipids extracted from spots A through F (lanes 2 to 7, respectively). Addition of lipids from spot B, but not from any other, resulted in processing of the Hh precursor protein.

Reproduced from Porter *et al.* Cholesterol modification of hedgehog signaling proteins in animal development. *Science,* 1996;**274**:255–259.

The identification of cholesterol as the covalent adduct of the Hedgehog protein, and related proteins in other organisms (Sonic hedgehog in mice, for example), had a huge impact on the field. Understanding of this molecular mechanism provided immediate perspective on defects associated with a number of mutant organisms, as well as on the effects of cholesterol levels on development and growth. The similarities between the Hedgehog autoprocessing domain and the inteins provide insights into the assembly of the Hedgehog signaling protein from evolutionarily ancient domains. Indeed, more extensive comparative genomic approaches suggest that the intein-related Hedgehog processing domain first came together with the signaling domain into a single gene as the first patterned multicellular animals evolved.

Find out more

Porter JA, Young KE, Beachy PA. Cholesterol modification of hedgehog signaling proteins in animal development. *Science*, 1996;**274**:255–259.

Related techniques

Coomassie staining; Section 19.7

Thin-layer chromatography; Section 19.7

14.5 GLYCOSYLATION OF PROTEINS

Having seen how a polypeptide may be modified through the attachment of a lipid molecule, we now consider a second widespread form of covalent modification, that of glycosylation.

Glycosylation can change the solubility of a protein and provide protein recognition sites

The most complex and diverse post-translational modifications to proteins are carbohydrate chains, referred to as **glycans**, which are attached through a process called **glycosylation**. Since carbohydrates are hydrophilic in nature, these modifications are usually found on the outside of the protein, where they have substantial effects on protein solubility. Almost all secreted and membrane-associated proteins in eukaryotic cells are modified with glycans. Collectively, the glycosylated proteins are referred to as **glycoproteins**.

Protein glycosylation serves a variety of functions. Inside the cell, the attachment of glycans generally increases the solubility of nascent glycoproteins and can prevent their aggregation. Small monosaccharides attached to proteins within the cell also are thought to have roles in intracellular signaling. Once the protein is secreted, glycosylation can provide protection from proteases and from non-specific interactions with other proteins. Glycosylated proteins on the cell surface can also be bound by other proteins that recognize specific glycans; this type of interaction commonly plays a role in the recognition of one cell by another. Important examples include the targeting of leukocytes to sites of inflammation and host cell recognition by viruses and bacteria. For example, the influenza virus binds to the

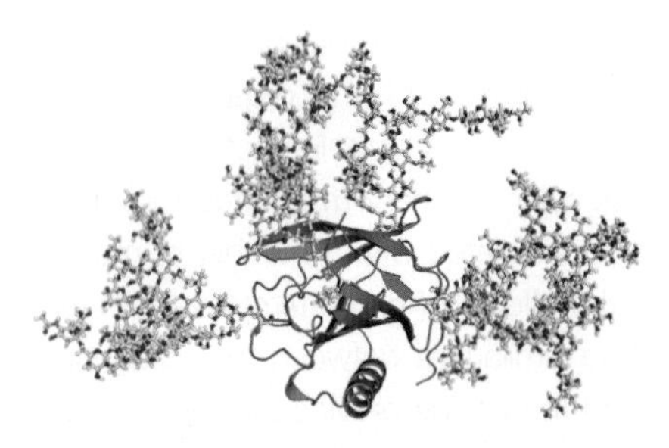

Figure 14.18 Volume added by carbohydrate addition. The *N*-linked glycan (light yellow, 3 kilodalton (kDa)), *O*-linked glycan (dark yellow, 1 kDa), and GPI anchor (orange, 1.5 kDa) only add a fraction of the mass to the CD59 protein, an immune defense protein (blue, 20 kDa). However, the carbohydrate additions which spread out in space, rather than folding inward as is typical for amino acid chains, more than double the volume of the molecule. PDB code 1CDR.

carbohydrate sialic acid that is attached to a glycoprotein on the mammalian host cell's surface. This binding results in internalization of the virus by the target cell, enabling the virus to enter and infect the host cell.

Glycans vary greatly in complexity: they range in size from a simple monosaccharide or just a few linked monosaccharides to many hundreds of linked saccharides. The nature of the glycan attached to any given protein also varies from species to species and often from one tissue to the next. Protein glycosylation is almost exclusively a eukaryotic property and the complexity of modifications increases as one proceeds up the evolutionary tree. Lower eukaryotes, like yeast, attach only a simple set of sugars to their proteins whereas mammals modify their proteins with highly branched oligosaccharides composed of a wide range of carbohydrates. The size of the attached carbohydrate can be significant, sometimes approaching 90% of the total mass of the glycoprotein. The hydrophilic nature of the glycan group prevents it from **condensing** into a compact structure and so the volume of a glycoprotein relative to its mass is generally large. The volume that can be added to a protein from glycan addition is illustrated in Figure 14.18

Carbohydrates are attached to proteins via nitrogen or oxygen

The diverse glycans that are attached to proteins are most often polymers built from a variety of hexose building blocks, examples of which are shown in Figure 14.19, and can be either linear or branched. Unlike many of the polymerization reactions that we have discussed throughout this book, the sequential attachment of hexoses depends on distinct enzymes to perform each addition and is specified without a guiding nucleic acid template. Carbohydrates are covalently attached to proteins in two different ways: through *N*-linked or *O*-linked glycosidic bonds. ***N*-linked carbohydrates** are covalently bonded to the amino nitrogen in asparagine side chains, whereas ***O*-linked carbohydrates** are covalently bonded to the oxygen of serine or threonine side chains; these differences are illustrated in Figure 14.20. The *N*-linked oligosaccharides are the most common, found on 90% of glycoproteins.

N-linked carbohydrates are usually more complex in nature than *O*-linked carbohydrates, and their addition can require 23 or more enzymatic steps. The process of protein glycosylation is summarized in Figure 14.21. N-linked glycosylation begins in the endoplasmic reticulum with the co-translational attachment of the precursor oligosaccharide to an asparagine residue found in a glycosylation signal sequence, Asn-X-Ser, Asn-X-Thr, or Asn-X-Cys, where X can be any amino acid except proline. The precursor oligosaccharide is composed of 14 monosaccharides

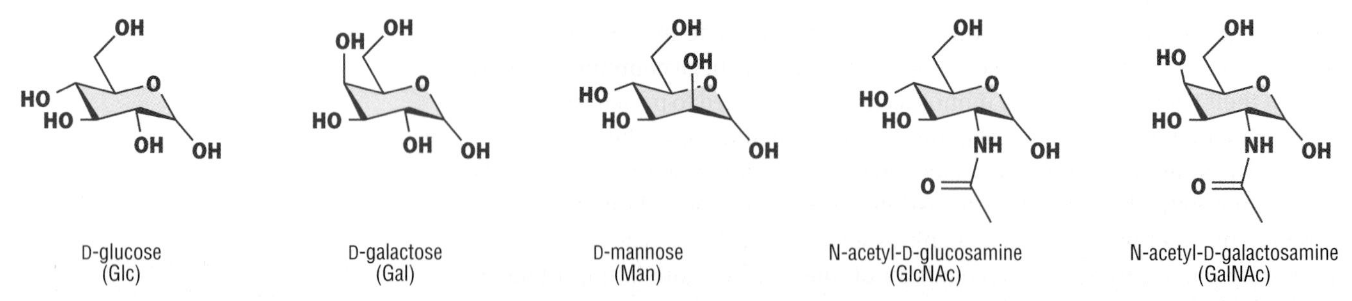

Figure 14.19 Chemical structure of some hexoses which are common building blocks for many oligosaccharides that are attached to proteins. The common abbreviation for each hexose is indicated in parentheses.

consisting of glucose, mannose, and *N*-acetylglucosamine arranged in a particular branched configuration that is common to most eukaryotes.

To produce the mature oligosaccharide, the *N*-glycosylation precursor is first trimmed by glucosidases and mannosidases in the endoplasmic reticulum. Additional sugars are then added when the protein is trafficked to the Golgi complex, giving rise to diverse complex glycans. The glycosyl transferase enzymes in this process all utilize sugars that are covalently linked to the nucleotide uridine diphosphate (UDP). As we will see in several more examples, most modifications that are added to proteins are coupled to nucleotide carriers. The breaking of the bond between the carrier and the modification group provides the energy to drive the modification reactions.

Many of the processing steps are not dependent on one another, that is, enzymes do not necessarily need to act in a specific order. Thus there can be significant heterogeneity in the final product due to the limited availability of individual enzymes or the varying efficiency of enzymatic steps, for example.

O-glycosylation is similar to *N*-glycosylation, in that it generally begins with the covalent attachment of a precursor sugar, *N*-acetylgalactosamine, which is subsequently derivatized with a variety of other saccharides to yield a number of final structures. Although these carbohydrate appendages are generally not as extravagant as their *N*-linked cousins, they are still diverse in structure.

The importance of these glycosylation steps in human development is underscored by the findings that mutations in any of the roughly 15 genes encoding proteins involved in *N*-glycosylation result in a variety of developmental disorders, including mental retardation and motor deficits.

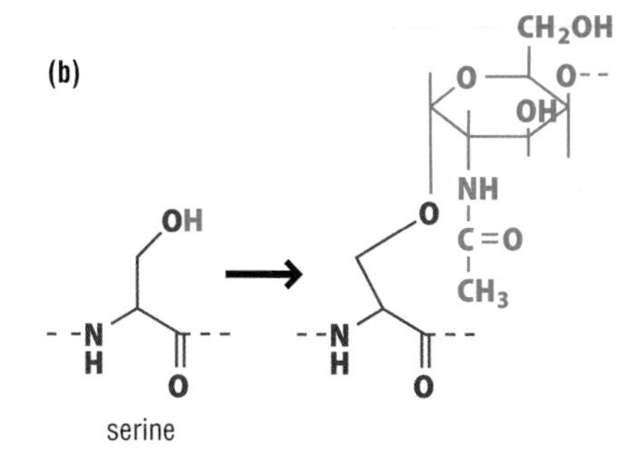

Figure 14.20 *N*-linked and *O*-linked glycosylation. Glycans are attached to (a) asparagine to give *N*-linked carbohydrates or (b) serine or threonine to give *O*-linked carbohydrates.

Reversible *O*-GlcNAc modification may serve a role in signaling

One form of *O*-glycosylation is different from the rather large, diverse set of carbohydrate modifications described above. The addition of a single *N*-acetylglucosamine

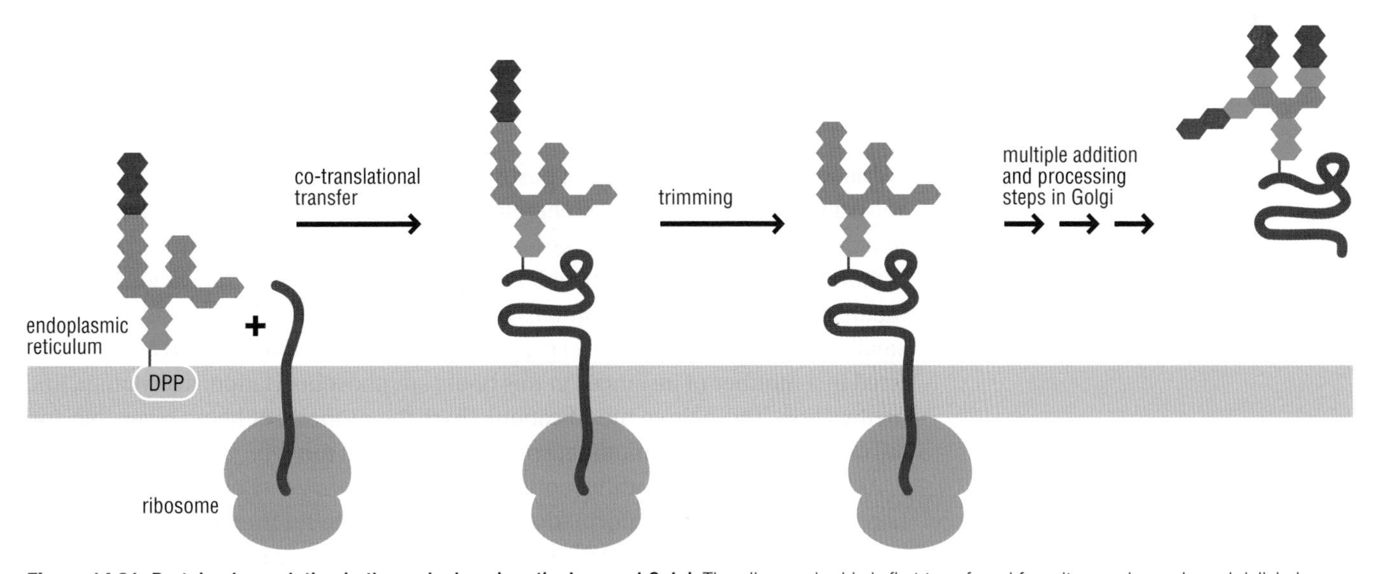

Figure 14.21 Protein glycosylation in the endoplasmic reticulum and Golgi. The oligosaccharide is first transferred from its membrane-bound dolichol pyrophosphate (DPP) carrier to a polypeptide chain while the latter is still being synthesized on the ribosome. As the polypeptide chain grows within the endoplasmic reticulum, monosaccharides may be cleaved from the ends of the oligosaccharide. After synthesis of the polypeptide chain is complete, the immature glycoprotein is transported to the Golgi apparatus, where further modification occurs. New monosaccharides may be added and others removed in a multi-step process involving many different enzymes. The completed glycoprotein is finally transported to its ultimate destination. Yellow hexagons represent N-acetyl-glucosamine, blue hexagons represent mannose, and red hexagons represent glucose.

residue to a serine or threonine residue, termed *O*-GlcNAc, is a reversible modification that is attached by *O*-GlcNAc transferase (OGT) and removed by *O*-acetylglucosaminidase (*O*-GlcNAcase). In contrast with other forms of protein glycosylation, which occur largely on membrane or secreted proteins, monoglycosylation by *O*-GlcNAc occurs on cytoplasmic and nuclear proteins.

The addition and removal of *O*-GlcNAc appears to play a role akin to that of other small, reversible modifications such as **phosphorylation** and acetylation, which we will discuss in Section 14.6. The sites of monoglycosylation are often identical to the sites of phosphorylation, suggesting that *O*-GlcNAc modification can be used to block phosphorylation and vice versa. The transcriptional regulator, c-Myc, contains a threonine residue that is either glycosylated or phosphorylated, depending on serum conditions. We learned in Section 8.2 that the C-terminal domain (CTD) of RNA polymerase II becomes phosphorylated at serine 2 and serine 5 of the heptapeptide repeats in the course of the transcription cycle. The CTD can also be monoglycosylated with GlcNAc at position 4 (threonine) of the heptapeptide. While the precise functional role of *O*-GlcNAc modification, other than its ability to block phosphorylation, is not well understood, there is evidence that monoglycosylation may play a role in nuclear localization and in modulating interactions between proteins.

The levels of *O*-GlcNAc modification are influenced by the availability of the GlcNAc donor, UDP-GlcNAc, whose levels are in turn controlled by the availability of sugars and other nutrients in the cell. *O*-GlcNAc modification thereby provides a way for the cell to respond to changes in nutrition. It is interesting to note that the carrier molecule for *O*-GlcNAc is also UDP. Other nucleotides are carriers for the reversible modifications that play a role in signaling, such as phosphorylation (via ATP), acetylation (via acetyl coenzyme A (acetyl-CoA)), and methylation (via *S*-adenosyl methionine), which we will discuss in the next section.

14.6 PROTEIN PHOSPHORYLATION, ACETYLATION, AND METHYLATION

Small reversible covalent modifications play central roles throughout biology

We have now explored a range of post-translational modifications, which mostly involve the addition of quite large chemical groups to a protein. However, the attached molecule does not need to be large for it to have a significant impact on the properties of the protein to which it is attached. There are a number of post-translational protein modifications utilized in all domains of life that involve the addition and removal of small chemical groups. The most prevalent of these, which we have already encountered in a number of instances in this book, are phosphorylation, acetylation, and methylation. These three modifications share the property of being reversible and are used very broadly to modulate protein function in the cell.

As with all protein modifications, phosphorylation, acetylation, and methylation can alter protein conformation (and hence protein function) or localization (and hence the site of function). The phosphate, acetyl, and methyl groups provide new binding surfaces and thus can recruit new binding partners. Many proteins possess specialized domains that can recognize side chains that have undergone phosphorylation, acetylation, or methylation, as illustrated in Figure 14.22. The reversible nature of these modifications enables them to serve as a type of chemical switch that can affect diverse processes such as transcription, chromatin

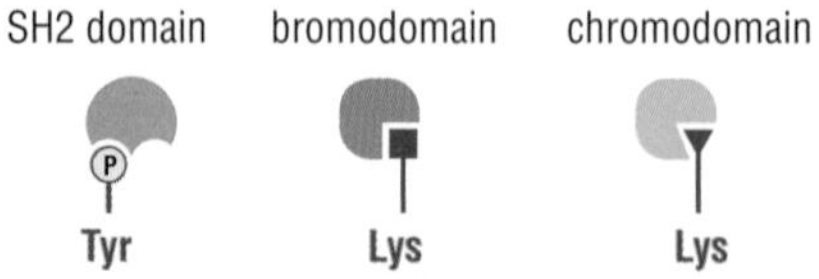

Figure 14.22 Specific modifications are recognized by specialized protein domains. Phosphorylated tyrosines are bound by SH2 domains, acetylated lysines are bound by bromodomains, and methylated lysines are bound by chromodomains.

structure, intracellular membrane traffic, and signaling cascades. Reversibility provides a means for the cell to respond rapidly to stimuli without the need to synthesize new proteins. In addition, since the same amino acid can sometimes be the target of more than one type of modification, the attachment of one modification may serve to block the addition of another modification. So, for example, acetylation of a particular residue might prevent methylation at the same site, hence inhibiting the event that would have been triggered had methylation occurred instead.

A common feature of phosphorylation, acetylation, and methylation is that the physical modification itself (a phosphate group, an acetyl group, or a methyl group) is transferred by an enzyme from a nucleotide donor molecule to a target protein. The identity of the nucleotide donor, and the amino acids targeted for modification, differ for each type of modification, as we shall discuss below. In each case, there are enzymes that reverse the modification.

Phosphorylation can alter protein function

Surely the best studied and perhaps the most common post-translational modification in cells is the attachment of a phosphate group. The source of the phosphate group that is added to a protein is the nucleoside triphosphate ATP, ATP has three phosphate groups: the third group (the 'terminal' or gamma phosphate) is transferred from ATP to the target protein through the action of enzymes known as **protein kinases** in a process referred to as **phosphorylation**, illustrated in Figure 14.23. Reversal of the modification is catalyzed by a different class of enzyme called **protein phosphatases**. In eukaryotes, the amino acid side chains that are usually phosphorylated are serine, threonine, and tyrosine, all amino acids containing a hydroxyl group to which the phosphate group is attached by an ester linkage (see Figure 14.23). Eukaryotic kinases are commonly named according to the amino acid they modify, such as serine/threonine kinase. In bacteria, phosphorylation on histidine and aspartate residues predominates (although some eukaryotic-like serine/threonine kinases and tyrosine kinases have been found in bacteria and some histidine kinases are present in yeast and plants).

The importance of phosphorylation in biology is underscored by the significant proportions of both eukaryotic and bacterial genomes that are devoted to phosphorylation. About 2% of all human genes encode kinases that phosphorylate serine, threonine, or tyrosine residues (representing the third most common protein domain) and 1.5% of all genes in *Escherichia coli* are histidine and aspartate kinases.

Phosphorylation, like other post-translational modifications, can affect the target protein in several ways. One possible effect is the creation of a new recognition

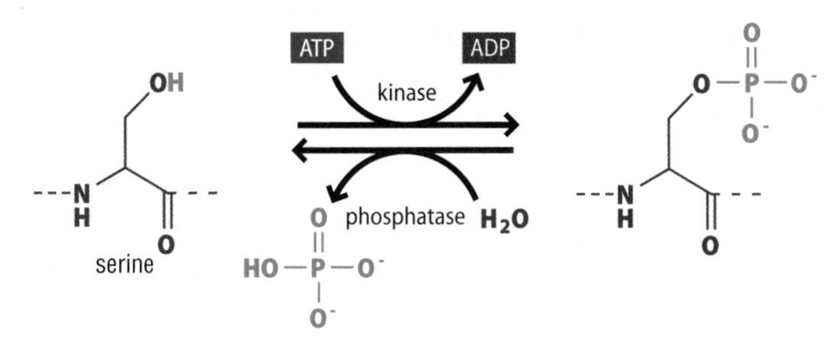

Figure 14.23 Reversible phosphorylation of serine side chains. Kinase enzymes catalyze the nucleophilic attack by the OH group of serine on the terminal γ-phosphate of ATP, yielding phosphoserine. The phosphate group is removed by phosphatases.

site to which another protein can bind. For example, as we saw in Section 8.4, phosphorylation of RNA polymerase II triggers recruitment of proteins required for transcriptional elongation, which bind to the phosphorylated CTD portion of the polymerase enzyme but not to the unphosphorylated version. There are specialized protein domains that bind specifically to particular types of phosphorylated side chains. For example, SH2 domains, which are found in some metazoans, bind to phosphotyrosine (a phosphorylated tyrosine residue).

Phosphorylation can also change the activity of the target protein itself, either considerably or subtly. This change in activity may come about from the added bulk and the negative charge of the phosphoryl group, or from the hydrogen-bonding potential of the phosphate oxygens. These physical properties of phosphorylated residues can sometimes trigger large conformational changes in the protein that can in turn have consequences for binding and catalysis. For example, in Section 12.1 we saw how phosphorylation of the translation initiation factor eIF2 modifies its interactions with its GTP exchange factor (eIF2B), thus down-regulating its activity in the translation initiation pathway.

Protein phosphorylation is a key event in transducing signals from the cell surface

The Jak–STAT signaling pathway (Jak stands for Janus kinase and STAT for signal transducers and activators of transcription) in eukaryotes provides an illuminating example of how binding of a signaling molecule to an extracellular receptor triggers a chain of phosphorylation events in the cell that lead to activation of certain genes. In this way, information is transmitted from the cell surface to the nucleus. The signaling molecule, α-interferon, is a protein secreted by cells in response to viral infection. The α-interferon protein binds to the surface of target cells, triggering a series of biochemical events that results in the transcription of genes encoding proteins that help the cell respond to the viral infection.

As shown in Figure 14.24, the cytokine receptor for α-interferon is a transmembrane protein containing an extracellular domain that binds α-interferon, and a cytoplasmic domain to which an intracellular tyrosine kinase is bound. The α-interferon

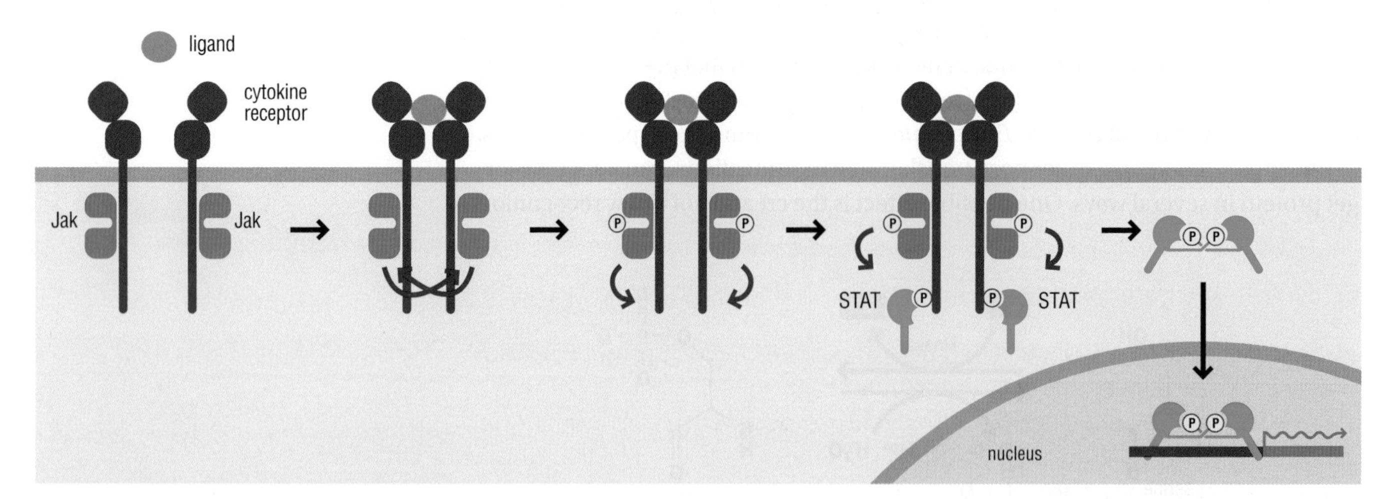

Figure 14.24 The Jak–STAT signaling pathway. Jak kinases associate with the cytoplasmic domains of cytokine receptors. Ligand binding to the extracellular domain induces cytokine receptor chain dimerization, bringing the two Jak kinases close enough to phosphorylate one another. The resulting activated Jak kinases phosphorylate key tyrosines in the cytoplasmic tail of the cytokine receptor. These sites attract STAT proteins, which contain SH2 domains (only STAT is shown). Bound STATs become phosphorylated, after which they dissociate from the receptor and dimerize via their phosphorylation sites and the SH2 domains. The STAT dimers enter the nucleus and activate transcription of specific target genes.

ligand associates with two extracellular domains, drawing them together in the membrane to form a receptor domain with a single α-interferon protein bound. This dimerization brings together the cytoplasmic domains of the two monomers, along with the tyrosine kinases to which they are bound. When brought together in this way, the tyrosine kinases phosphorylate one another. This phosphorylation activates the kinases, causing them to phosphorylate tyrosine residues in the cytoplasmic domain of the receptor. The phosphorylated tyrosine of the receptor can then recruit STAT, a DNA-binding protein bearing an SH2 domain, which recognizes the phosphotyrosine residue (see Figure 14.24). The STAT protein itself is then phosphorylated at a tyrosine residue, which allows it to dissociate from the receptor and form a dimer that is stabilized by the binding of the SH2 domain of one STAT monomer to the phosphotyrosine residue on the opposing monomer. The STAT dimer can now translocate to the nucleus, where it binds DNA and activates transcription. Thus a signal from outside the cell is transformed into a signal that activates transcription in the nucleus, all based on a series of interconnected post-translational modification events.

In bacteria, signals from the environment are also converted into changes in gene expression through the activation of kinases at the cell membrane (termed sensor kinases), which ultimately leads to the phosphorylation of transcription factors (termed response regulators). As we learned in Section 9.3 and saw in Figure 9.17, in these cases, the sensor kinases autophosphorylate in the presence of the inducing signal. Subsequently, the phosphate groups are transferred to another protein; either directly to the response regulator, which often regulates transcription, or to intermediary proteins, from which the phosphate groups are then transferred to the response regulators. Experimental approach 14.2 describes the discovery of this class of proteins and how comparative genomics was used to identify interacting pairs of sensor kinases and response regulators.

Proteins can be acetylated on side chains and at their amino termini

Another small, reversible post-translational modification is acetylation, which plays an important role in transcriptional regulation. The acetyl group derives from acetyl-CoA, a high-energy nucleotide species, and is attached to target proteins by enzymes known as **acetyl transferases**. This process, known as acetylation, results in covalent attachment of the acetyl group to an amine at either the epsilon position of lysine side chains or at the N-terminus of proteins; this attachment is illustrated in Figure 14.25a. It has been estimated that more than one-third of all yeast proteins may be acetylated at the N-terminus, and upwards of 80% of proteins in higher eukaryotes. This modification is, however, rare in bacteria, where just three ribosomal proteins are known to be acetylated in this way.

➔ We discuss the role of acetylation in transcriptional regulation in Chapter 9.

As with phosphorylation, the acetyl group can readily be removed from lysine side chains by **deacetylase** enzymes, which accomplish this by one of two different chemical pathways. As depicted in Figure 14.25a, one pathway is catalyzed by hydrolases that use simple hydrolysis to remove the acetyl group, releasing acetate and restoring the lysine side chain to its unmodified state. In a second pathway shown in Figure 14.25b, proteins known as sirtuins (named for their similarity to the yeast Sir2 enzyme), use nicotinamide adenosine dinucleotide (NAD) as a co-substrate in the deacetylation reaction. Sirtuins catalyze the cleavage of the oxidized form of NAD (NAD^+) to release nicotinamide and transfer the acetyl group from the lysine side chain to the remaining portion of the NAD^+ molecule, yielding 2′ and 3′-*O*-acetyl-adenosine diphosphate (ADP)-ribose. The biological advantage of this surprising use of NAD^+ as a cofactor in place of the simpler

14.2 EXPERIMENTAL APPROACH

Two-component systems in bacteria: how to interact with your partner

Histidine kinases activate response regulators by phosphorylation

Two-component signal transduction systems are ubiquitous in bacteria, where they help the cells respond to their environment. These systems are typically composed of a sensor kinase (usually a membrane protein), which is activated by an environmental signal and in turn phosphorylates a response regulator (usually a transcription factor). These kinases and their substrates were first identified in genetic screens that searched for mutants defective in responding to environmental factors. For example, bacterial variants defective in responding to nitrogen starvation led to the identification of three genetic loci important for directing transcription from the nitrogen-regulated *glnA* P2 promoter: *NtrC* (also called *NRI* and now understood to be the response regulator), *NtrB* (also call *NRII* and now understood to be the sensor kinase) and a specific alternate sigma factor σ^{54} (σ^{N}).

In a landmark study that followed, Alexander Ninfa and Boris Magasanik carried out *in vitro* transcription assays with purified NRI and NRII2302 (a mutant form of NRII that does not depend on nitrogen) proteins and RNA polymerase to examine the requirements for NRI activation of the *glnA* P2 promoter. In the course of these experiments, the investigators found that incubation of NRI with NRII2302 in the presence of mixtures of nucleotides, resulted in increased levels of transcription. By mixing NRI and NRII2302 with different combinations of nucleotides (where adenosine (A) + cytidine (C) + guanosine (G), A+C, A+G and A alone worked but C+G did not), the investigators determined *glnA* activation by NRI and NRII requires the presence of ATP. Their results are shown in Figure 1.

Using radiolabeled ATP, the investigators next demonstrated that NRII2302 catalyzes the transfer of the γ phosphate of ATP to NRI (see data in Figure 2), and that this modification is required for the increased ability of NRI to activate transcription. These data established the underpinnings of our understanding of the relationship between sensor kinases and response regulators, and their roles in activating transcriptional responses.

As the genes for additional cellular responses to environmental stimuli were characterized (e.g., for chemotaxis, cell envelope stress, and sporulation), it became apparent that cellular responses were quite often dictated by two proteins whose genes were adjacent to one another in the genome. Although it is hard to appreciate today how limited the sequence databases were in 1986, sequence comparisons revealed that these 'two-component' regulators, important for a variety of responses in a wide range of bacterial species, were related and thus must have evolved from a common ancestor.

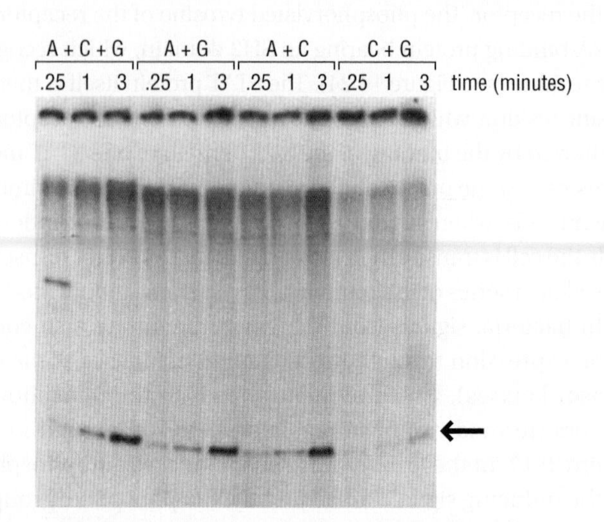

Figure 1 Effect of nucleotides on NRII2302 and NRI activation of transcription from the *glnA* P2 promoter. In each case NRI was pretreated with NRII2302 for 20 minutes at 37°C in the presence of the indicated nucleotides. These mixtures were then incubated with RNA polymerase and σ^{54} for the times listed above the gel before uridine triphosphate (UTP) and heparin were added to start the transcription reaction. The arrow indicates the transcription product from the *glnA* P2 promoter.

Reproduced from Ninfa and Magasanik, Covalent modification of the glnG product, NRI, by the glnL product, NRII, regulates the transcription of the glnALG operon in *Escherichia coli. Proceedings of the National Academy of Sciences of the U S A*, 1986;**83**:5909–5913.

Specificity between histidine kinase and response regulator pairs is determined by contacts between co-varying amino acid residues

As more and more of these homologous sensor kinase and response regulator proteins were identified, with some organisms containing as many as 200–300 pairs, the question of how specificity is maintained within the cell arose. Using batteries of purified kinase and response regulator proteins, Michael Laub and colleagues recently showed that most of the kinases are in fact quite specific, and only phosphorylate their own cognate response regulator. They surmised that this specificity must result from very specific interactions between the cognate proteins, and that the amino acids involved in this interaction (on the two surfaces) would co-vary. Taking advantage of 1300 sequences that were available in 2008, the investigators identified two clusters of co-varying residues, as shown in Figure 3a. By substituting specific residues in the EnvZ kinase with those found in the RstB kinase, they were able to

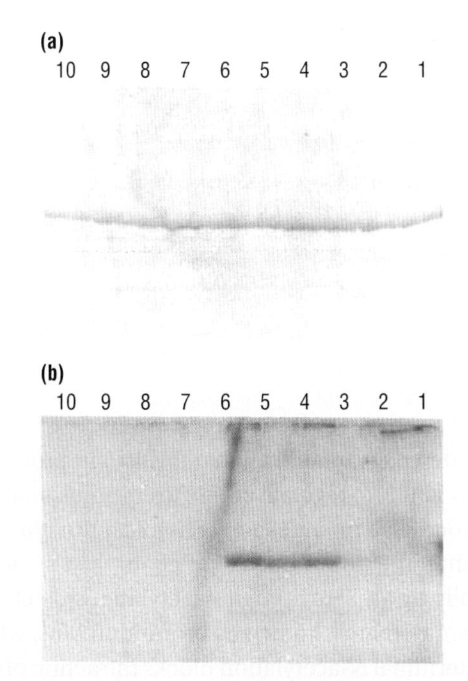

Figure 2 Covalent modification of NRI by NRII2302. NRI was incubated with [γ-^{32}P]-ATP for various times in the absence (lanes 6–10) or presence (lanes 1–5) of NRII2302. (a) Proteins separated by SDS-polyacrylamide gel electrophoresis (SDS-PAGE) stained with Coomassie blue and (b) an autoradiograph of the gel in (a).

Reproduced from Ninfa and Magasanik, Covalent modification of the glnG product, NRI, by the glnL product, NRII, regulates the transcription of the glnALG operon in *Escherichia coli. Proceedings of the National Academy of Sciences of the U S A*, 1986;**83**:5909–5913

convert EnvZ from a protein that transfers phosphate to OmpR to one that transfers phosphate to the RstA response regulator (see Figure 3b), thus effectively 'rewiring the specificity' of the system.

We previously discussed such a specificity swap experiment for an RNA (as applied to the identification of the Shine–Dalgarno interaction in Experimental approach 11.2), and we see here how very general and powerful the approach is for defining interactions between a range of biologically interesting molecules.

Find out more

Ninfa AJ, Magasanik B. Covalent modification of the *glnG* product, NR_I, by the glnL product, NR_{II}, regulates the transcription of the *glnALG* operon in *Escherichia coli*. *Proceedings of the National Academy of Sciences of the U S A*, 1986;**83**:5909–5913.

Nixon BT, Ronson CW, Ausubel FM. Two-component regulatory systems responsive to environmental stimuli share strongly conserved domains with the nitrogen assimilation regulatory genes *ntrB* and *ntrC*. *Proceedings of the National Academy of Sciences of the U S A*, 1986;**83**:7850–7854.

Skerker JM, Perchuk BS, Siryaporn A, *et al* Rewiring the specificity of two-component signal transduction systems *Cell*, 2008;**133**:1043–1054.

Related techniques

In vitro transcription assays; Section 19.8

Radiolabeling; Section 19.6

Genome editing; Section 19.5

SDS-PAGE; Section 19.7

Sequence alignment; Section 19.8

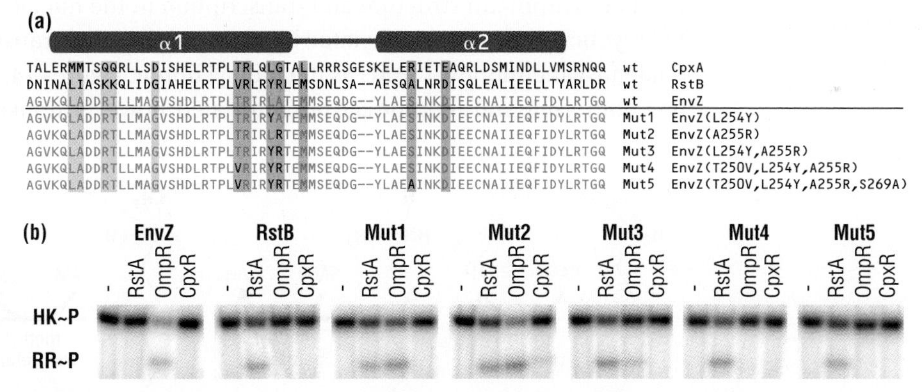

Figure 3 Primary sequence alignment of the histidine kinases EnvZ, RstB, and CpxA. (a) Residues showing the strongest covariation with residues in the cognate response regulator are highlighted in green and pink. Sequences of point mutants constructed in which amino acids with high inter-protein mutual information scores were mutated in EnvZ to match the corresponding amino acids in RstB. (b) Effects of mutations on ability of EnvZ histidine kinase (HK~P) to transfer phosphate to its cognate response regulator (RR~P) OmpR, the RstB cognate RstA, or the CpxA cognate CpxR. In these assays, the purified wild-type or mutant kinase protein was autophosphorylated and then incubated alone or examined for phosphotransfer to purified RstA, OmpR, or CpxR. The proteins were separated by SDS-PAGE, and phosphorylated proteins were detected by autoradiography.

hydrolysis reaction remains to be understood. In contrast with the acetylation of lysine side chains, *N*-acetylation of the N-terminus is usually not reversed.

What is the consequence of acetylation for the target protein? As with phosphorylation, acetylation affects the overall charge of the protein, in this case effectively neutralizing the positive charge of the lysine side chain. This neutralization of positive charge on lysine could, for example, reduce the affinity of proteins for negatively

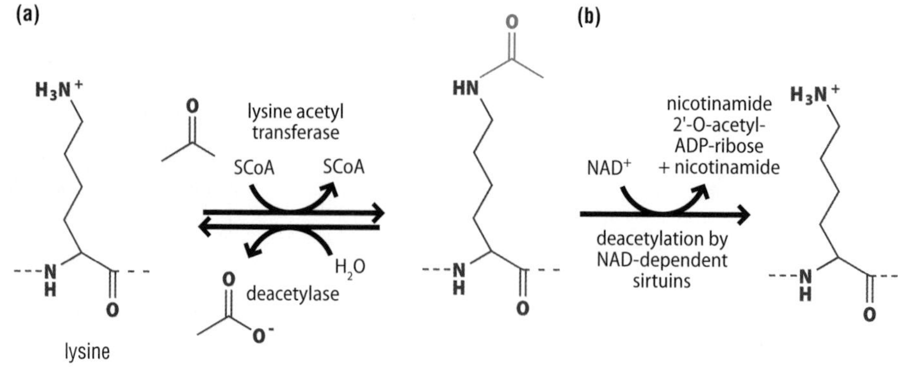

Figure 14.25 Reversible acetylation of lysine side chains. Acetyl transferase enzymes catalyzed the transfer of an acetyl group from acetyl-CoA to the amino group of lysine (for histones, acetyl transferases are denoted HATs). The acetyl group can be removed in two different ways by (a) deacetylases (for histones, deacetylases are denoted HDACs) or by (b) NAD-dependent sirtuins.

charged nucleic acids. Lysine side chain modifications on histone proteins have been shown to be critical for their ability to regulate access to chromosomal DNA.

The presence of the acetyl group also provides new chemical functionality for forming different interactions than the unmodified side chain. Just as phosphotyrosine is recognized specifically by an SH2 domain, acetylated side chains are recognized by specialized protein domains, known as bromodomains, which we learned about in Section 4.5. N-terminal *N*-acetylation blocks the action of aminopeptidases that degrade proteins, thus altering the lifetime of proteins in the cell.

A variety of side chains can be methylated

Methylation involves the addition of a methyl group, CH_3, to a target protein. A variety of proteins in both bacteria and eukaryotes have been found to be methylated. It is a common modification of histone proteins, which, like acetylation, regulates chromatin structure and transcription in the nuclei of eukaryotic cells. Eukaryotic proteins are methylated by a variety of **methyl transferases**, which use *S*-adenosylmethionine (SAM) as the methyl donor. Figures 14.26 and 14.27 show how both lysine and arginine can be methylated; however, unlike acetylation, more

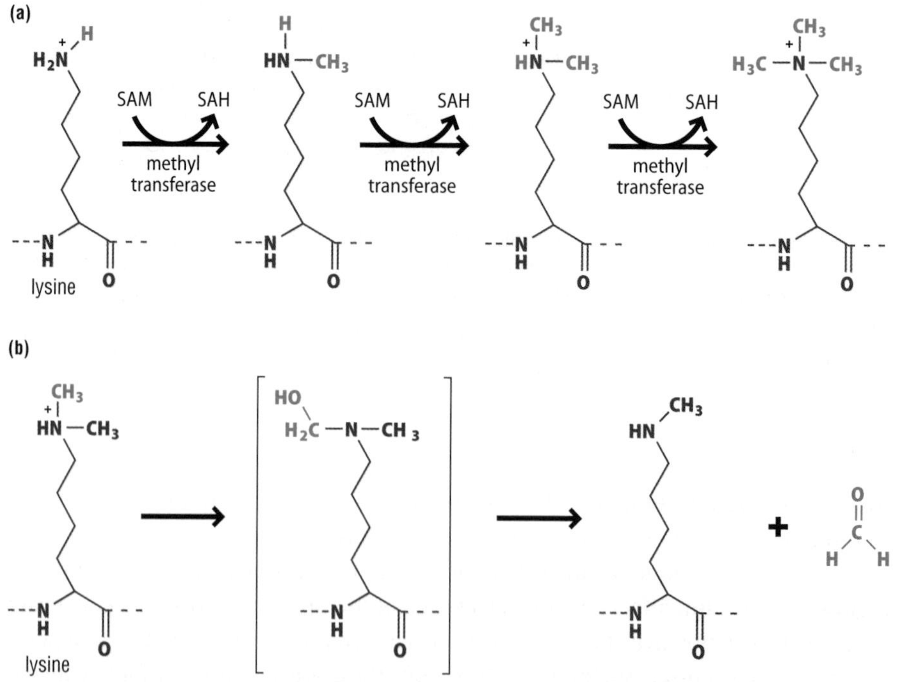

Figure 14.26 Reversible methylation of lysine residues. (a) Lysine may be mono-, di-, or tri-methylated by methyl transferases. The methyl donor is SAM resulting in the production of *S*-adenosylhomocysteine (SAH). (b) Methylated lysines can be demethylated by LSD1 or Jumonji family enzymes in reactions that involve a hydroxylated methyl intermediate and result in formaldehyde release.

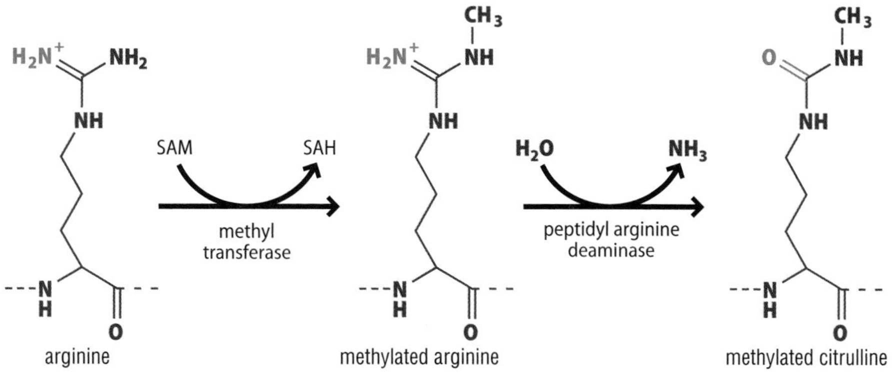

Figure 14.27 **Reversible acetylation of arginine residues.** Unmethylated and mono-methylated arginine can be converted to citrulline or methylated citrulline by the peptidyl arginine deaminase PADI4.

than one methyl group can be added to a given side chain. In addition, only some forms of methylation alter the overall electrostatic charge of the side chain.

Methylated lysine and arginine groups are chemically very stable, and it is only relatively recently that enzymes that remove the methyl group have been identified. One enzyme, LSD1, is an amine oxidase that demethylates mono- or di-methyl-lysine, releasing formaldehyde, as shown in Figure 14.26b. Another class of demethylases known as Jumonji domain-containing (JmjC) enzymes can remove up to three methyl groups from a lysine, also releasing formaldehyde. An intermediate in the demethylation reaction catalyzed by both of these types of enzymes is a hydroxylated methyl group. Methylarginine can be converted to citrulline, a non-canonical amino acid, by peptidyl arginine deaminase (PADI4), as shown in Figure 14.27. While this is not formally a reversal of arginine methylation, it does result in a chemical change that is proposed to play a role in transcriptional regulation.

A number of other side chains can also be methylated. Methyl groups can be covalently attached to oxygen atoms in aspartic and glutamic acid side chains, yielding a methyl ester form of these amino acids. This process is referred to as *O*-methylation. This modification is readily reversed because of the labile nature of an ester linkage. During bacterial chemotaxis, the cytoplasmic domain of a key transmembrane receptor becomes *O*-methylated at several glutamate residues by the methyl transferase, CheR. The modification is fully reversible by CheB, a glutamate ester demethylase. Other less common side chain methylations include *N*-methylation of histidine, glutamine, and asparagine and *S*-methylation of cysteine.

There are several different types of protein domains that recognize methylated lysine or arginine residues. As we saw in previous chapters, methyl-lysine found on histones can be recognized by chromodomains. Another well-characterized protein domain, the tudor domain, recognizes di-methylated lysines, whereas proteins with the WD40 domain recognize di- and tri-methylated lysine. Proteins that recognize methylated side chains often selectively bind to methylated side chains within a particular protein, suggesting that they recognize the modified side chain within the context of surrounding amino acid residues.

14.7 PROTEIN MODIFICATION BY NUCLEOTIDES

In the previous section, we focused on the reversible modifications phosphorylation, acetylation, and methylation, and the enzymes that catalyze the addition and removal of these small adducts. While these appear to be the most important and are certainly the best understood, it is important to bear in mind that other small reversible modifications have been identified, whose roles continue to be explored. For example, there are other acyl modifications such as butyryl, propionyl, crotonyl, succinyl, and malonyl groups that can also be transferred from coenzyme A (CoA) to lysines and whose functions are as yet poorly understood. Modern mass spectrometry methods have made it possible to identify these modifications in proteins, spurring investigations into their biological roles. As we will discuss next, another class of modifications that is receiving increasing attention is the addition of nucleotides to protein side chains.

Amino acids can be modified by one or more ADP ribose moieties

We saw in the previous section how the nucleotide, NAD^+, can serve as a co-substrate in certain types of deacetylation reactions. NAD^+ plays a very different role in a class of post-translational modification known as ADP ribosylation, in which NAD^+ is cleaved to release nicotinamide, and the remaining ADP ribose moiety is covalently attached to a side chain, most commonly glutamic acid, aspartic acid, or arginine. Enzymes that catalyze this reaction are generically known as ADP ribosyltransferases. A few enzymes that remove the modification have also been identified.

The attachment of a single ADP ribose as shown in Figure 14.28a, termed mono-ADP ribosylation, has diverse roles in biology, including the regulation of vesicular trafficking and transcription as well as the action of bacterial toxins. As a result, there are very different classes of enzymes that carry out ADP ribosylation on substrate proteins. Cholera toxin, for example, is an ADP ribosyltransferase that attaches ADP ribose to the G protein that regulates adenylate cyclase, which is responsible for synthesizing cyclic adenosine monophosphate (cAMP). Dysregulation of adenylate cyclase disrupts the proper functioning of selected ion channels and leads to the severe diarrhea characteristic of cholera.

In another form of ADP ribosylation, multiple ADP ribose units are attached to a side chain in the form of a linear or branched polymer in which one ADP ribose is attached to the next, as shown in Figure 14.28b. This modification, termed poly-ADP ribosylation, plays an important role in the response to DNA damage in mammalian cells as well as in telomere maintenance. Poly-ADP ribose is synthesized by poly-ADP ribose polymerase (PARP) enzymes, which both build up the poly-ADP ribose polymers and attach them to acidic side chains, and is broken down by poly-ADP ribose glycohydrolase (PARG) enzymes. Because of the intimate connection between efficient DNA repair and cancer, PARP inhibitors are being actively pursued as potential chemotherapeutic agents.

➔ We learn more about DNA damage and its repair in Chapter 15 and Chapter 16.

Modification of proteins with nucleotides is a ubiquitous mechanism used by pathogens to modify the host environment

We learned that ADP ribosylation is the mechanism by which cholera toxin modifies adenylate cyclase activity, resulting in uncontrolled efflux of ions. Bacteria have exploited the strategy of modifying host proteins, particularly regulatory proteins, to manipulate the host cell environment in a wide range of ways. The intracellular pathogen, *Legionella pneumophila,* provides another example of how the attachment of nucleotides to a GTP-binding protein changes the eukaryotic cell. As shown in Figure 14.29, once inside the host, *Legionella* secretes enzymes that sequentially catalyze the addition and removal of an adenosine monophosphate (AMP; denoted AMPylation) on the Rab1 protein, which regulates membrane trafficking. By reversibly converting the Rab1 protein into a constitutively active form, *Legionella* effectively diverts membranes to specialized organelles that surround and protect the bacterium.

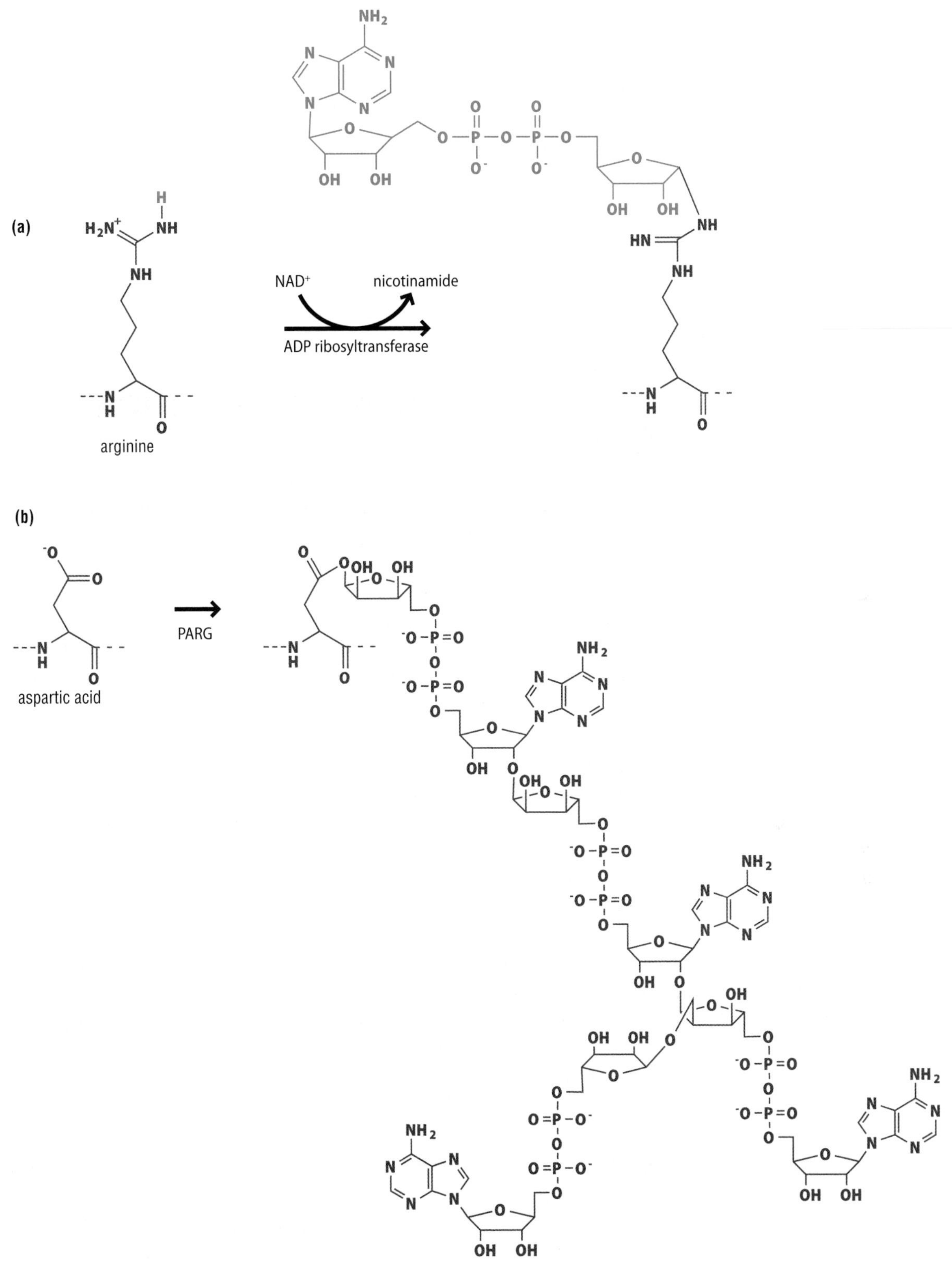

Figure 14.28 Reversible ADP ribosylation. (a) A single ADP ribose can be attached to a side chain by ADP ribosyltransferases. (b) Multiple ADP ribose units can also be attached to a side chain by poly-ADP ribose polymerase (PARP) enzymes. More than one type of linkage can occur, giving rise to linear and branched polymers.

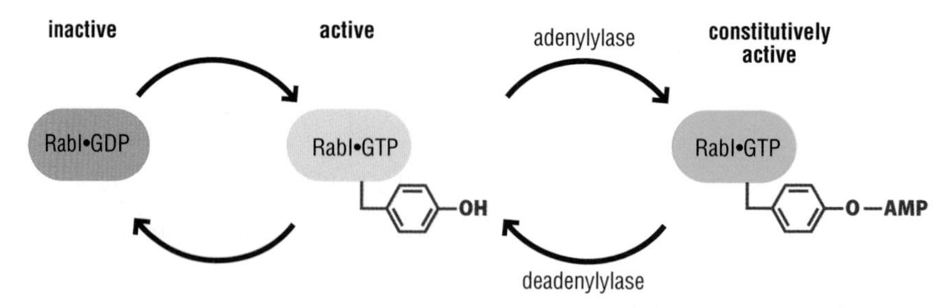

Figure 14.29 Reversible AMPylation. Adenylase enzymes secreted by the pathogen *L. pneumophila* modulate membrane trafficking by attaching an AMP to a tyrosine residue of the GTP-loaded Rab1 GTPase. This leads to constitutively active Rab1, because the GTP cannot be hydrolyzed until the AMP group is removed by deadenylase secreted by the bacteria later in infection.

14.8 DIRECT CHEMICAL MODIFICATION OF PROTEINS

Direct chemical modification of some amino acids can have both regulatory and detrimental effects

In addition to modifications that are catalyzed by specific enzymes, as discussed in Section 14.7, there are modifications that occur through direct chemical modification of specific amino acids without the help of an enzyme. These covalent changes may or may not be reversible. In some cases, the modifications are used to regulate protein activity whereas, in others, the covalent changes are detrimental and constitute a form of protein damage. Here we will discuss two of these chemical modifications - oxidation, involving the addition of oxygen, and nitration, involving the addition of nitrogen. We will also explore some consequences of these modifications.

Sulfur-containing amino acids are particularly sensitive to various forms of oxidation

A property of molecular oxygen, which contributes to its chemical reactivity, is that it can lose electrons to give rise to reactive oxygen species such as superoxide (O_2^-), hydroxyl radical (•OH), and peroxides (ROOH). (It is something of a paradox that we depend on oxygen for our survival, yet its chemical reactivity can make it toxic to the human body.) These reactive oxygen species can react with amino acids to covalently modify specific side chains. Cells possess a number of enzymes capable of eliminating reactive oxygen species, thus protecting the cell from oxidative damage. However, a variety of conditions, such as the body's response to infection, can give rise to oxidative stress whereby higher than normal levels of reactive oxygen species are produced in tissues, leading to oxidative damage of proteins and other biological molecules. The amount of oxidative damage generally increases in aging cells and is thought to contribute to the progression of age-related diseases.

Several examples of oxidatively modified amino acids are given in Figure 14.30. The chemical properties of the cysteine residue make it particularly well suited to facilitating chemical reactions, as well as binding to transition metals, as we learned in Chapter 2. These same properties render the thiol side chain of cysteine vulnerable to oxidation. As we saw in Section 14.1, the formation of disulfide bonds between

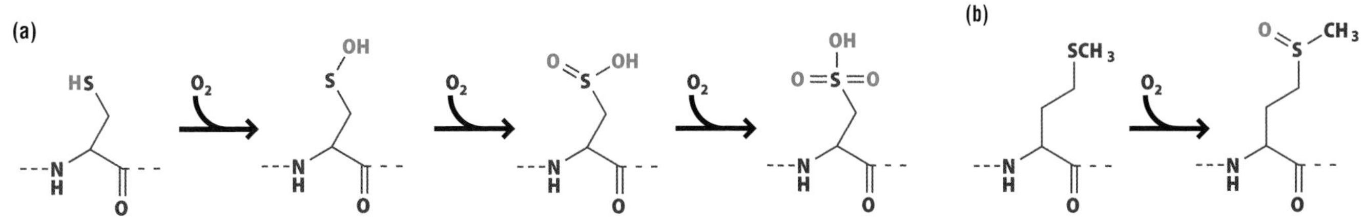

(a) cysteine → O_2 → sulfenic acid → O_2 → sulfinic acid → O_2 → sulfonic acid

(b) methionine → O_2 → methionine sulfoxide

Figure 14.30 Examples of amino acid oxidation. Formation of (a) cysteine sulfinic acid, cysteine sulfenic acid, cysteine sulfonic acid, and (b) methionine sulfoxide.

cysteine side chains is a positive consequence of oxidation: the formation of disulfide bonds between cysteine residues is critical to the proper folding of some proteins. Cysteine residues also can be directly oxidized by the covalent addition of one, two, or three oxygen atoms, producing sulfenic, sulfinic, and sulfonic acid, respectively (see Figure 14.30a). The tripeptide glutathione (γ-glutamylcysteinylglycine), which helps to maintain the reducing environment inside many cells, can also form adducts with single cysteine side chains through a disulfide bond, in a process called glutathionylation. All these modifications can alter the activity of the protein.

Glutathionylation and sulfenic acid formation can be reversed by thiol reductase enzymes or by other thiol-containing molecules (with reduced sulfur groups) in the cell. The enzyme, sulfinic reductase, specifically reverses the two-oxygen modification, while oxidation to a sulfonic acid is thought to be irreversible.

Two examples involving cysteine oxidation to sulfenic acid illustrate the varied effects of this modification. On the one hand, the oxidation of a unique cysteine in a bacterial transcriptional regulator is required for the protein to bind to DNA and activate transcription of a gene encoding an oxidative stress defense gene. On the other hand, oxidation of an active-site cysteine to sulfenic acid completely inactivates certain protein tyrosine phosphatases in eukaryotes. Thus, when cells are exposed to elevated levels of oxidants, pathways that depend upon the enzyme to remove phosphate groups from target proteins are disrupted. Inactivation of phosphatase enzymes by this oxidation event can be reversed by reduction of the active-site cysteine, restoring the original thiol group.

The other sulfur-containing amino acid, methionine, can also be oxidized by oxygen derivatives, giving rise to methionine sulfoxide, as shown in Figure 14.30b). Enzymes that reverse methionine oxidation, called methionine sulfoxide reductases, are widespread in biology, indicating a broad need for these enzymes. As a specific example, part of the pathology of emphysema results from oxidation of a critical methionine residue on a proteinase inhibitor. The methionine sulfoxide modification greatly lowers the affinity of the proteinase inhibitor for target proteases, such as elastase, which are found in lung tissue. The resulting increase in elastase activity causes the lung to deteriorate.

Proteins can be modified by reaction with nitric oxide

Side chains can also be covalently modified by reactive nitrogen species in the cell, resulting in ***S*-nitrosylation** and *O*-nitration. Reactive nitrogen species arise from the synthesis of the reactive gas, nitric oxide (NO), which is produced by specialized enzymes in eukaryotes, where it acts as a second messenger. Nitric oxide and other reactive nitrogen species can also be produced as a side product of nitrogen metabolism in bacteria.

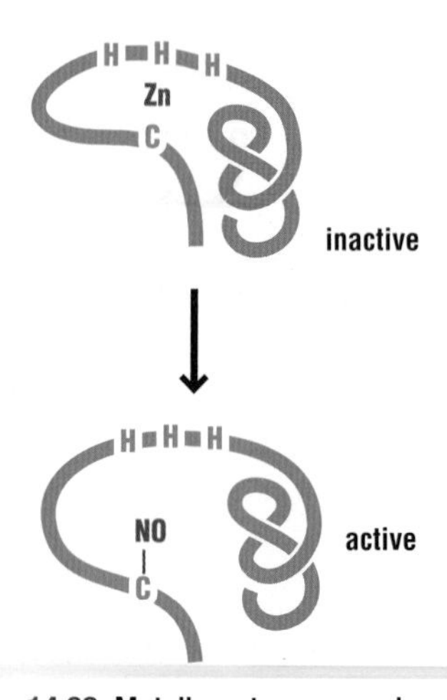

Figure 14.32 Metalloproteases can be activated by cysteine *S*-nitrosylation. The coordination of a zinc by three histidine residues and one cysteine residue blocks the activity of some metalloproteases. If the cysteine becomes S-nitrosylated, it can no longer coordinate the zinc resulting in activation of the enzyme. The enzymes also can be activated by proteolytic cleavage of the region encompassing the coordinating cysteine.

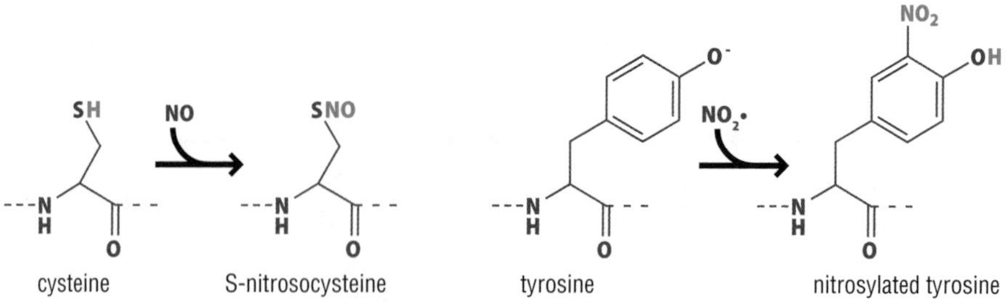

SH NO SNO O^- $NO_2\cdot$ NO_2 OH
cysteine S-nitrosocysteine tyrosine nitrosylated tyrosine

Figure 14.31 Examples of amino acid nitration. Formation of *S*-nitrosylcysteine and *O*-nitrated tyrosine.

Reactive nitrogen species have chemical properties similar to reactive oxygen species and can therefore also react with proteins, giving rise to a form of covalent modifications. Some of the same amino acids that are oxidized can also be subject to modification by reactive nitrogen species. As has been found for oxidation, these modifications can have a range of consequences. The reactive cysteines of many proteins can be modified directly by NO, thereby becoming *S*-nitrosylated. Tyrosine can also react with NO-derived species such as nitrogen dioxide, becoming *O*-nitrated. These modifications are shown in Figure 14.31. *S*-nitrosylation is chemically reversible by free thiols, although the mechanism by which this modification is removed in cells is not known in most cases.

As we have seen for other side chain modifications, protein activity can be regulated by the adducts derived from reactive nitrogen species. An example can be found in zinc-containing metalloproteases, whose active site is generally blocked by a zinc atom coordinated by three histidine side chains and one cysteine side chain. As illustrated in Figure 14.32, if the cysteine becomes *S*-nitrosylated, the zinc–cysteine coordination is disrupted, and the protease becomes activated.

It is easy to see similarities among the different post-translational modifications that we describe here and in earlier sections of this chapter. The distinguishing feature of the oxidation and nitration reactions is that these modifications are connected to the redox state of the cell, that is, the balance of reducing and oxidizing equivalents, allowing another key feature of cellular homeostasis to be monitored.

14.9 UBIQUITINATION AND SUMOYLATION OF PROTEINS

In the previous sections, we discussed relatively small post-translational modifications that are broadly used by the cell to modulate protein function. As we saw in our discussion of protein glycosylation, however, post-translational modifications can also be quite large and elaborate. Here we discuss another set of post-translational modifications that involves the attachment of an entire small protein to a polypeptide chain.

Covalent modification by ubiquitin and related proteins is required for a variety of cellular functions

A distinct set of post-translational modifications involves the covalent attachment of a small protein to lysines on a target protein. The most common and best-characterized of these proteins is ubiquitin, a compact, 76-amino acid protein found in all eukaryotes. Ubiquitin is covalently attached to lysine side chains in

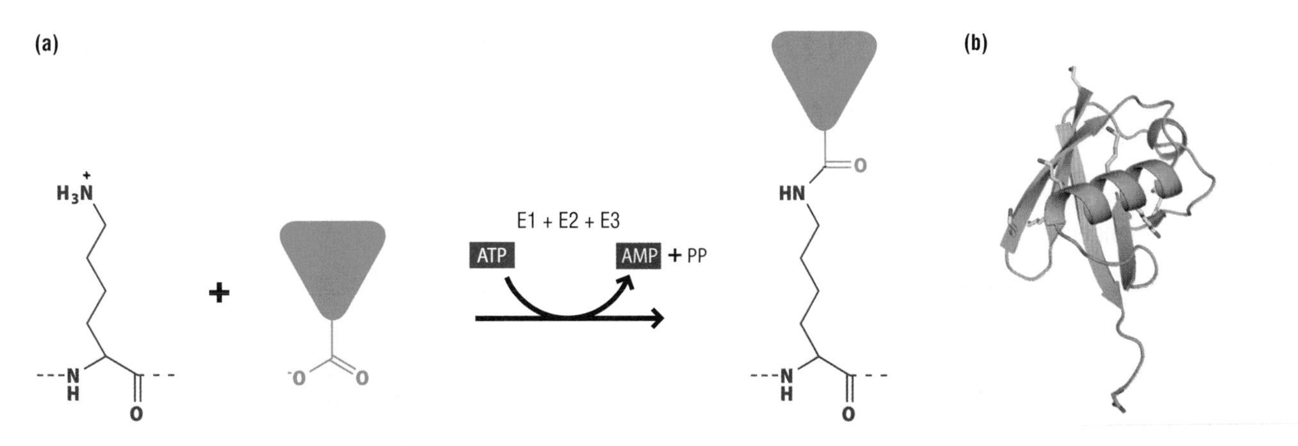

Figure 14.33 Ubiquitination. (a) Isopeptide bond between ubiquitin and a lysine side chain. (b) Three-dimensional structure of ubiquitin. The seven lysines in ubiquitin are shown in stick representation. PDB code 1UBQ.

a process known as **ubiquitination** (also denoted ubiquitylation), in which the C-terminus of ubiquitin is covalently attached to the lysine ε-amine by a peptide-like bond known as an isopeptide linkage. This process is depicted in Figure 14.33. This modification adds about 8500 Da for each ubiquitin attached, which is more than 100 times the molecular weight of a single phosphate group.

Ubiquitin has seven lysine residues which themselves can be ubiquitinated, giving rise to a polyubiquitin chain. There are different types of polyubiquitin chains, each distinguished by the ubiquitin lysine through which one ubiquitin is attached to the next. The C-terminus of one ubiquitin can also be conjugated to the N-terminal amine of the next, resulting in a linear polymer of multiple ubiquitins. The completely different configurations of polyubiquitin chains linked to lysine 48 (K48-linked) and lysine 63 (K63-linked) can be seen in Figure 14.34.

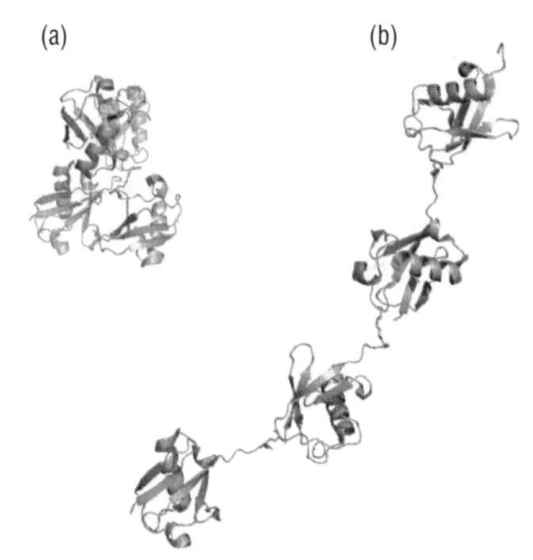

Figure 14.34 Structures of K48- and K63-linked polyubiquitin chains. The structures of (a) K48-linked (PDB code 2O6V) and (b) K63-linked (PDB code 3HM3) tetra-ubiquitin chains illustrate the significantly different conformations that result from different linkages in polyubiquitin chains.

The attachment of a single ubiquitin protein or any one of the different types of polyubiquitin chains can each specify distinct biological outcomes. Ubiquitin was initially identified as the modification that marked proteins for proteolytic degradation by a large multiprotein complex known as the **proteasome** (see Experimental approach 14.3). However, as will be discussed below, subsequent studies revealed that ubiquitin also plays non-degradative roles in a variety of processes including transcription regulation, the DNA damage response, and the subcellular localization of biological molecules. The precise chemical nature of the ubiquitin modification (position on the protein, length and type of ubiquitin polymer) determines the functional consequences and the fate of the modified protein.

Ubiquitin attachment and chain elongation occur through a series of enzymatic steps

Ubiquitin is attached to proteins in a cascade of reactions catalyzed by enzymes known generically as E1 (ubiquitin activating), E2 (ubiquitin conjugating), and E3 (ubiquitin ligase) enzymes, as shown schematically in Figure 14.35. In the first step, the ubiquitin C-terminus is covalently attached by a thioester bond to the active-site cysteine of the E1 enzyme. This reaction is powered by ATP hydrolysis using a two-step mechanism similar to that described for transfer RNA (tRNA) synthetases in Section 11.2. The E1–ubiquitin conjugate then interacts with an E2 enzyme and catalyzes the transfer of the ubiquitin onto the E2 active-site cysteine. This step is a transthiolation (or transesterification) reaction, since the final product is also a

14.3 EXPERIMENTAL APPROACH

Discovery of ubiquitin-dependent proteolysis

Fractionation of cell extracts led to the discovery of ubiquitin and the ATP-dependent proteosome

Aaron Ciechanover, Avram Hershko, Irwin Rose, and colleagues set out to address the basic question of why protein degradation in mammalian cells requires energy. This path of exploration led them to discover that degradation depended on the covalent attachment of a 76-amino acid protein known as ubiquitin, and to elucidate a pathway of central importance in eukaryotic cell biology. These scientists took a biochemical approach and simply examined proteolysis in cellular lysates (extracts) made from reticulocyte cells (immature red blood cells). They fractionated the lysate by column chromatography and identified two important fractions – a low-molecular weight fraction (fraction I), which contained a small heat-stable protein (initially named APF-1, but later found to be ubiquitin), and a second, higher-molecular weight fraction (fraction II), which in retrospect probably contained the proteasome and enzymes required for ubiquitination – that when combined yielded ATP-dependent proteolysis *in vitro*.

Ciechanover *et al*. next asked whether APF-1 directly associated with components in the system by incubating ^{125}I-labeled APF-1 with fraction II. Their experiments showed that APF-1 did in fact associate with proteins in the high-molecular weight fraction, in an ATP-dependent manner. The association was strikingly resistant to high pH, suggesting that the interaction was covalent. In a subsequent paper, Hershko *et al*. reported that denatured substrates, such as lysozyme, were covalently labeled with multiple APF-1 monomers to generate a discernible ladder (see Figure 1, bands C_1–C_5). Since the levels of the APF-1-labeled form of the protein decreased over time, Hershko *et al*. proposed that the ATP-dependent linkage of APF-1 to the substrate is followed by proteolytic breakdown. So, in answer to the original question, ATP is required both for the activation of ubiquitin (so that it can be transferred to targeted proteins), and for the proteasome that uses the energy to catalyze proteolysis.

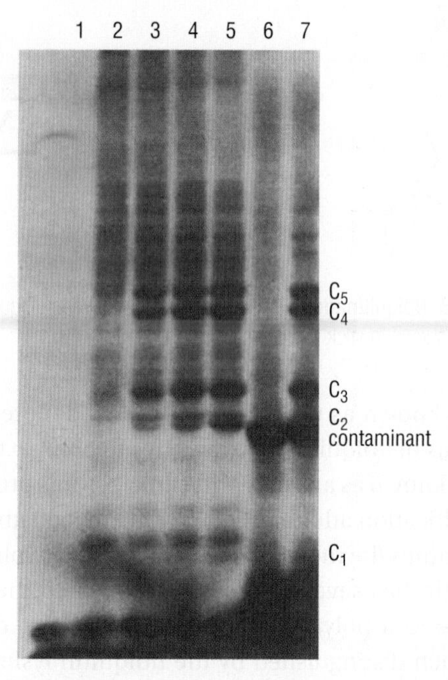

Figure 1 Formation of covalent compounds between APF-1 and lysozyme in an ATP-dependent reaction. [^{125}I]-APF-1 was incubated together with fraction II without ATP (lane 1), with ATP (lane 2), and with ATP and 5 µg (lane 3), 10 µg (lane 4), or 25 µg (lane 5) of unlabeled lysozyme. In addition [^{125}I]-lysozyme was incubated with unlabeled APF-1 without ATP (lane 6) or with ATP (lane 7). The same five bands, C1–C5, are observed when either protein is labeled and incubated with the other unlabeled protein in the presence of ATP.

Reproduced from Hershko *et al.* Proposed role of ATP in protein breakdown: conjugation of protein with multiple chains of the polypeptide of ATP-dependent proteolysis. *Proceedings of the National Academy of Sciences of the U S A*, 1980;**77**:1783–1786.

The SILAC mass spectrometric approach was used to identify proteins linked to lysine 11 of ubiquitin

Since the initial discovery that ubiquitin became covalently attached to lysozyme, hundreds of other proteins that are modified by ubiquitin (or other ubiquitin-like proteins) have been identified. Moreover, it has become clear that ubiquitin can be attached via any one of its seven lysine residues, and that the specific residue that is utilized impacts the fate of the ubiquitinated protein. Many questions arise from this diversity, including what determines the specificity for ligation and what are the functions of the different conjugations.

In one study, Mark Hochstrasser, Daniel Finley, Junmin Peng, and colleagues set out to identify lysine 11-specific linked proteins in *Saccharomyces cerevisiae*, knowing these linkages to be abundant *in vivo*, but poorly characterized. The investigators took a high throughput approach and profiled the total yeast proteins and ubiquitinated proteins in a wild-type strain and in a mutant strain in which lysine 11 of ubiquitin was mutated to arginine. The authors used a quantitative mass spectrometric approach termed SILAC (stable isotope labeling with amino acid in cell culture, discussed in Section 19.11) wherein wild-type and mutant cells were labeled with heavy and light stable isotopes, respectively, and then mixed. Total proteins from cell lysates, and those purified on the basis of their conjugation to ubiquitin, were separated on SDS gels. The separated proteins were then excised from the gel, digested with trypsin and analyzed by mass spectrometry; the results are shown in Figure 2. The peptide fragment masses detected by mass spectrometry allowed the different proteins to be identified, while the relative levels of the various fragments (and thus intact protein amounts) were assessed by comparison of the two isotopes.

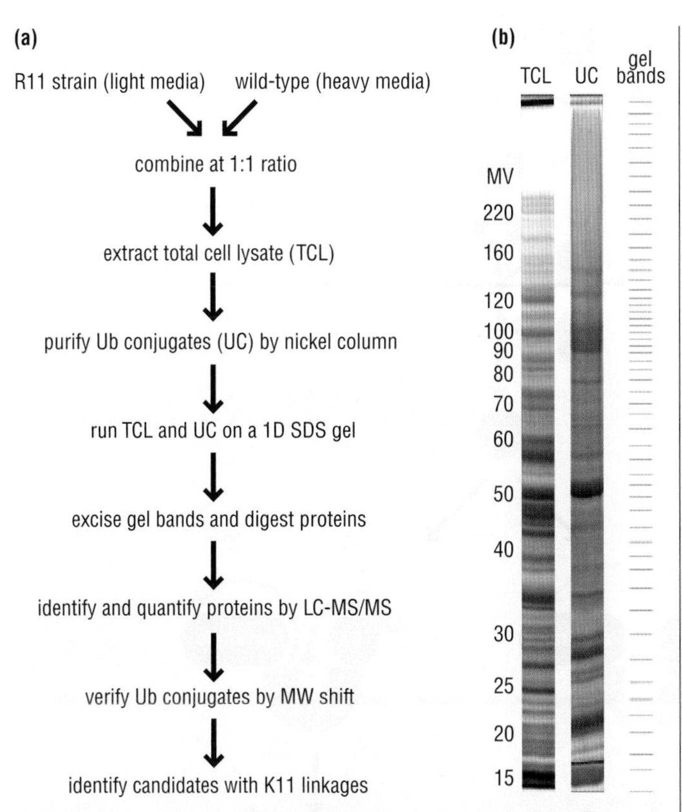

Figure 2 Use of SILAC for proteomic analysis of wild-type and mutant yeast strains. (a) Outline of the SILAC method for comparing total cell lysate (TCL) and purified ubiquitin conjugates (UC) from wild-type and mutant strains. (b) Comparison of the TCL and UC by SDS-PAGE showing the great number of proteins analyzed. The gel was excised into ~50 gel bands.

Xu *et al*. anticipated that potential lysine 11-ubiquitin-linked proteins would be present at higher levels in the mutant strain as a result of reduced rates of degradation in the absence of lysine 11-specific ubiquitin conjugation. A number of candidate ubiquitinated proteins were identified (347 to be exact), and a few of these were subsequently confirmed through independent assays to be modified with lysine 11-linked polyubiquitin. Although a fairly broad set of proteins appears to be targeted by this modification, one of the most intriguing candidates identified was Ubc6 (see Figure 3), itself an E2 ubiquitin-conjugating enzyme with known roles in the endoplasmic reticulum-associated degradation (ERAD) pathway. This lead was supported by subsequent experiments showing hypersensitivity of the mutant strain to reagents that induce stress in the endoplasmic reticulum. Like more traditional genetic approaches, the excitement for high-throughput approaches comes from their potential to identify new connections in our quest to understand biology.

Find out more

Ciechanover A, Heller H, Elias S, Haas AL, Hershko A. ATP-dependent conjugation of reticulocyte proteins with the polypeptide required for protein degradation. *Proceedings of the National Academy of Sciences of the U S A*, 1980;**77**:1365–1368.

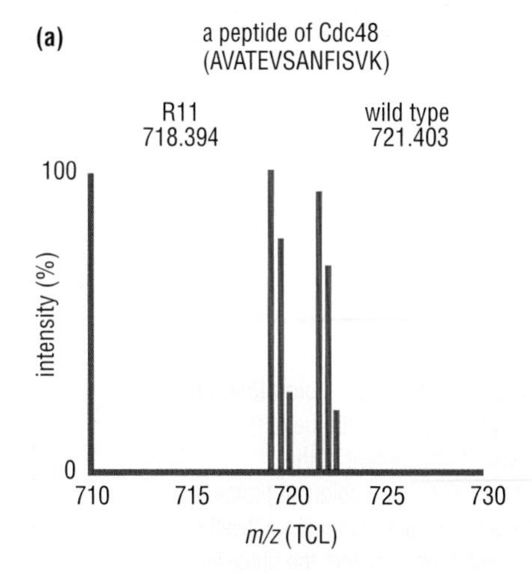

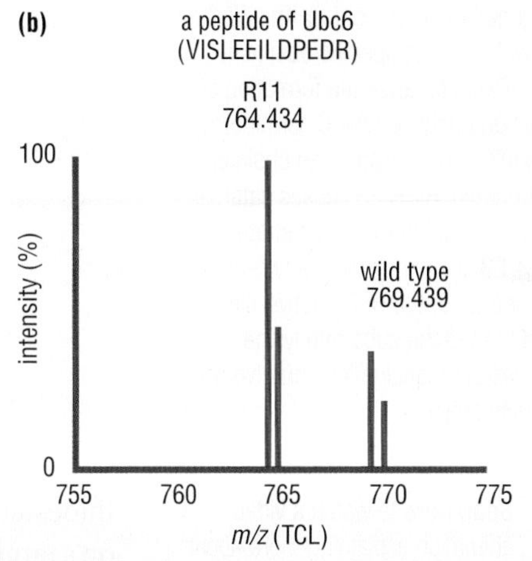

Figure 3 Isotope-labeled peptide pairs of Cdc48 (a) and Ubc6 (b) in the total cell lysate. The light- and heavy-labeled peptides were distinguished by their different mass-to-charge (m/z) ratios. While the intensity of the peak for Cdc48 is roughly equivalent in the two strains, the intensity of the Ubc6 peptide is markedly different, consistent with the idea that ubiquitin-dependent degradation is reduced in the mutant strain.

Hershko A, Ciechanover A, Heller H, Haas AL, Rose IA. Proposed role of ATP in protein breakdown: conjugation of protein with multiple chains of the polypeptide of ATP-dependent proteolysis. *Proceedings of the National Academy of Sciences of the U S A*, 1980;**77**:1783–1786.

Xu P, Duong DM, Seyfried NT, *et al.* Quantitative proteomics reveals the function of unconventional ubiquitin chains in proteasomal degradation. *Cell*, 2009;**137**:133–145.

Related techniques

Extract fractionation; Section 19.7

Protein radiolabeling; Section 19.6

SDS-PAGE; Section 19.7

S. cerevisiae (as a model organism); Section 19.1

SILAC; Section 19.11

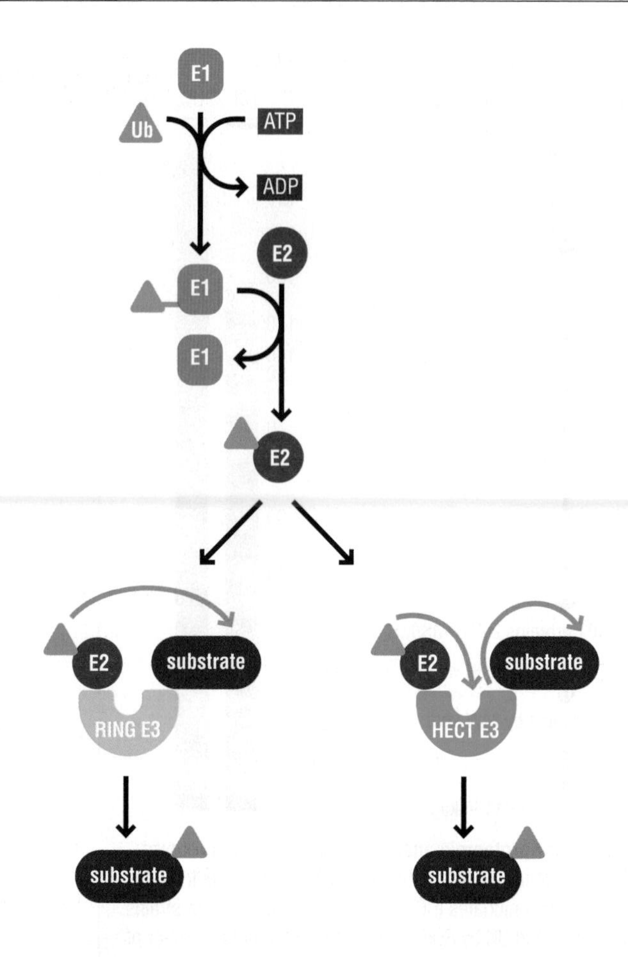

Figure 14.35 Steps in protein ubiquitination. Protein ubiquitination begins with the ATP-dependent covalent attachment of the C-terminus of the 76-residue ubiquitin protein to the active-site cysteine of an ubiquitin-activating enzyme (E1), which then transfers the ubiquitin to the active-site cysteine of one of a family of ubiquitin-conjugating enzymes (E2). The E2 then collaborates with a target-specific ubiquitin-protein ligase (E3) to catalyze the formation of an isopeptide bond between the C-terminal glycine of ubiquitin and a lysine side chain on the substrate protein. RING E3 ligases catalyze direct transfer of ubiquitin to the substrate, whereas HECT E3 ligases mediate a two-step transfer from the E2 to the HECT active-site cysteines and then to the substrate lysine. Additional cycles of ubiquitination can give rise to a polyubiquitin chain.

Scan here to watch a video animation explaining more about the process of ubiquitin-mediated protein degradation, or find it via the **Online Resource Center** at **www.oxfordtextbooks.co.uk/orc/craig2e/**

thioester-linked ubiquitin. The final step is the transfer of the ubiquitin to a lysine on a target protein. This step depends on the E3 ligase, which helps catalyze transfer of ubiquitin from the active-site cysteine of the E2 to the amino group of a lysine side chain in the target protein (see Figure 14.35).

Some E3 ligases (the RING class) promote the direct attack of the substrate lysine on the E2-ubiquitin thioester, whereas others (such as the HECT class) contain an active-site cysteine and catalyze a second transthiolation reaction followed by conjugation to the substrate lysine. Another E3 ligase of this type, called LUBAC, catalyzes attachment of ubiquitin to the N-terminal amine of another ubiquitin, a reaction that is chemically identical to lysine ubiquitination. It is thought that ubiquitination of particular substrate lysines is directed by the combined action of both the E2 and the E3 ligase, although the precise determinants of substrate specificity, or even the specificity of a given E2–E3 pair, remain to be elucidated.

Polyubiquitin chains consisting of one ubiquitin C-terminus linked to a lysine residue in the next are assembled in the same E1, E2, E3 enzymatic cascade. Polyubiquitin chains usually have a uniform linkage type, meaning that each ubiquitin is joined to the same lysine residue in the next ubiquitin. For example, K48-linked polyubiquitin chains are formed by linking the C-terminus of one ubiquitin to lysine 48 in the next, while K63-linked chains contain bonds between the C-terminal residue and lysine 63. This linkage specificity is thought to be governed primarily by the E2 enzyme. In the case of K63-linked polyubiquitin chains, there is also an additional subunit (called a ubiquitin E2 variant, or UEV) that binds to the E2 enzyme and positions the growing chain such that the next ubiquitin monomer is attached to lysine 63 only.

Although most eukaryotic cells contain a single type of E1 enzyme, they typically have a dozen or more different E2 enzymes and hundreds of E3 enzymes. The human genome contains around 30 E2 enzymes and over 500 E3 enzymes. It is not surprising that only a single E1 enzyme is needed for the uniform task of chemically activating the ubiquitin protein in an ATP-dependent reaction that is independent of the diversity of downstream targets. E2 enzymes are generally well conserved in overall structure and function, with some variation in sequence that affects their ability to synthesize polyubiquitin chains and interact with particular E3 ligases. The large number of E3 enzymes and the great diversity in the different domains they contain in addition to E2-binding regions reflects the need of the cell to very specifically target particular protein substrates for ubiquitination, as well as its need to interact with a variety of E2 enzymes.

Ubiquitination is reversed by deubiquitinating enzymes

Deubiquitinating enzymes are a diverse group of specialized proteases that hydrolyze the isopeptide bond between ubiquitin and lysine. These enzymes vary greatly in their substrate specificity and function: some do not appear to discriminate among different isopeptide linkages or substrates, whereas others are highly specific for breaking down specific types of polyubiquitin chains into monomers or for removing ubiquitin from particular substrate proteins. The different functions of deubiquitinating enzymes are reflected in the different types of catalytic domains they contain - there are four classes of cysteine proteases and one class of metalloprotease - as well as in additional domains or binding partners that govern their activity and specificity. The human genome encodes over 90 deubiquitinating enzymes, reflecting the diverse roles these enzymes play in specific pathways.

These opposing actions of ubiquitin conjugating and deubiquitinating enzymes enables ubiquitin to serve as a reversible signal, much in the way that covalent modifications such as phosphorylation and acetylation play a regulatory role in the cell. The number of eukaryotic genes devoted to ubiquitination and deubiquitination is on the same order as the number of genes devoted to phosphorylation and dephosphorylation, underscoring the comparable importance of both processes. Distant homologs of ubiquitin-modification enzymes that catalyze chemically similar sulfur-transfer reactions have been found in some bacteria, reflecting the ancient origin of the ubiquitin enzyme system.

Different types of ubiquitin modifications have distinct functional consequences

The diverse ubiquitin modifications - monoubiquitin or one of the eight types of polyubiquitin chains - can each play a different role in the cell (see Figure 14.36). Perhaps the best understood is the targeting of proteins for degradation by covalent attachment of a K48-linked polyubiquitin chain. The K48-linked chain is recognized by the proteasome, which proteolyzes the protein into short peptides. We shall learn more about the mechanism of protein degradation by the proteasome in the next section.

In contrast to modification by a K48-linked polyubiquitin chain, modification with K63-linked polyubiquitin chains does not lead to degradation of the modified protein. Instead, K63-linked chains modulate the activities of proteins with critical roles in the DNA damage response described in Section 15.9, as well as

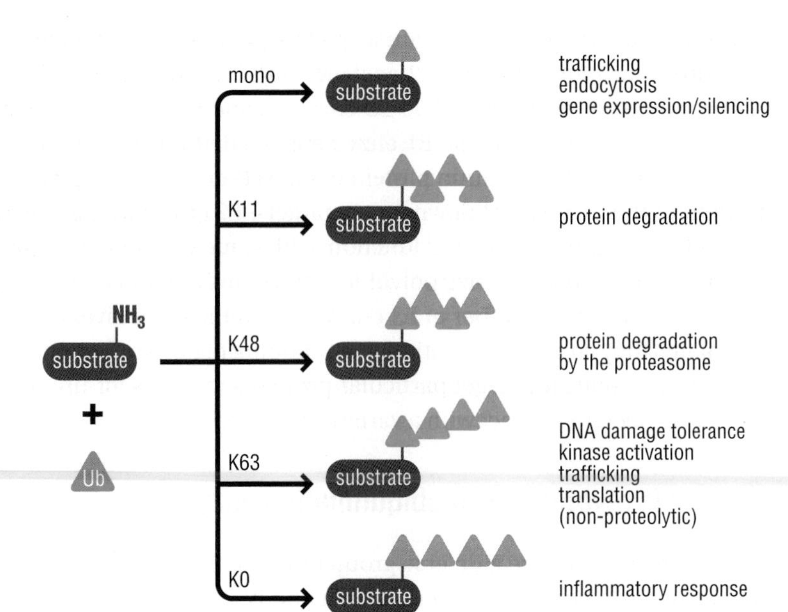

Figure 14.36 Functional consequences of different ubiquitin modifications. Each type of polyubiquitin chain is classified according to the lysine side chain in ubiquitin to which the C-terminus of the next ubiquitin is covalently attached.

Adapted from Pickart, CM, and Fushman, D. Polyubiquitin chains: polymeric protein signals. *Current Opinion in Chemical Biology*, 2004;**8**:610–616.

in the cascade of cytoplasmic events in the inflammatory response that lead to translocation of the NF-κB transcription factor into the nucleus described in Section 9.8. In the case of the DNA damage response, double-strand breaks trigger a chain of protein modifications at the break site, including attachment of K63-linked polyubiquitin to histones. The K63-linked chain recruits a protein complex containing BRCA1 (breast cancer type 1 susceptibility protein), an E3 ligase that plays a role in DNA repair by homologous recombination.

The attachment of a single ubiquitin to a substrate can also play a signaling role that does not involve proteasomal degradation. We learned in Section 4.5 that histone monoubiquitination plays an important role in nucleosome structure and transcriptional regulation. Again, ubiquitin exhibits a non-degradative role in this process.

Ubiquitin can act as a signal because there are protein domains that specifically recognize the different types of ubiquitin modifications. Several domains that interact with mono- or diubiquitin have been identified. These include the UIM (ubiquitin interacting motif), CUE (coupling of ubiquitination to endoplasmic reticulum degradation), and UBA (ubiquitin-associated) domains. In the example above, BRCA1 is found in a complex with a protein called RAP80 that binds selectively to K63-linked polyubiquitin with its UIM. In some cases, proteins with more than one ubiquitin recognition domain have been found to bind to longer polyubiquitin chains. It seems clear that recognition of specific ubiquitinated proteins is a relatively complex process that depends on recognition of the structure of the protein surrounding the modification, the length and type of ubiquitin polymer, and perhaps other post-translational modifications found on the protein.

Ubiquitin-like proteins and their functions

Ubiquitin-like protein	Function
SUMO	transcription, DNA damage
NEDD8	modifies CULLEN E3 ligases
ISG15	antiviral immune response
FAT10	unknown
UFM1	tRNA modification
ATG8	autophagy
ATG12	autophagy; is conjugated to a lipid

Figure 14.37 Ubiquitin-like proteins and their functions.

Modification by distinct ubiquitin-like proteins regulates protein activity

There are other small proteins that are similar to ubiquitin in their structure and in the enzymatic machinery used to attach them to substrates. Called ubiquitin-like proteins (UBLs), they serve signaling functions that are distinct from ubiquitin, as indicated in Figure 14.37. SUMO (which stands for Small Ubiquitin-related

Modifier) is one example of a ubiquitin-like protein; it is covalently attached to lysine side chains in a process known as **sumoylation**. The attachment of SUMO to a substrate relies on a dedicated set of E1, E2, and E3 enzymes that are similar in structure and in mechanism to those that ubiquitinate substrates.

Sumoylation plays an important role in gene expression and in targeting proteins to the nuclear pore complex. SUMO may also control protein degradation by competing with ubiquitin for common lysine residues. For example, attachment of SUMO to a lysine side chain can prevent attachment of a polyubiquitin chain and hence protect the protein from degradation through the ubiquitin–proteasome pathway. Although polymeric SUMO has been detected in cells, the biological function of this modification remains to be determined.

14.10 PROTEIN DEGRADATION

Cells must balance requirements for protein synthesis and degradation

Throughout this chapter, we have discussed the post-translational events that influence a protein's life following its release from the ribosome. The folding pathways and the post-translational modifications described in the previous sections illustrate the complexity of the processes that dictate the localization and activity of a protein molecule. The last step in the cellular life of a protein is its degradation. The cell must balance the requirements for the production of fully folded functional proteins with the need to eliminate protein activity when their function is no longer needed. Here we discuss the machineries in the cell that are dedicated to the destruction of proteins. Not surprisingly, these systems are tightly regulated to meet the ever-changing needs of the cell and to prevent unwarranted destruction of proteins that are still required. In contrast to many of the other processes that we have discussed in this chapter (primarily post-translational modifications), the process of degradation is not reversible; such finality requires this process to be strenuously controlled.

The bacterial and eukaryotic machineries for energy-dependent protein destruction are highly related

Related molecular machines power the energy-dependent unfolding and degradation of most proteins in bacteria and eukaryotes. These proteolytic machines are composed of two basic functionalities: an ATPase and a peptidase, which can reside either on the same or on different polypeptide chains. The polypeptide chains oligomerize to form a barrel-like structure with ATPase activities and small pores at both ends, and a protected internal cavity that contains the active sites of the degradative protease. The best-known eukaryotic degradation machine, referred to as the proteasome, is composed of about 30 different polypeptide chains. This structure is depicted in schematic form in Figure 14.38. The bacterial versions of this machine tend to be somewhat less complex. For example, the FtsH and Lon proteases are composed of just a single type of monomer (containing both the ATPase and proteolytic functionalities). While there are at least five ATP-dependent proteases in bacteria, there are probably many more in eukaryotes. Here we will focus our discussion on the well-characterized proteasome.

Figure 14.38 Proteasome structure. Proteins targeted for destruction (gray) are fed into a multiprotein complex where they are degraded in the tunnel-like enzymatic core. Bacterial ATP-dependent proteases have a similar structure although they lack the cap (dark blue) at either or both ends. PDB code 1PMA.

Adapted from Petsko and Ringe. *Protein Structure and Function*. Oxford: University Press, 2004.

The structure of the protease has obvious functional consequences

Even the smallest folded proteins have dimensions that make them unable to enter the cavity of the degradation proteases, thus providing a clear answer to the question of how most proteins in the cell escape the actions of roving proteolytic machines. The ATPase domains surround the entry portal, an ideal location that allows them to select desirable proteins and then mechanically denature and thread them into the protected central cavity. The amount of ATP that is needed to denature a range of substrates (of varying stability) is quite variable and, for the ClpXP protease, can range from 30 to 80 ATPs hydrolyzed per about 100 amino acids of protein sequence.

Once substrates are translocated into the central cavity, the effective concentration of protease active sites in the small chamber is extremely high (on the order of hundreds of mM), thus resulting in rapid degradation of the protein into peptide fragments ranging from five to 20 amino acids in length. Smaller fragments are thought to readily diffuse out of the chamber whereas larger fragments may depend on further ATP-dependent processes to be released.

Ubiquitin, molecular recognition sequences, and adaptor proteins collaborate to identify proteins that need to be degraded

The finality of protein degradation demands that this cellular event be highly selective. How then are the proper proteins recognized? In eukaryotes, selectivity is in part provided by the polyubiquitin chains, which we discussed in the preceding section. For example, Figure 14.39a shows how proteins tagged with K48-linked polyubiquitin chains are targeted to the proteasome, which has a subunit that binds specifically to K48-linked polyubiquitin chains. As the ubiquitinated protein is processed by the proteasome, associated deubiquitinating enzymes remove the ubiquitin monomers and thereby help to maintain the pool of available monoubiquitin in the cell.

In bacteria, substrate recognition by the proteases appears to be mediated by a diverse set of unstructured peptide signals, as depicted in Figure 14.39b. Such diversity in the recognition sequences helps to explain how these machines can target disparate classes of proteins. Not surprisingly, the primary recognition sequences tend to be located at both the N- and C-termini of proteins, where they are less likely to be sequestered in the core of the molecule. In general, there appears to be one sequence that acts as a primary degradation

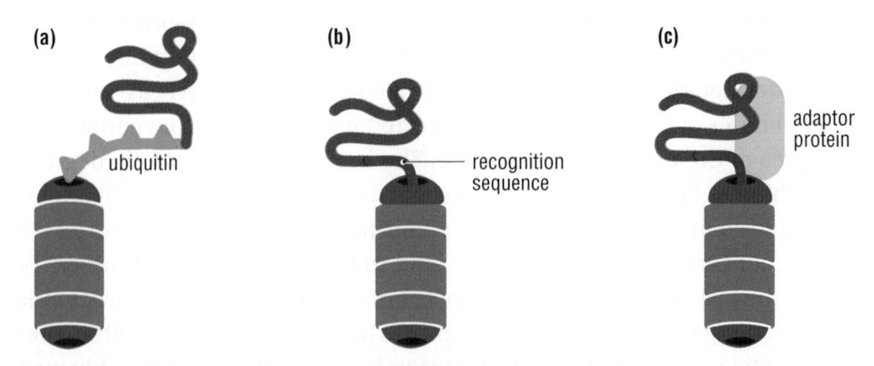

Figure 14.39 Protein recognition by ATP-dependent proteolytic complexes. Proteins to be degraded are recognized (a) by the ubiquitin tag in eukaryotes or (b) by an exposed N- or C-terminus in bacteria. (c) Bacterial substrates can also be brought to the ATP-dependent proteases by adaptor proteins.

tag, whereas other sequence elements contribute to the binding in more subtle ways.

Other important players in targeting the appropriate proteins for degradation are adaptor proteins (see Figure 14.39c), which bind to both the protease and the substrate protein to increase the interaction between these molecules.

Stress response programs maintain the balance between protein synthesis and degradation

A number of stresses to which cells can be exposed (including elevated temperature as well as exposure to various chemical reagents) ultimately result in the increased presence of aberrantly folded proteins. These unfolded proteins are detrimental to cell function and are dealt with by the cell in a number of general ways - by a decrease in overall protein production, or by an increase in chaperone or degradative activity. Both bacterial and eukaryotic cells have evolved systems that very specifically respond to such stresses by turning on transcriptional (or sometimes translational) programs that increase or decrease expression of the requisite genes that help to rebalance the protein status of the cell.

In bacteria, a pathway regulated by σ^E, an alternative sigma factor (alternative sigma factors are discussed in Section 8.3), is responsible for regulating a transcriptional program that is turned on in response to unfolded proteins in the envelope of cells. In this system, which is illustrated in Figure 14.40a, a membrane-spanning protein, RseA, has a periplasmic domain, which is important for the recognition of unfolded proteins in the membrane and periplasm, and a cytoplasmic domain that is bound by σ^E. Recognition of unfolded proteins initiates a proteolytic cascade that ultimately leads to the release of σ^E from RseA so that it can bind to RNA polymerase and stimulate transcription of the genes in response to the insult. The genes expressed include those encoding proteases, which degrade the unfolded

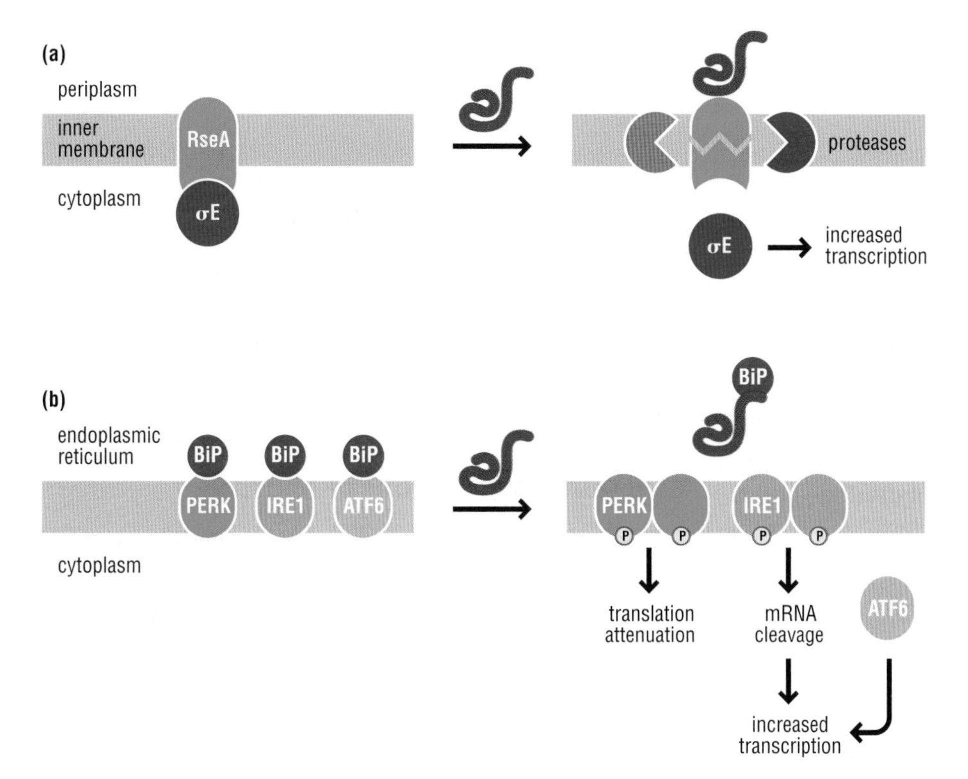

Figure 14.40 Responses to unfolded proteins. (a) Bacterial membrane stress response. The membrane-spanning protein, RseA (orange), has a periplasmic domain, which is important for the recognition of unfolded proteins in the membrane and periplasm, and a cytoplasmic domain that is bound by σ^E (blue). Recognition of unfolded proteins by RseA activates a proteolytic cascade that ultimately leads to RseA release of σ^E, which in turn can bind to RNA polymerase and stimulate transcription of the genes that help to reduce the number of unfolded proteins. (b) Eukaryotic Unfolded Protein Response (UPR). In the absence of stress, three membrane-spanning proteins (IRE1, PERK, and ATF6, orange) interact with the chaperone protein BiP (red) located in the endoplasmic reticulum. Upon stress, BiP instead preferentially binds unfolded proteins, thus releasing IRE1, PERK, and ATF6. IRE1, PERK and ATF6 can now become active, which includes the dimerization and phosphorylation of IRE1 and PERK. The activation of IRE1, PERK, and ATF6 leads to numerous protective downstream responses.

proteins, and non-coding RNAs, which repress the translation of abundant outer membrane proteins.

A eukaryotic system that is roughly equivalent to the RseA-σ^E system is known as the **unfolded protein response (UPR)** (see Figure 14.40b). In this response, cells sense unfolded proteins in the endoplasmic reticulum (the approximate equivalent of the periplasm in bacteria) that have been generated under various stress conditions. The UPR pathway in mammalian cells features three membrane-spanning proteins (IRE1, PERK, and ATF6) that respond to stress in the endoplasmic reticulum and communicate to other parts of the cell via their cytoplasmic domains. In the absence of stress, each of these proteins interacts with the chaperone protein BiP, located in the endoplasmic reticulum. Upon stress, BiP preferentially binds unfolded proteins rather than IRE1, PERK, or ATF6. Consequently, IRE1, PERK, and ATF6 are released from sequestration by BiP and become active, resulting in the dimerization and phosphorylation of IRE1 and PERK. The activation of IRE1, PERK, and ATF6 leads to numerous downstream responses, including mRNA cleavage to remove an intron of the transcript encoding a key transcription activator (activated IRE1 is an endoribonuclease), translational shut-down of most other genes (activated PERK directly regulates the activity of the translation initiation factor eIF2a), and the transcriptional activation of multiple genes (activated ATF6 is a transcription activator).

The intricate regulation described above, in addition to other regulatory responses to unfolded or misfolded proteins not described here, is not surprising given the burden imposed on the cell by these inactive proteins. Not only do non-functional proteins represent a waste of resources, but these proteins may also block the action of other proteins or have detrimental activities, such as the formation of the protein aggregates associated with many neurodegenerative diseases.

In subsequent chapters where we examine how cells cope with changes to the genome, we will see even more examples of how the protein modifications impact all aspects of cell metabolism.

- Most polypeptide chains are not functional upon release from the ribosome.

PROTEIN FOLDING

- Protein chaperones assist the proper folding and unfolding of proteins by interacting with hydrophobic regions of the protein.
- Some proteins require the formation of disulfide bonds for their correct folding.

PROTEIN TARGETING

- Proteins are targeted to specific subcellular locations by amino acid sorting motifs as well as by the attachment of lipids such as fatty acyl groups, isoprenoid groups, or GPI anchors.

POST-TRANSLATIONAL CLEAVAGE

- Some proteins need to be cleaved by site-specific proteases before they are active.
- Some proteins contain self-excising inteins, which must be removed from the protein product for it to become active.

COVALENT MODIFICATIONS OF PROTEINS

- Lipid modifications target proteins to membranes.
- The attachment of diverse carbohydrate chains increases protein solubility and can serve as recognition motifs or help protect proteins from degradation.
- Phosphorylation, acetylation, methylation, ADP ribosylation, and other small molecule additions reversibly regulate protein activity.

- Some non-enzymatic modifications such as oxidation and nitrosylation can be regulatory or detrimental depending on the adduct and the amino acid modified.
- The addition of monomeric or polymeric ubiquitin protein moieties by a series of reactions catalyzed by E1, E2, and E3 enzymes can have a variety of consequences including targeting proteins for proteasomal degradation, regulation of gene expression and mediating the cellular response to DNA damage.
- Many modifications added to proteins including sugar, methyl, acetyl, and ubiquitin groups, are coupled to nucleotide carriers that activate the groups for transfer to the protein.

PROTEIN DEGRADATION

- A barrel-like ATP-dependent proteolytic complex, termed the proteasome in eukaryotes, degrades most proteins.
- The substrate protein is threaded into the inside of the barrel, where the protease active sites are located, in a process powered by ATP hydrolysis.

QUESTIONS

14.1 CHAPERONE-ASSISTED PROTEIN FOLDING

1. Which of the following is NOT a function of chaperone proteins?
   a. Assist in the unfolding of specific proteins
   b. Guide appropriately folded proteins to their correct cellular location
   c. Prevent early folding
   d. Promote folding at the appropriate time
   e. Promote correct folding of a misfolded protein

2. Why might it be an evolutionary advantage to have chaperone proteins that physically interact with the ribosome?

Challenge question

3. α-Synuclein is a protein that has been implicated in the development of Parkinson's disease. In neurons, this protein is typically found as a protein monomer. Under oxidative stress, however, the protein changes shape and forms oligomers that can become toxic to the cell and ultimately lead to neurodegeneration. Additionally, specific polymorphisms in the α-synuclein protein have been associated with increased oligomer formation, insoluble aggregate formation, and Parkinson's disease. Heat-shock proteins (HSPs) have been studied to determine their potential as a therapeutic target for the prevention of Parkinson's disease.
   a. As a therapeutic target, would you want to increase or decrease the activity of HSPs? Explain your answer in terms of the role of HSPs in normal cells.
   b. Describe an experimental design that would test the above prediction.

14.2 TARGETING OF PROTEINS THROUGHOUT THE CELL

1. Describe the mechanism by which Ran-GTP acts as a molecular switch during nuclear import and export.

Challenge question

2. In order to form a functional ribosome, ribosomal proteins are translated in the cytoplasm, transported into the nucleus to become associated with rRNA, and then exported back to the cytoplasm. Propose a mechanism by which the ribosomal protein can be appropriately recognized for both import into and export out of the nucleus.

14.3 POST-TRANSLATIONAL CLEAVAGE OF THE POLYPEPTIDE CHAIN

Challenge question

3. The *proopiomelanocortin* (*POMC*) gene product undergoes extensive post-translational processing that produces several protein products with a variety of tissue-specific expression patterns, cellular locations, and functions. Mutations that affect POMC expression, sorting, processing, and function in specific neurons have been linked to altered feeding habits and obesity in humans. Propose a mechanism by which:
   a. The protein products of POMC can be tissue specific.
   b. POMC can be processed and targeted to secretory vesicles.
   c. The function of a specific POMC product can be different depending of tissue and cellular conditions.

14.4 LIPID MODIFICATION OF PROTEIN

1. In what way is the lipid modification associated with the GPI anchor different from the other types of lipid modification discussed?

Challenge question

2. Hutchinson–Gilford progeria syndrome (HGPS) is a rare genetic disorder characterized by accelerated aging. It is caused by mutations in the *LMNA* gene, which encodes the A-type nuclear lamins. At the carboxy-terminus, prelamin A contains a farnesylation site and an endoprotease recognition sequence. In order to form active lamin A, the endoprotease ZMPSTE24 cleaves the prelamin A protein and removes the carboxy-terminus containing the farnesylated amino acid. Mutations that cause HGPS lead to an in-frame deletion of 50 amino acids that includes the endoprotease recognition sequence but not the farnesylated amino acid.
   a. What is the function of lamin A?
   b. Propose a mechanism by which the HGPS mutation may lead to accelerated aging.

14.5 GLYCOSYLATION OF PROTEINS

1. Compare the purpose of lipid modification and glycosylation in terms of the changes to the general characteristics of the protein.

2. The glycosylation of one particular protein may vary greatly when compared between species, members of the same species, and even between tissues. What advantage does this variation serve?

3. Compare and contrast the *O*-GlcNAc glycosylation to other forms of *O*-linked glycosylations.

Challenge question

4. The wild-type form of multidrug resistance-associated protein 8 (also known as ABCC11) is an *N*-linked glycosylated protein found in intracellular granules, large vacuoles and the luminal membrane of some secretory cells, including the mammary glands. A single nucleotide polymorphism (SNP) variant of ABCC11 changes one amino acid and eliminates the *N*-glycosylation. In some forms of breast cancer, the ABCC11 gene is overexpressed. Patients with overexpression of which form of the ABCC11 protein would be candidates for chemotherapy? Explain your answer.

14.6 PROTEIN PHOSPHORYLATION, ACETYLATION, AND METHYLATION

1. Why are phosphorylation, acetylation, and methylation of proteins considered to be molecular switches, and what is their significance in terms of the cellular response to stimuli?

2. The main proteins involved in the Jak-STAT pathway are the ligand, the cytokine receptor, Jak, and STAT. Which of these proteins becomes phosphorylated, and how does the phosphorylation change the activity of each?

Challenge question

3. Chemotaxis is a process by which bacteria control their movement according to the concentration of specific molecules in the environment. Chemotaxis is controlled by both phosphorylation and acetylation of the response regulator protein CheY. The phosphorylation of CheY reduces its affinity for the signal receptor and increases its affinity for FliM, the switch protein that promotes the clockwise rotation of the flagellum. In order to determine the function of the acetylation of CheY, investigators compared the affinities of acetylated and deacetylated CheY for FliM. The results indicated that acetylated CheY had a lower affinity for FliM.

a. What do these results suggest as the role for acetylation in chemotaxis?
b. In terms of glucose concentrations in the environment, propose a reason why the dual regulation of the chemotaxis proteins would serve as an advantage and therefore be maintained in bacteria.

14.8 DIRECT CHEMICAL MODIFICATION OF PROTEINS

1. How are the modifications discussed in this section different from those discussed in Section 14.6?

2. Describe one example in which oxidation is helpful and one where it is harmful to the cell.

Challenge question

3. Oxidative stress and the formation of insoluble aggregates of the protein α-synuclein are known contributors to the onset of Parkinson's disease. α-Synuclein contains amino acids that are potential substrates for modification by methionine oxidation and tyrosine nitration. In an effort to determine if any of these modifications would promote the formation of a-synuclein aggregates, candidate amino acid residues were modified by treatment with hydrogen peroxide (oxidation) or tetranitromethane (nitration). Each sample was applied to a size exclusion chromatography column. The instrument produced a graph of absorbance versus time, and a peak in absorbance appeared as the amino acid residues exited the column. The results of the experiment are shown in Figure A.

a. Which modification(s) promoted the formation of aggregates? Explain your answer in terms of the data presented.
b. Based on what you know about the consequences of these modifications, suggest possible reasons that those modifications that produced an effect would lead to the formation of aggregates.

14.9 UBIQUITINATION AND SUMOYLATION OF PROTEINS

1. In what way does ubiquitination differ from the other types of modification discussed thus far?

2. What is the significance of the fact that most cells contain one E1 enzyme, several E2 enzymes, and many E3 enzymes?

3. Give examples of the functions of different types of mono- and polyubiquitin modifications.

14.10 PROTEIN DEGRADATION

1. Describe the basic structure of the proteasome and explain why its structure is uniquely tailored for its role in degrading proteins.

2. The UPR system is an example of a complex pattern of regulation involving most steps in the flow of genetic information from DNA to functional protein. Explain how each step is used in regulating the response to unfolded proteins.

a. Transcription
b. RNA processing
c. Translation
d. Protein localization
e. Protein modification
f. Protein degradation

EXPERIMENTAL APPROACH 14.1 – THE DISCOVERY OF CHOLESTEROL MODIFICATION FOR MEMBRANE TETHERING DURING SIGNALLING

1. What is the Hh family of proteins, and what characteristic of the Hh protein was studied in the Experimental approach?

2. What is DTT, and why was it used in the experiment that produced the results shown in Figure 1?

Challenge questions

3. Figure 1 demonstrates the lipid stimulation of Hh autoprocessing. Compare the results shown in lanes 1 and 2, and explain how they support the notion that Hh autoprocessing is stimulated by lipids.

4. Compare lanes 2 and 3 in Figure 1. What conclusions can you draw?

5. Compare lanes 3 and 4 in Figure 1 and describe the conclusion that could be made from the figure.

6. What is the relationship between the two figures that make up Figure 2?

7. What conclusion was drawn based upon these results?

Figure A

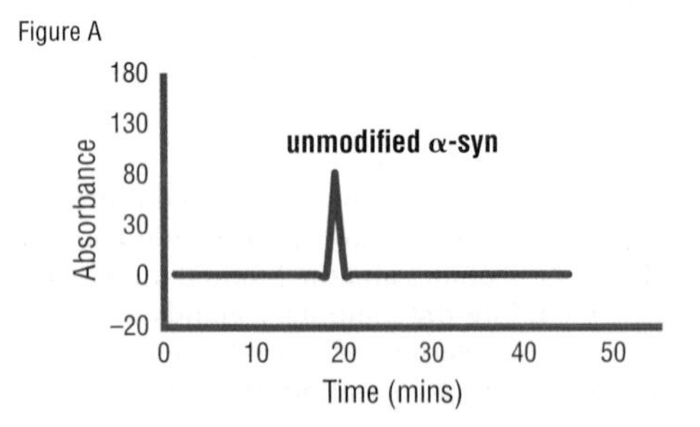

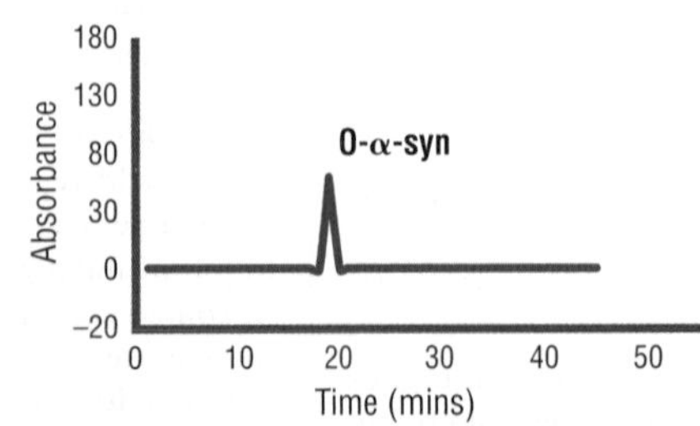

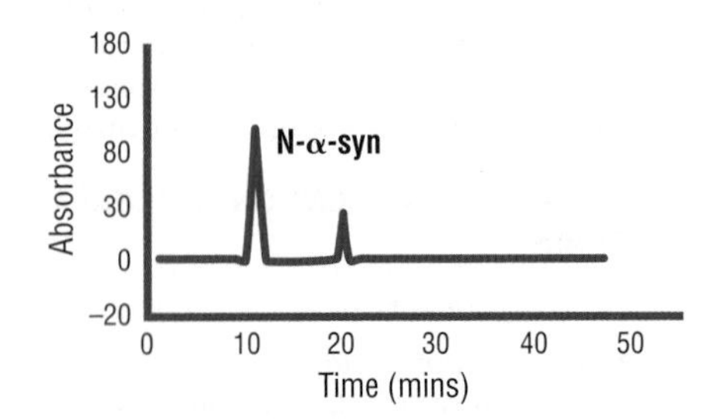

EXPERIMENTAL APPROACH 14.2 – TWO-COMPONENT SYSTEMS IN BACTERIA: HOW TO INTERACT WITH YOUR PARTNER

1. What are NRI, NRII, and NRII2302 in terms of the two-component regulatory system?
2. Why was the *glnA* P2 promoter activity studied in the experiments by Ninfa and Magasanik?
3. In the experiment that produced the results for Figure 1, what was the purpose of the pre-incubation vs the incubation with UTP and heparin for the times indicated on the gel?
4. What was the result of the experiment and what conclusion could be made based upon these results?
5. In Figure 2, what is the difference between (a) and (b) in terms of the type of information you can get from the procedure?
6. What conclusion can be drawn from the results of Figure 2?
7. What is covariation, and how was it used to determine how the specificity of kinase–response regulator pairs is maintained?

EXPERIMENTAL APPROACH 14.3 – DISCOVERY OF UBIQUITIN-DEPENDENT PROTEOLYSIS

1. Initial experiments to identify components of the ATP-dependent proteolysis process identified two important cellular fractions. What important components were later identified in those two fractions?
2. What additional knowledge regarding ubiquitin-dependent proteolysis has been added to the literature due to the availability of high-throughput technology?

Challenge questions

3. What are the main differences between lane 1, lanes 2–5, and lanes 6–7 in the gel in Figure 1?
4. Based on the results for each lane of the gel described in Question 3, what conclusions can be made regarding proteolysis?

FURTHER READING

14.1 CHAPERONE-ASSISTED PROTEIN FOLDING

Hartl FU, Bracher A, Hayer-Hartl M. Molecular chaperones in protein folding and proteostasis. *Nature*, 2011;**475**:324–332.

Mamathambika BS, Bardwell JC. Disulfide-linked protein folding pathways. *Annual Review of Cell and Developmental Biology*, 2008;**24**:211–235.

Preissler S, Deuerling E. Ribosome-associated chaperones as key players in proteostasis. *Trends in Biochemical Sciences*, 2012;**37**:274–283.

Tyedmers J, Mogk A, Bukau B. Cellular strategies for controlling protein aggregation. *Nature Reviews Molecular Cell Biology*, 2010;**11**: 777–788.

14.2 TARGETING OF PROTEINS THROUGHOUT THE CELL

Bonifacino JS, Traub LM. Signals for sorting of transmembrane proteins to endosomes and lysosomes. *Annual Review of Biochemistry*, 2003;**72**:395–447.

Stewart M. Molecular mechanism of the nuclear protein import cycle. *Nature Reviews Molecular Cell Biology*, 2007;**8**:195–208.

Wente SR, Rout MP. The nuclear pore complex and nuclear transport. *Cold Spring Harbor Perspectives in Biology*, 2010;**2**:a000562.

Wickner W, Schekman R. Protein translocation across biological membranes. *Science*, 2005;**310**:1452–1456.

14.3 POST-TRANSLATIONAL CLEAVAGE OF THE POLYPEPTIDE CHAIN

Elleuche S, Pöggeler S.Inteins, valuable genetic elements in molecular biology and biotechnology. *Applied Microbiology and Biotechnology*, 2010;**87**:479–489.

Jiang J, Hui CC. Hedgehog signaling in development and cancer. *Developmental Cell* 2008;**15**:801–812.

Wlodawer A, Vondrasek J. Inhibitors of HIV-1 protease: a major success of structure-assisted drug design. *Annual Review of Biophysics and Biomolecular Structure*, 1998;**27**:249–284.

14.4 LIPID MODIFICATION OF PROTEINS

Fujita M, Kinoshita T. GPI-anchor remodeling: potential functions of GPI-anchors in intracellular trafficking and membrane dynamics. *Biochimica et Biophysica Acta*, 2012;**1821**:1050–1058.

Mann RK, Beachy PA. Novel lipid modifications of secreted protein signals. *Annual Review of Biochemistry*, 2004;**73**:891–923.

Paulick MG, Bertozzi CR. The glycosylphosphatidylinositol anchor: a complex membrane-anchoring structure for proteins. *Biochemistry*, 2008;**47**:6991–7000.

Resh MD. Trafficking and signaling by fatty-acylated and prenylated proteins. *Nature Chemical Biology*, 2006;**2**:584–590.

Rocks O, Peyker A, Bastiaens PI. Spatio-temporal segregation of Ras signals: one ship, three anchors, many harbors. *Current Opinion in Cell Biology*, 2006;**18**:351–357.

14.5 GLYCOSYLATION OF PROTEINS

Hart GW, Housley MP, Slawson C. Cycling of *O*-linked beta-*N*-acetylglucosamine on nucleocytoplasmic proteins. *Nature*, 2007;**446**:1017–1022.

Imperiali B, O'Connor SE. Effect of *N*-linked glycosylation on glycopeptide and glycoprotein structure. *Current Opinion in Chemical Biology*, 1999;**3**:643–649.

Wells L, Vosseller K, Hart GW. Glycosylation of nucleoplasmic proteins: signal transduction and *O*-GlcNac. *Science*, 2001;**291**: 2376–2378.

14.6 PROTEIN PHOSPHORYLATION, ACETYLATION, AND METHYLATION

Aaronson DS, Horvath CM. A road map for those who don't know JAK-STAT. *Science*, 2002;**296**:1653–1655.

Hunter T. Why nature chose phosphate to modify proteins. *Philosophical Transactions of the Royal Society of London Series B Biological Sciences*, 2012;**367**:2513–2516.

Kouzarides T. Chromatin modifications and their function. *Cell*, 2007;**128**:693–705.

Laub MT, Goulian M. Specificity in two-component signal transduction pathways. *Annual Review of Genetics*, 2007;**41**:121–145.

14.7 PROTEIN MODIFICATION BY NUCLEOTIDES

Aktories K. Bacterial protein toxins that modify host regulatory GTPases. *Nature Reviews Microbiology*, 2011;**9**:487–498.

Messner S, Hottiger MO. Histone ADP-ribosylation in DNA repair, replication and transcription. *Trends in Cell Biology*, 2011;**21**:534–542.

Ribet D, Cossart P. Pathogen-mediated posttranslational modifications: A re-emerging field. *Cell*, 2010;**143**:694–702.

14.8 DIRECT CHEMICAL MODIFICATION OF PROTEINS

Hancock JT. The role of redox mechanisms in cell signaling. *Molecular Biotechnology*, 2009;**43**:162–166.

Hess DT, Stamler JS. Regulation by S-nitrosylation of protein post-translational modification. *The Journal of Biological Chemistry*, 2012;**287**:4411–4418.

Jacob C, Battaglia E, Burkholz T, Peng D, Bagrel D, Montenarh M. Control of oxidative posttranslational cysteine modifications: from intricate chemistry to widespread biological and medical applications. *Chemical Research in Toxicology*, 2012;**25**:588–604.

14.9 UBIQUITINATION AND SUMOYLATION OF PROTEINS

Elsasser S, Finley D. Delivery of ubiquitinated substrates to protein-unfolding machines. *Nature Cell Biology*, 2005;**7**:742–749.

Komander D, Clague MJ, Urbé S. Breaking the chains: structure and function of the deubiquitinases. *Nature Reviews Molecular Cell Biology*, 2009;**10**:550–563.

Komander D, Rape M. The ubiquitin code. *Annual Review of Biochemistry*, 2012;**81**:203–229.

Ravid T, Hochstrasser M. Diversity of degradation signals in the ubiquitin-proteasome system. *Nature Reviews Molecular Cell Biology*, 2008;**9**:679–690.

14.10 PROTEIN DEGRADATION

Ades SE, Regulation by destruction: design of the sigmaE envelope stress response *Current Opinion in Microbiology*, 2008;**11**:535–540.

Gardner BM, Pincus D, Gotthardt K, Gallagher CM, Walter P. Endoplasmic reticulum stress sensing in the unfolded protein response *Cold Spring Harbor Perspectives in Biology*, 2013;**5**:a013169.

Sauer RT, Baker TA. AAA+ proteases: ATP-fueled machines of protein destruction. *Annual Review of Biochemistry*, 2011;**80**:587–612.

Cellular responses to DNA damage

INTRODUCTION

The preservation of genetic information from one generation to the next requires the DNA sequence of the cell's genome to be maintained without alteration. Yet the integrity of the DNA sequence is under constant threat of damage as a result of errors in normal cellular processes such as DNA replication, and from reactive metabolites and environmental agents. DNA damage from such errors can result in changes in base sequence or even in chromosome structure. Despite the sophisticated mechanisms that have evolved to ensure accuracy of DNA replication, errors do occur and are one of the causes of sequence change.

➔ We learn about the process of DNA replication in Chapter 6.

If errors remain uncorrected, changes in DNA are propagated as mutations. If damage is encountered by the replication machinery in the course of DNA replication, chromosomes may be broken or replicated incompletely and fail to segregate properly during cell division.

The effects of DNA damage are greatly reduced by specialized processes of **DNA repair**. There are many pathways for the detection and repair of damaged DNA. As a result, most damaged DNA is repaired and only a small fraction of DNA damage – likely less than 1 in 1000 incidents – results in a permanent change in DNA sequence or a failure of chromosome replication and segregation.

In this chapter and the next, we discuss the principal ways in which damage and change are introduced, and the mechanisms that have evolved to repair damage and limit changes to the genome. Many strategies for repair rely on the double-stranded structure of DNA to provide the sequence information on which correction of the damaged region can be based: these are the repair pathways we describe in detail in this chapter. Where both strands of the DNA molecule are broken and there is no intact strand to serve as a template for repair, different pathways are required, and these are the subject of Chapter 16.

For some types of DNA damage, repair proteins are always present at levels that can rapidly recognize and repair DNA damage. In other cases, however, especially where DNA damage has arrested DNA replication, synthesis of repair proteins is increased in a global response to DNA damage in both bacteria and eukaryotes.

Before we embark on our exploration of DNA repair, a word on the history of repair protein nomenclature may help to make the terminology less confusing. As with many fundamental biological processes that were first elucidated through genetic analysis, the nomenclature of the proteins contributing to DNA repair pathways often reflects the way they were first discovered rather than their biochemical functions or structural relationships. For example, many repair proteins were discovered through the analysis of bacterial or yeast mutants showing abnormal sensitivity to DNA damaging agents, such as x-rays or ultraviolet (UV) light, and are consequently called Rad proteins or Uvr proteins. Human homologs of some of these proteins, discovered later, have sometimes inherited this nomenclature, and we shall encounter human Rad proteins in the course of this chapter. Many bacterial repair proteins play a part in recombination reactions and are consequently known as Rec proteins.

In humans, defects in repair proteins often lead to the development of diseases that are characterized by increased incidence of cancer, and repair proteins identified in this way are named after the diseases they cause when they are defective. For example, the XP proteins we shall meet in this chapter were identified in studies on the human disease xeroderma pigmentosum, in which there is increased sensitivity to DNA damage.

We begin the chapter by describing the types of DNA damage that can occur.

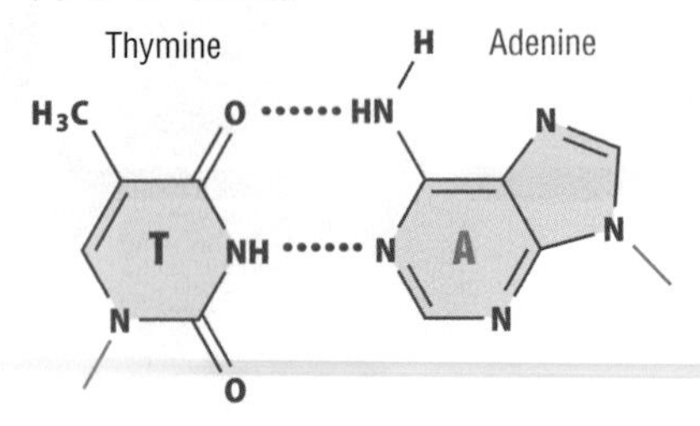

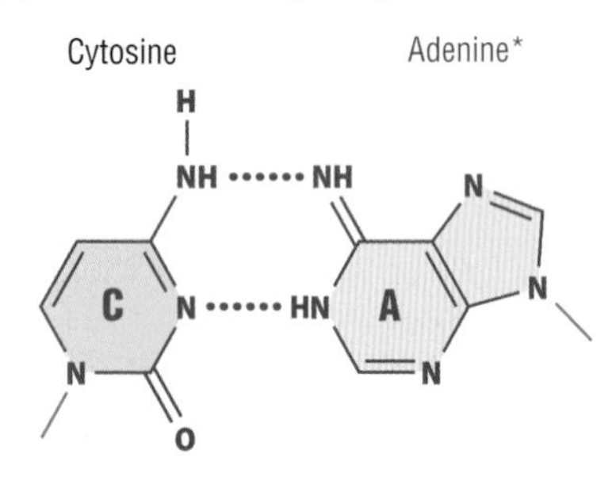

Figure 15.1 Mispairing due to tautomeric forms of the DNA bases. (a) The normal base-pairing of thymine and adenine. (b) In contrast, an imino tautomer of adenine (adenine*) mispairs with cytosine.

➔ We discuss tautomers in more detail in Section 2.4.

15.1 TYPES OF DNA DAMAGE

Misincorporation of nucleotides occurs during DNA synthesis

The integrity of genetic information in cells depends on the fidelity of DNA replication and the stability of the DNA molecule itself. The accuracy of DNA replication depends critically on the selection of the correct nucleotide to add to the growing DNA strand. Such accurate selection depends on correct base-pairing between the template strand of the DNA and the incoming nucleotide. The fidelity of DNA polymerization, which we discuss in Section 6.3, is increased by the proofreading activities of exonucleases associated with the replisome. The exonucleases excise inappropriate nucleotides and allow resynthesis of DNA with a high probability of incorporating the correct nucleotide. Despite this, however, DNA polymerization is not error-free: mispaired bases do occur in newly synthesized DNA every 10^5 to 10^6 base pairs.

One reason for such errors is the occasional transient formation of tautomeric isomers, or tautomers, of the four DNA nucleotides, in which some of the hydrogen atoms that mediate base-pairing have changed position.

Such tautomers can form non-Watson–Crick base pairs, as illustrated in Figure 15.1, so that an incorrect nucleotide pairs with the template and is incorporated into the new strand. The strand with the incorrect base will then be faithfully copied, leading to a mutation; this situation is depicted in Figure 15.2. If one purine base (adenine or guanine), is replaced by the other purine base, or a pyrimidine base (thymidine or cytosine) is replaced by the other pyrimidine, the mutation is called a **transition mutation**. If a purine base is replaced by a pyrimidine base or a pyrimidine base is replaced by a purine base, the mutation is called a **transversion mutation**.

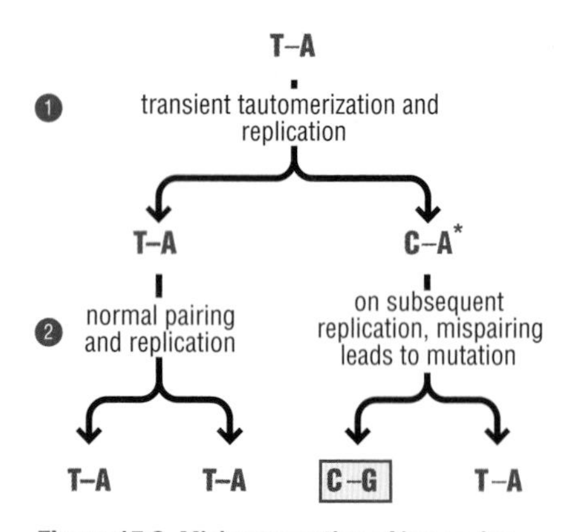

Figure 15.2 Misincorporation of bases due to tautomerization. A normal T–A base pair is shown at the top. (1) Tautomerization of adenine (adenine*) leads to mispairing with cytosine in the first round of replication. (2) In the second round of replication, the mispairing gives rise to a mutant C–G pair in one of the two copies of the DNA instead of the original T–A pair.

DNA is subject to spontaneous and induced damage that can lead to mispairing and mutation

Chemical alterations to DNA caused by both intracellular and extracellular agents disrupt the accurate base-pairing that is essential to accurate DNA replication, and hence can cause mutations. Bulky alterations such as pyrimidine dimers, which can be intrastrand cyclobutane dimers or 6,4 photoproducts that cross-link the

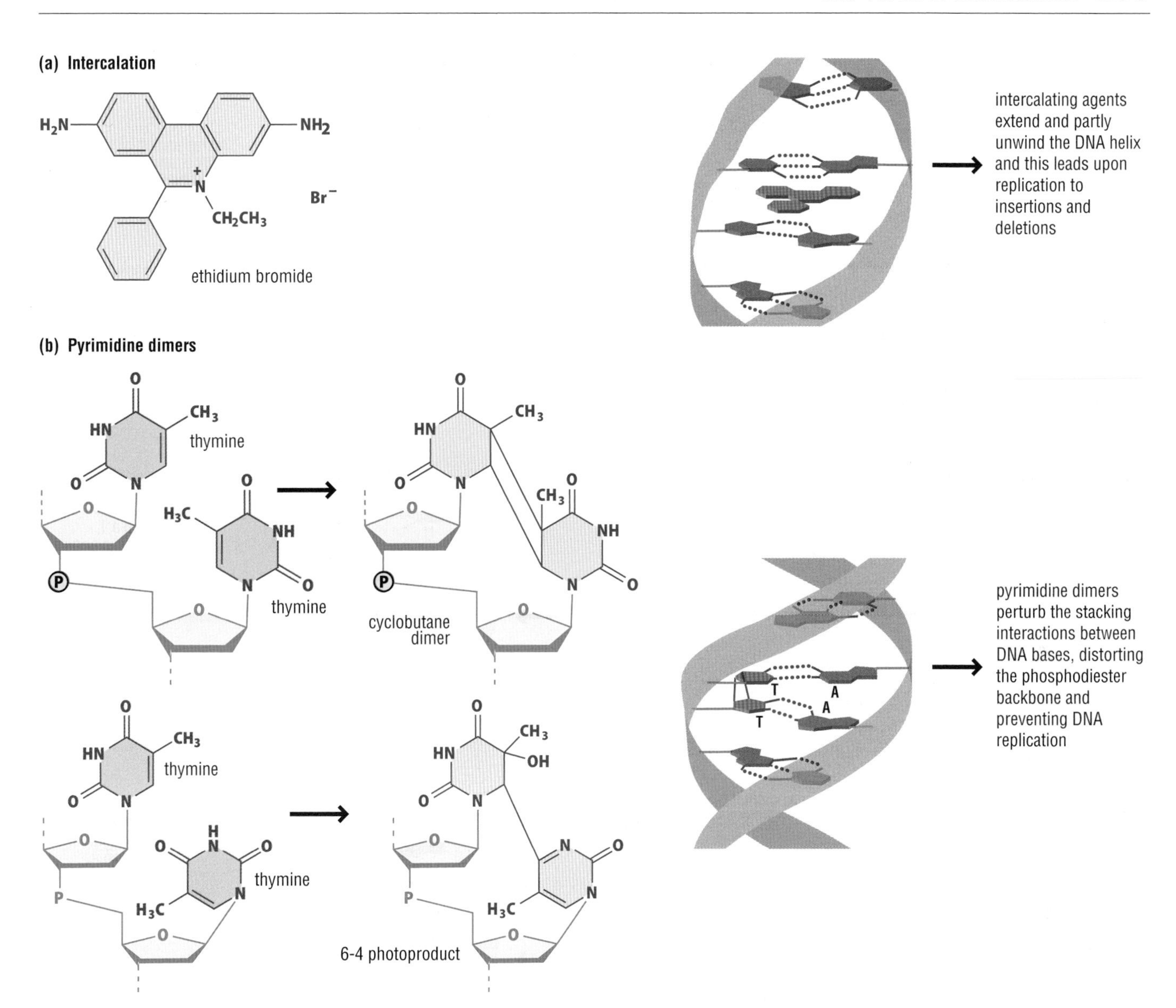

Figure 15.3 Agents that distort DNA. (a) Intercalating agents such as ethidium bromide insert themselves between adjacent bases, extending and partly unwinding the DNA helix, leading to insertions and deletions on DNA replication. (b) Irradiation by UV light generates pyrimidine dimers, which cannot fit into the active site of the replicative polymerase, preventing replication of the DNA. These dimers can exhibit various structures, including the cyclobutane dimer and 6-4 photoproduct structures shown here.

DNA strands, can also block DNA replication (see Figure 15.3b). In this section, we describe the different sources of such damage and their mechanisms of disrupting DNA. In the next sections, we explain how this damaged DNA is repaired.

Although DNA is highly stable, spontaneous damage does occur under physiological conditions. Hydrolytic **depurination**, a process illustrated in Figure 15.4a, results in loss of purine bases; hydrolytic **depyrimidation** results in the loss of pyrimidine bases; and **deamination** – illustrated in Figure 15.4b – results in changes in base structure. In mammalian cells, for example, about 18 000 bases per cell are lost by depurination each day, while deamination affects around 500 bases per cell per day. Such changes can also occur to free nucleotides, thus poisoning the pool of DNA precursors.

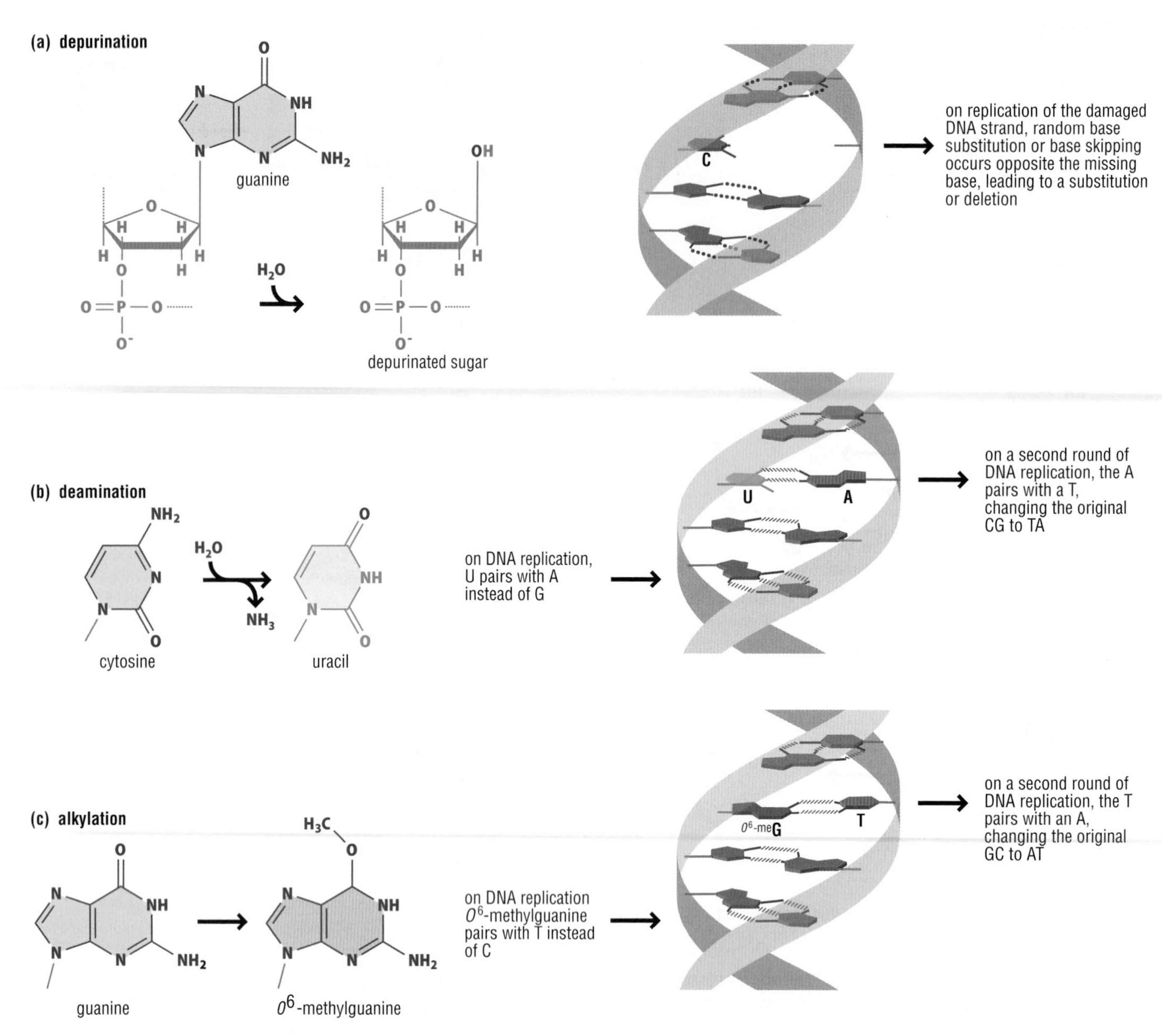

Figure 15.4 Some of the most common causes of base damage. (a) Possibly the most frequent of all types of DNA damage is the hydrolytic breakage of the glycosidic bond between a purine and the sugar–phosphate backbone of DNA, leaving a gap in the structure that causes substitutions or deletions when the DNA is replicated. (b) Hydrolytic deamination of cytosine or adenine substitutes an oxygen for the amino group. In the case of cytosine this generates uracil, which pairs with adenine, leading to the substitution of an A–T base pair for C–G; deamination of adenine (not shown) generates hypoxanthine, which pairs with cytosine, leading to the substitution of a G–C base pair for A–T. (c) Alkylating agents add methyl or ethyl groups to bases: O^6-methylguanine mispairs with thymine, leading to the substitution of an A–T base pair for C–G. Altered bases are shown in blue; mutant bases with which they mispair are shown in red.

A second important cause of DNA damage is the production of reactive metabolites in the course of normal intracellular metabolism. For example, several cellular processes, including energy production via electron transport chains, produce reactive oxygen species. These species include superoxides, peroxides, and other oxygen ions that cause oxidative damage. Although cells have enzymes such as superoxide dismutase and catalase that specifically target and destroy such reactive species, they are not destroyed completely and remain at sufficient levels to cause significant DNA damage.

Bases, sugars, and the phosphodiester backbone are all subject to oxidation. In fact, oxidative damage to the DNA backbone is the most dangerous type of DNA damage. Why is this? If the damage is not repaired, it can result in potentially lethal double-strand breaks that can lead to genome instability as a consequence of subsequent errors in chromosome replication and segregation.

Other natural products of metabolism can also damage DNA. Among the most important of these are **alkylating agents**, many of which are found in microorganisms (although some are man-made). Alkylating agents are potent modifiers of the structure of DNA bases, adding methyl groups (as illustrated in Figure 15.4c) or ethyl groups, which disrupt normal base-pairing interactions. Common reactions of this type include the formation of O^6-methylguanine, which pairs with thymine instead of cytosine, and 7-ethylguanine, which pairs with thymine instead of cytosine. Alkylating agents can also cause interstrand cross links that block strand separation.

Both natural products and human-made chemicals in the environment can interact with and damage DNA. Indeed, the vast majority of carcinogens work by interacting with DNA and causing mutations. For example, nitrosamines, which are present in some foods, can act as alkylating agents. Degradation products of nitrosamines can also oxidize adenine and cytosine. DNA damage resulting from the formation of benzo[a]pyrene adducts with guanine bases is an important consequence of smoking, leading to mutations and, potentially, cancer.

Most of the lesions we have discussed so far affect single nucleotides and, if unrepaired, cause single-nucleotide changes, or **point mutations**, in the DNA sequence. Many DNA damaging agents have more widespread effects, however. Among these are **intercalating agents**, which are aromatic or flat hydrophobic molecules that insert between the stacked bases in the DNA double helix. The benzo[a]pyrene adduct we have just mentioned is an example of an aromatic molecule that undergoes intercalation.

The insertion of an intercalating agent pushes the bases apart and distorts the helical structure, which becomes partly unwound and extended (as we see in Figure 15.3a). When such DNA is replicated, base pair insertions or deletions are generated in the newly synthesized DNA. In protein-coding regions, DNA insertions or deletions result in **frameshifting** during protein synthesis, such that essential proteins are not expressed. Dividing cells exposed to intercalating agents therefore often die. For this reason, intercalating agents have sometimes been used in cancer therapy, to target rapidly dividing cancer cells.

DNA is also subject to damage from physical sources. One of these is UV light, which promotes the formation of intrastrand and interstrand **pyrimidine dimers**; examples of such dimers are shown in Figure 15.3b. The coupled nucleotides formed from such dimerization cannot be accommodated in the active site of highly accurate replicative polymerases, and insertions and deletions are generated in the newly synthesized DNA.

Another source of DNA damage is ionizing radiation, such as x-rays, which can introduce single- and double-strand breaks into DNA.

The cell has at its disposal a number of different strategies for repairing damaged DNA, with the precise strategy adopted depending on the nature of the damage. In the next five sections, we will consider several repair strategies in turn, before examining how the cell detects the damage in the first place, prior to eliciting an appropriate response. We begin by considering the strategy of post-replication **mismatch repair**.

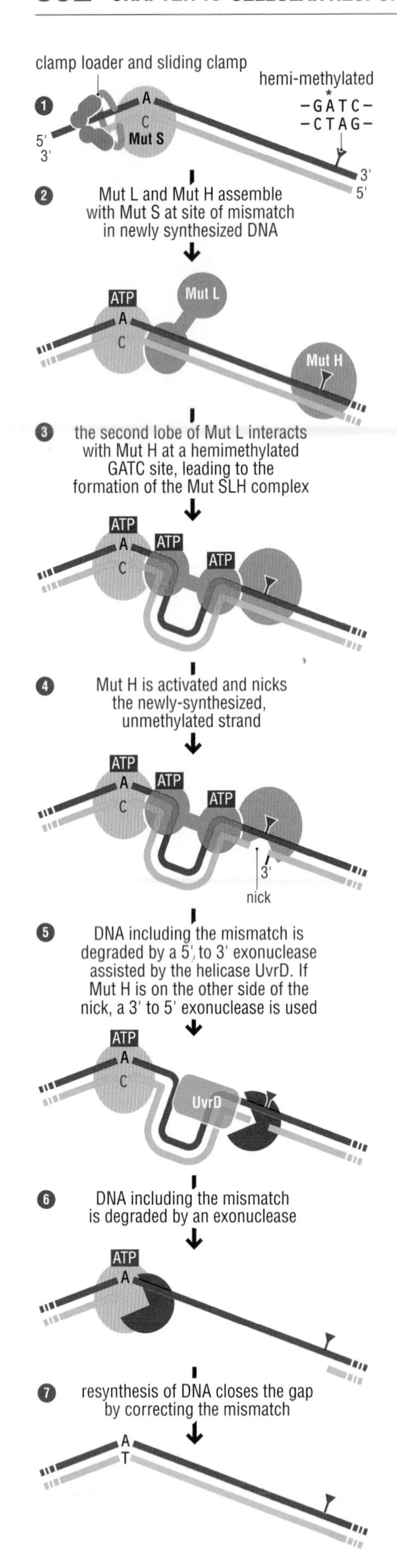

clamp loader and sliding clamp
hemi-methylated
-GATC-
-CTAG-
Mut S
A
C
5'
3'
3'
5'
Mut L and Mut H assemble with Mut S at site of mismatch in newly synthesized DNA
ATP
Mut L
Mut H
the second lobe of Mut L interacts with Mut H at a hemimethylated GATC site, leading to the formation of the Mut SLH complex
Mut H is activated and nicks the newly-synthesized, unmethylated strand
nick
DNA including the mismatch is degraded by a 5' to 3' exonuclease assisted by the helicase UvrD. If Mut H is on the other side of the nick, a 3' to 5' exonuclease is used
UvrD
DNA including the mismatch is degraded by an exonuclease
resynthesis of DNA closes the gap by correcting the mismatch
T

15.2 POST-REPLICATION MISMATCH REPAIR

Post-replication mismatch repair restores mismatched base pairs to their parental state

The *in vivo* mutation rate of about 10^{-9} to 10^{-11} per base pair is considerably lower than the error rate of DNA polymerase of about 10^{-5} to 10^{-6} per base pair. This low *in vivo* mutation rate is due to the activity of the polymerase proofreading exonuclease as well as a post-replication system that can detect and correct remaining mismatches immediately after they occur, a process known as **mismatch repair**.

Mismatch repair, like other repair systems described in this chapter, takes advantage of the fact that DNA is a duplex and contains a complete copy of its genetic information on both strands. Thus, if a base on one strand of a duplex is incorrect or damaged, the other strand can serve as a template to restore the original sequence.

The key steps in mismatch repair - recognition of the mismatched base pair, removal of the incorrect nucleotide on the newly synthesized strand, and subsequent resynthesis using the parental DNA strand as a template - are the same in prokaryotes and eukaryotes. Strategies for recognition of the newly synthesized strand and cleavage around the incorrect nucleotide, however, differ among organisms.

Inactivation of mismatch repair proteins in organisms from *Escherichia coli* to humans results in a high mutation rate, that is, a mutator phenotype. Therefore, when first identified in *E.coli*, these proteins were called Mut proteins, where Mut stands for mutator.

Mismatch repair in *E. coli* is mediated by three Mut proteins: MutS, MutL, and MutH. MutS specifically recognizes and binds mismatched base pairs, apparently sensing the way that such mismatches alter the bending properties of DNA. MutS can recognize virtually all mismatched base pairs. MutS protein also interacts with the sliding clamp components of the replisome, which thus serves to focus the repair machinery on newly synthesized regions of DNA. MutS also recruits MutL and MutH, which assemble at the site of mismatch, as illustrated in Figure 15.5.

MutH is an endonuclease that can specifically nick the newly synthesized daughter strand, targeting repair to this strand. (This targeting is important to ensure that the new, incorrect nucleotide, and not the original, correct, one is replaced.) MutH recognizes newly replicated DNA because of the way it contains hemi-methylated **GATC sites** in which only the adenine residue of the parental strand is methylated. MutH specifically binds to GATC sites (Figure 15.5, step 2), sometimes hundreds of base pairs away from the site of damage.

If the DNA is hemi-methylated and a mismatch MutS–MutL complex is present, a MutSLH complex forms (step 3) and MutH introduces a nick into the newly synthesized unmethylated DNA strand (step 4). MutL then recruits helicase II (MutU/

Figure 15.5 The MutSLH repair pathway of *E. coli*. A newly replicated DNA duplex with a parental strand (gray) and newly synthesized strand (green) that contains an A–C mismatch base pair formed during DNA synthesis is shown. The mismatch base pair is bound by MutS, which is recruited to the site of replication errors both by the mismatch base pair and by the sliding clamp (blue ring), which is an essential component of the replisome. Because there is a delay in methylation at GATC sites of newly synthesized DNA, only the parental DNA strand in the region of DNA closest to the replication fork is methylated (blue T), and the new strand (green) can thus be distinguished when mispairing occurs.

UvrD) that displaces the nicked strand from the duplex (step 5); exonucleases then degrade the newly synthesized DNA between the nick and the mismatch (step 6). Depending on the position of the nick relative to the mismatch, degradation occurs via a 5′ to 3′ exonuclease (as shown) or a different 3′ to 5′ exonuclease. New DNA synthesis fills the gap, replacing the mismatch with the correct base (step 7).

In eukaryotes, multiple MutS and MutL proteins interact to form heterodimers. As in the *E. coli* system, the eukaryotic MutS proteins recognize mismatched base pairs. Although DNA methylation occurs in many eukaryotes, it is not confined to newly replicated DNA, and there is no known eukaryotic homolog of MutH. Eukaryotic MutLα does have a latent endonuclease activity that can be activated by eukaryotic MutS bound to a mismatch. It is likely that the eukaryotic mismatch repair system distinguishes template from newly synthesized strands by recognition of discontinuities in the newly synthesized strand.

We discuss the sliding clamp and replisome in more detail in Section 6.9.

We learn more about the hemi-methylation of newly synthesized DNA in Section 6.13.

The mismatch repair system can also correct insertions and deletions

Mismatched base pairs are not the only errors that can occur during DNA synthesis. Repetitive DNA can form temporary hairpin structures while the DNA is in single-stranded form during DNA replication. Such structures can result in the deletion or expansion of repeated sequences. Figure 15.6 depicts how DNA polymerase can skip past these hairpins, so that the new strand will contain a deletion of one or more repeats. Hairpin formation can also occur in the new DNA strand as it is being synthesized, leading to repeated replication of the same template region, which results in the insertion of one or more extra repeats in the newly synthesized DNA. Taken together, these processes are often called **slipped-strand misreplication**.

The hairpins are detected by mismatch repair systems and their effects are eliminated by degradation of the newly synthesized strand and subsequent resynthesis. In the case of a hairpin in the template strand, for example, degradation of the newly synthesized strand leads to unfolding of the hairpin and accurate resynthesis of the new strand, thus restoring the original sequence. In eukaryotes, different MutS variants mediate correction of base-pair mismatches and repair of hairpins due to slipped-strand misreplication.

Slipped-strand replication in repeat regions (for example, triplet nucleotide repeats present in some human genes that encode proteins important for brain function) underlies some neurodegenerative diseases. It also provides an important mechanism for promoting the variable expression of bacterial cell surface proteins in a process called phase variation, which allows bacteria to evade immune responses. The genes whose expression varies contain short runs of the same base pair at the beginning of their open reading frames. Slipped-strand misreplication results in frameshifting, variably turning off and on the expression of different genes.

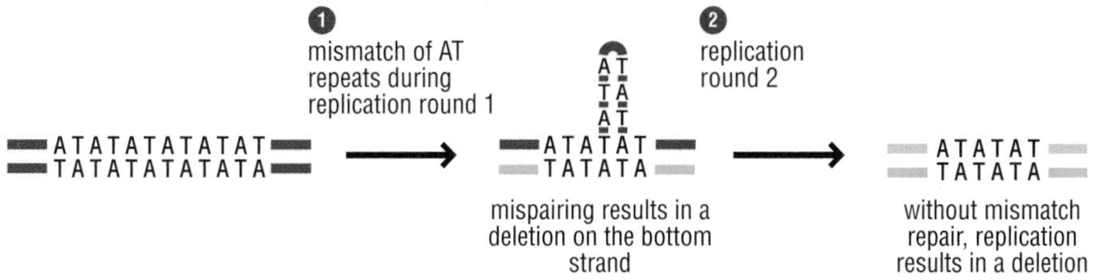

1 mismatch of AT repeats during replication round 1

2 replication round 2

ATATATATATAT
TATATATATATA

ATATAT
TATATA

mispairing results in a deletion on the bottom strand

ATATAT
TATATA

without mismatch repair, replication results in a deletion

Figure 15.6 Slipped-strand misreplication. A region of DNA consisting of AT repeats is shown.

Defects in mismatch repair result in human disease

Defects in mismatch repair in humans also lead to an increased rate of mutagenesis and are known to cause an increased predisposition to various types of cancer, arising because of the increased probability of mutation in genes that control cell growth and division. For example, the mutation of human MUT genes is the cause of many cases of hereditary non-polyposis colon cancer, one of the most common hereditary diseases.

The post-replication mismatch repair described in this section is really a means of correcting errors in newly synthesized DNA, rather than repairing damage that has been inflicted by some environmental factor - more akin to correcting typographical errors on the page of a book than to repairing the damaged binding of that book. In the next sections, however, we consider strategies that are adopted by the cell to repair physical damage.

15.3 REPAIR OF DNA DAMAGE BY DIRECT REVERSAL

Damaged regions of DNA are repaired either by reversal or by excision and resynthesis

Lesions to bases are repaired in two main ways, direct reversal of particular types of damage or by recognition of DNA damage, followed by damage excision (removal of a damaged site) and DNA resynthesis. We will consider several types of repair by resynthesis in the following sections. In this section, however, we focus on the strategy of direct reversal.

In the case of some modifications, specialized enzymes recognize particular modified bases and reverse the modification, thus restoring the original state without cutting the DNA. For example, in organisms other than placental mammals, pyrimidine dimers can be recognized and repaired by enzymes known as **photolyases**, which use energy from near-UV light to split the dimer and thus reverse the damage, restoring the DNA to its original condition. Pyrimidine dimers can also be repaired by the nucleotide excision pathway described in Section 15.5.

Base methylation can be reversed by specialized methyltransferases

DNA methylation at either adenine or cytosine, as promoted by cellular methylases, is an important process in normal development. DNA damage, however, can also result from DNA methylation. Alkylating agents can add methyl groups to bases (see Figure 15.4c) or to the DNA backbone. The modification to bases frequently occurs on guanine, which is particularly susceptible to alkylating agents.

➔ We learn more about the methylation of chromosomal DNA in Section 4.7.

Methylguanine, and methyl groups on the DNA backbone, are recognized by specialized repair enzymes called alkyltransferases, which reverse the damage by removing the methyl group from the base or the backbone and binding it covalently to a cysteine residue in the alkyltransferase active site. This is suicidal for the repair protein because the methylation is irreversible, inactivating the alkyltransferase activity.

Mice in which the gene encoding the repair alkyltransferase has been deleted are more susceptible to developing tumors after treatment with alkylating agents, revealing the importance of this enzyme in preventing the tumors that may arise as a result of this type of DNA damage. In *E. coli*, with remarkable evolutionary ingenuity, when the alkyltransferase, Ada, acquires a methyl group from the DNA backbone, it becomes an active, sequence-specific DNA-binding protein, which activates its own expression, thereby ensuring its replacement. This process is illustrated in Figure 15.7. Methylated Ada also activates the expression of a glycosylase enzyme (AlkA), which specifically removes methylated guanine bases from DNA by cleaving the linkage between the base and the sugar. This is the first step in the **base excision repair** pathway, as described below, which thus provides an alternative repair pathway for alkylated bases in *E. coli*.

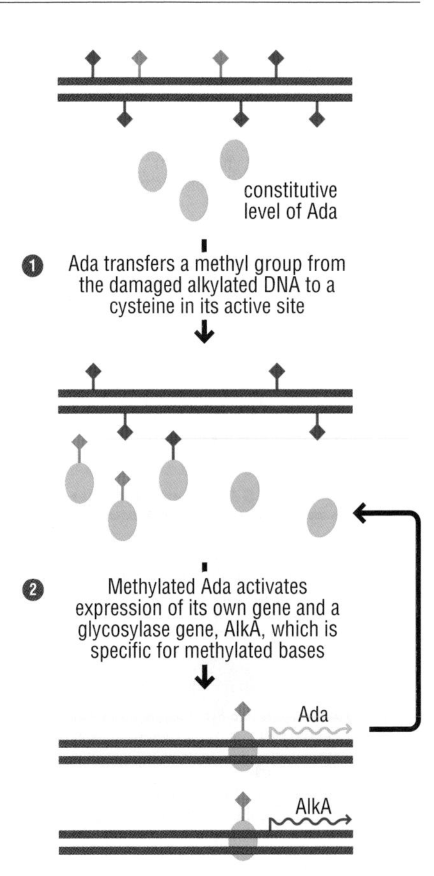
constitutive level of Ada
1 Ada transfers a methyl group from the damaged alkylated DNA to a cysteine in its active site
2 Methylated Ada activates expression of its own gene and a glycosylase gene, AlkA, which is specific for methylated bases
Ada
AlkA

Figure 15.7 Damage reversal by Ada. A DNA duplex that has been alkylated on the O^6 position of guanine (light blue squares) and the DNA backbone (dark blue squares) is shown. A low level of Ada (green ovals) is constitutively present.

Specialized proteins remove damaged nucleotides from the free nucleotide pool

Like DNA itself, the nucleotide precursors of DNA are also subject to damage and may be another source of errors in DNA if allowed to remain in the precursor pool once damaged. This is prevented by hydrolases that recognize damaged nucleoside triphosphates and convert them to monophosphates, which cannot be incorporated into DNA. For example, the *E. coli* hydrolase MutT promotes the destruction of nucleoside triphosphates containing the oxidized purine 8-oxo-guanine. The important role of MutT in removing damaged nucleotide precursors is illustrated by the way that inactivation of the human homolog of mouse MutT results in an increased incidence of cancer.

15.4 REPAIR OF DNA DAMAGE BY BASE EXCISION REPAIR

Most DNA repair occurs by excision of the damaged DNA and resynthesis

While some DNA damage is repaired by the direct reversal processes described in Section 15.3, in *most* cases DNA is repaired by excision of the damaged region and its replacement through synthesis of replacement DNA using the complementary DNA strand as a template. This process is called **excision repair**. Excision repair follows the same principles as in mismatch repair, but is catalyzed by different proteins: specialized enzymes detect and remove the damaged DNA, and specialized DNA repair polymerases replace the excised region.

There are two types of excision repair: **base excision repair** in which a single damaged nucleotide requires replacement; and nucleotide excision repair, acting on more bulky lesions, requiring a short stretch of nucleotides to be excised and replaced. In both these repair pathways, the complementary undamaged strand serves as a template for resynthesis of the correct sequence on the damaged strand.

In this section, we will describe base excision repair. We will discuss nucleotide excision repair in the next section.

Damaged DNA bases and abasic sites are repaired by base excision and DNA resynthesis

One of the simplest ways in which DNA can be damaged is through the breakage of the glycosidic bond linking a base to the sugar–phosphate backbone

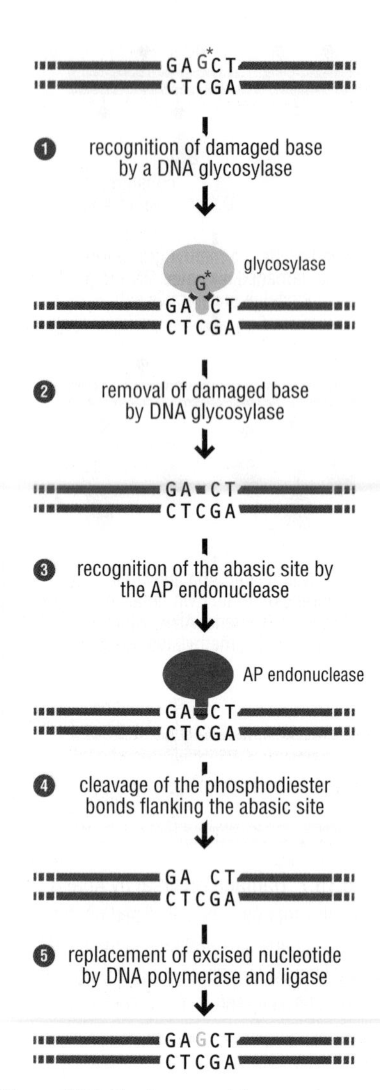

Figure 15.8 The base excision repair pathway. A DNA duplex containing a damaged base (pink) is shown.

of the DNA molecule, thus creating an **abasic site** at which the base is missing and there is a gap in the DNA. (An example of depurination is shown in Figure 15.4a.) Such lesions occur frequently and their repair is initiated by an enzyme called the **apurinic/apyrimidinic endonuclease (AP endonuclease)**. This enzyme recognizes the abasic site and cleaves the phosphodiester backbone of the DNA strand on either side of the abasic site, as illustrated in Figure 15.8, step 4. The gap is then filled in by repair DNA polymerases discussed in Section 6.2, such as Pol I in *E. coli* and Pol β in mammals, using the undamaged strand as a template.

The same pathway is used in the repair of damaged bases. Many of the mutagenic processes described in Section 15.1 result in the alteration of bases, with consequent mispairing at DNA replication that may result in point mutations (see Figure 15.4b, c). These lesions are repaired by base excision repair.

Cells contain a number of **DNA glycosylases**, enzymes that recognize and remove a particular damaged base. Some of these enzymes flip the base out of the DNA as illustrated in Figure 15.9, after which it can be excised via hydrolysis of the glycosidic bond by the gylcosylase (see Figure 15.8). The resulting abasic site is then recognized by the AP endonuclease, which cleaves the phosphodiester backbone as described above, excising the nucleotide and sometimes a few additional flanking nucleotides. The gap is filled in by a repair DNA polymerase and sealed by DNA ligase.

15.5 NUCLEOTIDE EXCISION REPAIR OF BULKY LESIONS

DNA damage that results in helix distortion is repaired by nucleotide excision repair

Some DNA damaging agents, including UV irradiation and some chemicals, produce bulky lesions that distort the DNA double helix (see Section 15.1). Although pyrimidine dimers can be specifically reversed by photolyases, these bulky lesions are usually repaired by the **nucleotide excision repair** pathway. In nucleotide excision repair, as with base excision repair, a region of the damaged strand is excised to allow accurate resynthesis of the original DNA sequence using the undamaged

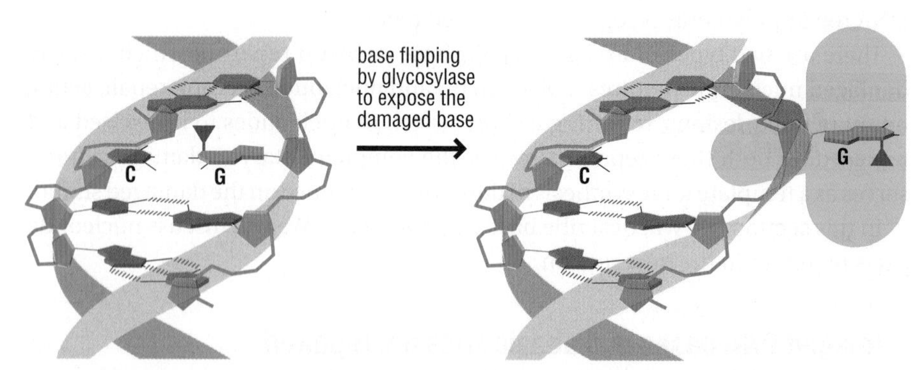

Figure 15.9 Schematic diagram of a repair glycosylase bound to DNA. A DNA duplex that contains a damaged base (blue triangle) that disrupts base-pairing is shown. The repair enzyme (green) has an alpha helix that extends into the groove of the DNA helix and flips out the damaged base (pink) that is to be cleaved.

strand as a template. In this case, however, an oligonucleotide of 10–30 nucleotides including the damaged site is excised, rather than just a single nucleotide.

As in base excision repair, the nucleotide excision repair machinery includes proteins that search for and identify damaged DNA, cleave phosphodiester bonds on either side of the site of damage, and cut away the damaged DNA strand, forming a single-strand gap. This gap is refilled by DNA synthesized by a DNA repair polymerase.

Although the proteins that mediate base excision repair are largely homologous between bacteria and eukaryotes, those that mediate nucleotide excision repair are not evolutionarily related, although they have the same functions and the pathways are similar. Nucleotide excision repair proteins in different eukaryotes are, however, homologous. As the bacterial pathway is much better understood, we will use it to illustrate the process.

Nucleotide excision repair enzymes excise and remove short DNA regions containing bulky lesions

In nucleotide excision repair, specialized proteins scan DNA and recognize most types of bulky lesions by sensing changes in the deformability of the damaged DNA. Following DNA strand separation when a lesion is recognized, a specific nuclease nicks the DNA backbone upstream and downstream of the lesion, releasing a damaged residue-containing oligonucleotide. DNA that is complementary to the undamaged DNA strand is then resynthesized.

Nucleotide excision repair in bacteria is mediated by the UvrA, UvrB, UvrC, and UvrD proteins. (The initial Uvr stands for 'ultraviolet resistant,' which causes the formation of pyrimidine dimers.) The nucleotide excision repair process occurs over multiple steps and is ATP-dependent; the process is schematically illustrated in Figure 15.10. A complex containing two UvrA and two UvrB proteins called $UvrA_2B_2$, whose formation requires ATP hydrolysis, scans the DNA looking for DNA damage. UvrA is the first damage sensor, and senses variations in DNA structure by detecting variation in DNA deformability (Figure 15.10, step 1).

The DNA strands at the site of damage are separated by the helicase activity of UvrB, which then stably associates with the DNA, while UvrA dissociates (Figure 15.10, step 2). UvrC, which has endonuclease activity, is then recruited to the UvrB–DNA complex, and cuts the phosphodiester backbone at either side of the lesion, first introducing a nick on the 3' side of the lesion and subsequently another nick on the 5' side of the lesion (Figure 15.10, step 3). The combined action of UvrD (DNA helicase II) and DNA polymerase I releases the oligonucleotide, UvrC, and UvrB (Figure 15.10, step 4), leaving a gap in the DNA that is then filled by repair synthesis (Figure 15.10, step 5).

The steps in nucleotide excision repair in eukaryotes occurs by the same steps as in bacteria, though the proteins involved are not evolutionarily related. The proteins that execute these steps in yeast and humans are listed alongside the functionally analogous bacterial proteins in Figure 15.11. The XP proteins were identified from studying the human disease XP, in which there is greatly increased sensitivity to DNA damage.

In eukaryotes, initial recognition of a DNA lesion is mediated by XPE, followed by interaction with XPC. Subsequent strand separation is mediated by the helicases XPB and XPD, which are part of the multi-subunit complex TFIIH, which also functions in transcription. XPG and XPF then generate incisions on each side of the lesion, resulting in a gap that is then repaired by DNA synthesis and ligation.

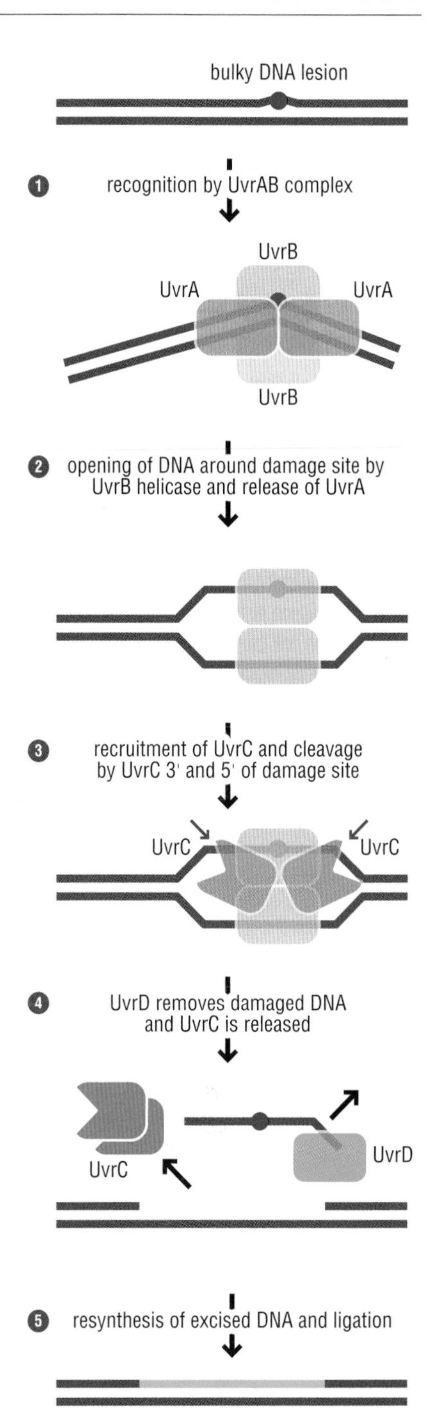

Figure 15.10 The *E. coli* nucleotide excision repair pathway. A DNA duplex containing a bulky lesion (red) is shown.

Some components of the nucleotide excision repair systems in bacteria and eukaryotes			
steps	***E.coli***	***S.cerevisiae***	**humans**
damage recognition (general)	UvrA, UvrB	Rad14, RPA1–3	XPC-RAD23B, XPA, RPA1–3
transcription-coupled repair	Mfd	Rad26, Rad28	CSA, CSB
DNA unwinding	UvrAB complex	Rad3, Ssl1–2, Tfb1–4	XPB, XPD
5′ excision of damaged site	UvrC	Rad1, Rad10	XPF, ERCC1
3′ excision of damaged site	UvrC	Rad2	XPG
displacement of damaged oligonucleotide	UvrD		
repair DNA polymerase	Pol I	Pol β	Pol δ, Pol ϵ

Figure 15.11 Some key components of the nucleotide excision repair machinery. Proteins involved in nucleotide excision repair from different organisms. Eukaryotic proteins are structurally related from yeast to humans. XP proteins were identified from humans with xeroderma pigmentosum. CSA and CSB were identified from humans with Cockayne syndrome. ERCC1 (excision repair cross-complementing) was identified from a mouse mutant with a defect in excision repair. Although they are functionally similar there is no sequence homology between the bacterial and eukaryotic nucleotide excision repair proteins.

Transcription-coupled nucleotide excision repair targets DNA repair to the template strand of genes that are being transcribed

In both bacteria and eukaryotes, nucleotide excision repair operates anywhere in the genome and on several types of damage; in this mode, it is often termed **global genomic repair (GGR)**. If the damage occurs in genes that are being transcribed, however, repair is preferentially targeted to these genes to remove stalled RNA polymerases that could disrupt gene expression. This process, **transcription-coupled repair (TCR)**, involves the additional recognition proteins listed in Figure 15.11.

TCR appears to be initiated by recognition of RNA polymerase molecules that have stalled as a result of encountering DNA damage. In *E. coli*, the DNA-binding protein, Mfd, interacts directly with stalled RNA polymerase complexes to dissociate them and remove them from the DNA. Mfd also recruits the nucleotide excision repair components to the site of damage by interaction with UvrA. Although not yet understood at the molecular level, the eukaryotic CSB proteins likely stimulate TCR via direct CSB-RNA polymerase interaction, which then leads to recruitment of the rest of the nucleotide excision machinery (except for XPE and XPC, which are involved in the GGR pathway).

Intriguingly, the CSB and analogous Rad26 proteins in eukaryotes belong to the SWI/SNF family of nucleosome-remodeling enzymes, which change the accessibility of DNA to proteins by altering chromatin structure. They may promote local TCR by facilitating the recruitment of the repair proteins to the damaged DNA by increasing their accessibility to it.

➔ We learn about nucleosome remodeling in Section 4.6.

Repair is then specifically targeted to the template DNA strand, that is, the DNA strand that is being transcribed into RNA. The stalled polymerase is removed along with the damaged DNA and the strand restored to its original state by the excision and DNA resynthesis steps outlined above. TCR ensures that the vital coding sequences in the genome are maintained in good repair, and also enables transcription of a damaged gene to proceed.

Defects in nucleotide excision repair result in mutator phenotypes and disease

Rare inherited defects in components of the nucleotide excision repair system cause diseases in which the underlying defect is an increased mutation rate. One of these diseases is xeroderma pigmentosum, in which individuals cannot repair pyrimidine dimers (and perhaps other adducts), which are formed in DNA on exposure to UV light. The result is a greatly increased incidence of skin cancer, caused by the greater likelihood of mutations in genes that control cell growth and division in skin cells, which are the cells most exposed to sunlight. Xeroderma pigmentosum can be caused by defects in any one of the XP proteins, the eukaryotic proteins functionally equivalent to UvrA, UvrB, and UvrC (see Figure 15.11).

A rare inherited disease linked to defective TCR is Cockayne syndrome, which is characterized by sensitivity to sunlight and the appearance of premature aging, among other symptoms, but not by an increased incidence of cancer. Patients with this disease have lost the function of either CSA or CSB, which are important for TCR (see Figure 15.11),

It is not yet understood why different molecular defects that all result in increased mutation frequency, such as defects in mismatch repair and in excision repair, have different phenotypic outcomes. For example, we don't yet know why we see a specific increase in the frequency of colon cancer occurring with defects in mismatch repair yet see premature aging with defects in transcriptional coupling of nucleotide excision repair.

15.6 TRANSLESION DNA SYNTHESIS

Error-prone polymerases can promote translesion DNA synthesis

The DNA repair pathways we have discussed above can restore damaged DNA to its pre-damage sequence. However, DNA lesions are not always repaired before they are encountered by a replication fork, especially in cells in which DNA has been heavily damaged, leading to stalled or even broken replication forks.

One possible response to a stalled replication fork is the recruitment of specialized 'translesion' synthesis (TLS) polymerases, many of which are members of the Y-family polymerases, a structurally distinct class of polymerase, whose members are summarized in Figure 15.12.

➔ We discuss polymerase families in Section 6.2.

Most TLS polymerases are much more error prone than the highly accurate replicative polymerases, and thus the cost of continued replication at an unrepaired lesion is the increased risk of mutagenesis. The occurrence of a mutation

Key subunits of some translesion polymerases

polymerase family	polymerase	domain	*E. coli*	*Sulfolobus solfataricus*	*S. cerevisiae*	mouse	human
Y	Pol IV	bacteria	DinB	-	-	-	-
Y	Pol V	bacteria	UmuC	-	-	-	-
Y	Pol IV	archaea	-	Dpo4	-	-	-
Y	Pol IV	archaea	-	Dbh	-	-	-
Y	REV1	eukaryotes	-	-	Rev1	Rev1	REV1
B	Pol zeta (ζ)	eukaryotes	-	-	Rev3	Rev3L/Polζ	REV3L/POLZ
Y	Pol eta (η)	eukaryotes	-	-	Rad30	Polh	RAD30A/XPV/POLH
Y	Pol iota (ι)	eukaryotes	-	-	-	Poli	RAD30B/POLI
Y	Pol kappa (κ)	eukaryotes	-	-		Polk	DINB1/POLK

Figure 15.12 Translesion DNA polymerases.

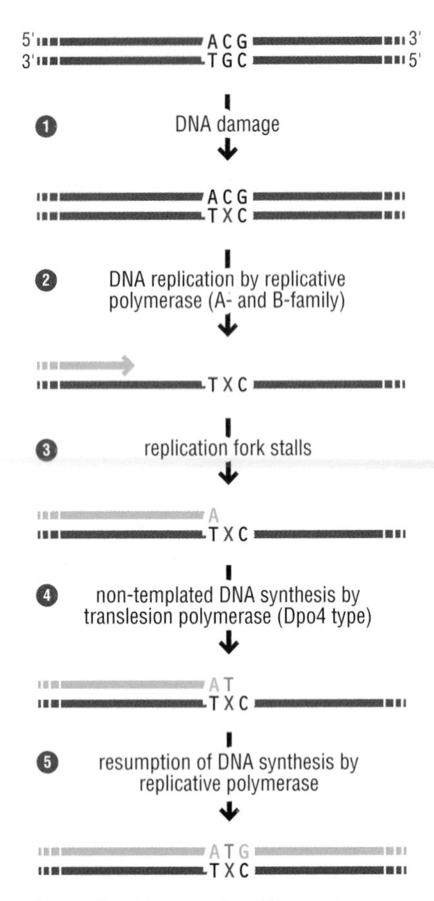

Figure 15.13 Translesion DNA synthesis. Undamaged DNA (gray) is shown.

event as the result of TLS is depicted in Figure 15.13. TLS may be more accurately considered as a 'damage tolerance' mechanism than a 'DNA repair' mechanism.

TLS polymerases display low fidelity of synthesis and lack 3′ to 5′ proofreading activity. They are able to bypass bulky lesions because their active sites are much more open and flexible than those of the highly accurate replicative polymerases such as *E. coli* DNA polymerase III (pol III) and the eukaryotic DNA polymerase δ (pol δ) and DNA polymerase ε (pol ε), as illustrated in Figure 15.14. Thus, altered base structures that are not compatible with the active sites of replicative polymerases can be accommodated in the active sites of TLS polymerases.

The error rates of these TLS polymerases however are 10^{-2} to 10^{-4} compared with error rates of 10^{-5} to 10^{-6} for replicative polymerases, and the cost of their flexibility is thus an increased rate of mutation. Indeed, it is the action of TLS polymerases that actually underlies most mutagenic events in response to DNA damage. Error-prone replication is, however, better than no replication at all.

Interactions between TLS polymerases and processivity clamps are central to TLS

To preserve the generally high fidelity of DNA replication, access of TLS polymerases to replication forks must be limited. Interaction of the TLS polymerases with processivity clamps, such as the *E. coli* sliding clamp and eukaryotic proliferative cell nuclear antigen (PCNA), mediates their recruitment to replication forks, as shown in Figure 15.15. All polymerases - including TLS polymerases - interact with the sliding clamp at the same site. It remains to be determined, however, how polymerase switching (from one to the other) is actually modulated.

An increase in the expression of some TLS polymerases occurs as part of a **DNA damage response**. As we shall see in Section 15.7, which explores the DNA damage response in more detail, expression of the *E. coli* TLS polymerases DinB and UmuCD is greatly increased in response to DNA damage. We

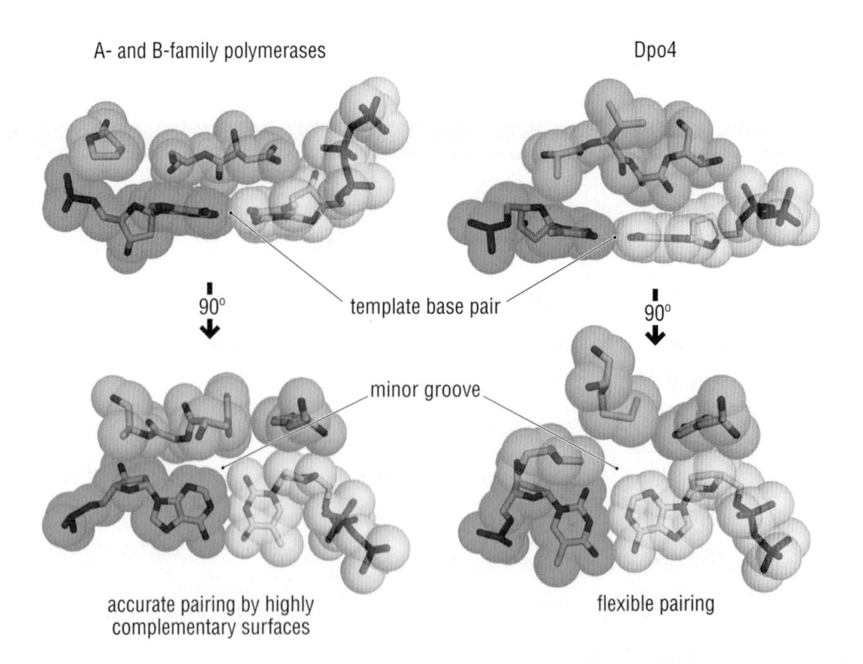

Figure 15.14 Comparison of replicative and translesion polymerases. Comparison of the interfaces between a replicating base pair (gray and yellow) and the protein surfaces (blue) of highly accurate replicative polymerases (A- and B-family) and the Dpo4 Y-family is shown.
From Yang, W. Portraits of a Y-family DNA polymerase. *FEBS Letters*, 2005;**79**:868–872.

shall also see in Section 15.9 how the modification of PCNA by ubiquitination controls a eukaryotic TLS polymerase as part of the eukaryotic DNA damage response.

We have considered a wide variety of mechanisms for the recognition, repair, and tolerance of DNA damage. We explore in Experimental approach 15.1 some of the methods used to identify genes involved in DNA repair.

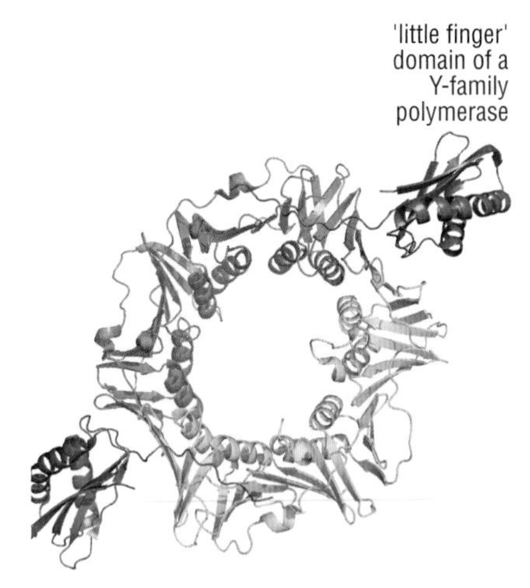

Figure 15.15 Structure of a translesion DNA polymerase and the beta clamp. The structure of a complex between the 'little finger' domain of the UmuCD Y-family polymerase (pink) is shown with the *E. coli* beta clamp (brown and yellow).
from Bunting, KA, Roe, SM, and Pearl, LH. Structural basis for recruitment of translesion DNA polymerase Pol IV/DinB to the a-clamp. *The EMBO Journal*, 2003;**22**:5883–5892.

15.7 THE DNA DAMAGE RESPONSE

Extensive damage to DNA induces mechanisms that preserve the viability of the organism

The repair mechanisms we have discussed so far depend on recognition of a specific type of DNA damage by a component of the repair pathway that recruits the proteins appropriate for its repair. Where damage is extensive, however - after exposure to UV light or ionizing radiation, for example - or where it interferes with DNA replication, a variety of cell-wide responses are invoked to amplify the recruitment of repair proteins and to halt the cell cycle while repair is completed. These responses are collectively known as the **DNA damage response**.

Although these responses have been studied largely in cells that have been experimentally subjected to large doses of radiation, it is likely that they are required continuously in dividing cells (especially those with large genomes) to prevent permanent damage that would otherwise result from unrepaired damage in the path of the polymerase during DNA replication.

Single-stranded DNA or double-strand breaks induce the damage response

There are two possible consequences if a polymerase encounters unrepaired DNA damage at a replication fork. Distorting lesions, such as unrepaired oxidative damage, pyrimidine dimers, or DNA adducts cause the polymerase to stall, and delay or prevent the completion of DNA synthesis, as depicted in Figure 15.16a. (As discussed in Section 15.6, translesion DNA synthesis by specialized polymerases can occur across a distorting lesion, but such synthesis is generally mutagenic.) Unequal movement of the polymerases at a stalled replication fork can lead to regions of single-stranded DNA.

Equally as hazardous as distorting lesions are nicks in the DNA, which can be caused by ionizing radiation but also occur through incomplete mismatch, base excision repair, or nucleotide excision repair, all of which involve nicking of the damaged strand by nucleases (as we have seen in the preceding sections). Single-strand nicks in replicating DNA not only cause the polymerase to stall but result in double-strand breaks that disconnect the arms of the replication fork and thus collapse the fork, as shown in Figure 15.16b.

Double-strand breaks in DNA, which can also be directly caused by ionizing radiation, are a special hazard because in this case there is no intact complementary strand for repair polymerases to copy. Thus double-strand breaks always elicit a global DNA damage response. The specialized repair mechanisms required to mend double-strand breaks are discussed in Chapter 16; here we are concerned with the DNA damage response pathways through which the cell cycle is arrested

15.1 EXPERIMENTAL APPROACH

Finding genes involved in the repair of damaged DNA

Insights into many cellular processes have come from the isolation of mutants with defects in the process. Indeed, this approach has been used with great success for the identification of proteins involved in the repair of DNA damage. When using such genetic approaches, it is generally easier to obtain or select for mutants that are *resistant* to a particular cellular insult, such as a DNA damaging agent. In such an experiment, cells are exposed to a condition that kills the wild-type cells, and survivors are directly isolated. However, some activities or genes are only identified by the isolation of mutants that lead to increased *sensitivity* to a particular condition. How can such mutants, for example, mutants that are sensitive to DNA damaging agents such as UV light or x-rays, be isolated?

One commonly used method has been to mutagenize bacterial or yeast cells, spread them on agar plates such that a population of single colonies is present, then replica-plate those same colonies to new agar plates where their growth can be monitored in the presence and absence of the DNA damaging agent. Colonies that do not carry mutations in DNA repair genes grow normally while colonies with mutations grow poorly. The challenge of this 'screening' approach is that the experimenter is looking for the rare single colony that will not grow among thousands of colonies that do. This is quite different from the direct isolation of mutants under a generally repressive set of growth conditions – scientists broadly refer to this approach as a genetic 'selection.'

How to select for mutants defective in DNA repair

Paul Howard-Flanders and Lee Theriot exploited a genetic selection approach in which only *Escherichia coli* cells defective in DNA repair would grow. These researchers took advantage of the fact that damage to the DNA of a lytic phage, caused by irradiation with UV light, for example, must be repaired for the phage to grow and kill the cell. Thus, after infection with a damaged lytic phage, those cells that repair the damaged phage DNA die and those that cannot repair the DNA damage live. Howard-Flanders and Theriot plated a population of mutagenized *E. coli* and exposed the cells to lytic phage that had been damaged by UV light. After infection with the phage, they identified colonies that were not killed by the damaged phage, that is, those that did not repair damaged phage DNA. Figure 1 shows that the 14 mutants isolated in this way were much more sensitive than wild type to killing by UV irradiation. These 14 mutations mapped to *uvrA*, *uvrB*, and *uvrC*, three genes that we now know mediate nucleotide excision repair (NER).

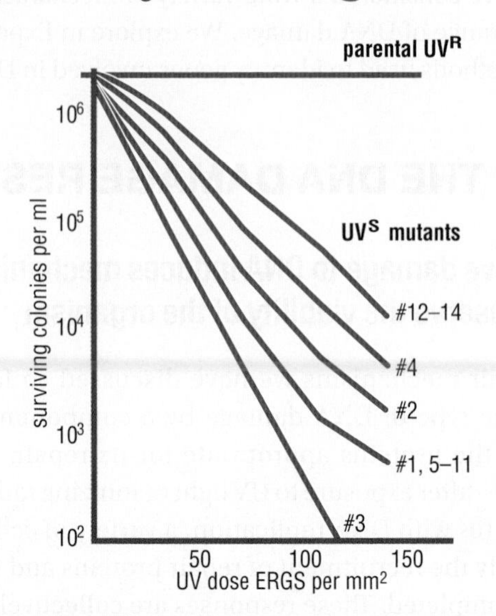

Figure 1 UV sensitivity of *E. coli* mutants defective in DNA repair. The ability of wild-type and mutant *E. coli* strains to grow in the presence of increasing doses of UV irradiation was measured. Cell survival was determined by the number of cells that can grow when plated on agar plates. Reproduced from Howard-Flanders and Theriot, A method for selecting radiation-sensitive mutants of *Escherichia coli*. *Genetics*, 1962;**47**:1219–1224.

Analysis of mutant collections in which all non-essential genes have been disrupted and tagged facilitates genome-wide hunts for genes that are involved in DNA repair

A powerful new genetic method exploits a collection of mutant strains in which each non-essential gene of the organism has been disrupted or 'knocked out', that is, where the coding region of the gene has been replaced by another marker, such as a drug resistance marker. In the process of constructing the disruption, one or two unique oligonucleotide sequences, called tags or barcodes, are introduced into each strain flanking the marker. These unique tags allow each mutant to be followed within the population of mutants by sequence analysis or hybridization to a microarray. We discuss these collections of disruption strains in Experimental approach 17.2 and Section 19.5.

Such 'knockout collections' have been constructed in several bacteria and the yeasts *Saccharomyces cerevisiae* and *Schizosaccharomyces pombe*. Large collections of knockout mice are also under construction.

Guri Giaver and co-workers used the *S. cerevisiae* knockout collection to identify DNA repair genes. To do this they compared a population of cells from the knockout collection grown with and without DNA damaging agents. Those genes that conferred sensitivity to DNA damage when knocked out were less abundant in the population after growth in the presence of DNA damaging agents.

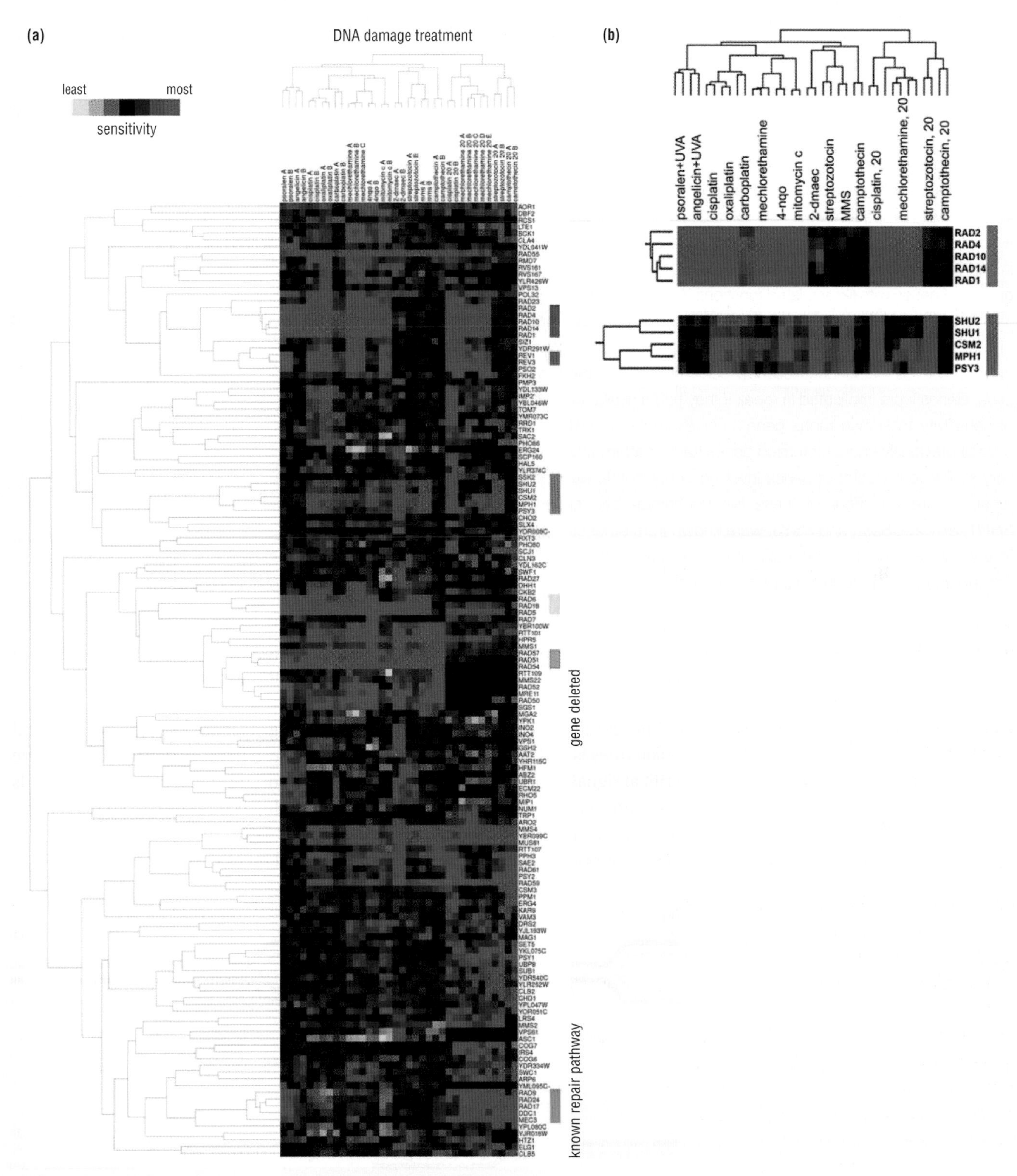

Figure 2 Relative growth of *S. cerevisiae* knockout strains in the presence of DNA damaging agents. (a) The yeast 'knockout' collection was grown in the presence of various DNA damaging agents (horizontal axis). The growth of individual mutants (vertical axis) was evaluated by isolating DNA from the population of cells, amplifying all of the tags (see Section 19.10), and hybridizing the amplified tags to a microarray. The amount of hybridization observed for a particular tag is considered to be proportional to the abundance of the corresponding mutant. The relative growth of different knockout strains is indicated by the colored bar where yellow indicates the least effect on growth and blue indicates the most inhibition of growth. The knockout strains with similar phenotypes are clustered on the vertical axis. Colored bars on the right represent gene clusters of known function, including NER (blue); error-prone translesion DNA synthesis (red); post-replication repair (yellow); homologous recombination (green); cell-cycle checkpoint control (orange); and a cluster of genes of unknown function shown in (b) (magenta). (b) Zoom view showing one cluster containing the class I NER genes and a second cluster containing several uncharacterized DNA-repair genes. Four of these five genes *(SHU1*, *SHU2*, *CSM2*, and *PSY3)* are known to encode proteins that physically interact.

Reproduced from Lee *et al.*, Genome-wide requirements for resistance to functionally distinct DNA-damaging agents. *PLoS Genetics*, 2005;**1**:e24.

Figure 2a shows the results of their genome-wide screen, using a cluster analysis in which the growth of each knockout strain in the presence of each DNA damaging agent is represented. Blue indicates knockout strains that are most sensitive (grow least well in the presence of DNA damaging agents), and yellow indicates knockout strains that are least affected by these agents. It is notable that mutations in particular groups of genes result in the same (or a very similar) pattern of sensitivity to different DNA damaging agents. Those with the most similar patterns are clustered together in the hierarchical cluster analysis shown in the figure, allowing groups of genes with likely related functions to be identified.

This screen re-identified some DNA repair genes known from previous studies, which served to validate the method. Importantly, however, it also identified new genes involved in the process. Genes were implicated in repair if they had a similar pattern of sensitivity to known repair genes. For example, a cluster of several previously uncharacterized genes had a pattern that was very similar to a cluster of genes involved in nucleotide excision repair (Figure 2b). Four of these five uncharacterized genes, *SHU1*, *SHU2*, *CSM2*, and *PSY3*, were known from other studies to encode proteins that physically interact with each other, suggesting that they are found in a protein complex.

While the actual function of these genes remains to be determined, this genome-wide functional analysis suggests that they are involved in repair, and this can now be tested directly for each of these genes. Knockout collections, which are being screened for phenotypes under hundreds of conditions, are providing important clues to the functions of many genes with previously unknown functions, but as in the case of *SHU1*, *SHU2*, *CSM2*, and *PSY3*, these clues often are most helpful when related clues come independently from other lines of experimentation.

Find out more

Howard-Flanders P, Theriot L. A method for selecting radiation-sensitive mutants of *Escherichia coli*. *Genetics*, 2008;**47**:1219–1224.

Lee W, St Onge RP, Proctor M, *et al*. Genome-wide requirements for resistance to functionally distinct DNA-damaging agents. *PLoS Genetics*, 2008;**1**:e24.

Related techniques

E. coli (as a model organism); Section 19.1

S. cerevisiae (as a model organism); Section 19.1

Mutant isolation; Section 19.5

Gene deletion by recombination; Section 19.5

Microarray analysis; Section 19.10

and repair proteins are recruited to the site of double-strand breaks or stalled replication forks.

The first step in the induction of the DNA damage response in both bacteria and eukaryotes is the recognition of damaged DNA by a damage-sensor protein. The critical signal in bacterial cells and the most important signal for eukaryotic cells is the presence of single-stranded DNA arising either from stalled DNA replication or from the actions of exonucleases on damaged DNA. In eukaryotic cells, damage sensors also directly recognize double-strand breaks.

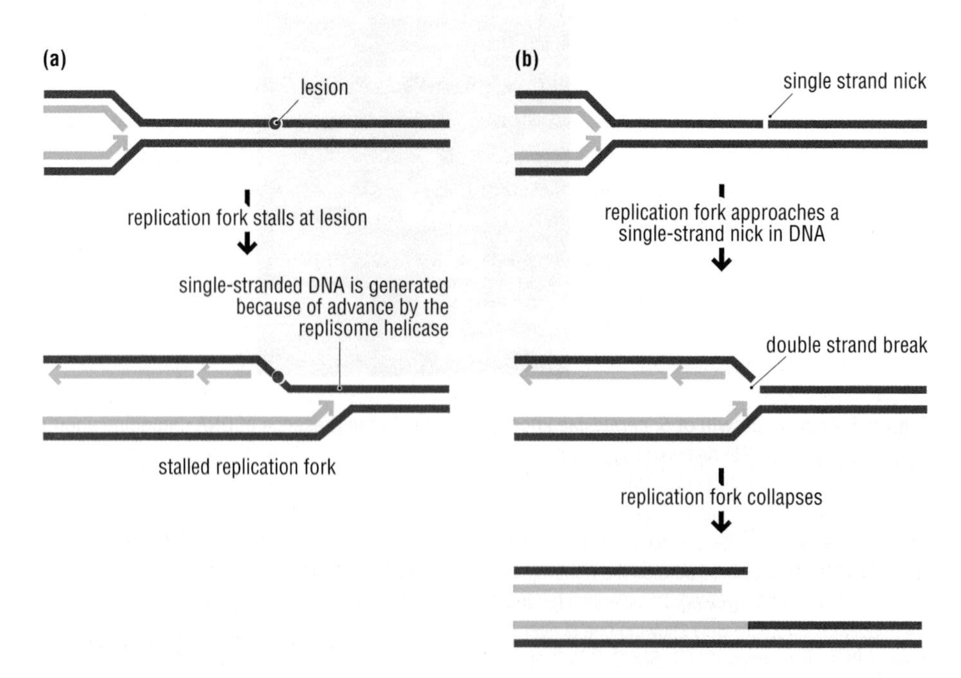

Figure 15.16 Effects of DNA damage at a replication fork. (a) Polymerase stalling, (b) Replication fork collapse.

DNA damage can result in programmed cell death

Single-celled organisms eventually resume division even when the damage cannot be repaired. In multicellular organisms, however, unrepaired DNA damage can ultimately trigger the programmed death of the damaged cell, preventing the propagation of mutations that may result in unregulated growth and thus to cancer, which can be lethal to the organism. The failure of these DNA damage response mechanisms underlies a variety of human diseases related by the common feature that they result in high sensitivity to DNA damaging agents and increased susceptibility to cancer.

15.8 THE DNA DAMAGE RESPONSE IN BACTERIA

The first identified and best understood example of a cell-wide response to DNA damage is the bacterial **SOS response**. We discuss this response here before turning later to the more complex responses of eukaryotic cells to DNA damage.

The bacterial SOS response depends on recognition of single-stranded DNA by the RecA protein

Many of the proteins involved in DNA repair in bacteria are expressed constitutively, but the presence of extensive damage to DNA leads to increased expression of repair genes to cope with the increased demand for repair. The SOS response is a multifaceted cellular response in which more than 40 genes, many of which are directly involved in repair, are induced when DNA is damaged. Central to the activation of the bacterial SOS response is the **RecA** protein, which binds to single-stranded DNA exposed by DNA damage.

Figure 15.17 illustrates how RecA forms a nucleoprotein filament on a single strand of DNA, which can act as a **co-protease**, stimulating the self-cleavage of several proteins that contribute to or inhibit the SOS response, and thereby activating or inactivating them. One of the proteins whose cleavage is promoted by RecA is LexA, a specific DNA-binding protein that acts as a repressor that blocks transcription of many of the SOS genes. In the presence of the RecA-single-stranded DNA filament, LexA cleaves itself, inactivating its repressor function by separating its DNA-binding domain from its oligomerization domain. This inactivation results in the induction of the genes it usually represses (see Figure 15.17). Many of these genes encode proteins that participate directly in DNA repair. These include some of the Uvr proteins, which act early in the SOS response to promote nucleotide excision repair (see Section 15.5).

The *recA* gene itself is repressed by LexA although a low level of RecA is synthesized even in the presence of LexA. Derepression of the *recA* gene upon LexA inactivation results in increased production of RecA protein itself, which also plays a central part in the repair of double-strand DNA breaks and in genetic recombination: the RecA–single-stranded DNA filament can also promote the specific pairing of its single strand with a complementary strand in a DNA duplex. We discuss the role of this RecA strand-pairing activity in the repair of double-strand breaks and homologous recombination in more detail in Chapter 16

Another SOS response that contributes to cell survival in the presence of DNA damage is the increased expression of the *sulA* gene, the product of which is a

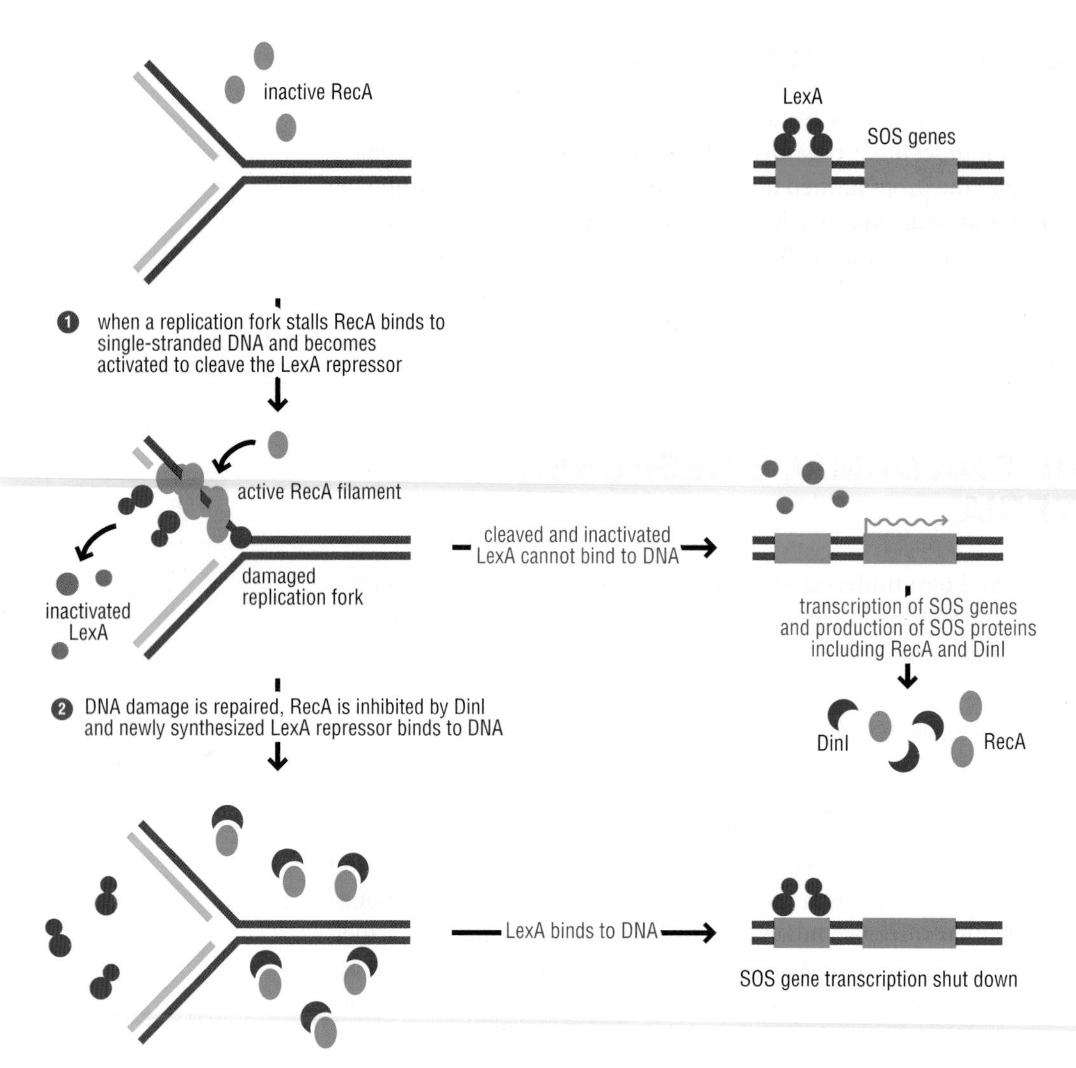

Figure 15.17 Activation and switching off of the SOS response. When DNA replication is stalled because of a damaged (red) replication fork, RecA protein binds to the exposed single-stranded DNA and forms a nucleoprotein filament that promotes the cleavage of the LexA repressor, thus releasing the SOS genes from repression. SOS proteins including RecA itself, other SOS proteins that promote DNA repair, and the regulator protein DinI (red) are produced.

➔ We discuss cell-cycle checkpoints in more detail in Section 5.6, and learn more about bacterial cell division – and the role played by FtsZ – in Section 7.8.

LexA-regulated inhibitor of cell division that increases the time window for DNA repair. This inhibition of part of the bacterial cell cycle is similar conceptually to the much more elaborate cell-cycle checkpoints of eukaryotes. SulA inhibits bacterial cell division by interacting with FtsZ, a key protein in the formation of the septum that divides a bacterium into two cells.

As discussed below, the synthesis of two TLS polymerases is also part of the SOS response. Other SOS-induced proteins include RuvA and RuvB, which we shall encounter in Chapter 16 when we describe the repair of double-strand breaks.

The SOS response is self-regulating. As DNA repair proceeds, the amount of single-stranded DNA decreases and less RecA filament assembles, reducing LexA cleavage. This, in turn, leads to re-repression of the SOS genes. Another SOS protein is DinI (damage inducible I) protein, which interacts with the RecA filament to inhibit the co-protease activity. The negatively charged surface of DinI helix probably acts as a DNA mimic, decreasing the amount of the RecA–single-stranded DNA filament. The structure of DinI is depicted in Figure 15.18. As a result of DinI activity, LexA is no longer degraded, and SOS genes are again repressed (see Figure 15.17c).

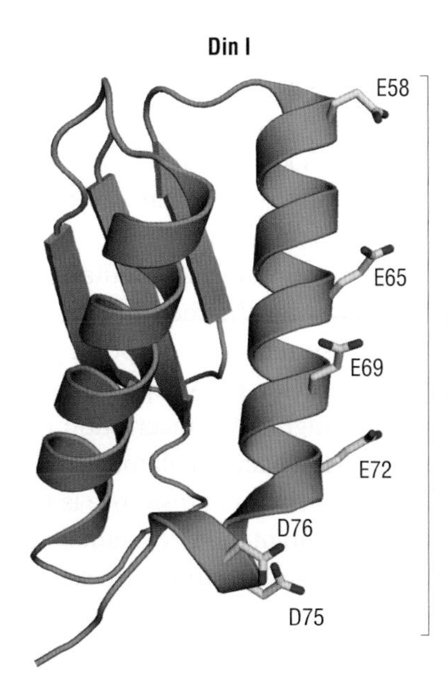
Din I
E58
E65
E69
E72
D76
D75
negatively-charged surface is a DNA mimic that binds to RecA, blocking its co-protease activity

Figure 15.18 DinI is a DNA mimic. The structure of DinI reveals a highly negatively charged alpha helix that can act as a mimic of DNA that can interact with RecA, blocking the interaction of single-stranded DNA with RecA that is required for the RecA co-protease activity. Protein Data Bank (PDB) code 1GHH.

from Voloshin, ON, Ramirez, BE, Bax, A, *et al.* A model for the abrogation of the SOS response by an SOS protein: a negatively charged helix in DinI mimics DNA in its interaction with RecA. *Genes and Development*, 2001;**15**: 415–427.

It is interesting to note that a number of the genes induced by the SOS response do not encode obvious DNA repair functions; moreover, DNA damage also results in the induction of many genes that are not regulated by LexA. The mechanisms of their contributions to cell survival and of their induction remain to be determined.

The SOS response includes increased expression of TLS polymerases and induced mutagenesis

A notable feature of the SOS response to DNA damaging agents is an increased frequency of mutagenesis. This mutagenesis is not a passive consequence of DNA damage but rather is dependent on RecA and LexA, revealing that the activity of SOS-induced proteins is required. The principal source of SOS-induced mutagenesis is a Y-family TLS polymerase composed of UmuC and UmuD; the expression of both UmuC and UmuD is regulated by LexA. However, the active TLS polymerase also requires proteolysis of UmuD, promoted by the co-protease activity of RecA bound to single-stranded DNA. This proteolysis generates the UmuD′ fragment that associates with UmuC to form the active UmuD$'_2$C complex. The active site of the polymerase lies in the UmuC protein.

Another Y-family TLS polymerase, DinB (Pol V), is also induced as part of the SOS response but its major role may be in mutagenesis induced by other stress conditions such as starvation stress.

15.9 THE DNA DAMAGE RESPONSE IN EUKARYOTES

The eukaryotic DNA damage response requires sensor, regulator, mediator, and effector proteins

Two key cellular events are perceived by the cell to be indicative of DNA damage: the sustained existence of single-stranded DNA, and the existence of double-strand DNA breaks. The eukaryotic DNA damage response, like the bacterial SOS

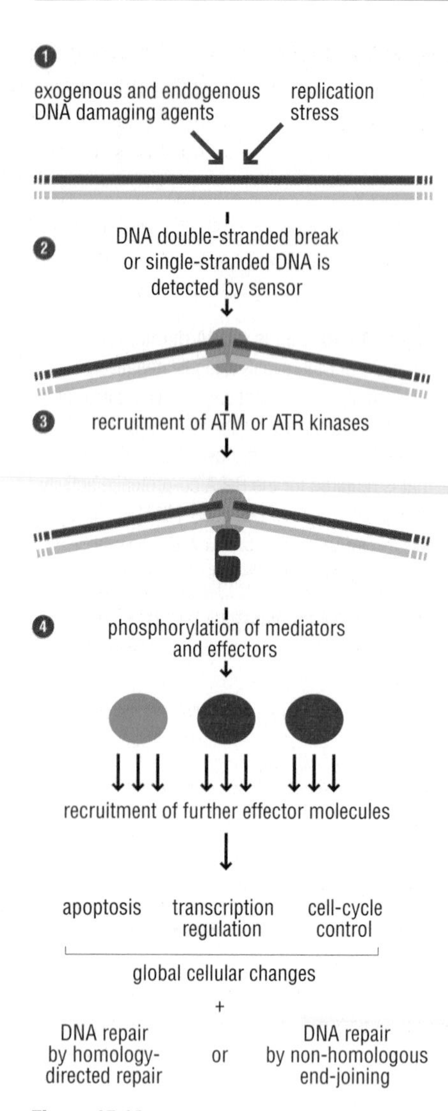

Figure 15.19 The eukaryotic DNA damage response. (1) DNA is subject to damage from exogenous and endogenous agents that result in DNA double-strand breaks and from replication stress that leads to the exposure of single-stranded DNA, for example at stalled replication forks. (2) DNA damage is detected by a sensor protein (such as RPA or MRN in Figure 15.21). (3) A regulator kinase, ATM or ATR (see Figure 15.21) interacts with the sensor protein at the site of DNA damage. (4) Phosphorylation of downstream mediator molecules (such as MDC1 and H2AX in Figure 15.25) that serve as a platform for the recruitment of effector proteins that control the cell cycle (Chk2 in Figure 15.26), control apoptosis (p53 as in Figure 15.27), or promote DNA repair by homology-directed repair or non-homologous end joining (see Chapter 16).

➔ We learn more about the activity of RPA in Section 6.5.

response, is initiated by the binding of **sensor proteins** at the site of such damage, as summarized in Figure 15.19. But instead of acting directly on the transcriptional regulation of repair proteins (as RecA does in bacterial cells) the sensor proteins of eukaryotic cells recruit specialized protein kinases that signal the presence of damage and coordinate a more complex response. The kinases are known as the **regulator kinases** of the eukaryotic damage response.

Regulator proteins phosphorylate and thereby activate **mediator** proteins that, when phosphorylated, provide binding sites for the recruitment of the **effector proteins** that repair DNA damage. They also recruit checkpoint protein kinases that halt the cell cycle to provide a window for DNA repair. Most DNA repair proteins are constitutively expressed, although the transcription of some genes encoding repair proteins is increased (as in the bacterial damage response). However, the mechanism of the eukaryotic transcriptional response is much less well understood. Let us now focus on the regulation of the DNA damage response; we will discuss the mechanisms of the repair of double-strand breaks in Chapter 16.

Figure 15.20 illustrates how many of the proteins and the processes of the eukaryotic DNA damage response are conserved from yeast to humans. A notable exception is the programmed cell death response that is elicited in multicellular eukaryotes when damage is not repaired, as we have already mentioned in Section 15.7. The specialized machinery that condemns such damaged cells to self-destruction is described in Section 15.10.

In addition to the phosphorylation cascade initiated by the regulator kinases, other post-translational protein modifications such as ubiquitination, sumoylation, methylation, acetylation, poly(ADP-ribosyl)ation and others also play important roles in the DNA damage response. Such protein modifications likely provide rapid and reversible protein activation and also function as essential signals for the recruitment of effector proteins to sites of DNA damage.

Let us now consider the mechanism of the eukaryotic DNA damage response in more detail. We begin by describing the eukaryotic damage sensors and the orchestration of the damage response by the regulator kinases.

Two major sensors initiate the DNA damage response in eukaryotic cells

The two major sensors of damaged DNA in human cells are the single-stranded binding protein, replication protein A (**RPA**), which binds DNA ahead of the polymerase at the replication fork, and **MRN**, which recognizes DNA double-strand breaks and is discussed below in more detail. The contrasting roles of RPA and MRN are depicted in Figure 15.21.

When RPA binds to single-stranded DNA ahead of DNA polymerase at a replication fork, it is normally rapidly removed as the replication machinery progresses. If the polymerase is stalled by DNA damage, however, the replicative helicase can disengage and continue to separate the two strands ahead of the rest of the replication machinery allowing RPA to accumulate and persist over extended regions of DNA (Figure 15.21a, step 1). This continued presence of RPA–single-stranded DNA acts as a signal to the cell of DNA damage, and results in the recruitment of a regulator kinase and the initiation of the damage response.

The other major sensor of DNA damage is MRN, which belongs to the SMC (structural maintenance of chromosomes) family of proteins. SMC proteins include the cohesins that are required for chromosome condensation and segregation during mitosis (see Section 7.2). MRN is so called because it contains MRE11 (hence M) protein, and the SMC protein RAD50 (hence R). These proteins are highly conserved in all domains of life. Mutations in either protein lead to genome instability as a result of radiation hypersensitivity, for example.

Mammalian MRN includes NBS1, a protein named for Nijmegen breakage syndrome, a human genetic disease that occurs when the protein is defective, and is characterized by radiation sensitivity, chromosomal instability and predisposition to cancer. NBS1 binds to MRE11 and, as we shall see shortly, provides the binding site for one of the two major regulator kinases, as well as other proteins involved in DNA double-strand break repair.

The analogous protein in yeast is called Xrs2, so named for the x-ray sensitive phenotype of cells in which it is defective; Xrs2 has weak sequence similarity to NBS1. We shall see in the next chapter that the mutant phenotypes of these proteins reflect the importance of MR complexes not only in activating the global DNA damage response, but also locally in the actual repair of individual double-strand breaks.

At a double-strand break, the MRE11 subunits interact directly with DNA on each side of the break and may also dimerize, forming a bridge between the two sides of the double-strand break (see Figure 15.21b). RAD50, like all eukaryotic SMC proteins, contains two half-site ATPase domains whose dimerization promotes interaction between the N-terminal and C-terminal domains of the protein.

Alternative names for DNA damage response components

S. cerevisiae	*S. pombe*	vertebrates
sensor regulator		
Mec1	Rad3	ATR
Tel1	Tel1	ATM
ATR regulatory subunit		
Ddc2/Lcd1	Rad26	ATRIP
effector kinases		
Chk1	Chk1/Rad27	CHK1
Rad53	Cds1	CHK2
MRN complex		
Mre11	Rad32	MRE11
Rad50	Rad50	RAD50
Xrs2	Nbs1	NBS1
9-1-1 (PCNA-like) complex		
Ddc1	Rad9	RAD9
Mec3	Hus1	HUS1
Rad17	Rad1	RAD1
Rad17–RFC complex		
Rad24	Rad17	RAD17
Rfc2-5	Rfc2-5	RFC2-5
adaptors and mediators		
Rad9	Crb2/Rhp1	BRCA1 53BP1 MDC1/NFBD1
Mrc1	Mrc1	CLASPIN

Figure 15.20 Some key components of the DNA damage response in different organisms.

from Morgan, DO. *The Cell Cycle: Principles of Control.* London: New Science Press, 2007.

➔ We discuss the SMC complex family in more detail in Chapter 7.

RPA and MRN recruit two distinct regulator kinases

The global regulator kinases that are recruited to the DNA damage sensors RPA and MRN, and which subsequently initiate the DNA damage response in mammals, are **ATR** and **ATM**. Both are serine-threonine phosphoinositide 3-kinase-related

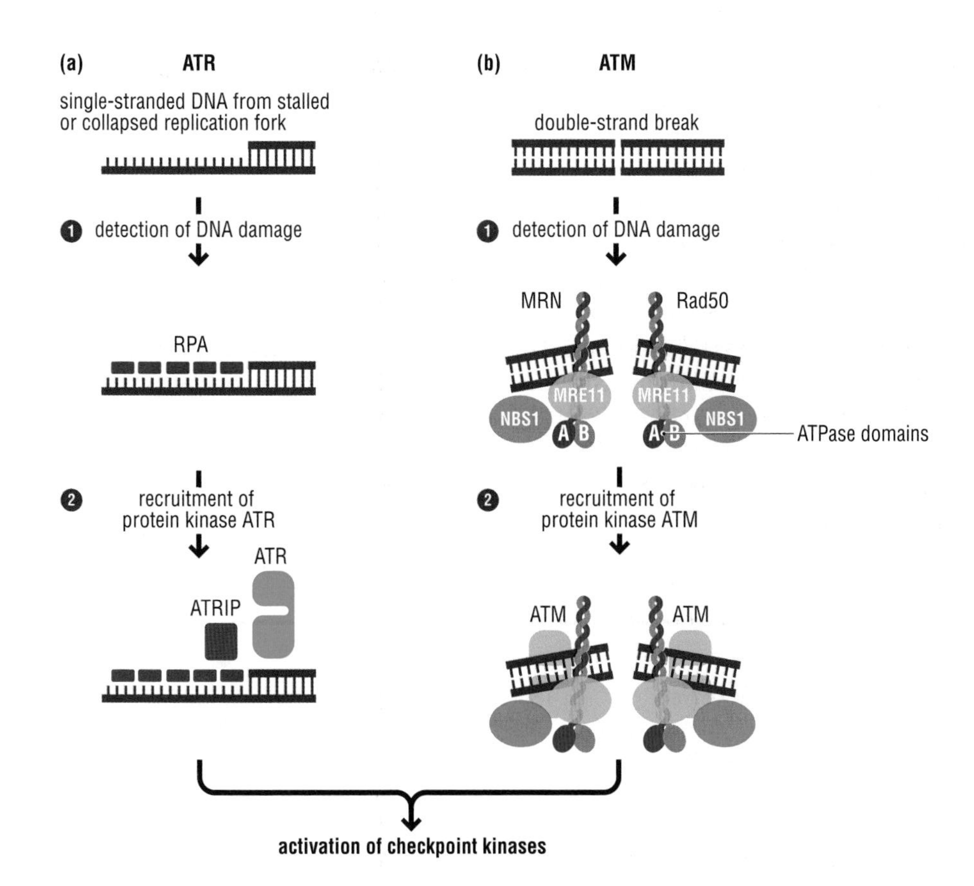

Figure 15.21 Recruitment of ATR and ATM to damaged DNA. The MRN complex recruits ATM through its NBS1 subunit (Xrs2 in yeast).

From Wyman, C, Warmerdam, DO, Kanaar, R. From DNA end chemistry to cell-cycle response: The importance of structure, even when it's broken. *Molecular Cell*, 2008;30:5–6.

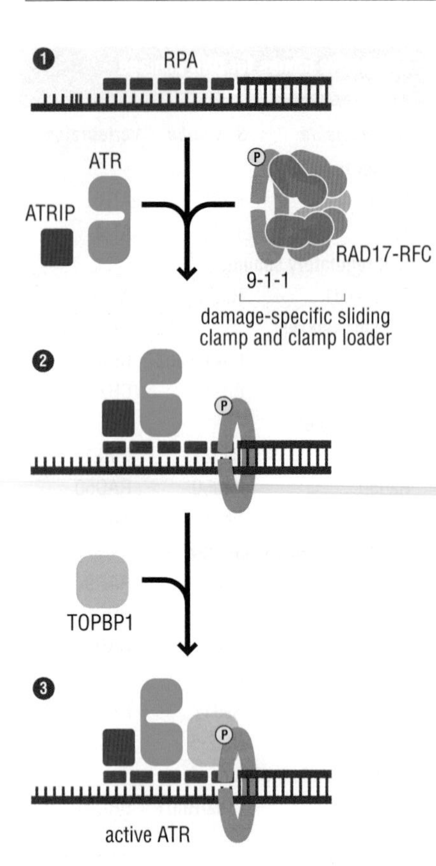

Figure 15.22 Recruitment of ATR to a single-stranded DNA.

From Crimpich, KA, and Cortez, D. ATR: an essential regulator of genome integrity. *Nature Reviews Molecular Cell Biology*, 2008;**9**: 616–627.

protein kinases (PIKKs) and are highly conserved, being present in organisms from yeast to humans. Each kinase initially responds to a different damage signal, and they are functionally non-redundant, although there is some overlap in their downstream phosphorylation targets. ATR is recruited to persistent single-stranded DNA by RPA; by contrast, ATM is recruited to double-strand breaks by MRN (Figure 15.21b, step 2). ATR stands for 'ataxia telangiectasia-related' while ATM stands for 'ataxia telangiectasia-mutated'.

Ataxia telangiectasia is a hereditary disease of humans that is characterized by, among other things, a greatly reduced ability to repair radiation-induced double-strand breaks and an increased risk of developing cancer: similar deficiencies, as we have seen, result from mutations in components of the MRN complex, reflecting the participation of these proteins in the same DNA damage response pathway. The single known mutation of human ATR causes a splicing variant, resulting in Seckel syndrome, which is characterized by growth retardation, small head size, and mental retardation; ATR-deficient mice are non-viable, and both human and mouse cells lacking ATR are highly sensitive to DNA damaging agents.

Stalled replication forks are far more common than double-strand breaks. The essential role of ATR likely reflects its role in the rapid repair of damaged forks, such that embryos lacking ATR do not survive.

Recognition of single-stranded DNA bound by RPA activates ATR

We saw earlier how RPA binds to single-stranded DNA, and the persistent accumulation of this protein–DNA complex acts as a signal of DNA damage. RPA at the DNA damage site does not directly recruit ATR, however. Instead, it recruits a protein called ATR-interacting protein (**ATRIP**) which acts as an adaptor, binding both to the single-strand-binding protein RPA and to ATR (Figure 15.21a step 2). The presence of RPA–single-stranded DNA also recruits another complex to the junction of single-stranded DNA and double-stranded DNA; this complex is a specialized damage repair-specific DNA sliding clamp-clamp loader complex, whose activity is summarized in Figure 15.22.

The damage-specific sliding clamp is the trimeric protein 9-1-1 (Rad9–Rad1–Hus1); the clamp loader is the Rad17–RFC2-5 complex. The function of 9-1-1 at sites of DNA damage is to recruit a third protein, TOPBP1. Interaction of TOPBP1 with ATR and 9-1-1 then leads to the activation of the ATR kinase. The initial targets of the ATR kinase are Rad17, 9-1-1, and TOPBP1 themselves at the break site: this increases their ability to activate ATR and amplifies the damage response.

➔ We discuss the sliding clamp-clamp loader complexes and their roles in DNA replication in Section 6.6.

Two important consequences of ATR activation are to control replication forks by phosphorylation of replication proteins; and to control cell-cycle transitions by phosphorylation of Chk1, as reflected in Figure 15.23. Phosphorylation of

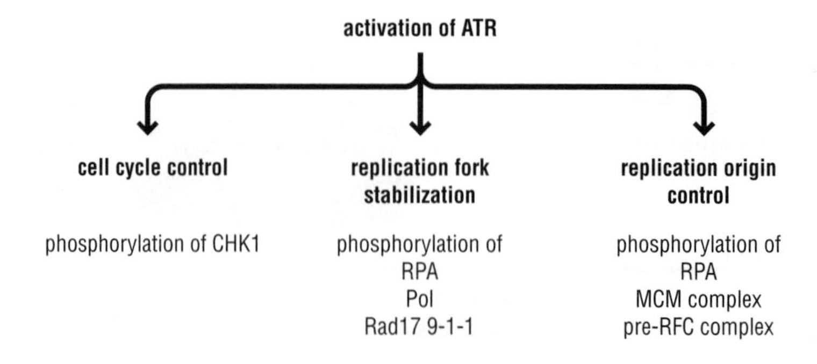

Figure 15.23 ATR activation modulates DNA replication. Activated ATR can modulate multiple aspects of DNA metabolism by phosphorylation of proteins that control the cell cycle, that act at replication forks and that control the firing of replication origins.

from Crimpich, KA, and Cortez, D. ATR: an essential regulator of genome integrity. *Nature Reviews Molecular Cell Biology*, 2008;**9**: 616–627.

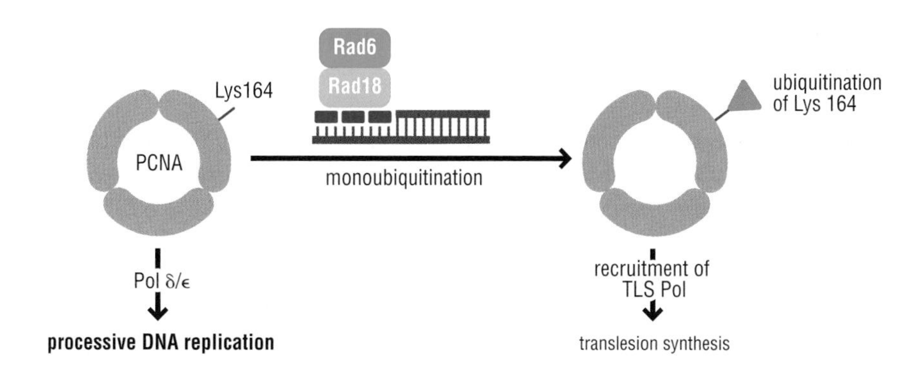

Figure 15.24 Ubiquitination of PCNA recruits translesion polymerases. TLS polymerases are recruited to PCNA by specific binding to ubiquitin and PCNA itself.

replication proteins slows the movement of replication forks, likely allowing more time for repair, and delays the firing of new origins of replication. We will discuss cell-cycle control by Chk1 in more detail below when we discuss how ATM activation also controls cell-cycle transitions.

Another response provoked by stressed replication forks is the recruitment of TLS polymerases, which facilitate the resumption of replication by allowing untemplated DNA synthesis at a fork-blocking lesion. In addition to the recruitment of the regulatory kinase ATR, RPA bound to single-stranded DNA also recruits the ubiquitination complex Rad6–Rad18, which results in the monoubiquitination of PCNA by Rad18. TLS polymerases are subsequently attracted to ubiquinated PCNA by their PCNA- and ubiquitin-binding domains and allow the resumption of DNA synthesis, as depicted in Figure 15.24. Although, as we saw in Section 15.6, such TLS DNA synthesis is error prone, it does at least allow progression of the replication fork. The molecular mechanisms by which the actual polymerase switch on the template strand occurs are not yet understood, however.

An additional error-free repair pathway, involving a template switching mechanism, also exists. In this pathway, monoubiquitylated PCNA is further modified with polyubiquitin chains linked to lysine 63 (K63-linked) through the action of the Rad5 ubiquitin ligase and the Ubc13–Mms2 conjugating enzyme. Polyubiquitylation of PCNA promotes homology-directed template switching, whereby the undamaged sister chromatid is used as a template for resumption of error-free DNA synthesis.

Ubiquitination of the yeast 9-1-1 clamp by Rad6–Rad18 also likely facilitates TLS polymerase activity. Intriguingly, recent studies in yeast suggest that ubiquitinated 9-1-1 may play an important role in the regulation of transcriptional induction genes encoding DNA repair functions, although the mechanism of induction remains to be determined.

ATM is activated by recruitment to double-strand breaks by MRN

In mammalian cells, several proteins essential to the DNA damage response interact directly with MRN. One is the regulatory kinase ATM, which is recruited to double-strand break sites though interaction with the NBS1 subunit of MRN (see Figure 15.21b). In the absence of DNA damage, ATM exists as an inactive dimer. However, the interaction of the ATM dimer with the MRN-DNA double-strand break complex results in activation of the ATM kinase activity through dissociation of the dimer and autophosphorylation of ATM. ATM-mediated phosphorylation of MRN also occurs.

Recruitment of repair proteins and chromatin modification are key to the DNA damage response

➡ We learn more about histone proteins in Section 4.3.

So what is the downstream effect of ATM? A key target of ATM phosphorylation is the histone variant H2AX, which makes up some 10–15% of the histone H2A to form a modified chromatin domain on each side of the double-strand break. This ATM phosphorylation is shown in Figure 15.25, step 1. Phosphorylated H2AX serves as a binding site for MDC1 (mediator of the DNA damage checkpoint 1), which is also a target of ATM phosphorylation (Figure 15.25, step 2). Phosphorylated MDC1, in turn, recruits additional MRN and ATM to the site of the double-strand break. MDC1 thus plays a critical role in *amplifying* the activation of ATM.

We examine the initial identification of ATM and its downstream targets in Experimental approach 15.2.

The formation of extended domains of phosphorylated H2AX and MDC1 provides a platform for the recruitment of large complexes of repair proteins to the sites of double-strand breaks to form subnuclear foci called ionizing-radiation-induced foci (IRIF). Indeed this protein recruitment to the sites of double-strand breaks can be considerable, extending to include a megabase of DNA around the damage.

MDC1 and H2AX both serve as mediators that recruit effector proteins (Figure 15.25 step 3). Key proteins that interact with MDC1 are the ubiquitin ligase RFN8, a second ubiquitin ligase, RNF168, and the ubiquitin conjugating enzyme Ubc13–Mms2. These enzymes build K63-linked polyubiquitin chains on histones that function as signals for recruitment of other repair proteins, including breast cancer type 1 susceptibility protein (BRCA1). BRCA1 is a breast cancer susceptibility gene that is mutated in some families with hereditary predisposition to breast and ovarian cancers. BRCA1 is also a target for ATM and ATR phosphorylation.

Histone ubiquitylation also affects the structure of chromatin surrounding double-strand breaks, thereby facilitating the recruitment of additional factors, including p53 binding protein 1 (53BP1), which provides a binding site for p53 and which we will discuss in detail below.

The repair process requires the recruitment of more proteins, including repair proteins and cohesins, which can link the broken DNA to a sister chromatid where the damaged DNA is undergoing replication; we shall see in the next chapter that one of the two pathways for repair of double-strand breaks relies on sister chromatids as a template for repair.

A third class of proteins that are recruited to DNA by MDC1 and H2AX are those that control the cell cycle. We will consider in Chapter 16 how these and other proteins recruited to DNA double-strand breaks promote the repair of these breaks.

Although it is a reasonable generalization that ATM is recruited in response to double-strand DNA breaks and ATR is recruited in response to stalled or collapsed replication forks (signaled by the presence of single-stranded DNA), this distinction is not absolute. As we shall see in the next chapter, DNA at a double-strand break that is not repaired will be resected – that is, one DNA strand will be preferentially degraded to generate single-stranded tails, which will be bound by RPA and will eventually recruit ATR. Conversely, stalled replication forks that are not rapidly repaired will result in double-strand breaks, with recruitment of ATM.

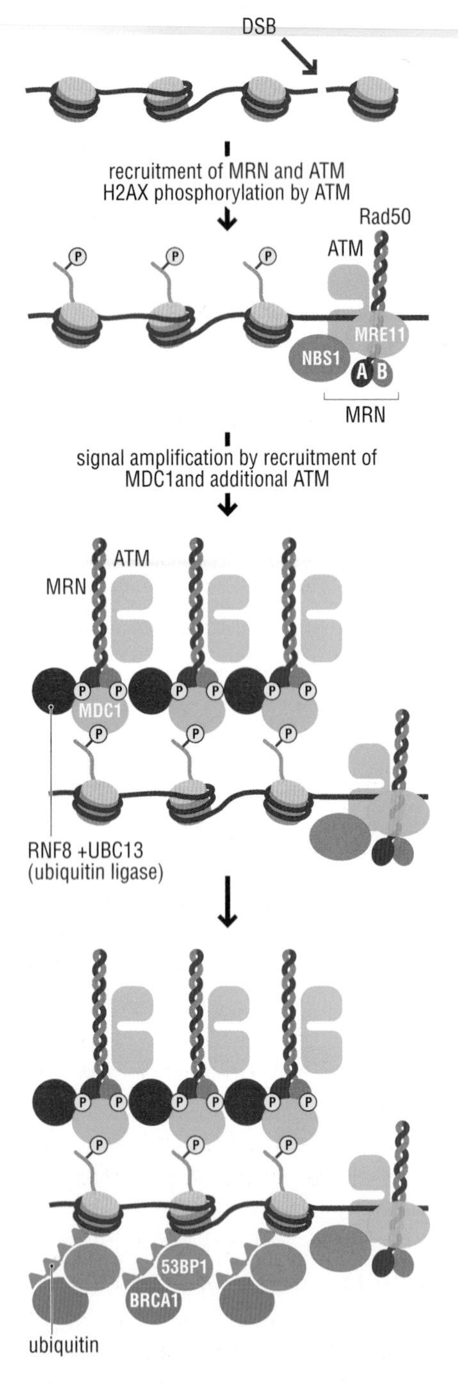

Figure 15.25 ATM activation results in chromatin modification at the site of a DNA double-strand break. The ubiquitin ligase RNF8 (pink) and its ubiquitin conjugating enzyme partner UBC13 ubiquitinate H2AX (as represented by orange triangles), providing binding sites for other proteins (not shown) that recruit BRCA1 (blue) and 53BP1 (orange).

15.2 EXPERIMENTAL APPROACH

Discovering the ataxia telangiectasia mutated gene and its functional characterization

One important goal of research in molecular biology is to identify the mutations that cause human disease and to understand why these mutations lead to the disorder. The mapping and characterization of the *ataxia telangiectasia mutated* (*ATM*) gene that is defective in ataxia telangiectasia illustrates how scientists can approach this problem. Ataxia telangiectasia is an autosomal recessive disease characterized by a variety of symptoms. It first manifests as a childhood neurological disorder with delayed motor skills, poor balance, and slurred speech, but individuals with ataxia telangiectasia also show a marked increase in radiation sensitivity and increase in cancer susceptibility. One percent of the population is heterozygous for ataxia telangiectasia, carrying a defect in one of the two *ATM* alleles.

By the mid-1990s, a variety of genetic studies, which included the mapping of the ataxia telangiectasia disease region in families through linkage analysis, had narrowed the location of the ataxia telangiectasia locus to a region of about 3 million base pairs (Mbp) of DNA on the q arm of chromosome 11. Linkage analysis is described in Section 18.10 and Section 19.15. As this mapping was performed well before the first draft sequence of the human genome was complete, how then could the scientists determine the actual identity of the *ATM* gene?

The ATM gene was isolated by positional cloning

In an example of broad collaborative effort, a consortium of research groups each focused on a defined segment of the ~3 Mbp 11q region for further characterization. Using positional cloning, which involves the systematic isolation and characterization of overlapping segments of the genome (similar to the cloning described in Experimental approach 4.3), the Shiloh group obtained additional markers that could be used in further linkage analysis. They initiated this analysis by isolating several large DNA segments (about 500 kilobase pairs (kb) in size; see Figure 1a) that were subsequently cloned into yeast artificial chromosomes (Figure 1b). Using this set of constructs, it was then possible to identify a polymorphism (a difference between individuals) on these segments, which could be followed to identify the DNA sequence that was most frequently associated with disease when family members who had the disease were compared with those who did not. We discuss the use of polymorphisms to carry out such linkage analysis in Section 18.10 and Section 19.15.

Eventually, the 500 kb fragments were subdivided into smaller 20 kb segments (Figure 1c), and the entire set of genes transcribed in this region was identified by isolating cDNAs. The transcribed regions were mapped (Figure 1d), and a 5.9 kb cDNA clone containing the 3′ end of the *ATM* gene was isolated. (With the availability of the complete human genome sequence today the isolation of overlapping clones and construction of transcription maps would no longer be necessary as this information is available online in the annotated sequence. When a new genome is sequenced, however, the construction of transcription maps is still an important phase of genome analysis.)

The 3′ end of the *ATM* gene was encoded by the 5.9 kb cDNA (Figure 1e) and had a predicted open reading frame (ORF) of 1708 amino acids; the entire *ATM* gene was later shown to encode a protein of 3056 amino acids. Sequencing of the exons of this gene in a number of ataxia telangiectasia patients revealed deletions and frameshift mutations, immediately indicating why the gene might not be functional in individuals with the disease (Figure 2). This sequence analysis also revealed that the protein contained a region related to protein kinases, suggesting that ATM is involved in signaling mediated by protein phosphorylation. Within a year of the identification of the human *ATM* gene, several groups made knockouts of the homologous genes in mice and observed radiation sensitivity similar to that which had been documented in human patients.

Although the path to identifying the *ATM* gene was laborious (largely because the complete genome sequence was not yet in hand), the same basic strategy of associating DNA changes with disease, or some other phenotype, is still required to establish that a gene is involved in a particular process.

Identification of proteins that are phosphorylated by ATM or ATR

The studies described above identified *ATM* as a gene that, when defective, results in radiation sensitivity and other phenotypes. What then does the gene product do? Many studies have shown that phosphorylation by the ATM kinase plays an important role in DNA damage signaling. In addition, a homologous protein that is also a kinase, termed ATR (ATM related) is also a key player in DNA damage response. Knowing a kinase is involved in regulation is an important step forward; however, knowing what substrates are phosphorylated gives even deeper insight into the mechanism of ATM function. A recent large-scale proteomic analysis of protein phosphorylation in irradiated cells by Elledge and colleagues set out to probe this question.

The basic strategy of the Elledge group was to identify proteins that were phosphorylated by ATM or ATR in response to irradiation. As these kinases share substrate specificity for serine–glutamine and threonine–glutamine motifs, they used antibodies that recognize these phosphorylated motifs to immunoprecipitate proteins

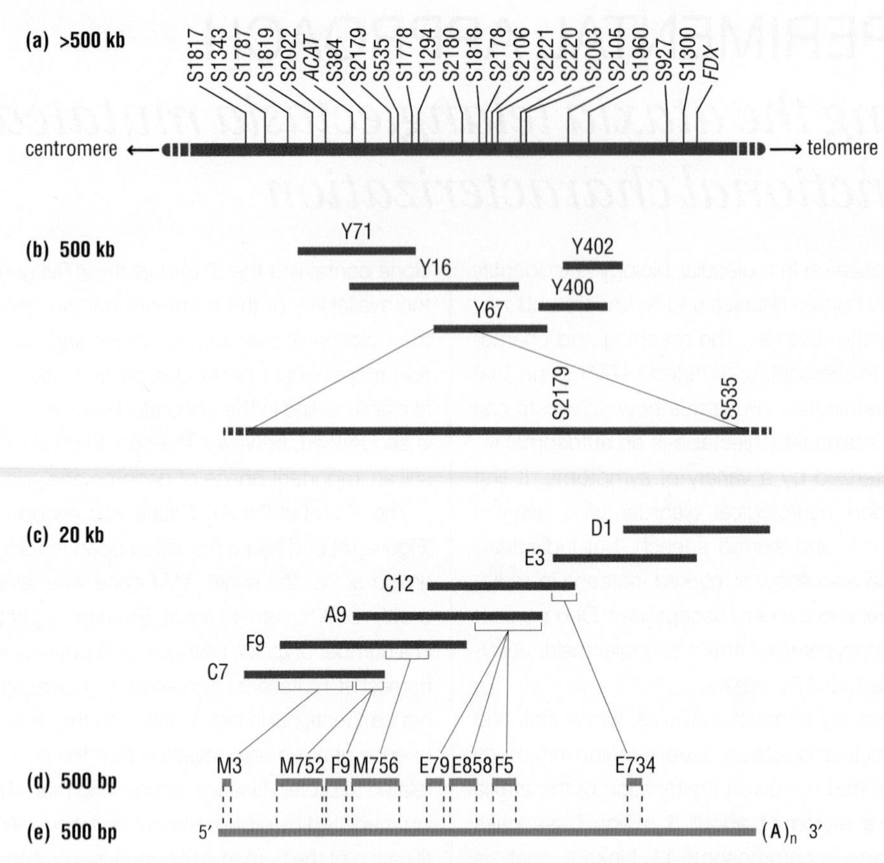

Figure 1 Positional cloning of the *ATM* gene. The general location of the ataxia telangiectasia locus mapped to human chromosome 11q. To identify the defect, the region needed to be narrowed down and the gene itself identified. Beginning with a set of markers for the 11q region (a), the Shiloh group cloned large >500 kb chromosome segments into yeast artificial chromosomes (b). These large fragments were used to identify smaller 20 kb fragments which were used to identify transcripts from the region (c). The fragments C7, F9, A9, and C12 were later all shown to contain part of the gene. The boxes in (d) show restriction fragments where mRNA transcripts (blue) were found. The C-terminus of ATM is encoded by the cDNA constructed from the transcribed exons (e).

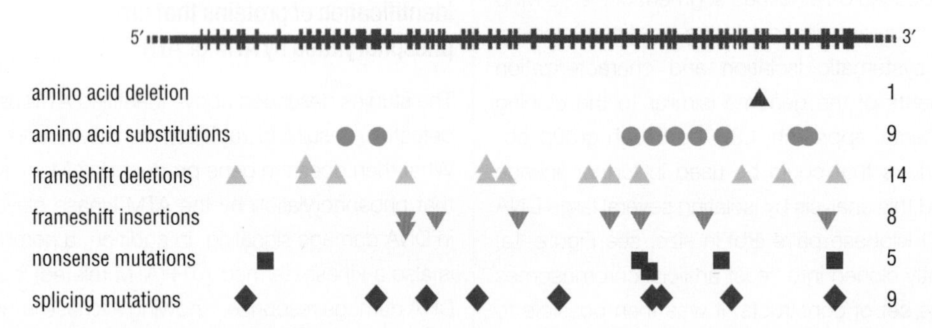

Figure 2 The *ATM* gene carries mutations in ataxia telangiectasia patients. The coding region of the entire *ATM* gene is shown at the top of the figure, with the locations of the mutations in 66 different individuals shown as hatched lines. Mutations in *ATM* are most frequently null mutations that result in complete loss of full-length protein. Deletions, insertions, nonsense mutations, and splice site mutations are also commonly found. Each type of mutation is illustrated below the diagram of the gene, and the location of the mutation is mapped onto where it occurs in the coding region. The numbers at the right indicate the total number of times the particular kind of mutation was found in this set of patients. For example, frameshift deletions were found in 14 of the 66 individuals analyzed.

of interest (in irradiated and untreated cells). These isolated proteins were then analyzed by SILAC (stable isotope labeling with amino acids in cell culture) following trypsin digestion. This mass spectrophotometric approach facilitates quantitative peptide analysis, allowing for direct comparisons of two related samples (as discussed in more detail in Experimental approach 14.3). In this manner, very many phosphorylated proteins were identified; indeed, more than 700 potential protein substrates were identified that displayed more than a fourfold increase in phosphorylation at ATM and ATR consensus phosphorylation sites. We learned about a related but different experimental strategy to identify kinase substrates in Experimental approach 5.1.

Testing the cellular role of ATM and ATR targets by decreasing their expression using siRNA.

The Elledge group then went on to show through direct biochemical methods that phosphorylation of a subset of the identified proteins did increase in response to irradiation. They experimentally tested whether 37 of the newly identified targets of phosphorylation were involved in the DNA damage response by using siRNAs to knockdown expression of the genes encoding the phosphorylated proteins (described in Section 19.5) and then measuring cellular responses to DNA damage. In one particular experiment (whose results are shown in Figure 3), the

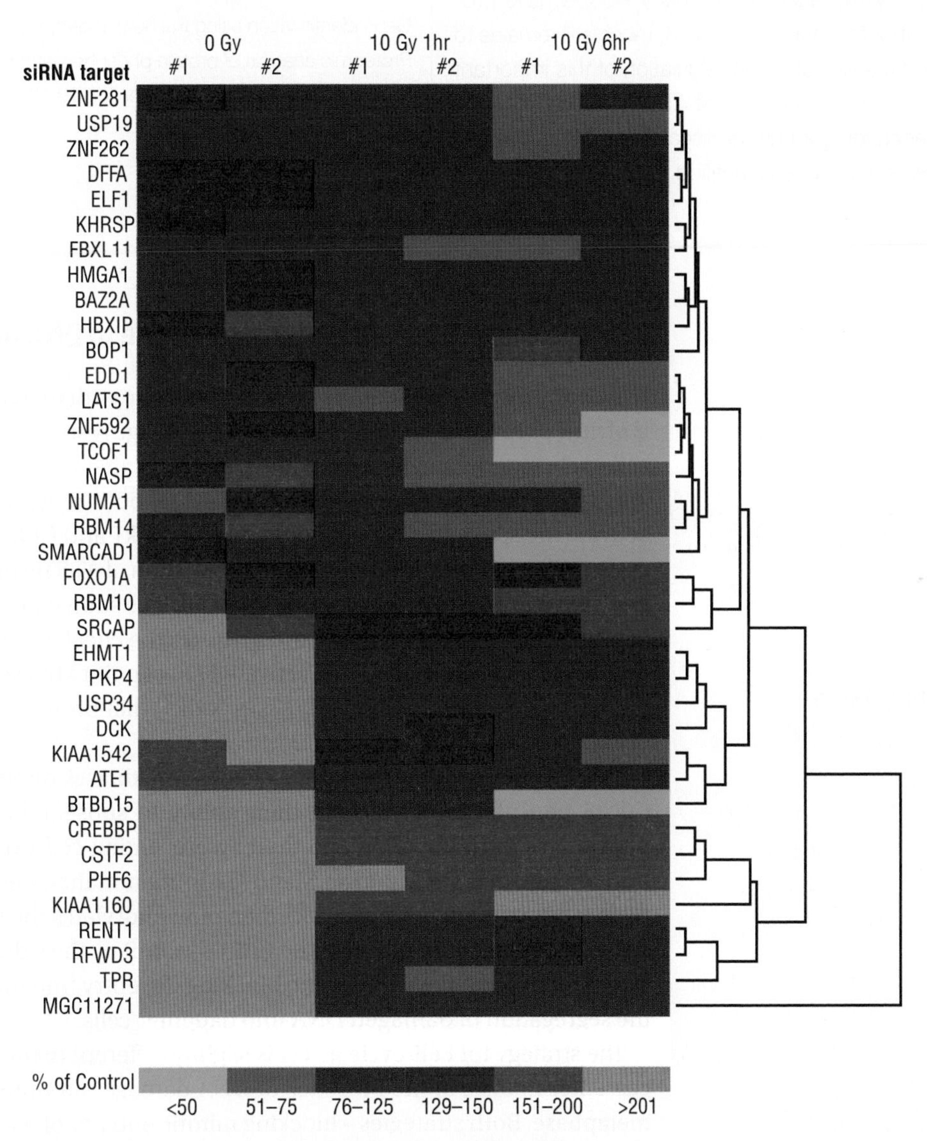

Figure 3 Analysis of the level of γ H2AX in cells treated with siRNA against specific genes. The levels of γ H2AX were measured by western blot assays using an antibody specific to γ H2AX. Cells were treated with the designated siRNA and then either not irradiated (0) or irradiated with 10 Gy and then protein levels measured one hour (#1) or six hours (#2) later. Each cell line treated with a specific siRNA had a characteristic response level at one hour and recovery at six hours. The colors indicate the percent change in level of γ H2AX compared with cells treated with a control unrelated siRNA. The color key is indicated at the bottom. The results are displayed as a hierarchical clustering in which cell lines giving similar responses are grouped next to one another. This clustering analysis helps to identify proteins that might be most functionally similar.

Reproduced from Matsuoaka *et al.* ATM and ATR substrate analysis reveals extensive protein networks responsive to DNA damage. *Science*, 2007;**316**:1160–1166.

group directly demonstrated increases in the levels of γ H2AX in response to γ irradiation in wild-type cells, a prediction that arose from the observation that γ H2AX levels appeared to be altered when genes of interest from the screen were knocked down. This experiment nicely confirmed the hypothesis that the newly identified target proteins are important in the DNA damage response.

Of the 700 target proteins identified, 31 had been previously shown to be involved in DNA damage; this study thus revealed a much wider range of proteins and cellular processes targeted by ATM than previously anticipated. ATM is now known to be a central regulator of the DNA damage response and thus has been extensively studied. While each of the many studies have provided information about the function of ATM, much still remains to be learned about the substrates and regulation of this important protein. This overview of the analysis of ATM thus illustrates the challenge of understanding a human disease, even after the defective gene or genes have been identified.

Find out more

Matsuoka S, Ballif BA, Smogorzewska A, *et al.* ATM and ATR substrate analysis reveals extensive protein networks responsive to DNA damage. *Science*, 2008;**316**:1160–1166.

Sandoval N, Platzer M, Rosenthal A, *et al.* Characterization of *ATM* gene mutations in 66 ataxia telangiectasia families. *Human Molecular Genetics*, 1999; **8**:69–79.

Savitsky K, Bar-Shira A, Gilad S, *et al.* A single ataxia telangiectasia gene with a product similar to PI-3 kinase. *Science*, 1999 **268**:1749–1753.

Related techniques

Linkage analysis; Section 19.15
Transcript identification using cloned cDNAs; Section 19.8
Gene identification using the human genome; Section 19.8
Proteomic analysis of protein phosphorylation; Section 19.8
Protein immunoprecipitation; Section 19.12
SILAC; Section 19.11
Use of siRNA; Section 19.5

Cell-cycle arrest is a short-term response to DNA damage

It is clearly disadvantageous for any cell to divide with damaged DNA. Therefore it is of little surprise that, as part of the DNA damage response, mechanisms to arrest cell-cycle progression become activated shortly after the DNA damage is detected. This response is mediated by ATR and ATM, both of which phosphorylate and activate the specialized checkpoint kinases, Chk1 and Chk2, which arrest the cell cycle to allow time for DNA repair. The activation of ATR results in phosphorylation and activation of the checkpoint kinase Chk1, as shown in Figure 15.26. Similarly, activated ATM phosphorylates and activates the checkpoint kinase Chk2.

➔ We discuss Cdks further in Section 5.2.

Once phosphorylated, these activated checkpoint kinases halt the cell cycle by inactivating cyclin-dependent kinases (Cdks) that promote transitions in the cell cycle.

This inactivation is indirect: the checkpoint kinases directly phosphorylate the Cdc25 phosphatases, blocking their ability to subsequently remove inhibitory phosphates from the Cdks and thereby preventing cell-cycle transitions. Detection of damaged DNA during S and G2 in mammalian cells, for example, leads to phosphorylation of Chk1. Chk1 then phosphorylates the Cdc25C phosphatase. Without dephosphorylation of the Cdk1-CyclinB by the Cdc25C phosphatase, the G2→M transition does not occur, blocking the entry into mitosis, and preventing the segregation of damaged DNA into daughter cells.

The strategy for cell-cycle arrest is slightly different in budding yeast cells: they can enter mitosis in the presence of DNA damage but mitosis is then blocked in metaphase. Both strategies - blocking mitotic entry or blocking chromosome segregation - allow increased opportunity for repair and decrease the possibility of permanent changes in DNA sequence.

This cell-cycle arrest is readily reversible in single-celled organisms, where each cell must divide in order for the organism to reproduce. In multicellular organisms, by contrast, cell-cycle arrest is generally irreversible because severe DNA damage in one cell can be a threat to the organism as a whole through the

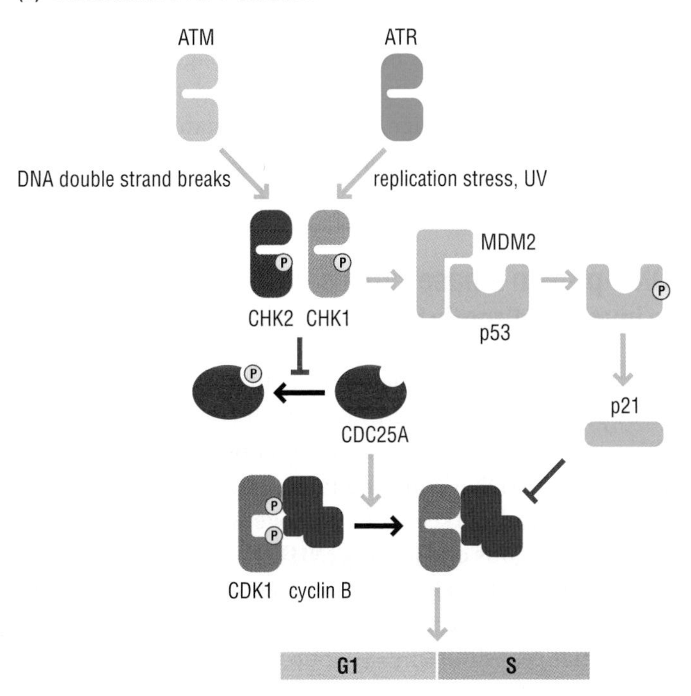

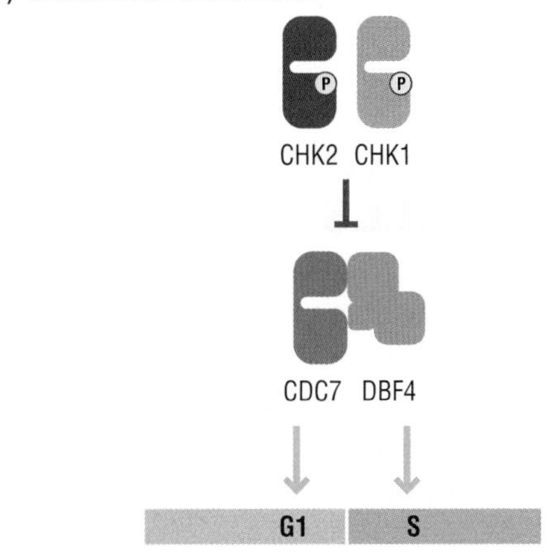

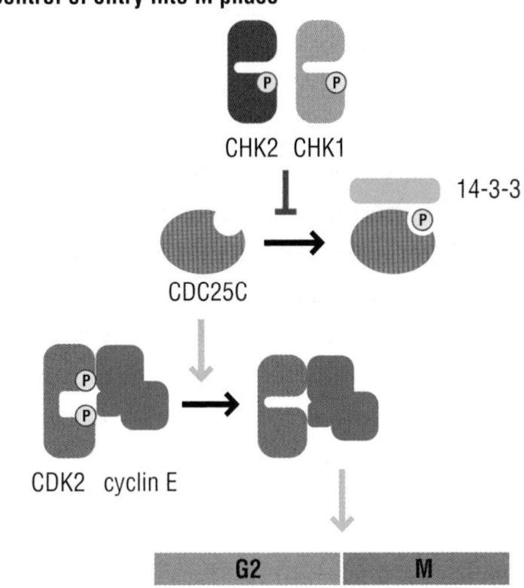

Figure 15.26 ATR and ATM control multiple cell cycle checkpoints. Activation of the DNA damage senor kinases ATM and ATR control transitions between multiple cell cycle stages. (a) ATM is activated by double-strand DNA breaks, and ATR is activated by single-stranded DNA breaks that result from replication fork stalling. Activation of ATM and ATR results in the phosphorylation and activation of the cell cycle regulators CHK2 and CHK1. These kinases then phosphorylate the CDC phosphatase, CDC25A such that there is no active CDC25A phosphatase to dephosphorylate cycCDK1 and promote the G1 to S transition. (b) Additional G1 and S controls. Activated CHK2 and CHK1 phosphorylate and block the action of CDC7 and DBF4 that are required at the initiation of S phase and within S phase. (c) Controlling entry into M phase. The active CDK2 and CHK1 kinases phosphorylate the phosphatase CDC25C, blocking its ability to dephosphorylate CDK2 and promote the G2 to S transition.

increased likelihood of cancerous changes in the progeny of the damaged cell. Multicellular organisms meet this threat with a specialized system for inducing the programmed death of cells with damaged DNA – the process of **apoptosis**. The key effector protein in this pathway is p53, a gene regulatory protein that is recruited to sites of damage by the ubiquitinated H2AX histone proteins that occur at DNA double-strand breaks. We will discuss in the next section the programmed cell death response to DNA damage that is mediated by p53.

DNA-protein kinase signals DNA damage and is involved directly in DNA repair

Another **PIKK** in mammalian cells that is activated by double-strand breaks is DNA-protein kinase (DNA-PK), which contains a catalytic subunit called DNA-PKcs and a double-strand break–binding subunit, the Ku70/80 heterodimer, that binds avidly to DNA ends. Interactions between the protein-bound ends of a broken DNA strand can keep the ends together. The ends are then rejoined through the **non-homologous end joining (NHEJ)** pathway, which we describe in Chapter 16.

Activation of the kinase by interaction with Ku bound to DNA is important in the actual repair step, possibly for autophosphorylation to promote conformational changes within the synaptic complex. However DNA-PK phosphorylation of downstream targets is also important for cell survival via pathways involving cell-cycle arrests and cell death. Yeast lacks DNA-PK but does carry out NHEJ that involves the Ku70/80 heterodimer.

15.10 DNA DAMAGE AND CELL DEATH IN MAMMALIAN CELLS

In the final section of this chapter, we consider the ways in which multicellular organisms cope with the wider impacts of DNA damage – damage that may require the organism to sacrifice some of its component cells to ensure its wellbeing. We focus in particular on the role of one protein, p53, which plays a major role in a multicellular organism's response to DNA damage.

p53 can block the proliferation of cells containing damaged DNA

Although the persistence of DNA damage in a single-cell organism can result in DNA change or death of that cell, the remaining cell population is not affected by the fate of the mutated cell. The consequences of DNA damage for a multicellular organism are quite different: if cells with damaged DNA can replicate within a multicellular organism, there is an increased possibility that some of those damaged cells will acquire genetic changes that lead to unregulated growth of that cell. Such unregulated growth underlies the development of cancers that can be fatal to the whole organism. Thus in multicellular organisms, there are special regulatory responses to unrepaired DNA damage: damaged cells are either permanently arrested in G1; or they are removed by **apoptosis**, a specialized process of programmed cell death. This response is orchestrated by the gene regulatory protein p53.

Importantly, p53 is the single most frequently mutated protein in human cancers, being mutated in at least half of all cases. In the absence of functional p53, the DNA damage response is deficient, mutations accumulate, and the chance of cancer is greatly increased.

Multicellular organisms activate p53 for long-term inhibition of cell proliferation

p53 is a specific DNA-binding protein that directly binds to the promoters of target genes and alters the rate at which they are transcribed. In most cases, expression of target genes is increased, resulting in the increased production of proteins that inhibit cell-cycle progression or stimulate apoptosis. p53 can also repress the transcription of some target genes, thereby reducing the levels of inhibitors of apoptosis. The result of p53 action is either cell-cycle arrest or cell death.

➔ We discuss the acetylation of histones in Section 4.5.

The level of p53 in the cell is controlled in a number of ways

Transmission of a DNA damage signal to p53 is mediated by the adaptor protein 53BP1, which binds to modified chromatin, that is, ubiquitinated H2AX, at the site of DNA damage, as illustrated in Figure 15.25. Because of the central importance of p53 in the response to DNA damage and other cellular stresses, and because its activation can cause the death of the cell, this protein is subject to an unusually large array of regulatory modifications, which, together, ensure that p53 is present and active only when necessary. Most of these modifications act to increase its concentration or its intrinsic gene regulatory activity, or both, when DNA damage occurs.

The major regulators of p53 include MDM2 and p300

The location of the functional domains of p53 are illustrated schematically in Figure 15.27. Like many transcriptional regulators, p53 has a DNA-binding domain and a separate activation region that interacts with the transcriptional machinery. This activation region, which is at the N-terminal end of the protein, along with a region at the C-terminal end, are the targets of an array of regulatory molecules that catalyze the ubiquitination, phosphorylation, and acetylation of specific amino acids - modifications that, collectively, control the activity of p53.

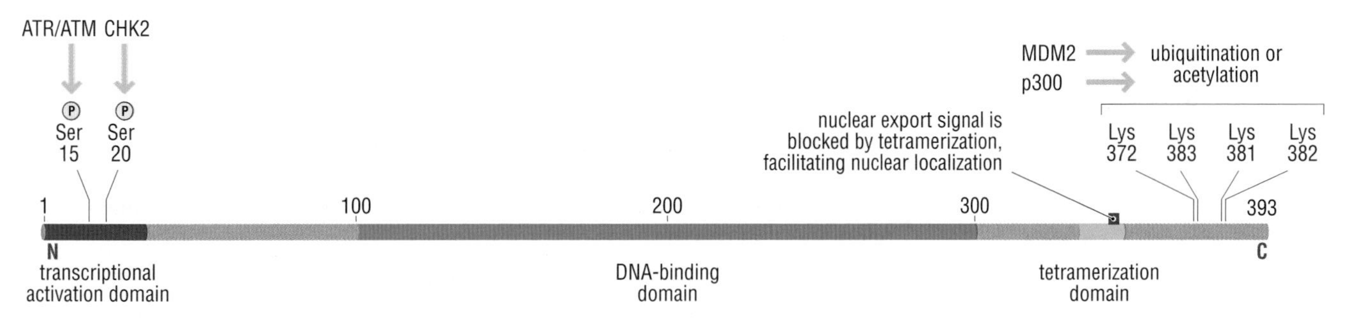

Figure 15.27 Domain structure and regulatory features of the human p53 protein. p53 contains several domains, including a large central region that interacts directly with the DNA target sequence and an N-terminal region that interacts with the transcriptional machinery to stimulate gene expression. When activated, p53 forms a tetramer, and a small region near the C-terminus is required for tetramerization. This region also contains a nuclear export signal that is blocked in the p53 tetramer, enhancing the nuclear localization of the activated protein. Additional regulatory regions are clustered near the termini of the protein. The key regulators MDM2 and p300 both interact with the N-terminal region and modify a group of lysines at the C-terminus, presumably because the termini are more closely apposed in the folded protein.

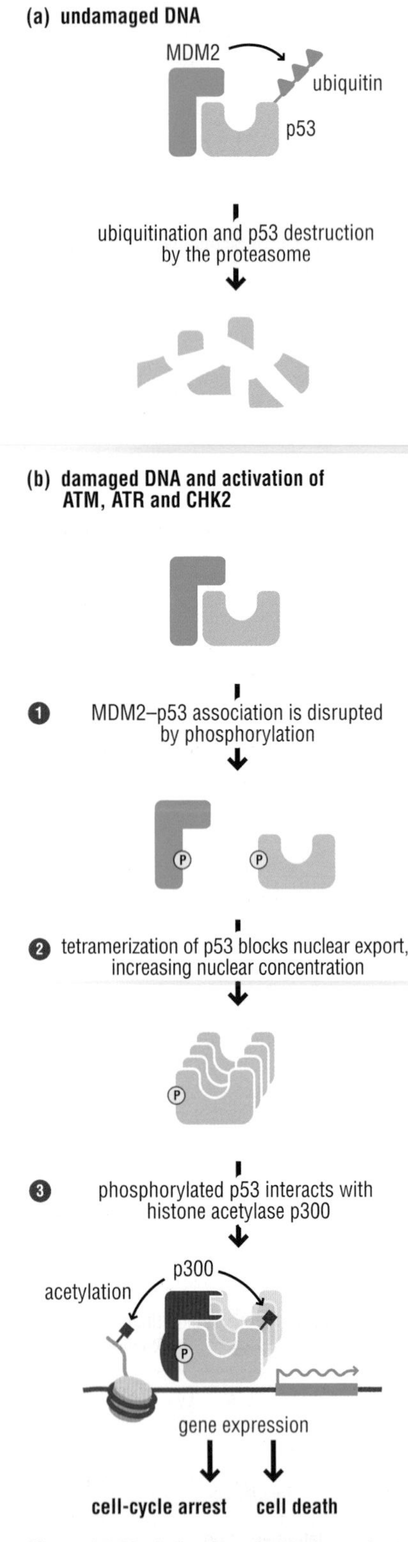
(a) undamaged DNA
MDM2
ubiquitin
p53
ubiquitination and p53 destruction by the proteasome
(b) damaged DNA and activation of ATM, ATR and CHK2
1 MDM2–p53 association is disrupted by phosphorylation
P
P
2 tetramerization of p53 blocks nuclear export, increasing nuclear concentration
P
3 phosphorylated p53 interacts with histone acetylase p300
p300
acetylation
P
gene expression
cell-cycle arrest
cell death

Figure 15.28 Activation of p53-dependent gene expression by DNA damage.
(a) Metabolism of p53 in the absence of DNA damage. (b) Metabolism of p53 in the presence of DNA damage and activation of ATM, ATR, and CHK2.

The major regulator of p53 is MDM2, an E3 ubiquitin ligase that ubiquitinates several lysine residues near the C-terminus of p53, thereby targeting it for destruction, as depicted in Figure 15.28a. Because it binds to the N-terminal region of p53, which contains the activation domain that is important in the stimulation of target-gene transcription, MDM2 is thought also to inhibit the gene regulatory activity of p53 until p53 can be destroyed. Mice lacking MDM2 die early in embryonic development, apparently because excessive amounts of p53 block the proliferation of many cell types. Normally, in the absence of DNA damage, MDM2 associates with p53 and keeps its activity at a minimum. When DNA damage occurs, however, the activity of MDM2 is suppressed, and p53 is stabilized, leading to expression of p53 target genes, a situation depicted in Figure 15.28b.

A second regulator, p300, which promotes the activity of p53, is a histone acetyltransferase. p300 associates with p53 during the DNA damage response and helps promote local gene expression by acetylating histones.

p300 also acetylates p53 itself, at the same lysines that are ubiquitinated by MDM2 in the absence of damage (see Figure 15.27). Acetylation of these lysines therefore blocks their ubiquitination, further ensuring the stabilization of p53 during the damage response.

p53 is activated in response to DNA damage

Early in the DNA damage response, MDM2 is phosphorylated, probably by ATM and perhaps also by ATR, inhibiting its association with p53 (see Figure 15.28b). The same kinases also phosphorylate p53 itself within its activation domain. This both inhibits MDM2 binding and increases p300 binding and acetylation, thereby increasing the stability and gene regulatory activity of p53. Finally, the effector kinase Chk2 (and probably Chk1 as well) phosphorylates p53 at serine 20, which also reduces MDM2 binding and helps stabilize p53.

Nuclear retention of p53, an indirect effect of stabilizing the protein, ensures its sustained activity during the damage response. In the absence of DNA damage, p53 is tagged for export from the nucleus by a nuclear export signal near the C-terminus (see Figure 15.27), and is thereby prevented from associating with target genes. When p53 is stabilized and activated after DNA damage, however, it forms a tetramer in which the nuclear export signal is blocked, ensuring that the active tetramer is retained in the nucleus.

So we see how the DNA damage response signal employs multiple overlapping mechanisms, including ubiquitination, phosphorylation, acetylation, and binding to inhibitor and activator proteins, to ensure the rapid and robust activation of p53-dependent gene expression.

Telomere degeneration promotes cell-cycle arrest

Another important pathway to p53-dependent cell-cycle arrest is initiated at the telomeres that cap the ends of chromosomes.

Telomerase, the enzyme primarily responsible for maintaining telomere length, is not expressed in many differentiated cells; in these cells, telomeres gradually decrease in length over many cell generations. This telomere shortening is associated with the loss of telomere function that is probably due to the deterioration of the protein cap. Since these chromosome ends now appear to resemble a double-break, ATM is activated, leading to p53-dependent cell-cycle arrest or apoptosis like that seen in response to double-strand breaks elsewhere in the genome.

In other cases, in the absence of telomere shortening, loss of the proteins that bind to telomeres can allow access of nucleases that generate single-stranded DNA tracts, which can also activate the damage response pathway.

➡ We describe the structure and function of telomeres in Section 4.11.

SUMMARY

DNA DAMAGE

- DNA damage threatens the integrity of the genome and can result in changes in DNA sequence (i.e., mutations).
- Errors in DNA replication result in mismatched base pairs.
- DNA can be damaged by both intracellular and extracellular agents, resulting in many different kinds of lesions.
- All components of DNA – the phosphodiester backbone, the deoxyribose sugar, and the bases – are subject to damage.

DNA REPAIR PATHWAYS

- There are multiple pathways for DNA repair in the cell.
- Because of DNA repair, only a small fraction of DNA damage events result in mutation.
- Defects in DNA damage repair pathways result in an increased rate of mutagenesis and in sensitivity to DNA damaging agents.
- Many genes encoding DNA repair proteins have been identified by the isolation and study of mutants with increased sensitivity to DNA damaging agents.
- Defects in DNA repair pathways underlie many heritable human diseases, which result in an increased incidence of cancer.
- Cancer can result from DNA repair failure in response to increased DNA damage or from a heritable decreased capacity for repair, both of which may lead to the mutation of growth control genes.
- Some repair pathways can recognize specific lesions and reverse them.
- Base methylation can be reversed by specialized methyltransferases.
- Specialized proteins sanitize the free nucleotide pool by removing damaged nucleotides.
- Multiple DNA repair pathways recognize and excise damaged DNA, and use the information on the undamaged strand to direct the resynthesis following damage excision, exploiting the fact that DNA is a duplex.

POST-REPLICATION MISMATCH REPAIR

- Although DNA replication is highly accurate because of proofreading by exonucleases, misincorporation does occur.
- Post-replication mismatch repair corrects DNA mismatches resulting from nucleotide misincorporation during DNA synthesis.
- The mismatch repair system recognizes mismatched base pairs, specifically removes the newly synthesized strand, and resynthesizes DNA to replace the removed base pair.
- Mismatch repair also corrects expansions and contractions that can occur at repeated sequences due to misalignment during DNA synthesis.
- Defects in human mismatch repair genes are associated with hereditary non-polyposis colon cancer.

BASE EXCISION REPAIR

- Base excision repair recognizes positions of base loss or damage, removes the damaged nucleotide or a few surrounding nucleotides, and resynthesizes DNA to replace the removed DNA using the complementary DNA strand as a template.
- Specialized glycosylases release the damaged base from its sugar.
- The AP endonuclease nicks the phosphodiester backbone at positions of base loss.

NUCLEOTIDE EXCISION REPAIR

- Nucleotide excision repair recognizes bulky lesions, removes an oligonucleotide containing the damage, and resynthesizes DNA to replace the removed oligonucleotide.
- Although the basic steps of excision repair are the same in bacteria and eukaryotes, there is no homology between the proteins involved.
- Nucleotide excision repair can be targeted to actively transcribed regions by interaction between the repair machinery and transcription machinery.

TRANSLESION DNA SYNTHESIS

- Lesions in DNA can result in stalling of replication of the replication fork.
- Specialized translesion DNA polymerases allow error-prone DNA synthesis through sites of DNA damage where other polymerases would be unable to function.
- Although mutagenic, error-prone DNA synthesis allows a chance of cell survival whereas failure to replicate is lethal.
- The active site regions of translesion polymerases are looser, allowing use of an imperfect template.
- Translesion polymerases can be induced in response to DNA damage.
- Nicks in DNA can lead to replication fork collapse.

THE DNA DAMAGE RESPONSE

- The presence of unrepaired DNA damage provokes cellular responses that lead to an increased chance for cell survival by inhibition of the cell cycle, and increased activity of DNA repair pathways.
- Single-stranded DNA and DNA double-strand breaks induce the DNA damage responses.
- In multicellular organisms, the continued presence of unrepaired DNA damage can lead to programmed cell death to avoid the multiplication of cells with decreased genome integrity.

THE BACTERIAL SOS RESPONSE

- The bacterial SOS response to DNA damage includes transcriptional induction of DNA repair genes and an increase in activity of translesion polymerases, which results in an increase in the rate of mutagenesis.
- Transcriptional derepression of the SOS genes results from the activation of a co-protease activity of a RecA protein–single-stranded DNA filament that cleaves and inactivates the LexA protein, a repressor of DNA repair genes.
- The RecA co-protease activates a translesion DNA polymerase proteolytic processing of an inactive precursor.

THE EUKARYOTIC DNA DAMAGE RESPONSE

- The DNA damage response in eukaryotes acts through phosphorylation signaling cascades and other protein modifications.
- ATR is the principal regulator kinase that responds to stalled replication forks and other kinds of damage that result in the accumulation of single-stranded DNA.
- The slowing of the DNA replication machinery to allow repair and an increase in the activity of error-prone translesion polymerases are responses to increased ATR activity.
- ATM is the principal regulator kinase that responds to double-strand breaks in DNA.
- Large complexes of repair proteins are recruited to the sites of double-strand breaks in DNA.
- Cell-cycle regulators are major phosphorylation targets of ATR and ATM, such that the cell cycle stops in the presence of DNA damage.
- Components of the DNA damage response pathway are defective in many human diseases.

DNA DAMAGE AND CELL DEATH

- DNA damage in multicellular organisms can lead to programmed death of damaged cells, avoiding the propagation of cells in which genome integrity has been compromised.
- The programmed cell death response is mediated by p53, a transcriptional regulator.
- The level of p53 is highly controlled, often by post-translation modification.
- Mutations in p53 occur in at least half of all human cancers.

15.1 TYPES OF DNA DAMAGE

1. Based on Figure 15.1, what is the main difference between adenine and the adenine tautomer, and what is the consequence if the shift to the tautomeric state occurs during DNA replication?

2. Compare and contrast the action of DNA damaging agents that lead to point mutations, frameshift mutations, and the formation of DNA dimers.

Challenge question

3. The Ames test is a simple test developed to determine the potential DNA damaging effects of various chemicals. The test uses a strain of bacteria that has a mutation in one of the genes required for the synthesis of histidine. The bacteria are spread on two agar plates without histidine, and then one plate is exposed to the chemical in question. The two plates are then observed for the ability of the mutant to grow in the absence of histidine.

a. What result would you expect on the agar plate that is not exposed to a test chemical?

b. What result would you expect on the plate that is exposed to a chemical that damages DNA? Explain your answer.

15.2 POST-REPLICATION MISMATCH REPAIR

1. Explain the discrepancy between the *in vivo* rate of mutation and the error rate of DNA polymerase.

2. What is the significance of DNA methylation in mismatch repair?

3. Match the following mismatch repair enzymes with the appropriate statements below.

A: MutH; B: MutL; C: MutS; D: MutU/uvrD

a. Interacts with hemi-methylated DNA
b. Interacts with the sliding clamp loader during DNA synthesis
c. Recognizes DNA base pair mismatches
d. Introduces a nick to the newly synthesized DNA strand
e. Recruits helicase to the site of a mismatch
f. Displaces the nicked, newly synthesized DNA strand

4. What is slipped-strand replication and how is it repaired?

15.3 REPAIR OF DNA DAMAGE BY DIRECT REVERSAL

1. In what fundamental way is the direct reversal repair process different from the other types of repair discussed in this chapter?

15.4 REPAIR OF DNA DAMAGE BY BASE EXCISION REPAIR

1. Which of the following enzymes functions in base excision repair?
a. Alkyltransferase
b. Glycosylase
c. MutS
d. Photolyase
e. UvrB

15.5 NUCLEOTIDE EXCISION REPAIR OF BULKY LESIONS

1. Compare the mechanisms for direct reversal base excision and nucleotide excision repair.

2. Compare and contrast GGR and TCR.

15.6 TRANSLESION DNA SYNTHESIS

1. Considering the high error rate of translesion DNA polymerases, speculate why they would provide an evolutionary advantage and be maintained in organisms.

15.7 THE DNA DAMAGE RESPONSE

1. What is recognized by the DNA damage sensor in both bacteria and eukaryotes, and what are the primary mechanisms by which the signal is produced?

15.8 THE DNA DAMAGE RESPONSE IN BACTERIA

1. By what mechanism does the RecA protein activate the bacterial SOS response system?

2. Describe the mechanism by which activation of the SOS response system can lead to the inhibition of the cell cycle and promote bacterial cell survival.

Challenge questions

3. Remember from Chapter 9 that the lambda repressor (cI) is a protein that binds to and represses transcription from the lambda promoters, which allow the prophage to switch from the lysogenic to the lytic stage. The cI protein has homology to LexA and is recognized by the RecA protein. What purpose might this homology serve to have provided an evolutionary advantage to the phage?

4. In terms of sensitivity to DNA damage, wild-type *E. coli* are most resistant, followed by *uvr* gene mutants, followed by *uvr* + *recA* mutants. Translesion polymerase mutants are most sensitive.
a. Name and briefly describe the pathways associated with the genes discussed in the question.
b. Give one possible explanation for the survivability of each combination of mutants in relationship to the other combinations.

15.9 THE DNA DAMAGE RESPONSE IN EUKARYOTES

1. Compare and contrast the eukaryotic DNA damage response and the bacterial DNA damage response.

2. By what mechanism is RPA able to distinguish between single-stranded DNA produced by normal DNA replication and the single-stranded DNA produced from DNA damage?

3. What are the two main targets of ATR kinase and what is the cellular consequence of the phosphorylation of each target?

4. What is H2AX and what is its role in DNA damage repair?

5. a. What is the function of the CDC25 phosphatases during cell cycle arrest?
b. By what mechanism do single-stranded breaks lead to the inactivation of the CDC25 phosphatases?

15.10 DNA DAMAGE AND CELL DEATH IN MAMMALIAN CELLS

1. Indicate if each of the following would (i) increase (ii) decrease or (iii) have no effect on the stability and activity of p53.
a. Persistent DNA damage
b. Mutation or deletion of MDM2
c. Ubiquitination of the C-terminal domain of p53
d. p53 association with p300
e. Phosphorylation of CDC25
f. Mutation of p53 nuclear export signal
g. Degradation of the telomeres

Challenge questions

2. Defects in both the mismatch repair pathway and the nucleotide excision repair in bacteria lead to a mutator phenotype. In humans, defects in these same pathways often lead to a predisposition to cancers such as hereditary non-polyposis colorectal carcinoma and xeroderma pigmentosum. Compare and contrast the consequences of defects in these repair pathways in single-celled and multicellular organisms.

3. The human *MLH1* gene is a homolog of the bacterial *MutL* gene. Mutations in this gene have been associated with hereditary non-polyposis colorectal carcinoma.
a. What does it mean to be a homolog? Based on this information, what is the likely function of *MLH1*?
b. Why might a mutation in *MLH1* lead to hereditary non-polyposis colorectal carcinoma?
c. In one study, eight hereditary non-polyposis colorectal carcinoma patients were found to have polymorphisms in the *MLH1* gene. In order to determine if the polymorphisms had an effect on mismatch repair and therefore likely contributed to the development of hereditary non-polyposis colorectal carcinoma, the investigators cloned the various *MLH1* genes and transfected them into ovarian cancer cells that already had inactive *MLH1* in addition to other mutations that had transformed the cell. These cells were then tested for their survivability after exposure to a DNA damaging methylating agent. Importantly, when the mismatch repair enzymes encounter DNA bases damaged by methylating agents, they normally trigger apoptosis.
i. What result would you expect from the ovarian cancer cell line with an inactive *MLH1* gene treated with a methylating agent compared to normal cells treated with a methylating agent?
ii. What result would you expect if these cells are transfected with normal *MLH1* and then treated with a methylating agent when compared to the untransfected cells treated with a methylating agent?
iii. If a patient with hereditary non-polyposis colorectal carcinoma had a polymorphism in the *MLH1* gene that contributed to the development of the disease and the sequence of that *MLH1* was transfected into the system described, what result would you expect?

EXPERIMENTAL APPROACH 15.1 – FINDING GENES INVOLVED IN THE REPAIR OF DAMAGED DNA

1. Define the terms forward and reverse genetics.

2. Which of the terms from Question 1 was used by Howard-Flanders and Theriot to isolate DNA damage repair genes? Explain the design of the experiment in order to explain your answer.

3. Which of the terms from Question 1 was used by Giaver and co-workers to identify *S. cerevisiae* DNA damage repair genes? Refer to the design of the experiment in order to explain your answer.

Challenge questions

4. Look at Experimental approach 15.1, Figure 1.
   a. What does the x-axis represent?
   b. What do the pink lines in the graph represent?
   c. Which phage were the most and least sensitive to UV light damage? Explain your answer in terms of the graph.

5. In bacteria carrying a mutation in the *umuC* gene the cells cannot be mutated when exposed to a mutagen such as UV light. Consequently, the cells that receive DNA damage either maintain their original phenotype or die.
   a. Describe one type of experimental design that would allow you to ***select for*** cells that have the umutable phenotype so that you could isolate genes like *umuC*.
   b. Based on the fact that *umuC* mutants cannot be mutated, *umuC* is most likely a gene involved in what type of repair? Explain your answer.

EXPERIMENTAL APPROACH 15.2 – DISCOVERING ATM AND ITS FUNCTIONAL CHARACTERIZATION

1. In general, how did scientists determine that the gene responsible for producing ataxia telangiectasia was located on 11q?

2. How has the identification of disease-causing genes changed since the sequencing of the human genome?

3. What is the function of the *ATM* gene, and how is it related to the characteristics of ataxia telangiectasia?

Challenge question

4 Look at Experimental approach 15.2, Figure 3.
   a. What was the purpose of the experiment shown in Figure 3?
   b. How did the investigators come up with the list of genes that were the targets of siRNA?
   c. Why was the level of γH2AX in cells the measure of the response to DNA damage?
   d. What do the bracket lines on the right of the figure represent, and how are they related to the conclusions based on the experiment?

FURTHER READING

INTRODUCTION

Friedberg E, Walker G, Siede W, *et al.* eds., *DNA Repair and Mutagenesis*. Washington DC: ASM Press, 2006.

Hakem R. DNA-damage repair; the good, the bad, and the ugly. *The EMBO Journal*, 2008;**27**:589–605.

15.1 TYPES OF DNA DAMAGE

Feuerhahn S, Egly JM. Tools to study DNA repair: what's in the box? *Trends in Genetics*, 2008;**24**:467–474.

Garinis GA, van der Horst GT, Vijg J, Hoeijmakers JH. DNA damage and ageing: new-age ideas for an age-old problem. *Nature Cell Biology*, 2008;**10**:1241–1247.

Thompson, LH. Recognition, signaling, and repair of DNA double-strand breaks produced by ionizing radiation in mammalian cells: the molecular choreography. *Mutation Research*, 2008;**75**:158–246.

Yang W. Structure and mechanism for DNA lesion recognition. *Cell Research*, 2008;**18**:184–197.

15.2 POST-REPLICATION MISMATCH REPAIR

Jiricny J. The multifaceted mismatch-repair system. *Nature Reviews Molecular Cell Biology*, 2008;**7**:335–346.

Li F, Mao G, Tong D, *et al.* The histone mark H3K36me3 regulates human DNA mismatch repair through its interaction with MutSα. *Cell*, 2008;**153**:590–600.

Li GM. Mechanisms and functions of DNA mismatch repair. *Cell Research*, 2008;**18**:85–98.

McMurray CT. Hijacking of the mismatch repair system to cause CAG expansion and cell death in neurodegenerative disease. *DNA Repair*, 2008;**7**:1121–1134.

Modrich P. Mechanisms in eukaryotic mismatch repair. *Journal of Biological Chemistry* 2008;**281**:30305–30309.

15.3 REPAIR OF DNA DAMAGE BY DIRECT REVERSAL

Eker AP, Quayle C, Chaves I, van der Horst GT. DNA repair in mammalian cells: Direct DNA damage reversal: elegant solutions for nasty problems. *Cellular and Molecular Life Sciences*, 2008;**66**:968–980.

Sabharwal A, Middleton MR. Exploiting the role of O6-methylguanine-DNA-methyltransferase (MGMT) in cancer therapy. *Current Opinion in Pharmacology*, 2008;**6**:355–363.

Sedgwick B, Bates PA, Paik J, *et al.* Repair of alkylated DNA: recent advances. *DNA Repair* 2008;**6**:429–442.

Tubbs JL, Pegg AE, Tainer JA. DNA binding, nucleotide flipping, and the helix-turn-helix motif in base repair by O6-alkylguanine-DNA alkyltransferase and its implications for cancer chemotherapy. *DNA Repair*, 2008;**6**:1100–1115.

15.4 REPAIR OF DNA DAMAGE BY BASE EXCISION REPAIR

Bosshard M, Markkanen E, van Loon B. Base excision repair in physiology and pathology of the central nervous system. *International Journal of Molecular Sciences*, 2008;**13**:16172–16222.

Fishel ML, Vasko MR, Kelley MR. DNA repair in neurons: so if they don't divide what's to repair? *Mutation Research*, 2008;**614**:24–36.

Hitomi K, Iwai S, Tainer JA. The intricate structural chemistry of base excision repair machinery: implications for DNA damage recognition, removal, and repair. *DNA Repair*, 2008;**6**:410–428.

Robertson AB, Klungland A, Rognes T, Leiros I. DNA repair in mammalian cells: Base excision repair: the long and short of it. *Cellular and Molecular Life Sciences*, 2008;**66**:981–993.

Wilson DM 3rd, Bohr VA. The mechanics of base excision repair, and its relationship to aging and disease. *DNA Repair*, 2008;**6**:544–559.

15.5 NUCLEOTIDE EXCISION REPAIR OF BULKY LESIONS

Bergoglio V, Magnaldo T. Nucleotide excision repair and related human diseases. *Genome Dynamics*, 2008;**1**:35–52.

Clement FC, Camenisch U, Fei J, Kaczmarek N, Mathieu N, Naegeli H. Dynamic two-stage mechanism of versatile DNA damage recognition by xeroderma pigmentosum group C protein. *Mutation Research*, 2008;**685**:21–28.

Goosen N, Moolenaar GF. Repair of UV damage in bacteria. *DNA Repair*, 2008;**7**:353–379.

Grollman AP. Aristolochic acid nephropathy: Harbinger of a global iatrogenic disease. *Environmental and Molecular Mutagenesis*. 2008;**54**:1–7.

Kuper J, Kisker C. Damage recognition in nucleotide excision DNA repair. *Current Opinion in Structural Biology*, 2008;**22**:88–93.

Schumacher B, Garinis GA, Hoeijmakers JH. Age to survive: DNA damage and aging. *Trends in Genetics*, 2008;**24**:77–85.

Tornaletti S. DNA repair in mammalian cells: Transcription-coupled DNA repair: directing your effort where it's most needed. *Cellular and Molecular Life Sciences*, 2008;**66**:1010–1020.

Truglio JJ, Croteau DL, Van Houten B, Kisker C. Prokaryotic nucleotide excision repair: the UvrABC system. *Chemical Reviews*, 2008;**106**:233–252.

15.6 TRANSLESION DNA SYNTHESIS

Broyde S, Wang L, Rechkoblit O, *et al.* Lesion processing: high-fidelity versus lesion-bypass DNA polymerases. *Trends in Biochemical Sciences*, 2008;**33**:209–219.

Lehmann AR. New functions for Y family polymerases. *Molecular Cell*, 2008;**24**:493–495.

Livneh Z, Ziv O, Shachar S. Multiple two-polymerase mechanisms in mammalian translesion DNA synthesis. *Cell Cycle*, 2008;**9**:729–735.

Waters LS, Minesinger BK, Wiltrout ME, *et al.* Eukaryotic translesion polymerases and their roles and regulation in DNA damage tolerance. *Microbiology and Molecular Biology Reviews*, 2008;**73**:134–154.

15.7 THE DNA DAMAGE RESPONSE

Ciccia A, Elledge SJ. The DNA damage response: making it safe to play with knives. *Molecular Cell*, 2008;**40**:179–204.

de Bruin RA, Wittenberg C. All eukaryotes: before turning off G1-S transcription, please check your DNA. *Cell Cycle*, 2008;**8**:214–217.

Harper JW, Elledge SJ. The DNA damage response: ten years after. *Molecular Cell*, 2008;**28**:739–745.

Jackson SP, Durocher D. Regulation of DNA damage responses by ubiquitin and SUMO. *Molecular Cell*, 2013;**49**:795–807.

Matic I, Taddei F, Radman M. Survival versus maintenance of genetic stability: a conflict of priorities during stress. *Research in Microbiology*, 2008;**155**:337–341.

15.8 THE DNA DAMAGE RESPONSE IN BACTERIA

Butala M, Zgur-Bertok D, Busby SJ. The bacterial LexA transcriptional repressor. *Cellular and Molecular Life Sciences*, 2008;**66**:82–93.

Erill I, Campoy S, Barbé J. Aeons of distress: an evolutionary perspective on the bacterial SOS response. *FEMS Microbiology Reviews*, 2008;**31**:637–656.

Galhardo RS, Hastings PJ, Rosenberg SM. Mutation as a stress response and the regulation of evolvability. *Critical Reviews in Biochemistry and Molecular Biology*, 2008;**42**:399–435.

Kelley WL. Lex marks the spot: the virulent side of SOS and a closer look at the LexA regulon. *Molecular Microbiology*, 2008;**62**: 1228–1238.

Yeeles JT, Marians KJ. The *Escherichia coli* replisome is inherently DNA damage tolerant. *Science*, 2008;**334**:235–238.

15.9 THE DNA DAMAGE RESPONSE IN EUKARYOTES

Bergink S, Jentsch S. Principles of ubiquitin and SUMO modifications in DNA repair. *Nature*, 2008;**458**:461–467.

Cimprich KA, Cortez D. ATR: an essential regulator of genome integrity. *Nature Reviews Molecular Cell Biology*, 2008;**9**:616–627.

Cohn MA, D'Andrea AD. Chromatin recruitment of DNA repair proteins: lessons from the Fanconi anemia and double-strand break repair pathways. *Molecular Cell*, 2008;**32**:306–312.

Hurley PJ, Bunz F. ATM and ATR: components of an integrated circuit. *Cell Cycle*, 2008;**6**:414–417.

Lavin MF. Ataxia-telangiectasia: from a rare disorder to a paradigm for cell signaling and cancer. *Nature Reviews Molecular Cell Biology*, 2008;**9**:759–769.

van Attikum H, Gasser SM. Crosstalk between histone modifications during the DNA damage response. *Trends in Cell Biology*, 2008;**19**:207–217.

Yang XH, Zou L. Dual functions of DNA replication forks in checkpoint signaling and PCNA ubiquitination. *Cell Cycle*, 2008;**8**:191–194.

15.10 DNA DAMAGE AND CELL DEATH IN MAMMALIAN CELLS

Halazonetis TD, Gorgoulis VG, Bartek J. An oncogene-induced DNA damage model for cancer development. *Science*, 2008;**319**: 1352–1355.

Kruse JP, Gu W. Modes of p53 regulation. *Cell*, 2008;**137**:609–622.

Vazquez A, Bond EE, Levine AJ, Bond GL. The genetics of the p53 pathway, apoptosis and cancer therapy. *Nature Reviews Drug Discovery*, 2008;**7**:979–987.

Wang W, El-Deiry WS. Restoration of p53 to limit tumor growth. *Current Opinion in Oncology*, 2008;**20**:90–96.

Repair of DNA double-strand breaks and homologous recombination

INTRODUCTION

In Chapter 15, we explored the strategies used by the cell as it detects and responds to damage that its genome may incur. Without such repair mechanisms, a cell would accumulate potentially deleterious mutations, with dire consequences for the cell or the organism it may be part of. The repair strategies outlined in Chapter 15 are those available to the cell when one of the two strands of a DNA duplex remains intact, whereby the strand can act as a template for the repair of its damaged complementary strand. But what happens if both strands of a duplex are damaged? How does the cell respond at times such as these?

In this chapter, we build on Chapter 15 to consider the repair of double-strand DNA breaks to complete our picture of the range of strategies adopted by the cell to maintain an intact genome. However, double-strand DNA breaks are not always due to unwanted damage; such breaks underpin the process of **homologous recombination**, through which the exchange of genetic material – and ultimately the generation of biological diversity – occurs.

➔ We see in Section 7.7 how homologous recombination is an essential process during meiosis.

We begin with a general overview of DNA double-strand break repair and homologous recombination to set out the key concepts that we will develop further during the course of the chapter. We will also learn more about homologous recombination toward the end of this chapter in Section 16.9.

16.1 AN OVERVIEW OF DNA DOUBLE-STRAND BREAK REPAIR AND HOMOLOGOUS RECOMBINATION

Double-strand breaks are repaired by two different mechanisms

Double-strand breaks in DNA, unlike other types of DNA damage, must be repaired in the absence of an undamaged template strand that could otherwise have been used to direct the repair process. There are two major mechanisms for repairing double-strand breaks, which have distinct requirements and different consequences (Figure 16.1). In the first of these, **non-homologous end joining (NHEJ)** (Figure 16.1a), the broken ends are simply rejoined. The second mechanism of repair, **homology-directed repair** (Figure 16.1b), involves (as its name implies) pairing between the damaged DNA and homologous sequences in another chromosome, which serve as the template for repair. Both mechanisms are deployed in eukaryotes and in most bacteria, although NHEJ does not occur in *Escherichia coli*.

Although NHEJ can occur at any time during the cell cycle, it occurs predominantly in non-dividing (or 'quiescent') cells. It is particularly important for bacteria and haploid yeast in stationary phase. Although NHEJ has the advantage that it does not require a template and can therefore occur during G0, G1, and early S phases of the cell cycle, when no sister chromatid is available, it has the disadvantage that it can cause mutations. Nucleotides at the breakage site are often lost through degradation by cellular nucleases before repair is initiated, and further loss frequently happens because of trimming that occurs during the repair process itself.

Homology-directed repair is the major mechanism for repairing double-strand breaks that occur during or after DNA replication. In eukaryotes, the majority of homology-directed repair occurs during late S phase or G2 phase of the cell cycle only, when closely apposed sister chromatids are available to serve as a template. Because new rounds of replication can take place before chromosome segregation, *Escherichia coli* often contains multiple copies of its genome, and thus double-strand breaks in *E. coli* can be repaired by homology-directed repair.

Homology-directed repair generally results in repair without mutation or loss of sequences because it involves copying intact information from an undamaged identical sister chromatid. More rarely, a homologous chromosome may provide the template instead of a sister chromatid; this allows homology-directed repair to also occur in G0, G1, and early S phase, but may not restore the original sequence to the damaged chromosome, because the sequences of the two homologous chromosomes may not be identical. In this case, the damaged region will acquire the sequence of the intact homologous chromosome, a phenomenon known as **gene conversion** or **loss of heterozygosity**. We shall see later that this loss of heterozygosity can have serious consequences if the sequence on the undamaged homolog contains a defective gene.

So what determines which repair strategy is adopted in cells in which both NHEJ and homology-directed repair are possible? The answer lies in the initial recognition of a double-strand break, which is key not only to the cellular response to double-strand breaks, as we saw in Chapter 15, but also to the choice of repair pathway. In eukaryotes, double-strand breaks resulting from damage to fully replicated DNA are quickly bound by either an MRN complex (see Section 15.9), or a heterodimer composed of two proteins called Ku70 and Ku80. Each of these complexes recruits repair proteins to the site of damage to determine the specific pathway adopted: we shall see in Section 16.4 how the MRN complex contributes to the cell-cycle restriction of the homology-directed repair pathway; the Ku70/Ku80 heterodimer contributes specifically to the NHEJ pathway.

In this chapter, we shall describe the steps that are required for each of these pathways, and some of what is known of the proteins that mediate them - many of which also play a role in DNA replication. We shall focus mainly on those proteins that have been identified through mutations that increase the susceptibility of single cells to radiation-induced damage, or confer genetic predisposition to cancer in humans and are therefore implicated in specialized roles in DNA repair.

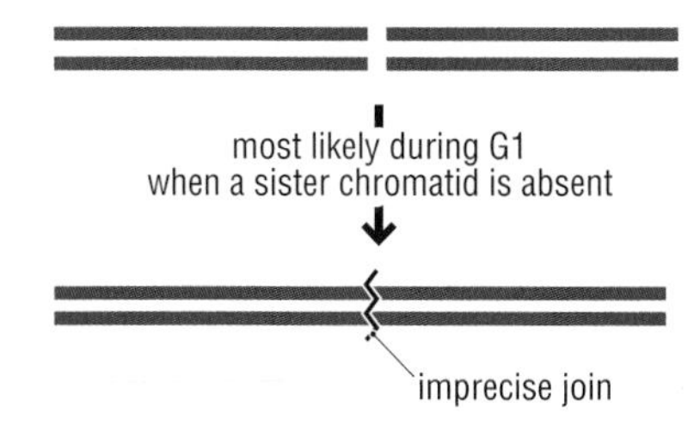

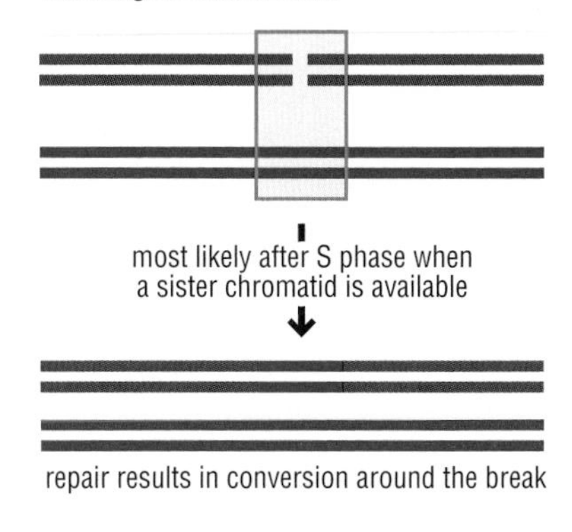

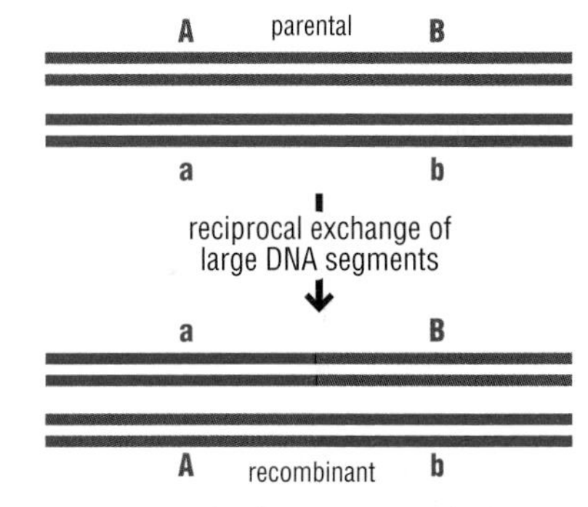

Figure 16.1 The major pathways of DNA double-strand break repair and their consequences. (a) NEHJ. DNA broken by a double-strand break is shown. Repair by this pathway can often be imprecise (jagged line). (b) Homology-directed repair using a homologous chromosome. A chromosome broken by a double-strand break and a homologous chromosome (pink) are shown. Repair results in the conversion of information in the region of the break (gray) to information from the homolog (pink). (c) Homologous recombination. Homologous recombination can exchange large segments of DNA, so that AB and ab markers on the parental duplexes (gray and pink) become Ab and aB on the recombinant daughter duplexes.

➔ We learn about proteins involved in DNA replication in Chapter 6.

The double-strand break repair machineries also underlie mechanisms to generate diversity in the immune system and mediate homologous recombination

The mechanistic principles that underpin double-strand break repair are also exploited in other biological contexts. The NHEJ pathway provides the means for the DNA joining steps in the programmed DNA rearrangement of specialized cells of the immune system. This rearrangement generates the highly variable proteins – of which antibodies are the most familiar – that protect us from infection. Defects in the NHEJ pathway can lead to immune deficiency as well as sensitivity to DNA damage that may result in cancer.

➜ We describe the programmed rearrangement of the DNA of specialized cells of the immune system in detail in Chapter 17.

The proteins of the homology-directed repair pathway have also been adapted to mediate the pairing interactions between homologous DNA molecules that underlie **homologous recombination** (Figure 16.1c). Homologous recombination is the reciprocal exchange of information between chromosomes that is intimately involved in the production of gametes during meiosis (see Chapter 7) and occurs occasionally during mitosis in eukaryotes. Homologous recombination also occurs in bacteria, often during **conjugation**, when DNA is transferred from one cell to another. In this case, the DNA from the donor cell can be integrated into the genome of the acceptor cell via homologous recombination.

Other paths of entry for DNA into bacteria include phage transduction, in which phage particles can carry cellular DNA, and transformation by naked DNA. Incorporation of these exogenous DNAs into the genome by homologous recombination is relatively common and is a significant source of the horizontal gene transfer that accounts for much bacterial diversification in the course of evolution. These foreign DNAs can also be the 'launch pads' for the discrete self-mobilizable elements that can insert into and diversify recipient genomes.

➜ We learn about mobile DNA segments in Chapter 17.

Thus, homologous recombination is universal, and provides a major mechanism for increasing the diversity of individuals in a population. Homologous recombination in meiosis is a highly regulated process involving homologous chromosomes, where exchange of DNA can occur without a net loss or gain of sequences. Homologous recombination can, however, occasionally occur between two stretches of homologous DNA located anywhere in the genome, potentially leading to **deletions** or **translocations**, both of which can have grave consequences for the cell. Such events are particularly likely in the genomes of higher vertebrates, which, as we have seen in Chapter 1, have a very high proportion of repetitive DNA, providing large regions of homology between non-homologous chromosomes, and between different regions of the same chromosome. We discuss such rearrangements in Section 16.9.

The conserved molecular components of the homology-directed repair and homologous recombination pathways highlight the close relationship between the mechanisms they adopt. We learn more about these commonalities during the course of this chapter.

16.2 DOUBLE-STRAND BREAK REPAIR BY NHEJ

Having now reviewed the key principles that underpin double-strand break repair, we begin our exploration of how this repair occurs in reality by considering the first of the two mechanisms for repair mentioned in Section 16.1, namely, NHEJ.

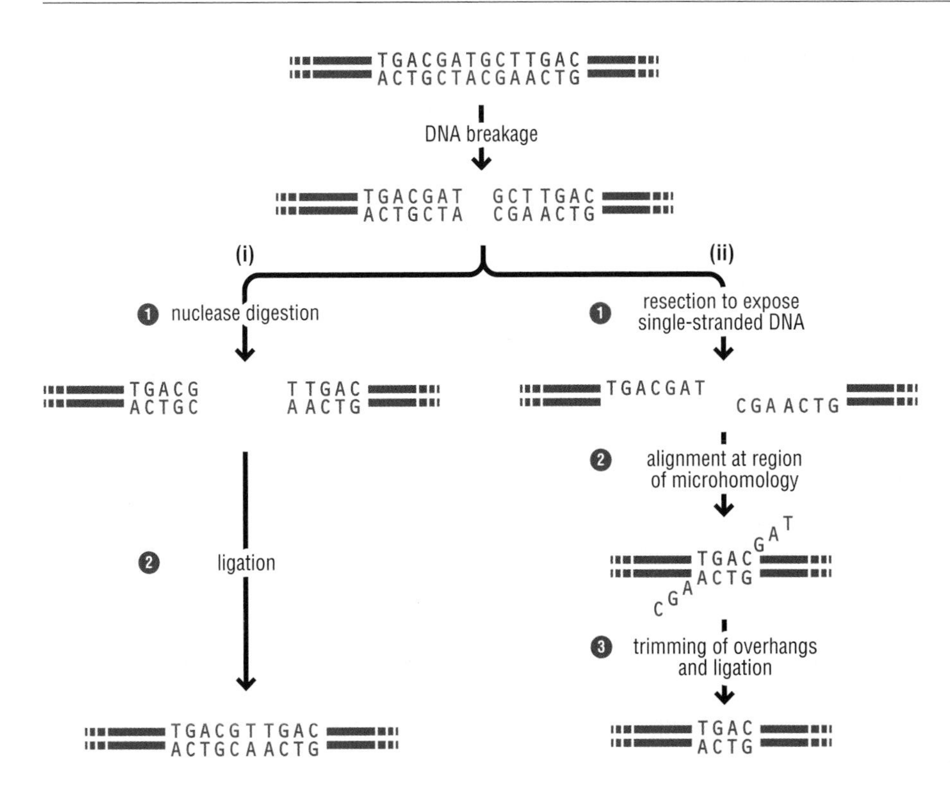

Figure 16.2 Pathways of NHEJ. A DNA broken by a double-strand break is shown, flanked by short regions of microhomology; rejoining usually results in the loss of sequence information. NHEJ pathway (i): (1) nucleases and/or polymerases remove a few base pairs at either side of the break to generate blunt ends; (2) blunt ends are aligned and joined by ligation. In the process, some sequence information is usually lost. NHEJ pathway (ii): (1) resection from the site of the break occurs to generate single-stranded tails; (2) base-pairing occurs between stretches of microhomology a few nucleotides long; (3) the tails are trimmed, and ligation occurs.

The NHEJ pathway is generally mutagenic

The process of NHEJ occurs by the rejoining of broken DNA ends and does not require another template. The exact nature of the repair reaction is highly influenced by the fact that different types of double-strand breaks result in different end structures. In some cases, the DNA ends have simple overhangs that can be simply rejoined by ligation with little or no sequence loss (Figure 16.2i). In other cases, particularly when radiation mediated breaks occur, the ends are rejoined only after considerable processing of the ends at the break (Figure 16.2ii).

NHEJ events usually occur with less than 10 base pairs (bp) of DNA degradation but can involve the degradation of more than 100 bp. In one pathway (Figure 16.2i), the ends at the break undergo only limited processing with perhaps the loss of a few base pairs, and two blunt ends are ligated together. In the other pathway (Figure 16.2ii), resection reveals short complementary regions (called microhomology) that occur by chance anywhere in the tails. These tails may then base-pair, after which the single-stranded tails are trimmed back to the base-paired DNA, the gaps are filled in by repair polymerases, and the ends are ligated. This process can involve considerable sequence loss before the trimmed ends are ligated to repair the break.

NHEJ is therefore often mutagenic and may be deleterious if, for example, it occurs in a region of functional DNA sequence such that important coding information is lost. However, such mutation is minor compared with the consequence of no repair happening at all, which could result in the loss of a chromosome arm, for example.

The NHEJ pathway depends on conserved end-binding and ligating proteins

Scan here to watch a video animation explaining more about NHEJ, or find it via the **Online Resource Center** at **www.oxfordtextbooks.co.uk/orc/craig2e/**.

The NHEJ process in bacteria and eukaryotes requires a protein, Ku, which binds to broken DNA ends and recruits the nucleases, polymerases, kinases,

Table of proteins that mediate NHEJ in different organisms

	bacteria	eukaryotes	
functional component		***Saccharomyces cerevisiae***	**multicellular eukaryotes**
DNA end binding	Ku (30–40 kDa)	Ku 70/80	Ku 70/80
polymerase	POL domain of LigD	Pol4	Pol μ and λ
nuclease	?	Rad50:Mre11:Xrs2	Artemis: DNA-PKcs
kinase/phosphatase	phosphodiesterase domain of LigD	Tpp1 and others	polynucleotide kinase/ phosphatase and others
ligase	LIG domain of LigD	Nej1:Lif1:Dnl4	Ligase IV -XRCC4 XLF/Cerunnos

Figure 16.3 Proteins and steps in NHEJ. Proteins that mediate NHEJ in different organisms are shown.

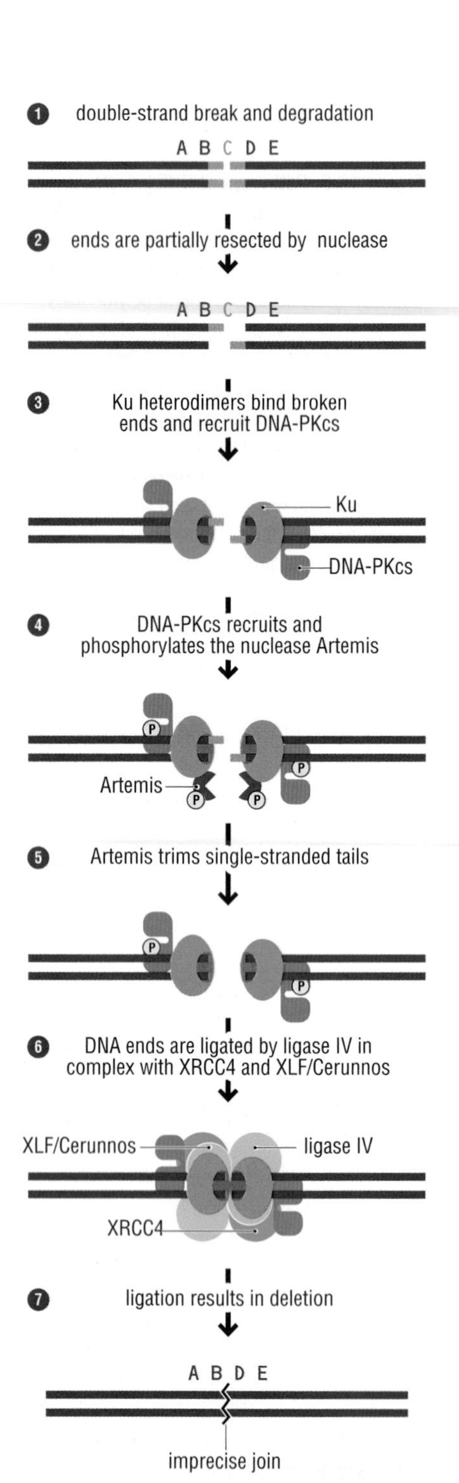

Figure 16.4 Steps in NHEJ in mammals. The C region of a DNA is broken by a double-strand break. Although intact after repair, the DNA was imprecisely repaired because resection and trimming of the ends leads to sequence loss.

phosphatases, and ligases that mediate the joining reaction (Figure 16.3). The principal known proteins in the mammalian NHEJ pathway, which is the best understood, are illustrated in Figure 16.4. What are the main steps in this pathway?

First, the Ku complexes slide onto each end of the double-strand DNA break and act as a scaffold for the recruitment of DNA repair enzymes, in particular a protein kinase subunit, DNA-PKcs, to form DNA-PK. Second, DNA-PKcs is activated by binding to Ku, resulting in autophosphorylation, and also phosphorylates and controls other important repair proteins. These include the overhang trimming nuclease Artemis, which is activated when phosphorylated by DNA-PK. Third, other DNA end-processing enzymes, including DNA polymerases and/or exonucleases and DNA kinases and/or phosphatases, further process the ends before ligation.

Finally, a specialized DNA ligase, DNA ligase IV, and its accessory proteins, XRCC4 and XLF/Cernunnos, form a complex that ligates the broken DNA ends. (XRCC stands for x-ray repair cross-complementing - a reference to the experimental analysis whereby the proteins were discovered.)

Ku is conserved from bacteria to mammals. By contrast, the kinase subunit DNA-PKcs seems to be a specialization found mainly in vertebrates and may play a part in the relatively high frequency of repair by NHEJ in mammals. DNA-PK can also activate cell cycle arrest and programmed cell death in mammals, like ATM (ataxia telangiectasia-mutated) and ATR (ataxia telangiectasia-related) (see Section 15.9).

The yeast MRN complex called MRX is involved in NHEJ

In budding yeast, the MRN complex, denoted MRX after its components Mre11, Rad50, and Xrs2, is known to also be involved in double-strand break repair by NHEJ, where the complex is thought to collaborate with Ku. As we have seen in Section 15.9, MRN complexes play a central part in eukaryotic DNA damage repair and may play a key role in holding broken DNA ends. As we shall see, the MRN and MRX complexes also participate in homology-directed repair of double-strand breaks. We learn more about their role in this pathway below.

16.3 HOMOLOGY-DIRECTED REPAIR OF DOUBLE-STRAND BREAKS

The NHEJ pathway allows the repair of double-strand breaks in the absence of a template, but at the cost of mutagenesis. By contrast, homology-directed repair exploits the availability of homologous sequences elsewhere, to provide a template for repair. The template most often used is a sister chromatid, which is identical in sequence to the damaged DNA, in which case the repair can be error-free. In the absence of a sister chromatid, homology-directed repair uses the homologous chromosome, which is closely related in sequence - but is not identical. In this section, and the two subsequent sections, we consider how this repair mechanism operates.

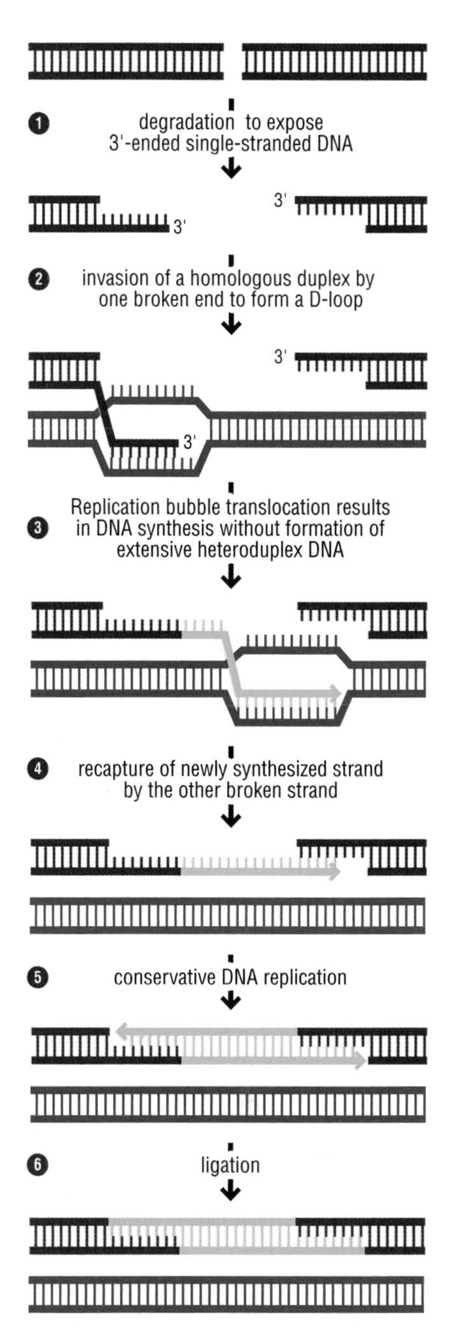

Figure 16.5 Homology-directed repair via SDSA. In this pathway, both strands of the repaired duplex are formed by new DNA synthesis (green). As DNA synthesis progresses, the newly synthesized strand is displaced from the template duplex such that the newly synthesized strand and the template strand do not remain paired as in semi-conservative DNA replication. Note that the gap in the gray duplex is replaced with newly synthesized (green) copies of the template (pink) duplex.

Accurate repair of double-strand DNA breaks involves pairing of homologous duplexes

Although the details of homology-directed repair vary between different organisms, the fundamental process is the same in all cases. The reactions that underlie homology-directed double-strand break repair occur within four main phases:

- generation of single-stranded DNA at the site of a double-strand break
- pairing by one of the single-stranded ends with the intact duplex by invasion of the duplex to form a **heteroduplex**
- repair of the damaged duplex by DNA synthesis using strands from the undamaged duplex as a template
- the separation of the two duplexes.

The formation of the heteroduplex is known as **synapsis**, and the steps that precede or follow it are referred to as **presynaptic** and **postsynaptic**. In this section, we will follow in detail what happens to the DNA during homology-directed repair. In subsequent sections, we discuss the specialized proteins that make these reactions possible, and how they can sometimes cause changes in genomic DNA.

Homology-directed repair requires long single-stranded DNA regions

The first step in homology-directed repair is the generation of single-stranded 3′ ends at the double-strand break by specialized helicases and exonucleases by degradation (also called resection) of the 5′ ends of the broken molecule (Figure 16.5, step 1). This process requires specialized helicases and exonucleases in both bacteria and eukaryotes. Homology-directed repair requires long regions of homology at the break for base-pairing to occur. Consequently, the single-stranded tails at the break are usually at least 50 nucleotides long. These long tails are needed to ensure accurate pairing with the template DNA in another chromosome, and thus to ensure error-free repair.

Formation of a D-loop containing heteroduplex DNA provides a template for DNA repair synthesis

In the second step in homology-directed repair, **strand invasion**, the single-stranded DNA from one end of the damaged duplex locates and pairs with its complementary strand in the related unbroken duplex to form a DNA **heteroduplex** - a region in which a single strand from one DNA molecule is base-paired with a single strand

Scan here to watch a video animation explaining more about the steps of homology-directed repair, or find it via the **Online Resource Center** at **www.oxfordtextbooks.co.uk/orc/craig2e/**.

from a different DNA molecule. This pairing displaces the DNA strand of the same polarity from the undamaged duplex with the formation of a three-stranded structure called a displacement loop (D-loop) (Figure 16.5, step 2). The degree of DNA sequence similarity required for efficient base-pairing and heteroduplex formation in these reactions is usually 95–100% identity extending over at least 20–30 base pairs. Thus, there is a very small likelihood of pairing between unrelated DNA molecules.

Homology-directed repair can proceed via synthesis-dependent strand annealing

Once specific heteroduplex pairing has occurred between the invading strand and the undamaged duplex to form the D-loop, homology-directed repair can proceed in any of several ways. The most frequent pathway is known as **synthesis-dependent strand annealing (SDSA)** (Figure 16.5, step 3) in which the strands in the repaired region are generated by DNA synthesis, As such, there is no extended region of heteroduplex DNA formed between parental and newly synthesized strands.

Repair by this pathway begins with DNA synthesis that is initiated in the D-loop using the 3′-OH end of the invading strand as a primer and the undamaged strand as a template (Figure 16.5, step 3). As synthesis proceeds, the newly synthesized strand is released from the template strand, and the D-loop translocates along the template strand.

When it is long enough (namely, past the site of the double-strand break), the newly synthesized strand pairs with the complementary strand of the damaged duplex on the other side of the break (Figure 16.5, step 4). Notably, the DNA that now bridges the double-strand gap is newly synthesized DNA, generated from the undamaged template duplex; there is no permanent formation of heteroduplex DNA between the damaged duplex and the template duplex. The damaged duplex now contains a region of single-stranded DNA rather than a double-strand gap. This gap is filled using the newly synthesized strand as a template (Figure 16.5, steps 5 and 6). As a result, in homology-dependent repair by SDSA, no DNA strand from the undamaged duplex (pink in Figure 16.5) is joined to or incorporated into the repaired duplex (gray in Figure 16.5), and the DNA synthesis is conservative: that is, both strands of the repaired duplex result from new DNA synthesis. This is sharply different from the semi-conservative DNA synthesis seen during DNA replication, during which the newly synthesized strand remains paired to the template strand.

➔ We learn about the semi-conservative nature of DNA replication in Section 6.1.

Having reviewed the overall events that are characteristic of homology-directed repair, we are now ready to consider the mechanism of this process in more detail. In the following sections, we describe the molecular machinery that operates to generate the single-stranded tails that invade the homologous duplex to generate the D-loop and to align the invading strand with the complementary strand of the intact homolog.

16.4 GENERATION OF SINGLE-STRANDED DNA BY HELICASES AND NUCLEASES

The RecBCD complex in bacteria and MRN complex in eukaryotes generate the single-stranded DNA needed for strand invasion

The process of generation of the long 3′ ended single-stranded tails that are required for homology-directed repair is called the presynaptic step. This step depends on both DNA helicases, which separate the two strands of DNA by unwinding (see

Section 6.5), and on nucleases that degrade one of the DNA strands at the broken DNA, namely the strands with the 5′ ends at the break. These nucleases include both exonucleases and endonucleases that expose single strands by degradation of one strand in a duplex.

In *E. coli*, there are two distinct presynaptic pathways by which 3′ ended single strands are generated: the **RecBCD** and the **RecF** pathways. The RecBCD pathway is the predominant presynaptic pathway and is mediated by a protein complex called by the same name: RecBCD (Figure 16.6). This complex, which has combined helicase and nuclease activities, attaches at a double-strand break (Figure 16.6, step 1) and translocates along the DNA, its helicase activity separating the two strands and its nuclease activity cleaving the unwound strands into small fragments (Figure 16.6, step 2). The nuclease activity of this enzyme is modulated by a specific 8 bp DNA sequence known as a **Chi sequence**, which occurs at frequent intervals of about 1 per 5 kilobase pair (kb) in the *E. coli* genome.

When the RecBCD complex arrives at a Chi sequence, the nuclease activity is reduced (Figure 16.6, step 3), and the enzyme now preferentially degrades the 5′-end strand, leaving a duplex DNA with a single-stranded 3′ tail containing the Chi sequence (Figure 16.6, step 4). Thus, the presence of a Chi site stimulates repair because it promotes formation of single-stranded 3′-end DNA. In Experimental approach 16.1, we explore how the study of single-molecule complexes of RecBCD and Chi-containing DNA defined how Chi modulates RecBCD activity.

RecBCD also actively facilitates the loading of the recombinase **RecA** protein onto the single-stranded 3′-end DNA generated at a Chi site. RecA then promotes homologous pairing between the single-stranded DNA and a DNA duplex that will provide the template for repair, preferentially recognizing the single-stranded 3′-DNA-end that participates in strand invasion to form a D-loop. Cells defective in RecBCD are very sensitive to DNA-damaging agents because they cannot repair DNA double-strand breaks.

In addition to RecBCD, the other presynaptic system in *E. coli* is known as the RecF pathway, which includes a helicase called RecQ and a nuclease called RecJ that, together, generate single-stranded DNA. This pathway appears to be used mainly in the repair of replication forks that have stalled as a result of single-strand lesions in the DNA, which we discuss in Section 16.7. Humans also have homologs of the bacterial RecQ helicases; defects in these genes are known to contribute to the hereditary diseases Bloom syndrome, Werner syndrome, and Rothmund–Thomson syndrome, which are associated with an increased risk of cancer as a result of increased genome instability.

In eukaryotes, the generation of the single-stranded DNA tails that are the substrates for strand invasion is carried out by several proteins (Figure 16.7). The activation of the specialized helicases and exonucleases required to produce these long single-stranded DNA tails is the decisive factor in determining whether a DNA break will be repaired by the homology-dependent repair pathway or by NHEJ.

The repair process is initiated by MRN, which, as we have seen, arrives early at the damage site. The MRN complex in yeast, MRX, associates with the exonuclease Sae2 to start the resection process. As mentioned earlier, homology-dependent repair is most efficient in the presence of the sister chromatid, so cells inhibit this process during G1 by preventing resection. This occurs, at least in part, by the requirement for the phosphorylation of Sae2, and its vertebrate homolog CtIP, by the cell-cycle regulator Cdk1 in the S and G2 phases of the cell cycle, to activate it for end degradation. One of two different proteins, EXOI, a 5′ to 3′ exonuclease, or exonuclease DNA2, then resect the 5′ ends at the break in conjunction with the 3′ to 5′ BLM RecQ family helicase to yield 3′ single-stranded tails.

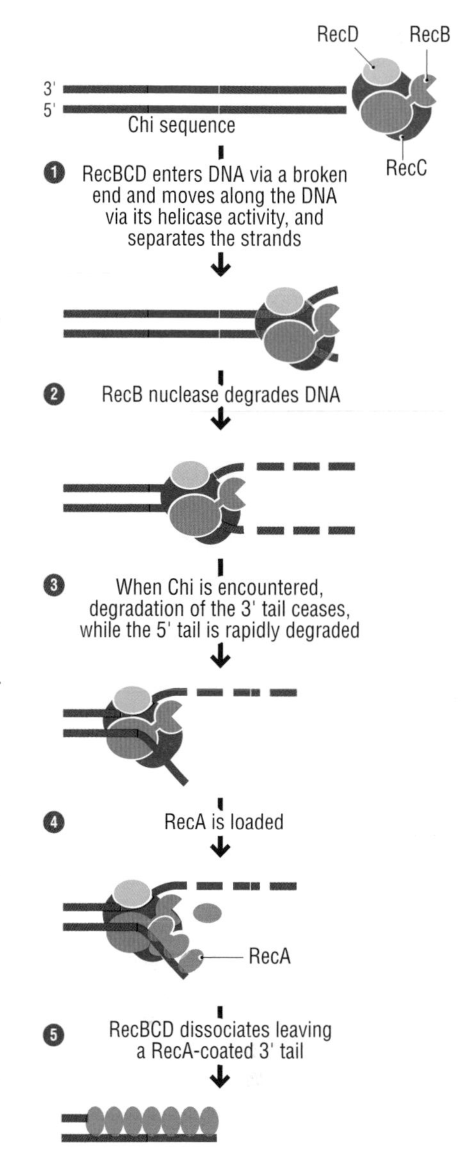

Figure 16.6 Schematic diagram of the generation of single-stranded tails at double-strand breaks by the RecBCD complex. The RecBCD protein complex loads onto DNA at a double-strand break, moves along DNA using its dual helicase and nuclease functions, and eventually dissociates leaving a nucleoprotein filament containing RecA bound to a 3′ single-stranded DNA end.

➔ We encounter the RecA protein briefly in Section 15.8.

➔ We discuss Cdk1 in Section 5.2.

16.1 EXPERIMENTAL APPROACH

DNA sequences that stimulate homologous recombination: Chi sites and RecBCD

The *E. coli* RecBCD complex has been known to play a role in recombination since *recB* and *recC* mutants that were defective in the process were isolated. However, the precise role of these proteins and how they might interact with DNA was not understood. Models for their mechanisms of action have been evolving and are still evolving from both careful observations of phage lambda and *E. coli* genetics and biochemistry.

A key finding in understanding the role of RecBCD was the recognition that the complex interacts directly with Chi sites in DNA. The initial biochemical studies of RecBC (the presence of the RecD subunit was a later discovery) identified it only as a DNA exonuclease. It was subsequently found that RecBC also had helicase activity that could interact with a duplex and expose single-stranded DNA. The fact that RecBC recombination activity was influenced by an 8 bp DNA sequence called Chi was discovered serendipitously in experiments looking at the growth of bacteriophage lambda, as discussed below.

Chi sites stimulate recombination

When the bacteriophage lambda grows lytically, it produces long, concatenated, linear DNA molecules that need to be packaged into the phage head. (See Section 9.4 for a review of the lambda life cycle.) However, if certain lambda genes (*red* and *gam*) are inactive, lambda replication produces only monomeric circles that are poorly packaged, and thus these lambda mutants do not grow well. If the circles are able to recombine, however, they will generate long, circular concatamers that can again be packaged, and the phage will grow well.

Several spontaneous lambda mutants were identified that resulted in higher recombination frequencies, as reflected by the production of more lytic phage. The production of more phage could be readily visualized by the presence of a larger plaque size (indicating a larger zone of bacterial killing), when the propagation of individual viruses was analyzed on a confluent lawn of bacteria, as depicted in Figure 1. In mapping these lambda mutations, Gerald Smith and colleagues found that the mutations did not affect proteins required for recombination, but were rather in DNA sites that stimulated recombination. These spontaneous mutations created 8 bp Chi sequences from natural sequences in the phage genome that were already almost consensus Chi sites (i.e., they already had seven of eight Chi nucleotides). Further work demonstrated explicitly that a Chi sequence was indeed a critical DNA site that stimulated *recBC*-dependent recombination. Although Chi was discovered because of its role in the phage lambda lifecycle, we now know that the *E. coli* chromosome has about one Chi site per 5 kb, and that these sites play a key role in stimulating homologous recombination in wild-type cells.

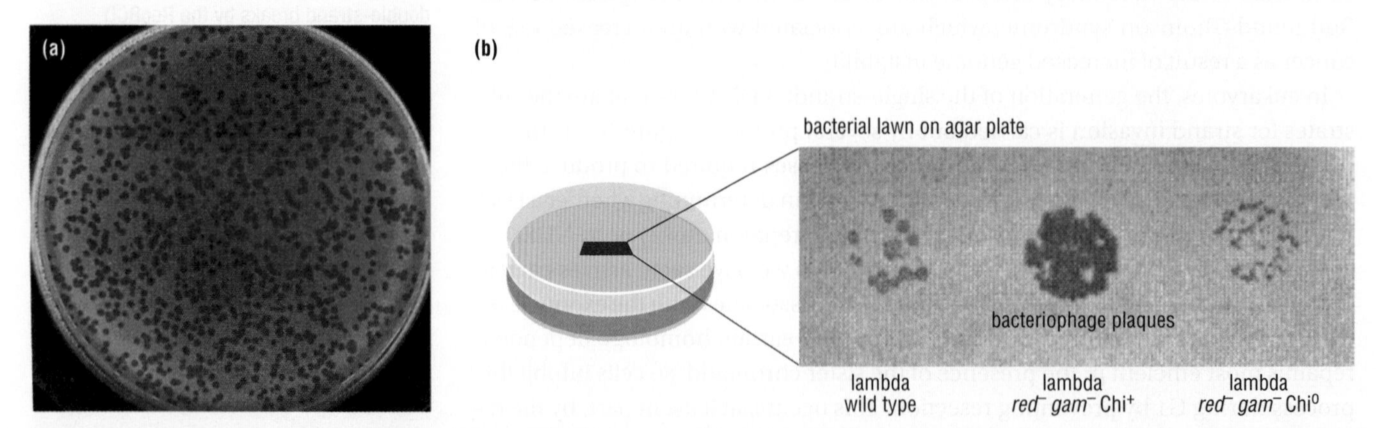

Figure 1 Growth of bacteriophage lambda on *E. coli*. (a) The ability of a lytic phage to grow can be monitored by the size of the plaque, or zone of bacterial killing, produced by the propagation of individual viruses on a confluent lawn of bacteria. (b) The effect of Chi on the growth of bacteriophage lambda: wild-type lambda lacks Chi sites and yields plaques of moderate size; *red⁻ gam⁻* lambda with Chi sites (Chi⁺) yields large plaques; and *red⁻ gam⁻* lambda without Chi sites (Chi⁰) yields small plaques.

Image courtesy of SK Amundsen.

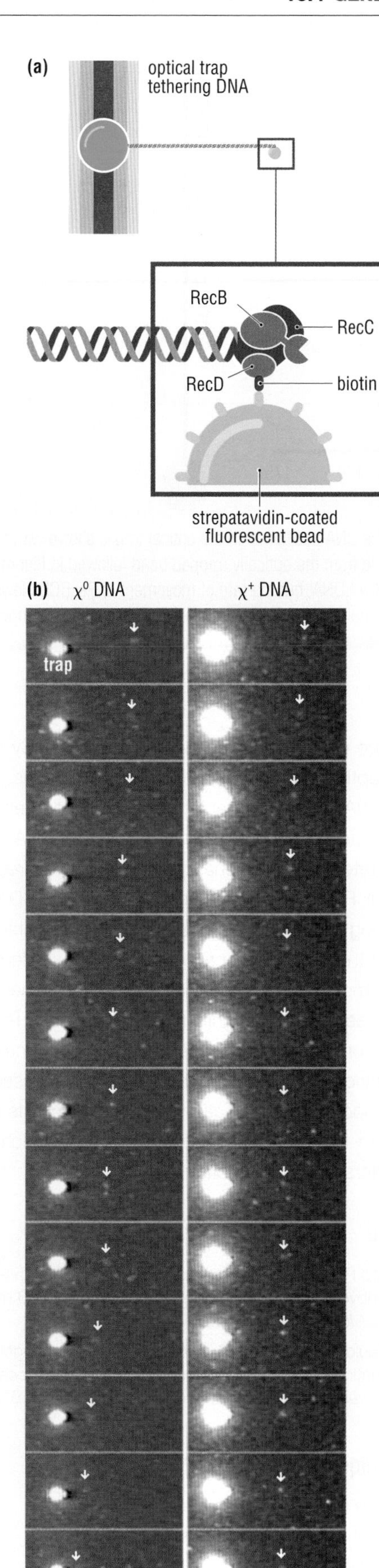

Chi sites do not stimulate RecBCD recombination in the absence of RecD

Biochemical studies of RecBC subsequently revealed that there was actually a third polypeptide present in complexes containing RecBC: this turned out to be the RecD protein. The *recD* gene was not identified in screens for recombination-defective mutants because *recD* mutants are hyper-recombinogenic, due to constitutive, unregulated activation of *recBC*-mediated recombination. Indeed, when researchers looked for revertants of the *recBC* recombination defects, many of those revertants that were hyper-recombinogenic turned out to be mutations in the *recD* gene. It was also observed that the presence of a Chi sequence did not increase recombination in the hyper-recombingenic *recD*⁻ mutants. This lack of Chi stimulation in the *recD*⁻ mutants suggested a direct link between the Chi sequence and modulation of RecBCD activity, and eventually led to the explicit suggestion that Chi activates recombination by inactivating RecD. As such, hyper-recombinogenic *recD*⁻ variants represent loss of function mutations in the RecD protein; in these variants, recombination is relieved of the inhibitory presence of RecD protein.

Chi stimulates 3′ single-stranded DNA production by RecBCD

How does Chi modulate the activity of RecBCD to stimulate recombination? As mentioned earlier, biochemical experiments had revealed that RecBCD has helicase and nuclease activity. This protein complex enters DNA at a double-strand break and moves through the duplex as a helicase/nuclease, degrading DNA as it proceeds. A key finding was that when RecBCD encountered a Chi site, the nuclease activity was decreased such that a 3′ single-stranded segment of DNA persisted. This single-stranded DNA is the substrate that RecA uses for recombination. Thus Chi stimulates homologous recombination by switching the activity of RecBCD from a DNA-degrading enzyme to one that makes 3′ single-stranded DNA.

Chi recognition does not result in ejection of RecD from RecBCD

What is the molecular mechanism by which Chi switches the activity of RecBCD? A popular (and entirely reasonable) hypothesis

Figure 2 Video images of the translocation of a single RecBCD molecule along a DNA. (a) Schematic of the experiment. Either χ^0 or χ^+ DNA is attached to a large bead (turquoise) that can be stably positioned in the optical trap (red), and the DNA is extended by flow to the right. The RecD subunit of RecBCD is labeled with biotin and thus associates with a small fluorescent nanoparticle covered with streptavidin (green). The nanoparticle tracks the movement of RecBCD along DNA towards the large optically 'trapped' sphere (b) Time lapse photography of the movement of the RecBCD bead along the DNA. The nanoparticle is at the tip of the arrow. The rate of RecBCD movement is plotted in Figure 3.

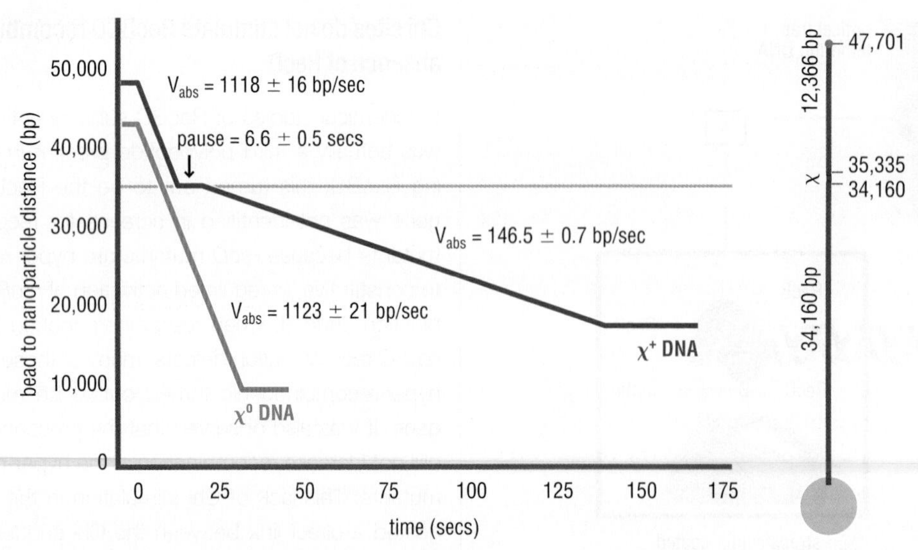

Figure 3 Plot of the translocation of a single RecBCD complex along DNA. (right) The size of the DNA positioned in the optical trap is shown with the positions of the Chi sites. (left) The position of the RecBCD molecules associated with the nanoparticle from the optically trapped bead followed in Figure 2 is plotted as a function of time. RecBCD initially moves at the same rate on χ^o (blue line) and χ^+ (pink line) DNA, but the rate of movement of RecBCD slows at the χ^+ site (downward arrow), reflecting changes in the RecBCD nuclease activity. Most notably, RecD remains associated with RecBC and continues to move along DNA, albeit at a slower rate.

was that when RecBCD encountered Chi as it moved along DNA, RecD was ejected from the complex, giving rise to a RecBC machine that had decreased degradation activity (while still having the helicase activity), thus resulting in the generation of 3′ single-stranded tails. Recent single molecule experiments by Stephen Kowalczykowski and colleagues have shown that this is not the case, however, since RecD remains associated with the RecBCD complex even after encountering Chi.

To follow RecBCD movement along single DNA molecules, Kowalczykowski's group used an 'optical trap' to hold a single DNA molecule in a fixed position such that the movement of proteins (the RecBCD complex) along the immobilized DNA could be analyzed. Long, linear, ~ 50 kb DNA molecules that either contained Chi sites (here called χ^+) or did not contain Chi sites (here called χ^o) were covalently linked to large beads, which were then positioned by 'optical tweezers' in a flow cell such that the DNA was extended by the flow to the right (Figure 2a); the Chi sites (in the χ^+ DNA) are found at sites proximal to the open right end of the DNA (see Figure 3). The movement of RecD can be specifically followed as it is tagged with a fluorescent nanoparticle (bead). The fluorescent complex enters the DNA at the exposed right duplex end and travels to the left (Figure 2b), towards the χ sites and the large 'optically trapped' bead. (The RecBCD complex is fluorescent because RecD forms part of it.) The positions of the beads were followed over time and those trajectories plotted as in Figure 3.

The initial rate of movement of RecBCD on both the χ^+ and χ^o DNA was virtually the same (Figure 3). Indeed, the speed of RecBCD movement along the χ^o DNA lacking the Chi sites did not change over the entire length of the DNA. By contrast, RecBCD abruptly changed speed at the Chi sites as it moved along the χ^+ DNA. Most importantly, however, the fluorescently labeled RecD protein remained associated with the DNA (and thus RecBC) even after the abrupt change in speed, thus directly demonstrating that RecD is not ejected from the RecBCD complex on encountering Chi. The researchers propose that this change in speed (and the associated production of free single-stranded DNA) might simply reflect conformational rearrangements in the complex that result from recognition of Chi by RecD. This example nicely demonstrates how single-molecule experiments allow direct visualization of molecular events and can be used to test very specific biochemical predictions. New experiments can now be designed to further probe the mechanism by which Chi site recognition alters RecBCD function.

Find out more

Handa N, Bianco PR, Baskin RJ, Kowalczykowski SC. Direct visualization of RecBCD movement reveals cotranslocation of the RecD motor after χ recognition. *Molecular Cell*, 2005; **17**: 745–750.

Sprague KU, Faulds DH, Smith GR. A single base-pair change creates a Chi recombinational hotspot in bacteriophage lambda. *Proceedings of the National Academy of Sciences of the U.S.A.*, 1978; **75**: 6182–6186.

Related techniques

Phage lambda (as a model organism); Section 19.1

E. coli (as a model organism); Section 19.1

Plaque assays; Section 19.2

Fluorescence microscopy; Section 19.13

16.5 THE MECHANISM OF DNA STRAND PAIRING AND EXCHANGE

A central process in homology-directed repair is the sequence-specific recognition that occurs between single-stranded DNA and DNA duplexes, and the exchange of the single strand with one of the strands of the duplex to form heteroduplex DNA. This step, as we have seen, is called synapsis. In this section, we describe the molecular components that are the key players in mediating synapsis.

DNA strand pairing is promoted by specialized recombinases

Synapsis involves adenosine triphosphate (ATP)-dependent, sequence-specific recognition events between single-stranded DNA and DNA duplexes, and the exchange of the single strand with one of the strands of the duplex to form heteroduplex DNA. These reactions are mediated by the **strand-exchange recombinases**. The best-understood of these are the **RecA** proteins of bacteria and the homologous **Rad51** protein in eukaryotes; homologous proteins also exist in archaea.

Bacterial cells can live without RecA, and yeast cells can live without Rad51, but they grow poorly and are highly sensitive to DNA-damaging agents. Mammals without RAD51 die *in utero*. These growth defects likely reflect the fact that homology-dependent repair reactions are required to reassemble broken replication forks resulting from DNA damage, which we will discuss in Section 16.7.

A second specialized eukaryotic recombinase, Dmc1, collaborates with Rad51 to promote strand exchange during meiotic recombination. The RecA, Rad51, and Dmc1 proteins are closely related structurally (Figure 16.8), and the mechanisms by which they promote strand exchange are the same. However, these mechanisms have been most extensively studied for the bacterial RecA strand-exchange recombinase, so we shall focus our discussion on RecA. (In addition, unlike RecA, which, as we saw in Section 15.8, promotes the self-cleavage of particular repressors, including LexA, thereby controlling the expression of DNA repair genes during the bacterial SOS response to DNA damage, Rad51 and Dmc1 do not act as co-proteases.)

The key step in the strand-exchange process is the identification of homology between an invading single strand and a DNA duplex. Strand exchange begins with the loading of the RecA monomers onto single-stranded DNA to form the **presynaptic filament** (Figure 16.9, step 1). This process requires ATP, which binds at the interface between the monomers of RecA and drives a conformational change in the RecA molecules that is essential for DNA binding by assembling RecA filament. DNA binding in turn promotes the assembly of further RecA monomers; therefore, the assembly of the presynaptic filament occurs through cooperative binding between RecA, ATP, and DNA.

Binding of the RecA nucleoprotein filament to the DNA in the invaded intact duplex distorts the duplex DNA molecule so that stretches of extended and untwisted DNA alternate with three-nucleotide segments that assume the conformation of normal B-DNA. Consequently, the whole molecule is extended by about 50% compared with standard B-form DNA (Figure 16.9, step 2). This arrangement is thought to provide the basis for accurate pairing with the recipient DNA duplex during strand exchange.

Untwisting of the extended duplex DNA is thought to enable the interrogation of the duplex for DNA sequence homology by the invading single strand, while the three-nucleotide segments of B-DNA promote correct base-pairing between

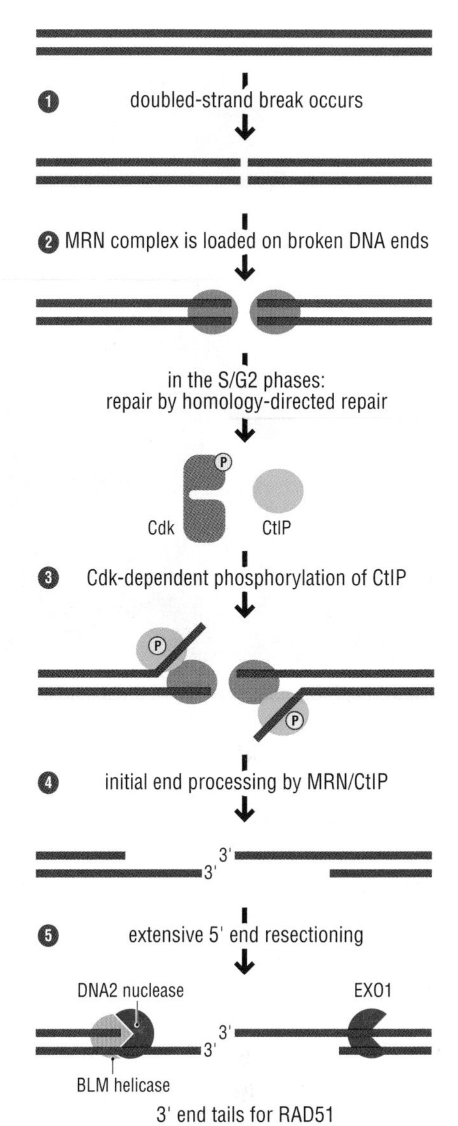

Figure 16.7 Schematic diagram of the generation of 3′ single-stranded tails in eukaryotes. A DNA duplex is broken by a double-strand break. In S/G2, CtIP is phosphorylated by Cdk1 and associates with MRN. The 5′ ends are further degraded to expose longer 3′ ends by an exonuclease (EXO1) or by a complex containing BLM, a RecQ helicase, and the nuclease DNA2

From Bernstein, K, and Rothstein, R. At loose ends: resecting a double-strand break. *Cell*, 2009; **137**: 807–809.

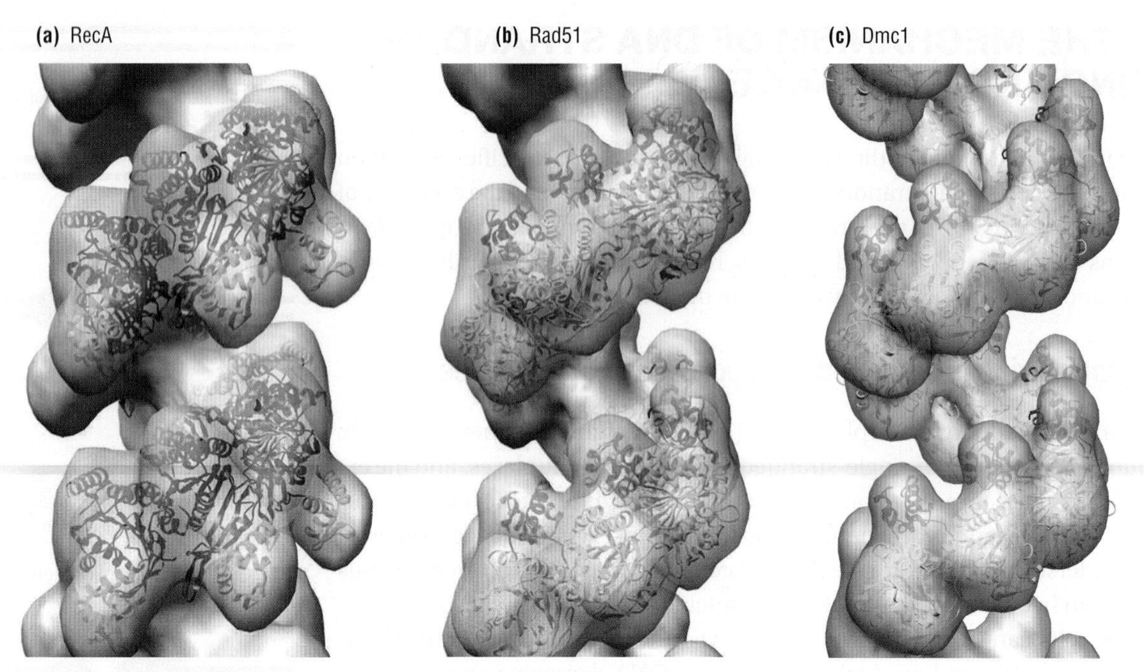

Figure 16.8 Structures of RecA, Rad5, and Dmc1 filaments. These models of the surfaces (pale gray) of *E. coli* RecA (a) and *S. cerevisiae* Rad51 (b) and Dmcl (c) filaments bound to DNA were generated by three-dimensional reconstruction of many electron micrograph images of these proteins bound to DNA. Structures of RecA and Rad51 determined by crystallography are shown within each model.

Structures provided by Edward Egelmann.

the invading and the recipient strands, since incorrect pairing would disrupt the stacking interactions of the DNA.

The RecA protein plays very little part in the sequence-specific binding interactions between the presynaptic filament and the recipient duplex. Instead, these interactions are largely dependent on the base-pairing interactions between the two DNA molecules. Thus binding is strongly disfavored by mismatched bases, which is thought to account for the requirement for very high sequence similarity between the interacting DNA molecules, and hence the accuracy of repair.

Once homologous pairing is established between the single strand of the nucleofilament and the complementary strand of the duplex DNA, strand exchange follows, and heteroduplex DNA is formed (Figure 16.9, step 3). This is accompanied by hydrolysis of ATP by the recombinase, the disassembly of the nucleoprotein filament, and the displacement of the unpaired strand of the original duplex DNA to form the D-loop, in which the 3′ end of the invading single strand can prime DNA synthesis.

Accessory proteins assist the strand-exchange recombinases

Although RecA and Rad51 recombinases can execute homologous pairing and strand exchange in the absence of other proteins *in vitro*, many other proteins are involved in recombination *in vivo*. For example, single-stranded DNA binding proteins prevent the formation of secondary DNA structures that would obstruct interaction of the single-stranded DNA with the strand-exchange recombinase; these proteins, however, must be removed to allow the recombinase protein to bind to form the presynaptic filament. When single-stranded DNA persists, it recruits accessory proteins that bind to the junction of single-stranded and double-stranded DNA. These proteins then catalyze the loading of the recombinase and the displacement of the single-strand binding protein.

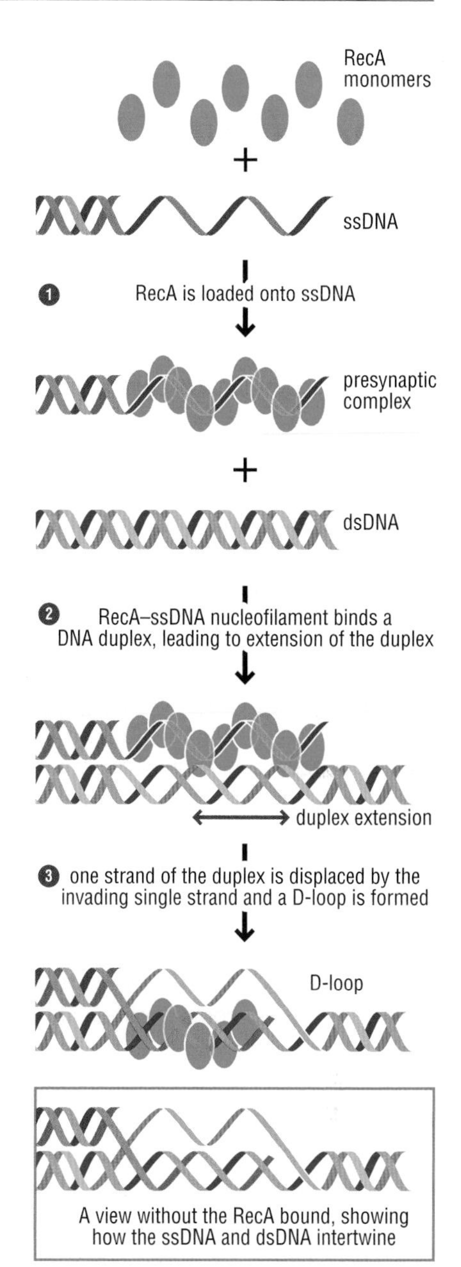

Figure 16.9 Strand exchange by the recombinase nucleoprotein filament. A broken DNA end with a 3′ single-stranded tail (gray) and RecA monomers (orange circles) are shown. RecA binds helically around the single-stranded DNA (ssDNA) to form a nucleoprotein filament called the presynaptic filament that can invade an intact duplex DNA (dsDNA; pink). Binding of the RecA–single-strand complex to the duplex extends the duplex such that the bases in the single strand and base pairs in the duplex have the same internucleotide spacing, forming a D-loop as shown in Figure 16.4.

Among the accessory proteins in mammals are breast cancer type 1 susceptibility protein (BRCA1), breast cancer type 2 susceptibility protein (BRCA2), and 'partner and localizer of BRCA2' (PALB2), a protein that mediates the interaction between BRCA1 and BRCA2. Strikingly, BRCA2 and BRCA1 were first identified as gene products that, when defective, predispose individuals to breast and ovarian cancers.

We now know that BRCA1, BRCA2, and PALB are all essential for homology-directed repair. Why mutations in these genes specifically cause predisposition to breast and ovarian cancer, rather than other kinds of cancer, is not known. However, it is not unusual for a protein with a very general role in the maintenance of genome stability to give rise to a surprisingly specific range of tumor types in families in which it is defective.

As we saw in Section 15.9, BRCA1 is one of the early proteins recruited to a double-strand break via an ATM response, thus recruiting PALB2 and BRCA2. Moreover, BRCA1 is a ubiquitin ligase and is thought to be important in the regulated recruitment of repair proteins to sites of DNA damage. This recruitment is driven by the addition of ubiquitin tags to molecules at the damage site that are recognized by components of repair pathways.

Several other ubiquitin ligases are also known to assemble at DNA double-strand breaks, although the exact role of each is not yet clear.

BRCA1 influences repair by directly interacting with CtIP, the enzyme involved in resection of 5′ ends to make their cognate 3′ single-ended tails, while BRCA2 interacts directly with RAD51. In Experimental approach 16.2, we explore how single-molecule studies demonstrate that BRCA2 stimulates RAD51 action by promoting the binding of RAD51 to single-stranded DNA and inhibiting its binding to double-stranded DNA.

➲ We see in Chapter 15 how the ubiquitination of the proliferative cell nuclear antigen (PCNA) sliding clamp in response to DNA damage facilitates the action of translesion polymerases at stalled replication forks.

16.6 GENE CONVERSION THROUGH HOMOLOGY-DIRECTED REPAIR

Homology-directed repair at a double-strand break can change gene sequence by replacing one version of a gene with another

We have just seen that, in order to be aligned by recombinases, the participating DNA molecules must be highly similar in sequence; but they need not be identical. Thus although the template for homology-directed repair of a double-strand break is *usually* the sister chromatid of the damaged DNA molecule, this is not always the case. For breaks that occur in G0, G1, and early S phases of the cell cycle, when sister chromatids are not available, a homologous chromosome may serve as the template.

Sister chromatids are exact replicas of each other (see Section 4.2), and if one serves as a template for repair of the other, the damaged chromatid will in effect be restored to its pre-damage sequence. Homologs, on the other hand, generally

16.2 EXPERIMENTAL APPROACH

Modulating the activities of RecA and RAD51

The function of a gene product is rarely understood from a single assay. Rather an ever-clearer picture emerges from a mosaic of insights that come from a variety of assays and approaches. This has been particularly true for the recombinases RecA and RAD51 (in bacteria and eukaryotes, respectively). While these proteins have been shown to have an astounding number of measurable activities, it has been an ongoing challenge to obtain a comprehensive picture of how these activities interface in the cell.

The *recA* gene was the first recombination gene to be identified. To search for genes involved in recombination, Alvin Clark and Ann Dee Margulies mutagenized cells and looked for variants in which recombination between the chromosome and a DNA fragment introduced into the recipient cell by conjugation was defective. Using this approach, they isolated several mutants, all of which were within the *recA* gene. These mutants also proved to be sensitive to a variety of DNA-damaging agents, establishing an early link between recombination and DNA repair. Subsequent work showed that the RecA protein also has functions in DNA strand exchange, signaling in the bacterial SOS response through proteolysis, and in mutagenic lesion bypass.

RecJ stimulates strand exchange by RecA

How can a single protein carry out so many different functions? Part of the answer comes from an understanding of its interactions with other proteins. Even though RecA is crucially involved in the DNA strand exchange event that plays a pivotal role in recombination, it does not act alone. Indeed, the type of genetic screens that led to the discovery of *recA* also resulted in the identification of other *rec* mutants that showed similar recombination defects. One of these genes encodes the RecJ protein, which was characterized by Susan Lovett and colleagues. Although RecJ protein has 5′ to 3′ exonuclease activity – which can generate the single-stranded DNA that is a substrate in recombination – Lovett considered the possibility that RecJ might also stimulate the strand-exchange process itself.

To test this idea, the group examined RecA-mediated strand exchange using purified components *in vitro*. In these assays, linear duplex DNA was mixed with complementary single-stranded circular DNA, and the mixture then incubated with RecA in the presence or absence of RecJ as shown in Figure 1. The extent of strand exchange was determined by measuring the amount of nicked circular duplex DNA formed during the incubation (resulting from the repartitioning of a strand of DNA from the linear duplex to the circular DNA). In the presence of RecJ, the band corresponding to the nicked circle duplex was detected within 15 minutes of starting the reaction, while in the absence of RecJ this band was not seen until 60 minutes later. The interpretation of this result is that RecJ stimulates strand exchange by degrading the strand that would otherwise compete with the exchange reaction by reannealing with the displaced strand.

The BRC4 domain BRCA2 stimulates strand exchange by RAD51

The mammalian homolog of RecA is RAD51. Like RecA, the monomer RAD51 assembles to form filaments along the DNA and promotes strand exchange between a single-stranded DNA and its complementary strand in a DNA duplex. As is true in bacteria, other proteins in eukaryotes are also required for homologous recombination, including BRCA2 in mammalian cells. BRCA2 was so named because it was first identified as a 'disease' gene associated with a predisposition to breast cancer. The finding that BRCA2 accumulated at double-strand DNA breaks *in vivo* suggested that it also plays a role in recombination.

What role does BRCA2 play in homologous recombination? The protein has two notable features: a DNA-binding domain and two regions that interact with RAD51 via short domains called BRC repeats. We now know that the BRC repeats are important because mutations in these repeats block the interaction of BRCA2 with RAD51 in families with a predisposition to breast cancer.

The large size of BRCA2 (3418 amino acids) made it difficult to study the protein activity *in vitro*, as it proved difficult to express and purify. Stephen Kowalczykowski and colleagues gained significant insight into the role of BRCA2 in recombination by studying RAD51-mediated reactions *in vitro* in the presence of a fragment of BRCA2 containing a BRC domain, BRC4. Kowalczykowski adapted the classic assay used to examine RecA-mediated strand exchange (see Figure 1) to measure the effects of BRCA2–BRC4 on RAD51-mediated strand exchange (Figure 2). As before, a labeled, linear, duplex DNA (asterisks) and complementary, single-stranded, circular DNA were incubated with RAD51 and ATP and the formation of joint molecules (formed as strand exchange begins) and a nicked circular duplex DNA (the final reannealed exchange product) were measured. The addition of the BRC4 protein domain stimulated RAD51-mediated strand exchange in these *in vitro* reactions, whereas no stimulation was seen when a mutant version of BRC4 (Δ1524–30) was used in the assay.

The BRC4 domain has several activities

Further analysis showed that BRC4 stimulates recombination through two distinct mechanisms. First it stabilizes the interaction of RAD51 with single-stranded DNA as evaluated in an electrophoretic gel mobility shift assay (EMSA).

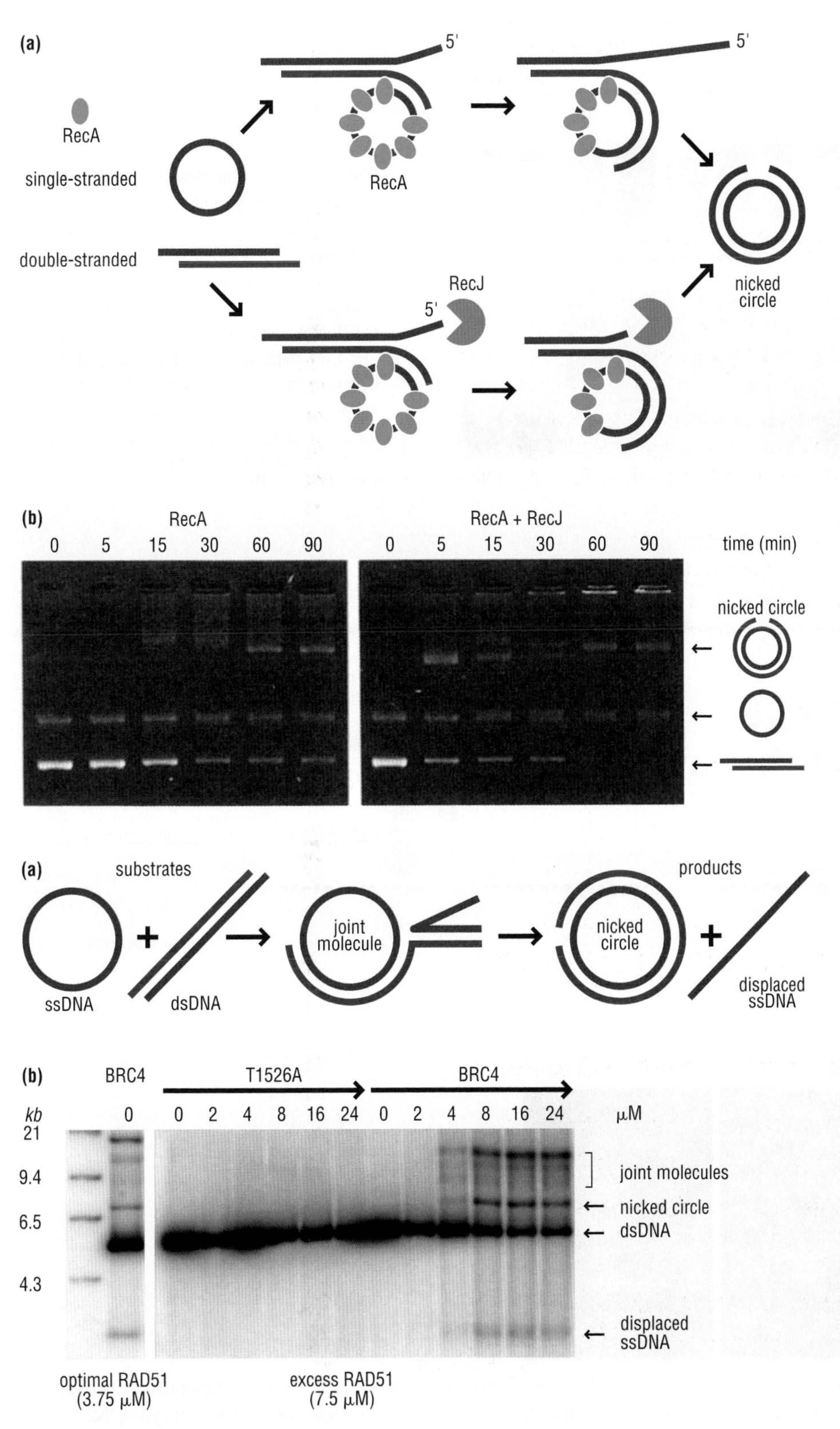

Figure 1 RecJ stimulates RecA-mediated strand exchange. (a) Schematic diagram of intermediates in the strand exchange assay in presence and absence of RecJ exonuclease. Single-stranded circular DNA (black circle) was mixed with double-stranded DNA (black and pink) in the presence of RecA (orange), with (bottom pathway) or without (top pathway) RecJ (blue). RecA faciliates the transfer of the pink strand from the double-stranded DNA to the circular DNA resulting in a nicked circle (black and pink). (b) *In vitro* assay for RecA strand exchange activity. A single-stranded circular DNA (gray) is incubated with a linear double-stranded DNA molecule, with one labeled strand (pink) and the other (gray) unlabeled. In the presence of RecA and with or without RecJ as described in panel (a). RecA facilitates strand exchange, measured as the accumulation of the nicked circle product, though the reaction proceeds more quickly in the presence of RecJ protein.

Reproduced from Corrette-Bennett and Lovett, Enhancement of RecA strand-transfer activity by the RecJ exonuclease of *Escherichia coli*. *Journal of Biological Chemistry*, 1995;**270**:6881–6885.

Figure 2 The BRC4 domain of BRCA2 stimulates RAD51-mediated strand exchange. (a) Schematic diagram of intermediates in the strand exchange assay. Incubation of an end-labeled double-stranded duplex (dsDNA, pink and black) with a complementary single-stranded circle (ssDNA) initially results in formation of joint molecule (JM) intermediates and culminates in the formation of labeled nicked circle (NC) and labeled displaced single-stranded linear DNA. (b) *In vitro* strand exchange assay wherein all lanes contain RAD51 and different concentrations of the mutant BRC4 domain T1526A or different concentration of the wild-type BRC4 domain. Impressive increases in the production of JM and NC, indicative of strand exchange are observed when the wild-type BRC4 domain is added.

Reproduced from Carreira *et al.*, The BRC repeats of BRCA2 modulate the DNA-binding selectivity of RAD51. *Cell*, 2009;**136**:1032–1043.

As we see, the addition of increasing amounts of BRC4 increases the extent of binding of RAD51 to the single-stranded DNA (Figure 3). A second role BRC4 plays is to inhibit the non-productive interaction of RAD51 with double-stranded DNA. Using a single-molecule approach, Kowalczykowski and colleagues directly examined the binding of RAD51 to double-stranded DNA with and without BRC4 (Figure 4). Fluorescently labeled RAD51 protein was added to a long (50 kb) double-stranded DNA attached to an optically trapped bead. As we saw in Experimental approach 16.1, the position of the bead can be manipulated by

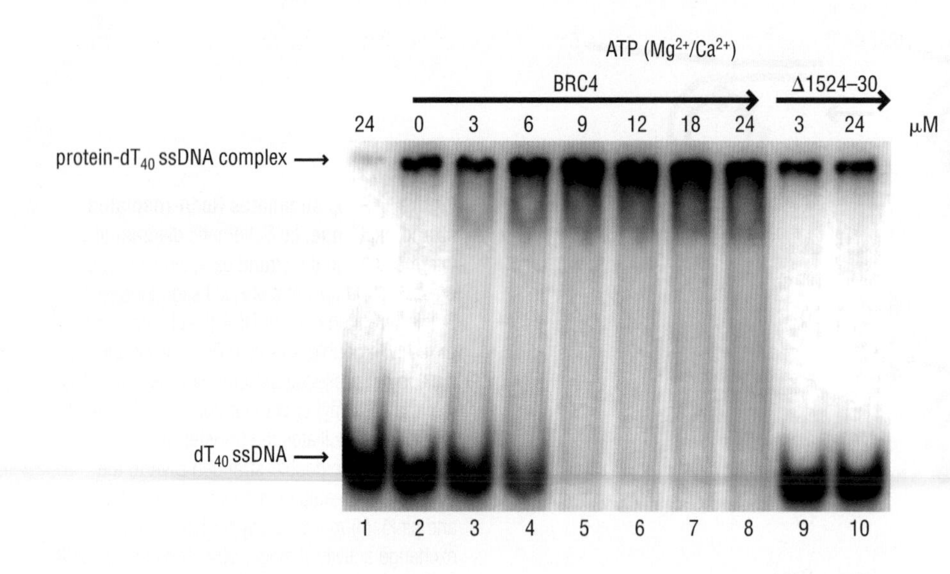

Figure 3 BRC4 promotes the binding of RAD51 to single-stranded DNA. Gel showing RAD51 incubated with either GST-BRC4 or GST-BRC4 Δ1524–30 prior to incubation with [^{32}P]-labeled single stranded poly-deoxythimidine DNA (dT_{40} ssDNA) for one hour in the presence of ATP, Mg^{2+}, and Ca^{2+}. In the presence of GST-BRC4, the majority of the dT_{40} single-stranded DNA is shifted to the top of the gel, while in the mutant (Δ 1524–30) less complex is formed.

Reproduced from Carreira *et al.*, The BRC repeats of BRCA2 modulate the DNA-binding selectivity of RAD51. *Cell*, 2009;**136**:1032–1043.

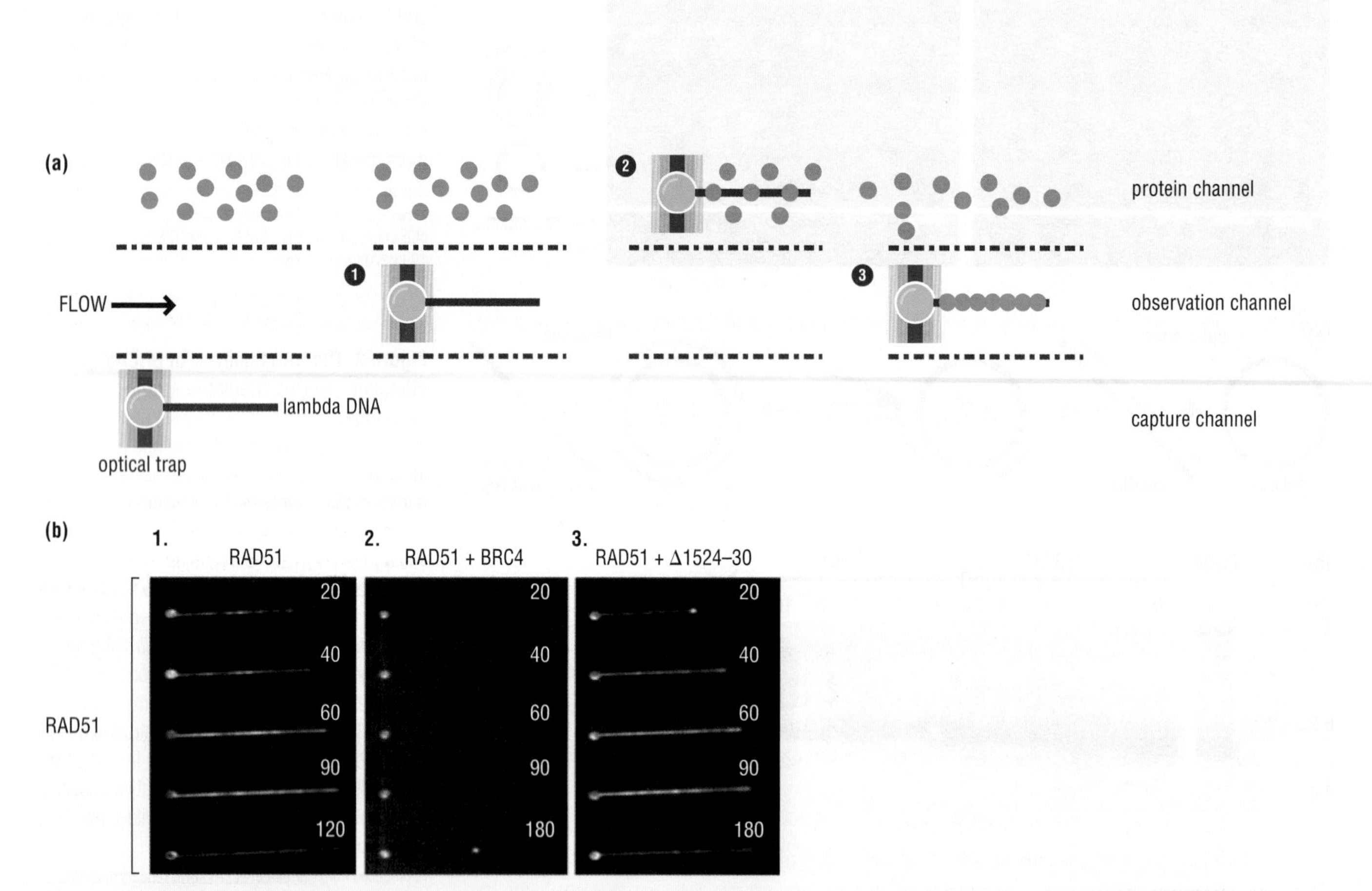

Figure 4 BRC4 inhibits the binding of RAD51 to double-stranded DNA. (a) Schematic of experimental design to examine binding of fluorescently labeled RAD51 to double-stranded DNA. In the capture channel (bottom row), a 50 kb double-stranded DNA, is tethered to a streptavidin-coated bead (green) that can be manipulated with optical tweezers (red). (1) One single-labeled DNA molecule is positioned in the observation channel of a flow cell (middle row); pictures are taken over time. (2) The DNA strand is moved into a protein channel containing fluorescently labeled RAD51 protein (orange) (top row). (3) The DNA is moved back to the observation channel, and a strong RAD51 specific fluorescent signal will be seen on the DNA if the RAD51 was able to bind the DNA. (b) Single molecule analysis of the effect of BRC4 on RAD51 binding to double-stranded DNA. (1) RAD51 binding to double-stranded DNA is imaged above over a time course from 20 to 120 minutes. (2) When BRC4 is added to the reaction containing RAD51 (as above), no RAD51 fluorescence coating of the DNA is observed. (3) When the mutant BRC4 fragment BRC4 Δ1524–30 is added, RAD51 coats the double-stranded DNA effectively.

Reproduced from Carreira *et al.*, The BRC repeats of BRCA2 modulate the DNA-binding selectivity of RAD51. *Cell*, 2009;**136**:1032–1043.

optical tweezers in a flow cell. In this experimental set-up, the optical bead and attached DNA was moved using the optical tweezers into a flow cell containing buffer (step 1). The DNA was then moved into a channel that contains fluorescent RAD51 (with a distinct fluorescent signature) in the presence or absence of the BRC4 domain (step 2). After incubation with RAD51, the DNA strand was then moved back into the observation channel, to see whether the RAD51 had bound to the DNA (step 3). RAD51 binding to double-stranded DNA was observed in the absence of BRC, but not when the BRC4 domain was included. As a control, the mutant protein, BRC4 Δ1524–30, fails to inhibit the binding of RAD51 to double-stranded DNA.

Together these experiments demonstrated that the BRC4 domain affects RAD51 activity in at least two ways. These experiments also highlight the challenges of any one experimental approach in yielding the 'mechanism' of action of a particular protein. Molecular insight comes most typically from the gradual accretion of clues that assemble bit by bit into an ever more detailed model.

Find out more

Carreira A, Hilario J, Amitani I, *et al*. The BRC repeats of BRCA2 modulate the DNA-binding selectivity of RAD51. *Cell*, 2009;**136**:1032–1043.

Clark AJ, Margulies AD. Isolation and characterization of recombination-deficient mutants of *Escherichia coli* K12. *Proceedings of the National Academy of Sciences of the U S A.*, 1965;**53**:451–459.

Corrette-Bennett SE, Lovett ST. Enhancement of RecA strand-transfer activity by the RecJ exonuclease of *Escherichia coli*. *Journal of Biological Chemistry*, 1995;**270**:6881–6885.

Related techniques

Agarose gel electrophoresis; Section 19.7
Radioactive labelling of DNA; Section 19.6
Mobility shift assays; Section 19.12
Fluorescence microscopy; Section 19.13

do not have the same DNA sequence throughout: most diploid organisms are likely to be heterozygous at many genetic loci. Thus, when DNA of a homologous chromosome serves as the repair template, the repaired chromosome will acquire a new sequence – that of the homolog. This process is called **gene conversion**: a sequence on one homologous chromosome is converted into the sequence on the other, which itself remains unchanged (Figure 16.10).

Note that this exchange of sequence is specifically targeted to a particular region that is at the site of the double-strand break that initiates homology-directed repair. For example, the 'B' marker on the parental gray duplex in Figure 16.10 in which the double-strand break occurred is converted to the 'b' marker as a consequence of repair. Thus, while one parental duplex contains the B marker and the other parental duplex the b marker, the two progeny duplexes each contain one b marker.

This conversion of markers is a signature of interaction between the two non-identical parental duplexes. However, note that the relationship and identity of the DNA segments that *flank* the double-strand break do not change: the A and C versions of these genes are still on one duplex and the a and c versions on the other.

In addition to its role in repairing damaged chromosomes, this double-strand break directed repair mechanism has been exploited in some organisms to carry out specific changes in gene structure and expression. Let us now consider several of these.

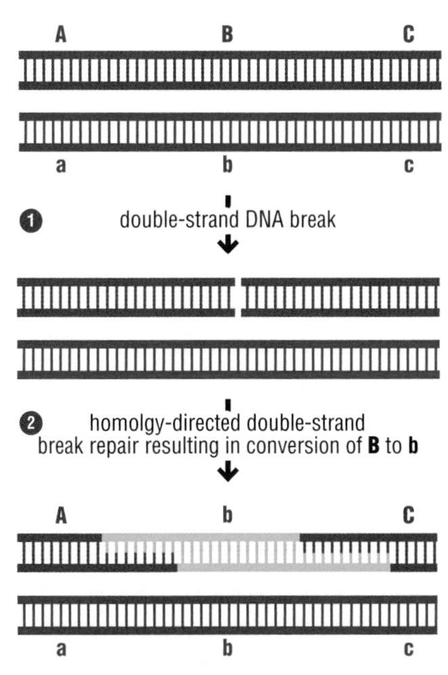

Figure 16.10 Homology-directed repair results in gene conversion. Two homologous parental DNA duplexes are shown, and a double-strand break occurs in the B region of the gray duplex. (2) The broken duplex is repaired by DNA synthesis (green) using the intact homolog (pink) as a template for repair as shown in Figure 16.5. Note that the information in the repaired region is converted from the parental B sequence to the template b sequence.

Yeast mating-type switching is the best-understood example of directed gene conversion

Haploid cells of the yeast *Saccharomyces cerevisiae* exist in two mating types, **a** and α, determined by the expression of either **a** or α mating-type transcriptional

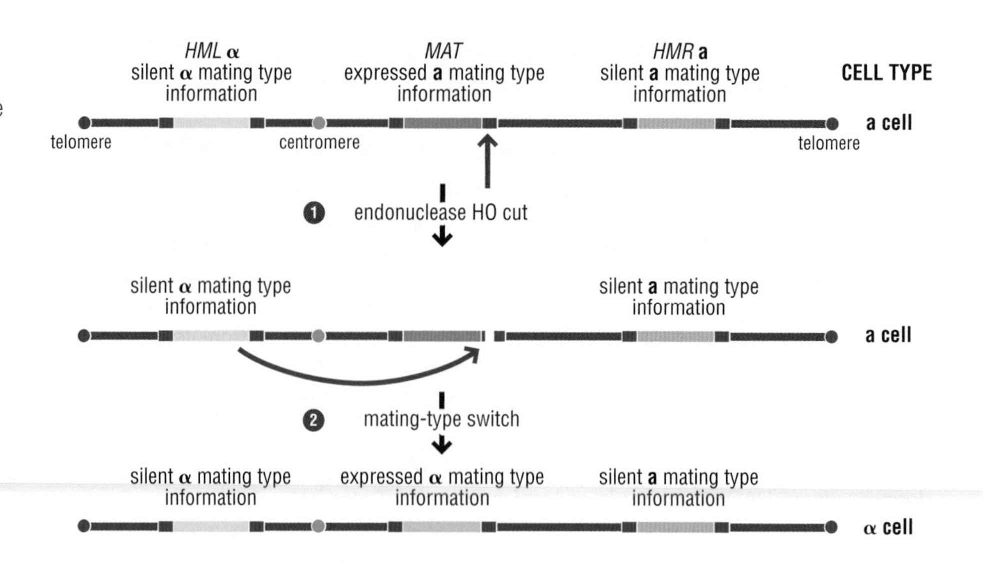

Figure 16.11 Yeast mating-type switching occurs via directional gene conversion. Three different loci on *S. cerevisiae* chromosome III encode the transcriptional regulators that determine whether a yeast cell is of mating type α or **a**. The center *MAT* site is the active site that expresses these transcriptional regulators. The silent flanking sites, *HML* (light green) and *HMR* (light blue) encode α mating regulators and **a** mating regulators, respectively. All the sites have regions of homology (dark blue). A switch in mating type is initiated when the HO endonuclease cleaves the DNA at the *MAT* site (pink arrow up), generating a double-strand break. (2) Initiating at the homologous sequences (dark blue) that bound the mating-type information, homology-directed repair copies α (green) information from the silent *HML* site into the MAT expression site in a gene conversion reaction. The source of information transfer into the expression site alternates between the silent α *HML* and the silent **a** *HMR* donor sites at each mating-type switch.

We learn more about the two mating types of *S. cerevisiae* in Section 9.7.

regulators at a site called *MAT* on chromosome III (Figure 16.11). In some strains of *S. cerevisiae,* the haploid cells can switch mating type by a process involving **directed gene conversion** by homology-directed repair.

Also on the left and right arms of chromosome III are transcriptionally-silenced copies of *MAT*α and *MAT***a**, at sites named *HML* and *HMR*, respectively (see Section 9.9). Mating-type switching between α and **a** cell types occurs by the transfer of information from a silent *HML* or *HMR* locus to the *MAT* expression site.

The transfer of DNA sequence information from the silent sites to the *MAT* expression site occurs by directed gene conversion, which is mediated by the homology-directed double-strand break repair pathway described in Section 16.3. The pathway is activated by the introduction of a double-strand break in DNA at the *MAT* locus by a specialized endonuclease called HO (Figure 16.11, step 1). Using the DNA at either the silent *HML* α or silent *HMR* **a** site as a template, homology-directed repair copies sequence information from one of these sites into the *MAT* site (Figure 16.11, step 2). The *HML*α and *HMR***a** silent loci serve alternately as donors, so that the expression of **a** or α cell-type information, and thus the mating type of the cell, alternates in a cell lineage.

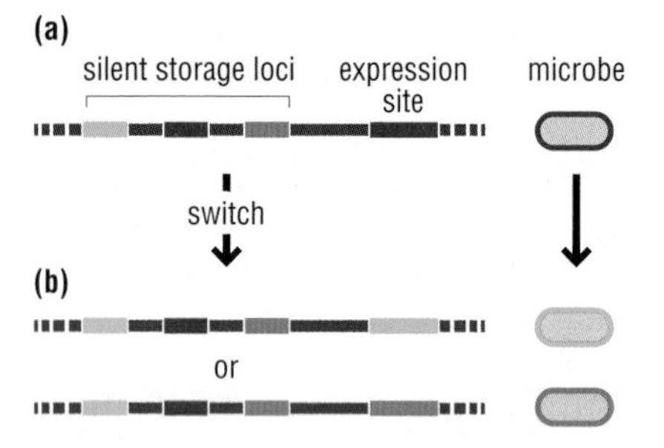

Figure 16.12 A schematic diagram of antigenic variation due to gene conversion. In some pathogens, the changes that occur in their surface protein antigens are due to directed gene conversion. (a) There is a single site from which the surface protein is expressed and a number of silent storage sites containing different versions of the surface protein genes. (b) Variation results from gene conversion of the sequence in the expression site using the sequence of one of the storage sites as template, changing the surface proteins of the organism.

Gene conversion is one cause of antigenic variation

Several pathogenic microorganisms, notably *Neisseria gonorrhoeae*, the bacterium that causes gonorrhea, *Borrelia burgdorferi*, the bacterial agent of Lyme disease, and the protozoan trypanosomes that cause sleeping sickness, can change the cell-surface antigens they express, enabling them to evade the host's immune response. The genomes of these organisms contain an active gene-expression site from which the cell-surface protein is synthesized and also contain many silent genes encoding variants of this protein. During the course of an infection the gene in the expression site can be replaced by gene conversion with a copy from one of the silent sites, resulting in expression of a different surface protein (Figure 16.12). The detailed mechanisms of these reactions remain to be established but likely in part resemble yeast mating-type switching.

16.7 REPAIR OF DAMAGED REPLICATION FORKS BY HOMOLOGY-DIRECTED REPAIR

DNA damage can collapse replication forks

We have already seen in Section 15.7 that, if a DNA replication fork encounters a nick in the DNA, the arm of the fork containing the nick breaks, generating a double-strand break that arrests replication. Abasic sites, base lesions or crosslinks, or defects in the replication complex itself can all block progression of the replication fork. In *E. coli*, it is estimated that 20–50% of the forks that initiate at OriC are blocked by such events. All these causes of the blocking of replication generate regions of single-stranded DNA at the fork that are targeted by repair proteins. Here, we describe how the homology-directed repair mechanisms operate at collapsed or blocked replication forks.

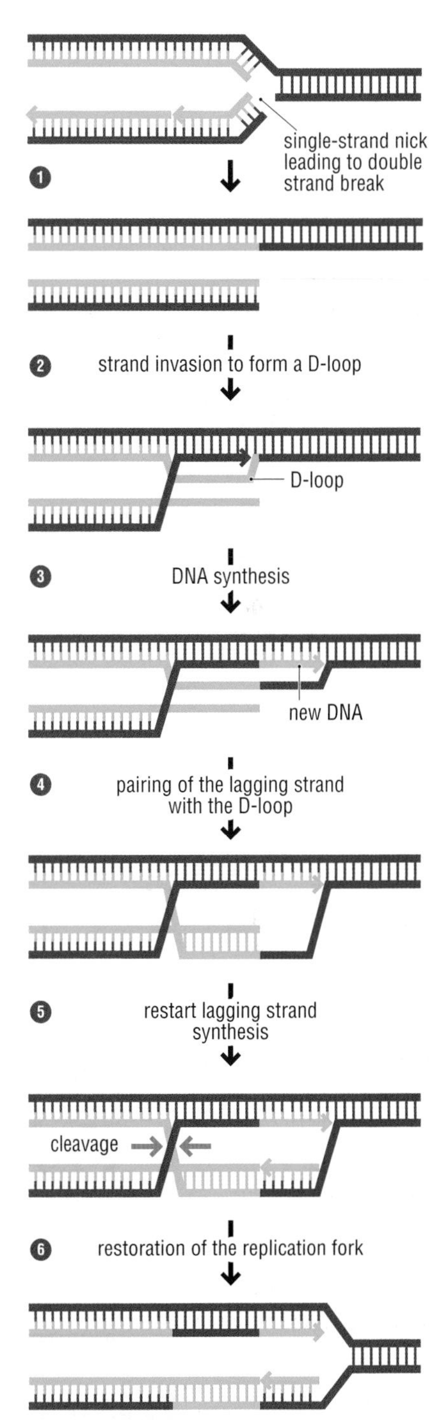

Figure 16.13 A DNA replication fork that has broken because of DNA damage can be rescued by homology-directed repair. A replication fork containing parental (gray), and newly synthesized (green) DNA strands is shown. The fork is broken by a nick in the gray parental strand that serves as the template for lagging strand synthesis.

Homology-directed repair can restore a collapsed replication fork

We saw in Section 15.7 how most collapsed replication forks occur during DNA replication as a consequence of nicks introduced into the phosphodiester backbone by reactive oxygen species that are a result of normal cellular metabolism. In such instances, interactions between the undamaged and damaged arms of the fork mediated by homology-directed repair can restore the fork (Figure 16.13).

So how does such repair unfold? Following degradation to expose 3′ single-stranded DNA, the broken arm of a replication fork (Figure 16.13, step 1) interacts with the undamaged arm through the invasion of a single DNA strand from the damaged arm into the undamaged arm to form a D-loop (Figure 16.13, step 2). This re-establishes a connection between the broken arm and the rest of the fork. This invading strand then serves as a primer for DNA synthesis (Figure 16.13, step 3).

Subsequently, the other single strand of the damaged duplex also specifically interacts with the undamaged arm by binding to the displaced strand of the D-loop (Figure 16.13, step 4). Thus single DNA strands from each parental duplex are reciprocally exchanged between the parental duplexes. A replisome can then be assembled (see below) with the resumption of DNA synthesis on the lagging strand (Figure 16.13, step 5). The exchanged strands that connect the duplexes can then be cleaved by a specialized nuclease called a resolvase (see below) that recognizes such reciprocally exchanged strands to separate the arms of the restored fork (Figure 16.13, step 6). Replication can then be restarted.

The resumption of replication requires reassembly of the replisome. Such **replication restart** may need to occur at many different sites (wherever fork collapse has occurred). Therefore, it doesn't require a specific template DNA sequence that binds a specific initiator protein (in contrast to replisome assembly at a specific site such as *OriC)*. Instead, in bacteria for example, a specialized protein called PriA can recognize a D-loop and promote loading of the replicative helicase and other replication components onto the D-loop, and thus promote initiation from a DNA intermediate assembled by homology-dependent interactions.

Replication fork reversal can repair a blocked replication fork

As we discussed in Section 15.6, translesion polymerases can promote untemplated error-prone DNA synthesis at positions of DNA damage at stalled replication

We discuss the blocking of replication in more detail in Section 15.7.

forks. However, other repair pathways related to the replication fork restart, including one related to the homology-directed repair pathway described above, can also facilitate the resumption of DNA synthesis.

The restoration of a damaged replication fork containing unrepaired lesions, such as an abasic site or pyrimidine dimer, can be achieved by a pathway that involves the pairing of the two newly synthesized strands of DNA with each other instead of with the template strands (Figure 16.14, step 1). In this case the fork

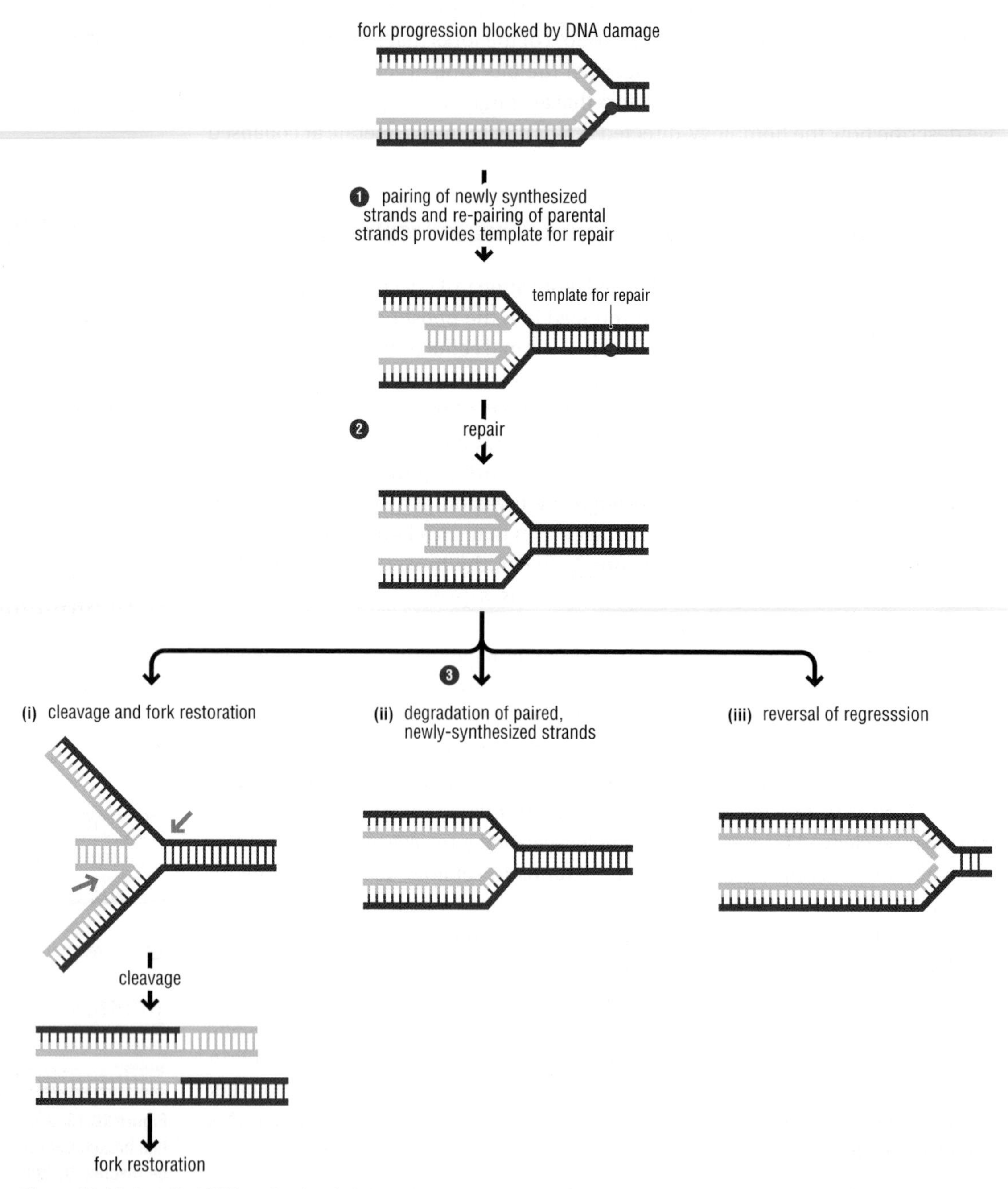

Figure 16.14 A stalled DNA replication fork that has stalled as the result of a lesion in DNA can be rescued by several mechanisms. Parental (gray) and newly synthesized (green) DNA strands at a replication fork are shown after the fork has stalled because of a lesion (red ball). Regression of the fork leads to pairing of the newly synthesized strands and re-pairing of the template strands such that the region containing the lesion is again paired to an intact strand and the parental strand is used as a template for repair.

reverses so that the damaged strand re-pairs with its complementary strand. This strand re-pairing provides another opportunity for lesion repair by systems such as base and nucleotide excision repair, which use the complementary DNA strand as a template (Figure 16.14, step 2).

Once repair has occurred, the replication fork may be restarted in several ways. First, the regressed fork may be cleaved to yield a collapsed fork (Figure 16.14, step 3, i) which may be repaired by homology-directed repair. Such repair allows reassembly of the replisome, as described above. The structure contains DNA strands that are reciprocally exchanged between duplexes, that can be specifically recognized and cleaved by the specialized resolvase nuclease (see later), with the subsequent reinitiation of replication. Second, the newly synthesized annealed strands may be degraded by a nuclease (Figure 16.14, step 3, ii) or third, the regression of the fork may be reversed (Figure 16.14, step 3, iii). In all cases, the replisome must again be assembled as above to enable replication to restart.

16.8 HOMOLOGOUS RECOMBINATION

Having now explored how the cell responds to repair damage to its genome in the form of double-strand breaks, we can now go on to consider how the cell uses the same double-strand breaks to its advantage - in the process of homologous recombination.

Large segments of DNA can be exchanged between two DNA molecules

Homology-dependent repair does not change the relative positions of the DNA segments flanking the initiating double-strand break although Figure 16.15a shows how local gene conversion can occur. However, the cellular machinery that carries out homology-directed repair has evolved into a major mechanism for generating new combinations of genetic information. This process is called **homologous recombination** and can result in the reciprocal exchange of large segments of DNA between homologous DNA duplexes (Figure 16.15b).

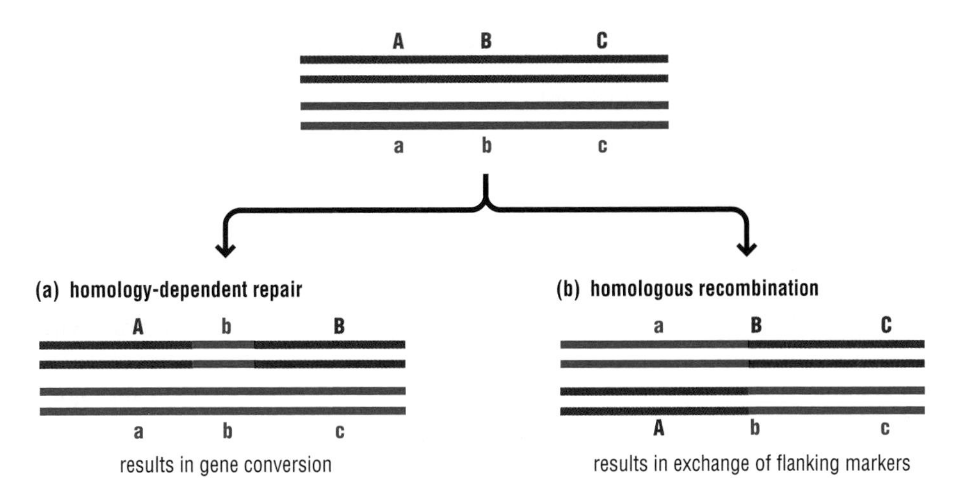

Figure 16.15 Homology-directed repair and homologous recombination differ in their effects of DNA. (a) Homology-directed repair at a double-strand break between A and C results in local gene conversion (B⟶b) but does not change the relationship of the flanking AC and ac markers. (b) Homologous recombination results in the reciprocal exchange of markers on the parental duplexes to yield progeny duplexes containing a segment from each parent; a and C now lie on the same duplex, and A and c are on the other duplex.

Scan here to watch a video animation explaining more about Holliday junction resolution, or find it via the Online Resource Center at www.oxfordtextbooks.co.uk/orc/craig2e/.

Following recombination, the linkage of the markers that were on the same parental duplexes in our hypothetical example in Figure 16.15 (A and C on one parent (gray) and a and c on the other (pink)), is changed in the progeny duplexes to give A and c on one duplex and a and C on the other (Figure 16.15b). Many enzymes involved in homology-directed DNA repair also play a role in homologous recombination.

So how is homologous recombination exploited by the cell? Homologous recombination in meiosis plays a key role in the segregation process that culminates in the production of haploid gametes and also provides another pathway for DNA repair; as such, it underlies most genetic variation in humans and most other eukaryotes. Homologous recombination also contributes to the diversification of bacterial genomes by mediating the incorporation of exogenous DNA during conjugation, transduction, and transformation. This is one way bacteria can acquire new metabolic functions or resistance to an antibiotic.

➜ We discuss the production of haploid gametes in Section 7.7.

Homologous recombination can occur between any DNA molecules with extensive regions of identical sequence (as with two sister chromatids), or very similar sequence (as with two homologs); no specialized recombination sequences are required. Such exchanges generate diversity by reshuffling segments of the genome to make new combinations of existing characteristics, rather than through the introduction of new random mutations or by the translocation of discrete DNA segments, as will be discussed in Chapter 17. Homologous recombination is an occasional accidental event in virtually all cells including the somatic cells of multicellular eukaryotes; however, it is a defined, programmed feature of meiosis.

The formation and resolution of Holliday junctions is the basis for homologous recombination

Homologous recombination occurs by the formation and **resolution** of **Holliday junctions** between homologous chromosomes. A Holliday junction is formed by the reciprocal exchange via D-loop formation by RecA-like recombinase mediated formation (Figure 16.16 step 1) of DNA strands between two duplexes to form regions of heteroduplex DNA (Figure 16.16, step 2). After formation, the two duplexes that have joined to form the Holliday junction are then separated, a step known as **resolution** (Figure 16.16, step 4). Resolution requires cleavage of the DNA strands at the Holliday junction at the point where the molecules are joined, and is promoted by nucleases called **resolvases.** The discovery of Holliday junction resolvases is described in Experimental approach 16.3.

Cleavage of Holliday junctions can result in permanent exchange of DNA between two duplexes

There are two ways in which Holliday junctions can be resolved; only one of these results in the permanent exchange of flanking DNA – that is, homologous recombination – between the participating DNA molecules. It is the location at which the Holliday junction is cut to separate the two molecules that will determine whether exchange occurs between the duplexes.

Holliday junctions adopt a planar, unfolded structure when they interact with recombination proteins (Figure 16.16, step 3). The Holliday junction has two axes of symmetry, one depicted here as horizontal and the other as vertical (Figure 16.16, step 4). If cleavage occurs in the horizontal direction (Figure 16.16, step 4i), the duplexes are separated without exchange of the DNA outside of the Holliday

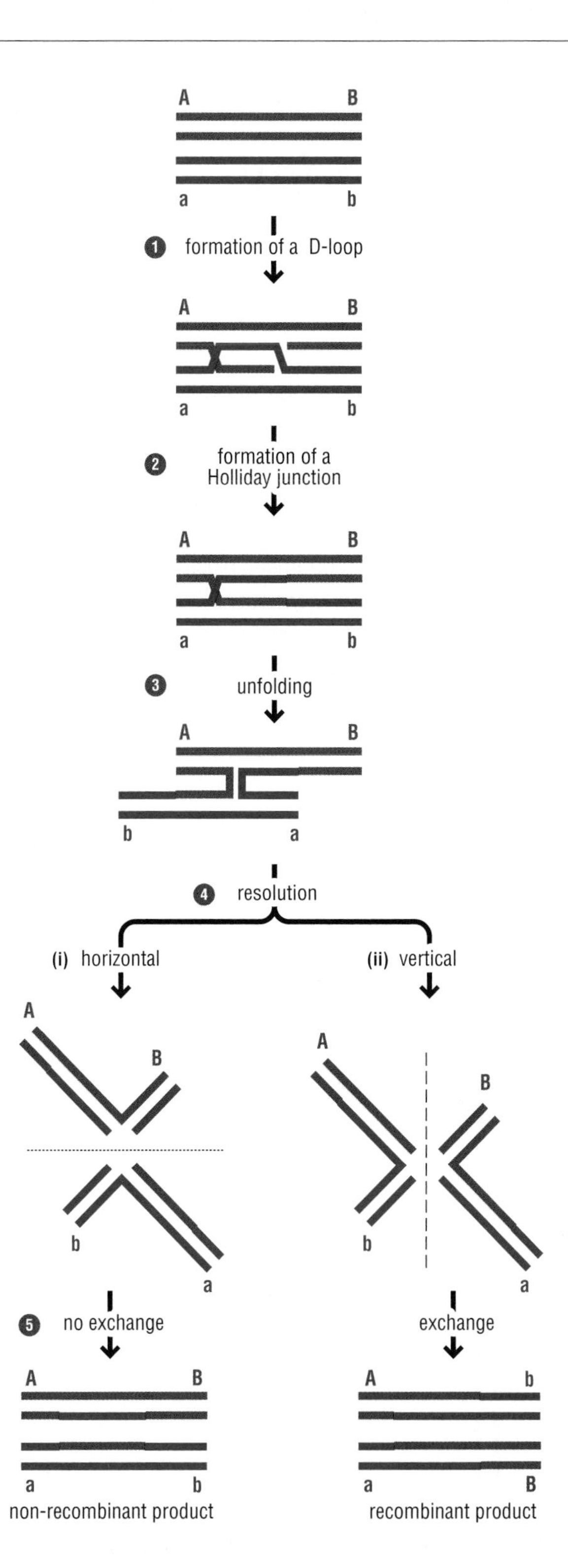

Figure 16.16 Cleavage of a Holliday junction can yield either of two products. A Holliday junction is formed by the reciprocal exchange of DNA strands between homologous duplexes. The position of the junction, and hence the amount of heteroduplex DNA involved, can change (see later). Rotation of the bottom double helix provides an alternative view of the Holliday junction. Note that both structures contain the same homoduplex and heteroduplex regions and that the connection between the duplexes involves four DNA strands. Reorientation of the junction emphasizes two axes of symmetry (lines) in the junction. The junction can be cleaved by the specialized endonucleases called resolvases in either of two planes, indicated by the dotted horizontal and dashed vertical lines, respectively.

junction. These products are called **non-recombinant** (Figure 16.16, step 5i). The parental segments – A and B, and a and b – are still linked to each other, and the only trace of interaction between the parental duplexes is the region of heteroduplex DNA containing one strand from each parental duplex formed by the exchange of strands to form the junction.

In contrast, if resolution occurs in the vertical direction (Figure 16.16, step 4ii), the DNA outside of the Holliday junction is exchanged or crossed over between the two parental duplexes. A and b are now on one duplex and a and B are on the

16.3 EXPERIMENTAL APPROACH

The difficulties in identifying Holliday junction resolvases

The idea that homologous recombination occurs through the formation and resolution of a DNA intermediate involving the reciprocal exchange of DNA strands was proposed entirely based on genetic observations. This intermediate, the so-called 'Holliday junction', was named after Robin Holliday who first proposed the model to explain certain products of fungal meiosis.

Holliday junctions can be cleaved *in vivo*

It only became possible to biochemically analyze Holliday junctions after the bacterial RecA protein was isolated, and Stephen West and colleagues showed using homologous plasmids that the protein promoted the formation of paired DNA molecules. Electron microscopy of such paired molecules confirmed that they had the anticipated structure of a Holliday junction. The West group also asked whether the paired molecules formed *in vitro* could give rise to recombinants when transformed into *E. coli*: could the Holliday junctions be cleaved (resolved)? Indeed, both dimeric and monomeric plasmids were recovered from *E. coli* following the transformation, indicating that resolution could occur in the two different orientations as first suggested by Holliday and as is illustrated in Figure 1a.

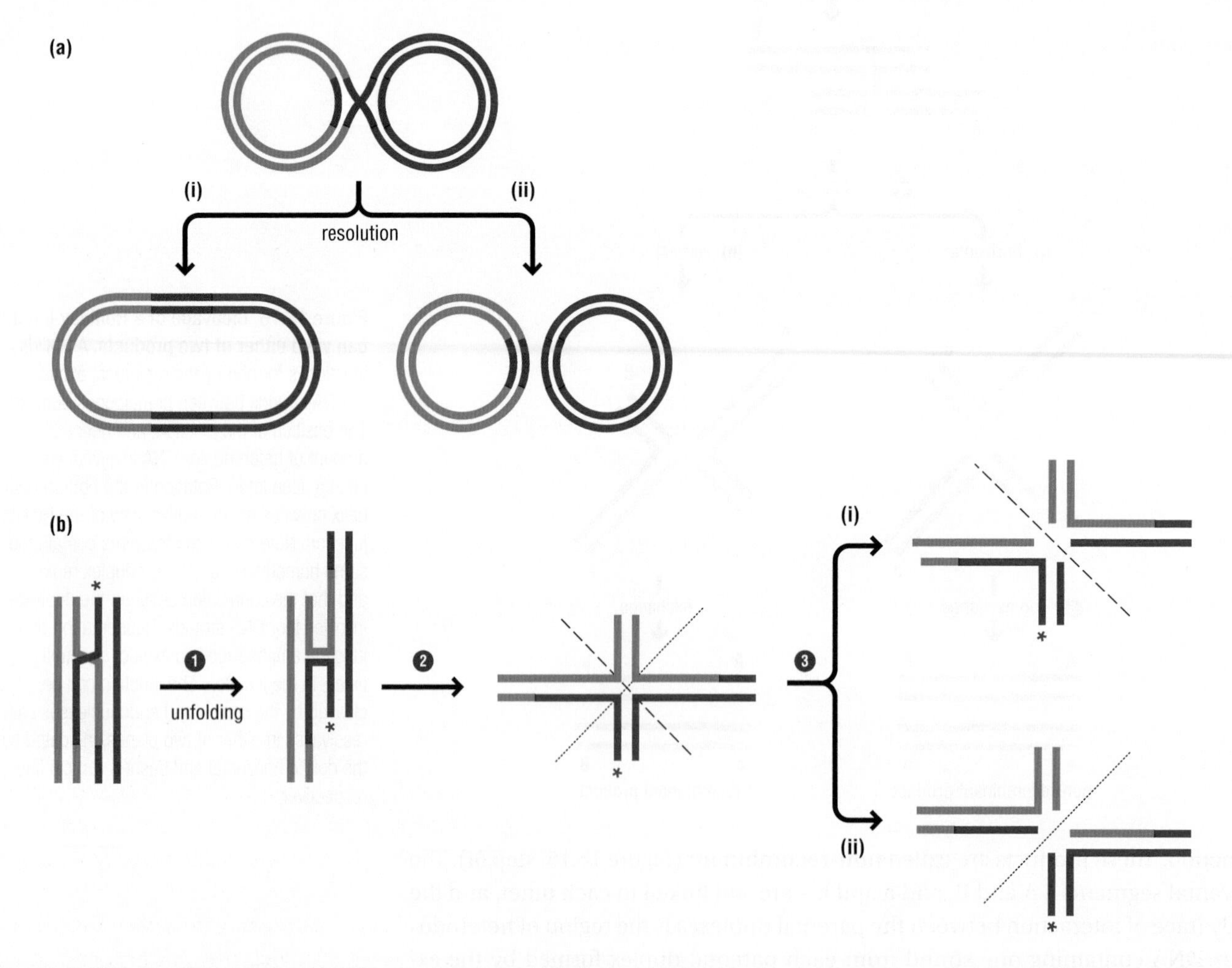

Figure 1 Holliday junction resolution. (a) Holliday junction resolution *in vivo*. A joined plasmid molecule is formed by purified RecA *in vitro*. When transformed into *E. coli*, the plasmid molecule is resolved to either (i) dimeric or (ii) monomeric form, indicating that resolution of Holliday junctions can occur *in vivo*. (b) Resolution of a synthetic Holliday junction *in vitro*. A Holliday junction can be made *in vitro* by annealing four oligonucleotides (see Figure 2b). (1) The junction is unfolded by rotation of one duplex. (2) The junction is reoriented to emphasize two axes of symmetry. This open 'four way' junction is the form of Holliday junctions *in vivo*. (3) Resolution of the junction can occur by the cleavage of two strands across either axis of symmetry.

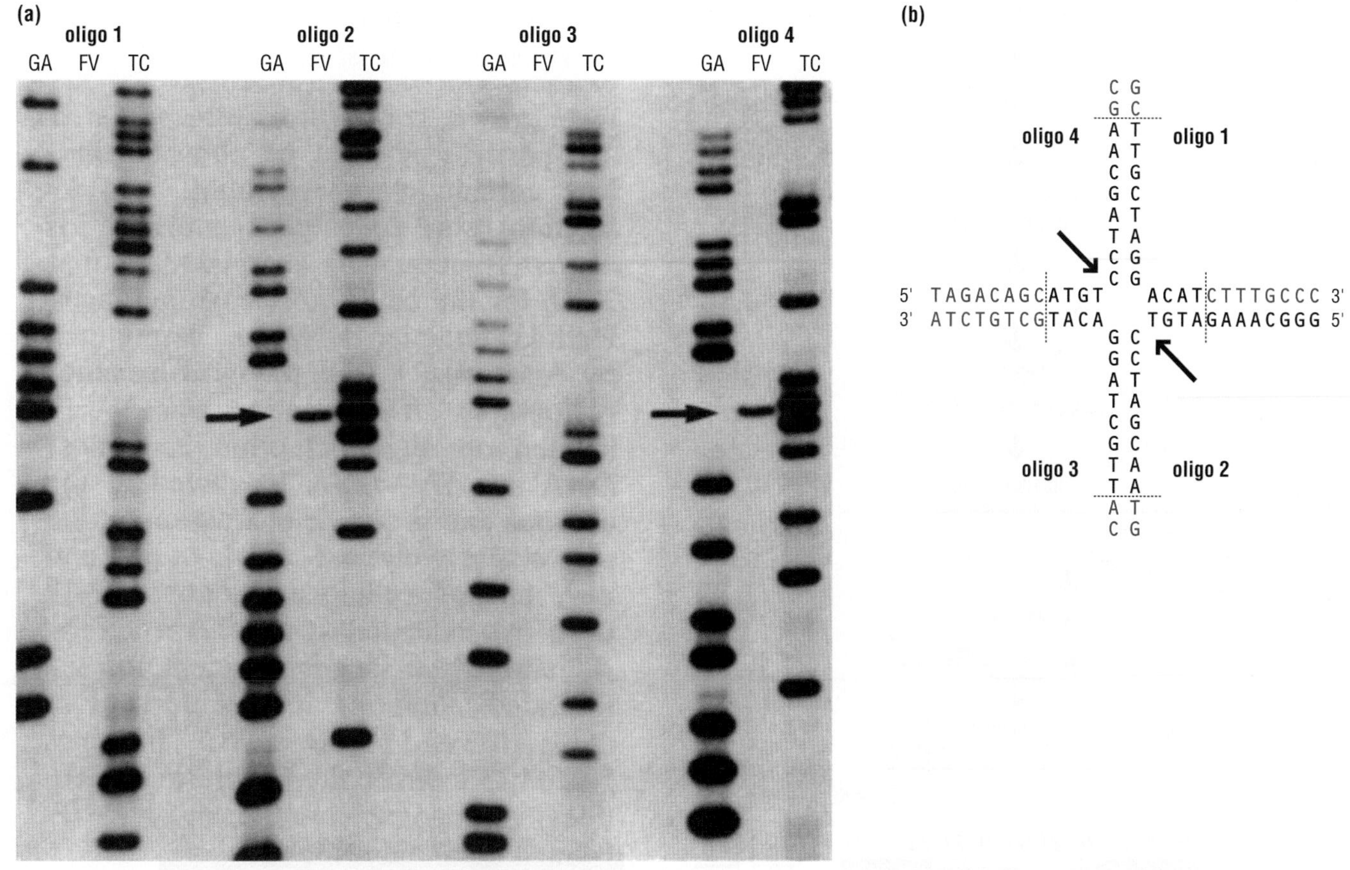

Figure 2 Resolution of a synthetic Holliday junction by RuvC resolvase. (a) Mapping of cleavage positions on a synthetic Holliday junction. A synthetic Holliday junction was constructed using oligonucleotides; four different forms were made in which a single oligonucleotide (1, 2, 3, 4) was labeled at its 5′ end. The Holliday junctions were incubated with purified RuvC (FV) and the products of the reaction displayed on a denaturing gel. GA and TC lanes serve as markers. Cleavage of oligonucleotides 2 and 4 is observed as indicated by the arrows. (b) Positions of RuvC cleavage on the synthetic Holliday junction. (The position of cleavage and the cleavage axis observed on such synthetic junctions are highly dependent on the sequence of the junction; symmetrical cleavage on both axes has been observed with other junctions.)

Reproduced from Connolly *et al.* Resolution of Holliday junctions in vitro requires the *E. coli ruvC* gene product. *Proceedings of the National Academy of Sciences U S A*, 1991;**88**:6063–6067.

Purification of a Holliday junction cleaving activity

The RecA protein promoted the formation of the Holliday junction in bacteria, but what protein was responsible for resolving the intermediate? Using *in vitro* generated Holliday intermediates made with RecA (Figure 1a) or by annealing of small oligonucleotide substrates (Figure 1b), West and colleagues were able to purify a resolvase activity from *E. coli* and establish that it was the product of the *ruvC* gene (Figure 2a). The *ruvC* gene had originally been identified genetically because of the way mutations in the gene rendered the bacteria extremely sensitive to killing by UV irradiation, a phenotype not inconsistent with resolvase activity. Analysis of RuvC resolution of a Holliday junction made from small oligonucleotides allowed the West group to further show that cleavage was symmetric about the junction (Figure 2b).

Isolation of eukaryotic resolvases by purification

Despite many subsequent genetic and biochemical studies in eukaryotes, it would be over 25 years before nuclear Holliday junction resolvases such as RuvC were isolated from these organisms. As we now know, the search for these enzymes was hampered by the little sequence homology between the resolvases identified in bacteria and eukaryotes. Moreover, no single mutations in yeast appear to give rise to cells that have the properties expected of cells defective in resolvase function. Identification of a human resolvase happened recently, emerging from biochemical assays that exploited synthetic Holliday junctions as substrates, as well as from sequence comparisons between several organisms (and the increasing sophistication of search algorithms).

The first human resolvase activity was isolated by West and colleagues in 2008, some 17 years after they identified the activity in

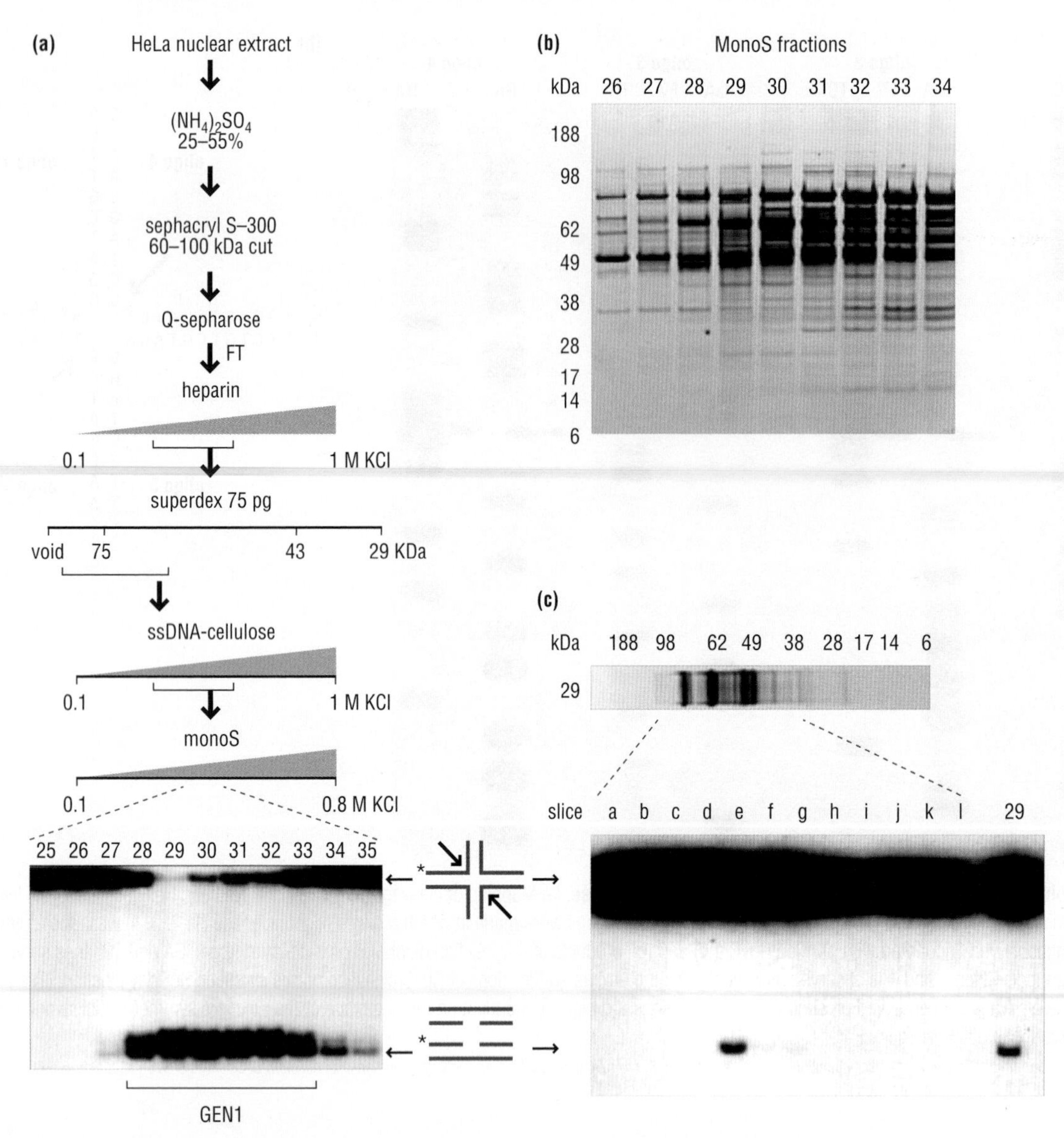

Figure 3 Isolation of the human Holliday junction resolvase GEN1. (a) Fractionation scheme for the purification of resolvase activity by multiple column chromatography steps. The presence of resolvase activity in the various fractions was monitored the ability to resolve a synthetic Holliday junction, as outlined at the bottom of the panel: the products of the reaction were displayed by gel electrophoresis. (b) Protein gel displaying polypeptides from fraction 29 of the last column in the purification, the MonoS column, which has the most enriched resolution activity. (c) Assay of gel slices from the gel lane containing proteins from MonoS fraction 29. Slice e contains the resolution activity. Mass spectrometry analysis of the polypeptides in this slice identified 20 polypeptides. One polypeptide, now called GEN1, has a nuclease motif related to the Rad2/XPG family.

Reproduced from Ip *et al.* Identification of Holliday junction resolvases from humans and yeast. *Nature*, 2008;**456**:357–361.

bacteria. In a brute force biochemical approach, the group fractionated large quantities of human tissue culture cells (200 L) and followed Holliday junction resolution activity through multiple protein purification steps; results are shown in Figure 3. Even after the many purification steps (Figure 3a), the final fraction had multiple polypeptides (Figure 3b), although final fractionation on a gel allowed the researchers to determine that a single gel slice had the resolvase activity (Figure 3c). Mass spectrometric analysis of the proteins located in that gel slice allowed them to identify 20 different proteins (from the defined polypeptide fragments). Of the identified target proteins, the most interesting candidate had homology to a known nuclease in human cells. In follow-up experiments, this protein, designated GEN1, was shown to cleave a synthetic Holliday junction symmetrically (the hallmark signature of the bacterial RuvC protein).

Exploiting sensitive sequence alignments to identify eukaryotic resolvases

Recently four different groups reported the identification of another Holliday junction resolvase in eukaryotes. In contrast to the extract fractionation strategy used by the West group, these

```
Hsap BTB012 1561 TPMPQYSIMETPVLKKELORFGVRPLP-KRQMVLKLKEIF...(152)...
Drer BTB012 1487 TPPPGFSDMETPFLKNRLNRFGVRPLP-KKQMVLKLKEIH...(101)...
Tcas MUS312  696 TPPANYDEMNTPQVCKELDKFGLKPLK-RSKGAKLLKYIY....(55)...
Dmel MUS312  924 TPKPOFATLPESEILQQLYKYGIKPLK-RKQAVKMLEFIY...(100)...
Cneo BSPI    608 RGPPOYOSWOVKALRLLIAOYGYRPIKOQSPLVQVAAECW....(64)...
Scer S1×4    620 KFCEIMMSQSMKELRQSLKTVGLKPHRTKVEIIQSLQTAS....(24)...

Hsap BTB012 TDEALRCYIRSKPAVYQKVVLYQPFELRELQAELRQNG----LRVSSRRLLDFLDTHCITF 1808
Drer BTB012 KLLAVRQFILSDPELYSRVLQYQPLPLAELRASDRAAG----IBLAAAKLLDFLDSQCITF 1682
Tcas MUS312 LHIAWHNLVMSNPKIREDILLYEPLQLENLHSMLKEQG----FRYNIQDLLTFLDKKCITI  846
Dmel MUS312 LHIAWHNLICANPQLHESVLTYEPIDLQAVYLHLKHMG----HRYDPKDLKTFFDRRCIIF 1119
Cneo BSPI   LHGQFQGMLTSDHDLYLRILRYEPIAFDELVSKAIASG--MTRRGWKKELKNYLDLQCVTY  769
Scer S1×4   IFDHLTELIEAFPDFLERIYTFEPIPLNELIEKLF-SAEPFVSQIDEMTIREWADVQSICL  743
```

Figure 4 Identification of SLX4 homologs through sequence alignment. Alignments of C-termini from proteins that were identified as homologs of yeast SLX4 derived from a BLAST search using the *Drosophila melanogaster* MUS312 protein as the starting sequence. In the alignment shown, sequences from widely divergent lineage are compared. Hsap is *Homo sapiens* (human), Drer is *Danio rerio* (zebra fish), Tcas is *Tribolium castaneum* (red flour beetle), Dm is *D. melanogaster*, Cneo is *Cryptococcus neoformans* (fungus), and Scer is *S. cerevisiae* (baker's yeast). The blue letters indicate positions conserved in at least four of the six species whereas the pink letters indicate positions conserved in all. The conserved predicted alpha helical structures are indicated with the bar underneath the sequence.

groups used a combination of sequence alignments, protein–protein interaction studies and biochemical assays. A complex of two yeast proteins, Slx1 and Slx4, was previously known to have some endonuclease activity. (An Slx1 homolog was known in humans, but an Slx4 homolog had not been identified.) Using sequence alignments between human, yeast, and *Drosophila* proteins known to be required for recombination, a candidate Slx4 homolog was identified based on similarities in sequence motifs in the C-terminal regions (see Figure 4). Biochemical purification and *in vitro* assays using both the yeast and human Slx1/Slx4 complexes showed this complex to have robust cleavage activity on Holliday junctions. Furthermore, inactivation of these genes in yeast and human cells resulted in a DNA damage response, consistent with the conclusion that these proteins encode a resolvase activity and that failure to properly resolve Holliday junctions will lead to DNA breaks.

Why have two different resolvases been identified in eukaryotes? It is possible that eukaryotic cells possess multiple resolvase activities that may function at different times or places, mirroring the way that *E. coli* has multiple resolvase activities. Further, it is not uncommon for different groups to isolate different proteins with the same activity or for scientists to come to different conclusions, driving the need for further experimentation. It is also not uncommon for several groups to publish similar results at the same time, especially given that research groups all strive to answer the next important biological question. Further studies will be required to understand these multiple eukaryotic resolvases.

Find out more

Andersen SL, Bergstralh DT, Kohl KP, *et al. Drosophila* MUS312 and the vertebrate ortholog BTBD12 interact with DNA structure-specific endonucleases in DNA repair and recombination. *Molecular Cell*, 2009; **35**:128–135.

Connolly B, Parsons CA, Benson FE, *et al.* Resolution of Holliday junctions in vitro requires the *Escherichia coli ruvC* gene product. *Proceedings of the National Academy of Sciences U S A*, 1991;**88**:6063–6067.

Dunderdale HJ, Benson FE, Parsons CA, *et al.* Formation and resolution of recombination intermediates by *E. coli* RecA and RuvC proteins. *Nature*, 1991; 354: 506–510.

Fekairi S, Scaglione S, Chahwan C, *et al.* Human SLX4 is a Holliday junction resolvase subunit that binds multiple DNA repair/recombination endonucleases. *Cell*, 2009;**138**:78–89.

Ip SC, Rass U, Blanco MG, *et al.* Identification of Holliday junction resolvases from humans and yeast. *Nature*, 2008;**456**:357–361.

Munoz IM, Hain K, Declais AC, *et al.* Coordination of structure-specific nucleases by human SLX4/BTBD12 is required for DNA repair. *Molecular Cell*, 2009;**35**:116–127.

Svendsen JM, Smogorzewska A, Sowa ME, *et al.* Mammalian BTBD12/SLX4 assembles a Holliday junction resolvase and is required for DNA repair. *Cell*, 2009;**138**:63–77.

Related techniques

Electron microscopy; Section 19.13

E. coli (as a model organism); Section 19.1

Denaturing sequencing gels; Section 19.8

Extract fractionation; Section 19.7

Protein purification; Section 19.7

Mass spectrometry; Section 19.8

Sequence alignment; Section 19.8

other. These DNA products are called **recombinant** (Figure 16.16, step 5ii): extensive genetic information has been exchanged between the parental duplexes.

Resolvases are nucleases that can cleave DNA at symmetrical positions in Holliday junctions and thereby separate the duplexes. The major *E. coli* resolvase is called **RuvC**. *E. coli* also contains another resolvase, RusA. In Experimental approach 16.3, we describe the identification of a new resolvase in humans (GEN1) that cleaves Holliday junctions in the same way that RuvC and RusA do. This eukaryotic resolvase is, however, structurally unrelated to RuvC; it belongs to the same family as the XP nucleases that participate in nucleotide excision repair. Another new type of resolvase, the eukaryotic SXL1/SLX4, is related to the UvrC endonucleases involved in bacterial nucleotide excision repair. Further investigation is required, however, to understand the precise roles of these enzymes in homologous recombination.

➔ We describe the process of nucleotide excision repair in Section 15.5.

Branch migration changes the position of Holliday junctions

When two identical duplexes are joined at a Holiday junction, (where sister chromatids become joined in the course of homologous recombination, for example), the junction can move along the DNA without any net change in base-pairing. Such movement is known as **branch migration** (Figure 16.17). Branch migration

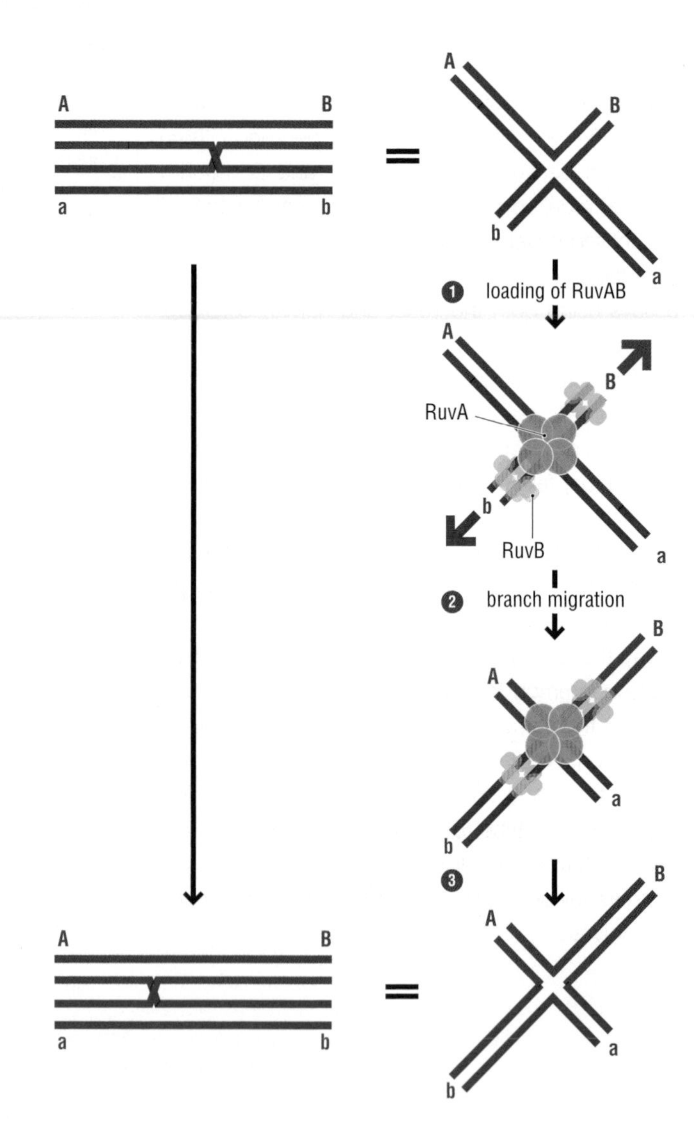

Figure 16.17 Schematic diagram of branch migration of a Holliday junction by the RuvAB complex. A Holliday junction formed by a reciprocal exchange of DNA strands between homologous duplexes is shown on the top left. The Holliday junction can move by a process called branch migration (left bottom). This reaction can occur spontaneously because the same number of paired bases are always present, the only change being which DNA strands are paired to each other (right). Branch migration is driven by RuvAB. All four arms of the Holliday junction are bound by a tetramer of RuvA, which converts it to the planar conformation. RuvB, which is a helicase, binds to two opposite arms. The RuvAB complex drives migration of the junction along the joined DNAs in the direction indicated by the arrows (pink and gray). The RuvAB complex recruits the resolvase RuvC to the junction, and it cleaves the DNA as shown in Figure 16.16.

can occur spontaneously in either direction over short distances. This can either increase or decrease the region of heteroduplex DNA between the linked duplexes. In the case of sister chromatids, where the duplexes are identical, branch migration makes no difference to the sequence of the two participating DNA molecules when they are separated.

However, branch migration can also occur between two homologous but non-identical DNA molecules, as in the case of two homologous chromosomes joined by Holliday junctions during meiosis. In this instance, migration requires specialized helicases and can influence the structure of the duplexes after resolution.

Specific enzymes drive branch migration

Although branch migration is a universal process, occurring both in eukaryotes and bacteria, the enzymatic machinery that drives it has been studied most extensively in bacteria and most of what is known to date has been learned from an enzyme complex called **RuvAB** in *E. coli*. As illustrated in Figure 16.17, step 1, RuvA is a tetramer that binds at the Holliday junction, keeping it in the unfolded and open conformation. RuvA also recruits RuvB to the junction. RuvB is a helicase that uses the energy of ATP hydrolysis to move the Holliday junction along the DNA, breaking and re-forming base pairs in one direction, and increasing or decreasing the region of heteroduplex in its wake, depending on the direction in which the junction is moved (Figure 16.17, step 2). The *E. coli* resolvase RuvC is recruited to Holliday junctions by its interaction with the RuvAB branch migration machine.

Thus, by changing the positions of the Holliday junction, namely how far it migrated from the original point of exchange, the helicases can determine the point at which the junction is resolved, and the point at which exchange between the participating chromosomes can occur.

In the above discussion, we have focused on the molecular processes associated with the resolution of a single Holliday junction. During homologous recombination, however, *two* Holliday junctions are formed and resolved. Let us now go on to consider the mechanism of homologous recombination in more detail. We begin by considering the initiation of the process of homologous recombination.

Homologous recombination is initiated by a double-strand break

In all organisms, from bacteria to yeast to humans, homologous recombination can initiate at a double-strand break. In meiotic recombination, this break is made by the programmed action of specialized nucleases. The break then undergoes reactions closely related to those underlying homology-directed repair as outlined in Figure 16.18, beginning with the formation of a D-loop between one of the broken duplexes and a homologous duplex.

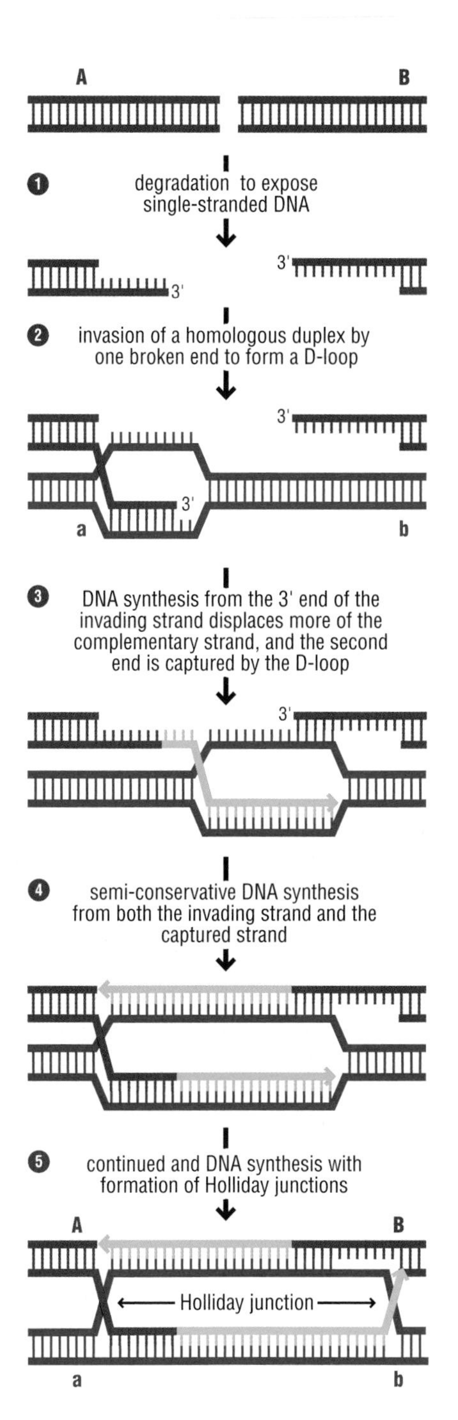

Figure 16.18 Homologous recombination initiates with a double-strand break. The 5′ ends of a broken DNA are resected to give 3′ tails, and one 3′ tail from the broken DNA invades the homologous intact duplex to make a D-loop. New DNA synthesis (green) initiates at the 3′ OH end of the invading strand. Extensive heteroduplex formation between the newly synthesized green DNA and the template duplex takes place, such that the pink displaced strand of the D-loop can pair with the gray DNA from the other side of the break. DNA synthesis continues from both ends of the broken DNA. In the region of gap repair, each duplex contains one strand of the template (pink) duplex and one newly synthesized strand (green); this DNA replication is thus semi-conservative. (5) Continued DNA synthesis and strand exchange results in the formation of two Holliday junctions linking the duplexes. The resolution of these linked duplexes is considered in Figure 16.20.

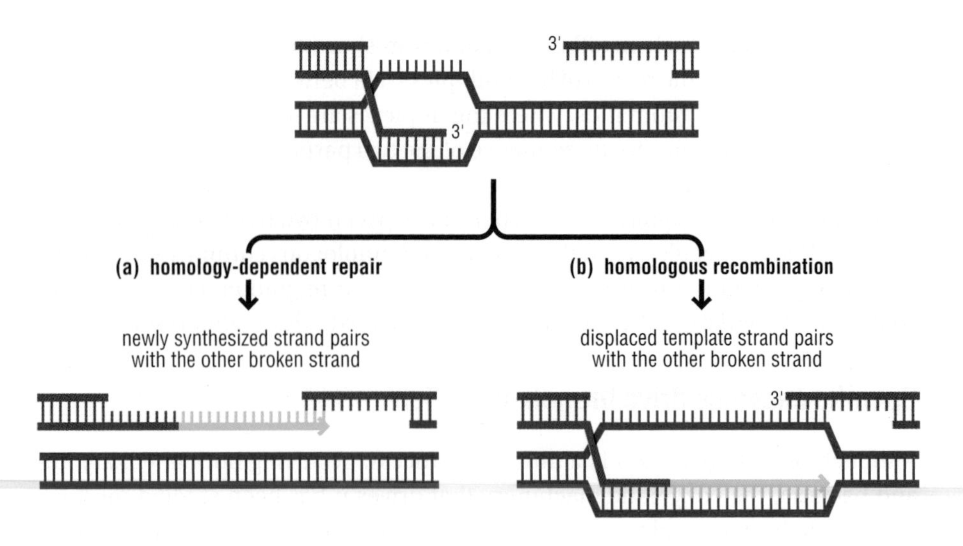

Figure 16.19 Homology-directed repair and homologous recombination differ in the processing of one of the double-strand break ends. Both homology-directed repair by SDSA, and homologous recombination initiate by formation of a D-loop and new DNA synthesis (green). (a) In homology-directed repair by SDSA, the newly synthesized strand (green) pairs with complementary sequences from the other broken end (gray). (b) In homologous recombination, the displaced strand from the template duplex (pink) pairs with the other broken end (gray).

The broken DNA ends are degraded to generate single-stranded 3′ tails (Figure 16.18, step 1). These ends are bound by strand-exchange recombinase molecules, RecA or Rad51, or the specialized meiotic recombinase, Dmc1, exactly as in SDSA, which we have already encountered in Section 16.3. The resulting nucleoprotein filament then invades the intact duplex (step 2). The next step is DNA synthesis, with the invading 3′ serving as a primer and its complementary strand in the participating homologous duplex as a template (step 3).

Subsequent events depart from those of SDSA (Figure 16.19a), however: instead of unwinding from the template strand, the newly synthesized strand remains paired to the template in the homologous duplex, and the D-loop is extended by continued DNA synthesis. This DNA synthesis displaces more of the non-template strand in the recipient duplex, and the displaced strand pairs with the complementary strand from the broken duplex (Figures 16.18, step 4, and 16.19b). In this way, the remaining broken strand on the other side of the break also has an intact template for repair. The gaps in the broken strands can be closed by repair polymerases using the free 3′-OH end of the cut strands as a primer (Figure 16.18, step 4). Note that DNA replication in this instance is semi-conservative: that is, the regions of DNA synthesis contain one newly synthesized and one old template strand.

The capture of DNA on the other side of the break by the displaced strand that results from D-loop extension is key to homologous recombination: without capture, the broken chromosome will be permanently fragmented. This failure to capture the other side of the broken DNA is also a hazard during homology-directed repair. We will see below that the uncaptured chromosome fragment can be replaced by continued DNA synthesis from the invading strand of the D-loop through the process of **break-induced replication (BIR)**.

The end result of the reactions described above is the formation of two intact DNA helices linked by two Holliday junctions (Figure 16.18, step 5). These junctions flank the position of D-loop formation by one end of the broken DNA, drawn on the left of Figure 16.18, step 2, and the position of capture of the other broken end by the displaced strand of the D-loop, drawn on the right of the same figure. Regions of repair synthesis, which will result in gene conversion (Figure 16.10), and heteroduplex DNA, as we discuss below, where gene conversion may occur by mismatch correction, lie between the two Holliday junctions. Once formed, however, the positions of the Holliday junctions and thus the distance between them may change by branch migration.

The nature of resolution of the double Holliday junction intermediate determines whether exchange of the DNA that flanks the junctions occurs

How does the resolution of two Holliday junctions affect the relationship of the duplexes joined by these junctions? Recall that, at each junction, resolution can occur in either of two ways, which differ according to which pair of strands is cut at each junction (Figure 16.16). Two of the four possible outcomes of the resolution of double Holliday junction are shown in Figure 16.20. In each case, the same pair of strands is cut on the left junction but different pairs are cut at the right junction. The consequence of the different cleavages at the right-hand junction is a change in the relationship of the DNA segments outside of the junctions.

Resolution of the right-hand junction in the horizontal direction (Figure 16.20, pathway a, step 2) results in non-recombinant progeny duplexes, that is, A and B on one duplex and a and b on the other, as in the parental duplexes. Note that even when resolution occurs without exchange of the flanking A,a and B,b markers, the DNA sequence between the junctions may change because of gene conversion, leaving a permanent reflection of the connection between the parental duplexes that did exist (see 'Holliday junction formation can lead to a change in the sequence of DNA following mismatch repair'). By contrast, resolution in the other direction (Figure 16.20, pathway iib, step 2), results in recombinant progeny duplexes, with one duplex containing A and b and the other a and B.

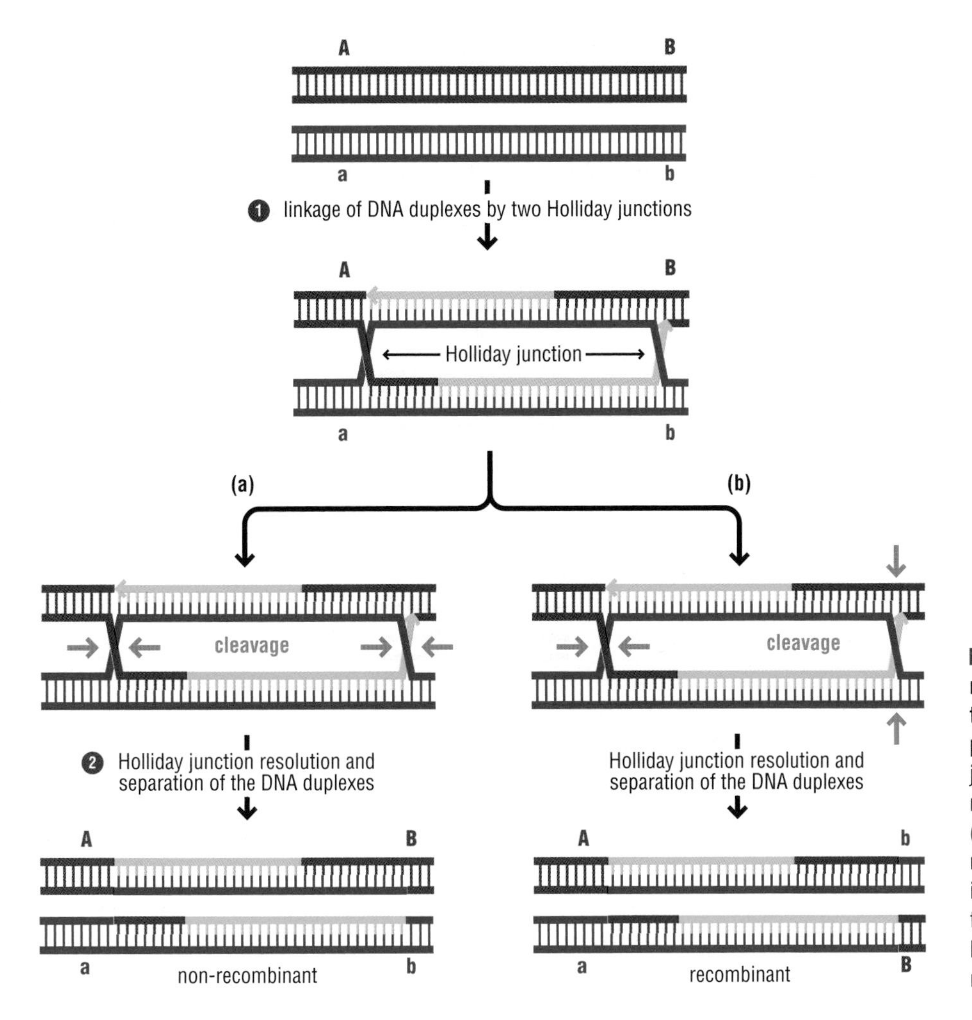

Figure 16.20 Resolution of a homologous recombination intermediate containing two Holliday junctions can yield multiple products. DNA duplexes joined by two Holliday junctions result from the initiation of homologous recombination by a double-strand break (Figure 16.17). Separation of the duplexes results from resolution of the Holliday junctions in alternative ways. In both pathways shown, the junction on the left is resolved along the horizontal axis, but resolution in the vertical axis may occur in either of two ways.

Thus recombination, a change in the spatial relationship of the DNA segments that flank a double-strand break, results from Holliday formation and resolution. By contrast, homology-directed repair does not involve Holliday junction formation, so no exchange of flanking DNAs is possible.

Holliday junction formation can lead to a change in the sequence of DNA following mismatch repair

Formation of a Holliday junction involves the pairing of strands from different DNA helices to form heteroduplex DNA, and branch migration influences the extent and location of this heteroduplex DNA. If heteroduplex DNA forms between strands from duplexes that are slightly different in sequence (as may occur between homologs, for example) the heteroduplex DNA will contain mismatches, which can be remedied by the process of mismatch correction, thus changing the sequence of DNA.

In post-replication mismatch repair (see Section 15.2; Figure 15.5), correction is guided by the parental DNA strand; that is, the newly synthesized strand is identified and the mismatched base specifically removed from the newly synthesized strand. In the case of mismatch correction of recombination intermediates, however, correction is not biased toward one strand or the other, and either DNA strand can serve as the template for repair. Figure 16.21 shows that if one parental duplex contains the gray marker B and the other the pink b marker, gene conversion by mismatch correction can result in two progeny duplexes with the B marker (step 2, i) or two progeny duplexes with the b marker (step 2, ii) without any change in the relationship of the markers, A and C, and a and c, that flank the region of gene conversion. Other possible patterns of correction are described in Figure 16.21.

So, we see how gene conversion of the B and b markers reveals that formation of heteroduplex DNA did occur between the parental duplexes although exchange of the flanking segments did not occur.

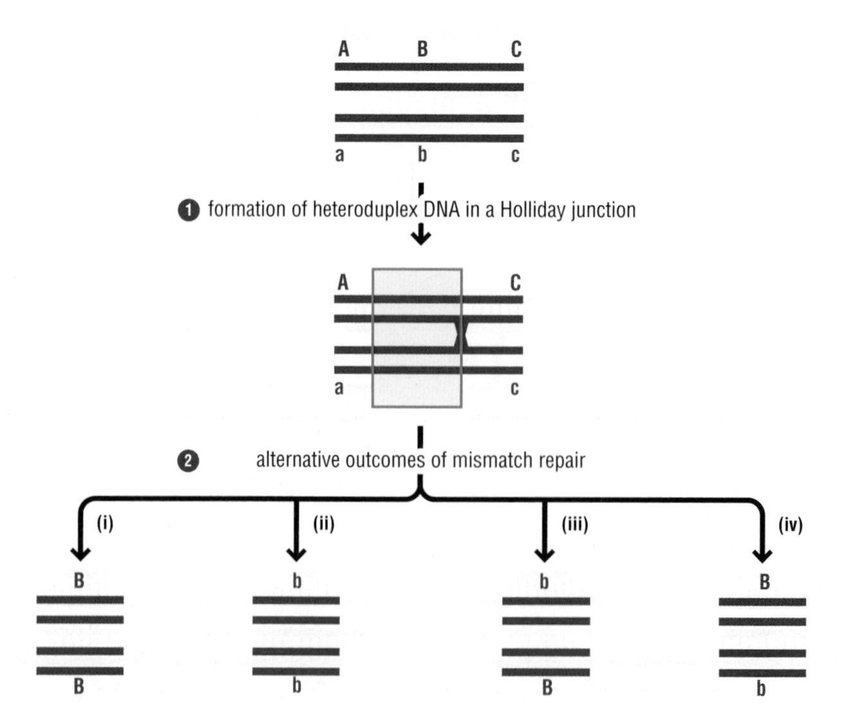

Figure 16.21 Mismatch correction of heteroduplex DNA can lead to gene conversion. Two parental homologous duplexes with different alleles at several positions are shown; these duplexes form a Holliday junction containing heteroduplex DNA. Mismatch correction of the heteroduplex (blue box) can have different outcomes, depending on which strand is used as the template for correction.

Although the mismatch correction described above *can* occur, it does not happen universally. Instead, DNA replication may happen before correction can take place; in this case, no gene conversion will occur. Mismatch correction can also occur within heteroduplex regions formed during homology-dependent repair.

Overall, we see that formation of a Holliday junction and branch migration can lead to significant changes in the sequence of a DNA molecule: the formation of heteroduplex DNA may lead to sequence changes in the parental duplexes by gene conversion through mismatch correction, and can also lead to the exchange of the DNA segments that flank the junction, depending on how the Holliday junction is resolved.

16.9 CHROMOSOME REARRANGEMENTS DURING ABERRANT REPAIR AND RECOMBINATION

Misdirected repair and recombination reactions can be damaging

The NHEJ reactions, and the recombination reactions that underlie homology-directed repair and homologous recombination at meiosis, are essential to life; but when they are misdirected, they can cause genetic changes that can be lethal to the cell, or can considerably increase the risk of cancer. Recombination in somatic cells, for example, can occur either in the course of double-strand break repair or through juxtaposition of chromatids from different chromosomes after DNA replication, with consequences that can be damaging. NHEJ reactions can join unrelated chromosomes, also with damaging consequences. In homologous recombination, mispairing interactions can cause deletions and **duplications** of DNA on the participating chromosomes. In the closing section of this chapter, we consider some of these aberrant processes, and explore their consequences to the cell.

Homologous recombination and homology-dependent repair in somatic cells can lead to loss of heterozygosity

One possible consequence of homologous recombination in somatic cells is a phenomenon known as **loss of heterozygosity**, to which we briefly alluded at the beginning of this chapter. Diploid organisms are heterozygous at many of their genes; that is, there are likely to be different alleles of any given gene on any two homologous chromosomes. In most cases, both alleles are functional, and loss of heterozygosity will have no adverse effects on the organism. However, when one of the alleles is defective, the heterozygous cell depends on the normal allele to grow and behave normally. Therefore, loss of the functional copy of the gene - that is, loss of heterozygosity - can have grave consequences. A particular hazard for multicellular animals is loss of the functional copy of a gene that normally suppresses cell proliferation and prevents the cell from giving rise to a cancer.

One mechanism by which loss of heterozygosity can occur as a result of homologous recombination is illustrated in Figure 16.22. After chromosomes have replicated (Figure 16.22, step 1), recombination may occur between two chromatids from different homologs, resulting in the reciprocal exchange of information between the two chromatids so that the defective gene (for example, the b marker) is now present in one chromatid of each homolog on the same template strand

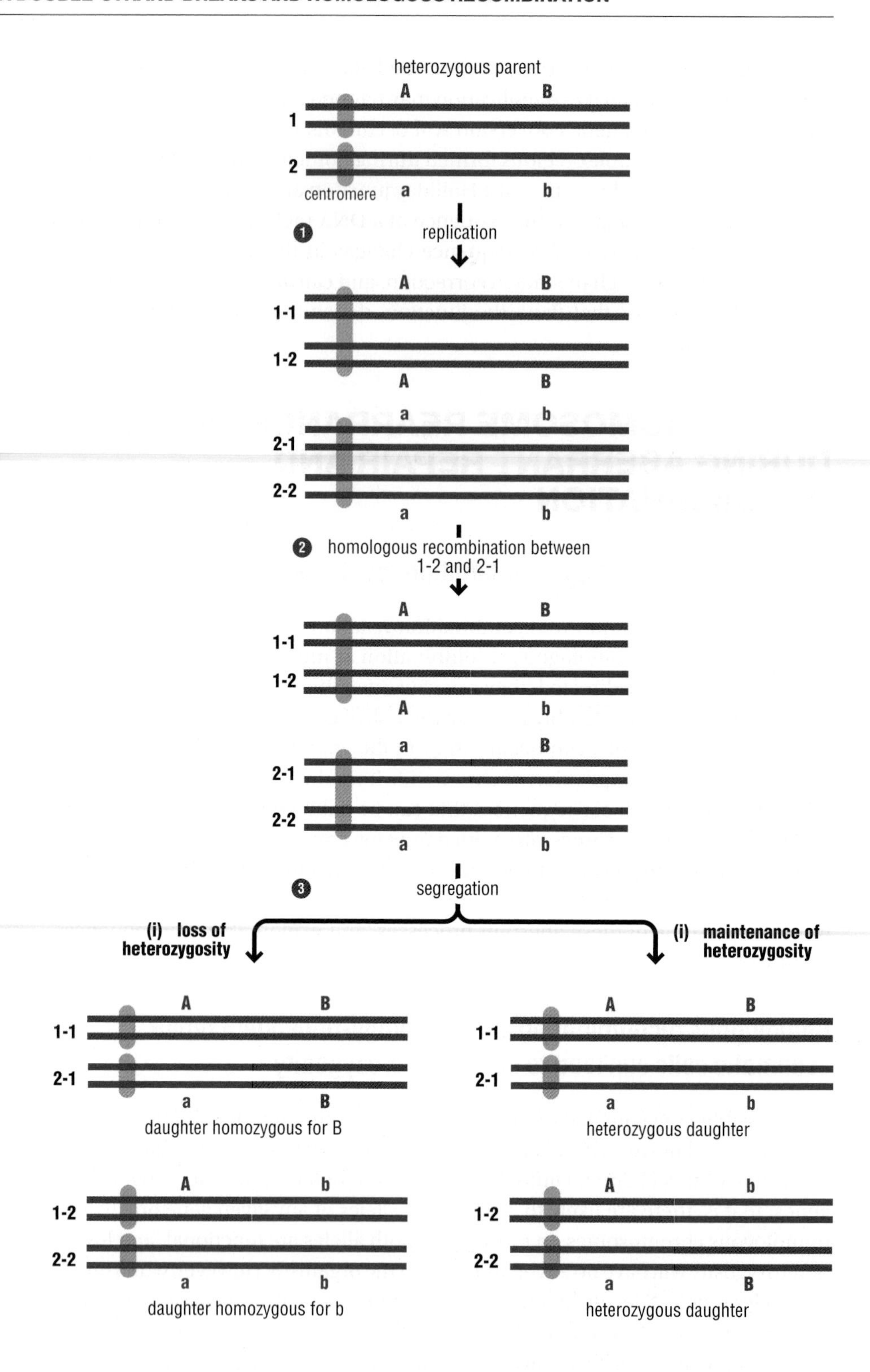

Figure 16.22 Loss of heterozygosity can occur by homologous recombination. Two homologs (1 = gray, 2 = pink) in a diploid that are heterologous at two loci, A/a and B/b, are shown. DNA replication generates two sister chromatids (1-1 and 1-2 for the gray chromosome; 2-1 and 2-2 for the pink chromosome) that are associated with one another at their centromeres (light gray oval). Recombination and segregation can result in loss of heterozygosity.

(Figure 16.22, step 2). After the chromatids separate and segregate at mitosis, one daughter cell could receive two copies of the normal gene while the other cell receives two copies of the defective gene (Figure 16.22, step 3).

Loss of heterozygosity can also occur through homology-directed repair that initiates within the chromosome and extends to the end of a chromosome, in a process called **break-induced replication (BIR)** (Figure 16.23). In BIR, a DNA end from one side of a double-strand break initiates repair by formation of a D-loop (Figure 16.23, step 2) and a new strand using a homolog is synthesized

(Figure 16.23, step 3). In BIR, however, the newly synthesized strand fails to pair with the other side of the double-strand break, and synthesis continues on the same template strand (Figure 16.23, step 4). New duplex DNA attached to only one side of the double-strand break is generated by the assembly of a replisome and initiation of lagging strand synthesis (Figure 16.23, step 5) that extends back to the side of the double-strand break that initiated D-loop formation (Figure 16.23, step 6). Notice that the newly synthesized chromosome segment is derived entirely from copying the intact duplex (Figure 16.23, step 7). Any markers that may have been different on the parental duplexes, for example B and b, are now the same on the progeny duplexes, b and b. Thus BIR results in loss of heterozygosity.

When loss of heterozygosity involves genes that regulate cell growth, such as the tumor suppressor genes described in Chapter 1, the consequences can be severe. For example, loss of heterozygosity underlies the greatly increased susceptibility to a retinal cancer called retinoblastoma in children who have inherited one defective gene for a protein known as Rb. The Rb protein suppresses the progression of cells from G1 to the S phase of the cell cycle (see Chapter 5). Growth control is normal in somatic cells heterozygous for Rb (Rb^+/Rb^-), because the undamaged gene is sufficient to maintain the essential functions of Rb. However, homologous recombination can result in a cell homozygous for the defective version of Rb (Rb^-/Rb^-). Such a cell does not properly observe the cell-cycle checkpoint that normally prevents DNA replication and cell division, and can proliferate to form a tumor.

Homologous recombination and homology-dependent repair are not the only ways in which loss of heterozygosity can occur. An entire member of a chromosome pair can be lost as a result of a malfunction of the cellular machinery that distributes copies of chromosomes evenly to the two daughter cells at mitosis (see Section 7.1).

BIR can restore telomeres

Telomerase mediates telomere maintenance in mammalian germline cells, as we discuss in Section 6.11. However, telomerase is not expressed in many differentiated cells; when telomeres in such cells become sufficiently shortened that shelterin, a complex of telomere-specific proteins that binds to and protects telomere ends (see Section 4.11), can no longer protect them, a DNA damage response can be provoked that results in apoptosis, that is, cell death. Thus the replicative potential of many differentiated cells is limited, a phenomenon termed 'replicative senescence'.

In a small fraction of cells, however, telomere length can be maintained by **'alternative lengthening of telomeres' (ALT)**. In yeast, the ALT pathway depends on BIR (Figure 16.23) in which the shortened ends can invade another chromosome end at similar sequences, thus priming DNA replication that extends to the end of the template chromosome and adding a new telomere to the invading chromosome end.

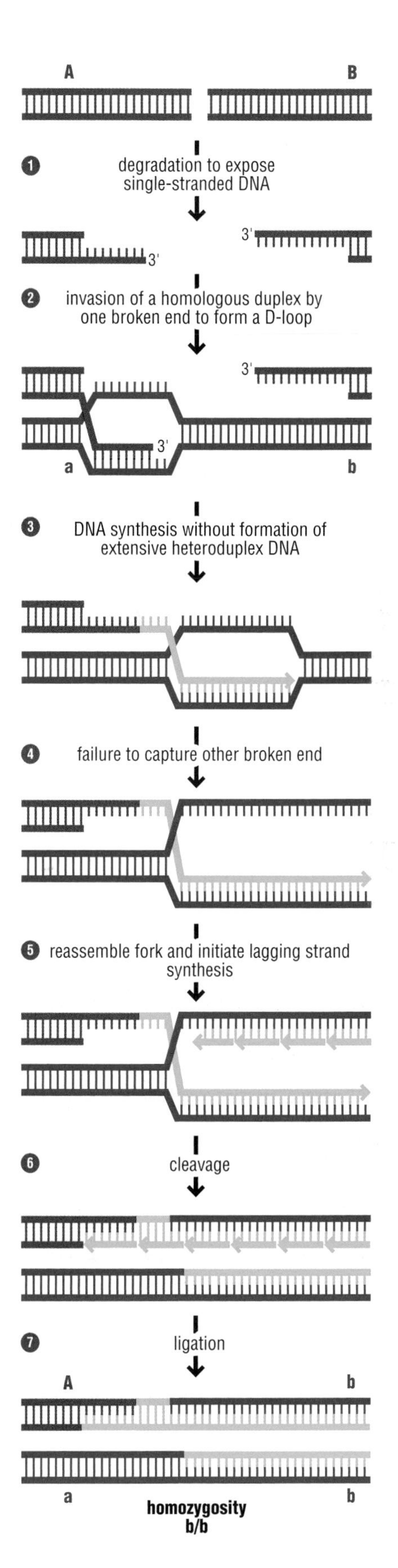

Figure 16.23 Loss of heterozygosity can occur by break-induced replication. When a double-strand break in DNA occurs, sometimes only the DNA at one side of the double-strand break forms a D-loop, resulting in loss of the DNA sequence information on the other side of the break, in contrast to the repair process shown in Fig 16.5.

A notable feature of tumor cells is that they can continue to divide and avoid the cell death response that usually results from the decreasing length of telomeres. In many of these cells, the expression of telomerase is reactivated, thus avoiding telomere shortening. Alternatively, induction of ALT is frequently observed in some types of tumor, which allows them to avoid the shortening of telomeres and resulting cell death.

Recombination between repeated sequences can result in chromosome rearrangements

Homologous recombination between repeated sequences is a particular hazard in multicellular eukaryotes, where such sequences are common. For example, the human genome contains about a million copies of a sequence called Alu, which is on average about 300 bp long.

➔ We learn more about Alu in Chapter 17.

The DNA rearrangements that result from recombination between repeated sequences depend on the relative position of the repeats: that is, whether they occur through intrachromosomal recombination, or through inappropriate interchromosomal recombination between repeats on sister chromatids, homologous chromosomes, or non-homologous chromosomes. Such rearrangements can include **deletions**, **duplications**, and **translocations**.

Recombination between direct repeats such as Alu on the same chromosome results in deletion of all the DNA between the repeats (see Figure 16.24a). When this DNA includes an essential gene, its loss can have severe consequences. Recombination between misaligned repeats on sister chromatids or on homologous chromosomes, often called **unequal crossing over**, can also occur, resulting in deletions and duplications on the different partners (Figure 16.24b and c). In a translocation, part of one chromosome is moved to another a non-homologous chromosome (Figure 16.24d).

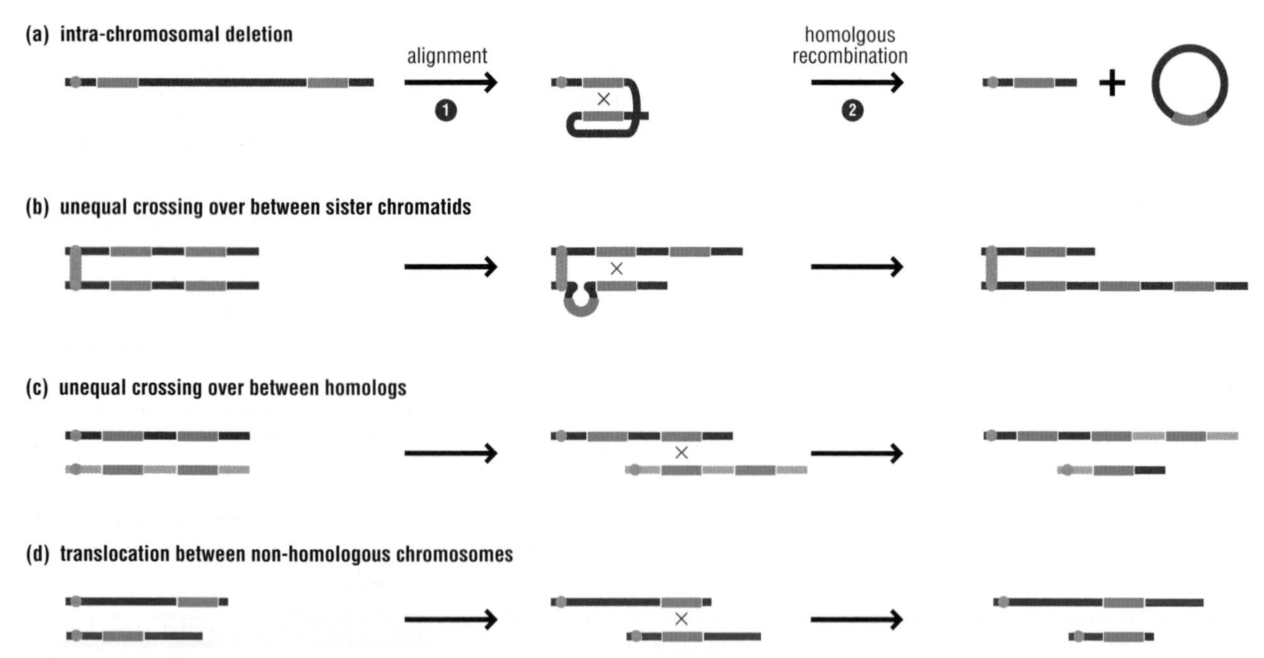

Figure 16.24 Homologous recombination between repeated sequences can lead to chromosome rearrangements. Deletions and translocations can result from recombination between homologous sequences at different positions in chromosomes.

Translocations and other chromosomal rearrangements are frequently found in cancer cells; some are specifically associated with particular cancers and are used to aid diagnosis and in defining cancer subtypes. Translocations that fuse two genes together, disrupting their function or regulation, are known to be directly responsible for several types of cancer, including chronic myelogenous leukemia and Burkitt lymphoma, both cancers of white blood cells.

As we will see in Chapter 17, Alu-mediated recombination occurring in germline cells at meiosis is the cause of about 0.3% of all cases of human genetic disease, including cases of insulin-resistant diabetes, immune deficiency due to the loss of a serum protein critical to resistance of bacterial infection, and familial hypercholesterolemia leading to heart disease.

Aberrant NHEJ can lead to chromosome rearrangements

Misdirected homologous recombination is not the only mechanism of chromosome rearrangement. If multiple double-strand breaks occur (for example, through environmental agents such as radiation, or as a result of DNA damage by oxidation, replication errors, or inappropriate cleavage by nucleases) the broken ends of the chromosomes may be incorrectly rejoined by NHEJ (see Section 16.1), such that a broken end is joined to the wrong partner. While such chromosomes can be faithfully transmitted during mitosis, they cause mis-segregation during meiosis because of errors in the homologous pairing necessary for accurate recognition and segregation of homologs. As a result, they are often lethal.

SUMMARY

DOUBLE-STRAND BREAKS

- DNA double-strand breaks are particularly dangerous lesions: failure to repair them can lead to chromosome fragmentation and cell death.
- Cells defective in their ability to repair double-strand breaks are very sensitive to DNA damage.
- There are two major strategies for double-stranded break repair: NHEJ and homology-directed repair.

NHEJ

- NHEJ rejoins the ends across a double-stranded break in the absence of a DNA template.
- NHEJ is often mutagenic because of nucleolytic processing of the ends prior to joining.

HOMOLOGY-DIRECTED REPAIR

- Homology-directed DNA synthesis can repair double-strand breaks by synthesizing new DNA across the break.
- The repair mediated by homology-directed repair is templated by a sister chromatid or homolog.
- Homology-directed repair changes the sequence of the broken duplex to that of the template duplex in the region of the break, resulting in gene conversion.
- Homology-directed repair occurs without changing the relationship of the DNA segments flanking the site of the double-strand break; that is, recombination or 'crossing over' does not occur.
- Identification of a template duplex occurs through the specific base-pairing of a single DNA strand from the broken double strand with its complement in a DNA duplex.

THE MECHANISM OF HOMOLOGY-DIRECTED REPAIR

- The base-pairing of single-stranded DNA from the broken DNA to a template strand forms a D-loop containing heteroduplex DNA.
- The single strand from the broken DNA duplex serves as a primer for DNA synthesis to replace the DNA lost at the break.
- The newly synthesized strand serves as the template to generate duplex DNA across the region of the break. This DNA synthesis is conservative.

- The ability to pair homologous DNA strands provides an important mechanism to reassemble replication forks that have broken or collapsed because of lesions such as nicks in DNA.
- Specialized proteins including helicases and nucleases generate the single-stranded DNA necessary for heteroduplex DNA formation.
- Strand-exchange recombinases, which include the bacterial RecA protein and eukaryotic Rad51 proteins, mediate the pairing reactions necessary for heteroduplex D-loop formation.
- The proteins that mediate homology-dependent repair have evolved to mediate homologous recombination.

HOMOLOGOUS RECOMBINATION

- Homologous recombination is the reciprocal exchange of large segments of DNA between homologous duplexes.
- Homologous recombination occurs during meiosis to generate gametes, and occurs between the bacterial chromosome and exogenous DNAs that enter the cell by conjugation, transformation, or in viruses.
- Homologous recombination can occur between any two duplexes that are nearly identical in sequence; it does not require a special site in either participating DNA.
- Homologous recombination is initiated by a double-strand break in one duplex.
- As in homology-dependent repair, a single-stranded DNA from the broken duplex forms a region of heteroduplex DNA in a homologous duplex via formation of a D-loop.
- The single strand from the broken duplex in the D-loop serves as a primer for DNA synthesis using the homologous chromosome as a template.
- The template strand that is displaced from the D-loop by semi-conservative DNA synthesis captures the other side of the broken DNA.
- This capture of the broken end by the template duplex is the key difference between homology-dependent repair and homologous recombination.

HOLLIDAY JUNCTIONS

- Holliday junctions are key recombination intermediates.
- In Holliday junctions, two single DNA strands are reciprocally exchanged between homologous DNA duplexes to form a branched four-way junction.
- The DNA duplexes that are linked by D-loop formation on one side of an initiating double-strand break, and end capture on the other side of the break, are joined by Holliday junctions, which result from the reciprocal exchange of single strands between two homologous duplexes.
- Holliday junctions bind regions of heteroduplex DNA and regions in which template-directed DNA synthesis has occurred.
- Gene conversion can result from mismatch repair of the heteroduplex regions and from template-directed DNA synthesis.
- A Holliday junction connecting two duplexes can be resolved (cleaved) in several ways in a process mediated by specialized enzymes called resolvases.
- Whether or not recombination occurs (i.e., whether a change in the relationship of the DNA segments flanking Holliday junctions occurs) is determined by the Holliday junction resolution pathway.

INAPPROPRIATE REPAIR AND RECOMBINATION

- Inappropriate homology-directed repair or homologous recombination between heterozygous duplexes can result in progeny that contain information from only one parental duplex.
- Such 'loss of heterozygosity' can be hazardous, most notably leading to cancer when a cell contains only mutant versions of growth control genes.
- Inappropriate homologous recombination between repeated sequences can lead to chromosome rearrangements such as deletions and translocations.

16.1 AN OVERVIEW OF DNA DOUBLE-STRAND BREAK REPAIR AND HOMOLOGOUS RECOMBINATION.

1. Which of the following statements correctly describes why unintended double-stranded breaks can be severely damaging?

a. They occur extremely frequently.
b. There is no intact complementary strand to use as a template for repair.
c. Double-stranded breaks propagate themselves and lead to the production of many more breaks.
d. They resemble the ends of chromosomes, so telomerase adds telomere sequences.

2. Briefly describe the two mechanisms that are used to repair double-strand breaks. Which is more likely to cause a mutation? Explain your answer.

3. Where does NHEJ normally occur?
   a. In meiotic cells
   b. In eukaryotes only
   c. In non-dividing cells
   d. In bacteria only
4. When is homology-directed repair most likely to be mutagenic?
   a. During G1, because repair is usually from a sister chromatid
   b. During G2, because repair is usually from a sister chromatid
   c. During G1, because repair is usually from a homologous chromosome
   d. During G2, because repair is usually from a homologous chromosome
5. Which key meiotic process involves a modified homology-directed repair process?

16.2 DOUBLE-STRAND BREAK REPAIR BY NHEJ

1. Briefly describe the way DNA ends are processed in the two main NHEJ pathways.
2. When a double-strand break is repaired by NHEJ, what protein binds the broken ends of the DNA strands? Which protein is subsequently recruited in mammals?

Challenge question

3. Discuss the significance of phosphorylation in NHEJ in mammals.

16.3 HOMOLOGY-DIRECTED REPAIR OF DOUBLE-STRAND BREAKS

1. Briefly describe the four main phases of homologous DNA repair.
2. What influences how much sequence will be resected when a double-strand break is repaired by the single-strand resection pathway?
3. What is a D-loop? Include the terms 'strand invasion' and 'heteroduplex' in your answer.
4. Which of the following correctly identifies when a D-loop usually forms?
   a. In homology-directed repair
   b. In NHEJ
   c. In M phase
   d. In G1 phase
5. During D-loop formation, which of the following is true?
   a. The template duplex does not base-pair with a damaged duplex.
   b. The non-template duplex does not base-pair with the damaged duplex.
   c. Both duplexes of the damaged strands base-pair with their complementary strands from the undamaged duplex
6. What causes the size of the D-loop to increase in size during the formation of a Holliday junction?

Challenge questions

7. Explain why the broken ends of the DNA are resected further in homology-directed repair of double-strand breaks than they are for non-homologous DNA repair
8. Explain SDSA. Include a diagram in your answer.

16.4 GENERATION OF SINGLE-STRANDED DNA BY HELICASE AND NUCLEASES

1. How do RecA monomers induce D-loop formation?
2. Name and describe the predominant pathway of single-strand DNA generation following double-strand breakage in *E. coli*.

16.5 THE MECHANISM OF DNA STRAND PAIRING AND EXCHANGE

1. Name the bacterial and eukaryotic strand-exchange recombinases, and briefly discuss how the bacterial recombinase mediates strand exchange.
2. Explain the purpose of single-stranded binding proteins in strand exchange.

16.6 GENE CONVERSION THROUGH HOMOLOGY-DIRECTED REPAIR

1. When is gene conversion least likely to occur?
   a. G0
   b. G1
   c. S
   d. G2

Challenge question

2. Using a diagram, indicate how gene conversion can occur during repair of a double-strand break.

16.7 REPAIR OF DAMAGED REPLICATION FORKS BY HOMOLOGY-DIRECTED REPAIR

1. Describe how homology-directed repair can repair a collapsed replication fork.
2. When a replication fork stalls due to a lesion in the DNA (such as a pyrimidine dimer), the fork can regress, causing the parental DNA to re-pair and the newly synthesized strands to pair up. Following repair of the lesion, DNA replication can recommence. What mechanisms can resolve the structure of the replication fork?

16.8 HOMOLOGOUS RECOMBINATION

1. What is homologous recombination, and what is required for its initiation?
2. What initiates homologous recombination, and what are the possible outcomes?
3. Briefly describe Holliday junctions and their formation.
4. Once a Holliday junction has formed, what affects the size of the region of heteroduplex? What enzymes can drive this process in *E. coli*, and what is the major enzyme is involved in Holliday junction resolution?

Challenge questions

5. Holliday junctions:
   a. On the following diagram of a Holliday junction, indicate where the DNA will be cleaved to produce:
      i. Recombinant DNA
      ii. Non-recombinant DNA

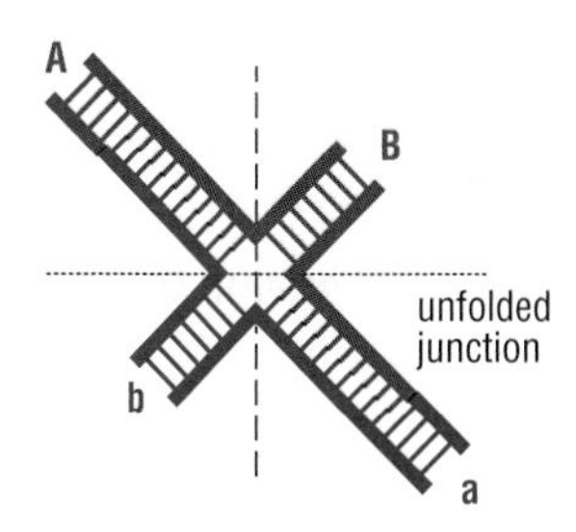

   b. Draw diagrams indicating the molecules that are produced from (a).
6. Compare and contrast meiotic recombination with SDSA.

16.9 CHROMOSOME REARRANGEMENTS DURING ABERRANT REPAIR AND RECOMBINATION

1. Why do double-strand breaks sometimes lead to a loss of heterozygosity?
2. Explain how cancerous cells are able to replicate indefinitely rather than experience 'replicative senescence'.
3. Homologous recombination can cause chromosomal rearrangements. Explain how this occurs.

EXPERIMENTAL APPROACH 16.1 – DNA SEQUENCES THAT STIMULATE HOMOLOGOUS RECOMBINATION: CHI SITES AND RECBCD

Challenge questions

1. Experimental approach 16.1, Experiment 1:
   a. Explain the significance of small and large plaque sizes in the first experiment detailed in Experimental approach 16.1.
   b. Using a similar approach to that in Experimental approach 16.1, Experiment 1, design an experiment to find out if Chi sites are the only sites upon which RecBCD can act.
2. In the second experiment described in Experimental approach 16.1, Kowalczykowski's groups were able to reject a prior hypothesis that RecD leaves the RecBCD complex when it encounters a Chi site, as shown in Figure 2b. With reference to the figure, explain what would have been observed if the hypothesis had in fact been correct.

EXPERIMENTAL APPROACH 16.2 – MODULATING THE ACTIVITIES OF RECA AND RAD51

Challenge questions

1. In Experimental approach 16.2, researchers used an electrophoretic gel mobility shift assay. Explain, with reference to Figure 3, how this kind of assay provides evidence of molecular interactions.
2. Design an experiment to determine which parts of the Rad51 protein interact with Brc4.

EXPERIMENTAL APPROACH 16.3 – THE DIFFICULTIES IN IDENTIFYING HOLLIDAY JUNCTION RESOLVASES

Challenge question

1. Although Holliday junctions were proposed in 1964, and a bacterial resolvase was identified from bacteria in 1991, resolvases from eukaryotes have only recently been identified. Explain two reasons why it was difficult to identify these proteins.

16.1 AN OVERVIEW OF DNA DOUBLE-STRAND BREAK REPAIR AND HOMOLOGOUS RECOMBINATION

Helleday T, Lo J, van Gent DC, Engelward BP. double-strand break repair: from mechanistic understanding to cancer treatment. *Repair*, 2007;**6**:923–935.

Pardo B, Gómez-González B, Aguilera A. repair in mammalian cells: DNA double-strand break repair: how to fix a broken relationship. *Cellular and Molecular Life Sciences*, 2009;**66**:1039–1056.

Wyman C, Kanaar R. DNA double-strand break repair: all's well that ends well. *Annual Review of Genetics*, 2006;**40**:363–383.

16.2 DOUBLE-STRAND BREAK REPAIR BY NHEJ

Haber JE. Alternative endings. *Proceedings of the National Academy of Sciences of the U S A*, 2008;**105**:405–406.

Lieber MR. The mechanism of human nonhomologous DNA end joining. *Journal of Biological Chemistry*, 2008;**283**:1–5.

Shuman S, Glickman MS. Bacterial DNA repair by non-homologous end joining. *Nature Reviews Microbiology*, 2007;**5**:852–861.

Weterings E, Chen DJ. The endless tale of non-homologous end-joining. *Cell Research*, 2008;**18**:114–124.

16.3 HOMOLOGY-DIRECTED REPAIR OF DOUBLE-STRAND BREAKS

Haber JE, Ira G, Malkova A, Sugawara N. Repairing a double-strand chromosome break by homologous recombination: revisiting Robin Holliday's model. *Philosophical Transactions of the Royal Society of London Series B, Biological Sciences*, 2004;**359**:79–86.

Hiom K. DNA repair: common approaches to fixing double-strand breaks. *Current Biology*, 2009;**19**:R523–525.

Li X, Heyer WD. Homologous recombination in DNA repair and DNA damage tolerance. *Cell Research*, 2008;**18**:99–113.

Malkova A, Haber JE. Mutations arising during repair of chromosome breaks. *Annual Reviews of Genetics*, 2012;**46**:455–473.

16.4 GENERATION OF SINGLE-STRANDED DNA BY HELICASES AND NUCLEASES

Bernstein KA, Rothstein R. At loose ends: resecting a double-strand break. *Cell*, 2009;**137**:807–810.

Dillingham MS, Kowalczykowski SC. RecBCD enzyme and the repair of double-stranded DNA breaks. *Microbiology and Molecular Biology Reviews*, 2008;**72**:642–671.

Kanaar R, Wyman C. DNA repair by the MRN complex: break it to make it. *Cell*, 2008;**135**:14–16.

Mimitou EP, Symington LS. DNA end resection-Unraveling the tail. *DNA Repair*, 2011;**10**:344–348.

Wigley DB. RecBCD: the supercar of DNA repair. *Cell*, 2007;**131**:651–653.

16.5 THE MECHANISM OF DNA STRAND PAIRING AND EXCHANGE

Antony E, Tomko EJ, Xiao Q, *et al.* Srs2 disassembles Rad51 filaments by a protein-protein interaction triggering ATP turnover and dissociation of Rad51 from DNA. *Molecular Cell*, 2009;**35**:105–115.

Chen Z, Yang H, Pavletich NP. Mechanism of homologous recombination from the RecA-ssDNA/dsDNA structures. *Nature*, 2008;**453**:489–494.

Cox MM. Motoring along with the bacterial RecA protein. *Nature Reviews Molecular Cell Biology*, 2007;**8**:127–138.

Finkelstein IJ, Greene EC. Single molecule studies of homologous recombination. *Molecular BioSystems*, 2008;**4**:1094–1104.

Kowalczykowski SC. Structural biology: snapshots of DNA repair. *Nature*, 2008;**453**:463–466.

Thorslund T, West SC. BRCA2: a universal recombinase regulator. *Oncogene*, 2007;**26**:7720–7730.

16.6 GENE CONVERSION THROUGH HOMOLOGY-DIRECTED REPAIR

Chen JM, Cooper DN, Chuzhanova N, Férec C, Patrinos GP. Gene conversion: mechanisms, evolution and human disease. *Nature Reviews Genetics*, 2007;**8**:762–775.

Deitsch KW, Lukehart SA, Stringer JR. Common strategies for antigenic variation by bacterial, fungal and protozoan pathogens. *Nature Reviews Microbiology*, 2009;**7**:493–503.

Palmer GH, Brayton KA. Gene conversion is a convergent strategy for pathogen antigenic variation. *Trends in Parasitology*, 2007;**23**:408–413.

16.7 REPAIR OF DAMAGED REPLICATION FORKS BY HOMOLOGY-DIRECTED REPAIR

Heller RC, Marians KJ. Replisome assembly and the direct restart of stalled replication forks. *Nature Reviews Molecular Cell Biology*, 2006;**7**:932–943.

Lambert S, Carr AM. Impediments to replication fork movement: stabilisation, reactivation and genome instability. *Chromosoma*, 2013;**122**:33–45.

Michel B, Boubakri H, Baharoglu Z, LeMasson M, Lestini R. Recombination proteins and rescue of arrested replication forks. *DNA Repair*, 2007;**6**:967–980.

Moldovan GL, Pfander B, Jentsch S. PCNA, the maestro of the replication fork. *Cell*, 2007;**129**:665–679.

Rudolph CJ, Dhillon P, Moore T, Lloyd RG. Avoiding and resolving conflicts between DNA replication and transcription. *DNA Repair*, 2007;**6**:981–993.

16.8 HOMOLOGOUS RECOMBINATION

Déclais AC, Lilley DM. New insight into the recognition of branched DNA structure by junction-resolving enzymes. *Current Opinion in Structural Biology*, 2008;**8**:86–95.

Nimonkar AV, Kowalczykowski SC. Second-end DNA capture in double-strand break repair: how to catch a DNA by its tail. *Cell Cycle*, 2009;**8**:1816–1817.

San Filippo J, Sung P, Klein H. Mechanism of eukaryotic homologous recombination. *Annual Review of Biochemistry*, 2008;**77**:229–257.

Schwartz EK, Heyer WD. Processing of joint molecule intermediates by structure-selective endonucleases during homologous recombination in eukaryotes. *Chromosoma*, 2011;**120**:109–127.

Walsh T, King MC. Ten genes for inherited breast cancer. *Cancer Cell*, 2007;**11**:103–105.

16.9 ABERRANT REPAIR AND RECOMBINATION AND CHROMOSOME REARRANGEMENTS

Haber JE. Transpositions and translocations induced by site-specific double-strand breaks in budding yeast. *DNA Repair*, 2006;**5**: 998–1009.

Jeggo PA. Genomic instability in cancer development. *Advances in Experimental Medicine and Biology*, 2005;**570**:175–197.

Llorente B, Smith CE, Symington LS. Break-induced replication: what is it and what is it for? *Cell Cycle*, 2008;**7**:859–864.

Weinstock DM, Richardson CA, Elliott B, Jasin M. Modeling oncogenic translocations: distinct roles for double-strand break repair pathways in translocation formation in mammalian cells. *DNA Repair*, 2006; **5**:1065–1074.

17 Mobile DNA

INTRODUCTION

In the preceding two chapters, we learned how the sequence of DNA can be changed by mutations, and by homologous recombination between large homologous regions of DNA. In this chapter, we will learn about two ways by which discrete segments of DNA can move from one region of the genome to another: by **transposition**, whereby **transposable elements** can insert in many different sites around the genome; and by **conservative site-specific recombination (CSSR)**, sometimes called site-specific recombination, whereby recombination occurs between pairs of specific related DNA sites to rearrange sequences bounded by those specific sites.

These recombination systems are widespread and have a profound impact on chromosome structure and function. Virtually all organisms contain transposable elements, and their mobility is a significant contributor to genetic variation. CSSR systems are found mostly in bacteria where they play important roles in bacterial chromosome segregation and the interactions of viruses and plasmids with their hosts. The mechanisms of DNA breakage and joining associated with transposition and CSSR, and the recombination enzymes that promote them, are different. However, they both involve recombination enzymes that act only at specific sequences, and form oligomeric complexes whose correct assembly with multiple DNA segments is required to coordinate and regulate the DNA breakage and joining reactions involved.

We begin the chapter with an overview of transposition before considering in more detail the mechanisms of action employed by different elements, and the strategies used to regulate these processes. We will then explore CSSR, including some specific well-characterized examples.

17.1 TRANSPOSABLE ELEMENTS: OVERVIEW

Transposable elements are DNA sequences that can move to different sites in the genome

Transposable elements, or **transposons**, are discrete mobile segments of DNA that can move around the genome and insert into different DNA **target sites** using a specialized type of recombination called transposition. Their ability to insert in many different sites reflects the fact that, unlike homologous recombination, transposition does not require any sequence homology between the transposon DNA and the target site.

Figure 17.1 shows how transposable elements move, or transpose, in two main ways. They are either (i) excised from one site and reinserted into a new site, or (ii) they replicate, with the duplicate copy being integrated at a new site in the

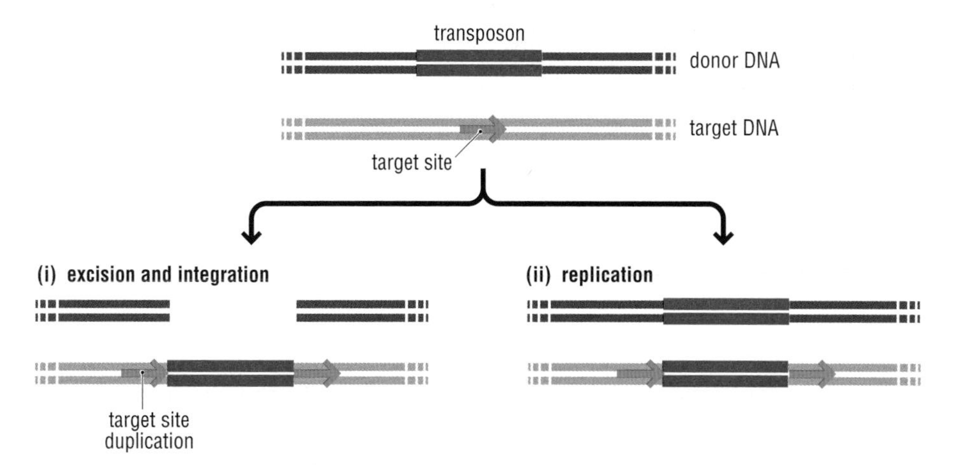

Figure 17.1 Transposons can move by excision and integration or by replication. Transposons move from a donor DNA (gray) to a target DNA (light gray) by either of two general pathways, and this movement is accompanied by duplication of a short sequence (blue arrow) at the target site.

genome. Most elements use only one of these pathways. A hallmark of transposition is that element insertion results in **target-site duplications**, that is, short direct repeats that flank the newly inserted transposon. (Such duplication is depicted in Figure 17.1) Transposable elements that encode the proteins necessary for their mobility are called **autonomous elements**. Elements that do not encode their own proteins but can be mobilized by proteins supplied from another element are called **non-autonomous elements**.

Transposable elements are found in all living organisms; indeed, they are often the largest component of the genomes of multicellular organisms. For example, about 45% of the human genome and 85% of the maize genome derives from transposable element sequences. Most transposable elements are confined to moving within the DNA of a single cell. Exceptions are viruses such as the bacteriophage Mu (discussed in Section 17.4) and the retroviruses (discussed in Section 17.7), which use multiple cycles of transposition to replicate their genomes during the viral lifecycle.

Transposable elements are powerful agents for genetic change

Because of their ability to move with a genome, transposable elements have great potential to generate changes in the genetic constitution of their host cells - and thus contribute to the genetic variation that drives evolution. For example, when a transposon inserts into the coding region of a gene, it acts to interrupt that coding region. As a result, the gene is usually inactivated because of the truncation of the coding region. Importantly, many transposable elements carry *cis*-acting regulatory sequences such as promoters, enhancers, terminators and assembly sites for modified chromatin. Therefore, when a transposable element inserts, it may have the sequence components necessary to influence the expression of nearby genes.

Moreover, transposons are often present in high copy numbers within a genome; the extensive regions of sequence homology resulting from their ubiquitous presence means that transposons can often provide substrates for homologous recombination. Such recombination may lead to intramolecular deletions and an increase in unequal sister chromatid exchange, ultimately causing a change in genome structure and function (see Section 16.9). Thus, transposable elements contribute to genetic variation in multiple ways.

The frequency at which transposable elements move and, hence, contribute to changes in host genotype (and, potentially, phenotype) varies considerably

	fraction of genome derived from transposable elements	fraction of *de novo* mutations resulting from transposable element insertion
human	45%	0.3%
mouse	39%	10%
Drosophila	5.5%	>50%

Figure 17.2 Frequency of *de novo* mutations resulting from transposable element insertion in different species.
From Belancio, VP, Hedges, DJ, and Deininger, P. Mammalian non-LTR re-trotransposons: For better or worse, in sickness and in health. *Genome Research*, 2008;**18**:343–358.

between elements and organisms, as summarized in Figure 17.2. Active mobile elements have been found in all model organisms, but transposition is generally rare, except when it is used as a mechanism for viral replication.

Experimental approach 17.1 describes the experiments that led to the identification of the first human mobile element, discovered because a *de novo* insertion resulted in hemophilia. There are now about 100 examples of human disease resulting from transposon insertion. Also described in Experimental approach 17.1 is the first analysis of mobile elements in a genome of a single human, allowing new information about the variability of transposable element populations between individuals to be revealed. From this work, it is estimated that in about one in every 20 human births, the genome of the newborn has a new transposition event compared to its parents' genomes.

Transposable elements have characteristic features

Most of the different transposons that exist can be divided, according to different criteria, into three other groups: the **DNA-only transposons**, the **long terminal repeat (LTR) elements**, and the **non-LTR** elements. These classes are distinguished by element structure, the type of recombination proteins they encode as shown in Figure 17.3, and the mechanism of transposition Figure 17.4 shows how the number and type of transposable elements present in the genome vary considerably between organisms. We consider each of these classes in the following sections.

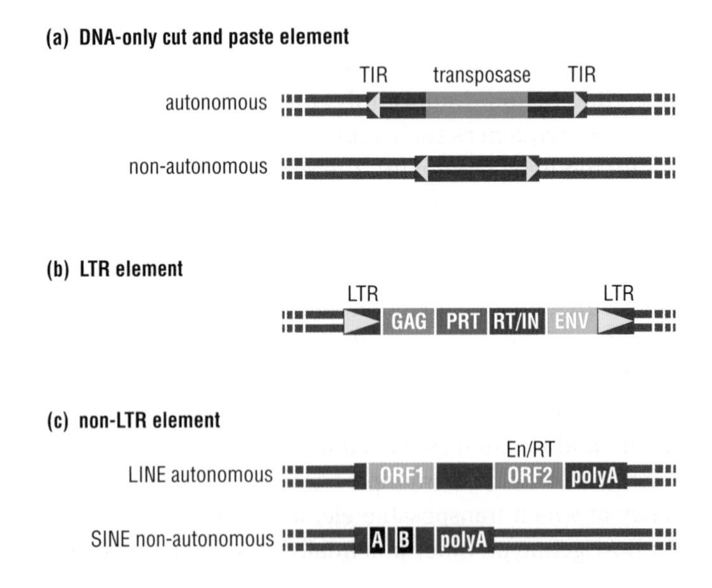

Figure 17.3 There are three major types of transposable elements. (a) DNA-only cut-and-paste transposons are bounded by terminal inverted repeat sequences (TIRs; yellow), which are transposase (orange) binding sites. (b) Long terminal repeat (LTR) elements are retroviruses or retroviral-like transposons, both of which transpose via an RNA intermediate. LTR element proteins include GAG, which interacts with the element's nucleic acid, protease (PRT), reverse transcriptase (RT), and integrase (IN), which is the transposase. Retroviral elements also have ENV, a coat protein important for cell exit and entry. (c) Non-LTR elements encode ORF1, which is an RNA chaperone protein important for making an RNA-protein complex (RNP), ORF2, which has reverse transcriptase activity (RT) and in some cases endonuclease (EN) activity, and a poly(A) tail. SINE elements are non-autonomous elements that can be mobilized by long interspersed element (LINE) proteins and have A- and B-box signals for DNA polymerase III.

17.1 EXPERIMENTAL APPROACH

Mobile DNA in humans

Transposable elements were first discovered by Barbara McClintock on the basis of the discernible phenotypes that they brought about upon their insertion into DNA. McClintock identified the DNA element Ac (for 'activator') in the late 1940s because it altered the pigment color of corn kernels. It was not until much later, in 1988, that Haig Kazazian and colleagues discovered a human disease, a form of hemophilia, similarly caused by a transposable element.

Transposable element insertion in human genes result in disease

Hemophilia is a group of hereditary diseases resulting from an inability to control blood clotting. Hemophilia A is a common form of the disease in which a protein required for clotting, called factor VIII, is missing. As the gene for this factor is on the X chromosome, hemophiliacs are usually male; in females, two X chromosomes are present, and having one wild-type copy of the gene is sufficient for effective clotting. Such heterozygous females are, however, carriers of the disorder and can transmit the defective gene to their progeny. Homozygous mutant females are rare, however, because relatively few males survive to reproductive age. Hemophilia A occurs in about 5000–10 000 male births; about 30% of these cases result from new spontaneous mutations.

To examine genomic changes associated with the disease, Kazazian and colleagues analyzed the DNA of 240 unrelated males with hemophilia A by restriction digestion and Southern blot analysis. The factor VIII gene is large, consisting of 186 kb, in which 26 exons contain about 9 kb of the total coding sequence. To analyze the locus, they digested genomic DNA with a restriction enzyme, fractionated the fragments on a gel, and examined the size of the factor VIII gene fragments by hybridization. Although the genes of 15 patients carried detectable deletions, the researchers were most interested to see that two patients yielded a *larger* fragment than expected for exon 14. In neither of these two cases were the DNA changes present in the DNA of the asymptomatic parents, suggesting that these insertions were new mutations that resulted in disease. The exon 14 DNA fragments from the two patients were isolated and sequenced (see Figure 1), revealing the presence of truncated forms of a human L1 retrotransposon.

There are many transposable elements in the human genome

Since this initial identification of mutations caused by transposition in humans, sequencing has provided a very powerful tool for in-depth analysis of transposable elements and their presence throughout the genome. An unanticipated finding from the initial sequencing of the human genome by the International Human Genome Sequencing Consortium (Eric Lander and colleagues, 2001) and Celera Genomics (Craig Venter and colleagues, 2001) was that transposable elements make up a very high fraction of the genome. The astounding prevalence of transposons in the genome, most significantly LTR and non-LTR retrotransposons, is discussed in Section 17.1.

While most of these transposable elements are no longer active due to the accumulation of mutations within them, there are important exceptions. Some elements do move, and both the recombination between dead elements and the movement of active transposable elements continue to contribute to structural

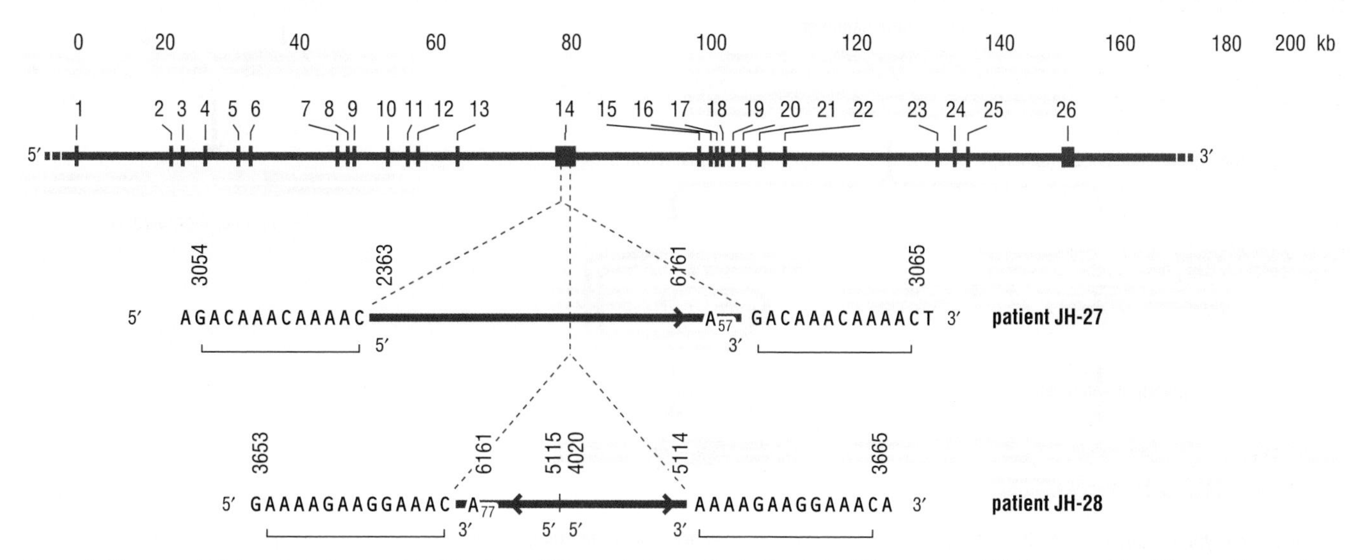

Figure 1 Diagram of L1 insertions in exon 14 of the factor VIII gene in two patients. The factor VIII gene spans 186 kb and contains 26 exons. The positions of insertion of segments of the human non-LTR element L1, a LINE element, into intron 14 are shown. The target site duplications are indicated by brackets.

variations between individuals. And, though most of the structural variations resulting from such events are selectively neutral, some insertions do have phenotypic consequences, as first described by Kazazian and colleagues (1988).

New transposable element insertion and recombination between elements contribute to structural variations between human individuals

The complete sequence obtained by the Human Genome Project in 2002 represented an assembly of DNA sequences that were derived from a few different people. To understand more fully variations *between* people it is necessary to compare the sequences of individuals, rather than use a composite sequence. Venter and colleagues (Levy *et al*. 2007) determined the complete DNA sequence of a single person, employing a new computer algorithm to correctly assign sequence information to each of the two homologous chromosomes. This complete diploid sequence of this individual is called the HuRef sequence. Having this sequence information from individuals allows for an understanding of the degree of heterozygosity between individuals and for fine structure mapping of the human genome, as discussed in Chapter 18.

The HuRef sequence also has proved to be very useful for understanding structural variations that arise from both the movement of 'live' transposable elements and recombination between inactive transposable elements, as alluded to above in 'There are many transposable elements in the human genome'.

To specifically examine the role of transposable elements in genomic change, Xing *et al*. (2009) took advantage of the complete HuRef sequence and compared it to the sequence from the Human Genome Project. This comparison identified a total of 643 992 sites at which either insertions or deletions (so called indels) occurred between the two sequences. A small fraction of these indels (8500 or 0.013%) were larger than 50 bp, and a subset of these larger indels was associated with transposable element sequences. In this manner, 846 'mobile element associated structural variants' (MASVs) were identified in the HuRef genome. Based on more detailed analysis of the sequences of the MASVs, Lynn Jorde and colleagues (Xing *et al*.) described four general mechanisms by which they were generated, as illustrated in Figure 2:

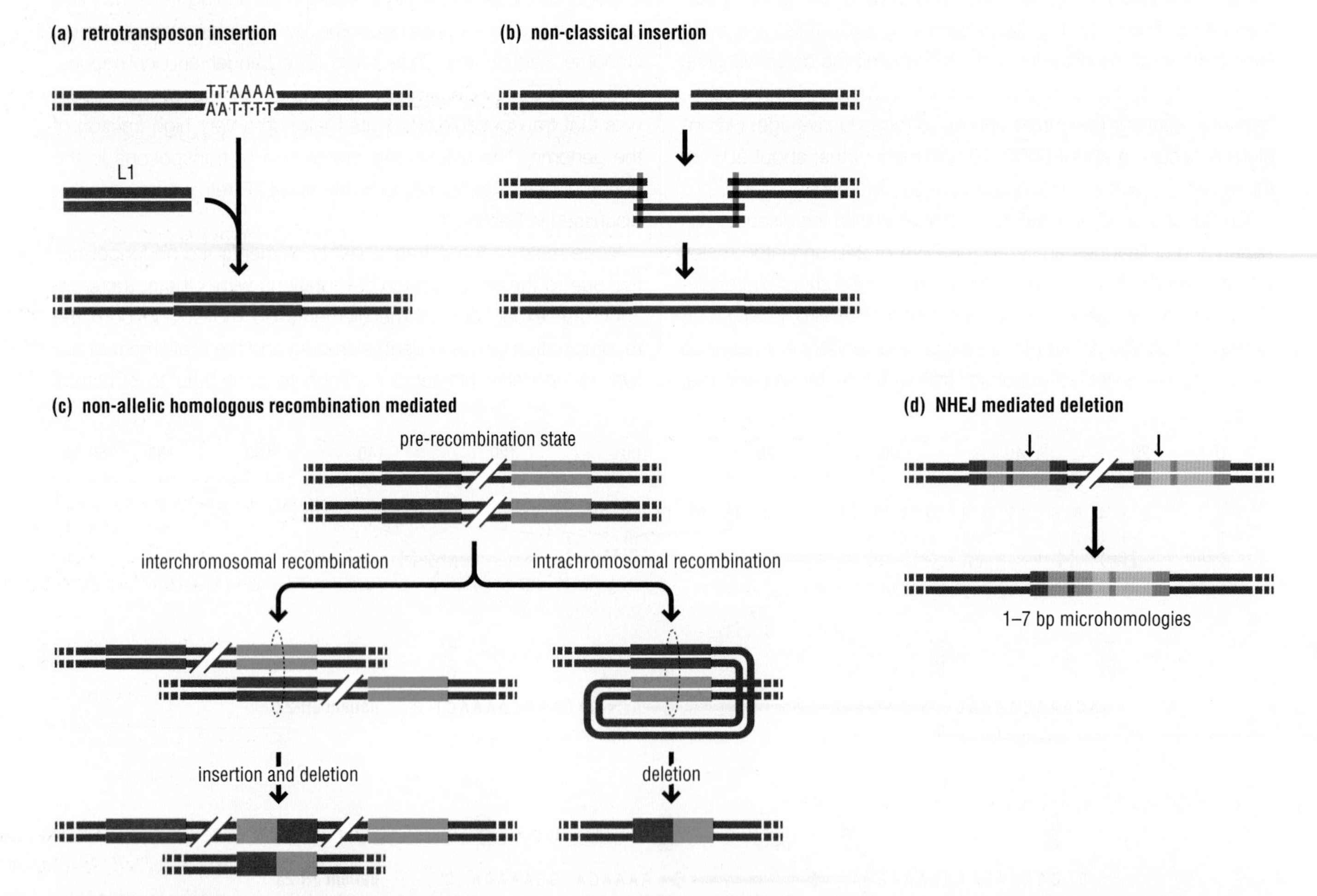

Figure 2 Four types of common MASVs in the HuRef genome and how they are thought to arise. (a) Insertion of non-LTR L1 (red) at a TTAAAA insertion site. (b) Non-classical insertion where a DNA fragment (red) is ligated into a double-strand break. (c) Non-allelic homologous recombination-mediated insertion/deletion; interchromosomal non-allelic recombination between homologous sequences resulting in a deletion and insertion; intrachromosomal recombination between homologous sequences resulting in a deletion. (d) NHEJ-mediated deletion. Rejoining at breaks in different chromosomal locations (dark blue and blue) resulting in deletion.

(a) retrotransposon insertions, (b) non-classical insertions, (c) non-allelic homologous recombination, and (d) non-homologous end joining (NHEJ)-mediated deletions. The latter three of these classes are mediated by recombination of repetitive DNA. Notably, only one of these classes, retrotransposition, requires active transposition, principally involving L1 and Alu retrotransposons. The role of transposons in genome evolution is discussed in Chapter 18.

At the end of this comparative analysis of HuRef and HGP sequences, retrotransposon insertions were the most frequent structural changes found (small indels are not considered in this calculation). These studies estimate an Alu transposition rate of about one in 21 births, and an L1 transposition rate of about one per 212 births. To date, 56 disease-causing mobile element insertions have been identified. Comparing this number to the total number of 'disease' mutations known suggests that 0.14% of disease-causing mutations are caused by mobile element insertions. Thus, although Haig Kazazian's initial finding of a retrotransposon causing hemophilia A was a rare find, mobile element associated structural variation has a significant role in disease and an even more significant role in genome evolution.

Find out more

Kazazian HH Jr, Wong C, Youssoufian H, *et al.* Haemophilia A resulting from de novo insertion of L1 sequences represents a novel mechanism for mutation in man. *Nature*, 1988;**332**:164–166.

Lander ES, Linton LM, Birren B, *et al.* Initial sequencing and analysis of the human genome. *Nature*, 2001;**409**:860–921.

Levy S, Sutton G, Ng PC, *et al.* The diploid genome sequence of an individual human. *PLoS Biology*, 2007;**5**:e 254.

Venter JC, Adams MD, Myers EW, *et al.* The sequence of the human genome. *Science*, 2001;**291**:1304–1351.

Xing J, Zhang Y, Han K, *et al.* Mobile elements create structural variation: Analysis of a complete human genome. *Genome Research*, 2009;**19**:1516–1526.

Related techniques

Restriction digests; Section 19.4

Southern blots; Section 19.9

DNA sequencing; Section 19.8

Human Genome Project; Chapter 19, 'Online resources for genomics and model organisms'

Transposable elements in various genomes

organism	common name	genome size (Mb)	number of ORFs	derived from TEs	% DNA TEs	% RNA TEs
bacteria						
Escherichia coli K-12 MG1655		4.6	4,300	1.0	100	0
archaea						
Mathanocaldococcus jannaschii		1.7	1,700	0.7	100	0
eukaryota						
Saccharomyces cerevisiae	budding yeast	12.1	6,300	3.1	0	100
Schizosaccharomyces pombe	fission yeast	14.0	4,800	1.1	0	100
Arabidopsis thaliana	(plant)	120.0	26,000	10	40	60
Oryza sativa	rice	420.0	23,300	30	85	15
Zea mays	maize	2300.0	32,500	85	11	89
Caenorhabditis elegans	worm	100.0	19,000	12	85	15
Drosophila melanogaster	fly	165.0	13,600	5.5	15	85
Mus musculus	mouse	2500.0	24,000	39	2	98
Homo sapiens	human	3280.0	25,000	45	7	93

Figure 17.4 Genome composition of a variety of organisms.

From Feschotte and Pritham, Computational analysis and paleogenomics of interspersed repeats in eukaryotes. In: Stojanovic, N, ed. *Computational Genomics: Current Methods*, Wyndmonham, Norfolk, UK: Horizon Bioscience, 2007.

DNA-only transposons move by DNA breakage and joining

As the name implies, all steps in the movement of **DNA-only transposons** involve DNA. Most of these elements move by simple excision and integration, as illustrated in Figure 17.5a; these elements are known as DNA-only **cut-and-paste** elements, and encode a specialized recombinase called a **transposase**, which recognizes particular sequences at the transposon ends and executes the DNA breakage and joining reactions that underlie transposition.

By contrast – and as we will discuss shortly – some DNA-only elements use a related nick-and-paste mechanism for replicative transposition.

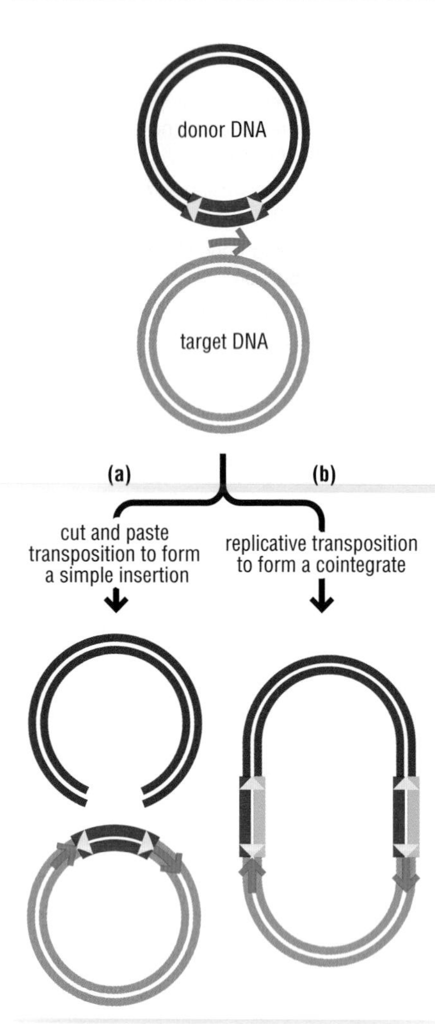

Figure 17.5 Non-replicative cut-and-paste and replicative nick-and-paste transposition by DNA-only elements. (a) In cut-and-paste transposition, the element excises by double-strand breaks and then integrates into a target. (b) In replicative nick-and-paste transposition, the transposon remains attached to the donor DNA and joins to the target DNA, followed by replication of the element and formation of a cointegrate. Each transposon copy contains one original strand (red) and one newly replicated strand (green). This structure is often eventually resolved to form two separate DNA molecules, each with a transposon.

➔ We learn about the synthesis of DNA from an RNA template in Section 6.11.

➔ We learn more about the regulation of transposition in Section 17.9.

DNA-only transposons are widespread in bacteria and eukaryotes. In bacteria, almost all transposable elements are autonomous elements (see Figure 17.3) whereas eukaryotic genomes, by contrast, often contain large numbers of non-autonomous elements.

Retrotransposons move by replicative mechanisms involving an RNA intermediate

The other two major classes of transposable elements, the **LTR elements** and **non-LTR** elements, move by replicative copy-and-paste mechanisms rather than through excision. With these elements, an RNA copy of the element is first generated. The RNA copy is then converted into DNA by an element-encoded reverse transcriptase. Because a reverse transcription step is involved (during which DNA is synthesized from RNA), these elements are also known as **retrotransposons**.

The LTR elements, which include retroviruses and retroviral-like elements, are characterized by long terminal repeats of hundreds of base pairs at each end of the element; the general structure of an LTR element is summarized in Figure 17.3b. These LTRs play a key role in reverse transcription, during which the RNA form of the element is converted to the double-stranded DNA form, a process mediated by the element-encoded reverse transcriptase. This DNA is then inserted into the target site by an element-encoded **integrase**. The mechanism of insertion is identical to that of the DNA-only cut-and-paste transposons. Thus, retroviral integrases are also transposases. To date, LTR elements have been identified only in eukaryotes.

Non-LTR elements also move via an RNA intermediate but lack long terminal repeats, as shown in Figure 17.3c. In eukaryotes, one autonomous family of non-LTR elements called **long interspersed elements (LINEs)**, which are prominent in the human genome, encode proteins that mediate their own transposition. These proteins include a reverse transcriptase, which makes a DNA version of the element at the target site using the RNA copy of the element as a template. Non-LTR elements are present in eukaryotic nuclei, mitochondria and chloroplasts, and in bacteria.

Host proteins play important roles in transposition

While most transposons encode the proteins required for their movement, host-encoded proteins are also involved in transposition reactions. Therefore, transposons cannot operate as completely independent entities: they rely on the cell whose genome they occupy for their ongoing propagation and movement. The target-site duplications flanking newly inserted elements (see Figure 17.1) result from repair by the host-cell's DNA repair proteins of single-strand gaps at each transposon end, which are formed upon element insertion. In some cases, host DNA-bending proteins may also promote the formation of the protein–DNA complexes that mediate DNA cut-and-paste transposition. Host proteins also play an important role in target-site selection for many elements and in the regulation of transposition. Much remains to be learned about the interesting interplay between the host and transposons during transposition.

Having now considered transposition in general terms, let us now examine in more detail the mechanism by which DNA-only 'cut-and-paste' transposons move, and also elements that move by a '**nick-and-paste**' mechanism that involves replication.

17.2 AN OVERVIEW OF DNA-ONLY TRANSPOSONS

DNA-only cut-and-paste elements move by DNA breakage and joining

As described in the previous section, the DNA-only cut-and-paste transposable **elements** move by DNA breakage reactions that separate the transposon ends from the donor DNA during excision, followed by DNA joining reactions that link the element to a new insertion site. In many cases, the element is completely excised from the donor site by double-stranded breaks before joining to the target site, as shown in Figure 17.5a. The product of this excision and integration reaction is called a simple insertion.

Unlike cut-and-paste elements, **nick-and-paste** elements move by a variation of the excision and integration mechanism, which results in copying of the DNA-only element. This mechanism is called replicative transposition, and results in a DNA structure called a **cointegrate**, as depicted in Figure 17.5b. A cointegrate contains two copies of the transposon linked by the donor and target DNA. To date, elements that move by this type of replicative transposition have only been found in bacteria. We will discuss the mechanism of this variation on cut-and-paste transposition below.

DNA cut-and-paste elements are widespread in both bacterial and eukaryotic genomes

DNA cut-and-paste elements have been found in virtually all organisms that have been sequenced. Although different terms are often used to describe elements in bacteria and eukaryotes, as detailed in Figure 17.6, their fundamental features - an element bounded by short terminal inverted repeats (TIRs) that are recognized by an element-encoded transposase - are the same.

One important difference between the DNA cut-and-paste elements found in eukaryotes and bacteria is that many bacterial transposons carry other genes in addition to those encoding their transposition functions. These include antibiotic resistance genes, determinants of bacterial pathogenicity such as toxin genes, or genes for metabolic functions that enable the bacterium to utilize unusual substrates. These additional genes are of great importance to both bacteria and humans. Indeed the rapid dissemination of antibiotic resistance among bacterial populations represents the ability of antibiotic resistance-encoding transposons to 'hitchhike' onto other DNAs, such as plasmids and viruses, using these DNAs as vehicles to move between cells. By contrast, in eukaryotes, other elements encoding non-transposition functions have not yet been identified.

For historical reasons, bacterial DNA cut-and-paste elements that encode only their own transposition functions are called **insertion sequences (IS elements)**. In some cases, a pair of IS elements flanks another gene (or genes) and the whole segment of DNA can act as a single transposable element; in such instances, the genes between the IS elements are mobilized along with the elements themselves, as illustrated in Figure 17.7. Such combinations of unrelated genes with flanking IS elements are sometimes called **transposons** or **compound transposons**. Transposon *Tn10*, for example, contains a tetracycline-resistance cassette bounded by two *IS10* elements. Coordinated breakage and joining at the outside ends of the two IS elements results in the movement of the tetracycline-resistance cassette along with the flanking *IS10*s.

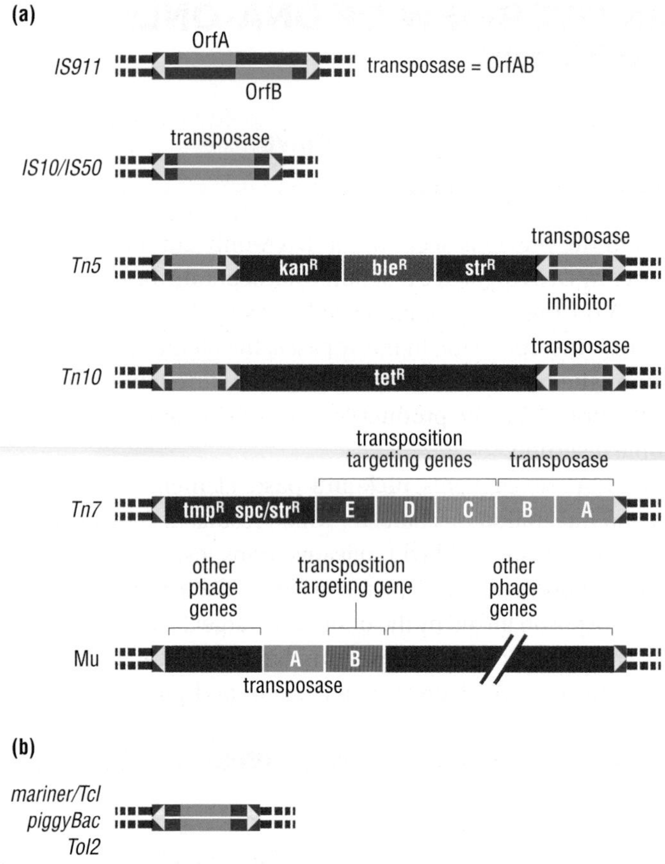

Figure 17.6 Some DNA-only transposable elements from bacteria and eukaryotes. Non-transposition genes including antibiotic resistance genes in bacterial elements are indicated in gray. In *IS911*, the transposase OrfAB is formed by translational frameshifting. IS elements can exist independently or as part of compound elements such as the transposons Tn5 and Tn10 in which IS50 and IS10, respectively, flank antibiotic resistance genes. Mu is a bacteriophage that inserts by transposition into the bacterial genome to form a lysogen and then replicates its DNA by replicative transposition during the lytic phase. (b) Eukaryotes. Eukaryotic elements encode only terminal inverted repeats and the transposase. Some transposases are encoded by a single exon, but others contain exons and introns.

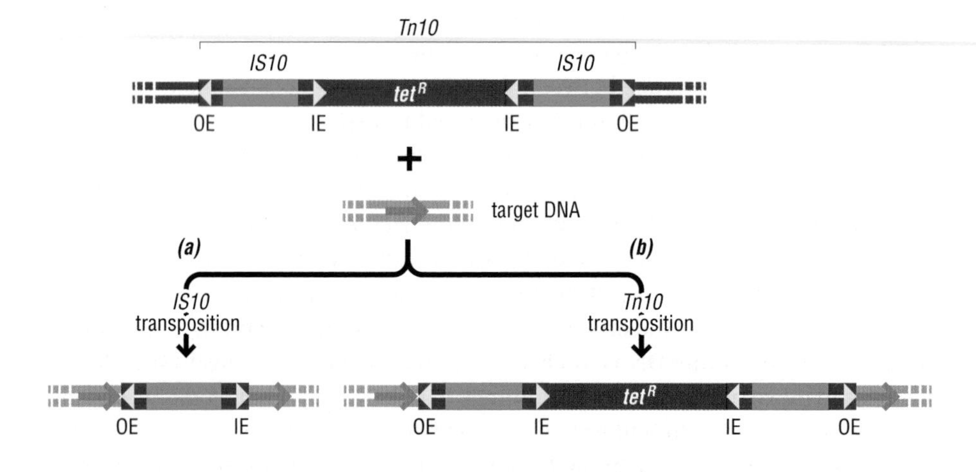

Figure 17.7 Comparison of *IS10* and *Tn10* transposition. *Tn10* is composed of two *IS10* elements flanking a tetracycline resistance cassette. Transposition of either *IS10* or *Tn10* can occur.

17.3 DNA-ONLY CUT-AND-PASTE TRANSPOSITION

We have seen in Section 17.2 how DNA-only cut-and-paste elements move around the genome, first by severing links to the host DNA on either side of the transposon, and then by forming new links to DNA at a target site elsewhere in the genome. But what is happening at the molecular level during this transposition process, and what molecular components are required to make it happen?

DNA breakage and joining reactions are coordinated within protein-DNA structures called transpososomes

The movement of all the DNA-only cut-and-paste elements that have been studied in bacteria and eukaryotes adopt the same fundamental mechanisms. Indeed, only slight variations on this mechanism mediate the replicative transposition of nick-and-paste elements such as bacteriophage Mu and the integration of the DNA versions of LTR transposons such as retroviruses.

Three DNA segments are involved in DNA cut-and-paste transposition: the two transposon ends and the target DNA. The key steps in moving the transposon from the donor site to the insertion site are:

- recognition of the transposon ends by the transposases
- the pairing of the transposon ends
- the excision of the transposon from the donor site
- interaction of the excised transposon with the target DNA
- integration of the transposon into the target DNA.

Transposition occurs within elaborate protein-DNA structures called **transpososomes**, which contain the transposon DNA, the transposase, and host proteins that may be required for transposition. The correct assembly of the transpososome coordinates the interaction of the transposon ends and target DNA and thereby controls the DNA breakage and joining reactions that underpin the transposition process. We first will consider the transpososomes that actually mediate transposition and then consider the chemistry of the DNA transactions in more detail.

Transpososomes mediate and control the progression of the DNA breakage and joining reactions

An overview of the way the transpososome mediates the DNA breakage and joining reactions that underpin DNA-only cut-and-paste transposition is shown in Figure 17.8. The initiation of transposition is marked by the recognition of the transposition substrate DNA by the transposase (Figure 17.8, step 1). This recognition involves the specific binding of the transposase to the TIRs at the transposon ends. The binding specificity of a transposase - such that it recognizes only specific sequences - ensures that it acts only on a transposable element with cognate binding sites. As such, we see how different transposases catalyze the transposition of only certain transposons.

In some cases, the formation of the complexes, in which transposase binding to the transposon ends occurs, is facilitated by host DNA-bending proteins such as the bacterial integration host factor (IHF), which bends DNA at specific sequences, or HU, a non-sequence-specific DNA-bending protein.

A key step in the formation of an active transpososome is the bringing together, or 'synapsis,' of the transposase molecules bound at the transposon ends through their oligomerization (Figure 17.8, step 2). Such synapsis results in the pairing of the transposon ends. In many instances, this oligomerization leads to a more stable complex, which activates the DNA breakage and joining activity of the transposase.

It is important to assure that no uncoupled DNA breakage occurs at a single transposon end that could lead to potentially lethal isolated double-strand breaks. A common strategy to assure the coupling of end synapsis and DNA cleavage is

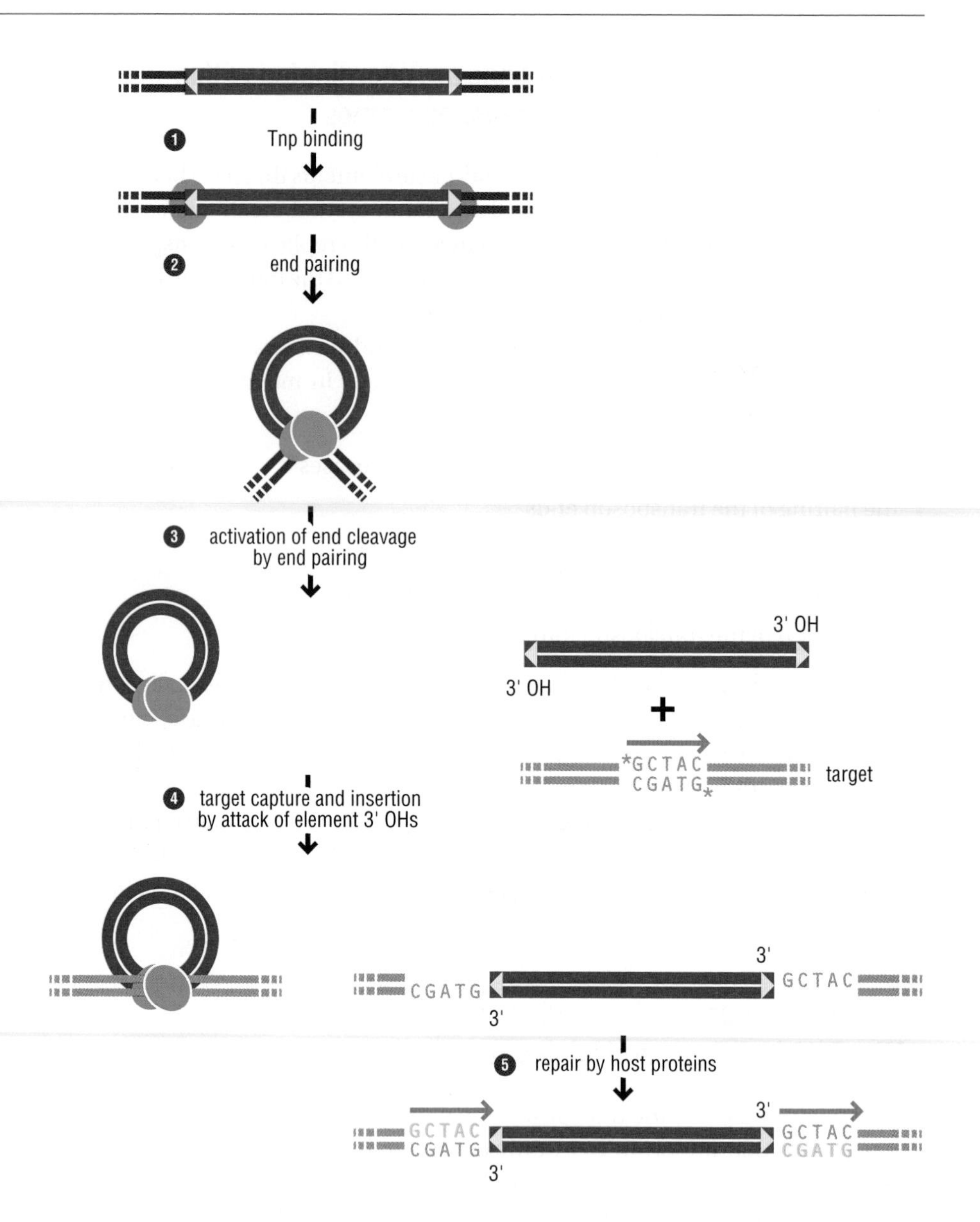

Figure 17.8 The transposon is excised by double stranded breaks at each end of the element and then integrated into the new target site by attack of the terminal 3'OHs at staggered positions (vertical arrows).

Scan here to watch a video animation explaining more about cut-and-paste transposition, or find it via the **Online Resource Center** at www.oxfordtextbooks.co.uk/orc/craig2e/.

cleavage *in trans* - literally, where the cleavage happens at the other transposon end from where the transposase is specifically bound. How does this happen? Let us consider *Tn5* as an example. In the case of this transposon, the active form of the transposase includes two monomers, which each bind one transposon end specifically through their binding domains, as shown in Figure 17.9. However, the active site domain (red spheres) of each monomer catalyzes cleavage at the opposite end of the transposon from where it is specifically bound. Thus, only when transposase binds to both ends and oligomerizes is there an active site domain for DNA breakage and joining positioned at *both* ends of the transposon.

Diverse DNA-only elements use the same chemical reactions for DNA breakage and rejoining

Following the interaction or synapsis of the transposon ends, the transposon is excised from the donor site by transposase-mediated double-strand breaks at both transposon ends (Figure 17.8, step 3. These double-strand breaks can be generated in a variety of ways as illustrated in Figure 17.10. For some elements, transposase

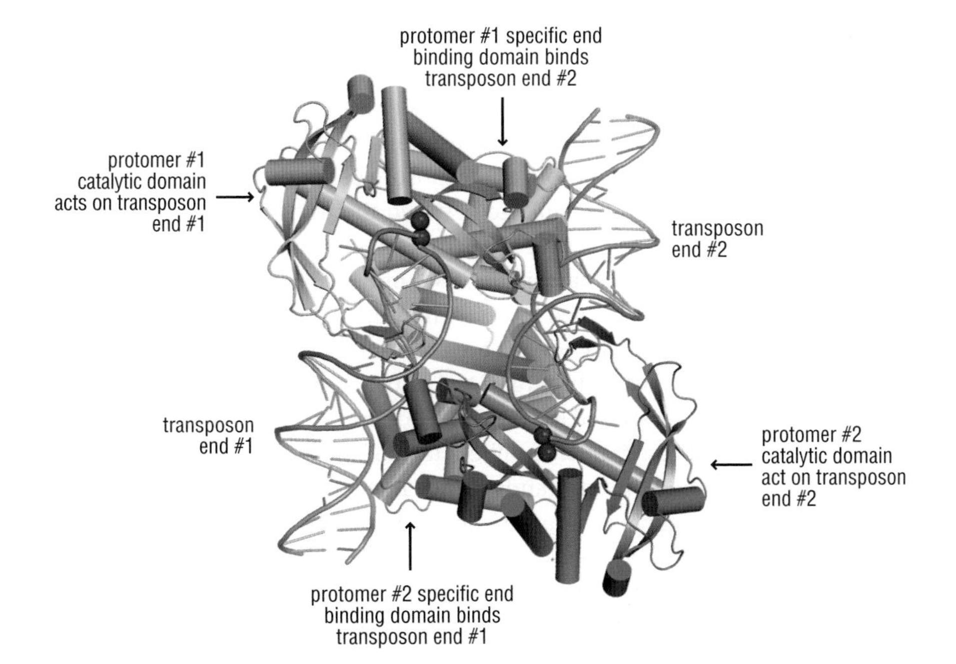

Figure 17.9 X-ray crystal structure of a *Tn5* transposase dimer bound to cleaved *Tn5* ends. The cleaved *Tn5* ends bind specifically to the N-terminal regions of each transposase. DNA breakage and joining is mediated by the catalytic domains (red spheres) *in trans* strategy, coupling end cleavage to end pairing. Protein Data Bank (PDB) code 1MUS.

From Steiniger-Whitem M, Rayment, I, and Reznikoff, WS. Structure/function insights into Tn5 transposition. *Current Opinion in Structural Biology*, 2004;**14**:50–57.

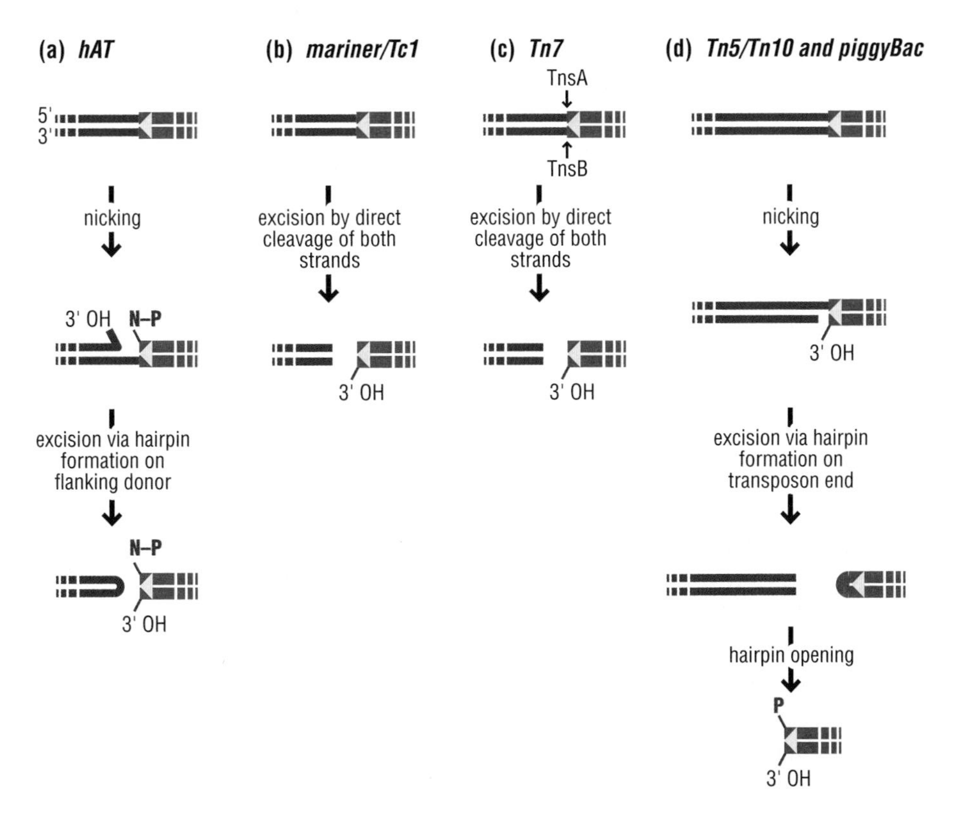

Figure 17.10 Strategies for transposon excision. The double-strand breaks that excise DNA-only elements can occur by a variety of combinations of nicking (hydrolysis) reactions, and hairpin formation reactions. The key result in all cases is the exposure of 3′ OHs at the transposon termini.

promotes a direct double-strand break by hydrolysis of phosphodiester bonds on both strands of the element (Figure 17.10b and c).

For other elements, the hydrolysis of one strand (whereby that strand is 'nicked') exposes a 3′ OH end, which can then attack the complementary strand by direct nucleophilic attack. This reaction – a transesterification reaction – forms a DNA hairpin in which either the top and bottom strands of the flanking donor DNA or the transposon end are covalently linked, and the transposon is thus excised. These two strategies are depicted in Figure 17.10a and d. If the hairpin forms on the

We discuss nucleophilic attack in general terms in Section 3.4.

transposon end (Figure 17.10d), transposase opens the hairpin. In all cases, the key outcome is the exposure of the 3′ OH ends of both strands of the transposon.

In contrast to the hydrolysis of a phosphodiester bond, such transesterification reactions do not change the number of phosphodiester linkages. Rather, there is now a covalent linkage in the hairpin and exposure of another 3′ OH end instead of there being a phosphodiester linkage between the transposon and the flanking donor DNA. As we see in Section 17.10, this direct transesterification strategy sharply contrasts with the mechanism of CSSR, in which the high energy of DNA phosphodiester bonds is preserved by covalent protein–DNA linkage.

So what happens once the 3′ OH ends are exposed? The transposase-bound exposed transposon ends engage the target DNA and insertion of the element occurs (Figure 17.8, step 4). The exposed transposon ends are joined to the target DNA by transposase-mediated direct nucleophilic attack of the 3′ OH ends at staggered positions on the top and bottom strands of the target DNA (Figure 17.8, step 5).

The length of this stagger is characteristic of a particular family of transposons. Transposons insert at many sites, although the choice of the insertion site is not entirely random. Instead, the choice of insertion site can be influenced by the DNA sequence that is recognized by the transposase, by non-transposase element-encoded proteins, or by host-cell proteins.

The nature of the attack of the 3′ OH ends on the target DNA is very similar to the reaction that occurs during DNA replication or transcription, when the 3′ OH end of the growing nucleic acid strand attacks the high-energy phosphodiester bond of the incoming nucleotide (see Figure 3.19). In both cases, an existing high-energy bond, in the transposon target DNA or within the incoming nucleotide, is converted into a high-energy bond that links the transposon end to the target DNA, or links the incoming nucleotide to another one in the growing chain.

As a consequence of the integration reaction, the 3′ ends of the transposon are covalently linked to the target DNA at staggered positions, whereas its 5′ ends are flanked by short gaps that result from the staggered positions of this targeted joining. These gaps are repaired by a host-cell polymerase, resulting in the characteristic target-site duplications that flank transposable elements (Figure 17.8, step 5). Thus, the regeneration of intact duplex DNA following transposition requires the formation of new phosphodiester bonds by replication and ligation.

Transposases and retroviral integrases share a common catalytic domain

The release of 3′ OH transposon ends and their direct attack on the target DNA is a highly conserved mechanism, being found in cut-and-paste elements from bacteria and eukaryotes but also in the integration of LTR elements, as we shall see later in the chapter. This mechanistic conservation is reflected in structural similarities in the active site regions of transposases from bacterial elements, eukaryotic DNA-only elements, and the integrases of retroviruses; this similarity is reflected in the structures shown in Figure 17.11.

The central feature of these active sites is the folding of the protein to juxtapose several beta sheet strands and alpha helices in an arrangement that was first identified in the enzyme RNase H; hence, it is called an RNase H fold. Positioned on these beta strands and alpha helices are several acidic amino acids, usually an aspartic acid–aspartic acid–glutamic acid (DDE) motif, to which the essential Mg^{2+} cofactor for transposition reactions is bound. These amino acids can be widely separated in the sequence of the transposase, but become juxtaposed at the active

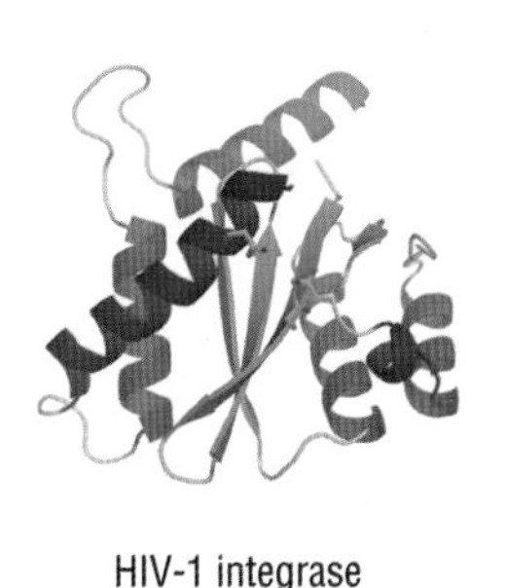

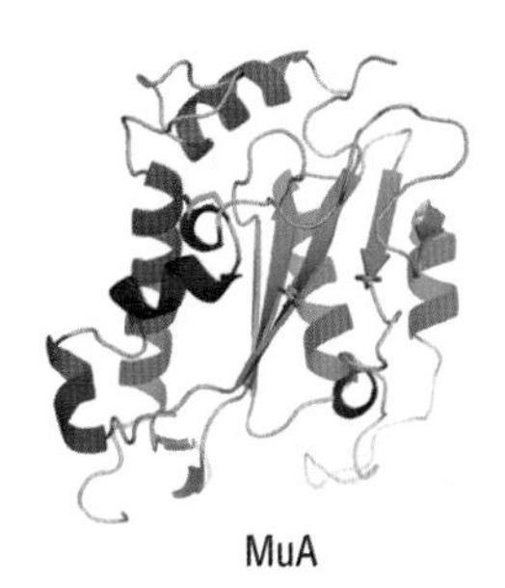

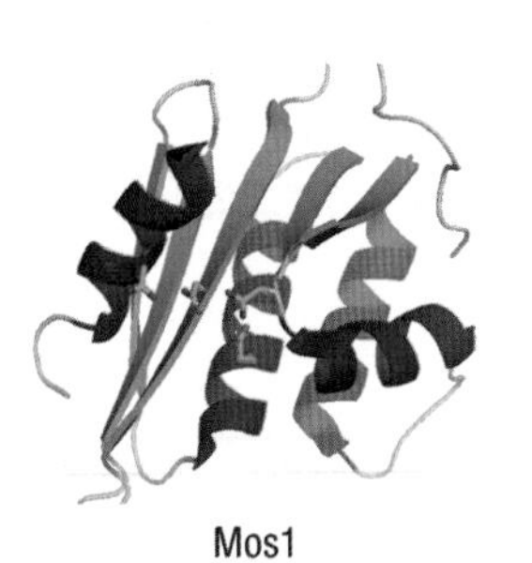

Figure 17.11 Similarity among transposases and a retroviral integrase. The structures of the human immunodeficiency virus (HIV)-1 integrase, the bacterial MuA transposase and the eukaryotic Mos1 transposase of the *mariner/ Tc1* family are aligned on their active site regions. The RNase H-like catalytic cores are shown as blue. The conserved acidic amino acids that are juxtaposed in the active site via the RNase H-like fold are shown in green.

From Hickman, AB, Chandler, M, Dyda, F. Integrating prokaryotes and eu karyotes: DNA transposases in light of structure. *Critical Reviews In Biochemistry and Molecular Biology*, 2010;**45**:50–69.

site by the RNase H fold. Moreover, although the amino acid sequences and secondary structures of transposases and other RNase H-related enzymes and polymerases are quite distinct, in all cases Mg^{2+} ions with similar spatial relationships form the active site.

Repair of the gapped donor site resulting from element excision is critical

The excision of a cut-and-paste transposon leaves a gap at the donor site that must be repaired. If a hairpin is left at the end of the donor DNA upon element excision, an important first step in repair is the opening of the hairpin by cellular nucleases. The repair often happens by non-homologous end joining (NHEJ) in eukaryotes, as illustrated in Figure 17.12a. Because of the target-site duplication that occurred upon element insertion into what is now the donor site, simple rejoining does not result in the restoration of this site to its original, pre-transposon sequence. Instead, 'footprints' that reflect the element's insertion will persist; the element is said to have undergone 'imprecise excision' (Figure 17.12a, step 2).

➲ We discuss NHEJ in more detail in Section 16.2, and learn about hairpin opening in Section 17.3.

If transposon insertion occurred within a protein-coding sequence, the proper reading frame and protein function will likely not be restored. 'Precise excision' that restores the donor site to its pre-transposon state may occasionally occur with nucleolytic processing during NHEJ.

Another strategy for repair of the gap at the donor site is to exploit homology-directed DNA repair using a sister chromatid or homolog as templates. As we shall see below, some transposons couple their transposition to DNA replication that may facilitate such repair. If transposon excision occurs and a sister chromatid still containing the transposon is present (as in Figure 17.12b), the transposon-containing sister chromatid can be used as the template for repair. The consequence of this repair will be to restore a copy of the transposon at the donor site. Note that, in this case, although the transposon moved by a cut-and-paste mechanism, transposition is effectively replicative: a copy of the transposon appears at the new insertion site and persists at the donor site from which excision occurred; it is also present in the sister chromatid.

➲ We describe homology-directed DNA repair in more detail in Section 16.8.

17.4 DNA-ONLY NICK-AND-PASTE TRANSPOSITION

Let us now go on to consider a related, but somewhat modified, version of DNA-only cut-and-paste transposition, which is exhibited by DNA-only nick-and-paste transposons.

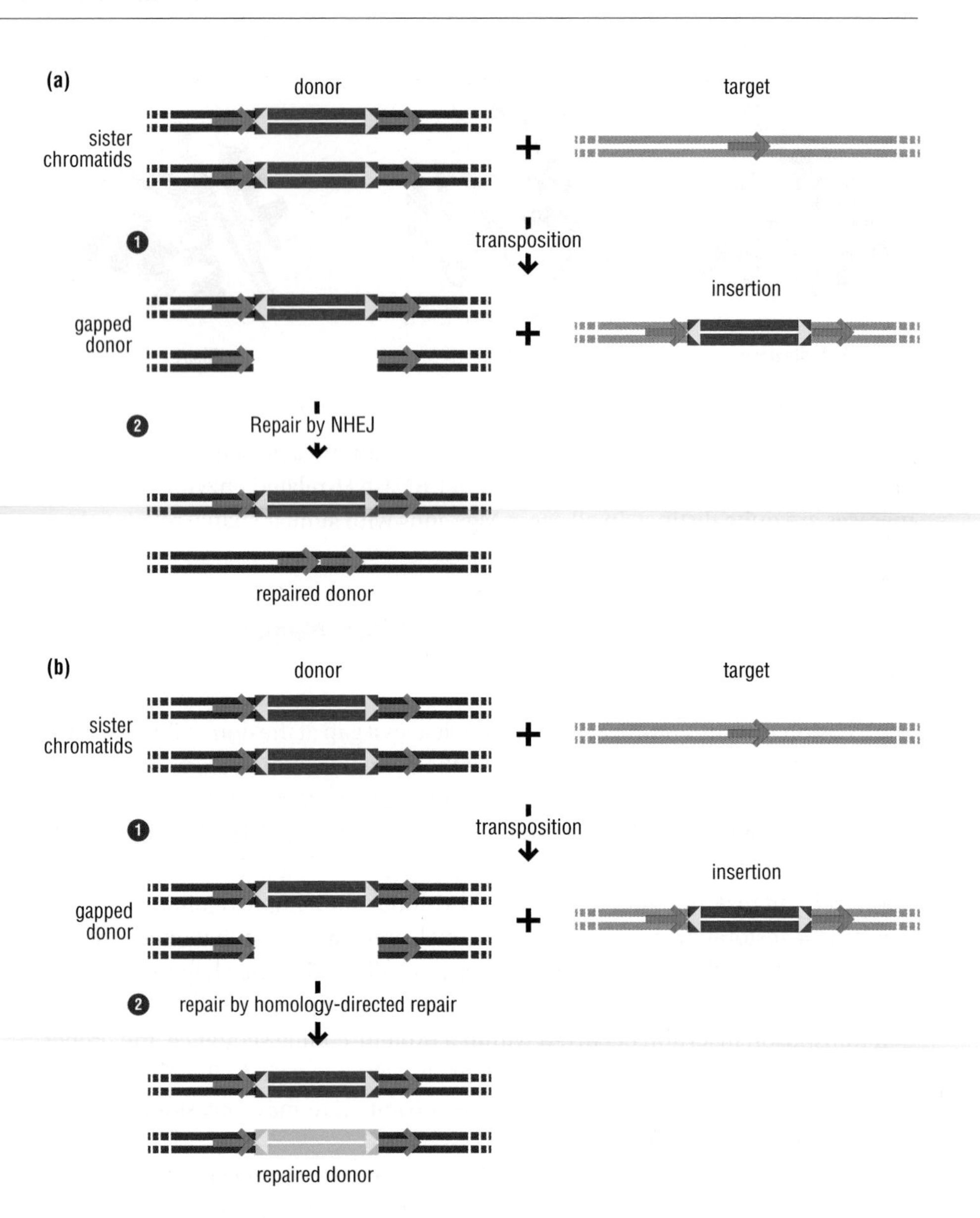

Figure 17.12 Transposition of a DNA cut-and-paste element may or may not change the donor site. Repair of the gap resulting from transposon excision can occur in several ways.

Bacteriophage Mu uses replicative nick-and-paste transposition to replicate its genome

A slight variation of the cut-and-paste mechanism described in the previous section is called the nick-and-paste mechanism, whereby the transposon is actually replicated, resulting in two copies of the transposon. This mechanism has been extensively studied for bacteriophage Mu, which can infect *Escherichia coli*. The bacteriophage has a double-stranded DNA genome and uses transposition at several key points in its lifecycle.

Within a virus particle, Mu is a linear DNA flanked at each end by several hundred base pairs of *E. coli* DNA. Upon infection, the Mu genome integrates randomly into the *E. coli* chromosome, using cut-and-paste transposition to excise from the flanking *E. coli* DNA that forms its 'donor site'.

There are two possible outcomes following Mu integration. On the one hand, integrated Mu can remain in a dormant state to form a lysogen. Alternatively, Mu can replicate its DNA for lytic growth by carrying out multiple rounds of replicative transposition, which result in the formation of cointegrates, the mechanism of which we discuss below. In addition to increasing Mu copy number, this replicative transposition destroys the *E. coli* chromosome. Following cointegrate formation, each Mu genome and several hundred base pairs of flanking *E. coli* DNA are packaged into a virus particle and the cell lyses to release the phage particles.

By contrast to Mu, another nick-and-paste element, *IS911*, uses an intramolecular nick-and-paste strategy to generate a circularized transposon species that is then opened to form a linear transposon that then inserts into a target site.

Nick-and-paste replicative transposition of bacteriophage Mu involves transposon end nicking and joining to the target DNA

Mu replicative transposition begins with the introduction of nicks in the donor site at the 3′ ends of the transposon by the MuA transposase, as illustrated in Figure 17.13, step 1. (This is in contrast to the double-strand breaks that excise a transposon from the donor site during cut-and-paste transposition (see Figure 17.5). The nick exposes the 3′ OH Mu ends, leaving the 5′ Mu ends still attached to the donor (dark gray) DNA. The 3′ OH Mu ends then attack the target DNA (light gray) at staggered positions, mirroring the insertion of an excised transposon into a target DNA (Figure 17.13, step 2). In the resulting fusion product, known as the strand-transfer product, the transposon is covalently linked to the target DNA (light gray) through its 3′ ends and to the donor DNA (dark gray) at its 5′ ends.

These joined molecules can then serve as templates for DNA replication, which initiates at the 3′ OH ends of the single-strand gaps flanking the Mu DNA on the target DNA (light gray) (Figure 17.13, step 3). Replication then proceeds across the entire element to form the cointegrate, in which the two copies of the transposon link the donor (dark gray) and target (light gray) DNAs.

Thus, although such a nick-and-join transposition reaction does not involve the complete excision of the element from the donor site, the key chemical steps in replicative transposition - exposure of the 3′ OH ends of the transposon and the direct attack of these ends to join the element to the target DNA - are the same as those underlying the transposition of an element that undergoes excision and integration.

Mu transposition proceeds within a series of nucleoprotein complexes

Mu has provided many key insights into the mechanism of transposition and its control. The MuA transposase is converted from an inactive monomer to a tetramer that executes breakage and joining through the multi-step assembly of the transpososome, as summarized in Figure 17.14. How does this assembly occur? First, monomeric MuA binds to multiple sites at the ends of the Mu genome. Pairing

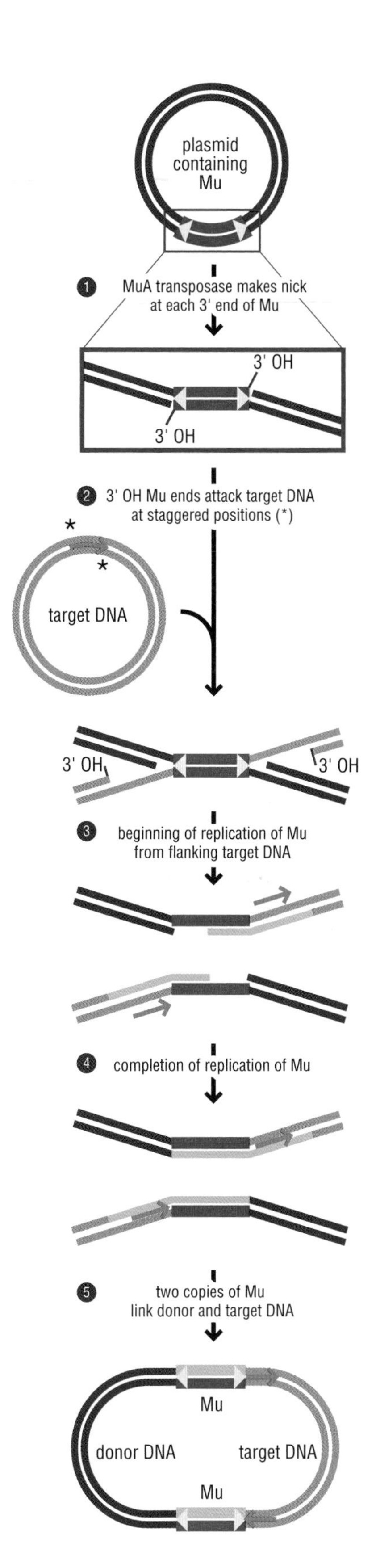

Figure 17.13 Mechanism of nick-and-paste replicative transposition to produce a cointegrate. The cointegrate contains both the donor and target DNA linked by two copies of Mu. Target-site duplications (blue) exist at the Mu-target DNA junctions. Note that each Mu copy contains one 'old' strand from the parent transposon (red) and one newly replicated strand (green).

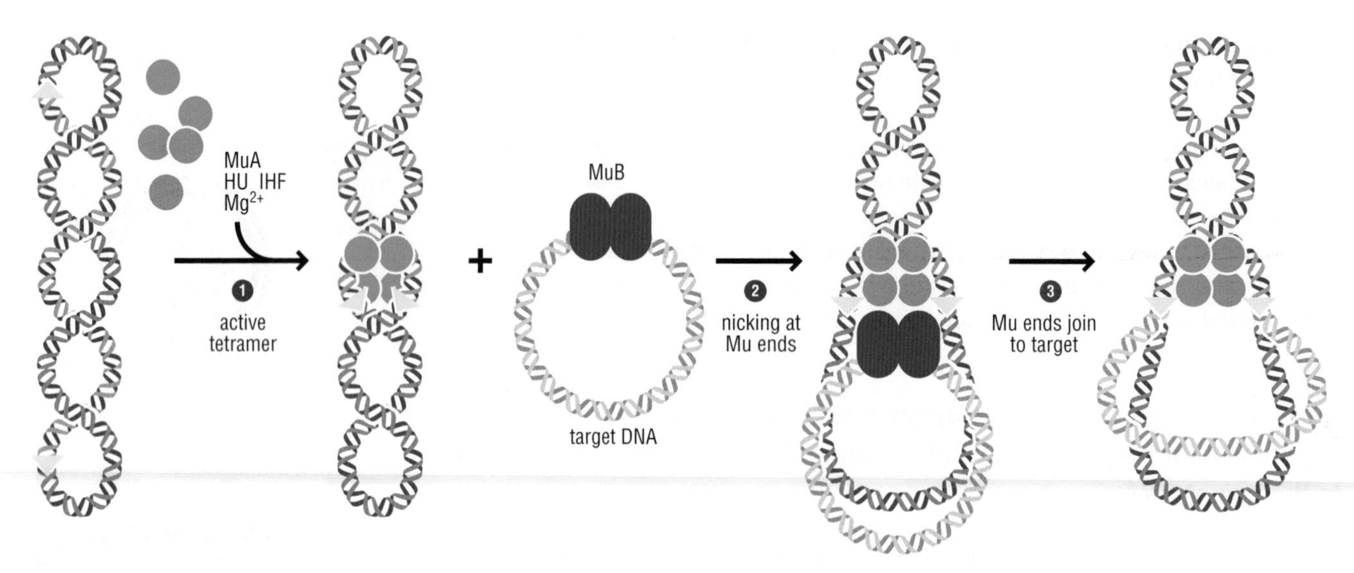

Figure 17.14 Mu transpososomes. Mu transposition proceeds via a series of protein–DNA complexes called transpososomes. The supercoiled donor DNA containing Mu (red) with its terminal TIRs (yellow) is shown. MuA transposase (orange) binds specifically to multiple sites in the transposon ends, and in the presence of the DNA-bending proteins HU, IHF, and the Mg^{2+} cofactor (not shown), a stable tetramer of MuA is formed that pairs the ends. MuB (dark blue) binds to the target DNA (light gray). In the strand transfer product, Mu remains covalently linked to the donor DNA through its 5′ ends and is covalently linked to the target DNA though its 3′ ends.

of the ends results in the formation of a MuA tetramer that can execute breakage and joining. Conversion of the MuA monomers to the active tetramer is facilitated by the host DNA-bending proteins HU and IHF, which also bind to the ends of Mu. Conversion is also facilitated by supercoiling, which also likely induces DNA bending. Another Mu-coded protein, MuB, which can also interact with MuA and target DNA, is involved in recruiting the target DNA to the MuA-bound ends (see Figure 17.14, step 2).

As discussed for the Tn5 transposase in Figure 17.9, Mu transposase executes end cleavage in *trans*: the MuA subunit that cleaves one end of the Mu sequence is bound specifically to the other end of the Mu DNA. This *trans* assembly strategy ensures that active enzymes can only cleave Mu when the DNA ends are brought together in the transpososome. The interaction of MuB with MuA is also important for MuA strand nicking activity and joining of the 3′ Mu ends to the target DNA (Figure 17.14, step 3).

Following the joining of the transposon ends to the target DNA, DNA replication must occur to repair the single-strand gaps that flank newly inserted transposon, and to copy the transposon, generating the cointegrate. Host chaperone proteins convert the highly stable MuA tetramer to a form that recruits the host replication machinery, which mediates this DNA synthesis.

17.5 DNA CUT-AND-PASTE TRANSPOSITION IN ADAPTIVE IMMUNITY

The transposition of DNA-only cut-and-paste elements is not just a nuisance event that must be tolerated by the cell without any real advantage. Instead, certain cell types have evolved to exploit such transposition reactions for their own benefit. In this section, we consider a particularly important example of this exploitation, the use of transposition systems in the generation of immune-system diversity.

Diverse transposition systems have been adopted to mediate a variety of host processes

Transposition proteins have been co-opted to perform a variety of cellular functions. There are, for example, multiple examples of transcription factors derived from DNA cut-and-paste transposases. Perhaps the most spectacular example of transposase domestication, however, is the use of a DNA cut-and-paste transposon derivative in the immune system to mediate the assembly of immunoglobulin superfamily proteins, the antibodies and antigen receptors produced by lymphocytes. This assembly mechanism allows for the generation of a vast diversity of proteins that are able to recognize any disease-causing microorganism. A schematic representation of an antibody molecule is shown in Figure 17.15.

The generation of diverse antibodies and antigen receptors stems from the way that the genes for these proteins are assembled in the individual immature lymphocytes from multiple gene segments, with each immature lymphocyte assembling a slightly different combination of gene segments. Consequently, each mature lymphocyte synthesizes a slightly different version of the protein.

Gene assembly occurs through recombination reactions that delete the DNA between gene segments and juxtapose the coding regions that remain, ready for subsequent rejoining. The generation of the DNA double-strand breaks that initiate this recombination is very similar in mechanism to the excision process of cut-and-paste DNA transposons. Indeed, the lymphocyte-specific recombinases that mediate this reaction are structurally related to transposases. The gene segments are then joined by the NHEJ system. This important immune-system recombination system probably evolved from a transposable element that was harnessed by the ancestors of vertebrates for this vital role.

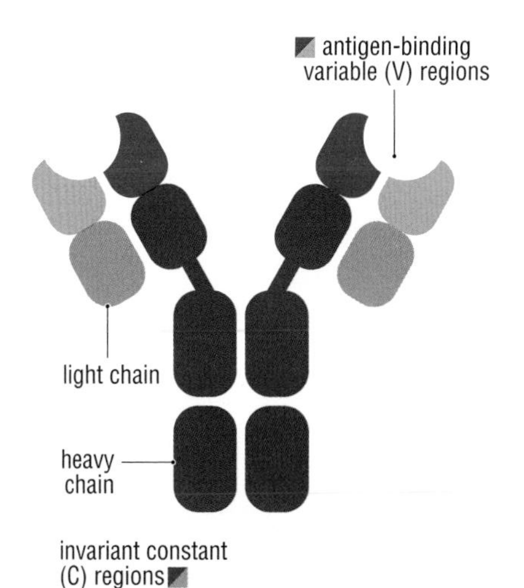

Figure 17.15 Schematic representation of an antibody molecule. Antibodies (immunoglobulins) are composed of two identical larger (heavy) chains and two identical smaller (light) chains, attached to each other by disulfide bonds (not shown). Each chain has a variable region (pink, light pink) that is formed by the assembly of different gene segments, and a constant region (gray, light gray). The variable regions form the antigen-binding site that recognizes foreign molecules.

Antibody and antigen-receptor genes are formed by joining gene segments together

The antibodies produced by B lymphocytes and the antigen receptors of T lymphocytes feature N-terminal regions called variable regions. These variable regions are encoded by several different loci that are each composed of multiple alternative protein-coding gene segments, as shown in Figure 17.16. Both these proteins

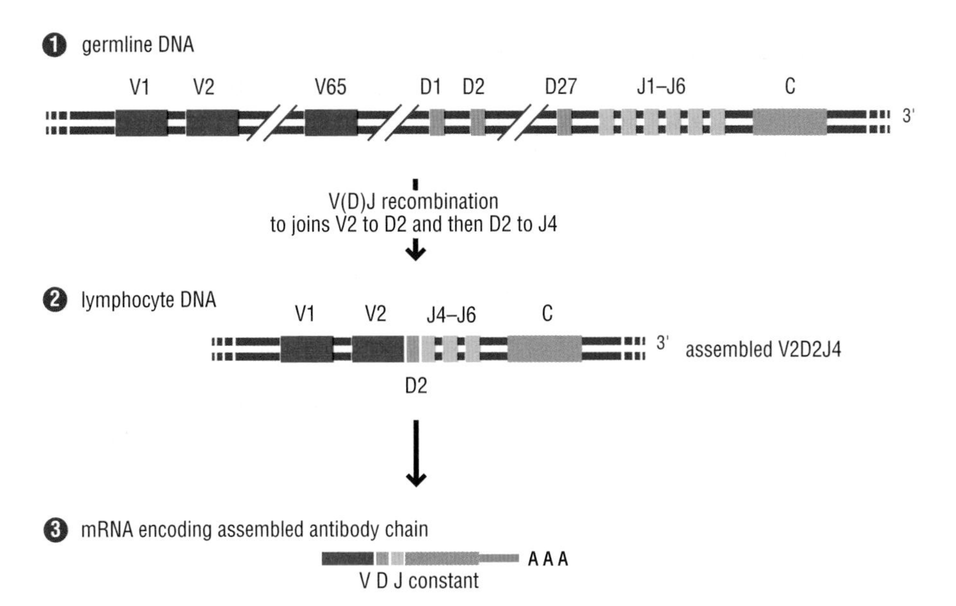

Figure 17.16 Assembly of immunoglobulin gene segments. For simplicity, only the segments encoding a heavy chain are shown. (1) The heavy chain is encoded in four separate gene segments, V, D, J, and C. In the human heavy chain gene locus schematically represented here, there are 65 V segments (V1–V65), 27 D segments (D1–D27), and six J segments (J1-J6) shown in different shades of red along with a constant-region segment shown in gray. (2) In the course of the differentiation of the lymphocytes that make antibodies, one each of the V, D, and J segments become joined by the host non-homologous end joining (NHEJ) system to encode a complete variable-region coding domain, here V2D2J4. (3) After this DNA rearrangement, the complete gene can be transcribed, and the variable region is joined to the constant region by RNA splicing.

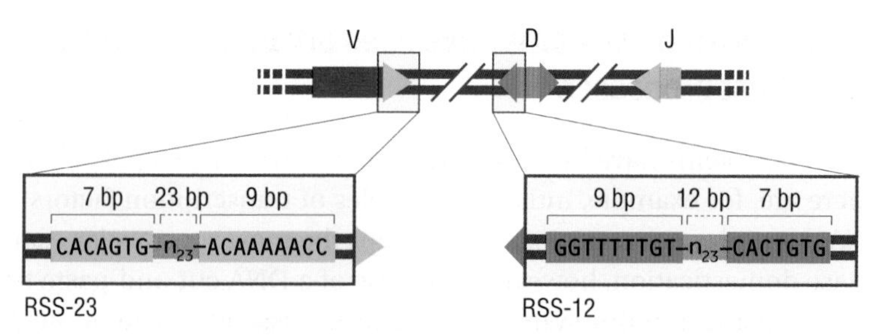

Figure 17.17 RSSs. The recombining segments of antibody genes are flanked by specialized recognition sites for the RAG recombinase that executes double-strand breaks at the edges of the gene segments. Each RSS comprises a seven-nucleotide (heptamer) sequence and a nine-nucleotide (nonamer) sequence, separated by either 23 (green RSS) or 12 (blue RSS) bp. Each type exists in a unique configuration with respect to the coding segments. Recombination requires one RSS with a 12 bp spacer and one RSS with a 23 bp spacer.

belong to the larger immunoglobulin superfamily, but they are the only ones whose genes undergo this programmed somatic recombination.

The assembly process is known as **V(D)J recombination** after the names of the gene segments that are joined together – the V (variable), D (diversity), and J (joining) segments. Each locus can include hundreds of different gene segments and spans hundreds of kilobases. Random selection of a gene segment of each type, and deletion of the DNA between them, generates novel combinations of gene segments in each B cell or T cell precursor. The gene segments that are assembled by recombination lie upstream of a C-terminal invariant region, called the constant region.

The immune-system recombinase RAG resembles a transposase

Antibody and T cell receptor gene segments are flanked by recombination signal sequences (RSSs), which are inverted repeats that contain conserved heptamer (7 base pairs (bp)) and nonamer (9 bp) segments, as illustrated in Figure 17.17. Thus, they are equivalent to the TIRs of transposable elements. Figure 17.17 shows how there are two kinds of RSS: one in which the conserved heptamer and nonamer are separated by 23 bp (to the left in Figure 17.17) and the other in which the heptamer and nonamer are separated by 12 bp (to the right in Figure 17.17); recombination requires one RSS of each type. RSSs are recognized and acted on by the lymphocyte-specific recombinase RAG in a process summarized in Figure 17.18. Thus the RSSs

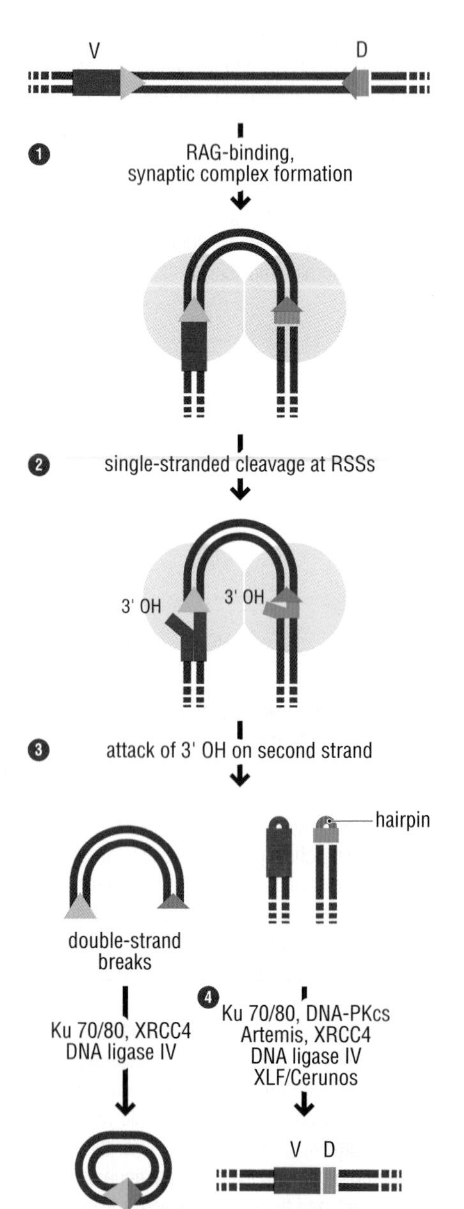

Figure 17.18 Breakage and joining by RAG recombinase. The DNA substrate for the joining of a V segment to D segment is shown. (1) An oligomer of RAG (pale blue) binds to the RSS-23 (green) flanking a coding segment; synapsis then occurs by capture of the RSS-12 (blue) of another coding segment (2) RAG then introduces a nick at the 5′ end of the RSS that abuts each coding segment. (3) The exposed 3′ OH ends of the coding segments then attack their complementary strands, forming hairpins at the end of each coding segment; the signal ends containing the RSSs are blunt ended. (4) Joining of the two coding segments (V and D) forms a coding joint within the chromosome, and joining of the excised RSS-bounded segment creates a circular DNA containing the signal joint. Joining is mediated by RAG in collaboration with universal DNA repair proteins that carry out NHEJ of double-strand breaks. These include Ku, composed of the Ku70 and Ku 80 subunits, which holds the broken ends near to each other and recruits the kinase DNA-PKcs. DNA-PKcs phosphorylates and activates the nuclease Artemis, which then opens the hairpins. Other proteins involved in forming signal joints and coding joints are DNA ligase IV and its partner XRCC4. XLF/Cerunos is involved in coding joint formation.

are functionally comparable to the terminal inverted repeats of cut-and-paste transposons whereas RAG is comparable with the transposase.

RAG is composed of two subunits, RAG1 and RAG2; RAG1 is the subunit that carries out the breakage and joining steps at the RSSs and is structurally similar to cut-and-paste transposases. In particular, it appears to have an RNase H fold on which essential conserved acidic amino acids are present. The RAG complex binds specifically to the RSSs (Figure 17.18, step 2) and promotes double-strand breaks that excise the RSS-bounded segment, mirroring the way a transposase recognizes and cleaves the ends of an integrated transposon, excising it from the DNA. This reaction exposes the ends of two gene-coding segments.

The detailed mechanism of double-strand break formation by RAG is mechanistically related to the excision of *hAT* transposons (transposon excision is described in Figure 17.10). RAG introduces a single-strand nick at the junction of the 5′ end of the RSS with the 3′ end of the adjacent coding gene segment (Figure 17.18, step 2). The exposed 3′ OH then launches an intramolecular attack on the un-nicked strand to covalently link the two complementary strands together to form a hairpin at the end of the coding segment, and leaves a blunt end on the RSS segment, which is now no longer attached to the gene segments it previously sat between (Figure 17.18, step 3).

The similarity of the RAG double-strand break mechanism to that of *hAT* and *Transib* transposons, and the close structural similarity to the *Transib* family transposases, suggests that the RAG system evolved from a transposon that invaded a common ancestor of the organisms which today display adaptive immunity.

Gene segments are joined together by the NHEJ DNA repair system

The DNA hairpins on the antibody and T cell receptor gene-coding segments exposed by the RAG-promoted double-strand breaks are cleaved by the nuclease Artemis and then joined together to form a functional protein-coding DNA by reactions that involve the universal NHEJ system of DNA repair (Figure 17.18, step 4). The joint formed between two coding segments is called a **coding joint**, while the joint formed by the exposed ends of the RSS segment is known as the **signal joint**.

➔ We learn more about NHEJ in Section 16.2.

A key step in the formation of the coding joint is the opening of the hairpins at the ends of the coding segments by the nuclease Artemis. Opening of the hairpin at its tip generates a blunt end. However, the hairpin is often opened at a position other than its tip, resulting in additional diversity being created in the sequence of the coding joint. Other nucleotides may also be added directly to the exposed ends of coding segments by the enzyme deoxynucleotidyl terminal transferase. Both blunt and filled ends can also be altered by the addition or deletion of nucleotides before the joint is finally sealed by the XRCC4/DNA ligase IV complex whose activity is modulated by XLF/Cernunnos.

The repair process also involves DNA-PK, a kinase that is apparently recruited to the ends through Ku, which itself binds to DNA ends. Mutant organisms lacking functions conferred by these proteins cannot repair other double-strand breaks in DNA, such as those generated by ionizing irradiation, and are thus both immunodeficient and radiation sensitive.

When the protein-coding DNA formed by the coding joint is fully assembled, transcription produces an RNA containing the assembled gene segments and the coding joints between them. RNA splicing then generates an mRNA, which is translated into a functional antibody chain or T cell receptor chain.

The ends of the excised RSS-bounded fragment are joined to form a circular DNA containing the signal joint. This DNA is lost from the cell because it lacks the structures, such as a centromere, that would enable its distribution into daughter cells.

The action of RAG can be dangerous and is tightly regulated

For the immune system to function properly, the sequence-specific recombination reactions promoted by RAG must occur only in the right genes in the right cell types at the appropriate developmental stage. One level of control ensures that RAG is only expressed in the precursors of B cells and T cells. Another ensures that recombination at RSS sequences only occurs in the presence of low-level transcription, which itself occurs only in lymphocytes at particular developmental stages. It is likely that this low-level transcription alters chromatin structure in the region of the RSSs, making them accessible to RAG.

Despite the strict controls on RAG action, it can sometimes be misdirected. A significant fraction of lymphoid cancers appears to result from rearrangements of chromosomal DNA that lead to the aberrant and unregulated expression of tumor suppressor or oncogenic proteins, for example, by the fusing of two different genes to encode a novel protein that displays oncogenic activity. These rearrangements are thought to be associated with aberrant RAG-mediated somatic recombination.

17.6 RETROTRANSPOSONS

Having considered the transposition of elements that use only DNA intermediates, we now consider the transposition of elements that have RNA intermediates – the retrotransposons.

Retrotransposons are numerous and widespread

Multiple types of transposable elements that move via an RNA intermediate have been found in both bacteria and eukaryotes. LTR elements include the retroviruses, which have an extracellular phase and are found only in vertebrates, and retroviral-like elements, the movements of which are strictly intracellular, and which are widespread in fungi, plants and animals. The human genome lacks endogenous active retroviruses; the pathogenic human immunodeficiency virus (HIV) which leads to acquired immunodeficiency syndrome (AIDS) is acquired by infection. The human genome however does contain multiple inactive retroviruses. It is generally thought that retroviruses evolved from the retroviral-like elements by acquiring the ability to exit and then re-enter cells.

Non-LTR elements are present in all domains of life. The human genome contains about 100 active autonomous non-LTR elements, and there are multiple examples of their transposition causing disease.

A notable aspect of retrotransposons is their success in colonizing many eukaryotic genomes. For example, more than 85% of the maize genome and more than 42% of the human genome is composed of retrotransposons. In some plant genomes, genes have been described as 'small islands surrounded by seas of LTR retrotransposons'. Non-LTR elements are particularly abundant in vertebrate genomes.

Reverse transcriptase mediates the transposition of LTR and non-LTR elements

Transposition of LTR and non-LTR elements begins by transcription of the element by host RNA polymerase II. These RNAs are converted to DNA by a reverse transcriptase, the DNA polymerase that can synthesize DNA using RNA as a template. This reverse transcriptase is usually element-encoded, as summarized in Figure 17.19. An RNase H activity, which can remove the RNA template strand to facilitate synthesis of the second DNA strand, is also required and is sometimes encoded by the element and in other cases is provided by the host.

Reverse transcriptases have two distinctive features: they are more error prone and less processive than replicative polymerases. Both of these properties are reflected in retrotransposon structure and function. The error-prone nature of reverse transcription means that mutations occur at a higher frequency in retroelements than in the course of chromosomal replication. This property is reflected by the rapid rate at which resistance to antiviral drugs can occur during treatment against HIV. Another impact of the nature of reverse transcriptase is the truncation of some elements during their replication because of the lower processivity of reverse transcriptase. Thus, these truncated elements lack their 5′ end promoters and are not transcribed, and therefore cannot transpose.

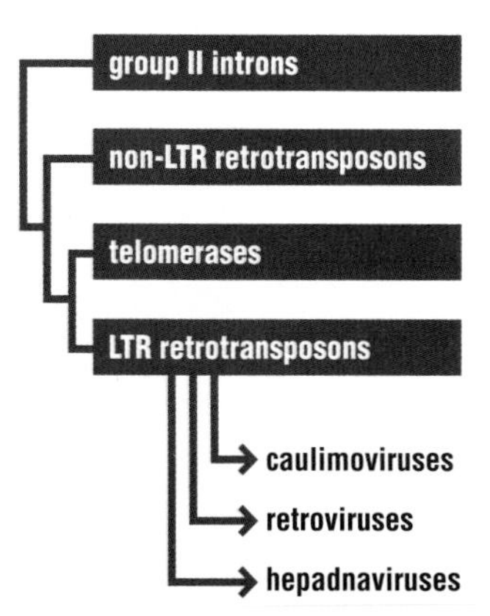

Figure 17.19 Groups of reverse transcriptases. All these elements encode a reverse transcriptase. The three arrows emerging from the LTR retrotransposons represent the independent origins of hepadnaviruses (mammalian viruses that can cause liver infections including hepatitis B in humans), retroviruses and caulimoviruses (plant viruses).

➲ We discuss the structure of reverse transcriptase and its role in chromosome replication in Chapter 6.

Retrotransposon transposition does not involve excision

A very significant difference between the movement of DNA-only cut-and-paste elements and retrotransposons is that retrotransposon transposition does not involve the excision of the element from the donor site. Retrotransposons move by copy-and-paste mechanisms that begin with synthesis of an RNA copy of the element. Thus, issues of donor site repair (see Figure 17.12) are not part of the retroelement lifecycle. However, changes to LTR elements can occur by homologous recombination between the several hundred base pair long LTRs that form the ends of the element, deleting the information between them and leaving a solo LTR end in the genome, as shown in Figure 17.20.

Thus solo LTR elements are not mobile but rather serve as markers of the prior presence of an LTR element.

➲ We discuss homologous recombination between repeated sequences in Section 16.7.

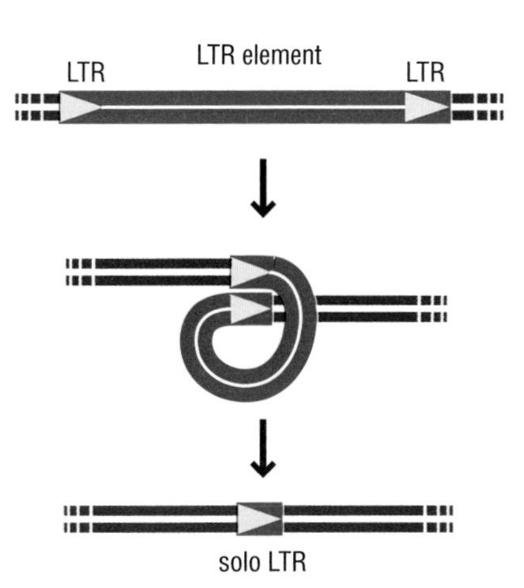

Figure 17.20 Homologous recombination can produce a 'solo LTR' from an LTR element. The proviral DNA of an LTR element with its several hundred base pair LTRs is shown. Homologous recombination between the LTRs 'excises' the element, leaving a single LTR copy.

17.7 LTR RETROTRANSPOSONS

Having now considered the nature of retrotransposons in overview, let us now turn our attention to the transposition mechanism of LTR elements. We will then discuss the non-LTR elements in the next section.

The LTR transposon lifecycle requires multiple element-encoded proteins

The lifecycle of an LTR transposon is more complex than that of a DNA-only transposon; this is reflected in the number and type of element-encoded proteins that are required for transposition. The structure of some well-studied retroviruses and retroviral-like elements is shown in Figure 17.21. All LTR elements encode four proteins: GAG, a polyprotein, which is cleaved after synthesis to give several structural

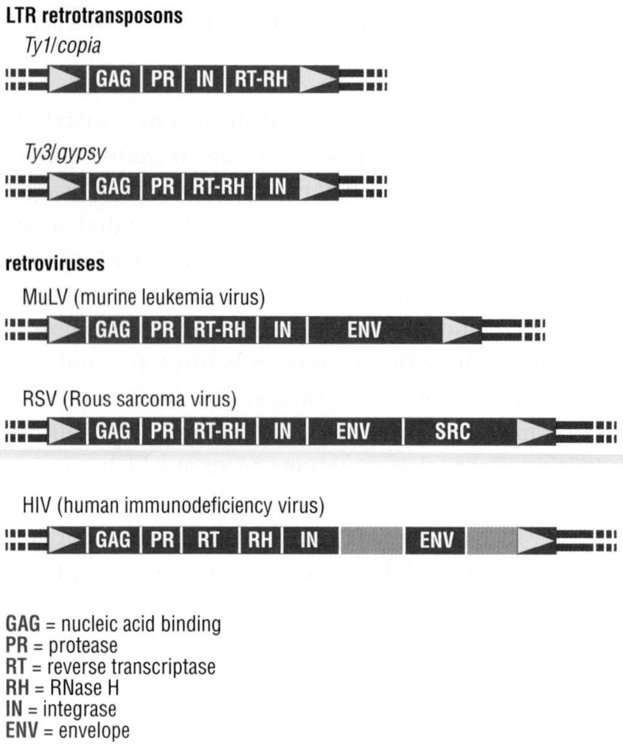

LTR retrotransposons

Ty1/copia

GAG | PR | IN | RT-RH

Ty3/gypsy

GAG | PR | RT-RH | IN

retroviruses

MuLV (murine leukemia virus)

GAG | PR | RT-RH | IN | ENV

RSV (Rous sarcoma virus)

GAG | PR | RT-RH | IN | ENV | SRC

HIV (human immunodeficiency virus)

GAG | PR | RT | RH | IN | ENV

GAG = nucleic acid binding
PR = protease
RT = reverse transcriptase
RH = RNase H
IN = integrase
ENV = envelope

Figure 17.21 The genetic structure of LTR elements. The terminal yellow arrows represent the long (several hundred base pair) terminal repeats (LTRs). *Ty1/copia* and *Ty3/gypsy* are families of LTR retrotransposons, of which the prototypes are the yeast *Ty1* and *Ty3* elements and the *Drosophila copia* and *gypsy* elements. Note the different orders of their IN (Integrase) and RT-RH (Reverse transcriptase and RNase H) domains.

proteins that bind DNA and RNA; PRO, a protease that processes GAG and other polyproteins; RT-RNase H, an enzyme with both reverse transcriptase (indicated by 'RT') and RNase (indicated by 'RNase H') activity that generates the DNA copy of the element from an RNA copy and IN, the integrase (transposase) that joins the ends of the retroviral DNA into the new insertion site.

IN binds to the extreme ends of the LTR element DNA, which in many LTR elements contain very short inverted repeats, like the termini of cut-and-paste transposons. As one would expect from the chemical identity of the LTR element and DNA-only element integration mechanisms as discussed below, the catalytic core of IN is closely related to that of a DNA-only element transposase (see Figure 17.8).

ENV (envelope) is found only in retroviruses, which can exit from one cell as an RNA virus and infect another. Env is a glycoprotein that enters the cell membrane and forms a lipid-protein envelope around the retrovirus nucleoprotein core, which enables the particle to be released from the cell. Interaction of Env with specific proteins on cell surfaces also enables the virus particles to infect new host cells.

LTR elements transpose via RNA and DNA intermediates

The life cycles of the retroviral-like element Ty1 and the retrovirus murine leukemia virus (MuLV) are compared in Figure 17.22. Transposition of both elements begins with transcription by the host-cell RNA polymerase of an integrated form of the LTR element called the **provirus** (Figure 17.22, step 1). These transcripts are exported from the nucleus (step 2) and initially serve as mRNAs for the synthesis of the element-encoded proteins that assemble with two copies of the element RNA into a large complex called a virus-like particle or virion (step 3). In the case of the

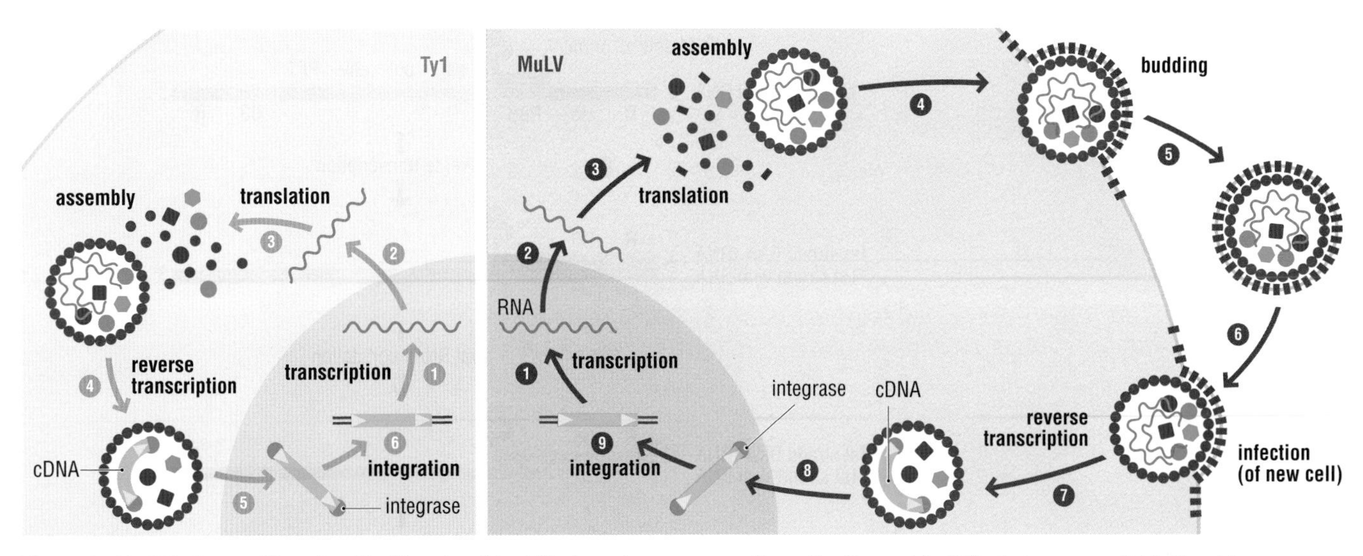

Figure 17.22 LTR element lifecycles. The lifecycles of two LTR elements are compared here: the *S. cerevisiae* LTR retrotransposon *Ty1* (left) and the mammalian retrovirus murine leukemia virus (MuLV) (right).

retroviral-like element, the RNA is converted to the double-stranded DNA form of the element by reverse transcription in the cytoplasm (step 4), transported into the nucleus (step 5), and integrated into the host genome (step 6). In the case of the retrovirus (right), the assembled virion interacts with the ENV protein in the membrane and buds from the cell (step 5); the infectious particle then infects another cell (step 6), and viral DNA is synthesized in the cytoplasm of the newly infected cell (step 7).

The **LTRs** are direct repeats several hundred base pairs long that each contain the U3, R, and U5 segments, and play key roles in reverse transcription as sites for the initiation of DNA synthesis as illustrated in Figure 17.23. Furthermore, their external 3′ termini are the sites at which joining to the target DNA occurs. The DNA form of the virus, which contains two LTRs, is generated through an elaborate series of reverse transcription and RNA degradation steps. Notice how the ends of the LTRs result from the 'jumping' of DNAs to a second viral template strand (Figure 17.23, steps 3 and 8), underscoring the importance of the packaging of two viral RNAs in each viral-like particle.

Once synthesized, the DNA form of the LTR element is integrated into the host genome by the element-encoded integrase (Figure 17.22, step 9). Integration of the LTR element DNA occurs by the same mechanism as that used by the DNA-only elements, as illustrated in Figure 17.24. The 3′ OH termini of the LTR element DNA directly attack the target DNA at staggered positions, giving rise to target-site duplications (Figure 17.24, step 4).

Like the DNA-only transposons, LTR elements insert at many different sites and can cause insertion mutations or an unwanted activation of adjacent host genes by gene regulatory signals, such as promoters, enhancers, and splice sites, which are present in the element.

Retroviruses do not usually actively destroy the cells they infect. Instead, they can both remain in the host genome and continue to generate infectious virus by transcription of the provirus and synthesis of viral proteins. Inhibitors of both the HIV protease and the reverse transcriptase are already used as drugs to block viral multiplication, and inhibitors of integrase that can block integration have been identified, which may also have therapeutic value. In Experimental approach 17.2 we consider how host proteins involved in HIV integration have been identified.

Scan here to watch a video animation explaining more about the LTR element lifecycle, or find it via the **Online Resource Center** at www.oxfordtextbooks.co.uk/orc/craig2e/.

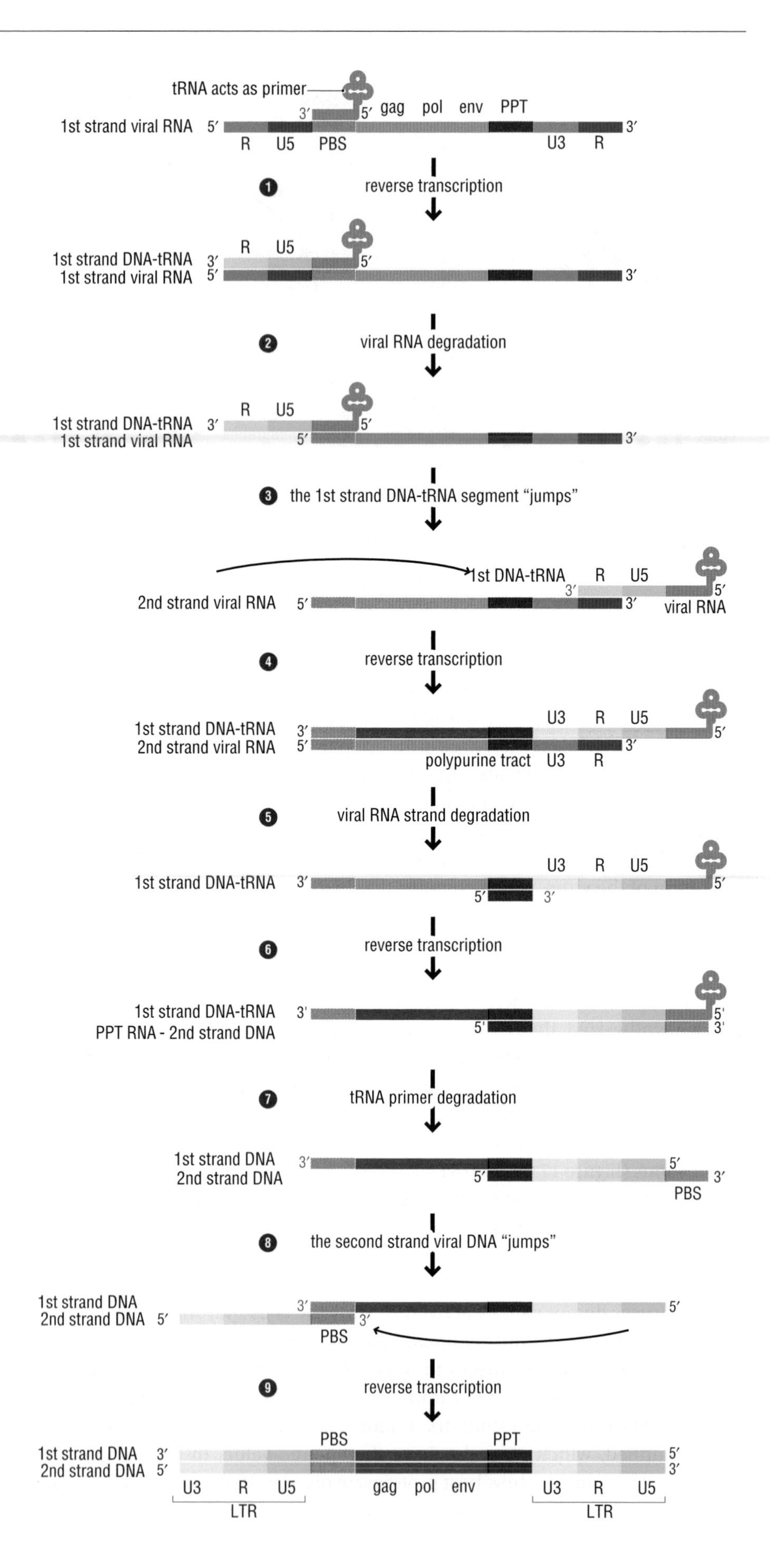

Figure 17.23 Synthesis of LTR element DNA from element RNAs. Host transcription of the integrated provirus generates viral RNA in which the protein-coding region (*GAG*, *POL*, *ENV*, (red)) is flanked by R, U5, U3 and R segments that play a key role in generating the U3-R-U5 LTRs that bind the DNA form of the element. PBS = primer binding site; PPT = polypurine tract. Multiple reverse transcription, RNA degradation, and template jumping steps are involved in element DNA production.

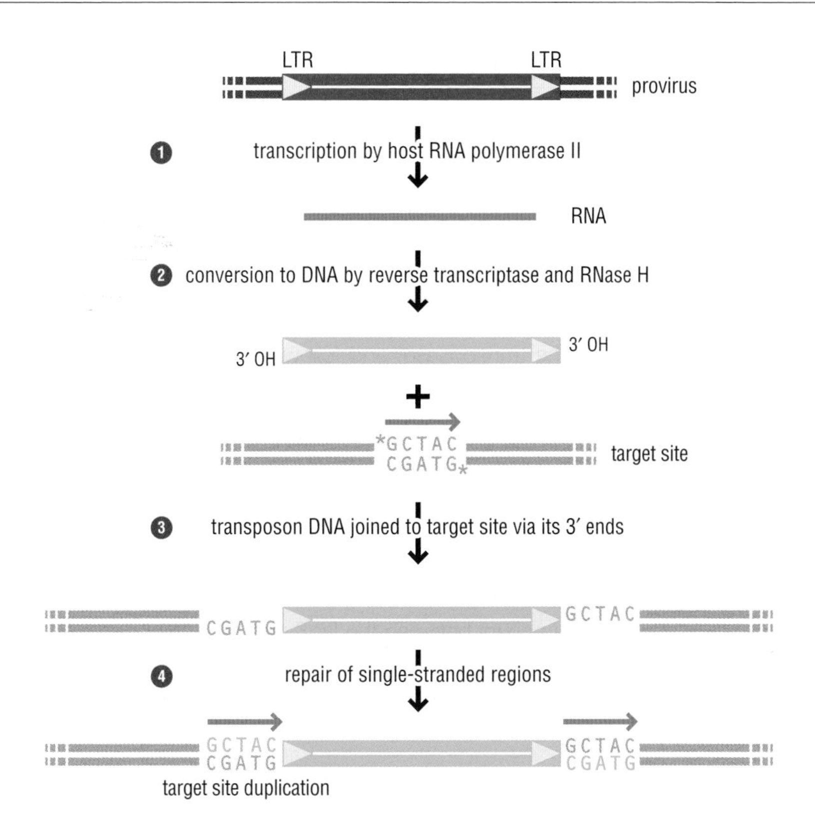

Figure 17.24 Integration mechanism of LTR elements. The proviral DNA of an LTR element in the donor site is shown. The element is transcribed by host RNA polymerase to generate the viral RNA (blue) form of the element, and this viral RNA is converted to the double-stranded DNA (green) form of the element by reverse transcription. The 3′ OH ends of the viral DNA attack the target DNA at staggered positions just as occurs during target joining of DNA-only elements (Figure 17.8). The newly inserted element is flanked by short gaps. Repair (green) of the gaps flanking the newly inserted element generates the target-site duplications (blue arrow).

17.8 NON-LTR RETROTRANSPOSONS

Scan here to watch a video animation explaining more about the integration of an LTR element, or find it via the Online Resource Center at **www.oxfordtextbooks.co.uk/orc/craig2e/**.

Non-LTR retrotransposons are the most common transposable elements in the human genome

Although DNA-only cut-and-paste elements and LTR elements are widespread, it is **non-LTR transposons** that make up a large proportion of the DNA of many eukaryotic genomes. Strikingly, these elements comprise about 34% of the human genome and are found in other animals and plants. A distinct group of non-LTR retrotransposons are the **mobile group II introns**, which are found in some bacteria and in mitochondria and chloroplasts of fungi, protists, and plants.

The processes of both cut-and-paste DNA-only and LTR element transposition involve the production of a 'free' DNA form that subsequently engages the target site. By contrast, when a non-LTR element transposes, an RNA copy of the element associates with the target site and is then used as a template for reverse transcription *in situ* at the site of insertion as we discuss below.

LINE elements are autonomous non-LTR retrotransposons

Several groups of non-LTR elements have been identified, and are depicted in Figure 17.25. One type of element found in mammalian genomes is a **LINE** element. These elements form about 21% of the human genome. Intact LINE elements are autonomous retrotransposons about 6 kilobases (kb) in length and encode several transposition proteins: ORF1 encodes an RNA-binding protein that forms a particle with the LINE RNA and is a nucleic acid chaperone; ORF2 encodes a protein with endonuclease and reverse transcriptase activities. These elements

17.2 EXPERIMENTAL APPROACH

Hunting for transposition host factors

Transposable elements depend on host-encoded processes and proteins to execute many stages in their lifecycle. These host factors can be identified by the isolation of host mutants that decrease or increase the frequency of the transposition of the element – experiments that, as we shall see here, are facilitated by visual assays for transposition.

Identification of transposition host factors in *E. coli*

To look for *E. coli* host factors that affect transposition of *IS903*, Keith Derbyshire and colleagues first created libraries of *E. coli* mutants containing 2000–5000 different *Tn5* insertions. The *Tn5* transposon inserts into many different sites (i.e., with little sequence specificity) so it was likely that many of the non-essential genes of the approximately 4600 genes in *E. coli* were mutated in the collection of cells represented in the libraries. Because *Tn5* makes mutations by insertion, each mutant gene is marked and the identity of the mutated gene can easily be determined. Derbyshire and colleagues next transformed the entire library of mutants with a plasmid encoding a specialized mini-*IS903* element and the transposase, and followed the transposition phenotype of thousands of transformants. The mini-*IS903* element contained the terminal inverted repeats (TIRs) necessary for transposition flanking a cryptic version of the *lacZ* gene that was not expressed because it did not contain a promoter or translational start site.

Utilization of this *lacZ* gene fragment allowed Derbyshire and colleagues to perform a 'promoter capture' assay: if the *lacZ* fragment is transposed into a gene such that an in-frame translational fusion occurs between the *lacZ* gene fragment and a host gene, *lacZ* will be expressed (and the cells will turn blue in the presence of the indicator dye) (Figure 1a). The frequency of transposition

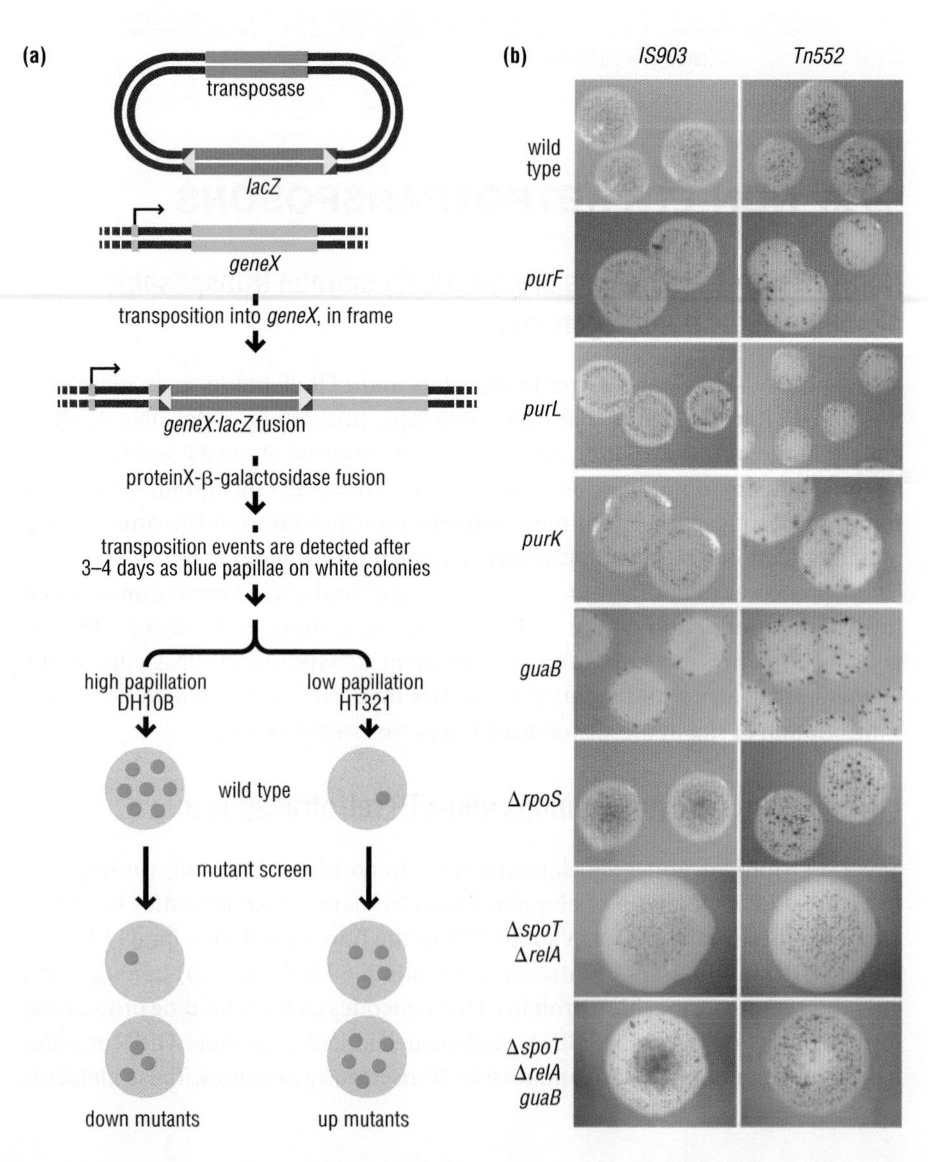

Figure 1 Assay for the isolation of *E. coli* mutants with altered *IS903* transposition. (a) The overall scheme of the assay was to look at the expression of a reporter ORF inside a transposon upon translational fusion to an ORF at a new insertion site. A donor plasmid contains both the transposase gene and a transposon containing a promoterless unexpressed *lacZ* reporter whose ORF extends to the end of the transposon terminal inverted repeat (red with yellow triangles). The target gene *X*, with an upstream promoter, is in the chromosome (X can be any non-essential gene). The host is *lacZ*$^-$. Transposon insertion into gene *X* results in an in-frame fusion of the transposon *lacZ* reporter with the N-terminus of gene *X*. The X-β-galactosidase fusion protein is synthesized. In the presence of X-gal (a chromogenic substrate for LacZ) in the growth media, transposition is detected as blue papillae (microcolonies) on the white colonies. In wild-type host cells (left), high frequency transposition is observed in a high copy number plasmid background; low frequency transposition is observed in a low copy number plasmid strain (right). Mutants in host genes can increase or decrease the frequency of transposition. (b) Examples of host transposition mutant phenotypes with *IS903* and the unrelated element *Tn552* to determine whether these elements use the same host factors.

From Twiss E, Coros AM, Tavakoli NP, Derbyshire KM. Transposition is modulated by a diverse set of host factors in *Escherichia coli* and is stimulated by nutritional stress. *Molecular Microbiology*, 2005;**57**:1593–1607

events can thus be simply followed as the frequency of blue papillae on otherwise white colonies on a plate. Derbyshire and colleagues used two different hosts, one in which the baseline transposition frequency was high so they could detect host mutants that *decreased* transposition, and another strain in which the baseline level was low so that host mutants that *increased* transposition could be detected. Over 100 different mutants that affected *IS903* transposition were isolated (Figure 1b).

Several mutants exhibited an interesting phenotype in which the *lacZ* color was enhanced in a ring in the colony (Figure 1b). Because colonies grow outward from the edge, this ring of color suggested that events at a particular stage during the growth of the colony were influencing transposition. When the corresponding *Tn5* insertions were cloned and sequenced (remember that the mutagenesis was implemented by a *Tn5* insertional strategy!), they were found in genes involved in guanosine triphosphate (GTP) synthesis: *purD purF*, and *purH*. Moreover, in follow-up experiments, Derbyshire *et al.* showed that the mutant phenotype was complemented (corrected) by plasmids carrying the wild-type versions of these genes. From these studies, it appeared that the GTP synthesis pathway plays a role in *IS903* transposition. Derbyshire and colleagues also showed that this pathway does not influence the transposition of the unrelated *Tn10* transposon.

Why would host mutants in GTP biosynthesis affect transposition and affect specifically one kind of transposon? One possibility is that GTP serves as a cofactor in *IS903* transposition. Such regulation by a cofactor is not unprecedented – the transposon-end pairing activity of the *Drosophila P* element is modulated by GTP, and adenosine triphosphate (ATP) plays a critical role in the assembly of Mu and *Tn7* target DNA-containing complexes, for example. Another interesting possibility is that (p)ppGpp (a global signal of nutrient availability in bacteria, see Section 12.1) modulates the expression of other host genes involved in transposition. Distinguishing among these models may depend on a fresh experimental approach.

Identification of host factors for retroviral replication in mammalian cells

In a conceptually similar study, now in eukaryotes, Sumit Chanda and colleagues screened large siRNA libraries to identify host factors required for HIV integration (König *et al.*, 2008). The life cycle of retroviruses including HIV is described in Section 17.7.

To follow HIV replication, the virus was engineered to contain a luciferase gene so that the presence of virus could be directly followed by luminescence, as depicted in Figure 2a. The genome-wide screen was then performed by independently targeting ~20 000 genes with siRNAs, infecting the cells arrayed in ~20 000 wells with the HIV derivative carrying the luciferase reporter, and looking for decreased luciferase signal resulting from reduced integration. As controls, parallel screens were performed on other viral vectors encoding luciferase, including murine leukemia (MuLV) and adeno-associated virus (AAV) constructs; general cellular toxicity was also evaluated (Figure 2b). It was reasoned that comparisons between 'hits' identified for all three viral vectors might prove useful in determining whether the identified host factors might play general or more specific roles in HIV pathogenesis.

As a technical note, siRNA-libraries are becoming more sophisticated, typically carrying more than one siRNA capable of targeting a specific gene in each well, increasing the likelihood

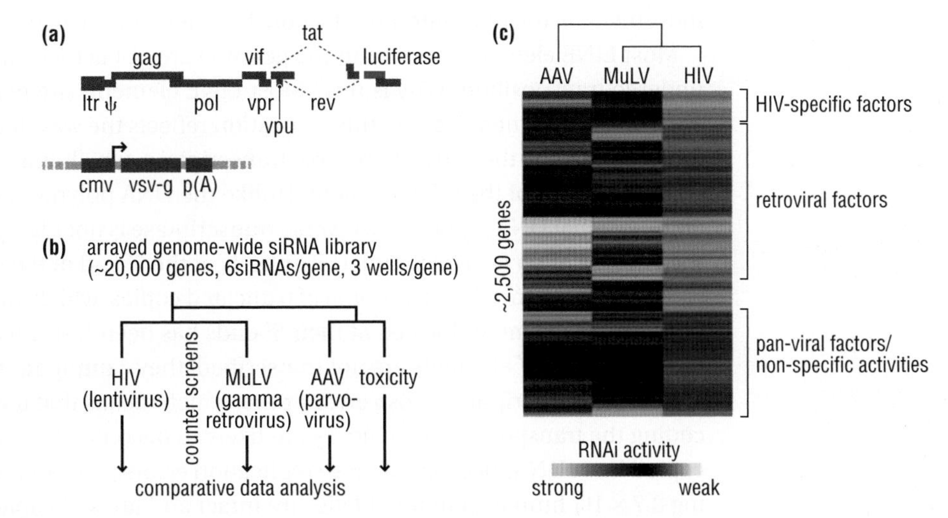

Figure 2 Reporter assay for siRNA-mediated screen for the identification of host factors important for HIV replication in human cells. (a) The HIV reporter virus contains a luciferase gene such that replication can be followed by luciferase activity. HIV was packaged in a different viral coat (vsv-g) to facilitate infection. (b) An arrayed library of human cells, each expressing a different siRNA, was infected with several different types of viruses, and viral growth was followed. (c) The effect of RNAi against about 2,500 different genes on viral growth is represented using the color scheme below: from blue (strong siRNA inhibition of viral growth) to yellow (weak inhibition of viral growth).

Konig *et al.* Global analysis of host-pathogen interactions that regulate early-stage HIV-1 replication. *Cell*, 2008;**135:**49–60

of efficient gene knock-down. Moreover, a given gene is typically targeted by several different sets of siRNAs (three in this study) in independent wells, thus allowing reproducibility to be evaluated. While it is possible that an individual siRNA will exhibit 'off target' effects that are misleading in the screen (knocking down another gene in the cell, for example), it is less likely that three different siRNA sets would all have the same secondary targets. Thus, similarities between responses of different siRNAs targeting the same gene should lead to increased confidence in the results.

The effects of specific gene knock-down on replication of the three different viruses is presented in a hierarchical cluster analysis that groups genes with strong inhibition profiles, those with intermediate profiles, and those with no effect (Figure 2c). In this case, the clustering was based on a number of different compiled criteria, not just the siRNA knock-down analysis. As we see, different genes display different effects for the three viral constructs: some genes are essentially HIV-specific, others affect both retroviruses (HIV and MuLV) while some affect all three viruses (HIV, MuLV, and AAV). Genes identified in this siRNA screen were further evaluated in a number of independent ways to determine which particular step in the HIV lifecycle is affected by the identified host factor. Host proteins that uniquely affected reverse transcription included a helicase and other nucleic acid binding proteins and ubiquitin–proteasome proteins. Nuclear importation proteins and proteins that could tether the integrase(transposase)/DNA complex to chromatin were also found to be important.

The use of libraries, whether a transposon, siRNA or chemical library, to screen for genes involved in a particular process has been extremely valuable for the identification and characterization of biological pathways, such as those involved in HIV biology. In all cases, the key to success is a reproducible phenotype or assay with which to screen for the desired effects. The availability of more and more high-throughput approaches with readily available reagents will expand the systems that can be studied and contribute to our understanding of the intricate workings of a cell.

Find out more

König R, Zhou Y, Elleder D, *et al*. Global analysis of host-pathogen interactions that regulate early-stage HIV-1 replication. *Cell*, 2008;**135**:49–60.

Twiss E, Coros AM, Tavakoli NP, Derbyshire KM. Transposition is modulated by a diverse set of host factors in *Escherichia coli* and is stimulated by nutritional stress. *Molecular Microbiology*, 2005;**57**:1593–1607.

Related techniques

E. coli (as a model organism); Section 19.1
Insertional mutagenesis; Section 19.5
Transformation; Section 19.4
Use of plasmids; Section 19.1
Complementation; Section 19.4
siRNA screens; Section 19.5
Microarray analysis; Section 19.10

also have a 3′ poly(A) tail. Beyond mediating their own movement, these autonomous LINE elements can also supply transposition proteins to, and promote the movement of, the non-autonomous non-LTR elements, which we discuss below.

Most LINE elements present in the genome are not active, that is, they cannot undergo transposition. Why is this? Most LINE elements are not full-length, but are truncated at their 5′ ends. This truncation reflects the way that these elements are generated by the action of reverse transcriptase, which initiates DNA synthesis at the 3′ end of the LINE element. Unlike the DNA polymerases that carry out chromosomal DNA replication, reverse transcriptase is not highly processive, and the enzyme often fails to complete synthesis of the 5′ end of the element. This results in the presence in the genome of truncated copies, which are not transcribed because the promoter located at their 5′ ends has been lost. Taking this one step further, if these 5′ elements are not transcribed, they cannot transpose.

Reverse transcription is also error prone, which means that the LINE genes encoding the transposition machinery are likely to become mutated when they are copied into cDNA. Because of these replication errors, only about 100 of the existing 8.7×10^5 human genomic LINEs are intact and are still capable of transposition. There are, however, multiple examples where gene disruption by a LINE has caused human disease.

Another type of non-LTR element, which is typified by the R2 elements of insects, has only one open reading frame (ORF) (see Figure 17.25b). The reverse transcriptase of these elements lies in the middle of the ORF while the C-terminus contains the endonuclease that cuts the target DNA. This endonuclease is highly site-specific, and insertion of these elements occurs only into ribosomal RNA genes.

So how does insertion of non-LTR elements occur?

Non-LTR elements insert by reverse transcription at the insertion site

The mechanism of transposition of non-LTR elements is summarized in Figure 17.26. Translocation of a LINE non-LTR retrotransposon begins with transcription of the element from an internal promoter for RNA polymerase II, the addition of a poly(A) tail to the 3′ end of the transcript, and export of the resulting mRNA to the cytoplasm (Figure 17.26, step 1). The LINE ORF1 and ORF2 proteins are synthesized from this mRNA, and preferentially associate with it; the resulting ribonucleoprotein complex then re-enters the nucleus. Nicking of the bottom strand of the target-site DNA by the ORF2 element-encoded endonuclease, which preferentially cleaves T-rich sequences, then occurs (Figure 17.26, step 2). In humans, the ORF2 endonuclease resembles an apurinic endonuclease (see Section 15.4) and can act at many sites. The role of ORF1 remains to be established.

The poly(A) tail of the LINE mRNA then associates with the target site by base-pairing with the cleaved bottom T-rich strand (Figure 17.26, step 3). The free 3′ OH end of the target DNA generated by endonuclease cleavage then acts as a primer for the synthesis of a new LINE DNA strand and extends to the end of the element RNA; this is catalyzed by the ORF2 reverse transcriptase using the LINE mRNA as template. Thus, a new LINE DNA strand that is covalently linked to the bottom strand of the insertion site is produced.

This mechanism is called **target-primed reverse transcription (TPRT)** because the target 3′ OH exposed by target nicking provides the primer for reverse transcription of the RNA copy of the non-LTR element. A consequence of TPRT is that a poly(A) sequence always becomes incorporated in the inserted element; this is a characteristic feature of non-LTR retrotransposons.

The mechanism by which the second top strand of LINE DNA is synthesized is not yet firmly established. However, one possibility is that, as the reverse transcriptase enzyme reaches the 5′ end of the LINE RNA template, a second cleavage event exposes a 3′ OH on the top strand of the target DNA (see Figure 17.26, step 4). The newly exposed target 3′OH then acts as a primer for synthesis of a second strand of LINE DNA using as a template the newly synthesized bottom strand LINE DNA. As with other transposons, target-site duplication occurs during non-LTR element insertion (Figure 17.26, step 6).

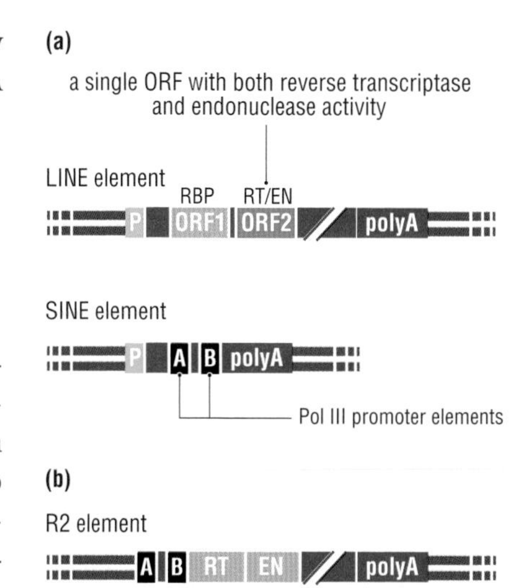

Figure 17.25 The structures of non-LTR elements. (a) LINEs and SINEs are transposable elements that move via target-primed reverse transcription. An intact LINE (about 6 kb) is an autonomous element that encodes its own transposition proteins in two open reading frames (ORFS), ORF1 and ORF2. ORF1 is an RNA chaperone, and ORF2 is an endonuclease-reverse transcriptase. Transcription initiates at P (green). SINEs are much shorter than LINES (at 100–300 bp), they do not encode any proteins, and are thought to translocate using LINE-encoded proteins. They do encode A- and B-boxes for RNA polymerase III. (b) R2 elements encode a single polypeptide that contains a reverse transcriptase domain and an endonuclease domain.

Short interspersed elements and pseudogenes are non-LTR elements generated from cellular RNAs

Short interspersed elements (SINEs) are another prominent class of non-LTR elements whose general structure is depicted in Figure 17.25. These elements derive from cellular RNAs whose initial integration and subsequent cycle of transposition are promoted by the LINE-encoded transposition proteins. Most SINEs are short (100–300 bp long), contain a DNA polymerase III (pol III) internal promoter and have a poly(A) tail. SINEs result from the integration of small host RNAs such as transfer RNAs (tRNAs) and 7SL RNA, which is the RNA component of the signal recognition particle involved in targeting proteins to membranes.

➲ We discuss the structure and function of the signal recognition particle in Section 14.2.

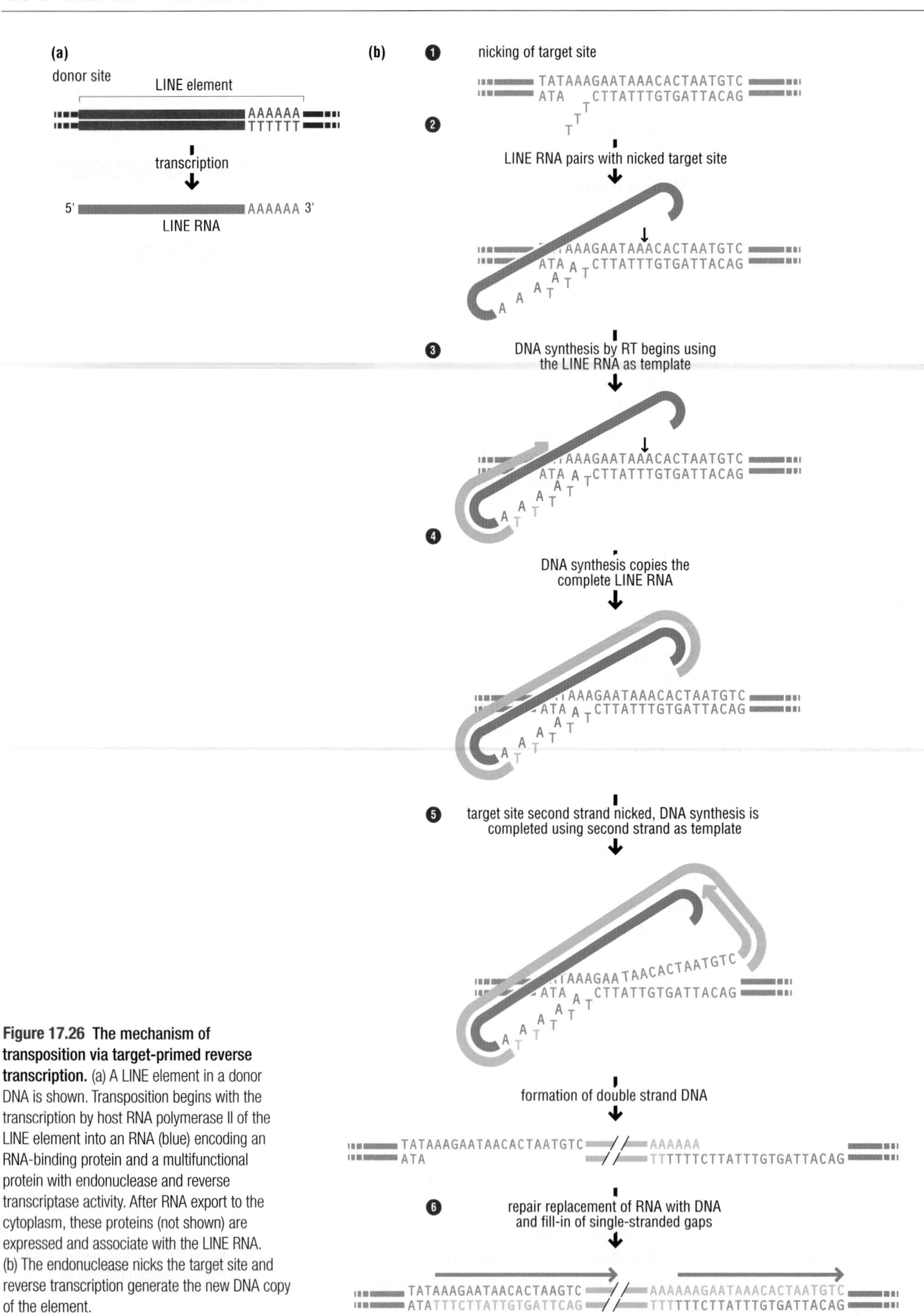

Figure 17.26 The mechanism of transposition via target-primed reverse transcription. (a) A LINE element in a donor DNA is shown. Transposition begins with the transcription by host RNA polymerase II of the LINE element into an RNA (blue) encoding an RNA-binding protein and a multifunctional protein with endonuclease and reverse transcriptase activity. After RNA export to the cytoplasm, these proteins (not shown) are expressed and associate with the LINE RNA. (b) The endonuclease nicks the target site and reverse transcription generate the new DNA copy of the element.

The 7SL RNA-derived SINEs are a prominent class of human SINEs. For historical reasons, these 7SL-derived elements are called **Alu elements**, because they include a target site for a restriction endonuclease called Alu.

There are about 1.6×10^6 SINEs in the human genome, forming about 13% of the genome (see Figure 17.4).

There are about 200 examples of human diseases resulting from gene disruption by LINEs and SINEs or and by deletions between elements by homologous recombination, examples of which are listed in Figure 17.27.

Pseudogenes are another type of element resulting from LINE protein-mediated TPRT and integration. Pseudogenes are protein-coding sequences that lack the introns associated with their active gene relatives; they also contain poly(A) tails, and are flanked by target-site duplications. Pseudogenes were probably formed by the LINE protein-mediated integration of a spliced mRNA. Pseudogenes are often not transcribed because they lack the promoters that lie upstream of the mRNA start site of their founding mRNA, and have generally been so mutated

Transposable elements and human disease

element	locus/protein affected	disease
insertions		
LINE L1	DMD	X-linked Duchenne muscular dystrophy
	F8	Hemophilia A
	F9	Hemophilia B
	RP2	X-linked retinitis pigmentosa
	APC	Colon cancer
	FKTN	Fukuyama-type congenital muscular dystrophy
	HBB	Beta-thalassemia
SINE Alu	F8	Hemophilia A
	F9	Hemophilia B
	IL2RG	X-linked severe combined immunodeficiency
	MSH	Hereditary non-polyposis colorectal cancer
	NT5C3	Chronic haemolytic anemia
	CFTR	Cystic fibrosis
	BRCA2	Breast cancer
	BRCA1	Breast cancer
	NF1	Neurofibromatosis
deletions by homologous recombination between elements		
L1 x L1	Collagen type IV	Alport's syndrome
	Cytochrome b-245, β polypeptide	Chronic granulomatous disease
	Phosphorylase kinase (PHKB)	Glycogen storage disease type II
Alu x Alu	Glycoprotein Ia	Glanzmann's thrombastheria
	Retinoblastoma gene (RB)	Glioma brain tumours (association to)
	MutL protein homolog 1	Hereditary non-polyopsis colorectal cancer
	β-globin cluster	Hereditary persistence of fetal hemoglobin
	Insulin receptor β	Insulin-dependent diabetes mellitus
	c-sis proto-oncogene	Meningioma
	Chondroitinase	Mucopolysaccharidosis type IVA
	Antithrombin	Thrombophilia
	β-globin cluster	β-thalassaemia
	β-globin cluster	γβδ-thalassaemia
	α-globin cluster	α-thalassaemia
	Von Willebrand factor	Von Willebrand's disease type II

Figure 17.27 Human diseases cause by transposable elements. Diseases have resulted from insertions of both L1 and Alu elements, and homologous recombination between non-LTR elements has caused human disease.

Belancio *et al.* Mammalian non-LTR retrotransposons: For better or worse, in sickness and in health. *Genome Research*, 2008;**18**:343–358; Deininger, PL, and Batzer, MA. Alu repeats and human disease. *Molecular Genetics and Metabolism*, 1999;**67**:183–193.

that they do not encode a gene functional product. The human genome has been estimated to contain about 8×10^3 pseudogenes.

LINEs and SINEs can contribute to genome instability as substrates for homologous recombination

Because LINEs and SINEs are highly repeated sequences that are dispersed throughout the human genome, homologous recombination between these elements can result in intramolecular and intermolecular deletions and duplications by unequal sister chromatid exchange and in translocations, as described in Section 16.9, Figure 16.24. There are more than 25 examples of human diseases that have resulted from homologous recombination between transposable element repeats; some of these rearrangements have occurred multiple times.

Mobile group II introns are present in the genomes of bacteria, mitochondria, and chloroplasts

The other major group of non-LTR retrotransposons are the **mobile group II introns**, which are found in bacteria and in mitochondria and chloroplasts of fungi, protists, and plants. Mobile group II introns are catalytic self-splicing RNAs – that is, they can excise themselves from an RNA transcript without the aid of proteins *in vitro*. Mobile group II introns move by the reverse splicing of their excised intron into the target DNA, followed by reverse transcription.

Group II introns are of particular interest because they are thought to be the progenitors of the nuclear spliceosomal introns of eukaryotic genes. An attractive hypothesis is that they arose in bacteria, and that, after eukaryotic cells acquired mitochondria and chloroplasts (which are of bacterial origin), the mobile group II introns invaded nuclear genomes. Once in the nuclear genome as spliceosomal introns, they can only be excised from mRNA with the aid of spliceosome, the large complex featuring both RNA and proteins.

➔ We learn more about the spliceosome in Section 10.6.

Mobile group II introns move by reverse splicing and TPRT of their excised RNA form

Mobile group II introns encode a conserved intron RNA structure that is essential for splicing and mobility; they also encode a protein called RME. RME comprises multiple functional domains including a maturase that assists in element splicing, a reverse transcriptase, and, in some cases, a DNA endonuclease. They can participate in two types of mobility reactions: **retrohoming**, in which they insert at high frequency into a specific site with considerable homology to the intron sequence; and **retrotransposition**, in which they insert at much lower frequency at an ectopic, non-specific site. The mechanism of retrohoming is summarized in Figure 17.28.

The first step in both retrohoming and retrotransposition is the splicing of the element from an mRNA to form a lariat (a looped structure), a reaction assisted by the maturase activity (Figure 17.28, step 3). In the retrohoming reaction, the excised RNA reverse splices into one of the DNA strands at a target site that is defined by base-pairing between the intron RNA and the target DNA (Figure 17.28, step 4). The intron-encoded endonuclease then cleaves the DNA strand just outside the insertion site, generating a 3′ OH that on the bottom strand can act as a primer for the reverse transcriptase to make a DNA copy of the intron RNA (Figure 17.28, step 5). The target site now contains an RNA copy of the element on the top strand and a DNA copy of the element on the bottom strand.

As is suggested for the synthesis of the second strand of LINE elements, the reverse transcriptase can then use this newly inserted intron DNA strand as a template to synthesize the second strand of the element at its new insertion site (Figure 17.28, step 6). The maturase reverse transcriptase, however, lacks an RNase H activity that can degrade the template RNA strand, and thus bacterial proteins play an important role in converting the spliced RNA–DNA form into duplex DNA (Figure 17.28 steps 7 and 8).

Despite differences in the origins of the 3′ OHs that prime reverse transcription, and differences in how the RNA template is brought to target DNA - by positioning by a protein or reverse splicing - the mobility of all LINEs, SINEs, and mobile group II introns is based on TPRT.

17.9 CONTROL OF TRANSPOSITION

In the previous sections, we have explored the range of different transposable elements that occur in the genomes of many organisms and have seen the various molecular processes that underpin transposition events. Very often, it may seem that the host cell is little more than an innocent bystander, powerless to stop transposition events from occurring within its genome. In this section, however, we see how this initial impression is far-removed from reality.

The frequency of transposition is usually tightly controlled

Much remains to be learned about the complex interrelationships between transposons and their hosts. In one view, transposable elements may be considered parasites that infest host genomes. Alternatively, the genetic diversity that can result from transposon mobility may be considered to reflect a more mutualistic relationship between transposons and their hosts. In any case, a high frequency of transposition is generally disadvantageous as it can result in a high frequency of deleterious mutations. An unacceptable burden of mutation (one that would be fatal to the host organism) due to transposition, however, is generally avoided because the transposition of most elements occurs very rarely.

There are many points at which the production of transposition proteins and the formation of protein–nucleic acid intermediates in transposition can be controlled. Although transposon-encoded proteins perform the central catalytic steps in element translocation, host proteins play essential roles in the movement of these elements. As such, intimate connections exist between transposons and their hosts. For example, while element-encoded transposases mediate breakage and joining at the ends of DNA-only cut-and-paste transposons, DNA-bending proteins, such as the sequence-specific protein IHF and the sequence non-specific proteins such as HU and HNS, participate directly in several transposition reactions by modulating the assembly of active transpososomes. As we saw in

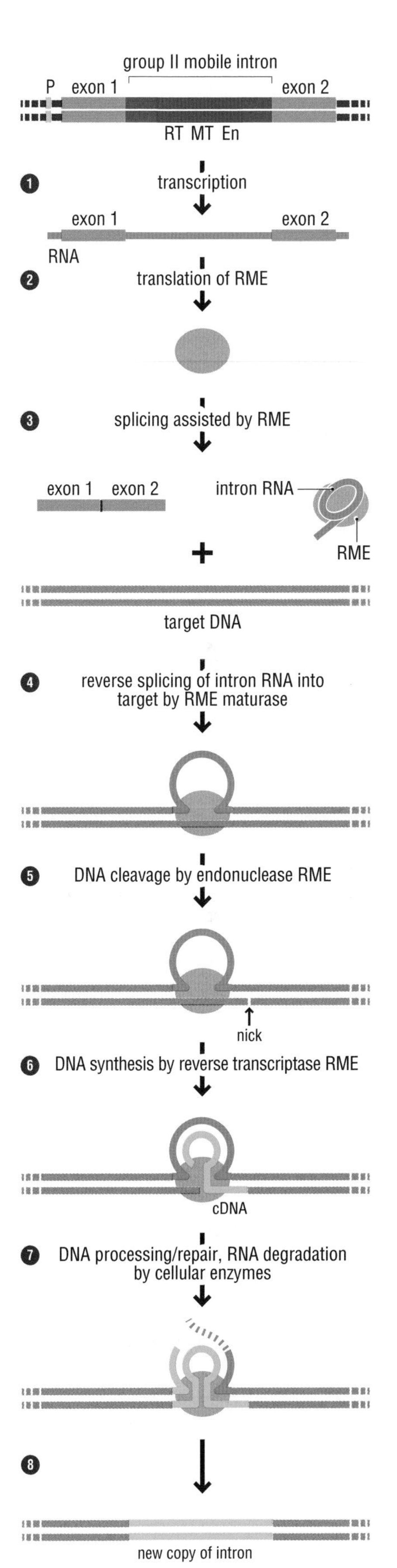

Figure 17.28 The mechanism of retrohoming by mobile group II introns. A mobile group II intron located within a gene is shown, flanked by exon 1 and exon 2 (orange). The mobile group II intron encodes the protein RME (dark blue) with reverse transcriptase (RT), maturase (MT) (which facilitates splicing), and endonuclease (En) activity. P is the gene's promoter. (1) Host gene transcription generates an RNA containing the host gene exons separated by the group II intron.

E. coli host factors involved in group II intron mobility

host factor	effect	identified function	putative effect on group II intron
Exo III	inhibits mobility	3′-5′ exonuclease	degrades nascent cDNA
ligase	stimulates mobility	DNA ligase	seals DNA nicks
MutD	stimulates mobility	3′-5′ exonuclease ϵ subunit of Pol III (*dnaQ*)	repairs second-strand cDNA synthesis
Pol I	stimulates mobility	3′-5′ exonuclease; removal of RNA primer from Okazaki fragments	removes intron RNA template
Pol II	stimulates mobility	repair polymerase (*polB*)	repair polymerization across DNA-RNA junctions
Pol III	stimulates mobility	replicative polymerase	second-strand cDNA synthesis
Pol IV	stimulates mobility	repair polymerase (*dinB*)	repair polymerization
Pol V	stimulates mobility	repair polymerase	repair polymerization
RecJ	stimulates mobility	5′-3′ exonuclease	5′-3′ resection of DNA
RNase E	inhibits mobility	ribonuclease; part of RNA degradasome	reduces half-life of intron RNA
RNase H	stimulates mobility	ribonuclease; cleaves RNA strand in RNA/DNA hybrid	removes intron RNA template
RNase I	inhibits mobility	ribonuclease	reduces half-life of intron RNA

Figure 17.29 ***E. coli*** **host factors involved in mobile group II intron transposition.** Multiple *E. coli.* proteins are involved in the transposition of a mobile group II intron.

From Beauregard, A, Curcio, MJ, and Belfort, M. The take and give between retrotransposable elements and their hosts. *Annual Review of Genetics*, 2008;**42**:587–617.

Experimental approach 17.2, many other types of proteins such as those involved in proteasome function and nuclear import are involved in retroviral integration.

The movement of bacterial group II introns and mammalian LINE non-LTR elements, in which element-encoded endonucleases and reverse transcriptases mediate target cleavage and reverse transcription, requires many host proteins such as exonucleases, ribonucleases, and repair polymerases. The host factors involved in mobile group II intron transposition in *E. coli* are summarized in Figure 17.29. However, a recent analysis of cellular factors required for HIV replication has revealed the involvement of even more types of host factors (see Experimental approach 17.2).

Transposase levels can be regulated in a variety of ways

The frequency of transposition is often directly related to the concentration of transposase; the synthesis of transposase, in turn, is frequently subject to elaborate control. For example, the transposition of the bacterial element *IS10* is kept to a low frequency by features of the element itself that result in production of a very low amount of transposase; these features are illustrated in Figure 17.30.

Figure 17.30 shows how the transposase is translated from an mRNA generated by the promoter P_{in}, which is a very weak promoter. The activity of P_{in} is also decreased by a promoter called P_{out} that promotes transcription from inside the element into the flanking (gray) DNA. The P_{in} transposase mRNA is also poorly translated, not only because of poor translation signals, but also because the P_{out} transcript acts as a regulatory RNA (as seen for the RyhB RNA in Section 13.2); this regulatory RNA forms a duplex RNA with the transposase mRNA that blocks its translation. The inhibition of P_{in} mRNA translation by P_{out} RNA is particularly important when IS10 copy number increases and P_{out} decreases the total amount of transposase synthesized.

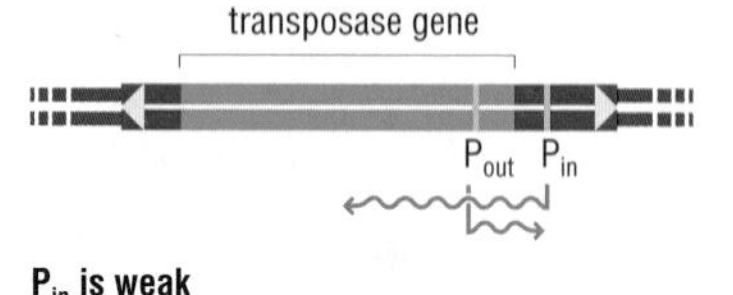

Figure 17.30 Mechanisms by which synthesis of the *IS10* transposase is controlled. The frequency of *IS10*/*Tn10* transposition is proportional to the amount of transposase present. Multiple levels of regulation act to keep transposase expression at a low level, making transposition infrequent.

Some elements only undergo transposition in certain cell types. This is the case for the *P* element of *Drosophila*: transposition of the *P* element can occur in germline cells but not in somatic cells. Although it may seem paradoxical to allow transposition in the germline because of the potential negative consequences of transposition in terms of mutation, only by transposition in the germline can the element be successfully propagated from generation to generation.

The production of active *P* element transposase requires the splicing of four exons to form the mRNA, as shown in Figure 17.31. A functional mRNA is only produced in the germline, however, because somatic cells contain several factors that specifically inhibit the splicing out of the last intron. A stop codon in this intron results in the synthesis of a truncated transposase that lacks the fourth exon, which encodes the catalytic domain of the protein. Consequently, the truncated transposase can bind to transposon ends, but it cannot promote transposition.

The transposition of some elements is directly influenced by cellular growth conditions. For example, transposition of the yeast *Ty1* transposon increases during filamentous growth (a response to starvation) as a result of increased transcription of *Ty1* and stabilization of the *Ty1* transposase in these conditions. It has been suggested that such an increased transposition rate could optimize cell survival in stressful environmental conditions by increasing genetic diversity.

Transposition can be controlled by DNA methylation

We see in Section 4.7 how cells use DNA methylation as a vital means of regulating gene expression. A further important role of DNA methylation is to prevent the movement of transposable elements, often by blocking transposase expression and therefore protecting the host. A particular hazard for DNA-only cut-and-paste transposons is that transposon excision results in a potentially lethal gap at the site of element excision. *IS10* has reduced this potential hazard by coupling its transposition to the passage of the replication fork, as illustrated in Figure 17.32. Thus, when *IS10* transposes, one arm of the replication fork will remain intact although

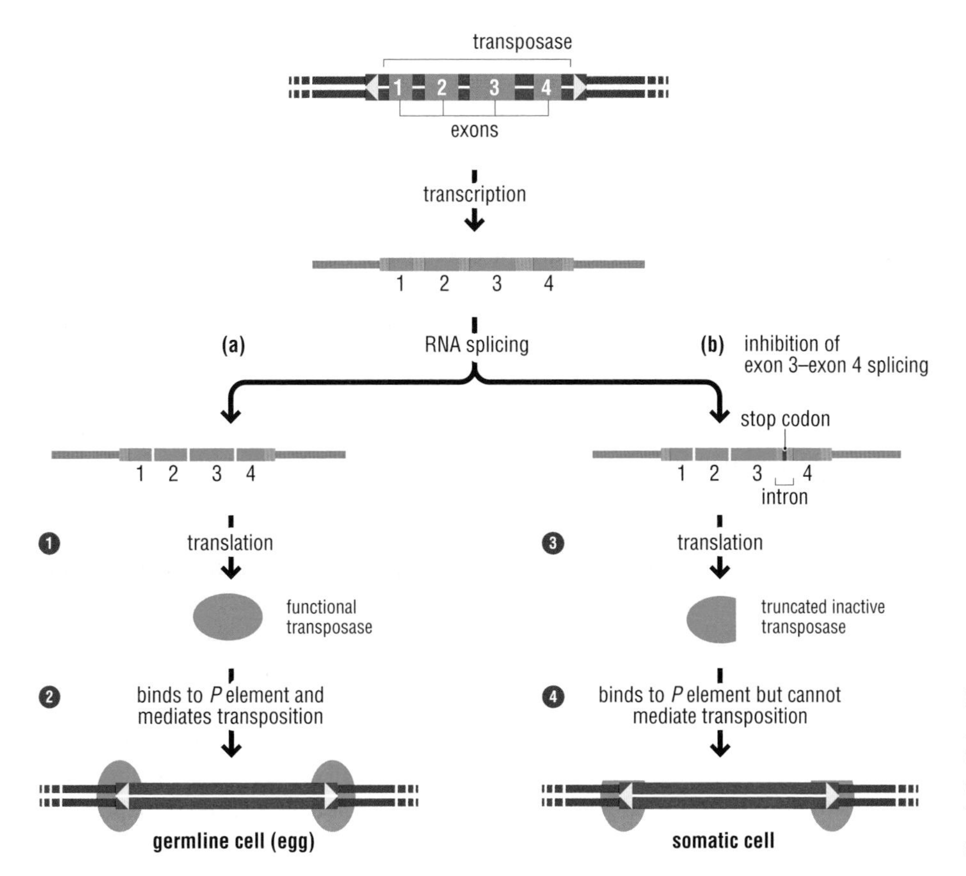

Figure 17.31 Inhibition of splicing of *Drosophila P* element mRNA blocks transposition in somatic cells. The *P* element transposase is encoded by four exons. Transcription forms the mRNA (blue) that can be alternately spliced.

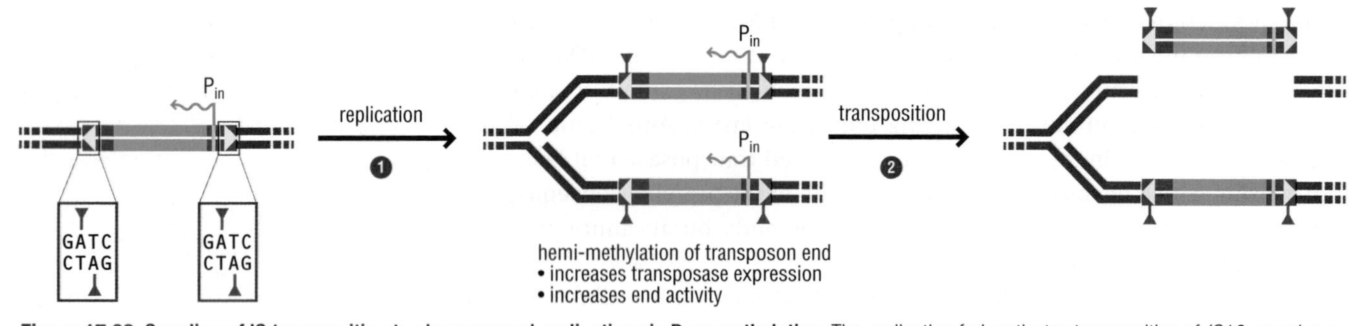

Figure 17.32 Coupling of IS transposition to chromosomal replication via Dam methylation. The replication fork activates transposition of *IS10*, assuring a single host chromosome does not undergo a potentially lethal double-strand break in the absence of a template for repair.

the other will have a gap. This gap may be repaired by homology-directed repair using the intact arm as a template as shown in Figure 17.12.

Figure 17.32 shows how the coupling of *IS10* transposition to replication is mediated by a Dam (GATC) methylation site in the end of the element. The methylation state of this element can affect transposase expression and the actual activity of the transposon end. As the replication fork passes through the *IS10* element, the two arms contain hemi-methylated *IS10*. Hemi-methylation of the P_{in} promoter stimulates transposase expression and the end activity of one of the hemi-methylated elements, presumably because of differential binding of transposase to the two asymmetric elements.

In eukaryotes, the methylation of transposon DNA also reduces transposition. As described below, this occurs via the RNA silencing mechanisms we learned about in Section 10.10.

Most eukaryotic organisms suppress transposition globally by gene silencing mechanisms

Most eukaryotic organisms possess a general means of blocking the movement of transposable elements throughout the genome. This suppression depends on mechanisms of gene silencing that are used against repeated DNA sequences throughout the genome. These defense systems act globally to decrease the transposition of all copies of an element by decreasing its transcription and downregulating the translation of any element mRNA that is synthesized.

Central to this general control is the production of a double-stranded RNA from the elements. As multiple copies of the element are located in many different positions in the genome, transcription impinging on the elements from external promoters will generate RNA copies of both strands of the element. These transcripts may then pair to form element-specific double-stranded RNA. Such double-stranded RNA species are specifically degraded to short RNAs by members of the Argonaute family of proteins and can then be targeted to other transposon-derived transcripts to induce mRNA degradation or block translation. Moreover, by recognizing their complementary DNA sequence, these short RNAs can recruit chromatin-remodeling complexes to sites of integrated transposon DNA. These particular chromatin-remodeling complexes modify histones to generate chromatin structure that represses gene expression, thus preventing transcription of the transposase gene. The short RNAs can also promote DNA methylation at their target sites by recruiting DNA methylases. It seems likely that these cellular responses to double-stranded RNA arose as specific defenses against transposable elements and viruses.

➔ We learn more about chromatin-remodeling complexes in Section 4.6.

Although successful transmission of an element to progeny requires element transposition in the germline, it is also important that transposition occurs at a low level to avoid the accumulation of deleterious mutations and maintain genome integrity. While transposition in somatic cells has no heritable effect *per se*, such transposition also is generally blocked to avoid reduced genome integrity, which could result in oncogenic transformation, for example. For some elements, including the *Drosophila P* element, there is an element-specific regulatory mechanism, the inhibition of transposase mRNA splicing, which blocks transposition in somatic cells.

A germline-specific mechanism for inhibition of transposition has recently been discovered in *Drosophila* and some other animals, including mammals. This mechanism uses a specialized class of small RNAs called piRNAs and specialized Argonaute proteins called Piwi, Aubergine, and Ago3. Several *Drosophila* genomic loci that have been long known to encode inhibitors of transposition are now known to encode clusters of inhibitory RNAs, not protein inhibitors, which result from transcription of clusters of elements or fragments of elements no longer capable of transposition; such clusters are depicted in Figure 17.33. These include *flamenco*, which inhibits the movement of retrotransposons such as *gypsy*, and X-TAS, which inhibits the movement of *P* elements. The primary transcripts from these regions are processed in as yet unknown ways to generate short piRNAs that interact with the Piwi, Aubergine, and Ago3 proteins. These piRNA–protein complexes then interact with transcripts from cognate target transposons, leading to the cleavage and inactivation of the transcripts (Figure 17.34). These interactions lead to amplification of the inhibitory piRNAs that can target transposon mRNAs for destruction.

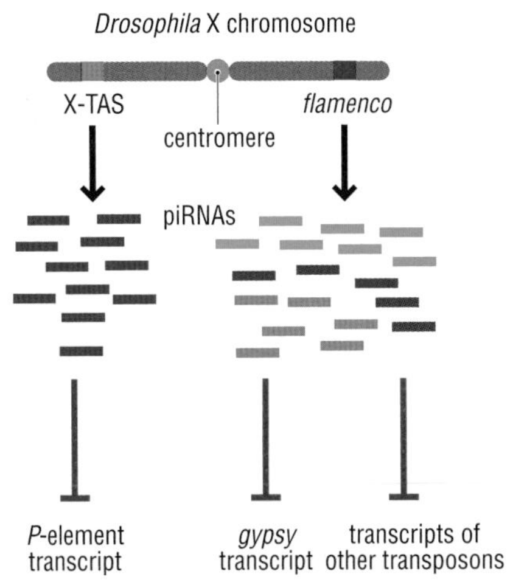

Figure 17.33 Specific loci in *Drosophila* encode inhibitors of transposition. The X-TAS and *flameco* loci on the *Drosophila* X chromosome encode small RNA transposition inhibitors of *P* element and of *gypsy* and several other types of retrotransposons, respectively. These regions do not encode protein inhibitors of transposition. Rather, they encode clusters of inactive transposable element copies whose RNA products are processed to make small RNA (piRNA) inhibitors that act upon the transcripts of active elements to decrease transposition (see Figure 17.34).

➔ We learn more about piRNAs in Section 13.4.

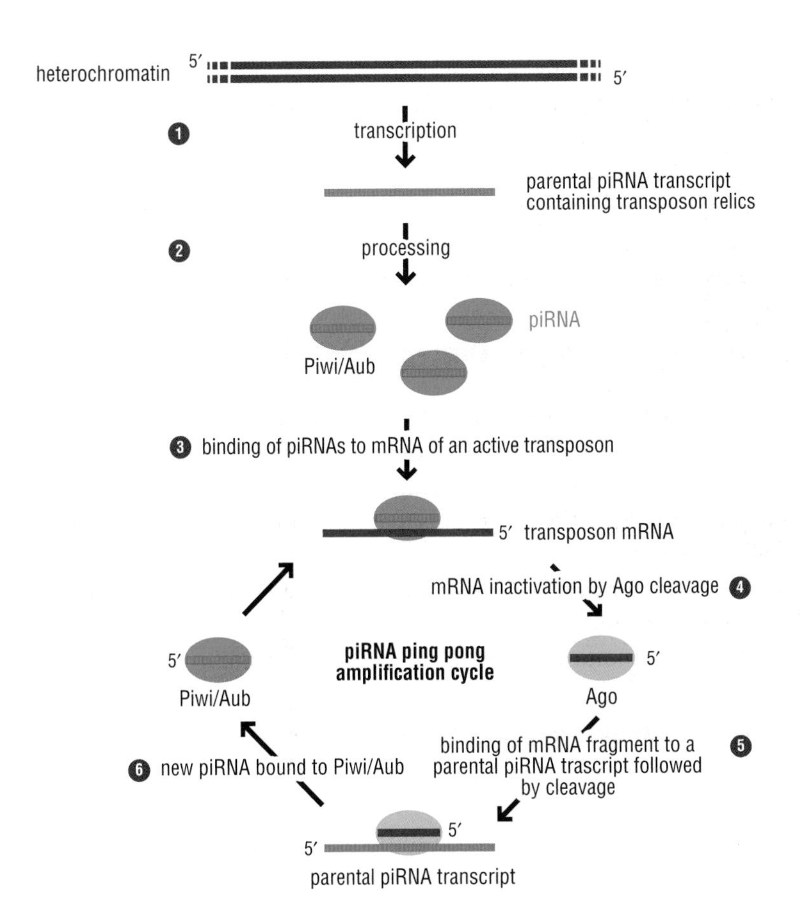

Figure 17.34 Inactivation of transposon mRNAs by piRNAs. A cycle of transcription of long RNAs containing transposon sequences, followed by processing to short piRNAs that can target transposon mRNAs for cleavage can lead to the amplification of piRNAs and increased inactivation of transposon mRNAs.

Adapted from Brennecke, J, Aravin, A, Stark, A, Dus, M, Kellis, M, Sachidanandam, R, Hannon, G. Discrete small RNA-generating loci as master regulators of transposon activity in *Drosophila*. *Cell*, 2007;**128**:1089–1103.

Transposition can be controlled through target-site selection

In some cases, the potentially deleterious effects of the movement of transposable elements can be limited by directing transposition away from gene-coding sequences. This means that, even when transposition does occur, the chances of inactivating a gene are small. Although it is true that most elements have the ability to insert into many different sequences, closer inspection reveals that target sites are generally not entirely random.

Target-site choice can be determined exclusively by the interaction of the transposase with the target DNA. In other cases, however, the insertion site is chosen by the interaction of the transposase with host-encoded proteins. The yeast LTR elements *Ty1*, *Ty3*, and *Ty5* have interesting mechanisms of target-site selection, all of which reduce the chance of potentially lethal insertions within genes. Both *Ty1* and *Ty3* inserts upstream of tRNA genes that are transcribed by RNA polymerase III by interacting with polymerase-specific transcription factors, thereby avoiding the actual coding regions. *Ty5* inserts into silent, heterochromatic regions of the genome, which include the regions around telomeres and the silent mating-type loci. In contrast, certain retroviruses insert preferentially into transcribed regions, a strategy that may favor element expression and, hence, activity.

Despite some restrictions on target-site selection, transposons can insert at many different sites in the genome. In the next sections, we will consider CSSR, a type of DNA movement that can only occur between particular related sites.

17.10 CSSR: OVERVIEW

The previous sections of this chapter have focused on the behavior of transposable elements – discrete portions of DNA that can insert into many different sites in DNA. However, it is not just the movement of defined transposable elements that can lead to the rearrangement of genetic material within the cell. In this section, we explore a quite different way in which DNA rearrangement can occur – the process of Conservative Site-specific Recombination (**CSSR**). This type of recombination is sometimes called site-specific recombination but as both CSSR and transposition involve specific DNA sites, we will use the term CSSR.

CSSR is required for diverse DNA rearrangements in the cell

CSSR can mediate a variety of different DNA rearrangements such as the integration and excision cycles of bacteriophages, the conversion of circular chromosomal DNA dimers to monomers to allow for accurate chromosome segregation to daughter cells, and the inversion of DNA segments that can control the alternative expression of a gene by changing the orientation of a promoter. Although most known examples of CSSR come from bacteria, it also occurs in eukaryotes and is known, for example, to control the replication of a yeast plasmid. We discuss the use of the bacterial CSSR system Cre-*lox* for genome engineering in mammalian cells in Section 19.5.

CSSR occurs between specific pairs of special sites that share a short region of homology

Unlike transposition, CSSR is reciprocal and conservative (that is, no DNA sequence is lost or gained during recombination), nor is a high-energy cofactor required to

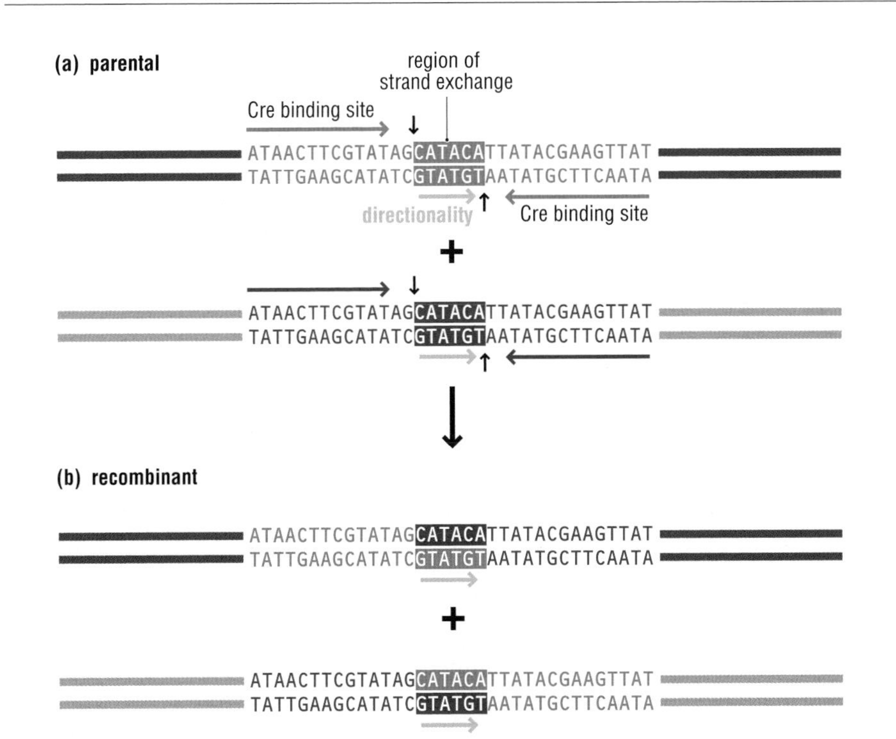

Figure 17.35 Schematic diagram of CSSR. Parental duplexes (gray and light gray) contain sites (blue and pink) for the Cre-*lox* recombination system of bacteriophage P1 (see also Section 19.5). (a) The parental *lox* sites share a region of homology (shaded boxes) called the spacer region that is flanked by binding sites for the Cre recombinase in inverted orientation (horizontal blue and pink arrows). The spacer sequence has directionality, that is, it is not a palindrome, as indicated by the green arrows. Each recombinase mediates breakage, exchange, and joining of one strand at each binding site (vertical arrows). (b) The recombinant duplexes contain one arm from each parent (gray and light gray), one Cre site from each parent (blue and pink) and a heteroduplex spacer region (shaded boxes) where strand exchange occurred.

regenerate intact DNA. Instead, DNA is simply exchanged by breakage and joining between the two recombining DNA sites in a short (2–8 bp) region of homology called the spacer that is shared between the sites. Figure 17.35 shows how the spacer sequence has directionality, and is flanked by recombinase-binding sites in inverted orientation, which position recombinases for action at the spacer region. Once positioned, one recombinase acts on the top strand and one acts on the bottom strand of each parental duplex. The DNA breakage and joining reactions that lead to recombination occur at the edge of these spacers.

Recombination begins by the binding of recombinases to each DNA site, after which the recombinases bind to each other to bring the DNA sites together. Only when the sites are paired do the recombinases execute DNA breakage and joining so that uncoupled DNA breaks, which may be harmful to the cell, do not occur.

Following pairing, the recombinases then break the parental duplexes of DNA at each spacer at staggered positions (see Figure 17.35a). Four recombinases bind to two duplexes, and each directs a single cleavage on one strand. Broken strands then are exchanged between the partner duplexes and rejoined to form recombinant duplexes (Figure 17.35b). Because the positions of strand breakage within each spacer are staggered, the DNAs within the exchanged and rejoined spacers are heteroduplex, that is, they contain one DNA strand from each parent between the positions of DNA strand exchange. This requirement for heteroduplex formation helps to enforce the accuracy of the recombination reaction, as only identical DNA sequences can form perfectly paired heteroduplex DNA. The recombinant duplexes contain DNA information from one parent on one side of the spacer and from the other parent on the other side of the spacer.

The outcome of CSSR reactions is determined by the relative positions of the recombining sites

CSSR can promote different kinds of DNA rearrangements, depending on the relative positions and orientations of the two recombining sites, with the orientation

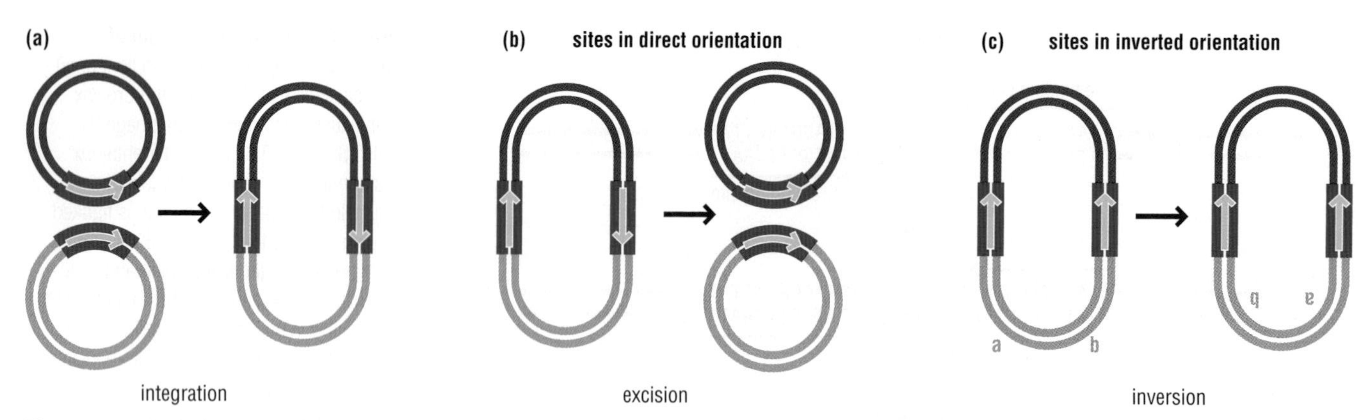

Figure 17.36 The outcome of CSSR is determined by the spatial relationship of the two recombination sites. (a) When recombination occurs between CSSR sites (blue boxes with green arrows) on different DNA molecules, the two molecules are joined together or integrated into one larger DNA molecule in which the spacer sequences (green) are oriented in the same direction. (b) Recombination between CSSR sites whose spacer sequences are in direct orientation results in 'excision', 'deletion', or 'resolution', depending on the biological context. (c) Recombination between CSSR sites whose spacer sequences are in an inverted orientation results in inversion of the segment between the CSSR sites.

of each recombining site being determined by the directionality of its spacer. The different rearrangements that are possible are illustrated in Figure 17.36. When the two recombination sites are located on different DNA molecules - for example, a bacteriophage genome and the host-cell genome - CSSR will integrate the phage genome into the host-cell genome (Figure 17.36a). For this reason, the phage-encoded recombinases that mediate such reactions are called integrases. As we will discuss in detail below, however, these bacteriophage integrases are a different type of enzyme from the integrases (transposases) of retroviruses and other LTR retroelements, and work by a quite different mechanism.

When the recombination sites are present on the same DNA molecule and their spacer sequences are oriented in the same direction, recombination between the two sites will separate the parental substrate DNA into two DNAs (see Figure 17.36b). This CSSR reaction is alternatively called 'excision', 'deletion', or 'resolution' depending on the biological process involved.

Finally, when two recombination sites are present on the same DNA molecule and their spacer sequences are in inverted orientation, recombination between them results in the inversion of the DNA segment that lies between them (Figure 17.36c). Inversion of a promoter region by this mechanism can be used to regulate gene expression, a process we explore further in Section 17.11.

It is important to note that not all CSSR systems can carry out all of these reactions. In some cases, they have sophisticated mechanisms to ensure that only particular reactions are carried out. For example, in the integration-excision cycles of bacteriophages, once integration occurs, excision cannot take place without the synthesis of additional phage-encoded proteins that are expressed only under excision conditions.

CSSR recombinases are topoisomerases

CSSR occurs by DNA breakage and rejoining. The recombinases that mediate these reactions are sequence-specific topoisomerases. As discussed in Section 2.5, a topoisomerase is a protein that can transiently break DNA, often resulting in a change in DNA topology such as the relaxation of supercoiling. Topoisomerases can break and rejoin DNA in the absence of a high-energy cofactor such as adenosine triphosphate (ATP) because DNA breakage is accompanied by a covalent linking of the topoisomerase to the DNA. This linkage conserves the high energy that was stored in the DNA phosphodiester bond. The high energy of the protein–DNA linkage is then used to rejoin DNA strands by re-forming a phosphodiester bond. Recombination by a topoisomerase activity is therefore quite distinct from transposition, during which DNA synthesis and ligation are required to generate intact duplex DNA.

The topoisomerases that carry out CSSR act specifically at recombination sites because they are sequence-specific DNA-binding proteins: they bind to their cognate sites, which flank the spacer regions of CSSR sites. Like transposases, the CSSR recombinases are usually encoded by genes adjacent to the DNA that undergoes recombination.

Two large families of topoisomerase-like recombinases carry out CSSR. Although the structures of these two families are unrelated, they both use the same fundamental mechanism to break DNA: the nucleophilic attack of a hydroxyl group of an amino acid, either a tyrosine or serine, to break DNA, with the concomitant formation of a protein-DNA linkage and release of an OH DNA end. Figure 17.37a shows how one recombinase family breaks DNA by forming a DNA-3′ P-tyrosine linkage and releasing a 5′ OH DNA end. These are called **tyrosine recombinases**. By contrast, recombinases of the other family break DNA by forming a serine-5′ P-DNA linkage and releasing a 3′ OH DNA end, and these are called **serine recombinases** (Figure 17.37b). Intact DNA strands can then be regenerated by the attack of the free DNA 3′ OH or the 5′ OH ends on the cognate covalent protein–DNA linkage, re-forming the high-energy phosphodiester bond in DNA and releasing the recombinase in its original state.

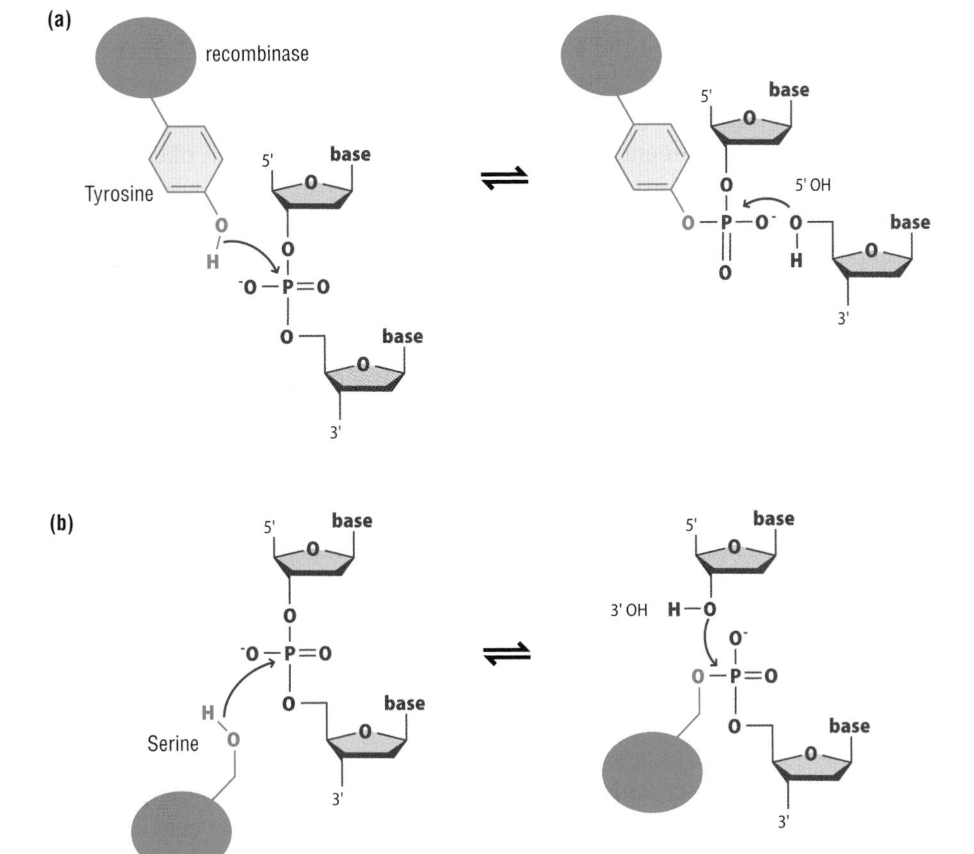

Figure 17.37 The chemistry of serine and tyrosine recombinases. (a) The OH of a tyrosine on a tyrosine recombinase attacks DNA to form a DNA-3′ P-tyrosine–protein linkage and a 5′ OH DNA end. The broken DNA strands can be rejoined by attack of the 5′ OH DNA end on the tyrosine–DNA linkage, regenerating the recombinase. (b) The OH of a serine on a serine recombinase attacks DNA to form a recombinase–serine–5′ P–DNA linkage and a 3′ OH DNA end. The broken DNA strands can be rejoined by attack of the 3′ OH DNA end on the serine–DNA linkage, regenerating the recombinase.

Adapted from Grindley, NDF, Whihteson, KL, and Rice, PA. Mechanisms of site-specific recombination. *Annual Review of Biochemistry*, 2006;**75**:567–605.

The serine and tyrosine recombinases differ from each other in that serine recombinases promote CSSR by simultaneously breaking and exchanging *both* strands in the two parental duplexes. By contrast, tyrosine recombinases first break and exchange one pair of strands from each duplex and then break and exchange the other pair in a subsequent step.

In some systems, the recombinase alone can carry out recombination. In other cases, accessory proteins which may be encoded by the recombination system (such as a phage) or by the host - often DNA-bending proteins that promote the formation of particular protein-DNA complex - are required. These accessory proteins can provide important avenues for the regulation of recombination.

17.11 CSSR SYSTEMS THAT CONTROL GENE EXPRESSION

Throughout this book, we have seen many ingenious ways in which biological systems mediate the regulation of gene expression. We begin this section by considering how CSSR mechanisms provide a further means of controlling the function of the genome.

CSSR deletion reactions can lead to the assembly of new genes

There are a few examples of CSSR-mediated deletion reactions in bacteria that lead to the assembly of active genes in specialized cell types. In *Bacillus subtilis*, for example, a serine recombinase CSSR deletion reaction generates the gene for a transcription factor required for sporulation by deleting a DNA segment that interrupts the gene in a specialized 'mother cell', which dies during spore formation. Similarly, a tyrosine recombinase CSSR deletion reaction assembles genes involved in the development of the specialized nitrogen-fixing cells of the cyanobacterium *Anabaena*. However, such deletion reactions are relatively rare as they are not readily reversible and thus represent terminal differentiation events, a state incompatible with bacterial growth.

Some microorganisms change their surface molecules at high frequency by inversion of a promoter-containing DNA segment

Given that the reactions are reversible, there are many examples of CSSR-mediated inversion reactions that control gene expression either by changing the orientation of a promoter with respect to a structural gene, or by changing the identity of the structural gene that is expressed. This type of alteration is most often seen with cell-surface proteins whose variation enables a microbe to rapidly respond to changes in its environment - for instance, the presence of antibodies - and thereby evade them. An important feature of such mechanisms is that they allow a high frequency of change at particular genes without putting the entire genome at risk by a high level of indiscriminate mutation.

While genes for surface antigens on pathogens are most likely to undergo variation, this also occurs in genes encoding virulence factors, as well as catabolic and biosynthetic systems in both pathogenic and symbiotic microorganisms.

Inversion of a promoter segment underlies the alternative expression of flagellar proteins in *Salmonella*

A well-studied example of surface variation by DNA inversion is the alternative expression of two types of flagellar protein by the bacterium *Salmonella typhimurium*. Flagella are long whip-like structures on the surface of the bacterium that enable it to move. The two flagellar proteins expressed by *Salmonella* are antigenically different; therefore, switching from one to the other helps *Salmonella* to circumvent an immune response against the protein expressed previously.

The two flagellar proteins are encoded by two genes, *fljB* and *fliC*, located some distance apart on the bacterial chromosome, as illustrated in Figure 17.38. The particular gene expressed is determined by the orientation of a promoter whose position can be changed by a chromosomal inversion. The FljB protein and the FljA protein (which represses the *fliC* gene) are encoded together in one region while *fliC* is located in a distal region of the chromosome.

The *fljB-fljA* region lies adjacent to an invertible DNA segment containing a promoter. When the invertible DNA is in one orientation, transcription from this promoter leads to expression of the flagellar protein FljB and the FljA repressor, which blocks expression of the *fliC* gene. When the promoter segment is in the other orientation, transcription from the promoter does not enter the *fljB-fljA* region so the flagellar FljB protein is not synthesized. In addition, since the repressor protein FljA is not produced, the flagellar FliC protein is now made.

DNA inversion occurs by CSSR between two sites called *hix* sites that lie in inverted orientations on either side of the DNA segment containing the promoter. Each *hix* site consists of the spacer region flanked by two inverted binding sites for the Hin serine recombinase, which is constitutively expressed from a gene in the invertible segment.

Hin inversion is carried out by a sophisticated nucleoprotein complex whose proper assembly is necessary to activate DNA breakage and joining. The mechanism of this inversion is depicted in Figure 17.39. Hin-mediated recombination also requires the accessory protein Factor for Inversion Stimulation (FIS). FIS is a DNA-bending protein and binds to specific sites on the invertible segment and is thought to provide a platform for the assembly of the Hin-bound *hix* sites. Assembly of the complex is also facilitated by the host DNA-bending protein HU.

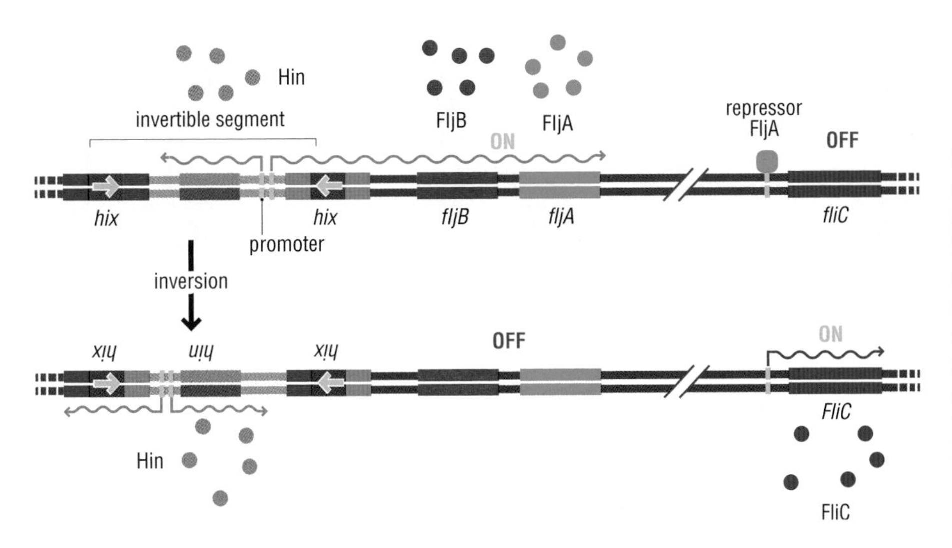

Figure 17.38 CCSR-mediated DNA inversion can change the expression of cell surface proteins. The alternative expression of different flagella surface proteins by inversion that changes the relative orientation of a promoter and directs transcription out of the invertible segment is shown. Inversion can occur by recombination between the *hix* (pink and blue) sites, mediated by the serine recombinase Hin that is encoded within the invertible segment. Inversion reorients the promoter such that the *fljB* and *fljA* genes are not expressed, resulting in *fliC* expression.

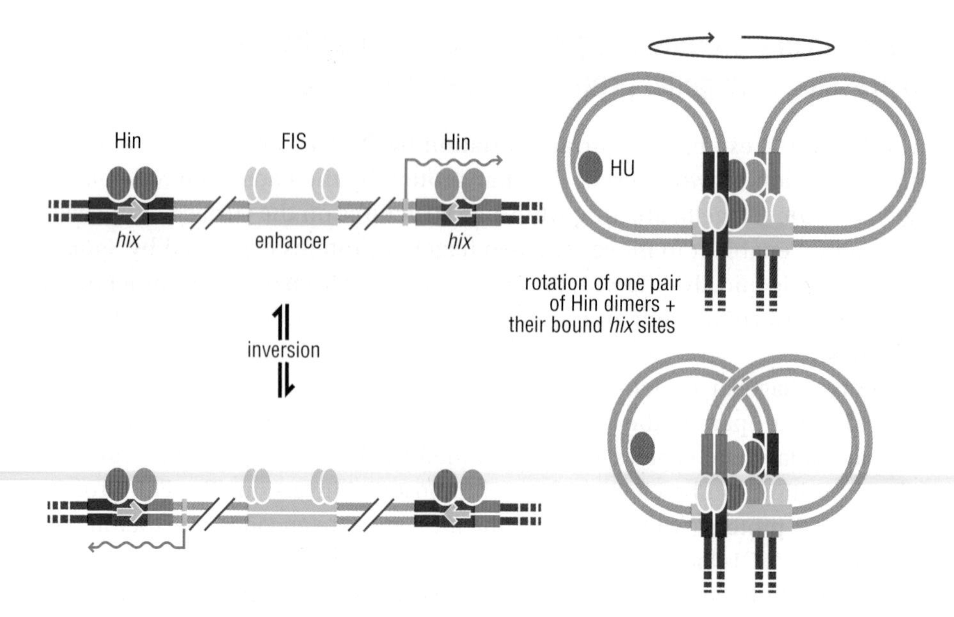

Figure 17.39 Mechanism of Hin-mediated recombination. The promoter-containing invertible segment (light gray) is bounded by *hix* sites and also contains the FIS-binding enhancer. Dimers of the recombinase Hin (pink and blue) bind to the *hix* sites and dimers of FIS (green) bind to two sites within the enhancer. The right panel shows an assembled 'invertasome' in which the enhancer bound by FIS forms a platform for the juxtaposition of the *hix* sites bound by the Hin dimers, facilitated by the sequence non-specific DNA-bending protein HU (gray). *hix* recombination occurs by the rotation of one pair of Hin dimers and their bound *hix* sites relative to the other.

Each *hix* site binds two monomers of Hin, which, mediate a cycle of double-strand breakage and joining to promote strand exchange, as illustrated in Figure 17.40. Once the Hin-bound *hix* sites are paired, the Hin molecules make a double-strand break in the *hix* site. In each strand, a serine from Hin attacks the phosphodiester backbone to form a protein–serine–5′ P-DNA linkage and exposes a free 3′ OH DNA end. The two Hins on each parental duplex thus introduce a double-strand break into each duplex, generating four 3′ OH DNA termini and four DNA-5′ P-serine recombinase termini. Rotation of the DNA bound Hin molecules then juxtaposes the cleaved strands from one duplex with the other cleaved strands of the other duplex. Attack of the 3′ OH ends on the protein–serine–5′ P-DNA linkages regenerates intact DNA in which the segment between the *hix* sites been inverted.

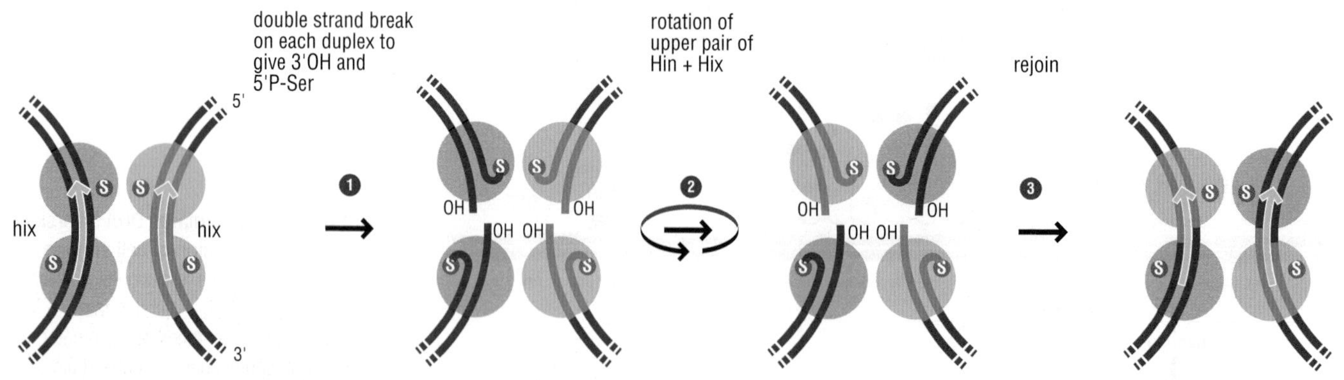

Figure 17.40 The serine recombinase Hin mediates double-strand breakage, exchange, and rejoining at the *hix* sites. Four molecules of Hin (blue and pink) bind to their recognition sequences in site I of the two *hix* sites and pair the parental duplexes. (1) The OHs of the Hin active site serines (circle S) attack the DNA, one per strand, making a covalent link between the recombinases and the DNA, and making a double-strand break in each duplex. (2) The rotation of one pair of Hins (left blue + pink) relative to the other juxtaposes the broken ends from one parental duplex to the broken ends of the other parental duplex. (3) The free 3′ OHs on the DNAs attack the 5′ phosphoserine linkages, rejoining the DNA backbones and releasing the Hin recombinases. These breakage, exchange, and rejoining events are highly coupled such that no DNA ends or Hin recombinases are released from the complex during this CSSR reaction.

17.12 CSSR CONVERSION OF DNA DIMERS TO MONOMERS

CSSR can convert DNA dimers to monomers

The conversion of a circular DNA to two circular DNA products is part of many cellular processes, and CSSR plays an important role in many such conversions. For example, with the circular genomes that occur in most bacteria and bacteriophages, accurate chromosome segregation to daughter cells is blocked by the formation of dimer chromosomes that result from homologous recombination between two monomer chromosomes, as depicted in Figure 17.41a. In response, many bacterial genomes, plasmids, and some viruses encode CSSR systems that can convert such dimers to monomers, a process illustrated in Figure 17.41b. Examples include a segregation system called Cre encoded by the bacteriophage P1 as well as the bacterial chromosomal monomerization system Xer.

Cre, a tyrosine recombinase, promotes CSSR by two cycles of single-strand exchange

As we see in Section 17.10, CSSR by a serine recombinase occurs by making and rejoining double-strand breaks in each parental duplex simultaneously. By contrast, CSSR promoted by tyrosine recombinases occurs by two cycles of single-strand

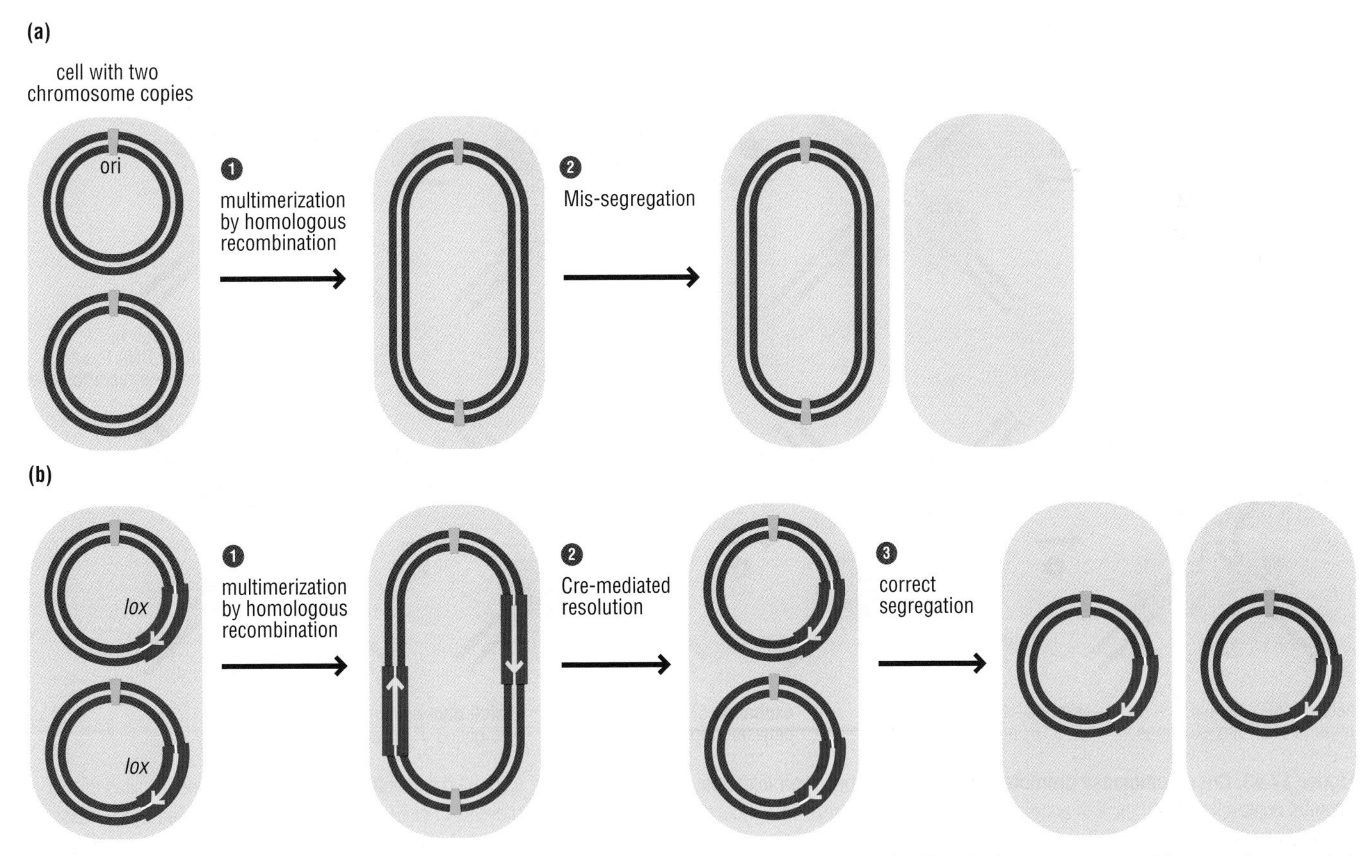

Figure 17.41 CSSR contributes to chromosome segregation by resolving dimeric chromosomes. (a) Dimeric chromosomes cannot be properly segregated to daughter cells. (b) Chromosome monomerization by CSSR allows segregation. The chromosome encodes a CSSR system that acts at special sites (blue box with yellow arrow), and converts dimers to monomers.

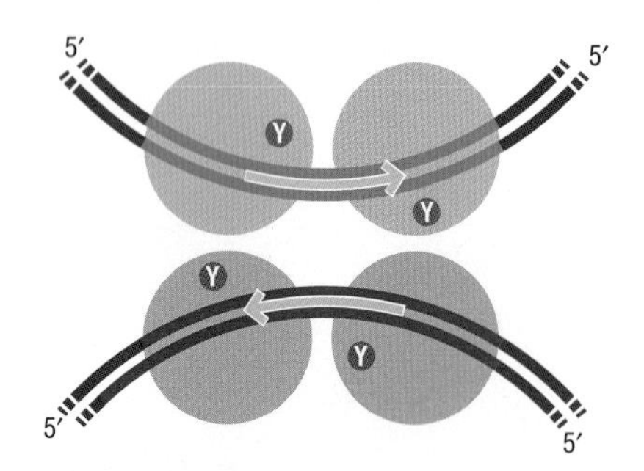

Figure 17.42 Four Cre molecules bound to two substrate *lox* sites with the spacer regions (green arrows) arranged in antiparallel fashion. The circled Ys represent the active site tyrosine in Cre.

Adapted from Chen, Y, and Rice, PA. New insight into site-specific recombination from FLP recombinase-DNA structures. *Annual Review of Biophysics and Biomolecular Structures*, 2003;**32**:135–159.

breakage, exchange and rejoining between the two parental duplexes. A well-studied example is the bacteriophage P1 Cre system, which acts at special DNA sites called *lox* sites. As in the Hin system, recombination begins with the binding of a dimer of Cre to a *lox* site. In two *lox* sites, four Cre molecules bind to inverted recognition sequences that flank the spacer region of two parental *lox* sites (see Figure 17.35). Two *lox* sites bound by Cre then pair in antiparallel alignment mediated by Cre–Cre interactions, resulting in a tetramer of Cre bound by two *lox* sites, as illustrated in Figure 17.42.

Instead of the concerted breakage of all four DNA strands seen with the serine recombinases, Figure 17.43 shows how tyrosine recombinases break and exchange only two strands at a time, one from each parental duplex. The sequence of the spacer region and the binding properties of Cre at each of its sites determines which pair of Cre molecules will execute the first strand exchanges. Following Cre binding and pairing, which juxtaposes the recombination sites, a tyrosine from one Cre protein on each duplex attacks the phosphodiester backbone of one strand to form a protein–tyrosine–3′ P-DNA linkage and expose a free 5′ OH DNA end (Figure 17.43, step 1). The 5′ OH ends from each duplex are then juxtaposed to the protein–tyrosine–3′ P-DNA linkages from the other duplex (Figure 17.43, step 2). DNA rejoining then occurs by the attacks of the 5′ OH DNA ends on the protein–tyrosine–3′ P-DNA linkages. This attack re-forms the DNA phosphodiester linkage and releases the protein (Figure 17.43, step 3).

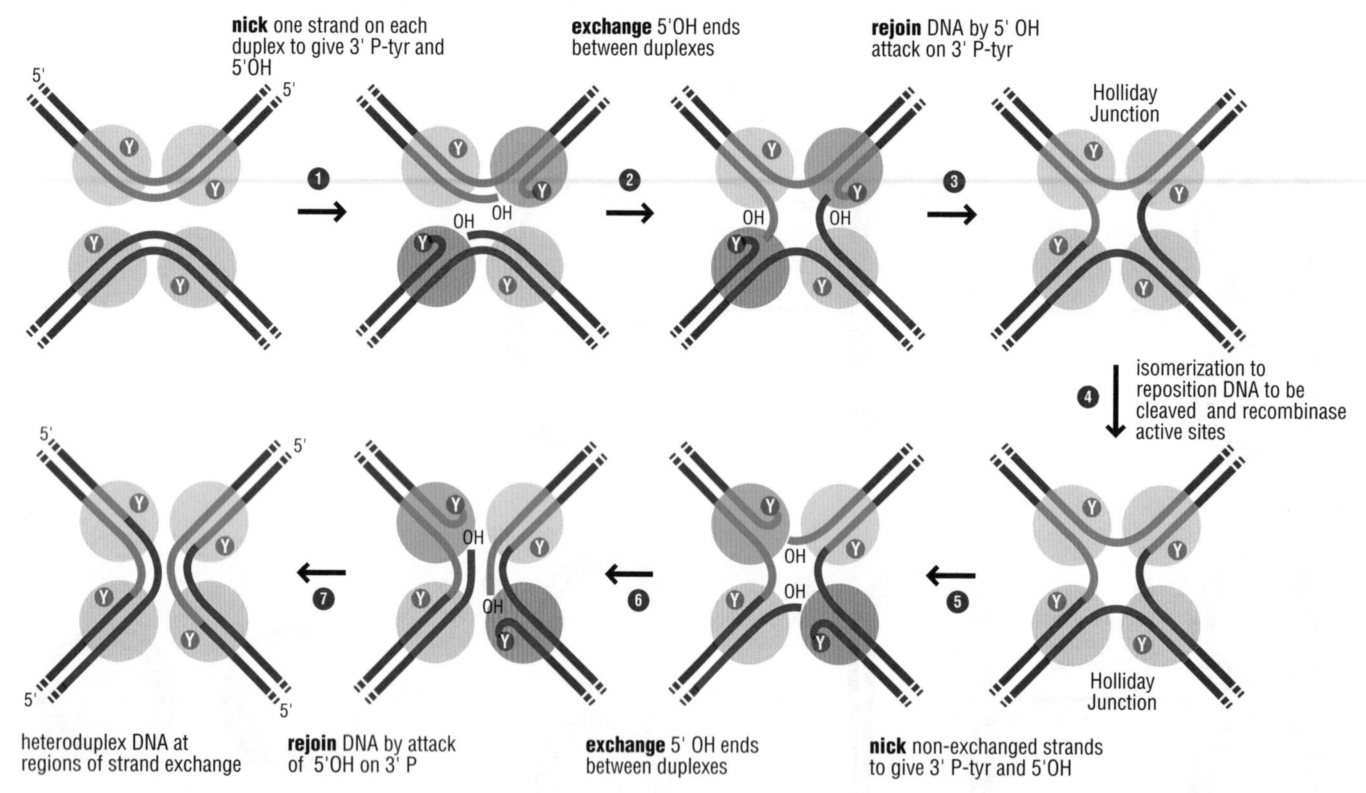

Figure 17.43 Cre recombinase promotes recombination between *lox* sites. Four Cre protomers (light blue and light pink) bind and pair the *lox* sites of two parental duplexes blue and pink in antiparallel fashion (see Figure 17.42). Recombination occurs by two cycles of single strand exchange mediated by breakage and joining at tyrosines. The darker shaded diagonally-opposed circles denote active protomers. One cycle of strand exchange results in formation of a Holliday junction whose changes slightly reposition the DNA strands and proteins for the second cycle of strand exchange (not shown here). The recombinant duplexes contain heteroduplex DNA in the spacer regions between the positions of strand exchange.

Adapted from Chen, Y, and Rice, PA. New insight into site-specific recombination from FLP recombinase-DNA structures. *Annual Review of Biophysics and Biomolecular Structures*, 2003;**32**:135–159.

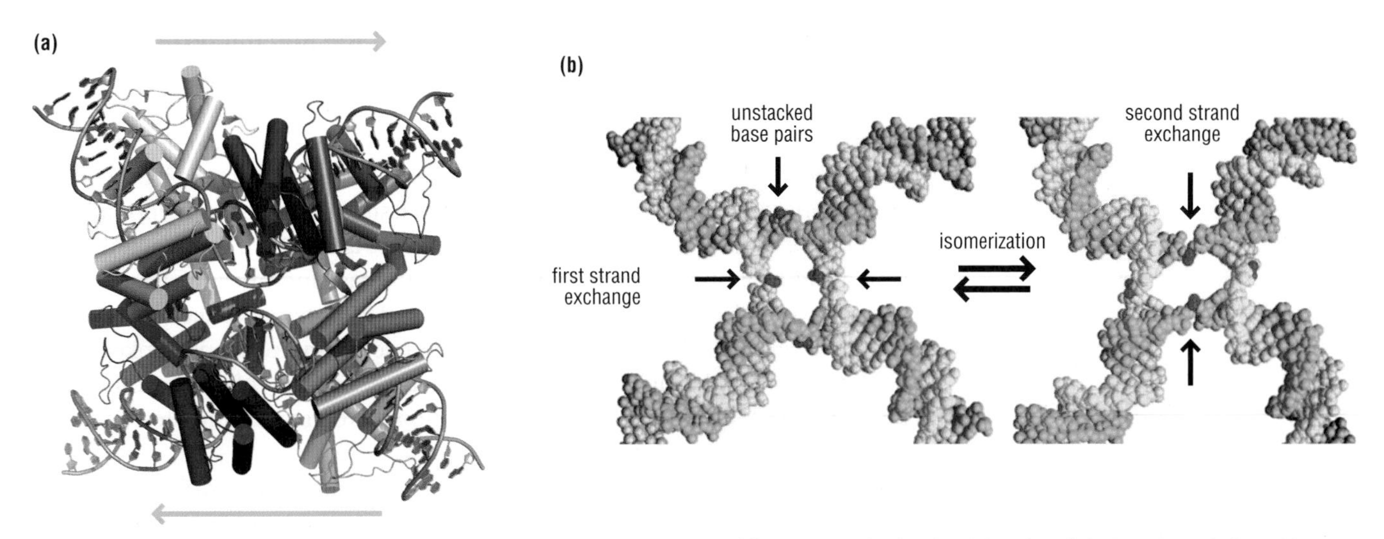

Figure 17.44 The structure of Cre-*lox* recombination complexes. (a) A tetramer of Cre pairs two *lox* sites (gray) in antiparallel orientation as indicated by the green arrows. The darker subunits will perform the first set of strand exchanges (step 1 in Figure 17.43). (b) During isomerization of the Holliday junction intermediate (step 4 of Figure 17.43), a modest local change in the structure of the junction DNA more closely juxtaposes the strand exchange positions (red) in the green strands, and this second pair of strands can undergo strand exchange. PDB codes 2HOI, 3CRK.

Panel (a) is From Ghosh, K, Guo, F, and Van Duyne, G. DNA: replication, repair, recombination, and chromosome dynamics: *Journal of Biological Chemistry*, 2007;**282**: 24004–240016; panel (b) is from Gopaul, DN, Guo, F, and Van Duyne. Structure of the Holliday junction intermediate in Cre-loxP site-specific recombination. *The EMBO Journal*, 1998;**17**:4175–4187.

The resulting CSSR intermediate formed, in which a single pair of strands is exchanged between the parental duplexes, is identical to the Holliday junction that is generated during homology-directed repair and homologous recombination.

➡ We discuss the structure of the Holliday junction in Section 16.8.

Figure 17.44a shows the protein–DNA structure of a Cre complex in which the parental duplexes are paired. The DNA in the Holliday junction intermediate in shown in Figure 17.44b. Following the isomerization of this intermediate indicated in Figure 17.44b, which positions the other non-exchanged strands to be cleaved, CSSR recombination is then completed by a second cycle of Cre-mediated DNA breakage and joining on the other side of the spacer region, using the recombinases on the other pair of strands (Figure 17.43, steps 5–7).

As will be discussed in Section 19.5, the Cre-*lox* system has been extensively used in genome engineering to allow the deletion of defined chromosomal regions at particular times.

17.13 BACTERIOPAGE LAMBDA INTEGRATION AND EXCISION

Bacteriophage lambda integration and excision are highly regulated CSSR events

The CSSR Cre-*lox* system is simple both in its components and directionality. Recombination requires only the recombinase Cre and the *lox* sites, comprising the spacer region and two flanking Cre binding sites. Integration, excision, and inversion can all occur with DNA substrates that possess appropriately oriented *lox* sites. By contrast the directionality of some other CSSR reactions is highly regulated. The integration and excision cycle of the bacteriophage lambda is an example of a highly regulated CSSR system that is mediated by elaborate protein–DNA complexes not unlike those that occur in DNA replication and transcription.

The lambda recombination sites on the phage and bacterial chromosomes are related but distinct

When the bacteriophage lambda infects *E. coli* and enters the lysogenic pathway as described in Section 9.4, CSSR occurs between a special site in the phage genome called *attachment site phage* (*attP*) and a special site in the bacterial genome called *attachment site bacteria* (*attB*), as summarized in Figure 17.45. This reaction is mediated by the lambda-encoded tyrosine recombinase Int (integrase), and several accessory proteins. Like the Cre system, *attP* and *attB* share a spacer region called the core region (*O*) in which recombination occurs and which is flanked by Int binding sites *C* and *C′* on *attP* and *B* and *B′* on *attB*. The arms of *attP* beyond the Int binding sites flanking the core contain additional protein-binding sites, and these regions are called the *P* and *P′* arms.

CSSR between *attP* and *attB* generates *attachment site left* (*attL*) and *attachment site right* (*attR*) which flank the integrated prophage; *attL* and *attR* are hybrid sites that contain sequences from both *attP* and *attB*. Prophage excision results from recombination between *attL* and *attR*, which regenerates *attP* and *attB*.

Figure 17.46 shows a higher resolution view of the *att* sites, shown in the antiparallel alignment in which recombination occurs. *attP* is large and complex and contains multiple binding sites for additional recombination proteins beyond the Int *C* and *C′* sites that flank the core, *O*; no additional protein-binding sites flank the Int *attB B* and *B′* sites. Thus each *att* site - *attP*, *attB*, *attR*, and *attL* - is distinct and has different protein-binding sites flanking the core.

The unique structure of each *att* site results in particular protein requirements for recombination with these different DNA substrates. Integration between *attP* and *attB* requires the phage-encoded integrase and IHF, a host-encoded DNA bending protein. Excision between *attL* and *attR*, however, requires Int, IHF, and two additional proteins: one encoded by the phage, called excisionase (Xis), and another host-encoded protein, FIS. FIS is the essential protein in the Hin inversion reaction discussed in Section 17.11, hence its name. Xis is not present during phage infection when integration occurs.

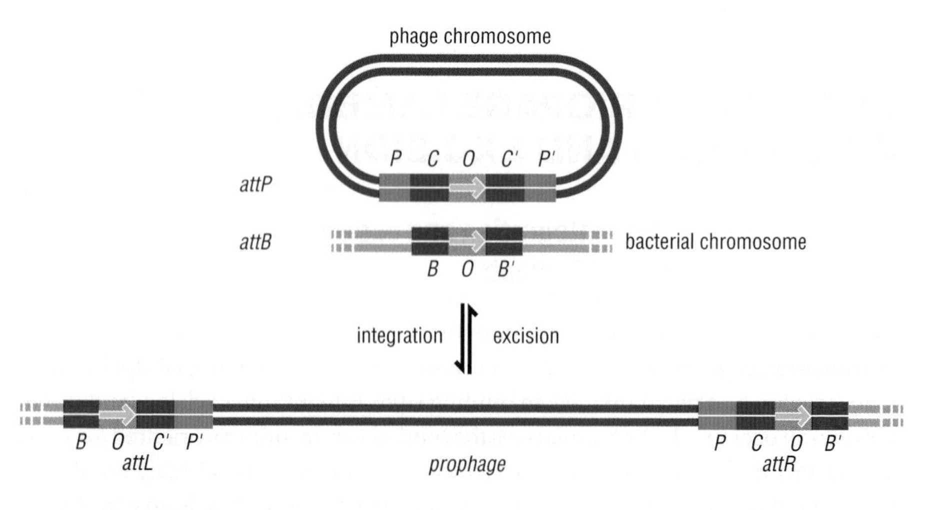

Figure 17.45 Schematic of the bacteriophage lambda integration and excision reactions. The viral chromosome (gray) contains *attP* (blue) which recombines with *attB* (red) in the bacterial chromosome (light gray) during lambda integration. This recombination event forms the prophage, which is flanked by *attL* and *attR*. Excision occurs between *attL* and *attR* and regenerates *attP* and *attB* when the prophage enters the lytic cycle.

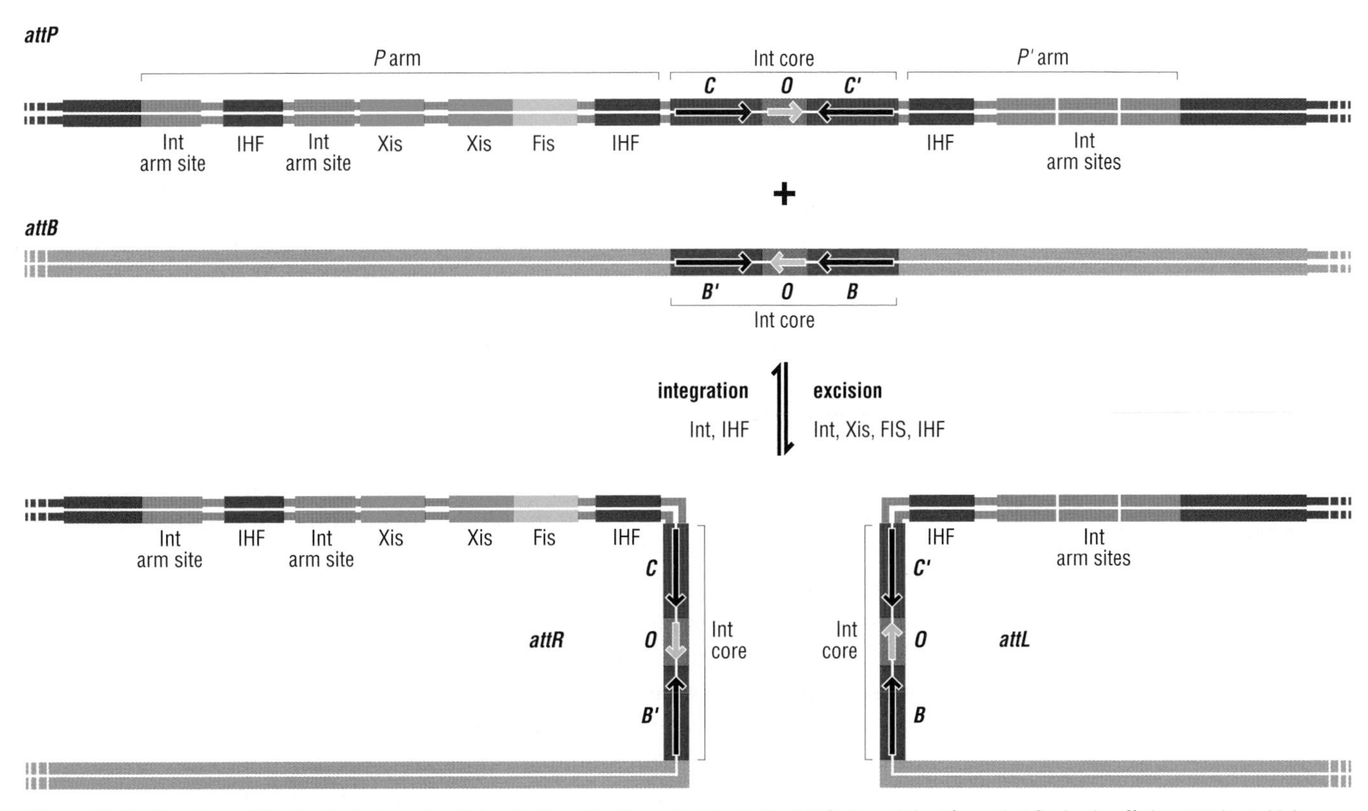

Figure 17.46 Structure of the bacteriophage lambda *att* sites. The *P* arm that flanks the Int *C* site and the *P′* arm that flanks the *C′* site contain multiple binding sites for accessory proteins. Both the *P* and *P′* arms contain Int arm binding sites (blue), whose sequences are distinct from the core *C* and *C′* sites and which are recognized by a different domain of Int (see Figure 17.47). The *P* and *P′* arms of *attP* contain binding sites for the host sequence-specific DNA-bending protein IHF (pink) and binding sites for two proteins required uniquely for excision, the phage-encoded Xis (excisionase; orange) and host-encoded FIS (green; see Figure 17.39). *attR* contains the *B′ O C* sites and the *P* arm that contains Int arm binding sites, binding sites for IHF and binding sites for Xis. *attL* contains the *B O C′* sites and the *P′* arm that contains Int arm binding sites and a binding site for IHF. Excision requires Int, IHF, Xis, and FIS and occurs between *attR* and *attL* to regenerate *attP* and *attB*.

Adapted from Grindley, NDF, Whiteson, KL, and Rice, PA. Mechanisms of site-specific recombination. *Annual Review of Biochemistry*, 2006;**75**:567–605.

Multiple recombination proteins are required for the assembly of active intasomes

Like Cre recombination at *lox* sites, Int recombination at *att* sites involves the assembly of a tetramer of Int that (i) binds to the four Int binding sites flanking the regions of homology, (ii) pairs the parental duplexes, and (iii) carries out strand exchange. The affinity between Cre and *lox* sites and between Cre protomers is sufficiently high, however, that Cre alone can execute recombination. By contrast, the affinities between Int monomers and the *att* core sites, as well as between the Int monomers, are too low to allow the formation of a stable tetramer that can pair the *att* DNA substrates in the presence of only Int; additional proteins are required to form active '**intasomes**'. These additional proteins promote DNA bending to facilitate the binding of the *att* sites to Int. Furthermore, the different arrangement of the accessory protein-binding sites on different *att* sites means that different proteins are required for recombination between different *att* sites. These differential requirements allow for the directional control of recombination, most notably that recombination between *attL* and *attR* (i.e., excision) requires Xis. Xis is only synthesized when the phage lytic functions are derepressed when cellular conditions prompt the phage to shift from the lysogenic state to lytic growth.

➔ We discuss the lytic and lysogenic growth of phage lambda in Section 9.4.

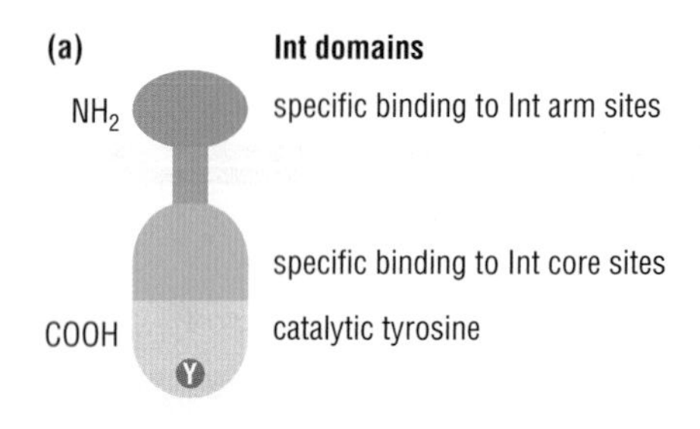

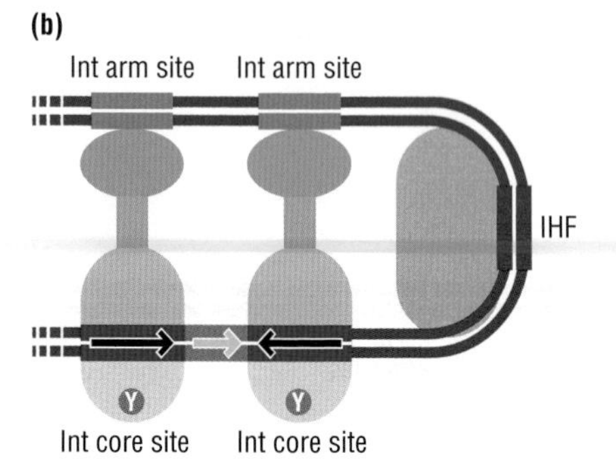

Int can bind two DNA segments

Figure 17.47 Int can bind to two different DNA sequences. (a) The domain structure of lambda integrase. Lambda integrase contains two domains: the NH2-terminal domain, which binds to specific sequences in the *P* and *P′* arms of *attP* and the COOH-terminal domain, which binds specifically to the Int sites that flank the core (*O*), *C*, *C′*, *B*, and *B′*, and contains the catalytic topoisomerase-like domain. (b) Because of its two different sequence-specific DNA-binding domains, the simultaneous binding of one Int protomer to two regions on a single DNA molecule can occur. The bending of the *att* DNA necessary for such bridging is facilitated by DNA-bending proteins IHF (as shown), Xis, and FIS.

Assembly of the *attP* intasome for integration requires IHF and DNA supercoiling

The *attP P* and *P′* arms contain multiple binding sites beyond the core *C* and *C′* sites. Why is the *attP* substrate so elaborate? We now know that the *P* and *P′* arms are required as an assembly platform to position an Int tetramer at the core sites of strand exchange of *attP* and *attB*. The active Int tetramer that mediates integration is initially assembled on *attP* alone, and naked *attB* DNA is then recruited into this complex.

Three features are key to the assembly of the Int tetramer in the *attP* intasome. First, Figure 17.47a shows how Int has two distinct DNA-binding domains that recognize two different DNA sequences – an N-terminal domain that binds to the Int arm sites and a C-terminal domain that binds the Int core sites; the C-terminal domain also contains the active site topoisomerase. Int actually binds with higher affinity to the Int arm sites. Thus the binding of four Int protomers to four Int arm binding sites in *attP* facilitates the recruitment of their C-terminal domains to the core regions.

Second, considerable bending of the *P* and *P′* arm DNA is required to allow the simultaneous binding of Int to the arm and core sites of *attP*, as shown in Figure 17.47b. The DNA-bending protein IHF is essential to the bending of *attP*.

Third, DNA supercoiling also plays a critical role in promoting the bending and wrapping of the *attP* DNA around the proteins of the *attP* intasome.

Once the *attP* intasome with its four catalytic domains is formed, as illustrated in Figure 17.48, *attB* DNA is thought to enter the reaction as naked DNA, with the *B′* and *B* core sites binding to the pair of unoccupied Int catalytic domains in the already-assembled Int tetramer. The Int catalytic domains then act on the *C*, *C′*, *B*, and *B′* sites of strand exchange in the paired *attP* and *attB* DNA duplexes, by the mechanism described earlier for Cre (see Figure 17.43). In Experimental approach 17.3 we consider how small molecules have been used as recombination inhibitors to identify intermediates in lambda recombination.

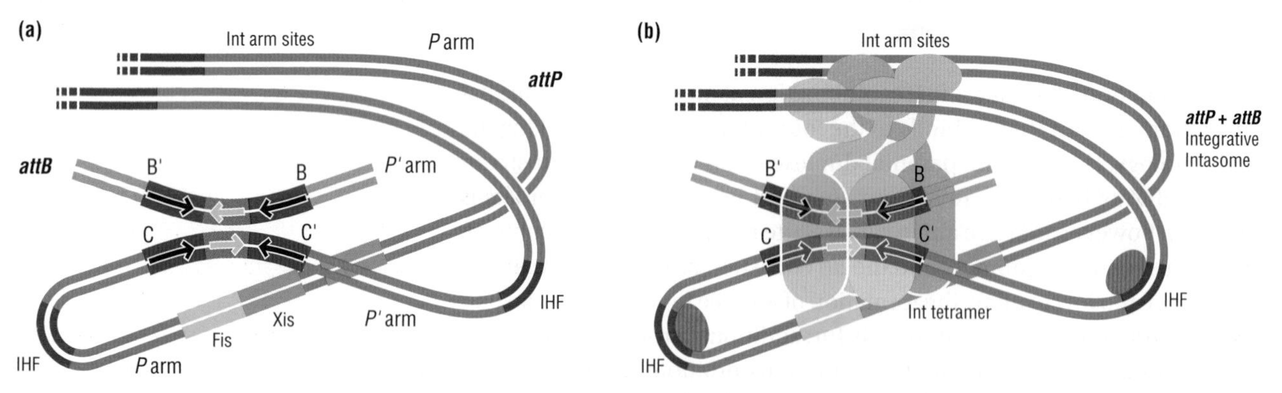

Figure 17.48 Structure of the bacteriophage lambda *attP-attB* intasome required for integration. Integrative recombination occurs in an elaborate nucleoprotein complex. (a) Path of the DNA in the integrative intasome. (b) The proteins in the integrative intasome. The core regions of *attP* (dark blue) and *attB* (red) bind to a tetramer of Int COOH domains (large blue ovals bound to *C* and *C′*, and large pink ovals bound to *B* and *B′*). The cores of *attB* and *attP* (green arrows) are oriented in antiparallel orientation. Formation of the Int tetramer bound to the core sites is facilitated by the Int NH2 domains (small blue and pink ovals) interacting with Int arm sites in the *P* and *P′* arms. Bending of the *attP* DNA is facilitated by the binding of the DNA-bending protein IHF (pink) to the *P* and *P′* arms of *attP*. (Which Int arm binding domains interact with which *attL* and *attR* Int arm binding sequences is not yet known in detail; thus this schematic diagram is meant to indicate only the general path of the *P* and *P′* arms.)

from van Duyne, GD. Lambda integrase: armed for recombination. *Current Biology*, 2005;**15**: R658–R660.

17.3 EXPERIMENTAL APPROACH

Identification of small molecule inhibitors of CSSR

The dissection of a reaction pathway requires identification and analysis of reaction intermediates as well as knowledge of the initial substrates and reaction products. Mechanistic understanding of a pathway can come from the analysis of mutant enzymes (that can only complete certain reaction steps) or the use of modified substrates (that stall the reaction at an intermediate stage). Another powerful strategy for studying reactions is the isolation of small molecule inhibitors that inhibit a process. For many years, chemists and chemical biologists have been developing methods to synthesize large libraries of molecules that can be used in screens to identify inhibitors of a specific biological process. The basic idea behind the use of so-called combinatorial libraries (called this because they are composed of chemical moieties in different combinations) is to apply complex mixtures of compounds to an assay that reports on a biologically interesting event.

For example, a screen for antibacterial compounds might start with a growth inhibition assay, and a screen for specific biochemical inhibitors might start with purified proteins performing a specific molecular event. The ultimate goal is to identify from within these complex mixtures a small molecule 'lead' compound that inhibits a specific molecular process. The inevitable challenge of such approaches is that the complex mixtures must be 'de-convoluted' to identify specific active molecular species within them, and this requires a careful synthetic plan.

A chemical screening approach was employed by Anca Segall and colleagues to study CSSR mediated by the integrase of bacteriophage lambda. Understanding of the integrase mechanism is incomplete in large part because several of the pathway intermediates are hard to isolate. The Segall group first established an assay that would allow them to screen complex mixtures of small peptides (hexapeptides) to identify inhibitors of the process that would result in the accumulation of the Holliday junction intermediates. By focusing on peptide inhibitors, the researchers reasoned that they might target protein surfaces that are involved in specific molecular interactions, and thus a specific molecular process that would result in the accumulation of the Holliday junction intermediates. As an assay, the CSSR reaction was visualized in a minimal system with radioactively labeled DNA substrate and two purified proteins (Int (the integrase) and IHF (a DNA-binding protein that promotes the bending of DNA)) using gel electrophoretic analysis, in which reaction substrates, products, and intermediates have characteristic mobilities; the results are shown in Figure 1. For example, Holliday junction complexes, which are reaction intermediates, (Figure 17.44b) migrate more slowly than the starting *attL* substrate or the recombinant reaction products.

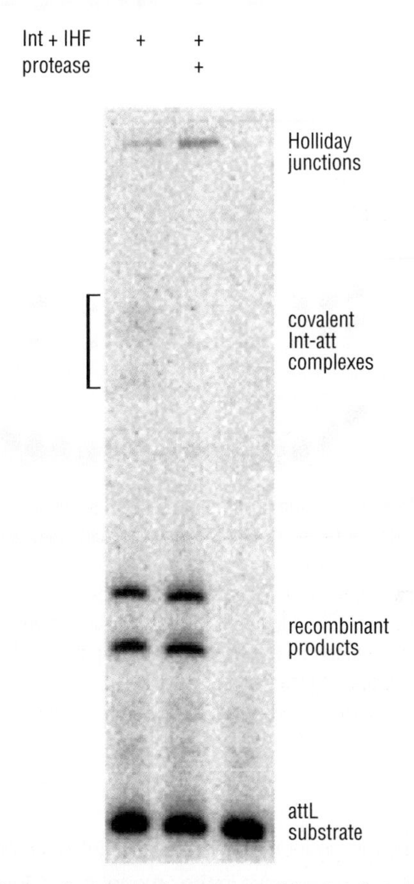

Figure 1 A biochemical assay for *attL* recombination for screening peptide libraries. A small radioactively labeled *attL* DNA substrate containing the *attL* sequence of a bacteriophage lambda lysogen and a larger unlabeled *attL* DNA substrate are incubated with purified Int and IHF proteins and the products of the reaction resolved on a polyacrylamide gel. The recombinant products, Holliday junction intermediates, and covalent Int–DNA complexes are seen.

Photo courtesy of Anca Segall (San Diego State University), of reactions like those in Cassell *et al.* *Journal of Molecular Biology*, 2000;**299:**1193–1202.

The next step was to ask whether compounds could be identified that resulted in the inhibition of the progress of the reaction or the accumulation of reaction intermediates, such as the Holliday junction. In a first pass, 120 combinatorial peptide libraries were prepared; in each library one of the 20 amino acids was fixed in a given position in the hexapeptide (6 positions × 20 amino acids = 120) and each contained about 2.5 million distinct molecular species (1 × 19 × 19 × 19 × 19 × 19, excluding cysteine from the randomized amino acid mixtures). Two different concentrations of the library were added to the CSSR reaction, and the progress of the reaction was followed by gel electrophoresis. In Figure 2, we see the analysis of seven different compound mixtures, most

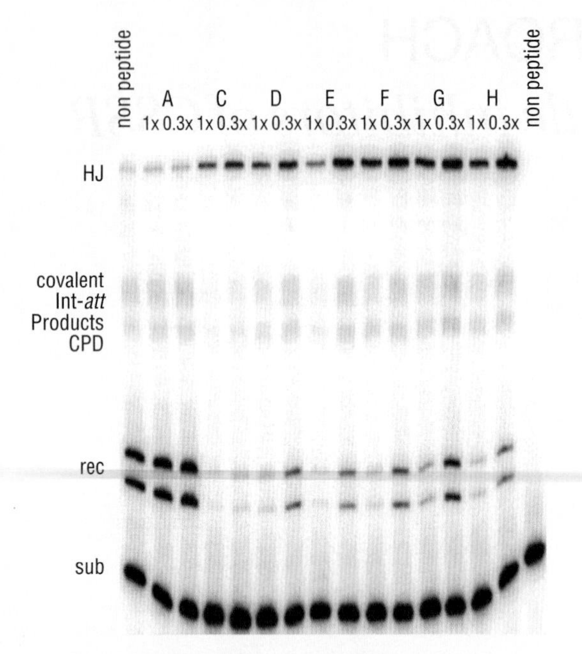

Figure 2 Effects of peptide libraries on a CSSR reaction. Peptide mixtures A, C, D, E, F, G, and H refer to the amino acid 'fixed' at position one of the hexamer (OXXXXX where O refers to a specific amino acid and X to random amino acids). Two different concentrations (1× and 0.3×) of the peptide mixture were added to the CSSR reaction. All mixtures except A had readily discernible effects on the reaction, increasing the amount of Holliday junctions, and decreasing the yield of products.

Cassell *et al.* Dissection of bacteriophage lambda site-specific recombination using synthetic peptide combinatorial libraries. *Journal of Molecular Biology*, 2000;**299**:1193–1202.

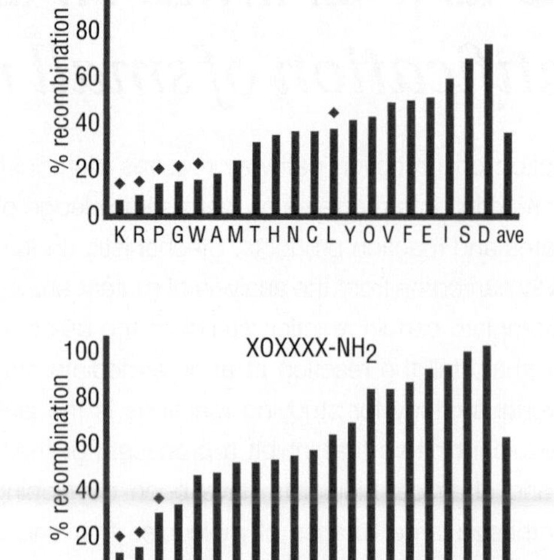

Figure 3 Inhibition of recombination by peptide libraries with fixed position 1 or 2. The amino acid residue in the fixed position is denoted along the x-axis. Values of percentage recombination were normalized to the extent of recombination in untreated reactions. (a) The most potent libraries carry K, R, P, G, or W at the first position. (b) The most potent libraries carry either W, R, or K at the second position. Subsequent libraries were designed with these positions fixed as indicated by these data.

of which result in an increase in the amount of Holliday junction intermediate and a decrease in the amount of product (mixture A, by contrast, has little effect). Promising mixtures were subjected to more careful dose-response titrations to determine which mixtures would be the most potent for subsequent analysis.

Following this initial screen, other positions in the hexapeptide were 'fixed' according to these initial leads, and new, less complex libraries synthesized around these fixed positions. In Figure 3, we see how the potency of a mixture in the assay can be compared to determine how to 'fix' a given position for synthesis of the next complex mixtures. As such a process is reiterated, the number of molecular species in the synthesized mixture becomes increasingly small, eventually allowing the scientist to identify the most promising leads in the library. The chemical structure of these lead compounds can then be systematically altered to increase the specificity and potency of the compound.

The isolation of these hexapeptide inhibitors has provided a new tool for studying CSSR and may prove useful as an antimicrobial agent if it shows specificity for the bacterial reactions. Such combinatorial library approaches for drug screening are quite powerful in that they allow millions of compounds to be tested simultaneously, thus decreasing the cost of the screening. By requiring the inhibitory activity to emerge from a complex mixture (where each compound is necessarily present at relatively low concentration), the potency of an initial lead must already be reasonably high, and thus worthy of pursuit.

Find out more

Cassell G, Klemm M, Pinilla C, Segall A. Dissection of bacteriophage lambda site-specific recombination using synthetic peptide combinatorial libraries. *Journal of Molecular Biology*, 2000;**299**:1193–1202.

Related techniques

Phage lambda (as a model organism); Section 19.1

Radioactive labeling; Section 19.6

Polyacrylamide gel electrophoresis; Section 19.7

Assembly of the *attL* and *attR* intasomes for excision requires additional accessory proteins

Excisive recombination requires the assembly of *attL* and *attR* intasomes and pairing between them. However, Int and IHF are not sufficient for the formation and interaction of these intasomes because neither *attL* nor *attR* have Int arm sites on both sides of the Int core sites to promote assembly of the tetramer on just one of these *att* sites. Instead, Figure 17.49 depicts how two additional accessory DNA-bending proteins, Xis and FIS, are required to promote the DNA bending necessary for Int to be able to bind to *both* core and arm sites on *attL* and *attR*.

As in integration, excision requires the assembly of an active Int tetramer and the wrapping of *attL* and *attR* around that tetramer. In contrast to integration, however, the *P* and *P′* arms that provide the platform for Int tetramer assembly do not lie on the same piece of DNA. The additional bending proteins Xis and FIS are required to appropriately orient the *P* arm in *attR* such that it can provide a platform for the correct assembly of the Int tetramer. Following the assembly of the intasome containing both *attR* and *attL*, in which the core Int binding sites are appropriately juxtaposed, strand exchange occurs.

The central feature of the lambda recombination directionality control is that recombination cannot occur between *attL* and *attR* in the absence of Xis. Xis (and FIS) are required to promote the bending of *attR*, which provides a DNA platform for the binding of the Int arm domains and, hence, facilitates the assembly of the active tetramer at the core regions. So, although the same DNA-binding sites are present in the *P* and *P′* arms of *attP*, the separation of the *P* and *P′* arms to the *attR* and *attL* DNAs imposes the requirement for Xis for the assembly of the paired, active *attR-attL* intasomes.

Now that we have learned about the myriad ways in which the sequence of a genome can be changed and how exquisitely gene expression and RNA and protein levels are regulated, we turn to a broader view of the genome as a whole. What do we now know about genomes and what do we have yet to learn? We next consider genome structure and evolution.

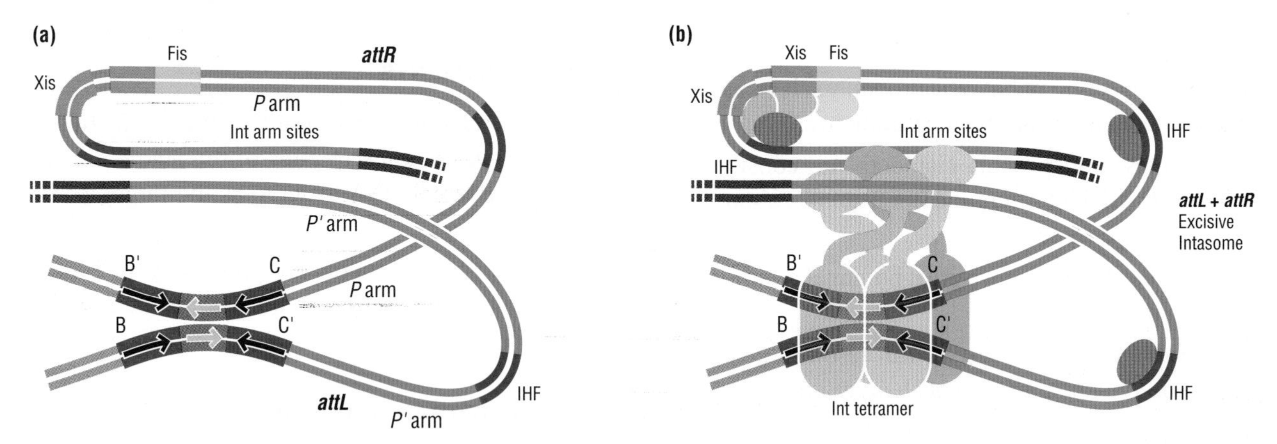

Figure 17.49 Structure of the bacteriophage lambda *attR*-*attL* intasome required for excision. Excision between *attL* and *attR* occurs within an elaborate nucleoprotein complex in which *attR* and *attL* are paired. (a) Path of the DNA. (b) The proteins of the excisive intasome. The core sites, B′OC in *attR* and BOC′ in *attL*, interact with the COOH domains (large blue and pink ovals) of a tetramer of Int. Formation of the tetramer is facilitated by the interaction of the four Int NH2 domains (small blue and pink ovals) with the *P* arm of *attR* and the *P′* arm of *attL*. The bending of the DNAs required for assembly of this complex is facilitated by the binding of the DNA-bending proteins IHF (pink), Xis (orange), and FIS (green) to the *P* arm of *attR* and of IHF to the *P′* arm of *attL*. Protein-protein interactions between Ints bound at both the core sites and the arm sites are important to pairing of the *attL* and *attR* DNAs. (Which Int arm binding domains interact with which *attL* and *attR* Int arm binding sequences is not yet known in detail; thus this schematic diagram is meant to indicate only the general path of the *P* and *P′* arms.)

from van Duyne, GD. Lambda integrase: armed for recombination. *Current Biology*, 2005;**15**: R658–R660.

SUMMARY

THE NATURE OF TRANSPOSITION AND CSSR

- Specialized recombination proteins acting at cognate special sequences can promote the rearrangement of defined DNA segments.
- Transposition promotes the movement of discrete DNA segments from place to place in the genome.
- CSSR promotes recombination between pairs of special sites, rearranging the DNA segments containing the sites.
- The mechanisms of DNA breakage and joining are different in transposition and CSSR.

THERE ARE MULTIPLE TYPES OF TRANSPOSONS

- Transposon insertion is accompanied by a target-site duplication, which results from the joining of the transposon to staggered positions of the top and bottom strands of the target DNA.
- DNA synthesis by host polymerases repairs the junction between the newly inserted transposon and the target DNA.
- The three main classes of transposable elements are DNA-only transposons, which move via a DNA intermediate, and LTR and non-LTR elements, which move via an RNA intermediate.

TRANSPOSONS IN NATURE

- Transposable elements are present in virtually all genomes.
- Transposons usually comprise a small fraction, <10%, of bacterial genomes.
- The rapid replication of bacteria probably selects against the accumulation of transposable elements.
- In eukaryotes, transposable elements can constitute a significant fraction of a genome (for example, about 45% of the human genome).
- Transposable elements have great potential to introduce genetic change.
- Transposable element insertion can result in gene disruption and the juxtaposition of element-encoded regulatory signals, such as enhancers, to host genes.
- Homologous recombination between transposable element copies can result in deletions, duplications, and translocations.
- There are multiple examples of human disease resulting from transposable element insertion and from chromosome rearrangements due to homologous recombination between element copies.

DNA-ONLY TRANSPOSONS

- Insertion of DNA-only transposons involves the direct nucleophilic attack of a 3′ OH of the transposon end on the target DNA.
- The transposases that catalyze these reactions share a common catalytic domain.
- The reactions used by DNA-only elements have been coopted in the immune system to mediate the assembly of antibody and antibody-receptor genes.
- The RAG recombinase that mediates these reactions is related to transposases.

LTR ELEMENTS

- LTR retrotransposons include retroviruses, and retroviral-like elements that do not leave the cell.
- LTR element transposition begins with the synthesis of an RNA copy of the element using an integrated DNA form as a template.
- The element RNA is the template for an element-encoded reverse transcriptase that generates a double-stranded DNA copy of the element.
- The 3′ OH ends of the DNA copy of an LTR element integrate into the target DNA by the same mechanism as the DNA-only transposons.
- The catalytic domain of the LTR integrase is related to those of the DNA-only transposases.

NON-LTR ELEMENTS

- Non-LTR elements insert into a new site by reverse transcription of an RNA copy of the element using a 3′ OH in the target DNA to serve as the target for reverse transcription.
- Active non-LTR elements are present in the human genome, and their transposition has resulted in disease-inducing mutations.

CONTROL OF TRANSPOSITION

- Given the potentially deleterious effects of transposon movement, transposition is tightly controlled at the levels of transposon expression and target-site selection.
- RNAi and DNA methylation are genome-wide defenses that have evolved to protect the genome against transposable elements.

CSSR

- CSSR between specific pairs of sequences can mediate a variety of DNA transactions.
- CSSR between sites on different DNAs allows for the joining (integration) of DNA molecules.
- CSSR between sites on the same molecule in direct orientation can result in excisions of the DNA between the sites or the resolution of DNA dimers to monomers.
- CSSR can mediate the integration and excision cycles of viruses.
- CSSR between sites on the same molecule in inverted orientation can result in inversion of the DNA between the sites and the regulation of gene expression by changing the orientation of a promoter.
- CSSR is reciprocal and occurs by DNA breakage and joining in the absence of DNA synthesis.
- CSSR is carried out by topoisomerases through a mechanism in which DNA breakage is accompanied by the covalent linkage between the protein and the DNA via either a serine or a tyrosine residue.

17.1 TRANSPOSABLE ELEMENTS: OVERVIEW

1. What is an autonomous element?
2. Euchromatin is gene poor and transposon rich; true or false?
3. What are target-site duplications? What causes them to be created?
4. What are the two pathways by which transposons can move?
5. Briefly describe the three main types of transposable elements and their mechanisms of transposition.

17.2 AN OVERVIEW OF DNA-ONLY TRANSPOSONS

1. Briefly, how do cut-and-paste elements differ from nick-and-paste elements?
2. Describe how DNA cut-and-paste transposons play a role in the development of antibiotic-resistant bacterial strains.
3. Explain why, when a transposase from one transposon is expressed, not all the transposons within the genome are affected.

17.3 DNA-ONLY CUT-AND-PASTE TRANSPOSITION

1. What are the five main steps in DNA cut-and-paste transposition?
2. How does the process of synapsis combat the formation of potentially lethal isolated double-stranded breaks?
3. Several mechanisms exist for transposon excision, but the final outcome at the cleaved end is the same in all cases. What is this outcome, and why is it important for insertion of the transposon into the recipient DNA site?
4. Although DNA cut-and-paste transposons do not replicate themselves, it is possible for the transposon to be reconstituted at the donor site after excision. Explain how this occurs.
5. Which of the following statements is not true?
   a. The active site of transposases and integrases are structurally similar.
   b. The amino acid sequences of transposases and integrases are highly conserved.
   c. Mg^{2+} ions are bound at the active site of all RNase H-related proteins.

Challenge question

6. Suggest one way in which the processes of DNA cut-and-paste transposition and spliceosomal intron splicing (Chapter 10) are similar and discuss why this is important for function.

17.4 DNA-ONLY NICK-AND-PASTE TRANSPOSITION

1. What is a strand-transfer product and during what process would it be produced?
2. Describe the process by how is it formed during Mu transposition.

17.5 A DNA CUT-AND-PASTE TRANSPOSITION IN ADAPTIVE IMMUNITY

1. For what purpose do lymphocytes use transposition? What is the name of the mechanism and how does it work?
2. How is RAG prevented from functioning at the incorrect stage of development?

17.6 RETROTRANSPOSONS

1. How do solo LTRs form in the genome? Are solo LTRs mobile?

17.7 LTR RETROTRANSPOSONS

1. Briefly discuss the roles of the following LTR retrotransposon-encoded proteins.
   a. GAG
   b. PRO
   c. RT-RNase H
   d. IN
2. What distinguishes a retrovirus from an LTR transposon, and how is this difference enabled?
3. What is a provirus?
4. Considering LTR element retrotransposons, why are two viral RNAs required in each virus-like particle?

17.8 NON-LTR RETROTRANSPOSONS

1. Briefly describe the following elements.
   a. LINE element
   b. SINE element
   c. Mobile group II intron
2. During transposition of which type of transposon would you expect to see a lariat?
   a. DNA-only transposons
   b. Retroviruses
   c. LTR transposons
   d. Mobile group II introns
3. How are pseudogenes created, and why are they not transcribed?
4. Explain how transposons can cause diseases.

Challenge question

5. Compare and contrast the mechanisms of replication and insertion in LTR retrotransposons with that in non-LTR retrotransposons.

17.9 CONTROL OF TRANSPOSITION

Challenge question

1. Uncontrolled transposition would rapidly lead to cell death. Discuss the various mechanisms that regulate the amount of transposition occurring.

17.10 CSSR: OVERVIEW

1. What is CSSR, and why is it considered to be conservative?

2. How does tyrosine recombinase CSSR differ from serine recombinase CSSR?

Challenge question

3.
   a. Briefly discuss the steps involved in CSSR.
   b. Which type of transposon interaction does this resemble? Explain your answer.

17.11 CSSR SYSTEMS THAT CONTROL GENE EXPRESSION

1. Discuss examples of the following ways in which CSSR systems affect gene expression.
   a. Deletion
   b. Promoter switching

Challenge question

2. Explain how mobile DNA bacteria has allowed rapid evolutionary responses to changing environmental conditions.

17.12 CSSR CONVERSION OF DNA DIMERS TO MONOMERS

1. How many Cre molecules are required to convert a dimeric plasmind containing two lox sites in direct orientation into two monomers.

2. Describe how Cre molecules recombinase mediate DNA breakage, strand exchange, and rejoining.

3. Which sites in the lambda and the *E. coli* genomes are used to complete CSSR during the insertion of lambda into the bacterial chromosome? What mediates the CSSR reaction? What recognition sites are used during lambda prophage excision?

EXPERIMENTAL APPROACH 17.1 – MOBILE DNA IN HUMANS

Challenge questions

1. Transposition in hemophilia.
   a. In the first experiment detailed in Experimental approach 17.1, researchers examined restriction fragment lengths. Explain the significance of the presence of larger sizes in some of the hemophilia patients and why this indicates transposition as opposed to other types of mutation.
   b. Sketch a diagram indicating how the difference in fragment sizes between DNA from a non-hemophiliac individual and that from a hemophiliac individual with transposition might look on an agarose gel. Explain your diagram.

2. Experimental approach 17.1 describes the work of Xing *et al.*, where the composite human genome reference sequence was compared to a single individual's sequence in order to map transposable elements. Design an experiment to ask whether these transposons are likely to be 'active'. Does your approach have any limitations, other than technical?

3. DNA cut-and-paste transposons excise themselves and are transferred to another region of the genome. Explain how it is possible for a copy of the transposon to remain at the original locus. Why is this more likely to occur during some parts of the cell cycle than others?

EXPERIMENTAL APPROACH 17.2 – HUNTING FOR TRANSPOSITION HOST FACTORS

Challenge questions

1. Considering Experimental approach 17.2, explain why it was important to use Tn5 insertions to generate random mutants rather than, for example, exposure to radiation.

2. In Experimental approach 17.2, the researchers used control strains with low and high rates of transposition. Design a control experiment to be sure that the observed blue color is due to transposition of the *lacZ* gene and not expression from the plasmid.

FURTHER READING

GENERAL

Craig NL, Craigie R, Gellert M, Lambowitz AM, eds. *Mobile DNA II.* Washington DC: American Society for Microbiology, 2002.

17.1 TRANSPOSABLE ELEMENTS: OVERVIEW

Chénais B, Caruso A, Hiard S, Casse N. The impact of transposable elements on eukaryotic genomes: from genome size increase to genetic adaptation to stressful environments. *Gene*, 2012;**509**:7-15

Cordaux, R, Batzer, MA. The impact of retrotransposons on human genome evolution. *Nature Reviews Genetics*, 2009;**10**:691-703

Curcio MJ, Derbyshire KM. The outs and ins of transposition: from Mu to kangaroo. *Nature Reviews Molecular Cell Biology*, 2003;**4**:865-877.

Fedoroff NV. Presidential address. Transposable elements, epigenetics, and genome evolution. *Science*, 2012;**338**:758-767.

Feschotte C, Pritham EJ. DNA transposons and the evolution of eukaryotic genomes. *Annual Review of Genetics*, 2007;**41**:331-368.

Hedges DJ, Batzer MA. From the margins of the genome: mobile elements shape primate evolution. *Bioessays*, 2005;**27**:785-794.

Jones RN. McClintock's controlling elements: the full story. *Cytogenetic and Genome Research*, 2005;**109**:90-103.

Kazazian HH, Jr., Mobile elements: drivers of genome evolution. *Science*, 2004;**303**:1626-1632.

Lander ES, Linton LM, Birren B, *et al.* International Human Genome Sequencing Consortium.Initial sequencing and analysis of the human genome. *Nature*, 2001;**409**:860-921.

Lisch D. How important are transposons for plant evolution? *Nature Reviews Genetics*, 2013;**14**:49-61.

Volff JN. Turning junk into gold: domestication of transposable elements and the creation of new genes in eukaryotes. *Bioessays*, 2006;**28**:913-922.

17.2 DNA-ONLY TRANSPOSONS

Claeys Bouuaert C, Chalmers RM. Gene therapy vectors: the prospects and potentials of the cut-and-paste transposons. *Genetica*, 2010;**138**:473-484.

Feschotte C, Pritham EJ. DNA transposons and the evolution of eukaryotic genomes. *Annual Review of Genetics*, 2007;**41**:331-368.

Ivics Z, Izsvák Z. The expanding universe of transposon technologies for gene and cell engineering. *Mobile DNA*, 2010;**1**:25.

William S. Reznikoff Transposon *Tn5*. *Annual Review of Genetics*, 2008;**42**:269-286.

17.3 MECHANISM OF DNA-ONLY CUT-AND-PASTE TRANSPOSITION REACTIONS

Dyda F, Chandler M, Hickman AB. The emerging diversity of transpososome architectures. *Quarterly Reviews of Biophysics*, 2012;**45**:493-521.

Gueguen E, Rousseau P, Duval-Valentin G, Chandler M. The transpososome: control of transposition at the level of catalysis. *Trends in Microbiology*, 2005;**13**:543–549.

Montaño SP, Pigli YZ, Rice PA. The Mu transpososome structure sheds light on DDE recombinase evolution. *Nature*. 2012; **491(7424)**:413-7.

17.4 MECHANISM OF DNA-ONLY NICK-AND-PASTE TRANSPOSITION REACTIONS

Harshey RM. The Mu story: how a maverick phage moved the field forward. *Mobile DNA*, 2012;**3**:21.

Harshey RM, Jayaram M. The Mu transpososome through a topological lens. *Critical Reviews in Biochemistry and Molecular Biology*, 2006;**41**:387–405.

17.5 CELLULAR DOMESTICATION OF A DNA CUT-AND-PASTE TRANSPOSASE IN ADAPTIVE IMMUNITY

Fugmann SD. The origins of the Rag genes—from transposition to V(D)J recombination. *Seminars in Immunology*, 2010;**22**:10–16.

Matthews AG, Oettinger MA. RAG: a recombinase diversified. *Nature Immunology*, 2009;**10**:817–821.

Sinzelle L, Izsvák Z, Ivics Z. Molecular domestication of transposable elements: from detrimental parasites to useful host genes. *Cellular and Molecular Life Sciences*, 2009;**66**:1073–1093.

17.6 RETROTRANSPOSONS

Cordaux R, Batzer MA. The impact of retrotransposons on human genome evolution. *Nature Reviews Genetics*, 2009;**10**:691–703.

Finnegan DJ. Retrotransposons. *Current Biology*, 2012;**22**:R432–R437.

Hancks DC, Kazazian HH, Jr., Active human retrotransposons: variation and disease. *Current Opinion in Genetics & Development*, 2012;**22**:191–203.

17.7 LTR RETROTRANSPOSONS

Arts EJ, Hazuda DJ. *HIV-1 Antiretroviral Drug Therapy. Cold Spring Harbor Perspectives in Medicine*, 2012;**2**:a 007161.

Craigie R, Bushman FD. HIV DNA Integration. *Cold Spring Harbor Perspectives in Medicine*. 2012;**2**:a 006890.

Montaño SP, Rice PA. Moving DNA around: DNA transposition and retroviral integration. *Current Opinion in Structural Biology*, 2011;**21**:370–378

17.8 NON-LTR RETROTRANSPOSONS

Beck CR, Garcia-Perez, JL, Badge, RM, Moran JV. LINE-1 elements in structural variation and disease. *Annual Review of Genomics and Human Genetics*, 2011;**12**:187–215.

Cordaux R, Batzer MA. The impact of retrotransposons on human genome evolution. *Nature Reviews Genetics*. 2009;**10**:691–703.

Edgell DR, Chalamcharla VR, Belfort M. Learning to live together: mutualism between self-splicing introns and their hosts. *BMC Biology*, 2011;**9**:22.

Lambowitz AM, Zimmerly S. Group II introns: mobile ribozymes that invade DNA. *Cold Spring Harbor Perspectives in Biology*, 2011;**3**:a 003616.

17.9 CONTROL OF TRANSPOSITION

Beauregard A, Curcio MJ, Belfort M. The take and give between retrotransposable elements and their hosts. *Annual Review of Genetics*, 2008;**42**:587–617.

Dooner HK, Weil CF. Give-and-take: interactions between DNA transposons and their host plant genomes. *Current Opinion in Genetics and Development*, 2007;**17**:486–492.

Goodier JL, Kazazian HH, Jr., Retrotransposons revisited: the restraint and rehabilitation of parasites. *Cell*, 2008;**135**:23–35.

Levin HL, Moran JV. Dynamic interactions between transposable elements and their hosts. *Nature Reviews Genetics*, 2011;**12**:615–627.

Lisch D, Slotkin RK. Strategies for silencing and escape: the ancient struggle between transposable elements and their hosts. *International Review of Cell and Molecular Biology*, 2011;**292**:119–152.

Nagy Z, Chandler M. Regulation of transposition in bacteria. *Research in Microbiology*, 2004;**155**:387–398.

Nakayashiki H. The Trickster in the genome: contribution and control of transposable elements. *Genes to Cells*, 2011;**16**:827–841.

Siomi MC, Sato K, Pezic D, Aravin AA. PIWI-interacting small RNAs: the vanguard of genome defence. *Nature Reviews Molecular Cell Biology*, 2011;**12**:246–258.

17.10 CSSR: OVERVIEW

Cambray G, Guerout AM, Mazel D. *Integrons. Annual Review of Genetics*, 2010;**44**:141–166.

Grindley ND, Whiteson KL, Rice PA. Mechanisms of site-specific recombination. *Annual Review of Biochemistry*, 2006;**75**:567–605.

Wozniak RA, Waldor MK. Integrative and conjugative elements: mosaic mobile genetic elements enabling dynamic lateral gene flow. *Nature Reviews Microbiology*. 2010;**8**:552–563.

17.11 CSSR SYSTEMS IN THE CONTROL OF GENE EXPRESSION

Carrasco CD, Holliday SD, Hansel A, Lindblad P, Golden JW. Heterocyst-specific excision of the Anabaena sp. strain PCC 7120 hupL element requires xisC. *Journal of Bacteriology*, 2005;**187**:6031–6038.

Cerdeño-Tarraga AM, Patrick S, Crossman LC, *et al*. Extensive DNA inversions in the *B. fragilis* genome control variable gene expression. *Science*, 2005;**307**:1463–1465.

Johnson RC, McLean MM. Recombining DNA by protein swivels. *Structure*, 2011;**19**:751–753.

van de Putte P, Goosen N. DNA inversions in phages and bacteria., *Trends in Genetics*. 1992;**8**:457–462.

17.12 CSSR CONVERSION OF DNA DIMERS TO MONOMERS

Marshall Stark W, Boocock MR, Olorunniji FJ, Rowland SJ. Intermediates in serine recombinase-mediated site-specific recombination. *Biochemical Society Transactions*. 2011;**39**:617–622.

van Duyne GD. A structural view of Cre-loxP site-specific recombination. *Annual Review of Biophysics and Biomolecular Structure*, 2001;**30**:87–104.

17.13 CSSR SYSTEMS REGULATED BY ACCESSORY PROTEINS

Lesterlin C, Barre FX, Cornet F. Genetic recombination and the cell cycle: what we have learned from chromosome dimers. *Molecular Microbiology*, 2004;**54**:1151–1160.

Radman-Livaja M, Biswas T, Ellenberger T, Landy A, Aihara H. DNA arms do the legwork to ensure the directionality of lambda site-specific recombination. *Current Opinion in Structural Biology*, 2006;**16**:42–50.

Rajeev L, Malanowska K, Gardner JF. Challenging a paradigm: the role of DNA homology in tyrosine recombinase reactions. *Microbiology and Molecular Biology Reviews*, 2009;**73**:300–309.

van Duyne GD. Lambda integrase: armed for recombination. *Current Biology*, 2005;**15**:R658–660.

van Duyne GD. Teaching Cre to follow directions. *Proceedings of the National Academy of Sciences of the U S A*, 2009;**106**:4–5.

Genomics and genetic variation

INTRODUCTION

Throughout this book we have explored the individual genes, proteins, functional RNA molecules, and metabolites that form the component parts of an organism, and that work together to bring the genetic information carried by that organism to life. Most of these components are similar in different species. Proteins that repair or replicate DNA and those that transcribe and translate mRNA are found in all species and carry out similar functions. Indeed, some of the most compelling evidence for evolution stems from the high degree of similarity in the molecules found among all species.

In this penultimate chapter, we look beyond the individual molecules, and the way they interact to carry out the instructions of our genetic material, to describe the entire collection of genes, proteins, and other functional molecules that act in concert to make up a living organism. The complete collection of genes in a species is known as its **genome**, so the branch of biology that studies genomes is called genomics. Genomics uses an organism's DNA sequence as a reference point for experimental analyses.

The field of genomics studies both the similarities and differences between the genomes of living organisms. With this in mind, this chapter has two key themes: the evolutionary conservation of genomes on the one hand and the variation that leads to differences between individuals and species on the other. Our first theme is an exploration of the *unity* of living things, that is, the features of the genome that are evolutionarily conserved. Because so many genes and gene products are shared within and between species, genomics illustrates the unity of living things and the descent of species from common ancestors.

Genomics also introduces us to the *differences* between species and indeed even within a species. Humans are not the same as fruit flies despite as many as three-quarters of human genes having an equivalent or similar gene in fruit flies. We are also not the same as chimpanzees; although we have nearly the same complement of genes, these genes differ by more than 1% at the nucleotide level. Furthermore, sequence variation means that even individuals within a species are not the same despite their component molecules being virtually identical.

Variation at the level of the genome drives variation at the level of the molecule – and of the organism of which the molecule is a component part. This variation links genomics and evolution and represents the second key theme of this chapter – the biological diversity between species and within a species that emerges from genetic variation.

We begin our exploration of the field of genomics by considering how we study genomes, and the kinds of information these studies can reveal.

18.1 GENOME SEQUENCES AND SEQUENCING PROJECTS

The functional information in a DNA molecule is carried by its nucleotide sequence. The nucleotide sequence encodes the properties of a functional RNA or the amino acid sequence of a protein product of a gene; it also provides the regulatory signals that direct gene expression. As it became possible to deduce the function of individual genes by determining their nucleotide sequences, the next step was to extend this study to entire genomes - to establish how all of the genes within an organism operate together as a unified whole.

The genome as a whole encapsulates the functional information for every gene - both the coding region and its regulatory region. If the nucleotide sequence of a single gene could suggest its function, so the nucleotide sequence of the whole genome could provide the 'parts list' for the entire organism - as well as many, if not all, of the assembly instructions for using those parts. One of the major challenges facing researchers today is to decipher this parts list, and establish how the different parts work together to generate the fully functional organism.

Many genomes have been sequenced and analyzed

Thousands of genomes have now been sequenced, from both bacteria and from eukaryotes. Indeed, all of the model organisms that are used for research purposes have had their genomes sequenced. Genome sequencing projects are also completed or underway for nearly every species that has medical, agricultural, or other economic importance. Other genome projects are focused on species of evolutionary and ecological importance. For example, the genome of the hemichordate *Ciona* was sequenced because this species represents an important evolutionary transition from invertebrates to vertebrates.

It is important to note that even genomes that are considered completely sequenced have sequences that are often not 100% complete or accurate, and the degree of completeness varies between organisms. A 'completed' sequence can have hundreds of small gaps where it has not been possible to amplify the DNA or assemble the sequence accurately. In addition, the completed sequence will contain some errors. For example, the exact sequence in regions containing many repeats is sometimes difficult to decipher. As it becomes more common to sequence more and more individuals within a species, these areas of error are diminishing. Nevertheless, even when the sequence is not 100% accurate, genome projects provide useful and biologically significant information.

Genome projects have revolutionized molecular biology

Having a complete genome sequence available for reference is an enormous benefit for researchers, since they can access the nucleotide and inferred amino acid sequence, the conserved regulatory elements and much more about an individual gene without having to obtain that information themselves. Experiments can be designed quickly on the basis of genome sequence information without having to perform some of the standard cloning and sequence analyses that were necessary even a few years ago.

The benefit of having sequenced genomes goes beyond exploring individual genes. It is possible to learn quickly about all of the related genes in an organism and in other organisms. This opens up a range of new experimental possibilities,

➡ Genome sequencing techniques are described in Section 19.15.

such as probing the function of the genes that are conserved, and identifying similar regions across different organisms. It allows investigators to analyze the organism as a species, with the knowledge of the full complement of genes potentially involved. This has profoundly changed the focus of molecular biology from the characterization of single genes to systematic multi-gene analysis.

Genome sequence information is also used to answer questions about genome structure. For example, what is the distribution of transposable elements and genes? What is the structure and function of the non-coding regions of the genome, and how are these regions distributed? Although these questions and many more could be approached by studying the process in individual species, genome projects offer deeper insights because they provide information about what has been conserved across species.

The genome sequences of individuals can be determined

Rapid advances in technology have made it possible to obtain the genome sequence of individual organisms. The examination of a collection of these sequences allows the variation among individuals to be studied and new questions to be posed. How frequent is a mutation and where do mutations occur? What are the rates of mutation in different regions of the genome? At a practical level, the sequence of the human genome is extremely valuable for the understanding, diagnosis, and treatment of human diseases. What mutations are associated with cystic fibrosis or breast cancer? What alleles are associated with detrimental side effects of certain drugs? As such, the availability of the genome sequence is revolutionizing medicine. However, the ability to obtain the genome sequence of an individual also raises ethical questions, as we discuss in Section 18.11.

Genome projects can be applied to ecological niches

One application of genomics is the field of metagenomics. Metagenomics involves obtaining collective sequence information, or the **metagenome**, for all of the organisms that occupy a particular ecological niche, particularly the microbes. This niche could be a particular pond, a soil habitat, a deep-sea vent, or even the microbes on the surface of human skin. The genome sequence information can be used to compile a profile of the population of a particular niche, similar to that performed with ecological population studies of animals, such as bird counts. This approach greatly expands genome analysis from individual organisms to their interactions with each other and their environment. It has been suggested that this approach, applied repeatedly over time, could indicate changes in a habitat perhaps due to climate change, or other ecological or industrial activity.

Metagenome sequencing data for microbes has revealed information about many more species and from a wider range of niches than other forms of population sampling used to study microbes. This is because sequence information can be obtained from organisms that are not possible to culture in the laboratory. Metagenomics bypasses the obstacle of culturing cells because it is not the organism itself but its DNA that is obtained and amplified. Organisms can be identified based on their DNA sequences; relationships between organisms can be analyzed; and comparisons between niches can be made through the sampling of whole populations of microbes from different environments.

Beyond telling us about ecological implications, metagenomics can offer new insights into human health and disease. For example, the population profile of the bacterial inhabitants of our digestive tract is large and diverse, but can be compiled by obtaining DNA sequences from these microbes. When applied to studies of microbes in humans, such metagenomics is often refered to as the human microbiome. Studies in mice and humans using these metagenomics profiles have found that some obese individuals have different populations of microbes living in the gut than some lean individuals. Furthermore, when the guts of lean mice are colonized with bacteria from the guts of obese mice, the lean mice rapidly gain weight. This indicates that, in addition to the diet and genotype of the individual, microbes in the gut can play a role in obesity, and possibly in several gastrointestinal diseases. This is a striking example of the application of genomics to a novel biological problem. Genome studies reveal the interacting parts in an organism; metagenome studies give insights into the organism as part of a system of interacting species.

Genomes have many different structural and functional elements

Much attention in genome sequencing projects is focused on the protein-coding genes, that is, the regions transcribed into mRNA and translated into the amino acid sequence of a polypeptide. However, for some species, including humans and plants, the protein-coding sequences are a small minority of the total genomic sequence, as illustrated in Figure 18.1. For example, only around 1.5% of the sequence in the human genome codes for proteins.

A major realization coming from genomic analysis is that a substantial fraction of the genome is in fact transcribed without the RNAs being subsequently translated into protein. For some of these transcripts, the functional end product is the RNA itself. Such functional RNA molecules include stable transcripts such as rRNAs, tRNAs, and X-chromosome silencing RNAs, among others. Also included are small regulatory RNA molecules, such as siRNAs, miRNAs, and piRNAs. However, the function of some of the untranslated RNAs is not known. Indeed, many may not even have a function.

➔ We learn about the processing and functions of the various regulatory RNAs in Sections 9.9, 10.10, and 10.15, and more extensively in Chapter 13.

In addition to protein-coding and RNA-coding genes, genome sequence analyses have allowed the fine structures of genomes to be examined, including repeated sequences which range from tandemly-repeated dinucleotides to repeats several kilobases in length. These repeated sequences vary in copy number from a few to tens of thousands per genome. Included among the repeated sequences are transposable elements or remnants of transposable elements, which are described in detail in Chapter 17. These elements are often found in different locations and in different copy numbers in different individuals.

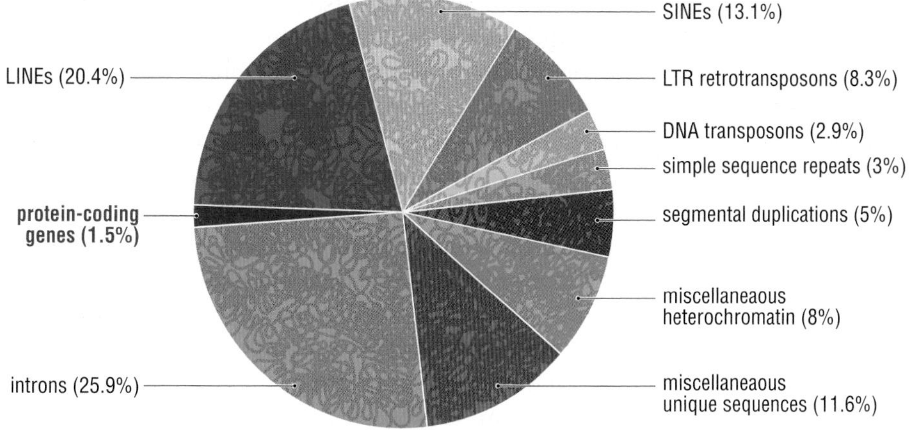

Figure 18.1 Less than 1.5% of the genome corresponds to protein-coding sequence. As shown by this pie chart, the majority of the human genome sequence consists of DNA encoding intronic sequences and the transposons and long and short interspersed nuclear elements (LINEs and SINEs) that are described in Chapter 17.

Figure 18.1 adapted from Gregory, RT. Synergy between sequence and size in large-scale genomics. *Nature Reviews Genetics*, 2005; **6**:699–708.

There are other sequence elements in a genome whose functions are not well understood, and which may not even have an identifiable function. Therefore, a significant challenge in genome analysis is how to identify the functional significance of the sequence elements within a genome, the topic we consider next.

18.2 FINDING FUNCTIONS IN A GENOME

Understanding the functions of genes and genomes is a monumental task. Two general approaches are being taken by investigators to elucidate the functions of genes and genomes. In one approach, called **comparative genomics**, insights into potential functions are obtained by comparing sequences. The comparisons can be carried out for a given gene or chromosomal region across a wide variety of species, or can even be carried out across whole genomes. The regions of sequence similarity between species highlight what has been conserved in evolution and can be used to infer function.

The second general approach, termed **functional genomics**, is based on genetic, molecular, biochemical, and cytological experiments, designed to study specific gene products or chromosomal regions to elucidate their functions. Initially, the methods were developed to explore the function of single genes or single protein complexes. However, the availability of complete genome sequences means these assays can now also be carried out on a genome-wide level.

In the next three sections, we explore what can be learned from both comparative and functional genomics, as together they provide the most complete understanding of the genome. In particular, we will explore the human ENCODE (Encyclopedia of DNA Elements) Project as an example of how comparative and functional genomics can be combined to provide a rich annotation of the genome sequence. But what do we mean by 'annotation'?

Genome annotation is an ongoing process that integrates comparative and functional genomics to create a map of the genome

The raw sequence of a genome is simply a string of adenosine, thymidine, guanosine, and cytosine nucleotides, albeit a string that is millions or billions of nucleotides in length. Having this raw sequence is of little use to biologists without some indication of where the genes start and stop and where control regions are. **Genome annotation** refers to the process of mapping genomic elements such as protein-coding regions, potential promoter elements, splice sites, and transposons (to name a few), to specific sequences in the genome. The DNA sequence signatures for some elements such as promoters, termination sequences, protein-coding sequences, and tRNAs are well defined. Indeed, sophisticated computer programs can search a genome sequence for specified features (and particularly the *combination* of features). Thus, a combination of sequence analysis within a species and conserved sequences from related species makes it possible to predict where genes and regulatory elements are located within long stretches of DNA sequence.

These predictions are not fool-proof, however, and often need to be complemented by another method. For example, some regions encoding very small proteins may be missed by computer programs since small protein genes are difficult to distinguish from random sequences that just happen not to contain stop codons. In some cases, these genes are shorter than the arbitrary size cut-off expected for proteins, but yet *are* indeed expressed as a protein. For this reason a robust genome annotation also needs to take into account information that comes from functional studies.

Functional genomics can experimentally validate the genetic elements that are predicted by the computer sequence analysis. For example, one can test whether a predicted splice site is indeed used, and whether the start and stop sites for transcription are correct, by sequencing the mRNA that corresponds to a region of the genome. This functional information is then integrated into the map of the location and structure of genes and other functional elements. Below, we describe some of the comparative and functional genomic approaches that allow a robust annotated map of the genome to be developed.

Sequence comparisons allow the identification of conserved sequences and can be used to infer function.

One of the main themes of this chapter is that the function of a genome sequence element can often be inferred by comparisons with other sequences, either in the same genome or in other sequenced genomes. After all, since the function of a genetic element is encoded in its nucleotide or amino acid sequence, similar sequences are likely to have similar functions. Sequences of related genes can be compared within a species, or the conservation across many species might be examined.

Let's consider a simple example of the comparison of protein sequences between species. A high degree of sequence identity can be enough to provide specific information about the function of the gene. For example, the amino acid sequence of the histone gene H4 differs by only seven amino acids (of 103 amino acids total) between the fungal species *Neurospora* and humans, as illustrated in Figure 18.2. This conservation of sequence implies a strong conservation of function. By knowing the function of a gene or gene product in one organism, we can infer the function of a similar gene or gene product in another often without conducting direct tests. Indeed, in many cases, a given human gene can successfully function in place of the equivalent gene in another species and vice versa.

Homo sapiens MSGRGKGGKGLGKGGAKRHRKVLRDNIQGITKPAIRRLARRGGVKRISGLIYEETRGVLKVFLENVIRDAVTYTEHAKRKTVTAMDVVYALKRQGRTLYGFGG
Mus musculus MSGRGKGGKGLGKGGAKRHRKVLRDNIQGITKPAIRRLARRGGVKRISGLIYEETRGVLKVFLENVIRDAVTYTEHAKRKTVTAMDVVYALKRQGRTLYGFGG
Bos taurus MSGRGKGGKGLGKGGAKRHRKVLRDNIQGITKPAIRRLARRGGVKRISGLIYEETRGVLKVFLENVIRDAVTYTEHAKRKTVTAMDVVYALKRQGRTLYGFGG
Danio rerio MSGRGKGGKGLGKGGAKRHRKVLRDNIQGITKPAIRRLARRGGVKRISGLIYEETRGVLKVFLENVIRDAVTYTEHAKRKTVTAMDVVYALKRQGRTLYGFGG
Monodelphis domestica MSGRGKGGKGLGKGGAKRHRKVLRDNIQGITKPAIRRLARRGGVKRISGLIYEETRGVLKVFLENVIRDAVTYTEHAKRKTVTAMDVVYALKRQGRTLYGFGG
Drosophila melanogaster MTGRGKGGKGLGKGGAKRHRKVLRDNIQGITKPAIRRLARRGGVKRISGLIYEETRGVLKVFLENVIRDAVTYTEHAKRKTVTAMDVVYALKRQGRTLYGFGG
Apis mellifera MTGRGKGGKGLGKGGAKRHRKVLRDNIQGITKPAIRRLARRGGVKRISGLIYEETRGVLKVFLENVIRDAVTYTEHAKRKTVTAMDVVYALKRQGRTLYGFGG
Anopheles gambiae MTGRGKGGKGLGKGGAKRHRKVLRDNIQGITKPAIRRLARRGGVKRISGLIYEETRGVLKVFLENVIRDAVTYTEHAKRKTVTAMDVVYALKRQGRTLYGFGG
Caenorhabditis elegans MSGRGKGGKGLGKGGAKRHRKVLRDNIQGITKPAIRRLARRGGVKRISGLIYEETRGVLKVFLENVIRDAVTYCEHAKRKTVTAMDVVYALKRQGRTLYGFGG
Caenorhabditis briggsae MSGRGKGGKGLGKGGAKRHRKVLRDNIQGITKPAIRRLARRGGVKRISGLIYEETRGVLKVFLENVIRDAVTYCEHAKRKTVTAMDVVYALKRQGRTLYGFGG
Arabidopsis thaliana MSGRGKGGKGLGKGGAKRHRKVLRDNIQGITKPAIRRLARRGGVKRISGLIYEETRGVLKIFLENVIRDAVTYTEHARRKTVTAMDVVYALKRQGRTLYGFGG
Oryza sativa MSGRGKGGKGLGKGGAKRHRKVLRDNIQGITKPAIRRLARRGGVKRISGLIYEETRGVLKIFLENVIRDAVTYTEHARRKTVTAMDVVYALKRQGRTLYGFGG
Zea mays MSGRGKGGKGLGKGGAKRHRKVLRDNIQGITKPAIRRLARRGGVKRISGLIYEETRGVLKIFLENVIRDAVTYTEHARRKTVTAMDVVYALKRQGRTLYGFGG
Vitis vinifera MSGRGKGGKGLGKGGAKRHRKVLRDNIQGITKPAIRRLARRGGVKRISGLIYEETRGVLKIFLENVIRDAVTYTEHARRKTVTAMDVVYALKRQGRTLYGFGG
Neurospora crassa MTGRGKGGKGLGKGGAKRHRKILRDNIQGITKPAIRRLARRGGVKRISAMIYEETRGVLKTFLEGVIRDAVTYTEHAKRKTVTSLDVVYALKRQGRTLYGFGG
Schizosaccharomyces pombe MSGRGKGGKGLGKGGAKRHRKILRDNIQGITKPAIRRLARRGGVKRISALVYEETRAVLKLFLENVIRDAVTYTEHAKRKTVTSLDVVYSLKRQGRTIYGFGG
Ustilago maydis MSGRGKGGKGLGKGGAKRHRKILRDNIQGITKPAIRRLARRGGVKRISGLIYDETRGVLKLFLESVIRDSVTYTEHAKRKTVTSLDVVYALKRQGRTLYGFGA
Candida glabrata MSGRGKGGKGLGKGGAKRHRKILRDNIQGITKPAIRRLARRGGVKRISGLIYEEVRAVLKSFLESVIRDAVTYTEHAKRKTVTSLDVVYALKRQGRTLYGFGG

Figure 18.2 Sequence comparisons allow functions to be inferred. The alignment of the human histone H4 protein sequence with those in various eukaryotes shows the extensive sequence conservation among these proteins; the only differences between them are indicated in pink. The high level of conservation suggests that all of these H4 proteins have a similar function.

Even without whole gene sequence similarity, the inferred amino acid sequence may provide some information about one part or domain of a protein, even when the entire protein is not conserved. If a small domain within the protein is conserved, one biochemical activity of a protein may be revealed. For example, there may be a domain that has homology to a DNA-binding region of another protein or a conserved kinase domain. This small region of homology will imply one activity of the protein, even if the homology does not extend throughout the entire protein. For instance, the zinc finger domain binds to nucleic acids (usually DNA) in a sequence-specific fashion. If the amino acid sequence of an untested protein contains zinc finger domains, we can predict from the presence of the domain that the untested protein functions as a DNA-binding protein. Although such sequence comparisons can *suggest* biological functions, additional genetic, molecular, biochemical approaches are needed to *confirm* the function.

➔ The method of sequence comparison used in the BLAST program is described in Section 19.8.

The quickest method by which we can infer the function of a gene is to compare its sequence with that of known genes of other species; this is commonly done using a **BLAST** program.

If the gene encodes a polypeptide, the most informative comparison is based on the inferred amino acid sequence encoded by the gene rather than the nucleotide sequence of the gene itself. Why is this? Comparisons of amino acids are the preferred method because an amino acid sequence comparison can ignore the changes in synonymous codons (different codons that encode the same amino acid) as illustrated in Figure 18.3. In addition, an amino acid sequence alignment can account for the nuances of the biochemical similarities and differences among the 20 amino acids – that is, it can regard a substitution of serine for threonine as more likely to be evolutionarily acceptable than a substitution of lysine for threonine because serine and threonine have similar chemical properties.

wild type	ATG Met	AAT Asn	ATT Ile	CGA Arg	GAT Asp
synonymous substitution	ATG Met	AAT Asn	ACT Ile	CGA Arg	GAT Asp
conservative substitution	ATG Met	AAT Asn	CTT Leu	CGA Arg	GAT Asp

Figure 18.3 Protein sequence comparisons can ignore synonymous changes, which do not change the amino acid sequence (but might impact the levels of expression), and account for conservative substitutions, which are less likely to interfere with the activity of the protein.

Alignments can identify conserved domains and sequence motifs, even when the functions of these domains and motifs are not yet known. Such an alignment, an example of which is shown in Figure 18.4, allows comparisons of amino acid sequences from proteins of two very distantly related organisms, such that the function of a protein in humans can be inferred from comparison with proteins in fruit flies, worms, budding yeast, and even *Escherichia coli*. (Figure 18.4)

```
                                                          Walker A
Walker A
NSF         220   INPDWDFQQMGIGGLDTEFSHIFRRAFASRVFPPEFIEQLGMKHV-RGILLFGPPGTGKTLMARQIGKMLNAREPKIVNGPQILDKYVGESESNVRKLFADAEE
Msp1p       86    INITFQ----DIGGLDPLISDL-HESVIYPLMMPEVYSNSPLLQAPSGVLLYGPPGTGKTMLAKALAKES-GANFISIRMSSIMDKWYGESNKIVDAMFSLANK
AAA ATPase  119   KKAKFS----EIIGQEEAKKKC-RIIMKY-LENPKLF---GEWAP-KNVLFYGPPGTGKTLMARALATET-NSSFILVKAPELIGEHVGDASKMIRELYQRASE
VPS4b       127   PNVKWS----DVAGLEGAKEAL-KEAVILPIKFPHLFT--GKRTPWRGILLFGPPGTGKSYLAKAVATEANNSTFFSISSSDLVSKWLGESEKLVKNLFQLARE
PEX6        653   PNVKWD----DVGGLEDVKTSI-LDTVQLPLLHKDLFSS-GLRKR-SGVLLYGPPGTGKTLLAKAVATEC-SLNFLSVKGPELINMYIGESEKNVRDIFEKARS
FtsH        147   IKTTFA----DVAGCDEAKEEV-AELVEY-LREPSRFQKLGGKIP-KGVLMVGPPGTGKTLLAKAIAGEA-KVPFFTISGSDFVEMFVGVGASRVRDMFEQAKK
Afg3p       283   IKISFK----NVAGCDEAKQEI-MEFVHF-LKNPGKYTKLGAKIP-RGAILSGPPGTGKTLLAKATAGEA-NVPFLSVSGSEFVEMFVGVGASRVRDLFTQARS
VCP         462   PNVHWD----DIGGLEDVKREI-KETVELPLLKPDVFKRLGIRPS-KGFLLYGPPGTGKTLLAKAVATES-NANFISIKGPEVLSKWVGESEKAIREIFKKAKQ
Rpt3        155   PDVSYA----DIGGMDMQKQEI-REAVELPLTHFELYKQIGIDPP-RGVLMYGPPGTGKTMLAKAVAHHT-TASFIRVVGSEFVQKYLGEGPRMVRDVFRLAKE

                        Walker B
Walker B
NSF               EWRRCGANSGLHIIIFDEIDAICKQRGS---MAGSSSVHD--TVVNQLLSKMDGV-EQLNNILVIGMTNRRDMIDEALLRPGRLEVQMEVSL     409(762)
Msp1p             --------LQPCIIFIDEIDSFLRER-----SSTDHEVTA--TLKAEFMTLWDGL-LNNGRVMIIGATNRINDIDDAFLR--RLPKRFLVSL     258(362)
AAA ATPase        --------SAPCIVFIDEIDAIGLSRE----YQSLRGDVS--EVVNALLTELDGI-KENEGVVTIAATNNPAMLDPAIRS--RFEEEIEFKL     286(371)
VPS4b             --------NKPSIIFIDEIDSLCGSR-----SENESEAAR--RIKTEFLVQMQGVGVDNDGILVLGATNIPWVLDSAIRR--RFEKRIYIPL     300(444)
PEX6              --------ARPCVIFFDEIDSLAPARGA---SGDSGGVMD--RVVSQMLAEIDGLSDSSQDLFIIGASNRPDLIDPALLRPGRFDKLLYVGV     828(941)
FtsH              --------AAPCIIFIDEIDAVGRQRGAG--LGGGHDERE--QTLNQMLVEMDGF-EGNEGIIVIAATNRPDVLDPALLRPGRFDRQVVVGL     321(644)
Afg3p             --------MAPSIIFIDEIDAIGKERGKGGALGGANDERE--ATLNQLLVEMDGF-TTSDQVVVLAGTNRPDVLDNALMRPGRFDRHIQIDS     459(761)
p97               --------VAPAIVFLDEIDSIAPRRG----TTSDSGVTE--RIVNQLLTSLDGI-EVMNGVVVIGATNRPDIMDPALLRAGRFDKLIYIPP     635(745)
Rpt3              --------NAPAIIFIDEIDAIATKRF----DAQTGADREVQRILLELLNQMDGF-DQTTNVKVIMATNRADTLDPALLRPGRLDRKIEFPL     330(413)
```

Figure 18.4 Protein sequence alignment. Amino acid sequence alignment for the ATPase domain of several distinct AAA+ ATPase proteins from different species (described in Section 6.6). The alignment shows how amino acids corresponding to the Walker A and Walker B boxes are conserved from bacteria and yeast to plants, flies, and mice, while much of the other sequence varies.

Nucleotide sequence comparisons can be used for closely related species

Amino acid sequence comparisons are nearly always more informative than nucleotide sequence comparisons for the protein-coding regions of the genome. However, nucleotide sequence comparisons are helpful when the functional element is the nucleotide sequence itself rather than its translated product. Among the functional elements of the genome that could be detected using nucleotide sequence comparisons are binding sites for transcription factors, structural elements of chromosomes, and the genes for regulatory RNAs. A nucleotide sequence comparison showing the promoter and coding sequence for a bacterial regulatory RNA is shown in Figure 18.5.

Although it may sound easy, defining regulatory regions and other functional DNA sequences can be challenging. Confounding factors include the shortness of many regulatory regions (on the order of eight to 12 nucleotides) and the requirements for conservation of a regulatory sequence often being less stringent than the conservation required to maintain protein function. If a protein sequence is changed there is a strong probability of a functional consequence. By contrast, the binding site for a transcriptional regulator can be degenerate, or multiple binding sites can replace one strong binding site.

Unlike amino acid sequences, in which informative comparisons can be made between two sequences in distantly related species, the most useful comparisons for functional nucleotide sequences require multiple sequences from more closely related species. For example, the genome sequences for 12 species of *Drosophila* were aligned for protein-coding genes based on the inferred amino acid sequences; the nucleotide sequences of the genes were then compared by a multiple alignment for conserved nucleotides, as described in Experimental approach 9.1. In this example, we explore how the comparison of *Drosophila* DNA sequences was used to predict RNA editing sites.

The assumption when carrying out sequence comparisons is that conserved sequences represent regions where selection has maintained sequence similarity or identity, possibly because the conserved sequence is the binding site for a regulatory protein, or regulatory RNA. However, conserved sequences may also be evidence of a sequence found in an ancestral species that has not yet diverged. Thus, not all conserved nucleotide sequences between two species are functionally important, but those conserved among multiple species are *more* likely to have a functional role. As more genome sequences from closely related species are available, the methods to detect functional sequence elements are becoming more sophisticated and successful.

```
Escherichia coli          TTTGCAA-AAAGTGTTGGACAAGTGCGAATGAGAATGATTATTATTGT-CTCGCG-ATCAGGAAGA----CCCTCGCGGAGAACCT------GAA
Salmonella enteric        TTTGCAA-AAAAAGCTAGACAAGTGCGAATGAGAATGATTATTATTGC-TTTGCG-TTCAGGG-GA----CCCTGACGGAAAACCT------GAA
Klebsiella pneumonia      TTTGCAA-AAAAAAGTAGACAAGTGCGAATGAGAATGATTATTATTGT-CCTGCA-TTCAGGGAGA----CCCTTGCGGAAAGCCT------GAA
Yersinia enterocolitica   TTCTCAACAAAGTGTTTG-CAACTGCTATTGATAACTATTATCATCTG-TTTGCTCTTCGAGACAAAGCTAGTTGTTTCAGGCTTTTGACAGAAA
Erwinia carotovora        TTCTCAACAAAGTACTTTACAGATGATACTGATAACTATTATCATTATTCTCGCACTTCA-GCAGTGG-CTCCTCGC--AGACACT--GAAGGAG
Invariant bases           TT..CAA.AAA....T...CA..TG..A.TGA.AA..ATTAT.AT.......GC...TC..G...........T.....A.....T.......A.

Escherichia coli          AGCACGACATTGCTCACATTGCTTCCAGTATTACTT-AGCCAGC-CGGGTGCTGGCTTTTTTTTTGATC
Salmonella enteric        AGCACGACATTGCTCACATTGCTTCCAGTATTACTTTAGCCAGC-CGGGTGCTGGCTTTTTTTTTGCTG
Klebsiella pneumonia      AGCACGACATTGCTCACATTGCTTCCAGTATTACTTTAGCCAGC-CGGGTGCTGGCTTTTTTTTTGCTG
Yersinia enterocolitica   AGCACGACATTGCTCACATTGCTTCCAGTGTTTTTTAAGCCAGCTCGGGTGCTGGCTTTTTTTTTGCTT
Erwinia carotovora        AGCACGACATTGCTCACATTGCTTCCAGTATTATTTTAGCCAGCCTACGTGCTGGCTTTTTTTT-GCCC
Invariant bases           AGCACGACATTGCTCACATTGCTTCCAGT.TT..TT.AGCCAGC....GTGCTGGCTTTTTTTT.G...
```

Figure 18.5 Nucleotide sequence alignment. DNA sequence alignment for the gene encoding the RyhB non-coding RNA (described in Section 13.2) for several different bacterial species is shown. The sequences encoding the non-coding RNA are pink. The regions encompassing the promoter (underlined), parts of the non-coding RNA, and the terminator (underlined) have a higher number of invariant residues.

As mentioned above, many regulatory regions such as protein binding sites are short, and often only some of the bases within such regions are essential for protein binding. This means that a potential binding site can be found many times in a complex genome simply by chance rather than because of any functional importance. This difficulty is compounded because the regulatory sequences for eukaryotic genes can be located in many different positions with respect to the coding region, and may not be the same distance away from the coding region for different genes.

All of these factors present challenges in inferring the functions of nucleotide sequences using computational comparisons of the sequences. For this reason, comparative genomics using sequence alignment is much more powerful when paired with assays that directly test function. In the next section, we will explore the information about genome function that functional genomics can provide.

18.3 FUNCTIONAL GENOMICS

The function of a gene or region of the genome can be tested by a wide variety of experimental manipulations. Historically, genetic approaches and the mapping of mutations were used to find regions of the genome that changed a phenotype when altered, thus implying a function for that DNA sequence. (As described in Chapter 1, a phenotype is a trait that can be measured in an organism.) For example, the function of the *white* gene for eye color in *Drosophila melanogaster* was inferred and mapped to a chromosome location before DNA was even identified as the genetic material.

Genetics has allowed the identification and mapping of thousands of genes in a wide variety of organisms from all domains of life. For many research organisms, the genome project grew directly out of the study of mutations, the creation of genetic maps, and the analysis of individual genes. Today genome-wide disruption studies combined with genome-wide expression and macromolecular interaction studies can accelerate the identification of gene function. These general techniques are being used to analyze the genomes of many different organisms.

Genome-wide gene disruption assays can uncover phenotypes that can give insights into gene function

Genome-wide gene disruption assays, in which every annotated gene in an organism is individually deleted or disrupted, allow the simultaneous analysis of the many independent mutations that may impact on a given phenotype, and the identification of the genes required for specific processes. A collection of mutants in which each mutant is individually identifiable by a barcode (a specific sequence that can be detected) is pooled and grown under a specific condition, for example in the presence of drug. If one mutant is sensitive to the drug and cannot grow, the barcode for that mutant will be missing at the end of the experiment, as illustrated in Figure 18.6a.

Let us consider an example of how this technique allows gene identification. A genome-wide set of *Saccharomyces cerevisiae* deletion mutants was screened for genes that when absent did not allow the budding yeast to form filaments. Several mutants were identified that blocked filamentation in *Saccharomyces cerevisiae*, and two of these genes then were further shown to be critical for the filamentation

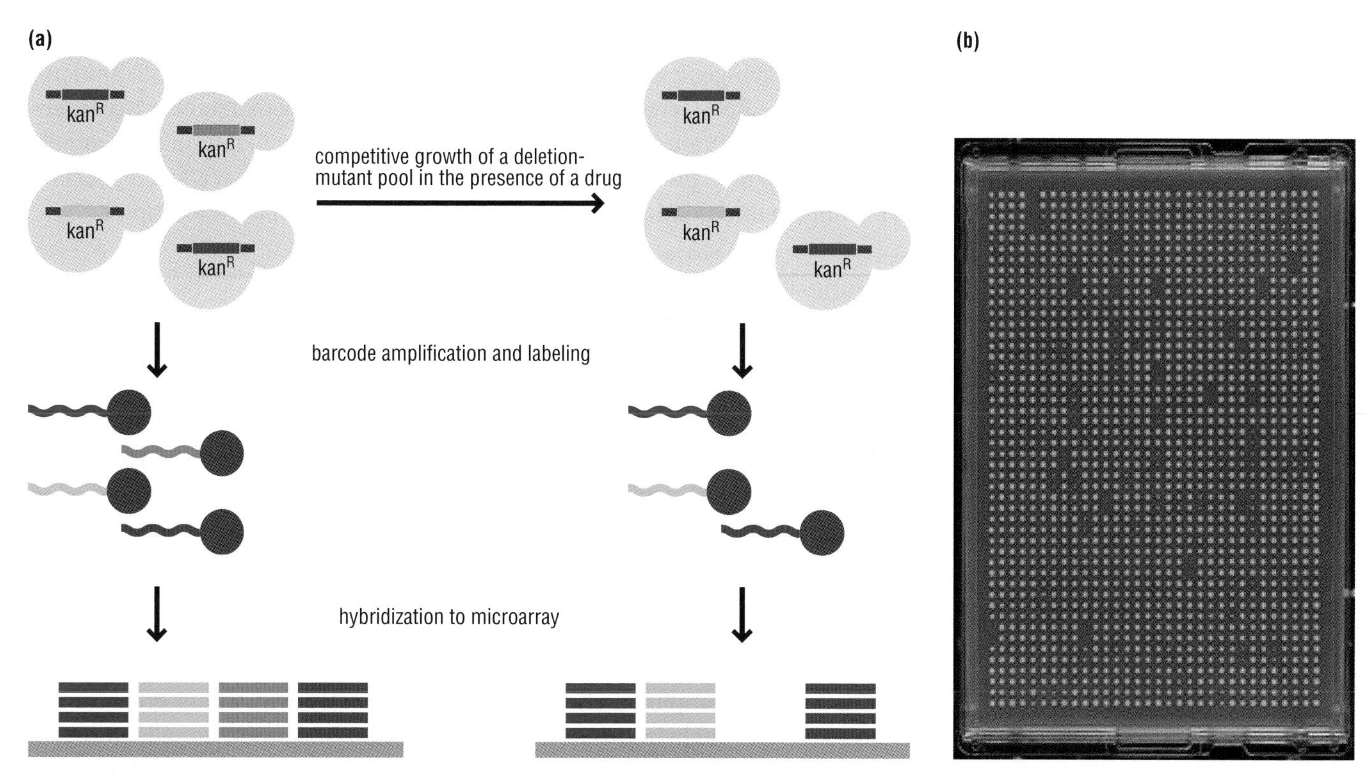

Figure 18.6 Genome-wide gene disruption assays. (a) By disrupting or replacing each gene by a unique tag or barcode, all mutants in a genome-wide disruption library (for example in *S. cerevisiae*) can be individually followed. If the collection of mutants is grown under some selective condition, those mutants that cannot survive under this condition are lost from the population. By amplifying the set of unique barcodes before and after selective growth and hybridizing the amplified DNA to a microarray, the mutants that are missing after selection can be identified because that unique barcode is missing. (b) In synthetic lethal screens, a particular mutation is moved into a library containing all other gene disruptions strains. In this case the double mutant yeast strain is unable to grow (a missing spot on the agar plate) while the corresponding single mutant strains are viable, the two mutations are considered to be synthetically lethal.
From Boone, C, Bussey, H, and Andrews, B. Exploring genetic interactions and networks with yeast. *Nature Reviews Genetics*, 2007; **8**:437–449.

of the pathogenic yeast *Candida albicans*, which adopts a filamentous form when it invades human tissues. The whole genome approach dramatically accelerated the discovery of the filamentation genes compared to classical genetic approaches. These genome-wide assays are now used in a variety of species to disrupt genes and rapidly identify which genes are involved in a specific biological process.

Gene disruption mutant collections also allow genetic interactions to be examined. Often the deletion of a single gene will have little or no effect. However, when two genes are deleted together, the cells die. Such an interaction is termed **synthetic lethal**: neither deletion alone is lethal but the combination of the deletions is. To assay for synthetic lethal effects, a particular mutation is moved into a **library** of cells that collectively contain single disruptions in all genes such that each cell ends up with two mutations. An example is shown in Figure 18.6b, in which the set of all double mutants is arrayed on an agarose plate. While most yeast cells with two mutations grow and form a colony on the plate, those cells that cannot grow are absent from the plate and can be seen as holes in the array. Since the identity of each mutation in the arrayed set of cells is known, it becomes immediately apparent which genes are synthetically lethal. Synthetic lethal interactions imply that the two genes are functionally redundant or play roles in separate pathways that ultimately affect the same fundamental process.

In addition to collections of deletion mutants, genome-wide RNAi screens are a powerful tool for identifying gene function, particularly in mammalian cells, in which the generation of gene disruptions is more difficult. In these assays,

➲ See Experimental approach 15.1 and Experimental approach 17.2 for examples of genome-wide disruption and RNAi screens.

synthetic RNAs or transcribed small RNAs are introduced into cells to block the expression of each gene in an organism individually. As with the gene disruptions, whole genome RNAi screens and whole genome synthetic lethal screens have identified proteins that are important in cellular processes, including mitotic spindle formation and RNA processing, and which act as host factors that affect replication of the human immunodeficiency virus (HIV)-1, to name but a few examples.

Genome-wide expression analysis informs which genes are co-expressed

The population of all RNAs that are expressed in a cell at a given time is called the **transcriptome** (a fusion of the words 'transcription' and 'genome'). Genome-wide RNA expression assays use methods such as microarray hybridization and deep sequencing (or 'RNA-seq') to examine what groups of genes are co-expressed in response to specific conditions or in specific tissues. The output of such assays is illustrated in Figure 18.7.

➲ See Experimental approach 9.3 for an example of genome-wide expression analysis.

➲ RNA microarrays and RNA-Seq are described in Section 19.10.

For example, to identify all of the genes whose expression changed when cells were grown on different carbon sources, total mRNA was isolated from *S. cerevisiae* cells grown in either glucose or galactose medium. The RNA from each sample was PCR amplified and the cDNA was labeled with either green or red fluorescent dye and hybridized to a microarray carrying probes for all annotated genes (see Figure 18.7a). The spots that are red represent genes only expressed in glucose, while those that are green are expressed only in galactose. Yellow spots represent hybridization of both red and green RNAs; they correspond to genes expressed in both conditions. Alternatively the amplified cDNA can be directly

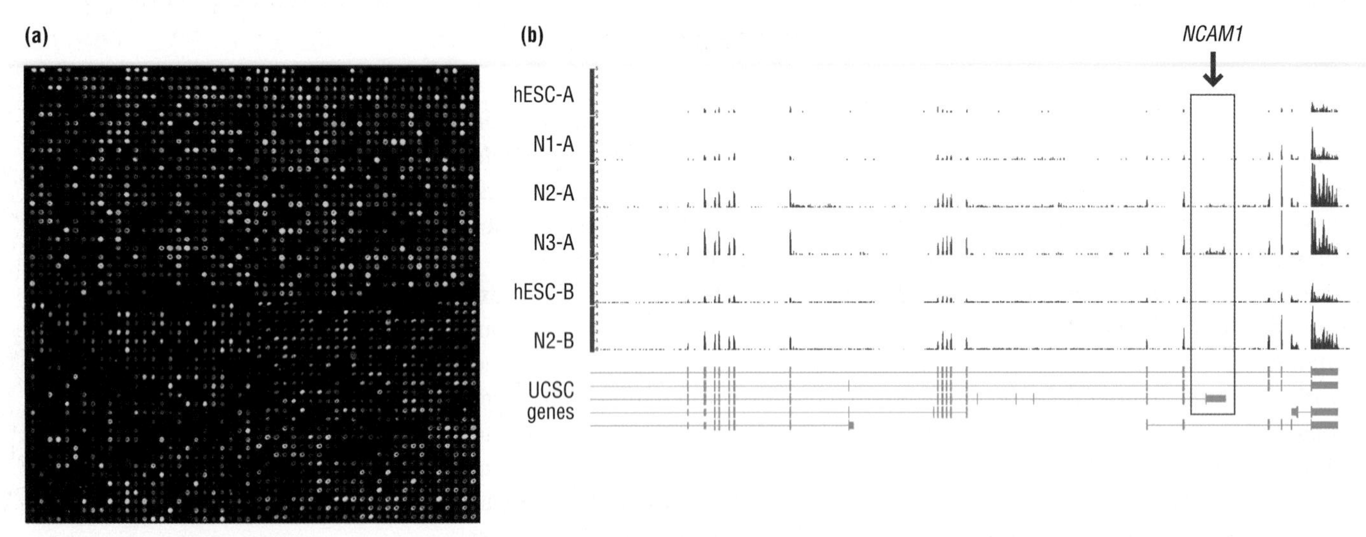

Figure 18.7 Genome-wide RNA expression assays. Two general types of gene expression assay can be performed to analyze the transcriptome. (a) RNA or a cDNA copy of the RNA can be hybridized to DNA probes on a microarray. The intensity of hybridization can be determined by labeling the cDNA prior to hybridization or by using labeled RNA or by detecting the DNA–RNA hybrid with labeled antibodies. If the cDNA prepared for two different samples is labeled with different fluorophores, the ratio between the labels can be used to determine whether a gene is upregulated (green), down-regulated (red) or unchanged (yellow) in one sample compared to the other sample. (b) The cDNA obtained for a sample can also be subject to deep sequencing (RNA-Seq) as is shown for samples isolated from neuronal cells at different stages of differentiation. All of the sequence reads are then aligned with the genomic region to which the sequence corresponds. The number of times a particular sequence is detected is represented above the reference genomic sequence. Genomic regions that are highly expressed are present in more copies in the cDNA population. Differences between the cell lines are highlighted by the gray box. Panel (a) in Lashkari *et al.* Yeast microarrays for genome wide parallel genetic and gene expression analysis. *Proceedings of the National Academy of Sciences U S A*, 1997;**94**:13057–13062. Panel (b) from Wu and Habegger *et al.* Dynamic transcriptomes during neural differentiation of human embryonic stem cells revealed by short, long, and paired-end sequencing. *Proceedings of the National Academy of Sciences U S A*, 2010;**107**:5254–5259.

sequenced (Figure 18.7b). The RNA expressed in different tissues or different points in development can be compared using similar techniques, to establish the tissue-specific or temporal pattern of gene regulation.

Since all genes are examined simultaneously in these experiments, groups of genes that are regulated together can be defined. Genes that are co-expressed under specific conditions are considered a 'signature' for that set of conditions. Such clusters of co-regulated genes can provide new insights into the function of unknown genes and suggest new functions for known genes based on the functions of other genes in the cluster. Future experiments can then focus on the genes in this signature to establish whether a cell is in a specific state. For example, stem cells that are differentiated into a particular lineage or stage have unique RNA expression profiles as shown for neuronal cells in Figure 18.7b. When a researcher is treating stem cells to allow them to differentiate into specific cell types, the expression profile of the treated cells can be compared to known expression profiles of specific cell types to identify whether correct differentiation has occurred.

Another genome-wide approach that is used to study transcription is the use of genome-wide reporter gene assays. In these studies, the promoters, and sometimes parts of the genes of interest, are fused to a gene encoding an easily assayable reporter protein (such as luciferase, which can be monitored by fluorescence). These reporter strains are then used to determine the time and location of gene expression, in cells or in whole organisms.

Proteomics extends genome-wide functional analysis to protein–protein interactions

Having the complete genome sequence facilitates **proteomics** (a fusion of the words 'protein' and 'genomics'), the large-scale study of proteins, protein complexes, and protein-protein interactions by allowing rapid protein identification and genome-wide protein-protein interaction assays. Rapid identification of proteins is made possible by the availability of completed genome sequences because the gene encoding a protein can be quickly identified from just a small amount of amino acid sequence. (That is, the sequence of the protein can be 'decoded' to identify of the gene that encodes the protein.) With genomic sequence available, only a short sequence of eight to nine amino acids is needed to uniquely identify a protein from the set of all known proteins sequences. As a result, when a protein complex is isolated, all of the associated proteins can be readily identified.

Understanding stable protein associations is a way of characterizing the function of a protein of interest (and, hence, the function of its corresponding gene) as represented schematically in Figure 18.8. For example, if a protein we would like to know more about is found to interact strongly with a complex of proteins that are already known to be subunits of a particular enzyme, we can infer that the query protein may be part of the same enzyme complex. Many of the proteins in the multisubunit RNA polymerase discussed in Chapter 8 were laboriously identified one by one over a 20-year period. By contrast, the availability of whole genome sequences means that the composition of such a large protein machine could now be analyzed in less than a month.

➔ See Experimental approach 14.3 and Experimental approach 15.2 for examples of large-scale protein identification.

In genome-wide protein-protein interaction assays, most or all of the proteins that interact with each other in a cell are identified. Each known protein can be individually tagged and purified using the tag, and all of the associated proteins identified. Alternatively, protein-based microarrays can be used to study the

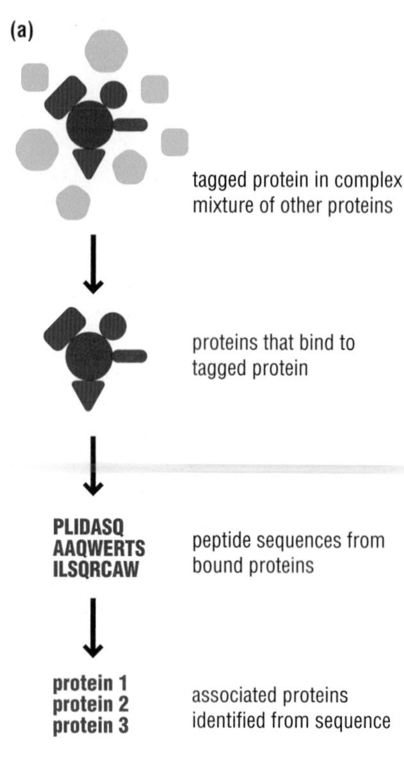

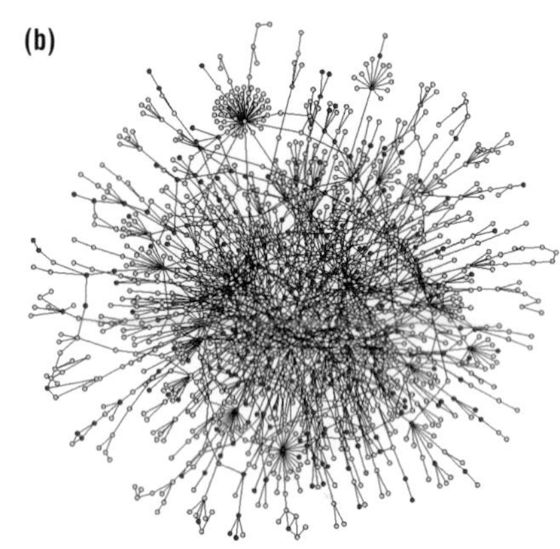

Figure 18.8 Proteomic approaches facilitated by the availability of genomic sequences. (a) To identify all of the proteins in a protein complex, one protein can be tagged (red bar) and used to 'pull out' any associated proteins by co-immunoprecipitation or co-purification as described in Chapter 19. All of the proteins isolated are then sequenced. With the availability of a complete genome sequence, a stretch of just eight to nine amino acids is needed to identify the gene that encodes the protein. (b) Using information from a variety of methods that detect protein–protein interactions, protein interaction networks can be displayed as scale-free networks with lines connecting interacting proteins represented as dots. Such a network shows the protein complexes that are associated with each other and may help point to biochemical pathways that the complex may participate in.

interactions. For example, all of the known proteins from the yeast genome have been produced and arrayed on a slide, such that the interactions between a protein of interest and all other proteins in the cell can be examined. The interactions between protein complexes, based on the presence of shared proteins, can be displayed as an interaction network, an example of which is shown in Figure 18.8b. The interactions between different complexes imply that these complexes may be in shared or related biochemical pathways, and thus give further insight about genome function. The set of protein interactions in a cell type is sometimes referred to as the protein **interactome**.

Immunoprecipitation assays allow genome-wide analysis of protein interactions with DNA and RNA

Genome-wide protein–DNA interaction assays allow protein binding to DNA across the entire genome to be examined. As we learned in Chapter 4, protein binding to DNA plays a major role in chromosome function and expression. To understand how sets of genes are co-regulated, protein–DNA interaction assays, such as those illustrated in Figure 18.9, can be carried out on a genome-wide level using chromatin immunoprecipitation (ChIP). Using this approach, proteins are cross-linked to the genomic DNA with which they are associated, the DNA is fragmented, and the fragments bound to the cross-linked proteins are isolated using immunoprecipitation. These DNA fragments are then identified by hybridizing to arrays (termed ChIP-chip), or more often are sequenced directly (termed ChIP-seq). The identification of such a set of binding sites allows a global view of all of the genes regulated by a specific transcription factor to be evaluated.

In addition to measuring the binding of site-specific binding proteins such as transcription factors, genome-wide DNA-binding assays can also be used to examine the modification state of the chromatin. As we saw in Chapter 9, the activity of a gene is influenced by the modification state of the histones that package the DNA. Antibodies that specifically detect the modification state of histones can be used in ChIP-seq experiments to determine the chromatin modification for the entire genome. For example, if an antibody to methylated H3K27 is used to probe the entire genome using ChIP-seq as shown in Figure 18.9b, all of the genes that have that modification under the conditions of the experiment will be identified and mapped. The ChIP-Seq can be repeated with cells grown under different conditions, and thus changes regions of the genome with H3K27 can be actively followed.

➲ ChIP-chip, ChIP-seq, and CLIP are described in more detail in Chapter 19 and Experimental approaches 4.1 and 16.2.

Protein binding to RNA is also critical to the function of many RNA molecules, as we learned in Chapters 11–16. Thus, the analysis of protein–RNA interactions is similarly being carried out on a genome-wide level using cross-linking immunoprecipitation (CLIP). This method stabilizes the interactions between proteins and RNA by cross-linking, after which the protein of interest is immunoprecipitated and the RNAs associated with the protein are identified by sequencing. Again, this type of analysis provides a global view of the RNAs bound by a particular protein.

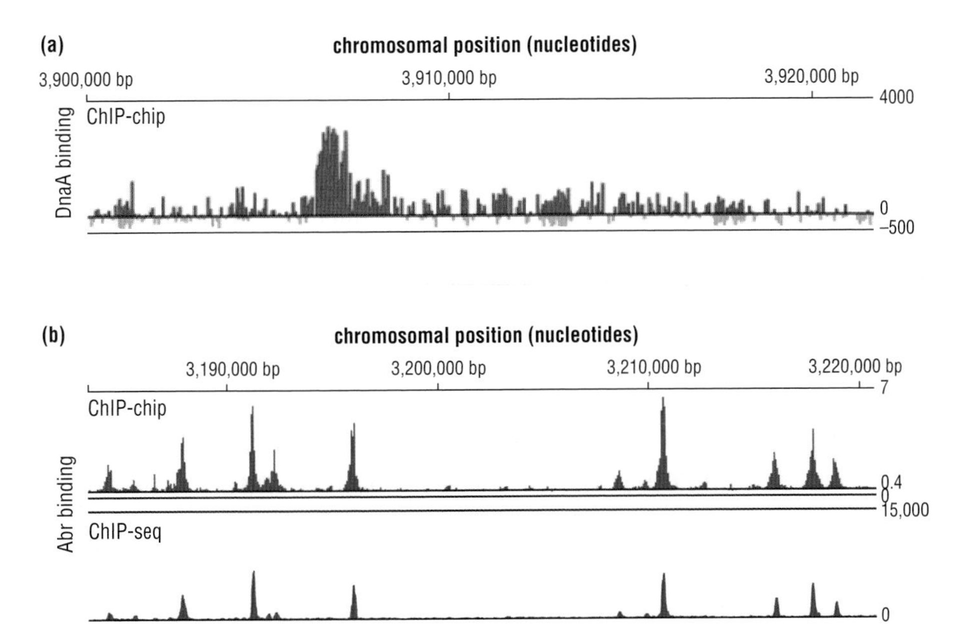

Figure 18.9 Identification of protein-binding sites across the genome. Whole genome sequences allow the genome-wide mapping of protein-binding sites on DNA. (a) In one example, the binding of the DnaA protein to the *Bacillus* genome is examined using genome-wide Chip-Chip analysis on microarrays. The majority of the DnaA protein is bound at the origin of replication, although there are some other binding sites nearby. The role of DnaA in the initiation of DNA replication in bacteria is discussed in Chapter 6. (b) In the second example, the genome-wide pattern of transcription factor ABr1 binding in *Bacillus* is compared using either Chip-Chip or ChIP-seq analysis. Fig 2b in Ishikawa et al. Distribution of stable DnaA-binding sites on the *Bacillus subtilis* genome detected using a modified ChIP-chip method. *DNA Research*. 2007;**14**:155–168.

Based on Chumsakul *et al.* High-resolution mapping of in vivo genomic transcription factor binding sites using in situ DNase I footprinting and ChIP-seq. *DNA Research*, 2013; **20**:325–338.

Genome-wide assays of DNase I hypersensitivity and DNA methylation can tell us about gene expression

Other genome-wide assays allow us to examine the state of DNA. DNase I hypersensitivity assays were developed many years ago and identified the regions in front of those individual promoters that are prepared for transcription initiation. DNase I nuclease cuts DNA independent of its sequence. However, if proteins such as nucleosomes are bound to the DNA, the nuclease will be unable to access and cut it. The identification of DNase I hypersensitivity sites across the genome thus allows a global view of regions that are actively being transcribed.

In contrast, DNA methylation is frequently associated with sites at which gene expression is silenced. As described in Section 4.7, DNA methylation in mammals involves the methylation of cytosine residues within CpG dinucleotides. The methylation of such residues within promoter regions is typically associated with transcriptional repression.

Sites of methylation across the genome can be assayed by several techniques including bisulfite sequencing, during which DNA is treated with the chemical bisulfite, which converts unmethylated cytosine (C) to uracil (U). The sequence of the DNA is then compared before and after the bisulfite treatment: the positions at which a C was not converted to a U indicate the C residue was methylated.

In the next section, we explore how the wealth of information provided by functional genomic approaches can be integrated with computational genomic approaches to give insights into the function of the human genome.

18.4 THE ENCODE PROJECT

We now turn to a recent large-scale functional analysis of the human genome termed the ENCODE project to illustrate how the integration of multiple computational and functional genomic approaches is helpful in understanding genome function. The ENCODE project is a collaboration between hundreds of research groups located all over the world. The data from different groups was assembled and analyzed to provide a comprehensive functional genomic annotation for the human genome. This project took advantage of previous genomic annotation, computational genomics and 1640 different experimental data sets generated by many independent labs to annotate specific functional elements in the human genome.

As we will learn at the end of this section, the analysis of this data is leading to the identification of new connections between molecules and the formulation of new hypotheses but is also raising fundamental questions - none more fundamental than how we define a gene.

The integration of all ENCODE functional genomic data gives a more complete picture of transcriptional control

The ENCODE project used many of the functional genomic methods outlined in Section 18.3 to gather data, including RNA expression, histone modification, and transcription factor binding. However, a central feature of the project was the development and use of sophisticated computational methods to integrate this data - to make sense of this data as a unified whole. Being able to draw out 'big picture' themes that emerged from the integration of the 1640 different data sets provided a much richer picture than the examination of any one kind of experiment in isolation could have yielded. For example, RNA-seq will tell you the RNA that is expressed and at what level. However, the ability to correlate expression data with transcription factor binding, DNase I hypersensitive sites, DNA methylation and histone modifications - all of which affect transcription - gives a much fuller, nuanced picture of transcriptional control.

As an example, the ENCODE project examined the binding of 119 different DNA-binding proteins, including 87 specific transcription regulators, RNA polymerase subunits, specifically modified histones, and histone modifying enzymes and chromatin remodeling complexes. In addition, the project mapped 205 109 specific DNase I hypersensitive sites. The vast majority of these hypersensitive sites overlapped with the sequences bound by specific regulatory transcription factors as expected from previous work. Taken together, these data showed how hypersensitive sites correlate with active genes. Across the genome, there also was a clear correlation between low levels of promoter methylation, transcription factor binding, and gene expression; this provided additional evidence that a lack of promoter methylation is correlated with gene expression as we discussed in Chapter 9.

One of the very powerful uses of having all of this data collected and coordinated in a genome-wide fashion is that information can be integrated across the genome, as illustrated in Figure 18.10. So, if we were interested in a particular region of chromosome 22, for example, the output from various functional assays performed on that region could be seen at once as different tracts of data. This makes a powerful visual image that is rich in information about gene function. For example, an active gene can be clearly identified as a region where the DNase 1 hypersensitive sites are aligned with binding of RNA polymerase II and transcription factors, and where active histone modification marks such as histone H3 lysine 4 trimethylation (H3K4me3) are also present.

If a scientist becomes interested in a particular gene, perhaps as a candidate gene that might be mutated in disease, this integrated functional map could provide a powerful starting point for determining whether the gene is transcribed

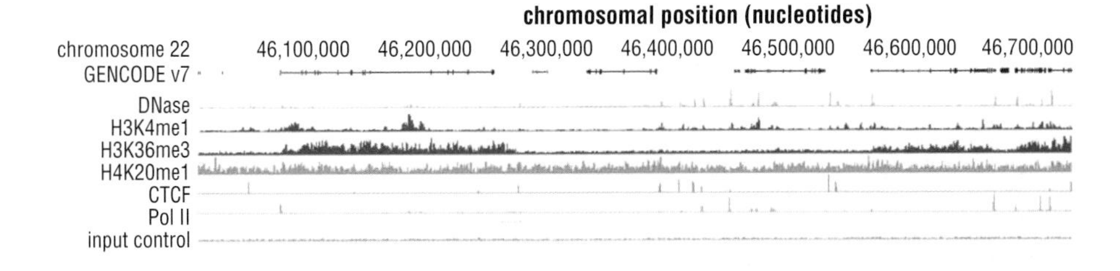

Figure 18.10 ENCODE data for chromatin accessibility, histone modification, and CTCF and RNA polymerase II binding across a region of human chromosome 22. Regions of open chromatin determined by DNase I sensitivity by two different groups are shown in blue; the signal for ChIP-seq data for listed histone modifications is shown in red; and the signal for ChIP-seq data for binding transcription factor CTCF and RNA polymerase II binding is shown in green. The signal for the ChIP-seq control is shown in mauve. Note that peaks of chromatin accessibility, histone H3 lysine 4 trimethylation (H3K4me3), and RNA polymerase binding generally are near the start of genes (genes are depicted by the colored line on the top line labeled GENCODE v7; the exons in the genes are denoted by the thicker bars within the lines), while the greatest signal for histone, H3 lysine 36 trimethylation (H3K36me3), spans the genes.

Fig 5A in The ENCODE Project Consortium, *Nature*, 2012;**489**:57–74.

and which transcription factors may regulate its expression. This wealth of information would previously have taken an individual investigator many years to uncover, yet it is now readily available online.

Genome-wide analysis allows the prediction of networks of gene, protein, and RNA associations

The need to analyze and synthesize the voluminous amounts of data being obtained from all of the genome-wide experiments is an ongoing challenge. When the expression data, DNA and RNA binding data, protein–protein interactions, and synthetic lethal interactions are analyzed as sets, either together or separately, networks of gene product interactions can be assembled. Clusters of interactions detected by these analyses often imply that the gene products function together in a larger complex or common pathway. Different complexes or pathways may share some components and have interactions with each other. The networks also can reveal which gene products are the major nodes, that is, those that have more interactions than others and so are probably the key players in certain biological pathways. For example, these nodes may represent key regulators whose functions should be further explored.

When specific genes cluster together in a network it may indicate they act together in a common functional pathway, and their clustering can help the function of genes to be predicted. An example of unexpected connections between transcription factors is again provide by the ENCODE project as shown in Figure 18.11.

The function of many genes remains unknown

Despite the successes of high throughput sequencing and computational and functional genomic approaches, the function of surprisingly many genes in each organism studied remains unknown. For example, dozens to hundreds of genes in every species have no detectable sequence similarity to any gene in another species, so a sequence comparison is not informative. Further, genome-wide mutant screens have found that many genes produce no mutant phenotype, at least under any of the many conditions that have been tested. This lack of phenotype could be because of functional redundancy - that is, multiple genes may carry out the same or similar function - or could be because a particular gene is not needed under the conditions tested.

Genomics has refined our definition of the gene

Comparative and functional genomics have reopened the decades-old debate, 'What is a gene?' When that question was asked during the second half of the twentieth century, a molecular biologist could answer with some confidence: a gene is

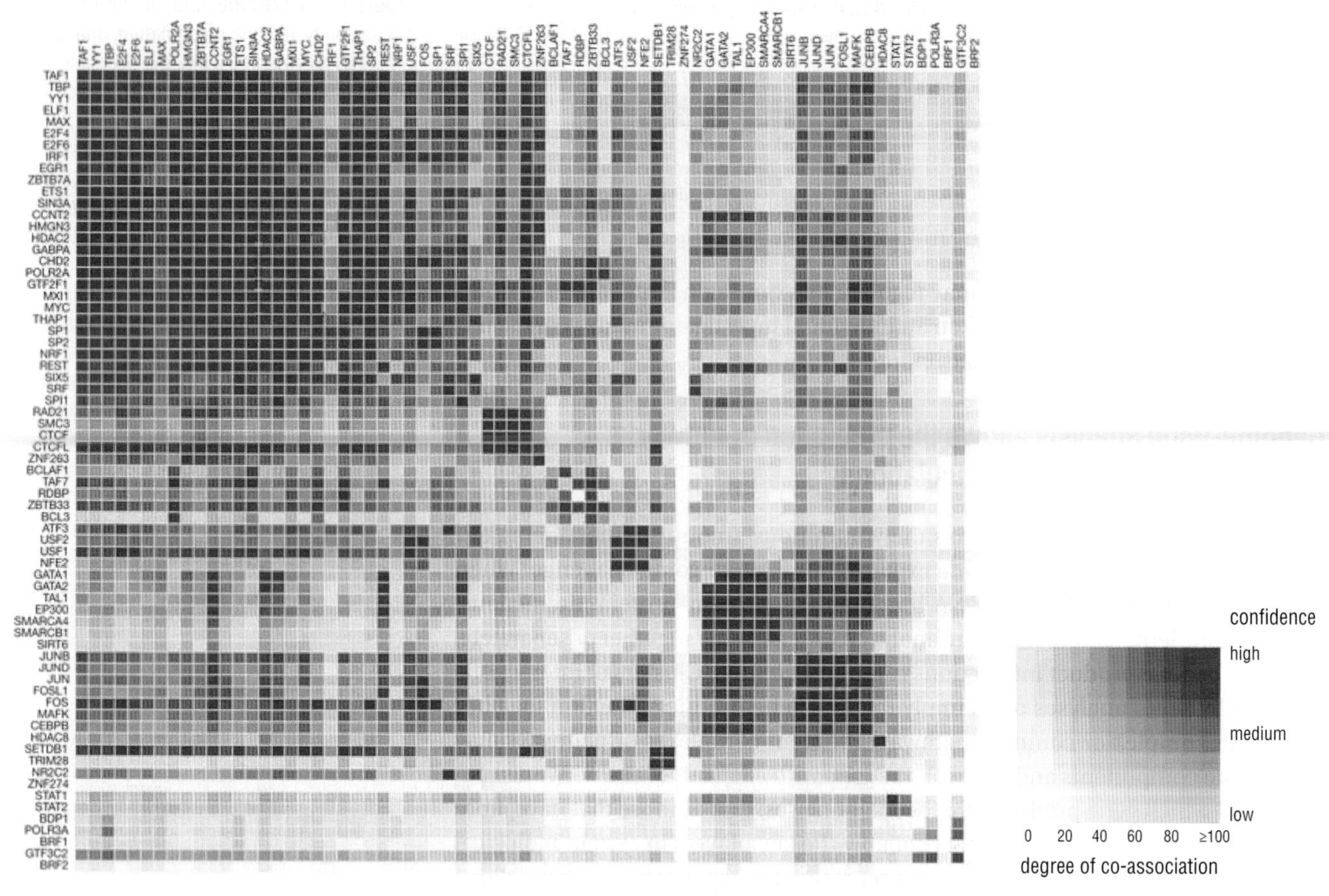

Figure 18.11 Co-association of binding sites for transcription factor pairs revealed by ENCODE data. The binding sites for each transcription factor were examined for co-association with the binding sites for each other transcription factor. The color corresponds to the strength of association, with red being strongest and yellow being weakest. (White indicates interactions where no conclusion could be made.)

Fig 4 in The ENCODE Project Consortium, *Nature*, 2012;**489**:57–74.

the nucleotide sequence information needed to make a functional RNA or polypeptide chain. Although that is still a good answer, we now realize it is incomplete. As we have seen, many genes encode hundreds of proteins due to processing to give alternative mRNA ends, alternatively spliced mRNAs, and alternatively processed proteins. In addition, many genes encode regulatory RNA molecules, perhaps even as many as the genes that encode polypeptide chains. Other sequence elements are conserved between species and are even transcribed but their functions are still unknown; should these also be classified as genes? Even as we learn more about the structure and function of genomes, the answer to this fundamental question appears to become more elusive.

18.5 THE EVOLVING GENOME: EVOLUTIONARY FORCES

The genome-wide comparative and functional genomic studies discussed in earlier sections enable us to characterize genomes as they currently exist: they give us a snapshot of the current genome sequence and the biological functions it mediates when expressed. However, genomes are not static, unchanging entities; they gradually change over time. Changes to DNA sequence provide the source material from which all evolutionary change arises.

Mutations are a critical source of genetic variation

The essential ingredient for evolution is genetic variation – a change in DNA sequence. Without variation at the genetic level, phenotypic variation would not occur. Changes in DNA sequence (that is, mutations) happen at a low frequency in all cells at all times at all sites in the genome. Small-scale mutational changes in the genome include the substitution, loss, or gain of one or a few nucleotides as described in Chapter 15 and shown in Figure 18.12. Most of these small-scale changes are the result of unrepaired errors in DNA replication, although a few come directly from DNA damage. By contrast, larger scale changes are the result of insertion and excision of transposable elements, recombination, and changes in the copy number of the whole genome or portions of the genome.

wild type	ATG	AAT	ATT	CGA	GAT
point mutation	ATG	AAT	ACT	CGA	GAT
deletion	ATG	AAT	A	CGA	GAT
insertion	ATG	AAT	ATT	ATTCGA	GAT

Figure 18.12 Possible mutations at a nucleotide level. Possible small-scale changes to genome sequence include point mutations as well as the deletion or insertion of one or a few nucleotides.

➔ We discuss larger scale DNA changes in more detail in Chapters 16 and 17.

Selection shapes the evolution of a gene and a population

Evolution occurs through the process of **natural selection**, which acts on the genetic variation in a population. Those individuals who are best suited to the environment will contribute more offspring than those who are less fit. As a consequence, the genes and specific alleles carried by the best-suited individuals will be selected for. Notice how selection occurs in a population of individuals; it can therefore be measured by examining the frequency of specific alleles in the population.

As illustrated in Figure 18.13, **positive selection** occurs when a new genetic variant is favored – when it puts an individual in a stronger position to survive and to pass its genes on to a subsequent generation. In contrast, **negative selection**

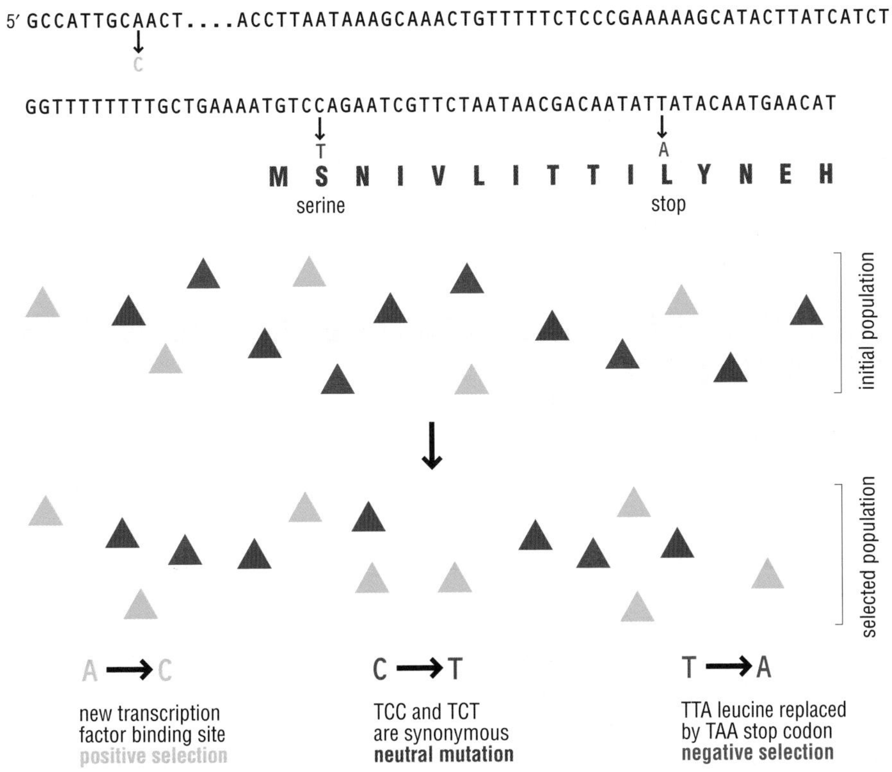

5′ GCCATTGCAACT....ACCTTAATAAAGCAAACTGTTTTTCTCCCGAAAAAGCATACTTATCATCT
↓
C

GGTTTTTTTTGCTGAAAATGTCCAGAATCGTTCTAATAACGACAATATTATACAATGAACAT
↓ T ↓ A
M S N I V L I T T I L Y N E H
serine stop

initial population

selected population

A ⟶ C
new transcription factor binding site
positive selection

C ⟶ T
TCC and TCT are synonymous
neutral mutation

T ⟶ A
TTA leucine replaced by TAA stop codon
negative selection

Figure 18.13 Different types of selection. Different kinds of genetic changes indicated in the DNA sequence on the top line are associated with positive and negative selection or neutral mutations. Initially, there may be several different genotypes (represented by the green, red, and blue triangles). If a mutation confers some selected advantage, the genotype (green) will predominate in the population after a period of time (positive selection). If a mutation confers a selected disadvantage, the genotype (red) will be reduced in the population after a period of time (negative selection). For a neutral mutation, the derived population (blue) will be similar to the original population.

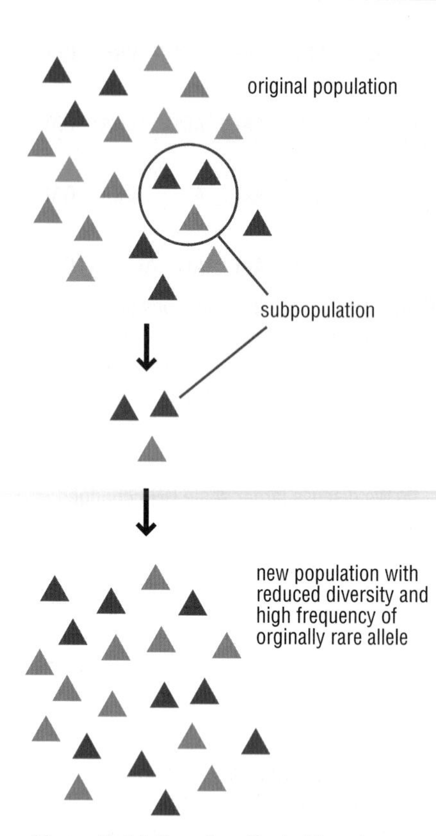

Figure 18.14 Founder effects. There is significant genetic variation in the original large population (represented by the blue, dark blue, orange, and pink triangles). If a subset of the individuals in the original population migrates to a new isolated location or if only a subset survives a disease, the genetic diversity in the new population is reduced.

occurs when a new genetic variant is disfavored, that is when it puts an individual in a weaker position to survive, such that it is less likely to pass its genes on to a subsequent generation. Many genetic variations - for example, some synonymous substitutions - are neither advantageous or disadvantageous and are thus considered **neutral mutations**.

Genetic drift also affects the evolution of populations and genomes

Another evolutionary force that shapes the genetic structure of a population is known as **genetic drift**. Genetic drift arises because both allele frequency in a population and population size can fluctuate, causing the alleles of a gene that happen to be predominant in a population to change randomly over a period of time.

One particularly important example of genetic drift is the **founder effect**. Imagine that a small group of individuals migrates and splits off from the main population to produce a new, isolated population. The new population is likely to have only a subset of the alleles that were present in the original population. These differences arise because the new population - the founding population - is much smaller than the original population and does not have all of the same variation as the original population. As shown in Figure 18.14, when a subset of the population is isolated, they represent a random subset of all possible alleles; thus in the next generations, allele frequencies are changed without selection simply due to random sampling in a population.

A similar outcome can arise from a *population bottleneck*, when many members of a large population suddenly die (due to disease or a natural disaster, for example), and the population re-grows from a much smaller group. The remaining population will probably have only a subset of the alleles that were present in the original population, so subsequent generations will possess a different range of genetic variations.

Although selection and genetic drift are probably the most significant forces that affect the structure of a population, they are not the only ones. The movement of individuals from one population to another (migration) and mating preferences among individuals in a population (non-random mating) also affect the amount and type of genetic variation in a population. These forces can often work in concert or in opposition with each other. Genetic differences arise as a result of mutation, but they persist or are lost over time largely because of the effects of selection and genetic drift.

18.6 THE EVOLVING GENOME: MECHANISMS OF VARIATION

Having learned about the forces that can shape evolution, let us now consider some of the mechanisms that underlie the generation of genetic variation in more detail.

Transposable elements are one source of genetic variation

Transposable elements do not move at a high frequency; a transposition rate of one movement per 1000 elements per cell generation is probably a realistic estimate. However, there are many transposable elements at different locations in the genome. In fact, they make up almost 45% of the human genome. Consequently, even a low frequency of transposition of any one element can have an impact on the

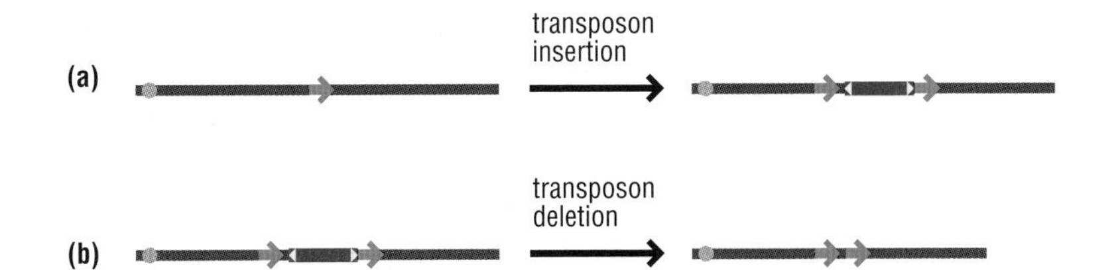

Figure 18.15 Transposons have a major role in rearranging the genome. Transposons can cause changes to the genome sequence by (a) inserting into or (b) excising out of the genome.

genetic structure of a population. Transposable elements frequently produce mutations at both the site of insertion and the site of excision, as shown in Figure 18.15. Not only does the genome of every organism include many transposable elements, but it also harbors evidence of former insertions and excisions.

Recombination changes gene number and location

In addition to causing mutations by insertion and excision, transposable elements have a major role in rearranging the structure of the genome as the most prevalent repeated DNA elements. The transposable element need not be active to serve as the substrate for recombination since the sequence similarity between elements, rather than their movement, provides the basis for rearrangement through recombination. As discussed in Chapter 17, recombination between transposable elements and also between other types of repeated sequences is probably one of the principal mechanisms by which genetic regions are rearranged, lost or duplicated. All of these rearrangements change the location of some genes, and two of them change the copy number of genes, as summarized in Figure 18.16.

➔ We describe the process of recombination in Chapter 16, and describe the structures of transposable elements and the mechanisms by which they move in Chapter 17.

Recombination between repeated sequences on two different chromosomes can also produce a *translocation*, whereby a region of one chromosome is swapped with a region from another chromosome (see Figure 18.16a). Rearrangements

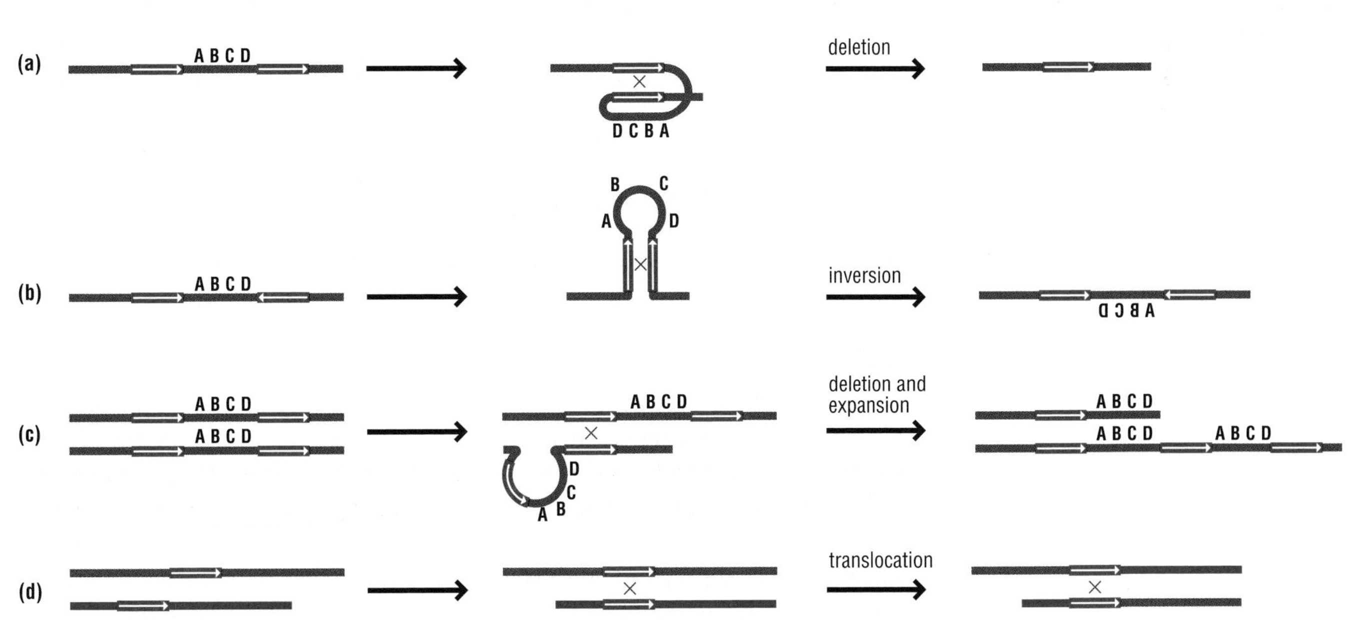

Figure 18.16 Recombination between repeated sequences, including transposable elements, can lead to genome rearrangements. (a) If the elements are oriented in the same direction (as indicated by black arrow in red repeated sequence region) looping and deletion of the genes A, B, C, and D can occur. (b) If the sequence elements are inverted, recombination between the repeated elements will lead to inversion of the order of the genes A, B, C and D. Such an inversion could separate a gene from a regulatory element. (c) If there are two homologs present (such as in eukaryotes), misalignment of the repeated regions can give rise to a deletion on one chromosome and expansion on the other. We discuss this mechanism further when we consider copy number variation. (d) Recombination between sequences repeats that are present on different chromosomes can lead to chromosomal translocations.

within chromosomes can lead to the movement of large blocks of DNA; such rearrangements frequently lead to changes in the copy number of genes.

Deletions represent the situation in which there is a loss of a region of the genome (Figure 18.16b). The deletion of a coding region may affect the fitness of the organism, and therefore may be selected against over the course of time. However, deletions involving non-functional regions may be selectively neutral and may persist. Likewise, deletion following polyploidization (as described below in 'Genome evolution involves changes in genome size between species'), or duplication of chromosomal regions is often tolerated and maintained: the multiple copies of genes generated by such a duplication event mean that, even if one copy is rendered inactive through a subsequent deletion event, the organism still has at least one functional copy left. In many, but not all cases, this remaining copy is sufficient for full function.

Duplications result in the presence of extra copies of specific regions of the genome (rather than the duplication of the entire genome, as described for polyploidy below). Gene duplications are exceedingly common, and nearly all eukaryotic genomes show evidence of regional duplication, as depicted in Figure 18.16c. Many of these so-called **copy number variants (CNVs)** in the human genome are associated with human disease.

Inversions, which are illustrated in Figure 18.16d, do not change the copy number of genes but do change the order in which they are found. In an inversion, a region of the chromosome A-B-C-D becomes switched to A-D-C-B. As a result, the regulation and expression of the genes within this region may be altered as they may have new promoter elements nearby or have lost their promoters as a consequence of the inversion.

Genome evolution involves changes in genome size between species

The small-scale changes in nucleotide sequence and regional rearrangements that we mention earlier in 'Recombination changes gene number and location' are dwarfed by the large-scale changes that have occurred in the overall size and structure of the genome during evolutionary time. One of the most significant of the large-scale evolutionary changes in genome size is **polyploidy**, an expansion in the number of genome copies (see Figure 1.18). Polyploidy happens as a regulated event in some cell types in some species as a way to increase gene copy number. But when it becomes fixed in the germline it leads to whole genome duplication that has evolutionary consequences. Autopolyploidy describes the situation in which there are multiple copies of all of the genes in a single genome. The frog *Xenopus laevis* has a genome twice the size of that of its close cousin *Xenopus tropicalis*; the banding pattern and structure of the chromosomes suggest that a complete duplication of chromosomes occurred in the evolutionary lineage leading to *X. laevis*. The budding yeast *Saccharomyces cerevisiae* also shows evidence of an evolutionarily ancient genome duplication followed by subsequent loss of many genes. Despite the losses, many genes and regions of the yeast genome are still present in the duplicate copies.

Polyploidy is also a characteristic of domesticated plant genomes. For example, bread wheat is a hexaploid produced from three different ancestral genomes, and bulgur wheat is a tetraploid arising from two different ancestral genomes.

The copies of genes resulting from polyploidy events may have acquired mutations over time, meaning that they are no longer exact duplicates. As such, they may have acquired distinct functions. Following polyploidization, some genes may also be lost entirely, so not all copies of the genome are complete replicates of the ancestral genomes.

Translocations and inversions are associated with speciation

As we described earlier in 'Recombination changes gene number and location', recombination between repeated sequences can lead to structural changes in chromosomes. Over evolutionary time, such structural changes can help drive the generation of new species. Because the pairing of homologous chromosomes is essential for chromosome segregation at meiosis in eukaryotes, two organisms with significantly different chromosome structures cannot interbreed even if the organisms can mate: they cannot undergo meiosis, and so cannot produce viable gametes.

➔ The steps in meiosis are described in Chapter 7.

Why is this the case? Recall that, during meiosis, each chromosome synapses along its length with its homologous chromosome. Figure 18.17 illustrates how, if one chromosome of the pair has an inversion or translocation, the rearranged homolog is unable to synapse with its unrearranged homolog (or, at best, it synapses very poorly). Inversion heterozygotes have to form a twisted loop in their chromosomes to synapse properly, and these structures are unstable at meiosis. Since organisms that cannot properly carry out meiosis are infertile, structural changes to chromosomes that prevent pairing can lead to the so-called 'reproductive isolation' of the organisms that carry them, leading eventually to **speciation**, the formation of a new species.

Speciation is also driven by karyotype evolution

Karyotype evolution is a special case in which speciation occurs because of changes in the number of chromosomes, rather than simply their structure. How can chromosome number change within a species? The mechanism that allows the stabilization of chromosome fusions, or the stabilization of broken chromosomes, leads to changes in chromosome number. This change in chromosome number may drive speciation due to the failure of meiosis as described above.

It is quite striking that, among vertebrates alone, the haploid chromosome number varies from as few as three to as many as 49, as illustrated in Figure 18.18a. Indeed, even two species of zebra that appear extremely similar to the eye have very different chromosome numbers: Grevy's zebra (*Equus grevyi*) has 23 chromosomes and Hartmann's mountain zebra (*Equus zebra hartmannae*)

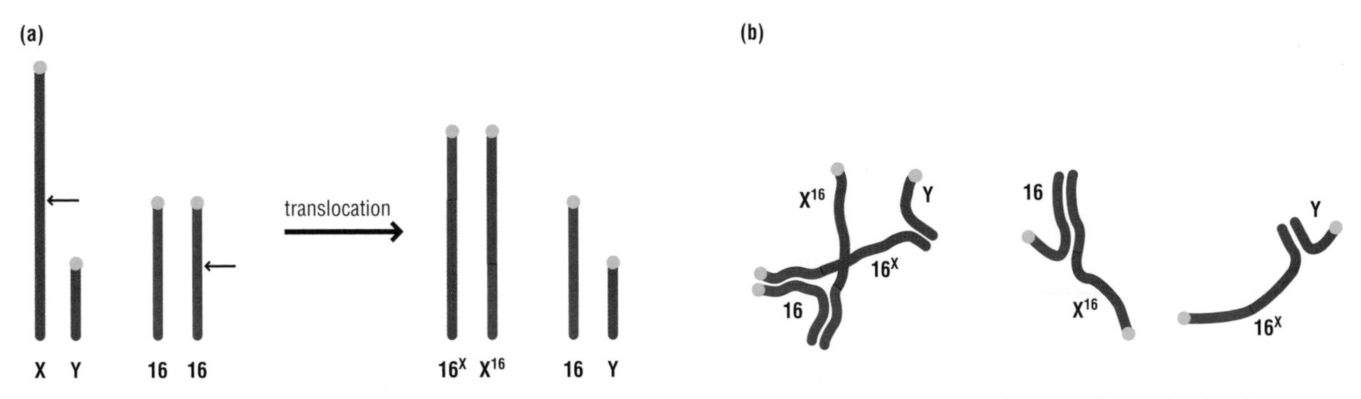

Figure 18.17 Rearranged chromosomes cause problems in meiosis. (a) On the left is the normal chromosome configuration with one X and one Y and two copies of chromosome 16. On the right is the translocation that has occurred between chromosome 16 and the X chromosome, denoted 16^X and X^{16}.
(b) Types of chromosome arrangements that are observed when cells attempt to synapse the translocated chromosomes during meiosis. Chromosome breaks and chromosome loss can occur when these poorly synapsed chromosomes are pulled apart during chromosome segregation.

From Turner JMA, Mahadevaiah, SK, Fernandez-Capetillo, O, Nussenzweig, A, Xu, X, Deng, C-X, and Burgoyne, PS. Silencing of unsynapsed meiotic chromosomes in the mouse. *Nature Genetics*, 2004;**37**:41–47.

(a) Variation in chromosome number in vertebrates

species	haploid chromosome number
human	23
mouse	20
chicken	39
carp	49
Indian muntjac	3
Chinese muntjac	23
Mongolian wild horse	33
Grevy's zebra	23
Hartmann's mountain zebra	16

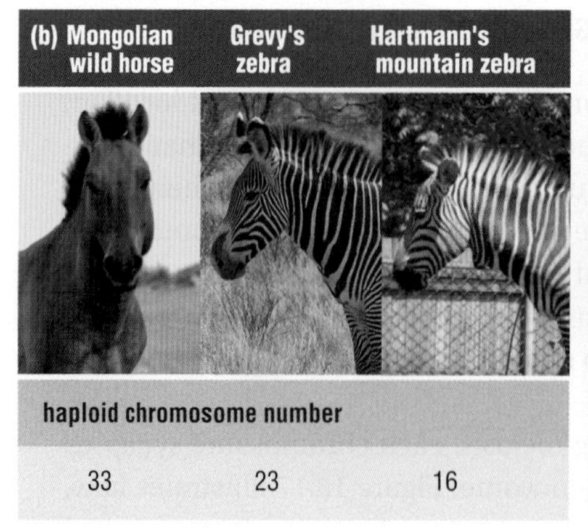

(b) Mongolian wild horse
Grevy's zebra
Hartmann's mountain zebra
haploid chromosome number
33
23
16

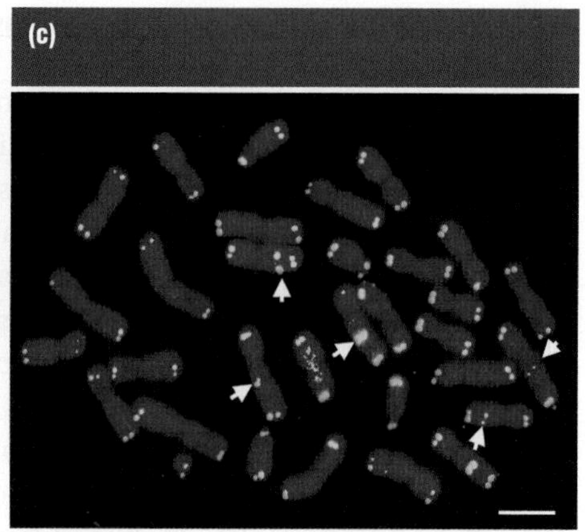

Figure 18.18 Karyotypes can evolve through changes in chromosome number. (a) The haploid number of chromosomes varies widely among vertebrates. Even closely related species of zebra that appear remarkably similar have very different chromosome numbers. (b) The most distantly related horse (the Mongolian wild horse) has 33 chromosomes, while two zebras have 23 (Grevy's) and 16 (Hartmann's). (c) The chromosomes of Hartmann's zebra show internal telomere repeats (arrows) indicative of reduction in chromosome number through chromosome end-to-end fusion.

has only 16. The oldest existent species of the same genus, the Mongolian wild horse (*Equus przewalskii*), has 33 chromosomes (see Figure 18.18b). The presence of telomere repeats at the junctions between ancestral blocks of DNA in Hartmann's mountain zebra, illustrated in Figure 18.18c, suggests that the decrease in chromosome number from the ancestral species occurred through chromosome fusion.

For a stable new chromosome to be generated from two or more ancestral chromosomes, the function of two telomeres and one centromere must have been lost, as shown in Figure 18.19a. Chromosome fission – the splitting of a chromosome – also contributes significantly to karyotype evolution; this process is depicted in Figure 18.19b. Sequence blocks that are found on a single chromosome in pigs, cows, and horses are located at the ends of human chromosomes 16 and 19. This implies that breakage of an ancestral chromosome occurred and new telomeres were added, probably by telomerase, in a germ cell early in vertebrate evolution. In addition, a new functional centromere must have been generated to allow segregation of the fragment that did not initially carry one.

We discussed in Chapter 4 how neocentromeres can form and become stabilized where there was no centromere before. Thus the ability to inactivate telomere and centromere function in the case of chromosome fusion, and the ability to *form* new telomeres and new centromeres in the case of chromosome fission, allows changes in chromosomes that can fuel karyotype evolution and drive speciation.

(a) chromosome fusion

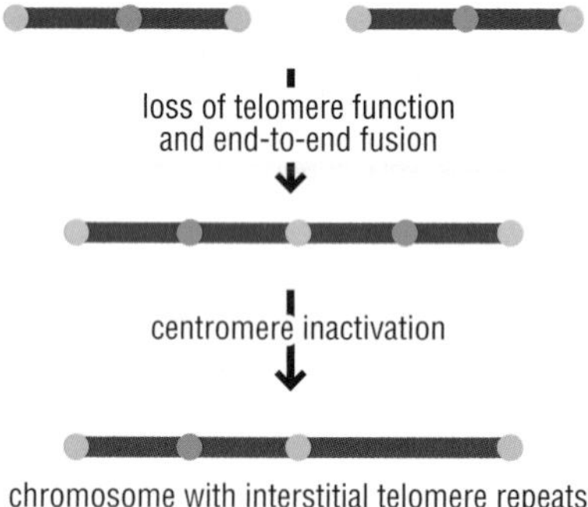

(b) chromosome fission

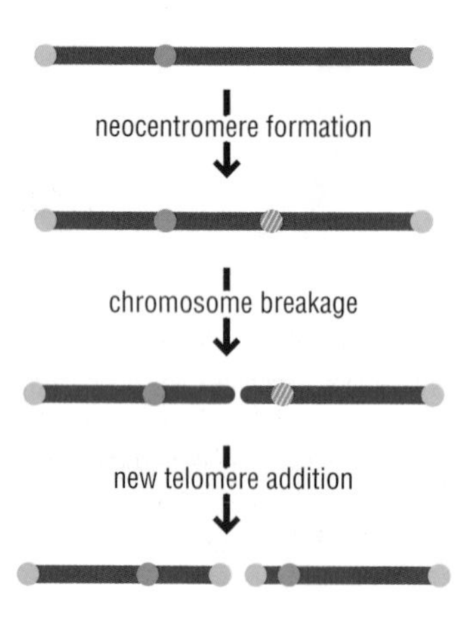

Figure 18.19 Mechanism for change in chromosome number. (a) Chromosome fusion. Two chromosomes can fuse if telomere function is lost. If one centromere also becomes inactivated, this chromosome will be a stable larger chromosome, and interstitial telomere repeats will be detectable at the point of fusion, as seen for the zebra chromosome in Figure 18.18. (b) Chromosome fission. Neocentromere formation may lead to chromosome breakage directly, or a break may occur at random on a chromosome with a neocentromere. If telomeres are added to the site of breakage by telomerase, two new stable chromosomes may result.

Completely novel sequences are introduced into a genome by horizontal gene transfer

The normal passage of genes from one generation to the next is referred to as **vertical gene transfer**. By contrast, yet one more source of genetic change is **horizontal gene transfer**, the insertion of a segment of the genome from one species into the genome of an unrelated species, as depicted in Figure 18.20. Horizontal gene transfer produces novel genetic functions in the genome, unlike duplications and changes in polyploidy, which duplicate existing functions. Many of these novel functions may be detrimental to the recipient organism, so the examples that we observe among living species are likely to only be a fraction of the horizontal gene transfers that have actually occurred during evolution.

Horizontal gene transfer is particularly evident between bacterial genomes but it can also occur between bacteria and their eukaryotic hosts. Horizontal gene transfer in bacterial genomes can occur due to the action of transposable elements or through other genetic exchange mechanisms such as conjugation (transfer of DNA via cell-to-cell contact), transformation (the uptake of DNA), and transduction (whereby viral particles incorporate up non-viral DNA). Horizontal gene transfer has also been inferred for some regions in plant genomes and in the genomes of some parasitic animals as well.

➔ We discuss centromere and telomere function, and the mechanisms that allow formation of neocentromeres, in Section 4.10.

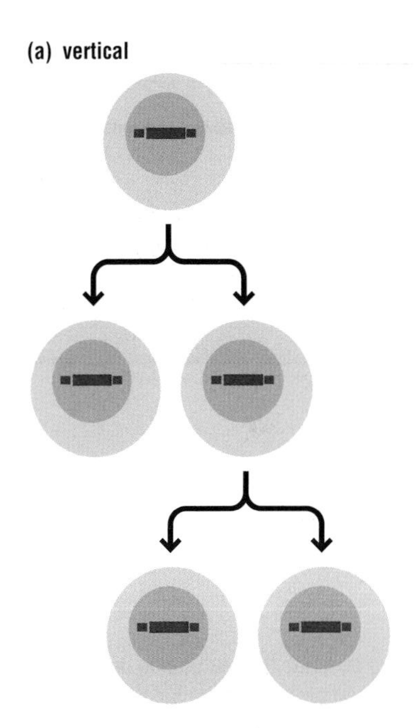

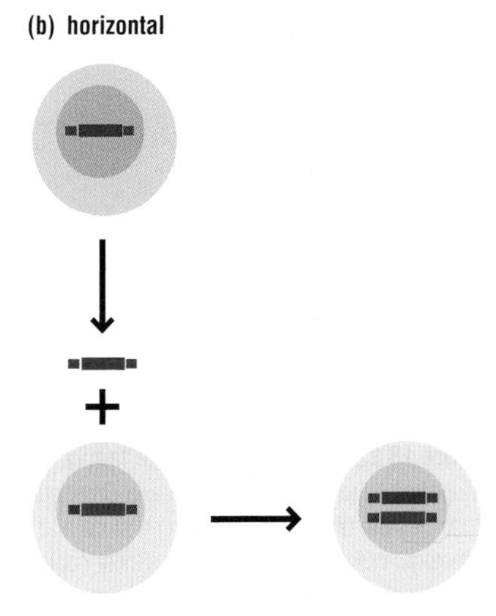

Figure 18.20 Vertical and horizontal gene transfer. (a) Normal passage of genes from one generation to the next is referred to as vertical gene transfer. (b) Integration of a segment of DNA from one species into the genome of another species is referred to as horizontal gene transfer.

18.7 GENE DUPLICATION AND DIVERGENCE OF GENE FUNCTION

Increases in the number of copies of a gene by genome duplication, or recombination that has important consequences for the evolution of gene function. Here we consider these consequences.

Changes in the number of genes drive the evolution of gene families

Once a gene has been duplicated, the copies of both the regulatory and coding sequences can accumulate different mutations. As a result, the functions of the duplicated genes may diverge over evolutionary time. The duplicate members of a gene or its protein are referred to as **gene** or **protein families**.

The α- and β-globin gene clusters provide an example of a duplicated region and a gene family. Hemoglobin is a dimer of α- and β-globin and is needed for carrying oxygen in the blood. In humans, there are three protein-coding genes at the α-globin locus and five at the β-globin locus as illustrated in Figure 18.21. In this case, the protein-coding regions of the genes have diverged only slightly in sequence and activity. However, significant divergence is seen in the time during development when the specific genes are expressed. During human embryogenesis, gestation, and postnatal growth, the expression of both the α- and the β-globin genes switch, such that a different gene is expressed at different times in development. The different globin proteins are not just markers of developmental stage, but have specialized functions in meeting differing oxygen transfer requirements (for example, placenta versus lungs) as the fetus matures and is born.

For other gene families, the divergence in function occurs when different family members are expressed in different tissues. The individual genes (like the globin genes) retain similar functions but the spatial pattern of expression changes. For example, the *hedgehog* gene family encodes important signaling molecules

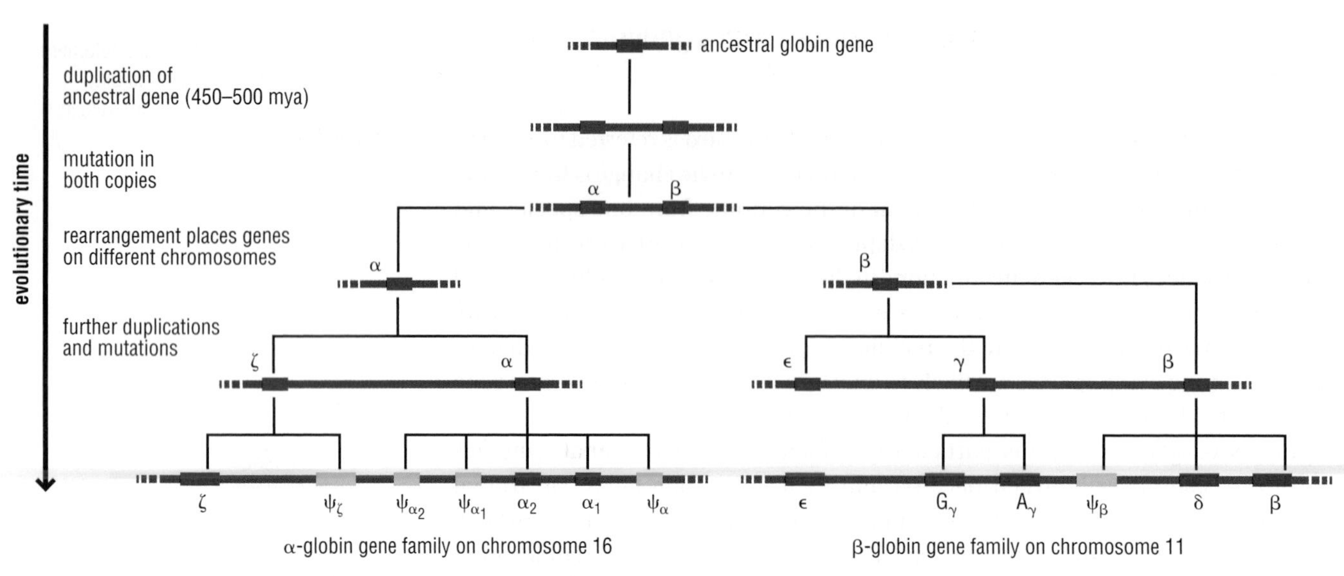

Figure 18.21 Globin gene family. The three α-globin and five β-globin genes in humans are all derived from an ancestral globin gene. The ancestral gene was duplicated 450–500 million years ago, and the two copies were translocated to different chromosomes. After translocation to these two new chromosomal locations, the genes further duplicated and accumulated additional sequence changes. Four copies of the α-gene family and one copy of the β-gene family were inactivated by the sequence changes. These are pseudogenes, shown here in green and denoted with a Ψ.

Modified from Campbell N. et al Biology 5th edition.

involved in spatial and developmental patterning during animal development. Figure 18.22 shows how mammals have three members of the *hedgehog* gene family whose expression patterns differ in different tissues. *Sonic hedgehog* has a widespread function in providing developmental patterning information for the brain, limbs, spinal cord, and many other tissues. By contrast, *Desert hedgehog* is involved in patterning the gonad, and *Indian hedgehog* affects cartilage and bone specification. The proteins are similar but not identical in sequence; the most important divergence has occurred in the location and time of gene expression.

➔ We describe the maturation of the Hedgehog protein in Chapter 14 and Experimental approach 14.1.

Gene families can be compared both within and between species

Gene families arising from duplication and divergence events are found in most eukaryotic genomes. Indeed and as already mentioned, there are often many related copies of a gene in a genome. Related copies of a gene in the same genome are referred to as **paralogs**. For example as shown in Figure 18.22, the *sonic hedgehog, desert hedgehog* and *desert hedgehog* genes in humans are paralogs of one another. The genes have similar functions but are expressed in different tissues.

Gene families can also be compared between species. Genes that are found in two different species and have arisen from a common ancestral gene are known

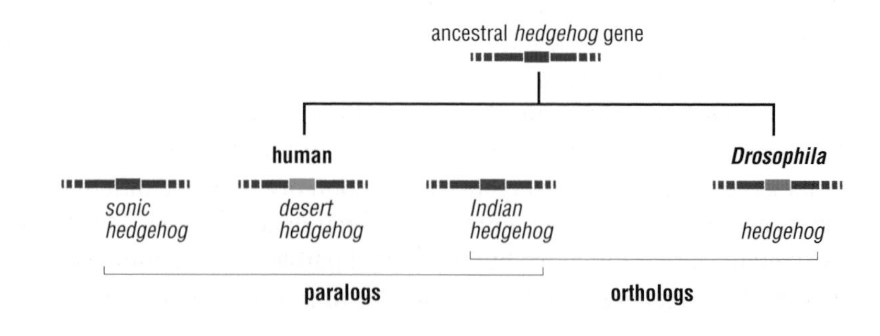

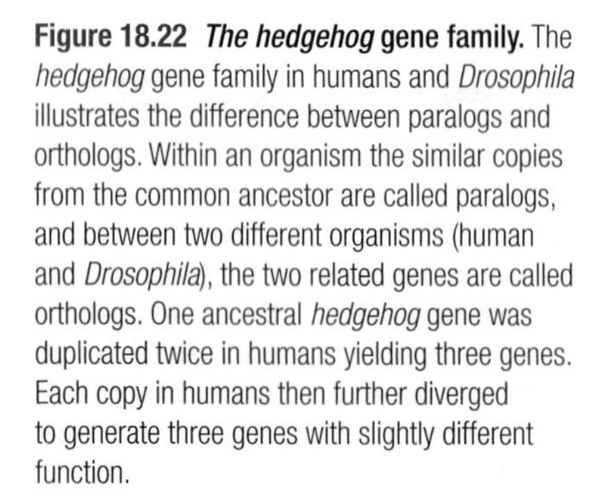

Figure 18.22 *The hedgehog* gene family. The *hedgehog* gene family in humans and *Drosophila* illustrates the difference between paralogs and orthologs. Within an organism the similar copies from the common ancestor are called paralogs, and between two different organisms (human and *Drosophila*), the two related genes are called orthologs. One ancestral *hedgehog* gene was duplicated twice in humans yielding three genes. Each copy in humans then further diverged to generate three genes with slightly different function.

as **orthologs**. For example, the first member of the *hedgehog* gene family was found as a single gene in *Drosophila melanogaster*, and the gene is named for the mutant phenotype it produces in flies. The mammalian *sonic hedgehog*, *desert hedgehog*, and *Indian hedgehog* are all orthologs of one *Drosophila hedgehog* gene (see Figure 18.22). In other words, the single gene found in insects became triplicated in mammals with the concomitant diversification of expression and function.

Both orthologs and paralogs are examples of the more general evolutionary principle of **homology**. Homology describes the way that structures or genes are descended from a common ancestor. It is a qualitative description of an historical event, so two genes are either **homologs**, and are descended from a common ancestor, or they are not. Homology is inferred from the sequence similarity between the two genes and the similarity of the corresponding gene products. The degree of sequence similarity can be quantitatively defined, which means that our inference about homology can be tested statistically. Thus, strictly speaking, we are inferring that the *Drosophila hedgehog* gene and the mammalian *sonic hedgehog* genes are homologs because of the high similarity of the amino acid sequences encoded by these genes. Likewise, we are inferring that *sonic hedgehog*, *desert hedgehog*, and *Indian hedgehog* genes are homologs of one another because of their similarity to each other and to the apparent ancestral *hedgehog* gene.

Duplicate genes experience different evolutionary pressures

Whenever a gene duplication occurs, the evolutionary pressure on the two resulting genes changes. After all, in a previous generation, one gene was sufficient for survival but, following the duplication, there are now two genes. Positive selection can occur for one or both duplicate copies of the gene so mutations that alter the time or amount of gene expression, or that slightly alter the activity of the protein, could be advantageous to the organism. This is a plausible explanation for the divergence in expression of the three mammalian *hedgehog* paralogs.

Mutations usually reduce or eliminate the function of the gene. However, because the selective pressure on the function of the gene has been relaxed following duplication (because the organism carrying the duplicated gene now has a 'back-up' if the activity of one of the two genes should become disrupted as a result of a mutation event), negative selection might not occur, and both the functional and non-functional copies of the gene may be successfully transmitted to the next generation. Thus, many gene families include non-functional gene sequences, termed **pseudogenes**.

In pseudogenes, mutations may have altered the promoter region, changed a splice junction, or introduced a stop codon that has eliminated the function of the gene. Once the gene has been rendered non-functional, further mutational events may occur, so that the pseudogene accumulates multiple additional changes. Thus, although the nucleotide sequence of the region resembles other genes in the family, the gene itself is non-functional.

Pseudogenes are common. For example, there are four of them in the α-globin and two of them in the β-globin gene cluster (see Figure 18.21). These probably arose by duplication and mutation to yield non-functional copies of each gene. Indeed, the genomes of all eukaryotes are littered with non-functional versions of once-useful genes. Pseudogenes can also arise from reverse transcription and the subsequent integration of the DNA copy of the processed mRNA. These 'processed' pseudogenes are easily recognized because they lack the introns that are present in the original gene.

Conserved sequences in a gene family arise from several different mechanisms

The process of mutation is continually at work to produce changes in the gene family members, and both the resulting similarities and differences between paralogs within a species or orthologs between species are informative. The similarity between two genes can be attributed to one or more of three possible events. First, gene family members that are highly similar in sequence might represent a very recent duplication event. Second, gene family members may be highly similar in sequence because there may be strong selection to maintain the sequence and function of both genes, even if they were duplicated a long time ago. Third, gene conversion may be occurring between gene family members, which will result in homogenizing the sequences.

As described more fully in Chapter 16, gene conversion is a repair process that uses one gene as a template to alter the sequence of a similar gene. By using one gene sequence to repair the sequence of a paralog, gene conversion keeps the sequences highly similar.

18.8 CHANGES IN CHROMOSOME STRUCTURE AND COPY NUMBER VARIATION

In the previous section, we examined the consequences of changes in the number of copies of genes and genome duplication. Let us now go on to consider the consequences of more global changes in genome structure.

Structural changes provide clues to genome history

As discussed above, chromosome rearrangements move entire regions of the genome from one location to another and can be involved in speciation. As a result, regions of a chromosome from one species can be aligned with the chromosomes of another species by using conserved genetic elements as landmarks for the alignment of the nucleotide sequences of two diverged genomes. Landmark genetic elements can include the relative orientation of two genes, the position of a repeated sequence or a transposable element, or some other defined sequence. This conservation of a block of sequence between chromosomes of different species is called **conserved synteny**, and the conserved regions are referred to as **syntenic blocks**. Examples of syntenic blocks between the human and the mouse genomes are shown in Figures 18.23 and 18.24. For mouse chromosome 14, note that a large sequence block shown in pink, which is much larger than a single gene, comes from human chromosome 13. This grouping of genes may be due to the evolutionary history, evolutionary pressure, and the amount of time elapsed since the rearrangement as discussed below.

Structural comparisons of genomes involve identifying as many syntenic blocks as possible, and then reassembling the jigsaw puzzle of blocks to observe the results of genome evolution. Specific sequences within syntenic blocks can be in the same orientation or an inverted orientation in different species. The orientation, size, and location of the syntenic regions allow an approximate reconstruction of the rearrangements that shaped the genomes. However, the reconstruction is approximate since there are usually hundreds if not thousands of syntenic blocks, and many of them are the product of more than one rearrangement event.

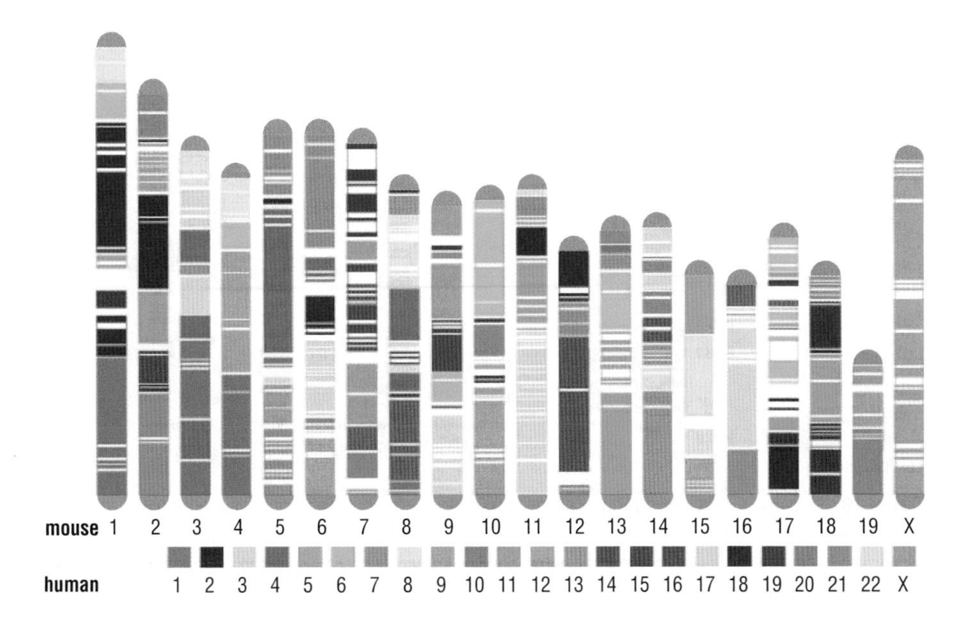

Figure 18.23 Mouse and human genomes show regions of synteny. The regions of the human genome that are homologous to mouse chromosomes are shown mapped onto the 20 mouse chromosomes. Each region of color corresponds to a block of human chromosome that is homologous to that region in the mouse. The colors representing each of the 23 human chromosomes are shown at the bottom.

from Waterston *et al.* Initial sequencing and comparative analysis of the mouse genome. *Nature*, 2002;**420**:520–562.

As a guiding principle, the greater the evolutionary distance, the fewer the number of syntenic regions that are shared between species. A comparison of syntenic maps of humans, chimpanzees and mice shows that there is greater similarity between humans and chimpanzees as compared to humans and mice.

The mammalian X chromosome is a special case of synteny

The most notable exception to the general rule about syntenic regions and divergence is the X chromosome in placental mammals. The X chromosome is almost completely syntenic between all mammalian species, although Figure 18.24b shows how intrachromosomal rearrangements such as duplications, deletions and inversions are seen. The synteny of the mammalian X chromosome was originally recognized because genes that exhibit X-linked inheritance in one mammalian species are nearly always X-linked in other mammals.

This conservation is likely a consequence of the X-chromosome inactivation that we learned about in Chapter 13. Since one X chromosome is inactivated in mammalian females, there is evolutionary pressure for the genes on the X chromosome to stay on this chromosome. If a rearrangement was to occur and a region of the X chromosome landed on an autosome, the transferred genes would be expressed in two doses, which cannot be tolerated for some X-chromosome genes. Conversely, if an autosomal region were placed on the X chromosome, only one copy would be expressed.

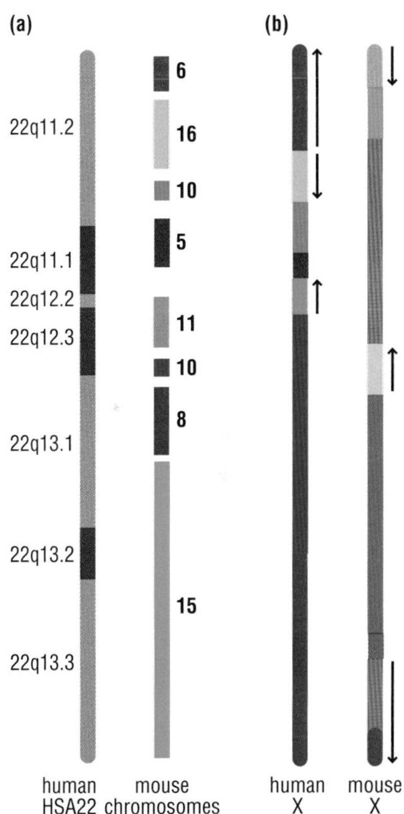

Figure 18.24 Synteny between autosomes is different from synteny between X chromosomes. (a) Human chromosome 22 has synteny with multiple mouse chromosomes. Blocks of mouse sequence that are homologous to sequences on human chromosome 22 are found on the seven different mouse chromosomes (indicated by the different colored blocks). (b) Blocks of mouse sequence that are homologous to sequences on human X chromosome are all found on the mouse X chromosome, although the order is changed.

Copy number variants alter genome structure and phenotype

In addition to the large-scale rearrangements of blocks of DNA over evolutionary time that result in conserved synteny between species, rearrangements are continually occurring *within* species. Rearrangements between repeated sequences during meiosis can result in gametes with different copy numbers of specific regions of the genome in different members of a family. Such changes are remarkably common in the human genome, as shown in Figure 18.25. These CNVs make each individual truly unique; even siblings can have large differences in copy number of certain regions of their genomes. CNVs can range in size from 50 base pairs (bp) to more than several megabases.

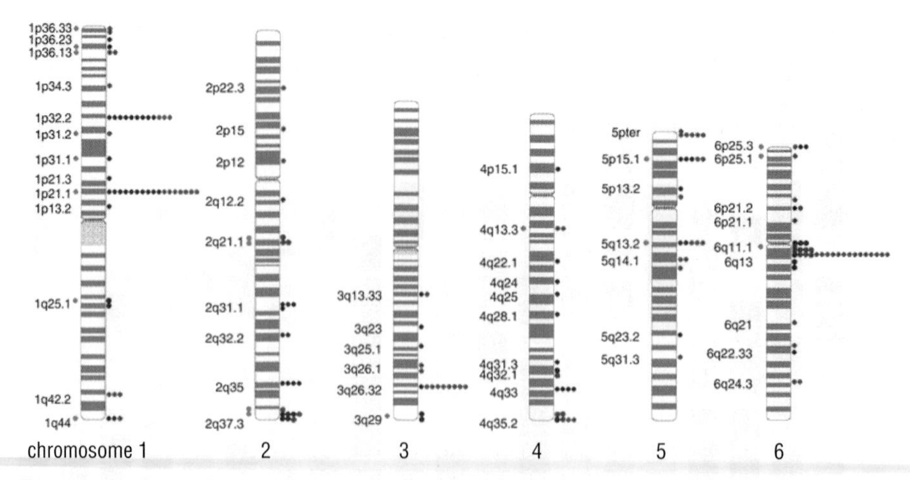

Figure 18.25 Copy number variation is surprisingly high. Circles to the right of each human chromosome indicate the number of individuals with gains (blue) and losses (red) for each region among 39 unrelated, healthy people. Green circles correspond to known genome sequence gaps within 100 kilobases (kb) of the variation, or segmental duplications known to overlap the variable region.

Based on Fig. 1 in Iafrate *et al.* Detection of large-scale variation in the human genome. *Nature Genetics*, 2004;**36**:949–951.

Whole genome sequencing has revealed the frequency of copy number changes that can occur. The genome sequences of six individual domesticated cows were compared, and 1265 unique CNVs were identified between the six genomes, with an average length of 50 kilobases. Thus, such altered regions make up 2.5% of the cow genome. Within the duplicated regions there were 413 complete genes that changed copy number between individuals and thus could contribute to individual variation in gene function. In addition, many of the changes that do not alter gene copy number might alter regulatory regions of specific genes.

While CNVs can occur at many places throughout the genome where repeats exist, they are found most frequently in regions of the genome that have segmental duplication – that is, regions where a block of DNA is duplicated as a tandem array, as illustrated in Figure 18.16c. In these regions, misalignment of the duplicated segment during recombination at meiosis can produce two gametes, one with one copy fewer and the other with an additional copy. The individual offspring from these gametes will then differ in the copy number of that region of DNA.

The effect of CNVs on phenotype has been documented in a number of mammals. In dogs, the duplication of a block of DNA containing four specific genes results in the hair ridge that is characteristic of the Ridgeback breed, while the wrinkled skin phenotype of the Chinese Shar-Pei breed is caused by a duplication of a region that is upstream of the HAS2 gene. In pigs, duplication in the region containing the KIT genes causes white coat color. Finally, in humans, CNVs have been associated with a number of diseases including autism, schizophrenia, neuroblastoma, Crohn's disease, and severe early-onset obesity.

18.9 EPIGENETICS AND IMPRINTING

Our discussion so far in this chapter has focused on changes in the DNA sequence that affect phenotype. However, sequence changes are not the only source of heritable variation. Epigenetic changes can alter phenotype and gene expression without altering the DNA sequence. Epigenetic changes typically are due to the modification of histones and the methylation of mammalian DNA. These modifications are heritable between cells and even between generations of individuals, and can affect the final phenotype that arises.

➲ The molecular bases for the inheritance of epigenetic changes are described in Chapter 4 and the consequences for gene expression are described in Chapter 9.

Epigenetics is the study of heritable changes in genome function that do not involve DNA sequence changes

One type of epigenetic change involves methylation of cytosines and is widespread among mammals. Most mammalian genes have clusters of CG dinucleotides (known as CpG islands) in their upstream regions, and methylation of cytosines is associated with transcriptional inactivation. The inactive X chromosome in female mammals is highly methylated, as are many individual genes that are transcriptionally silent on other chromosomes. A failure to methylate CpG islands, and the inappropriate gene activation that results, is the underlying molecular defect in the neurological disorder Rett syndrome. Loss of methylation is also frequently found in a variety of human cancers.

Histone protein modification is another important epigenetic change that is important in all eukaryotic organisms. As described in detail in chapter 4, histone proteins that package chromosomal DNA are subject to post-translational modifications, such as phosphorylation, methylation, and acetylation, among others. These modifications occur on specific amino acids on specific histones, and different modifications produce different effects on chromosome structure and gene expression.

➲ We discuss the post-translational modification of histones in Section 4.5.

Changes to histone or DNA modification underly epigenetic inheritance. This mode of inheritance is called *epi*genetic because DNA methylation and histone modifications can persist through many cell divisions (as described in Chapter 6), yet there is no underlying DNA sequence change. Thus, these changes in the chromatin structure produce changes in the heritable state of genes without actually altering the gene sequence or structure.

Epigenetic states can have consequences for disease

Genetic imprinting, the situation in which the expression of genes at a particular chromosomal regions differs according to whether the region is inherited from the mother or the father, is a special case of an epigenetic effect. A classic example of the detrimental consequences of genetic imprinting is Prader–Willi syndrome. This disease, which results in behavioral changes and obesity, is associated with abnormal expression of a block of genes on chromosome 15 as depicted in Figure 18.26. These genes are normally turned off (or 'silenced') by chromatin changes in the chromosome inherited from the mother. In contrast, they are normally expressed from the chromosome inherited from the father. The disease can arise if the genes are deleted in

healthy individual chromosome 15

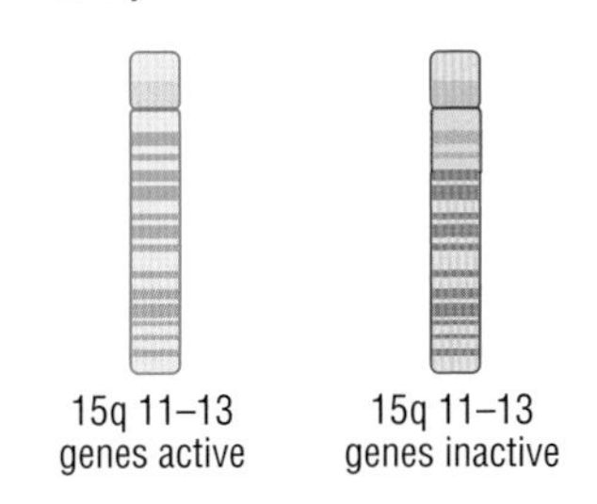

Prader-Willi syndrome chromosome 15

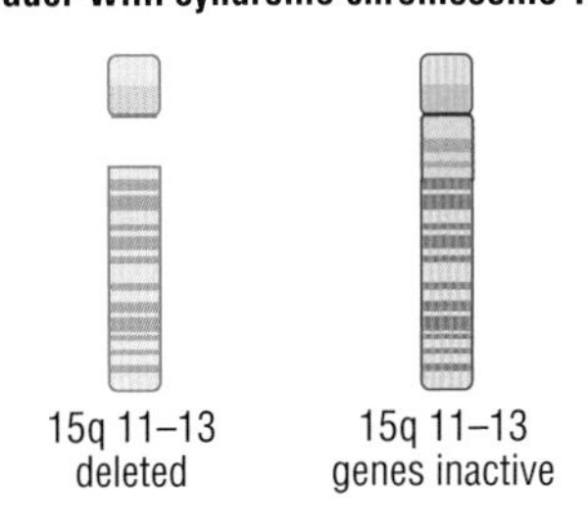

Figure 18.26 Imprinting can predispose humans to disease. Imprinting of a region of chromosome 15 inherited from the mother can lead to disease if there is also a deletion on the copy inherited from the father. Some genes in the 15d11–13 region of chromosome 15 are normally turned off (or 'silenced') by chromatin changes in the chromosome inherited from the mother (pink chromosome). Thus, if there is a deletion in this region of the chromosome inherited from the father (blue chromosome), there will be no expression of these important genes, giving rise to Prader–Willi syndrome. If, instead, there is a deletion in the chromosome coming from the mother, the syndrome will not occur because the genes inherited from the father are expressed.

➔ We discuss genetic imprinting in Section 4.7.

the chromosome inherited from the father, resulting in no expression from either gene in the offspring inheriting these chromosomes. Thus, we see how Prader–Willi syndrome can be caused by epigenetic silencing of the genes inherited from the mother.

18.10 HUMAN GENETIC DISEASES: FINDING DISEASE LOCI

We will now turn to humans to illustrate the effects of genome evolution and the information that can be revealed through genomics, especially as it relates to human health. This has become a significant focus of modern medicine, using information obtained from the Human Genome Project.

Diseases arising from single genes are the best-researched areas of human genetics

As illustrated in Figure 18.27, inherited risk for disease can come from a single mutant gene with simple Mendelian inheritance, multiple individual genes, or the combination of multiple loci throughout the genome and their interaction with the environment. A genetic disease may be highly penetrant (meaning that almost all people who inherit this mutant allele will develop the disease), or the mutation might simply confirm an increased risk of getting a particular disease, a risk that can be impacted by environmental factors.

By understanding the mutations that are causative of disease, risk can be better assessed, and in some, but not all cases potential life-saving changes for the individuals at risk may be implemented. For example, children born with phenylketonuria suffer mental retardation and other serious problems if they consume the amino acid phenylalanine. If it is known that a child carries a mutant gene, they can be put on a strict phenylalanine-free diet to avoid the devastating effects of their disease. Similar behavioral changes can also lead to better outcomes in other forms of genetic disease.

In addition, understanding the biochemical defect in a genetic disease is a starting point for the development of drugs for disease treatment. For this reason, much effort has been devoted to mapping and identifying mutant human genes that cause disease. In this section, we will discuss both single genes that cause disease, the so-called Mendelian traits, as well as complex diseases.

The field of human genetics has been revolutionized by the human genome sequence, the ability to sequence large genomes rapidly, and the many tools such as ENCODE described in Section 18.4 that are now available to researchers. We first describe how disease genes are mapped using specific genetic markers, and

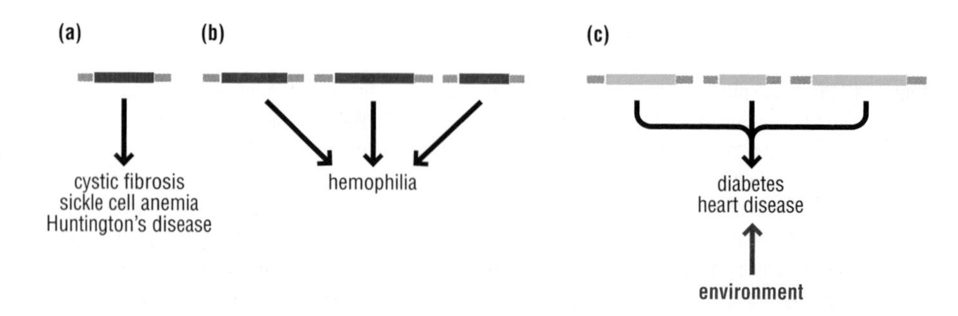

Figure 18.27 Phenotypic differences can be due to changes in single genes or in multiple genes. Some human diseases are caused (a) by mutations in one specific gene or (b) by mutations in one of several different genes (monogenic). The inheritance of these phenotypes follows classical Mendelian genetics. (c) Other human diseases are the outcome of genetic differences in multiple different genes (polygenic). Often these traits reflect both genetic differences and environmental variation (multifactorial).

then discuss new methods of sequencing and how those accelerate the discovery of disease genes.

Genetic markers allow disease gene mapping

Historically, the identification of disease-causing alleles followed classic genetic mapping procedures like those used to map a locus in any organism: markers are identified all over the genome and the frequency with which each of these markers is found in association with the trait of interest is then scored. The higher the frequency of association, the more likely it is that the gene of interest is close to the marker on the chromosome. However, rapid and inexpensive genome sequencing, which allows the identification of specific nucleotide changes, is bringing about changes to this approach.

Single nucleotide polymorphisms and microsatellites are used as markers in genome analysis

The human population has DNA sequence variation at many different sites. For example, some individuals have a cytidine at a certain position of a given site while others have a thymidine, as shown in Figure 18.28a. Such sites that differ at only one position are known as **single nucleotide polymorphisms (SNPs** - pronounced 'snips'). SNPs are the most frequently-used marker for gene mapping. As illustrated in Figure 18.28, SNP loci typically have two alleles (out of the possible four different nucleotides), corresponding to the two variant bases. To be useful as a marker for gene mapping, a SNP should be present at a frequency of 1% in the population. Below that, so few people would have base variations at that locus that the utility of the SNP would be too limited for most studies. With a 1% frequency, the SNP site itself is not expected to be the alteration that causes the rare disease; rather it is used as a marker for a region of the genome associated with a disease-causing mutation. The genomic locations of SNPs are known and are listed in the database called dbSNP.

➔ The dbsSNP database can be found at http://www.ncbi.nlm.nih.gov/SNP/index.html.

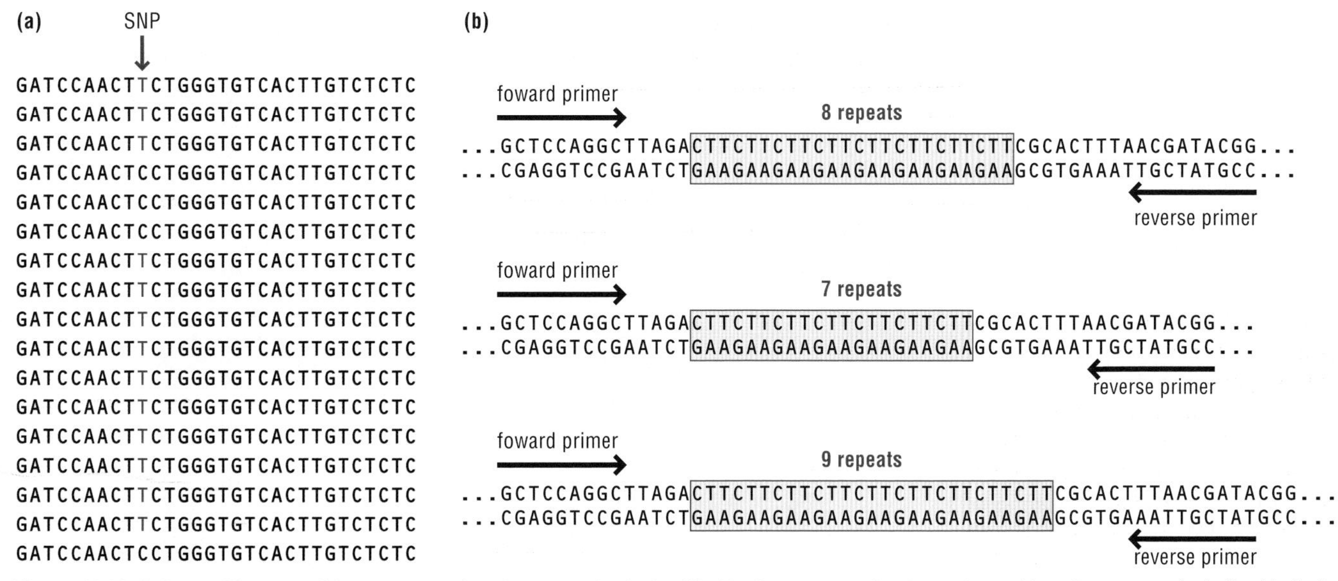

Figure 18.28 Polymorphisms used for gene mapping. Gene mapping is simplified by the presence of various polymorphisms in a genome including (a) single nucleotide polymorphisms and (b) microsatellite repeats. These variable length repeat tracts can be amplified with specific forward and reverse primers as shown in the figure. The different length of the PCR product represents the specific allele of the microsatellite.

A second type of variable region in the human genome are called **microsatellite repeats** - simple tandem repeats of two or a few nucleotides such as CTTCTT in the example shown in Figure 18.28b. These sequence repeats vary in length between individuals in the human population. As described above for SNPs, the locations of microsatellites are known and so can be used for mapping. To identify a given variant, primer sequences are used for polymerase chain reaction (PCR) amplification across the repeat region as shown in Figure 18.28b. The sequences to make these primers are provided in a database called UniSTS. To determine the length of a particular microsatellite repeat for an individual, a PCR reaction is performed and the size of the product analyzed on a gel.

The UniSTS website can be found at https://www.ncbi.nlm.nih.gov/unists.

SNPs and microsatellites are used as genetic markers to map disease-causing alleles for both single-gene Mendelian traits and for complex traits. When SNPs or microsatellites are used in mapping, their sequence or size for individuals with the disease are first determined, and then compared to individuals without disease. One common method to detect SNPs is high-density microarray analysis (referred to as SNP Chips). The SNP Chip has arrayed on it thousands of different 25-base oligonucleotides, corresponding to sequences of individual SNPs. To analyze the SNPs present in an individual, genomic DNA is fragmented and amplified by PCR. The fragments are then hybridized to the SNP Chip under conditions in which hybridization will only occur if a fragment is a perfect match with a given spot; consequently, hybridization to a spot on the chip uniquely identifies which of two possible sequences is present, as depicted in Figure 18.29. While direct genomic sequencing can be used to identify SNPs, SNP Chips are still a less expensive and rapid way of associating disease with a particular sequence.

Haplotype blocks reduce the number of markers needed to find a gene association

As SNPs from many individuals were characterized, it was discovered that SNPs are found across many unrelated individuals in characteristic sets called haplotype

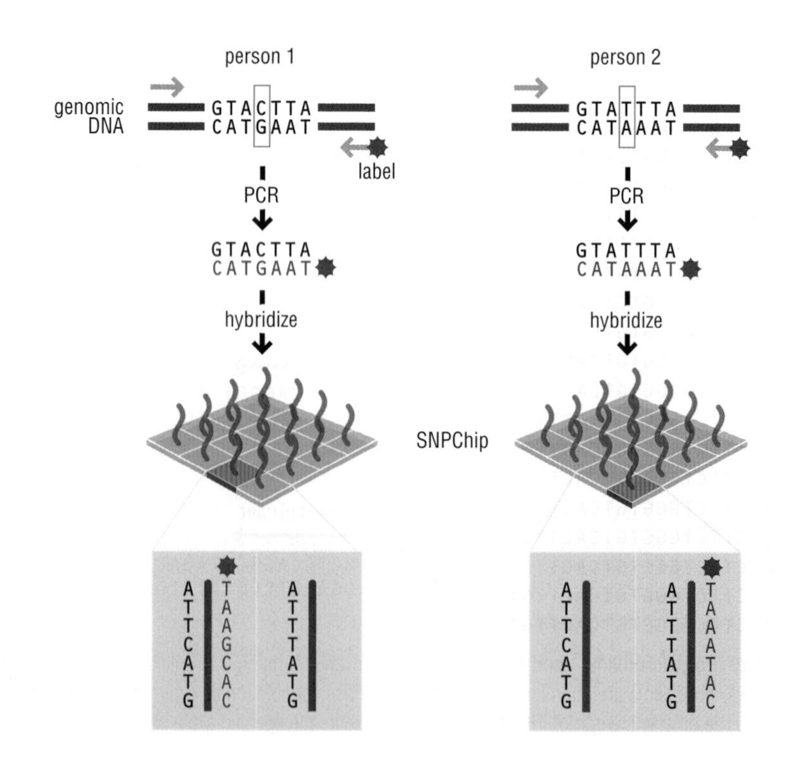

Figure 18.29 SNP Chips are used to detect single base changes at specific loci in the human genome. DNA from a region of a known SNP is first amplified by PCR. Different people will have a different sequence at the SNP; two people are shown in this example. The PCR products are fluorescently tagged – for example, with a red dye. The fluorescently tagged fragmented DNA (red) is hybridized to a chip carrying many thousands of different 25-nucleotide single-stranded DNA molecules (gray lines) arrayed in a known pattern under conditions where a single base mismatch will *not* hybridize. If an individual is a homozygote for a particular SNP, the DNA will only hybridize to one spot; DNA from a different individual may hybridize to the other spot containing the alternative SNP. If the person is heterozygous, both spots will be labeled (not shown).

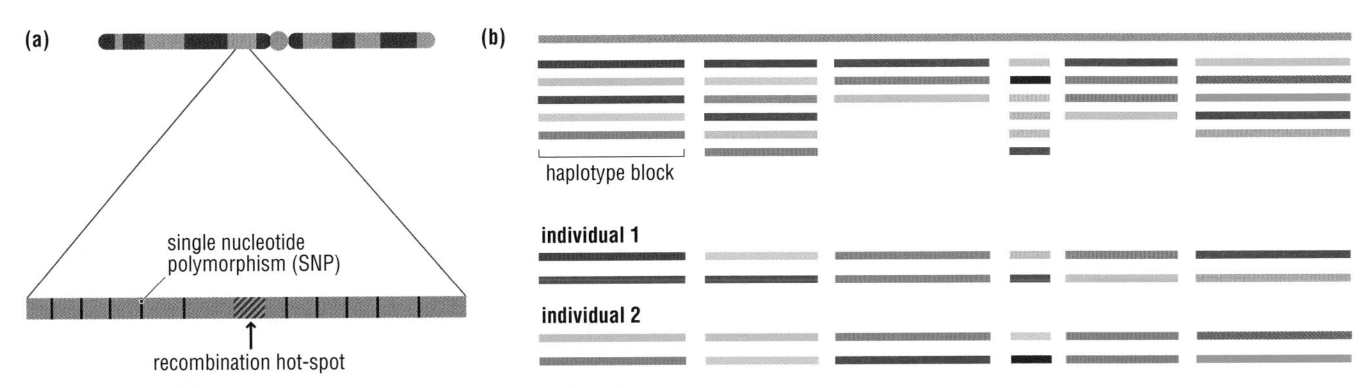

Figure 18.30 A block of polymorphisms that are frequently inherited together is called a haplotype. (a) Haplotypes arise because of the presence of recombination hotspots, positions in a genome where recombination is more likely to occur. (b) Knowing the structure of haplotype blocks reduces the number of markers needed to associate a phenotype with a particular DNA region. In the top part of panel (b), which depicts a theoretical chromosomal region, the gray line represents the DNA and the colored bars below represent the set of known haplotypes that are present for the haplotype blocks found on this chromosome. In the bottom part of the panel, we see a representation of two chromosomes from two individuals who each have different combinations of the possible haplotype blocks.

blocks. Of course, in related individuals, SNPs that are linked will tend to be inherited together. However, it was striking that some SNPs are consistently linked to other nearby SNPs, even among unrelated individuals in large populations. The characteristic size of haplotype blocks is different in different populations. For example, the blocks have a size of about 7.3 kilobases (kb) in African populations and 16.3 kb in European populations.

The presence of haplotype blocks primarily reflects the fact that there has been insufficient evolutionary time for recombination to bring the genome into equilibrium; in this regard, the difference in block size between different populations reflects the different number of generations (and therefore recombination events) since the founding of the population. Another contributing factor is that recombination occurs predominantly at certain characteristic spots in the genome, and so markers between those recombination spots tend to be inherited together (see Figure 18.30). Practically, this has enormous consequences for gene mapping: if you follow one marker in this block, the others will almost always move with it.

We can think of the human genome as a series of independent blocks. Knowing about these haplotype blocks in the genome simplifies both linkage analysis and association studies that are discussed later, because one needs only to look at one defining marker within the haplotype block to identify it. There is no need to use ten markers that are already known to associate with each other within a block, as just one informative SNP is sufficient to define a specific haplotype block.

Linkage analysis and association studies are two major approaches to disease gene identification

To identify specific genes, or gene variants that are associated with human disease, we must start with a population of individuals who have the disease and compare their DNA to a matched population of people who do not. DNA is collected from each individual, and the DNA from each person is analyzed for the presence of specific markers, such as SNPs. We will discuss two different approaches to mapping. **Linkage analysis** can be used to trace specific disease alleles within families. By contrast, **association studies** are used to identify disease genes by studying large groups of unrelated individuals.

The method chosen to identify a disease-related gene variant will depend in part on the population to be studied, and the availability of samples from people who have the disease of interest. If the disease is known to segregate in families, and DNA is available from many individuals in several such families, then linkage analysis will be an effective way to identify a potential disease gene. If DNA samples from related individuals are not available, association studies on large numbers of unrelated people sharing the disease can be employed.

Linkage analysis can map disease genes in families

Linkage analysis is a robust method for identifying single-gene Mendelian traits in families, if DNA from a sufficient number of family members is available. As detailed below, however, this mapping approach is rapidly being replaced with direct genomic sequencing. That said, linkage analysis has illuminated some of the structure of the human genome and so bears consideration.

Linkage analysis uses the pattern of inheritance of a marker compared with inheritance of disease to identify the location of a disease gene within a genome. If a marker is close to a disease gene on a chromosome, the marker and disease phenotype will typically be inherited together. By contrast, if a marker is relatively distant, there is an increased likelihood of recombination, reducing the chance of co-inheritance. Using a large number of markers, such as SNPs, one can accurately determine which SNP is most often associated with disease. Since the location of the markers in the genome is known, the region of the genome that causes the disease can be identified.

To determine how tightly a marker is linked to a particular trait, statistical analysis referred to as the **LOD score** (log of the odds score) is used. This is the score assigned to a marker to describe the likelihood that a marker is linked to a trait. The LOD score is calculated for each marker on every chromosome across the genome, as shown in Figure 18.31. A LOD score of over 5 is typically considered evidence for linkage.

The larger the family that can be analyzed, the more powerful is the approach of linkage analysis to identify disease genes. For example, for a dominant mutation,

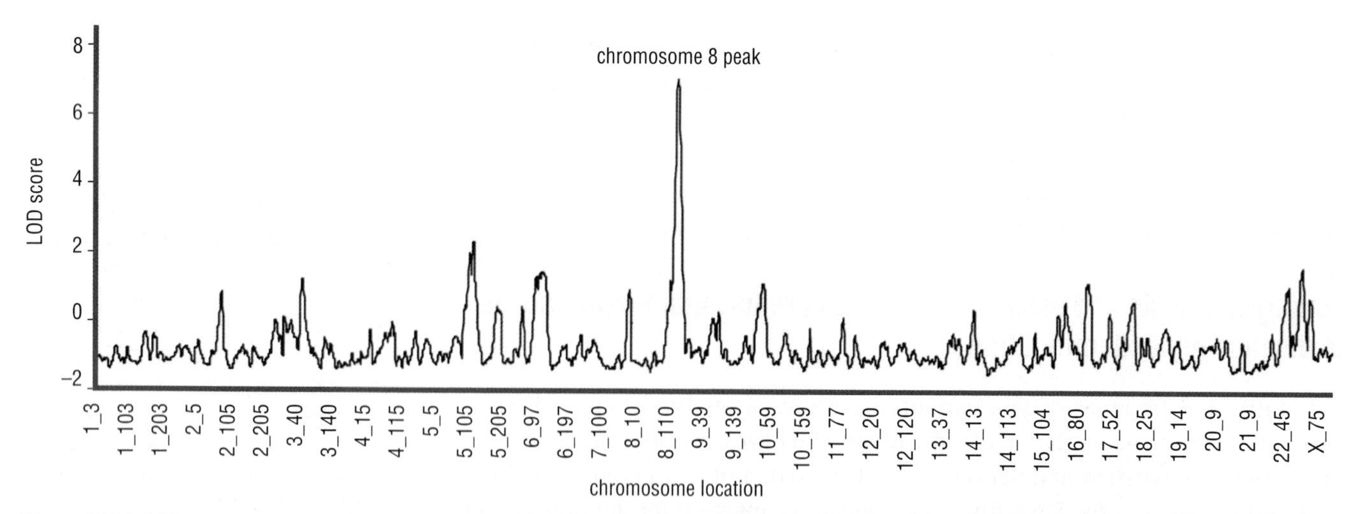

Figure 18.31 LOD scores are a quantitative description of the likelihood that a marker is associated with a trait. A whole genome scan was carried out using SNP Chips to identify genes associated with prostate cancer. The LOD score representing the likelihood of each SNP association with disease is plotted on the y-axis. The location of each of the SNPs assayed in the human genome is plotted on the x-axis. This analysis revealed that there was a set of SNPs located on chromosome 8 at the map position 8q24 that was tightly associated with disease.

Admixture mapping identifies 8q24 as a prostate cancer risk locus in African-American men. Freeman et al. Proc Natl Acad Sci USA. 006 Sep 19;**103(38)**:14068–73

an affected parent and affected child share half their genomes. Simply by genotyping these two, approximately half the genome (the proportion found in the child but not the parent) can be excluded as being a candidate for containing the disease locus. A second affected sibling in the same family would exclude a further half of the remaining half, narrowing the potential location of the disease gene to one-quarter of the genome. In this way, each child of an affected parent narrows the genomic region in which the potential disease gene may lie.

Genome-wide association studies are powerful tools to understand complex traits

Most of the diseases caused by single genes - the so-called **Mendelian diseases** - exhibit their phenotypes early in life. Many other traits and diseases 'run in families' but are not inherited in a simple Mendelian fashion. Examples of such complex traits include many types of cancer, schizophrenia, bipolar illness, diabetes, multiple sclerosis, and coronary heart disease, though there are many others. All of these traits have both genetic and environmental factors that contribute to susceptibility and so are called multifactorial or complex traits.

Genome-wide association studies (GWAS) have been widely used for identifying genes that contribute to complex traits. In these complex diseases, each of the genes that are identified may have only a small effect when considered in isolation, but their concerted action can be significant. GWAS use the same conceptual framework as linkage analysis, employing LOD scores as a statistical tool to determine the probability that a given marker associates with a trait. Unlike linkage analysis, however, a much larger set of unrelated people are studied in GWAS experiments.

GWAS experiments compare the association of markers with disease in a collection of people who have the trait with the markers in a collection of people in which the trait is absent. If a marker is enriched in the pool of people with the trait, it may be linked to the trait, as illustrated in Figure 18.32. Because the people examined in the studies are unrelated, however, the likelihood of recombination

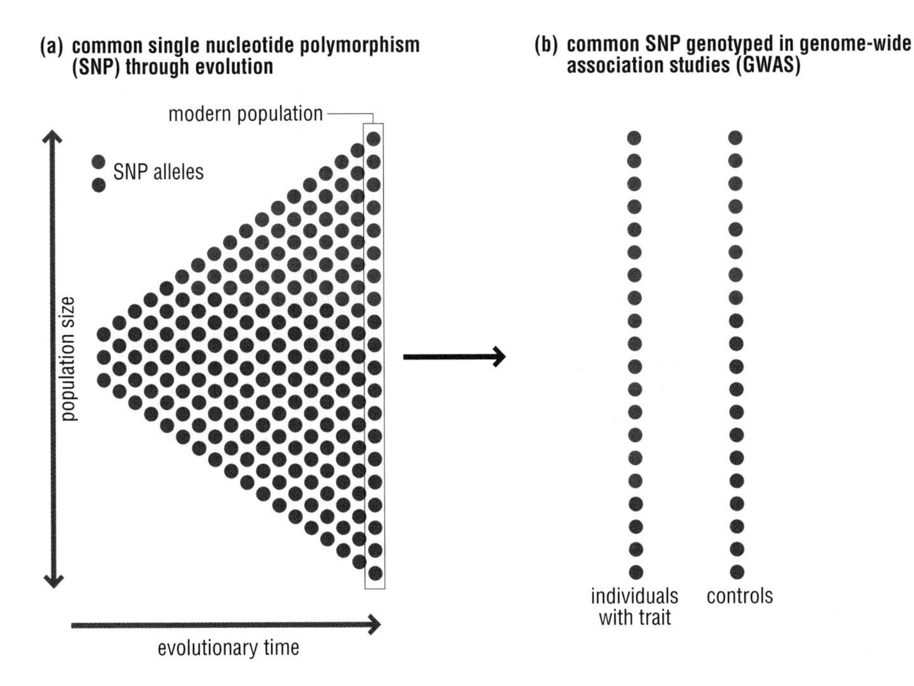

Figure 18.32 GWASs can be used to identify genes that contribute to a complex trait. (a) The allele frequency for any SNP in the population reflects the distribution that developed over evolutionary time. The two different SNPs are indicated in pink and gray. (b) The frequency of the two SNPs in the population carrying a specific trait is compared to a control population. In the example shown, the pink allele is more frequent than the gray allele for people having the trait being studied, suggesting that the SNP may be linked to the disease. This kind of analysis is done for all 500 000 SNPs to find the one or several SNPs that are most highly enriched for the population carrying the trait.

occurring between a given marker and the disease is higher and so a much larger number of people and many more markers must be used to find disease-associated loci by this method.

One of the complicating factors in GWAS studies is that the frequency of the disease gene allele in a population may be very low. GWAS is successful for those diseases that fit the 'common disease-common variant' hypothesis. This hypothesis states that if a disease is common in the population because of historic population bottlenecks in human history, there will be a small number of alleles for the gene that causes that disease. And if there is a common variant it makes it more likely that a causal gene would give a significant LOD score in an association study. (This is because more people will have that variant, generating a larger signal in the statistical test.) However, while some disease genes were found using this approach, many GWAS studies were not able to find variants that significantly associated with disease. It now seems that many complex diseases are cause by rare variants that are not tracked in GWAS studies. Instead, whole genome sequences strategies are now be used to try to identify these rare variants.

Whole genome sequencing can rapidly identify Mendelian disease genes

While linkage to markers was the most common method used to identify single-gene traits in the past, sequence analysis can now be used to identify these sequence changes more rapidly. For linkage analysis, the likelihood of identifying a gene is dependent on having a large family or several large families with multiple affected members to narrow the search to a small region of the genome. Today, however, sequence analysis can be used to identify disease-causing genes by analyzing just three family members - two affected and one unaffected. Typically, though, the sequence analysis of additional family members is used to corroborate a finding or increase the probability of finding the correct gene.

Genome sequence analysis initially focused on 'exome' sequencing. (The exome refers to all of the exons in a genome, that is, the regions that together encode all of the proteins.) As noted above, only 1.5% of the human genome encodes proteins, so exome sequencing was designed to reduce the amount of sequence information required to find mutations that reside in protein-coding regions. Exome sequencing does, however, miss mutations that are in regulatory regions, introns, and intergenic regions, and recent evidence from the ENCODE project indicates that there are many sequence variants that alter gene expression that are outside of the exon. Therefore, the sequencing of whole genomes, and not just exons, is now being used more often to ensure all potential variants are captured.

➔ Methods employed in carrying out genome sequencing and exome sequencing are described in Chapter 19.

The key to disease gene identification by whole genome sequencing is in the analysis of the data. The identification of causal mutations in whole genome data is complicated by the large numbers of rare variants that exist in each individual. After sequencing the complete genome of family members who are affected by the disease and at least one unaffected member, computer programs using sophisticated analysis algorithms are used to identify all of sequence variants. The computer program compares each genome sequence with the established human genome reference sequence, which represents a compilation of the sequences of three individuals; all sequence comparisons are currently made to this defined reference sequence. A variant is defined as a deviation from this reference genome.

The nucleotide positions at which each person varies from the reference genome sets is then listed by the program. Typically at this stage over 1000 variants might be found. The variants present in the affected and unaffected family members are then compared to remove those variants present in both. For each variant present only in affected family members, a statistical test is done to determine the likelihood that that mutation has arisen by chance, the result of which is a p value (much like the score given in the BLAST sequence alignment program). The variants that are found in all affected family members are then compared against a database of known common polymorphisms that occur in the human genome.

The 1000 Genomes Project, further described later, has as its aim the collection of a list of normal human variations. If the variant is in *not* found in the list of normal human variations, it will be deemed a candidate disease-causing variant. Typically on the order of 100 genes might be in this final list of candidate variants. As the number of completed genomes continues to grow, our power to distinguish rare polymorphisms from disease-causing variants will increase.

➔ You can find out more about the 1000 Genomes Project at http://www.1000genomes.org.

Sifting out the causal gene from a list of around 100 requires biological insight. Consequently, there are other computer programs that score the likelihood that a given mutation is deleterious to protein function. For example, if the change introduces a stop codon in a protein, or alters a conserved kinase domain, it would have a high likelihood of being a deleterious mutation. In contrast, a variant that results in a synonymous change in a protein is less likely to be a causal mutation.

While it might be possible to predict the deleterious effects of changes in protein-coding sequence, it is more difficult to determine the effect of sequence variants that are not in a coding region. Examining the ENCODE and other genome annotation sets can also be very useful. If the mutation is in regions of the genome that encodes a non-coding RNA, that RNA might play a role in disease; if the variant were in a known binding site for a transcription factor we would have a clue as to its function.

Whole genome sequencing can also be used to identify rare variants that contribute to complex disease

While whole genome sequencing has most commonly been used to study Mendelian traits in families, a number of new approaches are being developed to use genome sequence analysis for complex traits. In the recent past, GWAS studies using SNPs were used to find regions of the genome involved in complex traits in which many different loci together contribute to disease. As described earlier, it is hard to use SNPs to identify rare variants that cause disease because they are present at a level of just 1% in the population. Instead, many groups are now developing approaches that use whole genome sequence analysis from many affected people to find those loci that might contribute to disease susceptibility but are rare in the population.

The use of whole genome sequence analysis in this way is still in its infancy, to such an extent that we cannot yet point to any successes. But the rate at which sequence analysis is advancing suggests that successes may well have been identified by the time this book is published.

But how is whole genome analysis applied in this context? The concept is analogous to the methods to find Mendelian genes that we describe earlier - though with significantly more powerful analysis. Rather than finding one variant that associates with disease, a set of many variations needs to be identified that are found together in affected people but are not found in the references genomes.

This approach brings numerous challenges and questions. Which variations are meaningful and should be included in the set of changes being followed? How do we verify that this constellation of changes is truly predictive of disease? What computational programs are needed to carry out these analyses? The experimental basis of science - predicated on enquiry, analysis, and refinement - will doubtless see questions such as these soon answered, resulting in advances in our understanding of the genetic basis of disease.

18.11 HUMAN GENETICS: IMPACTS AND IMPLICATIONS

We see throughout this chapter how advances in gene sequencing technologies are revolutionizing our understanding of the structure and function of the human genome, and the intrinsic link between the nature of that genome and human disease. But to what real-world uses can these advances be applied, and what issues surround such applications? We explore these questions in this final section of the chapter.

Databases of human sequence variation allow rapid analysis of genetic disease.

A number of online databases provide valuable information to scientists about diseases caused by single genes; new databases will be an essential tool in understanding complex human disease. While these databases will change over time, let us now focus on four of the current major databases to provide an example of their importance.

One of the first compilations of human Mendelian disease genes was established in the 1950s by Victor McKusick and was named 'Mendelian Inheritance in Man.' This catalog is now maintained and known as **OMIM** (Online Mendelian Inheritance in Man). This database is the best source of information about Mendelian genetic disease, including inheritance, gene identification and function, animal models, known disease variants, and even links to support groups for afflicted families.

➡ You can find out more about OMIM at http://www.ncbi.nlm.nih.gov/omim.

To characterize the sequence diversity among humans, an international consortium called the **1000 Genomes Project** has set out to sequence complete genomes from a wide cross-section of people around the world, providing insight into the variety of human genetic variation. As of 2012, the project had sequenced 1092 complete genomes from 14 different populations drawn from Europe, East Asia, sub-Saharan Africa, and the Americas.

➡ Find out more about 1000 Genomes at www.1000genomes.org.

With the information now in hand, there are 38 million characterized SNPs, 1.4 million sites where a small insertion or deletion has occurred, and 14 000 sites with larger deletions. By comparing these changes across the 14 different populations examined, it is clear that some variations are more prevalent within a given population. Strikingly, the study found that each person in the population carries up to 20 heterozygous loss-of-function alleles that are unique to that individual.

The value of this database to scientists trying to find a disease-causing allele cannot be overstated. Imagine that 100 sequence variants were found in the process of trying to identify a disease gene for a rare disorder. Those nucleotide changes that represent common variants can be discounted as the causal changes in the disease. This can significantly reduce the number of the possible disease-causing genes that the scientists will need to further investigate.

As described in Section 18.8, point mutations aren't the only source of disease-causing mutations. CNVs change the copy number and arrangement of genes and also play a major role in human disease. To be able to probe the many alterations that have been found, a **human structural variation** database catalogs CNVs that have been identified.

➔ Explore the human structural variation website at http://www.genome.gov/25521748.

Projects to understand changes in cancer cells have allowed the establishment of a database of tumor DNA sequences and, in many cases, the normal DNA from the same individual. This sequence information can be mined using sophisticated computational methods to help identify those genetic changes in a tumor that most likely lead to disease. Indeed, these approaches have already led to the identification of specific genes that, when mutated, lead to cancer. The sequence data obtained from such sequencing projects are collected by **The Cancer Genome Atlas** (cleverly called TCGA, representing the four bases of DNA). By collecting this data in a central location, researchers can have access to large data sets and can compare sequences they have obtained in their own studies to sequences that have already been published.

➔ View The Cancer Genome Atlas at http://cancergenome.nih.gov.

These large databases accelerate the discovery of changes that may affect human health by orders of magnitude. No one researcher can accumulate all of the information needed to understand complex diseases on their own. Instead, they can draw upon the vast amounts of information collected by a worldwide community of scientists and thus make discoveries more rapidly.

The ability to easily obtain DNA sequence rapidly makes individualized medicine possible

While the 1000 Genomes Project illustrates the feasibility of cataloging the set of genome alterations within and between human populations, it says nothing specific about a given individual who was not sequenced. However, this vast knowledge of normal human variation makes it easier to determine what might not be normal. Today, **individualized medicine** is a plan to take advantage of the genome sequences of patients to help plan medical treatment on an individual level.

One example of such individualized medicine is the field of **pharmacogenomics**, which integrates information about a person's response to a particular drug with an understanding of their unique genetic makeup, as represented in Figure 18.33. Pharmacogenomics has yielded some striking successes in recent years, particularly in relation to the use of a patient's genotype to avoid treatment with a drug that carries a potential risk to the individual. For example, about 5% of Caucasians have at least one copy of a specific variant of an immune response protein, human leukocyte antigen (HLA)-B. Individuals carrying this specific variant are at significant risk of experiencing fatal hypersensitivity to the antiretroviral drug abacavir (Ziagen), which is used in the treatment of individuals with HIV infection. Genetic analysis can determine whether an individual carries this specific variant, and therefore indicate whether this drug should be avoided in their treatment regime.

Another example of the power of pharmacogenomics is the use of the blood-thinning drug warfarin (Coumadin), which is prescribed hundreds of thousands of times annually in the United States for people with blood clots or at risk of clotting after surgery. However, approximately 1% of patients develop life-threatening side effects from warfarin treatment. Genetic analysis has shown that specific variants at one of two different loci can predict who is at most risk from such side effects, and the dose of drug more carefully monitored for those individuals carrying the known genetic variant.

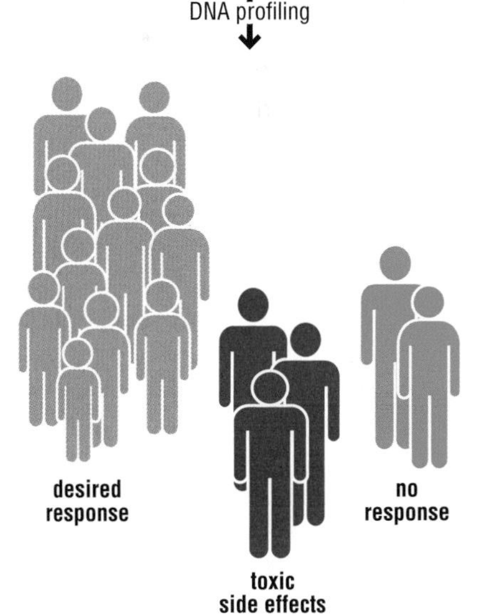

Figure 18.33 Individualized medicine is possible if a genomic sequence is known. Genomic sequence information for a population of individuals with the same disease can aid the development of a treatment regimen by identifying individuals who may not respond to the drug or who may even experience a toxic side effect.

While the avoidance of drug side effects is important, it only reflects part of the potential for individualized medicine. The complete genome sequence from a patient with a given disease may help doctors to determine the best overall course of treatment for that patient. Indeed, hospitals are now beginning projects to sequence the DNA from patients with specific diseases, and comparing their sequence information with that of both other patients with a similar disease and with the normal population in the 1000 Genome Project. As more and more information becomes available on diseases and genetic changes, the ability to determine associations will grow.

Obtaining your own genome sequence is powerful and controversial

The examples described above illustrate the value of individualized medicine in the future of patient care. Yet, hospitals aren't the only organizations now using DNA sequence information to understand how sequence changes may impact disease: direct-to-consumer companies are also now marketing genome sequencing to individuals, outside a medical setting. These companies sequence either the complete genome, the exome, or a subset of genes that may affect susceptibility to multifactorial disease for a fee, and provide some analysis of that sequence information.

In principle, such individual genomic information might allow recognition of a susceptibility to a disease before symptoms have appeared, and the affected individual can then make informed choices and lifestyle changes. However, while we have a great ability to obtain DNA sequences, except for mendelian genetic diseases, we have much less of an understanding of what sequence variations actually mean. We are just beginning to understand the complex interplay of multiple genetic changes, and the interaction of these changes with the environment.

Many people have raised doubts about the efficacy and ethics of direct-to-consumer genomic testing because the interpretation that might be provided to a person about their DNA sequence is variable. The industry is unregulated, and companies offering this service in the United States and most of Europe do not currently have to be certified or licensed and no standards have emerged for how risk factors are assigned to specific sequence variants. Further, studies indicate that many individual consumers have a poor understanding of the concept of risk factors and so are not well-placed to interpret the information presented to them as a result of their genome sequencing – yet the follow-up counseling that could make such a difference in this regard is often not available. Consequently, the information on which people are basing life decisions or lifestyle changes could be flawed. In fact, in December 2013 the US food and drug administration blocked these direct-to-consumer sequencing companies from providing any interpretation with the DNA sequence provided, because the medical implications are unclear.

Ethical issues arise with the availability of individual genome sequences

As the ability to obtain DNA sequences has grown, so issues about how genetic information is used in our society have come to the fore. Prenatal genetic testing has already been used for years to allow parents to make informed reproductive decisions; genetic testing can be sought when a previous child was affected with a disease, other family members have a disease, or there is an increased risk of disease because of environment, or maternal age. With the increased speed of sequencing today, such genetic testing may become both more comprehensive and more common. There are, however, significant ethical issues about prenatal testing and 'selective reproduction.' Is it ethical for parents to use genetic testing to 'select for' children who may be more likely to have a 'desired trait'? Is it ethical to 'select against' children who may have an increased risk of an 'undesirable trait,' rather than a significantly debilitating disease? And who decides what is desirable?

The answers to these questions depend heavily on the context, of course. A devastating disease like Tay Sachs is routinely screened for in the Ashkenazi Jewish population because infants born with this disease die in their first year. But what about a couple who want a child with red hair? If we soon can identify this trait at earliest embryonic stage, is it ethical to use genetics to favor such a benign trait? These kinds of questions are best discussed broadly in public discourse so that as a society we can make informed decisions.

Genetic privacy is also a consideration. There is considerable disagreement about who should have the ability to request individual genetic information. Many observers have raised the possibility of genetic discrimination when it comes to individuals trying to obtain insurance. Indeed, recent regulations in the United States ban the use of genetic information for discrimination for health insurance purposes, but it recently became clear that life insurance companies do not feel they are bound by this legislation.

Genetic privacy considerations extend to ethical issues related to inheritance and consent, since information derived from genetic testing of one family member may directly imply the genetic status of a relative who may not want to be tested or who are minors and have no legal right to refuse consent. For example, a young woman may choose to have her *BRCA1* and *BRCA2* genes tested for mutations associated with an increased risk of breast cancer. If the woman is found to carry a mutant *BRCA2* allele, this has implications for her mother and her sister who themselves may not wish to be tested. The issue is even more complex in cases where the presence of a particular gene allele always results in severe disease – for example, with the neurodegenerative disorder Huntington's disease, whose symptoms usually only begin at middle age, typically after the decision to have children has been made.

Human history is told in the sequences of our genomes

Finally, in addition to providing critical information about diseases and disease treatment, advances in human genome sequence analysis are providing insights into human history. For much of human history, we lived in small populations, so genetic drift exerted significant effects on human genetic diversity. Although we have recognizable genetic diversity between populations from different geographical regions, most of our genetic variation is found among individuals of the same population. Thus, for example, Swedes and Italians (to take two groups) do not look the same. However, the amount of genetic variation between a Swede and an Italian may be much less than the amount of variation found by comparing two Swedes or by comparing two Italians.

As our ancestors lived in small populations, haplotypes are a powerful way to trace our evolutionary past. For example, the native populations of the Americas have haplotypes that are also found in populations that arose in east-central Asia, consistent with migration across Siberia and the Bering Strait and south along the west coast of the Americas, as illustrated in Figure 18.34. Multiple prehistoric

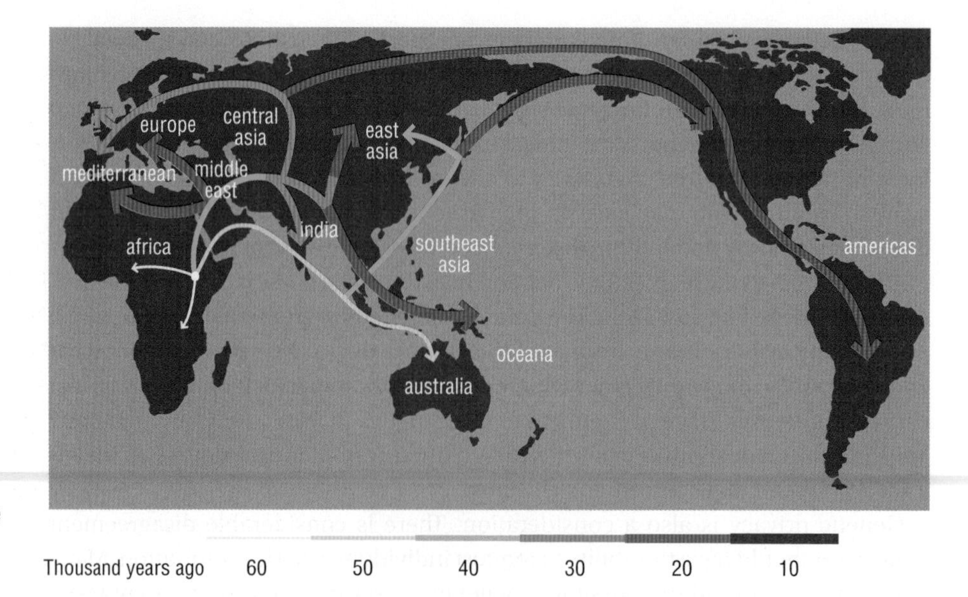

Figure 18.34 Haplotype analyses confirm maps of human migration. Because our ancestors lived in small populations, genetic analyses are a powerful way to trace human migration. The native populations of the Americas have haplotypes that are also found in populations that arose in east-central Asia, consistent with waves of migration across Siberia and the Bering Strait and south along the west coast of the Americas.

Courtesy of National Geographic's Genographic Project.

waves of migration can be inferred, the first being found in Africa and the most recent in western and northern Canada. These waves of migration are consistent with a regular southward and eastward movement of the immigrants. The prehistoric migrations across the islands of Polynesia and the general northwesterly settlement of Europe can also be reconstructed from examining the patterns of haplotypes.

Genome analysis supports what we learn from recorded history as well. For example, the presence of the Vikings in Ireland, Phoenicians in port cities around the Mediterranean Sea, and Spanish soldiers in Colombia are also evident from analysis of DNA polymorphisms and haplotypes.

The ancestral nature of some mutations and the effects of population migration are also apparent from some genetic diseases. One example is cystic fibrosis, the most common genetic disease among people of European descent. The gene responsible for cystic fibrosis is named *CFTR,* and the disease is inherited as an autosomal recessive trait. Although more than 125 different mutations in the gene *CFTR* have been identified, more than 75% of all cystic fibrosis cases are due to the same molecular lesion, a 3 bp deletion of a phenylalanine at position 508 of the *CFTR* protein (referred to as ΔF508). Analyses of polymorphisms near the ΔF508 mutation have revealed a common haplotype or a group of closely related haplotypes, suggesting that the causative mutation in almost 75% of cystic fibrosis cases arose in, and spread from, one particular individual.

SUMMARY

In his book *'On the Origin of Species'*, Darwin ended by musing on the 'endless forms most beautiful' as he contemplated the diversity of living things. Darwin saw the life of organisms and species in their environments; he knew little or nothing of cells, genomes, genes, RNA, and proteins. We now recognize that the unity and diversity of all living things that inspired Darwin is apparent at all levels of biology, from ecosystems to molecules.

Genomics both focuses and reflects all of biology. We may not yet have the insights necessary to understand the subtle interrelationships between molecules or between species, but genomics provides both the tools and perspective that we need to approach these investigations. It gives us molecular data to explore human history and our interaction with our environment. It can shape our understanding of disease, and behavior. Yet the power of genomics also brings responsibility. We recognize that humans are but one species among millions, and each of us is but one individual among billions. We understand how we are different, but we also recognize how we are similar.

Throughout this book we have explored the molecular components that cooperate in a series of processes to bring the information in our genome to life - what we call 'molecular biology'. However, while these components and processes remain remarkably conserved throughout the natural world, the end results of genome function - the organisms from which the natural world is formed - are remarkably diverse. Molecular biology paints only part of the picture of the intricacies of life. Only by considering molecular biology in a wider context - alongside fields of biology such as ecology, physiology, and evolution - can we really begin to appreciate just how intricate - and spectacular - the process of life really is.

PRINCIPLES OF GENOMICS

- The ability to obtain the complete genome sequence of thousands of different species, different individuals of one species (genomics) or a collection of species (metagenomics) has revolutionized biology.
- The genome includes protein-coding genes, RNA-coding genes, regulatory sequences, transposable elements, other repeat sequences, and many sequences of unknown function.
- Genome annotation involves using sequence information and knowledge obtained from other experimental approaches to assign functions to coding and regulatory sequences.
- Despite a wealth of information from molecular biology, the function of many genes and much of the intergenic sequence remains unknown.

EVOLUTIONARY FORCES

- Selection and genetic drift work on genetic variation to shape natural populations.
- Evolutionary differences between species involves changes in size of the genomes, the structure of genomes, and the number of genes, as well as the sequence of genes.
- Translocations and inversions, and large-scale chromosome rearrangements, including fusions, are associated with speciation.

ANALYZING GENOMIC SEQUENCE

- Many genes are members of gene families with related sequences and functions.
- Amino acid sequence alignments provide functional information about genes from distantly related species, whereas nucleotide sequence alignments provide functional information about sequences from more closely related species.
- Sequence polymorphisms are inherited in linkage blocks known as haplotypes. These blocks, which can be followed by SNPs or microsatellite repeats, are extremely important for the mapping of mammalian genes.

PHENOTYPIC DIVERSITY

- Biological complexity arises from both genetic and environmental sources, as well as the interactions between these sources.

HUMAN GENOMICS

- Our rapidly expanding knowledge of the sequence and haplotype structure of the human genome allows us to trace our history, as well as to identify genes associated with many different traits and important diseases, but also carries with it the responsibility to use the information wisely.

QUESTIONS

18.1 GENOME SEQUENCES AND SEQUENCING PROJECTS

1. What is a genome?

2. What is genomics?

3. What is a genome sequence?

4. Even genomes that are considered to be fully sequenced are often not 100% complete or 100% accurate. Explain why.

5. Explain why the study of evolution is enhanced by the comparison of complete genome sequences as opposed to comparing the sequences of just one or two genes.

6. Which of the following statements correctly describes a metagenome?

a. The genome sequences of all the organisms in an ecological niche.
b. The genome sequences of all the individuals in a species.
c. The sequence of a very large genome.
d. An individual genome, sequenced many hundreds of times.

7. Why does metagenomics provide more information than sequencing the genomes of organisms cultured in a laboratory?

8. How does genomics expand on the original notion of a gene, as described by central dogma (DNA encodes RNA which encodes protein)?

18.2 FINDING FUNCTIONS IN A GENOME

1. What is the main goal of comparative genomics?

2. What is the main goal of functional genomics?

3. Scientists use PCR to copy DNA. Is this copied DNA usually the equivalent of a genome? Explain your answer.

4. Which of the following describes a means of performing genome annotation?

a. The process of obtaining the DNA sequence of a whole genome
b. Assigning function to genes by mutating them and examining phenotype
c. Comparing the genes from one chromosome in an organism to those on another chromosome
d. Using similarity of DNA sequence among organisms to suggest gene function.

5. Give specific examples of short sequence strings that could be used to identify the following features.

a. The start of a gene
b. The end of a gene
c. A splice site

Challenge questions

6. Genome annotation:

a. Genomes are often annotated by comparing sequences among organisms. How might genome annotations be further enhanced and improved?
b. Computationally determined gene structures like introns and exons are often not completely correct. Explain why this is the case. How would you check whether an annotation was correct?
c. Many predicted genes do not have clear direct counterparts in other organisms. What other features can be used to make predictions regarding gene function?

7. DNA barcoding can be used for identification of species using characteristic regions of sequences. For fungi, a sequence termed ITS is used. The ITS sequence encompasses an rRNA gene flanked by two intergenic regions. Part of an ITS alignment is shown in Figure A. Indicate the rRNA gene region and an intergenic region. Explain your choices.

8. You have the sequence of a gene of interest from a known organism, and wish to find the sequence of the equivalent gene from two other organisms, organism A and organism B. There is a genome sequence available for organism A, but not for organism B. Describe the steps you would take to determine the sequence of your gene of interest in each of the two organisms.

18.3 FUNCTIONAL GENOMICS

1. Briefly explain the following terms.

a. Transcriptome
b. Proteomics
c. Genome-wide disruption assay
d. ChIP-seq

2. Which of the following statements correctly describes how genomics differs from proteomics?

a. Genomics looks at the whole genome, whereas proteomics examines the activity of just one gene.
b. These two terms have identical meanings.
c. Proteomics examines the output of a genome, which changes depending on circumstances. The genome does not change depending on circumstance.
d. Genomics is the study of genomes from different species, whereas proteomics only makes comparisons within a single species.

3. Explain one advantage of using deep sequencing rather than microarray analysis, to examine gene transcription.

Challenge questions

4. Functional genomics and experimental design:

a. Design an experiment to discover all the genes in *Saccharomyces cerevisiae* that encode proteins that are essential for sporulation. Discuss any difficulties or constraints you envision with your approach.
b. Modify the experiment to discover genes encoding proteins that are needed for sporulation, but for which there may be functional redundancy. Discuss any further difficulties.

Figure A

```
organism 2 --ATA-----------------------------CAGCTCTTCTGA--AATGTAT-TGAA 163
organism 3 --TTGG------------------------ATTGCAGCTCTTCTGA--AATGCAT-TGAA 163
organism 6 GATTGA------------------------TTAGCAGCTCTTCTGA--AATGCAT-TGAA 163
organism 5 GATTGAAGTCAGAGATTACTCTCTGATGAATTAGCAGCTCTTCTGA--AATGCAT-TGAA 163
organism 4 ----GA------------------------TTAGCAGCTCTTCTGA--AATGCAT-TGAA 163
organism 1 ---CGC------------------------CCGCCGGCCGAGCGTAGTAGTTTACATGAA 163
                                             * **    *  *  * *  *  ****

organism 2 ATGCGATAACTAATGTGAATTGCAGAATTCAGTGAATCATCGAGTCTTTGAACGCACCTT 223
organism 3 ATGCGATAACTAATGTGAATTGCAGAATTCAGTGAATCATCGAGTCTTTGAACGCACCTT 223
organism 6 ATGCGATAAGTAATGTGAATTGCAGAATTCAGTGAATCATCGAATCTTTGAACGCACCTT 223
organism 5 ATGCGATAAGTAATGTGAATTGCAGAATTCAGTGAATCATCGAATCTTTGAACGCACCTT 223
organism 4 ATGCGATACGTAATGTGAATTGCAGAATTCAGTGAATCATCGAATCTTTGAACGCACCTT 223
organism 1 ATGCGATAAGTAATGTGAATTGCAGAATTCAGTGAATCATCGAATCTTTGAACGCACATT 223
           ********  ******************************** ************** **

organism 2 GCGCCCT-TTGG--------------CTTTGCCTTCTCTTGAAGAG--AGATAACTATGC 266
organism 3 GCGCCCT-TTGGTACTTCACGGCTCCCTTTGTGTTCTCTCCTAGGG--AGATACCTATGC 280
organism 6 GCGCTCC-TTGG-------------------TATTCC----GAGGA--GCATGCCTGTTT 257
organism 5 GCGCTCCACCCGC------------------TGAACT----TAAGC--ATATCAATAAGC 259
organism 4 GCGCCCT-TTGGCA--------------TTCCGACCTCAAATCGGGTGAGACTACCCGCT 268
organism 1 GCGCCCG-CCAGTA-------------TTAAAAGCCTTATTTTACCCAAGGTTGACCT-C 268
           **** *     *                       *
```

5. Figure B indicates two ways in which gene expression levels can be analyzed.

Figure B

(a)

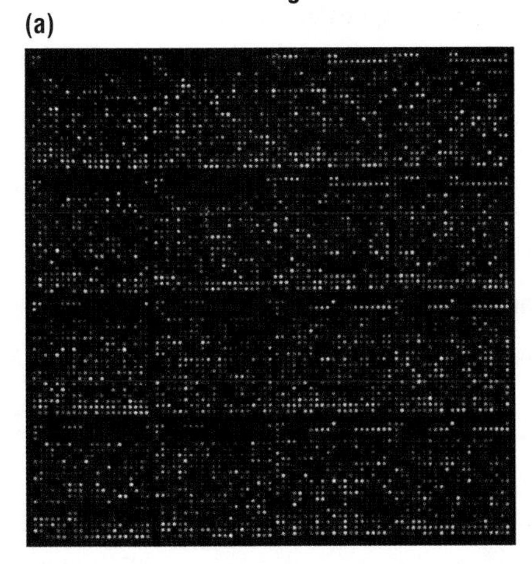

(b)

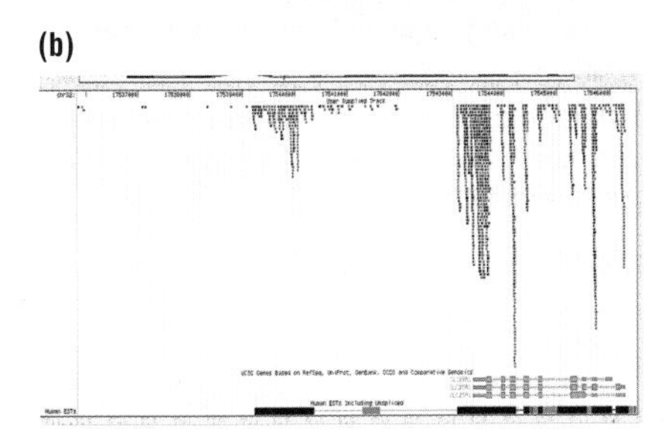

a. Assuming that this experiment used red dye for the control gene expression condition and green dye for the experimental condition indicate the parts of Figure B, panel (a) that indicate:
   i. A gene with unchanged expression
   ii. A gene with downregulated expression
   iii. A gene with upregulated expression
b. Which of the three expression possibilities in part (a) is most common, and how do you know?
c. What color would you expect to see if there is no gene expression?
d. Indicate the parts of Figure B panel (b) that indicate:
   i. No gene expression
   ii. A high level of gene expression
e. Do you think the gene structure predictions at the bottom right of Figure B panel (b) are correct? Explain your answer. (Gene predictions are represented by the blue boxes and lines at the bottom right of the figure, where a box represents an exon and a line represents an intron.)

6. Design an experiment to discover proteins that interact at the kinetochore. Assume you already know the sequence of one kinetochore protein and have the genome sequence of your experimental organism.

7. Which of these scientific questions would you use genomic approaches to address, and which would you address by investigating a single gene or protein? Explain your answers.
a. How does a particular gene mutation, which predisposes individuals to a certain type of cancer, affect transcription of other genes in a tumor?
b. Is a particular protein localized to a cell membrane?
c. Which amino acids in a particular protein are responsible for binding to its binding partner?
d. A new mushroom species has been discovered. Where should it be placed in the phylogenetic tree?

18.4 THE ENCODE PROJECT

1. What is the ENCODE project, and how does it differ from a single genomic experiment?

18.5 THE EVOLVING GENOME: EVOLUTIONARY FORCES

1. Explain why mutations in somatic cells have a minimal impact on evolution.

2. Mutations that change serine to threonine in an encoded protein tend to be neutral. Give another example of a neutral change, and explain your choice.

3. Evolution is a consequence of genetic mutation and selection. Explain each of the following terms and their evolutionary consequences.
a. Positive selection
b. Negative selection
c. Synonymous mutation
d. Genetic drift
e. Founder effect

18.6 THE EVOLVING GENOME: MECHANISMS OF VARIATION

1. Explain two ways in which transposable elements can shape evolution.

2. Briefly define the following terms in the context of genome evolution
a. Translocation
b. Deletion
c. Duplication
d. Inversion

3. Explain why chromosomal rearrangements may not be harmful for an individual yet may still lead to infertility.

4. What are horizontal gene transfer and vertical gene transfer, how do they differ, and which is more frequent?

18.7 GENE DUPLICATION AND DIVERGENCE OF GENE FUNCTION

1. Explain how genome duplication can lead to the development of gene families and pseudogenes.

2. Explain why it is preferable to compare amino acid sequences rather than nucleotide sequences when investigating the evolution of protein-encoding genes.

Challenge question

3. Homology:
a. Explain the terms homolog, paralog, and ortholog and discuss how these terms relate to one another.
b. Explain why it is incorrect to refer to 'percentage homology' when describing genes. What term would be used instead and why?

18.8 CHANGES IN CHROMOSOME STRUCTURE AND COPY NUMBER VARIATION

1. Examining synteny is a way to compare co-inherited blocks of DNA on chromosomes in different species. Use Figure 18.23 to answer the following questions.
a. Which human chromosome has the highest synteny with mouse chromosome 12?
b. Which mouse chromosome has the highest synteny with human chromosome 8?
c. Which chromosome is most syntenic between mouse and human?
d. Explain why it makes sense that your answer to (c) maintained a high degree of synteny during evolution.

18.9 EPIGENETICS AND IMPRINTING

1. What is epigenetic change, and how does it differ from a genetic change?

2. Describe the two major types of epigenetic change.

18.10 HUMAN GENETIC DISEASES: FINDING DISEASE LOCI

1. SNPs and microsatellites are types of sequence variation within a genome that can be used in disease mapping. Explain how these differ.

2. What are haplotype blocks, and why do they reduce the number of genetic markers that need to be used for association studies?

3. Compare and contrast linkage analysis and association studies.

4. Explain what is meant by a Mendelian disease, and contrast it to a multigenic or multifactorial disease. Use examples in your explanation.

Challenge questions

5. Haplotype blocks exist because recombination has not occurred within them. Recall chromosomal segregation (Chapter 7) and homologous recombination (Chapter 16) to help you answer these questions.

   a. Which two of these chromosomal markers are most likely to be linked? Explain your answer.

   b. Which two of these chromosomal markers are most likely to be linked? Explain your answer.

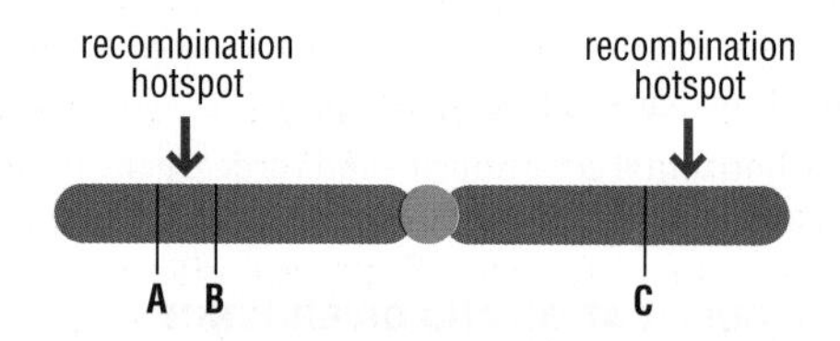

6. A patient has a number of cases of a particular disease in her close family. In the general population, the disease has been linked to known mutations in protein-coding gene X.

You sequence gene X in your patient. She does not have any of the previously characterized mutations, but does have three nucleotides that differ from the reference human genome sequence. The intron/exon structure of the gene is shown below, with the changes in the locations shown.

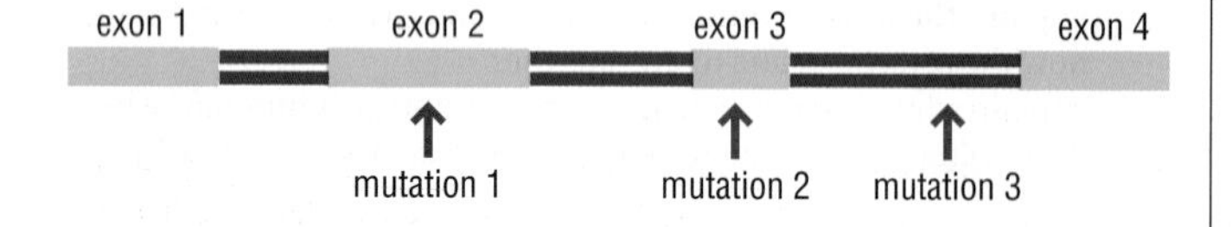

   a. Which change is least likely to be causing your patients' disease? Why?

   b. Describe a situation in which the change that you chose as your answer for (a) could be deleterious. Hint: think about splicing.

   You decide to investigate mutations 1 and 2 further. Alterations are shown in pink below.

part of exon 2:

normal 5' GAG ATG CAT TTA TAC CTG 3'

patient 5' GAG ATG CAT GTA TAC CTG 3'

part of exon 3:

normal 5' ATA GAG CGA TCT TAT ATC 3'

patient 5' ATA CAG CGA TCT TAT ATC 3'

   c. What effect does each nucleotide change have on the protein sequence?

   d. Which is more likely to be deleterious, mutation 1 or 2? Explain your answer.

   You then decide to compare the amino acid sequences of the normal human gene with those from other mammals. Your results are shown below.

part of exon 2:

human GLU MET HIS LEU TYR LEU
mouse GLU MET HIS LEU TYR LEU
chimp GLU MET HIS LEU TYR LEU

part of exon 3:

human ILE GLU ARG SER TYR ILE
mouse SER TYR MET ARG PHE TRP
chimp TRP TRP GLN MET TRP ASN

   e. Do you still agree with your answer to (d)? Explain why/why not.

   f. When performing a microarray hybridization for detection of SNPs, you realize that you made a mistake and your hybridization buffer had ten times more salt than the recipe required. Are the data resulting from your hybridization experiment still valid? Explain your answer.

18.11 HUMAN GENETICS: IMPACTS AND IMPLICATIONS

1. Discuss the 1000 Genomes Project.

2. Describe individualized medicine, and discuss how the development of cheaper and faster genome sequencing technologies will impact its development.

FURTHER READING

18.1 GENOME SEQUENCES AND SEQUENCING PROJECTS

Collins FS, Morgan M, Partings A. The Human Genome Project: lessons from large-scale biology. *Science*, 2003;**300**:286–290.

Kau AL, Ahern PP, Griffin NW, Goodman AL, Gordon JI. Human nutrition, the gut microbiome and the immune system. *Nature*, 2011;**474**:327–336.

Lander ES. Initial impact of the sequencing of the human genome. *Nature*, 2011;**470**:187–197.

Ley RE, Turnbaugh PJ, Klein S, Gordon JI. Human gut microbes linked to obesity. *Nature*, 2006;**444**:1022–1023.

Olson MV. The Human Genome Project: a player's perspective. *Journal of Molecular Biology*, 2002;**319**:931–942.

Tringe SG, Rubin EM. Metagenomics: DNA sequencing of environmental samples. *Nature Reviews Genetics*, 2005;**6**:805–814.

18.2 FINDING FUNCTIONS IN A GENOME: COMPARATIVE GENOMICS

Baetz K, Measday V, Andrews B. Revealing hidden relationships among yeast genes involved in chromosome segregation using systematic synthetic lethal and synthetic dosage lethal screens. *Cell Cycle*, 2006;**5**:592–595.

Boffelli D, Nobrega MA, Rubin EM. Comparative genomics at the vertebrate extremes. *Nature Reviews Genetics*, 2004;**5**:456–465.

Dunham I, Kundaje A, Aldred SF *et al.* An integrated encyclopedia of DNA elements in the human genome. *Nature*, 2012;**489**:57–74.

Friedman A, Perrimon N. Genome-wide high-throughput screens in functional genomics. *Current Opinion in Genetics and Development*, 2004;**14**:470–476.

Gerstein MB, Bruce C, Rozowsky JS, *et al.* What is a gene, post-ENCODE? History and updated definition. *Genome Research*, 2007;**17**:669–681.

Henson J, Tischler G, Ning, Z. Next-generation sequencing and large genome assemblies. *Pharmacogenomics*, 2012;**13**:901–915.

Mardis ER. A decade's perspective on DNA sequencing technology. *Nature*, 2011;**470**:198–203.

Jones SJ. Prediction of genomic functional elements. *Annual Review of Genomics and Human Genetics*, 2006;**7**:315–338.

Piskur J, Langkjaer RB. Yeast genome sequencing: the power of comparative genomics. *Molecular Microbiology*, 2004;**53**:381–389.

18.5 THE EVOLVING GENOME: EVOLUTIONARY FORCES

Bohne A, Brunet F, Galiana-Arnoux D, Schultheis C, Volff JN. Transposable elements as drivers of genomic and biological diversity in vertebrates. *Chromosome Research*, 2008;**16**:203–215.

Dujon B. Yeasts illustrate the molecular mechanisms of eukaryotic genome evolution. *Trends in Genetics*, 2006;**22**:375–387.

Eichler EE, Sankoff D. Structural dynamics of eukaryotic chromosome evolution. *Science*, 2003;**301**:793–797.

Frazer KA, Murray SS, Schork NJ, Topol EJ. Human genetic variation and its contribution to complex traits. *Nature Reviews Genetics*, 2009;**10**:241–251.

Nkongolo KK, Mehes-Smith M. Karyotype evolution in the Pinaceae: implication with molecular phylogeny. *Genome*, 2012; **55**:735–753.

Vinogradov AE. Evolution of genome size: multilevel selection, mutation bias or dynamical chaos? *Current Opinion in Genetics and Development*, 2004;**14**:620–626.

18.6 THE EVOLVING GENOME: MECHANISMS OF VARIATION

Coe BP, Girirajan S, Eichler EE. (2012). The genetic variability and commonality of neurodevelopmental disease. *American Journal of Medical Genetics Part C, Seminars in Medical Genetics*, 2012;**160**: 118–129.

Conrad B, Antonarakis SE. Gene duplication: a drive for phenotypic diversity and cause of human disease. *Annual Review of Genomics and Human Genetics*, 2007;**8**:17–35.

Fryxell KJ. The coevolution of gene family trees. *Trends in Genetics*, 1996;**12**:364–369.

Gogarten JP, Olendzenski L. Orthologs, paralogs and genome comparisons. *Current Opinion in Genetics and Development*, 1999;**9**:630–636.

18.7 GENE DUPLICATION AND DIVERGANCE OF GENE FUNCTION

Bailey JA, Eichler, E.E. Primate segmental duplications: crucibles of evolution, diversity and disease. *Nature Reviews Genetics*, 2006;**7**:552–564.

Blanchette M. Computation and analysis of genomic multi-sequence alignments. *Annual Review of Genomics and Human Genetics*, 2007;**8**:193–213.

Feil R, Fraga MF. Epigenetics and the environment: emerging patterns and implications. *Nature Reviews Genetics*, 2011;**13**:97–109.

Ferguson-Smith AC. (2011). Genomic imprinting: the emergence of an epigenetic paradigm. *Nature Reviews Genetics*, 2011;**12**: 565–575.

Ferguson-Smith MA, Trifonov V. Mammalian karyotype evolution. *Nature Reviews Genetics*, 2007;**8**:950–962.

Kumar S, Filipski A. Multiple sequence alignment: in pursuit of homologous DNA positions. *Genome Research*, 2007;**17**:127–135.

Liti G, Louis EJ. Yeast evolution and comparative genomics. *Annual Review of Microbiology*, 2005;**59**:135–153.

Pei J. Multiple protein sequence alignment. *Current Opinion in Structural Biology*, 2008;**18**:382–386.

Stanyon R, Rocchi M, Capozzi O, Roberto R, Misceo D, Ventura M, Cardone MF, Bigoni F, Archidiacono N. Primate chromosome evolution: ancestral karyotypes, marker order and neocentromeres. *Chromosome Research*, 2008;**16**:17–39.

18.10 HUMAN GENETIC DISEASE: FINDING DISEASE LOCI

Abecasis GR, Auton A, Brooks LD, DePristo MA, Durbin RM, Handsaker RE, Kang HM, Marth GT, McVean GA. An integrated map of genetic variation from 1,092 human genomes. *Nature*, 2012;**491**:56–65.

Altshuler D, Daly MJ, Lander ES. Genetic mapping in human disease. *Science*, 2008;**322**:881–888.

Daly AK. Genome-wide association studies in pharmacogenomics. *Nature Reviews Genetics*, 2010;**11**:241–246.

Evans WE, Relling MV. Moving towards individualized medicine with pharmacogenomics. *Nature*, 2004;**429**:464–468.

Gonzaga-Jauregui C, Lupski JR, Gibbs RA. Human genome sequencing in health and disease. *Annual Review of Medicine*, 2012;**63**:35–61.

Kamali F, Wynne H. Pharmacogenetics of warfarin. *Annual Review of Medicine*, 2010;**61**:63–75.

Kaye J. The regulation of direct-to-consumer genetic tests. *Human Molecular Genetics*, 2008;**17**: R180–183.

Kiezun A, Garimella K, Do R, *et al.* Exome sequencing and the genetic basis of complex traits. *Nature Genetics*, 2012;**44**,623–630.

McCarthy MI, Abecasis GR, Cardon LR, *et al.* Genome-wide association studies for complex traits: consensus, uncertainty and challenges. *Nature Reviews Genetics*, 2008;**9**:356–369.

Renegar G, Webster CJ, Stuerzebecher S, *et al.* Returning genetic research results to individuals: points-to-consider. *Bioethics*, 2006;20:24–36.

Yandell M, Huff C, Hu H, Singleton M, Moore B, Xing J, Jorde LB, Reese MG. A probabilistic disease-gene finder for personal genomes. *Genome Research*, 2011;**21**:1529–1542.

18.12 GENOME SEQUENCE REVALS THE PATH OF HUMAN EVOLUTION

Bustamante CD, Henn BM. Human origins: Shadows of early migrations. *Nature*, 2010;**468**:1044–1045.

Garrigan D, Hammer MF. Reconstructing human origins in the genomic era. *Nature Reviews Genetics*, 2006;**7**:669–680.

Tishkoff SA, Verrelli BC. Patterns of human genetic diversity: implications for human evolutionary history and disease. *Annual Review of Genomics and Human Genetics*, 2003;**4**:293–340.

USEFUL WEBSITES

- **National Human Genome Research Institute:** http://www.genome.gov/Education
- **OMIM:** http://www.ncbi.nlm.nih.gov/omim
- **Encode:** http://www.nature.com/encode/#/threads
- **The Cancer Genome Atlas:** http://cancergenome.nih.gov
- **The 1000 Genomes Project:** http://www.genome.gov/27528684
- **Survey of Human Structural Variation:** http://www.genome.gov/25521748

Tools and techniques in molecular biology

INTRODUCTION

Scientific discovery is driven by individuals who are able to look at a problem in a new way and have the right tools at their disposal. The Experimental approach panels throughout this book describe the process of discovery that has illuminated our understanding of some of the molecular processes and components described in the preceding chapters. Here we describe the tools and experimental methods that are used by biologists today in their continuing quest to understand how molecules in each cell carry out their many amazing functions.

Molecular biology became defined as a discipline around the middle of the twentieth century, when genetics and biochemistry were converging to provide an explanation for how genetic information was stored and harnessed by the cell. Although the use of model organisms in genetics was long established, we have since learned that many fundamental processes are so highly conserved from single-celled bacteria to multicellular eukaryotes that study of these model organisms yields information relevant to all domains of life. With the development of techniques to manipulate genetic material and reintroduce it into cells, it has become possible to study biochemical pathways as never before.

The term 'molecular biology' was coined to place emphasis on the molecules that carry out biological processes. To understand how something takes place in the cell, one must identify the molecules that participate in the process and determine their functions. A powerful approach to determining the components required for a given process, as well as the function of each macromolecule, is to replicate the process with purified components ***in vitro***, meaning in an artificial vessel (literally 'within the glass'). An mRNA copy of a gene, for example, can be readily synthesized in the test tube by mixing the DNA with purified RNA polymerase, nucleoside triphosphates (NTPs), magnesium, and the appropriate buffers. The fact that RNA synthesis will now occur without any other cellular components present tells us that the molecules in the test tube mixture comprise the minimal requirements for RNA synthesis in the cell. *In vitro* assays also make possible quantitative studies of binding constants, reaction rates, and the way in which particular mutations affect one or more steps in the process.

However, *in vitro* studies have their limitations: when a particular process is reconstituted in the test tube, some of the proteins that help to regulate the process in the cell might still be missing, or an important post-translational modification might not be found in a recombinant protein used in the experiment. Despite these caveats, much of what is known in molecular biology derives from *in vitro* experiments.

The advent of whole genome sequencing, and the resulting explosion of information about the nucleic acids and proteins in each organism, has made possible new types of experiments that exploit both cutting-edge techniques and the availability of vast amounts of information from thousands of organisms. The process of discovery has also been aided by other technological advances, including the use of robotics, the printing of slides containing thousands of different DNA or protein molecules, and the improvement in microscopy-based techniques. These methods have greatly increased the pace of discovery and have yielded profound insights into the inner workings of the cell. It is hard to believe that it was less than 75 years ago that DNA was first shown to be the carrier of genetic information in cells.

We begin this chapter by describing the organisms that have served as models for the study of biology in all types of cells and explore how cells and viruses can be grown in culture. From there, we explore the amplification and cloning of nucleic acids and how such cloning, in the context of **recombinant DNA** technology, can be used to both identify new genes and construct modified genes and chromosomes. We then present a large array of techniques used to study individual molecules, beginning with their purification from whole cells, through modern methods for identifying particular proteins and nucleic acids, before considering how we study their interactions with one another. The techniques used to image whole cells and study processes ***in vivo*** (in the living organism) on the one hand, and individual macromolecules on the other hand, are then explained. The chapter concludes with an explanation of how whole genomes are sequenced and how that information can be used to identify genes implicated in human disease.

19.1 MODEL ORGANISMS

Although millions of species exist in nature, only a few dozen, called model organisms, are extensively used in research. Many model organisms are studied not for their intrinsic importance to the economy or because they cause disease, but because they are easy to propagate and yet still can provide general insights into key molecular processes and components – insights which can be extrapolated to other organisms, including humans and human disease. This is because many molecules and molecular processes are evolutionarily conserved. For example, *Arabidopsis* is not a horticulturally important flowering plant, and *Caenorhabditis elegans* is not an agriculturally or medically important roundworm, but both organisms serve as laboratory models for the understanding of basic biological processes. Likewise, most strains of *Escherichia coli* are not human pathogens, yet they are similar in many ways to bacteria that are.

Model organisms have certain shared properties

The most widely used model organisms share several biological properties that have made them the standards for investigation. A model organism must have fairly simple and, ideally, very defined nutritional requirements, as well as a well-understood life cycle that allows it to be easily and safely grown in large quantities in the laboratory. A short generation time, small size and the ability to store or maintain stocks of the organisms for years are also advantages.

Nearly all important model organisms have been extensively studied by genetic analysis. This means that mutations which reduce or alter the function of a gene can be readily introduced, propagated, identified, and characterized. Since the advent of recombinant DNA technology, a further requirement is the ability to introduce modified genes into the organism, so that the effect of any type of mutation can be studied. The genomes of most model organisms have been sequenced, and the large communities of investigators studying these organisms have established databases containing genome sequence information, descriptions of mutant strains and their phenotypes, and the latest information about gene expression, enzyme activity, and protein-protein interactions. The availability of these data has greatly accelerated the pace of discovery and made possible new approaches to studying biology, as we will see throughout this chapter.

It is important to note that the nomenclature for gene and protein names varies from organism to organism. Refer back to page ii for a summary of the nomenclature rules adopted for some of the main model organisms discussed in this book.

Different model organisms have distinct advantages

Model organisms come from different kingdoms and phyla, and represent different evolutionary and biological properties, as depicted in Figure 19.1. Well-studied models include bacteria, single-celled eukaryotes, small animals, flowering plants, and lower vertebrates and mammals. How, then, does a scientist choose an experimental model in which to study a biological process? The answer depends on the question being asked, as well as the experimental tools that are available at the time. For example, much of our knowledge about DNA replication, transcription, and translation was originally gained from studies of bacteria. In part, this is because bacteria were among the first model organisms available, since it is easy to obtain large quantities of bacterial cells and they are easy to mutate. We now know that many aspects of the processes characterized in bacteria are universal to all organisms, validating the use of this model.

Phenomena that are unique to eukaryotes must, of course, be studied in eukaryotic cells. The inner workings of the nucleus of a cell can be readily studied in single-celled eukaryotes such as yeast, while the mechanisms by which a multicellular organism develops from a fertilized egg must be studied in a higher eukaryote, such as the fruit fly or the nematode. Similarly, phenomena unique to vertebrates or to mammals must be studied in a representative of that class of

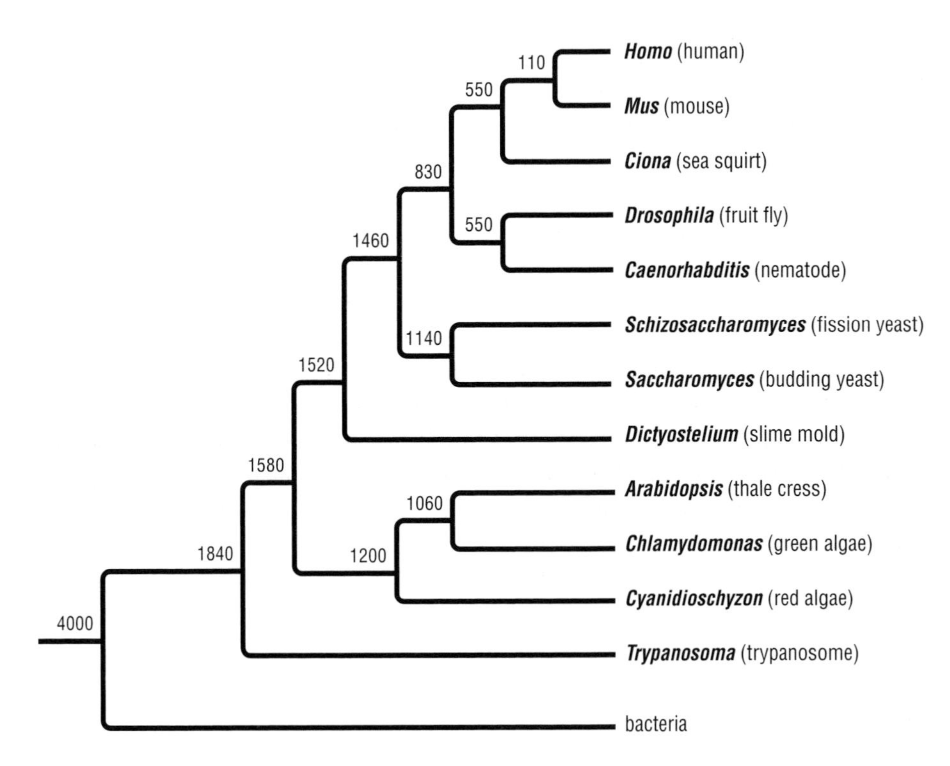

Figure 19.1 Phylogenetic tree of model organisms. The time of divergence, in millions of years, is indicated at the nodes of the tree (note that branch length is not proportional to time). For each organism, both the scientific name (italicized) and common name are provided.

Adapted from Hedges SB, *Nature Reviews Genetics*, 2002; **3**:838–849.

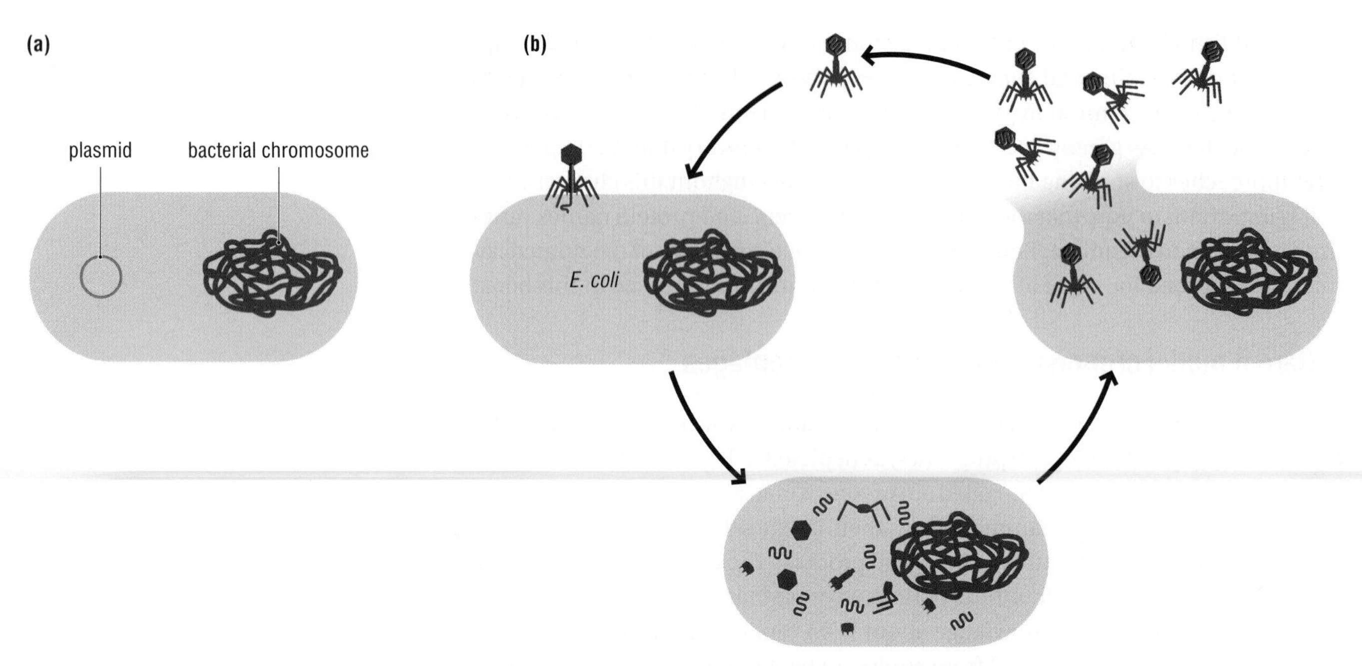

Figure 19.2 Bacterial cell with plasmid and phage. (a) Bacterial cell harboring a plasmid (in blue), a circular double-stranded DNA molecule that is separate from the bacterial chromosome. Plasmids are typically much smaller than the chromosome. (b) The lifecycle of bacteriophage T2: A T2 bacteriophage infects an *E. coli* bacterium by injecting its DNA (pink) into the bacterium. The bacteriophage genes are expressed in the bacterium, generating the various bacteriophage proteins and allowing the bacteriophage to make multiple copies of itself. Finally, the new bacteriophage particles are released by cell lysis, after which they can infect new cells.

organisms. Often, the choice of the appropriate model organism has been instrumental in making rapid experimental progress. The organisms mentioned in the following text are just a selected subset of the important and informative model organisms that have been used.

E. coli, *Caulobacter crescentus*, and *Bacillus subtilis* are commonly studied bacteria

The gut bacterium, *E. coli,* is the best-studied bacterium, and quite possibly the best-studied species. Many of the basic principles of molecular biology that are described throughout this book were first investigated in *E. coli.* This bacterium can be grown easily in the laboratory, and divides about every 20–30 minutes under optimal conditions. Importantly, genes can be readily introduced into *E. coli* using extrachromosomal DNA elements known as **plasmids**, which are illustrated in Figure 19.2a. These double-stranded, circular DNA molecules typically contain a DNA sequence that allows them to replicate, and hence be stably maintained, inside the bacterium. A plasmid can harbor multiple genes and be readily introduced into bacterial cells. Plasmids can also be easily manipulated by genetic engineering, making it straightforward to introduce mutant genes into bacteria and subsequently to study their effects. As we shall see below, plasmids can be used in eukaryotes as well.

E. coli cells can also be infected by a variety of viruses such as bacteriophage T2 and lambda, as depicted in Figure 19.2b. These bacteriophages provide another means by which to introduce genes into the cell. The dependence of bacteriophages on host cell function for their propagation has helped provide insights into basic cellular processes.

Many other bacteria have interesting or important biological properties that make them useful as model organisms. *B. subtilis* can form stress-resistant spores, and its sporulation program is a thoroughly studied example of a developmental process. Sporulation involves transcriptional activation of certain key genes, including a change in the sigma factor subunit of RNA polymerase. Thus, *B. subtilis* has given us insights into the regulation of transcription initiation.

The bacterium *C. crescentus* is another example of a bacterium with an interesting developmental process: it can adopt either of two different morphological forms. One form is the highly motile swarming cell that moves by means of a flagellum. The other form is a stalk cell, a stationary cell with a very different morphology. The switch between swarming and stalk formation is an example of a bi-stable biological switch. As we have seen in Chapters 5 and 7, *Caulobacter* has also been used to study bacterial cell division.

Saccharomyces cerevisiae, *Schizosaccharomyces pombe*, and *Dictyostelium discoideum* are commonly studied single-celled eukaryotes

Despite the enormous differences evident between highly complex eukaryotes, such as humans and single-celled eukaryotes, most fundamental pathways are remarkably conserved across all eukaryotes. Single-celled eukaryotes are therefore very attractive model systems for the study of eukaryotic molecular biology because of the ease with which they can be grown in culture and genetically manipulated. In addition, their short doubling time - about 90 minutes per generation for budding yeast - greatly enhances the rate at which studies of the effects of mutations can be performed.

The budding yeast *Saccharomyces cerevisiae*, which is of widespread practical use in brewing, baking, and production of biofuels, is also the best-understood model for a eukaryotic cell. Budding yeast has been a key model organism for understanding the fundamental processes of transcription, meiosis, recombination, cell division, and many other cell and molecular biology processes. Yeast has a fairly compact genome, with about 6000 genes in a genome of 12 million base pairs (Mbp). This type of yeast is distinct from other eukaryotes, and even from some other yeasts, in its manner of cell division: rather than simply dividing into two cells of equal size, budding yeast, as its name implies, reproduces by budding, in which the daughter cell first appears as a small bud on the mother cell, as depicted in Figure 19.3a. This small bud eventually grows into a daughter cell that separates from the mother cell and generates a bud of its own. Each *S. cerevisiae* cell can bud 20–30 times, depending on the strain.

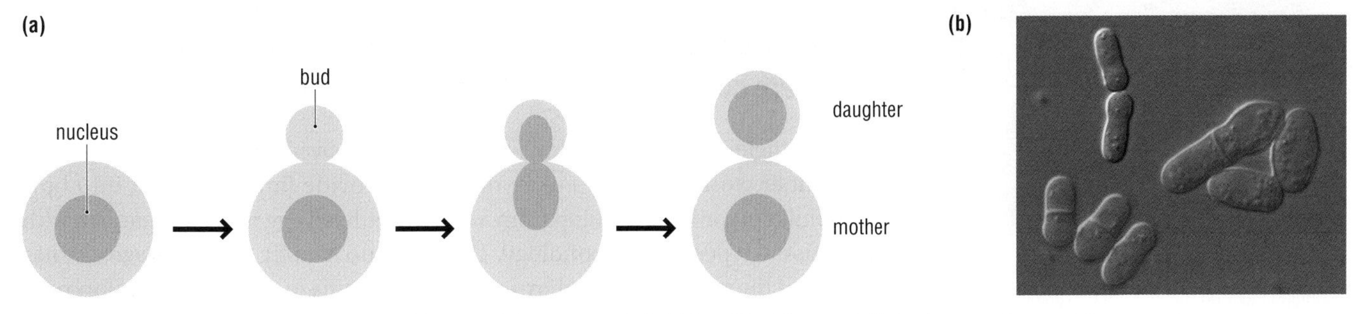

Figure 19.3 Budding and fission yeasts. (a) Diagram of the asymmetrical division of budding yeast. (b) A microscopy image of fission yeast.

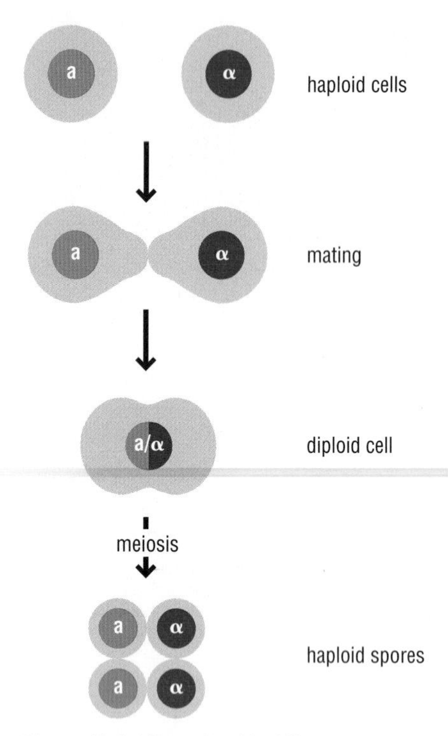

Figure 19.4 Life cycle of budding yeast. Yeast can exist as either haploid or diploid cells. Two haploid cells of opposite mating type, **a** and α, can mate with one another and fuse to form an **a**/α diploid cell. Under conditions of starvation, the diploid cell undergoes meiosis to produce four haploid spores: two **a** and two α.

The budding yeast life cycle, which is illustrated in Figure 19.4, lends itself to rapid genetic manipulation. As we learned in Chapter 8, this yeast can exist in both haploid and diploid form. There are two haploid yeast cell types: **a** cells and α cells. A haploid yeast cell can only mate with a cell of opposite mating type to produce an **a**/α diploid cell. Both haploid and diploid cells can be propagated indefinitely but, under conditions of starvation, diploid cells undergo meiosis to form four haploid cells: two **a** cells and two α cells.

The ability to work with a haploid organism makes the analysis of mutant phenotypes much easier, as only one gene copy has to be mutated (unlike the situation in diploid cells, where both gene copies have to be mutated for recessive phenotypes to emerge). Moreover, different mutations can be combined by mating, followed by meiosis and sporulation. Finally, homologous recombination in budding yeast is extremely efficient, allowing the generation of chromosomal gene fusions and gene deletions with relative ease. For this reason, there are commercially available yeast collections with all genes fused to a useful tag such as the green fluorescent protein (GFP), or where each of the non-essential genes is deleted.

Some phenomena common to higher eukaryotes that are absent from budding yeast can be studied in its evolutionarily distant cousin, *Schizosaccharomyces pombe*. This yeast reproduces by symmetrical cell division and is therefore called a fission yeast. A microscopy image of a fission yeast is shown in Figure 19.3b. Like budding yeast, fission yeast can be either haploid or diploid. However, DNA silencing and DNA structures of the centromeres in fission yeast are more similar to those found in multicellular eukaryotes than those of budding yeast. Moreover, because this yeast divides by fission, it has been used extensively in studies to elucidate how spatial cues, such as the cell's middle, are formed. The similarities and contrasts between the two yeasts are often used to provide a broader perspective on the fundamental properties of eukaryotic cells.

Most single-celled yeasts lack extensive extracellular signaling or multicellular organization. In contrast, the slime mold *D. discoidum* is often used to provide insights into such intercellular communication and signaling. *Dictyostelium* cells live as motile solitary cells for much of their life cycle. However, in response to a secreted signal from one cell, the nearby cells stream together to form a colony of functionally differentiated cells called a slug. The slug moves by the coordinated action of individual cells and can attach itself to a surface, whereupon it differentiates into a complex cellular structure called the fruiting body. Figure 19.5 illustrates how the fruiting body is composed of a stalk and a structure containing spores.

The secretion of the signal and the response of the cells to a signaling **gradient** for slug and stalk formation have been important in understanding the role of calcium in extracellular signaling in a relatively simple eukaryote.

Figure 19.5 Slime mold *D. discoidum*. Scanning electron micrograph of fruiting bodies formed by a colony of *D. discoidum cells.*
From David Scharf/Science Photo Library.

Drosophila melanogaster, *Caenorhabditis elegans*, and *Arabidopsis thaliana* are commonly studied multicellular eukaryotes

Studies of single-celled eukaryotes, while tremendously important for understanding the inner workings of all eukaryotic cells, provide limited insight into how multicellular organisms develop from a single fertilized egg into an organism with distinct tissue types and morphology. The common fruit fly, *D. melanogaster*, has been one of the most important experimental systems for studying development since it was first established as a model organism over 100 years ago. Indeed, *Drosophila* mutants, such as the one shown in Figure 19.6 a, have been instrumental in studying almost every process in animals.

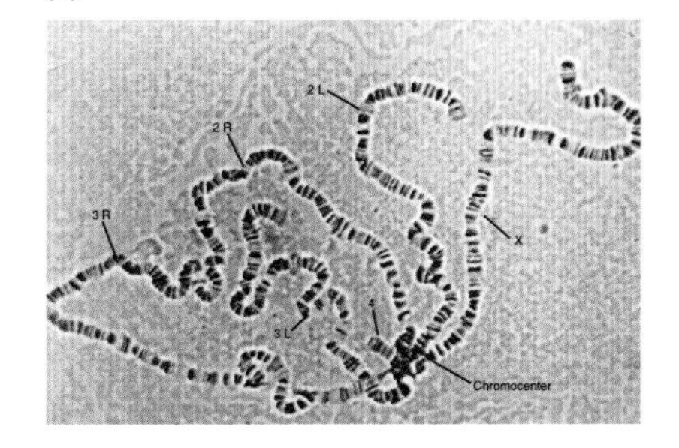

Figure 19.6 The fruit fly. (a) Wild-type (upper left) and mutant (lower right) fruit flies. The mutation in the mutant fly causes the third thoracic segment to develop wings instead of halteres. (b) Polytene chromosomes from salivary glands. The light and dark bands correspond to transcribed and non-transcribed regions, respectively. The numbers refer to the different chromosomes (L and R refer to the left and right arms, respectively, of chromosomes 2 and 3).

The entire development cycle in *Drosophila* is very rapid, with a fertilized egg taking just about a week to develop into a mature fly. A single female can lay thousands of eggs, thereby producing an enormous number of progeny. Random mutations can be induced by feeding flies chemicals that induce mutations or by subjecting flies to ionizing radiation, after which the many progeny can be screened for new and interesting phenotypes. Indeed, modern recombinant DNA techniques make it possible to engineer specific mutations into flies. By analyzing the effects of mutations on the mature organism, as well as on the different stages of development, genes that control development and differentiation as well as behavior can be identified. A number of genes that control development in all metazoans were first identified in the fruit fly.

A particularly interesting feature of the fly is its salivary glands. In order to produce large amounts of the material secreted by these glands, cells of the salivary gland undergo multiple rounds of DNA replication without undergoing mitosis. This results in the formation of the polytene chromosomes shown in Figure 19.6b, which have been useful in studying chromosome structure and gene expression.

➲ See Experimental approach 9.3 to learn more about the fruit fly salivary glands.

The nematode (or roundworm) *C. elegans* was developed as a genetic model organism because they have many fewer cells than other multicellular animals despite carrying out complex biological processes and behaviors. An adult worm has only 909 cells (excluding the germline), of which 302 are neurons. In addition, these cells arise from a pattern of precise and largely invariant cell divisions, which can be tracked by virtue of the worm's transparency. The cell lineage patterns, which were established by the early 1980s, were instrumental in understanding the important role of apoptosis in organismal growth and development. The *C. elegans* life cycle, which takes about three days, also has unique features that make it useful for genetic studies: the worms are either hermaphrodites or males and can reproduce by self-fertilization (hermaphrodites, as shown in Figure 19.7) or by mating (a hermaphrodite with a male).

An unanticipated benefit of developing *C. elegans* as an experimental model was the relatively recent discovery of the RNA interference (RNAi) machinery, which grew out of attempts to manipulate gene expression in worms using antisense RNA. Once it was realized that double-stranded RNA was causing targeted genes to be turned off, standard methods of worm genetics were rapidly used to identify components of the RNAi machinery. In fact, genes in *C. elegans* can be turned off simply by feeding the worm bacteria expressing the desired double-stranded RNA.

➲ See Experimental approach 13.2 to learn more about the discovery of RNAi.

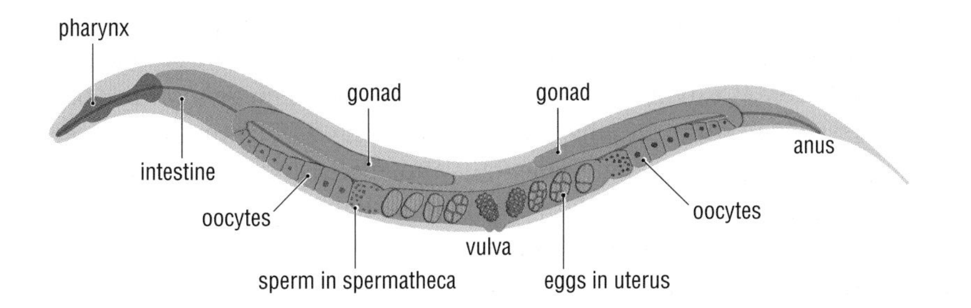

Figure 19.7 The *C. elegans* hermaphrodite. The hermaphrodite worm has two gonads that meet at a single vulva. The worm first produces sperm, which is stored in the two spermatheca, and then switches to producing oocytes. The oocytes are fertilized as they pass through the spermatheca.

Figure 19.8 ***Arabidopsis*** **as a genetic model system.** (a) Wild-type *A. thaliana*. (b) An *Arabidopsis* carrying a mutation that causes the plant to become bigger. Mutations such as this may be useful in the development of commercially valuable plants.

Image from Fig. 6a from Tominaga *et al.* Arabidopsis CAPRICE-LIKE MYB 3 (CPL3) controls endoreduplication and flowering development in addition to trichome and root hair formation. *Development*, 2008,**135**, 1335–1345..

The most prominent model organism from the plant world is the flowering plant, *Arabidopsis thaliana*, the wild type of which is shown in Figure 19.8a. Many other plants have served as model organisms, including maize, pea, tobacco, petunias, and snapdragons, but *Arabidopsis* is much smaller and has a shorter generation time. It is also a diploid and has a much smaller genome than agriculturally or horticulturally important plants (for example, some wheat cultivars are either tetraploid or hexaploid), making the isolation of mutants relatively straightforward. As we will see below, genetic manipulation of *Arabidopsis* also is quite straightforward since foreign genes can be introduced by infecting the plants with a soil bacterium, *Agrobacterium tumefaciens*. This soil bacterium harbors a tumor-inducing plasmid (Ti) into which genes can be cloned and introduced into the plant cells.

The zebrafish, frog, and mouse are model vertebrates

Several organisms serve as model vertebrates - that is, animals having a backbone and spinal column; these include reptiles, birds, and mammals. A small fish commonly found in pet stores known as the zebrafish (*Danio rerio*) has many features of other model organisms that make it amenable to study. Zebrafish are easy to raise and breed in the laboratory, with each female producing hundreds of eggs in a single clutch. Although the generation time of zebrafish is similar to that of the mouse (three to four months), the embryos develop in a matter of days. Importantly, embryo development occurs externally to the mother and the embryos are transparent, making it possible to observe the different stages of vertebrate embryogenesis and organ formation. A variety of techniques are available for genetic manipulations in zebrafish, including the ability to generate transgenic fish, as shown in Figure 19.9a. Such tools make it possible to even conduct large-scale screens in this organism.

Another vertebrate, *Xenopus laevis*, the African clawed frog, is very widely used in molecular biological and biochemical studies. Since oocytes and eggs of this frog are very large, it is possible to inject DNA and other molecules (and even whole nuclei) into them and examine the expression and function of these components. In addition, *Xenopus* eggs are a rich source for biochemical studies because the large quantities of concentrated cytoplasm can be easily isolated for biochemical reactions, as depicted in Figure 19.9b. The extracts can be fractionated to identify the important components for a given reaction. Cell cycle and DNA replication studies, in particular, have used such *Xenopus* extracts to identify many biological molecules. *Xenopus* is not, however, extensively used for genetic analysis because it is a partial tetraploid. The genome of the closely related but diploid frog, *Xenopus tropicalis*, has been sequenced, allowing genetic approaches to be more readily used in the study of this important vertebrate model system.

Some biological processes in humans can only be analyzed in other mammals. For example, the house mouse, *Mus musculus*, has provided invaluable insights into human biology and disease. It is often the experimental organism of choice for comparisons with humans because it too is a mammal, and yet can be cultivated for laboratory research. The mouse genome is similar to the human genome in size, and there is extensive sequence homology between genes. Mice offer an excellent system to study the effects of genes associated with human disease; they can be used to study both single gene traits and, increasingly, complex traits. Our understanding of the complexities of the mammalian immune system, for example, was largely derived from experimental studies in mice.

(a)

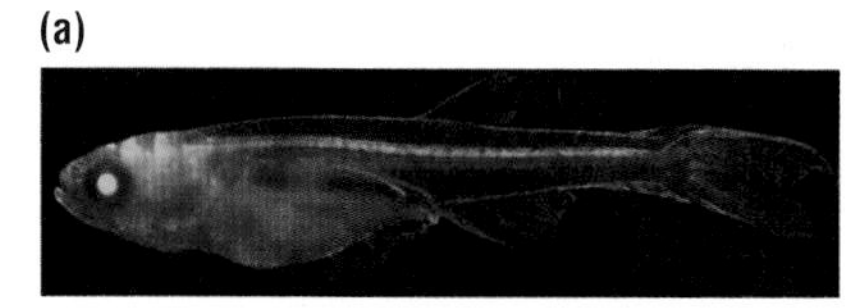

(b)

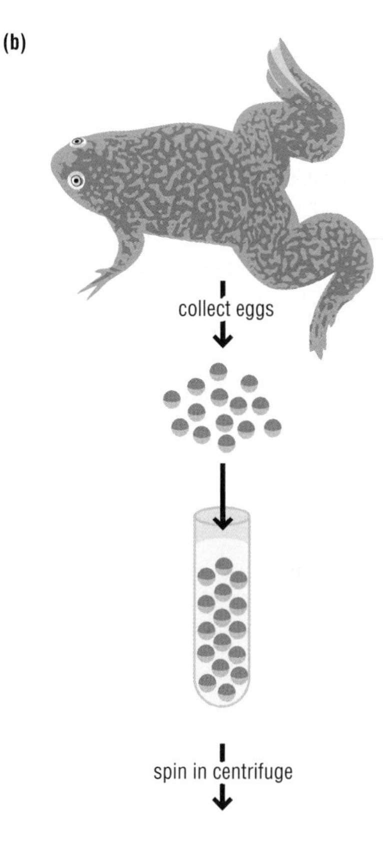

(c)

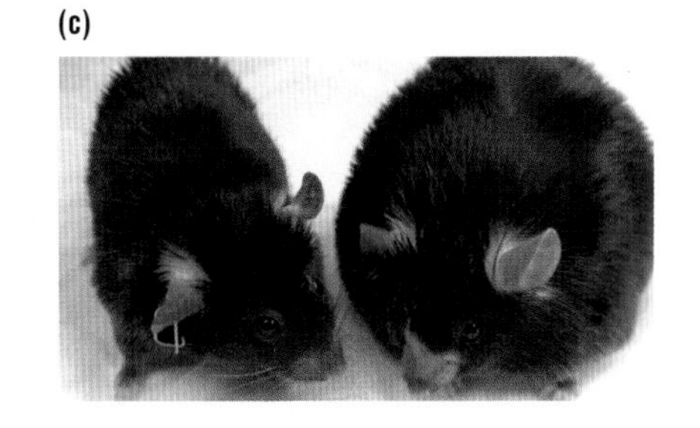

Figure 19.9 Vertebrate model organisms. (a) Zebrafish. The image shows a transgenic zebrafish expressing GFP under the control of regulatory elements of the *inhibitor of differentiation 1* gene (*id1*), which is expressed in the bones, skin, retina, pineal, optic tectum, and cerebellum. From http://www.sars.no/research/BeckerGrp.php. (b) Preparation of *Xenopus* egg extract. Each frog can lay hundreds of eggs. The eggs are collected in a centrifuge tube and spun to crush the eggs and separate the various egg components. The cytoplasmic fraction is often used for biochemical studies. (c) A mouse model for studying obesity. The mouse of the left is wild-type while the mouse on the right carries a mutation in the leptin gene that causes it to be obese.

Panel (a) courtesy of Mary Laplante. See also Kikuta *et al. Genome Research* **17**: 545–555, 2007; panel (c) courtesy of Shannon Reilly and the University of Michigan

There are multiple approaches that use mice to study those genes implicated in human disease. For example, it is possible to generate genetically modified mice by manipulating the genes in mouse embryos. The techniques for generating genetically modified mice are discussed in Section 19.5. The ability to disrupt single genes, thereby creating a 'knockout mouse', has been an invaluable tool in studying gene function in mammals and in understanding the molecular basis of an increasing number of genetic human diseases. Figure 19.9c illustrates just one such example, in which a mouse model was used to study obesity.

Some genes cannot be studied in this way because embryos in which both copies of the gene (on the two homologous chromosomes) have been knocked out fail to complete development and reach adulthood. In many of these cases, **conditional** mutants are used. These mutants make it possible to generate animals in which essential genes are deleted in only certain tissues. As the gene is not deleted in the entire mouse, embryonic development can occur. It is also possible to produce strains of **transgenic mice**, that is, mice whose cells express a foreign or mutant form of a gene.

Rats (*Rattus norvegicus*) are also used for studying molecular mechanisms of disease, but the genetic tools for rat research are less developed than in the mouse.

Tetrahymena is an example of a model organism studied for its special properties

Many of the model organisms described earlier in this section are representatives of particular phylogenic branches and together give a general overview of biological systems. Some model organisms, however, are chosen not for their common properties, but rather for some unique trait that can give insights into specific biological processes. A prime example is the ciliated protozoan, *Tetrahymena thermophila*, shown in Figure 19.10a. *Tetrahymena* is a single-celled eukaryote that is found in fresh water ponds. It is classified as a ciliate because it is covered with cilia, small appendages that allow it to move through the water. This single-celled eukaryote is unusual in that it has two nuclei: a diploid micronucleus, which serves as a germline 'vessel', and a polyploid somatic macronucleus, from which gene expression occurs.

The macronucleus is derived from a copy of the diploid micronucleus. This takes place in a developmental process that involves both gene amplification and the rearrangement of some regions of DNA and the elimination of other sections, as depicted in Figure 19.10b. As a result of the amplification, there is a very high copy number of small chromosomes, with many additional copies of both ribosomal RNA genes. These features led to the discovery of both self-splicing rRNA

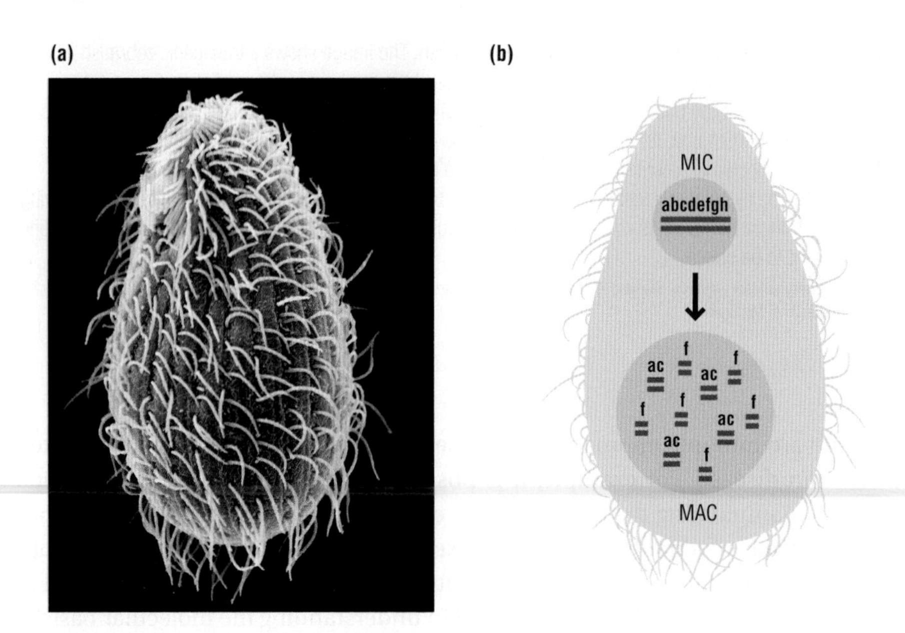

Figure 19.10 The ciliated protozoan *T. thermophila*. (a) Electron micrograph of *T. thermophila*. (b) *Tetrahymena* has two nuclei, a diploid micronucleus (MIC) that serves as a germline, and a polyploid somatic macronucleus (MAC). Gene expression occurs from the macronucleus, which is derived from a copy of the diploid micronucleus in a developmental process involving multiple DNA rearrangements.
Panel (a) courtesy of Aaron J. Bell.

and telomeres. The presence of both a transcriptionally silent micronucleus and a transcriptionally active macronucleus in the same cell allowed investigation of the role of histone modification in transcriptional regulation. These discoveries are described in Experimental approaches 4.2 and 10.2.

Even though its unique genome gymnastics make it seem unusual, studies in *Tetrahymena* have uncovered important processes and principles that are broadly applicable. Given the conservation of fundamental mechanisms in biology, what is learned in *Tetrahymena* and other unusual organisms often sheds light on processes across biological systems. For this reason, the exploration of diverse organisms to study specific biological processes makes it possible to gain new fundamental insights.

19.2 CULTURED CELLS AND VIRUSES

Detailed biochemical studies require samples of macromolecules that have been isolated from a cell. It is generally important to purify the molecule or molecules of interest from a homogeneous population of cells, in order to avoid variations in sequence or post-translational modifications that may be found among different strains or tissue types. Although proteins and nucleic acids can be isolated from relatively homogeneous large organs, such as a bovine heart or liver, most of the experiments underlying the information in this book were carried out with proteins and nucleic acids isolated from a single type of cell grown while suspended in liquid or immobilized on agar, plastic, or glass surfaces. We refer to this type of cell growth as **cell culture**. Some of the different types of cells that are grown in culture and the advantages and limitations of cell culture are the focus of this section.

Single-celled organisms can be grown in liquid culture or on agar plates

Single-celled organisms are most readily grown in culture. Model organisms such as the bacterium *E. coli*, and the budding yeast, *S. cerevisiae*, can be grown in liquid medium consisting of a mixture of nutrients that is buffered at a pH necessary for cell growth; such a culture is depicted in Figure 19.11a. The cells can also be grown on solid medium containing agar. (Agar is a substance added to culture medium that is liquid at high temperatures, but which solidifies at lower temperatures to form a gel on which microorganisms can grow.) When cells are spread on agar at low density, each cell grows and divides, giving rise to single colonies, as seen in Figure 19.11b. When cells are spread at high density, the proliferating cells coat the surface of the agar, giving rise to a lawn.

An advantage of culturing cells is that conditions under which the cells are growing can be controlled. For example, the cells can be continuously supplied with air by bubbling in gas or by vigorous stirring of liquid medium, or they can be grown completely in the absence of oxygen. Similarly, the cells can be maintained at different temperatures or grown with different nutrient sources.

Many species of bacteria, algae, fungi, and some archaea can also be grown in liquid culture or on solid medium. Other organisms, such as those that live in unusual environments, and even some that grow in soil, cannot be grown in the laboratory because we do not yet know how to create culture conditions under which they can grow and thrive. Our detailed knowledge of biochemical processes within cells has therefore been biased toward organisms that can be grown in culture.

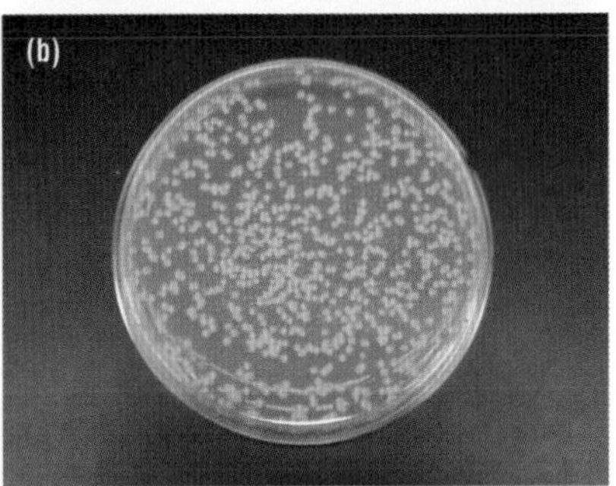

Figure 19.11 **Culturing of single-celled organisms.** (a) Growth of the yeast, *S. cerevisiae* in a glass tube. The solution is turbid because of the presence of yeast cells at about 10^8 cells/mL. The nutrient solution itself is yellow. (b) Growth of the yeast *S. cerevisiae* on an agar plate. A solution containing several hundred yeast cells was spread on the surface of a nutrient-containing agar plate; the agar is yellow because of the nutrients. On incubation, each cell gives rise to a single colony containing several million progeny cells.

Photos courtesy of Susan Michaelis, Johns Hopkins University School of Medicine.

Many differentiated cells from multicellular organisms need to undergo modifications to be maintained in culture

Eukaryotic cells from multicellular organisms can also be grown in culture. These cells, which can be derived by dissociating cells from tissues and are known as **primary cells**, are typically grown in a plastic culture bottle, as shown in Figure 19.12a, or on a plate under conditions that allow the cell to adhere to the surface. There are some cell types, such as blood cells, that can be grown for short periods while suspended in liquid culture.

The growth of cells in tissue culture is often limited by a number of factors. When grown on a solid surface, contact between cells may trigger cell cycle arrest so that cells cease to grow, a phenomenon known as **contact inhibition**. In addition, most primary cells can only divide a limited number of times due to certain intrinsic properties. For example, growth is limited in many cases by telomere shortening. Since telomeres shorten at every division, exceedingly short telomeres eventually cause **senescence**, during which cell division ceases. To avoid these problems, researchers often use special strains of cells, such as cancer cells, which can bypass senescence. These cultured cells, referred to **cell lines**, are often derived from tumors that actively express telomerase and have undergone other genetic changes that remove the normal inhibitions to continued cell division.

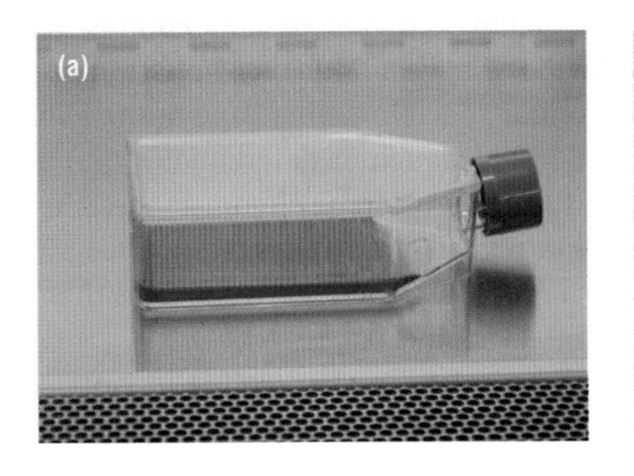

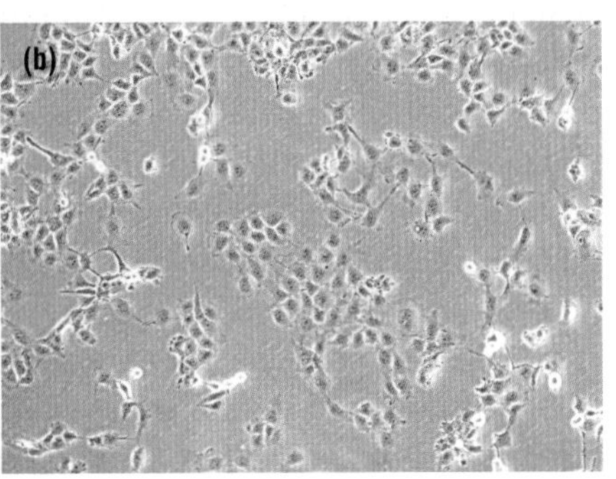

Figure 19.12 **Eukaryotic cells grown in tissue culture.** (a) Plastic flask containing mammalian cells in culture. The growth media is pink because of a pH indicator dye. (b) Layer of single tissue culture cells, as viewed by light microscopy.

Image kindly provided by Sin Urban, Johns Hopkins University.

Some commonly used cell lines		
cell line	**cell type**	**origin**
3T3	fibroblast	mouse
BHK21	fibroblast	Syrian hamster
MDCK	epithelial cell	dog
HeLa	epithelial cell	human
PtK1	epithelial cell	rat kangaroo
L6	myoblast	rat
PC12	adrenal gland cell	rat
SP2	plasma cell	mouse
COS	kidney cell	monkey
293	kidney cell, transformed with adenovirus	human
CHO	ovary	chinese hamster
DT40	lymphoma cell	chick
R1	embryonic stem cell	mouse
E14.1	embryonic stem cell	mouse
H1, H9	embryonic stem cell	human
S2	late embryonic cell	*Drosophila*
BY2	undifferentiated meristematic cell	tobacco

Figure 19.13 Commonly used cell lines. Adapted from Alberts *et al. The Molecular Biology of the Cell*, 4th edition. New York: Garland Science, 2002.

Cells can also be made to grow continuously in culture by infection with a cancer-causing virus.

Some commonly used cell lines are listed in Figure 19.13. While these cell lines have been central to the study of eukaryotic cells, the very changes that permit their continuous growth in culture also mean that these cells differ in fundamental ways from cells in intact tissues. These differences can affect the outcome of certain experiments.

Stem cells are the precursor cells to all cell types

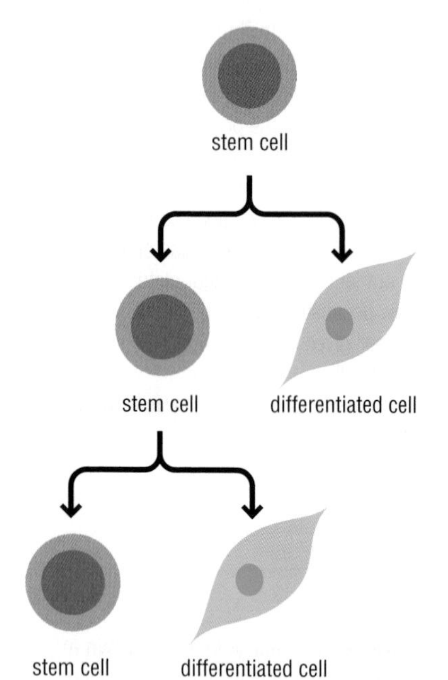

Figure 19.14 Stem cells are capable of self-renewal and differentiation. Stem cells divide in a special manner, yielding one daughter cell that is a stem cell and another daughter that can differentiate into a particular cell type.

Stem cells are a type of cell that can give rise to many different cell lineages. **Stem cells**, such as **embryonic stem cells (ES cells)**, are derived from early embryos and are considered to be **totipotent**, meaning that they have the potential to give rise to every cell type of the body. Other stem cells can generate a more limited repertoire of cell types; these are called **pluripotent cells**, as they give rise to a restricted set of cell lineages. Such pluripotent stem cells are present during development and are also found in fully developed animals. Adult stem cells, which are derived from a fully mature adult organism, are usually pluripotent, and the number of different cell types they can generate depends on the specific cell type. Stem cells from skin, hair follicles, and the intestine have been studied, but the best-characterized stem cells are those from blood.

Stem cells typically have an unlimited capacity to divide, in stark contrast to differentiated cells, which either do not divide or have a limited cell division capacity. When stem cells divide, they do so in a special manner: one of the two daughters forms a new stem cell while the other daughter differentiates into a particular cell type, as illustrated in Figure 19.14. Under some circumstances, the differentiated daughter can also give rise to other cell types and is referred to as a progenitor cell. Progenitor cells may have a more restricted set of cell fates than stem cells, and often have a more limited cell division capacity.

The best way to demonstrate that a cell is indeed a stem cell is to isolate the cell and implant it into a recipient animal. A stem cell will give rise to the

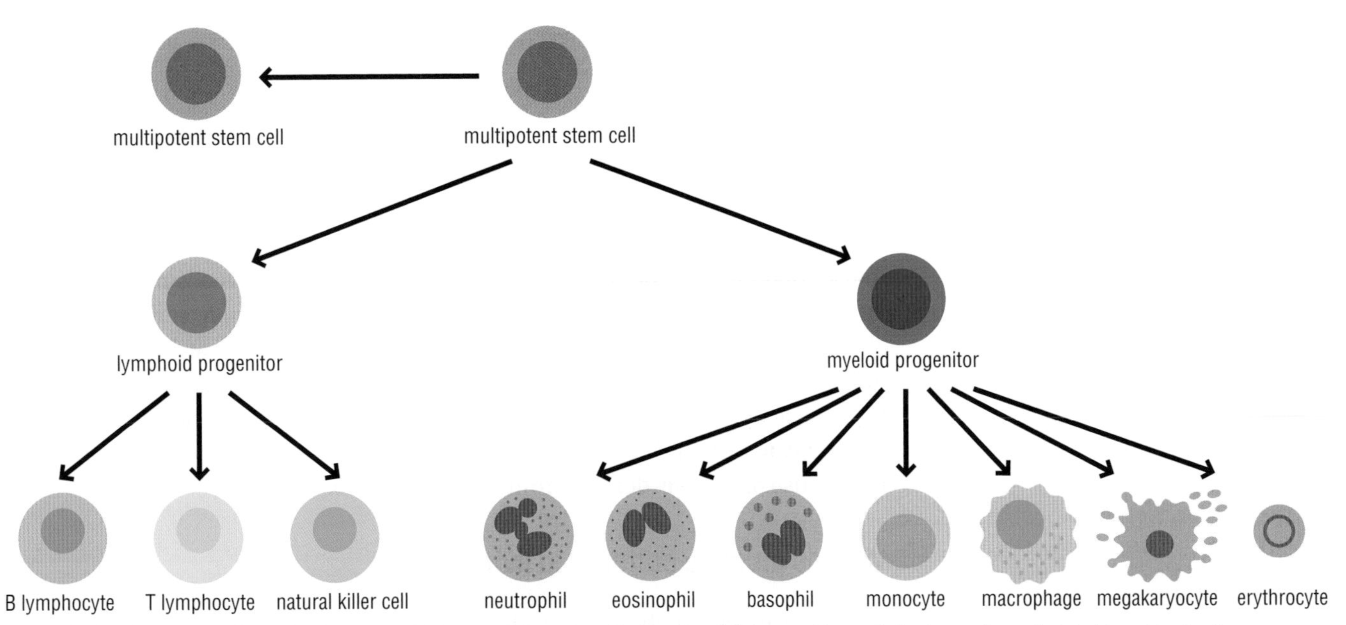

Figure 19.15 The differentiation of hemopoietic stem cells into specific blood cell lineages. Hemopoietic stem cells are isolated from blood or bone marrow. They are capable of self-renewal (that is, they can produce more stem cells) and also produce progenitor cells that can differentiate into many other cell types.

entire set of differentiated cells in this new setting. This type of transplantation is also the basis of bone marrow transplants that are carried out for the treatment of blood diseases and cancer. Blood cells have a limited lifespan, with some types lasting only a day. Blood cells must therefore be continually renewed throughout the life of the organism. In bone marrow transplants, stem cells that generate blood cells are isolated from the bone marrow, where they normally are found, and transferred to another individual. These blood precursor cells are referred to as **hematopoietic stem cells**, as **hematopoiesis** is the process of regenerating blood cells that occurs every day in adults. The differentiation of hematopoietic stem cells into specific blood cell lineages is one of the best-understood hierarchies of stem cell differentiation, and is illustrated in Figure 19.15.

ES cells can give rise to cell types needed to treat disease

ES cells have received a great deal of attention for their potential medical usage. These totipotent cells have been manipulated in the laboratory to generate specific cell lineages that might be used to treat specific diseases. For example, ES cells treated with certain growth factors in culture can cause >99% of the cells to differentiate into neurons. It is hoped that these neurons could be transplanted into patients with Parkinson's disease as a way to treat the loss of function of specific neurons, a characteristic of this disease.

One of the dangers of the use of ES cells, however, is the potential to generate tumors. For example, if less than 100% of the cells fully differentiate, the remaining undifferentiated cells could continue to grow as ES cells and later differentiate randomly to form a tumor. Indeed, direct injection of mouse ES cells into mice gives rise to a tumor known as a teratoma, which comprises a mixture of many different differentiated cell types.

iPS cells are a special form of stem cells derived from differentiated cells

In recent years, a new method has been developed to form totipotent cells for medical research. These cells are termed iPS cells: induced pluripotent stem cells. Researchers found that introducing four specific transcription factors into fully differentiated skin cells would 're-program' the skin cells, turning them into stem cells; that is, the transcription factors induce the expression of a set of genes that is needed in early embryonic development. iPS cells seem to be in many ways equivalent to ES cells.

Importantly, this new approach makes it possible to treat an individual with differentiated cells derived from their own tissue. Called autologous transplants, these cells are less likely to be rejected by the host's immune system. Moreover, the ability to form iPS cells from adult tissue bypasses the ethical issues surrounding the use of ES cells derived from human embryos.

Viruses must be cultured in living cells

As we saw in earlier chapters, studies of viruses that infect bacteria or eukaryotic cells have provided important insights into basic cellular mechanisms. Viruses are also a valuable means of introducing genes into cells. For example, the introduction of the specific transcription factors required to form iPS cells is often done with viral vectors. In some viruses, the genome is encoded by a DNA molecule, while other viruses (such as retroviruses) use an RNA molecule as their genome. Regardless, however, none of the viruses are free-living life forms, and they must be propagated within cells.

➔ See Section 9.4 for a description of the retrovirus life cycle.

Most viruses can infect just a few types of cells at most, so the ability to grow a particular virus depends on the ability to culture the type of cell it grows in. Perhaps the most important advance in developing a vaccine against the crippling disease polio was the development of methods to grow poliovirus in cultured human cells.

Viruses that have been propagated in cultured cells can be isolated in several ways. Since viruses infect a cell and harness the cell's machinery to produce many virus particles, one isolation method is to infect cultured cells with virus and harvest the virus-producing cells that result. The cell membrane can then be disrupted with detergents to release the virus particles, which can be separated from the cell debris by a variety of methods. With lytic viruses, which cause the cell membrane to burst in order to release the virus particles, cells can be infected and grown until the cells burst and release virus particles into the culture medium, or onto the surface of a Petri dish, where lysis of cells in a lawn produces a clear circle called a plaque as depicted in Figure 19.16. The virus particles present in a plaque or shed into the culture medium can then be purified.

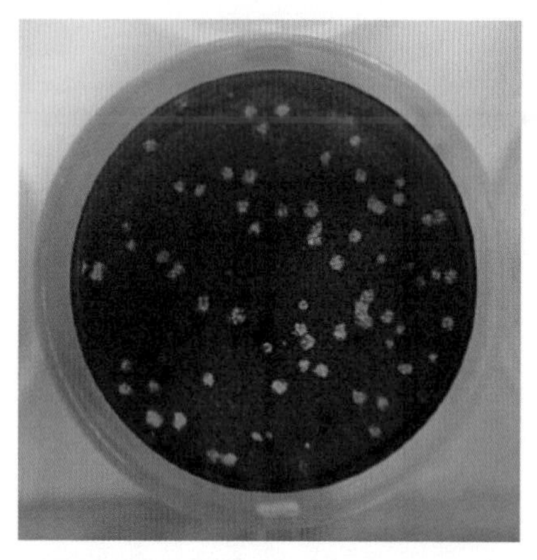

Figure 19.16 Viral plaque assay. To determine the concentration of viruses in a sample, appropriate cells are grown on a plate and then infected with the virus. As a virus infects a cell, it produces many virus copies, which after cell lysis infect neighboring cells. As a result, a zone of dead cells is formed. In the example shown, live cells take up a purple dye, and the white areas are plaques where cells died due to viral infection. The number of plaques is representative of the number of live viruses in the solution.

19.3 AMPLIFICATION OF DNA AND RNA SEQUENCES

The ability to manipulate DNA to generate many copies, or clones, of wild-type and mutant derivatives, and to determine the sequence of a particular DNA or RNA fragment, has been central to the advancement of molecular biology. Both molecular cloning and sequencing of DNA and RNA rely upon the generation of many copies of the DNA or RNA of interest from a small amount of starting material. In this section, we discuss commonly used methods for manipulating and amplifying DNA and RNA sequences.

The polymerase chain reaction can be used to amplify a specific DNA sequence

The polymerase chain reaction, commonly referred to as PCR, can be used to generate thousands or even millions of copies of a particular sequence of double-stranded DNA, beginning with just a few copies of starting material. The method relies upon repeated cycles of DNA strand separation followed by replication of each DNA strand. This doubling of the number of DNA molecules with every cycle leads to rapid amplification of the DNA sequence. Virtually any DNA segment can be amplified by PCR.

The boundaries of the DNA sequence to be amplified are set by two synthetic DNA oligonucleotides, each complementary to the 3′ end of one of the DNA strands in the DNA fragment of interest. PCR itself then proceeds as illustrated schematically in Figure 19.17. In the first cycle of PCR, the two strands of a DNA template are first separated by heat denaturation, followed by cooling of the sample. As the temperature drops, the two oligonucleotides (also called primers) hybridize to the denatured DNA, each complementary to a different 3′ end of the sequence to be amplified. DNA polymerase, which is present in the reaction along with dNTPs, utilizes the oligonucleotides as primers to replicate the complementary strand. This procedure will produce two copies of the desired DNA sequence. The cycle is then repeated, as once again the DNA is denatured so that primers can hybridize and be extended.

With each cycle, the number of template strands doubles. It is this amplification that makes PCR such a powerful experimental tool: it allows over one million copies of a single double-stranded template to be generated in just 20 cycles. PCR is usually done in an automated manner in a machine that cycles between different temperatures, using a DNA polymerase that is stable at high temperatures so that repeated cycles of heat denaturation do not inactivate the enzyme.

Although the PCR method is extremely powerful, a few caveats should be noted. Oligonucleotide primers can sometimes hybridize to sites other than those intended due to partial sequence complementarity. When this happens, the wrong DNA sequence could be amplified. Thus, primers need to be designed with care, and there are specialized computer programs to assist with this. In addition, misincorporation of bases by the polymerase can introduce random mutations during DNA synthesis. If such a mutation is introduced in an early cycle, the mutation will be amplified at each cycle. Finally, amplification of long DNA segments, typically over 4000–5000 base pairs (bp), may require specialized DNA polymerases and conditions.

While PCR is most commonly used to produce faithful copies of a DNA sequence, it can also be used to introduce random mutations by carrying out the PCR reaction under special conditions. For example, altering the salt concentration affects the fidelity of the DNA polymerase, and altering the ratios between the four deoxyribonucleotides in the reaction favors misincorporation of bases. The resulting population of DNA fragments containing random mutants can then be introduced into cells through a variety of methods and used in genetic screens for altered protein function.

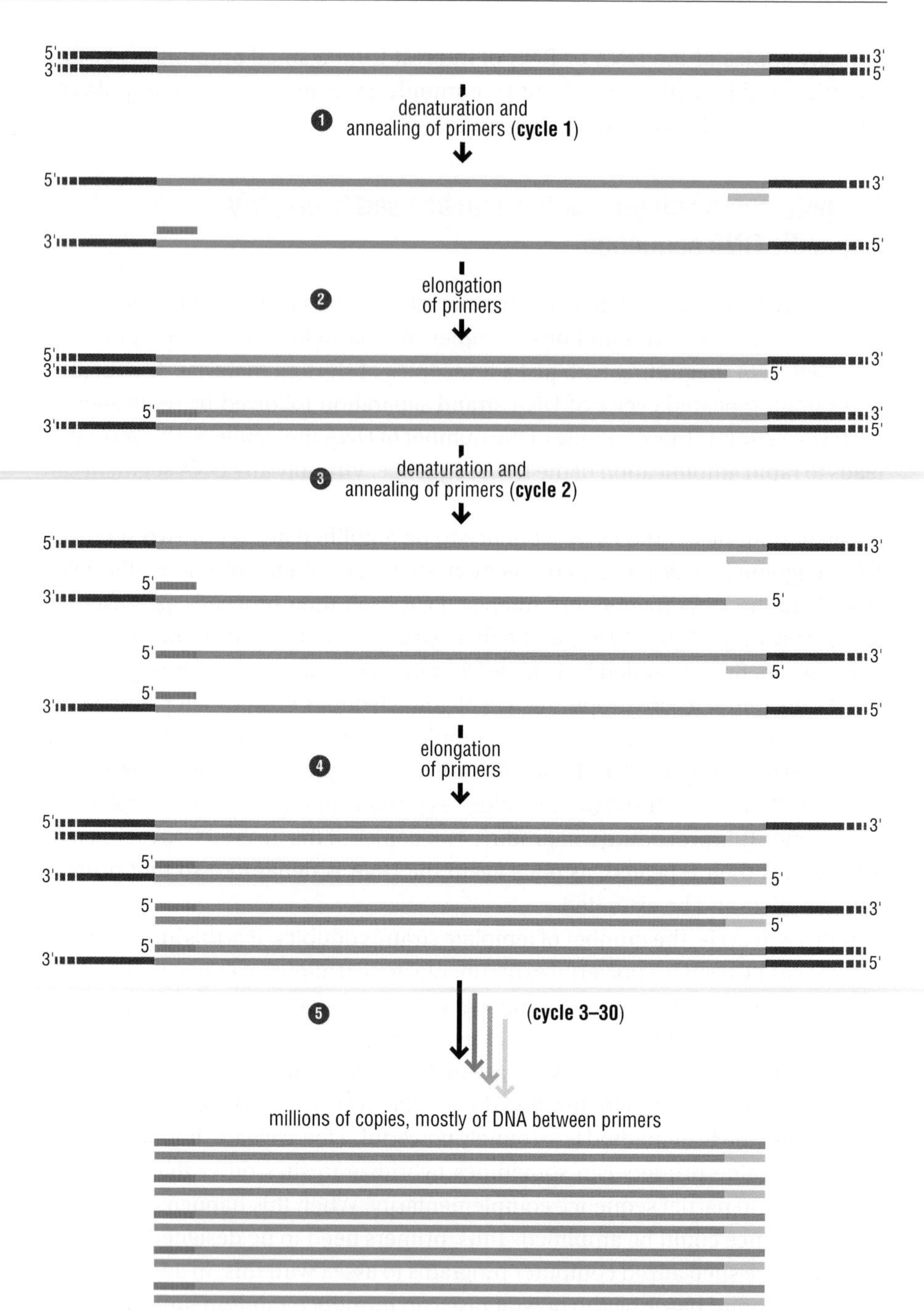

Figure 19.17 The use of PCR to amplify a specific DNA fragment, using multiple rounds of primer annealing and extension. Oligonucleotides (in blue and green) are designed to be complementary to each end of the fragment that is to be amplified (in orange). Such oligonucleotides, also known as primers, are typically around 20 bases in length. In the first PCR cycle, the DNA is denatured and the primers anneal to the complementary site on the single-stranded DNA (step 1). These primers then serve to prime DNA replication, usually by a heat-resistant DNA polymerase from the bacterium *Thermus aquaticus* (step 2). After the DNA is fully replicated, a second PCR cycle begins where the DNA is again denatured and a second round of primer annealing and extension takes place (steps 3 and 4). With every successive round (step 5), the amount of new DNA present in the reaction mixture is doubled such that the DNA region between the primers is preferentially amplified.

PCR can be used to add sequences that are not in the template DNA

There are a number of different applications of PCR . For example, it is sometimes useful to add additional nucleotides that are not found in the template DNA to either end of the amplified DNA fragment. These additional sequences are particularly useful for subsequent steps in cloning (Section 19.4), as well as in a variety of other applications. Additional nucleotides can be added to the amplified DNA fragment by synthesizing a primer that contains additional nucleotides of the desired sequence at its 5′ end. As a result, the 5′ end of the primer is not complementary to the template DNA sequence. However, as the DNA is amplified over successive PCR cycles, the sequences of the primer become part of the new DNA template

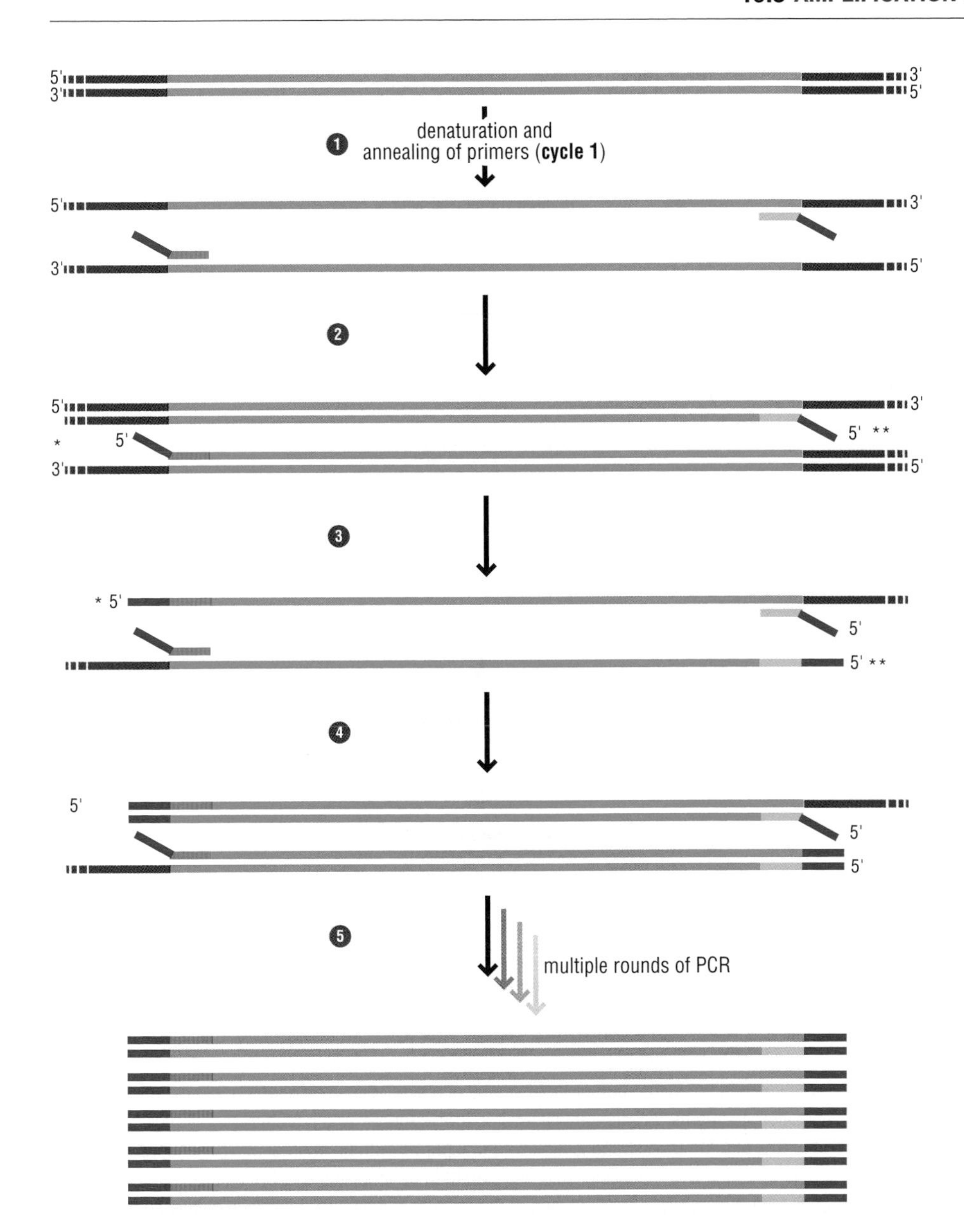

Figure 19.18 Adding specific sequences to DNA that is amplified by PCR. To add DNA sequences to the ends of a DNA fragment, one of both primers have unique sequences on the 5′ ends (purple) in addition to a region of complementarity to the template DNA (green or blue). During the first round of PCR only the complementary sequences will anneal to the template DNA (step 1). However, once these primers are extended by DNA polymerase (step 2), the new sequences become part of the new template DNA (* or **). In the next round of amplification (steps 3 and 4; for simplicity only the newly synthesized DNA is shown), the template DNA that was made in the first round will contain the added sequences, which will be incorporated into the amplified DNA in each successive amplification cycle (step 5).

and are incorporated into the PCR products, as illustrated in Figure 19.18. The additional DNA that has been added to the amplified fragment can then be used in subsequent manipulations, as we shall see in Section 19.4.

RNA amplification depends on a DNA intermediate

In some cases, investigators wish to generate a particular RNA molecule or obtain the full-length sequence of an RNA transcript. Since RNA cannot be directly cloned, a complementary DNA (cDNA) copy must be generated and then used in rounds of DNA amplification similar to PCR. In contrast with PCR amplification of a DNA sequence, though, the ends of a given RNA molecule of interest may not be known, for example, because of alternative splicing or there being multiple possible promoters from which the RNA may be transcribed.

This problem is overcome by a modification of the PCR technique for DNA cloning called RACE (for rapid amplification of cDNA ends) that makes it possible

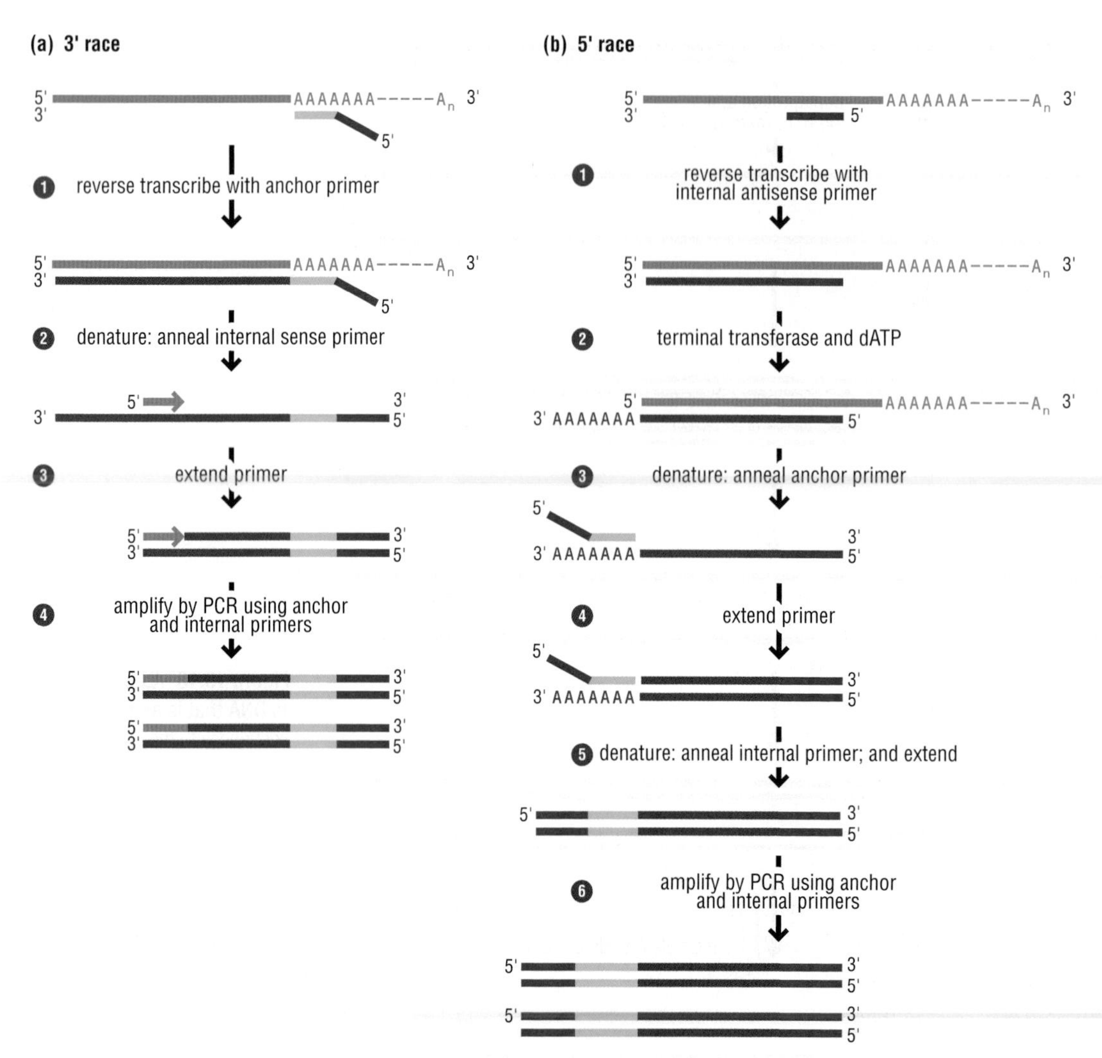

Figure 19.19 3′-and 5′-RACE. (a) 3′-RACE. The mRNA (in blue) is annealed to an anchor primer that is made of a sequence complementary to the mRNA's poly(A) (poly(T), in green) and a sequence of choice (purple). Once annealed, the primer is extended by reverse transcriptase, which creates a DNA molecule that is complementary to the mRNA (step 1). The DNA–RNA hybrid is denatured, and the DNA strand is annealed to a primer with an internal mRNA sequence (in blue). The primer is extended by a DNA polymerase (step 2) to create a double-stranded DNA molecule. This DNA can then be amplified by PCR (step 3) that includes the anchor primer (purple) and the internal primer (blue). The end result (step 4) is a DNA molecule, one of the strands of which is identical in sequence to the original mRNA. (b) 5′-RACE. This reaction is similar in principle to 3′-RACE, except that the anchor primer is on the 5′ end. The reaction begins with an internal primer that is complementary to the RNA sequence, which extended by reverse transcriptase (step 1). The 3′ end of the newly synthesized DNA is then extended by terminal transferase, which adds a poly(A) tail (step 2). This DNA strand can then be copied using an anchor primer (step 3) and further amplified by PCR (steps 4–6) as in 3′-RACE.

to amplify an RNA sequence even when its precise 3′ and 5′ boundaries are not known. The RACE method can be used to obtain the nucleotide sequence of an RNA transcript from a short known sequence within the transcript to either the 3′ or 5′ end. By combining these two approaches, known as 3′-RACE and 5′-RACE, one can determine the complete sequence of the RNA transcript.

For the purpose of this discussion, we will consider the amplification of an mRNA molecule. Since mRNAs typically have a poly(A) tail at their 3′ end, the presence of this characteristic stretch of nucleotides can be exploited in 3′-RACE, as illustrated in Figure 19.19a. A primer is designed such that its 3′ end is a poly(T) sequence – a sequence that is complementary to the poly(A) tail. Additional nucleotides may be added to the 5′ end of the primer to facilitate subsequent manipulations, as described above for DNA PCR.

In the first step, a cDNA copy of the RNA is made using a retroviral DNA polymerase called reverse transcriptase (Section 6.4), which uses the RNA molecule as a template to synthesize a complementary DNA strand. The poly(T) primer, called an anchor primer, is annealed to the RNA and then extended by reverse transcriptase to form an RNA–DNA hybrid. If a mixture of mRNAs is used as template, as would be the case if the mRNA is purified from a cell, this process will result in many different RNA–DNA hybrids, one for each type of polyadenylated RNA present. To convert the desired mRNA into cDNA, a second primer is used, this time using a sequence that is internal to the mRNA of interest (which would be complementary to the newly synthesized DNA). The RNA–DNA hybrid is denatured, the internal primer is allowed to anneal to the DNA strand, and the reaction now proceeds in a manner similar to PCR. Multiple rounds of amplification will generate a cDNA product, one strand of which is identical to the portion of the original RNA molecule that lies between the two primers used in these reactions.

The process is somewhat more complicated if one wants to convert the 5′ end of the RNA to DNA. In this method, called 5′-RACE, the sequence at the 5′ end of the RNA molecule is unknown. To overcome this problem, an artificial 5′ end is generated. As illustrated in Figure 19.19b, the first step of 5′-RACE is to anneal a primer that is complementary to an internal sequence within the RNA. This antisense primer is extended by reverse transcriptase, which synthesizes a DNA copy of the RNA molecule from the primer all the way to the 5′ end of the RNA molecule. A poly(A) tail is then added to the 3′end of the complementary DNA strand using an enzyme called terminal transferase. The subsequent steps are similar to 3′-RACE: the DNA–RNA hybrid is denatured, and an anchor primer complementary to the poly(A) sequence is used to prime synthesis of the cDNA. With two complementary DNA strands, one a copy of the original RNA and one the antisense strand, the DNA is amplified using the original primer to the internal sequence and the primer complementary to the added poly(A) tail.

19.4 DNA CLONING

The ability to propagate and re-engineer specific DNA fragments in cells has been central to the study of molecular biology in the past few decades. Thanks to these methods, scientists are able to change the sequence and expression of any gene or DNA region of interest. In this section, we discuss methods for cloning DNA sequences.

Specific DNA fragments can be amplified and modified by cloning into plasmid vectors

The process of isolating and propagating specific DNA regions is called **DNA cloning** because many copies of the same DNA sequence are generated from the original copy. The first step in cloning is to introduce the DNA fragment into an appropriate DNA **vector** that can be used to propagate the desired DNA fragment independently of the host cell chromosome. Since the final product is a DNA molecule containing both the inserted fragment and the vector DNA, the methods for generating and manipulating the cloned gene are referred to as **recombinant DNA** techniques.

Vectors for cloning can be plasmids, which are circular double stranded DNA molecules that can replicate independently of the chromosome, or viruses. These vectors contain specific elements that make them suitable for cloning. For example, a vector may contain a specific DNA sequence that allows it to replicate within its target host: plasmids that are propagated in *E. coli* contain an origin of replication that is recognized by the bacterium's DNA replication machinery. These vectors usually also carry a gene known as a **selectable marker** that allows only the host cells containing the vector to grow under the specific culture conditions. For example, vectors can carry an antibiotic-resistance gene that allows only the vector-containing cells to grow in the presence of an antibiotic drug that would otherwise kill the host, or a gene encoding a metabolic enzyme that allows the cells to grow in specific medium that would otherwise not sustain cell growth. DNA cloning is often carried out using bacteria, although other host organisms such as yeast or cultured cells are also used.

Once a DNA segment is cloned, it can be further manipulated according to the investigator's needs; a cloned DNA fragment is more easily manipulated than the equivalent DNA sequence situated in its 'native' chromosomal location. For example, a cloned gene can be introduced into a cell containing a mutation or deletion of a particular gene, and one can then test whether the introduced gene complements the mutation (that is, restores a particular gene activity lost by a given mutation). It is also possible to fuse the gene of interest to a **reporter gene**, such as a gene coding for GFP (see Section 19.10) or to a short peptide that facilitates detection or purification of the protein. Finally, by placing the gene next to a strong promoter, it is possible to direct high levels of gene expression, thus making it easier to purify the protein of interest.

The first step in cloning typically involves using PCR, as described in Section 19.3, to make many copies of the DNA fragment of interest. Once the desired DNA is amplified, there are two main strategies for cloning a DNA fragment:

- ligation – in which the two ends of the DNA fragment are cut with restriction enzymes and the fragment is then 'glued' to the vector DNA by a DNA ligase;
- recombination – in which the DNA fragment is inserted into the vector by recombination enzymes by virtue of sequence homology.

We begin by describing the use of restriction enzymes, which are an important tool for cloning by ligation.

Restriction enzymes cleave DNA at specific sites

One of the earliest tools developed for manipulating DNA was the use of bacterial proteins known as restriction enzymes that recognize and cleave double-stranded DNA at specific sequences, as illustrated in Figure 19.20a. These sequences, known as restriction sites, are typically palindromic sequences of 4–8 bp in length. Bacteria use these enzymes as part of a defense mechanism against foreign DNA, but scientists have adapted them as tools for manipulating DNA in the test tube.

There are hundreds of restriction enzymes that differ from one another in the DNA sequence that they recognize, and in the location of the cleavage sites relative to the restriction site. Depending upon the restriction enzyme, the DNA cleavage reaction may leave an end that is blunt or that has a small region of 5′ or 3′ overhanging bases, as shown in Figure 19.20b. The overhangs are useful during cloning, as we will see next. Restriction enzymes are also used for diagnostic purposes (see Section 19.9).

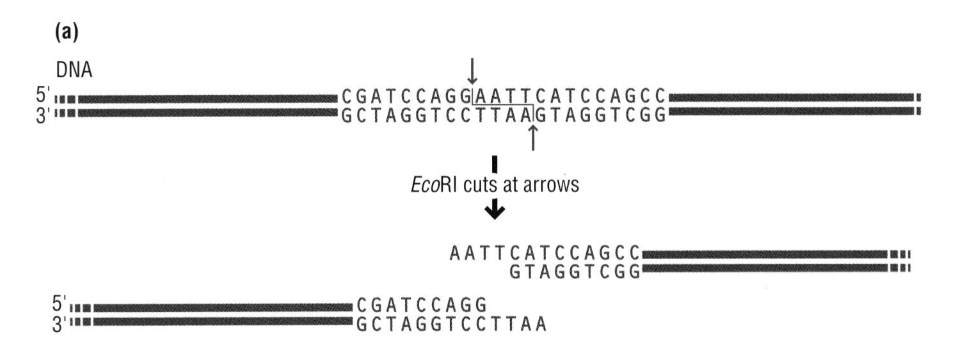

(b)

microorganisms	restriction enzymes	cleavage sites	cleavage products
Escherichia coli	*EcoRI*	5'-GAATTC-3' 3'-CTTAAG-5'	5'-G AATTC-3' 5'-CTTAA G-3'
Haemophilus influenzae	*HindIII*	5'-AAGCTT-3' 3'-TTCGAA-5'	5'-A GACTT-3' 5'-TTCGA A-3'
Haemophilus parainfluenzae	*HpaI*	5'-GTTAAC-3' 3'-CAATTG-5'	5'-GTT AAC-3' 5'-CAA TTG-3'
Klebsiella pneumoniae	*KpnI*	5'-GGTACC-3' 3'-CCATGG-5'	5'-GGTAC C-3' 5'-C CATGG-3'

Figure 19.20 Restriction enzymes. (a) Restriction enzymes recognize a specific DNA sequence (i.e., specific combinations of adenosine (A), cytidine (C), guanosine (G), and thymidine (T)) to cleave the double-stranded DNA. In the example shown, the restriction enzyme EcoRI recognizes the GAATTC sequence and cleaves after the G on both DNA strands. As a result, the new DNA ends now each contain a 5′ overhang. Since restriction enzymes of this type recognize palindromic sequences, the overhangs are complementary to each other and can be used in a reverse reaction, in which these ends are ligated (see Figure 19.21). (b) Examples of restriction enzymes produced by different bacteria. Note that these enzymes recognize different sequences and leave different types of overhangs.

From Stem Cell Information, 2001. http://stemcells.nih.gov/info/scireport/chapter5.asp.

Cloning by ligation requires restriction enzymes and DNA ligase

In cloning by ligation, a linear DNA fragment of interest is enzymatically joined with a larger DNA vector to yield one double-stranded DNA molecule. As we learned earlier, the DNA vector contains a number of different sequences or genes that make it possible to propagate the vector in the desired host organism. The inserted DNA fragment, which we will refer to as the insert, is typically amplified by PCR, as described in Section 19.3, or it may be excised from a larger region of DNA using a restriction enzyme, as illustrated in Figure 19.21a.

A PCR-amplified DNA fragment may have been obtained using PCR primers engineered to include particular restriction sites at both its 3′ and 5′ ends. Figure 19.21b illustrates how the insert and vector are prepared. The insert is prepared by digesting it with the appropriate restriction enzymes, which generates single-stranded overhanging ends. The vector is prepared by digesting it with the same restriction enzymes to generate overhanging ends that are complementary to the overhangs of the insert. The DNA insert and vector DNA are then incubated together so that the compatible ends of the insert and vector can base-pair with one another. A DNA ligase is used to covalently join the adjacent 5′ and 3′ ends, thereby connecting the DNA backbone.

The ligated plasmid is introduced into bacteria using a process called **transformation** (see Figure 19.21a) in which the receiving cells usually need to be treated with salts or exposed to an electric field before they are competent to take up the DNA. (The cells able to take up DNA are referred to as competent cells.) The presence of a selectable marker in the plasmid - for example, a gene that confers antibiotic resistance - is then used to select for cells that have received the plasmid.

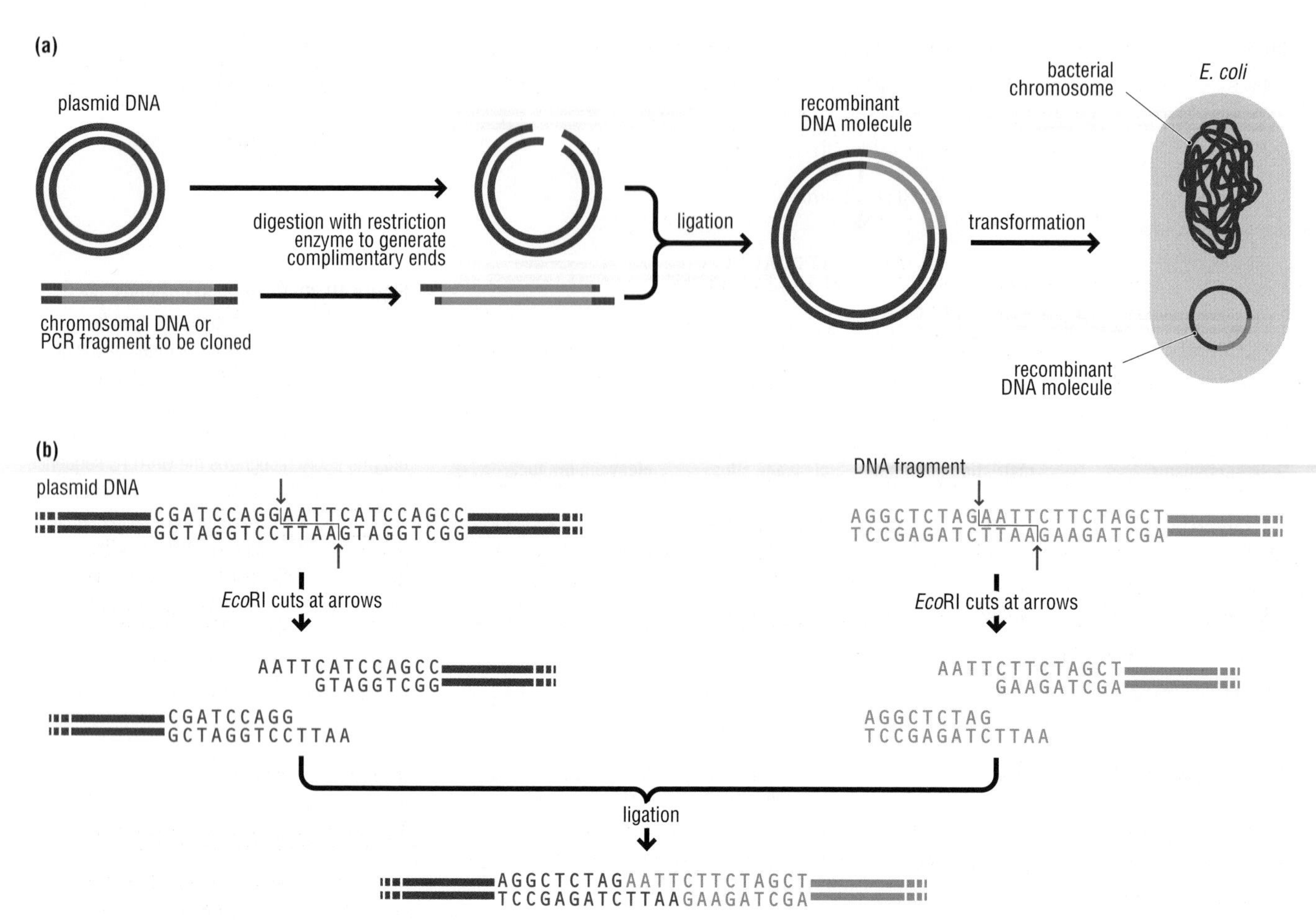

Figure 19.21 Cloning using restriction endonucleases. (a) In this cloning scheme, a plasmid is cut with a restriction endonuclease to create two ends (the restriction sites on both plasmid and fragment DNA are shown in blue). The fragment DNA to be cloned is cut with the same restriction endonuclease, making the single-stranded ends of the plasmid and the DNA fragment complementary, such that they can anneal. To clone the fragment DNA into the plasmid, the cut plasmid and the DNA fragment are mixed, allowed to anneal, and then the nicks are sealed by DNA ligase. The new plasmid is then transformed into a host, most commonly *E. coli*, where it can be propagated. (b) An example of DNA annealing following cleavage by the restriction endonuclease EcoRI is depicted.

This is done by taking cells that have undergone the transformation reaction and plating them on solid growth medium containing that antibiotic, so that only the cells that contain the ligated plasmid will grow.

Vectors used for cloning have a range of useful properties

The bacterium *E. coli* is the organism of choice for most routine cloning. A wide range of vectors can be used to propagate genes in *E. coli*; the vector chosen depends on the experiment to be undertaken. Vectors can accommodate inserts of different size; typical bacterial plasmids might have inserts of up to 15 kilobases (kb), whereas a specialized kind of bacterial vector, called a bacterial artificial chromosome (BAC), can have inserts that are hundreds of kb in size.

There are many vectors that can replicate in several different organisms and are therefore called shuttle vectors. These vectors contain two or more origins of replication: one for replicating in *E. coli* and the others for replicating in a different organism of choice. These vectors may include two or more selectable markers: one for *E. coli* and the other(s) for another organism. In this way, the cloning steps can be done in *E. coli*, after which the vector can be introduced into cells of the organism of choice.

The copy number of the gene in the specific organism is determined by the specific type of vector used. Some vectors are present in one or two copies per cell, whereas others are present in more than 50 copies. This is important because the expression of a gene is often proportional to the number of copies in the cell.

Gibson cloning allows for ligation of multiple DNA fragments

A variation on cloning by ligation can be used to join several different DNA fragments in a single ligation reaction. In this approach, called Gibson cloning, which is depicted in Figure 19.22, adjacent DNA fragments to be combined share regions of homology that can be introduced by PCR, as described above. The DNA fragments, typically a vector and one or more inserts, are mixed in a test tube and treated with an exonuclease that leaves a 5′ overhang at each DNA end. Complementary single-stranded regions base-pair, thus directing the order in which the DNA fragments and vector assemble. DNA polymerase fills in the gapped single-stranded regions, and DNA ligase seals the nicks. The mixture is then introduced into bacteria, which are plated on media with antibiotics to select for cells with complete plasmids.

An advantage of this method is that it does not require restriction enzymes or the presence of restriction sites at desired locations.

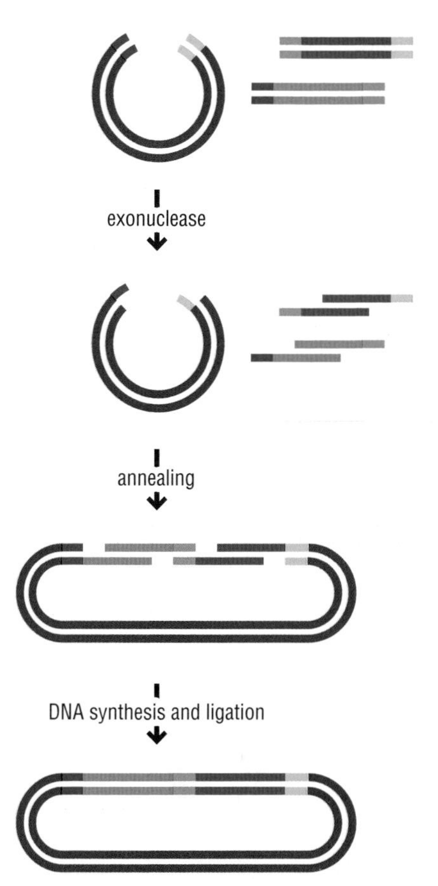

Figure 19.22 Gibson cloning. The DNA fragments to be cloned are designed to be flanked by DNA sequences of 20–40 bp in length that are identical to either the ends of the linearized vector (red and green regions) or to the ends of the other fragment (yellow regions). All DNA fragments are mixed and subjected to an exonuclease that digests DNA from 3′ to 5′, leaving a 5′ overhang. The DNA fragments are then allowed to anneal to each other, leaving single-stranded gaps. These gaps are filled in by a DNA polymerase and then sealed by ligase. Intact plasmids are selected by transformation into bacteria, as in the ligation mediated cloning described in Figure 19.21.

Cloning by recombination requires regions of DNA homology at the ends of the DNA fragment

The ability to amplify DNA by PCR and to introduce desired additional sequences to the ends of the DNA fragment has allowed investigators to introduce fragments into vectors by recombination. Recombination can occur when there is sufficient homology between the ends of the linear DNA and the target site. This method was initially used for research in yeast, where homologous recombination between a linear DNA fragment and either the chromosome or a plasmid is very efficient. More recently, methods have been developed to use recombination to insert a gene into bacterial chromosomes or plasmids, or in a test tube using purified recombination enzymes. Such an approach is illustrated schematically in Figure 19.23. The advantage of these methods is that it bypasses the need for restriction sites at specific locations. To clone a fragment of DNA by recombination, the DNA is first amplified with primers that introduce an additional sequence at each end of the amplified fragment that is homologous to the target sites in the vector (see Figure 19.23a). If the recombination reaction is to happen *in vivo*, a linearized vector and the fragment are transformed into cells, and the subsequent recombination reaction produces a recombinant plasmid. The plasmid will also code for a selectable marker (not shown) that makes it possible to select for cells in which the recombination event took place.

Certain types of recombination reactions can also be performed in a test tube. In this case, the DNA insert and target vector are incubated *in vitro* with purified recombination enzymes so that the DNA fragment becomes incorporated into the vector. Bacterial cells are then transformed with the plasmid as in ligation cloning.

➜ The modification of the chromosome by recombination is discussed further in Section 19.5.

Specific mutations can be introduced by site-directed mutagenesis or PCR

In the examples described above, the purpose of cloning was to generate exact copies of the DNA or RNA of interest. However, it is also often helpful to change the

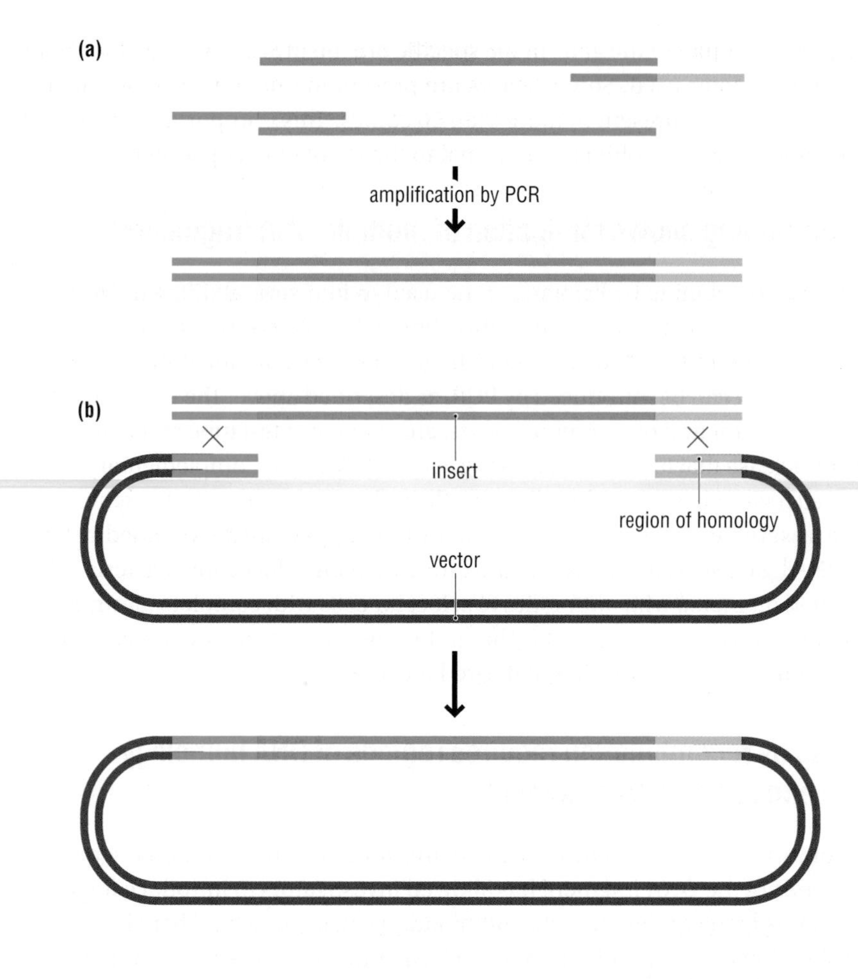

Figure 19.23 Cloning by recombination. (a) To clone a piece of DNA by recombination, the DNA can be amplified with a set of primers that contain, at their 5′ ends, sequences that are homologous to the vector (in turquoise and green), as illustrated in Figure 19.19. (b) The recombination reaction can take place in cells, such as yeast cells or bacteria, via a process in which a linearized vector and the fragment are transformed into the cells and the recombination reaction (indicated by X) occurs *in vivo* to produce a recombinant plasmid. The plasmid will also code for a selectable marker (not shown), allowing the identification of cells in which the recombination event took place. Certain types of recombination reactions can also be performed in a test tube using purified recombination proteins.

function of a gene or inactivate it by introducing mutations at specific sites. These mutations can be introduced after cloning, in methods known as **site-directed mutagenesis**, or they can be introduced by a specialized PCR process, known as PCR sewing, and then cloned into the vector of choice.

In site-directed mutagenesis of a cloned gene, sense and antisense oligonucleotides complementary to the DNA sequence of interest, with the exception of one or more nucleotide changes corresponding to the desired mutation, are synthesized, as depicted in Figure 19.24. The plasmid is denatured and each oligonucleotide is hybridized to one of the two strands of the denatured plasmid. These oligonucleotides serve as a primer for DNA polymerase, which synthesizes the remaining DNA strand. The process is then repeated; some plasmids will contain a strand of template DNA and a newly synthesized strand, while some will have two new strands. The reaction mixture will also include some of the original DNA.

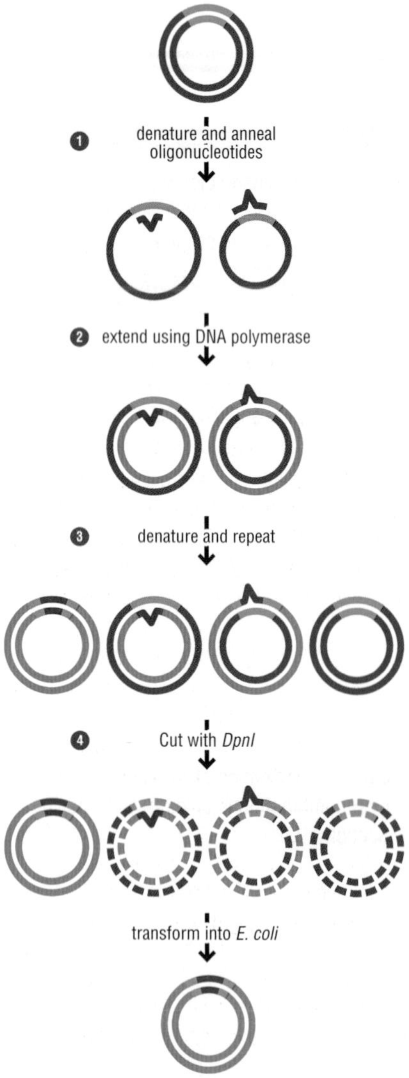

1 denature and anneal oligonucleotides
2 extend using DNA polymerase
3 denature and repeat
4 Cut with *DpnI*
transform into *E. coli*

Figure 19.24 Introduction of specific mutations by site-directed mutagenesis. In this scheme, oligonucleotides are synthesized whose sequence contains the desired mutations (depicted as pink Vs in the DNA) but are otherwise complementary to the flanking codons (step 1). These oligonucleotides are annealed to the denatured plasmid and serve as primers for DNA replication, which extends the primers and replicates the entire plasmid (step 2, new strands are shown in blue). This process is then repeated, to generate a population of plasmids containing the mutation on both strands (step 3). The plasmids are then digested with the enzyme DpnI, which cleaves plasmids containing one or both strands of the parental DNA (in black), which are methylated at the sequence GATC (step 4). When bacteria are then transformed with the reaction mixture, only the circular plasmid will preferentially promote colony formation on selective media (step 5).

To eliminate plasmids that do not contain the mutation on both strands, the mixture is incubated with an enzyme that cleaves plasmids containing one or both of the original DNA strands (that is, strands that were not synthesized in the test tube). The original DNA strands are methylated at the adenines of GATC sites, thanks to an enzyme present in many *E. coli* strains. This methylation is absent from the newly synthesized strands because the methylase responsible for this modification is not present in the reaction mixture. Plasmids containing these methylated bases can then be digested by the restriction enzyme DpnI, which specifically cleaves DNA that contains GATC sequences in which the adenine is methylated on either one or both strands. After DpnI digestion the only intact plasmids are the ones in which both strands are newly synthesized and so contain the mutation. Since circular, uncut plasmids transform bacteria with high efficiency, the circular plasmids containing the mutation preferentially will be propagated.

PCR can also be used to introduce a specific change into the amplified DNA, as illustrated in Figure 19.25. In this instance, the PCR primer carries the desired mutation

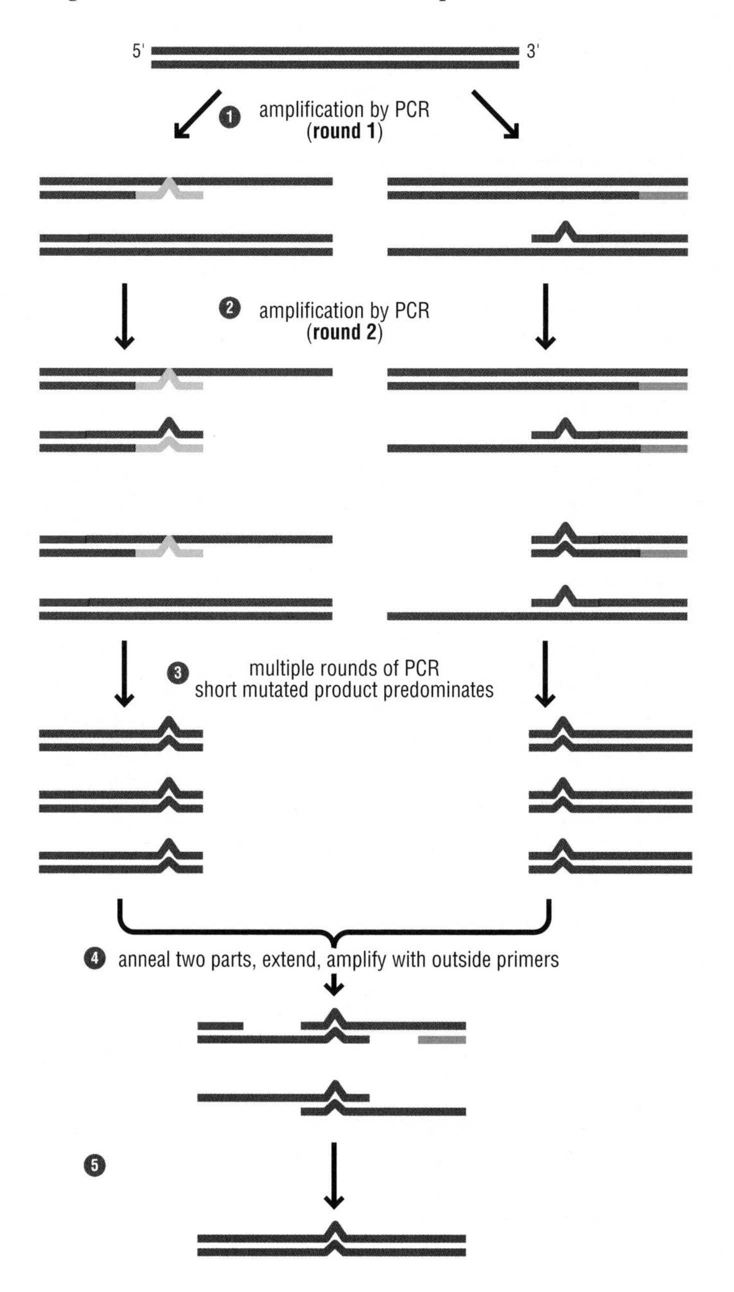

Figure 19.25 Introduction of specific mutations by PCR. In this scheme, the mutagenesis reaction takes place as part of a PCR amplification process. The first step of the mutagenesis reaction takes place in two PCR reactions, one amplifying the left half of the DNA fragment using a terminal (pink) primer and an interior mutagenic primer (green) and the other amplifying the right half of the DNA fragment using an interior mutagenic primer (dark blue) and terminal (orange) primer (steps 1 through 3). The green and blue primers are complementary to each other and carry a mutation in the same base pair. By incorporating the mutation into the oligonucleotide used in the PCR reaction, the majority of the amplified products will, eventually, carry the mutation in both strands, similar to adding sequences to the end of PCR products described in Figure 19.18. Once the two half-reactions are completed, the products are used in a 'sewing' PCR reaction, as shown (step 4): because the blue and green oligonucleotides were complementary to each other, the ends of the two PCR products can anneal as shown, and the DNA can be amplified using the terminal primers (orange and pink). The end result is a DNA fragment that is like the starting material, except that it carries a mutation at the desired position.

such that the products of the PCR will also predominantly carry the mutation. The resulting product can be used in a subsequent PCR reaction in which two PCR fragments are 'sewn' together by annealing the ends to each other. The products can then be cloned. In all cases, the resulting clones are sequenced to verify that the desired change has been introduced. We discuss methods for sequencing in Section 19.8.

A library of clones can be made from genomic sequences, cDNA copies of mRNAs, or synthetic DNA sequences

In molecular biology, a **library** is a collection of vectors (usually plasmids or viruses), each containing a different cloned sequence. For example, one can create a library of plasmids containing DNA fragments that, together, span the entire genome of an organism. Libraries are often used in genetic studies: for example, one could isolate a mutant yeast strain with an interesting phenotype without knowing which gene is mutated. By transforming the mutant strain with a library containing all the yeast genes, one can then identify a cloned gene that reverses the phenotype, thus revealing the identity of the altered gene in the mutant yeast strain. Complementation of a mutant phenotype by a library was used in Experimental approach 4.3 and Experimental approach 5.1.

A library can be made of total genomic DNA, of expressed genes only, or of a subset of expressed genes (for example, genes that are expressed only under a certain condition). For a genomic library, a genome is fragmented (for example, by mechanical sheering or by restriction enzyme digestion), and each fragment is independently ligated into the plasmid vector so that each plasmid carries a different fragment of the genome.

For a library made of expressed genes, a complementary DNA fragment of each mRNA, known as cDNA, is first made using the reverse transcriptase enzyme as described earlier in this section. A **cDNA library** featuring all mRNAs expressed in a certain cell type or tissue can be produced; such libraries make it possible to identify and study genes expressed in certain biological contexts, for example, from a particular human tissue.

More recently, investigators have designed libraries that express small hairpin RNAs (shRNAs), whose function is to inactivate gene expression in mammalian cells via RNAi pathways. These libraries contain hundreds of thousands of clones, each expressing a short RNA that can fold onto itself and direct the RNAi machinery to degrade the complementary mRNA. These libraries are generated using many different short palindromic synthetic oligonucleotides (of 40–50 bases) that are cloned into plasmids or viral vectors. Special computer programs are used to ensure that these libraries cover a certain portion of the genome, and the libraries typically contain several different clones for each gene.

➔ We learn more about RNAi pathways in Chapter 13.

Using this approach, researchers have screened for shRNAs that, when present, lead to a particular phenotype – for example, resistance to HIV infection. The researcher can then identify the sequence on the clone that conferred resistance, which is then used to identify the gene that was targeted by the clone. The use of this kind of library is described in Experimental approach 17.2.

19.5 GENOME MANIPULATION

In the preceding section, we learned how mutations are introduced into cloned regions of DNA. However, when a gene is expressed from a plasmid or a virus, its expression level is often different from its normal level. In addition, to eliminate gene function one must inactive the chromosomal copy (or copies, in a diploid organism) of the gene. Scientists therefore often want to introduce mutations into a genome.

Mutations in the genome were traditionally identified in an approach now referred to as **forward genetics**, in which the researcher generates a mutant that is detective in a particular process (or uses an existing mutant) and then seeks to identify and understand the gene(s) responsible for the observed phenotype or disease. For example, to study the process of DNA repair, researchers predicted that mutants defective in this process would be sensitive to radiation. Cells were treated in a way that produced random mutations in the genome and screened for mutants that were sensitive to radiation. Once mutants with the desired phenotype are in hand, one of a variety of methods could be used to identify the mutated genes, for example, by complementing the phenotype with a genomic library.

➔ We discuss genome libraries in Section 19.4.

Another approach to studying gene function is **reverse genetics**. In this case, the gene is already in hand, and the researcher seeks to understand its function by altering the gene's sequence and analyzing the phenotypic changes that result when the altered sequence is expressed *in vivo*. For example, a previously unstudied protein may be identified through its physical interaction with a known protein. To study the new protein, a researcher can delete the gene coding for this protein and examine how this affects cell function.

Sometimes it is also desirable to introduce a specific gene or specific mutations into a genome in a non-random fashion. The introduction of specific genes into the genome of an organism is denoted **transgenesis** and results in the generation of a **transgenic organism**.

In this section, we will discuss several different methods for genome manipulation including chemical mutagenesis, transposition, homologous recombination, site-specific editing, and conservative site-specific recombination (CSSR). For some organisms, however, it is difficult or even impossible to introduce specific mutations into the genome. In these organisms, scientists can turn to approaches such as RNAi and chemical libraries to block the expression or function of gene products, as we describe at the end of the section.

Mutations can be introduced by radiation or chemicals that damage DNA

As discussed in Chapter 15, mutations can be introduced more or less at random by the action of radiation or DNA-damaging chemicals. Thus, one of the simplest ways of introducing mutations is to expose the organism to agents such as ethyl methanesulfonate or ethylnitrosourea, or to irradiate with ultraviolet (UV) light. The advantage of these types of mutagens is that they often introduce point mutations that can change the function of the gene product instead of simply inactivating the gene (for example, due to a large deletion), which could result in cell death.

This type of mutagenesis brings with it the challenge of identifying the DNA sequence change in the context of a large genome (for example, that of the mouse). However, this limitation is rapidly being overcome by advances in DNA sequencing technology that make it possible to identify the mutation by sequencing the whole genome, making it likely that chemical mutagenesis will remain a widely used tool in genetic screens.

Transposons can be exploited for mutagenesis and transgenesis

Transposons were discovered because of the way their movement resulted in mutagenesis, since the insertion of a transposon into a coding sequence disrupts gene

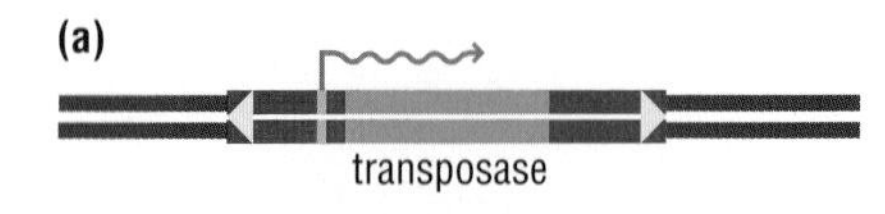

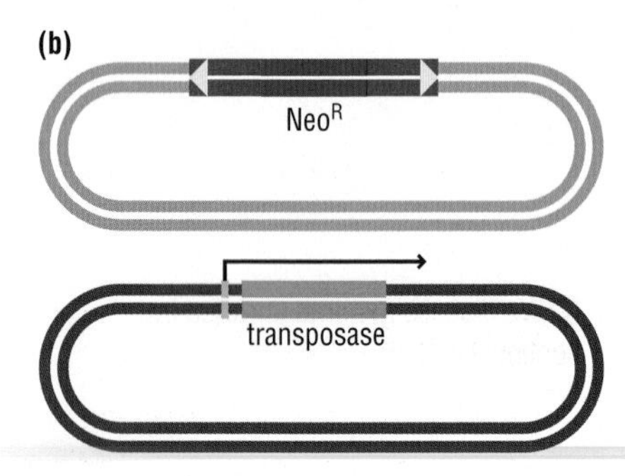

Figure 19.26 Two-component transposon system. (a) The structure of a transposon as found in nature, with its terminal recombination sequences (yellow triangles) and transposase gene (orange). (b) In a two-component transposition system, one component is a 'mini' transposon made of transposon ends flanking a cargo DNA segment, here the gene encoding Neo^R (which provides resistance to the drug neomycin). The other component is the transposase gene under the control of a regulatable promoter. When a cell contains both components, induction of transposase expression mobilizes the mini transposon.

We explore mutagenesis of the human genome by an endogenous transposable element in Experimental approach 17.1.

function. This property of transposons has been exploited in several genetically tractable model organisms to carry out insertional mutagenesis.

Transposon mutagenesis is most often carried out using a two-component system, where one component is a regulated source of transposase and the other component is a 'mini' transposon consisting of a selectable marker gene flanked by the transposon end sequences necessary for transposition, as illustrated in Figure 19.26. These elements are typically placed on separate plasmids that lack sequence elements necessary for efficient DNA replication or segregation and are thus eventually lost over multiple generations. Once the transposable element inserts into the target DNA, the plasmid carrying the transposase is eventually lost, at which point the inserted element is stably integrated and cannot transpose again.

The plasmids or viruses containing the transposon are introduced into the cell by transformation, or infection. Cells containing the transposon are isolated by screening for the selectable marker encoded by the integrated transposon, which may be an antibiotic-resistance gene or a visual marker such as a fluorescent protein.

Transposons are a particularly useful experimental tool because the DNA that flanks the transposon can be isolated and sequenced, with the transposon providing a molecular tag to identify the site of the insertion mutation. Some of the transposons that are the most widely used for mutagenesis are listed in Figure 19.27.

In addition to generating mutations, transposon insertions can be used to obtain information about the activity of specific promoters or enhancers, or to artificially activate a gene. For example, a transposon can contain a reporter gene whose expression can easily be assayed, as depicted in Figure 19.28. When the transposon is inserted in the vicinity of a promoter or enhancer element, the expression of the reporter gene is under the control of the neighboring promoter or enhancer. By assaying the relative levels of the reporter gene product, it is possible to obtain a measure of the level at which the endogenous gene is expressed. This approach is sometimes referred to as an **enhancer trap**.

A converse strategy is to activate endogenous genes using a transposon that contains a strong promoter. When the transposon containing the promoter is inserted into the chromosome, transcription of nearby genes can be controlled by the inserted promoter rather than by the endogenous promoter.

Transposons can be used not only to disrupt genes, as we have just discussed, but also as vectors for inserting genes of choice into cells or organisms. Transposons are thus being tested as vehicles for so-called 'gene therapy', in which a wild-type copy of a protein is introduced into the host chromosome in an effort to cure disease.

Some transposons commonly used as tools

organism	commonly used transposons
bacteria	*Tn5*
plants	*Ac*
insects	P element, *piggyBac*, *hobo*
fish	*Tol2*
mammals	*Sleeping Beauty*, *Tol2*, *piggyBac*, *Minos*

Figure 19.27 Transposons commonly used as tools.

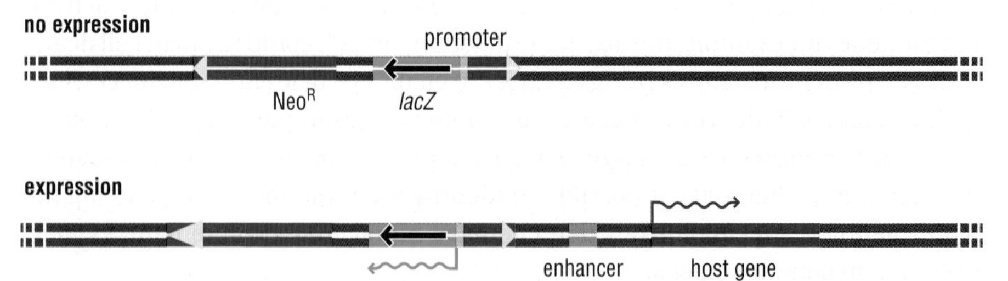

Figure 19.28 An enhancer-sniffer transposon. The transposon contains a selectable marker (to confirm transposon insertion into the genome) and a reporter gene (*lacZ*) with only a weak basal promoter such that it is not expressed in the absence of an enhancer. If the transposon inserts at a chromosomal region lacking a nearby enhancer (top panel), the *lacZ* reporter will not be expressed. If, however, the transposon inserts adjacent to the enhancer of a host gene (bottom panel), the reporter is expressed similarly to the host gene.

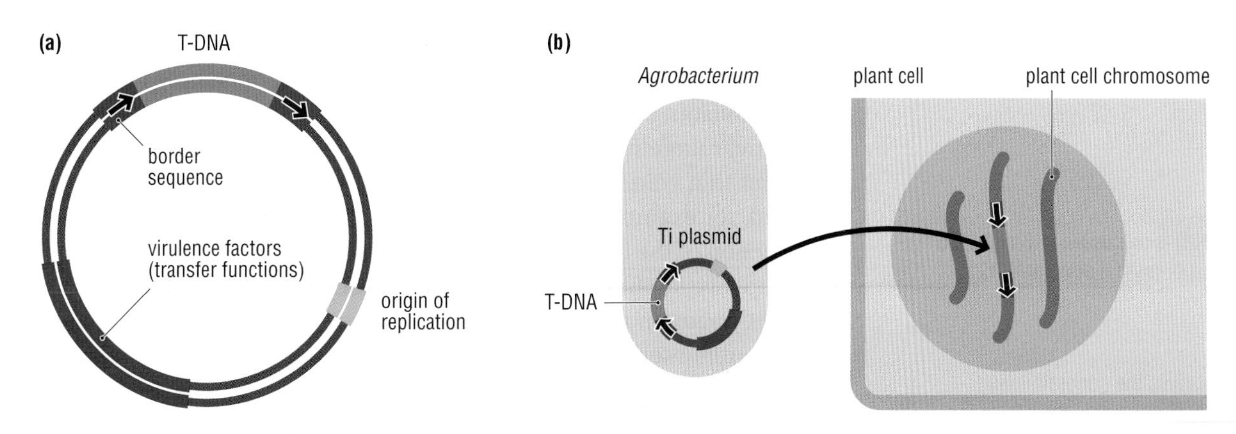

Figure 19.29 Use of T-DNA for modification of plant genomes. (a) The Ti plasmid from *A. tumefaciens* includes a bacterial origin of replication (green), the T-DNA, which is composed of two border sequences (dark blue) and the DNA that lies between (blue), and the virulence region (pink) that codes for proteins that copy and move the T-DNA to the plant cell. (b) Insertion of the T-DNA segment into the plant genome occurs when the proteins coded by the bacterial virulence region copy the T-DNA segment and move it to the plant cell, where it is integrated into the genome. When used for creating transgenic plants, the natural T-DNA segment is replaced with the DNA of interest.

Plant genomes can also be modified by mobile genetic elements

Mobile DNA elements have been the major tool for manipulating the genomes of plants which can be infected by the bacterium *A. tumefaciens*, resulting in the transfer of a DNA segment called the T-DNA from a specialized Ti plasmid into the plant genome, as shown in Figure 19.29. The virulence region of the Ti plasmid encodes the proteins that act on sequences at the ends of the T-DNA segment to transfer the DNA from the bacterium to the plant.

In nature, the T-segment encodes plant hormones that stimulate the plant cells to grow in unregulated fashion as 'crown gall tumors'; these 'tumors' produce compounds that the bacterium can metabolize for growth. For genome engineering, the tumor-inducing genes inside the T-DNA are replaced with a selectable marker, such as an antibiotic-resistance gene, and any other DNA of interest (for example, a gene expressing a plant protein fused to GFP). Another useful strategy is to use the T-DNA to deliver a transposon into the plant genomic DNA, which can be subsequently remobilized for further mutagenesis, or to express small inhibitory RNAs (siRNAs) to inhibit targeted genes.

Specific sequences or gene disruptions can be introduced by homologous recombination

In addition to randomly mutagenizing a genome, researchers often want to modify the genome in a precise manner. For example, an experiment might call for a genomic sequence to be replaced, as in the case of gene deletion, or for an exogenously supplied piece of DNA to be added, such as when creating gene fusions. These types of manipulations can be accomplished by homologous recombination.

➔ We discuss homologous recombination in more detail in Chapter 16 and encountered its use when we discussed cloning in Section 19.4.

To alter a genome using this method, a linear fragment of DNA that has homology to a genomic sequence on both of its ends is introduced into cells and becomes incorporated into the host cell genome by homologous recombination. This process is illustrated in Figure 19.30. This process may be mediated by proteins that are normally found in the host cell, or the cell may be engineered to increase the efficiency of the recombination reaction, for example by introducing components of the TALEN or CRISPR system, as described later in this section. In some organisms, such as *B. subtilis*, *S. cerevisiae*, and mammalian cells, the ends of the linear targeting DNA are sufficiently recombinogenic that recombination can occur in a wild-type strain. By contrast, bacteria such as *E. coli* need to be modified, for example by decreasing the nuclease activity of RecBCD to reduce the resection of the linear DNA, and introducing the recombination system from bacteriophage lambda to increase the efficiency of recombination.

RecBCD is discussed in Chapter 16 and in Experimental approach 16.1.

A DNA fragment that is incorporated by recombination can be engineered to contain unique sequences that make it easy to identify cells containing that particular gene replacement. For example, recombination can be used in yeast to replace a gene with a selectable marker. Researchers have taken advantage of this approach to substitute each non-essential yeast gene with a DNA fragment containing both a selectable marker gene and a unique sequence, often referred to as bar code, as illustrated in Figure 19.31. These barcodes make it possible to identify the different knockout strains uniquely in a pool of mutants, either by DNA sequencing or by hybridizing the chromosomal DNA to microarrays containing the corresponding probes.

We discuss the generation and use of knockout strain collections in Experimental approach 16.1.

(a) replacement

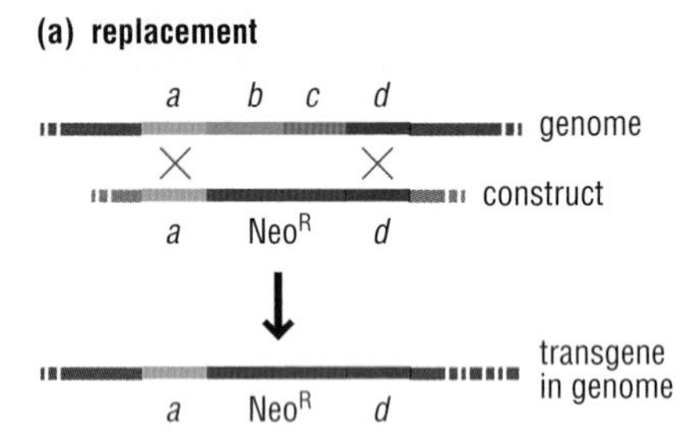

(b) addition

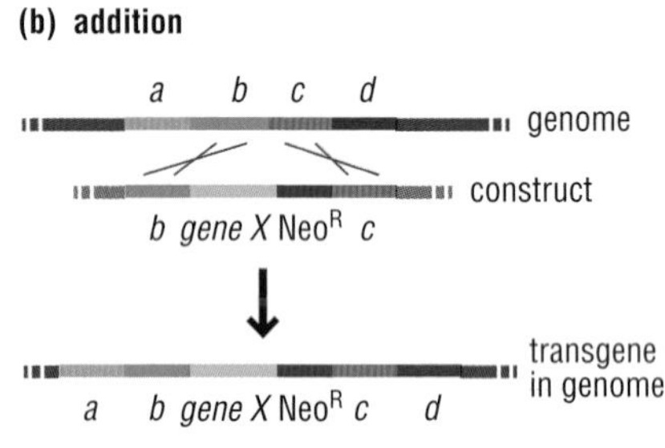

Figure 19.30 Gene replacement and addition by homologous recombination. (a) Gene replacement by homologous recombination. To delete the sequences between the a and d DNA segments, a linear DNA containing the gene encoding Neo^R (conferring resistance to the drug neomycin) is flanked by sequences that are identical to a on one side, and d on the other. When this linear DNA is introduced into cells it recombines with the genomic a and d sequences, resulting in the substitution of the genomic b and c regions with the Neo^R gene (b) Gene addition by homologous recombination. Let us assume that a researcher wants to express Gene X in a particular cell type, without removing any cellular DNA sequences. To do so, a linear DNA segment containing Gene X and a selectable marker (Neo^R) is flanked by sequences homologous to the genomic insertion site (in this case, sequences b and c). Homologous recombination between the b and c regions of the linear DNA and the b and c regions of the genomic DNA results in the insertion of Gene X and Neo^R into the genome.

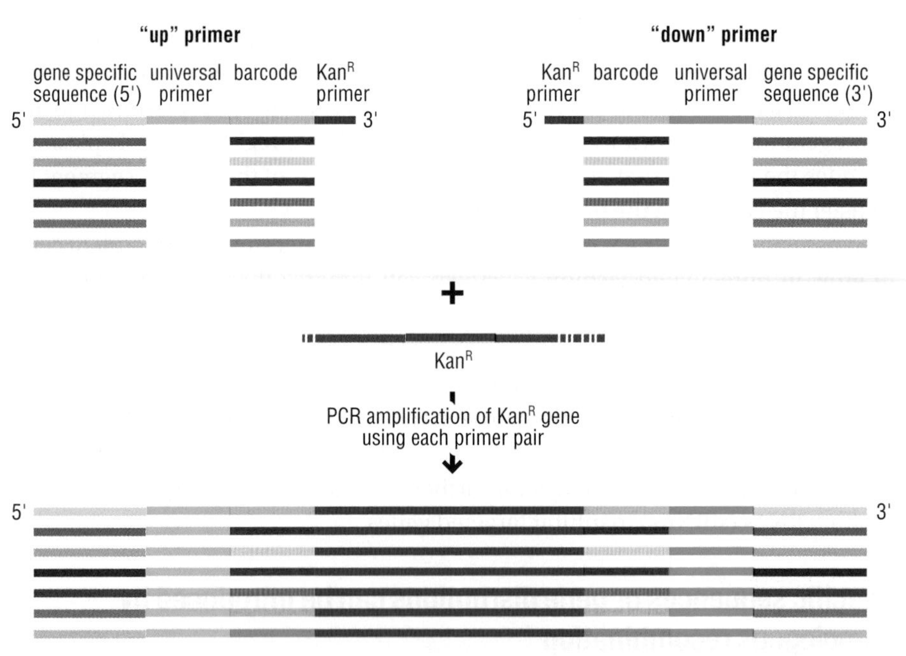

Figure 19.31 Gene disruptions (knockouts) of *S. cerevisiae* genes. The disruption of each gene in the yeast *S. cerevisiae* deletion collection requires a pair of DNA primers, referred to here is an 'up' primer and a 'down' primer. Each primer contains (starting from the 5′ end): (a) a gene-specific sequence that is immediately upstream (for the up primer) or downstream (for the down primer) of the coding region (shades of gray), (b) a universal primer, one for all 'up' primers and another for all 'down' primers (light green and light blue), (c) a bar code sequence that is unique to each primer (shades of pink), and (d) primers to amplify the *kanR* gene (in light and dark blue). Each pair of primers is used in a PCR reaction to amplify the *kanR* gene (conferring resistance to kanamycin), creating a collection of DNA fragments, each homologous at its ends to the DNA regions flanking the coding sequence of a specific gene. Homologous recombination between these gene-specific sequences and the chromosomal sequences results in the replacement of the gene with a *kanR* fragment flanked by universal primers and two barcodes specific to the targeted gene. The strains of interest (for example, those that are able to survive a particular treatment) can then be identified by virtue of the unique barcodes associated with their *kanR* cassette.

From Meneely, P. *Advanced Genetic Analysis: Genes, Genomes, and Networks in Eukaryotes*, Oxford: Oxford University Press, 2009.

Homologous gene replacements have also been quite useful in mammalian cells in culture. In these cells, DNA can integrate by non-homologous end joining (NHEJ), and it is thus important to distinguish between cells in which targeted integration has occurred and those in which NHEJ has occurred. A useful strategy is to include two markers on the targeting DNA, one that will be retained and another that will be lost when the desired homologous recombination occurs; such a strategy is depicted in Figure 19.32.

Knockout mice are generated using homologous recombination

Mice in which one or more genes have been disrupted, known as knockout mice, have become important tools for studying the role of particular genes in mammals.

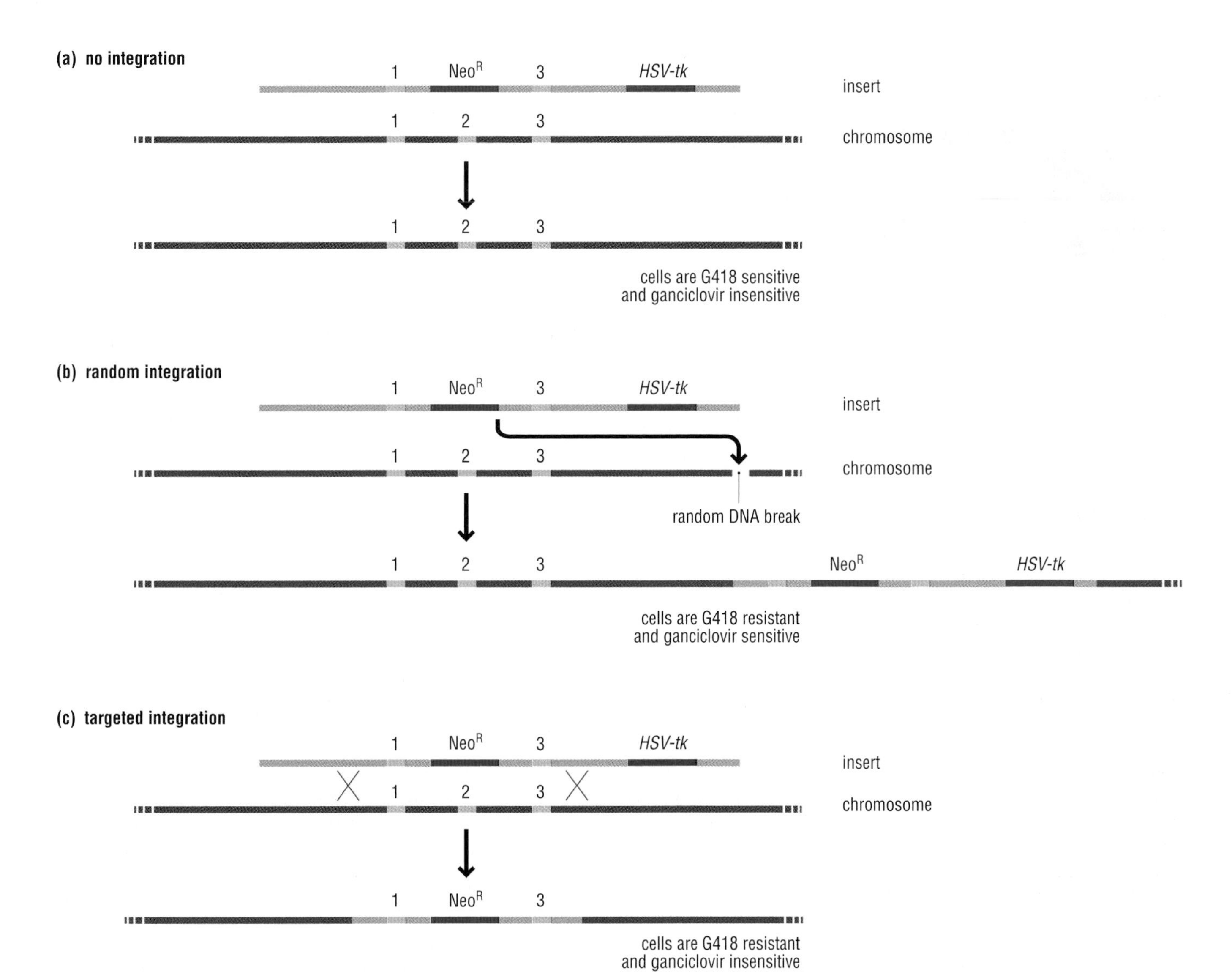

Figure 19.32 Characterization of gene integration in mammalian cells. The insert used in this example contains the gene for Neo^R, which confers resistance to G418, and *HSV-tk*, which confers sensitivity to the drug ganciclovir. The purpose here is to replace chromosomal DNA segment '2' with the Neo^R gene. (a) When the construct is not integrated, cells remain sensitive to neomycin and insensitive to ganciclovir. (b) Random integration of the insert by NHEJ results in cells that are neomycin resistant and sensitive to ganciclovir. (c) Integration by homologous recombination at the desired site results in cells that are resistant to neomycin and insensitive to ganciclovir.

From Meneely, P. *Advanced Genetic Analysis: Genes, Genomes, and Networks in Eukaryotes*, Oxford: Oxford University Press, 2009.

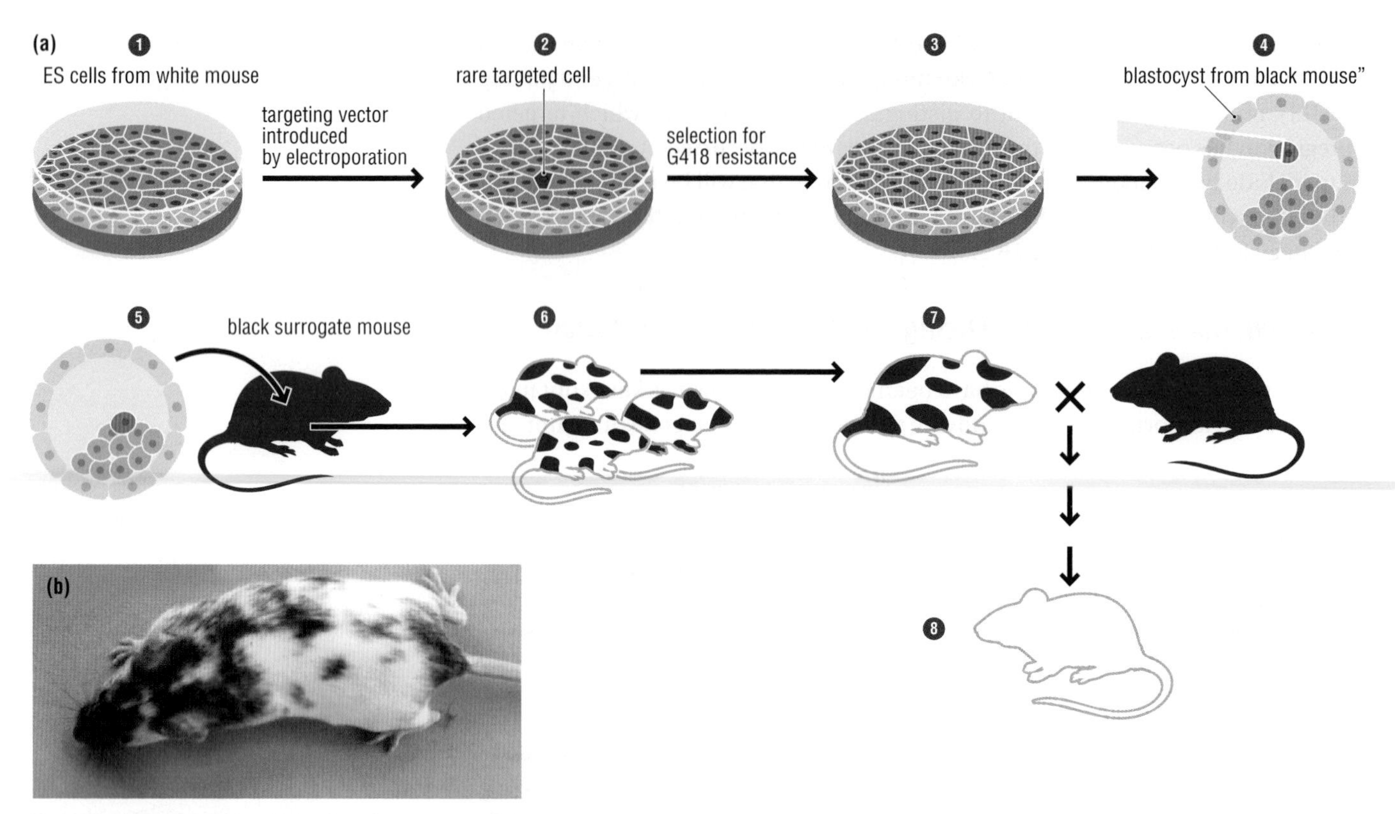

Figure 19.33 Generation of a transgenic mouse. (a) Scheme for generating a transgenic mouse: (1) ES cells from a mouse with white coat color are grown in culture. (2) The targeting vector is introduced into the ES cells by selection for a marker that confers resistance to the drug G418 (pink cell). (3) Cells in which the targeting vector has been correctly incorporated are grown as a pure population by selection in media containing G418. (4) The ES cells are introduced into a blastocyst from a mouse with a black coat color. (5) The blastocyst containing the targeted ES cells is injected into a female surrogate black mouse with black coat color. (6) Chimeric pups that contain white fur from the ES cells and black coat color from the mother are born. (7) Chimeric mice are backcrossed to a mouse with black fur in an effort to identify mice in which the ES cells gave rise to the germ line. (8) White coat color mice derived entirely from the ES cells are obtained by crossing chimeric pups to each other. (b) Photograph of a chimeric mouse, with patches of both white and black fur. From Meneely, P. *Advanced Genetic Analysis: Genes, Genomes, and Networks in Eukaryotes*, Oxford: Oxford University Press, 2009.

Knockout mice are also generated through a procedure that relies on homologous recombination to re-engineer mouse embryos, as illustrated in Figure 19.33a. The gene of interest is first replaced with a selectable marker by homologous recombination in mouse ES cell in culture, as described above.

For this example, we will assume the ES cell is derived from a white mouse. The ES cells are placed in a selective medium so that only the cells containing a correctly targeted selectable marker will survive. The transformed cells are then injected into a mouse embryo from a black-coated mouse, which is at the blastocyst stage of development and hence contains several hundred cells. The ES cells become incorporated into the blastocyst and will contribute to the tissues of the animal that develops. The blastocyst becomes implanted in the uterus of a black female mouse. Blastocysts that are naturally in the mother will give rise to black mice, while transplanted blastocysts that contain engineered ES cells will give rise to chimeric mice with a mixed black-and-white coat color, depending on which cells in the injected blastocyst contributed to a particular skin patch. Such a chimeric mouse is shown in Figure 19.33b.

The use of mice with different coat colors for generating the ES cells and as a source for the blastocysts makes it possible to identify progeny mice in which the ES cells contribute to the tissues. If the engineered ES cells contribute to the germ line, these mice can be crossed, first to a wild-type mouse, to generate heterozygous mice, and then to themselves, to generate homozygous mice in which all cells contained the gene knockout.

Conservative site-specific recombination (CSSR) allows the expression of mutations in specific cells

If a gene plays a critical role during development, it may not be possible to study its function in the mature animal by generating a knockout in an ES cell. An example is a gene knockout that causes the embryo to die or cease developing, thereby making it impossible to assess the function of the gene in the mature organism. In this case, a useful alternative strategy is to make a **conditional** allele of the gene of interest, which allows the gene to be deleted at a chosen time or in a specific cell type. A conditional allele can also be useful in studying the effects of a knockout in just a subset of tissue types, rather than in the whole organism.

A common method for producing conditional knockouts relies upon the site-specific Cre recombinase. As we learned in Section 17.13, the Cre recombinase mediates recombination between DNA sequences called *loxP* sites. When the *loxP* sites are in a direct orientation (meaning oriented in the same way in the DNA), recombination between these sites results in deletion of DNA that lies between the *loxP* sites as depicted in Figure 19.34a. By controlling whether or

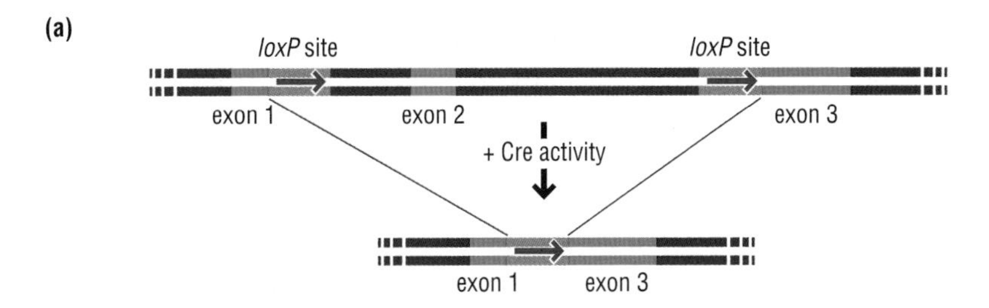

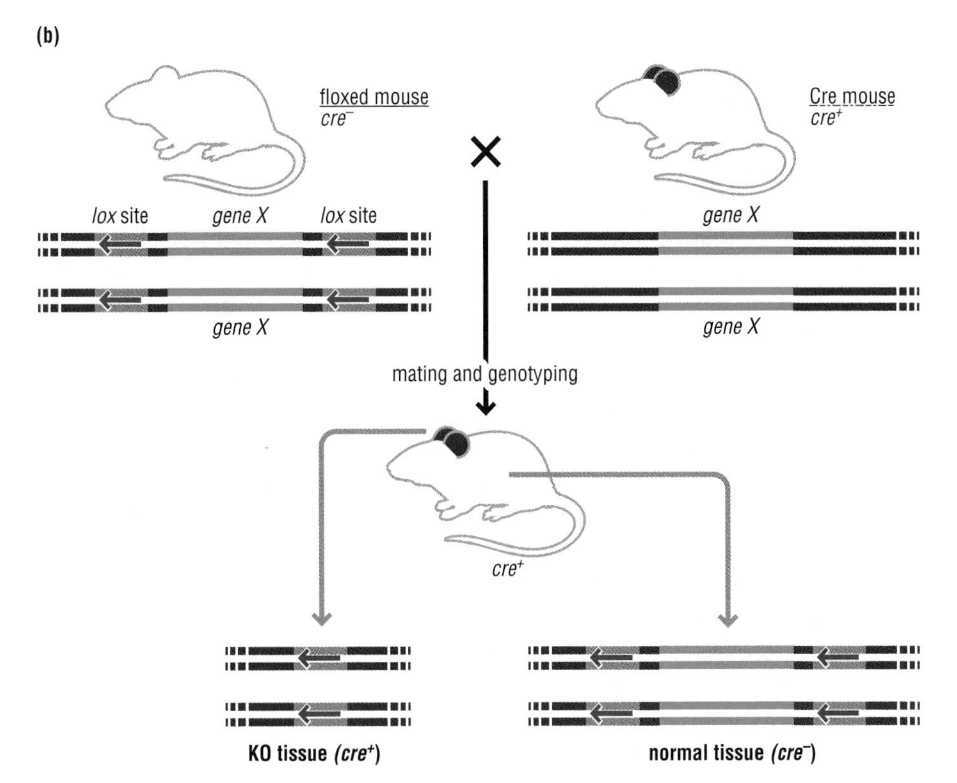

Figure 19.34 Using Cre-*lox* recombination. (a) The principles of Cre-*lox* recombination. The substrate DNA contains a *loxP* site (pink arrows) in the intron between exons 1 and 2 and another *loxP* site in the intron between exons 2 and 3. Cre-mediated recombination results in deletion of exon 2. (b) Scheme for making a conditional deletion using Cre-*lox* recombination. A homozygous transgenic mouse carrying gene X flanked by *loxP* sites (floxed gene X) but not expressing the *Cre* gene is crossed to a mouse expressing Cre in a tissue-specific manner (in the ears in the example shown) but carrying wild-type copies of gene X. After a couple of rounds of mating, a mouse homozygous for floxed gene X and Cre is generated. Note that this mouse will have intact copies of gene X throughout its body except where Cre is expressed (in the ears). In this way, it is possible to study the role of gene X in a specific tissue while the rest of the mouse has a wild-type genotype. Adapted from Meneely, P. *Advanced Genetic Analysis: Genes, Genomes, and Networks in Eukaryotes*, Oxford: Oxford University Press, 2009.

not Cre recombinase is expressed at a particular time or in particular tissues it is possible to control when and where the gene is deleted.

To generate a conditional knockout mouse, it is necessary to generate two different transgenic mice: one bearing the gene of interested flanked by *lox* sites (this allele is referred to as 'floxed,' for 'flanked by lox') and the second carrying the *Cre* gene under an inducible or a tissue-specific promoter. First, a transgenic mouse carrying the floxed allele in the desired position is constructed as described previously (see Figure 19.33). The mouse that is homozygous for the floxed allele is then crossed with a transgenic mouse carrying the *Cre* gene under an inducible promoter (for example a heat-shock promoter) or under a tissue-specific promoter; this strategy is depicted in Figure 19.33b. The floxed gene will thus be deleted only in tissues expressing Cre recombinase.

The use of the Cre-*loxP* system is not limited to mice and has been used in other model organisms such as *Drosophila* and *C. elegans*.

Genome editing using TALEN and CRISPR

While homologous recombination can, in theory, be used to modify any segment of DNA, in practice the efficiency of homologous recombination is extremely low in most organisms. One way to increase homologous recombination is by generating a double-stranded DNA break at the chromosomal locus of interest, but for this one needs to have a way of targeting an endonuclease to that particular chromosomal locus. Scientists have recently adapted processes from bacteria to accomplish just that.

Transcription activator-like effector nucleases, or TALENs, make use of DNA-binding proteins called transcription activator-like effectors (TALE) that are secreted by *Xanthomonas*, a bacterial plant pathogen. Each TALE protein contains multiple repeats, each 33–35 amino acids in length, that are nearly identical to each other in structure and sequence, except at residues in each repeat that govern the DNA-binding specificity to a single nucleotide. Since there is a one-to-one correspondence between TALE repeats and successive base pairs in the DNA recognition sequence, it is possible to engineer 'designer' TALE proteins containing a number of repeats that bind a particular DNA sequence (typically 15–30 nucleotides long) in the chromosome.

By fusing the catalytic domain of a nuclease called FokI to the designed TALE protein, the resulting TALEN can be targeted to specific sites in the DNA. The fact that the FokI nuclease is active as a dimer is further exploited to target TALENS very precisely to sites in the genome: by designing a heterodimeric TALEN in which each TALE DNA-binding domain recognizes a distinct sequence to either side of the desired chromosomal locus, one can induce a double-strand cleavage at a specific site, as shown in Figure 19.35a. In the absence of any added DNA, the double-strand break will be repaired by NHEJ. This process often involves the loss of a few nucleotides at the break site, thus generating mutations at a desired location. If, however, linear DNA with homology to sequences on either side of the break is present, repair may take place by homologous recombination, thereby resulting in the insertion of the DNA sequence of choice at a specific chromosomal location.

Another approach to introducing targeted DNA double-strand breaks takes advantage of a nuclease and DNA targeting mechanism from the CRISPR (for clustered regulatory interspaced short palindromic repeats) systems that are part of the cellular defense against foreign DNA in bacteria and archaea. Short foreign

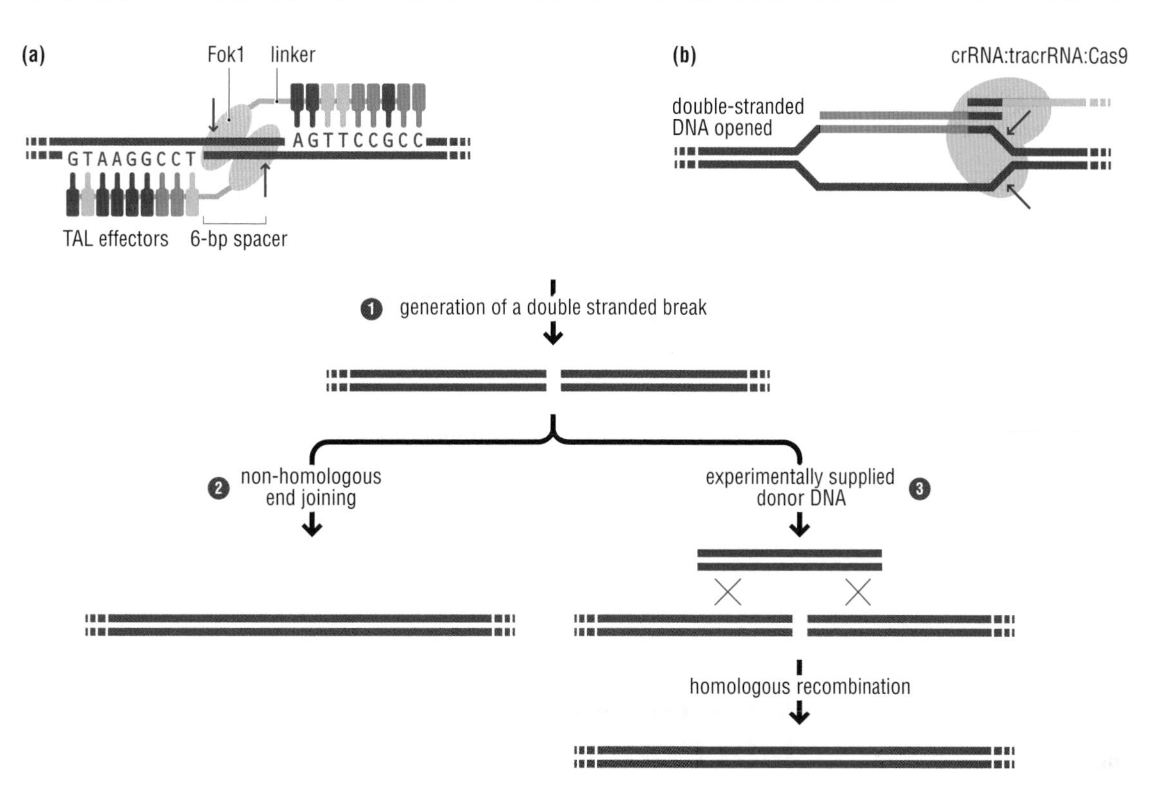

Figure 19.35 Using TALEN and CRISPR to generate double-strand DNA breaks. (a) The TALEN system. The TALEN proteins contain multiple TAL sequences (or effectors), shown in red, green, purple, and blue, each recognizing a specific nucleotide (for example, the 'red' TALE recognizes G, the 'green' TALE recognizes T, etc). The proteins also contain the catalytic domain of the FokI nuclease (pink). The two TALENs bind on opposite sides of the target site, bringing the two nucleases together to generate a double-strand DNA break. (b) The CRISPR system. A double-strand DNA break is caused by the Cas9 nuclease (light pink), which is targeted to a specific DNA site by the crRNA (dark pink), after it was processed by the tracrRNA (in green). The double-strand DNA break (step 1) generated by either the TALEN or CRISPR systems can be repaired by either NHEJ (step 2), which is likely mutagenic, or by homologous recombination (indicated by the two Xs), if a DNA fragment with homology to either side of the break is present (step 3).

DNA segments, or spacers, are inserted between the repeats and serve as a form of acquired memory of prior encounters with foreign DNA. When these segments are transcribed, the resulting RNA associates with a specific nuclease and targets it to foreign DNA that shares homology with the spacer sequences.

Researchers took advantage of this system to design a sequence-specific nuclease. Components of the *Streptococcus pyogenes* CRISPR system, including the Cas9 nuclease and two types of non-coding RNAs (an RNA that contains the space sequences needed for targeting, called crRNA, and a trans-activating RNA, called tracrRNA, that is required for processing the crRNA) are expressed in mammalian cells. As shown in Figure 19.35b, when these three components form a complex that binds to the DNA that is recognized by the crRNA, Cas9 generates a double-stranded DNA break. By designing the spacer sequences to be homologous to the gene of interest, the nuclease is targeted to this DNA sequence. As with the TALEN system, this break can be repaired by NHEJ, or, if homologous DNA is present, by homologous recombination.

➔ We learn about the CRISPR system in Section 13.7.

The expression of genes can be disrupted by siRNAs

Rather than directly disrupting a gene to study its function, the RNAi pathway discussed in Chapter 13 can be exploited to target particular mRNAs for degradation, thereby reducing expression of the targeted gene. In Section 19.4 we discussed the use of an shRNA library to disrupt gene function. shRNAs are small RNAs that form

➔ We learn about the double-stranded RNA fragments called siRNAs (21–22 bp), which can trigger the degradation of mRNA containing a region of complementarity to the siRNA, in Chapter 13.

hairpins and are processed by the cell's RNAi machinery. It is also possible to use synthetic siRNAs to accomplish the same goal. While shRNAs can be engineered to be stably expressed within a cell or an organism, siRNAs are mostly used in tissue culture, where they are transfected into cells and transiently inhibit expression. Both siRNA and shRNA libraries are now commercially available for every gene in the human and mouse genome.

A caveat for using RNAi to study gene function is the possibility of 'off target' effects, whereby genes that are related in sequence are inhibited by the same RNAi construct. If the expression of more than one gene is affected by a given RNAi construct, it can be difficult to determine which inactivated gene contributes to the observed phenotype. Sequences used for siRNA have to be designed carefully to avoid regions that are homologous with other genes. By using several different siRNAs that target the same gene, the chance of targeting a second, unrelated gene is much reduced. Finally, if the phenotype is due to down-regulation of the targeted gene, an siRNA 'resistant' version of the gene, in which the RNA sequence that was targeted by the siRNA is altered without changing the protein sequence, should complement the observed phenotype.

19.6 DETECTION OF BIOLOGICAL MOLECULES

In order to study or purify cellular components, methods are needed to detect the presence of specific biological molecules and distinguish them from others in the mixture. Some detection methods take advantage of the intrinsic properties of a molecule, such as its biochemical activity or its intrinsic spectroscopic properties. Molecules can also be detected by attaching to them moieties with particular spectral, chemical, or fluorescent properties that are then readily detectable. Many of these detection methods can also be used to determine the *amount* of a molecule that is present. Such measurements are essential for quantitative studies of the thermodynamic and kinetic properties of molecules. In this section, we review some of the basic methods currently in use for detecting the presence of certain biological molecules. Methods aimed at detecting specific DNA and RNA sequences or specific proteins are discussed in the coming sections.

We note that molecules are most generally studied and followed as a population of molecules and less often as single molecules. Thus when we refer to the properties of 'a protein' or 'a DNA fragment', what we most typically mean is the properties of a whole population of proteins or DNA fragments of a certain type.

Cellular components can be monitored by their spectroscopic properties

Biological molecules absorb electromagnetic radiation over a range of wavelengths specific to the molecule of interest; this property can be exploited to identify and quantify molecules in solution. The amount of radiation absorbed as a function of wavelength is called the **absorption spectrum**. Several key biological molecules have spectroscopic properties that allow their presence to be readily monitored by virtue of the characteristic absorption spectra they generate. For example, Figure 19.36a shows how DNA and proteins have distinct absorption spectra in the UV range. A spectrum typically has one or more peaks of absorbance that can be used as a chemical 'signature' to monitor the presence of a particular species. DNA

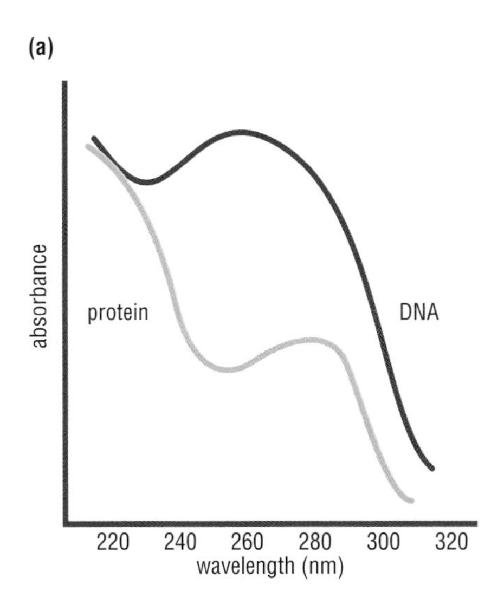

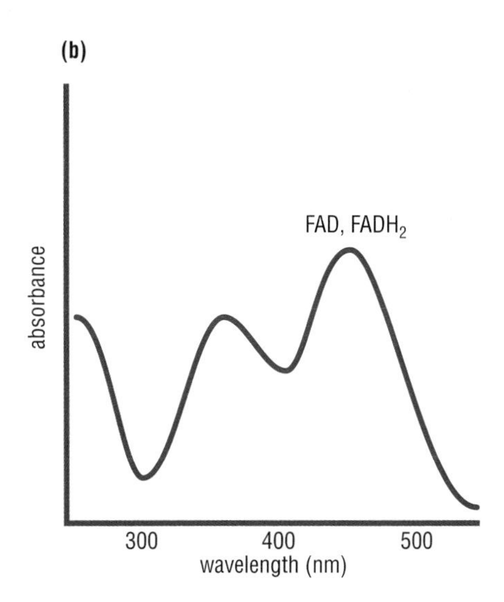

Figure 19.36 Representative absorption spectra. (a) Absorption spectra of DNA and protein at equal concentrations. At a wavelength of 260 nm, or A_{260}, a 50 μg/mL solution of double-stranded DNA or a 40 μg/mL solution of single-stranded RNA has an absorbance of ~1.0. At A_{280}, a 1 mg/mL of an 'average' protein has an absorbance of ~1.0, although this value depends on amino acid composition. (b) The absorption spectrum of FAD and $FADH_2$. Both FAD and $FADH_2$ exhibit two absorption peaks, at 380 and 450 nm.

From Harm, W. *Biological Effects of Ultraviolet Radiation*, Cambridge: Cambridge University Press, 1980.

and RNA have an absorbance maximum at ~260 nm that reflects the presence of heterocyclic bases (A, C, G, and T/U), whereas a typical protein with aromatic residues (tryptophan, tyrosine, or phenylalanine) usually has an absorbance peak at ~280 nm. Other peaks in the DNA, RNA, and protein absorption spectra at shorter wavelengths reflect the properties of the peptide and nucleic acid backbones.

Many small molecules in the cell also have characteristic absorption spectra that make them relatively easy to visualize (and identify). As shown in Figure 19.36b, the nucleotide cofactor flavin adenine dinucleotide (FAD) has a characteristic absorption spectrum with peaks at 380 and 450 nm. Proteins that are complexed with FAD (or its reduced derivative, $FADH_2$) can therefore be directly monitored by measuring their absorbance at these wavelengths. Cofactors that cause proteins to absorb light in the visible portion of the electromagnetic spectrum (from 400 nm to 700 nm wavelength) cause otherwise colorless proteins to appear colored, and are therefore called **chromophores** (from the Greek 'chromo', which means color, and 'phore', which means carrying). For example, the red color of hemoglobin, which gives blood its characteristic color, comes from the oxygenated heme prosthetic group that is complexed with globin peptide chains.

Optical absorbance is measured in an instrument called a **spectrophotometer**. Typically, a liquid sample is placed in a transparent vessel called a cuvette (illustrated in Figure 19.37), which is placed in the spectrophotometer. The intensity of a light beam that has passed through the sample is compared to the intensity of a beam that has passed through air or through a sample containing water or buffer. The relative amount of light absorbed by the sample is determined by three things: the concentration of the molecule, the distance the light traverses through the liquid, and the type of molecule the sample contains.

Each molecule has intrinsic properties that affect the absorbance of light and are indicated by a parameter called the **extinction coefficient**, denoted by the Greek letter ε. The extinction coefficient of a particular protein at 280 nm, for example, is dictated primarily by the total number of tryptophan and tyrosine residues in the protein. The light absorbed by the sample at a given wavelength λ, denoted A_λ, will then be the product of the extinction coefficient (ε_λ), the concentration of the molecule (c), and the distance the light traverses through the fluid (l). This relationship, known as the Beer-Lambert Law, can be represented by the following equation: $A_\lambda = \varepsilon_\lambda l c$.

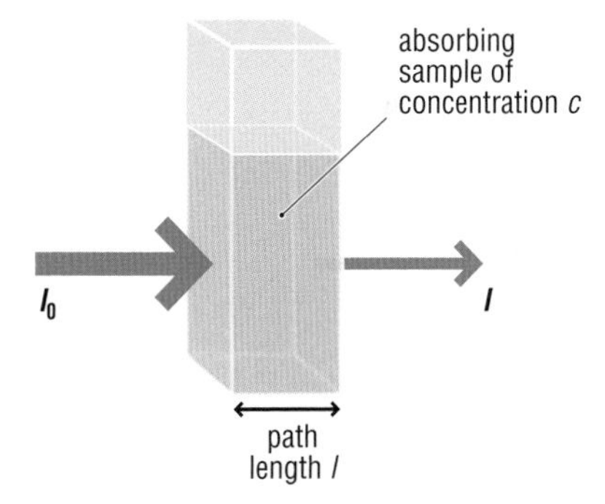

Figure 19.37 Absorption spectroscopy. The relative amount of light absorbed by a sample (I/I_0) is determined by the type of molecule it contains, the concentration of the molecule, and the distance the light traverses through the liquid. These parameters are related to one another according to the Beer-Lambert Law: $A_\lambda = \varepsilon_\lambda l c$, where A is the measured absorbance at wavelength λ, ε is the molar absorptivity or extinction coefficient (constant for each substance, units of M^{-1} cm^{-1}, where M is molar), l is the path length (in cm), and c is the concentration (in M). The key term is ε, which varies with wavelength and is a property of the particular molecule being evaluated.

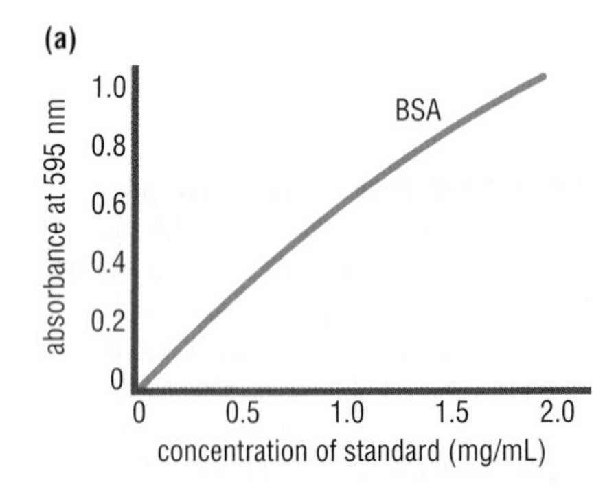

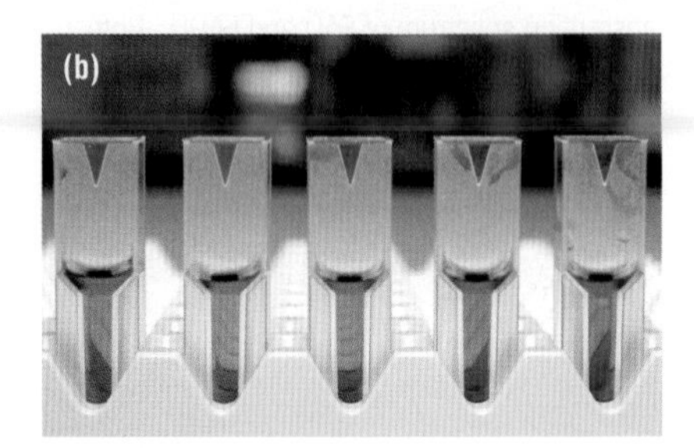

Figure 19.38 Typical standard curve for a Bradford assay. (a) A standard curve is constructed by performing the assay with known quantities of protein. Note that the assay becomes somewhat non-linear at higher protein concentrations. Samples are typically diluted until they fall within the linear range of the assay. (b) Visual examples of Bradford assay cuvettes with increasing concentrations of the protein bovine serum albumin (BSA). From http://www.eiroforum.org/media/photo_galleries/embl/index.html

When the extinction coefficient of the molecule under study is known, the absorbance of a sample can be used to determine the concentration of a given molecule (particularly if no other species are present that could contribute to the overall absorbance at that wavelength). In addition, the purity of a sample can sometimes be evaluated by asking whether the ratios of absorbance at different wavelengths match those of a pure sample. For example, the ratio of the absorbance at 260 nm and 280 nm is typically ~2 for a DNA population, whereas the ratio is ~0.6 for a typical protein preparation. Therefore, if the A_{260}/A_{280} ratio for a DNA sample is significantly less than 2, it is likely that it is contaminated with protein.

Cellular components can be monitored by binding or reaction with fluorochromes

In addition to using their intrinsic spectral properties, molecules can be detected by their binding or attachment to a specific molecule that either fluoresces or has a distinct color or spectroscopic property. The detectable molecules can be naturally occurring, as in the FAD/$FAHD_2$ example mentioned earlier, or they can be developed specifically for the purpose of detecting cellular components. For example, a method for determining protein concentration, known as the Bradford assay, relies on the binding of a dye called Coomassie Brilliant Blue to proteins. On its own, the dye has very little absorbance at 595 nm. However, when bound to protein, the dye changes conformation and absorbs light at 595 nm. Since the amount of bound dye is proportional to the amount of protein in a sample, the absorbance of the protein-dye mixture at 595 nm can be used to determine the concentration of the protein in the sample.

To use this assay, a standard curve is first generated by incubating a series of proteins standards of known concentration with dye and measuring their absorbance at 595 nm; an example of a standard curve is shown in Figure 19.38a. The protein sample in question is then similarly incubated with the dye and its absorbance compared to the standard curve to estimate the protein concentration in the sample (see Figure 19.38b). A drawback to this assay is that it assumes that the unknown protein binds about as much dye per mg of protein as the protein used as the standard, which is not always the case. Coomassie Brilliant Blue is also commonly used to stain proteins immobilized in gels after separation by electrophoresis, as we shall see in Section 19.7.

A commonly used molecule for the detection of DNA, and sometimes RNA, is ethidium bromide. This molecule inserts (intercalates) between the DNA or RNA bases, and the intercalated molecules fluoresce when irradiated with ultraviolet light. Methylene blue dye is also commonly used to stain nucleic acids. Although not as sensitive as ethidium bromide, methylene blue is less toxic and is less likely to interfere with subsequent manipulations (such as restriction enzyme digestion or annealing to complementary sequences) because it does not intercalate in the nucleic acid. Both reagents are commonly used to stain nucleic acid samples that have been resolved on acrylamide or agarose gels.

➔ We learn more about the use of acrylamide and agarose gels when we discuss electrophoresis in Section 19.7.

Radioactive labeling can be used to detect molecules

A powerful method for detecting very small amounts of a molecule is to incorporate radioactive atoms into them. Isotopes such as ^{32}P or ^{33}P, ^{35}S, ^{14}C, or ^{3}H are unstable, unlike their more common isotopes, ^{31}P, ^{32}S, ^{12}C, or ^{1}H. This instability means that they break down over time, emitting ionizing radiation that can be detected by a number of different methods.

Methods for detecting radioactivity
Geiger-Muller survey meter (Geiger counter)
Typically handheld gas-filled monitor that detects ions formed upon radioactive decay.
liquid scintillation counting
A scintillation counter converts radioactive emissions into visible light. Liquid scintillation fluid is an organic solvent containing a fluor (a compound that fluoresces in the presence of radioactivity). Scintillation fluid is mixed with the radioactive sample (which may be a small amount of fluid, a gel slice or a small piece of filter paper) in a vial which is then placed inside the counter in the dark so that the light emissions can be detected and counted.
film autoradiography
Photographic emulsions contain silver halide crystals that form silver atoms in the presence of radioactivity, leaving an image on the film after processing. X-ray film containing silver halide crystals is the format usually used in the laboratory and is exposed by pressing the film flat against a gel or paper that contains the radioactivity. Gel separation and autoradiography will be discussed further in section 19.7.
phosphorimager autoradiography
A phosphorimager screen is coated with a phosphor compound whose atoms are excited when they encounter radioactivity. When the screen is exposed to radioactivity, for example by being pressed against a gel, a latent image of excited atoms forms in the screen. When the screen is scanned with a laser, the stored energy in the excited atoms is released in the form of light, which can be measured.

Figure 19.39 Methods for detection of radioactivity.

Various approaches are available for incorporating radioactive isotopes into larger molecules, including proteins, nucleic acids, sugars, or lipids. Radioactive isotopes can be incorporated biosynthetically, for example, by growing cells in the presence of ^{35}S-labeled methionine, which becomes incorporated into newly translated proteins. Molecules can be labeled *in vitro* using enzymes that transfer a radiolabeled moiety to either nucleic acids or proteins. For example, adenosine triphosphate (ATP) synthesized with ^{32}P at the γ phosphate (the outermost phosphate) can be used as a donor to transfer the radioactive phosphate to a different molecule. To do this, T4 polynucleotide kinase is used to transfer the γ phosphate of ATP to a nucleotide substrate (for example, the 5′ end of a strand of DNA or RNA). In an analogous fashion, protein kinases typically transfer the γ phosphate of ATP to particular serine, threonine, or tyrosine residues in a protein.

Radioactively labeled molecules can be detected by a range of techniques, as shown in Figure 19.39. In many cases, it is possible to measure the amount of radioactivity present and thereby determine the quantity of radioactively labeled molecular species present in a sample.

Cellular components can be monitored by activity assays

In order to detect the presence of an enzyme and study how the rate of the catalyzed reaction changes under different conditions or as the result of mutation, it is necessary to have a method to monitor the particular activity of the protein. For example, a protein with kinase activity can be assayed by monitoring the addition of a radioactive phosphate to the target of the kinase. The advantage of an activity assay is that it can, in many cases, be sensitive enough to detect minute quantities of the species that might not be detectable by any other means, especially if the sample in question is impure. This is in part because product of an enzyme-catalyzed reaction may be easier to detect in small quantities than the enzyme itself, or because enough product may accumulate such that it becomes more abundant than the enzyme itself.

➔ Experimental approach 14.2 provides an example of an activity assay.

While the products of some enzymatic reaction cannot be easily monitored directly, in some cases, it is both possible and desirable to couple the reaction of interest to a second enzymatic reaction whose products can be monitored more readily. In the example shown in Figure 19.40, the reaction of interest is the hydrolysis of ATP by topoisomerase II, which yields ADP and inorganic phosphate. Although methods are available to measure ATP and ADP levels, a simple approach is use a product of the reaction – in this case ADP – to drive additional enzymatic reactions whose products can be easily detected using a spectrophotometer.

In the example depicted in Figure 19.40a, the coupled reactions take advantage of the fact that NADH has a peak absorption at 340 nm while NAD^+ does not absorb light at this wavelength. NADH and two additional enzymes are added to the reaction along with ATP and topoisomerase II. As ADP from the topoisomerase II reaction accumulates, it is used by pyruvate kinase to generate pyruvate, while regenerating ATP. The pyruvate is then converted into lactate by lactate dehydrogenase in a reaction that consumes NADH and releases NAD^+. Since one molecule of NADH is consumed for every molecule of ATP hydrolyzed by topoisomerase II, one can follow the progress of the topoisomerase reaction by using a **spectrophotometer** to monitor the absorbance of NADH at 340 nm. As the reaction progresses, NADH is converted to NAD^+, and the absorbance at 340 nm decreases, as depicted in Figure 19.40c.

Activity assays also make it possible to look for and isolate a biological molecule whose existence is suspected, but whose identity is unknown. For example, researchers were able to carry out RNA polymerase II-dependent transcription in a test tube using a nuclear extract when the identities of the particular proteins in the mixture that were required to facilitate transcription by RNA polymerase II were unknown. These proteins were identified by fractionating nuclear extracts

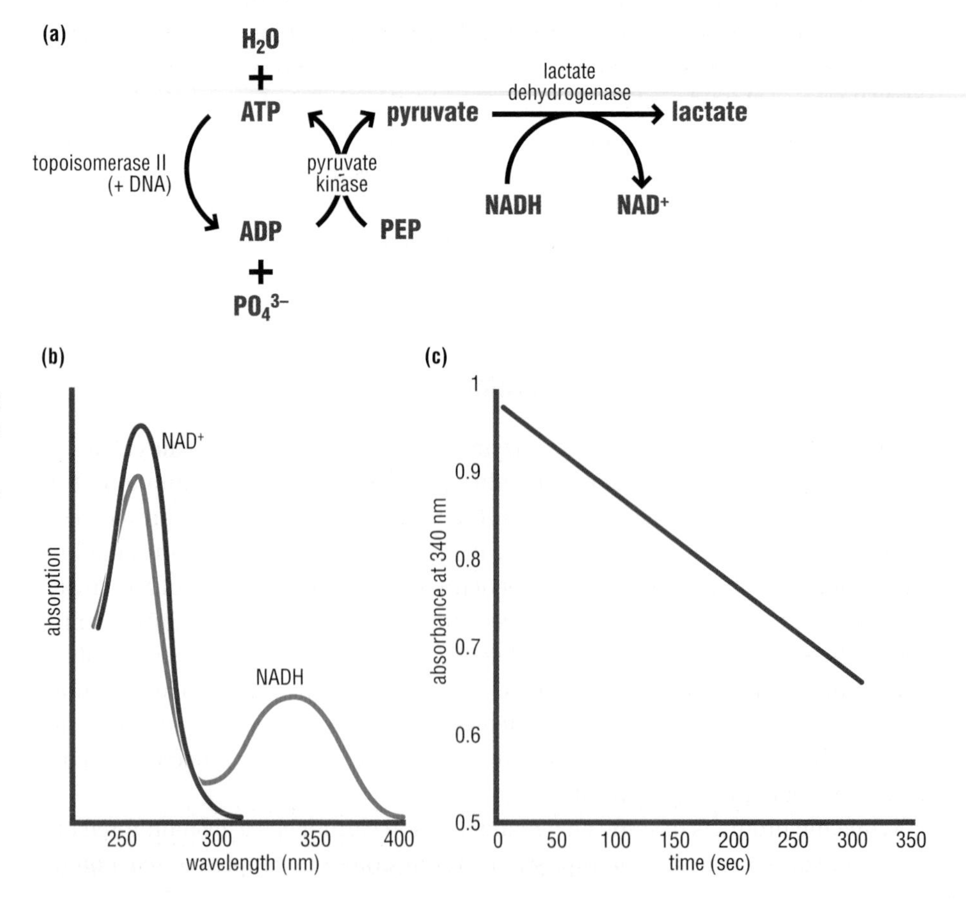

Figure 19.40 Measurement of enzymatic activity by coupling several reactions. ATP hydrolysis by DNA topoisomerase II can be followed by coupling the hydrolysis reaction to the oxidation of the reduced form of nicotinamide adenosine dinucleotide (NADH) to its oxidized form (NAD^+). (a) The reaction scheme. Topoisomerase II hydrolyzes ATP to adenosine diphosphate (ADP). This hydrolysis can be measured indirectly using two subsequent coupled reactions. First, the ADP that was generated by topoisomerase is converted back to ATP by pyruvate kinase, which uses phosphoenolpyruvate (PEP) as a phosphate donor. This converts PEP to pyruvate. Pyruvate, in turn, is a substrate for lactate dehydrogenase, which converts pyruvate into lactate, using NADH as a cofactor: for each pyruvate that is converted to lactate, one NADH molecule is converted to NAD^+. Thus, NADH concentration is directly correlated with ATP hydrolysis by topoisomerase. (b) Absorption spectra of NAD^+ and NADH. The spectra of these molecules are markedly different: NADH absorbs light at 340 nm whereas NAD^+ does not. (c) When ATP is hydrolyzed by topoisomerase, the concentration of NADH decreases as it is converted to NAD^+ and the absorption at 340 nm decreases in tandem.

and assaying the different fractions for their ability to promote transcription *in vitro*. Once a sufficiently pure sample with the desired activity was obtained, the proteins could be identified through a variety of methods, many of which will be described in the following sections.

19.7 SEPARATION AND ISOLATION OF BIOLOGICAL MOLECULES

Much of the information in this book derives from experiments done in the test tube using purified components. In order to study the behavior and properties of a specific molecule or complex in solution, one must be able to separate it from other cellular components. This can be done by taking advantage of the different physical and chemical characteristics that distinguish one molecule from another - for example, mass, shape, overall charge, or the ability to bind to a specific reagent. It is also possible to introduce a tag with known properties into a protein or nucleic acid molecule that can then be used to isolate the tagged molecule. In this section we will discuss some of the ways for isolating cellular structures and molecules.

Organelles and molecules can be separated by centrifugation

The first step in purifying biological molecules from cells often involves isolating the different organelles, or separating soluble and membrane-associated molecules from one another. This process is known as **cell fractionation** and can be done with whole tissues or with cells grown in culture. To begin this process, the intact cells must be disrupted to release their contents - but the treatment must not be so harsh as to disrupt the organelle or the molecular structures being studied. Figure 19.41 illustrates a variety of methods that can be used to disrupt cells; these include the use of a device similar to a mortar and pestle, or the use of ultrasonic waves, high pressure, or detergents.

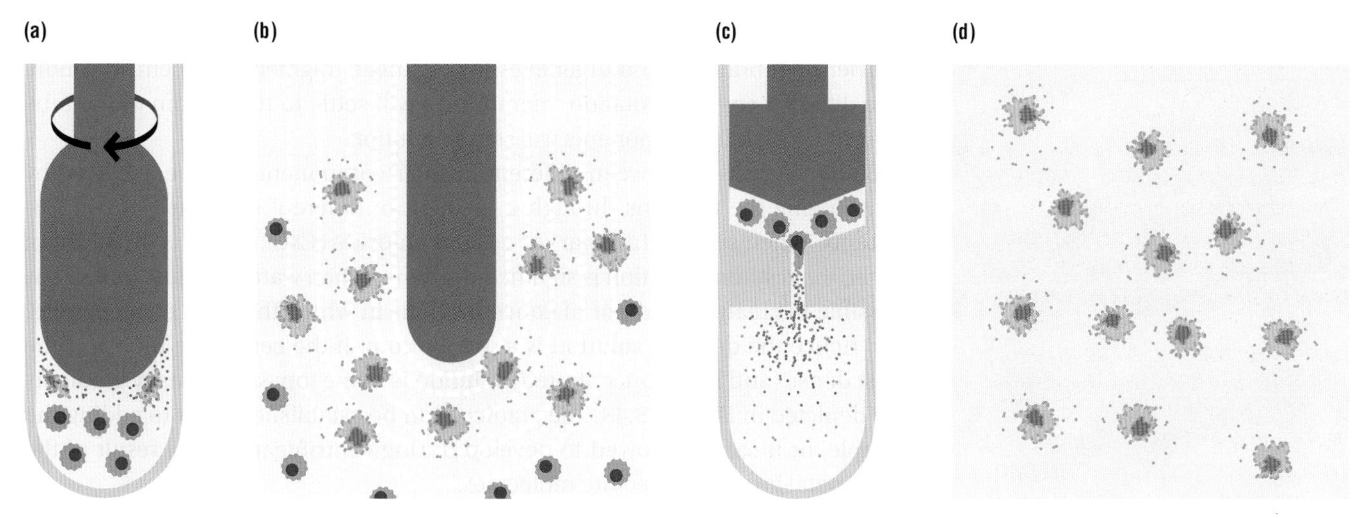

Figure 19.41 Methods for disrupting cells. Various methods are used to disrupt cells, including: (a) grinding cells with a pestle that fits tightly in a test tube, (b) subjecting cells to ultrasonic waves using an instrument called a sonicator (the tip of the sonicator is shown), (c) passing cells through a small opening by applying very high pressure, and (d) treatment of cells with mild detergents that generate holes in membranes.

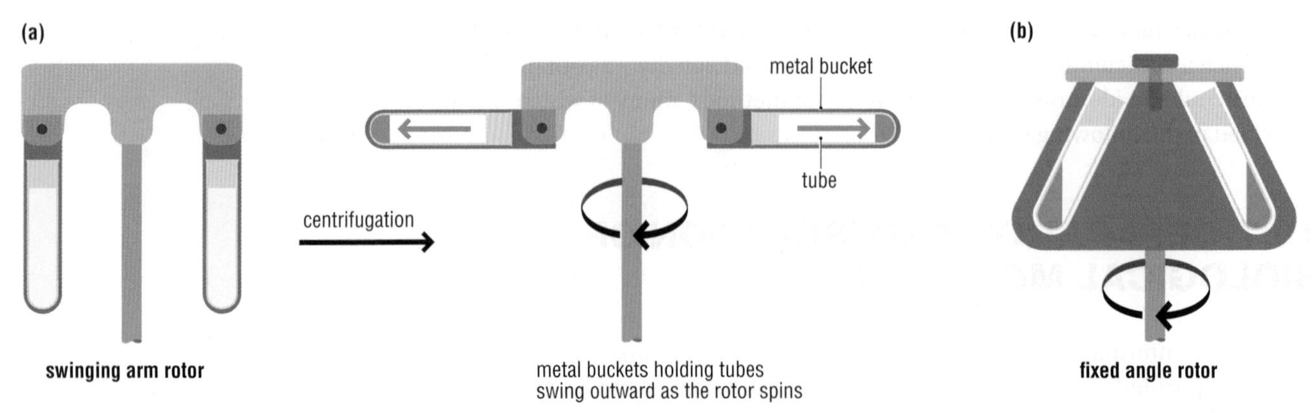

Figure 19.42 Centrifugation. Solutions can be subjected to centripetal force in a centrifuge rotor, which is spun at very high speeds (on the order of tens of thousands of revolutions per minute). This causes larger complexes and organelles to sediment to the bottom of the test tube. (a) In a swinging bucket rotor, the test tubes are placed in metal buckets that can swing outward as the rotor begins to spin. The centripetal force causes large complexes and organelles to sediment and collect in the bottom of the tube. (b) In a fixed-angle rotor, the test tube bottoms slant away from the central axis of rotation and remain at a fixed angle. The centripetal force acts perpendicular to the axis of rotation. As the rotor slows and stops, the layers of increasing density reorient so that they align along the vertical axis of the tube (that is, the densest layer is at the bottom and least dense layer is at the top).

Once cells have been disrupted, the individual components can be separated. This is commonly done using a **centrifuge**, an instrument used to spin samples in test tubes at high speed. The centrifuge creates centripetal force perpendicular to the rotation axis. Test tubes (or other vessels) are placed in a central holder in the centrifuge called a **rotor**, which either allows the tubes to swing outward as the rotor spins, or holds the test tubes at a fixed angle; these differences are illustrated in Figure 19.42. In either case, the centripetal force that results when samples are accelerated at upwards of 100 000 *g* (where *g* denotes the acceleration due to gravity) causes organelles or large molecular complexes to move toward the bottom of the tube and form a pellet.

The rate at which a particle descends through the tube depends upon its size and the speed (angular velocity) at which the centrifuge rotates: the faster the rotor turns, the greater the acceleration and thus the force that is exerted on the particle. By choosing a particular speed and amount of time, the desired particle can be collected at the bottom of a test tube (see Figure 19.42). For example, the cytosolic components of a cell that has been broken or lysed can be readily separated from the heavier membrane-bound organelles (or organelle fragments) by centrifugation. At a different speed of rotation, membranes will settle to the bottom of the tube while the cytosolic components will stay in solution.

Better separation between different cellular components can be achieved by centrifuging the fractions through a gradient of different sugar or salt concentrations. When sucrose (a sugar) or cesium chloride (CsCl; a salt) is dissolved in water, the resulting solution is significantly denser than water or dilute buffer. It is possible to create a **gradient** of concentrations in which the most concentrated, and hence the densest, solution is at the bottom of the centrifuge tube and the least dense (and least concentrated) solution is at the top; such density gradients are depicted in Figure 19.43. The gradient can be established prior to adding the sample, or it can be allowed to develop during centrifugation as a result of the centripetal force on the solute molecules.

A density gradient can separate different cellular components in one of two ways. In velocity sedimentation (Figure 19.43a), the sample is layered on the top of the gradient, and the sample is centrifuged for a fixed amount of time. The rates at which different molecules or organelles migrate toward the bottom of the tube depend on both their size and shape. After centrifuging the sample for a fixed amount of time, samples are collected from either the top or the bottom of the tube. (This can be done by punching a hole in the bottom of the tube and allowing the sample to drip out.) In equilibrium sedimentation, the sample is spun for a long period of time until the molecules reach equilibrium and are no longer moving within the gradient (Figure 19.43b). A molecule or complex will stop moving through the gradient when it reaches a position in the tube where the density of the gradient matches the density of the molecule.

Information about the size and shape of a macromolecule or complex can be obtained by a more sophisticated instrument known as an analytical ultracentrifuge, which can both spin samples at very high speeds and simultaneously record the concentration of the macromolecule at different radial positions in the ultracentrifuge vessel (known as a cell). In this method, there is standard buffer rather than a salt gradient. The rate at which a molecule or complex of molecules sediments in the centrifuge cell is denoted by a unit called a Svedberg (or S). The S value of a given complex is governed by both its molecular weight and its shape. The different subunits of the ribosome, which we learned about in Chapter 11, are denoted by the S value that characterizes their sedimentation behavior.

➔ We see an example of the use of CsCl gradient centrifugation to separate DNA of different densities in Experimental approach 6.1.

(a) velocity sedimentation

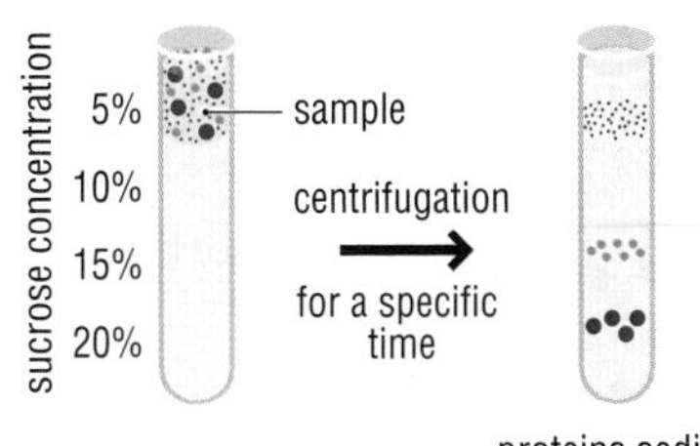

(b) equilibrium sedimentation

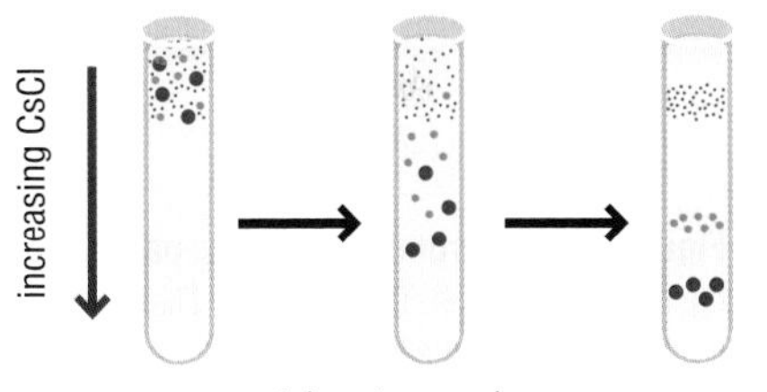

Figure 19.43 Density gradient sedimentation. Cellular components can be separated from one another by centrifugation through a solution whose density is highest at the bottom of the test tube and lowest at the top. (a) In a sedimentation velocity experiment, the gradient (for example, of sucrose) is first created by layering solutions of decreasing density in the test tube and then introducing the sample at the top of the tube. The rate at which particles sediment through this gradient is determined by their mass and shape. In the example shown, the red particles are larger and denser than the dark blue and orange particles, and thus, during centrifugation, they move faster through the gradient and end up at a more dense part of the gradient. (b) In an equilibrium sedimentation experiment, particles are separated by density. A solution containing the sample and a high concentration of a substance such as the salt CsCl is centrifuged at high speeds for long periods. Eventually, a density gradient forms under the centripetal force. The different particles in the sample will be found at a position in the tube where the solution density matches the particle density.

Macromolecules can be separated based on their solubility

The initial extract prepared from disrupted cells is often very complex, containing thousands of different proteins and complexes. These macromolecules differ in size and shape, as well as in other properties including overall charge and in the relative distribution of hydrophobic and polar residues on the surface. Each of these physical properties distinguishes one macromolecule or complex from the next and can be exploited to separate molecules.

A simple method for separating macromolecules from one another is to fractionate them based on their solubility, typically in a salt solution such as ammonium sulfate. Most proteins can be induced to become insoluble and precipitate at sufficiently high concentrations of ammonium sulfate, although the precise concentration at which a protein becomes insoluble varies widely among macromolecular species. Thus, by using the appropriate salt concentration, one can separate the protein of interest (and other proteins or complexes with similar solubility properties) from many other proteins. Either the solution containing the soluble protein of interest can be isolated, or the precipitate containing the protein of interest can be resuspended in buffer at a low-salt concentration, thus enabling the protein to re-enter the solution. Other types of precipitating agents, for example polyethylene imine, are also used to precipitate both proteins and nucleic acids.

Proteins and protein complexes can be purified by column chromatography

A common method for purifying proteins is column chromatography, which uses a glass or plastic tube filled with material (the matrix) that separates proteins based on their physical properties as they flow through the column. The protein sample is first applied to the top of the column as shown in Figure 19.44. As a buffer solution is flowed continuously through the column, proteins in the sample migrate through the column at different rates, depending on the nature of the column matrix and

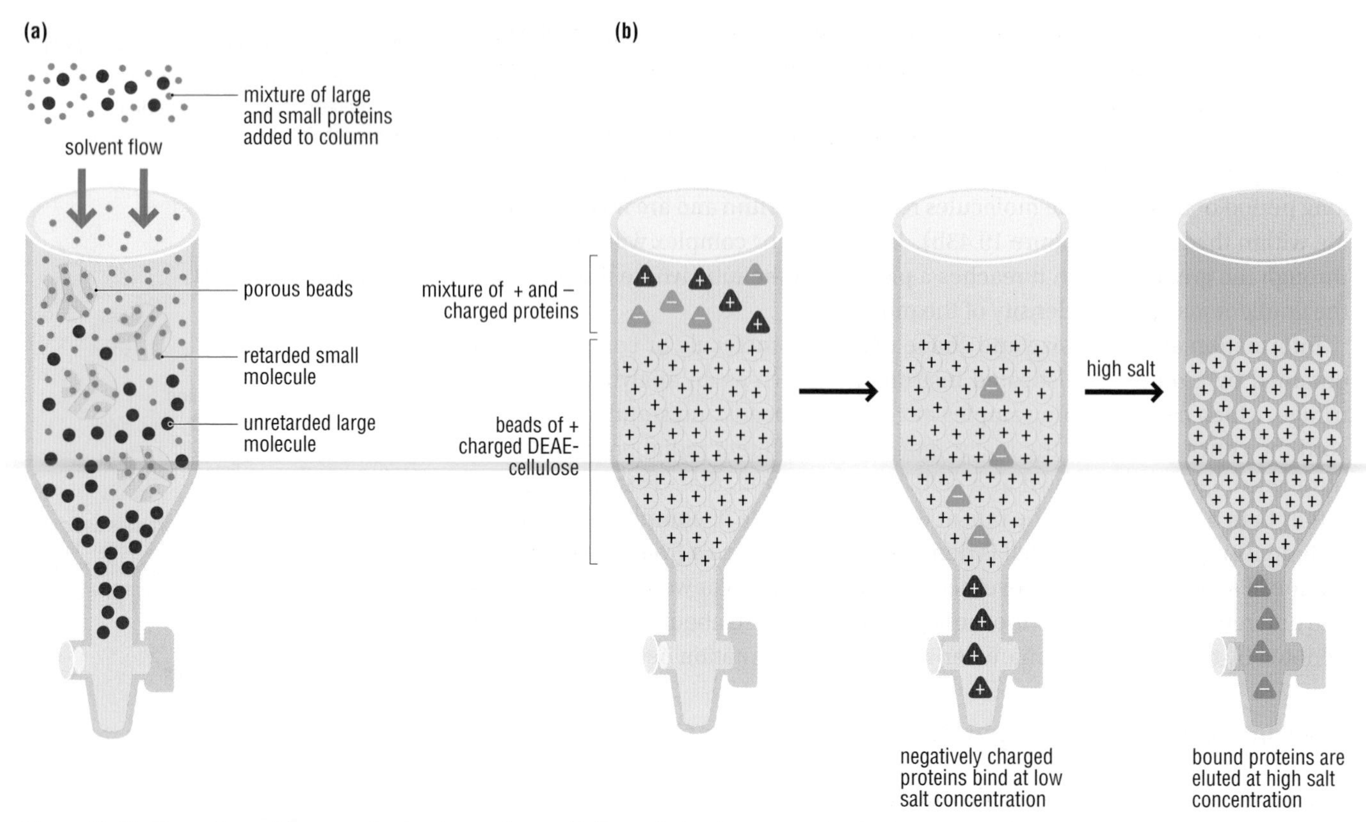

Figure 19.44 Chromatographic separation by size or charge. (a) Size exclusion chromatography, also known as gel filtration, separates molecules based on their size and shape. The column is filled with porous beads (in yellow). Large molecules (red) that cannot fit into the pores in the beads pass quickly between the beads, while smaller molecules (orange) that can enter the beads take more time to pass through the column. (b) Ion exchange chromatography separates molecules based on charge. If the beads in the column are positively charged, negatively charged proteins (green) will bind to the column matrix at low-salt concentrations, as a result of ionic interactions, while positively charged proteins (blue) will flow through the column. The negatively charged proteins can be induced to dissociate from the beads by increasing the salt concentration.

the physical and chemical properties of the proteins. Different types of columns are used to separate proteins based on size, charge, hydrophobicity, and other properties.

In **size exclusion chromatography**, protein molecules and complexes can be separated on the basis of size and shape on a **gel filtration** column (see Figure 19.44a). The column matrix consists of microscopic beads, each containing many tiny pores. Small molecules are more likely to go through the pores of the matrix and thus travel a greater distance and move through the column more slowly. Larger molecules, on the other hand, are excluded from entering the pores in the beads and can therefore only pass through the spaces between the beads. As a result they travel a shorter distance overall and pass through the column more quickly.

Ion exchange chromatography columns separate proteins on the basis of surface charge (see Figure 19.44b). These columns contain a matrix bearing either positively or negatively charged chemical groups. In low-salt solutions, proteins with a negative surface charge will bind to the positively charged matrix in **anion exchange** columns, while proteins with a positive surface charge will bind to the negatively charged matrix in **cation exchange** columns. The bound proteins can then be washed off (or 'eluted') in buffer containing higher salt concentrations, which competes for the charged chemical groups on the matrix.

Another property that can be used to purify proteins is the ability of the protein to bind selectively to a particular molecule or chemical group. This approach is called **affinity chromatography**, since it relies on the protein binding to a molecule for which it has specific affinity. For example, the heparin polymer mimics the sugar-phosphate backbone of DNA and RNA and, when attached to a matrix, can be used to purify proteins that bind nucleic acids. The dye, Cibacron blue, is used in an analogous way to purify proteins that bind to NAD^+.

Not all proteins have known a known substance to which they bind, however. Therefore, we can use cloning and recombinant DNA techniques (Section 19.4) to add a protein 'tag' that binds a known substance, as we discuss next.

Proteins and nucleic acid can be engineered to contain tags that facilitate purification

Instead of simply relying on the native properties of a protein to separate it from other proteins in the cell, the recombinant DNA technology described in Section 19.4 can be used to add an affinity tag that can be exploited to purify a protein using affinity chromatography. The tag generally consists of additional amino acids added to the N- or C-terminus, thus extending the polypeptide chain (see Figure 19.45). Some commonly used tags are a histidine-rich peptide that binds tightly to a matrix containing nickel ions; a peptide that binds to a specific antibody immobilized on a column; or an entire protein domain such as glutathione-*S*-transferase (GST) that binds to a matrix containing immobilized glutathione. More than one affinity tag can be incorporated into a single polypeptide chain, making it possible to use different types of affinity purification sequentially.

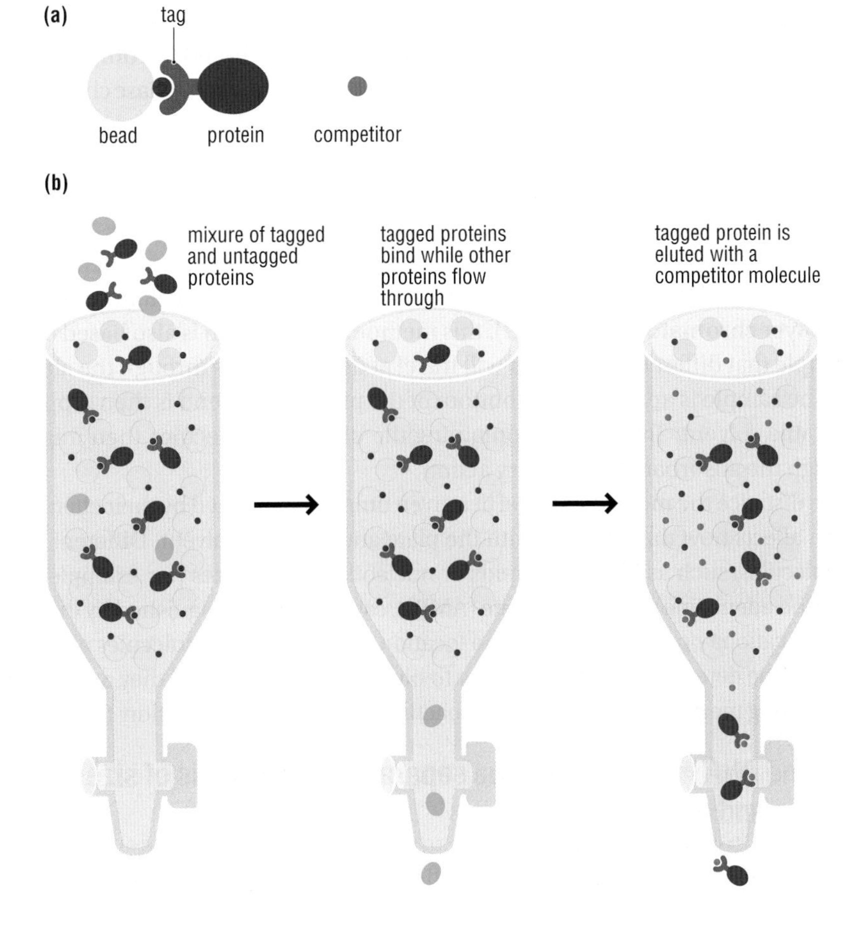

Figure 19.45 Using tags to purify proteins. (a) A fusion protein containing an additional protein tag can be engineered. The tag can bind to a specific molecule (pink) on the beads in the column. The competitor (light blue) has the same or similar structure to that of the resin-bound molecule. (b) Affinity purification of a tagged protein. The protein tag binds to the molecule that is immobilized on a column, causing the fusion protein (dark blue) to be retained while other proteins (green) flow through. The protein can be eluted by adding a molecule (light blue) that competes with the fusion domain for binding to the column matrix.

How are tagged proteins purified? The engineered protein of interest fused to the tag is expressed from a plasmid or the chromosome. The fusion protein is then isolated by lysing the cells and passing the resulting extract on the appropriate affinity column, as illustrated in Figure 19.45. The fusion protein binds to the column while the other components in the extract pass through. The fusion protein is then eluted from the column with a molecule that competes with the fusion protein's interaction with the matrix. Imidazole, for example, is used to elute histidine-tagged proteins from a nickel column (since it competes with the imidazole ring of histidine side chains, which coordinate the metal), whereas glutathione is used to elute proteins from a GST column.

An important caveat is that affinity tags have the potential to alter the activity of a protein. For systems where the activity of the protein can be assayed *in vivo* or *in vitro*, it is important to check that the tag has not altered the activity. Tagged proteins can also be engineered to contain intein sequences, which we learned about in Chapter 14, or a protease recognition site between the tag and the protein sequence of interest. These can then be exploited to remove the tag, either by changing the solution conditions to promote intein self-splicing or by adding a protease.

DNA or RNAs of interest similarly can be fused to tags that allow them to be separated from a pool of other molecules by affinity chromatography. Examples of such tags include sequences that are tightly bound by specific DNA or RNA binding proteins. These proteins can be attached to a solid matrix (such as in affinity purification) or be isolated using one of the methodologies described above.

High-performance liquid chromatography and thin-layer chromatography can be used to separate small molecules

The method most commonly used to separate small peptides, nucleic acids, and small molecules is called **high-pressure** or **high-performance liquid chromatography (HPLC)**. In a common application of this method known as reverse-phase chromatography, the column matrix consists of small beads derivatized with hydrophobic alkyl chains (such as $C_{18}H_{37}$ or C_8H_{17}). Small molecules bind to the beads via hydrophobic interactions and are then eluted with increasing concentrations of organic solvent. The liquid is passed through the column at high pressure, which increases the resolution.

Instead of using a column matrix, small molecules can also be separated on a flat support covered by a thin layer of an absorbant such as silica or cellulose. Termed **thin-layer chromatography (TLC)**, this method of separation is also based on the principle that different molecules will travel at different rates through the absorbant. The sample is applied to the bottom of the plate, whose end is then dipped in a suitable solvent. The solvent containing the different molecules then migrates through the absorbant by capillary action.

The distance the molecules travel in a given time is determined by their chemistry, which affects how they interact with the plate and with the solvent. Different types of molecules, such as particular lipids or nucleotide phosphates (for example, ATP, ADP, and adenosine monophosphate (AMP)), can then be distinguished by their different locations on the TLC plate. The location of the different molecules along the plate can be determined by their ability to bind a specific dye, or, if they are radiolabeled, using methods for detecting radioactivity as described in Section 19.6.

RNA and DNA molecules can be separated on the basis of size on agarose and acrylamide gels

As we recall from Chapter 2, DNA and RNA are negatively charged. Thus, when nucleic acids are subjected to an electrical field, they migrate toward the positive

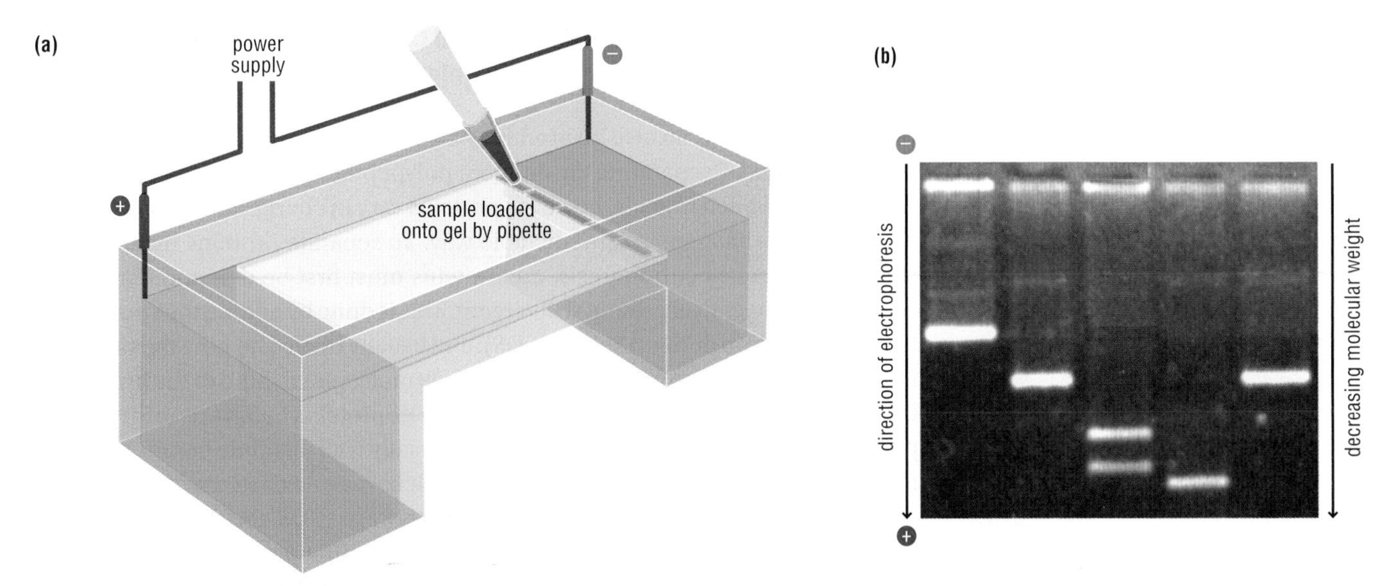

Figure 19.46 Agarose gel electrophoresis. (a) Apparatus showing an agarose slab gel (in light blue) immersed in buffer and connected to a power supply. The DNA (or RNA) sample (in pink) is introduced into the slots (gray) at the negative end of the gel, and a voltage is applied, causing the DNA fragments to migrate toward the positive pole. (b) Example of DNA fragments separated on an agarose gels. After electrophoresis, the gel was immersed in a solution containing a dye that binds to DNA (ethidium bromide). The DNA bands (in white) were detected by exposing the gel to UV light.

pole. This property is used to separate DNA and RNA molecules of different sizes by a method called **gel electrophoresis**, an apparatus for which is depicted in Figure 19.46. This method relies on a porous gel made of agarose or polyacrylamide through which nucleic acids migrate when the gel slab is immersed in an aqueous solution and an electric field is applied.

The gel slab is prepared by heating a solution containing agarose, thus causing it to liquefy, and then allowing it to solidify in a mold, or by polymerization of acrylamide (which can be induced by adding a cross-linking chemical) between two panes of glass or plastic, as shown in Figure 19.47. The pore size in the gel can be adjusted by altering the concentration of agarose or acrylamide. The nucleic acid sample is applied to the electropositive end of the gel, as illustrated in Figure 19.46a, and the sample migrates through the gel toward the positive pole at the opposite end. The rate at which nucleic acids migrate through the gel depends on their length, with smaller molecules migrating faster than longer ones. The DNA or RNA molecules can then be visualized by staining with dye or by autoradiography, as we learned in Section 19.6.

DNA molecules of longer than approximately 20 000 bases are not well resolved or separated by conventional electrophoresis on agarose gels, which are typically used to separate DNA fragments of more than a few hundred base pairs in length. Instead, chromosomes and large DNA fragments must be separated using a variant of the gel method, termed **pulsed-field gel electrophoresis**. In these gels, electric fields are applied in a series of orthogonally oriented pulses (meaning at right angles to one another). Every time the orientation of the electric field is changed, the DNA molecules are reoriented. Larger DNA molecules move more slowly through the gel because they take longer to reorient each time the electrical field is shifted, allowing for separation of very large fragments on the basis of size.

A variant of this method is the **field inversion gel (FIG)**, in which the charges at the two ends of the gel are periodically reversed in pulses of several seconds. This switching or inversion of the electric field serves the same purpose as the pulsed-field gels: to reorient the large DNA molecules. A FIG gel is simpler to manufacture and use, although the separation by FIG is typically not as good as it is for pulsed-field gels.

Proteins can be separated by size and charge on one- and two-dimensional poly acrylamide gels

Gel electrophoresis can also be used to analyze the composition of a protein mixture or complex by separating the different polypeptides based on their molecular weight. Unlike nucleic acids, however, proteins vary in charge and shape and thus do not migrate in a gel in a predictable way. To separate proteins by molecular weight using gel electrophoresis, the proteins must first be denatured and then coated with a negatively charged detergent. This is done by adding the strong ionic detergent, sodium dodecyl sulfate (SDS), and usually by also heating the sample to induce the protein to unfold and allow the SDS molecules to bind to the hydrophobic regions of the protein. In the presence of SDS, proteins behave as unstructured polymers coated with a negative charge, especially if all disulfide bonds are eliminated by adding a reducing agent. Proteins that have been coated with SDS can be separated by SDS-polyacrylamide gel electrophoresis (**SDS-PAGE**).

SDS-PAGE gels differ from the gels used to separate nucleic acids in that they are usually discontinuous. The sample first migrates through a small, slightly acidic 'stacking' gel with a low acrylamide concentration, followed by a longer, mildly basic 'resolving' gel containing a higher concentration of acrylamide; the stacking and resolving gels are depicted in Figure 19.47a.

The running buffer contains glycine, a zwitterion that is neutral in the acidic stacking gel and becomes negatively charged in the basic resolving gel. This change in ionization, along with the differing acrylamide concentrations in the stacking and resolving gel, helps the protein sample to become focused into a tight band in the stacking gel. Once the proteins reach the resolving gel, they migrate according to their molecular weights, with small proteins separating into faster migrating bands and larger proteins migrating more slowly. The mixture of proteins is thus separated into discrete bands according to size. The percentage of acrylamide in the resolving

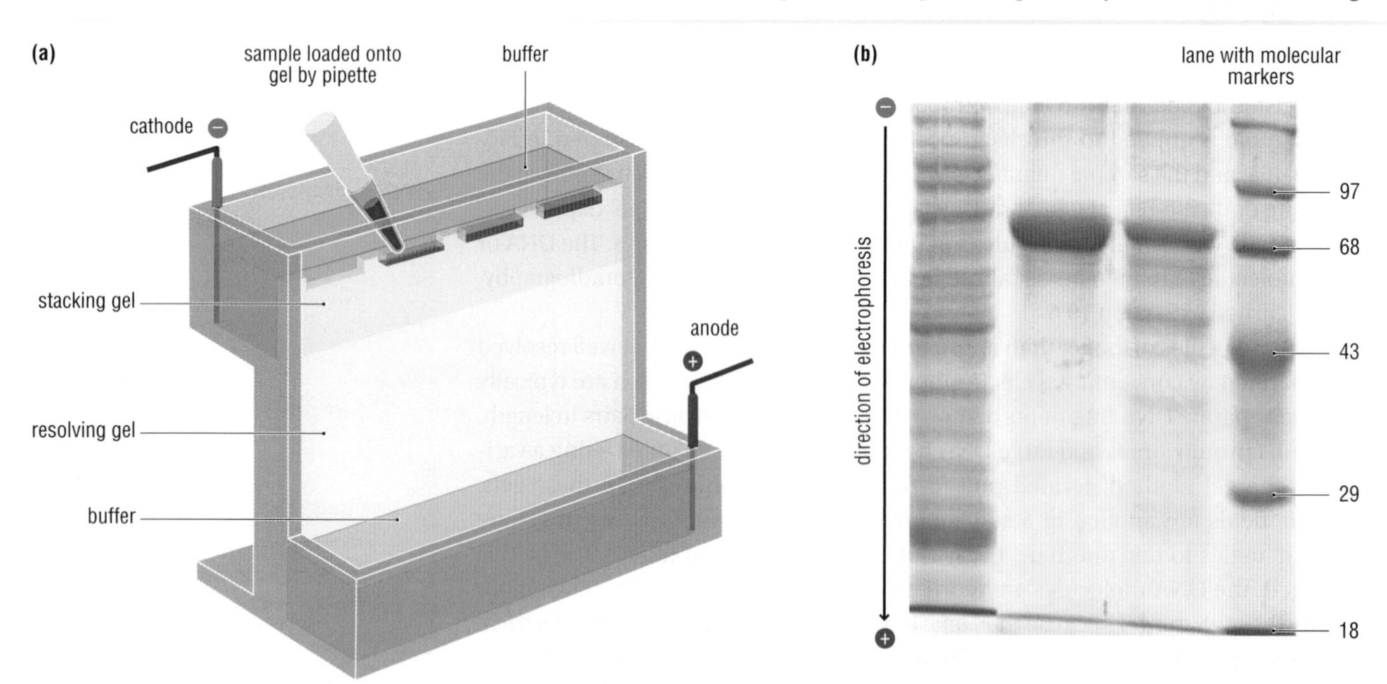

Figure 19.47 SDS-PAGE. (a) This type of gel is used to separate proteins by molecular weight after they have first been denatured in the presence of the detergent, SDS. The gel is discontinuous, with the sample first migrating through a stacking gel (in yellow), which helps to compact the sample into a single, tight band, followed by separation on a resolving gel, which separates proteins according to their size. (b) Depiction of an SDS-PAGE gel that has been stained with Coomassie Brilliant Blue to visualize the different proteins. The right lane contains a mixture of proteins of known mass, to indicate the approximate position of proteins of similar mass on the gel (numbers are in kilodaltons (kDa)).

gel can be adjusted to maximize the resolution of individual bands over a desired molecular weight range. The proteins can then be visualized by a variety of means, including the binding of the dye, Coomassie Brilliant Blue, as described in Section 19.6 and shown in Figure 19.47b, or by silver staining, whereby silver ions bind to the amino acid side chains (in particular the sulfhydryl and carboxyl groups) and are then chemically reduced to metallic silver, which is easily visualized.

Sometimes it is useful to separate proteins on the basis of other properties, such as their isoelectric point. The net charge of a protein depends upon its amino acid composition and relative number of acidic and basic residues. The isoelectric point, pI, is the pH at which a protein has no overall charge. Proteins are positively charged in solutions at a pH below the pI and are negatively charged at a pH above the pI.

An **isoelectric focusing (IEF) gel** is a type of polyacrylamide gel that has a stable pH gradient across the gel, as depicted in Figure 19.48a. When a native protein is loaded onto the high pH end of the gel and an electric field is applied, the protein will begin to migrate toward the positive pole of the gel because the protein has an overall negative charge (assuming the pI is lower than the highest pH in the IEF gel, which is typically around 9 or 10). The protein will continue to migrate through the gradient of decreasing pH until it reaches a zone in the gel whose pH corresponds to the pI of the protein. At this point, the protein is no longer charged and ceases to migrate in the gel.

IEF can be used in combination with SDS-PAGE to separate proteins by both pI and molecular weight. This method is called **two-dimensional gel electrophoresis** and is depicted in Figure 19.48b. Two-dimensional gels are particularly useful for separating proteins that have a similar size or pI and are thus difficult to separate by SDS-PAGE or IEF alone. Proteins can first be separated on the basis of pI using IEF, and separated subsequently on the basis of size using SDS-PAGE (Figure 19.48c).

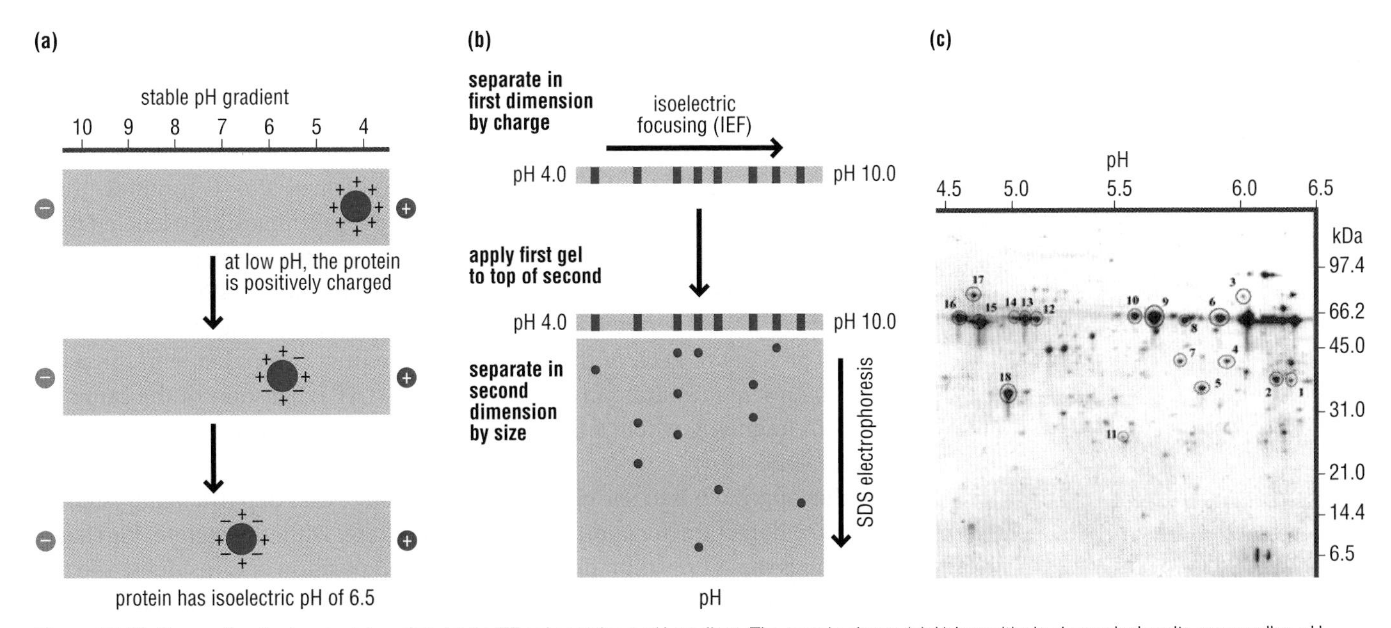

Figure 19.48 Separation by isoelectric point. (a) An IEF gel contains a pH gradient. The protein shown (pink) is positively charged when its surrounding pH is acidic. When an electrical field is applied, the protein migrates through the gel until it reaches a pH that matches its pI, at which point it is no longer charged and therefore stops migrating. (b) Two-dimensional gel electrophoresis. Proteins are first separated along one dimension by their pI, and then along the second dimension by molecular weight. In practice, the first separation by IEF is done in a thin tube of polymerized acrylamide containing the desired pH gradient. After the proteins reach their pI, the gel is removed from the tube and placed across the top of an acrylamide gel as shown Figure 19.47. (c) Example of a two-dimensional gel of *D. melanogaster* sperm. The purified sperm proteins are run on a pH 4–7 isoelectric focusing strip and run in the second dimension on a 12.5% SDS-PAGE gel. This gel demonstrates the resulting separation of proteins, which allows one to excise selected proteins from the gel (circled) for further analysis. From Karr, TL. Application of proteomics to ecology and population biology. *Heredity*, 2008;**100**:200–206.

19.8 IDENTIFYING THE COMPOSITION OF BIOLOGICAL MOLECULES

Once a cellular component of interest has been purified away from other components, it is usually desirable to determine its composition. For example, investigators often wish to determine the nucleotide sequence of a DNA or RNA molecule, or the amino acid sequence of a protein. In this section, we will discuss various methods for determining the sequence of these biological molecules. Many of the technologies described in this section are not sensitive enough to determine the sequence of a single molecule, and thus require the isolation of sufficient material, which can then be successfully analyzed.

The sequence of RNA polymers can be determined by reaction with specific nucleases or chemicals

The sequence of an RNA molecule can be determined with the aid of enzymes or chemical reagents that cleave RNA molecules after particular nucleotides. By carrying out these experiments under conditions in which a partial cleavage reaction is performed, a population of identical RNA molecules are cleaved at different places. The entire sequence can then be reconstructed based on the distance of each cleavage site from the end of the molecule. For this, the cleaved molecules need to be visualized.

One approach is to label either the 5′ or 3′ end of the unknown RNA so that it can be visualized when separated on a polyacrylamide gel (see Section 19.7). RNA can be easily labeled at the 5′ end with radioactive phosphate using T4 polynucleotide kinase and γ-[^{32}P]-ATP. Alternatively, the 3′ end can be labeled with [^{32}P]-pCp (cytidine-3′, 5′-bis-phosphate) using T4 RNA ligase. The end-labeled samples are then exposed to sequence-specific nucleases, or certain sequence-specific chemical modification reagents, which result in strand cleavage. These fragments can then be resolved on denaturing polyacrylamide gels to directly read out the sequence of the polymer. By comparing the length of the cleaved molecules, it is possible to obtain information about the entire sequence, as illustrated in Figure 19.49a.

For example, the ribonuclease T1 preferentially cleaves after G residues. Thus, the positions of G residues can be determined by partially digesting a labeled RNA molecule with ribonuclease T1, which produces a variety of fragments each ending in a G. Likewise, ribonuclease U2 cleaves after A residues, RNase PhyM cuts after U and A, RNase *B. cereus* cuts after U and C, and RNase CL cuts after C. The positions on a gel of each of these fragments are then compared with the positions of fragments resulting from a partial base (OH^-) digestion of the same labeled RNA fragment, where a band is seen for each consecutive nucleotide (see Figure 19.49a).

Another approach is to use chemicals that modify bases at particular positions, and then to detect the location of the modified base. Dimethylsulfate, for example, modifies the N1 position of A residues, the N3 position of C residues, and the N7 position of G residues. When a G residue is modified at the N7 position, treatment of the RNA strand with the chemical aniline will cause strand cleavage at the modified base. The fragments can then be separated based on size, as described above for the enzymatically cleaved RNA.

Another method of sequence analysis using chemical modification is primer extension. The RNA to be analyzed is annealed to a radiolabeled primer, which is then extended by reverse transcriptase. In the case of the modified A and C

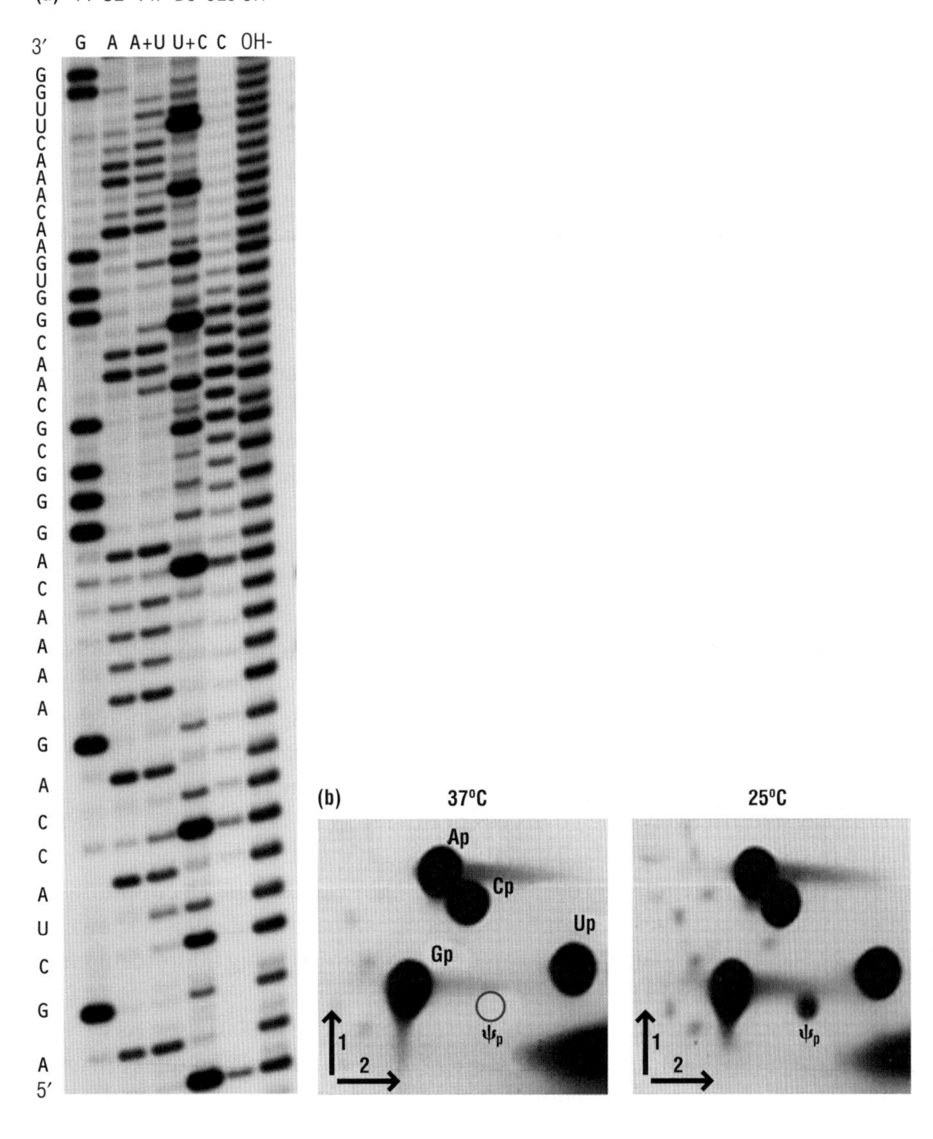

Figure 19.49 Sequencing RNA polymers. (a) Nuclease and chemical digestion of RNA. 5′ end-labeled RNA was prepared by *in vitro* transcription, subjected to partial digestion with ribonucleases T1, U2, PhyM (Ph), *B. cereus* (BC), CL3 or with base (OH⁻), and analyzed by denaturing polyacrylamide gel electrophoresis. The treatment with base generates a ladder of products resulting from cleavage at each nucleotide position, whereas T1 only cleaves after guanosine (thus the incomplete banding pattern), U2 cuts after A, PhyM cuts after U and A, *B. cereus* after U and C, and CL3 cuts after C. The sequence of this RNA, as interpreted from these data, is shown on the left side (b) Detection of a modified nucleotide by two-dimensional TLC. *In vivo* labeled ^{32}P pre-rRNA was isolated at 37°C or 25°C from yeast containing a temperature-sensitive allele in the pathway for RNA pseudouridylation. Thus, pseudouridylation is expected to happen at 25°C but not at 37°C. The RNA was digested with RNase T2 and analyzed by two-dimensional cellulose TLC: the samples were run on a TLC plate with one solvent mixture, and the plate then rotated in 90° and run in the second dimension with another solvent mixture. Arrows indicate the two directions of chromatography. Spots corresponding to the Ap, Gp, Cp, and Up residues are indicated. Pseudouridine 3′-monophosphate, whose expected position is indicated in each panel (Ψp), is not detectable in the heat-treated population (37°C, on left).

Panel (a) from Barrick, JE, Sudarsan, N, Weinberg, Z, Ruzzo, WL, and Breaker, RR. 6S RNA is a widespread regulator of eubacterial RNA polymerase that resembles an open promoter. *RNA*, 2005;**11**: 774–784; panel (b) from Bousquet-Antonelli, C, Henry, Y, Gélugne, J-P, Caizergues-Ferrer, M, and Kiss, T. A small nucleolar RNP protein is required for pseudouridylation of eukaryotic ribosomal RNAs. *The EMBO Journal*, 1997;**16**: 4770–4776.

residues, where the modification happens on the face of the nucleotide involved in Watson–Crick base-pairing, the reverse transcriptase cannot place a nucleotide opposite the modified base and extension will be aborted. Since different RNA molecules will be modified at different places, the lengths of the reverse transcriptase products, as analyzed by gel electrophoresis, will indicate the positions of the A and C residues. Other chemical reagents typically used for RNA modification include kethoxal (G specific), 1-cyclohexyl-(2-morpholinoethyl) carbodiimide metho-p-toluene sulfonate (CMCT) (U specific) and *N*-methylisotoic anhydride (NMIA) (acylation at the ribose 2′ OH position).

➲ The application of chemical probing to ribosome structural analysis is described in Experimental approach 11.1.

RNA molecules may contain both standard and modified ribonucleotides. Because post-transcriptionally modified nucleotides often behave like standard nucleotides in typical sequencing methods, their detection requires the use of other methods, such as TLC (Section 19.7). This approach, for example, can be used to detect pseudouridine in an RNA molecule, as illustrated in Figure 19.49b.

The sequence of both RNA and DNA molecules can be determined by primer extension in the presence of dideoxynucleotides (ddNTPs)

RNA and DNA molecules can both be sequenced by generating DNA copies of the molecules of interest in the presence of chain-terminating nucleotides in a procedure referred to as **chain-termination dideoxy sequencing**, or Sanger sequencing. In this approach, illustrated in Figure 19.50, an oligonucleotide primer that specifically hybridizes to the RNA or DNA is annealed. (If the sequence of the RNA or DNA is completely unknown, the RNA can be converted first to cDNA as described in Section 19.3, and oligonucleotides of known sequence can be ligated to the ends of cDNA or DNA molecule of interest.) The primer is then extended with an enzyme: a DNA polymerase in the case of DNA or reverse transcriptase in the case of RNA. If all four deoxynucleotide triphosphates (dNTPs) are supplied, a full-length DNA copy of the template strand will be synthesized. If, however, a small amount of ddNTPs (which are nucleotides that lack a 3′ hydroxyl group) is added, chain elongation terminates when that nucleotide is incorporated. This occurs because the chain cannot be extended by the polymerase when there is no 3′ OH group on the last added nucleotide.

For the sequencing reactions, four independent chain elongation reactions are performed, each containing the four dNTPs together with a small amount of a single ddNTP (ddA, ddC, ddG, or ddT). When supplied in an appropriate dNTP to ddNTP ratio, the ddNTP will be incorporated at random throughout the synthesized DNA strand, generating a pool of extension products terminating at each position that a given nucleotide occurs. The sequence of the DNA being made can be read by comparing the pattern of extension products of the four different reactions.

Typically, radioactively labeled dNTPs or fluorescently labeled ddNTPs are included in the reactions to make it possible to detect the extension products when they are resolved by regular gel electrophoresis or by capillary gel electrophoresis. An example of regular gel electrophoresis using radioactive labeling is shown in Figure 19.50c; an example of capillary gel electrophoresis using fluorescent labeling is shown in Figure 19.50d.

A single reaction can typically reveal the sequence of up to 900 bases. As we will see in Section 19.15, automated DNA sequencing and newer techniques are used when determining complete genome sequences of species and even individual organisms.

The N-terminal sequence of a protein can be determined by derivatization and cleavage

The N-terminal sequences of a protein can be determined by sequential derivatization, release, and identification of the N-terminal amino acid in a process called **Edman degradation**, a process depicted in Figure 19.51. In this procedure, the most N-terminal amino acid is specifically modified by phenylisothiocyanate (PITC). This modified amino acid is then specifically cleaved from the protein by mild acid treatment (which does not affect the remainder of the protein). The released derivatized amino acid is then identified by column chromatography, such as ion exchange chromatography (Section 19.7), in which each amino acid has a characteristic elution profile. The cycle is then repeated with the next amino acid, and so on.

There are two limitations of this method: (i) some modified residues are resistant to PITC modification or acid hydrolysis, and (ii) the ability to identify an amino acid decreases with each round of cleavage (due to limitations in efficiency as the process is iterated). In practice, only a limited number of N-terminal amino

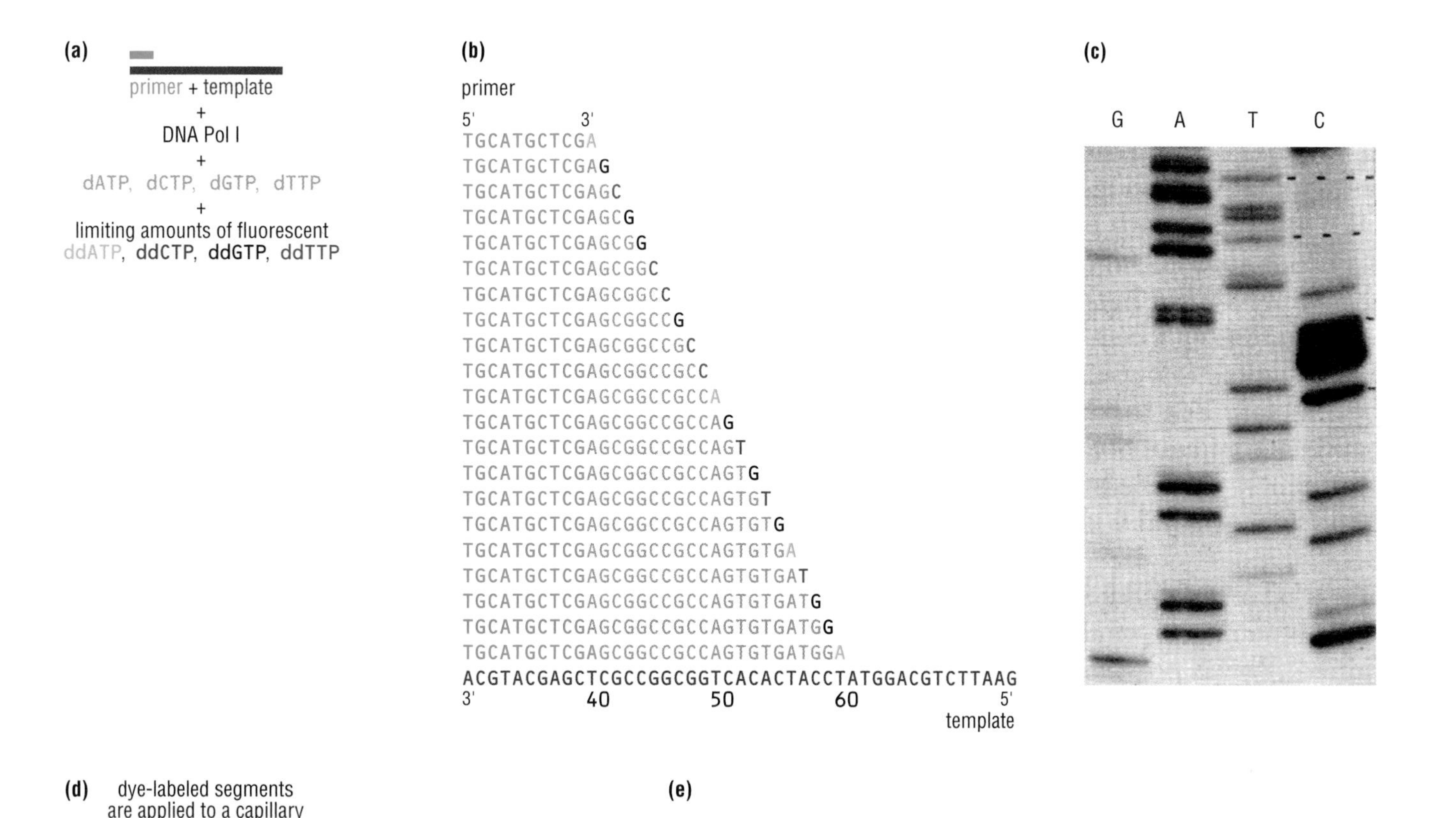

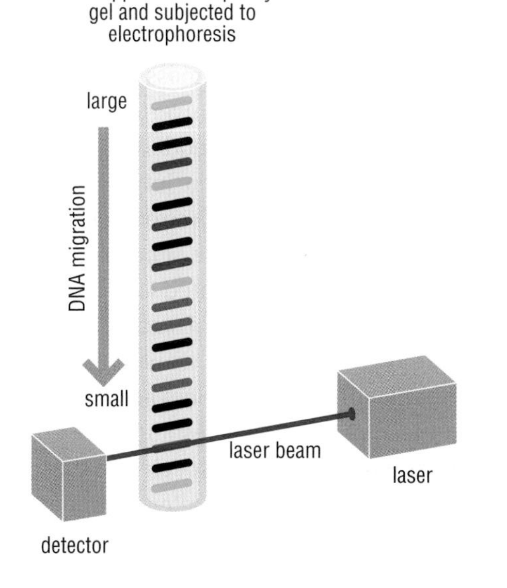

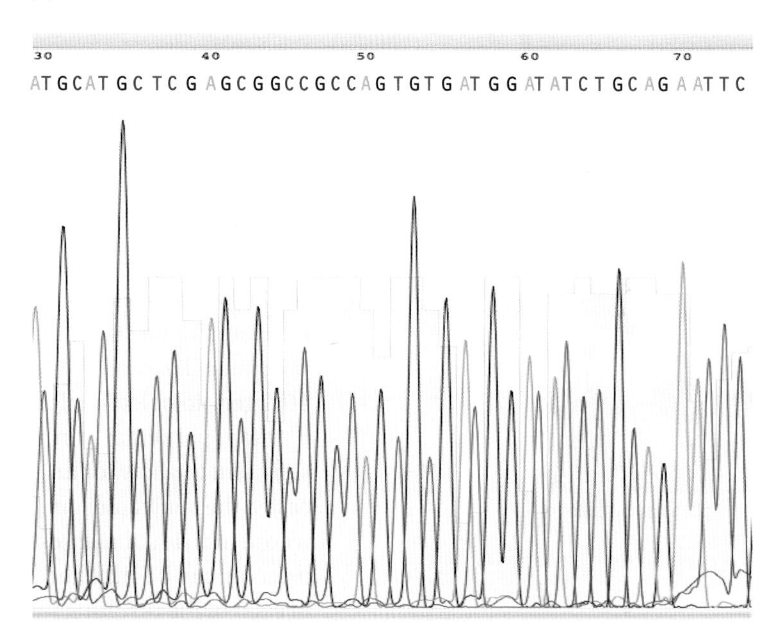

Figure 19.50 Chain-termination DNA sequencing using ddNTPs. (a) A DNA primer (turquoise) is annealed to one of the DNA strands whose sequence is being determined; DNA synthesis is carried out by DNA polymerase I (pol I; or reverse transcriptase, if RNA is being sequenced) together with dNTPs in the presence of limiting amounts of ddNTPs. If radioactive detection is used (as in panel c), one or more of the dNTPs is radioactively labeled. If detection is by fluorescence (as in panel d), each ddNTP is uniquely fluorescently labeled. (b) Products of DNA synthesis, each terminated by one of the ddNTPs. (c) An example of radioactive reaction products separated on a polyacrylamide gel and detected by autoradiography. (Note that this sequence is unrelated to the example shown in (b)). When radioactivity is used, each sequencing reaction contains a different ddNTP, and each reaction is loaded on a separate lane. The sequence can then be read from bottom to top (GCACAATGTCA....). (d) The products of DNA synthesis using fluorescently labeled ddNTPs are separated by capillary gel electrophoresis, which is similar to regular gel electrophoresis (Section 19.7) except that the gel is made in a capillary and it separates the molecules of a single sample. When fluorescently labeled ddNTPs are used, a single sequencing reaction can contain all four ddNTPs. The color of the fluorescent label at the 3′ end of the terminating nucleotide is detected as the corresponding fragment migrates past the laser. (e) An example of a sequencing reaction using fluorescent ddNTPs displayed on a chromatogram that reveals the sequence of the newly synthesized strand.

Panel (c) is Fig 2 from Sanger *et al.* DNA sequencing with chain-terminating inhibitors. *Proceedings of the National Academy of Sciences U S A*, 1977;**74**:5463–5467; credit: MRC Laboratory of Molecular Biology.

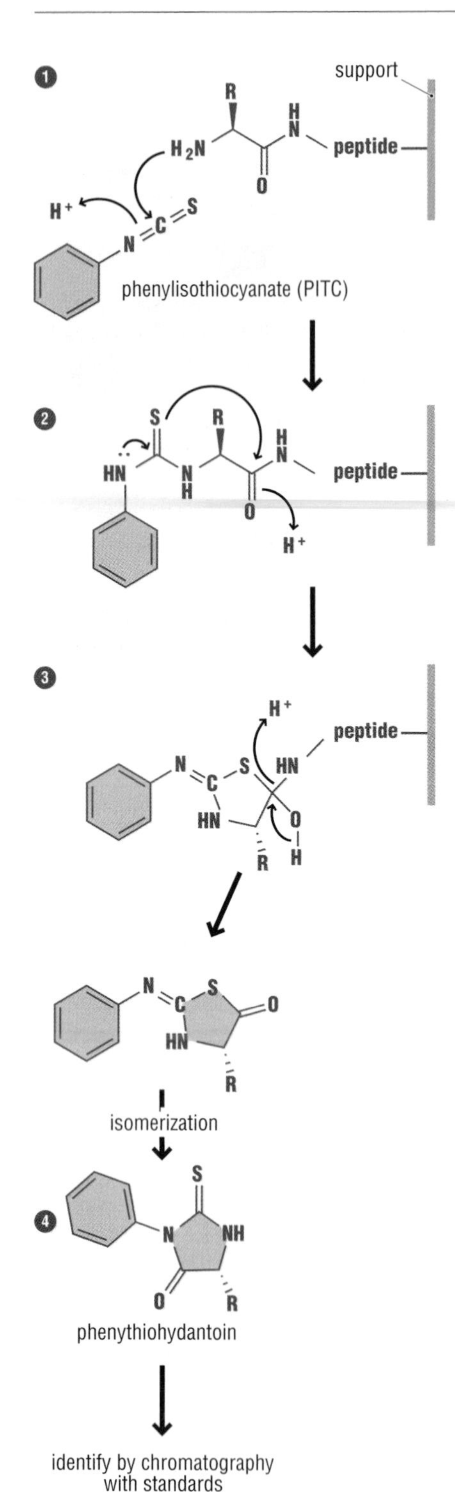

Figure 19.51 Steps in Edman protein degradation. The peptide is adsorbed to a solid support via its C-terminus and sequenced using sequential Edman degradation steps, in which one residue is removed from the N-terminus per cycle. Each cycle has the following steps: (1) PITC is reacted with the primary amine function of the peptide. (2) Under mildly alkaline conditions, a phenylthiocarbamoyl derivative is formed. (3) Mild acid treatment causes the terminal amino acid to be released, freeing the N-terminus of the peptide for another cycle of PITC reaction. (4) The released amino acid isomerizes to form a phenylthiohydantoin species that can be identified by chromatography or mass spectrometry.

acids (around 50–60 residues) can be identified through this approach, though in many cases the number of residues correctly identified may be sufficient for protein identification, especially in organisms whose entire genome sequence is known and protein sequences can be predicted. There are also methods for determining the C-terminal sequence of a protein based on the same principles as the N-terminal sequencing, but these approaches are more problematic and are much less widely used.

Mass spectrometry can be used to identify most molecules

Mass spectrometry allows the identification of the widest spectrum of molecules, even if the molecules of interest are only present in small amounts. Mass spectrometry is used extensively to identify proteins following their separation in and excision from a polyacrylamide gel, as illustrated in Figure 19.52; it can also be used to identify carbohydrates, nucleic acids, and small molecules.

The method of mass spectrometry makes it possible to determine the molecular weight of a molecule with very high accuracy by measuring its mass-to-charge ratio. A small amount of the material of interest must first be converted to gas-phase ions. This can be achieved by bombarding a sample embedded in a specific chemical matrix with a laser (matrix-assisted laser desorption/ionization or MALDI) or by generating a spray of the sample in the presence of a strong electric field (electrospray ionization or ESI). The ionized molecules are separated according to their mass-to-charge ratios, and the mass of the separated ions is then measured upon striking a detector (see Figure 19.52a). The mass of small biological molecules can be determined to within a single dalton (Da), while the mass of larger biological molecules can be determined with somewhat less accuracy.

Often, particularly in the case of larger molecules, the mass resolution is not sufficient to allow for unambiguous identification of a molecular species. In these cases, the molecules are usually digested or hydrolyzed into smaller fragments whose individual masses can be determined with greater accuracy. For example, proteins can be digested with specific proteases, such as trypsin, and the masses of each of the resulting fragments determined (see Figure 19.52b).

For organisms whose entire genome sequence is known, this information is usually sufficient to identify the protein, based on the predicted coding regions. If identification is not possible based on the fragment masses alone, however, the proteolytic fragments can be broken down further to the individual amino acids, whose masses can then be determined. This is often referred to as tandem mass spectrometry (or MS/MS): first the mass of the proteolytic fragment is determined, and then the masses of the amino acids in the fragment are determined, thereby identifying the exact amino acid composition of the fragment.

Mass spectrometry has become an important tool in studying protein biology. Using this approach, it is possible to determine not only the sequence of a protein, but also whether amino acids are modified, for example by phosphorylation.

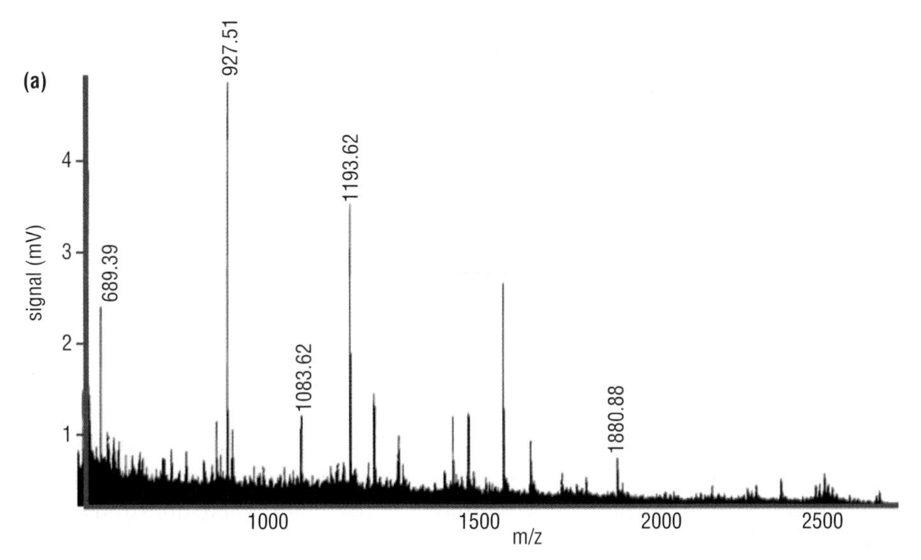

(b) Calculated Mass

amino acid	calculated mass (Daltons)	sequence
212–217	688.37	AWSVAR
137–143	926.49	YLYEIAR
1–10	1192.59	DTHKSEIAHR
323–335	1566.74	DAFLGSFLYEYSR
484–499	1879.91	RPCFSALTPDETYVPK

Figure 19.52 Representative mass spectrometric analysis. (a) BSA was digested with trypsin, and a MALDI spectrum was generated. The mass/charge (*m/z*) ratios are indicated on the x-axis and relative intensity is plotted on the y-axis. (b) Table showing the tryptic fragments of BSA that were detected and their calculated mass (accounting for the relative abundance of different amino acids).

From Shevchenko, A, Wilm, M, Vorm, O, and Mann, M. Mass spectrometric sequencing of proteins from silver-stained polyacrylamide gels. *Analytical Chemistry*, 1996;**68**: 850–858.

Proteins can be identified not only after purification, but also in complex mixtures containing hundreds of proteins. For example, one can use mass spectrometry to determine the composition of a large multi-protein complex, or even how different growth conditions affect the abundance of different cellular proteins. This use of mass spectrometry will be discussed further in Section 19.11.

BLAST Sequence alignment can be used to assign functions to genes

Once the sequence of DNA, RNA, or protein is obtained, it is possible to get some information about its function by examining regions of homology in other proteins or genes. As we discussed in Chapter 18, sequence homology implies that two sequences were derived at some point from a common ancestor. As such, one of the most common and powerful methods to infer the function of an unknown gene is to compare its sequence to that of genes with known functions.

The comparison of two or more sequences is called an **alignment**, in which the nucleotide or amino acid query sequence is compared with the nucleotide or amino acid sequence of known genes or proteins. After the alignment has been performed it is assigned a score, based on whether the residues are identical, similar, or different. Such alignments make it possible to identify other proteins or RNA molecules that are homologs with a common evolutionary origin, and therefore are likely to carry out similar functions or activities. The region of sequence similarity may encompass an entire coding sequence or it may be restricted to one or more protein domains (for example, kinase domains, zinc finger DNA-binding domains, or carbohydrate binding domains). The presence of an identifiable domain whose function in other proteins is known often provides clues about the function of an uncharacterized protein.

The most widely used method for sequence alignments is a suite of related programs known collectively as **BLAST** (Basic Local Alignment Sequence Tool), which are available from the National Center for Biotechnology Information (NCBI) website and on many other web servers worldwide.

➔ You can find BLAST on the NCBI website at http://blast.ncbi.nlm.nih.gov/.

The simplest BLAST alignments are done as pairwise alignments (that is, comparisons between a pair of sequences) using the query sequence and each sequence in a chosen database. BLAST algorithms can align either DNA sequences (BLASTN) or protein sequences (BLASTP). For proteins, the program starts with a string of three amino acids and searches for homology; it then moves in a predefined 'window' to examine adjacent sequences for additional homology. For DNA, an 11-nucleotide string is the initial sequence length that is probed.

After the alignment is optimized, it is given a score that provides a measure of how closely two sequences match one another. The score depends on the amount of similarity between the two sequences and the length of sequence that is similar. Gaps may be introduced into one sequence or the other to maximize the alignment of two sequences, although a penalty in the score is assigned for each gap. The score of the homology is expressed as an E value, which is a measure of the probability that the sequence alignment observed could have occurred by chance. The lower the E value, the better the homology, as the alignment in question is less likely to have occurred by chance.

An example of the output from a BLASTP search is given in Figure 19.53, which shows the results of a search for proteins similar in sequence to the Hermes transposase from the housefly, *Musca domestica* (the query sequence). Aside from finding other version of Hermes transposase in the database, the BLAST search also found homology with the *hobo* transposase, which is an orthologous protein from *Drosophila*.

While BLAST searches are the most commonly used searches for initially examining sequence alignments, there are a wealth of other programs that compare sequences, help identify specific domains, sites of protein modification, protein interaction domains, and other features. An increasing number of protein database entries incorporate multiple parameters including, for example, structural information.

```
Alignment of Hermes from Musca domestica (house fly) with hobo from Drosophila

>pir||A39652 Hobo element transposase HFL1 - fruit fly (Drosophila melanogaster)
Length=658
Score =  662 bits (1709),  Expect = 0.0, Method: Compositional matrix adjust.
Identities = 337/606 (55%), Positives = 448/606 (73%), Gaps = 15/606 (2%)

Query  9    VKAKINQGLYKITPRHKGTSFIWNVLADIQKEDDTLVEGWVFCRKCEKVLKYTTRQTSNL  68
            VK KIN G Y +  +HKG S IW++L DI KED+T+++GW+FCR+C+KVLK+  + TSNL
Sbjct  57   VKNKINNGTYSVANKHKGKSVIWSILCDILKEDETVLDGWLFCRQCQKVLKFLHKNTSNL  116


Query  69   CRHKCCASLKQSRELKTVSADCKKEAIEKCAQWVVRDCRPFSAVSGSGFIDMIKFFIKVK  128
             RHKCC +L++  ELK VS + KK AIEKC QWVV+DCRPFSAV+G+GF +++KFF+++
Sbjct  117  SRHKCCLTLRRPTELKIVSENDKKVAIEKCTQWVVQDCRPFSAVTGAGFKNLVKFFLQIG  176

Query  129  AEYGEHVNVEELLPSPITLSRKVTSDAKEKKALISREIKSAVEKDGASATIDLWTDNYIK  188
            A YGE V+V++LLP P TLSRK  SDA+EK++LIS EIK AV+   ASAT+D+WTD Y++
Sbjct  177  AIYGEQVDVDDLLPDPTTLSRKAKSDAEEKRSLISSEIKKAVDSGRASATVDMWTDQYVQ  236

Query  189  RNFLGVTLHYHENNELRDLILGLKSLDFERSTAENIYKKLKAIFSQFNVEDLSSIKFVTD  248
            RNFLG+T HY +  +L D+ILGLKS++F++STAENI  K+K +FS+FNVE++ ++KFVTD
Sbjct  237  RNFLGITFHYEKEFKLCDMILGLKSMNFQKSTAENILMKIKGLFSEFNVENIDNVKFVTD  296

Query  249  RGANVVKSLANNIRINCSSHLLSNVLENSFEETPELNMPILACKNIVKYFKKANLQHRLR  308
            RGAN+ K+L  N R+NCSSHLLSNVLE SF E  EL    + +CK IVKY KK+NLQH L
Sbjct  297  RGANIKKALEGNTRLNCSSHLLSNVLEKSFNEANELKKIVKSCKKIVKYCKKSNLQHTLE  356

Query  309  SSLKSECPTRWNSTYTMLRSILDNWESVIQILSEAGETQRIVHINKSIIQTMVNILDGFE  368
            ++LKS CPTRWNS Y M++SILDNW SV +IL   GE    V  NKS ++ +V+IL  FE
Sbjct  357  TTLKSACPTRWNSNYKMMKSILDNWRSVDKIL---GEADIHVDFNKSSLKVVVDILGDFE  413

Query  369  RIFKELQTCSSPSLCFVVPSILKVKEICSPDVGDVADIAKLKVNIIKNVRIIWEENLSIW  428
            RIFK+LQT SSPS+CFV+PSI K+ E+C P++ D++  A LK  I++N+R IW  NLSIW
Sbjct  414  RIFKKLQTSSSPSICFVLPSISKILELCEPNILDLSAAALLKERILENIRKIWMANLSIW  473

Query  429  HYTAFFFYPPALHMQQEKVAQIKEFCLSKMEDLELINRMSSFNELSATQLNQSDSNSHNS  488
            H  AFF YPPA H+Q+E + +IK FC+S+++          +S    L +T+  ++
Sbjct  474  HKAAFFLYPPAAHLQEEDILEIKVFCISQIQV-----PISYTLSLESTETPRTPETPETP  528

Query  489  IDLTS------HSKDIST-TSFFFPQLTQNNSREPPVCPSDEFEFYRKEIVILSEDFKVM  541
              L S      +K IS+   FFFP+L   ++        P DE E Y ++ V LS++F+V+
Sbjct  529  ESLESPNLFPKKNKTISSENEFFFPKLVTESNSNFNESPLDEIERYIRQRVPLSQNFEVI  588

Query  542  EWWNLNSKKYPKLSKLALSLLSIPASSAASERTFSLAGNIITEKRNRIGQQTVDSLLFLN  601
            EWW  N+  YP+LSKLAL LLSIPASSAA+ER FSLAGNIITEKRNR+  ++VDSLLFL+
Sbjct  589  EWWKNNANLYPQLSKLALKLLSIPASSAAAERVFSLAGNIITEKRNRLCPKSVDSLLFLH  648

Query  602  SFYKNF  607
            S+YKN
Sbjct  649  SYYKNL  654
```

Figure 19.53 BLASTP search using Hermes transposase as a query. BLAST alignment of the protein sequence (amino acid one letter code) between the Hermes transposase (Query) and the closely related Hobo transposase from *Drosophila* (Sbjct = Subject). The image shown was taken from the BLAST website. The alignment starts with amino acid number 9 within the Hermes sequence and with amino acid 57 in Hobo, as shown by the numbers 9 and 57 to the left of the aligned sequences. The numbers at each row to the right indicate the position in the sequence of the last amino acid in the row to allow orientation of the sequences over the many lines. The amino acids that are identical between the two sequences are indicated by the identical letter placed at that position between the two sequences, and positions where the two proteins have similar amino acids are indicated with a plus sign. When gaps have to be introduced in one of the two sequences to optimize the alignment, dashed lines are shown between amino acids.

19.9 DETECTION OF SPECIFIC DNA SEQUENCES

The base-pairing ability inherent in DNA sequence can be exploited to detect a specific DNA sequence amid a vast excess of unrelated sequences. In the previous section, we discussed how the sequence of a specific DNA segment is determined; in this section, we describe a number of methods by which a known DNA segment can be identified and distinguished from other DNA sequences in the context of many DNA fragments, a pool of DNA clones, or even chromosomes within a cell.

DNA molecules can be distinguished by their restriction enzyme digestion patterns

As discussed in Section 19.4, restriction enzymes are DNA endonucleases that cleave DNA at specific sequences, and different restriction enzymes recognize different sequences, as shown in Figure 19.20. These sequences are often palindromic and are between 4–8 base pairs in length. Restriction enzymes usually cleave DNA either at the restriction site or at a fixed distance from it. By choosing the right restriction enzyme (or enzymes), it is possible to distinguish between different DNA molecules and to learn something about their structure.

Consider, for example, a bacterial plasmid called pBR322, shown in Figure 19.54a: The sequence of this plasmid contains restriction sites for dozens of different restriction enzymes, of which only two are shown. Some restriction enzymes, such as EcoRI, have a single restriction site in the pBR322 sequence; some enzymes, such as AclI, have multiple restrictions sites; and some enzymes have none. The size of the DNA fragments generated by restriction enzyme digest can be determined by separating the fragments on an agarose gel (described in Section 19.7), as shown in Figure 19.54b. As such, restriction enzymes can be used to verify whether a cloning procedure worked as planned. In the example shown in Figure 19.54b, digestion by AclI was used to confirm that a 1 kb fragment was inserted at the EcoRI site of pBR322, resulting in an increase in size in one of the AclI fragments.

Specific DNA fragments can be detected by Southern blot hybridization

One method to detect a specific DNA fragment among many DNA fragments is called **Southern blot** analysis, which is illustrated in Figure 19.55. This method is named after its inventor, Edwin Southern, and hence the name carries an uppercase 'S'. (Note that in the names of 'northern blot' and '**western blot**' techniques, which we will learn about in Sections 19.10 and 19.11, respectively, the 'n' and 'w' are not capitalized because neither term is derived from a person's name; rather, they are whimsical names given to these techniques as they are similar to the Southern blot technique.)

In overview, Southern blot analysis proceeds as follows. DNA fragments cut by specific restriction enzymes are separated on an acrylamide, agarose, or pulse-field gel as described in Section 19.7. Without fragmentation, the DNA would be too large to enter the gel, and it would not be possible to analyze specific genomic regions. The DNA fragments separated in the gel are denatured with alkali to expose the single strands for hybridization and are then transferred to a positively

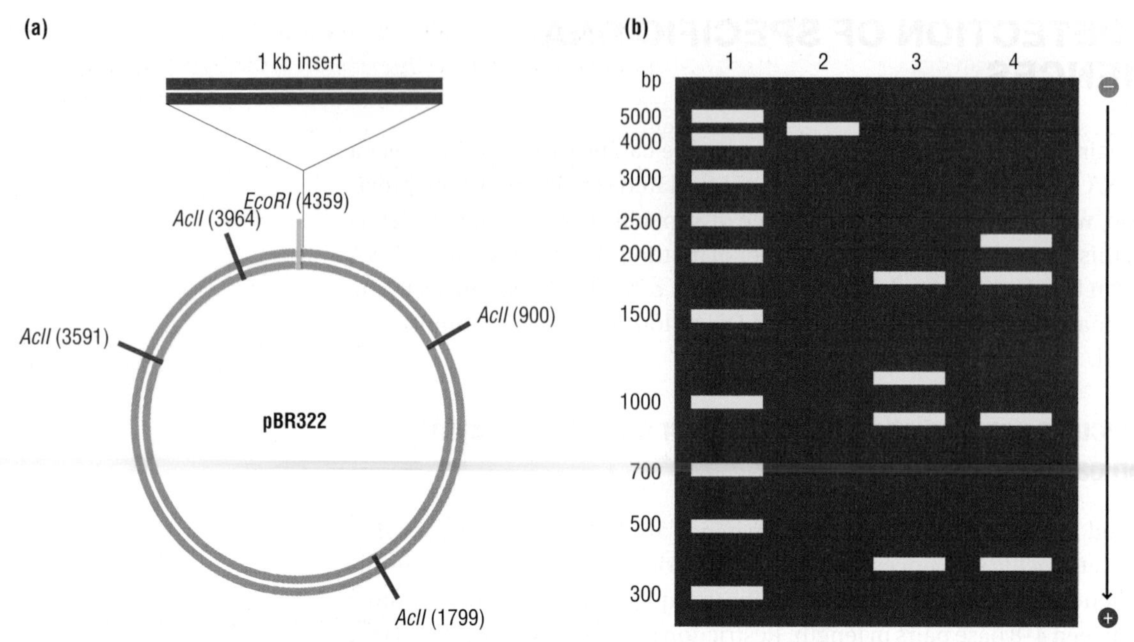

Figure 19.54 DNA analysis using restriction enzymes. (a) Plasmid pBR322 is a circular DNA molecule that is 4361 bp in length. Shown is a map of the plasmid with the locations of the restriction sites for EcoRI (green) and AclI (red). The numbers in parentheses indicate the position of the first base pair of the restriction site. The middle of the EcoRI site was designated as position 1. Also shown is the position of a 1 kb insert examined in panel b. (b) An illustration of an agarose gel used for separating different DNA molecules, which migrate according to size toward the positive pole. The gel is typically stained with a DNA-binding fluorescent dye (such as ethidium bromide) and visualized by exposing the gel to UV light. As a result, the DNA molecules appear as bright bands on a dark background. Lane 1 was used to separate a DNA ladder, which is a mixture of DNA fragments of known sizes (shown on the left). Plasmid pBR322 was digested with EcoRI (lane 2) to generate one DNA fragment, or with AclI (lane 3) to generate four DNA fragments. Lane 4 shows the results of an AclI digest of a pBR322 plasmid with a 1 kb fragment inserted at the EcoRI site. Notice that the band in lane 3 of ~1300 bp, corresponding to the fragment between the AclI sites at 3964 and 900 bp, is missing from lane 4, and instead there is a band of ~2300 bp, consistent with a 1 kb insert at the EcoRI site.

charged membrane. The membrane is incubated with a single-stranded DNA probe that is complementary to the sequence to be identified and is labeled with either radioactivity, fluorescence, or a chemical tag. The labeled DNA anneals to its complementary strand, and the location of the target DNA fragment is identified by exposing the membrane to a device that can detect the probe (an x-ray film in the case of radioactivity, or to a machine that can detect fluorescence or color in the case of fluorescent or chemical probes).

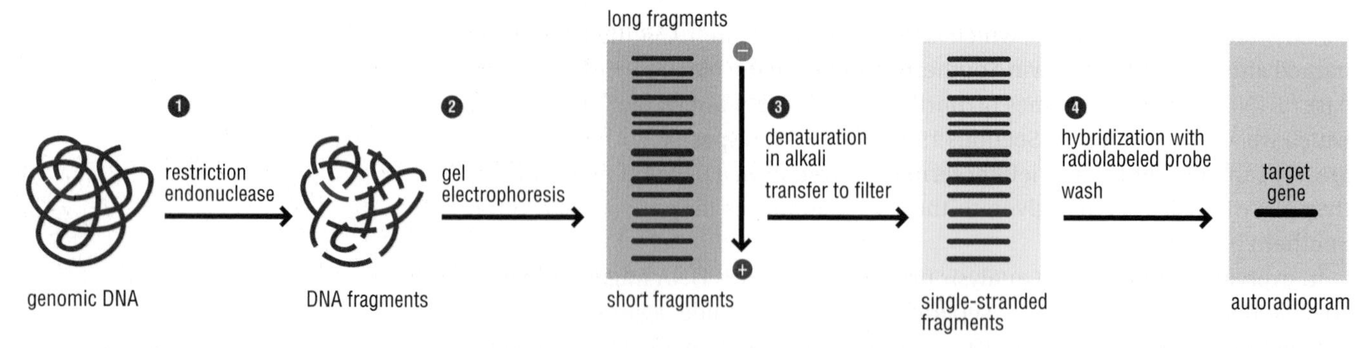

Figure 19.55 Southern blot analysis. In the example shown, the goal is to detect a region of interest within genomic DNA (shown in pink). The genomic DNA is first digested with restriction enzymes (step 1) and separated by gel electrophoresis (step 2). The DNA is then denatured and transferred to a filter (in blue, step 3). In parallel, a DNA fragment complementary to the region of interest is labeled, for example, with a radioactive isotope. The radioactive DNA probe is then hybridized to the filter (step 4). The probe will hybridize to its complementary DNA on the filter, which can be detected by exposing the filter to radiation-sensitive film.

Using this method, one can determine whether a particular chromosomal locus is altered by the insertion or deletion of sequences, as the DNA fragment size will be either larger or smaller than expected.

➔ Experimental approach 17.1 describes how Southern blot analysis was used in the detection of a transposon inserted into the factor VIII gene that causes hemophilia.

Specific DNA clones can be identified in libraries by colony or plaque hybridization

A similar hybridization method can be used to identify a specific DNA clone in a library of many different clones. If the library is in a bacterial plasmid or BAC vector, colony hybridization can be used. Alternatively, if the library is represented in a bacteriophage vector, plaque hybridization can be used.

➔ We discuss the construction of libraries in Section 19.4.

The process of colony hybridization is depicted in Figure 19.56. During this process, a collection of bacterial colonies, each of which arises from multiple divisions of a single cell containing one DNA clone, is transferred to a positively charged membrane, generating a copy of the plate on the membrane. The cells are then lysed *in situ* on the membrane. In an analogous method called plaque hybridization, the bacteriophages that formed plaques on bacterial lawns are transferred to the membrane. The DNA from the colony or plaque is denatured with alkali and probed in the same way as for the Southern blots described above. Because the membrane is a copy of the original plate, the researcher can go back to the plate once a positive signal is obtained and isolate the bacterial colony or bacteriophage that gave rise to the positive signal indicative of the desired DNA fragment. The colony or bacteriophage can then be propagated. These hybridization approaches make it possible to identify and isolate a single clone that contains the specific sequence of interest from among thousands of clones in a library.

Chromosome abnormalities can be detected by chromosome spreads

Sometimes it is also of interest to detect larger regions of DNA, including whole chromosomes. The complete chromosome complement of an organism, known as a karyotype, is determined by visualizing stained chromosomes isolated from mitotic cells. One stain that is commonly used for this purpose is Giemsa stain, which generates a characteristic banding pattern by binding to gene poor, A/T rich regions, as shown in Figure 19.57. The stained chromosomes are then examined under the microscope to detect the presence of additional chromosomes, missing chromosomes, and gross deletions or rearrangements. This approach of visual inspection of chromosomes is called karyotype analysis.

Karyotype analysis is routinely used in medical diagnosis. Aneuploidy (extra or missing chromosomes) and other chromosomal aberrations, such as deletions and translocations, are frequently the underlying cause of birth defects, intellectual disability, miscarriages, and other syndromes. Thus, direct visualization of

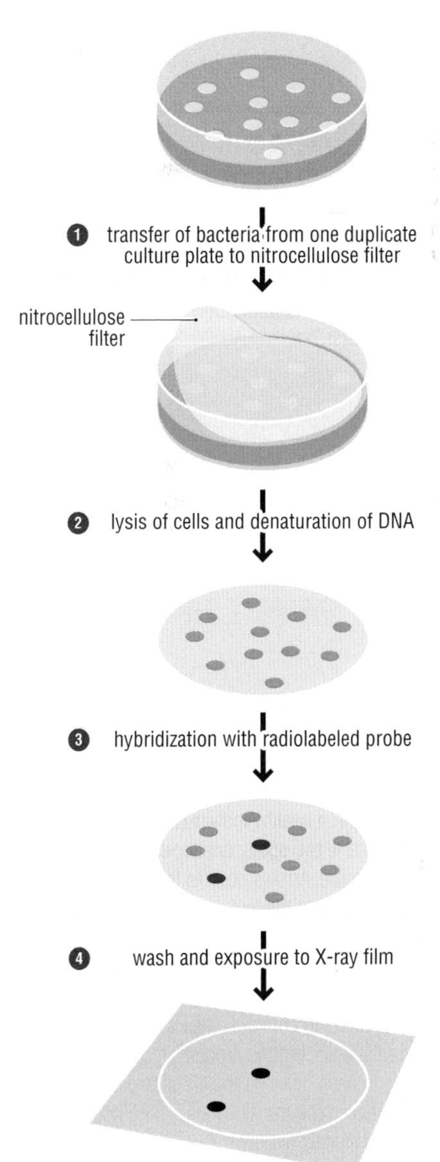

Figure 19.56 Colony hybridization. An outline of the method for probing bacterial colonies, each of which contains a unique plasmid, for the presence of a plasmid containing a DNA sequence of interest. The colonies are first transferred onto a filter that is pressed on the surface of the plate (step 1). The filter is processed in much the same way as that described for Southern blot analysis (step 2), then hybridized with a labeled DNA probe to the DNA region of interest (step 3). After washing, the filter hybridized with a radioactively labeled probe is exposed to film (step 4). If the colony carrying the relevant plasmid was present on the original plate, there will be a spot corresponding to its location on the film. The investigator can then isolate the corresponding colony from the original plate.

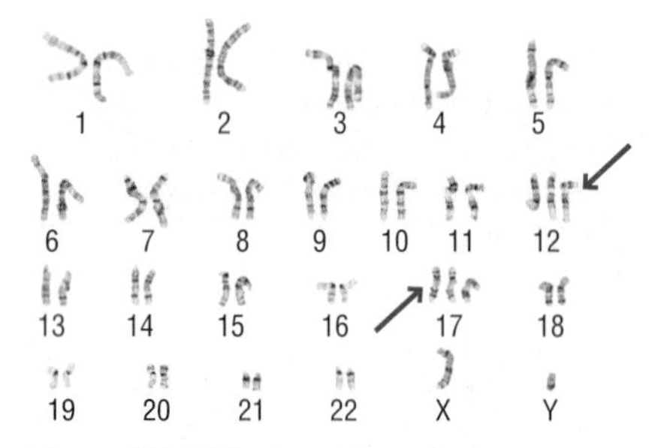

Figure 19.57 The karyotype of a human stem cell showing trisomy 12 and trisomy 17. The karyotype is determined by staining chromosomes of metaphase cells with Giemsa stain and identifying chromosomes based on their size and banding pattern. In the example shown, the cell has one X chromosome, one Y chromosome, and two copies of each autosome, with the exception of chromosomes 12 and 17, which are present in three copies each (marked by red arrows).

Adapted from Meisner and Johnson, *Methods*, 2008;**45**: 133–141.

chromosomes can be used in prenatal screening procedures, such as amniocentesis. In this procedure, fluid containing cells shed from the fetus is collected from the amniotic sac. The cells are then cultured, and used for karyotypic analysis. Chromosome abnormalities are similarly diagnosed in the case of malignancies and other diseases that may involve chromosomal rearrangements.

DNA sequences can be localized along a chromosome by fluorescent *in situ* hybridization

The examination of the overall structure of a chromosome makes it possible to detect gross chromosomal changes. However, it is often helpful to locate a specific DNA sequence *within* a chromosome. The method used to accomplish this is called **fluorescent *in situ* hybridization (FISH)**, as illustrated in Figure 19.58. This approach is also useful in diagnosing genetic diseases associated with chromosome abnormalities.

In this method, cells are fixed to a slide, permeabilized, and their DNA is hybridized with a fluorescent DNA probe in a manner similar to the Southern blots described above. If more than one probe is used, molecules that fluoresce at different wavelengths are attached to the different probes. The chromosomes are also stained with a fluorescent dye, 4′-6′-diamidino-2-phenylindole (DAPI), described in Section 19.13, which binds specifically to DNA. The image that appears in a fluorescent microscope shows the DAPI-stained chromosomes in blue, with the regions of interest highlighted by the fluorescent probe or probes. In Figure 19.58, the probes are against regions on two different chromosomes, as shown in the example from a healthy individual; Figure 19.58a. In a certain type of leukemia, a translocation occurs that bring these two regions in close proximity. As a result, the two probes overlap, as shown in Figure 19.58b.

➔ The use of FISH to study chromosome segregation is described in Experimental approach 7.1.

Two modifications of the FISH procedure, called **chromosome painting** and **spectral karyotyping (SKY)**, can be used to examine whole chromosomes. These methods are used to identify specific chromosomes or chromosomal rearrangements. For chromosome painting, the DNA for one entire chromosome is purified using flow cytometry, in which chromosomes are separated on the basis of size and base composition. The isolated chromosomes are then labeled with a fluorescent dye. SKY is similar to chromosome painting except that, in this case, every

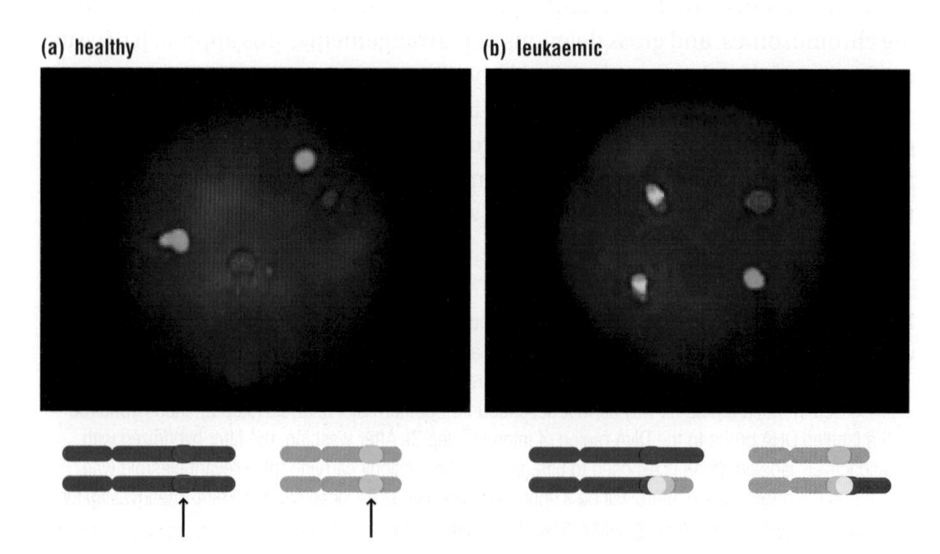

Figure 19.58 FISH. In the example shown, the assay was used to detect a reciprocal translocation between chromosomes 4 and 11, using probes that flank the breakpoint on both sides. (a) In a healthy individual, each probe gives rise to two spots, one for each homologous chromosome. Arrows point to the breakpoints that occurred in the leukemia patient shown in panel b. (b) In a leukemia patient, there was a reciprocal translocation between one chromosome 4 and one chromosome 11. The unaffected chromosomes still give green or red signals, but the probes become very close to each other on the chromosomes that experienced the translocation, giving rise to merged signals (yellow).

Photo courtesy of Sabine Strehl, Children's Cancer Research Institute.

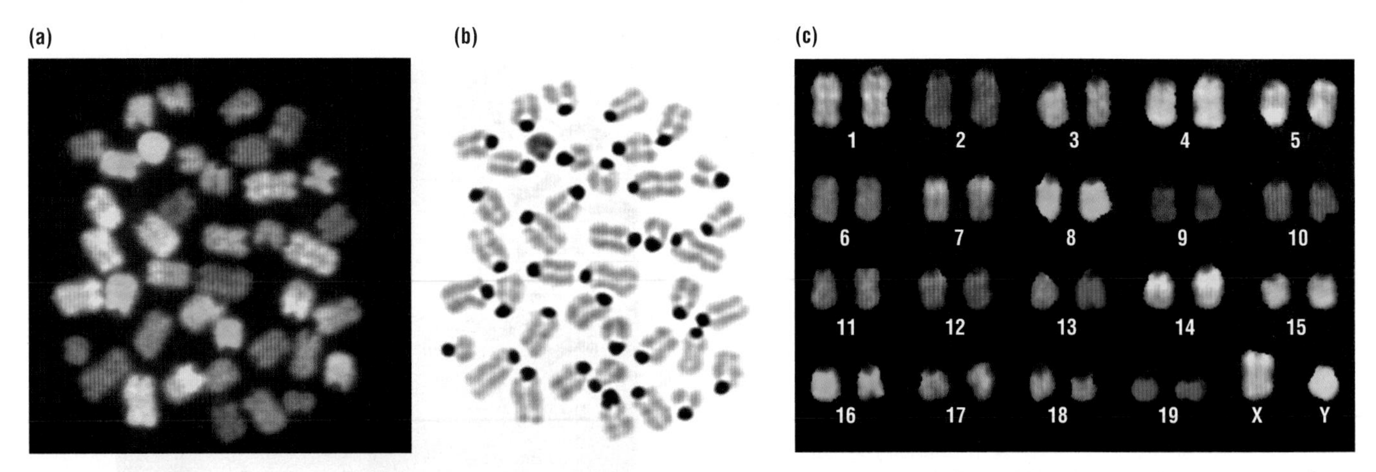

Figure 19.59 SKY. (a) DNA probes are generated for each individual chromosome by creating a unique combination of fluorophores, making each chromosome appear as a different color when these probes are hybridized to mitotic chromosome spreads. (b) In parallel, the chromosomes are also stained with a DNA dye that reveals the specific banding pattern of each chromosome. (c) An arranged SKY karyotype of the chromosomes can be generated from the chromosomes shown in (a).

Kindly provided by Margaret Strong, Johns Hopkins School of Medicine, unpublished.

chromosome has been separately purified by flow cytometry, and each is labeled with a different and spectrally distinct ratio of five different fluorescent dyes. This set of labeled chromosome probes is then pooled, denatured, and used to probe a metaphase spread of chromosomes as described above for FISH. Figure 19.59 illustrates that each pair of chromosomes in a normal karyotype has a unique color. If chromosome translocations or rearrangements have occurred, an abnormal karyotype will be seen. The chromosomes involved in the translocation are then known because the color uniquely identifies each chromosome.

Alterations in DNA from patients can be detected by array comparative genomic hybridization

Although FISH and SKY are very useful, they are not sensitive enough to detect small deletions or amplifications. In the case of FISH, we also need to have prior knowledge of which DNA region is to be analyzed. Array comparative genomic hybridization (aCGH) is a quantitative method for detecting changes in copy number (deletion, amplification) on a genome-wide basis.

This method is based on the abundance of the DNA sequence in a test cell population being compared to the abundance of the same sequence in normal cells. In this way, it is possible to determine whether certain sequences are missing or amplified. This method involves labeling genomic DNA from both normal cells and those from a patient with two different fluorophores. For example, one pool of DNA can be labeled with fluorescein, which emits green light when excited, while the other pool is labeled with rhodamine, which emits red light when excited. Instead of hybridizing these two labeled DNA pools to chromosomes, the labeled DNA is hybridized to a **DNA microarray** that contains DNA fragments from the entire genome, spotted in rows and columns on a microchip, as depicted in Figure 19.60a. If the copy number of a particular region is the same in the patient and in the control, the signal from the two fluorophores will be equal (and will appear yellow because the equal amounts of red and green fluorochromes generate a yellow light). However, if the patient has a deletion or duplication, the signal will be red or green, respectively. (In this instance, the excess of one

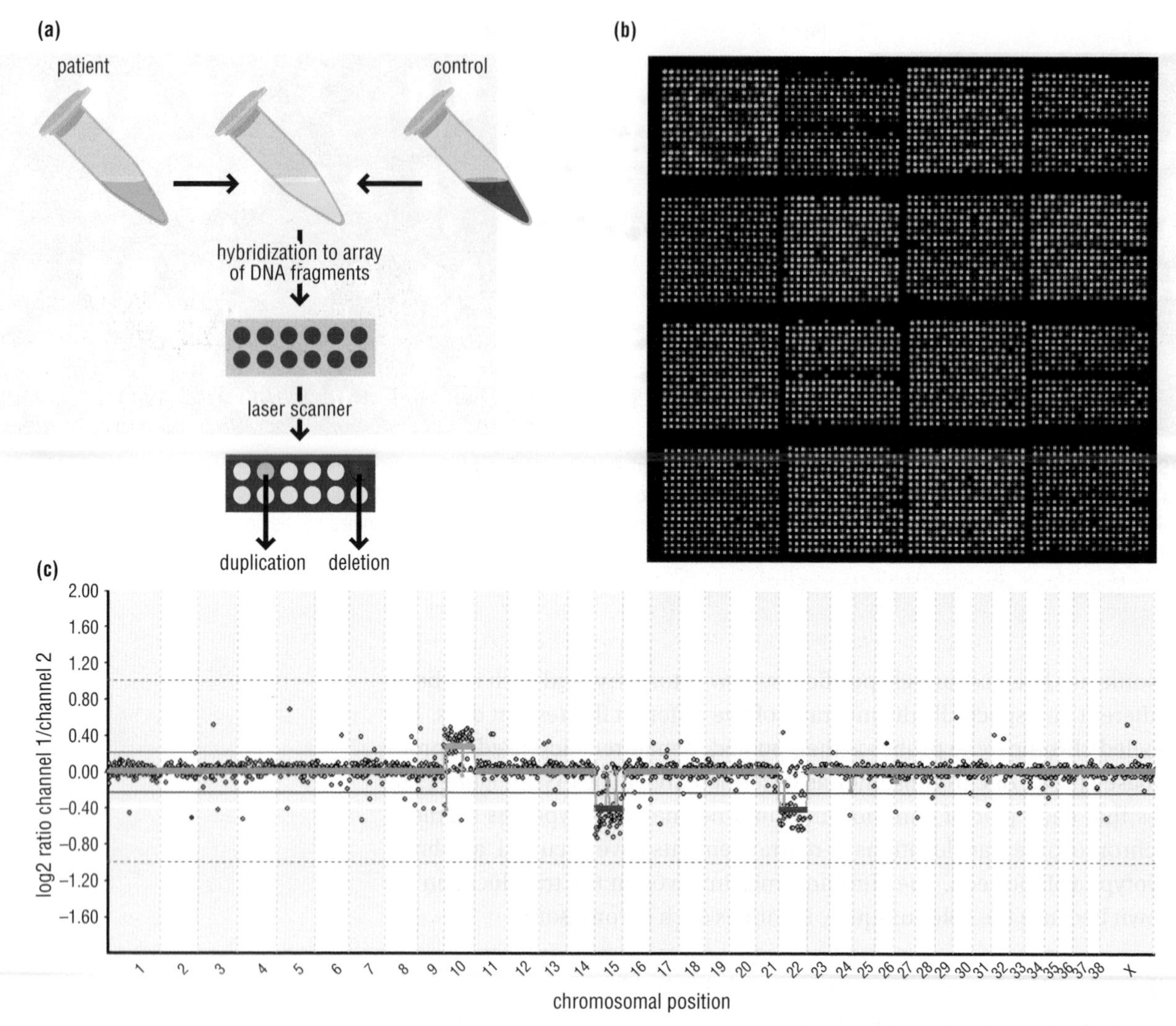

Figure 19.60 aCGH. (a) DNA from the sample to be tested is labeled with a green fluorescent dye and reference DNA is labeled with a red fluorescent dye. The two samples are mixed and co-hybridized to an array containing genomic DNA fragments that have been spotted on a glass slide. (b) The resulting ratio of fluorescence intensities is proportional to the ratio of the copy numbers of DNA sequences in the test and reference genomes. The green spots indicate extra chromosomal material (duplication) in the test sample at that particular region, while red spots indicate relatively less of the test DNA (deletion). The example shown is a small part of the entire array from a study on canine cancer. (c) The information generated by fluorescence intensities, such as is panel b, is converted to the abundance of different chromosomal regions. In the example shown, the tested canine cell line had an extra chromosome 10 and only one copy of chromosomes 15 and 22. Note that this type of analysis can uncover deletions and amplification of short chromosomal domains.

Panel (a) from Shinawi, M, and Cheung, SW. *Drug Discovery Today*, 2008; **13**:760; images in panels (b) and (c) are from http://www.breenlab.org/array.html, copyright Matthew Breen.

➲ We discuss the use of DNA microarrays to detect RNAs on a genome-wide level in Section 19.10.

fluorophore will generate a signal that exceeds the signal from the second, less abundant fluorophore.) The microarray is scanned by a special fluorescent scanner that measures the signal from each microarray spot and plots it with respect to each chromosome, as illustrated in Figure 19.60b.

19.10 DETECTION OF SPECIFIC RNA MOLECULES

As we learned in Chapter 9, many genes are regulated at the level of transcription. It is therefore often desirable to study the expression of an individual gene or a collection of genes by assaying the amount of RNA that is present. As described in this

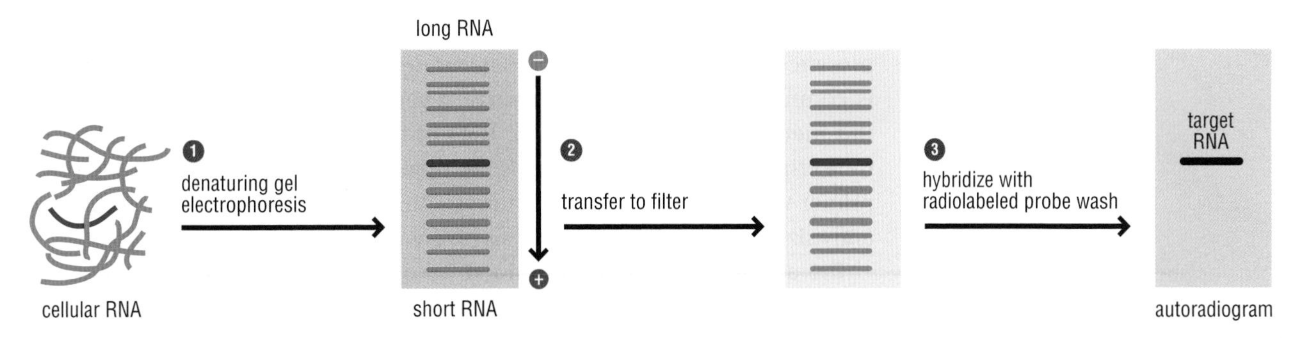

Figure 19.61 Analyzing RNA abundance by northern blot analysis. RNA is isolated from cells and separated by size on a denaturing gel (step 1). The RNA of interest is shown in pink. The RNA is then transferred to a filter (blue, step 2) and probed with a DNA or RNA probe that is coupled to a detectable molecule (for example, a radiolabeled isotope, step 3). The filter is then analyzed to determine the positions on the gel where the probe has hybridized (black band on the autoradiogram). In the case of a radioactive probe, the filter can be exposed to an x-ray film or scanned on a phosphorimager, which can quantify the amount of radioactive signal.

section, many methods that are used to examine RNA levels rely on hybridization to specific probes, although the use of reporter genes can also provide information about timing and patterns of expression. It is also possible to survey the entire population of mRNAs that are expressed or translated in a given cells type or tissue, as we will discuss toward the end of this section.

Northern blot analysis, quantitative reverse transcription-PCR, and nuclease protection can be used to evaluate the expression of individual genes

Just as we learned in Section 19.9 that specific DNA fragments can be detected by Southern blot, we can detect RNA through an analogous approach known as a northern blot. Specific RNA transcripts are generally detected by hybridization to specific DNA or RNA probes that have been labeled in some way.

So how is a northern blot performed? The procedure is summarized in Figure 19.61. First, RNAs are separated by size on denaturing acrylamide or agarose gels. (The denaturing step eliminates secondary structures in the RNA that would otherwise alter its migration in the gel and make it impossible to assess its true size.) The RNA is then transferred directly to a positively charged membrane. Unlike a Southern blot, where the DNA is fragmented prior to electrophoresis, RNA molecules are relatively small and thus will enter the gel and transfer efficiently to the filter without prior fragmentation. Also, RNA can be probed directly because it is single-stranded, and therefore does not need to be further denatured once it is on the filter (unlike double-stranded DNA). At this stage, the membrane is 'probed' with a radioactively or fluorescently labeled DNA (or RNA) probe, the unbound probe is washed off, and the hybridized labeled probe is detected.

Northern blot analysis is the least biased approach for RNA analysis as it directly follows the signal of the actual transcript in the sample, in contrast to other techniques we describe later. As such, the size of a full-length transcript can be determined by comparing the position of the signal in the membrane to the positions of marker bands of known size. Moreover, if there are multiple RNA species that hybridize to the selected probe (for example, due to alternatively processed versions of the transcript), each will be observed as an independent signal on the blot. Northern blot analysis can also be used to determine the relative abundance of a specific transcript under different growth conditions by comparing the intensity of the signals from samples isolated under different conditions.

The PCR method described in Section 19.3 can also be exploited to detect the presence of a specific RNA and determine its relative quantities. This approach is commonly called the quantitative reverse transcription-PCR (**qRT-PCR**). In this method, a cDNA copy of the RNA of interest is first synthesized by reverse transcription. The cDNA products are then amplified by PCR, utilizing a second primer within the transcript of interest. This step is designed to yield a conveniently sized PCR product for analysis on an agarose gel.

The amount of product obtained by qRT-PCR will be proportional to the amount of RNA present in the sample as long as every sample being compared is within the exponential phase of PCR amplification (that is, before one of the components in the reaction becomes limiting and the rate of polymerization decreases). Machines that automatically quantify the amount of PCR products after each cycle in 'real time' (through binding to a dye) have greatly simplified these assays, though proper controls must always be included to ensure that the amount of the PCR product is correlated with that of the input RNA. A distinct advantage of qRT-PCR is that its extreme sensitivity allows very small amounts of RNA to be detected. A disadvantage of this approach is that only a limited segment of the RNA molecule corresponding to the region between the chosen primers will be detected. As a result, if, for example, the RNA exists in multiple forms within the sample, this type of information may be missed in the analysis.

Another approach that has traditionally been used to determine the amount of an RNA molecule is a **nuclease protection assay**, which is illustrated schematically in Figure 19.62. In this approach, a radioactively labeled DNA or RNA probe of a defined size is hybridized to the transcript of interest in solution. Nucleases such as S1 nuclease, which cleave single-stranded nucleic acids but not double-stranded DNA–RNA or RNA–RNA hybrids, are then used to digest the unhybridized RNA and probe molecules. These digestion products are then resolved by gel electrophoresis and visualized; the signal of the probe is proportional to the amount of target RNA.

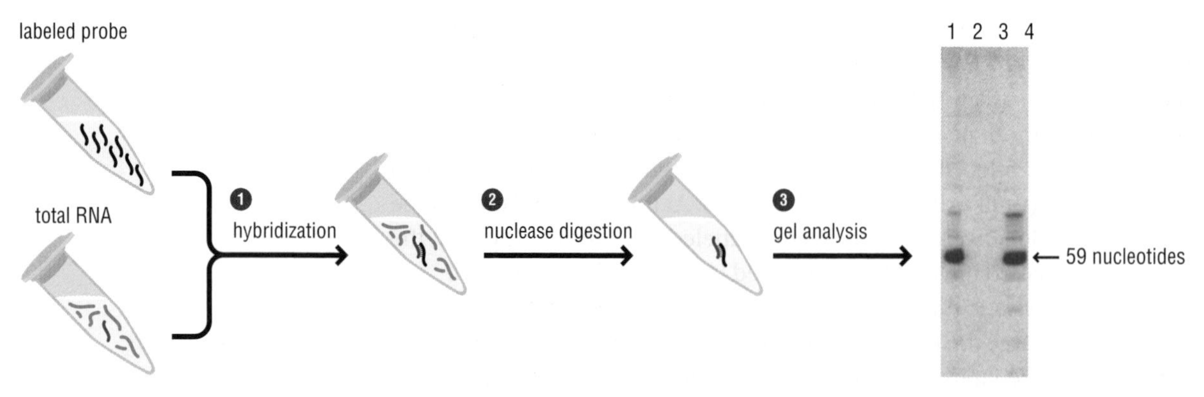

Figure 19.62 Analyzing RNA abundance by nuclease protection assay. RNA is isolated from cells and is hybridized in solution with a radioactive probe (DNA or RNA) targeting a specific sequence (step 1, the cellular sequence of interest is shown in pink). The mixture is then treated with nucleases that digest the non-target RNA and any probe that is not annealed, but not double-stranded RNA or DNA–RNA hybrids. Both the probe and the RNA are protected from nuclease digestion wherever the two are complementary to one another. The resulting mixture is subjected to gel electrophoresis; the amount of undigested probe visible in the gel is proportional to the abundance of the RNA of interest. In the example shown, mRNA was prepared from four cell lines (lanes 1–4), and the 59 bp probe used was against a segment of murine terminal transferase mRNA. Cell lines 1 and 4 express terminal transferase, whereas cell lines 2 and 3 do not.

Image adapted from Carey *et al. Cold Spring Harbor Protocols*, 2013; doi:10.1101/pdb.prot071910: 276–285.

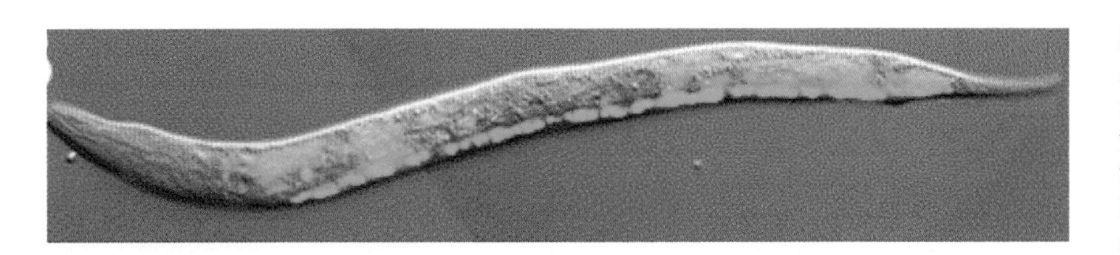

Figure 19.63 The use of reporter genes *in vivo*. Expression of a cyclin E:GFP reporter in *C. elegans*. GFP fused to the promoter of the cyclin E gene (cye-1::GFP) reports on its transcriptional activity. Shown is a larval stage L1 animal with strong expression of the *cye-1* reporter gene in a row of dividing cells (P-lineage descendants) in the ventral nerve cord (the line of cells in the middle of the animal), as well as in other tissues.

From Brodigan, TM, Liu, J, Park, M. Kipreos, ET, and Krause, M. Cyclin E expression during development in *C. elegans*. *Developmental Biology*, 2003;**254**: 102–115.

Reporter genes can simplify the detection of specific RNAs

A common method for assaying the expression of a particular gene is to fuse its regulatory region to a **reporter gene**, which encodes a gene product that can be readily detected. Products of some commonly used reporter genes include β-galactosidase (encoded by the *E. coli lacZ* gene), which breaks down particular galactose derivatives to give a colored product; luciferase, which catalyzes a chemical reaction that emits light; and GFP from jellyfish, which has intrinsic fluorescence. The availability of easily detected indicators of expression makes it possible to screen thousands of microbial colonies for those containing the coding and/or regulatory regions of a particular gene (as seen in Experimental approach 14.2).

Reporter genes can be quickly and repeatedly assayed, making it possible to study the way gene expression changes with time. Reporter genes also can be used to monitor gene expression that might be limited to certain tissues within the whole organism, as illustrated in Figure 19.63. The reporter can be fused to the promoter of the gene to reflect the transcriptional regulation in operation within the organism. Alternatively, the reporter can be fused directly to an open reading frame (ORF) of interest such that the reporter reflects both the transcriptional and translational regulation of the gene.

The specific ends of RNA molecules can be determined by nuclease protection, primer extension, or 5′- or 3′-RACE

Sometimes it is desirable to determine the 5′ or 3′ end of an RNA molecule - for example, to identify the promoter of a gene. All methods for identifying end regions rely on the hybridization of DNA or RNA probes in the vicinity of the ends. For example, the nuclease protection assays we have just described can be used to map the ends of an RNA. Such end identification is performed by using a probe that extends beyond the end of the RNA of interest on one side and overlaps the end of the RNA on the other side, as shown in Figure 19.64a. During nuclease digestion, the part of the probe that extends beyond the transcript will be digested, and the length of the nuclease-resistant probe will reveal the 5′ or 3′ end of the RNA of interest.

Primer extension assays can also be used to identify the 5′ end of an RNA molecule. As shown in Figure 19.64b, a labeled oligonucleotide complementary to sequences near the 5′ end of an RNA is extended by reverse transcriptase. Since the reverse transcriptase enzyme falls off when it reaches the 5′ end of the transcript, it is possible to determine the 5′ end precisely by comparing the extended product with a sequencing ladder.

Another approach for identifying RNA ends is called **5′- or 3′-RACE**. We encountered this amplification method when we discussed cloning of RNA molecules in Section 19.3 (see Figure 19.19). To identify a 5′ or 3′ end, an oligonucleotide of known sequence is ligated to the respective end of the RNA before the RNA is

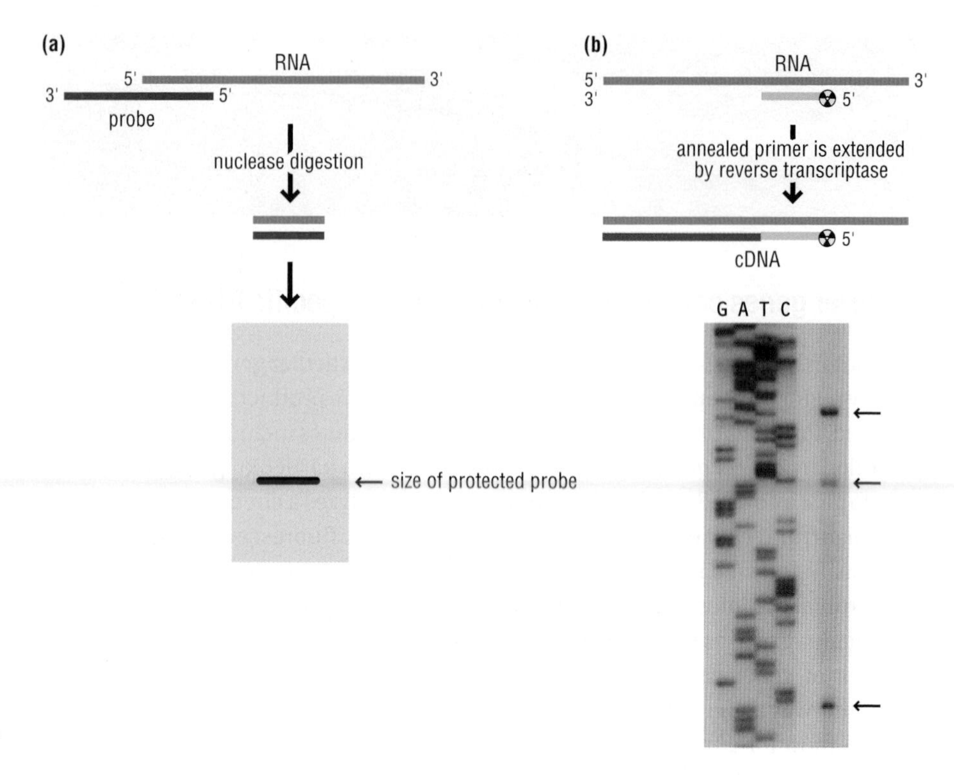

Figure 19.64 Methods to determine the identity of the ends of RNA molecules by nuclease protection and primer extension. (a) Identification of an RNA end by nuclease protection. The probe (in red) is designed to extend beyond the putative 5′ end of the RNA. After nuclease digestion, which removes any single-stranded nucleic acid, the 3′ end of the probe will lie at the 5′ end of the RNA. Determining the size of the protected probe will reveal where the transcript begins, relative to the 5′ end of the probe. (b) Identification of an RNA end by primer extension. A radioactive oligonucleotide (green) is hybridized to the RNA (blue) and then extended using reverse transcriptase to produce single-stranded cDNA (black). The resulting cDNA is subsequently analyzed on a denaturing polyacrylamide gel (by PAGE; bottom panel). A manual sequencing reaction is run next to the primer extension product to determine the size. In the example shown, there are three RNA forms: 59, 90, and 105 bp long (indicated by arrows).

From Opdyke, JA, Kang, J-G, and Storz, G. GadY, a small-RNA regulator of acid response genes in *Escherichia coli*. *Journal of Bacteriology*, 2004;**186**:6698–6705.

reverse transcribed (or sometimes at the end of the reverse transcriptase product, in the case of a 5′ end). Consequently, the resulting cDNA product has an additional string of nucleotides (whose sequence is known) at one end, which can then be used as a primer site for amplification by PCR. The products of PCR amplification are then sequenced, revealing the complete sequence of the original RNA molecule through the unknown 5′ or 3′ portion.

DNA microarray analysis is used to determine genome-wide expression patterns

The availability of whole genome sequences has made it possible to develop microarrays containing nucleotide fragments for all or much of the genome, which can then be used to detect specific RNA sequences on a genome-wide scale. As we encountered for aCGH in Section 19.9, the microarray contains many different DNA fragments that are immobilized on the surface of a specialized slide or chip, as depicted in Figure 19.65. The DNA molecules can be either synthetic oligonucleotides or larger fragments that have been generated by PCR. Microarrays can be constructed, for example, so that each spot contains a DNA fragment corresponding to a portion of an ORF. These microarrays can then be used to monitor transcription in a population of cells by isolating the pool of RNA, generating fluorescently labeled cDNA copies of this population of RNA molecules, and then hybridizing the mixture of cDNAs to the microarray. A labeled cDNA will hybridize to a spot in the microarray that contains a complementary DNA fragment. The relative intensity of each spot on the microarray reflects the level of transcription of the corresponding gene.

Expression of the same set of genes under different conditions (say, in the wild-type organism as compared with a mutant) can be compared by labeling each cDNA pool with a different fluorescent dye. Both samples are then used to probe

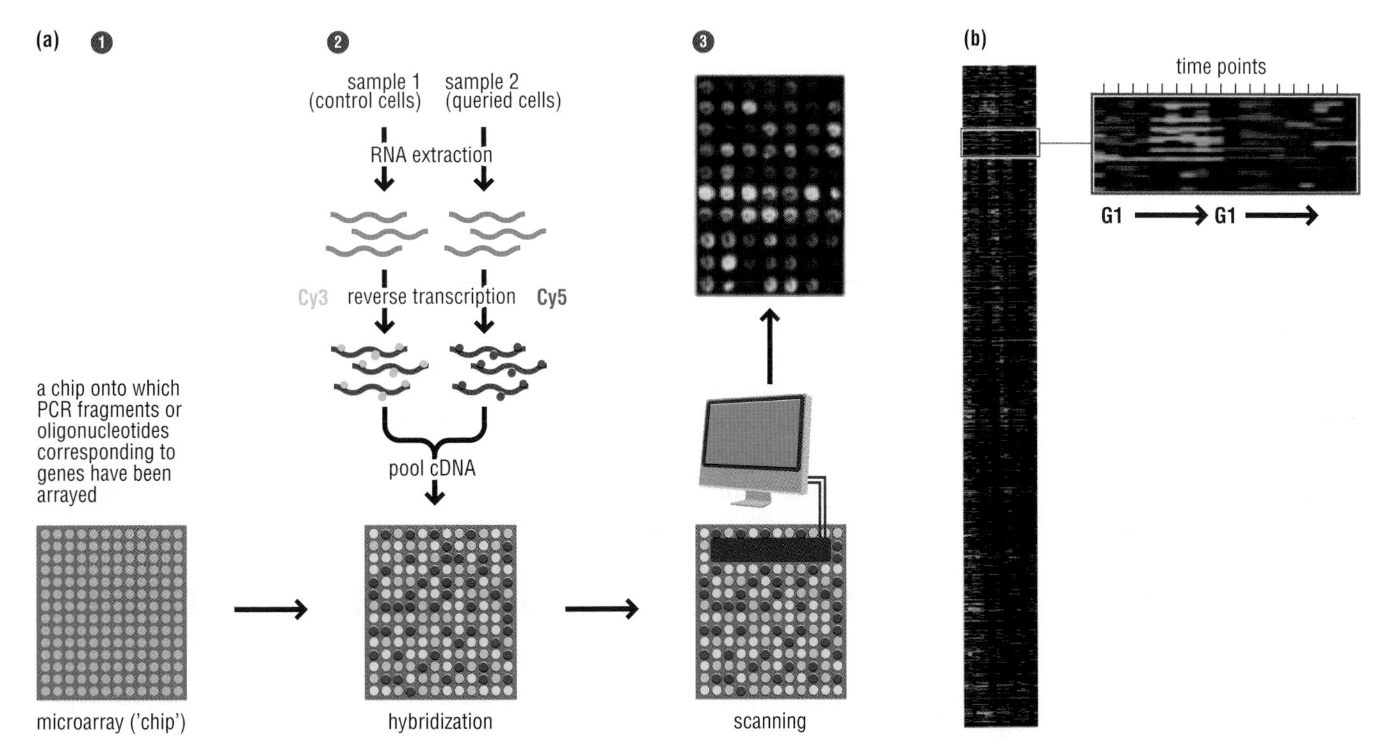

Figure 19.65 Microarray analysis. (a) Steps involved in microarray analysis. A chip is prepared that contains sequences representative of a select group of genes or an entire genome (step 1). RNA is isolated from two sets of cells: the control cells and the queried cells (for example, cells grown under a special growth condition, or a mutant strain vs a wild-type control). The RNA is reversed transcribed to generate cDNA, and each cDNA pool is labeled with a different fluorophore (in this case, Cy3, which emits green light, and Cy5, which emits red light). The labeled RNAs are pooled and hybridized to the microarray (step 2). Each spot on the chip is excited with a laser (at the Cy3 and Cy5 excitation wavelengths), and the emission is captured by a camera (step 3). If a particular transcript is expressed more robustly in the control than the non-control cells, the spot will appear green, while higher expression in the queried culture will yield a red spot. Equal expression gives a yellow signal. (b) Genome-wide microarray analysis of cell cycle dependent gene expression in *S. cerevisiae*. Cells were treated such that they progressed synchronously through the cell cycle, and the expression of every gene was examined at fixed time intervals. Each row represents the expression pattern of a single gene, and each column represents a time point. At the left-most time point, the cells are found in G1, and the cells proceed through two full cell cycles as the time course progresses. Red indicates genes with increased transcription relative to the control (RNA from asynchronous cells), while green indicates genes with decreased transcription. The data are displayed in the form of clusters, meaning that genes that have a similar expression pattern are placed next to each other in the presentation. For example, in the right panel, which is an enlargement of a small section from the left panel, the expression pattern in the top four rows is similar.

Adapted from Spellman, PT, Sherlock, G, Zhang, MQ, Iyer, VR, Anders, K, Eisen, MB, Brown, PO, Botstein, D, and Futcher, B. Comprehensive identification of cell cycle-regulated genes of the yeast *Saccharomyces cerevisiae* by microarray hybridization. *Molecular Biology of the Cell*, 1998;**9**: 3273–3297.

the same microarray and the relative intensities of each dye on each spot can be determined as described above. If red and green dyes are used, genes that are expressed at equal levels under both conditions will yield a yellow spot on the array, whereas differences in the ratio of expression will give rise to a spot that is either green or red in color. This approach can be used to assess relative gene expression under two different conditions, such as in the presence and absence of a stress such as heat shock, as discussed in Experimental approach 9.3.

Deep sequencing is used to determine genome-wide expression patterns

A final approach of increasing importance is the simple direct sequencing of cDNAs generated from an RNA population of interest, known as RNA-Seq. As sequencing methods have improved, it is now possible to sequence vast numbers of cDNAs. If enough sequences are read, the data from such an experiment can be used to evaluate actual amounts of a species in a population, in particular when directly compared with a related population. This approach is referred to as deep sequencing and will be described in more detail in Section 19.15. Deep sequencing can effectively be applied to different subpopulations of RNAs, such as RNAs below a certain size or RNAs that are processed in a specific way, to learn more about the transcripts. For example, to identify unprocessed RNAs with a 5′ triphosphate, the total RNA population can be treated with a terminator exonuclease enzyme that will degrade all processed RNAs with a 5′ cap or monophosphate before being subject to deep sequencing.

One caveat to deep sequencing, as well as the microarray approaches, for examining genome-wide RNA levels is that the need to generate a cDNA copy of the RNA can bias the population that is recovered for sequencing: for example, an RNA will be missed if reverse transcriptase cannot make a copy of the transcript due to the presence of inhibitory secondary structures or nucleotide modifications.

19.11 DETECTION OF SPECIFIC PROTEINS

It is often desirable to follow the fate of a specific protein. Proteins can be followed in a cell lysate or in their native location in the cell. In either case, there are two main approaches for detecting proteins: the use of antibodies directed against the protein or against a tag added to the protein by cloning. In this section, we discuss the detection of proteins in cell lysates and also consider an approach for evaluating the relative levels of specific proteins in two different samples. The detection of proteins inside cells will be discussed in Section 19.13.

Western blots rely on antibodies to detect proteins

A powerful method for detecting specific proteins takes advantage of the immune system's ability to generate antibodies that bind to proteins with very high affinity and specificity. Antibodies can be obtained by exposing the appropriate animal (often mouse, rabbit, goat or chicken) to either a purified protein or a short peptide corresponding to a region of the protein. Generally, two classes of antibodies are used: polyclonal antibodies, which are a collection of different antibodies that often react with different epitopes of the same protein, and monoclonal antibodies, which are obtained from a clonal population of immune cells that express only one type of antibody and thus react with only one epitope of the protein.

The western blot, or **immunoblot**, detects a specific protein immobilized on a membrane using antibodies directed against the protein (or a tag fused to the protein, as will be discussed next). In this method, illustrated in Figure 19.66, proteins are separated by SDS-PAGE, and transferred from the gel to a hydrophobic or positively charged membrane. The membrane is then incubated with specific antibodies in the presence of a protein-rich blocking agent such as a solution containing BSA, which reduces non-specific binding of the antibody. After the membrane is washed, the bound antibody can be detected in a variety of ways, most of which rely on 'secondary' antibodies that bind to the 'primary' antibody. These 'secondary' antibodies themselves can be directly labeled with radioactivity, coupled to a fluorescent molecule, or linked to an enzyme such as horseradish peroxidase, which produces a colored or fluorescent band at the location of the protein of interest when incubated with the appropriate substrate.

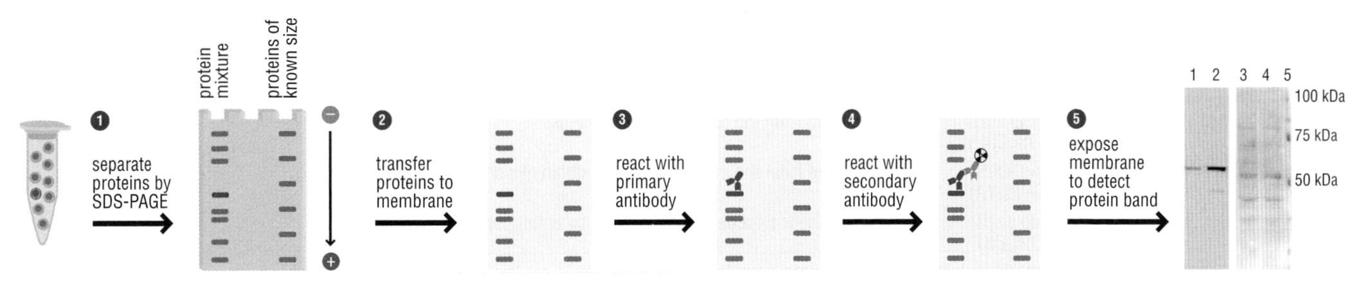

Figure 19.66 Western blot analysis. A protein mixture containing a protein of interest (pink) is loaded on a polyacrylamide gel and separated by electrophoresis, as described in Figure 19.47 (step 1). Proteins of known size that serve as molecular weight markers can also be loaded. The proteins are then transferred to the filter in a special chamber, using an electric field (step 2). The filter can then be stained with dye, such as Ponceau S Red, which reveals the proteins on the filter and shows the positions of the molecular weight markers (see pink-stained filter on the far right). The filter is then exposed to the primary antibody that recognizes the protein of interest (step 3). This reaction takes place in a buffer that minimizes non-specific binding of the antibody to irrelevant proteins (known as cross-reactivity). The filter with the bound primary antibody is washed several times to remove unbound antibody and then exposed to a secondary antibody that is conjugated to a detectable molecule, such as a radioactive isotope or a fluorophore (step 4). The filter is again washed to remove unbound secondary antibody and then processed to detect the secondary antibody using autoradiography (to detect radioactivity) or a phosphorimager (to detect fluorescence) (step 5). In the example shown, two samples containing a mixture of proteins were run on an SDS-PAGE gel and transferred to a filter. The filter was then stained with Ponceau S Red (on the right), revealing the proteins in both lanes that were transferred to the filter (lanes marked 3 and 4: note the presence of many protein bands). The Ponceau S staining also shows the locations of the molecular weight markers (lane 5). The same filter was then treated with antibodies against the protein of interest followed by secondary antibodies that recognize the primary antibody. The filter was then exposed to reveal the location of the band corresponding to the protein of interest (lanes marked 1 and 2).

The accuracy of this method depends upon the specificity of the primary antibody: some antibodies (especially polyclonal ones that may recognize several epitopes) may react with other proteins in addition to the intended target. It is thus important to include the appropriate controls (such as a lysate missing the protein of interest) to confirm the specificity of the primary antibody.

The addition of specific tags can be used to detect proteins

For some proteins, it is extremely difficult to generate antibodies. An alternative approach for detecting a specific protein is to introduce a protein tag that is fused to either the N- or C-terminus of the protein of interest, and then use an antibody that binds specifically to that tag. A summary of commonly used tags is given in Figure 19.67. Some of the tags comprise only a few amino acids, while other tags correspond to an entire small protein. Very specific antibodies have been developed for most of the tags, allowing for sensitive detection of proteins carrying these tags. As described in Section 19.7, several of the tags also bind to specific matrices and can be used to purify the tagged proteins.

The relative levels of proteins in two different samples can be determined by isotope labeling followed by mass spectroscopy

As we have seen for DNA and RNA, sometimes it desirable to detect a collection of protein molecules. A recent innovation in the area of mass spectroscopy has made it possible to compare the relative amounts of many protein species in two samples (collected under different conditions for example). This quantitative proteomic approach is referred to as stable isotope labeling with amino acids in culture (SILAC; Figure 19.68). The basic idea behind the approach is to grow cells under two different conditions in the presence of an amino acid containing one of two different isotopes, such as arginine labeled with ^{13}C atoms instead of the usual ^{12}C. One isotope is used under one set of conditions while the other isotope

Commonly used protein tags

tag name	size	source and comments
Myc	10 amino acids (EQKLISEEDL)	derived from the human c-Myc proteins
hemagglutinin epitope (HA)	9 amino acids (YPYDVPDYA)	derived from the influenza hemagglutinin glycoprotein
FLAG	8 amino acids (DYKDDDDK)	a synthetic peptide, can be cleaved by enteropeptidase
hexa histidine-tag (6 × His)	6 (or more) amino acids (HHHHHH)	a synthetic peptide that binds to a resin containing nickel (Ni^{2+})
vesicular stomatitis virus glycoprotein (VSV-G)	11 mino acids (YTDIEMNRLGK)	derived from a viral glycoprotein needed for the budding of vesicular stomatitis virus
V5	14 amino acids (GKPIPNPLLGLDST)	derived from a small epitope present on the P and V proteins of the paramyxovirus of simian virus 5 (SV5)
calmodulin binding peptide (CBP)	26 amino acids	derived from muscle myosin light-chain kinase, binds to resin containing calmodulin
glutathione-S-transferase (GST)	220 amino acids	naturally occurring protein that binds to resin containing glutathione
green fluorescent protein (GFP)	238 amino acids	a naturally flurorescent protein from the jellyfish, Aequorea victoria, commonly used in applications such as light microscopy to visualize directly the location of the tagged protein
maltose binding protein (MBP)	371 amino acids	E. coli protein, binds to resin containing amylose

Figure 19.67 Commonly used protein tags. For short tags, the amino acid sequence is shown. A 'synthetic tag' means that it was devised by scientists, rather than being derived from a naturally occurring protein.

is used in a population of cells subjected to different conditions. (For example, one population of cells might be treated with heat to stimulate a heat-shock response.) As a result of this isotopic labeling, any protein fragment containing an arginine will have a mass 6 Da heavier when growth in the presence of ^{13}C-labeled arginine compared to the equivalent fragment from the culture grown with ^{12}C-labeled arginine.

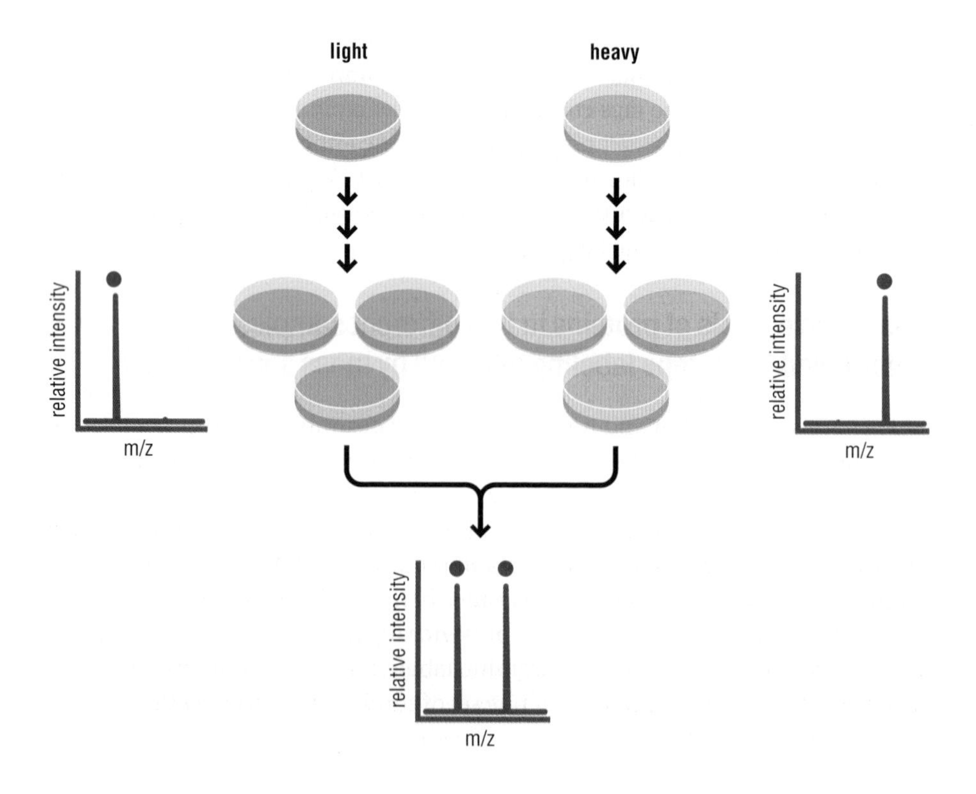

Figure 19.68 The principle of SILAC. To compare the protein composition of cells from two different sources, each cell type is labeled by its growth in medium containing an amino acid labeled with one of two different isotopes – for example, ^{12}C vs ^{13}C. In the example shown, the cells on the left were grown in 'light' arginine, containing ^{12}C (purple) while the cells on the right were grown in 'heavy' arginine, containing ^{13}C (light red). The incorporation of the ^{13}C labeled arginine into proteins results in a mass shift of the corresponding peptides relative to the same peptides from cells grown in the presence of ^{12}C arginine, and this shift can be detected by a mass spectrometer. The samples are combined, and the relative amounts of each peptide are compared. In the example shown, the peptide is present at similar levels in both samples.

Following whole cell labeling, protein extracts are prepared and the two samples digested with proteases in preparation for analysis by mass spectrometry, which we discussed in Section 19.8. The samples are then injected simultaneously into the mass spectrometer, where the ratio of peak intensities can provide information on the relative abundance of the peptides under different conditions.

Although not all proteins in the cell can yet be resolved and identified with this approach, a substantial number can be followed in a single experiment, yielding significant insights into global protein levels in the face of a given insult to the cell. As the technology is further developed, this method will undoubtedly become increasingly sophisticated.

19.12 DETECTION OF INTERACTIONS BETWEEN MOLECULES

The goal of the many techniques we have discussed in earlier sections of this chapter is to isolate and identify a single biological molecule of interest, such as a specific RNA or a specific protein. However, since no biological molecule acts alone, we are frequently interested in examining the interaction between molecules - especially interactions between proteins, and between proteins and nucleic acids. Thus we next consider some of the commonly used approaches to study binding between biological molecules.

Co-immunoprecipitation and co-purification can be used to identify and study interacting proteins

Among the simplest methods used to study interactions between macromolecules are co-immunoprecipitation and co-purification. If certain molecules of interest remain associated with one protein (for instance, during many purification steps) the molecules are said to co-purify. Thus, co-purification can be exploited to identify molecules that interact with one another in the cell.

Co-immunoprecipitation is a method for isolating protein complexes using an antibody that binds to just one specific protein in the complex. An antibody against a protein of interest is added to a mixture of proteins or to a cell extract. The antibody–protein complex is then precipitated from solution by adding microscopic beads that are covered with proteins that bind tightly to *any* antibody. The beads are relatively heavy and readily fall to the bottom of the test tube, bringing with them the protein that binds to the immobilized antibody. If any other proteins are bound to the protein of interest, they, too, will precipitate along with the rest of the complex. The binding partners that are precipitated in this way can then be identified by the method of mass spectrometry as described in Section 19.8.

An alternative to using antibodies relies instead on generating a fusion protein containing an affinity tag (as we learned about in Section 19.7 and Section 19.11), which can be used to precipitate the tagged protein along with other macromolecules that may bind to it. For example, a fusion protein containing GST can be precipitated by beads coated with glutathione. The interaction can be confirmed by performing it in reverse, in which the binding partner is precipitated and the association of the original protein is examined.

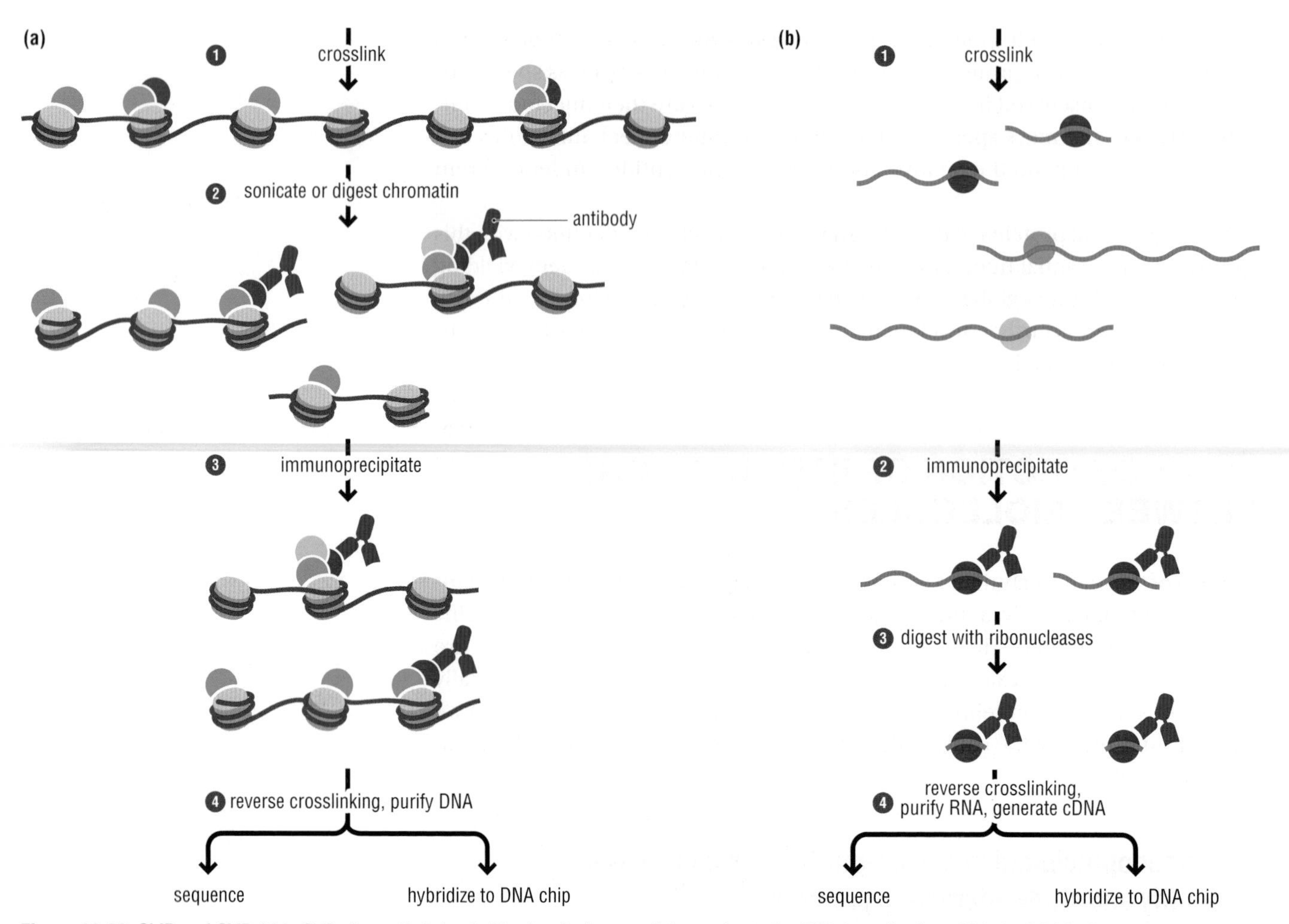

Figure 19.69 ChIP and CLIP. (a) In ChIP, chromatin is treated to chemically cross-link proteins to the DNA (step 1), after which the DNA is digested by restriction enzymes or mechanically sheared (step 2). The DNA fragments and their associated proteins are then subjected to immunoprecipitation, using an antibody (in dark blue) against a protein of interest (pink), which isolates only those DNA fragments that are bound to this protein (step 3). The cross-links are then removed, and the DNA is isolated, linkers are added to the DNA ends, and the DNA is amplified by PCR (step 4). (b) In CLIP, cross-linking is used to stabilize protein–RNA interactions (step 1) after which a protein of interest is immunoprecipitated (step 2). RNA not protected by the protein is then digested with ribonucleases (step 3). After the cross-links are reversed, the RNA is isolated, and cDNA is generated and amplified (step 4). For both ChIP and CLIP, the amplified DNA is analyzed by one of a variety of methods, including hybridization to a DNA chip or sequencing (step 5).

Protein binding sites in DNA and RNA can be identified by co-immunoprecipitation and co-purification

Co-immunoprecipitation or co-purification can also be used to isolate DNA or RNA fragments that bind to a protein that is, in turn, precipitated by the antibody or by an affinity tag. The DNA or RNA can then be identified using the microarray approaches we learned about in Sections 19.9 and 19.10 or the sequencing methods we learned about in Section 19.8.

An application of immunoprecipitation used to identify proteins that are bound to a particular region of chromosomal DNA in the cell is called **chromatin immunoprecipitation (ChIP)** and is illustrated schematically in Figure 19.69a. In this method, cells are treated for a limited period with a chemical reagent that covalently cross-links proteins to DNA. Once the bound proteins are immobilized in this way, the chromosomal DNA is sheared into relatively small fragments. Antibodies against a DNA-binding protein of interest are then added to the mix and used to immunoprecipitate the protein, along with the piece of DNA to which it is cross-linked. A reagent is then added that reverses the cross-links, releasing the DNA into solution.

All of the sites in a genome to which a particular protein binds can then be identified by hybridization to a microarray, sometimes referred to as 'ChIP on chip' or 'ChIP-chip'. As DNA sequencing becomes ever more available and cheaper, it is becoming common to sequence the individual precipitated DNA fragments directly in a method known as '**ChIP-Seq**'. PCR can also be used to amplify just a particular sequence of interest by using appropriately designed primers, as described in Section 19.3.

An approach similar to ChIP-Seq can be used to identify RNA bound to a particular protein, as illustrated in Figure 19.69b. In this method, known as cross-linking and immunoprecipitation (CLIP), proteins are cross-linked to their target RNAs in living cells using short-wave UV irradiation, thus generating a covalent bond between protein and RNA. The particular protein of interest can be immunoprecipitated along with its bound RNA, as in the case of ChIP. After the complex is precipitated, extraneous RNA that extends beyond the interaction of interest can be digested with ribonucleases, leaving the RNA fragment that contacts the protein and is thus protected from digestion.

In a variant of this CLIP method, the ribonucleotide 4-thiouridine is incorporated biosynthetically in the RNA by introducing it into cells. RNA containing this nucleotide can be cross-linked to proteins using lower energy UV light. This method is referred to as photoactivatable-ribonucleoside-enhanced cross-linking and immunoprecipitation (PAR-CLIP).

In all cases, the cross-linking is reversed, the RNA is isolated and cDNA is generated. The cDNA can either be sequenced using the methods described in Section 19.8, or it can be hybridized to a DNA microarray, thus making it possible to identify the sites in the RNA to which various proteins bind.

Co-purification followed by deep sequencing allows the mechanism of transcription and translation to be probed on a genome-wide level

Extensions of the co-immunoprecipitation and co-purification approaches are now being exploited to interrogate the processes of transcription and translation on a genome-wide level. For example, the mRNA regions that are bound to ribosomes and are thus protected from nuclease digestions and can be isolated and identified in an approach termed **ribosome profiling**. This process, which is summarized in Figure 19.70, begins with the preparation of a cell lysate that is treated with RNase. Ribosomes that are engaged in translation will protect a 30 nucleotide region on the mRNA from nuclease digestion. However, other RNA-associated proteins may also be present in the lysate. Therefore, the next step is the purification of ribosomes along with the protected mRNA segment to which the ribosomes are bound. This is achieved by using a sucrose gradient, which separates ribosomes from other cellular proteins. The RNA bound to the ribosome is then isolated, and linkers are attached to its ends to generate cDNA. This cDNA is then used for sequencing to reveal the genes that were being translated. Note that the relative abundance of the sequences obtained per gene is proportional to the expression level of that gene.

Global profiling approaches can provide many insights missed by studies of individual genes. For instance, ribosome profiling allows the quantification of the mRNAs that are being translated and the identification of protein-coding genes that might have been missed by standard annotation because they are short. The

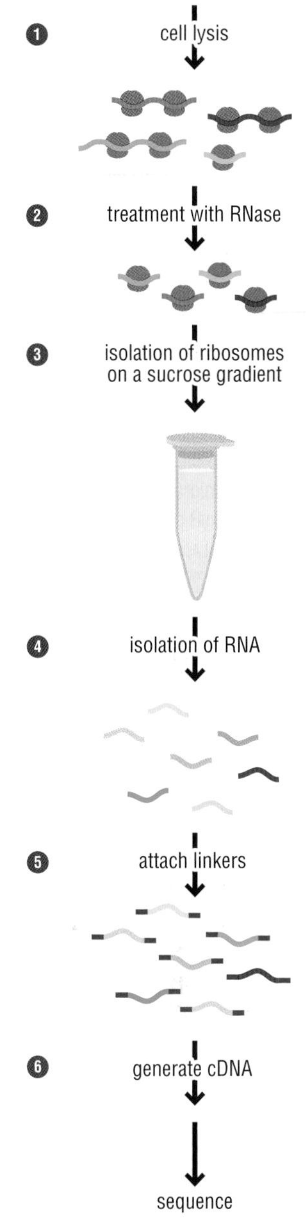

Figure 19.70 Ribosome profiling. Cell are lysed (step 1) and treated with RNase (step 2) to remove all RNAs that are not protein bound. Ribosomes are then purified by separating them on a sucrose gradient (step 3), after which the RNA that was bound to the ribosome is isolated (step 4). This RNA is then converted to cDNA by ligating it to primers (step 5) and converting it to cDNA with reverse transcriptase (step 6). These cDNA fragments are then sequenced and aligned to the genomic sequence to reveal the mRNAs that were being translated in the cells. The number of sequences recovered for each gene is proportional to the number of ribosomes that were engaged in translating the gene's mRNA.

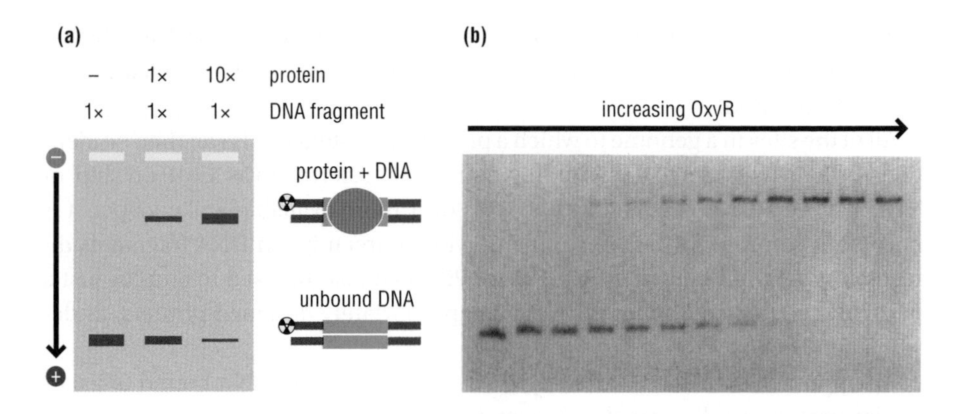

Figure 19.71 EMSA. A non-denaturing gel is used to assay binding of a protein to a DNA fragment. The first lane on the left shows the migration of the DNA fragment (dark gray band) on its own. In these experiments, the DNA is typically labeled with radioactivity or a fluorophore. The middle lane shows the migration of the DNA fragment when protein is present; the protein–DNA complex is larger than the DNA fragment alone and therefore migrates more slowly through the gel than the free DNA. The right lane shows the migration of the DNA fragment in the presence of ten times more protein than in the middle lane. This leads to a higher proportion of the DNA fragments binding to the protein, and thus more of the DNA fragment migrates slowly. (b) An example of an EMSA, showing how a radiolabeled DNA fragment bound to the OxyR protein is shifted to a more slowly migrating form as increasing amounts of the OxyR protein are added.

Adapted from Tartaglia, LA, Gimeno, CJ, Storz, G, and Ames, B. Multidegenerate DNA recognition by the OxyR transcriptional regulator. *Journal of Biological Chemistry*, 1992;**267**: 2038–2045.

approach can also be carried out for strains with mutations in components of the translation apparatus to identify the contributions of the components to steps in translation. Similar purification of RNA polymerase from wild-type and mutant strains allows the steps in transcription to be investigated for all transcribed genes.

A shift in DNA or RNA mobility in gels can be used to measure binding affinity

The rate at which a given macromolecule migrates in an electrophoretic gel depends upon its size. If a molecule forms a stable complex with one or more additional macromolecules, the complex will migrate more slowly in a non-denaturing gel than the individual molecule alone. This approach, termed an **electrophoretic mobility shift assay (EMSA)**, is frequently used to measure the strength of protein binding to a DNA fragment (an example of which is given in Experimental approach 16.2) or an RNA species.

In an EMSA of protein binding to DNA, a radiolabeled DNA fragment is incubated with a protein and then loaded onto a non-denaturing gel, to preserve protein-DNA interactions, as depicted in Figure 19.71. DNA with bound protein will migrate more slowly in the gel than the free DNA fragment. The relative amounts of bound and free DNA in each resulting band can be quantified, thus giving a measure of how much complex is formed at that concentration of binding partners. If this is repeated for a series of samples that each contains a different concentration of protein, a binding curve can be constructed, and an equilibrium binding constant can be determined.

➔ We discuss binding equilibria in more detail in Section 3.2.

Precise binding sites on DNA and RNA can be identified by footprinting

The precise site on a DNA fragment or RNA molecule to which a protein binds can be identified using **footprinting**. The underlying principle of this method, illustrated in Figure 19.72, is that a protein bound to DNA or RNA can protect the region of contact from chemicals or enzymes that cleave nucleic acids.

To carry out a footprinting experiment, many copies of a particular DNA or RNA molecule are labeled at one end with a radioactive atom (such as ^{32}P) or a fluorescent tag, and then treated with a chemical or endonuclease that cleaves the nucleic acid. The reaction is performed under conditions in which each DNA or RNA molecule in the solution is cleaved only once on average. This generates a set of fragments, with each 1 bp longer than the next. The DNA is denatured to separate

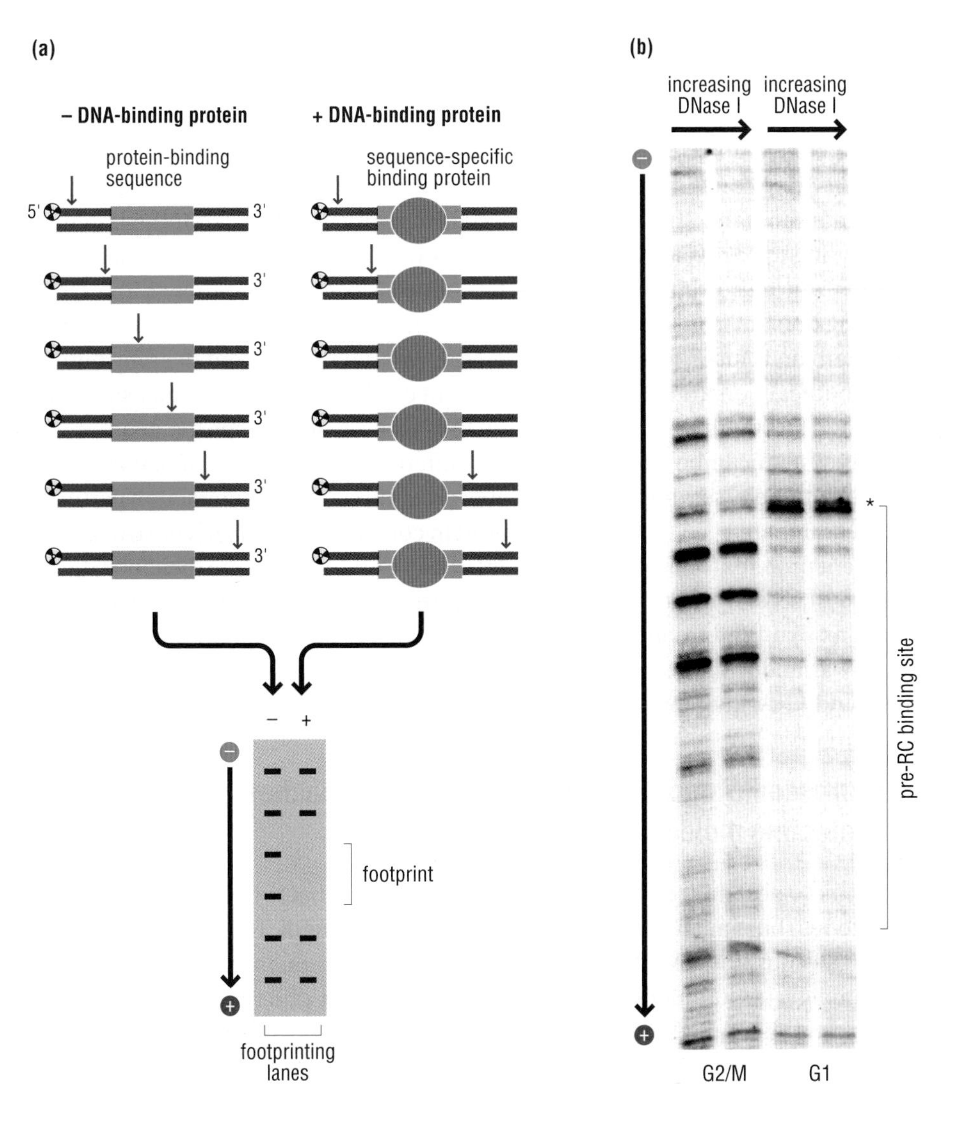

Figure 19.72 DNA footprinting assay. (a) DNA is labeled at one end and subjected to limited digestion, producing a ladder of fragments when electrophoresed on a gel, with smaller DNA fragments migrating further than larger fragments. If a protein is bound to the DNA during the digestion, it will protect a portion of the DNA from cleavage. See text for a more complete description. (b) An example of footprinting in a whole cell lysate. The region that is spanned by the black bar contains an *S. cerevisiae* origin of replication. When the cells are in G2/M, the origin is not bound by the prereplicative complex (preRC, see Chapter 6) and is thus exposed to DNase I digestion, generating the observed bands. In G1, however, the origin is bound by the preRC, protecting it from DNase I activity and generating a footprint that reveals the protected region. The dark band (indicated by an *) illustrates the paradoxical effect of protein binding sometimes enhancing endonuclease digestion at certain positions in the DNA.

Adapted from Labib, K, Kearsey, SE, and Diffley, JFX. MCM2-7 proteins are essential components of prereplicative complexes that accumulate cooperatively in the nucleus during G1-phase and are required to establish, but not maintain, the S-phase checkpoint. *Molecular Biology of the Cell*, 2001;**12**:3658–3667.

the strands, the resulting fragments are separated on a polyacrylamide gel, and the gel is autoradiographed.

If no protein is bound, a ladder of bands corresponding to the collection of cleavage products will appear. If, however, a protein is bound to the DNA or RNA before it is incubated with the enzyme, the nucleic acid will be cleaved everywhere except where it is protected by the protein. The resulting ladder on the gel will show fainter bands or no bands in the protected region. Base-pairing interactions between two different RNA molecules can also be examined by variations of the footprinting technique.

Two-hybrid analysis is a genetic approach for detecting interactions between proteins

A common method for studying protein interactions that is carried out *in vivo* and lends itself to large-scale screening is called the **two-hybrid assay**. This method is generally carried out in budding yeast, although it can be done in bacteria (*E. coli*) as well.

The yeast two-hybrid assay is carried out using a specially constructed strain of yeast containing a reporter gene that must be activated by a transcription factor

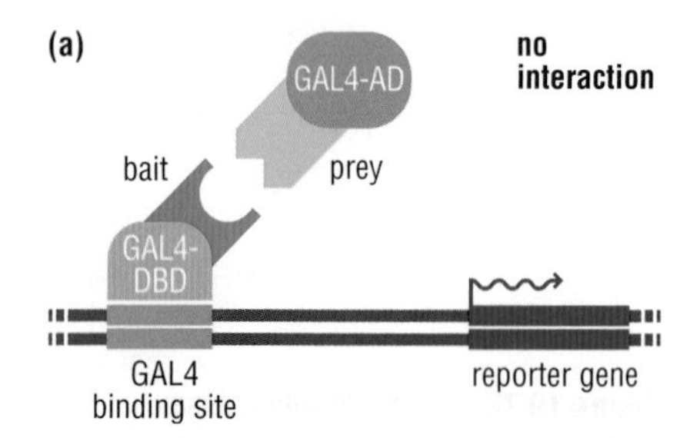

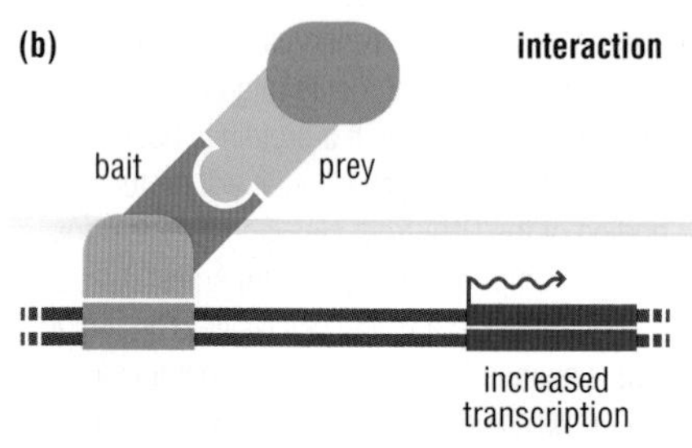

Figure 19.73 Yeast two-hybrid assay. A protein or domain of interest (the 'bait', in blue) is fused to the GAL4 DNA-binding domain (orange), while a different protein fragment, or a collection of protein fragments, the 'prey' (in green), is fused to the activation domain (orange). (a) If the bait and prey do not interact, the reporter gene will not be expressed. (b) Transcription of the reporter gene is only activated if the 'bait' and 'prey' bind to one another.

such as GAL4, which contains distinct DNA-binding and activation domains. The reporter gene can confer growth under a particular condition or encode an enzyme such as β-galactosidase that allows visual detection of gene activation. As illustrated in Figure 19.73, GAL4 is re-engineered to consist of two separate polypeptides: one containing the DNA-binding domain and the other containing the activation domain. Thus, the reporter gene can only be activated if the fragment containing the activation domain associates with the DNA-binding domain.

How can these separate proteins be induced to form a complex? The DNA-binding domain is fused to an additional domain called the 'bait,' and the activation domain is fused to a domain called the 'prey.' If the bait and the prey bind to one another, then the DNA-binding fragment can bind to the regulatory site in the DNA and recruit the activation domain, thus turning on the reporter gene.

This system can be used to test many different combinations of bait and prey proteins, thereby serving as a convenient and rapid screen for interaction partners.

Recombinant DNA techniques are used to construct a library of constructs with different proteins (or protein fragments) as bait or prey. These are then introduced into yeast cells, which can be assayed for particular combinations of bait and prey proteins that bind to one another, thus stimulating transcription of the reporter gene. Variations of the two-hybrid approach also can be used to detect protein–RNA and RNA–RNA interactions.

Spectroscopic signals provide sensitive approaches for detecting molecular interactions

Many biological molecules are fluorescent, meaning that they absorb light at one wavelength but emit light at a different (longer) wavelength. The chemical environment can substantially affect the absorption and emission properties of a fluorescent moiety; tryptophan side chains, for example, have different emission spectra depending on whether they are buried in the hydrophobic core of a protein or lie on the surface, where they are exposed to solvent. Tryptophan fluorescence can therefore be monitored to follow folding and unfolding of a purified protein in solution.

The fluorescence of a molecule can either increase or decrease when another species binds to it. Changes in fluorescence intensity can therefore be monitored to measure the binding of one molecule to another. Even if a molecule lacks intrinsic fluorescence, a fluorescent dye can be covalently attached and its fluorescence monitored. Proteins can also be engineered to have tryptophan side chains on the surface, whose fluorescence can then be monitored to follow a binding interaction. The instrument used to measure fluorescence is called a fluorimeter.

Another way in which fluorescence can be used to monitor binding interactions takes advantage of the fact that a fluorophore that is excited by polarized light will also emit polarized light, as depicted in Figure 19.74. If the molecules containing the fluorophore are tumbling rapidly in solution, after excitation the fluorophores in solution all reorient in a random fashion, and the sum of the fluorescent emissions no longer has a maximum in any particular direction. If, however, a larger molecule binds to some of the fluorescently tagged species and slows their rate of tumbling, some of the emitted light will still be polarized along the same direction because the molecules did not all have time to reorient in a random way. This method is referred to as **fluorescence anisotropy** (meaning not the same in all directions). By plotting the intensity of emitted polarized light as a function of increasing concentration of the unlabeled partner, it is possible to determine the rate constant for the binding reaction (the binding constant).

➔ We learn more about binding constants in Section 3.2.

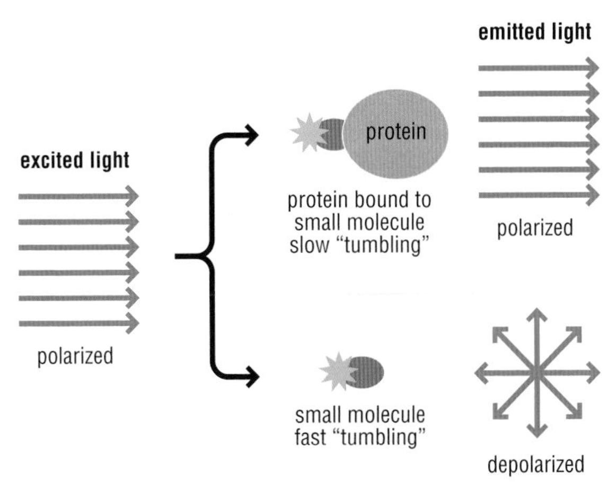

Figure 19.74 Monitoring binding interactions by fluorescence anisotropy. A fluorophore that is excited by polarized light will emit light that is polarized along the same direction. However, if the molecules are tumbling very rapidly (bottom), they will orient randomly, and thus each will emit light in random directions. The result is that the total emitted light is unpolarized. If, however, a large molecule binds to the fluorescently tagged molecule and slows down its rate of tumbling, the molecules will not have time to randomly orient after being excited and will thus emit light that is polarized along the same direction.

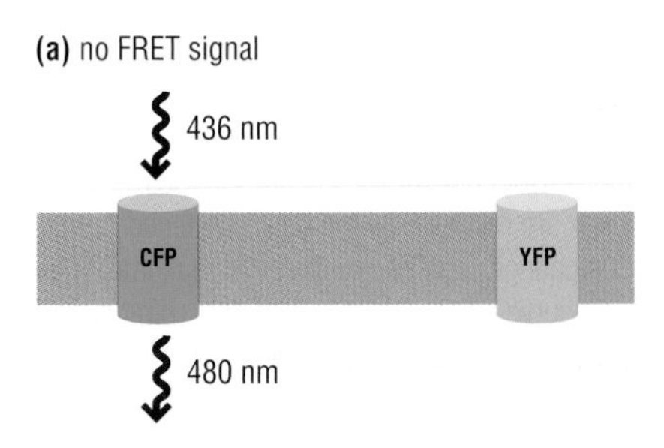

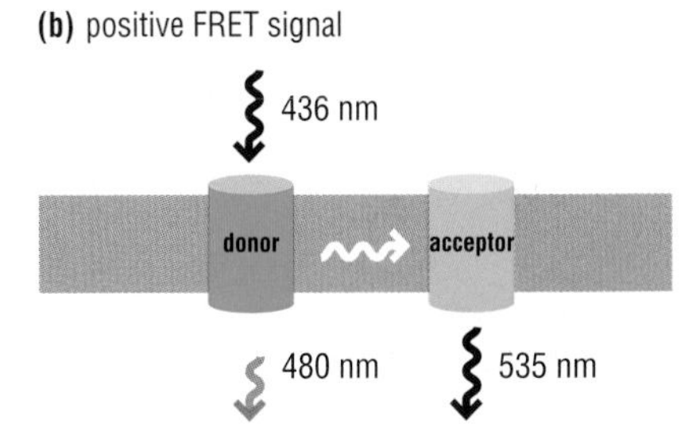

Figure 19.75 Principle of FRET. (a) The fluorescent proteins CFP (cyan fluorescent protein) and YFP (yellow fluorescent protein) have different absorption and emission spectra. Light with wavelength 436 nm will excite CFP, but because of its different absorption spectrum, YFP does not absorb light of wavelength 436 nm and therefore does not fluoresce. (b) If the two fluorescent molecules are sufficiently close to one another, following excitation by the light with wavelength 436 nm, some of the energy of the excited CFP (the FRET donor) can be transferred to the YFP molecule (the FRET acceptor). This transfer of energy excites YFP, which emits light at the YFP emission wavelength (535 nm). The donor and acceptor must be close to one another for energy transfer to occur, thus making measurements of acceptor fluorescence a sensitive measure of distance between the two molecules.

Another approach that also takes advantage of fluorescently labeled molecules is **Förster resonance energy transfer** (also called **fluorescence resonance energy transfer** or **FRET**). This method can be used to monitor interactions between two molecules, as well as to measure the distance between them. A FRET experiment requires two different fluorophores with distinct absorption and emission spectra, as depicted in Figure 19.75. If the molecules are sufficiently close to one another (within 5 nm), the energy emitted by the excited donor (that is, after exposure to the excitation wavelength) can be absorbed by the acceptor, causing the acceptor to fluoresce. By exciting the donor molecule and then measuring the fluorescence of the acceptor, it is possible to monitor interactions between the two molecules with a high degree of accuracy. Since the efficiency of the energy transfer depends strongly on distance (varying as $1/r^6$, where r is the distance between the donor and acceptor), the intensity of the emitted light can even be used to measure the distance between the donor and acceptor. We shall see in Section 19.13 how FRET measurements can be used in whole cells to detect molecular interactions *in vivo*.

Surface plasmon resonance can be used to monitor the kinetics of binding interactions

Another sensitive technique for studying interactions between two molecules relies on surface plasmon resonance (SPR)-based technology. This method is used to measure binding interactions where one of the interacting species has been immobilized on a surface. Like the fluorescence experiments we have just described, this approach can yield quantitative information on molecular interactions. A key difference is that, rather than directly measuring the equilibrium binding constant, SPR is used to measure both the association and dissociation rates, from which the K_d can then be calculated. The ability to monitor binding kinetics in real time provides additional information that cannot be deduced from the equilibrium binding constant alone.

➔ We learn more about association and dissociation rates in Section 3.2.

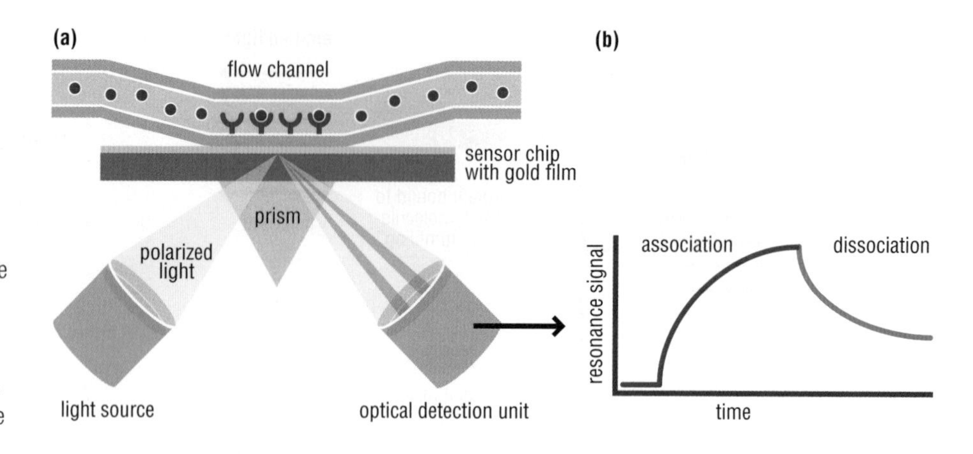

Figure 19.76 Surface plasmon resonance. (a) A glass sensor chip contains a thin layer of gold on which proteins or other molecules (shown in dark blue) can be immobilized. A solution containing a binding partner (pink spheres) flows across the upper surface. At the same time, a light is shone at an angle on the lower glass surface and the refracted light is detected. As molecules in the solution bind to the immobilized protein, the refractive index of the glass changes (due to a physical phenomenon known as surface plasmon resonance). This alters the angle at which the refracted light exits the sensor chip. (b) The changes in the refractive index are recorded by the optical detection unit. The changes can be monitored in real time. The pink curve represents the change in refractive index as more protein is bound, while the blue curve represents the change in refractive index as the binding partner is being washed away.

From Cooper, M. Optical biosensors in drug discovery. *Nature Reviews Drug Discovery*, 2002;**1**:515–528.

As illustrated in Figure 19.76a, one of the molecular species of interest (the 'bait') is immobilized on a thin gold film that lies on top of a glass surface. Buffer containing the other molecular species of interest (the 'prey') is then flowed across the surface. If the prey has an affinity for the immobilized bait, more and more prey will bind to the bait, until the surface is saturated. If fresh buffer is then flowed across the surface, the prey molecules will dissociate at a characteristic rate until no more molecules are bound.

The binding of molecules to the bait immobilized on the surface and their subsequent dissociation produces changes in the refractive index in the immediate vicinity of the surface layer. (The physical principles underlying this phenomenon are beyond the scope of this book.) As a consequence of the refractive index change, the angle at which the incident light is refracted changes. A plot of the change in the resonance angle as a function of time, known as a sensorgram (an example of which is depicted in Figure 19.76b), is then used to evaluate binding.

The thermodynamic parameters of a binding reaction can be measured using isometric titration calorimetry

The equilibrium binding constant gives a direct measure of the free energy of binding as given by the equation, $\Delta G = -RT \ln K$. While we have seen different methods for determining the equilibrium binding constant, from which the free energy of binding can be calculated, it is also possible to measure the free energy directly using **isothermal titration calorimetry**, or **ITC**. This method can be used to determine all three thermodynamic parameters that characterize a binding interaction: the enthalpy (ΔH), entropy (ΔS), and Gibbs free energy (ΔGo), which are related by the Gibbs free energy equation, $\Delta G = \Delta H - T\Delta S$.

➔ We learned about the thermodynamics of binding equilibria in Chapter 3.

While information about the strength of a binding reaction is given by ΔG, the relative contribution of entropy versus enthalpy provides valuable additional information. For example, an interaction that depends on a conformational change will have larger changes in entropy. Since ITC is performed in solution, it is free of potential artifacts that may arise with methods that require one or both molecules to be labeled or immobilized. A disadvantage is that relatively large amounts of pure material are needed for each experiment.

The apparatus used for ITC is depicted in Figure 19.77a. There are two cells: one containing a reference buffer and one containing a sample solution with one of

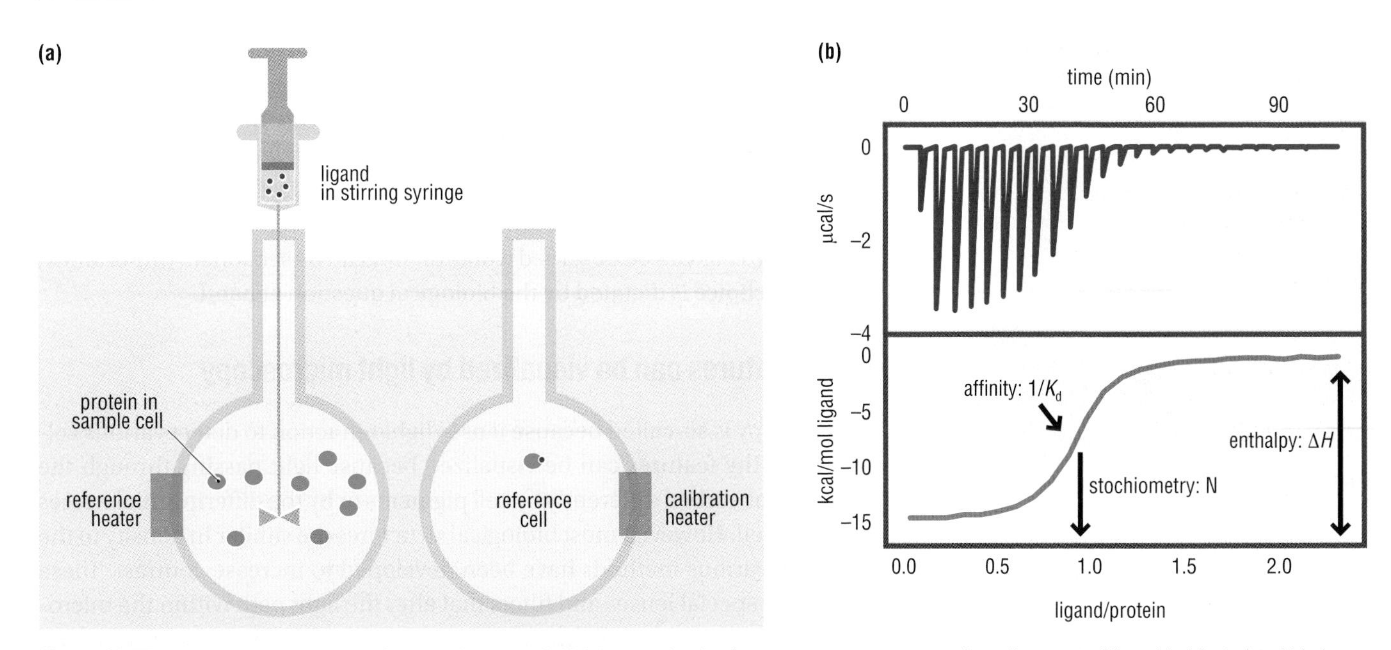

Figure 19.77 ITC. (a) The sample cell (left) and reference cell (right) are maintained at a constant temperature. A small amount of ligand is injected and binds to the protein, which causes the temperature to change slightly as heat is either absorbed or released by the binding reaction. The reference heater adjusts the sample cell temperature to return to that of the reference cell, and the resulting change in heat is recorded as in panel (b). (b) Top panel shows heat absorbed with each successive injection. As the binding sites on the proteins in the sample cell become occupied, less and less ligand is bound as its concentration exceeds that of the protein. Lower panel shows the area for each peak (or trough) can be used to plot the heat for each injection and to determine the binding constant and enthalpy of the binding reaction.

the molecules of interest. A sensitive temperature detection system ensures that the sample cell remains at the same temperature of the reference cell. Any temperature change in the sample cell will trigger either an increase or decrease in current going to the cell, depending upon whether the sample cell needs to be heated or cooled slightly so that its temperature once again matches that of the reference cell. This configuration is then used to measure the heat that is either released or absorbed when a ligand is introduced into the sample cell.

If a small amount of ligand is injected into the sample cell and the ligand binds to the molecule of interest, there will be a very slight change in the temperature of the sample that will result in a temporary trough or spike in current as the temperature is brought back to that of the reference cell (see Figure 19.77b). After repeated injections, each one adding more ligand to the sample cell, saturation will be reached as the concentration of ligand exceeds the K_d and little of the injected ligand binds. By repeatedly injecting small amounts of ligand, the relative change in heat for each injection can be measured and plotted as a function of ligand to determine the equilibrium constant.

19.13 IMAGING CELLS AND MOLECULES

The ability to visualize cells and to localize specific cellular components has been fundamental to our understanding of many aspects of biology. In this section, we discuss a range of microscopy techniques that are used to image cells and molecules. Microscopy makes it possible to determine the subcellular localization of particular proteins or cellular structures, to follow the dynamics of cellular components over time, to examine the expression patterns of a gene of interest, and to determine whether two proteins interact *in vivo*. In addition, the analysis of single

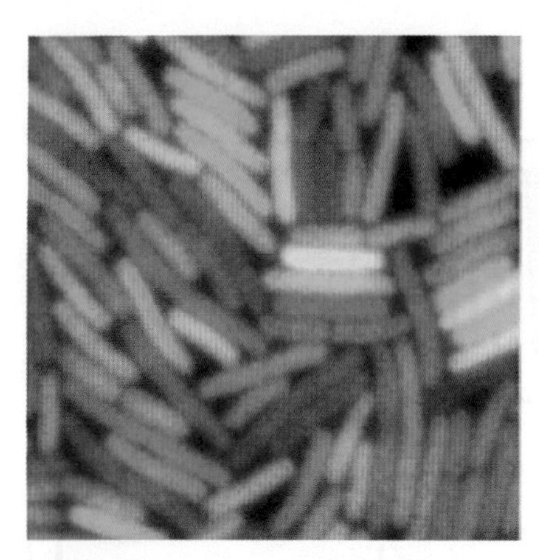

Figure 19.78 Cell-to-cell variation in gene expression. Each of these *E. coli* cells is expressing two proteins, one fused to CFP and one to YFP. The promoters of the CFP and YFP genes are the same. Although these cells are genetically identical and they are all growing in the same environment, different cells exhibit different levels of CFP (in green) and YFP (in red). This observation was made possible because the expression pattern was monitored on a cell-by-cell basis.

From Elowitz, MB, Levine, AJ, Siggia, ED, and Swain, PS. Stochastic gene expression in a single cell. *Science*, 2002;**297**:1183.

cells reveals information that is masked when examining an entire population. For example, Figure 19.78 shows the variation in gene expression observed between different *E. coli* cells, all of which are genetically identical. These cell-to-cell differences cannot be observed when assessing the protein levels in a population by western blot analysis, for example. Cells can be visualized live or treated with chemical cross-linkers (a process known as fixation) to preserve intracellular structures. Moreover, cells can be observed whole or they can be sectioned. Importantly, the method of choice is dictated by the biological question at hand.

Many cell features can be visualized by light microscopy

Light microscopy is so-called because it uses light refraction to detect various cellular features. The features can be visualized because light passing through the sample will be absorbed differently by cell pigments or by the differing thicknesses of parts of the cell. However, most biological structures are similar in density to the cytoplasm, so various methods have been developed to increase contrast. These usually involve special lenses and filters that alter the light path within the microscope. Two common types of light microscopy are **phase contrast** and **differential interference contrast (DIC)**, images from which are shown in Figure 19.79. Phase contrast often provides a penetrating image of the cell while the DIC method provides a three-dimensional-like image of the cell. Light microscopy can be used in combination with various strains or dyes that further increase the contrast and highlight particular cellular structures.

Intracellular structures can be seen by fluorescence microscopy in both fixed and live samples

Fluorescence microscopy makes use of the light emitted by certain dyes or proteins when they are illuminated with light of a shorter wavelength (known as the excitation wavelength, as depicted in Figure 19.80a). The characteristic excitation and emission wavelengths depend upon the chemical structure of the fluorescent molecule; scientists often use several different fluorescent molecules concomitantly, to highlight multiple cellular structures at once.

A variety of approaches have been developed to visualize specific cellular components using fluorescent molecules. These fall into two main categories: the use of fluorescent dyes or fluorophore-conjugated antibodies that are added to the cells, and the expression of fluorescent protein tags, such as **GFP**, fused to the protein of interest. The former approach is usually used on fixed samples, because

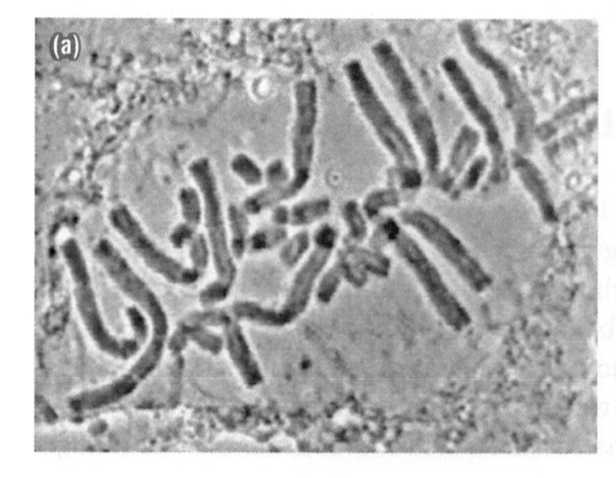

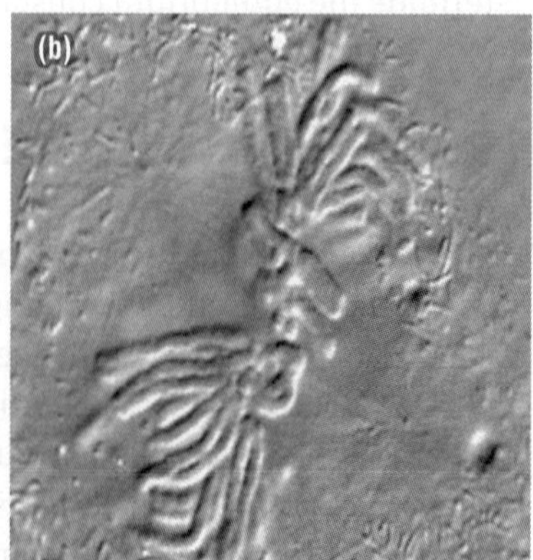

Figure 19.79 Phase contrast and DIC images of metaphase chromosomes. (a) A phase image of metaphase chromosomes from a flattened endosperm cell of the African blood lily *Haemanthus katherinae*. (b) A DIC image of metaphase chromosomes from newt lung cells. Note that the chromosomes in the phase image have light and dark regions; by contrast, DIC creates a three-dimensional image, but the chromosomes look uniform in composition.

Figure 19.79a from Inoué, S, and Oldenbourg, R. Microtubule dynamics in mitotic spindle displayed by polarized light microscopy. *Molecular Biology of the Cell*, 1998;**9**:1603–1607; Figure 19.79b courtesy of Alexey Khodjakov and Conly Rieder.

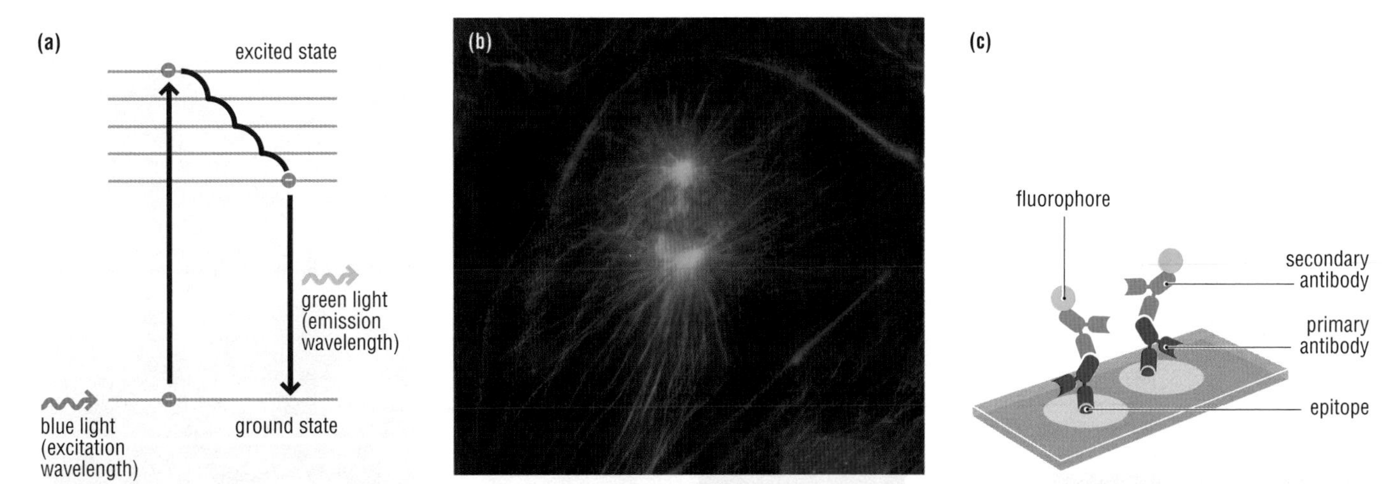

Figure 19.80 Fluorescence microscopy. (a) Each fluorophore or fluorescent protein has a specific excitation wavelength and emission wavelength. When the fluorophore is illuminated with high energy light at the excitation wavelength, an electron moves to an excited state. On its return to the ground state, energy is released in the form of light at the emission wavelength. In fluorescence microscopy, the specimen is placed on a slide on the microscope stage and is illuminated at the excitation wavelength. The light emitted by the fluorescent molecule can be visualized through the microscope's eyepiece or captured by a specialized digital camera. (b) Multiple fluorescent dyes or fluorophores can be used at once. In the image shown, the blue signal corresponds to DNA, the green signal corresponds to microtubules, and the red signal corresponds to actin filaments. These structures can be distinguished from one another because they are labeled with fluorophores that have different excitation and emission wavelengths. (c) Indirect immunofluorescence. To detect proteins for which antibodies are available, the cells are placed on a slide, permeabilized and treated with a primary antibody (dark blue) that recognizes the protein of interest (in red). The sample is then treated with a secondary antibody (light blue) that recognizes the primary antibody and is conjugated to a fluorophore. Thus, upon exposure to the appropriate excitation wavelength, the fluorophore associated with a secondary antibody will reveal the location of the protein of interest. This method is called indirect immunofluorescence because the fluorophore is not directly associated with the protein. Figure 19.80b provided courtesy Wadsworth Center, New York State Department of Health.

cells often need to be permeabilized to allow the dyes or antibodies to penetrate the cell. The fluorescent protein fusion approach can be used with either live or fixed cells.

A common example of a fluorescent dye is DAPI, which associates with double-stranded DNA (Figure 19.80b). Other fluorescent dyes bind to different subcellular structures, such as actin filaments, membranes or mitochondria. However, most cellular proteins or structures cannot be stained with a fluorescent chemical, and instead specific antibodies are used. Antibodies are generated in a host organism, such as mouse or rabbit, by injecting a protein or a peptide into the host. This, in turn, induced the immune system, which recognized these proteins as foreign, to produce antibodies against certain protein domains. These antibodies, referred to as 'primary antibodies,' recognize and bind to the protein of interest through the antibody's variable, or Fab, domains. To detect the primary antibody, secondary antibodies against the primary antibody's constant, or Fc, domains are made in a different host (for example, goat or donkey). The secondary antibodies can be conjugated to a fluorophore.

To detect the protein of interest, cells are fixed, placed on a microscope slide, permeabilized, and exposed first to the primary antibody. The slide is then exposed to the fluorophore-conjugated secondary antibody, as shown in Figure 19.80. The fluorescent signal from the secondary antibody can then be visualized under the microscope to reveal the location of the target protein. This method is called **indirect immunofluorescence**. As with any experiment using antibodies, it is important to establish that the primary antibody is specific to the protein of interest and does not cross-react with other proteins.

Since a highly specific antibody against the protein of interest is not always available, scientists sometimes create fusion proteins between the protein and an

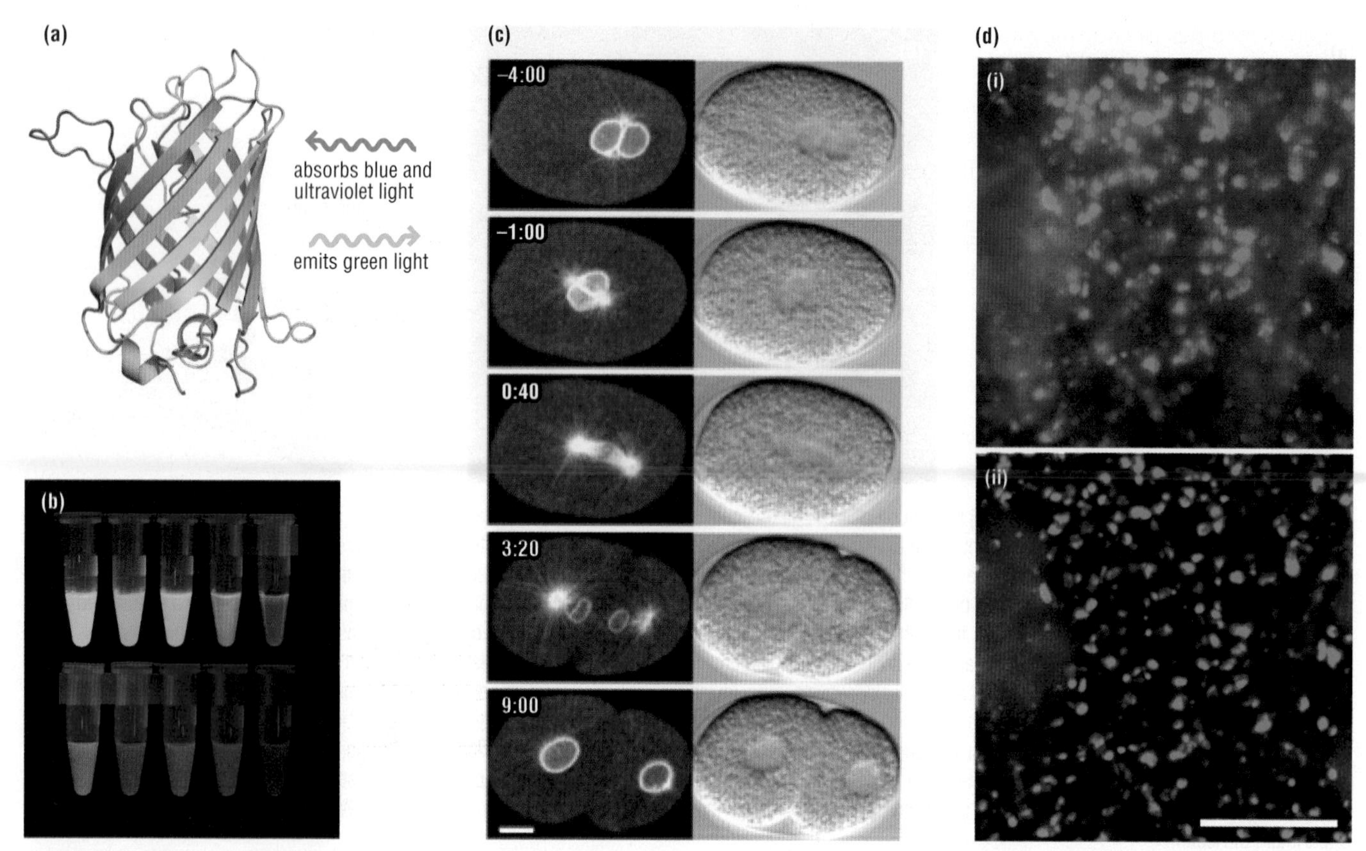

Figure 19.81 Fluorescent proteins. (a) The structure of GFP (Protein Data Bank (PDB) code 1EMA). (b) A series of fluorescent proteins, engineered from naturally occurring proteins, that emit at different wavelengths. (c) A series of images (fluorescence on the left, DIC on the right) taken of a *C. elegans* embryo expressing both GFP-lamin, which outlines the nuclear envelope, seen exclusively in time point 9:00 and GFP tubulin, which is visible on its own at time point 0:40, after nuclear envelope breakdown. These types of images are typically taken using a camera that detects light intensity rather than color, resulting in a black-and-white image. These images can be pseudocolored, as was done in Figure 19.82. The times (minutes:seconds) are relative to anaphase onset. (d) A comparison of the images produced by conventional fluorescence microscopy (i) and confocal microscopy (ii). Notice the reduction in out-of-focus fluorescence in (ii). Both images show axon boutons found in human neurons.

Panel (b) from http://www.tsienlab.ucsd.edu/Images.htm; panel (c) reproduced with permission from Franz C, Askjaer P, Antonin W, *et al. The EMBO Journal*, 2005;**24**:3519–3531; panel (d) reproduced from Sweet R, Fish K, and Lewis D. Mapping synaptic pathology within cerebral cortical circuits in subjects with schizophrenia. *Front. Hum. Neurosci.*, 2010; 4: 44. Copyright © 2010 Sweet, Fish and Lewis.

epitope for which there are good antibodies. Examples of such epitopes are the protein tags described in Section 19.11. The fusion protein is then expressed in the cell so that its location can be detected by indirect immunofluorescence, using primary antibodies against the tag. While this is a powerful method, a potential pitfall is that the fusion protein may not fully mimic the endogenous protein.

Significant information about the localization of a molecule can be obtained by immunofluorescence, but it is a methodology that cannot be used on live cells because it requires cell fixation and permeabilization. Consequently, other methods are used to monitor the localization of a molecule for a period of time in live cells. In many cases proteins are fused to the jellyfish **GFP** whose structure is depicted in Figure 19.81a; other naturally fluorescent proteins and their derivatives that emit and fluoresce at different wavelengths can also be used, as shown in Figure 19.81b. The localization of proteins tagged with these fluorescent proteins can be followed over time by recording a series of images (such as those shown in Figure 19.81c). In these types of experiments, researchers generate a gene fusion between their protein of interest and a fluorescent protein, such as GFP, and then express them *in vivo*. As is the case with immunofluorescence, these cells are then illuminated on the microscope stage with the appropriate excitation wavelength, and images are acquired using a sensitive digital camera.

Different types of microscopes can be used to detect fluorescence in biological specimens. In a conventional microscope, the specimen is flooded with light at the appropriate wavelength, resulting in the excitation of the fluorophores throughout the exposed area. Some of these molecules are in the microscope's plane of focus; others are either above or below this plane. Such out-of-focus fluorescence can result in a blurry image. Confocal microscopy overcomes this problem by using lasers and a special microscope configuration that minimizes the exposed area to a point while blocking out-of-focus fluorescence. To obtain an image, the laser rapidly scans the specimen, and the data from all the points are collected and assembled into an image. Figure 19.81d illustrates the difference between conventional fluorescence microscopy and scanning confocal microscopy.

Co-localization between two proteins can be detected by FRET

When several different cellular components are detected by distinct dyes, antibodies, or tags, we can determine if all of these components are found in the same location, a situation referred to as **co-localization**. It is important to note, however, that co-localization does not necessarily mean that these components physically interact (for example, that they are part of the same complex). FRET, which we described in Section 19.12, can also be applied to determine the distance between molecules within the cell. To do so, two proteins are expressed as a fusion to a FRET-compatible pair of fluorescent proteins, such as YFP and CFP. If the two proteins are close to each other (namely 10–100 angstroms (Å)), the energy generated by the excitation of the CFP will be transferred to the YFP, which will get excited and emit light that is detectable by microscopy. If the proteins are further away from each other, no such energy transfer will occur.

It is also possible to use FRET to sort out the organization of a multi-protein complex. For example, the nuclear pore complex consists of dozens of proteins. To determine how these proteins are organized relative to each other, researchers have measured the distances between pairs of proteins and created a map of interactions among the individual components of the nuclear pore, as illustrated in Figure 19.82.

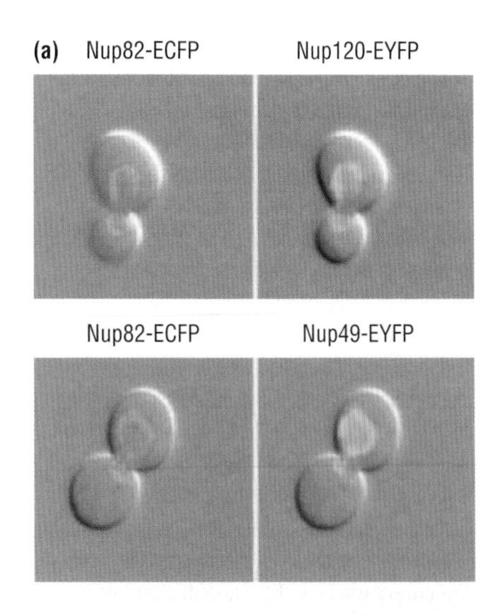

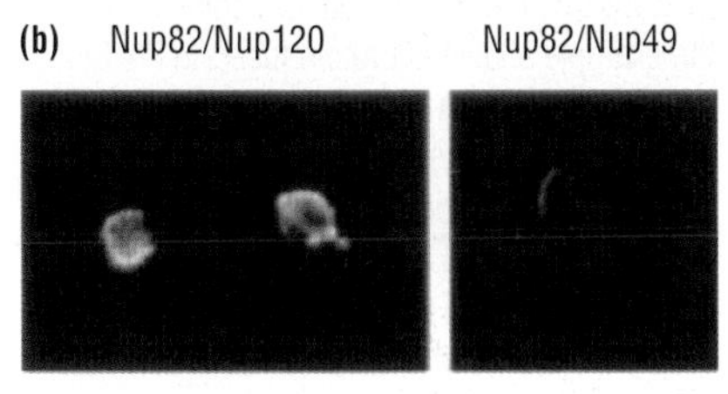

Figure 19.82 Analysis of yeast nuclear pore complex architecture by FRET. (a) Yeast cells co-expressing the nuclear pore complex protein Nup82 fused to enhanced cyan fluorescence protein (ECFP) and Nup120 fused to enhanced yellow fluorescent protein (EYFP) (top panel) or yeast cells expressing Nup82-ECFP and Nup49-EYFP (bottom panel) were excited with either the ECFP or EYFP excitation wavelengths, and the emission of either protein was recorded. The fluorescence images are superimposed on DIC images of the same cells. The images show that Nup82 co-localizes with Nup120 and Nup49 since the distributions of the proteins is similar. (b) When these cells were analyzed by FRET (by being excited with only the ECFP wavelength and the EYFP emission then being recorded), a FRET signal could be seen between Nup82 and Nup120 but not between Nup82 and Nup49, indicating that in the nuclear pore complex, Nup120 is relatively close to Nup82, but Nup49 is not.

Adapted from Damelin and Silver . In situ analysis of spatial relationships between proteins of the nuclear pore complex. *Biophysical Journal*, 2002;**83**: 3626–3636.

The movement of molecules can be followed using photobleaching or photoactivatable/photoswitchable GFP

Researchers can follow the movement of proteins and other cellular structures using live cell imaging. In the case of protein movement, the protein of interest is often tagged with a fluorescent protein, such as GFP, that can be visualized by microscopy. However, when dealing with an abundant protein that uniformly fills a particular cellular compartment, it is difficult to determine whether the protein moves within that compartment. Such information can be useful because lack of movement can suggest that the protein is tethered to an immobile structure. Several techniques that take advantage of properties of fluorescent proteins (or other fluorophores) can be used to examine movement within live cells.

Photobleaching occurs when a fluorophore is exposed to light of sufficiently high intensity that it destroys its ability to fluoresce. This can be a problem when taking repeated images of a cell, such as when acquiring time-lapse images, since the resulting photobleaching causes the fluorescence of a fluorescent protein to diminish over time. However, this irreversible loss of fluorescence can be exploited to assess the movement or diffusion of the protein within a cellular compartment.

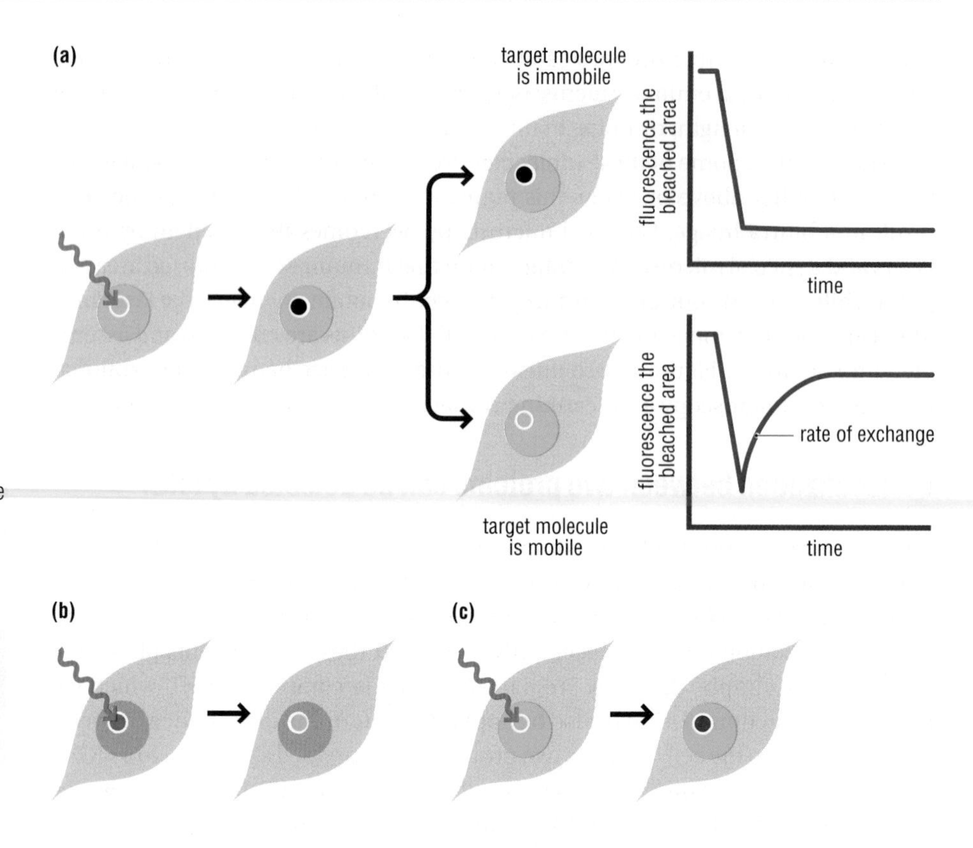

Figure 19.83 Following protein movement in live cells with FRAP and photoactivatable or photoswitchable proteins. (a) FRAP (fluorescence recovery after photobleaching). The cells in the diagram are expressing a fluorescent protein (for example, GFP) in the area indicated in green. A small area (circled in white) is rapidly bleached such that the fluorescent proteins in this area lose the ability to fluoresce (shown as a black circle). The recovery of fluorescence in this region is monitored over time. If the fluorescent proteins are immobile (i.e., fixed to their location and not free to diffuse), there will be no fluorescence recovery in the bleached area. If, however, the fluorescent proteins are mobile, the fluorescence in the bleached area will recover, and the rate of recovery will be indicative of the protein's mobility. (b) PA-GFP (photoactivatable-GFP). In the example shown, a protein fused to PA-GFP localizes to a particular cellular compartment (shown in dark gray) but is not visible. Following exposure to the appropriate wavelength, the PA-GFP in the exposed region becomes fluorescent (in green). (c) Photoswitchable proteins. In the example shown, the photoswitchable GFP fills a particular cellular compartment and is visible by fluorescence microscopy (shown in green). Exposure to the appropriate wavelength will cause the protein in the exposed region to switch from emitting light in one wavelength to another (in this case, red).

In this approach, called **fluorescence recovery after photobleaching (FRAP)**, a small region of interest (smaller than the total area occupied by the fluorescent protein) is illuminated repeatedly over a short period of time with high intensity light to bleach the fluorescent protein in that region, as shown in Figure 19.83a. If the protein is mobile within the compartment, the photobleached protein will diffuse away from the exposed area and be replaced by undamaged fluorescent protein molecules, resulting in the recovery of fluorescence signal in that illuminated zone. However, if the fluorescent protein is immobile within that cellular compartment, it will neither leave the bleached area nor be replaced by unaffected protein. Consequently, the bleached area will remain dark. By measuring the rate of recovery of fluorescent signal, we can estimate the rate of protein movement or diffusion in the cell.

A complementary method is **fluorescence loss in photobleaching (FLIP)**, where a small region is repeatedly bleached. Unlike the FRAP method, where the photobleaching period is brief, the photobleaching in FLIP occurs over the entire course of the experiment. This method is useful, for example, when a fluorescent protein occupies two distinct cellular compartments and the investigator wants to determine if the proteins can exchange freely between the two compartments. If there is an exchange, repeated photobleaching of an area within one pool will lead to loss of fluorescence in the other pool. If, however, there is no exchange, the protein in the compartment that is not being photobleached will retain its fluorescence.

While the FRAP method can be used to determine whether a protein is mobile, it does not allow the investigator to follow the protein as it moves. To do so, one can use photoactivatable or photoswitchable fluorescence proteins. A photoactivatable protein, as its name implies, is a protein that fluoresces only after being activated by exposure to an appropriate wavelength, as depicted in Figure 19.83b. One of the commonly used photoactivatable proteins is photoactivatable GFP (PA-GFP), a derivative of GFP that normally has very low fluorescence when exposed to light at 450-550 nm. However, after exposure to the activating wavelength, PA-GFP undergoes a conformational change that causes its fluorescence to increase 100-fold when illuminated with light with a wavelength of 504 nm.

PA-GFP can facilitate the elucidation of protein dynamics by being fused to the protein of interest. Exposing a small region to the activating wavelength will cause a subpopulation of the PA-GFP-fusion proteins to fluoresce, allowing the investigator to follow their movement in a cellular compartment that may contain many additional such proteins. In addition to protein movement, PA-GFP also allows the measurement of protein turnover: with a GFP-fusion protein, it is not possible to observe protein turnover as new protein is being continuously synthesized. However, activation of PA-GFP distinguished this subpopulation of fusion proteins from proteins that are synthetized after activation has taken place.

A disadvantage of PA-GFP and other photoactivatable fluorophores is that the molecule is not visible prior to activation, and thus it is sometimes difficult to identify the cellular region where the protein of interest initially resides. Photoswitchable proteins, which switch from fluorescing at one wavelength to another after being exposed to the appropriate wavelength, solve this problem (see Figure 19.83c). A commonly used, naturally occurring, photoswitchable protein is called Dendra, after the soft coral *Dendronephthya* from which it was isolated. One can use photoswitchable proteins fused to a protein of interest to identify the location of the fusion protein in the cell, and then photoswitch a subpopulation that can be followed over time, as described for PA-GFP.

Super-resolution microscopy overcomes the resolution limits of light microscopy

The resolution limit of conventional fluorescence microcopy discussed earlier in this section is around 200 nm, meaning that two objects that are closer than 200 nm will appear as one. This is due to the diffraction of light creating an image that is larger than the object itself, a phenomenon known as the point spread function. Although confocal fluorescence microscopy significantly improved imaging capabilities (as Figure 19.81d depicts), the overall resolution is still limited. Super-resolution microscopy overcomes this limitation.

There are several approaches for obtaining super-resolution with biological specimens, which reach resolutions of just tens of nanometers or below (hence 'super-resolution'). Some methodologies minimize the point spread function, while others use techniques that activate one fluorescent molecule at a time, allowing the location of the excited fluorescent molecule to be pinpointed. We will next discuss two super-resolution methods that utilize these approaches.

The **stimulated emission depletion (STED)** method was developed to minimize the point spread function. This approach utilizes a photoactivable fluorophore that can be excited and then turned off. The sample is illuminated rapidly by two successive beams: one that excites the fluorophore, and one doughnut-shaped beam at the so-called depletion wavelength that quenches it (shown as a STED beam in Figure 19.84a). Consequently, the molecules that are in the center of the STED beam remain excited, while molecules further away from the center return to the non-excited state, resulting in a more confined point spread function. The beams scan the specimen to generate a high-resolution image, as shown in 19.84b.

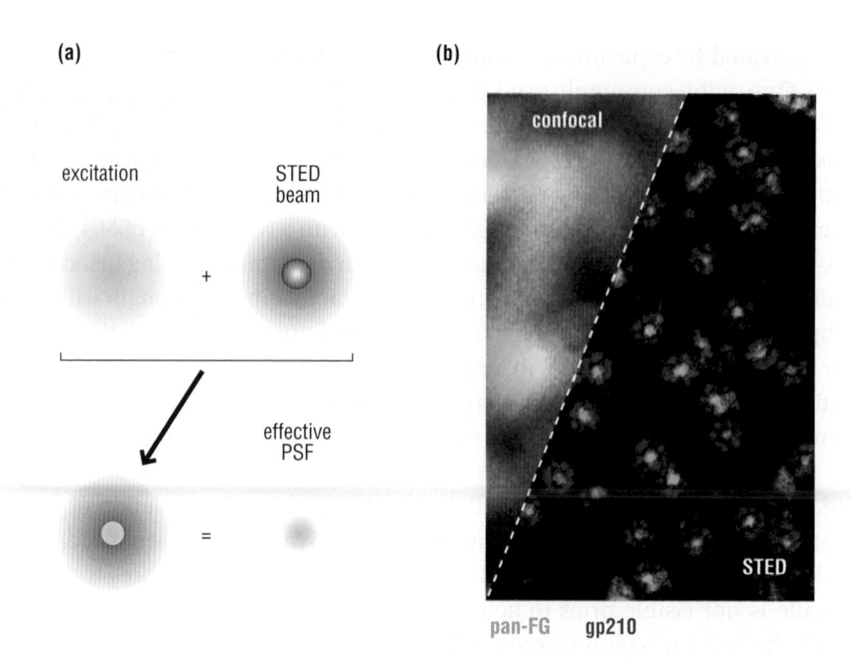

Figure 19.84 Super-resolution microscopy using STED. (a) STED limits the point spread function by using two beams. The excitation beam is shown in green, and the depletion beam is shown in red. The two beams illuminate the specimen in rapid succession, resulting in the point spread function being reduced in size. (b) Confocal fluorescence microscopy vs STED showing a side-by-side comparison of the gp210 protein subunit (red) and pan-FG (green) of the nuclear pore complex of a *Xenopus laevis* epithelial cell line. Note the much higher resolution (i.e., level of visible detail) revealed by STED.

Panel (b) Reproduced with kind permission from Stefan W. Hell.

A different approach for achieving super-resolution is **photoactivated localization microscopy (PALM)**. Unlike STED, where the excitation of the photoactivable fluorophore is reversible, in PALM the excitation is irreversible. The specimen is imaged multiple times using a very weak beam, with only a small percentage of photoactivatable fluorophores being excited each time, as depicted in Figure 19.85a. The fluorescent signal obtained from each molecule will be larger than the molecule itself, due to the point spread function. However, since each signal originates from a single molecule, it is possible to mathematically determine where within the signal the fluorescent molecule is localized. Thus, each time the specimen is imaged, the excited molecules are localized; they are then photobleached to prevent the same molecules from being resampled. To get the complete image, this process is repeated multiple times until enough data are collected. The locations of the fluorescent molecules from all the images are combined to yield a super-resolution image, as shown in Figure 19.85b.

Electron microscopy produces images at very high resolution

The resolution of light microscopy images is limited by the wavelength of light. Although high-resolution microscopy can improve imaging resolution, a much higher resolution can be obtained using electrons rather than light. **Transmission electron microscopy (TEM)** operates on the same basic principles as the light microscope, but electrons rather than light pass through the sample. Cells are fixed, stained with chemicals that bind different types of structures (for example, membranes), and the sample is cut into thin slices to allow the electrons to pass through. Objects whose dimensions are on the order of a few angstroms (10^{-10} m) can be detected, such that details can be visualized at near atomic levels.

In a second type of electron microscopy, **scanning electron microscopy (SEM)**, the scattering of an electron beam from a surface is monitored. The advantage of this method is the large depth of field, which allows much of the sample to be in focus at one time. This method mostly reveals the shape of the object of interest. Yeast cells imaged by both types of electron microscopy are shown in Figure 19.86. While electron microscopy can reveal structures not visible by any other imaging

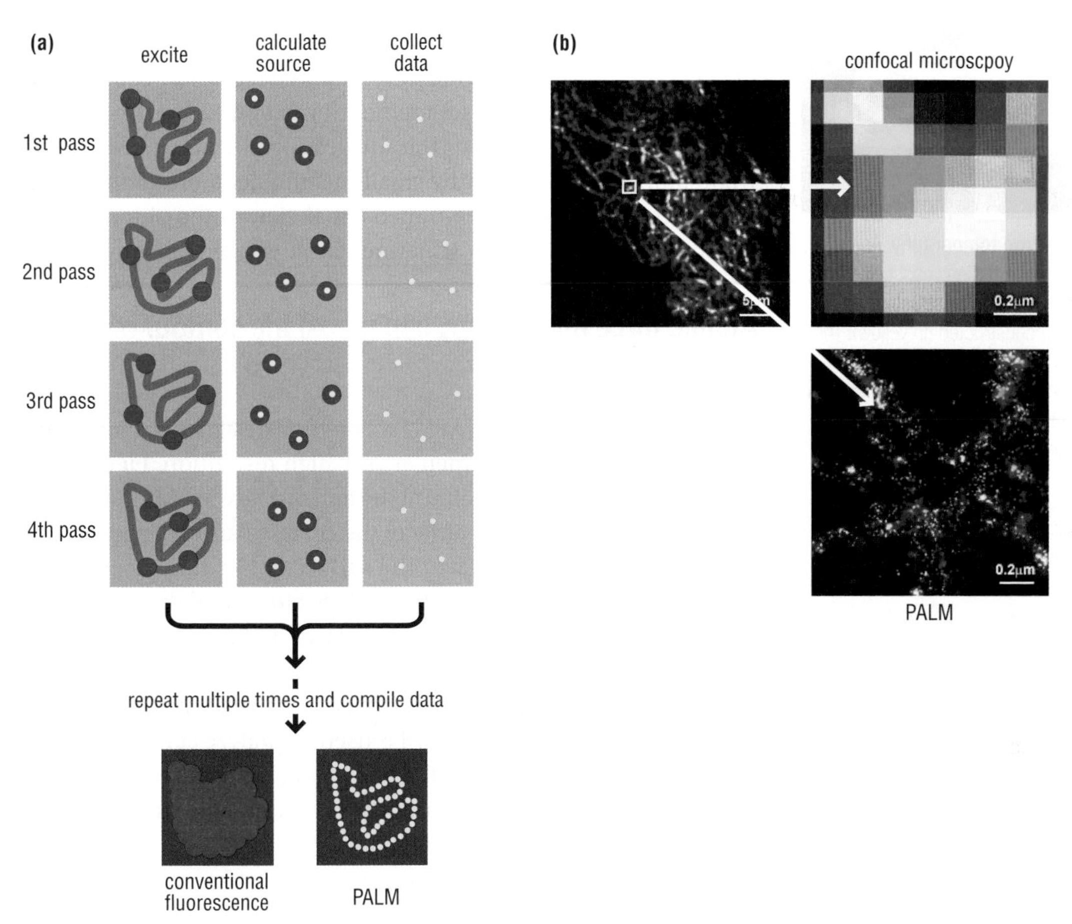

Figure 19.85 Super-resolution microscopy using PALM. (a) In the illustration shown, a cellular structure (dark gray) is expressing a photoactivable fluorescent protein throughout its surface. The structure is imaged multiple times; only a small number of fluorescent molecules are excited each time (shown in red). The location of the fluorescent molecule within the signal is calculated and registered as a point (shown in yellow). This process is repeated multiple times, after which the raw fluorescence data and the calculated locations of the fluorescent molecules from each image are combined. While conventional or confocal fluorescence imaging results in a blurry image of the structure, the imaged based on the calculated locations of the fluorescing molecules gives a more precise depiction of the imaged structure. (b) Fluorescently labeled cytokeratin (a cytoplasmic intermediate filament) was imaged by confocal fluorescence microscopy (left panel). The region shown in the white box was magnified; the top panel shows the image obtained with confocal fluorescence microscopy, while the bottom panel shown the image of the same region, obtained with PALM.

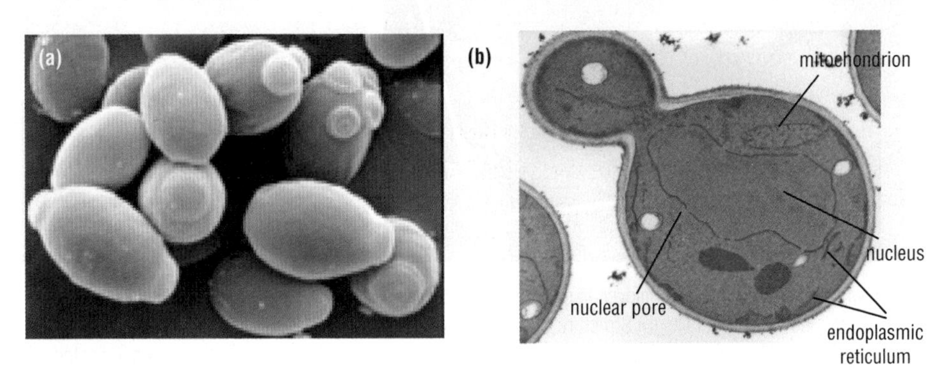

Figure 19.86 Electron microscopy. Budding yeast were examined by (a) SEM and (b) TEM. Note that SEM is a powerful method for examining shape while TEM reveals intracellular structures.

Figure 19.86a from http://www.micron.ac.uk/organisms/sacch.html; Figure 19.86b from http://www.medresearch.utoronto.ca/mil_transmission.html.

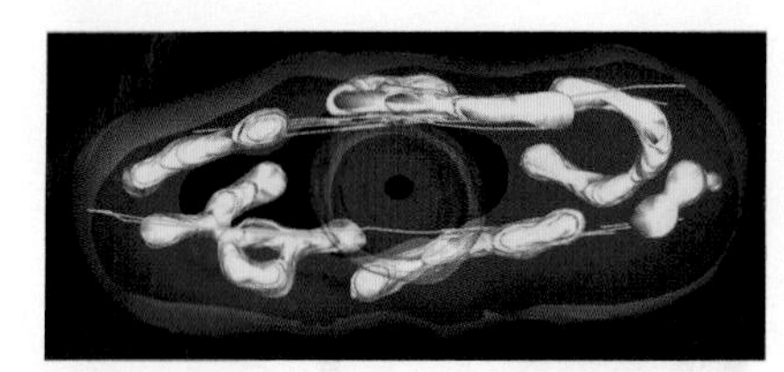

Figure 19.87 Electron tomography. An electron tomograph of a fission yeast cell, in which the three-dimensional structure of mitochondria was reconstructed (in yellow). The image also shows microtubules in light green, and the reconstruction reveals that mitochondria align with microtubules. The cell surface is shown in dark green and the nucleus in urple.

Adapted from Hoog JL, Schwartz C, Noon AT, O'Toole ET, Mastronarde DN, McIntosh JR, and Antony C. Organization of interphase microtubules in fission yeast analyzed by electron tomography. *Developmental Cell*, 2007;**12**:349–361.

methodology, it is limited to fixed samples. So, fluorescent microscopy remains the best approach for imaging live samples despite its lower resolution.

Electron tomography is a method for generating a three-dimensional image using electron microscopy. The images for electron tomography are collected while tilting the specimen by small increments around an axis. The images are then aligned to give a three-dimensional view of the object. This method is well suited for studying cellular structures, as illustrated in Figure 19.87.

Atomic force microscopy can reveal the contour of a cell surface or molecule

Atomic force microscopy (AFM) is a method designed to characterize and analyze the surface area of a specimen at very high resolution. Figure 19.88a shows how AFM utilizes a fine tip, situated at the end of a flexible cantilever, which tracks very closely to the specimen's surface; the tip rises and falls as the cantilever moves across the specimen's surface. The surface either attracts or repels the tip due to various forces acting between the surface atoms and those of the tip, and changes in the tip's position are recorded by a laser that is reflected off the cantilever. The result is a topological representation of the specimen, which not only provides a detailed image of the specimen's shape but also allows precise measurements of the specimen's dimensions. AFM is used to analyze complex biological molecules and structures, such as proteins bound to fragments of DNA, the topography of membranes and membrane-bound proteins, and the structure of mitotic chromosomes, as shown in Figure 19.88b.

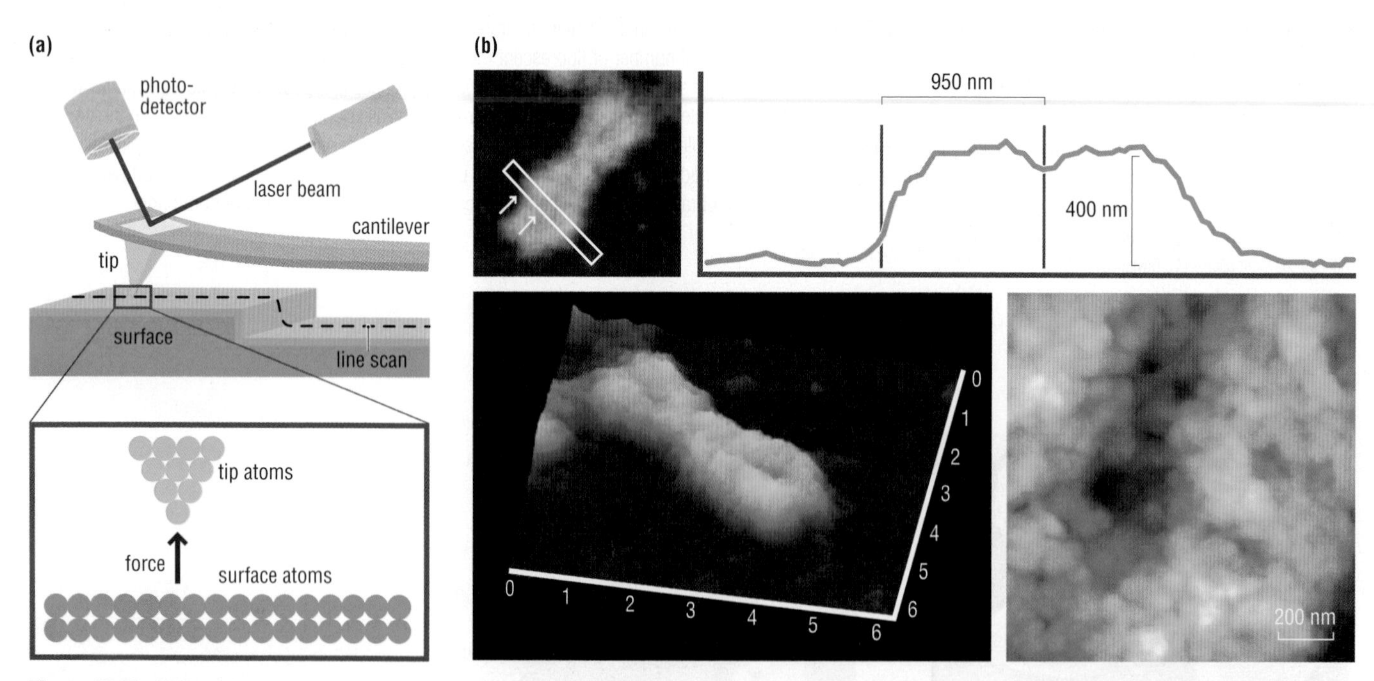

Figure 19.88 AFM. (a) General principles of AFM. The tip, which is attached to a cantilever, is moved across the surface of a sample while the laser is used to measure the relative displacement of the tip from the surface. See text for details. (b) Structure of a metaphase chromosome as determined by AFM. The chromosomes were isolated form the human cell line BALL-1. The profile (top right) is produced by calculating the average of the height in the region indicated by the box in the micrograph at left. The bottom left panel shows a three-dimensional representation of the chromosome, while the bottom right panel is a closer view of a part of the chromosome showing an aggregation of globular or fibrous structures.

Adapted from Ushiki T, and Hoshi O. Atomic force microscopy for imaging human metaphase chromosomes. *Chromosome Research*, 2008;16:383–396.

19.14 MOLECULAR STRUCTURE DETERMINATION

Much insight into biological processes has come from information on the three-dimensional structures of biological molecules. Nearly all of the macromolecular structures depicted in this book were determined by a method known as **x-ray crystallography**, which can be used to define the structure of a molecule by examining the way in which a crystal of the molecule of interest scatters x-rays. As compared with other techniques, x-ray crystallography can produce the most accurate and detailed model of a macromolecule. Other approaches, such as nuclear magnetic resonance (NMR) spectroscopy and cryoelectron microscopy, can provide complementary information on protein structure and dynamics as well as on large macromolecular assemblies. They are also used when it is not possible to obtain a crystal for analysis by x-ray crystallography.

(a)

(b)

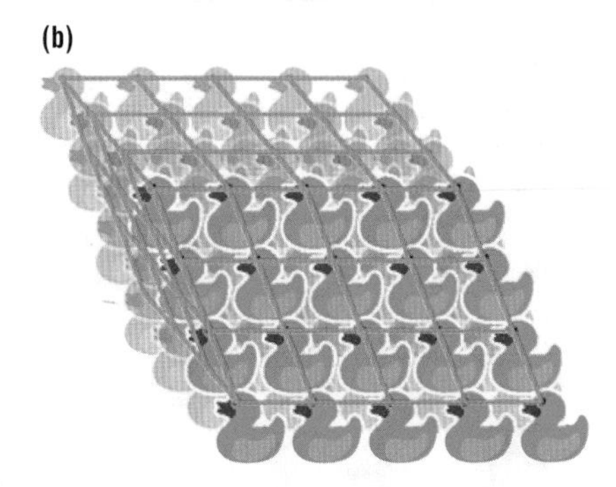

Figure 19.89 A crystal lattice. (a) A two-dimensional lattice of ducks. The parallelogram is the fundamental repeat unit of the crystal, called the unit cell (shown are three repeats of the unit). The lattice is generated by adding many identical copies of the unit cell, as indicated, along each of the two unit cell axes. Note that each unit cell contains a complete duck, although it is made up of bits and pieces of several intact ducks. (b) A three-dimensional crystal lattice contains many repeats of the unit cell, replicated along the three unit-cell axes. (The use of ducks to illustrate crystal lattices is an established, albeit obscure, tradition in protein crystallography.)

A macromolecular structure can be determined from the diffraction of x-rays

X-ray crystallography is used to determine the three-dimensional structure of macromolecules at a level of detail that includes the position of each atom. This methodology takes advantage of the fact that the way in which molecules scatter x-rays is determined by their shape. The wavelength of x-rays used in the laboratory is around 1–2 Å, making it most appropriate for visualizing distances equivalent to the length of a covalent bond (1.2–1.5 Å). There are fundamental technical barriers to constructing an x-ray microscope that could produce a magnified image of a molecule using x-rays as the 'light', so the technique of x-ray crystallography was developed instead.

Instead of scattering x-ray waves from a single molecule, x-ray crystallography uses scattering from a crystal, which consists of molecules lined up in a repeating three-dimensional array, as depicted in Figure 19.89. Purified proteins and nucleic acids can be induced to form a crystal by adjusting solution conditions, for example, by adding certain salts or organic solvents that reduce protein solubility, potentially causing the macromolecule to fall out of solution and form crystals. A trial-and-error approach is used to screen hundreds of different conditions that can potentially induce a molecule to crystallize.

When a crystal is placed in an x-ray beam, the x-ray waves are scattered by the electrons of each atom in the molecule, as shown in Figure 19.90a. We refer to this scattering as **diffraction**. Some of the scattered waves add together to produce a wave of high amplitude whereas other waves interfere with one another and produce a wave of low amplitude. As a result, the diffraction that is recorded on an x-ray detector appears as a pattern of light and dark spots, with the intensity of each spot determined by the three-dimensional structure of the crystallized molecule. It is from many pictures like this, taken from different angles, that the information about the molecular structure is extracted.

Using sophisticated calculations, it is possible to use the intensity of each spot in the diffraction pattern to deduce the three-dimensional structure of the protein or nucleic acid molecule that has been crystallized. Additional information about the scattered waves – referred to as the phase – is needed to perform this calculation. This phase information can be obtained by a number of methods that are beyond the scope of this book. Once the amplitudes and phases of all the scattered

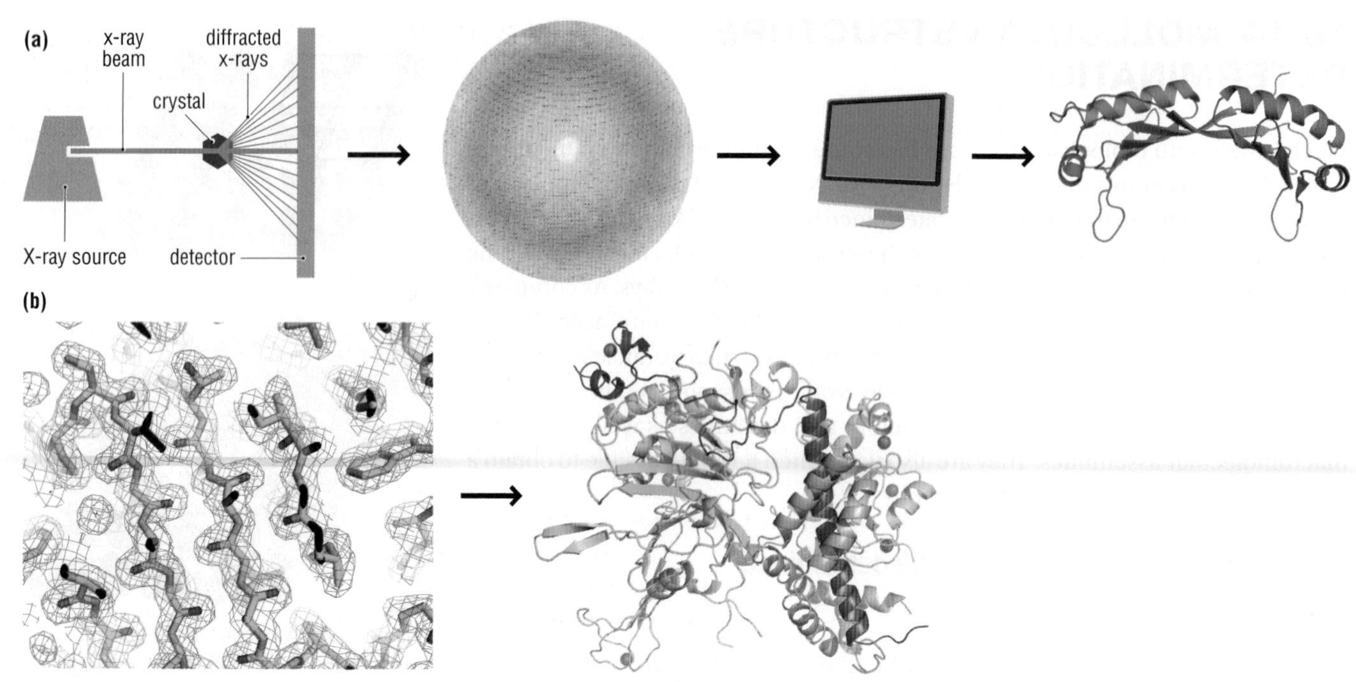

Figure 19.90 X-ray crystal structure determination. (a) An x-ray source is used to irradiate a crystal with x-rays, and the scattered x-rays are recorded on a detector. The intensity of each individual spot contains information about the structure of the molecule that has been crystallized. These intensities, together with additional information about the phase of the scattered wave, can be used in computer programs to calculate an electron density map, which is then used to construct a model of the macromolecules. The protein structure shown is a ribbon depiction of TATA-binding protein (TBP). (b) An electron density map (blue lattice) of Ubp8, a subunit of the yeast SAGA deubiquitinating module. The molecular structure that was constructed to fit the electron density map is shown in green (carbon), blue (nitrogen), and red (oxygen). The ribbon model on the right shows four proteins in the complex: Ubp8 (green), Sgf11 (pink), Sus1 (blue), and Sgf73 (salmon); orange spheres are bound zinc atoms. PDB codes 1TBP, 3MHH.

Figure 19.90b courtesy of Cynthia Wolberger.

waves are known, it is possible to mathematically reconstruct the shape of the molecule that has been crystallized.

The result of experiments to record x-ray diffraction intensities and obtain phase information is an experimental **electron density map**, which shows the outline of all the visible atoms; an example is shown in Figure 19.90b. For nearly all protein and nucleic acid structures, the hydrogen atoms cannot be visualized, and crystal structures in the PDB (Protein Data Bank) (and throughout this book) typically lack hydrogen atoms. A model of the protein is then constructed to best fit the electron density map.

The accuracy of the model will depend on the resolution of the data, which is a function of how far from the center of the image diffraction spots can be recorded. A structure determined at 1.8 Å resolution shows the positions of atoms with high accuracy, with all side chains clearly defined, whereas the side chains in a structure determined at 3.5 Å resolution are not well defined. Experimental approach 2.1 describes the application of x-ray crystallography to determine the structure of the DNA double helix. It is important to note that portions of a macromolecule that do not adopt a fixed conformation – perhaps a flexible loop, or a small mobile domain – do not appear in the electron density map irrespective of resolution. These portions of the molecule or complex are thus not present in the resulting structure.

The model for a protein (or nucleic acid) and its bound ligands is stored in a data file as a list of atoms along with their corresponding positions in space, which are given by x, y, and z coordinates. A single macromolecular structure may have several thousand atoms. The coordinates for protein and nucleic acid structures, including those depicted throughout this book, can be obtained from the PDB website and displayed using a personal computer.

➔ You can find the PDB at http://www.wwpdb.org.

NMR spectroscopy gives structural information on macromolecules in solution

Protein and RNA structures can be determined in solution by **nuclear magnetic resonance spectroscopy**, commonly referred to as **NMR**. The distinct advantage of this technique is that it does not require a crystal; only a concentrated solution of soluble protein. In addition to yielding information on the three-dimensional structure of a macromolecule, it is also possible to study structural fluctuations - known as protein dynamics - by NMR. For this reason, NMR spectroscopy can provide information on proteins that complements the static picture of a protein structure provided by x-ray crystallography. Although NMR was originally best suited to determining structures of small proteins, recent advances have made it possible to extend the technique to the study of proteins containing hundreds of amino acids.

NMR takes advantage of the fact that certain atomic nuclei behave as small magnets and can be aligned in a magnetic field. Any nucleus with an odd number of nuclear particles - protons plus neutrons - has a property known as spin that causes it to behave like a small magnet. Thus, hydrogen atoms, ^{1}H, have spin, as do the isotopes of carbon and nitrogen, ^{13}C and ^{15}N. When placed in a magnetic field, these nuclei behave like small gyroscopes, with their axes precessing around the direction of the magnetic field. If a pulse of electromagnetic radiation at radio frequencies is applied in a direction perpendicular to the magnetic field, the spinning nuclei are realigned in the perpendicular direction. Once the pulse is removed, the nuclei gradually return to their original orientations along the magnetic field, emitting energy as they do so. The energy emitted by individual nuclei and the rates at which they return to their original orientation are influenced by neighboring atoms in the molecule. By measuring these parameters, one can derive information about the bonding arrangement of the atoms with spin and their relationship to their neighbors. By applying a sequence of radiofrequency pulses of varying duration and frequency, different properties of individual atoms can be probed. NMR spectroscopists have developed numerous pulse sequences that give information on interatomic distances and bond angles.

In contrast with x-ray crystallography, which yields a set of atomic coordinates that constitute the best fit to an x-ray data set, the result of an NMR structure determination is a set of constraints on the interatomic distances and bond angles in a polypeptide chain. Together with the amino acid sequence of the protein, one can construct a model that provides a best fit to these constraints. It is generally possible to construct a number of different models that are consistent with the NMR data yet differ from one another in some way. Figure 19.91 shows an example of the set of models that provide a best fit to an NMR data set on a particular protein. Portions of the protein that are well ordered and have many constraints are nearly identical in the different models, while regions that are either mobile or have few distance constraints vary in the different models.

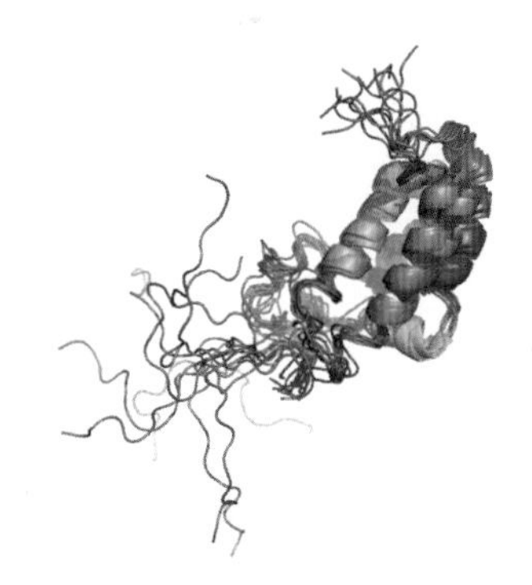

Figure 19.91 Structure determined by NMR spectroscopy. Shown is a set of superimposed models for the structure of a bromodomain of the transcriptional co-activator CREB-binding protein (CBP) as determined by solution NMR spectroscopy. The models shown are the ten best fits to the distance and angle restraints derived from the NMR measurements. The regions shown in green, orange, and red are the same in all models and their structure fixed, whereas the other regions are more mobile. (PDB code 1JSP).

Cryoelectron microscopy can be used to image large macromolecular assemblies

Some macromolecular complexes are so large and flexible that their structures cannot be determined by either NMR spectroscopy or by x-ray crystallography. In these cases, it is possible to use electron microscopy to determine their structures. A homogeneous preparation of the complex of interest is laid down on a special grid and flash-frozen. An electron micrograph is then recorded, yielding an image

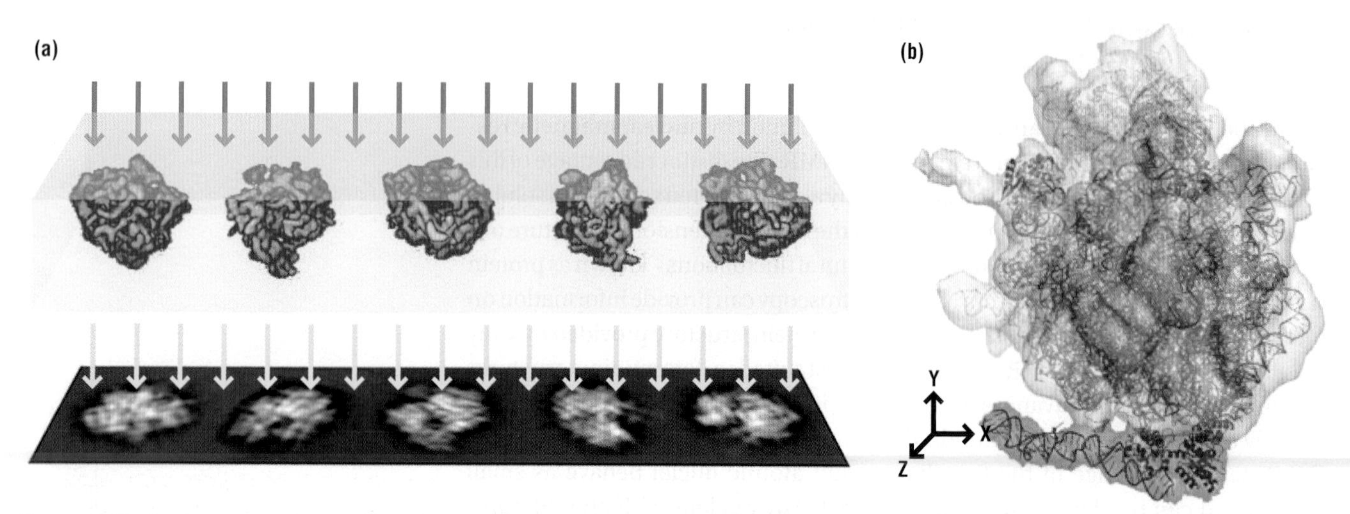

Figure 19.92 Cryoelectron microscopy: single molecule reconstruction. (a) The arrows symbolize the electron beam impinging on the specimen, which contains many randomly oriented molecules embedded in a thin layer of ice. The resulting images show the different projections of the particles. (b) Structure of the *E. coli* signal recognition particle (SRP) bound to the ribosome. Ribosomal subunits are shown in yellow (30S) and blue (50S); the SRP is shown in red. The surface depiction shows the structure obtained by cryoelectron microscopy. Crystal structures of the individual subunits are superimposed on the cryoelectron microscopy structure.

Figure 19.92a from Frank Single particle imaging of macromolecules by cryo-electron microscopy. *Annual Review of Biophysics and Biomolecular Structure*, 2002;**31**: 303–319; Figure 19.92b from Schaffitzel *et al.* Structure of the *E coli* signal recognition particle bound to a translating ribosome. *Nature*, 2206;**444**: 503–506.

of the randomly oriented particles, as illustrated in Figure 19.92a. The image of each individual particle represents a view down a particular axis: a projection of the structure along that axis. Thousands of these projections are collected, combined, and averaged, yielding an image of the particle with all its contours, in three dimensions (see Figure 19.92b). A structure determined by this method typically does not provide sufficient detail to distinguish individual atoms or residues. However, it provides the outlines of the complex to which is often possible to fit models of individual proteins and nucleic acid structures (Figure 19.92b).

19.15 OBTAINING AND ANALYZING A COMPLETE GENOME SEQUENCE

The availability of complete genome sequences has revolutionized molecular biology. Obtaining and using sequence information requires methods for both high-throughput DNA sequencing as well as computational analyses of the sequences obtained. The first genome to be completely sequenced, that of *Haemophilus influenzae*, was completed in 1995. Less than 20 years later, the sequences of thousands of genomes have been reported. The pace at which new genomes are sequenced continues to accelerate, driven by advances in DNA sequencing technology as well as in the methods for their analysis. Here we explore the rapidly evolving methods used to generate genome sequence information.

The sequence of a whole genome can be determined by analyzing many overlapping DNA fragments

How does one obtain the complete sequence of a genome? We lack the technology to sequence an entire chromosome from one end to the other, so other approaches must be used. Sequencing methods are rapidly evolving, but they are typically

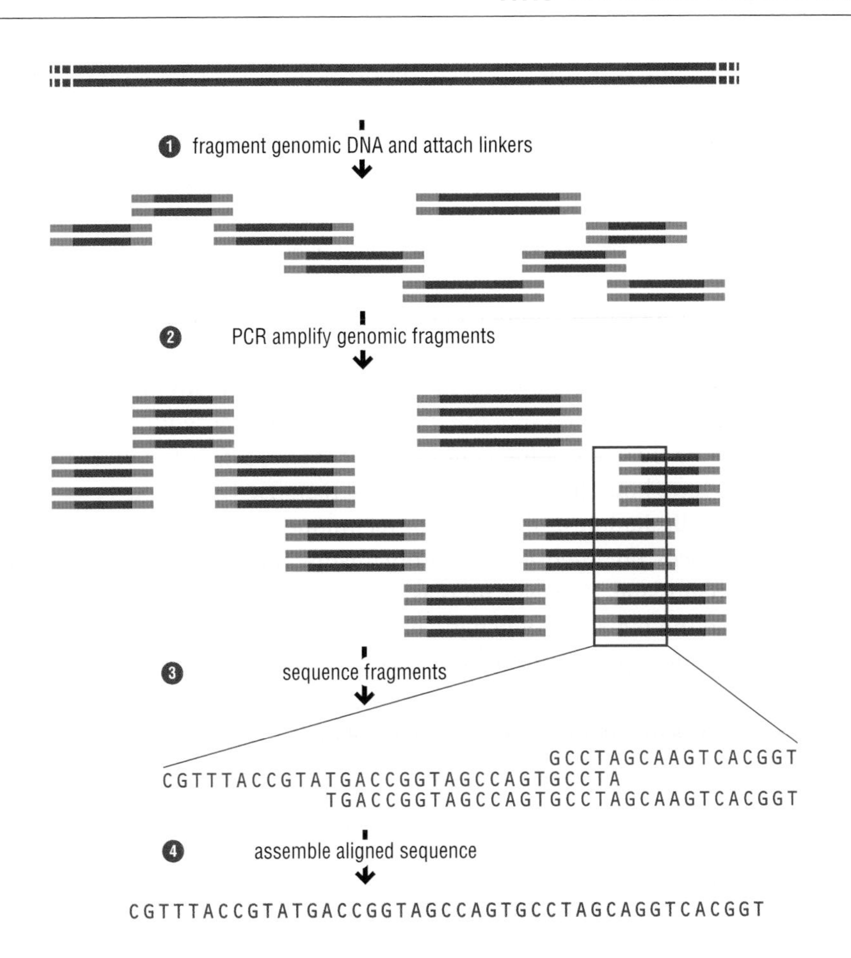

Figure 19.93 Assembly of a genome sequence from the sequences of multiple overlapping fragments. Genomic DNA is first fragmented by shearing or restriction enzyme digestion to generate a collection of thousands or millions of random overlapping fragments, after which linkers (blue) are ligated to all of the ends (step 1). The genomic fragments are amplified by PCR using primers complementary to the attached linkers (step 2). After sequencing of all of the fragments (step 3), computer programs can remove the linker sequence and assemble a complete genomic sequence by detecting overlaps between the many random fragments (step 4).

unified by a common underlying approach. The general strategy of whole genome sequencing is to obtain a collection of many small overlapping DNA fragments that together represent the entire genome. As we will learn below, there are different methods for generating these DNA fragments; the key point is that, at the start, it is not necessary to know where each fragment came from in the genome. By determining the base sequence of each fragment and then comparing the sequences of many overlapping regions, the complete sequence of the genome can be assembled in the correct order with the aid of a computer, as illustrated in Figure 19.93. This strategy is known as 'shotgun' sequencing because it begins with random fragmentation of the genome. When the human genome was first sequenced, the small overlapping DNA fragments were cloned into a library that was grown in *E.coli* to generate sufficient DNA for sequencing. Today, most protocols use PCR amplification to generate DNA for sequencing.

Genome sequencing requires over-sequencing

In order to obtain a complete genome sequence, it is necessary to sequence many more base pairs than the total length of the genome. There are two reasons for this. First, since DNA fragments are sequenced randomly, many different overlapping fragments must be sequenced to ensure that all regions of the genome have been sequenced at least once. In theory, clones corresponding to eight genome equivalents need to be sequenced to ensure coverage of 99.9% of the genome. For example, to obtain a nearly complete sequence of the 4.6 Mbp *E. coli* chromosome using shotgun sequencing, it is necessary to sequence about 37 Mbp of DNA (eight genome equivalents). In practice, even higher coverage is often sought, as it has the additional benefit of increasing the accuracy of the final sequence. This is because an error in the sequence of a single fragment can be recognized and corrected when many independent clones corresponding to the same region are sequenced.

The second reason that a complete genome sequence requires over-sequencing is that a sufficient number of random overlapping DNA segments are needed in order to determine the correct order in which overlapping sequences must be aligned to assemble the complete genome. The presence of repetitive DNA in multiple genomic locations complicates the task because they make it difficult to uniquely align the fragments containing identical repetitive sequences. This task becomes nearly impossible if the regions of the genome containing many repeats are longer that the sequenced shotgun fragments. Computer sequence alignment will not assemble this part of the genome from overlapping sequences because the repeat will have homology in many different regions.

This inherent limitation in the shotgun technique is the reason that the first genome sequences obtained for many organisms, including humans, had regions that were missing. However, increasingly sophisticated computer assembly algorithms and the inclusion of genetic linkage data are facilitating assembly of regions containing repetitive DNA sequences. In addition, new methods of sequencing are making it possible to sequence longer regions and thus connect unique sequences that flank these repetitive regions.

DNA sequences can be obtained by a variety of methods

For many years, the primary method used to sequence whole genomes was an approach called chain-termination sequencing, which uses ddNTPs as chain terminators, as described in Section 19.8. This DNA sequencing method was originally performed by hand and analyzed by gel electrophoresis. Later, machines were developed that automated the sequencing and detection process.

A single high-throughput automated chain-termination sequencing machine can process about 1000 reactions each day, generating about 0.5 Mbp of total sequence. (Each sequencing reaction is termed a 'read'.) This process would have to be repeated about 75 times to sequence eight genome equivalents of the 4.6 Mbp *E. coli* genome (37 Mbp). To sequence the human genome (3×10^9 bp), about 70 million chain-termination sequencing reactions, each generating 500 bp of sequence, were needed to generate ten genome equivalents of sequence, or around 35×10^9 bp. This immense effort required 100 automatic sequencing machines generating 1000 sequencing reactions per day for about two years.

Recent advances in sequencing technology have greatly increased the speed of DNA sequencing, while at the same time dramatically reducing the cost. As described in Chapter 18, these new methods are collectively called NextGen sequencing or, more commonly, deep sequencing. These methods share certain features, most importantly that they are 'massively parallel', meaning that huge numbers of unique DNA molecules can be sequenced simultaneously, thus generating a large number of unique reads that can then be aligned. In current deep sequencing methods, DNA fragments generated by PCR directly from genomic DNA are sequenced. New so-called single molecule methods are currently being developed that will make it possible to sequence individual DNA molecules directly, eliminating the need for PCR and thereby avoiding the errors that can be introduced during PCR amplification.

While the pace of innovation in sequencing technology exceeds the pace of textbook writing, we will describe the details for one method, known as Illumina sequencing (after the company that developed the method) that is currently widely used. This method utilizes fluorescently labeled nucleotides that are chemically blocked at the 3′ end, so that they can be added by a polymerase one at a time to a growing DNA strand.

In the Illumina sequencing method, illustrated schematically in Figure 19.94, the genomic DNA is fragmented and oligonucleotide adapter segments are ligated to the ends of the resulting DNA fragments. The DNA strands are then separated, and the single strands are allowed to hybridize to a surface that is densely coated with a mixture of immobilized oligonucleotides corresponding to a single strand of one of the adapters. Both ends of each DNA fragment hybridize with a complementary adapter immobilized on the surface, thus generating a bridge (see Figure 19.94).

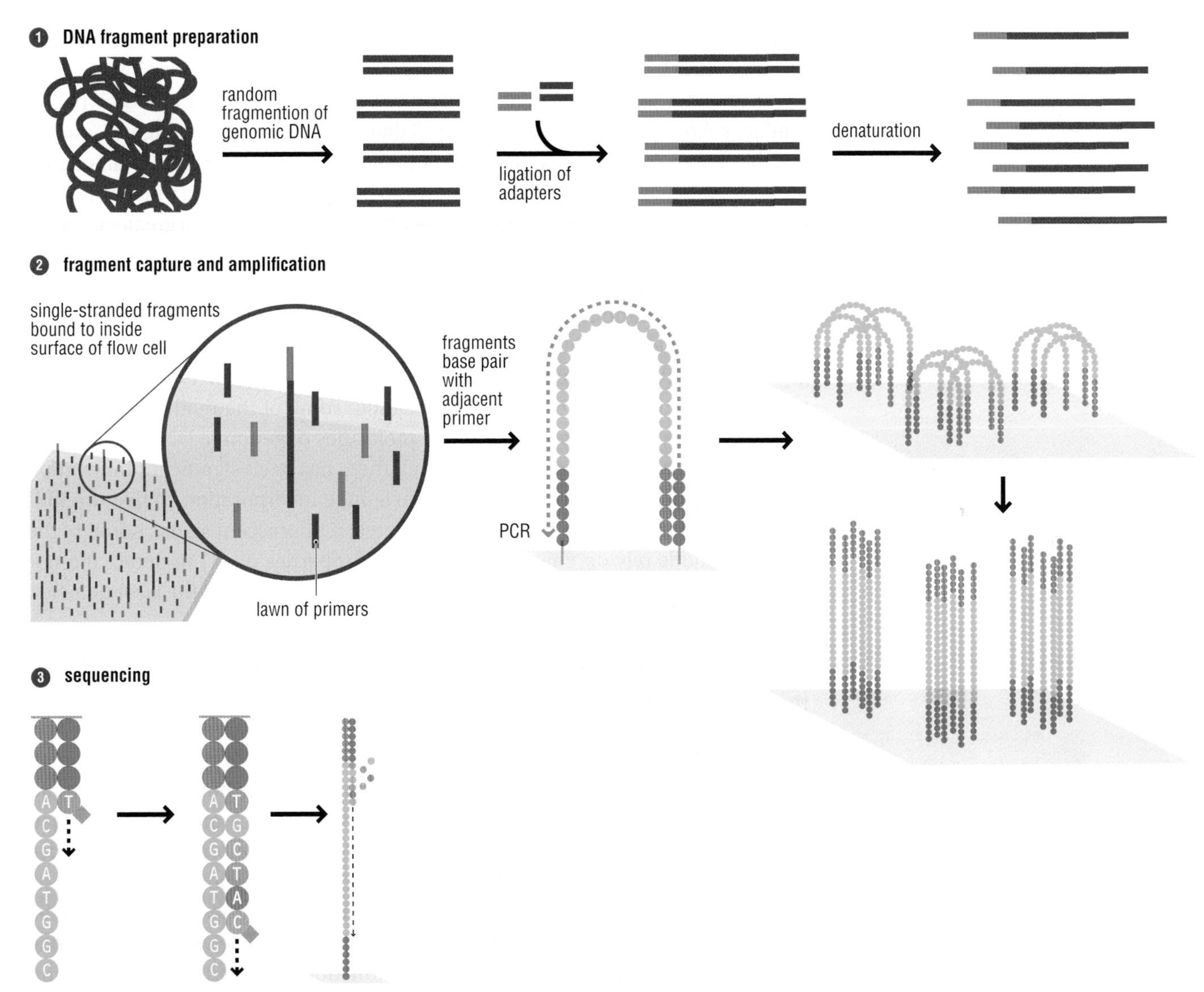

Figure 19.94 Illumina sequencing. (a) DNA fragment preparation. The DNA to be sequenced is broken into fragments by shearing or restriction digestion (step 1). Adapters are ligated to the ends of the fragments (step 2), after which the DNA is denatured (step 3). (b) Fragment capture and amplification. The fragments are attached to a surface decorated with a dense lawn of primers complementary to the adaptors (step 1). The individual fragments are then amplified by unlabeled nucleotides using a nearby primer to form a double-strand molecular bridge (step 2). Finally the DNAs are denatured to yield clusters of identical fragments that are sequenced in parallel (step 3). (c) Sequencing showing a single fragment. A specific primer and 4 3′ blocked nucleotides, each with a different fluorescent label, are added. After washing away the unincorporated nucleotides, laser excitation and optical detection identify the added nucleotide at each cluster position. The 3′ blocks are removed, and the cycle is repeated.

From Mardis. Next-generation DNA sequencing methods. *Annual Review of Genomics and Human Genetics*, 2008;**9**:387–402.

After PCR amplification, each spot on the array will contain a cluster of identical sequences. This high local concentration of each unique sequence is needed to produce a sufficiently robust fluorescent signal in the steps that follow.

The DNA in each spot is denatured to make it single-stranded in preparation for sequencing. Fluorescently labeled nucleotides (A, T, G, and C, each labeled with a differently colored fluorophore to allow them to be distinguished) are then added, along with a primer and DNA polymerase. Since the nucleotides are also chemically modified such that the 3′ end is blocked, the polymerase can add only one base at a time. A laser excites the fluorophore while a camera monitors whether the color of the emitted fluorescence corresponding to the addition of an A, T, G, or C base. The DNA is then chemically treated to remove the chemical block and the fluorophore, thus exposing the 3′ end and making it available for the next round of fluorescent nucleotide incorporation.

By repeating this process for many cycles, hundreds of nucleotides can be rapidly sequenced.

The Illumina sequencing method, like other deep sequencing methods, typically produces short reads of 75 to 150 bp. Recall that, in whole genome shotgun sequencing, the genome sequence is assembled by aligning overlapping regions among individual sequence reads. The short reads from the Illumina method thus compound the problem of sequence assembly, and make it necessary to have a greater depth of coverage of the genome to allow correct assembly of the final sequence.

At the time of writing, the Illumina method is the most popular because it is easy to automate and is highly accurate. However, many other methods are rapidly being developed. New Methods to generate read lengths of several kilobases great simplify the alignment and assembly of genome sequence. Unlike the Illumina method that relies on PCR amplification, some of the methods under development can directly sequence single molecules of genomic DNA without the need for PCR, as noted earlier. In addition, there has been significant innovation in the methods used to detect specific nucleotide incorporation. While the Illumina method uses a laser and camera to detect a fluorescent signal, some of the newer methods rely on changes in electrical properties or pH to detect the nucleotide added at a particular position, thus reducing the cost of the laser detector.

Exome sequencing reduces the amount of sequence information that needs to be analyzed

As we learned in in Chapter 1, protein-coding genes represent only 1.5% of the human genome. With the assumption that many functional changes in human disease would be a consequence of changes within the protein-coding genes, an approach termed 'exome sequencing' was developed. This method seeks to determine the sequence of all protein-coding exons in the genome, rather than the genome sequence in its entirety. Since sifting through trillions of base pairs to find those few that may be altered requires extensive computational analysis, focusing solely on protein-coding genes significantly reduces the sequence information that needs to be sorted, and narrows the search for meaningful changes that may cause human disease. By only sequencing exons that are thought to be expressed as proteins, however, this method misses mutations in promoters and splice sites that might affect gene function. However, the advantages of only having to sequence and interpret nucleotide changes for 1.5 % of the genome in many cases outweighs the potential disadvantage of missing a functionally important base change in non-coding regions. Exome sequencing has therefore become a common technique in recent years.

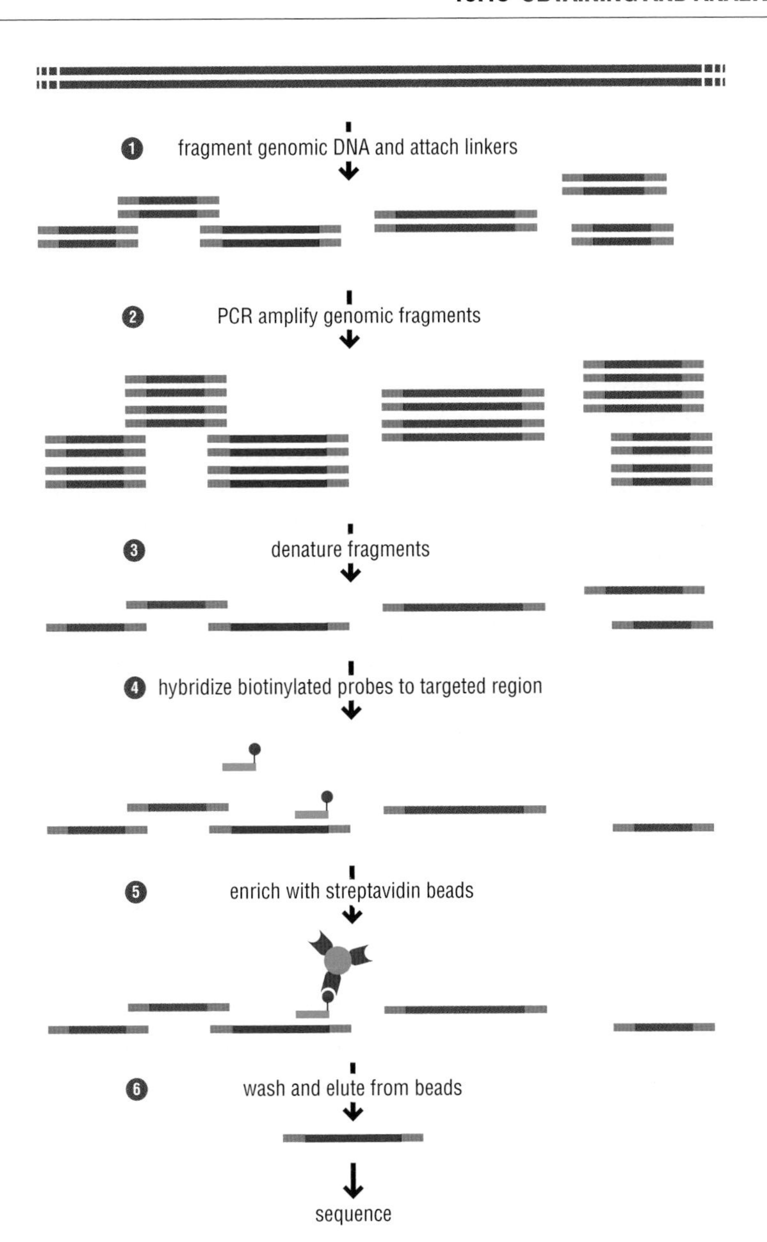

Figure 19.95 Exome sequencing. As for the sequencing of complete genomes, linkers are attached to fragmented genomic DNA (step 1), and the fragments are amplified by PCR (step 2). Subsequently the fragments are denatured (step 3) and incubated with biotinylated RNA probes corresponding to all known exons shown in purple (step 4). The DNA fragments hybridizing to the biotinylated probes can be captured by precipitating with streptavidin-coated beads, which strongly bind the biotin at the end of the RNA probes (step 5). After the unbound DNA is washed away, the remaining exon-containing fragments can be released from the RNA probes (step 6) and sequenced.

The sequencing step for exome sequencing is the same as for the deep sequencing methods we have just described, with one additional element: the enrichment for protein-coding genes. After genomic DNA fragments are amplified, the double-stranded DNA is denatured and hybridized with a set of specific RNA molecules that correspond to all known exons in the human genome. These RNA sequence probes are biotinylated, which allows them to be captured by magnetic beads coated with streptavidin as shown in Figure 19.95. Since the RNA probes correspond only to exons, only DNA fragments containing protein-coding genes will hybridize with the RNA. Once the RNA–DNA hybrids are formed and precipitated with the streptavidin-coated beads, the remaining 98.5 % of the genomic DNA is washed away. The purified DNA is then eluted and sequenced by the Illumina or other deep sequencing methods.

As the cost and speed of whole genome re-sequencing continues to decrease and the ability to analyze complex data sets increases, the use of exome sequencing is likely to decline in favor of whole genome sequencing.

Deep sequencing methods facilitate new kinds of experiments

The ability to rapidly sequence large amounts of DNA has led to the emergence of new experimental approaches. For example, deep sequencing technology can be applied to cDNA copies of RNA to monitor gene expression. This method, termed RNA-Seq, can be used to monitor transcript prevalence in particular cells or tissues. RNA-Seq can also be used to identify non-coding RNAs. Another application of deep sequencing replaces DNA microarrays as a method to identify DNA sequences bound by proteins that are isolated in ChIP experiments, a method termed ChIP-Seq. Similarly, deep sequencing after isolating ribosomes in the ribosome profiling approach allows all mRNAs being translated to be identified.

➔ We discuss RNA-Seq, ChIP-Seq and ribosome profiling in Sections 19.10 and 19.12 as well as in Experimental approach 12.1.

As we described in Chapter 18, deep sequencing is a powerful tool for identifying sequence variations in the genome that may play a role in human disease. Deep sequencing can also be used to rapidly re-sequence entire genomes and thereby identify naturally occurring genetic variation. In such applications, a reference genome sequence is used as a template to align short sequence reads into an already assembled complete genome, greatly reducing the difficulties in assembly. As sequencing becomes even less expensive, whole genome sequencing has become a powerful means of identifying disease-associated mutations.

SUMMARY

Advances in our understanding of molecular biology and genomics rely on ongoing experimentation, drawing on the kinds of tools and techniques we have introduced in this chapter. It is not only our understanding of molecular biology as a discipline that continues to move forward, but also the techniques used to probe genome function. As existing techniques become enhanced, and new methodologies become available, new and exciting approaches can be adopted to probe areas of interest, illuminating our understanding in new – and often unexpected – ways.

The principles of genome function presented throughout this book will doubtless continue to be refined as investigators around the world continue their research; so, too, will the tools and techniques used to explore these principles. It is this dynamic interplay between what we know and the experimental approaches used to further our understanding that makes molecular biology so fascinating, as we continue to uncover new questions that remain to be explored.

AN OVERVIEW OF MOLECULAR BIOLOGY METHODS

- Molecular biology methods range from genome-wide analysis to the dissection of interactions between isolated, single molecules.
- A key aspect of molecular biology is the study of processes both *in vivo* and *in vitro*.
- Most molecular components and processes are studied in a relatively small range of model organisms.
- Studies of the molecular components of a wide variety of genetically tractable organisms have proved Jacques Monod's prediction: 'What is true for *E. coli* is true for the elephant'.

CELL CULTURE

- Both bacterial and eukaryotic cells can be grown in culture.
- Individual molecules can be purified from cultures of these relatively homogenous populations of cells.
- ES cells and iPS stem cells can be grown in culture and manipulated to give rise to differentiated cells.

GENE AND GENOME MANIPULATION

- The ability to manipulate genes and genomes is central to molecular biology.
- DNA cloning, whereby specific DNA regions are isolated and propagated, is often carried out in bacteria.
- A library of clones may represent total genomic DNA, expressed genes only, or just a subset of expressed genes.

- The targeted introduction of mutations through DNA engineering (mutagenesis) allows for the generation of variant DNAs, RNAs, and proteins for study.
- Transposons and recombination are commonly used to introduce specific genes or mutations into the genome of an organism.
- Increasing use is being made of siRNAs for genome manipulation and analysis.

THE ISOLATION AND CHARACTERIZATION OF BIOLOGICAL MOLECULES

- The isolation and characterization of molecules *in vitro* is a key method in molecular biology.
- A variety of methods can be used both to identify and quantify molecules that are present in a sample.
- Many such detection techniques are spectroscopic in nature, and exploit the characteristic ways in which molecules interact with electromagnetic radiation.
- Chromatography techniques exploit the different physical characteristics of molecules to separate them from one another.
- Various approaches exist for determining the sequence (primary structure) of a biological molecule.
- Specific DNA sequences and RNA molecules can be detected using hybridization-based methods.
- Proteins can be detected by virtue of their specific interaction with antibodies.
- Genome sequence information and DNA manipulation allow for the addition of tags to specific proteins and RNA, facilitating their isolation without prior knowledge of their function.

CHARACTERIZATION OF INTERACTIONS BETWEEN MOLECULES

- The identification of components that interact with each other has enhanced our understanding of protein and RNA function.
- Advances in sequencing have facilitated the genome-wide analysis of interactions between protein and protein complexes with DNA and RNA.
- A range of biophysical techniques allows equilibrium and rate constants to be determined for the interactions between molecules.

CELLULAR IMAGING AND MOLECULAR STRUCTURE DETERMINATION

- Advances in imaging methods, including the use of reporter proteins such as GFP, have provided important new tools for studying cells in both live and fixed samples.
- Imaging can reveal information about both molecular localization and movement within the cellular environment.
- The three-dimensional structure of molecules is most widely determined by x-ray crystallography.

GENOME SEQUENCING

- Advances in genome sequencing have enabled us to obtain genome sequence information with increased speed and accuracy, and at lower cost.
- The study of human molecular biology has been facilitated and expanded by the availability of complete human genome sequences.

ONLINE RESOURCES FOR GENOMICS AND MODEL ORGANISMS

- National Institutes for Health
  http://www.genome.gov/
  http://www.ncbi.nlm.nih.gov/Genomes/
- US Department of Energy Office of Science
  http://genomics.energy.gov
- Major genomics centers
  J. Craig Venter Institute: http://www.jcvi.org/
  The Sanger Institute: http://www.sanger.ac.uk/
  Broad Institute of MIT and Harvard: http://www.broadinstitute.org/
  The Genome Center at Washington University: http://genome.wustl.edu/
  Baylor College of Medicine, Human Genome Sequencing Center: http://www.hgsc.bcm.tmc.edu/
- Microbes
  http://www.microbes.info/resources/General_Microbiology/Databases/Genetic/index.htm
- *E. coli*
  http://www.ecolicommunity.org
- *S. cerevisiae* (budding yeast)
  http://www.yeastgenome.org/
- *S. pombe* (fission yeast)
  http://www.genedb.org/genedb/pombe/

- *D. discoideum*
  http://dictybase.org/
- *C. elegans* (worm)
  http://wormbase.org/
  http://www.ncbi.nlm.nih.gov/projects/genome/guide/nematode/
- *Drosophila* (fruit fly)
  http://flybase.org/
  http://www.ncbi.nlm.nih.gov/projects/genome/guide/fly/
- *Arabidopsis*
  http://www.arabidopsis.org/
- *Xenopus* (frog)
  http://www.ncbi.nlm.nih.gov/genome/guide/frog/
- *D. rerio* (zebrafish)
  http://zfin.org/cgi-bin/webdriver?MIval=aa-ZDB_home.apg
  http://www.ncbi.nlm.nih.gov/projects/genome/guide/zebrafish/
- Mouse
  http://www.informatics.jax.org/
- Human
  http://www.genome.ucsc.edu/
- The Human Genome Project
  http://www.genome.gov/10001772

QUESTIONS

19.1 MODEL ORGANISMS

1. In the use of genetic model organisms, explain the significance of the fact that each of the following is well understood.

a. Life cycle
b. Genetics
c. DNA technology

2. Different model organisms are chosen to study particular biological processes because they have distinct properties that facilitate the study. For each of the following model organisms, name one or more biological processes given in the text as an example of the type of process studied in the organisms, and explain the unique characteristics that made the organism a good choice for the study.

a. *E. coli*
b. *B. subtilis*
c. *D. discoideum*
d. *C. elegans*
e. *M. musculus*

19.2 CULTURED CELLS AND VIRUSES

1. What is the significance of the use of cell culture in the study of proteins and nucleic acids?

2. Describe one advantage and one disadvantage of using eukaryotic cell lines for study.

19.3 AMPLIFICATION OF DNA AND RNA SEQUENCES

1. At the end of four cycles of PCR, you have 48 double-stranded DNA molecules containing the target sequence. How many double-stranded DNA molecules containing this sequence did you have before starting the PCR reaction?

a. 1
b. 3
c. 6
d. 12

2. Which of the following sequences would you find during first-strand cDNA synthesis?

a. 5′ AGTCGATGCTAGT 3′
b. 5′ AGTCGATGCTAGT 3′
   3′ TCAGCTACGATCA 5′
c. 5′ AGTCGATGCTAGT 3′
   3′ UCAGCUACGAUCA 5′

Challenge question

3. Quantitative PCR is a way to detect relative amounts of starting material (i.e., the PCR template) in a PCR reaction. After each cycle, the amount of PCR product is detected using fluorescence. Typical results are plotted in Figure A, with each line on the graph resulting from a single PCR reaction.

Which curve represents the sample with most abundant starting PCR template? Explain your answer.

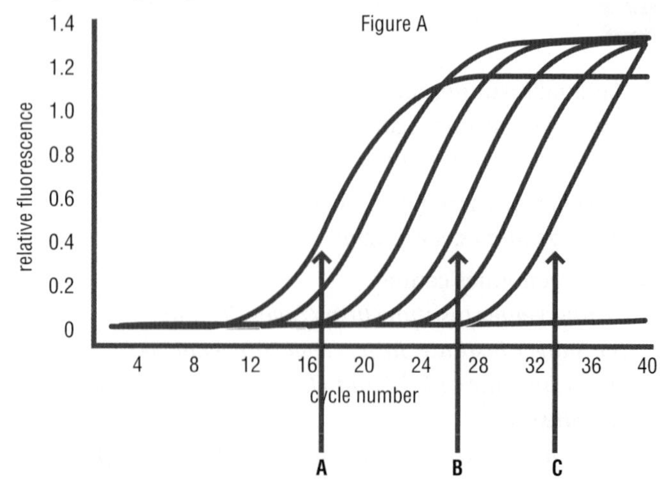

19.4 DNA CLONING

1. DNA cloning has been absolutely essential in the study of genes and their function.

a. What is DNA cloning?
b. Why is DNA cloning essential to the study of genes?
c. What is a DNA vector?
d. Why are DNA vectors important?

2. Site-directed mutagenesis is a technique that has significantly enhanced the ability to study gene function.

a. What is site-directed mutagenesis?
b. Why is this technique so significant in the study of gene function?

3. You extract some genomic DNA and cut it with MseI. What do you estimate the average size of the DNA fragments will be? Enzyme recognition sequences can be found easily online.

a. 1 bp
b. ~16 bp
c. ~250 bp
d. ~4000 bp

4. Compare and contrast the methods that would be used to clone a gene of interest as opposed to identifying a gene causing a particular phenotype.

Challenge question

5. Which of the following molecules will ligate to this sticky end:

5′ C-T-G-C-A
3′ G

a. 5′ C-T-G-C-A
3′ G
b. 5′ G
3′ C-T-G-C-A
c. 3′ G
5′ A-C-G-T-C
d. 5′ A-C-G-T-C
3′ G

19.5 GENOME MANIPULATION

1. Compare and contrast the use of transposons and the use of homologous recombination in genome manipulation.

2. Generating a knockout mouse.

a. What is a knockout mouse?
b. The first step in the process is to create a targeting vector containing a portion of the gene of interest, the Neo^R gene, and the *HSV-tk* gene. What is the purpose of each of these components?
c. The original ES cells for recombination are taken from a gray mouse, the cells that successfully underwent homologous recombination are injected into a white mouse blastocyst, and the blastocyst is implanted in a black mouse surrogate parent. Why are the three different colored mice necessary? Explain what possible colored pups could arise, which is desirable and why.
d. In the final step, the chimeric mice are mated with a white mouse, and the offspring are analyzed. Why is this step necessary? What are the possible outcomes of this cross?

19.6 DETECTION OF BIOLOGICAL MOLECULES

1. Which three amino acids absorb UV light and thus allow proteins to be assayed spectroscopically?

a. Glycine, tyrosine, proline
b. Proline, tryptophan, lysine
c. Tyrosine, proline, phenyalanine
d. Tryptophan, tyrosine, phenylalanine

19.7 SEPARATION AND ISOLATION OF BIOLOGICAL MOLECULES

1. In order to be able to study the particular characteristics of a biological molecule or molecules that are part of a biological process, it is first necessary to isolate and purify them away from other cellular components. Briefly describe the method and the general process by which one would isolate each of the following classes of cellular components.

a. Cellular organelles
b. Proteins
c. Nucleic acids

Challenge questions

2. Proteins often have domains of distinct functionality within a single polypeptide, or on different peptides within a multi-subunit protein.

a. Describe how you would use gel electrophoresis to distinguish between a protein that has domain A and domain B on a single subunit, and a protein that has domain A and domain B on two subunits. What results would you expect?
b. After performing the gel electrophoresis experiment described in (a), you note that the results for the two proteins look identical and realize you made a mistake in preparing the gel. What was omitted from the gel, and what does this tell you about the two-subunit protein and its likely cellular location?

3. This diagram represents a piece of genomic DNA. The red line indicates where a radioactively labeled probe hybridizes. Three samples of the genomic DNA are cut, one with PstI (P), one with BamHI (B), and one with both enzymes (PB). The digested samples are run on an agarose gel, blotted, and probed. Draw the results you expect to see from a Southern blot and explain your answer.

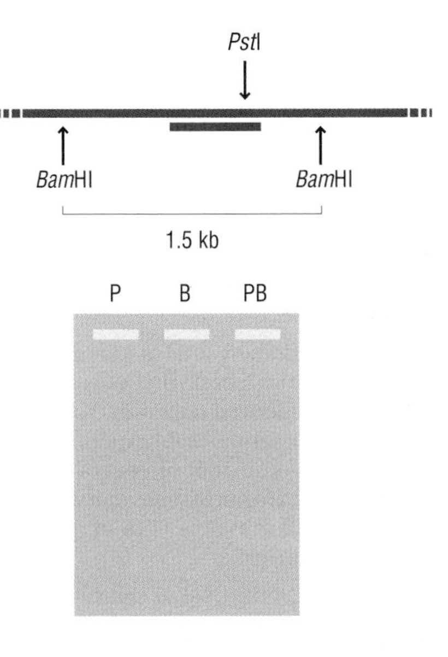

19.8 IDENTIFYING THE COMPOSITION OF BIOLOGICAL MOLECULES

1. You are preparing an 'A' reaction for radioactive Sanger sequencing. You have added the buffer, DNA template, DNA polymerase, labeled primer, and the four dNTPs to the tube. Which of the molecules in Figure B (page 876) must you also add?

2. Which common technique would you use a ddNTP for?

a. Affinity chromatography
b. Agarose gel electrophoresis
c. SDS-PAGE
d. DNA sequencing

Figure B

(a)

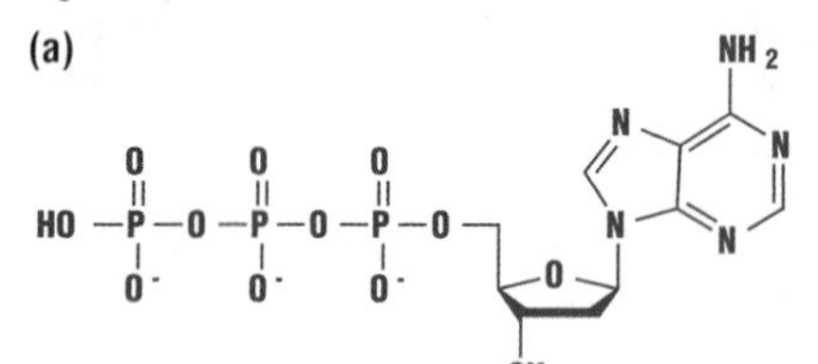

(b)

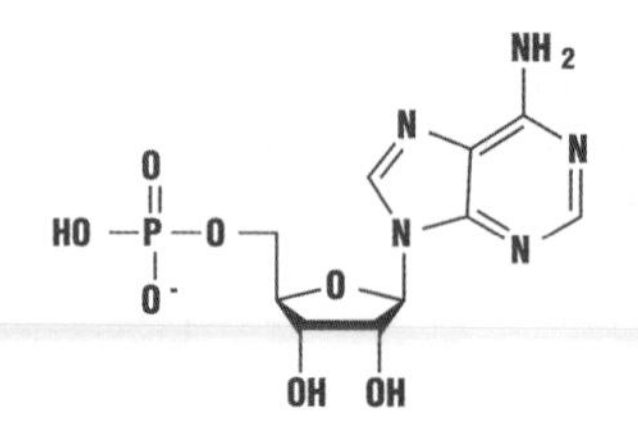

(c)

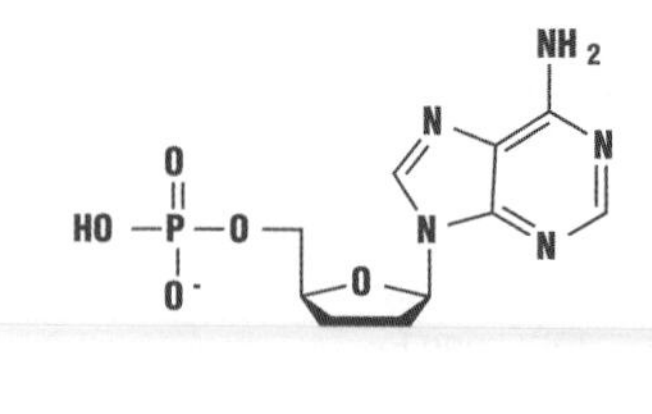

(d)

(e)

(f)

19.9 DETECTION OF SPECIFIC DNA SEQUENCES

1. Each of the following techniques can be used to examine whole genomes for insertions, deletions, and translocations. Briefly describe each method in terms of how it is unique from the other methods.
- a. Karyotyping
- b. FISH
- c. aCGH

2. When would colony hybridization be useful?
- a. To screen a genomic library
- b. To test for protein production
- c. To check for antibiotic resistance in a plasmid
- d. To investigate RNA:RNA interactions

3. What would be the most appropriate technique to look at the difference in global gene expression in mouse liver before and after a drug treatment?
- a. RT-PCR
- b. Northern analysis
- c. Microarray analysis
- d. Southern analysis

Challenge questions

4. You have been given an extraterrestrial species of bacterium isolated from an asteroid, and you wish to check if its DNA replicates in the same way as DNA from an Earth-derived organism. You decide to use the same approach as in the Meselson–Stahl experiment (Experimental approach 6.1). You have unfortunately run out of heavy nitrogen, but instead you have some modified nucleotides that will behave normally in replication but which will cause the DNA to migrate more slowly when you run it on an agarose gel.

First, you allow the bacterial DNA to replicate in the presence of the 'slow' nucleotide for several generations, until all of its DNA is 'slow'.

You extract DNA: this is SAMPLE A.

Second, you remove the 'slow' nucleotides, replace them with regular nucleotides, and allow the bacteria to undergo one round of replication.

You extract DNA: this is SAMPLE B.

Third, you allow one more round of replication with the normal nucleotides.

You extract DNA: this is SAMPLE C.

After running your three samples on an agarose gel, you determine that this species has dispersive replication. (Replication is discussed in Chapter 6.)

Draw a fully labeled diagram of your gel with samples A, B, and C, showing the relative positions of the DNA bands. Explain in detail why these DNA positions mean that replication is dispersive. Remember to label the electrodes of the gel.

5. You have transformed your organism with a gene for GFP, which should integrate randomly into the genome. You want to work further with the most stable transformants. You extract DNA from several different transformants and perform Southern analysis with a probe for the *GFP* gene. Which of the transformed strains (1–4) would you work with further and why?

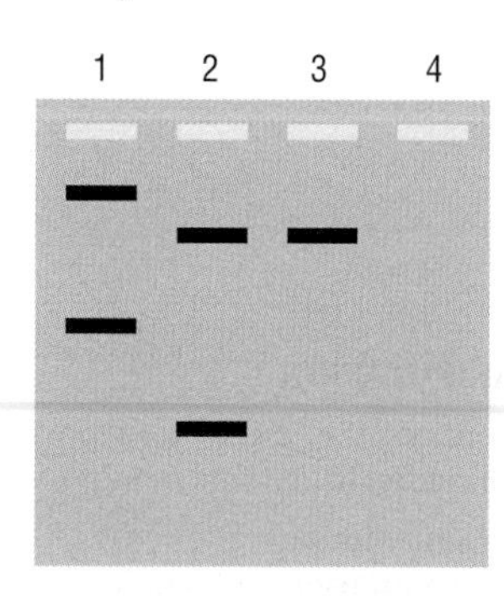

19.10 DETECTION OF SPECIFIC RNA MOLECULES

Challenge question

1. Analysis of the sequence for a gene of interest suggests that the four exons contained within the gene may be alternatively spliced to produce up to three alternative transcripts consisting of (1) all four exons, (2) exons 1, 2, and 4, and (3) exons 1, 3, and 4. Researchers performed a northern blot analysis to determine if the three possible transcripts were actually produced.
- a. Why did the researchers choose a northern blot?
- b. To which exon should the researchers create a probe? Why?
- c. Based upon the exon sized indicated in the figure below, draw a figure that shows the expected result of the northern blot if only (1) and (3) transcripts are produced.

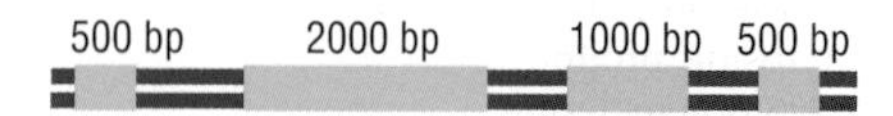

19.11 DETECTION OF SPECIFIC PROTEINS

1. Which of these techniques requires use of an antibody specific for a protein of interest?
- a. DNase footprinting
- b. Northern analysis
- c. Restriction enzymes
- d. ChIP

2. How can one detect a specific protein if there is no antibody available for that protein?

3. How is ChIP-chip or ChIP-Seq different from just ChIP?

19.12 DETECTION OF INTERACTIONS BETWEEN MOLECULES

1. What is ChIP used for?
   a. To test protein:protein interactions
   b. To test DNA:RNA interactions
   c. To test DNA:protein interactions
   d. To test DNA:DNA interactions

Challenge questions

2. Histone acetyltransferases were initially discovered in an elegant experiment that detected activity of the protein in a polyacrylamide gel infused with histones. (Experimental approach 4.2, Figure 1). Starting with the gel shown in that figure, design an experiment to determine the sequence of the gene encoding the protein.

3. The data in the image reproduced in Figure C show the results of a yeast two-hybrid experiment, with the parts of the proteins tested indicated with numbers. From these data, which protein binds only to *part* of one of its interactors? What does this suggest regarding how the proteins bind to one another? Explain your answer in terms of how yeast two-hybrid works and with reference to the data shown.

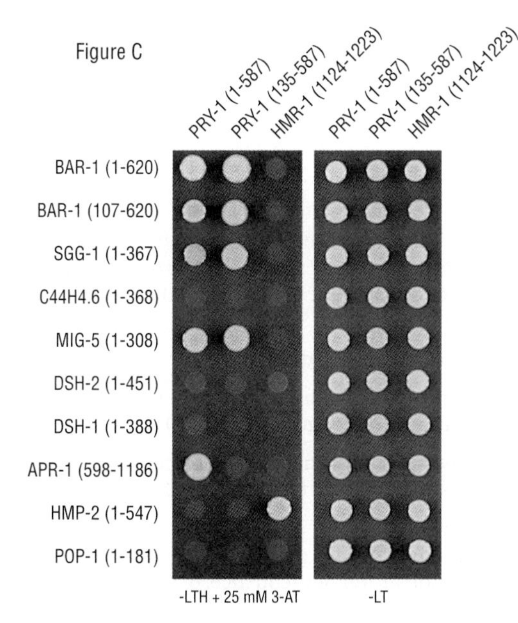

19.13 IMAGING CELLS AND MOLECULES

1. Figure 19.78 is an image that depicts one advantage to the use of microscopy in the study of biological processes.
   a. What does the figure represent?
   b. Why is this figure significant?
   c. Describe two additional advantages to the use of microscopy in the study of biological processes.

Challenge question

2. The graph in Figure D shows data from a FRAP experiment. Which of the proteins has the slowest movement? Explain your answer in terms of how FRAP works and with reference to the data shown.

Figure D

Fraction Recovery
Time (s)
[PSI⁻]
[PSI⁺] Gdn-0h
[PSI⁺] Gdn-1h
[PSI⁺] Gdn-5h

19.14 MOLECULAR STRUCTURE DETERMINATION

1. Provide one advantage and one disadvantage of each of the following techniques for determining the three-dimensional structure of a molecule.
   a. X-ray crystallography
   b. NMR
   c. Cryoelectron microscopy

19.15 OBTAINING AND ANALYZING A COMPLETE GENOME SEQUENCE

1. Compare and contrast classic Sanger sequencing methods to the next-generation sequencing methods that are currently available.

2. Once a genome is sequenced, what steps are taken to give meaning to the sequence?

GLOSSARY

(p)ppGpp ('magic spot'): a pentaphosphate guanosine analog that plays an important role in signaling the stringent response in bacterial systems.

–10: promoter sequence located ten nucleotides to the 5′ of the start of transcription and comprising part of the sequence directing bacterial RNA polymerase to bind and initiate transcription; also called the Pribnow box.

3′ CCA tail: the three terminal nucleotides of all tRNAs and the site of attachment for amino acids.

3′-RACE: RACE stands for rapid amplification of cDNA ends and 3′ indicates that the 3′ end of the RNA is being amplified. In 3′-RACE the 3′ end of the RNA is known (such as the poly-A tail of an mRNA) and is used to prime the synthesis of A DNA complementary strand. The resulting cDNA can then be amplified by PCR.

3′ splice site: consensus dinucleotide (AG) sequence found at the 3′ end of the intron that signals the intron–3′ exon boundary.

–35: promoter sequence located 35 nucleotides to the 5′ side of the start of transcription and comprising part of the sequence directing bacterial RNA polymerase to bind and initiate transcription.

454 sequencing: massively parallel sequencing technique in which individual sequence reads can be up to 250 bp.

4E-BP: class of protein that bind directly to the cap-binding protein eIF-4E thus preventing functional interactions with eIF-4G.

5′ cap: a special structure at the 5′ end of eukaryotic mRNAs containing a 7-methylguanosine residue attached via an unusual 5′–5′ triphosphate linkage.

5′ capping: process by which an mRNA is modified with the cap structure.

5′-RACE: RACE stands for rapid amplification of cDNA ends and 5′ indicates that the 5′ end of the RNA is being amplified. In 5′-RACE the initial cDNA reaction uses an internal primer that directs synthesis of a DNA strand, copying the RNA to it 5′ end. This is followed by terminal transferase, which adds a poly-A tail. This is then used as a template, using a 5′ primer that is complementary to the newly added poly-A tail. The resulting cDNA can then be amplified by PCR.

5′ splice site: consensus dinucleotide (GU) sequence found at the 5′ end of the intron that signals the 5′ exon–intron boundary.

β protein: subunit of the sliding clamp protein in bacteria; it dimerizes to generate a symmetrical ring structure with six similar domains.

A

A site (aminoacyl site): position on the small and large ribosomal subunits of the ribosome where the aminoacyl-tRNA binds.

AAA+ family (ATPases associated with a variety of cellular activities): family of proteins with a conserved region that includes the Walker A and Walker B ATPase motifs. Many replication proteins are members of this family, including the clamp loader and helicases. AAA+ family members are also involved in cellular activities other than DNA replication.

abasic site: site in DNA at which a base has been lost from a nucleotide residue.

abortive initiation: non-productive initiation of transcription leading to the production and release of short RNAs rather than elongation of the RNA transcript.

absorption spectrum: the amount of light absorbed by a sample as a function of wavelength.

acceptor stem: one of four helical elements in the tRNA that brings together the 5′ and the 3′ end of the molecule via base-pairing interactions.

accommodation: the step in tRNA selection that follows guanosine triphosphate (GTP) hydrolysis, which occurs when a tRNA is finally accepted by the ribosome and is thus committed to peptide bond formation.

acetylation: covalent addition of an acetyl group from acetyl-CoA to a nitrogen atom at either the N-terminus of a polypeptide chain or in a lysine side chain. The reaction is catalyzed by *N*-acetyltransferase. In the case of histone acetylation, acetyl groups are added to lysine residues in the histone tails.

acetyl transferase: enzyme that adds acetyl groups to side chains on proteins. A subgroup is the histone acetyl transferases (HATs).

acid: molecule that releases H^+ into solution.

activate: in transcription, to increase the number of RNA transcripts of a given gene.

activation energy: the energy required to bring a species in a chemical reaction from the ground state to a state of higher free energy, from which it can transform spontaneously to another low-energy species.

activator: sequence-specific DNA-binding protein that leads to increased transcription.

active site: pocket on or near the surface of a macromolecule that promotes chemical catalysis when the appropriate ligand (substrate) binds.

actomyosin ring: The contractile ring involved in cytokinesis. This ring attaches to the cell membrane at the cleavage furrow and its contraction leads to the formation of a membrane that separates the cytoplasms of the daughter cells.

ADAR: acronym for 'adenosine deaminase that acts on RNA'.

A-DNA: right-handed DNA double helix with 11 bp per turn; A-DNA has a deeper and narrower major groove than B-DNA.

affinity chromatography: purification of a molecule, for example, a protein, based on its ability to bind to a particular compound immobilized on a support.

alignment: comparison of each position in one sequence with the corresponding position in another sequence, or group of sequences.

alkylating agent: mutation-causing chemical that modifies bases in DNA with a methyl or ethyl group.

allele: different version of a gene, such as a wild-type or a mutant version.

allosteric effector: molecule that bind to a protein and induces a conformational change, thereby changing the activity of the protein.

alpha helix: an element of protein secondary structure in which a portion of the polypeptide chain forms a right-handed spiral, with the backbone NH group of every residue *i* forming a hydrogen bond with the C=O group of every residue *i*+4.

alpha satellite: highly repetitive DNA found at the centromere and elsewhere on human chromosomes. It consists of tandem repeats of a 171 bp sequence.

alternative lengthening of telomeres (ALT): break-induced replication (BIR) mechanism for the addition of telomeres to chromosome ends that have lost telomere sequences, usually because of the inactivation of telomerase.

alternative sigma (q) factor: sigma factor that recognizes promoters of specialized sets of genes and is expressed or activated under specific conditions, when it associates with the core RNA polymerase and enables the specialized genes to be expressed.

Alu element: genomic sequence derived from 7SL RNA, the RNA component of the signal recognition particle that targets protein to membranes. There are more than one million copies of this sequence in the human genome.

amino acid: chemical building block of proteins containing an amino group, a carboxyl group, and one of 20 possible side chains.

amino acid residue: the atoms that remain after an amino acid is incorporated into a polypeptide chain.

aminoacyl site (A site): position on the small and large ribosomal subunits of the ribosome where the aminoacyl-tRNA binds.

aminoacylation: process by which an amino acid is covalently attached to the terminal adenosine of a tRNA species.

aminoacyl-tRNA: a tRNA species carrying an esterified amino acid covalently bound to its terminal adenosine.

aminoacyl-tRNA synthetase: one of ~20 different ubiquitous enzymes, each of which is specific for one type of amino acid, which catalyzes the covalent attachment of an amino acid to all cognate tRNA species.

A-minor motif: a commonly seen structural motif in RNA, where adenosines found in distant regions interact via their minor-groove face with the minor groove of a neighboring helix, often with their 2′ OH groups. Various versions of the A-minor motif function similarly to bring together distant regions of the RNA.

amphipathic: having both polar and non-polar character. Synonym for amphiphilic.

amphipathic alpha helix: alpha helix with a hydrophilic side and a hydrophobic side.

amphitelic attachment: *See* **bi-orientation**.

anaphase: the stage of mitosis in which the sister chromatids are segregated by the mitotic spindle. In most species, anaphase is divided into anaphase A, the initial movement of chromosomes to the spindle poles, and anaphase B, the movement of spindle poles away from each other.

anaphase A: the movement of chromosomes after sister chromatid separation towards the spindle poles.

anaphase B: the elongation of the spindle during anaphase, resulting in an increase in distance between the two centrosomes that anchor the spindle poles.

anaphase-promoting complex (APC): large, multisubunit ubiquitin-protein ligase (also known as an E3 enzyme) that catalyzes the attachment of ubiquitin to mitotic proteins such as M cyclin and securin, thus promoting their ubiquitin-dependent proteolysis. It is also called the cyclosome or APC/C.

aneuploidy: a deviation in chromosome number from the normal complement. For example, having 3 copies of a chromosome in a cell that is normally diploid.

anomer: cyclic sugar isomer with a characteristic orientation of a hydroxyl group relative to the carbon atom.

antibiotics: small molecules that kill or slow the growth of bacteria or fungi by targeting essential cellular processes.

anticodon: three nucleotides in the anticodon loop of the tRNA that are responsible for directly interacting with the mRNA and decoding the genetic information.

anticodon loop: the single-stranded loop region of the tRNA that presents the anticodon nucleotides to interact with the mRNA.

antiparallel: arrangement of two DNA strands with the 5′ to 3′ polarity of one strand opposite to that of the complementary strand.

antiparallel beta sheet: a beta sheet, often formed from contiguous regions of the polypeptide chain, in which each strand runs in the opposite direction from its immediate neighbors.

anti-Shine–Dalgarno sequence: polypyrimidine sequence at the 3′ end of the 16S rRNA that directly interacts by base-pairing with the upstream leader sequence of mRNAs to position the AUG start codon in the P site.

anti-sigma factor: protein that binds a sigma factor and inhibits its function.

anti-termination: promotion of elongation over termination.

apoptosis: programmed cell death using a special cell machinery that occurs in response to unrepaired DNA damage.

apurinic/apyrimidinic endonuclease (AP endonuclease): an enzyme involved in base excision repair that recognizes the absence of a base (an abasic site) in DNA.

architectural DNA-binding protein: protein whose role is to induce a significant bend in the DNA, thereby either promoting or inhibiting interactions between proteins bound on either side of the bend.

Argonaute: key protein component of complexes (RISC and miRNP) that binds small RNAs and mediates RNAi or miRNA-mediated repression.

association study: method used to find the DNA marker closely associated with a disease gene or other trait when DNA is available from a large number of unrelated people who have the disease and a large number who do not.

ATM (ataxia telangiectasia mutated): a regulator kinase of the DNA damage response of multicellular eukaryotes first identified in patients with ataxia telangiectasia that interacts with MRN, which is a sensor of DNA damage.

Atomic force microscopy (AFM): a high-resolution technique for examining the surface of a specimen. It uses a nanometer-sized tip associated with a cantilever that scans the specimen surface and records changes in height that result from forces between the tip and surface features.

ATP-dependent nucleosome-remodeling complex: protein complex that uses the energy released by ATP hydrolysis to alter the positions of nucleosomes on the DNA.

ATR (ATM related): a regulator kinase of the DNA damage response of multicellular eukaryotes identified by its homology to ATM.

ATRIP (ATR interacting protein): a regulator of the eukaryotic DNA damage response that interacts with RPA bound to single-stranded DNA.

attenuation: mechanism for regulating transcription termination that exploits competition between ribosome binding and formation of a stem–loop structure in the transcribed RNA.

autocrine loop: when a secreted mitogen acts directly back on the cell that secreted it, increasing cell division of the secreting cell.

autonomous element: transposable element that encodes the transposase necessary for its mobility and also retains the sequences at which the transposase acts.

autoregulation: process by which a protein controls its own synthesis by activating or repressing transcription of the gene that encodes it.

autosome: a chromosome that is not a sex chromosome. Males and females have equal numbers inherited from the mother and the father.

B

backtrack: when an RNA polymerase enzyme stalls and then slides backwards along the template, causing the most recently added ribonucleotides to dissociate from the template.

bacteriophage: a virus that infects bacteria. Also referred to as phage.

basal transcription: level of transcription in the absence of activators.

base: molecule that reduces the concentration of H^+ in solution.

base (in a nucleic acid): aromatic purine or pyrimidine group attached to the sugar of a nucleotide.

base excision repair: mechanism of DNA repair in which a nucleotide with a damaged or missing base is excised from the DNA and replaced with an undamaged one, using the undamaged DNA strand as template.

base stacking: the stacking of bases in parallel planes along a nucleic acid strand. Base stacking, which is stabilized by van der Waals interactions, can occur in double- or single-stranded DNA and RNA.

base triple: three interacting, coplanar bases typically consisting of a Watson–Crick base pair that hydrogen bonds with a third base via its remaining unpaired chemical groups.

B-DNA: right-handed DNA double helix that predominates in cells, with 10.5 bp per turn and a rise of 3.4 Å per base pair.

beta sheet: secondary-structure element composed of multiple beta strands that interact through backbone hydrogen bonding interactions.

beta strand: an individual single strand of extended polypeptide chain in a beta sheet.

binding curve: a curve in which the amount of complex formed is plotted as a function of the concentration of the other component.

bi-orientation: the correct binding of kinetochores of sister chromatids to microtubules emanating from opposite spindle poles. Also known as **amphitelic attachment**.

BLAST: Basic Local Alignment Sequence Tool, a sequence alignment algorithm available on the web (http://blast.ncbi.nlm.nih.gov/Blast.cgi) for aligning either DNA or protein, in which sequence comparisons are based on local, that is regional, rather than global, comparisons.

boundary elements: DNA sequences that specify the transition between euchromatin and heterochromatin. Moving a boundary element sequence moves the transition point of chromatin structure.

branch migration: a process by which Holliday junctions move along DNA, increasing or decreasing the region of heteroduplex.

branch-point nucleotide: the adenosine residue that is directly involved in the nucleophilic attack on the exon–intron boundary and that is thus located at the branch site of the intron lariat structure.

break-induced replication (BIR): replication from the site of a double-strand break to the end of the chromosome using a homologous chromosome as a template.

bromodomain: A protein domain that binds to acetylated lysine side chains – usually specific positions on a histone protein (for example, acetylated lysine 14 on histone H3).

C

carbohydrate: organic molecule composed of one or more monosaccharides.

catalysis: the increase in the rate of a chemical reaction that is brought about by a substance known as a catalyst. In biological reactions, catalysts are typically referred to as enzymes.

catalyze: to increase the rate of a chemical reaction by decreasing the activation energy.

catastrophe (in microtubules): sudden shrinkage that occurs when all the tubulin at the microtubule end is in a GDP-bound form.

Cdk-activating kinase (CAK): protein kinase that activates cyclin-dependent kinases by phosphorylating a threonine residue (Thr 160 in human Cdk2).

cDNA: complementary DNA that is synthesized from mature (i.e. spliced) mRNA template by reverse transcriptase.

cDNA library: collection of clones carrying sequences of spliced mRNAs expressed under specific conditions.

cell culture: the process by which cells are grown under controlled conditions.

cell cycle: the highly regulated sequence of events that take a cell from its formation to its own division. In eukaryotic cells it is divided into four main phases: G1, S, G2, and mitosis (M).

cell fractionation: the process of breaking open cells and separating distinct organelles.

cell line: a particular strain of eukaryotic cells that will grow continuously in culture.

centrifuge: an instrument that uses the centripetal force generated by rotation at high speed to cause cells or molecules suspended in liquid to sediment to the bottom of the tube.

centriole: a cylindrical structure composed of short microtubule arrays present at the core of many types of centrosomes. Centrosomes typically contain a pair of centrioles at right angles to each other. Centrioles duplicate one per cell cycle and direct the duplication of centrosomes.

centromere: a region of heterochromatin in each eukaryotic chromosome through which the chromosome becomes attached to the mitotic spindle, allowing daughter chromosomes to segregate correctly at each cell division.

centromere inactivation: an epigenetic process by which a centromere loses its ability to promote chromosome segregation, presumably due to changes in chromatin structure.

centrosome: a cytoplasmic structure that nucleates microtubules. Also referred to as a microtubule-organizing center. Centrosomes typically contain a pair of centrioles surrounded by pericentriolar material.

chain-termination dideoxy sequencing: a method for sequencing DNA in which dideoxynucleotides, which cannot prime further DNA synthesis during primer extension because they lack a 3′ OH, are present in limiting amounts such that termination occurs at some frequency at each nucleotide position. Reaction products are analyzed by display following denaturing electrophoresis.

chaperone: a protein that aids in the folding of another protein by preventing the unwanted association of the unfolded or partially folded forms of that protein with itself or with others.

checkpoints: regulatory pathways that monitor certain cellular processes and can arrest cell cycle progression at specific transition points in the cell cycle when these processes fail.

chiasmata (*singular* chiasma): X-shaped chromosomal structures that are seen in the microscope at sites of crossing-over between homologous chromosomes at the end of meiotic prophase.

Chi sequence: the 5′-GCTGGTGG sequence that stimulates the formation of 3′ ended single-stranded DNA by reducing the nuclease activity of *E. coli*RecBCD

Chi (a) sequence: asymmetric, 8 bp DNA sequence that changes the activity of the bacterial RecBCD–protein complex.

ChIP-seq: stands for chromatin immunoprecipitation followed by deep sequencing. Often used when trying to identify *all* the DNA fragments that immunoprecipitated with a particular DNA binding protein.

chloroplast: organelle found in plant cells that houses the machinery for harvesting energy from sunlight for the production of sugars within the cell.

cholesterol: a lipid containing conjugated hydrocarbon rings that is a steroid.

chromatid: one of the two daughter copies of a chromosome after replication but before segregation at mitosis.

chromodomain: A protein domain that binds to methylated lysine and arginine side chains, usually on histone proteins (for example, lysine 9 on H3).

chromatin: the complex of DNA bound by proteins and assembled into the higher-order structure known as a chromosome.

chromatin assembly factor (Caf1): the protein that binds the H3–H4 tetramer and facilitates its incorporation into nucleosomes. Caf1 interacts with the sliding clamp protein, PCNA, to facilitate histone loading.

chromatin immunoprecipitation (ChIP): a method for identifying DNA sequences bound to a particular protein by cross-linking the DNA to the protein, shearing the DNA to produce fragments, and then immunoprecipitating particular protein-DNA complexes with a specific antibody against the protein. The associated DNA is then analyzed by microarray hybridization or sequencing (ChIP-seq).

chromophore: small molecule bound to a protein that causes it to absorb visible light and therefore have a color.

chromosome: a structure composed of DNA complexed with specific proteins, which is the form in which the genetic traits encoded in DNA are maintained and transmitted from one generation to the next. Generally, organisms have multiple, independent linear chromosomes or a single, circular chromosome.

chromosome condensation: the structural changes that result in compaction of the chromosomes into smaller, more compact structures before cell division.

chromosome cycle: variation in chromosome condensation state during the cell cycle, which changes from relatively decondensed in interphase to highly condensed during mitosis.

chromosome painting: method used to detect one unique chromosome in the set of all metaphase chromosomes.

chromosome segregation: the distribution of chromosomes to daughter cells at mitosis or meiosis.

chromosome territory: the unique space occupied by a chromosome in the nucleus.

***cis* splicing:** the standard form of splicing in which exons from the same RNA transcript are joined in a series of events coordinated most typically by the spliceosome.

clamp loader: five-subunit protein that loads the sliding clamp onto the DNA during DNA replication.

class 1 release factor: any of the protein factors responsible for recognizing the three stop codons in the A site of the ribosome and for facilitating the hydrolytic reaction that releases the growing peptide chain from the peptidyl-tRNA.

class 2 release factor: the GTPase protein factor involved in promoting different facets of the termination process in bacteria and eukaryotes.

Class I transposonable element: an element whose transposition pathway contains an RNA intermediate.

Class II transposonable element: an element whose transposition pathway contains only DNA intermediates.

cleavage furrow: during cytokinesis, the ingression of the cell membrane that is associated with the contractile ring and that eventually divides the cell in two.

closed complex: RNA polymerase bound to promoter DNA before the DNA strands are separated.

closed loop complex: complex of mRNA and eukaryotic translation initiation factors (minimally eIF4E, eIF4G, and PABP) thought to form a closed loop that is competent for translation initiation.

co-activator: protein or complex of proteins recruited to DNA by sequence-specific DNA-binding proteins that activates transcription.

coaxial stacking: end-to-end stacking of RNA helices commonly seen in folded RNA molecules.

coding joint: chromosomal fusion between various V, D, and J coding segments that occurs during RAG-mediated V(D)J recombination and results in the assembly of diverse antibody and antigen receptor genes in lymphocytes.

coding strand: the DNA strand that has the same base sequence as the RNA transcript.

coding, non-template, or sense strand: DNA strand that bears the same sequence as the RNA.

codon: three consecutive nucleotides (i.e. a triplet) in an mRNA that correspond to an amino acid or stop and are decoded by tRNAs, release factors, and the ribosome.

coenzyme: a cofactor that is an organic or organometallic molecule that assists in enzyme catalysis.

cofactor: a non-protein compound that binds to a protein and is required for its biological activities, including for example enzymatic transformations.

cognate interaction: when the codon of the mRNA and the anticodon of the tRNA are perfectly matched – these pairing interactions involve Watson–Crick interactions at the first and second positions of the codon and either a Watson–Crick or 'wobble' at the third position (typically a G–U interaction).

cohesin: protein complex containing a heterodimer of structural maintenance of chromatin (SMC) proteins, a kleisin, and other subunits, that links sister chromatids together after S phase.

cohesion: association between sister chromatids that is established immediately after DNA replication and is dissolved during M phase.

coiled coil: pair of alpha helices that coil around one another and associate via hydrophobic interactions to form a dimerization interaction.

cointegrate: DNA product generated by the replicative transposition of some DNA-only transposons, in which the transposon has been replicated to produce two new copies that link the donor and target DNA.

co-localization: degree to which distinct signals obtained from two or more fluorescent signals, each emitting at a different wavelength, overlap in the same spatial position within cells examined by microscopy.

combinatorial control: the use of several different proteins to regulate a single gene.

comparative genomics: comparison of genomic sequences to gain insights into genes and genome function.

competitive inhibitor: species that competes with substrate for binding to the active site of an enzyme and thus inhibits catalytic activity (affecting the apparent K_M of the reaction).

complementary (of DNA): containing the base pair partner(s) predicted by the Watson–Crick base pairs.

compound transposon: bacterial transposon in which two insertion sequence (IS) elements flank a segment of DNA. Coordinated movement of the IS elements results in the movement of the interior segment.

condensation: compaction of chromosomes.

condensin: protein complex containing a heterodimer of structural maintenance of chromatin (SMC) proteins and other subunits that mediates chromosome condensation.

conformation: three-dimensional arrangement of the atoms in a molecule.

conjugation: bacterial mating in which a plasmid is transferred from a donor strain to a recipient strain

conservative site-specific recombination (CSSR): exchange of DNA at specific sites with short regions of homology. It can result in integration, excision, recombination, or inversion depending on whether the sites are on different DNA molecules or the same one, and on their relative orientations.

conserved synteny: the regions of each chromosome that are very similar in gene composition and order to regions in the other species.

contact inhibition: contact between cells that stimulates cell cycle arrest, causing cells to cease dividing.

contractile ring: ring of proteins, including contractile assemblies of actin and myosin, which forms at the site of cleavage in dividing animal cells. Its gradual contraction pinches the cell in two.

convergent evolution: evolution of a common structure or function in macromolecules or organisms that do not share a common ancestor.

cooperative binding: occurs when binding of one ligand alters the affinity for the other ligand.

co-protease: protein that stimulates the self-cleavage activity of another protein.

copy number variants (CNV): a DNA sequence variation where the copy number of some segments of DNA at a specific position in the genome differs between members of a species.

core enzyme: RNA polymerase enzyme that is capable of DNA-dependent RNA synthesis *in vitro* but is incapable of specific promoter recognition. Consists of β, β', and two α subunits in bacteria, and more than ten subunits in eukaryotes and archaea.

co-repressor: protein or complex of proteins recruited to DNA by sequence-specific DNA-binding proteins that represses transcription.

core promoter elements: short DNA sequences found within eukaryotic promoter regions that serve as binding sites for proteins in the pre-initiation complex. Also called promoter proximal elements.

covalent bond: a chemical bond that results from sharing of electrons between two atoms.

covariation: concerted changes in sequences observed throughout evolution that provide evidence for interaction between two elements.

CpG and CpXpG: nomenclature designating the sequence that reads 5′ to 3′ CG or C (any nucleotide) G. These are the sequences in mammalian DNA that most often contain methyl cytosine.

cryptic splice site: sequence elements that are similar to the canonical splice sites but do not define biologically utilized exons.

CRISPR: Clustered Regularly Interspaced Short Palindromic Repeats found in bacteria and archaea that direct the degradation of foreign DNA and RNA.

CSSR (conservative site-specific recombination): exchange of DNA at specific sites with short regions of homology. It can result in integration, excision, recombination, or inversion depending on whether the sites are on different DNA molecules or the same one, and on their relative orientations.

CTD: term used for both the carboxy (C)-terminus domain of the bacterial α subunit and the carboxyl (C)-terminus domain of the eukaryotic RNA polymerase II Rpb1 subunit. There is no structural similarity between the bacterial and eukaryotic CTDs.

C-terminus (carboxyl terminus, also called carboxy terminus): end of a polypeptide chain with an exposed carboxylic acid.

cullin: a protein found in a core subunit of some ubiquitin-protein ligases, including SCF and the APC.

cut-and-paste: a transposition reaction in which a DNA transposable element moves by being excised from a donor site, and then being integrated into a new target site to yield a simple insertion

cyclin: regulatory subunit that binds and activates a cyclin-dependent kinase (Cdk). Cyclin levels oscillate in the cell cycle and different cyclins are present at different cell cycle stages.

cyclin box: a stretch of about 100 amino acids that is conserved between cyclin proteins and that identifies these proteins as members of the cyclin family.

cyclin-dependent kinase (Cdk): protein kinase whose catalytic activity depends on an associated cyclin subunit. Cdks are key components of the cell cycle machinery.

cyclin kinase inhibitor (CKI): protein that interacts with Cdks or Cdk–cyclin complexes to block activity, usually during G1 or in response to inhibitory signals from the environment or damaged DNA.

cysteine desulfurase: an enzyme belonging to the family of transferases which transfer sulfur-containing groups from one species to another.

cytokinesis: process following the completion of chromosome segregation in mitosis by which the parent cell divides, giving rise to two daughter cells.

cytoplasm (meaning 'cell substance'): contents of the cell contained within the cell membrane and, in eukaryotic cells, outside of the nucleus.

deacetylase: enzyme that specifically removes acetyl groups from the side chains on proteins. Histone deacetylases (HDACs) represent a subgroup.

deamination: loss of an amino group. If this occurs on cytosine the base is then uracil; if it occurs on 5′-methyl cytosine the base is then thymine.

decatenation: separation and unlinking of the two DNA molecules that result from replication of a circular DNA molecule.

decoding: process of interpreting the genetic code by pairing aminoacyl-tRNAs with cognate codons; this process is facilitated by an elongation factor known as EFTu (bacteria) or eEF1A (eukaryotes).

degradosome: complex of proteins in bacteria containing the endonuclease RNase E, which initiates the degradation of many bacterial RNAs, as well as a helicase and a 3′ to 5′ exonuclease.

deletion: in relation to chromosomes, the loss of part of a chromosome or of a whole chromosome.

deoxyribonucleotide: deoxyribonucleoside with attached phosphates.

deoxyribose: five-carbon sugar with H in place of OH at the 2′ position of ribose that is a component of the nucleic acid DNA.

depurination: removal of a purine base from a nucleotide in DNA.

depyrimidation: removal of a pyrimidine base from a nucleotide in DNA.

Dicer: RNAse III-type enzyme that carries out the cleavage of double-stranded RNA to generate siRNAs.

differential interference contrast (DIC): illumination technique in microscopy that increases the contrast of the specimen. Sometimes also referred to as Nomarski microscopy, which is a particular type of DIC. This illumination highlights the edges of the specimen, generating an illusion of a three-dimensional image.

diffraction: scattering of waves (such as X-rays) in different directions when they encounter a group of atoms or molecules.

diploid: containing two sets of chromosomes.

dipole: separation of positive and negative charge.

directed gene conversion: targeting of gene conversion to a specific position by a double-strand break in DNA.

disaccharide: two sugars covalently linked by a glycosidic bond.

distributive: refers to a type of DNA replication in which the polymerase adds only a small number of nucleotides to the growing DNA strand before dissociating from the template. The template is fully copied by repeated rounds of polymerase association, elongation, dissociation, and reassociation.

disulfide bond: a covalent bond joining two sulfur atoms, which can form by a chemical reaction between the sulfhydryl groups of two cysteine side chains.

divergent evolution: evolution from a common ancestor.

D loop (DNA): a three-stranded DNA structure in which one segment of a duplex is displaced by pairing with another complementary strand.

D loop (RNA): single-stranded region of tRNA that contains conserved dihydrouridine residues and participates in critical interactions with the T-loop to stabilize the elbow region of the tRNA.

DnaB: replicative DNA in *Escherichia coli* comprising six identical subunits.

DNA-binding domain: protein domain that binds to DNA.

DNA-binding motif: protein fold that binds to DNA. It may be a discrete, globular domain or a portion of a protein domain.

DNA cloning: generation of multiple copies of a particular DNA fragment. Often refers to the insertion of a DNA region into a plasmid or virus, which can then be propagated and amplified.

DNA damage response: cellular response to extensive DNA damage, in which cell cycle progression is delayed and DNA repair is promoted.

DNA glycosylase: enzyme involved in base excision repair of DNA that removes damaged bases by hydrolyzing the linkage between the base and the sugar. Different glycosylases are specific for different types of damaged bases.

DNA helicase: protein that moves along DNA, unwinding the double helix as it goes. DNA helicases use the energy of ATP hydrolysis to move along the DNA. The helicases involved in DNA replication in bacteria and eukaryotes are six-subunit ring structures.

DNA ligase: enzyme that links the ends of two strands of DNA by catalyzing the formation of a phosphodiester bond between the 3′ hydroxyl group at the end of one fragment and the 5′ phosphate at the end of the other.

DNA methylases: another name for DNA methyltransferases.

DNA methylation: addition of a methyl group to a base in DNA. Most often cytosine is methylated, although adenine is methylated in some organisms.

DNA methyltransferases: enzymes that catalyze the addition of methyl groups to bases in DNA.

DNA microarray: also known as a chip, because of its resemblance to a computer chip, a microarray contains a series of microscopic dots, each containing oligonucleotide sequence that corresponds to a specific chromosomal locus or gene. The DNA (or cDNA) under study is labeled with a fluorophore and hybridized to the microarray. A special scanner is used to detect the fluorescence at positions in the array to which labeled DNA has hybridized.

DNA-only cut-and-paste transposable element: DNA-only transposable element that moves by DNA breakage and joining reactions in which the transposon is cut out of the donor site and then inserted into a new site.

DNA-only transposon: transposable element whose transposition involves DNA intermediates only.

DNA polymerase: enzyme that synthesizes new DNA by copying a single-stranded DNA template. The polymerase moves along the template and synthesizes a new strand of complementary DNA sequence by adding nucleotides, one at a time, to the 3′ OH end of the new strand.

DNA repair: biochemical process that reconstitutes an intact DNA molecule after DNA damage.

domain: compact unit of protein structure that is usually capable of folding stably as an independent entity in solution. Domains do not need to comprise a contiguous segment of peptide chain, although this is often the case.

dominant mutation: change in DNA sequence that has phenotypic consequences even in the presence of the wild-type gene.

donor site: the site of a transposable element before transposition.

double bond: two covalent bonds joining two atoms, the result of sharing two pairs of electrons.

double helix: the helix formed by two base-paired DNA strands that wind around one another in a helical fashion.

double-stranded RNA-binding domain (dsRBD): RNA-binding motif consisting of a fold containing alpha helices and beta sheets, which can contact the major and minor grooves of the RNA helix.

downstream: to the 3′ side of any given reference point, such as a transcription start site or a coding sequence.

duplication: in relation to chromosomes, a duplication of part of a chromosome.

dynamic instability: tendency of microtubules to switch between states of rapid growth and rapid shrinkage.

E

E site (exit site): the position on the small and large ribosomal subunits which the deacylated tRNA occupies before completely exiting the ribosome following involvement in a round of elongation.

editing: process by which an aminoacyl-tRNA synthetase catalyzes the removal from the tRNA of amino acids that have been inappropriately loaded onto a given tRNA species.

editing site: site on the aminoacyl-tRNA synthetase that catalyzes a hydrolytic reaction on misactivated or aminoacylated amino acids.

Edman degradation: method for protein sequencing in which the N-terminal amino acid residue is removed by repeated cycles of degradation, each of which generates a derivatized amino acid that can be identified by comparison to known standards.

effectors (or effector protein): proteins that execute the checkpoint pathway response.

electron density map: depiction in three dimensions of the possible locations of all the electrons in a molecule, as is produced using the method of x-ray crystallography. The electron density map shows the outlines of the molecule, which can be used to construct a three-dimensional atomic model.

electronegative: partial negative charge on an atom that results from its power to attract electrons to itself.

electron tomography: based on transmission electron microscopy, electron tomography allows the generation of 3D reconstructions of intracellular structures. The images are obtained by tilting the specimen and subjecting it to transmission electron microscopy imaging, thereby collecting images of cellular structures from many different angles.

electrophile: reagent that is attracted to centers of negative charge and can accept electrons to form a chemical bond.

electrophoretic mobility shift assay (EMSA): method for examining the interaction of a protein with a nucleic acid by comparing the mobility in an electrophoretic gel of a DNA fragment on its own and in complex with a protein.

electropositive: partial positive charge on an atom that results from its tendency to donate electrons to other atoms.

elongation (transcription): iterative process of addition of ribonucleotides complementary to the DNA template to the nascent transcript; catalyzed by RNA polymerase.

elongation (translation): iterative process of addition of an individual amino acid to the growing polypeptide chain and the advancement of the mRNA–tRNA complex in the 3′ direction; catalyzed by the ribosome.

elongation factor (transcription): protein that promotes transcription elongation over pausing or termination.

elongation factor (translation): group of proteins that collectively is important for facilitating the various steps of the elongation cycle; best known are EFTu (eEF1A) and EFG (eEF2).

embryonic stem cells: stem cells isolated from very early stage embryos.

endonuclease: enzyme that cleaves a nucleic acid internally, rather than degrading it from the ends as an exonuclease does. There are both DNA endonucleases and RNA endonucleases.

endoplasmic reticulum: organelle found in eukaryotic cells comprising a network of tubules wherein lipids are synthesized and secreted proteins are translated, folded, modified and then sent on their way.

endosymbiont: organism living within the cell or body of another organism.

endothermic: a process or reaction in which the system absorbs energy from its surrounding in the form of heat.

end-replication problem: describes the problem of fully replicating the end of a linear DNA molecule, which would lead to progressive sequence loss without the addition of telomeric DNA by telomerase.

enhanceosome: complex consisting of a variety of proteins bound to an enhancer sequence.

enhancer: DNA sequences recognized by gene-specific regulators that can function efficiently at large distances from the promoter elements, both upstream and downstream of the promoter and in either orientation.

enhancer trap: using an exogenous gene (for example, an enhance-less reporter gene with a weak promoter, placed on a transposon) to identify and monitor the activity of a putative enhancer: the expression of the reporter gene will be very low unless the transposon is inserted next to an enhancer.

enthalpy: form of energy that can be released or absorbed as heat at constant pressure.

entropy: a measure of the disorder or randomness of a molecule or system.

enzymatic terminator: terminator sequence in the DNA that requires the action of additional enzymes or other proteins to cause RNA polymerase to stop transcribing and dissociate from the DNA.

enzyme: biological molecule that catalyzes a chemical reaction.

enzyme kinetics: rates of the different steps in an enzymatic reaction.

epidermal growth factor (EGF): small peptide that is a mitogen and stimulates cell division in response to receptor binding.

epigenetics: A heritable change in gene expression and phenotype that does not come from a change in DNA sequence. Epigenetic changes often result from histone modifications or DNA methylation.

equilibrium: state in which the rate of the forward reaction and the rate of the reverse reaction in a chemical transformation (or binding reaction) are equal. At equilibrium, the relative concentrations of reactants and products no longer change, although the reaction continues to proceed in both directions.

equilibrium association constant (K_a): quantitative measure of binding affinity for the association reaction $A + B \rightleftharpoons AB$. K_a is equal to a ratio of the concentration of free components and complex at equilibrium ($[AB]/[A][B]$) or of the rate constants for the forward and reverse reactions (k_{on}/k_{off}). The higher the association constant, the higher the affinity of the interaction.

equilibrium constant (K_{eq}): ratio of the forward and reverse rate constants for a reaction; also equal to the ratio of the product of the concentrations of reaction products to the product of the concentration of the reactants. For the reaction $A + B \rightleftharpoons C + D$, $K_{eq} = [C][D]/[A][B]$.

equilibrium dissociation constant (K_d): quantitative measure of binding affinity for the dissociation reaction $AB \rightleftharpoons A + B$; K_d is equal to the inverse of the equilibrium constant for the association reaction ($1/K_a$ or $[A][B]/[AB]$ or k_{off}/k_{on}). The lower the dissociation constant, the higher the affinity of the interaction.

euchromatin: active chromatin. The DNA that is packaged as euchromatin is accessible to many DNA-binding proteins that are involved in transcription, replication, recombination, and repair.

eukaryote (meaning 'well kernel'): organism having a nuclear compartment that contains its genetic information.

exit (E) site: the position on the small and large ribosomal subunits which the deacylated tRNA occupies before completely exiting the ribosome following involvement in a round of elongation.

excision repair: DNA repair reaction in which the damaged portion on one strand is excised and then repaired by DNA synthesis using the complementary strand as a template

exon: RNA sequence that is interrupted by introns in a pre-RNA and are brought together by splicing.

exon definition: the hypothesis that exons, rather than introns, are defined by proteins that bind to the exon sequence and by interactions between the 3′ and 5′ splice site spliceosome components across the defined exon.

exonic enhancer sequence: a sequence in an exon that increases the likelihood of that exon being included in the spliced mRNA product.

exonic silencer sequence: a sequence in an exon that decreases the likelihood of that exon being included in the spliced mRNA product.

exon junction complex (EJC): group of proteins deposited at exon–exon junctions during the splicing reaction that 'marks' the transcript as processed at that site.

exon shuffling: process by which the DNA sequences encoding exons are exchange and reordered through genetic recombination between the DNA sequences encoding introns.

exonuclease: enzyme activity that cleaves nucleotides from the ends of a nucleic acid molecule. There are DNA exonucleases and RNA exonucleases.

exosome: complex of proteins in eukaryotes and archaea, related to the bacterial degradosome, and responsible for the 3′ to 5′ degradation of many RNAs.

exothermic: a process or reaction in which the system releases energy into its surrounding in the form of heat.

extein: the sequences flanking an intein, which are religated after intein excision to form the functional protein.

extended –10 region: additional bases to the 5′ side of some –10 sequences that are bound by domain 3 of the sigma subunit of bacterial RNA polymerase.

extinction coefficient: coefficient that indicates how much light is absorbed by a particular molecular species at a given wavelength.

F

fatty acid: hydrocarbon chain with a carboxylic acid at one end.

fidelity: the accuracy of synthesis (by DNA or RNA polymerases or the ribosome). All polymerases occasionally make an error and add the wrong building block during chain extension. A polymerase that makes very few errors is said to have high fidelity.

first law of thermodynamics: states that energy can be converted from one form to another, but is neither created nor destroyed.

fluorescence loss in photobleaching (FLIP): a method for examining the movement of fluorescent molecules in a cell. It is based on the repeated photobleaching of a specific locus of the cell and determining whether the fluorescence in other loci is diminished.

fluorescence recovery after photobleaching (FRAP): a method for quantifying the diffusion rate of fluorescent molecules, such as green fluorescent protein (GFP)–protein fusions, in a cell, based on the rate at which fluorescence is recovered in a particular region that was previously photobleached.

fluorescent *in situ* hybridization (FISH): method for detecting a specific DNA sequence in a chromosome *in situ*. FISH typically involves placing cells on a slide, denaturing the DNA and hybridization to a DNA probe that can be detected by fluorescence microscopy.

fold: three-dimensional shape adopted by an RNA molecule or protein.

footprinting: method for identifying interacting macromolecular surfaces, for example the region of DNA that is bound by a specific transcription factor, by comparison of the consequences of treating the DNA or the protein–DNA complex with a reagent that modifies the DNA. The reactions are typically analyzed by gel electrophoresis.

forward genetics: random mutagenesis of a population of organisms followed by isolation of organisms with a given phenotype. An example is the screening for mutant yeast that cannot synthesize uracil, with the goal of identifying the genes involved in uracil biosynthesis.

founder effect: type of genetic drift that occurs when a new population is established by a very small number of individuals.

frameshifting: the process by which a ribosome shifts along the mRNA in a forward or reverse direction – thus changing the 'frame' of reading the triplet genetic code.

FRET- Forster (or fluorescence) resonance energy transfer: this occurs when the energy of a donor fluorophore is transferred to an acceptor fluorophore, resulting in the excitation of the acceptor. Because this type of energy transfer decays rapidly as a function of distance, FRET is used to examine the physical association between proteins.

functional genomics: genetic, molecular, biochemical, and cytological studies carried out on a genome-wide scale to gain insights into gene and genome function.

furanose: five-membered ring that forms when the C-2 keto group of a sugar reacts with a C-5 hydroxyl group.

G

G0: a prolonged non-dividing state that is reached from G1 when cells are exposed to extracellular conditions that arrest cell proliferation.

G1: first gap phase of the cell cycle. It is a growth stage that follows cell division (M phase) and precedes DNA replication (S phase).

G2: second gap phase of the cell cycle. This stage follows DNA replication (S phase) and precedes mitosis and cell division (M phase).

gamete: specialized haploid reproductive cell, for example, egg and sperm in animals, which is produced through meiosis.

GATC sites: DNA sequences that are usually methylated on the A residue in both strands in bacterial DNA.

gel electrophoresis: separation of protein or nucleic acid mixtures using an electric field to promote their movement in liquid through an agarose or polyacrylamide gel.

gel filtration: chromatographic method for separating molecules based on their size, as well as shape. Also known as size exclusion chromatography.

gene: region of DNA that controls a discrete hereditary characteristic.

gene conversion: replacement of a gene sequence with another version of that gene.

gene expression: utilization of the information provided by a gene to generate a functional product.

gene family: duplicate members of a gene.

general transcription factors: proteins required for promoter recognition and transcription initiation in eukaryotes and archaea. Also known as basal transcription factors.

gene regulation: mechanisms through which the expression of a gene product is controlled in time and space.

genetic code: set of three nucleotide 'words' that allow for the translation of the DNA sequence into proteins composed of amino acids.

genetic drift: change in the frequency with which a gene occurs in a population, resulting from the fact that genes in offspring are a random sample of those in the parents, and because chance determines whether a given individual survives and reproduces.

genome: the complete DNA sequence of an organism, which serves as its instructional blueprint. It should be noted that the genome of some viruses consists of RNA.

genome annotation: the process of annotating the functional sequence elements in a genome sequence.

genome-wide association study: an examination of how a particular trait is linked to various markers throughout the genomes of many unrelated individuals.

genotype: the collective DNA sequence of an organism.

germline cells: cells in multicellular eukaryotes that are involved in the production of the progeny.

GFP (green fluorescent protein): protein normally found in the jellyfish *Aequorea victoria* that exhibits green fluorescence when excited by blue light. Derivatives of this protein have been made that fluoresce at other wavelengths.

Gibbs free energy (*G*): the total energy change resulting from a chemical process, which is related to the entropy and enthalpy changes.

global genomic repair (GGR): the basic pathway of nucleotide excision repair that can act on many different types of damage at any site in the genome.

globular: roughly spherical in shape.

glycan: a complex carbohydrate constructed from monosaccharide units in either a linear or branched arrangement.

glycolipid: lipid molecule with covalently attached carbohydrate.

glycoprotein: protein with covalently attached carbohydrate.

glycosidic bond: chemical bond that joins a carbohydrate to another group. This term is used to describe the linkage between the heterocyclic bases of nucleic acids and the ribose or deoxyribose sugars.

glycosylation: the post-translational covalent addition of carbohydrate molecules to asparagine, serine, or threonine residues on a protein molecule. Glycosylation can add a single sugar or a chain of sugars at a given site and is usually catalyzed by an enzyme.

glycosylphosphatidylinositol (GPI) anchor: complex structure involving both lipids and carbohydrate molecules that is reversibly attached to some proteins to target them to the cell membrane.

Golgi apparatus: organelle found in most eukaryotic cells that processes proteins and modifies them with sugars before packaging them for export.

gradient: solution in which the concentration of a solute varies continuously from one value to another.

green fluorescent protein (GFP): protein normally found in the jellyfish *Aequorea victoria* that exhibits green fluorescence when excited by blue light. Derivatives of this protein have been made that fluoresce at other wavelengths.

G-rich strand: the DNA strand in telomere DNA that is rich in G residues; the end of this strand forms a short single-stranded extension at the end of the telomere.

group I intron: class of introns with a conserved secondary and tertiary structure, which catalyzes two sequential transesterification reactions initiated by the 3′ OH of an exogenous guanosine nucleotide.

group II intron: class of introns with a conserved secondary and tertiary structure that catalyzes two sequential transesterification reactions initiated by the 2′ OH of an internal adenosine nucleotide (the branch site).

group II intron transposable element: a mobile element that excises by RNA splicing from an mRNA transcript and then reverse-splices into its target DNA and is converted to DNA by an element-encoded reverse transcriptase.

growth factor: extracellular factor that stimulates cell growth (an increase in cell mass). This term is sometimes used incorrectly to describe a factor that stimulates cell division, for which the correct term is mitogen.

GTPase-activating protein (GAP): proteins that interact with G-proteins (GTPases) and accelerate their rate of GTP hydrolysis.

guanine-nucleotide exchange factor (GEF): proteins that catalyze the exchange of a GDP bound to a specific G-protein (GTPase) for GTP.

guide RNA: RNAs that direct modifications of RNAs by base-pairing with the target RNAs.

H

hairpin: RNA strand that folds over on itself to form a double helix separated by a short single-stranded region referred to as a loop.

haploid: containing one copy of each chromosome.

haplotype: a block of single nucleotide polymorphisms, copy number variation, and other sequence variations that are inherited together.

haplotype block: block of DNA along a chromosome that is usually linked together within which recombination rarely occurs.

HapMap: map of all of the known haplotype blocks in the human genome. The generation of the HapMap is ongoing as more and more individual human genomes are sequenced.

head group: hydrophilic end of a lipid molecule.

helix-turn-helix: DNA-binding motif consisting of a pair of α helices, the second of which fits in the major groove of the DNA (the recognition helix).

hematopoiesis: the process by which blood cells are made.

hematopoietic stem cells: multipotent stem cells that are in the bone marrow and give rise to all other types of blood cells.

hemi-methylated: DNA sequence that is methylated on one strand. Usually a site of methylation is symmetrical and both strands are methylated at the equivalent position.

heterochromatin: tightly packed chromatin that is less accessible to nuclease digestion than euchromatin and is associated with different chromosomal proteins. Regions packaged as heterochromatin are usually silenced.

heteroduplex: duplex containing complementary DNA strands from homologous parental duplexes, but which may not themselves be identical.

hexose: six-carbon monosaccharide.

high-pressure or high-performance liquid chromatography (HPLC): form of column chromatography in which the liquid phase, usually at high pressure, is pumped through the column.

histone: the core component of the nucleosome, which packages DNA in eukaryotic chromosomes.

histone acetyltransferases (HATs): enzymes that covalently attach acetyl groups to lysine residues in histones, as well as other proteins. Also known as histone acetylases or lysine acetyltransferases.

histone deacetylases (HDACs): enzymes that remove acetyl groups from lysine residues in histones and proteins.

histone methyltransferase (HMT): enzyme that catalyzes the covalent attachment of methyl groups to lysine or arginine residues in histones.

histone modification: covalent chemical alteration of amino acid side chains in histone proteins.

histone octamer: complex of eight histone proteins made up of two copies each of the histones H2A, H2B, H3, and H4.

histone protein: any one of the highly conserved histone family of proteins that bind and package DNA in eukaryotes. The family comprises the very highly conserved proteins of the histone octamer, H2A, H2B, H3, and H4, a number of histone variants, and linker histones such as H1.

histone variants: proteins that are related in sequence and in structure to the highly core histones, and that play specific roles in chromosome function.

holocentric: chromosome that lacks a defined centromere; instead, the mitotic spindle attaches to multiple points along the length of the chromosome.

holoenzyme: bacterial RNA polymerase consisting of the core enzyme (the β, β', and two α subunits), together with a σ subunit that is capable of specific promoter recognition.

Holliday junction: DNA structure in which two homologous DNA duplexes are linked by the reciprocal exchange of two complementary strands to yield a four-stranded junction

homeodomain: eukaryotic member of the helix-turn-helix family of DNA-binding proteins.

homologous chromosomes: two chromosomes with identical, or similar, gene composition and organization, one inherited from each parent. Homologous chromosomes are very similar, but not identical, in sequence.

homologous recombination: exchange of large DNA segments between homologous chromosomes that is mediated by strand exchange recombinases such as RecA and Rad51.

homolog (gene): gene descended from a common ancestor. Two different types of homolog are orthologs and paralogs.

homology: sequence similarity reflecting descent from a common ancestor.

homology-directed repair: mechanism for accurately repairing double-strand breaks in DNA in which a homologous DNA molecule provides the template for repair synthesis.

horizontal gene transfer: insertion of a segment of the genome from one species into the genome of an unrelated species.

hybrid states model: model for the elongation cycle wherein the tRNAs move distinctly with respect to the large and small subunits of the ribosome following peptidyl transfer.

hydrogen bond: non-covalent interaction between the donor atom, which is bound to a positively polarized hydrogen atom, and the acceptor atom, which is negatively polarized.

hydrogen bond acceptor: partially negatively charged atom participating in a hydrogen bond.

hydrogen bond donor: partially positively charged hydrogen atom in a hydrogen bond.

hydrophilic: tending to interact with water. Hydrophilic molecules are polar or charged and, as a consequence, are very soluble in water.

hydrophobic: tending to avoid water. Hydrophobic molecules are nonpolar and uncharged and, as a consequence, are relatively insoluble in water.

hyperacetylated: containing many acetylated side chains.

hypoacetylated: containing few acetylated side chains.

I

identity element: the group of nucleotides in a given tRNA species that confers specificity for the appropriate aminoacyl-tRNA synthetase.

immunoblot: an analytical method in which a protein is detected by the addition of a specific antibody.

imprinting: the inactivation of a gene or group of genes on one chromosome of a homologous pair, but not the other. The imprinted gene is typically methylated and is packaged as heterochromatin.

indirect immunofluorescence: method for detecting specific cellular molecules (for example, a specific protein) by microscopy using two antibodies: a primary antibody that recognizes the molecule of interest, and a secondary antibody that is fused to a fluorescent molecule and recognizes the primary one.

individualized medicine: The concept of being able to predict reactions to certain drugs or treatments based on the genome sequence of a given person. The thought is that more effective treatments can be given if the treatment is specialized to the individual and their unique genetic makeup.

initial selection: early steps in tRNA selection including codon recognition and subsequent GTP hydrolysis.

initiation (transcription): steps by which RNA polymerase begins transcription.

initiation (translation): process by which a ribosome is loaded with the initiating methionyl-tRNA species in preparation for the elongation phase of protein synthesis.

initiation codon: AUG codon (or sometimes UUG or GUG) that is the starting point for an open reading frame that encodes a protein product.

initiation factor: any one of a number of proteins that is critical in the initiation process.

initiator protein: any of a complex of proteins that bind to origins of replication in DNA and initiate the unwinding of the double helix in preparation for DNA replication.

initiator tRNA: a distinct class of tRNAs with particular structural and functional features that allow them to be specifically deposited in the P site to decode the initiation codon.

insertion sequence (IS element): small bacterial transposable element encoding only the functions required for its transposition.

insulator: thought to be a boundary element that helps define the limits of specific forms of chromatin, including preventing the spreading of heterochromatin. An insulator inserted between a gene and a gene-regulatory element preceding it can also block activation of the gene.

intasome: elaborate nucleoprotein complexes in which CSSR integrases carry out recombination.

integral membrane protein: membrane protein in which the protein chain is largely embedded in the membrane.

integrase: protein that mediates the integration of a retroviral provirus or long terminal repeat (LTR) retrotransposon into DNA.

intein: internal portion of a protein sequence that is post-translationally excised in an autocatalytic reaction while the flanking regions are spliced together to making a new protein product; a protein intron (intervening sequence).

interactome: complete network of the protein–protein interactions in a cell type.

intercalating agent: chemical such as ethidium bromide that becomes inserted between adjacent base pairs in DNA and sometimes gives rise to mutations.

intergenic DNA: DNA that lies between coding genes on a chromosome. These regions can contain regulatory elements, as well as transposable elements and other repeated sequences.

internal ribosome entry site (IRES): RNA element found in the 5′ untranslated region (UTR) of viral (and cellular) transcripts (in eukaryotes) that identifies the AUG start site for translation initiation independent of the standard cap-dependent scanning mechanism.

interphase: period between two M phases that includes G1, S, and G2.

intrinsic or simple terminator: DNA sequence that, when transcribed into mRNA, causes RNA polymerase to terminate transcription and dissociate from the DNA template.

intron: portion of RNA that is included in the primary transcript but is removed by splicing.

intron definition: hypothesis that introns, rather than exons, are defined through interactions between sequential 5′ and 3′ splice site bound factors.

intronic enhancer sequence: a sequence in an intron that increases the likelihood that neighboring exons will be included in the spliced RNA product.

intronic silencer sequence: a sequence in an intron that decreases the likelihood that neighboring exons will be included in the spliced RNA product.

***in vitro*:** in an artificial environment, such as a test tube (literally 'in glass').

***in vivo*:** in the living organism.

ion: an atom or molecule that is positively or negatively charged.

ion exchange chromatography: a type chromatography that separates molecules based on their charge, often used during protein purification. In this type of chromatography, charged molecules bind to a resin of the opposite charge. These molecules can then be eluted using a high salt buffer, which competes for their binding to the resin.

ionized: the charged state of a previously uncharged atom or molecule.

IRES-transacting factor (ITAF): protein factor that interacts with IRESs to facilitate the non-standard translation initiation pathway in eukaryotes.

isoelectric focusing (IEF) gel: a method for separating proteins based on their isoelectric point, which is the pH at which the protein has no net charge.

ITC: isothermal titration calorimetry. A method for examining physical association between molecules and can be used to determine the binding constant, enthalpy, entropy and Gibbs free energy of an interaction. This method is often used to determine the thermal changes that result form the association of two molecules.

K

karyotype: set of condensed metaphase chromosomes in eukaryotes stained so that each chromosome can be uniquely identified. The chromosomes are usually displayed by arranging them in order from the largest, designated chromosome 1, to the smallest.

kinetochore: a complex of proteins that attaches to the centromere and mediates the attachment of the chromosome to the spindle during mitosis.

K_M: the Michaelis constant, which is equal to the substrate concentration at which the reaction rate is 1/2 of V_{max}.

Kozak sequence: the weak consensus sequence of nucleotides surrounding the AUG start site in eukaryotic genes that favors efficient start site recognition.

L

lagging strand: in the replication of DNA, the new DNA strand that is synthesized on the template strand that runs from 5′ to 3′ into a replication fork. It is synthesized discontinuously as a series of short DNA fragments that are subsequently joined together to form a continuous DNA strand.

lariat: the circular intron structure generated during the process of splicing by the spliceosome or the group II introns.

lateral elements: a protein structure that is part of the symanptonemal complex, which associates two homologous chromosomes during meiosis. The lateral element organizes the paired sister chromatids of each of the homologous chromosomes into loops; lateral elements of homologous chromosomes are connected to each other via the transverse filament.

leading strand: in the replication of DNA, the new DNA strand that is synthesized on the template strand that runs from 3′ to 5′ into a replication fork. It is synthesized continuously as the fork moves forward.

lectin: protein that binds specifically to particular polysaccharides or other carbohydrate structures.

library: collection of clones, plasmids, or viruses that contain different DNA inserts, often representing a particular cell type. This could be a genomic library representing the entire genome or a cDNA library representing the set of expressed genes.

linkage analysis: method used to find disease genes in family pedigrees when DNA samples from a number of people in several large families is available (also denoted linkage study).

Linking number: in double-stranded DNA, the number of time one strand crosses the other (denoted Lk).

lipid: term used to describe a range of molecules that are all strongly hydrophobic, including fatty acids and cholesterol.

lipid bilayer: the structure of cellular membranes, formed when the hydrophobic tails from two sheets of lipid molecules pack against each other to form the interior of the sandwich while their polar head groups cover the outside.

locus control region: eukaryotic regulatory regions containing a combination of enhancer and insulator elements.

LOD score: a score that indicates the probability that two genetic markers or loci are linked to one another, and therefore likely to be inherited together; given by the logarithm of the odds ratio for linkage.

long interspersed element (LINE): genomic sequence derived from the duplication and transposition of retrotransposons that move by target-primed reverse transcription.

loss of function: a mutation that eliminates the function of a gene

LTR elements: class of transposable elements that are characterized by LTRs and move via an RNA intermediate. They include the LTR retrotransposons and the retroviruses.

long terminal repeat (LTR) elements: direct sequence repeats of hundreds of base pairs at the ends of retroviral genomes and some types of retrotransposons.

loss of heterozygosity (LOH): loss of one allele of a gene in a cell that is heterozygous for that gene (Aa), causing the cell to become effectively homozygous for the other allele (by virtue of becoming haploid). The term is usually used in the context of genes containing deleterious mutations, when the loss of the normal copy results in effective homozygosity for a deleterious recessive mutation.

LTR retrotransposon: type of long terminal repeat (LTR) element that does not form infectious particles and thus can only move within a single cell.

lysogen: bacterial cell containing a copy of the bacteriophage chromosome integrated into the host cell chromosome.

lysogeny: type of viral growth in which the bacteriophage chromosome is integrated into the host cell chromosome.

lytic growth: type of viral growth in which the bacteriophage reproduces within the infected host cell, ending with lysis (bursting) of the host cell to release the phage particles.

M

M phase: mitosis, the cell cycle phase during which the duplicated chromosomes are segregated and packaged into daughter nuclei.

macromolecule: large molecule; it is usually also a polymer.

major groove: one of two grooves running along the helix of double-stranded DNA, in which the N7 of purine bases is exposed. In B-DNA, this is the wider of the two grooves.

marker: a unique DNA sequence that can be followed in individual people to determine if it is linked to a disease gene or other trait. Sequence tagged sites (STSs) and single nucleotide polymorphisms (SNPs) are two common kinds of DNA marker.

mass spectrometry: analytical method for determining the composition of a molecule by generating ions and determining their mass to charge ratio.

mating: in yeast, the process by which two haploid cells, one **a** and one α cell, fuse to form a diploid **a**/α cell.

mating type: cell type of a yeast cell **a** or α for the yeast, *Schizosaccharomyces cerevisiae.*

mating-type cassette: transcriptionally silent copy of the yeast mating-type genes for either a cells (MATa) or α cells (MATα).

***MAT* locus:** location in *S. cerevisiae* chromosome III of the genes that are expressed in particular cell types. **a** cells contain the **a** cassette at this location, which encodes the **a**1 protein, while α cells contain the α cassette encoding α1 and α2.

mature RNA: fully processed and functional RNA.

MCM: replicative DNA helicase in eukaryotes and archaea comprising six different proteins.

MEC1: the ATR regulator kinase of *S. cerevisiae.*

Mediator: complex composed of 22 proteins required by RNA polymerase II for appropriate transcription and response to transcriptional activators.

meiosis: specialized division that occurs during formation of the gametes through two successive rounds of chromosome segregation. The products of meiosis have half the ploidy of the parent cell.

Mendelian disease: a disease trait that is inherited in a simple fashion from parent to child, usually involving a single gene.

messenger RNA: RNA copy of DNA coding strand that conveys genetic information to the ribosome, which uses the information in the RNA base sequence to guide synthesis of a polypeptide.

metagenome: genomic sequence information for all of the organisms that occupy a particular ecological niche.

metaphase: stage in mitosis in which the sister chromatids are fully attached to the spindle and await the signal to separate.

methylation: the covalent attachment of a methyl group to a molecule. In the case of histone methylation, methyl groups are added to lysine and arginine residues. Some bases on DNA and RNA can also be methylated.

methyl transferase: enzyme that adds methyl groups to side chains on proteins (lysine or arginine methyl transferase) or to DNA bases (DNA methyl transferase).

metalloenzyme: an enzyme that contains a metal ion cofactor that often participates in the chemical reaction catalyzed by the enzyme.

Michaelis–Menten equation: equation that expresses the rate of an enzyme-catalyzed reaction as a function of substrate concentration and is characterized by constants, V_{max} and K_M.

microsatellite repeats: short repeated stretches of DNA, such as CACACACA, which are variable in length between individuals and can be used as markers in genetic linkage and association studies.

microtubule: long hollow polymer of tubulin heterodimers with two distinct ends, a plus end and a minus end, which display different polymerization behaviors.

microtubule flux: flow of tubulin within microtubules from the microtubules plus end to the spindle poles, as a result of loss of tubulin from the minus ends at the poles. It is usually accompanied by addition of tubulin at the plus ends.

minor groove: one of two grooves running along the helix of double-stranded DNA, in which the N3 of purine bases is exposed. In B-DNA, this is the narrower of the two grooves.

miRNAs (microRNA): 21–24 nucleotide RNAs, encoded by distinct genes and specifically excised from characteristic hairpin structures, which generally modulate translation or RNA degradation.

miscoding agent: small molecule antibiotic that binds in the decoding region of the small subunit of the ribosome and induces the misincorporation of amino acids during the tRNA selection process.

mismatch repair: repair of nucleotide mismatches that are generated during DNA replication.

missense mutation: single nucleotide change in the DNA sequence that changes a codon encoding a particular amino acid to one encoding a different one.

mitochondria (*singular* mitochondrion): organelles found in eukaryotic cells that contain the macromolecular machinery responsible for deriving chemical energy from food precursors (known as the power houses of the cell).

mitogen: extracellular molecule that stimulates cell proliferation.

mitosis: process by which the duplicated chromosomes, the sister chromatids, are segregated by the mitotic spindle and packaged into daughter nuclei.

mitosis (M) phase: The stage of the cell cycle when two daughter chromosomes are segregated to opposite poles prior to cell division.

mitotic spindle: A spindle-shaped structure made of microtubules that serves to physically segregate chromosomes during mitosis.

mobile group II intron: type of group II intron that, in addition to catalyzing its own excision from RNA, encodes a maturase protein with reverse transcriptase activity that enables the element to insert itself at other sites.

model organisms: organisms that are extensively studied to gain insights into biological processes that occur in many other organisms, and that are easy to manipulate under laboratory conditions.

molecule: two or more atoms joined by covalent bonds.

monogenic: phenotypes or diseases that result from variation in a single gene.

monosaccharide: simple sugar with formula $(CH_2O)_n$; the building block of carbohydrates.

motor protein: proteins that are capable of moving along microtubules powered by ATP hydrolysis. Some motor proteins have a cargo-binding domain that allows them to move certain types of cargo along microtubules.

MRN: complex in multicellular eukaryotes containing MRE11-RAD50-NBS1 that interacts with DNA and plays several important roles in the DNA damage response.

MRX: complex in *S. cerevisiae* containing Mre11-Rad50-Xrs2 that plays an important role in the DNA damage response and in DNA transaction in homology-dependent repair, homologous recombination and non-homologous end-joining.

multifactorial trait: phenotype determined by both the genetic composition of an individual and environmental influences.

multifork replication: In bacterial cells, where a new round of DNA replication begins before the chromosome is fully replicated from the previous round.

mutator phenotype: phenotype characterized by an increased mutation rate as the result of genetic defects in DNA repair processes.

N

natural selection: process by which a heritable trait that causes an organism to produce more offspring in each generation eventually becomes more common than other genetic variants after multiple generations.

near-cognate interaction: when the codon of the mRNA and the anticodon of the tRNA are only appropriately matched at two of the three positions – these tRNAs are usually rejected by the ribosome.

negative selection: situation where a new genetic variant is disfavored in a population over time (also called purifying selection).

neocentromere: new centromere that forms at any position on the chromosome other than the normal centromere position.

neutral mutation: genetic variant that is neither favored nor disfavored.

NextGen (next generation) sequencing: massively parallel sequencing methods that have the capacity to generate hundred of megabases of sequence information.

nick-and-paste: a transposition reaction which initiates with a nick at the 3′ ends of a DNA transposable element, followed by the joining of those 3′ ends to a new target site. The 5′ ends of the transposon remain linked to the donor site such that replication of the element yields a cointegrate.

nitrosylation: modification of a cysteine or tyrosine residue by addition of nitric oxide (NO). The nitrosylation of cysteine is called *S*-nitrosylation.

***N*-linked carbohydrate:** sugar covalently linked to the nitrogen of an asparagine side chain.

NMR: nuclear magnetic resonance, a property of atomic nuclei in a high magnetic field that causes them to absorb energy when reoriented by an electromagnetic pulse, and then emit energy again as they return to their original orientation (a process called relaxation). This phenomenon is exploited by the technique of NMR spectroscopy, which can be used to obtain information on the structure and dynamics of biological molecules.

non-autonomous element: transposable element lacking a transposase gene, but which can be mobilized by a transposase from another similar element.

non-canonical base pair: base pair that does not form a standard Watson–Crick base-pairing interaction, but which generally is stabilized by some number of hydrogen bonds.

non-coding strand: *see*template strand

non-coding, template, or antisense strand: DNA strand complementary to the coding strand and which is the template for synthesizing an RNA copy of the coding strand.

non-cognate: a term used to describe a lack of correspondence between an amino acid and a tRNA isoacceptor species (or a codon) that will result in faithful interpretation of the genetic code.

non-cognate interaction: when the codon of the mRNA and the anticodon of the tRNA are mismatched at two or three of the potential three positions - these tRNAs are very effectively rejected by the ribosome.

non-competitive inhibitor: molecule that lowers the apparent maximal rate (V_{max}) of an enzymatic reaction by binding to the enzyme-substrate complex.

non-covalent interactions: interactions between atoms or molecules that do not depend on formation of covalent bonds.

non-homologous end joining (NHEJ): mechanism for repairing double-stranded breaks in DNA in which the broken ends are rejoined directly, usually with the loss of nucleotides at the joint.

non-LTR retrotransposon: transposable element that has an RNA intermediate and inserts into a new insertion site via target-primed reverse transcription. LTR, long terminal repeat.

non-LTR transposon: a retrotransposon that lacks long terminal repeats and inserts into a new site by reverse transcription of an RNA copy using a 3′ OH in the target site as the primer.

non-polar: (of a bond or molecule) containing an even distribution of electric charge.

non-recombinant: (section of DNA) produced by a homologous-recombination event between two DNA duplexes in which the DNA segments flanking the position of interaction between the duplexes that initiated recombination have not been exchanged.

nonsense codon (stop codon): a nucleotide triplet sequence that specifies a stop in the translation of the genetic information; there are usually three such triplets (UAA, UAG, and UGA).

nonsense mediated decay (NMD): process in eukaryotic cells that identifies RNA transcripts containing premature termination codons and targets them for degradation (thus preventing the production of truncated proteins).

nonsense mutation: change in the DNA sequence that introduces a premature stop signal such that only truncated gene products are made.

non-specific binding: lower affinity binding that occurs at random as opposed to relatively tight binding between correct partners.

northern blot: method for analyzing RNA in which RNA is extracted from the cell, fractionated by size by gel electrophoresis, and detected by hybridization to a labeled complementary nucleic acid probe (DNA or RNA).

N-terminus: end of a polypeptide chain with an exposed amino group.

nuclear envelope: A structure that defines the boundaries of the nucleus, made of a double membrane (inner and outer nuclear membranes) perforated by nuclear pore complexes (which allow selective passage of material between the nuclear and cytoplasm). In animal cells (but not in plant or fungi) underlying the inner nuclear membrane is a network of intermediate filaments called the nuclear lamina, which is also part of the nuclear envelope.

nuclear magnetic resonance spectroscopy: commonly referred to as NMR spectroscopy. It relies on the magnetic properties of certain atoms, which transfer energy when placed in a magnetic field in a manner that depends on its neighboring atoms. Thus, this method is used to obtain high resolution information on molecular structures.

nuclear receptor protein: member of a conserved family of proteins containing characteristic DNA-binding and ligand-binding domains. These proteins regulate transcription in response to the binding of hormones and other small molecules.

nuclease sensitivity: a way of differentiating euchromatin from heterochromatin experimentally by its greater resistance to digestion by the endonuclease, DNase I.

nucleic acid: polymer consisting of nucleotides joined by phosphodiester linkages.

nucleoid: region of bacterial and archaeal cells in which the chromosomal DNA is sequestered and packaged with protein.

nucleoid occlusion: a regulatory mechanism that prevents cell division in bacteria when the nucleoid lies near the division ring.

nucleolus: The structure in the nucleus where transcription of rRNAs and assembly of the ribosomes and other RNA-containing complexes occurs.

nucleophile: a reagent that is attracted to centers of positive charge and can donate electrons to form a chemical bond.

nucleoside: base joined to ribose or deoxyribose by a glycosidic bond.

nucleosome: basic unit of DNA packaging in eukaryotes, consisting of a protein core composed of a histone octamer, with DNA wrapped around the outside.

nucleotide: base joined to ribose or deoxyribose and one or more phosphate groups. Nucleotides form the repeating unit of the nucleic-acid polymers DNA and RNA.

nucleotide excision repair: DNA repair pathway in which major structural defects, such as thymine dimers, are excised along with a stretch of 10–30 nucleotides surrounding the site of damage. The now excised strand is resynthesized using the undamaged strand as the template.

nucleus: membrane-bound compartment found in eukaryotic cells where the chromosomes are contained.

null or loss of function mutation: change in DNA sequence that eliminates the function of a gene. These types of mutation typically are associated either with a change in the product that completely eliminates function (for example, by unfolding the protein) or from a large disruption in the DNA sequence that grossly perturbs gene expression.

O

odds ratio: odds of linkage of a marker to disease in individuals who have the disease divided by the odds in controls who do not. If the odds ratio is significantly greater than one, then the marker or genetic variant is associated with the disease.

Okazaki fragment maturation: the process by which the short DNA fragments, the Okazaki fragments, which are synthesized on the lagging strand during DNA replication, have their primers removed and are joined together to make a continuous DNA strand.

Okazaki fragments: short fragments of DNA that are made during lagging strand synthesis. They are initiated with RNA primers at the replication fork, synthesized in the direction away from the direction of fork movement and subsequently joined together by DNA ligase after removal of the RNA.

oligosaccharide: polymer composed of multiple (up to around ten) monosaccharide subunits.

***O*-linked carbohydrate:** sugar covalently linked to the oxygen of a serine or threonine side chain.

oncogene: a gene whose protein product promotes cell division that leads to cancer. Generally mutations or rearrangements in the normal gene have resulted in the oncogene version of the protein that is either overactive or overproduced.

open complex: complex formed by RNA polymerase and the promoter DNA in which the two DNA strands are separated.

operator: bacterial regulatory DNA sequence recognized by a repressor protein.

operon: a set of genes with a common promoter and operator that are transcribed as a single, polycistronic mRNA.

organic: containing the element carbon.

origin of replication (*ori*): a site on a chromosome at which DNA replication can begin. It contains a DNA sequence that is recognized by proteins that unwind the DNA and initiate DNA replication.

origin recognition complex (ORC): protein complex that binds to replication origins in eukaryotes and recruits other proteins to unwind DNA and initiate DNA replication.

ortholog: related copy of a gene in the genome of different species.

oxidation: loss of electrons from an atom or molecule.

P

p53: gene regulatory protein that orchestrates the long-term DNA damage response in multicellular organisms by activating a cell cycle checkpoint and/or apoptosis in response to DNA damage or to DNA replication occurring in an incorrect phase of the cell cycle.

parallel beta sheet: beta sheet formed from non-contiguous regions of the polypeptide chain in which each strand runs in the same direction.

paralog: related copies of a gene in the same genome.

parental histone segregation: the random redistribution of the old modified histones on the parental strand onto the two new daughter strands. Half the old nucleosomes go to one daughter and half to the other.

PCNA (proliferative cell nuclear antigen): subunit of the sliding clamp protein in eukaryotes; it forms a trimer to make a symmetrical ring structure with six similar domains.

penetrance: the percentage of individuals carrying a particular mutant genotype that exhibit the mutant phenotype.

pentose: five-carbon monosaccharide.

peptide backbone: the regularly repeating atoms in a polypeptide, typically referring to the atoms NH, Cα, and C=O, and excluding the side chain R group.

peptide bond: another name for amide bond, a chemical bond formed when a carboxylic acid condenses with an amino group with the loss of a water molecule. The term peptide bond is used only when the reactive groups come from amino acids.

peptidoglycan: rigid polymer of polysaccharides joined by peptide cross-links that is a major structural element of most types of bacterial cell walls.

peptidyl (P) site: position on the small and large ribosomal subunits which the peptidyl-tRNA occupies prior to peptide bond formation.

peptidyl transferase: activity of the enzyme responsible for catalyzing the formation of a peptide bond between the growing peptide chain and the incoming aminoacyl-tRNA.

peripheral membrane protein: protein associated with membranes but which does not contain a complete transmembrane region.

pH: negative log (base 10) of the H^+ ion concentration.

phage: virus that infects bacteria

pharmacogenomics: the use of an individual's genetic and genomic information to predict their response to specific drugs.

phase contrast: illumination technique in microscopy that amplifies small changes in the phase of light waves as they pass through a specimen, effectively increasing contrast.

phase variation: alternative expression of a cell surface protein on a microbe that facilitates evasion of recognition by the immune system.

phenotype: visual features and properties of an organism.

phospholipid: lipid containing a phosphate group on the head group.

phosphorylation: covalent attachment of a phosphate group to a molecule. In the case of proteins, the phosphate is added to a serine, threonine, tyrosine, or histidine residue.

photolyase: a DNA repair enzyme present in organisms other than placental mammals, which splits pyrimidine dimers formed in DNA by UV light.

phragmoplast: an organelle in a dividing plant cell on which the new cell membranes and cell walls between the two daughter cells are constructed.

phylogenetic analysis: a method for predicting interacting regions in RNA molecules by comparing homologous RNA molecules from different species and identifying covariations indicative of conserved interactions.

phylogenetic tree: representation in a quantitative branched model of the extent of relatedness between various species.

PIKK: ATM, ATR and DNA-PK are phosphoinositide-3-kinase-related protein kinases.

piRNAs (Piwi-interacting RNAs): 26–31 nucleotide long RNAs that associate with the Piwi subfamily of Argonaute proteins in RISC complexes and primarily act to silence transposons in the germlines of animals.

pK_a: negative log (base 10) of the acid dissociation constant (K_a) for an ionizable group.

plasma membrane: membrane that encloses a cell.

plasmid: DNA molecule, usually circular and containing an origin or replication, distinct from the chromosomal genome of the host.

ploidy: number of homologous sets of chromosomes in an organism.

P loop motif: weak amino acid consensus motif found in a variety of NTP binding proteins; gross conformational changes are observed in this motif following triphosphate hydrolysis.

pluripotent cells: stem cells that can give rise to a limited set of cell lineages.

point centromere: type of centromere that covers a very small region of chromosome (~100 bp of DNA).

point mutation: a mutation that changes a single base pair

polar: bond or molecule containing an uneven distribution of electric charge, with positive charge at one end and negative at the other.

poly(A) tail (polyadenosine tail): tail of approximately 200 adenosine residues added to the 3′ ends of translated mRNAs.

polyadenosine (poly(A)) tail: tail of approximately 200 adenosine residues added to the 3′ ends of translated mRNAs.

polyadenylation: process by which a poly(A) tail is added to mRNAs.

polyadenylation site: specific sequences responsible for recruiting the polyadenylation machinery.

polycistronic message: RNA transcript containing the coding sequences for several genes; typically found in bacteria.

polygenic: phenotype or disease in which multiple independent genes play a role.

polymer: molecule composed of many copies of a simple building block covalently linked to one another.

polymerase chain reaction (PCR): method to amplify a specific region of DNA with known sequence primers. The DNA is denatured and copied using primers and a DNA polymerase. Many cycles of denaturation and primer extension are carried out resulting in an exponential amplification of the sequence between the primers.

polymerase III holoenzyme: complex of proteins that carries out the elongation phase of DNA replication in bacteria, moving along with the replication fork and synthesizing new DNA on both template strands simultaneously. It is composed of two molecules of DNA polymerase III and the clamp loader with a bound sliding clamp. Also called the replisome.

polymerase switching: in eukaryotes, the switch that occurs from DNA synthesis by the polymerase α-primase complex to DNA synthesis by the processive replicative polymerase that carries out the bulk of DNA elongation.

polymorphism: a site in a genome that has more than one common sequence in a population.

polypeptide: a polymer of amino acids joined together by peptide bonds.

polyploid: Having two homologous copies of all chromosomes. One homolog is inherited from each parent.

polyploidy: expansion in the number of genome copies.

polyprotein: expressed polypeptide containing several discrete protein domains that must be separated by proteolytic cleavage before they become active.

polypyrimidine tract: pyrimidine-rich sequence found just upstream of the 3′ splice site recognized by U2AF65 during the early steps of splicing.

polysaccharide: large carbohydrate composed of a large number of sugars; can be a linear or branched polymer.

polytene chromosome: giant chromosomes that are formed by multiple rounds of DNA replication without separation of sister chromatids.

position-effect variegation: variation in expression of a gene as a result of changes in its relative position on the chromosome. The variegation refers to the visible effects of different degrees of silencing of a given gene by encroachment of heterochromatin when the gene is moved into a region adjacent to heterochromatin.

positive cooperativity: cooperative interaction that makes binding of more than one ligand energetically more favorable.

positive selection: situation where a new genetic variant is favored in a population over time.

post-transcriptional: refers to events that take place following the process known as transcription.

post-transcriptional gene silencing (PTGS): mechanism by which short fragments of double-stranded RNA lead to mRNA degradation and thus lead to the silencing of RNA derived from foreign genes.

post-translational modifications: changes such as cleavage, or the addition of lipids, sugars or small molecules, to a polypeptide chain that occur after protein synthesis.

postsynaptic: the steps after the formation of heteroduplex DNA during strand exchange, mediated by RecA-like recombinases

precursor RNA (pre-RNA): full-length unprocessed RNA transcript synthesized by RNA polymerase.

pre-initiation complex: complex formed by the general transcription factors and the RNA polymerase core enzyme, which is capable of transcription initiation.

premature termination codon (PTC): stop (nonsense) codon that is found in the transcript 5′ to the authentic stop codon for the gene.

prenylation: the irreversible attachment of either a farnesyl or geranylgeranyl group to a protein via a thioether linkage.

prereplication complex (preRC): large complex of proteins, including the origin recognition complex and its associated proteins, which assembles at replication origins in late mitosis and G1 and is activated to initiate DNA replication at the beginning of S phase.

presynaptic: the steps before the formation of heteroduplex DNA during strand exchange, mediated by RecA-like recombinases

presynaptic filament: nucleoprotein complex formed by the binding of RecA/Rad51-like protein to single-stranded DNA and that can initiate a search for homology in duplex DNA.

primary cells: cells derived from differentiated tissue.

primary sigma factor: sigma factor that is present in all normal growth conditions and associates with the core RNA polymerase to bind the promoters of many highly expressed genes. Each bacterial species has only one primary sigma factor.

primary structure: the amino-acid sequence of a polypeptide chain.

primase: specialized RNA polymerase that synthesizes a short stretch of primer RNA on the template at the beginning of DNA replication, thus generating a 3′ end to which nucleotides can be added by DNA polymerases.

primer: in DNA synthesis, a short stretch of RNA synthesized by a primase on a DNA template, and from which the new DNA strand is elongated by DNA polymerase.

processivity: ability of a DNA polymerase to remain associated with the template while synthesizing DNA. A highly processive enzyme will stay bound to the template and polymerize thousands of nucleotides before it dissociates.

progenote: a putative common original life form from which the entire diversity of modern-day life derives, often referred to as the last universal common ancestor (LUCA).

prokaryote: (meaning 'before the kernel' or pre-nuclear) unicellular species, including members of the bacterial and the archaeal kingdoms, which do not have internal membranes within their cells.

prometaphase: second stage of mitosis, in which the nuclear envelope breaks down and the sister chromatids become attached to the spindle.

promoter: DNA sequence needed for RNA polymerase to bind to the DNA and initiate transcription.

promoter clearance (promoter escape): step in transcription in which RNA polymerase moves past the promoter and enters into productive elongation

promoter escape (promoter clearance): step in transcription in which RNA polymerase moves past the promoter and enters into productive elongation.

promoter proximal pausing: pausing of RNA polymerase II somewhat downstream of the promoter after transcribing around 50 base pairs.

proofreading (DNA synthesis): when DNA polymerase recognizes that an incorrect nucleotide has been added and resynthesizes the region following the targeted action of an associated exonuclease function.

proofreading (protein synthesis): a phase in the tRNA selection process (a requisite step in elongation) that follows GTP hydrolysis and includes the alternative pathways of accommodation and rejection.

prophage: copy of the phage chromosome that is integrated into the bacterial host cell chromosome.

prophase: first stage of mitosis, when chromosome condensation, centrosome separation, and spindle assembly begin.

proteasome: large multisubunit enzyme complex that degrades proteins tagged with polyubiquitin chains into short peptides.

protein family: proteins that are related in structure or amino acid sequence and thus are presumably descended from a common ancestor.

protein fold: overall structure of a protein, as defined by its shape, secondary structure elements and topology.

protein folding: process by which a polypeptide chain collapses to form a compact, folded structure.

protein kinase: enzyme that adds phosphate groups to side chains on proteins.

protein phosphatase: enzyme that removes phosphate groups from phosphorylated serines, threonines or tyrosines on proteins.

proteomics: the study of proteins, protein complexes and protein interactions that is aided by having complete genome sequence of all predicted proteins.

protofilament: a protein structure formed by the head to tail attachments of tubulin dimers. Microtubules are made of 13 protofilaments that are aligned side by side in a cylindrical arrangement.

provirus: DNA copy of a retrovirus genome integrated into the cellular DNA.

pseudogene: gene that has a sequence similar to an active gene, but that is not expressed or does not encode a functional product.

pseudoknot: RNA structure minimally composed of two helical segments connected by single-stranded regions or loops; typically, the bases in the loop of a hairpin form another stem through interactions with a distinct set of bases.

pseudouridylation: conversion of uridine to pseudouridine.

P site (peptidyl site): position on the small and large ribosomal subunits which the peptidyl-tRNA occupies prior to peptide bond formation.

purifying selection: *See* **negative selection**.

purine: nitrogenous base found in nucleosides and nucleotides and consisting of a double ring structure.

pyranose: six-membered ring formed when the C-1 aldehyde of a sugar is attacked by C-5 hydroxyl.

pyrimidine: nitrogenous base found in nucleosides and nucleotides and consisting of single ring structure.

pyrimidine dimer: structure formed by the covalent linkage of two adjacent pyrimidine bases in a DNA strand, typically caused by UV light. It cannot be replicated correctly and thus causes a mutation.

pyrophosphate: two covalently linked inorganic phosphate groups, also denoted PP_i.

qRT-PCR: stands for quantitate reverse transcription couples with polymerase chain reaction. Also referred to as "real time" PCR. Typically used in the quantification of the amount of a particular RNA. This reaction can be followed in real time through the incorporation of the fluorescent nucleotide into the PCR product.

quaternary structure: subunit structure of a protein.

quiescence: a cell cycle phase, also referred to as G0, where a cell exits the cell cycle and stops dividing.

R

***rad* 3^{+}:** the ATR regulator kinase of *S. pombe*.

Rad51: eukaryotic strand-exchange recombinase.

***ram* (ribosomal ambiguity):** term for a type of ribosome mutation that results in decreased fidelity during protein synthesis.

Ramachandran plot: a two-dimensional plot of the values of the backbone torsion angles phi (φ) and psi (ψ), with allowed regions indicated for conformations where there is no steric interference between neighboring atoms.

rasiRNA: Repeat associated small interfering RNAs that down-regulate transcription from repetitive regions of the genome such as centromeres and transposon-rich regions.

rate constant: proportionality constant that describes how the rate of a reaction varies as a function of reactant concentrations.

reaction intermediate: chemical species that forms transiently in the course of a chemical reaction.

RecA: bacterial strand-exchange recombinase.

recessive mutation: change in DNA sequence that is compensated at the phenotypic level by the wild-type version of the gene.

recoding: process by which the genetic code is reinterpreted to allow for the encoding of different information.

recognition helix: the α helix in DNA-binding proteins that is used to recognize specific DNA sequences by forming side chain contacts with base pairs.

recombinant: DNA produced by a homologous-recombination event between two parental DNA duplexes in which the DNA segments flanking the position of interaction between the duplexes that initiated recombination have been exchanged.

recombinant DNA: DNA molecule that does not occur naturally but is a product of artificial manipulation, such as insertion of DNA sequences into plasmid or virus DNA.

recombinase: enzyme that recognizes specific DNA sequences and catalyzes breakage and joining reactions that result in their rearrangement.

recombination crossover: in meiosis, a homologous recombination event that results in the reciprocal exchange of DNA between two homologs.

redox reaction: reaction in which both oxidation and reduction occur.

reduction: gain of electrons by an atom or molecule.

regional centromere: a type of centromere that covers a large area of the chromosome (from hundreds to many thousands of base pairs of DNA).

regulator kinase: the ATM or ATR kinases that signal the presence of DNA damage

regulatory sequence: DNA sequence to which one or more transcriptional regulators bind and regulate a gene or set of genes.

rejection: the final step during the tRNA selection process by which an aminoacyl-tRNA can be discarded based on inherent binding stabilities for the ribosome.

relaxed DNA: DNA that has no supercoils.

replication: the process wherein a complementary copy of each strand of DNA is generated by DNA polymerase.

replication bubble: open structure formed in DNA undergoing bidirectional replication from an origin of replication. The bubble is generated after the DNA at the replication origin has been unwound and the two replication forks at either end move away from each other.

replication complex: large complex of proteins assembled at the eukaryotic replication fork and functionally equivalent to the bacterial polymerase III holoenzyme.

replication-dependent histone (S phase histone): histones H2A, H2B, H3, and H4, which are synthesized in large amounts at the same time as DNA is synthesized (during the S phase of the cell cycle) so that new nucleosomes can be assembled on the new DNA.

replication factor C (RFC): clamp loader in eukaryotes consisting of five different protein subunits.

replication fork: site at which DNA strands are separated and new DNA is synthesized. It is a Y-shaped structure and moves away from the site of replication initiation. Both strands of the DNA are copied at the replication fork.

replication fork barrier: site in eukaryotic DNA that prevents the progression of a replication fork coming from one direction by means of proteins that bind specifically to the site.

replication protein A (RPA): the single-stranded DNA-binding protein in eukaryotes which is a trimer of three proteins Rpal, Rpa2, and Rpa3.

replication restart: resumption of DNA replication at a site where the fork has collapsed because of a DNA lesion.

replicative transposition: transposition pathway in which DNA-only transposons are copied during transposition to form a cointegrate.

replisome: alternative name for the polymerase III holoenzyme.

reporter constructs: genetic elements introduced into various cell types to "report" on a given biological process. Often, these reporters encode proteins with easily visualized properties to allow for rapid and direct analysis of the biological activities of the cell (for example, fluorescent or luminescent proteins).

reporter gene: gene that is used to report on the expression of another gene by making a transcription fusion. The reporter gene product may produce a change in color or fluorescence, as in the case of green fluorescent protein (GFP), have an enzymatic activity that is easily detected, or may be detected by existing reagents such as nucleic acid probes or antibodies.

repressor: sequence-specific DNA-binding protein that leads to decreased transcription.

rescue (in microtubules): sudden shift from shrinkage to growth that occurs when a GTP cap forms at the microtubule end.

resolution: parameter that governs the limit of accuracy of a molecular structure and determines how well individual atoms can be distinguished from one another; related to how a crystal diffracts x-rays.

resolution (of Holliday junctions): the symmetric cleavage of a Holliday junction to yield two unconnected product duplexes

resolvase: an endonuclease that symmetrically cleaves Holliday junctions to yield two unconnected product duplexes

resonance: delocalization of bonding electrons over more than one chemical bond in a molecule. Resonance greatly increases the stability of a molecule. It can be represented, conceptually, as if the properties of the molecule were an average of several structures in which the chemical bonds differ.

response regulator: protein that transduces a specific response on phosphate transfer from a sensor kinase to a specific aspartate residue.

restriction endonuclease: enzyme that recognizes a specific sequence in DNA and cleaves both DNA strands at defined locations.

restriction point: term used in mammalian cells to describe the same regulatory points as Start. It is the point in the cell cycle beyond which cells are committed to complete the cell cycle.

restrictive: term for a type of ribosome mutation that results in high-fidelity protein synthesis.

retrohoming: the movement of a Group II mobile intron into a highly homologous target site by reverse splicing

retrotransposon: transposable element whose transposition involves an RNA intermediate and a reverse transcription step.

retrovirus: RNA virus whose genome is reverse transcribed into DNA and integrated into the cell's DNA as part of the viral life cycle. The viral genome is characterized by long terminal repeats (LTRs) similar to those present in the LTR retrotransposons.

reverse genetics: analysis of the consequences to the organism of directly inactivating a particular gene.

reverse transcriptase: DNA polymerase that synthesizes DNA from an RNA template.

Rho-dependent terminator: terminator sequence in bacteria that requires the additional action of Rho protein for transcription termination.

Rho-independent terminator: intrinsic terminator sequence in bacteria that does not require the Rho protein for transcription termination.

ribonuclease: nuclease enzymes that act on RNA.

ribonucleoprotein (RNP): complex of RNAs and proteins.

ribonucleoprotein (RNP) domain: *See* **RNA-recognition motif (RRM) domain.**

ribonucleoside: base covalently linked to ribose via a glycosidic bond with the C1′ atom of the sugar.

ribonucleotide: ribonucleoside with one or more attached phosphates.

ribose: five-carbon sugar that is a component of the nucleic acid RNA.

ribose 2′-*O*-methylation: methylation of the 2′ oxygen of ribose.

ribosome profiling: a method used to identify the RNA molecules that are associated with ribosomes under a particular condition. This method is useful in identifying *all* the RNAs that are being translated under that condition (including RNAs of genes that have yet to be annotated) and for studying the roles of translation-associated proteins in regulating translation and ribosome progression.

ribosomal ambiguity (*ram*): term for a type of ribosome mutation that results in decreased fidelity during protein synthesis.

ribosomal protein (r-protein): one of the ~55 (bacteria) or ~80 (eukaryotes) different types of protein present in the corresponding ribosome.

ribosomal RNA (rRNA): one of the three (bacteria) or four (eukaryotes) species of RNA present in the corresponding ribosome.

ribosome: macromolecular ribonucleoprotein machine responsible for the translation of the genetic information in the mRNA into the encoded polypeptide.

ribosome recycling: process by which the small and large subunits of the ribosome are separated and the tRNAs and mRNA released after termination of translation at stop codons in preparation for the next cycle of protein synthesis.

riboswitch: RNA element typically found in the 5′ untranslated region (UTR) of transcripts that, on binding to small molecule metabolites, change their conformation in a way that affects transcription or translation.

ribozyme: enzyme composed of RNA.

RISC (RNA-induced silencing complex): molecular complex which through the specificity of bound siRNAs acts as the effector for RNA interference or silencing.

RNA-binding domain (RBD): *See* **RNA-recognition motif (RRM) domain.**

RNA-binding motif: protein fold that binds to RNA.

RNA editing: process by which individual nucleotides in an RNA are modified, removed, or inserted.

RNA half-life: time in which half of the initial amount of an RNA is degraded.

RNA interference (RNAi): mechanism by which short fragments of double-stranded RNA lead to mRNA degradation.

RNA polymerase: the molecular machinery responsible for copying the DNA into RNA.

RNA processing: modifications that can occur to RNA including cleavage, chemical modification of individual nucleotides, as well as the addition and deletion of nucleotides.

RNA-recognition motif (RRM) domain: RNA-binding motif consisting of an alpha-beta sandwich structure in which the beta strands form a platform for RNA recognition. Also called RNA-binding domain (RBD) or ribonucleoprotein (RNP) domain.

RNA splicing: process by which introns are excised from a pre-RNA.

S

***S*-acylation:** reversible attachment of a fatty-acid group to a protein via a thioester linkage; palmitoylation is an example of *S*-acylation.

salt bridge: a non-covalent interaction in which both donor and acceptor atoms are fully charged.

saturated (fatty acid): containing no double bonds in the hydrocarbon chain.

scanning electron microscopy: a type of electron microscopy that uses an electron beam to scan the surface of structures in the sample. The result is a 3D-like image that shows the outline of the structure being examined.

SDS-PAGE: sodium dodecyl sulfate – polyacrylamide gel electrophoresis, a method for separating proteins based primarily on their molecular weight by denaturing proteins in the presence of SDS and then comparing their electrophoretic mobility in a polyacrylamide gel.

secondary structure: folded segments of a polypeptide chain with repeating, characteristic phi (φ) and psi (ψ) backbone torsion angles, that are stabilized by a regular pattern of hydrogen bonds between the peptide –N–H and C=O groups of different residues.

second law of thermodynamics: states that the entropy of an isolated system can only increase.

securin: inhibitor of separase protease activity. Securin is inactivated by degradation following ubiquitination by the anaphase-promoting complex (APC).

segregation: process through which the replicated DNA is divided into two chromosome clusters, in order for progeny cells to obtain a full complement of chromosomes.

selectable marker: a gene that can be used to provide a growth advantage to cells that express it. For example, a gene coding for a certain antibiotic resistance allows only cells that express it to grow in the presence of that antibiotic drug.

selenocysteine insertion sequence (SECIS): RNA sequence element that specifies the insertion of selenocysteine at UGA stop codons.

self-splicing intron: intron that is able to catalyze its own excision from the primary RNA transcript in the absence of any protein cofactors; such an intron is typically referred to as a ribozyme.

semi-conservative: a term applied to DNA replication to indicate that a new DNA double helix has one old strand of DNA and one new strand. Half of the parental DNA is conserved in each daughter helix.

senescence: when cells lose the ability to divide.

sense codon: a nucleotide triplet sequence that specifies a particular amino acid.

sensor kinase: protein that detects a signal which causes the protein to autophosphorylate a histidine residue.

sensors (or sensor protein): proteins that recognize a cellular lesion, such as breaks in DNA, and activate a checkpoint pathway, for example during the DNA damage response

separase: protease necessary for sister chromatid separation. Separase cleaves the kleisin subunit of cohesin.

sequence tagged site (STS): site that is uniquely identified in the human genome and for which PCR primer sequence are available to allow detection of this site.

sex chromosome: a chromosome that determines sex in a given species. The number of sex chromosomes often varies between males and females. In humans, females have two X chromosomes and males have one X and one Y chromosome.

Shine–Dalgarno sequence (ribosome binding site): polypurine sequence in the upstream leader of an mRNA that base pairs with a polypyrimidine tract in the 16S rRNA to help in the identification of the start site of the open reading frame.

short interspersed nuclear element (SINE): genomic sequence derived from small cellular RNAs such as tRNAs or 7SL RNA that spreads non-autonomously in the genome by target-primed reverse transcription mediated by LINE-encoded recombination proteins.

sigma factor: bacterial RNA polymerase subunit required for promoter recognition.

sigmoidal curve: s-shaped binding curve characteristic of cooperative binding interactions.

signal joint: fusion between the ends of the nonamer and heptamer bounded segments that are recognized by RAG and excised when the V, D, and J coding segments joined during RAG-mediated V(D)J recombination.

silencer sequence: *see* **intronic enhancer** sequence and **exonic enhancer sequence**.

silencing: establishment of a heritable state of chromatin – heterochromatin – in which genes are not transcribed, origins of replication are inactivated or replicate late, and recombination does not take place.

silent mutation: change in the DNA sequence that does not result in obvious phenotypic change.

simple sequence length polymorphism (SSLP): sites in the DNA where there are repeated sequences, usually di- or tri- nucleotide repeats, in which the number of repeats at a given site varies from person to person. Thus a region of the genome can be marked and followed by determining which version of the SSLP is present in association with disease.

simple terminator (or intrinsic terminator): DNA sequence that, when transcribed into mRNA, causes RNA polymerase to terminate transcription and dissociate from the DNA template.

single bond: one covalent bond joining two atoms, the result of sharing a pair of electrons.

single nucleotide polymorphism (SNP): DNA sequence variation where a single nucleotide at a specific position in the genome differs between members of a species.

single-stranded binding protein (SSB protein): protein that binds specifically to single-stranded DNA that is exposed when the double helix is unwound or damaged.

siRNAs (small interfering RNA): 21–24 nucleotide RNAs typically generated through non-specific excision from double-stranded RNA of either external or endogenous origin, which direct RNA interference and transcriptional gene silencing.

sister chromatid: one of the two daughter chromosomes generated after DNA replication. The two chromatids contain identical DNA sequences and remain attached to each other until chromosome segregation takes place.

sister chromatid cohesion: linkages that hold sister chromatids together between S phase and anaphase.

site-directed mutagenesis: generation of one or more mutations at a particular location in a gene.

size exclusion chromatography: method of chromatography that separates molecules based on size, as well as shape.

sliding clamp: ring-shaped protein that moves along the DNA with the DNA polymerase in DNA replication and thus increases its processivity. The ring is made of two copies of the β protein in *Escherichia coli* and three copies of PCNA in eukaryotes.

slipped-strand misreplication: errors in DNA replication that can occur in repetitive sequence regions, when intrastrand base-pairing of either the template strand or newly synthesized strand causes the replication machinery to slip on the template.

SL RNA: short, defined RNA sequences carrying a methylated cap structure, which are spliced onto the 5′ end of distinct transcripts in a *trans*-splicing reaction.

small nuclear ribonucleoprotein (snRNP): small nuclear ribonucleoprotein; ribonucleoprotein complexes composed of an snRNA and a variety of proteins that assemble to form the spliceosome.

small nuclear RNA (snRNA): RNA component of a snRNP.

small nucleolar RNA (snoRNA): small nucleolar RNA that guides ribose methylation and pseudouridylation.

snoRNP: snoRNA-containing RNP that catalyzes ribose methylation and pseudouridylation.

snRNA: RNA component of a snRNP.

snRNP: small nuclear ribonucleoprotein; ribonucleoprotein complexes composed of an snRNA and a variety of proteins that assemble to form the spliceosome.

Solexa sequencing: massively parallel sequencing technique in which individual sequence reads can be up to 40 bp in length.

somatic cells: cells in multicellular eukaryotes that constitute the building blocks of the body of an organism (such as liver cells) but are not involved in the production of progeny.

sorting motif: sequence identifiers within proteins that target the proteins to specific subcellular compartments (also denoted sorting signals or signal sequences in some cases).

SOS response: response of bacterial cells to maintain viability in cases of extensive DNA damage.

Southern blot analysis: method for detecting specific DNA sequences. The DNA fragments are typically separated on a gel, denatured, transferred to a filter, and probed by hybridization with a radioactive or otherwise labeled complementary DNA fragment.

species: reproductively isolated group of organisms.

specific binding: relatively tight binding between correct partners as opposed to lower affinity 'non-specific binding' that occurs at random.

spectral karyotyping (SKY): a method for detecting abnormalities in chromosome structure, by hybridizing chromosome-specific probes, each with a different fluorescent signature.

spectrophotometer: instrument for measuring the amount of light absorbed by a sample.

S phase: the stage of the cell cycle during which the DNA is replicated, resulting in chromosome duplication.

spindle checkpoint pathway: also known as the spindle assembly checkpoint pathway. A regulatory mechanism that inhibits progression through mitosis, by inactivating the Anaphase Promoting Complex, until all chromosomes have formed proper attachments (bi-orientation) to the spindle.

standard free energy change: the change in free energy that a system undergoes (at standard state) that expresses the thermodynamic potential of the system.

standard state: a reference point for a given material used to calculate its properties under different conditions; for compounds in solution, a 1 M concentration is standard state while for gases, 1 atmosphere of pressure is standard state.

start: major regulatory transition at the entry into the cell cycle in mid to late G1, also called the restriction point (in animal cells). It is the point in the cell cycle beyond which cells are committed to complete the cell cycle.

stem-loop: structure that forms in a single strand of RNA (and sometimes DNA) when two complementary regions base pair with one another, separated by several unpaired nucleotides.

stem cells: cells that have an unlimited capacity to divide and can give rise to many different lineages.

stop codon (nonsense codon): a nucleotide triplet sequence that specifies a stop in the translation of the genetic information; there are usually three such triplets (UAA, UAG, and UGA).

strand-exchange recombinase: recombination protein that identifies regions of homology shared between a single-stranded DNA and double-stranded DNA and promotes the formation of heteroduplex DNA containing the parental single strand and its complement in the DNA duplex.

strand invasion: the process by which a single-stranded DNA pairs with its complementary strand in a homologous duplex

structural maintenance of chromatin (SMC) protein: structural protein involved in cohesion, condensation, and DNA repair.

substrate: molecule that is transformed in a reaction.

sugar–phosphate backbone: the regularly repeating part of the DNA polymer, which consists of deoxyribose sugar joined by phosphodiester linkages between the 3′ oxygen of one sugar and the 5′ oxygen of the next.

sugar pucker: conformation assumed by the non-planar sugar in nucleic acid (ribose in RNA and deoxyribose in DNA).

sulfhydryl: –SH group, which is found in the side chain of cysteine. Also known as a thiol.

sumoylation: attachment of SUMO (small ubiquitin-like modifier) to a protein.

supercoil: in the case of DNA, the axis of the double helix follows a helical path. Supercoiled DNA is under tension.

synapsis: recognition of regions of homology between a single strand of one DNA molecule and an intact duplex, usually promoted by a strand exchange protein.

synaptonemal complex: protein structure that links a pair of homologous chromosomes along their length in early meiosis.

synonymous changes: changes to a coding sequence that do not change the encoded amino acid due to the redundancy of the genetic code.

syntenic blocks: conserved regions that contain the same genes in the same order between chromosomes of different species.

synthesis-dependent strand annealing (SDSA): a pathway for double-stranded DNA break repair using non-conservative DNA replication in which the newly synthesized DNA that restores a break is displaced from the template duplex and transferred to the broken duplex.

synthetic lethal: situation when the combination of two gene disruptions or mutations is lethal while the individual disruptions or mutations do not affect viability.

T

TAF (TBP-associated factor): polypeptide that associates with TBP to form the TFIID complex and includes proteins that are required for activation of transcription.

TagSNP: specific single nucleotide polymorphism that uniquely identifies a **haplotype block**.

T antigen: protein encoded by the mammalian SV40 virus. SV40 is a tumor-causing virus and T antigen stands for tumor antigen. T antigen binds the origin of replication in viral DNA to initiate replication and also serves as the helicase that unwinds the DNA ahead of the replication fork as well as a transcription factor.

target-primed reverse transcription (TPRT): insertion mechanism by which a DNA copy of a RNA copy of a non-long terminal repeat (LTR) element is copied *in situ* at the new insertion site.

target site: the site in DNA into which a moving transposable element is inserted.

target-site duplication: short direct duplication of the target-site sequence that is generated on either side of an inserted transposon by the insertion process.

TATA-binding protein (TBP): present in the TFIID fraction. It binds to the TATA box and directs transcription.

TATA box: AT-rich core promoter element located approximately 25–30 nucleotides upstream of the transcription start of many archaeal and eukaryotic promoters. It is bound by the general transcription factor TATA-binding protein (TBP).

tautomer: structural isomer of a molecule that can change spontaneously from one form to the other, usually by a change in position of a particular hydrogen atom or proton, and exists in equilibrium with alternative isomers.

tautomerism: the ability of a molecule to adopt isomers that differ only in the position of a hydrogen atom and a double bond.

TBP (TATA-binding protein): present in the TFIID fraction. It binds to the TATA box and directs transcription.

TBP-associated factor (TAF): polypeptide that associates with TBP to form the TFIID complex and includes proteins that are required for activation of transcription.

***tel 1$^+$*:** the ATM regulator kinase of *S. pombe*.

TEL1: the ATM regulator kinase of *S. cerevisiae*.

telomerase: enzyme that synthesizes the terminal telomere repeats on chromosome ends.

telomere: specialized structure at the end of a eukaryotic chromosome that enables the DNA to be fully replicated and also maintains the integrity of the chromosome. It contains repetitive DNA sequences packaged into heterochromatin.

telophase: final stage of mitosis, in which the spindle is disassembled, the chromosomes decondense, and the nuclear envelope re-forms.

template: parental nucleic acid strand that is copied by a polymerase, for example during replication or transcription. Each base is read by the appropriate polymerase and the complementary nucleotide is added to the new nucleic acid strand. For example, a G in the template specifies that a C be added to the new strand.

template strand: the DNA strand that is used as a template to synthesize an RNA copy of the coding strand

termination (transcription): steps by which RNA polymerase stops transcription.

termination (translation): process by which a stop codon in the mRNA is recognized and the growing polypeptide chain is released from the ribosome.

terminator: DNA sequence that signals RNA polymerase to stop transcription and be released from the DNA along with the RNA transcript.

terminus region: region on the *Escherichia coli* chromosome at which DNA replication terminates and the two interlinked daughter DNAs are resolved.

tertiary structure: folded conformation of a protein.

tetraploid: containing four complete sets of chromosomes.

thin-layer chromatography (TLC): type of chromatography in which a sheet of glass or plastic is coated with an absorbent material (such as polyethyleneimine or silica) and a solvent containing a mixture of molecules moves across the plate by capillary action, separating the components of the mixture based on their varying interactions with the absorbent material.

three domains of life: three distinct groups (the bacteria, archaea, and eukaryotes), or branches, into which all organisms can be divided.

TIR: the terminal inverted repeats at the ends of transposons that interact with transposase.

t-loop (telomere): structure formed at human telomeres when the G-rich single-stranded region loops around and base-pairs with the nearby double-stranded telomeric DNA.

T loop (tRNA): single-stranded region of tRNA that contains ribothymidine and participates in critical interactions with the D loop to stabilize the elbow region of the tRNA.

tmRNA: bacterial RNA species known to function both as a tRNA and an mRNA to recover stalled ribosomes.

topoisomerase: enzyme that changes the supercoiling of DNA.

TOR: protein kinase that regulates cell growth through the stimulation of protein synthesis.

totipotent cells: stem cells that can give rise to every cell type of the body.

transamidase: the enzyme responsible for catalyzing the amination of a carboxylic acid, for example to transform aspartate and glutamate into asparagine and glutamine.

transcript: RNA copy of a gene.

transcript cleavage factor: protein that promotes cleavage of the most recently added ribonucleotides that have separated from the template strand in a backtracked RNA polymerase complex.

transcription: process of copying the DNA sequence of interest into the corresponding RNA.

transcriptional arrest: cessation of RNA transcription.

transcriptional pausing: a temporary cessation of transcript elongation.

transcription bubble: unpaired region of DNA associated with RNA polymerase.

transcription-coupled repair (TCR): nucleotide excision repair that is preferentially targeted to DNA damage in genes that are being actively transcribed.

transcription unit: the RNA copy of a region of the genome, which may include one or more genes along with non-coding regulatory sequences.

transcriptome: levels of all of the transcripts detected for a genome.

transducers: proteins that are activated by sensors during a checkpoint response and activate various cellular processes, via protein effectors. Transducers are sometimes proteins kinases.

transesterification: energetically neutral reaction involving the exchange of phosphodiester linkages.

transfer RNA (tRNA): bifunctional adaptor molecule that interprets the genetic code and participates in the synthesis of proteins by the ribosome.

transformation: process of introducing DNA into cells. Cells are often treated with chemicals to make them competent to take up the DNA.

transgene: specific gene that has been added to a genome to make a transgenic organism.

transgenesis: the introduction of a specific gene, mutation or regulatory element into the genome of an organism.

transgenic organism: organism containing a transgene.

transition mutation: when a purine base, A or G, is replaced by the other purine base, A or G, or a pyrimidine base, C or T, is replaced by the other pyrimidine base, C or T.

transition state: species of highest free energy either in a reaction or a step of a reaction.

translation: process wherein the nucleotide sequence of the mRNA is deciphered in three-base triplets into the corresponding amino acids to form the encoded protein via the actions of the ribosome and auxiliary protein factors.

translational fusion: direct fusion of the open reading frame (ORF) of a reporter gene to the ORF of another gene.

translation factors: host of proteins that facilitate the various steps in the translation cycle.

translocation (chromosome): movement of a segment of a chromosome to another chromosome.

translocation (translation): step in elongation that follows peptidyl transfer that is defined by the directional 3-nucleotide movement of the mRNA–tRNA complex.

transmembrane helix: alpha helix spanning a cell membrane.

transposable element: a discrete mobile DNA segment that can move between non-homologous positions in a genome.

transposase: type of recombinase required for the movement of a transposable element and which is usually encoded by the element. It usually recognizes sequences at the termini of the transposon and executes the DNA cleavage and target-joining reactions that enable the element to move.

transposition: movement of a **transposable element**.

transposon: *See* **transposable element**.

transpososome: elaborate nucleoprotein complexes in which transposition occurs.

***trans* splicing:** form of splicing in which exons from two separate RNA transcripts are joined in a reaction that is otherwise analogous to that of *cis* splicing mediated by the spliceosome.

transverse filament: a protein structure that is part of the symanptonemal complex, which associates two homologous chromosomes during meiosis. The transverse filament connects two homologous chromosomes to each other.

transversion mutation: when a purine base, A or G, is replaced by a pyrimidine base, C or T, or a pyrimidine base, C or T, is replaced by a purine base, A or G.

tRNA: bifunctional adaptor molecule that interprets the genetic code and participates in the synthesis of proteins by the ribosome.

tumor suppressor gene: gene that encodes a protein that normally restrains cell proliferation or tumorigenesis, such that loss of the gene increases the likelihood of cancer formation.

Twist: number of turns about the local DNA helix axis (denoted Tw). Related to the quantities Linking number (Lk) and Writhe (Wr) by the equation $Lk = Tw + Wr$.

two-component signal transduction pathways: signal transduction cascades common in bacteria that signal extracellular changes and in the simplest cases comprise two proteins, a histidine kinase and a response regulator.

two-dimensional gel electrophoresis: separation method in which gel electrophoresis is carried out first in one direction, separating molecules based on one property, such as size, and then subjecting the gel to a second electrophoretic separation at a 90° angle to the first, using a different basis of separation such as pH.

two-hybrid assay: an *in vivo* technique for identifying proteins or protein domains that interact with one another.

ubiquitin: small protein that is covalently attached to lysine side chains in substrate proteins, either singly or in multiple copies as a polyubiquitin chain. Ubiquitin modification serves multiple signaling roles, notably targeting proteins to the proteasome for degradation, as well as for transcriptional regulation, the DNA damage response, and endocytosis.

ubiquitination: attachment of ubiquitin to a protein; also known as ubiquitylation.

unequal crossing over: homologous recombination between non-allelic repeat sequences of homologous chromosomes that result in chromosomal deletions or expansions.

unfolded protein response: multilayered response in eukaryotic cells when unfolded proteins are detected in the endoplasmic reticulum.

unsaturated (fatty acid): fatty acid containing one or more double bonds linking carbon atoms.

untranslated region (UTR) (5′ or 3′): region of an mRNA that precedes the start site of translation is known as the 5′ UTR whereas the region that follows the termination site is known as the 3′ UTR.

UP element: AT-rich sequence located just to the 5′ side of the –35 sequence and bound by the α subunit of RNA polymerase.

UPF: a group of proteins initially identified through genetic approaches for their ability, when mutated, to increase frameshifting (up frameshifting). These proteins are involved in recognizing exon-junction complexes and in signalling to the translation machinery that a particular stop codon is premature.

upstream: to the 5′ side of any given reference point, such as a transcription start site or a coding sequence.

upstream ORFs (uORFs): small open reading frames (ORFs) located 5′ of the principal large ORF in the transcript that in eukaryotes can play important roles in regulating translation.

van der Waals interaction: a weak attractive force between two atoms or groups of atoms, arising from the fluctuations in electron distribution around the nuclei.

van der Waals radius: effective radius of an atom or chemical group.

V(D)J recombination: cellular process for producing antibodies and immune system receptors by recombination of independent gene segments, selected at random from the V, D, and J clusters, to assemble a gene encoding an immune system protein. This process generates a diverse repertoire of antibodies and receptors that can recognize a variety of antigens.

vector: DNA molecule that is used in cloning. Often contains an origin of replication, a multiple cloning site and a selectable marker. A DNA fragment of interest is cloned into the vector, which can then be introduced into cells and propagated.

vertical gene transfer: normal passage of genes from one generation to the next.

viruses: particles composed of a nucleic acid-based genome surrounded by a protein coat. Although viruses contain their own genomes, they cannot replicate on their own and must replicate within another organism, where they usurp components of the host cell machinery.

W

Watson–Crick base pairs: characteristic pairing of guanine with cytosine and adenine with thymine that is stabilized by hydrogen bonds.

western blot: an analytical method in which an antibody is used to detect a protein of interest that has been separated by gel electrophoresis and transferred to a filter.

wild-type: the typical or most common form of an organism or gene in nature.

wobble pairing: non-optimal non-Watson–Crick interactions (most commonly G–U interactions); such pairings are best known for being accepted as cognate at the third nucleotide of the codon (and the first nucleotide of the anticodon) by the translational machinery.

Writhe: supercoiling. In a closed, circular plasmid, the numerical value of writhe is related to Twist and Linking number (Lk) by the equation $Lk = Tw + Wr$.

X

X inactivation: the process in mammals that generates equal expression of genes on the X chromosome in males and females. In females one of the two X chromosomes is permanently inactivated.

x-ray crystallography: a method for determining three-dimensional molecular structures from x-rays scattered from crystals.

Y

Y-family polymerases: class of specialized DNA polymerases that can replicate DNA in an error-prone fashion past a site of DNA damage.

Z

Z-DNA: left-handed DNA double helix that can form in alternating purine–pyrimidine sequences under conditions of high salt or torsional stress.

zinc finger: a protein domain whose single alpha helix and two antiparallel beta strands are held together by a central zinc ion.

Z ring: the contractile ring in bacteria that is made of the tubulin-like protein, FtsZ. The contraction of the Z ring in the driving force of bacterial cell division.

INDEX

Page numbers in *italics* refer to figures and tables; those in **bold**, to pages of the 'Experimental approach' sections.

C

F

G

Q

R

S

T

U

V

W

X

Y

Z